AF304943

Handbook Assisted and Automated Driving

Hermann Winner · Klaus C. J. Dietmayer · Lutz Eckstein · Meike Jipp ·
Markus Maurer · Christoph Stiller

Editors

Handbook Assisted and Automated Driving

Springer

Editors
Hermann Winner
Institute of Automotive Engineering
Technical University of Darmstadt
Darmstadt, Hessen, Germany

Lutz Eckstein
Institute for Automotive Engineering
RWTH Aachen University
Aachen, Nordrhein-Westfalen, Germany

Markus Maurer
Institute of Control Engineering
Technische Universität Braunschweig
Braunschweig, Niedersachsen, Germany

With Contributed by
Tobias Homolla
Riedstadt, Germany

Klaus C. J. Dietmayer
Institute for Measurement, Control
and Microtechnology
University of Ulm
Ulm, Baden-Württemberg, Germany

Meike Jipp
Divisional Board Member for Energy
and Transport
Deutsches Zentrum Luft and Raumfahrt eV
Berlin, Germany

Christoph Stiller
Institute of Measurement
and Control Systems
Karlsruhe Institute of Technology (KIT)
Karlsruhe, Baden-Württemberg, Germany

ISBN 978-3-658-45275-9   ISBN 978-3-658-45276-6   (eBook)
https://doi.org/10.1007/978-3-658-45276-6

Translation from the German language edition: "Handbuch Assistiertes und Automatisiertes Fahren" by Hermann Winner et al., © Springer Fachmedien Wiesbaden GmbH, ein Teil von Springer Nature 2024. Published by Springer Fachmedien Wiesbaden. All Rights Reserved.

Planung/Lektorat: Stella Müller

Daimler Truck AG Funders: Daimler Trucks Holding AG, Hans Turck GmbH & Co. KG

© The Editor(s) (if applicable) and The Author(s) 2026. This book is an open access publication.

Open Access This book is licensed under the terms of the Creative Commons Attribution-NonCommercial-No Derivatives 4.0 International License (► http://creativecommons.org/licenses/by-nc-nd/4.0/), which permits any noncommercial use, sharing, distribution and reproduction in any medium or format, as long as you give appropriate credit to the original author(s) and the source, provide a link to the Creative Commons license and indicate if you modified the licensed material. You do not have permission under this license to share adapted material derived from this book or parts of it.
The images or other third party material in this book are included in the book's Creative Commons license, unless indicated otherwise in a credit line to the material. If material is not included in the book's Creative Commons license and your intended use is not permitted by statutory regulation or exceeds the permitted use, you will need to obtain permission directly from the copyright holder.
This work is subject to copyright. All commercial rights are reserved by the author(s), whether the whole or part of the material is concerned, specifically the rights of translation, reprinting, reuse of illustrations, recitation, broadcasting, reproduction on microfilms or in any other physical way, and transmission or information storage and retrieval, electronic adaptation, computer software, or by similar or dissimilar methodology now known or hereafter developed. Regarding these commercial rights a non-exclusive license has been granted to the publisher.
The use of general descriptive names, registered names, trademarks, service marks, etc. in this publication does not imply, even in the absence of a specific statement, that such names are exempt from the relevant protective laws and regulations and therefore free for general use.
The publisher, the authors and the editors are safe to assume that the advice and information in this book are believed to be true and accurate at the date of publication. Neither the publisher nor the authors or the editors give a warranty, expressed or implied, with respect to the material contained herein or for any errors or omissions that may have been made. The publisher remains neutral with regard to jurisdictional claims in published maps and institutional affiliations.

Die Abbildung auf dem Umschlag ist im Rahmen des vom BMBF geförderten Projekts UNICAR*agil* entstanden.

This Springer imprint is published by the registered company Springer Fachmedien Wiesbaden GmbH, part of Springer Nature.
The registered company address is: Abraham-Lincoln-Str. 46, 65189 Wiesbaden, Germany

If disposing of this product, please recycle the paper.

Preface

The Handbook on Assisted and Automated Driving supersedes the Handbook on Driver Assistance Systems, which has so far been published in three editions in German and one edition in English. On the one hand, it follows in the tradition of this handbook, but, on the other hand, it also represents a clear transition in focus toward automated driving. Driver assistance systems have shown a breathtaking pace of innovation over the past 25 years and are now successfully established in the market, even in inexpensive vehicles. This great success is accompanied by a somewhat slower pace of innovation at present. On the other hand, we are at the beginning of the innovation wave of automated driving. Many findings and components of driver assistance can be found here, but most of them have to fulfill completely new requirements. With the selection of topics, the editorial team wants to represent both worlds. In particular, "value-stable" chapters of previous editions of the Driver Assistance Systems Handbook were not continued if they still correctly reflect the state of the art. This made it possible to add a larger number of new topics on automated driving. However, this also means that the older editions must also be consulted for an overview of driver assistance.

But it is not only the thematic change that characterizes this handbook, but also the change in the editorial team. The team now includes six professors who have been working together in Uni-DAS e.V. for many years. In addition to the expansion of competencies and the division of labor, this also creates a broader personnel basis for the publication of future editions.

Many people have contributed to the handbook. On the one hand, there are the 132 authors who, often in addition to their normal working hours, have written the contributions and have also taken on the tiresome review and proof recursions. For their part, the editors were supported by persons of the respective institute, especially in organizational matters, for which we would like to express our sincere thanks at this point. Mr. Tobias Homolla from the Institute of Automotive Engineering at the Technical University of Darmstadt deserves special mention because he was the organizational center of the book project for several years and always ensured an oversight.

We would like to thank the publishing house Springer Vieweg for their willingness to publish this handbook and their flexibility in solving the many contractual aspects for the authors and editors in a conciliatory manner. We would like to thank Mr. Markus Braun as Executive Editor for the trustful cooperation.

This handbook, published in English, largely reflects the content of the parallel German edition, which is also distributed commercially as a printed publisher's book. We are very grateful to the publisher for making it possible for the English edition to be published as an open-access work that is freely accessible electronically in order to reach a wider international readership.

<table>
<tr><td>Darmstadt, Germany</td><td>Hermann Winner</td></tr>
<tr><td>Ulm, Germany</td><td>Klaus C. J. Dietmayer</td></tr>
<tr><td>Aachen, Germany</td><td>Lutz Eckstein</td></tr>
<tr><td>Braunschweig, Germany</td><td>Meike Jipp</td></tr>
<tr><td>Braunschweig, Germany</td><td>Markus Maurer</td></tr>
<tr><td>Karlsruhe, Germany</td><td>Christoph Stiller</td></tr>
</table>

Contents

Part X   Automated Driving

Supplementary Information

Fundamentals of Driver Assistance Development

Human Performance in Vehicle Driving

Bettina Abendroth and Philip Joisten

Contents

© The Author(s) 2026
H. Winner et al. (eds.), *Handbook Assisted and Automated Driving*,
https://doi.org/10.1007/978-3-658-45276-6_1

1.1 Relevance of Human Performance for Vehicle Driving

Human performance in driving is reflected in the behaviour of the driver. It is determined by the human offer of performance and the requirements resulting from the driving task (see ◘ Fig. 1.1). The human offer of performance is composed of the general performance capacity and the current physiological and psychological willingness to perform, which is influenced by fatigue, emotion or illness, among other factors. The performance capacity in vehicle driving is determined by individual characteristics of the driver, for example, age and driving experience. Requirements from the driving task result from the driver's tasks in the vehicle itself (e.g. controlling the vehicles' longitudinal and lateral motion, other operational tasks) and from the driving environment (e.g. other road users, route).

The high relevance of human performance in vehicle driving results from the implication that humans, as vehicle drivers in manual driving (SAE Level 0, [1]), perform a control task in which information is translated into actions and responses and which requires continuous information processing by the driver. Therefore, human behaviour has a significant impact on the safety of the driver–vehicle–environment system.

From SAE Level 1, the use of advanced driver assistance systems causes a shift of tasks from the human driver to the vehicle, respectively, the advanced driver assistance system. In low automation levels (SAE Levels 1 and 2), the vehicle takes over longitudinal and/or lateral guidance independently, while the driver still has to monitor the driving environment and has to be ready to take over control at any time. In order for transitions from the vehicle to the driver to be safe, it is important that the driver's situational and mode awareness are maintained during assisted driving. Whereas situational awareness describes knowledge about what is currently happening in the driving environment and the actions that what will happen next [2], mode awareness refers to knowledge about the distribution of responsibility between the driver and the advanced driver assistance system, respectively, driving automation system [3].

Regarding the design of the driving task of humans in interaction with the use of advanced driver assistance systems and the resulting distribution of tasks between human and assistance system, it is particularly important to consider the process of human information processing as well as the interrelated factors of drivers' individual performance capacity and willingness to perform.

The simple system model shown in ◘ Fig. 1.2 (cf. [4]) serves as a description of the relationships between driver, vehicle and driving environment. It consists of the elements, driver and vehicle. The driving environment functions as an input variable for the two elements of the system. In addition, confounding variables, such as passenger distraction, can occur. The output variable from this system can be described by the system performances such as mobility, safety and comfort.

1.2 Human Information Processing

In order to describe human information processing, there are a variety of models (e.g. [5–7]) that specify general intake during which a signal entering the receptors (stimulus) is transformed into a cognitive representation and a human reaction (response). A distinction can be made between sequential and resource models. Sequential models allege that the transformation from stimulus to response occurs in a strictly sequential manner, meaning that the next step can only be performed once the previous one is completed. Resource models are based on the assumption that the capacity available for various activities is limited and must be shared between all simultaneously performed tasks. The theory of multiple resources extends this view; according to this theory, the degree of interference between two tasks depends on whether they demand the same resources [8]. In this model, simultaneous processing of visual, spatial image information (e.g. navigation displays) and auditory, verbal information (telephone calls, news on the radio) would be free from interference, since they use different sensory channels and different regions of the working memory. However, experimental studies have shown that this freedom from interference is not absolute (e.g. [9, 10]).

Human information processing can be explained through a combined sequential and resource model (see ◘ Fig. 1.2). This is based on the following processing steps: information intake (perception), information processing in the narrower sense (cognition) and information output (motor function) [11]. Additionally, it must be remembered that the available capacity of resources is limited since the capacity of resources determines the distribution of attention.

The efficiency of these three levels of the information processing system is influenced by available processing resources and requires the application of attention. This leads to targeted selection of information which is intended to be the content of conscious processing. The constant oversupply of information exceeds human processing capacity, so that it is impossible for a human being to consciously perceive everything that reaches the level of the sensory receptors. Human beings can distribute their entire attention to varying degrees among the three levels of the in-

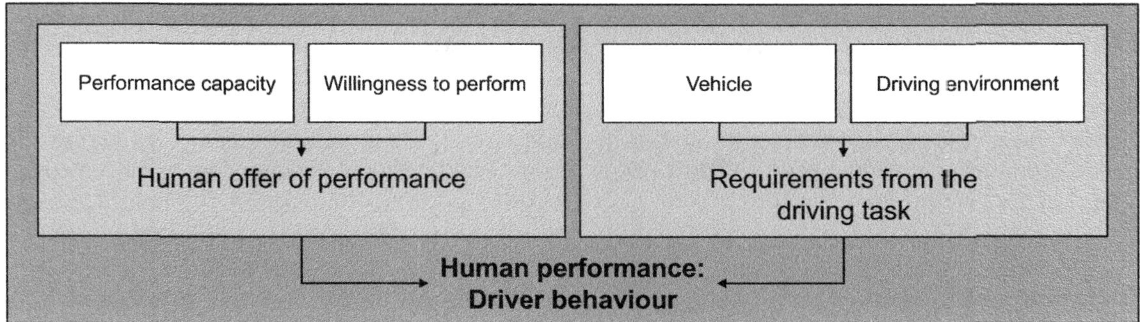

Fig. 1.1   Composition of human performance in vehicle driving

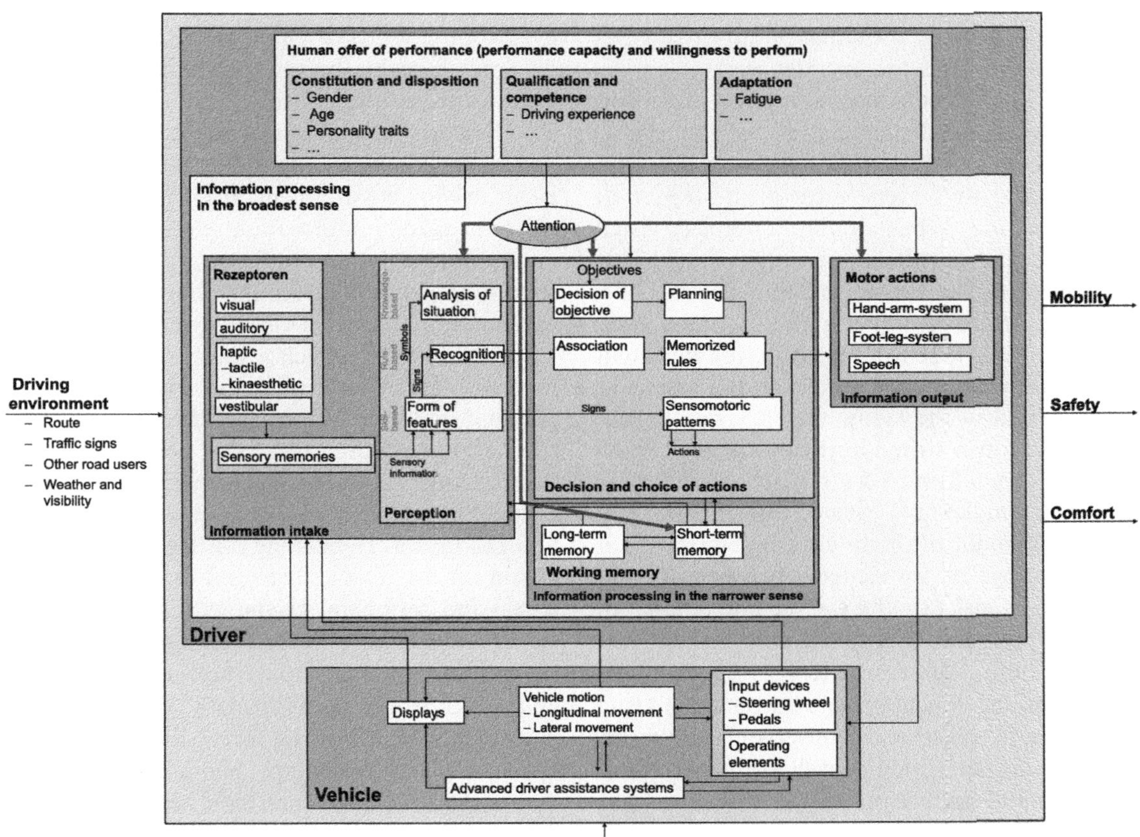

Fig. 1.2   Driver–vehicle–environment system model (cf. [4])

formation processing system in order to select relevant sources of information and further process information from these sources. For each operational task, it is possible to learn an efficient distribution of attention, while in extreme cases, a poor distribution of attention can cause human error.

On a theoretical level, varying forms of attention can be divided into the dimensions of selectivity and intensity. Selective use of attention describes the fact that human beings must decide between different competing sources of information, which can lead to task-relevant information being ignored. Related to this is cognitive tunnelling, which is often referred to in connection with head-up displays. It is the effect that humans tend to focus attention on information from certain areas and exclude information that is outside these areas. Furthermore, errors in the visual perception of information can be attributed to inattentional blindness. This term is used to describe the effect of not consciously perceiving a relevant but unexpected object, even though the gaze has been directed in the direction of the object or has even fixated the object.

1

Intensity of attention concerns the level of activation. In this context, a low proportion of relevant stimuli in the sense of underload can cause reduced vigilance and thus alertness problems, while a high proportion of relevant stimuli in the sense of overload can result in mental fatigue. Common to both cases is the effect of reduced attention performance.

1.2.1 Information Intake

Information intake includes all processes relating to the discovery and recognition of information. The procedure of internally representing the environment will be referred to as perception. This internal image of the environment is influenced by the situation in which a person finds themselves and the experiences at their disposal. Information intake occurs through the sensory organs. The specific performance range of the sensory organs influences the quantity and quality of information absorbed, and thus every subsequent step of information processing. While driving a vehicle, visual, acoustic, haptic and vestibular perception are of the highest significance. The sensory memory register (also called ultra-short-term memory) is also classified under information intake. The sensory memory register exclusively stores physically coded information. Visual information is stored in the iconic memory register and acoustic information is stored in the echoic register for a period of between 0.2 and 1.5 s, respectively [11].

During the intake of visual information, the eye reacts to stimuli through electromagnetic radiation within a range of wavelengths between 400 and 720 nm. The basic tasks of the eye include adaptation (adjusting the eye's sensitivity to the current light density), accommodation (alignment to varying visual distances) and fixation (directing the eyes onto the visual object so that both visual axes converge). Colour, object and movement perception as well as the perception of spatial depth and size are possible via the visual modality.

Periodical air pressure fluctuations within the frequency range between 20 Hz and 20 kHz are perceived auditorily. When receiving auditory information, the ear performs three basic functions: adaptation (increasing the auditory threshold necessary for distinguishing the hearing process); auditory pattern recognition (necessary for language and noise identification) and acoustic spatial orientation, which is accomplished by binaural (two-ear) hearing.

Haptic information intake involves the tactile and/or kinaesthetic perception channel. The tactile perception system enables recognition of distortions to the skin. Receptors (lamellar corpuscles and Meissner's cutaneous receptors) impart sensations of pressure, touch and vibration in and beneath the skin. The kinaesthetic perception system recognizes the stretching of muscles and the movement of joints. Various types of receptors located on the muscle spindles, joint regions and tendons enable the perception of body movements and the relative positions of parts of the body with respect to one another. The perception threshold in the range of the fingers is at deflections >2.5°.

In human beings, spatial orientation is accomplished in human beings by the vestibular perception system. The vestibular apparatus, located in the inner ear, functions as a receptor. This apparatus also has the task of relaying information about maintaining balance and triggering positional reflexes for normal head and eye posture. While driving, the vestibular sensory channel contributes to the perception of speed and acceleration of the vehicle.

The basis for a driver's correct choice of action is a preferably complete internal representation of the relevant driving environment. It is also important for the driver to take in information relevant to driving from a great distance, leaving sufficient time to react to this information. It results from this that most traffic-relevant information is taken in visually when driving a car [12].

The area from which a driver can take in visual information is determined by the driver's visual field. Depending on the movements of the head and the eyes, a distinction regarding the visual field is made. The spatial coverage of these visual fields as well as the lines of sight as reference values can be found in ◘ Table 1.1 and ◘ Fig. 1.3. Depending on the object's mapping location on the retina, there is a differentiation between foveal and peripheral vision: with foveal vision, the object is mapped onto the central depression of the retina (fovea); in this region, only objects within an aperture angle of 2° can be seen distinctly. The farther the image is located away from the fovea, the less distinct it will appear. In the peripheral field of vision, movements and changes of brightness can be perceived. Since peripheral vision is essential for planning shifts of attention and gaze, it is considered to be of an important role in the perception of the environment.

In order for the sensory information to be accessible to further cognitive processing, it must be given attention. The channelling of attention to relevant visual stimuli can be explained by the interaction of top-down and bottom-up processes. Through top-down processes, attention is deliberately shifted towards elements of the environment that are central to an action and thereby expected changes in the environment are captured based on experience, whereas bottom-up processes describe the fact that the property of the stimulus itself causes attention to be shifted towards an object.

□ **Table 1.1**   Extension of the visual field in accordance with [13, 14]

Term	Definition	Range—vertical	Range—horizontal
Visual field I (without eye and head rotation)	Angle of deviation from the visual axis as the limit of perception of visual stimuli (differences in luminance or colour) with the head at rest and the eyes not moving	Target: ±15° Boundary: ±45° (in relation to the visual axis)	Target (binocular): ±15° Boundary (binocular): ±60° Boundary (monocular): ±15°
Visual field II (eye rotation)	Total of all points that can be fixated when the head is at rest and the eyes are moving	Target: +15° to −45° For inclinations of the normal eye-related visual axis ≤25°: +25° to −35° (in relation to the visual axis)	Target (binocular): ±30° Boundary (binocular): ±60° Boundary (monocular): ±110°
Visual field III (head and eye rotation)	Total of all points that can be fixated by head and eye movements when the body is at rest	Target: +40° to −65° For inclinations of the normal eye-related line of sight ≤25°: +50° to −55° (in relation to the horizontal)	(binocular): ±110° Boundary (monocular): ±160°

1.2.2 Information Processing

Signals from the driving environment (e.g. road conditions, other vehicles, weather and visual conditions) and the vehicle (e.g. displays, input devices and vehicle motion) are taken in by human receptors, adapted and then further processed at the level of information processing in the narrower sense (cognition). Here, a decision is made about whether information should lead to an action (active case) or is merely endured (passive case). This decision is principally influenced by the driver's individual characteristics. The range of choices and actions can be explained by three levels of behaviour which build upon one another, which [15] explains as skill based, rule based and knowledge based. A distinction is made between the extent to which cognitive resources are used in different tasks that can range from everyday routine situations to unexpected challenges to rare complex demanding situations. The level of behaviour on which information processing occurs depends on the type of the task to be performed as well as the driver's individual characteristics, in particular, the driver's experience with the demands.

Sensorimotor actions that occur without conscious regulation as automated, uniform and highly integrated behaviour patterns are assigned to the skill-based level. This enables a high processing speed and thus a quick and flexible reaction to situational changes. These are automated processes that hardly require any attention. Rule-based behaviour occurs at more cognitively demanding levels and is determined by simple decision-making processes based on memories of defined rules. These rules are gathered through empirical experience, communication or written instructions. An association between the characteristics of the stored rules and the environmental characteristics takes place. In unknown situations that are new to humans and for which no rules exist, behaviour occurs at the knowledge-based level. Here, the goal is set based on a situation analysis and personal preferences. Alternative plans are developed and the most effective plan in terms of the defined objective is selected. In contrast to the processes that take place at the skill-based level, the processes assigned to the rule-based and knowledge-based levels are referred to as controlled and require more attention.

Memory plays a central role in the cognitive processing of information. With the help of memory, sensory impressions are compared with learned and stored structures of thought and judgement. According to the classic three-level storage model, memory consists of the sensory register (ultra-short-term memory), short-term memory and long-term memory [16]. Information is actively processed in short- and long-term memory. In a continuous process, stored information is summoned

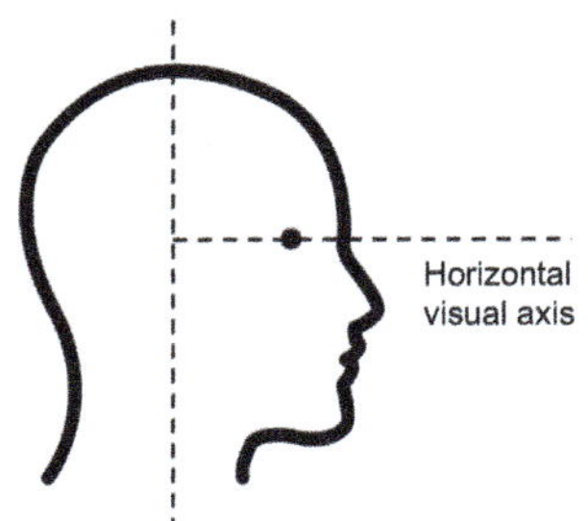
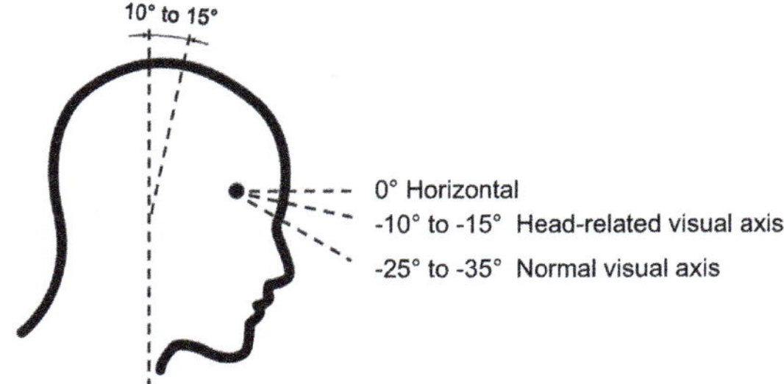

Fig. 1.3 Visual axes in accordance with [13, 14]

from long- and short-term memory and compared with the characteristic traits absorbed by the senses.

One significant aspect of the decision-making process within information processing is the fact that the selected action promises the greatest utility under varying external circumstances with regard to associated risk. The concept of risk is defined in various ways. Often it is interpreted as the probability that an undesirable outcome will occur. Thus, for example, [17] define risk as the relation between variables which describe the negative consequences of outcomes and variables which characterize the probability of the arising of conditions under which these consequences represent a possibility.

Countless models have been developed to explain drivers' risk perception. Among the most well-known are the "zero risk" model [18] and the "risk homoeostasis" model [19]. According to the zero risk model, humans behave in ways in which their subjective risk amounts to zero; this model is based on the individual motivations influencing road driving behaviour and adaptation to risks perceived in road traffic. The theory of risk homoeostasis asserts that, upon reduction of objective risk (e.g. through technical measures), humans alter their behaviour so far towards the direction of "more dangerous" that their subjective appraisal of risk takes on the same distance to their personally accepted risk as prior to the introduction of the measure.

1.2.3 Information Output

On the third level of the information processing system, decisions reached at the level of information processing in the narrow sense are transformed into actions. For driving, these actions encompass motor movements of the hand-arm system for the execution of steering movements and the use of controls or even gestures for additional functions inside the vehicle, movements of the foot-leg system for the control of the vehicle's longitudinal movement via the pedals, as well as speech for the operation of vehicle functions via voice input.

The actions serve to guide the vehicle longitudinally and laterally through the use of the steering wheel and pedals. In addition, other controls that may become relevant to vehicle driving because of necessity due to specific driving environments or because of the operation of advanced driver assistance systems are used. Comfort and entertainment functions unrelated to the driving task itself are controlled via various controls. Besides operation via conventional control elements in the form of buttons, switches and levers, touch displays and voice input systems are now well established. Alternative control options using hand gestures are also available on the market.

1.3 Determinants of Human Offer of Performance

Human performance in driving is determined firstly by the interaction with the driving environment and the vehicle and secondly by the performance offered by the human being, which is shaped by his or her performance capacity and willingness to perform (see **Fig. 1.1**). The human offer of performance is subject to inter- and intra-individual variations: not all individuals fulfil the same task equally well, but even individuals can demonstrate performance variability if performance of the same task is measured at different points in time.

This variability can be attributed to individual human characteristics and thus to the mutually influencing constitutional and dispositional characteristics, qualification and competence characteristics as well as adaptation characteristics of the driver [11]. Constitutional and dispositional characteristics are invariable or variable in a life cycle, but difficult to influence directly. Gender (see [20]), age and personality traits are often discussed as characteristics relevant to driving in the area of constitution and disposition. In contrast, qualification and competence characteristics are variable through short-, medium- and long-term processes. For the task of vehicle driving, the qualification and competence trait of driving experience is particularly important. Adaptation characteristics, on the other hand, can be changed in the short term through intervention. Here, fatigue is a relevant characteristic.

1.3.1   Age

The human capacity to orient oneself with the senses, process received information and carry out motor actions alters with age, during which human organs undergo changes. Increasing functional deficits can be at least partially compensated for by the fact that older drivers generally have more driving experience. Older drivers often use compensatory strategies in order to compensate for deficits in information processing; for example, older drivers are more likely to refrain from driving at night or at times of high traffic density and under complex traffic conditions [21]. In addition, many older drivers drive at lower speeds [22] and higher safety margins [23] and change lanes less frequently [24]. Overall, the functional changes associated with the ageing process are subject to significant inter-individual variations, making it difficult to name a limit beyond which someone belongs to the "elderly" [25]. Often, calendar or chronological age is used as an orientation point according to which people are considered older after their 60th or 65th year of life, even.

With increasing age, the ability of receptors diminishes, which leads to overall limitation in information intake. The changes within the visual system due to the ageing process [26, 27], the resulting effects on visual abilities and the relevance for vehicle driving are summarized in ▪ Table 1.2.

Changes in hearing during ageing include a reduction of the auditory amplitude, particularly at higher frequencies [26]. Difficulties in discriminating the frequency and intensity of tones as well as in recognizing complex noises such as language under difficult perceptive conditions (e.g. background noise, distortion) and partially impeded directional hearing are further changes.

As age increases, there is also a loss of sensitivity of tactile perception [28].

The sense of equilibrium is best formed in 20–30-year-olds and decreases strongly after the 40th year of life, reducing by half at the ages of 60 and 70 [29].

Sensory storage works less efficiently as age increases. Acoustic signals demonstrate a higher speed of decay in echoic storage, while visual signals remain inside the iconic storage for longer periods of time [30]. During the preparation of information relevant to driving, this leads to acoustic information only being available for processing for a temporally shortened span, while visual stimuli can only be received to a limited extent due to blocking of the iconic storage register [31].

Among the individual areas of attention, older people exhibit reduced performance, which in their additive effect results in an overall decrease in attention capacity and thus lower situational awareness. Consequently, older drivers must decide on their actions based on a relatively smaller pool of environmental information than younger drivers, since they do not have access to all potentially important information [26].

Especially in complex and novel situations that require fast action, difficulties can arise for older drivers. For instance, it has been demonstrated that the situational awareness of drivers deteriorates in complex traffic

▪ **Table 1.2**   Changes to the visual system which occur with age (↑ increase; ↓ decrease)

Impact	Cause and influence variables	Significance for vehicle driving
↓ Range of accommodation	↓ Fluid in tissue	Recognition of displays inside the vehicle
↓ Static visual acuity	Lighting conditions ↓ Density of photoreceptors in the fovea	Discovery and recognition of objects and events in the driving environment (other road users, traffic signs)
↓ Dynamic visual acuity	↓ Speed of accommodation ↑ Dullness of sensory cells	
↑ Sensitivity to bright light	↑ Functional disruptions in the retina ↑ Adaptation time	Discovery and recognition of objects and events in the driving environment in case of disturbance due to glare (sunlight, oncoming vehicles)
↓ Contrast sensitivity		Recognition of displays inside the vehicle
↑ Erforderliche Leuchtdichte	↑ Deterioration of cornea, lens and vitreous body	
↑ Limitation of the visual field		Discovery of dynamic objects (other road users), as the movement of relevant objects is initially observable in the peripheral field of vision

1

situations, but this effect is even more pronounced among older drivers [32]. Additional impediments occur in the form of limitations to information intake, which result in partially delayed sensorial preparation of relevant information, leaving less time for older drivers to process information relevant to driving and act accordingly.

1.3.2 Personality Traits

Various personality traits can also influence a driver's behaviour. Thus, correlations have been identified between a driver's willingness to take risks and their driving speed and use of traction [33]. Drivers who are emotionally unstable, impulsive and incapable of teamwork have a higher risk of accidents than people who are adaptable and emotionally stable [34].

1.3.3 Driving Experience

Driving experience can have varying effects on the risk of accidents. With increased driving experience, driving skills as well as situational awareness are improved, along with recognition and judgement of risks [33, 35]. Improvement in driving skills can be attributed to the fact that the number of varying driving situations experienced grows with an increase in distance driven, thus enabling the formation of routine actions. While vehicle control becomes better with increased driving experience, experience in other areas leads to the formation of errors and bad habits, for example, not looking in the mirror, braking late and tailgating [36]. When it comes to skills that reflect a driver's control over the vehicle, beginners have shown to perform more poorly than experienced drivers. This is evident in late acceleration, poor and inconsistent steering motions and slow gear shifting [36]. Inexperienced drivers make more steering motions than experienced drivers [37]. Inexperienced drivers' eye behaviour is often described as less efficient, since they fixate on points in close proximity too frequently [38]. Thus, young, inexperienced drivers recognize distant accident risks relatively poorly compared to experienced drivers [39]; however, there is no difference between the two groups when it comes to recognizing nearby dangers. With increasing experience, drivers learn to recognize dangerous objects and events based on certain parts of the traffic system. This also corresponds with the fact that visual fixation and search patterns differ between inexperienced and experienced drivers [17].

1.3.4 Fatigue

Fatigue is a consequence of continuous task-related stress and is thus a reversible condition that manifests itself physically and mentally depending on the cause and type of stress [40] 34. Task-related fatigue must be distinguished from day-related fatigue, also known as sleepiness/drowsiness, which is due to the circadian rhythm of humans. Both overload and underload can cause task-related fatigue. Overload while driving can be induced by high demands due to, for example, poor weather and visibility conditions, heavy traffic and complex routing. Overload causes active task-related fatigue [41]. In contrast, passive task-related fatigue is caused by persistent underload, which can be attributed to monotonous routing, low traffic density or to increasing automation in the vehicle through advanced driver assistance systems [42].

As a result of fatigue, reception, perception and coordination disorders as well as a decrease in concentration and attention can occur [11]. This can affect all steps of human information processing and the safety in the driver–vehicle–environment system (see ◘ Fig. 1.2). More than 10% of all road traffic accidents can be attributed, at least in part, to driver fatigue [42].

1.4 Demands on Vehicle Drivers in the Driver–Vehicle–Environment System

The demands on the driver result from the task of vehicle driving, which is jointly determined by factors of the driving environment. The complexity of the situation to be managed by the driver is influenced by the characteristics of the route, the dynamic behaviour of other road users and the operation of the vehicle, especially the advanced driver assistance systems. How the driver manages these challenges depends on both the driver's individual characteristics and the driver support offered by the vehicle's advanced driver assistance systems along with the resulting division of tasks between humans and advanced driver assistance systems. Depending on the level and duration of stress, bottlenecks occur in the driver's information processing system, which can lead to deviation from so-called "normal behaviour" all the way to critical traffic situations and even accidents, based on the continuum of traffic behaviour as outlined by [43]. In order to identify these bottlenecks, the following section will compile the subtasks involved in vehicle driving, and the demands resulting from these tasks.

1.4.1 Subtasks Involved in Vehicle Driving

According to the importance for the fulfilment of the driving purpose, a distinction is made between primary, secondary and tertiary tasks in the vehicle [44]. The actual driving process, i.e. keeping the vehicle on the road, is referred to as the primary task and involves guiding the vehicle longitudinally and laterally using the steering wheel, accelerator and brake pedal. Secondary tasks include other tasks relevant to vehicle driving that arise due to other road users, weather and visibility conditions and the chosen route, as well as the operation of advanced driver assistance systems. Various controls are used for this purpose, for example, to (de)activate lights, indicators and windscreen wipers. Operations to increase comfort and entertainment during driving are not related to the driving task, they are classified as tertiary tasks. These include, for example, climate control and infotainment applications.

The three-level model proposed by [45] describes a hierarchy of primary driver tasks using the following activities:

- navigation (choosing the route for a journey),
- guidance (determining target lane and target speed), and
- stabilization (adjusting vehicle motion to the designated driving variables).

This hierarchy also reflects the temporal margin available for the execution of a particular task, along with the tolerance for error. While a delayed decision or error at the navigation level generally does not lead to a critical situation, thoroughly critical driving situations or even accidents can arise at the stabilization level.

With regard to the tasks to be performed at different levels of automation by the human vehicle user and the vehicle itself (differentiating between the driving automation system and other vehicle systems and components), the dynamic driving task comprises all operational and tactical tasks necessary for the operation of a vehicle in road traffic, according to SAE [1]. Strategic tasks, such as route planning, are excluded from this. The following subtasks of the dynamic driving task are distinguished:

- Steering control of the vehicle's lateral motion (operational).
- Steering the longitudinal motion of the vehicle by accelerating and decelerating (operational).
- Object and event detection, recognition, classification and response preparation to monitor the driving environment (operational and tactical).
- Execution of reactions to objects and events (operational and tactical).
- Planning of manoeuvres (tactical).
- Increasing conspicuity through lighting, signalling, etc. (tactical)

1.4.2 Requirements for Vehicle Driving

Generally, the requirements involved in an activity stem from the tasks to be performed. With regard to task—nonspecific, situational operating conditions, stressors arise which can be described objectively. Among these situational factors, the duration and temporal composition of these requirements and influences from the operational environment are significant.

The requirements for drivers derived in this section, which result from the subtasks of vehicle driving, are detailed according to the classifications of [46–48] (cf. [20]) and supplemented by the tasks listed in [1] for automated driving, taking into account automation levels 0 (no driving automation) to 2 (partial driving automation). Requirements are assigned to the categories of information sources/sensory and perceptual processes, acts of judgements, decision-making and cognitive processes, and vehicle operation according to human information processing (see ◘ Fig. 1.2, ◘ Table 1.3).

1.5 Evaluating the Requirements for Driving with Respect to Human Offer of Performance

In conclusion, the requirement areas delineated above will be evaluated with respect to human performance capacity in order to demonstrate meaningful areas for technical driver support and design requirements.

1.5.1 Information Sources, Sensory and Perceptual Processes

Perception of those information sources relevant to fulfilling the task of vehicle driving is of the utmost importance for the driver to achieve sufficient situational awareness. With the help of this information, the driver creates an internal image of the current state of the vehicle and its driving environment and projects this into the near future. This image serves as a basis for the driver's decisions and actions.

The basis for the formation of situational awareness is the perception of the status, characteristics and movements of relevant elements in the environment. Through a synthesis of the elements which are, at this point only, perceived and disjointed, a holistic picture is formed, an understanding is created and meaning is attributed to the elements. Based on information perception and integration, the highest level of situational awareness represents a projection of the current state into the near future [36]. Mode awareness can be seen as a component of situational awareness. It refers to the awareness of the distribution of responsibility in interaction with driver assistance systems and automated systems. In addition to knowledge about the functionality, availability and usefulness of different modes of an automated system, mode awareness includes, in particular, knowledge about the currently active mode as well as future states of the automated system [3].

Table 1.3 Systematization of requirements for vehicle drivers (automation levels 0–2)

	Demands on vehicle drivers	
Information Intake	Information sources, sensory and perceptual processes	**Driver's vehicle movement** Speed, lateral acceleration, longitudinal acceleration and deceleration of the driver's vehicle **Other road users** e.g. position, direction and speed of movement of vehicles, cyclists, pedestrians **Characteristic of the route** e.g. transversal and longitudinal course of the route, junctions, road width, number of lanes **Traffic signs** e.g. speed limits, right of way and direction signs **Condition of the road surface, weather and visual conditions** e.g. moisture, dirt, snow, black ice; light in the eyes, fog **Visual displays in the vehicle** e.g. instruments (e.g. status of driver assistance systems, speedometer), positioning of input devices (e.g. climate control) **Acoustic information** e.g. warning and information sounds of the driver assistance systems, speech output from the navigation system **Secondary acoustic information** e.g. radio, conversations with passengers or on the telephone, sirens on emergency and rescue vehicles
Information processing in the narrower sense	Acts of judgement	**Longitudinal distances from or between other road users or objects** e.g. from vehicles in the direct surrounding area, from pedestrians, cyclists and obstacles **Speed, longitudinal and lateral acceleration of the driver's vehicle and other vehicles or road users** **Adequacy of the activities carried out by the driver assistance systems** e.g. distance to vehicles ahead, speed, acceleration, deceleration **Anticipation of critical traffic situations** e.g. sudden vehicle merging, disregard of right of way by others, a child running into the street
	Decision-making and cognitive processes	**Selecting suitable actions for navigating the vehicle** e.g. deciding which route to choose, deciding which direction to take at intersections **Selecting suitable actions for vehicle guidance** e.g. deciding what speed to drive at and what distance to maintain from other vehicles, passing manoeuvres, choosing a lane **Decision to take over control from or overrule the advanced driver assistance system**
Information output	Vehicle operation	**Controlling the vehicle's longitudinal and horizontal movement in order to stabilize the vehicle** Accelerating, braking, steering **Controlling of functions relevant to driving** e.g. driver assistance systems, lights, windscreen wipers **Controlling of comfort and entertainment functions** e.g. infotainment, climate

From this, the following requirement can be concluded: namely, that relevant situation-dependent information about the vehicle and its environment must also be perceivable, i.e. detectable and recognizable, by the driver. Relevant factors for detectability include the positioning of information in the vehicle as well as in the driving environment and the perception thresholds. When positioning displays, the human visual field (see ◘ Table 1.1) must be taken into account. The provision of additional information via head-up displays is particularly advantageous for older people and novice drivers, as they have trouble looking away from the road. Furthermore, it must also be ensured that the information presented is comprehensible to all drivers. This especially impacts information newly accrued for the driver through the use of advanced driver assistance systems, but also warnings and information that appear only rarely. Due to the requirements regarding the perceptibility of information, there is a need for systems that compensate for driver's possible information deficits resulting from the driving environment. In terms of comprehensibility, there is a need for advanced driver assistance systems to offer information about the current status of the system as well as currently executed and planned actions when driving, so that drivers are able to monitor the subtask performed by the advanced driver assistance system.

The process of human perception is limited by perception thresholds and the necessary application of attention. Perception thresholds vary from individual to individual, so, for example, age is a significant influencing factor, and they are also dependent on the environment. Since driving occurs in radically differing environments, it is important to ensure that information presented in the vehicle falls above the perception threshold. Relevant information from the environment should be presented by driver information systems (e.g. traffic sign assistant) in case it cannot be perceived under certain conditions. While driving, visual, acoustic and tactile information play a particularly important role and must be configured according to their environment. Light conditions can vary between full brightness and strong glare all the way to full darkness. Differences in acoustic environment can be equally great. Inside the vehicle, there are possible scenes without background noise or input of exterior sounds, all the way up to conversations or loud music. Haptic information within the vehicle, which is often used as a warning display, must also be adapted to possible vibrations transmitted by the vehicle or road. When configuring visual information within the vehicle in particular, it is important to note that human beings can only sharply distinguish objects when they are mapped on the fovea up to an aperture angle of 2°, hence it is necessary for the driver's vision to navigate far from the vehicle's external environment in order to take in complex information within the vehicle. This goes beyond simply coded signals, leading the driver to be visually distracted from the actual task of vehicle driving.

Whether the driver perceives (or does not perceive) relevant information depends largely on whether the driver pays attention to it in the first place. This application of attention is strongly influenced by the overall driver–vehicle–environment. Here, the amount and type of mutually competing information from the vehicle and its environment, the mental and/or emotional preoccupation of the driver with concerns irrelevant to driving as well as the driver's personal experience all play a role. With the increasing amount of information in the vehicle due to advanced driver assistance systems and due to the limited human ability to perceive information in parallel, care must be taken to ensure that information is coordinated with each other in the sense of "information management" regarding the time of information presentation and depending on relevance, urgency and the complexity of the driving environment. Another attentional problem associated with advanced driver assistance systems, which cause a shift in tasks for humans from "executing" to "monitoring", can lead to relevant information not being perceived due to reduced vigilance.

1.5.2 Acts of Judgement

Acts of judgement are required in order for the driver to evaluate distances, speeds, potentially critical situations and the appropriateness of the tasks carried out by advanced driver assistance systems. These judgements are carried out on the basis of information sources, sensory and perceptual processes.

Since judging absolute distances is difficult for humans, drivers use various information as variables for judging longitudinal distance. Perspective speed, which is calculated by the size of the vehicle in front as well as the difference in speed and absolute distance to that vehicle, provides drivers with an indication of how the distance to a vehicle in front is changing. Similarly, the time to collision (TTC), which takes into account absolute distance to the vehicle in front as well as difference in speed, is often identified as a relevant judgement variable for the driver [49].

The human threshold for perceiving motion while driving under ideal visual conditions lies between 3 and 10×10^{-4} rad/s. However, length of observation also influences the threshold for perceiving distances and differences in speed from vehicles in front [50]. With a decreasing difference in speed and decreasing length of observation, the distance from which a difference in speed can be registered also decreases. In general, it has been shown that drivers tend to leave a larger safety margin than necessary at lower speeds, while they fall short of this at higher speeds.

Acoustic information can also contribute to judging distance from other vehicles; however, subjective evaluation errors can result, for example, underestimating the distance of a very quiet truck.

Anticipation of critical situations is influenced by the driver's experience with a given potential critical situation. Depending on which situations a driver has already experienced and committed to long-term memory, the driver will classify a critical situation as such based on the situation's distinguishing characteristics and react accordingly. In terms of the ability to anticipate changes in the driving environment and to assess the development of critical situations, humans are superior to today's advanced driver assistance systems due to their situational awareness which is based on a comprehensive picture of the environment composed of individual elements.

In order to be able to assess the adequacy of subtasks of vehicle driving performed by advanced driver assistance systems, it is important that drivers possess a high level of situational awareness. This leads to the recommendation that objects and events in a driving environment that advanced driver assistance systems recognize and taken into account in their choice of action support situational awareness when represented suitably. In addition, information on performed and planned actions of advanced driver assistance systems should be offered.

1.5.3 Decision-Making and Cognitive Processes

While fulfilling the tasks of navigation and guidance, drivers must choose an action appropriate to a given situation on the basis of decision-making and cognitive processes. Under the condition that humans are given sufficient time to make a necessary decision based on an external traffic situation, humans will choose more successfully than technological systems. This can be attributed to the fact that a driver has access to a more complete representation of the driving environment, although this will be less precise in particular aspects, and that, with increased driving performance, a driver can draw on increased experience with identical and similar situations.

In addition to individual motives and preferences, a high level of situational and mode awareness is important for drivers when it comes to making decisions that are adapted to respective situations as to whether a subtask of vehicle control carried out by an advanced driver assistance system should continue to be left to the advanced driver assistance system, whether the system should be overridden or whether the task should be carried out by the driver themself. Especially the mental model a person has of how the system works and what the system's boundaries are, as well as its available system modes, influences the decision to keep a system activated, to override it or to deactivate it.

1.5.4 Vehicle Operation

Operating a vehicle in order to control the longitudinal and lateral movement of the vehicle normally does not present a problem for drivers with sufficient driving experience. The control processes occur on a skill-based level for the driver, which means they are automatic processes that rarely require any attention. This allows drivers to react quickly and flexibly to situational changes.

The analysis of driver reaction times during braking (time between occurrence of the action-triggering stimulus and execution of the movement) shows mean values of 0.7 s for expected situations such as an approach manoeuvre, 1.25 s for unexpected but common situations (e.g. braking of a vehicle in front) and up to 1.5 s for surprising situations [51]. The more critical a situation is, the faster the driver reacts. Overall, it must be taken into account that there is a large variation in braking reaction times, as the influencing factors from the individual area and driving environment are very diverse.

The operation of driving-relevant functions as well as comfort and entertainment functions in the vehicle do not pose a high demand and only require low attentional resources, provided that these activities occur frequently, have already been carried out by the driver correspondingly often and have thus been well learned. When new operating functions are added, e.g. for advanced driver assistance systems, these can require greater attention resources from the driver, at least within the learning period, in order to find the corresponding control element and the appropriate setting.

The allocation of attentional resources to displays and controls inside the vehicle inevitably means that fewer attentional resources are available for observing the driving environment, which can lead to critical situations. This is already addressed in today's vehicles by the functionalities offered by advanced driver assistance systems which warn of obstacles in certain areas and intervene in vehicle control if necessary.

References

1. SAE.: Taxonomy and Definitions for Terms Related to Driving Automation Systems for On-Road Motor Vehicles. SAE International J3016_202104. ► https://www.sae.org/standards/content/j3016_202104/. Zugegriffen: 29.07.2021 (2021). ► https://doi.org/10.4271/J3016_202104

2. Endsley, M.R.: Toward a theory of situation awareness in dynamic systems. Hum. Factors 37(1), 32–64 (1995). ► https://doi.org/10.1518/001872095779049543

3. Sarter, N.B., Woods, D.D.: How in the world did we ever get into that mode? mode error and awareness in supervisory control. Hum. Factors 37(1), 5–19 (1995). ► https://doi.org/10.1518/001872095779049516

4. Abendroth, B.: Gestaltungspotentiale für ein PKW-Abstandsregelsystem unter Berücksichtigung verschiedener Fahrertypen. Ergonomia, Stuttgart (2001)

5. Hollnagel, E., Woods, D.D.: Joint Cognitive Systems. CRC Press, Boca Raton, Foundations of Cognitive Systems Engineering (2005)

6. Welford, A.: The measurement of sensory-motor performance. Survey and reappraisal of twelve years' progress. Ergonomics 3(3):189–230. ► https://doi.org/10.1080/00140136008930484

7. Wickens, C.D.: Multiple resources and performance prediction. Theor. Issues Ergon. Sci. 3(2), 159–177 (2002). ► https://doi.org/10.1080/14639220210123806

8. Wickens, C.D., Hollands, J.G., Banbury, S., Parasuraman, R.: Engineering Psychology and Human Performance. Psychology Press, New York (2015)

9. Alm, H., Nilsson, L.: The effects of a mobile telephone task on driver behaviour in a car following situation. Accid. Anal. Prev. 27(5), 707–715 (1995). ► https://doi.org/10.1016/0001-4575(95)00026-V

10. Banbury, S.P., Macken, W.J., Tremblay, S., Jones, D.M.: Auditory distraction and short-term memory: phenomena and practical implications. Hum. Factors 43(1), 12–29 (2001). ► https://doi.org/10.1518/001872001775992462

11. Schlick, C., Bruder, R., Luczak, H.: Arbeitswissenschaft. Springer, Berlin (2018)

12. Rockwell, T.: Skills, judgment and information acquisition in driving. In: Forbes TW (Hrsg.): Human Factors in Highway Traffic Safety Research, S. 133–164. Wiley, New York (1972)

13. DIN 33414-1:1985-04 Ergonomische Gestaltung von Warten (zurückgezogen)

14. Landau, K., Stübler, E.: Die Arbeit im Dienstleistungsbetrieb. Grundzüge einer Arbeitswissenschaft personenbezogener Dienstleistung. Ulmer, Stuttgart (1992)

15. Rasmussen, J.: Skills, rules, and knowledge: signals, signs, and symbols, and other distinctions in human performance models. IEE Trans. Syst. Man Cybern. SMC-13(3):257–266 (1983). ► https://doi.org/10.1109/TSMC.1983.6313160.

16. Pastötter, B., Oberauer, K., Bäuml, K.-H.: Gedächtnis und Wissen. In: Kiesel, A., Spada, H. (Hrsg.): Lehrbuch Allgemeine Psychologie, S. 121–195. Hogrefe, Göttingen (2018)

17. Brown, I.D., Groeger, J.A.: Risk perception and decision taking during the transition between novice and experienced driver status. Ergonomics 31(4), 585–597 (1988). ► https://doi.org/10.1080/00140138808966701

18. Näätänen, R., Summala, H.: Road-user behaviour and traffic accidents. North-Holland Publishing, Amsterdam/Oxford (1976)

19. Wilde, G.J.S.: The theory of risk homeostasis: implications for safety and health. Risk Anal. 2, 209–225 (1982). ► https://doi.org/10.1111/j.1539-6924.1982.tb01384.x

20. Abendroth, B., Bruder, R.: Die Leistungsfähigkeit des Menschen für die Fahrzeugführung. In: Winner, H., Hakuli, S., Lotz, F., Singer, C. (Hrsg.): Handbuch Fahrerassistenzsysteme, S. 3–15. Springer Vieweg, Wiesbaden (2015)

21. Meng, A., Siren, A.: Older drivers' reasons for reducing the overall amount of their driving and for avoiding selected driving situations. J. Appl. Gerontol. 34(3), 62–82 (2012). ► https://doi.org/10.1177/0733464812463433

22. Trick, L.M., Toxopeus, R., Wilson, D.: The effects of visibility conditions, traffic density, and navigational challenge on speed compensation and driving performance in older adults. Accid. Anal. Prev. 42(6), 1661–1671 (2010). ► https://doi.org/10.1016/j.aap.2010.04.005

23. Andrews, E.C., Westerman, S.J.: Age differences in simulated driving performance: compensatory processes. Accid. Anal. Prev. 45, 660–668 (2012). ► https://doi.org/10.1016/j.aap.2011.09.047

24. Reimer, B., Donmez, B., Lavallière, M., Mehler, B., Coughlin, J.F., Teasdale, N.: Impact of age and cognitive demand on lane choice and changing under actual highway conditions. Accid. Anal. Prev. 52, 12–132 (2013). ► https://doi.org/10.1016/j.aap.2012.12.008

25. Ellinghaus, D., Schlag, B.: Alter und Autofahren. Eine zukunftsorientierte Studie über ältere Autofahrer. Uniroyal Verkehrsuntersuchung, Ifaplan, Köln (1984)

26. Ellinghaus, D., Schlag, B., Steinbrecher, J.: Leistungsfähigkeit und Fahrverhalten älterer Kraftfahrer. In: Bundesanstalt für Straßenverkehr (Hrsg) Unfall- und Sicherheitsforschung im Straßenverkehr, Heft 80, Bergisch Gladbach (1990)

27. Shaheen, S.A., Niemeier, D.A.: Integrating vehicle design and human factors: minimizing elderly driving constraints. Trans Res Part C Emerg Technol 9(3), 155–174 (2001). ► https://doi.org/10.1016/S0968-090X(99)00027-3

28. Geiser, G.: Ältere Benutzer und neuere Informationssysteme im Personenverkehr. In: VDI-Gesellschaft Fahrzeug- und Verkehrstechnik (Hrsg.): Der Mensch im Verkehr, München, S. 245–259 (1997)

29. Seo, T., Takamiya, S.: Improving road environments for the aging society. IATSS Res. 20(1), 21–28 (1996)

30. Oswald, W.D.: Gedächtnis. In: Oswald, W.D., Lehr, U., Sieber, C., Kornhuber, J. (Hrsg.): Gerontologie. Medizinische, psychologische und sozialwissenschaftliche Grundbegriffe, S. 178–182. Kohlhammer, Stuttgart (2006)

31. Metker, T., Gelau, C., Tränkle, U.: Altersbedingte kognitive Veränderungen. In: Tränkle, U. (Hrsg.): Autofahren im Alter. TÜV Rheinland, Köln, S. 99–119 (1994)

32. Kaber, D., Zhang, Y., Jin, S., Mosaly, P., Garner, M.: Effects of hazard exposure and roadway complexity on young and older driver situation awareness and performance. Transp. Res F: Traffic Psychol. Behav. 15(5), 600–611 (2012). ► https://doi.org/10.1016/j.trf.2012.06.002

33. Borowksy, A., Shinar, D., Oron-Gilad, T.: Age, skill, and hazard perception in driving. Accid. Anal. Prev. 42, 1240–1249 (2010). ► https://doi.org/10.1016/j.aap.2010.02.001

34. Kobayashi, T.: Human factors in driving. JSAE Rev. 6–7, 68–76 (1986)

35. Jackson, L., Chapman, P., Crundall, D.: What happens next? predicting other road users' behaviour as a function of driving experience and processing time. Ergonomics 52(2), 154–164 (2009). ► https://doi.org/10.1080/00140130802030714

36. Duncan, J., Williams, P., Brown, I.: Components of driving skill: experience does not mean expertise. Ergonomics 34(7), 919–937 (1991). ► https://doi.org/10.1080/00140139108964835

37. Miltenburg, P., Kuiken, M.: Driving performance and observational strategies of novice and experienced drivers. In: Queinnec, Y., Daniellou, F. (Hrsg.): Designing for Everyone. Proceedings of the 11th Congress of the International Human Factors Association, Paris. Taylor & Francis, London (1991)

38. Nabatilan, L.B., Aghazadeh, F., Nimbarte, A.D., Harvey, C.C., Chowdhury, S.K.: Effect of driving experience on visual behavior and driving performance under different driving conditions. Cogn. Tech. Work 14, 355–363 (2012). ► https://doi.org/10.1007/s10111-011-0184-5

39. Brown, I.D.: Exposure and experience are a confounded nuisance on driver behaviour. Accid. Anal. Prev. 14(5), 345–352 (1982). ► https://doi.org/10.1016/0001-4575(82)90012-4

40. Luczak, H.: Ermüdung. In: Rohmert, W., Rutenfranz, J. (Hrsg.): Praktische Arbeitsphysiologie, 3. Aufl. Georg Thieme Verlag, Stuttgart, S. 71–86 (1983)

1

41. May, J.F., Baldwin, C.L.: Driver fatigue: the importance of identifying causal factors of fatigue when considering detection and countermeasure technologies. Transp. Res. F: Traffic Psychol. Behav. **12**(3), 218–224 (2009). ▶ https://doi.org/10.1016/j.trf.2008.11.005

42. Bier, L., Emele, M., Gut, K., Kulenovic, J., Rzany, D., Peter, M., Abendroth, B.: Preventing the risks of monotony related fatigue while driving through gamification. Eur. Transp. Res. Rev. **11**(1), 1–19 (2019). ▶ https://doi.org/10.1186/s12544-019-0382-4

43. von Klebelsberg, D.: Verkehrspsychologie. Springer, Berlin (1982)

44. Bubb, H.: Der Fahrprozess, Informationsverarbeitung durch den Fahrer. In: VDA (Hrsg.): 4. Technischer Kongress, Stuttgart, S. 19–31 (2020)

45. Donges, E.: Aspekte der Aktiven Sicherheit bei der Führung von Personenkraftwagen. Automob. Ind. **27**(2), 183–190 (1982)

46. Fastenmeier, W.: Die Verkehrssituation als Analyseeinheit im Verkehrssystem. In: Fastenmeier, W. (ed.): Autofahrer und Verkehrssituation. Neue Wege zur Bewertung von Sicherheit und Zuverlässigkeit moderner Straßenverkehrssysteme. TÜV Rheinland, Köln, S 27–78 (1995)

47. Frieling, E., Graf Hoyos, C.: Fragebogen zur Arbeitsanalyse – FAA. Huber, Bern (1978)

48. Hacker, W., Iwanowa, A., Richter, P.: Tätigkeitsbewertungssystem – TBS. Handanweisung. Psychodiagnostisches Zentrum, Berlin (1983)

49. Färber, B.: Abstandswahrnehmung und Bremsverhalten von Kraftfahrern im fließenden Verkehr. Z Verkehrssicherheit **32**(1), 9–13 (1986)

50. Todosiev, E.P.: The action point model of the driver-vehicle-system. Dissertation, Ohio State University (1963)

51. Green, M.: How long does it take to stop? methodological analysis of driver perception-brake times. Transp. Hum. Factors **2**(3), 195–216 (2000). ▶ https://doi.org/10.1207/STHF0203_1

Open Access This chapter is licensed under the terms of the Creative Commons Attribution-NonCommercial-NoDerivatives 4.0 International License (▶ http://creativecommons.org/licenses/by-nc-nd/4.0/), which permits any noncommercial use, sharing, distribution and reproduction in any medium or format, as long as you give appropriate credit to the original author(s) and the source, provide a link to the Creative Commons license and indicate if you modified the licensed material. You do not have permission under this license to share adapted material derived from this chapter or parts of it.

The images or other third party material in this chapter are included in the chapter's Creative Commons license, unless indicated otherwise in a credit line to the material. If material is not included in the chapter's Creative Commons license and your intended use is not permitted by statutory regulation or exceeds the permitted use, you will need to obtain permission directly from the copyright holder.

2

Classification of Automated Driving Functions

Elisabeth Shi and Tom M. Gasser

Contents

© The Author(s) 2026
H. Winner et al. (eds.), *Handbook Assisted and Automated Driving*,
https://doi.org/10.1007/978-3-658-45276-6_2

2.1 Introduction

The report "Legal consequences of an increase in vehicle automation" from 2013 [6] (German version published in 2012 [7]) assessed the legal situation of increasing vehicle automation in public road traffic under German law for the first time. In this context, the scenarios of sustained automation emerged as a subject for the legal assessment. Five levels from Level 0 ("driver only") to Level 4 ("full automation") were distinguished. Based on this classification concept, SAE International published its first version of the J3016 standard 2014 [8], which established internationally and, with the 2016 version, has again come significantly closer to the BASt's preliminary work. Today, it is available in its most current version of 2021 [3]. In contrast to the original legal consequences report [7] SAE Standard J3016 has always distinguished between six levels of sustained driving automation from Level 0 ("no driving automation") to Level 5 ("full driving automation"). As can already be seen from the nomenclature applied to the highest level, the change from five to six automation levels also results in an inconsistency with regard to linguistic nomenclature in relation to level numbering. ◘ Table 2.1 compares the linguistic nomenclature of both classifications [3, 7].

The inconsistency in the linguistic nomenclature of the levels caused misunderstandings and confusion on international level not only in the media landscape, but also in the context of research and professional circles. The consequences of such misunderstandings can be misguided expectations of drivers, which in the worst case might lead to accidents. For this purpose, user-centered communication concepts have been developed by SAE and BASt, which are explained at the end of this chapter. Among experts, too, it is important to avoid misunderstandings due to incorrect assumptions regarding automation levels, in order not to talk past each other in the (interdisciplinary) discourse in the context of research and development as well as regulation. The goal must be to enable clear communication.

For this purpose, clear common understanding of the current classification concepts is essential. The concepts of user-centered communication are also suitable for an initial overview of classification.

This chapter helps to resolve misunderstandings by presenting the current classification concepts used among experts, and approaches to user-centered communication in detail. ▶ Chapter 3 deals with the legal framework for the driving functions described in this chapter. The internationally applicable technical vehicle regulations are described in more detail in ▶ Chap. 4.

2.2 What is Classified?

The generic term "automated driving functions" includes those functions that enable assisted and automated as well as autonomous driving as mentioned in the title of this book. The term "automated driving functions" refers to all driving functions that take effect on vehicle guidance [2]. In the following, the scope of the classification will be explained first. This is followed by a detailed description of the classification concept.

2.2.1 Distinction Between Vehicles, Systems, and Functions

The terms vehicle, system, and function must be distinguished from each other in the context of classification. The *vehicle* forms the structural entity in which automated systems and functions are contained, e.g., a passenger car, a delivery robot, or a truck. A *system* is the purchasable product of a manufacturer that is composed of one or more *(driving automation) functions.* For example, a system may comprise a lane change assist and a lane keeping assist. The lane change assist informs the driver, if a vehicle is in the adjacent lane or warns the driver in case the direction indicator is simultaneously activated. The lane departure warning intervenes if the driver leaves the lane in the direction of the

◘ **Table 2.1** Comparison of the linguistic nomenclature from [7] (Engl. [6]) and [3]

BASt level [6, 7]	Nomenclature according to [6, 7]	Differentiation criterion	Nomenclature according to [3]	SAE Level [3]
Level 0	Driver only	BASt and SAE differentiate according to the extent of automation applied within the task of driving	No driving automation	Level 0
Level 1	Assisted		Driver assistance	Level 1
Level 2	Partial automation		Partial driving automation	Level 2
Level 3	High automation		Conditional driving automation	Level 3
Level 4	Full automation	Only SAE differentiates Levels 4 and 5 according to the scope of the domain	High driving automation	Level 4
			Full driving automation	Level 5

occupied lane. In this example, lane change assist and lane keeping assist are the functions that make up the system. A function may be part of a system, as in the aforementioned example, or it may constitute a system of its own. In the latter case, the purchasable product (e.g., under the product name "Blind Spot Assist" by Mercedes-Benz) would, for example, only consist of a lane change assist.

2.2.2 Distinction Between Navigation, Stabilization, and Vehicle Guidance

According to Donges' three-level hierarchy of the driving task [2], the driving task can be divided into the navigation layer, stabilization layer, and vehicle guidance layer.

For the layer of *navigation,* the driving route is determined considering the existing infrastructure and travel time. It might be updated considering current traffic information and conditions. Navigation systems are an example of systems that support drivers with information relevant for this task.

For the layer of *vehicle guidance,* driving speed and distances to other road users as well as obstacles are selected and adapted to current conditions. For example, adaptive cruise control (ACC) maintains the driving speed and the distance to the vehicle in front as set by the driver.

At the *stabilization* layer, deviations to input variables that were previously defined within the layer of vehicle guidance are corrected by respective functions. For example, the anti-lock braking system (ABS) prevents the wheels from locking when braking. If a driver steers when the wheels are locked, the actual vehicle movement does not follow the driver's steering inputs due to reduced frictional force (sliding friction). The intervention of the anti-lock braking system helps maintain the frictional force between the tires and the road surface (rolling friction). This allows to bring the actual vehicle movement closer to the steering inputs of the driver in heavy braking situations.

2.2.3 Scope of the Classifications

The classifications described in this chapter refer to functions that have an effect within the vehicle guidance layer. Functions that take effect within the navigation layer (e.g., navigation systems) and functions that take effect within the stabilization layer (e.g., ABS), are to be classified as such, i.e., as functions of the navigation or stabilization layer. So far, no established classification concept is known for these two layers. Also, the present classification concept for vehicle guidance layer is not proposed for application to stabilization or navigation layer. Furthermore, it should be emphasized that classifications do not refer to entire vehicles, but individual functions are classified. That is, combination of terms such as "Level 3 vehicles" is not within the scope of the classification concept, since such combinations apply the classification of a single function (here, a "Level 3" function) to the entire vehicle which, however, may have many more classifiable driving automation functions than a Level 3 function (e.g., functions of Principle of Operation A and C, as well as Level 0 to 2 functions outside the domain of the Level 3 function). Another example is the combination of terms in "Level 5 vehicles". This combination is also not within scope of the classification concept, since the classification of a single function (here a Level 5 function) is again applied to the entire vehicle. Levels 4 and 5 allow for a specificity in the design of vehicles in the sense that vehicles with Level 4 or Level 5 automation can also be designed without control elements for human drivers. At the same time, the mere technical equipment of a vehicle with a Level 4 or Level 5 function does not fundamentally exclude the possibility of manual vehicle control (so-called *"ADS-equipped dual-mode vehicle"* [3]). If the term combination "Level 5 vehicles" is intended to refer to the special design of a vehicle without control elements for human drivers, it is recommended to focus on this very design instead of the driving automation function.

2.3 How is It Classified?

The central criterion is to check the effect of the function. In the first step, it is checked whether the function of interest is a function with effect within the *vehicle guidance layer.* If this is not the case, the function is to be described as a function taking effect within the stabilization or navigation layer. If the effect within the vehicle guidance layer applies, the function can be differentiated in the second step with regard to its *Principle of Operation* [4]. A total of three Principles of Operation are distinguished. Principle of Operation A (▶ Sect. 2.3.1) comprises functions that achieve their effect on vehicle guidance indirectly via information and warnings to the driver (e.g., lane departure warning, traffic sign recognition). Principle of Operation B (▶ Sect. 2.3.2) comprises functions with a direct and sustained effect on vehicle guidance (e.g., adaptive cruise control). Principle of Operation C (▶ Sect. 2.3.3) comprises functions that achieve their direct effect on vehicle guidance via temporary interventions (e.g., automatic emergency braking). This section explains the three Principles of Operation (◘ Table 2.2) and then their interaction and competitive relationship (▶ Sect. 2.3.4). For a deeper

Table 2.2 Overview of the effects of automated driving functions

	Principle of Operation A	Principle of Operation B	Principle of Operation C
Effect	Informs and warns the driver	Operates on a sustained basis	Intervenes in emergency situations
Examples	Traffic sign recognition	Cruise control	Automatic emergency braking
Further differentiation	Further differentiation into information, warning in case of abstract hazard, warning in case of concrete hazard	Further differentiation according to SAE Standard J3016 in Level 0 to Level 5	Further differentiation according to abstract and concrete hazard as well as according to the extent of control assumed

understanding of the classification concepts described here, please refer to [3] and [4]. For a former classification approach that distinguished between functions providing information and sustained automation among other functions, refer to [9]. For an extension of [9] that adds functions that aim at collision avoidance and mitigation, refer to [10].

2.3.1 Principle of Operation A: Informing and Warning Functions

Principle of Operation A includes all functions that provide information or warnings to the driver. This information or warning is provided to the driver via a human–machine interface. Thus, the functions of Principle of Operation A achieve their effect on the vehicle guidance only indirectly, mediated by the driver's actions. This indirectness is a central characteristic of Principle of Operation A functions. For example, a lane departure warning system warns the driver that the vehicle is leaving the lane. However, the action on vehicle guidance level, i.e., keeping the vehicle in its own lane, must be carried out by the drivers themselves.

Functions of Principle of Operation A can be further differentiated according to "information", "warning in case of concrete hazard", and "warning in case of abstract hazard". The distinction between concrete and abstract hazard is also made in Principle of Operation C and is explained in more detail there. In terms of time, functions of Principle of Operation A become active earlier than intervening functions of Principle of Operation C in the case of abstract and concrete hazard.

2.3.2 Principle of Operation B: Sustained Driving Automation Functions

Principle of Operation B comprises all functions that have a direct and sustained effect on the vehicle guidance level. They exert a direct influence on the vehicle

control over longer periods of time or travel sections. Sustained and direct influence on the vehicle guidance characterizes functions of Principle of Operation B.

As described in the beginning (▶ Sect. 2.1), a finer differentiation of the functions of Principle of Operation B along Levels 0 to 4 was first described in 2012 by the BASt project group "Legal consequences of an increase in vehicle automation" [6]. Based on this, SAE Standard J3016 established internationally in 2014 [8], and has been regularly updated since then. The current version from 2021 was developed for the first time together with ISO [3].

2.3.2.1 SAE International Standard J3016

▶ Table 2.3 shows the classification according to SAE Standard J3016 [3]. It represents a further subdivision of the functions of Principle of Operation B. The six levels are described in more detail below. For a more in-depth understanding, please refer to [3].

Level 0 describes the absence of any sustained driving automation function. In this case, only the driver controls the vehicle. A common misunderstanding is that Level 0 also implies the presence or absence of other functions of Principles of Operation A and C, or even functions operating on the stabilization or navigation layer. For this reason, it shall be explicitly stated here that the Level 0-category represents a subcategory for functions with a direct and sustained effect on the vehicle guidance layer. The Level 0-category alone does not provide any indication of the presence or absence of other functions, such as automated emergency braking of Principle of Operation C. These require separate mention.

Level 1 describes a driving automation function that supports the driver on a sustained basis in executing longitudinal or lateral control of the vehicle. The function can neither reliably detect the environment nor reliably react to it. The driver must permanently supervise the driving automation function and the traffic and correct the Level 1 function if necessary.

Level 2 describes a driving automation function that supports the driver on a sustained basis in execut-

Table 2.3 Six levels of sustained driving automation (positive description)

SAE level numbering	Description	Example
Level 5	Like Level 4 but with unlimited range of operation (domain)	Application, e.g., for trucks and buses
Level 4	Sustained automation of longitudinal and lateral vehicle guidance as well as object and event detection and response, with a limited range of operation (domain). If the driver does not take over when a functional limit is reached or if no driver is present (driverless design), the function initiates a minimal risk condition	Level 4 function operating on a fixed route; application, e.g., for delivery robots within a specific city
Level 3	Sustained automation of longitudinal and lateral vehicle guidance as well as object and event detection and response, with a limited range of operation (domain) and with the expectation that a driver will take over control after being prompted by the function without undue delay	Level 3 functions operating in a traffic jam on a motorway
Level 2	Sustained driving automation of longitudinal and lateral vehicle guidance, with a limited range of application (domain) and with the expectation that a driver will immediately correct the function if necessary	Combination of ACC and lane keeping assist
Level 1	Sustained driving automation of longitudinal or lateral vehicle guidance, with a limited range of use (domain) and with the expectation that a driver will immediately correct the function if necessary	Cruise control
Level 0	No sustained driving automation	No sustained driving automation function available

ing longitudinal and lateral control of the vehicle. The function can neither reliably detect the environment nor reliably react to it. The driver must constantly supervise the driving automation function, monitor the traffic, and correct the Level 2 function if necessary. In contrast to Level 1, a Level 2 function supports both longitudinal and lateral vehicle guidance, whereas a Level 1 function supports either longitudinal or lateral guidance.

Level 3 describes a driving automation function that completely takes over the driving task on a sustained basis within the domain for which it was designed and under the condition that the driver takes over the driving task again at the end of the domain at the latest. In contrast to Level 2, a Level 3 function does reliably detect the environment and reliably respond to it. When approaching a functional limit, the function issues a request to the user to resume the driving task with sufficient lead time.

Level 4 describes a driving automation function that fully performs the driving task on a sustained basis within its domain for which it was designed without the expectation that a driver will resume the driving task at the end of the domain. When the Level 4 function is active, all occupants are passengers. To drive outside the domain, a driver is required to take over the driving task. In contrast to Level 3, the Level 4 function sets itself to a minimal risk condition at the end of the domain if the driver does not continue the journey outside the domain on his or her own.

Level 5 describes a driving automation function that fully performs the driving task on a sustained basis on roads and under all conditions that can also be mastered by a human driver. Level 5 is distinguished from Level 4 by its domain-independent scope of action.

For clear orientation, it is recommended to use the numbering of the levels, as this is internationally congruent and avoids misunderstandings in contrast to the linguistic-level nomenclature (▶ Sect. 2.1).

2.3.3 Principle of Operation C: Temporary Intervention in Accident-Prone Situations

Principle of Operation C comprises functions that intervene temporarily in accident-prone situations. Functions of Principle of Operation C intervene when a near-collision or potentially near-collision situation arises to which the primary controller does not react or can only react with a time delay. Due to the proximity of a collision, the situation is no longer within the control of the primary controller, e.g., because of sudden inability to act or human reaction time. The primary controller can be the human driver as well as a driving automation function of Principle of Operation B. Principle of Operation C functions aim at avoiding an accident or mitigating its consequences. A classic example is automated emergency braking, which triggers emergency braking if a collision is otherwise unavoidable

[11]. Emergency stop assistants, which bring the vehicle to a halt on the hard shoulder if the driver suddenly becomes unable to act, also fall into the category of Principle of Operation C [12].

These examples also illustrate the difference between concrete and abstract hazard. Please note that the term "hazard" is used according to the Principles of Operation Framework [4]. A concrete hazard exists when an imminent collision has been detected and at the same time the primary controller can no longer handle the situation, e.g., due to functional performance limits (i.e., related to an active primary controller of Principle of Operation B) or human performance limits (e.g., reaction times). The direct intervention of the Principle of Operation C function is intended to avoid collisions or mitigate consequences.

An abstract hazard exists if the primary controller does not act according to expectations and/or rules or is unavailable for any reason. If the operational primary controller is unavailable, the driving task remains uncontrolled and thus constitutes a hazard to road safety, even if no immediate collision can be detected.

In the further differentiation of Principle of Operation C, the functions that intervene in the case of an abstract hazard are assigned with index I (Roman one), and the functions that intervene in the case of a concrete hazard are assigned with index II (Roman two). In addition, differentiation is made along the extent and duration of the intervening control assumption (level α, β, γ). For details on the differentiation, please refer to [4].

2.3.4 Interaction and Competition

The *classification* presented differentiates functions with regard to their Principle of Operation on vehicle guidance. Within each Principle of Operation, different levels are distinguished. However, the classification does not impose any limitation with regard to combinations of functions of different Principles of Operation and subordinate levels. This means that a system can consist of several functions that belong to the same or different Principles of Operation.

For example, a system can consist of a forward collision warning system (Principle of Operation A—warning in case of concrete hazard) and an autonomous emergency braking (Principle of Operation C—level α_{II}). In this example, two functions with different Principles of Operation have been combined into one system. Functions of the same Principle of Operation can also be combined into one system, e.g., a lane departure warning and a blind spot detection function that warns when changing lanes if another vehicle is in the blind spot.

In addition to this interaction of Principles of Operation, a competitive relationship between Principle of Operation C and Principle of Operation B can also be discussed. With increasing duration of an intervention, the question can be discussed at which point an intervention (Principle of Operation C, e.g., level γ) can be considered sustained control (Principle of Operation B).

2.4 Who is Addressed by This Classification?

The classification described so far is used among experts, e.g., in the context of research, development, regulation, legislation, and many more. At the same time, the development of driving automation functions aims at the application in road traffic. To enable the functions to be used as intended, users must be informed about their respective responsibilities when using the functions. Fulfilling this information requirement becomes increasingly important, as the sustained automation of vehicle motion control in particular changes the role of humans in the vehicle. As a result, concepts have been developed for *user-centered communication* of vehicle automation.

2.4.1 Approaches to User-Centered Communication

With the aim of increasing comprehensibility for users, technical principles have been developed by SAE International and BASt [1, 5]. Similarities and differences between the two concepts are briefly discussed below.

In order to simplify the J3016 standard in a user-centered way, SAE International combines Level 0, Level 1, and Level 2 into one category of "driver support features" [5] Level 3, Level 4, and Level 5 into a second category of "automated driving features" [5]. The category of "driver support features" does not completely relieve the driver of the driving task, but also provides functional support, i.e., the role of the driver does not change. In contrast, "automated driving features" take over the driving task completely, at least in some domains. Here, however, Level 3 represents a special case. Its central characteristic is that the user forms the fallback level for the function after it has been activated. This means that the user must necessarily take over the driving task when the function prompts him/her to do so. SAE International recognizes this characteristic of Level 3 as relevant knowledge for the user by selectively assigning this characteristic to the first category (Level 0–Level 2). The remaining Level 3 characteristics remain assigned to the second category (Level 3–Level 5). In this way, SAE International achieves a simplification to two approximately distinct categories [5].

◼ Table 2.4 Basis for user-oriented communication of vehicle automation

SAE Level	Role of the human being according to mode
Level 1	Driver role in assisted mode
Level 2	
Level 3	User role in automated mode
Level 4	Passenger role in autonomous mode
Level 5	

The German Federal Highway Research Institute (BASt) distinguishes between three different *roles* for the human (in the driver's seat): the role of the driver, the role of the user, and the role of the passenger (◼ Table 2.4) [1]. In the driver role, the person in the driver's seat performs the driving task. He or she can be supported by Level 1 and 2 functions. In these cases of redundant-parallel vehicle motion control, the driver must continue to perform the driving task manually even after the functions have been switched on. In addition, the driver must supervise the activated driver assistance functions to determine whether they react appropriately to the driving situation and, if not, take corrective action. At Level 3, the function and the driver alternate in performing the driving task. However, while the driving function carries out the driving task, the driver changes into the role of the user. As such, the user must be available as a fallback for the function. As long as this is fulfilled, the user can turn to other activities until the user is requested by the function to take over the driving task. Upon this request, the user must take over the driving task again and continue the journey as a driver. The user's availability for the active driving automation function is no longer necessary from Level 4 onward. When the Level 4 or Level 5 function is active, the on-board human is merely a passenger and no longer performs any driving-related tasks. BASt therefore achieves simplification to three distinct categories taking specifics of Level 3 into due account.

Both concepts for user-centered communication have in common that they are based on the roles of humans for the categories formed. In this context, BASt distinguishes a total of three distinct modes and SAE International two approximately distinct categories. A central difference is that the BASt concept attributes a specific role to Level 3 (the role of the user) in order to emphasize the special role of the fallback-ready user. SAE International highlights this by specifically assigning this feature to the first category of "driver support features" (Level 0 to Level 2 features). This softens the distinction between the two categories.

References

1. Bundesanstalt für Straßenwesen.: User communication: What does autonomous driving actually mean? (2020). Retrieved from ► https://www.bast.de/BASt_2017/EN/Automotive_Engineering/Subjects/f4-user-communication.html?nn=1844934
2. Donges, E.: Fahrerverhaltensmodelle. In: Winner, H. (ed.) ATZ/MTZ-Fachbuch. Handbuch Fahrerassistenzsysteme: Grundlagen, Komponenten und Systeme für aktive Sicherheit und Komfort, 3rd ed., vol. 78, pp. 15–23. Springer Vieweg, Wiesbaden (2015). ► https://doi.org/10.1007/978-3-8348-9977-4_3
3. SAE International/ISO.: Taxonomy and Definitions for Terms Related to Driving Automation Systems for On-Road Motor Vehicles. SAE International (J3016). 400 Commonwealth Drive, Warrendale, PA, United States (2021)
4. Shi, E., Gasser, T.M., Seeck, A., Auerswald, R.: The principles of operation framework: a comprehensive classification concept for automated driving functions. SAE Int. J. Connect. Autom. Veh. **3**(1) (2020). ► https://doi.org/10.4271/12-03-01-0003
5. Shuttleworth, J.: SAE Standards News: J3016 automated-driving graphic update (2019). Retrieved from ► https://www.sae.org/news/2019/01/sae-updates-j3016-automated-driving-graphic
6. Gasser, T.M., et al.: Legal consequences of an increase in vehicle automation: consolidated final report of the project group; Report on research project F 1100.5409013.01. Berichte der Bundesanstalt für Straßenwesen F, Fahrzeugtechnik: Vol. 83. Bremerhaven: Wirtschaftsverl. NW Verl. für neue Wiss (2013). Retrieved from ► https://bast.opus.hbz-nrw.de/opus45-bast/frontdoor/deliver/index/docId/689/file/Legal_consequences_of_an_increase_in_vehicle_automation.pdf
7. Gasser, T.M., et al.: Rechtsfolgen zunehmender Fahrzeugautomatisierung: Gemeinsamer Schlussbericht der Projektgruppe; Bericht zum Forschungsprojekt F 1100.5409013.01. Berichte der Bundesanstalt für Straßenwesen F, Fahrzeugtechnik: Vol. 83. Bremerhaven: Wirtschaftsverl. NW Verl. für neue Wiss (2012). Retrieved from ► http://bast.opus.hbz-nrw.de/volltexte/2012/587/
8. SAE International.: Taxonomy and Definitions for Terms Related to On-Road Motor Vehicle Automated Driving Systems. SAE International, (J3016). 400 Commonwealth Drive, Warrendale, PA, United States (2014)
9. Naab, K., Reichart, G.: Grundlagen der Fahrerassistenz und Anforderungen aus Nutzersicht. Seminar "Fahrerassistenzsysteme", Haus der Technik, Essen, 16./17.11.1998, S. 1–13 (1998)
10. Wörsdörfer, K.-F., Maurer, M.: Vortrag: Fahrerassistenzsysteme zur Erhöhung der Sicherheit. Driver Assistance Systems for Collision Mitigation and Collision Avoidance. Tagungsunterlagen: 5. Internationales Stuttgarter Symposium. Kraftfahrwesen und Verbrennungsmotoren. 18–20 Februar 2003 (2023)
11. Mimura, Y., Ando, R., Higuchi, K., Yang, J.: Recognition on trigger condition of autonomous emergency braking system. J. Saf. Res. **72**, 239–247 (2020). ► https://doi.org/10.1016/j.jsr.2019.12.018
12. Ardelt, M., Waldmann, P., Homm, F., Kaempchen, N.: Strategic decision-making process in advanced driver assistance systems. IFAC Proc. Vol. **43**(7), 566–571 (2010). ► https://doi.org/10.3182/20100712-3-DE-2013.00006

Open Access This chapter is licensed under the terms of the Creative Commons Attribution-NonCommercial-NoDerivatives 4.0 International License (▶ http://creativecommons.org/licenses/by-nc-nd/4.0/), which permits any noncommercial use, sharing, distribution and reproduction in any medium or format, as long as you give appropriate credit to the original author(s) and the source, provide a link to the Creative Commons license and indicate if you modified the licensed material. You do not have permission under this license to share adapted material derived from this chapter or parts of it.

The images or other third party material in this chapter are included in the chapter's Creative Commons license, unless indicated otherwise in a credit line to the material. If material is not included in the chapter's Creative Commons license and your intended use is not permitted by statutory regulation or exceeds the permitted use, you will need to obtain permission directly from the copyright holder.

The Regulation of Autonomous Driving in California

Stephen Wu

Contents

© The Author(s) 2026
H. Winner et al. (eds.), *Handbook Assisted and Automated Driving*,
https://doi.org/10.1007/978-3-658-45276-6_3

3.1 Introduction [1]

According to one popular saying, "As California goes, so goes the nation." In terms of social and political changes, California has been a pioneer within the United States for many decades. Likewise, Silicon Valley in California has led the nation in technological changes since the semiconductor revolution began over 60 years ago. The same holds true for law. Legal developments in California frequently set a trend within the United States. For instance, strict product liability—liability of a manufacturer without the need to prove the manufacturer's fault—originated from a 1962 California Supreme Court case.[1] Now, most states in the United States recognize the doctrine of strict liability.

California has also been on the forefront of both technological developments supporting autonomous vehicles (AVs) and the regulation of autonomous driving in law. In particular, Silicon Valley has been a hotbed of activity for AV technology. Silicon Valley has a number of advantages facilitating its strong support for businesses in the technology field, from which the AV sector benefitted. Examples of Silicon Valley's strengths include the following:

- The concentration of talented and diverse technology professionals from all over the globe.
- The abundance of venture capital firms with a long history of supporting technology startups.
- The proximity of world-class research universities that support spin-offs of technology developed in the lab.
- The attractiveness of living in Northern California with its warm weather, the nearby Pacific Ocean beaches, the nearby city of San Francisco, proximate skiing areas, and natural beauty.
- Laws protecting employee mobility that reduce the risk of new business formation and incentivize entrepreneurial activity.

Companies in the AV field have taken advantage of these strengths. Also, the dry and moderate climate is conducive to testing AVs without the need to deal with harsh weather conditions in other areas of the country. These strengths, combined with a forward-looking legislature and executive in California, made Silicon Valley an ideal test bed for AVs.

This chapter focuses on the regulation of AVs in California. The sections in this chapter discuss the sources of laws that apply to AVs in California, California's path-breaking AV statute, California's regulations governing testing and deployment of AVs, federal laws that apply to AVs, and the legal system's regulation of AVs via product liability law. The businesses most affected by the laws covered in this chapter are manufacturers, distributors, and sellers of AVs.

3.2 Sources of Law

When considering the regulation of autonomous driving in California, the first consideration is what sources of law may apply to AVs. The United States has a federal system of governance. Legal authority is split between federal laws at the national level and state laws for each of the 50 states and the District of Columbia. The United States Constitution is the supreme law of the land in the United States and therefore stands at the pinnacle of legal authority in the country. Indeed, any federal (national) law takes precedence over state law.[2] Below the U.S. Constitution are federal statutes, laws passed by both chambers of the legislature (the House of Representatives and the Senate) and either signed by the President, passed despite a presindent's veto, or, through the President's inaction, allowed to become law. Federal laws are codified in the United States Code. Federal agencies also enact regulations, which are subordinate to the statutes that authorized them. Regulations appear in the Code of Federal Regulations.

The California system of laws parallels the federal system. California has its own constitution, which is the highest source of state law. Below the California Constitution are state statutes, laws passed by the two chambers of the state legislature (the California Assembly and the California Senate) and either signed by California's governor 1 passed despite a governor's veto, or, through the Governor's inaction, allowed to become law. California codifies statutes into a series of codes for each main subject, such as the California Vehicle Code. Like federal agencies, state agencies can enact subordinate regulations via the authority granted to them by statute. Regulations in California appear in the California Code of Regulations.

In addition to these parallel forms of lawmaking, California authorizes counties and cities to exercise legislative authority that is consistent with the general laws of the state. Counties and cities create ordinances to regulate matters of local concern to a county or city. These ordinances are codified in county and city codes.

As discussed in greater detail below, the main California laws relating to autonomous driving are the California statute enacted by Senate Bill (SB) 1298 and regulations enacted under those statutes. AV suppliers and operators must also consider federal automotive standards and more general federal and state statutes on fair trade practices and data protection, as well as state cases and statutes governing product liability. The regulation of autonomous driving in California occurs through an interplay of these federal and state laws.

---

1   The information in this chapter was current as of june 2022.

Court cases play a role in defining California law. Law in the United States, with the exception of Louisiana, is based on a common law tradition. In common law jurisdictions, judges deciding cases and writing legal opinions create doctrines of law that create binding precedents. By contrast, in civil law jurisdictions, written codified codes of law are the key sources of binding law, rather than decisions of courts. Historically, the enactment of statutes at the federal and state levels in the United States supplement and sometimes supplant the common law derived from court cases. Regarding AVs, much of product liability law is derived from court cases, while the laws directly addressing AVs are statutes and regulations.

Regulating AVs through legislation and regulations is not the only model for governing AVs. At the time of this writing, governors in 11 states have issued executive orders governing AVs. Executive orders issued unilaterally by a president or governor have the force of law. Nonetheless, later presidents and governors can undo them simply by issuing new executive orders. As of 2020, all but 10 states had either legislation or executive orders governing AVs.[3]

3.3 California's Autonomous Vehicles Statute

As mentioned above, California's main source of law governing AVs is the bill known as SB 1298, which was codified at Section 38750 of the California Vehicle Code. SB 1298 was not the first AV-enabling law, but it has had considerable impact given that many of the companies testing AVs are located in California. Governor Jerry Brown signed SB 1298 into law on September 25, 2012. The legislature foresaw the promise of AV technology, stating that AVs "offer significant potential safety, mobility, and commercial benefits for individuals and businesses in the state and elsewhere."[4]

Section 38750 defines "autonomous technology" as "technology that has the capability to drive a vehicle without the active physical control or monitoring by a human operator."[5] An "autonomous vehicle" is "any vehicle equipped with autonomous technology that has been integrated into that vehicle."[6] Vehicles do not become "autonomous vehicles" simply because they have advanced driver assistance systems, such as "electronic blind spot assistance, automated emergency braking systems, park assist, adaptive cruise control, lane keep assist, lane departure warning, traffic jam and queuing assist, or other similar systems that enhance safety or provide driver assistance."[7] The definition of "autonomous vehicles" under the California statute excludes vehicles with such technologies when they "are not capable, collectively or singularly, of driving the vehicle without the active control or monitoring of a human operator."[8]

The statute is broad enough to cover both vehicles originally equipped with autonomous technology or those retrofitted to include autonomous technology. "Manufacturers" of autonomous vehicle technology can be the business or individual "that originally manufactures a vehicle and equips autonomous technology on the originally completed vehicle or, in the case of a vehicle not originally equipped with autonomous technology by the vehicle manufacturer, the person that modifies the vehicle by installing autonomous technology to convert it to an autonomous vehicle after the vehicle was originally manufactured."[9]

The California statute authorizes testing of AVs on public roads. As a condition of testing, the manufacturer must ensure.

- the AV is being operated by a driver who possesses the proper class of driver's license for the vehicle being operated;
- the operator is an employee, contractor, or other designee of the manufacturer;
- the manufacturer has insurance, a bond, or self-insurance in the amount of $5 million and has proved such coverage to the California Department of Motor Vehicles (DMV); and
- the DMV has approved the manufacturer's application to test the AV.[10]

The California statute also enables the operation of AVs on public roads, separate from testing. In the application to allow a manufacturer's AVs to operate on public roads, the manufacturer must certify.

- the AV has a mechanism to engage or disengage the autonomous technology, which is easily accessible by the operator;
- the AV has a visual indicator inside the cabin to indicate when the autonomous technology is engaged;
- the AV has a system to alert the operator of failures of the autonomous technology while it is engaged, which either requires the operator to take control or causes the vehicle to come to a complete stop;
- the AV allows the operator to take control of the AV in multiple manners via the brake, accelerator, steering wheel, or other means, and alerts the operator that the autonomous technology has been disengaged;
- the AV's autonomous technology meets Federal Motor Vehicle Safety Standards for the vehicle's model year and all other safety standards and performance requirements in state or federal laws or regulations;
- the AV has a separate mechanism to capture and store sensor data for at least 30 s before a collision occurs between the AV and another vehicle, object, or individual while the autonomous technology is engaged; the data must be captured and stored on a read-only format until the data is extracted by an

external device; and the manufacturer is required to preserve accident data for 3 years after the date of collision;

- the manufacturer has tested the autonomous technology on public roads in accordance with Department of Motor Vehicle regulations; and
- the manufacturer has $5 million available under insurance, a surety bond, or self-insurance under Department of Motor Vehicle regulations to cover liability claims.[11]

The DMV may establish an application fee imposed on manufacturers wishing to submit an application.[12]

The California statute calls for the DMV to issue regulations governing insurance requirements, application processes, safety standards, and any other requirements that the Department determines are necessary for the safe operation of AVs on public roads after consulting with the California Highway Patrol.[13] Regulations for safe operation may include limits on the aggregate number of AVs on public roads; special registration rules; new license requirements for operators; and rules for revocation, suspension, or denial of any license or approvals under the statute.[14]

The DMV must approve a manufacturer's application for allowing operation of its AVs if it finds that the applicant manufacturer submitted all required information, completed safety testing, and complied with all registration requirements in the statute.[15] Nonetheless, the DMV may impose additional requirements on manufacturers seeking approval for AVs capable of operating without the presence of a driver inside the vehicle, including requiring the presence of a driver in the driver's seat if the DMV believes such a requirement is necessary to ensure safe operation on public roads.[16]

The statute includes two other important sections. First, in accordance with Article VI of the U.S. Constitution, the statute recognizes that federal regulations issued by the National Highway Traffic Safety Administration will supersede with any state law or regulation.[17] Federal law preempts conflicting state laws under Article VI, and the statute recognizes this principle. Second, AV manufacturers must provide a written disclosure to any purchaser of an AV that describes what information is collected by the vehicle.[18]

3.4 California's Regulations Governing Autonomous Vehicles

In February 2018, California's Office of Administrative Law approved regulations issued by the Department of Motor Vehicles governing the testing and deployment of AVs.[19] These regulations implement, interpret, and provide detail to expand upon the California autonomous vehicle statute. The regulations forbid AVs from operating in autonomous mode on public roads unless operations are permitted under the California statute and regulations.[20] Definitions in the regulations provide additional detail about what is "autonomous mode" and the "dynamic driving task" being performed by AV technology.[21] Significantly, a manufacturer cannot claim that a test vehicle with autonomous technology is not an "autonomous test vehicle" just because the manufacturer placed a driver in the vehicle for safety; the presence of a human in the vehicle has no bearing on whether the vehicle is an "autonomous test vehicle."[22]

California's regulations cover both testing of AVs and "deployment" of AVs on public roads in California. "Deployment" refers to selling AVs for use on public roads for normal, non-testing purposes. ▶ Section 3.4.1 covers requirements for testing AVs. ▶ Section 3.4.2 discusses requirements for deploying AVs.

3.4.1 California's Regulations Governing Testing of Autonomous Vehicles

The regulations place four requirements on testing AVs on California public roads:

- The business conducting the testing must be the manufacturer.
- The operator must be working for the manufacturer as an employee, contractor, or other person designated by the manufacturer who has been certified by the manufacturer as competent and authorized to operate the vehicle.
- The manufacturer must have $5 million in coverage via insurance, a bond, or a certificate of self-insurance.
- The manufacturer must have received a testing permit from the DMV.[23]

The regulations provide additional detail concerning requirements for satisfactory financial responsibility, insurance, surety bonds, and self-insurance.[24] Manufacturers must specifically identify each AV being tested to the DMV.[25]

The regulations contain special rules for manufacturers wishing to test AVs "capable of operating without the presence of a driver inside the vehicle on public roads in California"—so-called "driverless vehicles." (Passengers may be present.) A manufacturer seeking to test driverless AVs must apply for a special permit for driverless vehicles.[26] Without such a permit and without testing the AV under controlled conditions that simulate public roads, manufacturers must not test driverless AVs.[27] Some vehicles are not eligible for testing: trucks, trailers, buses, motorcycles, and vehicles with a gross weight of 10,001 or more pounds.[28] Light-duty trucks are an exception; they may be tested.[29] The regulations permit the DMV to refuse, suspend, or revoke a permit for AV or driverless AV testing.[30]

To be qualified as a test driver, an individual must meet these requirements:

- The manufacturer must have identified the individual to the DMV.
- The manufacturer must have certified to the DMV that the individual has been licensed as a driver for 3 years who has not had more than one violation point; was not at fault in an injury accident; and was not convicted, suspended, or revoked because of driving while under the influence of alcohol or drugs.
- The driver must have completed the manufacturer's test driver training program.[31]

While driving, test drivers must meet the following requirements:

- They must be in the immediate physical control of the AV or actively monitoring the AV's operations and taking over immediate physical control (which seems to exclude remote control).
- They must be employees, contractors, or designees of manufacturers.
- They must obey the California Vehicle Code and local regulations, whether the AV is in autonomous mode or conventional mode, except when necessary for safety.
- They must be capable of safely operating the AV in all conditions, including by knowledge of the AV's limitations.[32]

To test driverless AVs, manufacturers must, among other things, meet the following conditions:

- The manufacturer must provide a notice to local officials with certain details about the testing.
- The manufacturer must provide a monitored two-way communications link between a remote operator and passengers in the AV.
- If the AV is involved in a collision, there must be a way to provide driver's license and insurance information;
- The AV must meet Federal Motor Vehicle Safety Standards.
- The manufacturer must certify that the AV meets the description of an Society of Automotive Engineers' (SAE) level 4 or 5 automated driving system.
- The manufacturer must inform the DMV of the intended operation design domains of the AV.
- The manufacturer must provide a copy of a plan that it will make available to law enforcement agencies and first responders on how to interact with the AV in emergency and traffic enforcement situations.
- The manufacturer must maintain a training program for remote operators and certifies that all remote operators have completed training and have an appropriate license for the type of vehicle.

- The manufacturer must have publicly disclosed an assessment demonstrating its approach to achieving safety, which must also be provided to the DMV.[33]

Notably, the regulation also calls for the manufacturer to "disclose to any passenger in the vehicle that is not an employee, contractor, or designee of the manufacturer what personal information, if any, that may be collected about the passenger and how it will be used."[34] This regulation addresses the privacy concerns of passengers.

The regulations impose reporting requirements on manufacturers testing AVs. In specific, manufacturers must report to DMV collisions on California public roads that result in bodily injury, death, or property damage within 10 days of the collision.[35] Manufacturers of driverless AVs must also provide the DMV with annual reports concerning incidents in which the autonomous mode disengaged and the circumstances of disengagement.[36]

3.4.2 California's Regulations Governing Deployment of Autonomous Vehicles

▶ Section 3.4.1 covers the testing of AVs. This section discussed the "deployment" of AVs on public roads in California.

» "Deployment" means the operation of an autonomous vehicle on public roads by members of the public who are not employees, contractors, or designees of a manufacturer or for purposes of sale, lease, providing transportation services or transporting property for a fee, or otherwise making commercially available outside of a testing program.[37]

California does not permit the deployment of AVs on California's public roads except as permitted under California's autonomous vehicles statute discussed in ▶ Sect. 3.3 and the regulations discussed in this section.[38]

The California autonomous vehicle statute requires manufacturers of vehicles operated on public roads to have $5 million to cover potential liability claims either through insurance, a surety bond, or self-insurance [39] The regulations provide additional detail. A manufacturer must demonstrate the same kind of financial responsibility as a manufacturer undertaking testing operations: either proof of a surety bond, insurance, or self-insurance meeting the same standards that a manufacturer testing AVs must meet.[40]

At the heart of the deployment regulations is the section covering the process to apply for a permit for post-testing deployment of AVs on public roads. California does not permit a manufacturer to allow deployment of its AVs on public roads in California until it has submitted, and the DMV has approved, an Application for a Permit to Deploy Autonomous Vehicles on Public Streets.[41] The application form requires manufacturers to.

- Identify the operational design domain in which AVs are designed to operate and certify that they cannot be used outside of that domain (which would be helpful for SAE level 4 vehicles).
- Identify weather and environmental conditions that the AVs are designed to be incapable of operating in or that would make operations unreliable.
- Describe how an AV would react when it moves outside of its operational design domain or encounters the weather or environmental conditions limiting its operations.
- Provide a filing fee.
- Certify that the AV has an autonomous technology data recorder that would record events of at least 30 s before a collision.
- Certify that the AV and the autonomous technology comply with all applicable Federal Motor Vehicle.
- Safety Standards or have an exemption.
- Certify that the AV detects and responds to roadway situations in compliance with the California Vehicle Code and local regulations, except where necessary to enhance the safety of occupants or other road users.
- Certify that it will make updates to the autonomous technology to keep up with changes in the Vehicle Code, local regulations, and mapping information, and will notify owners of the availability of the updates as well as instructions on accessing updates.
- Certify that the AV meets appropriate and applicable industry standards to defend against, detect, and respond to cyberattacks, unauthorized intrusions, or false vehicle control commands.
- Certify that the manufacturer has completed testing and is satisfied that the AVs are safe for deployment on public roads in California.[42]

For manufacturers of driverless AVs, manufacturers must also certify:

- A communications link is present between the AV and any remote operator to provide assistance if the AV experiences failures that would endanger the passengers or other road users.
- The AV has the ability to convey owner or operator information if a crash occurs or it is needed by a law enforcement offer.
- The AV complies with Federal Motor Vehicle Safety Standards or is exempt from them if the AV has no manual controls (i.e., a steering wheel, brake, and accelerator).[43]

Manufacturers will also need to explain their plan for educating consumers or end users on the operation of the AV, restrictions on use, responsibilities of the operator and manufacturer, and how the manufacturer will provide additional education. They will also need to provide a copy of educational materials to first responders.[44] In addition, manufacturers will have to tell DMV how the AV meets the SAE definition of level 4 or level 5. For level 3 vehicles, the manufacturer will have to explain how the vehicle will come to a safe stop if the driver does not or is unable to take control when there is a technology failure.[45]

Finally, manufacturers' applications must:

- Include a law enforcement interaction plan similar to the one needed for manufacturers testing AVs;
- Include a copy of the manufacturer's privacy notice;
- Certify that the AV satisfies all of the safety requirements in the California autonomous vehicle statute in Vehicle Code Section 3750(c)(1);
- Certify that it has complied with its responsibility to register with the National Highway Traffic Safety Administration and that it is aware of its responsibility to comply with federal motor vehicle safety requirements;
- Include a summary of its testing program; and
- Include a copy of any assessment demonstrating its approach to achieving safety.[46]

Manufacturers must report safety-related defects in their autonomous technology in compliance with federal law.[47]

The DMV is entitled to reject any application that violates the law if the manufacturer creates a safety risk.[48] Likewise, the DMV may suspend or revoke a permit for various violations, such as the failure to maintain financial responsibility, incorrect or misleading information on an application, failure to report changes in information to the DMV, other violations of the deployment regulations, or safety concerns.[49] Once an application is revoked or suspended, manufacturers must notify all owners and cease further deployments until the issues causing the revocation or suspension are resolved.[50]

The deployment regulations also include two consumer protection sections. First, the manufacturer must provide a written privacy disclosure to an operator of an AV or passengers in a driverless of AV describing the personal information collected by the autonomous technology that is not necessary for safe operations of the AV and how the personal information will be used. Alternatively, the manufacturer can opt to anonymize all such personal information.[51] If such personal information is not anonymized, the manufacturer must obtain written consent of an owner or lessee to its privacy practices.[52] A manufacturer must not deny use of an AV to any person just because the person does not provide such approval.[53] Presumably, a manufacturer must not collect personal information from a non-consenting owner or lessee or anonymize the personal information.

Second, the regulations address the possibility of manufacturers overselling their technology and deceiving consumers about their vehicles' capabilities. This section would cover manufacturers saying that their vehicles are autonomous when in fact they do not meet California standards. The regulations forbid a manufacturer from advertising a vehicle as autonomous unless it meets California's statutory and regulatory standards.[54]

3.5 Federal Laws Affecting Autonomous Driving

Because the U.S. is a federal republic, the law that applies to any given situation can be some combination of U.S. federal law and state law. Accordingly, manufacturers testing and facilitating operations of AVs on public roads in California need to consider federal as well as state laws. The initial approach toward AVs in the Department of Transportation involved issuing (non-binding) guidance to manufacturers. In September 2016, the Obama Administration's Department of Transportation (DOT) leadership and the National Highway Traffic Safety Administration (NHTSA) issued the nation's first automated vehicle guidelines.

DOT, under the Trump Administration, updated the 2016 guidance with three additional guidance documents in 2017, 2018, and 2020.[55] Before the Trump Administration ended, the DOT issued a comprehensive plan for AVs.[56] In general, the Trump-era guidance documents placed a larger emphasis on voluntary safety measures, in contrast to the 2016 guidance, which suggested a greater willingness to consider binding requirements at the appropriate time.

With the onset of the Biden Administration in January 2021, however, we can anticipate another revision of policy, just as the Trump Administration changed the Obama Administration's approach. As of this writing, the Biden Administration's DOT has not yet issued a revamped set of guidance. Moreover, Congress has yet to pass AV legislation, despite bipartisan negotiations.

The Federal Motor Vehicle Safety Standards (FMVSS)[57] are and will be a key source of requirements for AV manufacturers. AVs, as vehicles, must meet the current general FMVSS that apply to all vehicles. Nonetheless, the DOT is now considering new FMVSS that would apply to AVs. On December 3, 2020, the DOT issued an advanced notice of proposed rulemaking, seeking comments from the public leading to a proposed framework for a specialized set of FMVSS that would apply to automated driving systems (AVs).[58] With the change of administration, the status of DOT's work on new FMVSS is still unclear.

3.6 California Product Liability Law

Product liability law is another method of "regulating" AVs. Product liability law does not create mandates for sellers[2] of products to follow. Rather, the law of product liability imposes financial liabilities for companies that sell defective products. Concerns about liability from lawsuits deter careless behavior by sellers and promote safe processes to design and manufacture products. To date, potential product liability is perceived as one of the greatest, if not the greatest, challenge to the development and sales of AVs.

Following an AV accident, an injured plaintiff could bring a product liability suit against any manufacturer, distributor, or seller of the vehicle. Typical claims for plaintiffs seeking damages for bodily injury or property damage from an accident are "strict product liability," "negligence," and "breach of warranty." Other possible claims for product defects include fraud and unfair or deceptive trade practices.

"Strict product liability," or "strict liability" for short, refers to a claim against a seller that does not require a plaintiff prove "fault" on the part of the defendant. Instead, the analysis focuses on whether the product was "defective" in some way. A plaintiff may assert that the product was defective in its design, the product was defective in the way it was manufactured, and/or that the defendant failed to provide adequate warnings or instructions to the users of the product. Of greatest concern for AV litigation are design defect and failure to warn claims.

For design defect cases, a "trier of fact" (jury or judge)[3] may find a defendant seller liable under one or both of the following tests:

- The consumer expectation test, which asks the trier of fact to determine if a product performed as safely as an ordinary consumer would have expected it to perform.
- The risk–benefit test, which asks the trier of fact to weigh the risks of the product's design against its benefits.

The tests are not mutually exclusive and both could apply to a given case, depending on the facts.

---

2   "Sellers" may be manufacturers, distributors, or sellers, which could in theory include any party in the chain starting from a raw material provider to a component part manufacturer, the manufacturer of the final product, the distributor, and finally the retailer that sold the product to the customer who operated or used it.

3   In California, plaintiffs have the right to a trial by a jury in product liability cases. Nonetheless, if the parties waive the right to a jury, the parties may elect to have a judge render a verdict instead of a jury in what is called a "bench trial." The "trier of fact" refers to a jury or, in the case of a bench trial, the trial judge.

To prevail under the consumer expectation test against an AV seller, the plaintiff must prove the following elements:

- The defendant manufactured/distributed/sold the AV or a component.
- The AV or component did not perform as safely as an ordinary consumer would have expected it to perform when used (or misused in an intended or reasonably foreseeable way).
- The plaintiff was harmed.
- The AV or component's failure to perform safely was a substantial factor in causing the plaintiff's harm.[59]

The consumer expectation focuses on what a consumer would expect. If a consumer purchased an AV, and, upon leaving the showroom, it immediately ran off the road for no apparent reason, such an accident would be an example of not performing as safely as expected.

The risk–benefit test focuses on the benefits and risks of a particular design. Under this test, the plaintiff can establish a claim by proving:

- The defendant manufactured/distributed/sold the AV or a component.
- The plaintiff was harmed.
- The AV or component's design was a substantial factor in causing the plaintiff's harm.[60]

Once these three facts are established, the burden shifts to the defendant to prove the benefits of the product's design outweigh the risks of the design. In making this decision, the trier of fact must analyze factors such as.

- The gravity of the potential harm resulting from the use of the product.
- The likelihood that this harm would occur.
- The feasibility of an alternative safer design at the time of manufacture.
- The disadvantages of the alternative design.[61]

This test may prove to be the more relevant one for determining design defects in AV cases because of the complexity of AV design and the unlikelihood of obvious dangerous AV behavior, such as running off the road for no apparent reason.

If a plaintiff asserts a strict liability claim based on a failure to warn a driver or remote operator, the plaintiff must prove.

- the defendant manufactured/distributed/sold the AV or a component;
- the AV had potential risks or side effects that were known or knowable in light of scientific knowledge that was generally accepted in the scientific community at the time of manufacture/distribution/sale;
- the potential risks or side effects presented a substantial danger when the AV is used or misused in an intended or reasonably foreseeable way;

- ordinary consumers would not have recognized the potential risk or side effect;
- the defendant failed to adequately warn or instruct of the potential risk or side effect;
- the plaintiff was harmed; and
- the lack of sufficient instructions or warnings was a substantial factor in causing the plaintiff's harm.[62]

A component part manufacturer is not liable for an accident in the final product if the part was not defective at the time it left the hands of the part manufacturer. Nonetheless, a manufacturer, distributor, or supplier of a component part may be liable if the component part was defective or the part manufacturer integrated the component into the design of the end product and the integration of the component caused the end product to be defective, thereby causing the harm.[63]

As an alternative to strict liability, a plaintiff may assert a claim for negligence. Negligence is a harder claim to prove because the plaintiff must prove a certain "fault" on the part of the defendant. A plaintiff asserting a negligence claim must prove.

- the defendant designed, manufactured, or supplied the AV;
- the defendant was "negligent" in designing, manufacturing, or supplying the AV;
- the plaintiff was harmed; and
- the defendant's negligence was a substantial factor in causing the plaintiff's harm.[64]

"Negligence" refers to the defendant failing to use the amount of care that a reasonably careful business "would use in similar circumstances to avoid exposing others to a foreseeable risk of harm."[65] A trier of fact should balance what the defendant "knew or should have known about the likelihood and severity of potential harm from the product against the burden of taking safety measures to reduce or avoid the harm."[66]

It is possible for a plaintiff to assert that a defendant's failure to warn was negligent. Following a failure to warn, the plaintiff would have to prove the defendant knew or reasonably should have known that the product was dangerous or was likely to be dangerous when used or misused in a reasonably foreseeable manner, and that the defendant knew or reasonably should have known that users would not realize the danger.[67]

Finally, a plaintiff may contend that a defendant seller breached a warranty, in which a product failure caused the plaintiff harm. Warranties may arise from written or oral statements made to the purchaser ("express warranties") or may be implied as a matter of law. Setting aside implied warranties based on a seller's assurance that a product is intended for a particular purpose, the most common implied warranty is the "implied warranty of merchantability."

To prevail in an express warranty claim involving personal injury or property damage, the plaintiff must prove.

- the defendant gave the plaintiff a written or oral warranty regarding the AV;
- the AV did not perform as stated or promised;
- the plaintiff was harmed; and
- the failure of the AV as represented was a substantial factor in causing the plaintiff harm.[68]

If a plaintiff asserts a violation of the implied warranty of merchantability, the plaintiff must prove.

- the plaintiff bought the AV from the defendant (or was in the chain of purchasers leading to the defendant);
- at the time of purchase, the defendant was in the business of selling AVs;
- the product was not of the same quality as those generally acceptable in the trade or was not fit for the ordinary purpose of AVs;
- the plaintiff was harmed; and
- the failure of the product to have the expected quality was a substantial factor in causing the plaintiff's harm.[69]

Note that for personal injury and property damage cases, strict liability has almost entirely superseded implied warranty claims.

Plaintiff purchasers can also recover financial losses arising from the diminution in the value of the AV caused by the breach of warranty, although they must provide notice to the defendant within a reasonable time that the product did not have the expected quality and, in the case of an express warranty, the plaintiff must prove that the defendant failed to repair the AV as required by the warranty.[70]

3.7 Conclusions

California, as a leader among the states in the U.S., has established a statute and regulations that both enable autonomous vehicle testing and operations on public roads. However, the statute and regulations establish conditions for testing and operations and establish vehicle safety requirements, driver and remote operator competence requirements, and reporting requirements. Notably the regulations also include a requirement of fair privacy practices regarding the personal information of AV drivers or passengers in driverless AVs. Federal law provides a base of Federal Motor Vehicle Safety Standards to provide basic safety measures. There may be more detailed standards governing AVs in the next several years.

In addition to enabling the testing and use of AVs, California law also deters dangerous or careless practices by manufacturers through product liability law. Product liability can arise from defects in design, defects in manufacturing, or a failure to warn drivers of important safety information. Manufacturers can limit and manage their risk through insurance programs.

Legal developments will likely continue in California. Other states will begin to develop their own AV laws. Finally, federal or uniform state laws may be necessary to provide uniformity among the states. If the states and federal government create a comprehensive set of laws to regulate AVs and establish a path toward eventual deployment, state and federal policy will create an environment to bring lifesaving AV technology to the market and, at the same time, protect the public.

■ **Notes**

1. *Greenman v. Yuba Power Products, Inc.*, 59 Cal.2d 57, 27 Cal. Rptr. 697, 377 P.2d 897 (1963).
2. *See* United States Constitution, Article VI.
3. *See* Ref. [1].
4. SB 1298 § 1(a) (2012), ► https://leginfo.legislature.ca.gov/faces/billNavClient.xhtml?bill_id=201120120SB1298.
5. Cal. Vehicle Code § 38750(a)(1).
6. *Ibid.* § 38750(a)(2)(A).
7. *Ibid.* § 38750(a)(2)(B).
8. *Ibid.*
9. *Ibid.* § 38750(a)(5).
10. *Ibid.* § 38750(b)–(c).
11. *Ibid.* § 38750(c)(1)–(3).
12. *Ibid.* § 38750(h).
13. *Ibid.* § 38750(d)(1)–(3).
14. *Ibid.* § 38750(d)(2).
15. *Ibid.* § 38750(e)(1).
16. *Ibid.* § 38750(e)(2).
17. *Ibid.* § 38750(g).
18. *Ibid.* § 38750(h).
19. Cal. Code of Regulations title 13, § 227.00(a).
20. *Ibid.* § 227.00(b).
21. *Ibid.* § 227.02.
22. *Ibid.* § 227.02(b)(3).
23. *Ibid.* §§ 227.04, 227.18, 227.26.
24. *Ibid.* §§ 227.06–227.14.
25. *Ibid.* § 227.16.
26. *Ibid.* § 227.38.
27. *Ibid.* § 227.18.
28. *Ibid.* § 227.28.
29. *Ibid.* § 227.28(a)(5).
30. *Ibid.* §§ 227.40–227.42.
31. *Ibid.* § 227.34.
32. *Ibid.* § 227.32.
33. *Ibid.* § 227.38.
34. *Ibid.* § 227.38(h).

35. *Ibid.* § 227.48.
36. *Ibid.* § 227.50.
37. Cal. Code of Regulations title 13, § 228.02(c).
38. *Ibid.* § 228.00(b).
39. Cal. Vehicle Code § 38,750(c)(3).
40. Cal. Code of Regulations title 13, § 228.04(a).
41. *Ibid.* § 228.06(a).
42. *Ibid.* § 228.06(a)(1)–(a)(11).
43. *Ibid.* § 228.06(b).
44. *Ibid.* § 228.06(c)(1).
45. *Ibid.* § 228.06(c)(2).
46. *Ibid.* § 228.06(c)(3)–(c)(7), (d).
47. *Ibid.* § 228.12.
48. *Ibid.* § 228.16.
49. *Ibid.* § 228.20(a)–(b).
50. *Ibid.* § 228.20(c)–(d).
51. *Ibid.* § 228.24(a).
52. *Ibid.* § 228.24(b).
53. *Ibid.* § 228.24(c).
54. *Ibid.* § 228.28.
55. For a timeline of the creation of these guidance documents, see Ref. [2].
56. See Ref. [3].
57. Code of Federal Regulations, title 47, part 571.
58. See Ref. [4].
59. Judicial Council of California, Civil Jury Instructions [CASI] No. 1203, at 718 (May 2020).
60. CASI No. 1204, at 723.
61. *Ibid.*
62. CASI No. 1205, at 728.
63. *See* CASI No. 1208, at 742.
64. CASI No. 1220, at 745.
65. CASI No. 1221, at 747.
66. *Ibid.*
67. CASI No. 1222, at 750.
68. CASI No. 1230, at 758.
69. CASI No. 1231, at 762.
70. CASI Nos. 1230, 1231, at 758, 762.

References

1. Autonomous Vehicles | Self-Driving Vehicles Enacted Legislation. National Conference of State Legislatures (2020). ► https://www.ncsl.org/research/transportation/autonomous-vehicles-self-driving-vehicles-enacted-legislation.aspx
2. NHTSA is dedicated to advancing the lifesaving potential of new vehicle technologies. U.S. Department of Transportation (2021). ► https://www.nhtsa.gov/technology-innovation/automated-vehicles-safety
3. Automated Vehicles Comprehensive Plan. U.S. Department of Transportation (2021). ► https://www.transportation.gov/sites/dot.gov/files/2021-01/USDOT_AVCP.pdf
4. National Highway Traffic Safety Administration: Framework for automated driving system safety. Fed. Reg. **85**, 78050–78075 (2020)

Open Access This chapter is licensed under the terms of the Creative Commons Attribution-NonCommercial-NoDerivatives 4.0 International License (► http://creativecommons.org/licenses/by-nc-nd/4.0/), which permits any noncommercial use, sharing, distribution and reproduction in any medium or format, as long as you give appropriate credit to the original author(s) and the source, provide a link to the Creative Commons license and indicate if you modified the licensed material. You do not have permission under this license to share adapted material derived from this chapter or parts of it.

The images or other third party material in this chapter are included in the chapter's Creative Commons license, unless indicated otherwise in a credit line to the material. If material is not included in the chapter's Creative Commons license and your intended use is not permitted by statutory regulation or exceeds the permitted use, you will need to obtain permission directly from the copyright holder.

Driver Assistance Requirements from Vehicle Regulations and Consumer Protection

Patrick Seiniger, Andre Seeck, and André Wiggerich

Contents

© The Author(s) 2026
H. Winner et al. (eds.), *Handbook Assisted and Automated Driving*,
https://doi.org/10.1007/978-3-658-45276-6_4

Driver assistance systems can have an impact on traffic safety—desirable impacts as well as undesirable ones. Therefore, boundary conditions exist that, on the one hand, restrict the possible functions (such as the prohibition of lane assist systems which enable hands-free driving), and, on the other hand, require a certain functionality (blind spot information systems, automated emergency braking systems).

These boundary conditions are usually defined as requirements, both from vehicle technology regulations, without whose observance no type approval is possible, and from the criteria of product assessments from consumer protection, for example, the European New Car Assessment Program (Euro NCAP).

The aim of this chapter is to highlight the concepts behind the definition of requirements and to discuss essential requirements with regard to driver assistance and first automated systems. Legal framework conditions for automated driving of higher automation levels are dealt with in a separate chapter (▶ Kap. ID102, ▶ Kap. ID108), but the concepts of requirements definition are basically the same.

In the following, the basics of the vehicle type approval system as well as the concept of vehicle assessments in consumer protection using Euro NCAP as an example are first briefly described. Following this the impact of these demands on explicit or implicit requirements will be derived. Finally an overview of different types of Driver Assistance Systems is given, for which requirements or boundary conditions already exist.

4.1 Integration into the Product Development Process

Usually the product development process is based on the V-model of product development [1]: Over the course of the development of a system, the degree of abstraction of the system properties first decreases (in the left branch for development) and then increases again (in the right branch for verification). At the same time, the objective shifts from questions of system design to validation of the intended system characteristics.

The test procedures in this case are defined in the horizontal levels of the V-model in each case by requirements to be tested for the respective driver assistance system. In the left area of ▢ Abb. 4.1, during product definition and design, test procedures can be used to detail requirements, but also for feasibility considerations. In the right-hand area, the focus is then on verifying the requirements or, at the highest level, on validating the product properties with the acceptance test. At the same level of the V-model, the requirements

for requirements definition (in the left-hand branch) and test procedures (in the right-hand branch) may therefore be similar or the same. Requirements from vehicle technical regulations or standards and guidelines can also be placed in this scheme. On the left-hand side of the V-model, the requirements become more and more specific from top to bottom; on the right-hand side, you will find the tests required to verify the requirements. ▢ Abb. 4.1 shows the V-model with system levels and assigns standards and guidelines.

The highest level shown in ▢ Abb. 4.1 is not part of the V-model. In this level, on the left-hand side, it is examined which type of driver assistance system offers a benefit for road safety and this is compared with any additional risk arising from the system. On the right-hand side, the actual benefit after series deployment of the system in public road traffic is then to be evaluated. The transitions between the stages are fluent: a clear separation is not always possible, especially because a risk assessment requires a concretization of the system design, which is derived from the defined requirements.

4.2 Origin of Boundary Conditions

4.2.1 Type Approval of Motor Vehicles

The type approval of motor vehicles is described in detail in the literature [2]. The purpose of this section is therefore to highlight the basic features of type-approval of high-volume vehicles to the extent necessary for understanding the chapter, without going into special cases.

In principle, all vehicles registered in Europe have a type approval according to the EU Type-Approval Directive 2007/46/EC [3] (Framework Directive until August 31, 2020) or the EU Type-Approval Regulation (EU) 2018/858 [4] (Framework Regulation from September 1, 2020) including their amendments, especially by the so-called General Safety Regulations (GSR) Regulation (EC) No. 661/2009 (GSR I) [5] or Regulation (EU) No. 2144/2019 (GSR II) [6].

However, the amended type-approval directive and regulation do not yet define any concrete technical requirements, but instead refer either to so-called delegated acts of the European Commission or to concretely designated versions of UN regulations (preferred, because requirements can thus be harmonized worldwide).

Ultimately, the amended documents then contain a list, for example, per vehicle category, of mandatory or optional requirements to be met for vehicle type approval. It is thus clear that the so-called UN regulations will play an increasingly important role for driver assis-

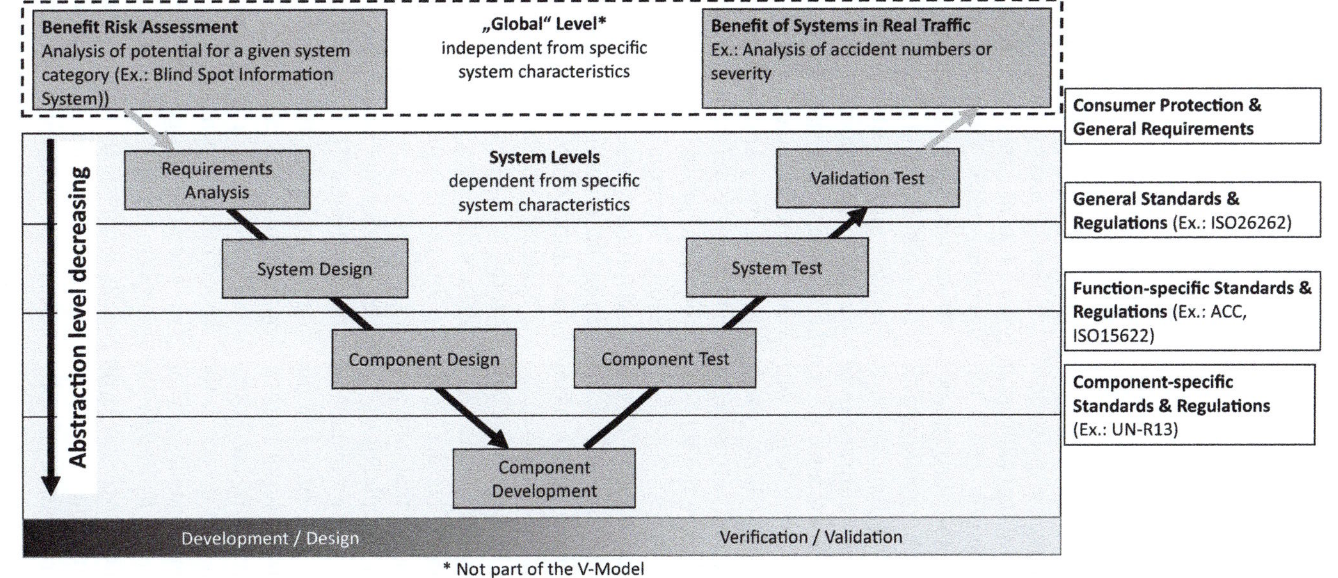

Abb. 4.1   Classification of standards and guidelines using a modified V-model

tance systems. Automated systems can serve as an example of optional regulations (so-called "if fitted" regulations).

The framework regulation also allows exemptions in certain cases for new technologies or new concepts (Artcle 39). In this context, deviations from the specifically defined requirements are made in order to promote innovations, provided that it can be demonstrated that road safety and environmental protection are not impaired.

4.2.2 Market Surveillance

Previously, under the Framework Directive 2007/46/EC, once a type approval had been granted by one member state of the EU, it could only be questioned and reviewed by another member state with high hurdles. The new Framework Regulation (EC) No. 2018/858 introduces the system of so-called market surveillance. It now allows any vehicle type-approved granted on the basis of the Framework Regulation to be inspected by the market surveillance authorities of the member states or the Joint Research Center of the European Commission with regard to compliance with the requirements. It is not yet fully clear how exactly market surveillance will be implemented. However, it has the potential to fundamentally change the type-approval system.

Without market surveillance, generically formulated test cases (e.g., "braking from any speed," as opposed to precisely specified test conditions "braking from X km/h with tolerance Y") are a disadvantage, because it is then up to only one technical service and one type-approval authority to verify compliance with the requirements, and this one technical service could possibly select parameter combinations that are particularly easy to meet without further specifications.

With market surveillance, generically formulated test cases are advantageous because not only the initial technical service and the approval-issuing authority check compliance with the requirements, but potentially many other member states can also do so on the basis of possibly completely different test cases. The vehicle manufacturer is therefore forced to design the system robustly. Optimization for the "one" test is no longer possible without the risk of sanctions.

This requires a rethink in the formulation of vehicle technology regulations toward the specification of generally valid requirements (especially in the case of driver assistance systems), which are not limited—as in the past—to a few requirements specified by concrete test cases. The generally valid requirements can be verified by more and generically defined test cases. Because artificial target objects often used in tests (see, for example, ▶ Kap. 9/0300) can by their very nature never

fully represent reality and can lead to malfunctions, this then also includes the creation of opportunities for retesting (in case of fails) on a limited scale with clear definition of criteria for test repetition.

4.2.3 Consumer Protection with Euro NCAP

With the requirements for type-approval, the legislator[1] defines minimum standards that must be met in order for a new vehicle model to gain access to the market. Accordingly, all new vehicles placed on the market must meet the legal requirements. However, these type-approval tests initially say nothing about the differences in safety levels between the various type-approved vehicle models that have gained access to the market by meeting the legal requirements. This is where the task of consumer organizations comes in. Independent tests, for example, crash tests and tests of driver assistance functions, are intended to determine the different safety levels of the type-approved vehicle models and to publish them in a differentiated manner as consumer information. Given this objective of the consumer organizations, it is clear that it would make little sense for a consumer protection test merely to repeat the type-approval tests with their requirements. The poorly differentiated result of such an approach for the approved test vehicles would be a "test passed" rating. Appropriate differentiation of approved products in terms of safety is often made possible by tightening both the test conditions and the evaluation criteria compared to the type-approval test. Furthermore, a consumer test must allow gradual differentiation of the products, whereas the type-approval test only allows a binary differentiation into "pass" or "fail."

In the European New Car Assessment Programme (Euro NCAP), a consumer protection organization under Belgian law as an example of consumer protection organizations [7], the gradual differentiation of test results is achieved by setting an upper and lower performance limit and calculating the rating within these limits by means of a linear interpolation ("sliding scale") based on the test result.

The strategic goal of consumer organizations is to supplement the manufacturer's internal requirements with additional requirements that consumers consider useful. This is done by assigning ratings that increase the value of a vehicle in the eyes of consumers or whose absence would reduce the value. These ratings thus give the abstract commodity of "road safety" a real market value that justifies investment on the

---

1   Vehicle engineering regulations are quite predominantly not laws enacted by the legislative branch, but regulations enacted by the executive branch.

part of the vehicle manufacturer. With the comparative product assessment, Euro NCAP thus forms a "marketplace for vehicle safety" in a figurative sense. Since this process has a strong economic significance, it is very important that the evaluation process is transparent and oriented toward road safety.

In order to promote the market entry of certain safety systems, Euro NCAP uses various instruments. As a matter of routine, new vehicles are evaluated for their safety in the so-called "rating" according to known criteria, which are usually amended and tightened every 2 years. For newly identified issues, Euro NCAP performs specific test campaigns that are based on a rating of their individual evaluation criteria and implementation and either remain a one-time campaign or are later converted into a fixed part of the rating. For example, other vehicle categories or new technical systems are examined that were not previously part of the rating.

In addition to the established "rating" system, a so-called "grading" system for continuous driver assistance and automation was introduced in 2018. This grading system will be successively expanded. It is based on the following three different degrees of automation (cf. on classification ▸ Kap. 2/ID0108):

- Assisted mode for combined lateral and longitudinal guidance (complies with SAE level 2).
- Automated mode (complies with SAE level 3).
- Autonomous mode (complies with SAE level 4 and 5).

In addition to this categorization of the degree of automation, the area of application, i.e., the Operational Design Domain (ODD), is also considered and suitable test and evaluation criteria are applied depending on whether assistance or automation is involved.

In order to be able to take new safety technologies into account as quickly as possible in terms of consumer information, Euro NCAP introduced the Beyond NCAP process in addition to the rating and grading process mentioned above. According to this process, a vehicle manufacturer can describe its new safety system by means of a standardized dossier. If this is positively assessed by a jury of experts, the manufacturer receives a "Euro NCAP Advanced Reward" for the safety technology, i.e., a safety award with which it can advertise the technology as having been tested by independent experts. In addition to the goal of informing consumers as early as possible about new safety technologies, Euro NCAP also pursues the goal itself of obtaining findings on new safety technologies as quickly as possible in order to accelerate the further development of the rating system.

When the ambitious Euro NCAP roadmaps are implemented, in which the requirements are to be tightened for the next 5 years and new test and assessment protocols are to be included for the rating process, the Beyond NCAP process with the "Euro NCAP Advanced Reward" will become less important.

4.3 Formulation of Requirements

There are two basic ways of specifying requirements, namely, the explicit specification of generally valid requirements, for example, in a regulation document, and the implicit specification via test definitions and pass criteria. While explicitly specified requirements apply to all cases that are not excluded in the definition (e.g., "Speed reduction always greater than 20 km/h when the road is dry"), requirements specified implicitly via test procedures initially apply only to the specific test case.

In practice, attempts are made to compensate for this limitation of the implicit specification by the fact that there are many different test cases and thus requirements apply to a wide range of driving situations.

However, it is still conceivable that driver assistance systems are optimized for specific tests and pass all tests, but show significantly poorer performance with parameters that deviate from the tests, so that it is not always possible to transfer the result to general traffic situations. This is referred to as "single-point" optimization, which is not aimed at real traffic and accidents, but only at passing the (for example) Euro NCAP test cases.

4.3.1 Implicit Specification of Requirements Through Precise Test Criteria

Requirements are implicit if the requirements are not specified in general terms, but result from the pass criteria of a limited test program. For this the precise description of the tests and the respective pass criteria is necessary.

For the selection of a method of the requirement definition it is important to know the boundary conditions. For the use of this method of implicit specification, the system or vehicle to be tested can be regarded as a "black box" whose internal decision-making procedures are unknown. Verification by means of tests is therefore possible in principle even without manufacturer knowledge, as it is applied in consumer protection, for example. Moreover, the method is easy to practice (tests are predefined and do not have to be adapted to the specific product).

For regions (and jurisdictions) where vehicles are type-approved, it is common practice to carry out type-approval tests, for which in the past quite predominantly implicit requirements were specified: Here, some (few) concretely described test cases with results to be achieved are specified, from which it is assumed that the test conditions represent the loads on vehicles or components occurring in practice. In most cases, worst-case conditions are used, as is justified in many cases, especially in classical mechanics (e.g., if the maximum trailer mass is borne by a connecting device, then—with a high degree of probability—all lower trailer masses will also be borne without damage).

However, this method does allow optimization for the test, as has become abundantly clear, especially as a result of the emissions scandal. Other examples are flap exhaust systems, such as those used in the past on motorized two-wheelers. Particularly in the case of driver assistance systems and all other "flexible" systems that are based on programming and that act in complex driving situations, it is difficult to define worst-case situations and create a representative test program with only a few individual tests, also because the worst case is not always clearly identifiable.

Consumer protection organizations such as Euro NCAP define a very large possible number of test cases, of which either all or individual, randomly selected tests are performed and evaluated. This allows a much wider range of product characteristics to be tested. However, the problem of determining the product's properties outside the specific test program remains.

A concrete example is the test definition for pedestrian emergency braking assistance systems by Euro NCAP, shown in ▢ Table 4.1. Test configurations with different impact points (25–75%), different pedestrian speeds (3 km/h as input test, 5 km/h, 8 km/h), a wide range of vehicle speeds (nominally 20 to 60 km/h with a step size of 5 km/h), and impact directions lateral and longitudinal implicitly define the parameter field in which the systems should work.

4.3.2 Explicit Specification of Requirements

The aim of vehicle technology regulations for assistance functions is to ensure effectiveness throughout the relevant operating range and not just for individual worst-case situations.

One approach that is becoming increasingly prevalent in more recent driver assistance regulation documents is the concrete, numerical definition of verifiable requirements. These requirements—unless further restricted—apply comprehensively (in the example: for all driving speeds, for all weather conditions, and so on).

In many cases, the function fulfillment (e.g., emergency brake assist function) is neither physically nor technically possible in every situation. Comprehensible restrictions are then specified (in the example: definition of speed reduction for dry road surfaces), in conjunction with the requirement that the corresponding function must not exhibit any unjustified switching of the control strategy even outside the restrictions (in the example: at least the same braking intervention times of an emergency brake assistance system on wet road surfaces as on dry road surfaces).

This definition of requirements is supplemented by fixation of (many) test cases, which in case of doubt can be supplemented by further test cases or the verification of all requirements defined in the document. Thus, the requirements for a system do not longer result from a single test with a pass criterion, but the par-

▢ **Table 4.1** Implicit requirement definition using the example of the Euro NCAP tests of pedestrian emergency braking assistance, 2018–2021 [8]: overview of the test configurations. A total of 52 tests are described, which are tested up to three times depending on the result

	CPFA-50	CPNA-25	CPNA-75	CPNC	CPLA-50	CPLA-25
Vehicle	20–60 km/h, 5 km/h steps	20–60 km/h, 5 km/h steps	20–60 km/h, 5 km/h steps	20–60 km/h, 5 km/h steps	20–60 km/h, 5 km/h steps	50–80 km/h, 5 km/h steps
VRU	8 km/h	5 km/h	5 km/h	5 km/h (child)	5 km/h	5 km/h
Obstruction				Yes (two vehicles)		
Impact location	Center	25% (right)	75% (left)	Center	Center	25% (right)
Braking or warning	Braking	Braking	Braking	Braking	Braking	Warning
VRU direction	Crossing, from the left (in cont. Europe)	Crossing, from the right	Crossing, from the right	Crossing, from the right	Longitudinal, same direction	Longitudinal, same direction
Day/Night	Day	Both	Both	Day	Both	Both

[Tabellenfußzeile - bitte überschreiben]

agraphs on test fulfillment then generally refer specifically to the paragraphs in which the requirements are defined.

However, the effort required to test the requirements can increase considerably when applying this method, and the evaluation of the extent to which the behavior outside the stated restrictions (in the example: on wet road surfaces) fulfills the requirements probably requires a scientifically based and thus costly individual case analysis in borderline cases.

In some cases, the tests that remain in the regulations are so vaguely defined that technical services, type-approval, and market surveillance authorities must have a deep understanding of possible system limits (for example, in the application of UN Regulation No. 157 on automated lane-keeping systems).

The use of this method of explicit requirements definition is therefore certainly not suitable for all vehicle technology regulations, but it should be used for sub-functions that are most relevant for road safety, and in particular for programmed systems.

Examples of regulation documents written in this way are UN Regulation No. 151 on turning assistance [9], No. 152 on emergency brake assist systems for passenger cars and light commercial vehicles [10], and No. 157 on automated lane-keeping systems [11].

In the quote below from the regulation on emergency braking assistance, a requirement definition consisting of three parts can be seen: first, the formulation of the ideal conditions for which the specified residual speed is to apply, then the paragraph stating that there must be no switching of the control strategy in the case of conditions that do not correspond to the ideal conditions, and finally the actual definition of the residual speed in form of a table, for which interpolation rules are also specified so that no gaps occur in the parameter space.

In sum, this results in a closed definition of requirements for pedestrian emergency braking assistance systems which apply comprehensively.

The following excerpt is paragraph 5.2.1.4 from UN Regulation No. 152:

> "5.2.1.4. Speed reduction by braking demand
> In absence of driver's input which would lead to interruption according to paragraph 5.3.2., the AEBS shall be able to achieve a relative impact speed that is less or equal to the maximum relative impact speed as shown in the following table:
>
> a. For collisions with unobstructed and constantly travelling or stationary targets;
> b. On flat, horizontal and dry roads;
> c. In maximum mass and mass in running order conditions;
> d. In situations where the vehicle longitudinal centre planes are displaced by not more than 0.2 m;
> e. In ambient illumination conditions of at least 1000 Lux without direct blinding of the sensors (e.g. direct blinding sunlight);
> f. In absence of weather conditions affecting the dynamic performance of the vehicle (e.g. no storm, not below 0 °C); and
> g. When driving straight with no curve, and not turning at an intersection.
>
> It is recognised that the performances required in this table may not be fully achieved in other conditions than those listed above. However, the system shall not deactivate or drastically change the control strategy in these other conditions. This shall be Annex 3 of this Regulation."

This paragraph thus precisely defines the expected speed reduction and the boundary conditions that apply to it. It is supplemented by ◘ Table 4.2.

◘ Table 4.2 Maximum relative impact speed (km/h) for M1 vehicle* (Quote from UN Regulation No. 152, paragraph 5.2.1.4. [10]). The requirements are given for two conditions: one condition is the vehicle in running order ("mass in running order"), the other condition is the vehicle laden up to its maximum permissible mass ("maximum mass")

Relative speed (km/h)	Stationary/Moving	
	Maximum mass	Mass in running order
10	0.00	0.00
15	0.00	0.00
20	0.00	0.00
25	0.00	0.00
30	0.00	0.00
35	0.00	0.00
40	0.00	0.00
42	10.00	0.00
45	15.00	15.00
50	25.00	25.00
55	30.00	30.00
60	35.00	35.00

4.4 Specific Requirements from Vehicle Regulations

4.4.1 UN-Regulations

The European Union and many other countries are contracting parties to the 1958 Agreement on the Harmonization of Vehicle Regulations [12] Known through the regulations developed under this agreement, the so-called UN regulations (numbered consecutively from UN R 1 to the latest regulation that has entered into force). These regulations, which currently number more than 150, are the result of transparent negotiations between representatives of the contracting states and non-governmental organizations (e.g., associations of the vehicle industry), which can be traced without any gaps on the United Nations homepage. The United Nations itself are involved only in an administrative capacity. ◘ Table 4.3 provides an overview of the regulations relevant to driver assistance systems.

In detail:

- Regulation 79 specifies requirements for steering systems. Until revision 2 of this regulation, it specified that steering interventions by assistance functions above a driving speed of 12 km/h were only permitted under certain circumstances. Continuously automated steering was prohibited. As of revision 3 of the regulation, requirements were therefore introduced into the regulation that specify how steering interventions are permitted.
- Regulation 130 and Regulation 131 were developed as part of the first General Safety Regulation in 2009; both regulations apply to heavy-duty vehicles. Regulation 130 defines requirements for lane departure warning systems largely along the lines of the relevant ISO Standard 17361:2007; Regulation 131, also identical as Delegated Act 347/2012 of the European Union, defines mandatory requirements for emergency brake assist systems for the first time. In the first version, the requirements are still implicitly specified by defining a test situation (first-person vehicle driving at 80 km/h toward a stationary or moving target) and test-related performance criteria (impact speed 0 km/h for moving target, 10–20 km/h for stationary target). An adaptation of the regulation to the state of the art is in progress.
- Regulation 140 implicitly defines requirements for electronic stability control based on a driving test with sinusoidal steering excitation with intermediate stops ("sine-with-dwell").
- Regulation 151 describes requirements for blind spot information systems to avoid accidents between trucks turning right and bicycles. It is an informing and warning system, i.e., without intervention in the driving dynamics of the truck. The requirements are still specified by pass criteria for test cases; however, the pass criteria are already generically described so that any test cases can be performed.
- Regulation 152 for emergency braking systems of passenger cars and light commercial vehicles is the first regulation in which the concept of direct specification of requirements (see Sect. 4.3) has been

◘ **Table 4.3** Overview of vehicle regulations that are relevant for advanced driver assistance functions

Regulation	Requirements for driver assistance systems	Comment
UN R79	Steering systems, driver assistance systems with steering interventions	Contains requirements for steering assistance function up to SAE level 2. Currently there are discussions whether it would be helpful to move those to a regulation of their own
UN R130	Lane departure warning systems	For vehicles of class M2/N2 and above (coaches/buses with more than 9 seats including driver's seat, goods vehicles with more than 3.5 tons of gross vehicle mass)
UN R131	Advanced emergency brake assist systems	For vehicles of class M2/N2 and above
UN R139	Conventional brake assist functions	For vehicles of class M1/N1, for conventional brake assist functions that judge situation criticality based on the brake pedal activation only
UN R140	Electronic stability control	
UN R149	Road illumination devices	Contains requirements for lighting assistance functions
UN R151	Blind spot information systems	For vehicles of class M2/N2 and above
UN R152	Advanced emergency brake assist systems Car-Car, Car-Pedestrian, Car-Cyclist	For vehicles of class M1/N1
UN R157	Automated lane-keeping systems	For vehicles of class M1/N1. Contains requirements for SAE Level 3

[Tabellenfußzeile - bitte überschreiben]

implemented. Vehicle manufacturers are therefore forced to design the system robustly. It is not possible to restrict the function when outside of the test conditions also defined in the regulation.

— Regulation 157 permits type-approval of automated steering systems for automation level 3 and higher; this regulates both longitudinal and lateral guidance. The requirements are also explicitly specified—as in Regulation 152—and the corresponding tests are only described in very abstract terms.

4.4.2 Delegated Acts of the European Commission

In cases where there is no pronounced interest in globally harmonized technical specifications outside the European Union or where the preparation of specifications would take too long, the European Commission prepares its own so-called delegated acts.

For capacity reasons, the drafting is usually carried out by a contractor. The draft regulations (mostly ordinances) are then discussed in working groups (e.g., the Motor Vehicle Working Group) and adopted by the EU member states in so-called comitology committees ("Technical Committee on Motor Vehicles" TCMV).

Examples of legal acts in progress are emergency lane-keeping systems, intelligent speed assistance, drowsiness detection, and warning systems.

4.5 Concrete Requirements from Consumer Protection

The European New Car Assessment Programme (Euro NCAP) assesses the driver assistance systems listed in ◘ Table 4.4 in the period 2020 to 2023 (extended assessment period due to the COVID-19 crisis). The eval-

uation criteria are developed by consensus in working groups and then confirmed by the Board. The working groups consist of representatives of the Euro NCAP members, the test laboratories, and representatives of the industry associations.

The individual elements of the vehicle assessment in the rating system at Euro NCAP are divided into four categories: Occupant Protection ("Adult Occupant Protection"), Child Occupant Protection ("Child Protection"), Vulnerable Road User Protection ("VRU Protection," formerly "Pedestrian Protection"), and Safety Assist ("Safety Assist"). The overall rating of a vehicle, zero to five stars, is determined on the basis of threshold values in the individual categories, so that a vehicle must demonstrate good results in each of the categories in order to receive a good rating. Currently, only in the categories of protection of vehicle occupants (adults and children) driver assistance systems are not relevant.

Detailed information on test procedures in Euro NCAP can be found in ▶ Chap. 9.

4.5.1 Grading of Comfort Assistance

Since 2018, Euro NCAP has also been evaluating potentially comfort-enhancing assistance functions that continuously support the driver in the execution of longitudinal and lateral guidance (cf. on classification ▶ Chap. 2) in terms of their safety. The evaluation of these functions is based on a balance principle that compares the degree of support of the assistance function and the driver engagement. In an optimal system, the amount of assistance provided by the system and the active engagement of the driver in the driving task are in balance. This is intended to counteract a false understanding as well as overreliance of the driver in the function. Since the purely supportive systems to be evaluated have system limits due to their

◘ **Table 4.4** Overview of requirements for driver assistance systems in the Euro NCAP "Rating "

Advanced driver assistance system	Introduced into Rating in	Assessment category
Automated emergency braking Systems (AEB) for Car-Car longitudinal situations	2014	Safety assist
AEB for pedestrians	2016	Vulnerable road users
AEB for cyclists	2018	Vulnerable road users
Lane departure warning systems	2014	Safety assist
Lane keeping assistance systems	2016	Safety assist
Junction assistance systems Car-Car	2020	Safety assist
Junction assistance systems Car-Pedestrian	2020	Vulnerable road users
Junction assistance systems Car-Cyclist	2020	Vulnerable road users
Dooring assist (Car-Cyclist)	2023	Vulnerable road users
AEB Car-PTW (powered two-wheelers)	2023	Vulnerable road users

principle, safety functions are also included in the evaluation. The overall evaluation from the three sub-areas of vehicle assistance, driver engagement, and safety functions are based on four grading levels—from "Entry" as the basic level over "Moderate" and "Good" to "Very Good."

4.6 Conclusion and Outlook

Type-approval regulations and consumer protection define the framework within which driver assistance systems can operate. Type-approval, which is regulated throughout Europe, is a sharp tool that regulates market access for driver assistance and automated driving functions, while the boundary conditions from consumer protection (e.g., Euro NCAP) are in principle voluntary for vehicle manufacturers, but a poor rating is nevertheless generally unacceptable.

The way in which vehicle regulations specify requirements is currently changing. In the past, individual, few, tests or inspections were defined here; for some time now, vehicle technology regulations have been developing toward the direct specification of requirements for driver assistance systems.

This is particularly advantageous in connection with market surveillance: vehicles from ongoing production can be retested against the requirement specifications, even in different test cases than those used for their type-approval—with severe penalties for violations. This system therefore forces vehicle manufacturers to develop robust systems that are adapted to individual test cases.

Consumer protection organizations evaluate products that are already on the market using a defined test catalog that contains a large number of test cases, so that a robust evaluation of systems over a wide range of applications is also possible. Here, the evaluation benchmarks are significantly severe than for type-approval, since vehicles on the market (have to) meet the type-approval criteria anyway.

The assessment criteria in consumer protection are generally adjusted every 2 years, with a tendency to include increasingly complex test scenarios and higher requirements, while a general overhaul of the requirements in the type-approval procedure (for Europe) takes place on a rotational basis approximately every 10 years (when a new "General Safety Regulation" is developed in the European Union).

Together, however, type-approval and consumer protection ensure that vehicle safety is continuously improved, always with the aim of reducing the number of road accident victims.

References

1. Der Beauftragte der Bundesregierung für Informationstechnik (2021) Das V-Modell XT. ► http://www.cio.bund.de/DE/Architekturen-und-Standards/V-Modell-XT/vmodell_xt_node.html. Zugegriffen: 19. Februar 2021
2. Siebert, N., Bahnert, J., Damm, R., Gaupp, W., Hoogen, M.: Das Typgenehmigungsverfahren für Kraftfahrzeuge. Kirschbaumverlag, Bonn (2019)
3. Europäische Union: Richtlinie 2007/46/EG des Europäischen Parlaments und des Rates vom 5. September 2007 zur Schaffung eines Rahmens für die Genehmigung von Kraftfahrzeugen und Kraftfahrzeuganhängern sowie von Systemen, Bauteilen und selbstständigen technischen Einheiten für diese Fahrzeuge (Rahmenrichtlinie) (2007). ► https://eur-lex.europa.eu/legal-content/DE/TXT/?uri=celex%3A32007L0046. Zugegriffen: 20. Januar 2021
4. Europäische Union: Verordnung (EU) 2018/858 des Europäischen Parlaments und des Rates vom 30. Mai 2018 über die Genehmigung und die Marktüberwachung von Kraftfahrzeugen und Kraftfahrzeuganhängern sowie von Systemen, Bauteilen und selbstständigen technischen Einheiten für diese Fahrzeuge, zur Änderung der Verordnungen (EG) Nr. 715/2007 und (EG) Nr. 595/2009 und zur Aufhebung der Richtlinie 2007/46/EG (2018). ► https://eur-lex.europa.eu/legal-content/DE/TXT/?uri=CELEX%3A32018R0858. Zugegriffen: 20. Januar 2021
5. Europäische Union: Verordnung (EG) Nr. 661/2009 des Europäischen Parlaments und des Rates vom 13. Juli 2009 über die Typgenehmigung von Kraftfahrzeugen, Kraftfahrzeuganhängern und von Systemen, Bauteilen und selbstständigen technischen Einheiten für diese Fahrzeuge hinsichtlich ihrer allgemeinen Sicherheit (2009). ► https://eur-lex.europa.eu/legal-content/DE/ALL/?uri=celex%3A32009R0661. Zugegriffen: 20. Januar 2021
6. Europäische Union: Verordnung (EU) 2019/2144 des Europäischen Parlaments und des Rates vom 27. November 2019 über die Typgenehmigung von Kraftfahrzeugen und Kraftfahrzeuganhängern sowie von Systemen, Bauteilen und selbstständigen technischen Einheiten für diese Fahrzeuge im Hinblick auf ihre allgemeine Sicherheit und den Schutz der Fahrzeuginsassen und von ungeschützten Verkehrsteilnehmern, zur Änderung der Verordnung (EU) 2018/858 des Europäischen Parlaments und des Rates und zur Aufhebung der Verordnungen (EG) Nr. 78/2009, (EG) Nr. 79/2009 und (EG) Nr. 661/2009 des Europäischen Parlaments und des Rates sowie der Verordnungen (EG) Nr. 631/2009, (EU) Nr. 406/2010, (EU) Nr. 672/2010, (EU) Nr. 1003/2010, (EU) Nr. 1005/2010, (EU) Nr. 1008/2010, (EU) Nr. 1009/2010, (EU) Nr. 19/2011, (EU) Nr. 109/2011, (EU) Nr. 458/2011, (EU) Nr. 65/2012, (EU) Nr. 130/2012, (EU) Nr. 347/2012, (EU) Nr. 351/2012, (EU) Nr. 1230/2012 und (EU) 2015/166 der Kommission (2019). ► https://eur-lex.europa.eu/legal-content/DE/TXT/?uri=CELEX%3A32019R2144. Zugegriffen: 20. Januar 2021
7. Euro NCAP: Informationen für Ingenieure (2021). ► https://www.euroncap.com/de/fuer-ingenieure/. Zugegriffen: 19. Februar 2021
8. Euro NCAP: Test Protocol—AEB VRU Systems (2019). Version 2.0.4. ► https://cdn.euroncap.com/media/43381/euro-ncap-aeb-vru-test-protocol-v204.pdf. Zugegriffen: 20. Januar 2021
9. Vereinte Nationen: UN-Regulation No. 151, Uniform provisions concerning the approval of motor vehicles with regard the the Blind Spot Information System for the Detection of bicycles (2019). ► https://www.unece.org/fileadmin/DAM/trans/main/wp29/wp29regs/2020/R151e.pdf, Zugegriffen: 20. Januar 2021

10. Vereinte Nationen: UN-Regulation No. 152, Uniform provisions concerning the approval of motor vehicles with regard to the Advanced Emergency Braking System (AEBS) for M1 and N1 vehicles (2020). ▶ https://www.unece.org/fileadmin/DAM/trans/main/wp29/wp29regs/2020/R152e.pdf. Zugegriffen: 20. Januar 2021

11. Vereinte Nationen: UN-Regulation No. 157, Uniform provisions concerning the approval of vehicles with regards to Automated Lane Keeping Systems Derzeit in Veröffentlichung, Link wird nachgereicht (2021)

12. Vereinte Nationen: Agreement concerning the Adoption of Harmonized Technical United Nations Regulations for Wheeled Vehicles, Equipment and Parts which can be Fitted and/or be Used on Wheeled Vehicles and the Conditions for Reciprocal Recognition of Approvals Granted on the Basis of these United Nations Regulations (Revision 3) (2017). ▶ https://unece.org/fileadmin/DAM/trans/main/wp29/wp29regs/2017/E-ECE-TRANS-505-Rev.3e.pdf. Zugegriffen: 19. Februar 2021

Open Access This chapter is licensed under the terms of the Creative Commons Attribution-NonCommercial-NoDerivatives 4.0 International License (▶ http://creativecommons.org/licenses/by-nc-nd/4.0/), which permits any noncommercial use, sharing, distribution and reproduction in any medium or format, as long as you give appropriate credit to the original author(s) and the source, provide a link to the Creative Commons license and indicate if you modified the licensed material. You do not have permission under this license to share adapted material derived from this chapter or parts of it.

The images or other third party material in this chapter are included in the chapter's Creative Commons license, unless indicated otherwise in a credit line to the material. If material is not included in the chapter's Creative Commons license and your intended use is not permitted by statutory regulation or exceeds the permitted use, you will need to obtain permission directly from the copyright holder.

Driver Assistance Systems and Automated Driving Functions—Accident Occurrence and Safety Benefits

Matthias Kühn, Jenö Bende, and Lars Hannawald

Contents

© The Author(s) 2026
H. Winner et al. (eds.), *Handbook Assisted and Automated Driving*,
https://doi.org/10.1007/978-3-658-45276-6_5

5.1 Accident Statistics

For a future-oriented statement on the effect of driver assistance systems (ADAS) and automated driving functions (AF) on road safety, it is essential to know and understand today's accident patterns. The accident patterns identified in this process should then be addressed by the systems through their specific functionality. To do this, it is necessary to move from a general, low-detail but representative view of a country's accident patterns into the details of the accidents. This, in turn, requires different qualities of accident data collections that result in special accident statistics. This field is spanned by the representative surveys of the Federal Statistical Office based on traffic accident reports at one end and the "in-depth" analyses of various accident research institutions in the context of their road safety work at the other end. In Germany, these are primarily the German In-Depth Accident Study (GIDAS) and the accident database of German Insurers (UDB). However, vehicle manufacturers, the ADAC and DEKRA, are also active here. With the exception of the official traffic accident statistics, the surveys of the organizations mentioned are not freely accessible. The individual surveys differ in their results due to the different databases available and the purpose pursued within and outside the organizations.

In terms of depth of detail and informative value, GIDAS comes first here. The interdisciplinary data collection at the accident site for a representative selection of traffic accidents in a defined region is a specialty. The joint research project of the Federal Highway Research Institute (BASt) and the Research Association for Automotive Technology (FAT) is therefore very well suited for the purposes of accident research. This also applies to the UDB, which is based on the insurers' claim data and depicts all damage events of the German motor insurers. The dataset is based on a representative selection of motor vehicle liability claims with a claim cost of at least 15,000 Euros and at least one personal injury. The UDB thus describes rather severe claims. For most questions—with the exception of single accidents with no involvement of third parties—it is comparable to the official traffic accident statistics or GIDAS.

The focus of vehicle manufacturers, on the other hand, is primarily on accidents involving vehicles of their own brand. When collecting the data, however, a great depth of detail similar to that of GIDAS is achieved. Air rescue operations mainly form the basis of the ADAC accident surveys. The accident data is then enriched with further information from the police, hospitals, and the fire brigade. Accident analyses by DEKRA are based on technical reports commissioned by third parties. These in turn still have to be enriched with medical data in order to obtain a more comprehensive picture of the accident. What both data collections have in common is that minor accidents are underrepresented here.

5.1.1 Accident Occurrence in Germany

If we look at the number of accidents in Germany over the last few decades, we can see an almost continuous decline in the number of fatalities (◘ Fig. 5.1).

While 21,332 people were killed on German roads (former East and West Germany) in 1970, the figure was 3,046 in 2019, with the number of accidents involving personal injury falling from 377,610 in 1970 to 300,143 in 2019 [1]. All of these figures should be viewed against the background of rising traffic performance and should be viewed even more positively. For example, the stock of all motor vehicles in Germany increased from 16.8 million in 1970 to 64.8 million in 2019, and the mileage traveled by motor vehicles nearly tripled from 251 billion km in 1970 to 738.8 billion km in 2019. Passenger cars are the dominant mode of transportation, with approximately 47.1 million registered vehicles in 2019 traveling 632.2 billion km, or about 86 percent of the total mileage traveled by all motor vehicles [2].

The distribution of fatalities for 2019 shows initial indications of areas for action to further increase road safety in Germany (◘ Fig. 5.2).

It is clear that almost 60% of fatalities occur on rural roads. A quarter of these fatalities resulted from accidents involving a tree. In urban areas, slightly more than 900 people were killed in road traffic accidents. These were mainly vulnerable road users such as pedestrians and cyclists. The remaining 356 fatalities occurred on motorways. An examination of fatalities by type of road use shows that in 2019, about 45% of the 3046 fatalities were passenger car occupants (1,364), about 18% were motorcycle occupants (542), and about 29% were non-motorized, vulnerable road users (417 pedestrians; 445 bicyclists). Although the latter group is smaller in absolute numbers, its relative share in all traffic fatalities has increased.

If we look at the situation for seriously injured road users in 2019, this picture changes (◘ Fig. 5.3). More than half of all seriously injured road users were involved in accidents in urban areas (52%), about 39% were involved in accidents on rural roads, and 9% on motorways.

The distribution of fatalities in 2019 is shown in ◘ Fig. 5.4. One can see the largest share of the number of fatalities being driving accidents with 35%, followed by accidents in longitudinal traffic with 26% and the group of turning accident and turning/crossing accident with 19%.

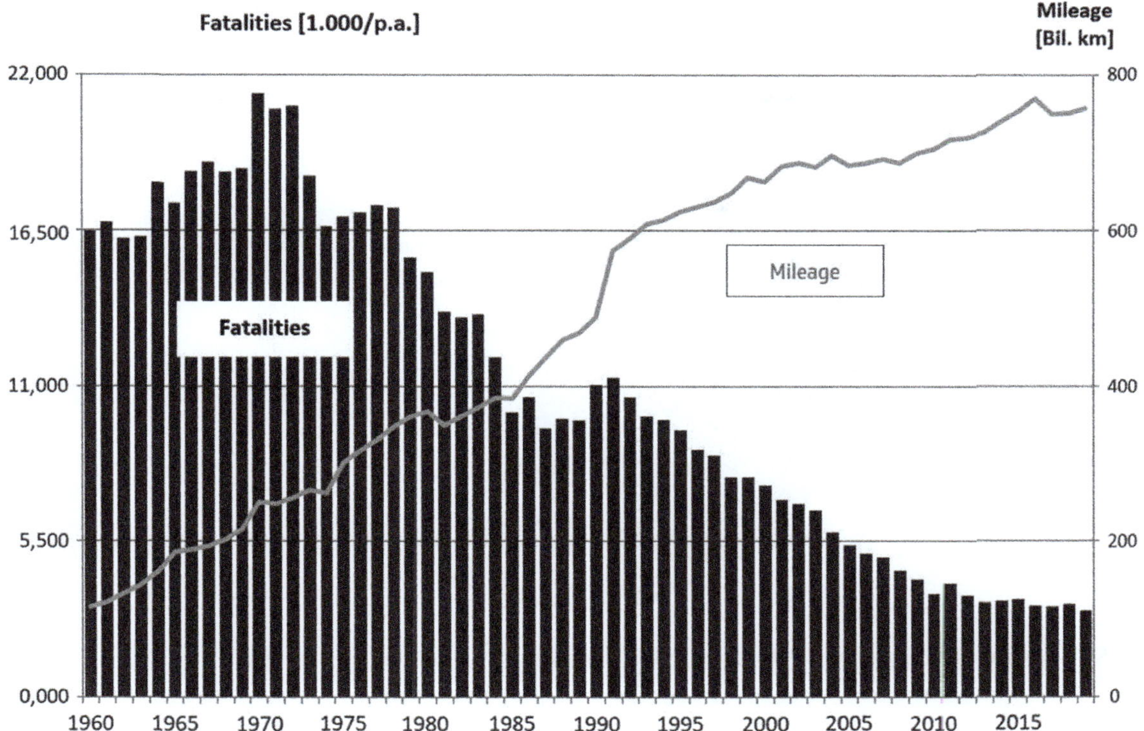

Fig. 5.1   Development of road safety based on the number of fatalities in road traffic in Germany [1]

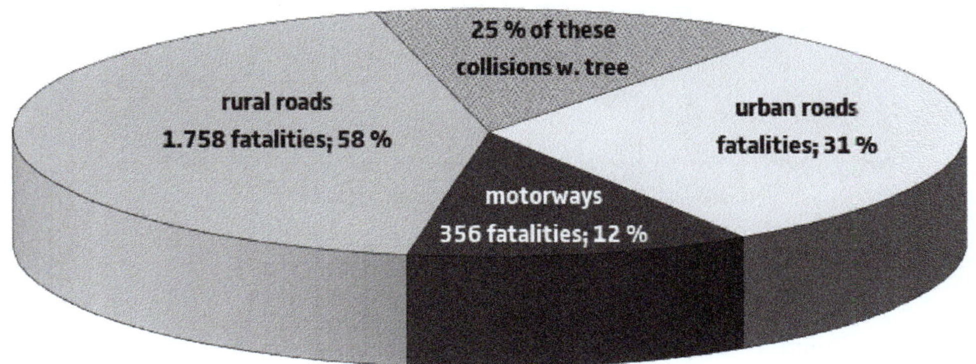

Fig. 5.2   Traffic fatalities in Germany by accident location in the year 2019 [1]

Especially in the case of a driving accident, which by definition is described as a loss of control over the vehicle because the speed was not adjusted according to the course, cross section, slope, or road conditions, one can already see the important role of the driving speed in the accident occurrence.

In addition to the location as an accident-specific parameter, various participant-specific parameters of the accident can also be evaluated, such as the influence of alcohol and drugs or the driver's distraction. Another important parameter is the age of the person who caused the accident. In relation to all accidents with personal injury in 2017, the risk of causing an accident with personal injury as a passenger car driver is as follows (Fig. 5.5).

It can be seen that both younger drivers aged 18–24 but also drivers aged 75 and older have a significantly increased risk of causing an accident with personal injury. The 18- to 20-year-old drivers cause 3.5 times more accidents with personal injury in relation to the distance driven. In fact, the curve continues to rise when the drivers older than 75 years are further differentiated. However, this was not shown here due to the small number of cases and the significantly larger confidence intervals. Young drivers are thus the "number 1" risk group in road traffic. However, over the last few years, older car drivers have also become increasingly apparent as a risk group in the accident statistics. Their driving performance-related risk of causing an

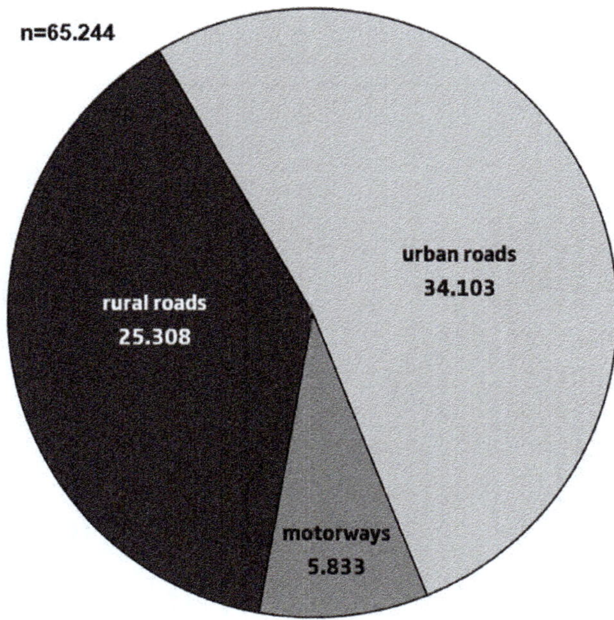

◘ Fig. 5.3   Seriously injured accident victims in Germany by location in 2019 [1]

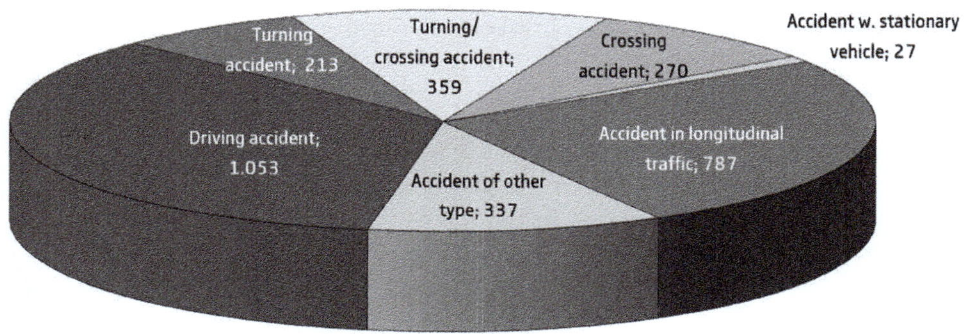

◘ Fig. 5.4   Traffic fatalities in Germany by accident type in 2019 [1]

accident with personal injury is now 1.8 and thus significantly higher than that of other adults. These figures justify discussing age-related measures for safe participation in road traffic in the future. For example, in the area of young drivers, accompanied driving in Germany is a successful model despite the still high figures in ◘ Fig. 5.5 [4].

In order to understand accidents even better and to target measures even more effectively, it is necessary to look not only at fatalities but also at the group of road users with serious injuries or life-threatening injuries. The current category of "seriously injured" according to German official statistics is based on the criteria of inpatient hospital treatment for at least 24 h and the criterion "did not die within 30 days as a result of the accident". This large group is therefore very heterogeneous and does not allow any statement to be made about the people who sustained life-threat-

ening injuries. Improved vehicle technology and a better rescue system, among other things, lead to a reduction in the number of fatalities, but possibly not in the number of survivors with serious injuries. Thus, the quality of road safety work should not be measured only by the reduction in fatalities. Study results suggest that about 10 percent of officially seriously injured road users suffer life-threatening injuries. For the year 2019 that would be about 6,500 polytraumatized persons [5].

So far, it has not been possible to establish a separate category for severely injured persons based on the Maximum Abbreviated Injury Scale (MAIS) in German traffic accident statistics. This category could then be used as a basis for targeted accident analysis, the development of subgroup-specific preventive measures, and a more accurate estimate of the economic costs of serious traffic accidents, among other things. The definition of

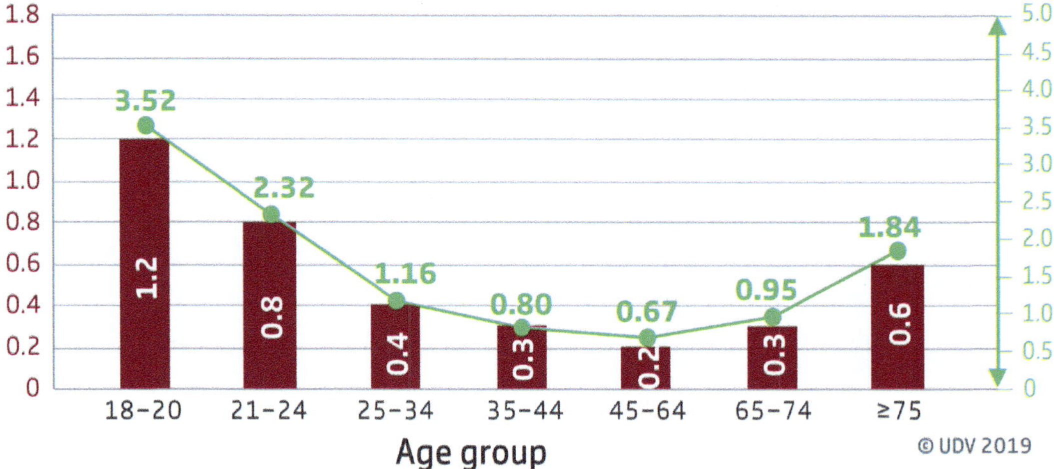

own calculations based on the datasets of infas, DLR, IVT and infas 360 (2018) as well as Statistisches Bundesamt (2018)

Fig. 5.5 Driving performance-related risk of causing an accident with personal injury as a passenger car driver by age group [3]

"seriously injured" road accident victims, which has been harmonized across Europe since 2015, aims in the same direction. In order to comply with the reporting obligation to the EU Commission, the figures provided on the most seriously injured in Germany are determined from an extrapolation algorithm based on accident data from GIDAS and the TraumaRegister DGU®. Based on this definition, 15,265 people were seriously injured on German roads in 2018 (MAIS3+). This is about a quarter of the category "seriously injured" on German roads [6].

5.1.2 Accident Occurrence by Vehicle Type

To further understand the accident occurrence of the individual vehicle types, it is helpful to change the consideration level and to use in-depth accident data. Table 5.1 shows an overview of kinds of accidents for different vehicles according to an analysis of the insurers' accident database (UDB). It becomes clear that even this simple distinction reveals special accident patterns for the individual vehicle types, which can often be explained by the different areas in which the vehicles are used. For example, the proportion of collisions with other vehicles driving or waiting ahead is significantly higher for trucks (28%) compared to the other vehicle types. This reflects, for example, the high level of motorway use and the typical rear-end collision. The collision with a pedestrian (11.2%) crystallizes as a more frequent event for the bus compared to the other vehicle types. The strikingly high proportion of the kind of accident ["accident of other kind" (27.4%)] for the bus is

predominantly due to passenger accidents (approx. 60%), in which bus passengers fall and are injured without a collision with another road user. For motorcycles, collisions with other vehicles turning or crossing are dominant over all other kinds of accident. Only the passenger car has a similarly high proportion of this kind of accident.

5.1.2.1 Passenger Car

Based on 6,451 claims in the UDB, passenger car accidents can be characterized by the parameter "kind of accident" and by the "accident cause" as defined in the official statistics [1] (see Table 5.1 and Fig. 5.6).

Thereby, more than 50% of the accidents involving passenger cars are described by the collision with another vehicle turning or crossing and the collision with another vehicle driving or waiting ahead.

The most frequent causes of accidents involving car drivers correspond to the kind of accidents just mentioned (Fig. 5.6). Failure to yield the right of way dominates, followed by turning, etc. as the cause of accidents, which is therefore typical for accidents involving crossing and turning vehicles. Inappropriate speed and insufficient safety distance then follow in roughly equal order. Here, too, is a connection to the kind of accident "collision with another vehicle which starts, stops or is stationary, is moving ahead or waiting".

5.1.2.2 Truck

For accidents involving trucks, the picture of the most frequent causes of accidents and the kind of accidents described in Table 5.1 is shown in Fig. 5.7.

Table 5.1 Distribution of the different kind of accidents depending on vehicle type on the basis of a UDB analysis

Accidents involving a:	Car (n=6,451; 100%) (%)	Truck (n=1,346; 100%) (%)	Bus (n=507; 100%) (%)	Motorcycle (n=2,603; 100%) (%)
Collision with another vehicle which starts, stops, or is stationary	4.9	6.5	5.5	3.4
Collision with another vehicle moving ahead or waiting	20.1	**28.0**	17.0	10.3
Collision with another vehicle moving laterally in the same direction	6.9	15.8	4.7	7.4
Collision with another oncoming vehicle	14.3	12.7	9.9	17.8
Collision with another vehicle which turns into or crosses a road	**36.8**	23.0	17.9	**37.9**
Collision between vehicle and pedestrian	8.7	4.8	**11.2**	7.1
Collision with an obstacle in the carriageway	0.3	0.6	0.0	0.2
Leaving the carriageway to the right	2.0	1.4	4.1	4.6
Leaving the carriageway to the left	2.2	2.1	2.2	2.8
Accident of another kind	3.8	5.2	**27.4**	8.5

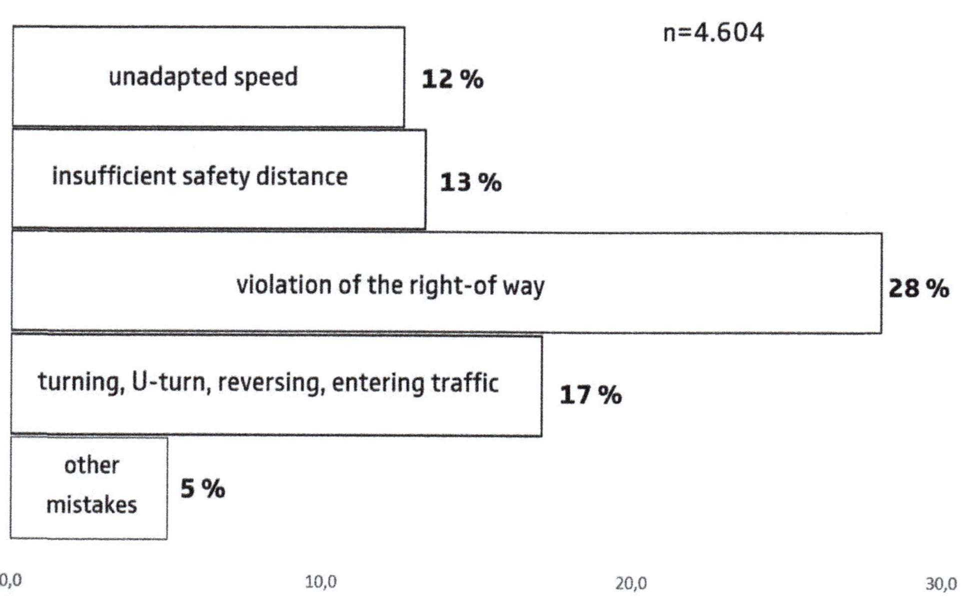

Fig. 5.6 The most common accident causes for passenger car accidents in the UDB

The analyses show that the collision with another vehicle moving in the same direction, either sideways or ahead, describes more than 40% of the accidents. Accidents with turning or crossing vehicles form the second typical accident pattern.

"Insufficient safety distance" dominates the causes of accidents. This in turn corresponds to the typical accident constellation of rear-end collisions. The other causes of accidents are found with about the same frequency (Fig. 5.7).

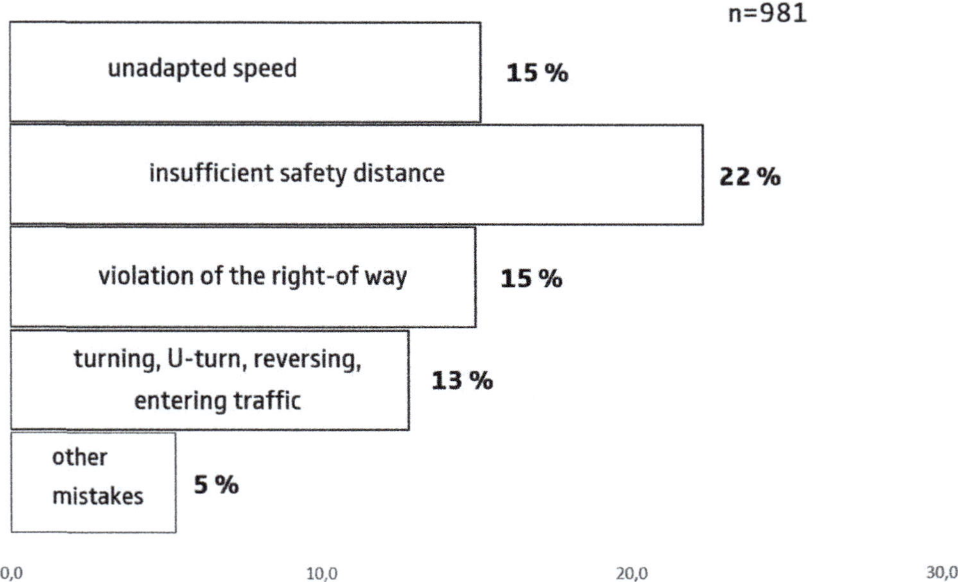

Fig. 5.7 The most common accident causes for truck accidents in the UDB

5.1.2.3 Bus

The analysis of accidents involving buses shows that the collision with a vehicle turning or crossing as well as with a vehicle driving ahead or waiting is equally ranked with about 17% each. The collision with a pedestrian is a more frequent event for the bus compared to the other vehicle types (■ Table 5.1).

The most frequent causes of accidents are shown in ■ Fig. 5.8. The statistics are dominated by the three accident causes: speed, safety distance, and right of way. However, knowledge of the causes in the dominant group of "other errors" can significantly influence the picture and thus the finding of measures.

This category includes, for example, all incidents involving bus passengers who fell without colliding with another road user. This group accounts for more than half of the cases here (▶ Sect. 5.1.2).

5.1.2.4 Motorized Two-Wheelers

Motorcyclists are among the most vulnerable road users. Acceleration, speed, narrow silhouette, misjudgment when driving a single-track vehicle and also by other road users are reasons for motorcycle accidents that often end with serious or fatal injuries for the rider. As shown in ■ Table 5.1, collisions with another vehicle turning or crossing dominate with almost 40% of the accidents. The collision with an oncoming vehicle is in second place with almost 18%.

If we look at the causes of accidents, it is noticeable that inappropriate speed was the cause of accidents in around one in four cases. Failure to yield the right of way and insufficient safety distance follow in roughly equal order (■ Fig. 5.9).

Even though motorcycling in groups accounts for a small share of motorcycle accidents (about 15%), a deviating accident pattern is evident [7]. These accidents mainly occur outside built-up areas. The conflict-triggering situation in these accidents is mainly the loss of control over the vehicle (driving accident) as well as a vehicle driving in the same or opposite direction (longitudinal traffic accident). Especially the collision of motorcycles within the group characterizes these accidents. Group dynamics and lack of driving skills in critical situations lead to overtaking or following with too little safety distance or too high speed, often involving group members as well.

5.2 Classification and Differentiation of Modern Driver Assistance and Automated Driving Functions

The discussion about assisted and automated driving and their contribution to road safety requires a clear understanding of the features and capabilities of the functions. The driving task can be divided into the navigation, path guidance, and stabilization levels [8]. According to this model, the navigation level describes route planning, the path guidance describes the driver's tactical task of forming target values for course and speed—i.e., dynamic driving—and the stabilization level describes the actual control and regulation in a closed loop in order to minimize the deviation between the target values formed on the path guidance level and the actual vehicle movement. Assistance and automation functions predominantly operate on the path con-

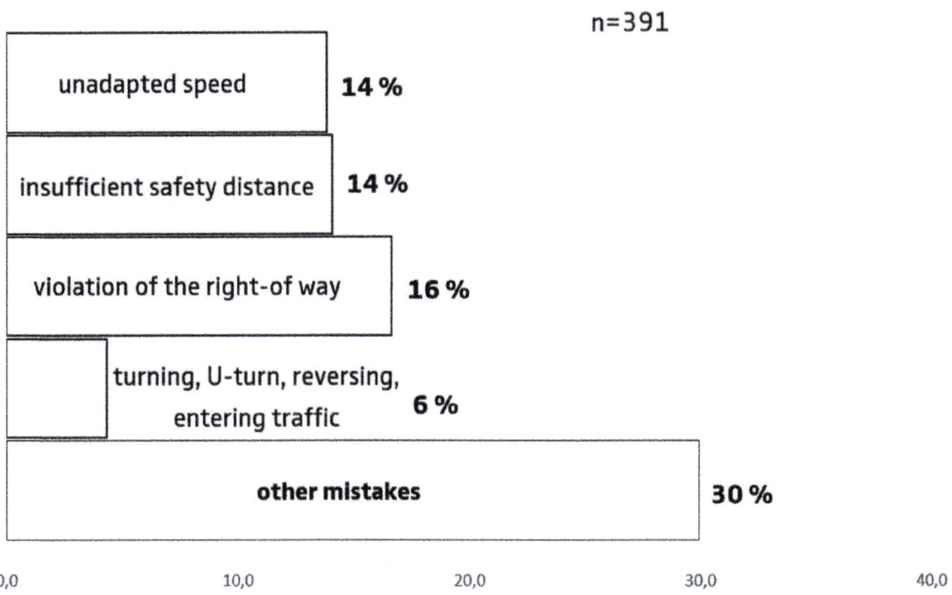

◘ Fig. 5.8 The most common accident causes for bus accidents in the UDB

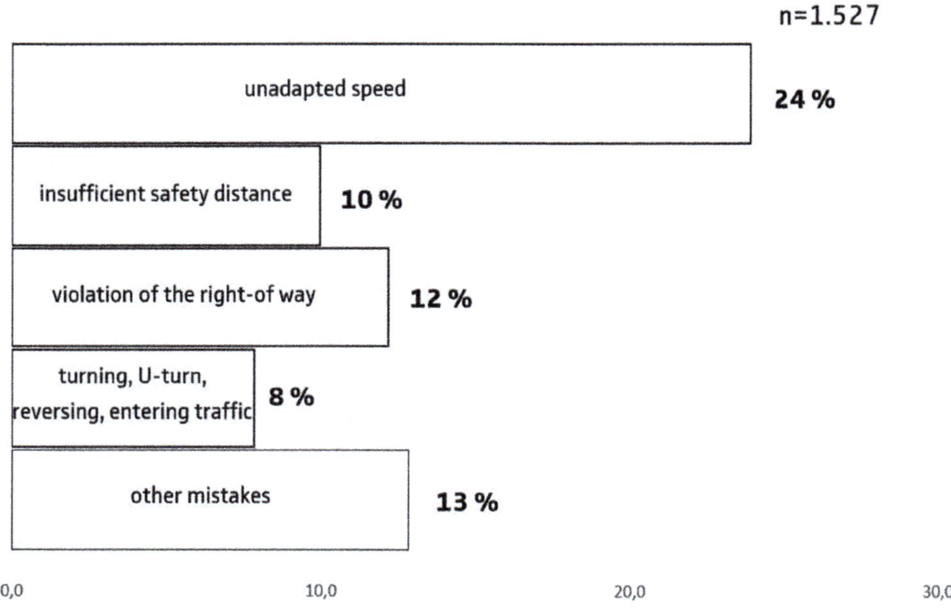

◘ Fig. 5.9 The most common accident causes for accidents with motorized two-wheelers in the UDB

trol level, while vehicle dynamics control systems such as the antilock braking system (ABS) and the electronic stability program (ESP) operate exclusively on the stabilization level. Basically, three different modes of action can be distinguished at the path guidance level [8]. These are the informing and warning functions (A), the continuously acting automating functions (B), and the temporarily intervening systems (C) (◘ Table 5.2).

This approach has the advantage that safety-oriented driver assistance systems and automated driving functions explicitly activated by the user to (at least partially) delegate the driving task become distinguishable.

Even within the driver assistance systems, visible differences can be clearly derived from their mode of action. Mode of action B describes the automation levels under discussion (◘ Table 5.3). Level 1 describes driver assistance systems that control longitudinal or lateral guidance thus enabling a more relaxed driver task. These include adaptive cruise control and lane departure warning. The lane departure warning system, on the other hand, is consequently assigned to mode of action A, since the driver carries out the driving task himself and only receives a safety-promoting instruction if necessary.

Table 5.2   Assistance and automation functions on the path guidance level [8]

Mode of action A **Informing and warning functions**	Mode of action B **Continuously acting automating functions**	Mode of action C **In accident-prone situations temporarily intervening systems**
Act exclusively and "indirectly" on the driver: 1. Status information, e.g., traffic sign recognition 2. Abstract warning, e.g., lane departure warning 3. Specific warning, e.g., • Blind spot assist • Forward collision warning	Have a direct influence on vehicle control can always be overridden	Preventive machine intervention in case of negative situation prognosis, e.g.: • Autonomous braking assist • Evasive steering support

Table 5.3   Levels of automation according to [8]

Nomenclature	Driving tasks of the driver according to the degree of automation
Level 5	The system takes over the driving task fully in all road types, speed ranges, and ambient conditions
Level 4	The system takes over lateral and longitudinal guidance completely in a defined use case *Driver does not monitor for this use case*
Level 3	The system takes over lateral and longitudinal guidance for a certain period of time in specific situations *Driver does not monitor for this period of time*
Level 2	The system performs the lateral and longitudinal guidance without taking the responsibility (for a certain period of time or/and in specific situations) *Driver monitors*
Level 1	The driver permanently performs either lateral or longitudinal guidance. The other driving task is performed by the system within certain limits, e.g., ACC, lane-keeping assistant
Level 0	Driver continuously (during the entire journey) performs longitudinal guidance (acceleration/deceleration) and lateral guidance (steering)

5.3 Safety Potential of Driver Assistance Systems

Driver assistance systems are electronic systems in the vehicle that are designed to assist the driver in driving. The aim of these systems is often to increase driving comfort, safety, or economy. This chapter will focus exclusively on the aspect of the influence of driver assistance systems on road safety. Against this background, the new technologies aim to address the main areas of accident occurrence and to have a positive effect.

Driver assistance systems counteract an escalation of the traffic situation in different phases (see Table 5.2) by helping the driver to.

- provide relief during normal driving (e.g., Adaptive Cruise Control (ACC), which ensures that sufficient distance is maintained);
- point out possible risks (e.g., information about the end of a traffic jam);
- indicate a need for action or issue a specific recommendation for action in the form of a warning (e.g., forward collision warning);
- assist with an action (e.g., brake assistant, which increases the driver's braking pressure);
- and, if necessary, intervene temporarily—after a warning has been issued—in the longitudinal or lateral guidance (e.g., emergency brake assistant, which triggers full deceleration if the driver fails to act, thus reducing the impact energy).

A clear distinction must be made between these and driving dynamics control systems, which can also be regarded as driver assistance systems, but which, unlike the systems discussed above, do not have any environment sensors or recognition. These systems therefore support the driver not at the path control level but at the stabilization level by supporting the driver's wish to decelerate or steer in such a way that the vehicle behaves dynamically as far as possible in line with expectations. For example, the antilock braking system aims to not only reduce the braking distance by controlling wheel slip, but also maintain the vehicle's steering ability. Based on the so-called single-track model, the Electronic Stability Program (ESP) calculates the transverse dynamic vehicle reaction expected by a normal user from the steering wheel angle, driving speed, and characteristic vehicle properties and minimizes the deviation of the actual vehicle yaw motion, i.e., rotation about the vertical axis, from the calculated target yaw speed by intervening in individual wheel brakes.

Due to this diversity and the wide range of effects of driver assistance systems, special methods are required to determine their safety potential.

5.3.1 Methods for Assessing the Safety Potential of ADAS

The safety potential of advanced driver assistance systems (ADAS) can be determined in different ways. For example, a retrospective comparison can be made between two accident groups: "vehicles with ADAS" vs. "vehicles without ADAS". Lie et al. applied this approach for the effectiveness demonstration of the Electronic Stability System (ESP) based on Swedish accident data [9]. Based on data from the German Federal Statistical Office for the years 1999–2001, Unselt et al. were able to demonstrate for the first time a more than 30% reduction in the probability of a driving accident for vehicles with ESP compared to otherwise identically constructed vehicles without ESP [10]. With increasing penetration of the systems in the vehicle fleet, this method is also applied to current ADAS such as the emergency braking system (AEB) [11]. It is also clear that only this method can describe the actual benefits of these systems. While this method describes the actual benefits of these systems, other methods only approximate it, more or less.

For the results presented in the following, the alternative method "What would be if…" was used [12]. Here, the course of the accident was considered as it occurred in reality and was compared with the calculated course of the accident with a generic driver assistance system. In this context, generic means a system that has freely compiled system properties and thus does not represent a product on the market. In this way, it is possible to determine what influence a specific driver assistance system would have on the accident if all vehicles were equipped with the system under consideration. For the implementation of this method, both the accident circumstances and the properties (functionalities) of the system to be investigated must be known or defined for a generic system. The multi-stage procedure used in this method distinguishes between two aspects: namely, whether the accident would have been avoidable or only positively influenceable or addressable. An accident is considered theoretically avoidable if it would have no longer occurred due to the influence of an ADAS. However, if the analysis shows that it would have happened anyway, but possibly with lighter accident consequences, it is considered to be positively influenceable or addressable.

The representation can be carried out much more precisely by means of simulation. Here, too, the accidents are examined in a prospective approach according to the method "What if…". The accident situation in the simulation environment, which can now be depicted in great detail, can also be used to evaluate the benefits of systems with considerably more complex functionality. With regard to warning and informing systems, the human reaction must also be represented with the help of a driver model. The definition of this

driver model is a great challenge, since an adequate reaction of the driver cannot always be assumed. This method was used as an example in the study "Equal Effectiveness for Pedestrian Safety" [13].

Alternatively, the Field Operational Test (FOT) can be used to analyze the safety potential. This is mainly used for the evaluation of new technologies, e.g., also ADAS [14]. For this purpose, the vehicle is equipped with extensive measurement technology. The driver is then instructed, for example, to drive for a period of time with the ADAS switched on or off. This type of behavioral observation has been made possible by rapid technical progress in terms of the collection, storage, and analysis of large amounts of data and the ever-smaller measurement instruments. Everything needed to explain and describe the driving behavior and the functionality of the ADAS is recorded: starting with the environment, vehicle movement (e.g., accelerations, speeds, direction, vehicle status, etc.) up to eye, head and hand movements as well as pedal actuations. This data includes information about the interactions between driver, vehicle, road, weather, and traffic not only under normal conditions but also in critical situations and even in the event of an accident. The biggest challenge here is evaluating the very large amounts of data.

5.3.2 Passenger Car

For passenger cars, an emergency braking assistant is the most promising driver assistance system, followed by the lane-keeping assist system and the blind spot warning system (◻ Table 5.4). The emergency braking assistant becomes even more important when it is able to address accidents involving pedestrians and cyclists. Then, up to 43.5% of all passenger car accidents in the database are avoidable [12].

Data from U.S. insurers can already quantify the real positive benefits of emergency braking assistants [11].

5.3.3 Truck

Based on the method described above, an emergency braking assistant was also identified as the driver assis-

◻ **Table 5.4** Safety potential of ADAS for passenger cars in relation to all passenger car accidents [12]

Theoretical safety potential	ADAS (%)
Emergency braking assist (a)	17.8
Lane-keeping assist (a)	4.4
Blind spot warning system (a)	1.7

a avoidable

Table 5.5 Safety potential of ADAS for trucks in relation to all truck accidents [12]

Theoretical safety potential	ADAS (%)
Emergency braking assist (a)	6.0
Emergency braking assist (detects stationary vehicles) (a)	11.9
Turning assist system (pedestrian detection) (a)	0.9
Turning assist system (cyclist detection) (a)	3.5
Lane-keeping assist (a)	1.8
Blind spot warning system (ad)	7.9

a avoidable, *ad* addressable

tance system with the greatest safety potential for trucks [12]. It was found that its potential doubles if it is also able to detect stationary vehicles in front of the truck. It is followed by the blind spot warning system and the turn-off assistant with cyclist detection, provided that the effect is related to all truck accidents (**Table 5.5**).

Accident data on a highly congested German highway section already show the positive effect of the emergency brake in the accident history of tractor-trailers. However, they also point to the need for further development [15].

Looking only at accidents between trucks and unprotected road users, the safety potential for a turning assist system with cyclist and pedestrian detection is very high. For the case analysis, a system was assumed to monitor the areas in front of and to the right of the truck and warn the truck driver if a pedestrian or bicyclist is in the critical area when starting or during a turn. It was assumed that the driver would react "optimally" to the warning.

For this turning assist, it was shown that approximately 43% of all truck accidents involving bicycles and pedestrians could be avoided and that approximately 31% of bicyclists and pedestrians killed in collisions with trucks could be prevented from death.

Such systems are now also available on the market. From 2022, they will become mandatory equipment [16]. Retrofit systems are also available and can be effective. They are subsidized by the state to quickly increase penetration in the vehicle fleet [17].

In this context, there is still a clear dependence of the truck body type and thus the intended use on the identified safety potential of the individual driver assistance systems (**Table 5.6**).

5.3.4 Bus

For buses and coaches, the emergency brake assistant is again the system with the highest safety potential (**Table 5.7**). It is followed by the blind spot warning system and the turn-off assistant, which detects pedestrians and cyclists. In the case of buses, too, there is a clear increase in potential if the emergency braking assistant can detect stationary vehicles.

Here, too, the safety potential of ADAS depends on the purpose of use and the type of bus (**Table 5.8**). The coach, for example, benefits significantly more from the blind spot warning system, whereas the turn-off assistant is more useful for the public bus.

5.4 Powered Two-Wheelers (PTW)

Riders of powered two-wheelers (PTW) are particularly at risk in road traffic. However, it has been shown that the technical possibilities for preventing or mitigating accidents involving PTWs have been very limited so far. Intelligent technical developments, on the other hand, can have a positive effect on the occurrence of accidents. In this context, experts talk about Intelligent Traffic Sys-

Table 5.6 Safety potential of ADAS for trucks broken down by body type [12]

ADAS	Theoretical safety potential for		
	Solo truck (%)	Truck with trailer (%)	Semi-trailer truck (%)
Emergency braking assist (a)	2.2	6.1	5.1
Emergency braking assist (detects stationary vehicles) (a)	7.9	10.7	9.5
Turning assist system (cyclist detection) (a)	4.2	0.6	2.9
Turning assist system (pedestrian detection) (a)	0.5	0.9	0.8
Blind spot warning system (ad)	6.8	5.2	6.4
Lane-keeping assist (a)	1.6	1.8	1.3

a avoidable, *ad* addressable

Table 5.7 Safety potential of ADAS for buses in relation to all bus accidents [12]

Theoretical safety potential	ADAS (%)
Emergency braking assist (ad)	8.9
Emergency braking assist (detection of stationary vehicles) (ad)	15.1
Turning assist system (cyclist and pedestrian detection) (a)	2.3
Lane-keeping assist (a)	0.5
Blind spot warning system (ad)	3.8

a avoidable, *ad* addressable

Table 5.8 Safety potential of ADAS for buses depending on purpose of use [12]

	Theoretical safety potential	
ADAS	Public bus (%)	Coach (%)
Emergency braking assist (ad)	11.9	4.5
Emergency braking assist (detects stationary vehicles) (ad)	16.6	17.3
Turning assist system (cyclist and pedestrian detection) (a)	3.4	–
Lane-keeping assist (a)	0.3	1.5
Blind spot warning system (ad)	0.2	14.6

a avoidable, *ad* addressable

tems (ITS), Intelligent Transport Systems (ITS), or Vehicle-to-X (V2X) systems. Among other things, these systems can exchange information with the environment in order to increase the safety of PTWs. The national research project simTD was able to prove that vehicle-to-x communication is practical (e.g., intersection and cross-traffic assistant) through a large-scale field test using FOT and involving the infrastructure as well as a wide variety of vehicles, including motorcycles. For example, the left-turn assistant to protect motorcyclists already showed the potential to prevent many accidents [18].

An analysis of accident data from the UDB [19] showsthat selected ITS systems for motorized two-wheelers would be able to positively influence safety (Fig. 5.10). The first four systems in the ranking address more than two-thirds of all accidents. These are the intersection and cross-traffic assistant, the curve warning system, the left-turn assistant, and the overtaking assistant.

When developing ITS systems for motorized two-wheelers, it is crucial to pay specific attention to the human–machine interface in order to avoid distraction, etc., and to increase driver acceptance. The spe-cial features of two-wheeled driving must be taken into account, since driving a single-track vehicle is significantly different from driving a passenger car.

5.5 Safety Potential of Automated Driving Functions in Cars

When vehicles are driven in automated driving mode (AD), some benefits for active safety and also consequences for passive safety can be expected. In the first development step, the functions of automation level 3 are offered for passenger cars and here for slow highway driving (congestion pilot up to 60 km/h). The extension to higher speed ranges (highway pilot up to 130 km/h) and other vehicle types (truck, bus) as well as road types will follow. In the context of automated driving functions and their design, we also speak of operational design domains (ODD). For passenger cars and the use case of the highway pilot (AD level 3), the highest safety potential of 21% of avoidable accidents caused by the passenger car is still expected from today's driver assistance systems (ADAS+). Here, ADAS+ is understood to mean the combination of an emergency brake assistant, a lane change assistant, a blind spot detection system, and an adaptive cruise control system (Table 5.9) [20].

An additional benefit of +5% can be expected for a level 3 automated driving function. This means that approximately 5% more highway accidents can be prevented if the vehicle is also equipped with a level 3 automated driving function instead of only ADAS+. The additional benefit of this system compared to ADAS+ systems is mainly to avoid more accidents due to lane changes [20].

However, it must be taken into account that a level 3 automated driving function may also have negative effects on road safety. These effects have not been quantified yet. However, studies show that these effects should not be underestimated and that they could reduce the additional positive benefits compared to ADAS+ [21–23]. Overall, under certain circumstances level 3 systems might not bring any additional positive safety effects at all.

It can already be predicted that passenger cars driving in level 3 automated mode will continue to cause accidents on highways. And they will also be involved in accidents through no fault of their own. In this context, those accidents caused by a lane change will account for the majority of unavoidable accidents (Table 5.10). And the driver will remain the critical part of a vehicle with a level 3 automated driving function in the future as well.

As can be seen from Table 5.10, only a Level 4 system can provide a high benefit in terms of an additional 18% avoidable accidents compared to a Level 3 system. This is because a Level 4 system will be able to

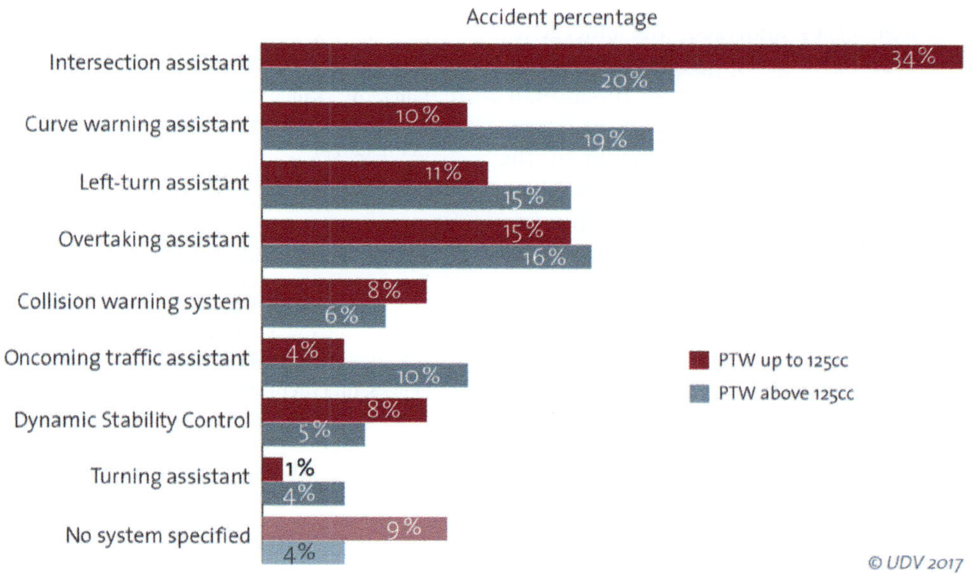

Fig. 5.10 Ranking of the most promising systems, divided by PTW type [21]

Table 5.9 Safety potentials of automated driving functions for accidents caused by passenger cars in the UDB [20]

Systems			Safety potentials in terms of avoidable accidents [%] that can be achieved in addition to the safety potentials of ADAS+			
			Motorway accidents involving a passenger car (n = 346) (%)	All accidents on motorways that were caused by a passenger car (n = 164) (%)	All accidents involving a passenger car (n = 3,029) (%)	All accidents that were caused by a passenger car (n = 1,834) (%)
Today	ADAS+		21	45	2.0	4.0
Future	Level 3	In addition to	+5	+11	+0.6	+1.0
	Level 4		+18	+38	+2.1	+3.4

Table 5.10 Safety potentials of automated driving functions related to the main accident scenarios in the UDB [20]

Accidents that were caused by a passenger car (n = 164)					
Accident scenarios	n	%	Avoidable accidents [%]		
			ADAS+	Level 3	Level 4
Rear-end accidents	83	51	80	87	98
Lane change accidents	60	37	10	27	97

handle almost all traffic situations within the intended application or operational design domain (ODD). Also important is the fact that the driver is eliminated from the operating area for almost the entire automated trip. Nevertheless, even with a level 4 system, a large proportion of highway accidents will remain unavoidable. Possible negative effects of mixed traffic from manually and automatically controlled vehicles have not even been taken into account yet.

5.6 Outlook

With the current state of research, it can be shown that driver assistance systems (ADAS) are and will remain the basis for increasing road safety—despite automated driving functions. Therefore, it is important to quickly increase their availability in the vehicle fleet. In addition to the efforts of the consumer protection organization EuroNCAP [24], the obligation to equip vehicles with ADAS, which is enshrined in law in Europe, will be vital for increasing road safety [16].

Across the vehicle types analyzed, the emergency brake emerges as the most safety-promoting ADAS. It can now be found in various forms and functionalities in almost all vehicle classes and, as current figures show, has its justification [11]. The expansion of the functionality of emergency braking systems includes, among other things, the detection of cross traffic and all forms of unprotected road users will increase the positive effect on accidents.

In addition to the emergency braking system, the turn-off assistant is particularly effective for trucks, especially in protecting cyclists and pedestrians. Against the background of the increase in the number of cyclists, its importance is increasing once again.

The next step in the development of the ADAS could be assisted or later automated evasive steering action in emergency situations. In this way, the effectiveness of pure emergency braking systems could be increased in certain accident-prone situations, since purely in terms of driving dynamics, an evasive maneuver can still take place at a later point in time compared to braking. The evaluation of such functionalities against the background of their influence on the increase of road safety is still pending, but it is absolutely necessary. However, this task places even higher demands on the quality of analysis methods and accident data.

In general, ADAS will benefit from the megatrend of automated driving. On the one hand, this will increase their robustness and availability. On the other hand, the boundary between comfort and pure safety systems will become blurred and the distinction between individual ADAS functionalities will also be increasingly eliminated. Road safety will benefit when it will no longer be necessary as a driver to develop an understanding of the individual ADAS functionalities in order to correctly interpret warnings and interventions. Rather, a smooth protection zone around the vehicle with a sophisticated human–machine interface will support the driver's natural behavior in critical situations to avoid accidents.

References

1. DESTATIS: Fachserie 8, Reihe 7. Statistisches Bundesamt, Wiesbaden (2020)
2. KBA, ► www.kba.de
3. Gehlert, T., Gaster, K.: Rückmeldefahrt für Senioren. Unfallforschung kompakt Nr. 93, Unfallforschung der Versicherer, Berlin (2019)
4. Schade, F.-D., Heinzmann, H.-J.: Sicherheitswirksamkeit des Begleiteten Fahrens ab 17. Summative Evaluation. Berichte der Bundesanstalt für Straßenwesen – Mensch und Sicherheit, Heft M 218, Bergisch Gladbach (2011)
5. Deutscher Bundestag: Bericht der Bundesregierung über Maßnahmen auf dem Gebiet der Unfallverhütung im Straßenverkehr 2018 und 2019., Unterrichtung durch die Bundesregierung, Drucksache 19/26135 (2021)
6. Malczyk, A.: Schwerstverletzungen bei Verkehrsunfällen. Fortschritt-Berichte VDI-Reihe 12 Nr. 722, VDI-Verlag, Düsseldorf, 2010
7. Lang, A., Kühn, M.: Motorrad fahren in Gruppen. Unfallforschung kompakt Nr. 103, Unfallforschung der Versicherer, Berlin (2020)
8. Shi, E., Gasser T.: Classification of automated driving functions. In: Handbook of driver assistent systems, 4th edn. Springer-Verlag, Wiesbaden; 2022SAE International: Taxonomy and Definitions for Terms Related to Driving Automation Systems for On-Road Motor Vehicles. SAE Standard J 3016, überarbeitete Version September (2016)
9. Lie, A., Tingvall, C., Krafft, M., Kullgren, A.: The effectiveness of ESC (Electronic Stability Control) in reducing real life crashes and injuries. In: 19th International technical Conference on the Enhanced Safety of Vehicles Conference (ESV), Washington D. C (2005)
10. Unselt, T.; Breuer, J.; Eckstein, L.; Frank, P.: Avoidance of "loss of control accidents" through the benefits of ESP. In: FISITA 2004 World Automotive Congress. Barcelona: Sociedad de Tecnicos de Automocion (2004)
11. HLDI: Compendium of HLDI collision avoidance research. Highway Loss Data Institute (HLDI), Bulletin, vol. 37, no. 12, Arlington, VA 22203, USA (2020)
12. Hummel, T., Kühn, M., Bende, J., Lang, A.: Fahrerassistenzsysteme - Ermittlung des Sicherheitspotentials auf Basis des Schadengeschehens der Deutschen Versicherer. Forschungsbericht FS 03, Unfallforschung der Versicherer, Berlin (2011)
13. Hannawald, L., Kauer, F.: Equal Effectivness Study on Pedestrian Protection, TU Dresden (2003)
14. Benmimoun, M., Kessler, C., Zlocki, A., Etemad, A.: euroFOT: Feldversuch und Wirkungsanalyse von Fahrerassistenzsystemen – erste Ergebnisse. 20. Aachener Kolloqium "Fahrzeug- und Motorentechnik" Aachen (2011)
15. Petersen, E., Scholze, C., Falke, C.: Notbremsassistenzsystemeim Lkw – eine Analyse niedersächsischer Autobahnunfälle in den Jahren, 2015 Bis 2019 und der Einfluss aktueller Systeme Zeitschrift für Verkehrssicherheit 4 (2020)
16. Verordnung (EU) 2019/2144 des europäischen Parlaments und des Rates vom 27. November 2019: Amtsblatt der Europäischen Union, L 325/1 (2019)
17. Kraftfahrtbundesamt: Aktion Abbiegeassistent - Freiwillige Aus- und Nachrüstung von Lastkraftwagen und Bussen mit Abbiegeassistenzsystemen. ► https://www.kba.de/DE/Typgenehmigung/Typgenehmigungen/Typgenehmigungserteilung/Abbiegeassistent/abbiegeassistent.html. Abgerufen am 22 Feb 2021
18. SimTD: Sichere Intelligente Mobilität - Testfeld Deutschland, Deliverable D5.5, TP5-Abschlussbericht – Teil A, Version 1.0 (2013)
19. Lindenau, M., Kühn, M.: Intelligente Systeme zur Verbesserung der Motorradsicherheit. Unfallforschung kompakt Nr. 73, Unfallforschung der Versicherer, Berlin (2017)
20. Kühn, M., Bende J.: Automated cars on motorways - Active and passive safety aspects. Unfallforschung kompakt Nr. 99, Unfallforschung der Versicherer, Berlin (2019)
21. Kühn, M.: Hochautomatisiertes Fahren im Mischverkehr auf der Autobahn. Unfallforschung kompakt Nr. 100, Unfallforschung der Versicherer, Berlin (2020)
22. Kühn, M., Vogelpohl, T., Vollrath, M: Übernahmezeiten beim hochautomatisierten Fahren. Unfallforschung kompakt Nr. 57, Unfallforschung der Versicherer, Berlin (2016)
23. Kühn, M., Vogelpohl, T., Vollrath, M.: Müdigkeit und hochautomatisiertes Fahren. Unfallforschung kompakt Nr. 70, Unfallforschung der Versicherer, Berlin (2017)
24. EuroNCAP: Fahrerassistenz-Tests (2020). ► https://www.euroncap.com/de/fahrzeugsicherheit/sicherheitskampagnen/2020-fahrerassistenz/. Recalled 22 Feb 2021

Open Access This chapter is licensed under the terms of the Creative Commons Attribution-NonCommercial-NoDerivatives 4.0 International License (▶ http://creativecommons.org/licenses/by-nc-nd/4.0/), which permits any noncommercial use, sharing, distribution and reproduction in any medium or format, as long as you give appropriate credit to the original author(s) and the source, provide a link to the Creative Commons license and indicate if you modified the licensed material. You do not have permission under this license to share adapted material derived from this chapter or parts of it.

The images or other third party material in this chapter are included in the chapter's Creative Commons license, unless indicated otherwise in a credit line to the material. If material is not included in the chapter's Creative Commons license and your intended use is not permitted by statutory regulation or exceeds the permitted use, you will need to obtain permission directly from the copyright holder.

Safety of Assisted and Automated Systems

Ulf Wilhelm and Susanne Ebel

Contents

© The Author(s) 2026
H. Winner et al. (eds.), *Handbook Assisted and Automated Driving*,
https://doi.org/10.1007/978-3-658-45276-6_6

6.1 Overview of Existing Safety Standards

Driver assistance and automated systems increasingly act and interact in and with the environment. Depending on the level of automation, increasingly complex environmental sensors (e.g., ultrasonic, radar, video, or laser sensors) are used to adequately perceive the environment. However, even with a surround view through the sensors, reality cannot be fully perceived. In addition, the behavior of road users must be predicted in order to evaluate possible courses of action. Both lead to uncertainties in the system behavior, which is generally described as functional insufficiency [1]. These uncertainties can lead to unintended system behavior. The same applies to hardware and software faults.

Regardless of the cause, life-threatening situations can quickly arise as soon as driver assistance and automated systems actively intervene in traffic events. This must be prevented, but a residual risk remains. Only after the remaining residual risk has been determined more precisely a decision can be made whether it is reasonable or not. In order to be able to make this decision objectively procedures from the field of product safety must be applied to product development.

According to the German Product Safety Act ProdSG[1] §3(2). [2] "*…a product may only be made available on the market if it does not endanger the safety and health of persons when used as intended or in a foreseeable manner.*" Despite all technical measures to avoid hazards, a residual risk may still exist. In this case, according to ProdSG §6 (1), it is necessary "*…to provide the consumer with the information he or she needs to assess and protect himself or herself against the risks associated with the consumer product during its normal and reasonably foreseeable period of use and which are not immediately apparent without appropriate information.*"

Standards and other technical specifications can be used as a basis for assessing whether a product is really safe (▶ Chap. 40). The standards for functional safety, safety of the intended functionality (SOTIF) and cybersecurity are relevant for product safety, cf. ◘ Fig. 6.1.

Here, safety is defined as "freedom from unreasonable risk," where, according to [2] risk is defined as the combination of the probability of occurrence of a hazard and the severity of the possible damage (▶ Chap. 40).

In the current standardization activities, various sub-areas have emerged for this comprehensive definition of safety, each of which makes different contributions to the safe product.

First and foremost is functional safety. Requirements for the functional safety of electrical, electronic, and programmable electronic systems (E/E systems) in general are summarized in the technical standard IEC/EN 61508 [3]. For the automotive industry, the relevant standard is ISO 26262, published in 2011 and updated in 2018 [4]. It addresses possible hazards caused by the malfunctioning behavior of safety-related E/E systems including the interaction of these systems and includes guidance to mitigate these risks by providing appropriate requirements and processes. The objective is to assess the hazard resulting from the malfunctioning behavior and implement appropriate measures to avoid systematic failures (e.g., faults in the software) and to control random hardware failures (e.g., faults in the electronic component) that may trigger the malfunction. In this way, the remaining residual risk is reduced to a reasonable level. Depending on the risk potential, the specified methodology identifies measures to reach a safe product within the framework of functional safety. In deriving the measures, the development team relies on models that describe the cause-effect relationships. Models are abstractions that allow conclusions about complex systems. If these conclusions can be automated, they are used by technical systems to calculate the appropriate responses to events. Model building is therefore essentially responsible for the design and evaluation of a system.

As a basic assumption of ISO 26262, fully described and sufficiently validated models of the system are assumed. Through the specified methodology and with the help of the assumed models a largely correct implementation is ensured. The more complex the environment to be considered becomes the more difficult it is to specify a consistent, fully described model for it. Especially in the area of driver assistance and automated systems, the development is based on incomplete models of the problem. The development team cannot derive a completely valid implementation from the incomplete models. The resulting gaps in the implementation, which are also described as functional insufficiencies, potentially lead to unintended system behavior to the same extent as software and hardware faults. However, it is often not possible to generate better solutions with reasonable effort. This contradicts the requirements of ISO 26262 in which no room for argumentation is provided for known functional insufficiencies that lead to potentially hazardous behavior.

If system reactions are of an informative nature only, no harm is usually triggered due to the incomplete models. As soon as driver assistance and automated systems also take over driving tasks there is a risk of harm to road users.

To fill this gap, another standard was introduced specifically for driver assistance and automated systems. In 2019, the specification (Publicly Available

---

1   Product Safety Act (German: Produktsicherheitsgesetz) only in German language available.

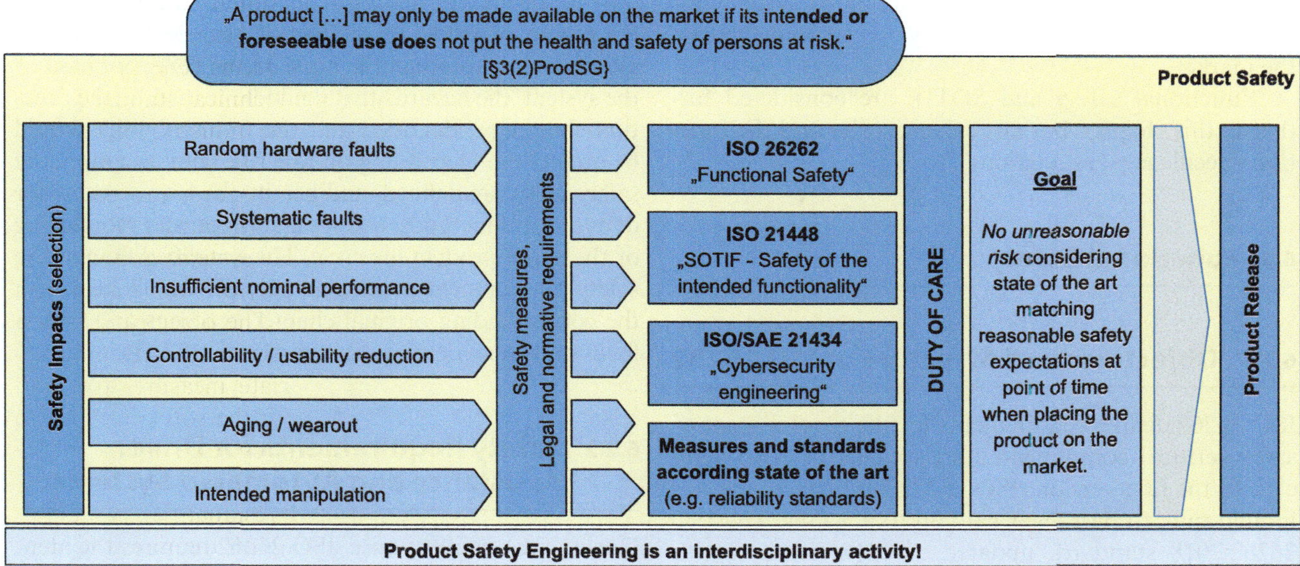

Fig. 6.1 Sub-areas of product safety

Specification, PAS) ISO/PAS 21448 [5] was published, which is commonly referred to as SOTIF for "Safety of the intended functionality". Based on this, the ISO 21448 "Road vehicles—Safety of the intended functionality" standard is currently being developed to replace the PAS. The ISO/FDIS 21448[2] [1] was published in February 2022. The focus here is on the safety of the intended functionality. Even without implementation faults, hazards to road users can occur during the intended use or misuse. These hazards occur due to functional insufficiencies or misuse of the function. Within the scope of the standard, reasonably foreseeable misuse must be covered but not intentional misuse. At the end of the development, the risk must be acceptable for intended use or reasonably foreseeable misuse. This also covers a large part of the safety in use, but there may also be aspects from the safety in use that go beyond the SOTIF standard.

In order to protect the system against external attacks the standard ISO/SAE 21434 "Road vehicles—Cybersecurity engineering" (as of June 2021) [6] is about to be published. The focus here is on protecting the safety-relevant system against threats from the system environment. The risk of external attacks that deliberately undermine safety measures or deliberately provoke previously unknown unsafe system behavior and thus have a negative impact on product safety must also be reasonable.

These three sub-areas of functional safety, SOTIF, and cybersecurity form an integral approach to ensure product safety. Specifically for automated systems, this integral approach is described in the technical report ISO/TR 4804 "Road vehicles—Safety and cybersecurity for automated driving systems—Design, verification and validation" published in 2020 [7]. Based on twelve guiding principles, the framework for the development of safe automated systems is defined. The next step will be to further concretize this as ISO Technical Specification TS 5083.

The cross-functional formulation also differentiates the standards from these three subfields to function-specific standards, such as ISO 15622:2018 for Adaptive Cruise Control (ACC) [8]. These describe the functional areas, scopes, and minimum requirements related to the respective driver assistance or automated function (▶ Chap. 40). They also describe test methods that can be used to verify that the requirements are met. As a rule, however, these function-specific standards only define the correct functional design, also considering social acceptance, but they do not contain any concrete safety argumentation.

In addition to standards, there are also UN/ECE regulations that are binding for vehicle approval. These are internationally harmonized technical regulations drawn up by the United Nations Economics Commission for Europe. UN/ECE regulations exist for different vehicle parts, such as UN/ECE 79 "Steering equipment [9] or functional specifications such as UN/ECE 152 "Advanced Emergency Braking Systems AEBS" [10]. UN/ECE 157 "Automated Lane Keeping System (ALKS)" was published in January 2021 specifically for highly automated driving [11]. This paves the way

---

2   The final standard ISO 21448 is about to be published. There will be no changes to the FDIS in terms of content. For this reason, the final standard will be named in this document.

in Germany for the first highly automated systems for traffic jam situations on highways to be launched on the market.

Functional safety and SOTIF are considered further in this chapter, but not cybersecurity and the function-specific legal regulations.

6.2 Functional Safety

6.2.1 Objectives and Structure of ISO 26262

ISO 26262 defines requirements for the development of safety–critical components and systems of road vehicles. In the first version, ISO 26262:2011, the scope was limited only to passenger cars up to 3.5 tons. The ISO 26262:2018 standard, updated in 2018, removes this weight restriction and the scope of the standard has been extended to other road vehicles such as heavy passenger cars, trucks, buses, and motorcycles.

The procedure described in ISO 26262 is based on the general V-Model of product development. At the beginning of the product life cycle, in the concept phase of the system, the hazards that could be caused by the function should be identified and the resulting risks should be quantified. Depending on this risk determination, the safety goals are defined, and hereby the requirements for development methods, quality assurance, and monitoring of the entire product life cycle. The simplified flow of the safety development process as per ISO 26262:2018 with the corresponding original chapter headings and with a focus on system development is shown in ◘ Fig. 6.2.

6.2.2 Safety Requirements for Driver Assistance and Automated Systems

During the concept phase, ISO 26262 requires the identification of hazards associated with an E/E system and the evaluation of the risk potential. This is done on the basis of hazard analysis and risk assessment (HARA),

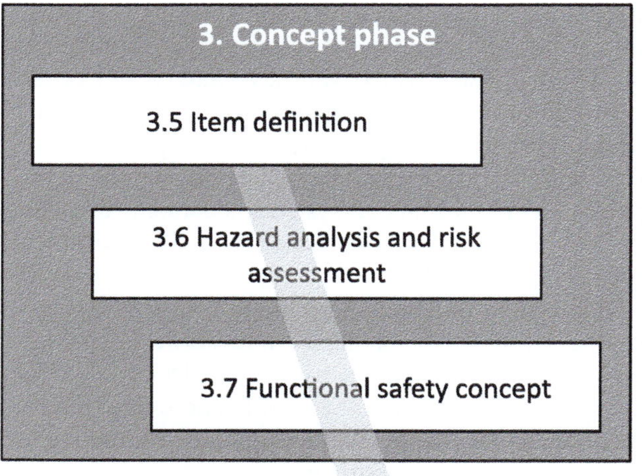
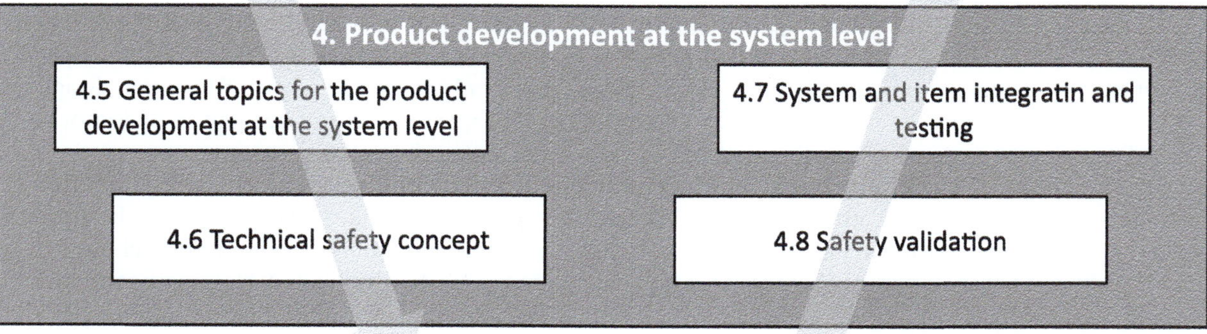

◘ **Fig. 6.2** Development process according to ISO 26262:2018 with original headings and with focus on system development

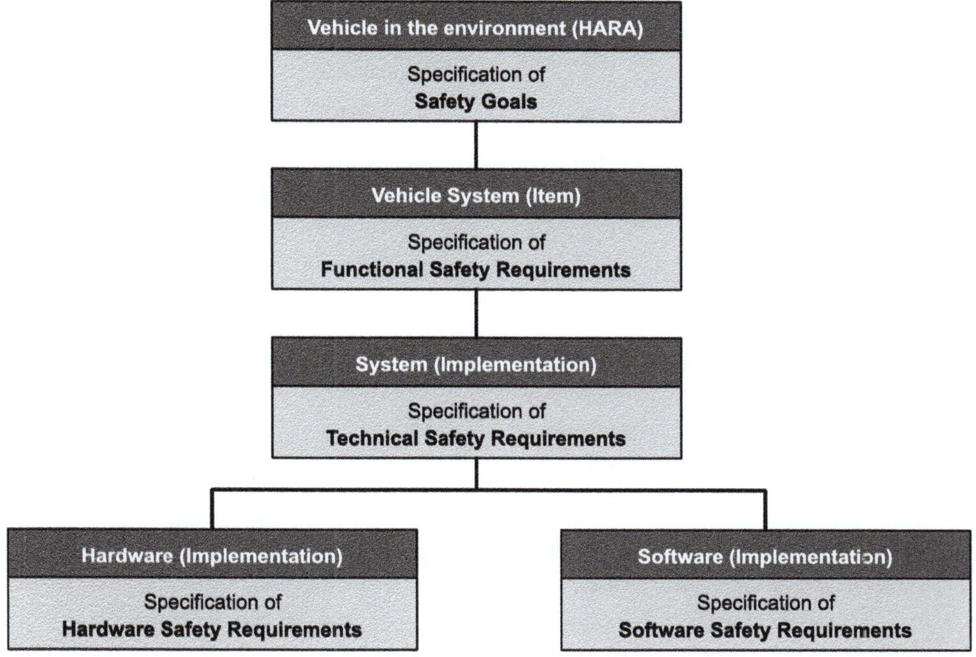

Fig. 6.3 Hierarchical structure of safety requirements

a method whose application is normatively prescribed in Part 3 of ISO 26262 (3–6 in ◘ Fig. 6.2). In the scope of HARA, potential hazards are analyzed at the vehicle level without considering the causes and the identified risks are classified. In this context, vehicle level means according to [12] the realization of one or more systems to which ISO 26262 is applied and thus denotes the highest level of abstraction in the overall "vehicle" system. To prevent the hazards defined in the scope of HARA, ISO 26262 requires the specification of safety goals. These provide the framework for the development of the safety concept. The hierarchical structure of safety requirements is shown in ◘ Fig. 6.3. The safety goals are defined at the vehicle level, taking into account the influencing factors and situations from the environment. In the vehicle system these safety goals are derived as functional safety requirements to the functions involved in the hazard (3–7 in ◘ Fig. 6.2). Together with the documentation of the problem solution, the functional safety requirements and their derivation form the functional safety concept. This derivation is still "solution-free" and thus independent of the actual implementation. The implementation of one or more functions involved is only done at the concrete system level, addressed by the technical safety requirements as part of the technical safety concept (4–6 in ◘ Fig. 6.2). Comparable to the functional safety concept, the technical safety concept is also composed of the technical safety requirements and their derivation, including the documentation of the problem solution. In contrast to functional safety requirements, technical safety requirements describe the implementation of the system and, in the next derivation step, hardware and software safety requirements are given in detail accordingly for implementing the hardware and software (5 and 6 in ◘ Fig. 6.2) are detailed.

In the following, the hierarchical structure of safety requirements is presented in more detail using the example of the driver assistance function "autonomous emergency braking" (AEB). The AEB is supposed to perform an automatic emergency braking in case of an imminent collision with the vehicle in front. Depending on the strength of the intervention, the function is designed for minimizing or preventing damage.

6.2.2.1 Specification of Safety Goals

Basic Principles

ISO 26262 specifies the methodology for the analysis and classification of hazards normatively. In a hazard analysis and risk assessment (HARA), the malfunctioning behavior of E/E systems in road vehicles is identified and the resulting risk is evaluated. This is done based on the following three risk parameters:

- Frequency of the driving situation (exposure): How often do the driving situations occur, in which a driver or other road users can pose a hazard?
- Controllability: How well are the driver or other road users able to control the hazardous event in this driving situation so that damage can be avoided?
- Severity: When damage occurs, what is its severity?

Table 6.1 ASIL determination matrix

Severity	Exposure	Controllability			
		C0 (general)	C1 (simple)	C2 (normal)	C3 (severe)
S0 (no injuries)	**E0–E4**	n/a	n/a		
S1 (light/moderate injuries)	**E0** (incredible)		n/a		
	E1 (very low)		QM	QM	QM
	E2 (low)		QM	QM	QM
	E3 (medium)		QM	QM	A
	E4 (high)		QM	A	B
S2 (severe and life-threatening injuries)	**E0** (incredible)		n/a		
	E1 (very low)		QM	QM	QM
	E2 (low)		QM	QM	A
	E3 (medium)		QM	A	B
	E4 (high)		A	B	C
S3 (life-threatening injuries, survival uncertain/ fatal injuries)	**E0** (incredible)		n/a		
	E1 (very low)		QM	QM	A
	E2 (low)		QM	A	B
	E3 (medium)		A	B	C
	E4 (high)		B	C	D

The product *risk* consists of the factors of *severity of damage* and *frequency*. In relation to the risk parameters mentioned above, the severity of damage is evaluated with the parameter of the same name. The frequency results from the *frequency of the driving situation* and the *controllability*. The higher the frequency of the driving situation, the higher is the resulting risk potential. The risk can be reduced only by means of high controllability, which will be present when the road users involved in the critical situation have the possibility to prevent the harm.

The result of the risk assessment is the risk potential, which is evaluated with an ASIL (Automotive Safety Integrity Level) in the classes QM, ASIL A, ASIL B, ASIL C, and ASIL D. Here, ASIL A corresponds to the lowest and ASIL D to the highest risk potential. If a hazard is classified only with QM (Quality Management), then the application of a certified quality management system is sufficient and the further application of ISO 26262 to the development process is not required.

The ASIL classification is defined concretely based on the evaluation of the individual risk parameters according to ◻ Table 6.1. Each risk parameter is divided into four to five classes. Each class corresponds to a different meaning. In the appendix of Part 3 of ISO 26262, there are informative tables with references and examples of which risk parameters are to be used to evaluate the severity of damage caused by ac-

cidents, driving situations, and controllability by the driver. Thus, driving situations that occur almost every time a car is driven (e.g., driving on a country road) are rated E4, and driving situations that occur at least once a month (e.g., driving with a trailer, traffic jams) are rated E3. Even rarer driving situations are then classified as E2 (e.g., driving on ice and snow) or E1 for very rare situations (vehicle on a chassis dynamometer). This example itself shows that an objective evaluation is not always possible, as driving situations can be evaluated differently based on the considered environment. For example, driving on ice and snow in the northern regions of Europe will certainly be evaluated with a higher exposure than in the southern regions.

Safety goals are defined for ASIL A–D classifications. Derived from the safety goal and the associated ASIL classification, the measures for the prevention and control of systematic failures and random hardware failures are specified normatively. According to ◻ Fig. 6.4 the risk is to be reduced so far that the remaining risk is below the residual risk according to application ISO 26262. In this context, to ensure safety more methods and measures must be applied at ASIL D than at ASIL A. A QM rating according to ISO 26262 expresses that a hazard and a risk basically exist ($E \geq E1$, $C \geq C1$, and $S \geq S1$). However, to reduce this risk at least to a reasonable level, the application of a standard QM system is sufficient. Only if the malfunction behavior can generally be controlled by the average

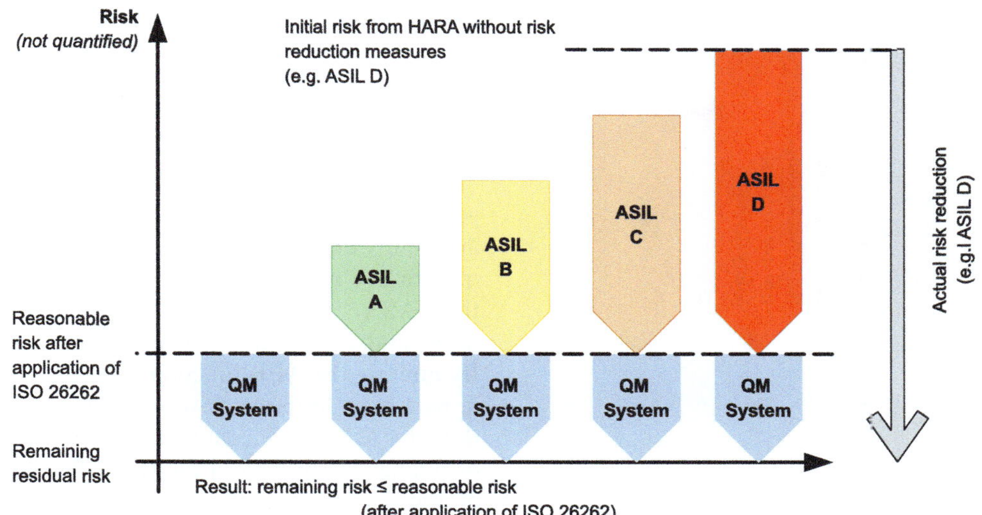

Fig. 6.4 Risk reduction according to ISO 26262

driver (C = C0), no injuries occur (S = S0) or the driving situation is classified as approximately improbable (E = E0), no requirements at all are derived in the scope of functional safety.

Example: Safety Goals for AEB

The AEB function controls the actuator "Brake". Based on the actuator function "braking," the following malfunctions can be determined as potential hazards:

a. Unintended automatic deceleration
b. Unintended increase of deceleration during driver braking
c. No deceleration despite deceleration request
d. Too low deceleration despite deceleration request
e. Unintentional holding force during standstill

In the first step, the risk potential of the hazards of the "brake function" is determined independently of the driver assistance function. This is done on the basis of the HARA with the risk parameters severity of damage (S), frequency of driving situation (E), and controllability (C) according to **Table 6.2. There, an ASIL C is determined for the hazards (a) and (b). The assessment is performed without taking safety measures into account (e.g., deactivation of braking by the driver or limitation of braking intervention).

The ASIL that is determined is used as a benchmark for risk reduction for the further development of the product. Here, safety goals are specified to avoid the hazard of "rear-end collision". As safety requirements are at the highest level of abstraction, these are the starting point for the development of the safety concept. In addition, the "safe state" has to be specified for each safety goal. For the example from **Table 6.2 the derived safety goals and the safe state are described in **Table 6.3.

Safety goals address all systems that can directly or indirectly access the actuator under consideration and thus have the potential to cause the hazards (a)–(e). The hierarchical structure of safety requirements from **Fig. 6.2 requires the specification of safety requirements from the vehicle level down to the hardware and software level in more detail. This safety requirements flow ensures the fulfillment of the given safety goals and thus the prevention of the identified hazards.

6.2.2.2 Specification of Safety Requirements

Basics

Safety requirements are specified starting from the safety goals at each level of the vehicle and system architecture, cf. **Fig. 6.3. At the vehicle level, the safe behavior of the function that can potentially violate the safety goals is specified as functional safety requirements. This is done for all systems involved and is part of the specifications of the individual system suppliers. The functional safety concept (3–7 in **Fig. 6.2) describes the assumptions and solutions used to derive the safety goals for functional safety requirements.

The specifications of the functional safety requirements are drawn up by the responsible for the vehicle system, this is usually the OEM. The system suppliers are then responsible for specifying and implementing the safety concept in such a way that the functional safety requirements for the system are not violated. The technical safety concept contains the documentation of the derivation of the functional safety requirements to technical safety requirements. This contains the concrete solution at the system level for avoiding the violation of the functional safety requirements and thus implicitly for avoiding the violation of the higher-level safety goals.

Table 6.2 HARA using the example of the vehicle function "Deceleration"

No	Malfunction	Situation	Hazard	S	Reason S	E	Reason E	C	Reason C	ASIL
(a) (b)	Unintended deceleration or unintended increase of the deceleration (assumption: vehicle remains stable)	City driving/country road/highway, heavy traffic with too short safety distance	Rear-end collision by the vehicle behind	2	Accident statistics show light to severe injuries	4	Heavy traffic in almost all driving situations	3	Difficult for following driver to control due to too short safety distance	C
⋮	⋮	⋮	⋮	⋮	⋮	⋮	⋮	⋮	⋮	⋮

Example: Safety Requirements for AEB

An example of safety requirements of the AEB function at the vehicle and at the system level is given in ◘ Fig. 6.5. The components of the brake actuator function are the "Driver Assistance System" (DAS) and the "Electronic Stability Control" (ESC) system. The safety goal "Avoid unintended deceleration that leads to a hazard" must now be distributed between the two systems in such a way that it can be fulfilled with ASIL C. For the DAS system, this means that an unjustified AEB request that could lead to a hazard has to be prevented. By limiting the AEB function in terms of duration and, in part, even in terms of strength, the risk potential and thus the ASIL classification can be reduced as a shorter or weaker braking has a positive influence on the extent of the damage as well as on the controllability. In the example in ◘ Fig. 6.5 the AEB function was limited to such an extent that the remaining risk potential for the driver assistance system can also be reduced to an ASIL A. However, the HARA described in ISO 26262 reaches its limits when it comes to concretely determining this risk reduction. As a rule, it is difficult to evaluate brake interventions with different braking profiles according to their risk potential due to the rough classification of risk parameters and the associated subjective assessment. In this case, objectified methods, such as simulations based on the equations of motion or the use of endurance data and accident statistics, can help to determine the ASIL classification of different brake profiles more precisely.

However, with the allocation of the functional safety requirement "Avoid unintended deceleration within the AEB limits" to the DAS system, the overarching safety goal "Avoid unintended deceleration that leads to a hazard" is not yet completely fulfilled. In addition, it has to be ensured with ASIL C that due to a fault, AEB deceleration is done outside the specified limits. If the DAS system is not capable of ensuring limitation with ASIL C, then this safety measure must be implemented in another ASIL C capable system. As a rule, the limitation of the AEB limits take place in the ESC system.

The technical safety requirements exemplify the concrete implementation of the functional safety requirement in the DAS and ESC system.

The example in ◘ Fig. 6.5 shows the allocation of a safety goal to the systems involved in the execution of AEB braking. The safety goal at the vehicle level was divided into two safety goals related to the specific safeguarding of the AEB function. With this split, a re-evaluation of the risk potential for deceleration within the AEB braking profile is performed. The limitation of the function is a very effective measure to reduce the risk potential and thus the ASIL classification for those systems that are not able to safeguard the initial risk. However, it must be noted that for at least one system a safety requirement with the ASIL rating must be assigned from the higher-level safety goal.

▫ Table 6.3   Safety goals using the example of the vehicle function "deceleration"

No	Safety goal	Safe state	ASIL
(a)	Avoid unintended automatic deceleration that leads to a hazard	Interrupt automatic deceleration	C
(b)	Avoid unintended increase of the deceleration that leads to a hazard	Interrupt increase of the deceleration	C

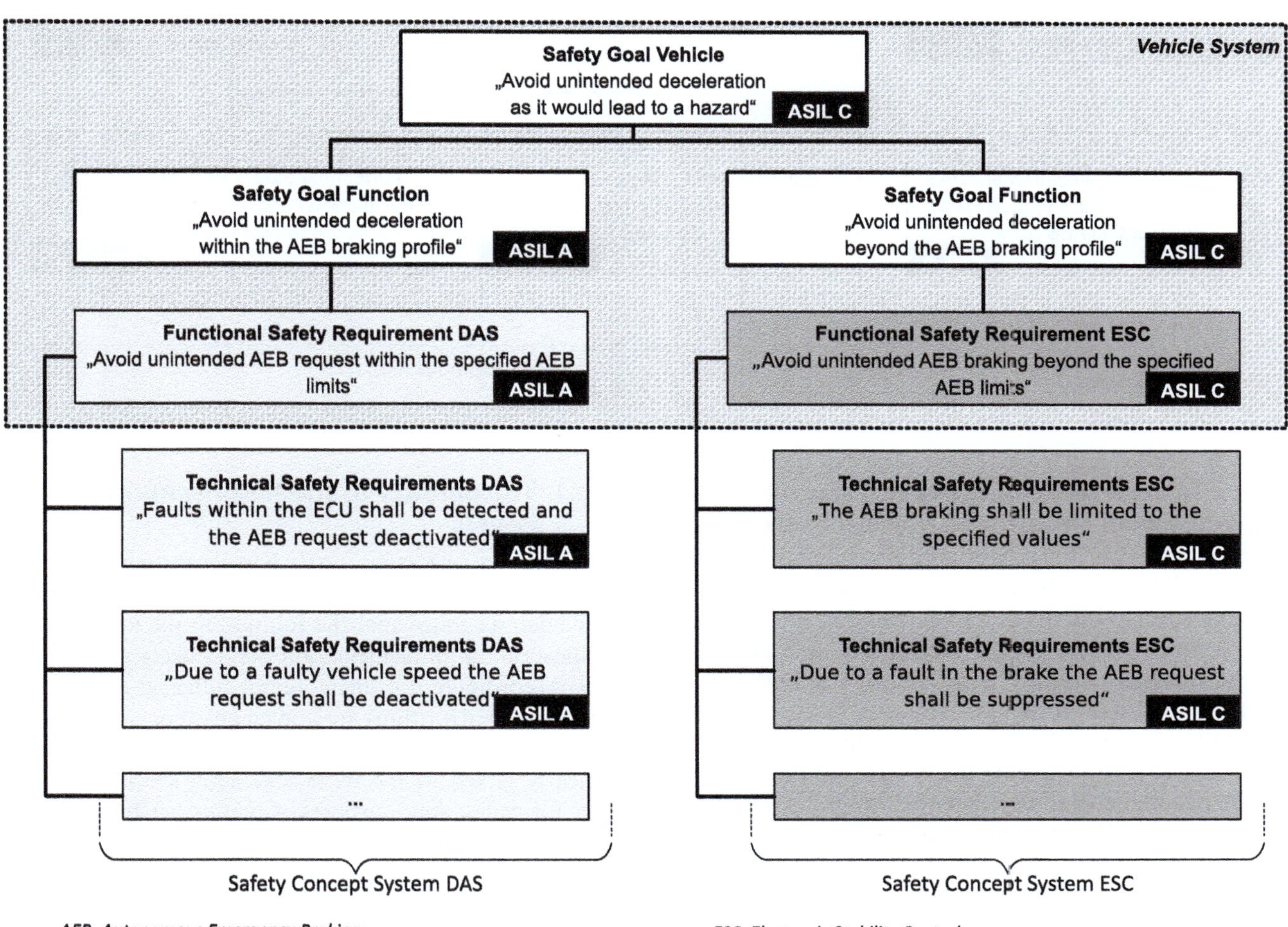

▫ Fig. 6.5   Hierarchical structure of safety requirements using the example of the AEB

ISO 26262 provides yet another way to reduce the ASIL classification to the systems involved. This method is described in the next chapter.

ASIL Decomposition

In addition to the allocation of safety requirements, in which the ASIL is assigned to at least one derived safety requirement, ISO 26262 also offers the possibility of decomposition. In this case, the ASIL of the derived safety requirements can be reduced. However, this is only possible if redundancy is present. According to ISO 26262:2018–9, Sect. 5.4, decomposition and thus reduction of the ASIL can only be performed if the safety requirement "before decomposition" (initial safety requirement) can be realized by at least two *sufficiently in-*

dependent elements or subsystems. Each derived safety requirement must therefore be able to fulfill the initial safety requirement "on its own". In this context, elements or subsystems are considered as "sufficiently independent" if, based on the analysis of dependent failures (see ISO 26262:2018–9, Chap. 7), no *Common Cause Failure* (CCF, see ISO 26262:2018-1, 3.18) or *Cascading Failure* (see ISO 26262:2018-1, 3.17) is identified that can lead to a violation of the initial safety requirement. Common cause failure in this context refers to failure due to a common reason, while a cascading failure causes further failures subsequently [12].

In ▫ Fig. 6.5 the ASIL is inherited by the systems involved. In this case, the ESC system receives the ASIL of the higher-level safety goal. The reduction of

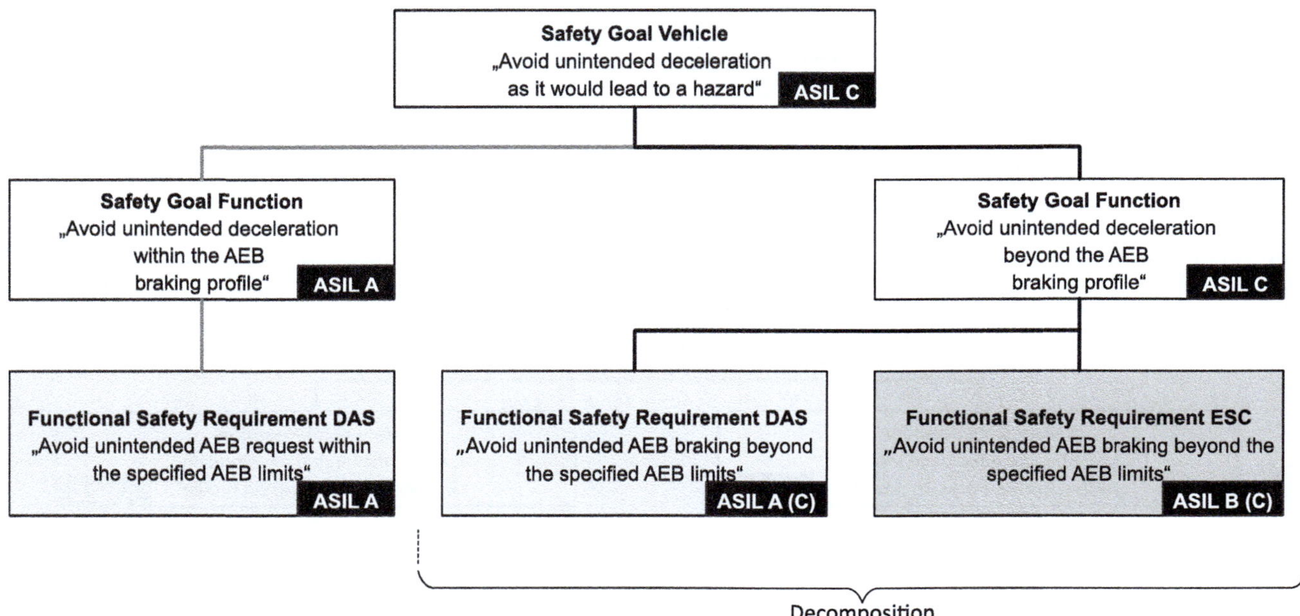

Fig. 6.6 Application of decomposition with the example of AEB

the ASIL in the case of the DAS system is not achieved by decomposition (there is no redundancy here) but by a reassessment of the risk potential with a limited braking function. Since the safety mechanism for safeguarding the specified limits is implemented in an independent control unit, it is perfectly permissible according to ISO 26262 to reduce the ASIL in the DAS system to the actual risk potential. However, if hypothetically no ASIL C can be implemented in the system ESC, it is conceivable to distribute the ASIL C to both systems by also safeguarding the specified limits in the system DAS. ISO 26262:2018 offers various options for reducing the ASIL rating in part 9, Chap. 5. However, the initial ASIL should always still be included in brackets. For example, an ASIL C can be decomposed into an ASIL A (C) and an ASIL B(C). For the safety requirements on the AEB function, the change would then result in **Fig. 6.6 result.

6.2.3 Compliance with Safety Requirements

After the risks emanating from a product have been systematically worked out with the aid of a HARA (3–6 in **Fig. 6.2) and safety requirements have been derived from the safety goals, ISO 26262 requires product development to ensure that these requirements are implemented.

According to the underlying V-Model [13] a complete requirements tree must first specify the product properties. This specification at all levels of detail and abstraction serves as the basis for the right side of the V-Model, the verification and validation of the product properties.

6.2.3.1 Traceability of the Requirement Levels

Assuming that the safety goals have been validated, according to ISO 26262 a product is safe precisely when it can be demonstrated that the solution in the form of the specification at implementation level meets the safety goals. Accordingly, it is therefore not sufficient to describe exactly which algorithms are implemented or which hardware is used and what properties it has. This solution must be clearly linked to the more abstractly specified safety goals for validation purposes. Creating such complete documentation is quite challenging for complex systems.

For example, in an initial attempt to specify a radar-based AEB system, the following requirements might end up in a specification document:

a. The electrical resistance of the heating wire for the radome must be xx Ohm.
b. The radar sensor must detect radar reflexes via a fixed-beam antenna.
c. AEB must not brake in situations without risk of an accident.
d. The driver must be able to override the AEB behavior at any time.
e. For object tracking a Kalman filter must be used.
f. The vehicle must not cause accidents through unwanted brake interventions.
g. On activation of the brake light switch the automated emergency braking must be canceled, and the deceleration must be adjusted to the value requested by the driver.
h. …

All sample requirements are valid product requirements. However, they are at very different levels of abstraction and detail. Requirement (f) corresponds to a general safety goal. Requirements (c) and (d) are abstract behavioral descriptions that do not yet define the concrete hardware and software solutions. Requirement (b) and (e) describe hardware components and software algorithms without anticipating implementation details. Requirement (a) in this example is the most concrete requirement that directly refers to a hardware implementation.

After deriving the functional safety concept (3–7 in Fig. 6.2), for example, requirement (d) was introduced as a safety requirement. The key question here is whether compliance with the internal resistance according to the requirement (a) is safety-relevant or "only" motivated by quality aspects. For this purpose, it is helpful to sort the requirements into a hierarchical requirements tree in a traceable manner.

The nodes of the tree connect general requirements with requirements that specify them: Fig. 6.7.

With each concretion, the solution space is further restricted. The tree can also be understood as a representation of nested solution spaces, see Fig. 6.8.

In this representation the benefit of traceability becomes particularly clear: All methods of ISO 26262 are designed to prove that the concretion and its implementation are subsets of the more abstract specifications. In the image of the subset, a solution is not allowed exactly when it leaves the solution space spanned by the abstract requirement.

In practice, it very often happens that hierarchy levels are inadvertently mixed. For example, the development team already requests a concrete algorithm at the behavioral description level of the FAS system. This leads to early, unnecessary constraints in the system design and also makes it harder to ensure traceability but is irrelevant for compliance with ISO 26262. ISO 26262 focuses on the seamless continuity of the requirements tree with respect to safety requirements.

It is a good idea to combine all the requirements belonging to a solution space into a system or subsystem description. For example, the requirements (c) and (d) are on the same description level and describe both the behavior of the function AEB. A complete set of such requirements is also called a model: It assigns an "output," the system response, to each "input," such as driver activity.

ISO 26262 roughly defines four hierarchy levels with the corresponding models:
(1) Abstract system level specified via the safety goals (requirement (f)).
(2) Solution-free product definition consisting of functional safety requirements (requirement (c), (d)).
(3) Requirements at the implementation level that may still affect several modules, i.e., do not yet relate to the smallest architectural element: technical safety requirements (requirements (b), (e), (g)).
(4) An implementation requirement affecting the smallest structural element: hardware/software safety requirements (requirement (a)) in this example only at the hardware level).

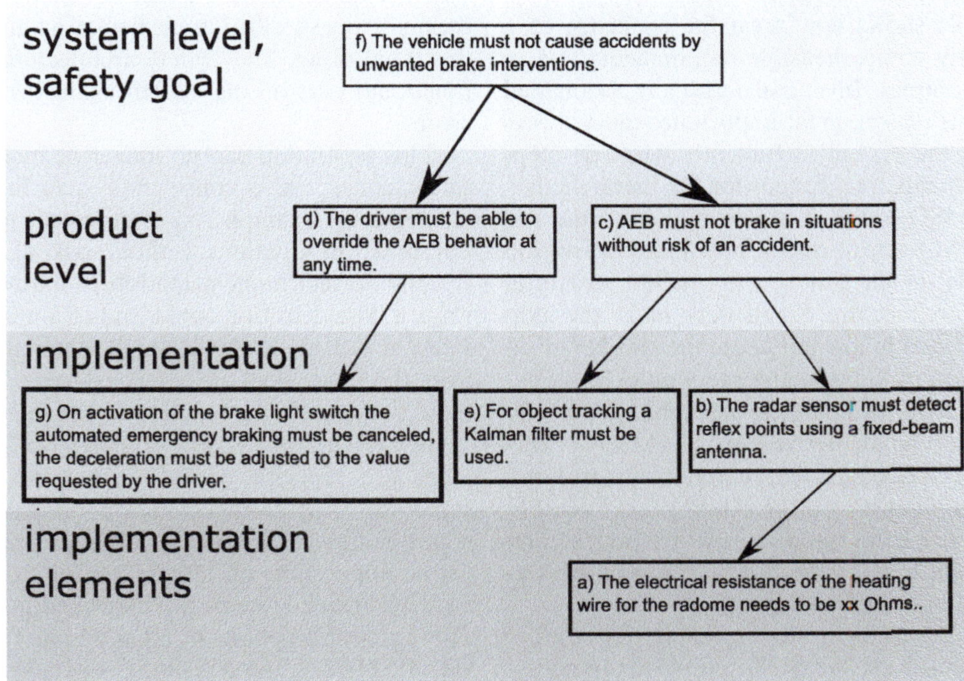

Fig. 6.7   The requirements can be structured in a requirements tree: This is the basis of traceability

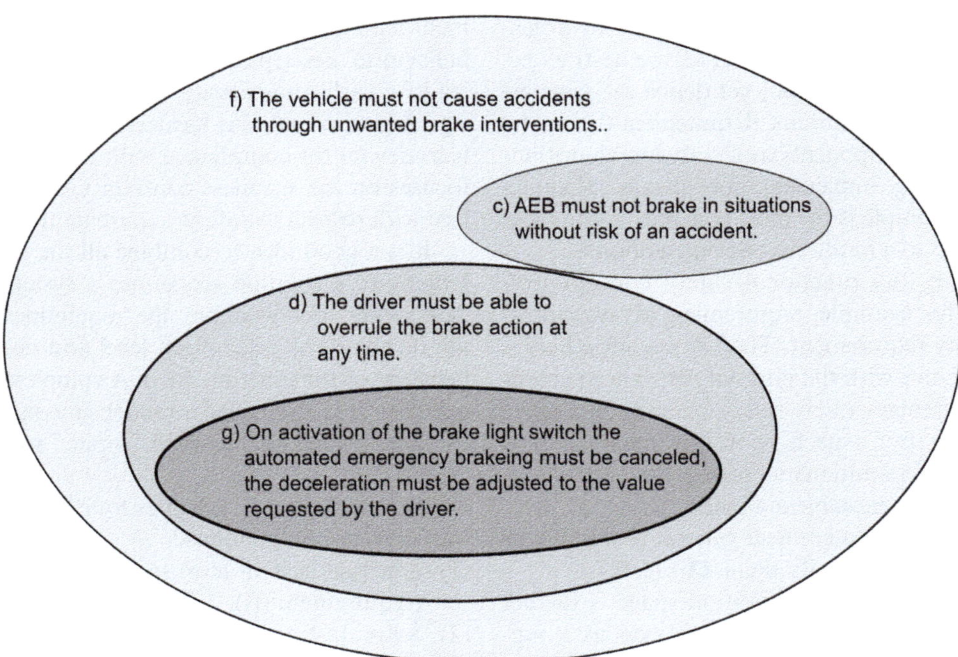

◘ Fig. 6.8 Alternatively, the requirement tree can also be represented in the form of nested solution spaces. In this representation, it is made clear that circular references are not allowed. The requirement "The system must not cause any accidents" defines a solution space that is restricted by the concrete solution "The driver must be able to override the system at any time"

The requirements tree sketched here as an example is incomplete in that no system can be developed using this simple "specification". In addition, however, the derivation steps are also too large: Is the heating wire property really not linked to requirement (f) (no unauthorized brake interventions)? This cannot be answered *unambiguously* from the few lines of text!

This example shows how great the challenge of a complete, clearly comprehensible requirements tree is, especially for complex driver assistance and automated systems. For this reason, great importance must always be attached to the system architecture when developing the requirements tree. Except for the hierarchy levels required by ISO 26262, it is generally up to the system development team to decide how many derivation steps to provide on the way to the smallest structural element of its architecture. Small steps make the individual step easier to follow but cause the number of elements of the specification to grow very quickly.

ISO 26262 requires complete traceability only regarding the safety goals derived in the HARA. However, since a well-documented requirements tree not only makes compliance with the safety goals traceable, but also generally helps product quality, the requirement for traceability is common in the automotive sector even without a safety reference.

In the example shown here, a functional requirement tree is created via the performance requirements of the AEB function and a separate requirement tree

from the safety requirement. ISO 26262 implicitly assumes in many places that a malfunction at the system level is always caused by a fault, i.e., a deviation from implementation requirements. In this case, the separation between safety architecture and functional architecture can be illustrated particularly well, since the safety requirements are essentially limited to the monitoring of the specified implementation and reactions to this monitoring. The "safety architecture" protects the functional part of the system against implementation errors.

This separation can no longer be maintained if the safety concept also considers system limits that have an effect, for example, on requirement (c) (AEB must not brake in situations without risk of an accident). Despite correct implementation in hardware and software, a false detection could lead to a misbehavior with safety relevance. In this case, all requirements resulting from the safety goal are safety relevant. Explicitly also those that specify the function of the subsystems.

In the current version of ISO 26262, such system limits are explicitly excluded from the scope of the standard: *"This document does not address the nominal performance of E/E systems."* The challenges this poses to the application of ISO 26262 for driver assistance and automated systems are described in [14] in detail. This publication points out that despite full compliance with ISO 26262, the remaining residual risk of systems with environmental sensors may still lie above the acceptable risk level (cf. ◘ Fig. 6.4). Thus, to reduce the residual risk to an acceptable level, additional measures must be implemented. The ISO 21448 provides the methods for deriving the necessary measures.

6.2.3.2 Verification

A prerequisite for robust verification is the complete system specification at the defined abstraction levels described in the previous section.

Verification aims at proving the input/output relation specified in the models on the corresponding elements of the product. This is the domain of requirements-based testing.

A derived model, for example, the concretion of requirements at the implementation level, is equivalent to the more abstract system specification. Accordingly, it would suffice to verify only the system-level model on the product. In practice, however, complete verification is not possible for more complex models. The more abstract the specification, the more input/output relations must be verified. In addition, the specification also becomes unambiguous only at the implementation level; only on the implementation level can a certain degree of completeness be practically achieved.

ISO 26262 attempts to mitigate this problem in two ways. On the one hand, verification is not only based on simple test methods or reviews, but, depending on the ASIL classification, recommends more advanced, more formalized methods that are known according to the state of the art (e.g., verification of software code by applying "abstract interpretation"). A more in-depth discussion of the different methods would go far beyond the scope of this overview. ISO 26262 itself also refers to established method descriptions. On the other hand, verification is repeated at all levels of detail. A simple software module can be tested much more completely than a complex, interacting system. In addition to verification at all levels of detail, there are also specific integration tests for each integration step identified in the design.

6.2.3.3 Validation

Verification, at its core, is concerned with proving that two equivalent models at the same level of abstraction—the specification and the implementation of that specification—are truly equivalent.

However, the safety goal derived from HARA is deliberately formulated in such an abstract and general way that design errors, i.e., incorrectly derived models, can also lead to a violation of these top-level requirements, cf. ◘ Fig. 6.9.

The phase 4–8 safety validation ("safety validation") of ISO 26262 (in ◘ Fig. 6.2) is specifically dedicated to the task of ensuring the validity of the derivations from more abstract requirements. In addition, the completeness and correctness of the actual safety goals are to be validated. This includes, for example, checking whether the system can be controlled by the driver in the "safe state". The required methods concentrate on system tests on the product and focus in particular on the robustness of the system with respect to "faults," i.e., implementation elements that fail to meet their specification due to errors.

In practice, according to [15] already 55% of all errors in the requirements and design phase occur during the concretion of abstract requirements to detailed technical models. These errors are particularly difficult to detect because all subsequent design steps and the generally more complete tests of the implementation levels cannot detect such errors.

ISO 26262 relies heavily on reviews by expert teams to detect such errors. A complete understanding of the design step by the expert teams according to the state

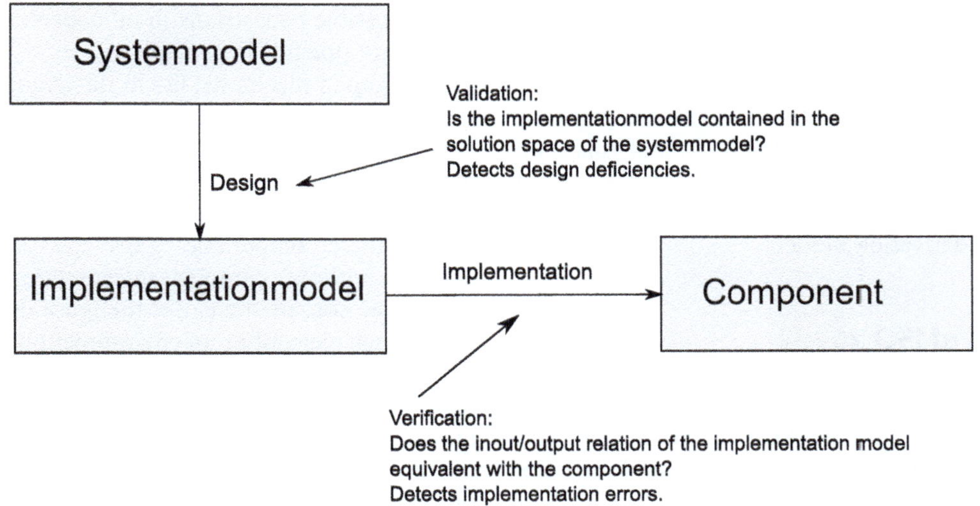

◘ **Fig. 6.9** Illustration of the term definition chosen here validation and verification. Validation focuses on the correctness of the design step: Vertical in the V-Model. Verification ensures the equivalence of implementation and implementation model: horizontally in the V-Model

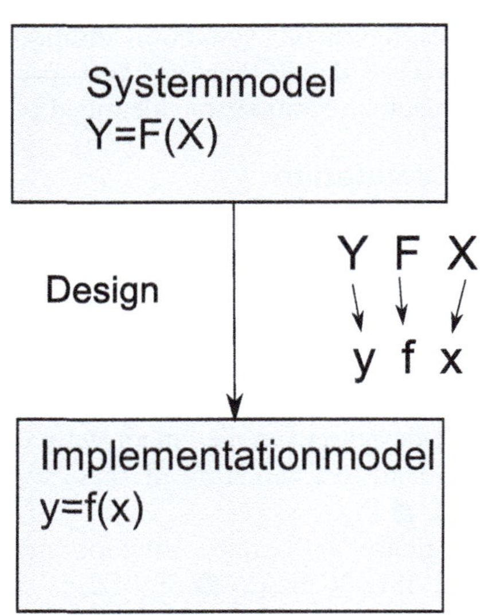

Fig. 6.10 The design combines the abstract with the concrete model. This is based on a "derivation model"

of the art is a prerequisite. Accordingly, ISO 26262 also emphasizes the value of simple designs as the basis for safe systems.

A design step that transforms an abstract system model into a concrete technical model is always based on a derivation model. Only with the help of such a derivation model is it possible to justify that the concrete model lies in the solution space of the abstract model. This justification is the basis of the validation of the concrete model, cf. **Fig. 6.10**.

If, for example, the design team chooses to fulfill requirement d) (**Fig. 6.8**: overdrive) an electromechanical brake light switch that is read into an ECU via an A/D converter, this decision is based on physical models of the operating principle of the switch and the A/D converter. These components are designed using the underlying derivative models such that, as long as the underlying hardware components meet their specifications, they will always perform their task by design.

ISO 26262 implicitly assumes that such a derivation model is accessible to the expert teams designing the system and to the teams reviewing it. Even if this is not explicitly required in the form of documents, a complete requirements tree can only be created with reference to such a derivation model.

6.2.4 Limits of ISO 26262

In the previous example, the safety goal "The driver must be able to override the AEB control at any time" can be achieved with a comparatively simple solution using the brake light switch.

But what if an emergency braking system designed to protect pedestrians has the safety goal of "never triggering in a situation where there is no risk of an accident"? In this case, the safety goal is equal to a system requirement that describes the actual function of the system. To separate a "safety architecture" from the core function is much more difficult, if not impossible. Traceability must accordingly be ensured along the functional path, which can lead to intrinsic gaps in complex, interpretive, and predictive systems.

Part of the system task is to predict the behavior of pedestrians in different situations: the system model thus maps all situations in which the behavior of the pedestrian, together with the behavior of the driver, will not lead to an accident to the system response "Do not react".

Is there a derivation model that makes it possible to derive an algorithm that maps the measured initial state of the situation to the predicted accident? This model would have to be able to predict the behavior of all possible pedestrians in all conceivable situations. This is not possible in general.

The gap in this model lies in the ignorance of the intention of the relevant traffic participants. This means that the prerequisites for the conventional validation methods, a complete derivation model according to the state of the art, are not given.

Are there comparable "gaps" also in conventional systems without complex interpretations or predictions? The core of the gap is the ignorance of the state of a system element or an environment property.

The validation discussed in the previous chapter relies on a complete understanding of the derivation model under the assumption that the hardware requirements underlying the design are met. The hardware properties are ensured as part of the manufacturing process and can be assumed for the validation argument if specified appropriately. It is more difficult to model the properties over time and under environmental conditions that are not precisely known.

The failure of a specific component after a known load leads to a principally predictable, systematic fault. After the failure has occurred, the cause and the failure process can be explained. However, for an accurate prediction, the exact nature of the device, possibly even at the atomic level, as well as the exact loads over the lifetime must be known.

Since this information cannot be known for all delivered products according to the current state of the art, individual failure appears to be random. In order to design the components reliably, statistical models are used in development. The ignorance of the individual properties of the component is thus averaged over a large quantity of components.

Like the statistical material model, the individual behavior of the pedestrian in a specific situation can also be statistically modeled in our example. The result is an algorithm which, over a larger set of similar situations, frequently estimates the situation correctly (in the example "true positive"), but also, with a certain probability, incorrectly (in the example "false positive"). An incorrect assessment that leads to undesirable behavior that contradicts the specification at the abstract system level is referred to below as an error. The probability of such a misbehavior is called "residual error probability".

This residual error probability can typically be tuned via model parameters in driver assistance and automated systems. In the case of a classifier with threshold, for example, the tuning parameter would be the threshold value. In contrast to clearing designs based on complete derivation models, "design-related" failure modes are known at the time of the product release. The probability of occurrence, and thus the relevance, are estimated with the help of system tests, if necessary, based on big data.

◘ Figure 6.11 shows the recognition performance of a classifier, which is adjustable by parameters. When misclassifications are reduced, i.e., "false positive" probability, the recognition performance, i.e., "true positive" probability, is reduced. Conversely, increased recognition performance is always accompanied by increased misclassification probability. The distance from the original straight line corresponds to the selectivity and thus the performance of the classifier. The maximum selectivity represents a compromise between recognition performance and misclassification probability.

For the design of such a system during the development time, residual error probabilities must be specified. This is also common for the design of hardware components. The common methods for modeling the fault frequency at the overall system level define this either quantitatively (Fault Tree Analysis, FTA) or qualitatively (Failure Mode and Effect Analysis, FMEA) via

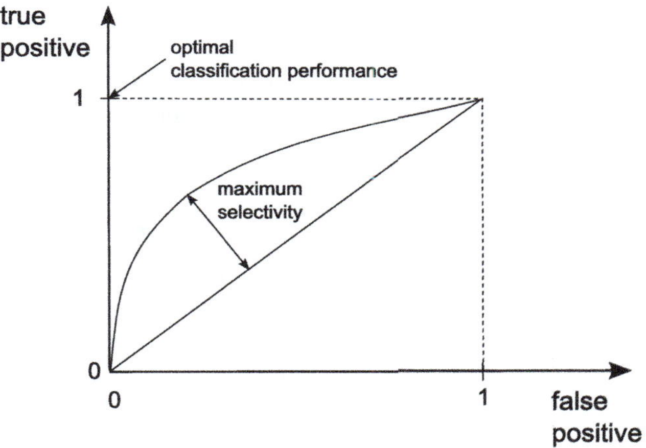

◘ **Fig. 6.11**   Recognition performance of a classifier

rough estimates. The design traceability gap described here is called a "functional insufficiency".

For systems with functional insufficiency, compliance with an upper limit [16] for the residual risk inherent in the system must be demonstrated. In the simplest case, a statistical "black box test" can be performed at the system level. This measures the probability of failure without knowledge of implementation details by performing tests on a statistically representative sample. For driver assistance and automated systems, this means a great many hours of validation under driving conditions that are representative of the function context.

Even for comparatively "uncritical" functions, depending on the derivation of the acceptable residual risk, residual error rates significantly smaller than 10^{-5} /h must be demonstrated. Unfortunately, the ISO 26262 does not provide any concrete information on the derivation of such residual error rates. For more critical functions with a higher degree of automation, the "black box test" can quickly become more expensive than the entire development or not realistically feasible within the time frame of a product development. It is also a particular challenge to prove that the test kilometers driven are sufficiently representative of the situations to be expected in the field. Which driving profiles, under which weather conditions, in which countries, with which drivers are to be carried out [16]? Statistical models of driver behavior are not always available to answer these questions easily. Here, it is often necessary to fall back on plausible expert assessments.

Accordingly, the statistical system tests are no substitute for a well-understood design derived as far as possible from robust models. During development, care must be taken to separate elements that can be fully derived from elements where gaps in the derivation model cannot be avoided. In the pedestrian safety example, imaging properties of a camera designed to detect the pedestrian do not need to be statistically modeled. Unambiguous physical models calculate an angle to the vehicle from the pixel assignment. The situation is different when the behavior of the individual pedestrian has to be predicted. Here, it is necessary to statistically model the degree of ignorance of the pedestrian's intent. The resulting algorithm will not correctly estimate intention in every situation. Residual error probabilities are the unavoidable consequence.

The described gaps in the derivation model are a potential safety risk in driver assistance and automated systems. However, ISO 26262 does not sufficiently address the safeguarding of the intended functionality, which is to be covered by the SOTIF standard ISO 21448 "Safety of the intended functionality".

6.3 Safety of the Intended Functionality (SOTIF)

6.3.1 Introduction to ISO 21448

The ISO 21448 standard provides a process model at an abstract level to ensure the safety of the intended function. Especially for driver assistance and automated systems, the validation methods provided by ISO 26262 are no longer sufficient, so SOTIF must be considered during development to implement a holistic safety concept. The standard provides a systematic approach to identify the risks arising from the target function and to develop countermeasures if necessary. Similar to the ISO 26262, the goal is to reduce the SOTIF risk to an acceptable level.

When classifying risks, according to ■ Fig. 6.12 four areas are to be considered:
1. Known not hazardous scenarios (Area 1)
2. Known hazardous scenarios (Area 2)
3. Unknown hazardous scenarios (Area 3)
4. Unknown not hazardous scenarios (Area 4).

The SOTIF standard defines a scenario as a combination of several scenes in a sequence of scenes with goals and values within a specified situation, influenced by actions and events. In this context, hazardous scenarios include concrete triggers from the environment that, together with the functional insufficiencies of the system, can lead to potentially hazardous system behavior. For example, driving in low sun is a known potential trigger for an unsafe scenario that can influence the video camera in sensing the environment (e.g., shadows are classified as obstacles that cannot be crossed). Such environmental conditions that may lead to a hazardous behavior due to the triggering system response are referred to as "triggering conditions" in the SOTIF standard. These

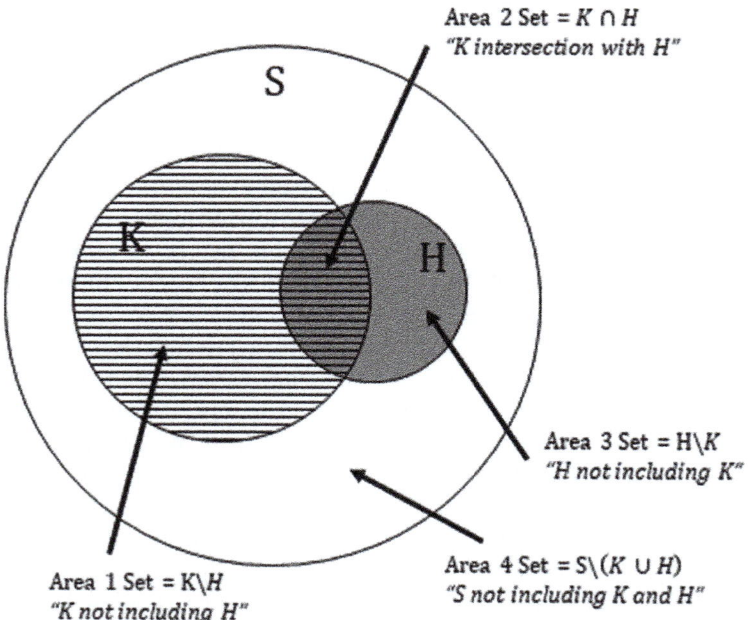

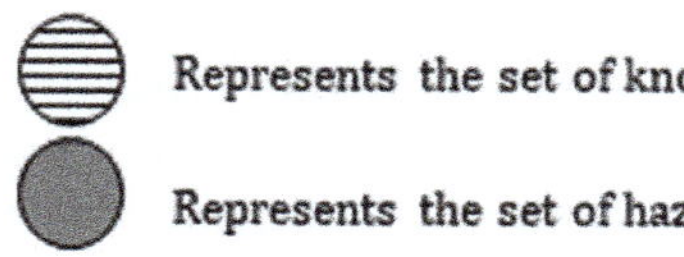

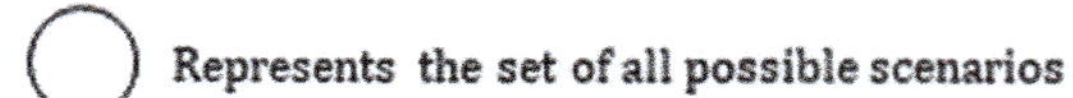

■ **Fig. 6.12** Classification of SOTIF Risks (© ISO 21448 Fig. 5)

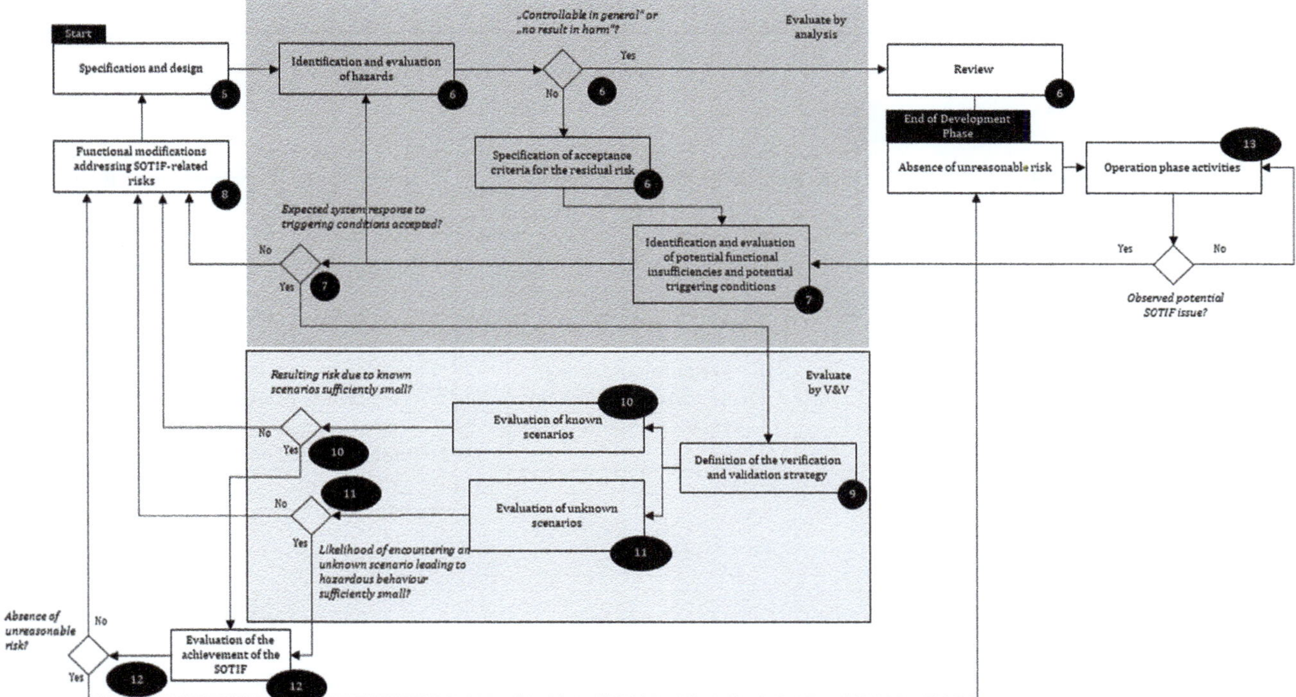

Fig. 6.14   SOTIF flowchart (© ISO 21448 Fig. 10)

cause-effect relationships are described in **Fig. 6.13** in the SOTIF cause-effect model from the triggering condition to the damage that occurs including the risk parameters from the HARA. In addition to the known and unknown hazardous scenarios containing triggering conditions, the reasonably foreseeable misuse of the function by the driver can also lead to hazardous behavior and thus to harm. ISO 21448 distinguishes between reasonably foreseeable direct misuse and reasonably foreseeable indirect misuse. The reasonably foreseeable direct misuse is treated as a trigger condition and can directly lead to a hazard, e.g., the driver activates a function on the country road, which is only intended for the highway. A reasonably foreseeable indirect misuse would be, for example, the driver inattentiveness while a driver assistance function is active, which can only lead to damage together with a malfunction from the system.

The development goal of SOTIF is to reduce the number of hazardous scenarios (areas 2 and 3) to an acceptable level. The known hazardous scenarios in area 2 are to be controlled by the system as best as possible. Of the unknown hazardous scenarios in Area 3, as many as possible should be detected through analysis and testing, evaluated, and addressed through additional measures so that they fall into one of the other areas. Unknown not hazardous scenarios (area 4) do not require any further measures.

The acceptance criteria for the reasonable residual risk of the known hazardous scenarios (Area 2 in **Fig. 6.12**) and the unknown hazardous scenar-ios (Area 3 in **Fig. 6.12**) must be defined at an early stage of development.

The following sections provide an overview of the SOTIF process model to achieve the development goal.

6.3.2   Overview SOTIF Process Model

The ISO 21448 describes normative activities in a total of 9 chapters. An activity is completed according to the standard when the achievement of the objectives in the respective first section of the chapter is documented using the listed artifacts. The methods for achieving the objectives must be selected depending on the problem.

An iterative approach is becoming increasingly common in the development of complex systems. Process steps are run through again and again until the set goal is reached. The standard also follows this logic. The normative activities are structured in a flow chart, cf. **Fig. 6.14**. The results of an activity are a prerequisite for the next activity in the flow chart.

Some activities (6,7,10,11,12,13) define decision points. Their result is a documented decision that determines which of the alternative paths must be pursued further. Depending on the system specification, not all paths need to be followed. The preceding decision point justifies which further activities are relevant.

A further iteration is necessary if the activities lead to a change in the original system specification or design. Regardless of the necessary iterations, the require-

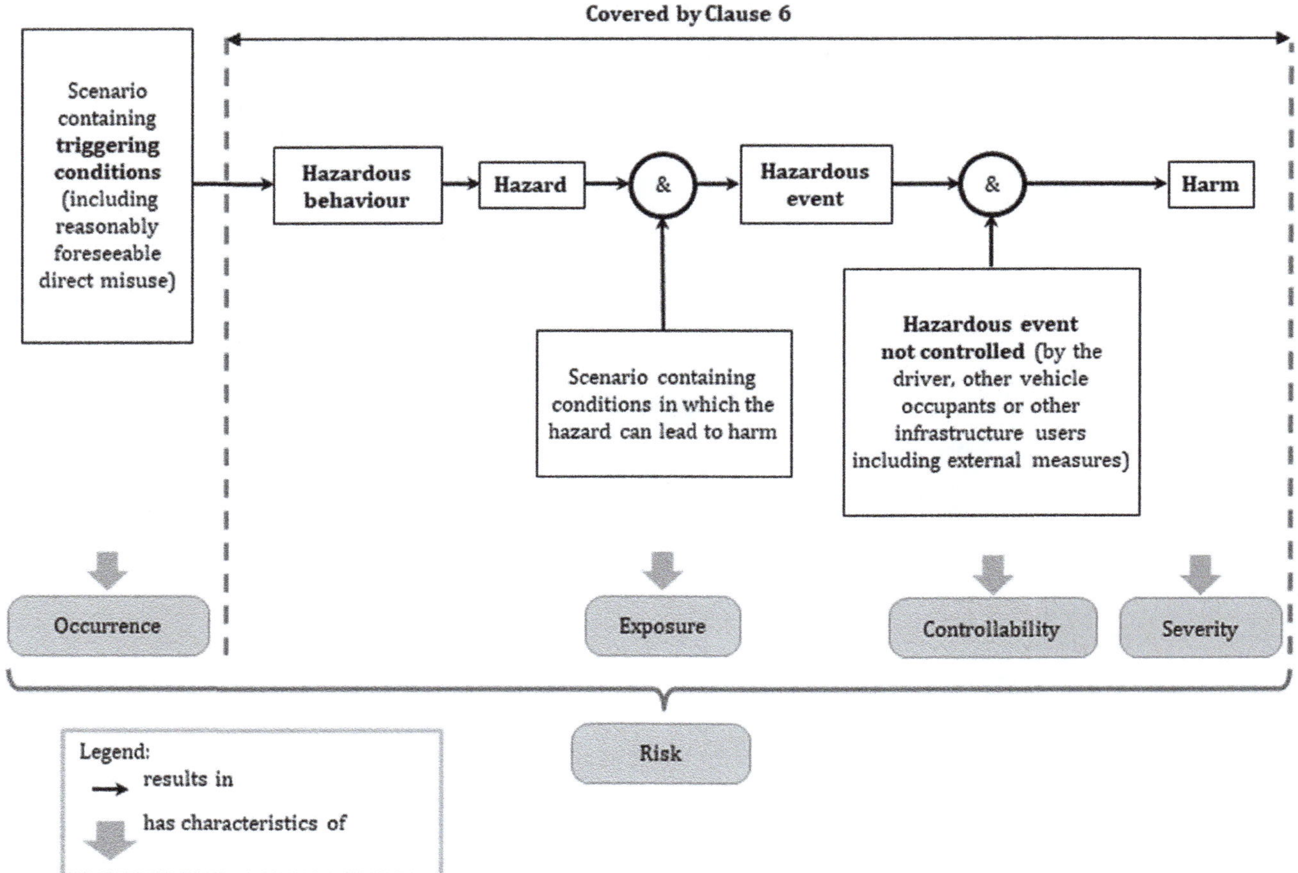

◘ Fig. 6.13 SOTIF cause-effect model linked to risk parameters from HARA (© ISO 21448 Fig. 12)

ment for placing a product on the market is achieved with the condition "acceptable risk" (see element "absence of unreasonable risk" in the flowchart).

The prescribed artifacts documenting the achievement of objectives in each activity then form a standards-based rationale arguing the absence of unreasonable risk.

The activities of ISO 21448 do not end when the product is placed on the market. As long as the product is in use, further activities may become mandatory in case of deviations from the SOTIF argument are observed in the field.

The chosen structuring of the activities provides a high degree of flexibility through the iterative approach and avoids unnecessary activities not needed, depending on product characteristics.

In order to guide the reader step by step through the process steps, we will discuss the steps using product examples and iteratively adapt the product specifications to bring more and more mandatory activities into play, to achieve the objectives. In the chapter headings, curly brackets indicate which chapter of ISO 21448 is covered in depth in each case.

The product specifications and argumentations are of course only exemplary. They do not claim to be complete or correct.

6.3.3 Function Without SOTIF Relevant Limitations {5, 6}

6.3.3.1 Specification and Design {5}

All SOTIF activities require a function specification with design documentation, including the characteristics of the implementation, as a starting point. In our example, a lane departure warning is described, which warns the driver in case of accidental lane departure by slightly shaking the steering wheel.

It is important that the documentation of the target function already contains a description of known functional deficiencies of the chosen design. In our example, it is known that lane detection will not work reliably in all situations.

This central documentation of the system properties is one of the prerequisites for the iterative approach of ISO 21448. Findings from subsequent SOTIF investigations regarding relevant limitations in the design, such as the intrinsic unreliability of lane detection mentioned above, lead to a revised version of the design documents. It is also planned to adapt the functional specification based on the evaluation results in order to arrive at a safe functional specification despite the known limitations in the design.

After such an adjustment, the process is run again to adjust the assessments to the new baseline.

Another aspect of the complete documentation of the target function referred to in Chap. 5 of SOTIF is the description of the driver's interaction with the system, including foreseeable cases of misuse.

Thus, in our example, not only is the functional goal "lane departure warning" specified, but also the functional shortcomings of the present design.

Specifically, this is the insufficiency of the lane estimation, but also the way the system warns the driver. In real systems, there will always be situations in which certain drivers do not react sufficiently quickly and appropriately to the warning. This, too, must be included as an insufficiency in the SOTIF argument.

For all deficiencies, there must be evidence about their extent, or the performance of the selected solution as a prerequisite for the next step of the process: the identification and evaluation of hazards.

This also applies to the evaluation of the unknown insufficiencies. For instance, without knowing exactly which events trigger an undesired behavior of the system, a residual probability can be estimated for example deduced from the known test coverage.

6.3.3.2 Hazard Identification and Assessment {6}

The first step in the SOTIF assessment is a hazard analysis and risk assessment familiar from ISO 26262 (HARA, s. 3–6 in ◰ Fig. 6.2). The assessment method itself is independent of the underlying causes, there are no differences between the two standards. The difference is concerning the events that cause the hazards. SOTIF exclusively focusses on hazards that can be caused by functional insufficiencies. Hazardous behavior that, according to the design, can only be caused by implementation-level errors is not considered. These errors denote deviations from the implementation specification. More concrete examples are a software implementation that does not behave as specified in the software requirements, or a defective hardware component.

The determining hazard scenario in our example is unwanted lane departure. This hazard is further detailed with the help of the present specification:

(1) The driver overreacts in case of incorrect warning. (Insufficiency of lane estimation, possibly in connection with misleading signaling of the warning).

(2) The driver reacts incorrectly when the warning is correct. (Due to misleading signaling of the warning).

In the first case, a relatively high probability of erroneous lane estimations is to be expected in the example system. However, the driving studies cited in the design documentation show that the erroneous warning can be controlled by the driver without any problems. This limitation does not lead to an unreasonable risk.

It can also be shown that the warning can be interpreted well. In the driving test, it could be validated that the potential insufficiency of the selected solution can be accepted: In no situation did dangerous overreactions of the drivers occur, all test persons were able to control the situation.

For both assessments, the standard requires that acceptance criteria are available that are part of the hazard assessments. This could be, for example, derived from the controllability, the rate of false reactions in defined driving tests over a representative selection of drivers.

Thus, the strength and nature of the steering wheel shake, together with the rate of false responses of the system, becomes relevant for the fulfillment of the acceptance criterion. These technical characteristics are to be treated as part of the SOTIF reasoning and thus as safety-related requirements.

In this example, the hazard analysis and risk assessment lead to the conclusion that the remaining residual risk is acceptable despite the existing functional insufficiency. From the SOTIF point of view, the system can be released after a review. Further process steps of ISO 21448 do not have to be applied.

In the example, the argument was essentially about controllability. This is a special case within the framework of ISO 21448: In the case of the classification controllability or damage severity equal to zero according to ◰ Table 6.1 the standard explicitly assumes that the risk is acceptable, and the classification only needs to be confirmed by a review. However, the measures to ensure controllability are well within the scope of the SOTIF standard.

In the general case, a validation target must be defined at this point.

The standard distinguishes between acceptance criterion and validation objective. The acceptance criterion specifies what should be achieved so that the remaining residual risk is reasonable. This can be controllability, as in our example, but generally also a concrete risk threshold. Unlike ISO 26262, no risk acceptance thresholds are defined (see ▶ Sect. 6.2.2.1). This is left to the user in the context of the SOTIF standard. Very general reference is made here to common reasoning such as GAMAB[3], ALARP[4] or MEM[5] (▶ Chap. 40).

---

3 **GAMAB (Globalement au moins aussi bons):** A new system should be at least as safe or low risk as any existing comparable system.

4 **ALARP (As low as reasonably practicable):** Risks should be reduced to a level that guarantees the highest degree of safety that is reasonably practicable (limiting the maximum expectation of harm).

5 **MEM (Minimum Endogenous Mortality):** MEM is a measure of the accepted (unavoidable) risk of death from the technology in question.

The validation target specifies which outcomes defined validation tests must have to argue the acceptance criterion. In our example, controllability must be demonstrated. This could be achieved by means of a driver study with representatively selected drivers on the controllability of the warning. The validation target would specify that at least 90% of drivers should react appropriately and that in the remaining cases the situation should remain controllable for other road users.

As we will see later, not only the definition of the acceptance criterion is left to the developer, but also the selection of the validation targets, i.e., the way in which the achievement of the target is demonstrated.

6.3.4 Iterative Adaptation of the Functional Specification {7, 8, 13}

6.3.4.1 Identification and Evaluation of Functional Deficiencies {7}

Let us consider an extension of the function: The system shall actively keep the lane, without intervention by the driver: When applying the risk analysis and hazard assessment, it quickly becomes clear that errors in the lane estimation can cause the vehicle to leave the lane. This behavior is not always controllable in the functional manifestation described and can potentially lead to serious accidents. An acceptably low risk cannot be argued based on the available data.

In this case, ISO 21448 calls for a more in-depth analysis of the functional deficiencies in the context of the underlying triggering conditions. One possible outcome of the analysis is acceptance criteria for the system derived from the HARA safety goals.

In our example, the responsible development team assesses the probability of the lane estimation errors as too high to keep the risk small enough without changing the functional design. If the targets defined in the risk assessment cannot be reached by further design measures trying to reduce the functional deficiencies, that is the probability of lane estimation errors, the initial risk cannot be accepted according to the SOTIF standard {7}.

In this case, the standard provides the possibility to adapt the system function specification {8} without further evaluating the functional deficiency. This can be, for example, a limitation of the situations in which the function can be activated. In our example, it is decided to ensure controllability by the driver through additional measures: The function may only be active when the driver has his or her hands on the steering wheel ("hands on" detection). In our case, additional studies show that with appropriate detection of whether the driver has his or her hands on the steering wheel and a limitation of the steering torque, the expected errors in lane guidance remain controllable by the driver at all times.

With this change in the function specification and design description, another iteration can be started. As with the lane departure warning, it can now be argued that the SOTIF objectives are met because of the controllability of the functional deficiency.

Alternatively, it would have been possible to modify the design to significantly reduce the extent of the functional deficiency, that is reduce the probability of lane estimation errors. For example, the inclusion of an electronic map significantly improves the performance of the lane estimation. With this improvement, the defined acceptance criterion could have been met without adjusting the functional behavior. The proof of this argument would require further steps within the framework of ISO 21448.

In practice, cost targets may stand in the way of such a design extension. With the way described above of adapting the functional scope, instead of a more costly reduction of the functional insufficiency on the perception side, the chosen design is nevertheless safe and consistent with SOTIF requirements.

Regardless of the path taken, a new iteration starts after each design or functional characteristic adjustment to revise all SOTIF relevant assessments accordingly.

6.3.4.2 Dealing with Events from Field Observation {13}

ISO 21448 requires that activities be set up to observe the correctness of the documented SOTIF safety argument in the field.

In our example, there are known cases of drivers deliberately overriding the "hands on" detection. The standard requires that the evaluation of functional deficiencies under ISO 21448 Chap. 7 be repeated based on any new findings that affect the SOTIF reasoning.

In the case of a deliberate overriding of safety measures by the person driving the vehicle, the standard speaks of "abuse". A further investigation of the resulting risk is thus no longer necessary according to SOTIF, because the deliberate abuse is explicitly not part of the SOTIF argument. This evaluation must be documented as part of the SOTIF argument. Further steps within the framework of ISO 21448 are not necessary.

If, on the other hand, the behavior is classified as "expected misuse," the risk assessment according to {6} must evaluate the new situation. In the simplest case, the additional risk is assessed as acceptable. Alternatively, the functional specification, for example, the execution of the "hands on" detection, can be adjusted. Once the respective specifications have been adapted, a complete iteration must again be performed over all necessary process steps until the target element "absence of unreasonable risk" is reached. In practice, this would lead to costly software updates, or in the worst case, retrofits of vehicles in the field.

6.3.5 Function with Accepted Functional Insufficiencies {7, 9, 10, 11}

6.3.5.1 Assessment of Functional Insufficiency and Triggering Conditions {7}

In the previous functional examples, the risk due to the functional deficiencies could be managed by controllability measures. If we consider an extension of the functional characteristic to "hands free," i.e., automated lane guidance without the need to keep the hands on the steering wheel, the argumentation becomes more complicated.

If, although the driver continuously monitors the vehicle, the controllability is not completely given, an acceptance criterion needs to be defined as the first step of the risk assessment. This defines an acceptable upper limit for the probability of occurrence and type of lane estimation errors. In addition to this acceptance criterion, however, typically additional criteria have to be defined, without which the remaining residual risk would not be acceptable.

These additional acceptance criteria could be:

(1) The effect of lane estimation errors must remain controllable by the driver despite the extended reaction time compared to "hands on":

   a. Activation of the function is limited to certain situations, for example on highways with wide lanes and low curve radii.

   b. If these conditions for activation are not met,

     i. the driver must be able to verify that the system is not active when attempting to activate it;

     ii. the driver must be able to safely take over the system again when the function is activated;

     iii. the driver must be able to easily correct misbehavior due to lane estimation errors without special training;

     iv. the driver's surveillance activity must be monitored by an interior camera due to expected misuse.

6.3.5.2 Definition of the Verification and Validation Strategy {9}

Special rules apply within the standard for the acceptance criteria derived from the previous activity. The verification and validation of these requirements is part of the SOTIF safety argument. Independent of other quality assurance measures, a separate verification and validation strategy must be defined for these requirements.

The strategy includes defining the necessary evidence, as well as the procedures to be used to obtain that evidence. In this context, an argumentation must be built as to why the selected methods for verification and validation are sufficient to achieve the acceptance criteria defined in the context of the hazard analysis and risk assessment. This is supported by a mandatory requirement in ISO 26262 for traceability of requirements (cf. ► Sect. 6.2.3.1).

A strict distinction is made between known scenarios and unknown scenarios with hazard potential.

If the design is based on a derivation model (see chapter "Validation"), scenarios with hazard potential can be found analytically. ► Chapter 10 of ISO 21448 deals with these known deficiencies.

For example, it is known that the video sensor can be limited in lane detection when the sun is low in the sky. Since the cause-effect relationships for this deficiency are known, the hazard potential can be specifically evaluated. Corner cases can be identified and specifically evaluated. Analytical models can be used to extrapolate or interpolate between specific test results.

In case of the analytical approach according to ► Sect. 6.3.4.1 the system behavior is not fully tested experimentally. Theoretical predictions based on the derivative model replace the system-level test. With increasing diversity of the relevant input states and with increasing complexity of the models, simulations complement the theoretical predictions. The simulations are again based on validated derivation models. If these are not demonstrably correct, the simulations must not be used as the basis for validation.

For many "open world" problems, however, no sufficiently expressive derivation models exist. In an "open world" problem, even the complete description of all relevant input states is no longer practical.

A typical example is video-based image segmentation. Here, each pixel of a video image is assigned to an object class. For example, the algorithm classifies whether a pixel belongs to a pedestrian, a vehicle or the road. The input vector consists of millions of integer values. Trying to solve this problem by defining explicitly derived filter kernels typically shows a very weak classification performance. For example, such a derivation model could try to model the property, that each pedestrian has a face that has certain symmetry properties and shapes. A complete derivation model is not feasible due to complexity. This is where machine learning methods come into play.

The stated goal of machine learning is to make the derivation model largely superfluous and to keep the explicit design effort to a minimum. The input–output correlation is "learned" directly. In the best case, a description of the desired behavior that is as simple as possible is sufficient and the system optimizes itself without the intervention of the developer. For individual, complex tasks, these approaches currently prove superior to explicit algorithms derived from a defined derivation model, using defined insights concerning the input–output correlation. The reason is that they can represent a significantly higher complexity than is practicable in explicit rule descriptions. Machine learning models are able to optimize millions of parameters.

Without a powerful derivation model, no analytical deduction of all expected, relevant deviations of the implementation from the system requirements is possible. The SOTIF standard speaks here of unknown, unsafe scenarios. How to deal with them is discussed in Chap. 11 of ISO 21448.

The ISO 21448 proposes in Chap. 11 methods to find possible unsafe scenarios by targeted experiments. This converts the unknown scenarios into known ones. These can then be handled as described in Chap. 10 (see also ▶ Sect. 6.3.5.3).

For many problems, however, this will never be completely successful. In this case, the focus moves from the complete identification of all unknown scenarios with risk potential to the argumentation of the sufficient completeness of the system tests. The more scenarios prove to be harmless in the test, the lower the residual probability of an unknown problem. Via an argumentation of the representativeness of the test data sets, the residual probability of an undesired behavior can be estimated. This corresponds to the methods (O) and (P) from Chap. 11, Table 11: "Exploration of scenarios under realistic conditions" and the "Functional decomposition and modeling of probabilities".

These considerations are driving the big data underlying validation efforts for increasingly autonomous systems.

The ignorance from the incomplete derivation models inevitably leads to probability estimates regarding the functional deficiencies. ISO 21448 recognizes the need to extend to data-driven reasoning for complex "open world" problems. As stated in the standard, data-driven arguments do not compete with analytical arguments in this regard. Rather, they complement each other in practice.

6.3.5.3 Evaluation of Known Scenarios with Hazard Potential {9, 10}

In order to be able to release the function, the achievement of the validation objectives and the correctness of the derivation of those objectives from the acceptance criteria must be demonstrated.

The basis for such an argument is a requirements tree as described in ▶ Sect. 6.2.3.1. All methods already described in ISO 26262 can be used for verification and validation.

In the driving test, it must be confirmed that the driver reacts correctly to the activation or deactivation of the function or, if applicable, that no foreseeable misuse is to be expected as a result of the specified measures.

The definition of all these tests requires that the so-called triggering conditions for a potentially dangerous situation are known. These known scenes lead to an explicit test catalog regarding the system behavior in the known scene.

Once one of the tests reveals specification or design gaps, another iteration must be undertaken after the system has been adapted.

However, for a release of the system in the sense of ISO 21448, the confirmation that the system acts within the set objectives in all known scenarios with hazard potential is not sufficient.

In our example, the system may only be active in scenarios with sufficiently wide lanes. This remains, even if map data is used, a classification problem based on environmental data. This is a limitation in the design where unknown triggering conditions cannot be excluded.

SOTIF requires a separate verification for this class of hazard potential scenarios.

6.3.5.4 Evaluation of the Hazard Potential of Unknown Scenarios {9, 11}

So far, the system in our example has been designed in such a way that the effects of the underlying limitation are acceptable. In the design selected in our example, an incorrect activation due to an incorrectly detected lane width, together with an incorrect lane estimation, can no longer be controlled by the driver if necessary.

Acceptance criteria in the form of residual error probabilities can be derived from the residual risk accepted in the hazard analysis for incorrect detection of the activation condition.

Falling below this reasonable residual error probability cannot be demonstrated by evaluating all known scenarios in which misclassification is expected. This could be accomplished by using the methods listed in Chap. 10 of ISO 21448. Furthermore, it must be argued that even unknown events do not jeopardize compliance with the threshold values for the residual error probability.

The reason for this is the nature of functional insufficiency. In the design of the environment sensor-based scenario classification and lane estimation, the hardware and software design are derived without a complete derivation model (cf. ▶ Sect. 6.2.4). Thus, it is not only clear that errors are to be expected despite correct implementation, but also that not all triggering conditions are known.

We are faced with the task of estimating a residual error probability without knowing the error mechanisms. The SOTIF standard only deals very superficially with a possible reasoning strategy here, with the help of which the required validation target can be derived.

Typically, a statistically representative test coverage is argued. In this context, representative means that all scenarios relevant for evaluation, including unknown ones, are likely to be covered in the test.

Statistically sufficient means that the weighting of the scenes in the test also corresponds to reality and that the scope of the test is sufficient so that rare, potentially dangerous scenes occur sufficiently frequently to be able to infer the remaining residual risk.

The test design should include all available knowledge regarding the insufficiency. Here, an appropriate design of the activation conditions can reduce the test scope and effort. If, for example, use in certain countries can be excluded by design specification, the test data set does not need to include scenes from the excluded countries.

Even for systems in which a relatively high residual error probability is acceptable, this procedure leads to a very high test effort.

Typically, testing and development iteration go hand in hand. If the residual failure probabilities are not achieved, the algorithms, and in the most complex case even the hardware, are adjusted until better results are achieved. Many models have free parameters that are adjusted as part of further optimization steps. If this adaptation takes place automatically, we are in the field of machine learning.

This explains why open world systems are developed data driven. Without enough data, the validation target can neither be achieved nor proven. Care must always be taken that the data for the optimization of the system, or reduction of the functional deficiency, is not mixed with the validation data in order to avoid so-called "overfitting".

The less responsibility lies with the driver, the more situations the system must master without intervention, and the more important the quality and quantity of the available data becomes. In some cases, these are expanded, "augmented",[6] by randomization techniques and models in order to be able to achieve the validation targets. The effective mastery of massively data-driven development processes now represents an essential part of the value chain in the development of driver assistance functions.

6.3.6 **Massive Data-Driven Development**

With the availability of increasingly powerful computer systems—both "on board" in the vehicles and in data centers—the implemented systems are becoming more and more complex.

For example, in the case of control systems such as the anti-lock braking system ABS or the electronic stability control ESC, the input variables are still limited to a few dimensions; in the case of environment-based systems—for example video—the number easily increases to several millions. The derivation of the mapping function, certainly not trivial even for ESC, can no longer be done explicitly for many environment interpretation problems.

This is where machine learning methods come into play. The functions are no longer explicitly designed and coded. The mapping functions are created by the automated optimization of an immense number of parameters. This automated optimization by parameter variation is also called training.

In most cases, no simplifying model of the function is available after this training. The development team cannot make predictions about the output of the function in concrete cases, let alone guarantee behavioral limits for corner cases.

Nevertheless, if these functions are part of a safety–critical system, a statement about the expected residual risk for a potentially dangerous system response must be argued.

In the context of machine learning, such statements are derived from observing the response of the trained system to statistically relevant large data sets.

The development effort thus shifts from explicit modeling to the definition of evaluation functions against which optimization is performed and the collection, evaluation, and management of very large amounts of data for training and validation of the function. ISO 21448 is the first to explicitly confirm the validity of such approaches.

Waymo, as the first company to commercially offer autonomous driving on a limited scale, based its release argument on more than 20 million miles of data from road traffic and, for statistical relevance, expanded it significantly through simulative augmentation [18].

This effort does not replace the "classical" model-based arguments applied in parallel but complements them.

Machine learning research attempts to compensate for this drawback by trying to subsequently derive simplified models of the trained functions, which may then be used for model-based validation reasoning. Alternatively, hybrid approaches are attempted where boundary value functions are guaranteed by explicit models and the complex training part has influence only within these fixed bounds set by those explicit models.

However, these considerations are currently still in the early stages of development.

To what extent this will limit the need for data, or whether the data will be available more cheaply soon, remains to be seen.

---

6 Augmentation uses existing data sets and transforms certain elements according to a verifiable model. For example, objects can be moved, or inserted into a scene. This results in new tests. The database is expanded.

6.4 UN/ECE in the Context of Safety Standards

The ISO 21448 deliberately gives room to newer, currently establishing, and continuously evolving procedures and system designs by not containing any concrete technical specifications.

Legislative initiatives such as the UN/ECE vehicle homologation regulations take a different approach. In the various technical addenda, very specific requirements are made in some cases regarding functional characteristics: "*UN/ECE R79 5.1.6.2.3.1: In the case of an ESF intervention on a road or a lane delimited with lane markings on one or both side(s), an automatic avoidance maneuver initiated by an ESF shall not lead the vehicle to cross a lane marking*" [9].[7]

With the background that currently no technical system can detect line markings without errors, a contextualization of the specification is missing. Which performance class of line detection is sufficient to achieve the legal specification? The UN/ECE ignores the issue of functional deficiencies.

Since the various requirements are not derived in a traceable manner, no context can be established that would allow the derivation of necessary performance classes. This is in contradiction to the approach of the ISO 21448. The lack of traceability is in contradiction to ISO 26262. Presumably, the manufacturer is to fill the gaps in his own development documentation. The UN/ECE requirements thereby represent restrictions that prohibit alternative solution options, potentially also improvement options.

This contradiction is also clear in the functional safety specifications: "*UN/ECE R152e Annex 3.1. a. The system is designed to operate, under non-fault and fault conditions, in such a way that it does not induce safety critical risks*" [10].

In contrast to the relevant safety standards, which require proof of an acceptable residual risk, absolute safety is demanded here, at least in a literal interpretation (▶ Chap. 40). This is not practically feasible, let alone verifiable. Accordingly, the test procedures defined in the UN/ECE are at best random tests and constitute an anachronism to the massively data-driven validation methods. It is left to the vehicle manufacturers to reconcile these two worlds.

6.5 Summary and Outlook

ISO 26262 and ISO 21448 pursue the same goals: To establish methods to ensure that the residual risk posed by the vehicle and its use can be reduced to an acceptable level.

ISO 26262 is limited to risks caused by systematic failures, e.g., in the software, and random failures in the hardware. It is assumed that the conceptual design is free of insufficiencies. This is generally not feasible for complex systems such as those required for driver assistance up to autonomous driving.

ISO 21448 complements ISO 26262 by providing methods for dealing with these functional insufficiencies.

For the design process of increasingly complex systems, iterative approaches with small and short change, evaluation and adaptation cycles are becoming more and more prevalent. The ISO 21448 reflects this trend in which the development of a safety argumentation follows an essentially iterative process model. This implies that compliance with the standard can also be implemented efficiently for simpler systems, because only parts of the SOTIF process model may have to be implemented, such as the controllability argumentation.

Within the framework of ISO 21448, a consistent safety argument is essentially required. Neither the acceptance criteria to be achieved nor the type of evidence of target achievement are normatively regulated. Neither qualitative nor quantitative validation or verification targets are specified, nor are development or verification methods.

The UN/ECE legislation takes a different approach in that very specific requirements are stipulated for certain functional characteristics without visible application of the established safety standards. This means that there is no reference to the safety argument from the safety standards presented. The manufacturer must resolve this contradiction.

The question of "how safe is safe enough" is thus essentially left to product development. This reflects the current state of the art. A standard should only specify objectives and procedures that are accepted in society and adapted to the target market. However, a social consensus on acceptance criteria and validation targets has yet to be reached. It is to be hoped that a societal discourse will lead to clear and assessable goals and not leave the derivation of these to the manufacturers (▶ Chap. 40). What is required, where possible, are quantifiable targets for the accepted risk, not the prescribing of technical solutions, as can be seen in some of the legislative initiatives. This would unnecessarily delay continuous progress. Nor will ignoring the problem in such legislative initiatives through specifications that are practically unworkable help safety. This is where academia could play a driving role because it can

---

7   ESF: Emergency Steering Function: Activate steering to avoid or mitigate collisions.

address the issue without conflicts of interest and bring a sound understanding of the problem [17].

Without this public alignment process, technology leaders like Waymo affirm the basic principles of existing standards but supplement them with their own considerations and development methods [18]. Concrete interpretations of ISO 21448 are emerging as part of these development projects, as well as extensions that will certainly influence an emerging state of the art.

Waymo also uses familiar methods but supplements them with the possibilities of massively data-driven approaches. These are at best mentioned and accepted in the standard, but not elaborated on. Unfortunately, the current ISO 21448 standard devotes the smallest part of the text to dealing with "unknown events," and thus the core of the present "open world" problems avoiding to practically support the key challenge of those systems. The future will show whether best practices will find their way into standardization and legislation in order to be able to implement the goals developed in a public discussion. Until then, we will see quite a few revisions.

References

1. ISO/FDIS 21448.: Road vehicles—safety of the intended functionality (2022)
2. Act on the Provision of Products on the Market (Product Safety Act—ProdSG) (2021). ▶ http://www.gesetze-im-internet.de/prodsg_2021/ProdSG.pdf
3. IEC/EN 61508 (2010) Functional Safety of Electrical/Electronic/Programmable Electronic Safety-Related Systems (E/E/PES), 2nd edn
4. ISO 26262 (2018) International Standard Road vehicles—Functional safety, 2nd edn
5. ISO/PAS 21448 (2019) Road vehicles—Safety of the intended functionality
6. ISO/SAE 21434 Road vehicles—Cybersecurity engineering
7. ISO/TR 4804 Road vehicles—Safety and cybersecurity for automated driving systems—Design, verification and validation
8. ISO 15622:2018 Intelligent transport systems—Adaptive cruise control systems—Performance requirements and test procedures
9. UN/ECE R79 Steering equipment
10. UN/ECE R152 Advanced Emergency Braking Systems (AEBS)
11. UN/ECE 157 Automated Lane Keeping System (ALKS)
12. Ross, H.-L.: Funktionale Sicherheit im Automobil. Carl-Hanser Verlag, Munich Vienna (2014)
13. V-Modell: Available at: ▶ https://www.cio.bund.de/Web/DE/Architekturen-und-Standards/V-Modell-XT/vmodell_xt_node.html
14. Spanfelner, B., Richter, D., Ebel, S, et al.: Challenges in applying the ISO 26262 for driver assistance systems. In: 5th Driver Assistance Conference, Munich (2012)
15. Balzert, H.: Lehrbuch der Softwaretechnik - Softwaremanagement, 2nd edn, p. 487. Spektrum Verlag (2008)
16. Ebel, S., Wilhelm, U., Grimm, A., et al.: Wie sicher ist sicher genug? Anforderungen an die funktionale Unzulänglichkeit von Fahrerassistenzsystemen in Anlehnung an das gesellschaftlich akzeptierte Risiko, 6. Workshop Fahrerassistenzsysteme, Löwenstein, 28–30. September (2009)
17. Weitzel, A., Winner, H., Cao, P., Geyer, S., Lotz, F., Sefati, M.: Absicherungsstrategien für Fahrerassistenzsysteme mit Umfeldwahrnehmung; Forschungsbericht der Bundesanstalt für Straßenwesen, Bereich Fahrzeugtechnik, Köln (2014)
18. Waymo.: Waymo's Safety Methodologies and Safety Readiness (2020). ▶ https://storage.googleapis.com/sdc-prod/v1/safety-report/Waymo-Safety-Methodologies-and-Readiness-Determinations.pdf

Open Access This chapter is licensed under the terms of the Creative Commons Attribution-NonCommercial-NoDerivatives 4.0 International License (▶ http://creativecommons.org/licenses/by-nc-nd/4.0/), which permits any noncommercial use, sharing, distribution and reproduction in any medium or format, as long as you give appropriate credit to the original author(s) and the source, provide a link to the Creative Commons license and indicate if you modified the licensed material. You do not have permission under this license to share adapted material derived from this chapter or parts of it.

The images or other third party material in this chapter are included in the chapter's Creative Commons license, unless indicated otherwise in a credit line to the material. If material is not included in the chapter's Creative Commons license and your intended use is not permitted by statutory regulation or exceeds the permitted use, you will need to obtain permission directly from the copyright holder.

Virtual Development and Test Environments for DAS

Virtual Integration

Michael Kochem and Stephan Hakuli

Contents

© The Author(s) 2026
H. Winner et al. (eds.), *Handbook Assisted and Automated Driving*,
https://doi.org/10.1007/978-3-658-45276-6_7

7.1 Consistent Testing and Evaluation in Virtual Test Drives

The guiding idea of the virtual road test is to transfer the real road test as realistically as possible into the virtual world. The primary aim is to benefit from the characteristic strengths of simulation in terms of reproducibility, flexibility, and reduction of effort. This establishes a testing and evaluation option for specifications and derived solutions early in the vehicle development process. The use of appropriate simulation methods enables more efficient design, development, and application of vehicles and vehicle components. They bridge and shorten the time until real vehicle prototypes are available. With real driving tests and the reliability of real test results as a template, the use of simulation techniques is an optimization task in which the modeling, parameterization, and simulation effort must be set in relation to the efficiency gained.

Like its real counterpart, the virtual road test consists of several components. The central role is played by a virtual vehicle prototype whose components are integrated as models, software code, or hardware, depending on the current state of the development process. Defined interfaces between the subcomponents of the virtual vehicle make it irrelevant for each component in which integration stage the respective other components are.

A virtual driver operates the virtual vehicle including its functions. The computer driver is parameterized via a behavior model, which can execute both open-loop and closed-loop maneuvers and which, like in the real driving test, is guided by maneuver step descriptions to execute the driving test based on distance, time, or a triggering event. As a consistent further development of signal-based testing, such a maneuver-step-based execution of the test order with a configurable driver behavior is a prerequisite for the transferability of the simulation results to the results of the real driving test used as a template.

The virtual driver moves the virtual test vehicle on a virtual roadway in a virtual environment that, for example, represents real road courses and their characteristics and in which he interacts with virtual road users.

While all elements of the test drive in the early concept phase are virtual, the development process subsequently passes through different integration stages, as illustrated in ◘ Table 7.1. Here, a gradual exchange of virtual components with the associated real test components takes places until the point at which a completely real test drive on a real road with real drivers and other real road users has entirely replaced the simulation.

However, the virtual road test not only includes the mapping of a test configuration into the virtual world but also the transfer of the evaluation and assessment methodology from the real road test into simulation. The continuous fulfillment of specifications at the system level and component level does not guarantee the desired behavior of the overall vehicle and thus does not guarantee the validity of a product in the sense of suitability in accordance with the product objectives. The achievement of the latter is only verified by the release test. The goal must therefore be to test partial solutions against the associated specifications during the development process and in the same context be able to test the entire vehicle concept for its ability to meet the desired properties of the overall solution. Hence, the real added value of the virtual road test is to consistently compare the maneuvers performed in the real vehicle during the release and the associated evaluation criteria with the specifications discussed in ► Sect. 7.3 using the example of the V-Model up to the beginning of the concept phase and to make them available already there. Ideally, design decisions can thus be checked at an early stage in a purely virtual test

◘ **Table 7.1** Gradual transition from the virtual to the real world

System/component	MiL	SiL	ECU-HIL	System HIL	Roller dynamometer	ViL	Road test
Function code	V	R	R	R	R	R	R
Control unit	V	V	R	R	R	R	R
System	V	V	V	R	R	R	R
Vehicle	V	V	V	V	R	R	R
Driver	V	V	V	V	V/R	V/R	R
Driving dynamics	V	V	V	V	V	R	R
Driving experience	V	V	V	V	V	R	R
Roadway	V	V	V	V	V	R	R
Traffic/environment	V	V	V	V	V	V	R

V: virtual, R: real

for their suitability for meeting the targets for the overall vehicle. In the subsequent course of development, this approach helps to avoid unnecessary disruptions in the test and evaluation chain between the integration stages and ensures that test results remain comparable across stages.

The use of virtual vehicle prototypes in virtual road tests thus makes it possible to evaluate design alternatives in the context of the overall vehicle early in the development process. Additionally, it helps to ensure that specifications for systems and components, as described in the following ▶ Sect. 7.2 can be checked for their suitability for achieving the overall vehicle goals even before the built of the first real prototype. ▶ Section 7.5 illustrates this aspect using the example of a function development along the V-Model through all integration steps.

7.2 Efficient Collaboration Between Manufacturer and Supplier by Means of an Integration and Testing Platform

Virtual road tests based on virtual vehicle prototypes and release maneuver catalogs mapped in the simulation, including the associated evaluation criteria, may also contribute to efficiency gains in the collaboration between vehicle manufacturers and system suppliers. The supplier benefits during the development and application of a component from the existence of a virtual vehicle prototype, as his organization becomes less dependent on real and usually rare prototype vehicles for the target achievement evaluation in the full vehicle context. Maneuver-based component test catalogs on the supplier side mainly focus on the verification and validation of contractually guaranteed system properties. By additionally applying maneuver catalogs as provided by the manufacturer, the supplier can examine his development results for their likely ability to meet the defined objectives in the overall vehicle context, even before this component is integrated into a real prototype and becomes verifiable in real-world test drives.

Such collaboration requires manufacturers and suppliers to use the same integration and test environment. Both vehicle and component models are exchanged, and release maneuver catalogs are implemented in the simulation with an integrated evaluation procedure according to defined evaluation criteria both at the vehicle and component levels. The manufacturer can hand over vehicle data in encrypted form as a "black box" optionally with an expiration date, which can be used in the simulation by the supplier but cannot be viewed or modified in detail. The assembly that the supplier develops is excluded from the encryption and can be in-

tegrated and tested by the supplier. In return, a supplier can provide component models and ECU software in a protected form, for example, via the FMI/FMU mechanism [1–3] independently of the authoring tool used to create the FMUs. This mutual exchange of protected models or model parameters enables the vehicle manufacturer as a system integrator to plan with the earlier provision of more mature components validated at an early stage in the overall vehicle context. This can significantly reduce the number of expensive and sometimes time-consuming iteration steps until the finalization of the implemented function.

While until some time ago the focus was essentially limited to the exchange or integration of individual components of the virtual prototypes, the description and exchange of the entire environment between the development partners is now added, especially in the ADAS and AV context. An increase in efficiency can be achieved if, in addition to the description of the actors ("agents") in the respective scenario, the entire process is also described independently from a specific authoring tool. Standards in the description of the environment, the agents, and their actions, such as OpenDrive, OpenScenario [5, 31] are meanwhile quite advanced. They contribute considerably to the establishment of scenario-based testing [6, 33, 34].

7.3 In-the-Loop Methods and Virtual Integration in the V-Model

Driver assistance functions are functionally mainly based on software. Therefore, it makes sense to use the V-Model known from software engineering [7] or its further development "V-Model" XT [8] as the development process for driver assistance functions. The V-Model basically represents a chronological development process. The sequence is not plotted linearly along the time axis, but rather in the form of the letter "V". There is a descending branch and an ascending branch. The descending branch contains the steps of the task analysis. The analysis results in step-by-step specifications for the components to be developed. It is essential that first the total product requirements (often also called customer requirements) are analyzed and converted afterward into a logical architecture. This is followed by the development of a technical architecture, which is subsequently decomposed and specified into systems and components. In parallel with each of these steps, test case specifications are created, which are later used to verify the development. The last step of the descending branch marks at the same time the first step in the ascending branch, which addresses the actual implementation and/or development of the specified components. The ascending branch contains all

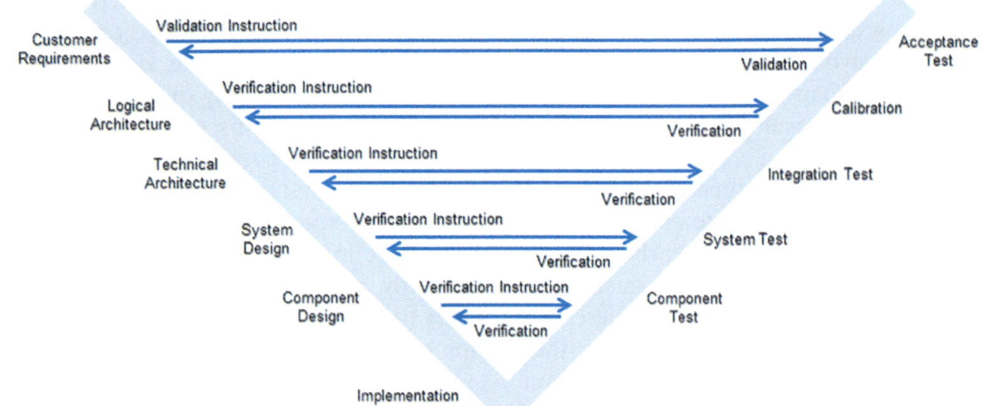

Fig. 7.1 Development process according to the V-Model

test and integration steps of the individual components over the entire system up to the acceptance test with the customer. It thus represents the integration and testing in the development process. Each step on the descending branch has a relationship to a step on the ascending branch. The connection corresponds to the verification of the subsystem created in the process step with the associated specification. The test cases used in each step are those which have been developed during the specification phase of the descending branch. The final step contains the validation, which is the verification of conformity to all customer requirements and acceptance tests. ▪ Figure 7.1 shows the general development process according to the V-Model.

Easy comprehensibility and the connection of development and quality management characterize the V-Model. This connection is achieved by using the test cases from the specification phase in the integration phase. However, a considerable disadvantage of the V-Model is that it is only possible to determine whether the associated specification is correct at the corresponding step of the integration phase. This becomes most relevant with regard to driver assistance functions when validating the overall functionality that was derived from initial customer requirements. Validation is only possible in the final step of the development process. If the specification of customer requirements is incomplete or even incorrect, all subsequent specifications and their realizations are affected. This can only be formally determined in the last development step according to the V-Model. In between, there are typically three years of development time in the automotive industry and often development costs of several million euros. A necessary change means an increase in development costs and an extension of the development time.

In order to reduce this risk, appropriate methodological additions should be made to the development process with the aim of being able to make a sufficiently reliable statement about the quality of the development at an early stage [9, 10]. This statement should be further substantiated along the development process in order to gradually arrive at a stable assessment. In order to maintain sufficient flexibility, it should also be possible to continuously adapt the specifications without significantly changing the previously created content. The methods used for this procedure originate predominantly from the repertoire of the development of embedded mechatronic systems. Considered are SiL, MiL, and HiL methods [11].

In this process, the models or real components available for the respective development step are coupled to a simulation of their real environment in order to obtain an overall system that can be evaluated. This replica is made available in virtual form by a simulation environment in which the real available models or mechatronic systems are embedded. Since there are no real components until the process step of implementation, the simulation environment must be able to offer virtual integration. ▪ Figure 7.2 shows the location of each method in the V-Model.

The Model-in-the-Loop (MiL) method can be used to confirm the specification of customer requirements up to the logical architecture step. In this methodology, algorithms are created that functionally correspond to the development objective. However, they are not yet related to the hardware in the target system. These algorithms are usually created in the form of model-based software. In order to validate these models, they are integrated into a corresponding simulation environment and tested in a virtual driving test. This means that all the necessary components (environment, driving route, driving dynamics, powertrain, sensors, driver model, etc.) are available in modular form. The created model is added to the simulation environment to experience the new function. The integration of

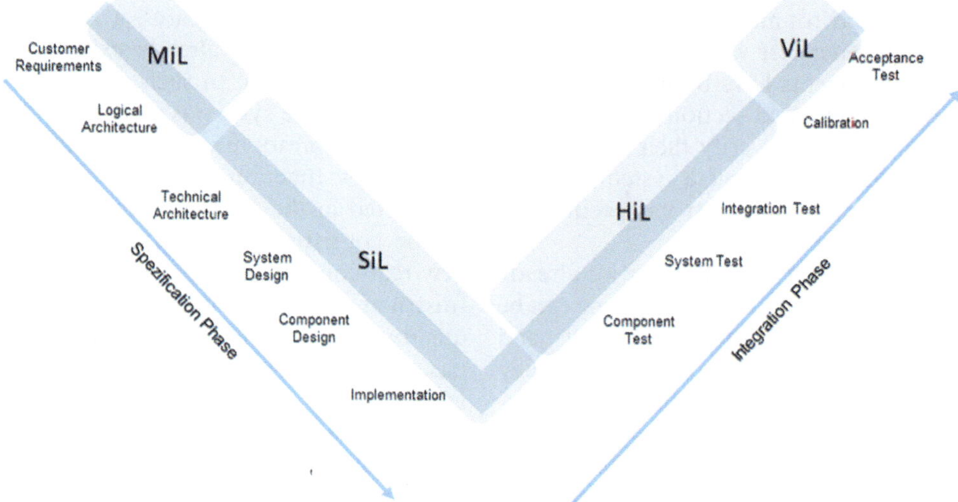

Ο Fig. 7.2 In-the-Loop methods in the V-Model

models into a virtual prototype and thus into the full vehicle results in more concrete requirement specifications and thus more concrete test specifications, which help to avoid possible surprises during validation. Ideally, the simulation environment is coupled with a driving simulator in which test drivers already can evaluate the effectiveness and user acceptance of the newly developed driver assistance function. Depending on the level of detail of this method, it is already possible to make far-reaching statements with regard to user acceptance. In terms of time, this step is well ahead of the corresponding step in the classic V-Model and thus enables a significant reduction in the development risk.

The Software-in-the-Loop (SiL) method allows validation down to the level of the individual components. This is achieved by transferring the previously created models into a simulation environment which is very similar to the technical characteristics of the target system in terms of computing power, real-time behavior, or resolution accuracy, but is still target hardware independent [11, 28]. Thus, the SiL method is the basis for the implementation and integration of software components. At the same time it represents a possibility to check and, if necessary, adapt the specifications of the individual components of a system prior to their implementation, if these software components are compiled for deployment to the target hardware.

If the development process is complemented according to the V-Model by the MiL and SiL methods in a powerful simulation environment, a virtual integration of the entire system is created at the end of the descending branch. Thus, before starting the integration phase, a virtual prototype is available that enables a complete testing and verification of each component and its functionality, individually and with regard to its interfaces. This virtual prototype can also be used for a virtual test drive as described in ▶ Sect. 7.1. This in turn results in the possibility of testing the effects of tolerances of each individual component on the driver assistance function. Since this can be done automatically and often faster than in real-time, the virtual prototype is a very powerful tool for checking the individual specifications and the entire system. In addition, if the virtual prototype is configured appropriately, the failure or misuse of individual components can be tested with respect to the overall system and its functionality. Nevertheless, once the real components are later available, similar tests should be executed with those real components. Tests with the virtual prototype, however, are much more flexible, faster, and less expensive. In addition, knowledge gained from this can be integrated into the specification of the individual components.

The described approach is not limited to the functional domain of driver assistance. However, this domain is predestined for such an approach for the following reasons:

There is a high degree of interaction between the vehicle dynamics, the powertrain, and the assistance systems on the one hand, and between the assistance systems and the human–machine interface on the other. Furthermore, there are high requirements for functional safety and its verification with increasing levels of automation. The use of novel sensor concepts and associated algorithms as well as little experience with user acceptance of a new function entail a high development risk.

In the next stage, the developed models are transferred from the SiL environment to the real components or replaced by them. The method is therefore referred to as Hardware-in-the-Loop (HiL). For distributed systems, this typically takes place over several stages. First, the individual components are tested independently against their respective specifications. Again, a simulation environment is used that provides the component under test with the necessary interfaces. Once all components have been verified using this method, they are integrated section by section using the same method (HiL) to verify their interaction. At the end of this stage, the complete system exists in real components and is tested against its specification up to the logical architecture level.

It is important to note that the test scenarios already used during the creation of the virtual prototype can be applied again. On the one hand, this reduces costs, and on the other, the test results can be compared directly between the real and the virtual components. This facilitates the identification of errors in case of discrepancies. If troubleshooting necessitates a change to a model of the virtual prototype, the effects of the change can be assessed again by means of a virtual road test.

Vehicle-in-the-loop (ViL) is another method to complement and improve development with the V-Model for driver assistance systems. It addresses the need for many driver assistance functions for a complex driving test and the high demand for functional safety. This group of driver assistance functions has increasingly gained in importance and scope. A major reason for this is the constantly growing number of vehicle variants in which driver assistance functions are offered and thus have to be safeguarded even as the degree of automation and networking continues to increase. The ViL method allows the real test vehicle to be operated in a virtual environment [12–14, 36, 37]. The coupling between real vehicles and virtual environments can be done in two ways. One option is replacing the real sensor system with an appropriate interface to the environment sensor system. At this interface, the simulation environment feeds in simulated sensor signals that correspond to the sensor response from a real environment. This is beneficial if real sensors cannot be stimulated by artificial signals with reasonable effort. The second option is to keep the real sensors and artificially stimulate them, as for example real ultrasonic sensors that are exposed to artificially generated response signals via ultrasonic transducers. In both variants, the real test vehicle responds to attributes and events in the virtual environment. Critical driving maneuvers with e.g., obstacles or objects on a collision course, can thus be tested safely and reproducibly The created interface can also be used to generate the sensor signals as they would occur due to a changed sensor mounting position in a vehicle variant or due to various tolerances. This method therefore offers the possibility to test these variants or tolerances with a single test vehicle. In addition to the much safer test operation, this allows effective testing and application of driver assistance functions. This results in considerable economic potential for driving tests of driver assistance systems.

Driver assistance functions are rarely developed from scratch. In most cases, a functionality that uses various components already exists. New features are added on this basis. In such cases, it is very helpful if the existing base system is already available in the form of a virtual prototype. Based on this, the specification of the new functionality can be efficiently integrated and tested in the existing structure of the virtual prototype consisting of MiL and SiL components. The new real component can also use the existing HiL infrastructure. This procedure represents a very helpful quality measure, since the changes to the already tested base system can be easily traced and verified. The existing test cases can be reused if the new function interacts with the existing base. Furthermore, this opens up an economic potential, since some development activities can be taken over with the existing infrastructure.

For complex driver assistance functions, the associated development process according to the V-Model cannot be regarded as the sole process. Driver assistance functions have strong interactions with functions from other domains in the vehicle. These interactions require a cross-domain concept for integration and testing. Typically, this is achieved today by agreeing on so-called synchronization points between the developers from the different domains in the ascending branch of the V-Model. These synchronization points represent the integration of all functions in the vehicle with a defined partial functionality. The development processes of the individual domains in the vehicle may well differ in terms of the sequence up to a synchronization point. Also, the set of realized subfunctions may differ from domain to domain. The example explained in ▶ Sect. 7.5 shows how virtual integration enables the availability of a virtual prototype at any point in the development process. Only the level of detail differs at the various points in time—from a rather coarse model with a few parameters at the top left of the V-Model to an ideally very detailed model toward the end of the development process. Therefore, coordinating the integration of functions is also useful already in the descending branch of the V-Model.

Despite the great potential of virtual development and integration of driver assistance functions, it will not be possible to completely replace real driving tests. One reason for this is that some test scenarios are discovered for the first time during real driving test, as any unexpected situations can occur here. If relevant, these test scenarios can then be transferred to the virtual driving test, provided that the implementation of relevant events and mechanisms in simulation is feasible. On the other hand, the subjective assessment of driver assistance functions is an aspect that cannot be fully adopted by a virtual driving test. Finally, the limited validity of simulation models restricts their use to certain applications [▶ Sect. 7.6.1].

7.4 Extension of the V-Model by Alternative Development Methods (Agile/Scrum)

If the development was to be carried out strictly in accordance with the traditional V-Model, it would only be possible to react slowly to the constantly and rapidly varying requirements in today's—primarily software-driven—development. It therefore makes sense to also apply processes from software development to modify the classic V-Model (e.g., Kanban, Scrum, etc.). Due to increasingly fast changing market requirements, agile methods in particular are gradually finding their way into well-established vehicle development, both at the OEM and at Tier1/2 suppliers. Scrum is a typical example of this [15–18]. Agile methods with continuous iteration cycles for adaptation enable the enrichment of the traditional V-Model by Micro-Vs, which are carried out in the form of small sprints—typical for Scrum (see ◘ Fig. 7.3) These Micro-Vs are not optional in the adapted V-cycle, but have become an integral part of the development process.

Such a Micro-V implementation always includes the steps planning/implementation, test and validation. The Micro-Vs, defined per section in the descending branch of the classic V-model for the areas MiL/SiL (compare ◘ Fig. 7.2), are executed several times until the created solution has reached the desired level of maturity.

The agile extensions in the development process do not contradict the traditional V-Model, but complement it in an ideal way. At the same time, this combination of V-Model and agile methods represents an analogy between the virtual prototype used for virtual integration, which is gradually refined, and the end result of a (software) development, which is refined step by step in the course of various sprints. The latter is done by creating one or more so-called Minimum Viable Products (MVPs)—individual software increments. Continuous verification of the MVPs within sprints, each of which usually lasts two to four weeks, prevents development from failing to meet the actual goals and integrates the responsible persons from the customer side into the product development process.

Agile working methods are intended to accelerate the development of executable products (MVPs) while simultaneously improving the quality of the work results. They facilitate increased customer satisfaction and support more effective communication between the involved project partners through improved transparency. Finally, cost savings can be achieved through increased efficiency based on the previously mentioned advantages [17, 19].

Such an agile approach requires not only an adaptation of the development processes, but also an adaptation of the organization in which the projects are carried out. Instead of one person ("chief engineer") making all the decisions, the decision-making and execution of the individual projects are distributed to and carried out by small multi-functional expert groups with the help of the sprints.

Further efficiency increase can also be achieved through the corresponding adjustments to the organization and the company culture toward agile working methods. The introduction of the corresponding information technology tools is partly a necessary prerequisite for the introduction of agile working methods. Here characteristic terms like Continuous Integration (CI), Continuous Testing (CT), and Continuous Deployment/Delivery (CD) should be mentioned. Ideally, not only the development areas but also the IT areas (IT operations) are considered. A significant increase in efficiency and quality can be achieved through high-level automation of the CI/CD tools, which in turn depends on the help of modified IT services (DevOps) [20].

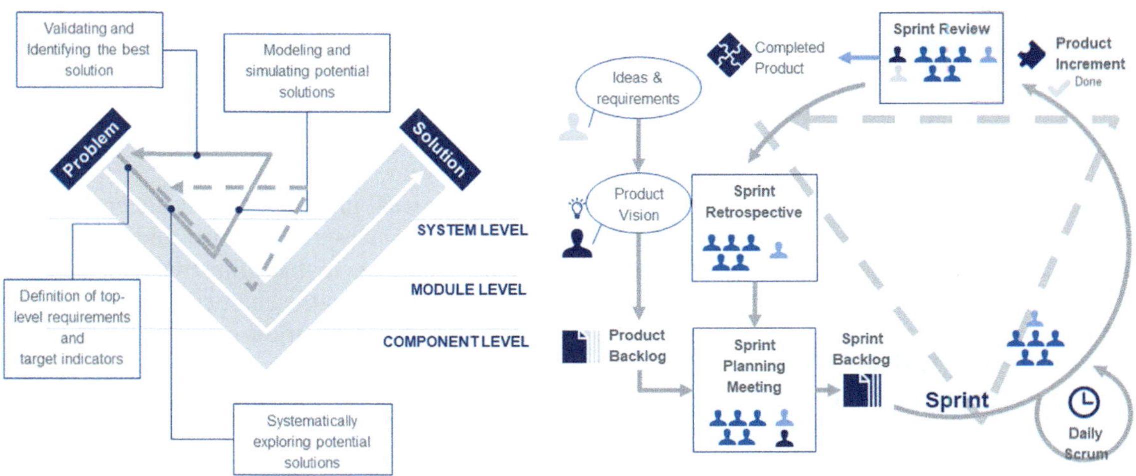

◘ **Fig. 7.3** Connection between traditional V-cycle and agile sprint cycle

The full potential of agile working methods in virtual integration can be exploited if all project partners involved apply agile methods and use the appropriate tools. Jointly conducted sprints very quickly reveal insights whether, for example, new functions are possible in the overall context and where there may be potential for improvement. New ideas and solution approaches can be entered and processed in a jointly managed backlog, and individual software increments (MVPs) can be quickly tested in a virtual prototype.

The approach described above is only possible with virtually integrated ECUs/virtually integrated software in a virtual prototype, which is simultaneously available to all project participants at anytime and anywhere. Thus, the described approach is primarily to be found in the left, descending branch of the V-Model.

7.5 Virtual Integration in the Development Process

The following is an example to show how virtual integration is part of the development process according to the V-Model. The description is divided into the specification phase and the integration phase. This corresponds to a division into the descending and ascending branches of the V-Model (cf. ◨ Figs. 7.1 and 7.2).

7.5.1 Specification by Means of Virtual Integration

The following example of a driver assistance function with respect to parking/maneuvering shows the extension of the development process according to the V-Model by virtual integration. First, a corresponding description is given for each step in the development process according to the V-Model. The description shows the perspective of a system supplier (Tier 1), is highly abbreviated, and is intended solely for the purpose of understanding the method. As a supplement, the respective activity for virtual integration and its result as added value in comparison to the classic process according to the V-Model are listed in three separate sections for each step. The description can be transferred to, for example, a software development unit of a vehicle manufacturer (OEM) without loss of general validity.

7.5.1.1 Customer Requirements

The customer requirement is formulated as: Avoidance of damage to the sides of the vehicle due to a collision with a stationary object during a parking maneuver. The maximum driving speed is 10 kph. Beyond this limit, the function is inactive. Typical driving maneuvers are described verbally and defined as test cases for further development.

In the simulation environment, driving maneuvers previously defined as test cases are configured as a virtual test drive to make the customer requirement more transparent. The sensors of the virtual test vehicle have an ideal behavior in terms of environmental perception. The simulation of the driving maneuvers gives the developers an important indication of the customer requirements' completeness and noteworthy details of the function. If a driving simulator is available, the responsible person from the customer side has the opportunity to experience and, if necessary, detail the functional requirement, which was initially formulated only verbally. If no driving simulator is available, the normal, ideally photorealistic animation of the simulation gives a first impression and the opportunity to adapt the requirements on this basis.

Virtual integration enables a much more transparent discussion of the functional requirements in this step. The development team thus gets an initial impression of the range of functions that require special attention. This also allows for an initial estimate of the specification of the required subcomponents and their feasibility. Most significant in this step is that the driver assistance function can at least be visualized so that the person responsible for the function and the development team have a common basis for discussion. This significantly reduces the risk of misunderstandings and resulting mistakes in the development process.

7.5.1.2 Logical Architecture

The logical architecture may be formulated as follows: Detection of stationary objects using the ultrasonic sensors attached to the front and rear bumpers. After the detection, the objects are further localized when exiting the visual range of the outermost sensors by means of object tracking, based on the movement of the vehicle. If the localization, based on the current speed and the current steering angle, yields an approach of an object to the side of the vehicle, a warning is issued. This warning is given by an acoustic signal. Corresponding test cases are formulated to stimulate this event.

In the virtual integration of this step, the previously simulated scenarios are specified and updated. This refers, for example, to the adjustment of the simulated sensors according to the characteristics of an ultrasonic sensor and the integration of an algorithm for object tracking, including the output of a warning within the simulation. The test cases that were developed for this step in the V-model are simulated.

At the end of this development step, the use of virtual integration results in a tested logical architecture of the customer requirements. This allows for a statement whether the customer requirements can be realized in terms of data flow and functional logic within the existing capabilities.

7.5.1.3 Technical Architecture

The technical architecture basically consists of the functional parts *Perception, Processing*, and *Output*. The scope of the perception refers to the detection of objects. For this purpose, the existing ultrasonic sensors and their interface to the vehicle bus are to be used. For processing, an additional control unit is integrated into the existing control unit network. The scope of information output is realized via a message on the same vehicle bus on which the sensor information is received. The processing of the message occurs as an acoustic warning via the infotainment system of the vehicle. The interfaces that are used for this purpose and not changed in the following development steps are described in their technical details. In addition, test cases are formulated in this development step, which are primarily used to check the interfaces between the defined functional parts.

Based on the previously created simulation models, the division of the previously modeled driver assistance function into the described functional parts results in a further detailing of the specifications. The interfaces between the functional parts are adapted to the actual technical conditions in terms of timing behavior and available bandwidth, and the test cases formulated for this step are conducted again in a simulation.

As a result of this step, virtual integration has made it possible to assess the impact on the driver assistance function of integrating it into an existing control unit network.

7.5.1.4 System Design

In this example, the system design focuses on the definition of the software architecture for implementing the driver assistance function. The required functionality is divided into different sub-tasks and their interfaces are defined. This is followed by assigning the tasks to the involved control units. The description of the tasks is formulated as a black box description. Required sensors or actuators are specified in a similar form. The test cases defined in this step essentially relate to a testing of the interfaces of the individual components. The interfaces can either be integrated into the control unit via a vehicle bus or via a direct hardware connection.

In virtual integration the overall functionality that is already included in the simulation is divided into subfunctions. The subfunctions correspond to the tasks defined in the system design or to the model refinement of sensors and actuators from an ideal to a real behav-

ior. This is done by respective modeling within the simulation environment. If an automatic code generation is planned for the tasks, it will depend on the standards arising from the used program. In a simulation, the previously formulated tests of the interfaces are carried out.

The system design is verified at the component interface level using virtual integration. This is a significant advantage over the traditional approach of the V-Model, which would not allow a verification in this early development phase. A subsequent change to the component interface would involve a significant change in its design. Such a change may additionally lead to changes in the system design.

7.5.1.5 Component Design

In this step, the black box descriptions from the system design are transferred into a detailed component specification. This specification describes the internal data and control flow of the respective tasks. Finally, a Whitebox description is available for each task. The test cases formulated for this purpose focus on testing the algorithms belonging to the respective tasks.

If necessary, an adaptation of the functionality previously created on the system level is performed in the virtual integration. Often, no significant adjustment is necessary as the component design is the result of the previous stages of development, all of which were already implemented in the virtual integration. The defined test cases are also conducted in a simulation.

Virtual integration means that tested virtual components are now available as the result of the function specification. For the following implementation, this has the advantage that there is already sufficient certainty about the correctness of the specification for implementation.

Interim Conclusion

The driver assistance function used in the example was developed by the authoring team using the virtual integration described. An integration and testing platform was used [21] as well as an authoring tool [22] for the actual function development. In fact, numerous minor and major specification errors were discovered as a result and corrected at an early stage. The errors mainly concerned missing specifications or faulty assumptions. For example, the geometric extent of the tracked objects, the necessary temporal behavior between object recognition and object tracking, or the behavior in the case of multiple simultaneous warnings were either initially undefined or incorrectly estimated. These errors may be considered as quite typical and can only be identified during the last two steps of the traditional development process according to the V-model. Therefore, the benefits of virtual integration could clearly be shown in a real project.

7.5.1.6 Implementation

In virtual integration, no explicit activity is required for this development step. The procedure during implementation is determined by the development and results in an integrable software component, taking into account all requirements for the target hardware.

7.5.2 Integration Using Virtual Integration

The following is a description of how the considered driver assistance function is progressively integrated according to the V-Model. The integration benefits from the previous work with a simplification of the process at higher quality.

7.5.2.1 Component Test

With the help of the HiL method in a white-box approach, each individual component is verified with regard to the behavior according to its specification.

The same test cases as in the component design step are used. For this, it is necessary that the simulation environment can communicate with the corresponding I/O measurement systems. In contrast to the testing of the component design, the interface to the component in this step is stimulated with real signals or read out respectively. This approach leads to results of the component tests that are directly comparable between the descending and the ascending branch of the V-model.

The reusability of the test cases and the comparability of the test results between the virtual and real implementation of each component enable the efficient investigation and evaluation of the reasons for the observed deviations.

7.5.2.2 System Test

The number of components tested together is increased step by step until all components of the tested system are integrated with real components. The HiL method and the test cases from the system design are also applied.

Similar to the previous step, the simulation environment can be used to execute the test cases. Due to the modular overall structure resulting from the V-Model, the number of real components can be gradually increased while at the same time reduced in the simulation environment.

The modular overall structure and its mapping as a virtual integration enable targeted and reproducible system testing. In addition, it is possible to perform variations in the sequence of the system test.

7.5.2.3 Integration Test

In the integration test, the new driver assistance functionality is combined with the overall system for the first time. The simulated scope may be reduced to the test components driver and environment. For the driver assistance function used in the example (parking assistant), integration with the infotainment system is carried out here and the test cases used in the technical architecture process step are applied. The HiL method is also used here. If there is a correspondingly high level of interaction with several functional domains in the vehicle, it may also be useful to introduce the Vehicle-in-the-Loop method as early as in this step.

The driver assistance function is verified in the complete vehicle after completion of the integration test. If the ViL method is used, the function can already be experienced.

7.5.2.4 Application/Calibration

The driver assistance function is applied to the entire real system. In the example of the described parking assistant, this refers, among other things, to the warning distance between the object and the driver's own vehicle. A basic calibration can be carried out purely virtually using a SiL or HiL environment. Ideally, the ViL method can be used for this step. Here, the real vehicle is integrated into a virtual environment. The reusability of the test scenarios, which are created and expanded step by step throughout the entire process so far, contributes to a significant increase in efficiency. Here, test variants resulting from the variation of vehicle and test cases, for example, different sized parking slots with and without obstacles, can be applied much more efficiently, since many environment properties can be set solely by parameterization.

The use of virtual integration in conjunction with the ViL method allows efficient calibration of the driver assistance function. The efficiency and reproducibility of the test cases required for this can thus be significantly increased. Depending on the application, however, the final calibration ("fine tuning") usually takes place in real tests. Due to this systematic virtual pre-calibration, the final tuning can start at a much higher level.

7.5.2.5 Acceptance Test

In the final development step according to the V-Model, the acceptance of the new function by the user or customer is tested. Ideally, this is done with a real vehicle in a real environment. The ViL method may still be installed to present variants or alternatives to the user that would otherwise mean a physical change to the target vehicle.

Virtual integration plays only a subordinate role in this step and may well be omitted completely.

Since the test cases used in this development step were already tested at the beginning of the development using virtual integration, the biggest disadvantage of the V-Model, namely the formally late validation of the customer requirement, is largely eliminated.

7.5.3 Application of Agile Methods and CI/CT/CD Tools During Virtual Integration

Without limiting the validity of the V-Model, efficiency increases can be achieved, if not only one development group works isolated in each of the individual phases of the V-Model and passes on the work result to the next group/department (e.g., from research to advance development and on to series development or to and from suppliers). Ideally, all involved development departments at the OEM and the supplier work simultaneously with the same virtually integrated model. In the example described above, this is initially characterized by ideal ultrasonic sensors, in later stages extended by increasingly realistic sub-models.

Regardless of the development methodology, whether according to the classic waterfall principle or to agile development methods, development can be carried out according to the V-Model. In the descending branch (specification phase), however, much faster throughput times with the help of Continuous Integration/Continuous Testing/Continuous Deployment (CI/CT/CD) tools [▶ Sect. 7.4] can be achieved. A considerable amount of development time can be saved while at the same time generating higher quality software. By using the CI/CT/CD tools for both the development of the driver assistance function as well as the test tools, executable increments of the final function are always generated in a short time (so-called Minimal Viable Products -MVPs), which can be tested directly against the specification. Ideally, the client of the function development is also part of the agile development team and can continuously test the respective created increments (MVPs) from the viewpoint of the end-customer. The immediate feedback directly refines the individual stages and thus directly improves the quality of the overall result (here in the example of the parking functionality) while at the same time reducing development time.

Starting with the application in the logical architecture, "living specifications" can be created by means of virtual integration. As soon as the first test cases, which can be simulated, are available (technical architecture / system design), the CI/CT/CD tool chain can be used to constantly perform automated checks in the course of the various specification stages and later the integration stages. In the spirit of agile development, multidisciplinary teams should work together on this. The main advantages of using CI/CD methods are constant, automated checking of all load cases created to date (e.g.,

overnight) to avoid the reintroduction of errors that have already been corrected and to check the effect of software changes on existing, already tested behavior.

In general, it can be stated that virtual integration is becoming increasingly important, particularly with the focus on software development, as the integration phase is primarily carried out independently of real hardware. In the future, function release tests will have to be performed purely virtually, since the availability of test facilities such as test sites, vehicles, and testing personnel is limited.

7.6 Limits of Virtual Integration

The exemplary virtual integration process described in this chapter is practically used with various extensions and modifications. Due to the highly interdependent component development in the individual vehicle domains, there are several central integration milestones within the vehicle development process, at which the respective integration statuses are checked against requirements at the system and full vehicle level. Even though the general trend is toward the "digital twin," i.e., the most exact possible replication of real development projects in virtual space, in practice not every vehicle component is virtualized to a degree and quality that would allow for a completely virtual (i.e., model based) verification of the development requirements. Limitations exist on the one hand due to the lack of validity of (partial) models, and on the other hand, there are purely practical limits that currently prevent a completely virtual validation.

7.6.1 Validity Limits

According to the British statistician George Box [23] "all models are wrong, but some are useful".

Viewed differently, this means that models used in the context of virtual validation must be carefully checked for their scope of validity. When categorized into failsafe tests, functional tests, and performance tests, different requirements may apply to the sub-models involved depending on the use case ("Purpose Driven Fidelity").

Vehicle dynamics models are comparatively easy to validate, sometimes complicated to set up, and complex to calculate. But they can be compared with physical measurements from real vehicles and the deviations quantified and evaluated using statistical methods. Sensor models with low-quality requirements can also be modeled phenomenologically with relevant properties and likewise evaluated with the aid of statistical methods in comparison with the characteristic values of the real sensors.

Physical sensor models, however, which emulate both the hardware and the physics, are "power hungry", i.e., they require a lot of computing power, have limited real-time capability, and are complicated to validate [4, 30]. On the one hand, the validity range of such models can be narrowed down in combination with a virtual vehicle in comparison with test data of real maneuvers. On the other hand, the quality of the implementation of the physical models can be evaluated against the expected behavior (e.g., multipath propagation, shadowing, strong reflections at gantries, etc. as in radar sensor models). Also, the combination of both approaches does not automatically lead to a quantification of the performance and a defined validity range of the sensor models.

Even more complex is the validation of environment models, which on the one hand provide the input signals for sensor and vehicle dynamics models, but on the other hand also have to react in an appropriate way to the behavior of the unit under test (UUT), depending on the application.

In the meantime, powerful environment models exist (often from the gaming sector) that not only provide visual input to camera systems but also, via reflection, diffraction, and attenuation models, useful input to virtual radar and lidar sensors. The physical sensor models described above are mostly computed with GPUs ("on the graphics card"), where they can effectively interact with the 3D environment models used.

However, the authors are not aware of any metric for quantitative quality assessment of an environment consisting of virtual vehicles, virtual sensors, and virtual environment objects. Numerous research projects (e.g., PEGASUS) [24, 25] approach this topic indirectly and propose procedures in which quality criteria for the individual parts of the chain of effects are discussed on the basis of application examples and exemplary tool chains [26].

In practice, this means in most cases that simulation results are regularly and systematically compared with real data. This is necessary to be able to use the main advantages of simulation (extreme parameter variation possibility and high availability) in a meaningful way.

7.6.2 Practical Limits

A few years ago, the available resources in terms of computing power and data storage were still considered the limiting factors on the way to a purely virtual development and validation environment. Today, however, state of the art is an almost unlimited scaling of high-performance computers that provide cloud-based computing power on demand.

These new possibilities have brought the term "DevOps" into focus, which describes the provision, operation and further development of an IT infrastruc-

ture that is well adapted to the use case [35, 38] (see ▶ Sect. 7.5.3).

Increasing standardization of test descriptions, virtual implementations (e.g., OpenDrive, OpenScenario, OSI, etc.) [5, 27, 31] and exchange formats for models incl. associated solvers (FMUs according to the FMI standard) [1, 3] facilitate reuse and simulation model exchange between OEMs and suppliers. But they do not yet cover all practical requirements and in some cases still clash with proven validated testing tool chains.

7.7 Conclusion

Virtual integration in general is not a completely new process for function development for vehicles. It uses established process models and methods and extends them using the metaphor of the virtual road test described in ▶ Sect. 7.1 above. Virtual integration is thus a methodology to develop complex, safety–critical, and highly networked functionalities for the vehicle. Driver assistance systems mostly possess these properties and thus benefit greatly from virtual integration.

In order to apply virtual integration, a powerful and flexible simulation environment is necessary. The necessary properties of this simulation environment go well beyond the simulation of a physical behavior and also include the connection to various real components. Therefore, it makes more sense, as described in ▶ Sect. 7.1, to speak of an integration environment or an integration platform for the virtual driving test. The simulation of physical behavior is one task of the integration platform, but it also includes the type of testing and the possibilities for more efficient cooperation between vehicle manufacturers and system suppliers (▶ Sects. 7.2 and 7.4).

The limits of virtual integration are essentially related to the validation of the environment of the virtual prototype. Since the computing power required, for example, for the simulation of complex sensor models can now be scaled almost arbitrarily, the methods and metrics to validate the environmental model and the sensor models define the limits.

Despite these limitations, driver assistance systems can be developed much more efficiently and with less risk by means of virtual integration and using the in-the-loop methods presented in ▶ Sect. 7.3. This was shown in ▶ Sect. 7.5 for the example of a function development.

Due to the constantly increasing and almost arbitrarily available computing power (HPC, cloud computing, etc.), as well as constantly improved simulation models and methods for virtual test driving, certain function developments in the domain of driver assistance systems can be carried out almost completely virtually today. This is necessary in order to cope with the ever-increasing variety of functions, especially in the field of highly automated driving.

Even though the development focus will increasingly shift from real driving tests to virtual driving tests, the real driving test will certainly remain an important part in the development process in the near future. In addition to final release tests, the necessary continuous extension of test cases that are classified as relevant in the real test operation as well as the subjective impression of experts and test subjects are still indispensable aspects of the development.

References

1. FMI Development Group.: Functional Mock-up Interface for Model Exchange and Co-Simulation (Version 2.0.3, Nov. 15th 2021) (2021). ► https://fmi-standard.org/
2. Schneider, S.-A., Frimberger, J., Folie, M.: Reduced validation effort for dynamic light functions. ATZ Elektronik 2014-9(2), 16–20 (2014)
3. FMI/FMU.: (2021). ► https://fmi-standard.org/. Accessed 10 Oct 2021
4. Roth, E., Dirndorfer, T., Knoll, A., v. Neumann-Cosel, K., Ganslmeier, T., Kern, A., Fischer, M.-O.: Analysis and validation of perception sensor models in an integrated vehicle and environment simulation. TUM Paper, 11–0301, 11–31 (2011). ► https://mediatum.ub.tum.de/doc/1287190. Accessed 4/2022
5. OpenScenario.: (2021). ► https://www.asam.net/standards/detail/openscenario/. Accessed 3 Oct 2021
6. Fremont et al. (2020). Formal scenario-based testing of autonomous vehicles: from simulation to the real world. In: 2020 IEEE 23rd International Conference on Intelligent Transportation Systems (ITSC), pp. 1–8 (2020). ► https://doi.org/10.1109/ITSC45102.2020.9294368
7. V-Modell.: ► http://de.wikipedia.org/wiki/V-Modell (09/2014) und ► https://de.wikipedia.org/wiki/V-Modell_(Entwicklungsstandard). Accessed 31 Oct 2021
8. V-Modell XT.: (2014). ► http://www.cio.bund.de/Web/DE/Architekturen-und-Standards/V-Modell-XT/vmodell_xt_node.html
9. Winner, H.: Challenges of automotive systems engineering for industry and academia. In: Maurer, M., Winner, H. (Hrsg.) Automotive Systems Engineering. Springer, Heidelberg (2013)
10. Palm, H., Holzmann, J., Schneider, S.-A., Koegeler, H.-M.: The future of car design—systems engineering based optimisation. ATZ Automobiltechnische Zeitschrift **115**(06), 42–47 (2013)
11. Schäuffele, J., Zurawka, T.: Automotive Software Engineering, 6th edn. Springer Vieweg, Wiesbaden (2016)
12. Miquet, C.: New test method for reproducible real-time tests of ADAS ECUs: "Vehicle-in-the-Loop" connects real-world vehicles with the virtual world. In: Pfeffer, P. (ed.): 5th International Munich Chassis Symposium 2014. Springer, Wiesbaden (2014)
13. Hakuli, S., Stölzl, S., Pithan, B.: Lab-on-Wheels – Ein Absicherungswerkzeug für den nahtlosen Übergang zwischen Simulation und Fahrzeug in Fahrerassistenzsysteme und automatisiertes Fahren 2018. VDI Verlag 2018 (2018). ► https://doi.org/10.51202/9783181023358-25. Accessed 30 Nov 2021
14. Berg, G.: Das Vehicle in the Loop - Ein Werkzeug für die Entwicklung und Evaluation von sicherheitskritischen Fahrerassistenzsystemen, Dissertation UniBW München (2014)
15. Agiles Manifest.: (2021). ► https://agilemanifesto.org/iso/de/manifesto.html. Accessed 7 Oct 2021
16. Scrum Guides.: (2021). ► https://scrumguides.org/. Accessed 7 Oct 2021
17. Schwaber, K., Beedle, M.: Agile Software Development with Scrum Prentice Hall Upper Saddle River (2002). ISBN 978-3-86645-643-3
18. Schwaber, K.: Agiles Projektmanagement mit Scrum. Microsoft Press, Redmond 2007 (2007). ISBN 978-3-86645-631-0 englisch: Agile Project Management with Scrum
19. Fowler, M.: Agile Software Guide (2019). ► https://www.martinfowler.com/agile.html. Accessed 6 Mar 2022
20. Niederbrucker, G., Kochem, M.: Clouds Ahead—The Transformation of Vehicle Development and Data Management Processes, ATZ-Tagung Automatisiertes Fahren, Wiesbaden (2020)
21. IPG Automotive GmbH.: CarMaker (2022). ► https://ipg-automotive.com/de/. Accessed Apr 2022
22. MathWorks.: MATLAB (2022). ► https://de.mathworks.com. Accessed Apr 2022
23. Box, G.: All models are wrong (but some are useful) (1978). ► https://en.wikipedia.org/wiki/All_models_are_wrong. Accessed 1 Nov 2021
24. Winner, H., Lemmer, K., Form, T., Mazzega, J.: PEGASUS—first steps for the safe introduction of automated driving. In: Meyer, G., Beiker, S. (eds.): Road Vehicle Automation 5. Lecture Notes in Mobility. Springer, Cham (2019). ► https://doi.org/10.1007/978-3-319-94896-6_16
25. PEGASUS-Projektseite. ► https://www.pegasusprojekt.de/de/. Accessed 13 Mar 2022
26. Bock, J., Krajewski, R., Eckstein, L., Klimke, J., Sauerbier, J., Zlocki, A.: Data Basis for Scenario-Based Validation of HAD on Highways. 27. Aachen Colloquium—Automobile and Engine Technology (2018)
27. Open Simulation Interface (OSI).: (2021). ► https://www.asam.net/standards/detail/osi/. Accessed 31 Oct 2021
28. Martinus, M., Deicke, M., Folie, M.: Virtual test driving—hardware independent integration of series software. ATZ Elektronik **8**(05), 16–21 (2013)
29. Schick, B., Schmidt, S.: Evaluation of video-based driver assistance systems with sensor data fusion by using virtual test driving. In: FISITA World Automotive Congress, Beijing, China (2012)
30. Roth, E., Calapoglu, T., Helmich, H., Neumann-Cosel, K., Fischer, M.-O., Kern, A., Heider, A., Knoll, A.: ADAS Testing using OptiX NVIDIA GTC, San Jose, USA (2012). ► https://on-demand.gputechconf.com/gtc/2012/presentations/S0319-Advanced-Driver-Assistance-System-Testing-Using-OptiX.pdf. Accessed 04/2022
31. OpenDrive.: (2021). ► https://www.asam.net/standards/detail/opendrive/. Accessed 3 Oct 2021
32. Infrastructure for Spatial Information in Europe.: (2021). ► https://inspire.ec.europa.eu/. Accessed 30 Oct 2021
33. Geyer, S., Baltzer, M., Franz, B., Hakuli, S., Kauer, M., Kienle, M., Meier, S., Weissgerber, T., Bengler, K., Bruder, R., Flemisch, F., Winner, H.: Concept and development of a unified ontology for generating test and use-case catalogues for assisted and automated vehicle guidance. IET Intell. Transp. Syst. **8**(3), 183–189 (2014). ► http://publications.rwth-aachen.de/record/446559. Accessed 19 Mar 2022
34. Weber, N., Frerichs, D., Eberle, U.: (2021) Sicherheitsrelevante Testszenarien für automatisierte Fahrfunktionen. ATZ Automobiltechnik Z **123**, 52–57 (2021). ► https://doi.org/10.1007/s35148-021-0748-5
35. DevOps.: (2021). ► https://azure.microsoft.com/de-de/overview/what-is-devops/oder ► https://aws.amazon.com/de/devops/what-is-devops/. Accessed 24 Oct 21
36. Horváth, M., Lu, Q., Tettamanti, T, Török, Á., Szalay, Z.: Vehicle-In-The-Loop (VIL) and Scenario-In-The-Loop (SCIL) automotive simulation concepts from the perspectives of traffic simulation and traffic control. Transp. Telecommun. J. **20**, 153–161 (2019). ► https://doi.org/10.2478/ttj-2019-0014

37. Bock, T.: Vehicle in the loop: Test- und Simulationsumgebung für Fahrerassistenzsysteme (2008, Audi-Dissertationsreihe; Bd. 10) (2008). ▶ https://d-nb.info/991137752
38. Halstenberg, J., Pfitzinger, B., Jestädt, Th.: DevOps – Ein Überblick; Springer-Vieweg (Heidelberg) (2020)
39. UNICARagil—Disruptive modular architectures for agile, automated vehicle concepts, 27. Aachen Colloquium Automobile and Engine Technology, Aachen. In: Woopen, T., Lampe, B., Böddeker, T., Eckstein, L. (eds.): Institute for Automotive Engineering, RWTH Aachen University, Aachen

7

Open Access This chapter is licensed under the terms of the Creative Commons Attribution-NonCommercial-NoDerivatives 4.0 International License (▶ http://creativecommons.org/licenses/by-nc-nd/4.0/), which permits any noncommercial use, sharing, distribution and reproduction in any medium or format, as long as you give appropriate credit to the original author(s) and the source, provide a link to the Creative Commons license and indicate if you modified the licensed material. You do not have permission under this license to share adapted material derived from this chapter or parts of it.

The images or other third party material in this chapter are included in the chapter's Creative Commons license, unless indicated otherwise in a credit line to the material. If material is not included in the chapter's Creative Commons license and your intended use is not permitted by statutory regulation or exceeds the permitted use, you will need to obtain permission directly from the copyright holder.

Dynamic Driving Simulators

Hans-Peter Schöner, Jens Häcker, and Katja Nagel

Contents

© The Author(s) 2026
H. Winner et al. (eds.), *Handbook Assisted and Automated Driving*,
https://doi.org/10.1007/978-3-658-45276-6_8

8.1 General Overview of Driving Simulators

8.1.1 Applications of Driving Simulators

Promoted by the increase of digitalization of development processes, driving simulators are used widely in the automotive industry and in automotive research institutions. Driving simulators are being used for many different use cases as listed in ◘ Table 8.1 (in the order of increasing requirements with respect to the realism of the driving experience in the virtual world).

The focus of all of these applications is on the *interaction of human drivers* with the technical systems of the vehicle, especially in challenging driving situations (with respect to other traffic participants, obstacles, dangers, …) and under variable and adverse driving conditions (road condition, weather, lighting, …). Depending upon the application there is a multiplicity of technical realizations of driving simulators: starting from a static seat with a flat screen and a simple steering wheel and pedals up to large dynamic simulators with perfected immersion technologies, in which complete vehicle cabins, systems for auditive, haptic and visual environment simulation as well as motion rendering mechanisms for providing a feeling of driving dynamics are implemented.

The common components of all simulator experiments are depicted in ◘ Fig. 8.1: The driver and his behavior in traffic situations are the focus of all (driver-in-the-loop) experiments. The driver operates his vehicle in a carefully planned and monitored experiment within a virtual world. The scenarios of interest are replicated on road networks with surrounding traffic, including the interaction between vehicles based on system functions using modeled sensor data, and applying model-based vehicle dynamics of the participating vehicles. Complex simulator software, executing on suitably matched simulator hardware, creates perceptions for the driver, which are presented to him via visual, acoustic, and haptic rendering of the virtual world in combination with a physical driver interface. The behavior of a large number of individual drivers can be analyzed in order to deduct robust evidence about the acceptance, usability, and controllability of driving functions in specific scenarios.

Related to testing and validation of driver assistance systems and automated driving functions, which are in focus here, the main advantages of driving simulators are:

- the precise tuning of situations and the high reproducibility of traffic conditions, by having control over factors that are out of control on roadways (traffic, signals, weather, etc.),
- exposing drivers to crash or near-crash situations without impacting their safety,
- simple and fast variation of vehicle and environment parameters.

Complementary to simulator experiments, various traffic situations with real vehicles need to be performed on test tracks and on public roads to cover effects, which cannot be modeled adequately in a simulator. Together with such real-world tests, driving simulators enable the study of cause and effect; thereby becoming an indispensable tool for the efficient and comprehensive testing of driver assistance and chassis systems.

An essential aspect of testing in driving simulators is to immerse the test persons into a virtual world which is perceived as being so close to the reality, that the driving behavior in this virtual world is representative of the behavior in the real world, so that the experimental results are valid and dependable. Three differ-

◘ **Table 8.1**   Driving simulator application areas and use cases

Driving simulator application areas	Use cases
Functional vehicle demonstrations, advertising new vehicle functions	Demos at auto shows
Investigations on interior concepts, display, and control concepts	Assessing reachability, clarity, and comprehensibility
Training for drivers	Emergency vehicles, transport vehicles for fuel saving driving style, race cars
Traffic safety research	Crash reconstructions, driver behavior analyses
Study of driver performance and development of driver models as basis for off-line simulations	Drowsiness, distraction, and impairment
Testing and verification of driver assistance systems and take-over situations in automated vehicles	Effectiveness, controllability and statistical analyses
Development of chassis systems and vehicle dynamics control systems	Analysis of variants and parameter tuning
Development and verification of driving functions for automated vehicles	Efficient software development, interaction of human drivers with automated vehicles

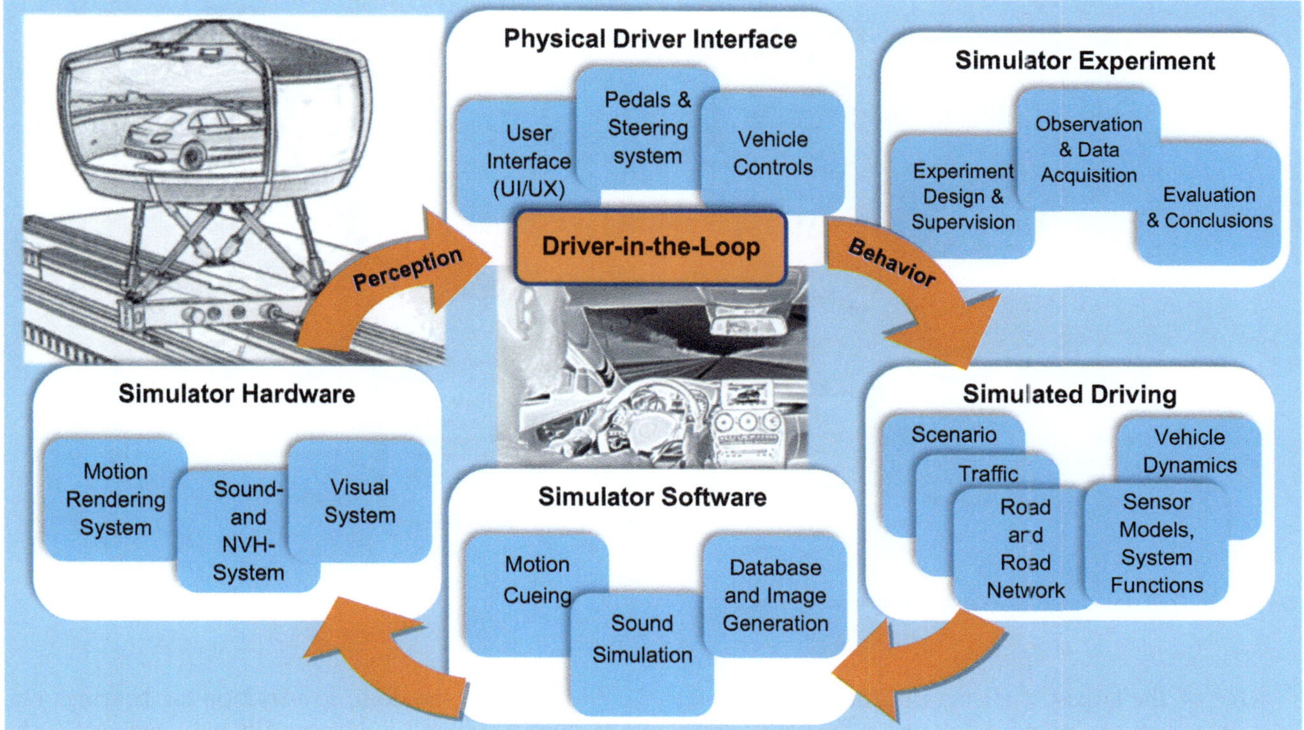

◘ Fig. 8.1   Driver-in-the-loop in virtual driving experiments

ent domains which need to be considered to reach this goal are in the focus of this chapter: hardware and software of the simulator, the design of the specific simulator experiments, and factors relating to the selection, observation, and questioning of the test persons (see ◘ Fig. 8.2).

8.1.2 Concepts for Dynamic Driving Simulators

An overview of the historical development of driving simulators has been published by Slob [44]. The first automotive driving simulator was realized in the decade of 1970 by Volkswagen with three degrees of freedom, for yaw, roll, and pitch motion. The VTI (Swedish National Road and Transport Research Institute) in Linköping, see Nordmark et al. [34], built a system, with a motion system likewise limited to three degrees of freedom, however using roll, pitch, and lateral motion, supplemented with vibration actuators for roll, pitch, longitudinal and vertical motion. In 1985 [7], Daimler-Benz in Berlin introduced a system based on flight simulator concepts, which was equipped with a hydraulic hexapod (*Stewart platform* with six degrees of freedom) and constituted the world-wide largest motion space available at that time. In the meantime, nearly every large car manufacturer and many tier-1

companies, as well as several large research institutes, have built and now run their own dynamic driving simulator in order to support the digital development and testing of driving functions. Depending on application focus and upon budget, different system concepts were selected, however, a hexapod, supplemented by one or two linear rails, is the most frequent design choice, see Mohajer et al. [27].

If a driving simulator is used as a development tool for driving dynamics investigations – e.g., for optimal tuning of chassis components for handling performance – the exact assessment of the lateral dynamic characteristics of the car is of greatest importance. With this in mind, during a revision of the Daimler Driving Simulator in Berlin in the year 1993, a transverse rail with a 6 m length was supplemented, which allowed precise replication of the vehicle lateral dynamics while performing a lane change maneuver [20]. For investigations with general test persons, the prevention of *kinetosis* (motion sickness) is important; crucial for this is an accurate coordination of visual and haptic motion cues. New low-friction hexapod actuators, a digital control system, and an increased field of view of the projection system were integrated with this goal into the Daimler Simulator in 2004. All these hardware components and precise coordination of the vision and the motion system contributed to the reduction of the kinetosis of the test persons to under 2% [21].

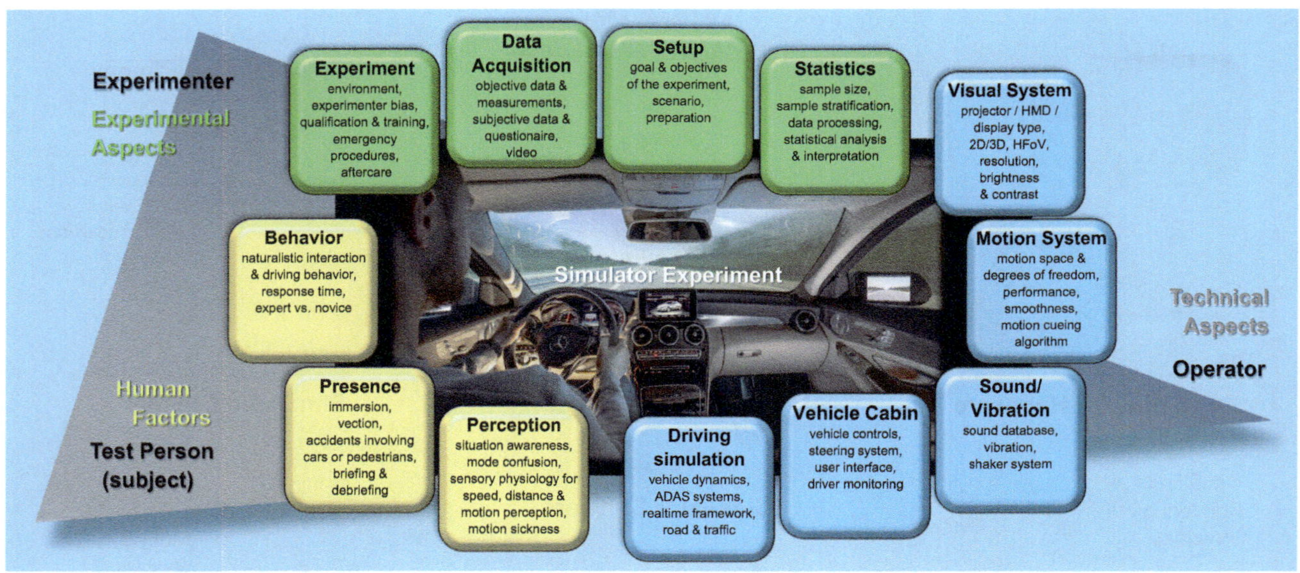

◘ Fig. 8.2 Different aspects of a driving simulator experiment

One of the largest challenges to the vision and the motion system of driving simulators is the realistic representation of city center scenarios, especially because of the frequent tight turning maneuvers in intersections. For an accurate motion perception in such situations, first, a large enough motion space must be available, which measures approximately the same size as the real-world maneuver. For this purpose in 2006, Toyota [29] built the up-to-now largest driving simulator (see ◘ Fig. 8.3) with a complete vehicle cabin and a motion space of 20 m × 35 m (hexapod on X–Y-rails, nearly identical to the NADS in Iowa from 2000). To support the accurate reproduction of tight turning maneuvers, both the vision system and the motion system should be able to render dynamic vehicle motions, especially during fast yawing motion, without perceivable jerks. Furthermore, both systems need to exhibit a low latency in order to provide a consistent motion perception to the test persons.

The necessary size for the representation of the turning maneuvers happens at the expense of the system dynamics; thus, such a large system is less suitable for the investigation of fast driving dynamics maneuvers. High efforts, cost, and technical risks of such large mechanical systems are not acceptable for many users; so, there are several new solutions under investigation to produce an acceptable motion perception with alternative motion systems. A variety of different design approaches should be mentioned here exemplarily (see ◘ Fig. 8.4):

— VI-grade DYNAMIC driving simulator DiM250, designed for the evaluation of driving comfort and handling dynamics of passenger cars and race vehicles [47], consisting of a hexapod platform with 9 degrees of freedom, low-friction air bearings on a base plate, and a dynamically actuated seat with active seat belts for rendering of longitudinal deceleration forces.

— Robot arm system of the company Kuka, with a realization at the Max Planck Institute in Tuebingen [33]; here the motion system is based on a conventional industrial production robot. An additional long linear motion axis is made available by a rail system.

— The concept of a "Wheeled Mobile Driving Simulator" is pursued in a research phase at the University of Darmstadt [5] and the University of Dresden [46]; the free-driving system is designed to be used on a sufficiently large even and open test area, so a large motion space would be available at relatively small cost.

— Coupled linear platform by AB Dynamics [1], with wedge-shaped couplings which provide a very stiff and space-efficient actuator configuration of high bandwidth.

8.2 Daimler's Dynamic Simulator as Example for a Driving Simulator Design

In 2010, Daimler opened a new dynamic driving simulator (see ◘ Fig. 8.5). It was designed on the basis of 30 years of experience with its predecessor, Daimler's first-generation simulator in Berlin [51]. The detailed motion system design was aimed at performance assessment of chassis concepts and driving dynam-

◻ Fig. 8.3 Toyota's driving simulator in Higashi Fuji [29]

ics control systems; variability and close to real world overall driving experience were the secondary design goals with respect to the evaluation of driver assistance systems with a large number of test persons. At its installation, this system was internationally benchmarked as the most dynamic and precise system. In the meantime, other simulators with similar properties, though each with a slightly different focus, have been designed and realized, see Mohajer et al. [27].

8.2.1 Motion System

The task of the motion system is to generate accelerations and rotations in all six degrees of freedom in order to provide a sensation of motion to the test person, which closely corresponds to the real vehicle motion in a specific driving situation. The driving simulator uses an electromechanically actuated hexapod which can move the vehicle cabin in its possible rigid body six degrees of freedom. Additionally, this hexapod is placed on an electrically propelled slider gliding upon a 12.5 m long linear rail, this way significantly expanding one degree of freedom. On top of the hexapod, there is a lightweight carbon fiber dome which contains the vehicle cabin. A rotating platform allows to reorient the cabin inside the dome by 90°around the vertical axis, so

that the linear rail can be configured for the representation of longitudinal or lateral motion. This way either longitudinal or lateral accelerations of app. 10 m/s^2 up to speeds of 10 m/s can be achieved.

The entire motion system was designed with the objective to minimize friction, in order to reach high dynamics, continuous motion and running smoothness; in fact, friction forces need to be compensated in order to achieve a precise motion perception. For this reason, the long linear slide is equipped with an air bearing system. This requires high precision in manufacturing and installation of the linear bearing system and regular conscientious system maintenance. Also, in the design and control of the electromechanical hexapod actuators friction reduction and compensation has been considered with extra care.

The design of the dynamics of the motion system is based on the requirements for vehicle dynamics investigations up to the stability limits of the tire-road contact. A second linear rail for a larger motion space in an x–y-rail configuration (in order to better replicate turning maneuvers in city driving) was omitted, as the larger inertia would have implicated too high restrictions for the system's dynamics. This fact needs to be taken into account in the conception of simulator experiments, e.g., by avoiding turning situations in traffic scenarios.

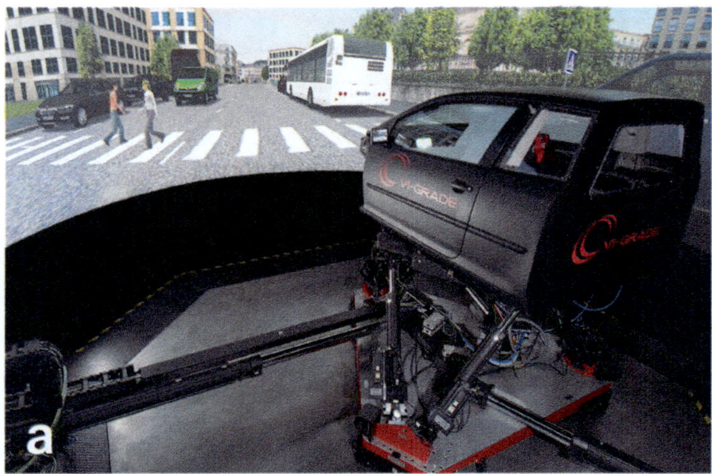

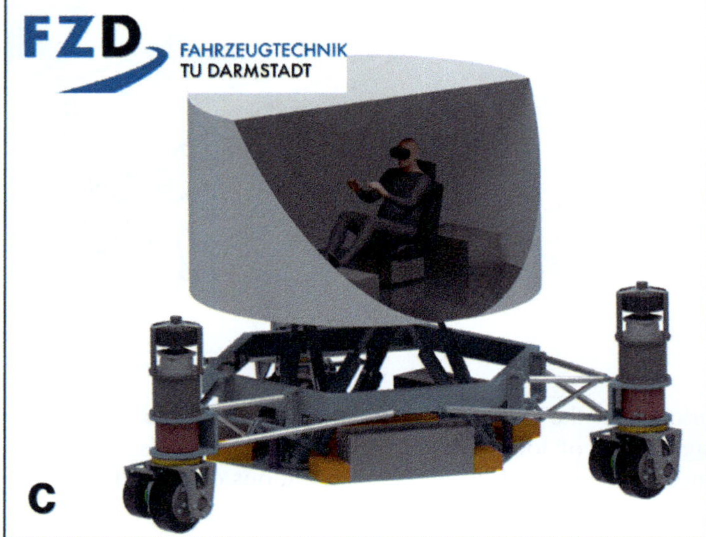

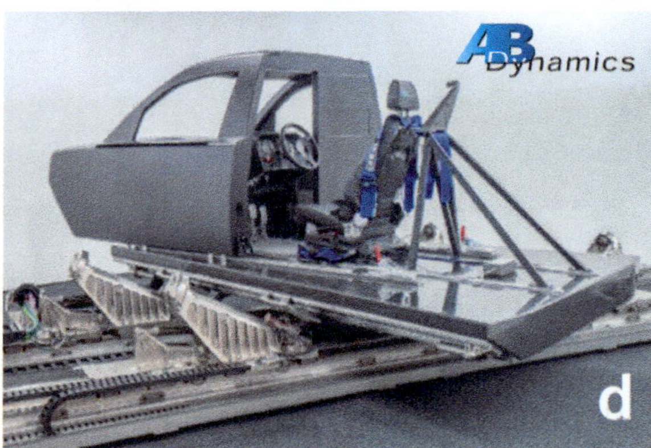

◘ Fig. 8.4 Driving simulators with alternative motion system concepts: **a** Vi-Grade DiM250 [47], **b** MPI Tübingen [33], **c** FZD TU-Darmstadt [5], **d** AB Dynamics [1]

In general, for any kind of motion system, the potentials and limitations resulting from the motion system configuration need to be considered when planning testing scenarios, because they influence how realistic the driving experience can be presented.

8.2.2 Physical Driver Interface

Simulator studies with test persons, which are essential for the conception and verification of driver assistant systems, require in particular, that the test persons should immerse in the virtual environment and thus feel completely like being in a real and natural driving situation: this results in realistic driver behavior. For this goal, the driver takes his seat in a vehicle cabin, originating from a real vehicle deprived of all unnecessary components of the powertrain and of many chassis components. On the other hand, actuators are installed to provide a realistic pedal feel and steering feedback, corresponding to vehicle dynamics and driving conditions.

In addition, the entrance to the dome of the driving simulator is arranged in such a way that the test person does not get to see the driving simulator exterior; instead, the person entering the dome will find a real vehicle standing on a road, the virtual scene already projected all around (see ◘ Fig. 8.6). All control elements of the vehicle and all other visible and audible items are set up for identical or at least close-to-reality experience. The test person will start the vehicle and engage it in driving with the same procedure as in a real-life vehicle; also, the instructions given to the test persons are chosen carefully to imply driving of a real vehicle and not a simulator.

Depending upon the type of investigation, different passenger car or truck cabins are used; a quick exchange of the cabins is facilitated with the help of a standardized mechanical and electrical interface system.

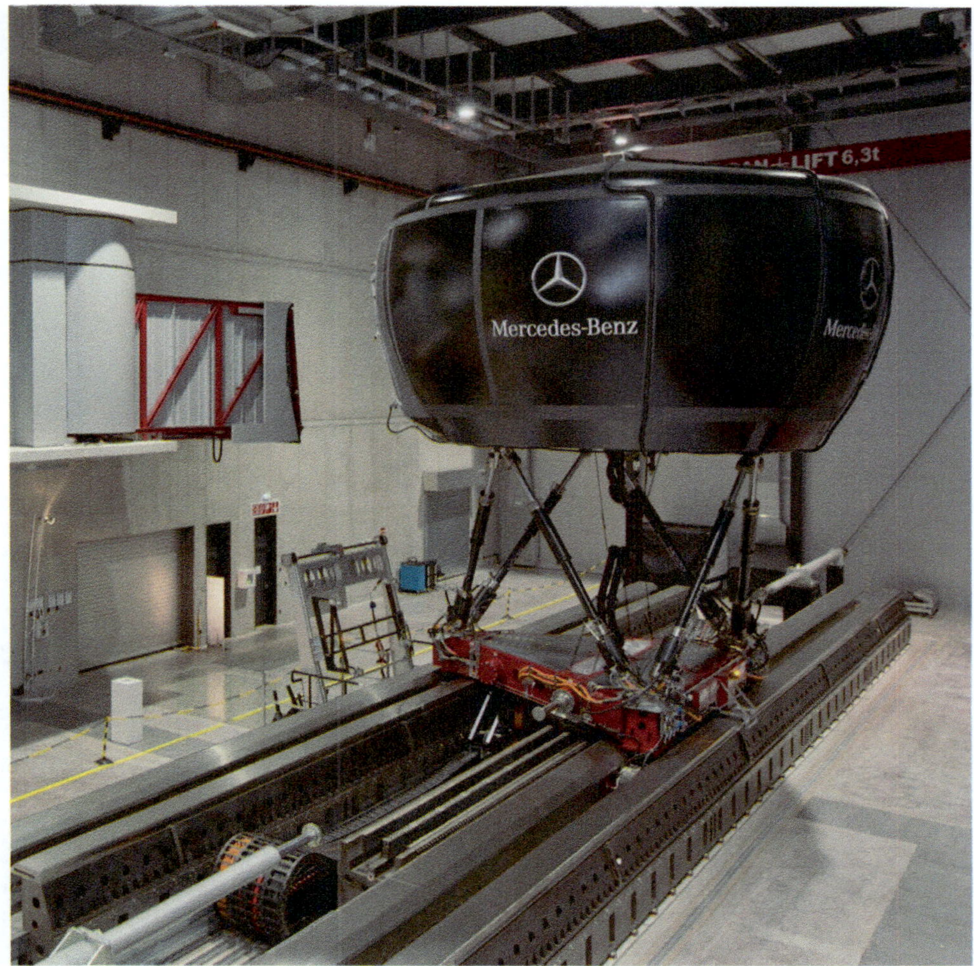

■ **Fig. 8.5** Daimler's dynamic simulator in Sindelfingen: motion system

8.2.3 **Visual System**

The visual system provides the dominant impression for the driver; it replicates the city or highway driving scene with a detailed rendering of the environment. While the motion system provides the correct feeling of acceleration forces to the test person, the visual system is responsible for the impression of speed, continuous movement, heading, and spatial orientation.

The visual system consists of image generation software, a database of the three-dimensional environment, the hardware for the calculation and transfer of the images into the simulator, and the projection of the images onto a screen. The actual projection system [41] is realized with 14 LED projectors with WQXGA resolution (2560×1600 pixel) installed on the dome's ceiling, projecting a seamless 360° scene for the driver onto the inner surface of the spherical dome. Additionally, the two outside mirrors of the vehicle cabin are replaced by displays; they provide the correct rear view as seen from the position of the driver. In combination with traffic simulation software and 3D picture rendering software, a close-to-reality representation of driving situations and driving maneuvers is provided; day and night scenarios and many weather conditions can be simulated.

For a realistic representation, apart from the level of detail of the displayed objects in the visual database, the resolution, shading, and surface reflection effects of elements in the pictures are important; there have been substantial improvements in rendering virtual worlds in simulators over the last decade. For high-speed driving scenarios, the frame rate of the picture generator and projection system is of vital importance: normal drivers in highway or city scenarios are well satisfied with a frame rate of 60 Hz; for high yaw rates, e.g., highly dynamic lane changes in vehicle dynamics experiments and for turns in intersections, a frame rate of 120 Hz leads to a much better perception of the image sharpness. For night driving conditions, the correct simulation of vehicle head lamps and street lights and their resulting illumination of the scenery are challenging.

Fig. 8.6 Daimler's dynamic simulator in Sindelfingen: vehicle in the dome

For highway driving, surrounding traffic consisting of vehicles of different categories (cars, trucks, motorbikes) are sufficient and may be represented as moving 3D objects with suitable models for driver behavior. In addition, for animated inner-city scenarios, with more naturally moving objects, scripted pedestrians, cyclists, traffic lights with changing light signals and high visual details become more important. The exact control and manipulation of position, timing, and behavior of traffic partners in relation to the test person's ego vehicle is essential for the design of critical traffic scenarios for the testing of driver assistance systems.

Visual systems with the option of stereoscopic projection and the use of VR headsets as head-mounted displays (HMD) are more and more used for driving simulation purposes. Their advantage should be to increase the immersion of the test persons in the virtual world, to improve distance and speed perception, and to make the complete driving experience even more realistic [12]. HMDs also have the advantage of being more portable and cost effective. Their drawbacks include decreased face validity (test person having to wear something artificial on their head) and lower visual performance in terms of resolution, frame rate, and motion latency.

8.2.4 Sound and NVH System

The sound of driving plays another substantial role in the immersion of the test person in the virtual driving situation. While motion systems are responsible for recreating acceleration cues, sound systems provide a sense of subtle changes in vehicle speed. It is calculated depending on engine power, engine speed, and driving speed. Engine, tire, and airflow noise are represented correctly by a sound system, that has to be correctly calibrated to the location of the driver's ears position in the vehicle; also, road roughness should be reproduced in the noise and additional vibrations when driving on an uneven road. For this end, recorded sound samples from road driving and test stands are adapted, mixed, and replayed according to a given driving condition, see Krebber [25]. Higher frequency excitation due to road roughness can be better rendered by additional shaker systems than by the hexapod actuators. Without such NVH and sound cues, it is much harder for test persons to control the speed of the vehicle. Spectral shifts (*Doppler effect*) in vehicle noise produced by passing other vehicles must be represented correctly, in order to make the test persons perceive the traffic situations close to reality.

8.2.5 Models of Vehicle Dynamics and of the Scenario

For the simulation of vehicle motion, vehicle dynamics models of varying complexity are in use. The overall vehicle motion must be evaluated in the driving simulator in a real-time simulation model, since the human driver is a substantial part of the control loop. For the creation of exact driving dynamics models of the different vehicles, an elaborate process of parametrization and validation is essential. Since simulations of passenger cars and commercial vehicles require very different vehicle models (including trailers, articulated busses, etc.), which in addition have been developed in different tool worlds, a flexible software interface for those real-time simulation models is realized within Daimler. Active chassis control systems and even autonomous vehicle driving

functions are integrated via the same interface. For the definition of road surface models, static environment models (road scenes), and traffic scenarios, open standards have been established for the flexible integration of tools and data sets from different partners (e.g., "Open CRG," "Open Drive," [14] and "Open Scenario" [2].

The simulation of driver assistant systems requires a further extended interface, which in addition covers the representation and behavior of other traffic participants. First, the surrounding traffic must be appropriately simulated. This requires behavioral models of the road users for all relevant situations, which need to be defined and integrated; the "Open Scenario" Standard intends to support this. In addition, the output signals of sensors (radars, cameras, lidars, etc.) must be provided, they have to be derived from the actual road scene and traffic scenario. For this end, suitable simulation models of how the sensors perceive the scene are necessary; their complexity depends strongly on the purpose of the investigation, see Schöner [43].

8.2.6 Mapping of Motion in a Limited Motion Space

Due to the physical relations and the limitations of the motion space of the actuators in a simulator, the exact motion of a vehicle on a real road cannot be reproduced in a simulator without trade-offs. However, one can take advantage of the fact that the human vestibular organ can perceive linear acceleration and rotation, but it cannot determine the absolute values accurately, and additionally it possesses perceptional thresholds [50], ▫ Table 8.2). Furthermore, visual, auditive and somato-sensorical cues (sensing abilities of skin, muscles, and joints) play an important part in sensing motion. Of fundamental importance is the presentation of overall consistent cues in all sensing channels in order to achieve a realistic perception resulting in a natural driving behavior.

In a driving simulator, a motion cueing algorithm (MCA) determines the actuator positions for motion rendering within the given motion space [16]. There are a large number of concepts for determining opti-

▫ **Table 8.2** Perception sensitivity thresholds (approximate values), from Zacharias [50]

Motion	Direction/axis	Acceleration	Speed	Frequency range with highest sensitivity (Hz)
Linear	Longitudinal	0.17 m/s^2	–	Approx. 1 Hz
	Lateral	0.17 m/s^2	–	Approx. 1 Hz
	Vertical	0.28 m/s^2	–	Approx. 1 Hz
Rotational	Roll	4–5°/s^2	Approx. 3.0°/s	Approx. 1–10 Hz
	Pitch	4–5°/s^2	Approx. 3.6°/s	Approx. 1–10 Hz
	Yaw	4–5°/s^2	Approx. 2.6°/s	Approx. 1–10 Hz

mal motion cues for the driver; with varying strategies each having their specific advantages and drawbacks. The latter include producing typical false cue effects, such as scaling errors or phase shifts. Since absolute values, e.g., accelerations, generally cannot be quantified by the driver, the physically correct signals might be reduced by a scale factor of 0.7 without significantly compromising the motion perception. A continuous and constant longitudinal acceleration or deceleration can be approximated by introducing a tilting angle of the dome, if the rotational velocity stays below the perceptional thresholds. Subsequently, a component of the gravitational force acts in the longitudinal direction of the vehicle and simulates longitudinal acceleration or deceleration. Providing a consistent screen picture of the vision system supports this motion illusion of the driver. This procedure is called "*tilt coordination.*"

Classically, the simulated vehicle movement is converted into coherent motion cues for the driver using *linear filter* functions in the motion rendering algorithm. In addition, after a dynamic maneuver, the driving simulator should return to the neutral position of its motion space in order to clear all traces of the motion rendering history; this is referred to as a *wash-out filter*, see Zacharias [50].

Advanced methods for generating simulator trajectories use *non-linear and adaptive filtering* methods which, depending on the situation, try to make better use of the available motion space, see, e.g., Fischer [16]. Modern methods of *model-based predictive control (MPC)* have also been implemented for the motion rendering of simulators. They use mathematical optimization procedures that run in a real-time context. These allow one to take into account human perception models and restrictions on motion space in real-time computing, see Bruschetta [8], as well as Lamprecht [26] and Cleij et al. [11].

It should be noted that the driver's perceptions are generally highly dependent on whether the test person is concentrating on the sensation of movement, is distracted, or even consciously manipulating the vehicle motion, see Nesti et al. [32]. For evaluations of chassis systems, which are always carried out by very sensitized and attentive test drivers, the tilt coordination and scaling processes are only used to a minimal extent in the Daimler driving simulator. Therefore, when selecting the driving maneuvers, specific attention must be paid that lateral movements during lane change and slalom maneuvers can be realistically represented within the available motion space.

8.2.7 Perception Conflicts and Kinetosis (Simulator Illness)

In artificial realities (e.g., VR systems, driving simulators) a phenomenon can be observed that can be described as a physiological reaction to an inconsistently perceived environment. This physiological reaction is not limited to artificial environments but can arise as the result of perceptional cues that are inconsistent to the brain in practically every situation, see Baer et al. [3]. During a journey, this phenomenon is called "motion sickness"; the more general term, "*kinetosis,*" see Schmäl [40], indicates an insoluble conflict in visual and kinetic perception. These effects can also arise if a presumed movement is perceived in a static simulator through optical stimuli. While the exact differentiation of those terms is vague, they generally describe the physiological response of the parasympathetic nervous system to an inconsistent perception of the environment. The parasympathetic nervous system, as part of the autonomic nervous system, controls a number of physiological processes in the body and, as part of the autonomic nervous system, can only be influenced indirectly.

If an inconsistent environmental perception occurs in the simulator, one also speaks of the so-called "simulator sickness," even if this is not a disease in the classic sense, but merely a healthy body reaction to a perceived inconsistency. During a simulator drive, the symptoms are mainly found in the area of visual perception (eye pressure/headache), circulatory and thermoregulation (dizziness, sweating), and the gastrointestinal tract (stomach pressure/awareness, belching, nausea, vomiting), but spatial orientation can also be affected (disorientation, dizziness). A common explanation for the occurrence of these symptoms in a driving simulator is the sensory-conflict theory, in which it is assumed that the symptoms mainly arise when the spatial orientation is disturbed, i.e., vestibular and visual sensory stimuli are not consistent, cf. Reason [36] and Johnson [19]. Since the occurrence of these symptoms depends very much on the situation experienced, the probability of occurrence can only be roughly estimated. It is assumed that around 5–10% of people are very sensitive and 5–15% are insensitive to it. Not only the current situation, but also previous experiences in the same or similar situation influence the "assessment" by the brain and thus the occurrence of the physiological symptoms.

To avoid "simulator sickness" – the undesirable physiological symptoms resulting from a perceived inconsistency – it is helpful to pay attention to a number of critical points influencing the subjects' perception, see also Kemeny e.a. [22]. Here are a few examples:

— *Certain driving situations* provide an unfavorable combination of visual and motion stimuli in a simulator (e.g., driving in a roundabout, i.e., very small curve radius). If possible, extremely stressful situations should be avoided or eased.

— *Sufficient adaptation time* at the beginning of the simulator ride supports the adaptation effect to technically unavoidable inconsistencies (e.g., limits of the motion rendering) in the perception of the driving situation.

— A *supportive environment* with a continuous slight airflow at temperatures below 22 °C and sufficient explanations by the experimenter enables a generally "positive" assessment of the experienced situation and thus a dampening of the "warning reactions" of the parasympathetic nervous system. The occurrence of symptoms such as nausea – which the brain uses to try to get out of the current situation immediately – can also be reduced in this way.

8.2.8 Preparatory Simulators

For the efficient use of Daimler's dynamic simulator, complex experiments are prepared in two static simulators, simultaneously to ongoing investigations in the dynamic simulator. They are not equipped with an expensive motion system, but else based on identical hardware and software. The environment is projected here with up to 6 channels onto a cylindrical screen around the vehicle cabin for a 220° field of view. In these simulators, scenarios can be optimized and a suitable operational sequence is verified before the experiments move to the real test in the dynamic simulator. If the experimental setup is designed accordingly, static simulators can also be used for complete investigations, e.g., if the motion perception is expected to have only a minor effect, such as in certain assessments of advanced man–machine interface concepts.

8.3 Design of Simulator Experiments

8.3.1 Goal of Test Person Experiments

During the development process of driver assistance systems, the development engineers test regularly new functions and systems (as "experts"). Supplementing tests with typical customers/drivers ("test persons") are conducted at different times in the process, in order to gain insights on effectiveness and acceptance of these functions and/or systems and their usability for later customers and users. This leads to the following investigation goals of test person experiments in driving simulators:

— Driver behavior on use of new vehicle systems, especially driver assistance systems, particularly in critical traffic situations
— Controllability of system limitations
— Optimization of innovative user machine interfaces
— Evaluation of the customer's benefit
— Analysis of the acceptance and the usage of new systems.

In comparison to testing on proving grounds and/or in road traffic driving, simulator experiments offer the following advantages:

— No risk for drivers, for other road users, and for the environment
— High reproducibility of the test situation
— Defined road status, weather, and visibility conditions
— Creation of situations, which are difficult or impossible to create in real-world traffic
— Use of the element of surprise
— Quick variation of driving conditions as well as vehicle and environment parameters.

Possible disadvantages of driving simulator experiments are a limited presence (lack of realistic feeling) of the driver in the simulator, a reduced perception of danger and consequences in critical traffic situations, and the high expenditure for these experiments. Due to the geographical location of the simulator, the variation within the test person collective is usually limited to the closer surroundings, so that cultural differences in the driving behavior of drivers from other regions cannot be well analyzed. Furthermore, in driving simulator experiments the focus is the analysis of a well-defined and reproducible situation to study cause and effect, while testing in road traffic intends to analyze a large number of different situations with a high variance of parameters.

8.3.2 Specification of Experiments

Tests on test persons in driving simulators follow the classic methods of empirical investigations, which requires to define and describe the test concept as precisely as possible at the beginning. This aspect is often underestimated in practice by new simulator users. It should be worked out jointly in an interdisciplinary team of test engineers and method experts. The following aspects are to be described:

— Determination of the aim of the experiment and delimitation from aspects not considered
— Exact description of the system to be examined
— Definition of the usage scenario under investigation (e.g., regular use, critical traffic situation, system limits, system failure)
— User group (e.g., unexperienced user, or previous experience with a system)
— Description of the variants to be examined (system versions/system parameters)
— Reference system (e.g., predecessor system)

From the test objectives specific hypotheses should be derived, from which an investigation concept is to be developed. In detail, this requires a definition of the examination process with the variables to be collected (including operationalization) and the sample of test persons (see ▶ Sect. 8.3.6). Since the technical preparation of these experiments is very complex, the investigation situation should also be described as precisely as possible here. The investigation scenario can be limited, for example, by the type of road (motorway, country road, city center), the time of day (day/night) and the road users involved (surrounding traffic, interactions).

The complete investigation concept generally consists of three parts: preliminary survey, simulator drive and follow-up survey. These three parts are divided into several subsections. The preliminary survey contains the formal aspects of the information and declarations of consent according to the General Data Protection Regulation GDPR of the EU [15]. At this point in time, the test persons receive general information on the content and procedure of the experiment and other experiment-specific information. Depending on the experimental concept, the actual preliminary survey takes the form of an interview or a questionnaire.

Due to the physiological limitations (see ▶ Sect. 8.2.7), for general research questions an average journey time in the simulator results to around 30–40 min. The basic structure of the simulator drive consists of a general familiarization with the simulator (5–10 min), a familiarization with the driving situation of at least 15 min, in which the test person builds trust in the virtual environment and gets to know the system to be examined in its regular functionality. In addition, the test person's attention level should change from that of an exam situation at the beginning to the normal level of a regular car journey. The investigation phase that follows seamlessly contains the specific test situations that are specified by the test content and vary in length. The behavior of the test person is observed and recorded by measurable variables. A survey while driving is possible and often supplements the final follow-up survey, which is carried out outside the simulator after the end of the journey.

Particular attention is paid to the design of the investigation situation, which – together with the results of the survey – serves to check the experimental hypotheses. It must be an imaginable situation in real road traffic and – depending on the test objective – focus on either the normal application of the system to be tested, its limits, or its failure scenarios. Furthermore, the criticality of the situation has to be chosen carefully between the extremes of *too easily manageable* and *completely uncontrollable situation* in such a way that an increase in knowledge can be achieved through the experiment. The unprepared behavior in critical situations can generally only be assessed once in an experiment. Once the test person has experienced such a situation, the anticipation of other, similarly structured situations results in increased attention and a biased behavior of the test person.

In ◪ Figs. 8.7 and 8.8, typical scenarios are shown as examples of defined criticality: the situations in ◪ Fig. 8.7 were investigated as part of the evaluation of automated highway driving functions (level 3) in commercial vehicles. During an automated drive on the motorway, the test persons performed a secondary activity which consisted, for example, of watching a video in the center display; related to a critical situation the driver was asked by the system to take-over the driving task. The critical driving situation was varied, e.g., poor visibility due to fog, a warning triangle appearing on the side, obstacles partially in the lane like a broken-down vehicle or lost cargo, roadworks on the side of the lane, or a narrowing road with reduced lane count. The subject of the investigation was the measured take-over time and the safe control of the sudden critical situation.

The investigated situation in ◪ Fig. 8.8 was used to examine the braking assistant BAS PLUS with intersection assistance. The test person is driving on a priority road in the city. He may be distracted by an event on the left side of the lane (pedestrians not shown in the picture). The vehicle crossing from the right disregards the test person's right of way and drives into the intersection. Here the collision frequency and severity of a group of test persons with and one without an assistance system were compared with each other.

To create the detailed test design, the following technical aspects must be considered taking into account the objective of the investigation:

- Is a dynamic or a stationary driving simulator needed for this investigation?
- When using a dynamic driving simulator, should the vehicle cabin be oriented longitudinally or transversely to the long axis of the motion system?
- Which cabin (vehicle type) should be used?
- Does the vehicle dynamics model represent the situations to be examined with sufficient quality or are refinements necessary?
- Are special installations necessary in the cabin (control systems, displays, operating elements, etc.)?
- Can the route be created from existing route elements, or do new elements have to be developed?
- What does the traffic (vehicles, pedestrians, etc.) look like in the scenario?
- Can the traffic simulation be implemented with existing maneuvers, or do new ones have to be developed?
- Which physical values and video recordings need to be recorded?

◘ Fig. 8.7 Critical situations for "Driver take-over from autonomous driving (Level 3)"

8.3.3 Preparation of the Experiment

Basis of each driving simulator experiment is a stringent project management including the definition of the responsibilities, the availability of the simulator, and a clear time line for the preparation and execution phases.

The operational preparation begins with the start of the development of new components like road elements or traffic maneuvers, which do not already exist in the tool box of standard elements. When these developments are completed and all new components have been tested, the integration of all components including hardware extensions in the cabin starts in a preparatory simulator (without motion system). Subsequently, the test sequences are optimized, first regarding technical aspects, afterward regarding answering the questions of the experiment. Here especially the learning phases during the routine drive and the critical situation are to be monitored and optimized. Then the test sequences for the different systems, system configurations, and/or parameterizations are derived and tested.

Now the vehicle cabin is moved from the preparatory to the dynamic driving simulator in which the experiment is performed. In a preliminary test with a smaller number of test persons, it will be finally verified that the test sequence is suitable for reaching the investigation goal, a time slot for final detail optimizations is reserved in the experiment's schedule.

Parallel to the described technical preparation, the survey (questioning concept) is developed, suitable test person groups are selected and invited, and supportive investigators are instructed.

8.3.4 Artificial Distractions

Crashes result frequently from missing attention and/ or driver distraction and a simultaneously arising unexpected critical traffic situation. In order to analyze the reaction of a driver in such a situation and/or to quantify the benefit of an assistance system, the situation in the simulator has to be represented reproducibly. Part of it is the application of reproducible artificial distractions. For this purpose, animated graphic elements are used, which are conceivable in real road traffic and which draw the attention of the driver, e.g., an unusual situation on the side of the lane. These graphic elements should arise surprisingly and must be precisely timed to the occurrence of a critical traffic situation – for example just when a vehicle coming from the crossroad drives without right of way into the intersection (see ◘ Fig. 8.8). Besides graphical elements outside of the vehicle, also close-to-reality control tasks inside the vehicle (finding a telephone number in a list, reading or texting a message, etc.) as well as synthetic distractions (pressing a key upon appearance of a visual signal in the field of view, etc.) may be used, depending on the task under investigation, see ◘ Fig. 8.9.

Fig. 8.8   Critical situation "crossing traffic from right side"

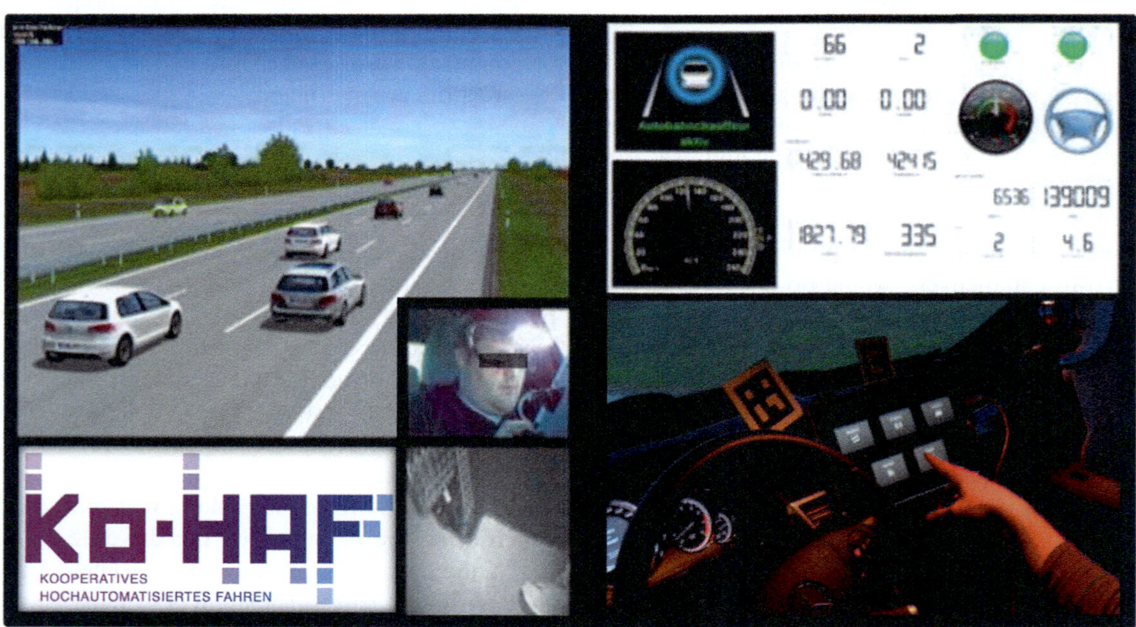

Fig. 8.9   Monitoring of experiments and driver distraction by additional tasks

8.3.5 Learning Effects

Usually, drivers are familiar with the assistance systems of their vehicle before they trap into a (rarely arising) critical traffic situation. This has to be considered when evaluating the benefit of an assistance system. In driving simulator experiments, however, the test persons often are confronted with future systems, which are unknown to them. A substantial aspect of the experiment design is the emulation of the learning process during the short test drive: becoming familiar with the system, its functions, and its limitations. System descriptions and briefings by the investigators are used before the test drive as well as experiencing the system during the routine drive. It is aimed to have the emulated learning process finished before the situation under investigation arises. Wrongly designed information and overemphasizing the system limitations can have a relevant influence on the test result: With too little information and learning on the system, the test person doesn't resemble a driver being familiar with the system. Too much information on system limitations, however, leads to unrealistically high attention of the test person waiting for the occurrence of a critical system performance.

8.3.6 Subject Sample Selection

The definition and stratification of the group of test persons is an essential pillar of the test design, see Knussmann [24]. The correct selection of the subject sample ensures the valid interpretation of the test results and, above all, the transferability of these results to the use of a system in a real vehicle. The relevant individual characteristics of the test persons and their distribution within the sample (see ◻ Table 8.3) must therefore consider the goal of the investigation and the defined test hypotheses, and this must be implemented in the sample in the best possible way.

The motorist population is made up of varying proportions of people in the total population and thus represents a sifting in the anthropological sense, see Nagel [30, 31]. Sifting means that the composition of this group can be relatively variable and only represents a snapshot of the current users. This selection of people therefore often does not correspond to the distribution of the total population. This aspect must be taken into account when defining a sample of test persons, in particular deviations between current and future user groups of a newly developed system must be taken into account. Above all, the aspect of competence or previous experience is important. Competence refers on the one hand to the general ability to drive a vehicle, but also to other individual domain-specific abilities and skills, e.g., in the operation of new technical systems.

The definition of a subject sample therefore includes general demographic criteria as well as characteristics of general driving skills and specific skills at the human–machine interface. The sample is often described on the one hand on the basis of minimum requirements, but also as a range representing the population variability.

To compare different systems, system characteristics or system parameterizations with each other and relative to a comparison basis, a sample size of 30–50 participants is necessary. The safeguarding of a system, on the other hand, requires significantly larger samples with more than 100 participants [9, 48].

8.3.7 Evaluation of Experiments with Test Persons

The hypotheses derived from the experiment goal are the basis of the analysis (▶ Sect. 8.3.2). The hypotheses can focus on purely objective data, e.g., the fuel consumption determined in the simulation (hypothesis: "With a new energy saving driving style program the consumption will be reduced."), on purely subjective data (hypothesis: "During the purchasing process of a new vehicle men focus more on engine power than women.") or on a mix of objective as well as subjective data (hypothesis: "The probability of purchase of a new energy saving program depends on the achievable consumption reduction.").

During the experiment the objective data are taken as measurements from the simulation, the subjective one by interviews and (possibly digital) questionnaires from the test persons. For the further analysis, programs such as *MatLab* are used for the objective data and *SPSS* for the subjective data. *SPSS* offers the possibility to integrate the results from *MatLab*. The validation of the hypotheses is done with the help of statistical tests (e.g., hypothesis tests or correlations).

An important question in collision-avoidance assistance systems is the evaluation of the effectiveness of the assistance systems. For this goal, a sufficiently high number of drivers is brought into accident-prone traffic situations and their behavior or the controllability

◻ **Table 8.3** Fictitious example of selection criteria for test participants

Selection criterion	Parameter	Stratification (example)
Privacy	Internal/external participants	Only company-internal participants
Demographical aspect 1	Age distribution	30% under 40 years/40% between 40 and 60 years/30% above 60 years
Demographical aspect 2	Gender distribution	50% male/50% female (approximately)
Driving experience	Distance driven per year	<5,000 km, 6,000 km up to 15,000 km, >16,000 km
System experience	Restrictions/preconditions	Previous experience with "adaptive cruise control" is a must

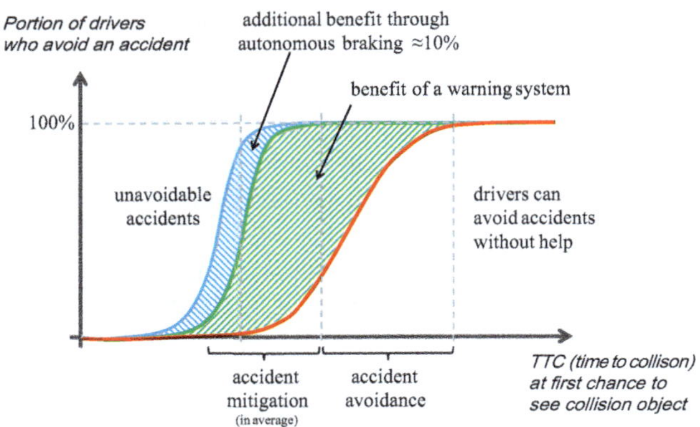

Fig. 8.10 Controllability of critical situations, from Schoener [42]

of the situations is recorded with various implementations of the assistance function. For this purpose, in particular, other traffic (oncoming, overtaking, crossing vehicles or cyclists or pedestrians) must be controlled in such a way that each test person experiences a comparable situation despite individually different driving behavior. The driver's attention plays a decisive role here: the majority of accidents happen because a driver notices a critical situation late or even too late due to fatigue, distraction, or inattention. A driving simulator test to evaluate collision-avoidance systems must therefore include a spectrum of different driver attentions.

In **Fig. 8.10**, according to Schoener [42], the principle of the assessment is shown: the distribution of the ability to avoid accidents by a driver without an assistance system depends on the time at which an obstacle first appeared in the lane (measured as TTC – "time to collision" assuming an unchanged driving speed); this is represented by the lowest curve in the figure. The interesting range of TTC for the tests in the driving simulator, in which accident avoidance can be influenced by a warning or autonomous braking system is around the steepest slope of this curve. The experiments must therefore be designed with regard to TTC in such a way that a sufficiently large number of test subjects can experience the situation and react accordingly.

Drivers who need a large TTC to avoid an accident are typically inattentive or distracted; these drivers can be supported particularly effectively by an assistance system, and accidents can be avoided. But even with attentive drivers with short necessary TTC times (through early warning or autonomous braking) a faster braking reaction and thus a statistically mean damage reduction is possible. In order to achieve the spread with regard to TTC, the critical situation is only triggered in the simulator after a relatively long journey; in addition, it may be necessary to distract the driver due to peculiarities in the street scene or secondary tasks in the vehicle. The quantitative determination of the hatched areas in the

figure from test persons allows an objective assessment of the effectiveness of assistance systems.

An example for the evaluation of driving simulator tests at the driver take-over time after a take-over request in autonomous vehicles (level 3) is published by Schittenhelm and Nöcker [38].

8.4 Transferability of Results

8.4.1 Validation of Driving Simulators

In the literature the distinction between absolute and relative validity, suggested by Blauuw (1982), is commonly used [6]: *Absolute validity* is the extent, with which the results of the simulation agree numerically accurate with real data. *Relative validity* describes the extent, with which a manipulation of a factor in a simulation has the same effect as in a real study, even if the numerical data does not agree accurately. For the investigation of driver behavior, absolute validity is not inevitably necessary, relative validity however is important and often sufficient. Most validation studies come to the conclusion that for professional driving simulators, relative validity can be assumed regarding the most important driving parameters (e.g., speed or lateral position), not however absolute validity, Mullen et al. [28]. For this reason, in all studies new systems are compared with an existing reference system for comparison (e.g., predecessor system).

With respect to validation studies, it has to be noted that validation can only refer to an exactly specified situation and only to one simulator. Generalizations over situations or simulators are generally of limited validity.

For validation, the data of a drive in the simulator can be compared with the data monitored during a naturalistic drive on a real road (not during an investigation situation) and/or an instructed drive on a real road (investigation situation, e.g., with an investigator in the vehicle), see Klüver et al. [22] and Wynne et al. [49]. Also interviews of test persons can elucidate differences between simulator drives and real drives.

Another approach of validation is to analyze the transferability of learning effects between a simulator and a real road drive, i.e., can driving abilities learned in a simulator be observed on a real road later.

8.4.2 Naturalistic Behavior and Risk Perception

In principle the following factors influence the validity of driving simulators:

- Technical restrictions, e.g., concerning the motion space or the visualization;
- Design of the experiment including the instruction of the test person;
- Influence by the investigator, like, e.g., the tendency of a test person to present itself positively;
- Other consequences than in the real traffic, as, e.g., with injuries from accidents or with tickets for exceeding the speed limit;
- Increased attention needed in the simulator, e.g., for lane keeping and speed control;
- Kinetosis.

According to statements of test persons, they get accustomed after the mentioned acclimatization phase of approximately 5 min. to the simulator and the new vehicle, so that they report and show to a large extent the impression to drive a real vehicle. When installing the vehicle cabin transversely to the linear axis of motion of the simulator, the driving and guidance behavior on the autobahn and rural highways are described as very close to reality; strong accelerating and braking, however, is perceived with unusual sensations. After a longer drive in the simulator, some test persons indicate increased stress on their eyes. On the described dynamic driving simulator only a few test persons report some early signs of kinetosis, in the fixed simulators used for the preparation of experiments this rate is significantly higher. City driving with tight curve radii and frequent start and stop situations are often found to be difficult in this regard. With modern control methods (*Model-based Predictive Control, MPC*), the range of motion of a driving simulator can be better used for many situations. This is the subject of intensive basic research, see Cleij [10].

From validation measurements a good agreement can be deduced with respect to reaction times in the simulator and on the road. There is a trend that the speed of the own car is estimated a little lower in the simulator, so that test persons tend to drive somewhat faster than intended, according to Tenkink [45]. The intention of the test persons to behave in the simulator correctly and therefore not to exceed the maximum speed limit may present a conflict. Furthermore, for tests in which it is important to estimate the distance to the driver in front at city traffic speeds, the validity of the test constellation must be verified, Klüver et al. [22], Schmieder et al. [41]; the distance in the simulator is generally estimated to be smaller, so that test subjects tend to set greater safety distances than in reality. Tracking and lane keeping in the simulator tends to be more difficult to maintain than in reality [12], see Blaauw [6].

The test persons in the driving simulator of the Daimler AG generally behave intuitively very close to reality, which shows up particularly clearly in potentially dangerous traffic situations: road departures, driving over curbs, or driving through guard rails typically do not occur. Before leaving the vehicle, the test persons convince themselves by a shoulder view whether they can open the door safely. In summary, in practice a very good relative validity can be observed for the investigation goals.

8.5 Summary and Outlook

A driving simulator in principle reproduces a driver cockpit with typical vehicle control elements like steering wheel and pedals. The sensory impressions during a drive are presented to the driver via the visual channel, showing the environment, road, and other traffic, and via the audible channel, producing appropriate vehicle and ambient noise. Dynamic simulators also present acceleration forces to the driver via a motion system. The arrangement of all those components can be quite different.

The main focus for driving simulators used in vehicle development lies on the analysis of the interaction of the vehicle's driver with the vehicle and/or with new vehicle systems, like driver assistance or vehicle dynamics control systems. In early development phases, new concepts can be assessed, and in later development phases the focus may be on optimization of the function and its user interface. For these analyses, selected test persons are familiarized with a new vehicle system in several road or traffic conditions; the test persons' behavior and reactions are monitored and evaluated by measurements, videos, and questionnaires. Usually, the behavior of several subject sample groups is compared to each other in order to assess the effectiveness of a function in different variations.

Driving simulators have been established as a valuable tool, especially for risk-free assessment of driver behavior in reproducible critical situations, allowing at the same time to quickly modify vehicle and environmental parameters.

In the automotive industry, there is an ongoing shift to further increase the validation activities based on simulations rather than road tests [35]. This requires a professionalization of the operations of driving simulators, turning them from exotic research tools to efficient service facilities. Key elements of investigations are shifting toward new applications, arising from questions around the field of automated driving and integral views of vehicle, communication, and entertainment functions, see Schöner [43]. At universities and research institutes as well as in supplier companies, further driving simulators are acquired for the development of driving functions in virtual traffic.

The technological advancement of driving simulator solutions will concentrate on the avoidance of today's deficits as in driving along curves with high yaw rates or as the limited distance judging capability in the near field, e.g., by the application of 3D visual systems, or on the further improvement of motion perception. In this field, the development of new simulator concepts as well as detailed optimizations of existing configurations is to be expected. While there will still be an increasing number of different mechanical simulator concepts, common tools as for the visual representation of the real world, the simulation of (active and interactive) traffic participants and complex traffic scenarios, as well as the interfaces of the different components will see more standardization. Especially as the standardization of interfaces will allow for an efficient share of efforts between suppliers and vehicle manufacturers, its role is becoming increasingly important. This applies to the simulation of vehicle components (especially sensor models, ASAM *OSI OpenSimulationInterface*) as well as to the description of traffic areas (ASAM OpenDRIVE) and traffic scenarios (ASAM OpenSCENARIO), see ASAM [2].

With the increasing spread of dynamic driving simulators in the development and validation of safety–critical vehicle systems, questions of the standardization of processes and the systematic documentation of tests are becoming increasingly important. It is particularly relevant to ensure that the results collected in a simulator will provide a representative and valid foundation. This will enable simulator studies to be compared with one another, results to be transferred, and systematic errors in the generation and evaluation of safety-relevant data to be excluded. With this aim in mind, an international initiative by research institutes and vehicle manufacturers to compile guidelines for carrying out driving simulator experiments was launched under the name "Operational Guidelines for Driving Simulator Experiments"; the resulting guidelines and best practice examples can be viewed on the website of the Driving Simulation Association DSA [13].

References

1. AB Dynamics.: (2020). ▶ https://www.abdynamics.com/en/products/simulation
2. ASAM.: (2020). ▶ https://www.asam.net
3. Bear, M., Connors, B., Paradiso, M.: Neuroscience: Exploring the Brain, 4th edn. Lippincott Williams & Wilkins/Wolters Kluwer (2015)
4. Betz, A.: Feasibility analysis and design of wheeled mobile driving simulators for urban traffic simulation. In: Fortschrittsberichte VDI: Reihe 12, Verkehrstechnik, Fahrzeugtechnik; 786, Düsseldorf, VDI-Verl., TU Darmstadt (2015)
5. Betz, A., Winner, H., Ancochea, M., Graupner, M.: Motion analysis of a wheeled mobile driving simulator for urban traffic situations. In: Driving Simulation Conference – Paris (2012). ▶ https://www.fzd.tu-darmstadt.de/forschung/research_projects_fzd/morpheus_1/index.en.jsp
6. Blaauw, G.J.: Driving experience and task demands in simulator and instrumented car: a validation study. Hum. Factors **24**(4), 473–486 (1982)
7. Breuer, J., Käding, W.: Contributions of driving simulators to enhance real world safety. In: Proceedings Driving Simulation Conference Asia/Pacific, Tsukuba (2006)
8. Bruschetta, M., et al.: A real-time, MPC-based motion cueing algorithm with look-ahead and driver characterization. Transp. Res. Part F (2017). ▶ https://doi.org/10.1016/j.trf.2017.04.023
9. Bubb, H.: Wie viele Probanden braucht man für allgemeine Erkenntnisse aus Fahrversuchen? In: Landau, K., Winner, H., Fahrversuche mit Probanden – Nutzwert und Risiko. Fortschr.-Ber. VDI Reihe 12 Nr. 557. Düsseldorf (2003)
10. Cleij, D.: Measuring, modelling and minimizing perceived motion incongruence for vehicle motion simulation. PhD thesis, TU Delft (2020)
11. Cleij, D., Venrooij, J., Pretto, P., Katliar, M., Bülthoff, H., Steffen, D., Hoffmeyer, F., Schöner, H.P.: Comparison between filter- and optimization-based motion cueing algorithms for driving simulation. In: Transportation Research Part F: Traffic Psychology and Behaviour (2019)
12. Schoener, H.P., Schmieder, H., Chardonnet, J.R., Colombet, F., Kemeny, A.: Verification of stereoscopic projection systems for quantitative distance and speed perception tasks. In: Proceedings Driving Simulation & Virtual Reality Conference, Strasbourg (2022)
13. Driving Simulation Association.: (2020). ▶ https://driving-simulation.com/site/driving-simulation-experience-sigdsep/
14. Dupuis, M., Strobl, M., Grezlikowski, H.: OpenDRIVE 2010 and beyond – status and future of the de facto standard for the description of road networks. In: Proceedinds Driving Simulation Conference DSC Europe, pp. 231–242, Paris (2010)
15. EU.: Verordnung 2016/679 des Europäischen Parlaments und des Rates vom 27. April 2016 zum Schutz natürlicher Personen bei der Verarbeitung personenbezogener Daten, zum freien Datenverkehr und zur Aufhebung der Richtlinie 95/46/EG (Datenschutz-Grundverordnung) (2016). ▶ https://eur-lex.europa.eu/legal-content/DE/TXT/?uri=CELEX:02016R0679-20160504
16. Fischer, M.: Motion-Cueing-Algorithmen für eine realitätsnahe Bewegungssimulation, Deutsches Zentrum für Luft- und Raumfahrt, Braunschweig (2009). ▶ http://www.digibib.tu-bs.de/?docid=00030516
17. Fisher, D.L., Rizzo, M., Caird, J., Lee, J.D.: Handbook of Driving Simulation for Engineering, Medicine, and Psychology. CRC Press, Boca Raton (2011)
18. Flanagan, M.B., May, J.G., Dobie, T.G.: Sex differences in tolerance to visually-induced motion sickness. Aviat. Space Env. Med. Aerosp. Med. Assoc. **76**, 642–646 (2005)
19. Johnson, D.M.: Introduction to and review of simulator sickness research. DTIC Document (2005)
20. Käding, W., Hoffmeyer, F.: The Advanced Daimler-Benz Driving Simulator. Society of Automobile Engineers, Warrendale, PA., SAE Technical Paper 950175 (1995)

21. Käding, W., Zeeb, E.: 25 years driving simulator research for active safety. In: Proceedings International Symposium on Advanced Vehicle Control (AVEC 2010), Loughborough (2010)

22. Kemeny, A., Chardonnet, J., Colombet, F.: Getting Rid of Cybersickness. Springer Nature (2020). ▶ https://doi.org/10.1007/978-3-030-59342-1

23. Klüver, M., Herrigel, C., Heinrich, C., Schöner, H.P., Hecht, H.: The behavioral validity of dual-task driving performance in fixed and moving base driving simulators. Transp. Res. F: Traffic Psychol. Behav. **37**, 78–96 (2016)

24. Knussmann, R.: Anthropologie. Handbuch der vergleichenden Biologie des Menschen. Band I, 1, Fischer, Stuttgart (1988)

25. Krebber, W., Sottek, R.: Interactive Vehicle Interior Sound Simulation. ISATA 00, Automotive & Transportation Technology, Dublin (2000)

26. Lamprecht, A., Haecker, J., Graichen, K.: Constrained Motion Cueing for Driving Simulators using a Real-Time Nonlinear MPC Scheme, IROS (2018)

27. Mohajer, N., Abdi, H., Nelson, K., Nahavandi, S.: Vehicle motion simulators, a key step towards road vehicle dynamics improvement. Veh. Syst. Dyn. **53**(8), 1204–1226 (2015). ▶ https://doi.org/10.1080/00423114.2015.103955

28. Mullen, N., Charlton, J., Devlin, A., Bedard, M.: Simulator validity: behaviors observed on the simulator and on the road. In: Fisher, D.L., Rizzo, M., Caird, J., Lee, J.D. (eds.): Handbook of Driving Simulation for Engineering, Medicine, and Psychology. CRC Press, Boca Raton (2011)

29. Murano, T., Yonekawa, T.: Aga, M., Nagiri, S.: Development of high-performance driving simulator. SAE Int. J. Passeng. Cars Mech. Syst. **2**(1), 661–669 (2009)

30. Nagel, K.: Praxisbezogene Anwendung gruppenspezifischer Anthropometrie. Tagung Arbeitsplatz Fahrerkabine, Haus der Technik, Berlin (2001a)

31. Nagel, K.: Gruppenspezifische Anthropometrie in der ergonomischen Gestaltung. Brandenburgische Umweltberichte **10**, 54–61 (2001)

32. Nesti, A., Masone, C., Barnett-Cowan, M., Robuffo Giordano, P., Bülthoff, H., Pretto, P.: Roll rate thresholds and perceived realism in driving simulation. In: Driving Simulation Conference, Paris (2012)

33. Nieuwenhuizen, F.M., Bülthoff, H.: The MPI cybermotion simulator: a novel research platform to investigate human control behaviour. J. Comput. Sci. Eng. **7**(2), 122–131 (2013)

34. Nordmark, S., Jansson, J., Lidström, M., Palmkvist, G.: A moving base driving simulator with wide angle visual system. In: The TRB Conference, Session on Simulation and Instrumentation for the 80s, Washington, D.C. (1985)

35. PEGASUS.: (2020). ▶ https://www.pegasusprojekt.de/files/tmpl/pdf/PEGASUS_Abschlussbericht_Gesamtprojekt.PDF

36. Reason, J.: Motion sickness adaptation: a neural mismatch model. J. Roy. Soc. Med. **71**, 819 (1978)

37. Reason, J.T., Brand, J.J.: Motion Sickness. Academic press, New York (1975)

38. Schittenhelm, H., Nöcker, G.: Übernahmequalität bei Level 3 Systemen (Drivers take-over reaction times and quality with Level 3 Systems). 8. Tagung Fahrerassistenz, München (2017)

39. Schlender, D.: Simulatorkrankheit in Fahrsimulatoren. Zeitschrift für Verkehrssicherheit **54**(2), 74–80 (2008)

40. Schmäl, F., Stoll, W.: Kinetosen. HNO 48, S. 346–356, Springer (2000). ▶ http://www.neuro24.de/schwind.htm

41. Schmieder, H.J., Nagel, K., Schoener, H.P.: Enhancing a driving simulator with a 3D-stereo projection system. In: Driving Simulation Conference – DSC 2017, Stuttgart (2017)

42. Schöener, H.P.: Erprobung und Absicherung im dynamischen Fahrsimulator. 17. VDI Congress SimVec - Simulation und Erprobung in der Fahrzeugentwicklung, Baden-Baden (2014)

43. Schöener, H.P.: Automotive needs and expectations towards next generation driving simulation. In: Driving Simulation Conference – DSC 2018 Europe VR, Antibes (2018)

44. Slob, J.J.: State-of-the-Art Driving Simulators, a Literature Survey. DCT 2008.107, DCT report, Eindhoven University of Technology (2008)

45. Tenkink, E., Van der Horst, A.R.A.: Effects of road width and curve characteristics on driving speed. Report IZF 1991 C-26. TNO Institute for Perception. Soesterberg (1991)

46. Tüschen, T.: Konzeptionierung eines hochimmersiven und selbstfahrenden Fahrsimulators, Schriftenreihe des Lehrstuhls Kraftfahrzeugtechnik, Universität Dresden (2018). ▶ https://tu-dresden.de/bu/verkehr/iad/kft/die-professur/ausstattung/fahrsimulator-1

47. VI-grade.: VI-grade DYNAMIC-Fahrsimulator DiM250 (2020). ▶ www.vi-grade.com

48. Weitzel, A., Winner, H.: Ansatz zur Kontrollierbarkeitsbewertung von Fahrerassistenzsystemen vor dem Hintergrund der ISO 26262. FAS 2012 – 8. Workshop Fahrerassistenzsysteme, Walting im Altmühltal (2012)

49. Wynne, R.A., Beanland, V., Salmon, P.M.: Systematic review of driving simulator validation studies. Saf. Sci. **117**, 138–151 (2019)

50. Zacharias, G.L.: Motion cue models for pilot-vehicle analysis. AMRL-TR-78-2 (1978)

51. Zeeb, E.: Daimler's new full-scale, high-dynamic driving simulator – a technical overview. In: Conference Proceedings of Driving Simulation Conference Europe, Paris (2010)

Open Access This chapter is licensed under the terms of the Creative Commons Attribution-NonCommercial-NoDerivatives 4.0 International License (▶ http://creativecommons.org/licenses/by-nc-nd/4.0/), which permits any noncommercial use, sharing, distribution and reproduction in any medium or format, as long as you give appropriate credit to the original author(s) and the source, provide a link to the Creative Commons license and indicate if you modified the licensed material. You do not have permission under this license to share adapted material derived from this chapter or parts of it.

The images or other third party material in this chapter are included in the chapter's Creative Commons license, unless indicated otherwise in a credit line to the material. If material is not included in the chapter's Creative Commons license and your intended use is not permitted by statutory regulation or exceeds the permitted use, you will need to obtain permission directly from the copyright holder.

Test Methods

Test Procedure for Consumer Protection and Vehicle Type Approval

Patrick Seiniger, Adrian Hellmann, and Andre Seeck

Contents

© The Author(s) 2026
H. Winner et al. (eds.), *Handbook Assisted and Automated Driving*,
https://doi.org/10.1007/978-3-658-45276-6_9

Test procedures serve as a tool for checking whether desired product properties are present, which of course also applies to the development of driver assistance systems. There are development-accompanying and release-accompanying tests, which are essentially carried out in-house by the vehicle or system manufacturer. There are also tests that are carried out by external, manufacturer-independent test organizations in the sense of independent product testing—whether for the approval of vehicle types for admission to the market, as part of the application of the functional safety standard (in both cases, for example, by technical services) or for customer information (by test institutes, e.g., Euro NCAP). Furthermore, tests are also carried out by authorities themselves as part of market surveillance.

Goal of this section is to provide an overview over those test procedures and tools that are currently used in type approval and consumer protection.

The term test procedure depicts a method according to which a test of a system for certain properties is to be carried out. For this purpose, required tools, aids, boundary conditions, and evaluation methods must also be specified.

The focus of this chapter is on these "external" test procedures. The adjacent area of tests during development is only described here for the purpose of delimitation; the tests required to demonstrate functional safety (although part of the development process, but also carried out by external organizations if necessary) can be found in ▶ Kap. 6/ID0105 and in the ISO 26262 standard [1].

9.1 Systematics of Test Procedures

A variety of different methods are used to evaluate driver assistance systems. The methods vary depending on the objective of the test, such as which system functionality is to be checked against which criteria, and—for development tests not considered here—the available development status of the respective system.

The methods used usually include a definition of the procedure, description of tools (example: artificial target objects to represent vehicles or persons in order to trigger emergency braking functions, measurement and control devices, and a definition of a suitable environment), description of the procedure for determining characteristic values based on the measurement data obtained (example: calculation of speed reduction and relative impact speed of an emergency braking assistance system from motion data of ego vehicle and target object including corresponding calculation rules) and often also contain an evaluation standard for deriving a test result from the characteristic values (example: weighting table depending on test speed

in Euro NCAP emergency braking assistance tests) or a reference thereto (example: reference to requirements ▶ Chap. 5 in test Chap. 6 of UN Regulation no. 152 [1]). If assessment standards are specified, this can be interpreted as a requirement for the corresponding system (see chapter 107).

Test procedures are often differentiated on the basis of characteristic properties. For the evaluation of driver assistance systems with environment perception, a relevant application situation must be generated for the system in each case. Since the systems often cannot be tested in public road traffic, an equivalent situation must therefore be generated in a test environment. In the case of tests for emergency braking systems, for example, this requires objects that represent stationary or moving vehicles. Distinguishing features of test procedures are then, for example, the type of these target objects, their motion systems, or also the degree of virtualization of the tests. A summary of the usual criteria for distinguishing test procedures is shown in ▪ Table 9.1; mixed forms are also possible. This scheme also includes characteristics that are irrelevant to vehicle regulations and consumer protection—common characteristics in this environment are therefore indicated in bold writing.

9.2 Test Procedures in Consumer Protection Using the Example of Euro NCAP

9.2.1 Concept

The approach of typical consumer protection organizations is described in ▶ Chap. 4, Sect. 4.2. In essence—for example, in the case of Euro NCAP, the European New Car Assessment Program—it is a matter of comparatively assessing products available on the market, which are thus fully developed and can be regarded as a black box, on the basis of their own criteria, most of which have already been published in advance, and presenting these assessments to the public. The requirements are only implicitly specified by evaluation criteria in tests. The test procedures are therefore of outstanding importance in consumer protection.

An assessment of the vehicles must therefore (as a rule) be possible without knowledge of the internal logic of the products and without the support of the vehicle manufacturer. This means that initially only test procedures for the complete vehicle can be considered, in which the vehicle is subjected to a critical situation or at least a situation that appears critical for the vehicle sensor system, and its reaction to this situation is measured and evaluated.

◾ Table 9.1 Systematics of test procedures

Properties	Possible characteristics (bold: relevant for regulations and consumer protection)
System category	Comfort systems, **safety systems**
Assessment level	Component, complete vehicle
Degree of virtualization	Virtual, partially virtual (development), **real**
Test environment	Simulation, laboratory, **test track, real road (closed off), real road**
Ego vehicle movement mode	Stationary, **self-propelled,** externally moved
Representation of the target object	Car, motorcycle, pedestrian, bicycle, other
Movement type of the target object	Stationary, self-propelled, externally moved
Collision tolerance of the target object	None, restricted for emergencies, **unrestricted** (within certain limits of collision speed)
Control of the ego vehicle	Test subjects, **professional test drivers, automated with supervisor, fully automated**
Valuation object	f. ex.: System properties (performance, false positives), overall vehicle performance, human–machine interface, biomechanical properties, false triggers for false positives
Evaluation method	f. ex.: **Questionnaires, vehicle metrics**
Valuation method	**Comparative assessment, absolute assessment**

The assessment criteria are set in such a way that system properties that are positive for road safety and go beyond the requirements of the technical vehicle regulations are also evaluated more positively. Since the aim of consumer protection organizations is a comparative evaluation of products, a gradual differentiation is necessary.

Euro NCAP is an association under Belgian law with currently 12 members (7 governments, mostly represented by the respective Ministry of Transport (Germany, France, Great Britain, Catalonia, Luxembourg, Netherlands, and Sweden) and 5 non-governmental organizations, such as automobile clubs (ACI, ADAC, FIA), an insurance organization (Thatcham Research) and the consumer protection organization ICRT. Furthermore, since 2020, there are also two associate members (insurance organization GDV and monitoring organization DEKRA). Euro NCAP evaluates new vehicles available on the market in the EU and the UK with regard to their vehicle safety. The Euro NCAP assessment is a comparative evaluation of different products (vehicles) with regard to assessment criteria relating to vehicle safety, which are specifically carried out in each case by test laboratories commissioned by the Euro NCAP members (example: Federal Highway and Transport Research Institute (Bundesanstalt für Straßen- und Verkehrswesen-BASt) as the test laboratory commissioned by the member Federal Ministry of Transport and Digital Infrastructure).

Euro NCAP products are the so-called "Rating" (safety assessment of vehicles, cardinal scale with 0 to 5 stars) and "Grading" (classification of automated driving functions according to the complexity and suitable safety in ordinal scale "Entry," "Moderate," "Good," "Very Good").

The evaluation concept in Euro NCAP's "Rating" is described below.

Various aspects of vehicle safety (example: passive occupant protection, protection of external road users, and safety assistance systems) are weighted for the overall assessment, with the individual aspects (example: speed reduction at various initial speeds for emergency brake assistance systems as a sub-aspect of "Safety Assistance") being assessed using a points scale. This point scale is in turn composed of the results of many different individual test cases (example: initial speed 10, 20 to 50 km/h, stationary target car, no lateral offset). In the individual tests, the evaluation is then often carried out by "sliding scale," i.e., quantification of a test result as a percentage between 0 and 100% by means of a linear equation.

The rating scales of the individual tests are defined by groups of experts and are largely based on the state of the art, so that the best products (vehicles) available on the market at the time the scheme is defined still remain well below the maximum rating (in example: the best systems available on the market achieve automatic accident avoidance up to 40 km/h in a particular test configuration → fixing the end point of the scale at 50 km/h).

In summary, the Euro NCAP assessment result of a vehicle in the "Rating" (safety assessment) consists of 0 to 5 stars, and in the "Grading" (assessment of vehicle automation in "Entry," "Moderate," "Good," "Very Good") consists of a variety of safety aspects, for which in turn different test procedures are used, each with different parameters.

Fig. 9.1   Driving robots from the company AB Dynamics in the test vehicle, as typically used in Euro NCAP

9.2.2 Boundary Conditions of Test Procedures at Euro NCAP

Individual test procedures of Euro NCAP (example: emergency braking assistance for passenger cars) consist of a basic test setup (in the example: longitudinal traffic situation, ego vehicle approaches target vehicle), for which different test cases are then created from the variation of parameters (in the example: driving speed of ego and target vehicle, overlap between ego and target vehicle). For each test case, a parameter characterizing the safety effect is usually calculated (in the example: relative impact speed), from which an evaluation is then determined for each test case and by weighting function for the entire test procedure.

In order to keep the results comparable between the eight EuroNCAP test laboratories distributed throughout Europe, the test conditions and the calculation steps for the evaluation result are standardized as far as possible and the permitted tolerances for the test performance are kept narrow. Driving tests are generally carried out under "ideal" environmental conditions (example: dry road surface with a sufficiently high coefficient of friction, wind within narrow limits, for tests during the day: ambient brightness of at least 2000 lx). Core documents are the so-called test protocols (in the example: "AEB Car-to-Car Test Protocol v.3.0.2" [2]). They precisely define test procedures, test objects, and test conditions that must be adhered to by the individual test laboratories.

The test technology to be used is not explicitly specified, but the tolerances are so tightly specified that a test execution requires driving robot systems that enable dynamic positioning of all objects with accuracy in the range of a few centimeters. These are devices that are mounted on the steering wheel and pedals and can completely take over the vehicle control by closed-loop control. Precise GPS-based inertial measurement technology is used to determine the actual position and speed. An example of the use of driving robot systems is shown in ▪ Fig. 9.1.

9.2.3 Test Procedures for Driver Assistance Systems

This section describes the individual test procedures for driver assistance systems at Euro NCAP in overview and structured according to the respective vehicle safety rating categories (there are also a large number of passive vehicle safety test procedures that are not considered here). Euro NCAP's "rating" assessment is divided into the following four categories, also called "boxes":

- Adult occupant protection (I). This category refers exclusively to vehicle safety that mitigates the consequences of accidents (Passive Safety).
- Child occupant protection (II). Again, no driver assistance systems are considered, only passive safety tests and criteria.

■ Table 9.2   Requirement definition of Euro NCAP tests of pedestrian emergency braking assistants, 2020–2023: Overview of test configurations for right-hand traffic

	CPFA-50	CPNA-25	CPNA-75	CPNC	CPLA-50	CPLA-25	CPTA	CPRA	
Vehicle initial speed and step width in km/h	↑10–60, 5				↑20–60, 5	↑50–80, 5	↗10, 15, 20 ↘10	↓-4, ↓-8	
Speed of the dummy in km/h	8 →	5 ←	5 ←	5 (child) ←	5 ↑	5 ↑	5 ↓[1]	5 ←r at the rear	0 at the rear
Impact location on the vehicle under test vut (braking of the vut not taken into account[2])	50%	25% (r)	75% (l)	50%	50%	25% (r)	50%	25, 50, 75%	50%
Obstruction	No			Yes	No				
Braking / Warning	B					W	B		
Day / Night	D	D+N	D+N	D	D+N	D+N	D	D	D

– Protection of pedestrians and cyclists (III). In addition to evaluating the passive safety of vehicles in terms of the safety of unprotected, external road users, appropriate emergency braking assistance functions are an important component. Thus, passive and active vehicle safety are evaluated together here in terms of integral safety.

– Safety assistance functions (IV). This aspect is evaluated using various test procedures for driver assistance systems: Occupant Monitoring Systems, Emergency Braking Assistance Systems, Lane Keeping Assistance Systems, and Driving Speed Assistance Systems.

A detailed description of all test procedures goes far beyond the possible scope of this chapter. The test procedures are therefore briefly outlined below in an introduction and with an overview table; for more details, please refer to the corresponding test protocol documents [2–4].

9.2.3.1 Pedestrian Advanced Emergency Brake Assistance

In the test procedure for pedestrian emergency brake assist systems, the vehicle under test driving straight ahead is guided on a collision course with crossing or longitudinally moving dummy pedestrians, except for the test configuration with a turning vehicle. The essential parameters for pedestrian emergency braking assistance system tests are given in ■ Table 9.2.

The designation of the individual test cases given in the table is made up of the collision partners (C—Car, passenger car, P—Pedestrian), the direction of movement N—Nearside, coming from the right in the case of right-hand traffic, F—Farside, coming from the left in the case of right-hand traffic. L—Longitudinal, driving longitudinally ahead, T—Turn, turning in an intersection, R—Reverse, driving backward), a more detailed designation of the target object (A—Adult, C—Child) and, if applicable, the collision point on the vehicle as a percentage (center corresponds to 50%). Other indices that are used: r—right, l—left, and arrows symbolizing directions of movement.

9.2.3.2 Bicycle Advanced Emergency Brake Assistance

In the test procedure for advanced emergency braking assistance systems for the protection of bicyclists, the vehicle under test traveling straight ahead is guided toward bicycle dummies crossing or moving longitudinally in the direction of travel. The parameters similar to the tests with pedestrians are given in ■ Table 9.3.

The designation of the individual test cases consists of the collision partners (C—Car, passenger car, B—Bicycle, bicyclists), the direction of movement N—Nearside, coming from the right in the case of right-hand traffic, F—Farside, coming from the left in the case of right-hand traffic, L—Longitudinal, driving

---

1   For both turning directions, the pedestrian moves in the opposite direction of the vehicle's initial driving direction before the turn starts.

2   When the ego vehicle brakes - which is the goal of the driving tests - the pedestrian dummy gains time to move on before the collision. The collision point therefore shifts.

152 P. Seiniger et al.

Table 9.3 Requirement definition of Euro NCAP tests of advanced emergency braking assistance systems for the protection of cyclists, 2020–2023: Overview of test configurations for right-hand traffic

	CBNA	CBNAO	CBFA	CBLA-50	CBLA-25
Vehicle initial speed and step width in km/h	↑10–60, 5			↑25–60, 5	↑50–80, 5
Speed of the dummy in km/h	15 ←	10 ←	20 →	15 ↑	20 ↑
Impact location	50% (Middle)				25% (right)
Obstruction		Yes			
Braking / Warning	B				W
Day / Night	D				

Table 9.4 Requirement definition of Euro NCAP tests of rear impact and intersection emergency brake assist systems, 2020–2023: Overview of test configurations for right-hand traffic

	CCRs AEB	CCRs FCW	CCRm AEB	CCRm FCW	CCRb	CCFtap
Vehicle output speed and step width in km/h	10–50, 5	30–80, 5	30–80, 5	50–80, 5	50, Initial distances 12 m and 40 m	10, 15, 20
Dummy vehicle speed in km/h	0		20		50	30, 45, 55
Dummy vehicle acceleration in m/s^2	0				-2 and -6	0
Overlap[3]	50%,75%, 100%, -75%, -50%				50%	50% (Vehicles stand at an angle)
Automatic (AB) or supported (SB) brake	AB	SB	AB	SB	AB and SB	AB[4]

longitudinally ahead), a more detailed designation of the target object (A—Adult) and, if applicable, the impact value in percent. Arrows symbolize directions of movement.

9.2.3.3 Advanced Emergency Brake Assistance Car-Car

In the test procedure for rear collision and intersection emergency brake assistants, the vehicle under test is subjected to situations that mimic inner-city, turning situations, or the end of a traffic jam. The essential parameters are given in **Table 9.4**. The ego vehicle moves predominantly straight ahead, in individual cases it turns off.

The designation of the individual test cases consists of the collision partners (C—car, example: car against car: CC), the direction of movement (R—rear, rear impact or coming from behind, F—front, oncoming), a more detailed designation of the test sequence (b—braking, s—stationary, stationary target object, m—moving, moving target object, tap—turn across path, turning).

9.2.3.4 Lane Support Systems

In the test procedure for lane departure warning systems, the vehicle is moved off course at a defined lateral speed and steered onto the edge of the lane. The test is designed in such a way that a possible steering reaction of the assistance system is not overridden (the steering robot therefore goes into neutral before the expected vehicle reaction). The lane edge consists of a real road edge (Real Road Edge), a lane marking, or a combination of both. The essential parameters for lane departure warning systems are given in **Tables 9.5 and**

---

3   Convention: Proportion of the width of the ego vehicle that is overlapped by the GVT. Exception: With 100% overlap, both vehicles collide centrally, even if the GVT could be narrower than the ego vehicle.

4   Since in these constellations an incomplete collision avoidance due to the shift of the point of impact could worsen the consequences of the accident, only complete avoidance is evaluated.

◻ Table 9.5 Requirement definition of the Euro NCAP tests of lane departure warning systems Part 1, 2020–2023: Overview of test configurations

	LDW Single Line	LKA Single Line	ELK Real Road Edge	ELK Single Line
System status after startup of the vehicle	Free selectable by vehicle manufacturer		Active by default	
Vehicle speed in km/h	72			
Lateral vehicle speed (in the direction of the lane marking), increment in m/s	0.2-0.5 , 0.1			
Lane marking	↑| or |↑	↑| or |↑	⁞↑| , ⁞↑⁞ , ↑| , ↑⁞ ,↑	↑| oder |↑
Roadside	No consideration	No consideration	True roadway edge (verge): gravel or companion greenery	No consideration
Marker type (solid — or gestrichelt ---)	— or ---	— or ---	— or ---	—
Evaluated system intervention	Warning	Vehicle steers back independently		

[Tabellenfußzeile—bitte überschreiben]

◻ Table 9.6 Requirement definition of the Euro NCAP tests of lane departure warning systems Part 2, 2020–2023: Overview of test configurations

	ELK Oncoming	ELK Overtaking unintentional	ELK Overtaking intentional
System status after ignition change	Active by default		
Vehicle under test-speed	72 km/h		
Lateral vehicle speed (in the direction of the lane marking), increment in m/s	0.3-0.6, 0.1		0.5 -0.7, 0.1
Lane marking	⁞ ⁞↑| or | ⁞↑|		
Direction blinker (turn signal)	Not operated		Active
Dummy vehicle-speed	72 km/h	A) 72 km/h and B) 80 km/h	
Dummy vehicle-movement	Oncoming ↑↓	↑↑ A) Parallel driving or B) Overtaking	
Evaluated system intervention	Vehicle steers back independently		

[Tabellenfußzeile—bitte überschreiben]

9.6. There are tests for pure warning systems (Lane Departure Warning, LDW) or automatically intervening systems (Lane Keeping Assist LKA, Emergency Lane Keeping ELK).

In addition, there are test procedures in which the criticality of the situation is aggravated by further oncoming or overtaking vehicles, sometimes also with actuation of the turn signal of the vehicle under test ("overtaking intentional").

9.2.3.5 Grading of Automated Driving Functions

In addition to the "rating," Euro NCAP introduced the so-called "grading" for assisted vehicle automation functions in 2018. The "grading" in relation to continuous assistance functions (according to SAE J3016 Level 2) is divided into two parts: It consists of both an assessment of the human–machine interface and an assessment of the driver assistance functions. The aim is to ensure that users do not develop unjustified overconfidence in the assistance function.

The assessment of the driver assistance function itself is essentially made up of modifications to the test procedures listed so far (example: higher driving speeds in tests of the emergency braking assistance function), supplemented by specific tests (example: testing of the maximum steering torque to be applied by the driver).

A detailed description of the "grading" goes far beyond the purpose of this chapter, please refer to the corresponding document [5].

For future automated driving functions (according to SAE J3016 Level 3), test procedures will be defined at a later stage.

9.3 Test Procedures Within the Framework of Vehicle Regulations

In contrast to the consumer protection assessment, which is not legally binding, a vehicle type approval is required for market entry and is therefore mandatory for the vehicle manufacturer. Requirements can and—in more recent times—are therefore directly specified and are generally verified using standardized test procedures, but verification of compliance with requirements using other tests is also possible at any time, for example in the event of doubtful results or suspected test optimizations. Therefore, in recent times, the requirements have become much more important than the test procedures and definitions themselves. Evaluation criteria in the individual test case can either be specifically written down or defined by reference to requirements.

In the concrete design of the test procedures, standardized tests from other areas are used where possible (example: advanced emergency brake assistance systems, which are essentially tested according to Euro NCAP methods). If this is not possible, own test procedures are specified (example: blind spot information systems). In some cases, the actual performance of the tests is deliberately left very free: In UN Regulation No. 157 on automated lane-keeping systems [4] for example, the concrete test definition for situations with objects in the driver's own lane looks like this:

4.2. Avoid a collision with a road user or object blocking the lane

4.2.1. The test shall demonstrate that the ALKS avoids a collision with a stationary vehicle, road user or fully or partially blocked lane up to the maximum specified speed of the system.

4.2.2. This test shall be executed at least:
- With a stationary passenger car target;
- With a stationary-powered two-wheeler target;
- With a stationary pedestrian target;
- With a pedestrian target crossing the lane with a speed of 5 km/h;
- With a target representing a blocked lane;
- With a target partially within the lane;
- With multiple consecutive obstacles blocking the lane (e.g., in the following order: ego vehicle—motorcycle—car);
- On a curved section of the road.

This new type of test definition assigns much more responsibility to the technical services to perform meaningful test criteria and setups themselves than was previously the case (with precisely defined tests). This requires, as defined in ▶ Chap. 4, the instrument of market surveillance to monitor this responsibility.

9.4 Properties of Test Tools

In tests of driver assistance systems, a critical driving situation is usually simulated for the vehicle under test using the appropriate tools, and the reaction to this is evaluated. Even if today's advanced emergency braking systems are designed in such a way that most test speeds and test parameters are completed collision-free, provisions must still be made for the event of a collision. Especially during development tests, the limits of the system are specifically determined, and a collision is very likely. For reasons of cost and time, it therefore makes sense to use a collision-tolerant target object.

Such target objects are usually lightweight and made of soft materials such as foam, in order to stress the front of the vehicle with the smallest possible forces that remain constant over the deformation in the event of a collision. However, the foam has fundamentally different properties than the target object to be represented, such as a vehicle, a bicycle, or a person walking.

The motion system for the target object must achieve the repeatability required within the test procedure while not affecting the collision properties and realism of the target object.

Essential components of test procedures are therefore target objects to represent other vehicles (passenger car advanced emergency braking assistance), people (passenger car to pedestrian advanced emergency braking assistance), and bicycles (passenger car to cyclist advanced emergency braking assistance) with suitable motion systems. The goal is to always prescribe properties of the target object rather than products or technical solutions. Thus, the properties are specified in such a way that they can be stably detected and unambiguously recognized by today's sensor technology. The sensors used in driver assistance systems are predominantly radar (▶ Chap. 15), camera (▶ Chap. 17), and lidar (▶ Chap. 16). The mockups are now defined to correspond exactly to the properties measured by the sensors of the object to be represented. In this context, the dummy and its carrier or drive unit are often considered together if this combination is detected together by the sensor system and an influence on the carrier or drive unit cannot be ruled out. Measures for matching the properties can be, for example, the application of metallic foil pieces (see ▶ Chap. 15) or the printing of the cover for the optical imitation of contours and structures.

Within this conflict of goals between realistic properties, collision tolerance, and mobility, it is necessary to find an optimal compromise corresponding to the respective requirements.

9.4.1 Car-Representing Targets and Movement Devices

The "Global Vehicle Target," abbreviated GVT, shown in ■ Fig. 9.2 is used to represent collision scenarios with a passenger car. It consists of a dummy and a self-propelled drive unit and is defined in ISO standard 19,206–3 [6]. The mockup imitates a Ford Fiesta (seventh generation, construction period 2009 to 2017) on a scale of 1:1 and is constructed as follows: foam sheets connected with Velcro strips to create a load-bearing structure form a framework. This is covered with a shell consisting of several parts, which is provided with metal fibers and foils to generate a representative radar reflection. The properties are tuned so that the GVT is reliably recognized as a passenger car from all sides.

The drive unit, or platform, can be moved manually, from a control station, or automatically traverse complex trajectories with the highest precision. During a collision, the GVT disintegrates into its lightweight individual parts, which are designed to be shock-and

abrasion-resistant and can then be quickly reassembled (■ Figs. 9.2 and 9.3).

The drive unit that can be driven over is designed in such a way that it can also withstand the test vehicle braking on it without damage.

9.4.2 Human-Representing Targets and Motion Devices

In the case of emergency braking systems designed for accidents with unprotected external road users, it is not to be expected, even in the distant future, that all test cases can be tested in a collision-avoiding manner. For ethical reasons, the question of using humans as target objects does not arise here. A target object (better: dummy) must appear as a human being to the usual sensor technology and be able to survive a collision. The pedestrian dummies used in Euro NCAP are specified in ISO standard 19,206–2 [7] and represent a typical adult and a child aged six years.

Platform solutions are now common, where a dummy is attached to a moving platform using magnets, for example. For this, the dimensional stability of the dummy and the abrasion behavior of the clothing are important—but such concepts just get by with simply constructed, portable platform solutions for movement compared to the previously used bridge solutions for hanging dummies. The platforms are available as

■ **Fig. 9.2**   Euro NCAP Vehicle Target (GVT) in assembled state

Fig. 9.3   Representation of the GVT after a collision

self-propelled or belt driven and guided. Dummies can also simulate the gait of a human being by moving extremities or the gaze by moving the head.

Static targets differ from pedestrians by the fact that they do not show any movement. This deficiency is still acceptable for optical sensors because, for example, when a real pedestrian is dressed in winter clothing, little movement of the arms and legs is visible. However, current radar sensors can use the movement of the extremities, which is always visible to them, as a classification feature [8]. Meanwhile, dummies that can at least move the legs are used quite predominantly accordingly.

There are also movement concepts in which the dummy is guided by ropes or bars from above. These concepts require complex setups and are therefore often not economical in the context of the test procedures described here—they may have their justification in the context of system development and validation.

Within the trade-off between collision tolerance, type of motion device, and complexity of the dummy, the available approaches can be classified as follows:

Belt-driven concepts, such as the "4a Surfboard" (see **Fig. 9.4**), allow in principle smaller overall heights and better repeatability, but are not very flexible in the representation of the pedestrian trajectory

because of the belt guidance—much more relevant: during tests in the longitudinal direction, the bearing of the belt could interfere with the radar sensor system. Removing the bearing means that the sled is only pulled in one direction and is no longer indirectly stabilized laterally. Compliance with the required tolerances must be ensured with great effort by adjusting the carriage and is very susceptible to crosswinds.

Because of the accuracy problems and the increasingly complex test procedures, nowadays—at least in consumer protection—mainly self-propelled platforms are used as carrier systems, as they were originally developed for bicycle targets, as explained in the following section.

9.4.3 Targets and Movement Devices Representing Bicycles and Motorcycles

Again, the same requirements for the bicycle and bicyclist combination as described in the previous section apply. The dummy bicycle must be able to survive collisions without unacceptable damage, support the bicyclist dummy, have high stability for bicyclist-typical speeds, have moving wheels or spokes for the radar sig-

◘ Fig. 9.4   Movement system "Surfboard" of the company 4a Active Systems during a night test

nature, and possibly replicate pedaling movements of the legs.

To overcome the disadvantages and limitations of the belt-driven concept (low flexibility, problems with lateral deviation when the sled is only pulled), the use of a self-propelled platform, as already used for passenger car emergency braking tests, is obvious. The high flexibility of the programmable trajectory, the high positional quality, and the fast set-up time have led to the development of a support system for pedestrians, cyclists, mopeds, and motorcyclists. Since the metallic spokes of the wheels have a significant contribution to the radar signature of the dummy, the platform "4a smallFB" (◘ Fig. 9.5) has a recess to allow the wheels of the bicycle to touch the ground and thus be rotated. However, to simulate motorized two-wheelers, the connection is rigid, and rotating triple mirrors are seated in the wheel hubs.

The physical properties of the target objects are specified in the standards ISO-19206–4 (bicycle) [9] and ISO-19206–5 (motorcycle, not yet published).

9.5 Outlook: Realism and Testing Effort

The development of driver assistance systems in recent years shows that the systems are supporting an increasing proportion of driving situations or even handling them (partially) automatically. To make this possible, the amount of information that has to be collected, processed, and evaluated by the systems regarding the driving situation is constantly increasing. Test procedures for these driver assistance systems must be able to represent all the information required in the respective situation or at least simulate it with sufficient realism.

This increases the effort required for the development and representation of real and relevant test situations for driver assistance systems. Likewise, it is also necessary to prove that these situations represent reality sufficiently accurately for a system response. For collision-tolerant target objects, this proof is particularly necessary with regard to the suitability for the sensor technology used and the driver's threat impression. For example, the requirement must be met that the objects

Fig. 9.5 Movement system "smallFB" of the company 4a Active Systems with cyclist dummy

used can not only be detected by the sensor technology, but also that they are to be considered representative of the collective of vehicles, pedestrians, or other road users encountered in real traffic. An example of such relevance studies for radar sensors can be found in [10].

If this proof is not provided, only little safety progress is achieved in real road traffic despite excellent functional and safety evaluations under artificial conditions.

The same applies to the selection of scenarios to be tested. With an increasing number of sensors and parameters included in the function of driver assistance systems with environment perception, the effort for the definition of scenarios, the execution of tests, and the proof of relevance for the field increases.

As the number of parameters included in the situation assessment increases, such as lighting conditions, weather, or road category, it must also be demonstrated with regard to the detection of objects and what proportion of real traffic or accident situations is addressed under these conditions. In the case of accident-avoiding systems, this can be done on the basis of accident databases, insofar as the data available there is sufficiently detailed. In the case of conditionally or highly automated systems, a driving situation collective with all relevant influencing parameters would have to be used for this purpose. A collective suitable for such

considerations is not currently known. As an alternative, a large number of test kilometers are therefore completed under a wide range of conditions in order to achieve sufficient coverage of driving situation constellations. However, the effort required for this is very high and the transferability is limited.

References

1. International Organization for Standardization: ISO 26262 Road Vehicles—Functional Safety (2018). ▶ https://www.iso.org/standard/68383.html. Access: 24. Februar 2021
2. Euro NCAP: Test Protocol—AEB Car-to-Car systems. Version 3.0.2 (2019). ▶ https://cdn.euroncap.com/media/56143/euro-ncap-aeb-c2c-test-protocol-v302.pdf . Zugegriffen: 24. Februar 2021
3. Euro NCAP: AEB VRU Test Protocol v3.0.3 (2020). ▶ https://cdn.euroncap.com/media/58226/euro-ncap-aeb-vru-test-protocol-v303.pdf. Access: 24. Februar 2021
4. Euro NCAP: LSS Test Protocol v3.0.2 (2019). ▶ https://cdn.euroncap.com/media/53143/euro-ncap-lss-test-protocol-v302.pdf. Access: 24. Februar 2021
5. Euro NCAP: System Grading (2020). ▶ https://www.euroncap.com/en/vehicle-safety/safety-campaigns/2020-assisted-driving-tests/gradings-explained/. Access: 24. Februar 2021
6. United Nations Organization: UN-Regulation No. 157, Uniform provisions concerning the approval of vehicles with regards to Automated Lane Keeping Systems (2021). ▶ https://unece.org/sites/default/files/2021-03/R157e.docx. Access: 20. September 2021

7. International Organization for Standardization: ISO/PRF 19206-3 Road vehicles—Test devices for target vehicles, vulnerable road users and other objects, for assessment of active safety functions—Part 3: Requirements for passenger vehicle 3D targets (2018). ▶ https://www.iso.org/standard/70133.html. Access: 24. Februar 2021

8. International Organization for Standardization: ISO/PRF 19206-2 Road vehicles — Test devices for target vehicles, vulnerable road users and other objects, for assessment of active safety functions—Part 2: Requirements for pedestrian targets (2018). ▶ https://www.iso.org/standard/63992.html. Access: 24. Februar 2021

9. International Organization for Standardization: ISO/PRF 19206-4 Road vehicles—Test devices for target vehicles, vulnerable road users and other objects, for assessment of active safety functions—Part 4: Requirements for bicyclist targets (2018). ▶ https://www.iso.org/standard/70134.html. Access: 24. Februar 2021

10. Heuel, S., Rohling, H.: Pedestrian Classification in Automotive Radar Systems. 19th International Radar Symposion, 23.-25.05.2012, Warschau, Polen

11. United Nations Organization: UN-Regulation No. 152, Uniform provisions concerning the approval of motor vehicles with regard to the Advanced Emergency Braking System (AEBS) for M1 and N1 vehicles (2020). ▶ https://www.unece.org/fileadmin/DAM/trans/main/wp29/wp29regs/2020/R152e.pdf. Access: 20. Januar 2021

12. Euro NCAP: Test Protocol—AEB VRU Systems. Version 2.0.4 (2019). ▶ https://cdn.euroncap.com/media/43381/euro-ncap-aeb-vru-test-protocol-v204.pdf. Access: 20. Januar 2021

Open Access This chapter is licensed under the terms of the Creative Commons Attribution-NonCommercial-NoDerivatives 4.0 International License (▶ http://creativecommons.org/licenses/by-nc-nd/4.0/), which permits any noncommercial use, sharing, distribution and reproduction in any medium or format, as long as you give appropriate credit to the original author(s) and the source, provide a link to the Creative Commons license and indicate if you modified the licensed material. You do not have permission under this license to share adapted material derived from this chapter or parts of it.

The images or other third party material in this chapter are included in the chapter's Creative Commons license, unless indicated otherwise in a credit line to the material. If material is not included in the chapter's Creative Commons license and your intended use is not permitted by statutory regulation or exceeds the permitted use, you will need to obtain permission directly from the copyright holder.

Human-Centred Assessment Methods for Assisted Driving Functions

Caroline Schießl

Contents

© The Author(s) 2026

H. Winner et al. (eds.), *Handbook Assisted and Automated Driving*,

https://doi.org/10.1007/978-3-658-45276-6_10

10.1 Why do we need human-centred assessment procedures for assisted driving functions?

The developments of assisted driving functions and consequently, in a broader sense, also for transport systems, are facing a number of challenges in current times. Both increasing urbanisation and a rise in individual, flexible and adaptive mobility are leading to an overall increase in mobility and therefore also to an increase in the volume of traffic. Growing technical and technological developments, such as artificial intelligence, automation, digitalisation and increased networking can, on the one hand, help to solve these challenges. On the other hand, however, they can lead to a further growth in mixed traffic—through the increasing establishment of assisted, automated and autonomous vehicles alongside conventionally driven vehicles. In cities, in particular, this is now leading to a changed urban traffic environment. Characteristics of today's urban traffic environments include a high degree of complexity, a large number of interacting traffic participants, complex driving tasks and varying, sometimes intersecting movement patterns of the traffic participants.

10.1.1 Methods and models of human-centred systems for driver assistance, automation and traffic

An essential prerequisite for a safe, efficient traffic environment that is therefore accepted by traffic participants is an understanding of the traffic environment, for example through its depiction by means of concepts and models. It is thereby essential for all participating agents to be considered, i.e. in addition to technology and techniques, the human being, namely the traffic participant and user of the driver assistance, automation and traffic systems, is also an intrinsic component of the concepts and models. If traffic participants understand the traffic environment, they are able to draw a comprehensive picture of the traffic environment and use this as a basis for, for example, predicting the intentions of other traffic participants and, consequently, for planning, communicating and executing their own safe movement through the traffic. This needs to be described.

With the help of a consistent, human-centred approach starting at an early stage of development, it is possible to develop driver assistance, automation and mobility systems which can later be deployed safely and adequately by the user, which can be interacted with safely, and which are characterised by a high degree of user acceptance and a high level of efficiency.

The human being forms the central focus of this observation. The approach described here is based on the assumption that by modelling and describing observable human behaviour in traffic, a basis can be created which enables the traffic environment to be understood and assessed. In addition, the interaction between different psychological methods and models with those of engineering and information sciences is addressed. This is considered essential for a human-centred assessment.

10.1.2 Acquisition, analysis, modelling and assessment

In order to be able to understand humans in their interaction with other traffic participants, not only models are required but also methods and techniques that enable the recording, analysis, modelling and also assessment of the human behaviour in traffic described above. As an example, parameters must be further developed that can be used to record, analyse, model and assess the behaviour of traffic participants. This includes, for example, parameters that describe the performance in lateral and longitudinal guidance, such as lane fluctuations or distances to preceding vehicles. The aforementioned challenges of increasing urbanisation and the associated growing complexity of the traffic environment thereby play an increasingly important role. Amongst other things, parameters must be defined by means of which aspects of interaction and communication between traffic participants can be described, or intersecting movement patterns of traffic participants can be mapped. In order to, for example, assess the traffic behaviour of traffic participants in an intersection in terms of safety, the observation of the speed or distance of individual vehicles is insufficient. Instead, the simultaneous speed and distance analysis of interacting traffic participants is relevant. Along with this, the development of new research methods and tools is also gaining in importance, with the help of which these new parameters can be established and recorded and which therefore also provide a contribution towards solving the aforementioned challenges. Approaches here are, for example, the networking of simulators or the application of differing methods and approaches within the framework of the analysis of complex data. For this, subjective observation data and categories and characteristics of the traffic environment extracted from video data (e.g. position of the observed traffic participants, relation to one another) are evaluated in combination with an inferential statistical analysis of objectively collected data (e.g. speeds driven) concerning vehicle and/or driver behaviour [1]. A comprehensive overview of this so-called mixed-methods approach can be found in [2]. The results are further compiled in the form of descriptive models.

Furthermore, methods and tools must be further developed so that the impact of new functions, technical systems and technologies on various categories such as safety, performance, efficiency, acceptance and trust can be mapped, as well as on aspects such as sustainability and resource conservation.

From the resulting descriptive models, knowledge and insights in the form of decision strategies, reaction times, etc. are passed on to the technology in order for the aforementioned goal of a safe, efficient and accepted traffic environment to be achieved.

10.2 Classical assessment methods and their limitations

In classical traffic psychology, various methods exist for recording the behaviour of traffic participants. Behaviour can thereby be considered, for example, as a function of the requirements that the driving situation places on people [3]. The requirements of the driving situation in turn result from the interaction of aspects of the driving environment, the vehicle and the surrounding traffic participants and can be subsumed under the actual driving task. According to [4], this can be further differentiated via the subtasks of navigation (route selection), path guidance (e.g. adaptation to traffic environment) and stabilisation (lateral and longitudinal dynamic control). Driving behaviour can be further regarded as an observable variable of human performance, the product of the driver's performance capability and the driving requirements of the driving task [5–7].

If the field of tension between performance capability and driving requirements is considered in relation to the methods and tools to be used for recording and evaluating driving behaviour, the classic investigation methods presented below can be derived.

If the driving requirements exceed the human performance capacity, the probability of critical situations or accidents increases. Accident analyses therefore provide information concerning scenarios and situations that can no longer be controlled. With the help of accident analyses, causes and contributing factors can be partially determined; the underlying causal relationships for the occurrence of accidents and actual triggering factors are, however, usually reconstructed in retrospect and are therefore often incomplete.

Natural driving studies ▶ Sect. 10.2.1 could provide information on this; critical situations and accidents are, however, relatively rare events. The expected data basis is therefore rather small or a very large number of driven kilometres are required in order to be able to make reliable statements. Natural driving data therefore provide information regarding normal driving, i.e. they generally depict behaviour in which the performance capacity corresponds to the requirements of the situation or is superior to them. Simulator studies ▶ Sect. 10.2.2 close the gap, i.e. in the simulation it is possible to systematically create critical situations as well as situations that represent normal driving. The methods mentioned above are described in more detail below.

10.2.1 Natural driving studies

Real natural driving data are collected in field studies (naturalistic driving studies NDS, field operational tests FOT) in which vehicle drivers are observed by means of instrumented vehicles over a longer period of time [8, 9]. Data recorded include driver behaviour (e.g. operator inputs, physiological measures, non-driving activities), vehicle behaviour data (e.g. speeds, transverse offset) and environmental data (e.g. surrounding vehicles, route characteristics). The data recorded in this way is characterised by very high external validity. External validity is a quality criterion for empirical studies and describes the generalisability and representativeness of the study results beyond the specific study situation, the study period and the subjects studied [10]. As it concerns natural driving, which is not influenced by experimental conditions, the external validity in terms of ecological validity, i.e. the representation of the situation under investigation, is high per se. Depending on the sample selected, population validity can also be very high, i.e. it is possible to fulfil the condition that, on the basis of a representatively selected sample, conclusions can be drawn concerning a population [11].

In addition to analysing normal driving behaviour, natural driving studies can also be used to investigate the effect of new technological developments in the field of driver assistance and automation systems. Classical driving behaviour parameters that are thereby recorded and analysed are parameters of longitudinal and lateral guidance, e.g. speeds driven, distances to vehicles in front and deviation from the centre of the road. According to [4], these primarily represent the path guidance level. In addition, however, the stabilisation level, i.e. stimulus-reaction mechanisms, are also mapped, e.g. via braking reaction times. Real driving data can also be the basis for the recognition of differing driving manoeuvres [12, 13]. These can be extracted and, in dependence on moderating factors (e.g. weather conditions, times of day, route characteristics, etc.) as well as different populations, can be described and analysed.

10.2.2 Simulator studies

Studies in driving simulators, i.e. studies in which the driver and driving behaviour of selected test persons is examined in a driving simulator, are characterised by a very high internal validity. Internal validity is another quality criterion for empirical studies. A high level of internal validity exists when the behaviour being investigated and measured can be clearly attributed to the study situation [10]. This is the case in experimental driving simulator studies, as driver behaviour is systematically analysed in strictly controlled and predefined scenarios or situations. Experimental studies consequently build on corresponding knowledge concerning existing dependencies between relevant variables to be investigated and examine these systematically. Experimental studies in the driving simulator also enable, for example, the analysis of underlying causal relationships of accidents. In addition, they also allow the limits of human performance to be approached by means of a systematically structured experimental design; as a result, not only can a statement be issued that an accident has occurred, but a prediction can also be made as to when and with what probability an accident will occur [14]. Furthermore, by means of experimental studies in the driving simulator, human driving behaviour can be recorded and analysed in situations which, e.g. due to low probabilities of occurrence or high associated criticalities, are covered neither by accident analyses nor by real driving tests.

Typical driving behaviour parameters that can be recorded and therefore analysed by means of simulator studies are again parameters of longitudinal and lateral guidance, e.g. speeds driven, distances to preceding vehicles and deviation from the centre of the lane. Parameters that map stimulus–response mechanisms, such as braking reaction times and steering intervention times, are, however, also relevant driving behaviour parameters. In addition to these behaviour-based driving behaviour parameters, physiological (e.g. heart rate, skin conductance, eye movements) and subjective data (assessment of current stress) of driver behaviour can also be collected in the simulative test environment, enabling constructs such as stress and exertion to be analysed [15, 16]. A construct is generally understood to be a circumstance which is not directly measurable or observable, but which can be measured via other measurable circumstances (indicators).

Nowadays, different types of simulators exist, which differ, for example, in their technical set-up, e.g. static laboratory set-up versus set-up with a motion system, simple control units (steering wheel and pedals) versus complex vehicle installation, as well as in their realistic visualisations and projections.

10.2.2.1 Dynamic driving simulator

Experimental studies in the dynamic driving simulator (◉ ▫ Fig. 10.1) are characterised by a very realistic design of the simulation and are therefore generally used for human-centred testing of assistance and automation functions and systems at an advanced stage of development. Realism is usually achieved through a motion system, high-quality projection and visualisation (e.g. visibility and simulated traffic environment ahead, to the sides, and in the rear-view and door mirrors) and the integration of real control elements and vehicle components (e.g. active steering wheel, active brake and accelerator pedals).

The time-synchronous recording of all data and systems, i.e. driver behaviour, vehicle and environment, enables the analysis of a holistic picture of the investigation scenario. As described above, the focus is primarily on the realistic analysis, testing and evaluation of systems and technological developments through the repeated, systematic and controlled production of scenarios and situations to be investigated. As a result, driving simulator studies are characterised by a very high internal validity, as already mentioned above [17].

In the Pegasus project (project for the establishment of generally accepted quality criteria, tools and methods as well as scenarios and situations for the approval of automated driving functions) funded by the BMWi [18], experimental studies were carried out in a dynamic driving simulator in order to analyse and define the limits of human performance in specific critical situations. For this purpose, driver behaviour was measured, analysed and modelled systematically and in a controlled manner in situations of differing criticality [5].

10.2.2.2 Virtual laboratory

In a virtual laboratory, as in the dynamic simulator, assistance and automation functions and systems can also be tested in a systematically controlled and repeatedly reproducible manner. In contrast to the dynamic driving simulator, however, the virtual laboratory does not have a movement system, i.e. the mock-ups of the integrated vehicles are permanently installed. The virtual laboratory is, however, characterised by a highly dynamic and scalable simulation environment and representation of the virtual world, including all components to be represented. The virtual laboratory is therefore utilised in particular for a human-centred development process characterised by many iterations in the early stages of development. This allows different layouts, designs and functions of assistance and automation systems to be implemented and tested in a time-efficient and cost-effective manner by means of experimental studies with test persons. The virtual laboratory can furthermore be used in the context of user stud-

◘ Fig. 10.1   Virtual laboratory

ies for the development of methods for the collection of user-state data [19, 20] as well as within the framework of design processes [21] for innovative assistance systems. An essential feature of the virtual laboratory is that as a result of the highly dynamic and scalable simulation environment and the representation of virtual components, it is particularly suitable for testing different design variants, representations and interpretations of assistance and automation components as well as human–machine interfaces. In addition to the aforementioned classic driving behaviour parameters, data is collected that can, for example, provide information concerning the use and acceptance of a new system.

10.3   Innovative new procedures

As described in ▶ Sect. 10.1, increasing urbanisation and the associated increase in mixed traffic (assisted, automated and autonomously driving vehicles alongside conventionally driving vehicles) represent a major challenge for the development and research of systems for assistance, automation and mobility. Characteristics of this urban traffic include a high number of in-

teracting traffic participants, complex driving tasks and different, sometimes intersecting movement patterns of traffic participants. If the goal is to understand this traffic environment and to use this in turn as a basis for the development of assistance, automation and mobility systems, then it is necessary to (further) develop, construct, test and validate the methods, tools and instruments which can map these aspects.

The level of consideration is consequently extended from one traffic participant interacting with one system to the joint consideration of several traffic participants interacting with each other. In addition to the still relevant constructs to be analysed, such as safety and acceptance, constructs such as interaction, communication and cooperation are gaining in importance. With these, the complexity of urban traffic, e.g. the intersecting and interacting movement patterns of traffic participants, is mapped and described. This also does justice to the dynamics that represent a central feature of urban scenarios, such as passing through a junction.

As already introduced in ▶ Sect. 10.1.2, in addition to the identification, development, modelling and analysis of e.g. interaction and cooperation behaviour, the development and construction of corresponding laboratory infrastructure and analysis tools is also essential. Networked simulations enable the mutual and simultaneous observation of several traffic participants.

The following sections describe the requirements and challenges for methods of data collection as well as data analysis and processing that arise from the use and application of networked simulations. In conclusion, two research examples provide an insight into the application of corresponding methods.

10.3.1 The networked simulation

In the urban traffic environment in particular, interaction, cooperation and communication are essential constructs that contribute to the description, analysis and modelling of the behaviour of traffic participants. The classical driving parameters described in ► Sect. 10.2 for describing driving behaviour can also be used here in order to describe how vehicle drivers react to the behaviour of vehicle drivers in ahead of them. Parameters such as braking intensity or time-to-collision (TTC) describe the behaviour in terms of safety or criticality [5, 7]. Driving manoeuvres can provide information regarding the quality of a behaviour and whether a behaviour is cooperative or not. For this purpose, however, it is essential that the selected parameters reflect not only the reaction of one traffic participant, but also the dynamics of an action and reaction of several traffic participants that may succeed each other several times.

The requirement for multi-directional consideration of the behaviour of traffic participants can also be fulfilled in experimental settings by means of networked simulators. The MoSAIC (Modular and Scalable Application Platform for ITS Components) is one such networked simulation at the German Aerospace Center in Braunschweig [17] (◉ ◘ Fig. 10.2).

The increasing complexity, particularly in urban areas, with the growing automation of traffic and the associated probable increase in interactions with mixed traffic represent the main challenges for the laboratory. Research topics in the area of interaction with weaker traffic participants are also becoming increasingly important. In order to be able to meet these challenges in the laboratory and test infrastructure, the laboratory has three interlinked vehicle simulators and two pedestrian simulators in virtual reality (VR) at its disposal. An additional VR bicycle simulator is currently under development and construction. The MoSAIC there-

◘ **Fig. 10.2**   Dynamic driving simulator

fore enables the scientific study of interactions, cooperation and communication between traffic participants through the creation, recording and analysis of widely differing interaction scenarios in a laboratory. The results represent human-centred models of observable behaviour of all traffic participants. Based on these, decision-making, communication or cooperation strategies are derived and made available to new assistance and automation systems. Furthermore, requirements for concepts, the design or the function of human–machine interfaces can be derived. Moreover, they also provide new, essential impulses for the further development of methods and parameters for the analysis of driver, vehicle and traffic participant behaviour in road traffic.

A detailed insight into the application of networked simulation as well as the requirements for the further development of the applied methods is provided below by means of two research examples.

10.3.2 Application and research examples

10.3.2.1 Analysis of interaction in urban space

Within the framework of a national research project[1] funded by the BMWi (German Federal Ministry for Economic Affairs and Energy), one of the project goals was to analyse and model urban traffic behaviour with a focus on the interaction behaviour of different traffic participants. Furthermore, methods and tools were developed which can empirically collect, record, analyse and simulate this interaction behaviour.

Amongst other aspects, the interaction behaviour of a vehicle driver equipped with a traffic-light assistant with surrounding non-equipped vehicle drivers was investigated. This was accompanied by the following research questions: What effects can be observed and analysed when vehicle drivers equipped with an assistance system meet non-equipped vehicle drivers? Do critical situations affecting safety or negative evaluations and emotions arise due to the fact that the behaviour of the equipped vehicle driver is not understood? Or, conversely, are there rather positive effects in the sense of increased safety, efficiency or optimisation of the traffic flow, as non-equipped vehicle drivers adapt their own

behaviour to that of the equipped vehicle drivers and therefore benefit from the assistance of the others?

In order to deal with the multidimensionality of the observation level of several interaction partners in the field of empirical and simulative methods and tools, the networked simulation MoSAIC described in ▶ Sect. 10.3.1 was implemented. In the course of the preparations for the empirical, experimental studies, however, it quickly became apparent that, in addition to the technical, simulative set-up of the laboratory, methods and approaches for human-centred assessment also needed to be further developed. Aspects of the experimental design, the creation of experimental scenarios and the instruction of test persons formed the main focus. Amongst other aspects, it was necessary to ensure that the test participants repeatedly met and interacted in the simulated world without their natural driving behaviour being too restricted. In the present case, this was realised by having test persons follow a vehicle from the beginning of the test scenario, which repeatedly came to a halt at traffic lights. The following vehicles were therefore able to come together again and again and to subsequently approach and cross the intersection in a group. Overtaking was not possible due to the traffic density. The test persons were therefore not restricted in their intended driving behaviour, but the advantages of the simulation to create systematic and controlled investigation scenarios were retained [22].

A further fundamental research focus was and still is related to the further development, definition and selection of suitable parameters for describing interaction behaviour. The analysis of the empirically collected data in the MoSAIC showed that classic parameters for describing driving behaviour, such as speeds and distances, are only partially sufficient to describe the interaction of several test participants. Solely the reaction of one vehicle driver to another can be described. The fact that the reaction of the latter in turn influences that of the first vehicle driver, i.e. the bidirectional, is not depicted.

As a result, two new parameters were established and applied that do justice to the dynamics of the study scenario and are therefore suitable for depicting the interaction behaviour. Firstly, the so-called "Difference in Distance" (DiD) was calculated, which relates the distances of three vehicles driving behind each other in a group [22] and allows statements to be made regarding the distance behaviour of the vehicle drivers in the group. The three-vehicle leaders show a more even distance behaviour in the group the smaller the difference is. Furthermore, the coherence between the speed curves was calculated in order to determine whether and to what extent the vehicle drivers behave similarly and one can therefore speak of an adaptation of the behaviour of non-equipped vehicle drivers to that

---

1 UR:BAN: Urbaner Raum: Nutzergerechte Assistenzsysteme und Netzmanagement Urban Space: (User-friendly assistance systems and network management), 2012–2015, funded by the BMWi, German Federal Ministry for Economic Affairs and Energy. Project goal UR:BAN: Making transport of the future safer and more efficient, thereby addressing in particular the challenges in the urban transport sector.

of equipped vehicle drivers. With both new parameters, a picture can be drawn of the interaction of several vehicles when driving through a junction and statements can be made regarding, for example, whether the vehicle drivers adapt their behaviour to that of a vehicle in front. To what extent do the vehicle drivers interact with each other, adjust distances and speeds and to what extent do they, for example, drive more uniformly through the junction, so that one can speak of an overall behavioural adjustment? [23–25].

In urban areas, not only the interaction between different vehicle drivers plays a role, but the interaction between vehicle drivers and weaker traffic participants is also increasingly gaining in importance. Here, too, the first steps have been taken towards an analysis of interactions. Observation studies at real junctions in inner-city areas have led to requirements for the further development of methods for creating urban interaction scenarios in the laboratory as well as methods for recording and analysing interaction behaviour between vehicle drivers and weaker traffic participants ▶ Chap. 29.

10.3.2.2 Analysis of interaction and communication as a possible basis for decision/behaviour strategies of automated vehicles

Automated driving in the city is one of the focal points of a further national research project @City [26]. Through the analysis of interaction, cooperation and communication between traffic participants and the modelling of traffic participants at the behavioural level (see ▶ Sect. 10.1.2), the urban traffic environment is to be better understood and this knowledge made available to automated driving functions in the form of behavioural or decision strategies.

The starting point is once again an urban traffic situation in which different traffic participants meet and interact with each other. Communication (e.g. explicitly via hand signals or implicitly via speed reduction) hereby plays an essential role for the understanding between the traffic participants and therefore also for their resulting behaviour. Understanding the different communication elements is therefore also a major challenge for automated vehicles. Furthermore, they have to understand the communication elements and messages, put them together to form an overall picture, and use this as a basis for predicting the intentions of the other traffic participants. Their own driving behaviour can be planned, communicated and carried out on the basis of this, so that safe coexistence becomes possible. Within the framework of the project, approaches, concepts and models for observing, recording and describing this communication were further developed and applied using the example of a right-turning vehicle driver interacting with a cyclist riding straight ahead, as well as the example of a vehicle turning left with oncoming traffic at a complex multi-lane junction. Amongst other things, speed and acceleration data ▶ Chap. 29 were analysed by means of annotation (addition, determination of explanatory, supplementary additional information) and analysis of metrologically recorded video and trajectory data at the DLR research junction AIM [27, 28].

In addition to examining the dynamics of the study scenario by means of analysed time series data, a combination of different empirical methods was deployed in order to obtain as accurate a picture as possible of the interaction and communication behaviour of the traffic participants. For example, the data were differentiated into encounters and conflicts depending on their criticality ▶ Chap. 29. In addition, different interaction patterns were determined descriptively, i.e. different spatial constellations in which vehicles and cyclists moved through the intersection. There were, for instance, constellations in which the vehicle driver was always behind or in front of the cyclist whilst in the more distant approach to the junction, in the general junction area and in the actual intersection point of the interaction partners, as well as constellations in which the spatial position of the partners alternated. Through further analysis of the data in dependence on the factors mentioned and through a combination of different methods (descriptive analysis, inferential statistical analysis, time series analysis), an increasingly detailed picture of the study scenario can be drawn. It is thereby essential that the human being and his or her behaviour are always the focal point and that the maximum amount of knowledge can be extracted from the collected and recorded data by means of the combination of different methods [29, 30].

10.4 Outlook/Significance for Automated Driving

In the previous sections, an overview of methods, tools and their further development in the context of human-centred development, design and evaluation of driver assistance, automation and mobility systems was provided. An approach was presented in which the modelling and description of observable human behaviour in traffic create a basis for understanding the traffic environment. Based on this, a variety of methods, tools and evaluation procedures, their limitations and further developments were outlined. Increasing urbanisation and the associated increase in mixed traffic (assisted, automated and autonomously driving vehicles alongside conventionally driving vehicles) represent one of the main challenges for the development and research of assistance, automation and mobility systems of the future.

Classical methods, approaches to the analysis of driving behaviour as well as evaluation procedures of assistance and automation systems are reaching their limits and must be further developed. Constructs such as interaction, cooperation and communication are gaining in importance and must be able to be mapped using experimental methodologies. Insights and knowledge regarding, for example, how traffic participants communicate with each other in urban areas, whether explicitly or implicitly, must also be available to automated vehicles and vehicle functions in order to enable them to interact safely with other traffic participants. Suitable parameters, e.g. for the quantification of communication in urban areas or between automated vehicles and traffic participants, will continue to be developed in the future. These can then be made available to automated driving functions, for example in the form of action strategies. In addition to providing the aforementioned action strategies, the approach described can also be used, for example, to provide the impetus for approaches to human-centred validation and verification of automated driving functions. In this context, models of observable behaviour and/or human performance can represent benchmarks for automated driving functions [5].

References

1. Dotzauer, M., Zhang, M., Junghans, M., Schießl, C.: Die Kunst der impliziten Kommunikation zwischen Auto- und Radfahrenden in Kreuzungen. 10. VDI-Fachtagung: Mensch-Maschine-Mobilität 2019, 05.-06. November 2019, Braunschweig (2019)
2. Creswell, J.W.: Research Design, Qualitative, Quantitative and Mixed Methods Approaches, 2nd edn. Sage Publications, Thousand Oaks (2003)
3. Fuller, R.: Towards a general theory of driver behaviour. Accid. Anal. Prev. **37**(3), 461–472 (2005)
4. Donges, E.: A conceptual framework for active safety in road traffic. Veh. Syst. Dyn. **32**(2–3), 113–128 (1999)
5. Preuk, K., Schießl, C.: Menschliche Leistungsfähigkeit als Gütekriterium für die Zulassung automatisierter Fahrzeuge: Methode zur Ermittlung der Grenzen menschlicher Leistungsfähigkeit. VDI-Fachtagung Der Fahrer im 21. Jahrhundert, 21.-22. Nov. 2017, Braunschweig (2017)
6. Rasmussen, J.: Skills, rules, and knowledge; signals, signs, and symbols, and other distinctions in human performance models. IEEE Trans. Syst. Man Cybern. **3**, 257–266 (1983)
7. Vollrath, M., Schießl, C.: Belastung und Beanspruchung im Fahrzeug - Anforderungen an Fahrerassistenz. In: Integrierte Sicherheit und Fahrerassistenzsysteme VDI-Berichte, 1864. VDI Verlag GmbH (2004). ISBN: 3-18-091864-0
8. Barnard, Y., Utesch, F., Nes, N.V., Eenink, R., Baumann, M., Barnard, Y.: The study design of UDRIVE: the naturalistic driving study across Europe for cars, trucks and scooters. Eur. Transp. Res. Rev. **8**(2), 1–10 (2016)
9. Lietz, H., Petzoldt, T., Henning, M., Haupt, J. et al.: Methodische und technische Aspekte einer Naturalistic Driving Study. FAT Schriftenreihe 229, Forschungsvereinigung Automobiltechnik (FAT) (2011)
10. Lexikon der Psychologie, Hogrefe AG—Dorsch - Lexikon der Psychologie
11. Bortz, J., Döring, N.: Forschungsmethoden und Evaluation für Human- und Sozialwissenschaftler, 3rd edn. Springer Verlag, Berlin, Heidelberg (2002)
12. Vollrath, M., Schießl, C., Altmüller, T.: Erkennung von Fahrmanövern als Indikator für die Belastung des Fahrers. In: Fahrer im 21. Jahrhundert VDI-Berichte, 1919: 103–112 (2005)
13. Schießl, C.: Modellierung von Belastung und Beanspruchung im Fahrkontext. Dissertation, Technische Universität Carolo-Wilhelmina zu Braunschweig (2009)
14. Quante, L., Zhang, M., Preuk, K., Schießl, C.: Human performance in critical scenarios as a benchmark for highly automated vehicles. Autom. Innov. **4**, 274–283 (2021). ► https://doi.org/10.1007/s42154-021-00152-2
15. Schießl, C.: Continuous subjective strain measurement. IET Intel. Transport Syst. **2**(2), 161–169 (2008)
16. Schießl, C.: Subjective strain estimation depending on driving manoeuvres and traffic situation. IET Intel. Transport Syst. **2**(4), 258–265 (2008)
17. Fischer, M., Richter, A., Schindler, J., Plättner, J. et al.: Modular and scalable driving simulator hardware and software for the development of future driver assistance and automation systems. driving simulator conference 2014. In: New Developments in Driving Simulation Design and Experiments, pp. 223–229 (2014), ISSN 0769–0266
18. Forschungsprojekt Pegasus Über PEGASUS - pegasus (pegasus-projekt.de)
19. Zhang, M., Ihme, K., Drewitz, U.: Discriminating drivers' emotions through the dimension of power: Evidence from facial infrared thermography and peripheral physiological measurements. Transp. Res. Part F Traffic Psychol. Behav. (63), 135–143 (2019), ISSN 1369–8478 ► https://doi.org/10.1016/j.trf.2019.04.003
20. Zhang, M., Ihme, K., Drewitz, U., Jipp, M.: Understanding the multidimensional and dynamic nature of facial expression based on Indicators for appraisal components as basis for measuring drivers' fear. Frontiers in Psychology (12), ISSN 1664–1078 (2021)
21. Schindler, J., Herbig, D., Lau, M., Oehl, M.: Communicating issues in automated driving to surrounding traffic—how should an automated vehicle communicate a minimum risk manoeuvre via eHMI and/or dHMI? In: HCI International 2020, Late Breaking Posters, Communication in Computer and Information Science, 1294: 619–626, Cham, Switzerland: Springer Nature (2020). ► https://doi.org/10.1007/978-3-030-60703-6_79
22. Oeltze, K., Schießl, C.: Benefits and challenges of multi-driver simulator studies. IET Intel. Transport Syst. **9**(6), 618–625 (2015)
23. Preuk, K., Stemmler, E.: Interdisziplinäre Bewertung der Auswirkung eines Fahrers mit Ampelassistenz auf nicht-ausgestattete Fahrer. In: Fahrer im 21. Jahrhundert VDI-Berichte, 2264 (2015)
24. Preuk, K., Stemmler, E., Schießl, C., Jipp, M.: Does assisted driving behavior lead to safety-critical encounters with unequipped vehicles' drivers? Accident Analysis and Prevention. Elsevier (2016). ISSN 0001–4575
25. Preuk, K.: Encountering Automated Cars—What are Consequences for Human Drivers? Dissertation, Technische Universität Carolo-Wilhelmina zu Braunschweig (2018)
26. Forschungsprojekt @City: Automatisiertes Fahren in der Stadt @CITY | Automatisiertes Fahren in der Stadt—Home (atcity-online.de)
27. Knake-Langhorst, S., Gimm, K.: AIM Reserach Intersection: Instrument for traffic detection and behavior assessment for a complex urban intersection. J. Large-Scale Res. Facilities JLSRF **2**(65) (2016)

28. Knake-Langhorst, S., Gimm, K., Frankiewicz, T., et al.: Test site AIM-toolbox and enabler for applied research and development in traffic mobility. Transp. Res. Procedia **14**, 2197–2206 (2016)

29. Zhang, M., Dotzauer, M., Schießl, C.: Analysis of implicit communication of motorist and cyclists in intersection using video and trajectory data. ICTCT Conference—International Co-operation on Theories and Concepts in Traffic Safety. 28.10. 29.10. Online (2021)

30. Quante, L., Junghans, M., Schießl, C.: Turning left at urban intersection: turning patterns and gap acceptance. ICTCT Conference—International Co-operation on Theories and Concepts in Traffic Safety. 28.10. 29.10. Online (2021)

Open Access This chapter is licensed under the terms of the Creative Commons Attribution-NonCommercial-NoDerivatives 4.0 International License (▶ http://creativecommons.org/licenses/by-nc-nd/4.0/), which permits any noncommercial use, sharing, distribution and reproduction in any medium or format, as long as you give appropriate credit to the original author(s) and the source, provide a link to the Creative Commons license and indicate if you modified the licensed material. You do not have permission under this license to share adapted material derived from this chapter or parts of it.

The images or other third party material in this chapter are included in the chapter's Creative Commons license, unless indicated otherwise in a credit line to the material. If material is not included in the chapter's Creative Commons license and your intended use is not permitted by statutory regulation or exceeds the permitted use, you will need to obtain permission directly from the copyright holder.

Wizard of Oz Vehicles

Alexander Frey and Meike Jipp

Contents

© The Author(s) 2026
H. Winner et al. (eds.), *Handbook Assisted and Automated Driving*,
https://doi.org/10.1007/978-3-658-45276-6_11

11.1 Introduction and Definitions

In 1900, Lyman Frank Baum published his narrative entitled "The Wonderful Wizard of Oz" [4], which after more than 120 years is not only world-famous as a literary classic, but also serves as the eponym for the research method presented in this chapter. In the story, the character "The Wizard of Oz" uses illusions and effects to feign his supposed magic skills to the protagonists in order to visibly impress them. In an analogous way, study participants can also be convinced of the high performance of technological systems [58]—although this is merely a (as valid as possible) deception and the systems are only conditionally functional. Whenever human behaviour in interaction with (modern) technology needs to be analysed—but the technology has not (yet) reached the final maturity for a convincing implementation—the "Wizard of Oz" principle (see [66]) offers a methodological solution. Kelley [39] first introduced the term "Wizard of Oz" after the parallels to the narrative became apparent to him [27]. Some of the research literature uses abbreviations such as "Oz principle" or "Woz"; in the following, "WoOz" will be used. The first applications of the then new psychological research method initially took place outside the automotive sector: According to [5] researchers had used "Wizard"-based illusions of supposed speech recognition and dialogue systems since the 1980s, for example [8]. Accordingly, Green and Wei-Haas [27] initially defined the method as "…an efficient way to examine user interaction with computers and facilitate rapid iterative development of dialogue wording and logic" ([27], p. 470). Later, these principles could also be applied to research in ergonomics of speech recognition and output in vehicles [22]. This laid the foundation for WoOz applications in the vehicle, with which researchers today have a powerful tool in their repertoire—especially for interaction-related questions from increasing vehicle automation: Specifically, so-called "WoOz vehicles" should be mentioned here, whose methodological basis will be dedicated accordingly in the following sections. WoOz vehicles allow the imitation of a ride in an automated or even autonomous vehicle. For this purpose, the "Wizard driver" takes over all or parts of the tasks that a corresponding automation function would take over if it were actually present in the vehicle. In this way, different SAE levels can be experienced by study participants without having to resort to a "real" vehicle with the corresponding technology.

Within the framework of this chapter, the WoOZ principle will first be presented in ▶ Sect. 11.2 in comparison to proven research methods. This is followed by a classification of the principle with regard to its methodological quality criteria on the basis of studies already performed (▶ Sect. 2.1). ▶ Section 2.2 deals with specific challenges in study planning and implementation. ▶ Section 11.3 deals with different WoOz vehicle concepts as well as implications from a covert control activity of the Wizard driver: In detail, Sects. 3.1.1 and 3.1.2 focus on the methodological application for SAE Level 3 by addressing both requirements for the Wizard driver and his workplace as well as prerequisites and limitations of the methodology as a whole. Based on this, ▶ Sect. 3.2 deals with the use of WoOz vehicles for SAE Level 4. Requirements for the vehicles, the Wizard drivers and the workplace are defined and the limitations and benefits of the research methodology are presented.

11.2 The Wizard of Oz Principle as a New Psychological Research Method in Automated Driving

The use of the WoOz method in a real vehicle to answer psychological questions in automated driving has become increasingly important in recent years. At the same time, the number of human–machine interaction-related research papers in this area as a whole must also be considered: These numbers have increased in the last 6 to 7 years and have seen exponential growth since then [16]. Still, the control of a real vehicle by a Wizard driver occupies a special position with regard to the research method used: The research method of choice for answering these questions was and still is experimental driving simulation—as prototypically demonstrated by the series of studies by Hohm et al. [34]. A meta-analysis of 118 studies on human–machine interaction-related issues in the field of automated driving showed that a driving simulator was used in more than 71% between the years 2010 and 2018. Meanwhile, the WoOz method played a subordinate role with only just under 3% [17]. This fact may be due in particular to the fact that many research institutions maintain a driving simulator as their primary test material. This has already been used in the past to answer various research questions in the area of manual driving [16]. With a high degree of experimental controllability, largely free of safety–critical concerns and relatively inexpensive, researchers thus find a reputable test environment. On the one hand, it is possible to investigate human interaction with functions of principle of operation A [20], e.g. the effect of different warning strategies on driver behaviour [72]. On the other hand, the driver's reaction in accident-prone situations can be estimated [54] and the controllability of the safety systems of principle of operation C [20] intervening in these emergencies can be evaluated [2]. Driving simulators show their methodological strength in particular when a high reliability (i.e. a high reproducibility of the measured values under the same conditions) is required and an adequate relative validity (corresponds to the representativeness of the measured values related to a specific criterion) can be achieved in comparison to real test scenarios [55]. Through appropriate programming adjustments and the generation of new test scenarios, the once purely "manually" operated driving simulator can be transformed into the simulation of a continuously automated or autonomous vehicle (principle of operation B; [20]) for any SAE level [59]. In the process, some hybrid concepts were implemented (partly transitionally) in which a so-called "Driving Wizard" (task: path guidance and switching between manual and automated driving) and a so-called "Interaction Wizard" (task: switching between manual and automated driving) were implemented. "Interaction Wizard" (task: replication of the vehicle-specific, automation-specific communication with the driver) mimic the automation functions in a manually operated driving simulator [31, 51]. However, the advantages of high internal validity (i.e. the ability to clearly attribute changes in the measured values—in the sense of high controllability—to the experimental manipulation) of experimental driving simulation are inevitably lost.

The higher the SAE level is, the greater the need is for successful validation of the simulation scenarios in reality, at best in public traffic, in order to achieve acceptable external validity [38]. In order for a credible presentation of the respective function to the participants, the corresponding performance levels must actually be available: After all, the use of a Level 3 functionality, for example, means that the research vehicle has the corresponding sensory performance and the technological reliability for automated driving, so that even if it is necessary to request the driver to take over manual driving, the function continues to act safely in a time window of several seconds. The requirements for the successful implementation and safeguarding of Level 3 in real road traffic are therefore high, and potent vehicles that can be operated in public traffic are still rare and expensive [31]. For Level 4, the technological hurdles are even higher. Another source of methodological problems arises from the implemented algorithms (partly trained via neural networks), whereby the functional behaviour is also subject to probabilistic influences [5]. Accordingly, it would be costly to design a controlled study in public traffic with real Level 3 or 4 vehicles, while benefiting from natural behaviours of manually driving surrounding traffic. Despite these challenges, there is an aspiration to research human behaviour and its interaction with these new technologies already in parallel with the development of functions, in order to be able to react early on to human performance limitations in dealing with the technology also in the development of the technologies (see, e.g. [47]).

These circumstances in particular motivate the development of WoOz vehicles, as they can perform the balancing act between high internal and external validity in real traffic.

Depending on the design and vehicle concept (cf. ► Sect. 11.3), WoOz vehicles can sometimes also be used on closed test tracks with less technical effort. This form of application additionally optimises the methodological quality by aspiring for the best possible combination of reproducibility and reality representation. In this way, the properties of the WoOz methodology can surpass even those of a highly immersive driving simulation. A test track provides a predefined study environment that can be varied according to the study-specific parameters. These include the density of the surrounding traffic (and thus the degree of monotony), specific superstructures (such as target or distractor stimuli), interactions with third-party vehicles, or the simulation of critical automation scenarios. The overriding credo is, of course, that a safe journey for occupants and the environment must be guaranteed at all times. The aforementioned variable and scenario adjustments can undoubtedly be made just as well or better in a driving simulator. However, study participants in a WoOz vehicle perceive the real driving dynamics of the car without latencies. This fact—the experience of congruent movement information while driving—provides an important indication of self-motion [29]. Furthermore, it is known that driver state measurement can be confounded by unrealistic lighting and temperature values [9, 61]—as sometimes prevalent in driving simulators. In general, visual cues are reduced in virtual environments [11, 40]—this concerns both disparity information (absence of motion parallax and stereopsis), which is important for depth perception, and texture information (Jamson, 2011). The aforementioned limitations basically do not exist in a WoOz vehicle. In parallel, study participants can thus imagine themselves in a "real" automated vehicle, develop a corresponding mental model and thus become aware of the immediacy of their actions (not least the associated safety relevance as a real entity).

The WoOz method offers a unique opportunity to answer research questions that deal with the external effects of automated or autonomous vehicles or even the interaction with pedestrians and cyclists in urban spaces: Without the need to merge several simulation environments (e.g. driving simulator and pedestrian simulator) [45], participants can be placed in real-life interaction and communication with a (particularly autonomous) vehicle [32, 48, 52]. Such studies aim at the ergonomic design of the "external HMI" of automated and autonomous vehicles [12]. In this context, the WoOz setting benefits, for example, from realistic perception speeds compared to simulation [18]. In the following section, the previously mentioned psychometric properties (objectivity, reliability and validity) of the WoOz methodology will be examined in more detail using the example of various studies.

11.2.1 Wizard of Oz Studies in the Context of Methodological Quality Criteria

Before using a WoOz vehicle in a study, fundamental questions arise with regard to the properties of the method. Analogous to a psychological test—although the WoOz method should not be called a test by definition (see Müller, Weinbeer, & Bengler, 2019)—high demands must be placed on the objectivity, reliability and validity of the principle. Müller et al. (2019) explained which methodological challenges arise specifically in WoOz studies: A minimum requirement is an objective study design, i.e. that different Wizard drivers reproduce the same driving style among themselves. In theory, it would therefore be advisable to use one and the same Wizard driver in one and the same study. In practice, however, the demands on the Wizard driver can sometimes be very high (cf. ▶ Sect. 3.1.2), so that several Wizard drivers take turns to enable a large sample to interact with a supposedly automated or autonomous vehicle in a short time. This inevitably involves a special need for coordination between the Wizard drivers and uniform driver training. If one and the same person not only acts as a Wizard driver, but also communicates with and instructs study participants, additional homologation is required. In principle, it should be considered that differences in performance between Wizard drivers are mainly reflected when the test scenario is complex and the driving speed is higher. In various studies, only one Wizard driver is used [3, 16, 35, 42, 71], while changes between Wizard drivers are found, for example, in [49].

If there is an acceptable level of objectivity, the next step is to aim for the best possible reliability: For this, the Wizard driver must be able to reproduce the same driving style at different points in time (Müller et al., 2019). It is important to consider all time points during a single trip as well as all time points resulting from a study execution over several days. In the case of a prolonged control activity of the Wizard, limitations related to driver state (development of fatigue) must be taken into account, which are well known from research on manual driving [15, 57, 62]. Current WoOz studies rarely use reliability criteria to assess Wizard's control quality (such as Standard Deviation of Lane Position (SDLP); [16]), although this would also facilitate comparisons between different studies. In addition to these challenges specific to WoOZ studies to maintain adequate reliability, general environmental factors that are inevitably encountered in field studies (also in real traffic) also play a role. These include, for example,

the weather conditions (especially temperature and precipitation), time of day, lighting and traffic density.

If one raises the question of the validity of a WoOz study, one is inevitably confronted with the complexity and the comprehensive claim of the WoOz principle: First, all study participants must be unquestionably convinced at all times that they are interacting with a real automated or autonomous vehicle (Müller et al., 2019). Secondly, it must be ensured that the behaviour of the vehicle actually corresponds to that of a real automated or autonomous vehicle (Müller et al., 2019). The first criterion places particular demands on the deception of the test subjects (cf. ▶ Sect. 2.2), which inevitably goes hand in hand with the vehicle set-up (cf. ▶ Sect. 11.3). In order for this validity criterion to be fulfilled in the best possible way, the study participants must be kept in the dark about the existence and function of the Wizard driver until the data collection is completed [35, 42]. The Wizard driver must not be visible, nor must the function of any visual cover be revealed to the subjects. Even the slightest speculation about human involvement must be excluded. After the study, a (subjective) query should be made as to whether the deception could be permanently maintained in a credible manner. This can be done with standardised metrics on acceptance [67] and trust in the automation function [36], or it can be clarified using an open question. "In what way do you think automated driving was technically implemented in this trial?" (translated from [16]) is just one possibility. After completion of the study, the participants must be informed about the true functionality of the technology—if necessary with a confidentiality agreement—in order to comply with valid ethical regulations (see, e.g. [14]; ▶ Sect. 2.2).

Studies that focus on the interaction of pedestrians or cyclists with an autonomous vehicle can usually implement these requirements satisfactorily (cf. [32, 48, 52]). However, once study participants take a seat inside the WoOz vehicle, the challenge for deception increases. For this reason, the WoOz principle is revealed in some studies: The test persons then know that a human driver is imitating the automated driving function, but the WoOz driver is either in a concealed position or even perceptible to the participants. A differentiated consideration of the topic of deception can be found in the following ▶ Sect. 2.2.

The second validity criterion to be fulfilled is a realistic implementation of the quasi-automated driving quality by the WoOz driver. This raises the fundamental question: How is "true to reality" defined? First of all, it must be considered that the future control quality of automation functions is per se unclear. Therefore, there is both, the risk of "overfitting" and the possibility of "underfitting" by assuming too many inaccuracies in the driving quality [5]. Here it is helpful to follow the classification of Scheiter (2021), which defines the specificity of Wizard's instruction: Variant 1 involves the Wizard driver imagining the control of the automation to be emulated, and implementing it to the best of his or her understanding. If no concrete information on this is disclosed in current studies, this handling can be assumed [6, 71]. In Variant 2, metaphorical instruction is used: For example, the Wizard driver is asked to imagine that he or she is working as a chauffeur for a famous person and does not exceed the target speed on motorways. Variant 3 gives concrete qualitative and quantitative recommendations. For example, the Wizard driver is instructed to always keep a sufficient safe distance from vehicles in front, to drive in the middle of the lane if possible and to use the direction indicator when changing lanes [35, 42]. It is important to note that all three variants require compliance with all traffic regulations and are aimed at comfort-oriented and defensive action, but this is communicated to the Wizard driver in different ways.

11.2.2 Method-Specific Challenges in Study Design and Implementation

In the context of any WoOz study, the initial question is whether study participants are made aware of the presence and activities of a Wizard driver. If so, there are only minor additional requirements associated with the disclosure of the methodology. In particular, it is recommended to completely conceal the control activities of the WoOz driver, in the best case the Wizard completely, so that the involvement of the study participants—and thus the external validity—can be increased. If the participants are kept unaware of the Wizard and its function, this requires more effort in study planning and implementation. The necessary condition for a valid deception of the study participants is an appropriate concealment of the Wizard and/or its control activity, which is mainly determined by the vehicle design (cf. ▶ Sect. 11.3). Therefore, the methodological limit regarding the maximum achievable involvement of the participants is basically already set at the WoOz vehicle design stage. If, on the other hand, a WoOz vehicle is only designed for a specific driving test or a single series of studies—which will probably rarely be the case due to the monetary costs—the degree of involvement of the test persons must already be defined here.

The basic psychological tool that must be used for deception schemes in WoOz studies is the deliberate misinformation within the instruction (the "cover story") [30]. Participants will eventually ask themselves, "Why is the study leader sitting next to me?" or "Why is there someone else in the back of the vehicle?" or "Why is there visual occlusion between me and the study leader?" These questions relate to covert control activity. Since WoOz vehicles often have experimental attachments and superstructures and thus resemble a research vehicle more than a production-ready automated vehicle, this issue must also be addressed with the cover story. The researcher should be aware that the content and quality of the cover story have a decisive influence on the involvement of the study participants and must fit the vehicle design. For example, if the participants use an obvious dummy steering wheel, this reality-based disadvantage must be credibly explained, or—if this is impossible—a partial or full explanation must be provided [71]. Alternatively, the presence of the Wizard driver can be justified by referring to it as an "engineer" (Sportillo et al., 2019) or as a "supervisor" of the automated driving function (Rittger et al., 2017). While these designations may serve as an explanation, they may also result in a lack of trust or misunderstanding regarding the driver role for the study participants. Therefore, if possible, the study driver should not be logically associated with the automated driving function, but should, for example, be responsible for recording the measurement data according to the cover story [16].

In any case, ethical aspects of good scientific practice must be maintained when using the WoOz method to deceive study participants [26]. After all, the participants are kept in the belief—possibly for several hours—that they are interacting with a highly technical system. Some participants may be looking forward to this in the run-up to the study or proudly telling their relatives and friends about it afterwards. The lie that has been planted methodically without any alternative must be uprooted again at the end of the study: All test persons must be informed accordingly. The recommendations of the German Psychological Society (DGPs) [14] or those of the American Psychological Association (APA) [1, 60] provide guidance here. As a consequence, this means that enlightened individuals cannot participate in a WoOz study again. The (sometimes regionally) interested group of participants, from which a renewed acquisition could take place, is thus reduced with each new study.

11.3 Use of Wizard of Oz Vehicles in Research on Human–machine Interaction in Continuous Vehicle Automation

The presumably first WoOz vehicle was presented by Petermann and Schlag [53] in the form of a converted VW Touran. The test persons take a seat at the conventional driver's workplace at the front left; the Wizard driver sits in the darkened rear, centrally and separated from the study participant by a heavily tinted glass pane. The Wizard driver can also see the traffic through the windscreen. In a validation using a real SAE Level 1 function, equivalent subjective judgements of participants could be measured in the WoOz vehicle [41]. A largely analogous vehicle design was implemented by the Federal Highway Research Institute (BASt) in a VW Caddy Maxi [50] and by BMW in a BMW X5 [24]. These WoOz vehicle concepts claim to achieve the maximum possible immersion quality by completely concealing the presence of the Wizard driver. In the BASt-WoOz vehicle, even the additional study guide is shielded from the study participant, whereas in the BMW this is located on the passenger seat. The WoOz vehicle presented by the Bosch company [49] does not separate the Wizard driver from the participants— here, the Wizard driver is located on the passenger seat. However, the Wizard driver can control the vehicle completely imperceptibly by means of a concealed joystick in the passenger door. Other study guides take a seat on the rear bench. A third WoOz variant was implemented by Audi in an Audi Q7 [71]: Here, all study participants are aware that they are interacting with a Wizard driver. In a right-hand drive car, the test persons sit on the left as usual and have a dummy steering wheel and additional displays in front of them— the Wizard driver steers the vehicle from the right-hand driver's seat, with a black curtain drawn between the two persons.

Although the immersion quality of the WoOz methodology is proportional to the degree of covering of the Wizard driver, and a high immersion quality is associated with high validity, a lower immersion quality can also be completely sufficient to answer a specific research question, because at least the construct validity is largely maintained [38]. For example, the operationalisation of Weinbeer et al. [71] allows to capture fatigue in an automated driving context (with certain validity restrictions, e.g. the influence of functional trust [16]), even if the study participants knew that the automation is simulated by a human.

All previously identified WoOz vehicle concepts were developed within the framework of the BMWI-funded project "Ko-HAF" (Cooperative, Highly Automated Driving). The aim was to have an alternative to driving simulation for SAE Level 3-typical questions. Accordingly, these vehicles can be used to simulate Level 3 automation scenarios, and some of them could also be used for Level 4 driving. Bengler et al. [5] already sorted possible vehicle setups. Regardless of the specific WoOz implementation in the vehicle, this involved structural changes to primary, safety-relevant functionalities in the vehicle, which usually ensure fail-safe (manual) vehicle control. The risk of failure of a control component in public road traffic must be kept as low as possible for safe operation. In the next

▶ Sect. 3.1, WoOz deployment at Level 3 is considered in more detail first, followed by Level 4.

11.3.1 Application for SAE Level 3

11.3.1.1 SAE Level 3 Implementation Concepts: Requirements for the Wizard Driver and the Driver's Workplace

An automated driving function according to SAE Level 3 is characterised, in addition to phases of continuous automated driving in a specific domain (e.g. on motorways), by the fact that before the function limit is reached, the vehicle control is handed back to the driver with a lead time of several seconds. Consequently, the Wizard driver in a Level 3 WoOz vehicle should be able to imitate these functional characteristics as best as possible. For this, it must be possible in a safely manner to a) take control of the vehicle after release by the study participant, b) complete a continuous automated ride and c) hand over control of the vehicle to the study participant. This results in specific requirements for the performance of the Wizard driver, which can only be met by fulfilling special requirements for his driver's workplace and can be supported with corresponding ergonomic design.

Basically, for a level 3 ride, a WoOz vehicle must be technically capable of switching between the control from the test person's seat and the Wizard's seat during the ride. The current driving or automation status must be clearly displayed to both drivers. The takeover time that elapses between manual and automated operation must also be visible to study participants. Depending on the technical implementation of the WoOz principle, different procedures of the Wizard driver are required (e.g. opening or closing a magnetic coupling for the steering device), which also take time to complete, so this should be taken into account when displaying the automation status to study participants. In any case, it must be unambiguously clear to both drivers at all times who has the responsibility for driving. A display concept that fulfils the aforementioned requirements can be found, for example, in the BASt-WoOz [42]. Appropriate displays are therefore essential for safe operation, regardless of where the Wizard driving position is located.

The arrangement of the Wizard seat [5], on the other hand, significantly determines the challenges and special features the Wizard driver is confronted with compared to conventional manual driving from the driver's seat at the front left. A Wizard driver will also have gained many years of driving experience in this position after acquiring his or her driving licence. However, the WoOz principle means that the Wizard seat is offset from this—possibly to the front right, or even in the rear from any position. In rear-driven WoOz vehicles, the Wizard's central, slightly elevated seating position, concealed by a darkened pane of glass, has proven its worth. Even a conventional steering wheel (because its dimensions are sometimes too large) may not find room for the hidden working Wizard driver—special steering devices such as a joystick may have to be used. If a second pedal set cannot be implemented, longitudinal guidance may also be implemented via a joystick. Thus, the need for a driving training of the Wizard driver is indicated, in order to familiarise himself on the one hand with the new seat position and the associated new alignment of the vehicle in the lane relative to the seat position. On the other hand, driving training should be used to learn the steering procedures, longitudinal and lateral guidance, the associated accelerations and how to deal with any visual covers that may be present. Driver assistance functions (according to SAE Level 1) can relieve the Wizard driver with regard to longitudinal and lateral guidance, or simplify lane centring. Adaptive Cruise Control (ACC) or a lane departure warning system are particularly worthy to mention here. In addition to primary vehicle control functions, secondary functionalities, such as the operation of the direction indicator, must also be possible from the Wizard position.

Visibility is particularly impaired if the Wizard driver is operating from a very concealed position, e.g. from the rear of the vehicle or from behind a curtain on the passenger side. The Wizard driver must then direct his or her view through a darkened glass pane, the enclosure of which may be accompanied by a widening of the B-pillars, which may further restrict the field of vision. It may be necessary to limit the Wizard driver's control to suitable lighting conditions. In addition, it must be possible to see the traffic behind with the same quality from the Wizard seat as is possible from the driver's seat in front on the left via the exterior mirrors. For this purpose, a so-called "camera monitor system" (KMS) can be used, whose technical specifications are based on UN ECE Regulation No. 46. Comparative studies between systems for indirect vision (KMS and conventional exterior mirrors) indicate an equivalent transmission of driver information if certain framework conditions are considered for KMS (e.g. monitor positioning) [46, 63]. In addition, the Wizard driver needs redundant displays about the vehicle status, especially the driving speed, as these can sometimes only be inadequately read from the Wizard position via the primary displays. Even if the Wizard driver operates from the passenger seat, additional (hidden) displays may be required for the Wizard driver [49]. Basic recommendations regarding the ergonomic design of the Wizard driver's workplace should be based on conventional driver's workplaces [7], whereby no restrictions in terms of interior design are necessary, but rather a high level of functionality can be focused on.

11.3.1.2 Requirements, Benefits and Limitations of the Methodology for SAE Level 3

The human–machine interaction-related issues can be very diverse at Level 3 [21] and can be assigned to the following topics, for example: The driver's state (specifically the driver's readiness to take over) [16], the assessment of the takeover time for manual driving [23, 42, 56, 68], situational awareness after takeover request [69], the role of the non-driving related task [35, 64, 73], the trust in the automated driving function [43], the development of kinetosis [65], and the avoidance of confusion between different levels of automated driving (the so-called "mode confusion") [13]. What methodological limitations must be expected in a WoOz vehicle on test tracks or in real traffic when answering research questions in these fields?

First of all, it must be remembered that a WoOz vehicle is intended to provide the best possible replication of a real automated ride, but can never fully replace it (although it may have other validity advantages—see ▶ Sect. 11.2). The biggest restriction results from the control changes between the study participant and the Wizard driver. The activation of the supposed Level 3 function by the participant serves as a first example: In order to secure the procedure, and to exclude any danger in traffic, the Wizard driver must first check whether the traffic situation is suitable for activation. The Wizard driver must have developed sufficient situation awareness (see [10]). Only then is it acceptable to let the Wizard carry out the vehicle control. The automation function is then switched "available" by Wizard for the study participant. As soon as the participant activates the function, he or she sees the following display in front of him or her, for example: "The automation is being prepared". This very short interval is needed for the Wizard to perform further processes, such as closing the magnetic coupling (see ▶ Sect. 3.1.1). When the control is returned from the Wizard driver to the participant, he or she receives, for example, the information "Deactivation is being prepared". Regardless of whether the participant requested control or the Wizard driver initiated a takeover request (and the participant confirmed it), the Wizard driver must ensure as best as possible that the participant is in a state in which he or she is capable of safe vehicle control. Only then may control of the vehicle be returned to the study participant. If this cannot be ensured, the Wizard driver must maintain control, inform the participant via communication link that deactivation is not possible at this moment or even initiate a state of minimal risk manoeuvre. It is possible that the participant is

11

still engaged in a non-driving related task, is still holding it in his or her hand, has not yet gained a sufficient overview of the driving situation, or his or her energetic level is insufficient for a safe continuation of the ride. In particular, the participant may be fatigued. It can be challenging for Wizard drivers to recognise an insufficient driver condition. Therein lies a residual risk (related to the degradation of the driver's state), which finally has to be assigned to the automation effects in need of investigation. This reveals a fundamental problem with WoOz studies in public traffic: If it is supposed to be part of the experimental manipulation to put the study participant in a critical state for the driving task, this can no longer guarantee safe driving in public spaces. Consequently, the study would have to be conducted on a closed test track.

Since the interaction concept between the study participants and the Wizard driver in Level 3 is very specialised in smooth control changes, it must be examined in each individual case whether a close-to-series differentiation (especially concerning displays) from Level 2 is possible per se. In this respect, the investigation of mode confusion could be limited. Alternatively, it is possible to use different types of instructions if, for example, a level 2 is to be compared with a level 3 [16].

Research questions that are less influenced by the takeover request, on the other hand, can very well be evaluated in public traffic. These include trust in continuous automation, first signs of the development of kinetosis or observations on the choice of the respective non-driving related task. In this context, gaze and head movement measurements are also possible—assuming a valid system. Even physiological parameters can be measured in WoOz vehicles. This should be taken into account when planning studies. Overall, it is clear that the demands on the Wizard driver for a Level 3 imitation can be high, and include additional tasks to be performed besides the actual vehicle control. Even the best trained automation imitator is, qua method, a human being who has performance limitations.

11.3.2 Application for SAE Level 4

11.3.2.1 SAE Level 4 Implementation Concepts: Requirements for the Wizard Driver and the Driver's Workplace

Compared to SAE Level 3, drivers do not need to intervene in the driving process in SAE Level 4. If, for instance, a vehicle offers Level 4 functionality on highways, its driver has to drive the vehicle to the highway manually or in assisted mode and activate the functionality as soon as it is available. Then, the Level 4 functionality takes over the lateral and longitudinal control of the vehicle until the exit or end of the highway is reached. During this Level 4 ride, the driver can do whatever he or she would like to do—including sleeping and working. Such Level 4 functionality can be evaluated with a Level 4 WoOz vehicle. The vehicle itself would resemble the implementation concept of the Level 3 WoOz vehicle. The only difference is that there is no need for the driver to take over the driving task, and the driver needs to be informed that the vehicle is a SAE Level 4 prototype. This information could be provided by corresponding instructions.

A Level 4 prototype could also be used for public transport so that there might be more than one passenger in the car. Consequently, the vehicle concept might be similar to a minivan (see, e.g. [19]). In such minivans, the passengers have no opportunity to take over control of the driving task, which is why the implementation concept is clearly different from the one of a Level 3 vehicle. The advantage of a Level 4 concept is that it is easier to hide the Wizard driver. One nearby option is to let the Wizard driver drive the minivan from his or her traditional seat and to cut off the driver area from the passenger area with a non-transparent separating wall. Such a wall has the disadvantage that the Wizard driver has difficulties supervising the surrounding traffic at the right in the direction of travel. This is why it is for reasons of safety recommended to implement a camera monitor system. This system should provide the driver with a live video-stream of the surrounding traffic at the right. The field of vision should correspond to the UN ECE R46, similar to the implementation concept of Level 3 vehicles.

Additional cameras should be placed in the passenger area so that the driver can react to safety–critical situations in that area as well. For instance, the driver should not initiate driving while a passenger is still standing or when the doors have not yet been closed. For emergencies, an intercom system should be implemented as well. Last, it should be ensured that the driver cannot be seen from outside the vehicle, for instance, by the passengers leaving or entering the vehicle. This is why the front and the side windows should be tinted, for instance, with darkened plastic sheets mounted onto the windows. Then, the vehicle needs, however, a special authorization for usage in road traffic. Typically, such authorizations are limited, temporarily or geographically. For instance, such vehicles are often not allowed to be used in darkness or during bad weather conditions. Moreover, logbooks are needed to be able to reconstruct who drove the vehicle at what time and was, thus, responsible, for the vehicle and its passengers (e.g. in case of speeding). Additional approvals are, of course, possible.

The implementation concept of a Level 4 WoOz vehicle has consequences for the Wizard driver, for the driving task, the passengers, and the study design: The Wizard driver needs—despite a valid driver's license for the relevant vehicle concept—a permission to transport passengers. The actual driving task is similar to driving a standard car. Yet, the supervision task needs to be expanded to the monitors providing the video streams of the surrounding traffic and the passenger area. This is why Wizard drivers should train to drive such vehicles. The duration of a ride should also be limited, because humans are per se not very good supervisors. From psychological studies (e.g. [33]), it is well known that the task load increased within the first 5–15 min during the conduct of a supervisory task (e.g. [70]) and the performance decreases during the same time period. The supervisory task is even more difficult if one considers darkened windows. These windows can provoke monotony, which can worsen the expected decline in human driving performance.

The implementation concept impacts the passengers. They can no longer see what is going on in front of the vehicle. This is why it is recommended to implement a camera streaming a video of the traffic situation in front of the vehicle into the passenger area. The video should be displayed at the wall separating the Wizard driver from the passenger area. The view sideways should also be guaranteed to avoid negative effects such as motion sickness (e.g. [25]). This is why the darkened plastic sheets should be placed only in the driver's area of the vehicle.

The study designs that can be implemented in such vehicles are limited. As already mentioned, such a vehicle needs a special authorization to be driven in public traffic. These authorizations are typically limited for a certain time period and for specific routes. If, for instance, one intends to investigate the acceptance of Level 4 minivans at night, it will hardly be possible to gain a special authorization for such a purpose and for a vehicle with darkened windows. It also has to be ensured that only participants enter the vehicle who have provided informed consent for their participation in the study and who will be informed about the inception related to the WoOz vehicle after the data acquisition. This prevents using such a vehicle concept as a public-transit bus, in which anybody could step in and out. Despite these disadvantages, a Level 4 WoOz vehicle has already demonstrated its benefits especially for identifying barriers preventing humans from using such functionality (e.g. [19]) and for implementing countermeasures at an early stage in the development process of the technology (e.g. [47]).

11.3.2.2 Requirements, Benefits, and Limitations of the Methodology for SAE Level 4

Wizard of Oz vehicles are especially valuable for investigating the impact of autonomous systems as a supplement for public transportation. In such settings, humans are passengers and no longer responsible for driving. WoOz vehicles can be drawn on to investigate a large number of research questions: Do passengers feel safe if the authority of a bus driver is no longer present? How can one deal with disruptions in the operating schedules and with failures of the technology? Which driving styles do passengers prefer for autonomous shuttles? How do passengers know that this is their bus? The relevance of these questions becomes apparent when one considers the decreasing intention to use autonomous vehicles in public transportation as soon as the driver leaves the driver's seat (e.g. [19, 28, 37]). It is surprising that the subjective feeling of safety is high with respect to the functional capability if there is no human driver (see [19]). However, the subjective feeling of safety decreases as soon as other passengers are present so that Grippenkoven et al. [28] concluded that humans are rather afraid of other humans than of technology. Still, the studies of Gade [19] demonstrated that the actual intention to use autonomous buses is low also for participants who do accept such systems. Experiences of technological errors do significantly reduce the acceptance and intention to use the systems (see [19]), which can be due to a reduced level of trust (see [44]). However, information systems for passengers might be able to absorb such effects. For instance, in the WoOz study of [19], it was found that only a few pieces of information help to avoid negative consequences. In this case, less information is better. If the participants of the Gade [19] study received more information, the effects were negative again. These results highlight the need to conduct studies with WoOz vehicles as early as possible in the development process. Still, the methodology should be used only if a WoOz vehicle is available with the required equipment and with the special authorization to be used in public traffic. The effort it takes to gain such special authorization is not small; the limitations for usage might be detrimental: Studies cannot take place during bad visibility conditions, which reduces the level of external validity (i.e. generalizability of the study results to other situations, other participants, other timings). By restricting access to previously selected participants, a bias of the sample might also reduce the generalizability of the results. Nevertheless, it is essential to execute such studies. It is one of the very few methods available with which the needs and requirements of passengers can be identified and, herewith, considered during the development process (see, e.g. [47]).

11.4 Conclusion and Final Discussion

Wizard of Oz studies belong to the psychological method toolbox (see, e.g. [5]). Within the scope of such studies, the technological functioning is played so that human reactions can be evaluated, and it can be checked whether or not system specifications need to be adapted (see [27]). The advantages of the methodology are obvious: Undesirable developments are seen at an early stage so that time can be saved and costs can be reduced. Consequently, the method is also used in the automotive sector for the development for automated and autonomous driving functionality (see [22]). The scope of this chapter was on an analysis of how WoOz vehicles can be used for the evaluation of automated and autonomous driving functionality.

A particular challenge in the use of the WoOz method lies in the development of adequate vehicle concepts and test conditions that, on the one hand, allow participants to believe that the technology actually controls the vehicle and, on the other hand, adequately reflect the subsequent intended use of the technology. The vehicles that can be used as WoOz for SAE Level 3 and Level 4 can be quite different. Level 3 vehicles need to be designed so that drivers can take over the driving task and that Wizard drivers can drive. Both drivers need to be informed at any time—in an unambiguous manner—who is responsible for conducting the driving task in a safe way. Moreover, it must be kept in mind that the driving task is challenging for the Wizard driver. First, he or she is not located on the driver's seat so that the routinized understanding of the own positioning on the road is no longer valid. Second, the lines-of-sight might be obstructed, which must be balanced by cameras and displays. Third, the vehicle is no longer controlled with a standard steering wheel but, for instance, with a joystick, which also must be trained. Last, the Wizard driver should mimic the driving behaviour of the automation, and the driver, which is the participant, should not notice that the Wizard driver is not only the contact person for the study but also actually driving the vehicle. Thus, it is essential to not only carefully design the vehicles but also carefully prepare such studies.

Level 4 WoOz vehicles can be designed in a similar way to Level 3 vehicles in case they have a similar intended use pattern. Then, it is the only difference that either the instructions or the interface need to communicate its higher level of automation. Level 4 WoOz vehicles can, however, also be designed in a totally different way, for instance, when their intended use case is for supplement of the public transport system. Then, the Level 4 vehicle needs to reflect a bigger vehicle concept, which driver cabin can be hidden from both, inside and outside of the vehicle. Again, this has consequences: Cameras need to enable supervision of the surrounding that is not visible and of the passenger area. The Wizard drivers need to supervise the relevant monitors regularly. Consequently, special challenges for the WoOz drivers and for designing the studies appear and need to be dealt with carefully.

Despite the difficulties, which are associated with the development of WoOz vehicles, the training of the WoOz drivers, and the design of the according studies, previous results (e.g. [19]) demonstrate the value of the WoOz method: They allow identifying what must be considered from the perspective of future users of automated and autonomous vehicles. The studies come along with a high level of external validity, because the data were captured in real situations. In similar situations, automated and autonomous vehicles will be used, so that difficulties in their acceptance and intention to use can be identified early and considered during their development process. It will hardly be possible to achieve the same level of external validity in driving simulators. It is thus desirable to see the WoOz method in more than nearly 3% of the studies that focus on automated driving (see [17]), if it has the highest validity under the previously described boundary conditions. With its early deployment, automated driving can become a reality on our roads sooner—not only as a deception of participants but with real functionality. Then, the title of the narrative from Lyman Frank Baum [4] would become reality: "The Wonderful Wizard of Oz".

References

1. American Psychological Association: Ethical principles of psychologists and code of conduct. Am. Psychol. **57**(12), 1060–1073 (2002)
2. Auerswald, R., Frey, A., Schneider, N.: Integrating different kinds of driver distraction in controllability validations. In: UR:BAN Human Factors in Traffic, pp. 495–519. Springer Vieweg, Wiesbaden (2018)
3. Baltonado, S., Sibi, S., Martelaro, N., Gowda, N., Ju, W.: The RRADS platform: a real road autonomous driving simulator. In: Proceedings of the 7th International Conference on Automotive User Interfaces and Interactive Vehicular Applications, pp. 281–288 (2015)
4. Baum, L.F.: The Wonderful Wizard of Oz. George M. Hill, Chicago (1900)
5. Bengler, K., Omozik, K., Müller, A.I.: The Renaissance of Wizard of Oz (WoOz)—Using the WoOz methodology to prototype automated vehicles. Proceedings of the Human Factors and Ergonomics Society Europe (2019)
6. Berghöfer, F.L., Purucker, C., Naujoks, F., Wiedemann, K., Marberger, C.: Prediction of take-over time demand in conditionally automated driving-results of a real world driving study. In: Proceedings of the Human Factors and Ergonomics Society Europe Chapter 2018 Annual Conference (2019)
7. Bubb, H., Grünen, R.E., Remlinger, W.: Anthropometrische Fahrzeuggestaltung [Anthropometric Design of Vehicles]. In: Bubb, H., Bengler, K., Grünen, R.E., Vollrath, M. (eds.) Automobilergonomie [Ergonomics of the Automobile], pp. 345–470. Springer Vieweg, Wiesbaden, Germany (2015)
8. Dahlbäck, N., Jönsson, A., Ahrenberg, L.: Wizard of Oz studies—why and how. Knowl.-Based Syst. **6**(4), 258–266 (1993)
9. Dorrian, J., Rogers, N.L., Dinges, D.F.: Psychomotor vigilance performance: Neurocognitive assay sensitive to sleep loss. In: Sleep Deprivation, pp. 67–98. CRC Press (2004)

10. Endsley, M.R.: Toward a theory of situation awareness in dynamic systems. Hum. Factors **37**(1), 32–64 (1995)

11. Espie, S., Gauriat, P., Duraz, M.: Driving simulators validation: The issue of transferability of results acquired on simulator. Orlando: Driving Simulator Conference North America (2005)

12. Faas, S.M., Mathis, L.A., Baumann, M.: External HMI for self-driving vehicles: which information shall be displayed? Transport. Res. F: Traffic Psychol. Behav. **68**, 171–186 (2020)

13. Feldhütter, A., Härtwig, N., Kurpiers, C., Hernandez, J.M., Bengler, K.: Effect on mode awareness when changing from conditionally to partially automated driving. In: Congress of the International Ergonomics Association, pp. 314–324. Springer, Cham (2018)

14. Fiedler, K., Elbert, T., Erdfelder, E., Freund, A.M., Kliegl, R., Stahl, C.: Empfehlungen der DGPs-Kommission "Qualität der psychologischen Forschung" [Recommendations oft he DGPs Commission "Quality of the Psychological Research"] (2015). Access from ▶ http://www.dgps.de/uploads/media/Empfehlungen-Qualitaet-der-Forschung-DGPs.pdf [2021–04–02]

15. Forsman, P.M., Vila, B.J., Short, R.A., Mott, C.G., Van Dongen, H.P.: Efficient driver drowsiness detection at moderate levels of drowsiness. Accid. Anal. Prev. **50**, 341–350 (2013)

16. Frey, A.T.: Zum Fahrerzustand beim automatisierten Fahren: Objektive Messung von Müdigkeit und ihre Einflussfaktoren [About the Driver's State during Automated Driving: Objective Measurement of Fatigue and Its Influencing Factors] (Dissertation [PhD Thesis], Technische Universität Braunschweig [Technical University Brunswik]) (2021)

17. Frison, A.K., Forster, Y., Wintersberger, P., Geisel, V., Riener, A.: Where we come from and where we are going: a systematic review of human factors research in driving automation. Appl. Sci. **10**(24), 8914 (2020)

18. Fuest, T., Schmidt, E., Bengler, K.: Comparison of methods to evaluate the influence of an automated vehicle's driving behavior on Pedestrians: Wizard of Oz, virtual reality, and video. Information **11**(6), 291 (2020)

19. Gade, K.: RAMONA (Realisierung Automatisierter Mobilitätskonzepte im Öffentlichen Nahverkehr)—Abschlussbericht [Realisation of Automated Mobility Concepts in Public Transport—Final Report]. Berlin, Germany: BMVI (2021)

20. Gasser, T.M., Frey, A.T., Seeck, A., Auerswald, R.: Comprehensive definitions for automated driving and ADAS. In: 25th International Technical Conference on the Enhanced Safety of Vehicles (ESV) National Highway Traffic Safety Administration (2017)

21. Gasser, T., Schmidt, E.A., Bengler, K., Chiellino, U., Diederichs, F., Eckstein, L., ... Hoyer, R.: Bericht zum Forschungsbedarf. Runder Tisch Automatisiertes Fahren—AG Forschung [Report about the Research Needs. Round Table Automated Driving—Working Group Research]. Bundesanstalt für Straßenwesen, Germany (2015)

22. Geutner, P., Steffens, F., Manstetten, D.: Design of the VICO spoken dialogue system: evaluation of user expectations by Wizard-of-Oz experiments. In: LREC (2002)

23. Gold, C.G.: Modeling of take-over performance in highly automated vehicle guidance (Dissertation [PhD Thesis], Technische Universität München [Technical University Munich]) (2016)

24. Gold, C., Meyer, M.L., Fischer, F.: Übernahmeleistung in einem Wizard of Oz Versuchsträger beim hochautomatisierten Fahren [Performance in Take over Control in a Wizard of Oz Vehicle for Automated Driving]. 3. Interdisziplinärer Expertendialog Aktive Sicherheit und automatisiertes Fahren [3rd Interdisciplinary Expert Dialogue Active Safety and Automated Driving], 187–199 (2017)

25. Golding, J.F.: Motion sickness susceptibility. Auton. Neurosci. **129**(1–2), 67–76 (2006)

26. Graziano, A.M., Raulin, M.L.: Research Methods, A Process of Inquiry, 7th edn. Pearson Education, Boston (2010)

27. Green, P., Wei-Haas, L.: The rapid development of user interfaces: experience with the Wizard of Oz method. In: Proceedings of the Human Factors Society Annual Meeting (Vol. 29, No. 5, pp. 470–474). Sage CA: Los Angeles, CA: SAGE Publications (1985)

28. Grippenkoven, J., Fassina, Z., König, A., Dressler, A.: Perceived safety: A necessary precondition for successful autonomous mobility services. In: de Waard, D. (Hrsg.), Proceedings of the Human Factors and Ergonomics Society Europe Chapter 2018 Annual Conference. Berlin: HFES Europe Chapter (2018)

29. Groen, E.L., Howard, I.P., Cheung, B.S.K.: Influence of body roll on visually induced sensations of self-tilt and rotation. Perception **28**, 287–297 (1999)

30. Gross, A.E., Fleming, I.: Twenty years of deception in social psychology. Pers. Soc. Psychol. Bull. **8**(3), 402–408 (1982)

31. Habibovic, A., Andersson, J., Nilsson, M., Lundgren, V.M., Nilsson, J.: Evaluating interactions with non-existing automated vehicles: three Wizard of Oz approaches. In: 2016 IEEE Intelligent Vehicles Symposium (IV), pp. 32–37. IEEE (2016)

32. Hensch, A.C., Neumann, I., Beggiato, M., Halama, J., Krems, J.F.: How should automated vehicles communicate?–Effects of a light-based communication approach in a Wizard-of-Oz study. In: International Conference on Applied Human Factors and Ergonomics, pp. 79–91. Springer, Cham (2019)

33. Helton, W.S., Hollander, T.D., Warm, J.S., Tripp, L.D., Parsons, K.S., Matthews, G., . . . Hancock, P.A.: The abbreviated vigilance task and cerebral hemodynamics. J. Clinical Exp. Neuropsychol. **29**, 545–552 (2007). ▶ https://doi.org/10.1080/13803390600814757

34. Hohm, A., Klejnowski, L., Skibinski, S., Bengler, K., Berger, S., Vetter, J., . . . Stürmer, T.: KO-HAF—Kooperatives Hochautomatisiertes Fahren: (Projektübergreifender Schlussbericht) [KO-HAF: Cooperative Highly Automated Driving (Cross-Project Final Report)] (2018). Available at ▶ https://www.ko-haf.de/fileadmin/user_upload/pro-jekt/19S14002_Ko-HAF_partner%C3%BCbergreifender-Schlussbericht_final.pdf [2021–03–13]

35. Jarosch, O., Paradies, S., Feiner, D., Bengler, K.: Effects of non-driving related tasks in prolonged conditional automated driving—A Wizard of Oz on-road approach in real traffic environment. Transport. Res. F: Traffic Psychol. Behav. **65**, 292–305 (2019)

36. Jian, J.Y., Bisantz, A.M., Drury, C.G.: Foundations for an empirically determined scale of trust in automated systems. Int. J. Cogn. Ergon. **4**(1), 53–71 (2000)

37. Jipp, M., Lemmer, K.: Moderne Mobilitätsformen und die Bedürfnisse der Gesellschaft [Modern mobility concepts and the needs of society]. In: Haux, R., Gahl, K., Jipp, M., Kruse, R., Richter, O. (eds.) Zusammenwirken von natürlicher und künstlicher Intelligenz [Cooperation of natural and artificial intelligence], pp. 97–115. Springer, Wiesbaden, Germany (2021)

38. Jipp, M., Schnieder, L.: Fahrtests unter Realbedingungen: Sicherheitsvalidierung nach ISO 26262 [Road Tests under Real Conditions: Safety Validation according to ISO 26262]. Springer Vieweg (2020)

39. Kelley, J.F.: Natural Language and computers: Six empirical steps for writing an easy-to-use computer application (Doctoral dissertation, Johns Hopkins University) (1983)

40. Kemeny, A., Panerai, F.: Evaluating perception in driving simulation experiments. Trends Cogn. Sci. **7**, 31–37 (2003)

41. Kiss, M., Schmidt, G., Babbel, E.: Das Wizard of Oz Fahrzeug: Rapid Prototyping und Usability Testing von zukünftigen Fahrerassistenzsystemen. VW Konzernforschung (2006)

42. Klamroth, A., Zerbe, A., Marx, T.: Transitionen bei Level 3-Automation: Einfluss der Verkehrsumgebung auf die Bewältigungsleistung des Fahrers während Realfahrten [Transitions at Level 3 Automation: Impact of the Traffic Situation for Managing Driving in Real Driving Situations]. FAT Schriftenreihe [FAT Mono-

graph], (323). Berlin, Germany: Forschungsvereinigung Automo-biltechnik e.V. (FAT) (2019)

43. Körber, M., Baseler, E., Bengler, K.: Introduction matters: Manipulating trust in automation and reliance in automated driving. Appl. Ergon. **66**, 18–31 (2018)

44. Lee, J.D., See, K.A.: Trust in automation: designing for appropriate reliance. Hum. Factors **46**(1), 50–80 (2004)

45. Lehsing, C., Feldstein, I.T.: Urban interaction–Getting vulnerable road users into driving simulation. In: UR: BAN Human Factors in Traffic, pp. 347–362. Springer Vieweg, Wiesbaden (2018)

46. Leitner, R., Oehme, A., de Silva, J., Blum, S., Berberich, J., Böhm, S.: Kamera-Monitor-Systeme als Fahrerinformationsquelle [Camera-monitor-systems as Driver Information Systems]. Berichte der Bundesanstalt für Straßenwesen, Unterreihe Fahrzeugtechnik, (F136) [Reports of the Federal Highway Research Institute]. Bergisch Gladbach: Bundesanstalt für Straßenwesen (BASt) (2021)

47. Lemmer, K., Jipp, M., Bubb, H., Vögel, H.-J., Jung, M., Laukart, G., Vorberg, T.: Mensch-Technik-Kooperation und Fahrzeuginnenraum [Human-Technology-Coopreation and Vehicle Interior]. In: Pischinger, S., Seiffert, U. (eds.) Vieweg Handbuch Kraftfahrzeugtechnik [Vieweg Handbook Automobile Technology], pp. 1146–1158. Springer, Wiesbaden (2021)

48. Liu, H., Hirayama, T., Morales, L.Y., Murase, H.: What is the gaze behavior of pedestrians in interactions with an automated vehicle when they do not understand its intentions? (2020). arXiv preprint ▶ arXiv:2001.01340

49. Marberger, C., Manstetten, D., Klöppel, C.: Highly automated driving in the real world-A Wizard-of-Oz study on user experience and behavior. VDI-Berichte [VDI Reports], (2360) (2019)

50. Marx, T., Frey, A.T.: BASt Wizard-of-Oz Vehicle (WoOz). Abschlusspräsentation des BMWi-Projekts Ko-HAF [Final presentation of the BMWi-funded project Ko-HAF], Rodgau-Dudenhofen, Germany (2018), 2018-09-19-2018–09–20

51. Mok, B.K.J., Sirkin, D., Sibi, S., Miller, D.B., Ju, W.: Understanding driver-automated vehicle interactions through Wizard of Oz design improvisation (2015)

52. Palmeiro, A.R., van der Kint, S., Vissers, L., Farah, H., de Winter, J.C., Hagenzieker, M.: Interaction between pedestrians and automated vehicles: a Wizard of Oz experiment. Transport. Res. F: Traffic Psychol. Behav. **58**, 1005–1020 (2018)

53. Petermann, I., Schlag, B.: Auswirkungen der Synthese von Assistenz und Automation auf das Fahrer-Fahrzeug-System [Impacts of the synthesis of assistance and automation for the driver-vehicle-system]. Proceedings of the AAET, 257–266 (2010)

54. Powelleit, M., Muhrer, E., Vollrath, M., Henze, R., Liesner, L., Pawellek, T.: Verhaltensbezogene Kennwerte zeitkritischer Fahrmanöver [Behaviour-based values for time-critical driving maneuvers]. Berichte der Bundesanstalt für Straßenwesen, Unterreihe Fahrzeugtechnik, (F100) [Reports of the Federal Highway Research Institute]. Bergisch Gladbach, Germany: Bundesanstalt für Straßenwesen (BASt) (2015)

55. Purucker, C., Schneider, N., Rüger, F., Frey, A.: Validity of research environments–comparing criticality perceptions across research environments. In: UR: BAN Human Factors in Traffic, pp. 423–446. Springer Vieweg, Wiesbaden (2018)

56. Radlmayr, J., Gold, C., Lorenz, L., Farid, M., Bengler, K.: How traffic situations and non-driving related tasks affect the takeover quality in highly automated driving. In: Proceedings of the Human Factors and Ergonomics Society Annual Meeting (Vol. 58, No. 1, pp. 2063–2067). Sage CA: Los Angeles, CA: Sage Publications (2014)

57. Reyner, L.A., Horne, J.A.: Falling asleep whilst driving: are drivers aware of prior sleepiness? Int. J. Legal Med. **111**(3), 120–123 (1998)

58. Riek, L.D.: Wizard of oz studies in HRI: a systematic review and new reporting guidelines. J. Human-Robot Interact. **1**(1), 119–136 (2012)

59. SAE: Taxonomy and Definitions for Terms Related to Driving Automation Systems for On-Road Motor Vehicles (Standard No. J3016). SAE International, April 2021 (2021)

60. Sales, B.D., Folkman, S.E.: Ethics in Research with Human Participants. American Psychological Association (2000)

61. Schmidt, E.: Effect of thermal stimuli on passive fatigue while driving. In: Effects of Thermal Stimulation During Passive Driver Fatigue, pp. 67–120. Springer Vieweg, Wiesbaden (2020)

62. Schmidt, E.A.: Die objektive Erfassung von Müdigkeit während monotoner Tagfahrten und deren verbale Selbsteinschätzung durch den Fahrer [The objective assessment of fatigue during monotonous day drives and their verbal self-assessments by the driver] (Dissertation [PhD thesis], Heinrich-Heine-Universität Düsseldorf, Germany) (2010)

63. Schmidt, E.A., Hoffmann, H., Krautscheid, R., Bierbach, M., Frey, A., Gail, J., Lotz-Keens, C.: Ersatz von Außenspiegeln durch Kamera-Monitor-Systeme bei Pkw und Lkw [Camera-monitor systems as a replacement for exterior mirrors in cars and trucks]. Berichte der Bundesanstalt für Straßenwesen, Unterreihe Fahrzeugtechnik, (F112) [Reports of the Federal Highway Research Institute]. Bergisch Gladbach, Germany: Bundesanstalt für Straßenwesen (BASt) (2016)

64. Shi, E., Frey, A.T.: Non-driving-related tasks during level 3 automated driving phases—measuring what users will be likely to do. Technology, Mind, and Behavior (2021)

65. Sivak, M., Schoettle, B.: Motion sickness in self-driving vehicles. University of Michigan, Ann Arbor, Transportation Research Institute (2015)

66. Steinfeld, A., Jenkins, O.C., Scassellati, B.: The oz of wizard: simulating the human for interaction research. In: Proceedings of the 4th ACM/IEEE International Conference on Human Robot Interaction, pp. 101–108 (2009)

67. Van Der Laan, J.D., Heino, A., De Waard, D.: A simple procedure for the assessment of acceptance of advanced transport telematics. Transp. Res. Part C Emerg. Technol. **5**(1), 1–10 (1997)

68. Vogelpohl, T., Vollrath, M., Kühn, M., Hummel, T., Gehlert, T.: Übergabe von hoch- automatisiertem Fahren zu manueller Steuerung. Teil 1: Review der Literatur und Stu-die zu Übernahmezeiten. Forschungsbericht/Unfallforschung der Versicherer (GDV), (39). Berlin: GDV (2016)

69. Walch, M., Mühl, K., Kraus, J., Stoll, T., Baumann, M., Weber, M.: From car-driver- handovers to cooperative interfaces: Visions for driver—vehicle interaction in auto-mated driving. In: Automotive User Interfaces, pp. 273–294. Springer, Cham (2017)

70. Warm, J.S., Finomore, V.S., Vidulich, M.A., Funke, M.E.: Vigilance: A perceptual challenge. In: Hoffman, R.R., Hancock, P.A., Scerbo, M.W., Parasuraman, R., Szalma, J.L. (eds.) The Cambridge handbook of applied perception research, pp. 241–283. Cambridge University Press, New York, NY (2015)

71. Weinbeer, V., Muhr, T., Bengler, K., Baur, C., Radlmayr, J., Bill, J.: Highly automated driving: How to get the driver drowsy and how does drowsiness influence various take-over-aspects?. In 8. Tagung Fahrerassistenz [Conference Driver Assistance] (2017)

72. Winkler, S., Powelleit, M., Kazazi, J., Vollrath, M., Krautter, W., Korthauer, A., ... Bendewald, L.: HMI strategy–Warnings and interventions. In: UR: BAN Human Factors in Traffic, pp. 75–103. Springer Vieweg, Wiesbaden (2018)

73. Wörle, J., Metz, B.: Driving with an L3—motorway chauffeur: how do drivers use their driving time? In: de Waard, D., Toffetti, A., Pietrantoni, L., Franke, T., Petiot, J.-F., Dumas, C., Botzer, A., Onnasch, L., Milleville, I., Mars, F. (Eds.), Proceedings of the Human Factors and Ergonomics Society Europe Chapter 2019 Annual Conference (2020)

Open Access This chapter is licensed under the terms of the Creative Commons Attribution-NonCommercial-NoDerivatives 4.0 International License (▶ http://creativecommons.org/licenses/by-nc-nd/4.0/), which permits any noncommercial use, sharing, distribution and reproduction in any medium or format, as long as you give appropriate credit to the original author(s) and the source, provide a link to the Creative Commons license and indicate if you modified the licensed material. You do not have permission under this license to share adapted material derived from this chapter or parts of it.

The images or other third party material in this chapter are included in the chapter's Creative Commons license, unless indicated otherwise in a credit line to the material. If material is not included in the chapter's Creative Commons license and your intended use is not permitted by statutory regulation or exceeds the permitted use, you will need to obtain permission directly from the copyright holder.

EVITA—The Procedure for Realistic Presentation of Rear-End Collision Critical Situations in Road Tests

Norbert Fecher, Jens Hoffmann, and Hermann Winner

Contents

System Description, Evaluation Methodology for Anti-Collision Systems and Application Examples in Real Driving Tests.

© The Author(s) 2026
H. Winner et al. (eds.), *Handbook Assisted and Automated Driving*,
https://doi.org/10.1007/978-3-658-45276-6_12

12.1 The Dummy Target EVITA

12.1.1 Motivation, Goals, and Requirements

The original reason for developing the EVITA (Experimental Vehicle for Unexpected Target Approach) tool presented here was the search for a suitable test procedure to investigate the effectiveness of anti-collision systems in longitudinal traffic. Since the overall system effectiveness depends on both the technical implementation of the system and the interaction of the driver with the system, a valid evaluation can only take place in real driving tests with test persons and activated system function.

The aim is to develop a dummy target that can be used to represent a rear-end collision-critical situation in a real driving test, that "looks real" both for the test person and for the sensor system in the vehicle under investigation, and that nevertheless does not pose any higher risks than are accepted in other common driving tests.

This objective results in numerous requirements for the system to be developed. Starting from a stationary subsequent journey, the dummy target must represent the motion variables of a vehicle in front in the event of an unexpected braking maneuver. In case of an insufficient deceleration of the following test vehicle for collision avoidance, the dummy target must automatically prevent a collision. If this is not possible or if there is a malfunction on the dummy target itself, there must still be no risk of injury to any of the persons involved. To achieve a valid test person reaction, the entire driving episode and in particular the subsequent drive and braking maneuver must be presented as realistically as possible, as it occurs in real road traffic. Furthermore, the dummy target must be identifiable as a vehicle for the usual vehicle sensors (e.g. radar, camera, lidar).

12.1.2 Concept

The realized concept consists of the combination of a towing vehicle with a rope-guided two-track dummy target (called EVITA), see ◉. ◘ Figure 12.1. The rear view is formed by the original vehicle rear end of an Opel Adam, which, in combination with the two tires in original track width, gives a deceptively real impression for the following vehicle to be examined and its occupants.

During a stationary follow-up run, EVITA brakes unexpectedly for the test person driving in the test vehicle. Regardless of whether the test person reacts to the maneuver in time or not, EVITA is actively pulled out of the collision area.

The high correspondence of the rear view with a conventional passenger car, combined with the ability to avoid rear-end collisions independently, offers the possibility of using EVITA in principle in all driving tests in which rear-end collision-critical scenarios have to be presented.

12.1.3 Structure

◉ ◘ Figure 12.2 shows an overview of the components of the dummy target.

In the rear of the towing vehicle, there is a cable winch with a friction-locked winch brake and an electric motor. The dummy target has its own power supply (1) and is connected to the towing vehicle only via the winch cable. The other end of the cable is attached to the Ackermann steering of the dummy target's front axle. The disc brakes of the dummy target are hydraulically operated by an electric motor via a hand brake lever (2). At the rear of the trailer is the rear of an Opel Adam. A radar sensor (3) is attached to this rear end. In the towing vehicle and in the dummy target there are computers (4), which are interconnected by radio modems (5). The basic frame for the dummy target is an aluminum truss frame from stage engineering with four individual wheel suspensions of a quad bike. The large caster of the front axle ensures smooth straight-ahead running. A moisture-proof housing houses the fanless computer together with the radio modem, power supply, and brake control. The rear-view brake lights are functional. The total mass of EVITA is approx. 200 kg.

12.1.4 Test Procedure

In the initial state, the trailer is short-coupled behind the towing vehicle. If a vehicle (target object) is detected at a predefined test distance interval by the rear-measuring radar mounted on the rear of the trailer, the entire system can be activated for a test run. A command from the operator in the towing vehicle opens the brake of the winch and applies the brakes of the trailer. During this process, the towing vehicle continues to drive at a constant speed under cruise control.

The braking of the dummy target causes the rope of the winch to unwind. While the trailer decelerates, the processing unit of the distance sensor permanently calculates the Time-To-Collision (TTC). The TTC is a quantity formed from distance and relative speed:

$$TTC = \frac{d}{v_{rel}}; [TTC] = \text{s} \tag{12.1}$$

◘ Fig. 12.1   EVITA (Experimental Vehicle for Unexpected Target Approach)

Here, d indicates the distance in m to the object in front and v_{rel} the relative speed in m/s. If the TTC falls below a specified value, the winch brake in the towing vehicle closes and the trailer accelerates to the towing vehicle traveling at constant initial speed. The acceleration of the trailer takes approx. 1 s at maximum differential speed. After completion of the test, the entire combination breaks to a standstill.

12.1.5 Hazards for Test Participants

A system FMEA (Failure Mode and Effects Analysis) was performed to determine potential system malfunctions and measures for safe operation were derived from this. Automated safety check routines run during each test execution. If a failure is detected, the system is transferred to a safe and stable state. The safety level can be further increased by automatically triggering emergency braking in the following test vehicle when a

TTC of 0.7 s is reached. The minimum achievable TTC by a collision-avoiding action of EVITA set for conducting the tests is 0.8 s (see ◉ ◘ Table 12.1). If a TTC smaller than 0.8 s is achieved, a malfunction of EVITA must be assumed. If a collision is unavoidable despite all precautions, no harm to the subjects is expected due to the low mass of the dummy target.

12.1.6 Performance Data

The performance data of EVITA are listed in ◉ ◘ Table 12.1.

12.2 Evaluation Methodology for Collision Avoidance Systems

With EVITA, the tool for generating rear-end collision critical situations in driving tests with test persons is available. In the following, one of the main evaluation

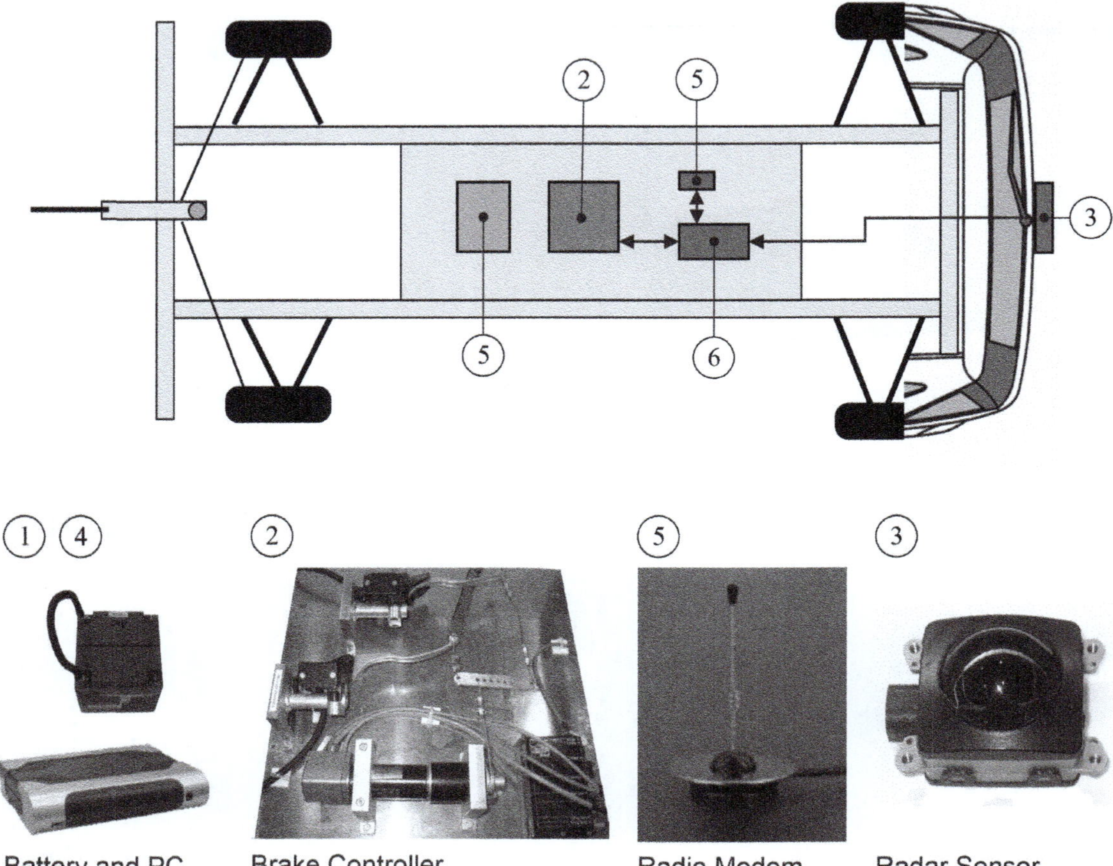

Fig. 12.2 Components of the dummy target

Table 12.1 EVITA performance data

Maximum differential velocity between vehicle approaching from behind and EVITA	**50 km/h**
Maximum braking deceleration of EVITA	9 m/s^2
Shortest TTC before end of test	0,8 s
Standard test velocity (initial velocity)	50–130 km/h

variables for assessing the quality of anti-collision systems is described.

12.2.1 Measurement Concept in the Test Vehicle

With the selected methodology, the measurement for the quality of the forward collision prevention is carried out independently of the EVITA tool. The measurement concept for determining the defined evaluation criteria is fully implemented in the test vehicle, which is equipped with an anti-collision system. The environment sensor system of the test vehicle classifies the dummy target EVITA driving ahead as a relevant target object. Object variables such as distance, relative velocity, and relative acceleration are measured to calculate the TTC. An operator interface is used by a test attendant to make settings for controlling the frontal collision countermeasures.

The vehicle has a measurement system for the combined and synchronized acquisition of CAN and camera data. Three cameras are used. The first camera is directed at the apron of the vehicle. In conjunction with the radar data, it enables reliable interpretation of the situation. The second camera is directed at the driver's face from the instrument cluster. This makes it possible, among other things, to assign the driver's line of vision. The third camera is focused on the vehicle's pedals. This makes it possible to analyze the driver's foot movements and determine action times, such as the change-over time from the accelerator pedal to the brake pedal. The repetition rate for each of the three individual images is 20 ms. The same measurement system records the CAN data so that a temporal assignment of images and signals is given. The CAN data available includes the usual vehicle data such as speed, lateral and longitudinal acceleration, data from the object in front, and data from the driver's operation such as steering wheel angle, brake pedal actuation, and others.

12.2.2 Vehicle-Independent Rapid Measuring Device

To evaluate systems without access to the CAN bus in the test vehicle, a quick-measurement device was developed and implemented that can be used in any vehicle without fixed installations. It consists of a compact measuring unit (trunk mounting), a radar setup (suction foot mounting vehicle roof), three cameras (fields of view: test person, footwell, instrument cluster with road) and a Correvit (trailer hitch). In addition, the radio-controlled remote triggering of EVITA is integrated. The power supply is provided by the on-board network of the test vehicle.

12.2.3 Effectiveness of an Anti-collision System

The reduction of the speed of the ego vehicle before the impact is used as an objective evaluation parameter for the effectiveness of an anti-collision system (especially of frontal collision countermeasures). This criterion maps the general goal of anti-collision systems to either reduce the impact speed, or to achieve complete avoidance of the impact. The higher the reduction in speed, the more effective the anti-collision system. In addition to the objective effectiveness, the subjective effectiveness assessed by the subjects is defined. This variable, determined by questionnaire, is defined as a comparison between different degrees of frontal collision countermeasures by forming a ranking.

12.2.4 Subject Trial

One finding from in-depth studies is that many vehicle drivers are distracted prior to a rear-end collision [1]. Therefore, the test subjects of the rear-end collision vehicle are induced by EVITA with a secondary task to avert their gaze for more than 2 s shortly before deceleration. The operator sitting in the test vehicle triggers the generation of the critical collision situation while the subject is averting his gaze. The subject is subsequently alerted when a predefined TTC threshold is reached, for example, by the warning elements of the anti-collision system. ◉ ◘ Figure 12.3 shows the idealized speed profile of the test vehicle over time. The distraction of the test person and the braking of the dummy target can be recognized. When the critical threshold is reached, for example, an alert to the driver or other intervention is triggered. Typically, this is followed by the subject turning his gaze to the situation in front of the ego vehicle and the start of braking.

For reasons of reproducibility, the test person is given the permissible distance to the EVITA driving ahead via a traffic light-like display at the rear of EVITA. If the distance is too great, a blue signal is displayed to the driver, and a red signal if the distance is too small. If the distance is within the range of.

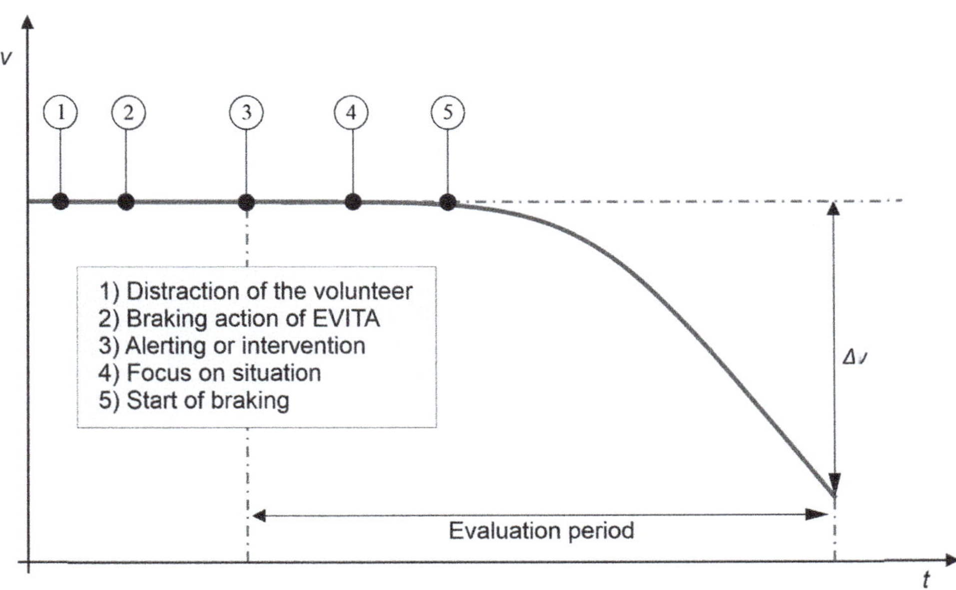

◘ **Fig. 12.3** Idealized test sequence as velocity curve of the test vehicle over time

12.2.5 Evaluation Criteria for Warning Front Collision Countermeasures

An evaluation period is set for the assessment of effectiveness. The period begins at the time of triggering a warning or vehicle intervention. It ends at the time of an imaginary, unbraked impact of the test vehicle on the dummy target driving ahead and braking continuously. This impact is "imagined" because a collision is automatically avoided by EVITA. The end time is determined as a function of the TTC algorithm and the trigger threshold in an unbraked calibration test without test subjects. For a typical warning with the TTC algorithm, the judgment period is 2 s. The warning threshold was defined as knowing warning times of known front collision countermeasures. Thus, warning elements can be compared with each other as well as with autonomous braking interventions.

The main evaluation parameter for the quality of frontal collision countermeasures is the reduction of the imaginary impact speed, this is called effectiveness. For this purpose, an evaluation period is defined from the time of the warning by the anti-collision system until the imaginary impact, at the end of which the reduced differential speed is determined. If it is a warning anti-collision system, the driver's reaction time and the amount of deceleration initiated are the most important components for a high decelerated differential speed. In the assessment period, the total reaction time is divided into different process steps.

In the literature, there are numerous data on the determination of driver behavior in hazardous situations, for which Bäumler [2] and Krause et al. [3] give an overview. The definition used in this context is based on the definition by Burckhardt, which is generally valid for the experimental conditions [4] or Zomotor [5] respectively. ◉ ◘ Figure 12.4 shows the temporal relationship of the reaction times, the assessment period, the typical speed course, and the effectiveness.

60 bar corresponds to a deceleration of 10 m/s^2 in the selected test vehicle and thus to the maximum deceleration at a high coefficient of friction of 1.0 between road surface and tire. During the assessment period, the criteria of the ◉ ◘ Table 12.2 are being assessed.

12.2.6 Comparison of Collision Avoidance Systems

The uniform evaluation procedure is the basis for comparing different versions of frontal collision countermeasures. For the evaluation, test drives are carried out with an appropriately divided collective of test subjects, taking into account different characteristics. The comparison of the distributions of the speed reductions in the assessment period across all test subjects reflects the effectiveness of the variants.

An assessment of the absolute effectiveness of an anti-collision system can be achieved by using a so-called baseline. Here, a part of the test subject collective is confronted with the critical situation without an intervention of the anti-collision system and, for example, the speed difference is determined.

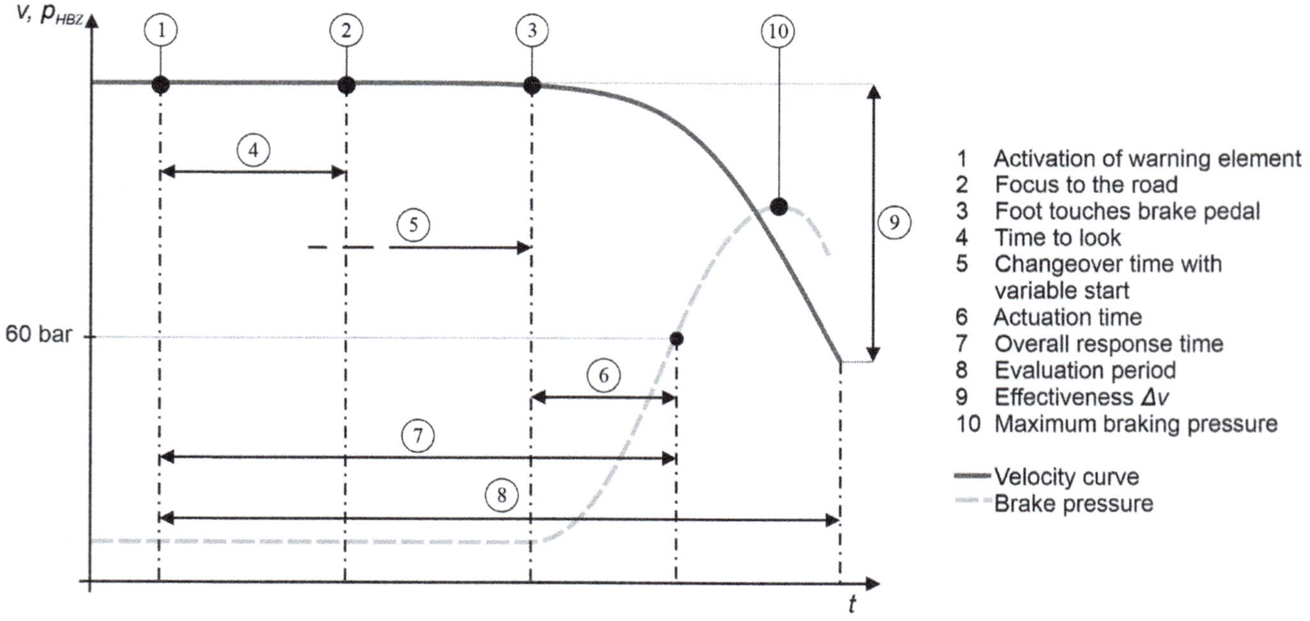

◘ **Fig. 12.4** Definitions of the temporal evaluation criteria

▫ Table 12.2   Evaluation criteria in the assessment period

Objective effectiveness	Speed change of the ego vehicle
Time to look	Time duration from the moment of the warning to the view of the road
Changeover time	Time duration from the first movement of the foot from the accelerator pedal to the first contact with the brake pedal
Actuation time	Time period from the first contact of the foot with the brake pedal until brake pedal pressure reaches 60 bar
Disturbance	Speed change of the ego vehicle from the beginning of a false warning without risk of collision
Subjective effectiveness	Subject-assessed measure of the level of a collision-avoiding effect of a warning element
Subjective forgiveness	Subject-assessed measure of the excusability of a warning element in the case of a false warning/unwarranted warning

For the evaluation of the effectiveness of the anti-collision system, only the first trial of the subject is an unbiased basis. In all further trials, the test person has understood the subject of a surprising emergency situation despite incomplete prior information about the actual purpose of the trials; he is considered to be biased. The evaluation of the acceptance by the driver is of great importance in the development of driver assistance systems [6]. The further tests after the first emergency situation are suitable for generating further findings, such as the handling of false warnings or the comparative subject assessments of variants of anti-collision systems. The assessment of test subjects on the experienced situation and on the evaluation of driver warning elements is collected with questionnaires. The evaluation of these questionnaires will provide information on the design of driver warning elements.

For the methodology according to Hoffmann presented here [7], extensive results are available, which are described in the 3rd edition of the "Handbuch Fahrerassistenzsysteme" (Winner et al. 2015, pp. 197–206).[8]

12.2.7   Usage in Further Studies

In numerous collaborations with OEMs and suppliers, EVITA was used in approximately 1,000 subject trials between 2010 and 2020, for example, to investigate the following questions:

- Investigation of different system characteristics (intervention times, intensity) of anti-collision systems.
- Development-accompanying investigation of different LED strip designs (luminous intensity, color, width, flash, etc.) as driver warning elements.

12.3   Presentation of Rear-End Collision Critical Situations in the Driving Test

In order to be able to achieve the original research objective of "evaluating collision avoidance systems in longitudinal traffic," EVITA was developed as a tool that has established itself to date in many research topics that go far beyond this issue. The use of EVITA is indicated wherever valid test results can only be obtained in real driving tests, e.g. due to complex interactions in the driver-vehicle environment system, and where at the same time rear-end collision-critical situations are the basis. It is irrelevant whether the critical situation itself is the focus of the investigation or is only required for plausibility checks of the system behavior for the test subjects.

Two applications in which EVITA has been successfully used are presented here as examples.

12.3.1   Autonomous Emergency Braking Systems for Motorcycles

The available findings from passenger car research in the area of anti-collision systems are not readily transferable to motorcycles. Motorcycles do not have restraint systems such as a seat belt. Motorcyclists must be preconditioned before warning, assisting, or automatic braking interventions. Automatic braking interventions require attention, situational awareness, and body tension. Therefore, a two-step approach is taken when investigating possible system characteristics. In a first step, possible deceleration limits are determined by experts. In a second step, these are applied in a test person study with normal drivers. In the real driving

□ Fig. 12.5 Investigation of emergency braking systems for motorcycles

test with EVITA, a total of 18 test subjects are confronted with a realistic emergency braking scenario (see ◉ □ Fig. 12.5). These experiments are used to analyze the reaction of unprepared participants to unexpected automatic braking maneuvers. The results are summarized in Merkel et al. [9].

12.3.2 Takeover Behavior with Partially Automated Driving (Level 3)

Research on automated driving also focuses on topics in which rear-end collision-critical situations are required in real-life driving tests. In a study conducted by the Federal Highway Research Institute, the question of whether drivers can react appropriately to a suddenly braking vehicle in front if they had to take back control of the vehicle only a short time before was investigated.

The results of the extensive EVITA study with a total of 45 subjects are described in Klamroth et al. [10]. The main result is that 80% of the test subjects only react to the suddenly braking front vehicle by braking. The remaining 20% additionally swerved. It is striking that evasive maneuvers are hardly secured visually.

12.4 Outlook and Future Usage

In addition to the studies on anti-collision systems and automated driving already described, the following individual issues were also investigated in cooperation with vehicle manufacturers:

- Investigation of driver actions/performance during a distracted driving episode in real subject driving test in human–machine interface design.
- Driver-adapted functional design of a brake assistant.

The use of the EVITA tool so far shows that the need for a system to represent rear-end collision critical situations in real driving tests is undiminished and goes far beyond the evaluation of anti-collision systems.

It is expected that further questions will arise in the investigation of Level 3 and Level 4 automation regarding, among other things, driver/passenger acceptance of system behavior in more critical situations.

In all of the more than 2,500 trips performed to date, EVITA was able to reliably prevent an imminent collision using the conservative collision avoidance strategy (see ◉ □ Table 12.1) reliably prevented an imminent collision. Although this confirms the high safety level of the tests, it can lead to EVITA being pulled out of the danger zone before it has been clearly decided whether test subjects could actually have handled the situation themselves if they had continued to brake. If necessary, it should be examined whether the conflict of goals between a sufficiently large safety reserve and the smallest possible residual distance can be shifted to even more critical maneuvers by a collision-tolerant design of the EVITA rear end.

References

1. NHTSA Report (2001)
2. Bäumler, H.: Reaktionszeiten im Straßenverkehr; VKU (Verkehrsunfall und Fahrzeugtechnik), Vieweg Verlag, Ausgaben 11/2007, 12/2007, 1/2008
3. Krause, R., de Vries, N., Friebel, W.-C.: Mensch und Bremse in Notbremssituationen. VKU (Verkehrsunfall und Fahrzeugtechnik) Juni (2007)
4. Burckhardt, M.: Reaktionszeiten bei Notbremsvorgängen Fahrzeugtechnische Schriftreihe. Verlag TÜV Rheinland, Köln (1985)
5. Zomotor, A.: Fahrwerktechnik – Fahrverhalten. Vogel, Würzburg (1987)
6. Bubb, H.: Fahrversuche mit Probanden – Nutzwert und Risiko, Darmstädter Kolloquium Mensch & Fahrzeug, Darmstadt (2003)
7. Hoffmann, J.: Das Darmstädter Verfahren (EVITA) zum Testen und Bewerten von Frontalkollisionsgegenmaßnahmen. Fortschritt-Berichte, VDI Reihe 12, Nr. 693, Düsseldorf (2008)
8. Winner, H., Hakuli, S., Lotz, F., Singer, Ch. (eds.): Handbuch Fahrerassistenzsysteme, 3. Auflage, 2015, Wiesbaden, Springer Vieweg, ▶ https://doi.org/10.1007/978-3-658-05734-3
9. Merkel, N., Pleß, R., Winner, H., Hammer, T., Schneider, N., Will, S.: Tolerability of Unexpected Autonomous Emergency Braking Maneuvers on Motorcycles – A Methodology for Experimental Investigation. Eindhoven, Conference on the Enhances Safety of Vehicles (ESV), Eindhoven, 10.06.19
10. Klamroth, A., Zerbe, A., Marx, T.: Transitionen bei Level-3-Automation: Einfluss der Verkehrsumgebung auf die Bewältigungsleistung des Fahrers während Realfahrten, Schlussbericht der Bundesanstalt für Straßenwesen, in: FAT-Schriftenreihe 323 (2019)

Open Access This chapter is licensed under the terms of the Creative Commons Attribution-NonCommercial-NoDerivatives 4.0 International License (▶ http://creativecommons.org/licenses/by-nc-nd/4.0/), which permits any noncommercial use, sharing, distribution and reproduction in any medium or format, as long as you give appropriate credit to the original author(s) and the source, provide a link to the Creative Commons license and indicate if you modified the licensed material. You do not have permission under this license to share adapted material derived from this chapter or parts of it.

The images or other third party material in this chapter are included in the chapter's Creative Commons license, unless indicated otherwise in a credit line to the material. If material is not included in the chapter's Creative Commons license and your intended use is not permitted by statutory regulation or exceeds the permitted use, you will need to obtain permission directly from the copyright holder.

Testing with Coordinated Automated Vehicles

Hans-Peter Schöner, Pim van der Jagt and Sebastian Werr

Contents

© The Author(s) 2026
H. Winner et al. (eds.), *Handbook Assisted and Automated Driving*,
https://doi.org/10.1007/978-3-658-45276-6_13

13.1 Motivation for Testing with Coordinated Vehicles

Driver assistance systems not only support the driver on long journeys doing routine tasks but also help the driver to react in time and correctly in critical situations. Assistance systems of the latest generation even react autonomously when an accident is becoming unavoidable, in cases where the driver does not react in time. For autonomous driving functions, even higher requirements are mandatory: they must recognize potentially critical situations early enough to avoid dangers by cautious and cooperative behavior proactively. For this kind of function, the systems must understand complex traffic situations and differentiate potentially critical situations from uncritical ones—this is also a challenge to the versatility and precision of the testing technology for verifying and validating such systems. In the former Daimler's (now Mercedes-Benz AG) Research Department in cooperation with AB Dynamics, a test methodology for Mercedes-Benz was developed [13, 17, 20, 21], which allows the testing of advanced driver assistance systems and autonomous driving functions precisely, reproducibly and safely. In the scope of the publicly funded PEGASUS project [16], this methodology has been established as a widely accepted standard for testing of autonomous driving functions.

For testing and validation of these systems, a large amount of physical testing of the integrated system is necessary—despite the increasing use of virtual development methods. The quantitative validation requires many tests changing in a wide range of parameters. The task of covering this space completely and as efficiently as possible is a challenge for the testing of these systems.

In comparison to the testing of vehicle dynamics systems—which react to vehicle-internal state variables—the testing of driver assistance systems requires the inclusion of additional state variables from *outside* the vehicle. For example, the relative position of the vehicle with respect to lane markings is important for testing lane departure warning and prevention systems, the relative speed and distance of several vehicles relative to each other is significant for adaptive cruise control systems. If the systems do not only warn but also react by supporting the driver's action, or if they even act autonomously upon the vehicle's motion, these control systems need to be tested and validated with respect to a large number of different driving situations and environmental conditions, in order to preclude hazards for the passengers and to reach certification of the system [15].

Systematic testing of such systems technically requires setting up precisely specific driving states of the vehicle on a given test track: typically, a predefined course with a specific speed profile must be controlled. If several vehicles are involved, all vehicles should move simultaneously in a coordinated manner.

Human drivers can control such conditions sufficiently in one single vehicle, however, controlling several vehicles simultaneously with respect to temporal and spatial specifications means asking too much from human drivers. Statistical variation of driving maneuvers might be acceptable or even necessary for some general testing tasks. Systematic and efficient testing according to precise specifications, objective comparison of different system variants and performing safety critical maneuvers, however, require a high precision and exact repeatability of tests is mandatory.

For specific assistance systems, testing methods have been developed in order to guarantee the comparability [10, 12] and/or to improve accuracy and repeatability [11, 18], see also ▶ Kap. 12); there are also some innovative solutions for reducing the risks for drivers in critical maneuvers [3]. An analysis of the general applicability of those systems to all relevant accident situations, which will have to be controlled by future systems, revealed that several tasks could still not be resolved with those systems. It remained the task to define and develop a testing system, which could manage potentially dangerous maneuvers with several participating vehicles in a precise, coordinated, and safe way. Testing with automated coordinated vehicles has proven to do this job as designed in daily application since 2009 [13].

Automation and—enabled by its precision—exact coordination of vehicles, allows achieving specific improvements in the following categories of driving maneuvers (see ◻ Fig. 13.1).

1. Maneuvers, which are hard to drive reproducibly for human drivers: Examples are cut-in and cut-out maneuvers with different speeds and distances between the individual vehicles but also the precise setting of lateral vehicle accelerations for sensor calibration and verification.
2. Risky maneuvers in which even small parameter changes can cause accidents: Examples are close passages, particularly used for the tuning and verification of the borderline between intervening and non-intervening of a collision avoidance system.
3. Dangerous maneuvers, which cannot be safely performed with human drivers: Examples are trajectories on a collision course with a high longitudinal difference in speed, especially with oncoming traffic, or even at relatively low speeds with vehicles crossing one another's trajectory in an intersection.

13.2 Requirements for Precision and Reproducibility

The accuracy requirements for automated vehicles are deduced from the analysis of assistance functions to be tested. Specifications for lateral and longitudinal accuracy are of a different type. For lateral accuracy, the

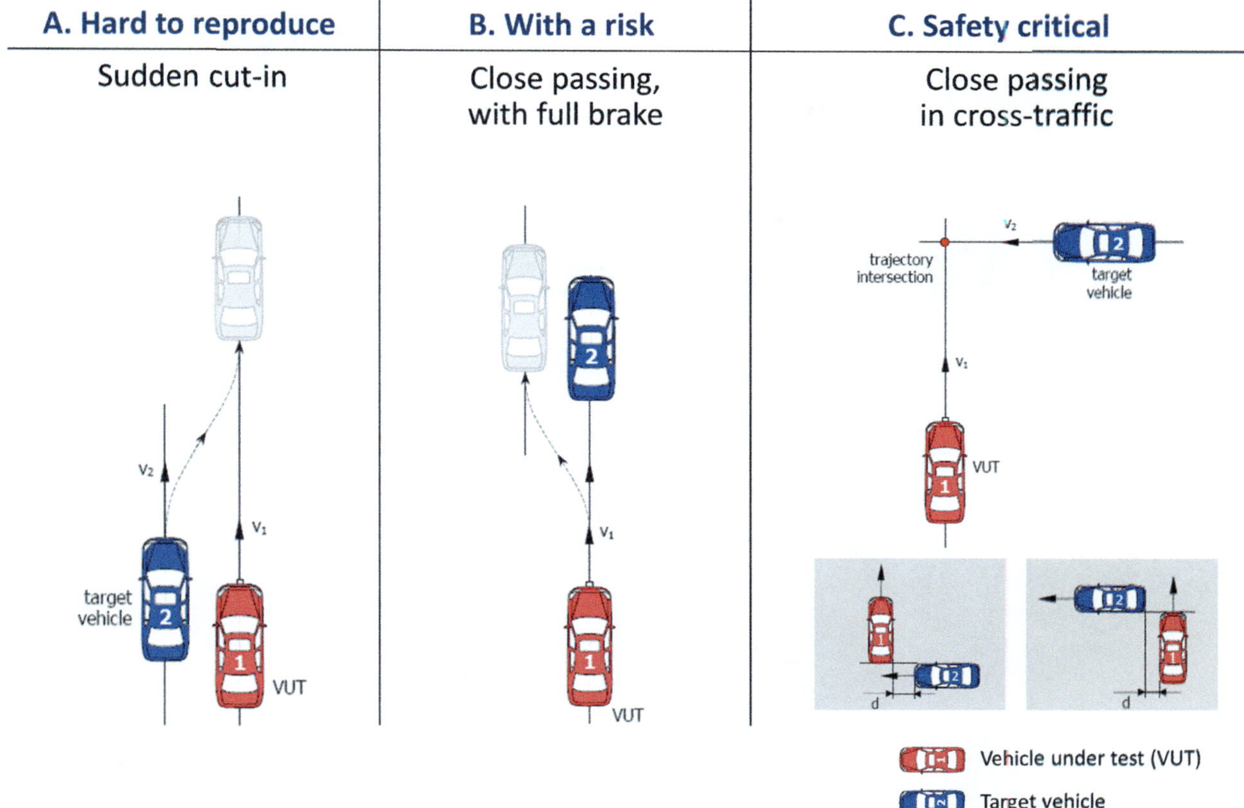

Fig. 13.1   Maneuver categories with significant improvements through automated driving

width of a lane marking (16 cm on rural roads in Germany) is a good reference. Passing an obstacle thus should be possible with a gap of less than 20 cm. For the precision of a vehicle driving in a lane, this translates into a lateral course tolerance of ± 10 cm. Some Euro-NCAP testing procedures require a maximum tolerance of 5 cm on straight and of 10 cm on curved trajectories.

Longitudinal accuracy requirements depend on the speed of the vehicle and are much more difficult to define generally for all tasks. In the specification, finally, it was required to reach a certain waypoint within a time tolerance of ± 20 ms from nominal timing. This translates into a spatial tolerance for a given time instance of, for example, 40 cm at a driving speed of 20 m/s(72 km/h).

Just by keeping these tolerances, a *single* vehicle can realize reproducible tracks and driving speed profiles. For the coordination of *several* vehicles a common time base for all vehicles with a tolerance in the range of 1 ms is an additional requirement.

13.3   Technical Realization

A flexible technical solution for this task was developed from 2007 on in cooperation between Daimler/Mercedes-Benz AG and AB Dynamics, which is the basis for the following description (see [17]; meanwhile similar solutions are available also from other manufacturers).

13.3.1   In the Vehicle: Steering Robot, Pedal Robot, Position Measurement, Safety Controller, Emergency Brake

As replacement for the driver, actuators are built into the vehicle under test. The actuators control throttle, brake pedal, and steering similar to human control actions (see **Fig. 13.2**); such robots have been in use for some time already, mostly for testing vehicles in driving cycles on dynamometers (pedal robots) or performing challenging steering maneuvers reproducibly (steering robots). For the use in development vehicles and autonomous vehicles, a direct electronic control of built-in steering, throttle, and brake actuators is possible, if they are realized with sufficiently high security. However, due to limitations in actuator speed and/or force, not all inputs can be implemented with the same intensity a human driver could reach.

For a precise control of the vehicle motion, with the goal to reach a certain location at a given time with a predefined speed and trajectory, the exact measurement of these values is essential. Therefore, an inertial navigation system (INS) is installed in the vehicle, which is

◘ Fig. 13.2   Robotic control of throttle, brake, and steering in the vehicle under test

supported by a differential GPS system. The measurement data are transferred to the vehicle with a rate of 100 Hz. The local differential GPS base station communicates a correction signal every 1 s; this is used to calculate the vehicle's position with an accuracy of typically ±2 cm at 100 Hz. A loss of the GPS signal will be compensated by the inertial navigation system with an assured precision of better than 10 cm for more than 30 s of GPS outage; this allows driving under bridges without GPS reception without problems, for example.

If it is necessary to test for longer periods without GPS connection, for example for tests in a multistory parking garage, ultra-wide band (UWB)-based indoor location systems can complement the GPS signals. They also reach an accuracy of ±2 cm, if the signal coverage is designed and verified with care. Inertial systems using both GPS and UWB correction can manage the transition between indoor and outdoor position sensing without interruption.

Apart from position, GPS also supplies a highly exact time reference signal. This signal is equally available on all vehicles and is used for synchronization of the maneuvers, with tolerances far below one millisecond. In particular, the maneuvers of all vehicles are started based on this reference signal at the same time and/or with a defined time delay.

The driving maneuvers are controlled by a real-time computer in each vehicle. If a driver is onboard the vehicle, he can operate the real-time controller by a connected tablet computer acting as an interface. This computer stores the test variants and the individual control parameters for the vehicle and the maneuvers, respectively. Drivers can start and stop single automatic driving sections while sitting in the vehicle; they can easily regain control of the vehicle at any time.

For driverless operation of the vehicles, an additional safety controller is used, which constantly monitors the normal function of all components in the vehicle as well as the position control of the vehicle and the communication to a base station. A spring-loaded emergency brake system and an additional cut-off switch for the engine of the vehicle complement the list of safety components; in case of an incident or a system failure, this will bring the vehicles to a quick and safe stop.

13.3.2   In the Base Station: Control Center, Visualization, Coordination, Safety, and Security

From the base station, the automated vehicle tests are started and supervised by a human operator supported

Fig. 13.3 Remote control of automated vehicles from the base station

by the *electronic control center* (■ Fig. 13.3). It contains wireless communication to all vehicles. The widely used WLAN provides a communication range with a radius of approximately 700 m. A safety controller, at the core of the control center, permanently exchanges a watchdog signal with the safety controllers in the vehicles. A first PC serves as remote control surface to the real-time computers in each vehicle, a second PC with a particular *"base station"* software (which is used to load the maneuvers and to control timing), and a control console for manual remote control (with steering wheel and pedals for rough positioning of the vehicle) are the essential components of the control center in the base station.

The control center can control more than 10 vehicles simultaneously on an enclosed test area. This allows controlling quite complex scenarios, which should cover all traffic situations to be tested according to current expectations.

From the base station, the vehicles and their systems are monitored, and test maneuvers are planned, the necessary information is sent to the vehicles and the vehicle maneuvers are started. When performing coordinated maneuvers, the *control center* guarantees that all vehicles use the same set of coordinated single maneuvers and assigns the coordinated starting time. During a test maneuver, the trajectories—which include spatial and temporal data of the track—of all vehicles are constantly monitored automatically and visualized for the operator. The

integrity and the communication of all systems are verified; if necessary, the test can be stopped at any time by software control via the "base station" software or by a hardware emergency stop button. The complete test procedure (including planned stopping trajectories after an emergency stop) is known to each vehicle at the start of the maneuver; this way the relatively unstable wireless communication is not within the safety-sensitive control loop and does not mitigate the safety of the test.

A single vehicle can be controlled remotely by hand from the base station: a video channel shows the view from the vehicle to the road ahead. The vehicle can be controlled via console with steering wheel (or possibly joystick) and pedals for throttle and brake. The operator can steer the vehicle at low speed, despite significant latency time of the communication links. This function allows to bring the vehicle to a suitable starting position for an automatic maneuver; it is especially handy after an interruption of a maneuver for any reason. For repetitively used positions, the planning software can propose collision-free routes (based on the known positions, orientation, and size of all known objects) on approved drivable areas. After verification and clearance from the base station, the vehicle can move automatically to the new starting position.

Another task of the operator at the base station is to ensure safety on the test track. A safety concept, considering accessibility of the test track, reliability of all technical components, and safety measures against errors while planning or operating the vehicle trajectories, is essential for the safe operation of such a system. The requirement for robust communication coverage is the limiting factor for the size of the test area for coordinated automated vehicles.

A WLAN communication with a single router has a sufficiently large range for the existing Euro-NCAP tests (at the time of writing), but for more complex scenarios over longer distances, a sufficiently large and uninterruptable communication network is mandatory. A WLAN mesh network with automatic frequency switching can provide a robust network over a distance of 1.2 km . Future traffic scenarios with even larger communication needs between a multitude of test participants and the base station, call for a robust network with a high data bandwidth. It is assumed, that the upcoming 5G-technology-based networks will provide suitable capabilities.

13.3.3 Other Systems: Data and Video Transmission, Data Synchronization, Aerial Photographs

Precise GPS time signals allow the data from all vehicles to be marked with synchronization time stamps, for both measurement data and video data, so that result evaluation can be done off-line after the tests are performed. Precisely timed maneuvers also offer the possibility of activating cameras or other triggered objects, like traffic lights or pedestrian and bicycle dummies, to operate with correct timing. Even documentation from the air with camera drones is easy to manage: the drone flies to the predefined position with optimal view on the point of interest and takes pictures at the exact time of interest. This is particularly useful for documentation and evaluation of close passage maneuvers.

13.4 Planning of Maneuvers

13.4.1 Planning of Individual Trajectories

For planning of test maneuvers, there are different possibilities:

The test vehicles can be driven by human drivers, even with the steering and pedal robots in place. In a learning mode, the control system can record the trajectory driven by the human driver and save it for further repetition in the maneuver catalog. Some parameters, like scaling of speed, lateral shift of the trajectory, and starting time, can be modified when the maneuver is retrieved for a new experiment.

For coordinated maneuvers of several vehicles, a planning with graphical planning tools in the maneuver editor is usually more efficient. The trajectory can be put together from predefined segments (straight tracks, curves, lane changes, spline curves, etc.) and placed onto a certain position of the mapped test track. It is also possible to import trajectories (sets of x-, y-, and time values) from simulations. The entire test sequence can be simulated in advance to check the limitations of vehicle dynamics for the trajectory. Of course, mathematical models in sufficient precision describing the properties of the vehicles used are a mandatory prerequisite (compare ▶ Sect. 13.2).

Planned trajectories of all single vehicles can be simulated separately or in combination and visualized in the "base station" software.

13.4.2 Planning and Examination of Coordinated Trajectories

The correctness of coordination of several vehicles should also be verified through simulation, especially to check for minimal vehicle distances. The distance between two vehicles in the relevant phase of the maneuver can be modified by variation of the starting delay between two vehicles. Critical maneuvers with several vehicles are firstly simulated for every single vehicle and subsequently verified by multiple test drives of the single vehicles; with the test drives the simulations are validated with respect to simulation accuracy and repeatability. The trajectories are then checked for unwanted collisions in combined simulations. In this way, even extremely close passage maneuvers (see ▶ Sect. 13.2) can be performed safely and reproducibly.

When testing with coordinated vehicles, it must be guaranteed that the test can be stopped at any time without creating safety critical situations. For example, a vehicle that brakes due to an emergency stop must not block the trajectory of any other vehicle. This can be tested by the simulation tools [13, 17, 21]. Each vehicle can be programmed to avoid such situations by predetermining a specific procedure when an emergency stop is triggered: instant braking, braking with a time delay, or even evasive acceleration with steering and final braking in a defined timing and magnitude are possible options. Such a sophisticated planning of safety procedures is typically only necessary for so-called "critical sections" (close encounters of the vehicles or path crossings), but they should not be neglected because of the short distances necessary for tests of collision avoidance systems.

13.4.3 Precision and Repeatability

The precision and reproducibility of the control system were verified with a position measurement system independent of the GPS measurement. The preci-

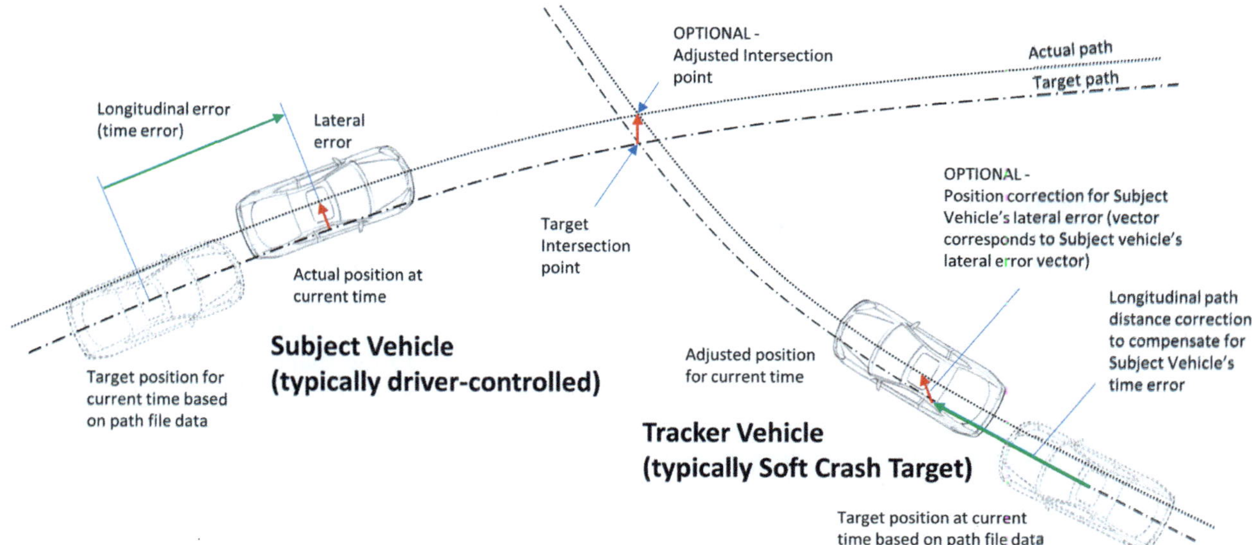

Fig. 13.4 Synchronization to a manually driven subject vehicle [2]

sion of differential GPS, which is around ± 2 cm, could be reproduced consistently for the closed-loop control of the trajectory, driving close by obstacles with a fixed position. Braking tests onto a target with high braking deceleration (approx. 7.5 m/s^2) show a reproducibility of the stopping point within ± 3 cm. Prerequisite for this precision is, however, that the vehicle controller is tuned to the specific vehicle type and that a sufficiently long settling time for dynamic deviation is contained in the trajectory. During acceleration and very dynamical steering maneuvers, temporary deviations in the range of decimeters were observed. However, these deviations are reproducible when driving the same maneuver repeatedly, and thus they can be accounted for.

Long-time stability of the position control should be checked by driving to reference points on the test track; with a differential GPS reference station on the proving ground, the specified stationary accuracy could be verified repeatedly by the authors—but it is not guaranteed by the system owners. The reproducibility of the control over several hours could be shown impressively by repeatedly driving along a pattern in snow for several hours [20]. In total, the required lateral accuracy of ± 10 cm can be reached easily. The repeatability of robot-driven trajectories is much better than those driven by human drivers.

The longitudinal accuracy depends on the dynamics of the trajectory and on the engine power of the vehicle. At fast changes of the controlled value of speed, a temporary deviation from the reference value is inevitable. The specified precision for reaching a certain waypoint within a time slot of ± 20 ms is achievable, under the condition of sufficiently long settling time within the trajectory. For safety critical situations, a sufficiently good repeatability should be ver-

ified by comparison of repetitive drives of the same trajectory. Changes in e.g. gear shift timing, which depend upon factors such as temperature, may be considered and avoided by choosing a suitable speed profile or shift limitations. The control system monitors and shows deviations constantly; it can trigger a shutdown of the test if spatial or timing deviations become too large. Aspects of reproducibility must be considered during planning of maneuvers with high precision requirements. The control software allows defining so-called "critical sections" of the trajectory, in which specifically small tolerances for position and timing error can be defined.

For certain tests, it can be necessary, that not all the scenario participants drive automatically on exact trajectories, but some are driven manually by a human driver. During such tests, it is impossible to reach the same precision of scenario constellations than with purely robot-driven maneuvers. For this kind of tests, AB Dynamics has developed a patented control software [2] called "synchro mode", which adapts the positions of all automated vehicles to the manually driven "subject vehicle". Figure 13.4 shows an example for the corrections of the automated "tracker vehicle". Longitudinal and lateral position of the target vehicle is controlled over time to reach, as far as possible, the desired geometrical constellation between the two vehicles at the point of collision.

13.4.4 Virtual Guide Rails

Coordinated automated vehicles allow the safe testing of software and hardware of autonomously acting driver assistance systems or driving functions during the development phase. For this purpose, it is important that in a given scenario there is some freedom for the system to control the vehicle; a trajectory precisely controlled by the robots would be counterproductive, since a steering or braking action of the robots could be interpreted as a corrective driver's intervention and thus lead to the abortion of the assistance function.

So-called "virtual guardrails" can specify a corridor around the nominal trajectory for the robot-controlled vehicle under test. Within this corridor, steering and/or braking intervention of the assistance systems is possible without interference of the robot controller. Only if the vehicle is about to leave this corridor due to too little or too strong intervention of the assistance systems, the robot system takes over control and brings the vehicle back on track or to a stop.

13.5 Self-driving Targets

Automated vehicles, i.e. vehicles equipped with driving robots as described above, are well suited for collision-free traffic scenarios. However, if a crash mitigation system is to be tested in situations that are not completely controllable, crashable objects should be used as targets. Most crash targets in common use until some years ago were either stationary objects, or they could drive only along a straight line, or they were coupled to a towing vehicle. Collision avoidance tests for intersection scenarios and especially with turning vehicles are not possible with such simple targets. Only a completely self-driving target vehicle can solve this problem.

In order to find a solution that can be seamlessly integrated into the rest of the testing environment, a collision target should have the following properties:

- fully automated, similar to automated real vehicles;
- no additional installations in the test environment;
- suitable for crashes from all directions;
- with a three-dimensional structure;
- a visual appearance from all sides as a real vehicle;
- and a sensor signature (for radar, lidar, ultrasonic sensors, etc.) equivalent to the signature of a real vehicle from any direction.

13.5.1 Soft Crash Target

A first target concept that fulfills these requirements is shown in ▪ Fig. 13.5, [19]. This crashable vehicle consists of a narrow chassis with an electric drive unit and a controller compatible to the robot vehicle control as described above, thus it can drive precisely along predefined trajectories. Its driving performance is sufficient for city driving scenarios. A maximum speed of more than 80 km/h is acchievable; accelerations of around 4 m/s^2 and decelerations up to 8 m/s^2 are possible.

The crash target is built from deformable air cushions, kept in shape by pressurized rubber hoses of 10 cm in diameter, which define the edges of the cushions. The enclosed space within the cushions is not pressurized; openings at the lower side allow air to escape when the inside pressure increases during a crash. This construction of the cushions causes a damper characteristic during crashes, and the differential speed of collision is reduced smoothly with an approximately constant force, providing a maximum "crumple-zone" (similar to passenger airbags). In addition, the shape of the cushion adapts easily to the shape of the colliding vehicle. In this way, the crash forces are distributed over a large area, and over the whole duration of the crash, and are thus minimized. Measurements show that this leads to substantially smaller peak crash forces compared to a conventional pressurized "balloon car" or a dummy car built from foam material, with less risk of damage to the vehicle under test. The pressurized rubber hoses make sure that the target reconstitutes its shape after a crash by itself. In this way, efforts for making the target ready for the next test are minimized.

The self-driving soft crash target is suited for tests of vehicles under test with a driving collision partner, especially during turns in intersections. The target is tuned for similar radar reflectivity of a real car by metal foils, and its shape is close to a real car. Thus, it can be used for radar- and camera-based assistance systems. The free choice of trajectories and the vehicle-like appearance from all directions make it suitable for many different traffic scenarios. Differential speeds during a crash of up to 50 km/h in longitudinal direction and up to 30 km/h in lateral direction are possible without risk of damage of the vehicle under test or harm to the target itself. The control electronics and the electrical system are set up and tested for high accelerations during such crashes.

13.5.2 Over-Drivable Target Carrier

If tests of the system vehicle are required with even higher differential speeds or if the risk of damage must be even lower, the concept of a self-driving over-drivable target carrier can be applied. Such a concept was developed (probably independently) by DSD in Austria [23], based on a suggestion by Daimler/Mercedes-Benz, and by Dynamic Research, Inc. in California [1, 24], (the latter is shown in ▪ Fig. 13.6).

The target carrier can be driven over if a collision takes place. The target itself, which is fixed on top of the target carrier, is pushed away by the colliding vehicle. Very high differential speeds are only relevant for longitudinal crashes. For such maneuvers, it may be sufficient to use only a front or rear segment, instead of a complete vehicle mockup on the target carrier, to

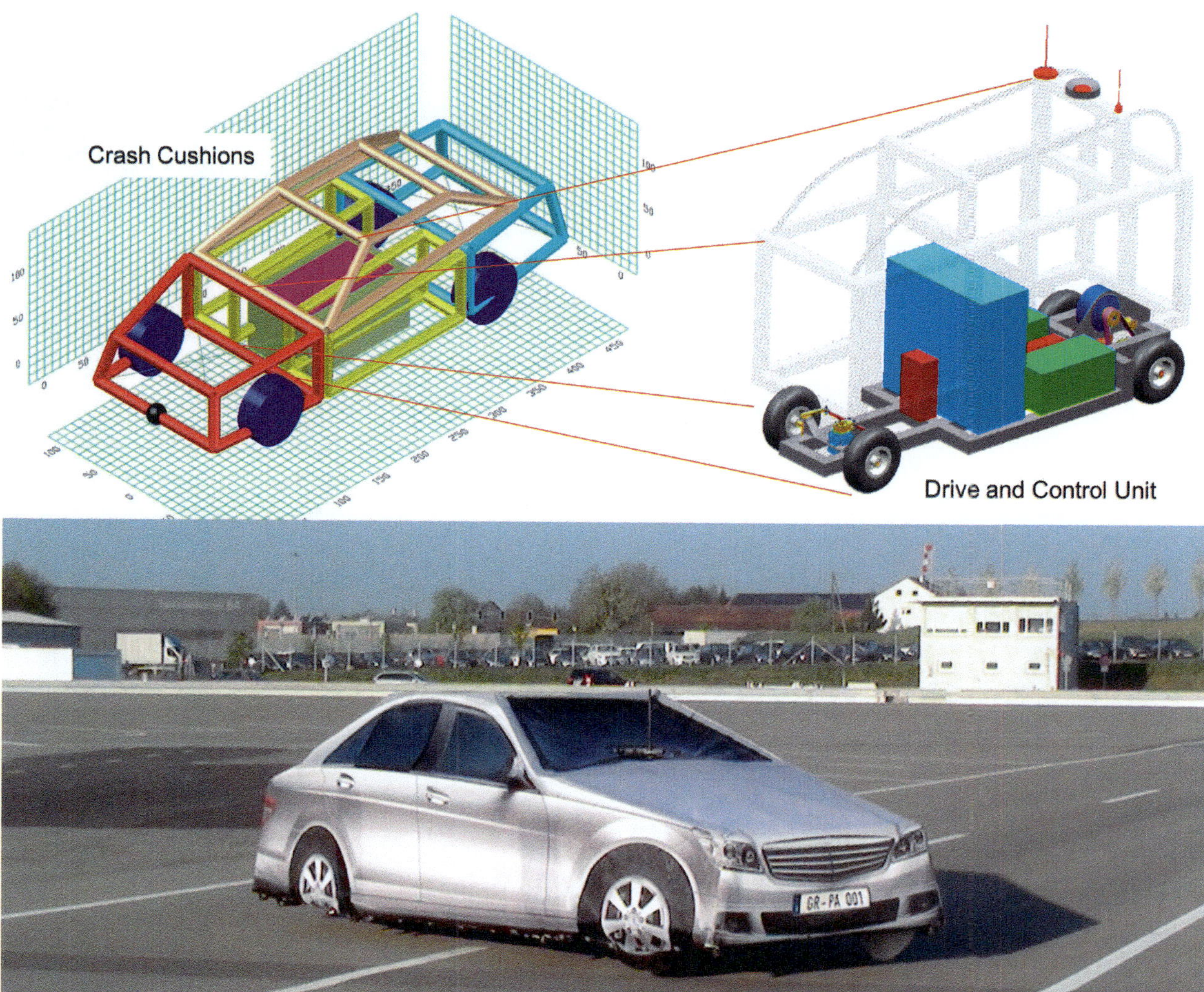

Fig. 13.5 Concept of the self-driving soft crash target

minimize the inertial mass involved. For example, the front or rear portion of the soft crash target can serve for this purpose. With this concept, differential speeds during crashes of more than 100 km/h have been performed without damage to the vehicle or the crash target. An important advantage of the over-drivable platform becomes evident for lateral collisions with relatively high possible collision speeds: a complete vehicle is needed as target for such constellations, but other concepts provide relatively short crumple-zones. The over-drivable target carrier concept is realized in several sizes with an appropriate turn radius, so it can be used for carrying pedestrian or bicycle dummies as targets.

Due to the low height of the target carrier vehicle, ground clearance and maximum driving power of the target carrier are limited. For this reason, the target carrier is less flexible for applications than the soft crash target. Nevertheless, driving speeds of more than 100 km/h can be reached with the target carrier. The concept of over-driving works even if the vehicle under

test reaches the collision point before the target and the target carrier. More care in coordinating the maneuver is needed if the test vehicle is stationary or has come to a stop in an intersection: the target carrier should not hit the *standing* car's wheels from the side. If the target carrier is used in conjunction with human-driven test vehicles for lateral crashes, such rare situations can be avoided by correcting the timing error of a human driver to ensure the correct collision configuration (see ▶ Sect. 13.4.3).

13.5.3 Certified Crashable Target

Based on the results of several global harmonization workshops, jointly organized by *EuroNCAP*, by *National Highway Traffic Safety Administration (NHTSA)* and by *Insurance Institute for Highway Safety (IIHS)*, a specification for a three-dimensional crash target vehicle has been established. This target is called *Global*

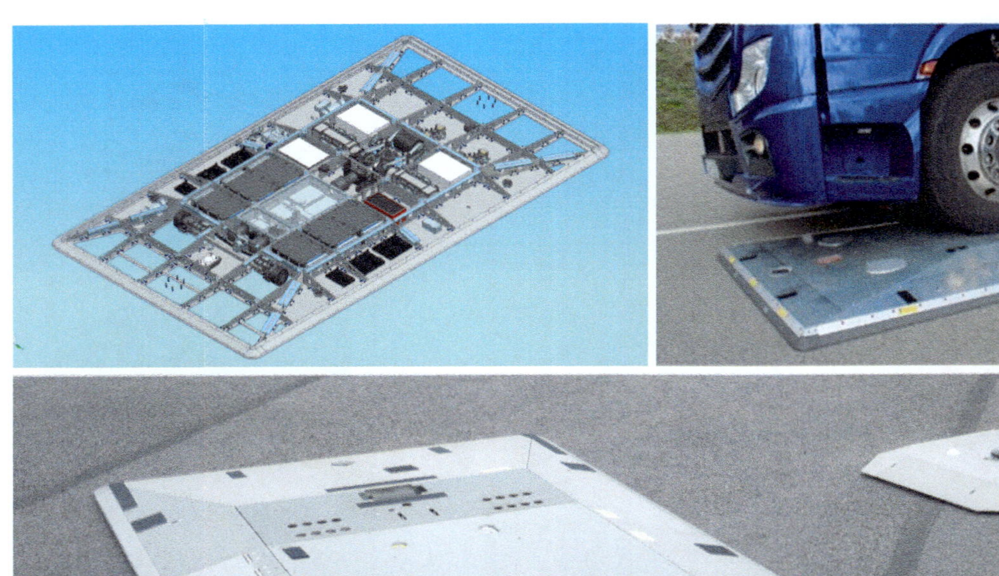

Fig. 13.6   Over-drivable, self-driving target carriers for targets of different sizes [1, 24]

Vehicle Target (GVT) (see Fig. 13.7), and its specification [7] is based mostly on pre-studies of Dynamic Research Inc [6]. EuroNCAP test protocols LSS [8] and AEB-C2C [9] require the use of a target according to this specification (see also ▶ Sect. 9.4.1).

Further specifications for a three-dimensional target vehicle have been established by *Society of Automotive Engineers SAE* [22] and ISO 19206–3 [14].

13.5.4   Targets for "Vulnerable Road Users" (VRU)

In addition to vehicle targets, *Vulnerable Road Users, VRU* play an increasing role for the capabilities (and thus for the testing and verification) of assisting and autonomous driving functions. Pedestrian targets and targets for single-track vehicles (bicycles, motorbikes) have been developed, which can be mobilized with an over-drivable target carrier or with a cable pulley. With suitable synchronization of the motion trigger, coordinated traffic scenarios and specific challenging constellations can be realized. Articulated arms and legs of pedestrian dummies make sure that radar signature and video images replicate important features of typical kinematic motion patterns of pedestrians. For a general description of such targets and specific assessment scenarios see ▶ Sects. 9.4.2 and 9.4.3

Most versions of pedestrian and single-track targets are well suited to replicate steady-state movement of a linear moving target, but in dynamic maneuvers, the motion patterns have an unrealistic appearance. A real pedestrian shows a significant forward or rearward inclination as a function of the longitudinal acceleration or deceleration, a single-track vehicle leans to the side as an early indicator for a change of direction. Most targets do not provide such motion clues; in contrast, they tend to lean to the opposite side due to inertial forces. The expected trajectory of VRU is much better predictable if longitudinal and lateral inclinations can be evaluated, especially trained motion predictors based on artificial intelligence require correct (even faint) motion indicators. Thus, the correct reproduction of VRU inclination should be an essential function of the target motion system. A first solution based on a cable robot system has been developed by Messring [4] and is operational on the Mercedes-Benz proving ground [5]; it coordinates body, leg, and arm motion of the dummy according to the dynamical properties of the target trajectory (see Fig. 13.8).

Fig. 13.7   Global vehicle target (GVT) on over-drivable target carrier

Fig. 13.8   Six-degrees-of-freedom target mover for "vulnerable road users" with articulated pedestrian target

13.6 Examples of Automated Driving Maneuvers

13.6.1 Automatic Maneuvers of Single Vehicles

Automatic maneuvers have advantages even for tests with a *single* vehicle:

Driving along curves and in circles: Some assistance systems are activated when reaching a specified lateral acceleration. An automatically driven maneuver with a predefined speed and track curvature can provide this lateral acceleration very precisely and (for example) with a continuously increasing value. Such maneuvers can be performed much more efficient than with a human driver.

Misuse and durability: For standardized testing of airbag systems, several situations with very high lateral and longitudinal accelerations are required; only under specific conditions an airbag is allowed to fire. For these tests, vehicles are, e.g. driven across a ramp at 70km/h, they have to perform a full brake while crossing a curb, or to ram a simulated boar. Durability testing is another example with extreme stress on the vehicle as well as on the test driver. Such maneuvers can be driven with automated vehicles—the test driver's health will not be stressed this way. The increase of efficiency of durability tests using automated vehicles (one active test driver in the base station controls several vehicles at the same time) is possible.

13.6.2 Coordinated Maneuvers with Several Driverless Vehicles

Longitudinal collision avoidance: In maneuvers with several participating vehicles, the coordinated automated vehicles show their full advantage. A large proportion of the testing of collision avoidance or collision mitigation systems is verifying the performance in situations close to the specified use case. In many of them, the assistance system has to show that it does not act unintentionally, in order to reliably avoid wrong emergency brake situations. For such situations, close passage is required, often with evasive maneuvers in the last moment.

With coordinated automated vehicles, such tests can be performed precisely, reproducibly, and safely for driver and vehicle. Parameters like initial speed, smallest distance, course angle, and more can be controlled easily. A new software version of the system or a new sensor variant can be tested with exactly the same maneuver as before, thus enabling efficient, reproducible, and comparable results of the tests.

Lateral collision avoidance: One of the biggest challenges for the coordination of two vehicles is the close crossing in an intersection. Mercedes-Benz vehicles equipped with "Driver Assistance Package", for example, include a cross-traffic assistance system since 2013. Crossing in intersections requires the highest accuracy concerning spatial *and* timing precision of the test. In comparison, in situations with longitudinal traffic like passing vehicles or oncoming traffic, spatial precision alone is sufficient for safe operation. In addition, if there is a fault in the test procedure, the risk for heavy damage is high. For these reasons, close passage maneuvers in cross-traffic situations may not be performed with human drivers on board. With coordinated automated vehicles cross-traffic situations with vehicles driving up to 70 km/h, and minimum distances of less than one meter could be performed repeatedly and safely.

Merging situations: Automated vehicles face combined longitudinal and lateral challenges in merging situations, and in addition high traffic density leads to temporarily reduced safety distances. Here relative positions and behavior of more than two vehicles are relevant for the situation and need to be coordinated precisely for meaningful testing scenarios. Realizing such test scenarios is a high challenge for coordinated automated vehicles and the testing team in charge, see ◘ Fig. 13.9.

13.6.3 Maneuvers with Driver, and with Triggered and/or Synchronized Targets

This is a testing variant with coordinated vehicles driving only part of the complete maneuver automatically, especially those parts of the maneuver after a specific trigger point, which need very precise execution. This is essential for tests with crashable targets that perform a critical maneuver, while the test vehicle is driven by a human driver to test the action and reaction of different drivers. For this variant, in the vehicle under test, no steering or pedal robot is required, but only exact position measurement and a communication link to the other vehicle or to the base station. Depending on the relative position, or other trigger conditions of the vehicle under test, for example, a lane change maneuver of the crash target can be initiated. Before the trigger point, the target drives along its trajectory on the adjacent lane (see ◘ Fig. 13.10).

Using this method, a coordinated self-driving crash target can be used with very little planning overhead for repeatable maneuvers with human drivers.

◘ Fig. 13.9 Challenging merging scenario with several coordinated vehicles and one vehicle target

13.7 Future Developments

Collision avoidance systems require the vehicle-under-test to be brought into critical situations, e.g. with close passages, as described in the examples given above. The methodology of coordinated automated vehicles has been developed for this purpose and works ideally for these tasks.

The systematic testing of specific critical situations and its variants with two or three vehicles—which can be done efficiently with this method—will be only *one* component of the safety assurance of autonomous driving functions. Testing of highly or fully automated vehicles on the proving ground will move more and more towards traffic scenarios, which represent the *early phase* of potentially critical situations. Without driver action, autonomous vehicles must provide evidence that they can cope with such scenarios—not only *by controlling* an immanently dangerous situation with an appropriate emergency action, but preferably, *by avoiding* such dangerous situations by proactive and cooperative driving behavior in the early phase. It can be anticipated that the scenarios to be tested will become more complex, and the coordination of *more than three* vehicles will be necessary. In addition to safety-related tests, there will be tests to ensure adequate and especially *cooperative* behavior for *fluent traffic* in high vehicle density scenarios. Probably, this will include the need to implement typical reactive and cooperative functions (with parameterized degree) also for the traffic vehicles.

A robust and safe operation of an *artificial surrounding traffic* with specific properties as testing environment for *driverless* autonomous vehicles on the proving ground (e.g. with variable density, dynamics, and cooperativeness) in order to provide a high density of challenging conditions with controllable parameters and in a wide variance might be a new challenge for testing environments. Testing scenarios for driverless autonomous vehicles in such artificial traffic under naturally changing environmental conditions will last not only a few seconds, but significantly longer—even to the scale of days and weeks. Coordinated automated vehicles provide the technological basis for this vision, while extensions in traffic control and interaction modes will be necessary.

An important future field is the interaction between the simulation of complex traffic scenarios (used for the development and first-stage verification of the driving functions) and the real-world verification and validation of vehicle behavior in "corner cases" (derived from design considerations during the development phase). The systematic, precise, and fast translation of a simulated traffic scenario to an executable test on the proving ground will be a significant building block for efficient testing of assisting and autonomous driving functions.

Fig. 13.10 Emergency braking scenario with coordinated subject vehicle and several targets

References

1. ABDynamics (2013). ▶ https://www.abdynamics.com/en/products/track-testing/adas-targets/guided-soft-target

2. ABDynamics: Path Control system. United States Patent No: US 10,214,212 B2, Feb. 26 2019 (2019)

3. Bock, T., Maurer, M., Färber, G.: Vehicle in the Loop (VIL)—a new simulator set-up for testing advanced driving assistance systems. Driving Simulation Conference, Iowa City, IA (USA) (2007)

4. Doric, I., Reitberger, A., Wittmann, S., Harrison, R., Brandmeier, T.: A novel approach for the test of active pedestrian safety systems. IEEE Trans. Intell. Transp. Syst. **18**(5), 1299–1312 (2017). ▶ https://doi.org/10.1109/TITS.2016.2606439

5. Doric, I., Arp, D., Werr, S., Schoener, H.P.: Test system for safe automated driving. January 2019, ATZ Automobiltechnische Zeitschrift 121(1), 58–63 (2019). ▶ https://doi.org/10.1007/s38311-018-0186-5

6. DynamicResearch (2020). ▶ http://www.dri-ats.com/soft-car-360/

7. EuroNCAP (2018) Technical Bulletin—TB025—Global Vehicle Target Specification for Euro NCAP. ▶ http://www.euroncap.com

8. EuroNCAP (2019). ▶ https://cdn.euroncap.com/media/53143/euro-ncap-lss-test-protocol-v302.pdf

9. EuroNCAP (2019). ▶ https://cdn.euroncap.com/media/56143/euro-ncap-aeb-c2c-test-protocol-v302.pdf

10. Gulde, D.: So testet AMS, Teil 9: Assistenzsysteme. Auto-Motor-Sport, 17–2010 (2010)

11. Hoffmann, J., Winner, H.: EVITA—das Untersuchungswerkzeug für Gefahrensituationen (EVITA—a tool for evaluation of dangerous traffic situations). 3. Tagung Aktive Sicherheit durch Fahrerassistenz, München (2008)

12. Huber, B., Resch, S.: Methods for Testing of Driver Assistance Systems. SAE Paper 2008–28–0020 (2008)

13. Hurich, W., Luther, J., Schöner, H.P.: Koordiniertes Automatisiertes Fahren zum Entwickeln, Prüfen und Absichern von Assistenzsystemen (Coordinated automated driving for development, testing and safeguarding of assistance systems). 10. Braunschweiger Symposium—AAET, Braunschweig (2009)

14. ISO 19206-3: 3D Car Target Specification (2020)

15. ISO 26262: bzw. DIN-ISO 61508: Road Vehicles—Functional Safety (2011)

16. PEGASUS: Securing automated driving effectively (2020). ▶ https://www.pegasusprojekt.de/en/pegasus-video

17. Pick, A., Hubbard, M., Neads, S.: Near-miss collisions using coordinated robot-controlled vehicles. Proceedings of the 10th International Symposium on Advanced Vehicle Control, pp. 634–639, Loughborough (2010)

18. Ploeg, J.: VeHIL—Vehicle Hardware-in-the-Loop. Testumgebung der Fa. TNO (2007): ▶ http://www.tno.nl/downloads/Ploeg%20-%20SUMMITS-VeHIL-RealWorldPilot.pdf

19. Schöner, H.P., Hurich, W., Haaf, D.: Selbstfahrendes Soft Crash Target zur Erprobung von Assistenzsystemen (Self-driving soft crash target for testing of assistance systems). 12. Braunschweiger Symposium—AAET, Braunschweig (2011)

20. Schöner, H.P., Hurich, W., Luther, J., Herrtwich, R.G.: Coordinated Automated Driving for the Testing of Assistance Systems. Automobiltechnische Zeitschrift ATZ, 01–2011 (2011)

21. Schretter, N., Sinz, W., Schöner, H.P.: Planung und Realisierung von automatisierten Fahrmanövern zur Erprobung von aktiven Sicherheitssystemen (Planning and realization of automated driving maneuvers for testing of active safety systems). 3. Grazer Symposium Virtuelles Fahrzeug, Graz (2009)

22. Society of Automotive Engineers SAE: Recommended Practice document J3122, "Test Target Correlation: Radar Characteristics" (2020)

23. Steffan, H., Moser, A., Ebner, J., Sinz, W.: UFO—ein neues System zur Evaluierung von Assistenzsystemen (UFO—a new system for evaluation of driver assistance systems). 13. Braunschweiger Symposium—AAET, Braunschweig (2012)

24. Zellner J W (2011) Guided Soft Target. ▶ http://www.dynres.com/resources-and-tools/#guided-soft-target

Open Access This chapter is licensed under the terms of the Creative Commons Attribution-NonCommercial-NoDerivatives 4.0 International License (▶ http://creativecommons.org/licenses/by-nc-nd/4.0/), which permits any noncommercial use, sharing, distribution and reproduction in any medium or format, as long as you give appropriate credit to the original author(s) and the source, provide a link to the Creative Commons license and indicate if you modified the licensed material. You do not have permission under this license to share adapted material derived from this chapter or parts of it.

The images or other third party material in this chapter are included in the chapter's Creative Commons license, unless indicated otherwise in a credit line to the material. If material is not included in the chapter's Creative Commons license and your intended use is not permitted by statutory regulation or exceeds the permitted use, you will need to obtain permission directly from the copyright holder.

Sensors for ADAS

Ultrasonic Sensors

Heinrich Gotzig

Contents

© The Author(s) 2026
H. Winner et al. (eds.), *Handbook Assisted and Automated Driving*,
https://doi.org/10.1007/978-3-658-45276-6_14

14.1 Introduction

Ultrasonic technology is used for many different areas of applications (sensing, welding, …). Sensors based specifically on this physical principle are used in many metrological application areas. The technology can be found in industry, medicine, measurement technology, material processing, and more up to cosmetics. A good overview of this can be found e.g. in [1–3]. Bats use ultrasound as a perfect localization system and this is where you can see the potential of this technology that has not yet been fully exploited. In general, ultrasound is regarded as a very robust and comparatively inexpensive technology that can be used as a simple distance-measuring device in a wide variety of media.

This chapter deals specifically with ultrasonic sensors. The basics, the underlying physical principles of environment detection sensors in automobiles are considered. Ultrasonic sensors were first introduced into driver assistance as a parking assistance system in 1991 with the BMW 7 series. The principle has now proven itself as a very mature system. It is robust against various environmental conditions and we now find ultrasonic sensors in almost every newly registered vehicle. But the technology is not finished yet. In this chapter, it will show how ultrasonic sensor technology is used for significantly more applications than the parking aid [4] and what potential (bat principle) this technology still has.

14.2 Basics of Ultrasonic Technology

14.2.1 Piezoelectric Effect

The piezoelectric effect, detected in 1880 by Jacques and Pierre Curie, consists of a linear electromechanical interaction between the mechanical and electrical states in a crystal. A mechanical deformation of the crystal produces an electrical charge shift (polarization) proportional to the deformation, which can be tapped as an electrical voltage (direct piezoelectric effect). Conversely, applying an electrical voltage to the crystal generates a mechanical deformation in it (reciprocal or inverse piezoelectric effect). Thus, piezoelectric materials are in principle suitable for generating mechanical vibrations or for producing deformations by applying electric fields. Conversely, they can serve as sensors for mechanical vibrations or deformations (Fig. 14.1). Since the direct and the inverse piezoelectric effect always occur together, piezoelectric transducers can be used both to emit and to receive sound.

14.2.2 Piezoelectric Ceramics

Nowadays ferroelectric ceramics are among the most widely used materials with piezoelectric effect. There are also organic materials that exhibit the piezoelectric effect. However, their use in motor vehicles has not played a role to date due to their lower robustness. Currently, the most important piezoelectric ceramic materials are based on the oxide solid solution system of lead zirconate and lead titanate, known as lead zirconate titanate (PZT). The specific properties of these ceramics, such as the high dielectric constant, depend on the molar ratio of lead zirconate to lead titanate as well as on substitution and doping with additional elements. This results in a wide range of modification possibilities for materials with a wide variety of properties. Below the so-called Curie temperature, an asymmetry in the distribution of positive and negative electrical charge occurs within the lattice of a cell. This results in a permanent electric dipole moment of the individual cell Ferroelectricity results from the formation of domains with uniform electric polarization, which are aligned by poling, i.e. the brief application of a strong electric DC field, while simultaneously heating the ceramic above the Curie temperature. This poling is associated with

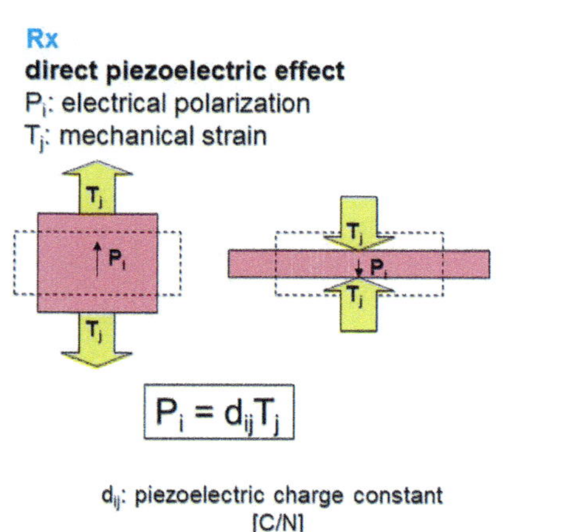

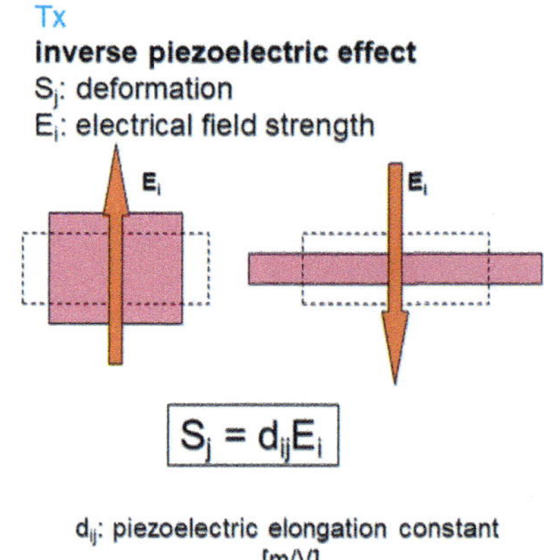

Fig. 14.1 Principle of the piezoelectric effect for generating ultrasonic waves

a change in the length of the ceramic (Fig. 14.1). In the course of efforts to eliminate the use of lead in motor vehicles as far as possible, intensive work is being carried out in the field of lead-free piezoelectric ceramics. However, on the basis of this research, no alternatives to the current ceramics are to be expected in the short term.

14.3 Ultrasonic Transducers

Since the ultrasonic transducer in the "driver assistance sensor" application case involves the transfer of mechanical kinetic energy from a solid body into the air, the efficiency is very low because of the different densities. It is significantly better with the well-known submarine sonar (**sound** and **na**vigation and **r**anging).

14.3.1 Basic Terms of Acoustics

In the following chapters, only a few terms of acoustics necessary for the basic understanding will be described.

14.3.1.1 Sound

The term sound is generally understood to mean elastodynamic oscillations and waves. In this context, sound fields generally presuppose solid, liquid, or gaseous matter.

14.3.1.2 Sound Pressure

Sound pressure p (pressure) refers to the pressure fluctuations of a compressible sound transmission medium (ultrasonic sensors in this case air) that occur during the propagation of sound. The sound pressure p is the alternating pressure that is superimposed on the static pressure p_0 (usually the air pressure). Here, the alternating sound pressure is

$$p = \frac{F}{A}$$

With the force F acting on the surface A per area of A.

Thus, the following applies to the total pressure p_{tot}:

$$p_{tot} = p_0 + p$$

The sound pressure p (acoustic pressure) is usually many orders of magnitude smaller than the static air pressure [5, 6].

14.3.1.3 Ultrasonic

Sound in a frequency range from 20 Hz to approx. 20,000 Hz is hearable to the human ear. If the frequency of the sound is below 20 Hz, it is called infrasound. When the frequency is above 20,000 Hz, it is called ultrasound. Animals (bats, whales, mice) can produce (and of course hear) ultrasound in a frequency range from 20 kHz up to several 100 kHz.

14.3.2 Construction and Functions

Figure 14.2 shows the principle construction of an ultrasonic sensor..

An ultrasonic sensor mainly consists of three main building blocks (Table 14.1)

14.3.2.1 Sensor Element

The sensor element is the most important and complex part of an ultrasonic sensor. It generates the energy field, which is the prerequisite for good performance as an environment detection sensor.

For driving assistance, the use of ultrasonic sensors for parking remains the most important one.

Figures 14.3, 14.4, and 14.5 show the requirements for the detection field as specified in ISO 17386—MALSO (Manoeuvring Aids for Low Speed

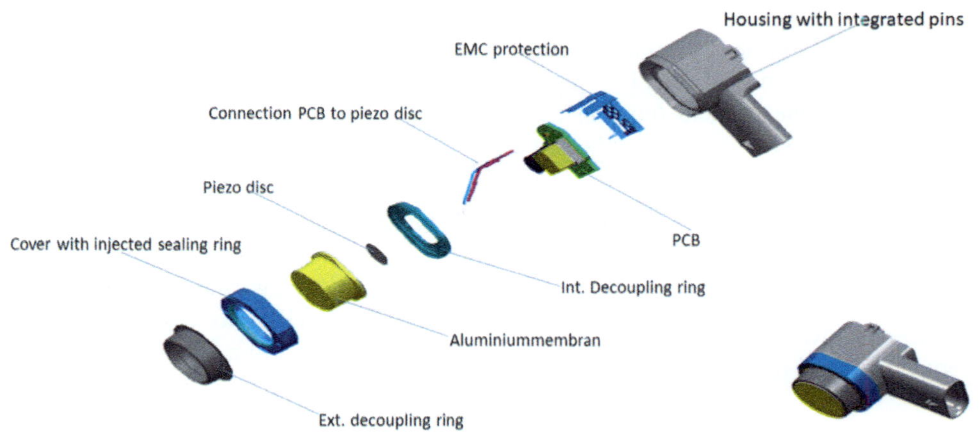

Fig. 14.2 Building blocks and their functions of an ultrasonic sensor

Table 14.1   Main building blocks of an ultrasonic sensor

Main block	Function, task
(1) Housing incl. connector	• Mechanical interface to the sensor holder • Mechanical connection to cable
(2) Elektronic/PCB	• Communication sensor ↔ electrical control unit • Electrical control of the piezo sensor element • Initial raw signal processing
(3) Sensor element (Aluminium membrane, damping, piezo disc, bonding)	• Generation of the necessary acoustic energy field • Conversion of electrical energy into mechanical energy and vice versa • Mechanical decoupling from bumper

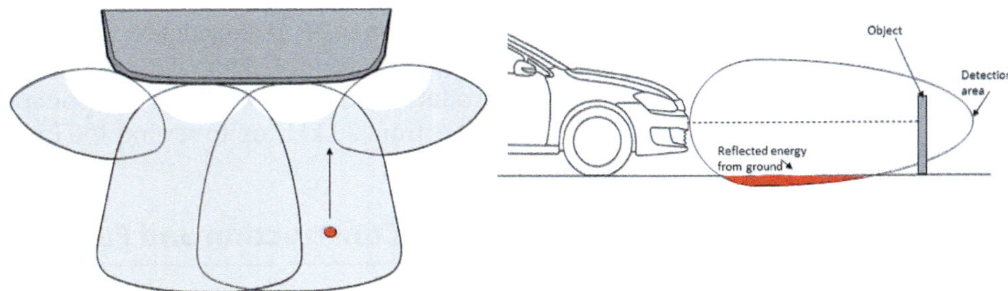

Fig. 14.3   Detection area for the application "near field sensing"

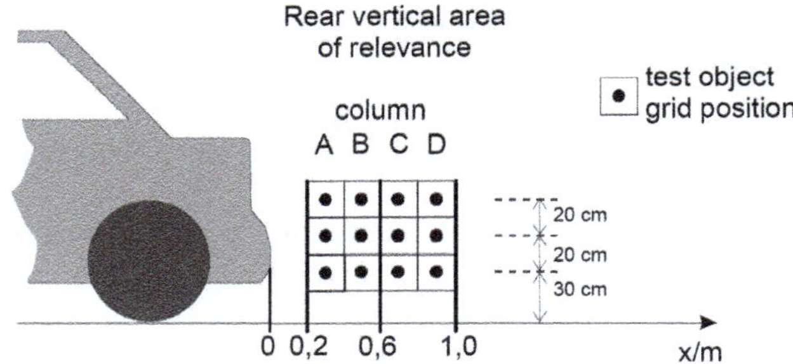

Fig. 14.4   Requirements for vertical detection of test objects (basis ISO 17386—MALSO)

Operation—parking aid) [7]. These requirements mean that a compromise must be found between the possible widest horizontal acoustic energy field (coverage and limiting the number of sensors) and a vertical field that detects objects with a height >15 cm. The best compromise is to use four sensors per bumper with a horizontal −3dB aperture angle between 60 and 90° and a vertical—3dB aperture angle between 35 and 60°.

The aperture angle Θ of the radiated energy for an ultrasonic transducer is inversely proportional to the transmission frequency f and the diameter of the radiating surface:

$$\Theta \sim \frac{1}{f * d}$$

For the design of the ultrasonic sensor membrane, in addition to the transmitter characteristic, the following other parameters are to consider:

— Matched impedance curve: the sensor element (see building block 3) and the control electronics must be matched to each other.
— Requirements for optimized transient and decay oscillation behavior of the ultrasonic sensor: Decay time as short as possible. The shorter the broader the mechanical resonance curve of the sensor and thus also the damping. Damping rate has a direct influence on both the amplitude and the mechanical resonance frequency. Operation in resonance is desired because of the highest possible quality/performance. In principle, the membrane could be excited at a higher, wider resonant frequency, but this is technically very difficult and would drive up costs.
— The sensor element should have an as low as possible shell vibration since energy is radiated laterally into the bumper. This energy can be reflected by the bumper and can thus lead to false signals.

Figure 14.6 shows the results of finite element methods "FEM simulations" as used for the design of the ultrasonic membrane. The mechanical resonance frequency of the ultrasonic membrane must ideally match the electrical frequency at which the membrane is excited. This operation in resonance is necessary for the highest possible quality/performance. Likewise, the re-

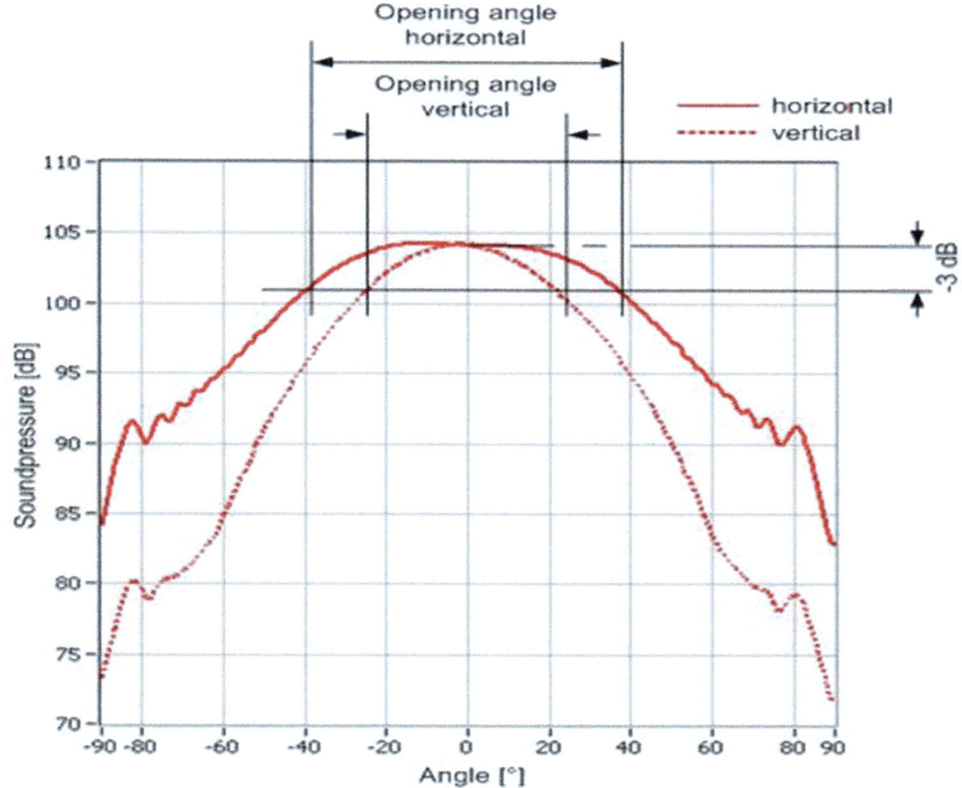

Fig. 14.5 Definition of the detection field (−3 dB-value) for an ultrasonic sensor

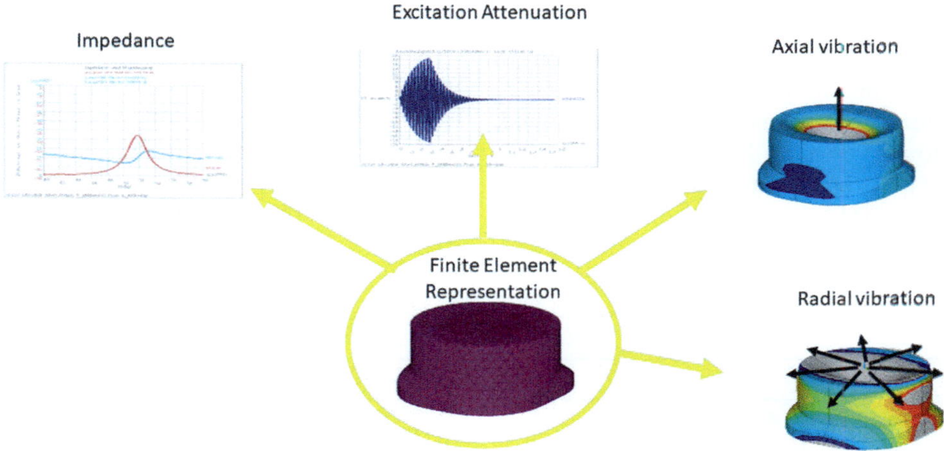

Fig. 14.6 Schematic results of a finite element method (FEM) for the design of the ultrasonic membrane

quired transient response of the damped diaphragm depends on the geometric design. To achieve good mechanical decoupling from the environment, vibrations on the shell surface must be kept as small as possible.

For the calculation of the impedance curve, the equivalent circuit diagram of a piezo transducer as shown in **Fig. 14.7** can be used with a very good approximation. A schematic representation can also be found in **Fig. 14.6**. The results of the measurement of the parameters C_0, Rs, Cs, Ls are used as control parameters in manufacturing. They can be used to determine quality of manufacturing steps such as bonding process of piezoceramics quite well.

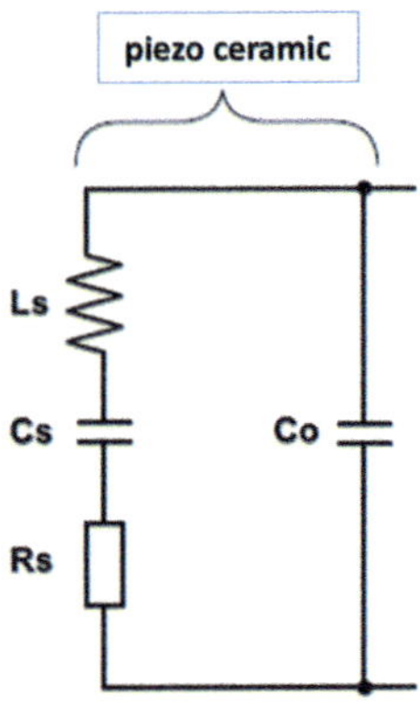

Fig. 14.7 Equivalent circuit diagram of an ultrasonic membrane

14.3.2.2 Electronic/PCB

Basically, a distinction is made between active and passive sensors for ultrasonic transducers. Passive means that the ultrasonic membrane is controlled and the signal is processed by a corresponding control unit. The sensor output is analog.

However, the vast majority of ultrasonic sensors used in automobiles today are active. Advantages include increased robustness against electrical interference (transmission of a digital signal from the sensor to the control unit) and a defined time reference is possible (direct and indirect measurement).

Figure 14.8 shows a typical block diagram of an ultrasonic sensor. The sensor shown here has three connections. In the block diagram on the left from top to bottom: Supply voltage, data line, connection to ground. Depending on the design of the electronics, variants with 2 lines (voltage and superimposed signal, connection to ground) or more lines are possible (dedicated vehicle buses). The control unit (ECU) sends a command, for example in the form of a pulse-wide modulated signal, which is evaluated in the logic. Depending on the length of the signal, the ultrasonic sensor jumps to different operating modes:

- Programming mode: sensor is specifically programmed
- Parallel transmit and receive mode: ultrasonic sensor starts with a transmit cycle, i.e. a voltage is applied to the piezo for a certain time via the transmit stage and at the same time the ultrasonic sensor is set to a receive mode
- Ultrasonic sensor is set to a receive-only mode.

Since the efficiency of ultrasonic sensors in automobiles is very low "transition from a solid medium to gaseous" must be excited with the highest possible voltage: order of magnitude—100 V. However, this voltage must be well below the Curie voltage (a correspondingly high electric field destroys the polarization). The voltage of the received signal is in the sub-μV range and is amplified accordingly for further signal processing. Afterwards, an analog/digital conversion takes place. There are basically two possibilities here: The direct comparison with a reference value and thus the digital information is either a 0 or 1, or the direct, e.g. 12-bit A/D conversion and corresponding transmission of the resulting values to the subsequent control unit. For cost reasons, and because it is sufficient for many applications in the vehicle, the simple 0/1 data transmission has been chosen until now. The latest generations of ultrasonic sensors transmit the complete sensor signal.

Each ultrasonic sensor usually has its own data line, so that the control of all sensors can take place simultaneously. Typically, different sequences are performed during the measurement: one sensor is put into the combined transmit-receive mode and the neighboring sensors into a pure receive mode.

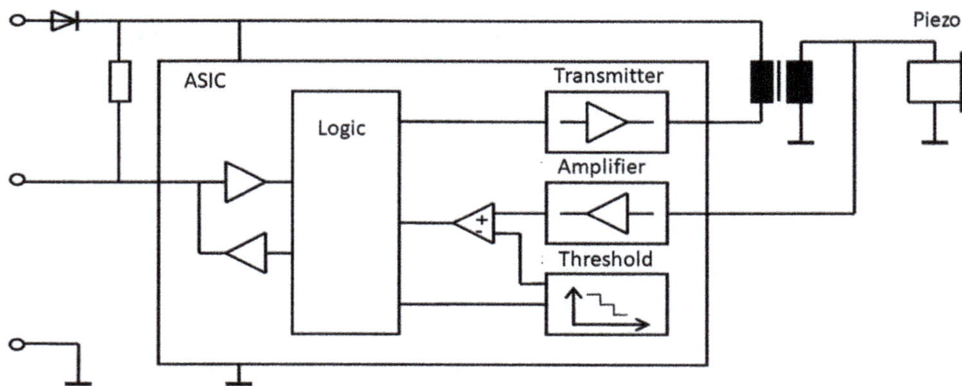

Fig. 14.8 Block diagram electronics of an ultrasonic sensor

14.4 Distance Measurement

Distance measurement by means of ultrasound is carried out according to the pulse transit time method. Since the speed of sound in dry air at 20 °C is relatively low at 343.2 m/s, the measurement of the pulse transit time is technically relatively easy to implement.

■ Figures 14.9, 14.10, and 14.11 show the basic principles of distance measurement by ultrasonic sensors in vehicles. At the time $t=0$, an AC voltage is applied to the piezoelectric disk for a few 100 µsec. (order of magnitude 100 V with a frequency between 40 and 70 kHz, depending on the manufacturer). This causes the sensor element to oscillate, and a pulse is emitted (red waves). After the oscillation has started, it takes another time t2 until the oscillation has decayed below a value necessary for the measurement. This is called the dead time of the ultrasonic sensor (approx. 1 to 1.5 ms). Objects in the energy field of the ultrasonic sensor reflect the emitted energy and are thus detected as echoes. However, a distinction must be made between relevant echoes, non-relevant echoes such as ground echoes, and other disturbances. This is done with the so-called threshold value comparison, as shown in ■ Fig. 14.11. The comparison results in digital information that is transmitted via the sensor's data line to the control unit, where it is used to calculate the distance:

$$r = \frac{t1 * Cs}{2}$$

$$Cs = 343{,}46 \text{ m/s}$$

r means the radial distance of the object to the sensor (■ Fig. 14.9); t1 corresponds to the echo transit time (■ Figs. 14.10 and 14.11) and Cs is the speed of sound. The given value of Cs refers to a temperature of 20 °C. For standard applications such as ultrasonic parking assistance, a constant value of Cs is used (20 °C); if higher precision is required, e.g. for ultrasonic parking space measurement, the temperature dependence of Cs is also taken into account.

Since the single ultrasonic sensor only provides radial distance information, but the exact location is required for most driver assistance applications, information from several sensors is combined. For more precise object localization, the principle of triangulation is used (see ■ Fig. 14.12). During a normal measurement cycle, the ultrasonic sensors are put into different modes. One ultrasonic sensor into a transmit/receive mode and the rest into a receive-only mode (see ▶ Sect. 3.2.2). If the distance of the sensors (X1,2) is known, the exact distance Y of the object to the vehicle (usually to the bumper) can be determined by trigonometric considerations as follows:

$$Y = \sqrt{r1^2 - \left(\frac{(X_{1,2}{}^2 + r1^2 - r2^2)}{4 * X_{1,2}{}^2}\right)^2}$$

A similar principle is used to measure the distance to more extended objects, such as a wall.

14.5 System Limitations of Ultrasonic Sensor

While 30 years ago ultrasonic-based driver assistance systems were only parking aids and had to be paid for separately as extra equipment, these systems are now mostly standard equipment. More and more applications are using ultrasonic technology. Over the years, mainly due to improved signal processing, the systems

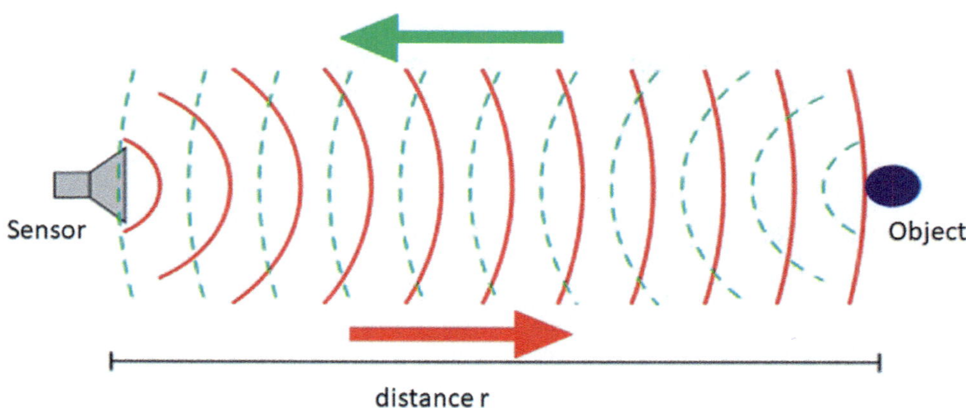

■ **Fig. 14.9** Distance measurement by means of pulse transit time method

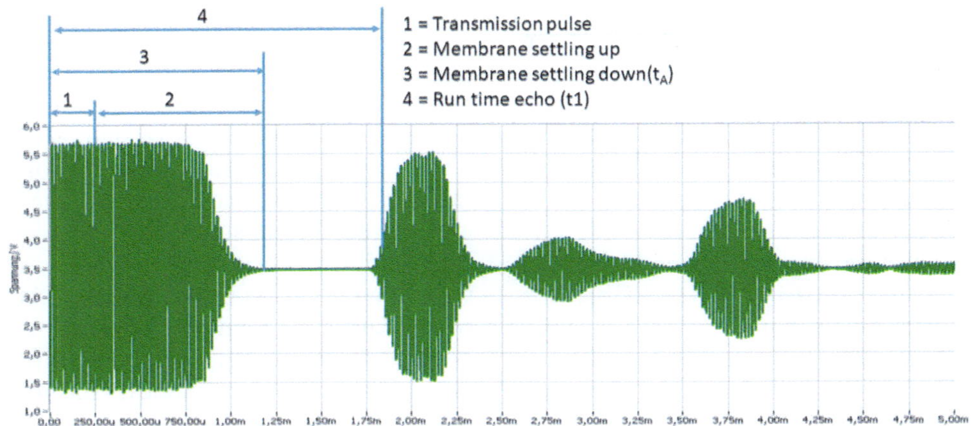

Fig. 14.10 Amplitude versus time of the ultrasonic sensor signal for the pulse transit time method

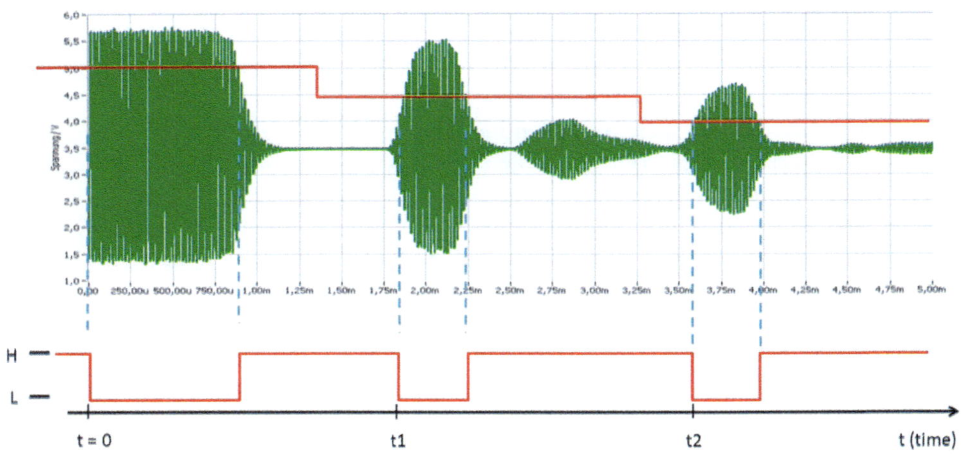

Fig. 14.11 Basic principle of time-of-flight measurement by A/D conversion using threshold comparison

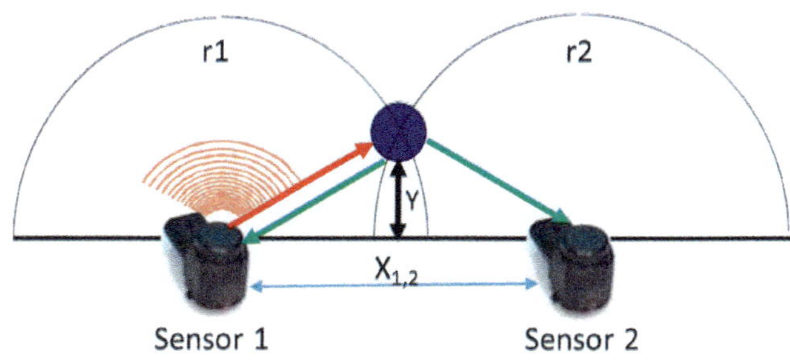

Fig. 14.12 Basic principle of triangulation

have improved, become more robust and less expensive. Ultrasonic sensors are still unbeatable in terms of cost/performance.

Pure parking assistance systems are purely informing systems: SAE automation level 0, while remote parking systems (ISO 20900: PAPS) [8] mainly based on ultrasonic sensors is a SAE level 2 system [9]. The most recent ISO document is even an SAE Level 4 system: valet parking (AVPS) [10]. Ultrasonic sensors are additionally used for perception. Accordingly, the requirements for system performance, reliability, SOTIF continue to increase, and the system boundaries of ultrasonic sensors must be pushed higher and higher. The system limits for ultrasonic sensors in automobiles are basically a function of the signal-to-noise ratio and the sensor signal processing. An optimum between the required performance and the required cost targets has to be achieved. The following list is intended to explain the individual influencing factors:

Influencing factors on the signal/noise ratio:
- Diameter of the aluminum diaphragm (piezo disk):
- the larger the piezo disk, the greater the signal. However, limits are given by the required opening angle (▶ Sect. 3.2.1, ◘ Fig. 14.4), installation in the bumper, costs.
- Mechanical damping of the sensor element
- the lower the damping, the greater the signal strength, but this extends the decay time and thus the shortest distance measurement.
- Voltage with which the piezo is actuated as well as material of the piezo disk.
- the higher the voltage, the greater the signal strength, but limits are given by the Curie voltage, lifetime, consumed energy.
- Coding of the transmitted signal and cross-correlation
- in principle a coding of the transmit signal and a following cross correlation (=mixing of the received signal (Rx) with the transmit code (Tx)) results in an improved signal to noise ratio, however typically an ultrasonic transducer is very narrowband, thus only very few codes are possible.
- Efficiency of energy transfer solid-state sensor element to air (and vice versa)
- The smaller the density difference, the better the coupling efficiency. Aluminum alloys have been shown to be the best compromise here.
- Matching of the oscillating circuit mechanics to that of the electronics (taking into account manufacturing tolerances, aging, and temperature effects)
- the better the matching, the more electrical energy is transferred into vibration energy. Here, too, a compromise must be found between the cost of the individual components (lower tolerances = higher costs) and the required overall performance.
- Distance, size, shape, and surface structure of the objects to be detected
- No optimization is possible here. The energy reflected at the object to be detected decreases with increasing distance. The law of reflection applies (angle of incidence = angle of reflection) and with increasing surface area, energy is absorbed (sonic surfaces reflect almost 100%, while materials such as powder snow absorb almost all energy).
- Attenuation of the energy field in the air (depending on temperature and humidity)
- And others.

Noise factors of ultrasonic sensors:
- Acoustic noise (sweeper, raindrops, ultrasonic sensors from other vehicles, …).
- Acoustic interferences.
- Mechanical vibrations of the vehicle
- Energy of the ultrasonic sensor is reflected by the bumper (insufficient mechanical decoupling).
- Electrical + magnetic noise (e.g. fluorescent lamp, induction loop, …).

The methods, which have been developed and tested for a long time, are based on plausibility considerations. In many cases, these are procedures for noise suppression that have been tried and tested over many years and constantly improved. Here, a lot of IP of the corresponding companies is behind it.

Current research (see also Outlook) benefits on the one hand from the knowledge in the field of artificial intelligence (ultrasonic sensor technology is biologically used by bats) and on the other hand from the constantly increasing and cheaper computing power.

14.6 Ultrasonic-Based Environmental Perception/Driving Assistance

14.6.1 Ultrasonic-Based Applications

Enclosed an overview of driver assistance applications that use ultrasonic sensors for perception:
- Ultrasonic Parking Aid—ISO 17386 Manoeuvring Aids for Low Speed Operation (MALSO).
- Assisted Parking—ISO 16787 Assisted Parking Systems (APS).
- Remote Parking—ISO 2090 Partially Automated Parking Systems (PAPS).
- Valet Parking—ISO DIS 2337 Automated Valet Parking System (AVPS) and new ISO 12768 Automated Valet Driving System (AVDS) (Start in 2021 Ultrasonic one of many sensors → sensor fusion).
- Ultraschall-based Blind Spot Detection System (BSD).
- Door Opening Assist.

14.6.2 Generic Ultrasonic System Architecture

◘ Fig. 14.13 shows a generic system architecture as realized in many vehicles. 2–18 sensors are installed in the outer skin (typically bumpers). The most common sensor variant is the 3-wire connection sensor ↔ control unit (separate data line, common ground, and supply voltage). The control unit is connected to the overall vehicle architecture via a data bus. Via this interface,

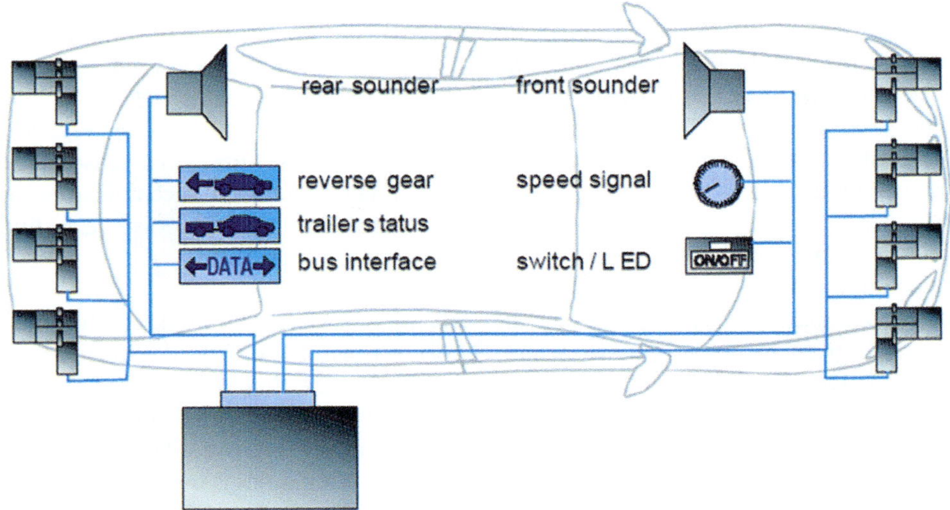

Fig. 14.13 Generic ultrasonic system architecture

the ECU receives necessary information such as temperature, speed, transmission status, trailer hitch, etc. It also controls the human interface, e.g. loudspeakers.

14.6.3 System Application of Ultrasonics

▶ Chapter 4 explained distance measurement in vehicles using ultrasonic sensors. For optimum performance, a system application must therefore be carried out depending on the ultrasonic sensor lobe and installation position in the vehicle. ■ Figure 14.3 shows schematically the task to be solved. Essentially, the task is to determine the sensor sensitivity parameters accordingly. In the first step, this is done by simulation.

■ Figure 14.14 shows an example of such a simulation. Reflection parameters are determined by preliminary tests, with which the reflected energy can be determined. The energy depends directly on the installation parameters. The simulation program then calculates the optimal value of thresholds (see ▶ Sect. 14.4) and with this the detection field can be calculated. Thus, it is directly possible to determine the detection performance depending on the mounting position.

14.6.4 Performance

In ▶ Chap. 5, influences on the system performance were explained. In principle, a compromise between cost and performance must be achieved for ultrasonic sensor systems in automotive applications. A continuous cost reduction with simultaneous performance increase is expected. This has been achieved in recent years primarily through a scaling effect, improved manufacturing processes. This currently results

in a distance measurement accuracy of 10% for simple applications (parking assistance) and a few cm for applications with higher requirements (assisted parking). The 10% is a result of the cost requirements for ultrasonic sensors and is mainly determined by the temperature dependency of the sound velocity and the size of the objects to be measured (a large object is detected closer with the threshold comparison technique—Fig. 14.11), in addition to design, manufacturing process, scattering of the individual components. Current sensors also provide only radial distance information, which can be improved by triangulation to an X, Y as described in ▶ Sect. 14.4. However, this requires that the reflection point is always known exactly, which is not always easy. Example: Measurement of the length of a parking space which is limited by 2 vehicles. The curvature of the bumper has a significant influence. Although by specification the ultrasonic sensors are used in a range from −40 °C to +90 °C, the optimum performance results in a range from −10 °C to +40 °C.

14.7 Summary and Outlook

14.7.1 Summary

Ultrasonic sensor technology in automotive applications has been around since the early 1990s. While in the beginning only a few vehicles of the upper class were equipped with a parking aid, this has changed significantly since the new millennium. The design of the vehicles in terms of optimizing the drag coefficient has as a consequence increasingly poor visibility, especially behind the vehicle. In the meantime, the technology has

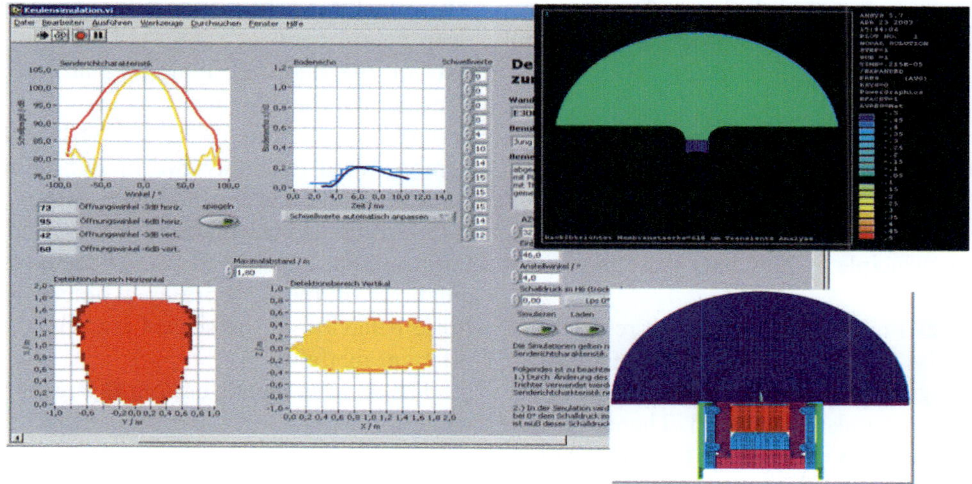

Fig. 14.14   Simulation ultrasonic detection field

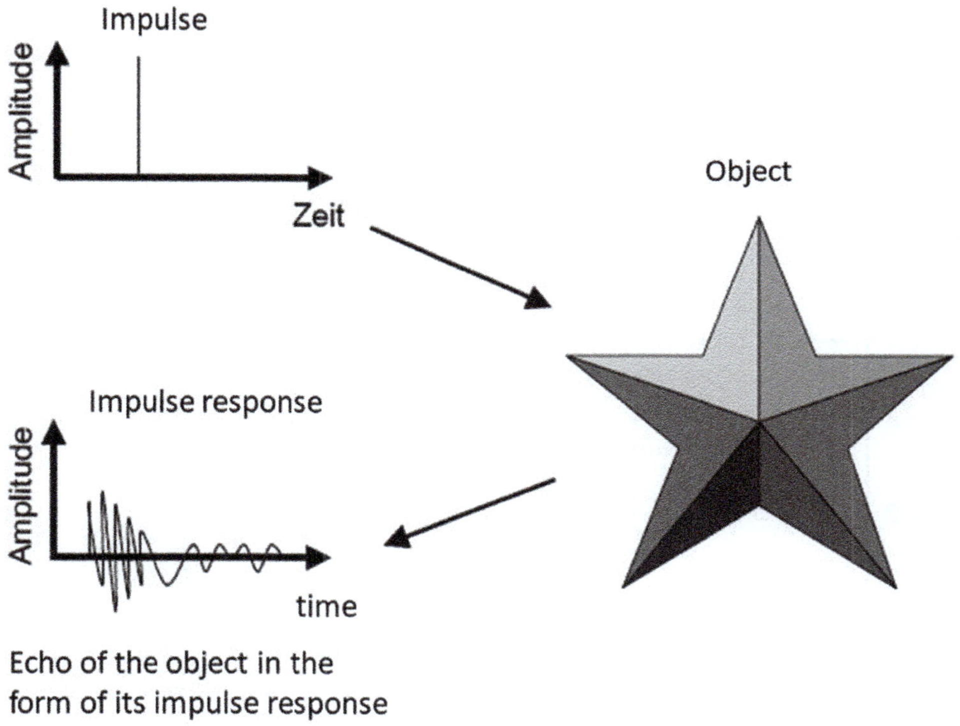

Fig. 14.15   Principle of the impulse response of a transmitted ultrasonic pulse

reached a very high level of maturity and the costs have been reduced more and more over the years. Currently, parking aids based on ultrasonic sensors can be found in almost every car, and in many, it is standard equipment. This has also led to the use of ultrasonic technology for other applications such as automated parking systems, blind spot detection, etc. The great advantage of ultrasonic sensors is that they can be used in a wide range of applications. The big advantage is that the basic architecture of a parking aid system can still be used. Only additional sensors, e.g. in the lateral area, need to be installed and the control unit requires a slightly larger processor with a little bit more memory. Advanced signal processing now even enables so-called "hidden integration". This means that the sensors can be installed behind the bumper. They then have slightly reduced performance, but can still be used very well as distance measurement sensors. Improvements are still ongoing.

14.7.2 Outlook

The next evolutionary stage of ultrasonic sensor technology uses the principle of how bats sense their environment.

As shown in ◼ Fig. 14.15, the impulse response of a transmitted pulse depends not only on the shape of the transmitted signal, but above all on the shape of the object from which the pulse is reflected. Therefore, artificial intelligence methods can be used for classification. For the required applications in the vehicle and taking into account, the total system costs, in a first step, three tasks were selected:

1. The echo originates from the ultrasonic sensor or from a source of noise → 2 classes
2. The echo is relevant or not relevant, whereby not relevant here means an obstacle that is easy to pass over (e.g. gravel) → 2 classes
3. Echo originates from the ultrasonic sensor and a height classification is to be made → e.g. 4 classes. (Lower than height 1, between height 1 and height 2, between height 2 u. height 3, higher than height 3) [11].

These tasks can be performed with a relatively small deep neural network (DNN). The input network consists of the digitized analog signal (see ◼ Fig. 14.10) considering the Nyquist criterion. For the AD converter, a resolution of 8-bit is sufficient. It could be demonstrated that only two hidden layers are suitable to meet the targets for the single output. A small µC is sufficient for the calculations. With this method, a clear improvement is achieved. This is demonstrated by calculations of the f1 score as a function of the signal/noise behavior of the sensor raw signal shown for the different methods.

The various signals used in ◼ Fig. 14.16 correspond to:

raw = unprocessed ultrasonic sensor signal.

corr = optimally correlated signal (see ► Sect. 14.5).

NN raw = neural network of unprocessed ultrasonic sensor signals.

NN corr = neural network with optimally correlated signals.

In the statistical analysis of binary classification, the f1 score is a measure of the accuracy of a test: how well the neural network does its job. It is the harmonious mean between accuracy and recognition.

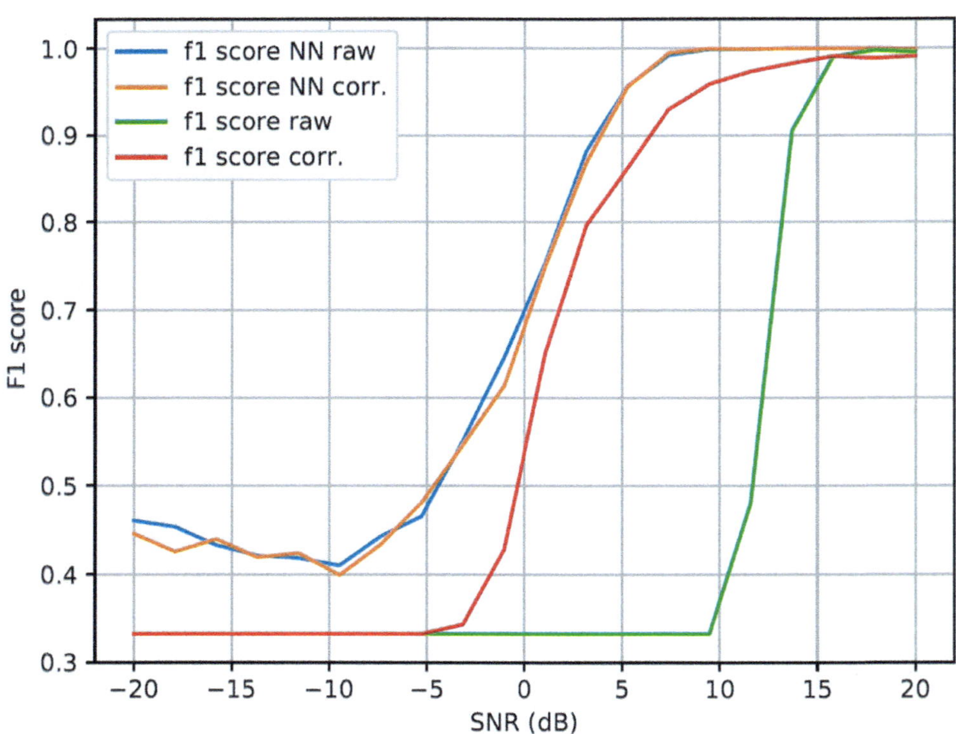

◼ **Fig. 14.16** F1 score for the distinction of relevant/irrelevant echo when driving over gravel

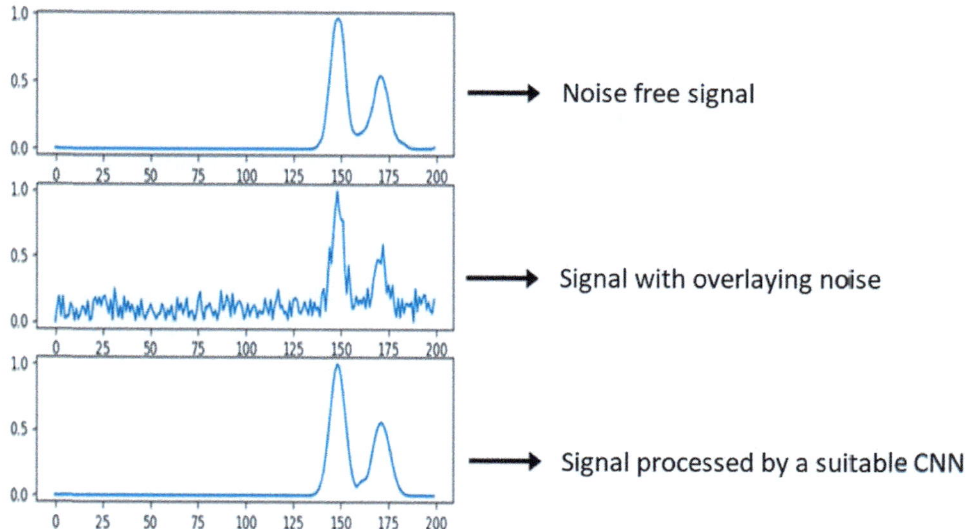

Fig. 14.17 Noise reduction by processing a noisy ultrasonic transducer signal via a trained Convolutional Neural Network (CNN)

$$f1 = 2 * \frac{precision * recall}{precision + recall} = \frac{t_P}{t_P + \frac{1}{2} * (f_P + f_N)}$$

t_P = number of true positives: signal is from the sensor and the classification is correct

f_P = number of false positives: signal is from sensor but classification is wrong (classified it as noise).

f_N = number of "false negatives"? Signal is not from sensor and classification is wrong (calculated it as signal from sensor).

Further improvement possibilities arise when using a CNN to extract noise from a signal. This is shown exemplarily with **Fig. 14.17**.

A Convolutional Neural Network (CNN) is an artificial neural network. It is a concept inspired by biological processes in the field of machine learning. CNNs are primarily used in machine processing of image or audio data.

Further improvements of ultrasonic sensors are expected in the coming years, mainly due to increasingly better signal processing, especially by using machine learning techniques.

References

1. Handbook of Acoustics. ▶ https://download.e-bookshelf.de/download/0000/0010/12/L-G-0000001012-0002340551.pdf
2. Ultrasonics: Data, equations and their practical uses. ▶ https://www.researchgate.net/publication/329736725_Ultrasonics_Data_equations_and_their_practical_uses
3. An Introduction to Acoustics. ▶ https://www.win.tue.nl/~sjoerdr/papers/boek.pdf
4. Sound pressure. ▶ https://en.wikipedia.org/wiki/Sound_pressure
5. The Master Handbook of Acoustics. ▶ http://tka4.org/downloads/audiosoft/BOOKS/1.%20sound/The%20Master%20Handbook%20Of%20Acoustics,%204th%20edition%20(F.%20Alton%20Everest,%20McGraw-Hill%202001).pdf
6. Particle velocity. ▶ https://en.wikipedia.org/wiki/Particle_velocity
7. ISO 17386:2010—Transport information and control systems—Manoeuvring Aids for Low Speed Operation (MALSO)—Performance requirements and test procedures. ▶ https://www.iso.org/standard/51448.html
8. ISO 20900:2019 Intelligent transport systems—Partially automated parking systems (PAPS)—Performance requirements and test procedures. ▶ https://www.iso.org/standard/69405.html
9. SAE Levels of Driving Automation. ▶ https://www.sae.org/blog/sae-j3016-update
10. ISO/DIS 23374-1 Intelligent transport systems—Automated valet parking systems (AVPS)—Part 1: System framework, requirements for automated driving, and communication interface. ▶ https://www.iso.org/standard/78420.html
11. A machine learning approach for ultrasonic noise classification and suppression. ▶ https://www.uni-das.de/images/pdf/fas-workshop/2020/FAS_2020_MOHAMED.pdf
12. Dissertation Petra Weißenbacher "Objekterkennung durch Echoortung und der Einfluss zeitlicher Iterationsmechanismen bei der Fledermaus Megaderma Lyra. ▶ https://edoc.ub.uni-muenchen.de/933/1/Weissenbacher_Petra.pdf

Open Access This chapter is licensed under the terms of the Creative Commons Attribution-NonCommercial-NoDerivatives 4.0 International License (▶ http://creativecommons.org/licenses/by-nc-nd/4.0/), which permits any noncommercial use, sharing, distribution and reproduction in any medium or format, as long as you give appropriate credit to the original author(s) and the source, provide a link to the Creative Commons license and indicate if you modified the licensed material. You do not have permission under this license to share adapted material derived from this chapter or parts of it.

The images or other third party material in this chapter are included in the chapter's Creative Commons license, unless indicated otherwise in a credit line to the material. If material is not included in the chapter's Creative Commons license and your intended use is not permitted by statutory regulation or exceeds the permitted use, you will need to obtain permission directly from the copyright holder.

Automotive Radar

Hermann Winner and Christian Waldschmidt

Contents

© The Author(s) 2026
H. Winner et al. (eds.), *Handbook Assisted and Automated Driving*,
https://doi.org/10.1007/978-3-658-45276-6_15

15.1 Introduction

Radar (Radio Detection and Ranging) has its origins in World War II military technology and remained tied to military applications for a long time. Its first use in traffic for a speed monitoring system had led to rather negative experiences for many drivers. However, applications perceived as useful for drivers were also considered early on, as evidenced by a magazine article [7] from the year 1955. In the seventies of the twentieth century, a major research project took place with the aim of developing radar sensors suitable for series production for rear-end collision protection. Although this project, which was sponsored by the German Federal Ministry of Education and Research, advanced radar development, the time was not yet mature for series production. It was not until 20 years later that the technical prerequisites were in place for radar to be used for driver assistance. In 1998, a vehicle with radar was available for the first time. Howztion was not rear-end collision warning, but rather adaptive cruise control ACC (see ▶ Chap. 32 and on history [24]), even though the collision warning was integrated as a functional part of this system. Other radar-based ACC systems followed at short intervals.

Radar technology received a further boost about 5 years later with the development of automatic emergency braking (cf. [21]) and lane change assistance (see ▶ Chap. 33 and [2]).

Three bands are currently available for use in road traffic (24,05 − 24,25 GHz, 76−77 GHz and 77−81 GHz). Currently, the dominant band is the 76,5 GHz range, which has been explicitly regulated for automotive radar and is available worldwide. The 24 GHz range is becoming less important, since the cost advantages over sensors that used to be prevalent with 77 GHz are no longer available today. Lastly, the 77 − 81 GHz band is increasingly used, especially when bandwidths above 1 GHz are required for high-range resolution.

The development of the first generation of radar for the automobile, as in comparable cases of innovation, also involved a lot of learning. They started with very heterogeneous technology concepts and were still very expensive for automotive sensors, even if already about two orders of magnitude cheaper than in the military sector. In the meantime, technology convergence is evident in many areas, and combined with the high number of units on the market, this leads to costs that are suitable for the mass market. The outdated technology concepts such as pulse and FMCW modulation (cf. [22] and [23]) are not described further in this edition. The automotive applications in the interior that go beyond the detection of traffic space, e.g. for seat occupancy or gesture recognition, are not considered in this chapter.

Radar cannot be understood without a basic knowledge of communications engineering. Nevertheless, an attempt is made here to present the theoretical considerations in such a way that they can be understood with minimal prior knowledge. Thus, readers with radar expertise may be surprised that the more condensed technical language and formula representation is not used. For the radar fundamentals and definitions used here, reference was made to standard works [11] and [16] in which much more extensive considerations of radar in general can be found. Due to the previous domain of radar in military as well as civil aviation and shipping, the subject area of automotive radar has hardly been addressed in the basic works so far. Therefore, this chapter specifically gives an overview of automotive radar technology, which comes up with significantly different solutions due to its requirements, which differ greatly from the above-mentioned areas of application (smaller ranges, smaller Doppler frequencies, high multi-target capability, small size, considerably lower costs).

15.2 Propagation and Reflection

The radar waves leave the sensor in a bundle manner and not as a spherical wave with uniform intensity in all spatial directions. This is ensured by the antenna (s. a. ▶ Sect. 15.4.1). The so-called directive antenna gain G_D describes the ratio between the intensity $P(\phi, \vartheta)_{\max}$ in the solid angle of the strongest radiation and the value $P_{\text{total}}/4\pi$ of a homogeneous omnidirectional radiator of the same total power. $P_{\text{total}} = \iint P(\phi, \vartheta)d\phi d\vartheta$. Here are ϕ the azimuth angle in the horizontal plane and ϑ the elevation angle in the vertical. The more the beams are focused, the greater is the antenna gain. The actual antenna gain G also takes into account the antenna losses, which are mostly conduction and dielectric losses. The Equivalent Isotropically Radiated Power (EIRP) resulting from the product of the total transmit power and the antenna gain is the decisive quantity for two criteria: Firstly, for the radio license, for which the power in the solid angle range of the maximum is important (specified in dBm (EIRP), where dBm refers to the basic power of 1 mW), and secondly, for the maximum detection range.

For the latter, however, other factors have to be taken into account. Obviously, the reflectivity of the radar target is one of them. This is specified as the so-called radar cross-section (RCS) σ representing an area and with the unit m^2. This area corresponds exactly to the central cross-sectionalarea πR^2 of a spherical reflector with the radius R. For the reflection to the receiver this quantity has to be multiplied by λ^2 because the effective area of the sphere reflected in the direction of reception decreases with decreasing wavelength and reflects outside this area past the receiver. The targets relevant for automotive applications in the middle and

Directional Reflection

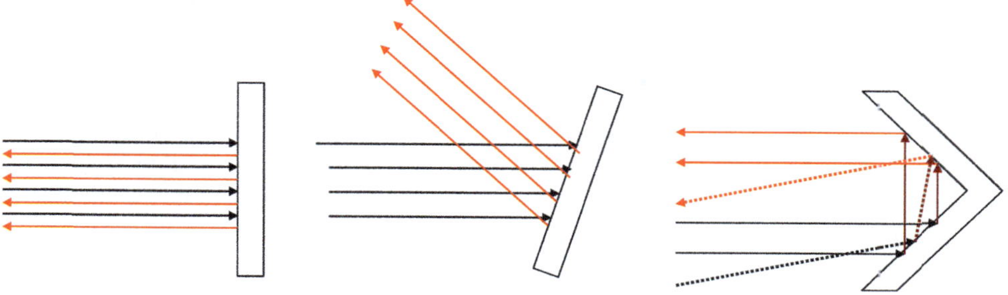

Fig. 15.1 Examples of directional reflection: **a** 90° reflection from a plate, **b** ≠90° reflection from a plate, **c** 9C° double mirror

far range have values of $\sigma = 1 \dots 10000\,\mathrm{m}^2$. If pedestrians are to be detected at close range, much smaller values ($\sigma = 0{,}01 \dots 0{,}1\,\mathrm{m}^2$) can be assumed. The scattering width depends on the material of the target, but even more on the geometry and orientation. A metal plate oriented perpendicular to the transmitting and receiving direction with the area A has at large ranges a backscatter cross-section of Eq. 15.1

$$\sigma_{\text{plate}} = 4\pi A^2/\lambda^2, \tag{15.1}$$

ref to [16]. At $A = 1\,\mathrm{m}^2$ and a frequency of $76.5\,\mathrm{GHz}$ ($\lambda \approx 4\,\mathrm{mm}$), the RCS is $\sigma \approx 0{,}8 \cdot 10^6\,\mathrm{m}^2$. Thus, a box car with a flat rear of $4\,\mathrm{m}^2$ can result in strong backscatter with an RCS of $12{,}5 \cdot 10^6\,\mathrm{m}^2$ (in the far range), but completely collapses when turned by one degree at a range of about 60 m, cf. ▪ Fig. 15.1 a and b. The remaining reflection then comes only from the edges or the axle components. An ideal retroreflector forms a so-called corner (cube) reflector with three right-angled triangular surfaces which are perpendicular to each other. For a perfectly oriented corner reflector, any incoming wave whose wavelength is significantly smaller than the dimensions will be reflected back in the direction from which the wave was sent, as shown in ▪ Fig. 15.1 c for the two-dimensional case. For a three-dimensional corner reflector which consists of three equal-sided, right-angled triangles lying perpendicular to each other with an edge length a and the diagonal dimension $L = \sqrt{2}a$ according to ▪ Fig. 15.2 is calculated according to [25] an RCS of

$$\sigma_{\text{CR}} = \pi L^4/3\lambda^2 \Leftrightarrow L = \sqrt[4]{3\sigma_{\text{CR}}\lambda^2/\pi} \tag{15.2}$$

With such a geometry, even small dimensions ($L = 35\,\mathrm{cm}$) can simulate a very strong reflection of $\sigma_{\text{CR}} \approx 1000\,\mathrm{m}^2$ corresponding to a strongly reflecting truck. For a passenger, car $100\,\mathrm{m}^2$ ($L \approx 20\,\mathrm{cm}$), for a motorcycle $10\,\mathrm{m}^2$ ($L \approx 11\,\mathrm{cm}$), and for a human $1\,\mathrm{m}^2$ ($L \approx 6{,}2\,\mathrm{cm}$) are considered as typical radar cross-sections. In the ISO standard for ACC and FSRA [18] a radar cross-section of $10 \pm 3\,\mathrm{m}^2$ is prescribed for

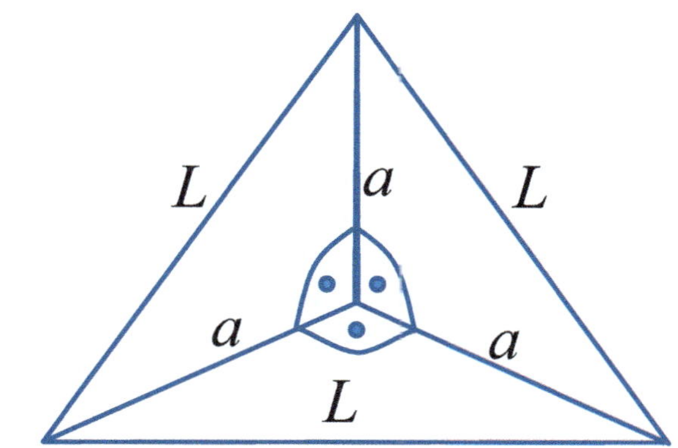

Fig. 15.2 Geometry of a corner reflector

the detection field measurements, whereby this is considered to be the lower limit for 95% the vehicles. Small values for radar cross-sections occur mainly on vehicles with path-reflecting flat or concave surfaces. Large values are mainly due to angular reflectors. For example, the support posts of crash barriers with their U-profile show quite high radar cross-sections, which leads to an appearance of a large number of these targets in the object list. Also, the entrance steps to truck cabs are so highly reflective as corner reflectors that they still bring sufficient signal power to the receiver even outside the main radar beam. On the one hand, the high dynamics of the radar cross-section over four to five orders of magnitude means that object classification via the radar cross-section must remain unsuccessful. On the other hand, the dynamics of the radar cross-section increases the dynamic demand on the receiving signal branch, which should therefore not be below 70 dB, which is nevertheless not safe from clipping.

In addition to the dynamics of the radar cross-section, the radial range r influences the signal strength at the receiver. As already considered, the power in a solid angle element remains constant, at least if absorption losses are not considered. The area of this angular segment increases with the range squared, the same applies to the reflected beam, so that for targets outside the close range a r^{-4}-decrease can be assumed. The absorption k (mostly given in dB/km) is only in a few cases so high that it has to be taken into account. At 76.5 GHz, the atmospheric attenuation is below $k < 1$ dB/km thus only 0.3 dB for the outward and return path to a 150 m distant target. At 60 GHz on the other hand, there is a maximum attenuation, with about $k \approx 15$ dB/km. Although this attenuation is significantly higher than the attenuation at 76.5 GHz, it is still small compared to the power drop with r^{-4}. The bands around 60 GHz are used for 5G mobile and short-range sensing applications in much of the world, therefore, the use of this band for automotive applications facing outward is not envisioned. Due to heavy rain, especially with large droplets reaching the order of the wavelength, attenuation is quite severe, resulting in a significant range loss, with the achievable visibility often exceeding that remaining for humans. In addition to the attenuation effect, heavy rain leads to an increased disturbance level (clutter). In most cases, it acts like an increased noise level and in this way lowers the signal-to-noise ratio (SNR) and thus the range. Another disturbing effect of a "water environment" occurs due to the guard (radome) of the beam output area. Due to the high dielectric constant, water has a high refractive effect on mm-waves, so that an uneven water coverage leads to unwanted "lens effects", which can distort the determination of the azimuth and/or elevation angle.

The last influencing factor on the received power mentioned here is multipath propagation. On the one hand, this concerns the vertical multipath propagation via the reflection at the road surface. Due to the consistent small grazing angles at longer ranges, the reflection occurs almost completely independent of polarization and road wetness [13]. Thus, the radar beams take different paths and arrive at the receiver with different phases. The difference is particularly strong if the superposition already acts at the reflector. At $\Delta r = (2n - 1)\lambda/2; n \in \mathbb{N}$ the incoming beams cancel each other out. At $\Delta r = n\lambda$, the reflection field strength is doubled, which means that both paths constructively overlap at the receiver if the return path is also parallel. In this case, four times the amplitude and thus sixteen times the power would be measured. However, the ideal condition is seldom singularly fulfilled, since within the resolvable range cells, see Eq. 15.27. Within the resolvable range cells, see also reflections with different phase positions are possible (e.g. vehicle rear with reflection surfaces of different ranges and heights in relation to the wavelength).

Depending on the installation height of the radar and the height of the reflection center above the road, cancellation occurs at certain intervals, causing the detection performance of the radar to fluctuate noticeably. For the most part, this is not problematic, since

even bouncing in and out of the target vehicle or one's own vehicle, as well as road irregularities, eliminate the superposition hole, and furthermore, with finite radial velocity, the associated cancellation spacing condition is quickly eliminated.

The vertical multipath reception thus manifests itself as a signal power "shaker" which can be described by a distribution function. V_σ so that a stochastically describable detection loss or drop-out rate is to be expected during detection. In the case of horizontal multipath propagation, reflection occurs on vertical surfaces lying approximately parallel to the direction of travel. In addition to walls, guard rails in particular can enable horizontal multipath propagation. The signal cancellation is less disturbing than the distortion of the azimuthal directional information.

The simple path difference between a direct wave and one reflected over the ground is [1]

$$\Delta r_{\mathrm{s}} = \frac{2 h_{\mathrm{S}} \cdot h_{\mathrm{T}}}{r}. \tag{15.3}$$

If one observes the course of the reception amplitude over a longer range, a transformation into the reciprocal range (i.e. $1/r$), a harmonic periodicity can be detected, the "frequency" of which indicates the product of sensor height and target height [5, 6], so that a bridge that can be driven under can be distinguished from an obstacle standing on the roadway. However, the standing obstacle over lateral reflectors (e.g. over guardrails) may produce similar patterns, since the product of the normal ranges to the reflector plane may be similar here as well.

As shown in ▪ Fig. 15.3, the power normalized to RCS, referred to as the r^{-4} normalized received power due to multipath reflection varies greatly. The roadway reflects almost completely at the low grazing angles, resulting in almost complete cancellation but also strong overshoot. On the one hand, the incident waves of the two paths overlap at the reflector. This leads to extinction when both wave strands overlap by $\Delta r = (2n - 1)\lambda, n \in \mathbb{N}$ distinguish. In the ideal case, constructive superposition results in a 16-fold increase in receiver power since the field strength amplitude is doubled on both the outward and return paths. However, this only applies to a small area at the reflector. On the other hand, reflections extending beyond this, e.g. at the rear of the vehicle, be it laterally or vertically offset, are only superimposed again at the receiver and cancel each other out there if $2\Delta r = (2n - 1)\lambda, n \in \mathbb{N}$ applies. The constructive superposition doubles the field strength amplitude at the receiver, corresponding to about four times the received power. Thus, the

---

1 Approximation for the case $h_{\mathrm{S}} + h_{\mathrm{T}} \ll r$.

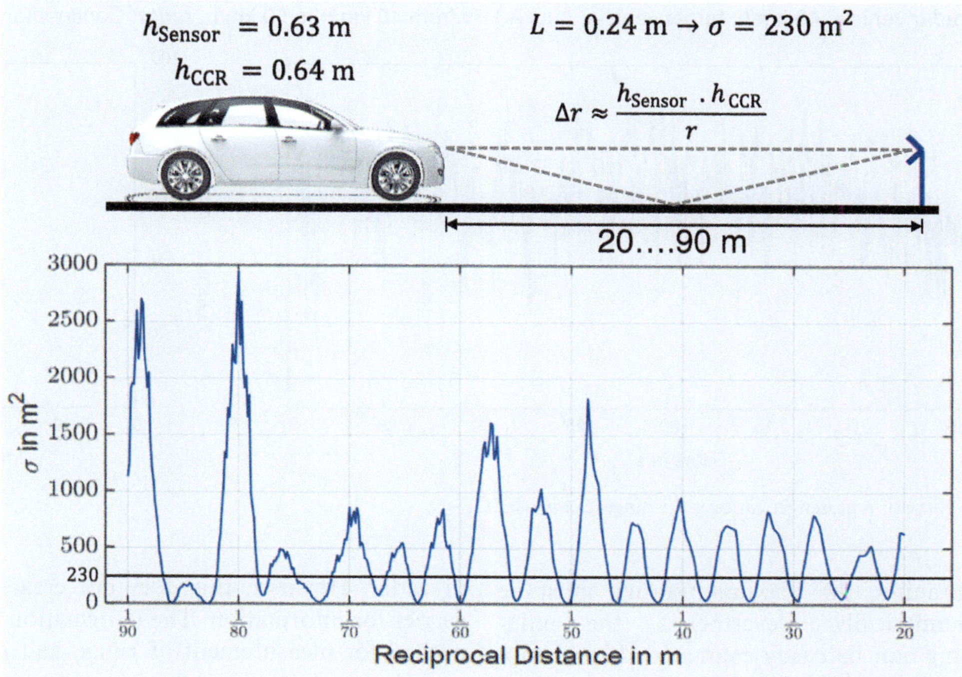

Fig. 15.3   RCS measurement of a corner reflector as a function of the reciprocal range

approach to a larger corner cube reflector shows in **Fig. 15.3**, that at large ranges (i.e. small $1/r$), in the picture on the left, the constructive superposition reaches almost to 16 times the CCR-RCS and the periodicity occupies larger ranges than at smaller ranges, when the dimensions of the CCR are no longer sufficient for a complete superposition at the reflector and both the maximum values correspond "only" to about four times the nominal value and change about twice as often.

In a scenario that is more realistic for road traffic applications, in this case, the departure of a passenger car with the radar vehicle moving by itself, the following picture is obtained, cf. **Fig. 15.4**. The RCS value (here plotted as a level value logarithmically in dBm^2) varies by about 10 dB peak-to-peak, i.e. by a factor of 10. As shown in [9] the measured RCS values can be assigned to an exponential distribution as well as to a log-normal distribution. The difference of the level values around a moving average over 10 s follows in a wide range the logarithm of a normal distribution with a standard deviation of $\Delta Q_{(\sigma)}$ so that the RCS value can be described as log-normally distributed in a good approximation, i.e.

$$V_\sigma \sim \mathcal{LN}\left(1, 10^{\Delta Q_{(\sigma)}}\right). \tag{15.4}$$

In the example **Fig. 15.4**, this way a $\Delta Q_{(\sigma)} = 3,7$ dB is determined. At greater range (25 s $< t <$ 40 s), a slight periodicity is recognizable, which is to be interpreted as a superposition pattern and shows single strongly decreasing RCS values, whose peak-to-peak even generates 20 dB fluctuations. A converted plot (σ vs. $1/r$) corresponding to **Fig. 15.3** also shows this periodicity, but much less pronounced than in the CCR experiment. A spectral evaluation and further investigations can be found in [9].

Summarizing the influencing factors described in this section, the maximum range for a detection can be derived. The power P_{Rx} of the received signal is calculated from the radar equation:

$$P_{\text{Rx}} = 10^{-2kr/1000} \cdot \sigma \lambda^2 \cdot G_{\text{Tx}} \cdot G_{\text{Rx}} \cdot V_\sigma \cdot P_{\text{Tx}}/4\pi r^4 \tag{15.5}$$

If the same antenna is used for transmission (Tx) and reception (Rx), the following applies $G_{\text{Tx}} = G_{\text{Rx}}$. For target detection, the power of the received signal must be sufficiently far from the noise. Depending on other signal evaluation for false target suppression, the threshold is a factor $SNR_{\text{threshold}}$ of about 6 to 10 dB above the noise (power P_{N}).

The achievable maximum range $r_{\max}$ is determined by multiplying the received power by Eq. 15.5 and set to the detection threshold $P_{\text{N}} \cdot SNR_{\text{th}}$. Neglecting the attenuation, i.e. $k = 0$ it can be calculated analytically:

$$r_{\max} = \sqrt[4]{\frac{\sigma \lambda^2 \cdot G_{\text{Tx}} \cdot G_{\text{Rx}} \cdot V_\sigma \cdot P_{\text{Tx}}}{4\pi P_{\text{N}} \cdot SNR_{\text{th}}}} \tag{15.6}$$

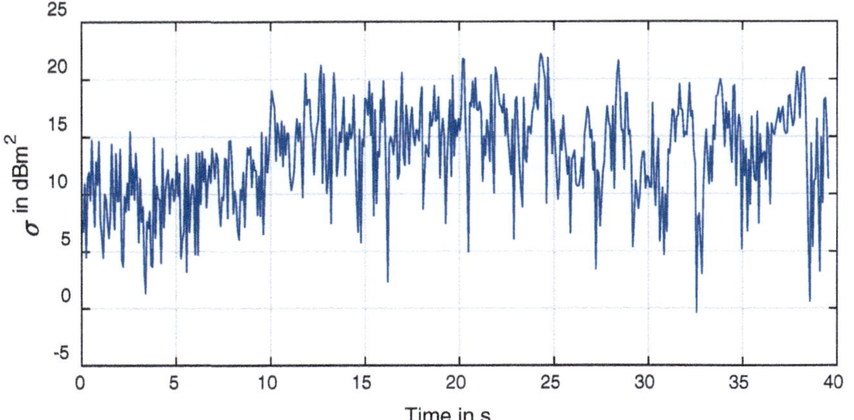
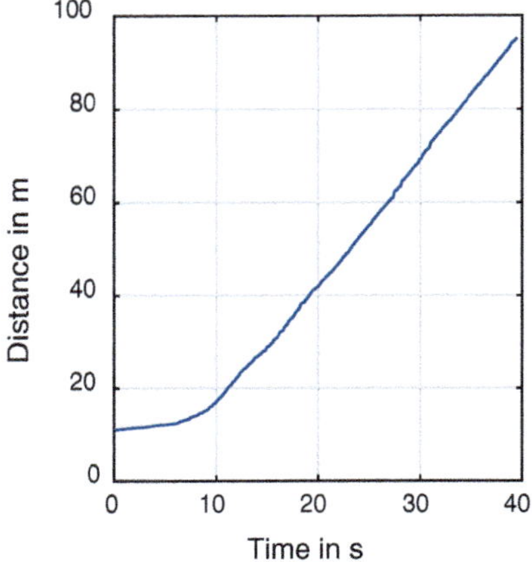

Fig. 15.4 RCS curve of a passenger car over the range. (*Source* FZD)

With finite damping, the maximum range must be determined numerically. Nevertheless, the influence of damping can be easily estimated. If a damping-free range of 200 m can be assumed, it is reduced at 21 dB/km to 140 m (^{2}), at 60 dB/km to 100 m and at 240 dB/km to 50 m. In principle, therefore, all the factors that determine the theoretical maximum range of the radar are known. In practical applications, however, further limits are imposed by signal processing, as described in the following section.

15.3 Fundamentals of Radar Signal Processing

To understand the operation of radar, we have to digress into the mathematics of telecommunications. The mathematical relations derived in the following sections are limited to the minimum from the authors' point of view and are presented in a rather popular notation. For an even deeper consideration of radar technology, reference is made to standard works on radar, such as the book by Skolnik [16].

15.3.1 Basic Principle of Modulation and Demodulation

15.3.1.1 Theoretical Foundations

The emission of electromagnetic waves and their reception is only a necessary condition for the function of radar. However, this does not create more than a carrier for information. The information itself, which is needed for measurement of range and radial velocity, must be modulated onto this carrier on the transmitting side and demodulated again on the receiving side. In summary, the transmitted wave train must be provided with an identifier for recognition and a time reference for measuring the transit time. This task is called modulation. The recognition and the determination of temporal relationships are performed by demodulation.

In a general form, the radiated wave can be described as a harmonic wave function:

$$u_{\text{Tx}}(t) = A_{\text{Tx}} \cdot \cos(\varphi_{\text{Tx}}(t)) \tag{15.7}$$

where the phase changes periodically with an instantaneous angular frequency $\omega_{\text{Tx}}(t) = 2\pi \cdot f_{\text{Tx}}(t)^3$ close to the carrier angular frequency $\omega_{\text{c}} = 2\pi \cdot f_c$.

$$\varphi_{\text{Tx}}(t) = \int_{t_0}^{t} \omega_{\text{Tx}}(t\prime)dt\prime \tag{15.8}$$

In principle, modulation of three variables amplitude A_{Tx}, instantaneous frequency f_{Tx}, and phase φ_{Tx} modulations can be performed. In the past, pulse modulation as a special form of amplitude modulation dominated, many radar application areas, but today this is no longer used in the current automotive radar developments. Therefore, the following calculations are limited to the phase $\varphi_{\text{Tx}}(t)$ and its derivative, and the instantaneous angular frequency $\omega_{\text{Tx}}(t)$ will be derived. The amplitude still plays the essential role for the signal strength but can be assumed to be constant in time in the course of a measurement cycle.

---

2   Calculation example $(200/140)^4 \triangleq 6\text{dB}$ times $(1\text{km}/(2 \cdot 140\text{m})) \triangleq 3{,}5\,\text{dB}) = 21\,\text{dB}$.

3   The index Tx refers to the transmitting, Rx is used for the receive signal.

For the signal to be received after reflection at a target, the same signal representation applies

$$u_{Rx}(t) = A_{Rx} \cdot \cos(\varphi_{Rx}(t)). \tag{15.9}$$

A_{Rx} depends according to the radar equation Eq. 15.5 from A_{Tx} where $A_{Rx}^2/A_{Tx}^2 = P_{Rx}/P_{Tx}$ holds. For the phase

$$\varphi_{Rx}(t) = \varphi_R + \varphi_{Tx}(t - \tau_{of}) = \int_{t_0}^{t-\tau_{of}} \omega_{Tx}(t\prime)dt\prime. \tag{15.10}$$

This means that the phase at the receiver, except for the phase jump[4] of φ_R is identical to the phase that was present at the transmitter a time-of-flight τ_{of} earlier. The time-of-flight in turn results from the propagation time from the transmitter to the reflector and from the reflector to the receiver. If the transmitter and receiver are close to each other, this duration is the same for the outward and the return paths. A relative movement of target and sensor changes the duration only by the ratio to the speed of light,[5] resulting in a negligible order of magnitude of 0.1 mm. The time-of-flight between reflector and sensor at the time of reflection $(t - \tau_{of}/2)$ is

$$\tau_{of}(t) = 2r(t - \tau_{of}/2)/c \cong 2r(t)/c, \tag{15.11}$$

with c as the speed of light (in air: $299{,}7 \cdot 10^6$ m/s) and can be calculated without relevant error (even under 0,1 mm) the range at the time of reception $r(t)$ represent. A range of 150m results in a time of flight of $1\mu s$. Equation 15.11 supplies also the basis for the determination of the radial velocity differentiated after the time $\dot{r}$:

$$\dot{\tau}_{of}(t) = 2\dot{r}(t - \tau_{of}/2)/c \cong 2\dot{r}(t)/c. \tag{15.12}$$

Again, the approximation errors are irrelevant.[6]

The decisive information for the measurement of the desired quantities range and radial velocity in radial direction lies in the phase shift between transmitted and received signal and its change. To make this information observable, the transmitted signal must be modulated in such a way that the phase change can be measured over the time-of-flight. Furthermore, the transmitted and received signals must be coupled in such a way that the information encoded in the phase becomes accessible. This is done by the process known in radio engineering as mixing. Mathematically, this process corresponds to a multiplication of two signals in the time domain. In the modulation principles presented here, it is mixed into the baseband. The multiplication of two harmonic signals in the time domain results according to the addition theorems

$$\cos x \cdot \cos y = \frac{1}{2}\{\cos(x - y) + \cos(x + y)\}. \tag{15.13}$$

a sum of signals with differential and sum phase. In the case of transmit and receive signals with a carrier frequency of e.g. 76,5 GHz, this means that on the one hand very high frequency signal components ($2 \cdot 76{,}5$ GHz $= 153$ GHz) are generated, which are very strongly attenuated in the following signal line without further measures. On the other hand, very low frequency difference signals, which are attenuated by a factor of 2, but still contain the complete information about the phase difference of transmit and receive signal. The output of the ideal mixer (according to Eq. 15.13) provides

$$\begin{aligned} u_M(t) = u_{Rx}(t) \cdot u_{Tx}(t) &= \frac{A_{Rx} \cdot A_{Tx}}{2} \cdot \cos(\varphi_{Rx}(t) - \varphi_{Tx}(t)) \\ &= A_M \cdot \cos(\varphi_M(t)). \end{aligned} \tag{15.14}$$

The phase of the mixed signal is the difference of the phase of the transmitter signal from the time of sending $t - \tau_{of}(t)$ to the time of reception t integrated:

$$\begin{aligned} \varphi_M(t) &= \varphi_R + [\varphi_{Tx}(t) - \varphi_{Tx}(t - \tau_{of}(t))] \\ &= \varphi_R + \int_{t-\tau_{of}(t)}^{t} \omega_{Tx}(t\prime)dt\prime \end{aligned} \tag{15.15}$$

A direct evaluation of the phase is not possible, since the phase difference is $2r/\lambda$ is far outside the uniqueness of 0 to 2π. On the contrary, the derivative, the difference angular frequency can be evaluated directly. However, this requires to know the temporal course of $\tau_{of}(t)$ and $\omega_{Tx}(t)$. For the assumption of within $\tau_{of}(t)$ time-linear range and instantaneous angular frequency curves, i.e.

$$\tau_{of}(t) = \tau_{of}(t_{Rx}) + (t - t_{Rx}) \cdot \dot{\tau}_{of}(t_{Rx}) \tag{15.16}$$

and

$$\omega_{Tx}(t) = \omega_{Tx}(t_{Rx}) + (t - t_{Rx}) \cdot \dot{\omega}_{Tx}(t), \tag{15.17}$$

results in the instantaneous angular frequency

$$\begin{aligned} \dot{\varphi}_M(t_{Rx}) &= \frac{d}{dt}(\varphi_{Tx}(t_{Rx} - \tau_{of}(t)) - \varphi_{Tx}(t_{Rx})) \\ &\cong \dot{\tau}_{of}(t_{Rx}) \cdot \dot{\omega}_{Tx}(t_{Rx}) + \dot{\tau}_{of}(t_{Rx}) \cdot \omega_{Tx}(t_{Rx}). \end{aligned} \tag{15.18}$$

The second term of Eq. 15.18 ($\dot{\tau}_{of}(t_{Rx}) \cdot \omega_{Tx}(t_{Rx})$) describes the Doppler shift, which already becomes visible with constant transmission frequency as measurable frequency, whereby the sign cannot be resolved without further measures. In order to determine the radial range over the transit time a change of the instantaneous frequency must necessarily take place. From Eq. 15.18 with Eqs. 15.11 and 15.12, the instantaneous

---

4   In the case of total reflection, as on metallic surfaces, $\varphi_R = \pi$.
5   The relative error is concretely $\dot{r}(t)/c \ll 10^{-6}$.
6   The absolute error is well below 10^{-4} m/s.

frequency $f_M(t)$ of the mixer signal can be described as a linear combination of range and radial velocity:

$$f_M(t) = \frac{2}{c}\left(r(t) \cdot \dot{f}_{Tx}(t) + \dot{r}(t) \cdot f_{Tx}(t)\right). \tag{15.19}$$

This equation is the **basic equation** for all radar sensors that use frequency modulation as the basic principle, as currently used.

15.3.1.2 Doppler Effect

Direct measurement of radial velocity with radar is a significant functional advantage over other measurement principles, providing resolutions up to two orders of magnitude that is better than range differentiation. The Doppler shift $f_D(\dot{r}, f_c)$ at a carrier frequency of $f_c = 76{,}5\,\text{GHz}$ provides for the radial velocity $\dot{r}$ according to Eq. 15.19 a Doppler shift of

$$f_D(\dot{r}, 76{,}5\,\text{GHz}) = -510\,\text{Hz} \cdot \dot{r}/(\text{m/s}) \tag{15.20}$$

or at the other frequency commonly used for driver assistance applications of 24 GHz about one third of this, namely

$$f_D(\dot{r}, 24\,\text{GHz}) = -161\,\text{Hz} \cdot \dot{r}/(\text{m/s}). \tag{15.21}$$

For conversion from m/s to km/h units, the values are to be divided by 3.6. With an assumed radial velocity of $\dot{r} = -70\,\text{m/s}(-252\,\text{km/h})$ when approaching a target, the maximum Doppler frequencies at 76.5 GHz carrier frequency are $f_{D,\max} = f_D(-70\,\text{m/s}, 76{,}5\,\text{GHz}) = 35{,}7\,\text{kHz}$ so that a measurement according to Nyquist theorem with a sampling rate of at least $f_{S,v} \geq 71{,}4\,\text{kHz}$ is required for an unambiguous determination.

Via the duration T_M of the Doppler shift measurement, the resolution can be calculated to

$$\Delta \dot{r} = c/2f_c T_M. \tag{15.22}$$

15.3.1.3 Mixer

As described up to this point, the mixing of signals is a simple and straightforward mathematical calculation. For the technical implementation, digital multiplication is out of feasibility, since affordable analog/digital converters are too slow for the radar frequencies used in automobiles.

Active mixers with so-called Gilbert cells, which are mainly used today, come quite close to the ideal multiplier. A Gilbert cell is shown in ▫ Fig. 15.5 figure. It consists of three differential amplifiers, here labelled A, B_1, and B_2. Differential amplifier means that these amplifiers respond only to a differential signal at the input; an in-phase component is not amplified. The high frequency signal U_S (corresponds in Eq. 15.14 $u_{Rx}(t)$) now alternately determines the current flow through transistors T_1 and T_2 given by the current source, depending on which half-wave of the high-frequency signal is

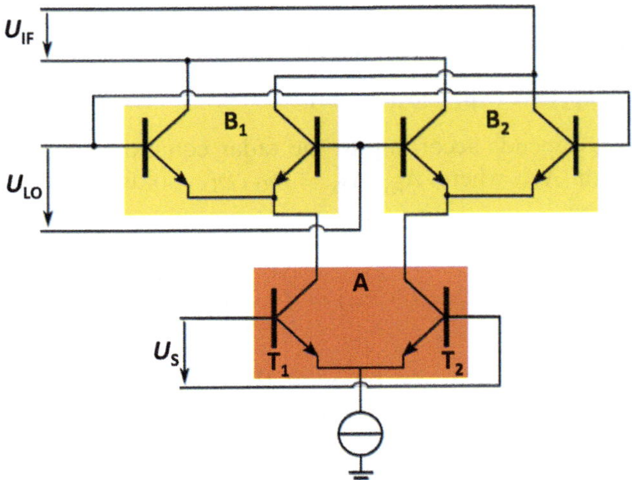

▫ **Fig. 15.5** Principle circuit diagram of a Gilbert cell (small-signal equivalent circuit with ideal current source and without peripheral circuitry)

applied to differential amplifier A. Thus, the current flow through the left or right half of the circuit alternates depending on the half-wave from U_S. The two differential amplifiers B_1 and B_2 are cross-connected, i.e. the two have the same gains but opposite signs, which are given by U_{LO} (corresponds to $u_{Tx}(t)$). Thus, the current flow is alternately crossed by U_{LO} determined. Thus, both differential amplifier stages result in a multiplication of the signals since the current flowing through the left and right sides of the circuit is given by both U_{LO} as well as U_S is controlled. The result is U_{IF} (corresponds to $u_M(t)$). For the implementation of the Gilbert cell, very fast transistors are needed.

Regardless of the frequency range 24 or 76 GHz, the two technologies, silicon germanium (SiGe) and CMOS, are available today as semiconductor technologies for realizing the very fast, high-frequency circuit components or transistors. Both technologies have different strengths and weaknesses, but in any case, they enable the high integration of the entire radar front-end into a Monolithic Microwave Integrated Circuit (MMIC). The technology that will prevail in the long term depends primarily on cost. Both technologies have very different cost structures in terms of development costs versus unit costs, with CMOS being the more cost-effective for large volume.

Modern MMICs contain the complete radar circuits for several transmit and receive channels. This includes not only the central component of the mixer but also all amplifiers, couplers, and filters, which are necessary for the preparation and post-processing of the signals before transmission or before A/D conversion. Currently, MMICs with 3 transmit and 4 receive channels, which are suitable for the design of low-cost radar sensors, dominate the market. If more powerful radar sensors with many channels are built, several MMICs

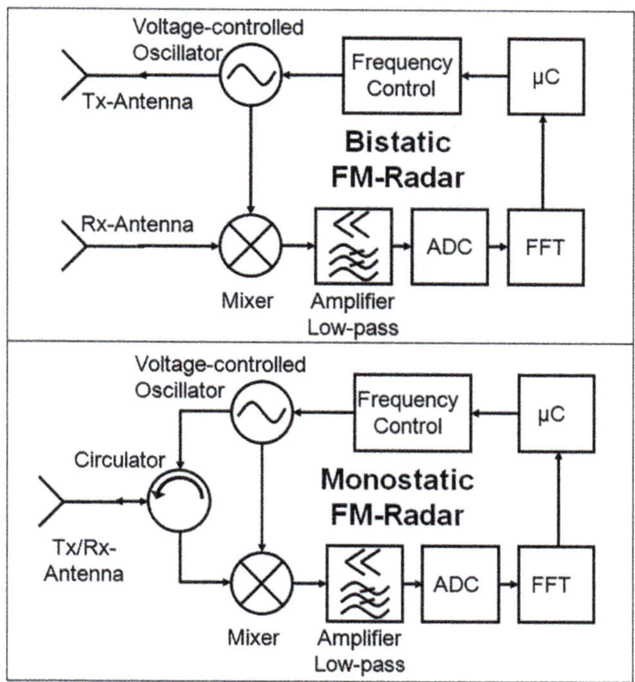

Fig. 15.6 Block diagram of a radar with frequency modulation. Top: in a bistatic design with separate antenna feeds for transmit and receive beams; bottom: in a monostatic design with circulator coupling

have to be coupled. In the research and start-up area, MMICs with considerably more channels have also been designed in the last two years, but these are not used in the mass market today.

15.3.2 Frequency Modulation (FM)

With frequency modulation, the transmission frequency $f_{\mathrm{Tx}}(t)$ is varied as a function of time, whereby it must be made clear that this is not an absolute and thus constant frequency, but an instantaneous frequency $f(t) = \omega(t)/2\pi$. In this chapter, frequency modulation is referred to all methods in which the information of the propagation time is achieved via the instantaneous frequency variation.

The basic structure of FM radar is shown in **Fig. 15.6**. Mandatory to its operation is the variation of frequency by means of a voltage-controlled oscillator, which provides the desired modulation directly or via a control loop (e.g. phase-locked loop, PLL). The received signal is mixed with the currently transmitted signal, filtered, sampled and converted. For signal separation of the transmitting branch and receiving branch, this could either be spatially separated feed lines (**Fig. 15.6** above) or special non-reciprocal[7] couplers (**Fig. 15.6** below), which couple directionally selectively.

15.3.2.1 Development of Modulation Types for Automotive Radar

Frequency modulation has completely displaced other modulation formats such as pulse modulation in the course of the last generations of automotive radar sensors. This was particularly due to the good integrability of radar circuits in silicon semiconductor technologies, since frequency-modulated radars have a constant low transmit power and do not require high peak pulse power.

Frequency modulations can be very different and consist, for example, of various fixed frequency steps (FSK—frequency shift keying) or so-called frequency ramps in which the frequency is continuously increased or decreased over time (FMCW—frequency modulated continuous wave). Above all, FMCW radar with a few approximately $1 \ldots 10\,\mathrm{ms}$ long ramps was used very frequently and turned out to be the dominant modulation format for automotive radar sensors before it was and is being replaced by chirp[8] sequence modulation with many, significantly shorter ramps.

Basically, the generation of the range information takes place with the FMCW modulation as with the chirp sequence modulation, as will be explained in the

---

7 Describes a direction-dependent component behavior.

8 Chirp sequence modulation also modulates the instantaneous frequency and is thus a variant of FMCW modulation. It is used here conceptually as an independent form, while FMCW refers only to the formerly widespread form with a few longer ramps.

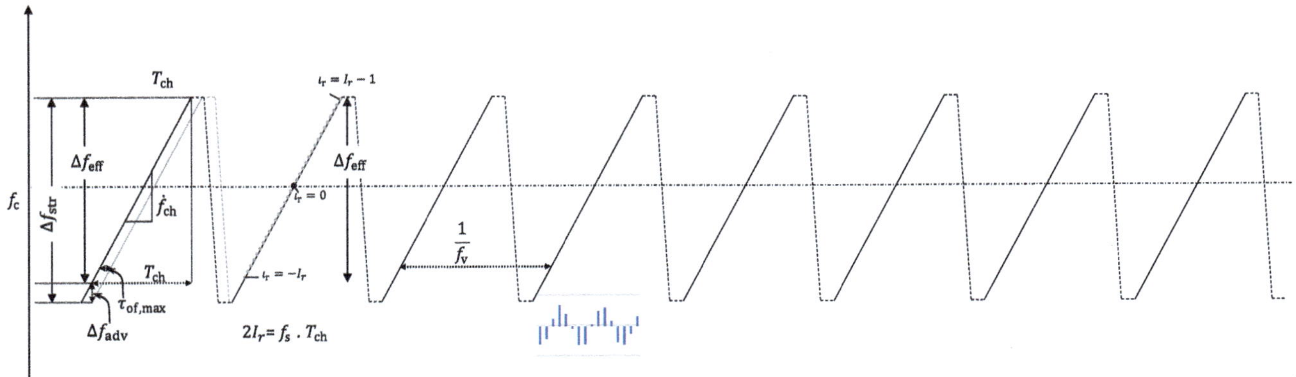

◘ Fig. 15.7 Chirp sequence modulation: instantaneous frequency graph and definition of the quantities, see text

following chapter. However, the Doppler shift, which occurs with moving targets and which is used for velocity measurement, is superimposed on the received signal or the range information. Consequently, the separation of range and velocity information for the individual targets is the major challenge in FMCW radar sensors, cf. [23]. Various methods have been developed for this purpose, but all are computationally expensive. With the replacement of FMCW by chirp sequence modulation, this problem no longer occurs.

15.3.2.2 Chirp Sequence Modulation (Multi-Chirp)

The modulation described below has several names. Here it is called chirp-sequence modulation because it consists of a sequence of equal linear frequency ramps, see ◘ Fig. 15.7. At short intervals $I_v \in \mathbb{N}$ identical linear frequency ramps are repeated at short intervals, which, if they were frequency-increasing (up-chirp), would sound like chirping of a bird or insect in the acoustic range.

15.3.2.3 Range

The instantaneous frequency of a chirp follows over the duration of the chirp T_{ch} the linear Eq. 15.17 with the constant rate of the instantaneous frequency $\dot{f}_{ch}$. The stroke of the ramps for the measurement determines the effective bandwidth $\Delta f_{eff} = T_{ch} \cdot \dot{f}_{ch}$ which is typically $100 \ldots 500\,\mathrm{MHz}$ for long range measurements (range > 100 m) and up to 2 GHz for short-range measurements (range < 100 m). For settling in, at the beginning of the ramp, a pre-run $\Delta f_{adv} = \tau_{of,max} \cdot \dot{f}_{ch}$ corresponding to the maximum time-of-flight must be performed at the beginning of the ramp, so that a total frequency deviation of $\Delta f_{str} = \Delta f_{eff} + \Delta f_{adv}$ follows.[9] During the effective duration of the ramp T_{ch}, the mixer signal $u_m(t)$ is sampled with the sampling frequency f_s whereby the $2I_r = f_s \cdot T_{ch}$ real discrete measured values

$$u_{\iota_r} = u_m(t_{\iota_r}) = u_m\left(t_{ch} + \frac{\iota_r}{f_s}\right); -I_r \leq \iota_r \leq -I_r - 1. \quad (15.23)$$

arise. The temporal reference point selected for the chirp t_{ch} lies here in the middle of the effective chirp, so that the data set of $(-I_r < \iota_r < I_r - 1)$ is indexed. At this reference point, the average frequency is also referred to as carrier frequency f_c. A Discrete Fourier Transform (DFT) is performed over the data set (mostly as a Fast Fourier Transform (FFT) to a radix-2 number $2I_r$). A DFT presupposes the periodic continuation of the signal, whereby with signals, which do not appear exactly as integer multiples of the base frequency T_{ch}^{-1} discontinuities are generated at this periodic continuation. This causes signal energy that is actually monofrequency to leak out into adjacent spectral bands, masking other, lower spectral lines. To reduce this effect, which is called leakage effect, a windowing is performed before the transformation as a multiplication with the window function W_{ι_r} (for some window functions this can also be done as convolution after the DFT). The choice of window function depends on what priorities are set. Window functions that suppress very strongly at the edge of the measurement window and even far from the edge allow a strong reduction of the leakage effect outside the vicinity of the spectral peak causing the leakage. The price one pays is the not inconsiderable broadening of this spectral peak. With these filters, the leakage is concentrated on the two or three neighboring spectral lines before and after the peak. For spectra with a high number of spectral lines, this priority is acceptable, for a small number and thus a small relative spectral resolution, this disadvantage is much more important.

As a result of the described process, the complex spectrum

$$\hat{u}_{\hat{\iota}_r} = \mathcal{F}\left(W_{\iota_r} \cdot u_{\iota_r}\right); -I_r \leq \hat{\iota}_r \leq I_r - 1 \quad (15.24)$$

---

9 Index derivation: adv→advance, ch→chirp, eff→effective, m→mixer, s→sampling, str→stroke.

is formed.[10] Since this spectrum is calculated from real values, the symmetry applies $\widehat{u}_{\widehat{\iota}_r} = \widehat{u}^*_{-\widehat{\iota}_r}$, so that the usual one-sided representation of $0 \leq \widehat{\iota}_r \leq I_r$ is sufficient. The DC component ($\widehat{\iota}_r = 0$) and the fraction of the Nyquist frequency ($\widehat{\iota}_r = I_r$) have almost no information value unless coverage or faults can be detected via the "zero line".

Each complex amplitude $\widehat{u}_{\widehat{\iota}_r}$ corresponds to the frequency $\widehat{\iota}_r/T_{\text{ch}}$ which according to the basic equation Eq. 15.19 is a linear combination of range and radial velocity:

$$\widehat{\iota}_r/T_{\text{ch}} = \frac{2}{c}\left(r \cdot \dot{f}_{\text{ch}} + \dot{r} \cdot f_{\text{c}}\right) \tag{15.25}$$

This results in the representation

$$\widehat{\iota}_r = \frac{2}{c}\left(r \cdot \dot{f}_{\text{ch}} \cdot T_{\text{ch}} + \dot{r} \cdot f_{\text{c}} \cdot T_{\text{ch}}\right)$$
$$= \frac{2}{c}(r \cdot \Delta f_{\text{eff}} + \dot{r} \cdot f_{\text{c}} \cdot T_{\text{ch}}) \Rightarrow r = \widehat{\iota}_r \cdot \Delta r - \dot{r}\frac{f_{\text{c}}}{\dot{f}_{\text{ch}}} \tag{15.26}$$

with

$$\Delta r = c/2\Delta f_{\text{eff}} \tag{15.27}$$

as range resolution. The Doppler component $-\dot{r} \cdot f_{\text{c}}/\dot{f}_{\text{ch}}$ causes a small bias, but this can also be seen as a temporal shift of the measurement by the duration $-f_{\text{c}}/\dot{f}_{\text{ch}}$. For representative values of $f_{\text{c}} = 76{,}5\,\text{GHz}; \dot{f}_{\text{ch}} = \pm 100\,\text{MHz}/\mu s$ this shift amounts to (depending on the sign) $\mp 0{,}765\,ms$ which is negligible in view of the measurement cycles of $T_{\text{cy}} > 10\text{ms}$. Related to the result of a single chirp, the spectrum can thus be $\widehat{u}_{\widehat{\iota}_r}$, which can be regarded as a range spectrum with a step size of Δr. In principle, the range spectrum would extend up to the range $I_r \cdot \Delta r$. However, there is a danger that over-reaches will be reflected in the spectrum as alias lines. To protect against this, an anti-aliasing low-pass filter is used, which cleans the mixer signal $u_{\text{m}}(t)$ which in turn leads to the fact that the theoretically possible bandwidth cannot be used completely, but only about 80...90%. For further processing the run variable $\widehat{\iota}_r$ can be limited to the used band, i.e. $0 \leq \widehat{\iota}_r \leq I_{ru}$; with $I_{ru}/I_r \sim 0{,}8 \ldots 0{,}9$. This results in the maximum range to provided that the received power is sufficient to exceed the detection threshold.

$$r_{\max} = I_{ru}\Delta r, \tag{15.28}$$

15.3.2.4 Radial Velocity

The sequence of chirps with the chirp number I_v is done with a constant repetition rate f_v over the measuring cycle duration $T_{\text{M}} = I_v/f_v$. Centered around the sequence reference time $t_{\text{cy},j}$ for each chirp $\iota_v; (-I_v/2 \leq \iota_v \leq I_v/2 - 1)$, the calculation process described above is carried out, whereby then Eq. 15.25 to

$$\widehat{\iota}_{r,\iota_v}/T_{\text{ch}} = \frac{2}{c}\left(r(t_{\text{ch},\iota_v}) \cdot \dot{f}_{\text{ch}} + \dot{r}(t_{\text{ch},\iota_v}) \cdot f_{\text{c}}\right) \tag{15.29}$$

is transformed. Therefore, the same spectra would be expected, which is largely true with respect to spectral power, provided the sequence is not spread too far apart. The complex amplitude $\widehat{u}_{\widehat{\iota}_r,\iota_v}$ to each spacer cell $\widehat{\iota}_r$ varies from chirp to chirp due to the phase change. For this, Eq. 15.15 can be used. For radially moving targets, the time-of-flight $\tau_{\text{of}}(t)$ changes by $\dot{\tau}_{\text{of}}(t)/f_v$ from chirp to chirp. This means that the over $\tau_{\text{of}}(t)$ integrated phase $\tau_{\text{of}}(t) \cdot \omega_{\text{TX}}$ changes accordingly from chirp to chirp.

$$\varphi_{\widehat{\iota}_r,\iota_v} - \varphi_{\widehat{\iota}_r,\iota_v-1} = \left(\dot{\tau}_{\text{of}}(t_{\text{ch},\iota_v})/f_v\right) \cdot \omega_{\text{c}} \tag{15.30}$$

If now over each range cell $\widehat{\iota}_r$ of all chirps of a sequence is Fourier transformed (DFT) including the windowing, a two-dimensional spectrum is obtained over the range ($\widehat{\iota}_r$) and radial velocity $\widehat{\iota}_v$:

$$\widehat{u}_{\widehat{\iota}_r,\widehat{\iota}_v} = \mathcal{F}\left(W_{\iota_v} \cdot \widehat{u}_{\widehat{\iota}_r,\iota_v}\right); \frac{-I_v}{2} \leq \widehat{\iota}_v \leq \frac{I_v}{2} - 1. \tag{15.31}$$

The cell width of the radial velocity $\dot{r}$ results, as already shown in Eq. 15.22, is derived as $\Delta \dot{r} = c/2f_{\text{c}}T_{\text{M}}$.

Aliasing cannot be prevented by an anti-aliasing filter because the velocity samples were generated discretely over the chirps. Therefore, suitable methods have to be applied in post-processing to resolve the ambiguity beyond the base interval of $v_{\text{uni}} = I_v \cdot \Delta v$. In many cases, it is even intended to allow the ambiguity in order to thus reduce the number of chirps. This allows a smaller number of measurement points to be collected per cycle and reduces the computational cost. The disadvantages are minor, since the velocity space is mostly sparse. Procedures to determine uniqueness consist of using the chirp rate f_v or the ramp center frequency from cycle to cycle. A difference of 10% allows an increment, i.e. 10 times the uniqueness interval. Although to avoid unfavorable overlaps, a threefold variation provides even more confidence for assignment to the correct velocity interval.

The sequence duration T_{M} determines, as Eq. 15.22 shows, the velocity resolution, which is why a long measurement time is advantageous. On the one hand, this is associated with a higher data volume and correspondingly more costly computation. On the other hand, the condition of a constant radial velocity can be violated to the extent that the Doppler frequency migrates into neighboring cells, thus giving away the resolution gained with the measurement time. With a relative acceleration $\ddot{r}$ during the measurement time T_{M} a velocity change of $\ddot{r} \cdot T_{\text{M}}$ can occur. This leads according to Eq. 15.32 to a change of

---

10 The run variables for the spectral lines have the same index of the corresponding run variables in the original area. They differ, like the transformed signals, by the circumflex ($\wedge$), e.g. ι_r and $\widehat{\iota}_r$.

$$\Delta\widehat{\imath_v} = \frac{\ddot{r} \cdot T_{\mathrm{M}} \cdot 2f_{\mathrm{c}} \cdot T_{\mathrm{M}}}{c} \Leftrightarrow T_{\mathrm{M}} = \sqrt{\frac{c\Delta\widehat{\imath_v}}{2\ddot{r}f_{\mathrm{c}}}} \qquad (15.32)$$

At the latest when two further lines are crossed over ($\Delta\widehat{\imath_v} = 2$) an extension of the measuring time is no longer reasonable. From this follows for a limit measuring time $T_{\mathrm{M,max}}$ of

$$T_{\mathrm{M,max}} = \sqrt{\frac{c}{\ddot{r}f_{\mathrm{c}}}} \approx 40\,\mathrm{ms} \,@f_{\mathrm{c}} = 76{,}5\,\mathrm{GHz}; \ddot{r} = 2{,}5\,\mathrm{m/s}^2 \qquad (15.33)$$

and a reasonable smallest velocity resolution of

$$\Delta\dot{r}_{\mathrm{min}} = \frac{c}{2f_{\mathrm{c}} \cdot T_{\mathrm{M,max}}} \approx 0{,}05\,\mathrm{m/s} \,@f_{\mathrm{c}} = 76{,}5\,\mathrm{GHz}; \ddot{r} = 2{,}5\,\mathrm{m/s}^2$$

$$(15.34)$$

The now determined complex two-dimensional spectrum $\widehat{u}_{\widehat{\imath_r},\widehat{\imath_v}}$ with $I_r \times I_v$ range and velocity cells contains the superposed received signals reflected from all targets but also amplifier noise and electromagnetic interference from the environment. Depending on the antenna concept, s. ▶ Sect. 15.4, such a spectrum is determined simultaneously for a large number of antenna elements. Here, too, it would be possible to determine one or more angles for each $\widehat{u}_{\widehat{\imath_r},\widehat{\imath_v}}$-cell, e.g. by a further Fourier transformation. But this is inefficient because already the two-dimensional space is occupied only very sparsely. For example, 60 targets occupy only one thousandth of a grid spanned with $I_v = 256 \times I_{ru} = 240$ grid. Therefore, the peaks are separated on the basis of the two-dimensional spectrum using mostly adaptive thresholding methods, which often use a CFAR (Constant False Alarm Rate) approach, ref. to ▶ Sect. 15.6.

Chirp sequence modulation achieves the best possible utilization of signal power, bandwidth, and measurement time. The quality of the measurement is determined not only by the noise of the receiver signal chain but also by the quality of the transmit signal. Non-linearity of the ramps, high phase noise and inaccuracies in the ramp repetition (time and frequency errors) lead to "leakage" of the detection peaks and degrade the resolutions and detection capability, especially at the edge of the detection field, i.e. at large ranges and radial velocities.

The disadvantage of chirp sequence modulation is the high sampling rate ($f_{\mathrm{s}} = 2I_r/T_{\mathrm{ch}}$) and the resulting large number of measured values $n_{\mathrm{s}} = 2I_r \cdot I_v$ per antenna for the two-dimensional Fourier transform.

15.3.3 Digital Modulation Methods

Due to the enormous progress in the performance of digital hardware in recent years, especially in digital/analog and analog/digital conversion, the realization of digital radars is possible today, at least in the development and research area. Thus, it can be assumed that such radars will be available on the market within a few years. In the following, digital radars will therefore be discussed briefly without going into more detail on the exact hardware designs and modulation layouts.

A digital radar is a radar that mixes the baseband signal from the digital/analog converter directly into the high-frequency (HF) position at 76 GHz and mixes the received signal from the HF position back into the baseband, in which it is directly analog/digital converted. Since the high-frequency front end of the radar sensor only performs a frequency shift, this is also referred to as software-defined radar. The modulations used in this process are so-called digital modulations, as they are used today in communications technology. In the scientific discussion today, two types of digital modulation are primarily considered:

- In Phase Modulated Continuous Wave Radar (PMCW), signals (so-called code sequences) are transmitted whose autocorrelation function is as low as possible. On the receiver side, the transmitted and received signals are correlated with each other. The time difference between the two signals can be read from the correlation function, which ultimately corresponds to twice the propagation time of the signal in the channel. The propagation time in the channel is proportional to the range between the targets. The velocity is measured by the Doppler shift, which can be determined by an evaluation over many code sequences, analogous to the use of many chirps in chirp sequence modulation.

- In Orthogonal Frequency Division Multiplex (OFDM) modulation, the RF bandwidth, e.g. $76\ldots77\,\mathrm{GHz}$, is divided into a very large number of subbands (typ. 1024–4096), with a frequency spacing of Δf_{sb} (typ. 1 MHz). In each sub-band, a signal with arbitrary data is transmitted at the sub-carrier frequency and received again after reflection at the target. In the simplest case, quasi-random data with a symbol duration of T_{sym} is used. The symbol duration T_{sym} results from the sub-carrier spacing $1/T_{\mathrm{sym}} = \Delta f_{\mathrm{sb}} \approx 1/1\,\mu\mathrm{s}$ (typ.), which makes all subcarriers orthogonal to each other, see ◘ Fig. 15.8. Such transmitted signals can be realized very easily by means of digital Fourier transform (Fourier synthesis). During evaluation, the channel information is first extracted from the random symbols. An evaluation via the subcarriers yields the range information, an evaluation via the symbol sequences yields the velocity information.

With both modulations, the high necessary sampling rates result in very large integration gains, which improve the SNR and relax the requirements on the dynamics of the analog/digital conversion. Analog fre-

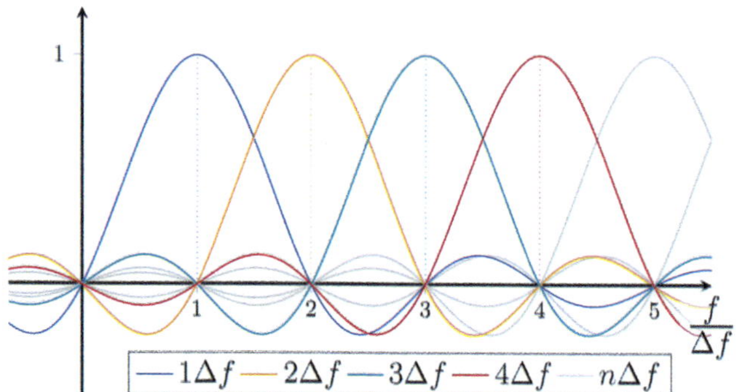

Fig. 15.8 Spectrum of an OFDM signal with subcarriers

quency filters cannot be used with these modulations, as is the case with FMCW (incl. chirp sequence modulation), which compensates for the range dependence of the received signal amplitude. An overview of modern modulation techniques can be found in [12]. It should be noted at this point that the parameters such as resolution or separability do not depend on the modulation format, but solely on the measurement duration (velocity) and bandwidth (range) of the measurement signal. Thus, these properties are the same for all modulations with the same measurement duration and bandwidth.

The disadvantage of digital radars lies in the very large signal bandwidths that have to be converted from digital-to-analog and analog-to-digital in an automotive radar. This results in enormous data rates that have to be processed and at least partially buffered. The advantage of digital radars is their great flexibility in signal processing on both the transmitter and receiver sides. This makes it conceivable, for example, to react more flexibly to interference scenarios (cf. ▶ Sect. 15.8) than is possible with analog radars. In addition, current research and development work shows ways to relax the requirements due to the large signal bandwidths in digital/analog conversion and vice versa. In PMCW, undersampling can be used at the expense of SNR, and in OFDM, variants such as stepped-OFDM [15] have emerged, which manage with a fraction of the bandwidth during conversion despite high range resolution.

15.4 Principle of Angle Measurement

15.4.1 Antenna Theoretical Considerations

Before describing the angle determination, needed fundamentals about the beam shape of radar sensors are introduced. The beam characteristic of the electric field strength $E(\phi, \vartheta)$ in the far field, i.e. at ranges much larger than the antenna dimensions, results (cf. [16]) as

inverse Fourier transform of the antenna current distribution function $A(x, y)$ where the azimuth angle ϕ corresponds to the $_Sx$-direction and the elevation angle ϑ corresponds to the $_Sy$-direction. The positive azimuth angle to the left ϕ is in the sensor horizontal plane of the sensor oriented in the $_Sz$-direction and the elevation angle describes the angle to the ϑ that describes the angle to the $_Sz - _Sx$-plane (upward positive). The origin of these spherical coordinates is located at the center of the antenna. For a plane antenna parallel to the $_Sx - _Sy$-plane, the electric field results according to [16]

$$E(\phi, \vartheta) = \iint A(x, y) e^{j\left(\frac{2\pi}{\lambda}(x \cdot \cos\vartheta \sin\phi + y \cdot \sin\vartheta)\right)} dx dy \quad (15.35)$$

This equation first describes the field strength distribution in the far field for a wave radiated with the current distribution function $A(x, y)$ but is valid in the same way for the reception. The far field is defined as the field at the location where a locally plane wave well approximates the radiated spherical wave, cf. [1]. This is always fulfilled if the range between the transmitting antenna or source point of a wave and the place of observation, e.g. where the receiving antenna or the target is located, is large enough. The range r_{ff} results from

$$r_{ff} = 2\frac{\ell_{max}^2}{\lambda}, \quad (15.36)$$

where ℓ_{max} describes the largest dimension of the antenna or target respectively. Frequently, this condition is not met in automotive radar, but this is usually ignored for practical reasons. Thus, the multiplication of the transmitting characteristic by the receiving characteristic applies to the angular dependence of a sensor. As long as the transmitting antenna is not far from the receiving antenna, the two-way characteristic can be described as the (generally complex) product of the one-way characteristics. If a monostatic antenna concept is used, i.e. if the transmit signal passes through the same antenna unit as the receive signal, the (in case of an occupancy function which is not mirror-symmetric about the center of the antenna, complex) square is given by $E^2(\phi, \vartheta)$.

For three simple, symmetric one-dimensional cases of the current distribution functions, the resulting antenna characteristics are shown in **Fig. 15.9**. The abscissa uses the normalized quantity $\Gamma_{\phi, A} \cdot q = (\ell_A/\lambda)\sin\phi$ and is thus given by the ratio ℓ_A/λ of aperture width (antenna aperture width) and wavelength.

These examples already show the conflict between the strongest possible concentration of the main lobe and the smallest possible[11] aperture width. As shown in a table at [16], a compromise that fits the angle eval-

---

11 This conflict is comparable to the interpretation of window functions, cf. ▶ Sect. 15.3.2.3.

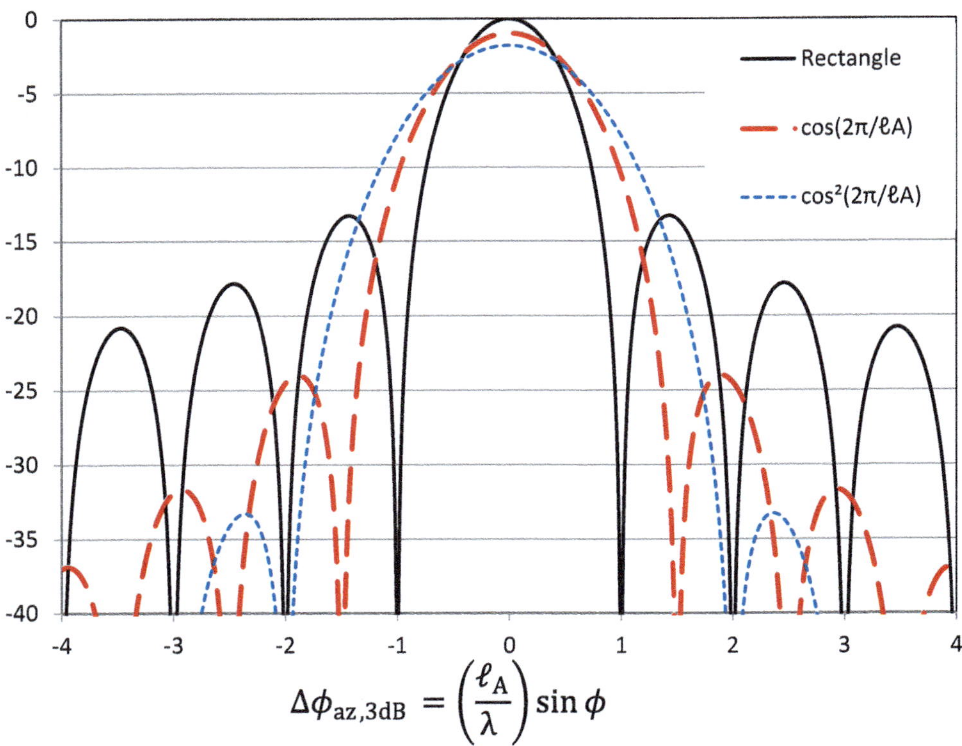

$$\Delta\phi_{az,3dB} = \left(\frac{\ell_A}{\lambda}\right)\sin\phi$$

Fig. 15.9 Calculated one-dimensional antenna characteristics for a rectangular occupancy function and a simple and a squared cosine half-bell, normalized to total power. The abscissa variable $\Gamma_{\phi,A} \cdot q = (l_A/\lambda)\sin\phi$ is normalized to the ratio l_A/λ of the aperture width to the wavelength

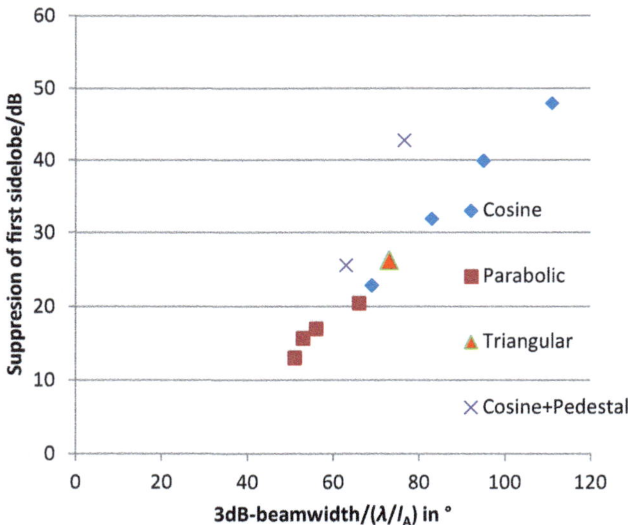

Fig. 15.10 Side lobe suppression versus width of main lobe (at -3 dB, one-way) according to [16]

tude prevailing in the center still remains at the edge. Despite such optimization strategies, the antennas must be about eighty times larger than the wavelength times the reciprocal of the main lobe width per degree, where one degree of main lobe width requires roughly $\ell_A = 80\lambda$ large aperture width, corresponding 32 cm at 1° and 76,5 GHz. However, modern antenna design concepts, such as the MIMO concept explained later, or high-resolution angle estimation techniques, allow a much smaller aperture width for the required angular separation capability.

Another undesirable side effect of high sidelobe suppression is a decrease in antenna gain (cf. **Fig. 15.10**) because the suppression is always caused by a current distribution function decreasing toward the edge of the antenna. Accordingly, the effective antenna area decreases, the main lobe becomes wider and thus the power is distributed over a wider area, which in turn leads to a decrease of the intensity in the center of the beam.

For long range radar application such as ACC (see ▶ Chap. 32), an angular range for the total coverage is $\Delta\phi_{max}$ of approx. 20° azimuth and 4° elevation is required. A separability with respect to elevation would be desirable for distinguishing a bridge from a station-

uation concept can be chosen depending on the current distribution function, cf. **Fig. 15.10**. A characteristic that is good for suppressing the first side lobe is the Hamming window, where 8 % of the ampli-

ary vehicle (height difference approx. 2 m), but this would require a separation capability in the far range of $1°(\approx 2\,\text{m}/116\,\text{m})$ and consequently an antenna of at least 30 cm would be required, which is unacceptable with respect to the available installation space. Thus, the angular evaluation in the far range is limited to the azimuth. For the application of radar in the close range, which has increased strongly in recent years, especially for the extension to low speeds of ACC (see ▸Chap. 32) or for collision avoidance systems, stationary obstacles must also be classified. Therefore, for these functions not only a significantly extended azimuthal range (30°...60°) is required, but also a resolution in elevation is desired. In particular, a planar antenna (see ▸ Sect. 15.4.4) with reasonable dimensions and thus distinguish underrideable (e.g. bridges) and overrideable (e.g. manhole covers, Coke cans) from potential collision objects.

15.4.2 Scanning

The simplest method of angle determination in terms of understanding is mechanical scanning, but this is no longer used in automobiles today. For this purpose, a beam deflection unit or a planar antenna is mechanically swiveled so fast that the entire azimuthal detection range is covered within one measurement and evaluation cycle (50 ... 200 ms). ▣ Fig. 15.11 shows the principle. Due to the dependence on aperture width described above, the radar lobe has at least 2° main lobe width, if the aperture width is not to exceed 15 cm. The lobe is measured in approx. 1°-steps over the measuring range. Instead of a discrete step control, a continuous scan movement is performed in order to avoid noise-generating accelerations[12] and to get by with lower positioning powers. The measured values are then nevertheless assigned to a discrete angular position, namely the center of the scan positions within a measurement window that is assigned to this angular segment. It is true that the blur resulting from the lobe width is additionally increased by a "motion blur". However, since the measurement data is windowed to avoid leakage effects, i.e. strongly suppressed at the beginning and end of the measurement interval, the effective motion blur is thereby reduced by approx. 30%. Furthermore, it is advantageous that the blurs add up geometrically approximately (or exactly in the case of a Gaussian characteristic of the antenna and window function), so that the loss of sharpness even if the lobe continues to move by its width during the measurement time is only about 10%. Of course, an even smaller step size can be selected and thus the motion blur can be

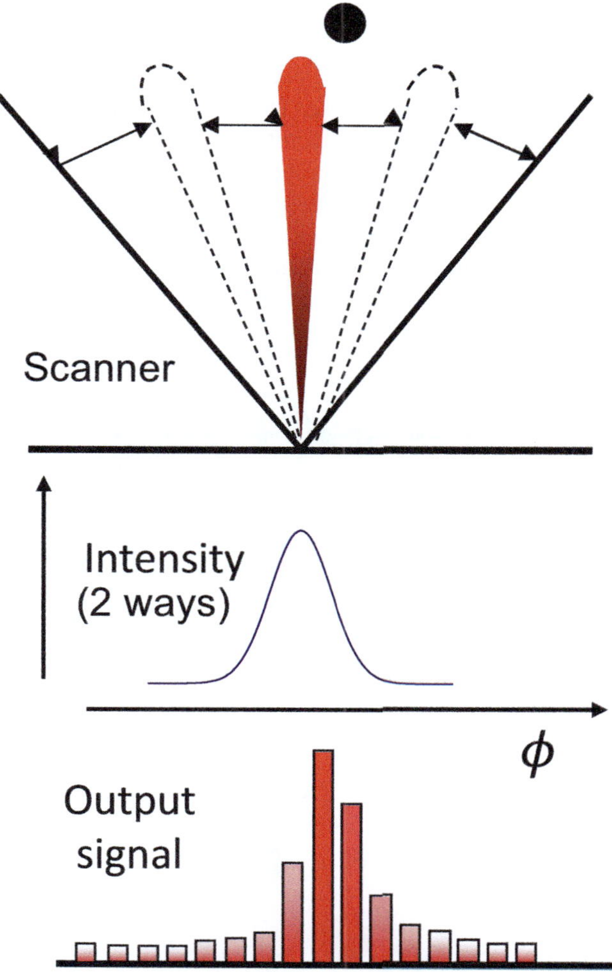

▣ **Fig. 15.11** Scanner principle for angle detection. Top: narrow focused beam scans the total detection area and detects the point target; middle: the azimuthal angle characteristic of the focused beam; bottom: Result for a point target

reduced. However, the division of the measuring time into many intervals assigned to the angular segments worsens the resolution for the Doppler evaluation. This also makes it clear that, in terms of radial velocity measurement, a mechanical scanner will be inferior in principle to a multi-beam arrangement measuring the same total duration.

Furthermore, it should be noted that the azimuthal evaluation range is effectively smaller than the scan range, because for a center of gravity determination, a drop must be detectable at least at the edge. Therefore, the actual angular range towards both edges is smaller than the scan range by about half a beam width. The angular resolution is basically limited by the lobe width.

However, something can still be achieved if the antenna characteristics are known, e.g. by measuring at the end of manufacture. With the help of deconvolution algorithms, both the values for resolution and the separability can be reduced by a factor of about ½ in the favorable case, cf. [6].

---

12 The high acceleration occurring while stepping causes considerable inertia forces, the support of them causes vibrations in and on the housing.

15.4.3 Monopulse

The monopulse method is based on a dual antenna arrangement, see ◘ Fig. 15.12. This is usually used for reception only, while the transmit beam is transmitted by means of a single separate antenna.

The (receiving) antennas can differ due to the beam characteristics or simply due to the position, which for azimuthal angle measurement is shifted horizontally by $\Delta y = \Gamma_\phi \cdot \lambda$. For two adjacent, otherwise equal antenna fields, one finds a phase difference of

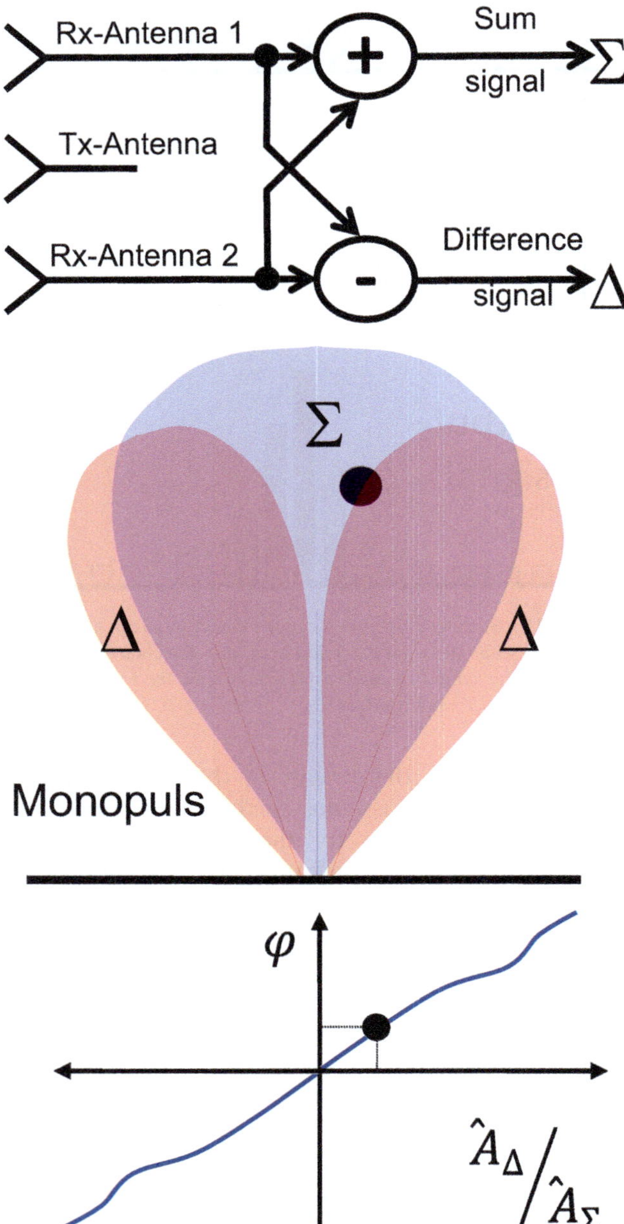

◘ **Fig. 15.12** Monopulse principle for angle determination. Top: formation of the sum and difference signals; middle: the azimuthal angle characteristic of the beams thus formed; bottom: typical characteristic azimuthal angle vs. quotient of the amplitude magnitudes of the difference and sum signals at smaller angles

$$\Delta\varphi = 2\pi\,\Gamma_\phi \sin\phi = 2\pi\,\Gamma_\phi q \qquad (15.37)$$

as a function of the sine of the azimuth angle $q = \sin\phi$. For the amplitudes of the difference signal, this means that instead of the original amplitudes A_1 and A_2 with $|A_1| = |A_2| = |A|$ for the difference and sum signal, an amount weighted with the sine and the cosine of the phase difference $\Delta\varphi$ is measured:

$$|A_\Delta| = |A_1 - A_2| = 2|A|\sin\frac{\Delta\varphi}{2};$$
$$|A_\Sigma| = |A_1 + A_2| = 2|A|\cos\frac{\Delta\varphi}{2} \qquad (15.38)$$

Thus, the sine of the azimuth angle can be determined from the ratio of the difference signal to the sum signal. $q = \sin\phi$ can be determined from the ratio of the difference signal to the sum signal without the need for a phase-sensitive measurement:

$$q = \frac{\arctan(|A_\Delta|/|A_\Sigma|)}{\pi\,\Gamma_\phi} \qquad (15.39)$$

However, because of the uniqueness, the restriction to phase differences of

$$\Delta\varphi = 2\pi\,\Gamma_\phi q_{\mathrm{uni}} < \pi \qquad (15.40)$$

required. From this follows the dimensioning rule of $2\Gamma_\phi q_{\mathrm{uni}} < 1$. With a (centered) maximum azimuth uniqueness interval of $\phi_{\mathrm{uni}} = 30°$, the antennas would be exactly 1 λ away, at 6° about 5λ.

The measurement procedure described here is accurate for single-point targets. However, even two targets can produce values without useful meaning in a way that is not detectable in the same measurement cycle. Therefore, when using this method, care must be taken to ensure that the probability of determining the azimuth of two or more targets is very low by using high range and/or radial velocity separation or by using mechanically or electronically scanning, highly focusing transmit antennas.

If the difference and the sum signal are measured simultaneously and a complex amplitude determination is possible, the uniqueness range of the phase evaluation can be increased to $\pm\pi$ because the signs of the complex amplitudes can also be used in the arctan-calculation. The monopulse method is no longer used as the sole method for determining the azimuth angle.

For the determination of the elevation angle, however, the monopulse principle is still used, e.g. by two vertically offset receiver arrays, see ▶ Sect. 15.4.4.

The monopulse principle can also be extended to more than two antenna elements, making angle determination multi-target capable and allowing higher resolutions. Here, the radiation characteristics of all antennas are recorded in a calibration process during the manufacture of the sensors and are stored in the sensor as a calibration data set. For angle estimation, the amplitude or phase ratios of the signals at the individual antennas are then compared with those of the calibration data, and the closest match corresponds to the estimated target angle. This type of angle estimation was commonly used in automotive radar in the past, especially when lens antennas were used.

15.4.4 Planar Antenna Arrays

Planar antennas have two positive properties that are relevant for practical applications:

- The construction depth of the sensors is significantly reduced to depths between about one and two centimeters.
- Arrays can be formed to control transmit and/or receive characteristics of the antenna.

The most common arrangement is a single or multiple transmit antenna patches consisting of commonly fed "patches" matched in size to the wavelength and having multiple (> 3) receive antenna patches (also consisting of a plurality of patches).

15.4.4.1 Structure and Design of the Antenna Elements

For the design of the antenna elements, on the one hand, a high occupations density, i.e. many antenna elements (patches) per unit area, is aimed at in order to achieve a high radiation power in accordance with Eq. 15.35 to achieve a high radiation power. On the other hand, increasing expansion is associated with greater directivity, which prevents broad radiation. However, if the ranges between the antenna elements are too large, this leads to angular ambiguities during transmission or reception and ultimately to an excessively large sensor size. In practice, a compromise must be found here to keep the number of channels and thus the number of antenna elements as small as possible for cost reasons.

The individual patches have a size of about half a wavelength. This causes resonance of the excitation current on the patches, which leads to radiation. If several patches are directly connected to each other with lines, attention must be paid to the phase shift between the patches, because the fields of the single patches can overlap constructively or destructively. By designing the phase shifts between directly connected patches, the transmission or reception characteristics can be designed very freely. For further information, please refer to the following literature [1].

15.4.4.2 Fundamentals of Digital Beamforming

Analogous to the consideration of the monopulse principle (▶ Sect. 15.4.3), the path difference is calculated as a function of the offset from the transmitting position. Simplifying, as shown in ◙ Fig. 15.13 for the horizontal plane, the transmitting antenna element is set to position 0 of the sensor transverse axis $_sx$ is set. The variable receive position $_sx = \Gamma_\phi \lambda_c$,[13] is calculated as the product of the carrier wavelength λ_c and the transverse position normalized to the wavelength Γ_ϕ, which is normalized to the wavelength. Analogously, for the elevation, $_sy = \Gamma_\vartheta \lambda_c$ would be defined. For a point target at a range r from the sensor center, the return path to the antenna position is given by $_sy = \Gamma_\vartheta \lambda_c$ from:

$$r(\Gamma_\phi) = \sqrt{r^2\cos^2\phi + \left(r\sin\phi - \lambda_c\Gamma_\phi\right)^2} \qquad (15.41)$$

which can be transformed to

$$r(\Gamma_\phi) = r\sqrt{1 - 2\sin\phi \cdot \Gamma_\phi\lambda_c/r + \left(\lambda_c\Gamma_\phi/r\right)^2} \quad (15.42)$$

The Taylor series expansion up to the quadratic member results in

$$r(\Gamma_\phi) - r \cong -\sin\phi \cdot \Gamma_\phi\lambda_c + \cos^2\phi\frac{\left(\lambda_c\Gamma_\phi\right)^2}{r} + \dots. \quad (15.43)$$

For more distant targets, the first term dominates, as it was also used for the monopulse principle. The spacing of the monopulse receiving antenna elements turns out to be small because of the necessary uniqueness condition, which is why the quadratic term must be disregarded.

For digital beamforming, it is assumed without limitation of generality that the antenna of width ℓ_ϕ has an even number of I_ϕ; $I_\phi/2 \in \mathbb{N}$ elements, which are symmetrically arranged on both sides around the transmitting antenna equidistantly $\Delta\Gamma_\phi \cdot \lambda = \ell_\phi/I_\phi$ arranged equidistantly, cf. ◙ Fig. 15.14.[14] The phase difference at the different antenna positions results in:

$$\begin{aligned}\varphi(\Gamma_\phi) = \frac{r(\Gamma_\phi) - r}{\lambda_c} &\cong -\sin\phi \cdot \Gamma_\phi \\ &+ \cos^2\phi\frac{\left(\lambda_c \cdot \Gamma_\phi\right)^2}{\lambda_c r} + \dots\end{aligned} \qquad (15.44)$$

---

13 As introduced in ▶ Sect. 15.4.1, index S at $_sx$ refers to the Cartesian sensor coordinate system, which is used for both the sensor geometry and the location of the locating objects. The sensor longitudinal axis points in the $_sz$-direction, $_sx$ in the azimuth plane, $_sy$ in the elevation plane. The spherical coordinate system (r, ϕ, ϑ) or the projection on the horizontal line (r, ϕ) as polar coordinate system use the same origin ($_sx = _sy = _sz = 0$). The solid angle ($\phi = 0, \vartheta = 0$) points into the $_sz$-direction.

14 Even if the distance between the centers of the outer elements is only $(l_\phi - 1)\Delta\Gamma_\phi$ the width is considered to be the period length, which also corresponds to the distance if half a $\Delta\Gamma_\phi$ is added on both sides.

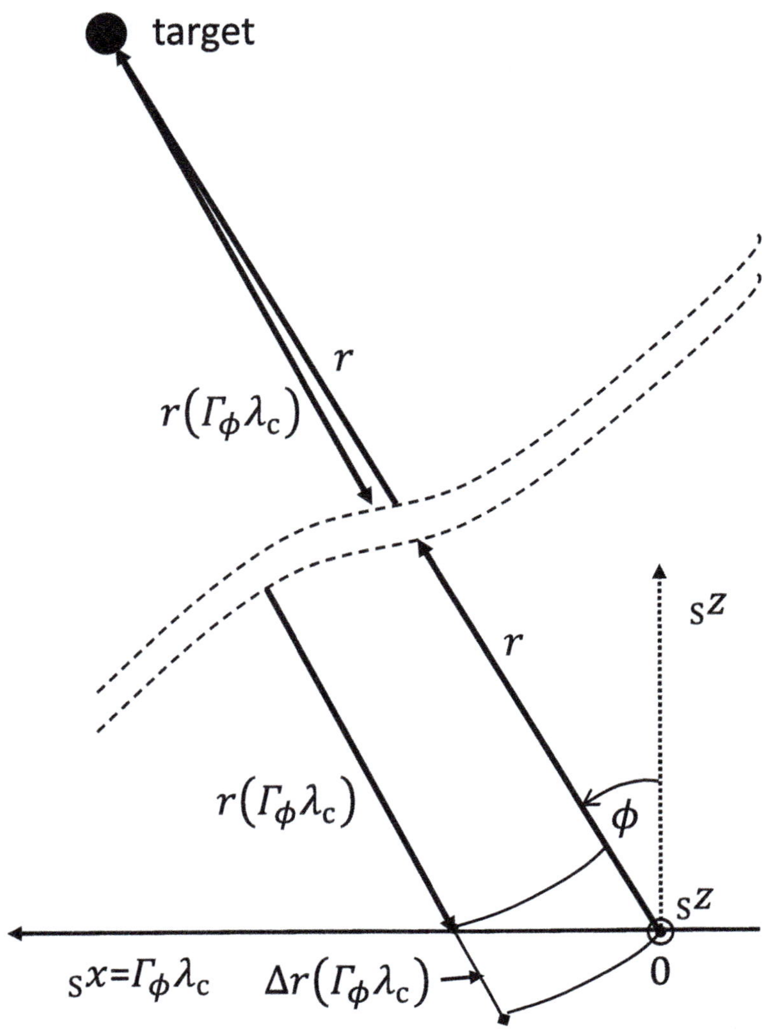

◻ Fig. 15.13 Definition for the calculation of the path difference depending on the receiving position

The quadratic term can be neglected if the far-field assumption is used, so that the complex amplitudes of each antenna element are ι_ϕ for the range/radial velocity cell $\widehat{u}_{\widehat{\iota_r},\widehat{\iota_v}}$ according to Eq. 15.31, i.e. $\widehat{u}_{\widehat{\iota_r},\widehat{\iota_v},\iota_\phi}$ (ideally) differ only by the phase

$$\Delta\varphi\left(\Gamma_{\phi,\iota_\phi}\right) = -\sin\phi \cdot \Gamma_{\phi,\iota_\phi}, \tag{15.45}$$

depending on the normalized lateral offset to the center of the transmitting antenna. With

$$\Gamma_{\phi,\iota_\phi} = \left(\iota_\phi - \frac{I_\phi - 1}{2}\right)\Delta\Gamma_\phi \tag{15.46}$$

the phase changes from antenna element to antenna element.

$$\Delta\varphi_{\iota_\phi} = -\sin\phi \cdot \iota_\phi \Delta\Gamma_\phi \tag{15.47}$$

A Fourier transformation

$$\widehat{u}_{\widehat{\iota_r},\widehat{\iota_v},\widehat{\iota_\phi}} = \mathcal{F}\left(W_{\iota_\phi} \cdot \widehat{u}_{\widehat{\iota_r},\widehat{\iota_v},\iota_\phi}\right); 1 \le \widehat{\iota_\phi} \le I_\phi \tag{15.48}$$

now yields the (complex) spectrum to $\sin\phi$ with a total of I_ϕ spectral lines. In the central area, this value may be converted into angles without major errors. Correctly, $\sin\phi$ corresponds to the lateral position in the unit circle, which in the following is denoted by $q = \sin\phi$, i.e. denoted by

$$_sx = \sin\phi \cdot r = q \cdot r \tag{15.49}$$

the lateral position in the Cartesian sensor coordinate system is also described. Therefore, the spectrum q is better described as a lateral spectrum than as an angular spectrum (◻ Fig. 15.15).

With Eq. 15.49 holds for the longitudinal position

$$_sz = \cos\phi \cdot r = \sqrt{\left(1 - q^2\right)} \cdot r. \tag{15.50}$$

Correctly should be r replaced by $r \cdot \cos\vartheta$ which, in the case of radar sensors used with elevation ranges of mostly no more than $|\vartheta| < 0.1\,\mathrm{rad} \approx \pm 6°$ leads to errors under 0,5‰.

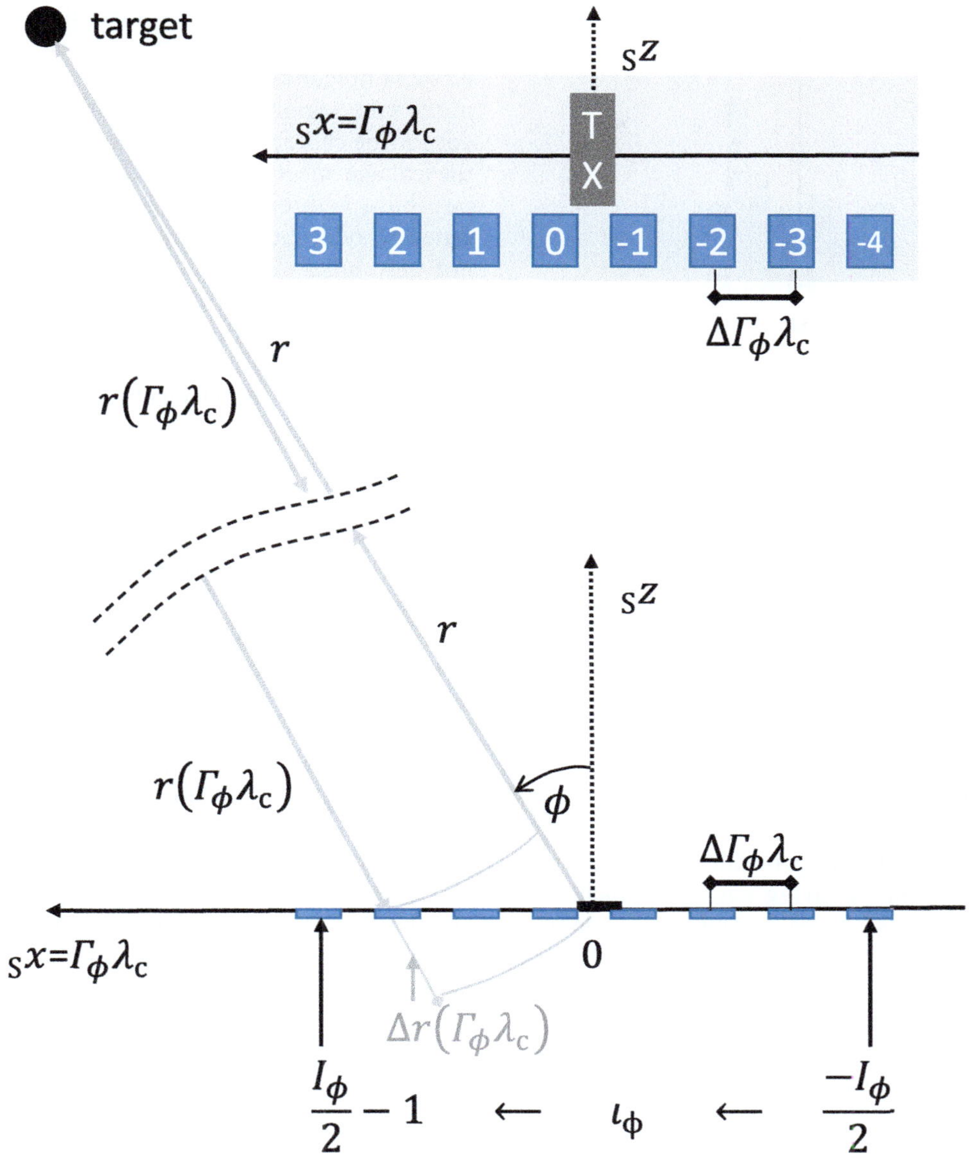

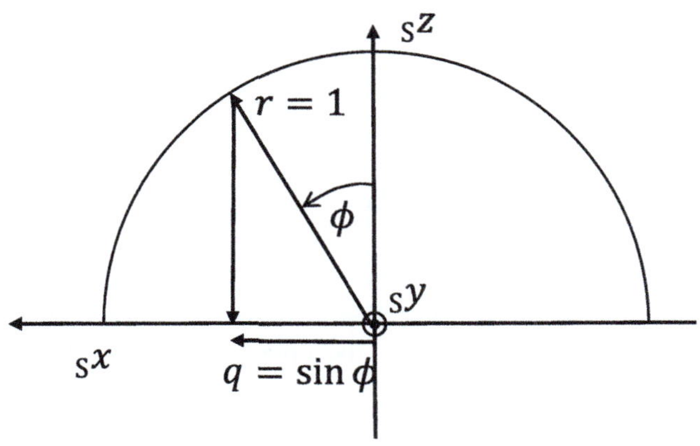

Fig. 15.14 Antenna arrangement with central transmitting antenna

Fig. 15.15 Graphical representation of the normalized lateral range q

The basic resolution of q is

$$\Delta q = \frac{1}{\Gamma_{\phi,\max}} = \frac{1}{I_\phi \Delta\Gamma_\phi} = \frac{\lambda_c}{\ell_\phi}. \tag{15.51}$$

As a calculation example an antenna width of 10 cm can be used. At $\lambda_c \cong 4\,\text{mm}$ corresponding to $f_c = 76{,}5\,\text{GHz}$ $\Delta q = 1/25$ is calculated. In the central area (best case), a resolution of $\Delta\phi \geq 2{,}3°$ can be derived.

The uniqueness range of the lateral spectrum depends exclusively on the normalized spacing of neighboring antenna elements

$$q_{\text{uni}} = \frac{1}{\Delta\Gamma_\phi} = I_\phi \Delta q \tag{15.52}$$

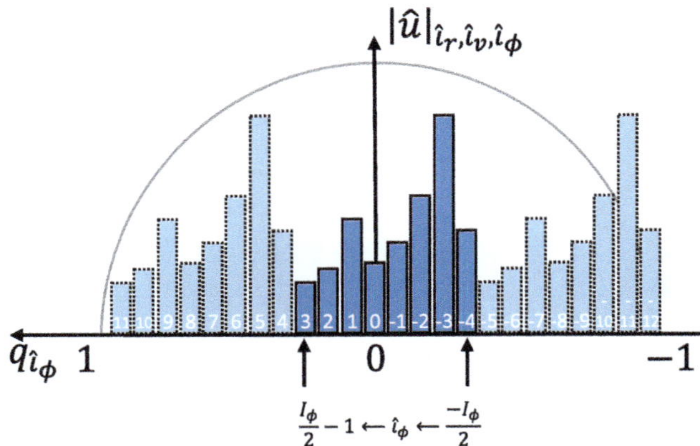

Fig. 15.16 Lateral spectrum of an eight-element receiving antenna with a uniqueness range of $q_{uni} = 2/3$

Should the aim be that the complete front area, i.e. from $-90°$ ($q = -1$) to $+90°$ ($q = +1$) is unambiguous, this would have to be $q_{uni} = 1/\Delta\Gamma_\phi = 2$ i.e. the space between the antenna elements should be exactly half of the wavelength λ_c. This leads, on the one hand, to a very high number of antenna elements, if at the same time a high lateral resolution is Δq is desired. On the other hand, there is little space left on the antenna for the individual elements, which reduces the antenna gain on the receiving side and thus the achievable range.

The example in **Fig. 15.16** illustrates the q-spectrum of an 8-element receiving antenna, which is designed for a total of three unambiguous domains $q_{uni} = 2/3 (\triangleq 39°$ in the middle segment). The basic resolution is thus $\Delta q = 2/24 (\triangleq 4,8°$ in the middle segment) corresponding to an antenna width of $\ell_\phi = 12\lambda_c = 4,8$ cm. The unambiguous angular range results in antenna elements with a normalized spacing of $\Delta\Gamma_\phi = 3/2\Delta y = 1,5\lambda_c = 6$ mm.[15]

Another exemplary compromise design with double the number of elements $I_\phi = 16$ in $2\lambda_c$ range resolutions ($\ell_\phi = 32\lambda_c = 12,8$ cm allows a $\Delta q = 2/32 \triangleq 1,8°$ can be achieved in an unambiguous angular range of $q_{uni} = 1/2 \triangleq 29°$ (degrees apply only to the center).

For a "dream antenna" with $\Delta q = 1/64 \triangleq 0,9°$ without ambiguity $q_{uni} = 2 \triangleq 180°$ would be 128 elements at a range of 2 mm would be required, which would increase the width to 25,6 cm.

If the ambiguity cannot be avoided, it must be resolved for equidistant antenna elements with subsequent procedures (mostly in object tracking). For this purpose, geometric relations to the radial velocity can be used.

Alternatively, the spacing of the antenna elements can be chosen aperiodically, at least in places, in order to assign individual targets to an unambiguous angular range. A DFT which works with higher "frequencies" $q > q_{uni}$ then yields a narrow peak only in the correct unambiguous region, while the corresponding lines in the false regions "leak out" with characteristic manner for the chosen aperiodicities in each case, similar to the leakage effects described in ▶ Sect. 15.3.2. Another possibility is through a staggered periodicity, where the series of odd and even antenna element numbers are not only offset by $\Delta\Gamma_\phi\lambda$ offset, but also by an offset preferably of $\lambda/2$. The false uniqueness interval manifests itself with the appearance of a "ghost target" with an offset of $(I_\phi/2)\Delta q$ to the true target, because the spacing between the elements is calculated "wrong". Only in the correct interval does the "ghost target" disappears again.

15.4.4.3 Combined Azimuth and Elevation Determination

Although the focus of the measurement of the solid angle is on azimuth angle resp. q., there are some scenarios where elevation information is needed, especially whether a stationary target can be moved over (example: manhole cover) or under (bridge) or should be interpreted as a potential collision object. A consistent implementation, both angles (or the sine; for elevation $p = sin\vartheta$) leads to a rectangular array with $I_\phi \times I_\vartheta$ antenna elements, which would multiply all signal chains (and therefore also the costs) by a factor of I_ϑ factor. Therefore, the currently used technology is mostly limited to the monopulse principle for elevation, which is why only $I_\vartheta = 2$ antenna element rows are needed, which are arranged in an offset of $_sy = \Delta\Gamma_\vartheta \cdot \lambda$ one above the other. $\Delta\Gamma_\vartheta$ must not be chosen too large, because there is hardly any possibility to resolve the ambiguities. Mostly the size of the uniqueness is adapted to the elevation characteristic of the transmitting lobe, so that outside the uniqueness range the antenna gain is strongly reduced. With a $\Delta\Gamma_\vartheta = 4$ a uniqueness range of just under 15° ($p_{uni} = 0,25$) is reached, which means that at $r = 20$ m range a vertical range of almost $_sz = 5$ m is available.

Without any additional antenna elements, an alternating row of elements in different heights($_sy$-direction) for the combined azimuth and elevation is determined, cf. **Fig. 15.17**. The evaluation is performed according to the approach presented above with a Fourier transform over ι_ϕ independent of the vertical arrangement.

In the case of $\vartheta = 0$, the result does not differ from a simple series because there are no elevation-related phase differences between the upper and the lower series. If the difference is π, the originally obtained line disappears in the lateral spectrum. Instead, a line of the same height appears in the spectrum offset by $(I_\phi/2)\Delta q$. Thus, the original line $\hat{\imath}_\phi$ can be interpreted as a sum signal of both series and $\hat{\imath}_\phi + I_\phi/2$ as a difference signal. Thus, analogous to the monopulse, only from the ratio

---

15 All non-standardized data refer to a carrier frequency of 76.5 GHz corresponding to a wavelength of about 4 mm.

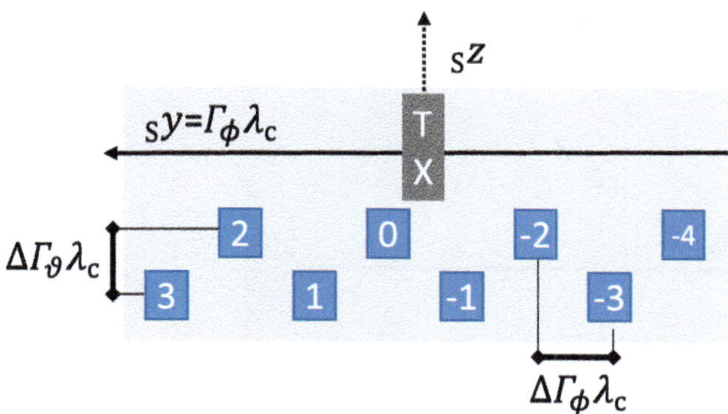

Fig. 15.17 Antenna arrangement for combined azimuth and elevation measurement (receiver elements only)

of the amounts can be concluded to p whereby only half of the uniqueness range is accessible. However, the signal can be reconstructed also in each case for the upper and lower row:

$$\widehat{u}_{\widehat{\iota}_r,\widehat{\iota}_v,\widehat{\iota}_\phi,\text{even}} = \widehat{u}_{\widehat{\iota}_r,\widehat{\iota}_v,\widehat{\iota}_\phi} + \widehat{u}_{\widehat{\iota}_r,\widehat{\iota}_v,(\widehat{\iota}_\phi+I_\phi/2)} \tag{15.53}$$

$$\widehat{u}_{\widehat{\iota}_r,\widehat{\iota}_v,\widehat{\iota}_\phi,\text{odd}} = \widehat{u}_{\widehat{\iota}_r,\widehat{\iota}_v,\widehat{\iota}_\phi} - \widehat{u}_{\widehat{\iota}_r,\widehat{\iota}_v,(\widehat{\iota}_\phi+I_\phi/2)} . \tag{15.54}$$

The indices even/odd refer to row order, which can be the upper or the lower one depending on the offset. From the phase difference of the reconstructed signals the elevation can be directly calculated as

$$\sin\vartheta = p = \pm\Delta\varphi_{\text{even}-\text{odd}}\Delta\Gamma_\vartheta \tag{15.55}$$

to the full extent of uniqueness $p_{\text{uni}} = 1/\Delta\Gamma_\vartheta$. The sign depends on whether the upper row or the lower row are occupied with even ι_ϕ are occupied. Further information can be found e.g. in [1].

15.4.4.4 Multiple-Input Multiple-Output (MIMO) Antennas

In the calculations in ▶ Sect. 15.4.4.2 and in ◻ Fig. 15.14, it was assumed for simplicity that the transmitting antenna element is located in the sensor coordinate origin. Furthermore, the receiver elements were arranged symmetrically to it in the $_s x$-direction.

If the transmitting antenna in ◻ Fig. 15.18 would instead be shifted laterally by $\Delta\Gamma_\phi \cdot \lambda \cdot I_\phi/2 = \ell_\phi/2$ thus at a range of $\Delta\Gamma_\phi \cdot \lambda/2$ outward from the last receiving antenna element, the (outward) path to the target changes by $q \cdot \ell_\phi/2$, i.e. the phase of the complex amplitude $\widehat{u}_{\widehat{\iota}_r,\widehat{\iota}_v,\widehat{\iota}_\phi}$ changes correspondingly also by $2\pi q \cdot \ell_\phi/2$. A displacement of the transmitting antenna in the opposite direction leads to a corresponding but negative phase shift. A multiple-input multiple-output (MIMO) antenna uses both arrangements, i.e. in addition to the receive antenna elements, it also has two transmit antenna elements on both the sides.

For the left- or right-hand transmitting antenna, neglecting the quadratic term in Eq. 15.43:

$$\Delta\varphi_l(\Gamma_\phi) = \frac{r(\Gamma_\phi) - r(\Delta\Gamma_\phi I_\phi/2)}{\lambda_c}$$
$$\cong -q \cdot (\Gamma_\phi + I_\phi\Delta\Gamma_\phi/2) \tag{15.56}$$

$$\Delta\varphi_r(\Gamma_\phi) = \frac{r(\Gamma_\phi) - r(-\Delta\Gamma_\phi I_\phi/2)}{\lambda_c}$$
$$\cong -q \cdot (\Gamma_\phi - I_\phi\Delta\Gamma_\phi/2) \tag{15.57}$$

For the discrete antenna elements ι_ϕ results in the first case (left)

$$\Delta\varphi_{\iota_\phi,l} = -q \cdot (\iota_\phi + I_\phi/2)\Delta\Gamma_\phi; \frac{-I_\phi}{2} \leq \widehat{\iota}_\phi < \frac{I_\phi}{2} \tag{15.58}$$

and in the second case (right)

$$\Delta\varphi_{\iota_\phi,r} = -q \cdot (\iota_\phi - I_\phi/2)\Delta\Gamma_\phi; \frac{-I_\phi}{2} \leq \widehat{\iota}_\phi < \frac{I_\phi}{2}. \tag{15.59}$$

whereas in the case of the central position of the transmitting antenna only $I_\phi/2$ antenna elements to one side, depending on the position of the transmitting antenna, there are I_ϕ antenna elements to one side and 0 to the other. In the case of operation alternating from chirp to chirp (time division multiplexing), the antenna side changes from measurement to measurement. If the receive amplitudes of the antenna elements of a measurement with a left-hand transmitting antenna are combined with those of a measurement with a right-hand transmitting antenna shortly afterwards, the number of measured values is doubled

$$\Delta\varphi_{\iota_\phi,\text{MIMO}} = \Delta\varphi_{\iota+I_\phi/2_{\phi,l}}; -I_\phi \leq \iota_\phi \leq -1; \Delta\varphi_{\iota_\phi}$$
$$= \Delta\varphi_{\iota-I_\phi/2_{\phi,l}}; 0 \leq \iota_\phi \leq I_\phi - 1 \tag{15.60}$$

and thus also twice the resolution.

$$\Delta q_{\text{eff}} = \frac{1}{2\Gamma_{\phi,\text{max}}} = \frac{\lambda_c}{2\ell_\phi} = \frac{\lambda_c}{\ell_{\phi,\text{eff}}}. \tag{15.61}$$

When evaluating both measurements, i.e. with both the left and right transmit antennas, the phase change occurring between the measurements due to radial range changes must be compensated. This MIMO approach results in $2I_\phi$ measured values for the Fourier transform with a corresponding number of lines in the lateral spectrum as a result.

The MIMO technique, as described above, makes it possible to improve the wavelength-related low solid angle resolution by a factor of about 2 with acceptable dimensions. However, the price to be paid for this doubling is that only every second chirp is available for Doppler evaluation, which halves the unambig-

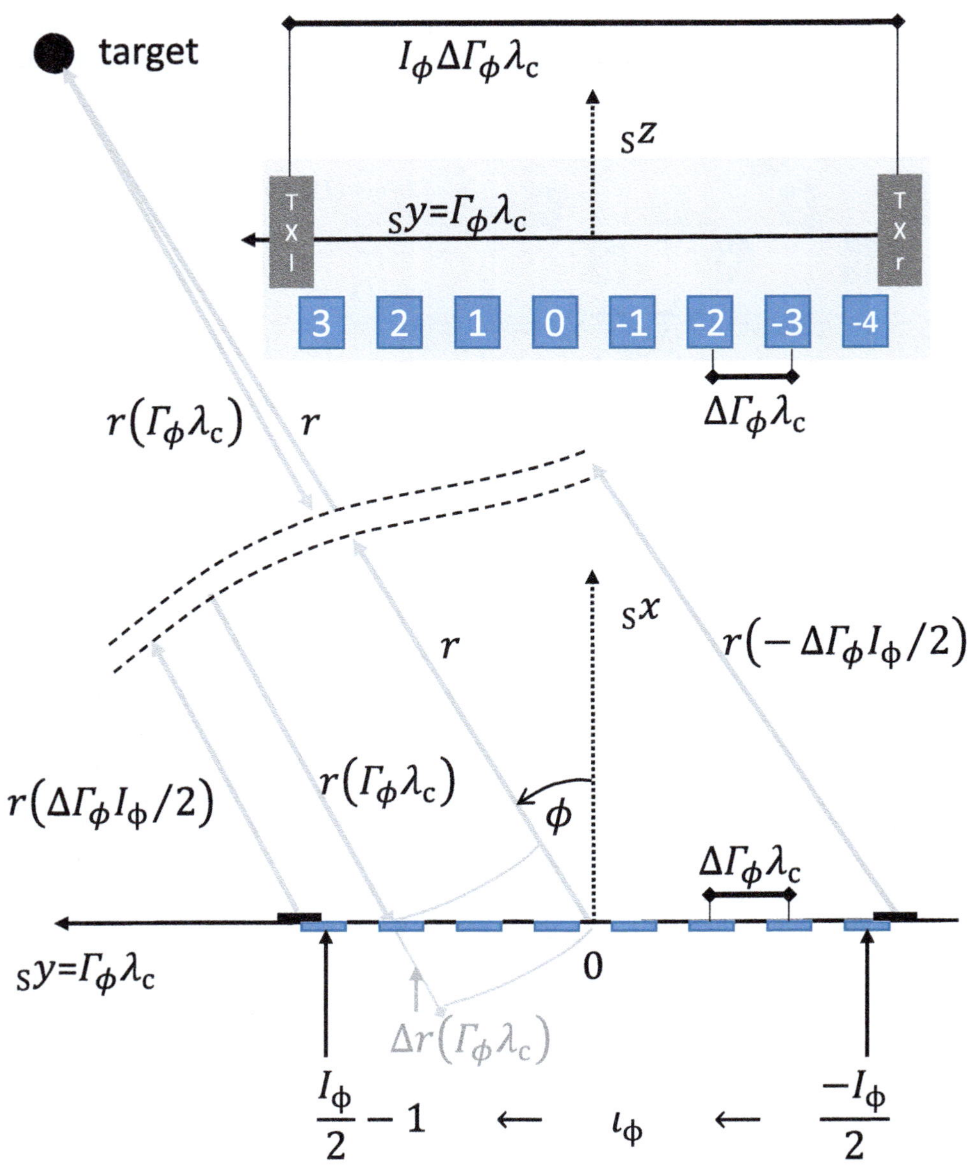

Fig. 15.18   MIMO array with 2 transmit antenna elements and 8 receive elements

uous range of the radial velocity determination. Alternatively, the transmit antennas can transmit in parallel if they are operated with slightly different carrier frequencies (frequency division multiplexing), but their difference is at least large enough that the spectra do not overlap in the result. The price here is a higher sampling rate for the receiver channels.

The statements on digital beamforming in the previous sections can be largely adopted for the MIMO arrangement as if the antenna had twice the width $\ell_{\phi,\text{eff}} = 2\ell_\phi$. The concepts for combined azimuth and elevation measurement as well as the resolution of ambiguity can be adopted unchanged.

In modern sensor designs, more than 2 transmitting or receiving antennas are used, as the examples in ▶ Sect. 15.9. For this general case, the so-called *virtual aperture*, i.e. the local distribution function of the antennas in the array, results from the convolution of the local distribution function of the transmit antennas with those of the receive antennas. This can virtually generate many more antenna positions with corresponding measured values than are physically present. This makes the angle estimation more robust and achieves higher resolution since the virtual aperture is larger than the real one. For a deeper understanding, please refer to references [17] and [19].

Also commonly used are sparse arrangements of the (receive) antenna elements, where the spacings from each other follow a rule that the spacings from each other are unequal, although this is constructed as a common multiple of one based on fractions of the wavelength, e.g. $\lambda/2$. In Sect.15.9.2, there is an example together with the antenna characteristic differences. In summary, the smallest space determines the uniqueness and the space between the outer elements determines the resolution.

15.5 Major Parameters of Performance

Even though the most important quantities of the performance result from the understanding of the function, especially the modulation and the angle evaluation, they are summarized here in a short overview.

15.5.1 Range

The performance of the range measurement is mainly given by the frequency bandwidth Δf_{eff} of the modulation, cf. B. Equation 15.19 and determines the range cell size

$$\Delta r \geq= \frac{c}{2\Delta f_{\text{eff}}} \tag{15.62}$$

and thus, the separability.

The measurement limit for the maximum (unique) range is determined for radar with frequency modulation by the sampling rate (cf. ▶ Sect. 15.3.1). However, the maximum range relative to a standard target depends not only on the modulation parameters but also on the transmit power, the antenna gain in main direction, and the signal-to-noise ratio of the receiver electronics, cf. Equation 15.6 in ▶ Sect. 15.2. It should be borne in mind that in practice the reflectivity of the objects varies by several powers of ten and, in addition, multipath superpositions make this limit appear anything but sharp.

The minimum range can be smaller than the separability interval only if the multi-target capability in the range is waived.

15.5.2 Radial Velocity

For the cell size $\Delta \dot{r}$ and thus for the separating ability as well as for the accuracy of the radial velocity, the uninterrupted measuring time T_{M} is decisive, cf. e.g. B. Equation 15.22. The maximum and minimum radial velocity are determined by the sampling rate of the Doppler effect. However, any ambiguity due to a sampling rate that is too low can certainly be compensated, as ▶ Sect. 15.3.2.4 is described.

15.5.3 Azimuth Angle

The separability of the azimuth angle determination of an array planar antenna depends on the ratio $\ell_{\phi,\text{eff}}/\lambda_{\text{c}}$

of the width of the (virtual) aperture $\ell_{\phi,\text{eff}}$ and the wavelength λ_{c}, cf. Equation 15.61, and provides the azimuth cell size relevant to the separability

$$\Delta q_{\text{eff}} = \frac{1}{2\Gamma_{\phi,\text{max}}} = \frac{\lambda_{\text{c}}}{2\ell_{\phi}} = \frac{\lambda_{\text{c}}}{\ell_{\phi,\text{eff}}}. \tag{15.63}$$

The uniqueness range is determined by the smallest space $_{\text{s}}\Delta x_{\text{min}}$ between (the center of) two adjacent (receive) antenna elements.

$$\Delta \sin\phi_{\text{uni}} = q_{\text{uni}} = 1/\Delta\Gamma_{\phi,\text{min}} = \frac{\lambda_{\text{c}}}{_{\text{s}}\Delta x_{\text{min}}} \tag{15.64}$$

However, it should be noted here that a variety of algorithms can be used to push both boundaries significantly. In addition to a significantly higher computational requirement compared to digital beamforming, they rely on certain assumptions, which in turn are not always met. In many cases, they also require a sequence of several measurements (multiple snapshots). In application cases favorable to these algorithms, resolutions are achieved that are almost an order of magnitude better than those specified in Eq. 15.63 given, cf. [17].

15.5.4 Elevation Angle

Since for the elevation determination usually only two antenna positions are used at a range of $_{\text{s}}\Delta y_{\text{n}} = \Delta\Gamma_{\vartheta} \cdot \lambda$, hence, there is no possibility of separation. In the end, this range exclusively determines the unambiguous angular range $p_{\text{uni}} = \sin\vartheta = \Delta\Gamma_{\vartheta}$ whereby a consideration must be made: If the unambiguous angular range is increased, the resolution deteriorates, which in turn also depends on the SNR.

15.5.5 Performance and Multi-Target Capability

A radar for automotive use as a surroundings sensor cannot do without a multi-target capability. This requires a separation capability in at least one of the dimensions range, radial velocity and azimuth angle. Depending on the concept, the separation capability is prioritized sometimes for range and sometimes for radial velocity. In a figurative sense, a "cell volume" as small as possible is aimed at for a multi-target capability that is high in practice, by which it is meant the product of the cell sizes of the three or four[16] dimensions, even if these have different units. Established conversion and thus weighting factors required for a volume consider-

---

16 Four if the antenna array allows elevation measurement.

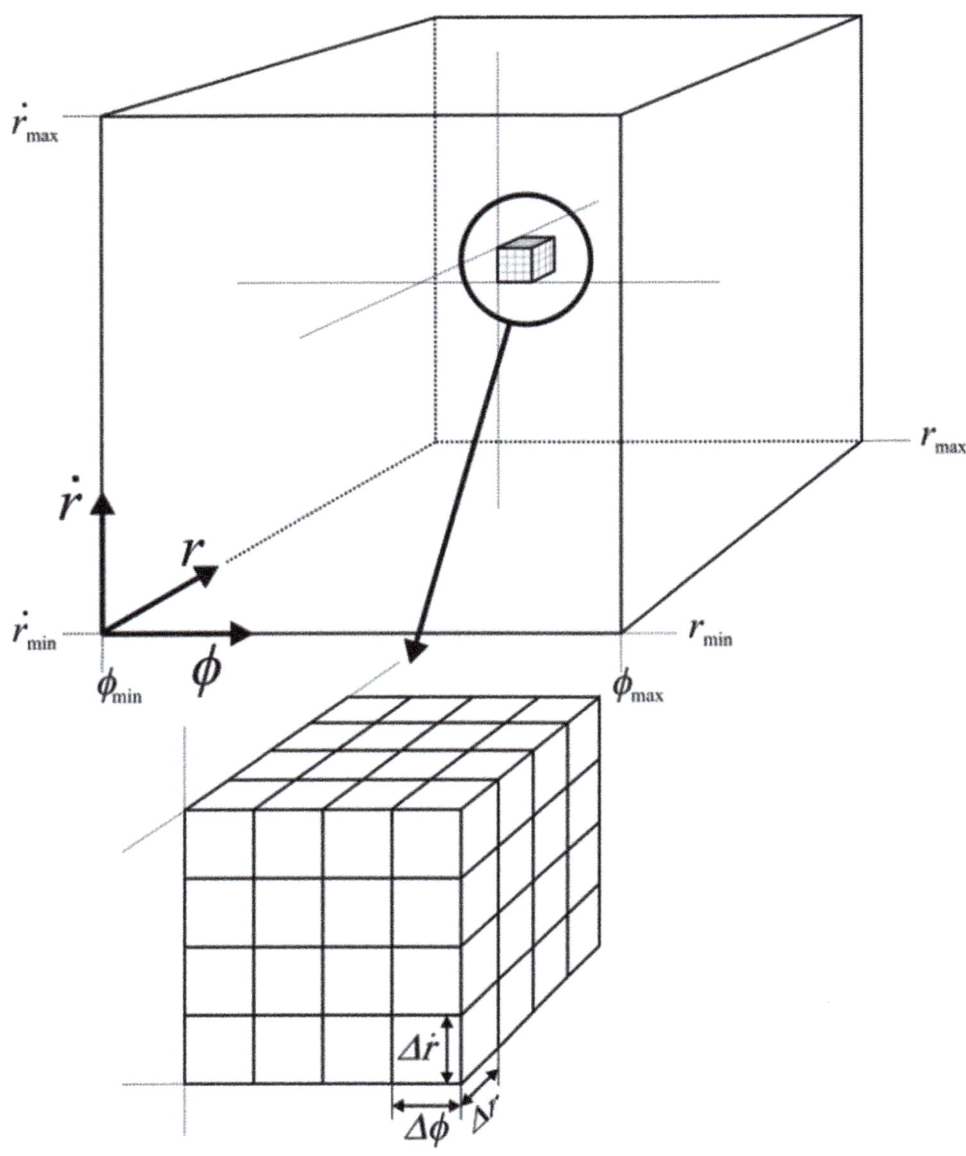

◘ Fig. 15.19   Visualization of separability as cell volume in the dimensions range, radial velocity and azimuth angle[17]

ation are not known and probably not always useful. This is especially true in cases that are far apart, for example, when a sensor that has uniformly small cell sizes is to be compared with a sensor that resolves only one dimension but can do so very accurately. Guideline values for the required cell size of a long-range radar are given below, which in themselves provide sufficient multi-target capability. Here, the longitudinal extent of a small passenger car is assumed to be at a 100 m range from the sensor. Further, for separation, it is assumed that a spacing of three cells is required. Theoretically, the range of two cells would also be sufficient, but the windowing and beam defocus hardly allow this.

$$\Delta r \approx 1,5\,\text{m}, \Delta\dot{r} \approx 0,1\,\frac{\text{m}}{\text{s}}, \Delta q \approx \sin(0,7°) \qquad (15.65)$$

Here we see that multi-target capability on an angular basis alone is not possible with installation-compatible antennas (effective aperture width would have to be $\ell_{\phi,\text{eff}} > 45$ cm) even when using MIMO techniques. Separation capability based on range alone reaches limits when multiple objects are at nearly equal range; separation capability based on radial velocity fails for stationary objects. Therefore, separation based on range and on radial velocity is sought. In ◘ Fig. 15.19, the domain cuboid $\{r_{min}\ldots r_{max}, \dot{r}_{min}\ldots\dot{r}_{max}, q_{min}\ldots q_{max}\}$ is schematically shown, which results from the minimum and maximum values, and which consists of the individual cell volumes $\{\Delta r, \Delta\dot{r}, \Delta q\}$. From this, the qualitative statement can be derived that the larger the range

---

17 In the picture was simplified.

volume and the smaller the cell volume, the higher the performance. However, this also requires at least

$$N_S \geq 2 \frac{r_{max} - r_{min}}{\Delta r} \cdot \frac{\dot{r}_{max} - \dot{r}_{min}}{\Delta \dot{r}} \cdot \frac{q_{max} - q_{min}}{\Delta q} \quad (15.66)$$

measurement points, which can quickly run into the order of magnitude of over 10^5/cycle and, with typical cycle times of 20 ms results in a raw data rate of well over 10 MB/s results.

However, it should be noted that other reasons for degradation exist in addition to the principle-related limitations. In particular, frequency generation and modulation can contribute to degradation. Both a non-constant amplitude over the modulation period and phase noise or nonlinearities of the high-frequency circuitry leads to broadening of the frequency peaks and reduces the separating capability.

In addition to the separation capability, the resolution plays a major role for the quality. For this purpose, neighboring cells are used for a peak centroid determination, which makes resolutions of approx. 1/10 of the cell width become possible. However, this again only applies to point targets. Real targets, on the other hand, cause dispersion considerably above this value. Changing reflection centers lead to jumps of several meters both longitudinally and laterally, but dispersion also occurs in the radial velocity when relative motions are detected such as moving parts or transported goods with a relative degree of freedom, e.g. cars on car trailers. This dispersion can be so strong that an object is detected in several cells. By mostly heuristic approaches, these cells then have to be clustered again to a common object.

In addition to the capability to detect objects, robustness against artifacts plays an important role. Thus, neither so-called ghost targets, i.e. object detections without an existing object, nor false values to existing objects are desired. Nonlinearities in the signal chain can lead to representatives of the first group, unresolved ambiguities, and interferences to the second group. Another expected capability is the suppression of targets that can be driven over or under, such as manhole covers or bridges at least in the area where the triggering of an emergency brake on a stationary target is no longer excluded, cf. [21]. Unfortunately, the addressed robustness cannot be predicted by specifying design parameters, since the countermeasures are "buried deep" in the evaluation algorithms.

15.5.6 24 GHz Versus 77 GHz

The frequency band of 24,05 ... 24,25 GHz allows radar use in road traffic, in addition to the 76 ... 77 GHz band and the 77 ... 81 GHz band. The advantages of the band from 24,05 to 24,25 GHz were the less expensive compo-

nents. However, this cost advantage has disappeared today. A disadvantage of 24 GHz is the increase of the radial velocity cell, because the Doppler frequency scales proportionally with the carrier frequency. The largest difference from the lower frequency results from the higher wavelength ($\lambda \approx 12$ mm), which in turn leads to a broadening of the beam pattern if the antenna size is to be maintained. The antenna gain is smaller, and the angular resolution deteriorates at 24 GHz compared to 77 GHz. Therefore, the use of 24 GHz for the mid-range up to 100 m was predestined. It is also well suited for the short range if a wide beam pattern is desired. However, due to the band limitation a detection below 0,5 m is hardly possible, so that a 24 GHz-radar cannot replace parking assistance sensors.

The ultra-wide band (UWB) technology at 24GHz, which enabled a large bandwidth and thus good range separation capabilities, was only temporarily tolerated. As of January 1, 2018, UWB radars were no longer allowed to be integrated into newly type-approved vehicles in the EU. In other regions, UWB regulation is also expiring. As a result, only the 77 ... 81 GHz band remains available when very wide bandwidths and thus high-range resolutions are required, although not all countries allow this band. Apart from the 76 ... 77 GHz band, global regulation for the automotive bands is heterogeneous and subject to change. The current status can only be researched via the country-specific and currently applicable regulatory provisions. Due to the advantages of even higher frequencies for spatial resolution (see 15.4.1) and for higher bandwidths, new frequency ranges above 100 GHz are also being discussed for frequency regulation.

15.6 Signal Processing and Tracking

Although there is an almost unmanageable diversity for signal processing and subsequent tracking, in this section, we attempt to organize this diversity into a basic structure (cf. ◘ Table 15.1) that applies to most procedures.

At the beginning, there is the **signal forming**. In all concepts, this includes signal modulation, e.g. ramp generation. If the antenna characteristics are also changed dynamically (e.g. by analog beamforming or scanning), this also counts as signal forming.

The first processing step with received signals consists of **preprocessing and digital data acquisition**. This step combines demodulation and digital data acquisition and often includes adjustment filters to compensate, for example, for the reduction in received power associated with range. After demodulation and amplification, the analog signals are sampled and converted to digital values. Both classic parallel converters and Σ-Δ converters can be used for this purpose. The latter are 1-bit converters with oversampling and downstream digital filter.

▫ Table 15.1   Generalized steps of radar signal processing

Processing step	Explanation
Signal shaping	Modulation (frequency steps or ramps, pulse generation), beam switching or forming
Pre-processing and digital data acquisition	Demodulation, amplification, digital data acquisition
Spectral analysis	Mostly one- or two-dimensional (fast) Fourier transformation of the digital data, where the frequency position and the complex amplitudes contain the information about range, velocity and azimuth angle
Detection	Detection of peaks in the spectrum, mostly by means of comparison with an adaptive threshold
Matching	Assignment of detected peaks to a target
Azimuth angle determination	Determination of the azimuth angle from the (complex) amplitudes of different receive branches
Clustering	Collection of detections that presumably belong to one object
Tracking	Associate current object data to previously known objects (= association) to obtain a temporal data trace (track), which is filtered and from which the object data is predicted for the next association

The amount of data corresponds to the number of cells according to Eq. 15.66, i.e. (at least) one measured value per cell. In all modern automotive radar sensors, spectral analysis performed by Fourier transform plays an important role in the preprocessing of signals. In simple terms, the Fourier transform is a conversion from the time domain to the frequency domain and vice versa. A sequence of measured values defined in time steps becomes a sequence of 'measured values' defined in frequency steps, which determines the frequency spectrum. Modern signal processors are powerful enough to perform this transformation in a few milliseconds even with many measurement points (order of magnitude 1000). However, this high transformation speed is only achieved if the number assumes certain values. With the classical Fast Fourier Transform (FFT) algorithm, it must be a power of 2 (e.g. 512, 1024, 2048).

It is common to use windowing in conjunction with spectral analysis to avoid artifacts caused by limiting the measurement window (so-called leakage errors). Even though different window functions can be used for this purpose, which are optimized on different criteria, this leads to an effective cell magnification of about 1.5 times in each dimension with a correspondingly reduced separation capability.

Detection is the search for special features in the measured data series. Often these are peaks in a spectrum, be it a frequency or a time-of-flight spectrum. Here, it is necessary to detect the reflection signals of individual objects and distinguish them from those of other objects. Due to the strongly varying signal strengths of the different objects, but also of the same object at different times, a threshold algorithm has to be found which not only finds all peaks originating from real objects as far as possible but also is insensitive to peaks caused by noise or interfering signals. Therefore, adaptive thresholds are mostly used, which are mainly chosen according to the principle of CFAR (Constant-False-Alarm-Rate). This principle is based on the fact that for a spectral point, the power values of neighboring points are used to form an adaptive threshold for the power of the selected spectral point, on the basis of which an overshoot is considered a detection and an undershoot leads to discrimination. The level of the threshold is determined, in addition to the spectral values of the neighborhood, by the specified probability of error (false alarm rate). The evaluation of the neighborhood can be done with different algorithms. An overview can be found in [8].

If systematic peaks occur that cannot be attributed to external reflections, these must be masked, as must any ground reflections. Unfortunately, strong reflections from a real object also complicate detection. On the one hand, they can mask weaker reflecting objects in neighboring frequency ranges, namely if the transmit frequency does not ideally follow the modulation course. The causes for this are the phase noise of the oscillator and linearity errors in FM processes.

As explained in chapter ▶ Sect. 15.4, **the angle determination of** modern sensors is done by an evaluation of the phase shift of the signals at the individual antennas in azimuth and partly also in elevation direction. In the simplest case Eq. 15.48 or alternatively, the complex signals at the antennas are compared with the calibration data. For MIMO sensors, the signal distribution on the virtual aperture must be determined beforehand, which is then used to estimate the angle using the above approaches. Today, more powerful and thus more complex angle estimation algorithms are primarily used when higher angular resolutions are required in individual fields of view.

Especially with high-resolution radar sensors, "too much" information is generated. For example, a small range cell from a truck often detects a large number of reflections or, at high resolution on velocity, relative movements of a coherent object (tractor, trailer, cargo or pedestrian limbs). Therefore, an attempt is made to combine the detections of the same object by means of heuristic **clustering** and to keep them as only one object in the measurement list. In the meantime, approaches based on machine learning methods are also being developed, which determine and classify objects analogously to the point clouds known from lidar sensors.

Tracking is the formation of the temporal context of individual measurement events into quasi-continuous "tracks" of individual objects. In this step, the transformation from the sensor system to the Cartesian vehicle coordinate system takes place at the beginning. Initially, the detection and the subsequent clustering leads to individual object hypotheses, which for the time being are valid only for present individual one cycle. In tracking, the assignment to hypotheses of previous cycles is attempted next (association). Usually, these object hypotheses are organized in lists and have Object Identifiers as "individual" identifiers. For association, the state variables of the previously known objects (e.g. range or lateral position) are predicted to the time of the current measurement. Then, the association of the current object hypotheses to the previous ones is performed, with a search window around the predicted values since both measurement and prediction errors are to be assumed. If a currently recognized object can be associated to a previous object in this search window, the trace is continued. At the same time, the object evidence is increased or remains at a high level. If objects of the current measurement remain, new objects are generated in the object list and initialized with the measurement data of the current measurement. However, this object starts its track with a low object evidence for existence, which is usually so low that one or more association in subsequent measurements are required before this object can be considered as a target object for an application (e.g. ACC).

If no hypothesis can be assigned for an existing object from the current measurement, the object evidence decreases. After several drop-outs, the evidence falls below a defined threshold, whereupon this object is removed from the object list. In addition to these basic cases, possible ambiguities must be taken into account, such as the fact that a current object falls within the search window of other objects or that several individual objects in the list belong to a single real object.

In addition to the association, mostly a very application-specific state data filtering is performed with the tracking, often connectable with the association as a Kalman filter, which already implicitly contains the prediction step necessary for the association, cf. ▶ Chap. 20. The state variables of objects detected with active sensors always contain the range in x- and y-direction (vehicle coordinate system), the relative velocity in these directions, the acceleration at least in longitudinal direction. If the state variables of the ego vehicle are added, the absolute object variables for velocities and accelerations can also be formed and carried along in the object list. This also makes it possible to distinguish between moving (in the same direction), stationary or oncoming objects. Since tracking provides the objects with a history, this can also be used to distinguish between stationary and "stopped" objects. These aforementioned distinctions represent the main classes of classification, albeit a simple one. The non-reaction of conventional ACC systems to stationary objects is based precisely on this classification and not on the often-mentioned but nevertheless false claim that stationary objects cannot be detected with radar.

In modern radar sensors, the aforementioned rough classification is no longer sufficient. Even if the reception amplitudes of a single measurement have hardly any significance, the observation of the reflection amplitudes over time, especially if the range changes strongly in the process, provides helpful information for the classification of the targets. As mentioned in ▶ Sect. 15.2, the variation due to multipath reflection can be used specifically to infer the height of the target.

Following the signal processing steps mentioned, the situation interpretation begins, which in the simplest way takes over the selection of a radar target from the object list. The target selection as well as a more extensive situation interpretation strongly depend on the application and are to be described as part of it. The target selection for ACC can be found accordingly in ▶ Sect. 32.4.

15.7 Installation and Adjustment

There are basically two concepts for installing the radar sensor: with or without an optical cover for the antenna (radome). An optical cover is often more design-friendly than direct visibility of the radar sensor, which means that more and more sensors are being concealed. It is important for the cover that the radar waves are only slightly attenuated and that the angle characteristic does not lead to any unexpected change. Plastics as a cover are rather problem-free. Depending on the thickness of the plastic, both the penetration and the angular characteristics change. If the waves propagating directly through the plastic superimpose with the waves reflected back and forth in the plastic, some of them several times, this is called constructive superimposition, which results in a low overall attenuation. If the partial waves superimpose destructively, this results in particularly high attenuation. In the case of a very large angular range, the phenomenon occurs that changes occur via the interfacial reflections of the layer then acting as a waveguide. Furthermore, the transmission at the surface itself depends on the angle (and the polarization).

Non-metallic paint tends to be unproblematic, but metallic paint can lead to considerable problems. Here, the repainting specification, which allows painting three times, is particularly problematic. Of course, metallic masking is completely unsuitable as long as the penetration depth is less than the material thickness. Very thin layers ($< 1\mu$m), however, can again be transparent for mm-waves without losing their metallic specular property for optical waves. This is exploited to reproduce metallic structures (radiator grille, brand logo) on plastic surfaces. Thus, a radome that is difficult to detect visually can be constructed.

Very often, radar sensors are either mounted directly in the bumper or on the body. A bracket acts as a coupling element to the body or chassis. The sensor can be adjusted in both azimuth ϕ as well as in elevation ϑ by screwing it to the bracket which allows alignment at the end of the vehicle production line or in the workshop.

In total, three sources of error for azimuth ϕ_{err} and elevationerr ϑ_{err} are to be considered:

- Error in the sensor-internal alignment ($\phi_{err,int}, \vartheta_{err,int}$). These are largely eliminated by calibrating the sensor, i.e. measuring the antenna pattern relative to the sensor housing.
- Error in alignment of the sensor on the vehicle ($\phi_{err,m}, \vartheta_{err,m}$). This can be checked and, if necessary, adjusted using optical mirrors on the sensor housing and a laser-based calibration system. Alternatively, metallic mirrors can be placed in front of the sensor at defined angular positions to the vehicle so that the sensor's own angle determination can be used for correction.
- Alignment error of the sensor carrier = ego vehicle due to a pitch angle deviating from the design position $\vartheta_{err,veh}$ or a vehicle slip angle that also occurs when driving straight ahead $\phi_{err,veh}$

With the radar sensors without elevation measurement capability, the misalignments in the elevation lead "only" to a reduction of the detection range or a reduced accuracy of the azimuth angles, but not to systematic measurement errors. This also means that no permanent checking of the alignment is required.

For the azimuth angle, an offset determination during operation is essential since the static vehicle slip angle only becomes apparent during travel if it has not been determined beforehand via a roller dynamometer. In addition, uncertainty still remains for the other angular errors. Because of the high sensitivity of the target selection to azimuth errors, azimuth offset estimation procedures are necessary. Basic information of these estimators are the averaged gradients of the measured lateral offset as a function of the longitudinal

range, corrected for the error caused by the rotation $\dot{\psi}$ of the ACC vehicle, but also the radial velocity of stationary targets, which are measured with

$$\dot{r}_{stat} = -v_{veh}\cos\phi \tag{15.67}$$

depends on the cosine of the azimuth angle. In the case of an azimuth angle offset, this leads to a vs. $\phi = 0$ asymmetric distribution of the radial velocities originating from stationary targets, which in turn can be evaluated for compensation.

If the radar sensor covers a wide azimuth range, fine adjustment at the factory or in the workshop is not necessary.

15.8 Interference

15.8.1 Occurrence and Effects

In the meantime, the order of 100 million radar sensors are installed in motor vehicles worldwide every year, and this trend is rising. The sensors are installed around the vehicles. Depending on their function, the field of view of the sensors is either oriented to the front, rear or side of the vehicle. Due to the ever increasing penetration rate of radar sensors, the probability of different radar sensors interfering with each other is growing at an increasing rate. While the probability of two FMCW or chirp sequence sensors interfering with each other is very low, interference between sensors has nevertheless become an actual and important issue.

Interference occurs in FMCW or chirp sequence sensors when the transmitting ramp of the interfering sensor is received in the receiver of another sensor. This can only happen if both ramps have almost the same frequency at one point in time, since the receiving sensor has a very narrow-band receiver in the analog modulation methods, see ◘ Fig. 15.6. This narrow bandwidth of the mixer output frequency means that the probability of interference is very low, and it is present only for a very short time, exactly as long as the interfering signal and the receiver use the same frequency. With different ramp slopes, this time is extremely short. However, if an interference occurs, it is often associated with a strong disturbance. This is because normal radar operation is designed to receive and process signals with channel attenuation that is inversely proportional to the fourth power of the target distance. However, interference signals between the transmitter and receiver are attenuated only to the second power of the target distance because they are not reflected off a target but propagate along the line of sight between the interfering transmitters and receiver. This then low channel attenuation leads to a strong short-time disturbance.

Although in the time domain the interference is only very short, the large energy input into the received signal leads to a massive degradation of the signal-to-noise ratio. The Fourier transform used to evaluate the signal causes the interference energy to be smeared across the entire received spectrum. This significantly reduces the sensitivity of the sensor and small targets are no longer detectable.

False targets could theoretically occur if the interfering ramp and the ramp of the interfered sensor have exactly the same slope and, as explained above, the interfering ramp falls into the receive filter. However since all sensors have different oscillators with different phase noise even with the same construction types, this case cannot occur in practice.

Interference can be detected in the sensor in the simplest case by a sudden power increase of the received signal. More complex methods can also be found in the literature, some of which allow the detection of weak interference. For example, the received signal can be predicted and compared with the actual signal. Sudden deviations then indicate interference.

15.8.2 Countermeasures

Measures against interference can be taken in the time and frequency domain. Furthermore, it is possible to mask out interferers in the angular range.

In the time domain, the simplest way to fight against interference is to cut out the disturbed signal component and set it to zero. This measure is usually already very effective, but leads to artifacts, because by cutting out a Sinc-function $(\sin x)/x$ convolves into the signal in the frequency domain. This can be prevented by not setting hard to zero, but by smoothing the cut-out signal part by windowing with smooth characteristic. In the time domain, very powerful reconstruction algorithms are also available that cut out the interfered signal portion and replace it with a reconstructed signal.

In the frequency domain, the measures focus on more or less intelligent frequency hopping schemes, i.e. changing the frequency position of the ramps. These measures are very effective as long as it is possible to move within the regulated bandwidth. It must also be taken into account that crossing in different areas of the ramps have different impacts depending on the analog filters in the radar.

The countermeasures that mask interferers in the angular domain are independent of the interferer's modulation scheme, which is the real strength of this type of measure. In the angular domain, it is possible to first estimate the direction of the interferer via digital beamforming and then, in a second measurement step, synthesize a receive directional characteristic of the antenna array via digital beamforming that

fades out exactly this direction. In this process, the target is usually not lost either, since other motor vehicles, for example, represent extended targets that have many reflection centers. Alternatively, so-called adaptive beamforming methods can be used, in which a criterion, such as the signal-to-interference ratio, is optimized without explicitly estimating the direction of the interferer. Both approaches are roughly equally powerful. An introduction to interference cancellation with extensive further reading can be found in [12].

Digital radar sensors are interfered within interference situations in the same way as analog radars. The interference mechanisms and the options for countermeasures are different, but interference occurs in the same way as with analog radar sensors. Since digitally modulated radar sensors have a transmit signal that instantaneously occupies a very large bandwidth over the entire time and the receivers receive over the entire time the maximum bandwidth, the probability of receiving interfering signals is much greater than with analog modulated radar sensors. However, the spectral power density is lower for digital radar sensors, so less interference energy is introduced per bandwidth. In identical interference situations, digitally and analog modulated radar sensors have exactly the same interference energy input [3]. The digital interference signals usually have the effect of noise in the disturbed sensor, both in the time and frequency domain. Beamforming is particularly effective as a countermeasure, since measures in the time and frequency domain often suffer from the fact that the digital signals interfere over long times and large bandwidths.

When digital radar sensors are interfered with by FMCW or chirp sequence sensors, the digital sensors can play to their strength that the interfered received signal is present at the digital radar as a baseband signal. This makes it very easy to filter out the ramps from the digital received signal. However, if analog radars are interfered with by digital radars, beamforming seems to be one of the few effective means according to the current state of research.

15.9 Design Examples

15.9.1 Bosch 4th Generation Radar Sensors

Since 2013, Bosch has produced their 4th generation of radar sensors. The technical details are described in [23] in detail. Like the previous generation, this is a 76.5GHz radar with an integrated electronic control unit. To address the increased amount of use cases and additional functions a flexible design kit was developed [10]. In addition to pure radar-based aspects, the complete system architecture must also be highly flexible in system integration aspects [4].

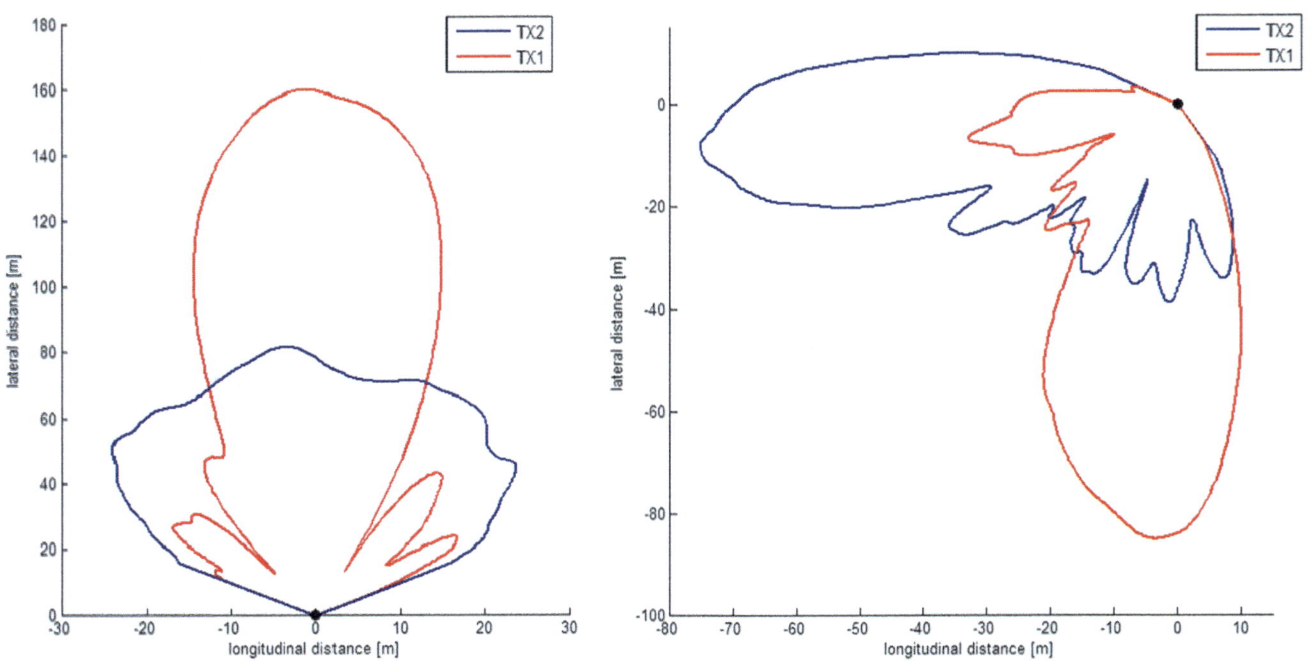

Fig. 15.20   Typical detection range of the MRR as front sensor (left) respectively as rear (right). Both transmitting antenna characteristics are shown. (*Source* Bosch)

The 4th Radar Generation consists of the following product variants: MRR, MRR Rear, and MRR Corner.

To cover the increasing spread of radar-based assistance systems, MRR was developed using a planar antenna system. To enable pedestrian protection, [14] the field of view can be switched to the 2nd transmitting antenna which has a wider field of view (■ Fig. 15.20, left). The MRR Rear and Corner variants also use different transmit antenna field of views. In these sensors, different main beam directions are realized. For a rear sensor, e.g. detection of approaching traffic up to 80 m is possible and additionally, cross traffic can be detected in the same range (■ Fig. 15.20, right). Additional use cases can be realized be adapting the antenna layout. The key performance indicators for the MRR can be found in ■ Table 15.2.

The angular estimation uses parametric deterministic maximum likelihood estimation. This is described in the chapter on monopulse principle (▶ Sect. 15.4.3). Based on this process, the performance of the radar's two-angle azimuth separability can be improved. The MRR has a separability down to 7°, and the front variant is also able to determine angles in elevation. For this, different antenna characteristics in elevation are used (see [23]).

15.9.2   Bosch 5th Generation Radar

The increasing requirements for driver assistance systems lead to the development of the 5th generation of Bosch radar sensors.

The use of a bandwidth of up to 1 GHz is standard, which has been made possible due to a change in legislation. Using the higher bandwidth allows better separability and accuracy in the distance domain. The RF front end even allows bandwidth greater than 2 GHz. Compared to Generation 4, the data acquisition time is significantly increased. This improvement is aided by the new Joint-Sequence FMCW modulation is developed.

The concept of a flexible design kit (see ■ Fig. 15.24) combined with an even higher integration of components form the base of the 5th Generation Radar (see ■ Fig. 15.21). The new MMIC consists of the RF components like VCO, PLL and mixer, as well as the digital control unit and the analog–digital converter. Controlling the MMIC is performed via SPI and the data transfer via LVDS. The benefit of this is the possibility to place all components on a single PCB and a significant reduction of the sensor depth to 19 mm. In addition, the MMIC has 3 transmit and 4 receive channels. With the use of MIMO (Multiple Input Multiple Output), up to 12 virtual channels can be used. The increased data rates are addressed using faster link technologies like CAN-FD and Ethernet.

The most significant performance jump compared to Generation 4 is reached due to the modulation concept change. This concept combines the advantages of Slow-FMCW and Chirp-Sequence modulation: direct resolution of range and velocity using 2D-FFT and efficient resource usage of FMCW. It allows for an effective modulation length four times higher as compared

■ **Table 15.2** Technical data of the Bosch 5th generation radar sensors

		Bosch Radar Generation 5
Product Variant		FR5CP: Front Radar (Plus) CR5CP/B: Rear Corner Radar (Plus/Base)
Start of production		2019 (FR5CP), 2020 (CR5CP), 2021 (CR5CB)
Dimensions (W x H x D)/mm³		72×63×19
Mass		75 g
Cycle duration		66 ms
Beam forming		Bistatic, patch arrays, incl. digital beam forming
Number of measurement ranges		Close range up to 125m and/or long range up to 210m
Transmit antenna beam width (3dB) azimuth/elevation		FR5CP: 75°/16°[*Near Range*]; 24°/16°[*Far Range*] CR5CP: 75°/16°[*Near Range*]; 24°/16°[*Far Range*] CR5CB:52°/16°
Radiated power (EIRPpeak/average)		FR5CP: 28.4/19.9 dBm (EIRP), CR5CP: 27.3/18.6 dBm (EIRP) CR5CB: 24.1/16.45 dBm (EIRP)
Frequency range		FR5CP, CR5CP, CR5CB: 76GHz–77 GHz
Modulation method		Joint Chirp Sequence
Effective modulation bandwidth/modulation bandwidth		FR5CP: 714MHz CR5CP: 652MHz CR5CB: 555MHz
Modulation details		32 chirps per TX sequence, up to 10 interleaved TX sequences per burst, chirp rate within a sequence:~1/525 µs (~1.9 kHz), Duration single chirp: 30us–61us, measuring time: 16.8 ms for all measurement ranges
Number of measurement ranges		FR5CP: 0.41 m–150 m / 0.5 m–180 m / 0.58 m–210 m CR5CP: 0.46 m–125 m/160 m CR5CB: 0.55 m–98 m
Azimuth		
	Angle measurement method	Sparse antenna array with additional virtual channels via MIMO 4 channels with phase evaluation
	Angle range	CR5CP, CR5CB:±75°, FR5CP:±60°
	Precision* point target	0,1° at 30 dB SNR
	Unambiguous range	[-65,+65]° for FR5CP, [-75,+75]° for CR5CP, CR5CB
	Resolution	0.2°, Separability FR5CP: 3°, CR5CP: 4°,CR5CB: 4°
Elevation		
	Angle measurement method	Thin antenna array
	Angle range	±15°
	Precision* point target	0,2° at 30 dB SNR, CR5CB without vertical angle measurement
	Unambiguous range	[−15,15]° for FR5CP, CR5CP CR5CB without vertical angle measurement
	Resolution	0.1°, separability FR5CP, CR5CP: 6° CR5CB without vertical angle measurement
Radial distance (range)		
	Distance range	For point target with RCS of 10 m²: FR5CP: 0.41 … 210 m, CR5CP: 0.47… 125.7 m, CR5CB: 0.55 m … 98,5 m
	Resolution	FR5CP: 0.21 m, CR5CP: 0.23 m, CR5CB: 0.27 m

(continued)

Table 15.2 (continued)		Bosch Radar Generation 5
	Separability	FR5CP: 0.41 m, CR5CP: 0.45 m, CR5CB: 0.55 m
	Precision* point target	FR5CP<0.11 m, CR5CP<0.09 m, CR5CB<0.14 m
Radial velocity (Range Rate)		
	Total range, of which unambiguous	CR5CP, CR5CB: -80 … +80 m/s, FR5CP: -110 … +55 m/s after parametric uniqueness resolution
	Resolution	0.115 m/s
	Separability	0.23 m/s
	Precision point target	<0.04 m/s
Special features		Online misalignment estimation, blindness detection, unambiguous single shot d/v-measurement, Euro NCAP ready

Fig. 15.21 Comparison of 4th and 5th generation exploded views. (*Source* Bosch)

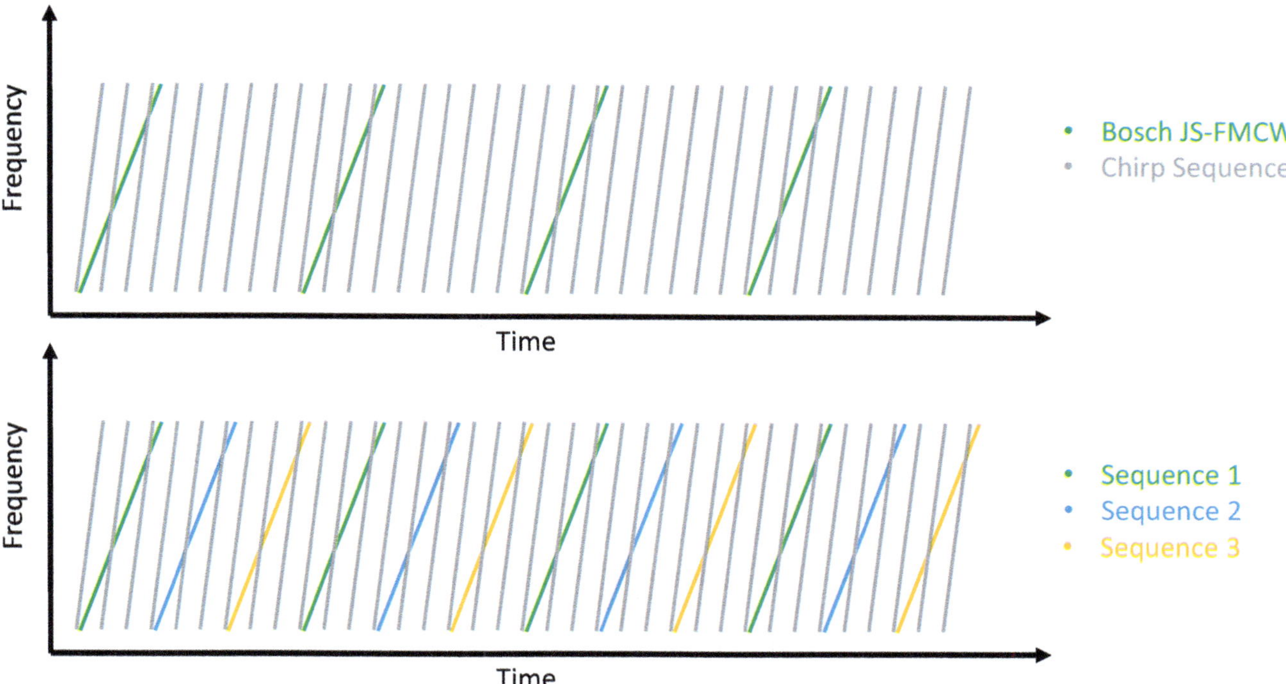

Fig. 15.22 Chirp Sequence versus JS-FMCW

to the previous generation. This leads to the same factor of improvement in the velocity domain.

The 2D-FFT enables a much higher number of radar reflections, allowing a better separation of standing and moving targets. To reduce the required computing resources, a specific subsampling in the velocity domain is used. A chirp length of ~50 μs is used, leading to a reduction of compute resources by a factor of 50 according to Bosch.

The velocity ambiguity (see ▶ Sect. 3.2.4) is resolved in BEP (Back-End Processing). It uses a model-based algorithm similar to the angular estimation algorithm. Using subsampling, the gaps between the more spread out chirps can be used for additional transmit channels. On simultaneous modulation cycles this allows the radar to cover different FoVs and MIMO (see ▪ Fig. 15.22). Additionally, the interleaving of all transmit channels allows for covering the full combined FoV in every cycle.

Using MIMO allows the radar to use a larger virtual aperture without increasing the size of the sensor, thus, leading to improvement in both the robustness and the precision of the angular estimation. As described, the MIMO (see ▪ Fig. 15.22), transmit channels are sampled in time domain multiplexing (see ▶ Sect. 15.4.4.4). Interleaving the channels leads to a minimal timing offset. Compared to the delta of ∼ 16ms (typical length of a JS-FMCW modulation), this is reduced to a few hundred microseconds. Due to this reduction, the environment is sampled nearly in-

stantaneously. Typically, the Bosch 5th Generation Radar uses 2 transmit and 3 or 4 receive channels simultaneously in azimuth, which leads to 6 to 8 virtual channels. In elevation, 2 transmitters and 2 receivers are used, therefore 4 virtual channels can be used. The antenna layout showing the transmit and receive antennas can be found in ▪ Fig. 15.23.

The microcontroller used also has a significant part in implementing the JS-FMCW modulation efficiently. It consists of a multi-core CPU architecture as well as dedicated DSP accelerators. The accelerators take care of FFT, CFAR, interference mitigation, and digital beam forming. In the back-end processing (BEP), velocity and angular estimation are done on a dedicated core. The additional cores are used for the radar perception and functional software.

In the Bosch 5th Generation Radar design kit, different variants of the microcontroller are available, which differ in the sizes of RAM and ROM and the number of cores and accelerators.

Due to the powerful HW and parallel processing, a fixed cycle time is possible. Synchronization of the radar to other ECUs is possible, allowing for the use of multiple radar sensors on one vehicle without mutual interference.

The design kit (see ▪ Fig. 15.24) makes it easy to develop different sensor types in Generation 5. Sensors are clustered in different families like Plus and Base, and these families differ in the number of RF channels and the microcontroller variant used.

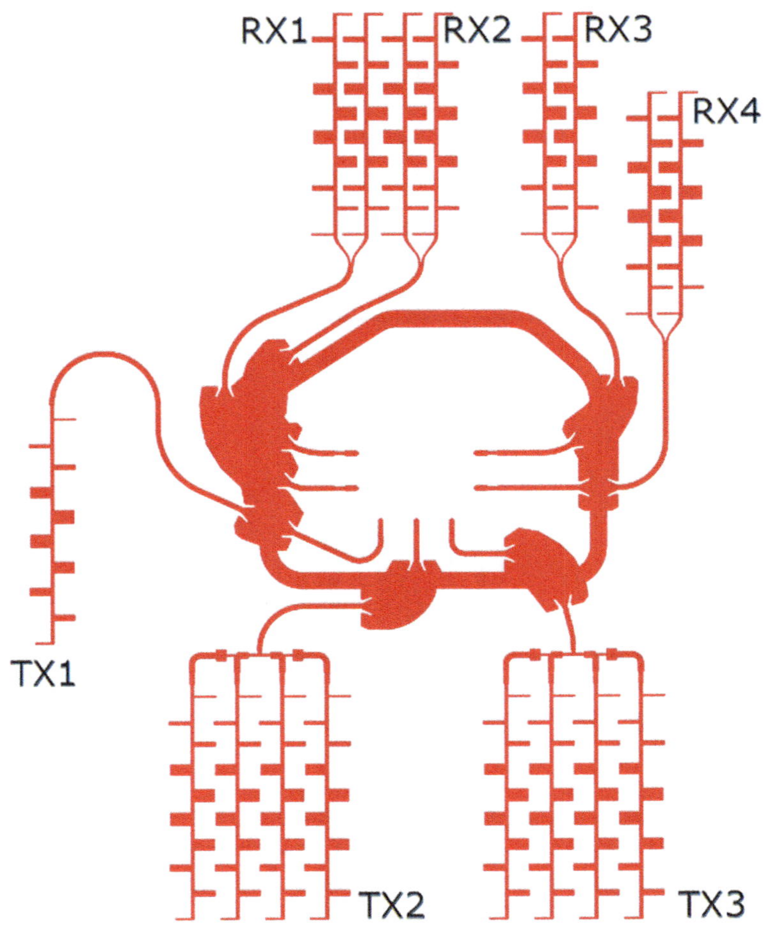

Fig. 15.23 FR5CP antenna layout (four receive antennas on top, the three transmit antennas below), *Source* Bosch

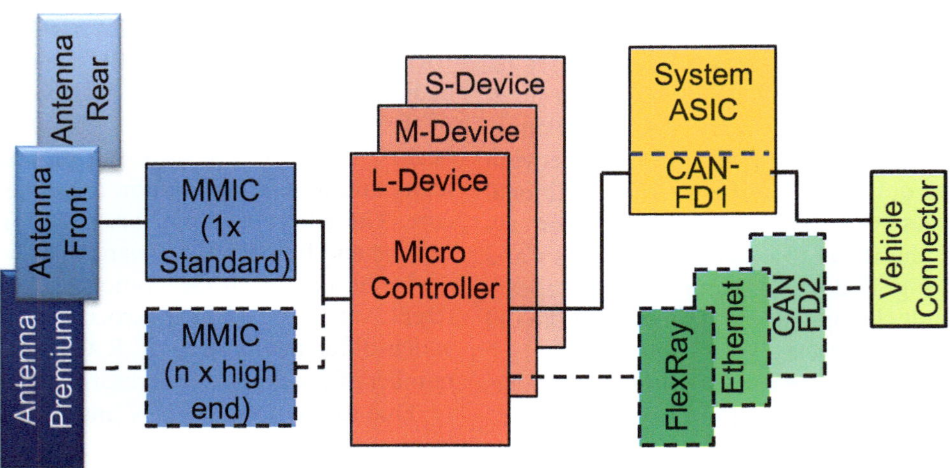

Fig. 15.24 Hardware architecture block diagram for 5. Generation Bosch design kit. (*Source* Bosch)

The front variant FR5CP has been available since 2019. Compared to the MRR 4th Generation, the azimuth aperture is increased by a factor of 2. The angular two target estimator performance is increased to separate targets down to 3.0°. It allows for separation of guardrail reflections in the far field from the relevant object, which especially improves ACC performance. To address the increased performance requirements for classifying standing objects, the number of elevation channels is increased from 2 to 4. As a result, a ro-

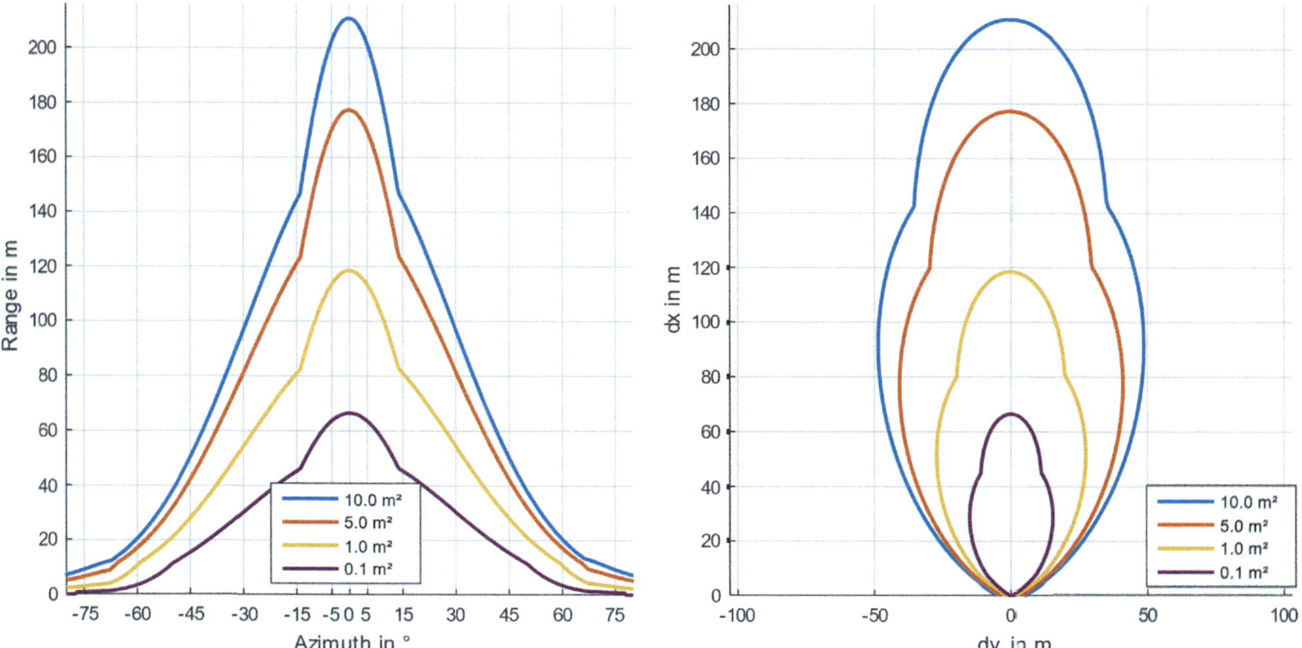

Fig. 15.25 Detection range Bosch 5th Generation Radar FR5CP, left in polar and right in cartesian origin

bust determination of over- and under-drivable objects is possible. In addition, this increase allows for calculation of quality factors and angular two target estimation in elevation for the first time.

The corner variant CR5CP (2020) replaces both Generation 4 variants MRR Rear and MRR Corner. This is possible due to the combination of two very different FoVs, and both are sampled simultaneously every cycle as described above. The high detection range of the focusing antenna allows for the implementation of a Lane Change Assist function that meets the requirements of LCA Type C. The Corner Plus variant also provides elevation angular estimation.

The Corner Base variant CR5CB is the cost optimized version of the CR5CP. Major changes are a smaller microcontroller and one less transmit channel to achieve the cost savings. The base variant addresses typical rear-radar functions like Blind Spot Detection (BSD) and Rear Cross Traffic Alert (RCTA), and additionally less performant LCA type A&B as well as front cross functionality can be offered. Unlike the Corner Plus, elevation angular estimation is not available.

All Bosch 5th Generation Radar sensors can be combined with communication interfaces in the processing chain. From classic ECUs with the full driver assistance system integrated, but also so-called "gateway units" (xGU). The available gateway unit variants are Locations (LGU), Sensor Objects (SGU), and Target Objects (TGU). The LGU provides locations based on radar reflections, the SGU delivers tracked objects, and the TGU delivers a subset of the SGU filter according to the needs of a defined function.

As an example of the Generation 5 radar performance, the detection characteristics of an FR5CP are shown in **Fig.** 15.25.

15.9.3 Continental Automotive Radar of 5th Generation

15.9.3.1 ARS 510 and SRR 520

Designed as a product family. Continental's 5th generation 77 GHz long- and short-range radars were developed in a successive evolutionary process from the SRR 200 (cf. [23]). The ARS 510 long-range radar and SRR 520 short-range radar differ mainly in the design of the antenna, the parameterization of the chirp sequence modulation, the associated signal processing, the realizable driving functions that can be displayed with the sensor and the vehicle interfaces. The block diagram of the long-range radar is shown in **Fig.** 15.26 and the corresponding exploded view in **Fig.** 15.27. The radar essentially consists of a RF board, which includes a planar antenna array with 3 Tx and 4 Rx antennas and an MMIC are located, and an LF board with a radar controller, the power supply, and the vehicle interfaces. The MMIC is programmed via an SPI interface and the ADC data are transmitted via CSI2

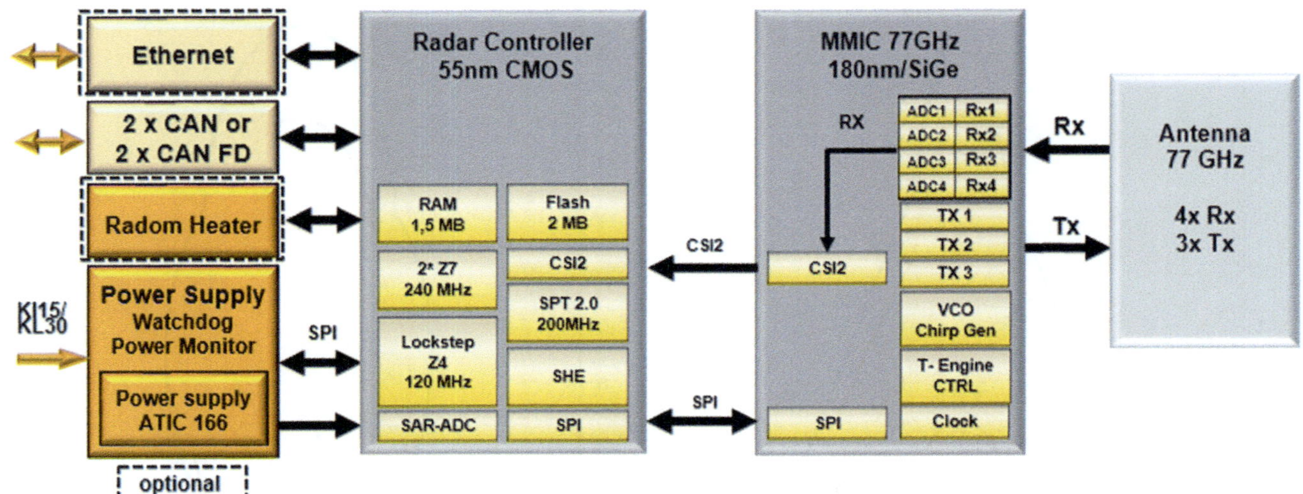

(SPT) **S**ignal **P**rocessing **T**oolbox

(CAN) **C**ontrol **A**rea **N**etwork

(SHE) **S**ecure **H**ardware **E**xtension

(SAR) **S**uccessive **A**pproximation **R**egister

(SD ADC) **S**ignal **D**ata **A**nalog **D**igital **C**onverter

(DAC) **D**igital **A**nalog **C**onverter

Fig. 15.26 Block diagram of the ARS 510 long-range radar (*Source* Continental)

Fig. 15.27 Exploded view of the ARS 510 (*Source* Continental)

Fig. 15.28 RF board of the ARS 510 with antenna layout and shielded MMIC. (*Source* Continental)

Fig. 15.29 RF board of the SRR 520 with antenna layout and shielded MMIC (*Source* Continental)

(Camera Serial Interface). **Fig. 15.28** shows the antenna arrangement of the ARS 510 with three external Tx antennas in two elevation planes and four centrally arranged Rx antennas, resulting in a virtual antenna array with 12 channels for digital beamforming. With the help of the two center Rx antennas, angular ambiguities can be resolved by means of monopulse techniques. **Fig. 15.29** shows the antenna layout of the SRR 520 with the 3 Tx antennas on the left and the 4 Rx antennas on the right. There are dummy structures between them which improve the symmetry of the single antennas and attenuate the propagation of surface waves.

The main innovations of the 5th generation are described in the following. One of them is higher integration. The complete RF front-end is integrated in an MMIC in SiGe technology including a PLL for ramp generation up to the ADCs as well as comprehensive online monitoring functionalities. Additionally, the entire processing is integrated in a radar controller with hardware accelerator for radar signal processing and three processor cores, one of which runs in lockstep mode to safeguard safety-relevant functions. Furthermore, in contrast to SRR 200, the 3 Tx and 4 Rx antennas are used in full parallel MIMO mode, i.e. transmission and reception take place simultaneously on all channels, which significantly improves the signal-to-noise ratio and the unambiguous speed range. To be able to separate the Tx antennas, each Tx signal has a separate phase modulation. The long-range radar has an elevation measurement feature using vertically displaced Tx antennas to support classification as overrideable, respectively, underrideable. To realize high range resolution and accuracy in the immediate vehicle environment, which is required especially for short-range radars, a near-range scan with high modulation bandwidth is implemented in addition to the far-range scan within one system cycle. In ARS 510, the bandwidth is varied in three steps depending on the ego-speed in order to adapt the range resolution to the required range (**Table 15.3**).

15.9.3.2 Continental ARS 540

The ARS 540 is a high-resolution premium long-range radar sensor with direct, independent measurement of four dimensions: range, velocity, azimuth, and elevation. Coherent operation of all four MMICs manufactured in 45 nm RFCMOS technology is realized by cascading them. One MMIC generates a radio frequency signal in master mode and distributes it to the other three MMICs operating in slave mode. This results in 12 Tx and 16 Rx channels, which together form a two-dimensional virtual antenna array of 192 channels. The antennas are arranged in a two-dimensional array that provides both precise angular measurement and angular separation in azimuth and elevation. The antenna, including the low-loss distribution network, is implemented in an innovative, plastic-based waveguide technology. Another innovation is the stepped chirp modulation method, which allows both high range separation capability and long maximum range without having to increase the number of range cells.

Table 15.3 Technical data of the continental ARS 510 and SRR 520 radar sensors

		Generation 5			
Product Variant		Continental ARS 510		Continental SRR 520	
Start of production		2019			
Dimensions (W x H x D)/mm³		83 × 69 × 20 (without mounting bracket and connector)		83 × 69 × 21 (without mounting bracket and connector)	
Mass		170 g		145 g	
Cycle duration		55 ms		50 ms	
Beam forming		Bistatic			
Number of measurement ranges		1		2 (far and near scan)	
Transmit antenna beam width (3dB) azimuth/elevation		48°/16.5°		123°/15.5°	
Radiated power (EIRPpeak/average)		average < = 25 dBm			
Frequency range		76–77 GHz			
Modulation method		Chirp Sequence			
Effective modulation bandwidth/modulation bandwidth		153 MHz 214 MHz 375 MHz	168 MHz @ v high 235 MHz @ v medium 408 MHz @ v low	238 MHz 750 MHz	266 MHz (FRS) 947 MHz (NRS)
Modulation details		256 ramps, ramp duration 84 µs, average repetition rate 104 µs		FRS: 256 ramps, ramp duration 64 µs, average repetition rate 82 µs NRS: 64 ramps, ramp duration 32 µs, average repetition rate 93 µs	
Number of measurement ranges		1		2 (far and near scan)	
Azimuth					
	Angle measurement method	Digital beam forming over 12 channels (3 Tx, 4 Rx, MIMO)		Digital beam forming over 12 channels (3 Tx, 4 Rx, MIMO)	
	Angle range	± 50° (plus ± 6° misalignment)		± 90°	
	Precision* point target	± 0.2° at SNR > 20 dB		± 1.3° at SNR > 20 dB	
	Unambiguous range	Unambiguous over complete detection range		± 90°	
	Resolution	2.4°		9.9°	
Elevation					
	Angle measurement method	Monopulse		–	
	Angle range	± 9° (plus ± 6° misalignment)		± 7.5° @ -6 dB ± 9.5° @ -10 dB	

(continued)

◻ Table 15.3 (continued)

		Generation 5	
	Precision* point target	<1° at SNR >20 dB	–
	Unambiguous range	±21°	–
	Resolution	20.6° (No separability)	–
Radial distance (range)			
	Distance range	0.2 m, 250 m with RCS of 10 m²	0.1 m, 100 m with RCS of 10 m²
	Resolution	0.4 / 0.7 / 0.98 m	0.63 m FRS / 0.2 m NRS
	Separability	0.8 / 1.4 / 2 m	1.3 m FRS / 0.4 m NRS
	Precision* point target	±0.13 m at SNR >20 dB	FRS: ±0.13 m at SNR >20 dB NRS: ±0.05 m at SNR >20 dB
Radial velocity (Range Rate)			
	Total range, of which unambiguous	-110 … + 55 m/s, unambiguous range 19 m/s in single cycle, unambiguous after 2 cycles	83- … + 83 m/s, unambiguous range 24 m/s in single cycle, unambiguous after 2 cycles
	Resolution	0.074 m/s	FRS: 0.093 m/s NRS: 0.33 m/s
	Separability	0.15 m/s	FRS: 0.19 m/s NRS: 0.66 m/s
	Precision point target	±0.03 m/s at SNR >20 dB	FRS: ±0.02 m/s at SNR >20 dB NRS: ±0.07 m/s at SNR >20 dB
Special features		Auto alignment in azimuth and elevation	Ability to self-align in azimuth

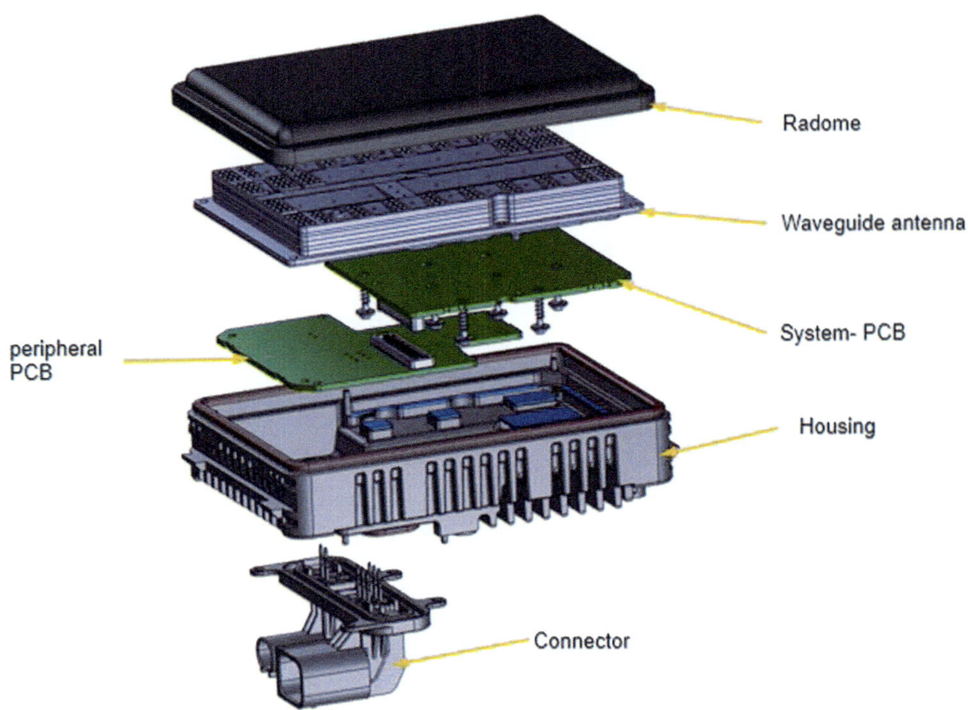

Fig. 15.30 Exploded view of the ARS 540

This modulation method is based on generating single ramps with a relatively low bandwidth whose starting frequency changes from ramp to ramp, resulting in a large bandwidth and thus a high range resolution over the entire ramp sequence.

Fig. 15.30 shows an exploded view of an ARS 540 sensor. It essentially consists of a system circuit board with the MMICs and a high-performance multiprocessor system implemented in a single chip, and a peripheral circuit board with the power supply, communication elements and external interfaces. The waveguide antenna is connected to the system PCB and covered with a radome (Table 15.4).

15.9.3.3 HELLA24 GHz and 77 GHz Corner Radar Sensors

In 2005, HELLA started the series production of its first generation 24 GHz narrowband corner radars which performed the following functions: "Lane Change Assist" and "Blind Spot Detection". Technological advances of integrated circuits and of the software algorithms helped to improve measurement performance and supported assistance functions of the following sensor generations (see Table 15.5).

While the first three sensor generations utilized slow chirp modulation, generation 4 uses fast chirps in order to achieve improved separation of objects in range and velocity (see ▶ Sect. 15.3.2.2). The precondition was the availability of a new, more powerful but cost-efficient microcontroller.

In addition, the development of a new, higher integrated, SiGe-BiCMOS-based MMIC allowed to realize the analog low-frequency and high-frequency circuits in a more compact way. Because of that, the overall sensor size and total cost were reduced further.

Continuous progress in RFCMOS technology and increasing radar sensor production volumes have motivated semiconductor manufacturers like NXP, Texas Instruments and Analog Devices to develop RFCMOS-based MMICs as an alternative to SiGe-based MMICs for radar sensors in the millimeter-wave frequency range [20]. Furthermore, various fabless start-ups were founded, some with very ambitious system concepts, outbidding RFCMOS technology and pushing forward state-of-the-art alternatives.

This new RFCMOS-based MMIC technology was used by HELLA in its radar sensor generation 5, which is in series production since 2020. The larger modulation bandwidth, compared to generation 4 (Table 15.6), in combination with a more powerful microcontroller, a new multi-modal modulation method and further developed algorithms resulted in a much improved radar measurement performance and allowed the realization of additional assistance functions.

15.9.3.4 Valeo MB79 Radar

The Valeo MB79 radar sensor is a short-range sensor with a detection range of up to 100 m. The sensor supports functions as Lane Change Assist (LCA), Cross Traffic Alert (CTA), Rear Collision Warning (RCW)

▢ Table 15.4 Technical data of the Continental ARS 540 radar sensors

		Generation 5	
Product Variant		Continental ARS 540	
Start of production		2021	
Dimensions (W x H x D)/mm^3		$130 \times 90 \times 42$ (without fasteners and connectors)	
Mass		530 g	
Cycle duration		50 ms	
Beam forming		Bistatic	
Number of measurement ranges		1	
Transmit antenna beam width (3dB) azimuth/elevation		67°/16°	
Radiated power (EIRPpeak/average)		Average < = 37 dBm	
Frequency range		76–77 GHz	
Modulation method		Stepped Chirp Modulation	
Effective modulation bandwidth/modulation bandwidth		690 MHz	958 MHz
Modulation details		512 Ramps, ramp duration 30 µs, average repetition rate 39 µs	
Number of measurement ranges		1	
Azimuth			
	Angle measurement method	Two-dimensional digital beam forming over 192 channels (12 Tx, 16 Rx, MIMO)	
	Angle range	± 60° (plus ± 6° misalignment)	
	Precision* point target	±0.1° at SNR > 20 dB	
	Unambiguous range	Clearly over total detection range	
	Resolution	1.1°	
Elevation			
	Angle measurement method	Two-dimensional digital beam forming over 192 channels (12 Tx, 16 Rx, MIMO)	
	Angle range	± 30° (plus ± 6° misalignment)	
	Precision* point target	±0.1° at SNR > 20 dB	
	Unambiguous range	Unambiguous over complete detection range	
	Resolution	1.6°	
Radial distance (range)			
	Distance range	0.2 m,, 300 m with RCS of 10 m^2	
	Resolution	0.22 m	
	Separability	0.44 m	
	Precision* point target	± 0.15 m at SNR > 20 dB	
Radial velocity (Range Rate)			
	Total range, of which unambiguous	−110 … + 55 m/s, unambiguous range 51 m/s in single cycle, unambiguous after 2 cycles	
	Resolution	0.1 m/s	
	Separability	0.2 m/s	
	Precision point target	± 0.03 m/s at SNR > 20 dB	
Special features		Enables highly automated driving, real altitude measurement, 4D imaging radar, multi-hypothesis tracking, automatic alignment	

Table 15.5 Evolution of HELLA corner radar sensors

Generation	1	2	3	4	5
Start of series production	2005	2011	2014	2017	2020
Vehicle integration	Rear corner				Front corner, side, rear corner
Assistance functions					
Lane change assistance	x	x	x	x	x
Blind spot detection	x	x	x	x	x
Pre-crash rear		x	x	x	x
Rear cross traffic alert			x	x	x
Exit assist				x	x
Front traffic assist					x
Trailer detection					x
Turn assist for trucks					x
Frequency range	24 GHz narrowband				77 GHz
RF components	Several discrete transistors and diodes	1 GaAs VCO MMIC and several discrete transistors and diodes	2 SiGe MMICs	1 SiGe-BiCMOS MMIC	1 RFCMOS MMIC
Antennas	1 Tx, 3 Rx patch antennas				3 Tx, 4 Rx patch antennas
Further details	[22]	[23]	[23]	Fig. 15.31 and Table 15.6, left	See Fig. 15.32 and Table 15.6, right

Table 15.6 Technical data on HELLA corner radar generations 4 and 5

		Generation 4	Generation 5
HELLA		RS4	RS5
Start of production		2017	2020
Dimensions (W x H x D)/mm^3		Size Housing: $82 \times 75 \times 22.5$ mm^3; Overall size: $109 \times 75 \times 24$ mm^3	$80 \times 68 \times 14.5$ mm^3
Mass		<130 g	<100 g
Cycle duration		50 ms	50 ms
Beamforming		bistatic	
Number of measurement ranges		1	2
Transmit antenna beam width (3dB) azimuth/elevation		$\pm 33°/\pm 7°$	$\pm 50°/\pm 6°$
Radiated power (EIRPpeak/average)		<20 dBm (peak) <12.7 dBm (average)	23 dBm (peak) 18 dBm (average)
Frequency range		24.05–24.25 GHz	76–77 GHz
Modulation method		Chirp sequence	
Effective modulation bandwidth		80 MHz	600 MHz (near), 250 MHz (mid)
Modulation details		Number of chirps: 256, Chirp duration: 60 µs	Near range: Number of chirps: 2×84 TDM MIMO, Chirp duration: 25.6 µs Mid-range: Number of chirps: 256, Chirp duration: 25.6 µs
Number of measurement ranges		1	2 [Near range (up to 24m) and mid-range (up to 91m)]
Azimuth			
	Angle measurement method	Phase monopulse	Phase monopulse
	Angle range	$\pm 75°$	$\pm 80°$
	Precision* point target	0.22° for 30 dB SNR	0.15° for 30 dB SNR
	Unambiguous range	$\pm 90°$	
	Resolution	0.11° (continuous measurements)	0.11° (continuous measurements)
Elevation		No elevation angle measurement	

(continued)

◾ **Table 15.6** (continued)

		Generation 4	Generation 5
	Angle measurement method		Phase monopulse
	Angle range		±10°
	Precision* point target		0.4 (near) 0.8 (mid) for 30 dB SNR
	Unambiguous range		±22°
	Resolution		0.18° (continuous measurements)
Distance (range)			
	Distance range	0.3 … 105 m for RCS of 10 m^2	0.15 … 24 m (near) 0.3 … 91m (mid) for RCS of 10 m^2
	Resolution	1.87 m	0.25 m (near) 0.6 m (mid)
	Separability	>3.5 m	0.5 m (near) 1.3 m (mid)
	Precision* point target	0.3 m	0.025 m (near) 0.06 m (mid)
Velocity (range rate)			
	Unambiguous range	±62 m/s	±106 m/s
	Resolution	0.24 m/s	0.22 m/s
	Separation	0.5 m/s	0.44 m/s
	Precision point target	0.18 m/s	0.022 m/s
Special features		Supports interleaving of chirps from several generation 4 radars in the same vehicle. Supports interference mitigation in frequency and time	Supports interleaving of chirps from several generation 5 radars in the same vehicle. Supports interference mitigation in frequency and time

*Precision here as accuracy without systematic errors such as bias or drifts

and Autonomous Emergency Braking (AEB). The sensors are installed in the right and left corners of the rear and/or front bumper (Fig. 15.33).

The Valeo radar sensor works according to the fast chirp sequence principle in the 77 and 79 GHz range. The sensor divides the detection area into a right and a left zone using two transmitting antennas that are separated by 60° from one another.

The target angles are measured on the receiving side with an electronic beamforming method using four receiving channels. The unique aspect of this design is the choice of the horizontal distances between the receiving elements. The distances are chosen according to the Golomb ruler of the fourth order in multiples of half the wavelength ($\lambda/2$), see. Fig. 15.34.

Compared to a linear $\lambda/2$ array, the Golomb array has higher side lobes, but a much smaller main lobe width, when keeping the number of elements constant, as shown in Fig. 15.36. The focus here is not on the requirement for angle separability but rather on the significantly higher accuracy of the angle measurement (Figs. 15.35, 15.36).

The sensor transmits a sequence of 128 frequency ramps. The duration of each frequency ramp is approximately 50 µs. The bandwidth of the frequency ramps can be set between 375 and 750 MHz in the 77 GHz range and up to 1.5 GHz in the 79 GHz band. Several bursts can be transmitted per measurement cycle, and it is possible to switch between the 77 GHz and 79 GHz range within one cycle.

Fig. 15.31 View of antenna side of Generation 4 corner radar. (*Source* HELLA)

Fig. 15.32 View of antenna side of Generation 5 corner radar. (*Source* HELLA)

Each of the two transmit antennas consists of a set of three columns, with each column having 12 patch elements. A set of three columns is used to focus the beam 90° in azimuth and to steer one transmit beam to the right side and the other beam to the left side.

The receiving antenna consists of four patch columns each with 12 patches. The receive signal of each of the four patch columns is processed separately by one receives channel.

The receive signal is mixed down, filtered and A/D converted. All further processing steps take place in a DSP.

The detection range for a range of target classes is illustrated in ◘ Fig. 15.35. The radar cross-section is assumed to be -5 dBm^2 for a pedestrian, 0 dBm^2 for a motorcycle, 5 dBm^2 for a motor vehicle and 10 dBm^2 for a delivery van (◘ Table 15.7, ◘ Fig. 15.37).

15.9.3.5 ZFMRGen21 and FRGen21

MRGen21

ZF's medium-range radar (MRR) is a high-performance 77 GHz front radar designed for 2020/22 Euro NCAP 5-Star Safety Ratings and L2/L2 + automated driving performance.

Compared to the previous MRR generation, it achieves increased speed resolution, higher resolution

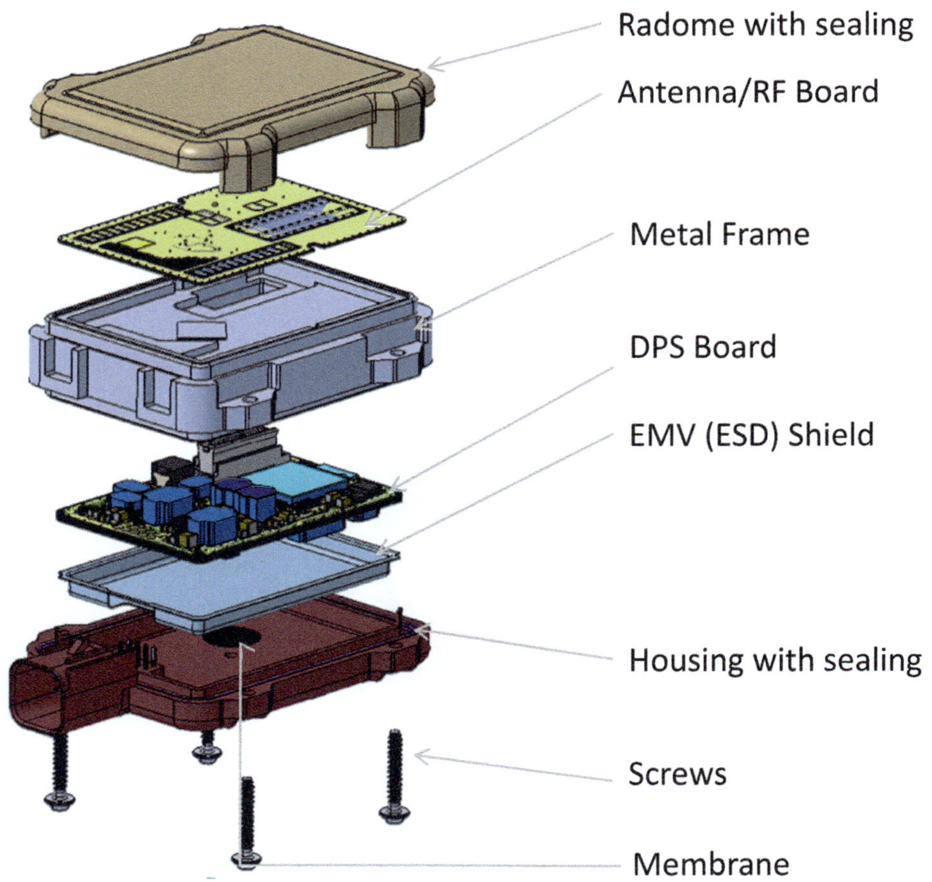

◘ **Fig. 15.34**   Exploded view of Valeo MB79 sensor. (*Source* Valeo)

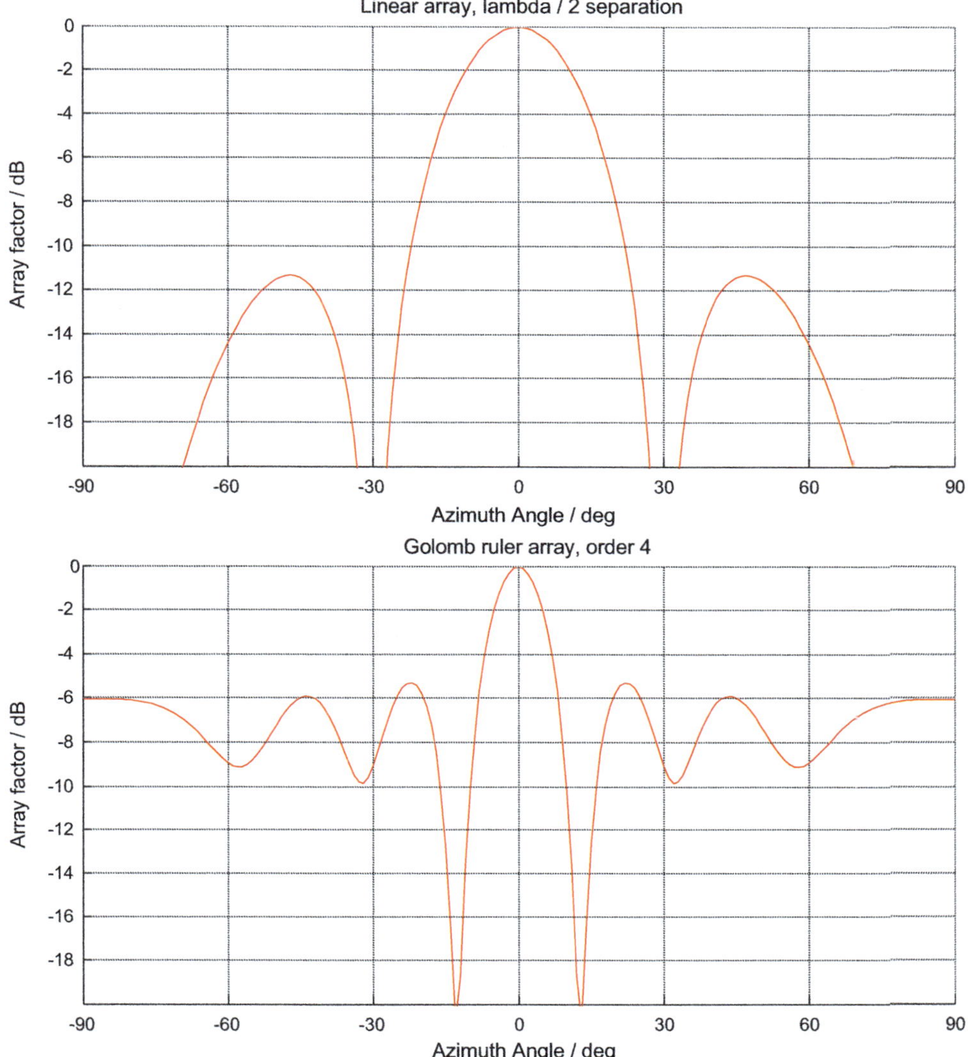

Fig. 15.35 Comparison of an equidistant 4-element array (top) and a thinned array according to the Golomb scale (4th order). (*Source* Valeo)

in azimuth and a high detection range of up to 220 m. It enables safety and convenience functions including forward collision warning, adaptive cruise control (even at high distances) and automatic emergency braking functionality.

The hardware components and dimensions are shown in Fig. 15.38.

A chirp sequence modulation with 256 frequency ramps on a MMIC with 3 Tx and 4 Rx antennas is employed with a planar antenna.

To measure distance, the sensor switches into one of three modes with differing frequency bandwidths depending on host velocity. The switch can be controlled by the sensor itself or via an external fusion unit. In the far-range mode at higher host speeds (> 30 m/s) the range cells are increased by reducing the bandwidth, effectively increasing the radar range to 220 m. At medium speeds (15...30 m/s) the cells are reduced, and the maximum range is scaled to 154 m. At lower speeds (< 15 m/s), the maximum range is

limited to 90 m to improve distance separation capabilities in complex urban environments. Depending on configuration, the speed mode can also be switched from one cycle to another by a central fusion unit (e.g. alternating long/short, etc.).

The azimuth measurement occurs via beamforming. In a two-step approach, two virtual antenna arrays (each after one MIMO demodulation) are observed. To measure the targets' azimuth position, data from the main array consisting of six channels are processed first. Next, the subarray with four channels is used to resolve ambiguities and enable a unique angular measurement with high resolution. Model-based algorithms improve the separation capabilities of the Gen21 radar sensors beyond the physical characteristics of the hardware by a factor of nearly two.

For this, the peaks in the spectrum in distance, doppler, and azimuth are considered. If the profile of a detection does not sufficiently match the sensor's refer-

▫ Table 15.7   Technical data of the Valeo radar sensor MB79

[Manufacturer] product variant		Valeo MB79		
Start of production		2018		
Dimensions (W x H x D)/mm^3		77 × 63 × 25		
Mass		200 g		
Cycle duration		50 ms		
Beam forming		Patch antenna on PCB, beam forming		
Number of measurement ranges		2 Tx Beams		
Transmit antenna beam width (3dB) azimuth/elevation		90°per beam / 12°		
Radiated power (EIRPpeak/average)		< = 23 dBm (EIRP)		
Frequency range		76–77 GHz, 77–81 GHz		
Modulation method		Chirp Sequence		
Effective modulation bandwidth/modulation bandwidth		375 MHz	750 MHz	1.5 GHz
Modulation details		128 chirps/50 µs chirp duration		
Number of measurement ranges		2 measurements per cycle (@ 50 ms cycle time)		
Azimuth				
	Angle measurement method	2 Tx × 4 Rx channel, non-uniform array beam forming (Golomb ruler separation)		
	Angle range	150°		
	Precision* point target	< 1° @ 20 dB SNR point target		
	Unambiguous range	180°		
	Resolution	<25° (3 dB width of beamforming beam)		
Elevation				
	Angle measurement method	–		
	Angle range	12°		
Radial distance (range)				
	Distance range	0.3 m–100 m	0.15 m–50 m	0.08 m–25 m
	Resolution	0.4 m	0.2 m	0.1 m
	Separability	0.8 m	0.4 m	0.2 m
	Precision point target	0.1 m	0.05 m	0.025 m
Radial velocity (Range Rate)				
	Total range, of which un-ambiguous	±68 m/s Unambiguous interval of single measurement:~25 m/s		
	Resolution	0.3 m/s		
	Separability	0.6 m/s		
	Precision point target	0.05 m/s (20 dB point target)		
Special features		Multi-mode 77 / 79 GHz Functions and multi-sensor tracking on radar DSP Auto alignment		

*Precision here as accuracy without systematic errors such as bias or drifts

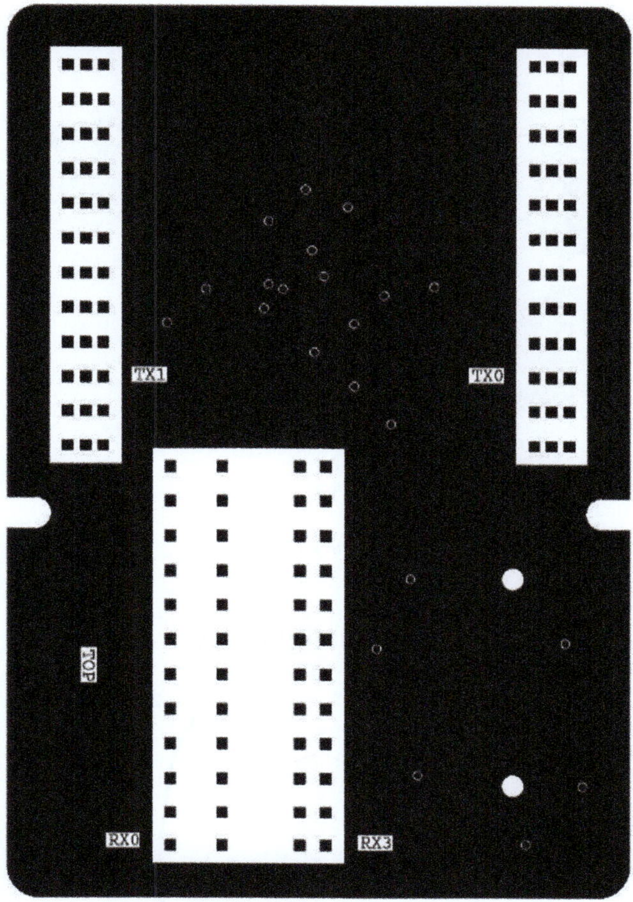

Fig. 15.36　Antenna arrangement of the Valeo MB79. (*Source* Valeo)

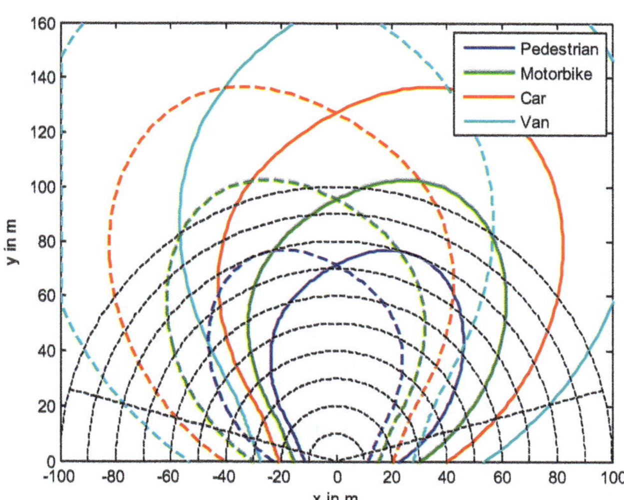

Fig. 15.37　Detection area for a selected range of target types, separately for each of both TX antennas. (*Source* Valeo)

ence model of a single target, a hypothesis is computed on how two targets would have to be placed within the spectrum to match the measured result. The dimension in which the deviation from the single target model is the largest is considered first while drawing up this hypothesis. In this dimension, the closest approximation to the measurement using the superposition of two targets is determined. If this two-target model achieves a better approximation of the measurement, the single detection is split up in two targets (■ Fig. 15.39).

This significantly improves the separability of the radar. A good compromise between high separability and the avoidance of undesired splitting of real single targets is the key to achieve optimal resolution in azimuth, range and Doppler. Using the azimuth angle as an example, the separability of 3,2° to 2,0° increased.

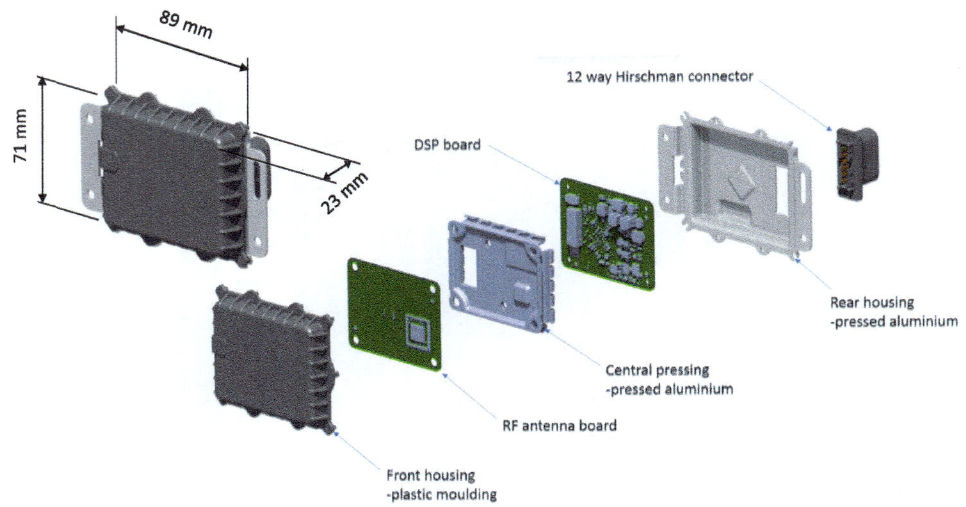

Fig. 15.38 Exploded view and dimensions of ZF MRGen21. (*Source* ZF)

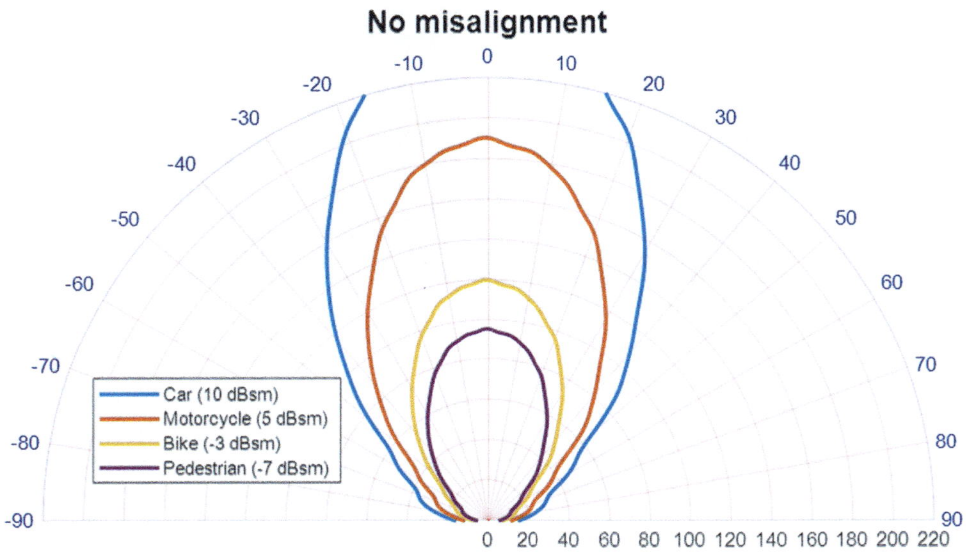

Fig. 15.39 Radar diagram of the ZF MRGen21 radar sensor (*Source* ZF)

FRGen21

The FRGen21 is a high-resolution 4D 77 GHz full-range radar. The FRGen21 is designed for driver assistance systems (L2+) as well as for Level 4 and above automated driving. In contrast to the MRGen21, the FRGen21 features a higher resolution in elevation. This is key for the classification of stationary targets (e.g. end of a traffic jam or bridge) and elevates the environment detection to a new level.

With 12 Tx × 16 Rx = 192 MIMO channels, the imaging radar FRR Gen21 enables a detailed picture of the environment in azimuth, elevation, distance, and radial velocity.

The radar uses a chirp sequence modulation with 512 frequency ramps. Four MMICs, each with 3 Tx and 4 Rx antennas, operate synchronously in a master/slave concept. As with the MRGen21, the antenna array for the FRR Gen21 is divided into two virtual arrays. The angular measurement is first carried out via the 128-channel main array with the 64-channel sub-array used to resolve ambiguities.

Due to its large bandwidth in combination with 512 range cells, the full-range radar can achieve a maximum detection range of up to 340 m while achieving a higher resolution compared to the MRGen21 across all radar modes.

2D-beamforming in azimuth and elevation achieves a true 3D representation of the environment, as seen in **Fig. 15.40** (**Table 15.8**).

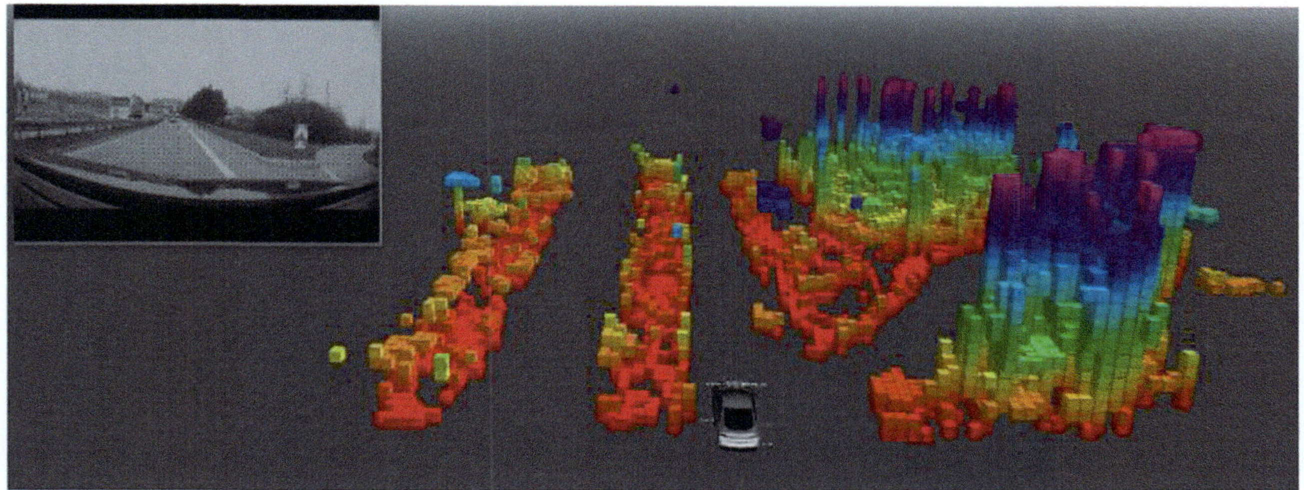

Fig. 15.40 Voxel-based visualization of the FRGen21 detections over time in a typical highway scenario (*Source* ZF)

Table 15.8 Technical data ZF radar sensors Gen 21

	Generation 5	
Product Variant	**FRR–FRGen21**	**MRR–MRGen21**
Start of production	2021	
Dimensions (W x H x D)/mm^3	120 × 120 × 30	71 × 86 × 23
Mass	590 g	160 g
Cycle duration	60 ms	60 ms
Beam forming	2D beamforming	1D beamforming
Number of measurement ranges	1	
Transmit antenna beam width (3dB) azimuth/elevation	50°/20°	50°/20°
Radiated power (EIRPpeak/average)	38 dBm (EIRP)	32 dBm (EIRP)
Frequency range	76–77 GHz	
Modulation method	Chirp Sequence	
Effective modulation bandwidth	215 MHz @ v high 375 MHz @ v medium 750 MHz @ v low	150 MHz @ v high 215 MHz @ v medium 365 MHz @ v low
Modulation details	512 chirps, duration 19 25 …ms	256 chirps, duration 17 … 2 4ms
Number of measurement ranges	1	
Azimuth		
Angle measurement method	2D Beamforming (16 × 8 channels; 16 in Azimuth and 8 in Elevation) with phase evaluation	Six channels with phase evaluation
Angle range	±60°	±°55
Precision* point target	±0.15° at SNR > 40 dB	±0.3° at SNR > 40 dB
Unambiguous range	±90°	±90°
Resolution	1.5°	3.2°
Elevation		
Angle measurement method	2D Beamforming (16 × 8 channels; 16 in Azimuth and 8 in Elevation) with phase evaluation	Three channels with phase evaluation

(continued)

■ **Table 15.8**   (continued)

		Generation 5	
Product Variant		**FRR–FRGen21**	**MRR–MRGen21**
	Angle range	±15°	±14°
	Precision* point target	±0.3° at SNR > 40dB	±0.5° at SNR > 40dB
	Unambiguous range	±17°	±45°
	Resolution	1.5°	NA
Radial distance (range)			
	Distance range	0.3 m…340 m for RCS of 10 m²	0.8 m…220 m for RCS of 10 m²
	Resolution	0.2 / 0.4 / 0.7 m	0.4 / 0.7 / 1 m
	Separability	0.1 m**	0.2 m**
	Precision* point target	0.02 / 0.04 / 0.07 m at SNR > 40dB	0.04 / 0.07 / 0.1 m at SNR > 30dB
Radial velocity (Range Rate)			
	Total range, of which unambiguous	70- … + 70 m/s, unambiguous range 0–53 m/s in single cycle	55- … + 83 m/s, unambiguous range 0–30 m/s in single cycle
	Resolution	0.1 m/s	0.1 m/s
	Separability	0.1 m/s**	0.1 m/s**
	Precision point target	±0.0 1m/s at SNR > 40 dB	±0.0 1m/s at SNR > 30 dB
Special features			

**: Via model-based algorithm "High-Resolution Techniques", see text

15.10 Summary and Outlook

Today's driver assistance systems and automated driving systems would be inconceivable without radar sensors. In particular, they show their strengths in terms of robustness, Doppler and range measurement.

The sensors have become a mass product, of which in the order of 100 million are installed in vehicles every year. Today, the frequency range around 77 GHz dominates, while 24 GHz is increasingly losing importance.

The radar sensors of different manufacturers differ less and less technically, at least in comparison to the previous years. Currently and in the coming years, chirp sequence is the predominant modulation method. The MIMO principle is used across the board and hardware integration is advancing steadily so that sensors today are built from very few highly integrated electronic components.

Automotive radar has thus assumed a technological pioneering role in the field of millimeter-wave radars. Many other application areas benefit from this, as can be seen, for example, in the numerous sensors for automation technology or the radar sensors in cell phones.

It remains to be seen how future developments will proceed. Digital modulation processes have been announced by individual manufacturers. However, it remains to be seen when these processes will become established. One of the current key topics of research is the question of how future systems will be partitioned. A variety of solutions are currently being developed for networking the sensors at the raw data level around the vehicle, as this would reduce the requirements on the individual sensors while increasing the overall detection performance of the network. Alternatively, increasingly powerful individual sensors could be used, which also achieve very good separation capabilities of less than 1 degree in the angular range.

Acknowledgements The authors would like to thank Raphael Hellinger from Bosch, Arnold Herb from Continental, Dr. Andreas John from HELLA, Dr. Urs Luebbert from Valeo and Sascha Heinrichs-Bartscher from ZF on behalf of their colleagues for their technical support and for providing the technical illustrations.

References

1. Balanis, C.A.: Antenna Theory: Analysis and Design, 4th Edition, Wiley, (2016). ISBN: 978–1–118–64206–1
2. Bartels, A., Meinecke, M.-M., Steinmeyer, S.: Fahrstreifenwechselassistenz. In Winner, H., Hakuli, S., Lotz, F., Singer, C., (Hrsg), Handbuch Fahrerassistenzsysteme, ATZ/MTZ-Fachbuch, (2015). ▶ https://doi.org/10.1007/978-3-658-05734-3_47, © Springer Fachmedien Wiesbaden 2015, S. 959–974
3. Bechter, J.: Eigenschaften und Unterdrückung von Interferenzen zwischen Kfz-Radarsensoren. Dissertation Universität Ulm, Open Access Repositorium der Universität Ulm (2019). ▶ https://doi.org/10.18725/OPARU-14428
4. Classen, T., Wilhelm, U., Kornhaas, R., Klar, M., Lucas, B.: Systemarchitektur für eine 360 Grad Fahrerassistenzsensorik, Uni-DAS Workshop Fahrerassistenzsysteme. Tagungsband, Bd 8 (2012)
5. Diewald, F., Klappstein, J., Sarholz, F., Dickmann, J., Dietmayer, K.: Radar-Interference-Based Bridge Identification for Collision Avoidance Systems. Proceedings IEEE Intelligent Vehicles Conference Baden-Baden, (2011)
6. Diewald, F.: Objektklassifikation und Freiraumdetektion auf Basis bildgebender Radarsensorik für die Fahrzeugumfelderfassung. Dissertation Universität Ulm (2013)
7. Fonck, K.-H.: Radar bremst bei Gefahr. Auto, Motor & Sport **22**, 30 (1955)
8. Gamba, J.: Radar Target Detection. In Gamba J Radar Signal Processing for Autonomous Driving (eds) Springer Singapore, (2019). ▶ https://doi.org/10.1007/978-981-13-9193-4
9. Holder, M.-F.: Synthetic generation of radar sensor data for virtual validation of autonomous driving. Dissertation Technische Universität Darmstadt (2012). ▶ https://doi.org/10.26083/tuprints-00017545
10. Hildebrandt, J., Kunert, M., Lucas, B., Classen, T.: Sensor Setups for future Driver Assistance and Automated Driving. Proceedings IWPC Frankfurt, Germany (2013)
11. Ludloff, A.: Praxiswissen Radar und Radarsignalverarbeitung, 4th edn. Vieweg + Teubner Verlag, Wiesbaden (2009)
12. Roos, F., Bechter, J., Knill, C., Schweizer, B., Waldschmidt, C.: Radar sensors for autonomous driving: modulation schemes and interference mitigation. IEEE Microwave Mag. **20**(9), 2019 (2019)
13. Schneider, R.: Modellierung der Wellenausbreitung für ein bildgebendes Radar, Dissertation Universität Karlsruhe (1998)
14. Schubert, E., Kunert, M., Menzel, W., Fortuny-Guasch, J., Chareau, J.-M.: Human RCS Measurements and Dummy Requirements for the Assessment of Radar Based Active Pedestrian Safety Systems. International Radar Symposium IRS. (2013)
15. Schweizer, B., Knill, C., Schindler, D., Waldschmidt, C.: Stepped-Carrier OFDM-Radar Processing Scheme to Retrieve High-Resolution Range-Velocity Profile at Low Sampling Rate. IEEE Trans. Microw. Theory Tech. **66**(3), 1610–1618 (2018)
16. Skolnik, M.: Radar Handbook. 3rd Ed McGraw-Hill Verlag, New York City (2008)
17. Sun, S., Petropulu, A.-P., Poor, H.-V.: MIMO Radar for Advanced Driver-Assistance Systems and Autonomous Driving - Advantages and challenges. IEEE Signal Process. Mag., p. 98 (2020)
18. TC204, WG14.: Intelligent transport systems—Adaptive cruise control systems—Performance requirements and test procedures. ISO **15622**, (2018)
19. Vasanelli, C., et al.: Calibration and Direction-of-Arrival Estimation of Millimeter-Wave Radars: A Practical Introduction. IEEE Antennas Propag. Mag. (2020). ▶ https://doi.org/10.1109/MAP.2020.2988528
20. Waldschmidt, Ch., Hasch, J., Menzel, W.: Automotive radar—from first efforts to future systems. IEEE J. Microw., **1**(1), p. 135 (2021)
21. Winner, H.: Fundamentals of collision protection systems. In Winner, H., Hakuli, S., Lotz, F., Singer, C. (eds.), Handbook of Driver Assistance Systems, Springer Reference 2015, pp. 1149–1176 (2016)
22. Winner, H.: Radarsensorik. In Winner H, Hakuli S, Wolf G (eds.) Handbuch Fahrerassistenzsysteme. Wiesbaden, Vieweg+Teubner Verlag, (2009). ISBN 978-3-8348-0287-3
23. Winner, H.: Automotive RADAR. In: Winner, H., Hakuli, S., Lotz, F., Singer, C. (eds.), Handbook of Driver Assistance Systems, Springer Reference 2015, pp. 325–404 (2016)
24. Winner, H., Schopper, M.: Adaptive cruise control. In Winner, H., Hakuli, S., Lotz, F., Singer, C. (eds.), Handbook of Driver Assistance Systems, Springer Reference 2015, pp. 1093–1148 (2016)
25. Wolff, C.: Radargrundlagen—Winkelreflektor.: (2020). ▶ http://www.radartutorial.eu/17.bauteile/bt47.de.html, Zugriff Nov. 2021

Open Access This chapter is licensed under the terms of the Creative Commons Attribution-NonCommercial-NoDerivatives 4.0 International License (▶ http://creativecommons.org/licenses/by-nc-nd/4.0/), which permits any noncommercial use, sharing, distribution and reproduction in any medium or format, as long as you give appropriate credit to the original author(s) and the source, provide a link to the Creative Commons license and indicate if you modified the licensed material. You do not have permission under this license to share adapted material derived from this chapter or parts of it.

The images or other third party material in this chapter are included in the chapter's Creative Commons license, unless indicated otherwise in a credit line to the material. If material is not included in the chapter's Creative Commons license and your intended use is not permitted by statutory regulation or exceeds the permitted use, you will need to obtain permission directly from the copyright holder.

LiDAR

Thorsten Beuth, Christoph Parl and Heinrich Gotzig

Contents

© The Author(s) 2026
H. Winner et al. (eds.), *Handbook Assisted and Automated Driving*,
https://doi.org/10.1007/978-3-658-45276-6_16

16.1 Definition

The acronym LiDAR stands for "Light Detection and Ranging". It describes all optical measurement systems measuring remotely object properties by scattering or reflection of electromagnetic waves of ultraviolet, visible or infrared wavelengths [21]. In broader sense, direct measurements of reflection and scattering properties, as well as distances, velocities and surface and material compositions are possible. The measurement of distances for reconstruction of point clouds is the most common use case in the automotive environment. It is common to assign a certain signal strength to each of these points which enables the object recognition for parts of the data into classes of lanes, cars, persons, trees or similar.

16.2 Introduction and History

The interest into distance measurement systems based on light was unbroken in the last decades. Already in the 1960s, the benefits of higher distances range and accuracy were recognized in comparison to RADAR (Radio Detection and Ranging) and used for mapping of see floors, landscapes and atmospheric research [5, 7, 8]. With the progress of the development of light sources and detectors, the effort and size of LiDAR systems decreased and their energy consumption was reduced. Thus, further areas of applications have been introduced in automation, entertainment and computer industry. In space flight, LiDAR supports autonomous flying [1].

While pursuing autonomous driving, the technology was introduced in the beginning of the 2000s in the automotive field. The first driver assistance systems were offered like traffic jam pilots [13, 23, 24]. The LiDAR supplies distance information of the surrounding of the car. In the upcoming sections, main components, measuring methods and applications of LiDAR-Systems are discussed.

16.3 Main Components

16.3.1 General Setup

Light sources are used on a broad bandwidth of components, which is based on the specific requirements of the measurement method. Details are discussed in ▶ Sect. 16.5.

Following the optical path from the laser source, a sending optic is used which relates to ◘ Fig. 16.1. The beam can be directed by mirrors or prisms. These elements can lead to angular changes via turning or oscillating movements, and thus, contain the essential part of scanning systems. On the other hand, flash systems illuminate the whole scene for each frame by diffusors or diffractive optical elements (DOE).

The sending optics are often refractive lenses, sometimes even DOEs, which are able to modify the illumination pattern and thus its intensity profile. Indirectly, the illumination can influence the resolution of the sensor because just the illuminated area contributes towards the measurement. The beam is influenced by the atmospheric conditions, scattered at the target and collected by the receiver optics. The receiving optic can be a lens or a DOE, as well. It influences the performance of the overall system by the aperture area and its ability to collect the scattered light, compare the LiDAR formula in Eq. 16.2. Furthermore, the beam can be refracted in the receiving path which is common for scanning systems. However, one must take care of unwanted reflections and the transmission of the used wavelength bandwidth. All optical elements which are not intended to reflect are typically antireflection-coated to ensure no crosstalk to other receiving channels. The transmission bandwidth has to be aligned with the laser source. Its properties influence the noise behavior and the sensitivity of the sensor. The receiver must match the wavelengths. It transforms the light into free charge carriers which work as a current. Additionally, a detector can

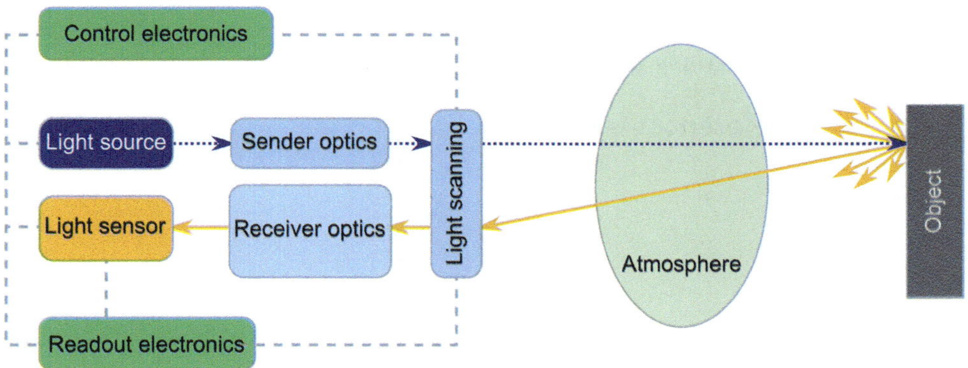

◘ **Fig. 16.1** Main components of a LiDAR system with biaxial architecture in simplified representation

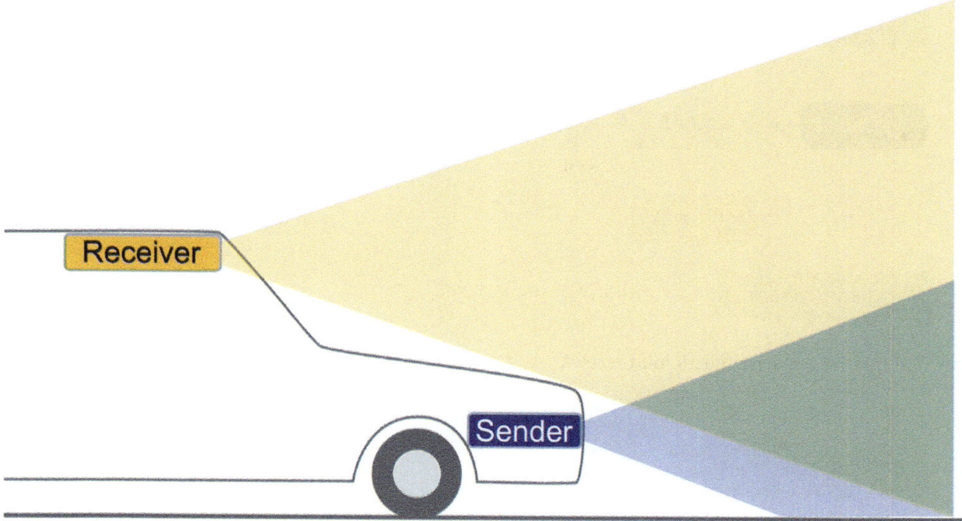

◘ Fig. 16.2 Example of the integration of a biaxial LiDAR system. A blind spot (blue) is created in front of the vehicle, since the transmitting (blue) and receiving (yellow) ranges do not overlap in this region. Only the green region can be resolved by the LiDAR system

consist of amplifying elements, e.g. in the case of APDs (Avalanche Photo Diodes).

The reading electronic has the task to digitalize the generated charge for data processing. Electronic amplifiers, analog-to-digital converters, time-to-digital converters and multiplexers are regularly used.

The readout electronics have the task of converting the charge generated by the light detectors for data processing, i.e. digitizing it. For this purpose, classic electronic amplifiers, analog-to-digital converters (ADC) or time-to-digital converters (TDC) and multiplexers are used. The readout electronics should also calculate the distance for each measuring point from the input signals. In addition, complex systems implement already defined vehicle functions on the sensor hardware. For example, measurement points belonging to an object are combined to a binding box and, if necessary, semantically classified. Possible functions are explained in more detail in ▶ Sect. 16.7, Application scenarios. Data is usually output via a standardized interface such as Ethernet, FlexRay or Camera Serial Interface (CSI).

Finally, control electronics are required to synchronize the various components with each other in terms of time. This is particularly important in order to ensure that the light sensor is activated only when a light signal has been emitted. Such a mechanism avoids unnecessary pickup of ambient light and thus reduces noise in the system. Likewise, the light scanning must also be controlled, so that the measurements can be assigned to the correct solid angle.

16.3.2 Variants of the Optical Path

In the previous subsection, the main components were discussed on the basis of the optical path. However, the optical path can have many different characteristics, which will be discussed in the next sections.

In system architecture, a distinction can be made between two broad categories based on the layout of the optical paths and the components used: a biaxial or bistatic and a monostatic setup. In biaxial systems, the transmitter and receiver paths are spatially separated. Thus, the optics for receiver and transmitter can be dimensioned differently. This architecture opens up a broad range of design options, especially for automotive applications. For example, the transmitter can be installed in the bumper or in the front headlights, and the receiver in the roof module. Likewise, several transmitters can be combined with one receiver. In this case, the time synchronization of transmitters and receivers must be realized with a precision in the sub-nanosecond range. However, it should be noted that different transmit and receive angles always generate parallax errors. This can result in incorrect angular assignment of the distance measurements. The measuring range of the system must always be covered by the transmitter and receiver. Otherwise, blind spots will occur, e.g. in front of the vehicle, see ◘ Fig. 16.2.

From efficiency point of view, the energy that is sent to blind spots is a loss and degrades the power budget of the LiDAR system. In addition, several components must be integrated into the vehicle and aligned with each other to operate properly. The overall installation space required will also be larger.

In contrast to the biaxial setup, the same optics can be used for transmitting and receiving as well. Thus, an overlap of transmit and receive angle as well as transmit and receive region is inherently given. ◘ Fig. 16.3 shows a schematic diagram of a coaxial LiDAR system. The advantage of coaxial systems is an intrinsic overlap between transmitting and receiving areas lead-

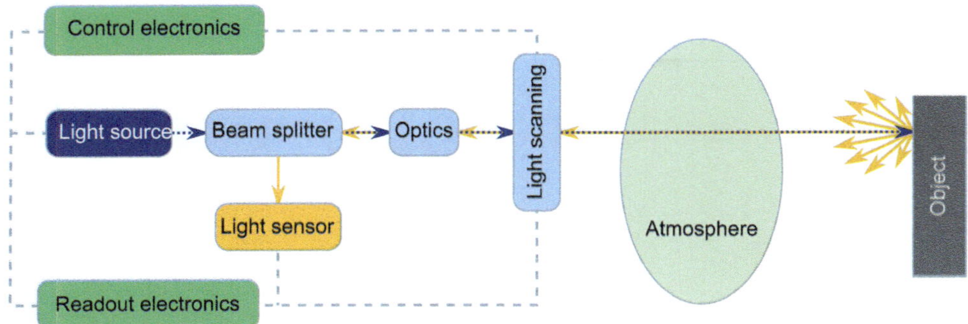

Fig. 16.3 Scheme of a monostatic system with joint receiving and transmitting optics

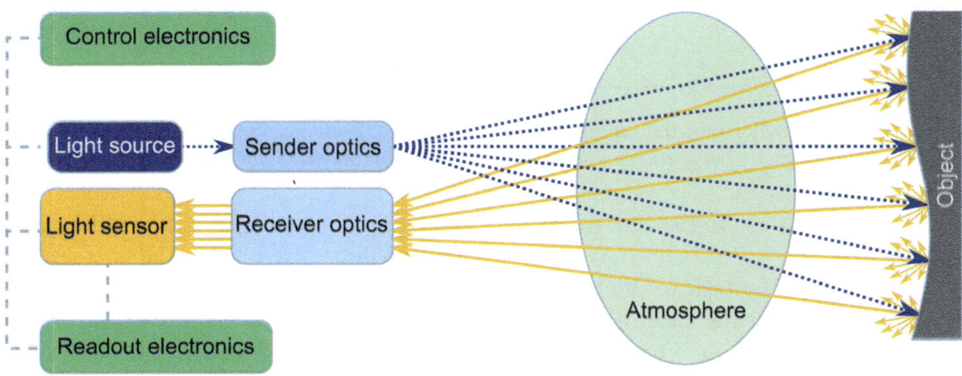

Fig. 16.4 Schematic setup of a flash LiDAR

ing to smaller mechanical tolerances. Still, it becomes difficult to separate crosstalk of the sending path to the receiving path, which might be wished for, e.g. in interfering measuring methods.

Besides of the biaxial and monostatic path design, one can distinguish LiDAR systems by their scanning method. In scanning systems, the FOV (Field of View) is subdivided into small areas, which are scanned one by another. The FOV specification differentiates between horizontal (hFOV) and vertical (vFOV). Similar to photography, the flash mode illuminates the whole FOV at once. Mixtures of the methods exist in which major parts of the FOV are illuminated subsequently, called sequential flashing.

Active beam steering is needed for a scanning LiDAR system. It guides at least one small beam to specific positions within the measuring area. The optical beam steering unit can be designed for horizontal (1D) and vertical (2D) direction. This functionality can be achieved by moving mirrors, moving system setups or comparable to RADAR with optical, phase controlled group antennas (Optical Phase Array, OPA). The biggest advantage of a scanning LiDAR system is the concentration of all light energy towards a very tiny measuring area enabling greater distance while keeping the energy consumption minimal.

For a flash architecture, the whole measuring area is illuminated. The gatherable light energy for each pixel is usually smaller than in scanning systems. Thus, the flash method is suited for small until medium distance of e.g. up to 50 m. Most flash LiDAR systems are based on one sender and a 2D receiver, similar to a common camera chip. The sender can consist out of several parallel controlled laser sources. **Fig. 16.4** shows a scheme of a flash LiDAR system.

Next to the capability of measuring all distances at the same time within a measuring area, flash systems usually have a lesser amount of optical components leading to lower costs.

16.4 Main Characteristics

16.4.1 System Range

One of the main characteristics of all distance measuring systems is the maximum range of the system. It is mainly determined by the signal–noise-ratio (SNR) of the received light. To achieve valid distance measurement, the receiving signal P_{RX} has to emerge from the amplitude of the noise. It can be argued that the higher the signal and the lower the system noise R_{SYS} and the noise of external sources, the higher the maximum range of the system is. The amplitude of the noise has to be minimized because the laser safety is capped, compare ▶ Sect. 16.6. The range of a system is determined by a root dependency towards the SNR.

$$SNR = \frac{P_{RX}}{\sqrt{R_{SYS}^2 + R_{EXT}^2}} \propto \frac{1}{r_{max}^2} \Rightarrow r_{max} \propto \frac{1}{\sqrt{SNR}} \quad (16.1)$$

The impact of the collected light is further discussed. To visualize the factors, we take a single circular receiving optic into account which covers the receiving path.

$$P_{RX} = P_{SRC} \cdot \rho \cdot \frac{1}{r^2} \cdot \frac{D^2}{4} \cdot \eta_{sys} \cdot \eta_{atm}^2 \quad (16.2)$$

The collected receiving power P_{RX} focused onto one receiving channel is a multiplication of the following characteristics.

P_{SRC}, the transmitted laser power of the source,

r, the range to the target,

ρ, the reflectivity of the object at the working wavelength of the system. The reflectivity of an object can vary strongly based on the spectrum of the laser source. It is dependent on the surface's composition, texture and temperature,

D, the diameter of the receiving optics,

η_{sys}, the transmission efficiency of the optical path at the working wavelength of the LiDAR system,

η_{atm}, the transmission of the atmosphere which is dependent on working wavelength, altitude, weather conditions [22].

Detailed information about signal determination of LiDAR system with more complex receiving and transmitting optics are describing by McManamon [21].

The noise of a LiDAR system has two main sources:

The first noise source is based on the light of external sources R_{EXT}, which is around the working spectrum of the LiDAR. The light can be directly or indirectly scattered or reflected from objects or the atmosphere towards the receiver.

Because the sun irradiance plays such a great role due to its common presence and its power spectrum in noise calculations, it is important to fit the working spectrum of the LiDAR in an area in which the sun's spectrum is low. ◘ Fig. 16.5 shows a typical power spectrum of the terrestrial sun's irradiance at sea level.

Following, most LiDAR systems for field application work in the near infrared (NIR)—specifically between 870 and 950 nm, because of the low power spectrum levels of the sun. Additionally, this spectral band is still covered by silicon-based detectors which are relatively cheap as a receiver material and offer a variety of semiconductor laser sources. Some manufactures offer LiDAR systems in the area of around 1060 nm and 1550 nm because their laser safety levels allow higher power output. Specific detectors and sources have to be used in these cases. More on the topic of laser safety can be found in ▶ Sect. 16.6.

The second source of noise is the noise of the system R_{SYS} which is mainly dependent on the noise characteristics of the receiver and the attached electronics.

16.4.2 Point Rate, Measuring Area and Resolution

Monostatic LiDAR systems as shown in ◘ Fig. 16.3 consist of a sending and a receiving channel. Thus, for each point in time, a distance measurement can be performed.

To achieve a first approximation of a possible maximum point rate per second, the maximum system

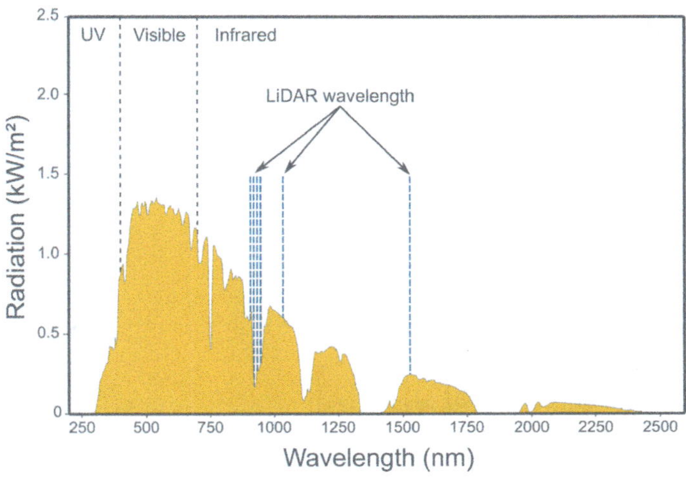

◘ **Fig. 16.5** Power spectrum of the sun at sea level [2]

range is taken into account. The time is calculated from which the light is send to a maximum ranged object and back at the speed of light $c \approx 3 \cdot 10^8$ m/s. For a range of 300 m, the time results in $300\,\text{m}/(3 \cdot 10^8\,\text{m/s}) \approx 2\mu s$ for a single distance measurement, hence the point rate comes up at $1/(2\mu s) = 500,000$ pt/s. Unfortunately,time for electronics reading, digitalization and dead time is not accounted for. If the point rate shall be increased, the maximum distance has to be decreased or the amount of channels has to be increased. It is possible to combine a one channel emitter with a multichannel receiver. The overall ideal point rate in this case is based on the point rate of the single channel multiplied by the amount of receiving channels.

To estimate the point rate per frame, the point rate per second has to be divided by the frame rate per second to estimate the point rate for each frame. One can then estimate the angular resolution through the overall FOV by dividing the maximum point rate by measuring cycle. In a system with 120° horizontal and 30° vertical FOV and a point rate of 30,000 pt/s, one can achieve a maximum resolution of $120° \cdot 30°/30.000 = 0,12°$ for 1 Hz and 1.2° for 10 Hz because the amount of points per frames would be just a tenth. An assumption in this case is that the FOV is covered without any gaps in between.

Another important characteristic is the coverage of the FOV, sometimes called point density. The FOV coverage is an amount of percentage of how much the FOV is actually covered with measurements. If the FOV is big and even if the angular resolution is small by each single measurement, gaps can occur in between neighbored measurements, which can be a multiple of the actual angular resolution. These gaps can already cause at middle ranged distances to miss objects of the size of a human being. Such low FOV coverages occur often in systems that are based on point-to-point measurements like flying-spot scanners.

16.5 Measuring Methods

To get an overview of the LiDAR measuring methods and their properties, one can distinguish two main categories: Direct and indirect Time-of-Flight measurements.

16.5.1 Direct Time-Of-Flight (dTOF)

The dTOF measuring principle is sometimes called pulsed Time-of-Flight (pTOF) or pulsed LiDAR. It is based on the direct measurement of needed time for transmitted light to hit a target, be scattered or reflected and be collected again by the receiver. Since the speed of light c within a medium with a constant refractive index is constant as well (for air $n \approx 1$), the distance d can be calculated directly from the passed time t ($d = c \cdot t/2$), compare ◌ Fig. 16.6.

The precision of the distance dTOF measurement is dependent on the quality of the time determination. In first approximation, a precision of the distance measurement of 0.15m requires one of 1ns. If the signal of the receiver is read directly, sensor systems of high signal bandwidth and sensitivity are used. Simultaneously, the reading electronics of high bandwidth has a high power consumption correspondingly creates heat. Thus, LiDAR systems based on such a principle consist of fewer channels, which means a smaller point rate.

An advantage of this method is that no coherent light is needed which improves the amount of usable components as a source. The maximum range of dTOF systems is dependent on the receiving light power within a certain time frame. To maximize the ranges, the power height has to be increased.

An advantage of this method is that no coherent (phase matching) light is needed increasing the amount of usable components as a light source. The maximum range of dTOF systems is mainly driven by the received light power. For higher ranges, shorter light pulses with higher peak powers are beneficial.

16.5.2 Indirect Time-Of-Flight (iTOF)

With the iTOF measuring principle, a distance is measured indirectly by e.g. a modulation of a transmitted signal. These modulations can be done either by amplitude or by frequency.

16.5.2.1 Amplitude Modulated LiDAR

LiDAR systems can be based on classic amplitude modulation (amplitude-modulated continuous-wave, AMCW) [3]. The transmitting signal (TX) is varied in

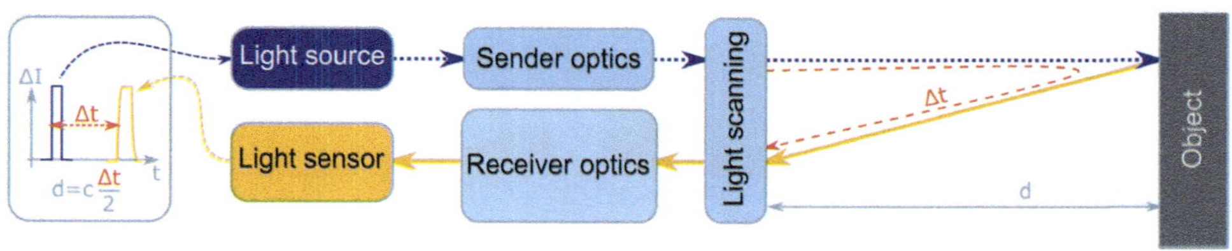

◌ **Fig. 16.6** Schematic example of a direct Time-of-Flight (dTOF) LiDARs

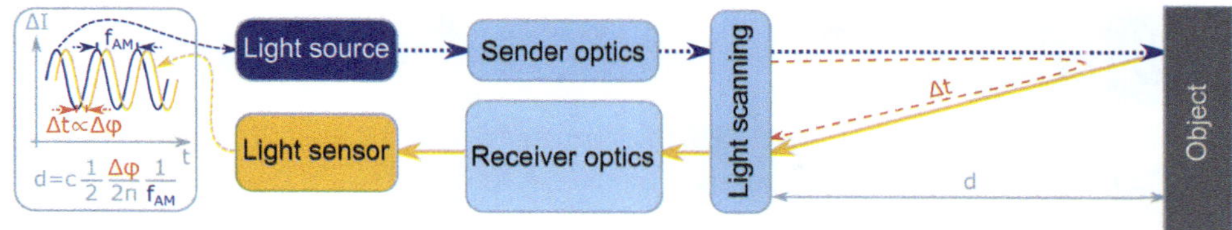

Fig. 16.7 Schematic depiction of an AMCW LiDAR

its intensity by a frequency of f_{AM}. The signal is reflected by an object and collected by the receiver. The distance to the object d is calculated by the phase shift $\Delta\varphi$ of the original TX to the RX signal. **Fig. 16.7** shows the schematic depiction of an AMCW LiDAR.

The maximum range of such a system is driven by two properties: The modulation frequency f_{AM} dictates the possible range in which distance phase information within one signal period of 2π can be determined and—as previously—the detected, reflected light within the background of the ambient light. Thus, there has to be enough contrast between the minima and the maxima of the collected light while keeping the electronics from being overloaded to avoid a disturbed signal curve.

Also, this method does not require a coherent light source.

In more advanced concepts, the amplitude can be modulated by the receiver, which enables a more integrated system with enhanced SNR. A multitude of channels can be realized with a minimal effort in this way with distance measurements of camera light resolutions, e.g. $640 \times 480 px$.

16.5.2.2 Frequency Modulated LiDAR

The frequency modulated measuring principle (frequency-modulated continuous-wave, FMCW) can be used with photons of the THz frequency band in analogy to the RADAR measuring method in the GHz frequency band. The distance is measured by the frequency difference of the received light to the transmitted light, which leads to an amplitude modulation (beat signal). **Fig. 16.8** shows a simple schematic depiction of the FMCW LiDAR based on classic linear rising and falling modulations.

In the FMCW approach, a majority of the light is sent to the objects but a small fraction is kept lo-

cal within the system to be coupled towards the receiver path. This so called local oscillator (LO) is mixed with the received light [20]. The receiver and the reading electronics must be able to resolute the interference signal with the beating frequency f_{BEAT} having a much smaller frequency than the received light or the LO. The range of a FMCW system is determined by the smallest detectable amount of received mixed light while the impact of ambient light and noise is R_{EXT} reduced heavily.

Because FMCW is based on frequency modulation, the light source is required to have a high coherence length with a multitude of the actual range of the system. The wavelength of the source needs to be controlled for the periodical shifting.

These complex requirements increase the cost of the source and its surrounding electronics.

16.5.3 Gated Imaging

The Gated-Imaging method is an additional measurement principle. It started in the submarine-based research and has been evaluated for automotive [9]. A distance measurement is a combination of several camera intensity pictures. For each of the pictures, which are called slices in this method, the system transmits light pulses and activates different time delays for the receiver. In this manner, a distance profile can be obtained by each distance. **Fig. 16.9** shows a schematic comparison with a conventional camera.

The distance to the object can be obtained by the time delays of the slices with the greatest intensity. To enhance the measuring precision, the light pattern can be modulated [18]. The precision is mainly determined by the amount of slices and the size of the delay steps. This dependency increases the acquisition time for each

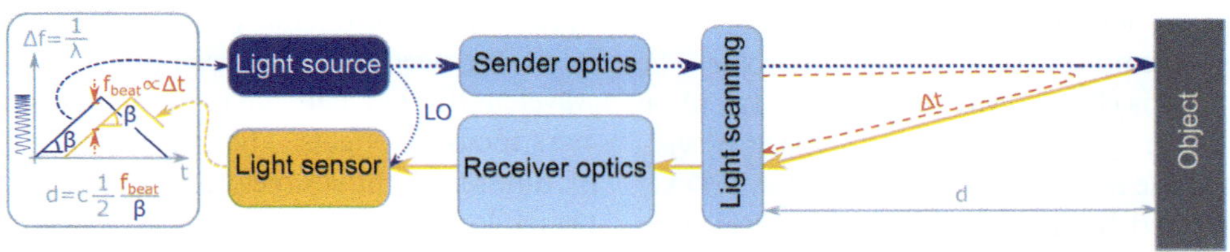

Fig. 16.8 Schematic of a FMCW LiDAR

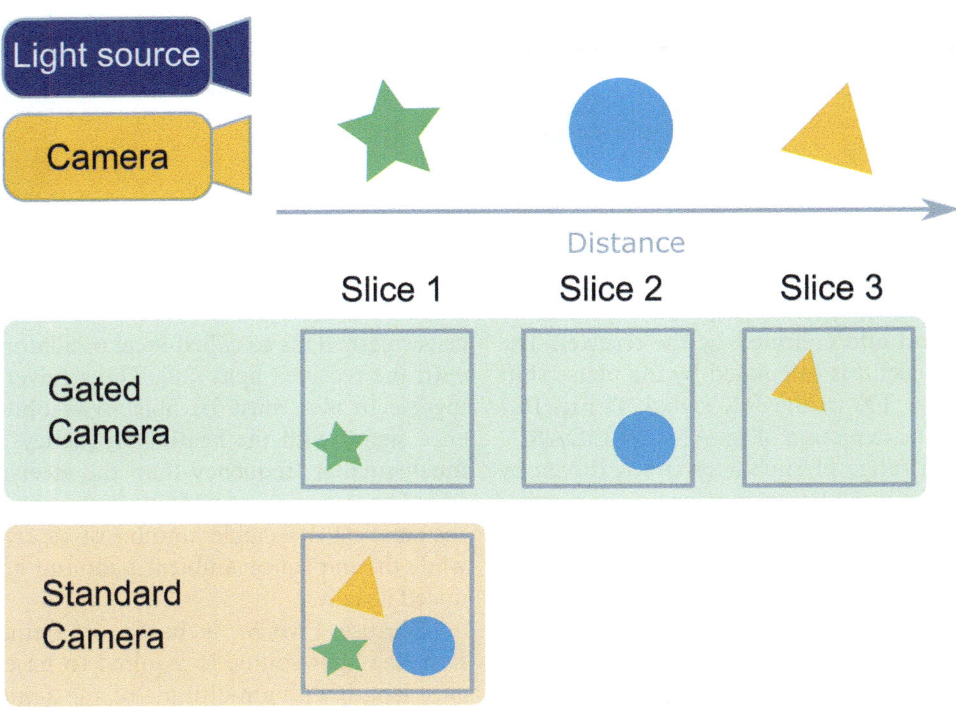

Fig. 16.9 Comparison between a conventional standard camera and pictures (slices) by a gated camera

distance measurement. This method is preferred in Flash LiDARs for enhancing the measuring rate and thus the information density, compare ▶ Sect. 3.2.

An advantage of Gated Imaging and the dTOF method is the possibility to mask certain areas across certain measurement distances. This enables a suppression of weather effects and semi-transparent objects like leaves on the height of the sensor. Additionally, the SNR is increased since the ambient light is just gathered in the relevant time frames. Machine learning algorithms further improve the Gated Imaging method [10]. Multi path occurrence can be suppressed and multiple reflections can be evaluated.

16.6   Laser Safety

Each laser product must be tested towards its safety aspects to avoid eye damage and in the worst offending scenario skin damage. The guidelines are based on the IEC standards 60825 [16], which exists in the third edition in the moment. Products are labeled by different laser classes, which categorize different hazard levels. The hazard level is depending on the wavelength, optical peak and average power and illumination profile. The only safe laser class is class 1. Already class 2 is based on assumptions like the eye-blink reflex enabled just by visible light. Because the active measurement of the LiDAR shall not be detected by human beings, the wavelength spectrum must be out of the area of 400 nm to 800 nm in the near infrared, typically even above 900 nm. In this case, a crossing of the laser safety class 1 results already at least in laser safety class 3R being potentially dangerous. By law, an operation of such devices is just allowed in specific laser safety areas in most countries.

Additionally, possible technical errors compromising the laser safety class have to be identified and surveilled, which equals the laser safety as an ASIL relevant function. Typical monitored attributes are shooting frequency or pulse train lengths. If the attributes' limits are violated, the laser is turned off.

The laser safety defines a physical limit for the LiDAR's performance. Thus, it is possible to determine the maximum performance of a system based on the derivation of its systematic attributes.

A common misunderstanding is that the main damage occurs at the eye. This is true for most cases in the area of visible or near infrared but there are other situations, as well. The infrared radiation with wavelengths above 1400 nm are absorbed mostly by water and hit the retina with a much reduced power level at the back of the eye. These water absorbing wavelengths fortunately are less harmful to the common known retina damage of the eye but the damage can unfortunately occur anywhere on the body by burning the skin. All in all, the needed energy level might be higher for these wavelengths but can harm skin independently on the eye's focusing effects. The overall risk of superposition of different light sources from different directions is in principle higher when more LiDARs are used. Also, long term damage can be generated, i.e. like the opacity of the eye's lens which is called cataract in medicine.

16.7 Application Scenarios

The use of LiDARs for the automotive industry is still under discussion. The complexity of this technology and its cost structure is reason for some OEMs to avoid this type of sensor [12]—others see here an important support to increase the redundancy which is necessary to enable safe autonomous driving and reduce the risk of crashes based on misinterpretations by other driver assistant systems [26]. In the following sections, typical application scenarios are discussed.

16.7.1 Point Cloud Generator

The LiDAR can be used as a simple measuring tool: The gathered data and three-dimensional point cloud can be delivered without any filters. In this configuration, the ECU can cover the interpretation and filtering of the data.

The use of LiDARs as a point cloud generator has some advantages for the OEMs: They have control over the data processing and can decide the algorithms and classifications as they please. The disadvantage of cause is that in this way, there is a much more processing effort to implement and fusion the data with other sensor outputs.

16.7.2 Smart Sensor

A smart sensor covers parts of the data processing. In contrast to the previous situation, data filtering for the elimination of noise, clustering of points, annotation or classifications in semantic objects can be used.

These objects can be fixed items or areas of significance for traffic maneuvering. For example, ground marking has the same distance as the road floor but can be distinguished by its high reflectivity value since these values are usually significantly higher. The smart sensor can recognize the road markings and divide the street into different lanes which it can describe as vector or surface information. A comparison between smart sensor and cloud point generator is given in ◘ Fig. 16.10. An incomplete list of functions is given in the following sections.

Object Classification

Parts of the point cloud can be recognized as certain objects. A set of point clouds can be given a meaning by determining the size and form of predetermined object classes. These object classes might be simple abstractions of the surrounding like person, bike, car or truck. As well, the classes can be even more abstract and be divided into subclasses: For example, a class of "moving objects" could contain all previously mentioned classes.

Object Tracking

There is the possibility to track objects and determine their moving vector after their classification. For example, a set of data points could be determined as the class "moving object" with the subclass "car" which is changing from one lane to another along a direction vector.

Weather conditions

Sensors measure the environment and can identify weather conditions. Even after applying the adaptive sensitivity of the LiDAR, it is triggered by statistical noise patterns. These patterns are so called false-positive measurements, which are depending on the weath-

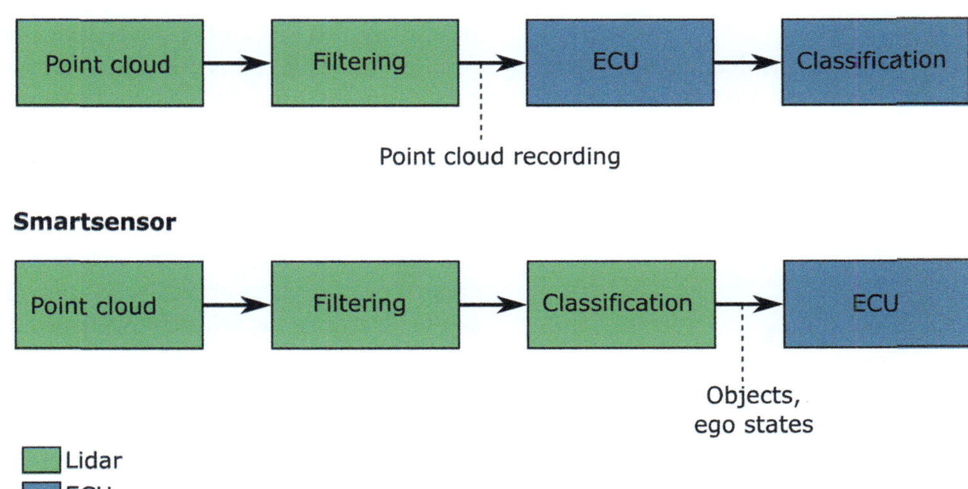

◘ **Fig. 16.10** Data flow of a LiDAR as a point cloud generator and a smart sensor

er's characteristics. Raindrops could cause such behavior when the laser beam hits them. The intensity of the rain can be analyzed by the strength and frequency of the noise pattern [14].

On the other hand, fog is so diffuse that is always causes backscattering while the range of the sensor is heavily decreased due to the levitating water particles in the air, which are scattering or absorbing the laser.

Road Conditions

Road conditions can be determined indirectly. A wet road condition will cause statistical noise by the tossed water drops behind driving vehicles. A similar noise pattern is caused by tossed sand, as well. It requires some amount of intuition and solid statistical mathematical analysis to distinguish the behaviors under each condition. The spatial limited noise phenomena is called clutter.

Another possibility to analyze the road condition is the smoothness of the road. Small bumps can be a recognized as derivations from an ideal course of smoothness in the point cloud. Depending on the wavelength, the behavior can be different. On a wavelength of e.g. around 1550 nm the high absorption would cause low strengths of reflected signals which can ultimately lead up to holes in the point cloud caused by targets. Thus, the point cloud characteristics is strongly based on wavelength [28].

Ego Motion

The knowledge about one's own movement within time and space is of utmost importance for autonomous driving. Many smart sensors can estimate their own position and movement and report the egomotion of the vehicle. This usually consist of velocity and direction of the movement.

Slam

If the sensor is able to determine its ego motion, SLAM can be executed. SLAM is an acronym for "Simultaneous Localization And Mapping" and allows the orientation and recording of unknown terrain. To achieve this, data is saved over time in an absolute map in which the vehicle is moving.

16.7.3 Functional Support

The LiDAR is useful to support several functions. In the following section, an incomplete list of possible support functions are stated.

Park Assistance

The park assistance is one of the most used comfort functions in the automotive sector. While this technology is usually based on ultrasonic sensors for distances and cameras for a higher resolution, LiDARs can offer both information. Also, LiDARs measure distance accurately for a good collision warning while maneuvering.

Automated Valet Parking

Valet parking is an advanced variant of the simpler park assistance function. The car is driving the last instance to the parking spot on its own, either short before the spot or at the beginning of a parking area. After an assignment, the car can find the rest of the way to the driver using the distance information of the LiDAR and a guidance system.

Collision Detection and Warning

The LiDAR monitors the area in front of the vehicle with high resolution and therefore allows reliable clearance estimation. Thus, it can warn of possible collisions while driving and if necessary, activate safety systems that prevent collisions or mitigate their impact. Since LiDAR provides active illumination, this function is not limited at night or in poor lighting conditions, e.g. when the vehicle approaches a tunnel.

Start-Off Clearance

A variant of collision avoidance is the start-off clearance. Here, the system analyzes whether the environment is clear at the initial start of the vehicle so that it can drive off without causing damage to itself or the environment.

Lane Detection and Lane Keeping

In case the LiDAR's FOV covers the closer ground in front of the vehicle, it is possible to perform lane detection as an input for the lane departure warning system and lane keeping assistance. Since LiDAR provides an intensity signal together with distance and location, the sensor can distinguish the higher reflective white or yellow markers from less reflective road surface. The precise location of lateral boundaries such as guardrails, delineators, cat's eyes or road studs provide valuable input to determine the vehicle's own lane situation.

(Partially) Automated Driving

LiDARs with a long distance range can be used as a sensor for Adaptive Cruise Control (ACC). Here, the necessary range performance depends on the speed of the vehicle. ACCs in urban areas are also much more complex than on the highway, as there are significantly more objects and more complex environmental situations to be expected. This is one reason why highway assistants are given priority when level three functions are introduced. This includes, above all, the traffic jam pilot, which is probably the simplest function to conceive, since low speeds and only one direction of travel are given.

Blind Spot Detection and Turning Assistant

LiDAR can precisely monitor the free space within the vehicle's blind spot. In contrast to ultrasonic and RADAR, LiDAR can classify surrounding objects at a very high resolution, the resolution also helps to suppresses false positives e.g. caused by rain. Furthermore, LiDAR provides precise information on non-metallic and sound-absorbing objects, unlike RADAR and ultrasonic.

16.7.4 Detection Zones, Applications and Requirements

The application scenario of a LiDAR determines the technical requirements. This includes, in particular, the question of which measurement ranges are reasonable and necessary. Although all object classifications according to ▶ Sect. 7.2 are possible and the functionalities of ▶ Sect. 7.3, the application scenarios depend on ranges and opening angles of the FOV. One example is parking support: parking typically uses low speeds and requires a very large field of view around the vehicle. A relevant maximum range for a LiDAR in this case would be in the order of 10 m. ACC for highway use requires exactly opposite properties. Here, a very long range in the order of 160 m is important to enable early and smooth breaking at high speeds. The required field of view is smaller, since freeways have a preferred direction of travel and no side traffic is expected. These different requirements can be transferred to zone models that try to cover the main use cases [24]. For example, a three-zone model in which different range requirements are required at different angular regions, see ◘ Fig. 16.11. This can apply not only to range but also to resolution. For example, in highway application the need for high resolution is usually greater in the central area than in the outer area. Exact description of the requirements for the corresponding use cases results in design margins for performance and cost optimization of the LiDAR systems. Some LiDAR systems offer to define variable measuring ranges with increased resolution (region-of-interest, ROI).

16.8 Use of Multiple LiDARs

Similar to ultrasonic sensors, the usage of multiple sensors might be necessary in applications to achieve the coverage of all functionality. In the following sections, you can find reasons and impacts for such use cases.

16.8.1 Improvement of Measurement Area and Redundancy

A combination of LiDAR sensors can increase the overall measurement area by the combination of their field of views. An overlapping of the FOVs usually does not increase the range since this is an intrinsic property of each of the LiDARs. Still, input signals can be analyzed independently from each other and within the overlapping area there is the chance to increase the redundancy of the system which increases the probability not to overlook objects.

16.8.2 Laser Class of the Car

Using a laser device in a vehicle leads to the necessity of classifying the vehicle within the laser classes. If just one LiDAR is used, the classification of this particular LiDAR is passed on to the vehicle. For the use of many LiDAR sensors, the situation becomes somewhat more difficult since the interaction of the sensors has to be analyzed, as well [4]. In principle, the laser beams can cross each other and brake the laser safety class 1 limits. Also, the LiDARs could hit the same spot at the same time leading to a similar situation. The vehicle itself would not be considered laser safe anymore although each LiDAR is laser safe. A detailed analysis by specialized experts and institutions is necessary to ensure the laser safety for the case of using several LiDAR sensors.

16.8.3 Extrinsic Calibration

Another aspect using multiple sensors in just one vehicle is the relative and absolute position of each other. Typical solutions are based upon mathematical transformations and/or mechanical mounting dealing with the relative and absolute positioning. Specifically, the yaw, pitch and roll angles have to be determined since already small differences can make large absolute errors on vertical heights on long distances. The absolute alignments of the LiDARs can be determined by an external reference frame, e.g. a specific pattern.

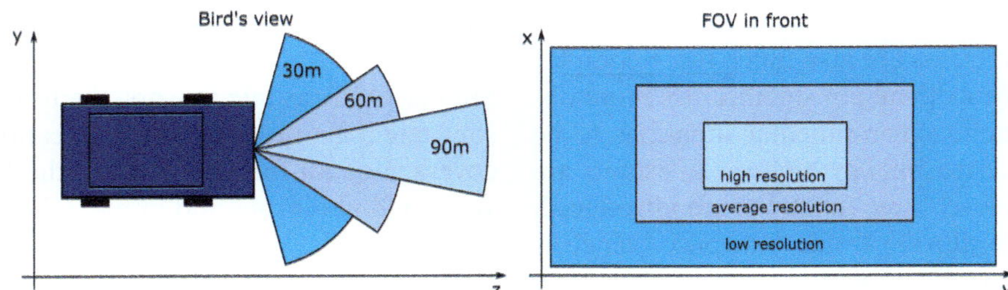

Fig. 16.11   Example zone model of a requirement description for a LiDAR

16.9   Current Series Examples

To get an insight into the variety of laser sensors, an incomplete list of LiDAR sensors is given in this section. The technical data can be found in ▪ Table 16.1. All LiDAR products mentioned here are laser class 1.

Valeo

Valeo is the first supplier to offer laser scanner technology in automotive mass production [11], which is today installed in series in several vehicle types. The second generation of the Valeo SCALA (Scanning Laser) has been installed in series production cars from 2021 onwards. The vertical FOV of the second version tripled compared to the first version to 16 layers, the vertical resolution improves to about 0.5°. The horizontal FOV of 133° is sampled at 0.125° in the central area and 0.250° in the outer area. The sampling rate in the central area doubled, so that possible gaps in the FOV coverage are closed, compared to the first generation.

Continental

Continental offers the HFL110, a flash LiDAR that provides a FOV of 120° horizontal and 30° in vertical direction. The wavelength of 1064 nm results from an Nd:YAG laser module, which illuminates the FOV in its entirety. The receiver consists of 128×32 pixels. This results in a resolution of about 1° per pixel.

Table 16.1   Comparison of products from serial and pre-serial models

	Valeo Scala 2	Ibeo NEXT	Continental HFL	Robosense M1
Picture	[27]	[17]	[6]	[25]
Range(m) @10%	~100	LR~140, SR~50		~150
Maximum range (m)	220	260		200
Wavelength (nm)	905	885	1064	905
Framerate (fps)	25	25	30	10
accuracy (cm)	<4	LR<16, SR<8		<4
Horizontal/vertical resolution	0.125° resp. 0.250° / 0.6°	LR 0.07°/ 0.09°, SR 0.47°/0.38°	~1°/~1°	0.2° / 0.2°
hFoV	133°	LR 11.2°, SR 30°	120°	120°
vFoV	10°	LR 5.6°, SR 60°	30°	25°
Power Consumption	<21 W	<10 W		<15 W
Technology	Scanning mirror	Sequential flash	Flash	MEMS mirror
Weight (g)	600	780	600	720

Ibeo

After developing the SCALA sensor together with Valeo, Ibeo focused with ZF on a LiDAR without moving parts. The IBEO Next is a combination of VCSEL (Vertical-Cavity Surface-Emitting Laser) and 2D-SPAD (2 Array. It has 80 rows and 128 columns of emitter and receiver pixels. The emitting and receiving lenses are available in different variants, allowing short range (SR) with large FOV and long range (LR) with small FOV [15]. The VCSELs can be driven in sections, allowing the data points to be generated in rows one after the other. This type of architecture is called sequential flash.

Robosense

The Robosense M1 is a LiDAR system that provides a horizontal FOV of 120° and 25° in vertical direction. The FOV is horizontally separated into five partially overlapping areas, which are scanned by MEMS mirrors.

16.10 Summary and Outlook

For automotive use, LiDAR is a versatile sensing technology that provides a large variety of functions based on superior angular resolution and distance range. However, the aspiration of OEMs to have one LiDAR for all use cases has not been fulfilled. The automotive sector recognizes more and more that this complex technology is more efficient if it is tailored to a specific set of use cases. Thus, different use cases call for different systems: e.g. parking and low-speed maneuvering can be covered by near field LiDARs, ACC and high-speed collision avoidance is based on long range LiDARs.

The long-term question is whether the competition with RADAR, camera and ultrasonic sensors exists, given that in principle all sensor types could be replaced by LiDARs. All these sensor types can be used for achieving certain driving assistance function up to SAE level 2. Nevertheless, achieving higher levels of SAE can only be achieved with a certain degree of redundancy, which gives space for leaving the competitive situation towards a general collaboration of all sensing technologies. By measuring principle, RADAR offers better performance in fog because the hovering water drops influence infrared light more than long wave radio frequencies. Most LiDAR systems are based on dTOF and it will be exciting to see if the transition towards iTOF (e.g. FMCW) will happen, since RADAR is using it because of the high efficiency. Due to the large-scale dynamics of the ongoing LiDAR development, a blanket statement about the system architecture of LiDAR is practically impossible and only the gathering of experience will show which application scenarios and which benefits are really desired by the OEMs for which price.

References

1. Amzajerdian, F., Pierrottet, D., Petway, L.B., Hines, G.D., Roback, V.E., Reisse, R.A.: Lidar sensors for autonomous landing and hazard avoidance. In: AIAA SPACE 2013 Conference and Exposition. (2013). ▶ https://doi.org/10.2514/6.2013-5312
2. ASTM.: G173–03: standard tables for reference solar spectral irradiances—direct normal and hemispherical on 37° tilted surface. ASTM International (2003)
3. Bamji, C.S., O'Connor, P., Elkhatib, T., Mehta, S., Thompson, B., Prather, L.A., Snow, D., Akkaya, O.C., Daniel, A., Payne, A.D., Perry, T., Fenton, M., Chan. V.H.: A 0.13 µm CMOS system-on-Chip for a 512 x 424 Time-of-Flight image sensor with Multi-Frequency Photo-Demodulation up to 130 MHz and 2 GS/s ADC. IEEE J. Solid-State Circuits **50**, 303–319 (2015). ▶ https://doi.org/10.1109/JSSC.2014.2364270
4. Beuth, T., Thiel, D., Erfurth, M.G.: The hazard of accommodation and scanning LIDARs. In: Proc. SPIE 10695, Optical Instrument Science, Technology, and Applications, (2018). 1069506. ▶ https://doi.org/10.1117/12.2312467
5. Collis, R.T.H.: Lidar observation of cloud. Sci. **149**(3687), 978–981 (1965). ▶ https://doi.org/10.1126/science.149.3687.978
6. Continental AG. Copyright (2020)
7. Gebel, R.K.H.: Optical radar and passive optoelectronic ranging. Ohio J. Sci., **66**(5), pp. 496–507. (1966). ▶ https://kb.osu.edu/handle/1811/5222. Zugegriffen: 31. Dezember 2020
8. Guenther, G.C., Thomas, R.W.L.: Depth measurement biases for an airborne laser bathymeter. In: Proceedings 4th Laser Hydrography Symposium, ERL-0193-SD. (1960)
9. Grauer, Y., Sonn, E.: Active gated imaging for automotive safety applications. In: SPIE Proceedings Volume 9407, Video Surveillance and Transportation Imaging Applications, 94070F. (2015). ▶ https://doi.org/10.1117/12.2078169
10. Gruber, T., Julca-Aguilar, F., Bijelic, M., Heide, F.: Gated2Depth: Real-Time dense lidar from gated images. In: Proceedings of the IEEE/CVF International Conference on Computer Vision (ICCV), pp. 1506–1516. (2019). ▶ https://doi.org/10.1109/ICCV.2019.00159
11. Hälker, J., Barth, H.: Lidar as a key technology for automated and autonomous driving. ATZ Worldwide **120**, 70–73 (2018). ▶ https://doi.org/10.1007/s38311-018-0102-z
12. Hawkins, A.J.: Elon Musk still doesn't think LIDAR is necessary for fully driverless cars. The Verge. (2018). ▶ https://www.theverge.com/2018/2/7/16988628/elon-musk-lidar-self-driving-car-tesla. Zugegriffen: 31. Dezember 2020
13. Hecht, J.: Lidar for self-driving cars. Opt. Photonics News **29**(1), 26–33 (2018)
14. Heinzler, R., Schindler, P., Seekircher, J., Ritter, W., Stork, W.: Weather Influence and Classification with Automotive Lidar Sensors External Link. IEEE Symposium on Intelligent Vehicle (2019). ▶ https://doi.org/10.1109/IVS.2019.8814205
15. Holzhüter, H.: IBEO next solid-state lidar. EPIC Online Technology Meeting on ADAS and Autonomous Driving. (2020). ▶ https://www.youtube.com/watch?v=-4pTHqQBNR4&t=1343. Zugegriffen: 31. Dezember 2020
16. IEC.: Safety of laser products—Part 1: Equipment classification and requirements. 60825–1 (2014)

17. Ibeo Automotive GmbH. Copyright (2020)
18. Kabashnikov, V., Kuntsevich, B.: Optimal conditions for distance determination by the range-intensity correlation methods using range-gated viewing system with nonrectangular pulses. In: SPIE Proceedings Volume 11160, Electro-Optical Remote Sensing XIII, 116000. (2019). ▶ https://doi.org/10.1117/12.2532293
19. Lesina, A.C., Goodwill, D., Bernier, E., Ramunno, L., Berini, P.: On the performance of optical phased array technology for beam steering: effect of pixel limitations. Opt. Express **28**, 31637–31657 (2020). ▶ https://doi.org/10.1364/OE.402894
20. Martin, A., Dodane, D., Leviandier, L., Dolfi, D., Naughton, A., O'Brien, P., Spuessens, T., Baets, R., Lepage, G., Verheyen, P., Heyn, P.D., Absil, P., Feneyrou, P., Bourderionnet,: Photonic integrated Circuit-Based FMCW coherent LiDAR. J. Lightwave Technol. **36**(19), 4640–4645 (2018). ▶ https://doi.org/10.1109/JLT.2018.2840223
21. McManamon, P.F.: Field Guide to Lidar. SPIE Press (2015)
22. McManamon, P.F.: LiDAR Technologies and Systems. SPIE Press (2019)
23. Editorial, N.: LiDAR drives forwards. Nature Photon **12**, 441 (2018). ▶ https://doi.org/10.1038/s41566-018-0235-z
24. Rasshofer, R.H., Gresser, K.: Automotive radar and LiDAR systems for next generation driver assistance functions. Adv. Radio Sci. **3**, 205–209 (2005)
25. Robosense EMEA GmbH. Copyright (2020)
26. Solon, O.: Lidar: the self-driving technology that could help Tesla avoid another tragedy. The Guardian. (2016). ▶ https://www.theguardian.com/technology/2016/jul/06/lidar-self-driving-technology-tesla-crash-elon-musk. Zugegriffen: 31. Dezember 2020
27. Valeo Schalter und Sensoren GmbH. Copyright (2020)

Open Access This chapter is licensed under the terms of the Creative Commons Attribution-NonCommercial-NoDerivatives 4.0 International License (▶ http://creativecommons.org/licenses/by-nc-nd/4.0/), which permits any noncommercial use, sharing, distribution and reproduction in any medium or format, as long as you give appropriate credit to the original author(s) and the source, provide a link to the Creative Commons license and indicate if you modified the licensed material. You do not have permission under this license to share adapted material derived from this chapter or parts of it.

The images or other third party material in this chapter are included in the chapter's Creative Commons license, unless indicated otherwise in a credit line to the material. If material is not included in the chapter's Creative Commons license and your intended use is not permitted by statutory regulation or exceeds the permitted use, you will need to obtain permission directly from the copyright holder.

Camera Sensors

Martin Punke, Boris Werthessen, Peter Geisler, Christian Mindescu, and Andreas Wagner

Contents

© The Author(s) 2026
H. Winner et al. (eds.), *Handbook Assisted and Automated Driving*,
https://doi.org/10.1007/978-3-658-45276-6_17

Camera systems are ubiquitous in today's world. They are used in a wide range of applications from consumer electronics to industrial environments. Cameras have also become indispensable in automotive applications.

Today's traffic environments such as traffic and information signs, road markings and vehicles are designed to be perceived by the human eye, even though initial approaches to automatic assessment by electronic sensor systems in the vehicle exist. This is done e.g., by different shapes, colors, or a temporal change of the signals.

Therefore, it is obvious to explore the environment by machine perception like the human eye. Camera systems are capable of this, since they offer comparable spectral, spatial, and temporal resolution. In addition to the "replica" of human vision, camera systems can offer additional functions, including images in other spectral ranges for night vision functions or distance measurement.

17.1 Applications

Due to their versatility, camera systems in automobiles are used both for surveillance of the interior of the car as well as the surroundings [1]. The following section discusses these applications and explains the specifics of using camera systems for these.

A first driver assistance system using a camera system was the so-called rear-view camera. The driver is supported by displaying a real-time video on a display system. Advanced functions using machine vision are used, for example, in the *automatic high beam* function and the *emergency brake assist*. In these systems, the video image is no longer displayed, but a function is derived directly from the camera image.

In addition, cameras are also used in the interior of the vehicle. Here, two functions are of particular importance. Firstly, the monitoring of the person driving the vehicle to detect the condition and intention, and secondly, the use of camera systems for interior monitoring and video functions.

17.1.1 Environmental Detection

The goal of automotive environment sensing is to detect all relevant road users, the road scene and traffic signs as completely as possible in order to draw appropriate conclusions. A variety of different sensor technologies can be considered, which in combination ensure both detection and reliability.

■ Figure 17.1 shows such a situation. It depicts the field of views of various sensors on the car. The sensors are designed such that the fields of view overlap, and different detection ranges can be achieved. In this example, the detection by camera systems as well as short and long range RADAR sensors is shown.

Camera systems for environment detection are implemented in two main ways. On the one hand, as a highly integrated "smart" camera (so-called front view camera), which takes care of both image acquisition and processing (see ■ Fig. 17.2a). Second, as an electronic control unit with integrated image processing and satellite camera heads for image acquisition (see ■ Fig. 17.2b). Both architectures will be explained in more detail later in ▸ ▸ Sect. 4.2.

17.1.1.1 Front View Cameras

Front view cameras are usually integrated behind the windshield of the automobile at the level of the rear-view mirror. This position has the great advantage of a wide field of view and protection by the windshield. In addition, the area in front of the camera is swept by the windshield wipers, thus ensuring a largely unobstructed view. Exceptions to this placement position are camera systems working in the far-infrared range. These systems are installed in the headlight or radiator area due to the low windshield transmission in this spectral range [2].

Cameras in the Visible Spectral Range

Most of the systems currently used are operating in the visible spectral region and thus similar to the human eye. As previously explained, this enables the detection of all relevant traffic signs, etc. by the camera systems that are also relevant for humans.

The *high-beam assist function* is used to automatically turn on and off the high beams of the car. More advanced systems also have a variable range control and independently dim certain areas. Such systems are enabled by segmented LED light-emitting diode headlamps. The various light functions are controlled by the camera that is analyzing the oncoming traffic. Important for this camera function is the ability to distinguish at least the color of the tail and front lights of a vehicle.

Therefore, color-sensitive camera systems are used (details in ▸ ▸ Sect. 3.3.5). Another requirement for this function is the need for a high dynamic range of the camera system, arising from the large differences of light intensities that occur at night (see also ▸ ▸ Sect. 2.1.4).

Traffic sign recognition detects all relevant traffic signs (e.g., speed limits, one-way street labels) and the information is made available to the driver. The traffic sign recognition feature requires a high-performance camera system [3]. The traffic signs must be recorded with a high resolution (>15 pixels per degree) so

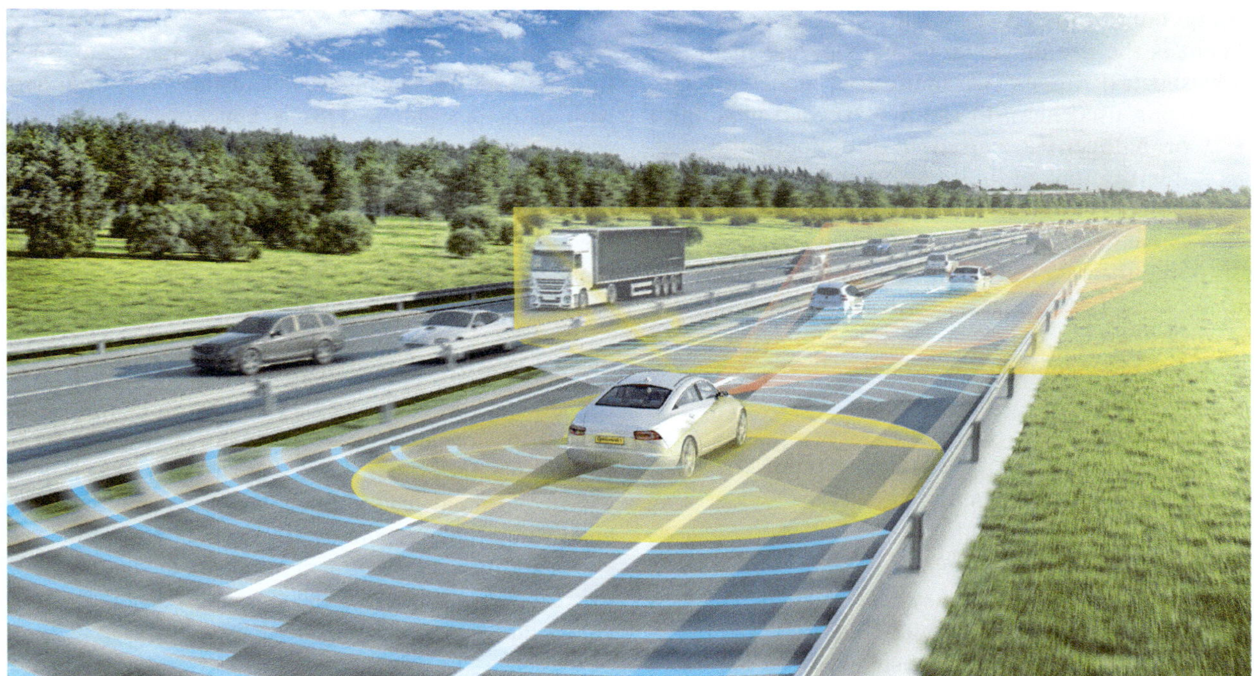

Fig. 17.1 Environmental detection by different sensor systems

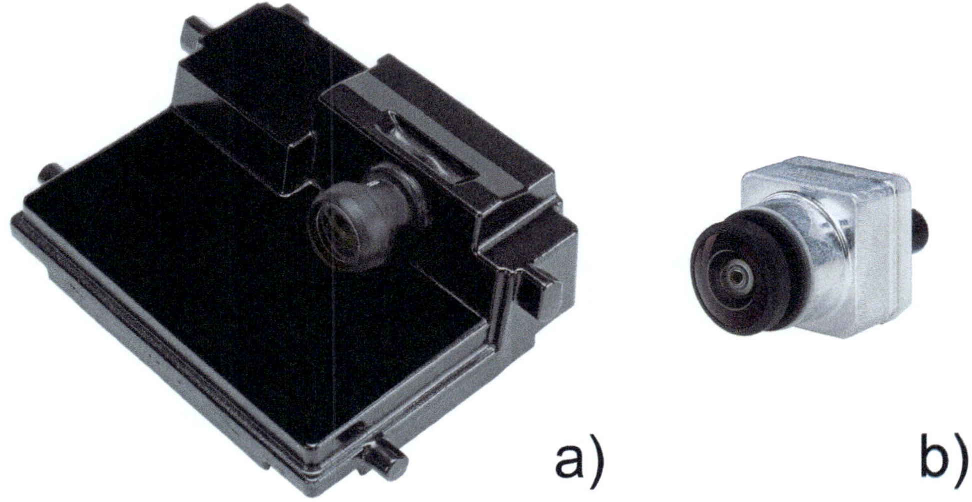

Fig. 17.2 Front view camera (**a**) and satellite camera head (**b**)

that the character recognition functions correctly. Since most traffic signs are placed at the edge of the road and the automobile is moving at high speed, a short (<15 ms) exposure time is necessary to avoid strong motion blur.

For safety reasons, many vehicles are equipped with a *lane detection* feature. Important for this function is a very high recognition rate even at nighttime and with poor road conditions. It is advantageous if the camera system can distinguish colors, since a detection of different colored markings on a street is enabled, for example, at construction sites (▶ ▶ Sect. 2.1.3).

In order to respond to other road users (e.g., vehicles, pedestrians, cyclists), a robust *object recognition* is necessary. Here, a variety of aspects are relevant for the camera system [4]. For a large detection range, e.g., for vehicle detection on highways, a high resolution is necessary. Pedestrian detection benefits from a large field of view. In general, a high sensitivity of the camera system is very important.

Cameras in the Infrared Spectral Range

One disadvantage of cameras operating in the visible spectral range is their limited sensitivity under poor lighting conditions. A possible alternative or supplement is the use of cameras that operate in the infrared spectral range. Two main approaches are currently being used here. In the first, special headlights (LED or halogen) are used to illuminate the traffic scene with infrared light. Another option is to use special cameras that are sensitive in the far infrared (FIR) range. These cameras can be used to directly capture the thermal radiation signature of pedestrians and animals, thus triggering an assistance function. The disadvantage, however, is the high cost, since these camera systems are not based on conventional image sensors [2].

17.1.1.2 Cameras for the Detection of the Surroundings of the Car

The class of environmental cameras can fulfill a wide variety of tasks, some of which overlap. On the one hand, there are human vision systems such as rear-view cameras, surround-view cameras, and mirror replacement systems. These camera systems show the video stream directly to the driver. However, they can also perform machine vision tasks, such as object recognition.

On the other hand, this camera class can also be purely used for environment detection, where displaying an image is not the focus of functionality. Such camera systems with a central image processing unit are growing in importance as they are needed for higher levels of automation (SAE L2–L5).

Rear-View Cameras

The camera module of rear-view cameras is usually integrated into the tailgate of the car (e.g., close to the license plate). The video stream is then displayed on a monitor on the instrument panel. Firstly, accidents by running over people behind the car can be avoided. On the other hand, advanced systems support the driver during parking by displaying a graphical overlay.

Surround View Camera Systems

Surround view systems are equipped with four or more cameras around the vehicle. The video information of the cameras is transmitted to an electronic control unit. Camera modules for such systems are usually equipped with the so-called fish-eye lenses, which allow a horizontal field of view of more than 180°. From the camera images, a 360° view of the environment is generated and provided to the driver as a parking aid on a monitor. In future, not only a parking support will be possible, but the camera images are also used for object detection and general environmental detection in addition to the front view camera.

Mirror Replacement

An approach that is likely to play a significant role in future is the replacement of normal exterior mirrors with camera systems. This is beneficial for fuel consumption (less drag) and opens up completely new design possibilities. Similar to surround view systems, a high dynamic range and a good color reproduction of the camera image are necessary for a good quality of the displayed video. Such camera systems are addressed in the international standard ISO/DIS16505 [5].

Environmental Detection

Cameras in this class can be mounted in a wide variety of positions on the vehicle, depending on the application. Commonly, they are integrated in the interior behind the windshield for detection in the front area of the vehicle and behind the rear window for the area behind the vehicle. In addition, these camera heads are also integrated in the rearview mirrors and the sides of the vehicle to provide complete coverage of the surrounding area.

17.1.2 Driver and Interior Monitoring

There are special requirements for the use of cameras in the vehicle interior that differ from those for environmental detection. The human being as an object that is captured by the camera is much closer to the camera than objects in the vehicle's environment. Artificial lighting is used to ensure that the system functions in all driving situations, e.g., at night or during rapid light changes. The light sources are in the near-infrared wavelength range, which is invisible to humans. The further development from pure observation of the driver towards full interior observation means that cameras with a larger field of view and higher resolution are needed.

17.1.2.1 Driver Monitoring and Gaze Control

Driver assistance systems are helpful and take the strain off the driver. At the same time, however, they also change his role: The human has more of a monitoring function than an active one. The goal of the development is a *holistic human machine interface (HMI),* in which information about the person driving the vehicle, their state and intention, is integrated into the situation-adapted interaction concept.

Starting from drowsiness detection, the application fields for a head-oriented interior camera have greatly expanded. Functions such as person identification e.g., for calling up individual profiles or preferences, adaptive warnings depending on head position and gaze behavior, and augmentation of the HMI (augmentation: an overlay of HMI information with objects from the vehicle environment), e.g., in the augmented reality head-up display, can be realized.

A scalable approach with one or more cameras makes it possible to define the range of detection and the accuracy of head position and gaze direction determination depending on the area of application. In ▫ Fig. 17.3a), a possible field of view for a mono camera driver monitoring system is shown.

17.1.2.2 Interior Monitoring

Installing a camera in a central location in the passenger compartment e.g., in the area of the roof control unit / rearview mirror (see ▫ Fig. 17.3b), makes it possible to monitor the passenger compartment as well as the person driving the vehicle. There are many ways of integrating a camera in the passenger compartment; the decisive factor is that the camera has a clear view of the objects to be observed. Sensors in both the infrared and visible spectral ranges are used. Currently, a fusion of the two sensor types is taking place. This leads to a reduction of the needed installation space and makes integration in the interior easier.

The image information obtained can be used to implement various functions. Algorithms allow the image content to be used to detect that people or objects are in the vehicle, the head pose can be recognized, and, with the appropriate resolution, people can also be identified. Deriving a body model of all occupants allows conclusions about posture, body size or the distinction between children and adults. This information can in turn be used to adjust the deployment of the airbags. Likewise, the proper use of seat belts can be verified. Gesture control is already one of the more familiar use cases. Images in the visible spectral range can be used for video telephony and for sharing photos/selfies.

17.2 Cameras for Driver Assistance Systems

As shown in the previous sections, the range of applications for camera systems in vehicles is very diverse. As a result, there are also a wide variety of camera system variants. In the following section, some aspects for the design will be discussed in more detail.

Basically, an object or scene is projected onto an image sensor by imaging optics. The pixels of the image sensor convert the photons into an electronic output signal that is processed by a processing unit. In the case of a direct presentation to the user, the output is displayed on a screen.

17.2.1 Criteria for the Design

For the design of a camera system, both the analysis of the individual parts and the complete system is necessary. Many performance parameters are influenced by several parts of the system.

The lens of the camera system is very important for a good overall performance. Among other components, it affects the possible resolution, field of view, depth of field, color reproduction, and the sensitivity of the system. Since an optical system never produces a perfect image (see ▸ ▸ Sect. 3.2), potential errors, such as a distortion, must be corrected.

The optical image is converted by the image sensor into digital values. Therefore, the design and adaptation of the optical system to the image sensor are crucial to image quality. The sensor mainly influences the resolution (number of pixels), field of view (number and arrangement of pixels), dynamic range, color reproduction, and especially the sensitivity (see ▸ ▸ Sect. 3.3).

▫ **Fig. 17.3** Field of view of interior cameras for **a** driver monitoring and **b** gesture control

In the next step of the processing chain, the image quality is affected by the image processing steps in the processor. In addition, the performance of the computer vision algorithms crucially depends on the performance of the processing unit.

17.2.1.1 Field of View

The *field of view (FOV)* of a camera plays an important role in the application and is essentially defined by the lens and the image sensor. One distinguishes the field of view in horizontal and vertical direction (HFOV, VFOV).

In front view camera systems, usually the horizontal field of view is the most important. However, the different assistance functions require differently wide horizontal fields of view. Relatively large HFOV values are needed in case of lane detection (in tight curve scenarios) and object detection (e.g., for the detection of crossing vehicles or pedestrians that are running onto the road). Values of more than 100° are meaningful for these applications (see ◻ Fig. 17.4a).

The vertical field of view is determined primarily by the mounting height and the minimum detection distance at close range. This distance is important for the lane detection function and when detecting objects, as the entire object may not appear in the field of view. An exemplary calculation for a passenger car results in an angle α of 18° below the horizon at a height h of 1.3 m and a distance d of 4 m (see ◻ Fig. 17.4b).

$$\alpha = \tan^{-1}(h/d) \tag{1}$$

For the vertical field of view above the horizon, it is decisive whether objects at short distances and large heights have to be detected in the application e.g., traffic lights or persons close to the vehicle.

Surround view systems almost exclusively use camera modules with a horizontal field of view of > 180°. This is due to the desired 360° representation of the vehicle environment. In order to calculate an overall image from the individual images, an overlap between the cameras' fields of view is necessary.

In the field of driver monitoring, the imaging of the head is important. Considering different anatomical requirements and installation situations, a necessary field of view of approx. 40°- 50° results. For gesture recognition, e.g., via a camera module in the roof function unit, larger (> 50°) fields of view are usually selected to allow users more freedom in gesture control. For interior monitoring, a horizontal field of view of at least 100° is required. If 'hands on the steering wheel' are to be detected or body models of the front occupants are to be created, horizontal fields of view of > 140° and vertical fields of view of > 100° are required.

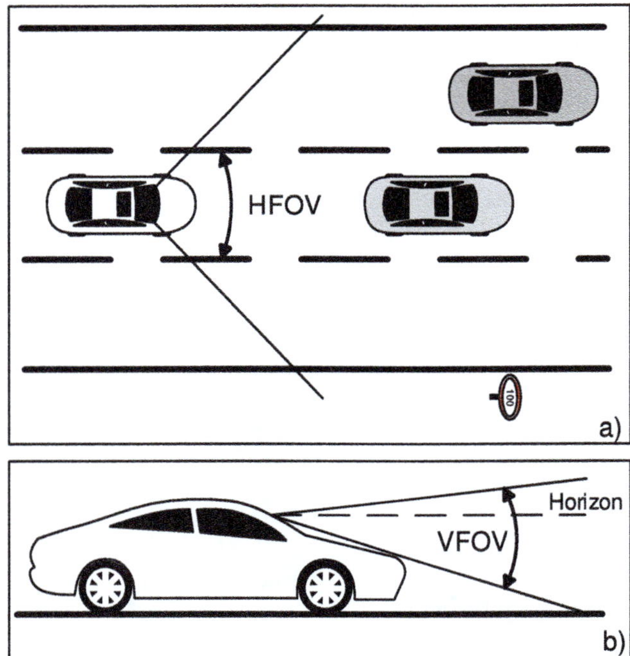

◻ **Fig. 17.4** Vertical **a** and horizontal **b** field of view of a front view camera

◘ Fig. 17.5 Effect of decreasing resolution using the example of a traffic sign ($480 \times 650/72 \times 96/36 \times 48/24 \times 32/18 \times 24/12 \times 16$)

17.2.1.2 Resolution

The possible resolution of a camera system is a complex interplay of the resolution of the lens and the image sensor as well as the image processing. For the design of a camera system, a purely theoretical analysis is necessary as a first step. Here, the object to be resolved (e.g., a vehicle 100 m away) is analyzed and the necessary resolution for the image processing defined (e.g., 10 pixels per vehicle width). Then, using the geometric relations, a necessary resolution in pixels per degree is calculated. Usual values are > 20 pixels per degree in the case of driver assistance functions in the field of environmental detection. In particular, traffic sign recognition has high demands toward the overall resolution (e.g., to recognize additional characters). In ◘ Fig. 17.5, the effect of different resolutions is shown. While the form and the warning symbol can be extracted from the right images, this will not work with pictogram and text in the supplementary sign. In the area of driver monitoring, higher resolutions may be necessary, e.g., to realize an eye tracking. An important aspect in choosing the optimal resolution is—in addition to the geometric requirements – the available computing power for image processing.

17.2.1.3 Color reproduction

In the area of advanced driver assistance systems (ADAS), the widely used CMOS image sensors are predominantly sensitive in the visible (VIS) and the near-infrared (NIR) spectral region. A separation in different color channels is realized via color filters on the image sensor (see ▸ ▸ Sect. 3.3).

How ▸ ▸ Sect. 17.1 describes, the color reproduction is of great advantage for applications in the field of front view and surround view cameras. While in surround view systems, a realistic representation of the camera image on the monitor is very important; in front view applications, the ability to distinguish between individual color channels is critical.

An example of the importance of color separation is shown in ◘ Fig. 17.6. While a camera system with color information (a) is clearly able to distinguish between white and yellow lane markings, in the case of a monochrome image (b), this is no longer possible.

17.2.1.4 Dynamic Range

The dynamic range (DR) of a camera system describes the ability to record both dark and bright areas in the image. In dark areas, the dynamic range is limited by the noise limit of the image sensor and in bright areas by the saturation limit of the image sensor. In addition to the image sensor, the dynamic range of the camera lens and the optical path defines the overall system dynamic range. The dynamic range of the lens is adversely affected by stray light. In addition, effects such as ghosting and flare can occur, which reduces the image quality especially in strong front light situations. Elements in the optical path such as the windshield can limit the total dynamic range of the system.

In the area of driver monitoring a design goal is to eliminate the portion of the unwanted light in the visible range as far as possible since this is determining the necessary dynamic range of the system. By using an artificial illumination in the near-infrared wavelength range and a band-pass filter in the lens of the camera this can be realized.

Traffic situations in the field of driver assistance systems are showing large differences in brightness levels. Examples are scenes with low-standing sun (see ◘ Fig. 17.7), entries/exits of tunnels and parking garages, or oncoming vehicles at night. A pavement marking at night can exhibit a luminance L of < 10 cd/ m², while the headlights of a vehicle in the same scene can have a luminance of up to 100,000 cd/m² [6]. Due to these issues, a very high dynamic range of the camera system and at the same time a high signal-to-noise ratio (SNR) is necessary. With an appropriate design (see also ▸ ▸ Sect. 3.3), one can achieve more than 120 dB dynamic range within imaging systems. The dynamic range is defined as follows:

$$DR(dB) = 20 \cdot \log_{10}(L_{MAX}/L_{MIN}) \tag{2}$$

Fig. 17.6 Importance of color separation for the detection of lane markers

Fig. 17.7 Illustration of a traffic scene with a high dynamic range

17.3 Camera Module

Camera modules can be very different in design. A camera module is defined here as the combination of lens, image sensor, electronics, and packaging technology. It is of course possible to accommodate more components within the camera module assembly, such as image processors.

17.3.1 Construction of a Camera Module

The main elements of the camera module are the lens and the image sensor which are held by a suitable mechanical structure. Additionally, electronic components and a connection to the image processor are necessary. As shown in **Fig. 17.8**, the single elements are assembled into a compact module using different packaging technologies.

The camera lens consists of multiple single lens elements that are built into a complete lens system. Part of the optical system is often an infrared cut-off filter (IRCF), which only transmits the non-infrared parts of the light spectrum. The lower part of the camera module exhibits the image sensor, the printed circuit board (PCB), and the electronic components.

Of course, the design parameters such as resolution and dynamic range are critical for the basic design of a module. For a robust design, in particular the environmental impacts during the lifetime of the system—such

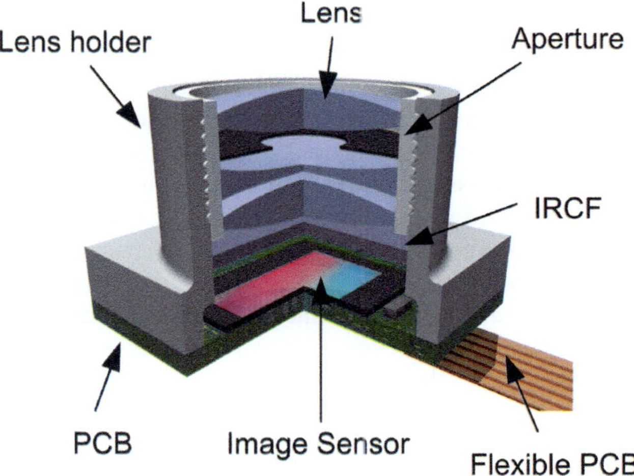

Fig. 17.8 Schematic structure of a camera module

as changes in temperature and humidity—are critical. The different applications lead to different requirements for the module design, since a camera in the area of surround view has direct contact with the external environment, while modules for front view and driver monitoring are only used inside the passenger compartment.

17.3.1.1 Electronics

Image sensors have a wide variety of analog and digital inputs and outputs. The most important are the power supply for the analog and digital parts of the image sensor, external timing signal (clock), digital control inputs and outputs, and interfaces for the configuration and image data transmission to the computing unit.

Analog voltage supplies that feed the pixel array or other functional blocks involved in image conversion require special attention. Fluctuations on these lines that exceed a certain amplitude are transferred directly to the image. At lower frequencies, this manifests itself as vertical wave motion in rolling shutter sensors see ▶ ▶ Section 17.3, and as pulsing of the entire image in global shutter sensors.

The transfer of settings (e.g., integration time) is realized via the configuration bus system with a low bandwidth. The transmission of the image data can be achieved via a parallel or serial interface. As data rates increase significantly with higher pixel counts and refresh rates, a trend is the usage of high-speed serial interfaces, such as the Camera Serial Interface (CSI) [7]. The data interface is then either directly connected to the image processing unit or the data is transmitted over longer distances via other interfaces (see also ▶ ▶ Sect. 17.4).

17.3.1.2 Packaging Technology

To ensure compliance with the requirements of the optical parameters over the lifetime of the camera module, the choice of suitable packaging technologies is very important. Firstly, the attachment of the image sensor to the PCB is of importance; on the other hand, the alignment of the lens system to the image sensor is critical. The image sensor and other electronic components can be assembled to rigid printed circuit boards, flexible printed circuit boards, or ceramic carriers. Image sensors are mounted onto the PCB in packaged or unpackaged form.

The alignment of the camera lens to the image sensor can be achieved using several mechanisms. Common are approaches that actively align the image sensor to the optical system via multi-axis adjustment and fastening with an adhesive bond. The biggest advantage is the precise adjustment. Alternatively, an adjustment only in the direction of the optical axis is possible, which is realized by means of a screw thread. However, a possible tilt of the optical axis to the image sensor plane will not be corrected with the latter method.

17.3.2 Optics

The camera lens typically comprises a stack of single lens elements, optical filter elements, and the overall lens housing. The applied lens designs in the automotive industry are a trade-off between optical performance, costs, and durability. This trade-off strongly influences the choice of materials of the single lens elements, as well as their number, and the choice of lens housing materials.

17.3.2.1 Design and manufacturing

Flint and crown lenses are the preferred optical materials for camera lenses. Plastic lenses are also used, but not for the entirety of the lenses in the lens system. The reason for this is the higher temperature dependence of the material properties of plastics and their lower resistance, for example, to scratches of the front lens in vehicle exteriors.

Glass lenses with spherical surfaces are usually produced by grinding methods. By means of alternative processes such as glass molding, lenses with aspherical surfaces can also be realized. Plastic lenses are manufactured by injection molding in spherical and aspherical shapes [8].

The lens barrel is generally made of metal or plastic and sealed against moisture penetration. To minimize image artifacts such as stray light, the housing is made of black material or is blackened. All optical surfaces are also coated with anti-reflective coatings, and an aperture ensures that no rays are scattered unintentionally at the edges. Ultraviolet and infrared filters are also used. This limits the light throughput to the visible spectral range. In the case of (near) infrared cameras, the filter is often narrow-band and matched to the wavelength of the active illumination.

17.3.2.2 Lens Design

The design of the optics is determined by the application. In particular, the characteristics such as field of view, light sensitivity, distortion, and sharpness play a role. The field of view of the optics is determined by the effective focal length. In the automotive environment, only lenses with fixed focal lengths are used. The focus position is also fixed and thus defines the available range of sharpness.

In most cases, a high **light sensitivity** of the camera system is required, so lenses with a small F-number, i.e., large aperture, are used to cope with poor lighting conditions and to achieve short exposure times. F-numbers around 2 and smaller are typical. More complex lens designs are required to achieve high light sensitivity while maintaining a good image quality.

The **sharpness** of a lens can be judged by the modulation transfer function (MTF). The MTF describes the contrast reproduction at different spatial frequencies in the image. In general, the detection of objects must be ensured without providing too much excess sharpness, as this is associated with additional lens elements and thus costs. The environment is not perfectly reproduced in the output image due to various influences. First, aberrations of the lenses in the lens limit the possible image sharpness, i.e., even an ideal point light source, such as a distant low beam at night, is not imaged as a point, but as an intensity distribution (point spread function—PSF) over a few micrometers [9]. Another influence is given by the division of the image into pixels, which discretize the PSF, or the overall image (sampling). Typically, one takes care in a camera design that the intensity distribution mentioned above covers approximately the area of a pixel.

An image through a lens always has a certain area along the optical axis where the image sharpness meets the requirements of the function. This so-called **depth of field** is always limited according to the low F-number. Since the focus is generally fixed, care must be taken to set it according to the area of application, e.g., to a greater distance for front cameras and to a short distance for surround view cameras. The camera system must have a large focus range to compensate for other effects such as temperature influences on the focal point.

The characteristics of a lens are negatively affected by **aberrations**. For example, spherical and chromatic aberrations, coma, astigmatism, and field curvature result in lower image sharpness because they together increase the PSF [10]. Chromatic aberrations such as lateral chromatic aberration affect color reproduction. The aberrations in optics can be minimized by optimized design and the use of aspherical lens elements [9]. In addition to aberrations, reflections and scattering from optical and mechanical surfaces in the optics can cause flare or ghosting [11].

Distortions in the range of a few percent can be tolerated because they are not relevant or can be easily corrected by software, and therefore distortion-free (and expensive) optics are not necessary. However, image distortions must be carefully compensated for when used the lens is used in a stereo camera system.

Optics for Front View Cameras

Older systems usually map a horizontal area around 50° in front of the vehicle. Due to increased safety standards and an expanded range of functions, newer systems rely on field angles of more than 100° HFOV. Depending on the vehicle equipment, these may be supported by one or two tele cameras with approx. 30° FOV in order to be able to detect objects with sufficient resolution even at high speeds and larger distances.

Optics for Surround View Cameras

In this area, a very large field of view is typically required. This is accompanied by greater image distortion, especially towards the edge of the image, which must be corrected. This correction is important because two or more cameras usually work together to generate an all-round view. For this panoramic view, correspondences must be found in the individual camera images to merge them correctly. Also, the decrease in brightness towards the edge of the image (vignetting) must be considered and compensated for. Typically, the first lens in the objective is exposed to direct environmental influences and must be designed to be extremely robust.

Optics for Driver Monitoring Cameras

Optical parameters for driver monitoring cameras are designed to image the person's head at a distance of about 40 cm to 100 cm from the camera. Since active lighting is used and a high depth of field is required, lenses with an F-number of approximately 2 are typically used.

To keep the amount of ambient light as low as possible, a narrow-band bandpass filter is built into the optics, which only transmits light in the wavelengths from 910 to 1000 nm. The active illumination of these systems also operates in this spectral range.

Optics for Interior Monitoring

The optical parameters of these cameras are designed so that they can image the entire interior of the vehicle, i.e., the front seats and the second row of seats. This means distances from 40 cm to approximately 170 cm. Since the interior cameras provide images in both the visible and near-infrared range, a dual bandpass filter is used to cover both spectral ranges.

17.3.3 Image Sensor

CMOS (Complementary Metal-Oxide Semiconductor) sensors, which are designed as APS (Active Pixel Sensor), are common today. "Active pixel" here means that each pixel converts its charges into a voltage via a source follower circuit. These sensors typically also include further amplifier circuits and analog-to-digital converters [12, 13].

CCD and CMOS-PPS sensors (PPS = Passive Pixel Sensors, without the above-described conversion in the pixel), which were also used in the past, are not considered further here.

In the following, the characteristics of sensitivity, resolution, increase of the dynamic range as well as the reproduction of color (by means of color filters) and shutter concepts of APS will be explained.

Since the image information is already digitally available to the sensor, it can be used both externally and internally, for example to create a histogram of the recorded scene, to implement automatic exposure control, to generate high dynamic range images or to guarantee functional safety.

A good basis for characterizing an image sensor is provided by standards such as EMVA 1288, which only consider the image sensor. For cameras, ISO standards, mostly from digital photography, such as the ISO 12231, 12232 and 12233 series (focus on exposure index, ISO sensitivity and resolution), ISO 14524 (OECF) or ISO 15739 (noise, signal-to-noise, and dynamic range) are suitable.

In the area of standardization for camera image quality assessment, the work of the *IEEE P2020*- Automotive Image Quality Working Group should be highlighted [14]. This working group has the goal to evaluate different aspects of a camera system in the automotive field and to make them comparable.

17.3.3.1 Shutter Concept

In digital cameras, there is usually no mechanical shutter that determines the integration time. An electronic "shutter" is used in which the pixel is reset to its initial value at the beginning of the desired integration time and the generated signal is read out at the end of the integration time. Global and rolling shutters are used today, which differ in their timing [15, 16].

Global shutter sensors integrate all pixels simultaneously and are therefore preferably used where active, pulsed illumination is applied, for example in driver monitoring. However, these sensors have the disadvantage of a more complex design, including complex sample and hold circuitry to store the information until it is read out.

In exterior applications, *rolling shutter* sensors are used almost exclusively. Here, the shutter "rolls" over the pixel array. The individual pixel lines are therefore recorded at different times, which leads to distortion of objects in the case of fast movements, or to different exposure situations within one recording in the case of high-frequency light sources. This disadvantage is accepted in favor of a simpler pixel design with better SNR compared to the global shutter and in addition more versatile possibilities to add HDR functions.

17.3.3.2 Sensitivity and Noise

The usability of the generated image data of the sensor is essentially influenced by its noise behavior, especially in dark areas of the recorded scene. Therefore, an important criterion in the selection of a sensor is its noise behavior over the required dynamic range and temperature range.

The **noise** of an image sensor is composed of temporal, photon shot and spatial noise [15–17].

At very low signal levels, temporal noise dominates, and is composed of reset noise, thermal noise, and quantization noise, among others.

Reset noise describes differences in the amount of charges at the start of integration and can be compensated by means of Correlated Double Sampling (CDS). Here, the reset level at the start of integration is also measured and subtracted from the actually generated signal.

Dark current noise is generated by charges created by thermal energy. With increasing temperature, this noise component increases nonlinearly, doubling at a temperature increase of about 7 °C.

At medium and high signal levels, *Photon Shot Noise* dominates, describing the statistical distribution of the number of incident photons. Photon Shot Noise is calculated from the square root of the generated signal and thus also determines the maximum signal-to-noise ratio. It is already generated outside the sensor and therefore cannot be corrected.

Quantization noise arises from the inaccuracy in the conversion of the electrical signal into a discrete digital signal. It can be minimized by a higher resolution, in the sense of dividing the same signal level into a higher number of digital values, of the analog-to-digital converter.

Spatial noise describes relative static differences in the offset and gain of individual pixels (fixed pattern noise—FPN) caused by variations in the current-to-voltage conversion of the active pixels, or as column FPN in the amplifier and A/D conversion circuits of the respective columns. As with dark current noise, there is a nonlinear dependence here on the temperature of the sensor [15, 16, 18].

Individual noise sources add quadratically to the variance. After applying the square root, the total noise is obtained. The result is that the largest noise source dominates, while others play only minor role. At higher signal levels, this is the photon shot noise mentioned above, which is therefore often used as a noise model for simplified calculations.

17.3.3.3 Resolution

In addition to spatial resolution, the quality of a digital video results from contrast resolution and temporal resolution. These describe the number of pixels onto which an object is mapped, the number of gray values into which a scene can be resolved, and the temporal distance between two images [16].

Spatial Resolution

To reconstruct a structure by a digital image sensor, it must be mapped to multiple pixels. The required number of pixels per angle is therefore determined by the smallest structure one wants to resolve. The total number of pixels is determined by the FOV and the desired resolution in pixels per angle ▶ ▶ Sect. 17.2.1.2.

To realize a high number of pixels on an image sensor, there are two options. The first is to keep the pixel pitch (distance between the centers of neighbored pixels) and increase the die size. The second approach is to lower the pixel pitch (and thus the pixel size) while keeping the die size. For cost reasons, the latter approach is chosen in most cases. From the perspective of the signal-to-noise ratio, keeping the pixel pitch is preferable, because with smaller pixels, also the filling factor further decreases. Furthermore, less temporal noise will be generated when pixel volume decreases, but it will not go down proportionally [16, 18]. Care must be taken to ensure that the disadvantages of decreasing the pixel size can be compensated by, e.g., new pixel design and improved manufacturing processes.

Contrast Resolution

For the detection of objects, the highest possible differentiation of object brightness is advantageous, for example to detect a dark-clothed person even at night. This is achieved by A/D conversion of the signal with a resolution (or bit depth) of 8–12 bits. HDR (High Dynamic Range) sensors often work internally with much higher bit depths, which are then compressed to a bit depth of typically 10–16 bits for simplified transmission, usually in piece wise linear (PWL) conversion.

During compression, care must be taken to keep the quantization noise to a minimum in relation to the noise of the uncompressed image. Here it is advantageous to let the compression factors increase in small increments over several linear sections.

With regard to the bit depth after compression, a trade-off must be made between the gain in quantization noise that results from a higher bit depth and the additional effort required for data transmission.

Temporal Resolution

The frame rate refers to the time interval between two images. A low frame rate poses the risk of not being able to react to events or reacting too late and it makes tracking of objects more difficult. A high frame rate increases the demands on the interface and further image processing. Typical values are around 20–60 frames per second. Temporarily, lower frame rates can be useful to allow for higher integration times. One example is parking scenarios with unfavorable lighting and low speed at the same time.

17.3.3.4 Dynamic Range and HDR Concepts

In the following, common HDR concepts are presented which are used in modern sensors in different combinations [19].

In the real world, situations with dynamic ranges of 120 dB corresponding to a contrast ratio of 1:1,000,000 can be expected. Linear sensors have a dynamic range of about 60–70 dB, so they often cannot represent the full dynamic range of a scene. High dynamic range HDR sensors achieve dynamic ranges of more than 120 dB by combining several linear images (hereafter referred to as partial integrations), which should happen simultaneously if possible.

The usable dynamic range of an image sensor is defined by the range of light intensities that can be digitally resolved, i.e., depending on the definition, from the clear distinction or equality of signal and noise to saturation (according to the EMVA1288 standard at $SNR = 1$). It is also crucial for the usability of the entire dynamic range that a sufficiently high SNR (transition SNR) is maintained in the transition areas from one partial integration to the next. Depending on the application, values between 20 and 30 dB are required here. This requires a large overlap range of two partial integrations, often at the expense of the maximum dynamic range.

Data sheet values for the maximum dynamic range of a sensor usually refer to individual noise sources (usually the readout noise) or even to the conversion of the internal bit depth into decibels (internally 24 bit $= 2^{24} = 16.78$ M $= 144$ dB), partly without taking the transition SNR into account and accordingly not necessarily representing a real application.

Ideally, a photodiode with a full-well capacitance (the measure of the maximum number of electrons a pixel can convert before going into saturation) of several million electrons would be able to directly generate a sufficiently high dynamic range. However, as of today, this is not (economically) possible. In the following, common HDR concepts are presented, which are used in modern sensors in different combinations. Here, the application and the technical implementation of the sensor determine which concept is most suitable.

Multi-Exposure

In this concept, several individual integrations with different sensitivities (influenced by integration time and gain) are carried out one after the other in time and this information is combined. In the case of moving objects, the multiple integration results in motion artifacts since they are located at different positions in the image during the individual integration times. In the case of pulsed objects (e.g., LED light sources), the individual partial integrations can be active during different switching states (on/off) and their respective determined gray values can be incorrectly interpreted and combined. Therefore, it is important for this method to have a small time difference between individual integrations and an appropriate combination scheme to form an HDR image [20].

The dynamic range of a single integration is in the range of 50–70 dB, the reasonably achievable dynamic range extension is about 12–25 dB per further exposure, limited by the minimum or maximum integration time as well as the desired quality of the signal in the overlapping area of two partial integrations (see introduction "Dynamic range and HDR concepts").

Split Pixel

In the split-pixel concept, one pixel is divided into two or more sub-pixels. These achieve different sensitivities through different sized photosensitive areas (typical ratios are 1:3 to 1:8), different gain levels, different integration times of the sub-pixels or the use of neutral density filters on individual sub-pixels. In most cases, a combination of several of these options is used to achieve the desired effect.

A major advantage of this concept is the temporally parallel recording of information in different dynamic ranges, which results in lower motion artifacts than is the case with multiple exposure methods. A disadvantage is more complex circuits for the two subpixels and thus a lower fill factor per subpixel, accompanied by weaker SNR values at low illumination.

In addition, the use of color filters often results in slightly different transmittances at different wavelengths for the larger versus the smaller subpixel, which can cause problems when combining the information.

Here, the dynamic range extension can be around 10–40 dB depending on the ratio between the large and small pixel, with very large extensions again causing problems in combining the information [21].

Dual Conversion Gain (DCG)

Dual conversion gain is often referred to as in-pixel gain and describes a process in which the generated charges are converted several times to a voltage by the pixel's source follower (SF). The pixel is internally switched between conversions [22].

This results in two possible conversion gains one for the darker image areas, another for brighter image areas.

With this method, the dynamic range of the linear pixel can be extended almost by the factor of the gains to each other. Realistically, about 20–30 dB can be expected, but often the sensor is designed for a much smaller extension of the dynamic range.

Full Well Enhance Capacitor (FWEC)

Here, during the integration of the pixel, charge is already diverted into an additional capacitor per pixel, for example via additional transistors in a lateral overflow mode.

In contrast to the classic lateral overflow, the charges are not "washed away" here, but are accessible for conversion into voltage and signal. Depending on the architecture of the pixel, the charges of photodiode and FWEC can be read out separately or combined.

Full well capacitances can reach a multiple of the full well of the actual photodiode and thus cancel the classical relation between pixel size and full well capacitance. However, they are also another source of noise, such as fixed pattern noise.

Pulsed Multiple Readout

In this procedure, the desired integration time is divided into several short partial integrations whose total time is less than the set one. For example, a total of 20 ms integration time is requested, which is divided into 20 partial integrations with 250 µs integration time each, in total 5 ms during the 20 ms total time.

This has advantages when capturing pulsed light sources, which are fully captured with a longer integration time, while a single integration of 5 ms can easily be in a pulse gap.

However, in the case of fast movements and strong contrasts, multiple edges occur because the individual integration times follow each other in time and an object moves in time. With a classic integration time, this would show up as motion blur.

Lateral Overflow

Here, only a certain level of charge in the pixel is permitted by means of partial resets (partial saturation), which is gradually increased in several steps over the course of the integration time, whereby the time difference for the renewed increase of the level becomes smaller and smaller [19]. For moving objects, motion artifacts result from the repeated partial integration (and superposition to form an overall image) when partial saturations are achieved.

The advantage of the method is that the highly dynamic information is available directly and therefore no information must be stored temporarily, which makes the method very suitable for global shutter sensors (see ▸ ▸ Sect. 3.2.2.4).

17.3.3.5 Color reproduction

A photosensitive element can only provide the information that electrons were generated by incident light. Information about the wavelength of the incident light is not directly available. Most digital image sensors are therefore in principle monochrome sensors and can only provide gray values.

In order to be able to assign color information to a pixel, various color filters (Color Filter Array—CFA) must therefore be inserted into the optical path so that a single pixel, in the case of the classic Bayer CFA (see ◘ Fig. 17.9), is sensitive only to the red (R), green (G), or blue (B) wavelength range. To give each pixel a color information, an interpolation with surrounding pixels with different color filters to color formats such as RGB, YUV or others takes place subsequently. However, the color filters also absorb a large part of the incident light power, which leads to a lower effective sensitivity. So, depending on whether sensitivity, sharpness or color fidelity are the focus of development, different combinations of the selected color filters are recommended.

Between a monochrome sensor for maximum sensitivity and sharpness to the classic RGB for maximum color fidelity in sufficient lighting conditions, the user has a variety of combinations of color filter materials, including the primary colors (R, G, B) and the complementary colors (C, M, Y).

The design of the blocking filters in the blue-UV and red-IR spectral transitions is also important. Most color filters lose their blocking effect in these areas, and, for example, a blue filter can let deep red light through to the pixel. However, since it is interpreted as "blue", this leads to problems with color reproduction. If color fidelity is the priority, an IR filter is selected accordingly so that these components are blocked. The opposite is true if IR components are desired, such as for active lighting in the passenger compartment.

17.3.3.6 Pulsed Light Sources and LED Flicker Mitigation (LFM)

If pulsed light sources are picked up by a sensor, two time-discrete systems meet, both with on and off times that are ideally synchronized. Typical frequencies for light sources are the mains frequencies of 50 Hz or 60 Hz, the legal minimum frequency for active traffic signs of 90 Hz (DIN EN 12,966) and typically multiples of 100 Hz. However, any other frequency can also occur in the recorded scene.

As a rule, even a large number of different frequencies within a scene must be covered. Thus, care must be taken in the design of the system to ensure a robust match of integration time and frame rate. For a known frequency of the light source, one chooses a frame period (1/fps) equal to that of the light source, or an integer multiple thereof.

The integration time is determined with the same methodology. Using the example of the 60 Hz mains frequency with a period of 16.6 ms, this ideally results in a frame rate of 16.6 or 33.3 ms corresponding to 60 or 30 fps with an integration time of 16.6 ms or 33.3 ms. For the European net with 50 Hz, this results accordingly in a refresh rate of 25 or 50 Hz and an integration time of 20 or 40 ms. The same applies to the known illumination by the own vehicle (reversing light, front headlights). This should be coordinated with the corresponding camera. If the integration time is too short, the active integration time of the sensor can fall into the pulse gap of the light source. The light source is accordingly not detected by the image sensor or only partially or only by parts of the image sensor that are actively integrating at this time (see also ► ► Sect. 3.1).

An arbitrary increase of the integration time is also not necessarily purposeful. If one sets a system to the minimum frequency of 90 Hz and determines the integration time as 11.11 ms, one will detect at least one pulse with a 100 Hz light source, but depending on the phase, a second pulse will be detected. In the image, this shows up as a pulsation in the brightness of the light source.

So-called LFM sensors try to provide the highest possible integration time even during daytime by increasing the dynamic range. The integration time should consist to the greatest possible extent of time-synchronous partial integrations (see also ► ► Sect. 3.2).

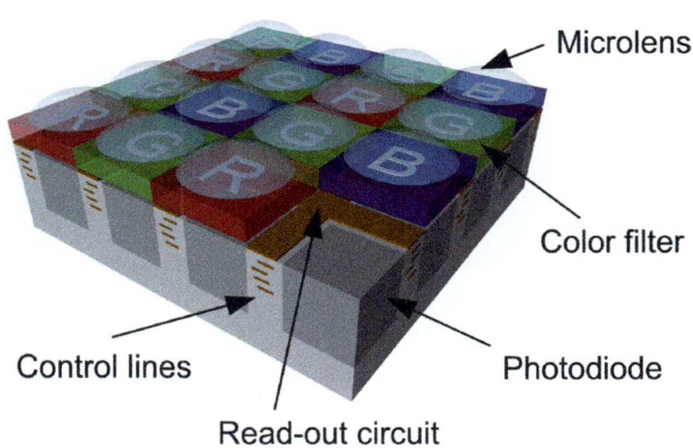

◘ **Fig. 17.9** CMOS image sensor with RGB color filters

This is illustrated by the following examples and Figures (Figs. 17.10, 17.11, 17.12, 17.13, 17.14, 17.15, 17.16). Figure 17.10 explains the principle, the other figures are examples of different situations with pulsed light sources.

A light source (yellow) pulses with a period P_LED. The duty cycle (DC) to P_LED in this example is about 1/5. Gray bars show the integration times of different lines of the image sensor, a longer integration time is represented by a longer bar.

For rolling shutter sensors the shifts of the bars among each other show the time offset of the image lines to each other and to the pulse of the LED (phase). The gray shades of the bars show the overlap of pixel integration time and the on-time of one or more pulses. For simplification only 5 states are shown here. A white bar (100%) indicates that two full pulses occurred during the integration of the pixel. Gray shades of 75, 50, and 25% indicate an overlap of integration time and LED pulses of 3/2 pulses, 1 pulse, and ½ pulse, respec-

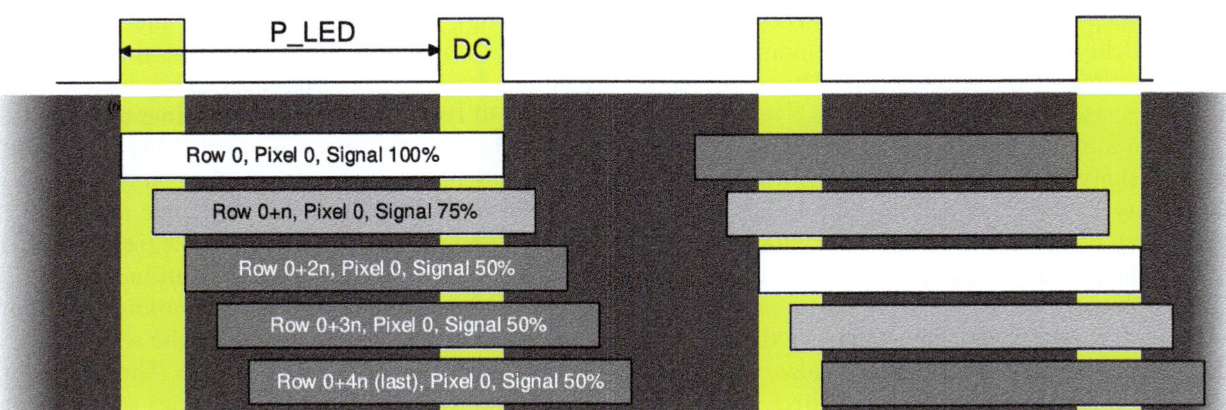

 Fig. 17.10 Schematic representation of different situations with flickering light sources

 Fig. 17.11 Example 1: Integration time longer than pulse period of LED and frame repetition period not equal to a multiple of pulse frequency ($T\_int > P\_LED$ and $P\_frame \neq n * P\_LED$)

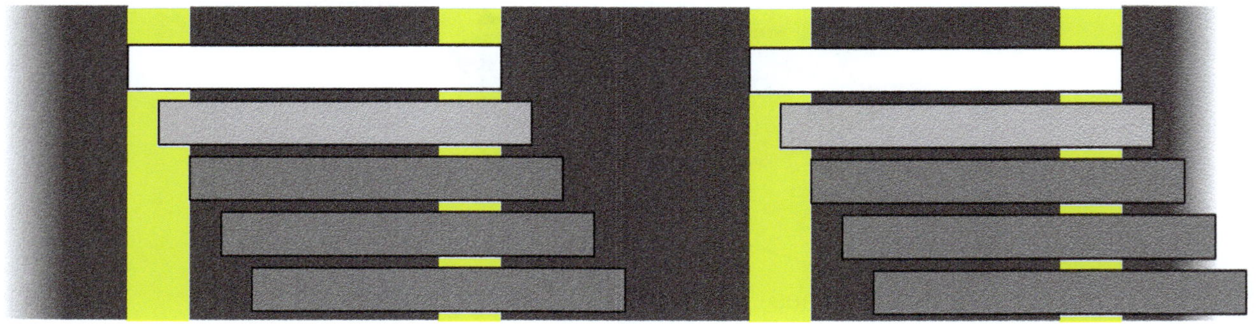

 Fig. 17.12 Example 2: Integration time longer than pulse period of LED and frame rate period equal to a multiple of pulse frequency ($T\_int > P\_LED$ and $P\_frame = n * P\_LED$)

Fig. 17.13 Example 3: Integration time shorter than pulse period of LED and frame period equal to a multiple of pulse frequency ($T\_int > P\_LED$ and $P\_frame = n * P\_LED$)

Fig. 17.14 Example 4: Integration time equal to the pulse period of the LED and frame repetition period not equal to a multiple of the pulse frequency ($T\_int = P\_LED$ and $P\_frame \neq n * P\_LED$)

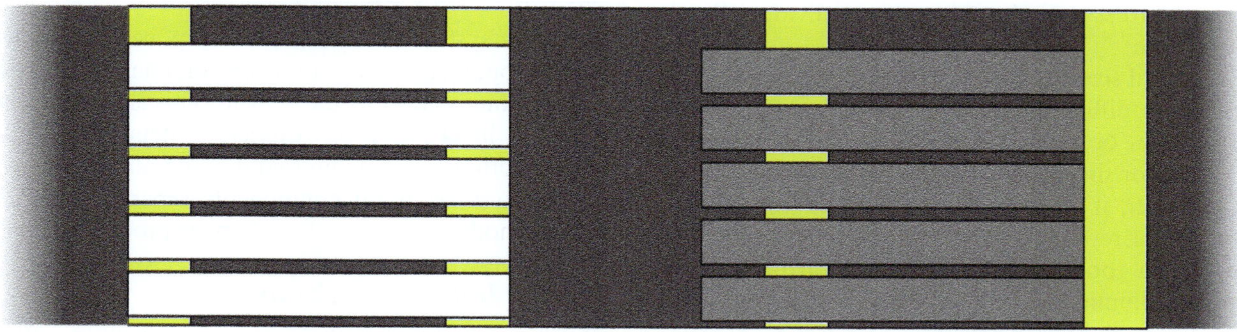

Fig. 17.15 Example 5: *Global-Shutter*

Fig. 17.16 Example 6: *Global-Shutter* with matched integration time

tively. With a black bar (0%), no LED pulse occurred during the integration time of the pixel. The graphs each show two consecutive shots of the sensor. The distance of the start points of two bars shows the frame period P_frame i.e., the reciprocal of the frame rate.

In Fig. 17.11a "wave" of different gray values appears instead of a uniform representation of the LED through all pixels. Since the periods of the image repetition and the LED pulse do not match either, this wave shifts from one image to the next.

Depending on the disproportion of the periods among each other, the result is a slow rolling of this gray value wave at nearly equal periods, while the effect jumps randomly at larger differences of the periods.

By tuning the pulse period and the frame repetition period, a "standing wave" of the gray values is achieved (see Fig. 17.12).

A too short integration time causes the sensor to integrate single lines between two LED pulses (black 0% bar). In this case, the sensor interprets this as if no pulse were present.

In the example (Fig. 17.13), the alignment of the refresh and pulse periods may even prove disadvantageous in this case since the sensor is permanently "blind" to this light source in the affected lines in this scenario.

Although the periods in Fig. 17.14 are not matched to each other, all lines appear with the same gray value, because here the integration time is equal to the frame repetition period. All lines see exactly one pulse regardless of what phase the individual line has to the pulse.

In a real scene, this is usually not achievable in this form since different light sources with different pulse periods can occur in a scene. The risk of integrating too short for single pulsed light sources is high.

However, if one knows the light source, as is the case with a reversing light, for example, then this approach is a good means of preventing artifacts. In this case, the illumination system can be matched with an appropriate rear view camera.

The *rolling shutter* effect is often held responsible as the sole reason for fluctuations in the image and a global shutter sensor is suggested as a solution to the problem. As example 5 shows (Fig. 17.15), however, in principle the same effects occur as in the previous examples with rolling shutter. The difference is that no gray value waves are formed here, but rather entire images change in intensity.

The integration time of the *global shutter* sensor must also match the pulse period of the LED (see Fig. 17.16).

17.4 System Architecture

To meet the demands for all the required functions, the system architecture requires the correct design of the hardware and software components as well as the image processing algorithms. In addition, there is the mechanical design of the system and the mechanical and electronic connection to the vehicle.

Since camera systems for driver assistance functions represent safety-relevant components in the vehicle (e.g., through brake intervention), the system must also meet the requirements of ISO standard 26,262 ("Road vehicles—Functional safety") [24]. Depending on the function, different ASIL levels (automotive safety integrity level) must be implemented.

17.4.1 System Overview

A camera system consists of the components for image acquisition, image acquisition control, image processing and communication to the vehicle. This is shown schematically in Fig. 17.17. Camera systems can be designed both as a single unit containing all components and as systems using components separately (e.g., a camera module with an external image processing unit).

17.4.1.1 Image Aquisition

Image acquisition is performed by one or more camera modules in the vehicle. The image data from the camera module is influenced by the image acquisition con-

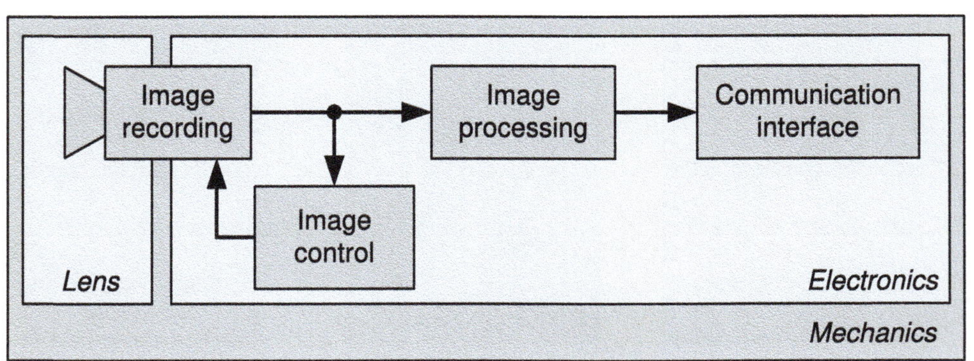

Fig. 17.17   System components of a driver assistance camera system

trol and then forwarded to the image processing system. Additional image (pre)processing functions can also be integrated directly on the image sensor. This is then referred to as a *System on Chip (SOC)* sensor. In this case, however, it must be weighed up whether the additional power requirement and associated heat generation, which leads to an increase in image sensor noise, makes the integration meaningful.

In the case of pre-processing on the image sensor or in the case of camera modules with an external processing unit, the image data is transmitted via special interfaces based on the LVDS standard (low-voltage differential signaling) or via Ethernet in compressed form (e.g., as MJPEG or in the H.264 standard). For future systems with high resolutions and frame rates, new standards such as MIPI A-Phy and Automotive Serial Alliance (ASA) will be applied [25, 26].

17.4.1.2 Image Acquisition Control

The control of the camera image sensor is necessary to set the optimal image parameters in all environmental situations. The two most important control systems are exposure control and white balance.

To adapt to different lighting situations, the exposure control is designed in such a way that structures can still be recognized in the dark areas of the image and the bright areas are not saturated. The prerequisite is a correspondingly large dynamic range of the camera module.

The white balance is controlled to achieve a constant color rendition with different color temperatures of the lighting of the scene. For example, white road markings must be recognized as white both in daylight and when driving through tunnels with partially yellowish lighting. For this purpose, the image content is analyzed, and the desired color rendering is set via a targeted weighting of the individual color channels.

17.4.1.3 Image Processing

In image pre-processing, the camera image is prepared, and initial processing steps are performed. If an RGB image sensor is used, a color image is reconstructed by a so-called demosaicing process. As a further step, a gamma correction is performed, i.e., the input values in the image are transformed into different output values via a non-linear transformation step. The background of this operation is either the adaptation to a certain reproduction system or also the improved image processing. In the first case, the display of the image, e.g., on a monitor, is optimized for the human eye by adjusting the image to the special characteristics of human vision (nonlinear perception of brightness differences).

When shown on a display, noise reduction, edge enhancement, and color corrections are also commonly applied [11, 27]. The image processing steps mentioned in this section (and others) are usually performed in a dedicated *Image Signal Processor* (ISP). An ISP is used to speed up the steps, since they are directly implemented in hardware functions and not done in software on a normal processor. A certain flexibility in the processing is nevertheless given by a possible parameterization or the omission of individual steps.

If the image data are used for machine vision tasks, distortion correction, optical flow calculation and, in the case of a stereo camera, rectification and disparity map generation are often performed. The desired information is extracted from the preprocessed images in image processing.

17.4.1.4 Communication

The data exchange with other controllers in the vehicle is realized via the communication interface of the camera system. Common vehicle bus systems are the *controller area network* (CAN) bus, the *FlexRay* bus, and the *Ethernet* standard. While CAN and FlexRay bus systems are only used to control the camera system and to transfer the output data in the form of, e.g., object lists, a transmission of raw image data is possible when using Ethernet because of the higher data rates.

17.4.1.5 Electronics

The design of the electronics follows the high standards in the automotive industry, among others, with regard to durability and electromagnetic immunity and compatibility. The large amounts of data that must be processed in real time using complex algorithms lead to a system design with several processors or multi-core processors [3]. The resulting amount of heat to be dissipated is a challenge in the vehicle installation space and must be considered already in the electronics and housing design.

17.4.1.6 Mechanics

The camera system housing forms the interface between the electronics and camera module and the vehicle. It must be thermally stable and easy to install. In addition, the housing usually forms the shielding of the electronics for better electromagnetic compatibility. In the case of the front-view camera, the housing is located behind the windshield. To avoid reflections at the windshield interfaces and on the housing, a straylight cover is often used between the camera module and the windshield. In camera systems where the camera module is in direct contact with the environment, the housing must also be sealed against moisture penetration.

17.4.2 Front View Camera Architecture

An exemplary mono camera architecture for a front view camera system is shown in ◼ Fig. 17.18. The image sensor is controlled via a communication bus and the image data are transmitted to the image processing unit via an interface. In most cases, a so-called *"system-on-chip"* (SOC) is used here. An SOC can combine all the necessary components for image acquisition, image processing and automated driving functions on a single chip. The SOC also handles exposure control, windshield heating control, communication on the vehicle bus, and other control and monitoring functions. In other systems, *digital signal processors* (DSPs), *field programmable gate arrays* (FPGAs) or dedicated *application specific integrated circuits* (ASICs) are also used [27]. The image processing unit is supported by fast memory chips, which are used for intermediate storage of processed data and for storing multiple images when tracking algorithms are used.

17.4.3 Satellite Camera System Architecture

In contrast to a highly integrated front view camera architecture, satellite camera systems are increasingly being used. The first systems of this type were surround view systems that acquire the input signals from four fisheye camera satellites in a central *electronic control unit* (ECU), process them and reproduce them in an output video. These systems are used primarily to support parking operations.

However, such centralized architectures are now becoming increasingly common in other areas. In some cases, front view cameras are being replaced by satellite cameras, and image processing is performed centrally. This has the advantage that fusion with other input data from cameras, radars, lidar systems and ultrasonic sensors can take place. This is particularly necessary for the automation of driving functions (SAE level 2–5).

Figure 17.19 shows the principle architecture of a satellite camera system. The central control unit includes one or more SOCs, microcontrollers, connections for data acquisition and communication, and power management units. Satellite cameras and other sensors are connected to the ECU via fast data links. In this case, the satellites are also supplied with power via the central processing unit, e.g., using "power-over-coax" technology. Since these architectures can be very different (e.g., an ISP function can be directly in the camera heads or centrally in the SOC), this is only shown here as an example. The structure of a central computer unit can be seen in Fig. 17.20.

17.4.4 Interior Camera Architecture

The interior camera represents a further development of the people monitoring camera. The first camera systems for people observation were characterized by a one-box design. This means that all functions and components are accommodated in a single housing. This unit is integrated into the steering column, for example, and has the advantage of looking directly into the face of the person driving the vehicle. This central arrangement leads to optimal results and high accuracy.

Since the available installation space inside the vehicle is always very limited, there is also the option of separating the camera head from the electronic control unit. This constellation is known as a two-box design. A cable provides the electrical connection between the two units. The ECU can be installed anywhere in the dashboard, thus increasing flexibility in the integration of the camera system. In addition, the actual camera head is much more compact, which allows it to be installed more inconspicuously.

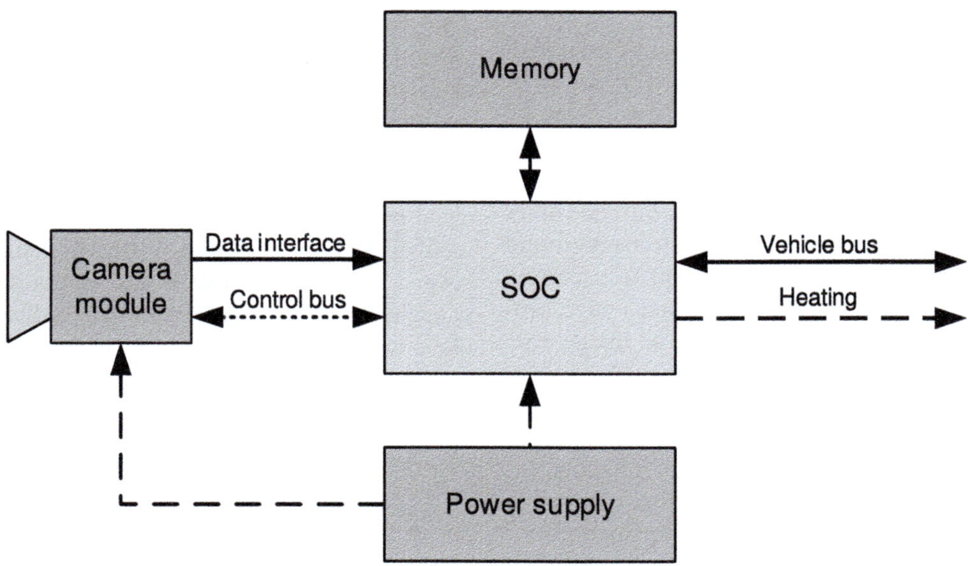

Fig. 17.18 Mono camera architecture

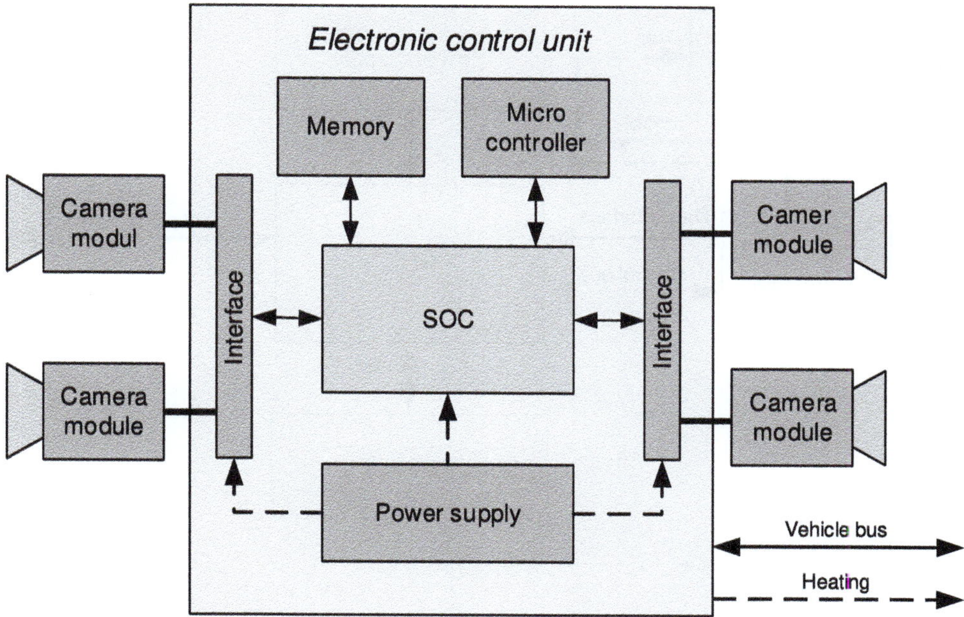

Fig. 17.19   Architecture of a satellite camera system

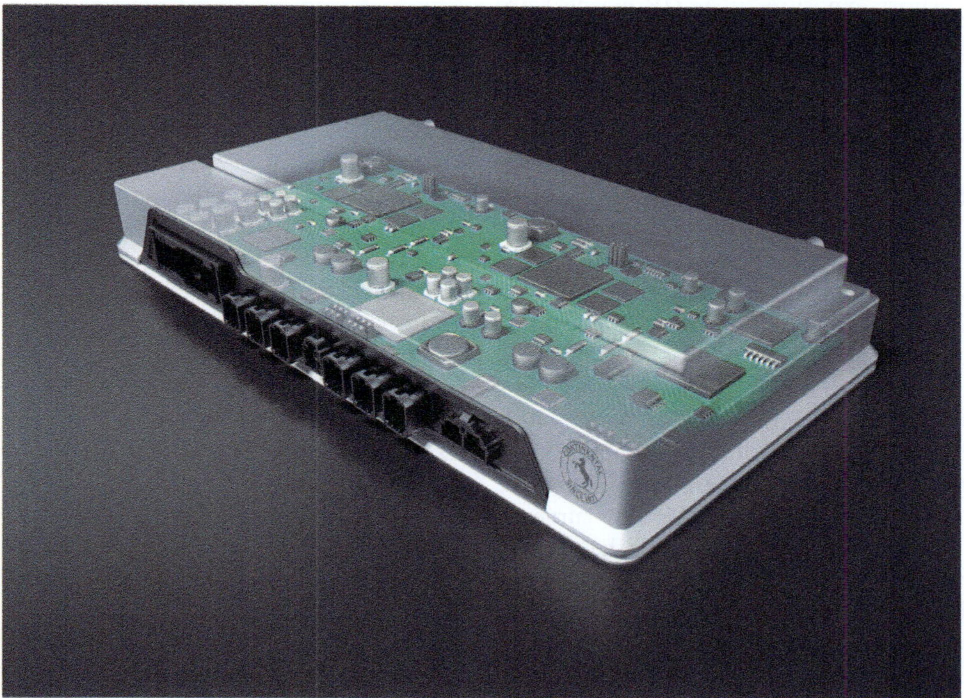

Fig. 17.20   Scheme of an electronic control unit

Alternatively, the camera system can also be integrated into other existing electronic systems. In this case, the instrument cluster is particularly suitable for a driver monitoring camera, while the central display, the roof control unit or the interior mirror housing is suitable for an interior camera. In these cases, it is referred to as an integrated solution. Ideally, the camera head fits seamlessly into the other component, and the evaluation electronics can also be integrated. Alternatively, the function of the camera ECU can also be taken over by an already installed high-performance computer.

Interior camera systems must function independently of existing natural ambient light. Therefore, camera sensors that operate in the infrared range are used in combination with artificial infrared illumination. The infrared radiation is generated by IR-LEDs

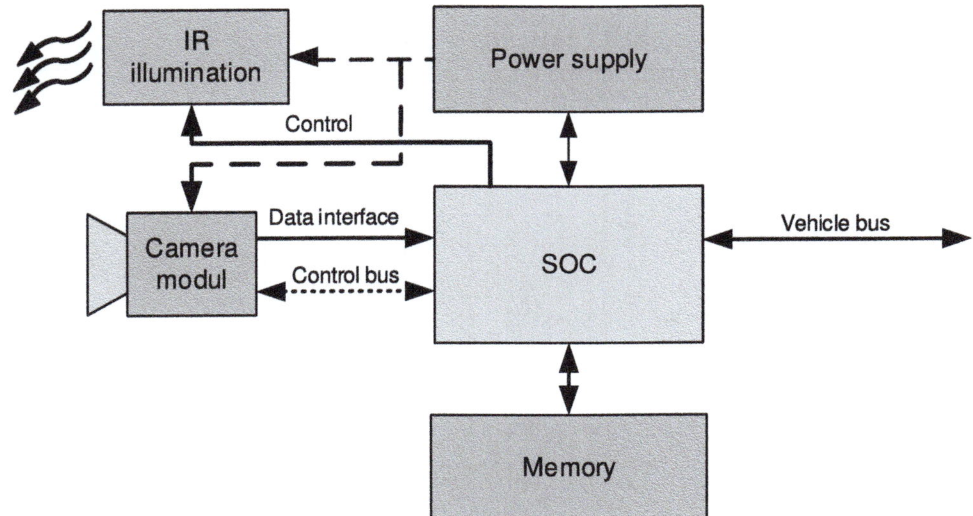

Fig. 17.21 Architecture of an interior camera

or IR lasers (e.g. in VCSEL technology), which are part of the camera head. Special attention is paid here to protecting the eyes from IR radiation exposure. Therefore, extremely short exposure times (IR flashes) are used (**Fig. 17.21**).

17.5 Calibration

To ensure that driver assistance functions like sign recognition, lane keeping assist, or head light control work in a reliable and correct manner, it is important to interpret images of the used camera system correctly. To this end, additional information about the images is necessary which can be determined by calibration algorithms. Such information, which we denote as calibration parameters, cannot only be used for better interpretation of the images but also for a compensation of deviations from a defined norm. One example might be the linearization of an image sensor's response curve.

This section provides a discussion about the calibration parameters that are typically determined for driver assistance systems. Furthermore, the section describes where a calibration normally takes place and how it can be conducted.

17.5.1 Calibration Parameter

Calibration parameters are variables of a model which is used to precisely describe the camera-based image acquisition system. For a specific system, the values of these parameters are found by using a calibration algorithm. One can distinguish between different classes of parameters, depending on the characteristics of the system to be modeled. The following list and **Fig. 17.22** give an overview on the most relevant parameters in the automotive field and their grouping:

Characterisation of the Camera Module
- OECF (*opto electronic conversion function*), image sensor response curve
- Noise due to dark current, defect pixels

Geometric Camera Calibration

Intrinsic Camera Parameter
- Focal length (f)
- Principal point (Cx, Cy)
- Distortion
- Pixel scale factors

Extrinsic Camera Parameter in Relation to Worlds Coordinate System
- Camera position
- Camera orientation

17.5.2 Calibration Environments and Calibration Procedures

Where and when a calibration is performed highly depends on the calibration parameters to be determined and thus the resulting requirements. In principle, a camera-based driver assistance system is calibrated already during the production process with respect to sensor characterization and intrinsic parameters. For doing this, target-based calibration setups are installed in production environments which enable measurements with a high reproducibility.

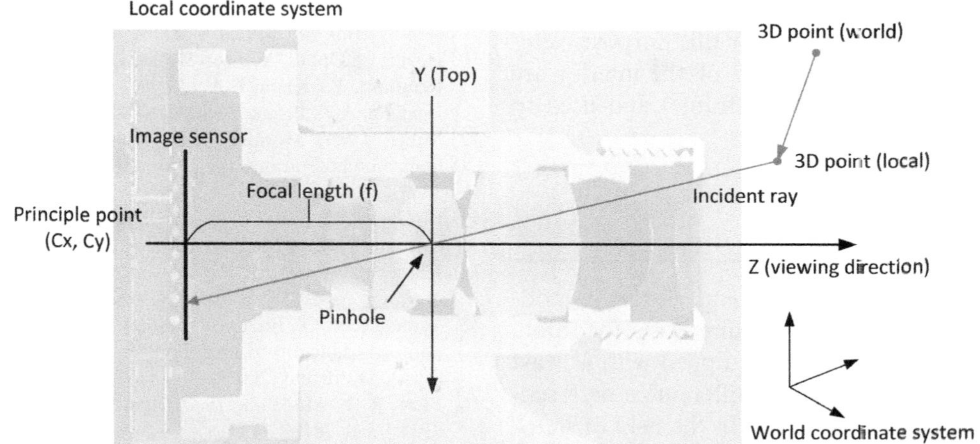

◻ **Fig. 17.22** Camera coordinate system

Camera Module Characterization—Camera Production Line

To determine the OECF, images with different but defined illumination intensities and constant exposure times are needed. Alternatively, the illumination intensity on the sensor remains constant, but the exposure time varies. Both ways have the same effect—the irradiation energy captured by the sensor is modified in a defined way such that the sensor response can be evaluated and plotted against the irradiation energy. With such a response curve, the dynamic behavior of the sensor is known. Potential deviations from the desired behavior can be compensated, and hence, fluctuations in the product quality are balanced out.

Intrinsic Camera Calibration—Camera Production Line

The calibration procedure during camera production can additionally be used to determine intrinsic parameters. Typically, a three-dimensional setup with checkerboard targets is used. With the known setup geometry and the positions of the checkerboard corners in the acquired images, the parameters of the model to describe the camera can be estimated. In most of the cases, a simple pinhole camera model plus a model for the lens distortion of low order is already sufficient. A well-known approach of determining these intrinsic (and additionally extrinsic) parameters is, e.g., the method described by Tsai [28].

In the case of a stereo camera module, the intrinsic parameters of both cameras are estimated in the production plant. Additionally, the position and orientation of the cameras to each other are computed. The transformation matrix from the right camera to the left camera is set so that feature correspondences extracted in the images satisfy the so-called epipolar condition.

Extrinsic Camera Calibration—Production Line of the Vehicle

During its production, the camera is not mounted at its final position in the vehicle. Hence, a calibration of the extrinsic parameters cannot be done at this point—but it can be done in the production line of the vehicle as soon as the camera is mounted. With a simple target which is placed at a known position with respect to the camera, the viewing orientation of the camera can be determined. The position of the camera is derived from construction data, or it is measured by using external devices.

Extrinsic Camera Calibration—During Vehicle Operation (Online Calibration)

Calibration in vehicle production is not always desirable for reasons of expense. Furthermore, the camera orientation is not constant relative to the world in the vehicle. Changing loading conditions of the vehicle can influence the orientation. However, the camera parameters must be exact even in this case. Therefore, the camera orientation is also calibrated during operation. Depending on the catch range of the process used and the installation tolerance, extrinsic calibration on the production line can thus be dispensed with completely. There are several ways to determine the calibration online. For example, it is convenient to use the results of functions that are already running on the driver assistance system, such as lane recognition, and calculate the calibration values from these.

In some vehicles, interior cameras are installed in positions that can be adjusted for design or integration reasons. For example, the driver monitoring camera on the steering column or the interior camera in the interior mirror. If the orientation of these cameras cannot be detected by (position) sensors, online calibration of the cameras must take place. For this purpose, vehicle-specific fixed points or contours of the interior are detected (e.g., B-pillar or window frame) and used to calculate the calibration values.

17.6 Outlook

Camera systems have become the main sensors in modern vehicles. Due to increased requirements for safety and comfort, all vehicles will be equipped with at least one camera in a few years. Many different camera sensors will be used in many vehicles. In the field of automated driving, the number of cameras will often exceed 10. This trend toward more and higher-quality cameras naturally also requires improvements in the area of camera sensor technology, signal transmission and image processing. Central processing units with high-performance processors will play a decisive role in the future.

In the application of camera systems, the merging of aspects of machine vision and human vison can be seen. This requires an additional focus on all aspects of the camera architecture to be able to enable these use cases in the best possible way.

References

1. Loce, R.P., Berna, l.E.A., Wu, W., Bala, R.: Computer vision in roadway transportation systems: a survey. J. Electron. Imaging **22**(4), 041121, (2013)
2. Källhammer, J.: Night Vision: Requirements and possible roadmap for FIR and NIR systems. Proc. SPIE **6198**, 61980F (2006)
3. Stein, G.P., Gat, I., Hayon, G.: Challenges and Solutions for Bundling Multiple DAS Applications on a Single Hardware platform. Israel Computer Vision Day, (2008)
4. Raphael, E., Kiefer, R., Reisman, P., Hayon, G.: Development of a camera-based forward collision alert system. SAE Int. J. Passeng. Cars – Mech. Syst. **4**(1), (2011)
5. ISO/DIS 16505: Road vehicles—Ergonomic and performance aspects of Camera-Monitor Systems—Requirements and test procedures
6. Hertel, D.: Extended use of incremental signal-to-noise ratio as reliability criterion for multiple-slope wide-dynamic-range image capture. J. Electron. Imaging **19**(1), 011007 (2010)
7. Homepage der MIPI-Alliance: ► https://www.mipi.org/specifications/camera-and-imaging, Accessed on 10 Jan 2021
8. Fischer, R.E.: Optical System Design. McGraw-Hill, (2008)
9. Sinha, P.K.: Image Aquisition and Preprocessing for Machine Vision Systems. SPIE Press, Washington (2012)
10. Hecht, E.: Optics. Addison Wesley Longman, (1998)
11. Reinhard, E., Khan, E.A., Akyüz, A.O., Johnson, G.M.: Color Imaging. A.K. Peters, Wellesley, (2008)
12. Miller, J.W.Y, Murphey, Y.L., Khairallah, F.: Camera performance considerations for automotive applications, Proc. SPIE, **5265**, (2004)
13. El Gamal, A.; Eltoukhy, H.: CMOS image sensors. IEEE Circuits and Devices Magazine, **21**(3), (2005)
14. Homepage der IEEE P2020 working group: ► https://site.ieee.org/sagroups-2020/, Accessed on 10 Jan 2021
15. Yadid-Pecht, O., Etienne-Cummings, R.: CMOS Imagers: From phototransduction to image processing. Kluwer Academic Publishers, Dordrecht (2004)
16. Fiete, R.D.: Modelling the Imaging Chain of Digital Cameras, SPIE Press, 2010
17. Holst, G.C., Lomheim, T.S.: CMOS/CCD Sensors and Camera Systems. SPIE Press, Washington (2011)
18. Theuwissen, A.J.P.: Course "Digital Camera Systems"—Hand out, CEI.se, Finspong, (2008)
19. Darmont, A.: High Dynamic Range Imaging. SPIE Press, Sensors and Architectures (2012)
20. Solhusvik, J., Yaghmai· S., Kimmels, A., Stephansen, C., Storm, A., Olsson, J Rosnes, A., Martinussen, T., Willassen,T., Pahr, P.O., Eikedal, S., Shaw, S., Bhamra, R., Velichko, S., Pates, D., Datar, S., Smith, S., Jiang, L., Wing, D., Chilumula, A.: A 1280x960 3.75um pixel CMOS imager with Triple Exposure HDR. Proc. of 2009 International Image Sensor Workshop, (2009)
21. Solhusvik, J., Kuang, J., Lin, Z, Manabe, S., Lyu, J., Rhodes, H.: A Comparison of High Dynamic Range CIS Technologies for Automotive Applications, Proc. of 2013 International Image Sensor Workshop, (2013)
22. Aptina Imaging Corporation. "Leveraging Dynamic Response Pixel Technology to Optimize Inter-scene Dynamic Range": ► https://www.photonstophotos.net/Aptina/DR-Pix_WhitePaper.pdf, Accessed on 31 Jan 2022
23. Baxter, D.: A line based HDR sensor simulator for motion artifact prediction. Proc. of SPIE **8653**, 86530F (2013)
24. ISO 26262: Road vehicles—Functional safety
25. Homepage der MIPI-Alliance: ► https://www.mipi.org/specifications/a-phy, Accessed on 10 Jan 2021
26. Homepage der Automotive SerDes Alliance: ► https://auto-serdes.org, Accessed on 10 Jan 2021
27. Nakamura, J.: Image Sensors and Signal Processing for Digital Still Cameras. CRC Press, (2006)
28. Tsai, R.Y.: A versatile camera calibration technique for high-accuracy 3D machine vision metrology using off-the-shelf tv cameras and lenses. IEEE J. Robot. Autom., **3**(4), (1987)
29. Civera, J., Bueno, D.R., Davison, A.J., Montiel, J.M.M.: Camera self-calibration for sequential bayesian structure from motion. IEEE International Conference on Robotics and Automation ICRA, (2009)
30. Zhang, Z.: Determining the Epipolar Geometry and its Uncertainty: A Review. International Journal of Computer Vision, **27**(2), (1998)

Open Access This chapter is licensed under the terms of the Creative Commons Attribution-NonCommercial-NoDerivatives 4.0 International License (▶ http://creativecommons.org/licenses/by-nc-nd/4.0/), which permits any noncommercial use, sharing, distribution and reproduction in any medium or format, as long as you give appropriate credit to the original author(s) and the source, provide a link to the Creative Commons license and indicate if you modified the licensed material. You do not have permission under this license to share adapted material derived from this chapter or parts of it.

The images or other third party material in this chapter are included in the chapter's Creative Commons license, unless indicated otherwise in a credit line to the material. If material is not included in the chapter's Creative Commons license and your intended use is not permitted by statutory regulation or exceeds the permitted use, you will need to obtain permission directly from the copyright holder.

Machine Vision

Christoph Stiller, Alexander Bachmann, and Ole Salscheider

Contents

© The Author(s) 2026
H. Winner et al. (eds.), *Handbook Assisted and Automated Driving*,
https://doi.org/10.1007/978-3-658-45276-6_18

18.1 Image Creation

18.1.1 Perspective Projection

The projection of most cameras can be described by the model of a pinhole camera shown in ◘ Fig. 18.1. Its aperture is assumed to be small enough to produce a sharp image in the plane of the radiation pickup. In practice, the pinhole is replaced by optics that produce a brighter image. The geometric description of the projection is conducted in the so called 3d camera coordinate system $X = (X, Y, Z)^T$, whose origin is placed in the pinhole and referred to as *optical center*. The Z-axis—also referred to as the *optical axis* in the following—is oriented perpendicular to the image plane. The X- and Y-axes are perpendicular to it and parallel to the line and column directions of the radiation pickup, respectively.

Instead of the image coordinates in the plane of the pickup, it is mathematically more elegant to choose an image plane parallel to it at unit distance in front of the pinhole. The image in this plane differs from the real camera image only by a scaling with the negative focal length f, so that the image no longer appears rotated by 180°. This virtual image defined in the image coordinates $x = (x, y)^T$, is called the *image of a calibrated camera*. From the image one can derive the projection equation

$$\lambda \cdot \begin{pmatrix} x \\ y \\ 1 \end{pmatrix} = \begin{pmatrix} X \\ Y \\ Z \end{pmatrix}; \quad \lambda \in \mathbb{R}, \qquad (18.1)$$

which—as usual in systems theory—is formulated as a numerical value equation freed from physical units. As the most important consequence of this projection, a camera can only determine ratios (i.e. angles) between 3d coordinates. Absolute distances can only be determined from camera images if the scale λ is known. Therefore, cameras are called *scale-blind*. In the automotive environment, the installation height of the moving camera or the distance in a stereo arrangement are often known for scale reconstruction.

Particularly elegantly one writes the projection equation after introduction of *homogeneous coordinates* $\tilde{x} = (x, y, 1)^T$ as

$$\tilde{x} \simeq X, \qquad (18.2)$$

where '$\simeq$' denotes *equality up to scale*, i.e. for a nonzero real number $\lambda : \lambda\tilde{x} = X$ holds. The resulting model is depicted in ◘ Fig. 18.2.

In image processing, *computer coordinates* $u = (u, v)^T$ are usually introduced, whose coordinate origin lies in the upper left corner of the image and which are scaled in such a way that the distance between neighboring pixels is 1, such that all pixels have integer computer coordinates (◘ Fig. 18.3). Image coordinates and computer coordinates can be directly transformed into each other by

scaling with the focal lengths $f_x = \frac{f}{\Delta x}, f_y = \frac{f}{\Delta y}$ in multiple of the pixel pitch Δx, Δy and displacement around the *principal point* $x_0 = (x_0, y_0)^T$. Such a mapping becomes linear in homogeneous coordinates $\tilde{u} = (u, v, 1)^T$

$$\tilde{u} = C\tilde{x} \quad \text{with } C = \begin{pmatrix} f_x & 0 & x_0 \\ 0 & f_y & y_0 \\ 0 & 0 & 1 \end{pmatrix}. \qquad (18.3)$$

The four parameters of the *focal lengths*, (f_x, f_y) measured in multiples of the pixel pitch and the *principal point* (x_0, y_0) thus determine the *intrinsic calibration matrix* C and are called *intrinsic camera parameters*. Knowing these parameters, one can determine the angle between two arbitrary viewing rays described by computer coordinates u_1, u_2.

Finally, we formulate the transformation of world to camera coordinates, where the world coordinates X_W are rotated by the rotation matrix R and shifted by the translation vector t yielding $X = RX_W + t$. In homogeneous coordinates $\tilde{X} = (X, Y, Z, 1)^T$, $\tilde{X}_W = (X_W, Y_W, Z_W, 1)^T$ this equation also becomes linear

$$\tilde{X} = \tilde{M}\tilde{X}_W \quad \text{mit } \tilde{M} = \begin{pmatrix} R & t \\ 0 & 1 \end{pmatrix}, \qquad (18.4)$$

where the *extrinsic calibration matrix* $\tilde{M}$ represents the six degrees of freedom of a rigid motion in 3d space. Summarizing the Eqs. (18.2)–(18.4), which are all linear in homogeneous coordinates, the mapping of a point in 3d world coordinates to 2d computer coordinates is given

$$\tilde{u} = P\tilde{X}_W \quad \text{mit } P = CM. \qquad (18.5)$$

Therein, $M = (R, t)$ includes the first three rows of the extrinsic calibration matrix $\tilde{M}$. The 3×4 matrix P is called the *projection matrix* [10]. The unequal dimension of the matrix in the row and column direction, respectively, shows the information loss of the perspective projection.

18.1.2 Image Representation

While the projection described in the previous section generates a signal that is continuous in space, time and amplitude, images are digitized by sampling and quantization. Here, the pixel raster on the image sensor performs the spatial sampling. Since natural images contain unlimited high spatial frequencies at sharp edges, strictly speaking, the sampling theorem is violated. However, the sensitive areas of the individual pixels, the optics and often the A/D converter act as a low-pass filter, so that aliasing effects are largely suppressed. The gray values of an image are classically quantized linearly with 8 bits. However, higher dynamic ranges are desirable in the automotive environment, which is why linear quantization with

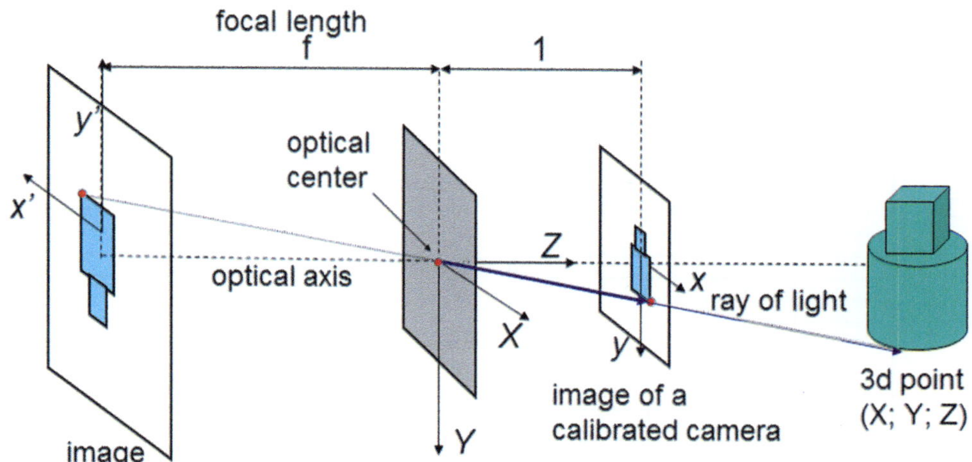

Fig. 18.1 Pinhole camera model for perspective projection

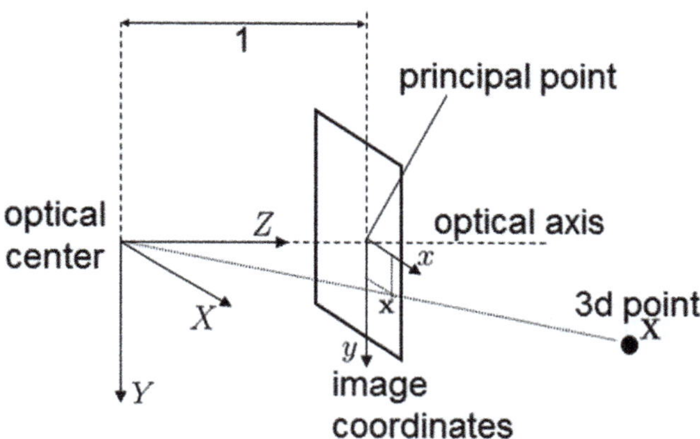

Fig. 18.2 Geometric camera model with perspective projection

up to 14 bits is used or non-linear quantization is applied. The value of a pixel can also represent other information, such as color, in addition to the gray value.

Moore's Law, according to which the computing power of control units doubles approximately every two years while complexity remains constant, seems to apply analogously to the number of pixels in image sensors. Accordingly, the cost relationship between camera and control unit would remain roughly constant. In any case, cameras already generate the highest data stream in today's automobiles. For example, a WSVGA image signal with 1024 × 600 pixels, a frame rate of 25 Hz and 8 bits per pixel in 3 color channels already produces a data rate of over 46 Mbyte/s. The development tends to increasing resolution of the cameras. From today's point of view, there is no physical limit that would stand in the way of catching up with the approximately 120 mega receptors on the human retina in the long term. In addition to programmable processors, digital logic modules and GPUs with a high degree of parallelization will gain in importance for processing these growing data volumes.

18.2 Image Processing

The term image processing refers to the preparation, analysis and interpretation of visual information contained in images. However, since the complexity of higher image processing processes, such as lane or object recognition algorithms, increases disproportionately with the amount of incoming data, these higher processes are usually preceded by large-scale image preprocessing and feature extraction. Image preprocessing aims to the reduce the noise contained in images and prepare the images for specific applications. Image signals can be manipulated in a targeted manner using a wide variety of filter operations or with the aid of transformations. Irrelevant or even disturbing information should be eliminated as far as possible. After successful preprocessing, relevant features can be extracted and fed to the respective higher evaluation process. ▶ Section 18.2.2 presents a selection of image features that are used in the context of automated driving.

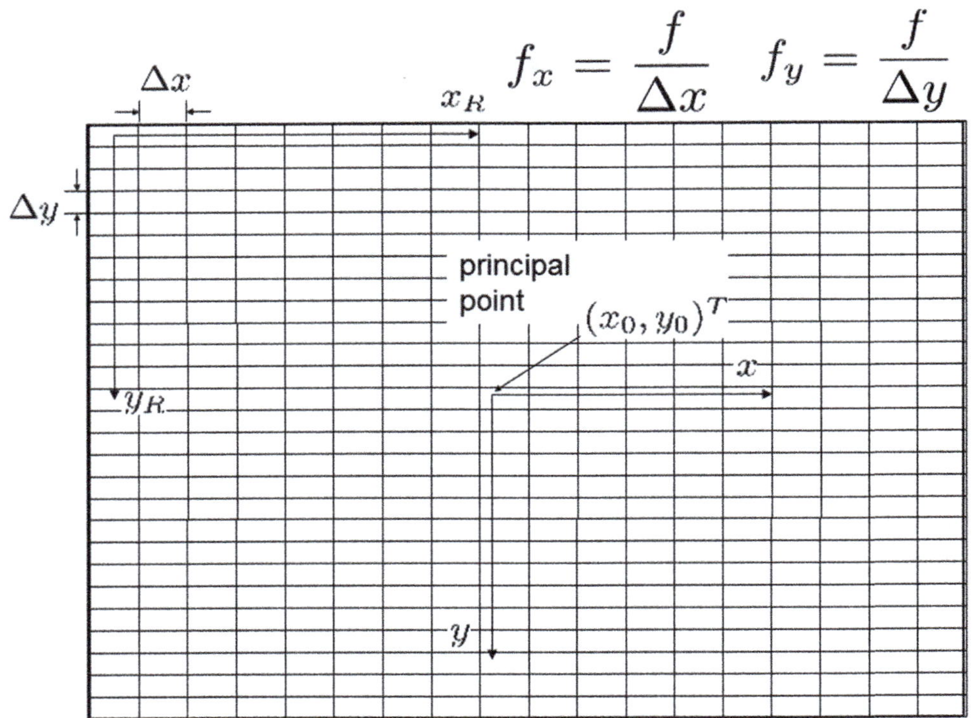

Fig. 18.3 Intrinsic parameters of a camera

18.2.1 Image Enhancement and Filtering

Image enhancement uses a wide variety of operations to enhance or extract relevant information from the recorded image signal and pass it on to the subsequent image processing process. The operations in image processing can be divided into three classes: Point operators, local operators and global operators. The classification is based on the number of pixels that influence an operation.

Point operators refer exclusively to an image point when processing the gray or color value of an image. Examples of this are histogram spreading, equalization or various thresholding methods. However, the spatial relationships of the gray values in the immediate vicinity are not captured here.

Local operators calculate a new color or gray value of a pixel based on a localized region around it. They are also called neighborhood operators and operate directly on the image signal. Examples of local operators are morphological filters, smoothing filters or gradient filters for highlighting specific textures. The important class of linear and shift invariant (LSI) operators can be mathematically described by the 2d convolution of the image $g(x, y)$ with a filter $h(x, y)$

$$f(u, v) = g(u, v) * h(u, v) = \sum_{i=-\infty}^{\infty} \sum_{j=-\infty}^{\infty} g(i, j)h(u - i, v - j).$$

$$(18.6)$$

In the case of local operators, the filter mask h has non-zero values only in a limited range. Since the impulse response then has a corresponding finite length, we speak of FIR (finite impulse response) filters.

An important local image operator is convolution with a smoothing filter. Gaussian or binomial filters are usually used for this purpose. The simplest binomial filter in one dimension is given by the convolution mask $[\frac{1}{2} \quad \frac{1}{2}]$. By repeated application of this convolution mask one yields higher order binomial filters. Thus, the sixth order binomial filter in one dimension is given by $\frac{1}{64}[1 \ 6 \ 15 \ 20 \ 15 \ 6 \ 1]$. For higher orders, binomial filters closely approximate Gaussian filters. The two-dimensional convolution masks necessary for application to images are not explicitly computed, since the filters are linearly separable. Instead, two-dimensional smoothing of an image with a binomial filter is implemented by successively applying a horizontal and vertical binomial filter in one dimension. A more detailed discussion of smoothing with Gaussian and binomial filters can be found in [24].

Global operators Global operators consider the entire image, as for example in the Fourier transform or the Hough transform. Geometric operators form a separate class of operators. They are responsible for the geometric manipulation of an image. Examples are displacement, rotation, scaling or mirroring. The intensity values of the image are preserved. ■ Figure 18.4 illustrates some examples of image operators.

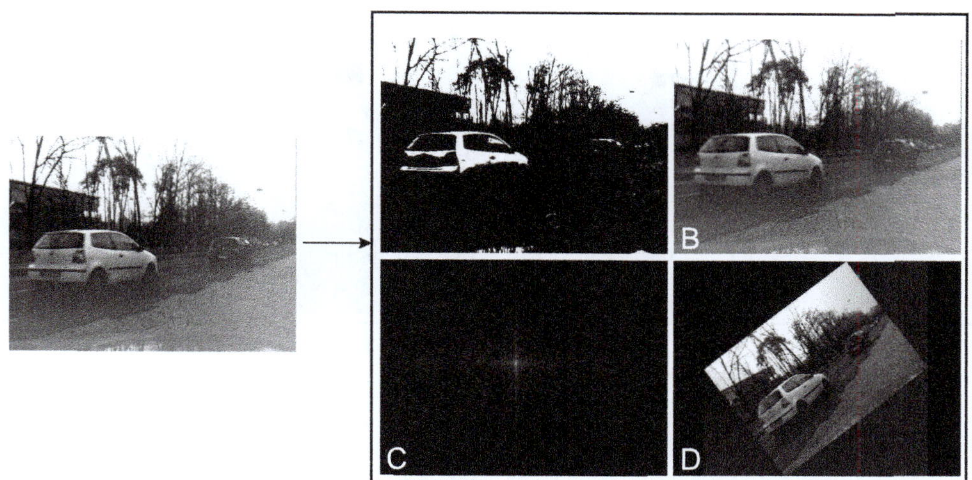

Fig. 18.4 Examples of image operators. Left: Original image; Right: **a** Binarization operator (point operator); **b** Binomial filter (local operator); **c** Fourier transform (global operator); **d** Rotation (geometric operator)

18.2.2 Feature Extraction

After the image signal has been corrected and processed by preprocessing, features can be extracted from the signal. Features are locally limited, meaningful parts of an image that provide a symbolic or empirical description of properties of the image or an object contained in the image. Features can be determined on various image primitives. For example, the gradient magnitude of the image signal is a feature for object contours. Object-typical distribution patterns of a measurable quantity provide an indication of image regions that depict the corresponding object. Thus, the goal of feature extraction is to highlight the structural properties relevant for the particular application from the extensive image information in a compact feature vector, while suppressing the information irrelevant for the application. This process leads to an enormous data reduction and finds its analogy in human visual perception. There, specific receptor fields are already formed by layers close to the retina, which extract edges, motion or local maxima, for example. Features can be roughly divided into the two groups of single image features and correspondence features: Single-frame features can be determined directly from the gray-scale pattern of a single image, while correspondence features relate the image positions of the projections of a real world point onto two or more images.

18.2.2.1 Single Image Features

Edges and corners contain essential image information. For example, people can already recognize image content excellently from line drawings that exclusively represent image edges. Accordingly, edges or corners are often extracted from a single image in image processing. Both features are characterized by an abrupt change of the image signal in one or two directions of the image plane. The gradient mathematically describes the local increase of the image signal. The most commonly used feature extraction algorithms are accordingly based on gradients. Since the image is only available in discrete spatial form, the partial derivatives of the gradient are approximated by FIR filters with usually few filter coefficients. A common approximation of the horizontal derivative is $\frac{1}{2}[1\ 0\ -1]$. The so-called *Sobel operator* uses this approximation and additionally smoothes with a filter $\frac{1}{4}[1\ 2\ 1]^{\mathrm{T}}$ in orthogonal direction. Overall, the Sobel operator thus approximates the derivative in the horizontal direction by convolution with the 3×3 filter mask

$$g_u(u, v)) = g(u, v) * h_u(u, v) = g(u, v) * \frac{1}{8} \begin{bmatrix} 1 & 0 & -1 \\ 2 & 0 & -2 \\ 1 & 0 & -1 \end{bmatrix}. \tag{18.7}$$

The corresponding image g_v filtered in vertical direction is obtained by convolution with the transposed filter mask, so that an approximation of gradient $\nabla g \approx (g_u, g_v)^{\mathrm{T}}$ is created. From the gradient image, edge detectors are used to extract connected line structures that locally exhibit maximum gradient magnitude. As a prominent representative the Canny edge detector is presented in the following [5]. First, a smoothing of the input image is performed for this purpose. This step serves to suppress sporadic noise-induced high gradients, which could lead to false positively detected edges. For this purpose, the image is convolved using, for example, the binomial mask described above. In the next step, the gradient of the smoothed image is approximated. From this, the local orientation α is calculated as the direction of a potentially existing edge in each pixel

Fig. 18.5   Left: Original image; Right: Result image of the **a** Canny edge detector; **b** Harris corner detector

$$\alpha(u, v) = \arctan\left(\frac{g_v(u, v)}{g_u(u, v)}\right) \qquad (18.8)$$

and quantized to multiples of $45°$. Furthermore, a measure for the edge strength—for example the magnitude of the gradient—is defined. As an intermediate step, a search for local maxima is now performed. Pixels are assigned an edge strength of zero if a neighboring pixel exists that is not in the edge direction and has a higher edge strength. This ensures that for the following processing, only edges that are not wider than one pixel are found. The calculated edge strengths of the remaining pixels are then compared with two predefined threshold values. If the edge strength is higher than the higher threshold value, the corresponding image point is marked as an edge. If the edge strength is lower than the lower threshold value, it is rejected as an edge. If the edge strength is between the two threshold values, then the pixel is detected as an edge if it is adjacent in the edge direction to a pixel previously classified as an edge pixel. In this way, hysteresis in edge detection is realized, resulting in a more stable detection of continuous edge features. The edges found can be examined in a next step, e.g. for specific shapes.

Corners are another primitive image feature whose localization is exemplified by the *Harris corner detector*. First, the gradient of the input image is approximated as described before. This is followed by the calculation of the *structure tensor*

$$\mathbf{S}(u, v) = \sum_i \sum_j w(i, j)$$

$$\begin{pmatrix} g_u^2(u+i, v+j) & g_u(u+i, v+j)g_v(u+i, v+j) \\ g_u(u+i, v+j)g_v(u+i, v+j) & g_v^2(u+i, v+j) \end{pmatrix}, \quad (18.9)$$

where w represents a weighting function. The eigenvector belonging to the larger eigenvalue of the structure tensor points in the direction of the local orientation and the eigenvalue represents a measure for the expression of this orientation. The eigenvector belonging to the smaller eigenvalue points accordingly in the direction perpendicular to it. Here, too, the eigenvalue is a measure of the orientation in this direction. With the help of the structure tensor, the image points whose structure tensor has two large eigenvalues can be detected as corners. For this the corner strength is calculated as $E(u, v) = |\mathbf{S}|(u, v) - \kappa \mathrm{sp}^2(\mathbf{S}(u, v))$ with $\kappa \in [0.04, 0.15]$. Pixels with a large corner strength are detected as corners. Alternatively to the calculation of the corner strength, the eigenvalues of the structure tensor can be examined directly. A corner is detected if both eigenvalues are sufficiently large. The detected corners can be used in a further processing step, for example, to search for correspondences between images. **Figure 18.5** shows results of the Canny edge and Harris corner detector. For a more detailed discussion on edge detection and the structure tensor, the reader is referred to [15, 24].

18.2.2.2 Correspondence Features

The knowledge of the projections of a real point in different images of a sequence of images or several cameras allows inference of its position in 3d space. Accordingly, the search for correspondences in several images is the basis for the reconstruction of the 3d information lost by the projection. The determination of correspondence features can be seen as a search task, where for an element in one view the corresponding element in the other view is determined [51, 60]. The search methods can be roughly divided into three classes: *Descriptor-based methods, gradient methods* and *matching methods*.

Detectors and Descriptors of Salient Pixels Although detectors and descriptors are now often learned using application-specific training data, processed descriptor-based methods such as SIFT [31], SURF [2], BRIEF [4], DAISY [59] and the learned descriptor DIRD [27] are still widely used and facilitate the understanding of corresponding search procedures. The search is typically performed in three successive steps: First, distinctive points such as corners [15] or distinctive regions (so-called blobs) [31] are detected in the image, which are likely found in other images of the same scene. For each of the selected points a descriptor is calculated in the second stage, which is as discriminative as possible and at the

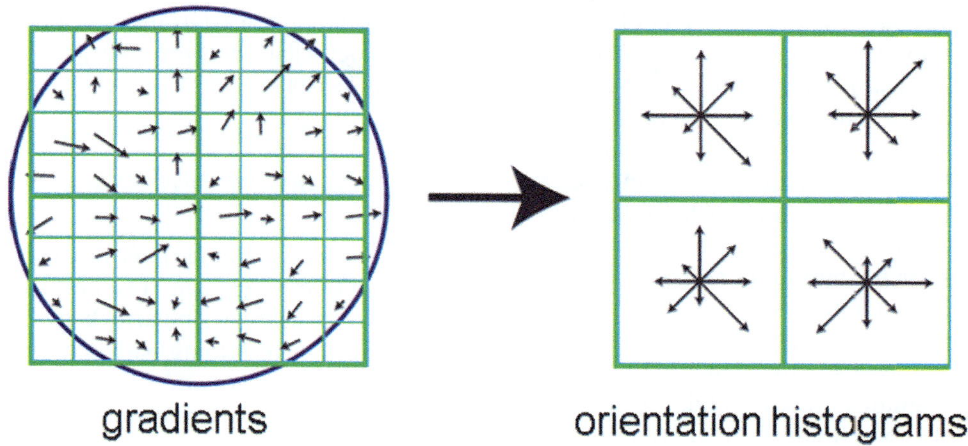

Fig. 18.6 SIFT descriptor of dimension $2 \times 2 \times 8$. Left: Gradients assigned to individual pixels and division into 2×2 squares; Right: Histograms of gradient orientations quantized to 8 directions

same time robust against illumination change, rotation, translation, change in scale as well as moderate distortions. The latter arise due to the perspective projection caused by different perspectives of the camera on a viewed object. Often, variants of the descriptors (U-SIFT, U-SURF) exist which exhibit only some of the invariances discussed. For example, in video sequences it is often useful to omit the rotation invariance altogether, since the rotation occurring between two time steps is usually negligible. With the DIRD descriptor, a method was proposed that learns a descriptor with an arbitrary selection of properties on the basis of given training examples. The third stage establishes correspondences between detected points in multiple images based on the similarity of their descriptors. A small Euclidean distance or a large scalar product of the corresponding descriptor vectors can be used as a criterion for a probable correspondence. While a simple method of assignment is to assign correspondences in descending order of their similarity measure, globally optimal methods such as the Kuhn-Munkres algorithm [26] (also known as the *Hungarian method*) may be used.

As a prominent representative for descriptor-based methods, the SIFT detector and descriptor is explained in more detail in the sequel. Other descriptors such as SURF, BRIEF, DAISY or DIRD follow a similar principle, but can offer computational time advantages through efficient calculations and approximations.

In the first step, the SIFT detector computes an image pyramid by successive low-pass filtering with a Gaussian filter and sub-sampling of the input image. Images of successive resolution levels of the Gaussian pyramid are finally subtracted so that each image of the pyramid represents the convolution of the original image with a Gaussian difference filter (DoG, difference of gaussians) with corresponding subsampling. Local minima and maxima are then determined at each level. Finally,

a so-called *non-maximum suppression* procedure reduces the number of points obtained in this way by requiring minimum distances. Subsequently, a descriptor is assigned to each remaining feature point. As shown in **Fig. 18.6** on the left, a local image area around the feature point is divided into equal squares for this purpose. For each pixel of these squares, the gradient is determined. A histogram of the gradient orientations is then generated for each square, with the directions quantized beforehand (**Fig. 18.6** right). To ensure extensive invariance to rotation, the histograms of the gray-scale gradients freed from the overall mean are used for the SIFT descriptor, i.e., the histograms are aligned relative to the dominant orientation of the gradients.

The resulting feature vector thus has the dimension of the product of the number of squares in horizontal and vertical direction with the number of quantization levels. In the example from **Fig. 18.6**, this is $2 \times 2 \times 8 = 32$ features. In his publication, Lowe [31] uses a vector with $4 \times 4 \times 8 = 128$ entries. In contrast to the direct comparison of intensities or gray value gradients, histograms guarantee a more robust feature description in the sense that it changes only slightly due to minor rotation, translation or distortion of the underlying image.

Gradient Methods In *gradient methods*, the intensity in the image is generally assumed to be constant along motion trajectories $\mathbf{u}(t) = (u(t), v(t))$. Then the continuous spatio-temporal function $g(t) = g(u(t), v(t), t)$ describes the intensity along such a motion trajectory and the total differential with respect to time must vanish. This gives the *continuity equation of optical flow*

$$0 = \frac{dg}{dt} = \frac{\partial g}{\partial u}\dot{u} + \frac{\partial g}{\partial v}\dot{v} + \frac{\partial g}{\partial t} = \nabla_{\mathbf{u}}^{\mathrm{T}}\dot{\mathbf{u}} + \frac{\partial g}{\partial t} \qquad (18.10)$$

where $\dot{\mathbf{u}}$ denotes the optical flow, $\nabla_{\mathbf{u}}$ is the spatial gradient and $\frac{\partial g}{\partial t}$ is the partial intensity derivative in time direction. This equation states that the image gradient vanishes along the motion trajectory. Equation 18.10 represents only a scalar constraint for the two sought parameters of the optical flow. Moreover, the partial derivatives in discretized image sequences can be determined only approximately. Therefore, for an unambiguous determination of correspondences with the help of Eq. (18.10), additional assumptions are needed. These consist mostly in simple models for the motion in an image region. In the simplest case it is assumed that the optical flow vector $\dot{\mathbf{u}}$ is constant in an image block. This assumption and the subsequent least square estimation of the optical flow leads to a gradient method [49], which is widely used and gives very good results for high frame rates and small motions. The fast and subpixel accurate correspondence determination of such methods is opposed by the disadvantage that Eq. (18.10) considers only infinitesimally small displacements. Gradient methods are therefore mostly used for determining the motion of points of temporally sequential image sequences in which a small optical flow can be assumed, as well as downstream of pixel-accurate correspondence methods, which may be based on the previously described descriptors or the matching methods described below.

Matching Methods In matching methods, correspondences are searched only within the image positions specified by the image grid. To find the correspondence to an image point, a small surrounding region from the first image is compared with corresponding regions placed around possible correspondence candidates in the second image. The correspondence is then given by the most similar region in terms of a given measure. Since the correspondence problem cannot be solved unambiguously within homogeneously or periodically textured image regions, it is useful to restrict the search space to distinctive image structures. In these so-called feature-based methods, the search space is restricted to a subset of features in the image. Features can be corner or line segments, among others. With the Harris corner detector presented above a frequently used feature detector has already been discussed. The region is often a symmetric window, for example a rectangular block, defined by the set of contained pixels in the reference image $\mathcal{B} = \{\mathbf{u}_1, \ldots, \mathbf{u}_n\}$. Corresponding methods are therefore often referred to as *block matching*. Let g_1 denote the reference image and g_2 the second image, respectively, and let the correspondence be described by the displacement vector $\mathbf{d} \in \mathbb{R}^2$. Then the *SAD* (sum of absolute differences) distance evaluates the absolute difference between two images blocks

$$J_{\mathrm{SAD}}(\mathbf{d}) = \sum_{\mathbf{u} \in \mathcal{B}} |g_1(\mathbf{u}) - g_2(\mathbf{u} + \mathbf{d})|. \qquad (18.11)$$

The SAD distance thus forms the L_1 norm of the *displacement-compensated frame difference*. Other norms

are also commonly used as distance measures. The correlation coefficient between the gray values of both image blocks can be used as a simple similarity measure

$$\rho(\mathbf{d}) = \frac{\sum_{\mathbf{u} \in \mathcal{B}}(g_1(\mathbf{u}) - \mu_1)(g_2(\mathbf{u} + \mathbf{d}) - \mu_2)}{\sqrt{\sum_{\mathbf{u} \in \mathcal{B}}(g_1(\mathbf{u}) - \mu_1)^2}\sqrt{\sum_{\mathbf{u} \in \mathcal{B}}(g_2(\mathbf{u} + \mathbf{d}) - \mu_2)^2}}. \qquad (18.12)$$

Here μ denotes the average gray value in the block of the image under consideration. The correlation coefficient thus forms a measure of similarity between image regions that is invariant to scaling with a constant factor and summation of a constant offset. Correlations obtained by maximizing this similarity measure are largely insensitive to changes in illumination.

18.3 Deep Learning

So far, classical image processing algorithms have been presented. Here, prior knowledge about the application domain and the problem to be solved is encoded in hand-engineered rules. For many applications, *artificial neural networks* (ANN) are an alternative that solves a given problem in a data-driven way: The model or structure of the neural network itself contains little prior knowledge about the application. Only by optimizing the numerous parameters of a model by representative training data, the necessary knowledge for solving the task is learned.

Model parameters $\theta = (\theta_1, \ldots, \theta_n)^{\mathrm{T}}$ are typically optimized by stochastic gradient descent or a variant thereof. For this purpose, a cost function is required that evaluates the accuracy of the model. During optimization, the model is continuously evaluated on training data and the costs are calculated using the cost function. Then the gradients of the cost function L with respect to the parameters θ are calculated and the parameters are updated by a step with step size η in the direction of the negative gradient.

$$\theta \leftarrow \theta - \eta \nabla L; \quad \nabla L = \left(\frac{\partial L}{\partial \theta_1}, \ldots, \frac{\partial L}{\partial \theta_n}\right)^{\mathrm{T}} \qquad (18.13)$$

For efficient computation of gradients, the backpropagation algorithm [44, 45] or generalized automatic differentiation in backward mode is typically used. A fundamental distinction of learning methods is made based on the required training data. If the cost function requires annotated training data to compute the cost, it is called *supervised learning*. *Self-supervised* and *unsupervised learning*, on the other hand, do not require any additional annotations.

The cost function is chosen according to the application. It can be very easy to calculate, like the mean square or absolute error, which are often used for simple regression tasks. On the other hand, complex cost functions are also common. In the case of *Generative Adversarial Net-*

works (GANs), the cost function is even given by a second neural network. The beginnings of artificial neural networks lie in the middle of the 20th century, when the perceptron was introduced [43]. It is the artificial counterpart to a biological neuron and computes the following generalized function:

$$f(x) = g(\mathbf{w}^{\mathrm{T}}\mathbf{x}) \tag{18.14}$$

The vector $\mathbf{x}$ contains the input data supplemented by a 1 as the last element to allow a bias. The vector $\mathbf{w}$ contains the weights to be optimized. The weighted sum then serves as input for a nonlinear *activation function* $g(\cdot)$. The original perceptron uses the Heaviside function for this purpose, but later ANNs use a variety of different functions.

If several perceptrons are arranged in layers where the input to the perceptrons of one layer are the output data of the perceptrons in the previous layer, a multi-layer perceptron (MLP) is obtained. This is visualized in ◘ Fig. 18.7. Single perceptrons or even a set of perceptrons arranged in a single layer cannot solve many tasks. For example, they cannot approximate the XOR function [32]. However, this is not true for MLPs: the *Universal Approximation Theorem* [7, 18] states that continuous forward neural networks with at least one hidden layer and enough neurons can approximate any continuous function arbitrarily well under mild conditions on the activation function. Thus the Universal Approximation Theorem motivates the use of KNNs. However, it is not constructive because it makes no statement about how many neurons are needed or how they should be arranged. It can be

empirically determined that a naive increase in the number of neurons, layers, and interconnections is often ineffective. The number of model parameters and the required computational power increases significantly, and the optimization problem becomes more difficult to solve. Therefore, a problem-specific model architecture is necessary, which has to be designed experimentally and with prior expert knowledge. This often stems from knowledge of the methods of classical image analysis.

Convolutional Neural Networks (CNN) are widely used in the field of image processing. As the name suggests, convolutions as introduced earlier in this chapter are performed in the layers of the CNN. Here, a neuron in one layer is not connected to all neurons in the previous layer, but only to those in a local neighborhood. The idea here, as already established by the FIR filters described earlier, is that information of the input signal is significantly localized in the image area. In the example of an object detector it is often sufficient to search different image sections for the desired object. It is not necessary to look at the whole image at once. In a CNN, moreover, all neurons of a layer share the corresponding weight vector. This dramatically reduces the number of parameters. The idea here is that many tasks are translation invariant or translation equivariant. In the object detector example, a neuron should respond to the presence of the object, regardless of its position in the input image.

In the last decade, the available computing power has increased by orders of magnitude, among other things due to GPGPU computing.[1] Together with architectural improvements, this has led to a breakthrough of ANNs in many areas of machine learning. In addition to the application field of image processing, for example, impressive results have also been achieved in the areas of Natural Language Processing and Reinforcement Learning, some of which exceed the capabilities of humans. In the field of machine vision, ANNs have become established for almost all applications and often define state-of-the-art results there.

A fundamental problem in image processing is image classification. Many CNN architectures exist that solve this problem. Architectures for other problems are usually derived from these architectures for image classification. The reason for this is that the classification problem lends itself to the search for new architectures: Several very large datasets are available and training the models is less computationally expensive than for many other applications. Well-known CNN architectures for image classification that achieve good results include various Inception architectures [54, 55], ResNet architectures [17], and MobileNet [19, 20, 47] as well as EfficientNet [56]. Most architectures consist of convolutional layers, activation functions, and normalization

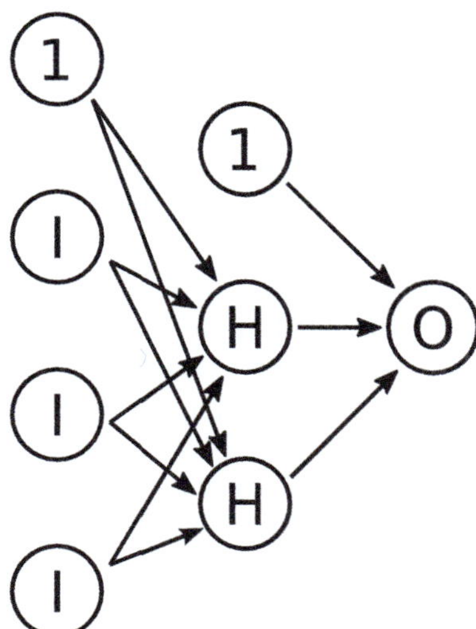

◘ **Fig. 18.7** Example of a multi-layer perceptron. Each node is a neuron and each edge is associated with a weight. The MLP contains input neurons I, bias neurons 1, hidden neurons H, and an output neuron O. The activation function (Eq. 18.14) is part of the hidden neurons and output neurons in this representation. The input neurons contain the input data

---

1 General Purpose Computing on Graphics Processing Units (GPUs) exploits their specialization for vector and matrix operations.

layers. However, they differ in the number, arrangement, and connectivity of these layers as well as their parameters. These hyperparameters have a decisive influence on the accuracy of the models as well as on the required computation time. Good hyperparameters close to the optimum have to be searched for experimentally, for example by a grid search.

CNN-based solutions exist for many other computer vision applications relevant to automated driving. They are frequently used for object detection, semantic segmentation of images or instance segmentation. Panoptic segmentation combines the last two tasks and provides an instance segmentation of all countable objects and a semantic segmentation of the rest of the scene. In object detection methods, a distinction is made between one- and two-stage detectors. One-stage detectors take an image as input and provide a list of object detections as output. Two-stage detectors on the other hand generate a list of object suggestions based on the input image in the first stage. In the second stage, these suggestions are classified and corrected again to obtain the final list of detections. Two-stage methods often achieve higher accuracy than single-stage methods, but also require more computing time. Well-known single-stage architectures include YOLO [39–41], SSD [29], RetinaNet [28] and EfficientDet [57]. Well-known two-stage architectures can be found in the R-CNN family [11, 12, 42]. Mask R-CNN [16] is also part of the R-CNN family. It extends the two-stage object detector with instance segmentation.

CNNs are also used for depth estimation. Both monoscopic and stereoscopic methods exist. Furthermore, CNNs can be used for optical flow estimation. Well-known works include MonoDepth [13, 14], GANet [63], FlowNet [9, 21], and PWC-Net [53].

There are countless other areas of application, although they are rarely used in the field of automated driving. However, some of the above-mentioned CNN architectures are in principle also applicable for structured lidar data or projections of point clouds into the Bird's Eye View. For unstructured point clouds, on the other hand, other ANN architectures exist, such as PointNet [37], which can handle the irregular data structure. ANN approaches to object tracking also exist, often integrated with the associated detectors. This allows the use of image information for object tracking, and thus a much less error-prone assignment of objects between the frames of a video. In this context, the learning of distance metrics with ANNs is also relevant. This can be used, for example, to assign image sections that contain the same object.

18.4 Reconstruction of the Scene Geometry

In many applications, cameras in driver assistance systems aim at the 3D reconstruction of the vehicle environment. With mono cameras, however, this is only possible to a limited extent due to the known *bearings-only ambiguity*. This fundamental property of projective imaging sensors means that by exclusively measuring directions from a moving sensor, it is only possible to determine a comparable motion of a tracked object down to one remaining degree of freedom. For example, for an automobile moving at constant velocity, this means that the position and speed of a likewise unaccelerated vehicle measured with a monoscopic camera will always have an indeterminable degree of freedom of one complete dimension. Although various heuristics can be used to determine this remaining uncertainty (e.g., driving over a plane with a known camera height), they hardly provide the necessary reliability for safety-critical driving functions.

In contrast, a stereo camera system directly provides depth information to almost all pixels and thus combines information about geometry and texture of a scene in one sensor. To describe a stereo camera system, the mathematical model of a monocular pinhole camera introduced in ▶ Sect. 18.1.1 is extended by a second camera. For both cameras, the intrinsic and extrinsic calibration described above is assumed to be known. In the following section the basics for stereoscopy are described. In the case of static scenes, stereo evaluation can also be performed from two images of a moving camera taken one after the other in time. The extension of the stereo evaluation for general temporally successively recorded images is presented in ▶ Sect. 18.4.2.

18.4.1 Stereoscopy

Stereoscopy[2] belongs to the passive methods of 3d scene reconstruction. Here, two or more images of the same scene are taken from different camera positions. From the position of a given scene point in at least two images, its spatial position can be determined given knowledge of the cameras' intrinsic and extrinsic calibration parameters. For the correspondence analysis itself, some procedures have already been described in the last section. In this section it will be shown that information about the calibration and position of the cameras to each other allows to impose further correspondence conditions, which allow an efficient stereo evaluation.

A stereo system consists of two cameras with optical centers O_l and O_r, which overlap in their field of view. Let X_W be a scene point that is imaged in both cameras. The indices l for left and r for right are used to distinguish the two cameras in the stereo setting. For practical reasons, the world coordinate system is often chosen to coincide with coincides with one of the two camera coordinate systems. In the further course we shall assume that the world coordinate system coincides with the camera

---

2   From Greek stereós: hard, firm, physical, spatial.

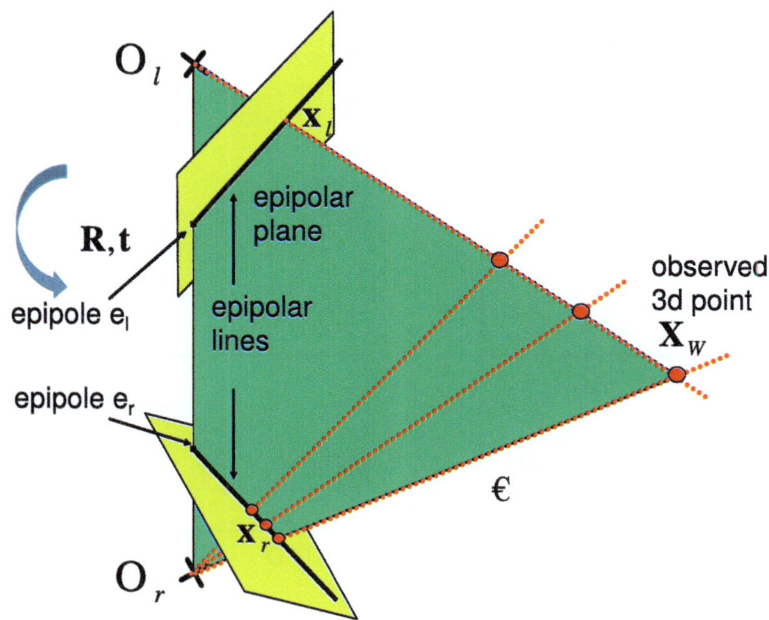

Fig. 18.8 Epipolar geometry

coordinate system of the right camera, i.e. the extrinsic parameters of the right camera result in $\mathbf{R}_r = \mathbf{I}$, $\mathbf{t}_r = \mathbf{0}$ and those of the left camera in $\mathbf{R}_l = \mathbf{R}$, $\mathbf{t}_l = \mathbf{t}$. The respective projection matrices are thus as follows

$$\mathbf{P}_r = \mathbf{C}_r[\mathbf{I}, \mathbf{0}]$$
$$\mathbf{P}_l = \mathbf{C}_l[\mathbf{R}, \mathbf{t}]. \tag{18.15}$$

The change in position and orientation of the left camera, relative to the world-fixed, right camera can thus be expressed by a rigid transformation. The transformation describes a transfer of the space point $\mathbf{X}_W$ from the right to the left camera coordinate system

$$\tilde{\mathbf{X}}_l = \tilde{\mathbf{M}}\tilde{\mathbf{X}}_r \quad \text{with } \tilde{\mathbf{X}}_r = \tilde{\mathbf{X}}_W, \tag{18.16}$$

where $\tilde{\mathbf{M}}$ denotes the *extrinsic calibration matrix* defined in Eq. (18.4).

The mapping of a point in 3d world coordinates $\mathbf{X}_W$ onto 2d computer coordinates of the left, respectively right camera $\mathbf{u}_l$, $\mathbf{u}_r$ is described in homogeneous coordinates with the respective projection matrix $\mathbf{P}$:

$$\tilde{\mathbf{u}}_r \simeq \mathbf{P}_r\tilde{\mathbf{X}}_W$$
$$\tilde{\mathbf{u}}_l \simeq \mathbf{P}_l\tilde{\mathbf{X}}_W \tag{18.17}$$

The three-dimensional displacement vector of the cameras is given as $\mathbf{t} = O_l - O_r$. It is often also referred to as the *base* with *base width* $b = |\mathbf{t}|$ and expresses the distance between the optical centers of the two cameras. In the choice of b, a compromise must be found between the highest possible depth resolution with large base widths and an increasingly difficult correspondence search, since occlusions occur more frequently with large

base widths and distortion effects are more pronounced in the images.

The intersections of $\mathbf{t}$ with the two image planes are called *epipoles* $\mathbf{e}_l$ and $\mathbf{e}_r$. The two optical centers O_l and O_r thus span a plane with $\mathbf{X}_W$, which is called the *epiploar plane*. Both image points $\mathbf{x}_l$ and $\mathbf{x}_r$ lie in this plane (**Fig. 18.8**). This geometric arrangement allows the subsequent formulation of the *epipolar constraint*, which reduces the search effort for stereo correspondences from the entire image plane to a line:

> **Epipolar Constraint**
> Let $\mathbf{x}_l$ and $\mathbf{x}_r$ be projections of the same point in space, then both lie on corresponding lines, the *epipolar lines*, which each run through the *epipole* in the same image and are assigned to each other by the *Essential Matrix* $\mathbf{E}$:
>
> $$\tilde{\mathbf{x}}_l^{\mathsf{T}}\mathbf{E}\tilde{\mathbf{x}}_r = 0 \quad \text{with } \mathbf{E} = \mathbf{t}_\times\mathbf{R}. \tag{18.18}$$
>
> Here, the antisymmetric matrix $\mathbf{t}_\times$ is defined in such a way that its multiplication corresponds to the cross product formation with the vector $\mathbf{t}$
>
> $$\mathbf{t}_\times = \begin{pmatrix} t_x \\ t_y \\ t_z \end{pmatrix}_\times = \begin{bmatrix} 0 & -t_z & t_y \\ t_z & 0 & -t_x \\ -t_y & t_x & 0 \end{bmatrix}. \tag{18.19}$$

Thus, for calibrated cameras, the essential matrix is completely determined by the position and orientation of the two cameras. It was first introduced by Longuet-Higgins [30].

For computer coordinates $\mathbf{u}_l$, $\mathbf{u}_r$ the epipolar constraint assumes a similar form with the *Fundamental*

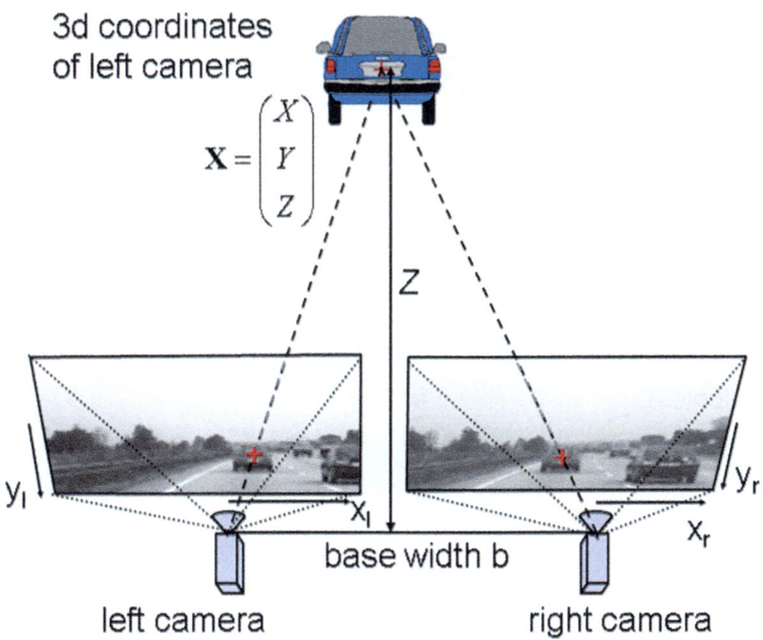

Fig. 18.9 Rectified stereo system: $\mathbf{R} = \mathbf{I}$, $\mathbf{t} = (b, 0, 0)^{\mathrm{T}}$

Matrix F

$$\tilde{\mathbf{u}}_l^{\mathrm{T}} \mathbf{F} \tilde{\mathbf{u}}_r = 0 \quad \text{with } \mathbf{F} = \mathbf{C}_l^{-\mathrm{T}} \mathbf{t}_\times \mathbf{R} \mathbf{C}_r^{-1}, \tag{18.20}$$

which contains both the intrinsic and extrinsic parameters of both cameras. For the determination of $\mathbf{E}$ or $\mathbf{F}$ from given pairs of correspondences, there is a plethora of linear and nonlinear approaches in the literature. Among the linear methods is the 8-point algorithm, which gives satisfactory results for sufficiently accurate point correspondences. However, it is shown in the literature that better results can be obtained by using classical nonlinear methods from numerical mathematics. For nonlinear estimation, for instance, the Gauss-Newton method or the Levenberg-Marquardt optimization can be mentioned [48]. The essential matrix has five degrees of freedom. These result from 3 rotation angles in $\mathbf{R}$ and 3 translations in $\mathbf{t}$, where the scale for the homogeneous matrix $\mathbf{E}$ is irrelevant and also not observable. Accordingly, a minimum of five point correspondences are required for its determination [36].

Equations (18.18) and (18.20), respectively, represent a necessary constraint for corresponding image point pairs, which reduces the search space to one dimension along the epipolar line. However, the generally oblique epipolar lines in the image plane present an unfavorable structure for the search. By a transformation called *rectification*, the mutually rotated stereo planes and camera coordinate systems are virtually aligned. Thus, after rectification, corresponding epipolar lines run horizontally and at the same height. As shown in ◘ Fig. 18.9 for rectified images of a stereo arrangement can be considered to be obtained from a stereo camera with $\mathbf{R} = \mathbf{I}, \mathbf{t} =$

$(b, 0, 0)^{\mathrm{T}}$. This is called an axis-parallel or a rectified stereo system. A detailed description of common rectification methods is given in [60].

Since correspondences are located in the same image row in an axis-parallel stereo system, the different perspective of the cameras with regard to the scene point leads to a purely horizontal shift in the image. This can be recognized directly by the following transformation of Eq. (18.2) for the spatial point $\mathbf{X} = \mathbf{X}_r = \mathbf{X}_l - \mathbf{t}$ in the respective image coordinates

$$Z \begin{pmatrix} x_r \\ y_r \\ 1 \end{pmatrix} = \begin{pmatrix} X \\ Y \\ Z \end{pmatrix} = \begin{pmatrix} X_l \\ Y_l \\ Z_l \end{pmatrix} - \begin{pmatrix} b \\ 0 \\ 0 \end{pmatrix} = Z \begin{pmatrix} x_l \\ y_l \\ 1 \end{pmatrix} - \begin{pmatrix} b \\ 0 \\ 0 \end{pmatrix}. \tag{18.21}$$

In it, the scale contained in Eq. (18.2) was determined directly from the third component of the vector equation. The second component of this equation forms the epipolar constraint for a rectified stereo arrangement. Finally, the first component allows the determination of distance

$$Z = \frac{b}{x_l - x_r}. \tag{18.22}$$

The displacement is called *disparity* after scaling with the referred focal length $\Delta = \frac{x_l - x_r}{f_x}$ and is given in the unit pixel. The 3d reconstruction of the searched point in space finally results as

$$\mathbf{X} = \frac{b f_x}{\Delta} \begin{pmatrix} x_r \\ y_r \\ 1 \end{pmatrix}. \tag{18.23}$$

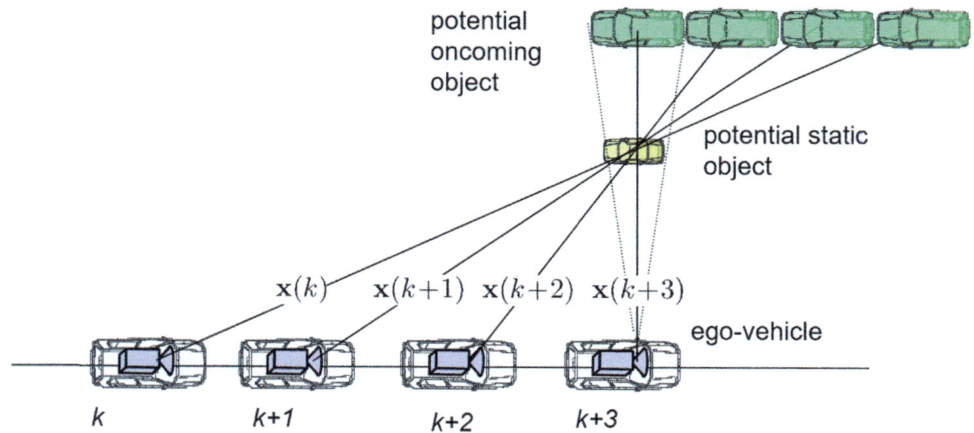

Fig. 18.10 The bearings-only ambiguity of a uniformly translationally moving monocamera: It is not decidable whether the observation sequence of the camera is caused by a static object or the exemplarily sketched moving object

Thus, the disparity is a measure for the spatial depth of the point $\mathbf{X}$ and is inversely proportional to it. For points at infinity in particular the disparity vanishes. By linear error propagation calculation the effect of small errors in the disparity estimate $d\Delta$ on the resulting error in the distance estimate can be approximated

$$dZ = \frac{dZ}{d\Delta}d\Delta = -\frac{Z^2}{bf_x}d\Delta. \qquad (18.24)$$

Thus for constant uncertainty in the disparity estimate, the uncertainty of the distance estimate grows quadratically with distance. Accordingly, the 3d reconstruction from disparity is highly accurate at close range, while it becomes useless for large distances.

18.4.2 Motion Stereo

In contrast to the classic stereo geometry, where two cameras are laterally shifted to each other, the so-called motion stereo method uses a single, moving camera. For non-moving environments, the 3d position of the corresponding points in space can be unambiguously determined from correspondences. However, if the requirement of motionless environments is not fulfilled, at least one ambiguity in one degree of freedom remains in the 3d reconstruction of position and motion in motion stereo techniques.

In the following we assume a coordinate system fixed to the camera. Due to the camera's own motion, a space point, which is assumed to be static, changes its position $\mathbf{X}(t)$ over the time t. Between time instances t and $t+1$ the space point moves by

$$\mathbf{D}(t+1) = \mathbf{X}(t+1) - \mathbf{X}(t). \qquad (18.25)$$

The 2d displacement of the corresponding image point on the image plane resulting from the camera movement is given by

$$\mathbf{d}(t+1) = \mathbf{x}(t+1) - \mathbf{x}(t). \qquad (18.26)$$

$\mathbf{d}$ denotes the displacement vector at position $\mathbf{x}$ in the image, caused by the movement of the camera relative to space point $\mathbf{X}$. For a camera system whose motion in space is described by the rotation matrix $\mathbf{R}$ and the translation vector $\mathbf{t}$, the trajectory of the point in space is given by

$$\mathbf{X}(t+1) = \mathbf{R}(t+1)\mathbf{X}(t) + \mathbf{t}(t+1). \qquad (18.27)$$

The epipolar constraint (18.18) restricts possible 2d displacement vectors of a rigid object to one dimension. Figure 18.10 shows the special case typical for automotive applications, where the rotational motion is negligible, making the rotation matrix an identity matrix: $\mathbf{R}(t+1) = \mathbf{I}$. Furthermore, it is assumed that the camera motion is oriented in the direction of the optical axis. The *bearings-only ambiguity* of monoscopic object detection becomes visible. From the image sequence of the monocamera it is not possible to decide whether a displacement trajectory $\mathbf{x}(t)$ has been created by a static object (Fig. 18.10 image center) or by an object moving uniformly parallel to its own motion (Fig. 18.10 drawn above). When estimating the position and motion of the object under consideration, a degree of freedom remains unobservable. Furthermore, for rigid objects with unknown motion, the epipolar constraint can only be effectively used to restrict the search space after estimating the epipolar matrix. As a consequence, corresponding points must first be searched for in the entire image space, which is associated with costly operations. In addition, it is not possible to determine a definite displacement vector $\mathbf{d}$ at many image points,

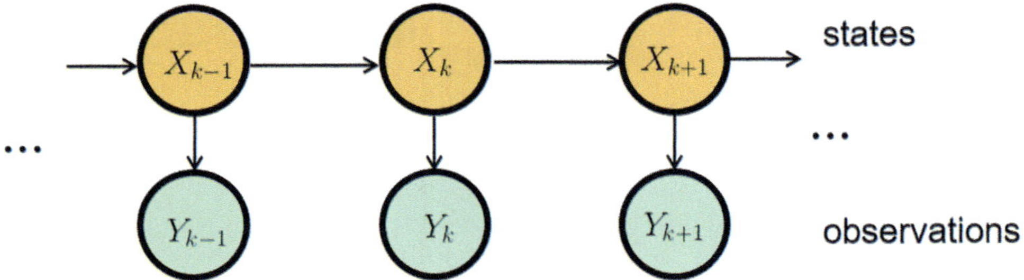

▫ Fig. 18.11   Directed dependency graph

since ambiguities caused by the aperture effect may not be resolved during projective mapping on the image plane. A decisive advantage of motion stereo compared to standard stereoscopy is the increasing base length with increasing time. The base length of stereo cameras, on the other hand, is limited for geometrical reasons. In motion stereo, the base length accumulates from the relative motion of the camera over time. This increases the accuracy and range of the sensor system with increasing time. Particularly attractive for automated driving is the simultaneous evaluation of stereoscopic disparity and motion stereo. While instantaneous stereoscopic 3d reconstruction in close range is possible, the motion stereo method accumulates displacement information with increasing time and thus achieves a high range [8].

18.4.3 Tracking

The loss of information associated with projective imaging can be partially compensated by temporal tracking of image features in 2d or 3d. Starting from the Bayes filter, which is the most general tracking method, the particle filter and the Kalman filter are described in this chapter as feasible approximations.

The task of temporal tracking is to estimate the quantities of interest $\mathbf{x}_k$ from observations $\mathbf{y}_k$. In general, both the observations and the quantities to be estimated are vector-valued. Observations and quantities to be estimated are time-varying and are acquired at discrete time steps $k = 0, 1, 2, \ldots$. Almost all filters for time tracking postulate the directed dependence graph shown in ▫ Fig. 18.11. In it, the sequence of states[3] $\mathbf{X}_k, k = 1, 2, \ldots$ forms a Markov chain with time index k and the observations $\mathbf{Y}_k$ are statistically independent of all other variables given knowledge of the state $\mathbf{X}_k$ at the same time. The relationship of the variables to be estimated with the given observations is then modeled in the state space.

$$\mathbf{x}_k = \mathbf{f}_k(\mathbf{x}_{k-1}, \mathbf{s}_k)$$
$$\mathbf{y}_k = \mathbf{g}_k(\mathbf{x}_k, \mathbf{v}_k) \tag{18.28}$$

The upper *system equation* describes the dynamics of the state $\mathbf{x}_k$. $\mathbf{s}_k$ is the realization of the stochastic system noise $\mathbf{S}_k$. The second so-called *observation equation* describes the emergence of observations $\mathbf{y}_k$ from the current state $\mathbf{x}_k$. Again, the equation includes uncertainty in the form of a realization $\mathbf{v}_k$ of the stochastic observation noise $\mathbf{V}_k$.

18.4.3.1 Bayes Filter

The Bayes filter represents an optimal state estimator, in the sense that at each time k the full probability density function $p(\mathbf{X}_k|\mathbf{Y}^k)$ of the state $\mathbf{X}_k$ is determined using all observations known up to that time[4] $\mathbf{Y}^k = (\mathbf{Y}_1, \mathbf{Y}_2, \ldots, \mathbf{Y}_k)$. This probability density function is often referred to in the literature as the *belief function* $\mathrm{bel}(\mathbf{X}_k) := p(\mathbf{X}_k|\mathbf{Y}^k)$ and written without explicit mention of the conditioning variables.

To derive the Bayes filter, one first formally rewrites the belief based on the definition of the conditional probability (Bayes' theorem)

$$p\left(\mathbf{X}_k|\mathbf{Y}^k\right) = \frac{p\left(\mathbf{X}_k, \mathbf{Y}_k|\mathbf{Y}^{k-1}\right)}{p\left(\mathbf{Y}_k|\mathbf{Y}^{k-1}\right)}$$
$$= \frac{p\left(\mathbf{Y}_k|\mathbf{X}_k, \mathbf{Y}^{k-1}\right) p\left(\mathbf{X}_k|\mathbf{Y}^{k-1}\right)}{p\left(\mathbf{Y}_k|\mathbf{Y}^{k-1}\right)}. \tag{18.29}$$

The denominator in Eq. (18.29) is independent of $\mathbf{X}_k$ and can therefore be taken as a normalization constant which ensures that the integral of the distribution density is normalized to 1. The second factor in the numerator of Eq. (18.29) is to be represented by a system model, i.e. by the system Eq. (18.28). For this, one first formally writes

$$p\left(\mathbf{X}_k|\mathbf{Y}^{k-1}\right) = \int p\left(\mathbf{X}_k, \mathbf{X}_{k-1}|\mathbf{Y}^{k-1}\right) d\mathbf{X}_{k-1}$$
$$= \int p\left(\mathbf{X}_k|\mathbf{X}_{k-1}, \mathbf{Y}^{k-1}\right) p\left(\mathbf{X}_{k-1}|\mathbf{Y}^{k-1}\right) d\mathbf{X}_{k-1}.$$
$$\tag{18.30}$$

---

3   In this section, capital letters denote random variables whose realizations are denoted by corresponding lowercase letters.

4   In the following, the high index.^{k} generally stands for the sequence of all corresponding quantities up to and including time k.

With the assumption expressed in the dependence graph of ◘ Fig. 18.11 that the state is closed, i.e., $\mathbf{X}_k$ is independent of all old observations $\mathbf{Y}^{k-1}$ for a given immediately preceding state $\mathbf{X}_{k-1}$ the *Chapman-Kolmogorov equation* is obtained

$$p\left(\mathbf{X}_k|\mathbf{Y}^{k-1}\right) = \int p\left(\mathbf{X}_k|\mathbf{X}_{k-1}\right) p\left(\mathbf{X}_{k-1}|\mathbf{Y}^{k-1}\right) d\mathbf{X}_{k-1}.$$
(18.31)

The first factor in the integral describes the actual system model.

The first factor in the numerator of Eq. (18.29) is described by an observation model, which forms the observation Eq. 18.28 of the state space model. In this, again the assumption is made that $\mathbf{X}_k$ is a closed system state. Mathematically formulated, the current observation $\mathbf{Y}_k$ is thus statistically independent of all previous observations $\mathbf{Y}^{k-1}$ for a given state $\mathbf{X}_k$

$$p\left(\mathbf{Y}_k|\mathbf{X}_k, \mathbf{Y}^{k-1}\right) = p\left(\mathbf{Y}_k|\mathbf{X}_k\right).$$
(18.32)

Substituting Eqs. (18.31) and (18.32) into (18.29) yields the recursive equation of the Bayes filter

$$p\left(\mathbf{X}_k|\mathbf{Y}^k\right) = \frac{p\left(\mathbf{Y}_k|\mathbf{X}_k\right) \int p\left(\mathbf{X}_k|\mathbf{X}_{k-1}\right) p\left(\mathbf{X}_{k-1}|\mathbf{Y}^{k-1}\right) d\mathbf{X}_{k-1}}{p\left(\mathbf{Y}_k|\mathbf{Y}^{k-1}\right)}.$$
(18.33)

The Bayes filter thus represents a recursive state estimator that starts from an initial estimate of the state $p\left(\mathbf{X}_0\right)$ and includes the following two updates at each time step
Prediction – In the prediction step, the estimate of the previous time step for $\mathbf{X}_{k-1}$ is projected to the current time. For this purpose, the Chapman-Kolmogorov integral in Eq. (18.31) is evaluated.

Innovation – In the subsequent innovation step, the prediction is improved by the current observation $\mathbf{Y}_k$, where the likelihood $p\left(\mathbf{Y}_k|\mathbf{X}_k\right)$ is evaluated before the integral in Eq. (18.33).

The Bayes filter is theoretically the best estimator for systems with closed state in the Wiener sense. However, the effort for the continuous estimation of full belief functions is often unjustifiably high. This is usually due to the large range of values of the state. Therefore, fast approximations of the Bayes filter are discussed in the following sections. The particle filter—also called Sequential Monte Carlo Sampling—approximates distributions by a set of particles. Assuming Gaussian distributed probability densities and a linear state space description, a special case of the Bayes filter is the Kalman filter, which allows for a particularly efficient implementation [1, 62].

Particle Filter To evaluate Eq. (18.33), particle filters approximate probability densities by a set of N particles (samples) $\mathbf{x}^{(i)}$ each assigned with a weight $w^{(i)}$ [22]

$$p(\mathbf{X}_k) \approx \sum_{i=1}^{N} w^{(i)} \delta\left(\mathbf{X}_k - \mathbf{x}^{(i)}\right), \quad \text{with} \sum_{i=1}^{N} w^{(i)} = 1,$$
(18.34)

where the particles are chosen randomly. Hence, not only the model but also the estimator is stochastic. The particles form 'typical' realizations for the distribution. Instead of the Dirac impulse, other kernel functions with pronounced maximum are used in different variations.

In the prediction step, each particle is individually predicted from time $k-1$ to the next Time step k propagated according to the Chapman-Kolmogorov Eq. (18.31). In that for each i a new sample is drawn from $p(\mathbf{X}_k|\mathbf{X}_{k-1} = \mathbf{x}_{k-1}^{(i)})$. If under the old particles $\mathbf{x}_{k-1}^{(i)}$ identical samples were included, they would generally be predicted differently by the system noise. The weights remain unchanged in the prediction step.

The innovation step follows the corresponding Bayes filter step from Eq. (18.33). The particles $\mathbf{x}^{(i)}$ remain unchanged in it, while the weights are adjusted to the new observation

$$w_k^{(i)} \propto p\left(\mathbf{Y}_k|\mathbf{X}_k = \mathbf{x}_k^{(i)}\right) w_{k-1}^{(i)}.$$
(18.35)

In order to keep the entropy of the approximated distribution as high as possible, eventually a third—so called resampling step—is introduced, in which particles with vanishing weight are removed and instead particles with very high weight are divided among several particles. This is done by randomly drawing N particles from the estimated distribution, which are then assigned identical weights.

Particle filters are usually considerably faster than Bayes filters and give almost the same results for a sufficiently large number of particles N. However, since the filter works stochastically, a strict convergence proof on the result of the Bayes filter is only possible for infinite particle number N.

Kalman Filter For the case where the system dynamics as well as the observation equation are linear and further the observation noise and the system noise can be considered as white and normally distributed, Eq. (18.33) can be easily and efficiently implemented in the so-called Kalman filter [62]. At each time k the resulting Gaussian distribution is completely represented by its mean $\hat{\mathbf{x}}_k$ and the covariance matrix $\mathbf{P}_k$.

In the *prediction step* of the Kalman filter, the estimate of the previous time step $\hat{\mathbf{x}}_{k-1}$, $\mathbf{P}_{k-1}$ is projected to the current time step using the linear system model

$$\hat{\mathbf{x}}_k^- = \mathbf{A}\hat{\mathbf{x}}_{k-1}$$
$$\mathbf{P}_k^- = \mathbf{A}\mathbf{P}_{k-1}\mathbf{A}^T + \mathbf{6}_\mathbf{R}$$
(18.36)

with the system matrix $\mathbf{A}$ and the covariance matrix of the system noise $\mathbf{6}_\mathbf{R}$. Finally, the subsequent *innovation step* accounts for the latest observation $\mathbf{y}_k$

$$\hat{\mathbf{x}}_k = \hat{\mathbf{x}}_k^- + \mathbf{K}\left(\mathbf{y}_k - \mathbf{C}\hat{\mathbf{x}}_k^-\right)$$
$$\mathbf{P}_k = (\mathbf{I} - \mathbf{KC})\mathbf{P}_k^- \qquad (18.37)$$
$$\mathbf{K} = \mathbf{P}_k^- \mathbf{C}^T (\mathbf{C}\mathbf{P}_k^- \mathbf{C}^T + \mathbf{6_S})^{-1}.$$

with the observation matrix $\mathbf{C}$ and covariance matrix of the observation noise $\mathbf{6_S}$. The matrix $\mathbf{K}$ is called the *Kalman gain*.

The conditions in the derivation shown above can at best be approximately guaranteed in practice. An obvious limitation is the descriptive power of the system and observation models used, which for reasons of computational manageability usually only approximate the real world. The resulting inaccuracies often lead to a diverging behavior of the filter, which can be prevented in practice by so-called parameter tuning. In the actual design of a filter, this step is an integral part of the development process. Another challenge in the implementation of the filter are numerical inaccuracies, especially in the matrix inversion. Numerically stable and efficient variants of the Kalman filter are based on the Cholesky or singular value decomposition.

18.5 Application Examples

The environment perception, consisting of sensor, sensor-related signal processing and data fusion, has the task of providing a generic and consistent image of the traffic scene, which is adapted to the respective driving function in the further processing layers scene interpretation and behavior decision [52]. In addition to the data fusion of different sensors, which already plays an important role today, the camera technology itself—due to the design of the traffic scene for human perception—offers an ideal platform for the provision and fusion or interpretation of different recognition results towards a consistent image of the scene in the environment of the vehicle. In order to represent the complex contents of a scene in a representative way, learning-based processes play a central role in today's development, which will be discussed in more detail below. Furthermore, the application of machine vision in future mobility concepts will be explained in more detail.

18.5.1 Panoptic Segmentation

In the past, a wide variety of methods were used for object detection and related tasks. These were based on manually designed image features and heuristics. For example, object hypotheses were generated by disparity, motion, or appearance features and subsequently plausibilized. In recent years, however, these approaches have been replaced almost completely by *Deep Learning*, which empirically achieves significantly higher detection quality and accuracy.

In this section, the application of *panoptic segmentation* [25] based on deep learning will be presented. Panoptic segmentation is the combination of several common tasks in the field of visual perception: The image is first semantically segmented to assign class labels to non-countable regions in the image. This may include, for example, the classes 'sky', 'road surface', or 'vegetation'. For countable objects, on the other hand, the individual instances are segmented. These include road users such as cars, cyclists, and pedestrians, as well as infrastructure elements such as traffic lights and traffic signs. Most approaches first detect the bounding boxes that belong to these instances. Then, a binary mask is estimated per bounding box, which segments the pixels of the image belonging to the object instance. In the last step of the panoptic segmentation, the semantic segmentation mask and the instance segmentation masks are fused. The result of the panoptic segmentation thus assigns an object class and, if possible, an object instance to each pixel of the input image. ◘ Figure 18.12 shows an example of the result of such a segmentation.

The result of the panoptic segmentation contains a compact representation of the scene, which is suitable for a variety of vehicle automation tasks. Object instances can be tracked in time so that their behavior can be predicted and taken into account in the planning modules. Object instances of infrastructure elements provide further information such as velocity constraints that influence the behavior of the ego vehicle. Semantic segmentation of the static world provides, among other things, the road surface and lane markings. Furthermore, information about selected semantic classes can be matched with a map. This can be used for localization or map validation, for example.

One architecture that achieves outstanding results in the area of panoptic segmentation is EfficientPS [33]. This architecture will be presented in the following. Many of its ideas can also be found in similar architectures for panoptic segmentation, as well as for object detection or semantic segmentation.

◘ Figure 18.13 illustrates the EfficientPS architecture. The input image is first processed by a slightly modified EfficientNet backbone [56] (red in ◘ Fig. 18.13). This backbone generates feature maps of different resolutions by repeated downscaling. These feature maps contain the image information in an abstract representation. Individual entries can, for example, describe the presence of objects or object parts as well as their properties, but in general the feature maps are not directly interpretable. The high-resolution feature maps contain local image structures with detailed spatial details, while the low-resolution feature maps describe global relationships.

The feature maps are then fused by a two-way FPN (purple, blue and green in ◘ Fig. 18.13). FPN stands for *Feature Pyramid Network*, a name derived from the fact that the feature maps of different resolutions from

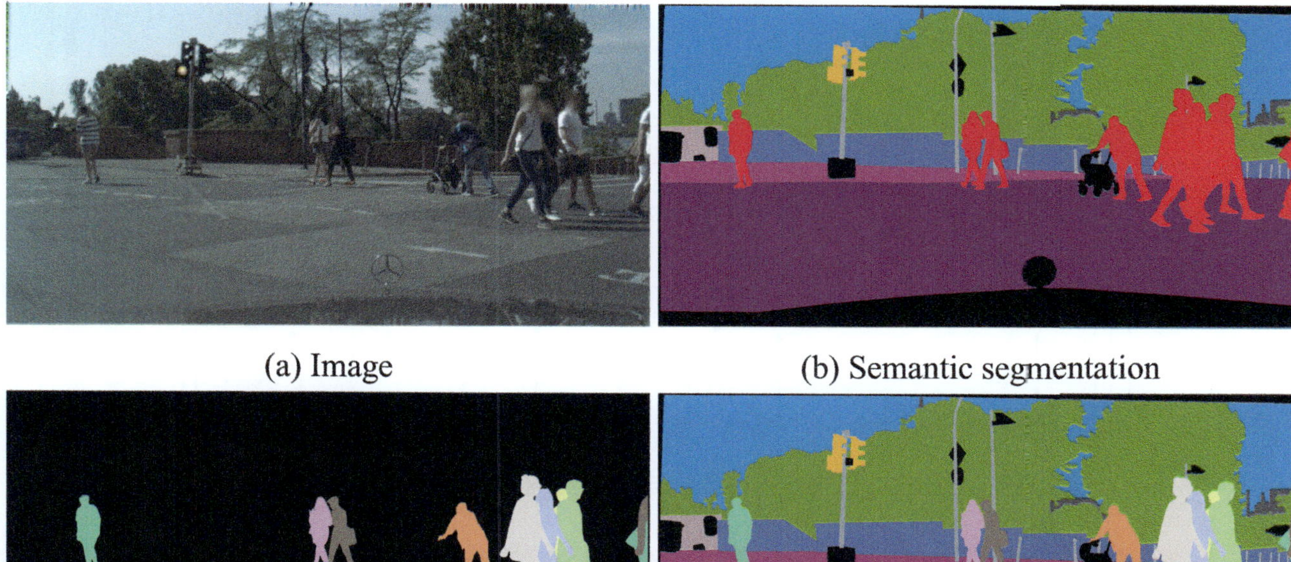

(a) Image

(b) Semantic segmentation

(c) Instance segmentation

(d) Panoptic segmentation

Fig. 18.12 Visualization of the panoptic segmentation on an input image. The different classes and object instances are encoded by colors. This image is taken from the Cityscapes dataset [6]

the backbone form a feature pyramid analogous to an image pyramid. In one branch, the FPN augments the low-resolution feature maps with local structures from higher-resolution feature maps. In the other branch, the FPN augments the information in high-resolution feature maps with global context from lower-resolution feature maps.

The output of the FPN is again a feature pyramid with the merged feature maps. These are now processed by separate network heads. A network head generates the semantic segmentation mask (yellow in **Fig. 18.13**). Another network head provides the instance segmentation (orange in **Fig. 18.13**). The network head for instance segmentation is based on Mask R-CNN [16]. A Region Proposal Network (RPN) generates a list of object proposals with associated bounding boxes. The 'ROI align' module crops the associated range of feature maps and scales the crops to a fixed size. Based on this, the following network layers estimate the object class, a more precise bounding box and an instance mask. The network head for semantic segmentation fuses the feature maps with different resolutions again and finally predicts the segmentation masks for the different semantic classes.

Subsequently, the outputs of the individual network heads are merged into the overall result. In a first step, all object instances are filtered. Among other things, instances with a low score or too large an overlap of the instance

masks are discarded. In a second step, the scores of the instance masks are offset against the class-specific scores of the semantic masks to obtain fused instance masks. Finally, the fused instance masks are transferred to an empty image and the free areas is filled with the information of the semantic segmentation. **Figure 18.14** visualizes an output of EfficientPS as an example.

18.5.2 Machine Vision in the Concept 'Mobility-as-a-Service'

To point out the important role of machine vision in modern and sustainable transportation systems, this section takes a closer look at the mobility concept 'Mobility-as-a-Service'. This term refers to autonomously driving vehicles[5] that can be used for both passenger and freight transportation. Experts have established the terms *people mover* for transporting people and *cargo mover* for transporting goods. For

---

5 Classified in the J3016 standard [46] for automated driving, this article refers to automation systems of development level 4. Such systems are to be understood as a collection of hardware and software components that work together in a structured manner to accomplish the driving task that is performed by the driver in conventional vehicles. The corresponding restriction to certain operating conditions defines the respective automation level.

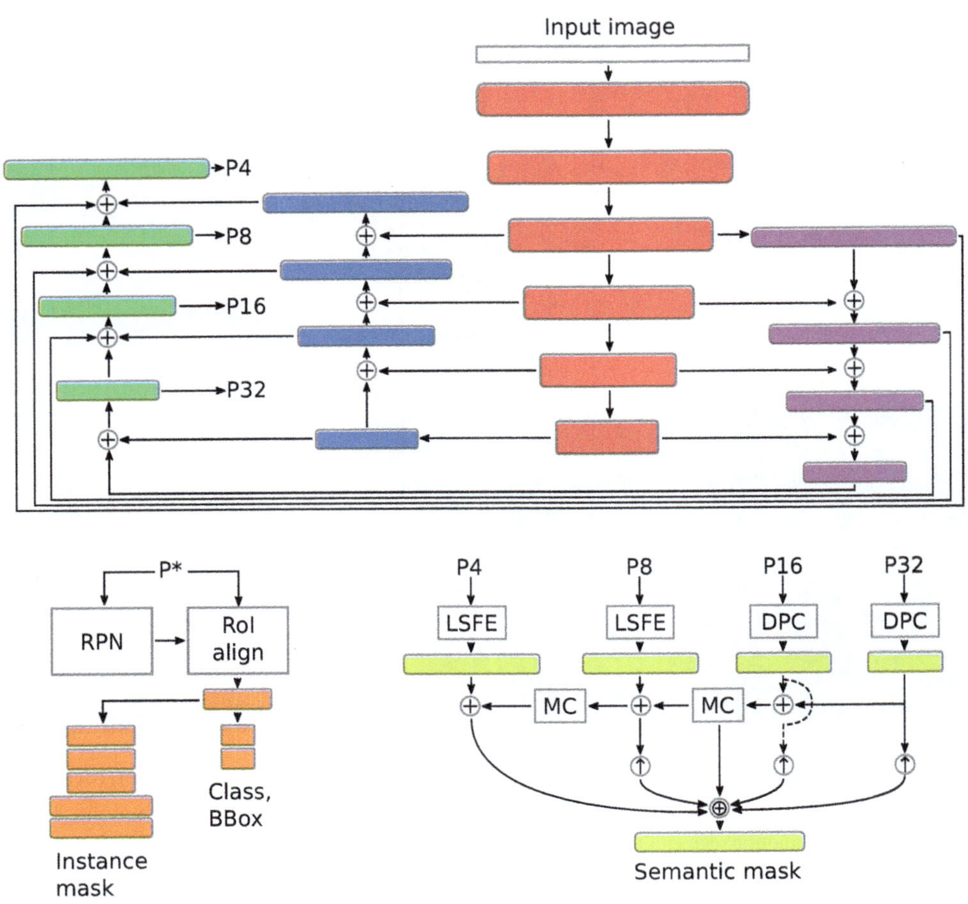

Fig. 18.13   Overview of the EfficientPS architecture

Fig. 18.14   Example of an output of EfficientPS

the mobility needs of tomorrow's urban society such systems represent a necessary supplement to today's predominantly individual traffic [3] and will probably become the signature of the new urban mobility.

Changing business models and customer structures must be taken into account in the development and production of such people and cargo movers. The competencies and unique selling propositions of successful automobile manufacturers are increasingly moving to the background: The driving dynamics and engine performance play a subordinate role in vehicles that will travel at a maximum speed of 70 km/h. Industrialization know-how in mass production is not absolutely necessary with potentially small production volumes. Today, we are in the midst of such a disruptive change in the automotive industry. It is being fueled by two technological trends: First, electromobility. The purely electric drive significantly reduces the complexity of vehicle propulsion and allows new suppliers to enter the market. Second, autonomous driving. This has evolved in part from the Advanced Driver Assistance Systems (ADAS) with the goal to achieve better safety ratings, for example EuroNCAP. The most important feature of such systems—classified as development level 1 to 3 in SAE standard J3016 [46]—is that the driver is in charge of the vehicle and must therefore be ready to take control of it at any time.

In contrast, automated driving for people and cargo movers requires solutions that do not require any intervention by the vehicle occupant. In the event of an error or a violation of the defined operating conditions, such an automated driving system (ADS) of level 4 must therefore be capable of transferring the vehicle to a state which rules out potential hazards as far as possible.[6]

A level 4 ADS is developed for a well-defined Operational Design Domain (ODD)[7] and plays a crucial role in terms of performance, integrity and security of the system. For example, an ODD can define the boundary conditions for embedding an ADS in an existing traffic infrastructure, i.e. the system is to be integrated into the prevailing traffic flow (mixed operation). This results in the need to react to static and moving road users and objects at any time. Typically, an *a priori* defined route set is also defined within which the mover system implements its

task. Light and weather conditions typical for the respective region are also specified precisely (◘ Fig. 18.15).

The system response required for an appropriate implementation of the driving task is the core task of the ADS (in the technical world this is referred to as Object and Event Detection and Response (OEDR)). This involves the detection of objects and other influencing factors in the vehicle environment and, derived from this, the determination of a suitable driving trajectory which the vehicle follows. Relevant objects in the area of the driving trajectory are all road users potentially occurring in an ODD and objects that have an influence on the safe operation of the vehicle. Furthermore, it must be ensured that the vehicle operates within the designated lane, follows the legally prescribed framework, and complies with the generally accepted etiquette in daily road traffic. For structuring and classification purposes, these tasks can be divided into core competencies [34]. Based on these competencies, requirements can be formulated to ensure the appropriate perception of the vehicle environment for a correct planning of the driving trajectory. Furthermore, an adequate measurement system must be defined to ensure the perception task, which provides all necessary information to successfully implement the driving task. In level 4 automation systems, this measurement system usually consists of several radar, lidar and camera sensors that survey the entire vehicle environment and ensure that all areas that could have an influence on the driving task are reliably measured.

The system architecture of an ADS usually follows this processing chain and is divided into three subsystems: Perception of the environment, behavior planning, and behavior execution [58]. The perception system plays a key role here: raw measurements of the measurement system are transformed into a representation that allows the calculation of a suitable behavior for the vehicle in the prevailing situation. Especially in the case of ADS, the perceptual system has to generate a deep, context-sensitive understanding of the world. It is not only about the reliable detection of object instances (and their geometric and kinematic attributes), but also about their object class, interaction and dependencies with or to other objects, and the consideration of environmental and situational conditions. Since the traffic infrastructure and the dynamic traffic events are designed for human perception, the visual interpretation of the environment determines the performance of a perception system. This results in a prominent position of camera technology and the associated signal processing methods. Established and common approaches in technical development can fulfill part of these requirements here. For example, visual odometry methods are used to localize an ADS and determine the vehicle position in a reference map based on learned feature vectors. Classification procedures help to determine object classes, both of road users and of traffic infrastructure such as traffic

---

6   The *minimum risk state* corresponds to the last fallback level of the vehicle and is assumed to be established independently by the ADS when a situation can no longer be controlled by the system or a fault occurs in the system. The minimum risk is defined as the state in which the vehicle no longer poses any significant danger. Depending on the situation, this can be the standstill itself or the safe transition to it.

7   The Operational Design Domain (ODD) defines the conditions under which an ADS operates safely and reliably. An ODD contains at least the following information: Road type, geographic area, speed range of the automated vehicle, and environmental conditions, such as weather and time of day, for the expected operational case.

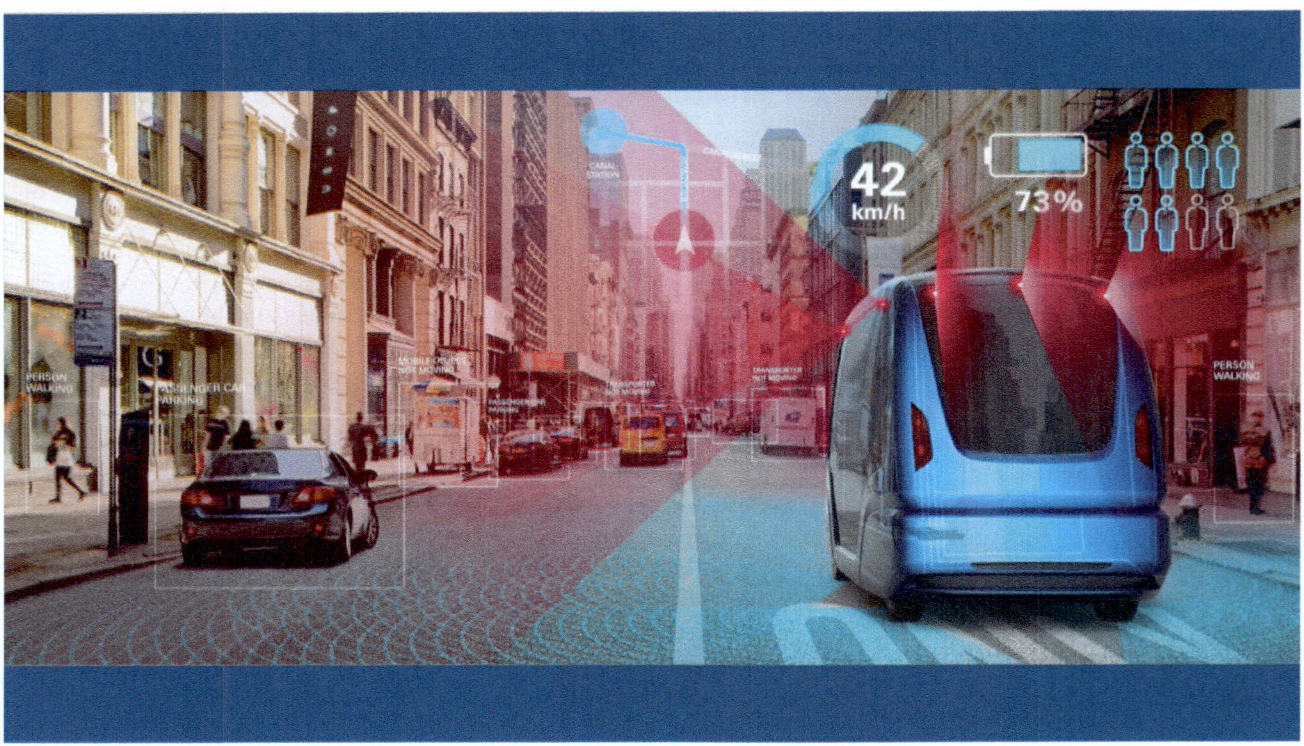

◘ Fig. 18.15   Machine vision represents an important building block for the use of people movers in the urban environment of the future

lights, signs or painted symbols. The optical flow of an image sequence can be used to draw conclusions about the motion signature of individual objects or the vehicle's own motion. By evaluating image pairs in temporal or spatial direction, a nearly dense 3D image of the environment can be generated, which in turn can be used to determine the position, extent and orientation of objects.

However, when it comes to the semantic interpretation of the situation in the vehicle environment (and also the monitoring of the vehicle interior), common methods reach their limits. However, this is of utmost importance for a successful implementation of the perception task, since only such an interpretation can ensure the consistency and plausibility of the computed environment model. Furthermore, only a visual interpretation of the scene can, with economically justifiable effort, determine how other road users interact with the ADS (keywords: collaborative vs. non-collaborative, ride hailing, human-machine interface, etc.) or whether they are even aware of the vehicle in its vicinity and potentially react appropriately. This is a challenge for mover systems in particular, since this type of transportation system is used in complex and highly dynamic environments with dense traffic and narrow buildings. To ensure user acceptance, the system must also react as robustly as possible to external factors such as changing weather and lighting conditions. The maximum possible availability is limited here by the reliable detection of the system limits and the associated transition of the vehicle to a safe state. Camera techno-

logy also plays an important role in this safety-relevant task.

While in the automation systems currently under development—and probably also in the first generation rolled out on the market—part of the safety concept and monitoring tasks are performed by humans. This must be fully automated for the following systems to be economically successful. In connection with this, open questions regarding development and operation of ADS have to be clarified. With respect to machine vision the following questions arise: what is the line of reasoning with respect to reproducible and deterministic behavior of algorithms that are based on training data? How can it be proven? In terms of system safety, design errors and functional misbehavior must be reduced to an acceptable minimum in a systematic manner during development. For example, it must be possible to explain comprehensively how the system handles objects that are not mapped in the training data [61]. In the recent past, the broader public has taken rather negative note of (partially) automated systems in which conceptual weaknesses in the design have led to fatal accidents [35]. Examples here are the collision of a Tesla vehicle with a truck in 2016 where the camera failed to detect the truck in backlight conditions and an error in the fusion logic led to a rejection of the object hypothesis. In 2018, a Tesla vehicle also collided with a lane separation element after the system failed to correctly detect lane markings. Also in 2018, a test vehicle operated by the company Uber collided at night with a person dressed in dark clothing who was pushing her bicycle as she crossed the road. Such incidents must remain isolated and regrettable exceptions to find social acceptance. For this purpose, a deep understanding of the system has to be developed to minimize the risk of functional insufficiencies and foreseeable misuse [23]. In connection with this, the question of how such procedures are validated must be clarified. Statistical validation, as in the case of level 2 and 3 automation systems, seems unrealistic due to the immensely higher validation effort.

Finally, attacks from outside must be averted during operation but also during development. For example, it must prevented that intentionally incorrectly trained networks may provoke unsafe behavior of the ADS. This is a sensitive point of attack that requires special attention for processes that interpret visual data and feed it into the driving decision. The computing power and bandwidth required for image processing methods in general and for machine learning in particular pose a further challenge. The design of a secure computing platform and the software integrated on it is already defined by various standards [50], but the requirements of an ADS and the methods required for sensor data interpretation define a new dimension here.

The actual task of image processing within the framework of the nominal function as described above is additionally supplemented by another important point: the monitoring of the ODD. As already mentioned, an ADS of automation level 4 is tuned to a specific ODD and designed for operation within the boundary conditions defined there. Therefore, monitoring of the ODD boundary parameters is essential for reliable operation of the safety-critical function. A visual evaluation of the prevailing conditions, such as visibility, lighting, etc., allows a gradual degradation of the driving task up to a safe standstill of the vehicle in case of inadequate ambient conditions.

18.6 Summary and Outlook

With machine vision systems, sensor systems are finding their way into our automobiles that come closest to the principle of human perception in terms of the visual information used. Benefiting from the continuing drop in the cost of cameras and evaluation hardware, image sensors are being used in a steadily growing number of applications. A decisive advantage of video sensors compared to other environmental sensors is their richness of information. At the same time, the analysis of the extensive image information is a great challenge for signal processing. In particular, the primary measurements of cameras consist only of brightness patterns. These can only be converted into 3d information and semantics using suitable image processing methods.

Image sensors have particularly high potential for use in driver assistance systems because:

- the passive measurement principle means that there are no legal restrictions with regard to approval for use in road traffic,
- the infrastructure and the traffic situation are designed for visual perception and can therefore only be fully captured by images,
- the information content of the sensor signal has a disproportionately higher degree of abstraction and depth than conventional environmental sensors,
- the proximity to human perception enables a high degree of transparency with regard to the function of video-based driver assistance systems.

This chapter has described the design and operation of cameras and associated control units for machine vision in vehicles. As a characterizing property of image acquisition, the loss of information by a whole dimension—resulting from the mapping of the 3d world onto a 2d image plane—was considered. With the introduction of homogeneous coordinates, such geometric and perspective transformations can be expressed in an elegant mathematically compact way and provide a linear description for a variety of the occurring mappings. The quantities necessary for the image description are given by the intrinsic and extrinsic calibration parameters of the camera.

There are a number of application-specific image processing methods for the interpretation and analysis of visual information. These are usually executed in a modular fashion, with initial preprocessing stages preparing and correcting the image signal itself. Finally, task-specific features are extracted by suitable operators and made available to higher-level processing steps. These then operate on the highly condensed feature information.

In the past decade, Deep Learning approaches have been used to train previously hand-designed image processing methods in a data-driven manner, resulting in significant performance advances. The development from simple network architectures such as convolutional neural networks to increasingly sophisticated network complexity suggests that this technology is only at the beginning of its development.

Point correspondences were introduced as particularly meaningful features for 3d reconstruction. These allow conclusions to be drawn about the 3D position and movement of the pixels under consideration. By means of temporal tracking methods, image information can be accumulated and stabilized step-by-step without having to store or even process past images in the long term. Based on the Bayes' filter, the particle filter and the Kalman filter have been described here as practicable realizations and supplemented by implementation notes.

Panoptic segmentation was presented as a practical application example from the Deep Learning domain. Finally, the immense importance of machine vision for the automation of our traffic was demonstrated using the example of a people mover application.

Acknowledgements The authors would like to thank Dr. Christian Duchow and Prof. Dr. Andreas Geiger for valuable contributions to early versions of this chapter.

References

1. Barker, A., Brown, D., Martin, W.: Bayesian estimation and the Kalman filter. Comput. Math. Appl. **30**(10), 55–77 (1995)
2. Bay, H., Ess, A., Tuytelaars, T., Gool, L.V.: Speeded-up robust features (surf). Comput. Vis. Image Understand. **110**(3), 346–359 (2008). ▶ https://doi.org/10.1016/j.cviu.2007.09.014. ▶ http://www.sciencedirect.com/science/article/pii/S1077314207001555. Similarity Matching in Computer Vision and Multimedia
3. BERYLLS: The revolution of urban mobility (2019). ▶ https://www.berylls.com/wp-content/uploads/2018/01/20171216_Studie_Mobilitaet.pdf
4. Calonder, M., Lepetit, V., Ozuysal, M., Trzcinski, T., Strecha, C., Fua, P.: Brief: Computing a local binary descriptor very fast. IEEE Trans. Pattern Anal. Mach. Intelli. **34** (2011). ▶ https://doi.org/10.1109/TPAMI.2011.222
5. Canny, J.F.: A computational approach to edge detection. IEEE Trans. Pattern Anal. Mach. Intell. **8**(6), 679–698 (1986)
6. Cordts, M., Omran, M., Ramos, S., Rehfeld, T., Enzweiler, M., Benenson, R., Franke, U., Roth, S., Schiele, B.: The cityscapes dataset for semantic urban scene understanding. In: Proceeding of the IEEE Conference on Computer Vision and Pattern Recognition (CVPR) (2016)
7. Cybenko, G.: Approximation by superpositions of a sigmoidal function. Math. Control, Signals Syst. **2**(4), 303–314 (1989). ▶ https://doi.org/10.1007/BF02551274
8. Dang, T., Hoffmann, C., Stiller, C.: Visuelle mobile wahrnehmung durch fusion von disparität und verschiebung. In: Maurer, M., Stiller, C. (eds.): Fahrerassistenzsysteme mit maschineller Wahrnehmung, chapter 2, pp. 21–42. Springer, Heidelberg (2005)
9. Dosovitskiy, A., Fischer, P., Ilg, E., Häusser, P., Hazirbas, C., Golkov, V., van der Smagt, P., Cremers, D., Brox, T.: Flownet: learning optical flow with convolutional networks. In: 2015 IEEE International Conference on Computer Vision (ICCV), pp. 2758–2766. IEEE Computer Society (2015). ▶ https://doi.org/10.1109/ICCV.2015.316
10. Faugeras, O.: Three-Dimensional Computer Vision: A Geometric Viewpoint. The MIT Press, Cambridge (1993)
11. Girshick, R.B.: Fast R-CNN. In: 2015 IEEE International Conference on Computer Vision (ICCV), pp. 1440–1448. IEEE Computer Society (2015). ▶ https://doi.org/10.1109/ICCV.2015.169
12. Girshick, R.B., Donahue, J., Darrell, T., Malik, J.: Rich feature hierarchies for accurate object detection and semantic segmentation. In: 2014 IEEE Conference on Computer Vision and Pattern Recognition (CVPR), pp. 580–587. IEEE Computer Society (2014). ▶ https://doi.org/10.1109/CVPR.2014.81
13. Godard, C., Aodha, O.M., Brostow, G.J.: Unsupervised monocular depth estimation with left-right consistency. In: 2017 IEEE Conference on Computer Vision and Pattern Recognition (CVPR), pp. 6602–6611. IEEE Computer Society (2017). ▶ https://doi.org/10.1109/CVPR.2017.699
14. Godard, C., Aodha, O.M., Firman, M., Brostow, G.: Digging into self-supervised monocular depth estimation (2019). ▶ https://arxiv.org/abs/1806.01260
15. Harris, C., Stephens, M.: A combined corner and edge detector. In: Proceedings of the 4th Alvey Vision Conference, pp. 147–151 (1988)
16. He, K., Gkioxari, G., Dollár, P., Girshick, R.B.: Mask R-CNN. In: IEEE International Conference on Computer Vision (ICCV) 2017, pp. 2980–2988. IEEE Computer Society (2017). ▶ https://doi.org/10.1109/ICCV.2017.322
17. He, K., Zhang, X., Ren, S., Sun, J.: Deep residual learning for image recognition. In: 2016 IEEE Conference on Computer Vision and Pattern Recognition (CVPR), pp. 770–778. IEEE Computer Society (2016). ▶ https://doi.org/10.1109/CVPR.2016.90
18. Hornik, K.: Approximation capabilities of multilayer feedforward networks. Neural Netw. **4**(2), 251–257 (1991). ▶ https://doi.org/10.1016/0893-6080(91)90009-T
19. Howard, A., Pang, R., Adam, H., Le, Q.V., Sandler, M., Chen, B., Wang, W., Chen, L., Tan, M., Chu, G., Vasudevan, V., Zhu, Y.: Searching for mobilenetv3. In: 2019 IEEE/CVF International Conference on Computer Vision (ICCV), pp. 1314–1324. IEEE (2019). ▶ https://doi.org/10.1109/ICCV.2019.00140
20. Howard, A.G., Zhu, M., Chen, B., Kalenichenko, D., Wang, W., Weyand, T., Andreetto, M., Adam, H.: Mobilenets: Efficient convolutional neural networks for mobile vision applications (2017). ▶ https://arxiv.org/abs/1704.04861
21. Ilg, E., Mayer, N., Saikia, T., Keuper, M., Dosovitskiy, A., Brox, T.: Flownet 2.0: Evolution of optical flow estimation with deep networks. In: 2017 IEEE Conference on Computer Vision and Pattern Recognition (CVPR), pp. 1647–1655. IEEE Computer Society (2017). ▶ https://doi.org/10.1109/CVPR.2017.179
22. Isard, M., Blake, A.: Condensation—conditional density propagation for visual tracking. Int. J. Comput. Vis. **29**, 5–28 (1998)
23. ISO International Organization for Standardization: Road vehicles—safety of the intended functionality. ISO/PAS 21448:2019 (2019). ▶ https://www.iso.org/standard/70939.html
24. Jähne, B.: Digitale Bildverarbeitung, 7th edn. Springer Vieweg, Berlin (2012)
25. Kirillov, A., Girshick, R.B., He, K., Dollár, P.: Panoptic feature pyramid networks. In: 2019 IEEE Conference on Computer Vision and Pattern Recognition (CVPR), pp. 6399–6408. Computer Vision Foundation/IEEE (2019). ▶ https://doi.org/10.1109/CVPR.2019.00656
26. Kuhn, H.W.: The Hungarian method for the assignment problem. Naval Res. Logist. Quart. **2**(1–2), 83–97 (1955). ▶ https://doi.org/10.1002/nav.3800020109
27. Lategahn, H., Beck, J., Stiller, C.: DIRD is an illumination robust descriptor. In: IEEE Intelligent Vehicles Symposium, pp. 756–761 (2014). ▶ https://doi.org/10.1109/IVS.2014.6856421
28. Lin, T., Goyal, P., Girshick, R.B., He, K., Dollár, P.: Focal loss for dense object detection. In: IEEE International Conference on Computer Vision (ICCV) 2017, pp. 2999–3007. IEEE Computer Society (2017). ▶ https://doi.org/10.1109/ICCV.2017.324
29. Liu, W., Anguelov, D., Erhan, D., Szegedy, C., Reed, S.E., Fu, C., Berg, A.C.: SSD: single shot multibox detector. In: B. Leibe, J. Matas, N. Sebe, M. Welling (eds.) Computer Vision—ECCV 2016—14th European Conference, pp. 21–37. Springer (2016). ▶ https://doi.org/10.1007/978-3-319-46448-0_2
30. Longuet Higgins, H.: A computer algorithm for reconstructing a scene from two projections. Nature **293** (1981)

31. Lowe, D.G.: Distinctive image features from scale-invariant keypoints. Int. J. Comput. Vis. **60**(2), 91–110 (2004). ► https://doi.org/10.1023/B:VISI.0000029664.99615.94

32. Minsky, M., Papert, S.A.: Perceptrons. MIT Press, Cambridge, MA, USA (1969)

33. Mohan, R., Valada, A.: Efficientps: Efficient panoptic segmentation (2020). ► https://arxiv.org/abs/2004.02307

34. NHTSA: Federal automated vehicles policy (2016). ► https://www.transportation.gov/AV/federal-automated-vehicles-policy-september-2016

35. NHTSA: 2018 fatal motor vehicle crashes: Overview. National Highway Traffic Safety Administration's Traffic Safety Facts Annual Report (2019). ► https://crashstats.nhtsa.dot.gov/Api/Public/ViewPublication/812826

36. Nistér, D.: An efficient solution to the five-point relative pose problem. In: CVPR (2), pp. 195–202. IEEE Computer Society (2003)

37. Qi, C.R., Su, H., Mo, K., Guibas, L.J.: Pointnet: deep learning on point sets for 3d classification and segmentation. In: 2017 IEEE Conference on Computer Vision and Pattern Recognition, CVPR 2017, Honolulu, HI, USA, July 21–26, pp. 77–85. IEEE Computer Society (2017). ► https://doi.org/10.1109/CVPR.2017.16

38. Ranft, B., Stiller, C.: The role of machine vision for intelligent vehicles. IEEE Trans. Intell. Veh. **1**(1), 8–19 (2016). ► https://doi.org/10.1109/TIV.2016.2551553

39. Redmon, J., Divvala, S.K., Girshick, R.B., Farhadi, A.: You only look once: unified, real-time object detection. In: 2016 IEEE Conference on Computer Vision and Pattern Recognition (CVPR), pp. 779–788. IEEE Computer Society (2016). ► https://doi.org/10.1109/CVPR.2016.91

40. Redmon, J., Farhadi, A.: YOLO9000: better, faster, stronger. In: 2017 IEEE Conference on Computer Vision and Pattern Recognition (CVPR), pp. 6517–6525. IEEE Computer Society (2017). doi:► https://doi.org/10.1109/CVPR.2017.690

41. Redmon, J., Farhadi, A.: Yolov3: An incremental improvement (2018). ► http://arxiv.org/abs/1804.02767

42. Ren, S., He, K., Girshick, R.B., Sun, J.: Faster R-CNN: towards real-time object detection with region proposal networks. In: C. Cortes, N.D. Lawrence, D.D. Lee, M. Sugiyama, R. Garnett (eds.) Advances in Neural Information Processing Systems 28: Annual Conference on Neural Information Processing Systems (NeurIPS), pp. 91–99. Curran Associates, Inc. (2015)

43. Rosenblatt, F.: The Perceptron, a Perceiving and Recognizing Automaton (Project Para). Cornell Aeronautical Laboratory, Buffalo, NY, USA (1957)

44. Rumelhart, D.E., Hinton, G.E., Williams, R.J.: Learning Internal Representations by Error Propagation, pp. 318–362. MIT Press, Cambridge, MA, USA (1986)

45. Rumelhart, D.E., Hinton, G.E., Williams, R.J.: Learning representations by back-propagating errors. Nature **323**(6088), 533–536 (1986). ► https://doi.org/10.1038/323533a0

46. SAE Society of Automotive Engineers: Taxonomy and definitions for terms related to driving automation systems for on-road motor vehicles. Society of Automotive Engineers (2018). ► https://www.sae.org/standards/content/j3016_201806/

47. Sandler, M., Howard, A.G., Zhu, M., Zhmoginov, A., Chen, L.: Mobilenetv2: Inverted residuals and linear bottlenecks. In: 2018 IEEE Conference on Computer Vision and Pattern Recognition (CVPR), pp. 4510–4520. IEEE Computer Society (2018). ► https://doi.org/10.1109/CVPR.2018.00474

48. Schreer, O.: Stereoanalyse und Bildsynthese. Springer, Berlin [u.a.] (2005). ► http://www.amazon.de/gp/search?index=books&linkCode=qs&keywords=354023439X

49. Shi, J., Tomasi, C.: Good features to track. IEEE Conf. Comput. Vis. Pattern Recogn. 593–600 (1994)

50. for Standardization, O.I.O.: Road Vehicles: Functional Safety. ISO (2018). ► https://www.iso.org/obp/ui/#iso:std:iso:26262:-2:dis:ed-2:v1:en

51. Stiller, C., Konrad, J.: Estimating motion in image sequences—a tutorial on modeling and computation in 2D motion. IEEE Signal Process. Mag. **7 & 9**, 70–91 & 116–117 (1999). (best paper award)

52. Stiller, C., Puente León, F., Kruse, M.: Information fusion for automotive applications—an overview. Inf. Fusion **12**(4), 244–252 (2011). ► https://doi.org/10.1016/j.inffus.2011.03.005

53. Sun, D., Yang, X., Liu, M., Kautz, J.: Pwc-net: CNNs for optical flow using pyramid, warping, and cost volume. In: 2018 IEEE Conference on Computer Vision and Pattern Recognition (CVPR), pp. 8934–8943. IEEE Computer Society (2018). ► https://doi.org/10.1109/CVPR.2018.00931

54. Szegedy, C., Liu, W., Jia, Y., Sermanet, P., Reed, S.E., Anguelov, D., Erhan, D., Vanhoucke, V., Rabinovich, A.: Going deeper with convolutions. In: 2015 IEEE Conference on Computer Vision and Pattern Recognition (CVPR), pp. 1–9. IEEE Computer Society (2015). ► https://doi.org/10.1109/CVPR.2015.7298594

55. Szegedy, C., Vanhoucke, V., Ioffe, S., Shlens, J., Wojna, Z.: Rethinking the inception architecture for computer vision. In: 2016 IEEE Conference on Computer Vision and Pattern Recognition (CVPR), pp. 2818–2826. IEEE Computer Society (2016). ► https://doi.org/10.1109/CVPR.2016.308

56. Tan, M., Le, Q.V.: Efficientnet: Rethinking model scaling for convolutional neural networks. In: K. Chaudhuri, R. Salakhutdinov (eds.) Proceedings of the 36th International Conference on Machine Learning (ICML), pp. 6105–6114. PMLR (2019)

57. Tan, M., Pang, R., Le, Q.V.: Efficientdet: scalable and efficient object detection. In: 2020 IEEE/CVF Conference on Computer Vision and Pattern Recognition (CVPR), pp. 10778–10787. IEEE Computer Society (2020). ► https://doi.org/10.1109/CVPR42600.2020.01079

58. Tas, Ö.S., Kuhnt, F., Zöllner, J.M., Stiller, C.: Functional system architectures towards fully automated driving. In: 2016 IEEE Intelligent Vehicles Symposium (IV), pp. 304–309 (2016). ► https://doi.org/10.1109/IVS.2016.7535402

59. Tola, E., Lepetit, V., Fua, P.: Daisy: an efficient dense descriptor applied to wide-baseline stereo. IEEE Trans. Pattern Anal. Mach. Intell. **32**(5), 815–830 (2010). ► https://doi.org/10.1109/TPAMI.2009.77

60. Trucco, E., Verri, A.: Introductory Techniques for 3-D Computer Vision. Prentice Hall, New York (1998)

61. Wang, W., Feiszli, M., Wang, H., Tran, D.: Unidentified video objects: a benchmark for dense, open-world segmentation. CoRR **abs/2104.04691** (2021). ► https://arxiv.org/abs/2104.04691

62. Welch, G., Bishop, G.: An introduction to the Kalman filter. Technical Report, USA (1995)

63. Zhang, F., Prisacariu, V.A., Yang, R., Torr, P.H.S.: Ganet: guided aggregation net for end-to-end stereo matching. In: IEEE Conference on Computer Vision and Pattern Recognition (CVPR), pp. 185–194. Computer Vision Foundation/IEEE (2019). ► https://doi.org/10.1109/CVPR.2019.00027

Open Access This chapter is licensed under the terms of the Creative Commons Attribution-NonCommercial-NoDerivatives 4.0 International License (▶ http://creativecommons.org/licenses/by-nc-nd/4.0/), which permits any noncommercial use, sharing, distribution and reproduction in any medium or format, as long as you give appropriate credit to the original author(s) and the source, provide a link to the Creative Commons license and indicate if you modified the licensed material. You do not have permission under this license to share adapted material derived from this chapter or parts of it.

The images or other third party material in this chapter are included in the chapter's Creative Commons license, unless indicated otherwise in a credit line to the material. If material is not included in the chapter's Creative Commons license and your intended use is not permitted by statutory regulation or exceeds the permitted use, you will need to obtain permission directly from the copyright holder.

Stereo Vision for ADAS

Stefan Gehrig and Uwe Franke

Contents

© The Author(s) 2026
H. Winner et al. (eds.), *Handbook Assisted and Automated Driving*,
https://doi.org/10.1007/978-3-658-45276-6_19

The first camera-based driver assistance systems were implemented as prototypes within the European project PROMETHEUS in the early 90's. Back then, only a few optimists envisioned this technology to play a vital role in the real world some 25 years later. Moreover, this technology now acts as an enabler for high-end safety systems such as autonomous braking for pedestrians. No other sensor besides the camera could benefit from recent technological advances so much.

The costs for a camera have dropped from well above 1000 $ in the beginning to a few dozen Dollars these days. The limited dynamic range of early CCD sensors was dramatically improved thanks to the modern CMOS-based consumer cameras. Back then, a presentation of in-vehicle Computer Vision algorithms was doomed when sunlight hit the camera. At the same time, the available computing power increased by 6 orders of magnitude, as predicted by Moore's law. Another important enabler is the FPGA (Field-Programmable Gate Array) technology, which is suitable for highly parallel image processing tasks on pixel level. In recent years, ASICs and GPUs are also deployed frequently for such tasks.

The enormous progress on the hardware side was complemented by innovations on the algorithmic level. This includes:

- The introduction of the Kalman filter in image sequence analysis by E.D. Dickmanns [1] which allows the integration of dynamic systems knowledge in the processing chain.
- The progress in the machine learning domain especially deep learning which enables the utilization of image statistics and large image databases.
- The step from local ad-hoc methods to global optimal approaches that use well-founded mathematical methods for finding optimal solutions.
- The introduction of stereo vision as the basis for robust image understanding.

In contrast to the monocular approaches using only one eye, stereo vision allows a full 3D reconstruction of the environment within a single image pair, independent of the motion state of the observer and the observed objects. This enables advanced driver assistance systems, e.g. autonomous braking for pedestrians or collision avoidance for crossing objects, and partially automated driving functions in traffic jams. Moreover, stereo vision also precisely measures the 3-dimensional shape of the road ahead, which allows an adaption of the active body control systems to compensate bumps. For these reasons, stereo cameras became available for Mercedes models in 2013.

In the literature, stereo vision is considered a largely solved problem. However, the outdoor traffic scenario and the usage in safety–critical driver assistance systems pose particularly hard requirements on current methods. These requirements include:

- Robustness: In traffic scenes, image disturbances occur due to illumination and weather circumstances that are not covered by most approaches in the literature. Artefacts are e.g. sun blinding, reflections, blurring caused by water on the windshield, spray water, partial occlusion due to the windshield wiper, snow, or darkness. ■ Fig. 19.1 shows four examples.
- Accuracy: The desired measurement range is rather large and ranges from the bumper (less than 2 m) to 50…80 m. Hence, the disparity estimate has to be sub-pixel accurate, which is not demanded in current stereo benchmarks.
- Real-time capability: In order to enable fast responses, the processing usually runs at 25…30 Hz. Many literature methods process one single image pair in seconds to minutes of runtime on high-end PCs or even high-end GPUs.
- Long-term stability: vision systems are expected to work for the entire life cycle of the vehicle. For a stereo system, this includes the capability to perform online-calibration in the field to maintain a well calibrated system under varying temperatures and aging effects.
- Power consumption: For cost reasons, a one-box solution hosting camera and processing unit is desirable. If the camera system is mounted behind the rearview mirror, the power consumption has to be rather low (without active cooling) to avoid overheating.

The organization of this chapter is following the processing pipeline used by successful stereo vision systems. In ▶ Sect. 19.1, the traditional local methods of disparity estimation are compared to modern global optimal stereo methods. Since the accuracy of the disparity estimation is vital for the success of downstream object formation methods, we dedicate ▶ Sect. 19.2 to this topic. Especially the demand for methods that also operate under adverse weather conditions is often overlooked in the literature. One reason for that is the lack of ground truth under adverse weather condition in outdoor scenarios.

Stereo vision delivers 3D measurements for every pixel. If the points are tracked over time, one can estimate the motion in 3D. This leads to the ▶ Sect. 19.3 on 6D-Vision with fusion in space and time for a powerful basis to quickly detect moving objects. Especially laterally moving objects such as crossing cars or kids running onto the street can be reliably detected this way.

Computer Vision in recent years has exhibited a trend towards using super pixels as opposed to pixels. With the goal to represent the traffic scene compactly in 3D despite the growing imager resolution and to offer efficient subsequent processing steps, the Stixel World has been developed. This efficient representation tailored for traffic scene is introduced in ▶ Sect. 19.4,

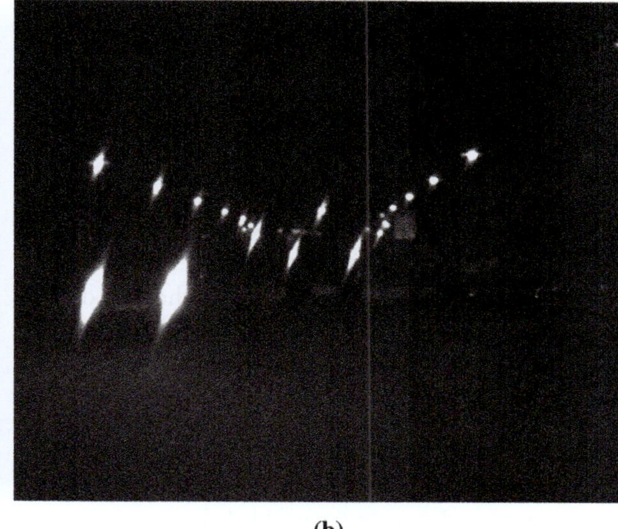

(a)

(b)

(c)

(d)

▫ Fig. 19.1 Safety–critical driver assistance applications pose high demands on the robustness of algorithms. **a** During rain, blinding and windshield smear can occur due to water on the windshield. **b** At night-time windshield wipers can generate disturbances with characteristic smear artefacts. **c** Snow fall and wet streets cause problems in disparity estimation. **d** Backlight can generate reflections on the windscreen that must not lead to erroneous obstacle detection. Figures courtesy of the HCI benchmark of D. Kondermann, University of Heidelberg [2]

as well as the subsequent steps for the final object detection. This concludes the transfer from pixels to objects. The chapter ends with a summary and remarks for further research.

19.1 Local and Global Disparity Estimation Methods

The basic steps of disparity estimation were introduced in ▶ Chap. 18. More similarity criteria common in the automotive field are described and several options for disparity estimation via correlation are detailed in this chapter. For a categorization of stereo methods and the different processing steps we refer to the seminal work of Scharstein and Szeliski [3]. According to the processing steps defined there, we distinguish between similarity criteria, disparity optimization, and sub-pixel estimation. We limit ourselves to spatially discrete methods that scan the disparity volume in discrete steps, since the continuous methods require significantly more computational resources to deal with the displacements present in traffic scenes. To compare published stereo methods, the Middlebury website (▶ http://vision.middlebury.edu/stereo/) was launched and currently ranks some 150 methods. A similar benchmark for the automotive field is the KITTI benchmark launched in 2012 [4]. This data set contains images with some typical problems for traffic scenes, such as reflections. Since some methods are very sensitive to such disturbances, differences in the ranking between both benchmarks occur.

For automotive purposes, a real-time computation is of uttermost importance. Thus, mostly local methods such as correlation were considered until recently.

19.1.1 Local Correlation Methods

As shown in ▶ Chap. 18, the classic method of disparity estimation works with correlation independently for every pixel. Independent refers to "independent of the results of neighboring pixels". For every pixel in the reference image (left image), the corresponding pixel in the search image (right image) that images the same world point is determined. For that purpose, the similarity of two pixels has to be established.

A large variety of criteria exists to determine the similarity of two pixels. We restrict ourselves to similarity metrics based on grey values. The simplest similarity metric is the difference of the two grey values in the left image (g_l) and the right image (g_r):

$$ABS(d) = |g_l(x; y) - g_r(x - d, y)| \tag{19.1}$$

This difference is computed line by line for all disparity hypotheses in the rectified image (s. ▶ Chap. 18). After rectification, the images are aligned in standard stereo geometry, i.e. corresponding points lie on the same image row. In the automotive field, disparities down to 0 have to be evaluated (corresponding to infinite distance). The maximum disparity is defined by the smallest distance to be measured.

Since even high dynamic range cameras hardly deliver more than 12 bits, this metric is always ambiguous. In addition, cameras cannot be equalized well enough for a common world point to be imaged with exactly the same grey value. This ambiguity can be reduced by using a window around the considered pixel. This was exemplified in ▶ Chap. 18 for the „Sum-of-Absolut-Differences" (SAD). Alternately, one can compute the "Sum-of-Squared-Differences (SSD)":

$$SSD(x, y, d) = \sum_{u,v \in B} \left(g_l(x + u, y + v) - g_r(x + u - d, y + v) \right)^2 \tag{19.2}$$

This computation scheme leads to strong deviations from the perfect similarity of 0, when the considered pixels differ by more than a few grey values. The authors prefer the similarity metric SAD that penalizes outlier less and leads to better results in practice. Independent of the chosen metric one can find wrong correspondences when radiometric deviations between the cameras occur. This problem can be alleviated by subtracting the respective means ($\overline{b_l}, \overline{b_r}$) of the image window B, exemplified with the absolute difference:

$$ZSAD(x, y, d) = \sum_{u,v \in B} \left| (g_l(x + u, y + v) - \overline{b_l}) \right.$$
$$\left. - (g_r(x + u - d, y + v) - \overline{b_r}) \right| \tag{19.3}$$

For all introduced similarity metrics one can find the optimal disparity value d^* for a pixel simply by finding the minimum over d:

$$d^*(x, y) = \min_d Z\,SAD(x, y, d) \tag{19.4}$$

A popular similarity metric, that also delivers a confidence value (measure of agreement) besides the optimal disparity is the previously introduced mean-free cross correlation function $\rho(d)$ (cf. ▶ Chap. 18). Its value ranges from −1 to 1. Good correspondences obtain values close to one, consequently the maximum of the metric is searched. Most of the time, correspondences with $\rho(d) > 0{,}7$ are accepted. Although the metric yields a good measure of similarity, one has to make sure that high-contrast regions are not matched with low-contrast regions that look the same after normalization performed by the score function.

The image window is typically chosen to be rectangular with a size of 3×3 to 9×9 pixels. Above metrics require that corresponding points lie on exactly same line. Similarity metrics on single points (i.e. 1×1 pixels) are particularly susceptible to small calibration errors. If the epipolar geometry is shifted by just one line, the corresponding point cannot be found since the search runs through the wrong image line. A larger image window reduces the sensitivity to calibration errors but induces higher computational load and causes the so-called foreground fattening: Since nearby object usually have higher contrast than the background, background pixels next to the foreground are often associated with the foreground disparity.

When rank statistics are used instead of grey value information, the similarity metric becomes less sensitive to small rectification errors [5]. The Hamming distance of the Census transform is a popular similarity metric that is more tolerant to such rectification errors and very efficient to compute. For every pixel within an image window, a bit is generated signifying whether the pixel's grey value is larger (1) or not (0) than the center pixel. This transform assigns a bitstring to every image window. The number of pixels within the considered window defines the bitstring length. The comparison of the two bitstrings from the reference image and the search image is straightforward. The Hamming distance provides an easy similarity metric for these two bitstrings. It just counts the number of different bits:

$$CENSUS(x, y, d) = HAM \left(T_{C(g_l, x, y)} - T_{C(g_r, x - d, y)} \right) \tag{19.5}$$

$T_{C(g_l,x,y)}$ represents the Census-transformed image at location x based on the grey values of the left image. The smallest Hamming distance is the best disparity estimation. The Census transform is invariant to linear transformations of the grey values in the left and right image, which occurs frequently due to sensitivity variations in cameras. Evaluations in [6] show that the Census metric is superior to other metrics, especially when used with global stereo algorithms.

Consistency Checks: The local methods introduced here cannot deliver meaningful results in texture-less areas. By applying a so-called interest operator, e.g. an edge filter (see Canny filter from ▶ Chap. 18), one can restrict the stereo analysis to areas with sufficient contrast. In addition, unreliable correspondences can be filtered out by computing the disparity image twice, swapping the reference and the match image: If pixel (x) in the left image obtains disparity d as a result, pixel (x-d) in the right image must have the disparity –d for the correspondence to be correct. This Right-Left-Check (RL-check) can be applied without computing the disparity map twice [7]. For in-vehicle applications, one has to perform such consistency checks to deal with windshield wiper occlusions, since the camera is usually mounted in the wiper area of the windshield. A windshield wiper is only visible in one image and the (wrong) disparities are removed with the RL check.

▪ Figure 19.2 shows the result of a local correlation method with ZSAD used as similarity metric. Unreliable matches were removed with a RL check. Despite this check occasional red (nearby) points remain in texture-less regions that are obviously wrong.

19.1.2 Global Stereo Methods

Correlation-based methods need sufficient image contrast to yield good results. This is not the case for traffic scenes with texture-less sky and pavements. Stereo analysis can be improved with additional constraints. In typical scenes, depth changes gradually and remains constant on planes parallel to the camera (fronto-parallel). The only depth discontinuities occur at object boundaries. Global stereo algorithms try to incorporate this so-called smoothness constraint in the disparity estimation process and achieve improved results. This algorithms can be subdivided into 1D-optimizing methods along an image line and 2D-optimizing methods across the image.

19.1.2.1 1D Optimization

With the assumption of piecewise constant depth, a new smoothness term appears in addition to the so-called data term represented by the similarity metric. The disparity optimization is interpreted as an energy optimization task minimizing the total energy. The total energy E_{total} hence consists of the similarity E_{data} and the smoothness energy $E_{smoothness}$, summed up for all pixels:

$$E_{total} = \sum_{x,y} (E_{data} + E_{smoothness}) \tag{19.6}$$

For modeling the smoothness energy the following principle has been established: Large depth discontinuities are penalized with a constant energy (P_2), small changes in disparity with a smaller energy P_1:

▪ **Fig. 19.2**   Color-coded disparity image (red = nearby … green = far) overlaid on the original image. The disparity image does not contain a result for every pixel (no color overlay). Isolated outliers (red points in the background) are visible

Fig. 19.3 2 Color-coded disparity image (red = nearby … green = far) obtained by scanline optimization overlaid on the original image. The row-wise optimization leads to streaking artefacts that are even worse in low-contrast regions

$$E_{smoothness} = \begin{cases} 0, if\, |d_1 - d_2| = 0 \\ P_1, if\, |d_1 - d_2| = 1 \\ P_2, if\, |d_1 - d_2| > 1 \end{cases} \qquad (19.7)$$

d_1 and d_2 denote disparities of adjacent pixels. The smaller energy P_1 is designed to correctly reconstruct slanted surfaces that exhibit gradual disparity changes, in most cases smaller than 1 pixel. The simpler Gibbs potential, often used in global stereo schemes, can be obtained by setting $P_1 = P_2$. Often, P_2 is adapted to the image gradient. If an intensity edge is present, P_2 is reduced because a depth discontinuity is more likely to coincide with intensity edges. Such an energy optimization can be efficiently executed along one direction, e.g. along an image row. This method considers every row independently and is known as "scanline optimization" in the literature. If the optimal disparity is found at the end of the row, backtracking is started to find the optimal disparity along the row. This method hence employs the principle of dynamic programming and is known as "dynamic programming stereo" [8]. Both methods can be efficiently implemented in embedded systems but the results exhibit streaking artefacts since the optimization takes place line-by-line independently. Fig. 19.3 shows a scanline optimization example with a 9×7 Census mask as data term.

19.1.2.2 2D Optimization

The streaking artefacts are avoided when the optimization takes all adjacent neighbors into account, i.e. when a two-dimensional optimization is carried out. There are several approaches to find the energy minimum. The most common ones are sketched in the following:

GraphCut: The optimization method GraphCut conducts an energy minimization on a graph. There, pixels are considered as nodes and the connections in between are considered edges. GraphCut is often used for foreground–background segmentation. For such a binary problem with 2 labels, GraphCut is guaranteed to find the optimal solution, i.e. the global minimum. The extension to more than 2 labels leads to an iterative scheme without optimality guarantee, usually giving very good results in practice. The labels in the stereo case are interpreted as discrete disparities [9]. Fig. 19.4. shows an example result using the 9×7 Census similarity metric as data term. The computation time depends on the number of pixels and disparity hypotheses. For typical image size and disparity ranges in driver assistance, computation times are prohibitively large. The optimization method "Belief Propagation" solves the same task as GraphCut and delivers similar results with similar computation times when comparable parameters are used [10].

Semi-Global Matching: With Semi-Global Matching (SGM), streaking artefacts are removed but the computational efficiency of 1D-optmization methods is maintained. SGM conducts a scanline optimization as described above, but this time in multiple directions (typically 8) and the results are summed up [11]. This method, proposed by Hirschmueller in 2005, approximates the 2D optimization by multiple independent 1D optimization steps. Similar results to GraphCut at a fraction of the GraphCut runtime are obtained.

Hirschmueller originally proposed SGM with Mutual Information as similarity metric, a pixel-based metric very sensitive to calibration errors. In driver as-

Fig. 19.4 2 Color-coded disparity image (red = nearby … green = far) obtained by GraphCut overlaid on the original image. A visually error-free disparity image is generated. However, computation time exceeds 10 s (in 2014)

Fig. 19.5 Color-coded disparity image obtained by SGM (red = nearby … green = far) overlaid on the original image. A visually error-free disparity image is generated, similar to the GraphCut result

sistance scenarios and for real-time implementations Census is preferred as similarity metric. Real-time implementations exist for Intel processors [12], GPU [13], and reconfigurable hardware (FPGA) [14]. This method is applied in the Daimler stereo camera. An example can be viewed in Fig. 19.5.

The key to success can be explained like this: Isolated pixels exhibit weak and ambiguous minima in low-textures areas. By comparing disparity hypotheses with the neighbors and the preference of smooth solutions, all pixels can find the correct disparity as a compatible solution, even when the disparity minima of the isolated pixels do not coincide.

An extension to color images is simple. For driver assistance, color-images for stereo matching have not been proven useful when contrasted with the additional computation effort. Using color information leads to worse results than robust similarity metrics such as ZSAD when color constancy is not perfectly fulfilled [15].

19.1.3 Stereo Methods Using Deep Learning

The enormous progress in the field of deep learning offers new options for disparity estimation. Since 2015, most of the newly developed stereo algorithms have utilized deep learning methods. The error metrics on the KITTI stereo benchmark were halved and most of the top 100 methods are based on deep learning.

One drawback of the methods described so far is the use of a manually engineered, fixed similarity criterion. Using machine learning this drawback can be removed and the similarity criterion can be learned.

The first successful example of Zbontar et al. [16] uses stereo training data from KITTI to train correspondences of 9×9px windows. The disparity estimation task is cast as a binary classification problem. Three disparities around the true disparity value are assigned to 1, all other disparities are assigned to 0. The cost volume is generated using a Siamese network architecture. This architecture first computes features using several CNN layers independently for the left and right image. Subsequently, these features are concatenated and connected via additional CNN layers. Using several post-processing steps such as regularization via Semi-Global Matching, the method achieved the top rank of the KITTI stereo benchmark in 2015. With this, a huge number of KITTI submissions using deep learning were triggered, and several follow-up papers used the base CNN architecture with improved post-processing steps.

The DispNet stereo method by Mayer et al. [17] has published a new network architecture as well as an artificial stereo training dataset that is used frequently for CNN trainings. The network architecture uses a hierarchical scheme via image pyramids to reduce compute time and improves convergence in the training step. The generic approach of directly regressing a disparity map from two images was shown and proven to work in general. However, an architecture that exploits the epipolar constraint and limits the disparity search range explicitly delivers better results. A significant improvement based on DispNet was obtained by Sakia et al. [18] with an automatic optimization of the network architecture as part of the training step.

Further improvements were achieved using explicit modelling of stereo processing steps from the classic stereo domain. Kendall et al. [19] uses 3D convolutions similar to the classic cost volume filtering to obtain a larger context region. The disparity estimation itself is trained end-to-end. Cost volume filtering and sub-pixel estimation are modelled in the network architecture, the sub-pixel part uses a differentiable ArgMin-loss.

Semantic information can also be used to robustify disparity estimation. Yang et al. [20] predicts both a disparity map and a semantic segmentation map simultaneously. The base network for semantic segmentation is ResNet50. A regularization loss based on semantics is applied which improves disparity estimation in ambiguous and low-contrast regions. This requires both ground truth for disparity and semantics. An unsupervised version of this algorithm can operate on disparity ground truth only, however, the results are slightly worse.

The Convolutional Spatial Propagation Network (CSPN) by Cheng et al. obtains currently the best results on the KITTi stereo benchmark. It applies ideas from several publications and optimizes them well. 3D CSPN combines correlations from the disparity space and the scale space to be robust against reflections while maintaining accuracy. The optimized spatial pyramid pooling method (SPP) concatenates information across scales, learns the appropriate context and the matching pyramid level. The original SPP method had a fixed connection between pyramid level and context by network design. Linear propagation models are applied to regularize the disparity results recurrently in a 3×3 window, similar to the global stereo methods from the previous section. All these ingredients contribute at similar levels to the excellent KITTi benchmark results. The CSPN consumes about 1 s runtime per image pair on a high-end GPU.

Despite the impressive results of CNN-based stereo algorithms on the KITTi benchmark no stereo systems deploying deep learning methods are available in products as of today. This is rooted on one hand to the high computational demands of such methods. On the other hand, no ground truth data for adverse weather conditions (rain, night, snow, fog) are available for training. This limits the generalization on such scenarios as can be seen in ☐ Fig. 19.6. An augmentation of night and fog effects in training based on real normal weather data is feasible, but the smearing artefacts during rain and snow are currently not sufficiently well modelled. Moreover, the functional validation of such disparity estimation schemes is more complex than with classical schemes.

19.2 Accuracy of Disparity Estimation

The global stereo methods described in the previous section deliver integer disparities. With the distance Z being inversely proportional to the disparity d via

$$Z = fB/d \tag{19.8}$$

with f being the focal length and B the baseline, integer disparities lead to undesired quantization artefacts of the estimated distance, that are only tolerable in the near-range. ☐ Fig. 19.7 illustrates these quantization effects.

These quantization effects can be reduced by using a larger focal length and/or a larger baseline. However, this is counterproductive to the goals of large field of view (i.e. small focal length) and design-friendly small baselines. Alternately, one can increase the imager resolution. Unfortunately, this increases computation and memory demands and reduces the imager sensitivity, when the imager size remains constant due to cost reasons.

Fig. 19.6 Disparity result (bottom) of the MC-CNN method [16] on a rainy highway scene with the windshield wiper in the image. Many false correspondences on the wiper are triangulated as being closeby

Driver assistance asks for large measurement range and high measurement accuracy, which makes sub-pixel estimation indispensable. At the same time, small decalibrations induce further systematic errors. The following section is devoted to these problems.

19.2.1 Sub-pixel Estimation

Deriving the disparity-distance relationship from Eq. 19.8 yields:

$$\frac{\partial Z}{\partial d} = -\frac{f \cdot B}{d^2} = -\frac{Z^2}{f \cdot B} \tag{19.9}$$

Accordingly, depth errors increase quadratically with the distance to the camera, a property of triangulation methods. This uncertainty as a function of distance Z leads to distance uncertainties of several meters in ranges above 50 m, given a stereo system with $f \cdot B = 250\ \mathrm{m} \cdot px$, roughly the stereo camera geometry used in Mercedes cars.

Independent of the applied stereo method, one can generate a sub-pixel estimation with little effort. For quadratic similarity metrics (e.g. SSD), the best known sub-pixel estimation conducts a second order Taylor approximation at the location of the minimum of the similarity function. The minimum of this quadratic polynomial marks the sub-pixel estimate. For linear similarity functions (e.g. SAD, Hamming distance of Census), the so-called equiangular fit has proven its usefulness [22]. In that case, the optimal disparity is determined by intersecting two lines defined by the minimum cost and its neighbors.

Shimizu [16] shows that the proposed sub-pixel disparity estimation leads to a preference of integer disparities. The effect depends on the similarity metric and on the contrast within the considered pixel window. In certain cases with high contrast, the error can amount up to 0.15 px disparity. With the right choice of interpolation function the effect is smaller than 0.1 px. More dramatic is the pixel-locking effect for global stereo methods that interpolate not only similarity costs but also smoothness costs. There, deviations of up to 0.3 px occur. (cf. ◻ Fig. 19.8).

The attainable accuracy of sub-pixel estimation is limited. While theoretical publications predict accuracies of 0.05 px, even perfectly calibrated systems generate accuracies of about 0.25 px on average for all measured points when using spatially discrete stereo methods.

With more computational effort, better estimates can be generated when necessary. In [23] for instance, the energy optimization is conducted on a sub-pixel level (e.g. in quarter pixel steps).

(a)

(b)

Fig. 19.7 A 3D reconstruction using SGM based on the disparity map of ◻ Fig. 19.5. At the top, the triangulated result without sub-pixel estimation, at the bottom the result with sub-pixel estimation is shown. Without sub-pixel estimation the oncoming car "decays" into two parts

19.2.2 **Effects of Decalibration**

So far, the sub-pixel analysis of disparity errors assumed an ideal camera system with perfect compensation of all lens errors and camera orientation errors.

Methods to determine these parameters, i.e. to perform a calibration, are introduced in ▶ Sect. 17.5.

All proposed lens models can only approximately represent the reality, and the calibration methods have residual errors, even when the system is designed and ana-

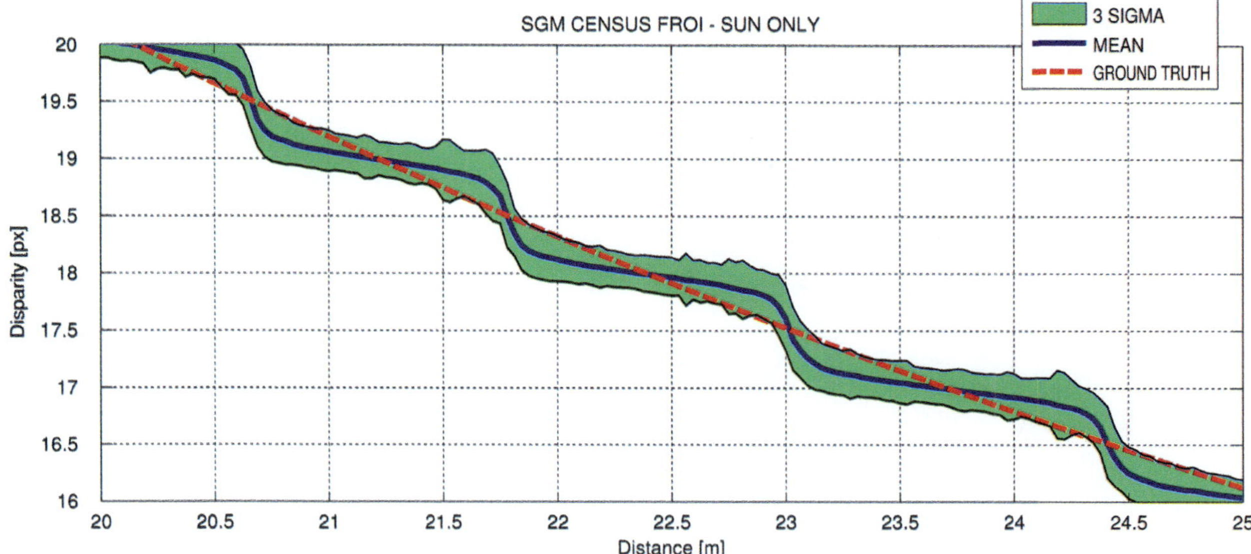

Fig. 19.8 Disparity progression approaching a planar obstacle. The reference disparity is generated by a rendered scene and the measured disparity using SGM are shown. Deviations of up to 0.3 px occur due to the smoothness term that implicitly favors integer disparities

lyzed very carefully. Residual errors in relative roll and pitch angle affect the correspondence analysis directly.

Roll and Pitch angle errors: As soon as the corresponding points do not lie on the same line after rectification, the correspondence search is affected. For mostly vertical structures such as poles, the problem is not apparent. The effect is amplified for mostly horizontal structures and leads to invalid or even worse, to wrong disparities. An extreme case with mostly horizontal structure is shown in ◘ Fig. 19.9. At the top, the correct result with perfect calibration is shown. At the bottom, the result with only 0.2px error in epipolar geometry is shown. The wall is triangulated significantly further away than it really is.

Squint angle error. Even more dramatic are small errors in squint angle or relative yaw angle of the camera system that generate a disparity offset Δd. If the object is to be found at distance Z with disparity d, one obtains an estimate $\widehat{Z}$ via:

$$\widehat{Z} = \frac{Z}{\left(1 + \frac{\Delta d}{d}\right)} \qquad (19.10)$$

◘ Figure 19.10 shows that this effect is not negligible for larger distances. The shown curves are valid for the stereo camera geometry described above. If one underestimates the disparity by one pixel at 60 m distance, the distance is overestimated by 20 m! If such an error is not detected, it can cause problems in a sensor fusion stage.

This effect is amplified when velocities are computed based on the distance change (cf. ▶ Sect. 19.3). Let the relative velocity be v_{rel}, then one obtains for the estimate $\widehat{v}_{rel}$:

$$\widehat{v}_{rel} = \frac{v_{rel}}{\left(1 + \frac{\Delta d}{d}\right)^2} \qquad (19.11)$$

Most of the time, the absolute velocity is searched. Subtracting the (known) ego-velocity on obtains:

$$\widehat{v}_f = v_{ego}\left(1 - 2\frac{\widehat{v}_{rel}}{v_{rel}}\right) \qquad (19.12)$$

If one follows a leading car with small relative velocity, this error is negligible. This is very different for oncoming cars as shown in ◘ Fig. 19.11. The estimated velocity is shown over the vehicle distance when both cars approach each other with 50 km/h. If the disparity offset is 1 px, velocities well above highway speeds are measured for trucks in urban scenarios.

This observation shows that squint angle errors are critical in stereo camera systems, especially for velocity estimation. The outer orientation of a stereo rig is changing due to aging and moreover due to temperature fluctuations. This makes an online calibration algorithm necessary for in-vehicle usage. By comparing with a reference sensor, it is possible to keep the disparity offset well below 0.1 px which is tolerable even for critical scenarios.

(a)

(b)

Fig. 19.9   Disparity map of a fronto-parallel wall with horizontal structure with correct calibration (left) and with small decalibration (0.2 px in relative tilt angle) which causes the wall to be a very distant obstacle

19.3 6D-Vision

Ideally, the disparity estimation delivers an unbiased estimate of the 3D position. Driver assistance systems, especially emergency brake systems, need a reliable detection of moving objects along with an accurate motion estimate and a detection confidence measure to assess the criticality of a traffic situation. The intuitive approach to extract objects from the disparity map and track them over time has not proven to be robust enough. The 3D object separation capability is very limited at larger distances due to the quadratically decreasing measurement accuracy. In an example shown in **Fig. 19.12 the car and the child cannot be separated and the criticality of the situation might be overlooked.

Hence, it is necessary to estimate the 3-dimensional motion of the image points directly in order to allow the detection of moving objects without error-prone prior segmentation. An example for a suitable algorithm is presented in the following.

19.3.1 The Principle

A substantial benefit of the camera sensor lies in the capability of finding corresponding points from the previous frame again, i.e. the points can be tracked. Optical flow methods and feature tracking methods have been widely researched for many years and are well understood. For every corresponding image point

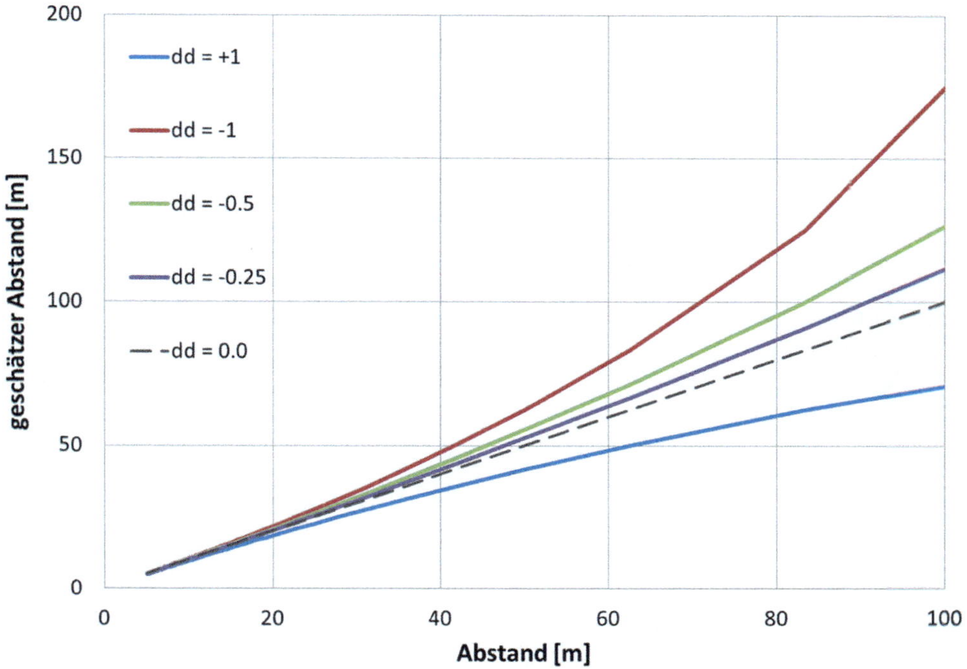

Fig. 19.10 Estimated distance over true distance as a function of disparity (squint angle) error Δd

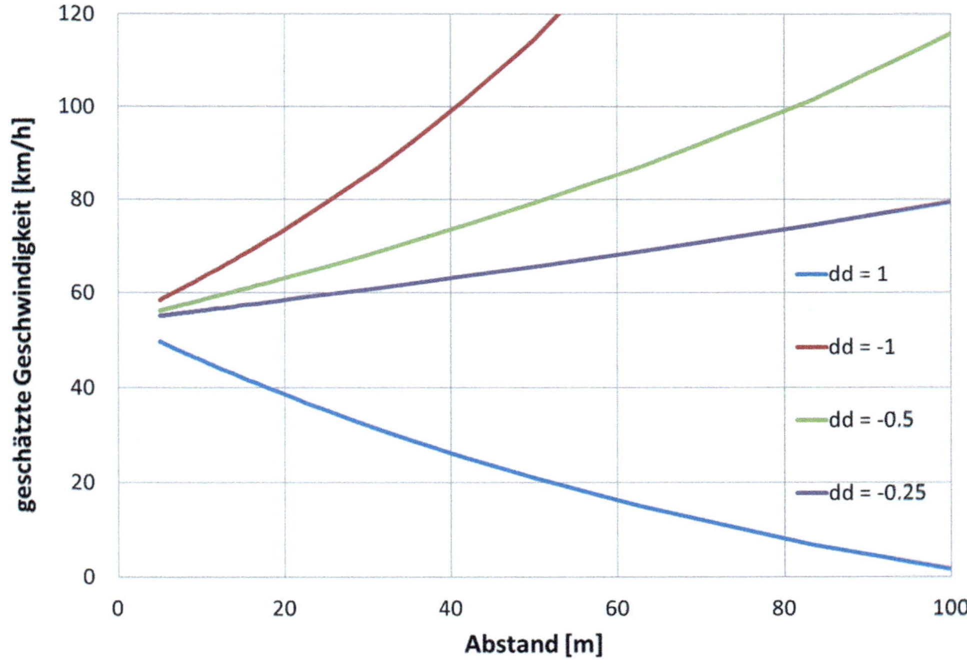

Fig. 19.11 Estimated velocity of an oncoming car driving 50 km/h with the observer moving at the same speed. At larger distances the errors increase dramatically

pair with 3D information, one could determine the motion vector by simple differentiation. However, due to the distance uncertainty of the considered measurements and the small time difference of typically 40 ms, the obtained motion estimations are extremely noisy. This behavior can be improved by increasing the time basis to larger time differences, but this is counterproductive for fast responses when objects appear suddenly.

The central idea of the 6D-Vision concept [18] is to track relevant image points over several frames and to reduce the measurement uncertainty by temporal inte-

◻ Fig. 19.12   Critical intersection scenario—the time-to-collision for the running kid behind the car is 1 s

gration. The goal is to deliver an optimal estimate of the motion for every frame. For that purpose, the pixels are modeled as objects with mass, which move with constant velocity through space.

This allows to estimate the pixel motion by means of the Kalman Filter (see ▶ Chap. 18). The 3D position $\vec{p} = (X, Y, Z)^T$ of an observed image point and its velocity vector $\vec{v} = (\dot{X}, \dot{Y}, \dot{Z})^T$ are stacked to a six-dimensional state vector $(X, Y, Z, \dot{X}, \dot{Y}, \dot{Z})^T$. After a time interval Δt, the new position at time k + 1 is:

$$\vec{p}_{k+1} = R\vec{p}_k + \vec{T} + \Delta t R \vec{v}_k \tag{19.13}$$

R denotes the rotation and T the translation of the scene, i.e. representing the inverse camera motion. For the velocity vector assuming constant motion one obtains:

$$\vec{v}_{k+1} = R\vec{v}_k \tag{19.14}$$

With that the time-discrete linear system of the Kalman Filter is

$$\vec{x}_k = A_k \vec{x}_{k-1} + B_k + \vec{\omega} \tag{19.15}$$

with the mean-free Gaussian noise term $\vec{\omega}$. The state transition matrix A_k is

$$A_k = \begin{bmatrix} R_k & \Delta t_k R_k \\ 0 & R_k \end{bmatrix} \tag{19.16}$$

and the control matrix is given by

$$B_k = \begin{bmatrix} \vec{T}_k \\ 0 \\ 0 \\ 0 \end{bmatrix} \tag{19.17}$$

The measurement vector $z = (u, v, d)^T$ consists of the tracker's current image position $(u, v)^T$ and the disparity measured by the stereo system. The easy-to-linearize measurement model with noise term $\overrightarrow{\omega_{mess}}$ reads as:

$$z = \begin{bmatrix} u \\ v \\ d \end{bmatrix} = \frac{1}{Z} \begin{bmatrix} Xf \\ Yf \\ bF \end{bmatrix} + \omega_{mess} \tag{19.18}$$

The compensation of the ego-motion causes a correct measurement of $v = 0$ for stationary points. If the vehicle is equipped with suitable inertial sensors, the sensor values for translation and rotation can be used directly. If a simple planar rotation is assumed relying solely on a yaw rate sensor, pitch motion is not measured, which results in a misinterpretation of stationary points being vertically moving. Alternately, these parameters can be estimated via visual odometry from the camera images. The authors rely successfully on the method developed by Badino [19].

The literature offers a good selection of feature tracking methods. For driver assistance, tracking large

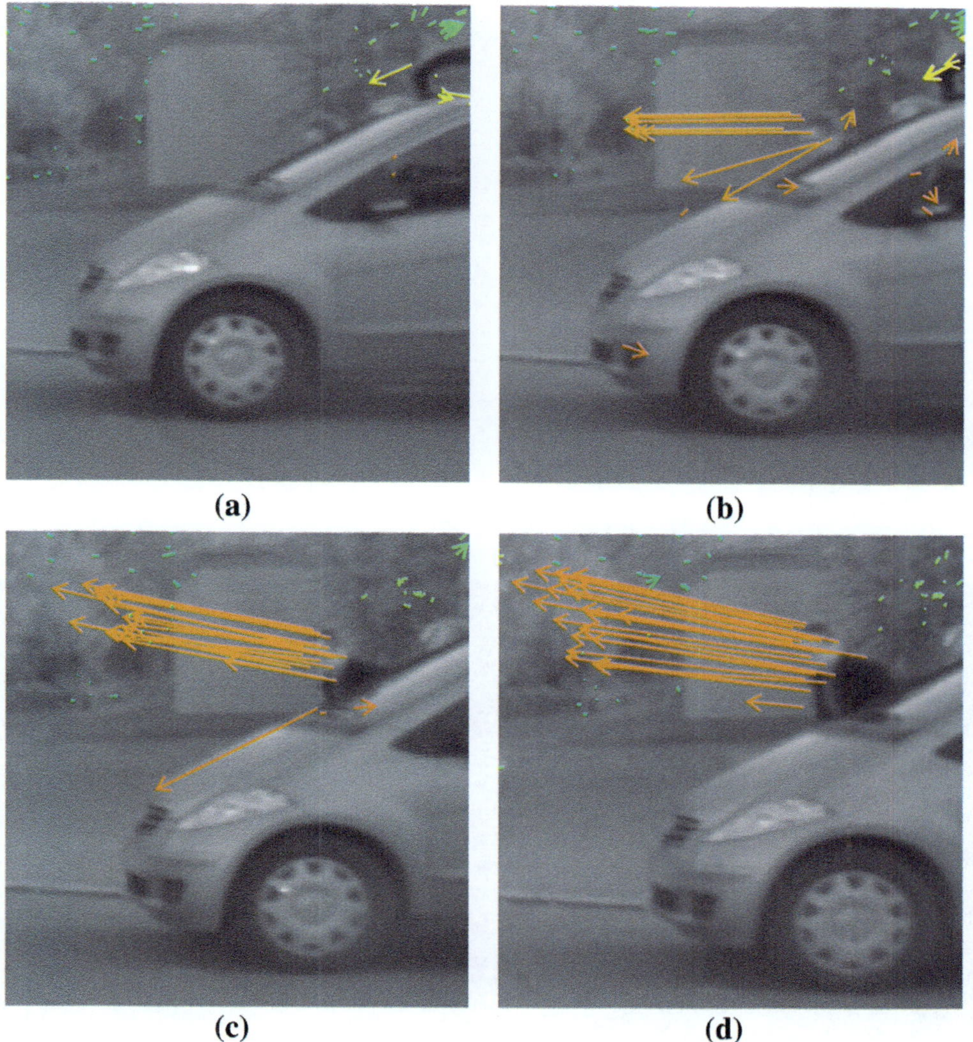

Fig. 19.13 Four enlarged image sections show the result of the 6D-Vision estimate for the situation from Fig. 19.12. The arrows point to the predicted position in 500 ms

displacements is vital, so descriptor-based feature trackers (cf. ▶ Chap. 18) are of interest. The authors use a real-time variant of the method proposed by Stein in 2004 [20] that is able to measure arbitrary large displacements with constant time and typically 10% of the image points matched. Imagery with oncoming cars or driving through tight curves generates displacements of more than 100 px/frame. Another vital property of feature tracking is the insensitivity to illumination changes, which often occur in practice. If the prediction from the 6D Kalman filter is used, one can also successfully work with robust variants of the popular Kanade-Lukas-Tomasi Tracker [21].

Figure 19.13 illustrates the performance of the described approach. The arrows depict the motion state of the considered pixels and point to the predicted position in 500 ms. For viewing reasons, every other frame is skipped so the time difference is 80 ms for the shown images. 200 ms are sufficient to robustly detect critical situations of that type, as proven by many experiments. The ego-vehicle from Fig. 19.13 drove about 30 km/h and would have collided with the pedestrian without intervention.

Another example is depicted in Fig. 19.14. The scenario shows a cyclist making a left turn in front of the approaching ego-vehicle. Here, the assumption of partial linear motion is confirmed: Although the cyclist turns (circular motion), the orientation of the estimated motion agrees well with reality.

The Kalman filter introduced above delivers position and velocity estimates of the tracked image point. For initialization, the position is readily available via stereo measurement. The initial velocity cannot be determined from a single position measurement. It must be computed with suitable velocity hypotheses and velocity variances. This aspect is covered by Franke et al. [18] in more detail.

(a)

(b)

Fig. 19.14 Turning cyclist scenario depicted at the top. At the bottom, the 6D-Vision interpretation shown in a 3D viewer. The arrows point to the predicted position in 500 ms. The colors encode distance (red = near … green = far)

19.3.2 **Dense6D**

The previous section introduced 6D-Vision for isolated image points that can be localized in space and time. For reasons of robustness and accuracy, a high measurement density is desirable. Ideally, a measurement of position and motion vector is available for all pixels. For that purpose, both optical flow and stereo methods should determine results for every pixel. The SGM method described above is able to deliver depth information for almost every pixel. Several dense optical flow schemes have been published recently. However, they were not used due to high computational demands and lack of robustness.

Classical methods to determine dense optical flow use the „Constant Brightness Assumption ", i.e. the intensity of corresponding points is assumed to be identical. From that, a data term is generated that minimizes the costs by varying the displacements. Similar to global stereo methods, a smoothness term is also introduced that penalizes changes of displacements of neighbor-

Fig. 19.15 Dense6D result for a crossing scenario. Colors encode the velocity magnitude (green=slow … red=fast). The vectors depict the predicted position in 500 ms

ing pixels. For computational reasons, smoothness deviations were penalized quadratically in older publications which resulted in high sensitivity to outliers. Especially the constant displacement assumption is not fulfilled for traffic situations and leads to oversmoothing.

Zach et al. [22] proposes a dense optical flow method that can be computed in real-time on high resolutions. They employed the computing power of modern graphics card (GPU). The implemented TV-L1 method uses total variation and penalizes smoothness deviations linearly, not quadratically. This results in more accurate flow fields, but is still error-prone to illumination changes. Müller et al. [24] shows that dense optical flow can be robustly computed under illumination changes by using the Census operator.

Despite this progress, the TV-L1 method tends to oversmooth the flow field and exhibits difficulties with large displacements. Müller [23] proposes two extensions to improve the method: Firstly, to estimate large displacements, measurements of a local flow method are used. Secondly, the smoothness constraint is modified to penalize the deviation from the expected flow field determined by stereo and ego-motion information under the assumption of a static world.

When optical flow and stereo information is available for almost every pixel, the 6D-Vision concept can be applied. Rabe et al. [25] presented a real-time variant under the name Dense6D in 2010. The Kalman filters are organized as iconic filters in 2D, similar to the image structure. By computing the flow field from the current to the previous frame for every pixel, one can identify the predecessor and continue the tracking chain.

Another type of method to determine both motion and depth for every pixel is called SceneFlow, where disparity change and optical flow is estimated simultaneously. From the disparity change and the original disparity, one can compute the velocity relative to the observer. As Rabe et al. [25] showed in their experiments, the velocity field obtained by these differential methods is significantly inferior to the Dense6D result. It seems that one gains more from the strong temporal smoothness of the 6D-approach than from the optimized spatial smoothness of the Scene Flow approach.

A Dense6D example result is shown in Fig. 19.15.

19.4 The Stixel World

Thanks to the real-time capability of dense stereo methods, subsequent processing steps can use about 2,000,000 3D points per frame (as of 2020). In upcoming years, the number of 3D points is expected to in-

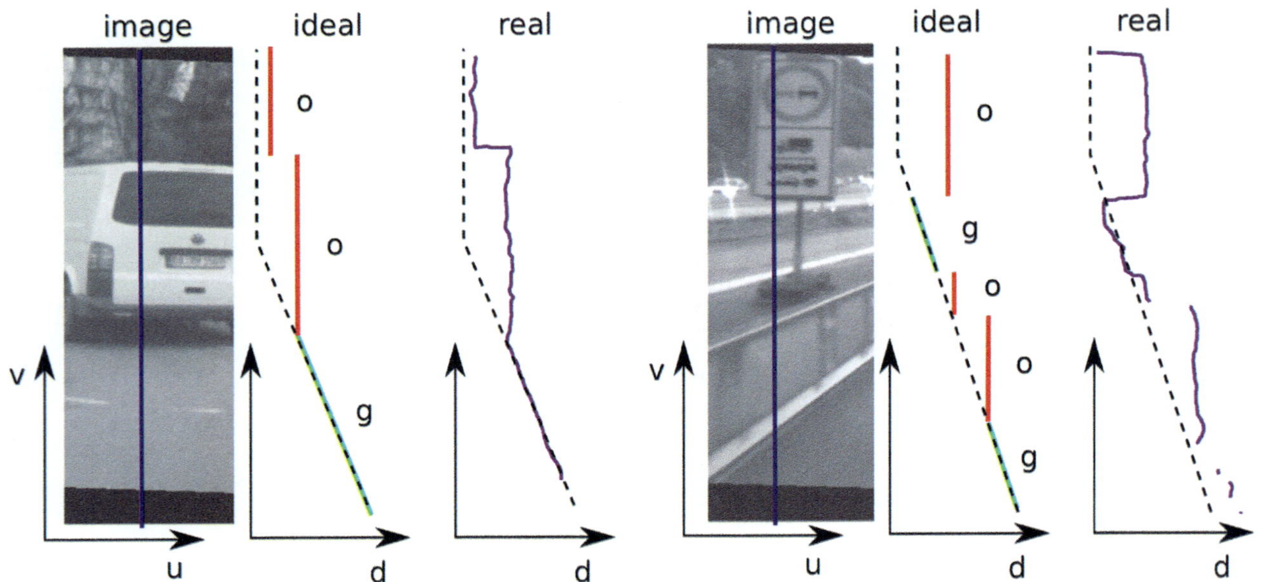

Fig. 19.16 Representation of the scene from **Fig. 19.5** with Stixels. Distances are color-coded, white rectangles define the Stixel outline. The white arrows indicate speed and motion direction

crease to 5–8 million points. At the same time, more detection modules for pedestrians, cyclists, vehicles, stationary obstacles, free space, road height profile etc. will use this information to improve their performance. This will lead to extreme demands on computation power and bandwidth.

This problem can be circumvented with a more compact representation that clusters image points to superpixels. A superpixel method suitable for traffic scenes is the Stixel World, proposed by Badino et al. [26] in 2009. As shown in **Fig. 19.16** the complete 3D information is approximated by a few hundred thin rectangular sticks defined by base point, distance and height. If these Stixels are tracked over time following the 6D-Vision concept, their motion state is available as well. Since the number of Stixels is small, moving object segmentation can be performed with a global optimal method. The steps from pixels via Stixels to objects are described in the following.

19.4.1 Optimal Computation of the Stixel-World

Traffic scenes are dominated by (approximately) horizontal and vertical planes. The most prominent plane is the street or ground plane. Objects such as cars, pedestrians and infrastructure stand on this ground plane. Only in rare cases, e.g. for bridges, objects do not touch the ground. The Stixel World exploits this property to generate a compact and robust representation of the scene.

The most powerful method to date to compute this representation was introduced by Pfeiffer et al. [27] in 2011. There, the computation was cast to a maximum-a-posteriori (MAP) problem solved by dynamic programming. For that purpose, small stripes of 5 to 9 pixels are considered independent of their neighbors, resulting in independent one-dimensional optimization tasks.

Figure 19.17 illustrates the basic idea of the Stixel World giving two examples. The blue line marks the considered image column/stripe. The measured disparities are shown in violet. A segmentation into the classes "object" and "ground" is sought. Pixels of an object have approximately the same disparity while pixels of the class "ground" have a linear disparity progression given by the camera geometry. The left case depicts the standard case of a dominating foreground object in front of a background object. For the right example, we expect a representation of 3 objects and two parts labeled as "ground".

The task of the optimization step is to find the most probable approximation of the disparity measurement vector D complying with the model. So we seek the most probable labeling L^*, formally:

$$L^* = \arg\max_{L} P(L|D) \tag{19.19}$$

Applying Bayes rule, the a-posteriori probability distribution can be converted to $P(L|D) = \frac{P(D|L)P(L)}{P(D)}$. With that, the optimization can be written as:

$$L^* = \arg\max_{L} P(D|L)P(L) \tag{19.20}$$

Fig. 19.17 Concept of the Stixel World: The purple disparity curve along a column is approximated by constant parts (objects, shown in red) and linear parts following the ground layout (ground, shown in green)

$P(D|L)$ represents the probability density of the observed disparity vector for any labeling L and can be considered the data term. The second term $P(L)$ accounts for the fact that not all possible object orderings are equally likely. This way, prior knowledge or statistics about the typical layout of traffic scenes can be incorporated. This represents the smoothness term of global stereo methods without limiting the term to smooth solutions. More constraints can be modeled with the following being the most powerful:

- **Bayes information criterion:** within a column only few objects are expected, solutions with few objects are preferred.
- **Gravitational constraint:** Hovering objects are unlikely. Objects with a base point close to the ground are preferred to touch the ground.
- **Ordering-Constraint:** The higher a Stixel in the image is, the further away it usually is. Bridges, trees and other objects violating this constraint are still approximated correctly when the data term supports it sufficiently.

A more detailed description of the prior term $P(L)$ can be found in [27]. Since the optimization task is a one-dimensional problem, dynamic programming can be efficiently used to solve it. For efficiency reasons, the disparity measurements are treated to be independent. The disparity distribution $P(D|L)$ is modeled as a Gaussian distribution overlaid with a uniform distribution to be robust against outliers.

Figure 19.16 shows the Stixel World for the scene introduced in ▶ Sect. 19.1. The distance is color-coded as before. White vertical rectangles define the outline of the foreground Stixels, the ground plane is shown in grey. The vehicles are clearly separated form the back-ground, the tree on the right is correctly modeled, despite violating the gravitational constraint.

In the example shown here, arrows on the ground depict velocity and motion direction of the Stixels. If the Stixels are considered "large" pixels, the 6D-Vision concept can be applied without changes. For the purpose of driver assistance, a 4 –dimensional state vector suffices since height and vertical motion are not important. With that, the pitch motion is not critical to estimate anymore. This "dynamic Stixel World" is the basis for subsequent high-level vision modules.

The Stixel World is an approximate representation of 3D data. It offers the following desired properties:

- **Robustness:** The implicit averaging of all disparities within a Stixel leads to disparity noise reduction. Thanks to the robust modeling, local disparity errors can be detected and eliminated. Larger occasional errors (e.g. from the windshield wiper) that are not consistent over time are removed by the temporal filtering.
- **Compactness:** The Stixel World is very compact. The 3D scene content of an image is represented by 300-600 Stixels, which can be encoded in a few kilobytes representing the full geometry of the scene including motion information.
- **Explicit Representation:** the representation extracts the scene content. If the Stixel World shown in ▶ Fig. 19.16 is viewed without the grey scale image, humans are still able to understand the scene. The implicit classification for street and object also encodes the drivable free space, which is necessary for the trajectory planning of swerve maneuvers.

The Stixel World models 3D data and is hence not limited to disparity maps. Also, data from high resolution

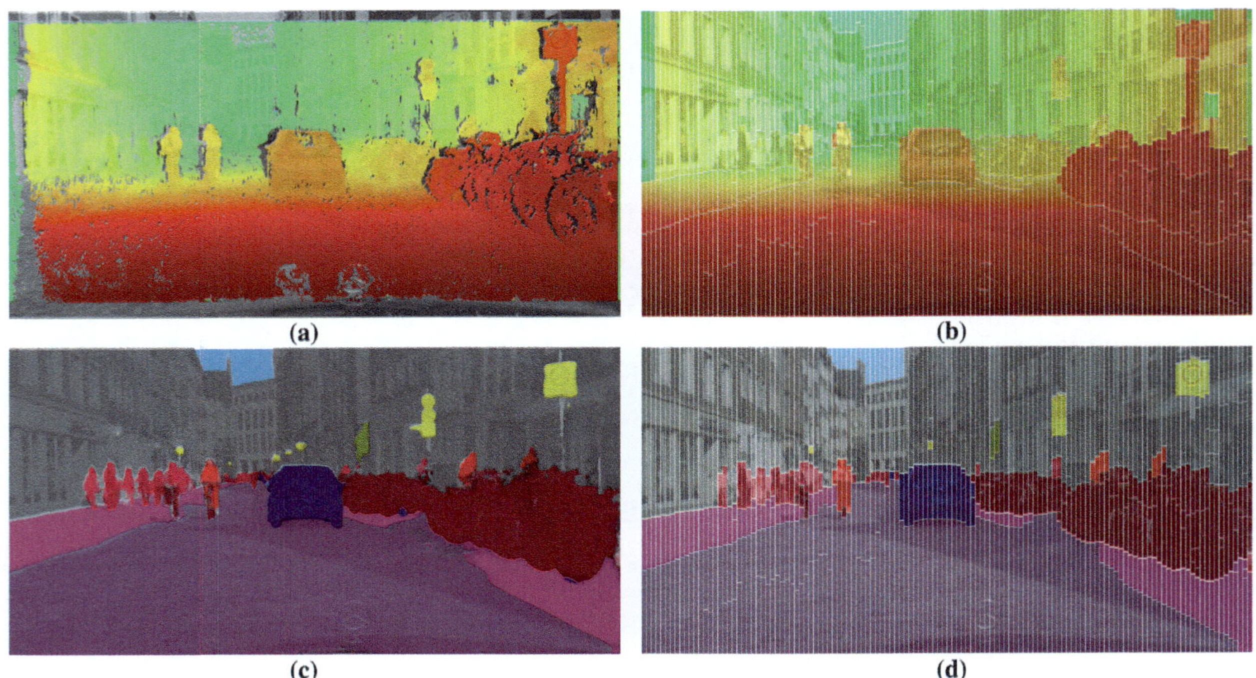

■ **Fig. 19.18**   **a** Disparity image and **b** the resulting Stixel optimization according to ▶ Sect. 4.1. **c** Result of semantic segmentation and **d** the resulting semantic Stixels obtained by fusion of semantics and disparity

Laser scanners, e.g. the Velodyne HD64, can be represented that way [34]. Such a unified representation exhibits clear advantages for the sensor fusion stage.

19.5 Semantic Stixels

The purely geometric representation of the scene ignores valuable information from the (monocular) images. Especially the reduction of the world into the classes road, obstacle, and sky does not match the requirements for automated driving. For this task, it is necessary to detect and classify all relevant traffic participants, e.g. separating a pedestrian leaning on a tree from the static tree, recognizing traffic signs and lights, separating sidewalk from road etc. Ideally every Stixel is assigned to a specific object class. We try to find a Stixel representation that includes both geometric as well as semantic information for subsequent processing steps.

Key ingredient for this goal is the semantic segmentation described in ▶ Chap. 44 which assigns a class label to every pixel in the image. Schneider et al. [35] introduce an elegant approach for fusing geometry and semantics. Instead of greedily assigning the majority class to every Stixel, they extend the optimization procedure described above. A new data term is introduced containing the class label probabilities per pixel. In the optimization step a solution with one class label per Stixel is preferred. Inconsistencies between stixel class

label and pixel-wise class label increase the overall energy.

■ Figure 19.18 shows the difference between the purely geometric Stixel world and the semantic Stixel world obtained by considering semantics. Sidewalk and road are clearly separated, the pedestrians on the left side are separated from the building wall behind them, traffic signs are classified in yellow, and the oncoming cyclists are separated in bicycle and rider Stixels.

There is an additional benefit of this joint optimization: One apparent weakness of the purely geometric analysis of the traffic scene are small ghost Stixels which might occur due to slight decalibrations on horizontal structures (stop lines or tram tracks) or due to adverse weather conditions (rain, wet road reflections). Such ghosts are almost completely eliminated by incorporating semantics. The classification result for the class "road" is extremely reliable and successfully votes against the presence of potential ghost Stixels. Hence, false braking events due to ghost Stixels are avoided, in contrast to early purely stereo-based approaches.

The label for "road" also helps to correctly model hilly roads which to do not comply with a planar world model. For extreme inclines known from San Francisco streets, a Stixel extension was proposed by Hernandez [36] which applies a piecewise linear approximation of the 3D road surface. For the different variants of Stixel world public implementations are available under ▶ https://github.com/gishi523/stixel-world and ▶ https://github.com/dhernandez0/stixels.

Fig. 19.19   Segmentation of the dynamic Stixel world in single, independently moving objects

19.5.1 Instances

Stixels represent the traffic scene geometrically in polar coordinates and contribute semantic and dynamic information in addition. They can seamlessly transferred into Cartesian grid representations as frequently used in fusion and behavior planning.

In many cases, a separation of the static background from the dynamic instances of traffic participants is demanded. In an early work by Erbs et al. Error! Reference source not found. This task is solved by graph optimization. Stixels are considered a node of a "Conditional Random Field" (CRF) and an energy minimization problem is formulated and solved via GraphCuts. If one is only interested in the separation from the stationary background, this binary problem can be solved optimally. Erbs also proposes an iterative approach that generates instances of an unknown number of moving objects step-by-step.

As depicted in ◘ Fig. 19.19 oncoming vehicles are segmented and their motion vector is estimated. It is recommended to apply an object-specific tracker on these Stixel groups. Barth et al. Error! Reference source not found. shows that this even allows to estimate the yaw rate of oncoming vehicles with such a tracker.

In a more recent approach, Hehn et al. [39] apply an image-based instance segmentation method proposed by Uhrig et al. [40] to separate independent traffic participants in the scene. Every pixel points to the geometric center of its corresponding instance. The computation of this instance image runs in parallel to the semantic segmentation using an extended CNN. An additional data term allows for an assignment of every Stixel to a specific instance ID after optimization. ◘ Fig. 19.20 illustrates the step from semantic Stixels to so-called instance Stixels. This instance information offers the possibility to separate adjacent Stixels of the same class when they belong to different objects.

19.6 Summary

The presented methods of stereo image processing are the result from years of research and have proven to be powerful in practice. Semi-Global Matching and the 6D-Vision concept became available in premium and midsize cars at Mercedes Benz in 2013. The applied classification methods benefit heavily from the dense stereo information. The false alarm rate is reduced by a factor of five at the same detection rate [33]. Here the advantage of a stereo-based approach compared to a monocular approach is clearly visible.

Also, the Stixel World and its subsequent processing steps have proven their effectiveness. The algorithms formed the basic scene understanding module within the Bertha-Benz autonomous drive in August 2013. There, a Mercedes S500 intelligent drive vehicle drove autonomously the famous route of Bertha Benz' first overland drive from Mannheim to Pforzheim, Germany with sensors close to serial production. Due to the compact representation of the environment, computation-intense methods can be run without runtime problems [34].

The vision of autonomous driving and the desire for more elaborate driver assistance systems will further drive the demands for more accuracy and robustness of stereo vision algorithms. Such stereo camera systems will further gain in importance as full redundancy to time-of-flight sensors thanks to their model-free 3D information. The CNN-based disparity approaches clearly represent state-of-the-art on public benchmarks such as KITTI. However, due to their high computational demand, these approaches are not yet available in commercial products.

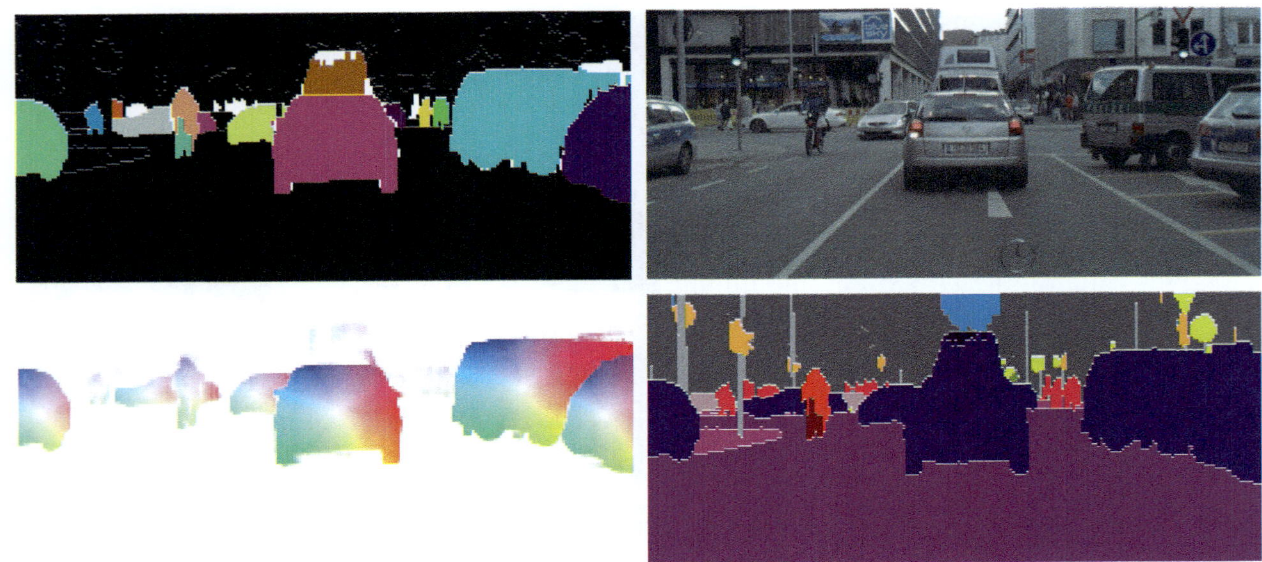

Fig. 19.20 Grouping of Stixels to instances: An independent instance segmentation supports this grouping (bottom left). The color encodes direction and distance to the instance center. The top right figure depicts the semantic Stixel result. On the bottom right, the resulting instance groups are shown in a random color coding

References

1. Dickmanns, E., Zapp, A.: A curvature-based scheme for improving road vehicle guidance by computer vision. SPIE **727**, 161–168 (1986)

2. Meister, S., Jähne, B., Kondermann, D.: Outdoor stereo camera system for the generation of real-world benchmark data sets, J. Opt. Eng., **51**(02), (2012)

3. Scharstein, D., Szeliski, R.: A Taxonomy and evaluation of dense two-frame stereo correspondence algorithms. Int. J. Comput. Vis. (IJCV), Kluwer, (2002)

4. Geiger, A., Lenz, P., Urtasun, R.: Are we ready for autonomous driving? The KITTI vision benchmark suite, computer vision and pattern recognition (CVPR), (2012)

5. Hirschmüller, H., Gehrig, S.: Stereo matching in the presence of sub-pixel calibration errors, computer vision and pattern recognition (CVPR), (2009)

6. Hirschmüller, H., Scharstein, D.: Evaluation of stereo matching costs on images with radiometric differences. IEEE Trans. Pattern Anal. Mach. Intell. **31**(9), 1582–1599 (2009)

7. Mühlmann, K., Maier, D., Hesser, J., Männer, R.: Calculating dense disparity maps from color stereo images, an efficient implementation, CVPR workshop on stereo and multi-baseline vision: pp. 30–36, (2001)

8. Belhumeur, N.: A Bayesian approach to binocular stereopsis. Int. J. Comput. Vis. **19**(3), 237–260 (1996)

9. Boykov, Y., Veksler, O., Zabih, R.: Fast approximate energy minimization via graph cuts. IEEE Trans. Pattern Anal. Mach. Intell. **23**(11), 1222–1239 (2001)

10. Tappen, M., Freeman, W.: Comparison of Graph Cuts with Belief Propagation for Stereo, using Identical MRF Parameters, International Conference on Computer Vision (ICCV), (2003)

11. Hirschmüller, H.: Stereo processing by semi-global matching and mutual information. IEEE Trans. Pattern Anal. Mach. Intell. **30**(2), 328–341 (2008)

12. Gehrig, S., Rabe, C.: Real-time Semi-Global Matching on the CPU. CVPR Embedded Computer Vision Workshop, San Francisco, CA (2010)

13. Banz, C., Blume, H., Pirsch, P.: Real-time semi-global matching disparity estimation on the GPU, ICCV workshop on mobile computer Vision, Barcelona, Spain, (2011)

14. Gehrig, S., Eberli, F., Meyer, T.: A Real-Time Low-Power Stereo Engine Using Semi-Global Matching, International Conference on Computer Vision Systems, Liege, Belgium, (2009)

15. Bleyer, M., Chambon, S.: Does Color Really Help in Dense Stereo Matching? International Symposium 3D Data Processing, Visualization and Transmission (3DPVT), pp. 1–8, Paris, France, (2010)

16. Zbontar, J., LeCun, Y.: Stereo matching by training a convolutional neural network to compare image patches. J. Mach. Learn. Res. **17**, 1–32 (2016)

17. Mayer, N., Ilg, E., Häusser, S., Fischer, P.;,Cremers, D., Dosovitskiy, D., Brox, T.: A large dataset to train convolutional networks for disparity, optical flow, and scene flow estimation. CVPR (2016)

18. Saikia, T., Marrakchi, Y., Zela, A., Hutter, F., Brox, T.: AutoDispNet: improving disparity estimation with AutoML. In: The IEEE International Conference on Computer Vision (ICCV) (2019)

19. Kendall, A., Martirosyan, H., Dasgupta, S., Henry, P., Kennedy, R., Bachrach, R., Bry, A.: End-to-End learning of geometry and context for deep stereo regression. In: Proceedings of the International Conference on Computer Vision (ICCV) (2017)

20. Yang, G., Zhao, H., Shi, J., Deng, Z., Jia, J.: SegStereo: exploiting semantic information for disparity estimation. ECCV (2018)

21. Cheng, X., Wang, P., Yang, R.: Learning depth with convolutional spatial propagation network. In: IEEE Transactions on Pattern Analysis and Machine Intelligence (T-PAMI) (2019)

22. Shimizu, M., Okutomi, M.: Precise sub-pixel estimation on area-based matching. In: International Conference on Computer Vision (ICCV), (2001)

23. Gehrig, S., Badino, H., Franke, U.: Improving sub-pixel accuracy for long range stereo. J. Comput. Vis. Image Underst. **116**(1), 16–24 (2012)

24. Franke, U., Rabe, C., Badino, H., Gehrig, S.: 6D-Vision: Fusion of Stereo and Motion for Robust Environment Perception, DAGM Symposium, Wien (2005)

25. Badino, H., Franke, U., Rabe, C., Gehrig, S.: Stereo Vision-Based Detection of Moving Objects under Strong Camera Motion. VisApp, Portugal (2006)
26. Stein, F.: Efficient Computation of OpticaL Flow, DAGM Symposium (2004)
27. Shi, J., Tomasi, C.: Good features to track, Computer Vision and Pattern Recognition (CVPR), Seattle. (1994)
28. Zach, C., Pock, T., Bischof, H.: A Duality Based Approach for Realtime TV-L1 Optical Flow, DAGM Symposium (2007)
29. Müller, T., Rabe, C., Rannacher, J., Franke, U., Mester, R.: Illumination-Robust Dense Optical Flow Using Census Signatures, DAGM Symposium (2011)
30. Müller, T., Rannacher, J., Rabe, C., Franke, U.: Feature- and Depth-Supported Modified Total Variation Optical Flow for 3D Motion Field Estimation in Real Scenes, Computer Vision and Pattern Recognition (CVPR), (2011)
31. Rabe, C., Müller, T., Wedel, A., Franke, U.: Dense, Robust, and Accurate Motion Field Estimation from Stereo Sequences in Real-time, European Conference on Computer Vision (ECCV), (2010)
32. Badino, H., Franke, U., Pfeiffer, D.: The Stixel World—A Compact Medium Level Representation of the 3D-World, DAGM Symposium (2009)
33. Pfeiffer, D. und Franke, U.: Towards a Global Optimal Multi-Layer Stixel Representation of Dense 3D Data, British Machine Vision Conference (BMVC), (2011)
34. Piewak F., Pinggera P., Enzweiler M., Pfeiffer D., Zöllner M.: Improved Semantic Stixels via Multimodal Sensor Fusion, German Conference on Pattern Recognition, Stuttgart, (2018)
35. Schneider, L., Cordts, M., Rehfeld, T., Pfeiffer, D., Enzweiler, M., Franke, U., Pollefeys, M., Roth, S.: Semantic stixels: depth is not enough In: IEEE Intelligent Vehicles Symposium, Göteburg (2016)
36. Hernandez-Juarez, D., Schneider, L., Cebrian, P., Espinosa, A., Vazquez, D., Lopez, A.M., Franke, U., Pollefeys, M., Moure J.C.: Slanted Stixels: A Way to Represent Steep Streets, International Journal on Computer Vision 127, (2019)
37. Erbs, F., Schwarz, B., Franke, U.: From stixels to objects, In: IEEE Intelligent Vehicles Symposium, Brisbane, (2013)
38. Barth, A.; Franke, U.: Estimating the Driving State of Oncoming Vehicles from a Moving Platform Using Stereo Vision, IEEE Transactions on ITS, (2009)
39. Hehn, T., Kooij, J.F.P., Gavrila, D.M.: Instance Stixels: Segmenting and Grouping Stixels into Objects. IEEE Intelligent Vehicles Symposium, Paris (2019)
40. Uhrig J., Cordts M., Franke U., Brox T.: Pixel-level encoding and depth layering for instance-level semantic labeling. In: German Conference on Pattern Recognition, Hannover, (2016)
41. Enzweiler, M., Gavrila, D.M.: A Multi-Level Mixture-of-Experts framework for pedestrian classification, IEEE Trans. on Image Processing, (2011)
42. Franke, U., Pfeiffer, D., Rabe, C., Knöppel, C., Enzweiler, M., Stein, F., Herrtwich, R.G.: Making bertha see, ICCV workshop computer vision for autonomous driving, Sydney, Australia, (2013)

Open Access This chapter is licensed under the terms of the Creative Commons Attribution-NonCommercial-NoDerivatives 4.0 International License (▶ http://creativecommons.org/licenses/by-nc-nd/4.0/), which permits any noncommercial use, sharing, distribution and reproduction in any medium or format, as long as you give appropriate credit to the original author(s) and the source, provide a link to the Creative Commons license and indicate if you modified the licensed material. You do not have permission under this license to share adapted material derived from this chapter or parts of it.

The images or other third party material in this chapter are included in the chapter's Creative Commons license, unless indicated otherwise in a credit line to the material. If material is not included in the chapter's Creative Commons license and your intended use is not permitted by statutory regulation or exceeds the permitted use, you will need to obtain permission directly from the copyright holder.

Data Fusion and Environment Representation

Representation of Fused Environment Data

Klaus Dietmayer, Fabian Gies, Dominik Nuß, Stephan Reuter, and Marcel Schreiber

Contents

© The Author(s) 2026
H. Winner et al. (eds.), *Handbook Assisted and Automated Driving*,
https://doi.org/10.1007/978-3-658-45276-6_20

20.1 Requirements of Fused Environment Data

A vehicle environment representation often also referred to as a vehicle environment model is a dynamic data structure in which all relevant objects, infrastructure elements, and contextual information in the proximity of the ego vehicle are contained. All elements must be represented in a common reference system as accurately as possible in terms of position and time. The detection and the temporal tracking of the objects and infrastructure elements are performed continuously using onboard sensors such as cameras, lidar, and radar (see ▶ Chap. 15, ▶ Chap. 16 and ▶ Chap. 18). In the future, more information from high precision attributed digital maps (see ▶ Chap. 22) and, if applicable, external information based on Car2x communication, such as the state of traffic lights on an independent information channel, will be available and can be incorporated. ◘ Figure 2.1 shows frequently used elements of a processing chain for automated and partially automated driving. The vehicle environment representation is the result of the perception layer.

The information sources and processing elements that are required for a vehicle environment representation depend to a large extent on the specified application and requirements of the automation. Hence, an essential requirement for the architecture and data structure used in the environment representation is their adaptability to different functional requirements. This can only be warranted through consistent modularity and largely generic interfaces to the sensors and their preprocessing, such as sensor-specific object detection and classification, as well as to the further processing steps of the situational interpretation. The focus of this chapter is to introduce the basic methods of a grid-based and an object-based representation of the vehicle environment.

In an object-based representation, all other traffic participants relevant for the representation, all relevant infrastructure elements, and the ego vehicle itself are described by their dynamic object model, e.g. a time-discrete state-space model. Its states such as position, velocity, yaw rate, acceleration, or 2D/3D object extent are updated continuously using sensor measurements and appropriate filtering processes (▶ Sect. 20.1). Since measurements are generally uncertain, these filtering processes should provide information on the current uncertainty of the vehicle environment representation to the further processing stages. This includes the uncertainty of the dynamic states, usually represented by covariance matrices, but also the existence probability as an indication of whether the object detected by the sensors exists or is caused by clutter measurements.

Since onboard sensors are preferred in environment perception, the object-based representation is commonly given relative to the ego vehicle's coordinate system. In principle, this description is sufficient for solving all driving tasks. It reaches its limits when context knowledge, for example from a high-precision digital map or through Car2x communication should be considered as well. In these cases, an absolute reference, i.e., a high-precision estimation of the global location of the ego vehicle, is necessary.

A grid-based representation is based on so-called grid maps. To generate them, the spatial environment is divided into fixed, normally identically dimensioned cells. The vehicle travels across this virtual grid and the onboard sensors provide information on whether a specified cell is free of objects or whether there is an obstacle in the cell. The measurements are temporally filtered with compensation of the ego vehicle's motion. This type of representation is suited primarily for representing static obstacles, but can also be used for the detection and modeling of moving cells up to the object generation from them, as shown below. It does not require any model hypotheses like object-based approaches and can therefore be classified as very robust regarding model errors.

The following chapters will address these two basic representations in more detail as well as their applicable algorithms. There are also very promising approaches that combine both representations. These will be briefly introduced at the end of the chapter. Regarding the architecture and required contents of environment representations, no official standards and guidelines exist yet, so the function modules presented here are only one possibility of implementation.

However, recent publications on system architectures for automated vehicles similarly define a central environment model in which all data from processed sensor measurements are collected to provide the data basis for further motion planning modules (see ◘ Fig. 20.1). As an example, the authors in [1] present a hierarchical modular system architecture that permits individual modules to be exchanged while maintaining overall functionality. The main functional modules are ego vehicle state estimation, dynamic occupancy grid map, road topology estimation, multi-object tracking, and the environment model. In the environment model, the consistency between the grid-based cell states and the object-based track states is checked. This enables for example the detection of falsely detected and missing objects.

In addition to a consistency check, each functional module should perform a self-assessment [2]. The system architecture is divided into different layers for this purpose, with each layer individually performing a self-assessment. This approach of self-assessment ena-

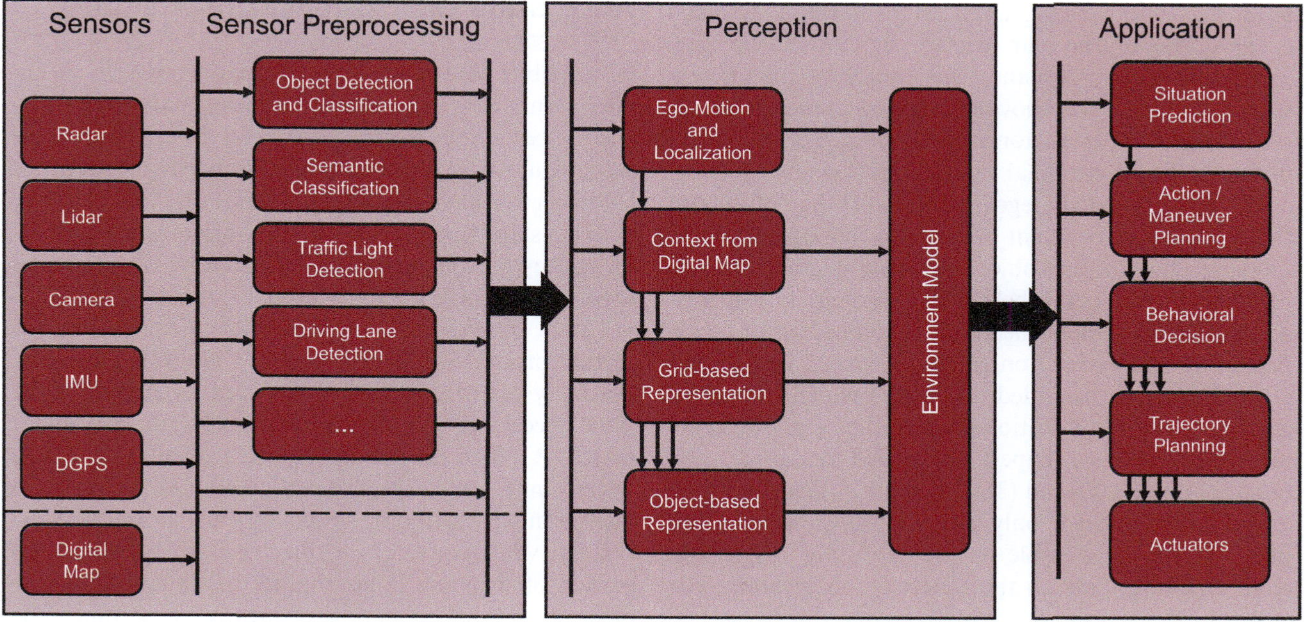

Fig. 20.1 Generic overview of the information processing chain and essential individual components as in automated and partially automated driving

bles ensuring system integrity and, consequently, avoids unintended behavior of the vehicle. For example, a fully comprehensive review of the necessary components of a system architecture for automated driving is presented in [3, 4]. Here too, the individual components are structured in a modular concept and the data levels are categorized into three different degrees of detail. As layers, the localization, environment perception, and planning are specified. Also, Schreier et al. [5] define a comprehensive environment model that combines all sources of information. The environment model abstracts the real world and provides a data basis for subsequent modules to perform the driving tasks. The environment model's tasks are to neglect irrelevant information, represent data in a compact and consistent format, and incorporate uncertainties. In the approach of [6], these concepts are utilized, and a hierarchical modular system architecture is presented using separate layers as well as a central environment model. Summarized, in the presented approaches, some core functions and common ideas regarding the systems architectures can be identified: Modular interchangeable functional modules, and the separation into different layers for a defined information flow are essential.

A completely different approach is the end-to-end system, as described, for example, by Waymo [7]. Thereby, all previously described concepts are bypassed, and a neural network processes all input data to predict the actuation signals for the motion controller. To demonstrate automated driving functions with such an end-to-end system, a significant amount of real and synthesized training data is required. A major disadvantage of these approaches is the missing transparency in the processing chain, which makes their security consideration and the possibilities for safety aspects significantly more difficult.

20.2 Object-Based Representation

20.2.1 Sensor-Specific Object Models and Coordinate Systems

Typically, the coordinate system of the moving vehicle is used for object-based representations. According to the conventions for vehicles, such a coordinate system is right-handed. The x-axis points in the direction of motion, the z-axis upwards, and the direction of the y-axis follows the definition of right-handed coordinate systems. Rotations around the x-axis are called rolling, the y-axis pitching, and the z-axis yawing. The center of mass of the vehicle is usually chosen as the coordinate origin for these kinetic models.

For environment perception, however, such complex vehicle dynamics models are usually not required, since only a comparatively simple estimation of the vehicle's motion is necessary. Therefore, at least for driving situations within dynamic limits, it is common to use a linear bicycle model neglecting wheel slip and slip angle for the motion estimation [8]. A suitable reference point for multi-sensor applications (fusion systems) is the center of the rear axle of the ego vehicle, projected onto the road plane. This choice has the advantage that, under the above-mentioned approximations, the vehicle velocity vector at the reference point always points to the longitudinal vehicle axis (x-axis).

In addition to the ego vehicle, all other objects in the vehicle environment must also be described by models. Due to the object tracking algorithms, explained in more detail below, their formulation is usually done as discrete-time state-space models for a planar motion. A distinction is made between point models and spatially extended models, in which the length and width (2D) or additionally also the height (3D) are modeled. The basic shape for extended models is a rectangle (2D) or a cuboid (3D). However, the use of such an extended model is only useful if the contour of the object is also observable by the capturing sensor system, which depends on the aspect of observation. For example, the length of a vehicle driving directly ahead is not directly observable, regardless of the sensor technology used, but only its width and height. This is also only the case if appropriate contour-resolving sensors such as (stereo) cameras, laser scanners, or high-resolution radar sensors are used. Due to this time-variant observability problem, the extent estimation is often performed in parallel to the state estimation in a separate filter algorithm.

For the state estimation, point models are modeled as so-called free-mass models, which means that when restricted to the plane motion, the motions in x-, y-direction, and the rotation around the vertical axis (yaw rate) are not coupled. Extended models (2D/3D) can be formulated as free-mass models or as bicycle models (see above) with corresponding coupled degrees of freedom of motion. The latter is usually the better choice for all wheel-bound objects such as vehicles or cyclists.

As state variables, the position in the plane, the velocity in the plane, and the yaw angle are considered minimally per object in the state vector. Such a model is called a "constant velocity, constant yaw angle" model. An extension of these models to include acceleration components as well as yaw rate or even yaw acceleration in the state vector is possible. However, this extension is only useful if sensors are available that can directly measure velocity components, such as radar sensors. Otherwise, the double differentiation of error-prone position measurements in the filter leads to such strong noise effects in the acceleration data that the overall estimation result tends to be worsened by this. The choice of the appropriate object model thus depends strongly on the available sensor configuration and its measurement capabilities. Details on the formulation of object models can be found for example in [9] or [8].

20.2.2 State and Existence Uncertainties

In an object-based representation of the vehicle environment, it is necessary to specify state and existence uncertainties for the individual objects to be able to safeguard action planning and behavior decisions based on the environment representation.

The state uncertainty of an object is usually described by a probability density function, which can be used to determine the most probable overall or individual state, as well as possible variations thereof with a certain probability. In the case of a multi-dimensional, normally distributed probability density function, the state uncertainty is fully represented by the covariance matrix P. When estimating static parameters, the state uncertainty can be progressively reduced by repeated measurements without systematic error, and the estimated value converges to the true value. In the estimation of dynamic states, due to the movement of the object between two successive measurement times, the convergence against the true value is no longer given. Therefore, when evaluating state estimation of dynamic systems, it is required that the expected value of the estimation error is zero and that the variance of the real estimation error is identical to the estimated variance. An estimator with these properties is called a consistent estimator for dynamic quantities.

For the realization of safety-relevant driver assistance functions, however, the existence uncertainty is at least as relevant as the state uncertainty. The existence uncertainty expresses the probability with which the modeled object in the driving environment representation corresponds to a real object. After all, emergency braking should only be triggered if the probability of the existence of the detected obstacle is very high. While the state-of-the-art estimation of state uncertainties is theoretically well-founded by methods of recursive Bayes estimation, in today's systems existence is usually inferred only based on a heuristic quality measure q(x) of an object hypothesis. An object is considered confirmed if the quality measure exceeds a sensor- and application-dependent threshold. For example, a quality measure is based on the number of successfully measured value associations since the object was initialized or the time interval between the initialization of the object and the current time. Often, the state uncertainty of the object or the quality measure of another system is also used for validation.

An alternative approach is the estimation of an object-specific existence probability. This first requires an application-dependent definition of object existence. While in some applications all real objects are considered as existing, the object's existence can also be restricted to the objects relevant in the current application. Furthermore, a restriction to the objects detectable with the current sensor setup is possible. In contrast to the threshold of the quality measure, the determination of the existence probability allows a probability-based interpretation possibility. For example, an existence probability of 90% means that the measurement history, as well as the motion pattern of the object, were generated by a real object with a probability of 90%. Consequently, assistance functions can use those objects whose existence probability exceeds an application-specific threshold value.

20.2.3 Fundamentals of Multi-object Tracking

The goal of object tracking is to estimate the time-variant state of all objects located in the field of view of the sensors. Changes in the object states result on the one hand through the motion of the objects as well as through the ego-motion of the observing vehicle. State-of-the-art object tracking is based on the recursive Bayes filter consisting of two parts: prediction and innovation. The prediction step models the movement of the object between two subsequent measurements in time based on an object-specific motion model (see Sect.2.1). Further, the ego-motion of the observing vehicle must be compensated in the prediction step. The ego-motion results in a change in the states (e.g. position) estimated relative to the ego vehicle as well as in an increased estimation uncertainty. In the subsequent innovation step, the predicted object state is updated based on the current sensor measurement, taking the measurement uncertainty into account.

In the following, the Bayes filter will first be introduced. Subsequently, the Kalman filter, which allows the analytic implementation of the Bayes filter for linear systems, will be explained and possibilities for its application to multi-object tracking are introduced. The Labeled Multi-Bernoulli filter in particular will be discussed. It realizes a probabilistic association of the obtained measurements to the currently tracked objects and estimates the objects' existence probability as called for in ▶ Sect. 2.2.

20.2.3.1 Bayes Filter

In the recursive Bayes filter [10] the estimated state of an object and the associated spatial uncertainty are represented by a probability density function (PDF)

$$p_{k+1}(x_{k+1}) = p_{k+1}(x_{k+1}|Z_{1:k+1}) \tag{20.1}$$

which is dependent on all measurements $Z_{1:k+1} = \{z_1, \ldots, z_{k+1}\}$ received up to point in time $k + 1$. The motion model of an object for the time between two subsequent measurements is given by

$$x_{k+1|k} = f(x_k) + v_k \tag{20.2}$$

where v_k is an additive noise process. Alternatively, the motion equation can also be described by a Markov transition density

$$f_{k+1|k}(x_{k+1}|x_k). \tag{20.3}$$

Applying the first-order Markov characteristic, the predicted state x_{k+1} of the object depends only on the state x_k since the latter implicitly contains the measurement history $Z_{1:k} = \{z_1, \ldots, z_k\}$. The prediction of the current object state x_k to the next measurement at time $k + 1$ is made based on the Chapman-Kolmogorov equation

$$p_{k+1|k}(x_{k+1}|x_k) = \int f_{k+1|k}(x_{k+1}|x_k)p_k(x_k)dx_k. \tag{20.4}$$

Subsequently, the predicted PDF of the object state is updated using the measurement . The measurement process of the sensor is described by the measurement equation

$$z_{k+1|k} = h_{k+1}(x_{k+1}) + w_{k+1}. \tag{20.5}$$

The stochastic noise w_{k+1} represents the error of the measurement model. The measurement equation transforms the state of an object into the measurement space of the sensor and therefore allows the innovation of the object state in the measurement space. The innovation in the measurement space is an advantage since the transformation of the measurement into state space generally is not possible due to the non-invertible measurement equation. An alternative representation of the measurement equation is the likelihood function

$$g(z_{k+1}|x_{k+1}) \tag{20.6}$$

which results from the measurement Eq. (20.5). The state innovation is subsequently performed using the Bayes equation

$$p_{k+1}(x_{k+1}|z_{k+1}) = \frac{g(z_{k+1}|x_{k+1})p_{k+1|k}(x_{k+1}|x_k)}{\int g(z_{k+1}|x_{k+1})p_{k+1|k}(x_{k+1}|x_k)dx_{k+1}}. \tag{20.7}$$

The recursive estimation process described by the prediction step (20.4) and the innovation step (20.7) is known as the Bayes filter. Besides the process and measurement equations, the process requires only an a priori PDF for the object state $p_0(x_0)$ at time $k = 0$.

20.2.3.2 Multi-instance Kalman Filter

Under the assumption of normally distributed signals as well as linear process and measurement models, the Kalman filter [11] allows the analytic implementation of the Bayes filter. Since a Gaussian distribution is completely described by its first two statistic moments, i.e., the mean $\hat{x}$ as well as the corresponding

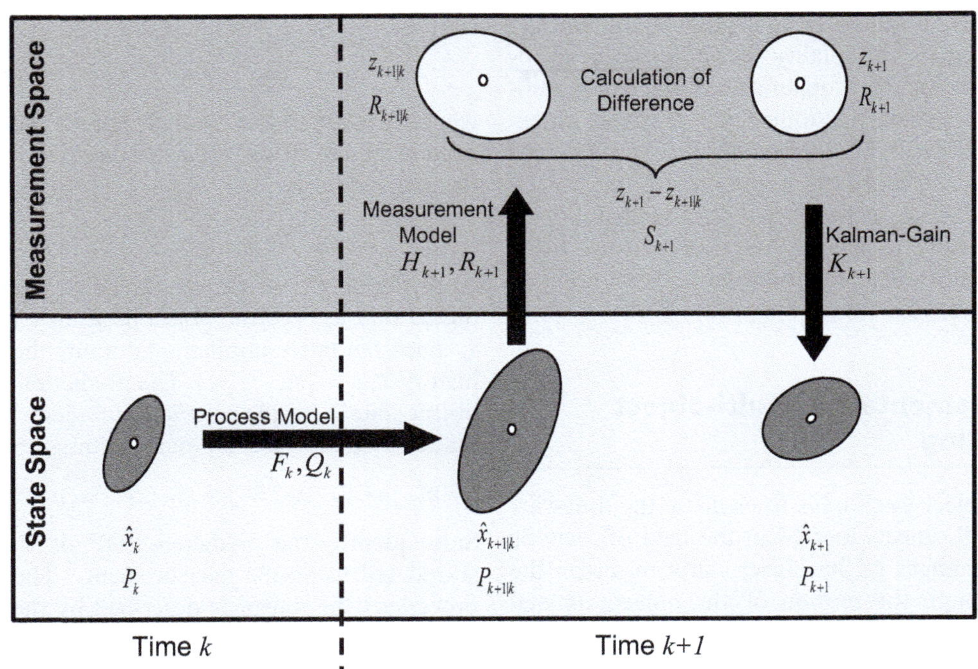

Fig. 20.2 Process of the Kalman filter: State prediction based on the process model, state innovation through transformation into measurement space, and the Kalman Gain

covariance matrix P, the temporal filtering of the moments presents an exact mathematical solution. It follows that given these assumptions, the Kalman filter is a Bayes-optimal state estimator. **Figure 20.2** illustrates the process of the Kalman filter, which will be described in detail in the following.

The initial state of an object in the Kalman filter is given by a multi-dimensional Gaussian distribution

$$N(x, \hat{x}_k, P_k) = \frac{1}{\sqrt{\det(2\pi P_k)}} \exp\left(-\frac{1}{2}\left(x - \hat{x}_k\right)^T P_k^{-1}\left(x - \hat{x}_k\right)\right) \tag{20.8}$$

with mean $\hat{x}_k$ and covariance P_k. For the case of linear process and measurement models, Eqs. (20.2) and (20.5) can be written as follows:

$$x_{k+1|k} = F_k x_k + v_k \tag{20.9}$$

$$z_{k+1|k} = H_{k+1} x_{k+1|k} + w_{k+1} \tag{20.10}$$

where F_k and H_{k+1} represent the system matrix and the measurement matrix for the current measurement. The process noise v_k and measurement noise w_{k+1} are assumed to be zero-mean, Gaussian noise and that the two noise processes are uncorrelated.

The prediction step of the Kalman filter is performed through the independent prediction of the expected value and the covariance:

$$\hat{x}_{k+1|k} = F_k \hat{x}_k, \tag{20.11}$$

$$P_{k+1|k} = F_k P_k F_k^T + Q_k. \tag{20.12}$$

The covariance matrix $Q_k = E\{v_k v_k^T\}$ of the process noise represents the uncertainty of the process model, for example, the maximum possible acceleration of an object when using a process model for constant velocity.

By applying the measurement matrix H_{k+1}, the predicted state $\hat{x}_{k+1|k}$ can be used to determine the predicted measurement

$$\hat{z}_{k+1|k} = H_{k+1} \hat{x}_{k+1|k} \tag{20.13}$$

as well as the corresponding covariance matrix

$$R_{k+1|k} = H_{k+1} P_{k+1|k} H_{k+1}^T \tag{20.14}$$

In the following innovation step, a measurement z_{k+1} with corresponding covariance $R_{k+1} = E\{w_{k+1} w_{k+1}^T\}$ is used to update the state. Applying the expected value of the predicted measurement $z_{k+1|k}$ as well as the actual measurement value z_{k+1} yields the measurement residual γ_{k+1} and the corresponding innovation covariance matrix S_{k+1}

$$\gamma_{k+1} = z_{k+1} - \hat{z}_{k+1|k}, \tag{20.15}$$

$$S_{k+1} = R_{k+1|k} + R_{k+1} = H_{k+1} P_{k+1|k} H_{k+1}^T + R_{k+1}. \tag{20.16}$$

The filter gain K_{k+1} can be calculated using the innovation covariance matrix S_{k+1}, the predicted state covariance $P_{k+1|k}$ as well as the measurement matrix H_{k+1}

$$K_{k+1} = P_{k+1|k} H_{k+1}^T S_{k+1}^{-1}. \tag{20.17}$$

The update of the object state estimate and corresponding covariance resulting from the current measurement is obtained using the following equations:

$$\hat{x}_{k+1} = \hat{x}_{k+1|k} + K_{k+1}(z_{k+1} - \hat{z}_{k+1|k}), \tag{20.18}$$

$$P_{k+1} = P_{k+1|k} - K_{k+1} S_{k+1} K_{k+1}^T. \tag{20.19}$$

The application of the Kalman filter to systems with nonlinear process or measurement equations can be realized by utilizing the Extended Kalman filter (EKF) [8] as well as the Unscented Kalman filter (UKF) [12]. While the EKF linearizes the process matrix F_k or measurement matrix H_{k+1} using a Taylor series approximation, the goal of the UKF is a stochastic approximation based on so-called sigma points.

The Kalman filter presents an optimal state estimator for an object and a measurement. Within the context of vehicle environment perception, however, it is necessary to simultaneously track multiple objects. In the literature, multi-object tracking is often realized using multi-instance Kalman filters as shown in ◘ Fig. 20.3 in which every object is tracked by an object specific Kalman filter. Since not every Kalman filter represents a relevant or an existing object, a subsequent classification and validation of the estimated tracks are necessary. To reduce the amount of processed data, object hypotheses are generated in the detection or segmentation step. One example is pedestrian detection in video images introduced in Chapter 18 as well as the deep neural network-based vehicle detectors for radar [13] and lidar [14] sensors. In the data association step, the obtained measurements are assigned to the available Kalman filters and the state of the objects are updated with the measurements, whereby the data association is ambiguous in many cases due to missed detections and false alarms.

The association process which requires the lowest computational effort is the Nearest Neighbor (NN) algorithm, which updates each object with the closest measurement relative to the state. The Mahalanobis distance [8] is commonly used to obtain the closest measurement. In scenarios with objects located closely together, the NN algorithm often leads to a single measurement being used to update multiple objects. However, this contradicts the assumption that a measurement is generated by at most one object. The Global Nearest Neighbor (GNN) algorithm guarantees compliance with this assumption through the calculation of an optimal association of all tracks and measurements. Both algorithms make a hard and possibly erroneous association decision at a given time step which cannot be reversed and in the event of a wrong decision often leads to the loss of a currently tracked object.

The basic idea of the probabilistic data association (PDA) [15] is therefore to perform a weighted update of the object state using all association hypotheses to avoid hard, possibly erroneous association decisions. Due to the probabilistic data association, the a posteriori probability density is not a Gaussian distribution anymore, since the superposition of several Gaussian distributions is typically a multi-modal distribution. Hence, the PDA algorithm requires an approximation of the a posteriori distribution by a Gaussian distribution to enable the application of the Kalman filter equations in the next innovation step.

20.2.3.3 Random Finite Set (RFS) Approaches and the Labeled Multi Bernoulli (LMB) Filter

An alternative to multi-object tracking with multi-instance Kalman filters is given by the multi-object Bayes filter introduced in [16] which models the multi-object states X as well as the measurements Z obtained at time $k + 1$ as random finite sets. A random finite set $X = \{x_1, \dots, x_n\}$ in this case is a random variable in which both the number of objects n (including $n = 0$) as well as the individual states x_i are random. Further, the states of a random finite set are unordered and hence permutation invariant. The prediction and update equation of the multi-object Bayes filter corresponds to Eqs. (20.4) and (20.7), however, the state vectors x and the measurements z are replaced by the random finite sets X and Z.

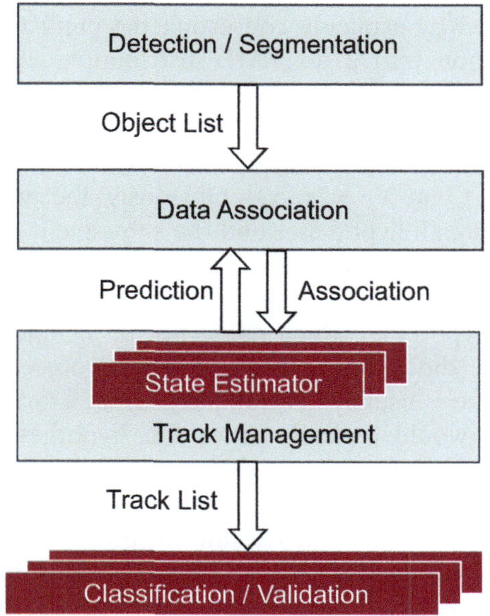

◘ **Fig. 20.3** Multi-instance Kalman filter

In the prediction step, the multi-object Bayes filter uses a multi-object Markov density which along with the motion of objects also accounts for the appearance and disappearance of objects. The representation of objects by a random finite set additionally allows the modeling of dependencies between the objects in the prediction step, while the use of a multi-instance Kalman filter presupposes the statistical independence of the objects. The update step of the multi-object Bayes filter is based on the multi-object likelihood function, which avoids an explicit data association by averaging over all possible association hypotheses. Besides the detection of an object, the multi-object likelihood explicitly models missed detections as well as false alarms or clutter measurements.

The multi-object Bayes filter can be implemented using Sequential Monte Carlo (SMC) methods. However, due to the high-dimensional state space, this is only applicable for a low number of objects. Based on the constant-gain approximation of the Bayes filter, an approximation of the a posteriori multi-object distribution by the first moment (PHD filter) is recommended in [10] to reduce the computational effort. While the first moment of a probability density function is given by the mean, the first moment of the multi-object distribution is given by an intensity function where the integral over the intensity function represents the estimated number of objects in the respective area. A disadvantage of the PHD filter is the considerable fluctuation in the estimated number of objects due to the low memory of the filter and the approximation of the cardinality distribution by a Poisson distribution. In the case of the Kalman filter, improved state estimation is obtained through the representation of the probability distribution through the first and second statistic moment. The Cardinalized Probability Hypothesis Density (CPHD) filter [16] provides a partial approximation of the second statistic moment of the multi-object distribution by propagating the intensity function and the cardinality distribution over time. Compared to the PHD filter, the CPHD filter distinguishes itself through the stable estimation of the number of objects but requires exact knowledge of the false alarm process beforehand which however is not available in the context of vehicle environment detection on account of the different environmental situations. A disadvantage of both, the PHD and the CPHD filter, is the absence of the estimate of the existence probability for the tracked objects.

In [17], the Delta Generalized Labeled Multi-Bernoulli (d-GLMB) filter is proposed for an analytical implementation of the multi-object Bayes filter. The d-GLMB filter represents the multi-object distribution using hypotheses and is based on labeled RFS, which assigns a unique label (or track ID) to each state. The Labeled Multi-Bernoulli (LMB) filter [18] is an efficient approximation of the d-GLMB filter which provides superior tracking results compared to the PHD and the CPHD filter and enables a real-time capable implementation.

The LMB filter is based on the labeled multi-Bernoulli distribution which represents M objects using M statistically independent Bernoulli distributions. With the existence probability r_i, the Bernoulli distribution of each object x_i is given by the single-object spatial distribution $p_k(x_{i,k}|Z_{1:k})$. With probability $1 - r_i$, the object x_i does not exist. By representing the objects' spatial distribution using a Gaussian mixture

$$p_k(x_{i,k}|Z_{1:k}) = \sum_{l=0}^{L_i} w_{i,k}^l N\left(x_{i,k}, \hat{x}_{i,k}^l, P_{i,k}^l\right), \qquad (20.20)$$

an analytical implementation of the LMB filter by using the Kalman filter equations is possible.

In the prediction step, each of the L_i mixture components of object x_i is predicted using the Kalman filter equations:

$$p_{k+1}(x_{i,k+1}|Z_{1:k}) = \sum_{l=0}^{L_i} w_{i,k}^l N\left(x_{i,k+1}, \hat{x}_{i,k+1|k}^l, P_{i,k+1|k}^l\right).$$

$$(20.21)$$

Note that the weight of the components is unchanged. The predicted existence probability of an object is given by

$$r_{i,k+1|k} = p_S\left(x_{i,k}\right)r_{i,k}, \qquad (20.22)$$

where $p_S(x_{i,k})$ represents the persistence probability of the object. Consequently, the object's disappearance probability is given by $1 - p_S(x_{i,k})$.

In [18], the innovation step of the LMB filter was computed by explicitly converting the predicted LMB distribution into a d-GLMB distribution which represents the multi-object distribution using hypotheses. For two objects x_1 and x_2, the d-GLMB distribution consists of four hypotheses: $X_0 = \varnothing$, $X_1 = \{x_1\}$, $X_2 = \{x_2\}$ und $X_3 = \{x_1, x_2\}$. Obviously, the process of generating all hypotheses and the subsequent measurement update is only feasible for a very small number of objects. Further, this approach generates a huge number of hypotheses which might not be significant. For example, the disappearance of several objects at the same time is usually very unlikely. Thus, a pruning algorithm would typically prune this hypothesis. However, using the knowledge that a set of objects has not been detected in the next measurement of a sensor would help to select the relevant hypotheses. In [19], an alternative implementation of the LMB filter based on a joint prediction and update step is proposed which will be introduced in the following.

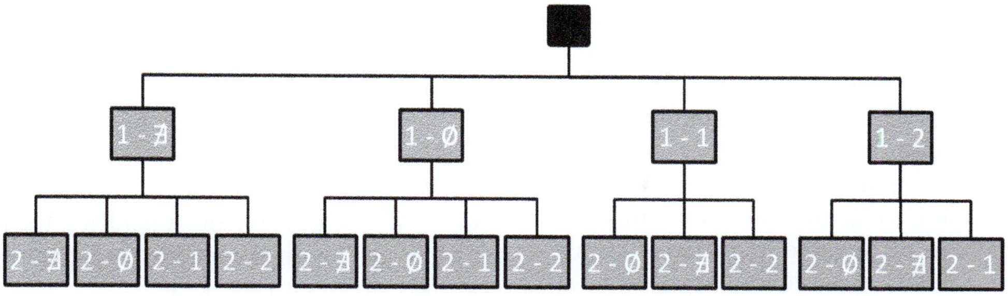

◻ Fig. 20.4 Association tree for two objects and two measurements. The first symbol of each node represents an element from the set of objects, the second symbol is an element from the set of measurements

In order to be able to represent all assignment hypotheses, it is first necessary to define the set of all measurements

$$Z = \{z_1, \ldots, z_m, \emptyset, \nexists\} \tag{20.23}$$

where the two pseudo-measurements $\emptyset$ and $\nexists$ represent the missed detection and the non-existence of an object, respectively. Further, set X consists of the n objects currently being tracked:

$$X = \{x_1, \ldots, x_n\}. \tag{20.24}$$

The assignment of an object to a measurement is therefore given by a pair $e = (x \in X, z \in Z)$. In a complete assignment hypothesis given by the set $E = \{e_1, \ldots, e_n\}$, the objects $x_1, \ldots, x_n$ must be assigned exactly once and the measurements $z_1, \ldots, z_m$ may be assigned at most once. The special elements $\emptyset$ and $\nexists$ may be used multiple times in an assignment hypothesis. The probability of an assignment hypothesis $E_l = \{e_1, \ldots, e_n\}$ can therefore be calculated using the product of the elementary assignment probabilities:

$$w(E_l) = \prod_{e \in E_l} \gamma(e). \tag{20.25}$$

In the following, the calculation rules are described for the three categories of elementary assignment hypotheses. The innovation of the spatial distribution of object x_i using measurement $z_{j,k+1}$ is given by

$$p_{k+1}\left(x_{i,k+1} | z_{j,k+1}\right) = \sum_{l=0}^{L_i} w_{i,k+1}^{j,l} N\left(x_{i,k+1}, \hat{x}_{i,k+1}^{j,l}, P_{i,k+1}^{j,l}\right). \tag{20.26}$$

The parameters $\hat{x}_{i,k+1}^{j,l}$ and $P_{i,k+1}^{j,l}$ of the Gaussian mixture components are calculated using the innovation step of the Kalman filter. In contrast to the prediction step, the weights of the mixture components need to be updated using

$$w_{i,k+1}^{j,l} = \frac{\frac{p_D}{\kappa(z_{j,k+1})} w_{i,k}^l N\left(z_{j,k+1}, z_{k+1|k}^l, S_{k+1}^l\right)}{\eta(e = \{x_i, z_j\})}, \tag{20.27}$$

where $N(z_{j,k+1}, z_{k+1|k}^l, S_{k+1}^l)$ denotes the measurement likelihood (20.6), p_D is the sensor specific detection probability, $\kappa(z_{j,k+1})$ is the Poisson distributed clutter density and

$$\eta(e = \{x_i, z_j\}) = \frac{p_D}{\kappa(z_{j,k+1})} \sum_{l=0}^{L_i} w_{i,k}^l N\left(z_{j,k+1}, z_{k+1|k}^l, S_{k+1}^l\right) \tag{20.28}$$

is the associated normalization constant. The normalization constant additionally represents the likelihood for the association of measurement j to the predicted object i in this case. Using the joint computation of prediction and update, the weight of the assignment is given by

$$\gamma\left(e = \{x_i, z_j\}\right) = r_{i,k+1|k} \eta\left(e = \{x_i, z_j\}\right), \tag{20.29}$$

since the assignment of the measurement requires the persistence of the object.

In addition to the association of one measurement, there is the possibility of missed detection for each object. In this case, the updated spatial distribution of the object equals the predicted spatial distribution (20.21) and the corresponding likelihood is given by

$$\eta(e = \{x_i, \emptyset\}) = 1 - p_D, \tag{20.30}$$

and

$$\gamma(e = \{x_i, \emptyset\}) = r_{i,k+1|k} \eta(e = \{x_i, \emptyset\}). \tag{20.31}$$

The third assignment option represents the disappearance of the object x_i with the according likelihood

$$\gamma(e = \{x_i, \nexists\}) = 1 - r_{i,k+1|k}. \tag{20.32}$$

An intuitive representation of all possible assignment hypotheses is possible using a hypothesis tree [20, 21]. ◻ Figure 20.4 shows exemplarily the hypothesis tree for a situation with two measurements, where each node of the tree represents an elementary assignment e. Every path from the root node to a leaf node represents a complete assignment hypothesis E_l.

Since the number of assignment hypotheses grows exponentially, the computation of all hypotheses is only possible for a limited number of measurements and objects. The number of possible assignment hypotheses can be reduced significantly by applying gating approaches which exclude very unlikely elementary assignments based on the innovation covariance S. The gating enables the grouping of objects described in [18]. The grouping procedure partitions the objects into groups that can be updated independently. A further reduction of the computational complexity can be achieved using the Murty algorithm [22] which enables to compute the K best assignment hypotheses in cubic complexity without the need to compute all hypotheses. Hence, the Murty algorithm allows for using a termination criterion which stops the computation of further hypotheses once the relative weight of the current hypotheses concerning the best hypotheses falls under a specified threshold. A very efficient approach for generating the hypotheses based on Gibbs sampling was introduced in [19] and features linear complexity. Starting from an initial assignment hypothesis, Gibbs sampling allows for drawing further hypotheses based on the conditional distribution

$$\pi_t\left(\gamma_t'|\gamma_{1:t-1}',\gamma_{t+1:n}\right). \tag{20.33}$$

Here, $\gamma_{1:t-1}'$ represents the already newly drawn assignments in the current run for the objects $1 : t - 1$ and $\gamma_{t+1:n}$ represents the assignments for the tracks $t + 1 : n$ from the previous solution. As an initial solution, three different assignment hypotheses may be used: the missed detection is associated with each object, each object is assumed to have disappeared, or the best assignment hypothesis (e.g., determined using the Hungarian method [23]) is used.

Based on the set of all assignment hypotheses, the existence probability and the spatial distribution of the objects can now be calculated. The a posteriori existence probability of an object x_i can be calculated by means of marginalization

$$r_{i,k+1} = \frac{\sum_{E\in E_i^{\exists}} w(E)}{\sum_E w(E)}, \tag{20.34}$$

where the set $E_i^{\exists}$ represents all hypotheses in which track x_i exists:

$$E \in E_i^{\exists} \Leftrightarrow (x_i, \not\exists) \cap E = \emptyset. \tag{20.35}$$

The calculation of the assignment weights

$$w_{ij} = \frac{\sum_{E\in E_{ij}^{TP}} w(E)}{\sum_{E_i^{\exists}} w(E)} \tag{20.36}$$

is also performed via marginalization. Here the set E_{ij}^{TP} represents all hypotheses which contain the assignment of measurement z_j to object x_i:

$$E \in E_{ij}^{TP} \Leftrightarrow (x_i, z_j) \in E. \tag{20.37}$$

Similarly, the assignment weight w_{i0} for the missed detection can be determined. Consequently, the updated spatial distribution results in

$$\begin{aligned} p_{k+1}\left(x_{i,k+1}|Z_{1:k+1}\right) = & w_{i0}p_{k+1}\left(x_{i,k+1}|Z_{1:k}\right) \\ & + \sum_{j=1}^{m} w_{ij}p_{k+1}\left(x_{i,k+1}|z_{j,k+1}\right). \end{aligned} \tag{20.38}$$

Since the innovation step significantly increases the number of mixture components, it is required to apply a post-processing step that merges similar components and prunes components with a very small weight.

The appearance of new objects is part of the prediction step of the LMB filter. The initialization of new objects can either be based on not associated measurements of the previous cycle or simply use a specified prior distribution. The existence probability of the new objects is typically set to a low value. In the innovation step, the newly spawned objects are added to the hypotheses tree-like already existing objects.

20.2.3.4 Extended Object Tracking

The multi-object tracking algorithms presented in the previous paragraphs assume that each object generates at most one measurement. Due to the increasing resolution and separability, this assumption is typically not fulfilled anymore for current sensor generations. To enable the usage of the presented multi-object tracking algorithms, preprocessing steps are required to cluster the raw measurements belonging to a single target. This is typically done by classical algorithms like Density-Based Spatial Clustering of Applications with Noise (DBSCAN) [24] or using state-of-the-art deep neural networks [25]. Since both approaches typically perform a hard decision for the association of the raw measurements to one cluster, the preprocessing leads to the same drawbacks as mentioned for the nearest neighbor approach.

An alternative approach is the so-called extended object tracking algorithm [26] which explicitly models multiple measurements per object using random finite sets. In addition to the association uncertainty described in the previous chapter, the Bayesian modeling of the innovation step requires the consideration of all possible partitioning of the raw measurements. Due to the combinatorial complexity, the set of all possible partitioning of the raw measurements is often approximated by applying a distance-based segmentation with different thresholds. The minimum threshold is chosen in such a way, that the event of combining the raw measurements of two objects into a single cluster is extremely rare. Consequently, this threshold is highly dependent on the measurement characteristics of the sensor. For the choice of the maximum threshold, the typical extent of the objects to be tracked plays an important role. Using an appropriate choice of the minimum and maximum threshold in combination with a suitable step size for the thresholds, one of the generated partitions typically provides the correct segmentation of the raw measurements. In contrast to this, a single threshold often results in over or under-segmentation in a large number of scenarios. An RFS based extended object tracking algorithm uses all generated partitioning of the raw measurements in the innovation step and creates a large number of hypotheses for the multi-object state.

For the estimation of the object extent, the approaches use either simple geometric shapes like rectangles and ellipses or for example star-convex shapes. The random matrix approach represents objects using a kinematic state vector and approximates the elliptical shape using an n-dimensional symmetric positive definite shape matrix. Hence, the random matrix approach is e.g. suitable for tracking pedestrians or cyclists. Further, it may also be applied for objects whose shape may not be estimated precisely due to a too-small number of measurements. In the case of a very precise measurement of the objects 'contour or surface, the temporal filtering of the shape using Gaussian processes (GP) or the Random Hypersurface Model (RHM) [26] is possible. Both approaches assume a star-convex shape of the objects and represent them using a kinematic state vector as a radial function. In [22], a representation of the object extent using B-splines is presented. In contrast to GP and RHM, the sampling points are represented in Cartesian coordinates. The B-spline approach and the GP approach achieve similar performance. However, an advantage of the B-spline approach in the context of vehicle environment perception is the more precise scaling of length and width due to the usage of the Cartesian sampling points.

20.2.4 Self-localization and Inclusion of Digital Maps

Machine perception for object recognition and object tracking, as described in the previous section, can be significantly improved by a-priori information, for example, the number and width of lanes, lane branches, turns, intersection topologies, or the position of traffic signs and traffic lights. Even though currently available commercial maps contain comparatively few such attributes,

automated driving in more complex environments will not be possible without such support. In addition, situational awareness and decision-making benefit from map knowledge, as recognized objects can be assessed in the context of the traffic environment.

Because digital maps are referenced in absolute terms, primarily in UTM or WGS84 coordinates (▶ Chap. 22) the use of their attributes requires highly accurate self-localization of the ego vehicle in the map. The accuracy of standard GNSS systems is not always sufficient for this purpose. In addition, reception in populated or forested areas is often severely impaired by multipath effects and satellite shadowing. However, since only the position on the map is decisive, self-localization can be achieved with the required accuracy based on landmarks that are recognizable by machine perception and recorded on the map. An easy-to-implement example is the determination of the exact lateral position in a lane that is also recorded on the map.

If this matching succeeds, all recognized dynamic objects from the object-based representation can be assigned to the context of the map. It is common to structure the map in layers in which, for example, the lower layer contains the road topology, higher layers then lanes and other static attributes, and a higher layer contains the detected dynamic objects. An overview with a focus on GNSS gives [27]. Special aspects are discussed in [28–31, 6].

20.2.5 Temporal Aspects

In addition to the correct spatial assignment, a no less significant challenge is to ensure the correct temporal reference of all elements of the driving environment representation. Finally, the latter should contain a temporally consistent image of reality.

Since the sensors for machine perception are generally not synchronized or cannot be synchronized and also have different latencies due to the different complexity of the pre-processing stages, a common global system time is required as a minimum, to which all measurements and other information refer. This ensures that the measurement time of the respective sensor or the external information can be assigned to other measurements. For the filtering methods based on recursive Bayesian filtering described in ▶ Sect. 2.3, for example, it is a basic requirement that measurements are introduced chronologically correctly. To ensure this, it may be necessary to wait for a slower sensor, i.e., a sensor with higher latency, before the measurements can be introduced into the filter. However, this procedure, known as buffering, has the disadvantage that the "slowest" sensor determines the latency of the environment representation. However, it is also clear from the explanations that such latency cannot be avoided, i.e.

the environment representation will always be behind the real situation in terms of time, which can be around 500 ms or more depending on the system configuration. This latency must be considered in the situation prediction and maneuver planning.

20.3 Grid Maps

20.3.1 Concept of Grid Maps

Grid maps partition the vehicle environment into cells. Each of these cells represents a location for which the respective cell contains information. Partitioning into cells is a spatial discretization of the vehicle environment. When the sensor data from different sensors allow inferences on the state of cells, then the mapping in grid maps corresponds to an indirect form of sensor data fusion.

Early publications on grid maps for environment modeling originated in the field of robotics and described two-dimensional maps generated in real-time and whose cells indicated whether the space they represented was occupied or free [9, 32]. These maps are known as occupancy grid maps. Their generation is done primarily using distance measurement sensors such as laser or radar sensors. Even though many different forms of grid maps have been introduced since then, occupancy grid maps remain an important foundation for many applications [33, 34].

Different grid map approaches can represent the space in a two-dimensional or three-dimensional way. Grid maps also differ in the type of information stored in the cells and the cell size and shape. Some approaches assume a static environment, while others explicitly model the movement of other objects. Those are often called dynamic grid maps. Both variations will be described in this chapter.

20.3.2 Ego-Motion Estimation

In order to generate a consistent grid map, it is necessary to account for the motion of the ego vehicle. This is required for grid maps assuming a static environment as well as dynamic grid maps. For the generation of a two-dimensional grid map, it is also necessary to estimate the pose (position and orientation) of the vehicle.

A simple possibility to estimate the absolute vehicle position results from the use of GNSS. The disadvantage of this method lies in the fact that reception is not guaranteed, and that the accuracy fluctuates greatly. For many applications, it is advisable to select an arbitrary relation between the grid map and a global coordinate system and only to consider the relative motion of the vehicle. This can be done using dead reckoning, which is described in detail in the following.

The vehicle pose at two points in time $(t_1, t_2), t_2 > t_1$ is shown in ▫ Fig. 20.5. At time t_1, let the vehicle pose $p_1 = \begin{bmatrix} x_1 y_1 \psi_1 \end{bmatrix}^T$ be known in the coordinate system of the grid map. Here, x and y correspond to the vehicle position and ψ to the orientation (yaw angle) of the vehicle. The goal is now to enter a measurement taken at time t_2 into the grid map. Hence, the pose p_2 at time t_2 must be determined. The vehicle velocity v and the yaw rate $\dot{\psi}$ are assumed measurable. In addition, it is as-

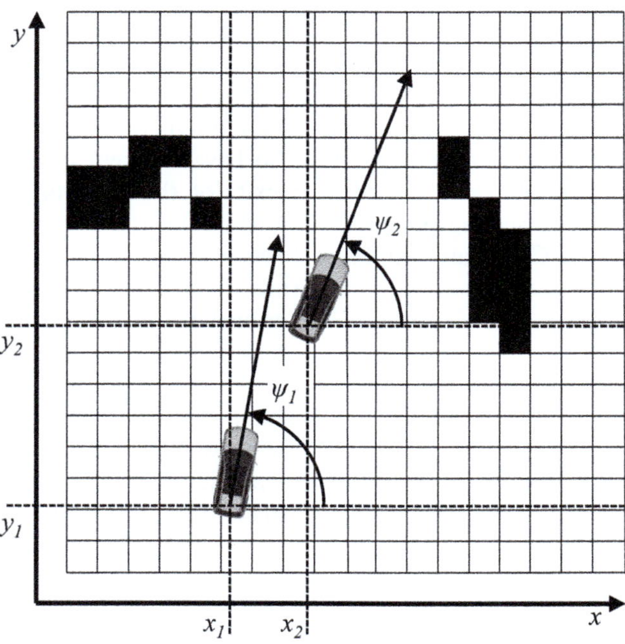

▫ **Fig. 20.5** Pose of the vehicle in the grid map measured at two different points in time

sumed that at every point in time, the vehicle is moving solely in the direction of the yaw angle ψ.

Then the change in the vehicle pose is given by

$$\begin{bmatrix} \dot{x} \\ \dot{y} \\ \dot{\psi} \end{bmatrix} = \begin{bmatrix} v\cos(\psi) \\ v\sin(\psi) \\ \dot{\psi} \end{bmatrix}, \tag{20.39}$$

so that vehicle pose p_2 can be calculated from

$$p_2 = p_1 + \int_{t_1}^{t_2} \begin{bmatrix} \dot{x}(t) \\ \dot{y}(t) \\ \dot{\psi}(t) \end{bmatrix} dt, \tag{20.40}$$

The assumption follows that the vehicle travels with constant velocity $\bar{v}$ and constant yaw rate $\bar{\dot{\psi}} \neq 0$ on a circular path during time interval $[t_1, t_2]$, with $\Delta t := t_2 - t_1$. For this case, the integral in Eq. (20.40) can be solved with

$$p_2 = p_1 + \begin{bmatrix} \frac{\bar{v}}{\bar{\dot{\psi}}}(\sin(\psi_1 + \bar{\dot{\psi}}\Delta t) - \sin(\psi_1)) \\ \frac{\bar{v}}{\bar{\dot{\psi}}}\left(-\cos(\psi_1 + \bar{\dot{\psi}}\Delta t) + \cos(\psi_1)\right) \\ \bar{\dot{\psi}}\Delta t \end{bmatrix} \tag{20.41}$$

There is a singularity in Gl. 41 for driving straight ahead with $\bar{\dot{\psi}} = 0$, which must be handled in the model. For vanishing yaw rates, the integration of Eq. (20.40) over the time interval $\Delta t := t_2 - t_1$ results in

$$p_2 = p_1 + \begin{bmatrix} \bar{v}\Delta t \cos(\psi_1) \\ \bar{v}\Delta t \sin(\psi_1) \\ 0 \end{bmatrix} \tag{20.42}$$

The estimate of the velocity $\bar{v}$ and the yaw rate $\bar{\dot{\psi}}$ is generally generated by measuring the wheel rates and via a yaw rate sensor, respectively.

For applications with highly dynamic maneuvers, it can be important to also consider the sideslip angle of the ego vehicle [35, 36]. Beyond the dead-reckoning technique, SLAM (simultaneous localization and mapping) approaches provide the possibility to include data of environment detecting sensors in the ego-motion estimation. Those approaches are computationally more expensive, although there are computationally efficient approximations as well [9].

20.3.3 Algorithms for Generating Occupancy Grid Maps

For many current and future driver assistance systems, the information on the drivable space in the vehicle environment is a prerequisite. Drivable space here means the area where the vehicle can drive without a collision occurring. Occupancy grid maps provide the possibility to depict in detail what area of the vehicle environment is occupied by an obstacle. In doing so, classical occupancy grid maps are especially suited to depicting static or slowly changing surroundings. Dynamic grid maps will be covered separately in ▶ Sect. 3.4.

Sensors which measure distance are best suited for generating occupancy grid maps because they allow a direct inference of the occupancy status of individual cells. The basic concept can be explained based on the following simplified example. ◻ Figure 20.6 schematically illustrates the inference of a measurement by a

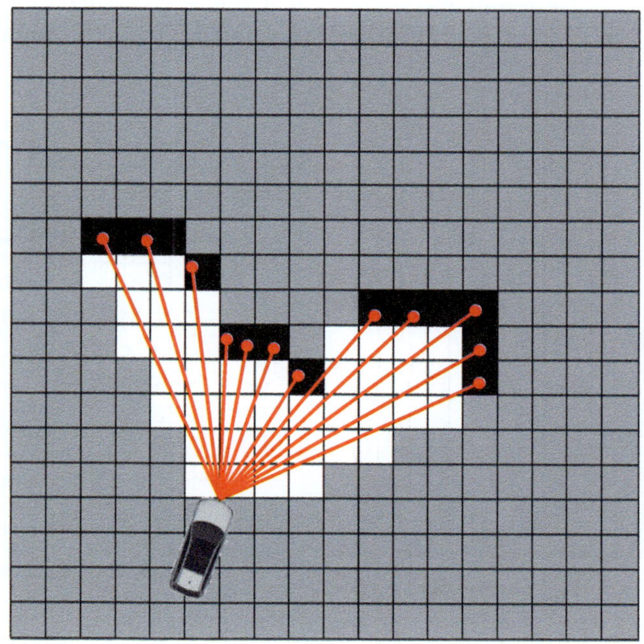

◻ **Fig. 20.6**   Simplified occupancy grid map based on measurements by a multi-beam lidar sensor

multi-beam lidar sensor onto a two-dimensional occupancy grid map (see Chap. 16). Lidar beams are reflected by obstacles, allowing the occupancy of corresponding cells to be inferred (shown in black). Cells that lie between the sensor and the reflection points are assumed to be free (shown in white). This measurement does not provide information on further cells (shown in gray). Such models are called inverse sensor models which in general, however, are far more complex. An example of a probabilistic inverse sensor model is described in the following.

20.3.4 Inverse Sensor Models

Sensor measurements are inherently afflicted with uncertainties. Therefore, an occupancy probability $p(o)$ is calculated for each cell. The event that the cell is occupied is described by o. The probability of the opposite event $f = \bar{o}$ (the cell is free) corresponds to the complementary probability $p(f) = 1 - p(o)$. This assumes that the occupancy state of a cell does not change over time; it is either occupied or empty.

A probabilistic inverse sensor model states which occupancy probability applies to a cell based on a sensor measurement. ◘ Figure 20.7 shows an example of a probabilistic inverse sensor model based on a single beam lidar distance measurement.

The left portion shows the occupancy probability $p(o)$ as a function of distance r from the sensor. The individual cells are modeled independently of one another. Cells around the reflection point receive an occupancy probability above 0.5. In contrast, the occupancy probability of the cells between the sensor and the reflection point is lower than 0.5. This area is often referred to as free space. The inverse sensor model accounts for the measurement accuracy of the sensor as well as the uncertainty of the vehicle pose estimate in the grid map. Therefore, besides the cell containing the reflection point, multiple other cells are assigned higher occupancy probabilities with $p(o) > 0.5$. In this example, the free space is assumed to be more certain the closer it lies to the sensor. This illustrates that the probability of erroneous measurements increases with distance from the sensor. The modeling in the azimuth direction is performed analogously. The right-hand portion of ◘ Fig. 20.7 shows the birds-eye view of the inverse sensor model.

In this example of an inverse sensor model, each lidar beam is modeled individually, which is not inherently the case. The modeling impacts the mapping algorithm, which calculates the occupancy probability based on the complete lidar measurement consisting of many individual beams. The fundamental goal in the design of the inverse sensor model is to reconstruct the characteristics of the sensor as precisely as possible while keeping the complexity and computational effort low. Some approaches also apply machine learning algorithms to derive inverse sensor models [9].

Inverse sensor models for radar sensors are different because radar reflections can appear behind line-of-sight occlusions via multipath propagation. This also leads to more clutter measurements (see ▶ Chap. 15). Free space functions for radar sensors are therefore more complex. Uncertainties in the azimuthal component are generally greater for radar sensors than for lidar sensors.

20.3.5 Combination Using Static Binary Bayes Filters

If no information is available yet on the occupancy of a cell, the occupancy probability is assumed to be $p(o) = 0.5$. This is the initial value for each cell. If more

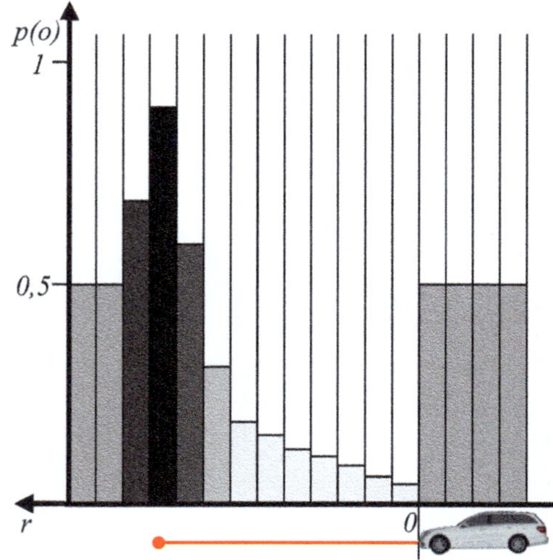
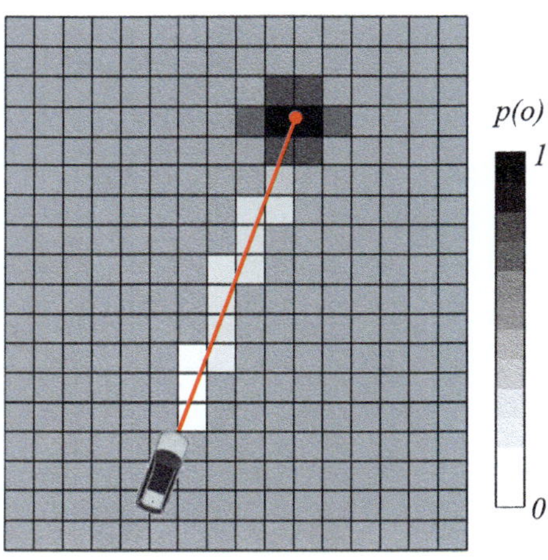

◘ **Fig. 20.7** Example of an inverse sensor model of a single lidar reflection showing the side view (left) and birds-eye view (right)

than one information source exists for a cell, then the information is combined. Depending on the inverse sensor model, this can occur if parts of a measurement are processed individually, e.g. as concerning the individual lidar beams in the preceding example. Dependent on the application, it is often the case that multiple successive measurements are incorporated into an occupancy grid map.

Under certain assumptions, multiple measurements z_i can be combined using the static binary Bayes filter:

Given are two conditional probabilities $p(o|z_1), p(o|z_2) \epsilon (0, 1)$. Let measurements z_1, z_2 be independent. Assuming the identical a priori probability $p(o) = p(\overline{o}) = 0.5$, then the combined probability is [9]:

$$p(o|z_1, z_2) = \frac{p(o|z_1)p(o|z_2)}{p(o|z_1)p(o|z_2) + (1 - p(o|z_1))(1 - p(o|z_2))}. \tag{20.43}$$

The combination rule (20.43) has the following characteristics:

$$p(o|z_1, z_2) > p(o|z_1) \Leftrightarrow p(o|z_2) > 0,5, \tag{20.44}$$

$$p(o|z_1, z_2) = p(o|z_1) \Leftrightarrow p(o|z_2) = 0,5, \tag{20.45}$$

$$p(o|z_1, z_2) = p(o|z_2, z_1), \tag{20.46}$$

An occupancy probability $p(o|z_2) > 0.5$ increases the previous occupancy probability. An occupancy probability $p(o|z_2) = 0.5$ corresponds to a neutral element and the sequence of the combination has no effect.

Measurements from different sensors or points in time can be combined using the combination rule (43). However, an important assumption for the practical realization of the grid map is the independence of individual cells. Dependencies between cells exist in the real world and can in principle be accounted for, but they increase the computational effort required for the creation of the grid map enormously.

For practical reasons, it is common to store the occupancy probability in logarithmic form as the log-odds ratio:

$$l(o) := \log \left(\frac{p(o)}{1 - p(o)} \right). \tag{20.47}$$

The advantage lies therein, that in logarithmic form the combination rule (20.43) can be performed as an addition:

$$l(o|z_1, z_2) = l(o|z_1) + l(o|z_2). \tag{20.48}$$

The same conditions apply as for Eq. (20.43). The reverse transformation is completed using

$$p(o) = 1 - \frac{1}{1 + e^{l(o)}} \tag{20.49}$$

20.3.6 Mapping Procedure

A sequence plan of the mapping algorithm for the presented inverse sensor model is shown in ◘ Fig. 20.8.

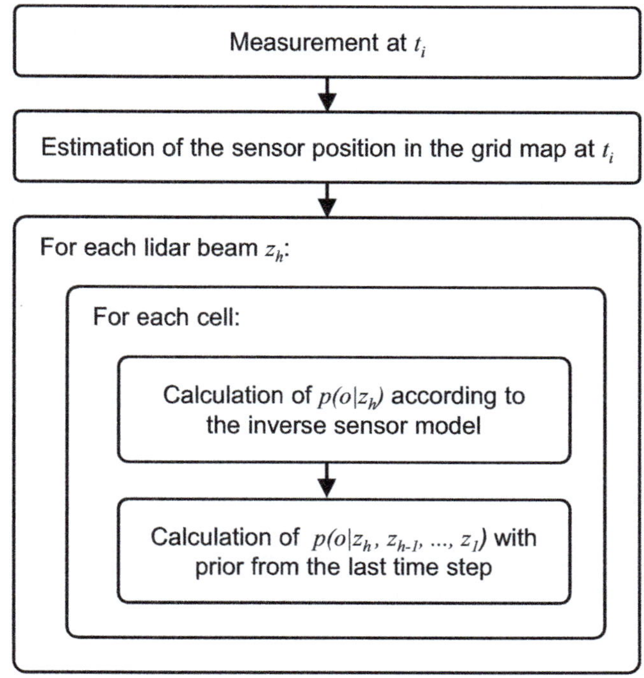

◘ **Fig. 20.8**   Sequence plan of a mapping algorithm

Fig. 20.9   Grid map of an actual vehicle environment

In this example, a lidar measurement consists of multiple individual beams. If a measurement that took place at time t_i is entered, the sensor pose within the grid map at time t_i is first estimated. Subsequently, each beam z_h is incorporated into the grid map individually. The occupancy probability $p(o|z_h)$ for each affected cell is first calculated according to the inverse sensor model and then combined with the occupancy probability $p(o|z_{h-1}, \ldots, z_1)$ resulting from the previous lidar beams. For each cell, the occupancy probability from the preceding time step is used as the prior value.

An occupancy grid map created based on a similar inverse sensor model is shown in **Fig. 20.9**. The sensor used was a multiple-beam lidar (see ▶ Chap. 16).

20.3.7   Dynamic Grid Maps

A prerequisite for the applicability of the static binary Bayes filter is the assumption that a cell does not change its occupancy state, which means that it remains either occupied or free. For the combination of individual lidar beams, which were all recorded almost simultaneously, this assumption is approximately true, since a cell does not change its occupancy state in this short time interval.

In general, and with regard to longer time intervals, this assumption for the vehicle environment does not apply but is violated by the presence of dynamic objects. In the literature, different approaches dealing with dynamic objects in grid maps are described. A distinction can be made between approaches in which dynamic objects are only to be filtered to avoid incorrect representations in the grid map, and approaches in which the grid map is used to detect dynamic movements and thus distinguish them from the static environment.

A simple method of the former type is the introduction of a forgetting factor. In this case, the inference of the occupancy probability of a cell depends not only on spatial but also on temporal conditions. The inference of the occupancy probability is therefore less certain the older the measurement is. More recent measurements are weighted stronger than older measurements when estimating the occupancy probability.

In many applications, grid maps are used explicitly to detect dynamic objects. This is generally done by determining the temporal consistency of the occupancy probability of cells. Cells that are consistently occupied or free are assigned to the static environment and cells for which the occupancy probability fluctuates greatly are marked as dynamically occupied areas. Especially occupancies occurring in areas previously detected as free are often treated as detections of dynamic objects. This often serves as a preprocessing step for tracking algorithms [34].

Dynamic occupancy grid maps offer the possibility of a self-contained representation of both the static and dynamic environment. These are characterized by providing an occupancy state and a velocity estimate for each grid cell and do not require a stationary environment. For this purpose, the Bayesian occupancy filter [37] uses a four-dimensional space consisting of two spatial coordinates and two coordinates for the velocity along each of the spatial coordinates. A cell is therefore associated with a two-dimensional location and a two-dimensional velocity vector. Thus, a Bayes filter consisting of prediction and update can be realized. This environment representation is also free of object assumptions. However, a Markov process that represents a state transition must be assumed for the prediction. This requires process assumptions, such as a constant velocity. The degree of abstraction of the Bayesian occupancy filter is therefore higher than for classic occupancy grid maps and the computational effort is also greater. As an alternative, especially particle filters have proven themselves for the estimation of a dynamic grid map. These methods have a significantly lower computational cost and can also be parallelized well. Particle filter approaches are described in more detail in the following chapter.

20.3.7.1 Particle Filter Approaches

A particle filter approach for dynamic grid maps was first described in [39]. Particle filters approximate the density functions by a large number of particles, i.e. possible realizations of, for example, the state vector. Therefore, the requirement of normally distributed quantities as in the Kalman filter is not needed anymore. Thus, it comes closer in its realization to the general recursive Bayes filter (▶ Sect. 2.2). Based on this approach, several variations emerged [40–42], differing mainly in the representation of the dynamic state. Generally, in all mentioned works, the occupancy of the grid map and the dynamic states of a grid cell are described by particles. Furthermore, the estimation is performed by an algorithm consisting of a prediction step and an update step as shown in ◘ Fig. 20.10. This common fundamental idea is explained in the following section.

State Representation

The dynamic state of a grid cell c is described by the set of n particles

$$S_c = \{x_i, w_i\}_{i=1}^{n} \tag{20.50}$$

which are located in the grid cell based on their position. A particle thus represents a hypothesis for the state and is a four-dimensional state vector

$$x_i = \left[p_{x,i} p_{y,i} v_{x,i} v_{y,i} \right]^T \tag{20.51}$$

consisting of a two-dimensional position and a two-dimensional velocity. It also has a particle weight w_i. The estimate of the velocity of a cell can be calculated by the average of the particle velocities:

$$\bar{v}_x = \sum_{i=1}^{n} w_i \cdot v_{x,i}, \tag{20.52}$$

$$\bar{v}_y = \sum_{i=1}^{n} w_i \cdot v_{y,i} \tag{20.53}$$

The occupancy probability $p(c)$ of a cell can be represented as the sum of the weights of the particles

$$p(o) = \sum_{i=1}^{n} w_i \tag{20.54}$$

Prediction Step

The calculation of the occupancy state of a grid cell through the prediction step and update step is based on the recursive Bayes filter (see ▶ Sect. 2.3.1). First, all particles are predicted independently based on their state and the chosen process model. A constant velocity and constant orientation model is generally used as process model. In the prediction step, particles can freely move between cells. Based on the distribution of the predicted particles among the grid cells, the predicted occupancies are obtained as shown in ◘ Fig. 20.10b).

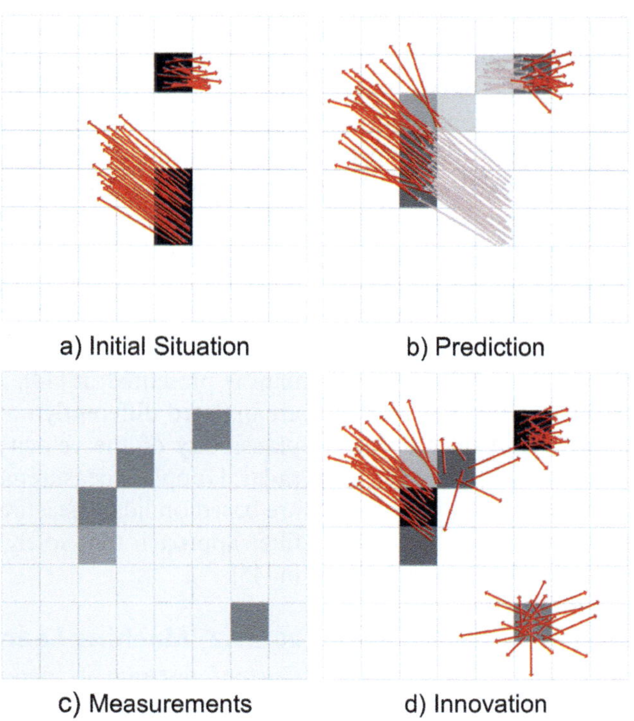

a) Initial Situation

b) Prediction

c) Measurements

d) Innovation

◘ **Fig. 20.10** Visualization of the particle filter-based algorithm for estimating a dynamic grid map

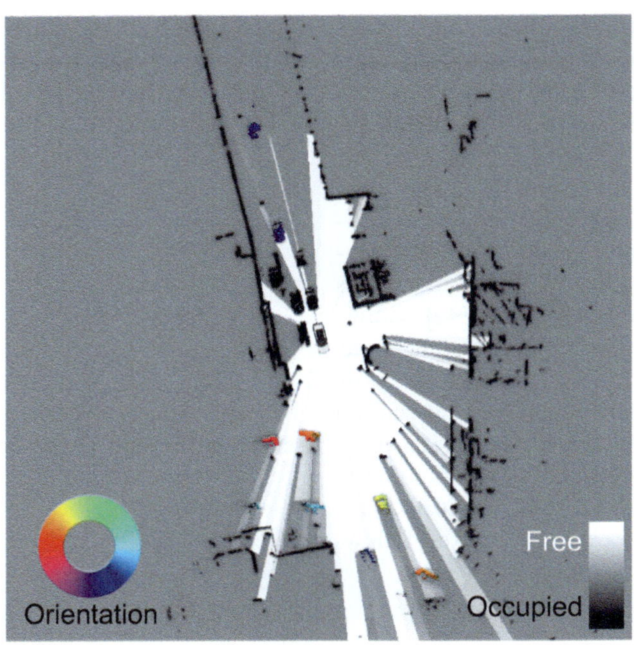

Fig. 20.11 Dynamic occupancy grid map in an urban environment. Dynamic cells are visualized in color according to the orientation of the velocity estimate

Update Step

In the update step, the predicted occupancies are updated with the current measurement by applying the Bayes formula (20.7). In practical applications, this often involves generating an occupancy grid map based on sensor measurements using an inverse sensor model. The occupancy probabilities of this grid map (see Fig. 20.10c) are interpreted as a measurement in the update step. Thus, assuming independent grid cells, the measurement and the predicted occupancy probabilities can be combined analogously to the binary Bayes filter (20.43) used in the static case. The combined occupancy probability is used to normalize the sum of the particle weights in a cell. So all particles of a cell are equally reweighted. As a result, particles predicted into a cell that is measured as occupied receive a high weight. Particles predicted into a cell measured as free receive a low weight. Subsequently, a resampling step is usually applied in which particles are removed or reproduced depending on their weight to prevent degeneration processes. Thereby the particles with high weight have a higher chance to be drawn again once or even several times. Similar to an evolutionary process, the particles which represent the movement and position of occupancies in the grid map most accurately by their state are retained. Similarly, particles that represent an incorrect hypothesis are lost. In addition, new particles are initialized in grid cells that have occupancy based on the current measurements to generate new hypotheses for the dynamic state.

A dynamic occupancy grid map generated with the algorithm in [42] is shown in Fig. 20.11.

Variations in Particle Filter Approaches

The variation in different particle filter approaches results mainly from the different state representations by particles. In [39, 42], the occupancy probability of a grid cell is given by the number of particles in a cell respectively by the sum of their weights. In contrast, in [41] the environment is divided into free space, static and dynamic occupancies, whereby only the dynamic occupancies are represented by a set of particles. This distinction into free, static, and dynamic is realized in [40] using the Dempster-Shafer theory [43]. Principally, in the approaches [40, 42, 44], in which the states are represented using the Dempster-Shafer theory, the combination rule of Dempster is used accordingly in the update step. A fusion of lidar and radar measurements for the generation of a dynamic occupancy grid map is presented in [44]. Thereby, the particle weights are updated differently even within a cell, based on the plausibility of the velocity of the particle state to the radar Doppler measurement. All mentioned methods are based on lidar measurements. In contrast, a particle filter approach that solely uses radar data is presented in [45].

20.3.7.2 Machine Learning Approaches

Current research is increasingly investigating how to use Deep Learning approaches in combination with a grid-based environment representation. Due to the image-like structure of grid maps, convolutional neural networks (CNNs) are mainly utilized. In several works, dynamic grid maps generated with particle filter approaches are used as input data for a neural network. Thus, in addition to the occupancies, i.e. the shape of objects in a top view, information about the dynamic states is also available to the neural network. The application of convolutional layers allows to use spatial context and thereby also overcomes the assumption of independence of individual cells assumed in the creation of grid maps. In [46], a neural network is used to distinguish dynamic and static areas. Here, the neural network somehow serves as a post-processing step to remove erroneous velocity estimates in the dynamic grid map. Further approaches investigate, for example, the prediction of future occupancies in the grid map based on the current dynamic grid map [47] and the detection of objects [48].

In the method introduced as deep tracking [49], a recurrent neural network is used to track static and dynamic objects in the environment of an automated vehicle. Recurrent neural networks have feedback loops, through which temporally encoded information can be stored and used for prediction. Thereby, the vehicle environment is represented as an occupancy grid map, which contains for each grid cell the binary occupancy state, free or occupied, and the observability of the grid cell at the current time. These occupancy grid maps used as input data are created based on lidar measurements of a single time step and thus represent only the currently observable area of the vehicle environment. By using a recurrent network, the observations of multiple time steps are captured in the internal memory states of the recurrent network layers. These internal memory states are able to capture not only the location of occupancies but also their dynamics. The network is able to estimate an occupancy grid map based on the internal states, which represents the occupancies generated by dynamic objects also in areas that are not observable. An extension of this approach demonstrates that in addition to occupancy, a semantic class can also be predicted. However, in contrast to the particle filter approaches, no velocity estimates are provided.

The estimation of a dynamic occupancy grid map using a learning-based approach is presented in [50, 51]. In these approaches, recurrent network layers are inserted into a convolutional neural network at multiple levels. This structure allows temporal filtering of learned features at different resolutions. Similar to the particle filter approaches, occupancy grid maps are calculated based on an inverse sensor model and lidar measurements are used as input data. The network output is a grid map with an occupancy probability and a two-dimensional velocity per grid cell. As a training label, the dynamic grid map of a particle filter approach, where velocities are replaced by the movement of manually annotated object boxes, can be used as shown in [51]. During the training process, sequences of occupancy grid maps are processed through the network and information is fed into the internal memory states to predict the dynamic grid map of the current time. The internal memory states replace, in a certain way, the function of the particles by using information based on measurements of several time steps to store the dynamics and the position of objects. In addition, the update of the internal states, based on the last state and the current input data, is a kind of temporal filtering. The advantages of a learning-based approach result from the independence of process models, which means that a diverse range of movements can be modeled. In addition, the use of convolutional neural networks enables the exploitation of spatial context, which for example can be useful for separating static and dynamic areas. The training process of deep neural networks in combination with recurrent network layers is comparatively computationally expensive and requires a large amount of memory due to the processing of input data sequences. However, in the implemented application, the hardware requirements are much lower compared to training. This is because the current prediction does not use an input data sequence, but is based only on the current input data and the memory states of the previous time step.

In some current research, similar to dynamic grid maps, motion is estimated based on lidar measurements and represented in a 2D grid map. In [52], lidar measurements from two consecutive time steps are transformed into an image-like representation and processed through a neural network. The network output provides an estimate of the scene flow, i.e. the movement between the current and the previous time step. In contrast, the approach in [53] uses recurrent network layers to aggregate information from five lidar measurements and estimate a velocity grid map. Both approaches use raw sensor measurements as input data, but compared to the dynamic occupancy grid maps, the free space is not modeled.

20.4 Fused Environment Modeling

Object-based and grid-based representations are different depictions of the vehicle environment that have specific advantages and disadvantages, respectively. Object-based representations depict the states of the individual objects such as the position, the velocity, and the extent of traffic participants as described in ▶ Sect. 20.2. This representation is based on process models which allow predictions for the near future. However, a disadvantage is that the assumptions made about the process model must coincide with reality to keep estimation errors as low as possible. Besides traffic participants, road topologies or traffic infrastructure elements such as traffic lights can also be represented by the object-based representation. Depending on the application, the road topology is detected while driving. This is possible especially in very structured environments with existing lane markings such as highways. In most cases, pre-recorded and highly accurate digital maps are used to navigate in complex areas with a lot of visual occlusions. A localization algorithm estimates the ego vehicle's position relative to the digital map, whereby all the map data is available with respect to the vehicle.

In contrast to the object-based representation, the generation of classic occupancy grid maps assumes a stationary environment as described in ▶ Sect. 20.3. Occupancy grid maps fuse and store information relative to location and not to objects and represent the environment by individual cells in the complete detection range of the sensors. Due to the temporal integration of the measurements, grid maps are also robust with respect to single erroneous sensor measurements. However, dynamic objects violate the assumption of a stationary environment, which is why dynamic occupancy grid maps (▶ Sect. 3.4) are increasingly used. Dynamic occupancy grid maps model dynamic cells, with which it is also possible to infer objects and their state values. A major advantage is the modeling without object assumptions, which allows to detect arbitrary object types and object shapes.

On account of their contrary calculation methods and different characteristics, it is an interesting question how these different domains for representing the environment could be advantageously combined. This section will address a selection of approaches.

20.4.1 Dynamic Occupancy Grid Map Object Extraction

A disadvantage of the dynamic occupancy grid maps is the missing associations between the cells and their corresponding object. To counteract this disadvantage, dynamic cells are clustered and transformed into an object-based representation [6, 54]. For this purpose, all cells with an occupancy probability $p(o) > 0.7$ are considered in [6]. Connected cells are identified using a clustering algorithm, e.g., Density-Based Spatial Clustering of Applications with Noise (DBSCAN) [24]. The DBSCAN algorithm determines a cluster consisting of several cells based on the Euclidean distance. Cells that are not included in the cluster either create a new cluster or are marked as noise points. In order to separate objects moving close to each other into different clusters, their velocity differences can be additionally evaluated. Finally, a cluster of cells is obtained that presumably belong to one object. These clusters include traffic participants as well as any other static obstacles in the vehicle's environment.

For transforming the cell cluster from the grid-based representation to the object-based representation, the object's position, velocity, orientation, length, and width are calculated based on the mean value of all individual cell states. Based on these states, the cell clusters' corresponding object type can additionally be classified by heuristically chosen parameters or, alternatively, using a machine learning model. As a classification model, a decision tree or a multi-layer perceptron (MLP) can be deployed. These trained models usually achieve higher accuracies than classification by heuristically chosen parameters. Since supervised learning requires a dataset, the cell clusters must be annotated as static or dynamic. The classification model uses the object states length, width, height, distance relative to the ego vehicle, absolute velocity, and the number of cells in a cluster as inputs. The model is then trained to distinguish between dynamic and static objects. This process is illustrated in ◘ Fig. 20.12 using an exemplary scene. To improve the estimation accuracy of the object states values and to receive state variances, the extracted object detections are additionally tracked by a multi-object tracking filter (refer to ▶ Sect. 2.3). Due to this temporal filtering, the occurrence of false or missing objects is reduced.

20.4.2 Environment Modeling and Object Fusion

As explained in the previous section, dynamic objects can also be extracted from the dynamic occupancy grid map and are then available in the object-based representation. So, besides the multi-object tracking from ▶ Sect. 2.3, a second algorithmically independent and redundant calculation method for objects and their states exists. These two sources of information can be used in environment modeling to create a more accurate and reliable representation of the environment. For example, both sources can be compared, and objects can be checked for consistency. Additionally, the accuracy of the state estimation can be increased by fusing both sources under the condition of statistical independence, which is briefly explained in the following.

A classic approach for environment modeling is object fusion on a track-level (Track-To-Track Fusion) to combine objects from multiple sources. First, presumably identical objects are associated with each other (Track-To-Track Association). For association, all objects from the first source are compared with the objects from the second source based on the Mahalanobis distance of their state vectors. The global best association to another object is determined using the Hungar-

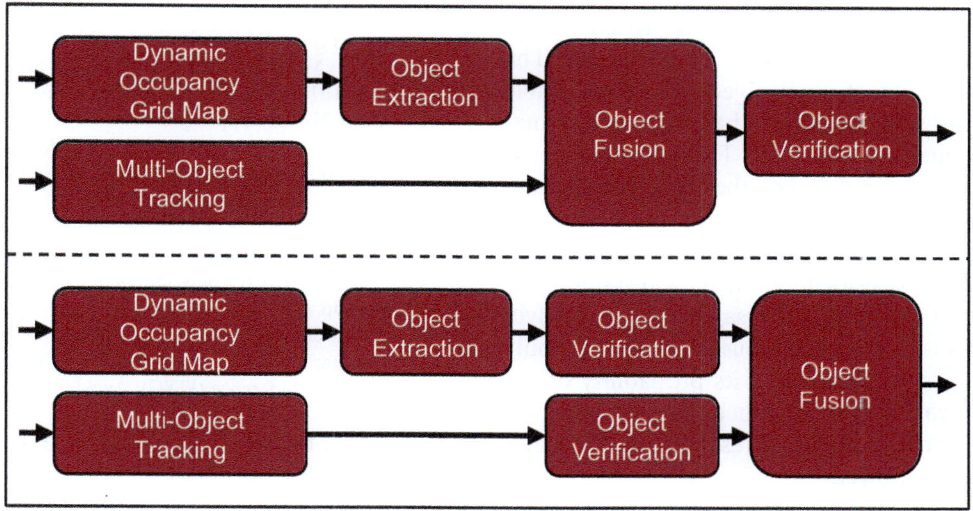

ian method [23]. After association, the fused state vector and covariance matrix are calculated by a fusion algorithm such as the Covariance Intersection [55]. The prerequisite for an accurate result is uncorrelated data. Objects without a successful association do not receive a state update. Finally, after object fusion, a redundant and complementary list of objects in the local environment is obtained.

Besides the object states and their uncertainties, the existence probability is an essential metric for evaluating whether the objects are real or falsely detected (false alarms). False alarms can occur due to incorrect sensor measurements over a time period, for example as reflections on glass facades. Therefore, a verification module decides on the existence of each object. In addition to the existence probability described in ▶ Sect. 2.2 the digital map can be used as a further source of information. In [56], road topologies and building outlines are utilized as digital map elements to compute an extended existence probability for object plausibility. This approach checks whether an object is located in the correct driving lane and is moving in the designated driving direction. If this is the case, the object's existence can be confirmed. If there is an object located inside a building due to incorrect sensor measurements, this object can be classified as a false alarm and removed from the list of existing objects. A probabilistic graphical model (Bayesian Network) defined by expert knowledge is used to determine the extended existence probability. When using digital maps, it is mandatory to consider the uncertainties of the map and the vehicle's localization result. Furthermore, an outdated map leads to incorrect assumptions being made. A validation of the map data helps to avoid these false assumptions [57].

◘ Figure 20.13 depicts two variants of a modular environment modeling. The upper variant shows

◘ **Fig. 20.12** Extraction of moving objects from a dynamic occupancy grid map

◘ **Fig. 20.13** Variants of a modular environment modeling structure with object extraction, object fusion, and object verification the process described here. First, the objects are extracted from the dynamic occupancy grid map (refer to ▶ Sect. 4.1) and fused with objects from the multi-object tracking. Finally, each object is verified individually, and a decision is made based on the existence probability whether the object is forwarded to subsequent modules. An alternative processing pipeline depicts the lower part from ◘ Fig. 20.13. Here, the object verification is calculated independently for both sources before the objects are fused. This has the advantage that falsely detected objects do not affect the fusion result negatively, but the associations may be ambiguous. The described modular environment modeling by fusing object-based and grid-based representations creates a consistent depiction of the environment and is thus an important component of the function system architecture from ▶ Sect. 20.1.

20.5 Summary

This chapter describes the basics of an object-based and a grid-based representation of the vehicle environment as well as necessary basic algorithms. While in the object-based representation each object is described by its dynamic state-space model, the grid-based methods are initially model-free, i.e. they do not require any physical model hypotheses. Thus, they are best suited to represent the static part of the environment such as drivable free space and boundaries. However, dynamic grid maps can be designed to also describe the movement of individual cells in a model-free manner. Applying post-processing steps such as clustering methods to dynamic grid maps, individual objects can also be extracted and dynamically modeled.

Object-based methods have the advantage that through the individual dynamic modeling of each object, it is also comparatively easy to infer its semantic meaning. However, the classification procedures required for this were not discussed in this chapter. The detection of the state of each object requires suitable filtering methods, which today are based on approximations of the recursive Bayes filter. The Kalman filter is the best-known representative of these approximations. Modern tracking methods are based on the random finite set theory and allow the formulation of multi-object tracking in a closed-form mathematical framework.

An important parameter for safety-relevant driver assistance systems and autonomous driving is not only the dynamic object state but also its probability of existence, i.e., a measure of confidence that an object appearing in the environment representation originates from a real existing object.

Since both, object-based and grid-based representations have specific advantages, the combination of both methods in hybrid architectures is obvious. In this context, a modular architecture with preferably generic interfaces to the sensor system and to the situation assessment and maneuver planning will gain importance. Furthermore, function modules based on machine learning methods will gain more importance in the future due to their very high performance. In consequence, these modules have also to be considered in a hybrid architecture. For object detection and object classification, they are nearly standard. Even for grid map generation, there are very promising approaches, as also shown in this chapter. However, an open research and development topic is still the validation of these AI-based methods for serial use.

References

1. Nuss, D.; Stuebler, M.; Dietmayer, K.: Consistent Environmental Modeling by Use of Occupancy Grid Maps, Digital Road Maps, and Multi-Object Tracking. IEEE Intelligent Vehicles Symposium (IV), 2014.
2. Tas, O. S.; Hormann, S. ;Schaufele, B.; Kuhnt, F.: Automated vehicle system architecture with performance assessment, IEEE 20th International Conference on Intelligent Transportation Systems (ITSC), 2017, ▶ https://doi.org/10.1109/ITSC.2017.8317862.
3. Ulbrich, S.; Reschka, A.; Rieken, J.; Ernst, S. Bagschik, G.; Dierkes, F.; Nolte, M.; Maurer, M.: Towards a Functional System Architecture for Automated Vehicles, 2017, ▶ http://arxiv.org/abs/1703.08557
4. Matthaei, R. Maurer, M.: Autonomous driving – a top-down-approach, De Gruyter, at – Automatisierungstechnik, 2015, Vol 63, pp. 155–167, ▶ https://doi.org/10.1515/auto-2014-1136
5. Schreier, M.: Environment representations for automated on-road vehicles, De Gruyter, at – Automatisierungstechnik, 2018, Vol. 66, pp. 107–118, ▶ https://doi.org/10.1515/auto-2017-0104
6. F. Gies, A. Danzer and K. Dietmayer, "Environment Perception Framework Fusing Multi-Object Tracking, Dynamic Occupancy Grid Maps and Digital Maps," 21st International Conference on Intelligent Transportation Systems (ITSC), 2018, pp. 3859–3865.
7. Bansal, M.; Krizhevsky, A.; Ogale, A.: ChauffeurNet: Learning to Drive by Imitating the Best and Synthesizing the Worst, 2018, ▶ http://arxiv.org/abs/1812.03079.
8. Bar-Shalom, Y., Fortmann, T.: Tracking and Data Association. Academic Press, Boston (1988)
9. Thrun, S.: Probabilistic Robotic. The MIT Press, (2005)
10. Mahler, R.: Multitarget Bayes filtering via first-order multitarget moments. IEEE Transactions on Aerospace and Electronic Systems, 2003, Vol. 39.4, pp. 1152–1178.
11. Kalman, R.: A new approach to linear filtering and prediction problems. Transactions of the ASME – Journal of Basic Engineering, 1960, Vol. 82, pp. 35–45.
12. Julier, S.; Uhlmann, J.; Durrant-Whyte, H.: A new method for the nonlinear transformation of means and covariances in filters and estimators. IEEE Transactions on Automatic Control, 2000, Vol. 45.3, pp. 477–482.
13. Danzer, A.; Griebel, T.; Bach, M.; Dietmayer, K.: 2D Car Detection in Radar Data with PointNets. IEEE Intelligent Transportation Systems Conference (ITSC), Auckland, New Zealand, 2019, pp. 61–66.
14. Herzog, M.; Dietmayer, K.: Training a Fast Object Detector for LiDAR Range Images Using Labeled Data from Sensors with Higher Resolution. IEEE Intelligent Transportation Systems Conference (ITSC), Auckland, New Zealand, 2019, pp. 2707–2713, ▶ https://doi.org/10.1109/ITSC.2019.8917011.

15. Bar-Shalom, Y., Tse, E.: Tracking in a cluttered environment with probabilistic data association. Automatica **11**, 451–460 (1975)

16. Mahler, R.: Statistical Multisource-Multitarget Information Fusion. Artech House, Boston (2007)

17. Vo, B.-T., Vo, B.-N.: Labeled Random Finite Sets and Multi-Object Conjugate Priors. IEEE Trans. Signal Process. **61**, 3460–3475 (2013)

18. Reuter, S.; Vo, B.-T.; Vo, B.-N.; Dietmayer, K.: The Labeled Multi-Bernoulli Filter. IEEE Transactions on Signal Processing, 2014, Vol. 62.12, pp. 3246–3260.

19. Reuter, S.; Danzer, A.; Stübler, M.; Scheel, A.; Granström, K.: A Fast Implementation of the Labeled Multi-Bernoulli Filter Using Gibbs Sampling. IEEE Intelligent Vehicles Symposium, 2017.

20. Munz, M.: Generisches Sensorfusionsframework zur gleichzeitigen Zustands- und Existenzschätzung für die Fahrzeugumfelderfassung. PhD Thesis, Ulm University, 2011.

21. Reuter, S.: Multi-Object Tracking Using Random Finite Sets, PhD Thesis, Ulm University, 2014.

22. Murty, K.: G: An Algorithm for Ranking All the Assignments in Order of Increasing Cost. Oper. Res.. Res. **16**(3), 682–687 (1968)

23. Kuhn, H.W.: The Hungarian method for the assignment problem. Nav. Res. Logist.Logist. **2**, 83–97 (1955)

24. Ester, M., Kriegel, H.-P., Sander, J., Xu, X., et al.: A Density-Based Algorithm for Discovering Clusters in Large Spatial Databases with Noise. Kdd **96**(34), 226–231 (1996)

25. Feng, D.; Haase-Schütz, C.; Rosenbaum, L.; Hertlein, H.; Gläser, C.; Timm, F.; Wiesbeck, W., Dietmayer, K.: Deep Multi-Modal Object Detection and Semantic Segmentation for Autonomous Driving: Datasets, Methods, and Challenges.IEEE Transactions on Intelligent Transportation Systems, 2020.

26. Granström, K.; Baum, M.; Reuter, S.: Extended Object Tracking: Introduction, Overview and Applications. Journal of Advances in Information Fusion, 2017, Vol. 12.2, pp. 175–188.

27. Skog, I.; Handel, P.: In-Car Positioning and Navigation Technologies: A Survey. IEEE Transactions on Intelligent Transportation Systems, 2009, Vol. 10.1, pp. 4–21.

28. Konrad, M.; Nuss, D.; Dietmayer, K.: Localization in Digital Maps for Road Course Estimation using Grid Maps. IEEE Intelligent Vehicles Symposium (IV), 2012, pp. 87–92.

29. Levinson, J.; Thrun, S.: Robust vehicle localization in urban environments using probabilistic maps. IEEE International Conference on Robotics and Automation (ICRA), 2010, pp. 4372–4378.

30. Mattern, N.; Schubert, R; Wanielik, G.: High-accurate vehicle localization using digital maps and coherency images. IEEE Intelligent Vehicles Symposium (IV), 2010, pp. 462–469.

31. Schindler, A.: Vehicle self-localization with high-precision digital maps. Intelligent Vehicles Symposium Workshops (IV Workshops), 2013, pp. 134–139.

32. Elfes, A.: Using occupancy grids for mobile robot perception and navigation. IEEE Comput., **22.6**, pp. 46–57 (1989)

33. Nuss, D.; Wilking, B.; Wiest, J.; Deusch, H.; Reuter, S.; Dietmayer, K.: Decision-Free True Positive Estimation with Grid Maps for Multi-Object Tracking. IEEE Conference on Intelligent Transportation Systems (ITSC), 2013, pp. 28–34.

34. Petrovskaya, A.; Perrollaz, M.; Oliveira, L.; Spinello, L.; Triebel, R.; Makris, A.; Yoder, J.-D.; Laugier, C.; Nunes, U.; Bessiere, P.: Awareness of road scene participants for autonomous driving. Handbook of Intelligent Vehicles. Ed. Springer, London, 2012, pp.1383–1432.

35. Scaramuzza, D.; Fraundorfer, F.: Visual Odometry: Part I - The First 30 Years and Fundamentals. IEEE Robotics and Automation Magazine, Volume 18, issue 4, 2011.

36. Mayr, R.: Regelungsstrategien für die automatische Fahrzeugführung. Springer, Berlin Heidelberg (2001)

37. Nuss, D.; Reuter, S.; Konrad, M.; Munz, M.; Dietmayer, K.: Using grid maps to reduce the number of false positive measurements in advanced driver assistance systems. IEEE Conference on Intelligent Transportation Systems (ITSC), 2012, pp.1509–1514

38. Coué, C.; Pradalier, C., Laugier, C., Fraichard, T., Bessiere, P.: Bayesian Occupancy Filtering for Multitarget Tracking: an Automotive Application. International Journal of Robotics Research, 2006, Vol. 25.1, pp.19–30

39. Danescu, R., Oniga, F., Nedevschi, S.: Modeling and Tracking the Driving Environment With a Particle-Based Occupancy Grid. IEEE Trans. Intell. Transp. Syst.Intell. Transp. Syst. **12**(4), 1331–1342 (2011)

40. Tanzmeister, G.; Thomas, J.; Wollherr, D.; Buss, M.: Grid-based Mapping and Tracking in Dynamic Environments using a Uniform Evidential Environment Representation. IEEE International Conference on Robotics and Automation (ICRA), 2014, pp. 6090–6095.

41. Nègre, A.; Rummelhard, L.; Laugier, C.: Hybrid sampling Bayesian Occupancy Filter. IEEE Intelligent Vehicles Symposium Proceedings (IV), 2014, pp. 1307–1312

42. Nuss, D., Reuter, S., Thom, M., Yuan, T., Krehl, G., Maile, M., Gern, A., Dietmayer, K.: A random finite set approach for dynamic occupancy grid maps with real-time application. The International Journal of Robotics Research **37**(8), 841–866 (2018)

43. Dempster, A. P.: A generalization of Bayesian inference. Journal of the Royal Statistical Society. Series B (Methodological), vol. 30, no. 2, pp. 205–247, 1968.

44. Nuss, D.; Yuan, T.; Krehl, G.; Stuebler, M.; Reuter, S.; Dietmayer, K.: Fusion of laser and radar sensor data with a sequential Monte Carlo Bayesian occupancy filter. IEEE Intelligent Vehicles Symposium (IV), 2015, pp. 1074–1081.

45. Diehl, C.; Feicho, E.; Schwambach, A.; Dammeier, T.; Mares, E.; Bertram, T.: Radar-based Dynamic Occupancy Grid Mapping and Object Detection. IEEE 23rd International Conference on Intelligent Transportation Systems (ITSC), 2020, pp. 1–6.

46. Piewak, F.; Rehfeld, T.; Weber, M.. Zöllner, J.M.: Fully convolutional neural networks for dynamic object detection in grid maps. IEEE Intelligent Vehicles Symposium (IV), 2017, pp. 392–398.

47. Hoermann, S.; Bach, M.; Dietmayer, K.: Dynamic Occupancy Grid Prediction for Urban Autonomous Driving: A Deep Learning Approach with Fully Automatic Labeling. IEEE International Conference on Robotics and Automation (ICRA), 2018, pp. 2056–2063.

48. Hoermann, S.; Henzler, P.; Bach, M.; Dietmayer, K.: Object Detection on Dynamic Occupancy Grid Maps Using Deep Learning and Automatic Label Generation. IEEE Intelligent Vehicles Symposium (IV), 2018, pp. 826–833.

49. Dequaire, J., Ondrúška, P., Rao, D., Wang, D., Posner, I.: Deep tracking in the wild: End-to-end tracking using recurrent neural networks. The International Journal of Robotics Research **37**(4–5), 492–512 (2018)

50. Schreiber, M.; Belagiannis, V.; Gläser, C.; Dietmayer, K.: Motion Estimation in Occupancy Grid Maps in Stationary Settings Using Recurrent Neural Networks. IEEE International Conference on Robotics and Automation (ICRA), 2020, pp. 8587–8593.

51. Schreiber, M.; Belagiannis, V.; Gläser, C.; Dietmayer, K: Dynamic Occupancy Grid Mapping with Recurrent Neural Networks. IEEE International Conference on Robotics and Automation (ICRA), 2021, pp.6717–6724.

52. Lee, K.; Kliemann, M.; Gaidon, A.; Li, J.; Fang, C.; Pillai, S.; Burgard, W.: PillarFlow: End-to-end Birds-eye-view Flow Estimation for Autonomous Driving. IEEE/RSJ International Conference on Intelligent Robots and Systems (IROS), 2020, pp. 2007–2013.

53. Filatov, A.; Rykov, A.; Murashkin, V.: Any Motion Detector: Learning Class-agnostic Scene Dynamics from a Sequence of LiDAR Point Clouds. IEEE International Conference on Robotics and Automation (ICRA), 2020, pp. 9498–9504.

54. Steyer, S., Lenk, C., Kellner, D., Tanzmeister, G., Wollherr, D.: Grid-Based Object Tracking With Nonlinear Dynamic State and Shape Estimation. IEEE Trans. Intell. Transp. Syst.Intell. Transp. Syst. **21**(7), 2874–2893 (2020)

55. Chang, K., Chong, C., Mori, S.: Analytical and Computational Evaluation of Scalable Distributed Fusion Algorithms. IEEE Trans. Aerosp. Electron. Syst.Aerosp. Electron. Syst. **46**(4), 2022–2034 (2010)

56. F. Gies, J. Posselt, M. Buchholz and K. Dietmayer, "ExtendedExistence Probability Using Digital Maps for Object Verification,"IEEE 23rd International Conference on Information Fusion(FUSION), 2020, pp. 1–7.

57. S. R. Bhavsar, A. Vatavu, T. Rehfeld and G. Krehl, "Sensor Fusion-based Online Map Validation for Autonomous Driving,"IEEE Intelligent Vehicles Symposium (IV), 2020, pp. 77–82.

58. Kaulbersch, H.; Honer, J.; Baum, M.: A Cartesian B-Spline VehicleModel for Extended Object Tracking, 21st InternationalConference on Information Fusion (FUSION),, Cambridge, UK (2018)

20

Open Access This chapter is licensed under the terms of the Creative Commons Attribution-NonCommercial-NoDerivatives 4.0 International License (▶ http://creativecommons.org/licenses/by-nc-nd/4.0/), which permits any noncommercial use, sharing, distribution and reproduction in any medium or format, as long as you give appropriate credit to the original author(s) and the source, provide a link to the Creative Commons license and indicate if you modified the licensed material. You do not have permission under this license to share adapted material derived from this chapter or parts of it.

The images or other third party material in this chapter are included in the chapter's Creative Commons license, unless indicated otherwise in a credit line to the material. If material is not included in the chapter's Creative Commons license and your intended use is not permitted by statutory regulation or exceeds the permitted use, you will need to obtain permission directly from the copyright holder.

SLAM and Map-Based Localization

Marc Sons, Mario Theers, and Isabell Hofstetter

Contents

© The Author(s) 2026
H. Winner et al. (eds.), *Handbook Assisted and Automated Driving*,
https://doi.org/10.1007/978-3-658-45276-6_21

21.1 Introduction

Automated driving functions such as route following assist or fully automated driving in urban environments are already being realized today in series and pilot projects [1, 2]. A prerequisite for such applications are HD maps which store static traffic elements and environmental data precisely localized. This information can be provided at runtime and support the automated vehicle efficiently in its driving function. Maps can provide information much earlier as on-board sensors, for example, due to occlusions or limited perceptual distance.

In order to use map data efficiently, it is necessary to precisely determine the vehicle ego position and orientation within the map. Especially in urban areas, this task cannot be achieved exclusively by global satellite navigation systems (GNSS) due to signal shadowing and multipath propagation [3]. Map-based localization methods enable a robust and cost-effective complement to pure GNSS-based localization. Within map-based localization, environmental measurements from onboard sensors are associated with landmarks from a map. Based on these associations, the position and orientation of the ego vehicle is precisely determined in the map frame. For this purpose, a mobile agent has a precomputed map available during runtime.

This chapter provides insight into both, map-based localization and mapping. Nearly all of the underlying concepts and methods for mapping and map-based localization emerge from the research field of "Simultaneous Localization and Mapping (SLAM)" [4–7]. Therefore, in this chapter, we also highlight and describe the basic concepts of SLAM. The focus of this chapter is, however, on map-based localization as it is of crucial importance for automated driving.

SLAM

The original SLAM problem addresses the issue of how to enable a mobile agent to find its way in an unknown environment. An additional challenge arises from the fact that the agent is initially in an unknown position with an unknown orientation. Based on measurements of its onboard sensors, the idea is to incrementally create a map in which the agent can localize itself. It moves from its starting point through the unknown environment and collects measurements of it continuously. By knowing about its ego motion, the agent can accumulate observed static and recognizable elements of the environment—so-called landmarks—into a map. When the agent repeatedly moves through already mapped territory, it can localize itself in the existing map and simultaneously improve the estimates of the already mapped landmarks in that area. This so-called loop closure optimization is one of the key concepts of SLAM [8].

Mapping

In contrast to the original SLAM problem, a map is computed in advance for automated driving and will not be updated during localization. For this purpose, one or more agents drive in the area to be mapped and record ego-relative sensor measurements. The data from all agents is then transferred to a server and aggregated into a common map. This server-side mapping can be understood as multi-session multi-agent SLAM, since this approach is mainly based on the original SLAM idea. The advantage is that the real-time and causality requirement of SLAM is removed for the mapping, since no real-time localization needs to be established during map generation [9]. This allows the map to be estimated more robustly and accurately. Also, agents require less computational resources since the computationally expensive mapping process is server-side. Likewise, the server-side mapping also allows for semi-automated quality checks of the created or updated map before it is distributed to the agents. In addition, maps for automated driving often still have to be manually post-processed, for example, to add additional semantic information.

Localization

For map-based localization, a mobile agent always has a previously created map available. In addition, the agent receives real-time measurements of the structure and appearance of its environment through its onboard sensors. The agent associates these measurements with landmarks from the map. Correspondences between measurements and landmarks allow the agent to precisely locate itself in the map. If the map is also precisely geodetically referenced, map-based localization can be supported by measurements from GNSS. In contrast to SLAM, localization only estimates the ego position and orientation without updating the map.

The terms landmark and sensor measurement are deliberately kept general at this point, since different landmark and measurement types from different sensors can be combined. The main exteroceptive sensors for map-based localization are camera and LiDAR systems. RADAR systems are also of fundamental importance for automated vehicles, however, they are less common in the SLAM domain due to the poor availability of semantic information [10]. In addition, SLAM applications are often supported by low-cost GNSS receivers, inertial measurement units (IMU), and wheel speed sensors. The aforementioned measurement systems will be addressed in this chapter with respect to their application in the SLAM context.

An important class of landmarks are simple geometric structures that have additional semantic meaning, so that they can be used for localization as well as for other purposes such as scene interpretation [11]. In this way, memory can be saved so that map information can

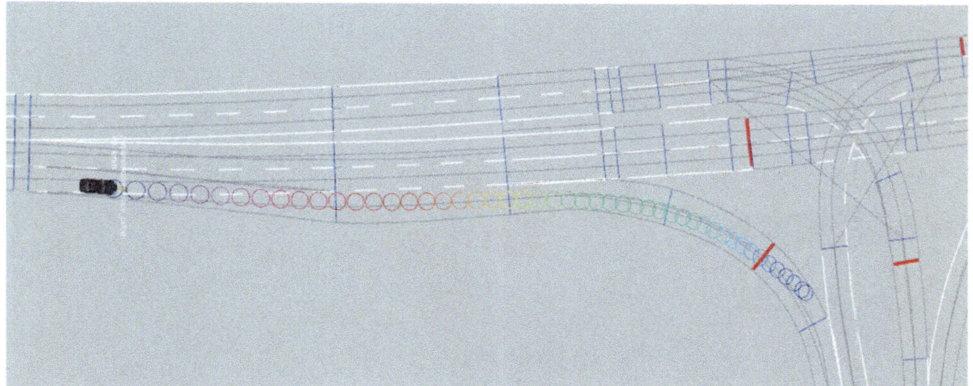

Fig. 21.1   Trajectory planning within a planning map

be exchanged between agent and server even over data connections with severely limited bandwidth. For this important class of landmarks, which includes, for example, lane markers, planar wall facades, and cylindrical sign posts, several localization approaches are presented in this chapter.

Requirements for map-based localization

By localizing an automated vehicle in an existing map, the information located in the map is put into a geometric relation to the current position of the agent. The accuracy requirements for localization depend largely on the application and the level of detail of the information stored in the map, as the following three examples illustrate:

Trajectory planning

Figure 21.1 shows an example of the planned trajectory of a vehicle within a turning lane in an intersection scenario. In this example, the depicted lane geometry and topology is stored in a map and has been made reliably available to the agent by an accurate map-based localization method. Thus, based solely on the map, the desired trajectory can be planned in advance while complying to lane restrictions and traffic regulations. Since the trajectory to be driven is recalculated at high frequency, localization inaccuracies of just a few centimeters lead to undesired and uncomfortable driving maneuvers. In particular, the localization signal for this application must be steady and smooth. Highly accurate localization methods often implement model-based regularizations for this purpose.

Navigation

An application that requires significantly lower localization accuracy compared to map-based trajectory planning is lane-based navigation. For example, if an automated vehicle is driving on a highway and wants to leave it at the next exit, the navigation system only needs to know on which lane it is and how far away the exit is, so that a lane change can be initiated if necessary. Neither longitudinal and lateral position estimation with centimeter accuracy nor orientation estimation are required for this function. For such a function, a precise geometric localization is not needed, since only information about the roadway topology and no information about the roadway geometry must be provided from the map.

3D-projection

Figure 21.2 shows the projection of two mapped traffic lights into a camera image. In this case, the map stores the 3D position of the traffic light. The localization enables the projection of the surrounding 3D boxes of the traffic lights into the image. A complex and error-prone detection of the traffic light from the image data is therefore not necessary. Especially in scenes with many visible traffic lights, the relevant traffic light signals can be robustly identified by using a map. In this case, however, only a very precise 3D localization can determine the correct image section in the camera image. A rotation error of a few tenths of a degree would already lead to a significant shift of the projected boxes in the camera image. Thus the correct determination of the traffic light state by an image classifier [12] would be disturbed or not possible.

In the field of automated driving, SLAM applications aim for the highest possible accuracy, since localization based on GNSS is usually sufficient for applications with low accuracy requirements. Therefore, SLAM is particularly relevant for higher levels of automation and complex environments, such as urban road traffic, since ad-hoc perception without map support often cannot detect all relevant scene information in real time.

In addition to localization accuracy, there are other important requirements:

- **Robustness**: A major challenge for localization arises from the dynamically changing environment. Varying weather conditions, season-based changes in vegetation, line of sight occlusions from surrounding vehicles and construction sites mean that sensor measurements can not be recognized and associated correctly

Fig. 21.2 Classification of signal status using a projection of mapped traffic lights into current camera images based on precise ego-localization

to the mapped landmarks. Too few or incorrect correspondences (so-called outliers) between landmarks and measurements lead to a significant degradation or failure of the localization. A real world localization system has to work robustly even under such conditions. A suitable choice of sensors and the use of robust and invariant feature and landmark types have an outstanding influence on the robustness of the localization system. A further increase in robustness can be achieved by a continuous and demand-driven update of the map. This involves updating landmarks that have already been mapped with sensor measurements from more recent surveys of the mapped area. In addition, obsolete landmarks can be removed and newly observed landmarks can be added so that the map continuously adapts to environmental changes.

- **Availability**: Compared to, for example, rail traffic, a vehicle's freedom of movement is much less restricted in road traffic. This means that vehicles can adopt a pose that differs greatly from vehicle poses during mapping. For example, lane changes or evasive maneuvers can lead to a significant change in orientation or position. Even under such influences, a correct association of landmarks from the map to the current measurements should be possible.

- **Resources and costs**: An essential aspect are the available computing and memory resources of the mobile system. While so-called robo-taxis are often equipped with a variety of laser scanners, cameras and radio detection and ranging (RADAR) systems, as well as a high-end computer, end consumer vehicles are equipped with much less and cheaper hardware for economic reasons. The available resources play a decisive role in determining the choice of localization method and the achievable localization quality under real-time conditions.

From the above it can be seen that the requirements concern both the localization and the map. The basic rule here is: The more precise, robust and efficient the map features and the map structure are, the better the requirements can be met. Map quality thus has a decisive influence on the performance of the overall system.

In addition, it should be noted that the maximum possible localization accuracy, robustness and availability are significantly influenced by the environmental structures of the mapped area. For example, in inner-city environments, there are significantly more clearly recognizable structures, such as street signs, facades, roadside and buildings than in rural areas with open forest and meadow areas.

21.2 Problem Formulation

SLAM problem

The starting point of the SLAM problem is a mobile agent, which is at an unknown pose[1] in an unknown environment and is supposed to map it systematically. While creating the map, the agent must continuously localize itself in this map. To do this, the agent moves through the unknown environment starting from its starting pose p_0 and records ego-relative measurements at discrete points in time using its onboard sensors. Features representing landmarks in the agent's environment are then extracted from the environmental measurements. The landmarks thus detected are aggregated in a map.

In a nutshell, the basic idea of SLAM is to minimize the sum of all squared residuals between ideal and real measurements. Here, the ideal measurements depend on the pose and the mapped landmark positions. Therefore,

---

1   A pose includes the position and orientation of the agent. A detailed description of the pose term can be found in the Appendix 21.9.

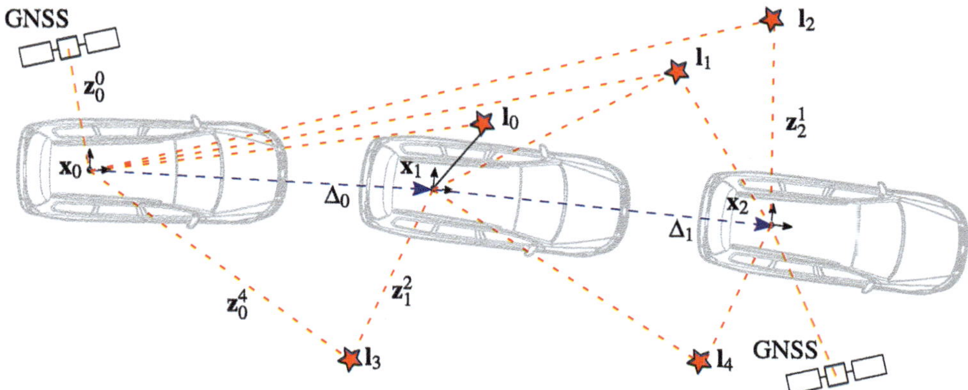

Fig. 21.3   Structure of the SLAM problem

map generation and localization are mutually dependent. The entire SLAM problem is formulated in discrete time for practical reasons. The agent pose at time step k or at time t_k is denoted by $\mathbf{p}_k$ and is modeled as a component of an agent state vector $\mathbf{x}_k \supseteq \mathbf{p}_k$. Often, in addition to the pose $\mathbf{p}_k$, dynamic quantities such as agent velocity or acceleration are modeled as components of the state vector $\mathbf{x}_k$. At time step k, the agent receives the measurement vector $\mathbf{z}_k = [\mathbf{z}_0^k, \ldots, \mathbf{z}_n^k]^T$ from its sensors. Concatenated in the measurement vector $\mathbf{z}_k$ are all measurements $\mathbf{z}_i^k$ recorded by the agent at time t_k, which can potentially correspond to landmarks. Additionally, the measurement vector $\mathbf{z}_k$ may include measurements that are independent of landmarks, such as GNSS measurements. A measurement $\mathbf{z}_i$ can also be the empty set if nothing was detected. This means that, in general, poses $\mathbf{p}_i$ can exist that are estimated only based on ego motion. The descriptive parameters of all landmarks detected up to time t_k are concatenated in the landmark vector $\mathbf{m} = [\mathbf{l}_1, \ldots, \mathbf{l}_L]^T$. The landmark vector $\mathbf{m}$ thus describes the map. If new area is explored, the landmark vector grows by the new landmarks detected in this region. The landmark vector $\mathbf{m}$ therefore also depends on the time step k, but this is suppressed in the notation hereinafter. Furthermore, the agent registers a control parameter vector $\mathbf{u}_k$, which may contain, for example, the velocity, and which allows, using a motion model to calculate the ego motion Δ_k between time t_k and t_{k+1}. The ego motion is formally defined in the Appendix 21.9.

Figure 21.3 illustrates all described quantities and relations.

So, in summary, the following input or measured quantities are given up to time t_k:

- The measurement history $\mathbf{z}_{0:k} = [\mathbf{z}_0, \ldots, \mathbf{z}_k]^T$, which includes the measurements at all measurement times $t_0, \ldots, t_k$.
- The history of the control parameter vector $\mathbf{u}_{0:k} = [\mathbf{u}_0, \ldots, \mathbf{u}_k]^T$. Additionally or alternatively, the ego-motion history $\Delta_{0:k-1} = [\Delta_0, \ldots, \Delta_{k-1}]^T$ can be used. It includes the ego-motion estima-

tes between all considered measurement times $t_0 \to t_1, \ldots, t_{k-1} \to t_k$.

In contrast, the following states are to be estimated up to time t_k:

- The agent's state history $\mathbf{x}_{0:k} = [\mathbf{x}_0, \ldots, \mathbf{x}_k]^T$, which includes all discrete agent states.
- The landmark vector $\mathbf{m}$ representing the map.

Since state estimation is always accompanied by estimation uncertainty, it is advantageous to formulate the SLAM problem probabilistically. A distinction is made between the full SLAM problem and the online SLAM problem. For the full SLAM problem the a-posteriori probability distribution

$$p(\mathbf{x}_{0:k}, \mathbf{m}|\mathbf{z}_{0:k}, \mathbf{u}_{0:k}) \qquad (21.1)$$

of all agent states $\mathbf{x}_{0:k}$ and all landmarks $\mathbf{m}$ is jointly estimated. In the online SLAM problem, on the other hand, at time step k only the probability distribution

$$p(\mathbf{x}_k, \mathbf{m}|\mathbf{z}_{0:k}, \mathbf{u}_{0:k}), \qquad (21.2)$$

of the current agent state $\mathbf{x}_k$ and all landmarks $\mathbf{m}$ are estimated.[2] In order to estimate the probability distribution (21.1) or (21.2) the relationship between the input variables and the state variables must be modeled. For this purpose, a probabilistic prediction model

$$p(\mathbf{x}_k|\mathbf{x}_{k-1}, \mathbf{u}_k) \qquad (21.3)$$

and a probabilistic measurement model

$$p(\mathbf{z}_k|\mathbf{x}_k, \mathbf{m}) \qquad (21.4)$$

---

2   In addition, so-called sliding window methods are also common, where $p(\mathbf{x}_{k-n:k}, \mathbf{m}|\mathbf{z}_{0:k}, \mathbf{u}_{0:k})$ is estimated. However, these will not be discussed further below.

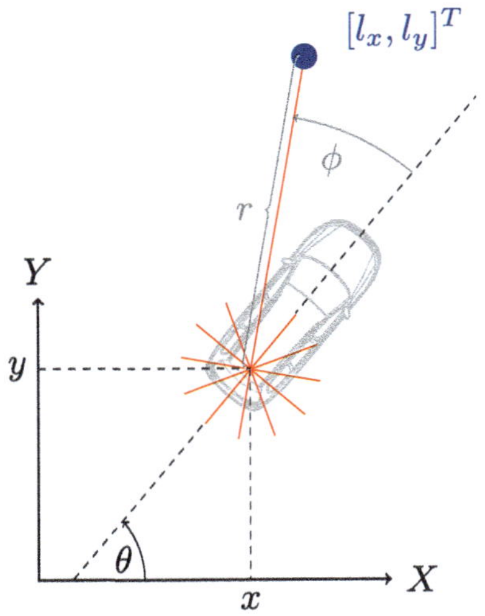

Fig. 21.4 Example: A vehicle with state $\mathbf{x} = \mathbf{p} = [x, y, \theta]^T$ is equipped with a LiDAR sensor. The red lines in the figure represent laser beams, one of which hits the blue landmark l with position $[l_x, l_y]^T$. The resulting measurement vector is $\mathbf{z} = [r, \phi]^T$, where r and ϕ denote the distance and relative angle to the landmark, respectively

are assumed to be known. The construction of these probabilistic models is usually based on a deterministic prediction model

$$\mathbf{x}_k = \mathbf{g}(\mathbf{x}_{k-1}, \mathbf{u}_k), \tag{21.5}$$

as well as a deterministic measurement model[3]

$$\mathbf{z}_k = \mathbf{h}(\mathbf{x}_k, \mathbf{m}). \tag{21.6}$$

To obtain probabilistic models, zero-mean Gaussian errors are typically added to the models (21.5) and (21.6). This is the case, for example, with the Kalman filter (see equations (21.31), (21.32), and (21.33)).

Example: Polar coordinate transformation

A two-dimensional geometric example for a deterministic measurement model (21.6) is

$$\mathbf{z} = \begin{bmatrix} r \\ \phi \end{bmatrix} = \mathbf{h}(\mathbf{x}, \mathbf{l}) = \begin{bmatrix} \sqrt{(l_x - x)^2 + (l_y - y)^2} \\ \arctan2(l_y - y, l_x - x) - \theta \end{bmatrix}. \tag{21.7}$$

In this model, a landmark position $\mathbf{l} = [l_x, l_y]^T$ is transformed into the agent's reference frame using the agent's pose $\mathbf{p} = [x, y, \theta]^T$ and then converted from Cartesian coordinates to polar coordinates (see **Fig. 21.4**). For example, a two-dimensional laser scanner could be described by this model (21.7).

The problem of map-based localization

The use case of map-based localization, which is especially important for automated driving, is formulated by a simplification of the SLAM problem. This simplification results from the fact that a static map with landmark vector $\mathbf{m}$ is available right from the start. In this case, only the pose of the agent has to be estimated, so that the full SLAM problem $p(\mathbf{x}_{0:k}, \mathbf{m}|\mathbf{z}_{0:k}, \mathbf{u}_{0:k})$ simplifies to $p(\mathbf{x}_{0:k}|\mathbf{z}_{0:k}, \mathbf{u}_{0:k}, \mathbf{m})$. In the case of the online problem, $p(\mathbf{x}_k|\mathbf{z}_{0:k}, \mathbf{u}_{0:k}, \mathbf{m})$ is determined instead of $p(\mathbf{x}_k, \mathbf{m}|\mathbf{z}_{0:k}, \mathbf{u}_{0:k})$.

21.3 Components of a SLAM System

For the solution of the SLAM or localization problem a multitude of process steps must be gone through. **Figure 21.5** represents the components of a typical SLAM system. Most approaches presented in the literature follow this structure.[4] Depending on the type of application, there can be significant differences in the detailed design of the individual components. Some components in **Fig. 21.5** are also not necessary for every application. For example, the place recognition module is not necessary if initial localization using GNSS is sufficient. As described in ▶ Sect. 21.2, landmark estimation is omitted for the localization problem.

Basically, the components for solving SLAM or map-based localization can be divided into the three process steps of association, prediction, and state estimation, which are briefly described in the following. Detailed explanations of the three process steps follow in ▶ Sects. 21.4, 21.5, and 21.6.

Association

The measurements $\mathbf{z}$ mentioned in the SLAM problem formulation cannot always be understood in the sense of a raw sensor measurement. Often, features are extracted from the raw sensor measurements in a first step, such as feature points from camera images (see **Fig. 21.6** and ▶ Sect. 21.4.1). These features are then understood as measurements $\mathbf{z}$. In a second step, the extracted features are associated to each other or to already mapped landmarks. Hence, there are two cases for the association:

1. **Sequential association:** Detected features at time step k are associated with the landmarks or features from the immediately preceding measurement cycle $k - 1$. This way features can be tracked over time. **Figure 21.6** shows, for example, the optical flow obtained from sequential association and tracking of point fea-

---

3 It should be noted that for some measurements, such as GNSS measurements, the model $\mathbf{h}$ in Eq. (21.6) does not depend on the landmarks, so $\mathbf{z}_k = \mathbf{h}(\mathbf{x}_k)$ holds.

4 However, there are approaches such as LSD-SLAM [13] whose structure is different.

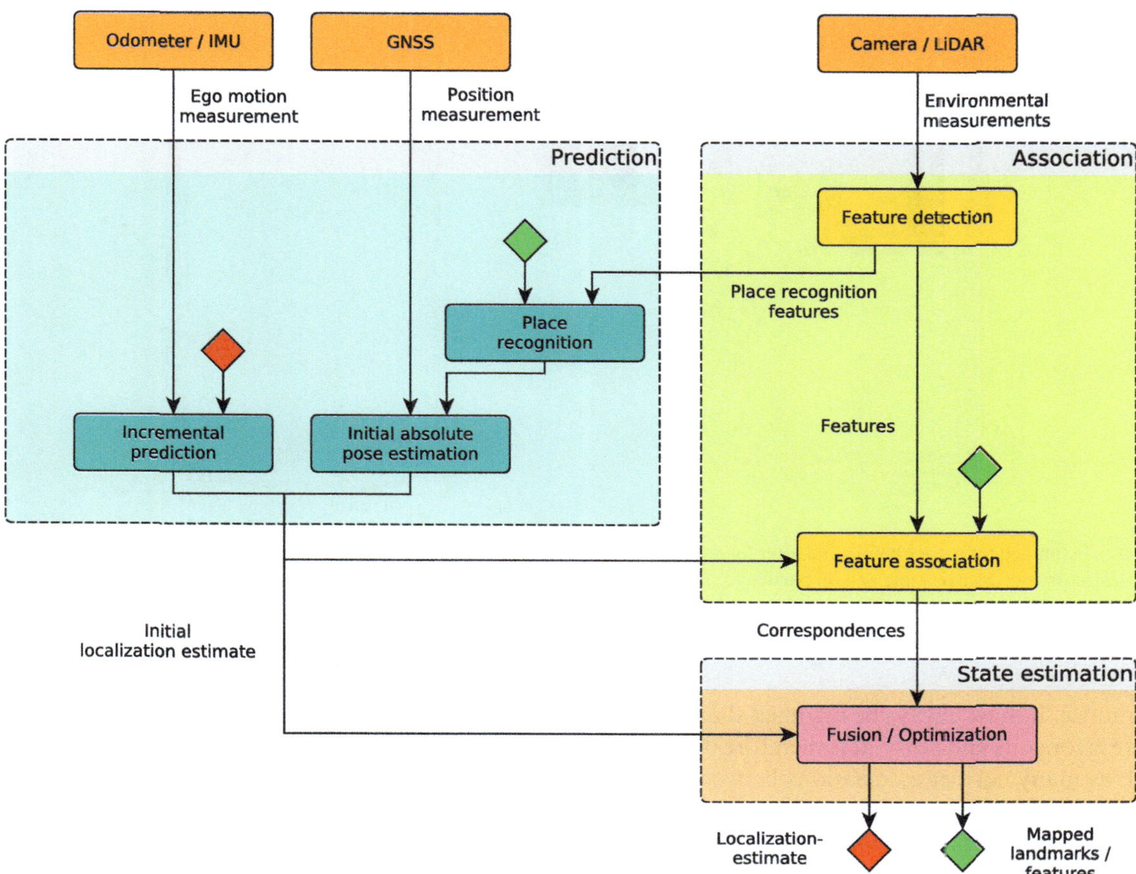

Fig. 21.5   Major components of a SLAM system

Fig. 21.6   Representation of optical flow by sequential association and tracking of point features from camera images. Here, each colored line connects the position of a feature in the previous time step to the position of the same feature in the current time step

tures from camera images. Sequential association is often used to estimate ego motion [14–16].

2. **Association with mapped landmarks:** If the agent revisits already mapped regions, detected features are associated to the existing map. This is always the case with map-based localization. In map generation by SLAM, it can take place during a new data-acquisition trip. However, it can also occur within the same trip. This important case is called loop closure, and it leads to a substantial increase in map consistency (see **Fig. 21.7**).

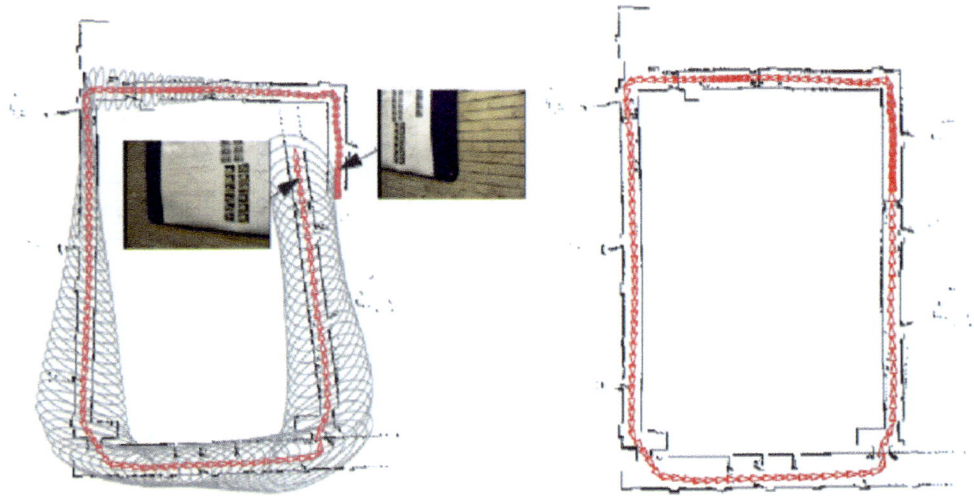

◻ Fig. 21.7   Estimated trajectory before and after loop closure. Due to an accumulation of small errors in sequential association, a drift in the estimated trajectory occurs over time and, in particular, an incorrect orientation is estimated. Once the agent realizes that it has already visited and mapped the current area, both the map and the estimated trajectory can be significantly improved. Figure taken from [17]

In practice, it must always be assumed that measurements are erroneous and noisy. It is therefore desirable to establish as many correct associations between features and landmarks as possible, so that measurement errors are averaged out in the parameter estimation, assuming a zero-mean error distribution. In addition, it is desirable to have a robust association method with unambiguously assignable features, so that, if possible, no incorrect landmark-feature correspondences (so-called outliers) are determined.

Prediction

For both robust feature association and state estimation, a prior state distribution is required in each process cycle. This can significantly narrow the search space in the map and avoid ambiguities, so that a reliable feature association is possible. In addition, the number of feature-landmark comparisons is reduced, so that the solution of the SLAM problem is possible with limited hardware resources. The more precisely the prior distribution is determined, the more efficient and robust is the landmark association and parameter estimation.[5] A distinction is made between two cases in prediction:

1. **Incremental prediction:** If the SLAM or localization system is already initialized, pose estimates from previous process cycles exist, so that ego-motion estimation can be used to predict the expected current pose. This predicted pose can also be used to initialize landmark parameters by triangulation, for example. The ego-motion estimation is done by the motion model (21.5) and/or by sequential association.

2. **Initial prediction:** A major challenge in map-based localization is the initialization phase, i.e. when the agent does not yet have prior knowledge about its current pose in the map. In this case, place recognition must be used to determine the closest map location. For automated driving, a GNSS system can be used here. For other domains, such as indoor SLAM, feature-based place recognition is needed (see ▶ Sect. 21.5.2).

State estimation

Using the correspondences determined in the association step, and based on the state prediction, the final state estimate is calculated. The updated and, if necessary, newly added state parameters are appended to the existing map database in the case of SLAM. In contrast, in the localization-only case, only the estimated agent pose $\mathbf{p}_k$ is returned.

Of particular importance for practical applications are, on the one hand, robust methods that either identify and remove unrecognized outliers, or strongly attenuate their influence on the state estimation. On the other hand, state initialization by prediction should be as precise as possible. Otherwise there is a risk that especially nonlinear optimization methods converge to an incorrect local minimum.

21.4 Sensors and Feature Association

The basic building blocks of any SLAM system are landmarks and features that are extracted from sensor measurements and associated with each other. Different fea-

---

5   In the ideal case, the localization prior is already so so precise that in the final state estimation, the prior state distribution is only slightly adjusted.

ture types are suitable depending on sensor configuration, application, and available computing and memory capacity. This section gives an overview of the most important sensors for SLAM and localization applications, shows different feature types and describes how they are detected, characterized and associated with each other. Camera and LiDAR (Light Detection And Ranging) sensors are of outstanding importance for environment perception. Both sensors measure light beams that are reflected from objects in the environment. Because a camera uses the available ambient light and has a comparatively high spatial sampling rate, a camera image has a high semantic information content. In contrast, a LiDAR sensor emits light pulses and precisely measures their travel time, enabling accurate distance determination to discrete points in space. The strengths and weaknesses of the two sensors are thus complementary, making a combination of these two sensors for SLAM applications advantageous. In addition, automated vehicles often have RADAR sensors, which can also be used for localization [18]. However, due to their measurement characteristics, radar sensors are less suitable and common for SLAM. Furthermore, the determination or measurement of ego motion is fundamental for SLAM. For the measurement of the acceleration, velocity and rotational speed wheel speed sensors or inertial measurement units (IMU) can be used. It is also possible to compute an odometry estimate using camera and LiDAR measurements [16, 19]. Since automated vehicles move mostly outdoors, the use of global navigation satellite systems (GNSS) is also particularly appropriate. In the following, the mentioned systems are described individually and the aspects relevant for SLAM are highlighted.

21.4.1 Camera

In a camera, an optical imaging system produces a two-dimensional projection, the camera image, of an observed scene. A digital camera image represents the local distribution of irradiance quantized and locally discretized. Due to the usually high resolution intensity distribution, an image provides a lot of semantic information, so that localization features can be detected and associated with each other comparatively reliably.

Projection

The imaging properties of a camera are defined by the projection function

$$\pi(\mathbf{l}) = \mathbf{z}, \tag{21.8}$$

which maps the spatial points $\mathbf{l} \in \mathbb{R}^3$ to image coordinates $\mathbf{z} \in \mathbb{R}^2$. The model parameters of the projection function (21.8) are determined by a calibration [20, 21].

Often it is assumed that the model has a unique projection center (single view point), so that this is defined to be the origin of the camera coordinate system. When projecting a landmark into the camera image using (21.8) the information of the distance of the landmark to the camera origin is lost. Therefore, equation (21.8) is not uniquely invertible. Due to the projection function

$$\kappa(\mathbf{z}) = \mathbf{r} \tag{21.9}$$

however, the view beam $\mathbf{r} \in \mathbb{R}^3$, $|\mathbf{r}| = 1$, corresponding to pixel $\mathbf{z}$ can be determined in the camera coordinate system.

Example: Spherical camera model

A common camera model is the spherical coordinate transformation

$$\mathbf{z} = \begin{bmatrix} u \\ v \end{bmatrix} = \pi([l_x, l_y, l_z]^T) = \begin{bmatrix} c_x - f \cdot \arctan 2(l_y, l_x) \\ c_y + f \cdot \dfrac{l_z}{\sqrt{l_x^2 + l_y^2 + l_z^2}} \end{bmatrix}, \tag{21.10}$$

where $\mathbf{l} = [l_x, l_y, l_z]^T$ represents the imaged spatial point in the camera coordinate system. The so-called image principal point $[c_x, c_y]^T$ and the focal length f represent the parameters of this camera model.

In addition, the visual ray projection function results in

$$\mathbf{r} = \begin{bmatrix} r_x \\ r_y \\ r_z \end{bmatrix} = \kappa([u, v]^T) = \begin{bmatrix} \sin(\frac{v-c_y}{f}) \cdot \cos(\frac{c_x-u}{f}) \\ \sin(\frac{v-c_y}{f}) \cdot \sin(\frac{c_x-u}{f}) \\ \cos(\frac{v-c_y}{f}) \end{bmatrix}. \tag{21.11}$$

3D-reconstruction

In order to reconstruct the depth information of a pixel $\mathbf{z}$, the spatial point corresponding to $\mathbf{z}$ must either have been observed from several different viewing positions or assumptions about the scene geometry must be fulfilled and known [22]. Several different viewing positions can be achieved, for example, by using a multi-camera system (stereo) or by changing the camera position of the same camera (motion stereo) [23]. If the images are recorded at different times, the observed spatial point must not change its position over the observation period.

Triangulation

An efficient method for determining the spatial point $\mathbf{l}$ given n pixels $\{\mathbf{z}_0, \ldots, \mathbf{z}_n\}$ and corresponding observation camera poses $\{\mathbf{p}_0, \ldots, \mathbf{p}_n\}$ is triangulation. Using the projection function (21.9) and the rotation component $\mathbf{R}_i \in SO(3)$ of $\mathbf{p}_i \in SE(3)$ the line of sight corresponding to image point $\mathbf{z}_i$

$$\mathbf{r}_i = \mathbf{R}_i \cdot \kappa_i(\mathbf{z}_i) \tag{21.12}$$

can be displayed in the reference coordinate system. The position component $\mathbf{t}_i \in \mathbb{R}^3$ of $\mathbf{p}_i$ represents under assumption of a unique projection center the receptor point of the line-of-sight ray in the reference coordinate system. Furthermore, it is assumed that the observation positions $\{\mathbf{t}_1, \ldots, \mathbf{t}_n\}$ are different and the corresponding sight rays $\{\mathbf{r}_1, \ldots, \mathbf{r}_N\}$ are therefore not collinear to each other. Theoretically, under these assumptions, the sight rays intersect exactly in $\mathbf{l}$. However, this is not fulfilled in practice due to measurement inaccuracies, so that for triangulation the square of the distance

$$\mathbf{d}_i = (\mathbf{l} - \mathbf{t}_i) - \mathbf{r}_i^T (\mathbf{l} - \mathbf{t}_i)\, \mathbf{r}_i \tag{21.13}$$

of the space point $\mathbf{l}$ to all view rays $\{\mathbf{r}_1, \ldots, \mathbf{r}_n\}$, is minimized. For $n \geq 2$ this results in an overdetermined system of equations, which can be solved in a closed form, resulting in the approximate solution

$$\hat{\mathbf{l}} = \left(\sum_{i=1}^{n} \mathbf{I} - \mathbf{r}_i \mathbf{r}_i^T \right)^{-1} \left(\sum_{i=1}^{n} (\mathbf{I} - \mathbf{r}_i \mathbf{r}_i^T) \mathbf{t}_i \right) \tag{21.14}$$

for the landmark position $\mathbf{l}$. The matrix $\mathbf{I} \in \mathbb{R}^{3\times 3}$ represents the identity matrix (Fig. 21.8).

Point features

One of the most widely used landmark and feature type for SLAM are point features. Here, in general, a landmark $\mathbf{l}$ corresponds to n pixels $\{\mathbf{z}_0, \ldots, \mathbf{z}_n\}$ from different camera images (see Fig. 21.9), which are associated to each other by local pixel descriptors.

Feature detection

The first step is to extract recognizable candidate pixels from the image. These pixels are determined using a detector that identifies distinctive patterns in the camera image. Often, this is done by convolving the camera image with a blob or corner mask and selecting the strongest filter responses as candidates [23]. For this step of the process, convolutional neural networks (CNN) are particularly useful [24], which learn the convolutional kernels in a data-driven way and thus adapt them to a specific application domain.

Feature-association

In order to decide whether two detected pixels $\mathbf{z}_a$ and $\mathbf{z}_b$ correspond to each other—i.e., depict the same landmark—usually the local image area around the pixels is analyzed. For this purpose, a local descriptor $\mathbf{v}$ is determined for the image point $\mathbf{z}$. A distance function

$$d = \delta(\mathbf{v}_a, \mathbf{v}_b) \tag{21.15}$$

can then be used to evaluate the similarity of pixel descriptor $\mathbf{v}_a$ and $\mathbf{v}_b$.

Particularly important in descriptor computation and association is invariance or robustness to perturbations such as partial occlusion, weather-related environmental changes, illumination changes, affine image transformations, and parallax errors [25]. Robustness to illumination changes is particularly important for localization (see Fig. 21.10), as the illumination of the mapped area is subject to many external influences. Important classes of local descriptors and distance functions are listed below:

— **Intensity comparison:** An intuitive method is to directly compare gray values using, for example, a normalized cross-correlation [26] or the sum of squared differences [27]. In this case, the descriptor vector directly represents the intensities from the incoming camera image or from an intermediate feature image. In the literature, direct SLAM methods in particular use this approach [13, 14].
— **Binary patterns:** Binary patterns are widely used descriptors in SLAM. These methods compare intensities of different pixel combinations from a local sampling pattern. The resulting binary vectors are compared, for example, with the Hamming distance and are therefore efficiently computable [28–30].
— **Gradient histogramming:** In this case, intensity gradients are locally discretized and accumulated in a histogram [31, 32]. All histograms together form a descriptor vector. A comparison is then possible, for example, by determining the Euclidean distance of two such vectors.
— **Machine-learned descriptors:** Learned features are a widely used variant. In particular, CNNs are suitable for this purpose, where feature computation and distance determination can be optimized in a data-driven manner [33–35].

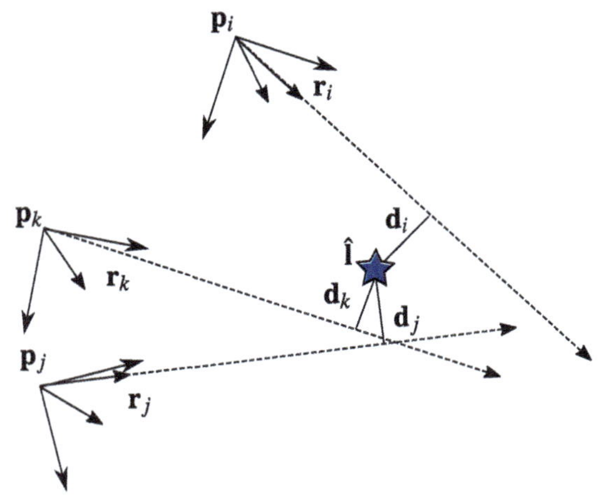

 Fig. 21.8 Triangulation by minimizing distance to line-of-sight rays

Fig. 21.9   Captures of corresponding pixels from different images from different perspectives

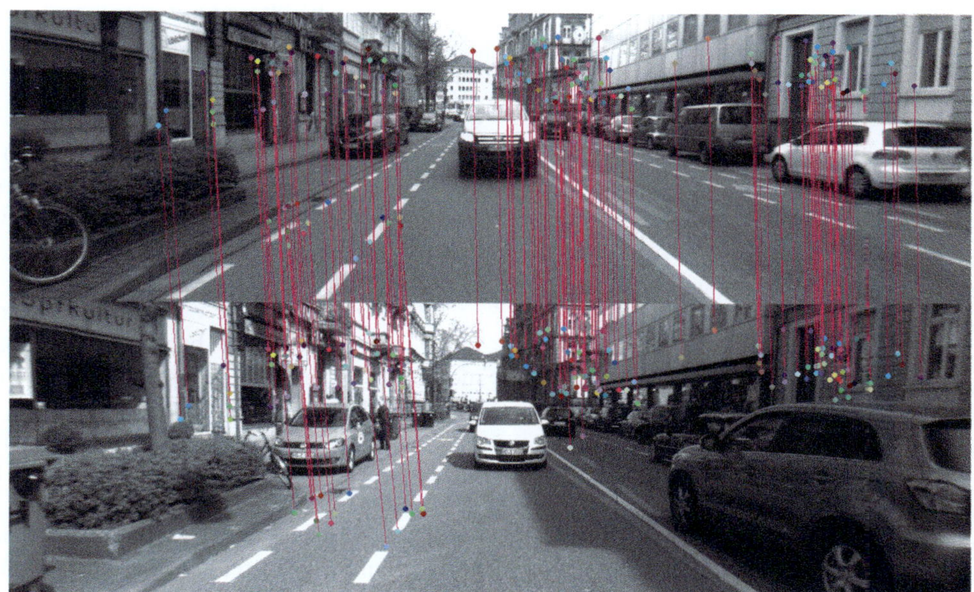

Fig. 21.10   Point features associated with each other between two camera images of the same scene recorded on different days

21.4.2 LiDAR

LiDAR (Light Detection And Ranging) is a measurement method for locating and measuring the distance of surrounding objects. A LiDAR system emits encoded light pulses and receives the light reflected from the surroundings. The distance d to objects in space can then be calculated from the signal travel time t. The distance

$$d = \frac{c \cdot t}{2} \tag{21.16}$$

between object point and sensor can be calculated using the speed of light c. Rotating LiDAR systems are commonly used in the field of automated driving. For this purpose, either a fan-shaped array of several laser signal emitters and receivers is used or a mirror that deflects the emitted light pulses and rotates them around its own axis at a constant frequency. Thus, a LiDAR produces a discrete ambient scan with constant azimuth angles. The diode fan also produces a scan with constant elevation angles. All measurement points together result in a so-called point cloud (see ◘ Fig. 21.11).

The LiDAR point clouds can either be used directly for association (scan association) [36, 37] or can be further processed for semantic feature extraction (see ▶ Sect. 21.4.3).

Scan association

When associating two point clouds, the optimal transformation Δ^* is determined that minimizes the distance between two point sets $\mathbf{S}_1$ and $\mathbf{S}_2$:

$$\Delta^* = \arg \min_{\Delta \in SE(N)} d(\Delta \cdot \mathbf{S}_1, \mathbf{S}_2), \qquad (21.17)$$

where the distance function d describes, for example, the Euclidean distance between each pair of points, and $N \in \{2, 3\}$. Here the transformation is applied to each point of the point set $\mathbf{S}_1$. Different methods can be used to determine this transformation Δ^*, such as the Iterative Closest Point (ICP) algorithm [38], which iteratively determines the underlying transformation parameters.

21.4.3 Semantic Features

For automated driving, known environmental structures such as lane markings, traffic lights, traffic signs, street lights, or even guardrails are suitable features. These so-called semantic features can be extracted from the measurements independently of the sensor and used for association. Compared to pure localization features such as point features or textured surface models [13], semantic features can not only be used for localization, but also provide important information for other tasks such as situation assessment and trajectory planning. The use of semantic features for vehicle localization has the significant advantage of not requiring a separate localization layer of the map, thus eliminating the need for pure localization maps with point features. Other advantages of semantic features are their ease of detection, their frequent occurrence, and their long validity span [11]. Moreover, semantic features can be efficiently described by geometric primitives, whereby a comparatively small map size can be achieved. However, there are also disadvantages compared to classical point features. First, feature uniqueness is lower, increasing the probability of misassociations. Furthermore, the existence of these features is required, which is often not guaranteed on small inner-city streets. In contrast, the only assumption for point features is that structure exists at all, regardless of the environment.

Feature detection

Methods for semantic feature detection have to be narrowed down depending on the type of features and the sensors used. In the following, three different types of semantic features and their detection using different sensors are considered as examples.

Lane markings

Lane markings represent commonly used semantic features. They can be extracted from camera images in different ways. On the one hand, it is possible to transform camera images into a bird's eye view in order to extract lane markings from them [39]. This has the advantage that features can be displayed regardless of perspective or camera coordinate system. After transforming the camera images, well-known blob or edge detectors [40, 41], or even neural networks [42] can be used to extract the features. LiDAR point clouds are also suitable for lane marking detection. Due to their color, lane markings have a higher reflectance than the adjacent asphalt, so this property is reflected in the intensity of the LiDAR measurement. By determining a suitable threshold, lane markings can be filtered out of intensity images [43].

Cylindrical objects

Cylindrical objects, such as traffic light posts, street lamps, or sign posts, frequently occur in road scenes and can be reliably detected with e.g. stereo cameras. By computing disparity images, hypotheses for cylindrical features can be generated based on the depth edges present [44].

Other methods for extracting cylindrical features from camera data are based on the use of learning methods [45] or on a combination with other sensors, such as LiDAR systems [46].

Cylindrical objects can be extracted from LiDAR measurement data even without cameras. The detection of such features is based on the definition of a cylinder as an upright structure with a limited maximum diameter [47]. Circular structures are extracted from the individual scan lines of the LiDAR that satisfy the radius requirements. A hypothesis for a cylindrical object is generated if a sufficient number of circular structures confirm the existence of a cylinder. A final least squares approximation of the cylinder parameters, i.e. coordinates of the center point as well as the radius, is performed considering all previously selected measurement points.

Planar surfaces

Planar surfaces, such as house facades, gantries, or guardrails, can be obtained from LiDAR point clouds in a similar manner to cylindrical objects [11] ◘ Fig. 21.12 shows a LiDAR point cloud including detected features, with cylinders in blue and planar surfaces in red.

Fig. 21.11   Extraction of road markings from camera images

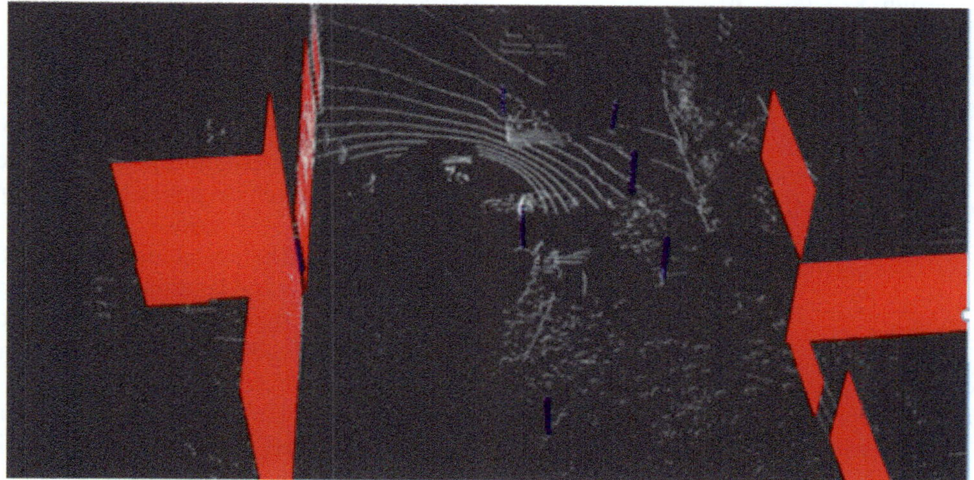

Fig. 21.12   Extraction of geometric primitives such as cylinders (blue) and planes (red) from LiDAR data

Feature association

Semantic feature association searches for correspondences between detected features $\mathbf{z}$ and landmarks $\mathbf{l}$ from a map. Unlike point features from the camera, semantic features often do not have unique descriptors, leading to the following challenges:

Global ambiguities

Semantic features such as lane markings, traffic light posts, or guardrails occur periodically in the world and in corresponding maps and cannot be distinguished from each other based on their geometry. This property leads to ambiguities that allow for a variety of hypotheses. To find the correct correspondence, the search space must be successively restricted using a localization prior.

Local ambiguities

Even with prior search space restriction, local ambiguities may occur that jeopardize correct association. For example, lane markings may also occur locally in periodic patterns. Due to such ambiguous feature patterns, it is sometimes impossible to establish correct correspondences without further knowledge.

Detection error

Further errors in the association of semantic features may result from incorrect feature detection. Here we distinguish between different types of errors:

- **Classification errors:** Due to classification errors features of different classes may be associated with each other. For example, dashed lane markings might be incorrectly associated with solid lane markings.
- **Coverage:** Especially for extended features such as guardrails or house facades, the start or end point is often not visible, as it is either not within sensor range or occluded. As a result, such features can be described incorrectly and associated incorrectly.
- **False positive (FP) detections:** Features that have been falsely detected when they do not actually exist, are caused e.g. by moving objects. Such a FP feature that is not recorded in the map can consequently only be falsely associated, unless it is rejected.

For these reasons, an ensemble of multiple detected features is often compared to the map and the best common match is sought. There are several methods for doing this. Three approaches are described below as examples:

Minimizing a cost function

The association problem can be solved by minimizing a cost function. The cost function $C(\mathbf{z}, \mathbf{l}, \mathbf{p})$ assigns a cost term to an association pair $(\mathbf{z}, \mathbf{l})$, consisting of a measurement $\mathbf{z}$ and a mapped landmark $\mathbf{l}$, depending on an estimated vehicle pose $\mathbf{p}$. This cost function should be chosen according to the representation of the semantic features present. The goal of the association is to minimize the sum of all costs for i association pairs regarding the pose $\mathbf{p}$:

$$\min_{\mathbf{p}} \sum_i C(\mathbf{z}_i, \mathbf{l}_i, \mathbf{p}).$$

In the following, cost functions that can be used for a wide variety of association problems, are discussed.

— **Cylindrical objects** are typically described by their position, radius and tilt angle. Thus, to associate two cylinders, all three parameters can be considered in the cost function $C_c(\mathbf{z}, \mathbf{l}, \mathbf{p})$ as follows:

$$C_c(\mathbf{z}, \mathbf{l}, \mathbf{p}) = \frac{d(\mathbf{z}, \mathbf{l}, \mathbf{p})}{d_{max}} + \omega_{c,1}\frac{\delta\alpha(\mathbf{z}, \mathbf{l}, \mathbf{p})}{\delta\alpha_{max}} + \omega_{c,2}\frac{\delta r(\mathbf{z}, \mathbf{l}, \mathbf{p})}{\delta r_{max}}. \tag{21.18}$$

The first term penalizes a large spatial distance d between an association pair $(\mathbf{z}, \mathbf{l})$, the second term considers a difference in tilt angle $\Delta\alpha$, while the third term considers the difference in radii Δr. All terms are respectively normalized by the maximum allowable distance d_{max} or tilt angle difference $\Delta\alpha_{max}$ or radius difference Δr_{max} between a pair of associations. The parameters $\omega_{c,1}$ and $\omega_{c,2}$ represent weighting parameters.

— **Straight line segments** can be associated if their mutual distance is small, their orientation matches, and if they have enough overlap. The cost function for the association of straight line segments is given by

$$C_l(\mathbf{z}, \mathbf{l}, \mathbf{p}) = \frac{d(\mathbf{z}, \mathbf{l}, \mathbf{p})}{d_{max}} + \omega_{l,1}\frac{\delta\alpha(\mathbf{z}, \mathbf{l}, \mathbf{p})}{\delta\alpha_{max}} + \omega_{l,2}\frac{o_{out}(\mathbf{z}, \mathbf{l}, \mathbf{p})}{o_{in}(\mathbf{z}, \mathbf{l}, \mathbf{p}) + o_{out}(\mathbf{z}, \mathbf{l}, \mathbf{p})}. \tag{21.19}$$

Figure 21.13 visualizes the influence of the three terms from equation (21.19). Figure 21.13a shows how the spatial distance d between two line segments can be defined by the distance between the center of the detection to the mapped line. This is the first term in equation (21.19). The second term accounts for differences $\Delta\alpha$ in the relative orientation of the straight line segments and the last term penalizes non-optimal feature overlap. To calculate the measure of overlap, the detected straight line segment is rotated around its center so that both straight lines are parallel. Then o_{in} describes the length of the overlapping part and o_{out} the length of the non-overlapping part. The parameters $\omega_{l,1}$ and $\omega_{l,2}$ allow the weighting of the individual terms.

— **Polylines** represent another type of semantic features. In Fig. 21.14 a description and association of lane markings is visualized using polylines. Uniform sampling along detections and minimization of distances between pairs of points can be the basis for determining associations here.

RANSAC (Random Sample Consensus)

The basis of the random sample consensus algorithm applied to the association problem is that there are significantly more pairs of associations than would be needed to compute a vehicle pose [49]. If this is the case, the RANSAC algorithm can be applied to determine association pairs $(\mathbf{z}, \mathbf{l})$ as follows:

1. Choose as many random association pairs, each from a detection $\mathbf{z}$ and a possible map element $\mathbf{l}$, as necessary to determine a position of the ego-vehicle.
2. Compute the vehicle pose $\mathbf{p}$ that supports these associations.
3. Determine other associations among the remaining measurements and landmarks that support this pose. This is called the *consensus set*. The cardinality of the consensus set indicates whether the computed vehicle pose is a good estimate.
4. Repeat steps 1. - 3. until a suitable estimate is found or a predefined maximum number of iterations has been undergone.

Due to its robustness to outliers, the RANSAC algorithm is also well suited when detection errors are to be expected.

Geometric hashing

Geometric hashing is based on the idea of encoding geometric feature patterns and using the resulting hash values for association [50, 51]. For localization, this is done a priori by creating a hash table based on the

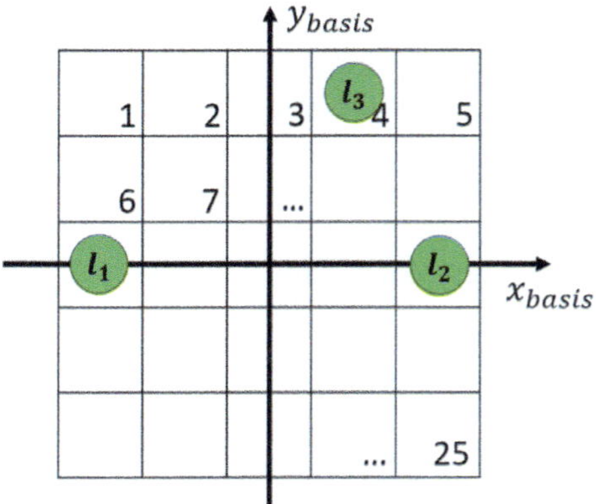

Fig. 21.15 Geometric Hashing: Hash function from equation (21.20), which assigns to each landmark the matching grid cell ID of a numbered grid. In this case, $H((\mathbf{l}_1, \mathbf{l}_2), \mathbf{l}_3) = 4$. The origin of the coordinate system associated to the base is calculated from the base pair $\mathbf{k} = (\mathbf{l}_1, \mathbf{l}_2)$ by $O_{\mathbf{k}} = \frac{\mathbf{l}_1 + \mathbf{l}_2}{2}$

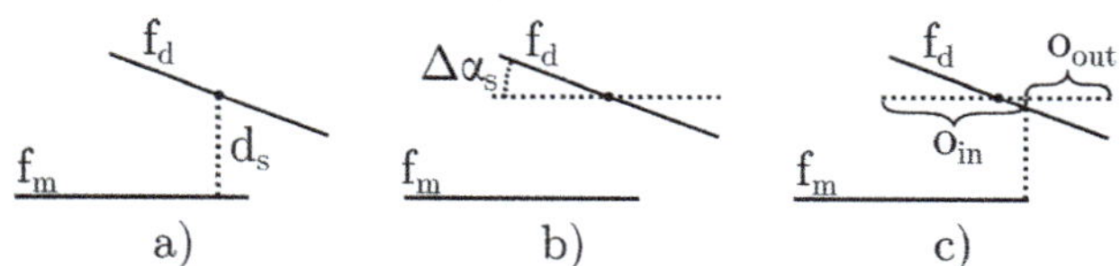

Fig. 21.13   Association of two straight line segments f_d and f_m by minimizing distance (**a**), angle (**b**) and non-overlap (**c**) [11]

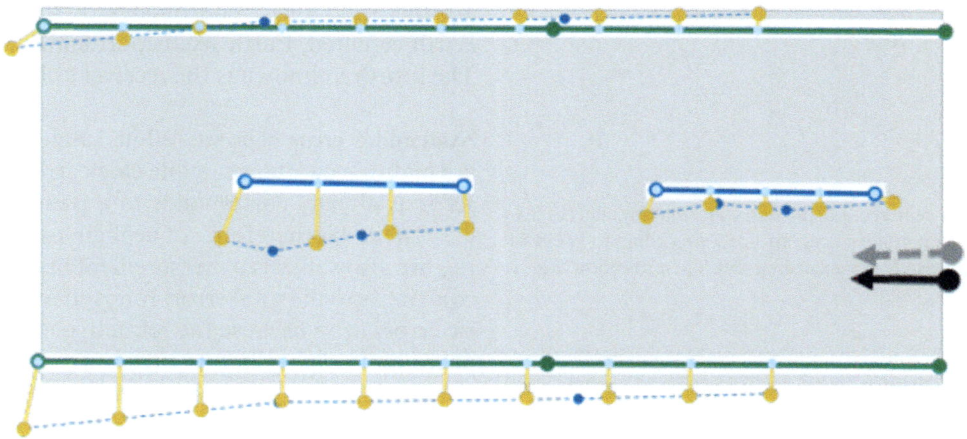

Fig. 21.14   Description and association of lane markings using polylines [48]

landmarks $\{l_1, l_2, \ldots, l_N\}$ contained in the map. Subsequently, this can then be used for association.

— **Generation of the hash table:** To calculate the hash values, which are stored in a table, local landmark patterns in the map are converted to new coordinate systems. It is important to note that the coordinate systems do not depend on the map or vehicle coordinate system, but are defined by the pattern itself. Thus, any combination of two landmarks $\mathbf{k} = (l_i, l_j)$, also called a base pair, represents a geometric base (see **Fig. 21.15**). Any other landmark in the vicinity can be represented relative to this base. Using a hash function, each landmark $\mathbf{l}$ is assigned a hash value h that refers to the coordinate system defined by $\mathbf{k}$:

$$H(\mathbf{k}, \mathbf{l}) = h.$$

The hash function itself is based on a discretization of the landmark positions, for example using a Cartesian grid (see **Fig. 21.15**). An example hash function that assigns the appropriate grid cell ID to each landmark $\mathbf{l} = (x, y)$ is given by

$$H(x, y) = \left\lceil \frac{G}{2} \right\rceil + \frac{1}{q} \cdot x - \frac{1}{q} \left(\frac{2R}{q} + 1 \right) \cdot y, \quad (21.20)$$

where G is the number of grid cells, q is the cell size, and $2R$ is the side length of the square grid. For a base $\mathbf{k}$, all hash values of the surrounding landmarks are then stored in the hash table.

— **Association using hash values:** To associate using these hash values, hash values for the detected features $\{z_1, z_2, \ldots, z_M\}$ are finally calculated in the same way as for hash table generation and compared with the hash table. For this purpose, an arbitrary detection pair $\mathbf{k}_d = (z_i, z_j)$ is chosen, which serves as a local basis. All further detected features z_k are now transformed into this newly generated coordinate system defined by $\mathbf{k}_d$. The hash values can now be calculated according to the chosen basis. By matching the hash values with the previously calculated hash table, association pairs can now be extracted.

21.4.4   GNSS—Global Navigation Satellite System

A GNSS receiver enables the determination of one's position with respect to an earth-fixed coordinate system. If a magnetometer is installed, the angle between the vehicle's longitudinal axis and compass north can also be measured.[6] GNSS receivers installed in production vehicles allow position determination with an error of about $1-10$ meters. Thus, the localization requirements described in ▶ Sect. 21.1 cannot be met with such a GNSS sensor alone.

Coordinate systems and map projections

---

6   An alternative to the magnetometer is the use of two offset GNSS receivers.

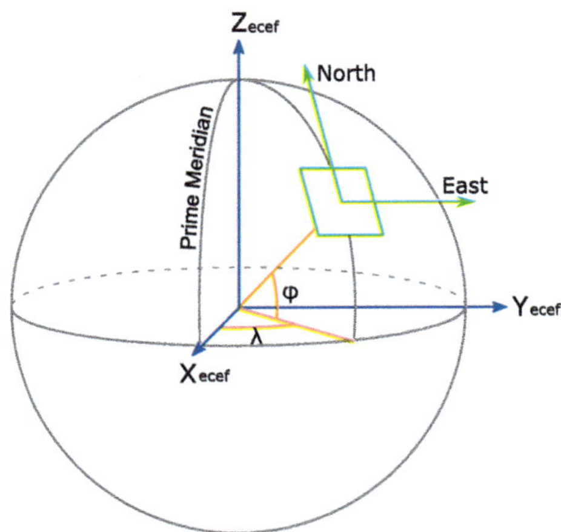

Fig. 21.16 Earth-centered, Earth-fixed coordinate system (ECEF) with axes X, Y, Z. Longitude φ and latitude λ according to WGS 84. East-North coordinates (E, N) according to UTM projection. Figure modified from [52]

In the context of satellite-based navigation, there are the following relevant coordinate systems:

World Geodetic System (WGS 84): The WGS84 coordinate system is based on the Earth-centered, Earth-fixed coordinate system (ECEF), which is a right-handed Cartesian coordinate system attached to the Earth (see **Fig. 21.16**). In addition to and equivalent to Cartesian coordinates, WGS84 defines the curvilinear coordinates longitude, latitude, and altitude. Intuitively—however not completely correctly—longitude and latitude are to be understood as angles in spherical coordinates, while the altitude corresponds to the distance to the earth's surface (see **Fig. 21.16**). In WGS 84 the exact definition of longitude, latitude and altitude is based on a modeling of the earth's surface by a reference ellipsoid on the one hand and an equipotential surface of the earth's gravitational field—the so-called geoid—on the other hand [53].

UTM coordinate system The UTM system divides the earth's surface into several zones, each of which is mapped to a Cartesian coordinate system using a transverse Mercator projection, which measures the coordinates E (East) and N (North) in meters [54]. A map which is generated using SLAM for automated driving is often georeferenced with respect to the UTM system.

Vehicle-fixed coordinate system (ISO 8855): Finally, since the GNSS receiver moves with the vehicle, it is important to mention the coordinate system attached to the vehicle (see **Fig. 21.17**).

GNSS positioning

The distance between the GNSS receiver and a satellite is estimated using the propagation time of the radio signal transmitted by the satellite, which includes satellite positions and timestamps. Since the clock of the receiver—unlike the synchronized atomic clocks of the satellites—is relatively inaccurate, the distance estimated in this way is subject to a high error and is therefore also referred to as pseudorange. Pseudoranges to at least four different satellites must be acquired in order to set up four equations for four unknowns. Three of the unknowns are the x, y, and z coordinates of the receiver in the Earth-centered, Earth-fixed coordinate system (ECEF). The fourth unknown is the receiver's clock error [55].

Systematic error sources: While satellite atomic clocks are highly accurate, a satellite clock error of $1 - 10$ nanoseconds already corresponds to a pseudorange error of 0.3–3 m and is therefore not negligible. Satellite trajectories are also subject to measurement inaccuracy and thus contribute to the total error. Ionospheric and tropospheric errors arise because the refractive index in the ionosphere and troposphere is subject to spatially and temporally dependent variations, which affect signal propagation times. All of these errors result in a position error that can be considered systematic because it varies only slowly in time. Such a systematic error can be estimated during localization, provided landmarks from the map are detected and associated with sufficient accuracy [56]. Additionally or alternatively, it is possible to use correction data from base stations to correct for the systematic error. Methods based on this approach are known as Differential GPS (DGPS) and Real-Time Kinematic (RTK) GPS [53, 55].

Error due to shadowing: Particularly serious errors result when the signal transmitted by the satellite is prevented from reaching the receiver by obstacles such as buildings or mountains. The fewer satellites are visible to the receiver, the less accurate the positioning becomes. If fewer than four satellites are visible, the position of the receiver can no longer be determined directly. This occurs especially when driving through tunnels.

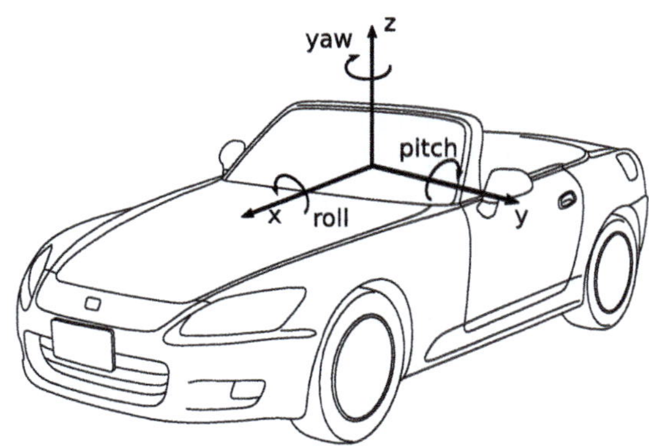

Fig. 21.17 Roll, pitch, and yaw angles, and x, y, z axes according to ISO8855. Image of the vehicle taken from Ref. [58]

Multipath errors: If satellite signals are reflected by buildings or mountains, this leads to multiple registrations of the same signal. This error, known as multipath reception, cannot be corrected by correction data as used in RTK or DGPS solutions [55] and, together with shadowing, is the major obstacle to highly accurate and robust positioning based solely on GNSS.

Tightly- and loosely-coupled integration: State estimation (see ▶ Sect. 21.6) is either based on the raw data from the GNSS receiver such as the pseudoranges (tightly coupled integration), or based on the processed data such as the WGS 84 coordinates (loosely-coupled integration). Loosely-coupled integration is easy to implement and requires little computational effort. The advantage of tightly-coupled integration, on the other hand, is the better accuracy in case of poor reception conditions, for example in an urban canyon [57], as well as the usability of GNSS measurements even when less than four satellites are visible.

21.4.5   IMU and Wheel Speed Sensor

While the sensors discussed so far take measurements, which were defined as environmental measurements $\mathbf{z}$ in ▶ Sect. 21.2, the following measurements are used to determine the control parameter vector $\mathbf{u}$, which is crucial for the motion model (Eq. 21.5).

- **Acceleration (x, y, z direction) and angular velocity (roll, pitch, and yaw direction):** These measurands are measured with angular rate and acceleration sensors, which are often combined in a so-called **IMU** (Inertial Measurement Unit). Since these measurands are typically integrated over time in the motion model, a zero point error of the sensor (also known as offset or bias) is particularly problematic and should be estimated.
- **Absolute velocity:** This is measured by a **wheel speed sensor**. If the wheel rotates once, the vehicle has traveled the distance πd_R, where d_R is the tire diameter. Note here that the tire diameter changes due to wear, which therefore leads to a scaling error in the speed. Furthermore, slip can occur, i.e. slippage between the tire and the road, which also leads to a scaling error.

For a localization in three spatial dimensions, all of the above measurands tend to be used, while for a localization in two dimensions the yaw rate, the velocity and optionally the accelerations in x- and y-direction are used.

21.5   Prediction

As described in ▶ Sect. 21.3, a localization prior is required for SLAM and map-based localization. If the agent is already initialized, i.e. the agent is active and has already been able to localize itself in the previous steps, the estimation of a prior by an odometry estimate or a motion model is preferred. A GNSS measurement or feature-based place recognition can be used for initialization. The different methods are described in this section.

21.5.1   Motion Model

An important foundation of SLAM and localization are motion models, which can be used to simulate the kinematics or dynamics of a mobile agent. In this context, all real motion models are subject to physical constraints that limit the mobility of the agent. The choice of an appropriate vehicle model depends on whether the modeling is done in two or three spatial dimensions, i.e. in $SE(2)$ or $SE(3)$. Furthermore, it must be decided whether dynamic effects, such as accelerations, are to be modeled. For SLAM considerations, a discrete-time deterministic motion model is usually used:

$$\mathbf{x}_k = \mathbf{g}(\mathbf{x}_{k-1}, \mathbf{u}_k). \tag{21.21}$$

As a simple example, the CTRV (constant turn rate and velocity) motion model is presented below.

CTRV model

In the CTRV model, the velocity $v \in \mathbb{R}$ and yaw rate $\omega \in \mathbb{R}$ of the vehicle are approximated as constant within one time step. The state includes the pose parameters $\mathbf{x} = \mathbf{p} = [x, y, \theta]^T \in SE(2)$ (see ◘ Fig. 21.18) and the control parameter vector is $\mathbf{u} = [v, \omega]^T$.

From the pose, the orientation vector of the vehicle is $[\cos(\theta), \sin(\theta)]^T$ and thus the velocity vector is $v[\cos(\theta), \sin(\theta)]^T$. The time derivative of the pose $\mathbf{p} = [x, y, \theta]^T$ is therefore

$$\frac{d}{dt} \begin{bmatrix} x \\ y \\ \theta \end{bmatrix} = \begin{bmatrix} v\cos(\theta) \\ v\sin(\theta) \\ \omega \end{bmatrix} \tag{21.22}$$

with constant v and ω. Equation (21.22) represents the continuous CRTV model, from which the discrete-time model that describes the transition from pose $\mathbf{p}_{k-1}$ to pose $\mathbf{p}_k$, can be determined by integration over time. First, integrating the equation $\frac{d}{dt}\theta = \omega$ from t_{k-1} to t gives the angle $\theta(t) = \theta_{k-1} + \omega(t - t_{k-1})$. In particular, $\theta_k = \theta(t_k) = \theta_{k-1} + \omega\Delta t$, where $\Delta t = t_k - t_{k-1}$ has been defined. The expression for $\theta(t)$ is substituted into the equations for $\frac{dx}{dt}$ and $\frac{dy}{dt}$, and these are integrated from t_{k-1} to t_k, yielding the discrete CTRV model

$$\begin{bmatrix} x_k \\ y_k \\ \theta_k \end{bmatrix} = \begin{bmatrix} x_{k-1} \\ y_{k-1} \\ \theta_{k-1} \end{bmatrix} + \begin{bmatrix} \frac{v}{\omega}(-\sin(\theta_{k-1}) + \sin(\theta_{k-1} + \omega\Delta t)) \\ \frac{v}{\omega}(\cos(\theta_{k-1}) - \cos(\theta_{k-1} + \omega\Delta t)) \\ \omega\Delta t \end{bmatrix}. \tag{21.23}$$

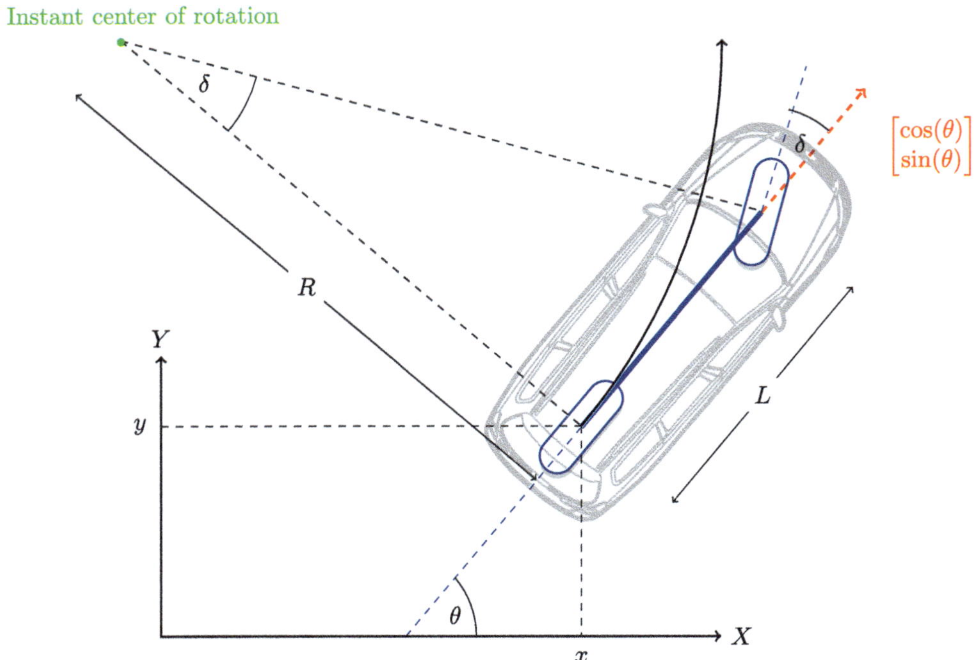

Fig. 21.18 The kinematic bicycle model is an extension of the CTRV model

Using a first-order Taylor expansion in $\omega \Delta t$, the above CTRV model simplifies to the so-called CV (constant velocity) model.[7] In the case of high accelerations, the CTRV model is inaccurate and should be replaced, for example, by the CTRA (constant turn rate and acceleration) model [59].

Finally, it should be mentioned that there are also motion models that include the wheel steer angle. In particular, the bicycle model should be mentioned here. The kinematic bicycle model extends the equations of the CTRV model by a relation between yaw rate, speed and wheel steer angle:

$$\omega = v \tan(\delta)/L. \tag{21.24}$$

Here L is the wheelbase, i.e. the distance between front and rear axle, and δ is the wheel steer angle (see **Fig. 21.18**).

21.5.2 Place Recognition

For the initial localization in an existing map and for loop closure it is necessary to detect that the agent is visiting a previously mapped area, and to determine the approximate pose within that area. If the map is georeferenced, a GNSS system can be used for initialization. Since GNSS systems are always available on the road, GNSS-based initialization is the tool of choice for automated driving. Since only rough positioning is required, the relatively poor accuracy and availability of GNSS systems in signal-shaded areas is usually still sufficient. As an alternative to GNSS, feature-based place recognition systems can be used, which are mentioned here for completeness.

For feature-based place recognition, additional place-recognition features are stored in the map. For real-time localization, it must be possible to provide and evaluate these features much faster and more efficiently than the landmarks and features of the iterative localization, since the entire map must always be searched. However, the biggest challenges in place recognition are environmental changes that alter the appearance of a mapped place (see **Fig. 21.19**). Camera-based approaches are much more suitable than LiDAR or RADAR-based approaches due to their high semantic information content.

Visual place recognition

In SLAM, for each camera image that was recorded for mapping, there is usually a map pose that can be associated with that image. In visual place recognition, a feature is then computed for a selection of map images that characterizes the map pose, i.e., the location of the agent at the time of recording. For place recognition, the currently recorded camera image is compared with the map images. The best association then represents the localization hypothesis. Basically, two variants of features can be distinguished:

---

7   In the numerical implementation of the CTRV model, note that Eq. (21.23) is unstable for small ω. Therefore, at each time step it should be checked if ω is small, and in this case the CV model should be used temporarily.

1. **Holistic features:** Holistic features evaluate the entire camera image and compute one descriptor per camera image. The map can then be represented by, for example, a $k-d$ search tree that stores the map descriptors [62].

2. **Local features:** Local features identify semantically representative features in the camera images. The assumption is that the spatial and semantic constellation of features per image is unique enough to avoid disambiguities [63].

A comprehensive review of current visual approaches can be found in [61]. It shows that CNN-based approaches currently have the highest performance [64]. Additionally, for robustness enhancement, localization hypotheses of consecutive localization cycles can be tracked to identify invalid sequences, that cannot be explained by consistent vehicle motion [60].

21.6   State Estimation

An essential process step of SLAM is the estimation of the pose or landmark parameters as well as the determination of an estimation uncertainty. The state estimation thus provides a solution to the SLAM problems introduced in ▶ Sect. 21.2.

21.6.1   Bayes Filter

To solve the online SLAM problem (21.2), the probability distribution

$$p(\mathbf{x}_k, \mathbf{m} | \mathbf{z}_{0:k}, \mathbf{u}_{0:k}) \tag{21.25}$$

must be determined using the prediction model $p(\mathbf{x}_k | \mathbf{x}_{k-1}, \mathbf{u}_k)$ and the measurement model $p(\mathbf{z}_k | \mathbf{x}_k, \mathbf{m})$. The probability distribution (21.25) results directly from the application of the recursive Bayesian filter [65]. To construct the recursive solution, it is assumed that $p(\mathbf{x}_{k-1}, \mathbf{m} | \mathbf{z}_{0:k-1}, \mathbf{u}_{0:k-1})$ is known. Using the prediction model $p(\mathbf{x}_k | \mathbf{x}_{k-1}, \mathbf{u}_k)$, the predictive distribution

$$p(\mathbf{x}_k, \mathbf{m} | \mathbf{z}_{0:k-1}, \mathbf{u}_{0:k}) = \int p(\mathbf{x}_k | \mathbf{x}_{k-1}, \mathbf{u}_k) \cdot p(\mathbf{x}_{k-1}, \mathbf{m} | \mathbf{z}_{0:k-1}, \mathbf{u}_{0:k-1}) d\mathbf{x}_{k-1} \tag{21.26}$$

can be calculated. Then, the measurement model $p(\mathbf{z}_k | \mathbf{x}_k, \mathbf{m})$ and Bayes' theorem

$$p(\mathbf{x}_k, \mathbf{m} | \mathbf{z}_{0:k}, \mathbf{u}_{0:k}) = \frac{p(\mathbf{z}_k | \mathbf{m}, \mathbf{x}_k) \cdot p(\mathbf{x}_k, \mathbf{m} | \mathbf{z}_{0:k-1} \mathbf{u}_{0:k})}{p(\mathbf{z}_{0:k}, \mathbf{u}_{0:k})}. \tag{21.27}$$

is applied. The denominator in equation (21.27) can be determined by normalizing the probability distribution. Thus, by applying Eqs. (21.26) and (21.27), $p(\mathbf{x}_k, \mathbf{m} | \mathbf{z}_{0:k}, \mathbf{u}_{0:k})$ can be calculated from $p(\mathbf{x}_{k-1}, \mathbf{m} | \mathbf{z}_{0:k-1}, \mathbf{u}_{0:k-1})$. In this context, Eq. (21.26) is

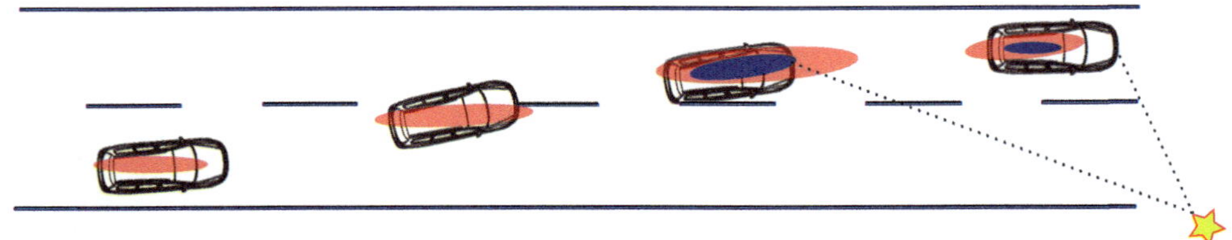

Fig. 21.20 Schematic illustration of the Kalman filter. The vehicle in the image moves to the right over four points in time t_1, t_2, t_3, t_4. The ellipses represent the estimated probability distribution of location $p(x, y)$ in terms of an error ellipse. The red ellipses correspond to a distribution obtained by a prediction step, while the blue ellipses represent a distribution after an observation step. Initially, only prediction steps take place in this sketch, since no landmark was perceived. Thereby the uncertainty increases, which is visualized by the size of the ellipse. At time points t_3 and t_4, the landmark represented by a star is detected by the vehicle sensors, causing the uncertainty to successively decrease

called the prediction step and Eq. (21.27) is called the observation or innovation step.

It is worth to mention that the Bayes filter should not be understood directly as an applicable algorithm, but rather as a theoretical framework from which approximate solutions for state distribution estimation can be derived. The first successful SLAM approaches such as EKF-SLAM [66] and fastSLAM [67] are based on the Bayes filter. However, most modern SLAM algorithms rely on numerical optimization and solve the full SLAM problem. This method is described in ▶ Sect. 21.6.3.

21.6.2 Bayes Filter for Map-Based Localization

For pure localization using a pre-generated map, Bayes filter-based algorithms are frequently applied in current approaches [48]. In this case, the map is fixed and the prediction and observation steps take the form

$$p(\mathbf{x}_k|\mathbf{z}_{0:k-1}, \mathbf{u}_{0:k}, \mathbf{m}) = \int p(\mathbf{x}_k|\mathbf{x}_{k-1}, \mathbf{u}_k)$$
$$\cdot p(\mathbf{x}_{k-1}|\mathbf{z}_{0:k-1}, \mathbf{u}_{0:k-1}, \mathbf{m}) d\mathbf{x}_{k-1}, \quad (21.28)$$
$$p(\mathbf{x}_k|\mathbf{z}_{0:k}, \mathbf{u}_{0:k}, \mathbf{m}) = \frac{p(\mathbf{z}_k|\mathbf{m}, \mathbf{x}_k) \cdot p(\mathbf{x}_k|\mathbf{z}_{0:k-1}, \mathbf{u}_{0:k}, \mathbf{m})}{p(\mathbf{z}_{0:k}, \mathbf{u}_{0:k})}. \quad (21.29)$$

In the following, we discuss two commonly used realizations of the Bayes filter in the context of map-based localization: The Kalman filter and the particle filter. To simplify notation, the dependence of the other quantities on the map $\mathbf{m}$ is not explicitly written in the following sections.

Kalman filter

In the Kalman filter [68], the probability distribution $p(\mathbf{x}_k|\mathbf{z}_{0:k}, \mathbf{u}_{0:k})$ is modeled by a Gaussian distribution

$$p(\mathbf{x}_k|\mathbf{z}_{0:k}, \mathbf{u}_{0:k}) = \mathcal{N}(\mathbf{x}_k, \boldsymbol{\mu}_k, \boldsymbol{\Sigma}_k), \quad (21.30)$$

where the abbreviation

$$\mathcal{N}(\mathbf{x}, \boldsymbol{\mu}, \boldsymbol{\Sigma}) = \det(2\pi\boldsymbol{\Sigma})^{-\frac{1}{2}} \exp\left(-\frac{1}{2}(\mathbf{x}-\boldsymbol{\mu})^T \boldsymbol{\Sigma}^{-1}(\mathbf{x}-\boldsymbol{\mu})\right) \quad (21.31)$$

is used. Since the Gaussian distribution is fully characterized by the mean $\boldsymbol{\mu}_k$ and the covariance $\boldsymbol{\Sigma}_k$, only these quantities need to be estimated over time. Based on a deterministic prediction model $\mathbf{g}$ (Eq. (21.5)) and a deterministic measurement model $\mathbf{h}$ (Eq. (21.6)), the following probabilistic prediction and measurement models are assumed:

$$p(\mathbf{x}_k|\mathbf{x}_{k-1}, \mathbf{u}_k) = \mathcal{N}(\mathbf{x}_k, \mathbf{g}(\mathbf{x}_{k-1}, \mathbf{u}_k), \mathbf{R}_k), \quad (21.32)$$
$$p(\mathbf{z}_k|\mathbf{x}_k) = \mathcal{N}(\mathbf{z}_k, \mathbf{h}(\mathbf{x}_k), \mathbf{Q}_k). \quad (21.33)$$

Here, the covariance matrices $\mathbf{R}_k$ and $\mathbf{Q}_k$ describe the uncertainty of the prediction and the observation step and should represent the actual uncertainties of the measurements and the prediction model. An appropriate choice of these covariance matrices is a challenging problem and is often coupled to the application requirements [69]. Substituting the equations (21.30), (21.32) and (21.33) into the prediction equation (21.28) and observation equation (21.29) of the Bayes filter, one obtains the Kalman filter equations describing how $(\boldsymbol{\mu}_k, \boldsymbol{\Sigma}_k)$ is obtained from $(\boldsymbol{\mu}_{k-1}, \boldsymbol{\Sigma}_{k-1})$ [70].

In the orginal Kalman filter, $\mathbf{g}$ and $\mathbf{h}$ are linear functions. In most applications, however, $\mathbf{g}$ and $\mathbf{h}$ are nonlinear. For this case, there are two prominent variants of the Kalman filter, namely the Extended Kalman Filter (EKF) and the Unscented Kalman Filter (UKF).

Extended Kalman Filter

In the EKF, the motion model $\mathbf{g}(\mathbf{x}, \mathbf{u})$ and the observation model $\mathbf{h}(\mathbf{x})$ are linearized by a Taylor expansion. Here, the Jacobian matrix of $\mathbf{g}(\mathbf{x}, \mathbf{u})$ with respect to $\mathbf{x}$ is denoted by $\mathbf{G}(\mathbf{x}, \mathbf{u})$, while the Jacobian matrix of $\mathbf{h}(\mathbf{x})$ with respect to $\mathbf{x}$ is denoted by $\mathbf{H}(\mathbf{x})$ (■ Fig. 21.20).

Example: Jacobian matrix of the CTRV motion model
For this example, the state includes only the agent pose $\mathbf{x} = [x, y, \theta]^T$. Let the motion model be given by the CTRV model (see Eq. 21.23). The Jacobian matrix in this case is given by

$$G(\mathbf{x}, \mathbf{u}) = \frac{\partial \mathbf{g}(\mathbf{x}, \mathbf{u})}{\partial \mathbf{x}} = \begin{bmatrix} 1 & 0 & \frac{v}{\omega}(-\cos(\theta) + \cos(\theta + \omega\Delta t)) \\ 0 & 1 & \frac{v}{\omega}(-\sin(\theta) + \sin(\theta + \omega\Delta t)) \\ 0 & 0 & 1 \end{bmatrix}.$$

$$(21.34)$$

The linearization $\mathbf{g}(\mathbf{x}_{k-1}, \mathbf{u}_k) = \mathbf{g}(\mu_{k-1}, \mathbf{u}_k) + \mathbf{G}(\mu_{k-1}, \mathbf{u}_k)(\mathbf{x}_{k-1} - \mu_{k-1})$ is inserted into Eq. (21.32) and the linearization $\mathbf{h}(\mathbf{x}_k) = \mathbf{h}(\mu_k) + \mathbf{H}(\mu_k)(\mathbf{x}_k - \mu_k)$ is inserted into Eq. (21.33). These equations are in turn substituted into the prediction equation (21.26) and observation equation (21.27) to obtain the linear update equations for μ_k and Σ_k [70]. This way one obtains the Kalman filter equations

$$\bar{\mu}_k = \mathbf{g}(\mu_{k-1}, \mathbf{u}_k) \tag{21.35}$$
$$\bar{\Sigma}_k = \mathbf{G}(\mu_{k-1}, \mathbf{u}_k)\Sigma_{k-1}\mathbf{G}(\mu_{k-1}, \mathbf{u}_k)^T + \mathbf{R}_k \tag{21.36}$$
$$\mathbf{K}_k = \bar{\Sigma}_k\mathbf{H}(\bar{\mu}_k)^T(\mathbf{H}(\bar{\mu}_k)\bar{\Sigma}_k\mathbf{H}(\bar{\mu}_k)^T + \mathbf{Q}_k)^{-1} \tag{21.37}$$
$$\mu_k = \bar{\mu}_k + \mathbf{K}_k(\mathbf{z}_k - \mathbf{h}(\bar{\mu}_k)) \tag{21.38}$$
$$\Sigma_k = (1 - \mathbf{K}_k\mathbf{H}(\bar{\mu}_k))\bar{\Sigma}_k. \tag{21.39}$$

The first two equations represent the prediction step of the EKF, while the following three equations are the observation step of the EKF.

To increase robustness against outliers, individual measurements $\mathbf{z}_k$ can be discarded if their squared Mahalanobis distance $(\mathbf{z}_k - \mathbf{h}(\bar{\mu}_k))^T\mathbf{S}^{-1}(\mathbf{z}_k - \mathbf{h}(\bar{\mu}_k))$ is larger than a threshold [71]. Here, $\mathbf{S} = \mathbf{H}(\bar{\mu}_k)\bar{\Sigma}_k\mathbf{H}(\bar{\mu}_k)^T + \mathbf{Q}_k$ is the innovation covariance. Moreover, to initialize the Kalman filter, an a priori distribution $p(\mathbf{x})$ must be determined—for automated driving, for example, by GNSS localization. If the nonlinearities of the measurement and motion model are not too large and a unimodal probability distribution is sufficient for estimation, the EKF is a good choice because of its simplicity and low computation time. If a multimodal distribution model is needed, the so-called MHEKF is an option, which uses a Gaussian mixture distribution [70].

Unscented Kalman Filter

The Unscented Kalman Filter uses a different approach to deal with the nonlinearity of the motion and measurement models in Eq. (21.28) and Eq. (21.29). Here, the Gaussian distribution is represented by a fixed number of deterministically chosen sample points [72]. While in the prediction step of the EKF a new mean is calculated by applying the motion model to the old mean (see Eq. (21.35)), in the UKF the motion model is applied to each sample point separately and a new mean is determined by a weighted average of the sample points.

While the EKF uses a 1st order Taylor approximation, the accuracy of the UKF corresponds to a 3rd order [72] approximation. Also empirically, the UKF was found to generally perform better than the EKF in nonlinear state estimation [70, 72]. Another advantage of the UKF is that the motion and measurement models do not need to be differentiable. The computational complexity of the UKF is identical to that of the EKF [72]. Thus, the computation time scales quadratically with the dimension of the state vector and approximately cubically with the dimension of the measurement vector. In practice, the UKF is usually somewhat slower than the EKF.

Particle filter

The particle filter is a derivative of the Bayes filter, where the state distribution is represented by a finite number of so-called particles [70]. Each particle has a state and a weight. Unlike the Kalman filter, in principle any probability distribution can be represented. The probability distribution does not have to be Gaussian and can also be multimodal. In the motion step, the state of each particle is updated using the motion model. Here, a perturbation $\delta\mathbf{u}$ is generated for each particle using a random number generator, which is added to the control parameter vector $\mathbf{u}$. In the observation step, weights are then assigned to particles: The closer the expected measurement $\mathbf{h}(\mathbf{x})$ is to the actual measurement $\mathbf{z}$, the higher the weight of a particle with state $\mathbf{x}$. As a result, N new particles are selected from the N particles. Particles with high weight have a higher probability to be selected (several times). The particle filter has some advantages. The algorithm is conceptually simple and easy to implement. As with the UKF, motion and measurement models that are not differentiable can be handled. The multimodality of the particle filter additionally allows to track different landmark associations simultaneously [73]. A disadvantage of the particle filter is the comparatively high computational cost [71], which scales exponentially with the dimension of the state vector.

21.6.3 Graph-SLAM

To solve the full SLAM problem (21.1), so-called Graph-SLAM methods have been established. For this, the SLAM problem is modeled as a graph (see ◻ Fig. 21.21). In this parameter graph, the state variables, i.e., the landmarks $\mathbf{m}$ and the agent states $\mathbf{x}_{0:k}$ are represented as nodes and the measurement quantities $\mathbf{z}_{0:k}$ and the ego-motion estimates $\Delta_{0:k}$ are interpreted as edges. Based on the underlying measurement models (21.6), residual functions $\mathbf{r}_i(\mathbf{z}_i, \mathbf{h}_i(\mathbf{s}))$ define costs that penalize differences between the true measurements $\mathbf{z}_i$ and the measurements $\mathbf{h}_i(\mathbf{s})$ predicted by the states. To simplify the notation in this chapter, the ego-motion estimates $\Delta_{0:k}$ are interpreted as measurements, i.e., $\Delta_{0:k} \subseteq \mathbf{z}_{0:k}$. The parameter graph thus links the measurements to the parameter vector $\mathbf{s} = [\mathbf{x}, \mathbf{m}]^T$, which in the case of SLAM is formed by the agent states $\mathbf{x}$ and the map landmarks $\mathbf{m}$. Due to the fact that usually a single parameter or parameter block is related to several measurements, the graph represents

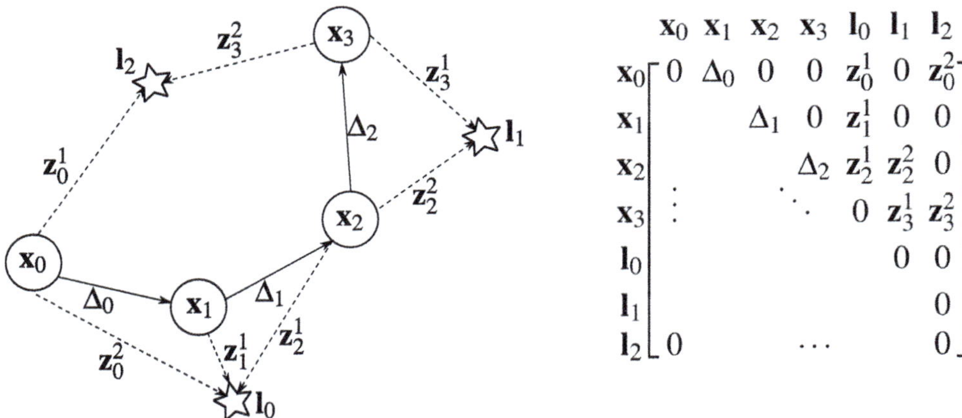

Fig. 21.21 Left: Example graph with agent states (round nodes), landmarks (star-shaped nodes), self-motion estimates (edges), and landmark observations (dashed edges). Right: adjacency matrix corresponding to the graph

an overdetermined—mostly nonlinear—system of equations. The goal of Graph-SLAM is to find that parameter configuration $\mathbf{s}^*$ which minimizes an overall cost function $c : \mathbb{R}^N \to \mathbb{R}$, which evaluates all residuals jointly. By solving the problem

$$\mathbf{s}^* = \arg \min_{\mathbf{s}} c(\mathbf{s}) \tag{21.40}$$

the optimal state vector $\mathbf{s}^*$ can then be determined, which describes the measurements $\mathbf{z}_{0:k}$ at best with respect to the model assumptions Eq. (21.5) and Eq. (21.6).

Least-Squares

One of the most widely used methods for solving Graph-SLAM is the Least-Squares method. Here, the cost function $c(\mathbf{s})$ is defined as a linear combination

$$c(\mathbf{s}) = \mathbf{r}^T(\mathbf{s})\mathbf{W}\mathbf{r}(\mathbf{s}) \tag{21.41}$$

of squared residual functions $\mathbf{r} : \mathbb{R}^N \to \mathbb{R}^{N_r}$ of the form

$$\mathbf{r}(\mathbf{s}) = \mathbf{h}(\mathbf{s}) - \mathbf{z} \tag{21.42}$$

which are weighted by a positive-semidefinite matrix $\mathbf{W} \in \mathbb{R}^{N_r \times N_r}$.[8] The formulation of the residual function (21.42) as difference is important, because in that way, assuming Gaussian distributed measurement noise, the optimal state vector $\mathbf{s}^*$ can be understood as the expected value of the maximum log-likelihood probability distribution[74]. Thus, in this case, the state vector $\mathbf{s}^*$ represents the expectation value of the maximum A-posteriori probability distribution (21.1) for the full SLAM problem. To get a unique solution, the resulting system of equations must not be underdetermined, i.e., the condition $N_r \geq N$ must be satisfied.

Linear Least-Squares

A particular special case of the least-Squares problem (21.41) arises when the residual vector

$$\mathbf{r}(\mathbf{s}) = \mathbf{H} \cdot \mathbf{s} - \mathbf{z}, \quad \mathbf{z} \in \mathbb{R}^{N_r}, \tag{21.43}$$

is affine in the parameters $\mathbf{s}$. In this case, the cost function $c(\mathbf{s})$ is convex, so that the optimal parameter vector $\mathbf{s}^*$ can be determined by

$$\mathbf{s}^* = (\mathbf{H}^T\mathbf{W}\mathbf{H})^{-1}\mathbf{H}^T\mathbf{W}\mathbf{z} \tag{21.44}$$

assuming that the local minimum $\mathbf{s}^*$ exists.

Numerical optimization

In the case of a non-linear cost function (21.41), in general several local cost minima exists. This defines a so-called Nonlinear Least-Squares problem (NLS). In this case, the optimal parameter configuration $\mathbf{s}^*$ can no longer be determined in a closed-form manner. Therefore, to determine $\mathbf{s}^*$, starting from an initialization state $\mathbf{s}_0$, the nearest minimum in the state space is determined by an iterative gradient descent procedure. Here it is crucial that the initialization state $\mathbf{s}_0$ is close to the global minimum. Otherwise there is a risk, that the gradient descent converges to a local minimum. For the gradient descent, the cost function (21.41) is linearized at the state $\mathbf{s}_i$

$$c(\mathbf{s}_i + \Delta\mathbf{s}) = \mathbf{r}^T(\mathbf{s}_i + \Delta\mathbf{s})\mathbf{W}\mathbf{r}(\mathbf{s}_i + \Delta\mathbf{s})$$
$$\approx \left(\mathbf{r}(\mathbf{s}_i) + \mathbf{J}_\mathbf{r}|_{\mathbf{s}=\mathbf{s}_i}\Delta\mathbf{s}\right)^T \mathbf{W}\left(\mathbf{r}(\mathbf{s}_i) + \mathbf{J}_\mathbf{r}|_{\mathbf{s}=\mathbf{s}_i}\Delta\mathbf{s}\right) = c_{GN}(\Delta\mathbf{s}, \mathbf{s}_i), \tag{21.45}$$

where $\mathbf{J}_\mathbf{r}$ is the Jacobian matrix of $\mathbf{r}(\mathbf{s})$. The resulting linear Least-Squares problem is affine in $\Delta\mathbf{s}$ and can be solved by using equation (21.44). The state variation $\Delta\mathbf{s}$ defines the initial state

$$\mathbf{s}_{i+1} = \mathbf{s}_i + \Delta\mathbf{s}, \tag{21.46}$$

for the next iteration $i+1$, where $c(\mathbf{s}_{i+1}) \leq c(\mathbf{s}_i)$ holds. The iteration ends when a termination criterion is reached, for

---

8   If $\mathbf{W} = diag(w_k), w_k \geq 0, 0 \leq k \leq N_r$ holds, equation (21.41) is called Weighted Least-Squares (WLS). If $\mathbf{W} = \mathbf{I}$, it is often referred to Ordinary Least-Squares (OLS).

example, when the change in cost becomes smaller than a defined threshold. This iterative method is also known in the literature as the Gauss-Newton algorithm [75].

A common problem when using the Gauss-Newton algorithm is its poor convergence behavior on gradient descent. To avoid this, a regularization term can be added to equation (21.45), which controls the step size on state update (21.46).

This variation of the Gauss-Newton algorithm is known as the Levenberg-Marquardt algorithm—one of the most widely used optimization algorithm for NLS problems [76].

Covariance estimation

Minimizing problem (21.41) yields the optimal parameter estimate $\mathbf{s}^*$, which is the maximum a-posteriori expected value of distribution (21.1) under the given model assumptions and measurements $\mathbf{z}_{0:k}$. An approximate estimate of the second-order central moment, called the covariance matrix, of distribution (21.1) is given by

$$\mathbf{C}_{\mathbf{s}^*} = ((\mathbf{J}_\mathbf{r}^*)^T \mathbf{W} \mathbf{J}_\mathbf{r}^*)^{-1}, \qquad (21.47)$$

where

$$\mathbf{J}_\mathbf{r}^* = \left.\frac{\partial \mathbf{r}}{\partial \mathbf{s}}\right|_{\mathbf{s}=\mathbf{s}^*} \qquad (21.48)$$

represents the Jacobian matrix of $\mathbf{r}(\mathbf{s})$ evaluated in the optimal estimation $\mathbf{s}^*$.

In this context, it is worth mentioning that the weighting matrix $\mathbf{W}$ can be treated as the Fisher information matrix of the measurements $\mathbf{z}_{0:k}$ [77]. If uncertainty information exists for these measurements in the form of a covariance matrix $\mathbf{C}$, the Fisher information $\mathbf{W} = \mathbf{C}^{-1}$ can be estimated by the inverse of the covariance matrix $\mathbf{C}$. Thus, the estimation uncertainty, represented by the covariance matrix (21.47), can be understood as linear propagation of the measurement uncertainty.

Manifolds

For the optimization algorithms described above, it is assumed that the parameter vector $\mathbf{s}$ is an element of $\mathbb{R}^N$. For SLAM and map-based localization, the orientation is estimated as part of the agent pose and thus represents a part of the state vector $\mathbf{s}$. However, in the three-dimensional case, the agent orientation is an element of the special orthogonal group $SO(3)$, which cannot be bijectively mapped to the $\mathbb{R}^3$.[9] To avoid regularizations, which are necessary for direct optimization in overparameterized representations by orthogonal matrices or quaternions which can lead to convergence problems, it can be exploited that $SO(3)$ is a manifold, and thus

for each element of $SO(3)$ there is a local neighborhood which is homeomorphic[10] to the $\mathbb{R}^3$ [74].

Sparse matrices

In practice, the full SLAM problem (21.1) is often composed of a large number of parameters and measurements. This means that the linear system of equations, which is used to solve the NLS problem (21.41) in each iteration, is very high dimensional and the computation with conventional methods is not feasible. However, in SLAM, a single measurement $\mathbf{z}_t^j \in \mathbf{z}_t$, $\mathbf{z}_t \in \mathbf{z}_{0:k}$ measured at time t sets only a single landmark $\mathbf{l} \in \mathbf{m}$ in relation to an agent state $\mathbf{x}_t \in \mathbf{x}_{0:k}$. All other landmarks $\mathbf{m} \setminus \mathbf{l}$ and agent states $\mathbf{x}_{0:k} \setminus \mathbf{x}_t$ have no direct relation to the measurement $\mathbf{z}_t^j$ (see ◘ Fig. 21.21 on the left). This means that the information matrix $\mathbf{J}_\mathbf{r}^T \mathbf{W} \mathbf{J}_\mathbf{r}$ to determine $\Delta\mathbf{s}$ for the state increment (21.46) is sparse (see ◘ Fig. 21.21 right) [76]. Therefore, on the one hand, so-called sparse matrix solvers can be used to solve the linearized cost function (21.45) [78, 79]. On the other hand, the sparse structure of the information matrix $\mathbf{J}_\mathbf{r}^T \mathbf{W} \mathbf{J}_\mathbf{r}$ allows to divide it into blocks, to solve the system of equations only for a part of the matrix blocks and to determine the total solution by recombination of the blocks. For this purpose, the so-called Schur complement trick is utilized commonly[75].

The exploitation of the sparse matrix structure for efficient inversion of the information matrix has ultimately led to the breakthrough of NLS-based approaches in the field of SLAM.

Robust Optimization

In addition to good initialization, the identification and removal of outliers is of fundamental importance in solving NLS problems. An outlier occurs, for example, when there is an accidental misassociation between a landmark and a measurement which usually leads to very large but incorrect residual values and thus significantly distorts the cost landscape. A relatively small fraction of outliers is already sufficient to significantly distort the optimal estimate $\mathbf{s}^*$. The following strategies have been established for dealing with outliers:

- **Identification and removal:** One way to counteract outliers is to identify and remove them before optimization. To do this, measurements that already have exceptionally high residuals initially can be removed before problem optimization. Another common method for outlier detection are randomized hypothesis testing methods such as the RANSAC algorithm [49], which, however, can only be used efficiently for low-dimensional state spaces.
- **Influence reduction:** Robust error functions are suitable as an alternative to outlier removal. By replacing

---

9   For the two-dimensional case—i.e., the agent orientation is an element of $SO(2)$—there is also no bijection $SO(2) \rightarrow \mathbb{R}^2$. For practical applications, however, this is not a problem, since a scalar rotation angle is sufficient to describe the rotation.

10   A mapping is homeomorphic if it is itself bijective and continuous and its inverse mapping is also continuous.

the quadrature $\|\mathbf{r}_i(\mathbf{s})\|^2$ of the residuals with a robust error function. $\rho(\|\mathbf{r}_i(\mathbf{s})\|) \in \mathbb{R}$, the influence of gradients of high residuals on the optimization is mitigated. It should be noted that the use of error functions can have a significant impact on convergence behavior. If the error function is discontinuous or constant in sections, convergence problems may occur.

Example: Common error functions

◘ Figure 21.22 shows characteristic progressions of common error functions:

$$\text{Huber}: \quad \rho(\|\mathbf{r}_i(\mathbf{s})\|) = \begin{cases} \frac{1}{2}\|\mathbf{r}_i(\mathbf{s})\|^2, & \|\mathbf{r}_i(\mathbf{s})\| < \alpha \\ \alpha\|\mathbf{r}_i(\mathbf{s})\| - \frac{1}{2}\alpha^2, & \text{else} \end{cases}$$

$$\text{Cauchy}: \quad \rho(\|\mathbf{r}_i(\mathbf{s})\|) = \frac{1}{2}\log(1 + \|\mathbf{r}_i(\mathbf{s})\|^2)$$

While in the standard quadrature of the residual norm the cost gradient increases linearly, the Huber error function results in a constant gradient for large errors. For the Cauchy error function, the gradient converges to zero for increasing costs.

21.7 Multi-session Mapping

This section describes an iterative mapping method that allows for continuous map updating and expansion. The idea is to continuously update and extend an existing map by adding further measurements of an already mapped area in order to guarantee a high localization quality over arbitrarily long periods of time. This approach is suitable both for the use of dedicated mapping vehicles[11] as well as for the use of vehicles of the application fleet and is also referred to in the literature as centralized multi-session SLAM without real-time requirements [7, 81].

Process

In centralized multi-session mapping, each vehicle in a fleet uses not only its sensor measurements for localization, but also records them and sends them to a central server. On the server side, the new data is integrated into the existing map, which means that the map can be updated and expanded on an area-by-area basis to include previously unmapped areas (see ◘ Fig. 21.23). After the map update is completed, it can be distributed to the agents of the fleet. The decision whether to integrate a newly recorded sequence into the map can be made based on the localization quality achieved. If this has deteriorated to a certain degree over time, for example due to seasonal environmental changes, an update is performed.

Cascaded graph optimization

If large areas such as entire cities or metropolitan regions are to be mapped using graph-based SLAM methods, it is inevitable to divide the entire area into local areas and optimize them separately. In a second step, the local areas must then be consistently combined (see ◘ Fig. 21.24). This procedure is realized differently in many so-called Smoothing And Mapping methods (SAM) and is a basic requirement for scalable mapping with limited hardware resources [82]. Thus, partial map merging is also necessary for scalable multi-session SLAM. In the following concepts, the goal is always to estimate the vehicle poses as robustly and precisely as possible, i.e., to reference the driven trajectories accurately to each other. Since in SLAM the estimation of vehicle poses and landmarks interact with each other, the landmarks are always estimated in the respective process steps, but are not used afterwards. Thus, the landmarks estimated during optimization are only intermediate and may differ from the final landmarks. When all optimization steps have been completed so that the new sequence has been finally referenced to the existing map, the final landmarks are determined. Thus, the basis of the map adjustment are always the estimated map poses.

Experimental configuration

In the following, the ideas and concepts for a cascaded multi-session SLAM variant are exemplified using point features, extracted from images of a camera system [60]. The focus is on a robust and scalable graph optimization that realizes the concepts presented in ▶ Sect. 21.6.3. Extensive research and criteria on whether and when a map section should be updated when using point features, can be found in [83]. For the following considerations, it is assumed that appropriate matches of point features have been determined from camera images which show the same scene from a similar perspective. All observations are three-dimensional, i.e. the vehicle and camera poses are elements of the Euclidean group $SE(3)$ and the landmarks are 3D points. Each camera in an agent's onboard camera system is both intrinsically and extrinsically calibrated, i.e., for each camera there exist the projection functions (21.8) and (21.9) as well as an extrinsic camera pose $\mathbf{c}$ which defines the transformation of the camera with respect to a vehicle reference point (usually the vehicle rear axle center point). The feature matches $Z = \{Z_1, \ldots, Z_M\}$ used for optimization result from matching and tracking of sequential correspondences as well as loop closure and localization correspondences, that link pixels from images at different times from different measurement sequences. Therefore, the correspondences contain different numbers of features, i.e., the landmark associated to $Z_i = \{\mathbf{z}_1, \ldots, \mathbf{z}_n\} \in Z$ associated landmark $\mathbf{l}_i \in \mathbb{R}^3$ was generally observed in n different camera images. ◘ Figure 21.25 (left) shows a histogram of n in urban areas. Correspondences exist

---

11 Vehicles, which are equipped with high-precision and usually high-cost measurement technology.

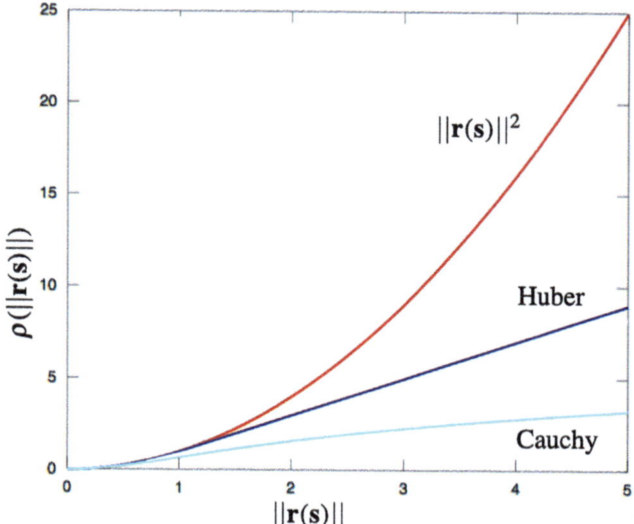

Fig. 21.22 Characteristic graphs of various common error functions. *Source* [80]

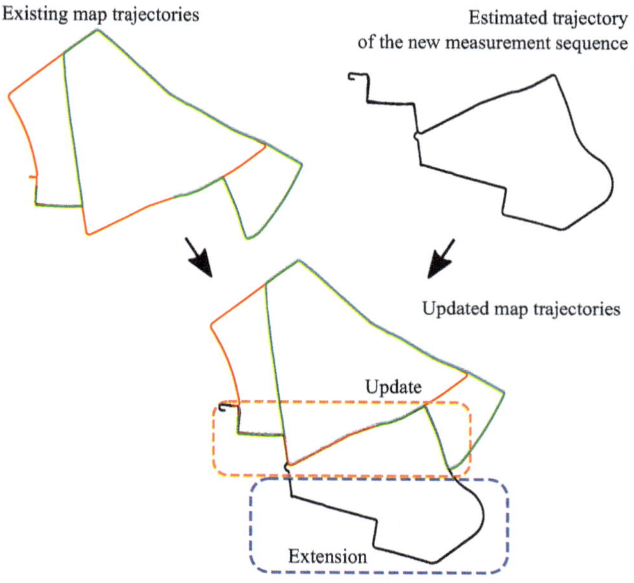

Fig. 21.23 Adding the trajectory of a newly recorded measurement sequence to existing map trajectories

where a landmark has been associated with up to 50 different feature points. These are often landmarks that are farther away from the camera poses, so that the parallax created by vehicle motion is low and an association was possible even if the vehicle moved much farther. However, there are typically significantly more landmarks with short observation periods (< 20 measurements in ◘ Fig. 21.25 (left)) because much of the visible structure is close to the vehicle and closer landmarks are only visible for short periods of time due to the high driving speed.

Sub-map optimization

One possibility for geometric delineation of individual sub-maps and consistent sub-map optimization are locally overlapping optimization windows (see ◘ Fig. 21.24). An optimization window can be understood as a spatially constrained area (for example, rectangular). The size of the window may vary depending on local properties (for example, number of poses, landmarks, and measurements) or available hardware resources. For each window, a so-called bundle block adjustment problem, which includes all vehicle poses and landmarks within the window area, is optimized. Due to the overlap, inconsistently estimated parameters due to truncated correspondence points at the edges of the optimization windows are avoided. In map windows which comprise already mapped areas, associated map features as well as the corresponding map landmarks are added as invariant parameters to the optimization. As a

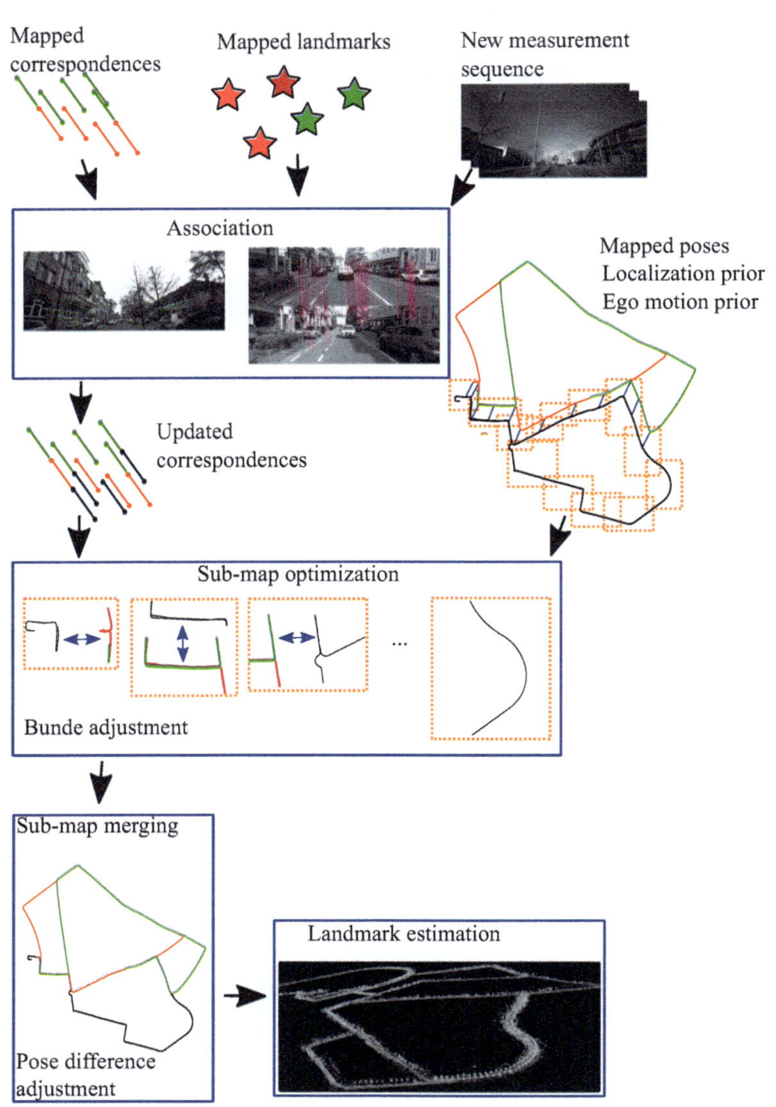

Fig. 21.24   Processing steps of the cascaded graph optimization

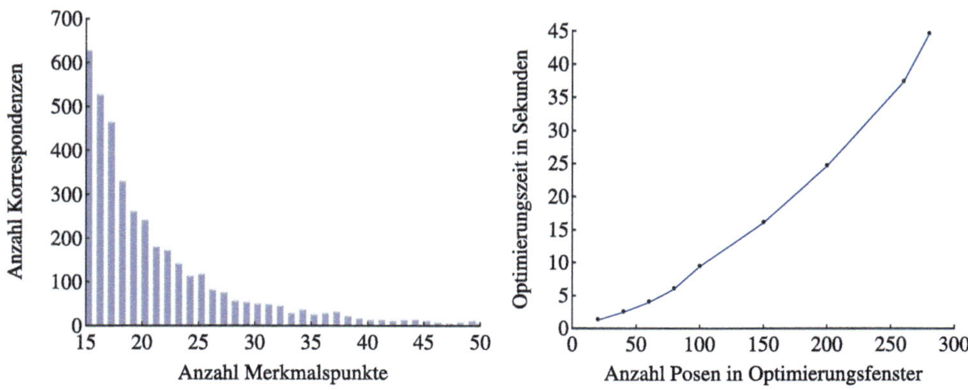

Fig. 21.25   Left: Typical distribution of correspondences. Right: Exemplary computation times of the bundle block adjustment problem for different optimization window sizes. The number of pose parameters is on average proportional to the number of landmarks involved

result, the poses to be estimated of the new measurement sequence automatically adapt to the existing map. In this context, each optimization window can be understood as a sub-map (see **Fig. 21.24**).

Bundle adjustment

The basic idea of bundle adjustment is to bundle line-of-sight rays that ideally meet in a point landmark. An error metric commonly used for this purpose is the so-called reprojection error

$$r_{\mathbf{z}} = \mathbf{z} - \pi_{\mathbf{z}}(\mathbf{c}_{\mathbf{z}}^{-1} \cdot \mathbf{p}_{\mathbf{z}}^{-1} \cdot \mathbf{l}), \qquad (21.49)$$

in which a point landmark $\mathbf{l}$ is transformed into the camera coordinate system using the agent pose $\mathbf{p}_{\mathbf{z}}$ and the extrinsic camera calibration pose $\mathbf{c}_{\mathbf{z}}$, and is subsequently mapped into the image plane using the projection function $\pi_{\mathbf{z}}$. The subscript $\mathbf{z}$ at $\mathbf{p}_{\mathbf{z}}$, $\mathbf{c}_{\mathbf{z}}$ as well as $\pi_{\mathbf{z}}$ indicates the agent pose and the camera which measured the observation $\mathbf{z}$. By using the reprojection error (21.49) within the optimization problem (21.41), one obtains the bundle adjustment problem

$$\hat{P}, \hat{L} = \arg\min_{P,L} \sum_{j=1}^{|Z|} \sum_{\mathbf{z} \in Z_j} \|r_{\mathbf{z}}\|^2, \qquad (21.50)$$

which estimates the poses $\hat{P}$ and landmarks $\hat{L}$ that best describes the involved image point correspondences Z.[12] In this context, the landmarks L relate to the correspondences Z ($|Z| = |L|$).

The computational effort to solve adjustment problems using numerical optimization methods depends essentially on the dimension of the information matrix, since this must be evaluated and inverted in each optimization step in the current state. In the case of bundle block adjustment, $|L| \gg |P|$ usually holds, i.e., there exist significantly more landmark parameters and landmark observations than agent poses, so the problem complexity of (21.50) is dominated by the landmarks. Moreover, the observation poses of an image point correspondence extend only over a limited range, so that bundle adjustment is particularly suitable for partial map optimization. In this way, the theoretically best local estimate can be determined while ensuring reasonable computation time (see ◘ Fig. 21.25 (right)).

Sub-map merge

After each sub-map has been optimized separately, the individual sub-maps must be adjusted consistently to each other and to the existing map. Pose difference adjustment can be used for this purpose. Within pose difference adjustment, the pose differences represent the measurements and can be derived from the poses of each previously computed sub-map. By that, the totality of the derived pose differences D from the sub-map optimizations are combined in the pose difference adjustment, whe-

reby the poses P^s of the entire new acquisition sequence can be estimated consistently to each other as well as to the existing map P^m. This means that the estimated poses from the partial map optimizations are indirectly used as measurements for the sub-map merge, so that the high estimation quality achieved by the bundle block adjustment within each sub-map optimization is directly transferred to the final adjustment. [60]. The advantage here is that the pose difference adjustment scales only with the number of poses to be estimated and not with the landmark count and thus much larger areas can be optimized together.

Pose difference adjustment

The pose difference residual

$$r_{\mathbf{d}_{i,j}} = \Psi\left(\Delta_{i \to j} \cdot \mathbf{p}_j^{-1} \cdot \mathbf{p}_i\right) \qquad (21.51)$$

evaluates the difference between the pose difference measurement $\Delta_{i \to j} \in D$ and the pose difference $\mathbf{p}_i^{-1} \cdot \mathbf{p}_j$ defined by the vehicle poses $\mathbf{p}_i$ and $\mathbf{p}_j$ to be estimated, where the function $\Psi : SE(3) \to \mathbb{R}^6$ transforms the resulting residual $(\Delta_{i \to j} \cdot \mathbf{p}_j^{-1} \cdot \mathbf{p}_i) \in SE(3)$ to an error vector. This vector includes the rotational error in an angular axis representation [84] as well as a three-dimensional translation error. Solving the pose difference adjustment

$$\hat{P} = \arg\min_{P} \sum_{\mathbf{d}_{i,j} \in D} \|r_{\mathbf{d}_{i,j}}\|^2, \qquad (21.52)$$

yields the joint estimate of the poses using the derived pose differences D from all sub-map optimizations. In the sub-map merging step, the set of poses $P = P^s \cup P^m$ is composed of the poses of the measurement sequence to be integrated P^s and the already existing map poses P^m. Since the existing map should not to be modified, the map poses P^m are modeled as invariant parameters in problem (21.52) so that the sequence poses P^s adjust to the map.

Georeferencing

A problem, which cannot be avoided even by robust and highly accurate reconstruction of the map parameters by bundle adjustment, is a position drift. This drift arises from small errors in calibration, model assumptions, and measurement and estimation inaccuracies. If the map is extended over sufficiently large areas, pose drift due to long accumulation leads to drastic errors in absolute pose accuracy. With a well-chosen graph topology using ring closure and consistent intersequence optimization, the drift can be reduced but not completely avoided.

Global position drift can therefore only be countered by using global and invariant reference measurements. For this purpose, GNSS measurements can be integrated

---

12   The landmarks L estimated within the context of partial map optimization are used only to optimize the partial map and, therefore, are not be primarily understood as part of the map landmarks $\mathbf{m}$.

into the pose graph optimization. Even if these arise from a low-cost GNSS receiver and therefore have a comparably poor local accuracy, the drift problem can be avoided without significantly disturbing the local pose and landmark accuracy due to the bundle block adjustment (see ■ Fig. 21.26).

A simple error term for integrating GNSS measurements is the position difference

$$r_{\mathbf{t}_{i,\text{GNSS}}} = \mathbf{t}_i - \mathbf{t}_{i,\text{GNSS}} \qquad (21.53)$$

between the translational component $\mathbf{t}_i \in \mathbb{R}^3$ of pose $\mathbf{p}_i$ and the GNSS position $\mathbf{t}_{i,\text{GNSS}}$ previously converted to Cartesian coordinates (e.g. UTM coordinates). Another possibility is to use the measured GNSS pseudoranges. This so-called "tightly-coupled" integration theoretically leads to better accuracies. However, the modeling of the error terms and the determination of the model hyperparameters must be very precise. Therefore, this variant is demanding from a practical point of view (see ► Sect. 21.4.4). The presented "loosely-coupled" variant, on the other hand, is much more robust and easier to handle and is usually sufficient for the purpose of global referencing. It is assumed here that the GNSS measurement was recorded synchronously with the recording time of $\mathbf{p}_i$. In general, a corresponding GNSS measurement does not exist for every pose $\mathbf{p} \in P$, since the measurement frequency of simple GNSS receivers is lower than that of cameras or laser scanners. However, this undersampling is not problematic for global referencing and is even advantageous for the overall accuracy, since it reduces the interference of the GNSS measurements on the local estimation accuracy.

Results

■ Figure 21.27 shows the trajectories of 14 measurement sequences that were consistently referenced to each other using the described methodology. Sequences were conducted within a period from November to April of the following year in inner-city and suburban areas of Karlsruhe, Germany. Six sequences were entered on the same circuit, except for lane changes. The other runs extend this circuit to include additional areas to be mapped.

In addition to synchronously recorded images from a multi-camera system, GNSS information was recorded from a near-series GPS receiver. Across all sequences, 194k of imagery and 18k of GNSS measurements were recorded and used as measurements in the mapping process. All sequences were mapped iteratively and fully automatically using commercially available hardware.

Local accuracy of map pose estimation can be illustrated using manually annotated structures of the environment. For this purpose, prominent structures were annotated with pixel precision in the images of a single randomly selected pose sequence and triangulated

with the estimated map poses. Based on the mutually adjusted vehicle poses, the triangulated points were projected into the images from other measurement sequences. ■ Figure 21.28 shows examples of some of the structures projected in this way. The images shown originate from different cameras at different times of the day and year and show the projections from different perspectives. It can be seen that regardless of the observation pose, the projections are located with high accuracy at the expected position in the image, which illustrates the high estimation quality by the bundle block adjustment method.[13] To further demonstrate accuracy, the point cloud of map landmarks is shown. ■ Figure 21.29 shows the landmarks, which were determined by triangulation using the final map poses.

21.8 Localization: Example Applications

After a system for map generation using SLAM was shown in the previous section, two exemplary localization systems are now presented. The point-feature-based approach from the previous section can also be used for localization, but requires a lot of compute and memory in comparison. The maps require memory of the order of one gigabyte per kilometer. This is acceptable for robotaxis, but not for ordinary production vehicles. In this application domain, the method from ► Sect. 21.7 can be used to determine highly accurate poses. In a second step, a low-memory map can then be created using the known poses (see ► Sect. 21.8.2).

This map is then stored on a server and made available to production vehicles, for example via the the internet.

The following two localization systems were selected as examples because they satisfy the real-time requirement even with comparatively low computing capacities. They also work with maps that require memory of the order of kilobytes per kilometer. The first system is simple, but satisfies the accuracy requirements (see Ref. [86]) of a localization on the highway. The second system, on the other hand, is more elaborate, achieves higher accuracy, and can therefore be used for localization in urban areas.

---

13   It is worth noting that a quantitative evaluation of the absolute accuracy and consistency of the georeferenced map pose by comparison with the estimation by a high accuracy GNSS measurement system is problematic in inner-city scenarios. In particular, when referencing multiple trajectories of the same route together at different times of the day and year, it is not a a high-accuracy GNSS system does not represent a reliable ground truth [3, 85].

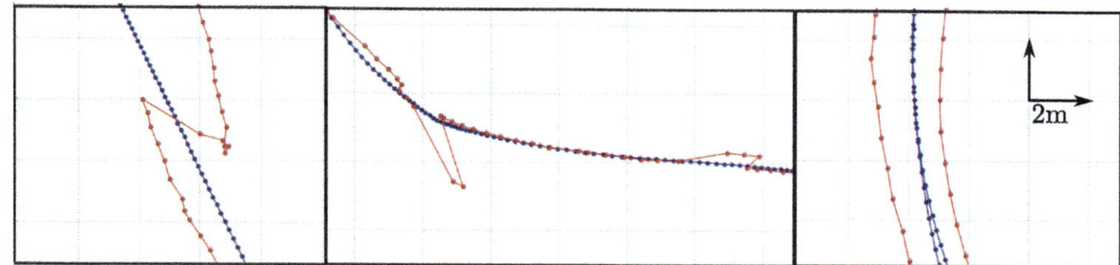

Fig. 21.26 Illustration of estimated vehicle trajectories (blue) and corresponding GNSS trajectories (red) from a near-series GNSS receiver. From left to right, different types of errors are visible in the GNSS trajectories, which, due to suitable optimization weights, have no significant influence on the local pose estimation accuracy (smoothness). Left: Position jump. Center: Outlier measurements Right: Erroneous systematic position offset in a ring closure scenario

Fig. 21.27 Overlay of estimated map poses from 14 acquisition sequences with an aerial photo reference. Each mapped sequence is shown in a different color. Aerial imagery: ©Microsoft/Here

21.8.1 Localization Using Low-Cost Sensors on the Highway

In Ref. [56], Tao et al. describe a system for localization based on sensors found in most current production vehicles:

- GNSS receiver (standard for navigation systems)
- A camera that detects lane markings (standard for lane departure warning systems)
- Yaw rate sensor (standard in ESP systems)
- Wheel speed sensor (standard in ABS systems)

Algorithm

For localization, an Extended Kalman Filter (EKF) with state $\mathbf{x} = \mathbf{p} = [x, y, \theta]^T \in SE(2)$ is used, where $[x, y]^T$ is the position of the center of the rear axle (see Fig. 21.30). The CTRV model (see Eq. (21.23)) is used as the prediction model. The position determined by a GNSS receiver is projected into the map coordinate system (for example UTM) to obtain a measurement $\mathbf{z}^{(GNSS)} = [x_{GNSS}, y_{GNSS}]^T$. The measurement model $\mathbf{h}^{(GNSS)}(\mathbf{x})$ calculates the measurement vector which is to be expected due to the state $\mathbf{x}$, and reads as follows

Fig. 21.28 Green: Manually annotated structures. Yellow: Projections into images of other acquisition sequences based on estimated map poses

Fig. 21.29 Jointly mapped point landmarks from 14 measurement sequences

$$\mathbf{h}^{(GNSS)}(x, y, \theta) = \begin{bmatrix} \cos(\theta)(g_x - t_{lag} \cdot v) - \sin(\theta)g_y + x \\ \sin(\theta)(g_x - t_{lag} \cdot v) + \cos(\theta)g_y + y \end{bmatrix}. \tag{21.54}$$

Here, v is the velocity of the vehicle, $[g_x, g_y]^T$ denotes the position of the GNSS receiver in the vehicle's reference frame, and the latency t_{lag} is the time that elapses from the reception of the GNSS signal to its processing in the EKF. Latency is a constant that can be determined once by data analysis [71]. The camera models the lane markers by a third degree polynomial, so that in the x_C, y_C coordinate system of the camera a lane marker is described by

$$y_C(x_C) = C_0 + C_1 x_C + C_2 x_C^2 + C_3 x_C^3. \tag{21.55}$$

The measurement vector for the lane markers detected by the camera is restricted to the first two coefficients C_0 and C_1 of the polynomial: $\mathbf{z}^{(C)} = [C_0, C_1]^T$. Here, C_0 corresponds to the lateral distance from the center of the front bumper to the lane marker, and C_1 measures the angle between the vehicle orientation and the lane marker orientation (see **Fig. 21.30**). The measurement

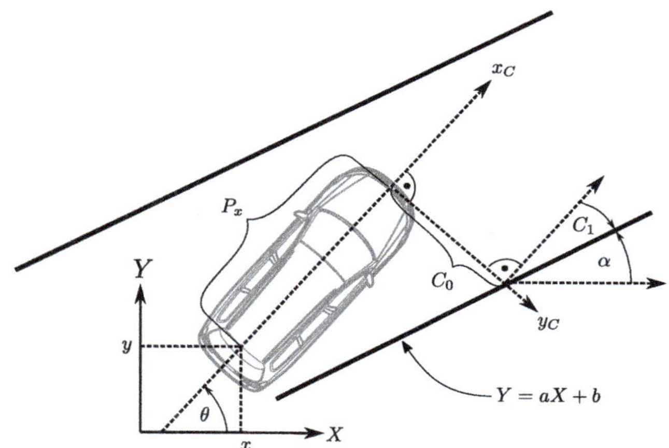

Fig. 21.30   Geometry of the measurement model for lane markers. The map coordinate system is labeled X, Y, while the camera coordinate system is labeled x_C, y_C and is pinned to the vehicle. The lane markers in the map are represented by polylines. The straight line equation of the relevant line segment is $Y = aX + b$

model is $\mathbf{h}^{(C)}(\mathbf{x}) = [\hat{C}_0(\mathbf{x}), \hat{C}_1(\mathbf{x})]^T$ with

$$\hat{C}_0(x, y, \theta) = \frac{P_x \sin(\theta) - aP_x \cos(\theta) - ax + y - b}{a \sin(\theta) + \cos(\theta)},$$

(21.56)

$$\hat{C}_1(x, y, \theta) = \alpha - \theta.$$

(21.57)

Here, P_x, α, a and b are defined in Fig. 21.30. Eq. (21.57) is valid within the small-angle approximation $\tan(x) \approx x$ with $x = \alpha - \theta$. To compute $\mathbf{h}(\mathbf{x}) = [\hat{C}_0(\mathbf{x}), \hat{C}_1(\mathbf{x})]^T$, the map which stores lane markers as polylines, is queried. The polylines can be stored in a spatial index, such as an R-tree, for quick access [71, 87]. The Jacobian matrices that are necessary for the EKF can be computed analytically from Eqs. (21.54), (21.56), and (21.57). A similar approach to the one described here was implemented in Ref. [88] using a particle filter and in Ref. [89] using an information filter.

Results

Empirically, a mean lateral position error of about 10cm was obtained, while the mean longitudinal error is 1m [56]. This is explained by the fact that lane markers yield a very good estimate for the lateral vehicle position, whereas their information content for longitudinal positioning is comparatively low—especially on straight roads. The accuracy achieved does not meet the requirements for automated driving in the city. For the highway, however, it is sufficient [86].

21.8.2   Localization Using Geometric Primitives in Urban Terrain

Since localization based on lane markings only is not always sufficient, especially in urban terrain, another localization approach is considered in this chapter, which uses cylindrical structures and planar surfaces as localization features [11]. Such features provide a high feature density in urban areas and can be efficiently described as geometric primitives (see ▶ Sect. 21.4.3).

Map

The type of localization features contained in a map has a large impact on the localization performance that can be achieved. In addition, properties such as the memory size and the up-to-dateness of the map data are important to be considered when choosing localization features. The use of geometric primitives as localization features offers the following advantages:

- **Cylindrical objects** such as sign posts, traffic lights, street lamps, or tree trunks occur frequently in street scenes. Thus, they not only provide broad coverage, but also represent long-lasting localization features that contribute to the timeliness of the map. Another advantage is their height, which makes them easy to detect from a long distance. The efficient description of such cylindrical objects consists of their intersection with the ground plane, their height as well as their radius and possibly a tilt angle.
- **Planar surfaces** are also frequently found in road scenes. In urban areas, house facades represent such planar surfaces. On the highway, bridges, guard rails, noise barriers or even gantries can be interpreted and described as planes. These features can also be detected from a great distance due to their size. The description of these planes is achieved by the intersection line with the ground plane and is stored as a 2D straight line through its start and end points in the map.
- **Lane markings** are often used for localization and are intended to support the localization procedure in scenes with few geometric primitives. Thanks to their color, lane markings stand out clearly from their

surroundings and are easy to detect. However, their periodic occurrence is problematic, which can lead to ambiguities in the association step.

To map these features, trajectory estimation of the mapping run is performed first. A SLAM approach can be used for this, as described for example in 21.7. Then, the detected features are transformed from the vehicle coordinate system to the existing map coordinate system. All detections that potentially describe the same object are then clustered. Finally, each cluster of detections can be transformed into a single map element.

Such a map with approximately 500 cylindrical features as well as 760 planar surfaces is shown in ◘ Fig. 21.31. It is a route of 3.8 km length in an urban area. The black dots represent the positions of the cylinders. The black lines indicate locations where planar surfaces were mapped.

Localization framework

The sensor technology used is limited to several LiDAR sensors, a stereo camera, an odometry unit, and a GNSS receiver. However, only the camera and LiDAR sensors are important for the detection of the localization features. Cylindrical features as well as planes are extracted from raw LiDAR data, while a stereo camera is used for the detection of lane markings. Details of feature extraction have already been described in ▶ Sect. 21.4.3. After feature extraction, the next step is to search for associations with the map. For this, the detections are first accumulated in a local map using odometry measurements. This is followed by the association with the map using a sampling procedure. The resulting associations as well as pose differences provide the basis for an optimization problem, which is solved in the final step to estimate the vehicle pose. ◘ Figure 21.32 presents a schematic overview of the localization framework.

Feature association

A drawback of geometric primitives is that they do not have a unique descriptor. Moreover, such features often occur in periodic patterns. This leads to ambiguities in the association step that must be resolved. One approach to do this is to accumulate multiple detections over a limited period of time and merge them into a local map. The resulting feature patterns then enable reliable association.

1. **Generation of a local map:** For generating a local pattern from multiple detected features, a limited time $\mathscr{S}$ is considered. Within this time window, all detections are transformed into a static coordinate system. To do so, the motion of the vehicle must be estimated. The more feature detections are included in this local map, the more unambiguous the resulting pattern becomes and the more reliable associations can

be made. However, odometry errors distort the resulting map if the time $\mathscr{S}$ is too long, so that no association is possible afterwards. If the time span $\mathscr{S}$ is too short, too few features might be accumulated in the local map. In this case, reliable association is also not possible. Thus, the goal of the local map is to combine as many features as possible in one map to form a pattern without introducing major odometry errors.

2. **Association with the global map:** To find associations between the local map with the global map, the vehicle pose that allows the best possible overlap of the two maps has to be found. A sampling approach is used for this purpose. Based on a previous pose estimation or GNSS measurement, poses can be sampled. Then, association costs are computed for each of these poses (see ▶ Sect. 21.4.3). The pose with the lowest cost is determined and the underlying association pairs are passed to the optimization for final pose estimation.

Optimization

An optimization problem is formulated to estimate the final vehicle pose. For this purpose, the final N vehicle poses are considered and these are optimized based on the feature associations as well as odometry measurements. First, the pose difference residuals are minimized as described in 21.7. Second, for each association pair $\mathbf{a}_i = (\mathbf{z}_i, \mathbf{l}_j)$ found, consisting of a detection $\mathbf{z}$ and the map element $\mathbf{l}$, a cost function C is minimized

$$C(\mathbf{z}, \mathbf{l}, \mathbf{p}) = \sum_{\mathbf{a}_i} C_f(\mathbf{z}_i, \mathbf{l}_j, p). \tag{21.58}$$

The corresponding cost functions for cylindrical as well as planar features can be found in ▶ Sect. 21.4.3. The optimization problem can then be solved to estimate the most current pose.

Results

The localization quality is determined based on the position error. Two runs are recorded to determine a reference solution: the localization run and the mapping run. A trajectory estimate can thus be performed for both trips together, registering the localization trip to the mapping trip. The estimated trajectory of the localization run serves as the reference solution.

◘ Table 21.1 illustrates the results on a test section of 3.8 km length in an approximately 10 minute trip. This corresponds to a drive along the mapped route in ◘ Fig. 21.31. For evaluation, we consider the mean errors in pose estimation using different features. A distinction is made between position error Δpos and orientation error Δyaw. The position error is further divided into the error in lateral (Δlat) and longitudinal (Δlon) direction.

When using all available features, i.e. all cylindrical objects, planes as well as road markings, mean errors below 10 cm are achieved. When using only one type of localization features, some of the localization accuracy is lost, which is due to the lower feature density.

Fig. 21.31 Localization map with about 500 cylindrical features and 760 planar surfaces along a 3.8km demo route in an urban terrain in Sindelfingen, Germany

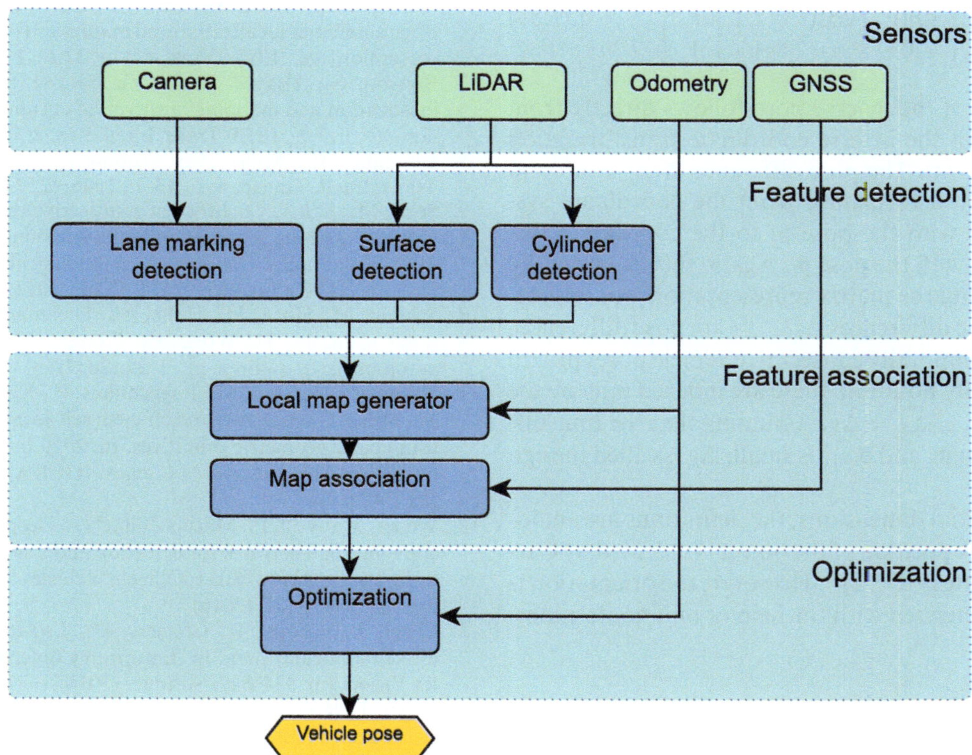

Fig. 21.32 Overview of the localization framework consisting of sensing, feature detection, feature association, and pose estimation

21.9 Appendix: The Concept of Pose

A pose $\mathbf{p}$ is the position and orientation of an object with respect to the world coordinate system. A coordinate system whose coordinate axes are pinned to the object is selected as the object coordinate system. The coordinate transformation from the object coordinate system to the world coordinate system is the coordinate transforma-tion[14] uniquely assigned to the pose $\mathbf{p}$. Vehicles and sensors appear as objects in this chapter. The axes of the object coordinate system then usually correspond to the forward, left, and up directions of the vehicle or sensor.

In two spatial dimensions, the pose can be represen-ted as $\mathbf{p} = [x, y, \theta]^T$, where $[x, y]^T$ specifies the position

14 This coordinate transformation is element of the special Euclidean group $SE(N)$.

◻ Table 21.1 Localization results: accuracies achieved using different semantic features [11]

Used features	Δlat [cm]	Δlon [cm]	Δpos [cm]	Δyaw [°]
All available features	3	6	8	0.14
Cylinders	7	8	12	0.33
Planes	4	11	12	0.18
Lane markings	10	29	34	0.47
GNSS	156	135	227	1.50

of the object by Cartesian coordinates, and $\theta \in [0, 2\pi)$ describes the rotation angle between the object and the world coordinate system. The transformation associated with the pose can be represented by a matrix, such that a point $[X_o, Y_o]^T$ is transformed from the object coordinate system to the world coordinate system as follows:

$$\begin{bmatrix} X_w \\ Y_w \\ 1 \end{bmatrix} = \begin{bmatrix} cos(\theta) & -\sin(\theta) & x \\ sin(\theta) & \cos(\theta) & y \\ 0 & 0 & 1 \end{bmatrix} \begin{bmatrix} X_o \\ Y_o \\ 1. \end{bmatrix} \qquad (21.59)$$

The definition of the inverse pose follows directly from the definition of the inverse coordinate transformation (or inverse matrix). The pose difference $\Delta_{i \to j} = \mathbf{p}_i^{-1} \cdot \mathbf{p}_j$ describes the transformation from the coordinate system associated with the pose $\mathbf{p}_j$ to the coordinate system associated with the pose $\mathbf{p}_i$. Again, this can be easily understood using the matrix representation. An important class of pose differences $\Delta_{k \to k+1}$ are pose differences between two temporally consecutive vehicle poses $\mathbf{p}_k$ and $\mathbf{p}_{k+1}$. To simplify notation, these are indexed only by the current time $\Delta_{k \to k+1} = \Delta_k$. Assuming that the time difference between $\mathbf{p}_k$ and $\mathbf{p}_{k+1}$ is small, Δ_k is called the ego motion.

In three spatial dimensions, the definitions are analogous. Again, the position of the object is specified by Cartesian coordinates $[x, y, z]^T$. However, the orientation is usually parameterized with the help of unit quaternions.

References

1. Ziegler, J., Bender, P., Schreiber, M., Lategahn, H., Strauss, T., Stiller, C., Dang, T., Franke, U., Appenrodt, N., Keller, C.G. et al.: Making bertha drive-an autonomous journey on a historic route. IEEE Intell. Transp. Syst. Mag. **6**(2), 8–20 (2014)
2. Furgale, P., Schwesinger, U., Rufli, M., Derendarz, W., Grimmett, H., Mühlfellner, P., Wonneberger, S., Timpner, J., Rottmann, S., Li, B. et al.: Toward automated driving in cities using close-to-market sensors: an overview of the v-charge project. In: 2013 IEEE Intelligent Vehicles Symposium (IV), pp. 809–816. IEEE (2013)
3. Sons, M., Lategahn, H., Keller, C.G., Stiller, C.: Multi trajectory pose adjustment for life-long mapping. In: 2015 IEEE Intelligent Vehicles Symposium (IV), pp. 901–906. IEEE (2015)
4. Durrant-Whyte, H., Bailey, T.: Simultaneous localization and mapping: part i. IEEE Tobot. Autom. Mag. **13**(2), 99–110 (2006)
5. Bailey, T., Durrant-Whyte, H.: Simultaneous localization and mapping (slam): Part ii. IEEE Robot. Autom. Mag. **13**(3), 108–117 (2006)
6. Cadena, C., Carlone, L., Carrillo, H., Latif, Y., Scaramuzza, D., Neira, J., Reid, I., Leonard, J.J.: Past, present, and future of simultaneous localization and mapping: Toward the robust-perception age. IEEE Trans. Robot. **32**(6), 1309–1332 (2016)
7. Bresson, G., Alsayed, Z., Li, Yu., Glaser, S.: Simultaneous localization and mapping: a survey of current trends in autonomous driving. IEEE Trans. Intell. Veh. **2**(3), 194–220 (2017)
8. Lategahn, H., Stiller, C.: Vision-only localization. IEEE Trans. Intell. Transp. Syst. **15**(3), 1246–1257 (2014)
9. Sons, M., Stiller, C.: Efficient multi-drive map optimization towards life-long localization using surround view. In: 2018 21st International Conference on Intelligent Transportation Systems (ITSC), pp. 2671–2677. IEEE (2018)
10. Schoen, M., Horn, M., Hahn, M., Dickmann, J.: Real-time radar slam. In: 11. Workshop Fahrerassistenzsysteme und automatisiertes Fahren, pp. 1–11 (2017)
11. Kümmerle, J., Sons, M., Poggenhans, F., Kühner, T., Lauer, M., Stiller, C.: Accurate and efficient self-localization on roads using basic geometric primitives. In: 2019 International Conference on Robotics and Automation (ICRA), pp. 5965–5971. IEEE (2019)
12. Weber, M., Wolf, P., Marius Zöllner, J.: Deeptlr: a single deep convolutional network for detection and classification of traffic lights. In: 2016 IEEE Intelligent Vehicles Symposium (IV), pp. 342–348. IEEE (2016)
13. Engel, J., Schöps, T., Cremers, D.: Lsd-slam: Large-scale direct monocular slam. In: European Conference on Computer Vision, pp. 834–849. Springer (2014)
14. Engel, J., Koltun, V., Cremers, D.: Direct sparse odometry. IEEE Trans. Pattern Anal. Mach. Intell. **40**(3), 611–625 (2017)
15. Zhang, J., Singh, S.: Visual-lidar odometry and mapping: Low-drift, robust, and fast. In: 2015 IEEE International Conference on Robotics and Automation (ICRA), pp. 2174–2181. IEEE (2015)
16. Graeter, J., Strauss, T., Lauer, M.: Momo: monocular motion estimation on manifolds. In: 2017 IEEE 20th International Conference on Intelligent Transportation Systems (ITSC), pp. 1–6. IEEE (2017)
17. Ho, K.L., Newman, P.: Detecting loop closure with scene sequences. Int. J. Comput. Vis. **74**(3), 261–286 (2007)
18. Ward, E., Folkesson, J.: Vehicle localization with low cost radar sensors. In: 2016 IEEE Intelligent Vehicles Symposium (IV), pp. 864–870. IEEE (2016)
19. Zhang, J., Singh, S.: Loam: Lidar odometry and mapping in real-time. In: Robotics: Science and Systems, vol. 2 (2014)

20. Strauß, T., Ziegler, J., Beck, J.: Calibrating multiple cameras with non-overlapping views using coded checkerboard targets. In: 17th International IEEE Conference on Intelligent Transportation Systems (ITSC), pp. 2623–2628. IEEE (2014)

21. Beck, J., Stiller, C.: Generalized b-spline camera model. In: 2018 IEEE Intelligent Vehicles Symposium (IV), pp. 2137–2142. IEEE (2018)

22. Gräter, J., Schwarze, T., Lauer, M.: Robust scale estimation for monocular visual odometry using structure from motion and vanishing points. In: 2015 IEEE Intelligent Vehicles Symposium (IV), pp. 475–480. IEEE (2015)

23. Geiger, A., Ziegler, J., Stiller, C.: Stereoscan: dense 3d reconstruction in real-time. In: 2011 IEEE Intelligent Vehicles Symposium (IV), pp. 963–968. IEEE (2011)

24. Deng, J., Dong, W., Socher, R., Li, L.-J., Li, K., Fei-Fei, L.: Imagenet: a large-scale hierarchical image database. In: 2009 IEEE Conference on Computer Vision and Pattern Recognition, pp. 248–255. IEEE (2009)

25. Zhang, F., Liu, F.: Parallax-tolerant image stitching. In: Proceedings of the IEEE Conference on Computer Vision and Pattern Recognition, pp. 3262–3269 (2014)

26. Zhao, F., Huang, Q., Gao, W.: Image matching by normalized cross-correlation. In: 2006 IEEE International Conference on Acoustics Speech and Signal Processing Proceedings, vol. 2, pp. II–II. IEEE (2006)

27. Hisham, M.B., Yaakob, S.N., Raof, R.A.A.: Nazren, A.B.A.: Template matching using sum of squared difference and normalized cross correlation. In: 2015 IEEE Student Conference on Research and Development (SCOReD), pp. 100–104. IEEE (2015)

28. Calonder, M., Lepetit, V., Strecha, C., Fua, P.: Brief: Binary robust independent elementary features. In: European Conference on Computer Vision, pp. 778–792. Springer (2010)

29. Leutenegger, S., Chli, M., Siegwart, R.Y.: Brisk: Binary robust invariant scalable keypoints. In: 2011 International Conference on Computer Vision, pp. 2548–2555. IEEE (2011)

30. Rublee, E., Rabaud, V., Konolige, K., Bradski, G.: Orb: an efficient alternative to sift or surf. In: 2011 International Conference on Computer Vision, pp. 2564–2571. IEEE (2011)

31. Lowe, G.: Sift-the scale invariant feature transform. Int. J. 2(91–110), 2 (2004)

32. Bay, H., Tuytelaars, T., Van Gool, L.: Surf: Speeded up robust features. In: European Conference on Computer Vision, pp. 404–417. Springer (2006)

33. Lategahn, H., Beck, J., Kitt, B., Stiller, C.: How to learn an illumination robust image feature for place recognition. In: 2013 IEEE Intelligent Vehicles Symposium (IV), pp. 285–291. IEEE (2013)

34. Zagoruyko, S., Komodakis, N.: Learning to compare image patches via convolutional neural networks. In: Proceedings of the IEEE Conference on Computer Vision and Pattern Recognition, pp. 4353–4361 (2015)

35. Sons, M., Kinzig, C., Zanker, D., Stiller, C.: An approach for cnn-based feature matching towards real-time slam. In: 2019 IEEE Intelligent Transportation Systems Conference (ITSC), pp. 1305–1310. IEEE (2019)

36. Bosse, M., Zlot, R.: Continuous 3d scan-matching with a spinning 2d laser. In: 2009 IEEE International Conference on Robotics and Automation, pp. 4312–4319. IEEE (2009)

37. Olson, E.B.: Real-time correlative scan matching. In: 2009 IEEE International Conference on Robotics and Automation, pp. 4387–4393. IEEE (2009)

38. Besl, P.J., McKay, N.D.: Method for registration of 3-d shapes. In: Sensor fusion IV: control paradigms and data structures, volume 1611, pp. 586–606. International Society for Optics and Photonics (1992)

39. Poggenhans, F., Schreiber, M., Stiller, C.: A universal approach to detect and classify road surface markings. In: 2015 IEEE 18th International Conference on Intelligent Transportation Systems, pp. 1915–1921. IEEE (2015)

40. Canny, J.: A computational approach to edge detection. IEEE Trans. Pattern Anal. Mach. Intell. 6, 679–698 (1986)

41. Duda, R.O., Hart, P.E.: Use of the hough transformation to detect lines and curves in pictures. Commun. ACM 15(1), 11–15 (1972)

42. Tian, Y., Gelernter, J., Wang, X., Chen, W., Gao, J., Zhang, Y., Li, X.: Lane marking detection via deep convolutional neural network. Neurocomputing 280, 46–55 (2018)

43. Hata, A., Wolf, D.: Road marking detection using lidar reflective intensity data and its application to vehicle localization. In: 17th International IEEE Conference on Intelligent Transportation Systems (ITSC), pp. 584–589. IEEE (2014)

44. Spangenberg, R., Goehring, D., Rojas, R.: Pole-based localization for autonomous vehicles in urban scenarios. In: 2016 IEEE/RSJ International Conference on Intelligent Robots and Systems (IROS), pp. 2161–2166. IEEE (2016)

45. Kampker, A., Hatzenbuehler, J., Klein, L., Sefati, M., Kreiskoether, K.D., Gert, D.: Concept study for vehicle self-localization using neural networks for detection of pole-like landmarks. In: International Conference on Intelligent Autonomous Systems, pp. 689–705. Springer (2018)

46. Sefati, M., Daum, M., Sondermann, B., Kreisköther, K.D., Kampker, A.: Improving vehicle localization using semantic and pole-like landmarks. In: 2017 IEEE Intelligent Vehicles Symposium (IV), pp. 13–19. IEEE (2017)

47. Brenner, C.: Global localization of vehicles using local pole patterns. In: Joint Pattern Recognition Symposium, pp. 61–70. Springer (2009)

48. Poggenhans, F., Salscheider, N.O., Stiller, C.: Precise localization in high-definition road maps for urban regions. In: 2018 IEEE/RSJ International Conference on Intelligent Robots and Systems (IROS), pp. 2167–2174. IEEE (2018)

49. Fischler, M.A., Bolles, R.C.: Random sample consensus: a paradigm for model fitting with applications to image analysis and automated cartography. Commun. ACM 24(6), 381–395 (1981)

50. Hofstetter, I., Sprunk, M., Ries, F., Haueis, M.: Reliable data association for feature-based vehicle localization using geometric hashing methods. In: 2020 IEEE International Conference on Robotics and Automation (ICRA), pp. 1322–1328. IEEE (2020)

51. Lamdan, Y., Wolfson, H.J.: Geometric hashing: a general and efficient model-based recognition scheme (1988)

52. Wikimedia Commons. File:ecef enu longitude latitude relationships.svg — wikimedia commons, the free media repository (2020). https://commons.wikimedia.org/w/index.php?title=File:ECEF_ENU_Longitude_Latitude_relationships.svg&oldid=479685496 [Online; accessed 14-May-2021]

53. Seeber, G.: Satellite geodesy: foundations, methods, and applications. Walter de gruyter (2008)

54. Lu, Z., Qu, Y., Qiao, S.: Geodesy. Springer (2014)

55. Zogg, J.M.: U-Blox. GPS: Essentials of Satellite Navigation. U-Blox (2009)

56. Tao, Z., Bonnifait, Ph., Fremont, V., Ibanez-Guzman, J.: Mapping and localization using gps, lane markings and proprioceptive sensors. In: 2013 IEEE/RSJ International Conference on Intelligent Robots and Systems, pp. 406–412. IEEE (2013)

57. Falco, G., Pini, M., Marucco, G.: Loose and tight gnss/ins integrations: comparison of performance assessed in real urban scenarios. Sensors 17(2), 255 (2017)

58. Free SVG: Honda s2000 outline (2020). https://freesvg.org/ivak-honda-s2000-outline [Online; accessed 14-May-2021]

59. Schubert, R., Richter, E., Wanielik, G.: Comparison and evaluation of advanced motion models for vehicle tracking. In: 2008 11th International Conference on Information Fusion, pp. 1–6. IEEE (2008)

60. Sons, M.: Automatische erzeugung langzeitverfügbarer punktmerkmalskarten zur robusten lokalisierung mit multi-kamerasystemen für automatisierte fahrzeuge (2020)

61. Lowry, S., Sünderhauf, N., Newman, P., Leonard, J.J., Cox, D., Corke, P., Milford, M.J.: Visual place recognition: a survey. IEEE Trans. Robot. 32(1), 1–19 (2015)

62. Bentley, J.L.: Multidimensional binary search trees used for associative searching. Commun. ACM 18(9), 509–517 (1975)

63. Zitnick, C.L., Dollár, P.: Edge boxes: Locating object proposals from edges. In: European Conference on Computer Vision, pp. 391–405. Springer (2014)

64. Sünderhauf, N., Shirazi, S., Jacobson, A., Dayoub, F., Pepperell, E., Upcroft, B., Milford, M.: Place recognition with convnet landmarks: Viewpoint-robust, condition-robust, training-free. Robotics: Science and Systems XI, pp. 1–10 (2015)

65. Särkkä, S.: Bayesian Filtering and Smoothing. Number 3. Cambridge University Press (2013)

66. Bailey, T.,Nieto, J., Guivant, J., Stevens, M., Nebot, E.: Consistency of the ekf-slam algorithm. In: 2006 IEEE/RSJ International Conference on Intelligent Robots and Systems, pp. 3562–3568. IEEE (2006)

67. Montemerlo, M., Thrun, S.: Fastslam 1.0. FastSLAM: a scalable method for the simultaneous localization and mapping problem in robotics, pp. 27–62 (2007)

68. Kalman, R.E.: A new approach to linear filtering and prediction problems. Trans. ASME J. Basic Eng. 82(Series D), 35–45 (1960)

69. Odelson, B.J., Rajamani, M.R., Rawlings, J.B.: A new autocovariance least-squares method for estimating noise covariances. Automatica 42(2), 303–308 (2006)

70. Thrun, S., Burgard, W., Fox, D.: Probabilistic Robotics (Intelligent Robotics and Autonomous Agents). The MIT Press (2005)

71. Stannartz, N., Theers, M., Llarena, A., Kuhn, M., Kind, O.M., Bertram, T, et al.: Efficient localization on highways employing public hd maps and series-production sensors. In: 21. Internationales Stuttgarter Symposium, pp. 395–409. Springer (2021)

72. Wan, E.A., Van Der Merwe, R.: The unscented kalman filter for nonlinear estimation. In: Proceedings of the IEEE 2000 Adaptive Systems for Signal Processing, Communications, and Control Symposium (Cat. No. 00EX373), pp. 153–158. IEEE (2000)

73. Thrun, S., Montemerlo, M., Koller, D., Wegbreit, B., Nieto, J., Nebot, E.: Fastslam: an efficient solution to the simultaneous localization and mapping problem with unknown data association. J. Mach. Learn. Res. 4(3), 380–407 (2004)

74. Grisetti, G., Kümmerle, R., Stachniss, C., Burgard, W.: A tutorial on graph-based slam. IEEE Intell. Transp. Syst. Mag. 2(4), 31–43 (2010)

75. Agarwal, S., Snavely, N., Seitz, S.M., Szeliski, R.: Bundle adjustment in the large. In: European Conference on Computer Vision, pp. 29–42. Springer (2010)

76. Grisetti, G., Kümmerle, R., Strasdat, H., Konolige, K.: g2o: a general framework for (hyper) graph optimization. In: Proceedings of the IEEE International Conference on Robotics and Automation (ICRA), Shanghai, China, pp. 9–13 (2011)

77. Hoeting, J.A., Madigan, D., Raftery, A.E., Volinsky, C.T.: Bayesian model averaging: a tutorial. Stat. Sci., 382–401 (1999)

78. Duff, I.S.: A survey of sparse matrix research. Proc. IEEE 65(4), 500–535 (1977)

79. Chen, Y., Davis, T.A., Hager, W.W., Rajamanickam, S.: Algorithm 887: Cholmod, supernodal sparse cholesky factorization and update/downdate. ACM Trans. Mathem. Software (TOMS) 35(3), 1–14 (2008)

80. Agarwal, S., Mierle, K. et al.: Ceres solver. http://ceres-solver.org

81. Schneider, T., Dymczyk, M., Fehr, M., Egger, K., Lynen, S., Gilitschenski, I., Siegwart, R.: maplab: an open framework for research in visual-inertial mapping and localization. IEEE Robot. Autom. Lett. 3(3), 1418–1425 (2018)

82. Ni, K., Steedly, D., Dellaert, F.: Tectonic sam: Exact, out-of-core, submap-based slam. In: Proceedings 2007 IEEE International Conference on Robotics and Automation, pp. 1678–1685. IEEE (2007)

83. Mühlfellner, P., Bürki, M., Bosse, M., Derendarz, W., Philippsen, R., Furgale, P.: Summary maps for lifelong visual localization. J. Field Robot. 33(5), 561–590 (2016)

84. Diebel, J.: Representing attitude: Euler angles, unit quaternions, and rotation vectors. Matrix 58(15–16), 1–35 (2006)

85. Sattler, T., Maddern, W., Toft, C., Torii, A., Hammarstrand, L., Stenborg, E., Safari, D., Okutomi, M., Pollefeys, M., Sivic, J. et al.: Benchmarking 6dof outdoor visual localization in changing conditions. In: Proceedings of the IEEE Conference on Computer Vision and Pattern Recognition, pp. 8601–8610 (2018)

86. Reid, T.G.R., Houts, S.E., Cammarata, R., Mills, G., Agarwal, S., Vora, A., Pandey, G.: Localization requirements for autonomous vehicles. arXiv preprint arXiv:1906.01061 (2019)

87. Poggenhans, F., Pauls, J.-H., Janosovits, J., Orf, S., Naumann, M., Kuhnt, F., Mayr, M.: Lanelet2: A high-definition map framework for the future of automated driving. In: 2018 21st International Conference on Intelligent Transportation Systems (ITSC), pp. 1672–1679. IEEE (2018)

88. Li, F., Bonnifait, P., Ibañez-Guzmán, J.: Map-aided dead-reckoning with lane-level maps and integrity monitoring. IEEE Trans. Intell. Veh. 3(1), 81–91 (2018)

89. Al Hage, J., Xu, P., Bonnifait, P.: High integrity localization with multi-lane camera measurements. In: 2019 IEEE Intelligent Vehicles Symposium (IV), pp. 1232–1238. IEEE (2019)

21

Open Access This chapter is licensed under the terms of the Creative Commons Attribution-NonCommercial-NoDerivatives 4.0 International License (▶ http://creativecommons.org/licenses/by-nc-nd/4.0/), which permits any noncommercial use, sharing, distribution and reproduction in any medium or format, as long as you give appropriate credit to the original author(s) and the source, provide a link to the Creative Commons license and indicate if you modified the licensed material. You do not have permission under this license to share adapted material derived from this chapter or parts of it.

The images or other third party material in this chapter are included in the chapter's Creative Commons license, unless indicated otherwise in a credit line to the material. If material is not included in the chapter's Creative Commons license and your intended use is not permitted by statutory regulation or exceeds the permitted use, you will need to obtain permission directly from the copyright holder.

Digital Infrastructure

Digital Maps in Navigation Data Standard Format

Ralph Behrens, Roland Homeier, and Katharina Jülge

Contents

© The Author(s) 2026
H. Winner et al. (eds.), *Handbook Assisted and Automated Driving*,
https://doi.org/10.1007/978-3-658-45276-6_22

Digital map data for navigation and driver assistance systems are not generated automatically during data acquisition. There are some companies that specialize in capturing geodata (e.g., roads, intersections and their geometries and properties) in this environment; for example, HERE, TomTom and NavInfo should be mentioned. In addition to commercial companies, there are official institutions that collect geodata, including the cadastral offices in Germany. Finally, there is freely available geospatial data collected by countless volunteers; The OpenStreetMap project is of particular importance in this area.

The data from the geodata acquisition are not used directly in the navigation and driver assistance systems. Before use, the data is compiled; meaning it is converted into a format so that the systems can process the digital map data efficiently. This includes data reduction, but also enriching the geodata with additional information from sensor data such as radar objects for localization. In the past, each navigation manufacturer designed its own data format, and the compiled navigation data was in fact not reused.

The standardization of map data for navigation systems has been an important topic in the automotive industry for many years, especially since the costs for creating navigable data are very high. As early as 2004, a group of companies formed an interest group to promote standardization of map data. This initiative resulted in the Navigation Data Standard (NDS) e.V. in 2009. NDS e.V. is a registered association with the aim of standardizing map data for navigation systems. The AutoDrive extension for automated driving has been in the works since 2014, and the first product release was released in 2018.

As part of the standardization, the requirements, definitions, and official versions of the NDS map format are specified. The licenses are available to association members free of charge [1].

The members of the association consist of vehicle manufacturers, suppliers to the automotive industry, map data suppliers and providers of telematics services. As of the end of 2020, 42 companies are listed as members [2]. The companies in the consortium come from Europe, America and Asia. The distribution shows that NDS has developed into a worldwide standardization.

22.1 Objectives of Standardization

The requirements for the standard arose almost automatically as a result of the increasing international demand for vehicles that are already equipped with a navigation system during series development. Initially, numerous navigation solutions were created locally, but the development and integration of regional solutions is complex and expensive.

In terms of economy and the expectation that the navigation data should always be as up to date as possible, the following objectives of the NDS standardization follow: [3].

1. Development of a globally valid map format
   Over the years, NDS has developed into a universal standard for navigable map data and has replaced the proprietary map formats of many manufacturers of navigation solutions. The aim is to transfer new requirements for navigation and driver assistance systems into map standardization.

2. Separation of application and data
   Many manufacturers of navigation programs also install a new version of the software when they update the navigation data. The aim of the NDS map format is a clear separation so that the navigation data can be updated independently of the navigation and driver assistance system software.

3. Data compatibility and interoperability
   With this standard, it should be possible to exchange the map data across the series of a vehicle and to keep the components of the digital map identical across vehicle manufacturers, with the NDS data map formats being designed to be downward compatible. The purpose of this is that a vehicle manufacturer only needs one NDS data set, which is supported by several different navigation programs. This simplifies logistics and makes it easier for users and garages to use the various navigation systems.

4. Defining a procedure for updating the data
   More and more systems, especially the driver assistance functions, require up-to-date map data for their function, which is why updating them is one of the consortium's goals.

5. Data compactness and application efficiency
   Navigation and map-based driver assistance systems that are ordered as original equipment and supplied with a vehicle purchase must ensure a service life of more than 10 years. This means that the systems not only have to be very robust, but they also have to work with changed data content and volumes. The aim of the standardization is to design the data format in such a way that the application of the systems continues to function with an acceptable performance even after the data has been updated several times.

6. Independence of the data format for the storage and transmission medium
   This means that the format of the map data must also be usable with different media. It must be possible to distribute the data not only on storage media, but also via a mobile broadband connection.

7. Extensibility of navigation data
   The map format must allow for the extension and integration of proprietary content. Every vehicle

manufacturer wants their vehicles to sell particularly well, which is achieved, among other things, through special functions. This goal shall to ensure that a manufacturer-specific extension of the data is possible without violating the goal of compatibility and interoperability.

8. Support of copy protection measures and protection against manipulation
   The map format must protect the data in terms of piracy and misuse, such as tampering. For this purpose, the Digital Rights Management (DRM) standard and integrity checks must be supported

22.2 Characteristics of the NDS-Standard

A special feature of the NDS map format is the organization of the data in so-called building blocks. A building block has uniform interfaces. Each building block can be replaced individually, with the possibility of varying the color and size of the building block. Translated, this means that NDS creates the possibility of combining functional module components for different driver assistance functions (e.g., functions for lane localization). It is also possible to add further new module blocks for future new functions or to store customer-specific adaptations of the data. Furthermore, the contents of the building blocks can be updated incrementally with changes in the real world.

The data of the individual building blocks are linked to one another via certain properties, which must follow a uniform format across all building blocks. The most important uniform properties are coordinates and names. For example, all the different building blocks of the NDS format support coordinates for georeferencing: therefore, all parts of the database can be searched for information in one place.

The data is preferably stored in a database based on SQLite. The database creates an efficient basis for incremental updates and searching in the various modules based on uniform parameters.

Right from the start of the standardization activities, attention was paid to extensibility for new functions. Particularly noteworthy are the new Building Blocks Detailed Lane Model, Landmarks and Obstacles for automated driving functions in addition to a special ADAS Building Block with a flexible approach that was already available at an early stage to optimally support driver assistance systems with map information.

These new building blocks for automated driving functions have very precise geometries or curve data, which means that large amounts of data are generated, which are only loaded along the route due to the need for up to data map data and are no longer kept completely on a data carrier in the car for a certain geographical area.

22.3 Data Structure Within an NDS Database

The data in the NDS database is initially divided into update regions (see ◉ ◘ Fig. 22.1). Within the regions, the data is organized into components, which in turn contain the building blocks; the building blocks in turn contain the smallest unit, so-called tiles.

This structure can be illustrated using the following example: An NDS database contains the data for a specific market (database coverage), e.g., B. Europe. This market is divided into update regions, e.g., France and Monaco, Spain and Portugal, Benelux, Germany etc. These regions can then be updated independently. The exact cut of the regions is not fixed, it can be reconfigured for each project.

22.4 NDS Building Blocks

NDS databases are organized into building blocks (see ◉ ◘ Fig. 22.2) [4]. This subchapter presents the most important components and their task for the driver assistance functions, whereby products can be assembled

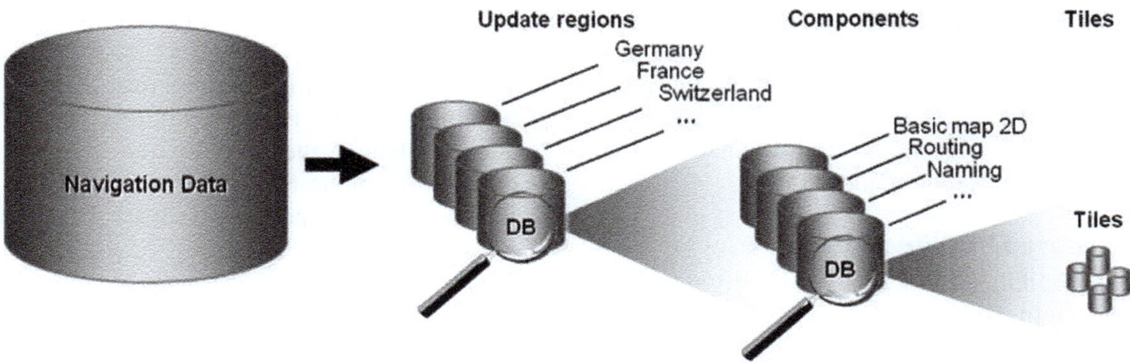

◘ **Fig. 22.1** Definition of Update-Regions (courtesy of NDS Association)

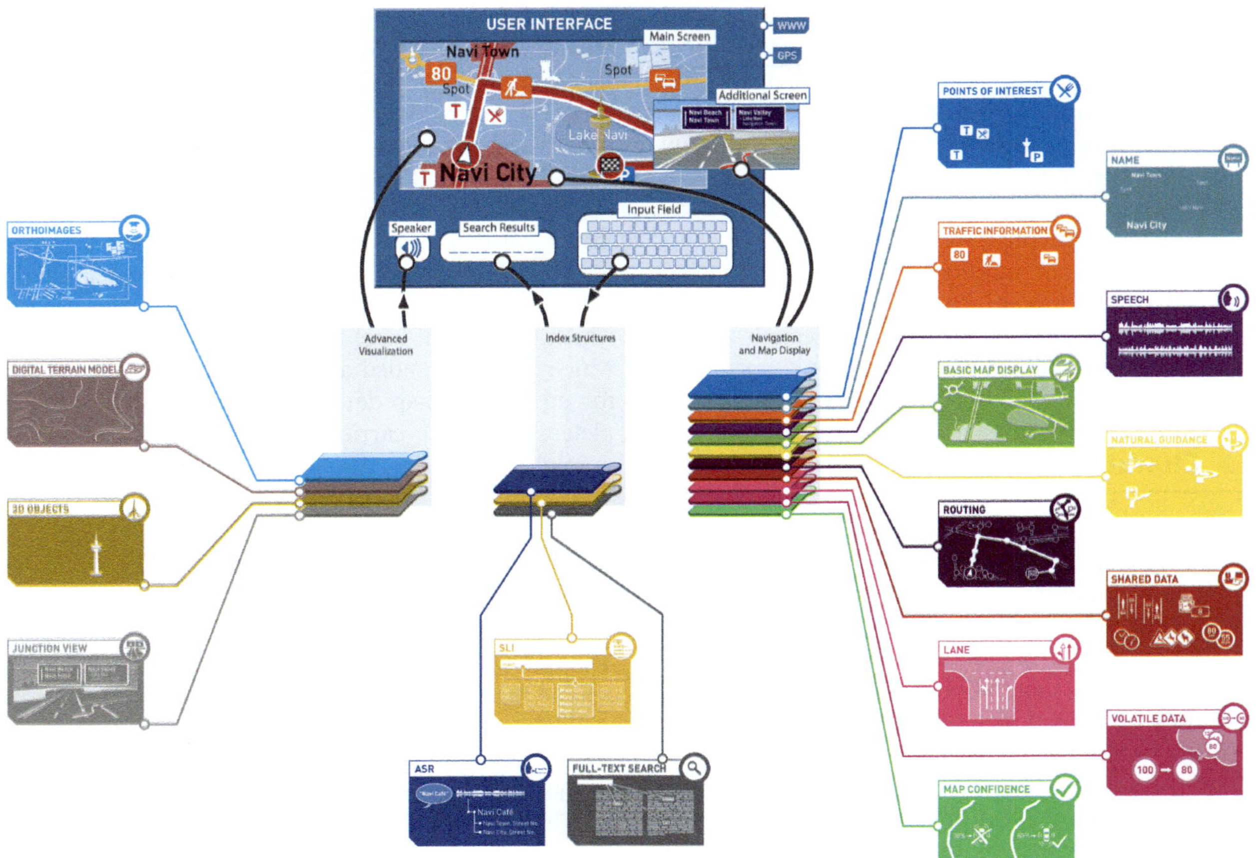

Fig. 22.2 NDS building block overview (courtesy of NDS Association)

from individual building blocks and divided into three groups as follows:

1. Navigation and map display for route calculation, route guidance and their visual representation
2. Index Structures for destination entry
3. Advanced visualization for realistic map representation, e.g., with building models

22.4.1 Navigation and Map Display

The various navigation and driver assistance function applications require different components of the map data, for example, the building blocks (see ◉ Fig. 22.3) used for localization in automated driving, such as shared, routing, lane, landmarks, obstacles, volatile data and map confidence are used without the building blocks for map display and destination input, whereas today's navigation requires them.

22.4.1.1 Shared Building Block

The shared building block is used to store variable, global metadata and regionally specific information that must be available as a constant component for the building blocks and their interpretation. As an ex-

ample, there are the region definitions with ISO identification for their state and federal state, the information whether there is left-hand or right-hand traffic, and the definition of the time zones valid in these regions.

22.4.1.2 Routing Building Block

The Routing Building Block contains the road network and is subdivided into different areas of application, which are stored as layers (e.g. Routing, Guidance, ADAS, Names, Volatile Locations (see Volatile Building Block ▸ Sect. 1.4.1.7) and Truck). It serves as a basis for various applications:

- Route Calculation
  The route calculation reads the road network topology and attributes from the Routing layer to calculate a route with the desired parameters using various cost functions. Special properties for trucks can be provided in a separate layer truck if needed.
- Route Guidance
  The route guidance accesses the topology and geometry of the road network in order to compare the current position with the road network and derive instructions for driving manoeuvres

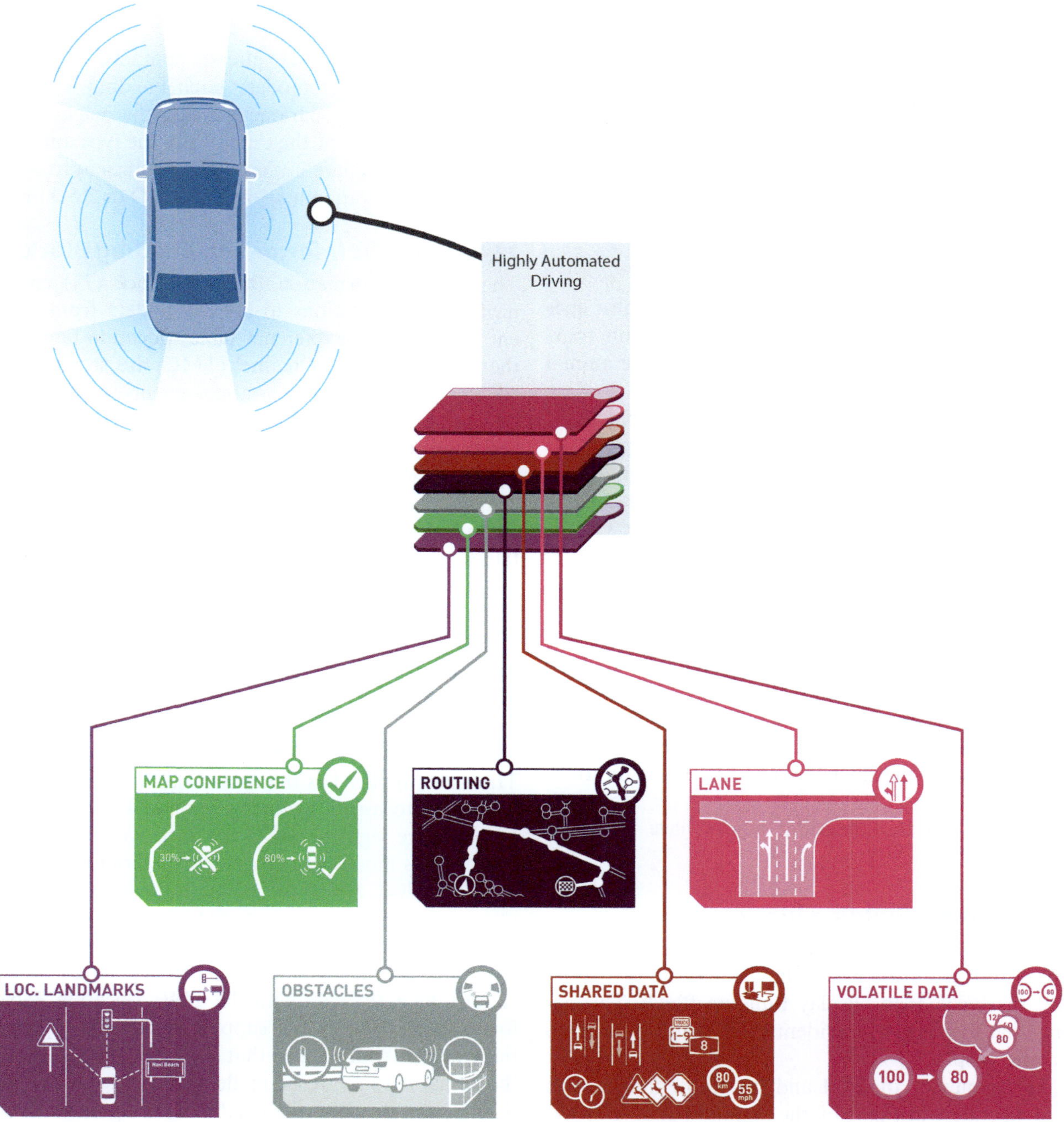

■ **Fig. 22.3** Building blocks for highly automated driving in NDS and their data

— **Map Matching**
  The map matching uses the road geometry from the Routing and Guidance layers to map the dead reckoning position to the road network.

— **ADAS**
  An ADAS layer was added to the routing building block specifically to support driver assistance systems, e.g. to record curve data, road width, etc.

22.4.1.3 POI Building Block

Points of Interest (POI) are distinctive or user-interesting points that are displayed on the navigation map and that must be navigable [5]. Examples would be: Eiffel Tower, train stations, hospitals, gas stations, etc. In addition to the location information, the POI can provide additional information, e.g., telephone numbers or opening hours.

The POI can be presented and used in different ways:
- Display of POI using an icon on the navigation map,
- Display of POI in list form, sorted by category, location or other rules,
- Display of additional information after selecting a POI: This can be supplemented with images, texts, payment information, etc. in the building block according to the needs of the application.

In addition to the point-based POIs, there are also lines or area-related POIs: These can e.g. B. certain tourist roads; an area-related POI could be a large animal park.

Regarding the POI, there are two building blocks:
- Integrated POI Building Block
  The integrated POI Building Block may contain direct references to other NDS building blocks. These references target route segments or intersections in the Routing Building Block, or they target names in the Naming Building Block. The direct links serve to improve search performance.
- Non-integrated POI Building Block
  The non-integrated POI Building Block is linked to the other building blocks exclusively via geo-coordinates. This means that it is only possible to search with geo-coordinates and determine whether POI information belongs to a specific map section or not.

Internally, both POI building blocks have the same data structure.

22.4.1.4 Name Building Block

The Name Building Block contains the street network names from the Routing Building Block and the names from the Basic Map Display Building Block. This ensures that names are used identically in a route list and in a map display.

The names of the POI and the traffic information are deliberately not part of the naming building block. This is due to the need to update this specific data more frequently. The corresponding names are therefore part of the POI or traffic building blocks.

22.4.1.5 Basic Map Display Building Block

The Basic Map Display Building Block (BMD) groups the essential data for a map display. The Basic Map Display Building Block obtains the names for the display from the Naming Building Block.

The contents of the BMD are points (including city centers, mountain peaks), lines (including roads, rivers, border lines, etc.), areas (including forests, bodies of water, building footprints, land use), icons and drawing styles and rules (including information on which points streets etc. can be labeled).

In addition, the BMD already contains the data for displaying 2.5D city models. Each building is characterized by different parameters, e.g., the height of the building, the color of the outer walls, the type and color of the roof. This data in the BMD is sufficient to ensure a complete navigation map display in 2D or 2.5D mode.

22.4.1.6 Traffic Information Building Block

The Traffic Information Building Block (TI) enables navigation applications to use traffic data from different standards. The reference tables (location tables) for the Traffic Message Channel (TMC) are stored in the TI; this allows the traffic announcements to be assigned to various languages. The reference tables are defined according to region, for example by the TMC forum. With regard to the traffic messages according to the Transport Protocol Expert Group (TPEG), the Traffic Information Building Block provides the reference tables so that the event information (e.g., congestion, complete blockage, disruption, etc.) of a TPEG message can be evaluated. The event information is transmitted as a number. The system can use this number to load the corresponding text in the respective language from the reference table.

22.4.1.7 Volatile Building Block

The data supplier has the option in the Routing Building Block to assign a unique Volatile Location ID to one or more segments of the road network in order to refer to this mostly long road section from the Volatile Building Block and then be able to continue to clearly reference these segments in the event of changes to the map. These IDs only change in the event of major structural changes and are not determined by organizations, as is the case with the TMC Location Tables, for example, but by the supplier. If dynamic properties (variable traffic signs) that are more up to date than the actual map are to be made available to the system, they can be stored in the volatile building block as different messages and transmitted to the system, with reference being made to previously define volatile location IDs. In order to also address sections of a volatile location, the messages are provided with start and end validity information. The messages can come from different suppliers, provided that an alignment has taken place about the assignment of the volatile location ID.

22.4.1.8 Lane Building Block

This new building block contains a lane model, which enables the creation of a localization and planning map for automated driving in the car for its environment. For this purpose, the 3D geometries of lanes, i.e., center lines, lane markings with their type, color, material, and lane boundaries (e.g., curbs) are stored. In addition to these, special lane-specific properties (connectivity, merging / dividing lanes) and original properties from the Routing Building Block can be added to the lanes. The individual markings are stored in sections for the lanes, and this is done in a serial, parallel order. To optimize storage space and ensure consistency, the inner lane markings of two adjacent lanes can only be stored once on the map. Due to the more detailed lane geometries, properties with a higher position accuracy can be achieved compared to the storage in the routing building block, since the properties are referenced to the center line geometry of the lane and not to the generalized road geometries. The adjacent lanes are combined into a lane group and assigned to one or more segments of the road network from the routing building block. Thus, there is a relationship to the properties of the road, which does not have to be stored more than once.

22.4.1.9 Landmark Building Block

In addition to lane markings, other localization objects with their 3D geometries next to the road are also used for localization by the vehicle's sensors. These are divided into signs, posts, walls, traffic lights, various markings on the road, and boundaries (e.g., crash barriers, jersey barriers). Landmarks are assigned to road network segments from the Routing Building Block where relevant.

22.4.1.10 Obstacle Building Block

Obstacles lie along the road and are stored as 2D polylines with a height indicator and are determined, for example, from point clouds. Obstacles are assigned to the segments of the road network from the Routing Building Block, where they are important for locating them.

22.4.1.11 Map Confidence Building Block

Confidence in the accuracy of the map data for individual properties is stored in this building block. Because different driver assistance systems require different levels of accuracy of map properties (e.g., absolute and relative position accuracy, positive / negative failure rate of occurrence), so that the function works reliably and safely in the system.

22.4.1.12 Speech Building Block

For navigation systems, there is also a Speech Building Block for the voice output of names via phonetic transcriptions or recorded speech fragments, as well as Natural Guidance for easier orientation when routing using noticeable objects (e.g., monuments, fountains).

22.4.2 Index Structures

Different index structures are available in different building blocks for the different use cases of destination input and their storage.

22.4.2.1 SQLite Location Input (SLI) Building Block

The SLI building block is used for destination input; this is now equipped with a variety of options. In addition to the classic hierarchical destination input, in which letter after letter can be entered, the SLI supports destination input via a speller using Next Valid Character (NVC) and First Letter Input (FLI), with the FLI being a mode specially designed for the Asian market; in FLI mode only the first letter of a syllable needs to be entered. Another entry mode is called one-shot destination entry. Similar to a search in an Internet-based search engine, this function allows the entire address to be entered at once before the search function is started. The advantage of the SLI-based destination input is that the search is not only based on one criterion—such as city, street, postal code or house number—but that the search is carried out simultaneously using several criteria.

22.4.2.2 Full Text Search (FTS) Building Block

To implement a free text search, the content must be indexed beforehand when creating the FTS: The POI building block, the naming building block and all other building blocks that contain names serve as source material for creating this index. An index is built with this information. This index is then filtered with the search text. The result of the search is then a list with links to the files that contain one or more of the words from the search text and point directly to the source data. With the help of this information, an input function can be implemented that is comparable to meta searches on well-known Internet platforms.

22.4.2.3 Automatic Speech Recognition Build Block

In addition to the following building blocks, the NDS Automatic Speech Recognition (ASR) building block can also be used for speech recognition, or a more proprietary data format from the speech recognizer manufacturer can be used in order to be able to use special manufacturer properties.

22.4.3 Advanced Visualisation

The advanced visualization summarizes the extended content in the building blocks described below for the Advanced Map Display (AMD) in order to be able to display the map as realistically as possible.

22.4.3.1 Digital Terrain Model Building Block

The building block stores the elevation model or the topography of the earth's surface (Digital Terrain Model—DTM). In addition to the height information of each area part, this part also contains the textures for displaying the earth's surface in case satellite or aerial photographs are not available. The polygon mesh (Batched Dynamic Adaptive Meshes—BDAM) is also stored in this part so that the representation of the earth's surface can be displayed without discontinuous transitions.

22.4.3.2 Orthoimages Building Block

The Orthoimages Building Block contains satellite or aerial photographs of the earth's surface. It must be possible to select the satellite or aerial photographs in relation to the viewing position over the map. Depending on the scale and viewing angle, the photographs are selected for display and used as a texture file for the BDAM.

22.4.3.3 3D Objects Building Block

The 3D Objects Building Block is used to store the 3D objects that are required to display the digital map as realistically as possible. A single 3D object is described by geometry information, a material description, textures, and names. Some of the 3D objects are shown in great detail, these are the 3D landmarks. These designate very well-known and very distinctive elements, e.g., the Eiffel Tower in Paris or the Hamburger Michel.

22.4.3.4 Junction View Building Block

For the representation of complex intersection situations, motorway exits or for example, display of roundabouts, images from the Junction View Building Block can be used. It is possible to generate the images from the road data at runtime, or pre-built images are used. The pre-built images are stored in the Junction View Building Block.

22.5 NDS-Database Structure/ Generalization

22.5.1 Database Structure and Partitioning

Some building blocks contain very finely granular data. It is therefore necessary to group the data and store it in a partitioned way in the database. NDS groups the data into building blocks, each of which is stored in its own SQLite database. Not all map data is always required for different functional use cases, so that some building blocks are furthermore divided into layers (e.g. truck, guidance, routing for route calculation) within

a database. Furthermore, the world surface is divided into tiles in which the map data is stored individually.

22.5.2 Generalization

A map that is supposed to show the whole of Germany on a display in a car or on a mobile device, for example, cannot do anything with street geometries of the lowest street categories from residential areas. In an overview display, often only roads of a higher class, e.g., motorways are loaded; so that the loading process and the display take place efficiently. This type of thinning is called generalization and is implemented in NDS using levels. It is comparable to the scales of maps. As a result, there are separate tiles for the different levels and the number of tiles decreases the more general the level is, which in turn expands the area of the tile.

22.6 Structure of the NDS Database

The NDS database preferably uses a custom SQLite engine to process the data. For example, the NDS committee has called for and implemented the addition of the "multiplexing" function to the SQLite engine. With this function it is possible to send several requests from different applications for parallel map processing to the database at the same time. The database itself consists of objects, attributes, and metadata. The structure of the data is described using DataScript files. A DataScript compiler translates these files into the appropriate programming language for the respective target system, such as C + + or Java. The application developer then only needs to know the generated access classes to access the data; the application developer cannot see the direct SQLite commands behind the access classes.

- Objects (features) in the NDS database: All real existing objects are represented in the NDS database using one or more features. A road segment between two intersections is represented as a link feature in the Routing building block.
- Attributes in the NDS database: The NDS database distinguishes between fixed and flexible attributes. The attributes describe the specific properties of the NDS features. For example, a fixed attribute of route section is the road type. This attribute is always part of a street segment. Optional information about a street segment is, for example, on which days of the week a street is open or passable. This information is called a flexible attribute of a road segment.
- Metadata in the NDS databases: The metadata contains all the necessary information to describe variable content and database properties. They relate to a

specific part of the NDS database, a building block, or to the entire database. For example, the static metadata indicates whether length data is stored metrically or not. Furthermore, the ISO country code is stored in the static metadata.

— If content consists of nested objects with lists and attributes (e.g., routing, lane), these are stored in the database itself as Binary Large Objects (BLOB) within database tables. For example, the Routing Building Block has multiple layers, so for each layer, a BLOB per tile identifier is stored in a single database table. Other map content (e.g., metadata, FTS), on the other hand, can be stored more easily directly in database tables and queried using SQLite statements.

The objects, attributes and metadata can be searched in the database. These values are stored in plain text in the database. The situation is different with the detailed map data, which is stored in binary form as a BLOB. Before the individual data can be used, a BLOB must be read from the database and interpreted. The BLOB structure was introduced so that the binary size of an NDS database corresponds to a size that is still manageable on an embedded device.

22.6.1 DataScript and RDS

The entire format of the NDS database is described using the formal description language DataScript. As briefly outlined in the last paragraph, this is divided into two parts: On the one hand there is the formal description of the database, on the other hand there is the language binding, i.e,. the representation of the data interface in a dedicated programming language. DataScript was originally developed by Godmar Back [6]. Binary formats, bit streams or file formats can be described with the help of DataScript. This enables the data formats required in NDS to be clearly described, and there is a reference implementation of the DataScript compiler for the Java programming language. This led to DataScript being selected in the NDS consortium after evaluating various database description methods. A dialect of DataScript emerged within the NDS consortium, which is referred to as Relational Data Script (RDS) [7].

22.6.2 NDS Format Extension

The NDS data format enables the user companies to make extensions with the aim of later standardization, adjustments and proprietary extensions. This freedom is necessary so that each NDS development can offer additional objects. An example of this would be the proprietary supplementation of ski areas with parking lots as a navigation destination and the representation of the ski lifts and slopes in the navigation map. Not everyone needs this feature, but some users may find it a useful addition. The companies that decide to adapt or extend the NDS objects must observe a strict set of rules: the NDS database must in any case fully support the standard, every modification must support the standard-compliant mechanism for updating the database and each modified NDS database must always fully meet the requirements for interoperability with pure NDS standard databases.

22.6.3 NDS Database Tools

The NDS standard supports development with a number of proprietary tools for validating the database, examining the content of an NDS database and checking the NDS format used. The developer has a validation suite, a database inspector with a map viewer, the RDS compiler, an adapted and optimized SQLite engine and various drivers at their disposal.

22.7 Map Learning

Keeping maps up-to-date is a complex and expensive process. In the past, dedicated measuring vehicles with high-quality sensor equipment were mainly used. It was correspondingly expensive to collect data in this way, and the update frequency was correspondingly low. In the meantime, crowdsourcing methods are increasingly being used, and the required data is taken from the vehicle sensors that are already installed. The lower accuracy of the measurement is compensated by the aggregation of many measurement results using machine learning methods.

22.7.1 SENSORIS as Transmission Protocol

The SENSORIS protocol (Sensor Interface Specification, [9]) creates an interoperable standard for the transmission of sensor data. This makes it possible to transmit data from vehicles from different manufacturers, which in turn use sensors and control units from different suppliers, to backends in a uniform way. The standard is managed by ERTICO [10], members are map suppliers, system and infrastructure manufacturers as well as vehicle manufacturers.

The protocol defines the request-response logic between client and server and the data types for the transmitted sensor information. Both client and server can control the data transfer. SENSORIS is designed to support vehicle-to-backend, vehicle-to-vehicle and backend-to-backend transmissions, however, the vehicle-to-backend transmission use case will be the most common. Google Protocol Buffers are used for encoding and decoding [11].

22.7.2 Aggregation in the Sensor Versus Aggregation in the Backend

When learning maps from sensor data, it is in principle conceivable to send completely unprocessed raw data from the respective sensor to the backend and to carry out all processing steps for learning (application of an appropriate algorithm on data from a single sensor to extract a map feature, cleaning of invalid data, application of an aggregation algorithm) on the backend. The advantage here is that less powerful hardware is required in the vehicle. Upgrading the backend is also usually possible without any problems, while the hardware in the car remains unchanged over the entire vehicle life cycle and software updates at least generate a logistical effort. Algorithms can also be more easily optimized and validated based on the complete sensor information available on the backend.

At the other end of the scale is aggregation in the sensor. Here, a relevant map feature is extracted from the raw data obtained, which can be assigned to a measurement in a single vehicle. Only this map feature is transmitted to the backend. The advantage here lies in the significantly lower bandwidth that is required compared to raw data transmission. The further processing steps are then carried out on the backend and the data of many measurements are aggregated. The disadvantage is that the raw data is not available.

In practice, the division of functions between vehicle and backend requires careful consideration.

22.7.3 Example Attribute (Slopes)

The slope can be mapped as a linear interpolated slope to geometry points on the road in the NDS. When the vehicle is driving, the vehicle sensor system determines the gradient values with their position for a defined resolution. This data is transmitted to the map data backend via the Telematic Control Unit (TCU) in the SENSORIS transmission protocol and aggregated there in order to make it available to the applications as an NDS map in accordance with customer requirements (resolution, accuracy, etc.).

22.8 Future of the NDS-Standard

In recent years, it has become increasingly important that digital map data is up-to-date and quickly reaches the vehicle from the cloud via its mobile data connectivity. This affects the required NDS data format. The application no longer has to rely on optical data carriers and systems with low performance and memory. This is currently leading to the development of NDS. Live with the approach of bringing the data as services via an API for navigation and driving functions from a backend into the car of tomorrow, taking into account aspects of functional safety and faster map updates [8].

References

1. Müller, T.M.: Navigation Data Standard (NDS): bald industriestandard? Automobil-Elektronik **6**, 30 (2010)
2. NDS-Association: NDS Association Members. ► https://nds-association.org/map-data-structure. Link visited: 29. Dezember 2020
3. NDS-Association: Objectives ► http://www.nds-association.org/the-nds-standard. Link visited: 16. April 2014
4. NDS-Association.: Navigation Data Standard Format Specification NDS Version 2.5.4, S 59 ff (2018)
5. Wikipedia: Point of Interest ► http://de.wikipedia.org/wiki/Point_of_Interest. Link visited: 16. April 2014
6. Back, G.: DataScript (2003). ► http://datascript.sourceforge.net. Link visited: 16. April 2014
7. Wellmann, H.: DataScript Tools (2013). ► http://sourceforge.net/projects/dstools/. Link visited: 18. April 2014
8. NDS-Association: NDS.Live: An NDS update for the future of navigation. ► https://nds-association.org/nds-update. Link visited: 10. Januar 2021
9. SENSORIS: Sensor Interface Specification ► https://sensoris.org/. Link visited: 14. Januar 2021.
10. ERTICO: European Road Transport Telematics Implementation Coordination Organisation-Intelligent Transport Systems & Services Europe ► https://ertico.com/. Link visited: 14. Januar 2021
11. Google Protocol Buffers ► https://developers.google.com/protocol-buffers. Link visited: 14. Januar 2021

Open Access This chapter is licensed under the terms of the Creative Commons Attribution-NonCommercial-NoDerivatives 4.0 International License (▶ http://creativecommons.org/licenses/by-nc-nd/4.0/), which permits any noncommercial use, sharing, distribution and reproduction in any medium or format, as long as you give appropriate credit to the original author(s) and the source, provide a link to the Creative Commons license and indicate if you modified the licensed material. You do not have permission under this license to share adapted material derived from this chapter or parts of it.

The images or other third party material in this chapter are included in the chapter's Creative Commons license, unless indicated otherwise in a credit line to the material. If material is not included in the chapter's Creative Commons license and your intended use is not permitted by statutory regulation or exceeds the permitted use, you will need to obtain permission directly from the copyright holder.

Vehicle-2-X

Frank Hofmann, Hendrik Fuchs, Ignacio Llatser, Christian Zimmermann and Kurt Eckert

Contents

© The Author(s) 2026
H. Winner et al. (eds.), *Handbook Assisted and Automated Driving*,
https://doi.org/10.1007/978-3-658-45276-6_23

23.1 Basic Principles and Motivation

Connectivity among vehicles and between vehicles and the infrastructure and other road users (pedestrians, cyclists, motorcycle riders) is becoming increasingly important. It forms the technological basis for future cooperative intelligent transportation systems (C-ITS). The ability of a vehicle to communicate with its immediate surroundings—that is to say, with other vehicles and road users and with the road infrastructure (e.g., traffic beacons, road sign gantries, and traffic lights) as well as with traffic control centers—makes it possible to provide many new or improved functions that lead to increased road safety, improved traffic efficiency, and greater personal comfort and convenience. In this context it has therefore become customary to use the term vehicle-to-X communication, where "X" refers to the respective communication partner. Generally, the terms "vehicle-to-X" and "vehicle-2-X" (V2X) or "car-to-X" and "car-2-X" (C2X) are used. V2X will be used throughout this chapter. The terms V2V (vehicle), V2I (infrastructure), V2P (pedestrian), and V2N (network) have also been introduced for referring to specific communication partners.

The topic of V2X has been investigated in various German, European, and international R&D projects already since the late 1990s. In the first phase, they centered on the development and testing of the underlying technologies. The second phase included system testing using individual demonstrators in vehicles and the infrastructure. As a result, it was possible to demonstrate basic feasibility. In the third phase, proof of suitability for practical use and proof of interoperability were demonstrated in large-scale field trials in real traffic (e.g., sim^{TD} in Germany, DRIVE C2X in the EU, and Safety Pilot in the United States). The use cases initially considered were essentially limited to warning and information functions for improving road safety and traffic efficiency. In the next step, the research and development activities have focused very strongly on supporting automated driving, particularly in terms of cooperative maneuver coordination between the vehicles involved (e.g., in the R&D projects AutoNet2030 [1], IMAGinE [2], and TransAID [3]).

These activities are supported by the Car-2-Car Communication Consortium (C2C-CC)—which has set itself the goal of further improving safety and efficiency in road traffic through cooperative ITS systems [4]—as well as by the 5G Automotive Association (5GAA) [5].

Since the period of relevance for information shared via V2X is generally limited in terms of time and location, a Wi-Fi-based method of direct communication optimized for the specific requirements was initially considered. In order to achieve similar transmission performance also by means of mobile communications, the inclusion of a so-called geocast server was considered in a first step. A geocast server manages all of the logged-in recipients within a certain area and so can relay a received message relevant to this area per multicast to all relevant recipients. New generations of mobile communications now also incorporate various options for direct communication called Cellular V2X (C-V2X). The coordination processes for this are being handled by the 5GAA. Research and development results have been achieved in various projects like 5G NetMobil [6].

The inclusion of communication as a vehicle sensor makes it possible to extend the horizon of what is perceivable past the range of the driver's vision as well as past the range of the vehicle's on-board sensors (e.g., radar, lidar, and video). The field of view of autonomous on-board systems is curtailed by specific sensor properties, such as range limitations or a line-of-sight requirement. Compared with this, V2X-based driver assistance systems enable significantly more comprehensive coverage: in particular, they provide a view around bends and through obstacles, such as buildings, terrain, and other vehicles. Further improvement can be achieved with so-called collective perception. Here, the vehicles share information (e.g., the objects their own sensors have detected) with one another, so that an improved model of the surroundings can be generated through a combination of this data. This firstly enables the effectiveness of existing driver assistance systems to be improved considerably and secondly allows many completely new functions to be realized. A disadvantage of the technology, however, is the dependency—to a greater or lesser extent—on the equipment installed in the vehicles involved and the infrastructure components. The equipment penetration rate has a particularly strong influence on the functions that can potentially be realized, and this must therefore be factored in when considering rollout scenarios. Detailed portrayals of the topic of vehicular networks can be found, for example, in [7] and [8].

A basic prerequisite for V2X systems is the standardization of all necessary components, which will be covered in the following section. This will then be followed by a consideration of different classes of use cases that have been conceived for a phased rollout. The next section will explain the underlying system concept, followed by a description of the data transmission technologies and the data formats required for this as well as a consideration of the necessary security aspects. The current status of market rollout as well as an outlook on further developments round off this chapter.

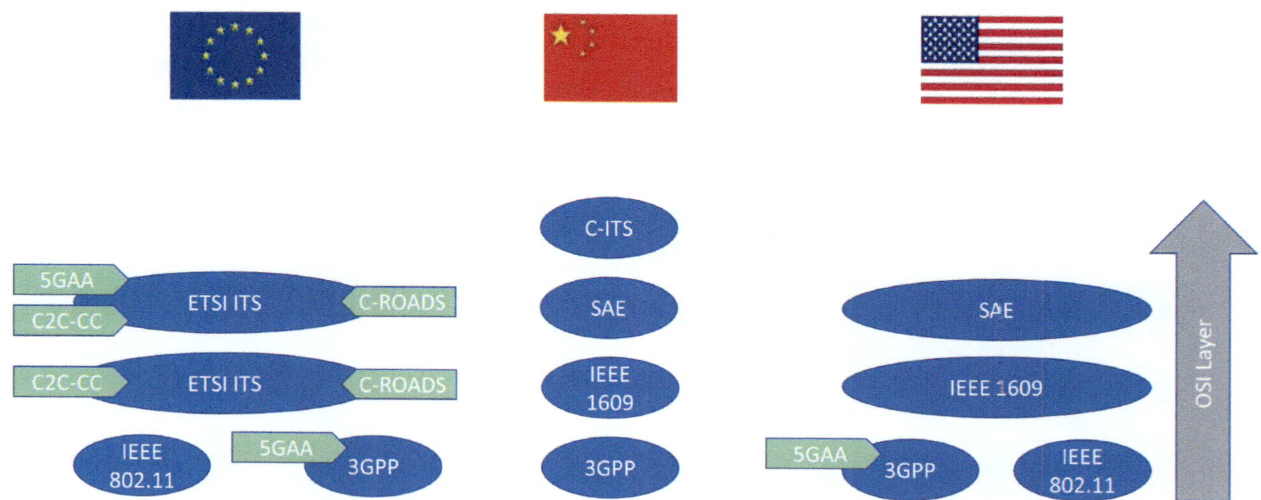

Fig. 23.1 Simplified view of the key standardization activities. (*Source* Bosch)

23.2 Standardization Bodies and Corresponding Committees

Many of the goals of V2X communication, such as increasing road safety and traffic efficiency and support for cooperative driving, make it essential to strive for a penetration rate that is as high as possible. To achieve this, it makes sense to ensure that interoperability between the various potential participants in the V2X ecosystem is as seamless as possible. Standardization on various levels of the OSI layer model is necessary for this. This standardization process is being spurred on in various bodies around the world. **Fig. 23.1** provides a simplified view of the key standardization activities.

Essentially two variants exist on the lower OSI layers. One variant is based on the Wi-Fi standard of the IEEE 802.11 family; the other variant is being pursued at the 3rd Generation Partnership Project (3GPP) on the basis of the mobile telecommunications technologies that are globally standardized there. On the application layer, the two variants then converge again to some extent, though they can possess different characteristics for different regions (e.g., EU, United States, and China). Different variants are therefore being standardized in different parts of the world. In Europe, the standardization process is mainly being handled at the European Telecommunications Standards Institute (ETSI) [9]. This is where adaptations of the existing IEEE 802.11p standard known as ETSI ITS-G5 as well as the definition of services on higher OSI layers all the way up to application protocols are being standardized. In the United States the lower layers are based on the 3GPP specification, named LTE-V2X, the higher layers are to some extent also standardized at the IEEE in the 1609 family, while the application protocols are specified at the Society of Automotive Engineers (SAE). In China, a local standardization body (C-ITS) is pushing forward a standard that on the lower layers is based on the 3GPP standard and on the higher layers follows DSRC though it also incorporates additional services.

Besides the standardization bodies themselves, there are also industry consortia that are primarily involved with the preparation of the actual standardization work. These include, as the most notable representatives, the Car-2-Car Communication Consortium (C2C-CC) [4], which first and foremost prepares the ETSI ITS-G5 standards, the 5G Automotive Association (5GAA) [5], which is involved in working on mobile telecommunication-based variants throughout the world, and C-Roads, where primarily adaptations for the use of V2X on the road infrastructure side are considered. Besides carrying out preliminary work on the standards, these bodies also specify so-called profiles. Profiles are used to define the parameters specified in the standards so that these parameters can be used by particular groups (e.g., carmakers or manufacturers of road infrastructure) as a basis for the products they develop. The C2C-CC has, for example, specified the so-called Basic System Profile [10] for this purpose.

23.3 V2X Use Cases

23.3.1 V2X Roadmaps

Communication systems are currently evolving at a very dynamic pace toward higher data rates, lower transmission latency, greater reliability, and increasing connectivity. The automotive industry is at the same time striving for higher levels of automation with the goal of realizing driverless vehicles. These two trends

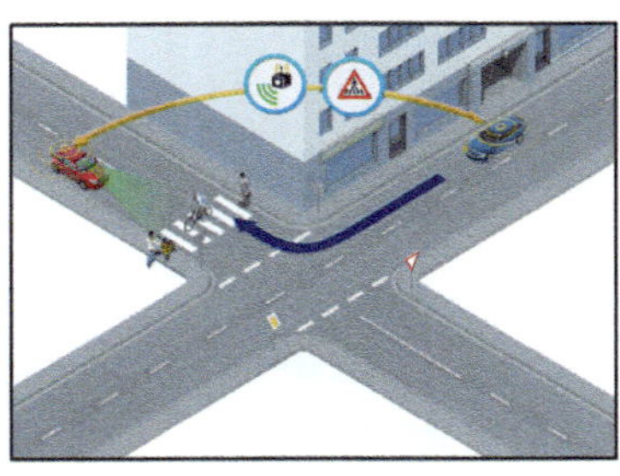

Fig. 23.2 C2C-CC phase model for the introduction of Vehicle-2-X systems. (*Source* Bosch, created using the C2C-CC Illustration Toolkit)

stimulate each other and, driven by research institutions and industry, lead to innovations in the area of cooperative intelligent transportation systems (C-ITS).

Based on intensive discussions among their members, the two consortia 5GAA and C2C-CC have developed and published roadmaps. The 5GAA roadmap [8] comprises various use cases, each of which is assigned to the two C-V2X technologies *direct communication* and *network communication*. A categorization in traffic efficiency, road safety, and automated driving functions is also made. The C2C-CC roadmap [11] divides the evolution into three phases (with associated examples of use cases and communication services): *Awareness Driving*, *Sensing Driving*, and *Cooperative Driving* (cf. **□** Fig. 23.2). The requirements placed on communications, the accuracy of position and time, the level of automation, the vehicle equipment, etc. increase in the roadmap from phase to phase. In the first phase *Awareness Driving*, essentially only status information and unexpected events are shared among vehicles and between vehicles and the road infrastructure. In the second phase *Sensing Driving*, the vehicle and the road infrastructure additionally exchange information about detected objects, such as pedestrians. The third phase, *Cooperative Driving*, is based on higher levels of vehicle automation that allow maneuver planning and implementation to be coordinated, thus increasing traffic flow and safety.

The information presented above shows that vehicular communication is not only suitable for information exchange but will in future also support automated driving. Particularly challenges like the identification of current traffic signal phases and switching times or the detection of objects like pedestrians that are not

in the field of vision of the vehicle sensors (e.g., radar or video) can be solved with V2X. The trustworthiness of messages must therefore in future be enhanced with methods from the area of functional safety for this.

The three development phases of the C2C-CC roadmap will be looked at in more detail below. Special data formats and data services are required for data communication, and these are explained in greater detail in **▶** Sect. 5.4. Here, a distinction is made between messages that are transmitted periodically with a certain frequency and messages that are transmitted on an event-driven basis.

23.3.2 **Day1 Development Phase**

The Day1 development phase (Awareness Driving) includes the Cooperative Awareness (CA) and Decentralized Environmental Notification (DEN) services, which enable transmission of local status information and warnings. The associated message formats are the Cooperative Awareness Message (CAM) [12], which is transmitted periodically with a frequency of up to 10 Hz, and the event-driven Decentralized Environmental Notification Message (DENM) [13]. For the exchange of data with road infrastructure components (e.g., traffic signal systems), there is the Signal Phase and Timing (SPaT) data format.

Potential use cases of the Day1 phase are simple warning functions like obstacle warning, forward collision warning, electronic emergency brake light warning, warning about broken-down vehicles, emergency vehicle warning, traffic jam ahead warning, hazardous weather warning, road construction warning, intersection collision warning, traffic signal status information, etc.

How these kinds of warning functions basically work can be described using two examples. The electronic emergency brake light warning function is an event-driven warning function whereby a vehicle that is braking hard will generate and send a DENM (including all the necessary data, like position, direction of travel, speed, brake deceleration, etc.) when a certain brake deceleration threshold is exceeded. The DENM can be received by all V2X-enabled vehicles located within a relevant area. The receiving party performs a relevancy check and may, if appropriate, issue a warning using visual and acoustic signals. This ensures that other vehicles that cannot see the braking vehicle are warned too.

An example of a CAM-based warning function is the intersection collision warning feature. The vehicles approaching the intersection and the ones at the intersection broadcast CAMs with sufficient frequency containing position and motion data (speed, direction of travel, etc.). At the same time, they receive such data from other vehicles too. The motion of one's own vehicle and that of the other vehicles is predicted and used to ascertain a collision risk. This is done by first determining a collision zone for each transmitting vehicle. This is essentially the location where the predicted driving trajectories of the vehicles intersect. If a potential collision zone exists, the system checks whether the vehicles involved will be within the zone at approximately the same time or whether one of the two vehicles will pass through the zone significantly earlier than the other vehicle. In the first case, the system in the vehicle that does not have the right of way calculates the time it will take for it to reach this zone; if this value is below a critical time threshold, the system warns the driver in good time about a potential collision. This calculation takes into account an analysis of the driver's intentions; so if, for instance, the driver has already begun to apply the brakes, no warning will be issued.

23.3.3 Day2 Development Phase

In the Day2 phase (Sensing Driving), vehicles and roadside units (RSUs) that are equipped with V2X and surround sensors (e.g., cameras, radar, lidar) can exchange information about detected objects. This functionality makes it possible for receiving vehicles to become aware of objects outside the range or field of vision of their own sensors (e.g., pedestrians or cyclists who are hidden behind another vehicle). On the basis of this, applications can be developed that not only generate a warning for the driver but also control the vehicle with some degree of automation. Examples of this are an automated emergency stop in front of a pedestrian and cooperative adaptive cruise control (C-ACC).

Potential use cases of the Day2 phase are, for example, cooperative overtaking warning, cooperative intersection collision warning, cooperative ACC, protection for vulnerable road users, and emergency vehicle prioritization.

An important functionality in this phase is collective perception, which uses Collective Perception Messages (CPM) for exchanging information about a sending party's perceived driving environment. The Collective Perception Service (CPS), which specifies the CPM, is currently going through the ETSI standardization process. The associated technical report [14] has already been published, though the specification has not yet been fully concluded. The CPM can be transmitted by a vehicle or a roadside unit (RSU) and contains information relating to the sender, its surround sensors, and the objects detected (for further details on this, see Sect. 5.4 "V2X message formats").

◉ Figure 23.3 shows a typical scenario for the application of collective perception. Vehicle A is approaching an intersection and, using its sensors, detects oncoming Motorcycle D. The information about the detected object is transmitted in a CPM, and this is received by Vehicle B. Alternatively, the RSU that is equipped with sensors can detect the motorcycle and can broadcast a CPM that is received by all vehicles. This means that Vehicle C also becomes aware of the concealed motorcycle that is approaching, thus enabling Vehicles B and C to take the motorcycle into account in their maneuver planning.

In addition to this, other infrastructure-based functionalities are also planned in the Day2 development phase, such as the prioritization of certain vehicle categories (e.g., local public transport) via Signal Request Extended Messages (SREM) and Signal request Status Extended Messages (SSEM) and the transmission of GNSS position corrections via Radio Technical Commission for Maritime services Extended Messages (RTCMEM).

A further important development is the inclusion of functional safety in V2X communication, particularly in order to enable partially automated driving functions. For this purpose, the C2C-CC has proposed an extension of the CAM and DENM message formats with safety containers that indicate which ASIL level is supported by the message elements.

23.3.4 Day3 + Development Phase

The Day3 + phase (Cooperative Driving) benefits from the increasing automation of driving functions. Cooperative automated and partially automated vehicles are able to plan and execute driving maneuvers and can

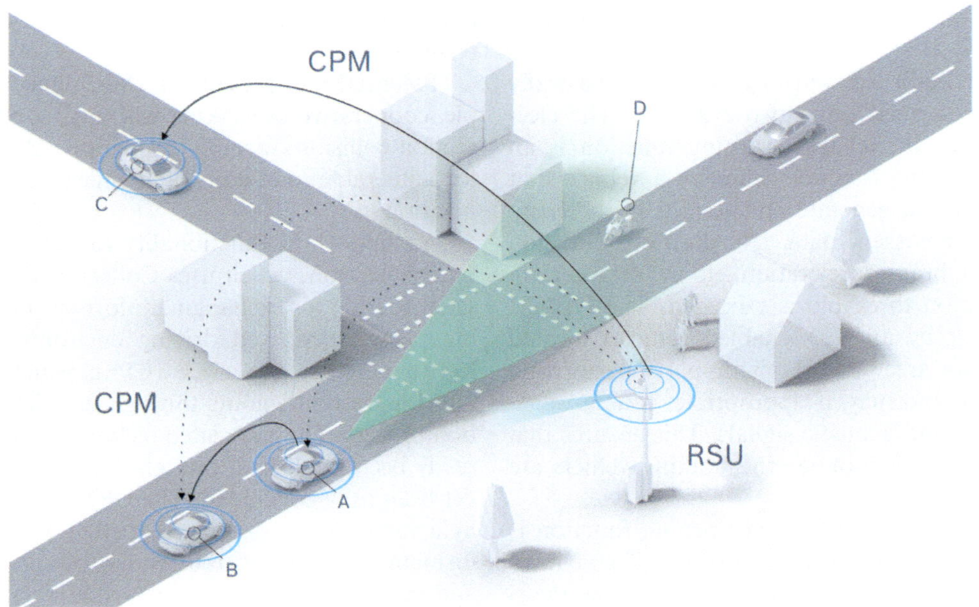

Fig. 23.3 Use of collective perception in an intersection scenario. (*Source* Bosch)

share the planned maneuvers with other road users. Several approaches are currently being examined, from a non-binding notice of intent all the way up to cooperative automated maneuver coordination.

Potential use cases of this phase are, for example, cooperative merging, cooperative lane changing, cooperative left or right turn, cooperative overtaking, and automated green wave (Automated Green Light Optimal Speed Advisory, GLOSA).

For maneuver coordination between vehicles, some projects propose the transmission of Maneuver Coordination Messages (MCM) [2, 3] that contain information about the possible future trajectories of one or more vehicles (see details in ▶ Sect. 5.4 "V2X message formats"). The necessity for maneuver coordination is identified when, for instance, the safety clearance for the planned trajectories of at least two vehicles is no longer maintained. Maneuver coordination aims to resolve the conflict and find a suitable maneuver for the vehicles involved. An example scenario is shown in Fig. 23.4. Vehicles A and B are driving toward the intersection and use V2V communication to periodically transmit MCMs (dashed lines) that contain their planned trajectories (solid lines). Since an overlap of trajectories is identified, the vehicles plan new trajectories (e.g., Vehicle A slows down) to enable a collision-free maneuver.

Alternatively or additionally, the cooperative maneuver coordination process can also be supported by the road infrastructure [3, 15], e.g., in order to improve the traffic flow at intersections. In the example scenario shown in Fig. 23.4, the RSU transmits speed recommendations using I2V communication to the vehicles involved in the maneuver. The RSU can additionally relay the MCMs from more distant vehicles (e.g., from Vehicle C) so they can be involved in maneuver coordination at an early stage.

Other aspects in the Day3+phase are the extension of the CAM and the In-Vehicle Information (IVI) message format in order to support automated driving functions.

23.4 V2X System Concept

V2X systems are made up of various entities that communicate with one another. In general, they can be divided into the following basic component classes.

ITS Central Station (ICS): These are components that provide transportation support for a larger geographic or logical area. Examples are traffic control centers. ICSs can be integrated into hierarchical topologies in which, for example, an ICS monitors and controls a very large area that in turn on the hierarchy level beneath it is made up of smaller sub-areas that are likewise monitored by ICSs.

Roadside Unit (RSU): On the lowest hierarchy level, the ICSs are typically connected to RSUs. RSUs are stations that serve a limited geographic area and are connected by cables or wirelessly to a number of monitoring or influencing devices, such as cameras, variable traffic signs, and traffic signals. The connection to

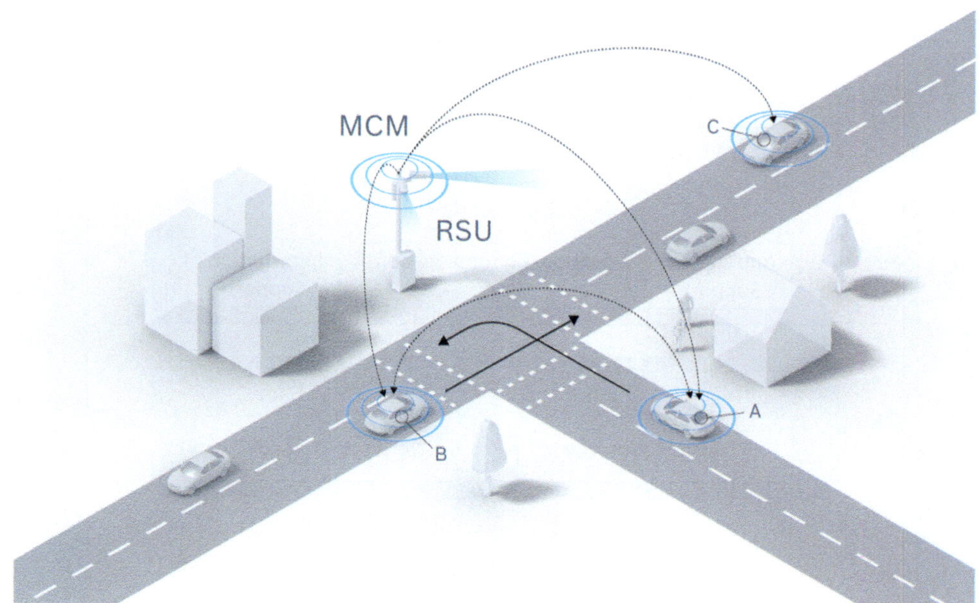

Fig. 23.4 Use of cooperative maneuver coordination in an intersection scenario. (*Source* Bosch)

the higher level Central Station is implemented through cables or wirelessly (e.g., cellular technology or directional radio). RSUs may also be equipped with a V2X communication module that supplies the vehicles within radio range of the module with various information, such as hazard warnings or congestion alerts, and can, in the opposite direction, also receive information from the vehicles.

ITS Vehicle Station (IVS): Vehicles that are equipped with V2X are each referred to as an IVS. In the sending direction, an IVS processes vehicle information (e.g., vehicle status, vehicle dynamics information) and from this generates V2X messages (e.g., CAM or DENM), which are sent to other ITS users (i.e., other IVSs, IPSs, RSUs). In the receiving direction, the IVS processes information from other ITS users and uses this, for instance, to inform the driver or for driver assistance functions. In addition to using direct communication, the IVS can also communicate via mobile networks and can relay the information exchanged over the networks or incorporate it into its own vehicle functions.

ITS Personal Station (IPS): Besides vehicles, other road users, such as pedestrians, can also participate in the exchange of messages. IPSs are very similar to the IVSs and are typically integrated in smartphones or e-bike systems.

Gateways: Gateways are components that establish connections between various subunits within an ITS station and perform tasks such as data transformations and data filtering. Routers are a special gateway category that establish connections between the subunits on the lower communication layers.

Data backends: For general, overall data exchange, there are servers that manage certain data and make it available to the users in the ITS network and also receive such data from them. Modern vehicles are typically connected to the backend of manufacturers over mobile networks and exchange information relating to operation, maintenance, and fault states. Software updates for the operating systems of modern vehicles are also carried out via these backends. Besides the OEM backends, there are also other data backends (e.g., the Mobility Data Marketplace (MDM) in Germany) that manage and supply data for specific applications.

Data brokers: Data brokers are servers that enable data exchange between data backends. The data brokers establish connections between the data backends. They perform data conversions between various interface formats, though they typically do not store data on a long-term basis.

The basic structure of all previously mentioned classes is similar and can be explained using the simplified diagram based on ETSI-ITS shown in Fig. 23.5 [17].

The structure of the actual communication stack is based on the OSI layer model and comprises a physical layer, an access layer, and a network layer. The latter

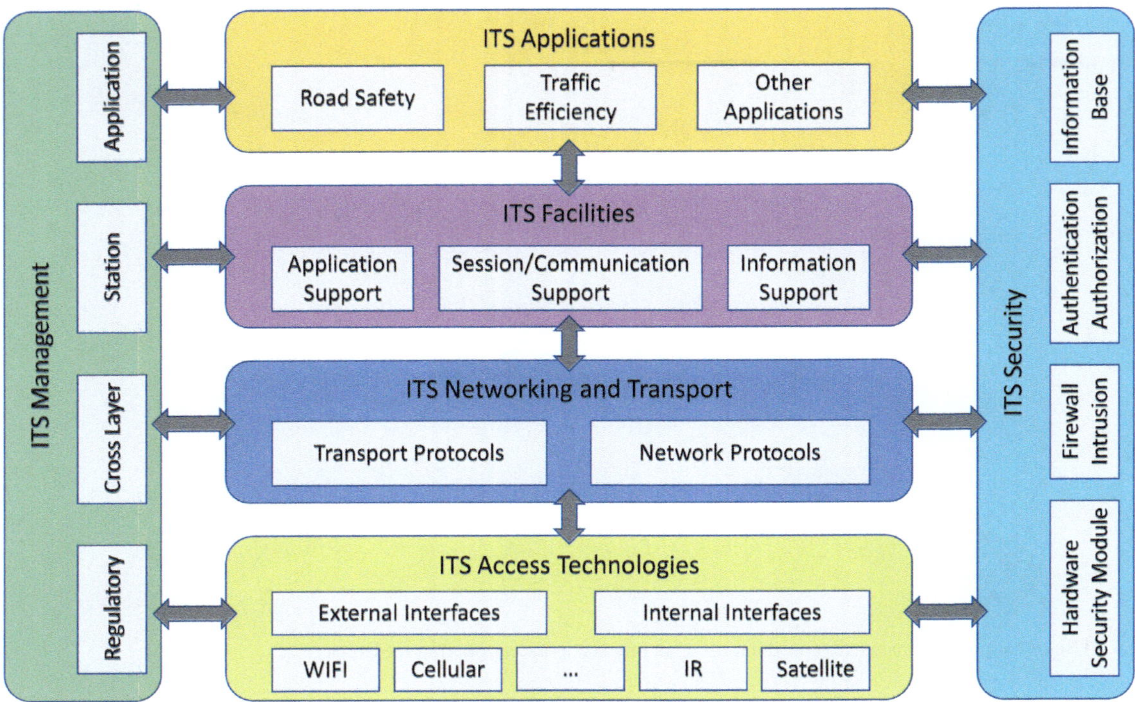

Fig. 23.5 Simplified diagram of an ITS station based on ETSI ITS. (*Source* Bosch)

can differ depending on the communication method, e.g., mobile communications (direct or via base stations), Wi-Fi, or satellite. These different variants then come together in the so-called facility layer. This is where the actual generic messages are generated and evaluated and then passed on to applications above it. In addition, management and security functions exist across the layers and operate across the communication layers. Their primary tasks are likewise depicted in ▪ Fig. 23.5. Interfaces exist between the individual components; the exchange of data and control information takes place via these interfaces.

Example configurations for various applications that use the above-mentioned component classes are shown in ▪ Fig. 23.6.

23.5 V2X Data Transmission Technologies

23.5.1 Overview of V2X Communication Technologies

A variety of system approaches have been pursued in recent years for the exchange of information among vehicles and between vehicles and road infrastructure components. An overview, which is by no means exhaustive, is shown in ▪ Fig. 23.7—though, it should be noted, the terminology in the literature is not always used consistently.

For information exchange between the road users, the cellular network can be used together with a message distribution service (mostly referred to as a geocast service), or a direct means of communication can be established. In the first case, the data is transmitted to the cellular network via the air interface between the terminal device and base station (referred to as the Uu interface). Here, the geocast server is responsible for the distribution of the messages to all relevant road users. The first implementations of this were realized in the CONVERGE project [16]. An advantage is that it works regardless of the distance between the users; a disadvantage is the dependency on operators of the mobile network and geocast server.

In contrast, when direct communication between the road users is used, data exchange takes place without a detour via the mobile network. Here, a distinction can be made between two methods. In network-assisted communication, the coordination of which user is allowed to send and when they can do so is handled centrally by the mobile network. Control data is exchanged between the users and the network, but the payload is exchanged directly between the users. Without network assistance, on the other hand, the coordination takes place between the users, and no mobile network infrastructure is required. Due to the technical benefits and current market development, the information below will refer to the latter system approach, which in turn is subdivided into two different transmission systems.

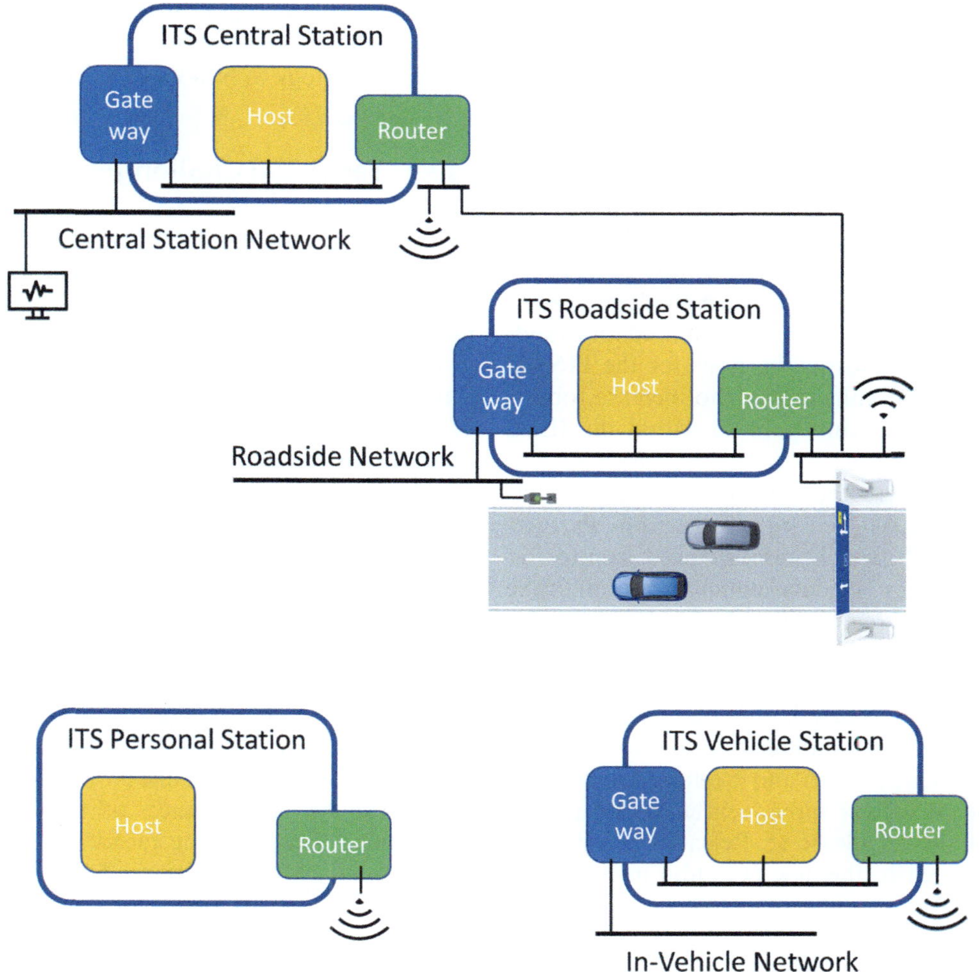

Fig. 23.6 Example configurations of an ITS station's component classes for various applications. (*Source* Bosch)

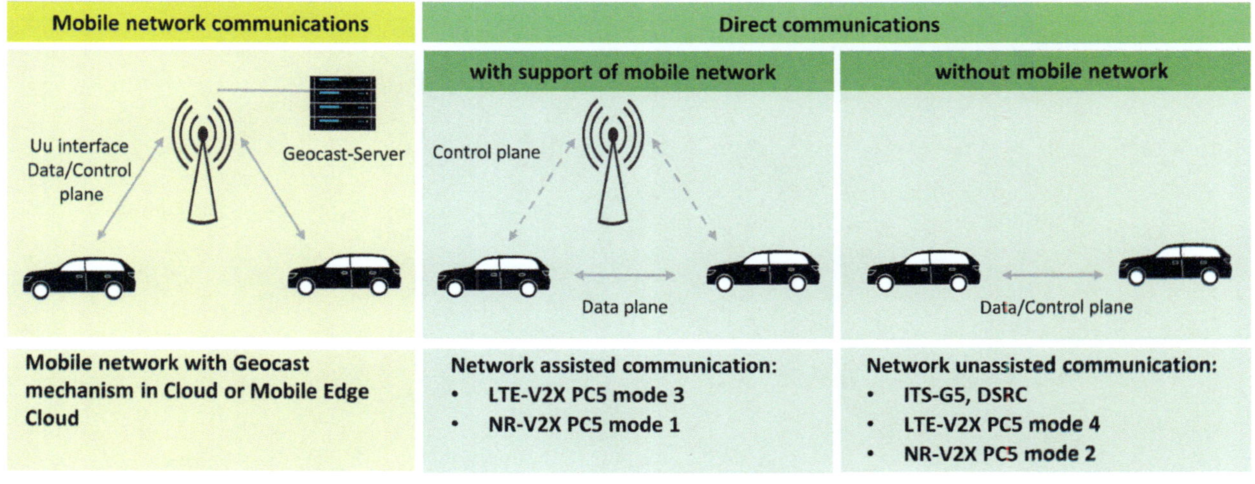

Fig. 23.7 Overview of the system approaches for V2X communication. (*Source* Bosch)

The IEEE Standards Association expanded the Wi-Fi standard for V2X ad hoc communication early on and published it in 2010 as IEEE 802.11p [18] for the PHY and MAC layers. On the basis of this, the **DSRC** (Digital Short Range Communication) system was established in the United States together with the IEEE 1609.x family of standards for the higher OSI layers and the message formats specified by the Society of Automotive Engineers (SAE). In Europe, the IEEE 802.11p standard was complemented with developments at ETSI and European Committee for Standardization (CEN) and forms the basis for the **ITS-G5** system. A compatible enhancement to increase robustness and efficiency is being pursued within the IEEE 802.11bd standardization.

The huge potential for V2X communication has also been recognized by 3rd Generation Partnership Project (3GPP), which has enhanced its Proximity Service (ProSe) for vehicular communication. ProSe was developed for direct data exchange between mobile communications terminal devices. The terminology used to describe this direct communication is "device-to-device" and "sidelink". With release 14 of 3GPP in 2017, the first standard for direct vehicular communication was achieved and is referred to as **LTE-V2X PC5 mode 4**. It was subsequently expanded with added functionality in release 15. Release 16 in 2020 introduced the switch to the new 5G radio interface, referred to as New Radio (NR), and this also resulted in a change in the designation of the vehicular communication technology to **NR-V2X PC5 mode 2** [8, 19].

23.5.2 Technical Challenges

Radio-based data exchange between vehicles among themselves and between vehicles and road infrastructure components imposes a wide range of requirements on the transmission system. These requirements involve the various layers (OSI layers), from the physical layer to the application layer—the most important of which are briefly explained here. The characteristics of the transmission channel are forever changing due to the mobility of the users in combination with multipath propagation (see ◘ Fig. 23.8). Doppler effects due to high vehicle speeds must furthermore be taken into consideration. Various channel models that consider these effects have been developed and are suitable for use in simulations and laboratory tests for evaluating transmission reliability.

The radio range is mainly impaired by shadowing effects due to obstacles blocking the direct signal path between the communication partners. Such obstacles can be, for instance, other vehicles—particularly trucks—and building facades in cities, especially in the vicinity of intersections. To ensure sufficient ra-

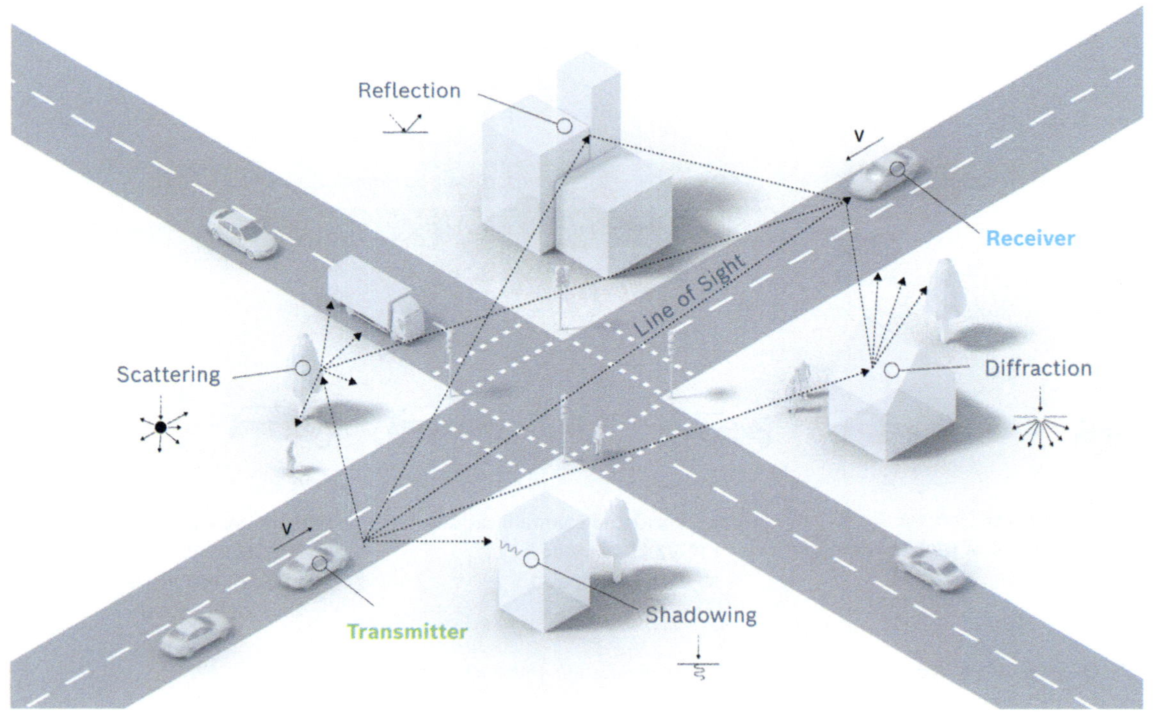

◘ **Fig. 23.8** Illustration of the principle of multipath propagation. (*Source* Bosch)

dio range also in these kinds of difficult situations, the system must have a transmitter with sufficient transmitting power, a suitable transmission system must be selected, and the system must possess dedicated receiver algorithms and antenna technology.

A further challenge is the channel access method, which controls which vehicle can occupy the transmission channel (i.e., the limited frequency resource), when it can do this, and for how long. The desire for efficient frequency utilization with high data throughput and simultaneous prevention of packet collisions during transmission between dynamic network nodes (i.e., between the vehicles) can only be achieved with special protocols. Here, the issues of fairness (i.e., that all vehicles have the same channel access) and prioritization of important time-critical messages must be taken into consideration.

23.5.3 Transmission Channel and Frequency Allocation

While the use of mobile GHz network frequencies is licensed by the operators, free frequencies in the 5.9 band are available in most countries around the world for ITS services with direct communication. In Europe, the European Commission specified a frequency range for ITS services already in 2008 and divided it into 10 MHz blocks. In 2020, this was expanded as shown in ◘ Fig. 23.9. The two lower frequency blocks are shared with other radio transmission systems, the middle four are available with priority for road safety services, and the upper block is shared with applications for rail vehicles. The EU has taken a technol-

ogy-neutral position and therefore allows the use of both systems (i.e., ITS-G5 and C-V2X PC5) in this frequency range. The maximum transmitting power is set at 33 dBm equivalent isotropically radiated power (EIRP).

In the United States, the Federal Communication Commission (FCC) already allocated frequencies in the 5.9 GHz band for DSRC in 1999. In 2021, the spectrum was reallocated to assign a larger frequency range to Wi-Fi. For ITS services, a switch was implemented from DSRC to C-V2X PC5, and the frequency range was limited to 5.895–5.925 GHz. Currently, however, there are still lawsuits pending against the FCC decision.

In China, the Ministry of Industry and Information Technology (MIIT) decided in 2016 to provisionally use the 5.905–5.925 GHz frequency range for C-V2X PC5.

In other countries, like Korea, Singapore, and Australia, a spectrum in the 5.9 GHz band was likewise allocated for ITS services. An exception is Japan with a frequency allocation in the 755.5–764.5 MHz range. Here, the 5.9 GHz band is used for toll applications.

The necessary radio range for the considered applications is around 50–100 m in urban areas at low vehicle speeds; on the freeway at high speed, it is around 100–300 m.

Depending on the transmitting power, antenna, receiver sensitivity, interference due to packet collisions, channel characteristics, and system parameters (like modulation and channel coding), these radio ranges are achieved and even greatly exceeded. The transmission channel plays an essential role here and is characterized by reflections, diffraction and scattering of the

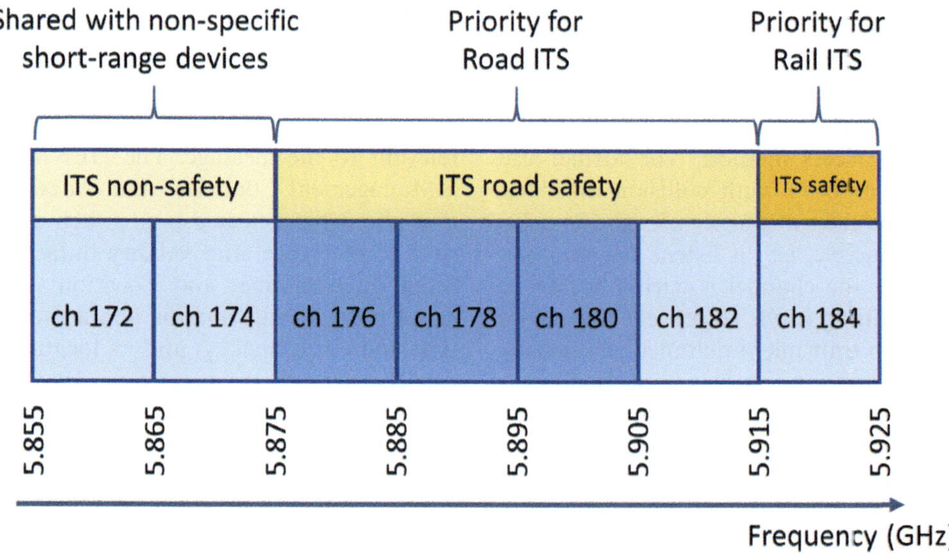

◘ **Fig. 23.9** Frequency range allocation for direct communication in Europe. (*Source* Bosch)

electromagnetic waves, and Doppler effects. Complex statistical channel models have therefore been developed to describe the short-term fading based on the interference caused by the superposition of the waves as well as the long-term fading caused by shadowing effects. These channel models can be used in simulations, but they can also be used with real transmitters and receivers to determine their performance.

In the section on the technical challenges and the characteristics of the transmission channel, it becomes clear that designing a capable transmission system for V2X is a challenging task. All components in the transmitter and receiver have been designed to meet the exacting requirements, but the further details below focus on frequency division multiplexing and channel access, since they are the essential elements.

Orthogonal frequency-division multiplexing (OFDM) was chosen for the IEEE 802.11p system. With OFDM, a large number of orthogonal carriers are used for data transmission, and each carrier is modulated separately. Since the number of carriers is reciprocal to the symbol duration, a very long symbol duration can be chosen. The result is that intersymbol interference caused by multipath propagation is reduced and can even be largely eliminated with the introduction of an additional guard interval at the beginning of each symbol. An advantage here is that the transmitter and receiver implementations can be realized with a low level of complexity on the basis of the fast Fourier transformation. A disadvantage, however, is the high peak-to-average power ratio. The LTE-V2X PC5 system uses single-carrier frequency division multiple access (SC-FDMA) as the multiplexing method. With this, the peak-to-average power ratio is more advantageous, but receiver implementations are more complex due to the necessary channel equalization and the performance in channels with strong multipath propagation is lower. With NR-V2X PC5, a switch to OFDM took place.

A key difference between IEEE 802.11p and LTE-V2X is the channel access method. The former uses carrier sense multiple access with collision avoidance (CSMA/CA) in which the transmitter uses the so-called listen-before-talk principle, i.e., it listens on the channel to check whether the channel is currently free before it begins transmitting itself. To prevent several stations from starting to transmit simultaneously, a backoff procedure with a random timer is used. In contrast to this, LTE-V2X PC5 uses semi-persistent scheduling (SPS), where the transmitter allocates a portion of the spectrum's frequency-time resources for a certain period and thus prevents packet collisions with other transmitters.

23.5.4 V2X Message Formats

The messages that are necessary for implementing V2X use cases have largely been standardized by ETSI or are currently in the process of being standardized. A fundamental function in the Day1 development phase is cooperative awareness, where vehicles transmit their own status data. This service has been standardized in Europe with the Cooperative Awareness Message (CAM) [12], which can be transmitted by vehicles and the road infrastructure. CAMs are transmitted with variable frequency depending on how the dynamic state of the sending party varies (i.e., current position, speed, and direction of travel). ◘ Fig. 23.10 shows an overview of the structure of the most important V2X message formats.

The CAM format begins—as do all other messages standardized by ETSI—with an ITS Protocol Data Unit (PDU) header, which contains the protocol version, message type, and station ID of the sender. That is followed by a Basic Container, which contains basic information about the transmitting ITS station (e.g., vehicle type and position), and a High Frequency Container that contains dynamic information, like the current speed and direction of travel. The PDU header, Basic Container, and High Frequency Container are included in every CAM. In addition to that, there are two optional containers that are transmitted with a frequency of up to 2 Hz: the Low Frequency Container, which contains less dynamic information relating to the sender (e.g., the status of the headlights); and the Special Vehicle Container, which contains information about vehicles that play a specific role in road traffic (e.g., local public transportation and emergency vehicles). The CAM is broadcast periodically in the immediate surrounding area with a variable frequency of between 1 and 10 Hz.

The Decentralized Environmental Notification Message (DENM) standard [13] specifies a protocol for the event-based (i.e., non-periodic) distribution of safety information within a geographic region that is relevant to the message. The DENM format includes a Management Container that contains basic information about the detected event, such as its position, period of relevance, and validity duration. The optional Situation Container and Location Container contain supplementary information about the event (e.g., event type and event history) and its location (e.g., waypoint lists that lead to the event as well as the event's speed and direction of travel). An optional À La Carte Container makes it possible to add application-specific data to the DENM.

The exchange of object data for collective perception is carried out using Collective Perception Messages (CPM). CPM is currently undergoing standardization at ETSI [14] to ensure interoperability between all equipped vehicles.

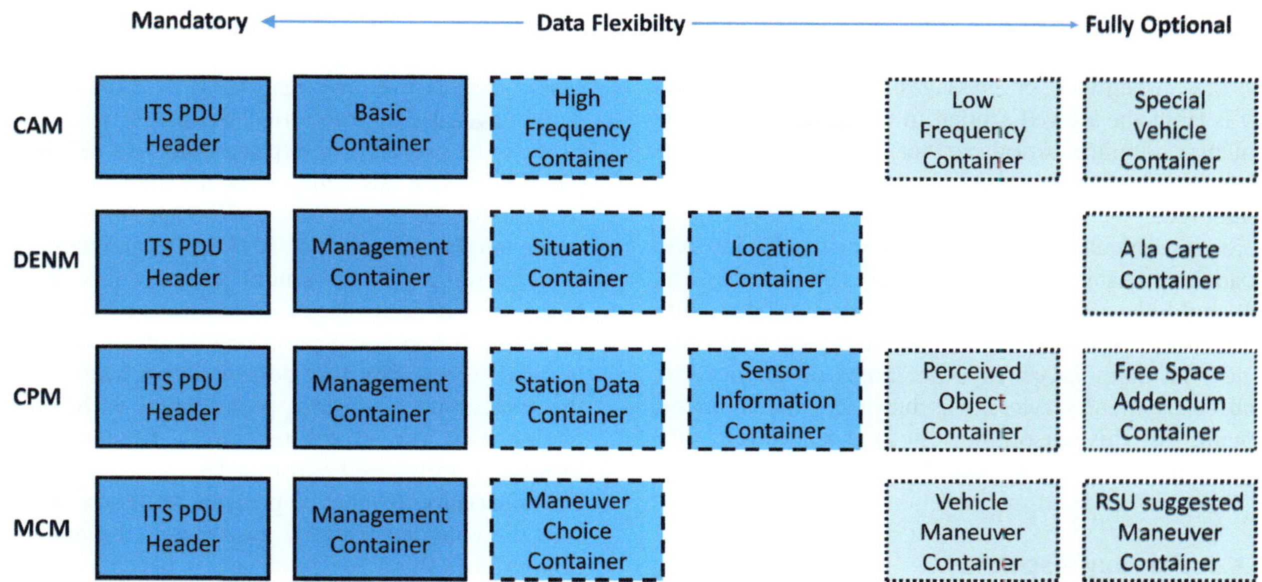

Fig. 23.10 Overview of the structure of the most important V2X message formats. (*Source* Bosch)

The CPM structure begins with the CPM Management Container and the Station Data Container, which provide information about the transmitting station, such as its position and direction of travel. The Sensor Information Container contains details about the on-board sensors in the transmitting station, such as their number (quantity), range, and angle of beam spread. The Perceived Object Container contains a list of the relevant objects that the transmitting station has detected, together with their relative positions, speeds, and dimensions as well as other data. The final Free Space Addendum Container contains detected free space areas that cannot be computed by potential receiving parties on the basis of the other containers.

Cooperative maneuver coordination is implemented with the periodic transmission of Maneuver Coordination Messages (MCM) that enable the exchange of trajectories between road users. In the IMAGinE project, the researchers developed a concept for cooperative maneuver coordination based on trajectory tuples weighted with cost values. The concept has been proposed for standardization at ETSI [20].

The proposed MCM format includes a Management Container that contains information like the automation status of the vehicle and its current position. The Trajectories Container contains the trajectory currently being driven along, optionally followed by at least one possible cooperation offer (alternative trajectory) or a cooperation request (request trajectory) for other vehicles. The trajectories are essentially represented as a vehicle path (position and direction of travel) relative to the current traffic lane over time. Each trajectory furthermore contains a cost value that represents the sender's relative preference for the transmitted trajectories.

23.6 Data Security and Privacy Protection

[1]Protection against unauthorized manipulation is, needless to say, extremely important in the context of V2X-based assistance systems and functionalities. Furthermore, the continuous transmission of data by vehicles, like their position and speed, also raises questions concerning data protection and the privacy of the vehicle occupants. This section will first examine the basic issues surrounding data security and privacy protection in vehicular communication. Afterwards, existing potential solutions and the current state of technology will be briefly outlined.

23.6.1 Security Issues

Without suitable protective measures, there is a risk in vehicular communication that fake messages can be sent and legitimate messages can be altered. Using such manipulation, attackers could, for instance, fake non-existent traffic congestion or a construction site with the aim of hindering or diverting other vehicles. With messages that fake extreme situations, like an emergency stop, they could also put other road users in danger. These and similar scenarios make it necessary

---

1   The section on data security has been largely adopted from the third edition, where it was written by Hans Löhr.

for manipulated messages and messages sent by illegitimate senders to be identified as being invalid. That is to say, the authenticity of messages (i.e., the message really is from the alleged sender, in this case, a legitimate vehicular communication partner) and the integrity of messages (i.e., the message was not manipulated) must be assured.

Since vehicular communication primarily involves broadcast messages that are of relevance to safety and that should be readable for everyone, the confidentiality of the message content (i.e., the secrecy of sent data) is not the main focus. In other areas of application, such as payment services, this may be quite different, though. For this reason, current approaches also provide for the capability of confidential communication.

23.6.2 Privacy Aspects

If messages sent within the context of vehicular communication could be associated with a vehicle or even its driver with little effort, this could have serious consequences for the privacy of the driver—and also for the privacy of other vehicle occupants. If vehicles are continuously sending messages, they can be received by any individuals or organizations within range and also stored. This could lead to critical scenarios arising, for instance, if it were easy to determine or even prove that a particular vehicle had been present at a potentially sensitive event, such as a political party convention or an association general meeting. The possibility that logged messages could be of assistance in the automated punishment of traffic offenses, like speeding, is an issue that may also be worthy of discussion.

To avoid seriously infringing on the privacy of vehicle occupants as a result of vehicular communication, an effort should be made to ensure no permanent identifiers that are easily associated with the vehicle are used in these messages. Furthermore, only the data that is actually necessary for the use cases should be transmitted (principle of data minimization). In particular, when devising a security solution, attention must be paid to ensuring that the cryptographic keys cannot be used to identify and trace vehicles easily or even verifiably.

23.6.3 Protection Objectives and Challenges

The following essential protection objectives for a security concept emerge from the above considerations:
- **Integrity:** The messages must be protected against manipulation.
- **Authenticity:** It must be ensured that only messages from legitimate parties are accepted. In special cases, such as the investigation of a serious crime, it may even be desirable that the origin of a message can be proven (non-repudiation) also to a third party, such as a law court, after the necessary order to do so has been issued by a sovereign authority.
- **Anonymity/pseudonymity:** To keep curtailment of users' privacy to a minimum, support for anonymous or pseudonymous communication should be provided. When using pseudonyms, attention must be paid to ensuring that de-pseudonymization (i.e., the association of various pseudonyms with a particular user) can at the most only be easily performed by authorized parties.
- **Confidentiality:** It must be possible to optionally ensure the confidentiality of messages if the application requires this.

In addition to this, the following aspects also play a crucial role in the security solution:
- **Performance:** The security mechanisms must perform sufficiently well so that communication is not impaired. In particular, steps must be taken to ensure that the realistically expected quantity of received messages can indeed be processed.
- **Costs/effort:** The costs and effort entailed in equipping vehicles and roadside units with security modules should be kept as low as possible. The same applies to the infrastructure required for certificate and key management.
- **Maintainability/futureproofing:** Since a system that has been rolled out in the automotive sector must function in the field for decades, it is essential that the aspects of maintainability and futureproofing are already considered before the system's introduction. This naturally also applies to the security solution, which should provide—at least as far as can be judged from today's perspective—the necessary security over corresponding periods of time.

23.6.4 Security Measures

To secure communication adequately from a cryptographic perspective, a system has been implemented in which all messages are digitally signed. The signatures generated by vehicles can be verified using pseudonymous certificates that change on a per vehicle basis and that are issued by a dedicated public key infrastructure (PKI). ◨ Fig. 23.11 provides an overview of this approach as defined by C2C-CC and ETSI [21, 22].

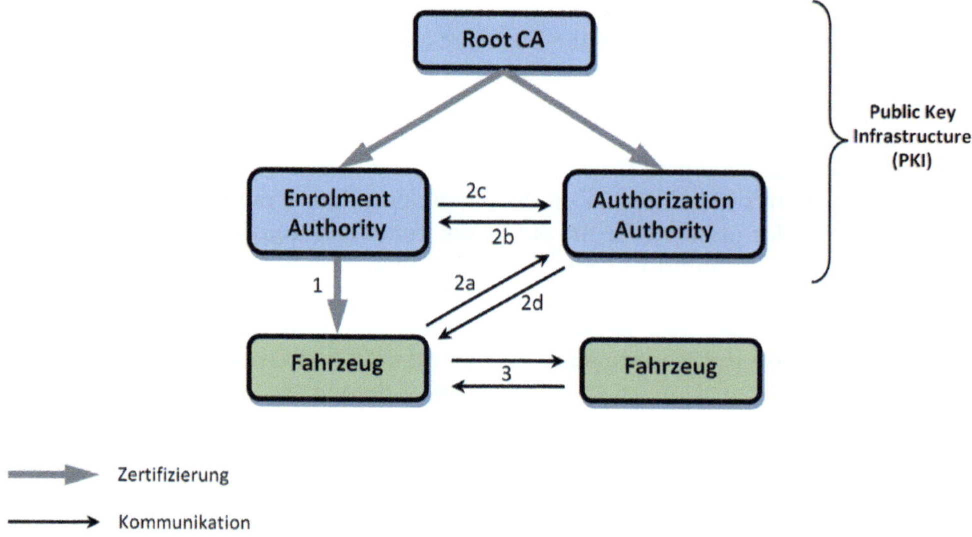

☐ Fig. 23.11 Securing vehicular communication. (*Source* Bosch)

The PKI comprises three different types of Certificate Authority (CA): the Root CA, the Enrollment Authority (EA, also called Long Term CA), and the Authorization Authority (AA, also called Pseudonym CA). The trust anchor in the security architecture is the Root CA, which is known to all system participants. The Root CA issues certificates for the EA and AA and monitors compliance with the policies governing the issuing of certificates. It is envisaged that there will be several instances of Enrollment and Authorization Authorities that are operated by different organizations.

A vehicle that wants to participate in the communication system first of all requires an Enrollment Credential (EC, also called Long Term Certificate) from an Enrollment Authority (see ☐ Fig. 23.11, Step 1).

After receiving an Enrollment Credential, the vehicle can generate pseudonyms (cryptographic key pairs) and have them certified by an Authorization Authority (see ☐ Fig. 23.11, Step 2). To do this, the vehicle encrypts its Enrollment Credential for the Enrollment Authority and sends the encrypted credential together with the newly generated public keys (the pseudonyms) to the Authorization Authority (Step 2a). The Authorization Authority then sends the encrypted credential to the responsible Enrollment Authority, which subsequently checks whether it is a proper certificate that qualifies for the issuance of pseudonymous certificates (Step 2b). After a positive response from the Enrollment Authority (Step 2c), the Authorization Authority certifies the public keys with a digital signature and generates so-called Authorization Tickets (ATs) that incorporate these certified keys. The ATs are now transmitted to the vehicle (Step 2d), which can then use them for pseudonymized communication.

To secure the messages for communication (see ☐ Fig. 23.11, Step 3) between vehicles—or also for communication between vehicles and infrastructure components—the messages are now signed digitally by the sender. A private key is used for this—the sender possesses a valid Authorization Ticket for the public counterpart to this private key. The recipient of the message can now verify that the message was generated by a legitimate communication partner and was received unaltered. To do this, the recipient must first check the signature using the public key from the sender's Authorization Ticket. Then the recipient must verify that the Authorization Ticket is valid (in particular that it has not expired) and that it was signed by a legitimate Authorization Authority. If the issuing Authorization Authority is not yet known to the recipient, the recipient must verify the AA's certificate that was signed by the Root CA in order to make sure the Authorization Authority in question is indeed legitimate. With this approach, secure communication can take place even between vehicles that are encountering each other for the first time and that only trust the same Root CA. If required, the system can also be expanded to integrate several Root CAs that certify one another through so-called cross-certification. Optionally, confidential messages can also be transmitted in encrypted form. Since this is not necessary for the main currently planned (safety-relevant) use cases, it will not be dealt with here in more detail.

Effective pseudonymization is made possible through the procedure described here on the premise that the vehicles replace their Authorization Tickets often enough to prevent the association between the AT and vehicle from being easily identified. In order to ascertain the identity of the sender (which is equivalent to the sender's Enrollment Credential) on the basis of received messages, the Authorization Authority (that issued the Authorization Tickets) and the Enrollment Authority (as the only CA that has seen the vehicle's Enrollment Credential in unencrypted form) must cooperate with each other. If this identification of the sender is to be made possible (for instance, for the purpose of revealing it to a sovereign authority), the AA and EA must keep a log of the issuance of the ATs and of the examination of the ECs respectively. By merging these logs, it is possible to revoke the pseudonymity of the sender of a message.

23.6.5 Current State of Technology and Implementation

The cryptographic mechanisms and data formats necessary for protection are already specified. The certificate formats and message formats are defined in the ETSI technical specification TS 103 097 (for Europe) [23] and in the IEEE 1609.2 standard (for North America) [24]. The Elliptic Curve Digital Signature Algorithm (ECDSA—see, for example, [25]) is used for the digital signatures. The plan is to use the Elliptic Curve Integrated Encryption Scheme (ECIES—see, for example, IEEE 1363a [26]) for optional encryption. Various instances of the PKI have been implemented and are in part already being used in vehicles on the market. Certain divergences between United States and European standards are the focus of ongoing advanced harmonization activities.

23.7 V2X Market Rollout and Outlook

After an extensive research phase, large-scale pilot projects in the United States, Europe, and China were initiated to support the market rollout. They were intended to not only produce scientific results but also provide information on real operation. What also became clear here is the strong dependency between vehicle equipment penetration rate and the benefits that can actually be experienced by vehicle users. This dependency can, however, be lessened through investment in the road infrastructure, which enables certain functionalities that allow all users of V2X-equipped vehicles to experience benefits regardless of the penetration rate. An example in this regard is the ITS Corridor project [27] in which construction site warning sign trailers are equipped with communications units and can inform vehicles in good time to help prevent collisions. The integration of traffic signals in the communication infrastructure, so they can inform vehicles in the vicinity about their

signal phase, is another important step toward enabling use cases like a green wave.

With the introduction of the Golf 8 in Europe at the end of 2019, a major breakthrough was achieved in the transition of V2X technology to the mass market [28]. The vehicle comes equipped with an ITS-G5 system and offers a range of V2V and V2I hazard warning functions. Further vehicle models from the new ID range have meanwhile come onto the market with V2X technology, and more are being planned. In the United States, the situation is difficult to predict owing to the FCC's decision to partially reallocate the spectrum—despite the fact that the first production vehicle equipped with DSRC, the Cadillac CTS, already went on sale in 2017 [29]. In the Asian region, China has taken on a pioneering role and, beginning with pilot projects in numerous metropolitan regions, is investing heavily in V2X infrastructure. A great many vehicle manufacturers have equipped their first vehicle models with LTE-V2X technology in 2021.

Another driver for the introduction of V2X systems in Europe is the announcement by Euro NCAP (European New Car Assessment Programme) that it would be including the first V2X-based functions in its rating system for vehicle assessments by 2024/25 [30]. Alongside this, its Chinese counterpart C-NCAP outlines an even more ambitious plan in terms of planned functionalities and time frame.

In the longer term, V2X technology will evolve from a system for increasing vehicle safety and traffic efficiency to an enabler for automated driving. The integration of the road infrastructure will be an essential element here, also for achieving the appropriate level of safety in critical and complex traffic situations, such as busy intersections, where the view may be obscured.

References

1. Weblink to the AutoNet 2030 research project. ▶ https://www.eucar.be/autonet-2030/
2. Weblink to the IMAGinE research project. ▶ https://imagine-online.de/
3. Weblink to the TransAID research project. ▶ https://www.transaid.eu/
4. Weblink to the Car-2-Car Communication Consortium (C2C-CC). ▶ https://www.car-2-car.org/
5. Weblink to the 5G Automotive Association (5GAA). ▶ https://5gaa.org/
6. Hofmann, F., Mahdi, A., Brahmi, N., Wegmann, B., Mühleisen, M., Petry, F., Mudriievskyi, S., El Assad, A., Jornod, G., Franchi, N., Fettweis, G.: "5G NetMobil: pathways towards tactile connected driving". In" 2019 IEEE 2nd 5G World Forum (5GWF), (2019)
7. Campolo, C., Molinaro, A., Scopigno, R.: Vehicular ad hoc Networks. Springer Verlag, (2015)
8. Fallgren, M., Dillinger, M., Mahmoodi, T., Svensson, T.: Cellular V2X for connected automated driving, Wiley, (2021)

9. Weblink to ETSI ITS: ► http://www.etsi.org/technologies-clusters/technologies/intelligent-transport

10. Weblink to C2C-CC Basic System Profile: ► https://www.car-2-car.org/documents/basic-system-profile/

11. C2C-CC White Book: Guidance for day 2 and beyond roadmap, (2021). ► https://www.car-2-car.org/fileadmin/documents/General_Documents/C2CCC_WP_2072_RoadmapDay2AndBeyond_V1.2.pdf

12. ETSI: EN 302 637–2, Intelligent Transport Systems (ITS); Vehicular Communications; Basic Set of Applications; Part 2: Specification of Cooperative Awareness Basic Service, European Telecommunications Standards Institute (ETSI). ► https://www.etsi.org/deliver/etsi_en/302600_302699/30263702/01.03.02_60/en_30263702v010302p.pdf

13. ETSI: EN 302 637–3, Intelligent Transport Systems (ITS); Vehicular Communications; Basic Set of Applications; Part 3: Specification of Decentralized Environmental Notification Basic Service, European Telecommunications Standards Institute (ETSI). ► https://www.etsi.org/deliver/etsi_en/302600_302699/30263703/01.02.02_60/en_30263703v010202p.pdf

14. ETSI: TR 103 562, Intelligent Transport Systems (ITS); Vehicular Communications; Basic Set of Applications; Analysis of the Collective Perception Service (CPS), European Telecommunications Standards Institute (ETSI). ► https://www.etsi.org/deliver/etsi_tr/103500_103599/103562/02.01.01_60/tr_103562v020101p.pdf

15. Weblink to the INFRAMIX research project: ► https://www.inframix.eu/

16. Weblink to the CONVERGE project: ► https://fgvt.htwsaar.de/site/converge-2012-2015/

17. ETSI: EN 302 665, ITS Communications Architecture, V1.1.1. 09/2010. ► http://www.etsi.org/deliver/etsi_en/302600_302699/302665/01.01.01_60/en_302665v010101p.pdf.

18. 11p-2010—IEEE Standard for Information Technology—Local and Metropolitan Area Networks—Specific Requirements—Part 11: Wireless LAN Medium Access Control (MAC) and Physical Layer (PHY) Specifications Amendment 6: Wireless Access in Vehicular Environments

19. Castañeda Garcia, M., Molina-Galan, A., Boban, M., Gozalvez, J., Coll-Perales, B., Şahin, T., Kousaridas, A.: A Tutorial on 5G NR V2X communications, IEEE communications surveys & tutorials, **23**(3), Q3/2021

20. Llatser, I., Michalke, T., Dolgov, M., Wildschütte, F., Fuchs, H.: Cooperative automated driving use cases for 5G V2X Communication. IEEE 5G World Forum (5GWF), Dresden, Germany, pp. 120–125, (2019). ► https://doi.org/10.1109/5GWF.2019.8911628

21. ETSI: TS 102 940, ITS communication security architecture and security management, V2.1.1. 07/2021. ► https://www.etsi.org/deliver/etsi_ts/102900_102999/102940/02.01.01_60/ts_102940v020101p.pdf

22. Bißmeyer, N., Stübing, H., Schoch, E., Götz, S., Stotz, J., Lonc, B.: A generic public key infrastructure for securing Car-to-X communication. The 18th World Congress on Intelligent Transport Systems, Orlando, Florida, USA, (2011)

23. ETSI: TS 103 097, Security header and certificate formats, V1.4.1. 10/2020. ► https://www.etsi.org/deliver/etsi_ts/103000_103099/103097/01.04.01_60/ts_103097v010401p.pdf

24. IEEE Standards Association: 1609.2–2016—IEEE Standard for Wireless Access in Vehicular Environments—Security Services for Applications and Management Messages. ► https://standards.ieee.org/standard/1609_2-2016.html

25. National Institute of Standards and Technology: The Digital Signature Standard (DSS). FIPS Publication 186–4, (2013)

26. IEEE Standards Association: 1363a-2004—IEEE Standard Specifications for Public-Key Cryptography. ► http://grouper.ieee.org/groups/1363/.

27. Weblink to Autobahn GmbH: ► https://www.autobahn.de/cits/aktuelles/detail/intelligente-mobilitaet-fuer-weniger-verkehrsunfaelle

28. Weblink to Volkswagen press release: ► https://www.volkswagen-newsroom.com/en/press-releases/world-premiere-for-the-new-golf-digitalised-connected-and-intelligent-5490

29. Weblink to Cadillac press release: ► https://media.cadillac.com/media/us/en/cadillac/news.detail.html/content/Pages/news/us/en/2017/mar/0309-v2v.html

30. Weblink to Euro NCAP 2025 roadmap: ► https://cdn.euroncap.com/media/30700/euroncap-roadmap-2025-v4.pdf

Open Access This chapter is licensed under the terms of the Creative Commons Attribution-NonCommercial-NoDerivatives 4.0 International License (► http://creativecommons.org/licenses/by-nc-nd/4.0/), which permits any noncommercial use, sharing, distribution and reproduction in any medium or format, as long as you give appropriate credit to the original author(s) and the source, provide a link to the Creative Commons license and indicate if you modified the licensed material. You do not have permission under this license to share adapted material derived from this chapter or parts of it.

The images or other third party material in this chapter are included in the chapter's Creative Commons license, unless indicated otherwise in a credit line to the material. If material is not included in the chapter's Creative Commons license and your intended use is not permitted by statutory regulation or exceeds the permitted use, you will need to obtain permission directly from the copyright holder.

Actuation for ADAS

Steering Systems for Passenger Cars and Commercial Vehicles

Thilo Bitzer, Heinz-Dieter Heitzer, Christoph Elbers, and Michael Scholand

Contents

© The Author(s) 2026
H. Winner et al. (eds.), *Handbook Assisted and Automated Driving*,
https://doi.org/10.1007/978-3-658-45276-6_24

24.1 Types of Steering Systems for Passenger Cars and Commercial Vehicles

Steering systems for passenger cars and trucks are intended to provide the driver with a steering device and to convert steering inputs via the steering device into a change in the direction of the vehicle.

In the course of the development of the automobile, steering wheels have established themselves as a device for steering control and the change in direction of the vehicle is caused by a swiveling motion of the two wheel carriers on the front axle of the vehicle. Steering wheel and wheel carrier are mechanically linked to each other in all conventional steering systems.

The steering gear is the core piece of the conventional steering system. It converts the rotary motion of the steering wheel into the translational motion of a steering linkage that controls the swiveling motion of the wheel carriers. Almost all modern steering systems use a hydraulic or mechatronic power amplifier to reduce the manual force required by humans. Steering systems with integrated steering booster are also referred to as power steering.

The correlation between the angle of rotation of the steering wheel and the swivel angle of the wheel carriers is referred to as the steering ratio. When driving at low driving speeds, for example when parking, a low or direct steering ratio is desirable in order to limit the steering effort of the passengers. However, the steering ratio must not be reduced too much, because the steering sensitivity of the vehicle becomes too high when driving at high driving speeds. The steering sensitivity describes the correlation between the steering input of the driver and the change in direction of the vehicle. Excessive steering sensitivity impairs the driver's ability to control the vehicle at high driving speeds.

Experience has shown that the lateral dynamics of a vehicle can be controlled considerably more easily at higher driving speeds, if a return force is effective which is perceived by the driver on the steering wheel as a haptic feedback to his steering inputs, when deflecting from the straight-ahead position in the steering system. The haptic feedback should also provide the driver with a feel for the road contact of the tires and be as free as possible from unwanted faults, such as those caused by road bumps or tire imbalances.

The steering system characteristics, consisting of steering sensitivity and haptic feedback, are often summarized under the term steering feel.

24.1.1 Pitman Arm Steering Gear

The Pitman arm steering gear converts the rotary motion of the steering wheel into the swiveling motion of a rotatably mounted lever, which is called the Pitman arm (Fig. 24.1). The circular segment that is run through by the free end of the Pitman arm is transferred into the translation movement of the steering linkage by means of a coupling rod.

Pitman arm steering gears are particularly suitable for vehicles with rigid front axles. However, with the exception of a few off-road vehicles, all passenger cars now have an independent suspension. For heavy trucks, a chassis with conductor frame design and rigid axles is generally still used. Accordingly, almost all heavy trucks are equipped with a Pitman arm steering gear.

Within the Pitman arm steering gear, the recirculating ball steering has been identified as the best solution. The input shaft is equipped with a spindle thread which transfers the rotary motion of the steering wheel to the axial displacement of a rack section. The rack section meshes with a gear segment that drives the axis of ro-

 Fig. 24.1 Pitman arm steering gear for heavy trucks (TRW)

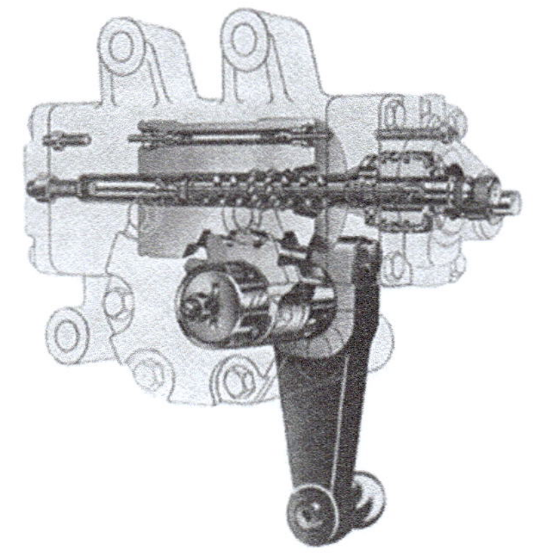

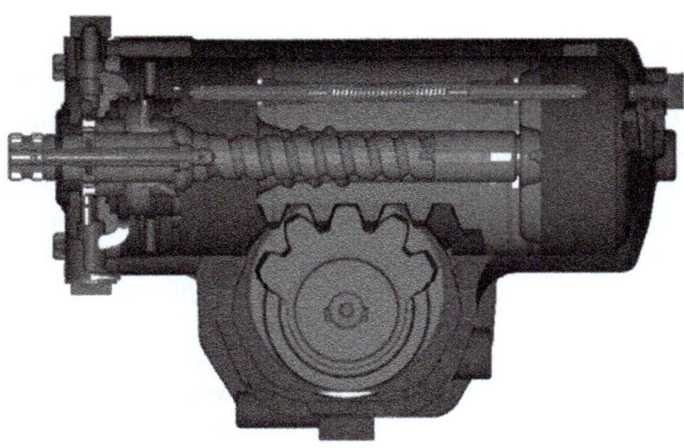

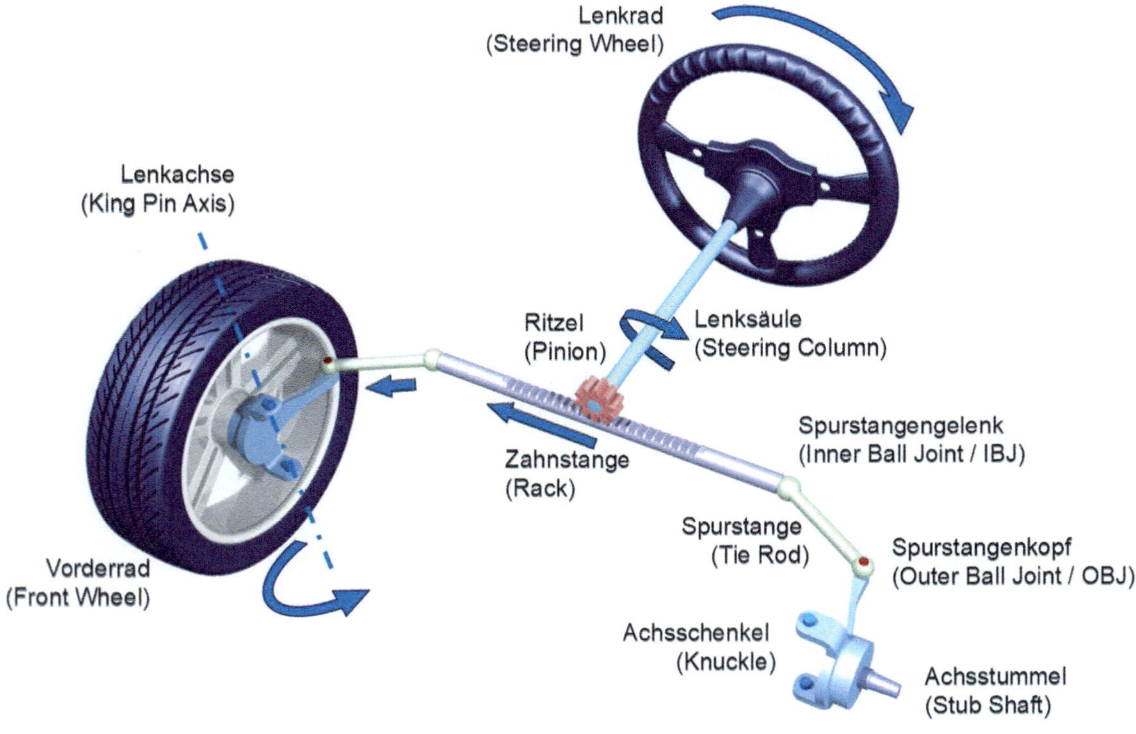

tation of the Pitman arm. To achieve the required efficiency of the steering gear, the spindle is designed as a ball thread spindle in which one or more rows of balls rotate in the thread between the spindle and the rack segment. This is called a recirculating ball steering gear.

The recirculating ball steering can be extended by a hydraulic booster by designing the rack section as a hydraulic piston which can be alternately pressurized with hydraulic pressure (■ Fig. 24.2). The direction and pressure of this assist force are determined by the valve of the steering wheel.

24.1.2 Rack and Pinion Steering Gear

The rack-and-pinion steering gear (■ Fig. 24.3) is available as a mechanical system (for manual steering or in conjunction with power assistance on the steering column, see ■ Fig. 24.4), or with either an integrated hydraulic booster or a mechatronic booster.

24.1.3 Steer-by-Wire

In contrast to conventional steering systems with hydraulic or mechatronic boosters, the steer-by-wire steering system does not have a mechanical connection between the steering wheel and the input shaft of the steering gear. Accordingly, a mechanical intermediate

■ **Fig. 24.2** Recirculating ball steering gear with integrated hydraulic booster for heavy trucks (TRW)

■ **Fig. 24.3** Principal of a rack-and-pinion steering gear in the vehicle (Wikipedia)

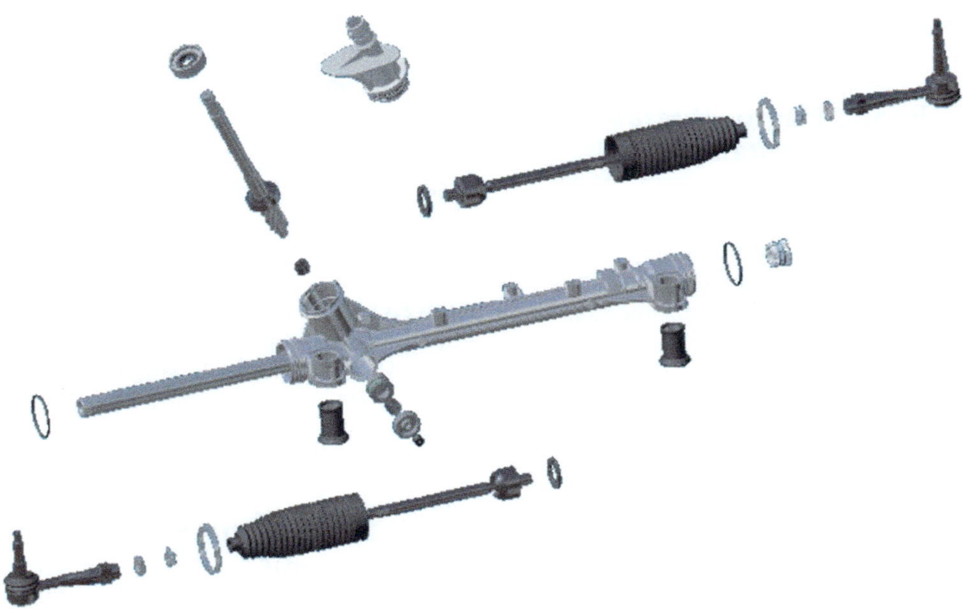

◘ Fig. 24.4   Example of a mechanical rack-and-pinion steering gear for passenger cars (ZF)

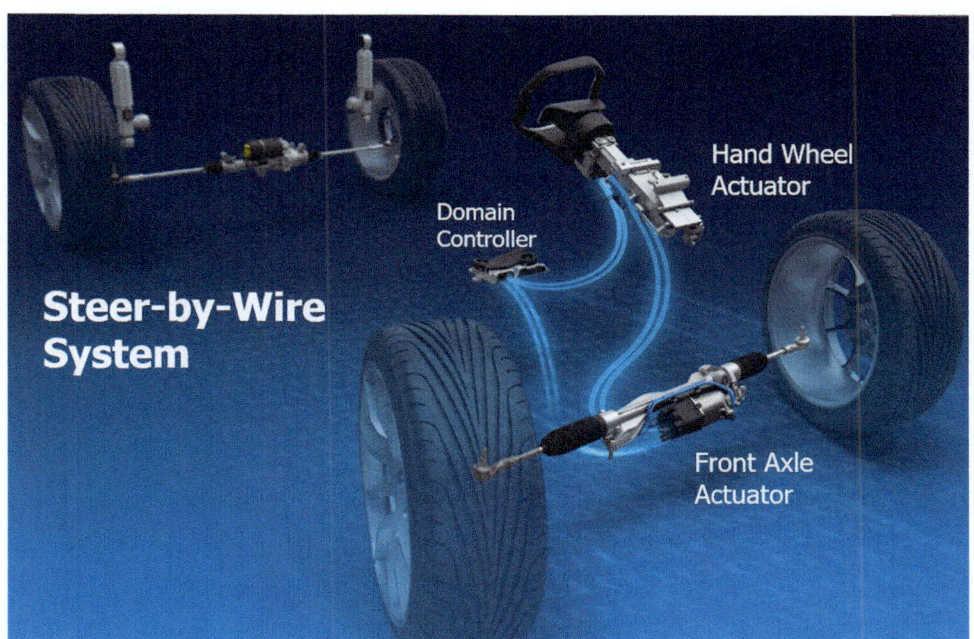

◘ Fig. 24.5   Steer-by-wire steering system (ZF)

shaft does not exist (◘ Fig. 24.5). The steering wheel rotation is measured by means of an electronic steering angle sensor. Depending on this information, an electronic control unit determines a nominal value for the actuating movement of the controlled actuator which is coupled to the steering linkage. At the same time, the control unit determines a nominal value for the actuating force of a second electronically controlled actuator which is coupled to the steering wheel and generates a haptic feedback signal based on the driver's steering inputs. This means, the steering gear functions including the steering booster of a steer-by-wire system are represented by at least two independent actuators on the steering linkage and the steering wheel, which are only connected to each other via electronic communication. Thus, the steering ratio and the haptic feedback can be optimized independently of each other and can for example be varied as a function of driving speed.

24.2 Power Steering Systems

24.2.1 Hydraulic and Electro-Hydraulic Power Steering Systems

Since its introduction in the earlys 1950 (Chrysler Imperial 1951), passenger car steering systems with hydraulic power support (HPS) have become the dominant steering technology. With the hydraulic rack-and-pinion steering, one end of the rack is inserted into the piston rod of a hydraulic cylinder, which can be alternately pressurized with hydraulic pressure. The direction of pressurization and its height are then controlled by the manual force the driver exerts on the steering wheel via a hydraulic steering valve which is arranged in the area of the input shaft of the steering gear (◘ Figs. 24.6 and 24.7).

Despite the very precise haptic feedback on the driving condition to the driver, hydraulic steering systems have some significant disadvantages:

- Due to limited parameterization, they are only suitable to a limited extent for driver assist functions.
- They are not suitable for HEV/FEV as the pump is driven by the combustion engine.
- Permanent operation leads to high power consumption and consequently high CO2 emission.
- The required hydraulic pipes result in installation space restrictions.

Since the end of the 1990s, electro-hydraulic steering systems (EPHS) have been used as a transitional technology which enables an even better haptic feedback of the driving condition compared to HPS. The controllable motor pump unit (MPA, see ◘ Fig. 24.8) used in these systems in combination with a hydrostatic steering gear with low-leakage rotary valve and steering position sensor enabled the implementation of power-on-demand systems. Compared to the purely hydraulic steering system, an EPHS shows substantial improvements in terms of efficiency and packaging and thus becomes independent of the combustion engine. However, the EPHS is also not suitable for driver assist functions that require an active steering intervention.

EPHS is still used in small volume production because of the comparably low application effort, but not in high-volume applications.

24.2.2 Mechatronic Power Steering Systems

Meanwhile, the different types of hydraulic rack-and-pinion steering systems have been almost completely replaced by rack-and-pinion steering systems with mechatronic power amplifier (EPS—Electric Power Steering). In the mechatronic booster, instead of the hydraulic cylinder, an electric motor is used which is attached to either the steering column or the rack via transmission elements (◘ Figs. 24.9 and 24.10).

Both the EPS steering system and the hydraulic booster control the direction and amount of power boost depending on the manual force the driver exerts on the steering wheel. The hydraulic steering valve is replaced by an electronic steering torque sensor. The output signals of the steering torque sensor are converted

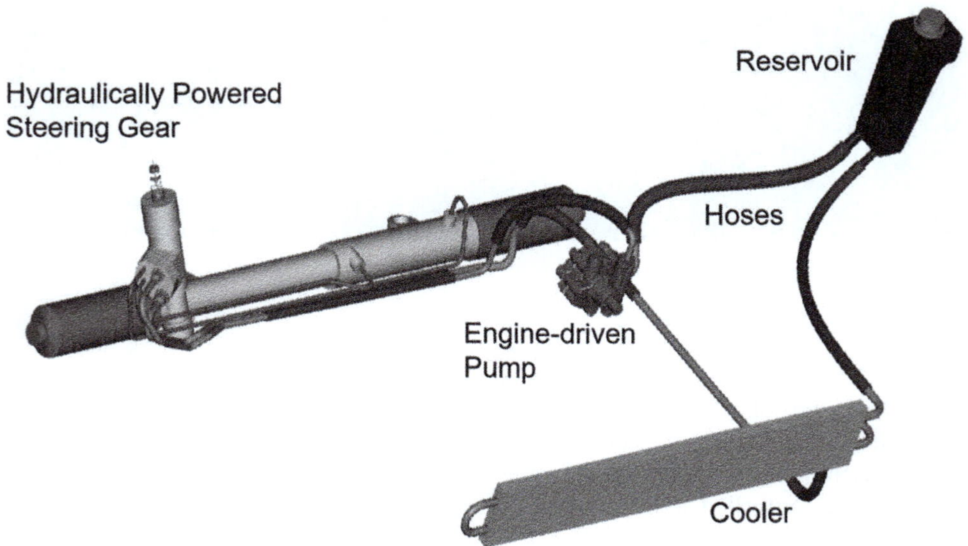

◘ **Fig. 24.6** HPS steering system (TRW) [7]

24

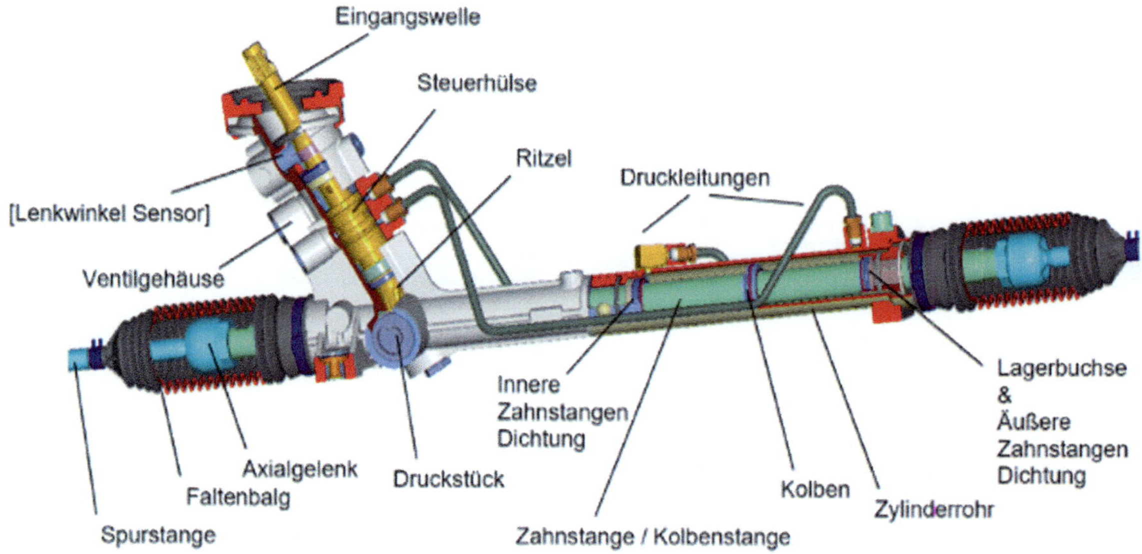

Fig. 24.7 Rack-and-pinion steering gear with integrated hydraulic booster (TRW)

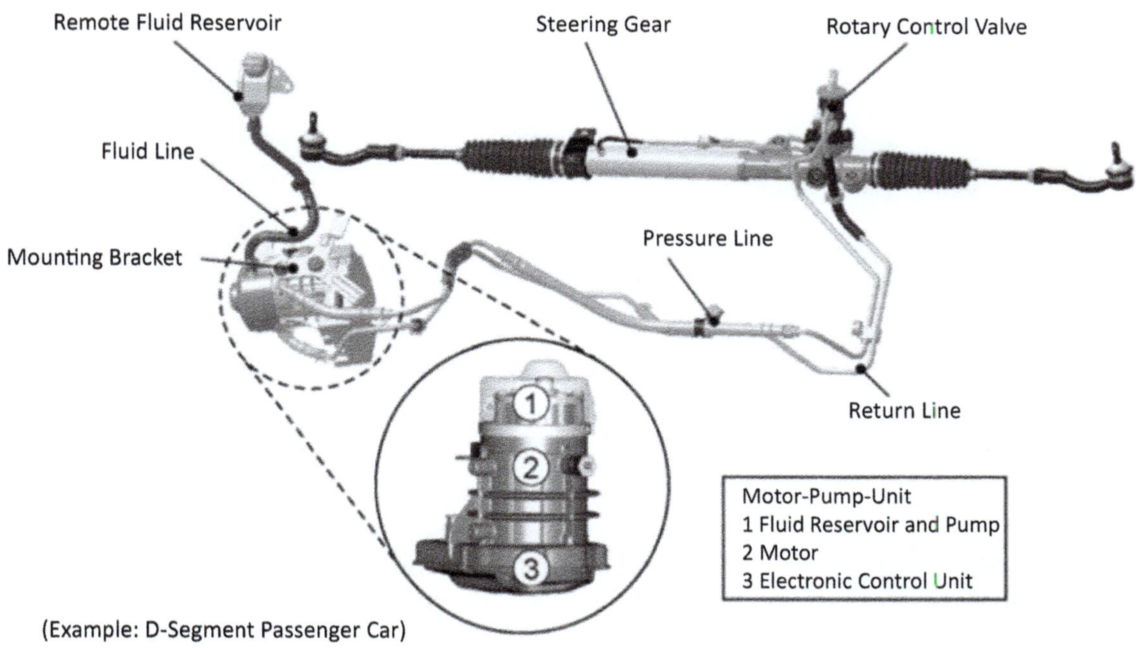

Fig. 24.8 Electro-hydraulic steering system (TRW) [3]

into control signals for the power electronics by an electronic control unit which can be used to continuously regulate the phase current and thus the output torque of the electric motor. Due to the high-power density and the low torque ripple, multi-phase permanent magnet motors with electronic commutation are generally used. Rack-and-pinion steering systems with mechatronic power amplifier are also referred to as electro-mechanical power steering, or abbreviated electric power steering.

24.2.3 Combined Hydraulic and Mechatronic Steering Systems

In heavy trucks, the mechanical power required to actuate the steering system at the desired speed cannot be supplied solely by the electric motor that is driven by the vehicle power supply, due to the high forces that need to be generated. As a consequence, a combination of a hydraulic and a mechatronic steering system is used in these vehicles, in order to facilitate the options

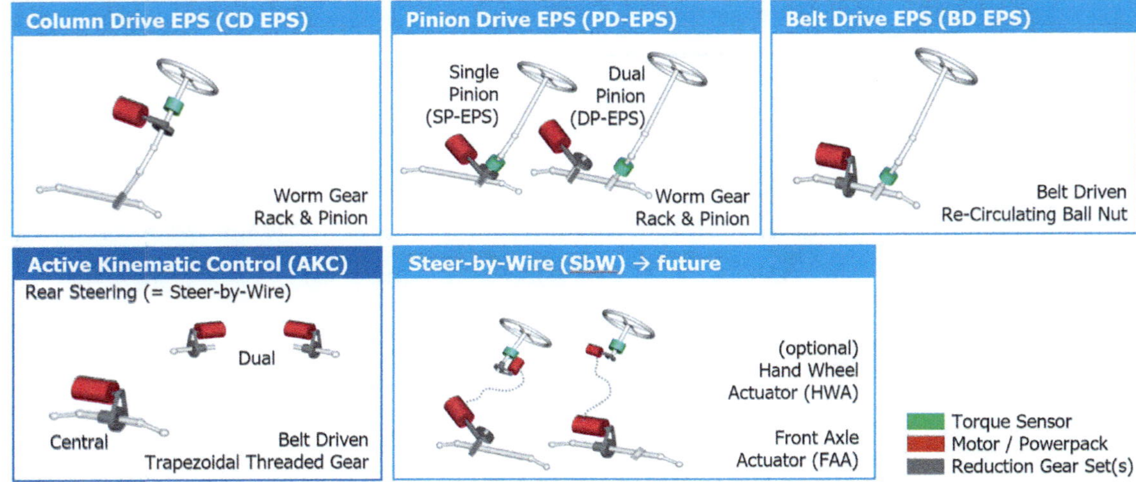

◼ Fig. 24.9 Topologies of different electro-mechanic steering systems (ZF)

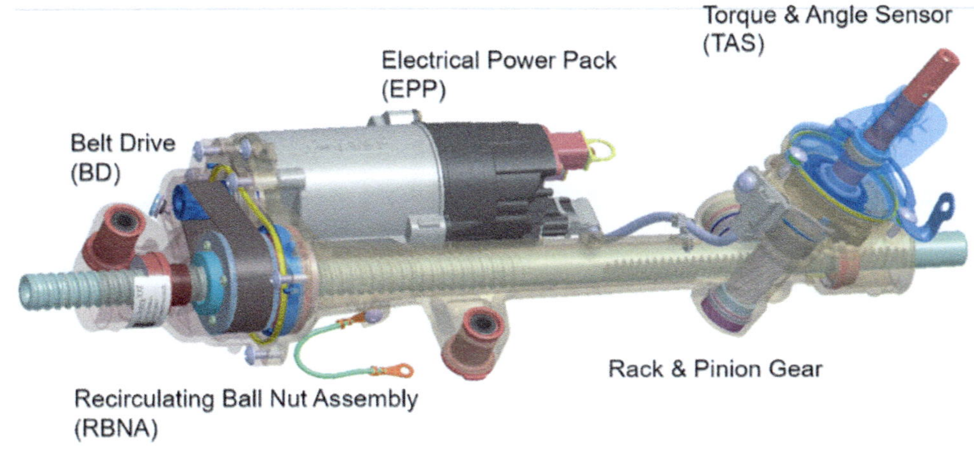

◼ Fig. 24.10 Rack-and-pinion steering gear with integrated mechatronic booster (TRW)

◼ Fig. 24.11 Parameterizable "ReAx" steering gear designed by ZF

share of the force amplification. An additional mechatronic force amplifier, consisting of an electric motor with transmission and a steering torque sensor, is installed close to the steering column or directly on the input shaft of the block steering system. As a result, the hydraulic steering valve is activated by the parameterizable output torque of the mechatronic force amplifier. ◼ Fig. 24.11 shows a version of this steering type with gear-mounted electric motor for heavy trucks.

24.2.4 Mechatronic Steer-By-Wire Systems

In the steer-by-wire system, the steering wheel rotation is recorded by means of an electronic steering angle sensor. Depending on this information, an electronic control unit determines a nominal value for the actuating movement of an electronically controlled actuator which is coupled to the steering linkage. The con-

described in the following chapters that result from the use of a mechatronic steering system. A conventional hydraulic block steering system takes over the major

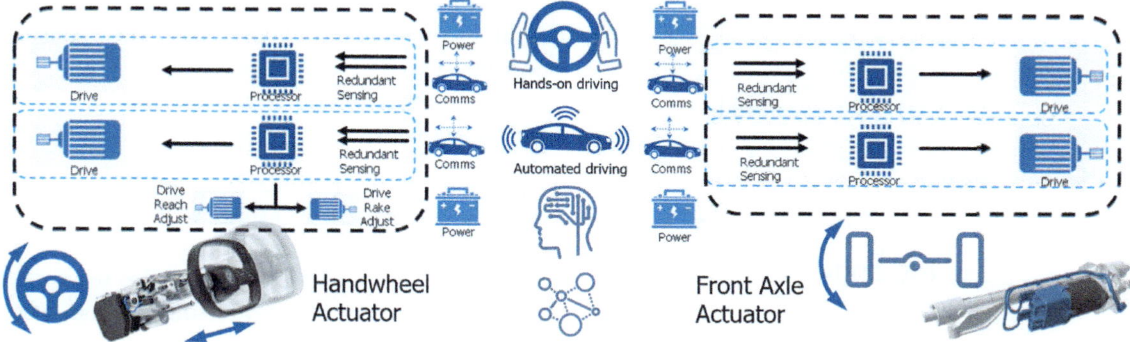

Fig. 24.12   Architecture of a steer-by-wire steering system (ZF)

trol unit simultaneously determines a nominal value for the actuating force of a second electronically controlled actuator which is coupled to the steering wheel and which generates a haptic response to the steering inputs of the passengers. In the steer-by-wire system, the steering gear functions, including the steering booster, are represented by two mutually independent actuators mounted on the steering linkage and the steering wheel, which are only connected to each other via electronic communication (**Fig. 24.12**). Thus, the steering ratio and the haptic feedback can be optimized independently of each other and can for example be varied as a function of driving speed.

24.3 Rear Axle Steering for Passenger Cars and Commercial Vehicles

First active rear axle steering concepts were already implemented in the beginning of the 1930s. These concepts focused mainly on increased maneuverability.

Rear axle steering systems for vehicles with double or triple axles are common in the commercial vehicle sector. The benefit of actuated double or triple axles is a significant reduction of tire wear depending on the utilization profile. The rear axle steering is actuated via the front axle steering system, either by a direct mechanical connection or by a hydraulic transfer via master and slave cylinders. In the new millennium, electronically controlled hydraulic systems were also introduced in the truck sector. These systems are parameterizable with regards to the rear axle steering angle in a closed loop control, i.e. without dependence on the front axle steering angle.

The introduction of ABS control units in the passenger car sector at the end of the 1970s initiated the success story of electronic systems. Starting with ABS, systems that influence the vehicle's behavior, especially in the longitudinal dynamic driving range, were developed. With the help of the Electronic Stability Control (ESC), it was then possible to actively influence the yaw motion of the vehicle by using the brakes for steering.

In parallel to the development of the ESC, the first four-wheel steering systems and active rear axle kinematics were developed. Although these systems were partly used in volume production vehicles, they did not prevail for reasons of costs and reduced functionality.

Compared to previous systems, the technological advancements of the rear-wheel steering system into an intelligent mechatronic system facilitate new target values. There is further potential to optimize the chassis system in terms of driving dynamics while improving driving comfort and thus significantly improving the spread. Furthermore, it is now possible to adjust the vehicle handling depending on the driving situation and link the mechatronic chassis actuators in the integrated network, in order to control the vehicle in longitudinal, lateral, and vertical dynamics at the optimal operating points in all driving situations.

Active Kinematics:

The design of modern wheel suspensions aims to combine precision and elasticity. Wheel suspensions should be sufficiently elastic to absorb jolts, but also allow to precisely guide the wheels and position them exactly in a suitable manner. The elasticity of the wheel suspension is used to achieve wheel movements that have a favorable influence on vehicle handling and driving stability. Wheel suspensions that meet the different requirements for kinematic and elastokinematic properties to a great extent require quite complex designs with a variety of possible influencing factors. However, there is always a trade-off between driving comfort and driving dynamics.

Regarding for example the toe angle change on the rear axle via the spring travel of a passive chassis, the toe angle change results from the kinematics via the wheel stroke and the superimposed elastokinematics. However, if the length of the steering angle of the toe link could be actively adjusted, the solution space

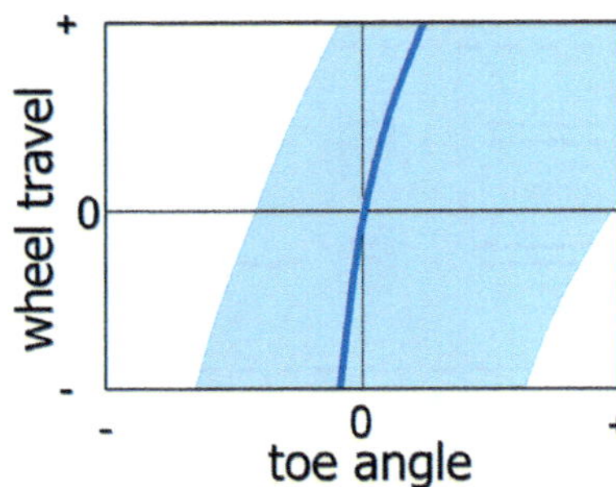

Fig. 24.13 Toe-in angle curve as function of wheel stroke versus active control with dual actuator (ZF)

CAN, while only the left actuator sends a response. Internal communication takes place via a master–slave connection [4].

Central Actuator

Depending on the requirements and available installation space, there are solutions with only one actuator in the center of the rear axle, also called "central actuator" (Fig. 24.15). These systems also work energy-efficiently according to the power-on-demand principle.

Characteristics of rear axle steering with central actuator:

- Toe adjustment via centrally arranged steering gear.
- Trapezoidal thread drive with self-locking device.
- Belt drive.
- Locking mechanism not required due to coupling left/right.

Even large vehicles can skillfully maneuver with more actuating force and higher actuating stroke. Future actuators allow for a higher actuating force of 11 kN (instead of 8 kN to date), enabling the application in vehicles up to 3.5 tons weight. This is primarily a benefit for battery-powered vehicles (BEV) with higher mass determined by the system. They generally also have a longer wheelbase because the energy accumulators are usually located between the axles. Without rear axle steering, these vehicles would be difficult to maneuver, especially in urban environments.

The steering assistance is provided by an electro-mechanical actuator without mechanical connection to the steering wheel. Just like the dual actuator, it is therefore a pure by-wire system, with the advantage that the intelligent actuators can be integrated into the active control network of the respective passenger car. It then supports the function of other active systems—for example in combination with the stabilization via the brake (ESC). If the rear axle steering system and the braking system are networked, stabilizing brake

would be significantly extended by the surface shown (Fig. 24.13).

The adjustment and regulation of the track is therefore an important task in chassis development. It has a major influence on optimum vehicle handling, because the toe-in angle in the chassis is responsible, among other things, for straight-ahead driving during braking and for the steering precision perceived by the driver, ref. to [2].

Dual Actuators

Dual actuators adjust the track via a variable-length control arm on each wheel of the rear axle. The individual actuators essentially consist of an electric motor, a toothed belt drive, and a trapezoidal threaded spindle. The trapezoidal threaded spindle translates the rotary motion of the drive into a longitudinal motion. Sensors record information on the movement and state of the actuator. The spindle is designed to be self-locking, i.e. it blocks under the influence of external loads. A locking mechanism can also lock the system. An integrated electronic control unit usually controls the actuators (Fig. 24.14).

The two actuators are connected virtually via a separate bus (private CAN) and "listen" to the vehicle

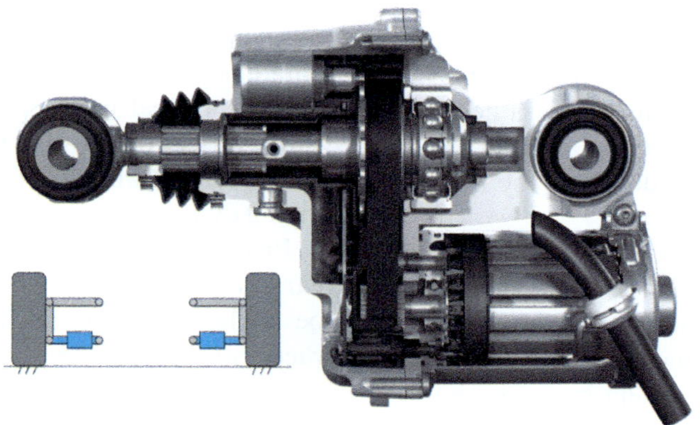

Fig. 24.14 Dual actuator (ZF)

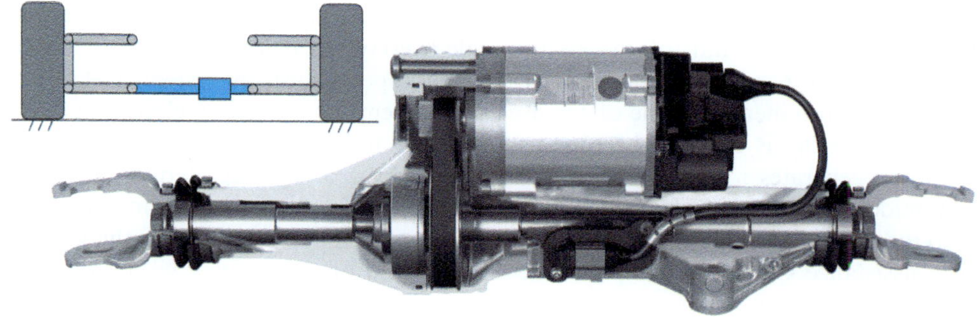

□ Fig. 24.15 Central actuator (ZF)

and rear axle interventions improve vehicle handling during deceleration. Thus, the system increases safety and driving dynamics at the same time. When braking on surfaces with different grips, the stopping distance is shortened. With the coupling of the two front axle and rear axle steering systems, safety can be increased even on slippery surfaces to prevent break-out during lane changes and overtaking. Networking the systems offers further potential for future partially or fully automated vehicles in order to be able to perform automated steering maneuvers.

In active rear wheel steering systems, in-phase and antiphase steering actuation can be realized. While steering in antiphase leads to increased maneuverability of the vehicle, steering in phase improves driving stability due to the apparent longer wheelbase (□ Fig. 24.16).

While steering the rear axle in phase, the current rotational pole of the overall vehicle moves backwards, which corresponds to the same pivot point of a conventional vehicle with a longer wheelbase [1].

The maximum angles on the rear axle range from 0.5° to 5.3° for the different system designs. Large steering angles are always required when steering in antiphase to minimize the turning circle diameter. Currently, systems are on the market which enable a steering angle of up to 10°.

Advantages of vehicle handling with rear axle steering include agility, smaller turning radius, and higher vehicle stability. With sports cars that have their center of gravity close to the rear axle, there is also a significantly more dynamic response to steering impulses.

24.4 Basic Steering Functions for ADAS and AD

In addition to the elementary functions of the steering system as explained in ► Sect. 25.1, i.e. converting the steering input of the driver via the steering control device into a change in direction of their vehicle and applying a haptic feedback on the steering control device, modern steering systems provide additional basic steering functions that are important for the implementation of driver assistance systems. In principle, a distinction must be made between steering functions that influence the steering torque applied by drivers on the steering wheel and steering functions that cause a controlled swiveling motion of the wheel carriers.

24.4.1 Steering Torque Overlay

Influencing the steering force is also referred to as the steering torque overlay, since the primary task is to modulate the steering torque to be applied by the drivers on the steering wheel in certain driving situations and to support the passengers in executing the driving task by means of the haptic feedback generated by it.

For this purpose, the electronic control unit of the steering system is equipped with an interface which communicates steering torques in amount and direction, for example by means of CAN bus messages. The steering torques requested by an advanced driver assist system via this interface are additionally linked to

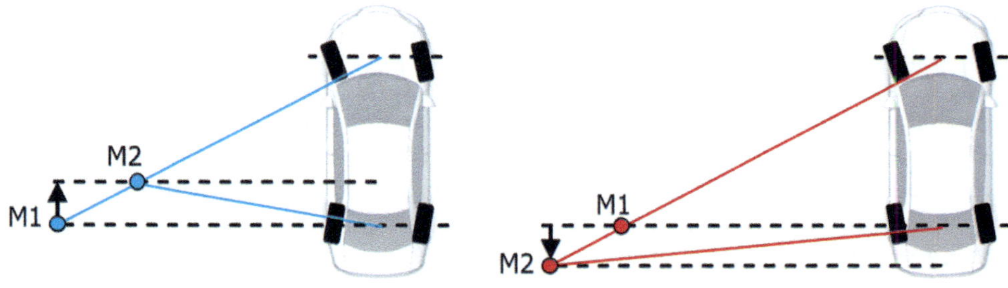

□ Fig. 24.16 Virtual wheelbase change during different steering maneuvers (Porsche)

the steering torques resulting from the current driving situation by the electronic control unit of the electric power steering and are thus perceptible to the driver via the steering wheel. A prominent example of the use of the steering torque overlay is the lane keeping assistance system, which generates a steering torque directed toward the middle of the lane when the vehicle comes too close to the edge of the lane.

In many driver assistance systems, a complete safeguarding of the steering torque requirements cannot be realized with reasonable effort. In this case the steering torque requirements are subjected to a gradient and amplitude limitation by the electronic control unit of the steering system, in order to ensure the functional safety of the driver assist system. As a result, faulty steering torque requirements are limited by the driver assistance system to a level that can still be safely mastered by drivers. A steering torque of ± 3 Nm has been established as a reference value for amplitude limitation. The final parameters of the gradient and amplitude limitation are usually determined after test drives with error activation.

24.4.2 Steering Angle Overlay

In conventional steering systems there is a fixed connection between the steering wheel rotation and the swiveling motion of the wheel carriers due to the mechanical coupling via the steering gear and the steering linkage, which is described by the steering ratio. As explained in ▶ Sect. 25.1, the fixed steering ratio represents a compromise between the optimal values for low and high driving speeds. In addition, a fixed steering ratio limits the implementation of driver assist functions, which are supposed to directly influence the vehicle's direction changes by means of interventions in the steering system. The steering wheel rotations associated with these functions must not be perceived as irritating to the passengers.

The desire to be able to design the steering ratio variably depending on the driving speed, or to enable corrections to the driving direction of the vehicle without steering wheel rotation by means of steering interventions by a driver assistance system, has led to the development of the steering angle overlay. Steering angle overlay is defined by a freely definable additional swivel angle which is dynamically superimposed on the swivel angle of the wheel carriers, which results from the steering wheel rotation with a fixed steering ratio.

In the steer-by-wire steering system, the steering angle overlay can be achieved as a software function. However, compared to conventional steering systems, steer-by-wire steering systems require a high technical effort to ensure fail-safe operation.

In a conventional steering system, the steering angle overlay can be realized with the help of a so-called superimposed steering system. The superimposed steering system consists of an electric motor with electronic control unit and a mechanical planetary gear. The planetary gear is integrated in the steering gear input shaft connected to the steering wheel. The rotational motion of the input shaft of the steering gear is then composed of the steering wheel rotation and the rotational motion of the electric motor. With the help of corresponding software functions controlling the electric motor, the rotational motion of the steering gear input shaft is thus kinematically decoupled from the steering wheel rotation without giving up the mechanical coupling. However, the kinematic decoupling is also limited because the power flow among the steering wheel, superimposed steering system, and the steering gear is permanently maintained by the mechanical coupling, and system dynamics-related irregularities in this power flow can impair the haptic feedback on the steering wheel.

The superimposed steering system can be completely deactivated in the event of an error without impairing the elementary steering functions of the conventional steering system, with a device that mechanically blocks the electric motor of the superimposed steering system automatically under certain conditions. Compared to a steer-by-wire steering system, the functional safety of the superimposed steering system is thus much easier to guarantee.

Apart from realizing a variable steering ratio as a function of the driving speed, the application of the steering angle superimposition for advanced driver assist systems has not played a major role until now. If the safeguard against faulty steering angle requirements on the part of a driver assist system can only be ensured by gradient and amplitude limitation on the electronic control unit of the superimposed steering system, this significantly limits the possible scope of the kinematic decoupling of the steering angle requirement from the steering wheel rotation and thus significantly reduces the benefit of such driving assist systems.

24.4.3 Steering Angle Control

The external control of the steering system by an advanced driver assist system without the driver being involved is referred to as steering angle control. The electronic control unit of the steering system is in this case equipped with a communication interface via which a steering angle can be specified in terms of amount and direction, for example by means of CAN bus messages. Depending on the definition agreement, the steering angle refers to the swivel angle of the wheel carriers or alternatively describes a corresponding rotational movement of the steering wheel.

The best-known example of a steering angle control application is parking assistance systems in which the steering system is automatically actuated during reverse parking.

Steering angle control is the most important steering function for automated driving. While a sudden failure of the steering angle control can still be tolerated in parking assist systems due to the very low driving speed, the safety analysis of applications for automated driving beyond the steering speed leads to high requirements with regard to the failure safety of the steering function. With all the characteristics of automated driving that enable drivers to take their hands off the steering wheel, these safety requirements can only be fulfilled by a fault-tolerant design of the electric drive and the electronic control unit of the steering system, including the communication interface for transmitting the steering angle requirement.

To avoid the risk of a malfunction of the control quality caused by manual steering interventions of the drivers, the steering angle control with steering force assistance is usually not active in parallel to the manual steering process, but only after a function change. Switching between the functions should not cause abrupt steering torque or steering angle changes on the steering wheel which could irritate the driver. During parking assist, the transition to steering angle control therefore only takes place when the vehicle is stationary and only when the driver has released the steering wheel for parking. During automated driving, the request to activate the steering angle control is only implemented once a straight-ahead driving with low steering forces is guaranteed. In both cases, deactivation occurs when the driver intervenes manually in the steering angle control and a previously defined steering torque threshold is exceeded.

24.4.4 Cooperative Steering Angle Control

In the course of the development of automated driving, the lane guidance assistance system has become established as a preliminary stage for highly automated driving. In contrast to the lane keeping assist system, which only generates a steering torque directed toward the middle of the lane, if the vehicle comes too close to the edge of the lane via steering torque superimposition, the lane keeping assist system offers complete automatic lane guidance via steering angle control.

The lane guidance assistance system differs from the automatic lateral guidance of the vehicle in highly automated driving mainly due to a considerably lower technical effort in terms of functional safety. These

simplifications are only possible because the safety concept of the lane guidance assistance system assumes that the drivers are always prepared for intervention. Within the scope of the UN-ECE approval guidelines, the lane guidance assistance system may therefore only be active while the driver does not release the steering wheel.

If drivers keep their hands on the steering wheel during automatic lane guidance maneuvers via steering angle control, this boundary condition must be taken into account in the control engineering design of the steering angle control, regardless of whether it is a lane guidance assist system or highly automated driving. The control processes within the steering angle control loop should not be perceived by the driver as a disturbing haptic feedback on the steering wheel and the intended manual interventions of the driver in the steering angle control should not lead to a possibly irritating counter reaction of the steering angle control to the driver. These requirements have led to the development of the so-called cooperative steering angle control.

The cooperative steering angle control does not require a function switch between manual steering with steering force assist and steering angle control. Steering movements of drivers and external steering angle requirements on the part of automated lateral guidance are linked to one another by means of control technology in such a way that the drivers can temporarily overlay the automated lateral guidance while maintaining natural haptic feedback.

The requirements that led to the development of the cooperative steering angle control for electric power steering systems also apply to steer-by-wire steering systems. Since there is no mechanical coupling between the steering actuator, which regulates the swivel angle of the wheel carriers, and the steering wheel, and therefore the haptic feedback on the steering wheel is instead generated by means of a separate actuator, completely different control engineering concepts are used here.

24.4.5 Supplementary Rear Axle Steering

The reaction of a vehicle to steering inputs from passengers at low driving speeds is mainly determined by geometric correlations, above all by the wheelbase and steering ratio. Other influencing factors are noticeable at higher driving speeds, such as the mass distribution of the vehicle body, the power transmission characteristics of the tires, as well as the kinematic and elastic characteristics of the wheel suspensions on the front and rear axles. The steering behavior of a vehicle in both operating ranges can be optimized by means of an additional steering device which actively controls the swiveling motion of the wheel carriers on the rear axle.

At low driving speeds, a pivoting motion in antiphase of the wheel carrier on the rear axle relative to that of the front axle reduces the turning radius of the vehicle and thus improves maneuverability. At higher driving speeds, however, a pivoting motion in phase makes it possible to build up tire side forces on the front and rear axle almost simultaneously, which contributes to improving driving stability, especially during fast lane changes and evasive maneuvers.

The control of the additional steering device on the rear axle corresponds to a steering angle control. However, the electronic control unit of the steering device is usually not supplied with steering angle specifications via the communication interface, but with the steering angle output by the steering system on the front axle and other signals available in the vehicle. An algorithm implemented on the electronic control unit of the steering device continuously checks the plausibility of this information and calculates the target value specification for the steering angle control on the rear axle. In the event of an external calculation of this target value specification by a driver assistance system or an integrated driving dynamics control, the plausibility check on the steering device control unit may not be possible due to a lack of sufficient information. The integrity of the target value specification via the communication interface must then be ensured in another way.

24.5 Safety Requirements for Steering Systems and Steering Functions

24.5.1 Type Approval Regulations

For many countries, the most important regulations for the homologation of steering systems are defined in the UN ECE R79. Some countries that have not ratified R79 have specified their own regulations, for example China in GB 17,675. In addition to a standard type registration for volume production vehicles, individual registrations are also possible, i.e. for vehicles for disabled persons. The homologation of steer-by-wire has been possible according to R79 since 2005. In China, release of steer-by-wire was not permitted in the releases of GB 17,675 up to 2020, but was made possible in the 2021 revision.

24.5.2 Fatigue Testing of Mechanical Components

As usual for safety–critical components in the vehicle, steering systems are designed for a specific requirement regarding service life and loads. In addition to strict ad-

herence to quality standards and requirement-based testing of mechatronic components and software contents, the complete inspection of mechanical components is also becoming increasingly important.

A test strategy developed for this purpose is based on three pillars:
- Basic requirements that steering manufacturers determine for modern steering systems and their safety
- Analyses based on an FMEA
- A DRBFM analysis for the respective application/design.

The requirements for steering systems are derived from market analyses, the experience of manufacturers from the development of steering systems, and a manufacturer-internal definition of the operating conditions of a steering system during a vehicle life (mission profile). The Failure Mode & Effect Analysis (FMEA) deals with possible error cases during the production and operation of a steering system and evaluates the risk potential for every single error case. A test method is defined for the error modes under consideration. The Design Review Based on Failure Mode (DRBFM) analysis considers the design differences between a specific application and already known and validated designs and derives an additional scope of validation from this. The resulting scopes of validation may only apply to individual components or subsystems.

The test strategy defines which validation tests have to be made on system level, on subsystem level, or on component level. It follows a superordinate development strategy that compares requirements and validation methods. This is manifested in the separation of the steering system into subsystems, which is either done based on the physical structure or on functional units. Often the initial focus is on the physical structure, and then standard assemblies from standard components are set up accordingly. These standard components and assemblies can then be validated individually and incorporate preliminary knowledge into each new validation. However, the problem with this test approach is that components and subsystems must meet the requirements of a range of applications and can therefore also be over-dimensioned for a specific application and as a result be too expensive. Furthermore, with this approach, the interfaces between the subsystems are only inadequately mapped so that aspects such as friction, wear behavior, and noise characteristics are only addressed at the system level. This can be represented in a better way by a functional distribution of the steering system, which enables clearances or lubrication properties to be taken into account in these functional groups. An example of a functional unit is the conversion of the steering wheel rotation movement into the linear positioning movement of the rack by the "pinion, rack, yoke" assembly.

The validation contents for the defined validation level results determine whether a physical or functional approach should be used to describe the requirements to be validated. In both cases, the requirements that are initially only available on system level are broken down to the subsystems and, if necessary, to the components in order to be able to define validation contents at the respective abstraction levels. Historically, the release was granted solely for component or system tests. The release via component tests is increasingly replaced by virtual tests on digital twins. For this purpose, the component characteristics, such as stiffness or the continuous vibration behavior, are required and these are used as a basis for the development of the computer models. The comparison between the computer and hardware model is repeated permanently from the component to the system level, thus creating the digital specimen, i.e. digital twin. Ideally, the virtual model runs in the background of the physical tests in order to show differences from both test runs.

Since additional functions and desired customer variation via the software have an enormous influence on the mechanical specimen, not all variations with the classic validation approach can be released within the given period. The release will now take place via a hybrid model, i.e. a validation is carried out according to the standard procedure combined with the use of the digital twin.

Initially this procedure requires a considerably higher amount of resources because, in addition to physical testing, the virtual model also needs to be developed and validated. At the same time, the development of the virtual model reduces the additional testing effort for function or parameter changes and accelerates the release time.

24.5.3 Functional Safety of Electric and Electronic Systems

ISO 26262 (cf. ▶ Chap. 6) is used as an integral element of the product evolution process for mechatronic passenger car and truck steering systems as described below.

The active safety systems for longitudinal and lateral dynamics, especially mechatronic steering systems as well as mechatronic braking systems, are generally classified as ASIL D systems.

24.5.4 ASIL Classification of Steering Functions

Derived from ISO 26262–3 (Automotive Safety Integration Level (ASIL) classification via the risk graph), the vehicle manufacturer evaluates the safety goals and breaks them down to the steering system requirements. This results in different requirements (safety goals) depending on the vehicle (size, architecture, system limits) and application (e.g. regional market requirements or regulations, vehicle design, and degree of automation). ◘ Table 24.1. shows, as an example and in a simplified form, an overview of typical steering system safety goals and corresponding ASIL requirements:

There are different classifications for the failure of power steering assist in EPS systems. The controllability of the vehicle during or after failure of the support in certain operating situations is primarily evaluated. In this case, the range of ASIL classifications reaches from QM to ASIL D and thus requires the availability of corresponding E/E architecture families in order to cover the respective vehicle-specific requirements. Typically, the large jump from ASIL B to C already partly

◘ **Table 24.1** Exemplary safety goals (ZF)

Steering system—safety goal (SIMPLIFIED DESCRIPTION) Avoidance while driving of	ASIL requirement
Unmanageable, unwanted steering actuation	ASIL D
Unmanageable heavy steering due to electric/electronic dampening of the motor	ASIL D
Loss of power assist of an EPS system	QM to ASIL D *
Sudden variation or interruption of power assist	QM to ASIL D *
Loss of wheel position control in a Steer-by-Wire system	ASIL D
Undesired removal of the (retractable) steering wheel from the grip area of the driver	ASIL D

* depending on application-specific boundary conditions

requires a 2-channel E/E architecture, for ASIL D additionally complete redundancy and potential diversity. The trend toward higher requirements is driven by more direct steering gear ratios, increasing vehicle weights, as well as ADAS and AD functions. The highest requirements apply to vehicles with automated driving functions that include driving at higher speeds as well as steer-by-wire steering systems.

24.5.5 Cyber Security

Further challenges arise for future networked systems within the development and production process. Safe software on a safe control unit cannot be analyzed externally or overwritten with development statuses—so-called fault injection. Future interfaces are potential intrusion gates for digital attacks. Cyber security plays a very important role in preventing external access. General requirements regarding security against external access to mechatronic systems are described in the ISO/SAE 21,434 and SAE J3061 standards.

Due to steering systems being classified as ASIL D, they have been subject to specific cyber security requirements for some time. With increasing networking of modern vehicle electronics and computer architectures as well as the requirement for functional over-the-air updates, the requirements for access to the steering system control units continue to increase (for example to avoid provoked failures of the power steering assist or unintentional steering movements). As a consequence of the high ASIL classification of the steering system (see ▶ Sect. 25.5.4), high Cyber Assurance Level (CAL) classifications usually result. A safeguard against attacks must be in place for the entire product life cycle.

Cyber security therefore not only describes the requirements for safeguarding the software, but also necessitates an additional, integrative development process with considerable safeguarding efforts in the disciplines of project management, systems engineering, testing, and production. In order to achieve a robust cyber security concept, all expert departments must be considered in their entirety and must be supported by quality-assuring processes/measures (analogous to Automotive Spice).

In addition to safeguarding the main interfaces between steering product (supplier) and vehicle/gateway (OEM), a tamper-proof steering system also prevents external accessibility (e.g. plug, wiring, radiation) as well as accessibility to internal components (e.g. ECU, microprocessor/software, sensors) (see ◙ Fig. 24.17). For steer-by-wire systems, one or more specific private communication links between the actuators or control units may be added which must be additionally secured (e.g. encryption of the data streams).

24.6 Safety Concepts for Mechatronic Steering Systems

24.6.1 Fail Safe and Fail Operational Concepts

Standard steering systems with hydraulic, electro-hydraulic, or electro-mechanical power steering support are usually designed to be "fail-safe," i.e. if a serious electronic error is detected, the system must be transferred to a safe state, for example by switching off the power steering assistance. Since the mechanical connection is retained, the vehicle can continue to be steered by the application of (increased) steering torques by the driver. The UN ECE R79 regulates the permissible steering forces on the steering wheel for certain driving maneuvers on vehicle level.

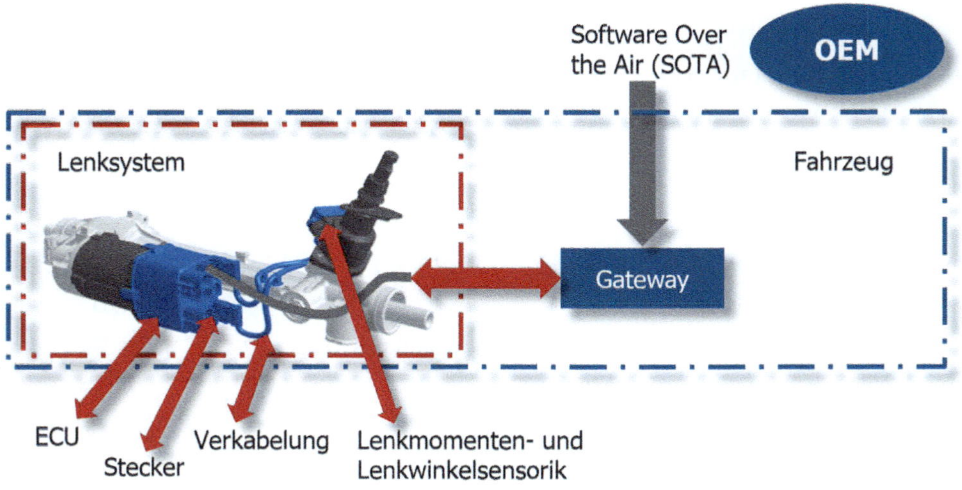

◙ **Fig. 24.17**   Possible access points that need to be secured in a steering system (ZF)

For steer-by-wire steering systems without mechanical fail-safe level (as well as for EPS steering systems that are designed for demanding SAE AD L4 use cases, if applicable), a fail-operational system architecture is required. Following an initial electronic fault, it must be ensured that steerability is maintained by drivers or ADAS/AD systems without violation of an essential safety goal. A temporary, restricted operation of the steering system must then be ensured in order to be able to transfer the vehicle into a safe state.

This fail-operational requirement also applies to the onboard infrastructure, such as redundant power supplies and data buses, and certain functions on domain or central control units.

24.6.2 Fault Tolerance

The requirements for the availability of power steering assistance have risen steadily. The driving factors for this are, among other things, increasing front axle loads (e.g. for SUV, BEV) which result in higher steering forces and more direct steering ratios. The forces required to rotate the steering wheel are thus increased in the event of a loss of power steering assistance making it more difficult to control the vehicle. As described in ▶ Sect. 25.5.4, the corresponding safety goals are therefore typically defined in the range of QM up to ASIL C.

In parallel to these increased availability requirements for certain vehicle types and modern chassis designs, availability is also influenced by the various automation applications. Requirements in the range from QM to ASIL C are common for SAE AD levels 0 to 2. Apart from SAE AD L3/4 use cases at low driving speeds, safety targets up to ASIL D classification are common for more demanding SAE AD use cases above L3, for the availability of the steering function (◘ Fig. 24.18). In addition to the improved mechatronic development processes, the availability of the steering system is also included in the volume production. ◘ Fig. 24.19 shows an example of a possible way to gradually increase fail-safety by adding redundant assemblies in different areas such as sensor system, microprocessor, engine control, and engine topology.

In order to cover the described range of steering system requirements and safety objectives, the development of a modular, scalable electronics architecture modular kit is recommended. ◘ Fig. 24.20 illustrates such an efficient technological solution approach. The redundancies for the key assemblies are increased in various stages, and in the full configuration level as a completely redundant electronics architecture. In addition to the electronic assemblies, the software is then implemented redundantly and requirement-based, also redundantly in terms of diversity.

24.6.3 Application Examples

Three application examples for fault-tolerant steering systems are described below.

24.6.3.1 Fault-Tolerant Electro-Mechanical Power Steering Systems

The column-drive electro-mechanical steering system shown is equipped with a multi-core microprocessor, redundant motor control, and a three-phase motor with two lanes. The torque sensor has a triple-redun-

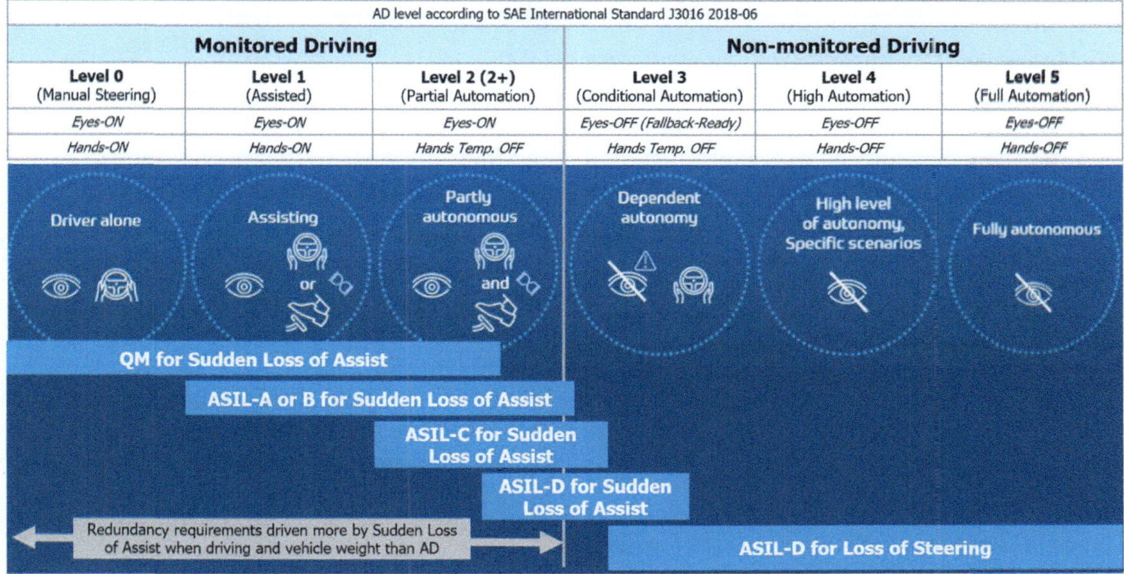

◘ **Fig. 24.18** Allocation of safety targets to SAE AD levels (ZF)

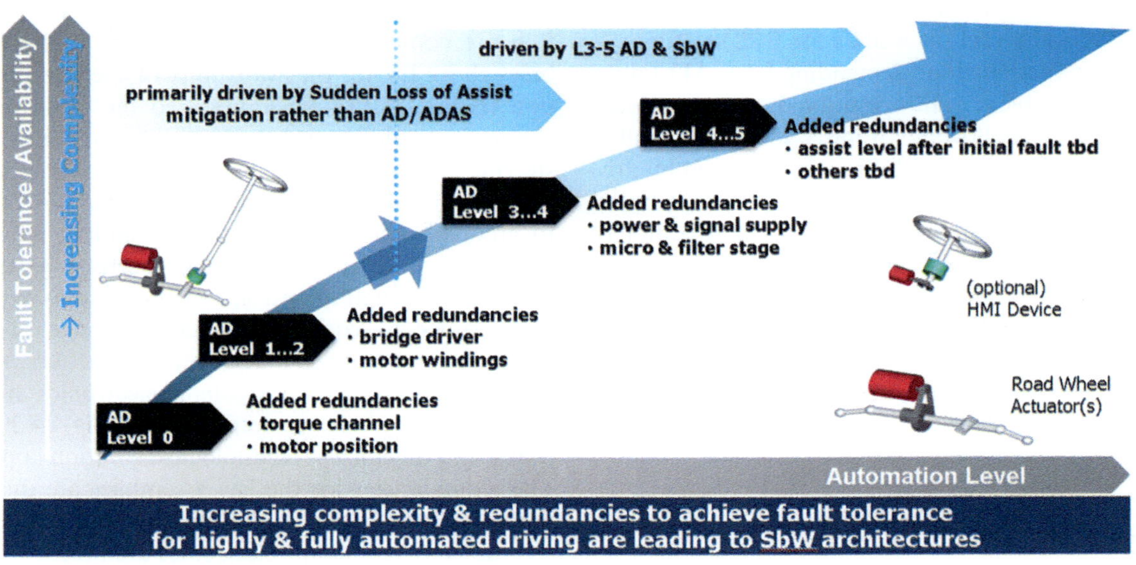

Fig. 24.19 Allocation of steering system redundancies to SAE AD levels (ZF)

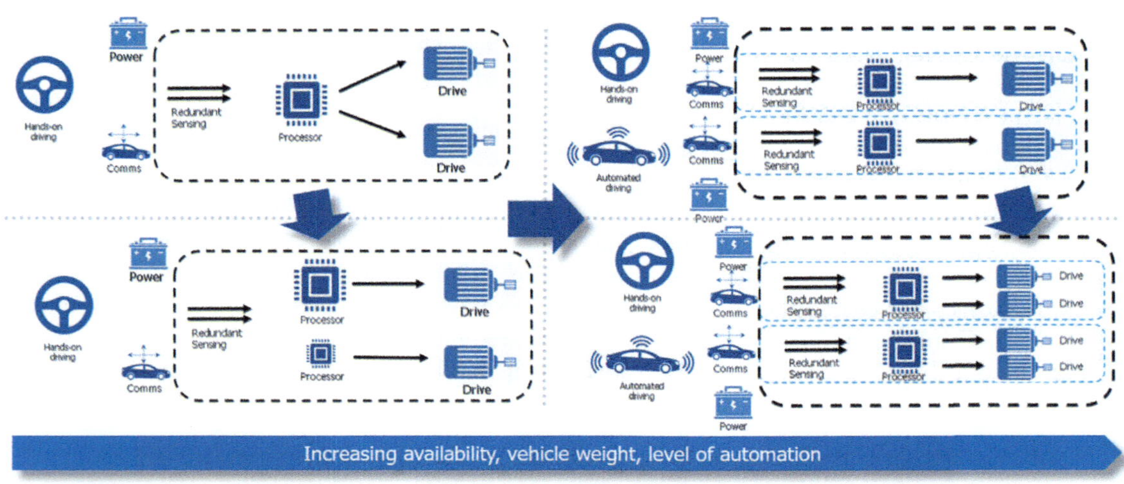

Fig. 24.20 Modular, scalable electronics architecture construction kit (ZF)

dant design and can therefore be operated safely and without functional restrictions after a single fault. Such a system fulfills, depending on the characteristics, QM, ASIL A, or ASIL B requirements. Two block diagrams (Figs. 24.21 and 24.22) of a column drive steering system are shown here, i.e. the functional interfaces of the steering system as well as the interfaces of the corresponding electronic drive module.

24.6.3.2 Steer-by-Wire System with Redundant Actuators

Various types of steer-by-wire steering systems will be used for the mobility of the future, especially for highly and fully automated vehicles according to SAE AD L3-5, in the following vehicle categories:

(1) driverless robot vehicles (SAE AD L4/5) without steering wheel actuation in regular operation.
(2) conventional passenger cars/trucks with steering wheel actuation.

In these intelligent mechatronic steering systems of the future, the driver steering input device for conventional passenger cars/trucks will at first continue to be designed as a steering wheel, but will increasingly be transformable, in particular for SAE AD L4/5 use cases, such as.

— retractable and/or stowable steering wheel

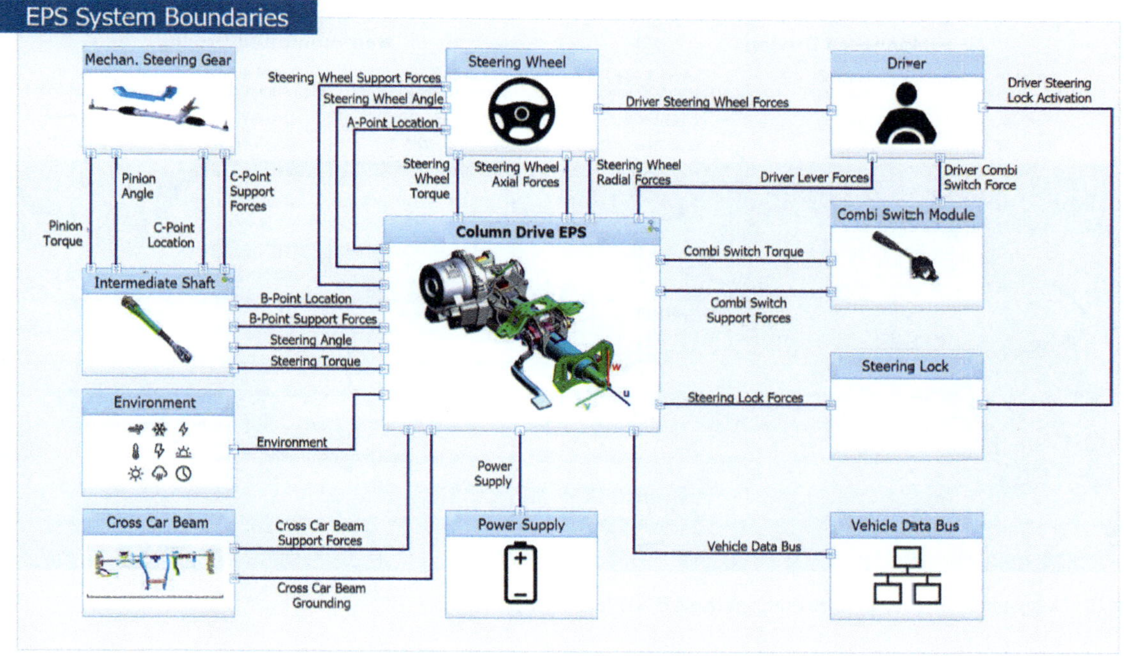

Fig. 24.21 Block diagram of a column-drive steering system (ZF)

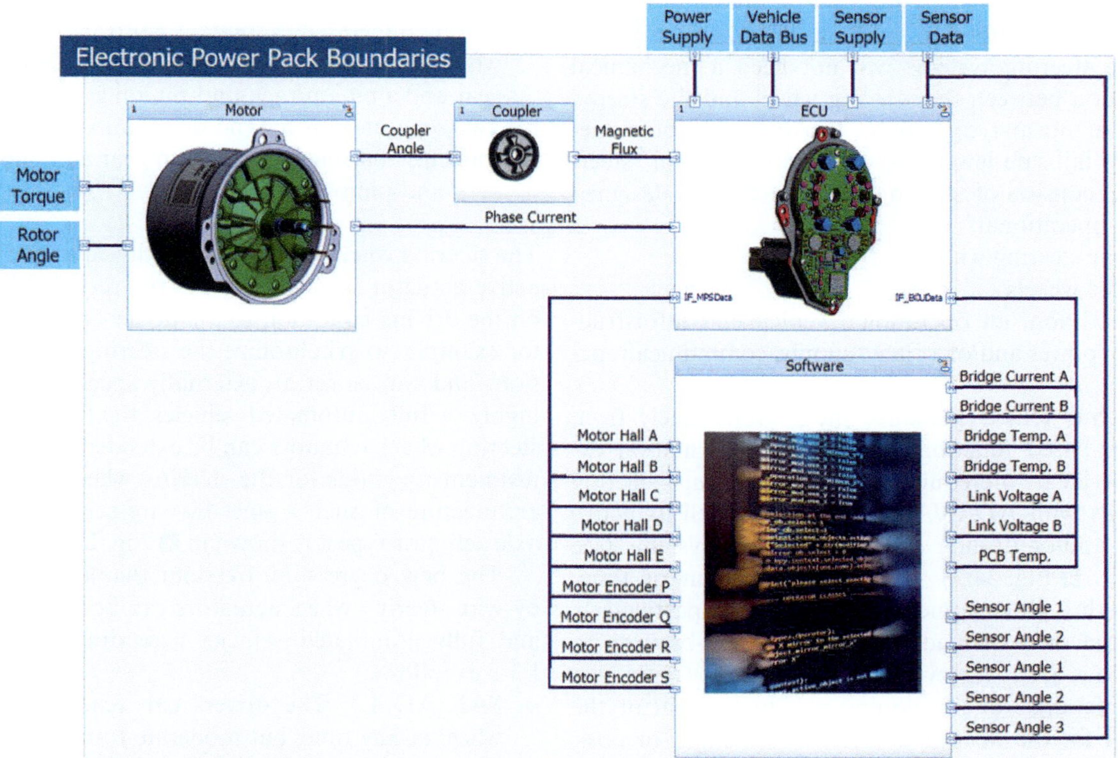

Fig. 24.22 Block diagram of an electronic drive module (ZF) [9]

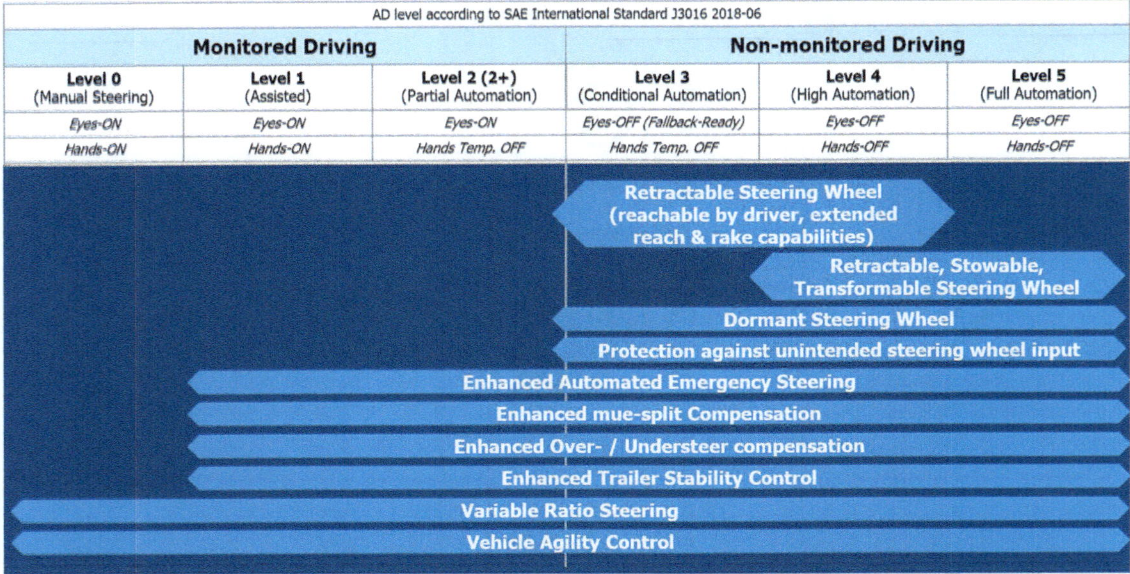

Level 0 (Manual Steering)	Level 1 (Assisted)	Level 2 (2+) (Partial Automation)	Level 3 (Conditional Automation)	Level 4 (High Automation)	Level 5 (Full Automation)
Eyes-ON	Eyes-ON	Eyes-ON	Eyes-OFF (Fallback-Ready)	Eyes-OFF	Eyes-OFF
Hands-ON	Hands-ON	Hands Temp. OFF	Hands Temp. OFF	Hands-OFF	Hands-OFF

Fig. 24.23 Assignment of steer-by-wire function to SAE AD levels (ZF)

- foldable or collapsible steering wheel
- alternative versions (top and/or bottom flattened or open steering wheel, joystick).

As already described in ▶ Sect. 25.1.3, future Steer-by-Wire steering systems will not need a mechanical connection between the steering wheel and the steered wheels in normal operation. The omission of a mechanical fail-safe level (also referred to as "real" steer-by-wire) consists of steering system actuators designed as "fail-operational":

- on the steering wheel,
- on the wheels,
- in addition, an operational vehicle-end infrastructure (power and/or voltage supply, communication).

The lateral vehicle guidance then results solely from software-based functions and digital information exchange—ideal prerequisites for integrating steering functions with ADAS/AD functions as well as active chassis and/or driving dynamics control systems (see also [1]). **■** Fig. 24.23 shows an assignment of functions with high customer value, which are particularly supported or even made possible by steer-by-wire, to the various SAE AD levels.

A safe angle or positioning position output of the actuator for the steering control, a safe angle, or positioning position control of one or more actuators on the wheels, as well as fast, secure data transmission are essential for steer-by-wire systems.

■ Fig. 24.24 e.g. shows two steer-by-wire system configurations which differ in the architecture of the front axle actuator.

The corresponding safety concept of the two front axle actuators shown differs in the characteristics of the redundancies in the area of the drive and the E/E architecture:

a. A fully integrated redundant control and drive unit with two transmission stages (in this case a worm gear and a pinion/rack-and-pinion gear)
b. Two separate control and drive units, each with two transmission stages (one worm gear and one pinion/rack-and-pinion gear).

The steering wheel actuator is usually equipped with an active actuator which transmits the required feedback on the driving condition to the driver and is also used, for example, to synchronize the steering actuator positions and can be set an externally specified angle. For highly or fully automated vehicles, the functionality of steering wheel actuators can be extended by electric adjustment functions for the steering wheel position. An architecture of such a steer-by-wire system with front axle actuator type a is shown in **■** Fig. 24.25.

The new degrees of freedom thanks to the steer-by-wire steering wheel actuators can be used for highly and fully automated vehicles according to SAE AD L3-5 as follows:

- SAE AD L3: The drivers can reach the steering wheel at any time, but moderate retraction and lifting of the steering wheel can offer the drivers increased freedom during certain use cases. For this purpose, the electrical adjustability of the steering column is appropriately enabled and the intended trajectory and positioning is adjusted accordingly via the steering wheel actuator.

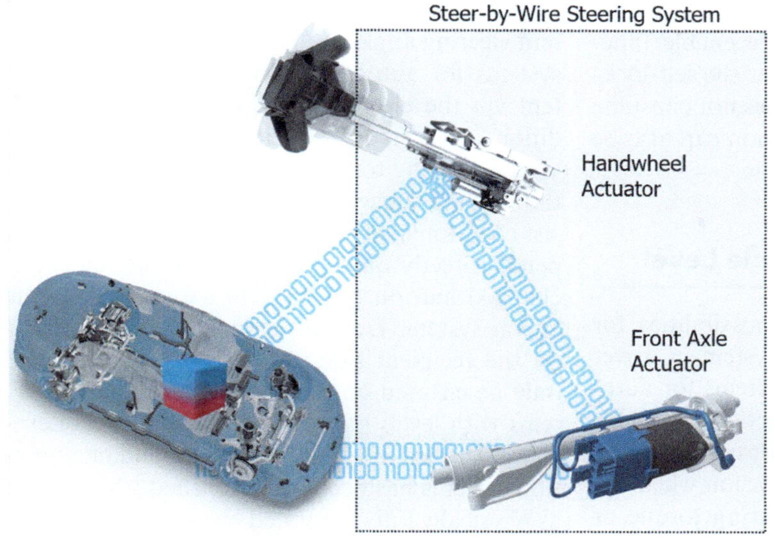
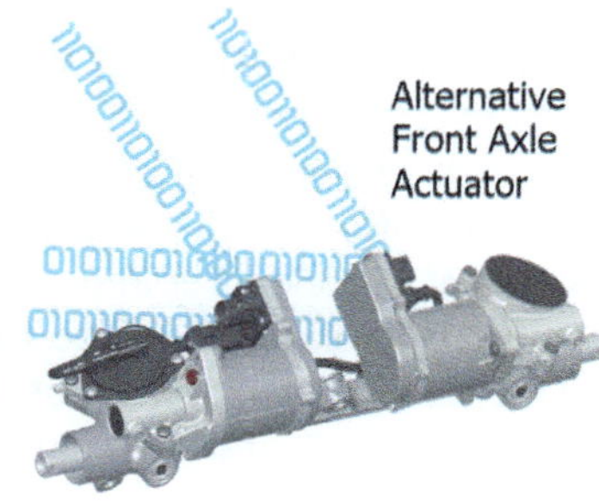

Fig. 24.24 Steer-by-wire system configuration (ZF)

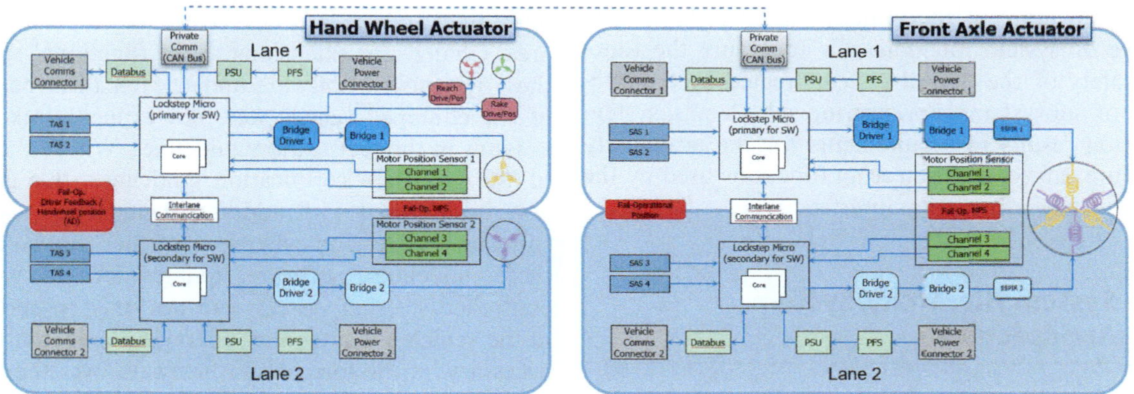

Fig. 24.25 Steer-by-wire system architecture of steering wheel actuator and front axle actuator (ZF)

— SAE AD L4/5: The drivers can no longer reach or use the steering wheel during certain use cases. It may even have to be ensured that drivers or passengers on board cannot intervene unintentionally or intentionally in vehicle control during certain use cases. The steering wheel rotation may optionally be configured to no longer follow the position of the front wheels, instead remaining stationary in the middle position or only providing the driver with an indication of the steering maneuvers initiated. Optionally, the steering wheel can be retracted electrically or even stowed in the cockpit to provide drivers with the best possible freedom.

A steer-by-wire system already established in volume production vehicles is the rear-wheel steering system described in ▶ Sect. 25.3. A modern application example of such a fault-tolerant system design is described in the following section.

24.6.3.3 Intrinsically Safe Rear Axle Actuator

The limits for the positioning and synchronization of mechatronic rear axle actuators are maintained to a very tight tolerance and the safety concept as well as the hazard and risk analysis are classified as ASIL D. The safe state must be reached independently of the available supply voltage, and unintended adjustment must be prevented under all circumstances for all operating points.

Reactions to system errors can be realized via a two-stage concept:

1st Stage: System has functional restrictions.
— System moves into zero position and remains there.

2nd Stage: System can no longer proceed.
— System switches off and remains in the current position.
— Vehicle handling remains predictable and manageable.

The electro-mechanical lock and the self-locking feature of the trapezoidal threaded spindle enable functional safety. The additional advantage of the self-locking feature is that the steering system does not consume energy in a static state because the position can also be kept passive under lateral force fluctuations.

24.7 System Integration at Vehicle Level

In ▶ Sect. 25.4, the most important possibilities for functional integration of the steering system in driver assist systems at vehicle level and in systems for automated driving were described. The functional integration primarily results in requirements for control accuracy, control dynamics, and fault suppression when implementing an externally requested steering torque or steering angle.

In contrast to functional integration, system integration is defined as the integration of electronically controlled steering systems into the electric and electronic architecture of the vehicle. The most important objective of system integration is to ensure the functional safety of the overall system created by the integration of subsystems. For steering subsystems, system integration results in requirements for the availability and failure safety of the steering functions used by the system and the information interface.

24.7.1 System Hierarchy/System Architecture

In the system hierarchy of the electronic and electrical architecture of the vehicle, electric power steering systems and steer-by-wire steering systems have some of the highest requirements regarding fail-safety of all the chassis systems and are currently being developed as complete subsystems. The system integration tasks are thus limited to the connection to the power supply via the electric vehicle power supply as well as the connection to the electronic communication system of the vehicle, such as CAN or Flexray Bus.

As long as the power supply remains active, the elementary steering functions are still available in the vehicle, even in case of a malfunction or after a failure of the electronic communication system. Vehicles with electric power steering which are insufficiently controllable in the event of a failure of the steering power boost, and vehicles with a steer-by-wire steering system must provide the steering system with a fail-safe energy supply. The fail-safety of the energy supply in the vehicle's electric vehicle power supply is sufficiently maintained by two independent voltage sources with completely separated supply lines to the steering system.

Advanced driver assistance systems receive the current steering angle of the vehicle at the vehicle level and systems for automated driving from the steering system via the electronic communication system, in addition to status information. In return, these systems send a steering torque or steering angle requirement to the steering subsystem. The functional safety of an assist or automated function implemented in this way depends directly on the integrity of the information exchanged and on the fail-safe nature of the communication system. Due to the lack of plausibility options on the recipient side, the integrity of the signals can only be ensured on the respective sender side. For systems with levels of automation from SAE Level 3 onwards, the use of at least simple redundant communication systems has become established which are realized as physically fully separated bus systems.

Due to the progressing digitalization in the automotive sector, software updates are becoming increasingly important. Historically, the approval guidelines for safety-relevant systems in the vehicle, such as the steering system and the braking system lagged behind in this area. Changes or extensions to the functional scope of these systems via software update after the type release of the equipped vehicle were not explicitly taken into account in the registration guidelines. Within the scope of the UN-ECE certification guidelines, this gap has now been closed with additional guidelines regarding software updates and cyber security. In compliance with these guidelines, software updates for electronically controlled steering systems can not only be performed in the vehicle manufacturer's service plants, but also, if necessary, in the form of a "firmware over the air update" (FOTA) via an application server in the vehicle.

The trend toward electric and electronic architectures in the vehicle, intended for the centralization of functional scopes on domain or zone control units (◻ Fig. 24.26), has led to the possibility for steering functions to also be implemented on an electronic control unit away from the steering system, with the electric power steering system or the subsystems of steer-by-wire steering systems then replaced by so-called smart actuators. The basic requirements are provided by the introduction of very powerful and fail-safe electronic communication systems in the vehicle, which are combined with the use of domain or zone control units.

The idea of dividing the steering system subsystem into functional software and smart actuators comes with the aim of reducing both the development and system costs of the steering system. The savings potential for development costs is difficult to quantify in advance, because the established development cycle for the complete steering subsystem is then no longer valid and must be replaced by a new development cycle, in which several development partners may be involved.

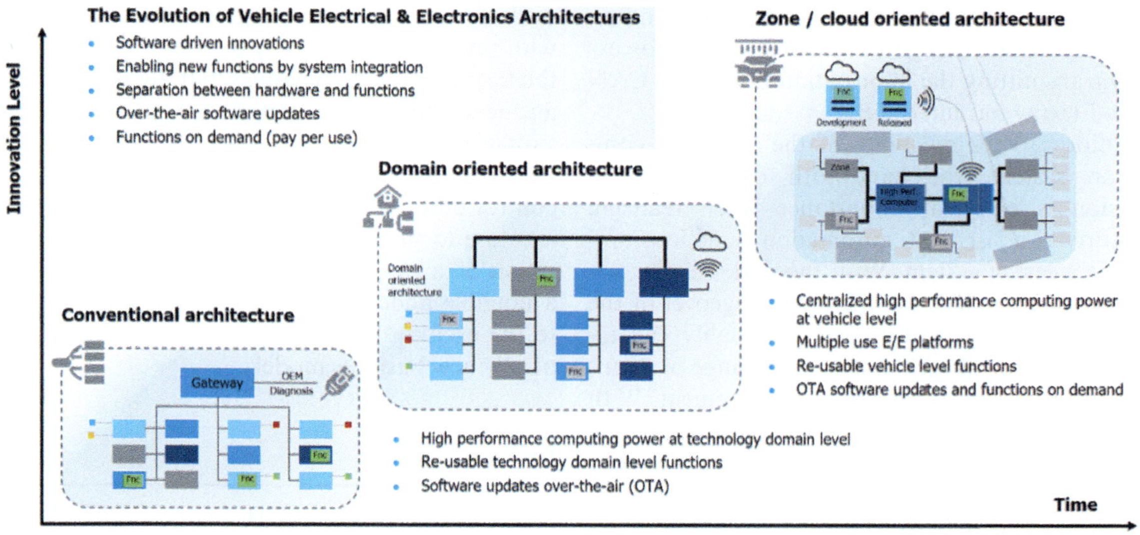

◘ Fig. 24.26 Evolution of the electric and electronic architecture in the vehicle (ZF)

The expected savings potential regarding system costs is based on the assumption that a smart actuator can be designed with a significantly lower cost microcontroller. This expectation will only be partially fulfillable, because the functional software only consumes a certain portion of the computing power of the electronic control unit of the steering system. The major part is used for hardware-specific services, such as the processing of sensor signals, the control of the electric motor, and the continuous fault monitoring of the electronic components, as well as services that ensure the compliant operation and safeguarding of the electronic communication interfaces.

An electronically controlled steering system with high levels of fail-safety requires a major technical effort in the form of redundant electrical and electronic components. As a result of the division of the steering subsystem into functional software and smart actuators, alternative solutions for maintaining the steering functions of the vehicle are increasingly being discussed. The solutions are based on the proposal to compensate the failure of the steering subsystem by means of a yaw rate control on chassis level. This kind of yaw rate control can theoretically be realized by means of a centrally coordinated wheel-specific distribution of drive and brake forces, which may be supported by the use of the additional steering device on the rear axle. With the steering systems and axle kinematics commonly used for front axles today, the wheel-specific distribution of the drive and brake forces in parallel to the yaw rate control must also stabilize the steering angle of the wheels on the front axle of the vehicle. So far, it has not been proven that the performance and reliability of this range of functions at chassis level would be sufficient to enable the elimination of fail-safe operation of the steering subsystem.

24.7.2 Steering System Interfaces

The interface of the steering system subsystem to the electrical and electronic architecture of the vehicle comprises the connection to the power supply via the electric vehicle power supply, and the connection to the electronic communication system. For systems that require fail-safe steering power assistance and steer-by-wire steering systems, the connection to the energy supply must be redundant. A redundant power supply requires two independent voltage sources. With the introduction of hybrid vehicle drives, the 12-V vehicle network typically forms the primary voltage source, and the second independent voltage source independently provided by a DC-DC converter via the vehicle drive's high-voltage system.

The steering system is usually connected to the electronic communication system of the vehicle via a CAN bus interface. Communication via a singular CAN bus interface can implemented in a manner that safeguards against the transmission of incorrect information by use of monitoring algorithms, but cannot protect against the complete absence of information. This must be taken into account in the safety analysis of the functions based on the information. For example, a speed-dependent steering power assistance automatically transitions into a non-critical mode in terms of vehicle handling when the information on the current driving speed is no longer available on the steering side.

If the steering system is supposed to receive steering torque or steering angle requirements that are part of driver assistance or automation functions with high functional safety requirements via the electronic communication system, the communication bus must be redundant. Depending on the results of the safety analysis, it may even be necessary to rule out a possible failure of the communication by a redundant and simultaneously diverse design of the connection, for example by transmitting the information both via a CAN bus and a Flexray bus interface.

In addition to the reliability of the electronic communication system, the bandwidth of information transmission is of great importance when realizing modern driver assist and automation functions, including the steering system. With the park steering assist system, which includes a function changeover of the steering system, a satisfactory control quality can already be achieved with cycle times in the range of 20 ms for the transfer of the steering angle requirement. With driver assist functions that have a direct influence on the haptic feedback on the steering wheel, such as the cooperative steering angle control, the requirements for the short cycle times increase considerably. Even higher requirements, up to virtual real-time capability of information transmission, will arise in the course of outsourcing the steering system's functional software to domain or zone control units, as well as the need to implement highly dynamic control loops in a network of control units and physically separate smart actuators. The same applies to the communication between the subsystems of a steer-by-wire steering system.

24.8 Summary and Prospects

The automotive megatrends *electrification, assisted and automated driving,* and *digitalization* are already having significant impacts on the development of modern steering systems in terms of availability, reliability, and mechanical robustness.

Higher AD levels require a stationary steering wheel during automated operation. Electrified vehicles have various redundant voltage supplies and thus enable the accelerated introduction of steer-by-wire steering systems.

In steer-by-wire systems, the steering characteristics can be freely adapted in new ways that were not previously possible. The new degrees of freedom associated with steer-by-wire not only offer exciting possibilities in the field of driving dynamics, but also for variant management and interior design due to the packaging advantages.

Over-the-air updates by means of powerful central control units, in combination with the power electronics close to the steering actuator, enable the implementation of a range of functionalities that will represent a tangible added value for end users and, in addition to the exciting technical possibilities, also enable completely new business models.

References

1. Elbers, C.: Mathematische Abbildung von Kinematik und Elastokinematik, in: Dissertation 2001 (2001)
2. Elbers, C.: Mechatronische Fahrwerksysteme in: Heißing B, Ersoy M, Gies, S (Hrsg.) Fahrwerkhandbuch, 1. Aufl. 2007, Vieweg + Teubner Verlag, Springer Fachmedien Wiesbaden GmbH (2007)
3. Harrer, M., Pfeffer, P.: (2013, Hrsg.) in: Pfeffer P, Harrer M (Hrsg.), Lenkungshandbuch; 2., überarbeitete und ergänzte Auflage; Vieweg + Teubner Verlag, Springer Fachmedien Wiesbaden GmbH (2013)
4. Lunkeit, D., Weichert, J.: Performance-oriented realization of a rear wheel steering system for the Porsche 911 Turbo, in: 4. Internationales Münchner Fahrwerk Symposium, 2013, München (2013)
5. Pruckner, A.: Nichtlineare Fahrzustandsbeobachtung und –regelung einer Pkw-Hinterradlenkung, in: Dissertation, 2001 (2001)
6. Rieth, P.E., Raste, T.: Integrationskonzepte der Zukunft. In: Winner, H., Hakuli, S., Lotz, F., Singer, C. (eds.) Handbuch Fahrerassistenzsysteme; 3, p. 1090. Auflage; Vieweg + Teubner Verlag, Springer Fachmedien Wiesbaden GmbH (2015)
7. Seewald, A., Heitzer, H.-D.: Leichtbau – eine Herausforderung an die Gestaltung von Fahrwerkskomponenten und Lenksystemen in: 7. Aachener Kolloquium Fahrzeug- und Motorentechnik (1998)
8. Wiertz, A., Williams, D.E.: Improved steering feel for heavy commercial vehicles by torque overlay, in: Steering.tech, München, 01.04.2008 (2008)
9. Williams, C., Cagatay, K., Jobanputra, M., Chaujar, S.: Reusable architectures for safety-critical smart-actuators in: Chassis. Tech Plus 2021 (2021)

Open Access This chapter is licensed under the terms of the Creative Commons Attribution-NonCommercial-NoDerivatives 4.0 International License (▶ http://creativecommons.org/licenses/by-nc-nd/4.0/), which permits any noncommercial use, sharing, distribution and reproduction in any medium or format, as long as you give appropriate credit to the original author(s) and the source, provide a link to the Creative Commons license and indicate if you modified the licensed material. You do not have permission under this license to share adapted material derived from this chapter or parts of it.

The images or other third party material in this chapter are included in the chapter's Creative Commons license, unless indicated otherwise in a credit line to the material. If material is not included in the chapter's Creative Commons license and your intended use is not permitted by statutory regulation or exceeds the permitted use, you will need to obtain permission directly from the copyright holder.

Electronic Brake Systems

Paul Linhoff, Stefan Arnold Drumm, Thorsten Ullrich,
and Bernhard Schmid

Contents

© The Author(s) 2026
H. Winner et al. (eds.), *Handbook Assisted and Automated Driving*,
https://doi.org/10.1007/978-3-658-45276-6_25

25.1 Passenger Vehicle Braking System

25.1.1 The Braking System as a Vehicle Subsystem

The braking system of a passenger car is a subsystem of the vehicle chassis, which itself constitutes a subsystem of the vehicle. In order to classify the braking system into a system hierarchy, the passenger car is defined as a four-wheeled road vehicle with its own propulsion, approved for use on public roads, for the primary purpose of passenger transport.

From a system-theoretical point of view, this vehicle comprises.

- Drivers and/or virtual drivers
- Passengers
- Load
- Vehicle body and
- Vehicle chassis.

The chassis is the aggregation of the vehicle elements used to control and perform the vehicle movement (■ Fig. 25.1).

25.1.2 The Braking System as One of the Vehicle Control Action Chains

In a traditional human-controlled vehicle, the chassis provides drivers with the three essential structures to control vehicle movement: powertrain, steering and brake.

To classify the technical evolution status of the braking system, the technologies currently used in these three control structures will now be compared.

In today's passenger cars, electric by-wire technology is used for the powertrain to convert a command, entered via the accelerator pedal into a drive action, while in historical vehicles a mechanical connection between the pedal and the engine was used. Steering still uses a mechanical connection between steering wheel and front wheel steering mechanism, but a steering support system reduces the steering wheel torque to be applied by the driver. Electromechanics has prevailed as a technology for this steering support system and has replaced the electrohydraulics used previously. The braking system is currently in a transitional phase. The conventional direct mechanical-hydraulic connection between the brake pedal and the wheel brakes, which was interrupted by the braking system only during anti-lock and stability control applications, is increasingly replaced by a permanent electric by-wire transmission of the driver brake command. In this context, "by-wire" means that signal transmission for the nominal function is done by electrical means at least at one point in the chain of actions. This by-wire braking technology has the advantage that no brake booster is needed anymore, that no irritating pedal effects occur during stability control applications and that the braking system is already equipped for automated driving (AD). In particular, by-wire braking technology enables brake pedal controlled recuperative braking, in which the desired deceleration of the vehicle is partly or fully achieved by operation of an electric drive motor as an electric generator, while the frictional braking force is automatically reduced accordingly, without causing a retroactive effect on the brake pedal.

25.1.3 Electrification of Vehicle Propulsion and Automated Driving

There is currently a clear technological trend toward a full-by-wire chassis. This trend is also intensified by the fact that the existence of by-wire technology is a prerequisite for automated driving. Another aspect, requiring a switching to by-wire technology is that, in addition to the vehicle functionality, which is covered by separate controls, enhanced functionality can be achieved by combined control of steering, drive, braking and suspension.

Future chassis technology will also use the well-known drive, steering and brake chains, but instead of being independent of each other, they interact. Regenerative braking is an already existing example for that.

A virtual driver is provided parallel to the human driver on the front side of the chassis impact chains. This computer-based virtual driver communicates with the chassis not via steering wheel and pedals, but via electronic signals. A chassis for future mobility includes sensors, interfaces, controllers and actuators for driving, brakes, steering and suspension.

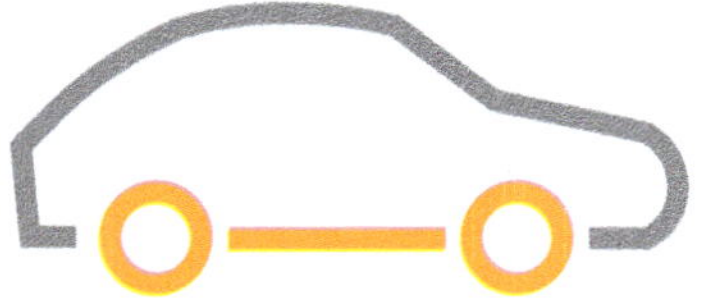

■ **Fig. 25.1** Vehicle body and chassis

25.1.4 Sense-Plan-Act-Structures

Most vehicle systems and subsystems at each hierarchical level can be presented as a set of functional blocks that interact according to the Sense-Plan-Act scheme. This is illustrated in ◻ Fig. 25.2 for the chassis control structures.

25.1.5 Main Functions of the Braking System

The task of the braking system is not only to slow down the vehicle and keep the vehicle stationary, but also to provide assistance functions which are essential in today's vehicles.

Main task of the braking system is to apply the wheel brakes in the event of a driver's brake command, with wheel-specific actuation values in accordance with a predetermined relational pattern.

Another task of the braking system is to keep the movement of the vehicle stable, so that the reaction of the vehicle to control commands remains controllable. Stabilization functions include anti-lock braking system or ABS function which prevents wheels from locking during braking, traction control system or TCS function which prevents wheels from spinning due to an acceleration command and an "active yawing control" or AYC function which prevents the vehicle from uncontrolled spinning around its vertical axis by means of targeted braking interventions.

25.1.6 Brake System Ancillary Functions

When stepping on the brake pedal, drivers expect the pedal to decline depending on the force of the pedal application, in addition to the main function of the vehicle deceleration. Drivers control the braking rate over the actuation degree of the brake pedal. In the classic hydraulic braking system with a brake booster amplifying the pedal force, the resulting vehicle deceleration is proportional to the pedal force. The assigned pedal travel is a result of the hydraulic volume absorption of the wheel brakes, on the basis of their characteristic absorption curves in the form of a monotone-growing, generally non-linear function. In by-wire brake systems, the pedal force over pedal travel characteristic curve is determined by means of a pedal travel simulator. A target value for the intensity of the brake application is calculated by means of a further characteristic curve from the pedal travel. In electrohydraulic brake-by-wire systems with central brake pressure provision, this is the brake system pressure.

The conventional brake system has pedal repercussions during both automated braking and stabilization procedures (ABS and ESC functionality). In the brake-by-wire brake there are no such pedal implications that could provoke irritations.

In the event of simultaneous and different commands, for acceleration or deceleration from the human driver and from a virtual driver, a generally accepted approach to pedal behavior has not yet been established.

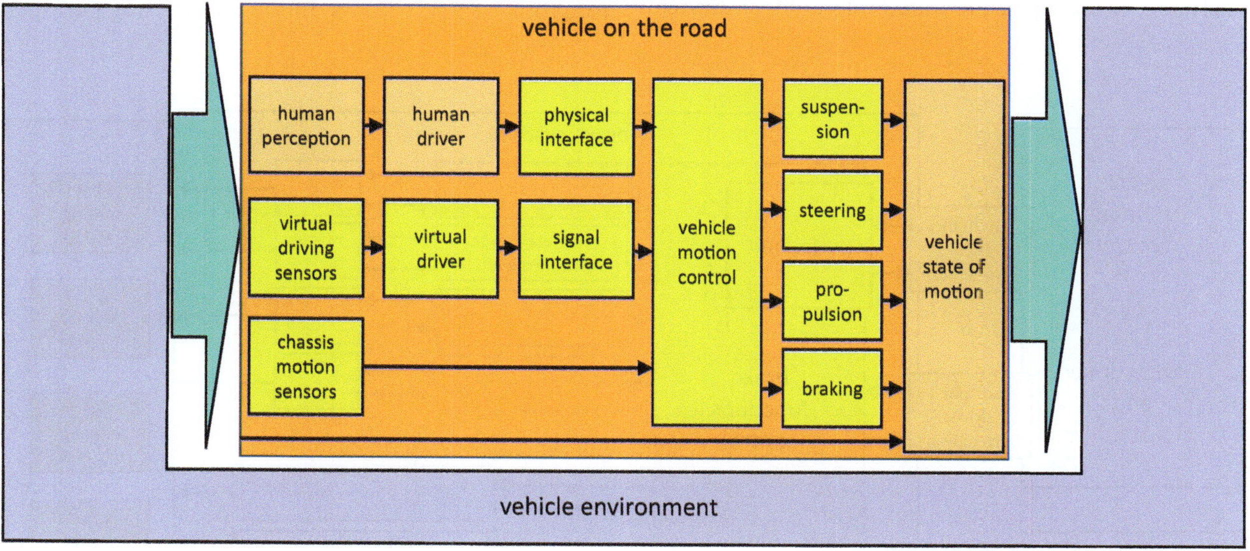

◻ **Fig. 25.2** Control chains of vehicle guidance

25

With regard to operation of parking brake, the transition to electric by-wire technology has already been made for many vehicles. The hand brake lever has been replaced by a switch and the wheel brakes of the rear axle have been equipped with gearbox electric motors, which generate the actuation forces necessary for parking braking.

Another interaction of the braking system with the driver is the indication of a possible malfunction by means of warning lamps. However, switching on a warning lamp does not have any effect when driving automated without the assistance of a human being. For this purpose, fault signaling must be developed by means of signal forwarding to the virtual driver.

25.1.7 System Scope of Braking System, Functional Blocks, System Inputs and Outputs

The term "braking system" originally refers to the whole of the chassis components in the active chain that control the wheel brakes.

The braking system as a chassis subsystem can be described by a set of functional blocks in accordance with the Sense-Plan-Act scheme. Since an electric vehicle can be braked by the electric drive, in addition to the friction brake, to a degree defined by the performance of the electric drive, this definition is extended to an interface between the brake and the drive system, so that the drive system can be used as an additional deceleration device.

The structure of the functional blocks shown in ◘ Fig. 25.3 needs not being identical to the physical structure. For example, the functional control unit can

be distributed decentrally to those control units, which are assigned to the wheel brakes for controlling the wheel brakes.

25.2 Evolutionary Steps of Passenger Car Braking Systems

25.2.1 Conventional Braking System with Vacuum Brake Booster

The braking system, most commonly used in today's passenger cars, is the one described in ◘ Fig. 25.4. It consists of a so-called vacuum booster, which is a vacuum-operated pedal force amplifier, acting on a tandem master cylinder TMC, an ESC module for refined wheel-specific hydraulic pressure control and a hydraulically driven friction brake per vehicle wheel.

The ESC module consists of an electronic control unit and a hydraulic control unit HCU. During driving or braking, the electronic control unit monitors the state of movement of the vehicle at all times. In dynamically simple driving situations, which do not require the intervention of a control system, the HCU only distributes the brake pressures supplied by the TMC to the wheel brakes. If the control unit detects signs of instability in the rotation of the vehicle wheels or the yaw movement of the body, the ESC module shall respond by generating appropriate individual brake pressures to stabilize them. The ESC module is designed to generate all required brake pressures below or above the TMC pressure. It is designed primarily for effective and therefore robust interventions. In the event of regular opera-

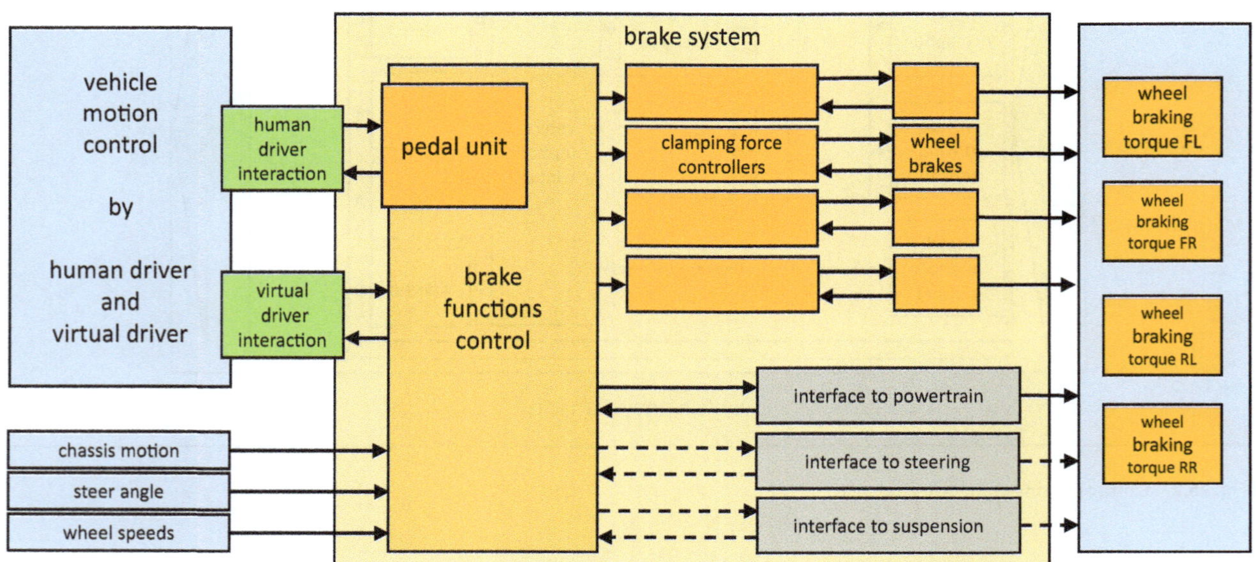

◘ **Fig. 25.3** Function blocks, system inputs and outputs

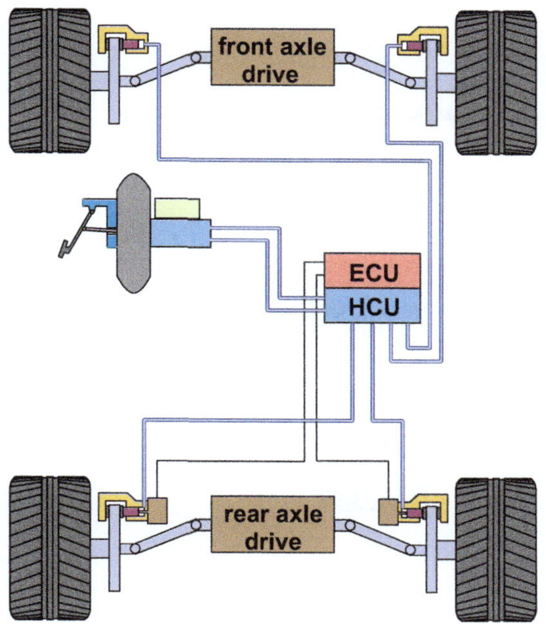

Fig. 25.4 Conventional brake system

tion, driving comfort has a lower priority than stabilizing the vehicle movement. This is, why for conventional braking systems it is accepted that drivers might experience impaired pedal performance, chassis jerking and vibration during the ESC operation.

In the past, vacuum power was used for the brake booster because petrol engines whose combustion air supply was controlled by a throttle valve provided the vacuum supply free of charge. In addition, this air intake throttle produces maximal underpressure, precisely when it is needed, namely when the accelerator pedal is released to transfer the foot to the brake pedal. However, there is a technological trend for internal combustion engines, to avoid reducing the air supply and to use variable fuel injection instead.

Therefore, additional vacuum pumps were introduced in order to continue the use of the well-known and inexpensive vacuum booster. Especially for diesel engines, vacuum pumps coupled to engine rotation are used. However, this does not solve the problem of lack of vacuum energy supply when the combustion engine is running at low speed or even comes to a standstill. This is why additional electrically driven vacuum pumps are installed in vehicles that are still equipped with this conventional braking system.

Vacuum pumps generate noise, even if there is no braking at the moment. Particularly in vehicles which are driven temporarily or permanently by almost noiseless electric motors, the noise of a vacuum pump is disturbing. In addition, it is a cumbersome method to convert electrical energy into pneumatic energy by means of a vacuum pump, then to convert it into mechanical energy in the vacuum booster, and then to generate hydraulic energy in the master brake cylinder for actuation of hydraulic wheel brakes, which finally act mechanically via the piston in the wheel brake. There are other options that apply the brakes in less steps, including those that do not use hydraulic forces.

25.2.2 Electric Booster-Based Braking System

A simple approach to avoiding vacuum technology is to replace vacuum pumps and vacuum brake boosters with an electric booster that is directly powered by electric energy. The rest of the braking system remains unchanged—cf. **Figs. 25.5 and 25.6.**

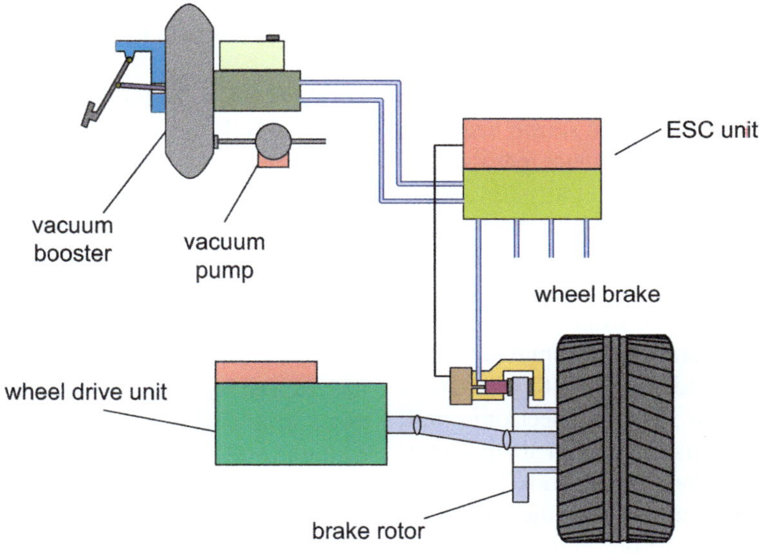

Fig. 25.5 Vacuum booster-based brake system

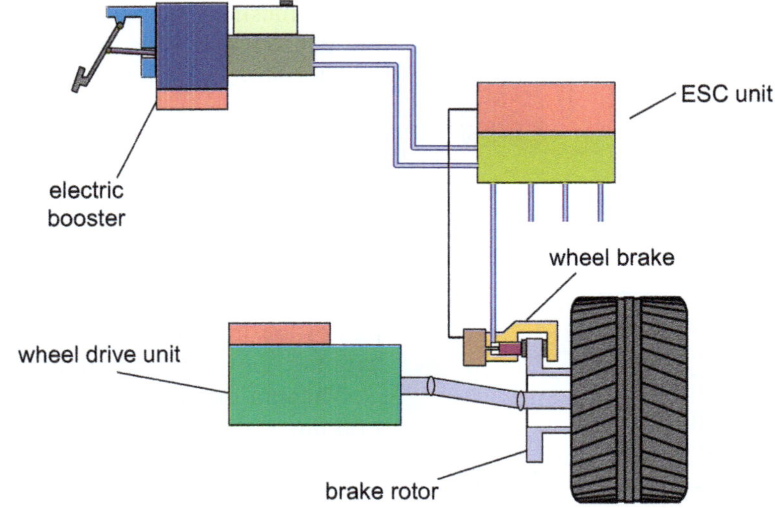

Fig. 25.6 Electric booster-based brake system

The electric booster system consists of a pedal actuation sensor, an electronic control unit, an electric motor and a transmission gear to combine driver pedal force and booster force. As an example, the Bosch second generation iBooster is depicted in **Fig. 25.7** and is described below.

As with a vacuum brake booster, the pedal movement is coupled to a pressure rod with a central stamp head in a symmetry axis, which pushes the pedal force to a circular area of a reaction disk. A second, electromechanically driven stamp head adds an amplification force by pressing it onto a ring-shaped surface of the reaction disk. On the opposite side of the reaction disk, a pressure piece exports the sum of pedal force and booster force to a master cylinder. Usage of such an elastomer reaction disk is known from the vacuum brake booster, and it is a proven concept that can be transferred to the electric booster as follows: The amplification force is controlled in a way to minimize the travel difference between the two above stamps. In the vacuum brake booster, this is done by pneumatic valves which are coupled mechanically to the stamp heads. In the Bosch iBooster Gen. 2, the difference path of the stamp heads is captured by a differential path sensor. An electronic unit receives its travel difference signal and uses it for controlling an electric motor that drives a spindle nut via a multi-stage gear transmission, with the spindle nut directly driving the annular force stamp. The electric motor is electronically commuted, has a three-phase stator winding and a rotor with permanent magnets. The forces addition unit, the electronic unit, the electric motor and the master cylinder form a unit. Unlike a vacuum booster, the electric booster can be used for automated braking without further technical modifications. It should be noted that due to the working principle of this booster, the brake pedal is involved in the process of automated braking. Resulting brake pedal travel without pedal application by a driver's foot can be problematic in terms of the possibility of the driver's foot being squeezed.

Brake-by-wire was introduced to overcome this, among other things.

25.2.3 Electrohydraulic By-Wire Braking System

Brake-by-wire means that somewhere in the chain of effects from the brake pedal to the wheel brakes, there is a passage of electrical signals. The brake pedal is connected to a pedal-feel simulator, the pedal actuation is detected by sensors, and the corresponding brake actuation is carried out by electric actuators.

Brake systems with hydraulic accumulators

Historic electrohydraulic brake-by-wire systems have used hydraulic accumulators as storage reservoir for hydraulic energy. By means of a high-pressure pump, the hydraulic accumulator was kept in a loaded state, at a pressure at least equal to the highest wheel brake pressure required when braking—typically around 200 bar. The required pressures, which are mostly considerably smaller, were derived from this supply pressure by means of electro-magnetic valves. In the event of failure, caused by an empty hydraulic accumulator, direct, not amplified hydraulic connection between the pedal cylinder and the wheel brakes was provided. Such electrohydraulic braking systems (EHB) are described in the Handbook of Driver Assistance Systems, first edition, ▶ Sect. 2.2 [1]. The hydraulic accumulators used in EHB brake systems store hydraulic energy by com-

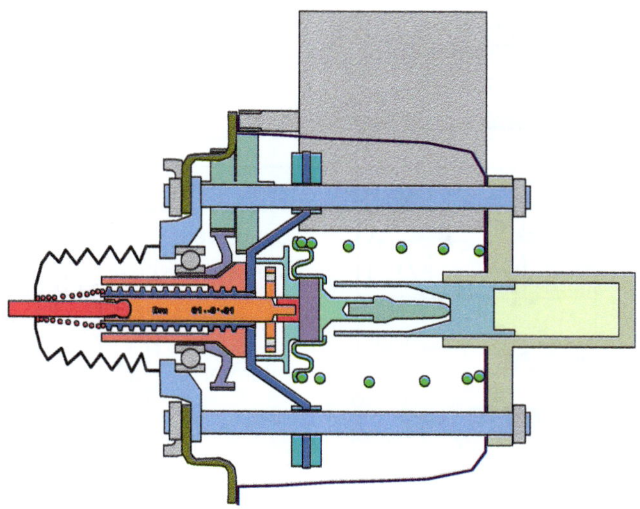

Fig. 25.7 Cross-sectional slice image of Bosch iBooster Gen. 2

pressing a gas volume, mostly nitrogen. It is therefore necessary to consider, when analyzing possible technical failures of the hydraulic reservoir, the fact that brake fluid could be contaminated by gas. This could lead not only to a loss of by-wire braking functionality but also to the loss of reserve functionality with direct hydraulic coupling. The technical effort required to hedge against this possible fault is another argument against the use of hydraulic accumulators.

Brake systems without hydraulic accumulators

To avoid an energy storage in the braking system, newer brake-by-wire systems use only the electrical power available from the on-board power supply at the current time. Therefore, other pressure providing concepts must be used. For charging a hydraulic accumulator, a pump powered by a simple electric brush motor with a capacity, sufficient to recharge the hydraulic accumulator during and after braking, was sufficient. For the design of an electrohydraulic by-wire braking system without hydraulic storage, at least similar requirements apply to rapid brake pressure build-up as to an e-booster and, due to the unavailability of the driver's contribution, requirements are even tougher than for an electric booster. Like there, an electronically commutated electric motor is used here, which has a stator with a three-phase coil arrangement, a rotor with permanent magnets and a rotary angular sensor. The electric motor drives a rotation to translation gear directly (MKC1, MKC2 from Continental) or via an intermediate rotation to rotation transmission (integrated braking system from Bosch), with the translational output moving a hydraulic piston to build a system pressure. This system pressure is only increased to the amount needed, which reduces energy consumption and average

pressure compared to hydrostorage systems. A pressure sensor is used to control system pressure. In addition, actuator piston travel and speed information are used by an appropriate control algorithm, with this information being derived from the rotor angular sensor data of the electric motor. Electronic unit, electric motor, transmission gear elements, hydraulic piston and hydraulic cylinder form an electrohydraulic actuator.

In these braking systems, a fallback operational mode provides direct hydraulic coupling between the brake pedal and the wheel brakes, which is however cut off by electro-magnetic valves in normal operation. In addition, in normal operation, the pedal cylinder is connected to a simulator which, by means of its hydraulic volume absorption, gives in the brake pedal a pedal travel dependent on the pedal force according to a characteristic curve. For detecting a driver's brake demand, the pedal path and the pressure in the pedal cylinder are captured by means of appropriate sensors. In normal operation, braking system pressure is set by the electrohydraulic actuator. Each wheel brake is connected to the system pressure by an electrohydraulic valve which is open when deenergized and can be connected to reservoir pressure via another electrohydraulic valve which is closed when deenergized. These valve pairs, already known from ABS hydraulic modules, can be used to individually reduce the wheel brake pressure to values below the system pressure.

The MK C1 electrohydraulic module, the hydraulic layout of which is given in Fig. 25.8, integrates the pedal control unit, simulator, electrohydraulic actuator, electro-magnetic valves and an electronic control unit into a unit to which the brake pedal is connected on an input side and the hydraulic lines to the wheel brakes are connected on an output side. Compared to the abovementioned brake systems, no addi-

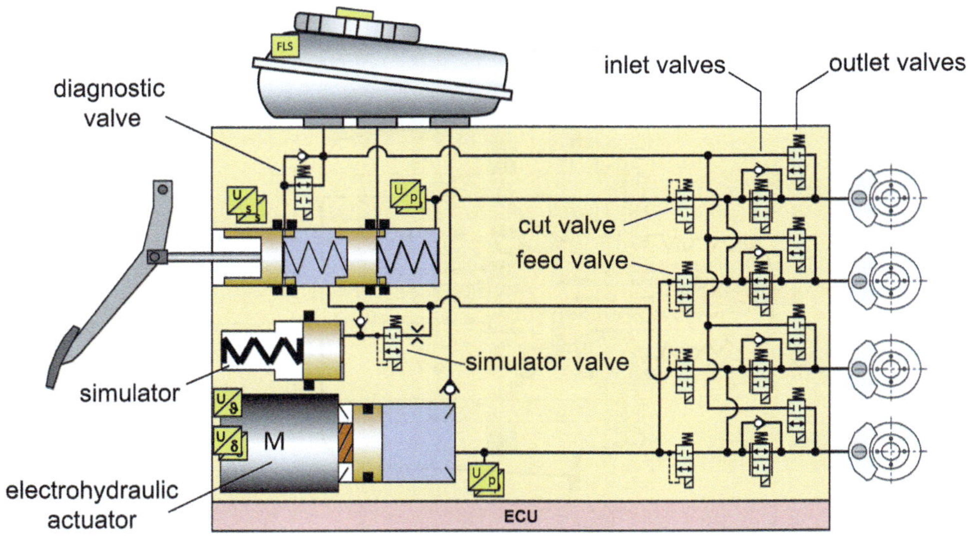

Fig. 25.8 MK C1—one-box electrohydraulic brake module

tional ESC module is needed for ABS, TCS and ESC functionality, and the installation space previously occupied by the ESC module is no longer required. MK C1 uses brake fluid and uses conventional hydraulic wheel brakes. In addition, MK C1 controls the currents for electric parking brakes EPB. Electromechanical wheel brake actuation is a proven technology for these parking brakes.

With the MK C2, there is a further development of this electrohydraulic one-box brake with focus on simplifications in the hydraulic layout.

25.2.4 Electromechanical By-Wire Braking System

In order to avoid the use of hydraulic fluids completely, so-called dry brake systems can be used. An electromechanical actuator replaces the hydraulic slave cylinder of a wheel brake. It is supplied with energy by electrical lines and controlled by signal lines. In this context, the acronym EMB refers to the entire "electromechanical braking system", not to a single electromechanical wheel brake.

The EMB system in the vehicle, which is schematically shown in **Fig. 25.9**, consists of an actuator module per wheel, a pedal module, communication lines and electrical power supply. As the intelligent actuators on the wheel brakes comprise electronic control units for brushless DC motor control, braking system-specific algorithms can also be processed there. Another possible EMB system architecture includes a control unit, located in the chassis to receive deceleration commands, calculate the braking force distribution and, if necessary, wheel-specific braking interventions to stabilize the vehicle movement and transmit the resulting commands to the wheel brake modules and the vehicle's propulsion system.

For every wheel, the EMB electromechanical braking system uses a friction brake, whose actuation force is generated by an electric actuator. For safety reasons, the service brakes shall automatically open when the actuator is not operated. This requirement meets the strict need to avoid overheating of the brakes during driving without brake application. In the field of system technology, this is a so-called "safety objective". On the other hand, an actuated parking brake must continue to hold the wheel, even if its power supply is switched off. An electromechanical service brake cannot be a simple modification of an EPB with increased dynamics because of this fundamental difference with parking brakes.

25.2.4.1 E-Caliper

At Continental, an E-Caliper wheel brake module is being developed, of which a design example is shown in **Fig. 25.10**. An electric motor and a transmission gear generate the clamping force for a floating brake caliper.

The Continental E-Caliper, whose internal structure is presented in **Fig. 25.11**, is a so-called "smart actuator" since the electric motor is mechanically and electrically combined with an electronic control device to form a unit. This enables the usage of a three-phase brushless electric motor with a rotary angle sensor and with stator coil and sensor wiring inside the module. The module is electrically connected to the vehicle body by a cable for electric power supply and for signal exchange.

The transmission consists of a gear which couples the rotation of the electric motor's rotor with the drive

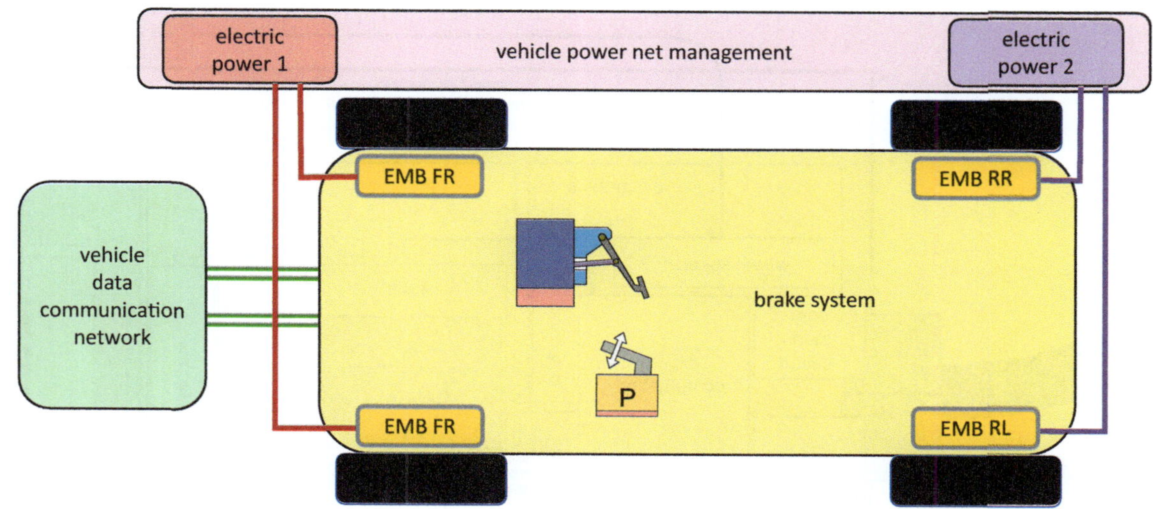

Fig. 25.9 Generic architecture of electromechanical brake system

Fig. 25.10 E-caliper—design example

side of a ball screw spindle. On its output side, the ball screw spindle causes a linear shift of a piston-shaped thrust piece. Like a hydraulic piston, the thrust piece acts on the inboard brake pad of a floating caliper type disk brake. The E-Caliper is therefore a conventional brake with the hydraulic cylinder being replaced by an electromechanical actuator.

The parking brake function may be operated using the same brake linings as the service brake application, as is the case with hydraulic combined brake saddles. However, due to the above safety objective for service brakes, the service braking drive of an electromechanical brake shall not be used to maintain the braking force during parking. The combination of service braking and parking braking functions in one device requires two different actuators, one self-releasing and one persisting, if deenergized, as illustrated in **Fig. 25.12**. In addition to the electromechanical service brake actuator, another actuator for parking braking functionality is needed. Besides the electromechanical service brake actuator, an electric parking brake actuator, similar to that, already known in combination

with hydraulic service brakes can be used. Another solution, available for the electromechanical braking systems, is the use of the service brake actuator to build both the service and parking braking forces, as well as a transmission lock, being controlled by a separate actuator to maintain the control force for parking purposes. In order to keep a parked vehicle stationary, this transmission lock shall remain locked even in the case of a powerless braking system.

25.2.4.2 Electric Drum Brake EDB

A drum brake requires less actuator effort than a disk brake to produce the same brake torque. This is due to a self-energizing effect within the drum brake. A smaller electric motor can therefore be used in an electromechanically operated drum brake. In addition to the E-Caliper, Continental is also developing an electric drum brake (EDB), which is shown in **Fig. 25.13**. An unfavorable characteristic of friction brakes with internal self-energizing effect is that friction coefficient variations have a greater impact on the braking torque produced, than with not self-energizing brakes. Usage of an additional control loop for the electronically controlled actuator allows for the compensation of this uncertainty. For this purpose, a quantity, representing the braking torque, is captured by an appropriate sensor. Systematic studies on this topic can be found in the dissertation by Christian Vey entitled "Entwicklung eines Verfahrens zur integrierten Bremsmomentmessung für Simplex-Trommelbremsen" [2]. This eliminates an existing disadvantage of drum brakes.

25.2.4.3 Decelerating Vehicles by Means of Electric Powertrain

Electric drive systems can not only accelerate a vehicle, but also decelerate it in a controlled manner. With

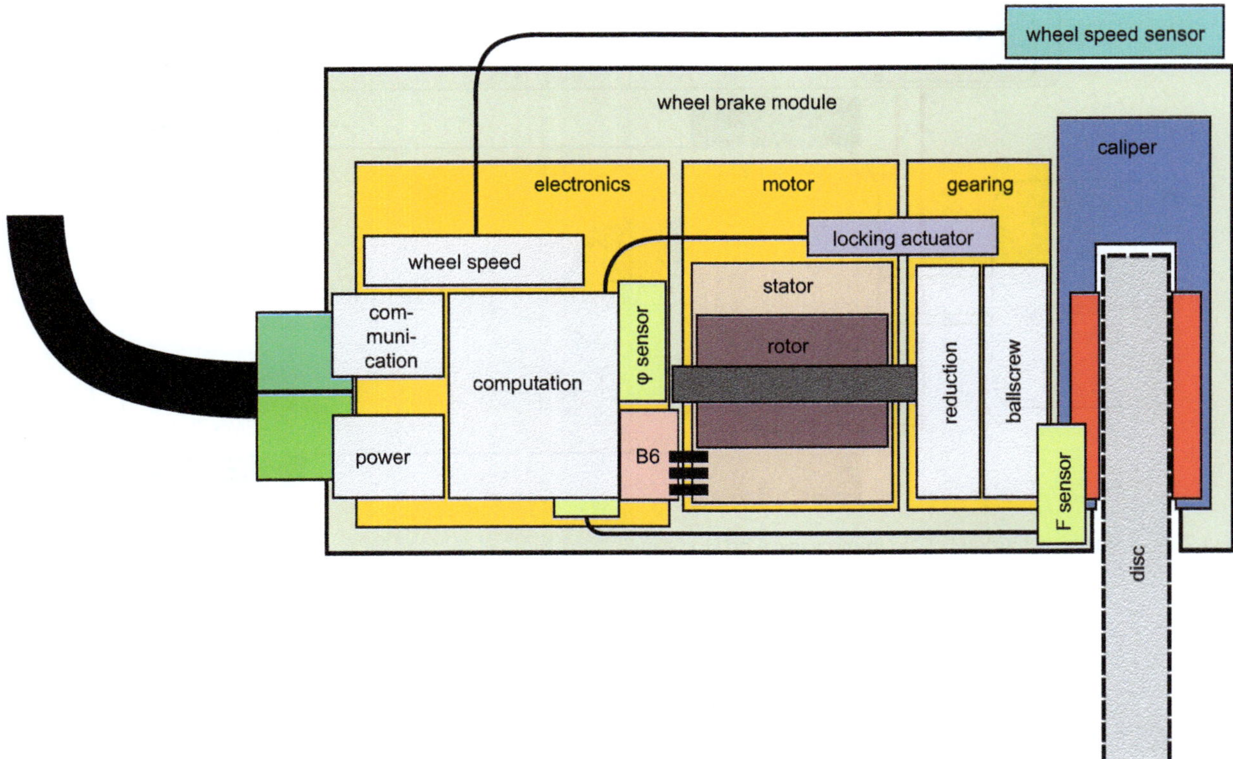

Fig. 25.11 E-caliper—internal structure

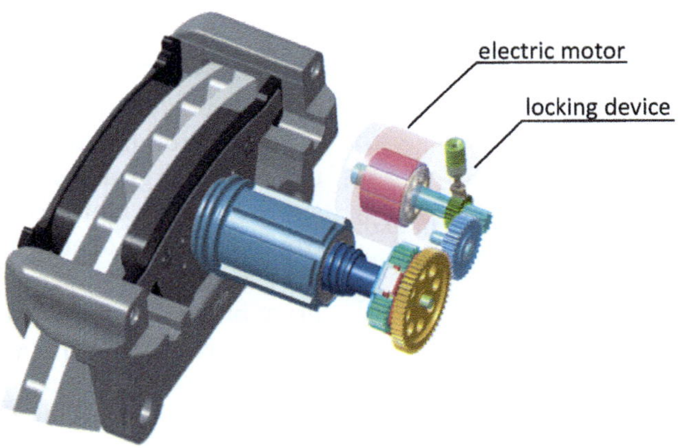

Fig. 25.12 E-caliper with locking actuator for parking brake function

regard to a balance of the energy flows, necessary for moving a vehicle, braking the vehicle means reducing its kinetic energy by a portion corresponding to the change in speed. For this purpose, electric drive motors can be operated in a generator mode. In a friction brake, the complete kinetic energy to be dissipated is converted into heat, whereas with recuperative braking, a significant part of this kinetic energy can be converted into electrical energy and can be used to charge an electrical energy storage device, as reflected in **Fig. 25.14**.

This recuperation braking process is therefore capable of recovering part of the energy, precedingly used for propulsion of the vehicle and, for example, increasing the range of a battery electric vehicle. Because the technical prerequisites for recuperative braking are present in a vehicle with electronically controlled electric drive engines per se, it is useful to perform braking as recuperative braking as far as possible. However, a friction brake cannot be dispensed within these vehicles for several reasons:

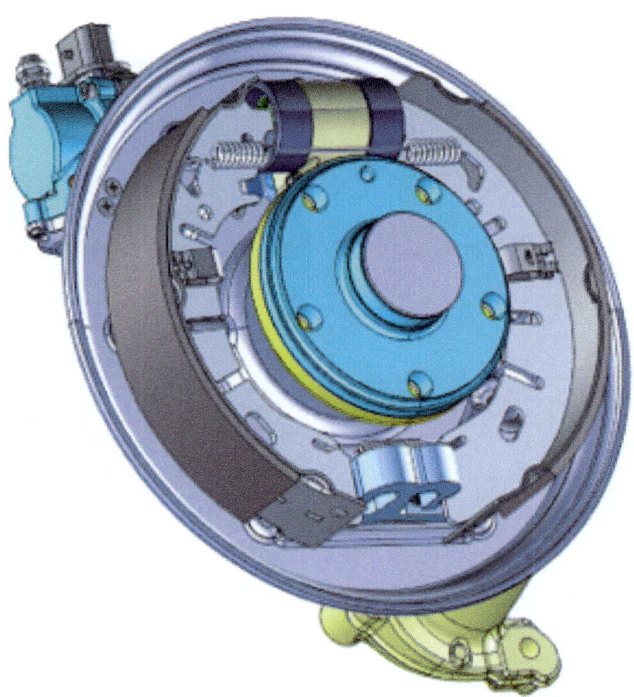

◘ Fig. 25.13   Electric drum brake—with the brake drum removed for an inside view

- The maximum permissible energy flow for electric propulsion systems is normally lower than the permissible heat flow in the friction brake.
- Unless each single vehicle wheel is driven by a separate electric motor, it is not possible to control the braking torques of wheels individually using only electric drive components.
- Also, in case that the recuperated energy cannot be fully used for battery charging—for example, if the battery is already fully charged or the permissible electric charging power is exceeded—the braking performance must be available at full height.

There are two main concepts for coordinating recuperative braking and frictional braking: Accelerator pedal-controlled braking and brake pedal-controlled braking. For the accelerator pedal-controlled recuperative braking, the friction brake system may remain unchanged. A "motor braking effect" can be dosed over a section of the pedal path. For the pedal-controlled recuperative braking, the braking system detects a request for braking on the basis of a pedal application, determines a recuperative part which is communicated to the vehicle drive system and controls the friction brake so that the in-sum resulting brake effect is adjusted to the requested one. Brake-by-wire systems are particularly suitable for carrying out this process.

25.3 Impact of Automated Driving on the Development of Braking Systems

25.3.1 Brake System Requirements Depending on the Degree of Automation

Individual mobility has always been characterized by the desire to overcome large distances as safely as possible in the shortest possible time with minimal effort, but with maximal comfort.

It was quickly realized that using human effort is not a practical way to achieve this and that animals such as horses are much more suitable for this purpose. Two significant additional properties of domesticated draft animals were thus implicitly available:

- In hazardous situations, the system tries autonomously to stop or to avoid the hazards, even without human intervention.
- The system will find its way autonomously.

However, for safety reasons, a redundant braking system was installed to ensure that the system can be secured in standstill and also to brake in case of emergency.

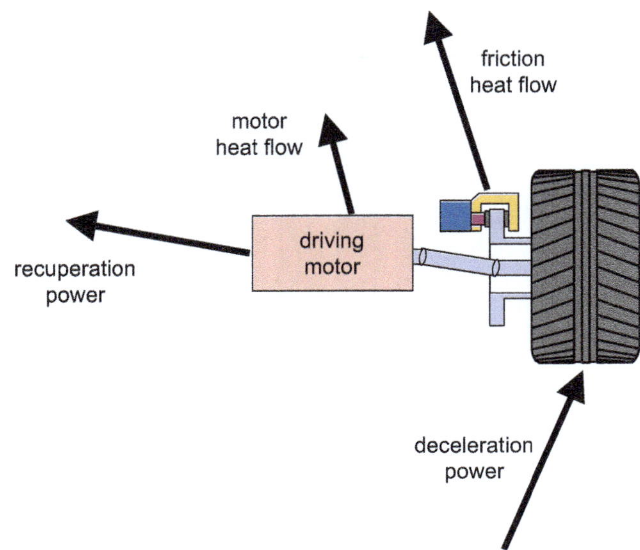

Fig. 25.14 Energy flow during combined frictional and recuperative braking

25.3.1.1 Basic Principles for Advanced Driver Assistance Systems

With the introduction of engine driven vehicles that can bridge larger distances in a shorter time, redundancy of vehicle control, which was implicitly present due to the draft animal, has been lost. In addition, the higher speeds significantly reduce the time slot remaining for an intervention in hazardous situations.

Drivers have therefore become an essential part of the basic safety concept in "car mobility". For harmonizing the scope for the driver, performing control of a vehicle in public areas, this scope was regulated in the Vienna Convention on Road Traffic [3] in 1968 and ratified by the acceding countries. The Federal Republic of Germany adopted the Convention in 1977 [4].

From this three key points for driving a car can be derived:

1. The vehicle shall have a driver
2. Driver of the vehicle shall be able to drive the vehicle safely
3. The driver must have control over the vehicle at all times.

Control of a vehicle is manifested by control of longitudinal guidance (acceleration and braking) and lateral guidance (steering) and is generally referred to as dynamic driving task (DDT).

In an annex to the Economic Commission for Europe of the United Nations ECE report of the sixty-eighth session of the Working Party on Road Traffic Safety [5], electronic systems were approved in 2014 to assist drivers in controlling the longitudinal and lateral movement of vehicles. However, a human driver must be able to override or deactivate the electronic system at any time. Responsibility for the vehicle's driving actions, including the use of the assistance system skills, remains with the human being. From this the following requirement is derived:

» To ensure controllability, a braking system must always make its function available to the driver with highest availability (fail-operational). The same applies to the steering systems of vehicles, being operated in public traffic.

The braking system is one of the most important system elements of the vehicle control chains for a vehicle equipped for automated driving Fig. 25.2. As a rule, modern braking systems have two channels to ensure high availability, so that a driver's braking command can be executed in the form of a vehicle deceleration. This is shown in Table 25.1, depending on the degree of automation. For this purpose, the classification according to automation levels as defined in standards document SAE J3016 [6] is used. In the case of driver-monitored systems, usage of system elements provided for SAE Levels 0 to 2 is required.

Importantly, unlike steering, braking has established itself as the ultimate measure in emergency control of the vehicle. After taking advantage of all the possibilities to maneuver the vehicle in safe zones, such as sidestrips, the vehicle's standstill is considered to be the risk minimal vehicle state.

25.3.1.2 Impact of Increasing Levels of Automation of Driver Assistance

As a result, the driver of the vehicle must have control over the vehicle at all times up to the SAE Level 2 and, if necessary, eliminate hazardous situations by immediate action. From Level 3, the driver may partially turn away from watching the traffic situation. However, in

Table 25.1

Brake system channel	Type 1 characteristics "Electrical hydraulic mechanical braking system"	Type 2 characteristics "Brake-by-wire brake system"	Suitability, depending on the degree of automation of driving
"Service brake" for system and driver intervention	Electronically controlled, electrically generated hydraulic brake pressure build-up	Electronically controlled pressure build-up	SAE levels 0 to 5
"Service brake" for system and driver intervention	Electronically controlled, electrically generated hydraulic brake pressure build-up	Electronically controlled pressure build-up	SAE levels 4 to 5
"Secondary brake" for system and driver intervention	Electronically controlled, electrically generated hydraulic brake pressure build-up	Electronically controlled pressure build-up	SAE level 3
"Secondary brake" for driver intervention	Hydraulic braking pressure build-up by brake pedal only		SAE levels 0 to 2
Braking action implementation	Two to four channel hydraulic braking system	Four channel hydraulic braking system	SAE levels 0 to 5

List of different types of braking systems with appropriate measures to ensure high availability in a SAE Automation Level 3 control system

the context of safety in use, it is necessary to examine more closely the influence of a temporary release of the steering wheel by the driver on the delay time, until the driver might intervene. This is done in ▸ Sect. 25.3.1.1.

25.3.1.3 Automated SAE Level 3 Driving

With automated driving at SAE Level 3, the passenger is granted the option to transfer responsibility and control of the vehicle to the automated driving system in a defined operational design domain (ODD).

However, when the ODD limits are reached or in the event of a system failure, the human driver is responsible for regaining control of the vehicle control within a certain period of time and subsequently deactivating the system. A ratified regulation for a first SAE Level 3 system is contained in ECE/TRANS/WP.29/2020/81 [7]. The deactivation phases described therein are shown in ◘ Fig. 25.15.

The regulation also gives the option of bringing the vehicle without a Phase I to a minimum risk condition in the event of a serious failure. In case a risk minimizing maneuver (MRM) has to be initiated, the regulation describes a maximum delay of $4\,\mathrm{m/s^2}$.

» This results in the requirement of providing a minimum deceleration being achievable by the secondary braking system of at least $4\,\mathrm{m/s^2}$.

In summary, the deactivation process, shown in ◘ Fig. 25.15, can be described as follows.

Phase I: Here, the vehicle automated driving system calls on the human driver to take over by sending optical, acoustic and haptic warnings, with an escalation of signals increasing over time. As far as the technical prerequisites, other road users and local conditions permit, the system tries to maneuver the vehicle by changing lanes onto a breakdown lane, a road shoulder, a holding bay or a parking lot. At the latest, the warning lights will then be activated. Meanwhile, depending on the reasons for the takeover and the traffic situation, the vehicle is being moderately and gradually decelerated. Deceleration values, referenced in ◘ Fig. 26.15 with δ and α and specified in $\mathrm{m/s^2}$, are within a typical range of 0 to 3.5. For δ, an exemplary value of zero is used, corresponding to the speed being constant in this section of the graph. Deceleration value α can be realized with a towing torque of the drive motor.

Phase II: If the human driver does not take control during Phase I, the "minimum risk maneuver" MRM is performed, which brakes the vehicle to a standstill and, if possible, maneuvers the vehicle onto a breakdown lane or a holding bay. For a SAE Level 4 or 5 system, the dotted line indicates that the system shall be able to maintain the emergency run until a stopping spot can be reached. If the human driver still does not react after stopping, an emergency call will be made by the system depending on the technical prerequisites. The deceleration value referenced in ◘ Fig. 26.15 with β and specified in $\mathrm{m/s^2}$ is typically greater than or equal to 4.

In these scenarios, the vehicle automated driving system as well as the human driver are responsible for avoiding accidents. However, in situations where other road users violate this maxim, a SAE Level 3 system shall also be able to perform emergency braking maneuvers to prevent an accident or minimize the

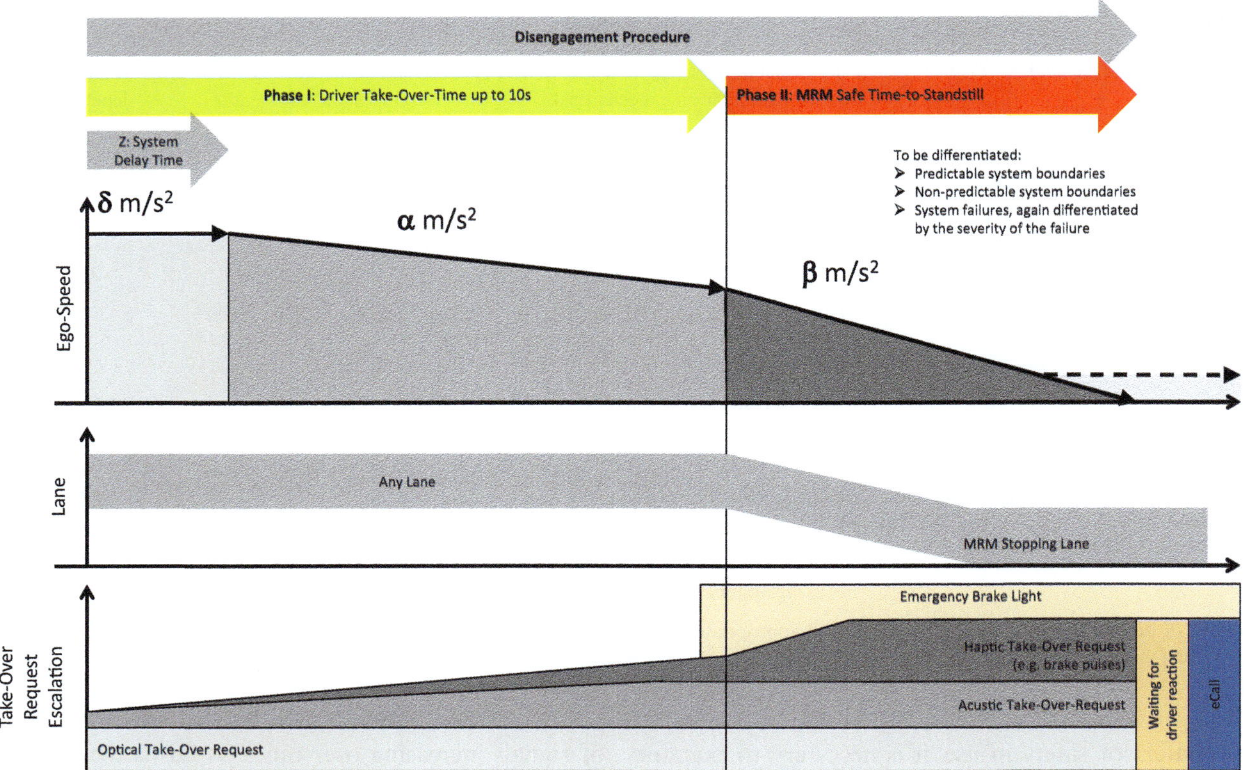

◘ Fig. 25.15 Automated vehicle guidance deactivation process with the aim of human takeover or achieving a minimum risk condition

severity of the consequences of the accident. These maneuvers require the ability of performing decelerations of at least 10 m/s², analogous to the emergency brake assist, provided that the friction coefficients between road and tires make this possible.

» This results in the requirement of providing a minimum deceleration being achievable by the service braking system of at least 10 m/s².[1]

Due to the need for controllability by the vehicle automated driving system, it is no longer possible to use an auxiliary braking system with mechanical-hydraulic pressure build-up. In this case, the system solutions described in ◘ Table 25.1 according to SAE Level 3 are necessary for the automation of driving. In the case of automated driving systems not permanently supervised by humans, the minimum requirement is to use the system elements provided for SAE Levels 3 to 5.

◘ Figure 25.16 exemplifies an overall system overview of a SAE Level 3 vehicle automated driving system, which illustrates the connection of the redundant brake system to the vehicle guidance system.

In the overall system overview, the connection with the outside world from the vehicle is realized via the Intelligent Antenna Module (IAM), the Telematic Control Unit (TCU), the Map-Processing Unit (MPU) and the Vehicle-to-X System (V2X). Furthermore, the first control system has displays and controls of the Human Machine Interface I (HMI I), a powerful Automated Driving Control Unit ($ADCU_{AD}$) and associated environment sensors such as front and surround view (SV) cameras, radar, lidar and ultrasound sensors (US), and the braking, steering and drive system. The first control system is mainly intended for automated driving but also includes an MRM function if the emergency system fails. The parking functions are also located there due to the connection of the SV cameras and the US.

The fallback emergency system consists of another $ADCU_{High}$ and a braking and steering system, redundant to the first control system. The fallback system includes both MRM and SAE Level 2 functions, such as adaptive cruise control (ACC) and lane keeping assistance (LKA), and safety functions such as emergency brake assist. The first control system and the fallback system are supplied by independent energy supplies (Battery I and II). Other redundant system elements, such as driving, braking and flashing lights, are not shown here.

---

1 State of the art for enabling comprehensive collision avoidance and mitigation of accident consequences, as required in accident test scenarios, e.g. Euro-NCAP (► https://www.euroncap.com/de).

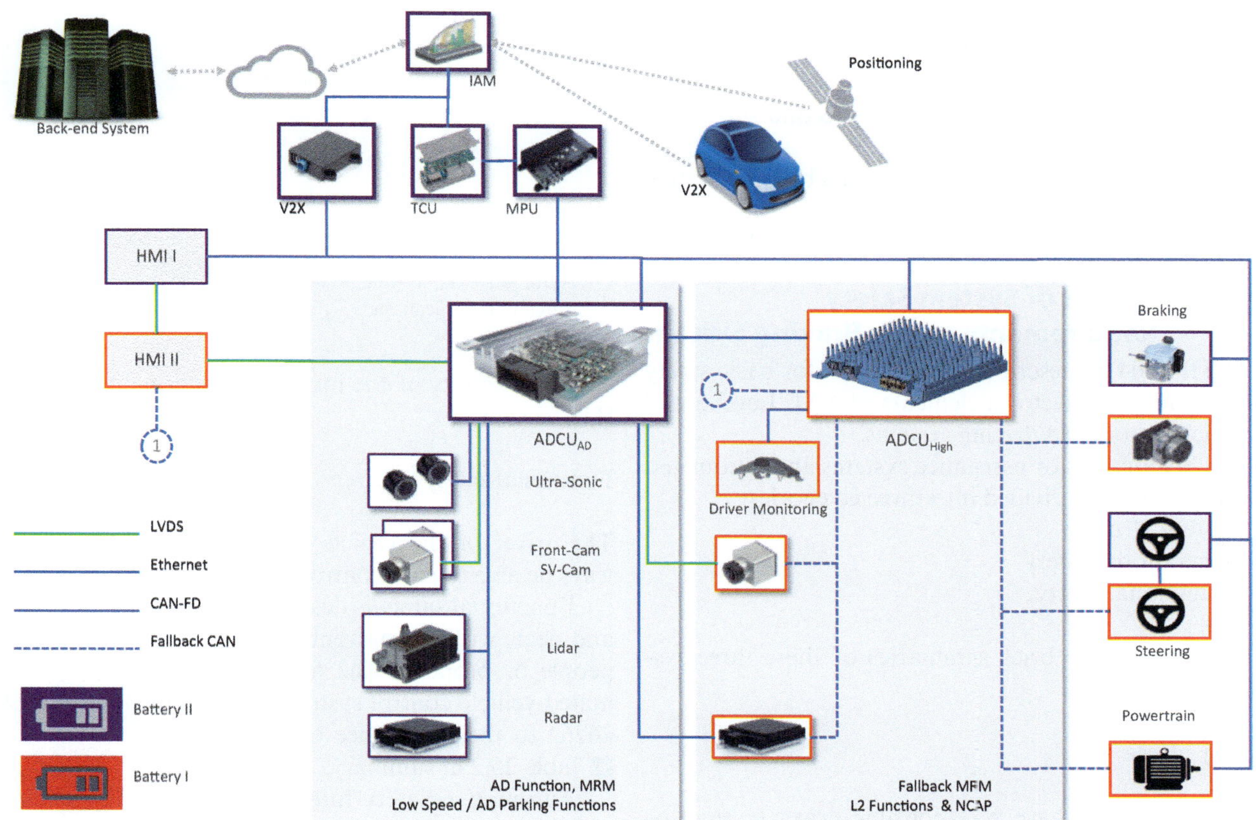

◻ Fig. 25.16 Continental AG configuration example for a SAE Level 3 vehicle guidance system, with an operational design domain covering speeds up to 130 km/h and being limited to freeways

25.3.1.4 Automated Driving at SAE Levels 4 and 5

As indicated in the previous chapter, a human driver of a vehicle with SAE automation Level 3 will be granted a transitional period for taking over control of the driving of the vehicle, after which he will be responsible solely.

This option is no longer available in a SAE Level 4 and 5 vehicle automated driving system. These systems must maintain control of driving until the end of the mission. System-initiated handover to a person is no longer planned.

Depending on the operational or business model of the operator of such a system, the termination of the mission may also be a stop at an appropriate parking space from which passengers are offered a service for onward travel.

Impact of Automated Level 4 Vehicle Driving Systems on Brake Systems

Such an automated vehicle driving system comprises two subsystems. If the first subsystem fails, the mission will continue in degraded condition. It is assumed that the mission can be completed within a few hours, for example by parking the vehicle in a safe parking lot.

To fulfill this requirement, the system shall comprise a fully functional first subsystem and a second subsystem, designed to carry the mission to its destination in the event of failure of the first subsystem.

In case of specific restrictions of the ODD, such as slow-moving driverless taxis or minibuses operating in the city center, immediate stopping at the roadside is also an option after the failure of the first subsystem. In this special case, a brake system configuration is possible as with a Level 3 system.

Such scenarios for Level 4 systems are discussed in more detail in ISO TR 4804 [9]. The deceleration requirements are those of the first subsystem and the emergency fallback system as indicated in ▶ Sect. 25.3.1.2.

The limitation of the operating time in degraded system conditions, e.g. direct access to the nearest parking lot at a very low speed, significantly reduces the probability of failure of the second subsystem, which is reflected in the hazard and risk analysis accord-

25

ing to ISO 26262 [10] (▸ para. **Fehler! Verweisquelle konnte nicht gefunden werden.**) and results in lower requirements to the availability of the second subsystem. Therefore, further redundancy to the second subsystem is not foreseen.

This results in configurations for the braking system as shown in ◘ Table 25.1.

25.3.1.5 Impact of System Safety Requirements on the Braking System

ISO TR 4804 [9] describes a framework for basic safety aspects and architectural elements of SAE Level 3 and 4 vehicle automated driving systems.

Safety analysis of assistance systems and automated driving systems is divided into three categories:
1. Safety of use
2. Operational safety
3. Functional safety.

The following are brief summaries of these three system safety aspects.

Safety of use

The vehicle shall have a responsible driver (system or person).

The driver assistance or automated driving system shall be designed to enable human drivers to assume responsibility for driving the vehicle, with their cognitive and physical skills.

The driver, the passengers and other involved parties shall not influence the system in such a way as to cause damage, in particular to persons or property, by abuse or by acting unwillingly. It is therefore necessary to make it clear to the persons concerned, what their responsibilities and tasks are in the different states of operation of the system.

The system shall ensure that foreseeable abuses or unintentional actions cannot cause a risky disruption of control over the vehicle and that its state of operation is clearly displayed to the driver.

Operational safety

The driver of a vehicle (system or person) shall have the ability to drive the vehicle in an accident avoiding manner. When a non-defective assisted or automated vehicle control system is activated and operated, it must not present a risk that goes beyond a socially responsible risk, particularly for people or property. This is discussed in detail in ▸ Chap. 40. SAE Level 3 and higher automated driving systems must certainly have at least the accident prevention capabilities as human drivers possess them. The requirements on such systems are currently being developed in ISO 21448 Road

◘ **Table 25.2**   .

Requirement type	ASIL classification			
	A	**B**	**C**	**D**
FIT[2]	10^{-6}/h	10^{-7}/h	10^{-7}/h	10^{-8}/h
SPFM[3]	–	>90%	>97%	>99%
LFM[4]	–	>60%	>80%	>90%

Automotive-Safety-Integrity-Level (ASIL) Requirements for different classifications

vehicles—Safety of the intended functionality (SOTIF) [11].

Functional safety

The driver of the vehicle (system or person) must always be able to gain control over the vehicle.

Functional safety is described in detail in ISO 26262 and deals with the prevention of damage, especially to people or property, due to faults in assisted or automated vehicle control systems. Requirements from ISO 26262 to the occurrence of system faults are listed in ◘ Table 25.2 in summary.

Automotive-Safety-Integrity-Level (ASIL) classifications can be decomposed via the system design, i.e. an ASIL D requirement can be met by two design elements in ASIL B(D) or ASIL A(D) and ASIL C(D). The "(D)" indicates that the development processes, tools and test procedures must meet the requests of ASIL D when merging the decomposed elements. The operational safety achieved by the design must not be compromised by random or systematic faults.

State of the art in driver assistance or automated driving systems up to SAE Level 2 is ASIL classification **D** for brake and steering controls, which must ensure for the human driver the possibility of overriding them. For the rest of the chain of action, the requirement can be reduced to ASIL B and lower.

Analyzing automated driving systems with SAE Level 3 and above leads to ASIL classification **D** being allocated to the whole system as there is no need for an overriding capability and for immediate transfer of control to a person. The emergency system is subject to a special consideration as its design depends on the duration of its operation and the possible maneu-

---

2   FIT: Failures-In-Time → Failures per time interval.

3   SPFM: Single-Point-Fault-Metric → Robustness of the system or impact of a system fault on the risks. These faults are detected in the system design and reduced by design. The higher the value, the more robust the system is.

4   LFM: Latent-Fault Metric → Robustness of the system or impact of a "sleeping" or hidden system fault on the risks that could not be foreseen by design.

vers, such as lane changes or a simple stop in the lane and depends on speed. For a simple emergency system with immediate stop on the lane, an ASIL classification "B" is required. The availability of the braking system for the automated driving control systems to achieve the risk minimal vehicle state is classified with ASIL D (maximum availability, lowest fault rate). Due to the technical challenges involved, braking systems are designed redundantly and each of the redundant channels receives an ASIL-B(D) rating for availability.

25.3.1.6 Hands-Free Driving with SAE Automation Level 2 and Above

Due to complexity of driver assistance systems being growing with increasing level of automation, the high costs for SAE Level 3 and higher systems are currently difficult to market. To reduce system costs by reducing monitoring and redundancy elements and to reduce the production costs of higher automation level system elements through economies of scale, so-called SAE Level 2 + (or "2.5") systems are currently being developed. They are available in some markets, which allow drivers to take their hands off the steering wheel for at least a limited period of time.

As mentioned in Section▸ **Fehler! Verweisquelle konnte nicht gefunden werden.**, requirements on automated driving systems up to SAE Level 2 depend on the intervention time, they are designed for. Intervention time is the time it takes for human driver to oversteer or shut down the vehicle automated driving system, i.e. the time a human driver needs to enforce his control responsibilities. In the case of hands-free driving, in addition to the so-called "reaction time" which is needed for cognitive capture of the situation, another time span needs to be considered: The time the human driver needs for putting his hands on the steering wheel and to establish control over the whole chain of effects,

spanning from his intention to change the direction of travel on the road through manual steering wheel interventions to the vehicle's lateral dynamic. The manageability of the event "acquisition of guidance by the driver" is classified to be C2, which is "Normally Controllable" in a "hazard and risk assessment" according to ISO 26262.

Studies by Zhang, de Winter, Varotto and Happee from 2019 [12] show a mean time of 2.14 s for establishing human steering control capability after a period of hands-free traveling. Studies with stricter rules to avoid distraction by Eckstein, Zlocki and Josten [13] resulted in an average response time of 1.19 s in 2016. Continental internal studies were conducted with trained drivers, who needed 2.7 s as an average. This immediately leads to requirements on operational safety and functional safety of the vehicle lateral guiding process.

No special requirements were seen for the longitudinal guiding as current ACC systems are already operated without direct contact of a foot with the brake pedal, and this is considered to be manageable by the driver due to field experience.

Therefore, additional requirements for the braking system by a SAE Level 2 + system with hands-free driving are not expected.

25.3.2 Redundancy Concepts for Braking Systems

◘ Figure 25.17 shows in an extrapolation of brake system development, which degree of redundancy is needed within the brake system to equip a car for automated driving.

The EU law regulation ECE R 13 [13] requires an achievable deceleration of 6.43 m/s^2 for a car with an

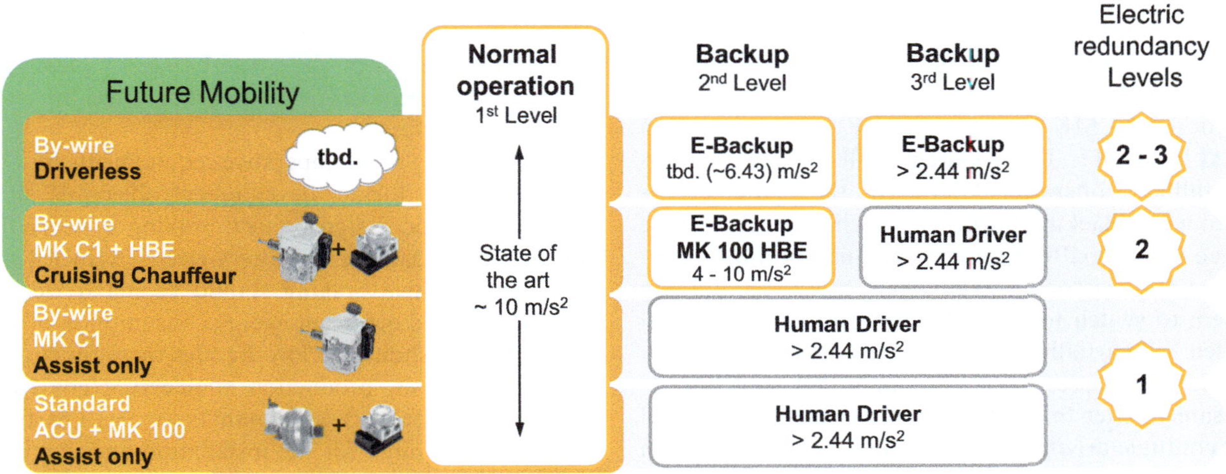

◘ **Fig. 25.17** Extrapolation of brake system development with the required redundant designs

intact braking system. However, a passenger car that could only reach this deceleration value would not be able to persuade a potential car buyer. Braking distances of below 38 m for a standardized brake application from 100 km/h to standstill on dry asphalt, which are described as acceptable in automobile magazines, can only be reached with deceleration values of more than 10 m/s^2.

After any possible initial fault, ECE R 13 requires an achievable deceleration at the lower level of 2.44 m/s^2. The state of the art, underlying ECE R 13, was originally based on a brake, operated exclusively by the driver's muscle force, acting on the brake pedal. In the course of brake systems development, the pedal force was complemented by an amplification force (auxiliary braking). Brake boosters are designed to be "fail-silent". This means that if the booster force fails, the wheel brakes can still be operated by means of human muscle force.

In an electrohydraulic brake-by-wire system, such as the Continental MK C1, the normal braking function of a fault-free braking system is performed with actuator power only (power brake). The work carried out by the driver when applying the pedal is not used for brake application. Pedal stroke and pedal force are used only to determine a target brake pressure. In the fallback operational mode, i.e. in the case of braking applications with an unavailable by-wire function, a pressure, produced by the driver using the brake pedal, is transmitted to the wheel brakes, so that in the fallback operational mode the actuation of the wheel brakes is applied by the driver's muscle force.

With the introduction of the automation Level 3, the driver muscle force is no longer available immediately, if needed. A possible fault in the braking system can therefore only be addressed by a redundant build-up of the braking system. One possible technical solution is the use of an additional hydraulic module with pumps and valves downstream of an electrohydraulic MK C1 module. This module is referred to as Hydraulic Brake Extension (HBE). An example of a realization of the double redundancy of the electric brake actuation in the form of a combination of a MK C1 and an HBE module is shown in ■ Fig. 25.17. These systems still include mechanical fallback operational level. In the figure, this is noted with "Backup 3rd Level". Such a vehicle may be driven in a partially automated manner until the detection of any electrical failure which causes the braking system to switch to the fallback operational level, intended for partially automated operation. Then, in a transitional phase, the vehicle quits automated driving safely. After that, a human driver can take control and continue driving safely for a defined time (Mission Completed). In the absence of any mechanical con-nection of the brake pedal to the braking device, for the actuators, provision could be made for a triple redundancy. Such triple redundancy of actuating devices would mean a significant additional effort for a corresponding braking system, compared to the state of the art with one or two actuators. Therefore, both the safety concept and the brake system technology to be used are currently being revised in the development departments of vehicle manufacturers and suppliers in line with the requests of future vehicle automated driving systems. It is not yet clear whether triple redundancy can be avoided or whether it is necessary to develop new, triple redundancy featuring concepts and to make them cost-effective.

25.4 Outlook

For safety reasons, a car which is suitable for automated driving has two independent electrical on-board power networks. Future braking systems can and must take advantage of this in order to maximize the availability of the braking function even in the event of a possible technical failure. This holds for braking commands of both driver and automated driving system. At present, it is not yet possible to predict the layout of future standard braking systems for passenger cars with automation Level 3, 4 or 5 in detail. However, it is clear that they will be more complex and more expensive than current brake systems. It is therefore expected that there will be competition between different future brake system concepts.

For automated driving, the braking action shall be controllable by signals and the braking system shall have, in addition to normal operation, reserve modes of operation which ensure this function even in the event of failures of any kind. Depending on the nature of the failure, the corresponding reserve mode may be accompanied by degradation of some functions, for example a lower deceleration capability. Reserve mode of operation in today's passenger car, whereby the driver takes over operation of the wheel brakes by means of a "direct hydraulic access from pedal to wheel brakes", can no longer be used.

In addition to these imperative requirements for automated driving, further development objectives must be pursued in the design of new braking systems: To improve energy balance of battery-electric vehicles, usage of recuperative braking should be maximized in driving mode and quiescent energy consumption of the braking system should be low. Besides these functional development goals, production engineering goals are also important for the vehicle manufacturer. In particular, the time required for the installation and commissioning of the braking system during the production of cars is to be reduced. In this context, there is a discussion of a waiver of brake fluid, i.e. the replacement of the electrohydraulic brake system by an electromechanic brake system.

For vehicles with automation Level 4, stowable or retractable pedals have already been shown at automobile fairs. They can only be realized with reasonable effort as so-called by-wire pedals. Such a pedal module only delivers signals and does not provide any mechanically usable output when applied. The conventional brake pedal with a pressure rod that presses a hydraulic cylinder, which is installed outside the cabin, will no longer exist.

References

1. Hermann, W., Stephan, H., Felix, L., Christina, S.: Editors Handbook of Driver Assistance Systems. Springer International Publishing Switzerland (2016)
2. Christian, V.: Entwicklung eines Verfahrens zur integrierten Bremsmomentmessung für Simplex-Trommelbremsen. Dissertation, Technische Universität Darmstadt (2021)
3. United Nations Economic and Social Council.: International treaty: convention on road traffic (1968)
4. Bundesgesetzblatt BGBl.: II S. 809, Text S. 811 Gesetz zu den Übereinkommen über den Straßenverkehr und über Straßenverkehrszeichen (1977)
5. United Nations Organisation UNO.: Report of the sixty-eighth session of the Working Party on Road Traffic Safety, ECE/TRANS/WP.1/145 (2014)
6. Society of Automotive Engineers SAE.: Norm SAE J3016, Taxonomy and Definitions for Terms Related to Driving Automation Systems for On-Road Motor Vehicles (2014)
7. United Nations Organisation UNO.: Consideration of proposals for new UN Regulations submitted by the Working Parties subsidiary to the World Forum, ECE/TRANS/WP.29/2020/81 (2020)
8. International Organization for Standardization ISO: Intelligent transport systems–forward vehicle collision mitigation systems–operation, performance, and verification requirements. ISO **22839**, 2013 (2013)
9. International Organization for Standardization ISO: Road vehicles–safety and cybersecurity for automated driving systems–design, verification and validation. ISO/TR **4804**, 2020 (2020)
10. International Organization for Standardization ISO.: Norm 26262, Road vehicles–functional safety (2018)
11. International Organization for Standardization ISO.: Norm 21448, Road vehicles–safety of the intended functionality SOTIF (2019)
12. Bo, Z., de Winter Joost, C.F., Varotto Silvia, F., Riender, H.: Determinants of Take-Over Time from Automated Driving: A Meta-Analysis of 129 Studies. Elsevier Ltd (2019)
13. Johanna, J., Adrian, Z., Lutz, E.: Untersuchung der Bewältigungsleistung des Fahrers von kurzfristig auftretenden Wiederübernahmesituationen nach teilautomatischem, freihändigem Fahren, Forschungsvereinigung Automobiltechnik (2016)
14. Regulation No 13 of the Economic Commission for Europe of the United Nations (UN/ECE) — Uniform provisions concerning the approval of vehicles of categories M, N and O with regard to braking [2016/194]

Open Access This chapter is licensed under the terms of the Creative Commons Attribution-NonCommercial-NoDerivatives 4.0 International License (▶ http://creativecommons.org/licenses/by-nc-nd/4.0/), which permits any noncommercial use, sharing, distribution and reproduction in any medium or format, as long as you give appropriate credit to the original author(s) and the source, provide a link to the Creative Commons license and indicate if you modified the licensed material. You do not have permission under this license to share adapted material derived from this chapter or parts of it.

The images or other third party material in this chapter are included in the chapter's Creative Commons license, unless indicated otherwise in a credit line to the material. If material is not included in the chapter's Creative Commons license and your intended use is not permitted by statutory regulation or exceeds the permitted use, you will need to obtain permission directly from the copyright holder.

Human-Machine-Interfaces for ADAS

User-Friendly Design of Human–machine Interaction of Driver Assistance Systems

Klaus Bengler and Lutz Eckstein

Contents

© The Author(s) 2026
H. Winner et al. (eds.), *Handbook Assisted and Automated Driving*,
https://doi.org/10.1007/978-3-658-45276-6_26

26.1 Introduction and Motivation

Both in the area of driver information systems and driver assistance systems, the number and complexity of functions continuously increase [1]. This is particularly true for systems that are designed to support the user in challenging driving situations, such as when changing lanes, i.e., primarily aiming to increase safety. In the case of functions to which the driver can increasingly delegate the driving task, a key challenge is to make the capabilities and system limits clear to the person so that he or she has a correct idea of the distribution of roles. In addition, a vehicle usually offers a large number of situational support systems and several additional comfort functions to which the driver can delegate a greater or lesser part of the driving task, depending on the functional area (Operational Design Domain ODD). This is why mode awareness, which refers to the awareness of the distribution of roles and tasks between the driver and the vehicle, also plays a role beyond directing attention.

The term assistance implies, above all, that humans and machines interact. In the case of the motor vehicle as a moving system, it is not enough to strive for efficient, effective, and accepted interaction compared to the design of screens and CE devices. In addition to these three dimensions of classical usability, there are the requirements of interaction safety and controllability, which are explained below.

The goal is therefore to achieve a quality of interaction between humans and machines that is efficient, effective as well as safe and thus generates a high degree of user acceptance. To this end, the human–machine interface must be suitably designed, i.e., according to Bubb [1, 2] the input technologies, the interaction logic as well as the output and display technologies. However, the design of the interaction also corresponds to requirements for the actual function, its performance limits, and the effects of system errors, since these must be recognized and controlled by humans, which is addressed by the term controllability.

The development and implementation of the functional scope of the HMI represents a complex process in the context of which the human–machine interaction must be designed and optimized and then implemented in the form of software and hardware. The result must measurably meet a wide variety of internal and external requirements. For information and communication systems, many of these requirements are very well structured in documents such as the European Statement of Principles on In-vehicle Information and Communication Systems, or ESoP [3] for short are formulated. For their verification, mostly experimental methods have been standardized. For driver assistance systems and automated driving functions, relevant criteria and methods are described in the Response Code of Practice [4], but they are by far not as widely harmonized and distributed internationally. Due to the large interindividual differences, user studies form the backbone of the methodological canon. To counter subjectivity in the decision-making process, quantification of a wide variety of metrics is a key tool for ensuring safe usability while driving. In addition, there are internal quality requirements that supplement the external minimum requirements with company-specific standards. They aim to ensure brand-specific imprints, consistency between different vehicle models of a brand and, for example, premium requirements for the interaction between man and vehicle.

Since driver assistance systems make up only a small part of the more than 1000 functions with which a human driver can interact in a modern vehicle, the activities of many people are intertwined in the development process [5]. In the end, these must result in a secure human–machine interface or a display and operating concept from a single source [6].

A corresponding development process should be based on the expectations and skills of the users, take into account the relevant external and internal requirements, and result in a coherent concept despite the diversity of technologies. Above all, the limitations of the users should be taken into account, which can lie in perception, cognition, and motor skills and should be compensated for by suitable assistance. This should not neglect the strengths of humans as vehicle drivers to bridge or control technical limitations. This consideration is initially oriented to the principle MABA-MABA (Paul Fitts) "man are better at—machines are better at". It is discussed in detail by Reichart [7]. However, this has to be fundamentally re-discussed against the background of automated vehicle guidance. Furthermore, intercultural aspects have to be taken into account throughout.

26.2 User and User-Friendly Design

The challenge of developing driver-vehicle interfaces is primarily to design a suitable interaction concept for a wide range of possible users and usage situations that allows the most error-free and efficient interaction possible with a wide range of functions right from the initial contact.

The requirements can be merged into several groups:

- Fundamental are legal requirements, the violation of which means that a vehicle cannot be registered in the respective market. Such requirements now also exist for driver assistance systems and automated driving functions.

- Requirements based on voluntary commitments by automobile manufacturers are almost as binding [8, 9]. This instrument makes sense in particular when a technology or system class is still relatively young, so that a legal regulation does not yet adequately address possible better solutions. For example, the ESoP already mentioned is not a law, but a European recommendation. Vehicle manufacturers and other stakeholders have committed themselves to compliance with the principles formulated therein order to reduce the distracting effect of information and communication systems in motor vehicles.
- Norms and standards define a state of the art accepted by several parties. As a manufacturer, you can deviate from this, provided that this can be justified in a comprehensible manner.
- In contrast, user-specific requirements for an interaction within the framework of the previously mentioned requirements are differentiating factors for competition. For example, the quality of the interaction in the initial contact, often described as a so-called rental car scenario, plays a central role. Conversely, it should also be defined how efficient the interaction with prioritized functions should be in the practiced state (continuous use). This can be done either in absolute terms in the form of necessary interaction steps or interaction times, or relative to the competition classified as relevant.
- In addition, brand-specific requirements often play a role, which in the simplest case affect the choice of font and the color palette of the graphic displays. If, on the other hand, a brand is to be positioned by almost completely dispensing with physical controls, analogous to smartphones, this can lead to a very distracting design of the interaction, which has even been criticized in court rulings.

- Finally, technical requirements and boundary conditions can have a major influence on interaction quality, especially if technical building blocks have to be built on top of existing technical building blocks. Ideally, the validated design of an interaction concept can be used to derive requirements for its technical implementation, e.g., to keep the latency between user input and system response low.

Simply meeting the legal requirements as well as the minimum requirements defined in voluntary commitments would certainly produce a usable HMI. However, the automobile represents one of the most complex consumer goods, as regular appearances by vehicle manufacturers at corresponding trade shows such as the Consumer Electronics Show (CES) in Las Vegas suggest. Consequently, different brands compete and try to differentiate themselves, especially through differences in the HMI concept. Brand-specific design should take into account that users compare the interaction concepts of vehicle manufacturers with the interaction concepts of their mobile devices and computers and expect similar or identical approaches in the automobile.

This initially understandable expectation should not or cannot be completely fulfilled, since one must sufficiently consider interaction safety and controllability as additional requirements when designing the human–machine interface in the automobile. ◘ Fig. 26.1 illustrates the difference between the interaction design of a classic consumer product, which must fulfill the criteria of usability [10], and the much more complex design of the human–machine interface in the automobile, which must pay particular attention to the aspect of safety and controllability.

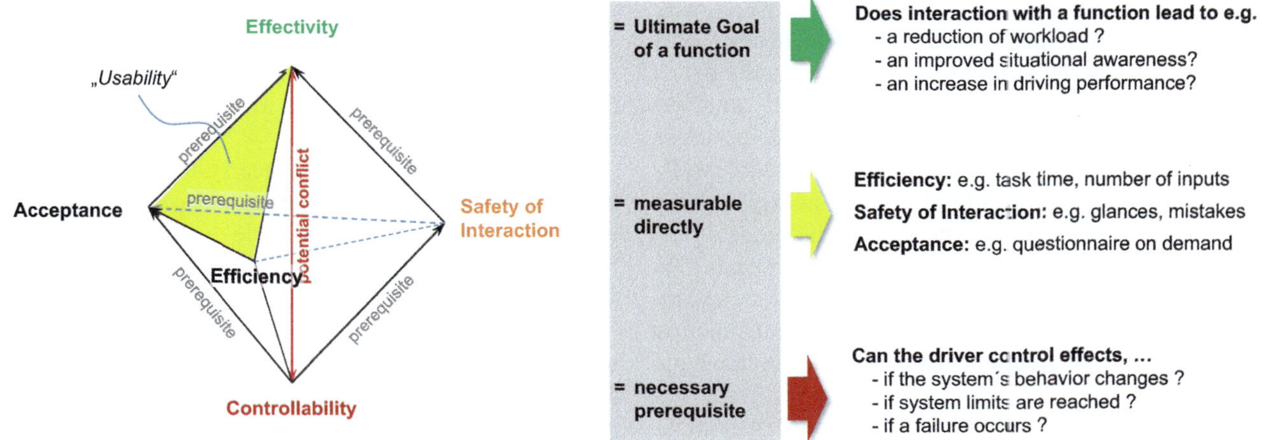

◘ Fig. 26.1   Additional requirements of interaction safety and controllability for the HMI design in the motor vehicle for usability

In order to achieve a high degree of usability, the design of an interaction should be fundamentally **efficient, i.e.,** require few operating steps and a short interaction duration, and achieve the desired **effect**, thus creating very good conditions for a high degree of **acceptance**.

In addition to these three classic usability criteria, **interaction safety is the** first factor to be considered in motor vehicles, since the human being must devote himself or herself primarily to the safe performance of the driving task and thus distribute attention appropriately. For this reason, long glances at a system display, for example, must be avoided just as much as unintentional distraction of attention, because an attractive graphic animation can become a fatal distraction while driving. For example, a graphic display of the current lateral acceleration must be critically scrutinized, as it can potentially lead to inadequate gaze behavior at high lateral accelerations or even to riskier driver behavior.

To be clearly distinguished from the design criterion of interaction safety is the criterion of **controllability**, which is of particular relevance for driver assistance systems and automated driving functions. Against the background of ISO 26262, functional safety and controllability are often reduced to the consideration of a system failure and possible effects. In the context of interaction design, however, the limits of a system function as well as the controllability of a function as such must also be considered, which is now addressed by the supplementary ISO 21448 on safety of intended function (SOTIF).

A challenge highlighted in ◘ Fig. 26.1 in interaction design lies in the possible **conflict of goals** between the effectiveness of a function on the one hand and its controllability at the system boundaries or in the event of an error. This can be explained by the example of an automatic distance control system (ACC). The effectiveness of this function in the sense of the intended relief of the human undoubtedly increases if the functional limits are extended: If the system, for example, decelerated twice as hard instead of only braking at 3 m/s^2 as in many vehicles today, the driver would have to intervene much less frequently and feel a greater relief effect. On the other hand, however, when the system limits are reached, a person would have to take over longitudinal control at significantly higher decelerations, who experiences this process only rarely and perhaps for the first time. This means that the conditions for good controllability are less favorable in the case of a technically very powerful function than in the case of frequent takeover at low vehicle dynamics.

Against the background of the complex requirements for the design of human–machine interaction

and the multitude of functions of modern motor vehicles, it is essential to adopt a **user-oriented approach**. This user-centered approach should ensure that, in addition to technical and legal requirements, the concerns of cognitive ergonomics regarding intuitiveness, learnability, and user experience are taken into account. Specifically, this means iteratively involving potential users in the evaluation of concepts as part of the development process. On the one hand, this requires assessable representations of the interaction concept or individual aspects, and, on the other hand, the use of suitable assessment methods. In [11] it is shown how the interaction concept of a luxury class sedan was developed and validated. A variety of methods were used, which on the one hand focus on different aspects of the interaction concept, such as the information representation or the interaction logic, and on the other hand abstract the driving task to different degrees; see ◘ Fig. 26.2.

The basic understanding of the information presented as well as of interaction sequences can be assessed using simple representations and simulations without a parallel driving task. However, if the question is how distracting the interaction with an information and communication system (IVIS) could be and—vice versa—how effective the guidance of attention by a driver assistance system is, it is necessary to present the driving task in parallel to the interaction with the system to be evaluated.

The strongest abstraction of the driving task is the so-called occlusion method (ISO 16673:2017). Study participants wear special glasses that can block or occlude the view of the testing system in an electronically controlled manner. The basic idea is that during the occlusion phase, the human observes the driving scene rather than the system being evaluated. While the Peripheral Detection Task (PDT), referred to in standardization as Detection Response Task (DRT) (ISO 17488:2016), and the so-called Lane Change Test (LCT, cf. [12]) are aimed at the question of the extent to which humans are still able to perceive and react to peripheral visual stimuli during the use of an IVIS, the driving task is mapped realistically in static and dynamic driving simulators so that the interaction with driver assistance systems can also be investigated and evaluated.

As soon as a driver assistance system actively influences the dynamic behavior of the vehicle, a sufficiently good representation of the vehicle dynamics is required to make valid statements, since otherwise vestibular and kinesthetic stimuli are missing.

The development of a viable interaction concept must therefore take into account a wide variety of user and usage aspects and repeatedly involve future users in the process in the spirit of participatory and iterative development ISO 9241 [13].

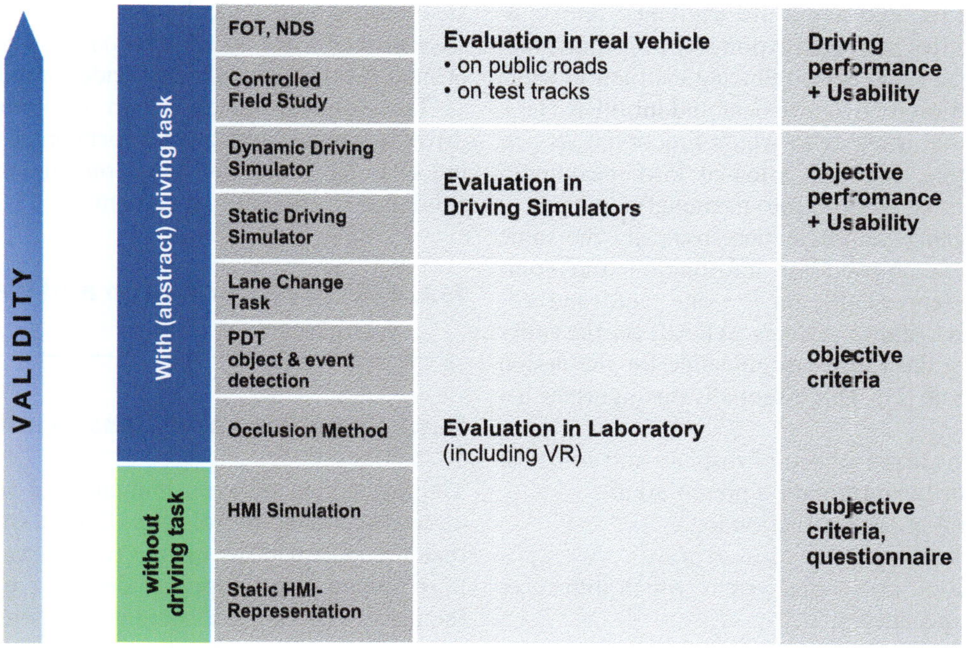

VALIDITY				
With (abstract) driving task	FOT, NDS	Evaluation in real vehicle • on public roads • on test tracks		Driving performance + Usability
	Controlled Field Study			
	Dynamic Driving Simulator	Evaluation in Driving Simulators		objective perfomance + Usability
	Static Driving Simulator			
	Lane Change Task	Evaluation in Laboratory (including VR)		objective criteria
	PDT object & event detection			
	Occlusion Method			
without driving task	HMI Simulation			subjective criteria, questionnaire
	Static HMI-Representation			

Fig. 26.2   Methods for the evaluation of the human–machine interface in motor vehicles

26.3 MMI for Driver Assistance Systems Compared with Information and Communication Systems

Over a long period of time, individual systems could be made usable for humans in the vehicle with specific interaction concepts, so that special controls and lights/displays were provided for operating the system and displaying the relevant information. Examples of this are the interaction with the first cruise control/tempomat systems, radio systems, or climate control functions.

This approach could no longer be pursued for several reasons [14]. First, the necessary space for displays and controls was no longer sufficient. Second, the need for learning increases with the number of different interaction paradigms, so that intuitive use of the many functions in the initial contact is practically ruled out. Third, end users simply expect many technical driver assistance functions to belong together. Therefore, in the UR:BAN project, an integration approach was developed for the integration of functionalities that assist in the execution of the driving task on the maneuver level and navigation level, which provides for an early and content/logic integrated presentation [15, 15].

For this purpose, all functions were listed and grouped with their area of application, their functional character (information, warning, intervention). In this way, functionalities that provide information regarding longitudinal guidance or even intervene in longitudinal guidance were coordinated with one another so that contradictory or superfluous instructions could be avoided. For example, a request from the traffic light assistant to decelerate was combined with a further recommendation from the ACC to decelerate to a vehicle ahead. Similarly, this was possible in the area of lateral guidance. Through this approach it was very possible to reduce the number of messages and outputs and to enable drivers to adopt a more anticipatory driving style.

However, with this strategy, users lose the direct association of an AD to a function, potentially making it more difficult to understand the individual function. It is therefore necessary to establish and visualize a **cross-functional metaphor**. In the case of the UR:BAN HMI strategy, this was the metaphor of protection and influence areas arranged around the vehicle, to which the functionalities can be assigned in combinations.

Similar designs were made by von Gijsel [16] or Lindberg [17] presented. The crucial factor is the fit of the metaphor to the users' imagination/expectations. Lindberg showed with his card sorting experiments that this spatial metaphor with different safety zones fits the mental models of the later users.

A similar development can be observed in the area of **infotainment functions**. A very component- or device-oriented interaction is increasingly being replaced by a function- or intention-oriented interaction. Whereas for a long time various devices (radio, MP3, Bluetooth, …) were activated in the area of audio functions in order to listen to specific information or music, users can increasingly express the desire to listen to a specific song or feature, whereby the source of playback is irrelevant. Similarly, comfort settings of air conditioning, lighting, and audio are combined into interior profiles. In this way, interaction steps for individual operation can be saved, thus reducing distraction while making interaction more convenient and intuitive. Here, too, the question arises as to which functional scopes, oriented to users, can be combined into meaningful units oriented to be grouped into meaningful units.

When **designing the interaction**, basically the same approach can be applied for information and communication systems (IVIS) on the one hand and advanced driver assistance systems (ADAS) on the other hand, following the European guideline for the design of IVIS (ESoP, cf. [3]). This essentially distinguishes between.

— the geometric arrangement of displays and controls,
— the design of the information presentation,
— the definition of the interaction logic,
— the system behavior, and
— communication about the system, which influences the user's expectations.

Nevertheless, the design rules defined in the ESoP cannot be transferred one-to-one **from IVIS to driver assistance systems** because the design goals are diametrically opposed in parts. This will be explained by means of ■ Fig. 26.3 as an example.

The difference becomes particularly obvious when considering the design of **information presentation**: while information and communication systems should not be designed in a distracting way, driver assistance systems and their outputs should *direct attention appropriately* so that the human recognizes the need for an action if possible and initiates it. The same applies to the **interaction logic**: IVIS should present only driver-initi-

ated information, while driver assistance systems should *issue warnings and information on their own initiative to* prompt the driver to take an accident-avoiding action.

Therefore, dedicated design recommendations for driver assistance systems are formulated in the following sections, which are based on the structure of the ESoP but clearly differ in content.

26.4 MMI for Information and Warning Functions

26.4.1 Definition and Classification

The interaction between humans and vehicles can be understood as a control loop. Driver assistance means that the system takes on the role of an assistant and thus prepares and supports decisions and actions of the driver. The human being is therefore the active element in the driver-vehicle environment control loop: he or she perceives the information directly and via displays and processes it in order to decide and act with confidence. Assistance systems support humans, for example, by displaying relevant information in a timely manner as well as evaluated information that can have a warning character. The system can be compared to the personal assistance of a manager, who prepares decisions but does not make the final decision [1, 1, 18].

Consequently, the aim of designing driver assistance systems is to support the human to perform his task of driving the vehicle and to fulfill his responsibility. Since decisions at the path guidance level, such as stopping at an intersection, must be made several sec-

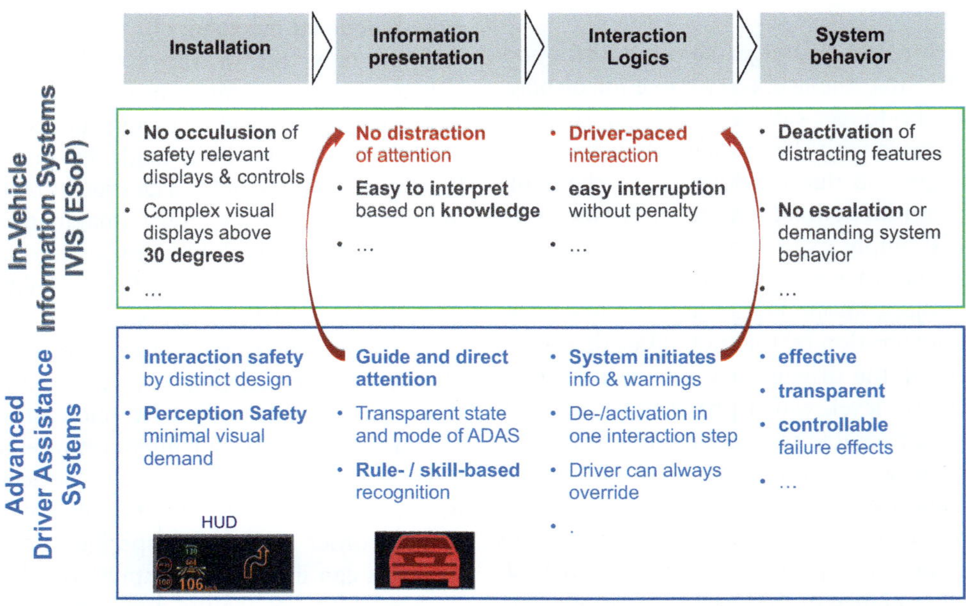

■ **Fig. 26.3** Comparison of exemplary design principles for IVIS and ADAS

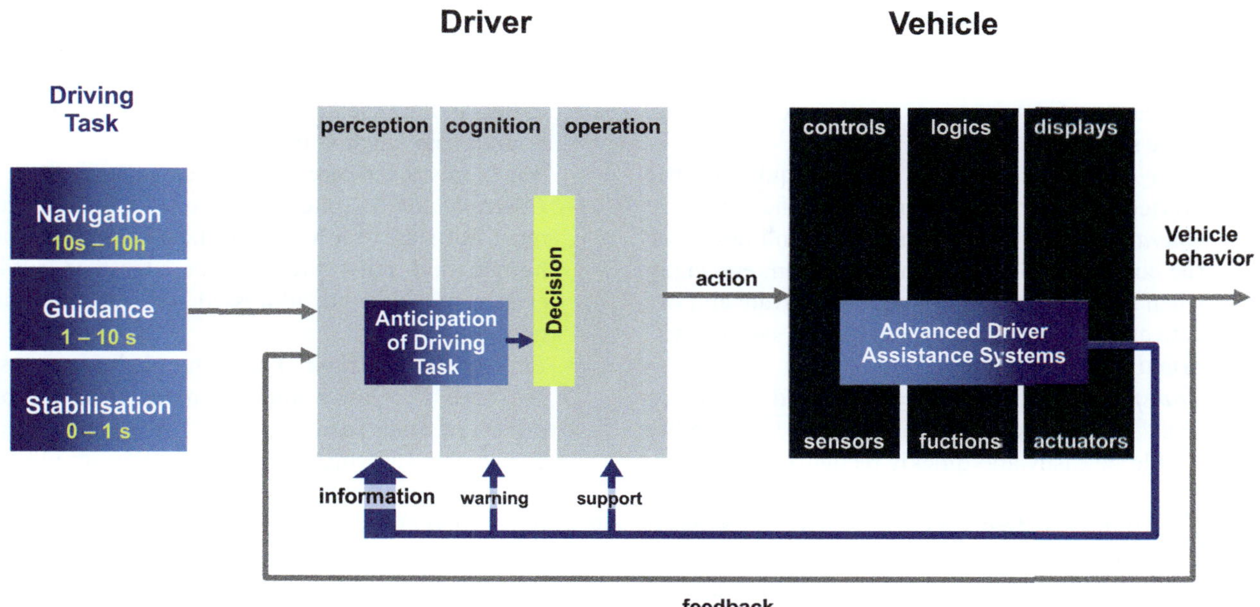

Fig. 26.4   Informing driver assistance in the control loop consisting of driver, vehicle, and environment [1]

onds before the vehicle comes to a standstill and the vehicle also reacts to driver inputs with a time delay, all relevant information must be displayed early and in full. On this basis, the human can anticipate the development of the traffic situation in order to decide the right action in time; cf. **Fig. 26.4**.

In order to decide on the choice of speed and a suitable course at the guidance level, **information** can be displayed, for example, on the course of the road ahead, speed limits, or events such as the end of traffic jams. Drivers must evaluate the relevance of the information themselves and take it into account in their decision.

The *classification of information* can be based on its relation to the elements of the control loop shown in **Fig. 26.4**: it can refer to other road users or objects (e.g., pedestrians, cyclists, animals) or provide information about the course of the road (e.g., curve, bump), regulations (e.g., speed limit), or weather conditions. Increasing attention is paid to information regarding the condition and behavior of the driver (e.g., fatigue). A clear distinction must be made between this and information regarding vehicle operation (e.g., step on the brake pedal to engage drive mode D). Information on the driving status (e.g., driving speed) and the status of the vehicle's systems (e.g., door open, dipped headlights on) has long been established and is partly regulated by law.

Warnings are the result of a machine evaluation of information with regard to its potential relevance. They indicate deviations of an actual state from a reference or target value, such as falling below the legally required minimum distance to the vehicle in front, but can also include concrete requests for an action, such

as braking. Warnings aim at directing people's attention in a targeted manner.

Warnings can preferably be classified in terms of content based on the type of action that is to be influenced. Important warnings concern longitudinal dynamics, such as the request to brake or stop. If the reason for the warning is displayed, it should be conveyed quickly and in a way that is easy to interpret, e.g., in the form of an internationally recognized symbol. With regard to lateral dynamics, warnings can be differentiated on the basis of the object to which attention is to be drawn. This can be the movement relative to other road users (e.g., vehicle in the "blind spot") or the own course relative to the course of the road (e.g., threatening to leave the lane or the roadway). Another class of warnings relates to one's own vehicle and its systems (e.g., refilling wiper fluid). Few warnings so far address driver behavior as such (e.g., wearing a seat belt).

Information and warnings can of course also be classified and differentiated in terms of their design. A good overview and introduction are given by standard works on system ergonomics [19].

26.4.2 Design Rules

The design of the human–machine interface for automated driving functions is detailed below in the form of concrete design principles for informing and warning driver assistance systems. These principles are structured analogously to the European Statement of Principles on Information and Communication Systems (ESoP), the revision of which is explained in [20].

Based on [18], these categories and selected design principles for driver assistance systems are presented below.

1. **Arrangement of** displays and controls for driver assistance systems
   - *Perceptual safety*: Reading should require as little averted gaze as possible from the traffic scene.
   - *Operating safety*: The arrangement and design of the controls should prevent confusion, especially when they are needed not only to activate and deactivate the function, but also during system use.
2. **Information presentation**
   - *Transparency* of assistance type and assistance level: system status (on/off) and availability should be easily and quickly recognizable.
   - *Perceptibility*: Information and warnings should be designed so that they can be easily perceived by the driver under all relevant conditions.
   - *Recognizability*: Information presentation should preferably use easily interpretable and/or internationally known/standardized symbols and metaphors to ensure easy and quick recognition of the meaning (e.g., steering wheel vibration when leaving lane resembles nail band rattles).
   - *Unambiguity*: The unambiguous assignment of information presentation and driving situation is a prerequisite for a very fast and correct reaction of the human at the level of skill-based behavior.
3. **Interaction with driver assistance systems**
   - *(De)activation*: Driver assistance systems should always be activatable and deactivatable by users. Exceptions may be appropriate, particularly in the case of chassis control systems such as ABS.
   - *Interruptibility*: The use of driver assistance systems should not require permanent interaction, but should allow driver-initiated interaction at any time.
   - *Stability*: All information output by driver assistance systems and the control of actuators should be designed in such a way that the stability of the driver-vehicle environment control loop is not negatively influenced. An important prerequisite for this is that drivers can clearly distinguish the system outputs from their setpoint inputs, which is not trivial, for example, when a steering wheel torque is applied unexpectedly.
   - *Override capability*: To meet his or her ultimate responsibility, the driver must have the ability to turn off, cancel, or override all functional outputs of driver assistance systems.
4. **System behavior**
   - *Effectiveness*: In addition to a function designed in accordance with the system's objective, the output of information and warnings must be designed effectively. In addition to the type of information presentation, the timing in particular

should be selected so that system outputs are provided neither too early nor too late.
   - *Traceability*: This requirement addresses reproducibility and transparency of system function and boundaries, which in turn requires a strong limitation of customer-perceivable system complexity.
   - *Controllability*: System errors and limitations must not lead to a loss of vehicle control by the human and must therefore be identified sufficiently quickly and addressed with appropriate measures.
   - *Lack of repercussions*: The system behavior must not have a negative impact on the behavior of drivers or other road users.
   - *Timeliness*: Information and warnings should be displayed in a timely manner, otherwise neither the desired effect nor a positive acceptance of the system will be achieved.

In addition, as in the ESoP, the communication and description of a driver assistance system must be designed in such a way that, for example, it does not raise the expectations of drivers too high, which the actual function does not meet. The design principles listed in the ESoP with regard to communication can be directly transferred to driver assistance systems in this respect.

26.4.3 Design Examples

Both information and warnings can be perceived through different sensory modalities.

Information from driver assistance systems is usually displayed visually, often in the instrument cluster or head-up display. Although an acoustic presentation of information is possible and widely used, for example, in parking assistance systems, it has the disadvantage that it is perceived by all vehicle occupants. Haptically, humans perceive information very quickly, such as a variation in steering wheel torque and ABS control via the pulsation of brake pedal movement. The interpretation of a haptic warning is always more difficult when the signal can also have other causes. For this reason, concepts such as the haptic accelerator pedal were designed. It is intended to guide the driver to drive more efficiently by increasing the actuation force and thus to use haptics for the explicit presentation of information from driver assistance systems, have not yet become established.

Warnings should be perceived and correctly interpreted within the shortest possible time, ideally independent of the person's current line of vision and body posture. For this reason, several sensory modalities are often addressed in combination, possibly staggered over time. For example, a visual display is com-

	MODALITIES			
	Optical (eye)	Acoustic (ear)	Haptic (force, path)	Vestibular (acceleration)
INFORMATION				
Environment				
Traffic	Vehicle in front			
Roadway	Sharp curve ahead			
Weather	Temperature below 3°C			
Driver				
Operate	Notes for action			
Behavior	Attention Assist: „Pause"			
Vehicle				
Driving state	Speed limit			
System state	System on / off / active			
WARNINGS				
Longitudinal guidance				
Small distance	Red vehicle icon			
Speed	Flashing speed limit			
Time to Collision	„BRAKE"	Intermittent warning tone	Increased accelerator pedal force	Partial deceleration or brake jerk
Lateral guidance				
Traffic	Flashing vehicle icon		Steering wheel torque or vibration	Reduction of the lateral approach
Lane	Flashing lane symbol		Steering wheel torque or vibration	Lateral course correction by brake
System				
System active	Flashing ESP control light			Course and speed correction
System error	Sensor dirty	Gong at system ejection		Reduction of the drive torque

Fig. 26.5 Suitable modalities for displaying information and warnings from driver assistance systems. Informing driver assistance in the control loop consisting of driver, vehicle and environment, according to [4]

bined with an acoustic warning to prompt the driver to stop, for example. In collision avoidance systems, a visual and acoustic warning is usually followed by partial deceleration, which simultaneously acts as a further warning stage.

Fig. 26.5 systematizes frequently used modalities for the transmission of information and warnings along the lines of the RESPONSE Code of practice for the design and evaluation of ADAS. Vehicle manufacturers offer an ever-growing number of individual driver assistance systems, which are generally harmonized only to a limited extent with respect to the design of information, warnings, and interventions. From a system ergonomics perspective, this can be seen as disadvantageous, since the human must develop an internal model for each function in order to understand and use the system outputs. Research and predevelopment therefore develop concepts that combine these numerous individual functions into just two systems by using metaphors. Fig. 26.6 shows a perspective representation of the driving-relevant information. In analogy to the established desktop metaphor [23] it is form familiar to the driver and is referred to as the roadview metaphor [24].

The illustration shows the currently active collision avoidance system, which supports the human during self-driving like a repulsive magnetic field. This virtual magnetic field acts in all directions, even when reversing and parking, and its strength can be adjusted. As an alternative to self-driving, the user can delegate the driving task to the second system; the magnetic field lines turn green and are used to set the desired distance.

Also in the project PRORETA 3 [25] 0 the approach was taken to combine the variety of individual driver assistance systems in only two functional modes: the mode for assisted self-driving is called "Safety Corridor", the one for delegation of the driving task "Cooperative Automation".

26.5 MMI for Longitudinal and Lateral Guidance Functions

As soon as the human can delegate the execution of the driving task to the vehicle to a large extent (Level 2) or completely (Level 3 and higher), this results in a whole series of relevant questions:

- How is the respective system activated and switched between possible modes?
- How is the distribution of roles and tasks sufficiently recognizable (mode awareness)?
- How does the driver take over the driving task again?
- Based on what other conditions is the driving task handed back to the human and how is this appropriately indicated?

26

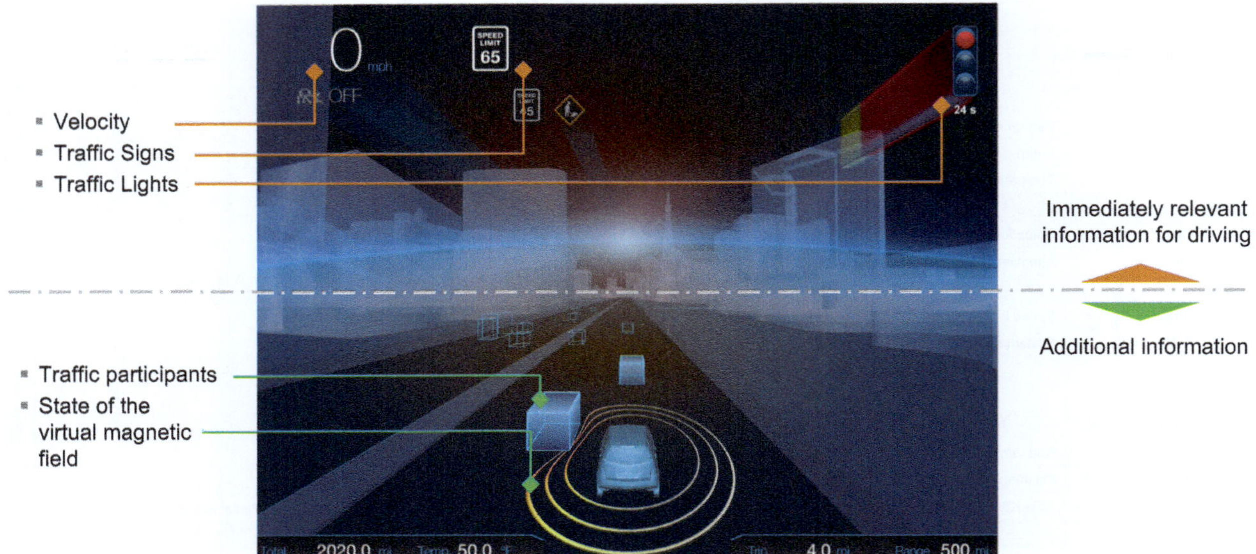

Fig. 26.6 Integration of all collision-avoiding functions by the annular metaphor of magnetic repulsion embedded in the roadview metaphor

— What requirements does the respective function place on the monitoring of the user or how is sufficient monitoring by the driver ensured?
— What should the system do if the human does not respond to multiple requests for action?

The complexity of the task illustrated by **Fig. 26.7** shows the state space of interaction with driver assistance systems on the basis of four cornerstones of Interaction with ADAS.

While the human is performing the driving task, he or she consciously divides the attention between the driving task in the lower left corner and a secondary task, which may be, for example, talking to the passenger or changing the radio station (upper left corner in **Fig. 26.7**). The situation to avoid is that the person can no longer consciously control attention but is distracted by an external influence (distraction). Furthermore, fatigue can lead to a situation where attention is no longer sufficiently focused on the driving task (attention @ limits).

If the driver activates a driver assistance system that takes over the execution of the driving task but not the responsibility for it (Level 2), then he or she must sufficiently monitor the system behavior. This means that the user's attention must be regularly focused on the driving task (lower right corner) and only temporarily turned to a secondary task (upper right corner). If the system detects by a so-called driver monitoring system that the user does not sufficiently monitor the execution of the driving task by the system (distracted or drowsy driver), the system must direct the user's attention back to the driving task by means of warnings (1st Call, 2nd Call). If this does not succeed, the system function should noticeably degrade by reducing the driving speed so that the human is once again challenged to take over the driving task and the risk of an accident is reduced. Independently of this, the system must detect when its own system limits are reached in order to hand over the execution of the driving task to the human in an orderly manner (TOR @ system limits).

The figure clearly shows that, in addition to a present representation of the system state, the **transitions** between states in the state space in particular must be carefully designed. Furthermore, it becomes clear that offering another additional mode such as a Level 3 function adds another dimension to the state space to be mastered, so that sufficient mode awareness of the user becomes even more important. From the perspective of system ergonomics, the priorities are therefore clear: the decisive criterion should not be what is technically possible, but the complexity that can be mastered by humans.

26.6 Outlook and Further Developments

At present, the differentiation of automated driving functions is increasing, often for technical reasons: in the U.S. and other countries, for example, Level 2 functions are on the market, in which driver monitoring is realized via cameras so that the human can take his hands off the steering wheel. Furthermore, we can

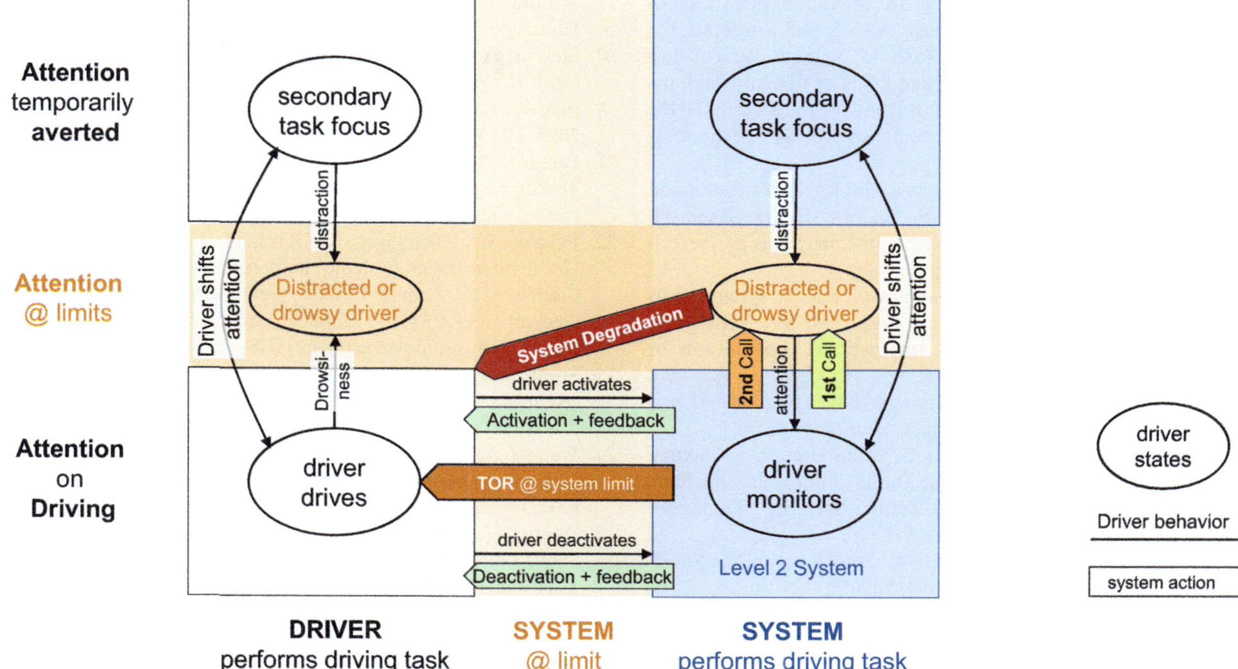

◘ **Fig. 26.7** State space of interaction with driver assistance systems based on four cornerstones

expect to see additional, manufacturer-specific Level 3 functions that will complement Level 2 functions in the same vehicle due to the limited ODD.

Since human perception and cognition develop evolutionarily on a time scale that is orders of magnitude longer than technological progress, the challenge arises on the one hand to design human–machine interaction in motor vehicles in such a way that humans benefit as much as possible from these developments and are not confronted with their technical complexity. On the other hand, this chapter makes it clear that the number and variety of driver assistance functions should be significantly reduced and that this is already possible on the basis of the concepts presented as examples.

While there are internationally harmonized guidelines and recommendations regarding the design of information and communication systems, this is only rudimentarily the case for driver assistance systems. Since the safety relevance of ADAS is to be classified higher than that of IVIS, this results in the recommendation that companies involved in the development and integration of these systems take special care to establish stable development processes that cover development, testing, and documentation and also include learning from field data and updating of functions.

References

1. Akamatsu, M., Green, P., Bengler, K.: Automotive technology and human factors research: past, present, and future. Int. J. Veh. Technol. (526180) (2013)
2. Bubb, H., Bengler, K., Grünen, R.E., Vollrath, M.: Automobilergonomie. Springer-Verlag (2015)
3. Commission of the European Communities: Commission Recommendation of 22 December 2006 on Safe and Efficient In-Vehicle Information and Communication Systems: Update of the European Statement of Principles on human machine interface. Brussels **22**(12), 2006 (2006)
4. RESPONSE Consortium: Code of practice for the design and evaluation of ADAS (V5.0): RESPONSE 3: a PReVENT Project (2009)
5. Bengler, K., Rettenmaier, M., Fritz, N., Feierle, A.: From HMI to HMIs: towards an HMI framework for automated driving. Information (Switzerland) **11** (2) (2020). ▶ https://doi.org/10.3390/info11020061
6. Bengler, K., Dietmayer, K., Färber, B., Maurer, M., Stiller, C., Winner, H.: Three decades of driver assistance systems: review and future perspectives. IEEE Intell. Transp. Syst. Mag. **6** (4), S. 6–22 (2014). ▶ https://doi.org/10.1109/MITS.2014.2336271.
7. Reichart, G.: Menschliche Zuverlässigkeit beim Führen von Kraftfahrzeugen. Fortschritt-Berichte VDI, Reihe 22 Mensch-Maschine-Systeme. VDI-Verlag, Düsseldorf (2001)
8. JAMA (Japan Automobile Manufacturers Association).: Guideline for In-vehicle Display Systems (2004)
9. AAM.: Principles on Human Machine Interface (HMI) for In-Vehicle Information and Communication Systems (Draft), Alliance of Automobile Manufacturers, Detroit (2002)

26

10. ISO 9241–11:2018–03. Ergonomie der Mensch-System-Interaktion - Teil 11: Gebrauchstauglichkeit: Begriffe und Konzepte

11. Niedermaier, B., Durach, S., Eckstein, L., Keinath, A.: The new BMW iDrive – applied processes and methods to assure high usability. In: Duffy, V.G. (eds) Digital Human Modeling. ICDHM 2009. Held of Part of HCI Internations 2009, San Diego, CA, USA (2009)

12. Mattes, S.: The lane-change-task as a tool for driver distraction evaluation. In: Strasser, H., Kluth, K., Rausch, H., Bubb, H. (eds.), Quality of Work and Products in Enterprises of the Future. Ergonomia, Stuttgart (2003)

13. ISO DIS 9241–210:2008.: Ergonomics of human system interaction-Part 210: Human-centered design for interactive systems (formerly known as 13407). International Organization for Standardization (ISO). Switzerland

14. Freymann, R.: HMI: a fascinating and challenging task. IEA—16th World Congress on Ergonomics. Maastricht (2006)

15. Bengler, K., Drüke, J., Hoffmann, S., Manstetten, D., Neukum, A.: UR: BAN Human Factors in Traffic. Approaches for Safe; Efficient and Stress-Free Urban Traffic. Springer: Wiesbaden, Germany (2018)

16. Gijssel, A.V., Brunner, G., Künzner, H.: Assistance for the Sovereign Vehicle Driver–A Driver-Vehicle Interface for Sustained Situation Expectancy and Anticipative Vehicle Control. Der Fahrer im 21. Jahrhundert, VDI-Bericht Nr. 2015 (S. 251–256). Düsseldorf: VDI-Gesellschaft Fahrzeug- und Verkehrstechnik (2007)

17. Lindberg, T.: Entwicklung einer ABK Metapher für gruppierte Fahrerassistenzsysteme. Dissertation an der technischen Universität Berlin (2012)

18. Eckstein, L.: Souveräne Interaktion mit Fahrerassistenzsystemen. In: Tagungsband, VDA Technischer Kongress 2008, Ludwigsburg (2008)

19. Schmidtke, H.: Ergonomie, 3. Auflage – München; Wien: Carl Hanser Verlag (1993)

20. Stevens, A., Hallen, A., Pauzie, A., Vezier, B., Gelau, C., Eckstein, L., Victor, T., Koenig, W., Moutal, V., Hoefs, W.: The European Statement of Principles on Human Machine Interaction 2005. ITS World Congress, San Francisco (2005)

21. Perelló, J.R., Gomila, A., García-Quinteiro, E.M., Miranda, M.: Testing new solutions for eco-driving: haptic gas pedals in electric vehicles, published by J. Transp. Technol. 7(1) (2017)

22. Bengler, K., Dietmayer, K., Eckstein, L., Stiller, C., Winner, H.: Fahrerassistenzsysteme und Automatisiertes Fahren. In: Pischinger, S., Seiffert, U. (eds) Vieweg Handbuch Kraftfahrzeugtechnik. ATZ/MTZ-Fachbuch. Springer Vieweg, Wiesbaden (2021). ▶ https://doi.org/10.1007/978-3-658-25557-2_8

23. Smith, D.C. et al.: The Star User Interface: An Overview. Proceedings of the AFIPS National Computer Conference, pp. 515–528 (1982)

24. Wagner, V., Bavendiek, J.: Metaphor based interaction safety. Aachener Kolloquium Fahrzeug- und Motorentechnik. Aachen (2017)

25. Winner, H., Lotz, F., Bauer, E., Konigorski, U., Schreier, M., Adamy J., Pfromm M., Bruder, R., Lüke, S., Cieler, S.: PRORETA 3–an integrated ADAS concept–Comprehensive driver assistance by safety corridor and cooperative automation. In: Tagungsband 1. Internationale ATZ-Fachtagung, 28.-29.04.2015, Frankfurt am Main)

26. Winner, H., Lotz, F., Bauer, E., Konigorski, U., Schreier, M., Adamy, J., Pfromm, M., Bruder, R., Lüke, S., Cieler, S.: PRORETA 3: comprehensive driver assistance by safety corridor and cooperative automation. In: Winner, H., Hakuli, S., Lotz, F., Singer, C. (eds.) Handbook of Driver Assistance Systems, pp. 1449–1470. Springer International Publishing Switzerland (2016)

Open Access This chapter is licensed under the terms of the Creative Commons Attribution-NonCommercial-NoDerivatives 4.0 International License (▶ http://creativecommons.org/licenses/by-nc-nd/4.0/), which permits any noncommercial use, sharing, distribution and reproduction in any medium or format, as long as you give appropriate credit to the original author(s) and the source, provide a link to the Creative Commons license and indicate if you modified the licensed material. You do not have permission under this license to share adapted material derived from this chapter or parts of it.

The images or other third party material in this chapter are included in the chapter's Creative Commons license, unless indicated otherwise in a credit line to the material. If material is not included in the chapter's Creative Commons license and your intended use is not permitted by statutory regulation or exceeds the permitted use, you will need to obtain permission directly from the copyright holder.

Control Elements for Driver Assistance Systems

Matthias Pfromm and Klaus Bengler

Contents

© The Author(s) 2026
H. Winner et al. (eds.), *Handbook Assisted and Automated Driving*,
https://doi.org/10.1007/978-3-658-45276-6_27

This chapter describes how to design ergonomic control elements for driver assistance systems. Since driver assistance systems are safety-relevant vehicle functions, distraction-free and intuitive operability is of utmost importance. For this reason, the control elements must be developed according to human requirements.

This chapter first describes the requirements that must be considered during development. The requirements are defined by the user's needs but also by the different characteristics of the assistance systems. Then, the general design recommendations are clarified by concrete implementation examples to facilitate access to the topic and to illustrate the now large variety of different control elements.

An operating element is a technical device at the interface between a human and a machine, with the aid of which a controlling or regulating influence is exerted on the technical process or the functional sequence. In this presentation, we do not limit ourselves to mechanical control elements, but also consider input options based on recognition technologies, such as gesture or voice input.

27.1 Control Element Development

Since the requirements for the control elements are derived not only from the human characteristics but also from the functions of the system to be operated, the first step is to categorize the assistance systems that support the driver in order to derive requirements for user-friendly design.

27.1.1 System-Specific Requirements

The requirements for control elements depend on their tasks. To specify the tasks, we divide assistance systems into two groups; see Gasser [1]:

1. Systems that *do not require constant interaction* with the driver. They are active in the background and require no or only occasional operating actions while driving. These systems can usually be activated and deactivated, and their characteristics can often be changed by parameter settings. For example, collision avoidance systems can be activated and deactivated and the warning threshold can be set.

2. Systems that *require constant interaction* with the driver. The driver must activate or deactivate the system depending on the situation, set parameters, or confirm suggestions from the system. The operating actions are usually time-critical. For example, when using Adaptive Cruise Control (ACC), the driver must change the desired speed, confirm speed limit suggestions, or activate and deactivate the system regularly.

Operating elements of both groups must meet fundamental requirements:

- The control element must be easy to find.
- The user must be able to recognize what the operating action does.
- The input status must be clear.
- Direction of movement must be compatible with the parameters to be set (e.g., tapping upwards increases value, tapping downwards decreases it).
- It must be possible to operate the device without having to look away for a long time.

Systems that require constant interaction with the driver (2nd group) have further requirements for the controls:

- Unerring, error-free operability without averting the driver's gaze.
- It must be possible to operate the functions quickly, safely, intuitively, and precisely without having to adopt uncomfortable postures.
- As far as possible, the hand should not have to reach around.

Various regulations exist for the operation of driver assistance systems, e.g., from the United Nations [2], consumer protection organizations such as NCAP [3], laws [4], and standards [5]. For example, they specify the minimum number of operating steps when deactivating a system, or they stipulate that a safety system that was previously switched off must be active again when the vehicle is restarted. These laws and guidelines must be taken into account when designing the operating elements but are not discussed further in this chapter.

27.1.2 Determining the Control Element Type

For operating driver assistance functions a number of control elements are available, which should be selected depending on the requirements and constraints. The currently common control elements for driver assistance systems are presented here and described for which tasks they are suitable.

For some years now, touchscreens, as shown in ◉ Figs. 27.5 and ◉ 27.6, have been becoming increasingly common in motor vehicles. In addition to "infotainment" content, they are also increasingly used to operate vehicle functions, which include driver assistance systems. In the corresponding menus, driver assistance functions (such as lane departure warning) can be switched on and off, and parameters (such as sensitivity) can be configured. Due to the large number of functions now available, it would not be possible to provide a hardware control element for every function for reasons of limited space and high complexity. The advantage of touchscreens is that many functions can be integrated into a small space, such that the control elements can be clearly identified, and that they can provide clear feedback on the function status. Two major disadvantages are that a longer gaze aversion is required for operation than with all other types of control elements, since the controls cannot be felt (even though in some vehicles the touch displays provide haptic feedback) [6] and, with today's common arrangement in the center console, the hand usually must be removed from the steering wheel and placed in a position that is uncomfortable in the long run. Therefore, this type of operation is not very suitable for systems that require constant/frequent and/or time-critical interaction with the driver, for reasons of safety and comfort. For settings that do not require constant or time-critical interaction, operation via touch-sensitive screens, especially in combination with haptic feedback, is a common and suitable solution.

In many vehicles, menu systems in the screen can be controlled with a central operating element (rotary pushbutton or touchpad, ◉ ◘ Fig. 27.1). This type of input has the advantage that the hand does not have to be placed in a forced position during operation and that a shorter gaze aversion is required than with touchscreens [6]. Recently, however, it has lost importance in favor of the touchscreens familiar from smartphones and tablet computers.

Operating driver assistance functions via hardware controls allows direct and immediate access to the respective function. This type of operation is particularly preferable for time-critical and frequent operating actions, such as ACC operation. With a suitable design, the operating element and the switch position can be felt by the user and thus "blind operation" is possible. Pushbuttons and switches, as well as rocker switches, are often used for discrete control tasks. Rotary actuators and levers are suitable for both discrete and continuous inputs. Several control elements are combined in integrated control elements, such as the ACC control element ◉ ◘ Fig. 27.4.

It is also possible to combine the hardware controls with on-screen operation. For example, a hardware element can be used to quickly jump to a specific menu (◉ ◘ Fig. 27.7).

Recently, the voice operability of functions has been increasing due to the improvement of recognition technologies. The advantage of this input is that it can be performed without averting the gaze from the road and without changing the posture of the hands. However, due to the duration of the input, it is not suitable for time-critical tasks, or those that require constant interaction with the vehicle. Even for safety–critical inputs, at least one confirmation by the technical system must be provided.

Input via freehand gestures is currently of secondary importance in the operation of vehicle functions. In

◘ **Fig. 27.1** Rotary pushbutton and touchpad (AUDI AG)

some vehicles, gestures can be used to navigate in the menu or make text entries.

27.1.3 Spatial Arrangement and Geometric Integration

During the ergonomic analysis, the vehicle interior is assessed in terms of accessibility and visibility. As described in DIN EN 894–3 [7], the importance and frequency of operation as well as the sequence of operation must be considered when placing and grouping controls. The more important a control element is for safe vehicle guidance, the more centrally it must be located in the optimum viewing and gripping range. Controls that are used for constant and/or time-critical interaction with a driver assistance system must therefore always be easy to reach and, if possible, easy to operate without having to reach around or change posture. Common locations for this are the spokes of the steering wheel or a lever behind the steering wheel. Controls that are used less frequently can also be positioned outside the optimal area. Here, for example, the center console or the console on the left and right behind the steering wheel are suitable locations.

When arranging the operating elements, not only the geometric position but also the relative assignment to other operating parts must be taken into account. To facilitate operating sequences (e.g., 1. activating, 2. parameterizing the ACC), the spatial proximity of the individual operating elements should prevent the user from reaching around.

The size of the operating elements and the distances between operating elements must be based on anthropometric conditions, whereby the values for finger sizes and gripping widths can be taken from corresponding tables [8]. To avoid unintentional actuation, care must be taken to ensure that sufficient distance is kept from other operating elements, such that they can be clearly distinguished and that they cannot be touched accidentally.

27.1.4 Feedback, Operating Direction, Haptics, and Labeling

For ergonomic operation, the user must receive feedback on the input made. In the case of hardware operating elements and touchscreens, haptic feedback should be provided at the location of operation. A successful input is haptically recognizable by a click of the button or a corresponding grid and resistances for rotary elements.

Controls that maintain their switching position can also provide the driver with feedback on the current system status via position coding (e.g., *ACC switched on* recognizable by the detent position of the lever). If

the feedback is optical, it must refer to the corresponding control element. In general, multimodal redundant coding that addresses several of the driver's sensory channels is recommended for feedback.

The direction of operation of controls should be obvious, i.e., oriented to the stereotypical movements (◉ ◘ Fig. 27.2) and match the intended direction of movement of the system, to ensure intuitive and quickly learnable operation and to minimize the risk of incorrect actions. For example, in the ACC controls (◉ ◘ Figs. 27.3 and ◉ 27.4), the desired speed is increased when the lever is tapped upward and decreased when it is tapped downward [9].

To ensure that the driver does not have to direct his or her gaze to the control element during use, haptic distinguishability should be ensured by the shape, size, position, and surface design of the buttons (for example, as in ◉ ◘ Figs. 27.3 and ◉ 27.4) as well as sufficient spatial separation. The less the spatial separation can be ensured, the more clearly the controls would be distinguishable to avoid operating errors.

The labeling of the operating elements should be based on international standards, such as the ISO standards [10] and common conventions, in order to ensure self-explanatory information.

27.2 Implementation of Control Elements

There are a wide variety of possible solutions for the design of control elements in accordance with the requirements described above. Common solutions are presented first, followed by future trends.

27.2.1 Common Solutions

27.2.1.1 Steering Wheel Controls

The steering wheel is an ideal location for controls of systems that require constant interaction with the driver, as it is in the primary field of vision and reach. The controls are easy to find and there is little or no need to reach around the hand to operate them. If the controls are form-coded and ideally a combination of different control types, such as buttons in combination with a lever, the controls can be felt without looking and blind operation is possible. Operation is only difficult or impossible when the steering wheel is turned sharply.

Frequently, systems for continuous automatic longitudinal and lateral guidance of the vehicle can be operated by controls in the spokes of the steering wheel. ◉ ◘ Figure 27.3 shows a control element in the current Mercedes S-Class. Various functions, such as activation

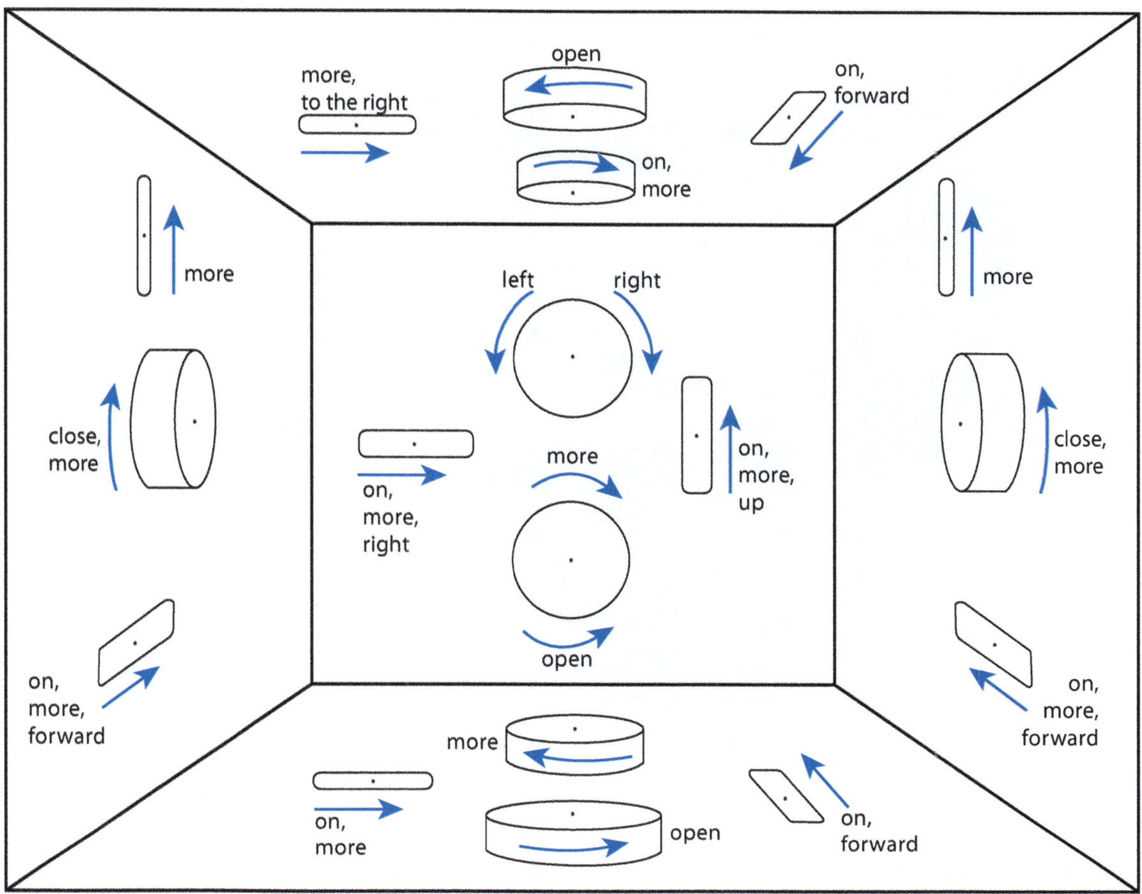

Fig. 27.2 Spatial alignment of rotary knobs and sliders with associated stereotypical movements acc. to [11]

Fig. 27.3 Control elements on steering wheel for driver assistance systems

27

◘ Fig. 27.4 Integrated control element for ACC operation

and deactivation of the system, a mode switch, and selection of the desired time gap for driving ahead, can be operated by pushbuttons.

27.2.1.2 Operating Levers

The control elements of longitudinal guidance systems, such as ACC or the Intelligent Speed Adaption (ISA), are often integrated into a control lever with attached pushbuttons and shift paddles mounted behind the steering wheel. The advantage of this control element, compared to the buttons on the steering wheel, is that the lever can be operated independently of the position of the steering wheel without having to look at it. The disadvantage is that it is not as easy to find as the steering wheel buttons.

◘ Figure 27.4 shows the control lever of Audi's Adaptive Cruise Assist. If the lever is pulled in the direction of the driver, the system is activated; if it is pushed away from the driver, it is deactivated. Pushing the lever upwards increases the speed, pushing it downwards decreases it. The operation is thus based on the movement stereotypes (◉ ◘ Fig. 27.2). The time gap to the car in front is set on the rocker at the top of the lever. Pressing to the right increases it, and pressing to the left makes it smaller. There is a button on the left side of the lever that is used to set the current speed as the desired speed.

27.2.1.3 Pushbuttons

Hardware buttons in locations other than the steering wheel are usually only used if frequent or time-critical operation of a specific system must be enabled. ◉ ◘ Figures 27.5 and ◉ 27.6 show various hardware buttons under the touchscreens to control functions, such as the parking systems. Hardware buttons are often combined with menu-based operation. The button shown in ◉ ◘ Fig. 27.7 is used to jump directly to the driver assistance menu, where various driver assistance systems can be operated. This saves several operating steps in the vehicle menu.

27.2.1.4 Menu-Based Operation (Touchscreen)

In current vehicles, the configuration of driver assistance functions is usually integrated into the menu structure of the infotainment system. The menu is shown on the center display or in the instrument cluster and can be operated via touch-sensitive screens or controls on the steering wheel or center console. Parameters of the systems can be changed according to the user's wishes, and driver assistance systems that do not require regular interaction with the driver can be switched on and off.

◘ Fig. 27.5   Touchscreen and buttons for direct access to driver assistance functions (AUDI AG)

Systems that require constant interaction with the driver during use are only operated via the screen in rare cases (e.g., Audi Trailer Assistant).

27.2.1.5 Gesture Operation

Just like various consumer electronics devices, the infotainment menu in the vehicle can increasingly be operated using gestures. In addition to gestures executed directly on the display or separate touchpad, some vehicles also allow inputs using freehand gestures. The gestures performed by the driver "in the air" are recognized by sensors. They are used to control infotainment functions. For example, gestures can be used to "push" content from the center console screen to the instrument cluster [12]. Studies have shown that gesture control reduces the driver's gaze deviation from the road compared to conventional control [13, 14].

In the context of human-vehicle interaction, a distinction can be made between the following gestures: Direction-inducing (kinematic) gestures (e.g., pointing or waving to the left/right), mimic gestures (e.g., imitating picking up/hanging up a telephone receiver), deictic gestures (e.g., pointing in the direction of the display), and symbolic gestures (e.g., a horizontal swipe for "cancel"). In general, it is important to use gestures that are as simple as possible and already familiar from consumer electronics so as not to overwhelm the driver with a complicated "interaction language" [13].

Fig. 27.6 Touchscreen in the Mercedes S-Class

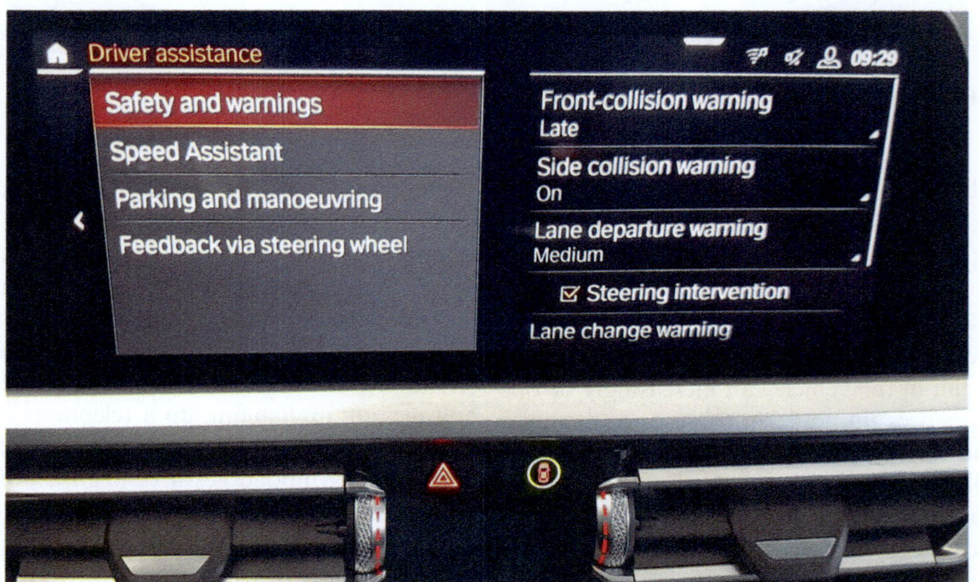

Fig. 27.7 Driver assistance menu and button for direct access

27.2.1.6 Voice Control

Recently, speech recognition in automobiles has made great progress, so that freely formulated commands can be used to control various functions. Combinations of onboard and off-board solutions are frequently used: If no mobile Internet is available, local speech recognition techniques are used; if the Internet is available, more powerful systems in the cloud can be used [15]. The control of infotainment functions by voice is now standard. Control of secondary functions, such as driver assistance settings, are partially possible and will become more widespread as recognition algorithms improve.

27.2.2 Future-Oriented Concepts

27.2.2.1 Brain-Computer Interface

Göhring et al. [16] investigated vehicle guidance using a Brain-Computer Interface (BCI). 16 EEG sensors were attached to the head measure potential differences on the scalp. The driver was to think of different movement patterns (e.g., left, right, push, pull), which were then detected and classified. Free ride and control of semi-automated system scenarios were investigated. A free ride, meaning that steering, throttle, and brake activities are directly controlled by the BCI, was possible but simply not accurate enough for practical use. Control of a semi-automated system with the BCI was, however, more promising. Here, the driver communicated the desired direction via the BCI at given decision points. The recognition accuracy was 90%.

27.2.2.2 Eye-Tracking Control

In a research project [17], the control of a vehicle by the driver's gaze was investigated. The driver wears an eye tracker system on his head, which simultaneously films the pupil and the surroundings and thus determines the direction of gaze. Two variants were implemented: In the "freeride" mode, the gaze directions are directly linked to the steering actuator, i.e., the vehicle drives in the direction in which the driver is looking. In "routing" mode, the vehicle drives largely automatically— only at certain decision points, such as traffic junctions, the driver selects the desired direction by directing his/her gaze [17]. Such a system is certainly still a long way from series production, but in the future, it could be an opportunity for people with impairments to operate a highly automated vehicle.

27.2.2.3 Holographic Devices

At CES2017, BMW presented a colored holographic display that floats freely in the interior of the vehicle. It is operated by finger gestures and confirms the commands with haptic feedback. Freely configurable buttons are displayed to the driver next to the steering wheel at the level of the center console. In this area, a camera detects the driver's hand movements and registers the position of the fingertips. As soon as these touch one of the virtual buttons, a haptic impulse is emitted by ultrasonic waves and the linked function is activated. In this way, invisible buttons can be positioned in space that provide haptic feedback to the user when operated [18, 19]. Future studies must show to what extent this type of interaction is suitable for the operation of driver assistance systems.

27.3 Conclusion

This chapter has provided an overview of the ergonomic design of controls for driver assistance systems. The first step was to categorize the characteristics that must be considered in the design of control elements of driver assistance systems. It was looked at what kind of operating actions must be performed: Is continuous interaction with the system necessary and do time-critical inputs have to be made while driving? Or is only occasional operation and change of settings necessary? Subsequently, hints were given for the selection of the control unit, the spatial arrangement in the vehicle interior, as well as for the design of the haptics and the operating directions. In the final step, current and future-oriented solutions were presented and described in terms of their advantages and disadvantages.

References

1. Gasser, T. (Hrsg.): Rechtsfolgen zunehmender Fahrzeugautomatisierung: gemeinsamer Schlussbericht der Projektgruppe ; Bericht zum Forschungsprojekt F 1100.5409013.01. Berichte der Bundesanstalt für Straßenwesen F, Fahrzeugtechnik 83. NW Verl. für neue Wissenschaft, Bremerhaven (2012)
2. UN Regulation No. 79: Steering Equipment. ► https://unece.org/transport/vehicle-regulations-wp29/standards/addenda-1958-agreement-regulations-61-80. Zugegriffen 29. November 2021 (2021)
3. Euro NCAP: TEST PROTOCOL – AEB Car-to-Car systems. ► https://cdn.euroncap.com/media/62794/euro-ncap-aeb-c2c-test-protocol-v303.pdf. Zugegriffen 29. November 2021 (2021)
4. The European Parliament And The Council: Regulation (EU) 2019/2144. Type-approval requirements for motor vehicles. ► https://eur-lex.europa.eu/eli/reg/2019/2144/oj Zugegriffen 29. November 2021 (2019)
5. ISO 9241–400: Ergonomics of human-system interaction-Part 400: Principles and requirements for physical input devices (2007)
6. Spies, R.: Entwicklung und Evaluierung eines Touchpadbedienkonzeptes mit adaptiv haptisch veränderlicher Oberfläche zur Menübedienung im Fahrzeug. Dissertation an der Technischen Universität München (2013)

7. DIN EN 894-3: Safety of machinery-Ergonomics requirements for the design of displays and control actuators-Part 3: Control actuators (2010)

8. Flügel, B., Greil H., Sommer, K.: Anthropologischer Atlas Alters- und Geschlechtsvariabilität des Menschen. Minerva, München (1986)

9. Bubb, H., Bengler, K., Breuninger, J., Gold, C., Helmbrecht, M.: Systemergonomie des Fahrzeugs. In: Bubb, H. (Hrsg.) Automobilergonomie. ATZ/MTZ-Fachbuch. Wiesbaden: Springer Vieweg, S. 260–340 (2015)

10. ISO 2575: Road vehicles—Symbols for controls, indicators and tell-tales (2021)

11. Woodson, W., Conover, D.: Human Engineering Guide for Equipment Designers, 2nd edn. University of California Press, Berkeley, Los Angeles, USA (1964)

12. Continental Magic User Interface. Verfügbar unter: ► http://www.continental-corporation.com/www/ presseportal_com_de/themen/pressemitteilungen/3_au- tomotive_group/interior/ press_releases/pr_2011_10_19_ magic_user_interface_de.html, Zugegriffen: 05. Juni 2014 (2011)

13. Geiger, M., Nieschulz, R., Zobl, M., Lang, M.: Bedienkonzept zur Gestenbasierten Interaktion mit Geräten im Automobil. In: Useware 2012 VDI-Bericht, Bd. 1678, VDI Verlag, Düsseldorf (2012)

14. Kreifeldt, R., Roth, H., Preissner, O.: Bediensysteme mit Gestensteuerung Stand der Technik und Zukunft. ATZelektronik 0(4), S. 248–253 (2012)

15. Haag, A.: Sprachsteuerung mit semantischer Spracherkennung. ATZelektronik (6), S 466–471 (2012)

16. Göhring, D., Latotzky, D., Wang, M., Rojas, R.: Semi-autonomous car control using brain computer interfaces Intelligent Autonomous Systems, Bd. 12. Springer, S. 393–408 (2013)

17. Car Steered with Driver's Eyes: ► http://www.fu-berlin.de/en/presse/informationen/fup/2010/fup_10_106/. Freie Universität Berlin. Zugegriffen 05. Juni 2014 (2010)

18. Mangold, M.: CES 2017 – HoloActive Touch. Neues HOLO Bedienkonzept von BWM fürs Auto. ► https://magic-holo.com/ces-2017-bwm-bringt-holo-bedienkonzept-ins-auto-holoactive-touch/. Zugegriffen 29. November 2021 (2017)

19. Kallweit, J.: BMW HoloActive Touch: Frei schwebende Anzeige im Interieur. ► https://www.automobil-produktion.de/specials/interieur-infotainment/bmw-holoactive-touch-frei-schwebende-anzeige-im-interieur-213.html. Zugegriffen 29. November 2021 (2016)

Open Access This chapter is licensed under the terms of the Creative Commons Attribution-NonCommercial-NoDerivatives 4.0 International License (► http://creativecommons.org/licenses/by-nc-nd/4.0/), which permits any noncommercial use, sharing, distribution and reproduction in any medium or format, as long as you give appropriate credit to the original author(s) and the source, provide a link to the Creative Commons license and indicate if you modified the licensed material. You do not have permission under this license to share adapted material derived from this chapter or parts of it.

The images or other third party material in this chapter are included in the chapter's Creative Commons license, unless indicated otherwise in a credit line to the material. If material is not included in the chapter's Creative Commons license and your intended use is not permitted by statutory regulation or exceeds the permitted use, you will need to obtain permission directly from the copyright holder.

Occupant State Detection

Stephan Cieler, Fabian Faller, Moritz Groh, and Manfred Wilck

Contents

© The Author(s) 2026
H. Winner et al. (eds.), *Handbook Assisted and Automated Driving*,
https://doi.org/10.1007/978-3-658-45276-6_28

A common description of driving is based on the assumption that a vehicle can only be driven safely if *driver*, *vehicle* and *environment* interact in the best possible way. Modeling the task of driving is often carried out with the aid of control loop models, which are supplemented by driver assistance or automation systems in the case of assisted or automated driving [1]. Typically, these systems understand the driver or user as a *controller*, who is responsible for continuous control or—depending on the degree of automation—for monitoring the transport system.

Whether it is speed, outside temperature or road geometry, many *conditions* that affect the vehicle and the environment are incorporated into the control strategy of an automated vehicle. Information on vehicle users and their physical and psychological states, however, receives significantly less attention—which is also due to the fact that many of these states can only be inferred indirectly.

28.1 Why Occupant State Detection?

When specifying and designing assistance and comfort systems, taking user states into account can make an important contribution to improving the driving experience and enhancing road safety. In Germany alone, for example, more than 2,000 accidents involving personal injuries were caused by driver fatigue in 2019 [2]. A large proportion of these could presumably have been prevented if a fatigue warning system had been in place.

In the past, the focus of user state detection was on simple behavioral observations that served as influencing variables for information and warning systems (Is the seatbelt fastened? How long ago did the driver take a break? etc.). Today, however, it is possible to also detect user states that cannot be directly observed (e.g. feelings, intentions or stress levels) with the aid of innovative sensor technology and modeling algorithms in order to optimize the functions of a variety of innovative safety and comfort applications according to the situation. By monitoring vital signs such as body temperature or respiratory rate, it is also possible to infer indications relating to drivers' or users' health and to initiate countermeasures in the event of impending driver incapacity.

While many of these applications are optional comfort or safety systems, the detection of different user states becomes indispensable in the context of automated driving and the transition from high levels of automation to manual control. Depending on the applicable automation level, delegation of the control task back to the driver can only take place reliably and without accidents once the driver's presence, alertness and focus on the traffic situation have first been confirmed (see ▸ Chap. 51 in this handbook).

This chapter provides a practical description of the technological processes of modeling user states, lists the relevant sensor systems and provides an overview of functions and applications currently under development or already launched on the market. Since the well-being of vehicle passengers also comes into focus in the context of automated driving—the problem of travel sickness (*motion sickness*) should to be mentioned in particular here (see, for example, [3])—the term *occupant state detection* is also used in this chapter, which includes the monitoring and analysis of states of the user or driver as well as other persons in the vehicle cab. The chapter also addresses the issue of infants mistakenly left in the vehicle and the associated health hazards (so-called *hyperthermia issue*).

28.2 Occupant State Detection in Practice—Seven Areas of Application

Since the advent of the first standard drowsiness detection systems, interest has grown steadily in detecting and evaluating user states and taking them into account as input variables for driver assistance, warning and comfort system functions.

While the recent focus of driver state monitoring has been on detecting levels of alertness and attentiveness, current research and development is additionally addressing many other physiological and psychological aspects of vehicle users/occupants. The detection of intentions to act, emotions, state of health and other user states makes innovative functions and applications possible that contribute to an enhanced driving experience, increase the safety of vehicle use and make it possible to switch between automation levels during driving in the first place. Safety systems for distraction, drowsiness and driving fitness detection will also find their way into Euro NCAP's assessment routines over the next few years under the term *driver monitoring systems* [4].

The potential and usefulness of occupant state detection are demonstrated based on seven areas of application below.

Regulation of alertness

According to official statistics, fatigue, microsleep and falling asleep at the wheel are among the common causes of accidents and have always been a focus of accident and driver state research. Early fatigue detection systems monitored the duration of driving periods and detected typical steering angle corrections that correlated with the subjectively perceived extent of fatigue above a threshold value [5]. Modern systems (additionally) estimate the fatigue level of users by analyzing eyelid closure behavior (e.g. PERCLOS indicator: proportion of time in which the eyelid covers 80% of the pupil) and other physiological predictors (see summaries in [6, 7]).

Depending on the detected fatigue level, *fatigue mitigation systems* issue warnings or break recommendations in different sensory modalities [8], change the lighting and ventilation settings in the car interior [9], adapt the music offerings according to the situation [10] or emit "refreshing" fragrances in the car interior [11]. These *fatigue management systems* aim to continuously measure states of alertness and prevent phases of fatigue and microsleep from occurring through continuous monitoring and intervention [12].

Attention regulation and distraction prevention

To safely drive a vehicle, it is important at low automation levels that drivers focus their attention on the relevant aspects of the traffic situation and driving task. Inattention manifests itself in incorrectly prioritized gaze directions, daydreaming or distracting *non-driving activities*. The increased risk of accidents due to distraction has been demonstrated in numerous studies (e.g. [13, 14]) and addressed in safety campaigns (e.g. ► www.nhtsa.gov/risky-driving/distracted-driving).

With the aid of *attention assistants*, it has become possible to continuously record, analyze and evaluate visual attention to the driving task or the degree of distraction. In the event of deviations from a target behavior, such as use of a cell phone during manual driving, warnings or notifications can be issued to direct the driver's attention back to driving. While the first generations of these systems operated by evaluating steering behavior and analyzing steering corrections (e.g. Mercedes Attention Assist; see [15]), gaze detection systems are used in modern systems, in a similar manner to drowsiness detection, to continuously evaluate the visual behavior of vehicle users (see, for example, [16]).

By analyzing the attention of vehicle users over a defined period of time, it is possible, with the aid of modeling assumptions, to map their insight into the current driving situation, or their *situation awareness*— an assessment that is considered, among other things, to be an important predictor for the successful resumption of the driving task following a phase of automated driving [17].

Regulation of affective states

The fact that emotions and moods influence driving behavior and road safety is in line with everyday experience and has been documented in many studies (see summary in [18]). Anger and frustration are in particular considered major causes of aggressive driving behavior (so-called *road rage*) and many road accidents [19, 20].

The detection and regulation of drivers' and users' moods and emotions have increasingly become the focus of research in recent years. To calm down or stimulate vehicle users, relaxation and breathing techniques [21], situation-adapted music playlists [22] and situation-dependent lighting and color concepts in the car interior [23] have been proposed, among other things. Here, the "measurement" and classification of affective states are mainly conducted using camera-based detection of facial expressions and AI-based classification (see summary in [24]), where individual emotions are frequently localized in a multidimensional space (see, for example, Russell's classification model based on the dimensions of *valence* and *arousal* [25]).

Going one step further, *empathetic assistants* "empathetically" adapt to the user's current mood and in this way achieve "natural human" communication [26]. Several studies have demonstrated the positive effects of empathetic assistance on driving behavior and the regulation of affective states (see summary in [18]) (◘ Fig. 28.1).

OLIVIA—An empathetic driving assistance concept from Continental

Continental's OLIVIA driving assistant is capable of assuming the role of a co-pilot, trainer or companion, depending on the traffic situation and the user's current need for support. OLIVIA, for example, reminds the user of speed limits (co-pilot role), rehearses the process of switching from automated to manual driving together with the user (trainer role) or suggests playing music during phases of boredom (companion role). Using camera-based facial expression detection, OLIVIA is able to respond to the user's current emotion or mood and behave differently—just like a real person—depending on whether the user is upset, sad, happy, stressed, etc. As an avatar in the vehicle cockpit, OLIVIA can express feelings by means of her facial expressions and thus empathetically respond to the user's emotions. OLIVIA's empathetic skills range from conversational content (e.g. expressing an encouraging thought) to facial expressions, gestures and different communication styles (intonation, tempo, sentence melody, etc.).

Detection of user intentions

Detecting user intentions and desires is a particularly innovative area of research in the field of occupant state detection. The prerequisite for these functions is

Fig. 28.1 Empathetic assistant OLIVIA in various driving situations

to be able to "look into the mind" of users, to reliably detect their desired actions and respond in a manner appropriate to the situation.

It is well known that many driving maneuvers, especially evasive, lane change and emergency braking maneuvers, are preceded by behaviors that are reflected in typical steering wheel and pedal operations as well as characteristic gaze behavior patterns (see, for example, [27]). If characteristic driver intentions can be deduced via these signals, assistance systems can be activated with a "lead time" and maneuver processing times significantly reduced. Studies have shown that reading neurophysiological brain signals can help determine specific driver intentions even earlier [28].

Detecting driver intentions also enables the situation-dependent design of assistance strategies. A *smart* assistance concept of this type would, for example, adjust the steering assistance of a lane-keeping system whenever it detects the intention to change lanes [29]. In the future, we can expect comfort systems that detect certain desires or intentions to act—for example, when searching for a parking space—and offer users appropriate assistance functions at the right moment.

Stress management and health applications

Under the title "automotive health", many research and development projects have been established to address the health and well-being of vehicle occupants (see summary in [30]). Based on *vital monitoring*, which includes data on heart rate and body temperature, the aim is to detect situations of physical or cognitive overload at an early stage and positively influence them. In order to "exit the vehicle healthier than you entered it"—a motto propagated by Mercedes a few years ago [31]—adaptive lighting concepts, seat massages and the targeted use of odors, for example, have been proposed. Other applications envisage stimulating the user in phases of underload and monotony in order to counteract fatigue and vigilance.

In addition to stress management and the well-being functions described above, future systems will use (non-invasive) derivations of physiological signals to detect symptoms of illness and anticipate critical health situations associated with impaired driving (heart attacks, hypoglycemia, epilepsies, strokes, etc.). The detection of sudden driving incapacity (*sudden sickness*) is one of the new safety requirements of the European New Car Assessment Program (Euro NCAP), which will be relevant for evaluations from 2023 [4].

Critical health conditions also need to be detected during highly automated driving. Otherwise, an unconscious occupant could be transported for many miles without assistance being called.

Presence detection and classification

As explained in the following section, driver presence detection plays an important role in automated driving when the driver needs to assume control (*take over* situation). Presence detection systems are also gaining importance in the area of passive safety. UNECE regulation 16, for example, prescribes warning systems for new vehicles that emit an alarm if *detected* occupants are not wearing a seat belt [32].

Another application example in this context relates to so-called *child presence detection*. The sad background to *child presence detection systems* is the observation that babies and small children can be unintentionally left behind in vehicles and come to harm when the interior overheats. Warning devices that are soon to be launched on the market detect such situations and inform the parents or driver about the left-behind child via a mobile app. Child presence detection systems place particular demands on detection sensor technology (► ► Sect. 1.6).

Transitions between automation levels

As detailed in the introduction, the detection of user states during the transition of driving tasks from the automation system to the driver plays a key role: a *takeover maneuver* can only work reliably if the driver is present (i.e. sitting in the driver's position), awake and mentally prepared for the upcoming driving task.

For this important prerequisite on the user side—the automation levels beneath highly automated driving (SAE level 4) are particularly relevant here—the notions of *ability to take over* [33] and *driver readiness* have become established [34]. Many studies have shown that the speed and quality of the takeover depends in particular on the characteristics of the previously performed non-driving activity [35].

Regardless of takeover situations, the switch between different levels of automation can lead to users being overwhelmed by the complexity of task distribution between human and machine, resulting in them not reacting appropriately in certain traffic situations. To address this issue, intensive research has for several years been conducted on driver state-based interaction and takeover strategies that adapt human–machine interactions to the relevant driver state. The EU-funded ADAS&ME project [36] (▶ www.adasandme.com) dealt, among other aspects, with the development of driver-state-based transitions between highly automated (SAE level 4) and manual driving (SAE level 0), taking into account the emotions, distraction moments and stress potentials of the drivers/users. Based on a combination of these sub-aspects of the driver state, pre-warning time, HMI modalities and escalation levels of transitions were individually adapted to the driver to ensure a safe and smooth transition. Launched in 2021, the German KARLI funding project goes one step further and explores the use of driver and occupant state detection as well as AI-based interaction concepts to permanently ensure driver behavior appropriate to the level of automation.

28.3 From User to User Model

Operationalization of user states

Fatigue, stress, anger and *intentions* in the context of occupant state detection are human characteristics that cannot be detected or observed directly. As such, they can be classified as *latent variables* or *theoretical constructs* inferred on the basis of observations. Tapping into dimensions that cannot be "measured" directly is familiar from everyday life. Based on frequent yawning, it is possible to conclude that a person is tired—albeit

with a margin of error. Based on a person's facial expressions, it is possible to arrive at a statement about their current mood.

Providing clearly defined measurement rules to quantify latent variables is referred to in the social sciences as *operationalization*. The term, which goes back to Bridgman [37], refers to empirically applicable rules that can be used to define and quantify theoretical constructs—in our case, various dimensions of user states. Such an operational definition requires a detailed semantic analysis and definition of the construct to be operationalized.

The mapping of a holistic and all-encompassing user state is not feasible due to the complexity and diversity of human behavior. Instead, sub-aspects of the human state and behavior are modeled that are significant for the operation and safe driving of a vehicle as well as the comfort of the person driving the vehicle and the vehicle occupants.

User models as an abstraction of reality.

> **»** Models are deliberately not true to the original. They emphasize certain features and omit others. The intended purpose of the model determines which properties are modeled and which calculation is particularly suitable for describing them [38].

When discussing user models, reference is made to specific safety and driving comfort-relevant sub-aspects (e.g. *fatigue, attention*) of human beings. User models are thus an abstraction of reality that fails to take into account subtleties of human behavior and inner processes. The input variables of the models are the measurement data based on operationalizations, which, when mathematical and statistical methods are applied, make it possible to estimate the user state in question.

Considering specific sub-aspects of the user, application-specific characteristics can be defined in the form of *states*, which make modeling possible in the first place. The sub-aspect *fatigue* can be assigned the modelable states *asleep, tired* and *awake*.

◉ ▣ Table 28.1 illustrates the relationship between sub-aspects (constructs) and proven operationalizations (measurement methods) using the seven areas of application from ▶ ▶ Sect. 1.2.

The challenges for state modeling relate in particular to the issues of.

- which sub-aspects of human behavior and experience are to be modeled in relation to a defined application in the vehicle,
- how a valid and technically feasible operationalization of the relevant sub-aspects can be defined,
- how measurement technology and modeling can specifically be implemented.

◻ Table 28.1   Sub-aspects of the user state to be modeled and corresponding proven model input variables

Sub-aspects (constructs) to be modeled	Model input variables—proven operationalization through determination …
Alertness/fatigue	Of eyelid closure [7]
Attention/distraction	Of movement and gaze patterns [16]
Affective states/emotions	Of facial characteristics [24]
User intentions/driving style	Of pedal positions, steering settings, driving data, gaze pattern [39]
Stress/state of health	Of vital signs [30]
Presence	Of vital signs and biometric characteristics [40]
Ability to take over	Of presence, alertness and attention (see therein)

28.4 Sensors and Measured Variables

The model input variables are determined by making measurements of the occupant using suitable sensor systems and subsequent processing of the raw sensor data. While it is quite common to use contact and invasive measurement methods (ECG recordings, blood glucose level measurements, skin resistance measurements, etc.) in research projects or for validation purposes, *contactlessness* and *unobtrusiveness* are important acceptance criteria for end customers in the case of market-ready systems. In this section, the focus is therefore on *camera* and *radar systems*, since these operate without contact and can be integrated unobtrusively into vehicles.

Contactless sensor systems and reference measurement technology

Contact-based sensor systems, such as chest straps for recording respiratory movements, usually provide the most accurate results and are particularly suitable for the *development* and *validation* of user models. Beyond research and development activities, however, these measurement methods are less suitable because they do not meet with the acceptance of end customers.

By contrast, contactless measurements are often accompanied by limitations in data availability and accuracy. In the case of a contactless respiratory rate measurement system integrated near the rearview mirror, for instance, an unobstructed view of all of the occupants' chests to observe their respiratory movements is not always available and simple occupant movements can make measurement even more difficult. Moreover, body movements can interfere with the respiratory movement being measured, distorting the measurement results. In the worst case, the movements can cause the upper body to be completely obscured, preventing the measurement altogether.

Thus, the challenge in selecting and developing a contactless sensor technology is to provide model input variables in a quality and availability that enable the implementation of the intended target applications while meeting the necessary requirements. The technologies explained in this section concern contactless systems based on electromagnetic waves (radar systems and camera systems). A more detailed overview of these measurement systems, which also covers sound waves and Wi-Fi sensing, is given by Leonhardt et al. [40].

Camera systems

Relevant camera-based indoor sensing systems include near-infrared cameras (NIR cameras), far-infrared cameras (FIR cameras, thermal imaging cameras), color cameras and time-of-flight cameras (ToF cameras). The main components of the camera systems include lens, image sensor, image processor, optical filter and, if necessary, an additional illumination unit.

Camera systems operate in different ranges of the electromagnetic spectrum. While color cameras operate in the wavelength of 380 nm to 780 nm using an infrared filter and FIR cameras detect thermal radiation emitted or reflected from surfaces between 8 µm and 14 µm, ToF and NIR cameras use additional illumination units based on near-infrared radiation. The illumination units can be light-emitting diodes or laser diodes/surface emitters (*vertical-cavity surface-emitting laser*, VCSEL) whose near-infrared radiation is in the wavelength range of 850 nm to 950 nm. The usage of a bandpass filter allows the NIR camera system only to sense light within the specified wavelength range. This assures homogeneous light conditions, even in case of extreme sun light.

Because the camera systems operate in different regions of the electromagnetic spectrum, they vary in their ability to determine different model input variables. For the detection of users' facial expressions, gaze directions and eyelid closure behavior, NIR cameras are considered to be the standard technology in the automotive industry. Typically, they are integrated in the instrument cluster to provide an optimal view of the driver's face.

◻ Table 28.2 Suitability of the various camera and radar systems for determining selected model inputs for occupant state detection under realistic in-vehicle conditions (including nighttime driving)

Measure-ment system	Place of presence	Body posture	Minimal body movements	Respiratory rate	Facial features	Gaze direction	Body temperature
NIR camera[a]	+	O	O	–	+	+	n.f
ToF camera[a]	+	+	O	O	O	O	n.f
Color camera[a,b]	+	O	–	–	+	–	n.f
FIR camera[a]	O	n.f	–	–	–	n.f	+
UWB radar	–	–	+	+	n.f	n.f	n.f
CW radar	–	–	O	O	n.f	n.f	n.f
FMCW radar	O	–	+	+	n.f	n.f	n.f

Key: + highly suitable, O suitable to some extent, –unsuitable, n.f. not feasible
[a] Direct visibility without occlusion is a necessary precondition for the functionality
[b] Functionality is only possible at night with suitable ambient illumination during automated driving

Where determining body positions is essential, ToF cameras provide the depth information relevant for an application in addition to an NIR grayscale image. If the driver's posture is of interest—e.g. in order to identify an activity unrelated to driving—the integration of a ToF camera near the rearview mirror is practical, as this allows the seating position to be detected easily.

If determination of the occupants' body temperature is relevant, e.g. for health and well-being applications, FIR camera systems are the method of choice. These systems are often integrated into the central area of the instrument panel so that the faces of the driver and front passenger can be easily observed.

Radar systems

Owing to the small number of antennas, radar systems achieve a lower spatial resolution of measurement results than camera systems. By contrast, the advantage of radar systems lies in the fact that, due to their properties in the microwave range, the slightest movements (*minimal body movements*) can be detected with millimeter precision and objects obscured by an obstacle (clothing, textiles, etc.) can be sensed. The unobtrusive integration of a radar system in the headliner enables monitoring of a large area of the car interior, including the footwells, and the reliable detection of the presence of occupants.

Radar systems proposed for use in car interiors are the *frequency-modulated continuous-wave radar* (FMCW radar), the *unmodulated continuous-wave radar* (CW radar) and the *ultra-wide band radar* (UWB radar). The main components of radar systems are transmitter and receiver unit, antennas, oscillator and mixer. A digital signal processor and a microcontroller are used to compute the signals.

Radar systems for in-vehicle applications are currently being used or developed in the frequency range of 3.1 GHz to 10.6 GHz (according to [41] in the USA, according to [42] in Europe), 24 GHz to 24.25 GHz (according to [43] worldwide), 57 GHz to 64 GHz (according to [44] in the USA, according to [45] in Europe) and above 120 GHz. Approval of frequency utilization is subject to national guidelines that regulate, among other things, the type of operation, frequency, duty cycle, radiated power and intended use of the applications. Based on the regulations, organizations such as the European Telecommunications Standards Institute (ETSI) issue the applicable recommendations and standards (see [46, 47]).

Selection of the sensor system

The camera and radar systems described are suitable to different extents for determining model input variables and have distinctive strengths and weaknesses. A rough estimate of the suitability of the various systems is made in ◉ ◻ Table 28.2. In addition aspects such as price, availability, maturity level of the technology, integration possibilities and customer acceptance in terms of privacy and health have to be taken into account for a final system design.

28.5 Occupant State Modeling

In the context of occupant state detection, models describe approximate abstractions of the user and represent very specific sub-aspects of the driver's state that are relevant in the driving context. Data from one or several sensor systems is employed for this purpose (▸ ▸ Sect. 1.4), which serves as input variables for a model. Based on these input variables, the model provides an estimate or, more precisely, an operationalized sub-aspect of an occupant state.

Model development distinguishes between *knowledge-based models* and *data-driven models*, both of which rely on a *ground truth*.

Ground truth

A ground truth (or gold standard) is a reference against which models are optimized and evaluated. The goal is to create a "truth" that is as close to reality as possible, that corresponds to the actual state being modeled or detected and that allows the model performance to be evaluated as objectively as possible. If a sub-aspect cannot be recorded through reference measurement techniques, test subject surveys using standardized questionnaires, expert assessments or a combination of the above may be employed.

For the modeling of human behavior or states, it is often difficult or even impossible to define an objective ground truth. The reasons for this include in particular the high interpersonal and intrapersonal variance in behavior and responses to stimuli. Moreover, the actual conditions are not readily observable and can only be determined, for instance, through interviews with test subjects. Here, the problem arises that answers sometimes vary widely between test subjects or are biased if the test subjects have knowledge of the background of a study or survey. Ahlstrom et al. [48] have shown, for example, that for fatigue detection during real driving, expert assessments based on video data and self-assessment of the test subjects diverge significantly and that the assessments of the experts involved also sometimes differed considerably from one another.

It should therefore be borne in mind that the development and evaluation of models that represent human behavior are subject to a certain "residual error", since no universally valid and absolute truth generally exists in this regard.

Knowledge-based models

Knowledge-based models, also often referred to as rule-based models or expert systems, draw on expert and domain knowledge in specific areas of experience and represent this knowledge in the form of fixed rules. Acquiring the a priori *knowledge* is not part of, but is in fact the basis for, model development. Russel and Norvig [49] describe this approach as using more powerful domain-specific knowledge that allows for logical derivation steps and can handle simple typical cases in limited areas of experience. Knowledge-based models more or less assume to already know the answer to a "difficult problem" [49].

Not infrequently, a priori knowledge is transferred from other application fields. This approach is used in particular when there are few, clearly identifiable indicators of a particular state that can be observed almost universally. This applies, for example, to writing a text on a smartphone. In addition to inputting the text using their fingers, users often repeatedly glance at the screen to check, among other things, whether the text input corresponds to their intended wording. A further example is that an increase in heart rate can usually be observed and measured in stressful situations caused by a sudden, unexpected event.

One of the greatest challenges in knowledge-based systems is the high interpersonal and intrapersonal variance that people exhibit in their behavior and responses to external and internal stimuli. It is therefore difficult to define universally applicable rules that allow adequate decisions for each individual at any given time. Defining fixed thresholds and decision rules is thus always a compromise between the required *generalizability* and the sufficiently individual *personalization* of model parameters. Knowledge-based models are therefore only suitable to a limited extent for adequately describing more complex sub-aspects of a driver state that are based on individual reactions or behaviors.

However, they do offer the advantage that decisions made by such a system are comprehensible, since they are based on established knowledge and experience, and their technical implementation is fully testable. The transparency of system decisions is particularly relevant for the risk minimization of safety–critical systems in motor vehicles for functional safety according to ISO 26262–9 [50]. Furthermore, knowledge-based systems can in many cases be implemented in embedded systems in a resource-efficient manner with regard to memory and processor utilization.

Seatbelt reminders as an example of knowledge-based models

Since 2019, seatbelt reminders have been mandatory as a warning system on all seats in newly registered vehicles within the EU [32]. This requires a seat occupancy check. Here, pressure-sensitive sensors under the seats, so-called "smart" pins or occupant classification mats, determine the approximate weight of the object on the seat [51]. If the measured weight is greater than or equal to a minimum weight predefined through a priori knowledge, a human is assumed to be present and a check is made to determine whether the seatbelt buckle is correctly fastened at the relevant seat position. At the same time, false detections caused by heavy objects such as a crate of water bottles or a bag are reliably prevented. If the seatbelt buckle is not fastened, a visual and audible seatbelt reminder is emitted. The entire set of rules for such a warning system can be implemented comparatively easily and reliably as a knowledge-based model.

Learning, data-driven models

In contrast to knowledge-based models, data-driven models do not usually make use of a priori knowledge, or do so only to a very limited extent. Instead, they are the result of an automated, computer-aided learning process that helps identify patterns and regularities in large data volumes and derive rules for the model. Data-driven models consequently acquire adequate knowledge by evaluating experiences (see [52]). The associated learning process is referred to as training. The data used for this purpose usually comprises measured values or derived statistical parameters that correlate with the sub-aspect being sought (see ◉ ◘ Table 28.1).

Data-driven methods enable the development of realistic as well as adaptive models that can be adjusted to the characteristics of individual drivers and occupants. Given a suitable data set, interpersonal and intrapersonal characteristics can be taken into account during model building. This is particularly an advantage over knowledge-based models, whose decision bases are much more generic.

One challenge in applying learning methods is finding or generating a suitable database and—depending on the learning method—a suitable representation of the ground truth used for training these models. A large database may be required to train a model whose complexity is appropriate to the sub-aspect of the occupant state being modeled. The generation of such large data sets for training, but also for the subsequent mandatory validation, can result in significant costs and technical effort. In addition to the recording of real data sets, the generation of synthetic, simulated data is therefore playing an increasingly important role. Particularly in the field of image processing, it is now possible to generate detailed, photorealistic scenes that can usefully supplement small real-world data sets in order, for example, to increase the variability of certain input variables contained within them. Compared to real data, synthetic data can be generated very quickly, cheaply, repeatably and safely.

Depending on the modeling approach used, the traceability of model decisions is not always transparent and may be difficult to interpret, which makes the use of these methods in a safety-relevant context difficult, requiring additional effort for verification. The verification of data-driven models continues to be a subject of research and is addressed in detail, for example, in the German funded project "KI Absicherung—Safe AI for Automated Driving" (see ▶ www.ki-absicherung-projekt.de). For the reliability of model output

variables in particular, it is necessary to take measurement uncertainties into account with regard to observations in the input data on the one hand and the uncertainties of the model itself on the other. Kendall and Gal [53] describe a Bayesian method that models these uncertainties and applies them to image processing tasks.

Differentiation of vehicle occupants according to age group

Modern occupant monitoring systems rely on contactless sensors for detection and also allow occupants to be differentiated by age group (for example, infant, small child, adult, pet). Such systems are used, among other things, to detect the presence of children. A camera system with near-infrared lighting, for example, is suitable as a sensor for ensuring that images of occupants can also be taken at night. The basis for further image processing are artificial neural networks from the area of object classification (see [54, 55]). For this purpose, the artificial neural network is trained on a data set of labeled (annotated) images, with each age group being assigned to a *class*. The result is a data-driven model in the form of an artificial neural network that can distinguish occupants according to age group with ideally high confidence. A specific product that operates on the basis of a radar system is described in ▶ Sect. 1.6.

28.6 Application Example: Child Presence Detection

» "Deathtrap back seat. Dozens of children die every year because their parents leave them in the car in hot weather" (Autobild magazine, 26.09.2019, p. 10).

On summer days, the power of the sun increases, and outdoor temperatures climb to high levels. If the sun shines directly onto the car, life-threatening heat can develop in the interior within a short time. Experimental measurements have shown that the interior temperature of vehicles can increase by 7 °C in five minutes, by 16 °C in ten minutes and by as much as 24 °C in 30 min [56] (for an overview of various studies, see [57]). If an infant or small child is left unattended in a vehicle, heat stroke or heat exhaustion can occur after only a few minutes. Accident statistics in the USA have revealed that hyperthermia and heat stroke are the leading causes of death in children under 15 years of age in the *non-crash-incidents* group, with an average of 19 cases per year [58]. As Hammett et al. [59] have shown in a recent survey study, most victims (78.2% of cases) are left in the vehicle inadvertently. If children are deliberately left in the car, reasons cited include work or school activities (19%), drug and alcohol use (13%) or the child sleeping (11% of responses).

Against this backdrop, a number of initiatives and programs have been launched, such as the US National Highway Traffic Safety Administration's "Where's Baby? Look Before You Lock" campaign (▶ https://www.nhtsa.gov/campaign/heatstroke) and the ▶ www.noheatstroke.org platform. Calls for technological solutions that immediately warn drivers or caregivers when children are left behind alone in the vehicle have also grown louder. Under the name "Hot Cars Act 2021", a law is currently being prepared in the USA that will require the installation of appropriate systems in new vehicles in the future [60]. Leading North American car manufacturers have agreed to install so-called *rear seat reminder systems* in new cars by 2025 [61]. Italy became the first country in Europe to make the installation of such systems mandatory in 2019. Euro NCAP wants *child presence detection* systems (CPD systems) to be considered in the safety assessment of new cars from 2023 [62].

The way CPD systems work is based on a simple principle: as soon as children left behind are detected in the vehicle cab, a warning cascade is triggered in which first the driver (via mobile app) and later the vehicle's surroundings (via audible warning) are informed that the child is in danger. Other intervention steps can include activating the air conditioning and sending an emergency call.

The reliable detection of children in the car interior places special demands on sensor technology. First, monitoring—preferably contactless—must be ensured for both rows of seats and footwells. Second, living beings (and not other objects) must be identified. Third, living beings must be detected even if they are under a blanket or concealed by a child-seat sunshade.

This requirement profile suggests the integration of a radar system. Its microwaves penetrate light textile fabrics, enabling the detection of even the slightest movements of living beings, such as registering chest movements during breathing.

CoSmA[1] is a UWB-based radar system from Continental that primarily implements a digital access control function. Thus, vehicle enablement takes place without a physical vehicle key, via the communication function of the UWB modules. For this purpose, UWB modules are located both in the vehicle body and in the passenger compartment, which communicate with the user's UWB-enabled smart device (e.g. smartphone). By using the UWB modules in sensing mode, further functionalities can be implemented. CoSmA, for instance, enables gesture recognition, through the use of the external modules, making a sensor-controlled tailgate mechanism to open the trunk possible (so-called kick sensor). Through use of the UWB modules in the car interior, the occupant state variables of the CPD function can be determined.

An alternative system for child presence detection, which is also under development at Continental, is FMCW radar. With an appropriate antenna design, improved spatial resolution is achieved, making it possible to determine the location of detected persons (see ◉ ▢ Table 28.2). This allows for additional applications such as the detection of occupied seats or the measurement of vital signs (e.g. breathing rate) of passengers in these seat positions.

Thanks to their compact installation dimensions, the radar systems and the UWB modules can be unobtrusively integrated into the vehicle headliner. They also allow NCAP-compliant implementation of the CPD functionality. Accurate differentiation between objects and living beings based on motion patterns can be achieved with high precision. However, it is difficult to distinguish between large children and small adults based on their movement patterns, which can lead to false positive detections in edge cases. There is thus an overlap in the body sizes and vital signs of small adults and large children, which highlights their similarity for the radar systems [63].

By pairing one of the radar systems with Continental's interior camera, system performance can be further improved.[2] When positioned in the center console or dashboard, the camera can view the car interior and be used to detect occupants and objects. However, since the camera system cannot "see" through visual obstacles (e.g. a blanket) or the sunshade of a restraint system for small children, NCAP-compliant child presence detection cannot be implemented using a camera alone. In conjunction with the radar, the camera "recognizes" the presence of persons on the seats and supports their classification into adults and children. For non-visible locations—such as behind the seats, in the footwells and behind visual obstacles—the results from the radar are used (▢ Fig. 28.2).

The example of *child presence detection* illustrates the goals and challenges of occupant state detection. Starting from a problem—in this case, the health risk to children—relevant constructs must be defined that can be modeled and made observable by means of suitable measurement instruments. Generally speaking, merging data from different sensor technologies in the car interior enhances the quality and accuracy of state

---

1   ▶ https://www.continental-automotive.com/en-gl/Passenger-Cars/Architecture-and-Networking/Comfort-Security/Access-Control-Systems/Smart-Device-Access.

2   ▶ https://www.continental.com/en/press/press-releases/20211013-cabin-sensing/.

Fig. 28.2 Radar-based child presence detection system. Explanations in the text

detection and increases the detection range. By applying different detection technologies, a CPD system can, for example, monitor several rows of seats and also the footwell areas in the car interior. Apart from questions relating to occupant privacy and the associated data protection aspects, the acceptance and market success of occupant status monitoring systems will depend above all on the extent to which the linked applications contribute to increased comfort and safety during driving. It has been shown that user state detection is of particular importance for the implementation of automated driving.

References

1. Hofauer, S.: Fahrer-Fahrzeug-Interaktion einer automatisierten, kraftstoffeffizienten Fahrzeuglängsführung. Dissertation. Technical University of Munich, Faculty of Mechanical Engineering, Munich (2017)
2. German Federal Statistical Office: Verkehr - Verkehrsunfälle 2019 (Subject-matter series 8, series 7). Federal Statistical Office, Wiesbaden (2021)
3. Mühlbacher, D., Tomzig, M., Reinmüller, K., Rittger, L.: Methodological considerations concerning motion sickness investigations during automated driving. Information **11**, 265 (2020). ► https://doi.org/10.3390/info11050265
4. Euro NCAP.: Assessment protocol–safety assist. Version 9.0.1. Euro NCAP, Brussels (2019)
5. Krajewski, J., Sommer, D., Trutschel, U., Edwards, D., Golz, M.: Steering wheel behavior based estimation of fatigue. Proceedings of the Fifth International Driving Symposium on Human Factors in Driver Assessment, Training and Vehicle Design, June 22–25, 2009, Iowa City, 118–124 (2009). ► https://doi.org/10.17077/drivingassessment.1311
6. Balkin, T., Horrey, W.J., Graeber, R., Czeisler, C., Dinges, D.F.: The challenges and opportunities of technological approaches to fatigue management. Accid. Anal. Prev. **43**(2), 565–572 (2011)
7. Barr, L., Popkin, S., Howarth, H.: An evaluation of emerging driver fatigue detection measures and technologies (Report No. FMCSA-RRR-09–005). U.S. Department of Transportation, Federal Motor Carrier Safety Administration, Washington, DC (2009)
8. Gaspar, J.G., Brown, T.L., Schwarz, C.W., Lee, J.D., Kang, J., Higgins, J.S.: Evaluating driver drowsiness countermeasures. Traffic Inj. Prev. **18**, S58–S63 (2017). ► https://doi.org/10.1080/15389588.2017.1303140
9. Schmidt, E., Decke, R., Rasshofer, R., Bullinger, A.C.: Psychophysiological responses to short-term cooling during a simulated monotonous driving task. Appl. Ergon. **62**, 9–18 (2017). ► https://doi.org/10.1016/j.apergo.2017.01.017
10. Li, R., Chen, Y.V., Zhang, L.: Effect of music tempo on long-distance driving: which tempo is the most effective at reducing fatigue? i-Perception **10**(4), 1–19 (2019). ► https://doi.org/10.1177/2041669519861982
11. Raudenbush, B., Grayhem, R., Sears, T., Wilson, I.: Effects of peppermint and cinnamon odor administration on simulated driving alertness, mood and workload. N. Am. J. Psychol. **11**(2), 245–256 (2009)
12. Wu, B.: Driver Fatigue Management System–Fatigue Inspector. Project Report. Eindhoven University of Technology, Eindhoven (2013)
13. Klauer, S.G., Dingus, T.A., Neale, V.L., Sudweeks, J.D., Ramsey, D.J.: The Impact of Driver Inattention on Near-Crash / Crash Risk: An Analysis Using the 100-Car Naturalistic Driving Study Data (Report DOT HS 810 594). National Highway Traffic Safety Administration, Washington, DC (2006)
14. Huemer, A.K., Vollrath, M.: Ablenkung durch fahrfremde Tätigkeiten – Machbarkeitsstudie (German Federal Highway Research Institute reports, issue M225). German Federal Highway Research Institute, Bergisch Gladbach (2012)
15. Yekhshatyan, L., Lee, J.D.: Changes in the correlation between eye and steering movements indicate driver distraction. IEEE Trans. Intell. Transp. Syst. **14**(1), 136–145 (2013)
16. Ahlström, C., Georgoulas, G., Kircher, K.: Towards a context-dependent multi-buffer driver distraction detection algorithm. IEEE Trans. Intell. Transp. Syst. 1–13 (2021). ► https://doi.org/10.1109/TITS.2021.3060163
17. Vlakveld, W., van Nes, N., de Bruin, J., Vissers, L., van der Kroft, M.: Situation awareness increases when drivers have more time to take over the wheel in a Level 3 automated car: a simulator study. Transport. Res. F Traffic Psychol. Behav. **58**, 917–929 (2018). ► https://doi.org/10.1016/j.trf.2018.07.025

18. Braun, M., Weber, F., Alt, F.: Affective automotive user interfaces–reviewing the state of emotion regulation in the car. Preprint. Under Review at ACM Computing Surveys (2020)

19. Galovski, T.E., Malta, L.S., Blanchard, E.B.: Road Rage: Assessment and Treatment of the Angry, Aggressive Driver. American Psychological Association, Washington, DC (2006). ▶ https://doi.org/10.1037/11297-000

20. James, L.: Road Rage and Aggressive Driving. Prometheus Books, Amherst, NY (2000)

21. Paredes, P.E., Zhou, Y., Hamdan, N.A., Balters, S., Murnane, E., Ju, W., Landay, J.A.: Just breathe: in-car interventions for guided slow breathing. In: Proceedings of the ACM on Interactive, Mobile, Wearable and Ubiquitous Technologies, 2 (1), Article 28 (2018). ▶ https://doi.org/10.1145/3191760

22. Çano, E., Coppola, R., Gargiulo, E., Marengo, M., Morisio, M.: Mood-based on-car music recommendations. In: Maglaras, L.A., Janicke, H., Jones, K. (eds.), Industrial Networks and Intelligent Systems, vol. 188 (pp. 154–163). Springer International Publishing, Cham (2017). ▶ https://doi.org/10.1007/978-3-319-52569-3_14

23. Coughlin, J.F, Reimer, B., Mehler, B.: Driver wellness, safety & the development of an awarecar. Massachusetts Institute of Technology, Center for Transportation & Logistics, AgeLab, Cambridge, MA (2009). ▶ http://web.mit.edu/reimer/www/pdfs/coughlin_wellness_2009.pdf

24. Marechal, C., Mikołajewski, D., Tyburek, K., Prokopowicz, P., Bougueroua, L., Ancourt, C., Węgrzyn-Wolska, K.: Survey on AI-based multimodal methods for emotion detection. In: Kołodziej, J., González-Vélez, H. (Eds.), High-Performance Modelling and Simulation for Big Data Applications. Lecture Notes in Computer Science, vol. 11400 (pp. 307–324). Springer, Cham (2019). ▶ https://doi.org/10.1007/978-3-030-16272-6_11

25. Russell, J.A.: A circumplex model of affect. J. Pers. Soc. Psychol. **39**(6), 1161–1178 (1980)

26. Sadler, P., Labouchère, P.: Das empathische Fahrzeug – Wenn das Auto Verhaltensmuster erkennt. ATZ extra, October 2020, 26–28 (2020)

27. Doshi, A., Trivedi, M.M.: On the roles of eye gaze and head dynamics in predicting driver's intent to change lanes. IEEE Trans. Intell. Transp. Syst. **10**(3), 453–462 (2009). ▶ https://doi.org/10.1109/TITS.2009.2026675

28. Welke, S.: Lenkmanöverprädiktion basierend auf einer Analyse der hirnelektrischen Aktivität des Fahrers. Dissertation. Technical University of Berlin, Faculty V Mechanical Engineering and Transport Systems, Berlin (2012)

29. Yan, Z., Yang, K., Wang, Z., Yang, B., Kaizuka, T., Nakano, K.: Intention-based lane changing and lane keeping haptic guidance steering system. IEEE Trans. Intell. Veh. (2020). ▶ https://doi.org/10.1109/TIV.2020.3044180

30. van Berck, J., Knye, M., Matusiewicz, D. (eds.).: Automotive health. Gesundheit im Auto im (Rück-) Spiegel der Kundenbedürfnisse. Springer Gabler, Wiesbaden (2019). ▶ https://doi.org/10.1007/978-3-658-27285-2

31. Köllner, C.: Wenn das Auto zum Gesundheitsmanager wird. Online article (2019). ▶ https://www.springerprofessional.de/ergonomie---hmi/gesundheitsmanagement/wenn-das-auto-zum-gesundheitsmanager-wird-/15507094

32. UNECE.: Regulation No 16 of the Economic Commission for Europe of the United Nations (UN/ECE)—Uniform provisions concerning the approval of: I. Safety-belts, restraint systems, child restraint systems and ISOFIX child restraint systems for occupants of power-driven vehicles II. Vehicles equipped with safety-belts, safety-belt reminder, restraint systems, child restraint systems and ISOFIX child restraint systems [2018/629]. Official Journal of the European Union, 27.04.2018 (2018)

33. Naujoks, F., Befelein, D., Neukum, A.: Welche Aspekte fahrfremder Tätigkeiten schränken die Übernahmefähigkeit beim hochautomatisierten Fahren ein? In: VDI Wissensforum GmbH (Ed.), Fahrerassistenz und automatisiertes Fahren (pp. 245–262). VDI Verlag, Dusseldorf (2016)

34. Braunagel, C., Rosenstiel, W., Kasneci, E.: Ready for take-over? A new driver assistance system for an automated classification of driver take-over readiness. IEEE Intell. Transp. Syst. Mag. **9**(4), 10–22 (2017). ▶ https://doi.org/10.1109/MITS.2017.2743165

35. Biedermann, A., Cieler, S.: Bedeutung und Herausforderungen der Fahrerzustandserfassung im Kontext des vollautomatisierten Fahrens. VDI reports **2311**, 91–106 (2017)

36. Willstrand, T.D., Anund, A., Strand, N., Nikolaou, S., Touliou, K., Gemou, M., Faller, F.: Driver/rider models, Use Cases and implementation scenarios. EU Horizon 2020 project "Adaptive ADAS to support incapacitated drivers Mitigate Effectively risks through tailor made HMI under automation" (ADAS&ME, Del.1.2). European Commission, Brussels (2017) . ▶ https://www.adasandme.com/wp-content/uploads/2017/06/ADAS-ME-20170619-D1.2-Driver-Rider-models-Use-Cases-and-implementation-scenarios.pdf

37. Bridgeman, P.W.: The Logic of Modern Physics. MacMillan, New York (1927)

38. Kastens, U., Kleine Büning, H.: Modellierung. Grundlagen und formale Methoden, 5th edn. Hanser, Munich (2021)

39. Schwehr, J., Luthardt, S., Dang, H., Henzel, M., Winner, H., Adamy, J., Fürnkranz, J., Willert, V., Lattke, B., Höpfl, M., Wannemacher, C.: The PRORETA 4 city assistant system. at - Automatisierungstechnik **67**(9), 783–798 (2019). ▶ https://doi.org/10.1515/auto-2019-0051

40. Leonhardt, S., Leicht, L., Teichmann, D.: Unobtrusive vital sign monitoring in automotive environments-a review. Sensors **18**(9), 3080 (2018). ▶ https://doi.org/10.3390/s18093080

41. Federal Communications Commission.: Technical Requirements for Indoor UWB Systems. Code of Federal Regulations, 47, Paragraph 15.517. Federal Communications Commission, National Archives and Records Administration, Washington, DC (2002)

42. European Commission.: Commission Implementing Decision 2019/785 of 14 May 2019 on the harmonisation of radio spectrum for equipment using ultra-wideband technology in the Union and repealing Decision 2007/131/EC, Section 3: UWB devices installed in motor and railway vehicles (file number C(2019) 3461). European Commission, Brussels (2019)

43. International Telecommunication Union.: Frequency ranges for global or regional harmonization of short-range devices. Annex 1: Frequency ranges for global harmonization of SRDs. (Recommendation ITU-R SM.1896–1, 09/2018). International Telecommunication Union, Geneva (2018)

44. Federal Communications Commission.: Operation within the band 57–64 GHz. Code of Federal Regulations, 47, Paragraph 15.255. Federal Communications Commission, National Archives and Records Administration, USA (2013)

45. European Commission.: Commission Implementing Decision 2019/1345 of 2 August 2019 amending Decision 2006/771/EC updating harmonised technical conditions in the area of radio spectrum use for short-range devices, vol. 74a (file number C(2019) 5660). European Commission, Brussels (2019)

46. Europäisches Institut für Telekommunikationsnormen.: Short Range Devices (SRD) using Ultra Wide Band (UWB); Part 3: Worldwide UWB regulations between 3,1 and 10,6 GHz (ETSI TR 103 181–3). ETSI, Sophia Antipolis (2019)

47. Europäisches Institut für Telekommunikationsnormen.: Short Range Devices (SRD) to be used in the 40 GHz to 260 GHz frequency range. Harmonised Standard for access to radio spectrum, part 4: vehicle application (ETSI EN 305 550–4). ETSI, Sophia Antipolis (2019)

48. Ahlstrom, C., Fors, C., Anund, A., Hallvig, D.: Video-based observer rated sleepiness versus self-reported subjective sleepiness in real road driving. Euopean Transp. Res. Rev. **7**, 38 (2015). ► https://doi.org/10.1007/s12544-015-0188-y

49. Russell, S., Norvig, P.: Künstliche Intelligenz: Ein Moderner Ansatz. Pearson Studium, Munich (2012)

50. International Organization for Standardization: Road Vehicles-Functional Safety-Part 9: Automotive Safety Integrity Level (ASIL)-Oriented And Safety-Oriented Analyses (ISO 26262–9). ISO, Geneva (2018)

51. Reif, K., Dietsche, K.-H.: Kraftfahrtechnisches Handbuch, 27th edn. Springer Vieweg, Wiesbaden (2011)

52. Paaß, G., Hecker, D.: Künstliche Intelligenz. Springer Vieweg, Wiesbaden (2021)

53. Kendall, A., Gal, Y.: What uncertainties do we need in Bayesian deep learning for computer vision? In Luxburg, U.v., Guyon, I., Bengio, S., Wallach, H., Fergus, R., (Eds.), Proceedings of the 31st International Conference on Neural Information Processing Systems (NIPS'17) (pp. 5580–5590). Curran, Red Hook, NY (2017)

54. Redmon, J., Divvala, S., Girshick, R., Farhadi, A.: You only look once: unified, real-time object detection. Proceedings of the IEEE Conference on Computer Vision and Pattern Recognition, 779–788 (2016)

55. Howard, A.G., Zhu, M., Chen, B., Kalenichenko, D., Wang, W., Weyand, T., Andreeto, M., Adam, H.: MobileNets: Efficient Convolutional Neural Networks for Mobile Vision Applications (2017). ► arXiv:1704.04861

56. Gibbs, L.I., Lawrence, D.W., Kohn, M.A.: Heat exposure in an enclosed automobile. J. La. State Med. Soc. **147**(12), 545–546 (1995)

57. Grundstein, A., Dowd, J., Meentemeyer, V.: Quantifying the heat-related hazard for children in motor vehicles. Bull. Am. Meteor. Soc. **91**(9), 1183–1192 (2010). ► https://doi.org/10.1175/2010BAMS2912.1

58. National Highway Traffic Safety Administration.: Not-in-traffic surveillance: non-crash fatalities and injuries (Research Note DOT HS 812 120). NHTSA'S National Center for Statistics and Analysis, Washington, DC (2015). ► https://crashstats.nhtsa.dot.gov/Api/Public/ViewPublication/812120

59. Hammett, D.L., Kennedy, T.M., Selbst, S.M., Rollins, A., Fennell, J.E.: Pediatric heatstroke fatalities caused by being left in motor vehicles. Pediatric Emergency Care (under review) (2020). ► https://doi.org/10.1097/PEC.0000000000002115

60. House of Representatives.: Hot Cars Act of 2021 (2021–05–12) (2021). ► https://www.congress.gov/bill/117th-congress/house-bill/3164/text

61. Alliance of Automobile Manufacturers.: Leading Automakers' Commitment to Implement Rear Seat Reminder Systems. Online Document (2019). ► https://www.autosinnovate.org/safety/heatstroke/Automakers%20Commit%20to%20Helping%20Combat%20Child%20Heatstroke.pdf

62. Euro NCAP: European New Car Assessment Programme (Euro NCAP). Test and assessment protccol–Child Presence Detection, Version 1.2 (2023). ► https://www.euroncap.com/media/79888/euro-ncap-cpd-test-and-assessment-protocol-v12.pdf

63. Fleming, S., Thompson, M., Stevens, R., Heneghan, C., Plüddemann, A., Maconochie, I., Tarassenko, L., Mant, D.: Normal ranges of heart rate and respiratory rate in children from birth to 18 years of age: a systematic review of observational studies. Lancet **377**(9770), 1011–1018 (2011). ► https://doi.org/10.1016/S0140-6736(10)62226-X

Open Access This chapter is licensed under the terms of the Creative Commons Attribution-NonCommercial-NoDerivatives 4.0 International License (► http://creativecommons.org/licenses/by-nc-nd/4.0/), which permits any noncommercial use, sharing, distribution and reproduction in any medium or format, as long as you give appropriate credit to the original author(s) and the source, provide a link to the Creative Commons license and indicate if you modified the licensed material. You do not have permission under this license to share adapted material derived from this chapter or parts of it.

The images or other third party material in this chapter are included in the chapter's Creative Commons license, unless indicated otherwise in a credit line to the material. If material is not included in the chapter's Creative Commons license and your intended use is not permitted by statutory regulation or exceeds the permitted use, you will need to obtain permission directly from the copyright holder.

Human Behaviour as Basis for Situation and Risk Assessment

Sascha Knake-Langhorst, Mandy Dotzauer, Kay Gimm, Marek Junghans, Hagen Saul, Caroline Schießl, and Meng Zhang

Contents

© The Author(s) 2026
H. Winner et al. (eds.), *Handbook Assisted and Automated Driving*,
https://doi.org/10.1007/978-3-658-45276-6_29

29.1 Introduction

29.1.1 Safety and Efficiency in Complex Traffic Spaces

In science and research, we face increasing individual mobility challenges, a growing range of new transportation options (e.g. small electric vehicles such as electric scooters and cargo bikes) and the accompanying increase in traffic. For some years now, the focus has increasingly been on looking at transport in urban areas, where different types of traffic with their respective requirements and specifics share the always limited road space. These investigations address, in particular, questions of traffic safety on the one hand and the highest possible efficiency of traffic on the other. These two aspects must always be reconciled.

In the relevant disciplines,[1] one of the ways in which the above challenges are addressed is through the definition and development of models of observable behaviour. The goal is to understand and analyse the urban traffic behaviour of all road users involved, to build up knowledge about it and to model behaviour. Building on this, requirements for automation and assistance functions and systems can be derived and traffic can be simulated in many ways under the most diverse of boundary conditions. Effective measures to improve traffic safety, efficiency and the negative environmental impact of traffic can thus be derived and evaluated.

29.1.2 From Driver to Road User

The fundamental thing here is to use these models to quantify the overall dynamics of traffic (i.e. the spatio-temporal behaviour of all road users in a traffic area) as an essential characteristic of urban traffic. Models must include factors, categories and criteria that reflect these dynamics, rather than simply describing a point-in-time snapshot of traffic behaviour. Parameters and variables to be developed (e.g. traffic flow-related and kinematic parameters, interaction, communication and cooperation variables) for describing and analysing behaviour as a basis for model building must also meet these requirements.

Analysing human behaviour in urban environments is far more complex than behaviour on motorways. In urban areas, a variety of traffic types and modes encounter one another with a wide range of speed and possible driving/moving manoeuvres. Characteristic driving manoeuvres on the motorway include following a vehicle, changing lanes, or entering and leaving the motorway. The urban space is characterised by changes of direction, vehicles parking and leaving parking spaces, pedestrians and cyclists crossing the road, lane changes and overtaking manoeuvres. An initial approach to the research area of urban transport and understanding interactions between different groups of road users began in the early 2010s. Observational studies on the interaction of cyclists and pedestrians with motorists were conducted to investigate aspects of urban traffic [1–3]. In parallel, technical research infrastructures were established (e.g. [4]) allowing for human behaviour to be recorded and used for the further development of methodologies, methods and parameters for traffic analysis. This is an essential step in making a lasting contribution to road safety and in quantifying human behaviour, considering a variety of situational factors.

Once such correlations have been worked out, they can be used to develop risk predictions which can then in turn serve as the basis for innovative technical system designs as automated support for road users in relevant traffic situations. One example of this chain of effects is provided by the use case "right-turning motorists interacting with crossing cyclists" at inner-city intersections, for which an infrastructure-based procedure for risk prediction was developed with the help of the outlined work structure [5]. In the course of the aforementioned work, it became clear that risk-prediction procedures must incorporate aspects of human situational awareness so that, for example, the timing of a warning generated and communicated by the infrastructure is tailored to the needs of the user. However, testing of such warning systems also shows that it is not as simply as to identify situational factors and consider aspects of perceptual psychology. Understanding human behaviour goes far beyond that. Models and algorithms need to be extended to include constructs such as interaction, cooperation and implicit communication—phenomena that are essential to the outcome of encounters in human interaction [6, 7].

29.2 Methods for the Technical Observation of Traffic Behaviour

Technical data capture offers enormous potential for the in-depth analysis of human behaviour in traffic. While a few years ago traffic observations were conducted by people at intersections noting individual aspects, technological advances now permit detailed microscopic data collection. Based on objectively collected kinematic measurement data in the form of

---

1   Traffic, engineering, industrial psychology and human factors in interaction with various engineering disciplines.

trajectories, analyses can be performed automatically and repeatably over long survey periods. This eliminates the subjective factor of human observation. In order to achieve this, the traffic flow to be recorded must be defined before trajectory data requirements are set and targeted recording systems are deployed.

29.2.1 Methods for Traffic-Flow Observation

Traffic is the realisation of mobility in terms of location changes of people or goods. Traffic flow is the spatio-temporal process of traffic and motion of people or goods (driving, accelerating/decelerating, stopping, waiting, crossing, turning, overtaking, parking) in the traffic areas. The sum of moving road users and goods is called traffic stream, and the movement itself is called traffic flow [8–10]. Traffic flow can be understood as a location- and time-dependent, non-stationary random process [11], characterised by individuality (free movement on one's own initiative), collectivity (restriction of free movement by imposed routes, mutual consideration, traffic and anticipation behaviour, traffic rules, weather conditions, etc.) and complexity [12].

Traffic flow can essentially be described by macroscopic and microscopic parameters, as well as behaviour-specific parameters. Macroscopic parameters refer to the traffic density, which is defined by the fundamental relation $Q = D \cdot V_m$ with traffic volume Q, traffic density D and average instantaneous speed V_m. Essentially, a distinction is made between free flow, synchronised traffic and congested traffic [10–14]. Microscopic parameters refer to road users [15], for example, speed, acceleration function, journey time, waiting time and loss time, but also to interaction behaviour. Behavioural metrics are used to refer to interaction between road users, for instance time gap, distance to the vehicle ahead or metrics of traffic-conflict technique.

Traffic data can be measured by suitable sensors in order to realise concrete transport-related applications. In the past, the focus was often on motorised vehicles (MV). This mainly concerns the areas of driver assistance, traffic-dependent traffic-light control, detection of traffic violations, toll collection and systems for parking-space management [9, 11, 16–19].

Traffic data can be derived directly or indirectly from the road-user trajectories in space-time diagrams (◘ Fig. 29.1). Thus, a differentiation is made between instantaneous (e.g. photograph), local (e.g. inductive loop detector), spatio-temporal (e.g. video camera) and moving observations both in the direction of traffic flow (e.g. floating car) and against the traffic flow (e.g. floating car observer). For example, traffic volume Q at location s_1 is a time-continuous and constant consideration of the count function $N(s_1,t)$ divided by the considered time interval Δt. The time gap t_L between

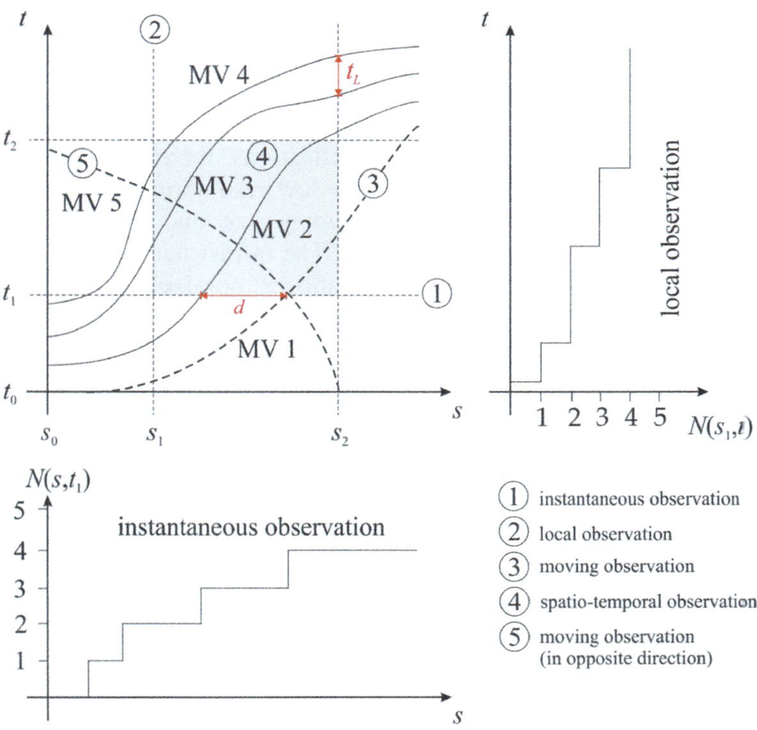

◘ **Fig. 29.1**　Characteristics of traffic flow using the example of path-time diagrams (adapted from [11])

vehicles 3 and 4 and distance d to the vehicle in front between vehicles 1 and 2 result directly from this.

29.2.2 Requirements for Technical Observation of Traffic Behaviour

Another application is the technical observation of traffic for traffic-related analysis in research. In order for traffic to be observed and traffic-flow data to be used in the form of trajectories, specific requirements of the technical observer must be met. The basic concern here is, that measurement does not feed back into traffic flow in order to prevent unwanted behavioural patterns. A broad bandwidth of additional aspects is presented in ◘ Table 29.1, including description and possible expression of artefacts in real data. The requirements are grouped, according to what they cover, into the categories of General, Detection, Tracking and Robustness to Environmental Influences.

29.2.3 Application of Sensor Technologies as a Reflection of Requirements

Various sensor technologies can be used to record road users. The choice of technology depends on the question to be answered and available resources. Advantages and disadvantages must be weighed up, as shown in ◘ Table 29.2.

The fusion of different sensors can compensate specific disadvantages. In principle, bringing together different sensors and sensor types can:

- increase accuracy or minimise errors, for example, when determining the position of objects;
- extend the measuring range, for instance for cameras with different orientations but overlapping field of view (similar to stitching [20] in image processing);
- extend the dimensionality, for example, by combining LiDAR and mono video camera data, where the camera image data (2D) is supplemented by the depth dimension given by LiDAR. However, different resolutions and resultant problems must be considered here:

Plausibility checks are another way to gain an information advantage. The processing of mono-camera data can be supported by depth information from LiDAR, for example.

Technical observation using video cameras offers the essential advantage for traffic analysis that the image can be interpreted directly by humans and can be used to verify automatically detected events. Fusion with additional technologies can improve the quality of recording.

Despite best efforts, errors inevitably occur during a measuring process. These can have a relevant negative influence on data quality. Data quality can be impacted in a wide variety of ways, as described in ◘ Table 29.3.

It is extremely important that the potential impacts on associated quality metrics are always considered and that data is scrutinised and adequately reviewed in this regard. This is particularly relevant because modelling, as a subsequent step in the processing chain, usually allows no or only limited conclusions to be drawn about potential problems in the underlying database.

29.2.4 Consideration of Known System Approaches for Traffic Monitoring

A range of different approaches is used for the technical detection of traffic. Induction loops offer one possibility for local, possibly lane-specific observations. This method allows traffic counts to be carried out and time gaps between vehicles to be measured. In the context of the latter, critical situations can be calculated on the basis of low gap distances. On the basis of the measurement principle, it is not possible to detect vulnerable road users or only possible to a very limited extent.

In naturalistic driving-data collections, such as [22], moving observation takes place dominated by the first-person perspective of the vehicle. The advantage of this survey method is the continuous monitoring of traffic events from the perspective of a road user. However, there is a limit to the in-depth analysis of first-person driving behaviour and interaction with other road users since the purely vehicle-based sensor set-up does not usually permit an overall view of the situation. Probe vehicle data offers another moving-observer possibility, which is not really suitable for microscopic investigations, for example, regarding criticality, due to the low resolution. The focus here is on determining journey times and macroscopic traffic variables.

The infrastructure-based observation of traffic is a spatio-temporal observation (see 4. in ◘ Fig. 29.2.1). This allows a detailed view of the traffic situation, as a complete picture of the traffic is given on a microscopic level. Detailed analyses of traffic behaviour in terms of interaction and criticality can be carried out based on this comprehensive technical observation and before-and-after comparisons can also be made. In research and development landscape, some efforts have been made in the past to develop adequate tools that encompass different solution approaches and the traffic areas addressed by each. For example, the Institute of Transportation Systems (TS) at the German Aerospace Center (DLR) applies the AIM test bed where it operates suitable large-scale research infrastructure

◘ Table 29.1 Requirements for technical observation of traffic behaviour

Requirement category	Short form	Description	Possible artefacts in real data
General	No feedback to traffic flow	Recording has no repercussions on the traffic observed	Heavy braking at sighting of measuring apparatus
	Verifiability of data	– In addition to trajectories, also capture scene videos to be able to check plausibility of traffic event and detection later – Make annotations that cannot be made automatically on the basis of the data material	
	Monitoring of data acquisition	System monitoring to ensure correct recording	
Detection	Coverage	Road users are recorded in spatially relevant areas making analyses possible. One example is the detection of cyclists in the arm of a junction in order to study interaction with turning traffic	– False-positive detection (ghost objects) – False-negative detection (object not detected) – Object detected late
		Road users are continuously recorded	– No time gaps for identical ID (no broken trajectory) – No tear-offs and reassignment of the ID
	Temporal resolution	Road users are recorded densely meaning that analyses are possible	Data are available with such a low temporal resolution that, for example, investigations in interaction can be carried out conditionally
	Accuracy	Position measurement corresponds to real position	Position strongly lags real position, so that calculated time gap is smaller and no conflict is present
		Classification corresponds to reality	Cyclists classified as pedestrians
		Dimensions correspond to real road users	Vehicle size estimates are too large, so that calculated gap between vehicles is smaller than actual gap
Tracking	No outliers	Smooth and steady position history corresponding to the course of the journey	Noise in speed data
	Maintenance of an object hypothesis	Road user "truck with trailer" is considered as one object	No splitting effect with large objects such as articulated buses
Robustness with regard to environmental influences	Weather conditions	Insignificant influence of normal weather conditions, rain, snow	Lower detection quality in rain
	Light conditions	Insignificant influence of normal lighting conditions, darkness	Traffic cannot be detected at all or with greater difficulty in darkness

□ Table 29.2   Overview of selected sensor technology and associated advantages and disadvantages

Sensor technology	Advantage	Disadvantage
Radar	Direct distance measurement, can also be used in poor light conditions	Evaluation by humans sometimes difficult (point cloud)
Lidar	Direct gap measurement (and speed measurement), High refresh rate (>100 Hz), Maximum coverage in the medium range (10–40 m)	Evaluation by humans sometimes difficult (point cloud)
Mono video camera	Image available for traffic analysis (easily interpretable by humans), High coverage range possible if resolution is high enough	Dependent on weather and light conditions
Stereo video camera	See mono video camera, additional depth information available	Dependent on weather and light conditions
Ultrasonic	No particular advantage, suitable for collision detection at close range (a few metres)	Low range
Time-of-flight cameras (TOF)	Direct gap measurement, complete image available immediately	Interference/mutual influence or disturbance possible when using several TOF cameras

□ Table 29.3   Known data quality metrics (see [21] and examples)

Data-quality metric	Remarks and examples
Correctness: conformity with reality	Object recognition: traffic and static objects must be correctly recognised, ghost objects (false-positive) or wrongly assigned objects (false-negative) reduce accuracy
Consistency: no contradictions	Object classification in multi-sensor systems: the sensors must not differ in their object classifications. Data fusion methods can reduce consistency problems
Reliability: traceable origin of the data	Object recognition: the detected objects originate from sensors on the infrastructure side, of which one camera failed for a certain time and did not provide any data. Sensor fusion techniques improve reliability
Completeness: all necessary attributes are included	Situation verification: LiDAR do not provide image-based information for the verification of identified situations, they are incomplete. They can be completed by data from imaging sensors
Accuracy: data available at the required level of precision	Determining the position of road users during tracking: video-based tracking methods provide the required information with accuracy (e.g. decimal places) dependent on distance. At greater distances, these are no longer fulfilled
Timeliness: data correspond to the current state of reality	Object recognition: sensors always provide object information with a certain latency. Latency of 1 s or higher may lead to data that no longer is up-to-date, for example, for tasks relevant to safety
No redundancy: no duplications	Object tracking: if information is available that can be derived from information contained in trajectories then there is redundancy in the data
Relevance: information content must meet information need	Object tracking: if information on the differential velocity of successive objects is not available based on trajectory data and cannot be derived, the criticality metric TTC cannot be determined
Consistency: uniform structure and presentation	Determining velocity in object tracking: if velocity information is alternately available in m/s and km/h, it is not uniform
Clarity: unambiguous interpretability	Object recognition: if there are multiple instances of a single object ID and all other data attributes are the same, it is likely that it is one and the same object
Comprehensibility: conformity with users' expectations	Object classification: objects are coded by sensor systems as, for example, "C" (car) and "T" (truck), whereby the information regarding an "unknown object"—coded as "U"—is not known. In this case, the data coded as "U" cannot be understood

such as the AIM Research Intersection [4] located at a busy intersection in the city of Braunschweig, as well as the data acquisition track which is part of the Test Bed Lower Saxony [23] along a section of the A39 motorway. Set up as semi-stationary camera masts, mobile versions enable individual measuring operations at level crossings [24], bottlenecks or shared spaces [25]. The systems have identical system architecture and each permits sustained traffic observations under a wide range of external environmental conditions. There is also a research intersection for permanent traffic monitoring and analysis set up in cooperation with Tongji University in Shanghai. Other research initiatives and institutions also have similar tools. The activity around the Digital test bed at the German A9 motorway as part of the Providentia project should be mentioned foremost [26], as well as the research intersection in Aschaffenburg (Ko-FAS research initiative [27]), Ingolstadt (SaveNoW [28]), Lund University, Sweden with the Ruba tool for recognising road users (EU project INdev [29]), and Purdue University's mobile camera masts for traffic monitoring in the USA [30]. Apart from infrastructural approaches, we also find aerial approaches using drones: "highway Drone" dataset (highD) [31] familiar from the PEGASUS or Datafromsky projects [32]. However, these recordings are time-limited, as constant traffic monitoring by drones is difficult to implement.

It must be noted that measurement systems demonstrate measurement errors. The artefacts in the data should be considered when conducting traffic analyses. It is advisable to perform a plausibility check and implement quality improvement measures before processing the data, for example, smoothing a signal using a Kalman filter or associating and merging broken trajectories [33].

With regard to framework conditions such as the installation and operation of infrastructural measuring equipment for recording traffic, a wide range of organisational aspects must be ensured. The effort required for installation and operation grows with the complexity of the system. The number of mounted sensors appropriate to the detection areas, artificial lighting for 24/7 operation, and associated data distribution and computer technology should be mentioned here, for example. Civil engineering work for data lines and power supply or the construction of a switch box to house computer technology may be necessary. Operating power must be provided by means of a generator or temporary or permanent metered supply provided by the energy supplier, depending on the duration of measurement campaigns. Permission from local authorities is required, as well as compliance

with data-protection aspects of the country concerned. Remote access permits installation and configuration work, perhaps by means of a monitoring system to be set up. Depending on the performance of computing power, the infrastructural data-acquisition system is able to provide data in approximately real time in order to pass on situational information and warnings in the context of automated and connected driving. Recorded contextual information—from a weather station, for example—such as visibility, rainfall, estimation of friction coefficient, emissions or switching states of traffic signals can be merged with traffic data for analysis purposes. In the context of processing sensor data and analysing trajectory data, high-performance backend systems are advantageous in order to automatically extract traffic events, over a period of several weeks if necessary.

What does this mean for traffic analyses? For the analysis of traffic behaviour, infrastructural data-acquisition approaches offer a very promising basis. On the one hand, they can provide simple low-cost solutions using a webcam, for example, and on the other hand complex, permanently installed productive systems. With increased cost and effort, more complex and in-depth analyses of traffic patterns become possible. These include aggregated analyses over long measurement periods in 24/7 operation. With an appropriately large detection area, it is possible to systematically and completely investigate the development of a traffic situation in order to gain knowledge regarding any behaviour of relevance to safety. One example is the interaction of cyclists with turning vehicles. Human behaviour can be analysed and quantified with the help of systematic surveys, such as those made possible by the use of traffic-conflict techniques (see ▶ Sect. 29.3). Real-time-capable systems for the infrastructural support of driver assistance and automation enable the implementation of the developed models in the form of algorithms (see ▶ Sect. 29.4) and to send risk-related warnings by means of wireless communication (see ▶ Sect. 29.5).

29.3 Methods for the Evaluation of Traffic Situations

The majority of traffic situations are undisturbed passages or controlled encounters between road users. Less frequent are traffic conflicts, near misses and crashes. In addition to these are atypical situations that deviate from normal traffic behaviour and generally involve non-critical behaviour (e.g. driving in a serpentine manner).

29.3.1 **Risk-Inhibiting and -Contributing Factors**

An overview of any (safety-related) behaviour is given in ◘ Fig. 29.2. Road traffic crashes are rare events, but in more than 90% of cases they are due to human error. In addition to the human factor, environmental factors and technical, vehicle-related or infrastructural factors are also identified as causes of crashes [34]. Based on this information, four major categories of factors influencing the course of a traffic situation can be derived: (1) infrastructure, (2) environment, (3) vehicle and (4) human.

In the category "infrastructure", aspects of traffic control play a role which enable or prevent certain traffic flows from encountering one another. A study by Sturm [36] shows the influence of signal control on crash occurrence. While 32% of crashes involving unprotected left turns without left-turn protection at signal-controlled intersections are turning crashes, at signal-controlled intersections with left-turn protection it is only 3%.

The "environment" category includes the volume of surrounding traffic, but also physical influences due to unfavourable weather and light conditions. When there is high traffic volume, such as at an urban intersection, more encounters occur that potentially evolve into a critical situation. In relation to vulnerable road users (e.g. cyclists), a phenomenon called Safety in Numbers (SiN) has been observed. It is postulated that a larger number of non-motorised road users will decrease the risk of individual non-motorised road users being involved in a crash [37, 38]. Karndacharuk et al. [39] also found that the number of pedestrians has an influence on the speed of motorists. The more pedestrians present in a shared space, the lower the speed of motorists. In contrast to these findings, there were studies in which the SiN effect could not be confirmed. For example, Lücken [40] found, when looking monthly at the individual risks of cyclists, that the individual risk of a crash actually increases as the traffic volume increases.

The "vehicle" (automobile) has been in constant development since the beginning of the twentieth century. For example, wing panels were fitted to cars early on to reduce the fatal consequences of collisions between pedestrians and motorists. These efforts have continued over the last century and into the current century, so that the modern vehicle is characterised by a range of active and passive safety features. Passive safety systems include, for example, the seat belt (including tensioners), airbags, safety headrests and the crumple zone, but also the airbag for cyclists [41]. The aim of passive safety systems is to minimise the consequences of a crash as far as possible. Active safety systems are designed to help prevent a crash. Where this is no longer possible, then active safety systems are intended to help reduce the consequences of a crash so that, if possible, fatalities no longer occur. The anti-lock braking system (ABS) and electronic stability program (ESP) are two of these systems which, according to EU regulation, must be fitted as standard in all new passenger cars. Despite advances in the development of the automobile, a vehicle can be a risk-enhancing factor. For example, a brake system that is not functioning faultlessly increases the probability of a collision, as the braking distance can be significantly increased.

The "human" factor is the main influencing one, since about 90% of all road traffic crashes are due to human error. In order to gain a better understanding of the causes, it is necessary to identify and analyse the

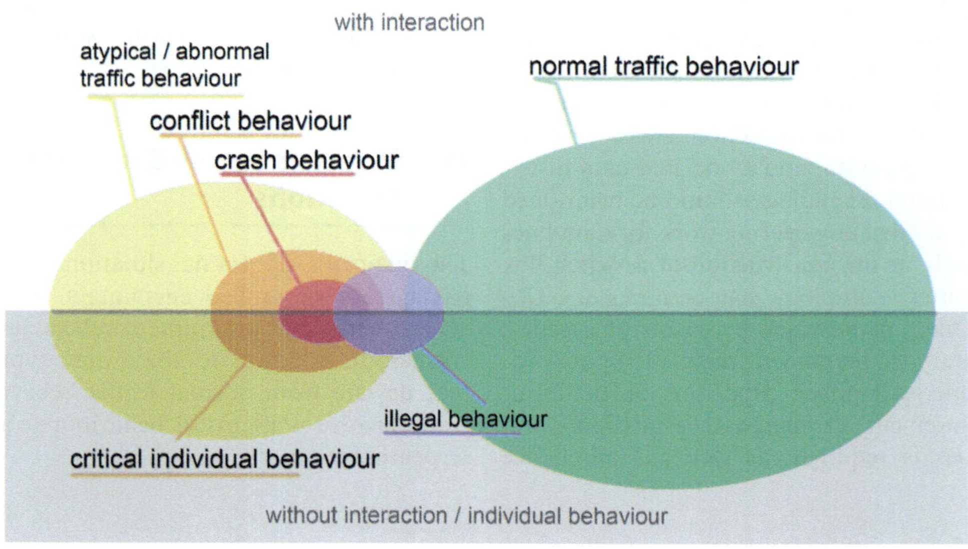

◘ **Fig. 29.2** Road traffic situations classified by road safety [35]

errors that are key to the occurrence of crashes. The error model provided by Rasmussen [42] offers an approach here. Rasmussen defines seven types of error:

1. Structural error (no driver action possible).
2. Information error (a relevant piece of information was perceived too late or not at all).
3. Diagnostic error (the situation was recognised but incorrectly evaluated).
4. Goal-setting error (the situation was correctly evaluated, but wrong intention was formed).
5. Method error (the wrong procedure was chosen to achieve the goal).
6. Action error (an error was made in the attempt to achieve the goal).
7. Operating error (an error was made in the operation of the vehicle).

An analysis of human error in near misses (data taken from the Austrian 100-Car Naturalistic Driving Study) showed that Rasmussen's error model was inadequate for determining the causes of near misses (e.g. [6]). Graab et al. [43] adapted Rasmussen's error model, with information error playing a central role and being further differentiated. This modified model describes five types of human error: (1) information access, (2) information intake, (3) information processing, (4) goal setting and (5) action. For example, road users may be distracted by operating a smartphone. This leads to an error in information access and reception. According to Klanner [44], carelessness, impaired visibility and incorrect assessment of the situation are the main causes of crashes at intersections and junctions. These causes can be attributed to the first three levels of the error model developed by Graab et al. [43]. Errors on the first three levels resulted in an action error. 42% of the drivers had not braked and 60% had not taken evasive action to prevent the crash or minimise the consequences [44].

These four categories of influencing factors cannot be considered individually. Infrastructure, vehicle (in a broader sense), the environment and human being interact to influence spatial and temporal encounter of two or more road users.

29.3.2 Well-Known Approaches from the Traffic-Conflict Technique

Relative to the total traffic volume, crashes are rare events that also occur in isolation. The sequence or course of crashes must be reconstructed and this is prone to error for several reasons. Reconstruction is often based on statements by witnesses and those involved, as these events are rarely recorded. Furthermore, assessment of the situation by police officers and experts can be biased. Near misses, conflicts, rule violations and encounters, in addition to crashes, are observable and measurable events in traffic that permit conclusions to be drawn regarding the hazard potential of a particular traffic area. Conflicts in are of particular importance for traffic safety. Conflicts can be systematically recorded using the traffic-conflict technique. Although they are also comparatively rare events, they occur more frequently than crashes, and compared to crashes, the sequence or process of occurrence and resolution is observable and thus measurable [45].

The traffic-conflict technique (TCT) is a tool for the preventive analysis of traffic situations (interactions, one-off situations) based on spatio-temporal proximity between at least one road user and another entity (road users, objects, traffic infrastructure) with the aim of reducing systematic negative influences on traffic safety and, if possible, lowering them to near zero. Efforts to establish a standardised tool go beyond European borders. Over the years, many national TCT methods have emerged: among others, for example, the Swedish [46, 47], German [45], Austrian [48], British [49], Belgian [50], Dutch [51], American [52], Canadian [53], Finnish [54], French [55] and the Czech [56].

What these TCT methods have in common is that traffic conflicts are considered as events in which at least one of the interacting entities must take evasive action to prevent a crash. In this context, the analysis of such situations is performed by trained personnel (tending to the subjective) on site and/or by an objective assessment based on imaging techniques (e.g. videos of the situation [4, 57, 58]) and the use of special analysis software (e.g. semi-automated analysis software such as T-Analyst; see [59]). In addition, traffic conflicts are categorised by (1) type in relation to the traffic area (e.g. in the direction of travel, against the direction of travel, crossing, turning), (2) severity (e.g. on a scale from minor to fatal) and (3) type of situation (intentional or controlled and uncontrolled situations). These categories are quantified by the spatio-temporal proximity of the interaction partners. Suitable metrics, known as Surrogate Measures of Safety (SMoS), are used. These include distance to conflict point (DCp), time to collision (TTC) or time to accident (TA), post encroachment time (PET) and probabilistic ones such as collision probability or risk (e.g. risk of collision—ROC), as well as the ratio between the number of crashes and the number of conflicts or the expected number of crashes (e.g. [60]).

The identification and analysis of traffic situations are carried out in part by trained personnel. To quantify conflicts, the metrics presented above are mostly used with different threshold values and, in some cases, different scaling and factor levels, especially as regards conflict severity. There are also differences in the application of certain thresholds depending on the region (urban vs. rural, involvement of pedestrians and cyclists). Which method is used to evaluate traffic situations depends on the question and the objective. The Swedish TCT, for example, focuses on the moment of the evasive manoeuvre. The method was primarily developed to evaluate interactions between motorised road users.

As a study by Jakobowsky et al. [25] shows, this method is not suitable for evaluating the criticality of interactions between motorised road users and pedestrians in a shared space. Methods (e.g. Pedestrian-Vehicle Conflicts Analysis [61]) specifically developed for this particular traffic space and the interaction partners are more promising here, as these methods take into account the characteristics of each road user. For example, Kaparias et al. [61] developed different categories of changes of velocities for pedestrians and motorised road users.

The Dutch TCT is used when considering the time period from the beginning to the end of the evasive action. It is based on the Swedish TCT, but has been specially adapted to local conditions. It is possible to evaluate interactions between motorised and non-motorised road users. The method takes into account, for example, that two cyclists on a collision course have a greater degree of freedom with regard to evasive actions than, for example, two cars on a collision course. In addition, the method is designed not only to determine the likelihood of a conflict or crash, but also to consider the degree of severity for individual road users [51]. Another possibility is to consider a traffic situation as a whole process from the beginning, for example, from the moment when spatio-temporal proximity between two entities starts and finally leads to an evasive action, until the end of that action (e.g. in the German TCT). The German TCT developed by Erke and Gstalter [45] enables the consideration of individual road users in interaction with objects/infrastructure. It is equally applicable to interactions between drivers. A modified observation method was developed to systematically observe and analyse interactions between car drivers and cyclists. Particular characteristics affecting cyclists were taken into account when drawing up the observation protocol. Erke & Gstalter divide interactions into encounters and conflicts. Encounters are defined by the spatial and temporal proximity of two or more road users, where the road users adjust to avoid each other in time. The driving/riding manoeuvres are controlled. Conflicts, on the other hand, are instances in which road users converge in space and time where an increasingly likely collision can only be avoided by taking evasive action. The manoeuvre involves a change in speed and/or direction, that is braking, accelerating and swerving. For cyclists, abrupt jerking of the handlebars and jumping off the bike are also defined as evasive manoeuvres. The severity of a conflict is determined by the distance between two road users, the speed and change in speed, intensity of acceleration and type of evasive manoeuvre, with the authors defining two levels of severity. A severity 1 conflict involves controlled braking/acceleration and/or swerving. Road users here have just enough time to consider other road users when choosing a manoeuvre. However, there is no longer enough time to communicate this intended manoeuvre. Severity level 2 is when heavy braking/acceleration and/or abrupt swerving is carried out at the last moment. There is then no longer sufficient time to perform a controlled manoeuvre, which means that road users can no longer take the situation of other road users into account when choosing a manoeuvre.

Erke and Gstalter [45] divide conflicts into: (1) origin and its conditions, (2) conflict occurrence, (3) resolution and its consequences. The origin of the conflict is determined by the collision course up to the onset of the critical manoeuvre. The type of involvement (triggering, sudden, controlled, aggressive and non-involved), speed, lines of movement and any rule violations by those concerned are taken into account. The resolution of the conflict extends from the onset of the manoeuvre to reorientation. The following factors are determined for analysing the resolution of the conflict: type of involvement (coping, reactive, cooperative, aggressive and non-involved), speed, movement and reference to rules. The behaviour of all participants is documented for the origin and resolution of the conflict. Furthermore, neighbouring spaces are included in addition to the actual observation space. Erke & Gstalter assume that encounters and conflicts are prepared for or continued outside the observation space. In the example of right-turning motorists and crossing cyclists at inner-city intersections, the observation space is not limited to the bicycle lane. The approach to the junction is also included.

Applying the TCT and the associated in-depth analysis of traffic events have helped in the effort to derive and put into practice measures to increase traffic safety, including for encounters between car drivers and cyclists. Examples of measures introduced with the aim of reducing the number and severity of crashes between motor vehicles and cyclists include: separated cycle lanes built and traffic-calmed areas introduced ([62–65]); signs and road markings installed [66]; intelligent systems for vehicles and traffic infrastructures developed to detect cyclists and warn of impending collisions [35, 67]. A look at current crash statistics suggests that these measures have the desired positive effect. The number of all traffic fatalities in Germany is expected to reach a new low in 2021. The number of injured road users has also decreased over the last three years. Looking at the statistics for cyclists in isolation, however, this positive trend is not apparent. The number of cyclists killed in Germany has been stagnant at a high level for several years (more than 400 per year). Approximately half of the fatalities occur at intersections as a consequence of a collision with a motorised vehicle [68]. Based on current crash statistics, the question remains: How can crashes between motor vehicles and cyclists with fatalities and serious injuries at inner-city intersections be prevented?

29.4 Understanding Traffic, Describing It and Determining Risks

29.4.1 Understanding and Describing Traffic

Understanding traffic and describing what effective measures can be taken to increase road safety for cyclists requires first an understanding of the causes of crashes and the situations that precede them (known as pre-crash situations). Drivers and cyclists are often equally responsible for crashes. The main reasons for the increase in cycling crashes are primarily behaviour which is misunderstood or unclear behaviour between cyclists and drivers, which can lead to an inappropriate sense of safety. While motor-vehicle drivers disregard cyclists' right of way or commit errors when turning, cyclists ride on the wrong side or disregard the right of way. The authors of the report Ordnungspartnerschaft Verkehrsunfallprävention (2009) cited by [69] conclude that lack of cooperation or unwillingness to cooperate is the cause. On the other hand, the majority of interactions between cyclists and motor-vehicle drivers do not lead to crashes with serious injuries or fatalities, but to crash-free, yet possibly conflictual situations that may require evasive manoeuvres. If the continuity of incidents with relevance for safety is assumed (see [45, 46, 70]), critical situations that do not lead to crashes occur far more frequently. Attempts have been made to quantify this relationship for some time (e.g. [71–76]). To understand how these critical situations arise, it is necessary to understand how road users encounter each other without conflict or, when a conflict does occur, how it arises.

29.4.1.1 Descriptive Analysis of Interactions

Determining, analysing and quantifying behaviour in traffic requires a comprehensive and detailed psycho-technical view of the dynamics of road users in traffic spaces in order to recognise road users' intentions and to identify, evaluate, prevent, but also predict potential risks. Observational studies allow us to explore these questions (see ▶ Sect. 29.3.2, e.g. [45]). Compared to traditional on-site observation, video observation studies, for example, offer the advantage that scene video and trajectory data can be collected over an extended period of time, allowing a variety of situations to be analysed. In order to achieve this aim, cameras are installed and directed at the areas of interest in the traffic space. The image data is processed into object data using object recognition and classification techniques to identify the road user types (e.g. car, bicycle, motorcycle). Association techniques are used to assign a unique ID to each object classified in the temporally sequential frames, so that the spatio-temporal history of each object is known. By means of tracking methods (e.g. by applying Kalman or particle filters), trajectories are created from the associated object data which are available either in the local camera coordinate system (coordinates of the pixels in the video images) or in the world coordinate system (coordinates of the traffic area). At a minimum, trajectories contain information on positions and time reference, but in some cases much more, for example, road-user type, speeds, accelerations, object sizes. If the trajectories are available in world coordinates, then these can be analysed (e.g. [77–81]). In a study of this kind, the study area can be divided into sections (e.g. approach, turn and distance areas when turning right at an intersection) and into scenarios (e.g. vehicles decelerating vs. not decelerating or accelerating at the stop line, see ▶ Sect. 29.3.2). For analysis, certain road-user data is collected (e.g. speed development, waiting or stopping times, spatial and temporal distances of interacting road users) which serves to answer the research questions. If certain data are available, they are usually compared with each other with regard to their statistical values (e.g. mean, median, standard deviation) (e.g. [7, 34]). These analyses take place post-hoc, so that the observers have the opportunity to run a situation several times, compare it with other situations and thus develop a deeper understanding of each situation. At the AIM Research Intersection in Braunschweig, video and trajectory data are recorded for this kind of analysis [4]. For example, video data of drivers turning right and cyclists crossing the intersection were recorded over a period of four weeks and analysed afterwards. Based on the results, an interaction matrix was created that provided initial indications of how motor-vehicle drivers and cyclists encounter each other at inner-city intersections in turning and conflict areas. In approximately 27% of the situations, motor-vehicle drivers had adjusted their speed in the turn and conflict areas, while cyclists travelled at a constant speed through both areas. In other situations (19%), motor-vehicle drivers had adjusted their speed in the turning area and then drove at a constant speed across the conflict area. As a rule, cyclists crossed the intersection before motor vehicles. In another 10% of the situations, cyclists had adjusted their speed in the turning area, while motor-vehicle drivers continued at a constant speed.

Here it was observed that motor-vehicle drivers passed through the conflict area before cyclists.

It became apparent that observation and assessment of encounter behaviour alone are not sufficient to understand the interactions between motor-vehicle drivers and cyclists. Other essential objective measures, such as velocity and acceleration profiles, must be included to achieve this aim. In addition, objective data, such as trajectory data, provide opportunities to derive criticality measures (see ▶ Sect. 29.2) or to develop new measures that can be used to quantify interactions.

29.4.1.2 Objective Analysis of Interaction

An interaction is understood as a process in which the interaction partners have a certain proximity in space and time and in which they coordinate their traffic behaviour. A critical interaction is a process whose (measurable) criticality increases as it progresses until the moment of maximum criticality (in the case of a crash the collision, in the case of a near miss the moment of minimum spatio-temporal proximity) and ends with criticality decreasing again (after evasive action by at least one of the interaction partners [45]). Some metrics from TCT (see ▶ Sect. 29.2) have proven suitable for this purpose, such as TTC, gap time, delta-V, which can be derived from trajectory data. The spatio-temporal relationship between interaction partners is established, located in a world coordinate system. The granularity and relevant error rates depend on the chosen measurement set-up, for example, camera location, resolution and viewing angle, distance of camera to scene [82]. This spatio-temporal localisation of the interaction partners also makes it possible to determine the criticalities of the interaction situations under consideration. The situations considered can be verified using the original video material.

Critical situations are rare events that cannot be fully captured by classical on-site observation studies, neither in their frequency nor in their development. Since the early 2000s, increasingly powerful digital video capture and computing technology has made it possible to conduct semi-automated surveys and interaction analyses of traffic situations (e.g. [59, 83]), long-term surveys of traffic spaces (e.g. [57, 78]), and infrequently occurring situations. The disadvantage of such extensive video material is that manual identification of situations relevant to the investigation is very costly and time-consuming [84]. A reduction of the material to relevant situations is required. The application of simple metrics such as TTC or PET is not sufficient for the selection of relevant situations (e.g. critical interactions between motor-vehicle drivers and cyclists). In the same way, simply exceeding or falling below defined thresholds of these types of metrics is not sufficient for classifying traffic situations as critical and non-crit-

ical. A combination of multiple triggers decreases the false-positive rate. In addition, a selection of recorded traffic situations based on such metrics is required for understanding traffic situations.

In order to understand and quantify complex interactions between road users, certain technical requirements must be met. There is thus a need for powerful object recognition and classification, based, for example, on specialised neural networks (e.g. [85–87], trajectory generation methods that enable the application of advanced object-tracking techniques (e.g. use of state estimators such as Kalman or particle filters [88, 89, 90, 91, 92]). If these are present, procedures for trajectory prediction and collision warning can be developed and applied (e.g. [35, 93, 94, 95]), and the interaction behaviour of road users in special situations or areas, such as in shared space (e.g. [25, 61]), when turning right [7, 69] or turning left [96, 97], can be studied. It is not only the TCT metrics processed for the interaction process which play a role here, but also kinematic and driving-dynamics parameters and situational factors, and those which can be derived from road-user trajectories and which are relevant to the traffic process. Research has shown that situational factors play a significant role in the development of critical situations. Dotzauer et al. found that in the use case of motor-vehicle drivers turning right and cyclists going straight ahead, their position relative to each other was a decisive factor for the development of conflicts [69]. When approaching the conflict point together, the conflict probability increased when cyclists were situated relatively behind motor-vehicle drivers. At the same time, the studies showed that conflicts were the exception and the majority of encounters between motor-vehicle drivers and cyclists were conflict-free. In order to better understand conflicts, the basic patterns of communication and cooperation must be understood. These findings, as they concern interaction behaviour and patterns, must be identified and quantified. They are essential building blocks for the further development and implementation of automated driving functions, especially in urban areas.

29.4.1.3 Objective Analysis of Communication and Cooperation

In certain traffic areas (e.g. roundabouts, bottlenecks, right-before-left controlled intersections), there is a need to understand the interaction of communication and traffic rules. Road users often pursue the goal of maximising time efficiency and convenience while maintaining a level of traffic safety assumed to be acceptable. For instance, road users might make a trade-off between compliance with traffic rules and personal efficiency, for example, crossing an empty street on red [98]. Studies at a roundabout showed that in more than 60% of cases, motor-vehicle drivers do not yield the right of way to cyclists when they leave the roundabout, although this right of way is governed by traffic rules [99]. On the other hand, it was observed that road users tend to balance compliance with traffic rules against safety as experienced. For example, motor-vehicle drivers may yield the right of way to pedestrians at locations not designated for crossing in order to ensure traffic safety [7]. Studies at zebra crossings showed that motor-vehicle drivers gave right of way to cyclists who used the zebra crossings to cross, even though this is not required by law [100, 101].

Road-user movement information plays an important role in communication. The positions and kinematic information of the vehicle and its traffic environment are recorded based on surveys [102], subjective observations (e.g. [103]), video capture (e.g. [104]) or traffic environment data captured using sensors mounted in the vehicle (e.g. [105]) or based on virtual reality [106] studies. This permits the description of different types of communication and cooperation behaviour between road users. The results of the survey study [102] show that pedestrians are more likely to use vehicle-based motion information to judge specific motion behaviour when crossing. For example, speed is often used for this purpose. If speed decreases, for example, the motor vehicle decelerates, then "giving way" is more likely to be understood [107, 108]. If, on the other hand, the speed of motor vehicle drivers remains high or increases, it cannot be expected that motor vehicle drivers are giving way to other road users [99, 109, 110]. The results show that intention recognition or behaviour prediction is not only based on traffic rules. Communication between road users plays an essential role in understanding intentions. Communicating and understanding intentions require explicit communication (e.g. hand signals) as well as implicit communication (e.g. position in lane, change in speed).

29.4.2 Prediction of Road-User Trajectories

Predicting road-user trajectories, meaning the future state of motion of all road users, is a prerequisite for determining collision risks in real-time systems (e.g. by means of collision probabilities and possible crash severities, see [35]. The TCT (see ▶ Sect. 29.2.2) forms the methodological basis here in the sense of identifying points, areas and times of conflict, although human factors and factors favouring and inhibiting collision risks are usually not taken into account or only to a limited extent, as for example in [111], where correlative relationships between cyclists and pedestrians were investigated in connection with winter conditions.

29.4.2.1 Observable and Unobservable Behaviour

The future distance travelled by road users depends on various (conditionally) observable as well as unobservable physical and environmental, traffic and human factors:

Human Factors Such as Intention and Cognition of Road Users (Not Observable): Traffic as a realisation of mobility is the consequence of the individual goals and intentions of all road users [12]. A distinction is made between strategic intention (e.g. moving from A to B), tactical intention (e.g. turning) and operational intention (e.g. turning the steering wheel, braking and giving way to oncoming cyclists). Road users perceive road users around them, the traffic infrastructure, static and moving objects more or less well and estimate their motion states in the traffic environment in order to adapt their own actions accordingly: road users anticipate. However, parts of the environment may be perceived poorly or not at all, so that the prediction is always subject to error and may influence individual traffic behaviour (e.g. safety distance to other road users). Factors that play a role include reaction times [112], age, state of health and other physical factors, as well as the current mental and emotional state of road users which make a risk-favouring or risk-inhibiting contribution to cognitive capacity and thus to traffic behaviour.

Physical and Environmental Factors (Observable): Adverse weather, visibility and light conditions affect the physical behaviour (e.g. braking ability) of road users [10]. The mass and inertia of road users in combination with speed at which they are travelling result in the kinetic energy that is released in the event of a collision with other road users and, in the worst case, can lead to fatal injuries (e.g. [113]). Traffic infrastructure (carriageways, cycle paths and footpaths, constructions and planting, etc.), in conjunction with the rules and prohibitions of the German Road Traffic Regulations (StVO), determines the routes and traffic regulations for road users and the areas to be avoided. While motor vehicles are rarely driven on footpaths and cannot make sudden, sharp changes in direction due to the laws of physics, the situation is different for cyclists and pedestrians. The prediction of their trajectories is therefore to be seen as particularly challenging.

Traffic Factors (Observable): Traffic volume and speed travelled have an impact on risk behaviour (e.g. [114]). Naturally, the type of road user also plays a role due to its mass, the protective effect (protected vs unprotected road users) and the degree of freedom; for example, motor vehicles tend to use the roadway and have limited agility, while pedestrians can be anywhere and can change their direction of movement instantaneously. Traffic rules and prohibitions, but in particular signalling in a traffic area, for example, at intersections with traffic lights, have on the one hand by definition a collision-risk-inhibiting influence on road users (e.g. [10]), whilst on the other hand, for example, conditionally compatible traffic-light switching and long signalling phase durations favour risk behaviour and thus collision risk (e.g. [114]). Road users usually follow the rules, so this knowledge can be used in trajectory prediction. In addition, static or other moving objects in the roadway, for example, can affect trajectory classification and prediction.

29.4.2.2 Trajectory Prediction Method

Currently, there is a variety of different methods for predicting trajectories and identifying points of spatio-temporal intersection while accounting in part for the above factors, namely, state estimators and machine-learning methods, and as a subset of these also what are known as prototypical trajectories. All of the above methods allow the prediction of trajectories and conflicts/collisions and thus form the basis for reliable risk prediction. Kalman, particle and other related linear and nonlinear filters, known as state estimators, are commonly used methods for predicting trajectories. Filters such as these usually have what is known as a predictor–corrector structure, in which one or more variables of interest (e.g. position, direction of a road user) are predicted one or more time steps into the future and model-derivable variables (e.g. speed, acceleration) are estimated simultaneously [89]. Deciding whether two road users are on a collision course depends on the extrapolation hypothesis, the underlying behavioural pattern-detection procedures, and the estimation procedures. Commonly used learning-based models include hidden Markov models [115], Gaussian mixed models [116] and dynamic Bayesian networks [117]. In recent years, more and more deep-learning models such as recurrent neural networks have been proposed, which are able to capture the underlying interactive behavioural patterns [116–119]. However, most of these approaches lack the consideration of physical constraints and boundary conditions, which can only be learned implicitly and not assumed as, for example, in [116]. These deep-learning methods provide good results in terms of average and final position error with a reliable prediction horizon of about three to five seconds.

One alternative approach which also makes use of machine-learning methods, amongst others, is the formation of so-called prototypical trajectories which can be used to model the typical behaviour patterns of road users (e.g. [80]). In doing so, this approach exploits the learned motion patterns of road users, since they usually move with corresponding intention in the traffic areas imposed by the infrastructure and thus follow typical routes (e.g. [78, 120]). If, for example, at an intersection only straight-ahead and left-turning traffic occur in separate lanes, each of these lanes can be assigned to exactly one prototypical trajectory resulting from the normal behaviour of a large number of road users. If, for instance, a left-turning vehicle and an oncoming straight-ahead vehicle are on a collision course, the position and speed values and the underlying prototypical trajectories can be used to check whether and when a collision is foreseeable. A similarity measure is required in order to map the observed trajectory to a prototype and thus determine a future trajectory. This approach enables collision predictions of more than five seconds to be made [80]. Spatio-temporal clustering and concatenated Gaussian distribution methods have been used to represent identified motion patterns of such prototypical trajectories (e.g. [121]). Saunier et al. [81] used unmodified road-user trajectories as prototypes. Wiest et al. [122], on the other hand, applied a probabilistic Gaussian-mixture-model approach in order to learn a motion model based on previously observed trajectories. This permitted the estimation of complete probability density functions that quantified the potential areas occupied by road users.

29.4.3 Derivation of Risk Potentials

In order to warn of an impending collision, its risk must be estimated first. According to [123], the collision risk R is defined as the product of the collision probability P and the expected severity S (also expected severity of damage), i.e. $R = P \cdot S$. Traffic conflicts are traffic situations that occur immediately before a collision, but do not necessarily lead to a collision if at least one interaction partner takes evasive action. The spatio-temporal proximity (determines the collision probability P to a significant extent) of the interacting road users is not the only factor of relevance for determining the risk R [123, 124]; other risk-favouring or -inhibiting factors such as weather-related influences are also relevant (see ▶ Sect. 29.2.1). Conflicts are usually calculated using methods and metrics from the traffic-conflict technique (see ▶ Sect. 29.2.2). SMoS and situation-dependent thresholds are used for this purpose, for example.

An SMoS metric frequently used in practice to determine collision probability P is the time to collision (TTC). It is defined as the time remaining until a collision if the road users maintain their current speed and directions of motion. For example, TTC is used as a measure for detecting rear-end conflicts and collisions and describes the collision (TTC = 0 s), but also severe conflicts ($1 \leq$ TTC ≤ 1.5 s and less severe conflict situations (e.g. $1.5 < \text{TTC} \leq 3.0$ s; see [46, 84, 125]). This is a continuous process variable as long as there is a relative speed at which the following vehicle approaches. In the case of intersecting traffic flows, there are limits to the applicability of TTC in that sometimes there is no collision course and the TTC cannot be calculated, even though two road users miss each other by a very narrow margin.

For the conflict assessment of intersecting traffic flows (e.g. right-turning motor vehicles with crossing-cyclists), for example, the SMoS metric post encroachment time (PET) is used. The PET is defined as the time interval by which two road users sharing a common conflict area miss each other. A collision exists at $\text{PET} = 0$ s and a critical situation exists at $\text{PET} < 1.5$ s [126], although here, varying thresholds are assumed. PET can be determined once the interaction situation has occurred. However, PET cannot be used in isolation to determine the conflict severity of an encounter. For example, if cyclists have to brake abruptly and come to a full stop to avoid a collision with right-turning motorists, it takes longer for cyclists to cross the common conflict point. In a case of this kind, it is very likely that the PET is much higher than 1.5 s. An apparently low PET can also mean that the behaviour of the two interaction partners is so well coordinated that they pass each other very efficiently and with a small time gap. Taking such external factors into account, PET lends itself to drawing conclusions about the nature of the encounter. However, since it is not a continuous process variable, it cannot be used to predict potential conflicts. Using the SMoS metric time advantage (T_{Adv}), alternatively known as gap time (GT) or predicted PET (pPET), the PET is regarded as a continuous process variable. It is assumed here that the road users continue to move at their current speed. A determination is then made as to how the two interaction partners sharing the same conflict area will miss each other. Consequently, low GT values are classified as critical. However, GT depends on the accuracy of the interaction partners' location and velocity as detected and that this detection is more reliable the closer both interaction partners are to the common conflict area. Thus, with a low GT value when both interaction partners are far from the conflict area and have low velocities, the value of the information is limited, whereas the information becomes more reliable when the interaction partners have high velocities and are close to the conflict point. There are also other metrics that are used in the design of assistance systems, such as emergency braking assistants, and that have a predictive collision-avoidance character. Examples include the time to last possible manoeuvre (TLM) or time to last-second braking (TLSB) [127]. These consider how long the vehicle can continue to travel at constant speed un-

til only braking with maximum deceleration will prevent a collision. If there is no response from the drivers, an ADAS can intervene to avoid a crash. These parameters can be relevant if the intervention strategy of the assistance function goes beyond informing or warning and the vehicle is meant to intervene autonomously to prevent a collision.

By definition, the transition from encounter to conflict lies in evasive action and the exceed of a defined threshold. There is still no consensus in the scientific community on which indicators should be preferred and which thresholds are appropriate [124]. However, it is agreed that a single SMoS metric and a single threshold are usually not sufficient to distinguish between critical, non-critical and controlled situations and that the indicators used (i.e. event triggers of single SMoS with predefined thresholds) are not suitable for measuring human-perceived risk [126]. This goes hand in hand with the need to verify traffic situations identified as critical or non-critical, for example, by expert evaluation of the original video footage. Furthermore, when surveying critical and non-critical traffic situations and distinguishing between them, the chosen survey set-up has an influence on the measurement results, for example, camera resolution, distance to scene, camera tilt angle [82, 124]. Thus, it has been shown that the majority of traffic situations determined as critical were in fact non-critical. In [82], the ratio of situations incorrectly determined as critical (based on a threshold) to actual critical situations ranged between 4:1 and 5:1. The cause of this was the propagation of measurement errors in the chosen survey set-up. Accordingly, a manual evaluation must always be used in the determination of conflicts, and must give similar results for a majority of users. However, study results show that users and experts evaluate situations differently. Goos [128] found that the assessments of situations based on video material differed between users and experts. A report by the UDV [129] shows that the immediate evaluation of situations involving cyclists and the post-hoc evaluation of the videos by experts did not show a high degree of agreement. Cyclists used elements of explicit (e.g. eye contact) as well as implicit (assurance of space and safety distances) communication [130] to assess the situation, which were not necessarily evident in the videos. Annotating videos is a first step towards laying a foundation for risk prediction. However, the results also show that real-world experiments are required for fine calibration (see [131]).

Post-hoc analysis of verified critical or non-critical situations enables the identification of thresholds as well as combinations of SMoS and, as a result, the detection of further relevant situations. This leads to an opportunity for collision/conflict prediction. In [132], TTC was used in combination with a classifier and vehicle movements detected based on a nonlinear state estimator to determine collision probabilities P between the interacting road users. In [81], P for interacting road users was calculated using a weighted TTC. This weighting resulted from matching the interacting trajectories with prototypical trajectories (see ▶ Sect. 29.3.2), which were learned from the spatial distribution of typical trajectories. In [133], a deep neural network (DNN) model was used to predict traffic conflicts in real time, using traffic process-related characteristics and SMoS as input parameters. The results of this study, with conflict classification accuracies of 94%, showed that risk prediction depends not only on TTC, but also on the speeds of the interacting partners, weather conditions and traffic flow. Furthermore, the results indicate that this type of method with a high classification quality can be used reliably in advanced driving assistant systems (ADAS) as a proactive safety management strategy. The algorithm was used as a means of evaluating interaction situations and not as a risk-prediction method.

Although the use of machine learning, deep learning and neural networks in the field of conflict research is increasing, few scholarly works have been identified to date. Some evidence of predictive characteristics and patterns of interaction that lead to conflict can be found in [134]. Collisions between cyclists and right-turning trucks were investigated. It was shown that in 2/3 of all cases, truck drivers and cyclists did not change their speed in the event of a collision. In most of these cases, the speed of the trucks was below 30 km/h and that of the cyclists below 20 km/h. Obstructed or poor visibility due to unfavourable weather conditions was not identified as an influencing factor.

For the timely warning of collisions, a technically mature and reliable procedure requires more than individual SMoS and threshold values. It requires the combination of metrics related to the traffic process with appropriate SMoS (derived from human behaviour) and primarily takes into account technically induced latencies in the detection systems deployed, individual human-reaction times and road-user safety distances in interaction with controlled or uncontrolled behaviour, as well as existing risk factors.

29.5 Use in the Area of Driver Assistance and Automation

The methods discussed in the previous sections for the technical observation, modelling and evaluation of traffic behaviour enable the development and use of measures aimed at increasing traffic safety and efficiency. The key question here is how knowledge and understanding of individual road user behaviour users and their interaction can be utilised, what requirements result from this and what problems need to be overcome.

29.5.1 Concepts to Increase Traffic Safety and Efficiency

On the one hand, this information can be used, for example, in the design and conception of traffic-engineering measures. The scope of action includes the design of the road space, traffic routing and the form and design of the signalling, which must always form a unit [135] in the way in which they interact. In addition, there is great potential for impact in the approach of using this information in the area of connected innovative system approaches as part of driver assistance or automation concepts.

One fundamental issue in the development of assistance or automation functions is always the design of the functionality, which in turn is based on model assumptions of a desired driving strategy or the understanding of an optimal form of manoeuvre design. The objective should be, by means of fundamentally driver-compatible design, to enable the most appropriate support system which reduces as far as possible the problems encountered in the interaction between humans and the system [136]. An important component here is the use of derived knowledge in the appropriate design of the technical systems. This includes both derived knowledge concerning necessary support for drivers in specific situations and reference points for the machine-based evaluation and prediction of traffic scenarios (see ▶ Sect. 29.3.2). The knowledge can be used at all levels of an automation or intervention strategy, for example, for the optimal dissemination of information to road users, as in the presentation of risk-related warnings, although strategies must be developed that include road users both with and without specific instrumentation or equipment.

In [131], SMoS metrics and traffic-related parameters (gap time, time-to-arrival, speed and distances of the interaction partners to the conflict point) were combined and evaluated by experts, for example, for situation-dependent risk prediction in the use case of a right-turning car vs a cyclist going straight ahead. A decision tree (see ◘ Fig. 29.3) was trained as a trigger for the outgoing risk warning, which was helpful for the interaction partners in about 2/3 of the situations studied. Different channels were combined to convey the warning to the road users involved: this included acoustic warning stimuli for instrumented motor vehicles as well as haptic feedback for cyclists.[2] A similar solution was developed and tested in the RealLabHH

---

2   The warning was conveyed by a vibration in the handlebars.

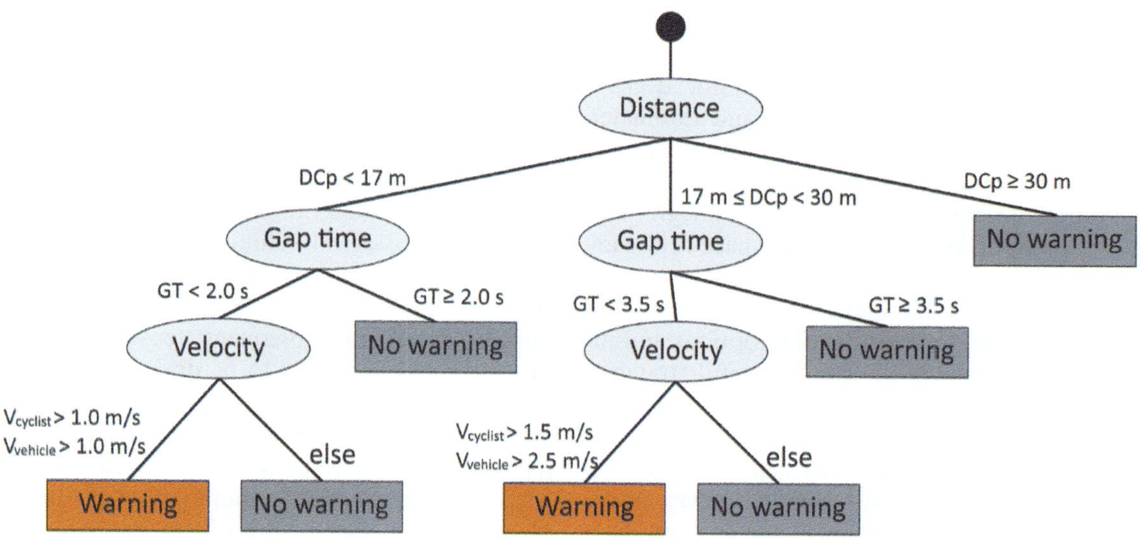

Fig. 29.3 Decision tree for triggering a risk warning as a function of distances to the conflict point (DCp), speeds travelled (velocity) and gap time (GT) (from [131])

project. The digital guardian angel (collision warning) warns right-turning drivers and vulnerable road users crossing the road of a possible collision. To achieve this, road-user motion data from within a defined space enriched with information from high-precision data, pedestrian and cyclist manoeuvre recognition, and traffic-light data, are processed in near-real time by means of machine learning and a residence probability of the objects is determined. The information is transmitted to the devices of the respective road users [137]. In addition, other work has successfully used flashing warning lights to include non-instrumented motorised road users in the warning strategy [138]. In a pilot test, the flashing warning light was installed at the AIM Research Intersection. The objective was to inform right-turning motorised traffic when they are about to encounter cyclists in the intersection area. In this event, the warning light lit up. When a conflict was predicted, the warning light flashed according to the severity of the conflict. If there was no encounter between motorists and cyclists, the warning lights remained off. In contrast to existing systems such as Bike-Flash [67], measures such as these have made it possible to avoid the habituation effects caused by static flashing in the presence of cyclists without any potential for conflict.

29.5.2 Architectural Concepts for Automated and Connected Driving in Complex Urban Spaces

Technical developments in digital solutions are advancing rapidly; they are driving things forward and paving the way for system and architecture approaches that go far beyond a single vehicle-autonomous design form.

The vision here lies in a cooperatively designed transportation system, such as the one described in [139]. The connection between relevant system structures is an essential core element of this kind of C-ITS system [140]. Current communication standards come into play here, such as V2X communication in accordance with the 802.11p standard, or approaches using mobile communications such as 5G. Communication can be purely vehicle-based in order to exchange information on the system's own status, detected traffic situations or other relevant information.

Although these system approaches allow fundamentally new fields of thought and functions, they do not fully exploit the possibilities of distributed information collection and processing. For example, vehicle-based perception approaches for the traffic environment reach their limits in case of more complex situations, such as those often observed at urban intersections. The reasons for this are obscuring effects and influencing factors that are difficult to detect and that render the recording of the relevant vehicle environment inadequately. Examples of this are, in addition to environmental influences, in particular road users or obstacles obscuring the sensor areas. This does not usually put a function itself into question, but these effects significantly reduce the possibilities for function designs with pronounced dynamics or speed ranges comparable to typical human driving behaviour. As a result, differences in road-user behaviour occur, which affect the homogeneity of the traffic flow. The lack of homogeneity in turn leads to poorer traffic performance for the system solution and possibly even to a detrimental effect on safety. Most approaches therefore pursue the incorporation of intelligent infrastructure, particularly in urban environments, as evidenced by the diverse and extensive body of work in the related research (e.g. [141–143]).

In addition to these technical issues, the decisive factor for the applicability of connected and automated system concepts in real-world operation will be whether and how the inclusion of a variety of future mixed-traffic situation conditions is successful in an integral system design. In urban traffic, the traffic space is already shared by road users who differ greatly from one another in terms of dynamics and motion characteristics as well as in terms of their automation options and instrumentation. One possibility for successful strategies to introduce for digitally designed and connected traffic concepts would be the strict separation of traffic space by creating separate reserved areas with restricted access for non-instrumented road users. However, under current conditions, and in particular due to the limited traffic spaces and their capacities, this solution only appears feasible if the entire transport process is subjected to disruptive changes. Assuming mixed-traffic concepts, the key question is: What does a system design look like that allows road users to be included as independently as possible from their specific configuration and instrumentation? The use of intelligent infrastructure in the context of distributed environment perception may provide an answer. The objective of these approaches is to completely capture all relevant traffic movements. If solutions such as these make it possible to reliably detect non-equipped road users even in complex traffic situations, this will open up a new chance of possible active and functional principles. Examples of this are provided by work so far in the area of R&D, as mentioned in this section, as well as the first product-oriented approaches, as shown, for example, in [144]. What they all have in common is that behavioural models are generated and used which enable and exploit the possibilities of situation evaluation, risk assessment and forecasting. While ongoing development in the respective fields is constantly pushing the limits of what is technically feasible, there is still a lack of standardised principles for scalable and operationally robust concept designs that cover the given requirements.

29.6 Conclusion and Outlook

Traffic is becoming increasingly complex, especially in urban areas, and is characterised by a variety of interaction and cooperation processes, meaning that an isolated view of individual road users falls short. Instead, all relevant entities and influencing variables must be integrated into an analysis process in order to provide the most comprehensive overall understanding of the respective traffic events. It can be stated that the combi-

nation and interweaving of different methods from the context of traffic psychology and engineering in particular opens up new possibilities which, in their respective individual disciplines, remain closed.

The basis for this is the technical observation of road-user behaviour by means of the development of trajectories. A wide range of requirements for the underlying metrological solutions must be considered to ensure that the data provides an adequate basis for the subsequent processing steps. This is particularly important, because these subsequent steps usually do not allow further conclusions to be drawn about metrological artefacts or error effects. The applicability of available metrics is determined by the specific situation and its constraints. The complexity of the situations always requires a critical examination of the question of whether the selected model depth is sufficient or requires additional steps, for example, the further development of several metrics or their clever combination.

The use of the resulting understanding can serve as valuable background knowledge, for example, in the appropriate design of road space and vehicular system designs. However, there is decisive added value, in particular in the possibility of a functional coupling of situation observation and analysis as they relate to infrastructure with other information sources to create innovative connected system solutions. It may be observed here that the newly created field brings together disciplines that previously had only limited points of contact. By merging transportation engineering domains (such as traffic safety) with vehicle engineering (e.g. vehicle-based perception), the classic separation between vehicle and infrastructure is eliminated and transferred to an overall level. This allows for comprehensive architectures and system approaches that hold great potential for addressing current and future issues in the area of assistance and automation development.

References

1. LaMondia, J.J., Duthie, J.C.: Analysis of factors influencing bicycle–vehicle interactions on urban road-ways by ordered probit regression. Transp. Res. Rec. **2314**, 81–88 (2012)
2. Twaddle, H., Schendzielorz, T., Fakler, O., Amini, S.: Use of automated video analysis for the evaluation of bicycle movement and interaction. Video Surveillance and Transportation Imaging Applications:90260U (2014)
3. Wu, C., Yao, L., Zhang, K.: The red-light running behavior of electric bike riders and cyclists at urban intersections in China: an observational study. Accid. Anal. Prev. **49**, 186–192 (2012)
4. Deutsches Zentrum für Luft- und Raumfahrt e.V., Institute of Transportation Systems.: AIM research intersection: instrument for traffic detection and behavior assessment for a complex urban intersection. JLSRF 2 (2016). ▶ https://doi.org/10.17815/jlsrf-2-122
5. Gimm, K., Knake-Langhorst, S., Dotzauer, M., Urban, U., Arndt, R.: Increasing cyclist safety with infrastructural sup-

ported cooperative ADAS in EU XCYCLE by extending test site AIM research intersection-concept & status. In: University of Bologna (Hrsg) ICSC 2016. Book of Abstracts, Bologna (2016)

6. Dotzauer, M., Preuk, K., Patz, D., Schießl, C.: Das autonome Fahrzeug oder der Mensch: Wer ist besser und leistungsfähiger? 34. VDI/VW-Gemeinschaftagung Fahrerassistenzsysteme und automatisiertes Fahren. VDI Wissensforum GmbH, S 299–314 (2018)

7. Zhang, M., Dotzauer, M., Schießl, C.: Analysis of implicit communication of motorists and cyclists in intersection using video and trajecto-ry data. Berlin (online) (2021)

8. Coers, H.G.: Über die Methodik der wissenschaftlichen Erforschung des Straßenverkehrsflusses; Beitrag zur Theorie des Verkehrsablaufs, Dresden (1969)

9. Kühne, R.D.: Hinweise zum Fundamentaldiagramm; Grundlagen und Anwendungen. Forschungsgesellschaft für Straßen- und Verkehrswesen (2005)

10. Schnabel, W., Lohse, D.: Grundlagen der Straßenverkehrstechnik und der Verkehrsplanung. Forschungsgesellschaft für Straßen- und Verkehrswesen e.V., Köln (1997)

11. Junghans, M.: Ein Beitrag zur Qualifizierung von Verkehrsdaten mit Bayesschen Netzen. TUDpress Verl. der Wiss, Dresden (2009)

12. Rose, M.: Modellbildung simulation von Autobahnverkehr. Dissertation, Hannover (2003)

13. Leich, A.: Ein Beitrag zur Realisierung der videobasierten weiträumigen Verkehrsbeobachtung. TUDpress Verl. der Wiss, Dresden (2006)

14. Leutzbach, W.: Introduction to the Theory of Traffic Flow. Springer, Berlin Heidelberg, Berlin, Heidelberg (1988)

15. Treiber, M., Kesting, A.: Verkehrsdynamik und -simulation; Daten, Modelle und Anwendungen der Verkehrsflussdynamik. Springer, Berlin (2010)

16. Beyerer, J., Puente León, F., Sommer, K.D. (eds.): Informationsfusion in der Mess- und Sensortechnik. Univ.-Verl, Karlsruhe (2006)

17. Forschungsgesellschaft für Straßen- und Verkehrswesen.: Hinweise zur Datenvervollständigung und Datenaufbereitung in verkehrstechnischen Anwendungen (2003)

18. Leich, A., Fliess, T., Jentschel, H.-J.: Bildverarbeitung im Straßenverkehr - Überblick über den Stand der Technik. Technische Universität Dresden, Zwischenbericht (2001)

19. Skutek, M., Linzmeier, D.: Fusion von Sensordaten am Beispiel von Sicherheitsanwendungen in der Automobiltechnik. Automatisierungstechnik **53**, 295–305 (2005)

20. Rankov, V., Locke, R.J., Edens, R.J., Barber, P.R., Vojnovic, B.: An Algorithm for image stitching and blending. In: Conchello, J.-A., Cogswell, C.J., Wilson, T. (Hrsg) Three-Dimensional and Multidimensional Microscopy: Image Acquisition and Processing XII. SPIE, S 190 (2005)

21. Turner, S.M.: Defining and Measuring Traffic Data Quality: White Paper (2002)

22. Eenink, et al.: UDRIVE: the European naturalistic driving study. In: Proceedings of transport research Arena. TRA, 14–17 Apr 2014, Paris, France. IFSTTAR (2014)

23. Köster, F., Mazzega, J., Knake-Langhorst, S.: Automatisierte und vernetzte Systeme–erprobt und evaluiert. ATZ extra, pp 20–23 (2018)

24. Grippenkoven, J., Gimm, K., Stamer, M., Naumann, A., Schnieder, L.: Untersuchung des Fehlverhaltens von Verkehrsteilnehmern an einem Bahnübergang mit Blinklichtsicherung. SIGNAL + DRAHT. DVV Media Group. ISSN 0037–4997 (2015)

25. Jakobowskya, C., Siebert, F., Junghans, M., Schießl, C., Dotzauer, M.: Why so serious?-comparing two traffic conflict techniques for assessing encounters in shared space. ToTS (2021). ► https://doi.org/10.5507/tots.2021.009

26. Providentia Proaktive Videobasierte Nutzung von Telekommunikationstechnologien in innovativen Autoverkehr-Szenarien. ► https://innovation-mobility.com/projekt-providentia. Zugegriffen: 06. Oktober 2021

27. Doll. K.: Forschungskreuzung. Labor für kooperative automatisierte Verkehrssysteme. ► https://www.th-ab.de/hochschule/organisation/organisationseinheiten/labor-fuer-kooperative-automatisierte-verkehrssysteme#forschungskreuzung-2011, Zugegriffen:02.07.2023

28. SAVeNoW Konsortium.: Digital Twin For Safe & Sustainable Mobility. ► https://savenow.de/de/, Zugegriffen:02.07.2023

29. Bahnsen, C.H., Madsen, T.K.O., Jensen, M.B.: The ruba watchdog video analysis tool. aalborg university denmark (2018). ► https://vbn.aau.dk/ws/portalfiles/portal/309152018/InDeV_deliverable_4.3.pdf., Zugegriffen:02.07.2023

30. Tarko, A.: Measuring road safety using surrogate events. Elsevier, Amsterdam (2020)

31. Krajewski, R., Bock, J,, Kloeker, L., et al.: HighD dataset - The highway drone dataset. institut für kraftfahrzeuge, RWTH Aachen. ► https://www.highd-dataset.com, Zugegriffen:02.07.2023

32. RCE systems s.r.o.: Data from Video Analysis - DataFromSky.► https://datafromsky.com., Zugegriffen:02.07.2023

33. Sun, X., Zhu, S., Jin, D., Liang, Z., Xu, G.: Tracklet association for object tracking 2016 Chinese Control and Decision Conference (CCDC). IEEE, S 107–112 (2016)

34. Chiellino, U., Winkle, T., Graab, B., Ernstberger, A., Donner, E., Nerlich, M.: Fahrerassistenzsysteme Was können Fahrerassistenzsysteme im Unfallgeschehen leisten?, Zeitschrift für Verkehrssicherheit : ZVS, Zeitschrift für Verkehrssicherheit : ZVS : Organ der DGVM Heidelberg : Organ der DGVP Berlin. Zeitschrift für Verkehrssicherheit : ZVS, Zeitschrift für Verkehrssicherheit : ZVS : Organ der DGVM Heidelberg : Organ der DGVP Berlin **56**, 131–138 (2010)

35. Berghaus, M., Ehlers, J., Hoffmann, R., Kalló, E., Leich, A., Saul, H., Wagner, P.: Ansätze zur datengetriebenen Verkehrssicherheit als Ergänzung zu Unfalldaten. Straßenverkehrstechnik 2021 (2021)

36. Sturm, P.: Verkehrssicherheit an plangleichen und teilweise planfreien Knotenpunkten von Ausserortsstrassen. Dissertation, Darmstadt (1989)

37. Jacobsen, P.L.: (21) Safety in numbers: more walkers and bicyclists, safer walking and bicycling. Injury Prevent. 271–275 (2015)

38. Robinson, D.L.: Safety in numbers in Australia: more walkers and bicyclists, safer walking and bicycling. Health Promot. J. Austr. **16**, 47–51 (2005)

39. Karndacharuk, A., Wilson, D.J., Dunn, R.C.M.: Analysis of pedestrian performance in shared-space environments. Transp. Res. Rec. **2393**, 1–11 (2013). ► https://doi.org/10.3141/2393-01

40. Lücken, L.: On the variation of the crash risk with the total number of bicyclists. Eur. Transp. Res. Rev. **10**(2018).► https://doi.org/10.1186/s12544-018-0305-9

41. Volvo Car Germany GmbH.: Fußgänger-Airbag. ► https://www.volvocars.com/de/support/manuals/v40/2018w17/sicherheit/airbags/fussganger-airbag. Zugegriffen: 30. Januar 2022 (2020)

42. Rasmussen, J.: Human errors; a taxonomy for describing human malfunction in industrial installations. J. Occup. Accid. **4**, 311–333 (1982)

43. Graab, B., Donner, E., Chiellino, U., Hoppe, M.: Analyse von Verkehrsunfällen hinsichtlich unterschiedlicher Fahrerpopulationen und daraus ableitbarer Ergebnisse für die Entwicklung adaptiver Fahrerassistenzsysteme. In: TU München & TÜV Süd Akademie GmbH (Hrsg) (2008)

44. Klanner, F.: Entwicklung eines kommunikationsbasierten Querverkehrsassistenten im Fahrzeug. Dissertation, Darmstadt (2008)

45. Erke, H., Gstalter, H., Zimolong, B.: Verkehrskonflikte im Innerortsbereich (1985)

46. Hydén, C.: The development of a method for traffic safety evaluation; The Swedish Traffic Conflict Technique, Lund, Schweden (1987)

47. Laureshyn, A., Várhelyi, A.: The Swedish traffic conflict technique; observer's manual (2018). ▶ http://www.tft.lth.se/fileadmin/tft/images/Update_2018/Swedish_TCT_Manual.pdf

48. Risser, R., Zuzan, W., Tamme, W., Steinbauer, J., Kaba, A.: Handbuch zur Erhebung von Verkehrskonflikten mit Anleitungen zur Beobachterschulung. Literas-Univ.-Verl, Wien (1991)

49. Baguley, C.J.: The British traffic conflict technique NATO advanced research workshop on international calibration study of traffic conflict techniques. Denmark, Copenhagen (1984)

50. Mortelmans, J., Vits, A., Venstermans, L., vanEssche, M., Boogaerts, M., Vergisson, E.: Analyse van de verkeersveiligheid met behulp van de Bijna-Ongevallen methode. Katholieke Universiteit Leuven, Departement Bouwkunde, Analysis of road safety by means of the near-accident method (1986)

51. Kraay, J., Horst A van der, Oppe, S.: Manual conflict observation technique DOCTOR. Foundation Road safety for all (2013)

52. Parker, M.R., Zegeer, C.V.: Traffic conflict techniques for safety and operation; FHWA-IP-88–027 (1989)

53. Sayed, T., Zein, S.: Traffic conflict standards for intersections. Transp. Plan. Technol. 22, 309–323 (1999). ▶ https://doi.org/10.1080/03081069908717634

54. Kulmala, R.: The Finnish traffic conflict technique NATO advanced research workshop on international calibration study of traffic conflict techniques. Denmark, Copenhagen (1984)

55. Muhlrad, N., Dupre, G.: The French conflict technique NATO advanced research workshop on international calibration study of traffic conflict techniques. Denmark, Copenhagen (1984)

56. Kocárková, D.: Traffic conflict techniques in Czech republic SIIV-5th International Congress - Sustainability of Road Infrastructures (2012)

57. Laureshyn, A.: Application of Automated Video Analysis to Road User Behaviour. Lund University (2010)

58. Saunier, N., Sayed, T., Ismail, K.: Large-scale automated analysis of vehicle interactions and collisions. Transp Res Record J Transp Res Board 42–50 (2010)

59. Johnsson, C., Norén, H., Laureshyn, A., Ivina, D.: InDeV project deliverable 6.1; T-Analyst - semi-automated tool for traffic conflict analysis (2018)

60. Asmussen, E.: International Calibration Study of Traffic Conflict Techniques. Springer, Berlin Heidelberg, Berlin, Heidelberg (1984)

61. Kaparias, I., Bell, M., Biagioli, T., Bellezza, L., Mount, B.: Behavioural analysis of interactions between pedestrians and vehicles in street designs with elements of shared space. Transport. Res. F: Traffic Psychol. Behav. 30, 115–127 (2015). ▶ https://doi.org/10.1016/j.trf.2015.02.009

62. Buehler, R., Pucher, J., Altshuler, A.: Vienna's path to sustainable transport. Int. J. Sustain. Transp. 11, 257–271 (2017). ▶ https://doi.org/10.1080/15568318.2016.1251997

63. Prati, G., Marín Puchades, V., Pietrantoni, L.: Cyclists as a minority group? Transport. Res. F: Traffic Psychol. Behav. 47, 34–41 (2017). ▶ https://doi.org/10.1016/j.trf.2017.04.008

64. Prati, G., Marín Puchades, V., de Angelis, M., Fraboni, F., Pietrantoni, L.: Factors contributing to bicycle–motorised vehicle collisions: a systematic literature review. Transp. Rev. 38, 184–208 (2018). ▶ https://doi.org/10.1080/01441647.2017.1314391

65. Pucher, J., Dill, J., Handy, S.: Infrastructure, programs, and policies to increase bicycling: an international review. Prev. Med. 50(Suppl 1), S106-25 (2010). ▶ https://doi.org/10.1016/j.ypmed.2009.07.028

66. Hurwitz, D., Monsere, C.: Towards Effective Design Treatment for Right Turns at Intersections with Bicycle Traffic, Portland (2015)

67. Weiss, T.: Dieses Warnsystem soll Radfahrer schützen; Vor Unfällen mit Lastwagen. Frankfurter Allgemeine Zeitung (2018)

68. Statistisches Bundesamt.: Verkehrsunfälle. ▶ https://www.destatis.de/DE/Themen/Gesellschaft-Umwelt/Verkehrsunfaelle/_inhalt.html. Zugegriffen: 19. Januar 2022 (2021)

69. Dotzauer, M., Junghans, M., Schnücker, G.: Cycling Through Intersections: Patterns Affecting Safety, Olomouc, Tschechien (2017)

70. Zheng, L., Sayed, T., Mannering, F.: Modeling traffic conflicts for use in road safety analysis: a review of analytic methods and future directions. Analytic Methods Accident Res. 29, 100142 (2021). ▶ https://doi.org/10.1016/j.amar.2020.100142

71. Dozza, M.: What is the relation between crashes from crash databases and near crashes from naturalistic data? J. Transp. Safety Secur. 12, 37–51 (2020). ▶ https://doi.org/10.1080/19439962.2019.1591553

72. Lu, G., Cheng, B., Kuzumaki, S., Mei, B.: Relationship between road traffic accidents and conflicts recorded by drive recorders. Traffic Inj. Prev. 12, 320–326 (2011). ▶ https://doi.org/10.1080/15389588.2011.565434

73. Migletz, D.J., Glauz, W.D., Bauer, K.M.: Relationships Between Traffic Conflicts and Accidents. Federal Highway Administration (1985)

74. Tarko, A.: Measuring Road Safety Using Surrogate Events. Elsevier (2020)

75. TU München & TÜV Süd Akademie GmbH (Hrsg) (2008)

76. Zheng, L., Ismail, K., Meng, X.: Shifted gamma-generalized pareto distribution model to map the safety continuum and estimate crashes. J. Safety Sci. 155–162 (2014)

77. Laureshyn, A., Ardö, H., Svensson, Å., Jonsson, T.: Application of automated video analysis for behavioural studies: concept and experience. IET Intell. Transp. Syst. 3, 345 (2009). ▶ https://doi.org/10.1049/iet-its.2008.0077

78. Ismail, K., Sayed, T., Saunier, N., Lim, C.: Automated analysis of pedestrian-vehicle conflicts using video data. Transp. Res. Rec. 2140, 44–54 (2009). ▶ https://doi.org/10.3141/2140-05

79. Twaddle, H.: Development of tactical and operational behaviour models for bicyclists based on automated video data analysis. Dissertation, München (2017)

80. Saul, H., Junghans, M., Dotzauer, M., Gimm, K.: Online risk estimation of critical and non-critical interactions between right-turning motorists and crossing cyclists by a decision tree international cycling safety conference (ICSC2021) (2021b)

81. Saunier, N., Sayed, T., Lim, C.: Probabilistic Collision Prediction for Vision-Based Automated Road Safety Analysis 2007 IEEE Intelligent Transportation Systems Conference. IEEE, S 872–878 (30.09.2007 - 03.10.2007)

82. Leich, A., Kendziorra, A., Saul, H., Hoffmann, R.: Calculation of Error Rates for Detection of Critical Situations in Road Traffic TRB Annual Meeting 2016. Transportation Research Board (2016)

83. Anderson, J.: Image-processing for analysis of road user behaviour: a trajectory-based solution. Dissertation, Lund, Schweden (2002)

84. Archer, J.: Indicators for traffic safety assessment and prediction and their application in micro-simulation modelling: A study of urban and suburban intersections. Dissertation, Stockholm (2005)

85. Bahnsen, C.H., Madsen, T.K.O., Jensen, M.B.: The RUBA Watchdog Video Analysis Tool. Aalborg University Denmark (2018). ▶ https://vbn.aau.dk/ws/portalfiles/portal/309152018/InDeV_deliverable_4.3.pdf

86. Girshick, R.: Fast R-CNN 2015 IEEE International Conference on Computer Vision (ICCV). IEEE, S 1440–1448 (07.12.2015 - 13.12.2015)

87. Szegedy, C., Toshev, A., Erhan, D.: Deep neural networks for object detection. Advances in Neural Information Processing Systems (NIPS) 2013

88. Bar-Shalom, Y., Li, X.-R., Kirubarajan, T.: Estimation with Applications to Tracking and Navigation. John Wiley & Sons Inc., New York, USA (2001)

89. Haykin, S.S. (ed.): Kalman Filtering and Neural Networks. Wiley, New York, NY (2001)

90. Meissner, D., Reuter, S., Wilking, B., Dietmayer, K.: Road user tracking using a dempster-shaferbased classifying multiple-model PHD Filter. In: 16th International IEEE Conference on Intelligent Transportation Systems (ITSC 2013). IEEE (06.10.2013 - 09.10.2013)

91. Ristic, B., Arulampalan, S., Gordon, N.: Beyond the Kalman Filter: Particle Filters for Tracking Applications. Artech House (2004)

92. Zhang, Z., Wu, J., Zhang, X., Zhang, C.: Multi-Target, Multi-Camera Tracking by Hierarchical Clustering: Recent Progress on DukeMTMC Project (2017)

93. Goldhammer, M., Gerhard, M., Zernetsch, S., Doll, K., Brunsmann, U.: Early prediction of a pedestrian's trajectory at intersections. In: 16th International IEEE Conference on Intelligent Transportation Systems (ITSC 2013). IEEE, S 237–242 (06.10.2013 - 09.10.2013)

94. Petrich, D., Dang, T., Kasper, D., Breuel, G., Stiller, C.: Map-based long term motion prediction for vehicles in traffic environments. In: 16th International IEEE Conference on Intelligent Transportation Systems (ITSC 2013). IEEE, S 2166–2172 (06.10.2013 - 09.10.2013)

95. Weidl, G., Breuel, G., Singhal, V.: Collision risk prediction and warning at road intersections using an object oriented Bayesian network. In: Terken J (Hrsg) Proceedings of the 5th International Conference on Automotive User Interfaces and Interactive Vehicular Applications - AutomotiveUI '13. ACM Press, New York, New York, USA, S 270–277 (2013)

96. Junghans, M., Krauns, F., Sonka, A., Böhm, M., Dotzauer, M.: Comparison of safety and kinematic patterns of automated vehicles turning left in interaction with oncoming manually driven vehicles. ToTS (2021). ▶ https://doi.org/10.5507/tots.2021.003

97. Quante, L., Junghans, M., Schießl, C.: Turning left at urban intersections: turning patterns and gap acceptance; accepted presentation (28. und 29. Oktober 2021)

98. Nygårdhs, S., Kircher, K., Johansson, B.J.E.: Trade-offs in traffic: does being mainly a car driver or a cyclist affect adaptive behaviour while driving and cycling? Eur. Transp. Res. Rev. 12 (2020).▶ https://doi.org/10.1186/s12544-020-0396-y

99. Silvano, A.P., Ma, X., Koutsopoulos, H.N.: When do drivers yield to cyclists at unsignalized roundabouts? Transp. Res. Rec. 2520, 25–31 (2015). ▶ https://doi.org/10.3141/2520-04

100. Bjørnskau, T.: The zebra crossing game–using game theory to explain a discrepancy between road user behaviour and traffic rules. Saf. Sci. 92, 298–301 (2017). ▶ https://doi.org/10.1016/j.ssci.2015.10.007

101. Färber, B.: Kommunikationsprobleme zwischen autonomen Fahrzeugen und menschlichen Fahrern. In: Maurer, M., Gerdes, J.C., Lenz, B., Winner, H. (Hrsg) Autonomes Fahren. Springer Berlin Heidelberg, Berlin, Heidelberg, S 127–146 (2015)

102. Lee, Y.M., Madigan, R., Giles, O., Garach-Morcillo, L., Markkula, G., Fox, C., Camara, F., Rothmueller, M., Vendelbo-Larsen, S.A., Rasmussen, P.H., Dietrich, A., Nathanael, D., Portouli, V., Schieben, A., Merat, N.: Road users rarely use explicit communication when interacting in today's traffic: implications for automated vehicles. Cogn. Technol. Work 23, 367–380 (2021). ▶ https://doi.org/10.1007/s10111-020-00635-y

103. Rettenmaier, M., Requena Witzig, C., Bengler, K.: Interaction at the bottleneck–a traffic observation. In: Ahram, T., Karwowski, W., Pickl, S., Taiar, R. (Hrsg) Human Systems Engineering and Design II. Springer International Publishing, Cham, S 243–249 (2020)

104. Risto, M., Emmenegger, C., Vinkhuyzen, E., Cefkin, M., Jim, H.: (Hrsg) Human-vehicle interfaces: the power of vehicle movement gestures in human road user coordination (2017)

105. Schneemann, F., Gohl, I.: Analyzing driver-pedestrian interaction at crosswalks: a contribution to autonomous driving in urban environments. In: 2016 IEEE Intelligent Vehicles Symposium (IV). IEEE, S 38–43 (19.06.2016 - 22.06.2016)

106. Feldstein, I., Dietrich, A., Milinkovic, S., Bengler, K.: A pedestrian simulator for urban crossing scenarios. IFAC-PapersOnLine 49(19), 239–244 (2016)

107. Ackermann, C., Beggiato, M., Bluhm, L.-F., Krems, J.F.: Vehicle movements as implicit communication signal between pedestrians and automated vehicles. 6th Humanist Conference, The Hague (2018)

108. Beggiato, M., Witzlack, C., Krems, J.F.: Gap acceptance and time-to-arrival estimates as basis for informal communication between pedestrians and vehicles. In: Boll, S., Pfleging, B., Donmez, B., Politis, I., Large, D. (Hrsg) Proceedings of the 9th international conference on automotive user interfaces and interactive vehicular applications, S 50–57. ACM, New York, NY, USA (2017)

109. Himanen, V., Kulmala, R.: An application of logit models in analysing the behaviour of pedestrians and car drivers on pedestrian crossings. Accid. Anal. Prev. 20, 187–197 (1988). ▶ https://doi.org/10.1016/0001-4575(88)90003-6

110. Šucha, M.: Road users'strategies and communication: driver-pedestrian interaction. Transport Research Arena 2014 (TRA 2014) (2014)

111. Bärwolff, M., Reinartz, A., Gerike, R.: Correlates of pedestrian and cyclist falls in snowy and Icy conditions. ToTS (2021). ▶ https://doi.org/10.5507/tots.2021.007

112. Green, M.: How long does it take to stop? Methodological analysis of driver perception-brake times. Transp. Human Factors 2, 195–216 (2000). ▶ https://doi.org/10.1207/STHF0203_1

113. Laureshyn, A., de Ceunynck, T., Karlsson, C., Svensson, Å., Daniels, S.: In search of the severity dimension of traffic events: extended delta-V as a traffic conflict indicator. Accid. Anal. Prevent. 98, 46–56 (2017). ▶ https://doi.org/10.1016/j.aap.2016.09.026

114. Chan, C.-Y.: Characterization of driving behaviors based on field observation of intersection left-turn across-path scenarios. IEEE Trans. Intell. Transport. Syst. 7, 322–331 (2006). ▶ https://doi.org/10.1109/TITS.2006.880638

115. Wang, W., Xi, J., Zhao, D.: Learning and inferring a driver's braking action in car-following scenarios. IEEE Trans. Veh. Technol. 67, 3887–3899 (2018). ▶ https://doi.org/10.1109/TVT.2018.2793889

116. Li, J., Ma, H., Zhan, W., Tomizuka, M.: Generic probabilistic interactive situation recognition and prediction: from virtual to real. In: 2018 21st International Conference on Intelligent Transportation Systems (ITSC). IEEE, S 3218–3224 (04.11.2018 - 07.11.2018)

117. Kasper, D., Weidl, G., Dang, T., Breuel, G., Tamke, A., Wedel, A., Rosenstiel, W.: Object-oriented bayesian networks for detection of lane change maneuvers. IEEE Intell. Transport. Syst. Mag. 4, 19–31 (2012). ▶ https://doi.org/10.1109/MITS.2012.2203229

118. Huang, Z., Wang, J., Pi, L., Song, X., Yang, L.: LSTM based trajectory prediction model for cyclist utilizing multiple interactions with environment. Pattern Recogn. 112, 107800 (2021). ▶ https://doi.org/10.1016/j.patcog.2020.107800

119. Park, S.H., Kim, B., Kang, C.M., Chung, C.C., Choi, J.W.: Sequence-to-sequence prediction of vehicle trajectory via LSTM encoder-decoder architecture. In: 2018 IEEE Intelligent Vehicles Symposium (IV). IEEE, S 1672–1678 (26.06.2018 - 30.06.2018)

120. Quehl, J., Hu, H., Wirges, S., Lauer, M.: An approach to vehicle trajectory prediction using automatically generated traffic maps. In: 2018 IEEE Intelligent Vehicles Symposium (IV). IEEE, S 544–549 (26.06.2018 - 30.06.2018)

121. Hu, W., Xiao, X., Fu, Z., Xie, D., Tan, T., Maybank, S.: A system for learning statistical motion patterns. IEEE Trans. Pattern Anal. Mach. Intell. **28**, 1450–1464 (2006). ► https://doi.org/10.1109/TPAMI.2006.176

122. Wiest, J., Hoffken, M., Kresel, U., Dietmayer, K.: Probabilistic trajectory prediction with Gaussian mixture models. In: 2012 IEEE Intelligent Vehicles Symposium. IEEE, S 141–146 (03.06.2012 - 07.06.2012)

123. Laureshyn, A., Svensson, A., Hydén, C.: Evaluation of traffic safety, based on micro-level behavioural data: theoretical framework and first implementation. Accident Anal. Prevent. **42**, 1637–1646 (2010). ► https://doi.org/10.1016/j.aap.2010.03.021

124. Mahmud, S.S., Ferreira, L., Hoque, M.S., Tavassoli, A.: Application of proximal surrogate indicators for safety evaluation: a review of recent developments and research needs. IATSS Res. **41**, 153–163 (2017). ► https://doi.org/10.1016/j.iatssr.2017.02.001

125. Svensson, A., Hydén, C.: Estimating the severity of safety related behaviour. Accid. Anal. Prev. **38**, 379–385 (2006). ► https://doi.org/10.1016/j.aap.2005.10.009

126. Johnsson, C.: Surrogate Measures of Safety with a Focus on Vulnerable Road Users : An exploration of theory, practice, exposure, and validity. Ph.D. thesis, Lund, Sweden (2020)

127. Zhang, Y., Antonsson, E.K., Grote, K.: A new threat assessment measure for collision avoidance systems. In: 2006 IEEE Intelligent Transportation Systems Conference (pp. 968–975). IEEE (2006)

128. Goos, K.: Risiko im Verkehr: Wie wird es wahrgenommen und welche Kritikalitätsphänomene werden berücksichtigt?. Technische Universität Berlin, Masterarbeit (2021)

129. Kolrep-Rometsch, H., Leitner, R., Platho, C., Richter, T., Schreiber, A., Schreiber, M.: Abbiegeunfälle Pkw/Lkw und Fahrrad. GDV, Berlin (2013)

130. Aldred, R., Woodcock, J., Goodman, A.: Does more cycling mean more diversity in cycling? Transp. Rev. **36**(1), 28–44 (2016)

131. Saul, H., Junghans, M., Dotzauer, M., Gimm, K.: Online risk estimation of critical and non-critical interactions between right-turning motorists and crossing cyclists by a decision tree. Accid. Anal. Prev. **163**, 106449 (2021). ► https://doi.org/10.1016/j.aap.2021.106449

132. Ward, J., Agamennoni, G., Worrall, S., Nebot, E.: Vehicle collision probability calculation for general traffic scenarios under uncertainty. In: 2014 IEEE Intelligent Vehicles Symposium Proceedings. IEEE, S 986–992 (08.06.2014 - 11.06.2014)

133. Formosa, N., Quddus, M., Ison, S., Abdel-Aty, M., Yuan, J.: Predicting real-time traffic conflicts using deep learning. Accident Anal. Prevent. **136**, 105429 (2020). ► https://doi.org/10.1016/j.aap.2019.105429

134. Seiniger P, Gail J, Schreck B (2015) Development of a Test Procedure for Driver Assist Systems Addressing Accidents Between Right Turning Trucks and Straight Driving Cyclists

135. Forschungsgesellschaft für Straßen- und Verkehrswesen; Forschungsgesellschaft für Straßen- und Verkehrswesen.: Richtlinien für Lichtsignalanlagen; RiLSA: Lichtzeichenanlagen für den Straßenverkehr (2010)

136. Löper, C.: Manöverbasierte kooperative Automation für teil- und hochautomatisiertes Fahren. Technische Universität Braunschweig (2019)

137. Agnesmeyer, L., Bieker, L., Blunck, F., Christ, P., Döler-Brede, L., Dotzauer, M. et al.: ReallabHH-Wir verändern Mobilität. Erkenntnisse des Reallabors Hamburg für eine digitale Mobilität von morgen. Hg. v. Reallabor Hamburg (RealLabHH). Hamburg. Online verfügbar unter ► https://reallab-hamburg.de/wp-content/uploads/2022/04/Reallabor-Hamburg-Abschlussbericht-5.pdf, zuletzt geprüft am 01.06.2022 (2022)

138. XCYCLE Konsortium (31.05.2018) XCYCLE D5.3; Integration of stationary and quasi-permanent demonstrator - Integration of stationary demonstrator

139. Lemmer, K. (Hrsg): Neue autoMobilität II; Kooperativer Straßenverkehr und intelligente Verkehrssteuerung für die Mobilität der Zukunft (2019)

140. Sjöberg, K., Andres, P., Buburuzan, T., Brakemeier, A.: C-ITS deployment in europe-current status and outlook. ArXiv abs/1609.03876 (2016)

141. Wertheimer, R. (Hrsg): Fahrerassistenz und präventive Sicherheit mittels kooperativer Perzeption; Partnerübergreifender Schlussbericht. http://ko-fas.de/files/19-S-9022_Ko-PER_partneruebergreifender-Schlussbericht.pdf. (16.06.2014)

142. ICT4CART consortium.: ICT infrastructure for connected and automated road transport; a connected future for automated driving. ► https://www.ict4cart.eu/assets/homeSliderImage/ICT4CART-General-presentation.pdf. Zugegriffen: 10. November 2021 (Juni 2019)

143. Kaul, R., Barthauer, M., Bläsche, J., Böhm, M., Dobmeier, S., Ehmen, G.: DK4.0 - Digitaler Knoten 4.0; Gestaltung und Regelung städtischer Knotenpunkte für sicheres und effizientes, automatisiertes Fahren im gemischten Verkehr: Verbundbericht des Gesamtvorhabens: Laufzeit des Vorhabens: 01.12.2016–31.05.2019, Berichtszeitraum: 01.12.2016–31.05.2019, Verbundbericht des Gesamtvorhabens - Digitaler Knoten 4.0, Final report - joined project efforts - Digitaler Knoten 4.0. Institut für Verkehrssystemtechnik (2019). ► https://www.tib.eu/de/suchen/id/TIBKAT%3A1733625313

144. Continental, A.G.: Intelligente Kreuzung. ► https://www.continental.com/de/presse/messen-events/techshow-2019/intelligente-kreuzung. Zugegriffen: 10. November 2021 (2019)

145. International Cycling Safety Conference (ICSC2021) (2021)

146. 09.2007 - 03.10.2007) IEEE Intelligent Transportation Systems Conference. IEEE (2007)

147. Goos, K.: Risiko im Verkehr: Wie wird es wahrgenommen und welche Kritikalitätsphänomene werden berücksichtigt? (2021)

148. Leich, A., Kendziorra, A., Saul, H., Hoffmann, R.: Calculation of Error Rates for Detection of Critical Situations in Road Traffic. Washington D.C, USA (2016)

149. Li, J., Ma, H., Zhang, Z., Tomizuka, M.: Social-WaGDAT: Interaction-aware Trajectory Prediction via Wasserstein Graph Double-Attention Network (2020). ► http://arxiv.org/pdf/2002.06241v1

150. MRS Mobile Road Safety GMBH (Hrsg). bike flash–das präventive Dialogdisplay. ► https://bike-flash.de/. Zugegriffen: 22. September 2021

151. Aldred, R.: Cycling near misses: their frequency, impact, and prevention. Transp. Res. Part A Policy Pract. **90**, 69–83 (2016). ► https://doi.org/10.1016/j.tra.2016.04.016

152. Statistisches Bundesamt (Hrsg).: Jeder siebte Mensch, der 2019 im Straßenverkehr ums Leben kam, war mit dem Fahrrad unterwegs; Mehr als die Hälfte aller getöteten Radfahrerinnen und Radfahrer war 65 Jahre oder älter. ► https://www.destatis.de/DE/Presse/Pressemitteilungen/2020/08/PD20_N049_46241.html. Zugegriffen: 31. März 2021 (19.08.2020)

153. T-Analyst. Lund University, Sweden

Open Access This chapter is licensed under the terms of the Creative Commons Attribution-NonCommercial-NoDerivatives 4.0 International License (▶ http://creativecommons.org/licenses/by-nc-nd/4.0/), which permits any noncommercial use, sharing, distribution and reproduction in any medium or format, as long as you give appropriate credit to the original author(s) and the source, provide a link to the Creative Commons license and indicate if you modified the licensed material. You do not have permission under this license to share adapted material derived from this chapter or parts of it.

The images or other third party material in this chapter are included in the chapter's Creative Commons license, unless indicated otherwise in a credit line to the material. If material is not included in the chapter's Creative Commons license and your intended use is not permitted by statutory regulation or exceeds the permitted use, you will need to obtain permission directly from the copyright holder.

Open Access This chapter is licensed under the terms of the Creative Commons Attribution-NonCommercial 4.0 International License (http://creativecommons.org/licenses/by-nc/4.0/), which permits any noncommercial use, sharing, distribution and reproduction in any medium or format, as long as you give appropriate credit to the original author(s) and the source, provide a link to the Creative Commons license and indicate if you modified the licensed material. You do not have permission under this license to share adapted material derived from this chapter or parts of it.

The images or other third party material in this chapter are included in the chapter's Creative Commons license, unless indicated otherwise in a credit line to the material. If material is not included in the chapter's Creative Commons license and your intended use is not permitted by statutory regulation or exceeds the permitted use, you will need to obtain permission directly from the copyright holder.

ADAS on Guidance- and Navigation Level

Visibility Enhancement Systems and Signaling Devices

Tran Quoc Khanh, Jonas Kobbert, and Timo Singer

Contents

© The Author(s) 2026
H. Winner et al. (eds.), *Handbook Assisted and Automated Driving*,
https://doi.org/10.1007/978-3-658-45276-6_30

The following chapter deals with lighting systems in vehicles that have introduced over the last couple of years to provide drivers with better visibility and orientation when driving at night. The chapter begins by presenting the fundamentals of lighting technology, perception, and glare. Then the currently available light sources are discussed. Starting from these, the currently available light assistance systems are described technically, and the advantages of these systems are illustrated by means of examples. This part is concluded by open questions and current problems of automotive lighting technology. The last section discusses possible new types of light signal systems for highly automated vehicles. These are classified in terms of their technological implementation and placed in relation to existing signal systems. In the outlook, the future role of lighting technology in motor vehicles is discussed.

30.1 Introduction

While most of the driver assistance systems described so far are so-called "active" assistance systems, since they actively intervene in the control of the vehicle, light-based driver assistance systems are passive assistants. This means that the systems support drivers without directly influencing driving behavior. Instead, the aim of light-based driver assistance systems is to support drivers, and to a certain extent other road users, in driving safely on the road by improving visibility or by means of special light signals.

The fact that light plays a decisive role in this respect is directly evident from the fact that the average driver takes at least 90% of all the information required for driving a vehicle in visually [1]. The fact that driver assistance systems are necessary at night is a direct result of the available light. While during the day the sun provides global illumination and enables drivers to perceive almost everything necessary for safe driving, at night drivers must rely on artificial lighting. To provide drivers with a similar level of safety when navigating the roads at night, vehicle and headlight manufacturers have been working for years on various technologies to increase visibility and thus safety on our roads.

The need for such technological development can be clearly demonstrated by comparing the occurrence of accidents during the day and at night. Various studies show that accidents with fatal or severe outcomes are significantly more likely at night than during the day [2, 2]. A detailed analysis of the accident by Sullivan [4] even shows that particularly accidents, such as collisions with roadside vehicles, deer, and pedestrians, where limited visibility while driving at night can be identified as the clear cause of the accident, occur more frequently by a factor of 4 than during the day. However, other types of accidents, such as running off the road, where driver visibility is further aided by guardrails and reflective delineators, are just as common at night as during the daytime [4]. This is shown in ◉ ◘ Fig. 30.1. The red curve shows the number of pedestrian fatalities in the United States. The blue curve represents those accidents in which the vehicle left the roadway. The left half of the graph shows accidents in light, the right side in darkness.

To make the data comparable between light and dark, *Sullivan* examined accidents four weeks before and after the time change from summer to winter time. Here, only times were considered that were shortly after sunrise during daylight saving time and shortly before sunrise during winter time. This is clearly shown again in ◉ ◘ Fig. 30.2.

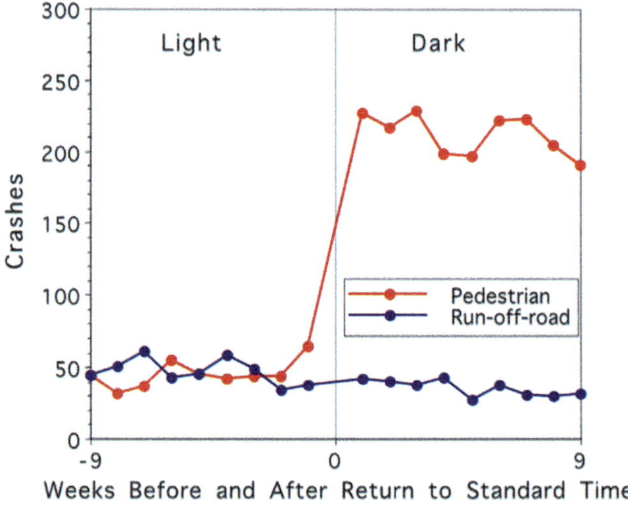

◘ **Fig. 30.1** Number of fatal accidents before and after the changeover from summer to winter time. Accidents in which pedestrians were killed are shown in red. Accidents in which the vehicle left the road and caused a fatal accident are shown in blue [4]

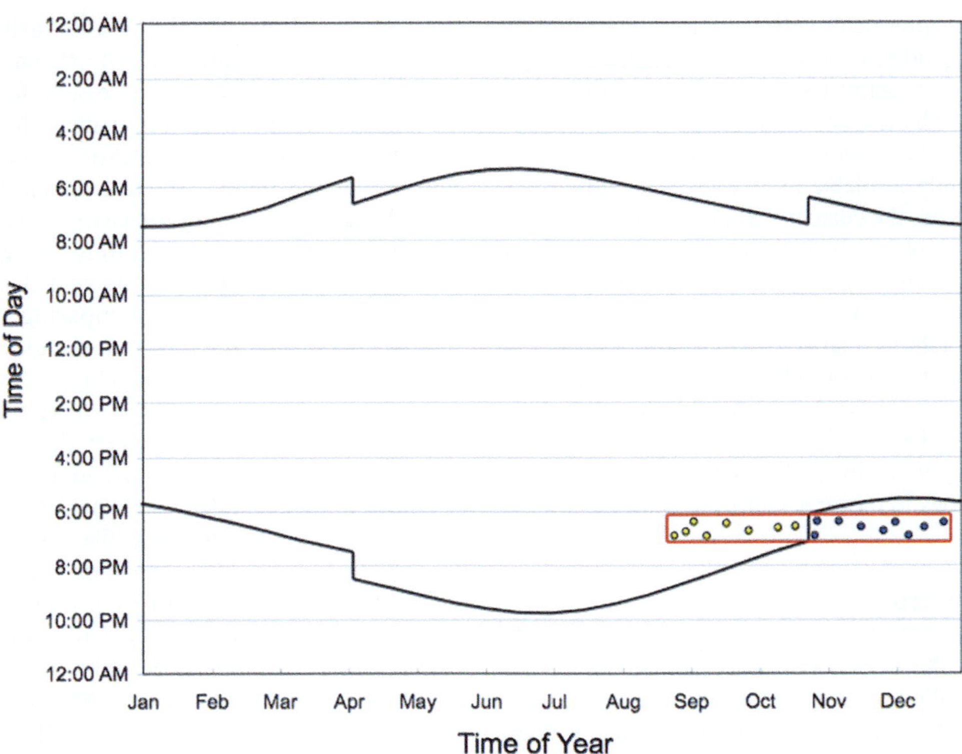

Fig. 30.2 Period considered for *Sullivan's* accident analysis. In yellow are marked the periods for accidents in the light and in blue are the periods for accidents in the dark [4]

For these data, accidents were evaluated over a period of more than 10 years in the U.S. [4]. In addition to this very objective rationale for why further development of headlamp technologies is essential, *Zydek* demonstrates [5] in a survey that a large majority of people, over 80%, want better light.

To begin with, the necessary lighting technology basics are covered. Then the various headlamp technologies that have been developed are presented and the advantages of each technology are substantiated with various studies. Finally, the new role of lighting technology in the age of highly automated vehicles is addressed and existing as well as novel light signal concepts are explained and discussed.

30.2 Basics

30.2.1 Lighting Conditions

As described at the outset, a person visually absorbs at least 90% of all the information needed to drive a vehicle safely. In ◉ ▢ Fig. 30.3 it becomes clear that the lighting conditions in road traffic cannot only be divided into day/night, but that local differences also occur. For this reason, the aim is to adjust headlamp light distributions as adaptively as possible to a given traf-

fic situation. ◉ ▢ Fig. 30.3 (left) shows the measured illuminance at the driver's eye during the day, subdivided into city (blue), country roads (red), and on highways (yellow). ◉ ▢ Fig. 30.3 (right) shows the data for nighttime driving on the same route with the same coding. This data also directly indicates the need for additional lighting at night, since the brightness here is several orders of magnitude lower than during the day [6].

Furthermore, it can be seen directly from the data that at night the lighting in urban areas reaches a good level and the requirements for headlights are different than on rural roads and highways. The influence of different lighting conditions on perception can be explained by the contrasts and their perception.

30.2.2 Contrasts and Contrast Perception

In lighting technology, a contrast always describes a difference between two surfaces or objects. A distinction is made here between color contrast, i.e., the ability to distinguish colors from one another, and brightness contrast. Since light-based driver assistance systems are necessary in twilight and night driving, this section only deals with brightness contrast. Contrast can be defined in different ways. The simplest contrast should be familiar to most readers from monitor or television data sheets. Here, the luminance, i.e., the bright-

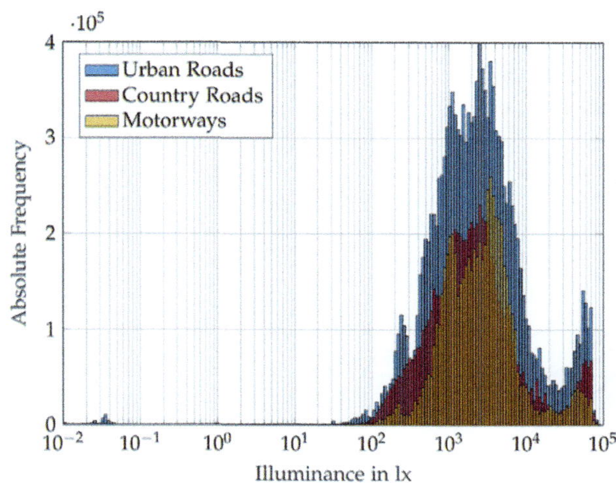 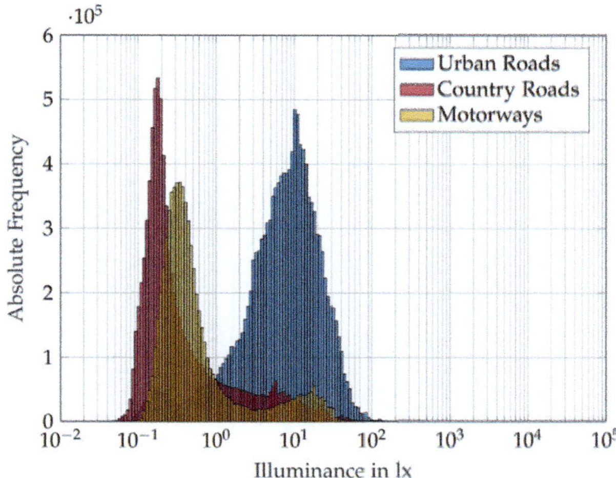

Fig. 30.3   Illuminance at the driver's eye during daytime driving (left). Data from urban areas are shown in blue, from rural roads in red, and from highways in yellow. The diagram on the right shows the same data for nighttime driving [6]

ness, of two surfaces is divided (light surface/dark surface). In automotive lighting technology, however, this contrast definition is hardly used. Here, the Weber (Eq. 30.1) or Michelson (Eq. 30.2) contrasts are usually evaluated.

$$K_{\text{Weber}} = \frac{L_{\max} - L_{\min}}{L_{\min}} = \frac{L_{\max}}{L_{\min}} - 1 \qquad (30.1)$$

$$K_{\text{Michelson}} = \frac{L_{\max} - L_{\min}}{L_{\max} + L_{\min}} \qquad (30.2)$$

Here, K stands for the contrast and $L\_\max$ (for the illuminated object) and $L\_\min$ (for the immediate surroundings around the object) for the brighter and darker area, respectively. Decisive for the perception, however, is that the contrast that can just be perceived, i.e., the so-called contrast threshold, changes with the existing ambient or adaptation luminance. The connection to this was discovered by Adrian [7] directly in relation to automotive lighting conditions. The finding is that a lower ambient luminance means that contrasts have to be increased in order to be perceivable. In other words, the darker it is, the greater the difference in brightness between an object, for example a person, and the surroundings must be for drivers to be able to perceive the object safely. This relationship is described in ⊙ ■ Fig. 30.4.

Here it can be seen that the threshold contrast depends on the object size, shown here in angular minutes. The larger the object, the easier it is for a person to recognize it. Furthermore, it is shown here that the threshold contrast increases with decreasing ambient luminance, starting from a constant value. These findings lead directly to the necessity of providing targeted lighting for drivers at night.

30.2.3   Glare

If vehicles come toward us while driving at night, it can lead to glare. With the introduction of new lighting technologies, such as Xenon headlights, LED headlights, and LED taillights, complaints about glare from other road users have increased. Glare itself can basically be divided into two different types. Physiological glare is directly responsible for a deterioration in visibility, while psychological glare leads to a discomfort. Both types of glare will be briefly explained below.

30.2.3.1   Psychological Glare

Psychological glare is purely a feeling of being dazzled. The visual performance is not necessarily impaired. In order to measure psychological glare, it is always necessary to interview the test subjects. Subjects are exposed to glare stimuli and subsequently assign the experienced stimuli to a glare scale. The most widely used glare scale is the *de Boer* scale. Here, the glare stimulus is assigned to a number on a 9-step scale [9]. A 1 corresponds to unbearable glare, 9 to imperceptible glare. A 5 therefore corresponds exactly to the borderline between dazzling and non-dazzling.

It should be particularly emphasized that it can be shown that the psychological evaluation of glare is strongly dependent on the lighting conditions experienced immediately beforehand. A study by Kobbert [10] shows, for example, that with the same test setup and test procedure, the same glare stimuli are evaluated differently only by shifting the glare intensity between two runs. This means that a stimulus that can lead to very strong glare in one case, for example, when approaching another vehicle for the first time after a long time at night, does not lead to any psychological glare

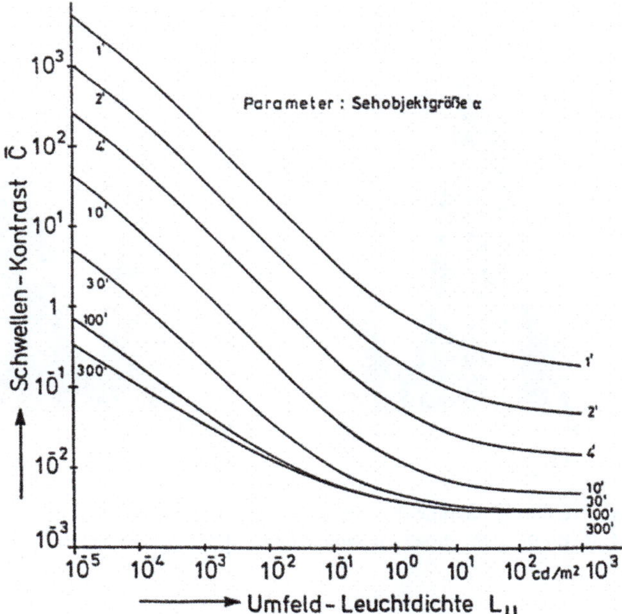

Fig. 30.4 Threshold contrast (Schwellen-Kontrast C) over ambient luminance (Umfeld-Leuchtdichte L_U) at different object sizes (Sehobjektgröße α) [8]

in another case, for example, because several vehicles have already approached before.

30.2.3.2 Physiological Glare

With physiological glare, however, people's visual performance is reduced. This means that the already low contrast perception at night is further reduced by glare. This can be vividly explained by looking into a glare source and trying to recognize an object next to the glare source. If the glare source appears to outshine the object, this is an extreme form of physiological glare.

Physiologically, the following happens here. The light emitted by the glare source is not perfectly imaged on the retina in the human eye. Rather, the vitreous body, the lens, and the cornea cause not only refraction and imaging of the light, but also reflects and deflects individual light bundles. In sum, this results in a kind of veil which, starting from the image of the glare source and sloping outward, covers the entire retina. This is illustrated in ◉ ▢ Fig. 30.5

Mathematically, the situation can be described further. Based on Eq. 30.1 the glare source causes an additional luminance to be applied to the retina: the so-called veil luminance L_Veil. Since this luminance is superimposed on the retina both over the surroundings of the object and over the object itself, to a first approximation also of equal intensity, this luminance can be added to Eq. 30.1 For this purpose, we assume that the surroundings have the lower luminance, and the object is brighter. Thus, L_O stands for the object luminance and L_U for the ambient luminance (see Eq. 30.3).

$$K_{\mathrm{Glare}} = \frac{(L_O + L_{\mathrm{Veil}}) - (L_U + L_{\mathrm{Veil}})}{(L_U + L_{\mathrm{Veil}})} = \frac{L_O - L_U}{(L_U + L_{\mathrm{Veil}})}$$

$$(30.3)$$

If we compare Eq. 30.1 with Eq. 30.3, we see that the effective contrast has become smaller due to the larger denominator. If one was already close to the threshold contrast without the glare source, the introduction of this glare source can lead to the effective contrast on the retina being below the contrast threshold and the object thus no longer being perceptible.

The veil luminance itself can also be calculated. The exact dependence is not important for further understanding. However, it should be mentioned that the veil luminance is dependent on the luminance of the glare source as well as reciprocal to the angle between the object and the glare source. Both physiological and psychological glare have been much discussed or reported in recent years as new light sources have been used in vehicles. Therefore, in the next section, all currently used light sources in vehicles will be described.

30.3 Functionality of Light Sources

One of the ways to increase safety in road traffic is the development and introduction of new light sources. The advantage of new light sources is normally either an increased luminous flux, increased efficiency, i.e., luminous efficacy, or significantly higher flexibility in

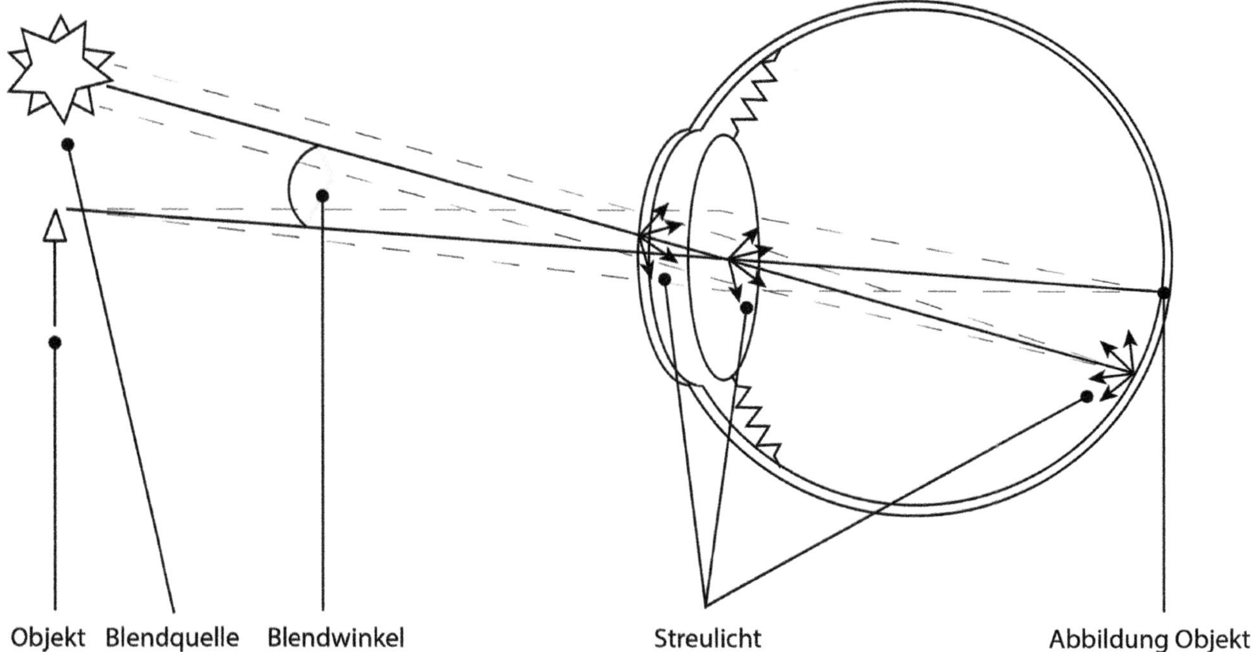

◘ Fig. 30.5   Schematic representation of the scattered light from a glare source on the retina

use. As there have been many more light sources used in the history of automotive lighting technology than are mentioned here, it should be noted that this section only deals with the light sources from recent years that are still relevant today.

30.3.1 Tungsten Halogen Lamps

The most widely used light source in the world is still the halogen incandescent lamp. Their simple design, low-cost production and decades of experience and continuous further development mean that halogen incandescent lamps are still used in low-priced vehicles in particular. In these headlamps, light is emitted via a wire (tungsten) that becomes hot due to the applied voltage and due to the converted thermal energy, and according to the principle of a Planckian radiator. The light produced by a tungsten halogen lamp is warm white (< 3200 K) and even the most modern tungsten halogen lamps achieve a luminous flux of currently 1.500 lm, (lumens) at an electrical power of 55 W.

The light is normally collected via a reflector and cast onto the road in a controlled manner. However, since halogen bulbs are becoming less important, they will not be discussed further here.

30.3.2 Xenon Lamps

The next big step in light source development was taken by the Xenon lamp. In contrast to the halogen incandescent lamp, light is generated here by the principle of gas discharge. The light produced is perceived as much colder than that of an incandescent lamp, with a color temperature similar to that of a halogen lamp of over 4.000 K. This makes it appear much brighter, even with the same luminous flux. In addition, a Xenon lamp with about 3.200 lm at only 35 W electrical power generates significantly more light with higher efficiency. Xenon lamps use a projection system consisting of lenses and shutters. The Xenon lamp was used in the 1990s and 2000s as a high-tech lighting system in automobiles. In the meantime, it is hardly used any more, which is why it is not considered further here.

30.3.3 Semiconductor LED Light Sources

The current variant for light generation in automotive front headlights is the LED. The first series-production headlight to be implemented exclusively with LEDs was installed in the AUDI R8 (Type 42) in 2008. Electroluminescence is used to generate the light. Light is

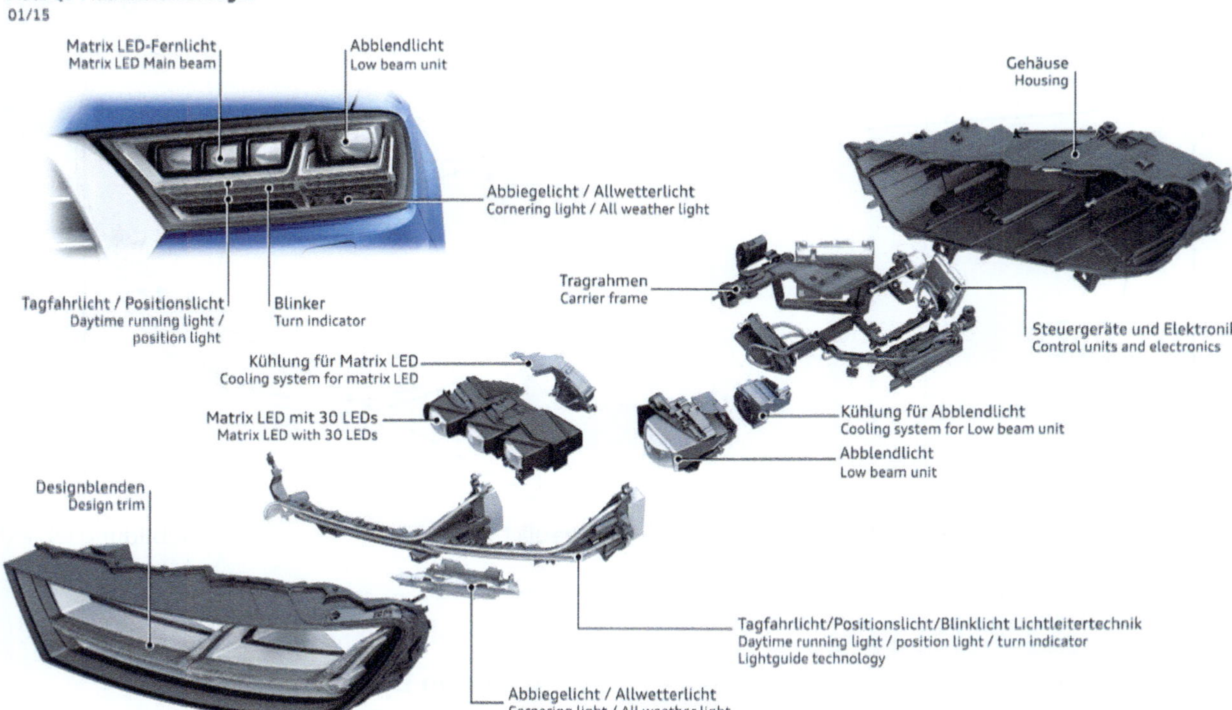

Fig. 30.6 Exploded view of the Audi Q7 Matrix LED headlight. All light functions, low beam, high beam/matrix beam, cornering light, daytime running light/position light as well as turn signal, are implemented by LEDs. *Source* Audi AG

generated by recombining electrons and "holes" in semiconductors. For further details on light generation, please refer to the relevant technical literature [11].

While halogen and Xenon light sources for headlights are standardized and thus values for luminous flux and power can be easily specified, LEDs are used differently. Since a single LED is not sufficient, at least efficiently, as the only light source for a headlight, a large number of individual and also different LEDs are used for a single headlight. As an example, here in ◉ **◘** Fig. 30.6 is an exploded view of a modern LED headlight in the Audi Q7.

In this example it can be seen directly that not only are LEDs used in a wide variety of functions, but also individual functions, for example the matrix beam, are used with a high number of LEDs, 30 in this case.

Furthermore, both reflection and projection systems can be set up, so that a wide variation of application types of LEDs occurs. However, this flexibility also leads to the possibility of efficiently implementing various new lighting functions. However, these will be discussed in more detail in the next section. These circumstances make it impossible to specify an exact luminous flux or specific power for LED headlights. Overall, however, it can be said that headlights are being developed with LEDs that can be significantly more energy efficient than halogen bulb or xenon headlights. At the same time, the compact form of LEDs makes it possible to get significantly higher luminous fluxes out of headlights than was possible even with the best xenon headlights. Furthermore, LEDs offer enormous design freedom precisely because they can be arranged almost anywhere.

30.3.4 Semiconductor-Laser

The use of lasers in headlights represents the logical and consistent further development of LEDs. The basic function of lasers will not be described here. From the point of view of lighting technology, it is important that a laser achieves a much higher luminance than an LED. This leads directly to higher visibility ranges for the driver. If one compares the widely used 1lux line, i.e., the positions at which a headlight can still generate 1lx (lux) of illuminance, an LED headlight with a laser as an auxiliary main beam reaches approx. 600m thus doubling the range of the main beam of "normal" LED headlights. However, since the 1lx line is not meaningful in terms of visibility, but only serves as a

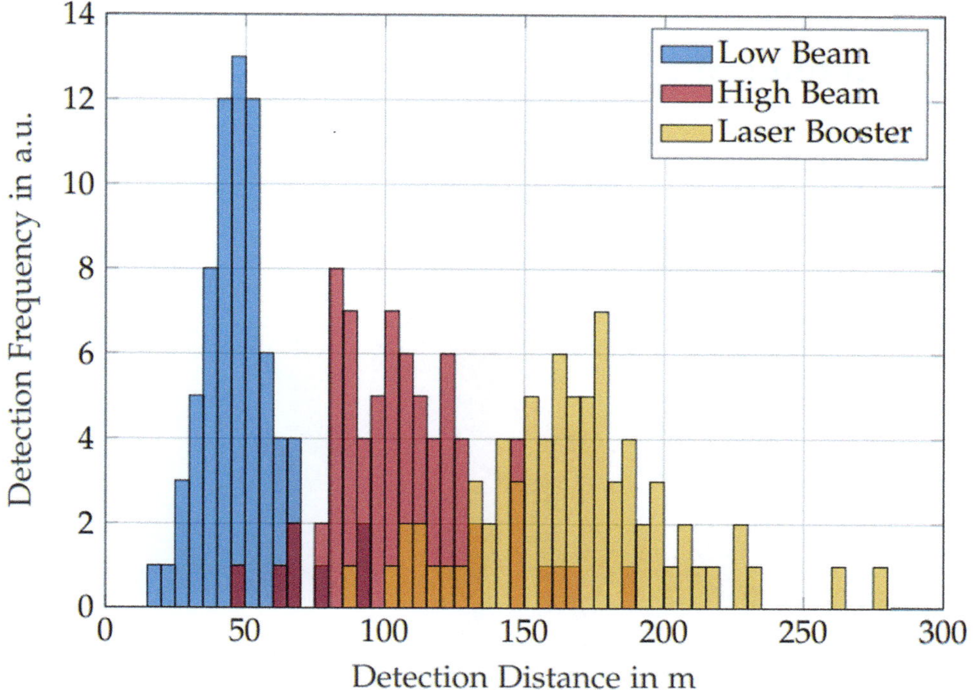

Fig. 30.7 Visibility ranges with LED low beam (blue), LED high beam (red), and laser high beam (yellow). On average, visibility ranges of 48,2 m for low beam, 107,0 m for LED high beam, and 167,4 m for laser high beam are achieved [6]

simple and descriptive comparison between different headlights, ◉ ▢ Fig. 30.7 shows the actual visibility range with LED low beam, LED high beam, and laser as additional high beam.

From this study, it can be seen that the additional laser high beam achieves a very significant gain in visibility range. The first production vehicle with a laser as an auxiliary high beam is the Audi R8 LMX (built in 2014). Several points are important here. First, the laser light is not cast directly onto the road as a laser. As indicated at the beginning, the laser light is the further development of the LED headlight. Just as with an LED, blue light is beamed onto a yellow phosphor. Only the mixed light appears white. With a laser, however, another important effect comes into play here. The parallel and coherent light is partially absorbed, scattered, and reflected. The characteristic that makes a laser so potentially dangerous in itself, namely the high parallelism in the light and the strong bundling, are directly canceled out by the phosphor. Nevertheless, other safety systems are built into an LED headlight with a laser module.

On the one hand, the laser is only used as an add-on to normal LED high beams and is also only activated at speeds of over 70 km/h; for example, it always remains inactive when flashing lights. Secondly, light traps and electrical shut-off devices are used in the event of a mechanical or electrical defect. This ensures that coherent laser radiation cannot escape from the headlight at any time.

30.3.5 OLED

Another well-known lighting technology is the OLED (organic light emitting diode). While OLEDs have long been used in smartphones and televisions in private life, they are very new in the automotive sector. Due to the different requirements for service life and durability, OLEDs have only been available as a combination rearlight variant in series production since the Audi TT RS (2016). The special thing about OLED combination rear lamps, apart from their extremely low thickness (an OLED segment is less than 1 mm thick), is that they are surface emitters. Particularly in the case of automotive combination rear lamps, where lower luminance levels are required, the use of OLEDs results in a homogeneity and contrast that cannot be achieved by any other technology. The basic structure of an OLED is shown in◉ ▢ Fig. 30.8 For more details, please refer to the corresponding source [11].

But the OLEDs are not just a design feature in the rear. By implementing individually controllable elements, safety-relevant functions can also be implemented. In the current Audi Q5 (model year 2020), for example, proximity detection is used for the first time

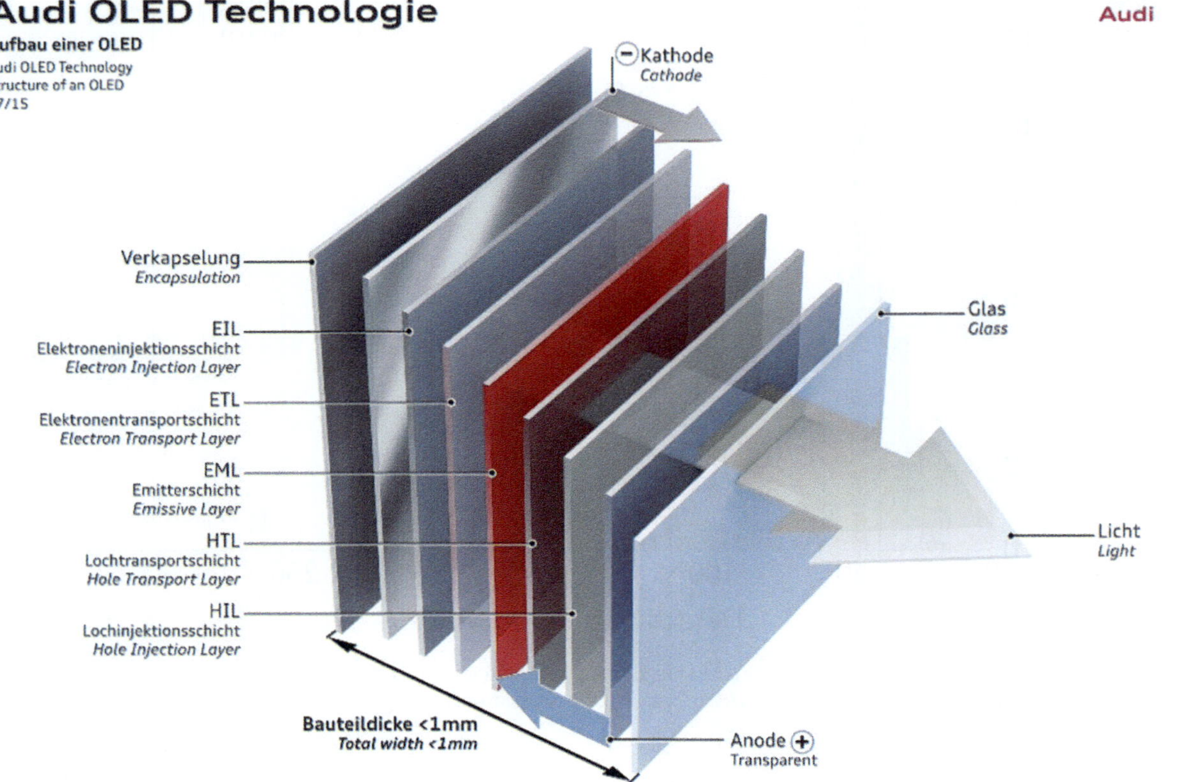

Fig. 30.8 Design of automotive-grade OLEDs with a total thickness of less than 1 mm. *Source* Audi AG

as an option in the combination rear light. If another road user, whether a car, cyclist, or pedestrian, approaches within 2 m, all OLED segments light up. A further development of this provides for different light signals to be displayed via the OLED combination rear lamp, depending on the situation, in order to communicate with other road users. Examples of this are shown in ◉ ◻ Fig. 30.9.

30.4 Light-Based Driver Assistance Systems

After the various light sources and light technologies have been explained in the previous sections, this section will present the currently available light-based driver assistance systems. These are explained as already clearly shown in ◉ ◻ Fig. 30.10.

30.4.1 Cornering Light/Static Cornering Light

The cornering light, also known as static cornering light, was introduced jointly by Audi and Hella in the Audi A8 in 2002. The aim of the cornering light is to provide enough light and corresponding visibility at intersections or very tight curves when the low beam does not illuminate these areas. The light distribution is shown schematically in comparison with a low beam distribution in ◉ ◻ Fig. 30.11.

The cornering light can achieve an illumination angle of up to 90°, as seen from the driver. The cornering light is normally implemented by an additional LED mounted on the inside of the headlamp. The light distribution is then generated by a reflector. Since the cornering light is only activated at very low speeds and braking distances and therefore also the required visibility ranges are low, a particularly high luminous flux is not required here. The aim of the cornering light is to illuminate pedestrians in the immediate vicinity of the vehicle and to detect them in good time before a turn is made.

30.4.2 Dynamic Bend Light

Dynamic bend lighting is a direct evolution of the static bend lighting. Here, the light cone of the headlamp on the inside of the bend is initially swiveled into the bend depending on speed and steering wheel angle.

Audi digitale OLED-Technologie
Audi digital OLED Technology
Personalisierung des Lichtdesigns und Einsatz bei Car-to-X-Kommunikation
Customizing lighting design and to be used for Car-to-X communication
08/19

Fig. 30.9 Display of various warning or info symbols in an OLED combination rear light by means of the individually controllable OLED segments. *Source* Audi AG

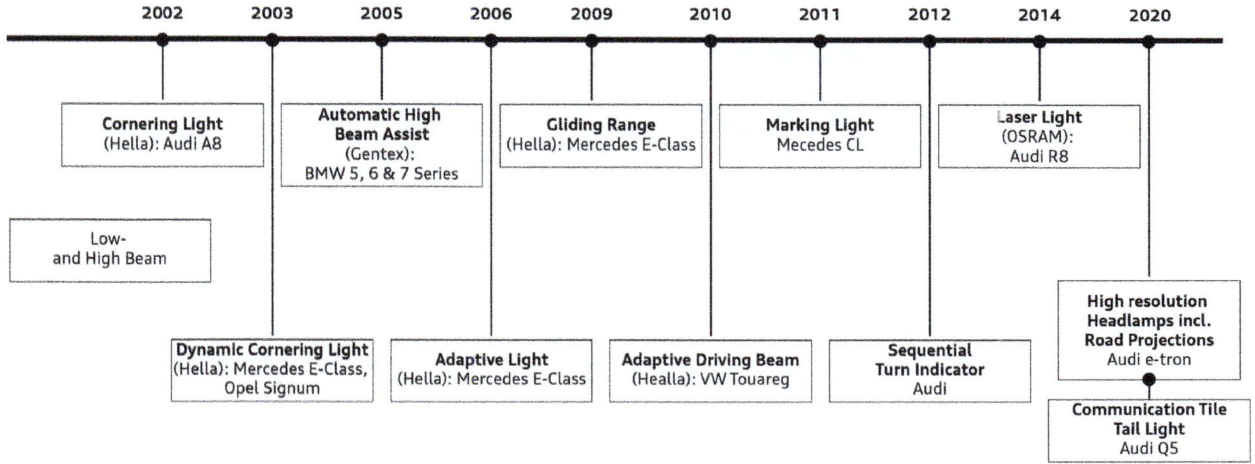

Fig. 30.10 Chronology of introduction of light-based driver assistance systems from 2002 to 2020. The dates always refer to the introduction in the UN-ECE countries. In the USA, for example, glare-free high beam is still not approvable

The aim here is to provide better visibility of the inside of the bend, since the low beam would always shine only tangentially to the passage through the bend. This is shown schematically in ● **Fig. 30.12**.

In the first implementations with projection systems (Xenon and LED), the complete low beam module was actually swiveled. In more modern LED headlights, additional LEDs are activated or at least the center of gravity of the light is shifted into the curve by controlling the LEDs differently, which has the same effect as swiveling the headlight module, but without moving parts in the headlight.

Since such a system, which reacts only to steering wheel angle and speed, can only ever respond to the

30

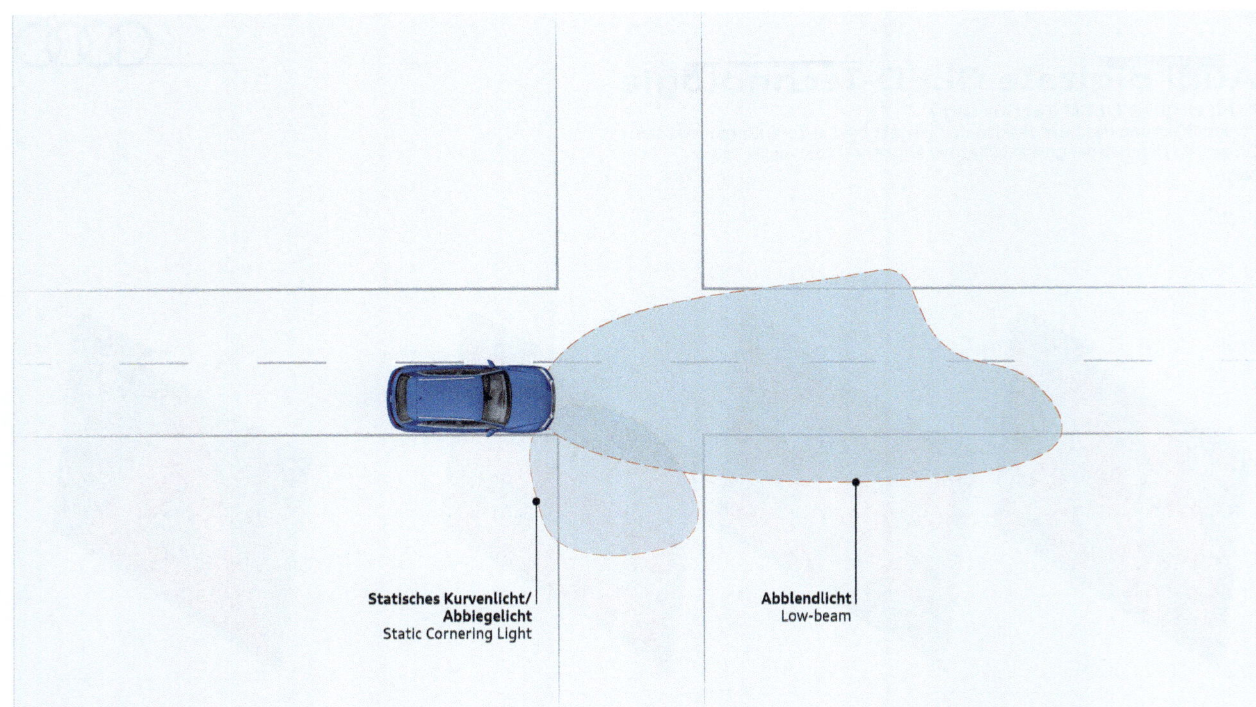

□ Fig. 30.11 Schematic representation of cornering light/static bend lighting in comparison to low beam. The cornering light can achieve an illumination angle (as seen from the driver) of 90° and therefore provides better visibility when turning or negotiating very narrow bends

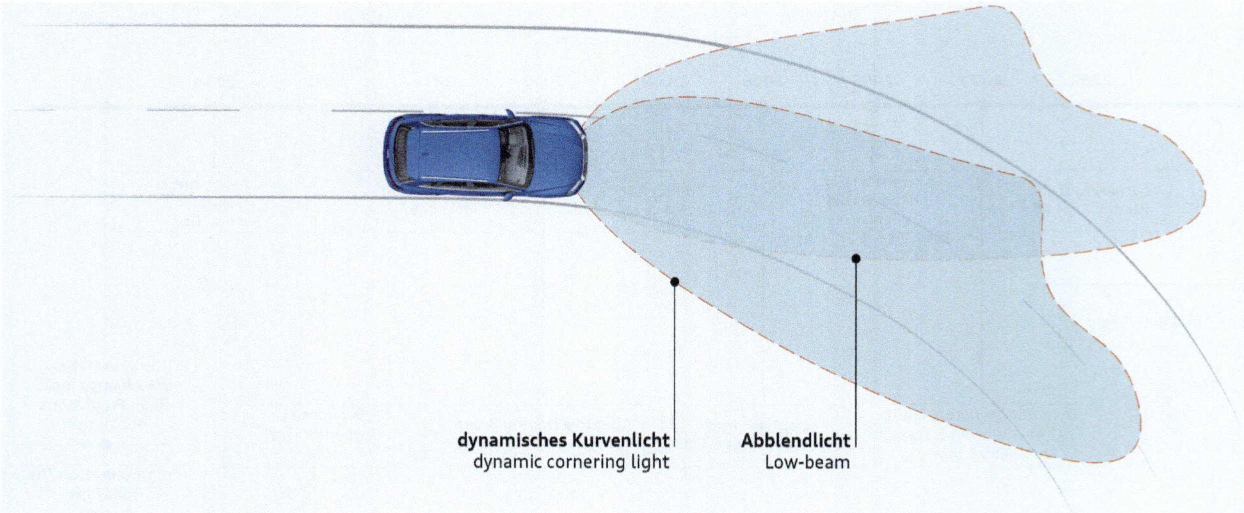

□ Fig. 30.12 Schematic representation of dynamic bend lighting compared to low beam. The bend lighting is swiveled in/controlled depending on speed and steering wheel angle. Modern systems also use navigation support and digital map data

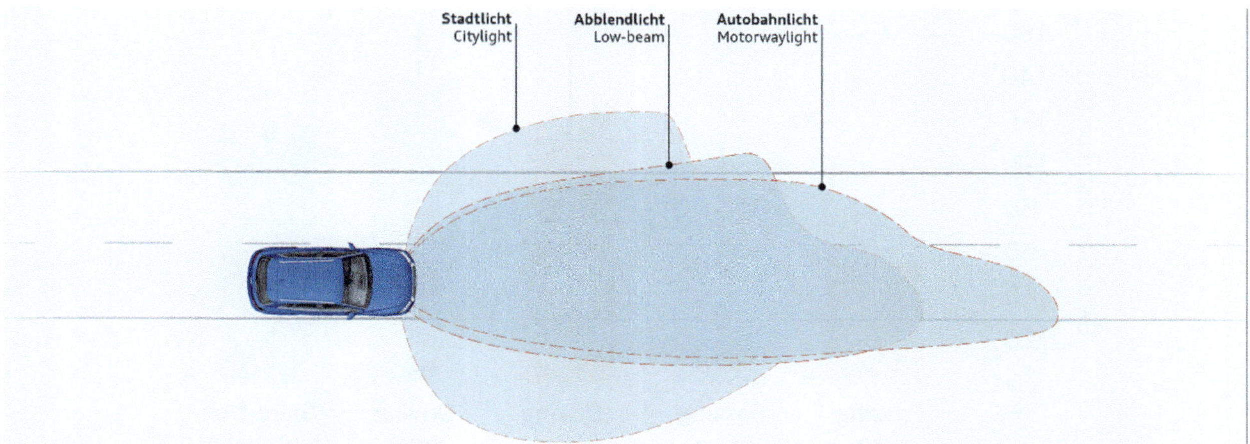

Fig. 30.13 Schematic representation of the different light distributions, city light, low beam, and highway light of an AFS headlamp

current vehicle status, the bend lighting therefore always swings into the bends with a corresponding time delay. Newer systems therefore use the support of the navigation system and digital map data to also be able to illuminate curves predictively.

30.4.3 High Beam Assist

While bend lighting improves illumination when driving through curves, there is still potential for improvement even when driving straight ahead. It has been shown, for example, that the high beam usage falls far short of the potential useful time [12]. Since the low beam, as shown in ⊙ ▫ Fig. 30.7, only provides an average visibility of about 50–60 m, at speeds above 60 km/h there is already insufficient visibility to stop in time in front of an obstacle. To remedy this shortcoming, a so-called high beam assistant is used very widely nowadays. A camera, usually installed behind the rearview mirror, registers other road users, such as oncoming vehicles or vehicles in front. If no other vehicle is detected within a distance of 400 m in front of you (legal minimum), the high beam assistant automatically switches to the high beam function. This directly ensures that only low beam is used when low beam must also be used in any case.

30.4.4 Adaptive Light Distributions

It should be quite easy to understand that the two light distributions, low beam and high beam, cannot provide optimum visibility or illumination of the road in all different traffic situations. For this reason, the so-called AFS (Adaptive Frontlighting System) was introduced by the legislator. This system refers to headlights that can adapt dynamically to the current traffic situation. Basically, the following classes are defined by the leg-

islator: Highway light, country road light (low beam), city light and bad weather light. City lights, country roads/dipped beam lights, and highway lights are shown schematically in ⊙ ▫ Fig. 30.13.

In built-up areas the driving speed is reduced. In most cases, the road is well illuminated by additional fixed lighting. This means that visibility no longer has to be provided by the low beam. On the other hand, pedestrians and cyclists are particularly vulnerable road users in the immediate vicinity of the vehicle. Accordingly, the light distribution in the city is extended to provide better visibility at the direct edge of the road.

When the road surface is wet or the weather is bad (fog, snow…), the road surface has a reflective effect due to wetness. On the one hand, it becomes darker for drivers and, on the other hand, the light can dazzle drivers of oncoming vehicles [13, 13]. The darker roadway can make it more difficult for drivers to orient themselves to traffic guidance lines. For this reason, the legislator allows the low beam to be swiveled outward, similar to the city light. This is to enable drivers to orient themselves better at the edge of the road by means of the additional lighting. If possible, the intensity of the inside roadway headlight can be reduced to provide less glare for oncoming traffic.

On highways, the requirements for the lighting system are significantly different. Since driving at high speeds is permitted here, an increased visibility range is necessary. The light cone of the headlamp may therefore be raised in steps depending on the speed currently being driven.

30.4.5 Matrix Beam/Glare-Free High Beam

The current expansion stage of an adaptive headlamp is the so-called glare-free high beam. With this light system, it is possible to drive permanently with high beam without dazzling other road users. The first such

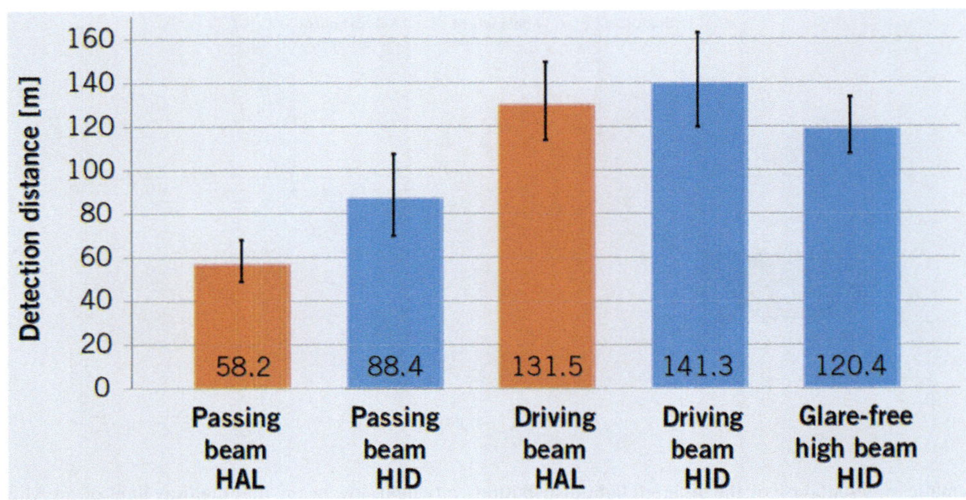

Fig. 30.14 Visibility with different light systems. From left to right: halogen low beam, HID (High Intensity Discharge/Xenon) low beam, halogen high beam, HID high beam, and the HID/Xenon implementation of glare-free high beam [5]

system was installed by Volkswagen in the VW Touareg in 2010. In these systems, movable shutters were used that were inserted into the beam path of the projection headlamp. These shutters create shadows in the light distribution and can be moved to the left or right to create a blanked out area in the high beam distribution. This enables drivers to achieve the same visibility as with high beam, despite other road users, while the light impact on oncoming road users is the same as with low beam. On the one hand, the advantage of the high-intensity xenon system over the conventional halogen system is directly achieved. On the other hand, it is directly shown that the visibility range with glare-free high beam is significantly closer to high beam than to low beam (◉ ▣ Fig. 30.14).

In ◉ ▣ Fig. 30.15 the measured psychological glare during the same experiment is shown. The arrangement of the systems corresponds to the representation of ◉ ▣ Fig. 30.14. In this figure, however, the psychological glare of oncoming motorists is plotted.

For this purpose, the *de Boer* scale mentioned at the beginning was used. It is important to note here that low numbers correspond to high glare and high numbers to low glare. It can be seen from the data that glare-free high beam systems certainly live up to their name. The glare is equivalent to the glare caused by low beam.

Modern LED headlights use individually controllable LEDs, each of which can be switched on or off separately. This type of glare-free high beam was first used in the Audi A8 in 2013. In turn, individual "shadows" in the light distribution can be generated by switching these LEDs off. By electrically controlling the LEDs, there is now no limit here to the number of gaps that can be generated if the number of LEDs available is correspondingly high enough. This system is shown schematically in ◉ ▣ Fig. 30.16.

Furthermore, it is possible to dim the LEDs in the high beam individually and thus, for example, illuminate traffic signs with reduced intensity. This ensures that drivers are not dazzled by the retroreflection of their own light. Such systems can already be found in all price classes. They differ greatly in the number of LEDs available. In the simplest case (example Mini Cooper), 4 LEDs are available per headlight. Since 2020 systems have also been available that use DMD (Digital Mirror Device) technology, a chip with over 1 million micromirrors (example: Audi e-tron Sportback). However, since these also have other functions, they are dealt with separately in ▶ ▶ Sect. 32.4.7.

Studies on glare-free high beams clearly show that they achieve the same visibility ranges as high beams even in situations with oncoming traffic. At the same time, both physiological and psychological glare are at the level of low beam [5].

30.4.6 Marking Light

An exactly opposite approach is taken with marker lights. An infrared camera, usually installed in the vehicle's grill, detects other road users, such as pedestrians or game on the side of the road. An additional, swiveling LED module then illuminates the corresponding object. This is shown schematically in ◉ ▣ Fig. 30.17.

Studies show that the spontaneous illumination of the marker light means that visibility ranges remain more or less the same as with high beam, but that reaction times until an obstacle is detected can be significantly reduced [15]. A disadvantage of the entire system is, however, that due to the lack of illumination of the "danger zone", an infrared camera must be used instead of a normal camera. Here, either passive cameras

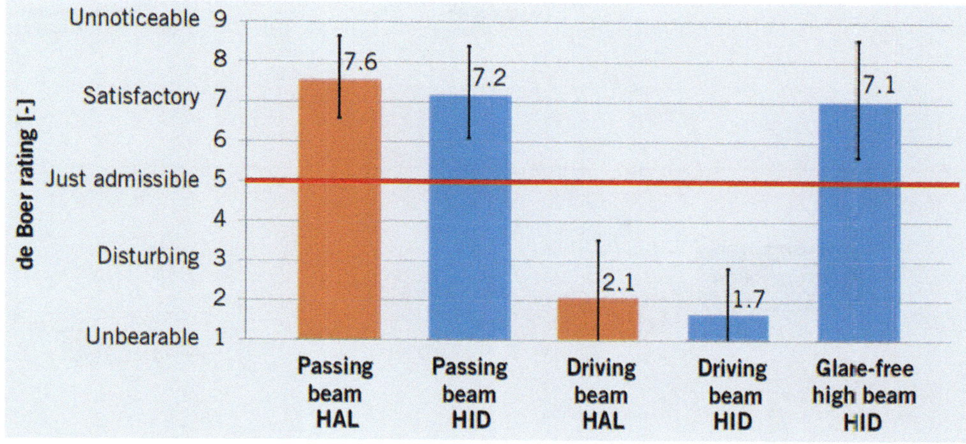

Fig. 30.15 Psychological glare of low beam, high beam, and glare-free high beam for halogen (HAL) and xenon (HID) headlights. High values correspond to low glare [5]

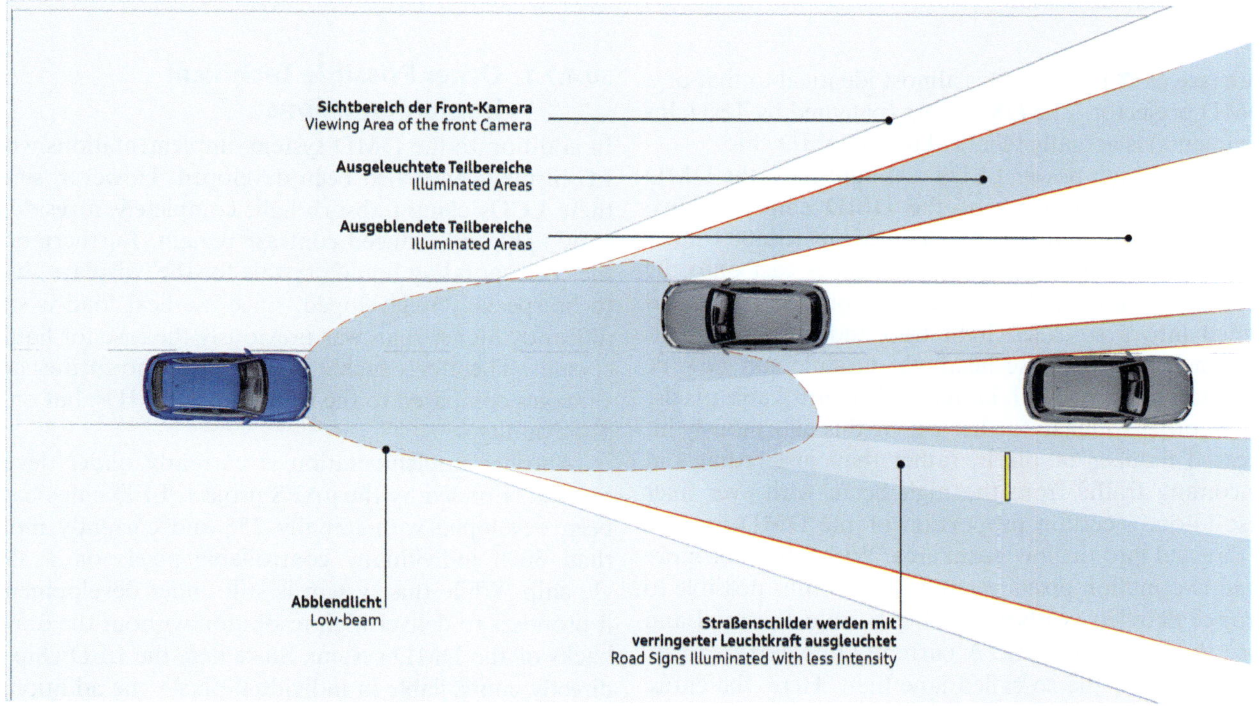

Fig. 30.16 Schematic representation of a glare-free high beam. Both the low beam, the camera's field of view, and the blanked areas are shown. In addition, such a system is also able to illuminate traffic signs with reduced intensity in order to avoid dazzling drivers with its own headlights

can be used, which then detect especially game well. In the case of clothed people, however, the same does not work so well, because the clothing has an insulating effect. Alternatively, an active system can be equipped with an infrared emitter. However, this is much less efficient than LEDs in the visible range and takes up additional space in the headlamp, which is already packed to the limit.

30.4.7 High-Resolution Headlights and Road Projections

As already described in ▶ Sect. 30.4.5 on glare-free high beam, the Audi e-tron Sportback is already on the market as a vehicle with 1.3 million individually addressable micromirrors per headlight using DMD (Digital Mirror Device) technology. The exact struc-

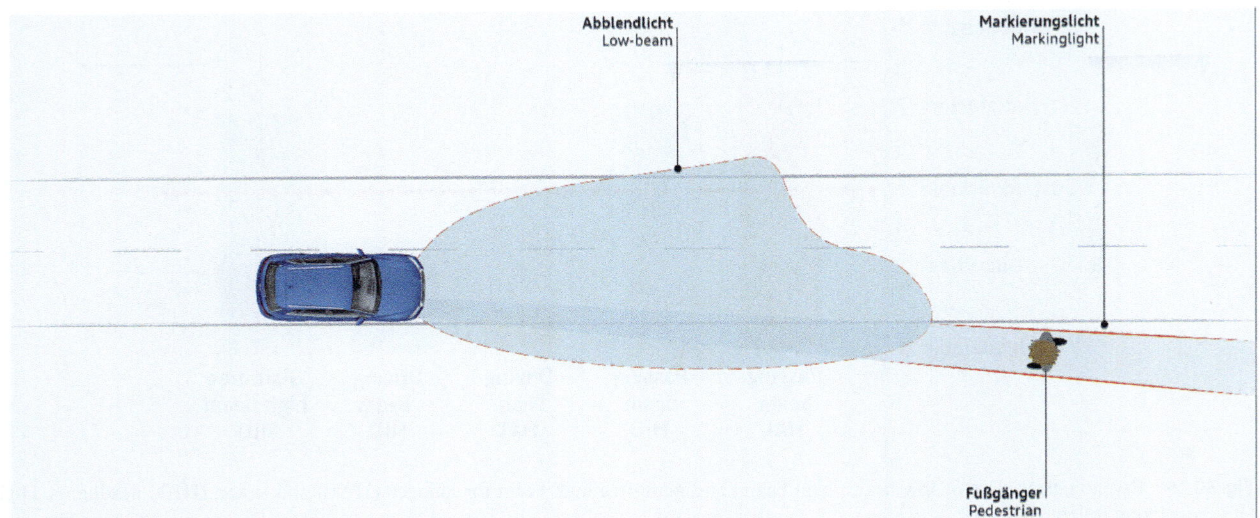

Fig. 30.17 Schematic representation of marking light. An additional light module is pivoted so that a road user, such as a pedestrian or game, is explicitly illuminated

ture (see ◉ **▫** Fig. 30.18) is almost identical to that of a DMD projector. The DMD chip (patented by Texas Instruments) is actually identical to the DMD chip.

Here, 3 high-power LEDs cast light onto the DMD chip. The micromirrors of the DMD chip have two fixed positions that can be changed up to 5000 times per second. At one position, the light is cast onto the road via optics. At the second position, the light is directed into a so-called light trap and absorbed. This system means that the headlight module can now be seen as a projector with individually controllable pixels.

In order to make further use of this enormously increased number of pixels, rather than just fading out oncoming traffic from the high beam with ever finer resolution, a certain proportion of the DMD module is directed into the low beam area. With a total of more than two million pixels per vehicle, it is thus possible to project detailed symbols or objects onto the road and into the low beam area. A currently implemented variant of this is the so-called lane light. Here, the entire width of the vehicle, including the side mirrors, is projected in front of the vehicle in the dipped beam distribution. Drivers can therefore always use this projection (shown schematically in ◉ **▫** Fig. 30.19) to orient themselves in their own lane.

A study by Hamm [16] showed that drivers feel much safer, especially when overtaking trucks or driving through bottlenecks, for example in construction zones. This was demonstrated by more constant acceleration and less brake activation. The projection of symbols on the road is currently still being discussed by the legislator (as of October 2021).

30.4.7.1 Other Possible Technical Implementations

In addition to the DMD system, implementations with LC displays have also been developed. However, since these LCDs cannot absorb light completely, a residual stray light and reduced contrast remain. Furthermore, the corresponding liquid crystals for the "displays" had to be specially redeveloped, since the heat load is significantly higher than was previously the case for liquid crystals. Therefore, such a system offers no further advantages compared to the solution via DMDs, but only disadvantages.

Another implementation is currently under development. Funded by the µAFS project, LED chips have been developed with initially 256 and currently more than 3000 individually controllable pixels on a single chip. While this system is still under development, it promises to deliver high resolution without the drawbacks of the DMD system. Since here the LED chip is directly controllable in individual pixels, the additional path via the micromirrors is not needed. The control of the DMD chip is omitted, and the overall system becomes much more compact and lighter.

30.5 Headlamp Rating Systems

Modern headlamps differ, as shown at the beginning, not only in the type of light source and the available functions, but also in the implementation of low beam and high beam. For this reason, the International Commission on Illumination (CIE) introduced its first

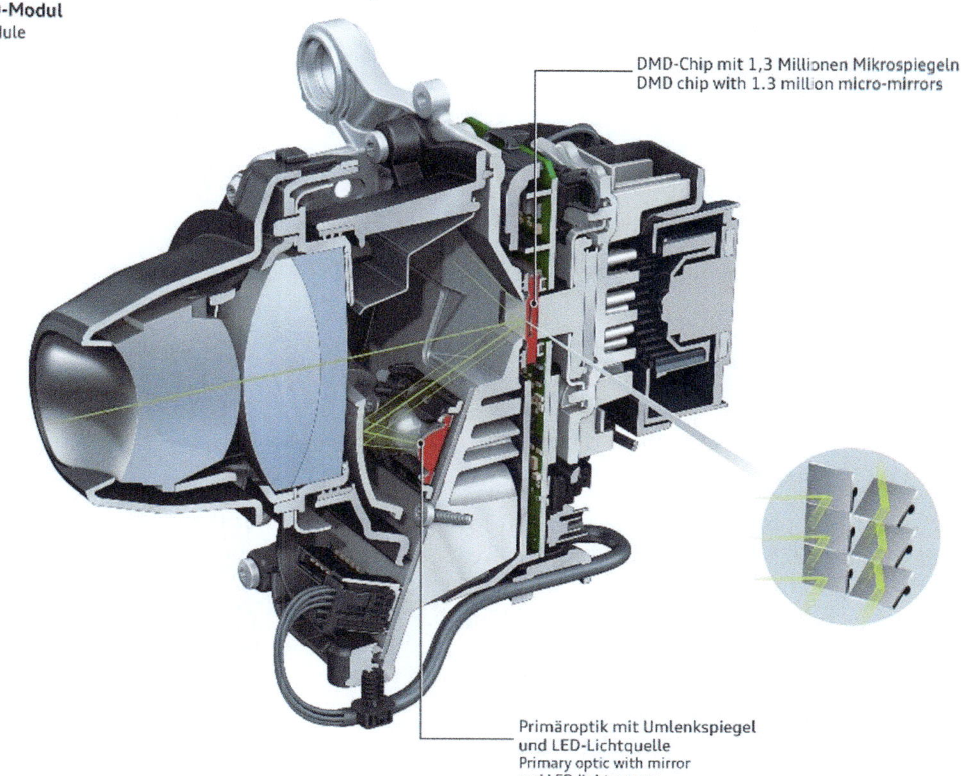

Fig. 30.18 Functionality of a high-resolution headlight as used in the Audi e-tron Sportback. The beam path is shown in yellow

headlamp rating system, the so-called TC4-45, back in 2011 [17]. Since then, this evaluation system has offered an established system for the assessment of low beam and high beam at automotive manufacturers and suppliers as well as in the general circle of technical experts. Due to the high complexity and the many measurement points and zones of the rating system, the ratings have not been made available to end users.

Additionally, in 2016, the Insurance Institute for Highway Safety (IIHS) in the U.S. came up with its own headlight rating system. The IIHS evaluates vehicles available on the U.S. market in terms of their road safety and makes recommendations in this regard. The two possible awards are the "Top Safety Pick" and the "Top Safety Pick Plus". These can sometimes be decisive for customers in the USA. In this context, the headlights are classified into one of the following categories: Poor, Marginal, Acceptable or Good. For a vehicle to receive one of the two awards, for the Top Safety Pick Plus, every headlight available for the tested vehicle must be rated at least Acceptable or Good. For the Top Safety Pick, it is sufficient for an optionally available headlamp to be rated Acceptable or Good. It is important to note that there are a certain number of bonus points, for example for an available high beam assistant. However, these bonus points are not related to the safety gain demonstrated in reality. An evaluation of glare-free high beam is not carried out, as these systems, although already in use in Europe since 2010, still do not receive approval in the USA. From a technical point of view, it is problematic here that the headlights are not correctly adjusted before the headlight test. As will be described in the following chapter, the optimization of the headlight setting is currently still a working point in the entire automotive industry. One point of criticism of this headlamp evaluation system, as Hinterwälder [8] shows, is that a misalignment of the headlamp by only 0.1° can already result in a devaluation from "Good" to "Acceptable" and, in the case of a misalignment of 0.2°, to "Marginal". It should be noted here that the interpersonal adjustment accuracy alone can account for up to 0.4° [18] (see also ▸ ▸ Sect. 32.6).

In China, a headlamp assessment system also exists within the framework of the China NCAP. Its structure is strongly based on the TC4-45 system. The headlamp can achieve a total of 10 points here. The headlamp rating is included in the overall vehicle rating. Adaptive lighting functions are also positively taken into account

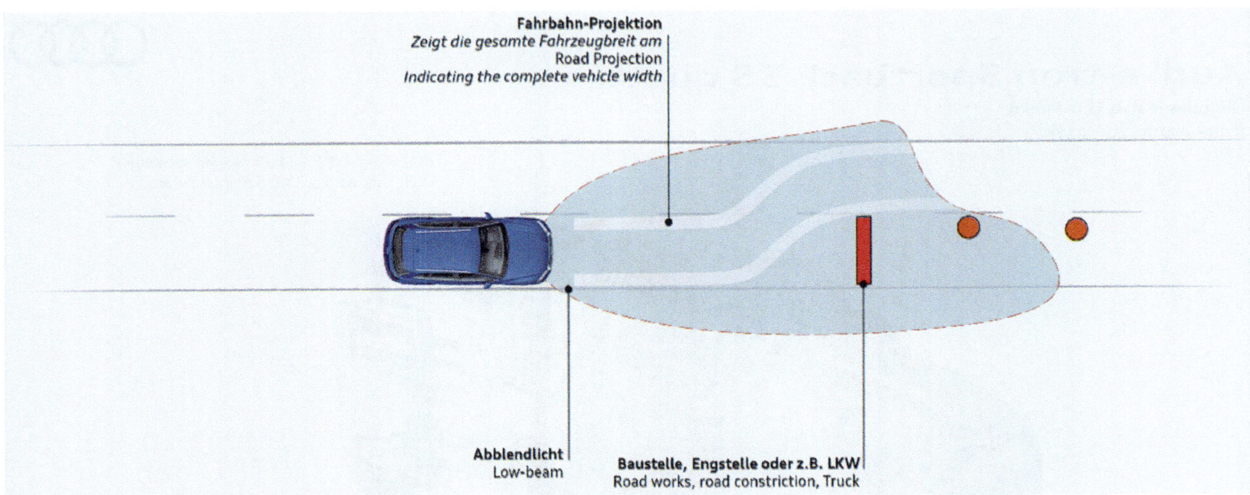

Fig. 30.19 Schematic representation of the roadway projection by a high-resolution headlight

here. In this context, 0.2 bonus points are awarded for a high beam assistant or 0.5 bonus points for glare-free high beam.

Since all of these systems do not evaluate the light assist systems described at the beginning, or only to a minor extent, a group of experts in automotive lighting technology is working on an extension of the TC4-45 evaluation system for the evaluation of glare-free high beam. In addition, the evaluation is to be simplified to make the system understandable for the end user [6, 19].

For a meaningful evaluation of the headlamps, only laboratory measurements of the headlamp system are used, as in TC4-45. This avoids influences on the headlamp setting in the vehicle and the evaluation takes place exclusively on the headlamp system. For a detailed description of the evaluation system, please refer to the original publication [17] and to the two articles [6, 19].

Briefly summarized, the system from TC4-45 is adopted for low beam (incl. glare) as well as high beam. Here, the individual evaluations are first calculated in the form of effective sight distances. For the evaluation of glare-free high beam, situations with oncoming traffic and traffic ahead are now additionally measured. Here, the high beam segments are switched off in such a way that the corresponding vehicles as well as a safety tunnel of 0.5° each on both sides of the vehicle are blanked out. Analogous to the evaluation of low beam and high beam, an effective visibility range is also calculated here. In order to combine the three assessments of low beam, high beam and glare-free high beam into a final assessment, the respective effective sight distances are offset against each other analogously to their respective service life (see ◉ **Table 30.1**).

As an example, a system from [19] is presented here in Eqs. 30.4 to 30.6. The system achieves an effective

Table 30.1 Utilization ratio of low beam, high beam, and glare-free high beam depending on degree of automation [6]

	Unit	Low beam	High beam	Glare free high beam
Manual	%	90	10	–
Automatic	%	70	30	–
Glare free high beam	%	30	30	40

range of 97 m in low beam, 186 m in high, and 133 m in glare free high beam mode.

Accordingly, a total effective range is calculated as follows:

If the headlamp system is only offered with low beam and high beam as well as manual high beam switching:

$$0{,}9 \cdot 97\,\text{m} + 0{,}1 \cdot 186\,\text{m} = 106\,\text{m} \tag{30.4}$$

Is the system now offered with automatic high beam assistant:

$$0{,}7 \cdot 97\,\text{m} + 0{,}3 \cdot 186\,\text{m} = 124\,\text{m} \tag{30.5}$$

If an ADB system is also used, the following results

$$0{,}3 \cdot 97\,\text{m} + 0{,}3 \cdot 186\,\text{m} + 0{,}4 \cdot 133\,\text{m} = 138\,\text{m} \tag{30.6}$$

These effective ranges are now assigned numerical values between 0 and 60 using an assignment table created from a large database of headlights that have already been evaluated. In order to be able to communicate these points to the outside world in a comprehensible and intuitive way, they are divided into the categories Standard, Good, Advanced, Excellent, Premium, and

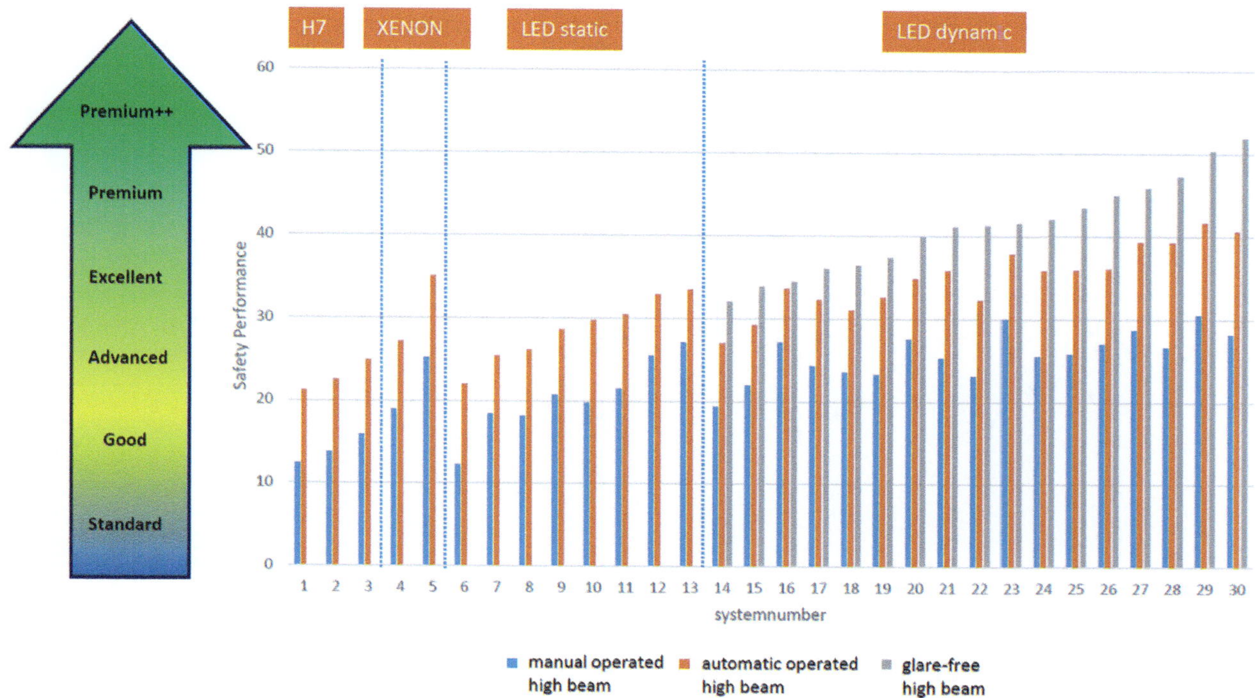

Fig. 30.20 Headlamp evaluation sorted by light source type and evaluation result

Premium +, each in steps of 10. This rating is shown for 30 headlamps already evaluated with the respective options, manual, automatic and, if available, also with glare-free high beam, in ◉ **Fig. 30.20**).

It can be seen that there is a mix between different lighting technologies—a good halogen or xenon headlight can be as good or even better than a simple LED headlight. At the same time, it can be seen that an automatic or even glare-free high beam (ADB mode) always brings a significant bonus.

The entire rating system is currently still being revised and finalized.

30.6 Current Challenges in Lighting Technology

Despite this rapid development of lighting systems, there are still a number of problems that automotive lighting technology has been dealing with for some time.

30.6.1 Headlight Aiming

The problem most prominently represented in the media is certainly the headlight aiming. The ADAC conducts an annual light test and publishes the results. Year after year, the ADAC shows that over 20% (28.8% in 2019) are defective. The ADAC understands defec-

tive to mean not only incorrect settings but also defective lamps. Similar studies by the Technical University of Darmstadt paint an even worse picture, with around 75% of headlights incorrectly adjusted. Incorrect headlight adjustment can have two consequences. If the headlight is set too low, this leads to reduced visibility for the driver. If the headlight is set too high, this can quickly lead to glare for other road users. If the horizontal setting is incorrect, this will result in reduced visibility at the edge of the road and, should the headlight be adjusted too far to the left, also in glare for other road users, as European headlights (at least in right-hand traffic) have an increase in the light distribution on the right-hand side to provide more visibility at the right-hand edge of the road.

Despite being set correctly after leaving the factory, headlights seem to adjust a lot within the first 10,000 km. The assumption is that mechanical stresses during transport or the first drive cause the headlights to settle. Tensions are released and components in the headlight become misaligned in such a way that an adjustment occurs. ◉ **Fig. 30.21** shows the settings of various headlamps measured by the lighting laboratory of Darmstadt Technical University.

Kosmas [20] and Hamm [18] systematically demonstrate the strength of various influencing parameters in their respective studies. Hamm investigated intra- and interindividual adjustment accuracy of skilled personnel, the stick–slip effect, changes in tire pressure, the influence of temperature, different driver weights, and the

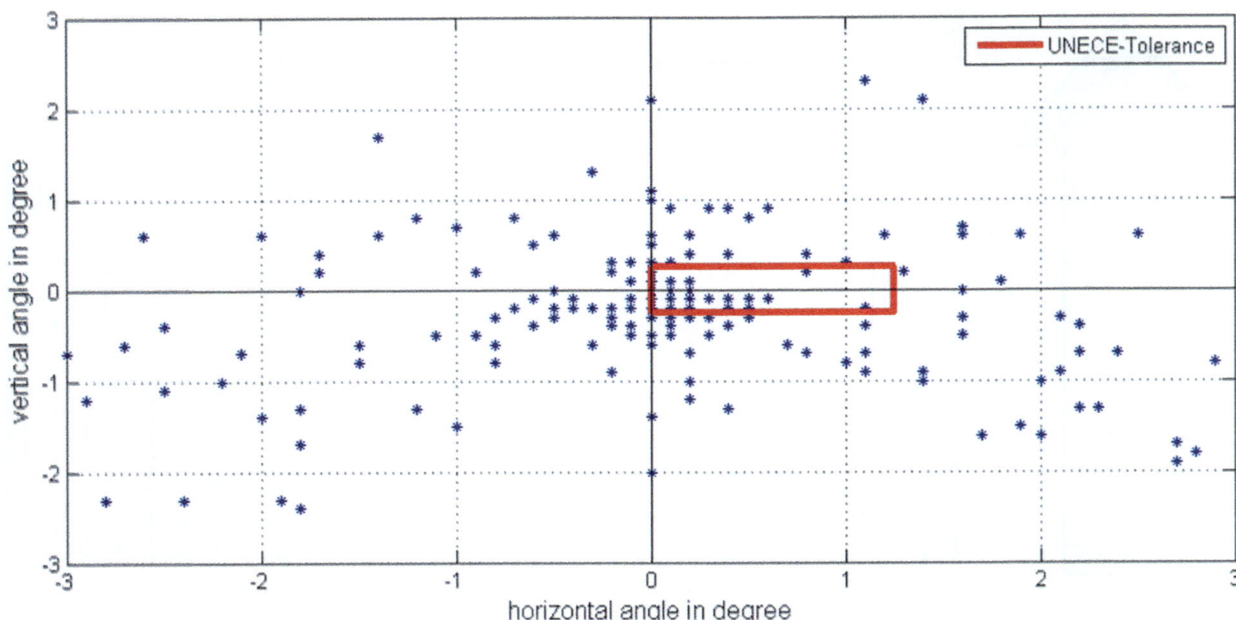

Fig. 30.21 Horizontal and vertical adjustment of automotive headlights. The red area marks the corridor for correctly adjusted headlights. Overall, only 24 of the 248 headlights were correctly adjusted

fill level of the tank. In summary, Hamm finds that in a worst-case scenario, these effects alone allow for a maximum adjustment accuracy of $+0,4°$ or $-0,5°$ Here, the greatest influences are due to the intra- and interindividual setting accuracy.

30.6.2 Headlamp Cleaning

Another topic of discussion is the headlight cleaning system. According to UN-ECE, this is mandatory on headlamps with light sources with a total luminous flux of over 2000 lm. With the introduction of xenon lamps, the fear was that dirt accumulating on the cover lens in combination with the increased luminous flux could cause stray light that could dazzle other road users.

Current investigations show that this problem certainly also occurs with current LED headlamps, but also with "low-light" halogen headlamps [21]. However, since headlamp cleaning systems cost space and weight, it is unfortunate that car manufacturers are happy to do without them. Some manufacturers still offer it as a paid extra, and Audi is one of the last manufacturers to still install the headlight cleaning system as standard equipment almost everywhere across all vehicles.

However, the introduction of the glare-free high beam described above is now causing new problems. While the legislator only requires that the low beam is cleaned, since at that time only the low beam was allowed to be active when other road users were registered in front of the vehicle, the glare-free high beam results in two new limits of high light intensity in the direct vicinity to other road users. The study by Schiller [21] also shows that there is still a need for additional solutions here.

30.7 Light Signals as Part of a Novel Human–Machine Interface for Highly Automated Vehicles

In the previous chapters, mainly light-based assistance systems were described that support a human being in driving safely through road traffic at night. As the level of automation continues to increase, the responsibility of the driving human will move more into the background. This becomes clear when looking at the 6 levels of automation. While in level 0 to 2 the control of the vehicle and the monitoring of the traffic area is still completely taken over by humans, this is taken over by a machine from levels 3 and 4 for some traffic situations and from level 5 completely.

In §1 of the road traffic regulations for Germany all road users are required to exercise "constant care and mutual respect" [22]. In order to be able to assess and anticipate the behavior of others, various intuitive ways of communication, such as eye contact or hand gestures are used for this purpose [23]. These communication channels cease to exist when humans are no longer responsible for the driving task. Accordingly, fully automated vehicles represent a new road user, especially in urban areas, which must be integrated into the existing traffic network. To achieve this, trust in this new

◼ Table 30.2 Signaling devices prescribed and permitted on the motor vehicle, based on the ECE regulations [26]

Signal function	Meaning	Mounting location	Color	Light modulation frequency in Hz	Luminous intensity in cd
Daytime running light	Better detection during the day/Better estimation of the distance to the vehicle	Front	White	Static	40 … 1200
Direction indicator	Direction indicator (left or right)	Front, tail, side	Yellow	1,5 ± 0,5	50...1000
Tail light	Detection of the vehicle from following traffic at night	Tail	Red	Static	4...42
Stop light	Warning of braking maneuvre	Tail	Red	Static/ 4 ± 1 during emergency brake	60...730
Rear fog lamp	Better visibility in adverse weather situations	Tail	Red	Static	150...840
Hazard warning light	Warning of potential dangerous situations	Front, tail, side	Yellow	1,5 ± 0,5	50...1000
Reverse light	Illumination of the rear lane/warning others of reversing maneuvers	Tail	White	Static	80...300
Side marker light	Detection of the vehicle to the side at night	Side	Yellow or Red	Static	4...25

technology must be built up, especially from a social perspective.

A 2018 study by the American Automobile Association (AAA) shows that 63% of 1014 surveyed drivers would feel more unsafe if they had to share the road with an automated vehicle while walking or cycling [24].

This could be remedied by novel, external light-based communication devices. The driving maneuver planning based on data from various sensor systems can thus be made visible to other road users externally.

30.7.1 Novel Light Signal Concepts for Highly Automated Vehicles

Before possible new light signal concepts are discussed in more detail, an overview of existing signaling functions is given. A precise overview is provided by ◉ ◼ Table 30.2. This shows that the meanings of the various signaling devices in the vehicle can be distinguished in terms of their color, light modulation frequency, or light intensity: For example, the tail light and stop light can be distinguished by the different brightness levels perceived by humans. The color can be used to draw conclusions about the direction in which the vehicle is driving. When white light is detected, the vehicle approaches the observer (white light functions at the front and as reverse light). If the detected light function is red, it can be assumed that the vehicle is moving away from the stationary observer. The direction indicator can be clearly identified not only by its color but also by its flashing frequency. Flashing light functions are used in more safety–critical situations to additionally increase attention (hazard warning lights, change of direction, hazard braking) [25]. Furthermore, it is evident that light functions that must also be recognizable during the day require higher luminous intensities.

With the introduction of highly automated driving functions, there is the possibility of displaying additional information about a planned maneuver or also additional warnings for road users who are in danger outside the vehicle. Possible information displayed by an automated vehicle could be, e.g., information about the driving state (manual or automated), information about future driving maneuvers (e.g., vehicle will brake or drive off), information about the perception of the environment (were pedestrians or other road users detected?), possibilities for cooperation such as requests to pedestrians to stop or walk [27].

In the following subsections, different novel signal installation concepts are highlighted and the advantages and disadvantages are discussed. From a lighting technology perspective, these can be divided into three categories: Additional signal lamps or light strips, displays, and road projections.

Fig. 30.22 Signal concept from Ford and Virginia Tech with a light strip above the windshield [28]

30.7.2 Additional Signal Lamps and Light Strips

Various concepts for this are shown in ◉ ▣ Fig. 30.22 and ◉ ▣ Fig. 30.23. It can be seen that light strips are installed between the windshield and the roof of the vehicle so that they are at the eye level of pedestrians who are in the immediate vicinity of the vehicle.

By using light strips, additional mounting around the entire vehicle can be realized, making detection possible from any position relative to the vehicle. With this technology, it can be assumed that light intensities similar to those of existing signaling devices can be generated to ensure detectability both during the day and at night. Limited coding options are available here for the transmission of information. For example, animations can be used to indicate driving maneuvers [30], or color can provide information about driving status [31].

30.7.3 Display-Based Signaling Devices

The use of displays offers a wide range of possibilities for displaying information to the outside world by means of symbols or symbol animations. Integration into the vehicle is more difficult here than with normal signal lights. The concepts in ◉ ▣ Fig. 30.24 und ◉ ▣ Fig. 30.25 show that displays tend to be attached to the front of the vehicle. There, the signals are easily visible in the event of a direct encounter with another road user.

Daytime recognizability is difficult to achieve with conventional display technologies whose luminances are in the range of $200\ldots1000\,\mathrm{cd\,m^2}$ ($50\ldots250$ cd for a display size of $50\cdot50\,\mathrm{cm^2}$) When using symbols or lettering, the size of the display and the resulting maximum distance of road users to the display must also be taken into account, which must not be exceeded for clear recognition of the signal (see [33]). In order to achieve the light intensities required for daytime detection, there are concepts for this purpose that do not use high-resolution displays but LED arrays with a lower resolution (see ◉ ▣ Fig. 30.25). Here, the number of symbols that can be displayed is limited, but the required luminance for recognition during day and night is guaranteed. Displays can also be used to indicate future driving states or cooperation possibilities [34].

■ **Fig. 30.23** Nissan IDS concept car with light strips above the doors [29]

■ **Fig. 30.24** The Smart Vision EQ fortwo concept car from Daimler with display in the area of the radiator grille [32]

Fig. 30.25   The AI:ME concept car from Audi with display in the area of the front headlight Source: Audi AG

30.7.4 Road-Projections

In addition to road projections as an additional information channel for drivers, there are concepts that make it possible to communicate with other road users via projections (see ◉ ▫ Fig. 30.26). High-resolution DMD systems in the headlights can also be used to generate various symbols that can be projected into the direct path of pedestrians. In this case, the light signal would even be recognizable if no direct view of the vehicle is possible.

Due to the low reflectance of road surfaces (< 10%) the light intensity of the projection system must be high enough to achieve the threshold contrast for detecting a symbol. This is only possible at night with common light intensities of current headlamp systems. Other disturbance variables are weather conditions such as rain or snow, which can have a high impact on the detectability of roadway projections. A possible area of application would therefore be more covered locations such as parking garages, since there is no influence of weather or direct sunlight here.

30.7.5 Summary and Discussion

From a photometric point of view, the three categories presented offer various advantages and disadvantages, which become important depending on the visibility or the signal to be displayed. The first step is to define whether and which additional messages an automated vehicle should display. From a regulatory perspective, messages that indicate driving status or information about future maneuvers are the easiest to implement. Prompts to other road users are more difficult to implement here, as it must always be ensured that the right person receives the message. However, current studies show that road users generally benefit from additional signals [35].

30.8 Outlook

In the development process of new vision enhancement systems, the focus in recent decades has always been on the driving human being. With the introduction of

Zukunftsausblick: digitale Matrix LED-Technologie

Future perspective: Digital matrix LED light
11/20

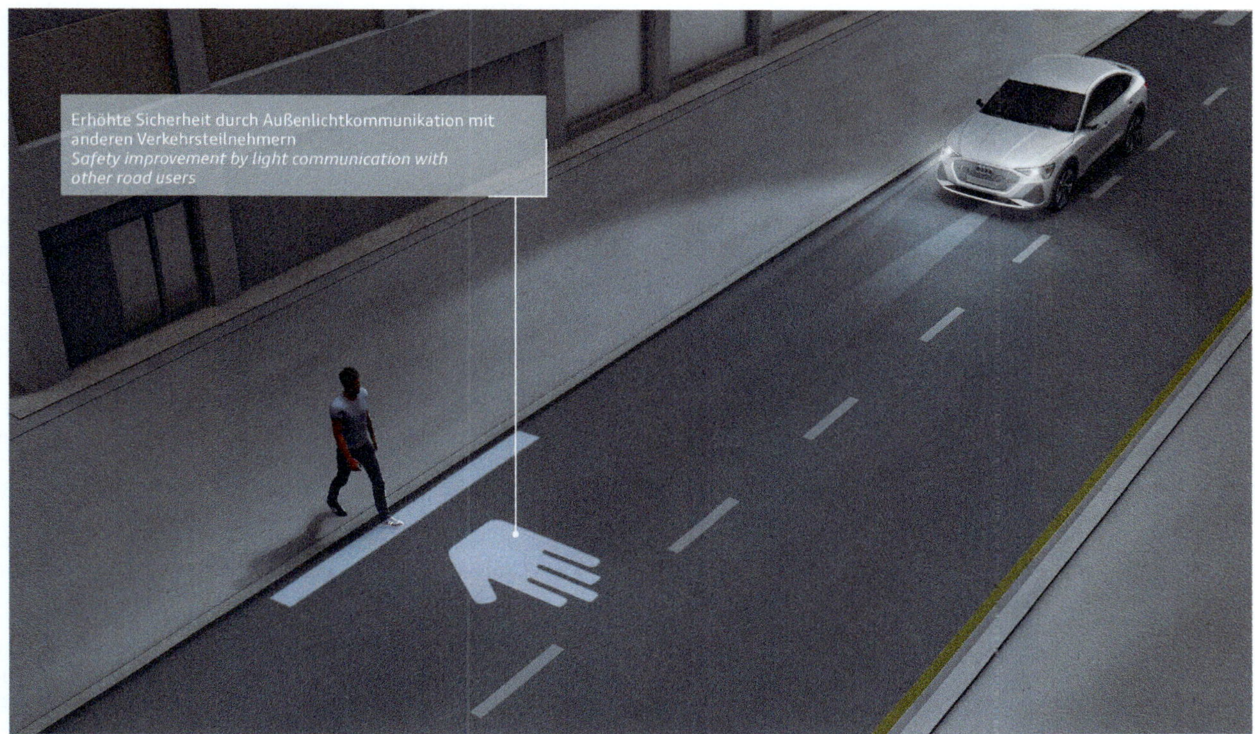

◘ Fig. 30.26 Symbol projection in the path of a pedestrian with a high-resolution DMD headlight system from Audi *Source* Audi AG

highly automated vehicles, the question arises whether this will fade into the background in the future. On the one hand, there will be a very long transition period during which highly automated vehicles will share the road with manually controlled vehicles. Consequently, conventional headlamp systems will not disappear from the production program. On the other hand, the number of sensors in vehicles continues to increase, and camera-based sensors, like humans, need enough light to detect objects on the road even in poor visibility conditions. This opens up a new field of research that focuses on optimizing lighting systems for sensors.

The research field of novel light signaling devices has been growing over the last years and the many new concepts promise to simplify the integration of automated vehicles into existing road traffic. However, from a regulatory point of view, it has not yet been clarified what kind of new signals should be implemented. Many questions can only be answered here once a large number of automated vehicles are on the road. However, early consideration of new signaling devices provides the basis for finding a uniform solution at the international level.

Another area of research not considered in detail here in this chapter is vehicle interior lighting. When drivers no longer have to concentrate solely on the driving task, there is time for other activities. Here, the focus shifts back to the human beings, and light can be used here to enhance their sense of well-being while driving.

References

1. Schlag, B., Petermann, I., Weller, G., Schulze, C.: Mehr Licht — mehr Sicht — mehr Sicherheit? VS Verlag für Sozialwissenschaften, Wiesbaden (2009)
2. Åkerstedt, T., Kecklund, G., Hörte, L.G.: Night driving, season, and the risk of highway accidents. Sleep, 24. Jg. Nr. **4**, 401–406 (2001)
3. Lerner, M., Albrecht, M., Evers, C.: Das Unfallgeschehen bei Nacht. Berichte der Bundesanstalt für Straßenwesen, Fahrzeugtechnik, Heft M 172 (2005)
4. Sullivan, J.M., Flannagan, M.J.: The role of ambient light level in fatal crashes: inferences from daylight saving time transitions. Accid. Anal. Prev. **34**, 487–498 (2002)
5. Zydek, B., Schiller, C., Polin, D., Khanh, T.Q.: Evaluation of headlamps with a glare-free high beam function. ATZ Worldw **116**, 46–51 (2014)

6. Kobbert, J.: Optimization of automotive light distributions for different real life traffic situations. Dissertation, Technische Universitaet Darmstadt (2019)

7. Adrian, W.: Die Unterschiedsempfindlichkeit des Auges und die Möglichkeit ihrer Berechnung. Lichttechnik, 21. Jg. Nr. **1**, 2A-7A (1969)

8. Manz, K.: Licht und Displaytechnik, Visuelle Leistungsfähigkeit, Vorlesungsunterlagen (2009). ► https://www.lti.kit.edu/rd_download/licht2008/Grundlagen_der_Lichttechnik-LTI-Basiskurs-05.pdf. Accessed 16 Nov 2021

9. De Boer, J.B.: Visual perception in road traffic and the field of vision of the motorist. Pub. Light. S. 11–96 (1967)

10. Kobbert, J., Löffler, W., Kosmas, K., Khanh, T.Q.: Société des Ingénieurs de l'Automobil (SIA) (2016): Estimation of glare load from small illuminance peaks in real life driving situations. VISION 2016, Vehicle and Infrastructure Safety Improvement in Adverse Conditions and Night Driving, Paris, 13.-14. Okotber 2016 (2016)

11. Brutting, W., Rieß, W.: Grundlagen der organischen Halbleiter. Physik Journal, 2008, 7. Jg. Nr. **5**, 33–38 (2008)

12. Sprute, J.H.: Entwicklung lichttechnischer Kriterien zur Blendungsminimierung von adaptiven Fernlichtsystemen. Technische Universitaet Darmstadt, Dissertation (2012)

13. Rosenhahn, E.O.: Entwicklung von lichttechnischen Anforderungen an Kraftfahrzeugscheinwerfer für Schlechtwetterbedingungen. Herbert Utz Verlag (2000)

14. Kosmas, K.: Optimierung von adaptiven Kfz-Scheinwerfertechnologien zur Blendungsbegrenzung unter dynamischen Bedingungen. Dissertation, Technische Universitaet Darmstadt (2019)

15. Schneider, D.: Markierungslicht-eine Scheinwerferlichtverteilung zur Aufmerksamkeitssteuerung und Wahrnehmungssteigerung von Fahrzeugführern. Herbert Utz Verlag (2011)

16. Hamm, M.: Real driving benefits and research findings with digital light functions. 13th International Symposium on Automotive Lighting (ISAL). Darmstadt, Germany, 23.09 – 25.09.2019 (2019). ISBN 978–3–8316–4817–7

17. Commission Internationale d'Éclairage (CIE).: Vehicle Headlighting Systems Photometric Performance–Method of Assessment. CIE 188/E (2011)

18. Hamm, M., Hinterwaelder, C.: Investigations on headlamp and car body tolerances in real life. SAE Technical Paper (2020)

19. Langhammer, G.: Ideas for including ADB functionality into the TC4–45 assessment system. 13th International Symposium on Automotive Lighting (ISAL), Darmstadt, Germany, 23.09 – 25.09.2019 (2019). ISBN 978–3–8316–4817–7

20. Kosmas, K., Kobbert, J., Khanh, T.Q.: Anforderungen an die dynamische Leuchtweitenregelung zur Vermeidung der Blendung entgegenkommender Verkehrsteilnehmer. Berichte der Bundesanstalt für Straßenwesen, Fahrzeugtechnik, Heft F 129 (2019)

21. Schiller, C.: Dirty headlamps-efficiency of headlamp cleaning devices and the impact on stray light–method and first results.12th International Symposium on Automotive Lighting (ISAL). Herbert Utz Verlag GmbH, München, S. **2017**, 145–154 (2017)

22. Straßenverkehrs-Ordnung (StVO), Date of issue: March 6, 2013

23. Sucha, M., Dostal, D., Risser, R.: Pedestrian-driver communication and decision strategies at marked crossings. Accid. Anal. Prev. **102**, 41–50 (2017)

24. American Automobile Association.: Vehicle Technology Survey–Phase IIIB: Fact Sheet, Heathrow, FL, USA (2018)

25. Boyce, P.R.: Human Factors in Lighting, 2nd edn. Taylor & Francis, London, New York (2003)

26. ECE R-48: Uniform Provisions Concerning the Approval of Vehicles With Regard to the Installation of Lighting and Light-Signalling Devices, Effective date: July 19, 2018

27. Schieben, A., Wilbrink, M., Kettwich, C., Madigan, R., Louw, T., Merat, N.: Designing the interaction of automated vehicles with other traffic participants: design considerations based on human needs and expectations. Cogn Tech Work **21**, 69–85 (2019)

28. Ford: Ford, Virginia Tech Go Undercover to Develop Signals That Enable Autonomous Vehicles to Communicate with People, ► https://media.ford.com/content/fordmedia/fna/us/en/news/2017/09/13/ford-virginia-tech-autonomous-vehicle-human-testing.html (Online) (2017). Accessed 16 Nov 16

29. Nissan: Nissan IDS Concept. ► https://www.nissan.co.uk/experience-nissan/concept-cars/ids-concept.html (2017). Accessed 16 Nov 2021

30. Habibovic, A., Lundgren, V.M., Andersson, J., Klingegård, M., Lagström, T., Sirkka, A., Fagerlönn, J., Edgren, C., Fredriksson, R., Krupenia, S., Saluäär, D., Larsson, P.: Communicating intent of automated vehicles to pedestrians. Front. Psychol. **9**, 1336 (2018)

31. Fuest, T., Feierle, A., Schmidt, E., Bengler, K.: Effects of marking automated vehicles on human drivers on highways. Information **11**, 286 (2020)

32. Daimler: Autonomous concept car smart vision EQ fortwo: Welcome to the future of car sharing, ► https://media.daimler.com/marsMediaSite/en/instance/picture/Autonomous-concept-car-smart-vision-EQ-fortwo-Welcome-to-the-future-of-car-sharing.xhtml?oid=29043146&ls=L2VuL2luc3RhbmNlL2tvLnhodG1s-P3JlbElkPTEwMDEmZnJvbU9pZD0yOTA0MjcyNSZyZXN1bHRJbmZvVHlwZUlkPTE3MiZib3JkZXJJzPXRydWUm-ZnJvbUluZm9UeXBlSWQ9NDA2MjYmb2lkPTI5MDQyN-zI1&rs=18 (2017). Accessed 16 Nov 2021

33. Graphical Symbols–Safety Colours and Safety Signs—Part 1: Observation Distances and Colorimetric and Photometric Requirements, DIN Deutsches Institut für Normung e. V., Berlin (2012)

34. Singer, T., Kobbert, J., Zandi, B., Khanh, T.Q.: Displaying the driving state of automated vehicles to other road users: an international, virtual reality-based study as a first step for the harmonized regulations of novel signaling devices. IEEE Trans. Intell. Transport. Syst 1–15 (2020)

35. Rouchitsas, A., Alm, H.: External human-machine interfaces for autonomous vehicle-to-pedestrian communication: a review of empirical work. Front. Psychol. **10**, 272 (2019)

36. Hinterwaelder, C., Hamm, M.: Analysis of human intra-and interpersonal aiming accuracy of cutoff lines using different adjustment methods. No. 2021–01–0849. SAE Technical Paper (2021)

37. Kobbert, J., Khanh, T.Q.: Objective assessment of the safety contribution of today's automotive headlamps. ATZ Worldw **122**, 66–71 (2020)

30

Open Access This chapter is licensed under the terms of the Creative Commons Attribution-NonCommercial-NoDerivatives 4.0 International License (▶ http://creativecommons.org/licenses/by-nc-nd/4.0/), which permits any noncommercial use, sharing, distribution and reproduction in any medium or format, as long as you give appropriate credit to the original author(s) and the source, provide a link to the Creative Commons license and indicate if you modified the licensed material. You do not have permission under this license to share adapted material derived from this chapter or parts of it.

The images or other third party material in this chapter are included in the chapter's Creative Commons license, unless indicated otherwise in a credit line to the material. If material is not included in the chapter's Creative Commons license and your intended use is not permitted by statutory regulation or exceeds the permitted use, you will need to obtain permission directly from the copyright holder.

Low Speed Assistance

Ahmed Benmimoun and Sebastian Klaudt

Contents

© The Author(s) 2026
H. Winner et al. (eds.), *Handbook Assisted and Automated Driving*,
https://doi.org/10.1007/978-3-658-45276-6_31

31.1 Introduction

Maneuvering in the low-speed range of up to approx. 15 km/h (parking, maneuvering, turning, etc.) is often a challenge for the driver, especially when reversing. The changed driving dynamics compared to driving forwards, the strenuous posture (e.g., looking backwards and simultaneously steering the vehicle with the steering wheel, perception by means of mirrors.) make this driving task more difficult. In addition, the duration of this driving task is short compared to the total driving time. As a result, the driver usually lacks practice and routine.

Due to strict requirements on the aerodynamics of the vehicle to minimize fuel consumption as well as design-related shaping of the vehicle geometry, the driver's view of the immediate surroundings is often restricted, and the dimensions of the vehicle cannot be accurately estimated.

Low-speed maneuvering also usually occurs in complex environments with easily overlooked road users and objects in the near vicinity of the vehicle (pedestrians and cyclists, posts, bicycles, flowerpots, etc.).

Many drivers therefore feel partially overwhelmed when maneuvering in the low speed range and try to avoid such driving tasks if possible. Reference [1] shows that the heart rate increases significantly during a parallel parking task compared to the phase before and after parking, indicating that the parking task is stressful and challenging.

The difficulty of this driving task is reflected in accident statistics by a high number of accidents in the low-speed range. For example, [2] shows that around 40 percent of all passenger car accidents with material damage are parking or maneuvering accidents. Around 80% of these accidents occur while reversing.

Reference [3] reports that in the USA about 290 people are killed each year in accidents involving reversing vehicles. The largest proportion of these accidents (about 75%) occur off the road (garage driveways, parking lots, etc.). Children under the age of five account for 44% of those killed in these car accidents, and people aged over 70 account for 33%. According to the study it is estimated that the number of people injured in reversing accidents is around 18 000 per year.

Driver assistance systems can make the driving task easier for drivers in these situations by assisting them in finding a parking space, taking over sub-tasks when parking and backing out, supporting perception of surrounding objects, and warning them of obstacles, thus helping to avoid accidents or minimize the damage caused by accidents [1, 4, 5].

31.2 Classification of Low-Speed Assistance and Automation Systems

There is currently a great variety of assistance and automation systems that aim to support the driver when parking and maneuvering. The type of assistance varies depending on the system and can be classified as follows:
- Informing and warning assistance systems:

In this category, the driver is supported by the provision of information in various ways (acoustically, visually, and haptically). The purpose of these systems is to present drivers with a simplified model of the environment by aggregating information, to direct their attention to relevant objects in the vehicle's surrounding, to point out objects outside their field of vision, or to make it easier for them to steer the vehicle by visualizing the vehicle geometry and the vehicle's predicted path.

In some cases, the system evaluates the information collected by environmental sensors. The system calculates the probability of a collision with other road users or objects in the surrounding of the vehicle. In the event of an impending collision, the driver is warned and prompted to take corrective action.
- Intervening assistance systems:

Assistance systems aiming at collision avoidance in particular assist drivers by additionally intervening in the vehicle actuation system if they have not previously reacted to indications or warnings. In some situations, drivers may not be able to avoid an impending collision despite a warning from the system, for example, if the collision is detected too late and they do not have enough time to react. Therefore, it is more effective if the assistance system itself intervenes by braking the vehicle and removing the drive torque, thus reducing the driving speed or bringing the vehicle to a standstill. Research is also being conducted on systems that, while maneuvering manually at low speeds, also intervene by means of steering torque in order to influence the trajectory of the vehicle [6].
- Partially assisted systems:

While parking and maneuvering, the driver has to perform a variety of actions (steering, braking, pressing the accelerator pedal, shifting gear, etc.) and carry out several tasks simultaneously (detecting obstacles, planning the trajectory, guiding the vehicle longitudinally and laterally, etc.). The basic idea of this group of assistance systems is to completely take over some of these tasks, thus enabling the driver to better concen-

trate on the remaining tasks. For example, in semi-assisted parking, the driver is responsible for monitoring the vehicle's surrounding and longitudinal control. Trajectory planning and lateral guidance are taken over completely by the assistance system.
— Fully assisted systems supervised by the driver:

In this form, the assistance system takes over environment detection and monitoring, trajectory planning and longitudinal and lateral guidance. The driver is therefore no longer involved in vehicle control. The driver's task is to activate and monitor the system and to intervene in the event of system errors or inappropriate system behavior. Since this is still L2 automation according to SAE, the system is not designed to manage all possible situations and to recognize all relevant objects and road users in all scenarios and under all conditions. The driver is still fully responsible for the driving task.

In the event of a system error (e.g., failure to detect an obstacle), the driver is required to act and prevent a collision by intervening by means of the vehicle actuators or by a mechanism, acting like a dead man's switch. In order to do this, the driver does not necessarily have to be sitting in the vehicle. In remote-controlled parking assistance systems, drivers are outside the vehicle and can use a remote control to bring the vehicle to a stop if they deem it necessary. Several vehicle manufacturers in series production vehicles already offer this type of system.
— Fully automatic systems without driver supervision:

This category describes systems that are classified as L4 automation according to SAE. Here, the driver no longer has to supervise the vehicle and can perform other tasks while the vehicle maneuvers automatically. In case of parking functions, the user no longer needs to remain near the vehicle once the system has been activated to park the vehicle.

The majority of these systems are still at the research and development stage. However, the first isolated systems (automated valet parking) are already available to end customers in some markets (China and Germany).

In addition to the type of assistance and type of automation, low-speed assistance/automation can also be classified according to the area of use and application.
— Parking

This group of assistance/automation systems (▶ Sect. 31.4) supports the driver when parking or pulling out of a parking spot. This also includes the search for and the identification of a suitable parking space. The assistance/automation provided ranges from the display

of obstacles in the area of the parking space, visual guidance of the driver when steering, taking over lateral and longitudinal guidance, up to automated and driverless maneuvering in the parking garage or in a parking lot.
— Reversing and maneuvering

A large number of assistance/automation systems (▶ Sect. 31.4.2) address reversing. The driver is shown the area behind the vehicle and is warned of road users or objects behind the vehicle or crossing road users. In case of an imminent collision with these, the systems intervene by means of braking or the entire reversing journey is carried out automatically.
— Driving with trailer

Driving with a trailer is particularly difficult for many drivers due to the unfamiliar kinematics of the vehicle-trailer-combination. This group of assistance systems (▶ Sect. 31.4.3) helps the driver, for example, to monitor the surroundings around the trailer, to reverse a vehicle with a trailer, and to position the vehicle appropriately for coupling a trailer.

31.3 Historical Development of Driver Assistance for the Low Speed Range

Passive maneuvering and parking aids were already introduced in the 1950s. These included flexible dipsticks or a corresponding design of the body (e.g., tail fins), which were intended to make it easier for the driver to assess the orientation and dimensions of the vehicle [8]. These dipsticks were attached to the front and rear fenders for this purpose. Later, these were concealed and extended automatically as soon as reverse gear was engaged. When parking, so-called "curb feelers" were supposed to assist. These were mounted on the side of the sills and produced a scraping sound when they touched the curb during parking [9].

In the 70s and 80s, active systems were developed that monitored the vehicle environment by means of sensors [7]. These were only used in isolated cases because, among other things, they were very expensive. These systems were not widely used until the beginning of the 1990s. ◨ Figure 31.1 provides an overview of assistance and automation systems for the low-speed range. These are classified according to the area of application (blue = parking, green = reversing and maneuvering, orange = driving with a trailer), type of assistance, and according to when they were introduced to the market. The naming and description of the systems may differ depending on the car manufacturer, but the scope of functions is generally comparable.

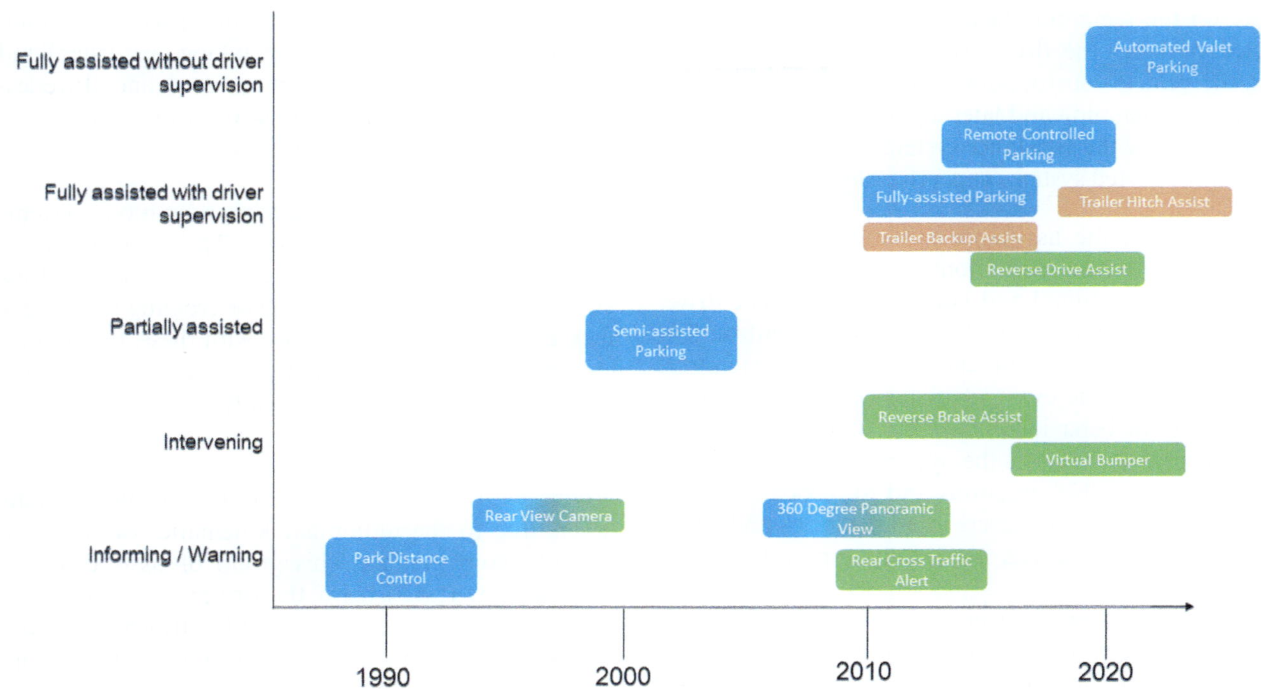

Fig. 31.1 Development of assistance and automation functions for the low-speed range

31.4 Technical Realization

31.4.1 Parking Applications

31.4.1.1 Parking Assistance/Park Distance Control

Parking aid (also known as Park Distance Control by most car manufacturers) uses ultrasonic sensors installed in the bumper to measure the distance to obstacles. In the simplest version, the driver is warned by an acoustic tone. When the system is activated, the tone starts to sound at a distance of approximately 150 cm. The frequency of the tone signal (tones per second) increases as the distance to the obstacle decreases (echo sounder principle). From a distance of 30 cm or 20 cm on, a continuous tone is emitted. In addition, many models also have a visual display, which shows the distance to obstacles schematically and color-coded on a display (**Fig. 31.2** right). In addition to activation by pressing a button, the system is usually activated automatically when reverse gear is engaged and deactivated at a speed of 15 to 20 km/h.

The number of ultrasonic sensors installed varies (two-, four-, and six-channel systems, number of sensors installed per bumper). The more sensors are installed, the larger the area around the vehicle that can be monitored. In addition, the quality of the measurement results improves.

To determine the distance, the sensors transmit ultrasonic pulses in combined transmit and receive mode. The signal reflected by an obstacle (echo) is received by the sensor, amplified, and further processed in the control unit. In the process, each sensor also receives echoes from signals transmitted by neighboring sensors. The distance to the obstacle is calculated from the propagation time of the signal from the sensor to the obstacle and back. The signal travels at the speed of sound.

Alternatively to ultrasonic sensors, short-range radars or ultra-short-range radars can also be used for this function. These can be installed in a concealed manner and are therefore not visible from the outside. One disadvantage is the high cost, which is only worthwhile when used in combination with other assistance functions.

Due to its simple design, Park Distance Control has many limitations that can lead to missing warnings or false warnings. For example, narrow objects (thin posts) or people wearing sound-absorbing clothing may be overlooked. For dynamic objects, the warning is often too late due to system latency. Air pressure, humidity, and noise sources affect the propagation of sound and therefore the accuracy of the distance measurement. Dust, water, and other substances on the sensor membrane are partially detected as objects at close distance.

Performance requirements and corresponding test procedures are described in the Maneuvering Aids for

Fig. 31.2   Left: View of rear view camera with integrated PDC, Right: View of stand-alone PDC

Low Speed Operation (MALSO) standard (ISO 17386) [10]. Further regulatory requirements can be found in ECE R158 (devices for reversing motion and motor vehicles with regard to the driver's awareness of vulnerable road users behind vehicles).

31.4.1.2   Rear View Camera

The rear view camera makes reversing easier for the driver because the uncomfortable posture caused by looking to the rear is not permanently required. In addition, the rear view camera allows drivers to see objects that they would otherwise not be able to see because they are obscured by the vehicle.

For this purpose, wide-angle cameras with a large horizontal aperture angle (approx. 180°) are used, which make it more difficult to detect objects in the far distance but allow a large area behind the vehicle to be covered. In addition, the camera is directed downwards so that even small objects (bollards, curbs, etc.) in the immediate vicinity of the vehicle can be detected.

The image shown to the driver is mirrored so that the driver has a familiar and consistent view compared to the rear view mirror. The image is usually shown on the central display in the vehicle. In some vehicles, the display is also integrated into the rear view mirror.

The rear view camera is activated automatically when reverse gear is engaged and is switched off automatically after a defined forward speed has been reached.

Both when reversing and parking, auxiliary lines are displayed in the camera image to facilitate the maneuver (**Fig. 31.2** left). The auxiliary lines visualize the current driving path (without steering angle) as well as the predicted driving path (when driving on with current steering wheel position). This enables drivers to better estimate how far they have to steer in order to come to a stop parallel to the curb, for example. In addition, further color-coded auxiliary lines are drawn to estimate the longitudinal distance. In combination with ultrasonic sensors, nearby objects in the image can also be highlighted in color.

Due to the large number of accidents while reversing and the high number of children involved in these accidents, rear view cameras have been mandatory for all passenger cars in the USA and Canada since May 2018. According to a study by the Insurance Institute for Highway Safety (IIHS), reversing accidents can be reduced by about 17% with the help of rear view cameras [11].

Regulatory requirements for rear view cameras can be found in ECE R158 (devices for reversing motion and motor vehicles with regard to the driver's awareness of vulnerable road users behind vehicles) and FMVSS 111 (Rear Visibility).

31.4.1.3   Degree Panoramic View

The rear view camera has proved to be very helpful since its market launch. As a consequence, further cameras have been installed in vehicles to expand the scope of visual support for the driver. Compared to the rear view camera systems, there are also systems with a wide-angle camera installed at the front of the vehicle. These support drivers in situations in which they have to detect obstacles very close in front of the vehicle or if the view of the flowing traffic is obscured by obstacles on the side of the vehicle (e.g., at a garage exit) (**Fig. 31.3**).

With the installation of additional side cameras, which are usually integrated in the exterior mirrors, a complete surround view can be achieved. The driver is thus able to monitor the entire vehicle environment, as each of the cameras has an aperture angle of over 180°.

The images from the four wide-angle cameras are read into a control unit and processed for the driver.

◘ Fig. 31.3 Split view using front wide-angle camera (*Source* Ford Motor Company)

For this purpose, the four individual camera images are stitched together. Since the camera images overlap each other at the horizontal edge, the camera images can be seamlessly connected with each other, thus creating a true 360° surround view. Further processing includes a calculation of a virtual camera view from a bird's eye view, which gives the driver a good overview of objects in the near vehicle environment (◘ Fig. 31.4). In addition, a 3D view of the vehicle environment can be displayed, in which the driver can position the virtual camera as desired to take a closer look at a specific area. However, since, the wide-angle camera has only a limited vertical aperture angle, the image has to be partially extrapolated. Thus, very close objects often do not appear in the correct proportions but are vertically distorted. The vehicle itself is only partially within the cameras' field of view and is animated by either an image (bird's eye view) or a 3D model.

Systems are currently being researched in which augmented reality is used to make the bonnet or trailer transparent so that the user can see what is directly in front of or underneath the vehicle or what the traffic situation is like behind the trailer (◘ Fig. 31.5). For this purpose, an appropriately positioned camera is used (at the front of the bumper looking downwards/at the trailer looking backwards), the displayed images of which are processed and shown to the driver by means of a head-up display or a display integrated in the rear view mirror. Thus, the camera image and the real view are superimposed, and the driver is given the impression of transparency.

31.4.1.4 Semi-assisted Parking

The broad introduction of electromechanical steering systems laid the foundation for assisted/automated lateral guidance. This also made active parking assistance possible. In semi-assisted parking, the driver first drives past the slot. Potential parking spaces are identified and measured by the vehicle's sensor system. Depending on the car manufacturer, this takes place automatically at speeds below about 40 km/h, or the driver must activate the function by pressing a button. The dimen-

Fig. 31.4 360-degree bird's eye view (*Source* Ford Motor Company)

Fig. 31.5 Left: Transparent bonnet, Right: Transparent trailer (*Source* Jaguar Land Rover Ltd.)

sions of the parking spaces detected are evaluated and displayed to the driver if they are large enough (in case of most manufactures: vehicle length + about 80 cm for parallel parking spaces, vehicle width + about 80 cm for perpendicular parking spaces). In some systems, parking spaces found are always communicated to drivers; in other systems, drivers must first select which side of the lane they want to park on and whether it is a parallel or perpendicular space. In partially assisted mode, they must first stop the vehicle at a position defined by the system and activate the partially assisted parking system, which is usually done by engaging reverse gear.

The system now takes over lateral control and steers the vehicle along a calculated trajectory. Meanwhile, drivers are responsible for longitudinal control (accelerator pedal, brake pedal and gear shifting). The system shows them when they should bring the vehicle to a halt and when they should change direction by engaging the appropriate gear. In doing so, they should drive slowly; as otherwise, the vehicle will not be able to follow the planned trajectory due to the limited steering wheel angular velocity. In addition, potential obstacles might not be detected in time. The system maneuvers the vehicle into the parking space, trying to achieve an ideal parking position. In the case of parallel gaps, this is achieved when the vehicle is parallel to the curb with a lateral distance of about 15 cm and is centered between the two bordering vehicles. In case of perpendicular gaps, the vehicle aligns itself centrally between the two adjacent vehicles in lateral and longitudinal position (◘ Fig. 31.6). There are also systems that are able to detect the parking lines and use these for the final alignment of the vehicle.

The system does not only assist with parking, but also with pulling out in case of parallel parking spaces. In this case, the vehicle is steered out of the parking space until the driver can take over and drive out of the space without any further steering maneuvers.

The first systems of this type used only ultrasonic sensors for the detection of parking spaces and the surrounding area. Later on, the environment detection was expanded to include radars and cameras, which also enabled the detection of parking lines, objects at a longer distance, etc. At first only parallel gaps were supported, later on also perpendicular gaps and some car manufacturers also offered parking forward or backwards into perpendicular gaps. However, the driver and the surrounding traffic perceive the trajectory driven when parking forward very unnatural, as with today's systems the vehicle must first have completely passed the parking space for scanning purposes.

For planning and tracking of the trajectory, the position of the vehicle relative to the parking space must be known at any time from the moment of the detection of the parking space. For this purpose, odometry is used, which is mainly based on the observation of wheel speeds. In order to keep the slip-induced error as small as possible, usually only the non-driven wheels are considered. The effective rolling radius is essential for determining the distance traveled and must therefore be estimated continuously (e.g., by means of suitable models), as it can change as a result of tire wear, tire changes, etc.

In general, these systems work very well and are especially helpful for inexperienced drivers. However, depending on the system design, there are some weaknesses causing the systems to function incorrectly. The quality of detection of parking spaces is highly dependent on the sensor technology used. In case of

◘ **Fig. 31.6** Semi-assisted parking for longitudinal and perpendicular slots (*Source* Ford Motor Company)

purely ultrasonic sensor-based systems, "unsuitable" gaps are often detected, since the system in principle classifies all suitable free spaces between two obstacles as parking slots. In addition, suitable gaps are sometimes rejected due to false detections (inaccurate measurement of width, ground reflections detected as obstacles, etc.). In some systems, the classification according to parallel and perpendicular slot is done automatically. In this case, it can happen that two adjacent free perpendicular slots are detected as parallel slot, since the ultrasonic sensors are not able to measure the parking space depth due to their limited range.

Before activating the system, the driver therefore needs to be aware of whether the selected parking space is actually suitable for parking. This is difficult because the driver often cannot assess where the system wants to park. Car manufactures are trying to solve this problem with suitable HMI concepts by converting the received sensor information into an environment model that is displayed to the driver. However, this environment model may be very rudimentary in case of purely ultrasound-based systems and deviates often significantly from the real environment.

Another weakness of these systems is that they often require more moves than a trained driver would perform. Firstly, this is due to the starting position, resulting from the need for the vehicle to drive completely past the parking space parallel to the road. Secondly, the gap is sometimes inaccurately detected and therefore the planned trajectory is not suitable. In addition, odometry errors can cause the driven trajectory to deviate from the planned one. However, this is only detected after the vehicle has partially driven into the slot, thus requiring corresponding corrective moves.

ECE-R79.03 (ACSF-A) regulates the most important framework conditions for assisted parking. It, e.g., defines that the automatic steering function during parking may only take place up to a maximum of 10 km/h (with a tolerance of 20%) [12]. This applies to all functions with an automatic steering function in the low-speed range (fully automatic parking, remote-controlled parking, Trailer Back Assist, and Trailer Hitch Assist).

ISO 16787 (Assisted parking system (APS)) defines minimum requirements for the functionality (detection of parking spaces, calculation of a trajectory, lateral guidance, detection of obstacles in the path) and performance of semi-assisted parking functions and corresponding test scenarios [13].

31.4.1.5 Fully Assisted Parking

Based on the semi-assisted parking, the fully assisted parking system takes over the longitudinal guidance in addition to the lateral guidance. The systems do not differ from each other in the slot detection phase. Once the slot has been detected, the driver must put the transmission in a position defined by the car manufacturer (reverse, neutral, or park) and apply the brake. Subsequently, drivers are released from active vehicle control and only have to perform a monitoring function. Depending on the manufacturer, they must press a button and keep it pressed (dead man's switch concept, where the vehicle is brought to a halt as soon as the drivers release the button) or simply observe the surroundings. They can override the system at any time and, for example, brake and bring the vehicle to a standstill.

After activation, the vehicle shifts to the appropriate gear (automatic transmission or automated transmission) and follows the planned trajectory, also adjusting the speed accordingly. Most systems move the vehicle at a speed of about 3 to 4 km/h. The system steers, changes the direction of travel and brakes in case of obstacles. When the target position is reached, the vehicle is brought to a standstill and secured against rolling away (parking position and/or electronic park brake).

Depending on the manufacturer, these systems can also park at slopes of up to about 15% and drive over curbs if the location of the parking space makes this necessary.

31.4.1.6 Remote-Controlled Parking

Remote-controlled parking differs from fully assisted parking mainly in that drivers are no longer in the vehicle and controls it from outside. This allows them to park in and park out from very tight parking spaces and garages, since they no longer have to get in or out of the vehicle while the vehicle is parked (◘ Fig. 31.7). However, in doing so, they must respect the drivers of the neighboring vehicles and park in such a way that the drivers of these vehicles are able to get into their vehicles.

Another advantage of remote-controlled parking is that users have a better overview from the outside and can position themselves accordingly around the vehicle and thus monitor the parking maneuver well.

To activate and control the system, the driver uses a remote control. Some car manufacturers use a smartphone for this purpose, while others use the vehicle key fob (sometimes equipped with a touch display). The driver can select the parking maneuver partially already in the vehicle, then get out, and activate the system with the remote control.

After activation, the vehicle drives along the planned trajectory and parks in the previously detected slot. After completion of the parking process, the vehicle is secured against rolling away, locked and the engine is turned off. After activation of pulling out, first, the engine is started, and then the vehicle drives out of the parking space and hands over to the driver (for most manufacturers with the engine turned off).

31

◙ Fig. 31.7   Remote-controlled parking in a narrow space (*Source* Ford Motor Company)

In addition to parking, most manufacturers also offer an operating mode that allows the driver to move the vehicle over a defined distance in one go. This is particularly useful in situations where the driver just wants to position the vehicle differently without intending to drive away (e.g., to reach objects in the garage, to get to the trunk, etc.). In this case, the vehicle drives straight ahead and avoids obstacles that protrude slightly into its path. If these obstacles are near the center of the driving path, the vehicle just stops. Many manufacturers also use this operating mode for parking and backing out. When parking, however, the vehicle must be positioned in front of the garage or parking space in such a way that it can drive into the parking position with only minor steering corrections and without requiring to change driving direction. However, this eliminates the need to scan and detect the gap.

The driver is still responsible for the parking maneuver and must continuously press a button on the remote control or perform an action on the remote control over the entire duration of the maneuver (dead man's switch concept). As soon as the driver stops the action, the vehicle is immediately stopped and secured against rolling away.

The vehicle also stops if the driver moves too far away from the vehicle. For this purpose, the distance from the vehicle to the driver is determined. As a rule, the key fob, which drivers must have with them, is localized for this purpose. In principle, drivers could also be localized via Bluetooth in smartphone-based systems, as these communicate with the vehicle. How-

ever, this is not precise enough and leads often to false alarms (drivers are indicated to be outside of valid range although they are inside valid range).

Remote-controlled parking poses additional challenges to the system. Since drivers are no longer in the vehicle, they would not be able to stop the vehicle in the event of a fault. For this reason, the actuators (especially the braking system) and the power supply must meet functional safety requirements according to ISO 26262 (ASIL-relevant safety goals). In addition, appropriate measures must be taken to prevent the accumulation of pollutant emissions in enclosed spaces when the engine is running for extended periods. In case of smartphone-based systems, it moreover has to be taken into account that they are not developed according to automotive standards and can therefore be prone to errors. To ensure that the driver can stop the vehicle at any time, appropriate measures must be taken to secure the necessary functions on the smartphone and the communication with the vehicle.

ECE-R79.03 (ACSF-A and RCP) stipulates that, for the purpose of system monitoring, the driver must not be more than six meters away from the vehicle during the entire maneuver and must "keep pressed" some kind of dead man's switch all the time. If one of these conditions is violated, the vehicle must be brought to a standstill as quickly as possible. In addition, the driver must not be allowed to exert any influence on the vehicle speed or lateral guidance by means of the remote control, apart from the request to come to a standstill (no remote-control function) [12].

31.4.2 Reversing and Maneuvering

31.4.2.1 Rear Cross Traffic Alert

When backing out of a parking space or reversing out of a driveway, the view on moving traffic as well as pedestrians and cyclists on the pavement is often obscured by obstacles (e.g., other vehicles, fences, walls, hedges, etc.) (◻ Fig. 31.8). The Rear Cross Traffic Alert assistance system support drivers in these situations by warning them of approaching crossing road users. Usually, an audible warning signal is combined with a visual one. Displays are also used to indicate the direction from which the road user is approaching.

For this purpose, the system uses the two radar sensors mounted right and left on the vehicle rear for Blind Spot Assist. These can monitor the area behind and to the side of the vehicle up to a distance of approximately 50 m. If a moving object is detected in this area, its movement trajectory is first anticipated and checked for overlap with the vehicle's path of the driver's own vehicle. If the probability of a collision exceeds a defined threshold value, a warning signal for the driver is triggered.

Newer systems also intervene autonomously and brake the vehicle to a standstill if the driver does not react or no longer have time to react.

The performance of Rear Cross Traffic Alert is partly tested according to the Euro NCAP test protocol (AEB VRU systems, emergency braking systems for pedestrians and cyclists) and is thus included in the overall vehicle safety assessment [15].

31.4.2.2 Reverse Brake Assist

Reverse Brake Assist supports the driver when reversing at speeds of up to about 12 km/h. In contrast to Rear Cross Traffic Alert, this system focuses on obstacles, which are already in the driving path of the driver's own vehicle and are sometimes overlooked by the driver (◻ Fig. 31.9). While the rear view camera does help the driver to detect obstacles behind the vehicle, the driver must observe his entire surroundings when reversing and ensure, for example, during steering maneuvers that the front of the vehicle that is pulling out does not collide laterally with objects in the near vicinity. Thus, it can happen that the driver overlooks persons or obstacles behind the vehicle despite the acti-

◻ **Fig. 31.8** Pulling out with crossing traffic (*Source* Ford Motor Company)

Fig. 31.9 Automatic braking due to an obstacle using Reverse Brake Assist (*Source* Ford Motor Company)

vated rear view camera, especially if these are only of a small size/height (children, posts, flower pots, etc.).

Reverse Brake Assist detects objects in the vehicle's driving path and determines the probability of a collision. If a defined threshold is exceeded, the driver is warned, and emergency braking is (additionally) initiated if the driver does not react in time. The system brings the vehicle to a standstill, allowing the driver to check the area behind the vehicle before continuing to drive again.

Depending on the manufacturer, Reverse Brake Assist uses ultrasonic sensors, rear radars, and the rear view camera for object detection. The image from the rear view camera is evaluated in an electronic control unit and obstacles are extracted from the camera image using digital image processing.

ISO 22840 (Devices to aid reverse maneuvers—Extended-range backing aid systems (ERBA)) standardizes the functionality and scope of reversing assistance systems. This standard only considers information and warning systems. Intervening systems are not included [14].

The performance of Reverse Brake Assist is partly tested according to the Euro NCAP test protocol (AEB VRU systems, emergency braking systems for pedestrians and cyclists) and is thus included in the overall assessment of vehicle safety [15].

31.4.2.3 Reverse Assist

The reversing assistant supports drivers in situations when they have to reverse over a longer distance. Especially, if the distance to the right and left is limited (e.g., walls, hedges, etc.) and not much space is available, the driver needs to steer very precisely and in time to keep the vehicle on track (e.g., narrow underground car park or narrow driveway with no turning space). The reversing assistant takes over lateral guidance during reversing. The user is responsible for longitudinal guidance and must monitor the system and intervene if necessary.

When driving forward at a speed of less than 35 km/h, the reversing assistant stores the trajectory driven for about the last 50 m. Once the vehicle is stationary, the reversing assistant can be activated, and the vehicle attempts to reverse exactly along the stored trajectory up to a maximum speed of approx. 10 km/h.

The first systems of this type do not take account of the vehicle's surroundings and therefore do not brake for obstacles. In the future, it is expected that these systems will monitor the environment by means of the sensors in the vehicle and will take over longitudinal guidance in addition to lateral guidance.

31.4.2.4 Virtual Bumper

Virtual Bumper is a kind of extension of the emergency braking functions (Reverse Brake Assist etc.). This system observes the entire vehicle environment and intervenes in the event of imminent collisions in the low-speed range by means of emergency braking, regardless of where the obstacle is located relative to the vehicle. Thus, front and rear as well as side collisions can be avoided.

In order to do this, the system must first detect obstacles. In addition to the direct detection of obstacles, this system also uses indirect methods (e.g., occupancy map (▶ Sect. 31.4.4), tracking and extrapolating the position of previously detected obstacles, etc.). Indirect detection is particularly relevant for systems of the first generation, as these only monitor the environment by means of ultrasonic sensors. These, however, cannot monitor the entire vehicle surrounding due to their mounting position. Thus, the detection range of the sensors does not cover a large part of the lateral space next to the vehicle. When passing this area, possible obstacles in it are detected and recorded in the environment model. The position of the obstacle remains valid as long as the obstacle does not change its movement vector in the meantime.

Versions of this function currently under development also use radars and cameras (surround-view cameras) in addition to ultrasonic sensors. This allows for the entire vehicle surrounding to be monitored without gaps, which increases the reliability of the function and prevents incorrect emergency braking.

31.4.3   Driving with Trailer

31.4.3.1   Trailer Backup Assist

Reversing with a vehicle-trailer combination is a challenge for inexperienced drivers, as the kinematics change significantly compared to a vehicle without a trailer. The rear of the vehicle for example must be moved to the right if the trailer is to be maneuvered to the left. The so-called Trailer Backup Assist supports the driver in such situations by taking over the lateral guidance. When the system is activated, the driver specifies the direction of movement of the combination (◘ Fig. 31.10 left) and takes over the longitudinal guidance. The system determines the appropriate steering angle required to drive in the specified direction, which is then implemented by means of the electromechanical steering actuator.

To do this, the system must know the relevant dimensions of the trailer (e.g., distance between the pivot point of the wheels and the trailer coupling) and in particular, the relative orientation of the trailer to the

vehicle (articulation angle) at all times. The kinematic parameters of the trailer can be specified by the user or determined automatically. For this purpose, some systems require drivers to perform certain driving maneuvers with the trailer (e.g., driving in a circle) before they can activate this function. The articulation angle can be determined in different ways. It can, for example, be measured by means of a sensor integrated in the trailer coupling (e.g., rotation angle sensor) or be determined with the aid of the rear view camera. For this purpose, a defined pattern (◘ Fig. 31.10 right) is applied to the trailer, which is tracked by means of the camera. The articulation angle can then be determined from the position of the pattern in the camera image. In principle, the articulation angle can also be determined without a pattern being applied on the trailer by using the rear view camera and neural networks. These networks must be trained appropriately to identify and track specific characteristic features of any trailer. The articulation angle can also be determined by means of model estimation. However, for this, the trailer must be equipped with a yaw rate sensor. From the yaw rates of the towing vehicle and the trailer, the articulation angle can then be estimated using a suitable model.

31.4.3.2   Trailer Hitch Assist

When coupling the trailer, the vehicle must be positioned in such a way that the coupling ball of the trailer coupling is situated directly below the coupling claw, otherwise coupling cannot take place and either the vehicle or the trailer must be positioned differently. Larger trailers are difficult to maneuver manually, the driver has no choice but to maneuver the vehicle until the appropriate position is reached. The use of the rear view camera makes this maneuvering much easier for drivers, because they no longer have to get out of the

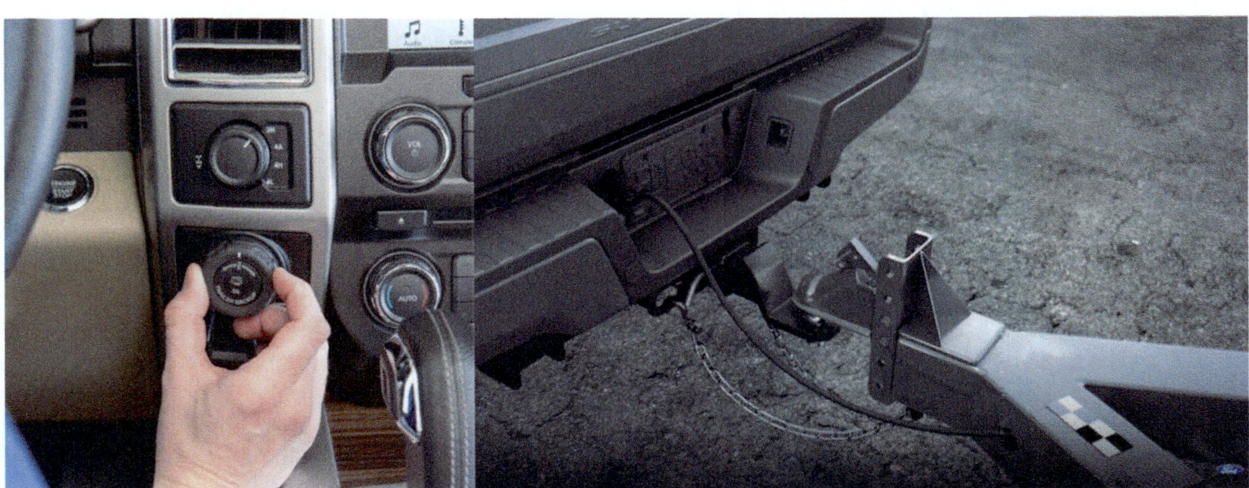

◘ **Fig. 31.10**   Left: Button for entering the direction of travel, Right: Trailer with pattern for articulation angle detection (*Source* Ford Motor Company)

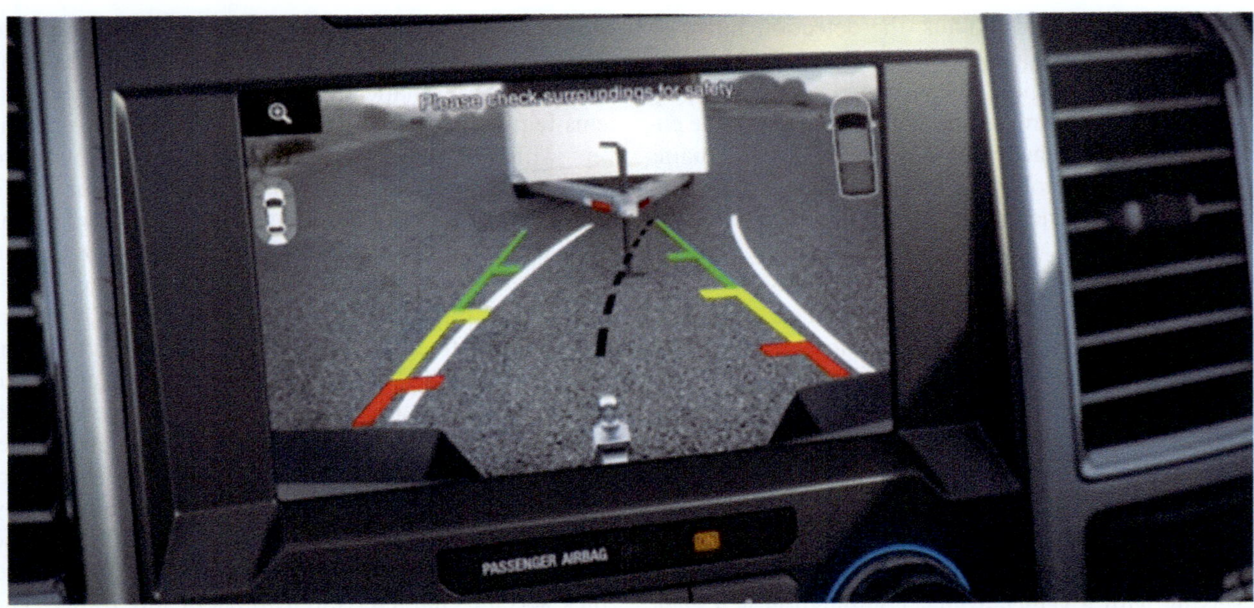

◘ Fig. 31.11   Trailer hitch assist—hitch ball trajectory in rear view camera (*Source* Ford Motor Company)

car to determine the position of the trailer coupling relative to the trailer claw.

Trailer Hitch Assist aims to make this maneuvering task even easier. For this purpose, an additional auxiliary line is displayed in the rear view camera image, which shows the driver on which path the coupling ball will move with the currently applied steering wheel angle (◘ Fig. 31.11). The driver can check whether the position of the coupling claw is on this path and, if necessary, readjust it with the steering wheel.

Systems are being developed which take over lateral and longitudinal guidance during such maneuvers and automatically steer the vehicle into the position where the trailer can be coupled. After activating the system, the driver has to monitor its operations and press a dead man's switch. The rear view camera is used in this case to detect trailer and trailer hitch, and to determine their position and orientation relative to the vehicle. Once the position is known, a trajectory is calculated to move the vehicle to the required position for hitching. The position of the hitch claw is continuously tracked so that the trajectory can be adjusted in case of deviation.

31.4.4 Environment Perception/ Environment Modeling

As described in the previous chapters, ultrasonic sensors, radar sensors, and camera systems are classically used for low-speed assistance systems. In addition to the pure visualization of the vehicle's surroundings and the display of objects in informative assistance systems, environment modeling forms the base framework for (partially) assisted systems. The described assistance systems are mainly used in rather unstructured environments. Grid-based methods for modeling the environment are particularly suitable for this task: in an occupancy grid map, the probability of occupancy can be modeled for each individual cell. Thus, it is very easy to query whether there is an obstacle at a certain location around the vehicle, whether the space is free or whether no corresponding information is available yet.

For collision checks, the vehicle geometry can be approximated, e.g., by polygonal trajectories or circular arcs, so that a collision between, e.g., the planned or predicted trajectory with the objects in the vehicle surrounding can then be determined very easily by means of location interrogation. However, it is disadvantageous that the accuracy also depends on the cell size, among other things. Since the field of application of the assistance and automation systems considered here is primarily in stationary traffic, the disadvantage of the classical occupancy maps in which moving objects cannot be represented very well, does not weigh so heavily. In addition, dynamic objects can also be covered with corresponding methods (e.g., dynamic occupancy maps) or these can be represented separately (e.g., trackers for dynamic objects).

A decisive advantage of occupancy maps, especially in the field of low-speed assistance, is that they can be generated independently of the sensor data source or allow for a fusion of data from different types of sensors, easily. Detections from short-range radar sensors, distance measurements from ultrasonic sensors or derived features from a 360° camera image as computed by a surround-view system are suitable for gen-

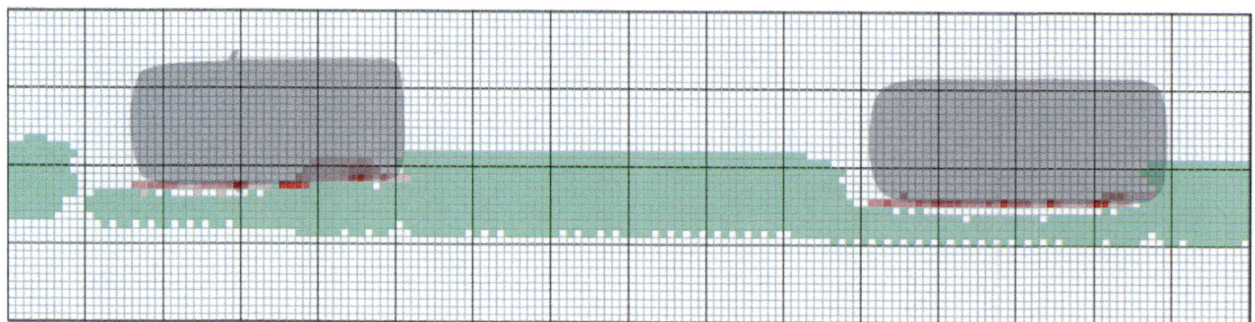

Fig. 31.12 Ultrasound-based occupancy map

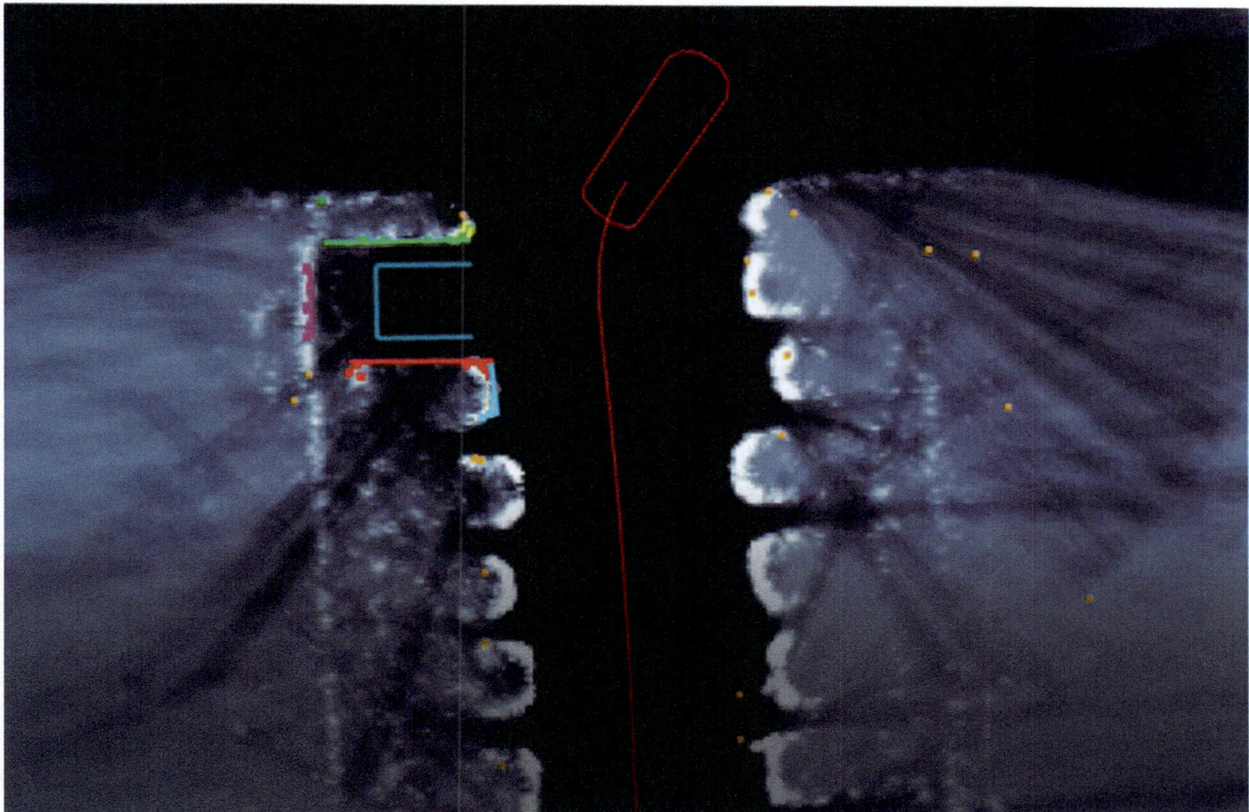

Fig. 31.13 Radar-based occupancy map (*Source* Ford Motor Company)

erating an occupancy grid map [16–18]. For this purpose, a point cloud is determined from the camera images using methods such as Structure from Motion, Monodepth, etc., which is suitable for describing the environment. Using the camera features, it would in principle be possible to create a three-dimensional occupancy map. However, since the other sensors provide only limited height information, two-dimensional occupancy maps are usually used for low-speed assistance. The height information of the camera is often evaluated separately, e.g., to determine whether a certain object (e.g., signage hanging from the ceiling, half-open garage door, etc.) can be driven under.

Figure 31.12 shows a section of an occupancy grid map generated by ultrasonic sensors. The map is the result of a vehicle with lateral ultrasonic sensors driving past parked vehicles to measure a parallel parking space. The occupancy of the cells was generated by the inverse sensor model of an ultrasonic sensor and then improved in several optimization steps. Free cells are marked in green and occupied cells in red. Unknown cells are colored gray. The limited range of the ultrasonic sensor is clearly visible, which means that this type of sensor is not able to detect the depth of the parking space, sufficiently. The vehicle edges, on the other hand, are very clearly visible in the map.

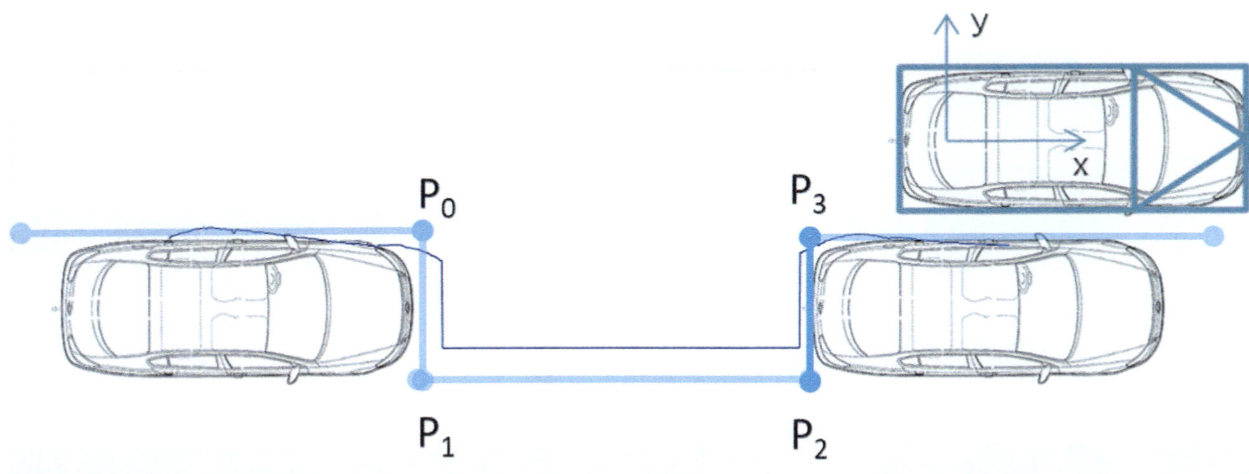

Fig. 31.14   Measurement and detection of a parking space

31

In comparison, Fig. 31.13 shows an occupancy map generated by radar sensors. The cells are color-coded according to the probability of occupancy (black = definitely not occupied, white = definitely occupied, shades of gray correspond to the probability of occupancy). You can clearly see the higher range of the radar sensors, which allow for a mapping a much larger area around the vehicle. The depth of the parking spaces is fully taken into account. It is also evident that the "resolution" of the radar-based occupancy map is higher compared to the ultrasonic version. The vehicle contours, for example, can be clearly recognized.

The measurement and the detection of parking spaces are the basis for the applications of (semi-) assisted parking systems. This can be done based on e.g., the displayed occupancy map or also classically by edge detection based on ultrasonic measurements. In this case, the distance measurement of a side-mounted ultrasonic sensor is plotted over the traveled path of the vehicle. This signal path is shown in Fig. 31.14. Based on this signal path, the edges of the objects bordering the parking space can be extracted. By doing so, the length or width of a parking space can then be determined. Due to the propagation characteristics of ultrasound (fanning out at a greater distance from the sensor origin) and different bumper shapes, it is particularly challenging to detect the exact position and length of the parking space with a single sensor. This can be improved by using additional sensors and applying the principle of trilateration, i.e., position determination by three distance measurements (► Sect. 16.6).

If a free parking space is only bounded by lines, image-processing methods are suitable for determining the parking space geometry. The abstracted environment can then be represented, for example, by the four corner points with the respective edges and be transferred to the path planning for boundary and target point determination.

31.4.5 Trajectory Planning

For systems to take over vehicle guidance, such as a (semi-)assisted parking assistant, they must be able to plan a path or trajectory based on the model of the environment presented in the previous chapter. Especially in path planning for parking systems, the nonholonomic constraints of a vehicle that can only be steered at the front axle and with a limited maximum steering angle play a crucial role: It cannot change its orientation without moving forward or backward at the same time. The orientation of the vehicle must therefore always be taken into account during planning. A path is thus always planned between two poses. A pose consists of the Cartesian coordinates x and y as well as the orientation θ relative to a reference point on the vehicle (typically the center of the rear axle).

According to Reeds-Shepp, the shortest path between two vehicle poses can be found in one of 48 (or 46) possible combinations of circular arcs with minimum curve radius (C) and straight lines (S) [19]. The minimum curve radius is constrained by the maximum steering angle. A path constructed in this way takes into account the kinematic constraints, but not the dynamics of the steering actuator, which would be necessary to implement such a planned vehicle movement. At the transition between circular arcs and straight lines or circular arcs with different signs, the vehicle would have to stop to realign the front wheels, or further optimization of the planned path in terms of curvature continuity would be necessary. Another possibility is to extend the procedure in such a way that curvature-continuous paths are created. This can be implemented, for example, in line with the CC-Steer method, by using clothoids at the transition between individual path elements [20]. However, a disadvantage of this method is it always generates a curvature-continuous path, so that no steering motion can take place even at points of direction change. Especially in parking scenarios with very narrow parking spaces, this is a disadvantage. Therefore, this method can be extended by the possibility of curvature discontinuity at the point of direction change [21]. At transitions without a change of direction, clothoids are generated assuming a maximum steering angular velocity and a maximum vehicle speed. The assumed values for these parameters must therefore be correspondingly achievable during the entire automated vehicle guidance.

Common to all methods is the assumption that they can only determine the shortest path in the absence of obstacles. Since this assumption is not valid or only valid to a very limited extent for parking processes and since the resulting paths do not always appear very natural in this procedure, an exemplary and schematic

Thus, for each point of the polygon, a circular path can also describe the position during a constant circular drive. In this way, the respective collision points during forward or backward moves can be determined and thus the respective change in orientation and the length of the circular arc element can be determined. If no collision with one of the edges takes place any longer, a geometric construction from a subset of the above-described methods (e.g., SCS, CCC) can be used to plan the remaining path to the actual starting pose. A similar procedure is described in detail in [22].

The result of a parking plan for a simple, single-move parking maneuver is shown as an example in ■ Fig. 31.16. In the case shown, the maneuver can be completed without any further change of direction. For this purpose, clothoids with different maximum veloci-

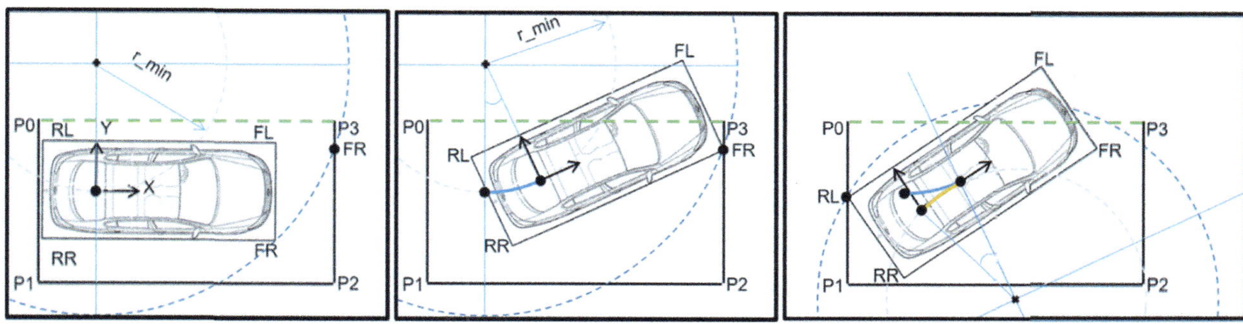

■ **Fig. 31.15** Parking strategy for multiple parking: end position (left), backward pull (middle), forward pull (right)

procedure for generating a parking trajectory for corresponding parking processes (longitudinal and transverse) is presented in the following. As an outlook, a more general planning framework is presented, which can be used to implement more complex scenarios, e.g., for automated valet parking systems.

31.4.5.1 Parking Planner

Parallel parking is usually performed at the side of the road in parking bays. A (partially) assisted parking procedure can help the driver, especially in narrow parking spaces, to position the vehicle ideally within the parking space, which is usually limited by two other vehicles or other obstacles, which cannot be driven over. The target criteria are a defined lateral distance to the curb, a central alignment to the bounding vehicles, and an equal distance to both sides within the gap. To plan the ideal parking trajectory, the calculation can also be performed from the target to the start pose and subsequently inverted. ■ Figure 31.15 shows such a procedure for very narrow parking spaces. In this example, the vehicle contour is assumed a simplified polygon.

ties (cf. ■ Fig. 31.16) were used at the beginning and at the end, as well as between the straight line and the circular arcs with R_{min} and R_2.

When parking in perpendicular gaps, the vehicle should be positioned in the middle between the two bounding objects. The parking maneuver can therefore also be planned as an exit maneuver and then be inverted accordingly (■ Fig. 31.17). The start should be an appropriate straight line. If the drivable space on the opposite side is limited by, for example, other parked vehicles, a similar iterative sweeping method can be used as for parallel parking. This ends as soon as a 90° change of orientation for parking out can be achieved without collision. The connection with the actual starting pose can then again be represented in a corresponding analogous manner by means of one of the above-described combinations.

31.4.5.2 Generalized Motion–Planning Scheme

For more general motion planning in the context of low-speed assistance such as automated valet parking

31

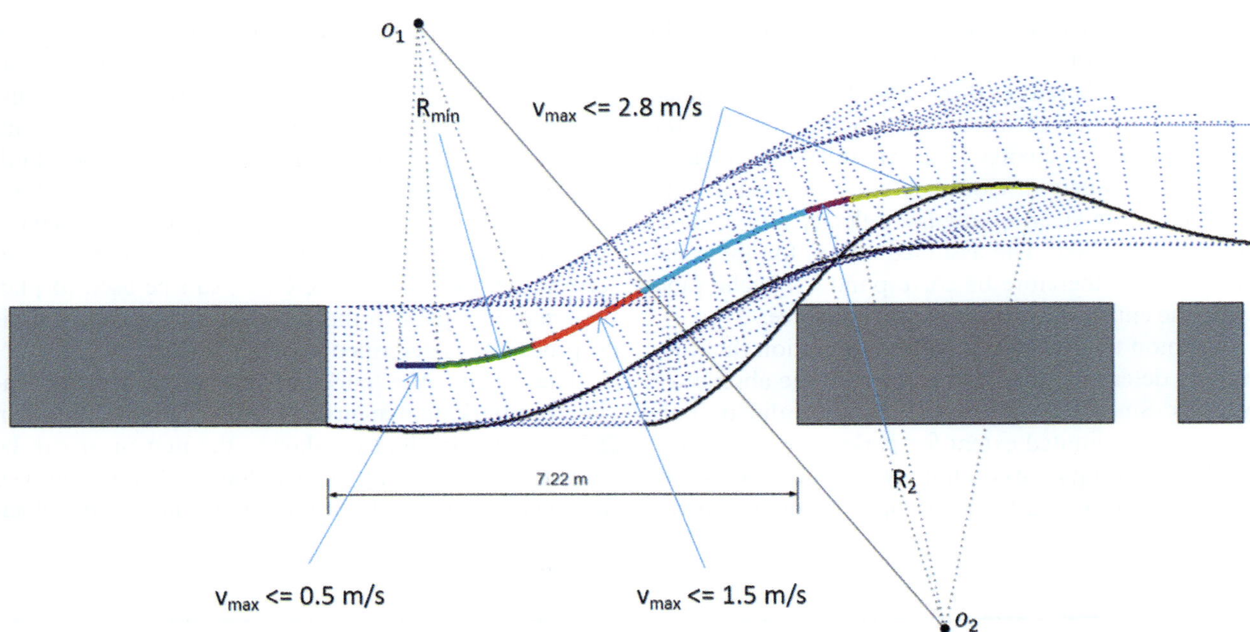

Fig. 31.16 Planned trajectory for a single parallel parking maneuver

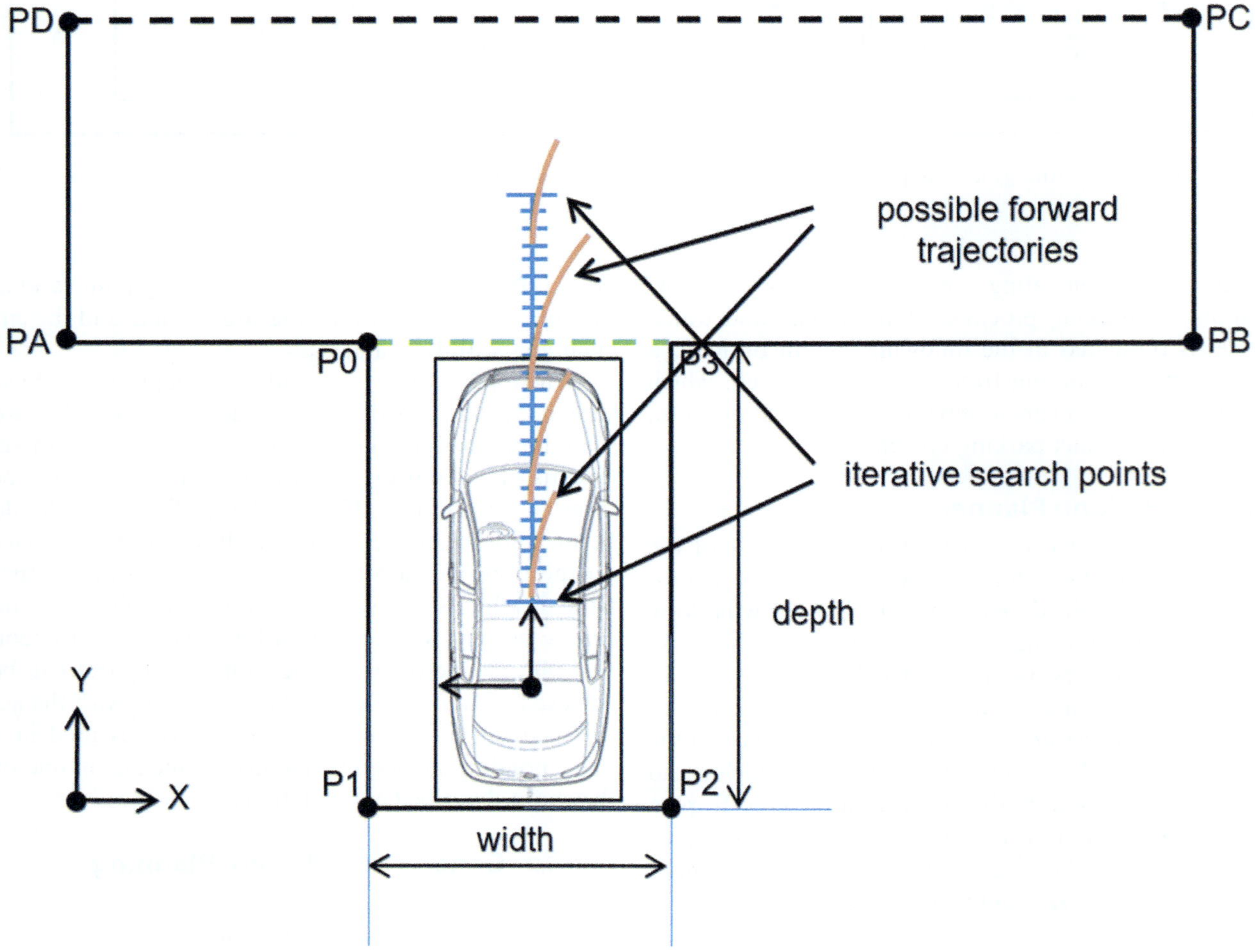

Fig. 31.17 Planning of a perpendicular parking maneuver

systems, the very problem-specific methods for path planning presented in the previous section are not sufficient. For this purpose, path planning algorithms are typically used which can solve a more general global optimization problem such as A*-based [23] or RRT methods [24]. In the hybrid A* method, for example, a graph is built by sampling motion primitives such as those mentioned above (circular arcs, straight lines, clothoids) or even control variables of a vehicle model (e.g., single-track kinematic model). Due to the sampling discretization, an exact hit of the target pose in the graph search cannot be guaranteed or only within the order of the discretization. In order to achieve an exact target pose (such as a target position within a parking space), the method must therefore be combined with one of the methods mentioned above (e.g., Reeds and Shepp) [25]. The advantage of these methods is that any number of additional optimization criteria (for example, the length of the path) can be defined. Thus, the methods are on one hand also suitable for planning parking procedures and, on the other hand, for following a reference route to a parking space while avoiding static obstacles for example. ◨ Figure 31.18 shows a parking maneuver planned with a hybrid A* method in green on the left side. It can also be seen that the contour of the vehicle for collision detection is approximated by circles and the obstacles are represented in an occupancy map. In the context of automated valet parking systems, where the planning of a trajectory, including a time horizon, may also be necessary, the combination of path planning with a local optimization method [26], is suitable, in order to also include dynamic obstacles in the (then multi-stage) planning. This extension is shown on the right side of ◨ Fig. 31.18. In addition to the planned path, which was planned around the parked vehicle, the result of a local trajectory optimization is shown in green. An appropriate vehicle model with constraints was used to generate the trajectory, so the underlying planned path does not necessarily have to consider these, thus sim-

plifying the planning. In addition, a drivable space is made available for the trajectory optimization (orange lines).

31.4.6 Execution

For the execution of the planned path, corresponding target values must be generated by the parking system. For this purpose, it is necessary to perform a continuous target/actual comparison between the planned path and the actual vehicle movement and to track the vehicle position relative to the start position. This is typically done based on odometry in conjunction with so-called dead reckoning. The vehicle motion is usually determined and integrated by evaluating the wheel speeds and inertial sensor data. Depending on the available environmental sensor technology, further information can also be included in the position estimation to correct the integration error, which is referred to as radar odometry or visual odometry. For example, in visual odometry the required distance information is usually generated by feature-based methods that extract image feature points and track them in the image sequence.

Depending on the system, replanning is also possible and useful if the position deviates too much or the environment model has changed significantly. ◨ Figure 31.19 schematically shows the target/actual comparison between the planned path and the vehicle position. On the one hand, the nominal position along the planned path can be determined, and on the other hand, the current lateral distance, as well as the orientation difference, which can then be used as input variables for the generation of target values. For the longitudinal guidance, this results in the maximum permitted speed and the stopping distance.

For active lateral guidance as part of a parking assistant, current electromechanical steering systems have a steering wheel angle interface. The calculation of the

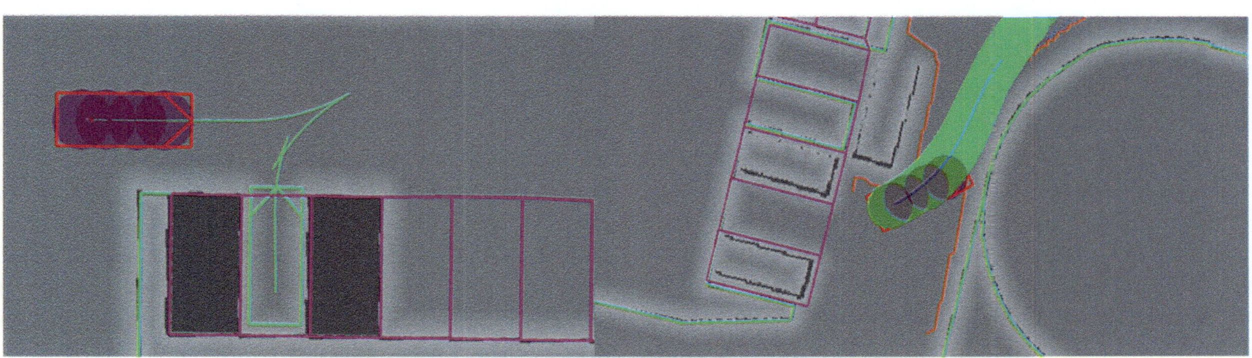

◨ **Fig. 31.18** Planned parking trajectories: Multi-move parking in a perpendicular parking space (left) and a planned obstacle avoidance (right)

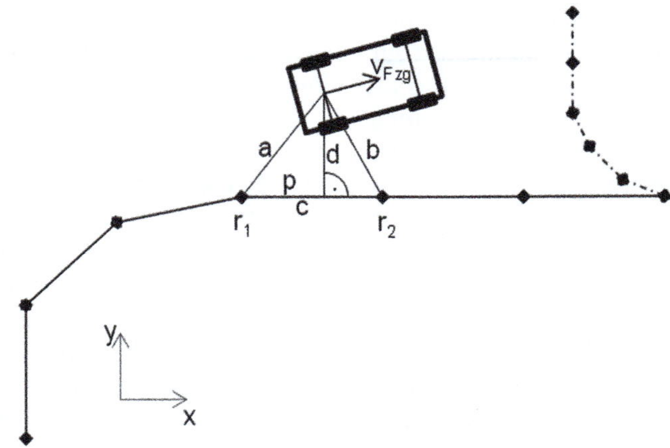

Fig. 31.19 Target/actual comparison between planned path and current vehicle position

target steering angle for following the planned path can be implemented as follows: The curvature at the current target position of the vehicle can be converted into a target steering angle with the help of the vehicle model, which then serves as feedforward control. Lateral distance and orientation difference can be used as feedback control input and serve to compensate for any system modeling errors and other noise factors.

In case of semi-assisted parking systems, drivers must take over longitudinal guidance themselves. The system only monitors compliance with the maximum speed defined at the planning level and continuously determines the remaining distance to the stopping or direction change point. In systems with active longitudinal guidance, the system itself controls compliance with the maximum speed and the stopping distance. In conventionally driven vehicles, this is done by controlling the two torque-generating actuators, the engine and the brake. Since the brake has a faster and more dynamic control behavior than an internal combustion engine drive, it is usually used to control the vehicle speed. On flat ground, the torque provided by the converter at idle speed is sufficient to provide the necessary propulsion. The brake then usually works against this converter torque, because the creep speed of the vehicle is higher than the targeted vehicle speed for parking. On slopes, or if a curb must be traversed, the internal combustion engine generates increased propulsion torque. Since most brake actuators have a limited control accuracy at low brake pressures, oscillating behavior often occurs because the exact brake pressure for controlling a constant parking speed cannot be set. In such a case, it can be helpful, for example, to have the engine generate additional propulsion torque in order to increase the required brake pressure for setting the parking speed. Shifting the operating range of the brake actuator to higher torques can improve its control accuracy. Especially in narrow parking spaces, where sometimes very short movements are required,

overshooting must be avoided as far as possible, as this can lead to a collision. For this purpose, the interaction of the actuators involved (engine, brake and transmission) and their characteristics must be taken into account for longitudinal control.

Another task during the execution of the parking process is the monitoring of the environment: In principle, drivers are still responsible for monitoring the environment in the systems described and they are supported for that by the informing systems. For fully assisted parking systems, however, the system must also react and stop the vehicle in case of suddenly appearing and detected obstacles in the planned path. This requires continuous monitoring of the planned or the predicted trajectory of the vehicle by means of collision checking with the environment (e.g., via the described occupancy map, tracking lists with dynamic objects).

31.5 Outlook

Current low-speed assistance systems on the market described here focus mainly on assistance in maneuvering operations or on taking over driving tasks when parking. These support the driver particularly in narrow, complex and sometimes unusual situations. The focus is on the one hand on avoiding collisions and minimizing resulting damage and on the other hand on automating the driving task with the goal to relieve the driver from complex and strenuous tasks. In the future, the focus will however also be on optimizing general efficiency with regard to parking space search and utilization through increased automation. With remote-controlled parking, there is already a functionality that allows for using narrow parking spaces or garages by means of automation, as there is no need to open the doors of the automated vehicle when it is parked.

This type of driverless automation represents a significant development path. Until now, the driver is still needed to monitor the system. In the future, this will no longer be necessary. This saves time for the driver, because the vehicle performs the parking process automatically without the driver's presence being required.

Home Zone Parking is an example of such a system [27]. Here, the vehicle is first taught by the driver to drive a certain route and park in a defined position. After the teaching process, the vehicle is able to perform this maneuver fully automatically without human supervision. The system follows the learned trajectory and reacts to obstacles, if necessary. This functionality is however, limited to private driveways and garages, which are known, do not change much, and most importantly do not require the system to interact extensively with other road users. If the system cannot proceed because the learned trajectory is blocked, it can stop and ask the user to take over.

Especially in urban areas, technologies, which enable the principle of valet parking, have the potential to make the use of and search for parking spaces much more efficient. The principle of valet parking is mainly known from the US: Drivers pull up the vehicle to a defined location (e.g., entrance of a hotel or restaurant), get out, and hand over the key to a third person who then takes care of parking the vehicle. When they return, they receive their vehicle back at the drop-off point. In addition to the gain in comfort for the driver, this procedure enables also efficient parking.

This principle is represented by the Automated Valet Parking (AVP) system: In this case, the system not only handles the actual parking process, but also the allocation and management of parking spaces. There are systems that operate in mixed traffic with other road users (vehicles, people, etc.) not controlled by the system, or systems which only operate in an exclusive area in which only system-controlled vehicles move (e.g., rental car terminal). For such systems, it is important that special handover areas are set up so that the driver can leave or take over the vehicle again not feeling the need to rush. Within the defined area, the system then takes over the driving task (Dynamic Driving Task) completely (L4 automation) and the driver does not have to be available as a fallback level. Within such areas, the driver can thus leave the vehicle, virtually hand over the key to the system, and start the process (for example via a smartphone app).

In principle, Automated Valet Parking can be implemented purely vehicle-based (type1) or supported by corresponding smart infrastructure (type2). A purely vehicle-based system has to be able to independently master all the necessary driving tasks, which is very challenging in some cases. The system for example needs a map of the parking lot to navigate, which can be provided by a third party or taught in advance by the driver. Alternatively, the vehicle can explore the parking lot or parking garage and search for a suitable parking space. The biggest challenge, however, is that there is no fallback level in case the vehicle cannot handle a situation on its own. In principle, drivers can be informed, but they are usually not directly on site and not always available.

In the so-called type 1 system, the infrastructure only takes over coordinating tasks such as the allocation of a parking space or an area. For this purpose, the infrastructure also provides extended information such as a highly accurate map. To further support the system, artificial landmarks can also be introduced into the infrastructure (☐ Fig. 31.20), which can then

☐ **Fig. 31.20** AVP infrastructure: artificial landmarks (left) and digital map (right)

be detected by the sensor system installed on the vehicle side, thus enabling position determination within the high-precision map. In addition to localization, also motion planning and collision avoidance, as well as maneuver execution, are performed on the vehicle itself. The advantage of this variant is that the effort required to integrate the system into the infrastructure is significantly lower than for type 2 systems. The disadvantage is that in particular the collision avoidance is only based on the sensors installed on the vehicle, which have to be designed accordingly considering all possible application scenarios.

Infrastructure support can significantly facilitate the driving task to be mastered by the vehicle but leads to a dependency on the availability of this infrastructure. Therefore, the function is not available everywhere where the driver wants to use it. Infrastructure-supported systems can be distinguished with regard to the division of tasks. Infrastructure of the so-called type 2 is equipped with sensor technology to monitor the environment. Allocation of an appropriate target parking space, as well as planning of the vehicle movement and collision avoidance takes place in a backend within the infrastructure. The vehicle only carries out the planned movement. Since the vehicle's position must also be determined via the infrastructure sensors, this type of system requires low-latency communication between the vehicle and the infrastructure. The advantage here is that there are no special requirements on the environment sensor technology installed in the vehicle. A disadvantage is that the complete operating range must be equipped with the corresponding sensor technology.

In addition, a third variant is conceivable (type 3), in which both systems share the responsibility for monitoring the surrounding of the vehicle appropriately. So far, all three types are still in a research and development phase. Only type 2 is already available in the Mercedes S-class. But because of the required infrastructure the feature can only be experienced in the parking structure of the Mercedes-Benz Museum in Untertürkheim and at the airport of Stuttgart [28].

Fundamental challenges arise primarily when such a system is to be introduced into the existing operation of, for example, a parking garage. For this purpose, either the system must be able to deal and negotiate with other road users or a dedicated area for the systems must be provided. Additional areas must also be designated for drivers to be able to leave or re-enter their cars. Especially in urban areas, the systems have the potential to reduce parking search traffic and, in future, to support e.g., garages in residential areas to which the vehicles can be automatically dispatched and called back by the residents instead of having to be parked at the roadside, as is the case today.

References

1. Reimer, B., Mehler, B., Coughlin, J.F.: Reductions in self-reported stress and anticipatory heart rate with the use of a semi-automated parallel parking system. Appl. Ergon. **52** (2016)
2. N.N.: Unfallforschung kompakt 61, Park-und Rangierunfälle, GDV (2016)
3. NTHSA: Fatalities and injuries in motor vehicle backing crashes (2008)
4. N.N.: Automatisiertes Fahren, Auswirkungen auf den Schadenaufwand bis 2035. GDV (2017)
5. Keßler, M.; Mangin, B.: Nutzerorientierte Auslegung von teilautomatisierten Einparkassistenzsystemen. In: Tagungsband zur 4. VDI-Tagung Fahrer im 21. Jahrhundert, Braunschweig (2007)
6. Benmimoun, A., Shah, J., Suermann, M.: Next generation active safety systems—collision avoidance by means of active intervention. VDI Bericht (2011)
7. N.N.: Automotive Engineering, SAE volume 90 (1982)
8. Katzwinkel, R. et al: Einparkassistenz. In: Handbuch Fahrerassistenzsysteme, ATZ/MTZ-Fachbuch (2009)
9. N.N.: Curb "Feelers" save tires, popular science **153**(3), 135 (1948)
10. N.N.: ISO 17386, transport information and control systems—Manoeuvring Aids for Low Speed Operation (MALSO)—performance requirements and test procedures, 2nd edn (2010)
11. Cicchino, J.B.: Effects of rearview cameras and rear parking sensors on police-reported backing crashes. Traffic Injury Prev. **18**(8) (2017)
12. N.N.: Regelung Nr. 79 der Wirtschaftskommission der Vereinten Nationen für Europa (UNECE)—Einheitliche Bedingungen für die Genehmigung der Fahrzeuge hinsichtlich der Lenkanlage (2018/1947)
13. N.N.: ISO 16787, Intelligent transport systems—Assisted parking system (APS)—Performance requirements and test procedures, 2nd edn (2017)
14. N.N.: ISO 22840, Intelligent transport systems—Devices to aid reverse manoeuvres—Extended range backing aid systems (ERBA), 1st edn (2010)
15. N.N.: European New Car Assessment Programme (Euro NCAP), Test protocol—AEB VRU systems (2020)
16. Will, D., Klaudt, S., Beizerov, E., Eckstein, L.: Automated parking applications—from environmental modeling to planning and tracking of a feasible trajectory. In: 24th Aachen Colloquium Automobile and Engine Technology (2015)
17. Dubé, R., Hahn, M., Schütz, M. Dickmann, J., Gingras, D.: Detection of parked vehicles from a radar based occupancy grid. In: IEEE Intelligent Vehicles Symposium (IV), Dearborn, Michigan, USA (2014)
18. Schwesinger, U., et. al: Automated valet parking and charging for e-mobility—results of the V-charge project
19. Reeds, J.A., Shepp, L.A.: Optimal paths for a car that goes both forward and backward. Pac. J. Math. **145**(2) (1990)
20. Fraichard, T., Scheuer, A.: From reeds and shepp's to continuous-curvature paths. IEEE Trans. Robot. **20**(6) (2004)
21. Banzhaf, H.; Palmieri, l.; Nienhüser, D; Schamm; T.; Knoop, S. Zöllner, J.: Hybrid curvature steer: a novel extend function for sampling-based nonholonomic motion planning in tight environments. In: IEEE 20th International Conference on Intelligent Transportation Systems (ITSC), Yokohama, Japan (2017)
22. Vorobieva, H., Glaser, S., Minoiu-Enache, N., Mammar, S.: Automatic parallel parking in tiny spots: path planning and control. IEEE Trans. Intell. Transp. Syst. **16**(1) (2015)
23. Dolgov, D., Thrun, S., Montemerlo, M., Diebel, J.: Path planning for autonomous vehicles in unknown semi-structured environments. Int. J. Robot. Res. (2010)

24. Banzaf, H.: Nonholonomic motion planning for automated vehicles in dense scenarios. Dissertation, Karlsruher Institut für Technologie (2020)

25. Klaudt, S., Zlocki, A. Eckstein, L.: A priori map information and path planning for automated valet-parking. In: 28th IEEE Intelligent Vehicles Symposium (IV) (2017)

26. Nietzschmann, C., Klaudt, S., Klas, C. Will, D., Eckstein, L.: Trajectory optimization for car-like vehicles in structured and semi-structured environments. In: 29th IEEE Intelligent Vehicles Symposium (IV) (2018)

27. Musabini, A., Bozbayir, E., Marcasuzaa, H., Islas Ramırez, O.: Park4U mate: Context-aware digital assistant for personalized autonomous parking. In: IEEE Intelligent Vehicles Symposium (2021)

28. Nimmo, M., Heß, F., Sommer, N.: Sicherheitskonzept für einen automatisierten Parkservice. ATZ Automobiltechnische Zeitschrift **122**, 26–31 (2020)

Open Access This chapter is licensed under the terms of the Creative Commons Attribution-NonCommercial-NoDerivatives 4.0 International License (▶ http://creativecommons.org/licenses/by-nc-nd/4.0/), which permits any noncommercial use, sharing, distribution and reproduction in any medium or format, as long as you give appropriate credit to the original author(s) and the source, provide a link to the Creative Commons license and indicate if you modified the licensed material. You do not have permission under this license to share adapted material derived from this chapter or parts of it.

The images or other third party material in this chapter are included in the chapter's Creative Commons license, unless indicated otherwise in a credit line to the material. If material is not included in the chapter's Creative Commons license and your intended use is not permitted by statutory regulation or exceeds the permitted use, you will need to obtain permission directly from the copyright holder.

Longitudinal Control

Hermann Winner and Jens Desens

Contents

© The Author(s) 2026
H. Winner et al. (eds.), *Handbook Assisted and Automated Driving*,
https://doi.org/10.1007/978-3-658-45276-6_32

32.1 Introduction

The assistance function Longitudinal Control adapts the driving speed to the traffic situation. It was introduced to the market with Adaptive Cruise Control, abbreviated to ACC, as a distinct function at the end of the 1990s.

Active cruise control, automatic distance control, automatic cruise control, or autonomous intelligent cruise control tends to be used as synonyms. DISTRONIC and automatic distance control (ADR, Automatische Distanz-Regelung) are registered trademarks.

ISO 15622 (Intelligent transport systems—Adaptive cruise control systems—Performance requirements and test procedures) [1] is available as an international reference. It describes both the functionality first introduced, later called Limited Speed Range-ACC (LSRA), and the extended functionality for the low-speed range, also called Full Speed Range-ACC (FSRA). This variant, which has been available on the market since 2005, has dominated for years because it significantly expands the use of ACC. Only in vehicles with manual transmissions does the functionality remain limited to the early standard LSRA feature set. In line with the market dominance of the FSRA variant, ACC is used in the following in the broader sense with FSRA functionality. For details on the limited LSRA function, please refer to the chapters of the earlier manual editions [2, 3]. They also contain a review of the beginnings of the development.

Parallel to the development of longitudinal control systems, systems for lateral control emerged, as described in the following chapter (▶ Chap. 33). Consequently, a combination of both functions is evident, in which the basic functions of longitudinal and lateral control can be loosely coupled together as level 2 systems (cf. ▶ Chap. 2). Hence, the basic features of longitudinal control of an independent ACC function are also valid for systems combined in this way. An Integrated Assistance system that plans longitudinal and lateral control together, on the other hand, is described in ▶ Chap. 34.

32.2 Functional Definition and Functional Requirements of ACC

32.2.1 Reference to the ISO Standard on ACC

In this chapter, reference is made to the international standard ISO 15622 [1] for definitions, system states and modes, and requirements. It represents the state of the art regarding the minimum requirements for an ACC system. No legally binding interpretations are derived from it. Conformity to the standard is neither obligatory nor sufficient for national approval. A deviation from the standard will oblige the manufacturer to explain why the deviation was chosen in the event of liability suits. If a reliable justification is available, nothing stands in the way of the deviation.

National homologation is governed by the respective legal framework of the country, which in individual cases may even refer to standards in a binding manner, but this does not apply to this standard. The catalogue of fines for traffic offences will already be observed by the manufacturer when interpreting the functional limits, such as the minimum following distance. This can prevent drivers from being misled into driving conditions that could result in driving bans.

32.2.2 Definitions

In ISO 15622 [1] ACC function is described as follows:

» The main system function of Adaptive Cruise Control (ACC) is to control vehicle speed adaptively to a forward vehicle by using information about: (1) distance to forward vehicles, (2) the motion of the subject (ACC equipped) vehicle and (3) driver commands. Based upon the information acquired, the controller sends commands to actuators for carrying out its longitudinal control strategy and it also sends status information to the driver.

ACC is derived from the long-standing driving speed control, referred to in English-speaking countries as cruise control which is widely used in North America and Japan (abbreviated to CC) or commonly in German-speaking countries as "Tempomat." Its role is to control a desired speed v_{set} set by the driver, and it is included as part of the ACC function (■ Fig. 32.1 top).

The main extension concerns adjusting the speed to the speed of the immediately preceding vehicle, here in addition to v_{to} (to: target object identified by ACC as target for control) (■ Fig. 32.1 center).

Although ISO 15622 leaves it open as to whether the brake is used for the control, the application of the brake to increase deceleration has become established as a de-facto standard. The appropriate distance mentioned in this standard is determined by τ, the time gap that is often colloquially referred to as distance in seconds. It is defined as:

» Time gap: "time gap calculated as clearance, *c, divided by* vehicle speed v by:
$$\tau = c/v."^{[1]}$$

---

1 For clearance (inter-vehicle distance), the letter d is used instead of c, since c is often used for the propagation velocity of waves.

Using temporal rather than spatial reference follows the basic idea, that is to prevent a rear collision. It is sufficient to have a distance, which is in accordance with the reaction time, assuming the same deceleration capability for both the preceding and the subject vehicle. Therefore, in the presence of a preceding vehicle that is moving slower than one's own desired speed, the control task of the ACC is to adapt one's own speed to that of the preceding vehicle to obtain compliance with a clearance that ensures a constant reaction time. However, as soon as the target leaves the immediate driving corridor and no other vehicle is designated as the target, ACC restores the set speed without further action required by the driver (⊡ Fig. 32.1 bottom).

32.2.3 Functional Requirements for FSRACC According to ISO 15622

The function definitions in ▶ Sect. 32.1 give rise to the following functional requirements:

- Under free cruising conditions:
    - Constant speed control and high control comfort, i.e., minimal longitudinal jerk, no swinging, and high control quality (with no obvious deviation from the set speed).
    - Cruise control with brake intervention in case of a lowered desired speed or downward slope.
- When following another vehicle:
    - Control throughout the entire speed range down to 0 km/h, particularly in the creep speed range (with increased requirements for the coordination of drivetrain and brakes).
    - Vehicle-following control with oscillation damping adjustment to the speed of the vehicle ahead so that the speed fluctuations are not copied.
    - Time-gap control to maintain the set time gap τ_{set} and gradual "falling back" when the interval is greatly shortened by vehicles cutting in, in line with typical driving behavior.
    - Control with the dynamics expected by the driver.
    - Convoy stability of the control when following other ACC vehicles.
    - Adequate acceleration capability for dynamic following.
    - Ability to decelerate for the majority of pursuit driving situations (>90%) in moving traffic.
    - Automatic target detection when vehicles in front approach or cut in and out within a defined distance range, i.e., determination of a target-seeking corridor.

- When approaching:
    - For slow approaches, prompt speed control to the desired distance.
    - For faster approaching, predictable deceleration behavior in order to facilitate an assessment by the driver of whether to intervene because of inadequate ACC deceleration.
    - If the vehicle has become closer than the desired clearance, "falling back" in a typical driving manner.
- When stopping:
    - Control of appropriate stopping distance (typical: 2...5m).
    - Greater deceleration capability at low speeds (⊡ Fig. 32.2).
    - Safe stopping with service brake in active system mode.
    - In the case of system shutdown to a standstill without driver intervention, transition into a safe holding state without a power supply is required.
- Functional limits
    - Driver is responsible to set desired speed v_{set} and desired time gap τ_{set}.
    - The time gap must not be less than $\tau_{min} = 0.8$s in the steady state. Priority of driver intervention, i.e. deactivation when the brake pedal is actuated and override when the accelerator pedal is actuated.
    - No free ride regulation below $v_{set,min} = 4.4$m/s ($\approx$ 16km/h Speedometer speed), therefore also no lower desired speed.
    - Suitable driver handover of the longitudinal control task in the event of system failure, especially if this occurs during a deceleration process.
    - The maximum deceleration[2] $D_{max}(v)$ is set above $v_{high,min} = 20$m/s to $D_{max}(v \geq v_{high}) = 3.5$m/s^2 and below $v_{low,max} = 5$m/s to $D_{max}(v \leq v_{low}) = 5.0$m/s^2 and is linearly interpolated between these two limits cf. ⊡ Fig. 32.2.
    - Analogously, the maximum acceleration is set to $a_{max}(v \geq v_{high}) = 2$m/s^2 as well as $a_{max}(v \leq v_{low}) = 4$m/s^2 with interpolation in between.
    - The build-up rate of deceleration (jerk) γ shall not exceed the jerk limit $\gamma_{max}(v \leq v_{low}) = 5$m/s^3 below $v_{low,max} = 5$m/s and the limit $\gamma_{max}(v \geq v_{high}) = 2,5$m/s^3 above $v_{high,min} = 20$m/s. In between, the limit is interpolated linearly.

---

2   In order to avoid confusion between sign and minimum/maximum, the symbol D (deceleration) is introduced ($D = -a$) and the negative number range for parameters to D and a is excluded.

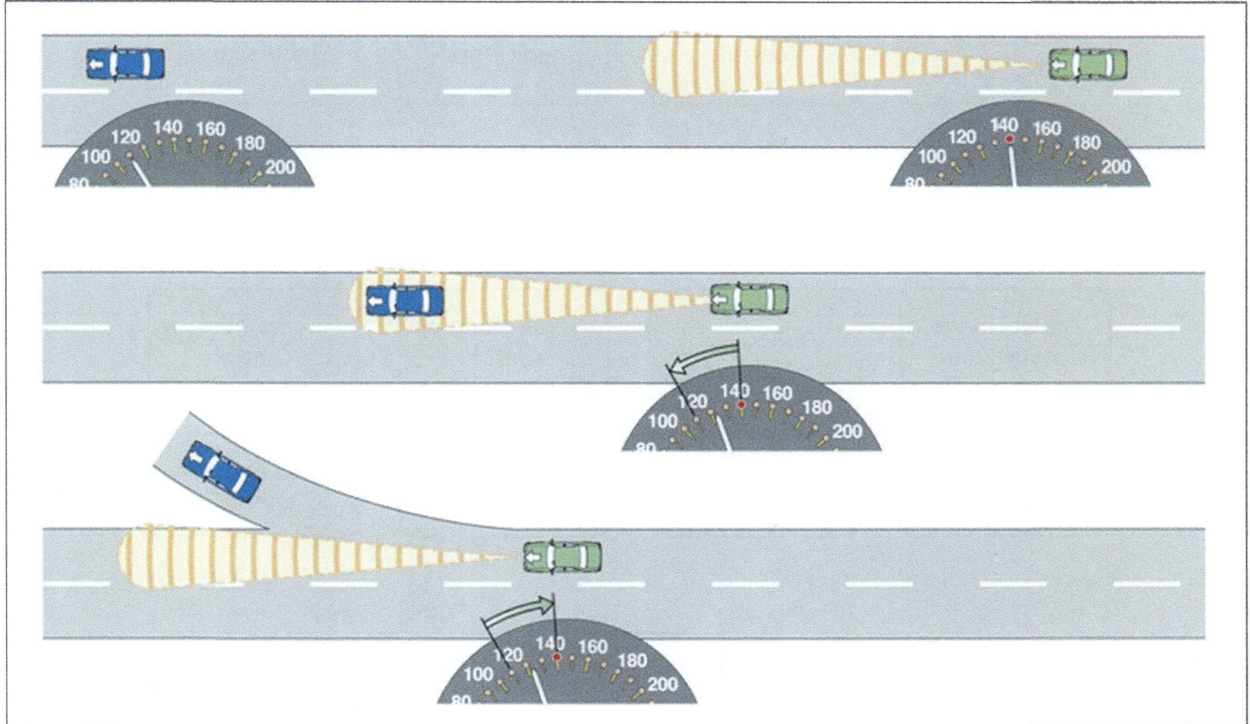

Fig. 32.1 Situation-adapted change from free driving to following and back (*Source* BOSCH)

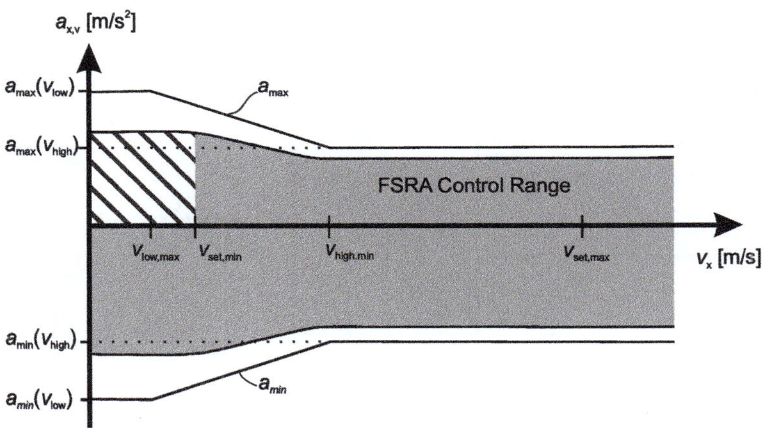

Fig. 32.2 Acceleration limits as a function of driving speed of FSR-ACC according to ISO 15622:2018 [1], definitions of parameters in the text

32.3 System Structure

The multitude of tasks of an ACC can be assigned to modules with the structure shown in Fig. 32.3. The modules themselves can in turn be further subdivided and assigned to different hardware units, see also the examples described later. The information interfaces between the modules can also vary considerably. This applies to the physical content as well as the data rate and bit representation. The four levels of this structure and their modules are described in the following Sects. 32.4, 32.5, 32.6, 32.7, 32.8 and 32.9. In such cases, the ACC-specific requirements for this component are listed.

32.3.1 Example: Mercedes-Benz DISTRONIC

Mercedes-Benz DISTRONIC concentrates the following modules from Fig. 32.3 in the driver assistance control unit called the Intelligent Drive Controller (see Fig. 32.4).

ACC state management, self-diagnosis, target object selection, course determination and prediction, control

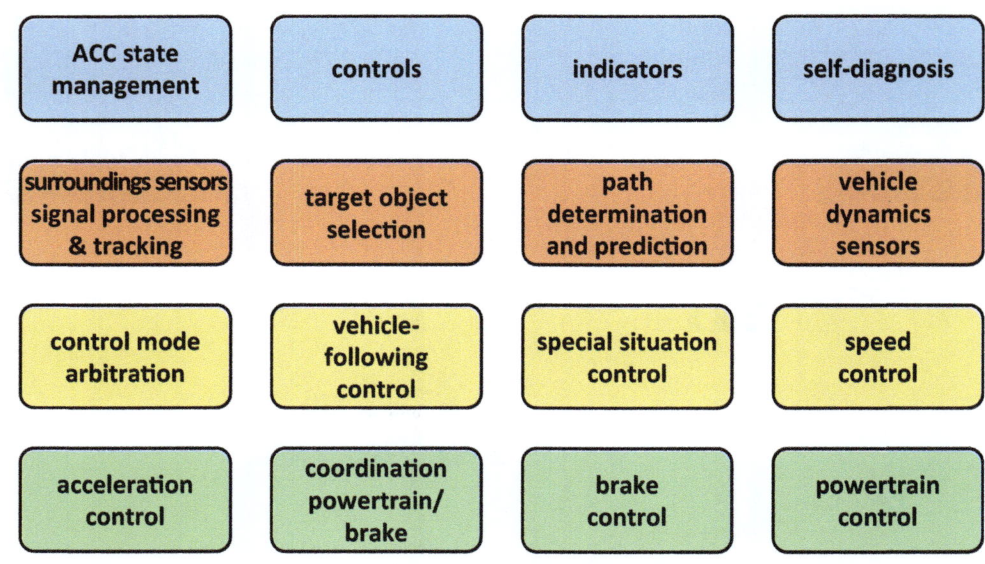

Fig. 32.3 Function modules of the ACC system

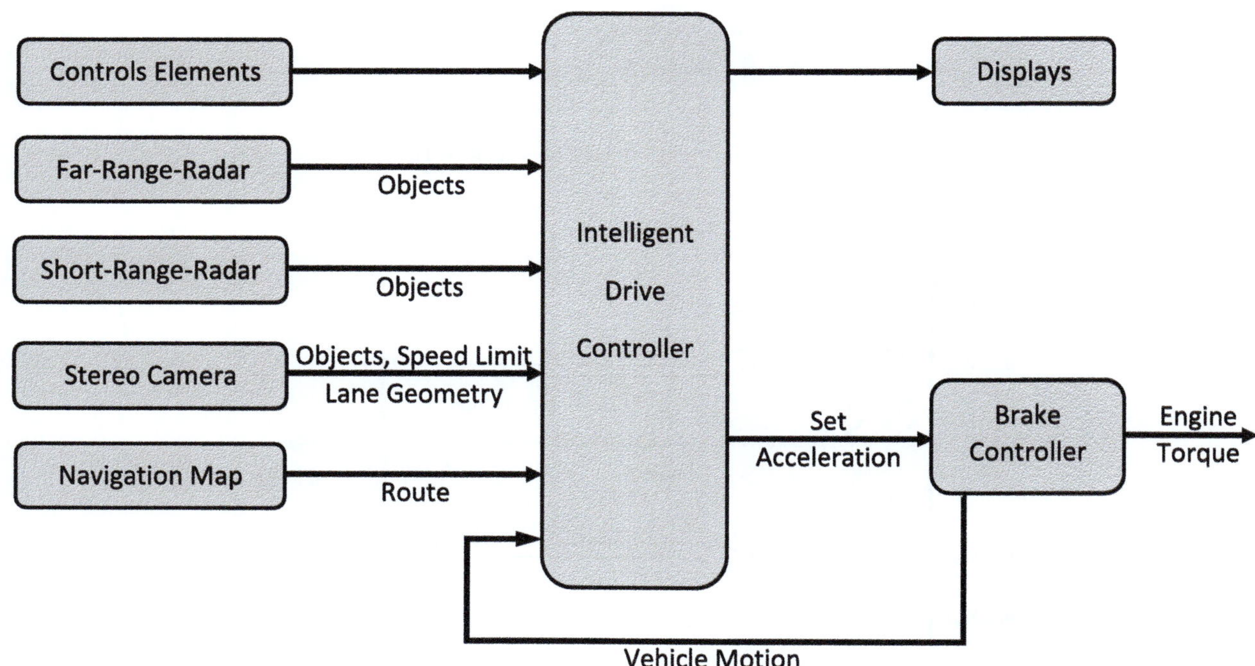

Fig. 32.4 Function modules of the DISTRONIC-system

mode arbitration, follow-up control, special situation control, speed control.

The acceleration control, driveline/brake coordination, and brake control modules are implemented in the Elecrtonic Stability Control (ESC) system, which thus plays an important role in the structure of this system example. It not only supplies, as in most known systems, the vehicle dynamics measured variables for the course prediction (see ▶ Sect. 32.6.1), but also takes over the monitoring of the ACC system and functions as a communication and coordination center for the drive control. This is in addition to the task of brake control, which is also often located there. The separation of sensor data generation from the downstream traffic scene analysis and longitudinal dynamics control enables the use of sensors from different suppliers. Furthermore, the adaptation to or extension by new sensor technology is also simplified since only the processing of the raw sensor data is performed in the sensor itself, cf. [4].

In addition to this fully comprehensive variant, which also includes the functional extensions described in ▶ Sect. 32.10 via the special situation control, there is also a less expensive variant that does without a camera, the short-range radars, and the driver assistance control unit. The modules of the Intelligent Drive Controller are then implemented in the long-range radar, which forms a so-called sensor and control unit.

32.3.2 Functional Gradations

ACC was the first system to be introduced that influences vehicle dynamics and thus, as a distributed system, loses its core functionality, if a part of the peripheral system fails. The residual functionality of a cruise control system (CC) without distance control capability could still be realized if all the systems required for cruise control, such as drive, brake, display, and control elements, were available. However, this is not advisable because it is not immediately apparent to the drivers, e.g., after a long period of free driving, that the familiar distance control function is not available when approaching an object. Such an option would therefore violate the principle of mode awareness, cf. ▶ Chap. 26. The vast majority of systems on offer, including DIS-TRONIC, therefore do not downgrade the ACC function to a cruise control function, even if the environment-detecting sensor system fails.

32.4 ACC State Management and Human–Machine Interface

32.4.1 System States and State Transition

The system states that an FSR-ACC can take on are illustrated in ◻ Fig. 32.5. The power-on state is the **ACC off** state, which, after a successful automatic function test, can be switched automatically to the **ACC standby** state or directly by the driver with a main switch. This standby state, insofar as the defined criteria for activation (see ◻ Table 32.1) are met, enables activation of the **ACC active** state.

If FSRA has been successfully activated, then two control states exist: **ACC speed control** in free-driving situations and **ACC following control** when following a vehicle in front that is traveling at a speed lower than the desired speed v_{set} speed. If this is not the case, the set speed is adjusted to v_{set} and controlled corresponding to ACC speed control. The transition between these control states takes place automatically, without driver intervention, solely from the detection of a target object and its distance and speed by the forward ACC sensor, as illustrated in ◻ Fig. 32.1.

Deactivation, i.e., the transition from **ACC active** to **ACC standby**, is usually initiated by brake pedal actuation or intentionally switching off via the control button. The various system variants available on the market have further deactivation criteria, which are listed in the right-hand column of ◻ Table 32.1 transition to the **FSRA off state** is caused by detected malfunctions and, if present, by the main switch.

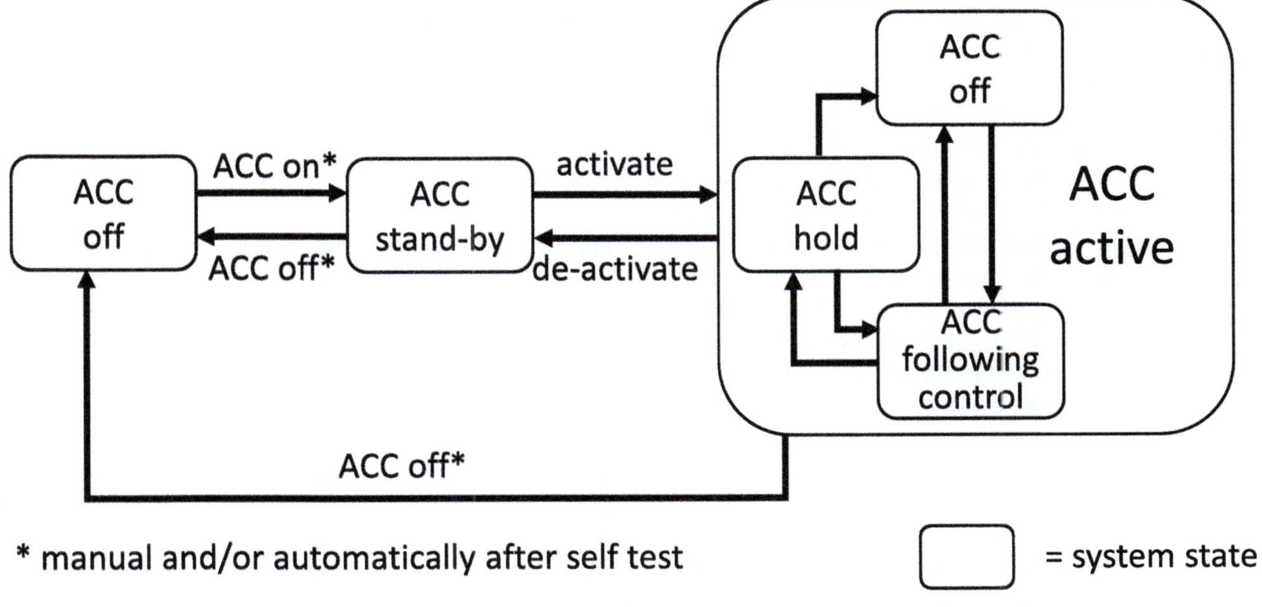

◻ **Fig. 32.5** States and transitions for FSR-ACC according to ISO15622:2018

◘ Table 32.1 Activation and deactivation conditions (for activation all conditions must be fulfilled, for deactivation one is sufficient)

Activation only if simultaneously (selection)	Deactivation, e.g., for
Drivetrain ready for operation	Deactivation via control switch
Transmission mode D engaged	Driver brakes at $v > 0$
ESC fully functional	Transmission mode invalid
Slip control not active	ESC disabled
Parking brake released	Parking brake activated
No ACC system failure	ACC system error
Driver 's door closed	$v = 0$ AND at least 2 of 3 signals active: door open, no seat belt, driver's seat empty
Driver seat belt is fastened (if available: driver's seat occupancy positive)	Slip control active for longer than a specified time (can vary depending on type, e.g., 300ms for yaw rate control, approx. 600...1000ms for drive slip control)
	Note: With ($v = 0$ AND driver brakes) alone no shutdown

In contrast to the early LSR-ACC, FSR-ACC contains another state, called **ACC hold**. The **ACC hold** state indicates the holding of the vehicle at a standstill by the ACC system. A transition from the **ACC speed control** state to **ACC hold** would only occur if a set speed of 0km/h would be possible. It makes sense, however, to limit the minimum set speed $v_{set,min}$ to a value > 0 such as $v_{set,min} = 30$km/h.

In the **ACC hold** state, some special features must be observed. Regardless, if a driver handover of the hold function is possible with appropriately signaled indications, it has proven effective to continue holding the vehicle and not to switch off the system. This is even when simply pressing the brake pedal, but rather continue holding the vehicle safely to avoid unintentional rolling away. Thus, preventing more critical system states.

The transition from the **ACC hold** state to one of the two driving states that can—except for very short stops—only takes place after the confirmation from the driver or optionally also automatically if the system can monitor the area directly in front of the vehicle with sufficient certainty.

Driver presence is also monitored, as this does not have to be ensured when the vehicle is at a standstill. If an intention to exit the vehicle is detected (e.g., via door, seatbelt, or seat occupancy detection), a suitable system shutdown takes place with a safe stopping condition—even in the event of a power failure—e.g., by activating an electromechanical parking brake. If this is not possible, a warning must be given before exiting the vehicle or the system must be switched off at such an early stage such that the passengers who have not yet left the vehicle, can secure themselves against the vehicle rolling away.

Once a standstill has been detected, the responsibility for a safe stop is transferred to the ESC system. In the process, it must ensure a safe stop for a short time by increasing the brake pressure and a permanent stop, in the case of a longer stop, by activating an electric parking brake (EPB) without consuming power.

32.4.2 Control Elements with Design Examples

The controls for ACC provide the state transitions and set the defaults for the control, namely the desired speed and the set time gap. In detail, the following elements are.

- Control element to switch from **ACC off** to **ACC stand-by.** There are two versions of this control:
  - Switch that only must be operated once and then remains permanently in the "On" position.
  - A push button which activates the controls once per ignition.
- Control element for activating the ACC system. This control is often also used in case of active control to increase the current set speed.
- Control element to reduce the current desired speed.
- Control element to activate the ACC system by using the last desired speed (resume).
- Control element for setting the desired time gap. Here again, there are two fundamentally different power-up states:
  - A constant initial state with a default setting, which usually corresponds to a time gap of $1.5 \ldots 2$s;
  - the last selected state.

The controls are frequently arranged in a group or integrated into remote control levers, as illustrated by the following example.

◘ Figure 32.6 shows the arrangement of the controls and the displays on the left. The Mercedes-Benz DISTRONIC PLUS is operated via the control cluster on the steering wheel shown on the right in ◘ Fig. 32.6. Pressing the SET+ or SET− buttons activates the system and initially adopts the current speed as the desired speed. By pressing one of the two buttons again, the desired speed is rounded to smooth 10km/h values and increased or decreased in 10km/h increments.

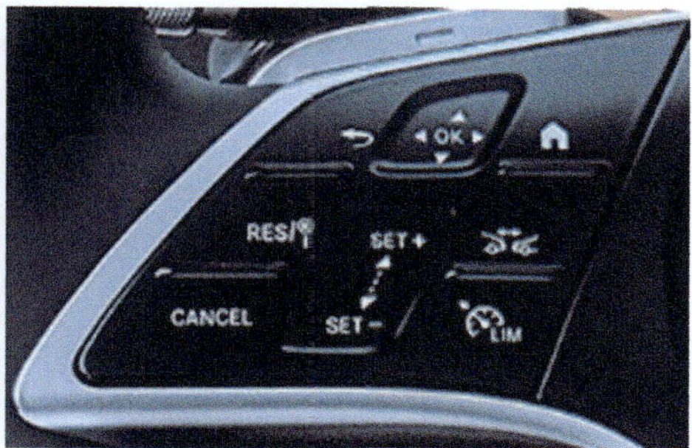

Fig. 32.6 Human–machine interface of DISTRONIC PLUS in the Mercedes-Benz S-Class (W223), right steering wheel button arrangement

If you swipe across the area of the touch surfaces area between the SET+ and SET− buttons up or down, the desired speed is increased or decreased in 1 km/h steps. When the RES button is pressed, the set speed is also activated, but the previously used set speed is resumed (so-called resume function). When activated for the first time via the RES button, the current speed is taken. This button can also be used to initiate the restart from standstill. When CANCEL is pressed, the system is deactivated. Pressing the button with the distance symbol calls up the four set time intervals offered in sequence.

In addition to this typical type of control interface integrated into the steering wheel, there are others that are implemented, such as via a lever. Despite more than two decades of experience, there is no evidence of convergence between the different types, such that today's drivers have to readjust to the operation and, as will be shown later, also to the display when they change vehicles.

32.4.3 Display Elements with Design Examples

Even though most ACC states can be determined from the current control behavior, clear feedback of the states is important for monitoring the system, especially during the state transition. However, the feedback of the current set speed and the set time gap are also essential for a user-friendly design of the operation. In the following, a distinction is made between permanent (p) and situational (s) displays. The latter are only displayed when a certain event occurs or is displayed for a certain time during driver operation. A display that is only active situationally has the advantage that it cannot only

share the display space with other situational display functions, but also that attention can be focused.

A further distinction concerns the importance (I) of the display, whereby this is made between the levels *essential* (e), *important* (w), and *helpful* (h). According to this grading, essential displays are found in all systems, important ones in most. Even though helpful displays are mostly found in some providers, they improve the learnability of system functions and allow drivers to make detailed predictions or improve their understanding of system responses. By displaying the distance and relative speed of the object just detected, drivers can also easily identify any false detections and plausibly assign the desired/required system responses.

As is often the case with the operating functions, the activation state and the set speed are also combined for the display. The **Table 32.2** shows the most common display functions and the technology (T) that can be reasonably used for the display, although only optical (o) or acoustic elements (a) are distinguished here. Haptic elements are not considered since no haptic display function is used for ACC except for the inherent kinesthetic feedback by the driving dynamics.

The corresponding display concepts are presented in **Fig. 32.7** in accordance with the operating examples presented in ▶ Sect. 32.4.2.

The display shown in **Fig. 32.7** contains the activation status of Mercedes-Benz DISTRONIC PLUS and the other driver assistance systems. From left to right, the currently permissible speed, the status of DISTRONIC, the currently driven speed, and the activation status of the lateral control systems are displayed. The symbol for the state of DISTRONIC contains information about the activation state (green speedometer symbol), the set speed, and whether a

Table 32.2 Common display functions for ACC (abbreviations for: **P: presentation**: p: permanent, s: situational; **I: importance**, e: essential, i: important, h: helpful; **T: technical**, a: acoustical, o: optical)

States	P	I	T
Activation	p	e	o
Relevant target object detected	p	i	o
Override by driver	s	h	o
Below a critical distance (e.g., during on/off operations)	s	i	o
Go possible	s	h	o
Transition automatic go ↔ driver-triggered go	s	h	o
System settings			
Desired speed (set speed)	p	e	o
Desired time gap (set time gap)	p, s	i	o
Speed of the vehicle in front	p	h	o
Actual distance to preceding vehicle and/or deviation between set time gap to actual time gap	p	h	o
State transitions			
ACC off ↔ ACC stand-by, if present	p	e	o
Takeover request when system limit is reached	s	i	a + o
System shutdown without the driver's intervention	s	e	a + o

Fig. 32.7 Display elements of DISTRONIC PLUS in the Mercedes-Benz S-Class (W223)

vehicle ahead is detected. In that case, the vehicle symbol would also be green. Not listed in the image is the display when the driver overrides the system (DISTRONIC PLUS passive).

The same information is presented in a head-up display, available as an option.

32.5 Target Object Detection for ACC

32.5.1 Requirements for the Surrounding Sensor System

The ACC functionality succeeds or fails with the detection of the relevant target vehicle, which is the basis for the control. The first prerequisite is a surrounding sensor system in order to detect and consequently decide whether and which of the vehicles/detected objects in the relevant area is to be selected as the target object. RADAR and LIDAR are successfully used as surrounding sensor technologies. The requirements listed below apply equally to both. A technology description of the sensors can be found in ▶ Chaps. 15 and 17.

32.5.2 Measuring Dimensions and Associated Requirements

32.5.2.1 Distance

According to subdivisions defined for ISO15622 [1] the standard ACC function requires that objects are detected from the minimum detection distance $d_{min} = 2m$, and beyond that from $d_{min1} = 4m$ the distance to the

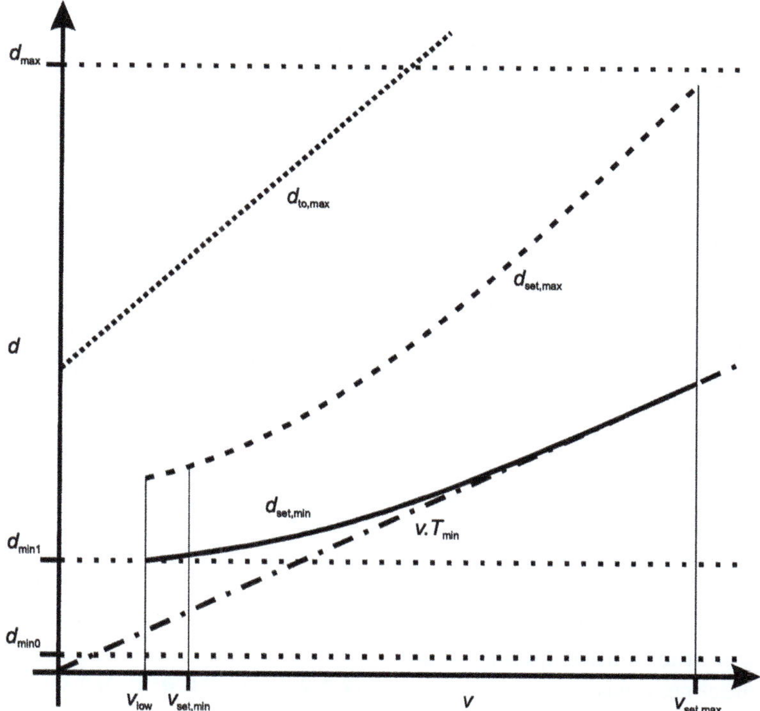

Fig. 32.8 Distance range requirements according to ISO15622 plus requirements as a function of driving speed, horizontal straight lines: Parameters for minimum distances and maximum range, $d_{to,max}$ exemplary limit curve for relevant targets, $d_{set,max}$ and $d_{set,min}$ Examples of distances for largest and smallest target time gap

object must be determined (**Fig. 32.8**). The reason why no distance measurement is required below this value is that the ACC control will in any case decelerate in this situation. With regard to falling below the threshold d_{min}, it can be assumed that the control process will be interrupted by the driver before such a small distance is reached. The same applies in the case of a cut-in in close proximity to the subject vehicle, for which drivers will not rely on the ACC function but will resolve the situation through their own brake operation.

The maximum required range d_{max} must allow control with the largest desired range, i.e., the range in the setting for the largest time gap at maximum desired speed $v_{set,max}$. It makes sense to add a reserve so that the control remains comfortable. Since at least one target time gap $\tau_{set} \geq 1,5s^3$ must be provided, the request by lowering the maximum time gap can only be made up to this limit.

The above requirements are minimum requirements and are only designed for stationary following. A greater distance is desirable for an approach, particularly if the speed difference is considerable. As shown in **Fig. 32.7**, the more difficult the target

selection is, the greater the distances involved; consequently, in many cases, a deceleration reaction is experienced as negative at a distance greater than 120 m even if the target selection is working without error. This is particularly the case if you intend to overtake the target vehicle. Overtaking is impeded by the ACC deceleration reaction before the lane change has started. In practice (see also [5]), a limitation of the reaction range, i.e., the range within which the reaction to relevant targets takes place, has proven effective. Especially in the lower and medium speed range, it makes little sense to use the entire detection range of the sensor, since objects at a greater distance have no influence on one's own vehicle. An exemplary limit curve $d_{to,max}$ is shown in **Fig. 32.8**. In another case, $d_{to,max} = \max\left(v \cdot 3.6\,s, d_{range,min}\right)$ is used, i.e., the same number of meters as km/h, but at least $d_{range,min}(\approx 80\,m)$.

No high demands are made on the accuracy of the distance measurement, since the control only reacts softly to distance deviations, as will be shown below. With the control loop gain given below, distance errors of $d_{err} = 1m$. (rms value in the frequency band of $0.1-2.0$Hz) propagate to acceleration amplitudes of maximum $a_{set,err} = 0.1$ m/s^2 and thus, remain below the perception threshold of 0.15 m/s^2 [6] for car following. Gain errors of the distance ε_d can, thus, be as high as 5% as it can be tolerated without noticeable ef-

---

3   For Germany, an interpretation according to the "half of speedometer" specification is necessary ($\tau_{set} \geq 1,8s$).

fects for the driver. However, the minimum target time gap should be selected with a corresponding reserve so that the minimum time gap defined for the stationary following process $\tau_{min} = 0.8s$ ([1]) is not undercut due[4] to the amplification error associated with the relative error.

32.5.2.2 Relative Velocity

The accuracy of the relative velocity must fulfill far higher requirements than that for distance. Any deviation of the relative velocity leads to a change of the acceleration (see ▶ Sect. 32.7). A static offset leads to a stationary deviation of the distance, whereas an offset of 1m/s leads to an approximately 5m large deviation of the distance. Fluctuations in the velocity from $v_{rel,err} = 0.25m/s$ (rms value in the frequency band of 0.1–2.0 Hz) can still be accepted, as the resulting subsequent acceleration fluctuations remain below the driver's sensitivity. While filtering the speed signal can reduce the fluctuations effectively, an excessive delay has to be avoided as otherwise, the control quality is adversely affected. As a guideline, delay times of a maximum of 0.2s can be used as guide values, which means that for a stable control with the smallest time gap of $\tau_{min} = 0.8s$, 0.6s remains for the control time constant and the actuator delay.

Relative error ε_{vrel} of the relative velocity measurement up to 5% are without problems for vehicle-following control, since the subsequent acceleration control with the control systems for brake and drivetrain generates similarly large deviations. Thus, the distortions of the control setpoints caused by the relative errors are hardly perceptible, cf. Sects. 32.7 and 32.9.

Rather higher demands on the accuracy of the relative velocity are made by the classification of the objects, whether they are moving in the same direction, at a standstill, or moving in the opposite direction. For this classification, tolerances smaller than 2m/s and 3% of v_{rel} are to be demanded. However, the relative errors can also be calibrated with standing objects because these are measured much more frequently and can therere be seen as an accumulation in a statistical recording. In this way, it is also possible to correct the errors in the determination of the driving speed, which are caused by the rolling circumference and is usually only known to a 2% accuracy of the rolling circumference.

32.5.2.3 Lateral Field of View

For lateral detection, on the one hand, it is required to be able to follow a preceding vehicle in a curve. ISO 15622 specifies a curve radius of $R_{min} = 500m$ (cor-responding to a curvature of $\kappa = 2 \cdot 10^{-3}/m$), while a lateral acceleration of $2\,m/s^2$ is provided for. This requirement can already be met with a field of view of $\pm 4°$ that can be achieved, see [2]. However, since many curves outside of highways have higher curvatures, a field of view of at least $\pm 10°$ is required, whereby the longitudinal range required in the outer angle range is significantly reduced compared to the values for d_{max} (rule of thumb $d_{max}(R) < R$), since often the visual obstruction inside the curve (hedges, buildings, ...) already prevents higher detection ranges.

On the other hand, the lateral field of view in the close range should be covered to such an extent that objects approaching are detected as soon as they enter the driving corridor. It is not specified at what distance from the front of the vehicle this condition must be met. Sensors currently in use (cf. ▶ Chap. 15, Radar) have visibility ranges of $\pm 45°$ or more in the near range, so that they completely cover the driving corridor from the 1m on completely from side to side.

32.5.2.4 Vertical Field of View

For the vertical field of view, the detection of all objects relevant for ACC (trucks, cars, motorcycles) is required. Since the objects are neither very high off the ground nor lower than the usual sensor installation heights, only the influencing variables of changes in gradient as well as static and dynamic pitching within the dynamics exploited by ACC are to be considered. Here, in practice, requirements of $\Delta\vartheta_{max} \geq 3°$ $(\pm 1, 5°)$ have resulted.

For the most part, elevation misalignment angles have only a minor negative effect since elevation as a measured variable is of little significance for the function of longitudinal control. Thus, it is essentially only necessary to prevent the available elevation range from being reduced by misalignment to such an extent that the above requirement is no longer guaranteed. Meanwhile, radar sensors also have the elevation measurement capability to meet the requirements of intervening emergency braking systems to distinguish stationary objects from targets that can be driven under or over. This allows better response to stationary objects in sub-areas without classifying them as "stopped" objects, cf. ▶ Sect. 32.7.

32.5.2.5 Multi-target Capability

Since several objects can be present in the sensor area, multi-target capability is very important. This means above all the separability between relevant objects on the own driving corridor and objects, e.g., on the neighboring lane. This separability can be achieved by a high separability of at least one measure and (distance, relative speed, or azimuth angle). However, the requirement for high separability must not be at the expense

---

4   Too low for Germany, as falling below "quarter of speedometer", i.e. $\tau < 0,9s$ leads to severe penalties.

of association problems in which the objects are repeatedly generated as new objects. This can be countered, as described in ▶ Chap. 15 (radar), by association windows in tracking that are adapted to the dynamics of the objects and the sensor-carrying ego-vehicle.

32.6 Target Selection

Target selection is of very great importance to the quality of the ACC, as both relevant objects can be overlooked, and false targets may be selected. In both cases, the user's expectation of this system is not met.

The following error consideration is based on the steps necessary for target selection, shown in ◘ Fig. 32.9 on the left. The measurement of the lateral position $Y_{u,i}$ of the object i is performed by the ACC sensor with an uncertainty of $Y_\varepsilon \approx \phi_\varepsilon \cdot d$, which results from the inaccuracy ϕ_ε of the angle determination.

32.6.1 Determination of the Course Curvature

The curvature κ describes the change in direction of a vehicle velocity as a function of the distance traveled. In the constant part of a curve, the curvature is reciprocal to the curve radius $\kappa = 1/R$. The curvature of the vehicle trajectory can be determined using various onboard sensors and under the assumption that they are used outside the dynamic vehicle limits. Therefore, the determination is not valid in skidding situations or in the presence of significant wheel slip.

Curvature calculated from the steering wheel angle

For the calculation of the curvature κ_s from the steering wheel angle δ_H, three vehicle parameters are required: the steering gear ratio i_{sg} the wheelbase ℓ and the characteristic speed v_{char}, which characterizes the understeering behavior in the linear driving dynamics range, i.e., at low lateral accelerations. Under typical ACC conditions of very good approximation, κ_s could be determined according to

$$\kappa_S = \delta_H / \left[\left(i_{sg} \ell \right) \left(1 + v_x^2 / v_{char}^2 \right) \right] \tag{32.1}$$

The uncertainty of this calculation is based on the time and velocity dependent change of the characteristic velocity, which is why this curvature calculation becomes problematic at velocities in the range of or above v_{char}.

Curvature calculated from yaw rate

For the calculation of the curvature κ_ψ from the yaw rate $\dot\psi$, the driving speed v_x is required and the vehicle slip angle rate is neglected:

$$\kappa_\psi = \dot\psi / v_x \tag{32.2}$$

Curvature calculated from lateral acceleration

Also, for the calculation of the curvature κ_{ay} from the lateral acceleration a_y, the driving speed v_x is required:

$$\kappa_{ay} = a_y / v_x^2 \tag{32.3}$$

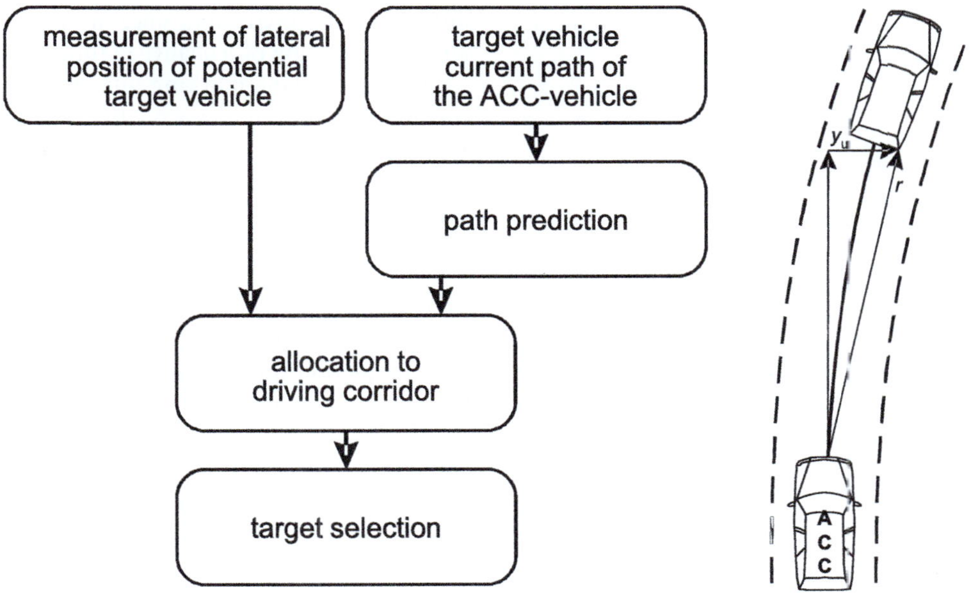

◘ **Fig. 32.9** Left: Steps for target selection; right: Definition of the quantities used in this section

Curvature calculated from the wheel speeds

For the curvature κ_v from the wheel speeds, the relative difference of the wheel speeds v/v_x and the track width w is required. In order to keep propulsive influences low, the difference $v = v_l - v_r$ and also the driving speed $v_x = (v_l + v_r)/2$ are determined at the non-driven axle.

$$\kappa_v = v/(v_x w) \tag{32.4}$$

Comparison of the curvature calculation methods

Although all the above methods can be used to determine curvature, they have different characteristics under different operating conditions. They differ particularly in crosswind, lateral road inclination, wheel radius tolerances, and in terms of sensitivity in different speed ranges.

As ◘ Table 32.3 shows, the curvature from the yaw rate is most suitable. However, a further improvement of the signal quality can be achieved if some or all signals are used for mutual comparison. This is especially possible because the ACC vehicle is equipped with ESC and thus, all of the above sensors are components of the system. When the vehicle is at a standstill, offset calibration of the yaw rate is an option, but this requires standstill phases that do not occur when driving on the highway without traffic jams. In this case, statistical averaging methods can be used, as the average of the yaw-rate sensor supplies the offset over long distances.

◘ **Table 32.3** Comparison of the different methods for curvature determination

	κ_s	κ_ψ	κ_{ay}	κ_v
Robustness against crosswind	– –	+	+	+
Robustness against lateral road inclination	– –	+	– –	+
Robustness against wheel radius tolerances	o	+	+	–
Measurement sensitivity at low speeds	+ +	o	– –	–
Measurement sensitivity at high speeds	–	o	+ +	–
Offset drift	+	– –	– –	+

32.6.2 Course Prediction

To predict the future path, the system needs to know the path (ahead) of the roadway and the future driving lane choice of the ACC vehicle and those of the potential target vehicles. If this information is not accessible, working hypotheses are used that employ simplifying assumptions.

One simple hypothesis is the assumption that the current curvature will be retained. This basic hypothesis is always used when no further information is available. This approach disregards inlets and outlets of curves to bends, changes in the lane markings, and also drivers' steering errors. If past lane marking assignments are available, the hypothesis that the objects and the ACC vehicle will remain within their lanes will be used. However, this is invalidated if objects cut in or out or if the driver changes lanes. Furthermore, it does not help with the initial assignment.

The compromise approach is to delay the object data by half of the time gap and to allocate it based on the then-current path curvature. This is very robust at the start and end of curves, because, due to the delay, the curve of the road between the objects and the ACC vehicle is used, and thereby good assignment is permitted even when curvatures change. This method does not, however, constitute a replacement for the first one in the case of initial assignment.

Additional options for course prediction are offered by GNSS-based navigation with access to the digital map and the curvature information stored there. Unfortunately, such maps are not always up-to-date, and roadworks are not marked. Also, the method of using roadside stationary targets for curvature determination is of little support in the absence of such objects but is presumably, nevertheless, included in most ACC path prediction algorithms. The lateral motion of vehicles ahead can also be used to improve course prediction, since in most cases it provides an early indication of a change in future curvature.

The use of lane marking information from the processing of camera images is obviously very promising. However, with the quality achieved today, hardly any improvement in the distance range above 100 m can be expected, since with conventional camera designs one pixel corresponds to about $0,05°$. This value can be achieved with a width of about 120 m and corresponds to a width of 10 cm and therefore, can hardly perform lane marking detection. Furthermore, image-based path prediction outside the headlight beam is impossible in darkness, especially when roads are wet.

The aforementioned algorithms are used in different ways and to different degrees by different manufacturers, but overall, they deliver as a starting value a predicted trajectory curvature κ_{pred}. The trajectory can then be extrapolated depending on the distance. Instead of the circular function, a parabolic approximation is sufficient in the case of the normal opening angle:

$$y_c(d) = \frac{\kappa_{\text{pred}}}{2}d^2 \qquad (32.5)$$

A correction offset can be added to the distance from the ACC sensor, which represents the position of the center of rotation in the longitudinal direction of the vehicle. The trajectory curve, determined according to Eq. 32.5 can now be used to calculate the lateral offset values $y_{i,c}$ as difference to the lateral position y_i of the object i determined by the ACC sensor.

$$y_{i,c} = y_i - y_c(d_i) \qquad (32.6)$$

Errors of the predicted curvature therefore increase quadratically with the object distance d. At high velocities ($v_x \geq 150\,\text{km/h}$), a curvature error κ_{err} of less than $10^{-4}/\text{m}$ is attainable, which means that at $100\,\text{m}$ distance an error of $y_{i,c,\text{err}} \approx 0.5\text{m}$ results. At $140\,\text{m}$ the error doubles due to the quadratic propagation. At low speeds ($\approx 50\text{km/h}$) the error of the curvature according to Eq. 32.2 about three times higher. Accordingly, the distance for $y_{i,c,\text{err}} \approx 0,5\text{m}$ is only $57\,\text{m}$ from which a reduction of the maximum target selection distance at low velocities is derived.

32.6.3 Driving Corridor

The driving corridor is the corridor used for ACC target selection. In its simplest form, it is defined by a distance-independent width w_{corr} with the predicted path as the centerline (Eq. 32.5). Initially, one might equate the driving corridor width with the lane width w_{lane}. However, this assumption has been found to be unsuitable.

The example in 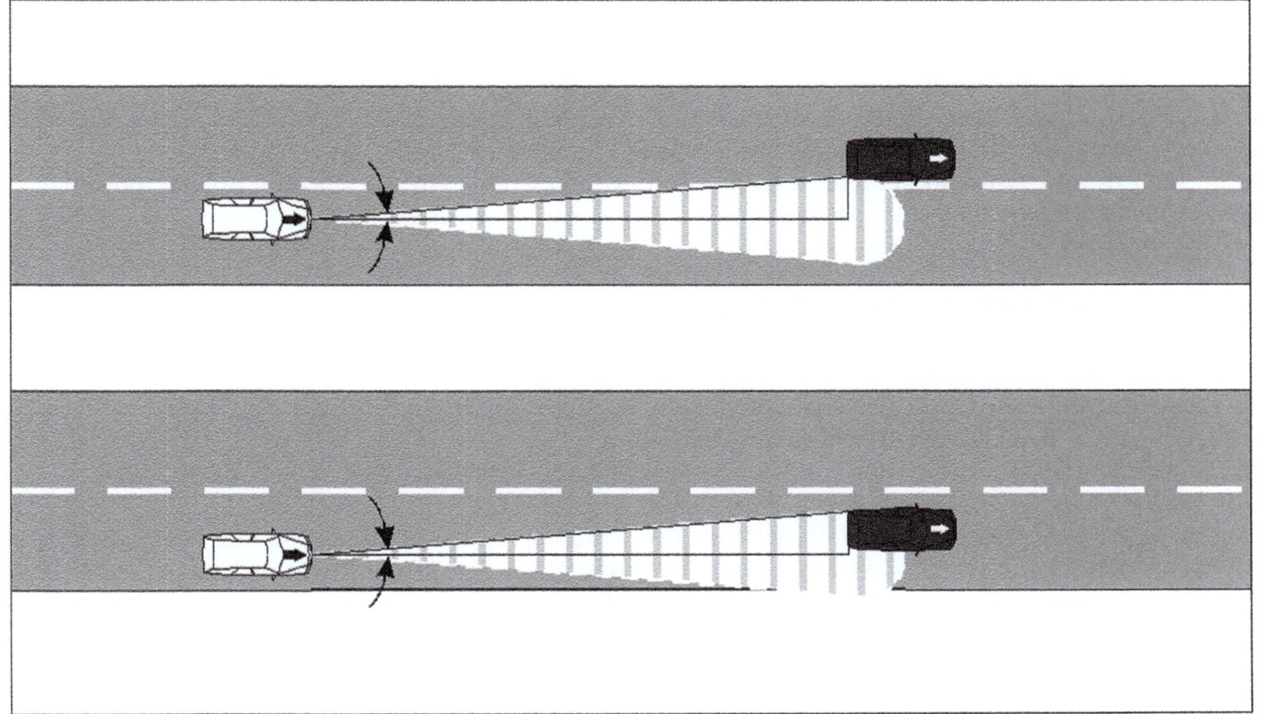 Fig. 32.10 shows that there is a range in which an unambiguous assignment based on the measured transverse position alone is not possible because it cannot be assumed that the measured lateral position of the object corresponds to the object center. Therefore, both the right and left object edges must be considered. Further uncertainty in the assignment arises during off-center travel, from both the ACC vehicle and the potential target vehicle. Therefore, the assignment to the own lane is only certain if the (without error) measured lateral position with respect to the (without error) predicted course center lies within approx. $\pm 1.2\,\text{m}$. The assignment of the object to the neighboring lane also succeeds reliably only if the position is at least $2,3\,\text{m}$ from the course center. The values refer to a lane width of $3.5\,\text{m}$.

The lack of unambiguous assignment in the transition areas, thus, leads to false assignment with both false negative assignment (not recognized as a target object, although lying in the travel corridor) and false positive (incorrectly recognized as a target object). The selection of the corridor width for the target selection

Fig. 32.10 Example of a different assignment despite the same relative data

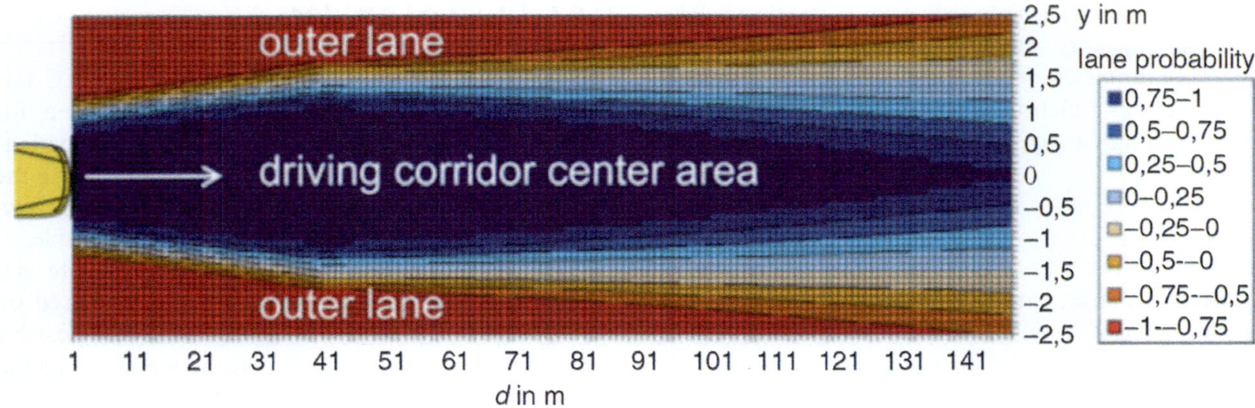

◘ Fig. 32.11   Fuzzy driving corridor contour to avoid assignment errors

can be used to weigh which type of false assignment is to be suppressed, but an error-free assignment solely from the object positions measured by the sensor is not successful, see [3] or in the original [7]. A dynamic corridor width via the detection of the number of lanes and the lane width (e.g., in road works) can further improve the compromise.

Another measure is local hysteresis, by which is meant that a wider travel corridor applies to an object marked as a control object rather than to all the other objects. Typical differences are about 1m, i.e., on both sides about 50 cm. On one hand, this prevents false detections of objects in the neighboring lane, especially in changing conditions (entering and exiting curves, unsteady steering), but on the other hand the target object can be kept stable in such situations.

Furthermore, a temporal hysteresis is applied, which means that the assignment is not performed in single time step alone, but the results of the local assignment must be confirmed over several measurement cycles, corresponding to the accumulation of increments per cycle. These can be weighted as in ◘ Fig. 32.11 by a fuzzy-like location function in which the uncertainty of the assignment is represented.

32.6.4 Further Criteria for Target Selection

It can make sense to use other criteria in addition to the lane assignment. The most significant criterion for target selection is the object speed. Oncoming vehicles are completely ignored for the control. Stationary objects that have already been detected as moving in the direction of travel (so-called "stopped objects") are especially relevant for the Full-Speed-Range ACC function in the same way as the objects moving in the same direction. Permanently static objects are often used for other functions, such as AEB systems, cf. ▶ Chap. 9, and therefore, are subjected to separate filters. In basic ACC functionality, however, only some systems

use permanently static objects as targets in certain situations. They are particularly difficult to take into account because there are numerous stationary objects in real traffic scenes and therefore there is a great risk for false reactions to objects on the side of the road.

Another simple but very effective approach is to limit the distance as a function of the driving speed (cf. ◘ Fig. 32.8). Thus, at a speed of 50 km/h a reaction to targets that are more than 80 m away is neither necessary nor useful because the frequency of incorrect assignment increases significantly with distance. Empirical values yield a distance value $d_{to,0} = 50$m and a slope of $\tau_{to} = 2$s.

$$d_{to,max} = d_{to,0} + v \cdot \tau_{to} \tag{32.7}$$

If several objects meet the criteria for a target object, the following decision criteria can be considered individually or in combination:
- the smallest longitudinal distance,
- the smallest distance to the center of the course (minimum $|y_{c,i}|$),
- the lowest target acceleration of the potential targets.

The last criterion, however, requires coupling to longitudinal control or a multi-target object interface, but also improves the transition in the case of leaving target objects.

32.6.5 Limits of the Target Selection

The approaches illustrated in the latter sections are highly effective and have reached a high-quality level. There are, however, situation-dependent constraints which require explanation as in the following two examples. The vehicles in ◘ Fig. 32.12 move identically at the moment shown, but the assignment as a "correct" object is done differently with the differing progress of

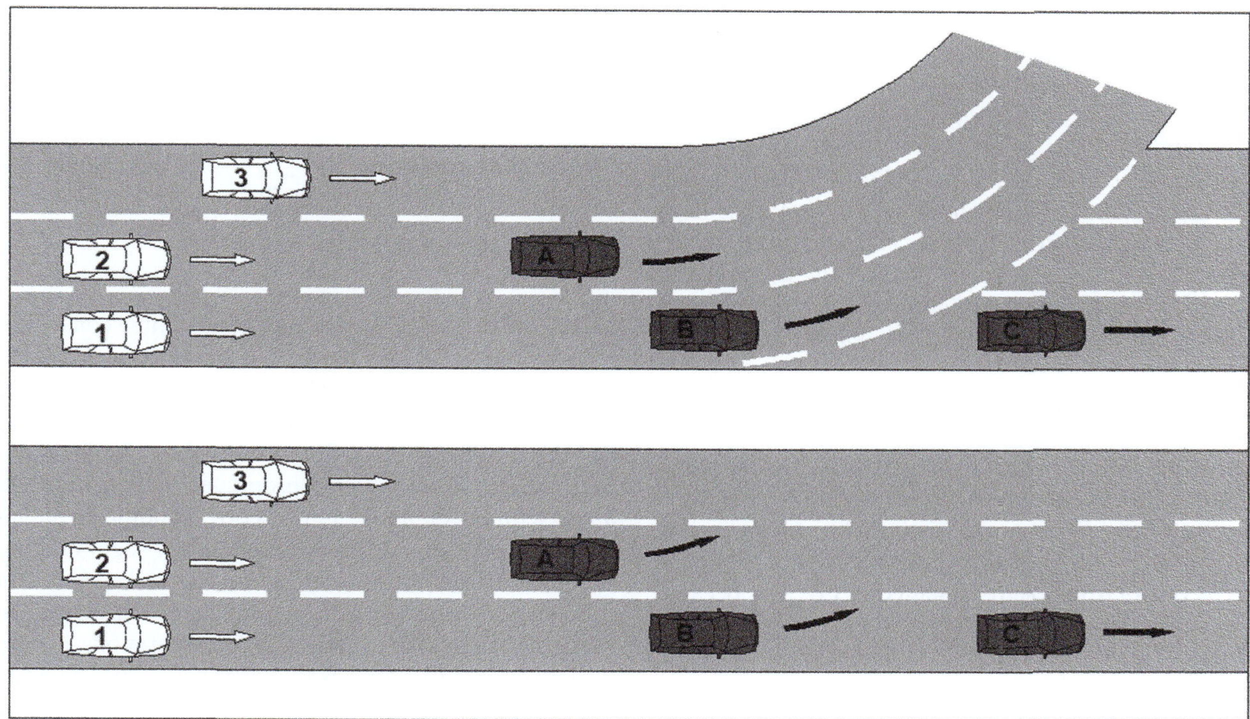

◘ Fig. 32.12 Situation example for ambiguous target assignment (vehicle positions and movements are identical in both pictures)

the road. Another example is the "overtaking dilemma" when approaching at high speed. For comfortable deceleration to follow behind a significantly slower moving vehicle, on the one hand, deceleration should already begin at a large differential. On the other hand, the probability that the vehicle in front is to be overtaken is particularly high at a large differential speed. However, early deceleration would significantly disrupt the overtaking process. Since the overtaking process is rarely indicated earlier than six seconds before reaching the vehicle to be overtaken [5], the dilemma arises between reacting too early for an undisturbed overtaking and reacting too late for a comfortable or even sufficient approach.

Another limit is the late detection of merging vehicles. On the one hand, the temporal and spatial hysteresis of the driving corridor assignment leads to a delayed reaction of about two seconds with respect to the moment when the merging vehicle is crossing the lane marking. Since the situation and the indication of the lane change by the direction indicator cause the driver to anticipate the lane change even before the vehicle crosses the marking, the late reaction is repeatedly criticized by the users. The same phenomenon occurs when the cut-out lane change is detected, although the target clearance objectively takes place correctly when the adjacent lane is fully reached.

A significant improvement of the on/off behavior of ACC can only be achieved by situation classifica-

tion, whereas the transparency of the ACC functionality may suffer under such "intelligent" functions.

Improving the target selection in the case of lane changing on the part of the ACC vehicle can be achieved by interpreting the direction indicator, resulting in a shift of the driving path to the indicated direction. The digital map in combination with the search and detect function also enables an adaptive driving corridor function.

All in all, modern ACC systems achieve an average duration of about one hour between two noticeable misclassifications; a value that is surprisingly good considering the many possibilities for errors and is difficult to improve upon, as detailed in [8].

32.7 Vehicle-Following Control

Although the ACC vehicle-following control is often described as a distance control, it is anything but a distance guided control. As a starting point for further considerations, it is assumed that the controller output converts the vehicle acceleration directly and without time delay and further that the ACC vehicle follows the target vehicle with the set time gap τ_{set}. If the ACC vehicle now copies the position of the target vehicle shifted by the time gap, the time gap is maintained regardless of the speed. Similarly, the speed and acceleration of the vehicle ahead are copied time-shifted. Thus, a simple control law can be derived for the steady-state case that even avoids feedback:

$$\ddot{x}_{\iota+1}(t) = \ddot{x}_{\iota}(t - \tau_{set}) \tag{32.8}$$

The index $\iota + 1$ stands for the ACC vehicle in a vehicle convoy with running index ι. The notation is introduced with regard to the consideration of the convoy stability, as whose measure is the quotient $V_K = \hat{\ddot{x}}_{\iota+1}(\omega)/\hat{\ddot{x}}_{\iota}(\omega)$ of the (complex) acceleration amplitudes. The convoy is stable if the condition

$$|V_K| = \left| \hat{\ddot{x}}_{\iota+1}(\omega)/\hat{\ddot{x}}_{\iota}(\omega) \right| \leq 1 \text{ for } \forall \omega \geq 0 \tag{32.9}$$

is fulfilled. From a small disturbance, the frequency components of the frequencies for which this condition is not fulfilled will be greater with each subsequent convoy position. Convoy stability clearly applies to the idealized control principle represented in Eq. 32.8 because

$$|V_K| = \left| e^{-j\omega\tau_{set}} \right| = 1, \tag{32.10}$$

even if semi-stable without damping. This approach is not suitable in practice, but it illustrates the basic controller design. Disadvantages of this approach are the numerically unfavorable determination of the acceleration of the vehicle in front (differentiation of the relative speed and the own driving speed, the necessary filtering again leads to phase delay) and that there is no correction option if the speeds do not match or in the case of deviations in the distance.

For this purpose, the following is a control design based on relative speed:

$$\ddot{x}_{\iota+1}(t) = (\dot{x}_{\iota}(t) - \dot{x}_{\iota+1}(t))/\tau_v = v_{rel,\iota}/\tau_v \tag{32.11}$$

or in the frequency range

$$\hat{\ddot{x}}_{\iota+1}(\omega) = \hat{\ddot{x}}_{\iota}(\omega)/(1 + j\omega\tau_v) \tag{32.12}$$

With a few steps, this approach can be transformed into an acceleration-led approach such as in Eq. 32.8, where the acceleration value of the vehicle ahead is not delayed by a fixed time, but is filtered in a PT1 element and thus, implicitly delayed by τ_v delayed. The approach (Eqs. 32.11 and 32.12, respectively) is obviously convoy-stable, but only when τ_v equals τ_{set} as it conforms to the rule requirement of constant time gap. Also, this control approach is not suitable to reduce a once existing distance deviation. For this purpose, the controller is extended by an additive correction part to the relative velocity, which is proportional to the difference between set and actual distance

$$\ddot{x}_{\iota+1}(t) = \left(v_{rel,\iota} - \frac{d_{set} - d_{\iota}}{\tau_d} \right)/\tau_v = \frac{\Delta v}{\tau_v} \tag{32.13}$$

or in the frequency range

$$\hat{\ddot{x}}_{\iota+1}(\omega) = \hat{\ddot{x}}_{\iota}(\omega)\frac{1 + j\omega\tau_d}{1 + j\omega(\tau_d + \tau_{set}) - \omega^2\tau_d\tau_v} \tag{32.14}$$

The stability condition $|V_K| \leq 1$ is now only fulfilled, if τ_v is chosen as sufficiently small:

$$\tau_v \leq \tau_{set}(1 + \tau_{set}/2\tau_d) \tag{32.15}$$

So far, the choice of the distance control time constant τ_d remains open. For this purpose, a reference scenario can be used, namely falling back in a cut-in situation. In this case, it is assumed that the merging vehicle cuts in without any speed difference at a distance which is 20 m is smaller than the set distance. An appropriate reaction would be a deceleration of about $1\,\text{m/s}^2$ corresponding to taking one's foot off the accelerator pedal or very slight braking. For such a reaction to occur according to Eq. 32.13 the product must amount to $\tau_v \cdot \tau_d = 20\,\text{s}^2$. This value will be used as a basis for the following considerations.

From Eq. 32.15 follows that the loop gain defined by τ_v^{-1} for the relative velocity must be the higher, the smaller the target time gap τ_{set} is. However, a high loop gain also means a low attenuation of velocity fluctuations of the target vehicle, as ◘ Fig. 32.13 for frequencies above 0.05 Hz.

Stability on the one hand and high decoupling from the speed fluctuations of preceding vehicles on the other cannot be achieved simultaneously. An implicit differentiation between cases offers itself as a way out of the conflicting goals. The high sensitivity of the driver to fluctuations occurs mainly when driving in a calm following situation characterized by small speed differences. The problems associated with the lack of convoy stability do not manifest themselves until large deviations from steady-state occur. Therefore, it is obvious to selectively design the control loop gain for this difference. In a very simple form, this can be realized by a characteristic curve with two kinks at $\pm \Delta v_{12}$ (◘ Fig. 32.14), although a rounded transition is also conceivable. With this approach, within the control differences of $\Delta v_{12} \approx 1\,\text{m/s}$ the velocity fluctuations can be damped with a large control time constant, but if larger dynamics are required, such as a larger delay stage, stability can be ensured at the large-signal level.

In the actual controller implementation, only the basic principle is adopted since other influencing variables require modification. This can be expressed via look-up tables or more complex mathematical functions. Furthermore, all other system delay times are neglected in the above consideration, which is justified neither for the environment sensor system nor for the subordinate acceleration control loop. As a guideline, the control loop time constants must be reduced by the delay times in order to fulfill the stability conditions.

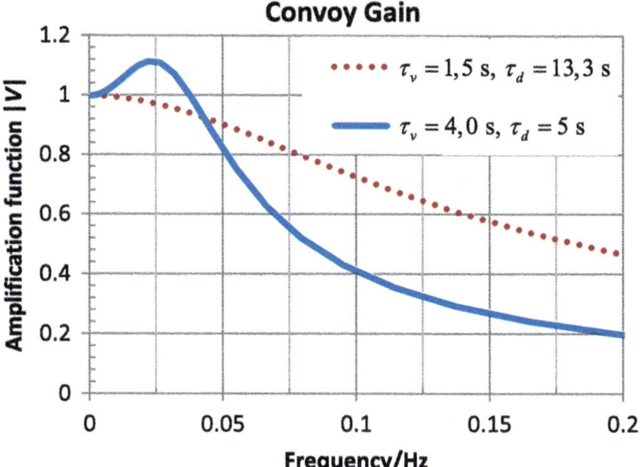

Fig. 32.13 Convoy gain for two different loop gains

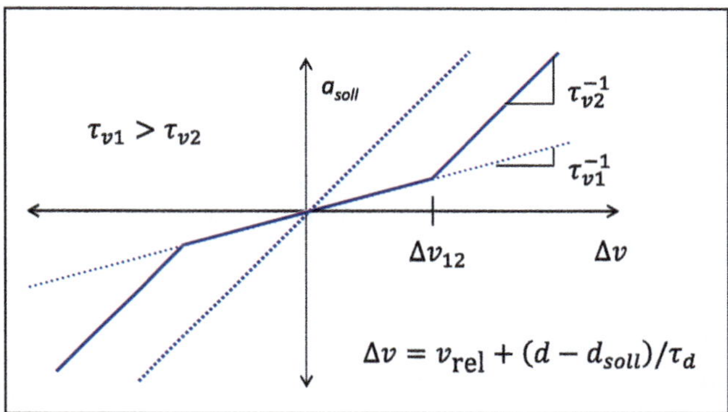

Fig. 32.14 Control loop gain characteristic from a nonlinear distance and relative speed controller (schematic), τ_v according to Eq. 32.11 as velocity control time constant in two expressions, Δv according to Eq. 32.13 as control speed residual

32.8 Controller Adaptations

32.8.1 Target-Loss Strategies and Curve Control

In ACC systems of the first generations, the angular ranges of the ambient sensors were quite limited, which is why target loss often occurred in curves. Approaches for a "blind flight control" to prevent acceleration after target loss are described in detail in previous editions of the handbook [2, 3]. Due to the dominance of FSRA over LSRA and the wider field of view required for it, the curve-related situations hardly occur anymore. Nevertheless, the curve curvatures influence the longitudinal control in order not to let the lateral acceleration become too large (in terms of magnitude). Also, a loss of target without a confirmation that the target was seen outside the driving corridor will not lead to an immediate acceleration. Additionally, the controller design is also jerk-limited, thus avoiding sudden acceleration changes. On the one hand, this avoids critical reactions in case of false target selection, and on the other hand, it supports the mental model for drivers that they remain responsible for fast reactions.

32.8.2 Approach Strategies

The approach capability is defined as the maximum velocity difference $v_{diff,appr}$ (= negative minimum relative speed $v_{rel,appr}$), which can be controlled by ACC with respect to a vehicle traveling at a constant speed before a critical distance $d_{appr,0}$ is exceeded. It depends on

- the distance $d_{appr,0}$ at the start of deceleration,
- on the assumed constant maximum increase of the deceleration (jerk) $\dot{D}_{max} = \gamma_{max}$ and
- the maximum deceleration $= D_{max}$.

$$v_{diff,appr} = \sqrt{2D_{max} \cdot \left(d_{appr,0} - d_{appr,min} + \frac{D_{max}^3}{6\gamma_{max}^2} \right)} - \frac{D_{max}^2}{2\gamma_{max}}$$

$$(32.16)$$

$$d_{appr,0} = d_{appr,min} - \frac{D_{max}^3}{6\gamma_{max}^2} + \left(v_{diff,appr} + \frac{D_{max}^2}{2\gamma_{max}}\right)^2 / 2D_{max}$$

$$(32.17)$$

The distance required for a noncritical approach grows approximately quadratically with the differential velocity and approximately reciprocally with the maximum deceleration. At a distance of 100m at $D_{max} = 2.5\,\text{m/s}^2$, around 20m/s (72 km/h) differential velocity can be compensated for, while for an approach capability of $v_{diff,appr} = 100$ km/h, $d_{appr,0} \approx 120$ m and $D_{max} \approx 3.5\,\text{m/s}^2$ are required.

The ramp in the buildup of deceleration reduces the approach capability, but as mentioned in the previous section, it also increases transparency for the driver.

In the case of dynamic approaches, it is clearly impossible to avoid values falling below the stationary set distance and/or the set time gap. Therefore, a significantly smaller reserve distance value can also be used as the set distance $d_{appr,min}$ for a successful approach. It should be noted that this undershooting is, however, recognized as "temporary", since the responsibility for the distance ultimately remains with the driver.

32.8.3 Overtaking Support

Following and overtaking conflict with each other. Thus, to support overtaking, the following control must be temporarily modified. If the overtaking action could be predicted exactly, the vehicle currently ahead could be ignored, and the vehicle could be driven as if there were no vehicle ahead. However, this is only the case if an "Active Lane Change Assistant" carries out the lane change automatically and thus, on a trajectory known to the system, as is also the case with integrated longitudinal and lateral control (cf. ▶ Chap. 34). In the case of manual lane changes, the direction indicator is not a clear indicator of either the actual intention to overtake or the desired or possible start of the maneuver. The first case occurs if the direction indicator signals the intention to turn left. Since this situation rarely occurs at high speeds, and also because overtaking is usually associated with high speed, a compromise solution is that overtaking assistance is only used at speeds in excess of 70 km/h.

It is not feasible to simply blank out the current target, because "signaling a left turn" is often also used to tell the driver in front to keep the overtaking lane free. However, because it is impossible to predict if and when this intention is going to be followed up, only a cautious reduction of the previous nominal distance remains for "gaining momentum". Within this phase, the overtaking process should be initiated with a detectable change in direction. The necessary rapid "letting go" of the previous target can be supported by a vehicle path

offset to the left. If the overtaking maneuver cannot be followed through as desired, the ACC reverts back to normal vehicle-following mode after a few seconds of the advance phase. However, this function is only suitable in countries with high relative dynamics, e.g., in Germany. In the US, on the other hand, there are often only slight differences in speed between the lanes, so that this function has to be offered in a much different form or omitted altogether. Alternatively, with corresponding sensor performance (especially azimuthal coverage range and multi-target capability), the speeds of vehicles in the target lane can be analyzed to form a basis for overtaking assistance.

32.8.4 Reaction to Static Targets

Static objects may be obstacles located in the driving corridor. But in many cases, they are targets that can be driven over, such as drain covers, or under, such as bridges or signs that extend above the roadway. Even at speeds of 70 km/h, deceleration of around $2.5\,\text{m/s}^2$ should be started at about 100m before the object. However, since the error rate in target selection is still very high in this case, a reaction to static targets is useful in exceptional cases and at low speeds only. The most important exception concerns the history of static objects. If these have been previously measured with an absolute velocity distinguishable from zero, they are classified as "stopped" objects and may also be treated as potential target objects. Otherwise, the conditions for reacting to static targets are restricted only in the close range up to approx. 50m. The reaction can be an "acceleration suppression", by preventing the increase of speed as long as the static object is located in the driving corridor, or a collision warning. If advanced classification methods are used for the surroundings sensors, e.g., image-based semantic segmentation (cf. ▶ Chaps. 18 and 19), all relevant obstacles can be included for the control, i.e., triggering a timely stop in front of them.

32.8.5 Stopping Control and Specifics of Low-Speed Control

In principle, it is not necessary to have a different control approach for low-speed control; however, the nominal values must give greater weightage to distance and speed deviations. For example, in a state controller structure, the corresponding controller gains can be adapted to the situation via a fuzzy estimate. This gives a greater degree of comfort and driver-like behavior. Since the distance to the preceding vehicle is short at low speeds, particular attention must be given to situations such as "vehicle cutting in close", "object too close", or "stopping" in this speed range. Due to the greater weightage of distance and speed deviations in detecting these situations, the controller reaction is quicker, which ensures appropriate dynamics to the given situation.

Compared with "normal" ACC operation, higher decelerations (up to 5.0 m/s^2 according to [1], cf. ▶ Sect. 32.2.3) are permitted in low-speed control. This also allows dynamic stopping. However, decelerations in excess of this are not useful, as this would make drivers think that the system can automatically master any situation, so that they cease to be aware of the function limits. An additional feature of low-speed control is traffic jam detection. If a traffic jam is detected on the basis of sensor data (e.g., due to the person in front repeatedly driving off at a low pace, low maximum speed, stopping shortly after setting off), the distance and speed deviations are given a lower weightage, to produce smoother controller behavior. This can, consequently, guarantee different behaviors with dynamics appropriate to the situation, e.g., a tailback on a motorway or a queue at traffic lights.

32.9 Acceleration Control and Actuators

32.9.1 Basic Structure and Coordination of Actuators

The acceleration control presents the challenge of converting adaptive speed control, namely, the ultimate set acceleration obtained from various individual controllers, into an actual acceleration. For this purpose, the sum forces (or sum torques) of respectively self-contained subordinate control circuits of the drivetrain and brake survey systems are adjusted so that the desired acceleration can be implemented. Even though simultaneous actuation of drive torque and brake torque is possible, it is generally avoided and the respective elements are controlled independently.

With respect to the design of harmonic transitions between drivetrain and brakes, it makes sense to choose a physical value with which both actuator subsystems can be controlled equally. Wheel torques (but also wheel forces) are an option. In this case, a summary consideration is sufficient, i.e., the sum of the wheel torques acting on all four wheels since ACC does not apply torques to individual wheels. In this way, coordination on the basis of the same physical signal can be affected as close to the actuators as possible, as shown in the following sections.

ACC requires the implemented actual sum wheel torque (and/or force) to calculate the driving characteristic equation and therefore, among other things, esti-

mate the road gradient. The drivetrain must also provide the current set maximum and minimum value as the sum wheel torque. In this case, specifically the minimum possible torque, i.e., the achievable torque in the current gearshift operation, is important, since the brake can only be activated when it is no longer possible to decelerate via the drivetrain.

As outlined in more detail in later sections, ACC control does not require absolute accuracy of the actuators since deviations from the required reference value can generally be well compensated for by the closed control loop. Only at the start of the control process and at transitions between the different actuators does good absolute accuracy simplify the control. In principle, however, good relative accuracy is necessary to achieve the desired control comfort.

32.9.2 Brake

For the relationship between deceleration D (= negative acceleration) and brake pressure p_{Br} applied:

$$\Delta D = \Delta p_{Br} \frac{A_K \mu_{Br} R_{Br}}{m_v R_{dyn}} \tag{32.18}$$

Key to symbols:
ΔD – Change in vehicle deceleration

Δp_{Br} – Change of brake pressure

A_K – Total area of brake pistons

μ_{Br} – Sliding friction coefficient between brake pad and brake disk

R_{Br} – Effective radius at the brake disks (mean value)

m_v – Vehicle mass

R_{dyn} – Dynamic radius of the wheels

$A_K \mu_{Br} R_{Br} = 70\,\text{Nm/bar}$ and $m_v = 2100\,\text{kg}$ and $R_{dyn} = 0.34\,\text{m}$ can serve as approximate values. Depending on the vehicle configuration, transmission ratio values of $\Delta D / \Delta p_{Br} = 0.07 \ldots 0.14\,\text{m/s}^2$ bar can result. This means that pressure changes have an effect with the factor $0.1\,\text{m/s}^2$ bar in the deceleration, i.e., pressure jumps of 1 bar are just below the notice threshold value of $0,15\,\text{m/s}^2$. The requirements on the dosing capability of the brake control are correspondingly high.

32.9.2.1 Acting Ranges
If a deceleration of 5 m/s^2 is requested by the longitudinal control system, even on a plane roadway, 50% of the maximum pressure required for good road conditions for emergency braking of 80bar is required. However, if the maximum payload, unbraked trailer loads, and downhill driving are taken into account, significantly higher values are obtained such as up to a maximum of 80% of the maximum pressure. In order to

have sufficient reserves for different friction conditions on the brake disk as well as fading, the permissible setting range must be extended upwards so that the entire setting range available must be used. Therefore, the safety monitoring of the brake intervention cannot take place with the amount of the requested braking torque but rather only via the set deceleration.

32.9.2.2 Actuator Dynamics

For comfort functions such as ACC, deceleration rates of up to $5\,\text{m/s}^3$ are permitted (see ▶ Sect. 32.2.3). This would result in a required pressure rate of $\dot{p}_{max} \geq 30 - 40\text{bar/s}$. However, in order to be able to follow the specified torque or pressure curves with sufficient dynamics, the brake system must be able to follow rates with up to $\dot{p}_{max} \approx 150\text{bar/s}$.

Dynamic following of the set value with sufficiently rapid pressure increases up to the start of braking, and as far as possible delay-free following, in the case of pressure modulations, is required. The maximum time delays in this case should remain below $\tau_B < 300\text{ms}$. The preconditions for this, along with a correspondingly dimensioned pump, are predominantly the de-throttling of the hydraulic system in the pump intake area to provide the required volume, largely irrespective of temperature. The control of the set value must be free from vibration and overshooting as this will be perceived by the driver as extremely unpleasant.

Together with the rapid following of dynamic set values, as far as possible stepless following of small or slowly changing set values is absolutely essential because precisely this type of control of small control differences is typical for ACC operation. Stationary deviations are also to be avoided as they turn into speed and distance errors and can, consequently, lead to limit cycle oscillations.

32.9.2.3 Control Comfort

As already described in the introduction, the vehicle responds very sensitively to pressure changes. To ensure that a sensitive driver perceives the pressure increase as continuous, the brake system must be able to handle braking increments of less than $\Delta p \approx 0.5\text{bar}$. As far as possible, braking pressure increase and decrease should be silent, harmonious, and continuous. Inadvertent pressure changes of more than 1 bar are to be avoided. Additional pump elements are beneficial for an even pressure increase, while continuously controlling valves are beneficial for pressure decrease. In terms of acoustics, a low pump speed is desirable, as are suitable location of the hydraulic unit and the suitable placement of brake lines in order to prevent the absorption of vibrations from the chassis. A complicating factor is that in the case of brake intervention, one of the substantial

noise emitters in the vehicle, the engine, is reduced to its acoustic minimum, the drag range.

For braking in the low-speed range, there are increased requirements for the acoustics of the brake control, primarily due to the lack of driving and engine noise. Braking noises such as squealing or judder must also be minimized. Due to the higher decelerations at lower speeds, this results in a modified braking behavior, whereby the pedal must not harden excessively.

32.9.2.4 Miscellaneous Requirements

- The brake light has to be actuated independently of the driver applying the brakes. In brake actuator systems with an active booster, this can be achieved at the pedal without changing the brake light switch. Whereas in brake actuating systems with hydraulic pumps, the brake light can be controlled by the control unit as a function of brake pressure and deceleration. Flickering of the brake light must be avoided by means of minimum actuation times or switching hysteresis.
- The distribution of brake pressure between the front/rear axles must be kept identical to normal brake application to prevent overloading of the brakes on one axle or unstable vehicle behavior. Additional brake circuit pressure sensors have proven useful in this context. During long periods of braking, potential leakage can to be detected and compensated for in one circuit.
- When the driver brakes during ACC braking, the pedal feedback should be kept to a minimum. In particular, vibrations or even shocks at the pedal must be avoided, and the transition into the normal brake pressure curve should be smooth.
- If the vehicle becomes unstable, the vehicle control (ABS, TSC, ESC) has priority, and transitions into skid control must be suitably designed.
- Safeguards: In the event of failures in the ESC system, the brake pressure must be reduced immediately. In the event of failures in the partner control devices, braking must be terminated, or the pressure gradually ramped down, depending on the severity. It is also necessary to ensure that all shut down signals (in addition to the brake light switch), such as the actuator elements, handbrake operation, incorrect gear, etc., are safely processed.

32.9.2.5 Feedback Information

The brake subsystem is the key source of internal vehicle state values, the most important of which are vehicle speed, yaw rate, steering angle, brake light switch, and slip control information. Furthermore, the current actual brake moment is fed back such that the ACC can carry out a gradient estimate. The ESC system provides binary state information (flags) (e.g., ABS active,

TSC active, ESC active) for an appropriate response to control states.

32.9.2.6 Additional Requirements for Standstill Management

Once it detects that the vehicle is at standstill, the FRSA hands over the responsibility for keeping the vehicle at standstill to the ESC, giving rise to the following tasks:

- Increasing (or even decreasing after sharp decelerations) the brake pressure for keeping the vehicle at standstill. Road gradient detection is advantageous for this.
- Permanent roll monitoring and, if necessary, increasing brake torque.
- Slip detection at very low friction condition, possibly releasing the brake, to maintain steerability.
- Safe transition to no energy holding (actuation of the electric parking brake, EPB) on recognizing the driver's intention to alight from the vehicle.
- Monitoring the temperature of the hydraulic system because of heating due to permanent valve currents and switching off with driver warning when overheating is expected.
- With engine start-stop systems, it is necessary to ensure that all necessary functions remain active during the voltage drop that accompanies engine start-up, in particular, the correct closing of the hydraulic valves that are responsible for maintaining the necessary brake pressure.

32.9.3 Drivetrain

In the following, the combination of combustion engine and automatic transmission is considered. The combination with manual transmission is considered as a special case, while combinations with hybrid drives are also conceivable. In principle, it should be noted that transitions between the electric motor and the combustion engine must be just as imperceptible for ACC as they are for the driver; for ACC, the drive continues to be merely a torque actuator, since it is irrelevant for the system function how these torques are generated. With regard to recuperative braking with an electric motor, care must be taken to ensure appropriate coordination with the braking system, which must take over the transition to the friction brake. Compared with the hybrid, the pure battery-electric drive fundamentally simplifies controllability, since there are no transitions to consider with the combustion engine. However, battery-electric drives are also simpler compared to the combustion engine drive alone since the transmission is usually designed with a constant ratio.

Regardless of the drivetrain concept, it has proven effective to view the engine and transmission as an entity from the point of view of longitudinal control and to directly specify sum-wheel setpoint torques. The drivetrain subsystem is then left to decide how these torques are to be set, either by changing the engine torque or by changing the transmission ratio.

Thus, a change in acceleration Δa has a proportional relationship with the change in the total wheel force $\Delta F_{R\Sigma}$ or the change in the sum of the wheel moments $\Delta M_{R\Sigma}$:

$$\Delta a = \frac{\Delta F_{R\Sigma}}{m_v} = \frac{\Delta M_{R\Sigma}}{m_v R_{dyn}} \tag{32.19}$$

Key to symbols:

Δa – Change of the vehicle acceleration

m_v – Vehicle mass

R_{dyn} – Wheels radius

While direct actuation of the engine via engine torque set value is possible, specific action is required to influence the transmission in order to maintain an adequate dynamic while avoiding unwanted gearshifts. Simply using a heavily gearshift-based characteristic as in CC is inadequate because the ACC-following controller must be far more dynamic in design than a vehicle speed controller that is purely designed for constant cruising.

Equally, a direct conversion of the engine torque set values into virtual driving pedal angles to control the transmission logic is not suitable because the ACC attempts to emulate a preset acceleration precisely. Additionally, other than with the driver, deviations are directly reflected in set value changes which at certain operating points could lead to toggling gearshifts.

32.9.3.1 Engine Control

As with the brake, the ACC requires access to the entire possible torque range for the necessary control range in order to cover all relevant driving situations. The required actuator dynamics corresponds to the dynamics required by the driver, which should not present a problem in most modern systems because the driver set values can also be transferred electronically; driver and ACC settings are therefore principally transferred via the same path.

The drivetrain optimally applies the required sum wheel torque of the ACC function (similar to the accelerator pedal) at the respective operating point. The states of engine, transmission, and auxiliary units will be considered for calculating the set value. Coordination is autonomous in the drivetrain system as far as possible. If this is not supported, conversion to the engine torque by the ACC control unit or a longitudinal dynamic module is required, for which the current transmission ratio must be known.

The ACC function differentiates between different operating modes for comfort purposes which relate to the coordination of the drivetrain's different actuation options (e.g., fuel cut-off, gearshifts, incorporation of ancillary units). Thereby, minor inconstancies in the torque application occur, for example, when activating the fuel cut-off, these inconstancies have to be avoided or tolerated. In addition, more serious inconstancies in the torque application occur, which for example are avoided or permitted during additional gear shifting in automated gearboxes.

Examples:

- Triggering the fuel cut-off, but no additional gearshifts (only rolling gearshifts), on approaching a slowly moving target object or on reducing the desired speed.
- Triggering the fuel cut-off and additional gearshifts during static downhill travel to support the brake system in downhill mode.
- Suspending the fuel cut-off during static downhill travel to cancel previously effected gearshifts. This prevents "fuel cut-off toggling" and allows the gearbox to cancel shift-in, if there is a change in the gradient during static downhill travel.

32.9.3.2 Transmission Control

ACC state control essentially requires information from the transmission that a valid (forward) gear is engaged as one of the activation conditions.

If engine torques are to be preset, the current torque amplification V_{DT} is required by the gear unit. This value is the ratio of force $F_{R\Sigma}$ at either one or both drive axles to the engine torque M_E and can be calculated by the product of converter gain μ_C, the transmission ratio i_{TR} of the current gear, and lastly, the final drive ratio i_A divided by the dynamic wheel radius R_{dyn}:

$$V_{DT} = F_{R\Sigma}/M_E = \mu_C \cdot i_{TR} \cdot i_A/R_{dyn} \qquad (32.20)$$

The converter gain is generally incorporated as a curve in a lookup table which may need to be temperature compensated.

FSRA may also use electronic gearshift systems as additional protection for stopping management. In this case, the parking lock is engaged on detection of the intent to leave the vehicle. This is sufficient in combination with a multistep driver early warning system which advises the driver of the responsibility to make the vehicle safe when it is stopped.

However, it is not sufficient as the only safety means for a fully automated stopping management system (without driver intervention), since the parking lock blocks only the drive axle. At corresponding µ-split conditions, the wheels could turn, and the vehicle could roll away. Also, in the case of a late request or of an error when the vehicle is already rolling, it is not possible to safely engage the parking lock above approx. 3km/h , while an EPB can function at any speed in principle.

32.10 Functional Extensions

32.10.1 Consideration of the Right-Hand Overtaking Ban[5]

In some countries, overtaking on the side of the main direction of traffic such as in Germany where overtaking on the right is prohibited at driving speeds of above 80km/h. At lower speeds, however, overtaking on the right is permitted at low relative speeds. The ban does not apply to lanes to different destinations, such as those found at exit lanes at highway interchanges or exits. In order to comply with these regulations, more knowledge is required than can be provided by a simple ACC system. The system must

- know in which country it is used and whether overtaking on the right is permitted there. This also includes knowing whether right-hand or left-hand traffic prevails in the country. The system can obtain this information from the navigation system map.
- know whether traffic is currently traveling in a lane that branches off to another connection opposite the adjacent lane. This can be realized by the system being able to locate itself lane by lane on a detailed map or by recognizing the lane marking types, which are often wider and/or differently dashed at pull-out lanes.
- detect the next vehicle in its own lane as well as in the neighboring lane by means of sensors.

Some ACC systems thus also realize compliance with the regulations mentioned at the beginning.

32.10.2 Adjustment of the Set Speed to the Maximum Permissible Speed

If a speed limit assistant is also installed in the ACC system vehicle, which displays the currently valid speed limit to the driver, it makes sense to support the driver in adjusting to the speed limit with the ACC system activated. This can be done semi-automatically by pressing a button while ACC is activated to adopt the speed

---

5 Related to right-hand traffic. Must be adapted accordingly for left-hand traffic.

limit displayed by the speed limit assistant as the new set speed. There are also ACC systems that perform this acquisition automatically when the detected speed limit changes. This function can be activated in a configuration menu and, for some systems, a desired deviation from the speed limit of, e.g., ±15km/h can be set.

32.10.3 Speed Adaptation to the Course of the Track

Most vehicles equipped with ACC are also equipped with a navigation system. This means that the ACC system is potentially aware of the route ahead. With this knowledge, some ACC systems reduce the driving speed before curves, intersections, and roundabouts. This keeps the lateral acceleration within the comfort corridor and prevents drivers from having to intervene by braking, thus deactivating the ACC system. If driving without activated route guidance, it will be more difficult to determine the route ahead and the deceleration via ACC will not always correspond to the route actually being driven by drivers.

32.11 Usage and Safety Philosophy

For the acceptance of the ACC system, a high transparency of the system functionality including the comprehensibility of the system reactions is essential. Only when users are quickly able to predict the system responses will they also use the system effectively. This presents the developer with the problem of making the control as simple as possible and sometimes leaving out features which an experienced user and of course the developer would value. As the driver gives up a part of the vehicle management task to the system when the ACC function is active, and needs only to monitor this, the transparency of the system plays an important role. Since the current ACC systems only take over a part, albeit a large part, of the longitudinal control, it is sensible and necessary to design the system limits when the system is used as intended so that they are reached or exceeded with a certain regularity. This ensures that drivers are aware of the system limits at all times and are trained to take over control from the system.

Adaptive cruise control is not a safety function. It is designed first and foremost to increase driving comfort. Sensibly, a comfort system should not pose any dangers; therefore, the ACC system must guarantee a level of safety corresponding to this requirement. Fault tree analyses have shown that dangerous situations can only occur if the driver does not employ intervention options. Three consequences can be derived from this:

1. The drivers must not be overburdened with the takeover. In particular, this means that they recognize the necessity of the takeover and choose the subsequent reaction in time and with the right course of action.
2. The driver takeover option must be designed to be fault-tolerant, so that these options, such as switching off the control, decelerating more strongly, or accelerating more strongly, may be blocked in only highly unlikely ways.
3. Takeovers must occur sufficiently often so that they can be considered practiced for drivers.

Prompt recognition of the need to take over the controls is derived from the driver's perception gained from past experience. In particular, too much confidence in the technology because only error-free functioning is experienced would be problematic because the driver would be unprepared both for the occurrence and also for the response. In the case of ACC, this difficulty is not an issue because, as explained above, it is impossible to perfect its function. This negative aspect, therefore, has the benefit that the driver is permanently trained for the error situation. The driver remains aware that he/she may have to resume control in the case of unwanted performance and is well-practiced in when and how this is carried out. A further elaboration of these aspects can be found in [3].

32.12 Special Features for Combined Longitudinal and Lateral Control

This section briefly discusses the loose coupling of longitudinal and lateral control. In this combination, the basic functions of longitudinal control and lateral control are performed in parallel, largely independently of each other. The main interactions between these two control functions relate to:

- Corridor adaptation during lane change, which in turn is started by the lateral control on triggering by the direction indicator, after the lane change guard has assessed the lateral gap as sufficiently large, see ▶ Chap. 33.
- The target trajectory of the lateral control can be used for target selection if following a staggered vehicle is considered more relevant than following the lanes.

With the transition from Level 1 (ACC alone) to Level 2 with lateral guidance, the focus of the safety design also shifts to lateral guidance, and particularly, to the monitoring of the hands-on condition, since unintended reactions in lateral guidance must be quickly intercepted by the drivers.

32.13 Outlook

Longitudinal control as a stand-alone ACC function is now very mature. It is available as an option in essentially all new vehicles, unless it has already been replaced by integrated longitudinal and lateral control assistance. ACC has helped many other assistance and safety systems to achieve a breakthrough. Further development, however, now depends more on the other objectives, such as intervening active safety systems with a higher scope of application than earlier, but also application for electromobility and emission avoidance. These will change slightly in terms of functionality when compared with what is known today, but the components, however, will undergo changes. In the long term, it will merge into Integrated Assistance or even into higher levels of automated driving.

References

1. Winner, H.: Die Aufklärung des Rätsels der ACC- Tagesform und daraus abgeleitete Schlussfolgerungen für die Entwicklerpraxis. Tagungsbeitrag Fahrerassistenzworkshop Walting (2005)

2. Winner, H., Schopper, M.: Adaptive cruise control. In: Winner, H., Hakuli, S., Lotz, F., Singer, C. (eds.) Handbook of Driver Assistance Systems, pp. 1093–1146. Springer International Publishing, Switzerland, (2016). ▶ https://doi.org/10.1007/978-3-319-12352-3_46

3. ISO TC204/WG14: ISO 15622:2018 Intelligent transport systems. In: Adaptive Cruise Control systems—Performance Requirements and Test Procedures, 3rd edn. ISO ICS, 03.220.20 (2018)

4. Winner, H., Olbrich, H.: Major design parameters of adaptive cruise control. In: AVEC'98. Nagoya, Paper 130 (1998)

5. Pasenau , T., Sauer, T., Ebeling, J.: Aktive Geschwindigkeitsregelung mit Stop&Go-Funktion im BMW 5er und 6er. ATZ 10, 900–908. Wiesbaden, Vieweg Verlag (2007)

6. Meyer-Gramcko, F.: Gehörsinn, Gleichgewichtssinn und andere Sinnesleistungen im Straßenverkehr. Verkehrsunfall und Fahrzeugtechnik 3(1990), 73–76 (1990)

7. Luh, S.: Untersuchung des Einflusses des horizontalen Sichtbereichs eines ACC-Sensors auf die System performance. Dissertation TU Darmstadt, Fortschritt-Berichte Reihe 12. VDI, Düsseldorf (2007)

8. Winner, H., Danner, B., Steinle, J.: Adaptive cruise control. In: Winner, H., Hakuli, S., Wolf, G. (Hrsg.) Handbuch Fahrerassistenzsysteme, ATZ/MTZ-Fachbuch, S. 478. © Vieweg+Teubner Verlag|Springer Fachmedien Wiesbaden GmbH, Wiesbaden (2012). ▶ https://doi.org/10.1007/978-3-8348-8619-4

Open Access This chapter is licensed under the terms of the Creative Commons Attribution-NonCommercial-NoDerivatives 4.0 International License (▶ http://creativecommons.org/licenses/by-nc-nd/4.0/), which permits any noncommercial use, sharing, distribution and reproduction in any medium or format, as long as you give appropriate credit to the original author(s) and the source, provide a link to the Creative Commons license and indicate if you modified the licensed material. You do not have permission under this license to share adapted material derived from this chapter or parts of it.

The images or other third party material in this chapter are included in the chapter's Creative Commons license, unless indicated otherwise in a credit line to the material. If material is not included in the chapter's Creative Commons license and your intended use is not permitted by statutory regulation or exceeds the permitted use, you will need to obtain permission directly from the copyright holder.

Lateral Guidance Assistance

Dirk Wisselmann, Arne Bartels, Felix Fahrenkrog and Gregor Nitz

Contents

© The Author(s) 2026
H. Winner et al. (eds.), *Handbook Assisted and Automated Driving*,
https://doi.org/10.1007/978-3-658-45276-6_33

33.1 Motivation

Lateral control of a vehicle is one of the primary tasks of drivers which they must perform continuously while driving. The goal is to keep the vehicle safely in the lane. Assistance systems for continuous support of lateral guidance offer a gain in comfort for drivers, especially on longer journeys, if they are designed as "lane-keeping assistance" (LKA) with active intervention in lateral vehicle guidance. "Lane departure warning systems" (LDW) are intended to warn drivers in good time of an unintentional departure from the lane or to prevent this by actively intervening in the lateral guidance and therefore address the safety gain.

In order to assess the potential of lateral guidance assistance systems for increasing road safety, a distinction must be made between the two crash scenarios of "leaving the lane" and "leaving the road". Crashes that occur due to unintentional lane departure are recorded in Germany as part of the so-called driving accidents. Approximately 37,000 such crashes were reported by the police in 2019. A comparison of the proportions of crashes involving unintentional lane departure in relation to the location and injury categories shows that this type of crash mainly occurs outside built-up areas and, at 29.4%, represents the largest proportion of all crashes recorded by the police outside built-up areas (◘ Fig. 33.1). LDW and LKA systems reduce the number of these crashes, which has been proven in many studies (▶ Sect. 33.7).

The relevance of lane departure warning systems for road safety is also taken into account by the European legislator, who has prescribed a lane departure warning system for new commercial vehicle types in classes M2, M3, N2 and N3 since 2013. For the year 2022, a corresponding requirement will also be extended to new passenger car vehicle types by means of the current General Safety Regulation (EU) 2019/2144, which must be equipped with an "Emergency Lane Keeping System" (ELKS) from this date (▶ Sect. 33.4).

33.2 Requirements

Assistance systems for the support of the lateral guidance of vehicles are intended to prevent unintentional departure from the lane by
- informing or warning the driver of this in good time
- steering the vehicle back into the lane if possible or
- actively supporting the drivers to keep the vehicle in the centre of the lane as a continuously controlling system.

If the system is only to warn against leaving the lane, it must be able to determine the vehicle's position relative to the lane's boundary. Depending on the type of road, the lane boundary is indicated by marking lines on one or both sides. The systems must detect at least one of these lines ("single line detection"). Current systems also detect the edge of the lane, which is mandatory for the ELKS systems to be installed from 2022 (▶ Sect. 33.4).

To steer back or keep the vehicle in the lane, the system must determine the vehicle position relative to the centre of the lane, the future direction of movement of the vehicle and the course of the lane ahead. In addition to suitable vehicle sensors, this also requires environment sensors with sufficient foresight and detection accuracy. If lane detection is based on the left and right lane marking lines, both lines must be detected by the system ("dual line detection"). Ideally, this line detection should take place on all roads in all countries, even under adverse environmental conditions.

Driver information or intervention should be avoided unless required by the vehicle's lateral position and orientation. Furthermore, information/interventions should be avoided in the case of a deliberate lane change, e.g. when overtaking, or in curvy areas where the driver deliberately "cuts" the curves.

The information about the unintentional lane departure should be clearly perceptible to the drivers, but it should not be "annoying". To this end, the de-

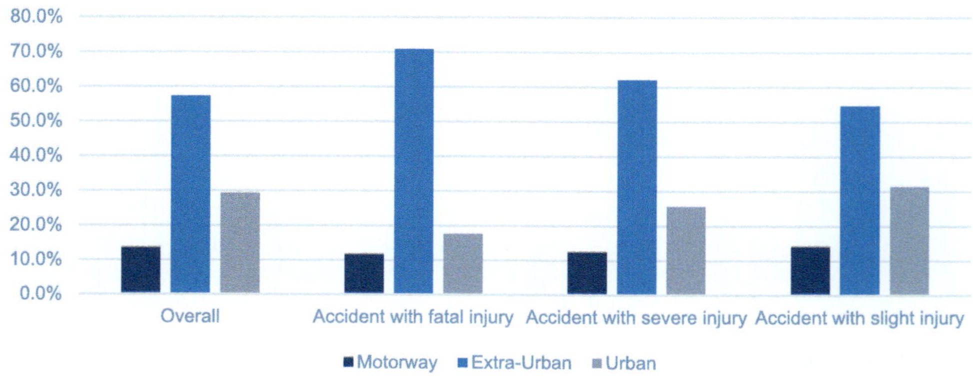

◘ **Fig. 33.1**   Proportion of crashes with lane departure related to the driving location [1]

sign must carefully balance visual, auditory and haptic driver information.

An active system intervention in the vehicle's lateral guidance is required to keep or to return the vehicle in the lane. This must be designed in such a way that the driver is always able to override the system. Therefore, steering interventions are often limited to a steering wheel torque of approx. 3 Nm.

When the vehicle is controlled to stay in the lane, the lateral guidance assistance should reproduce steering behaviour that is as natural as possible, e.g. no high-frequency continuous steering movements should appear.

It should be clearly and transparently indicated to the driver whether the system is switched on and active. In addition, intuitive operation of the system by the driver should be ensured.

The system shall support drivers in holding the vehicle in the centre of the lane without completely relieving them of this task. The responsibility for guiding the vehicle in the lane remains with the drivers and they must not turn away from this task. This applies in particular to systems with active intervention in vehicle lateral guidance. Therefore, drivers should usually be continuously involved in the execution of the vehicle's steering.

33.3 Classification

◘ Table 33.1 shows the technical classification of LDW and LKA systems. LDW systems inform the driver haptically, visually and/or audibly about the imminent, unintentional departure from the lane. LKA systems support drivers by actively intervening in the lateral guidance when holding the lane. This can be done in two different ways: Type I first prevents lane departure in the best possible way by actively intervening in the vehicle's lateral guidance, more or less as an extension of an LDW system. If the crossing of the marking cannot be prevented despite active steering support, information in the sense of an LDW is issued to the driv-

ers. Type II supports drivers when keeping the vehicle in the middle of the lane by continuously controlling the vehicle's lateral guidance and informs the drivers if necessary (see LDW). This differentiation leads to different goals of the functions: While type I contributes to vehicle safety, type II additionally addresses comfort aspects.

The safety-enhancing LDW and LKA Type I systems are typically active when the vehicle is restarted and provide support when needed usually above a speed of 60-70 km/h. LKA Type II systems are activated as comfort systems by the user and can be used in the entire speed range with most manufacturers.

LDW or LKA type I systems can be assigned to automation level 0 "No Automation" according to SAE standard J3016 (▶ Chap. 2), as they only warn the driver or prevent the vehicle from leaving the lane by corrective steering intervention. LKA type II systems can be assigned to automation level 1 "Driver Assistance", as they support the driver in lateral control of the vehicle without relieving him of the driving task. LKA Type II systems in combination with longitudinal guidance assistance systems can then be assigned to automation level 2 "Partial Automation", as they support the driver in both longitudinal and lateral guidance of the vehicle, without again relieving him of the driving task. Examples of SAE Level 1 and Level 2 systems can be found in ▶ Sect. 33.6.

Lane-centred lateral guidance without driver involvement, as required, e.g. for automated driving functions (SAE Level 3, 4 and 5), can be built purely technically on the LKA system components described below. However, it is not part of this chapter on lateral guidance assistance.

33.4 Regulations, Standards and Tests

In the following, some important regulations and standards with direct reference to lateral guidance assistance systems are mentioned by way of example, from which requirements for corresponding systems re-

◘ **Table 33.1** Classification of the systems for lateral guidance assistance

LDW	Lane departure warning	
	Informs the driver haptically, visually, acoustically about unintentional lane departure	
LKA	Lane Keeping Assistance	
	Assists the driver in keeping the vehicle in lane by actively intervening in lateral control	
	Type I	Prevents the vehicle from leaving the lane as far as possible by actively intervening in the vehicle's lateral guidance and informing the driver if necessary (see LDW), safety-oriented function. ELKS are subsumed here
	Typ II	Supports the driver while keeping the vehicle in the lane centre by actively assisting the vehicle's lateral guidance. System informs the driver if necessary (see LDW), comfort-oriented function

sult. In addition to limit values, requirements for override capability and HMI (human machine interface) concepts are specified.

ISO 17361 "Lane departure warning systems—Performance requirements and test procedures" [2] and ISO DIS 11270 "Lane keeping assistance systems (LKAS)—Performance requirements and test procedures" [3] specify, among other things, minimum functional requirements, basic HMI elements and test methods for LDW and LKA systems, in each case for passenger cars, commercial vehicles and buses on motorways and equivalent roads. ISO DIS 11270 does not differentiate between Type I and Type II systems. ISO standards are good references of the state of the art and thus form an important basis for system design. Beyond that, however, they have no approval-related legally binding character.

There are different international legal bases for placing steering driver assistance systems on the market. For example, the requirements for steering systems for the signatory states to the UN Convention of 1958 are specified in UN Regulation 79 on steering equipment [4]. In addition to requirements for the mechanical design of the steering system, this also contains technical specifications for corrective steering functions (CSF) and automatically commanded steering functions (ACSF). For both categories, the driver remains primarily responsible for driving the vehicle at all times.

For LDW systems that fall under the CSF category, there are specifications in particular with regard to the HMI: an intervention must be visually indicated to the driver, and an acoustic signal is also required from a certain intervention duration. The minimum display duration increases with an increasing number of interventions in a rolling time interval.

LKA systems fall under ACSF category B1. In addition to requirements for the HMI, there are specifications in particular regarding the maximum permissible lateral acceleration of 3 m/s^2 caused by the system, the ability of the driver to override the steering wheel torque applied by the system, and the system behaviour in the absence of driver activity. If the driver leaves the steering wheel out of his hands for more than 15 s, a visual warning must be given, and after a further 15 s an acoustic warning must also be issued. After 60 s at the latest, the function must be deactivated.

The last General Safety Regulation (GSR) (EU) 661/2009 already obliges the installation of LDW systems for commercial vehicles of the classes M2, M3, N2 and N3, since 1.11.2013 for all new vehicle types and since 1.11.2015 for all newly registered vehicles of the corresponding class. In a separate regulation 351/2012, the EU specifies type approval requirements for LDW systems. For further information on lateral guidance assistance for commercial vehicles, please refer to ▶ Chap. 36.

With the decision of the new GSR (EU) 2019/2144 by the European legislator, LDW/LKA type 1 systems will also become mandatory for passenger cars of classes M1 and N1 from 1.7.2022 for new approval types and from 1.7.2024 for all new registrations in Europe. These functions, referred to in the GSR as ELKS—"Emergency Lane Keeping Systems", must have both a purely warning (LDWS—"Lane Departure Warning System") and a direction-correcting (CDCF—"Corrective Directional Control Function") component (LKA Type 1). The requirements for ELKS include specifications regarding the HMI, limited deactivation with automatic reactivation after vehicle restart, a minimum speed range of 65 km/h to 130 km/h to be covered and the latest possible warning time, depending on speed, 30 cm beyond contact with a lane marking (LDWS) or active prevention of further drifting away from the lane (CDCF).

In the USA, driver assistance systems up to and including SAE Level 2 are currently not subject to any special legal requirements. Market introduction is possible here without specific technical requirements. In addition, however, the strict product liability law and, if applicable, the behavioural requirements imposed by individual states (e.g. the ban in the State of New York on taking one's hands off the steering wheel) limit the technical scope for design.

In addition to legal requirements, consumer protection tests accelerate market penetration and drive a certain level of performance of active safety functions. For example, LDW systems on the European market have been assessed in the Euro NCAP rating since 2014.

33.5 System Components

◘ Figure 33.2 shows the generic block diagram of a lane departure warning system (LDW) and a lane-keeping system (LKA) with the required system components. In addition to the function blocks of a LDW system, a LKA system requires an actuator and the corresponding extensions in the "function module" (shaded grey).

The environment sensors (mainly cameras) generate measurement data from which the subsequent signal processing extracts the required environment features (e.g. position and course of marking lines). By means of sensor data fusion, it is possible to include further environmental features in the signal processing. Within the function module (◘ Fig. 33.3), a warning algorithm determines the need for driver information, the output of which is decided by a state machine depending on vehicle and system status. The HMI contains the driver information as well as the output of the system status. The driver can use controls to switch the system on and off and to configure it (e.g. time for driver information).

LKA systems also require a controller for lateral guidance. This controller calculates manipulated vari-

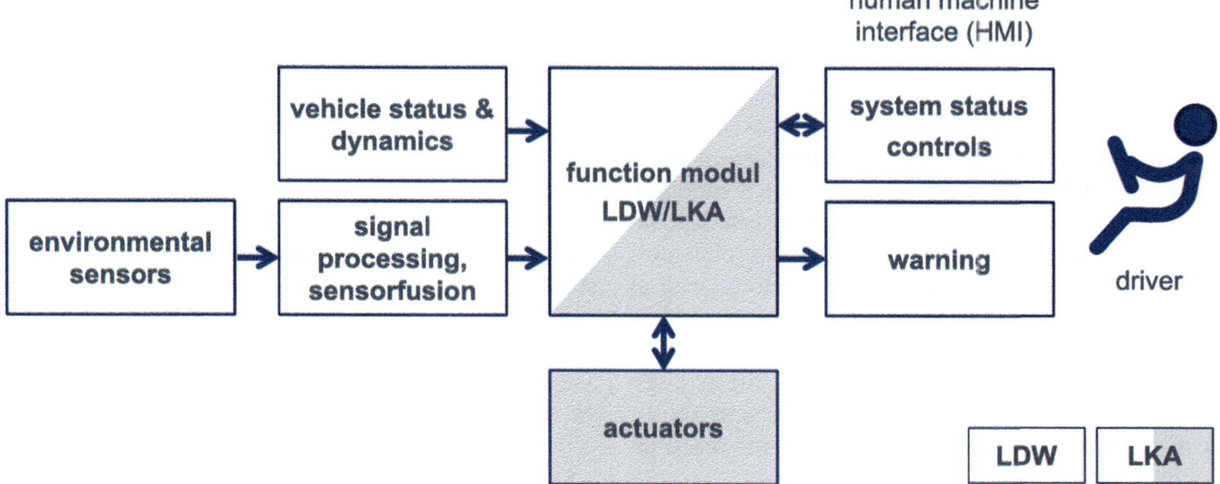

Fig. 33.2 System components for lateral guidance assistance

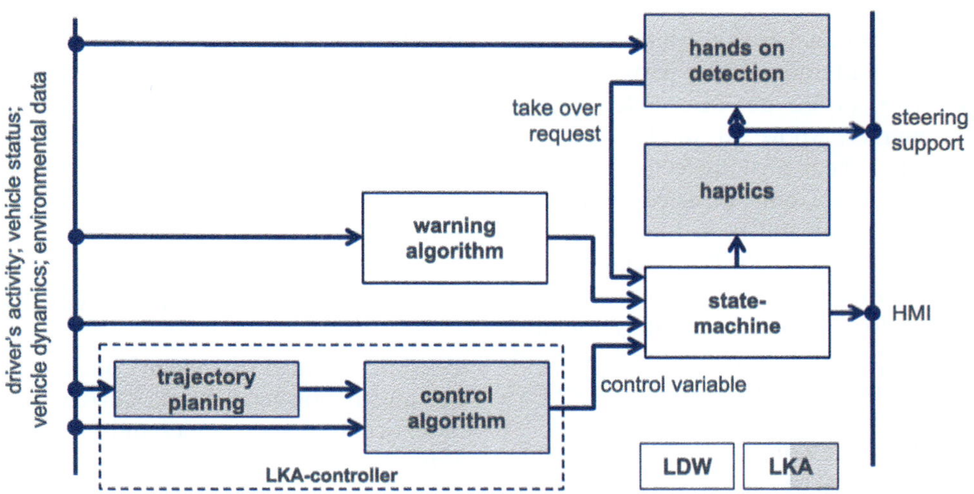

Fig. 33.3 Function module LDW/LKA

ables (corrective yaw moment, which is usually implemented by intervening in the steering or by braking individual wheels) and sends them to an actuator (steering or brakes) for the purpose of suitably influencing lateral control, which can ideally be perceived by the driver as haptic feedback in case of a steering torque intervention or kinaesthetic feedback in case of wheel individual braking. A detection of hands-free driving should prevent permanent vehicle guidance without hands on the steering wheel.

The individual components of a lateral guidance assistance system are described in detail below.

33.5.1 **Environmental Sensors**

In addition to the mono cameras that are mainly used, stereo or trifocal cameras constitute an alternative to detect the lane markings. The cameras are installed behind the windscreen at the level of the interior mirror, looking in the direction of travel and invisible to the driver. In principle, infrared diodes and laser scanners may also be used in addition to cameras.

The requirement for the field of view of the cameras results from the look-ahead time of the LKA systems required for a "smooth" vehicle control, for which approx. 1 s can be given as a reference value. At 210 km/h, this corresponds to a look-ahead distance of approx. 60 m, which today's cameras with viewing distances > 150 m clearly exceed. The required field of view is also covered with aperture angles of approx. 50°(horizontal) and + 27/−21°(vertical). A resolution of approx. 2.6 MP and a "multi-path" image processing (combined use of different image processing algorithms) enable the detection of marking lines with the required accuracy. The data is taken as an example from [5]. Furthermore, camera-based systems offer the potential for multiple use of the sensor, e.g. for traffic sign recognition, rain detection or for high beam assistance.

Increasingly, colour cameras are being used so that the system remains available even in construction zones with yellow and white marking lines, since ambiguities can now be resolved. Optimised exposure controls and a high dynamic range (HDR) increase robustness against extreme changes in lighting conditions, such as those that can occur at tunnel entrances/exits or when driving through an avenue with trees in summer.

In vehicles with a very high number of driver assistance systems and a corresponding "all-round sensor system" (cameras, radar sensors and, in some cases, ultrasonic sensors), further information about the vehicle environment such as static and dynamic objects (e.g. concrete barriers in construction zones or vehicles approaching from the side) is included in the algorithms of lane keeping systems [6].

Lane marking detection based on infrared diodes installed near the ground and looking perpendicularly at the lane (◘ Fig. 33.6) has not been able to establish itself on the market. Above all, the lack of look-ahead distance makes them unsuitable for LKA systems, as marking lines are only detected shortly before they are crossed. Ambiguities in road works can only be resolved very late, if at all. Both delay the output of driver information. Furthermore, the detection of so-called "botts'-dots", which are mainly found in the USA, is not guaranteed. Compared to cameras, infrared diodes are exposed to more pollution. However, due to their perpendicular viewing direction, they are insensitive to backlight and rain compared to cameras.

Location-based approaches using highly accurate, digital road maps have so far only been used by GM in the USA ("Supercruise", [7]). Laser scanners for lane detection have so far only been used in research projects [8], as have infrastructure-based solutions using magnetic nails or guide cables.

For more detailed information on environment sensors, please refer to Part D of the handbook "Sensors for Driver Assistance Systems".

33.5.2 Signal Processing

Image processing algorithms significantly influence the quality of camera-based LDW and LKA systems. Their central task is the detection of marking lines and the edge of the road. An algorithm for this is presented as an example in ▶ Chap. 18. General requirements for lane detection are:

- available on as many infrastructural conditions as possible,
- robust against adverse environmental influences.

The diversity of the road infrastructure is a challenge here. White (Europe) and yellow (USA, Canada) marking lines on dark asphalt or light concrete must be recognised, as well as marking pins in construction sites or "botts' dots" in the USA. Line lengths, gap lengths and line widths vary widely around the world (see ISO 17361 Annex A). In addition to well-maintained marking lines on motorways, worn or weathered marking lines on minor roads should also be recognised. Line structures from bitumen joints, tar seams or crash barriers, on the other hand, must be classified as irrelevant, as well as skidmarks and snow tracks on the road.

The performance of camera systems in adverse weather conditions has extremely improved in recent years. This has been achieved through a combination of different measures such as increased sensitivity through new pixel technology or an increased dynamic range of the cameras (split pixel, super capacitor). In the case of cameras installed externally, such as parking cameras, which are increasingly being used to detect lane markings, an increase in availability is achieved through cleaning and heating systems, as well as through the use of dirt-repellent coatings.

Short-term interruptions in line detection due to the above-mentioned impairments can be bridged by suitable algorithms such as Kalman or particle filters.

33.5.3 Function Module LDW/LKA

Based on the data of the environment recognition and the information available in current vehicles on vehicle status, driving dynamics and driver activity, a LDW system with driver information can already be implemented. If additional suitable actuators are available in the vehicle, such as an electromechanical power steering system (EPS "Electric Power Steering"), then type I and II LKA systems can also be realised. The central element here is the so-called function module.

◘ Figure 33.3 shows an example of the structure of such a software component with a central state machine and warning algorithm as well as the controller required for the LKA lateral control, a haptic calculation and hands-on detection. These components are described below.

33.5.3.1 State Machine

The state machine is a central component of LDW and LKA systems (◘ Fig. 33.3). It checks whether all boundary conditions for a driver information or an intervention in the lateral guidance of the vehicle are fulfilled. The respective system must be switched on (→on/off button) and ready for operation (→self-diagnosis); among other things, the sensors must neither be defective nor dirty, the direction indicator must not be set and the vehicle speed must be within the activation limits (→vehicle status). To avoid constant activation/deactivation, a hysteresis may be added to the speed thresholds. In the case of an LKA system, an additional check is made as to whether drivers are actively steering or keeping their hands on the steering wheel or allowing themselves to be driven by the system (→hands-free driving detection). In the state machine, a coupling with other driver assistance systems can be realised. When combined with lane change assistance systems, the LDW or LKA system can, for example, issue a driver warning if the driver announces his intention to change lanes by activating the direction indication, while the neighbouring lane is occupied.

33.5.3.2 Warning Algorithm

The so-called "Distance-to-Line-Crossing" (DLC) d_{LC} is the simplest criterion for a driver information about the imminent leaving of the lane. It denotes the lateral distance between a certain part of the vehicle and the lane boundary. By defining a minimum and maximum DLC, a warning zone is created that begins shortly before and ends shortly after the lane boundary (◨ Fig. 33.4 left). If the vehicle enters this warning zone, the driver is informed or warned. If the vehicle leaves the warning zone, the driver information ends. The DLC may be determined with simple sensors with low look-ahead distance, such as infrared diodes. However, the approach of using the DLC to infer an action-relevant situation can also be disadvantageous: For example, if a vehicle is driving very closely parallel to the lane marking, driver information is given even though the vehicle is not about to leave the lane.

The so-called "Time-to-Line-Crossing" (TLC) t_{LC} is more suitable as a criterion for a lane departure warning, as it can predict the departure from the lane, thus preventing unnecessary driver information as described above for the DLC. TLC refers to the amount of time after which a vehicle is expected to cross the lane boundary based on the vehicle's position and movement. It is calculated in the simplest case to be ($t_{LC} = d_{LC}/v \bullet \sin(\Psi)$), where $v \bullet \sin(\Psi)$ is the approaching speed to the edge of the lane with the vehicle's longitudinal speed v and the lane-related orientation of the vehicle Ψ (◨ Fig. 33.4 right). A generally valid approach to calculating TLC should also take into account the curvatures of the vehicle's trajectory and the lane. Calculations for this can be found, e.g. in [9, 10]. In the simplest case, driver information is given as soon as the TLC falls below a threshold value.

It is desirable that drivers have the option of setting this threshold value. Depending on driving style and type of road, it may make sense for the driver information to be given either shortly before, during or even shortly after the lane limit is exceeded.

If a driver deliberately cuts corners on a curvy road, drives closely to the edge of the lane on a narrow road, or does not use the direction indicator during overtaking manoeuvres, driver information may be judged to be inappropriate or annoying. This unwanted driver information can be avoided with the help of driver intention recognition (see ▶ Chap. 51). By evaluating additional environmental and contextual information such as, e.g. vehicle acceleration, accelerator pedal position, steering wheel angle, yaw rate, lane curvature as well as lane marking types on the left and right, intended curve cutting and overtaking manoeuvres can be detected in many cases and unnecessary driver information can be suppressed, or on narrow roads the time for driver information can be postponed.

If driver state detection is available, it seems reasonable to adapt the driver information times according to driver activity, e.g. to warn distracted, tired or inattentive drivers earlier or to improve the acceptance of active drivers by later information times.

33.5.3.3 Lateral Control

Initially, comparatively simple path-following controllers were used for lateral control in LKA systems of type I and II. In the meantime, approaches with trajectory planning and downstream control have become established. These approaches are described in detail in ▶ Chap. 33, so they will not be discussed further here. The decisive factor in these approaches is a valid representation of the driving environment in a model that

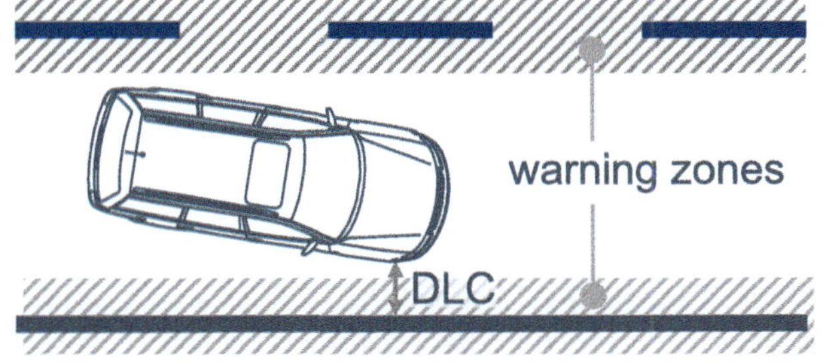

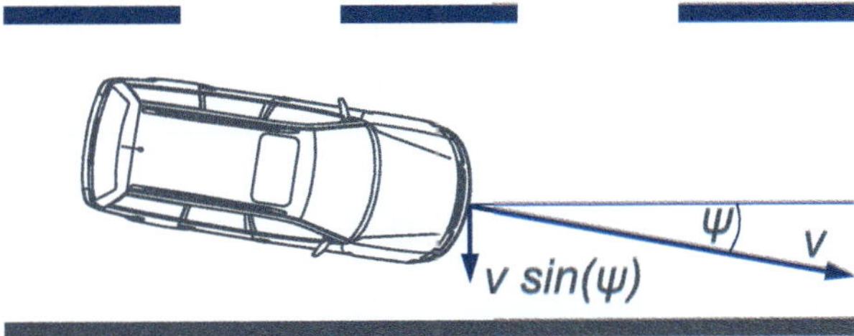

◨ **Fig. 33.4**   Left: DLC and warning zones right: Lateral vehicle speed to determine TLC

takes into account the roadway, boundary development and dynamic objects. With this information, an adapted reference path can be defined, which is then implemented in compliance with the manufacturer-dependent comfort, dynamics and safety targets.

33.5.3.4 Haptics LKA

Haptics and controllability of an LKA intervening in the vehicle lateral guidance are essential design criteria for the system. In the example presented here, they are considered part of the LKA function module, although their actual implementation in the architectures of current LKA systems may well be in other components (e.g. as a module in the control unit of EPS or ESC in LKA systems with course-correcting brake intervention).

Due to the direct and regular (LKA type I) or even permanent (LKA type II) interaction with the driver via the lateral guidance actuator, the haptics have a direct influence on the perception and acceptance of the system function. Since LKA lateral guidance may sometimes be not perfectly in line with the driver's expectation (e.g. active central guidance against a driver's wish to move to the edge of the lane), the haptics function module especially in LKA type II systems uses vehicle status, driver activity and environment data to determine when and with what noticeability the controller's requirements are to be implemented by the actuator system.

For the immediate driver activity, there are two decisive situations for the haptics: During countersteering, the driver's action is opposite to the controller demand, thus the driver perceives the controller demand as resistance on the steering wheel. The haptics module reduces the demand on the actuator proportionally depending on the desired steering support of the LKA system. This reduces the warning character and control quality, but at the same time the system appears more comfortable, less "disturbing". In the case of co-steering, driver activity and controller demand are directed in the same direction; drivers therefore perceive the controller demand as an unusually strong vehicle re-

action. This is considered rather unpleasant by many drivers.

Since the warning character and control quality of the LKA compete with the steering activities of active drivers, especially during countersteering, the haptics module shown here as an example also takes into account the position of the vehicle within the lane: If the vehicle is at the edge of the lane, the haptics system provides strong steering support during countersteering. If, however, the driver drives further in the middle, the module reduces the steering assistance and thus decreases the controller demand more when the driver is active.

◘ Figure 33.5 illustrates this relationship graphically using two possible characteristic curves as examples: With LKA type I, the warning character at the edge of the lane is in the foreground, the steering assistance thus remains strong—in the central area, there is no assistance with LKA type I. With LKA type II, it is also important to implement a clear warning characteristic at the edge of the lane, and steering support is high here. In the central area, however, areas with lower steering assistance can be set through system application or driver adjustment in order to improve the comfort of the system.

The controllability of actively intervening assistance systems must be considered from two points of view. On the one hand, a risk analysis is required within functional safety according to ISO 26262, which is based on an assessment of the controllability in the event of malfunctions in various driving situations in order to conclude a risk together with an exposure and a severity of the effect (cf. ► Chap. 6). An evaluation of the controllability of the steering torque applied by the system is explained, for example, in [11]. Here it is pointed out that, in addition to a maximum amplitude of the steering torque, its gradient is particularly relevant for the controllability by the driver.

On the other hand, there are requirements from directives and regulations (see ► Sect. 33.4). LKA systems in particular are subject to regulations and standards that have an impact on the implementation of per-

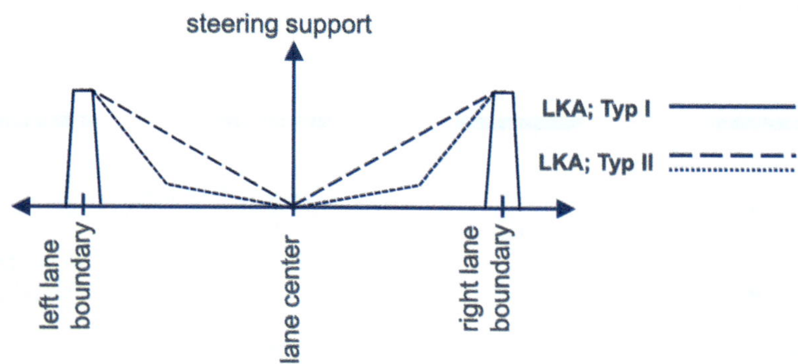

◘ **Fig. 33.5** Steering assistance for LKA type I and type II as a function of the lateral lane shift

missible limit values that ensure controllability. The implementation of these requirements takes place in the LDW/LKA function module in the interaction of the haptics module and state machine.

33.5.3.5 Hands-Free Driving Detection

Current SAE Level 1 and 2 LKA systems do not serve an automated driving, drivers must therefore monitor the system function at all times. To comply with legal regulations and standards (see ▶ Sect. 33.4), detection of this driver activity (currently generally realised by detection of hands-free driving) is required.

In the electromechanical power steering systems fitted as standard in almost all of today's passenger vehicles (see ▶ Chap. 24), the necessary data for evaluating driver activity is already available from the steering's integrated sensors. By analysing the steering activity, the driver's steering wheel movement can be distinguished from steering influences caused, for example, by road unevenness. For example, the different frequency of the various forms of stimulation can be used for this purpose. As the differentiation becomes increasingly difficult with very low steering activity, e.g. when driving straight ahead for a long time, a hands-free driving detection may occur in these cases although the driver is driving "hands on". Various manufacturers therefore use additional methods for driver activity detection, such as driver observation cameras or capacitive or pressure-sensitive sensors in the steering wheel.

If the system cannot detect sufficient steering activity or operating actions on steering wheel buttons (depending on manufacturer and vehicle), this indicates that the drivers may no longer have their hands on the steering wheel, whereupon they are prompted in an appropriate manner (auditory, visual) to resume lateral control of the vehicle. If a driver does not comply with this request, the system is switched off after an appropriate waiting time or—depending on the manufacturer—the vehicle in the lane is braked to a standstill and the hazard warning lights and an emergency call are activated.

33.5.4 Driver Information

According to ISO 17361:2017 "… an easily perceived haptic and/or audible warning shall be provided". … If the haptic and/or audible warning is not designed to indicate a direction, then a visual cue may be used to supplement the warning" [2].

General requirements for driver information for LDW and LKA systems include.

- clear, so that they can be easily perceived by inattentive drivers, for example,

- intuitive, so that the type of driver information favours the intended driver reaction,
- exclusive, so that drivers can react quickly without having to think twice,
- side-selective, so that drivers can infer where to steer,
- perceptible only by drivers, so that other vehicle occupants do not notice the driver information,
- inexpensive, requiring as few additional components as possible.

Haptic driver information can be provided, for example, by steering wheel vibration, seat vibration or a seat belt tensioner. The steering wheel vibration can be technically generated by vibration motors integrated in the steering wheel or alternatively as a functionality of the electromechanical steering support (EPS). Interventions in the vehicle lateral guidance can also be used as haptic driver information, e.g. with the help of the EPS steering actuator, which should then have a steering torque characteristic similar to ◘ Fig. 33.5. Alternatively, a clear, course-correcting brake intervention of the ESC system can also act as driver information.

Acoustic driver information can be provided by so-called "auditory icons", such as a specific information tone or the sound of driving on a rumble strip. These can be emitted, for example, via the stereo loudspeakers of the audio system, which must then, however, fulfil additional safety requirements. In the case of driver information, music and voice output must then be suppressed. Alternatively, the drivers can be informed via buzzer or gong of the instrument cluster. Obviously, such an output cannot fulfil the above-mentioned requirement that only the driver perceives the information.

Visual information should be in the primary field of vision of the driver, e.g. as an image or symbol in the instrument cluster or head-up display (HUD). LDW and LKA symbols are standardised in ISO 7000:2682 and 7000:3128 respectively.

An evaluation of these different possibilities of driver information for LDW and LKA systems according to the criteria mentioned above is shown in ◘ Table 33.2.

It becomes obvious: A clear perception of the driver information is possible with almost all variants. Only the seat vibration may not be detected when wearing thick clothing in winter. With exclusively visual driver information, it cannot be ruled out that a distracted driver will overlook the symbols in the instrument cluster and HUD, which is why, according to ISO 17361, it may only supplement but not replace haptic or auditory driver information.

Intuitive driver information that favours the intended driver response can be provided by lateral guidance intervention and steering wheel vibration. The sound of driving on a rumble strip and a sym-

◘ Table 33.2 Evaluation of driver information for LDW and LKA systems

Driver information			Clearly	Intuitive	Exclusive	Side-selective	Driver only
Perception	Action	Actuator					
Haptic	Lateral intervention	EPS-steering	+	+	+	+	0
		ESC	+	+	+	+	−
	Steering-vibration	EPS-steering	+	+	+	−	+
		Vibrator	+	+	+	−	+
	Belt jerk	Belt tensioner	+	0	+[1] −[2]	−	+
	Seat-vibration	Vibrator	0	0	+	+	+
Auditory	"Rumble strip sound"	Loud speaker	+	+	+	+[3] −[4]	−
	Spec. info sound		+	0	+	+[3] −[4]	−
	Gong, summer	Instrument cluster	+	0	−	−	−
Visual	Image, icon	Instrument cluster	−	+	+	+	+
		Head-up display	−	+	+	+	+

Belt tensioner exclusiv für LDW: [1]yes, [2]no; stereospeaker present: [3]yes, [4]no

bolic or pictorial representation in the HUD or instrument cluster are also suitable for this purpose in principle. This is not easily possible with seat belt tensioners, vibration information in the seat, information tone and instrument cluster buzzer or chime. These types of driver information do not necessarily indicate the need for steering intervention on the part of the driver.

All haptic driver information is exclusive to the sensory channel used. The drivers can therefore clearly and without much thought assign it to an LDW or LKA system, which favours a quick driver reaction. For these reasons, steering wheel vibration and steering back in lanes are now used by most vehicle manufacturers. In the case of the seatbelt tensioner, fast, confusion-free assignment is only possible if it is not used by other applications (then "+"). Specific information tones and the sound of driving on a rumble strip also allow clear assignment, as do images and symbols in the instrument cluster and HUD, if they are easily perceived and understood.

Side-selective driver information can only be provided by lateral guidance intervention, seat vibration and via a suitable symbol or image in the instrument cluster or HUD. Stereo speakers are also suitable for this (+), but mono speakers are not (−).

Belt jerk, steering wheel and seat vibration as well as visual driver information can only be perceived by the driver. The lateral guidance intervention by means of EPS steering can be designed to be moderate and almost imperceptible for passengers, because the recommendation for action for the driver comes from the auxiliary steering torque and not from the vehicle movement. A lateral steering intervention by means of ESC, on the other hand, should be strongly pronounced so that the driver can clearly perceive the vehicle movement and derive a recommendation for action from it. The ESC intervention, like acoustic driver information, is therefore clearly perceived by all vehicle occupants, which can have an effect on system acceptance. Frequently unnecessary misinformation is therefore particularly disturbing for these types of driver information and may contribute to a reduction in acceptance.

Usually, driver information is cost-effective if the required components are already installed in the vehicle as standard.

For the acceptance of driver information by LKA systems, influences from culture and society are relevant in addition to subjective evaluation criteria of the vehicle drivers. While Asian vehicle manufacturers tend to emphasise auditory driver information, European vehicle manufacturers focus on haptic and visual information channels.

33.5.5 **Actuators**

In passenger cars, electromechanical steering systems (EPS) are usually used as lateral control actuators for LKA systems, as described in detail in ▶ Chap. 24. Their steering torque influence can be directly experienced by the driver at the steering wheel as haptic feedback, whereby informative steering wheel vibrations can be realised as well as additional steering torques as a recommendation for action for the driver. This is just as impossible with hydraulic power steering systems as with superimposed steering without additional actuators.

The targeted braking of individual wheels can also suitably influence the lateral guidance of the vehicle. This effect is used by the ESC system to stabilise the vehicle in the driving dynamic limit range (see ▶ Chap. 25). However, in consideration of fuel consumption and brake wear, course-correcting brake interventions should not be continuous but only temporary. Thus, they are mainly suitable for type I LKA systems with lane feedback and less for type II systems with lane centre control.

33.5.6 **Status Indication and Operating Elements**

The system status indication should inform the driver about the current status of the LDW or LKA system in a way that is easily perceivable, but unobtrusive and easy to understand. This information is usually provided visually. In the simplest case, the readiness of the system is indicated to the driver by an illuminated LED in the system's on/off button. A more elaborate solution is shown in ◘ Fig. 33.7c: The display of the instrument cluster shows an image with the detected lines, the driver's own vehicle, and the position of the driver's own vehicle relative to these lines. In series systems, additional variants of these combinations are used.

The transition from "ready for use" to "not ready for use" is signalised to the driver, for example, by the LED in the on/off button going out or by a colour change of the lines in the image in the display of the instrument cluster. There is usually no audible information about this status change.

All vehicle manufacturers offer the possibility of switching LDW or LKA systems on and off via push-button or menu settings. Optionally, drivers are offered a possibility to configure the system so that they can adjust thresholds for driver information, switch certain driver information on and off, and choose between a type I and type II LKA system (menu item in the instrument cluster).

33.6 **Exemplary Implementations**

LDW systems were first used for commercial vehicles in Europe in 2000 and shortly afterwards also in the USA. For passenger cars, they were available in Japan from 2001, in North America from 2004 and in Europe from 2005. LKA systems were first offered in Japan in 2002 and in Europe in 2006.

The first LDW system in Europe was offered by Citroën under the name "AFIL" ("Alerte de Franchissement Involontaire de Ligne"). The lane markings were detected by means of infrared diodes installed on the underbody and a warning was issued to the driver via a side-coded vibration in the left or right seat cheek [12] (◘ Fig. 33.6).

Toyota launched the first Type II LKA system in Europe in 2006 in the Lexus LS 460. LDW driver information was provided via an audio/visual warning and steering back into the lane. A stereo camera was used to detect lane markings [13].

Mercedes' steering assistant (LKA Type II) was first introduced in the S-Class in 2013. Using a stereo camera (windshield top) and radar sensors at the front, side, and rear, the vehicle was able to orient itself not only to the lane markings but also to surrounding vehicles as a world first. The steering assistant was available on all road types. In the current system generation, Mercedes has coupled the steering assistant with a variety of other functions (avoidance of side collisions with stat. and dyn. objects, support for the formation of an emergency lane, braking in the lane in the absence of driver activity, etc.) [6]. In the current S-Class (W-223), Mercedes' LDW uses a steering wheel vibration for warning and steers back into the lane via the EPS. All other models use a wheel-individual braking to steer back into the lane.

Tesla has offered its LKA Type II function, coupled with ACC, under the name "Autopilot" since 2015. Characteristic of Tesla is that all vehicles are equipped with environmental sensors as standard and the function software can be installed remotely [14]. From 2014–2016, the vehicles were equipped with the hardware package I (mono camera in the upper area of the windshield, mid-range radar in the front of the vehicle, 12 ultrasonic sensors with a range of about 5 m) and from 2016 to the present with the hardware package II (8 mono cameras installed all around, mid-range radar in the front of the vehicle and 12 ultrasonic sensors with a range of about 8 m). By installing the sensors in each vehicle, Tesla can use its entire fleet of vehicles to collect relevant driving situations as input for its machine learning algorithms [15]. The use of the name "Autopilot" for a Level 2 function has been strongly criticised from a consumer protection perspective and has been banned by a court of first instance in Germany, for example, in July 2020 [16].

In 2017, GM introduced the LKA Type II function "Supercruise", also coupled with ACC, in the Cadillac CT6 on the American market, and from 2020 it was also available in Canada. A world first, it allows permanent "hands-free driving" on GM-approved, structurally segregated roads [7]. "Supercruise" is designed as a Level 2 feature that requires continuous monitoring by drivers. Sensors include a long-range radar and stereo camera (forward-facing), two front/side short-range radars, two rear/side short-range radars, and two rear-facing short-range radars [17]. A driver-observation camera detects whether drivers are looking ahead and monitoring traffic. If this is not the case, drivers are requested to take over the driving task. If they do not comply, the vehicle brakes slowly in the lane, activates the hazard warning lights and initiate an emergency call [18].

Since their introduction, the systems have undergone significant technical development and have been introduced also in lower vehicle classes. The use of cameras and haptic driver information via steering wheel vibration and steering back into the lane via the EPS has become established as a quasi-standard. In Sept. 2020, the German automobil association ADAC evaluated Euro NCAP tests of 71 vehicles (manufactured in 2018/2019) offered in Germany and assessed the LDW and LKA systems with the following criteria [19]:

- Active after every restart,
- Roadside without marking,
- Protects against collision with oncoming traffic,
- lane keeping dashed line,
- lane keeping solid line,
- Blind spot warning,
- Active Blind Spot Assist.

In addition to technical improvements (detection of the edge of the lane, detection of the line type, detection of oncoming traffic, etc.), ADAC thus also evaluates the coupling with other systems (blind spot warning, etc.) that require additional side and rear sensors.

By way of example, the systems from BMW and Volkswagen are discussed in more detail below with regard to operation, scope of functions and functional limits.

33.6.1 BMW "Driving Assistant Professional"

BMW generally offers its LDW and LKA functions in the two function packages "Driving Assistant" and "Driving Assistant Professional", which differ in the sensor equipment and thus the range of functions.

BMW's LDW is designed as a Type I LKA in both packages and informs the driver before unintentionally leaving the lane by means of a steering wheel vibration and active steering back into the lane. The system activates automatically when the vehicle is started and is ready for use from a speed of 70 km/h. Setting the direction indicator before or while crossing a lane marking temporarily deactivates the system, as the vehicle assumes an intentional lane change. Steering back into the lane can be stopped at any time by a deliberate oversteer by the driver. Deactivation takes place automatically below approx. 70 km/h or by the driver via the configuration menu in the control system (◼ Fig. 33.7a). The system status is displayed in the instrument cluster via symbols with status-dependent colour. The haptic driver information can be extensively configured in the menu; warning time, vibration and/or steering intervention can be changed or selected. The detection of the lane markings or the edge of the road is carried out via cameras (mono camera or trifocal camera depending on the model and installed options).

BMW offers the Level 2 function "Steering and Lane Guidance Assist" in the function package "Driving Assistant Professional" [20]. When this function is activated, the vehicle steers at speeds of up to 210 km/h. In doing so, it orients itself to the lane markings, the vehicles driving to the side and in front, as well as the lane edge with speed-dependent weighting. The function is available on all road types. A capacitive "hands-on" detection prevents abusive continuous hands-free driving or, in the event of a permanently absent signal, serves to initiate a braking manoeuvre until the vehicle comes to a standstill and to switch on the hazard warning lights. When the function is activated, ACC is switched on at the same time ◼ Fig. 33.7b The common function symbol is shown in ◼ Fig. 33.7c. BMW combines further comfort and safety functions with the "Steering and Lane Guidance Assist", such as assisted lane change at the driver's request, or a side collision warning, if the vehicle approaches the edge of the road too closely, e.g. in a construction zone, or if a collision with a vehicle driving on the side is imminent. In traffic jams, the system supports the formation of an emergency lane. All functions can be individually configured or activated/deactivated via the menu.

33.6.2 Volkswagen "Lane Assist"

The lane departure warning system "Lane Assist" from Volkswagen, as installed in the Golf 8, uses a camera to detect lane markings as well as other lane boundaries, such as clear transitions from the road to the grass verge. Taking into account driving dynamics data, the system calculates the danger of leaving the lane. If this becomes imminent, it gently countersteers by means of a corrective intervention in order to prevent the vehicle leaving within the given system limits (◼ Fig. 33.8a). The driver is warned during this intervention visually (◼ Fig. 33.8b) and haptically by means of a steering

Fig. 33.6 AFIL from Citroën as of 2005. field of view of the infrared diodes (*Source* Citroën)

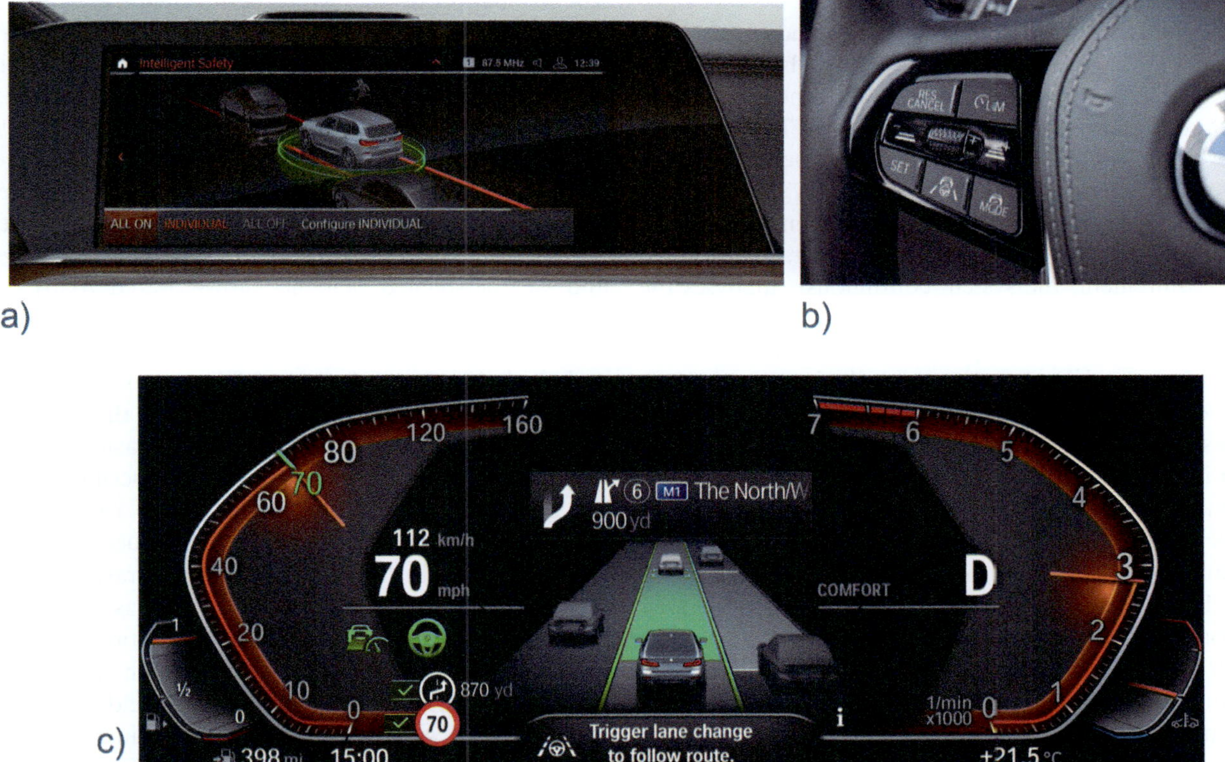

Fig. 33.7 BMW configuration menu (**a**) Steering wheel button LKA (**b**) BMW LKA display (**c**) [*Source* BMW AG]

◘ Fig. 33.8 "Lane Assist" from Volkswagen **a** Steering intervention **b** Multifunction display—illustration shows a Golf 7 [*Source* Volkswagen]

wheel vibration that can be switched on and off, if a lane limit is crossed despite the steering intervention of the system.

"Lane Assist" is designed for use on motorways and well-maintained country and federal roads. The system is available at speeds above 60 km/h, and deactivates when the speed drops below 55 km/h. The assistant also works in the dark and in bad weather conditions. Drivers can override "Lane Assist" at any time with little use of force. This does not relieve them of their responsibility to consciously drive the car. To monitor this requirement, the system continuously analyses the steering activity of the driver to register whether he is steering or relies on the system. If the latter is detected, the driver receives an acoustic and visual request to take over. If he does not react, the system switches off. The system also switches to passive mode, if the direction indicator is activated, if the driver brakes or steers heavily during system intervention, if no lane limits are detected or if the ESC is deactivated. „Lane Assist" is offered by Volkswagen in almost all vehicle models.

„Lane Assist" and ACC merge in Emergency Assist to form a new assistance system [21]. If the system detects that the driver is no longer reacting, he or she is prompted to take over control of the vehicle by acoustic and visual signals as well as an uncomfortable brake jerk. If the user still does not react to this, then the system subsequently initiates an emergency stop. The hazard warning lights are switched on automatically. In addition, the vehicle performs slight steering manoeuvres to make other road users aware of the dangerous situation. In this situation, "Lane Assist" keeps the vehicle in the centre of the lane as well as possible, and ACC prevents the vehicle from colliding with the traffic ahead. Finally, the system brakes the car continuously until it comes to a standstill.

33.7 Safety Performance

The evaluation of the safety performance of a driver assistance system generally consists of two factors. Firstly, the field of action, which quantifies the potentially addressed crashes, and secondly, the effectiveness of the system within this field of action.

The identification of the field of action first requires a description of the system's mode of action as well as the system-related (e.g. restriction due to the speed range), infrastructural (e.g. presence of lane markings) and weather-related (e.g. no snow-covered road) operating conditions. Based on the accident database of the German Insurance Association from 2002 to 2006 and the basic mode of action of lane departure warning systems, Hummel et al. [22] quantify their field of action as 17,848 (13.0%) from 136,954 crashes. Taking into account a generic functional description and the associated restrictions as well as a case-by-case consideration, the field of action is reduced to 4.4% (6005 crashes). In terms of fatal traffic crashes, Hummel et al. calculate the field of action as 10.1%. In other countries, however, the crash situation is different. For Sweden—a country with a high number of rural roads—Sternlund quantifies the field of action of LDW / LKA systems as 23.9–27.5% in relation to 138 fatal crashes in 2010 [23]. For the USA, Cicchino calculates the field of action as 5% in relation to all crashes and 20% in relation to fatal crashes on the basis of crashes recorded by the police [24]. It should be noted that in the USA in 2019, 24.4% of the 36,096 road users killed died in rollover crashes [25]. This type of crash can be easily associated with unintentional road departure. However, the dynamics in the accident sequence determine whether this crash can be avoided by a system.

Proving the effectiveness of the systems in their field of action is more difficult because retrospective accident analyses require sufficient market penetration of the systems and ADAS are often offered in combinations. Therefore, the number of studies related to the effectiveness of LKA systems is currently still rather small. For LDW there are several studies, e.g. by Hickmann et al. (48% lower crash risk for equipped trucks) [26], Sternlund et al. (53% lower risk for LKA/LDW vehicles in the field of action) [27] as well as Cicchino (11% lower crash rate for vehicles with LDW; 21% for crashes with injuries) [24].

Overall, the results show comparatively large differences. This is partly due to the above-mentioned challenges in the analysis. However, system-related effects must also be taken into account here. For example, it should be noted for warning systems (LDW) that they only work in interaction with the driver. In contrast, for intervening systems (LKA) it must be taken into account that a system design always represents a compromise between crash avoidance, the legal requirements and ensuring functional safety and safety in use. An example of this is ensuring that steering interventions can be overridden, which limits the strength of corrective interventions in critical driving situations.

33.8 System Assessment by Trade Press and Consumer Protection

The first systems of lateral guidance assistance were controversially discussed. The potential safety gain was contrasted above all with paternalism and too frequent unjustified warnings. As late as 2013, the German trade magazine Auto-Motor-Sport wrote in a review article on driver assistance systems: "Ambivalent: Lane Keeping Assist" [28]. However, the criticism has been steadily weakening since the systems were introduced. In 2019, for example, the same magazine argued in a comparative test of systems in upper mid-range vehicles with those from the small car segment: "Better simple than not at all" [29].

The now high acceptance of the systems is also reflected in the equipment rate. According to an evaluation by Bosch, a "lane assistance system" was installed in 32% of newly registered vehicles in Germany in 2016 [30]. This represents a doubling of the number of systems sold compared to 2015.

The main reason for this improved perception was the continuous revision and technical improvement of the systems on the part of manufacturers and suppliers, which also took into account suggestions from the trade press and consumer protection organisations. In 2012, for example, the ADAC derived the requirements for an ideal system with the help of a study of test per- sons [31]. Many of the points found to be positive in this study, such as the adjustable steering intervention, have since been implemented. Since 2014, Euro NCAP has also been evaluating LKA and ELK lane assist systems, thus confirming the high safety significance of these systems [32].

33.9 Outlook

Due to the legal requirement to install an ELKS system from 2022 to 2024, it is foreseeable that in the medium term "independent" LDW functions will no longer be offered as optional equipment in Europe. Today's steadily increasing integration of functions even in the lower vehicle classes is likely to continue, depending on which sensor equipment will be fitted as standard in the lower vehicle classes in the future.

From a technological point of view, the recognition of lanes and lane edges has today reached a level that already approaches the recognition ability of humans. The technical improvement of the cameras is therefore not the sole goal of further development. A significant increase in the performance of LKA systems is more likely to be achieved through the extended inclusion of "context information" (surrounding vehicles, static objects in or on the driving lane, traffic information, navigation data, etc.). By combining all available information into a consistent environment model, it is possible to warn or intervene earlier and in a more targeted manner, and the system behaviour can be more closely matched to human driving behaviour.

Comfort and safety systems will continue to be combined and, from the driver's point of view, merge with each other. The development of highly and fully automated vehicles is leading the way in this respect. From a technological point of view, it is therefore to be expected that many approaches from the L3/L4-development will benefit the Level 2 functions. Examples include earlier detection of potentially demanding situations in Level 2 operation (road works detection, traffic light detection/warning, etc.) and thus an even more comprehensible takeover request at system boundaries.

References

1. DeStatis (2019)
2. ISO 17361:2017: Intelligent transport systems—lane departure warning systems—performance requirements and test procedures (2017)
3. ISO 11270:2014: Intelligent transport systems—lane keeping assistance systems (LKAS)—performance requirements and test procedures (2014)
4. ▶ http://www.unece.org/fileadmin/DAM/trans/main/wp29/wp-29regs/2018/R079r4e.pdf
5. Multifunktionskamera (bosch-mobility-solutions.com)

6. Die Technik der neuen S-Klasse—Mercedes me media (mercedes-benz.com)
7. Super cruise—hands free driving|Cadillac ownership
8. Homm, F., Kaempchen, N., Burschka, D.: Fusion of laser scanner and video based lane marking detection for robust lateral vehicle control and lane change maneuvers. In: IEEE Intelligent Vehicles Symposium (IV), pp. 969–974 (2011)
9. van Winsum, W., Brookhuis, K.A., de Waard, D.: A comparison of different ways to approximate time-to-line crossing (TLC) during car driving. Accid. Anal. Prev. **32**, 47–56 (2000)
10. Mammar, S., Glaser, S., Netto, M.: Time to line crossing for lane departure avoidance: a theoretical study and an experimental setting. IEEE Trans. Intell. Transp. Syst. **7**(2), 226–241 (2006)
11. Schmidt, G.: Haptische Signale in der Lenkung: Controllability zusätzlicher Lenkmomente. In: Berichte aus dem DLR-Institut für Verkehrssystemtechnik – Band 7, Dissertation (2009)
12. Avoiding accidents—PSA Peugeot Citroën—sustainable development (archive.org)
13. Lane departure warning system—Wikipedia
14. Autopilot|Tesla
15. Autopilot AI|Tesla
16. LG München I, Endurteil v. 14.07.2020–33 O 14041/19—Bürgerservice (gesetze-bayern.de)
17. GM enhanced super cruise review: hands-free driving adds auto lane-change—SlashGear
18. 2021 Cadillac super cruise convenience and personalization guide
19. Test: Können Spurhalteassistenten Unfälle verhindern?|ADAC
20. BMW Medieninformation 06/2018: Der neue BMW X5
21. ▶ https://land-der-ideen.de/wettbewerbe/deutscher-mobilitaetspreis/preistraeger/best-practice-2017/emergency-assist
22. Hummel, T.; Kühn, M.; Bende, J.; Lang, A.: Ermittlung des Sicherheitspotenzials auf Basis des Schadengeschehensder Deutschen Versicherer. Gesamtverband der Deutschen Versicherungswirtschaft e. V. Forschungsbericht FS 03. Erschienen: 09/2011
23. Sternlund, S.: The safety potential of lane departure warning systems—a descriptive real-world study of fatal lane departure passenger car crashes in Sweden. Traffic Inj. Prev. **18**(sup1), S18–S23 (2017). ▶ https://doi.org/10.1080/15389588.2017.1313413
24. Cicchino, J.B.: Effects of lane departure warning on police-reported crash rates. J. Saf. Res. **66**, 61–70 (2018)
25. NTHSA: Traffic safety facts—early estimates of motor vehicle traffic fatalities and fatality rate by sub-categories through June 2020. National Highway Traffic Safety Administration, DOT HS 813 054 (2020)
26. Hickman, J.S., Guo, F., Camden, M.C., Hanowski, R.J., Medina, A., Mabry, J.E.: Efficacy of roll stability control and lane departure warning systems using carrier-collected data. J. Saf. Res. **52**, 59–63 (2015)
27. Sternlund, S., Strandroth, J., Rizzi, M., Lie, A., Tingvall, C.: The effectiveness of lane departure warning systems—a reduction in real-world passenger car injury crashes. Traffic Inj. Prev. **18**(2), 225–229 (2017). ▶ https://doi.org/10.1080/15389588.2016.1230672
28. Assistenzsysteme im Praxistest: so arbeiten die Helfer im Alltag|auto motor und sport (auto-motor-und-sport.de)
29. Assistenzsysteme im Test: Teuer gegen günstig|auto motor und sport (auto-motor-und-sport.de)
30. Bosch-Auswertung: Fahrerassistenzsysteme sind weiter stark auf dem Vormarsch—Bosch Media Service (bosch-presse.de)
31. ADAC Probandenstudie: Warnsignale von Assistenzsystemen (2012)
32. Euro NCAP|Spurassistent

Open Access This chapter is licensed under the terms of the Creative Commons Attribution-NonCommercial-NoDerivatives 4.0 International License (▶ http://creativecommons.org/licenses/by-nc-nd/4.0/), which permits any noncommercial use, sharing, distribution and reproduction in any medium or format, as long as you give appropriate credit to the original author(s) and the source, provide a link to the Creative Commons license and indicate if you modified the licensed material. You do not have permission under this license to share adapted material derived from this chapter or parts of it.

The images or other third party material in this chapter are included in the chapter's Creative Commons license, unless indicated otherwise in a credit line to the material. If material is not included in the chapter's Creative Commons license and your intended use is not permitted by statutory regulation or exceeds the permitted use, you will need to obtain permission directly from the copyright holder.

Integrated Longitudinal and Lateral Control

Christian Rathgeber, Dirk Odenthal, and Norbert Nitzsche

Contents

© The Author(s) 2026
H. Winner et al. (eds.), *Handbook Assisted and Automated Driving*,
https://doi.org/10.1007/978-3-658-45276-6_34

34.1 Introduction

On the way towards autonomous driving, driver assistance functions have been developed in recent years that support the driver in distinct driving situations with a limited range of functionality. Based on these distinctions, the functions can be imagined as being detached from one another in a criteria space that is spanned by the following axes, among others:

— operating range for longitudinal speed
— operating range for curvature
— operating range for road adhesion utilization
— operating range for degree of cooperation
— primary motivation (comfort or safety)
— direction of intervention (longitudinal, lateral, combined)

■ Figure 34.1 shows a three-dimensional cross-section through this criteria space and the location of some exemplary driver assistance functions.

In the past, simultaneous lateral and longitudinal guidance was often realized by simultaneously activating a longitudinal and a lateral guidance function. The fact that the individual driver assistance functions were so clearly differentiated from one another is related to the way in which they have been developed sequentially and in separate organizational units. Individual solutions were created in terms of control technology implementation, safety concepts and regarding the functional and network architecture, including the interfaces between OEMs and suppliers. Initially, these function-specific solutions were advantageous for industrializing the functions at a manageable cost in terms of function development, testing and validation, and calibration. By now, however, feature-specific solutions represent, on the one hand, a problem of mastering complexity and, on the other hand, prevent the development of functions that contiguously cover even larger areas in the criteria space outlined above. Such contiguous coverage is, however, a prerequisite for highly and fully automated driving.

Degree of Cooperation

When merging the operating ranges, the aspect of driver interaction is of particular importance. Therefore, regardless of the automation levels as defined by SAE [16] and BASt [7], we introduce the concept of *degree of cooperation*. The degree of cooperation describes to which extent the driver and the driver assistance system guide the vehicle together in a cooperative manner. From a control and system dynamics perspective, this is more relevant than the automation level. ■ Figure 34.2 shows the qualitatively well distinguishable degrees of cooperation using lateral guidance as an example.

— *Manual Driving*: The driver assistance system is switched off, the driver guides the vehicle with regard to

a specific driving task as shown in the upper left of ■ Fig. 34.2.
— *Autonomous driving*: The driver assistance system guides the vehicle with regard to a specific driving task. The driver is mechanically decoupled (*hands off, feet off*) as shown on the top right in ■ Fig. 34.2.
— *Cooperative driving*: Driver assistance system and driver cooperatively guide the vehicle. Both the driver and the assistant should be able to assume the leading role in cooperative vehicle guidance. The other partner estimates/observes the intention of the leader and adapts to it. We can distinguish two cases here:
  – First, the driver is guided specifically and can be interpreted in a very reduced way as a mechanical impedance passively connected to the steering wheel or the (possibly force-reflecting) accelerator or brake pedal, as shown in the lower left of ■ Fig. 34.2.
  – Or second, the driver takes the lead with a assistance system that is appropriately compliant in this case, as shown on the bottom right in ■ Fig. 34.2.

In robotics, such cooperation between humans and robots is referred to as *Human Robot Collaboration HRC* [4], which includes all perceptual modalities. For the design of cooperative aspects of assistance functions, the haptic modality is of particular importance from a control engineering perspective, since the mechanical coupling of the driver and the interface elements must be considered here.

For many driver assistance systems, the degree of cooperation is not a constant, nor does it necessarily correlate with the automation level of the system. On the one hand, even an SAE level 2 function with the driver remaining permanently responsible for monitoring must be capable, for a short time span, of guiding the vehicle with an *autonomous* degree of cooperation, i.e., without driver intervention. On the other hand, functions with SAE level 3 and higher must also work safely while being influenced by driver interaction. Furthermore, switching degrees of cooperation necessarily occur during transitions between individual functions as well as during activation and deactivation. In addition, for assistance functions with simultaneous longitudinal and lateral guidance, the current cooperation levels of longitudinal and lateral guidance may differ. For example, in the case of Adaptive Cruise Control (ACC) with simultaneous lane keeping assistance, the longitudinal guidance is autonomous for longer periods of time, while the lateral guidance can be *hands on* and thus cooperative.

Objective of the Contribution—Requirements for an Integrated Approach

The purpose of this contribution is to demonstrate a way to realize driver assistance systems from different

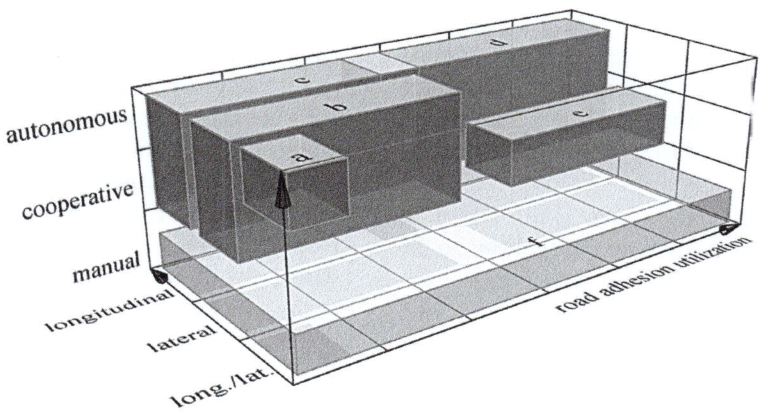

Fig. 34.1   Some driver assistance functions and their operating ranges (qualitative) in terms of road adhesion utilization, direction of intervention, and degrees of cooperation; **a** autonomous parking, **b** lane keeping assistance, **c** ACC, **d** emergency braking assistance, **e** evasion assistance, **f** manual driving without assistance

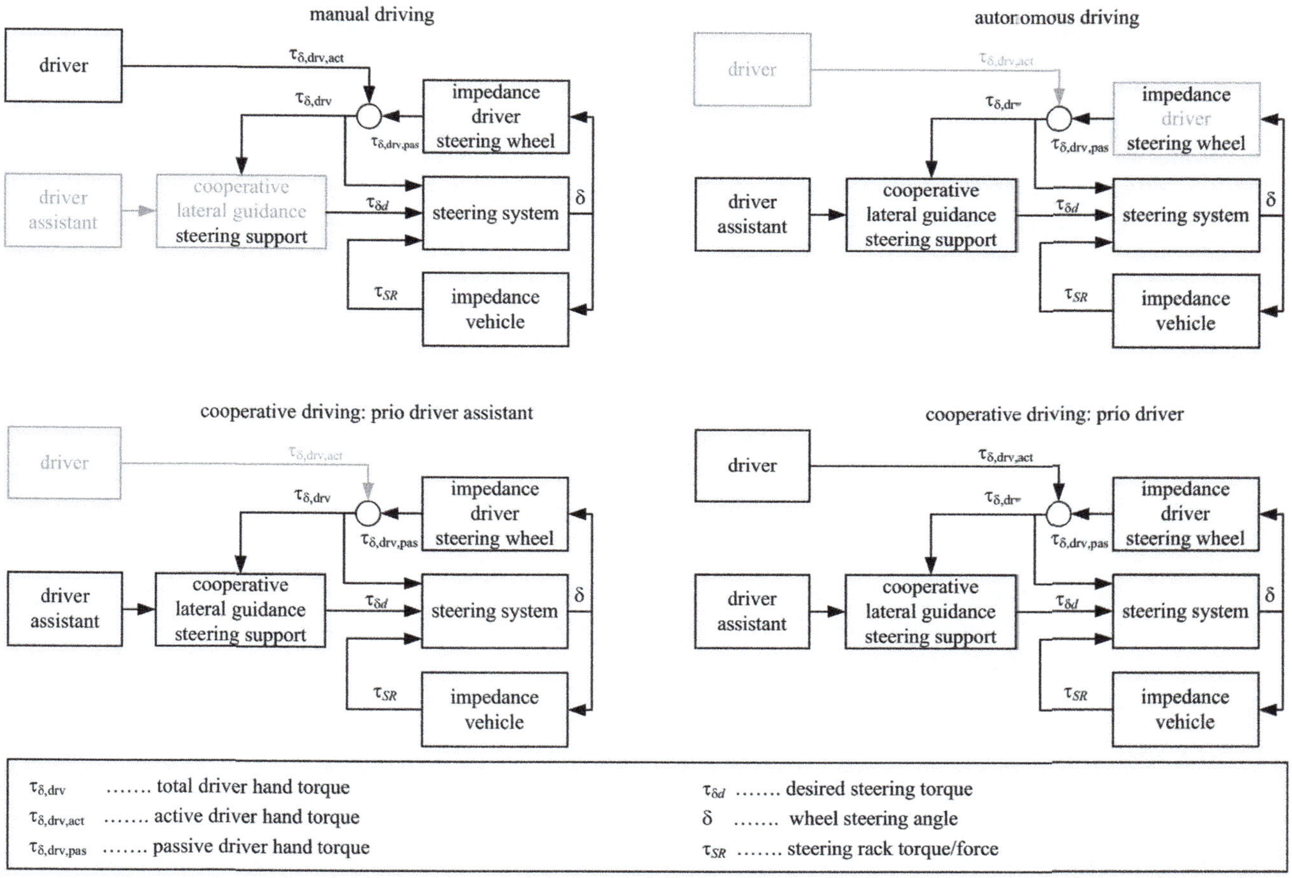

Fig. 34.2   Degrees of cooperation illustrated on the basis of lateral guidance with steering actions

regions of the criteria space, see ❑ Fig. 34.1, with a unified functional architecture and, embedded therein, identical controllers for all assistance systems. The proposed approach directly results in a reduction of development, validation, and calibration effort. Furthermore, it will, in the future, enable new features that cover increasingly larger portions and finally the entire criteria space.

In the sequel, we will derive exemplarily the requirements for such a control engineering approach based on existing and future driver assistance functions. Among these requirements, the generic control requirements *stability, good guidance behavior, good disturbance rejec-*

tion and *robustness* are of foremost importance. Meeting them is the subject of the following sections. In addition to these generic requirements, the context of integrated longitudinal and lateral guidance implies a number of specific requirements that need to be considered in more detail.

- *Balance between adjustability and simplicity*: The approach should be able to cover functions such as autonomous parking, evasion assistance, lane keeping assistance, ACC, emergency braking and emergency stop assistance, as well as future functions and all transitions between different operating ranges. Autonomous parking, for example, requires the most accurate guidance and disturbance attenuation and must also be able to compensate for driver interventions up to the switch-off threshold. A lane keeping system, on the other hand, must be able to be overruled even with low driver steering torques. Similar prioritization differences in control objectives can be found between longitudinal functions such as ACC and emergency brake assistance. To satisfy such partially conflicting requirements with a unique controller structure calls for a high degree of adjustability, which, however, must not be at the expense of simplicity. In particular, it should be avoided to switch between different controllers depending on the degree of cooperation or the driving situation.
- *Scalability*: The approach should work as much as possible with generalized control commands of the longitudinal and lateral guidance (target acceleration and target curvature). These generalized control commands are then distributed via *control allocation* to the actuators actually present in the vehicle, and, thus, concretized to actuator specific control commands, e.g., motor and brake torques, steering angles or differential torques. On the one hand, this allows different vehicle variants to be operated with almost identical functional logic, and on the other hand, safety and availability can potentially be enhanced in the event of actuator failures.
- *Accounting for lateral/longitudinal coupling when approaching the handling limits*: Using a single unified structure also for realizing assistance systems close to or at the vehicle's handling limits requires that the coupling of longitudinal and lateral dynamics is taken into account. When approaching the adhesion limit of the tires (Kamm's circle), this coupling is no longer negligible. Then, in analogy to manual driving, the anticipatory planning of maneuvers gains decisive importance. Such planning must take into account the current driving dynamics potential (actuator limits, road adhesion coefficient, etc.) and its uncertainties (model uncertainties, estimation errors, etc.) and can permit trajectories close to or beyond the handling limits only after careful balancing of all risks, i.e., to avoid an otherwise unavoidable collision vs. increased

slip, for example. If the planning is thus conservative, the longitudinal/lateral coupling caused by limit situations must not be taken into account when implementing the maneuvers in the case of *cooperative* driving. However, the maneuvers must be implemented in such a way that the existing stabilization functions (ESC, ABS) can also do their work with the input from an automated longitudinal and lateral guidance with only minimal adjustments. *Autonomous* driving at the handling limits requires additional effort on the controllers which is not the subject of this chapter.

34.2　Three-Level Model of Cooperative and Autonomous Driving

Inspired by [11] and [5], we separate the driving task into the *navigation level*, the *trajectory guidance level*, and the *vehicle guidance level*, see ◘ Fig. 34.3.

The tasks of the *navigation level* include all discrete decisions from planning a route to the destination of a journey to selecting a lane, a parking space and, for example, deciding whether to decelerate in the lane or to change lanes.

At the *trajectory guidance level*, the driving task specified by the navigation level is translated into a continuous trajectory. The generation of a continuous trajectory is realized in two stages: trajectory planning and trajectory tracking control. The trajectory guidance level is *vehicle parameter free*: target and actual vehicle motion are described by mere kinematic quantities. Essentially, trajectory planning must find a trajectory that satisfies the driving task, is collision-free, and does not exceed the limits of feasibility at the vehicle guidance level. To ensure collision-free driving, knowledge of the immediate vehicle environment and the geometric dimensions of the vehicle is necessary.

In order to plan feasible trajectories by accounting for parameters specific to the vehicle and to a given driving situation, the vehicle guidance level provides a potential vector. The potential vector comprises all limits resulting from the limited road adhesion, from the actuator limits and any control command limits introduced to ensure functional safety and safety of the intended functionality (SOTIF) as well as limits introduced to tune the functional characteristics. Since the planned trajectory does not exceed the handling limits in the nominal case, the trajectory tracking controllers can be implemented separately for longitudinal and lateral motion. Their outputs are *vehicle-independent* quantities, i.e., the desired longitudinal acceleration and the desired curvature, which are transmitted as reference values to the vehicle guidance level. With such an interface, the trajectory guidance level remains abstract and the vehicle guidance level behaves like a *Smart Actuator* for acceleration and curvature. The concrete actuator technology used remains encapsulated.

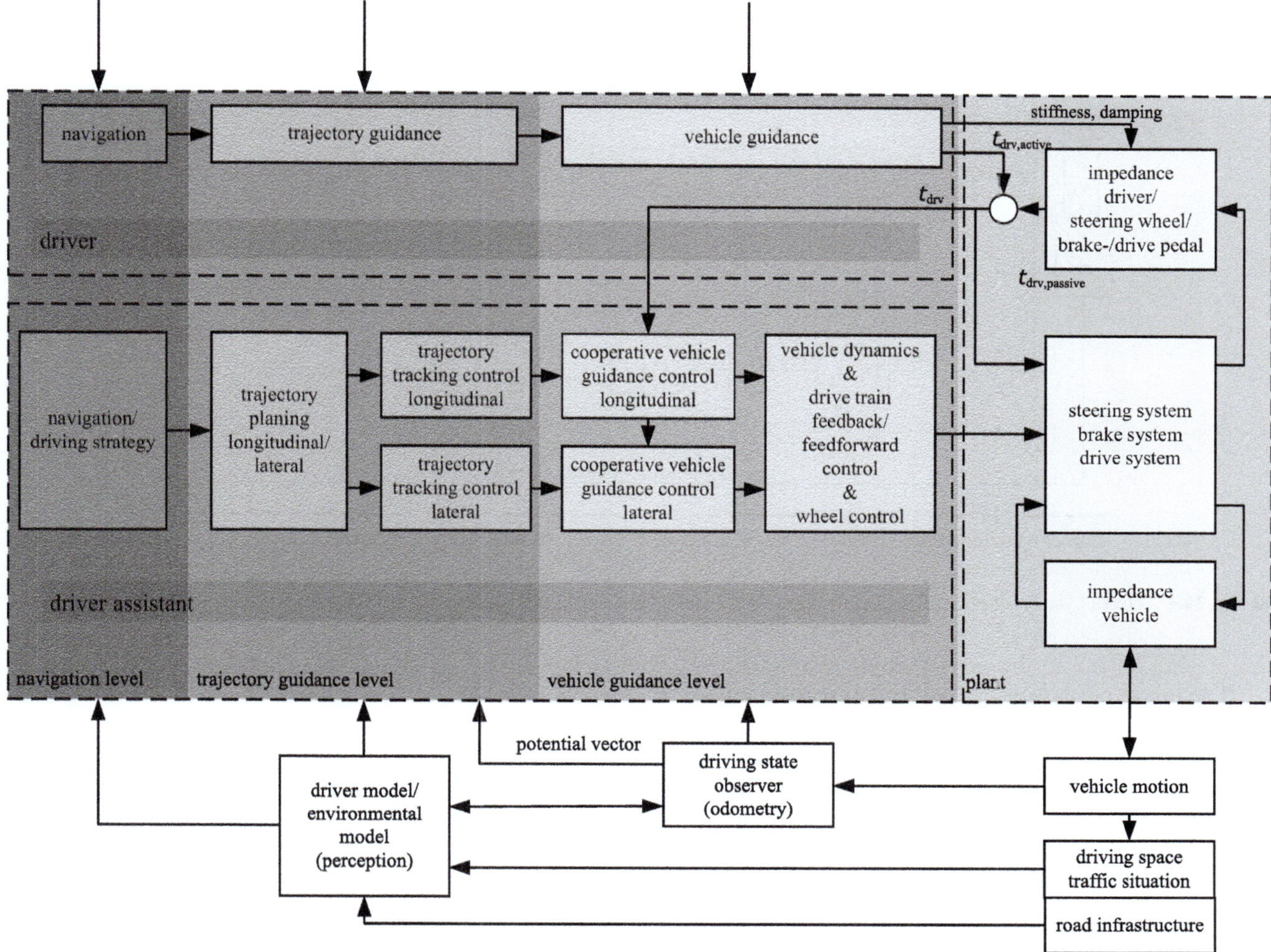

Fig. 34.3 Three-level model for integrated lateral and longitudinal control

For example, a desired curvature could be realized at the vehicle guidance level without adjusting the navigation and trajectory guidance levels by means of either or a combination of front-axle steering, rear-axle steering, unilateral braking, or torque vectoring.

The *vehicle guidance level* performs the task of controlling the desired longitudinal acceleration and curvature as specified by the trajectory guidance level by actuating the available actuators. Secondly it provides the available potential of the vehicle dynamics to the trajectory guidance level. When determining the vehicle dynamic limits, the dependency on the road adhesion coefficient in particular is still subject of research. At the present time, rather conservative approximations are applied for this purpose. The control of the desired motion in terms of longitudinal acceleration and curvature is subject to vehicle- and function-specific requirements. The vehicle dynamics properties are incorporated into the controller design by means of actuator and vehicle dynamics models. Thereby, the characteristics regarding the control transmission and disturbance attenuation behavior as well as the

collaboration with the driver have to be adjustable at real time. This ensures that these three properties can be adapted to the respective active driver assistance function or that transitions between the functions can be performed.

Figure 34.4 shows the interaction of the hardware and software components involved in more detail. The figure serves as a reference for where to find the subsequently used signals and transfer functions in the overall signal flow.

The trajectory planning (TP) generates the target trajectory $\left[x\ y\ \theta\ \kappa\ v_x\ a_x \right]^T_{\text{traj}}$ based on the demands of the navigation level (not shown). The subsequent trajectory tracking controllers, TC_{lat} for the lateral motion and TC_{long} for the longitudinal motion, calculate from the target trajectory on the one hand and the actual quantities position, track angle and velocity $\left[x\ y\ \theta\ v_x \right]$ on the other hand the compensating control commands target curvature κ_d and target longitudinal acceleration a_{xd}. All quantities refer to a reference point which is fixed to the vehicle geometry (e.g. center of rear axle, front axle or nominal vehicle center of gravity).

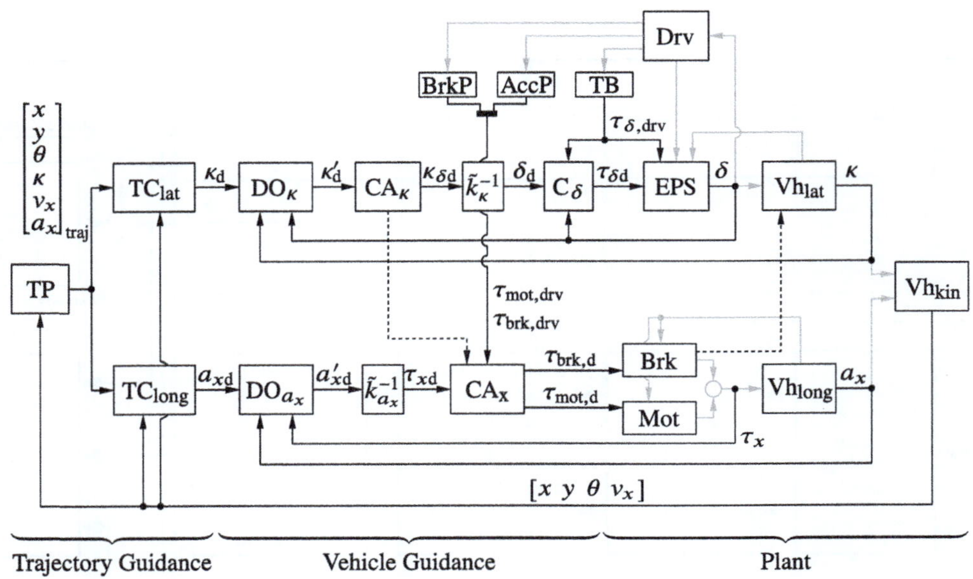

Fig. 34.4 Overall view of the cooperative integrated lateral and longitudinal control system

The target curvature κ_d and target longitudinal acceleration $a_{x\mathrm{d}}$ are augmented by the two disturbance observers, DO$_\kappa$ and DO$_{a_x}$, such that observed disturbances are compensated for steady state situations. The adapted target curvature κ_d' can now be distributed, by means of a control allocation CA$_\kappa$, to lateral actuators. In most cases, this will be the front wheel steering or a steering system with coordinated front and rear wheel steering angles. Unilateral braking and direct control of the rear-axle steering can also be considered here. The proportion κ_δ of the target curvature, which was assigned to the front wheel steering, is then mapped to a target steering angle δ_d. The adapted target longitudinal acceleration $a_{x\mathrm{d}}'$ is converted to a torque command $\tau_{x\mathrm{d}}$, which represents the sum of all positive and negative torques for all wheels, and is then allocated to the motor and brake system (CA$_\mathrm{x}$). This allocation also includes the desired torques communicated by the driver (Drv) via the brake pedal (BrkP) and accelerator pedal (AccP).

The braking and motor target torques τ_{brk} and τ_{mot} thus calculated are transmitted to the appropriate actuator control units, which will process them essentially without distinction between manual or autonomous driving (Brk and Mot; the blocks represent both the control unit and the actuator). In contrast, an additional controller, C$_\delta$, is needed to control the steering angle δ to match δ_d. This controller is distinctly more complex than the one used for manual driving.

Motor (Mot), brake (Brk) and steering system (EPS) generate the vehicle motion according to the longitudinal and lateral dynamics of the vehicle (Vh$_{\mathrm{lat}}$ and Vh$_{\mathrm{long}}$). Vehicle motion is expressed by the actual curvature κ and the actual longitudinal acceleration a_x. The physical feedback on the actuators caused by the vehicle motion is indicated in gray.

The outer loop of the trajectory tracking control is closed via the vehicle kinematics (Vh$_{\mathrm{kin}}$), which integrates κ and a_x to $[x\ y\ \theta\ v_x]$.

34.3 Models for Lateral and Longitudinal Vehicle Dynamics

This section describes the mathematical models of the vehicle's lateral and longitudinal dynamics as well as the mathematical models of the actuators, which later provide the basis for the controller design. The employed controller structure, which is robust with respect to model errors, allows the use of very simple linear models. The comparisons between model predictions and measurements, as shown examplarily in the following sections, are primarily intended to demonstrate that not prediction quality but simplicity and low application effort are the essential criteria. With the selected approach, it is in most cases sufficient to estimate only one single set of parameters for the mathematical models, which covers a whole vehicle series and to use this single set in production software.

34.3.1 Steering System

Two models capturing the behavior of the front axle steering system are prerequisite for the lateral control algorithms introduced later in this chapter. Steering model 1 describes how the front axle steering angle δ changes as a result of the commanded input $\tau_{\delta\mathrm{d}}$, see **Fig. 34.4**.

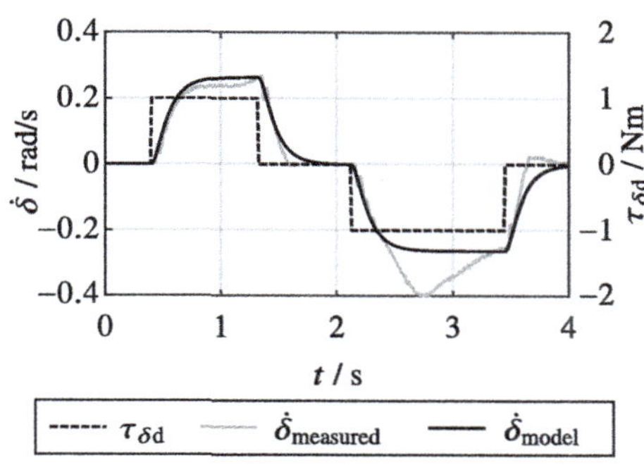

Fig. 34.5 Identification of steering behavior torque to steering angle speed, comparison measurement and model

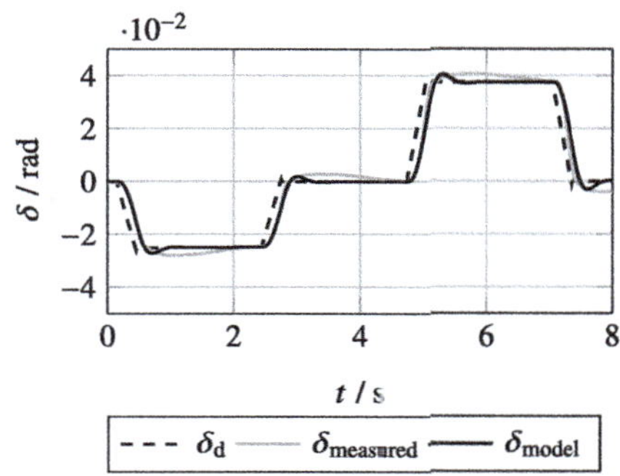

Fig. 34.6 Identification of steering behavior from desired steering angle to actual steering angle, comparison of measurement and model

Steering model 2 describes the behavior of the steering system when steering angle control is activated. The input of model 2 is the desired steering angle δ_d, the output is the actual front axle steering angle δ, see **Fig. 34.4**.

Steering Model 1: Steering Torque to Steering Angle Speed

The steering model 1 is greatly simplified, linear model of the front axle steering with the steering torque $\tau_{\delta d}$ as the input and the steering angular velocity at the front axle $\dot{\delta}$ as the output. $\tau_{\delta d}$ is the torque that is requested at the steering actuator in addition to the torques of the power steering assist functions. Written as a transfer function, the model has the form

$$G_{\tau\delta}(s) = \frac{1}{m_\delta s + b_\delta} \frac{1}{T_\delta s + 1} \tag{34.1}$$

with inertia m_δ, a damping coefficient b_δ, and a time constant T_δ. $G_{\tau\delta}$ is used as an internal model for the disturbance observer within the steering angle controller presented in ▶ Sect. 34.5.6. The identification of the parameters is done with power assist activated, whose effects on the steering behavior are thus captured, too. The vehicle is on a lifting platform during this process to eliminate the feedback effects of lateral tire forces or bore torques. **Figure 34.5** shows, as an example, the comparison of the measured and the model-predicted curves of the steering angle speed. The significant deviations cannot be avoided with the simple model (34.1). However, as explained in ▶ Sect. 34.5.1, the disturbance observer is robust against these deviations and enforces, in the lower frequency range, the linear behavior imposed by the model.

Steering Model 2: Desired Steering Angle to Actual Steering Angle

The steering model 2 is a simple linear model which subsumes all aspects of the behavior of the steering system including the steering angle control and thus represents the dynamic transfer behavior between desired steering angle δ_d and actual steering angle δ. This model is used as an internal model for the curvature disturbance observer presented in ▶ Sect. 34.5.4, which compensates for disturbances on the lateral dynamics. Furthermore, the steering model 2 is used in the dynamic feedforward controller of the trajectory tracking controller for the lateral motion to compensate the undesireable delay effects of the steering system. Written as a transfer function, the model has the form

$$G_\delta^*(s) = \underbrace{\left(\frac{\omega_\delta^2}{s^2 + 2D_\delta\omega_\delta s + \omega_\delta^2} \right) e^{-T_{D,\delta}s}}_{G_\delta} \tag{34.2}$$

with the parameters ω_δ, D_δ und $T_{D,\delta}$. The identification is done by driving tests with steering angle ramps δ_d at various vehicle speeds, see **Fig. 34.6**.

34.3.2 Single Track Model

The linear single track model for the lateral and yaw dynamics of the vehicle [15] serves as a plant model for the synthesis of the lateral trajectory tracking and vehicle guidance control. It can be described in a simple form by the transfer functions between the wheel steering angle input and the yaw rate and the sideslip angle output based on [1]

$$G_{\dot{\psi}}(s) = k_{\dot{\psi}} \frac{(T_{\dot{\psi}}s + 1)\omega_{\psi\beta}^2}{s^2 + 2D_{\psi\beta}\omega_{\psi\beta}s + \omega_{\psi\beta}^2},$$
$$G_\beta(s) = k_\beta \frac{(T_\beta s + 1)\omega_{\psi\beta}^2}{s^2 + 2D_{\psi\beta}\omega_{\psi\beta}s + \omega_{\psi\beta}^2}. \tag{34.3}$$

Furthermore, the curvature of the vehicle motion is $\kappa = (\dot{\psi} + \dot{\beta})/v_x$. Accordingly, the transfer function from steering angle to curvature can be written as

$$G_\kappa(s) = [G_{\dot{\psi}}(s) + s \cdot G_\beta(s)]/v_x$$
$$= k_\kappa \frac{s^2 + 2D_\kappa\omega_\kappa s + \omega_\kappa^2}{s^2 + 2D_{\psi\beta}\omega_{\psi\beta}s + \omega_{\psi\beta}^2} \frac{\omega_{\psi\beta}^2}{\omega_\kappa^2}. \tag{34.4}$$

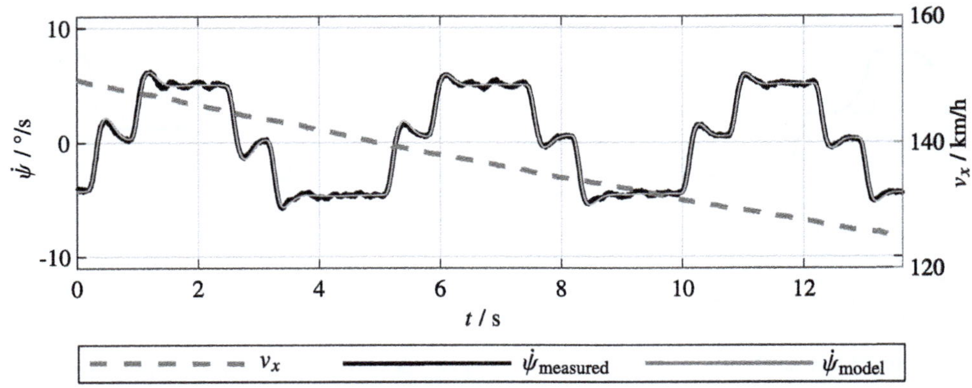

◘ Fig. 34.7 Comparison between single track model and measured yaw rate for bidirectional steering angle steps at descending driving speed between 150 and 120 km/h

Here, the variables $k_{\dot{\psi}}$, k_β and k_κ describe the steady-state yaw rate, sideslip angle and curvature gain, $T_{\dot{\psi}}$ and T_β the derivative time in the dynamic transfer behavior from steering angle to yaw rate and sideslip angle, respectively. The variable $\omega_{\psi\beta}$ is the eigenfrequency (also natural frequency or bandwidth) and $D_{\psi\beta}$ the damping ratio of the yaw and lateral dynamics. The curvature transfer function (34.4) has a conjugate complex, i.e., oscillatory pair of zeros with natural frequency $\omega_\kappa = 2\pi f_\kappa$ and damping ratio D_κ. All variables result from the equations of motion of the single track model linearized with respect to straight-line driving and under the assumption of constant driving speed.

The vehicle parameters and characteristic variables of the single track model and thus the dynamic transfer behavior depend to a large extent on the vehicle speed, the uncertain vehicle payload and the friction-dependent tire-road contact and thus on the road condition and the properties of the tire. Payload, tires and weather-related road adhesion coefficient cause a change in the steady-state and dynamic properties, which are captured in the six independent velocity-sensitive parameters $k_{\dot{\psi}}, k_\beta, T_{\dot{\psi}}, T_\beta, \omega_{\psi\beta}$ and $D_{\psi\beta}$. As a function of these six, the variables k_κ, ω_κ and D_κ are computed variables. For control design, it is essential to adapt the control laws to all measurable, estimable, or observable varying parameters and to be sufficiently robust to non-measurable, non-estimable, or non-observable varying, i. e. uncertain, parameters. An adaptation of the controller functions, for example, to the driving speed—a varying but known parameter of the single track model and at the same time an observed driving state—via classical *gain scheduling* is commonly used.

The single track model is identified in road tests on the basis of bidirectional steering angle steps in the linear range of lateral acceleration ($3 - 4\,\text{m/s}^2$) throughout the entire speed range. This process, repeated for different payloads and tires, can be used as a basis for determining the parameters of the single track model and for describing the uncertainties to be considered (◘ Fig. 34.7).

34.3.3 Longitudinal Dynamics

In today's vehicle architectures, *sum wheel torque* is an established interface for controlling the motor and the brake. This is sufficient for any part of a driver assistance system that does not take responsibility for stabilizing the vehicle at the handling limits but relies on an inner stabilization loop. Then, the outer control structures do not need to consider wheel individual torques. Such an architecture offers the advantage that both motor and brake can be driven by the same physical quantity.

In the course of the chapter, the transfer behavior from the desired motor torque $\tau_{\text{mot,d}}$ and from the desired braking torque $\tau_{\text{brk,d}}$ to the resulting wheel torque τ_x is modelled as a delayed second-order system with the parameters ω_{τ_x}, D_{τ_x} and T_{D,τ_x}:

$$G^*_{\tau_x}(s) = \underbrace{\left(\frac{\omega^2_{\tau_x}}{s^2 + 2D_{\tau_x}\omega_{\tau_x}s + \omega^2_{\tau_x}} \right)}_{G_{\tau_x}} e^{-T_{D,\tau_x}s} \qquad (34.5)$$

Similar to steering model 2, identification in driving tests is based on target wheel torque steps at various driving speeds. The model parameters are chosen to average over a wide range of actuator variants from different vehicle derivatives and at various operating points. ◘ Figure 34.8 shows the result of such an identification. As with the steering models, the existing deviations cannot be avoided with the simple approach. However, as shown later, the model quality is sufficient for the selected control design.

The transfer behavior from sum wheel torque τ_x to the longitudinal vehicle acceleration a_x is modelled, again, in a highly simplified way as first order transfer behavior. Using the assumed vehicle mass m and the wheel radius r, the longitudinal acceleration is obtained from the transfer function

$$G_{a_x}(s) = \frac{1/(mr)}{T_{a_x}s + 1}. \qquad (34.6)$$

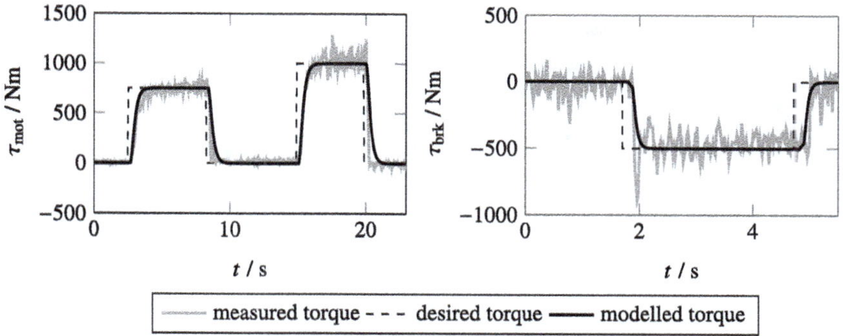

◻ Fig. 34.8 Comparison between modelled and measured cumulative wheel motor torque (left) and cumulative wheel braking torque (right)

34.3.4 Kinematics

The global position of the vehicle, described by x and y, results from the absolute value v and the direction θ of the vehicle velocity $\mathbf{v}$:

$$\dot{x} = v\cos\theta \quad \text{und} \quad \dot{y} = v\sin\theta. \tag{34.7}$$

θ describes the course angle of the vehicle and represents the angle of the velocity vector relative to the global coordinate system. It is calculated as the sum of the yaw angle ψ and the sideslip angle β.

$$\theta = \psi + \beta. \tag{34.8}$$

34.4 Trajectory Guidance

The task of the trajectory guidance level is to provide the target values for the realization of various maneuvers required by the respective driver assistance function. For example, maneuvers such as lane keeping, avoiding collisions, following a vehicle, or holding a set speed must be realized. In the past, these maneuvers have often been implemented by purely control-based approaches, e.g. [2]. In this case, the controller has to be adapted to the respective driving task or driving situation or it has to be switched between different control approaches. In contrast, trajectory-based approaches on the trajectory guidance level represent a very generic solution that can be used for various driver assistance functions.

34.4.1 Trajectory Planning

The realization of the respective maneuvers with a trajectory planning has become a widely used method (compare [19]). The trajectories to be calculated have to meet various requirements. First of all, safety has to be guaranteed. The computed motion must always be collision-free with other road users and obstacles. In addition, quality aspects such as maximum jerks and accelerations must be adhered to or minimized. This is not only necessary for comfort aspects, but also to be able to realize the motion by the used actuators. For this purpose, the potential of the engine or steering system and the available friction value must be taken into account. These quantities are combined in the potential vector.

Trajectories can be calculated using a wide variety of methods. Optimization-based approaches are particularly suitable for the realization of the aforementioned requirements, since trajectory planning can be formulated as an optimal control problem, compare [9]. The task is to compute the control commands minimizing a given cost function subject to constraints. For comfort-oriented systems such as ACC or lane keeping assistance, cost functions that maximize driving comfort are suitable; for functions that aim to maximize driving dynamics, time optimal cost functions can be used; and for functions that are intended to minimize fuel consumption, energy optimal approaches are a possible solution. Safety must always be ensured by suitable constraints.

The driver assistance function to be realized must define the driving target: For the longitudinal vehicle motion, the target state is defined by a set speed and the target distances to be maintained (time gaps). For the lateral vehicle motion, the specification of a reference path is suitable. For lane keeping assistance this can be the center of the lane and for lane change assistance or evasion assistance the neighboring lane. The driver assistance function can be adjusted by parameterizing the cost function. By considering the potential vector as an additional restriction in the optimization problem, it can be ensured that the calculated trajectory can be realized by the underlying control. The formulation of the trajectory planning problem as an optimization task has been handled in many publications. Chapter 45 describes the implementation using different optimization methods in more detail.

The planned trajectory is always represented as a sequence of states and actuating variables as a function of time. In the course of the chapter it is represented in the form

$$\mathbf{x}_{\text{traj}}(t) = [x(t), y(t), \theta(t), \kappa(t), v_x(t), a_x(t)]_{\text{traj}}^{T} \tag{34.9}$$

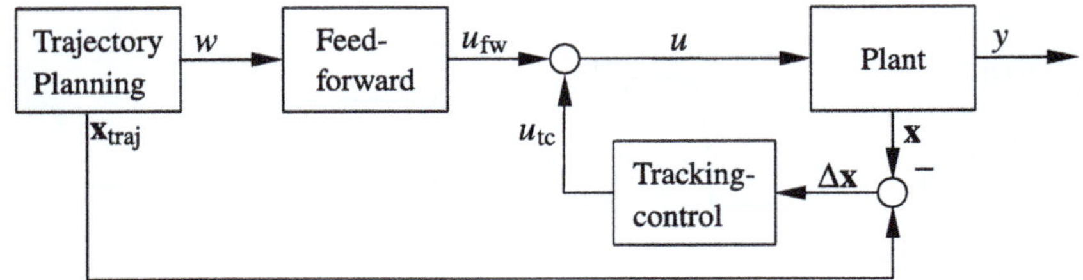

Fig. 34.9   Two-degree-of-freedom control structure consisting of model-based feedforward and output control

x_{traj} and y_{traj} describe the planned position in global coordinates with the planned track angle θ_{traj}. Furthermore, the planned curvature κ_{traj} as well as the planned longitudinal velocity $v_{x,\text{traj}}$ and longitudinal acceleration $a_{x,\text{traj}}$ are provided.

Feeding back the measured vehicle states in each call of the trajectory planning and thus closing the control loop with the trajectory planning, is referred to as model predictive control (MPC, compare e.g. [14]). However, solving the optimization problem can take a considerable amount of time, depending on the constraints to be considered. Therefore an implementation with high replanning rates on a production ECU is often difficult to realize. For this reason, trajectories are planned in a slower computation cycle without feeding back the measured state and stabilization is left to the underlying trajectory tracking control according to the two-degree-of-freedom structure [8]. Due to its simple structure, the latter can be executed at a much higher computational rate. With the separation into two degrees of freedom, disturbance rejection behavior and guidance transfer behavior can be designed separately. This is particularly advantageous if different driver assistance functions are to be implemented with the same control structure. The guidance behavior of the function is achieved by setting the trajectory planning. In contrast, the trajectory tracking control can be designed independently of the respective assistance function. Thus, no compromise has to be made in the design of the control or the control structure has to be adapted situationally.

Since the trajectory planning takes into account the driving dynamic limits and the limitations of the actuators by considering the potential vector, the underlying control can be divided into a lateral and longitudinal part.

34.4.2 Trajectory Tracking Control

The goal of the trajectory tracking control is the realization of the planned trajectory $x_{\text{traj}}(t)$ with the trajectory guidance control commands κ_{d} and $a_{x\text{d}}$. For that, the two-degree-of-freedom structure [8] is used. First, it will be briefly introduced and afterwards applied to lateral and longitudinal control.

The two-degree-of-freedom structure is a commonly used method for solving trajectory tracking problems. Figure 34.9 illustrates the common implementation. Both the trajectory x_{traj} and the measured state x are available. The control command u is calculated by

$$u = u_{\text{fw}} + u_{\text{tc}} \tag{34.10}$$

from the dominant feedforward part u_{fw} and the feedback part u_{tc}. The task of feedforward control is to approximately invert the estimated dynamics of the underlying system and to realize the control objective w in the absence of disturbances:

$$u_{\text{fw}} = \tilde{G}^{-1} w \tag{34.11}$$

Since the trajectory is represented as a sequence of states over time, known time delays can be compensated directly by evaluating the trajectory at the time $t + T_{\text{D}}$ to calculate the reference variable w.

The control objective is mainly achieved by feedforward control. Remaining deviations from the planned trajectory, caused by modeling errors or external disturbances, must be compensated by the tracking control. It can be implemented comparably simple and also a linearization of the system along the reference trajectory is possible. Thus, the control command u_{tc} is often obtained by linear feedback of the control error $\Delta x = x_{\text{traj}} - x$ by using the gain factors $\mathbf{k}$:

$$u_{\text{tc}} = \mathbf{k} \Delta \mathbf{x} \tag{34.12}$$

Figure 34.10 illustrates the states of the planned trajectory as well as the measured vehicle states. In the lateral direction, the control deviations from the planned trajectory are described by the lateral deviation Δd and the angular deviation $\Delta \theta$, and in the longitudinal direction by the longitudinal position error Δs and the velocity control error Δv_x:

$$\begin{bmatrix} \Delta s \\ \Delta d \\ \Delta \theta \\ \Delta v_x \end{bmatrix} := \begin{bmatrix} \begin{bmatrix} \cos \theta_{\text{traj}} & \sin \theta_{\text{traj}} \\ -\sin \theta_{\text{traj}} & \cos \theta_{\text{traj}} \end{bmatrix} \begin{bmatrix} x_{\text{traj}} - x \\ y_{\text{traj}} - y \end{bmatrix} \\ \theta_{\text{traj}} - \theta \\ v_{x,\text{traj}} - v_x \end{bmatrix} \tag{34.13}$$

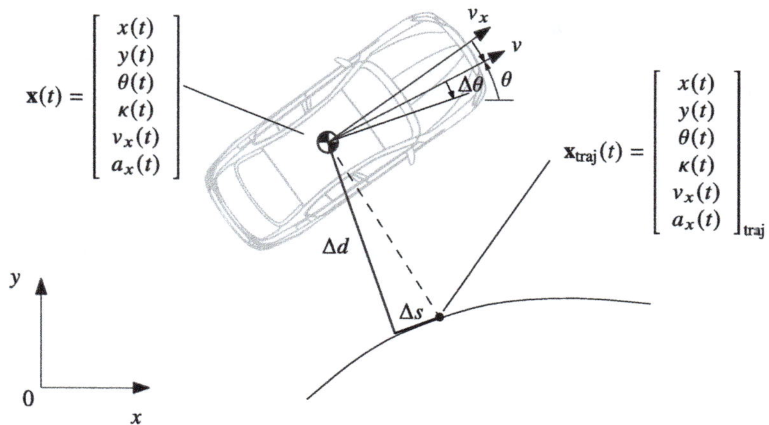

$$\mathbf{x}(t) = \begin{bmatrix} x(t) \\ y(t) \\ \theta(t) \\ \kappa(t) \\ v_x(t) \\ a_x(t) \end{bmatrix}$$

$$\mathbf{x}_{\mathrm{traj}}(t) = \begin{bmatrix} x(t) \\ y(t) \\ \theta(t) \\ \kappa(t) \\ v_x(t) \\ a_x(t) \end{bmatrix}_{\mathrm{traj}}$$

■ **Fig. 34.10**   Illustration of the states of the trajectory tracking control

The track angle θ is the sum of the yaw angle ψ and the sideslip angle β. The error dynamics is calculated with $v = \frac{v_x}{\cos\beta}$ as

$$\Delta\dot{s} = v_{x,\mathrm{traj}}\kappa_{\mathrm{traj}}\Delta d + v_{x,\mathrm{traj}} - v\cos\Delta\theta \quad (34.14a)$$

$$\Delta\dot{d} = -v_{x,\mathrm{traj}}\kappa_{\mathrm{traj}}\Delta s + v\sin\Delta\theta \quad (34.14b)$$

$$\Delta\dot{v}_x = a_{x,\mathrm{traj}} - a_x \quad (34.14c)$$

$$\Delta\dot{\theta} = v_{x,\mathrm{traj}}\kappa_{\mathrm{traj}} - v\kappa. \quad (34.14d)$$

Ignoring the lateral errors in the calculation of the longitudinal errors and vice versa and considering small deviations, we obtain the simplified linear equations

$$\Delta\dot{s} = \Delta v_x \quad (34.15a)$$

$$\Delta\dot{d} = v_x\Delta\theta. \quad (34.15b)$$

These often serve as the basis for the design of linear path tracking controllers (see e.g. [2]).

34.4.2.1 Lateral Trajectory Tracking Control

According to the three-level model presented in ▶ Sect. 34.2, the desired curvature κ_{d} is defined as the control command of the lateral control. The dynamics of the lower-level control is given by $\tilde{G}_{\kappa_{\mathrm{d}}\to\kappa}$ and the time delay $T_{\mathrm{D},\kappa}$. According to (34.11), the feedforward of the lateral trajectory tracking control is calculated with

$$\kappa_{\mathrm{fw}} = \tilde{G}_{\kappa_{\mathrm{d}}\to\kappa}^{-1}\kappa_{\mathrm{traj}}. \quad (34.16)$$

The inversion of the transfer function can be realized with common methods.[1] To compensate for the time delay of the underlying plant, the curvature of the trajectory at time $t + T_{\mathrm{D},\kappa}$ is used.

The relevant control errors Δd and $\Delta\theta$ can be calculated with (34.13) using the measured and planned states. Thus, all necessary information is available and the control command of the lateral control is calculated to be

$$\kappa_{\mathrm{d}} = \kappa_{\mathrm{fw}} + \kappa_{\mathrm{tc}} = \tilde{G}_{\kappa_{\mathrm{d}}\to\kappa}^{-1}\kappa_{\mathrm{traj}} + k_\theta\Delta\theta + k_d\Delta d \quad (34.17)$$

with the gains k_θ and k_d. These are derived using the Eqs. (34.14d) and (34.15b): The goal of κ_{tc} is to asymptotically suppress lateral deviations from the target trajectory: $\lim_{t\to\infty}\Delta d(t) = 0$. Neglecting the underlying dynamics, the control law directly results from the time derivative of the differential equation (34.15b):

$$\Delta\ddot{d} = \dot{v}_x\Delta\theta + v_x\Delta\dot{\theta} = a_x\Delta\theta - v_x^2\kappa_{\mathrm{tc}} \quad (34.18)$$

Demanding a second order transient response with a desired time constant $T_{\mathrm{tc},\kappa}$

$$T_{\mathrm{tc},\kappa}^2\Delta\ddot{d} + 2T_{\mathrm{tc},\kappa}\Delta\dot{d} + \Delta d = 0 \quad (34.19)$$

and solving for the control variable κ_{tc}, results in

$$\kappa_{\mathrm{tc}} = \Delta\theta\underbrace{\frac{a_xT_{\mathrm{tc},\kappa}^2 + 2v_xT_{\mathrm{tc},\kappa}}{T_{\mathrm{tc},\kappa}^2v_x^2}}_{k_\theta} + \Delta d\underbrace{\frac{1}{T_{\mathrm{tc},\kappa}^2v_x^2}}_{k_d}. \quad (34.20)$$

This control law guarantees the asymptotic decay of lateral position errors.[2]

34.4.2.2 Longitudinal Trajectory Tracking Control

The desired acceleration $a_{x\mathrm{d}}$ is defined as the control command of the longitudinal trajectory tracking control.

---

[1]   If the computed trajectory meets sufficient continuity requirements and thus can provide not only the command variable $\kappa_{\mathrm{traj}}(t)$ but also its derivatives, no filter is necessary for the inversion of the plant's dynamics.

[2]   In order to achieve steady-state accuracy, the track angle must be correctly determined. Since the included sideslip angle is usually determined by an observer, an additional observer on the trajectory guidance level may be necessary if the quality is not sufficient, cf. [13].

The dynamics of the lower-level control is assumed to be $\tilde{G}_{a_{xd}\to a_x}$ with a time delay T_{D,a_x}. The longitudinal control errors Δs and Δv_x can be calculated according to (34.13). This results in the control law

$$a_{xd} = a_{x,\text{fw}} + a_{x,\text{tc}} = \tilde{G}_{a_{xd}\to a_x}^{-1} a_{x,\text{traj}} + k_v \Delta v_x + k_s \Delta s \tag{34.21}$$

with the gains k_v and k_s. These are determined analogously to the lateral control by demanding an ideally damped second order disturbance rejection behavior with the time constant T_{tc,a_x}. Thus, we obtain the control law

$$a_{x,\text{tc}} = \Delta v_x \underbrace{\frac{2}{T_{\text{tc},a_x}}}_{k_v} + \Delta s \underbrace{\frac{1}{T_{\text{tc},a_x}^2}}_{k_s} . \tag{34.22}$$

34.5 Vehicle Guidance Level Lateral/Longitudinal

In the previous section, it was shown how the trajectory controllers guide the vehicle along the planned trajectory by commanding a desired curvature κ_d and a desired longitudinal acceleration a_{xd}. This section details the control structures of the vehicle guidance level that track desired values by controlling the actuators. During the design of these controllers, a number of requirements have to be taken into account, which result from the control task introduced in ▶ Sect. 34.1. In particular, the vehicle guidance level is responsible for disturbance rejection and robustness with respect to vehicle and actuator characteristics. In addition, the direct interaction with the driver takes place at this level, where the various degrees of cooperation have to be accounted for.

34.5.1 Disturbance Observer—Basic Structure

Vehicle motion is, in both the longitudinal and lateral direction, subject to non negligible disturbances. These must be compensated by the controllers, however, whether this compensation must be complete or not, depends on the desired degree of autonomy. A classical Integral-action in the controller is capable of completely rejecting certain stationary disturbances, but does not offer the necessary degrees of freedom to adapt its disturbance attenuation behavior to the desired degree of autonomy.

Therefore, *disturbance observers* (DO, see [10]) with simultaneous compensation are used, whose basic structure is shown in ◘ Fig. 34.11. The structure is obviously related to IMC (*internal model control*, see [6]), which is shown in ◘ Fig. 34.12. Both structures provide very good disturbance rejection if the transfer function $Q(s)$ is chosen appropriately. DO also offers intuitively understan-

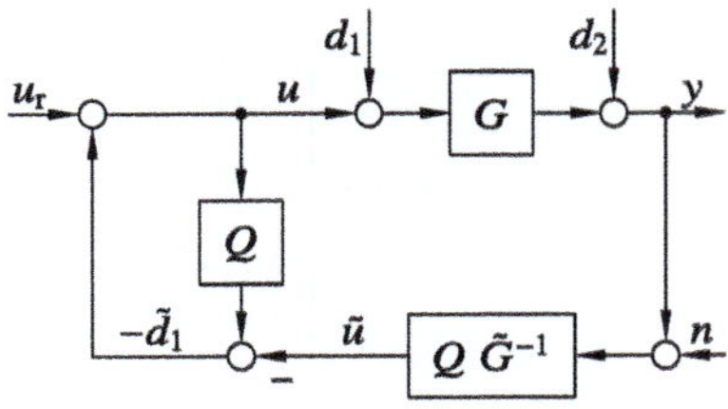

◘ **Fig. 34.11**   Basic structure of the *disturbance observer* (DO) with compensation

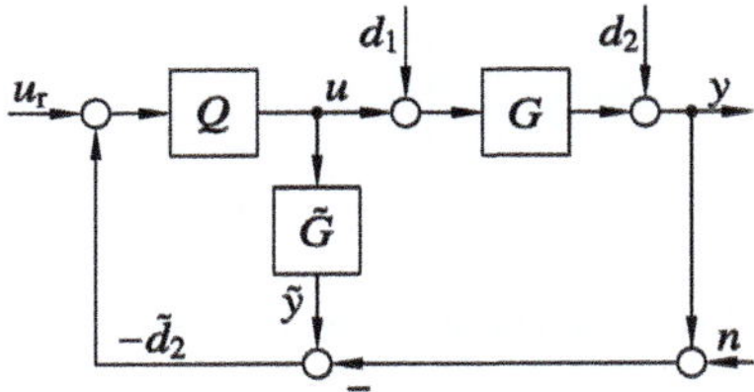

◘ **Fig. 34.12**   Basic structure of the *internal model control* (IMC)

dable manipulation options in the context of the desired adjustability of the disturbance attenuation. Moreover, assuming $G \approx \tilde{G}$, the command tracking behavior of DO does not depend on Q and can be adjusted separately via a superimposed controller.

In ◘ Fig. 34.11 G represents the dynamic behavior of the plant to be controlled, with $\tilde{G}$ being an approximation of its dynamics. By inverting $\tilde{G}$ in the feedback path, the output y is converted to a virtual input $\tilde{u}$, which is then compared to the actual input u. Since the inversion of $\tilde{G}$ is usually improper, a low-pass filter Q is introduced to make $Q\tilde{G}$ proper. The delay of $\tilde{u}$ with respect to u is taken into account by also applying the low-pass filter Q to the control input u. According to this interpretation, the difference $Qu - \tilde{u}$ provides an estimate $\tilde{d}_1$ of the input perturbation d_1, and at the same time takes into account the effect of d_2, translated to the control input u. By feeding $\tilde{d}_1$ back to the input, the disturbances d_1 and d_2 are compensated.

The transfer functions of the disturbances d_1 and d_2 to the output y are

$$G_{d_1\to y} = \left.\frac{G\tilde{G}(1-Q)}{\tilde{G}(1-Q) + GQ}\right|_{Q=1} = 0$$

$$\tag{34.23}$$

$$G_{d_2\to y} = \left.\frac{\tilde{G}(1-Q)}{\tilde{G}(1-Q) + GQ}\right|_{Q=1} = 0.$$

Both disturbance transfer functions in (34.23) become 0 for constant inputs when the steady state gain of Q is chosen to be 1. The trade-off between dynamic behavior of the disturbance attenuation and measurement noise cancellation can be adjusted by the choice of the poles of Q.

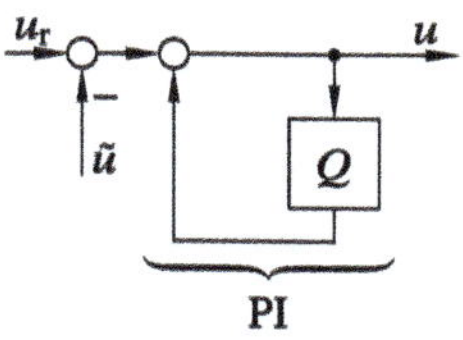

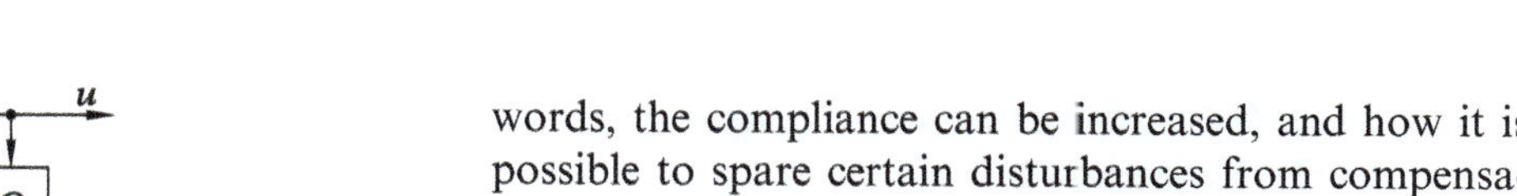

◘ Fig. 34.13 Feedback of the input u in the basic structure of the disturbance observer results in Proportional-Integral behavior

The close relationship between the disturbance observer used here and classical Integral-action controller becomes obvious when the left feedback loop in the signal flow of ◘ Fig. 34.11 is rearranged as shown in ◘ Fig. 34.13.

If $Q(s)$ is a low-pass filter of the form

$$Q(s) = \frac{1}{1 + \lambda_1 s + \lambda_2 s^2 + \cdots} \tag{34.24}$$

then the transfer function from $(u_r - \tilde{u})$ to u is given by

$$G_{\mathrm{PI}}(s) = 1 + \frac{1}{s(\lambda_1 + \lambda_2 s + \cdots)}, \tag{34.25}$$

i.e., a combination of proportional behavior with gain 1 and a more or less delayed integral behavior which is a function of the coefficients λ_i. Consequently, u closely tracks u_r in the upper frequency range, whereas, in the lower frequency range, u will converge such that $u_r - \tilde{u} = 0$.

In addition to the perturbation attenuating effect of the disturbance observer presented thus far, another property of this structure is of great advantage, which becomes evident by studying the transfer behavior from u_r to y for $Q = 1$.

$$G_{u_r \to y} = \left. \frac{G\tilde{G}}{\tilde{G}(1 - Q) + GQ} \right|_{Q=1} = \tilde{G} \tag{34.26}$$

Equation (34.26) implies that the disturbance observer behaves like $\tilde{G}$. This gives rise to two consequences: First, the disturbance observer is robust even in the presence of larger deviations between the real system behavior and $\tilde{G}$, and second, this property can be used as a design degree of freedom to use $\tilde{G}$ to specify the behavior of the disturbance compensated system.

To quantify the amount of disturbance attenuation, it is convenient to define the *stiffness* of a control system, following mechanical systems, as the ratio of disturbance magnitude (force) to the thus caused deviation from nominal behavior (deformation). The disturbance observer in its standard from compensates constant disturbances completely and has, therefore, infinite stiffness or a compliance of 0.

In the following two sections, we will show how this high stiffness can intentionally be reduced or, in other words, the compliance can be increased, and how it is possible to spare certain disturbances from compensation creating selective disturbance attenuation.

34.5.2 **Adjustable Steady State Accuracy**

Because the disturbance observer explicitly determines the disturbances before compensating them by subtraction, it is very easy to make the degree of compensation adjustable. Essentially, the structure from ◘ Fig. 34.11 is augmented for this purpose by the feature of scaling and limiting the disturbance to be compensated as shown in ◘ Fig. 34.14.

With the scaling k the disturbance transfer functions (34.23) become

$$G_{d_1 \to y} = \frac{G\tilde{G}(1-kQ)}{\tilde{G}(1-kQ)+kGQ} \quad G_{d_2 \to y} = \frac{\tilde{G}(1-kQ)}{\tilde{G}(1-kQ)+kGQ}. \tag{34.27}$$

The effect of k can be reasonably estimated under the assumptions $Q = 1$ and $G \approx \tilde{G}$. (34.27) then simplifies to

$$\begin{aligned} G_{d_1 \to y}\big|_{Q=1,\tilde{G}\approx G} &= G(1 - k) \\ G_{d_2 \to y}\big|_{Q=1,\tilde{G}\approx G} &= (1 - k). \end{aligned} \tag{34.28}$$

This means that from the original effect of the disturbances on the output y (Gd_1 in case of d_1 and d_2 in case of d_2) a fraction $(1 - k)$ remains or only a fraction k of the disturbances is compensated when the disturbance observer is used with such scaling.

Similar estimates can be used to make the effect of a limitation in the compensation path tangible. Furthermore, the limitation is an effective anti-wind-up feature, which is necessary if the actuator in the real system modelled by G is limited.

▶ Section 34.5.6 shows in detail how scaling and limiting are used to intentionally make the disturbance observer compliant in steering angle control.

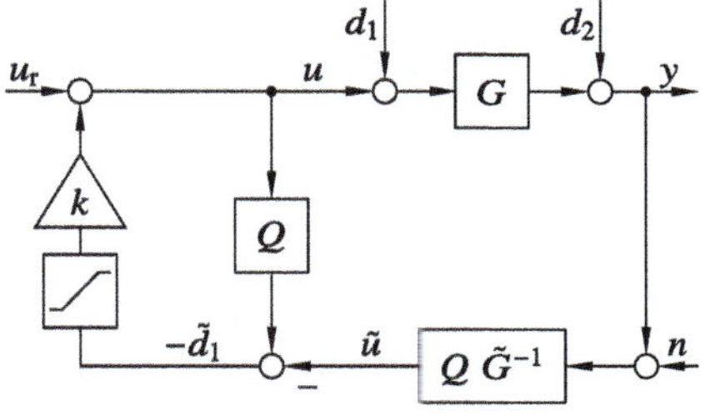

◘ Fig. 34.14 Disturbance observer with scaling and limitation of the disturbance

34.5.3 Selective Compensation of Disturbances

In this section, we present another adaptation of the basic structure of the disturbance observer from ▶ Sect. 34.5.1, which blinds the disturbance observer to selected disturbances, i.e., leaves the transfer behavior of these disturbances to the output unaffected. This selectivity can be achieved under the condition that an additional *auxilliary* signal y_1 can be measured at the plant, e.g. at the output of the actuator, and that this signal reflects only the disturbances that are to be ignored.

The plant output y contains all disturbances, see ▶ Fig. 34.15. Comparing y_1 and the result of mapping y back to the input by the inverse of the subsection G_2 of the plant now yields an estimate $\tilde{d}_{12}$ only for the disturbances d_{12} and d_{22} at the plant subsection G_2. Multiplying this estimate by G_1^{-1} yields a value $\tilde{d}$, which is a version of the disturbance estimate converted to the input of the plant. Subtracting $\tilde{d}$ form u_r compensates the disturbances d_{12} and d_{22}, but ignores the disturbances d_{11} and d_{21}. As in the case of the basic structure of the disturbance observer, low-pass filters Q_1 and Q_2 with steady-state-gains of 1 are needed to make the inversion of the transfer functions proper and to tune the dynamic behavior of the observer.

With $Q_1 = 1$ and $Q_2 = 1$ (stationary and lower frequency range), the disturbance transfer functions for all four disturbances are

$$G_{d11} = G_1 G_2 \tilde{G}_1 \tilde{G}_2 \cdot D^{-1}$$
$$G_{d12} = G_2 \tilde{G}_2 (\tilde{G}_1 - G_1) \cdot D^{-1}$$
$$G_{d21} = G_2 \tilde{G}_1 \tilde{G}_2 \cdot D^{-1} \qquad (34.29)$$
$$G_{d22} = \tilde{G}_2 (\tilde{G}_1 - G_1) \cdot D^{-1}$$

with the common denominator $D = \tilde{G}_2(\tilde{G}_1 - G_1) + G_1 G_2$. It is straightforward to see that the condition $\tilde{G}_1 = G_1$ must be satisfied in the relevant frequency ranges for the disturbances d_{12} and d_{22} to be fully compensated there. With this condition (34.29) simplifies to

$$G_{d11}\big|_{\tilde{G}_1=G_1} = G_1 G_2 \quad G_{d12}\big|_{\tilde{G}_1=G_1} = 0$$
$$G_{d21}\big|_{\tilde{G}_1=G_1} = \tilde{G}_2 \quad G_{d22}\big|_{\tilde{G}_1=G_1} = 0. \qquad (34.30)$$

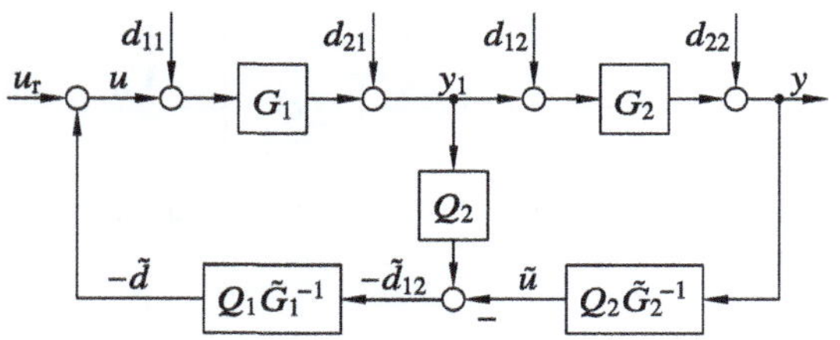

Thus, d_{11} and d_{21} have essentially the same transfer behavior to the output y as they had without the disturbance observer. d_{12} and d_{22} are compensated.

The transfer function $G_{u_r \to y}$ is equal to G_{d11} and thus also equals $G_1 \tilde{G}_2$. As a consequence, with this selective form of the disturbance observer, too, the behavior of the plant can, at least partially, be specified by the choice of $\tilde{G}_2$, or, in other words, also this form of the disturbance observer is robust against deviations between $\tilde{G}_2$ and G_2. $\tilde{G}_1$, on the other hand, must match G_1 very well.

34.5.4 Curvature Disturbance Observer

The objective of the curvature disturbance observer, DO_κ in ◻ Fig. 34.4, is to track the desired curvature κ_d from the trajectory guidance level with high steady-state accuracy. Therefore, all lateral force disturbances acting on the vehicle must be compensated. Such forces can be caused by crosswinds, inclined roadways, etc. The challenge is, in the case of cooperative assistance functions, not to counteract steering interventions by the driver as if they were disturbances, thus creating a compliant controller behavior. As mentioned in the previous section, the structure of the selective disturbance observer can resolve these conflicting goals by feeding back of an auxiliary variable.

In the sequel, only front axle steering is considered for the control the lateral dynamics. Therefore, control allocation CA_κ can be ignored. Using the assumed steady-state gain $\tilde{k}_\kappa$, the target curvature $\kappa_{\delta d}$ can be converted into a target steering angle δ_d:

$$\delta_d = \tilde{k}_\kappa^{-1} \kappa_{\delta d} \qquad (34.31)$$

The Eqs. (34.2) and (34.4) serve as models for the design of the disturbance observer. Accordingly, the curvature is a consequence of the transfer behavior of the front axle steering with steering angle control G_δ and the single-track model G_κ. The measured steering angle δ is fed back as the auxiliary variable. It subsumes both the applied torque of the subordinate steering angle control and the driver steering torque $\tau_{\delta,\mathrm{drv}}$. In contrast, the lateral force disturbances z_κ which need to be compensated are only detectable in the measured curvature κ. ◻ Figure 34.16 shows the implementation of the curvature disturbance observer as a block diagram. Its output (curvature to be compensated) is given by:

$$\kappa_{do} = \tilde{k}_\kappa Q_\delta \tilde{G}_\delta^{-1} \left(Q_\kappa \delta - Q_\kappa \tilde{G}_\kappa^{-1} \kappa \right) \qquad (34.32)$$

The overall output κ_d' is calculated as the sum of κ_d and κ_{do}. To render the inversions of the assumed transfer functions $\tilde{G}_\delta$ and $\tilde{G}_\kappa$ proper, lowpass filters Q_δ and Q_κ

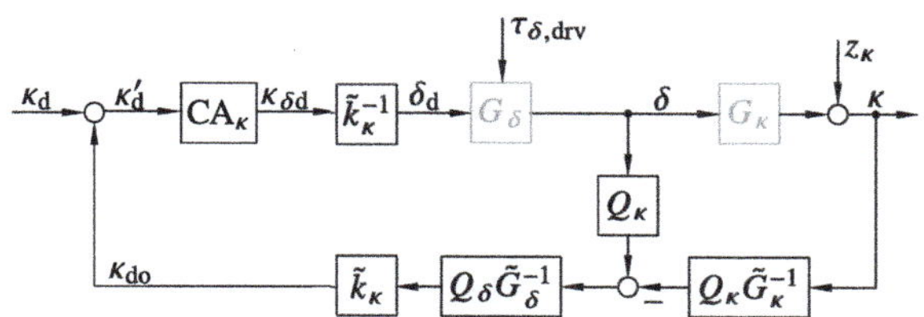

Fig. 34.16 Selective disturbance observer on curvature level for compensating the disturbance z_κ without compensating the driver's hand torque $\tau_{\delta,\mathrm{drv}}$

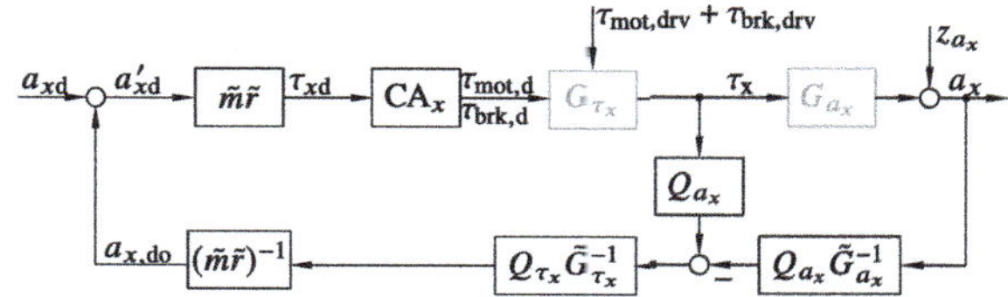

Fig. 34.17 Selective disturbance observer on acceleration level for compensation of disturbance z_{a_x} without compensation of driver torque $\tau_{\mathrm{mot,drv}}$ and $\tau_{\mathrm{brk,drv}}$

are necessary. Disturbances z_κ are compensated completely in steady-state with this structure, whereas driver interactions are not interpreted as disturbances and, therefore, not compensated. Furthermore, anti-wind-up measures [3] against the effects of actuator limitations are not necessary, since the measured steering angle δ already reflects the steering angle limits in contrast to the command input δ_d. The control loop closed by the disturbance observer is characterized approximately by the nominal transfer behavior

$$\tilde{G}_{\kappa_\mathrm{d}\to\kappa} \approx \tilde{k}_\kappa^{-1} \tilde{G}_\delta \tilde{G}_\kappa. \qquad (34.33)$$

Parameter uncertainties can thus be handled and the transfer function can be used for the design of the trajectory following control within the trajectory guidance level.

34.5.5 Longitudinal Acceleration Disturbance Observer

The longitudinal control is designed analogously to the curvature control: By means of a disturbance observer, the steady-state accurate tracking of the target acceleration $a_{x\mathrm{d}}$ is ensured. Here, too, an auxiliary variable, the resulting sum wheel torque τ_x, is fed back to ensure that driver interactions are not compensated as disturbances. The sum wheel torque is a result of the driver torques commanded by the accelerator and brake pedals and the inputs from the assistance system. The assumed transmission behavior according to Eqs. (34.5) and (34.6) serve as models for the design. The commanded target acceleration $a_{x\mathrm{d}}$ is converted into a target wheel torque using the assumed vehicle parameters $\tilde{m}$ and $\tilde{r}$:

$$\tau_{x\mathrm{d}} = a_{x\mathrm{d}} \underbrace{\tilde{m}\tilde{r}}_{\tilde{k}_{a_x}^{-1}} \qquad (34.34)$$

$\tilde{m}$ and $\tilde{r}$ are available as estimated values at runtime. The control allocation logic for the engine and brake (CA_x) has approximately the transfer behavior 1 due to the logic presented in the next section and is therefore ignored in the design of the disturbance observer.

Figure 34.17 shows the implementation of the selective disturbance observer to compensate the disturbances z_{a_x}. Its output (the acceleration to be compensated) is calculated as

$$a_{x,\mathrm{do}} = Q_{\tau_x}\left(\tilde{m}\tilde{r}\tilde{G}_{\tau_x}\right)^{-1}\left(Q_{c_x}\tau_{a_x} - Q_{a_x}\tilde{G}_{a_x}^{-1}a_x\right). \qquad (34.35)$$

To render the inversions of the assumed transfer functions $\tilde{G}_{\tau_x}$ and $\tilde{G}_{a_x}$ proper, two filters Q_{τ_x} and Q_{a_x} are necessary. The control loop thus closed behaves approximately like

$$\tilde{G}_{a_{x,\mathrm{d}}\to a_x} \approx \tilde{m}\tilde{r}\tilde{G}_{\tau_x}\tilde{G}_{a_x}. \qquad (34.36)$$

Disturbances z_{a_x} are compensated, whereas the driver's commanded motor torque $\tau_{\mathrm{mot,drv}}$ and braking torque $\tau_{\mathrm{brk,drv}}$ are not compensated.

34.5.6 Cooperative Steering Angle Control with Disturbance Observer

The steering angle controller, C_δ in **Fig. 34.4**, is responsible for tracking the commanded steering angle δ_d as part of the vehicle lateral control. Depending on the driver assistance system in action, it should react more or less stiffly to driver torques $\tau_{\delta,\mathrm{drv}}$ as well as to external and internal forces acting on the steering system.

The challenge of assigning a defined stiffness or compliance to a controlled system is known from robotics, where it is addressed by various approaches to impedance control. Following loosely [17], four basic structures can be identified for this purpose:

1. realization of the desired impedance by the choice of the controller gains (no force sensing necessary),
2. stiff position control with superimposed admittance model of the desired system dynamics (admittance architecture),
3. stiff force control with superimposed impedance model of the desired system dynamics (impedance architecture), and
4. combined position/force control as a type of multivariable control.

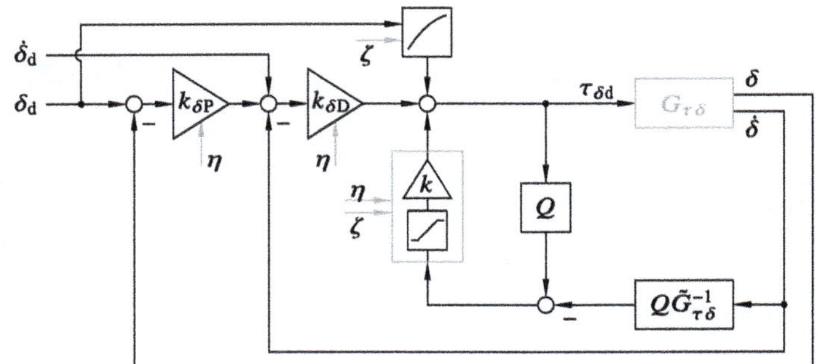

Fig. 34.18 Cooperative steering angle controller with disturbance observer; stiffness and steady-state accuracy are adjustable by the signals η and ζ

The steering angle controller proposed here makes use of these ideas to meet the requirement for adjustable stiffness and steady-state accuracy. In this context, the steering system, which interacts with the driver via the torsion bar, takes the place of a robot, which interacts with its environment via a force sensor.

Essentially, the steering angle controller consists of two cascaded P controllers, $k_{\delta P}$ and $k_{\delta D}$ with additional feedforward control, whose plant is the disturbance compensated steering system according to ▸ Sect. 34.5.2, see **Fig. 34.18**.

The steering model 1 $(G_{\tau\delta})$ introduced in ▸ Sect. 34.3.1 is used as the desired plant model $\tilde{G}$ in the disturbance observer.

The controller parameters affecting stiffness and steady-state accuracy can be manipulated externally by the η (stiffness) and ζ (steady-state accuracy) signals. This manipulation of the controller parameters essentially corresponds to the idea of impedance control 1. However, due to the chosen controller structure, it is possible, with the controller from **Fig. 34.18**, to adjust stiffness and steady-state accuracy *independently* within certain limits.

The approaches for impedance control with force feedback generally require very high controller bandwidths, since the transfer behavior from the commanded torques to the measured reaction forces typically show high bandwidths. Depending on the partitioning of the controller, these controller bandwidths may not be realized in the vehicle. Nevertheless, a feedback of the driver's hand torque $\tau_{\delta,\text{drv}}$, which is partly a passive reaction, is necessary if the controller shall behave stiff with respect to all disturbances on the one hand, but compliant with respect to the driver on the other hand. The solution proposed here decouples the actuating torque $\tau_{\delta d}$ and the reaction torque $\tau_{\delta,\text{drv}}$ (driver's hand torque) in the relevant frequency ranges by reducing the gains of the various components of the controller depending on $\tau_{\delta,\text{drv}}$ but with a time delay. The control thresholds and the extent of these gain reductions again depend on the desired stiffness η.

34.5.7 Allocation to the Motor and Brake System

In contrast to lateral control, no additional controller is required for actuating the motor and brake system. The strategy for (function-independent) distribution of the target sum wheel torque τ_{xd} to the two actuators will, at first, be explained for the case without driver interaction:

For the target motor torque $\tau_{\text{mot,d}}$ the target torque τ_{xd} is used directly:

$$\tau_{\text{mot,d}} = \tau_{xd}. \tag{34.37}$$

However, for the control of the brake, a case-by-case analysis is necessary with respect to the available deceleration torque $\tau_{\text{mot,min}}$ reported by the motor:

$$\tau_{\text{brk,d}} = \begin{cases} 0 & \text{if } \tau_{xd} \geq \tau_{\text{mot,min}} \\ \tau_{xd} - \tau_{\text{mot,min}} & \text{if } \tau_{xd} < \tau_{\text{mot,min}} \end{cases} \tag{34.38}$$

Thus, the entire operating range of the motor can be utilized. By means of the drag torque in case of an ICE or by means of generator mode in case of an electric drive, the motor is able to generate decelerations. The case-by-case approach avoids negative target accelerations to be automatically generated by the brake.

To include the driver in the actuation of the motor, the maximum of the commanded driver torque and the assistance system torque is usually prioritized, so that drivers can always override the system. In the case of braking commands, on the other hand, automated longitudinal guidance is switched off for comfort functions such as ACC as soon as a driver braking torque is requested. Only in the case of safety functions such as emergency brake assist, the minimum of driver braking request $\tau_{\text{brk,drv}}$ and the driver assistance function request $\tau_{\text{brk,d}}$ is prioritized. This ensures that the higher deceleration request is always executed.

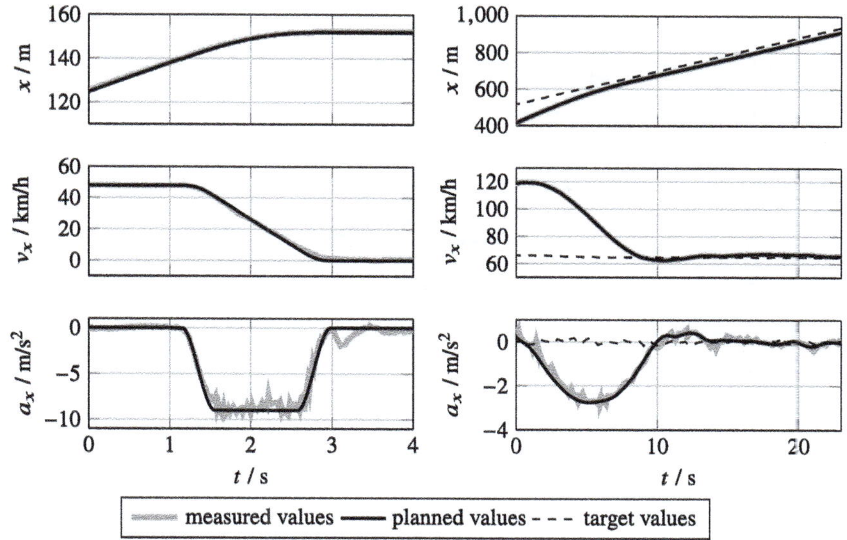

Fig. 34.19   Emergency braking maneuver from an initial speed of 50 km/h on a dry road (left) and braking towards a constantly moving vehicle in front in ACC mode (right)

34.6   Calibration and Experimental Validation

In the following, selected aspects regarding the calibration properties of the control structure will be discussed and illustrated by means of driving tests with various driver assistance systems. The driving tests were performed with a production vehicle of the BMW 3 series. Essentially, the adjustment of the control structure can be divided into three categories:

- functional behavior
- tracking behavior and disturbance attenuation
- cooperativity and transition behavior

34.6.1   Functional Behavior

The functional behavior of any driver assistance system implemented according to the proposed architecture is primarily shaped by the properties of trajectory planning and is not a result of changes in the underlying control structures.

Figure 34.19 shows an example of planned and driven longitudinal trajectories during emergency braking (left) and following a front vehicle in ACC mode (right). Both trajectories have been calculated by the same algorithm for trajectory planning according to [12]. Different parameterization of the cost function and different constraints (e.g. safety distances) in the trajectory planning create widely divers maneuvers.

34.6.2   Tracking Behavior and Disturbance Attenuation

In contrast to trajectory planning, the underlying trajectory tracking and vehicle guidance control are parameterized independently of the respective driver assistance system. They affect how well the planned trajectory is tracked and how quickly external disturbances are compensated. Tuning is usually first done for the case of autonomous, i.e., *hands-off* guidance with the primary goal to achieve good disturbance attenuation and tracking behavior, while, at the same time to apply only moderate actuation. For this purpose, the assumed transfer functions from ▶ Sect. 34.3 are roughly identified and then the time constants in the filters of the disturbance observers and the trajectory tracking control are chosen.

Figure 34.20 demonstrates the behavior of the overall system tuned in this way during a combined lateral and longitudinal maneuver. The trajectory planning requests a dynamic deceleration with simultaneous lane change. The maneuver is performed without any driver intervention. From an initial velocity of 50 km/h, the vehicle is braking to standstill while deviating laterally by 3 m from the starting position. Despite the high dynamics, the planned trajectory is tracked by the control system with very low control errors.

External disturbances acting on the vehicle (such as crosswinds or inclined roadways) are robustly attenuated. As an example, this will be illustrated using a lane keeping assistance function. Figure 34.21 demonstrates the results of a driving test. The goal is to follow a straight lane at a speed of 50 km/h. One-sided braking interventions are used to reproducibly emulate lateral force disturbances. For this purpose, a braking torque

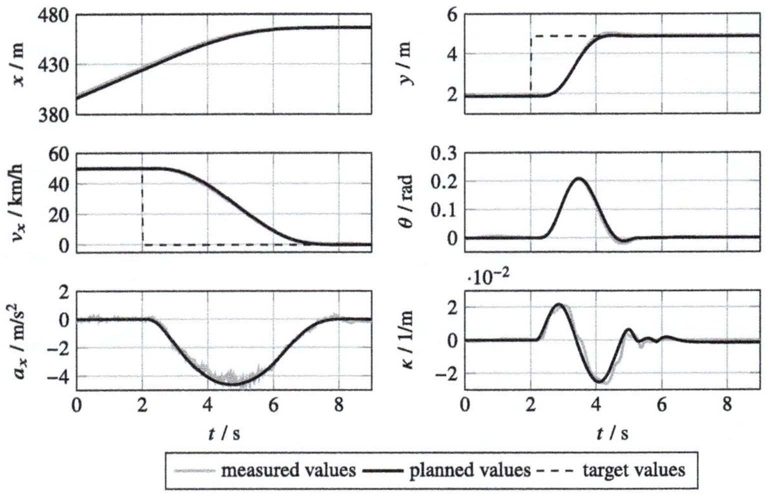

Fig. 34.20 Coupled lateral-longitudinal maneuver for evasion with simultaneous braking to standstill

of 500 Nm is applied to the right wheel of each axle. The disturbance observer compensates the disturbance without building up a significant lateral position error Δd. As shown in ▶ Sect. 34.5.4, however, steering interventions by drivers are not compensated in a steady-state manner. This is again illustrated in a second experiment. ▪ Figure 34.22 shows the results of a measurement with driver interaction. The vehicle travels along a straight reference with a speed of 50 km/h. The driver intervenes the lateral guidance (clearly visible from the driver's hand wheel torque $\tau_{\delta,\mathrm{drv}}$) and steers the vehicle a few meters away from the planned trajectory. In this case, ordinary tracking controllers with integral behavior would constantly increase the command input thus working against the driver. By using the presented control structure with a selective disturbance observer, cooperative driving behavior, such as is required for lane keeping assistance, can be realized. The output of the disturbance observer κ_{do} is almost 0 over the entire time of the driver's intervention, since no disturbance other than the torque caused by the driver is acting on the vehicle. Only the control law of the trajectory tracking control works against the driver. As stated previously, this is desired to provide feedback to drivers about the goal of the implemented driver assistance system. After the manual torque drops back to zero (end of driver intervention), the assistance system guides the vehicle back to reference.

34.6.3 Cooperativity and Transition Behavior

The realization of cooperative control behavior is particularly relevant for lateral guidance, since the driver feels the steering torque directly via the steering wheel. The adjustment of the counter torque, which the driver feels as resistance when steering contrary to the system, is mainly

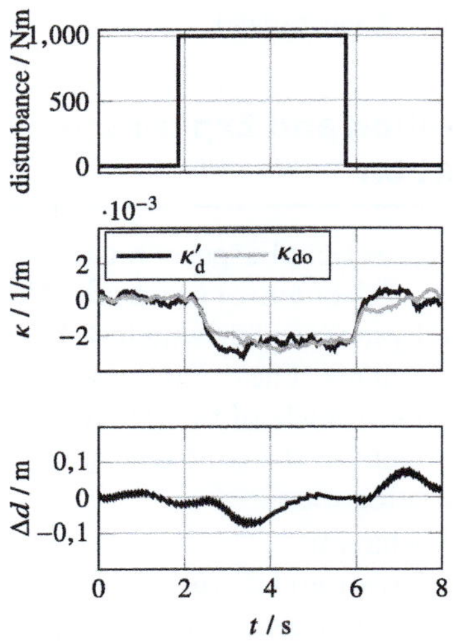

Fig. 34.21 Attenuation of lateral force disturbances

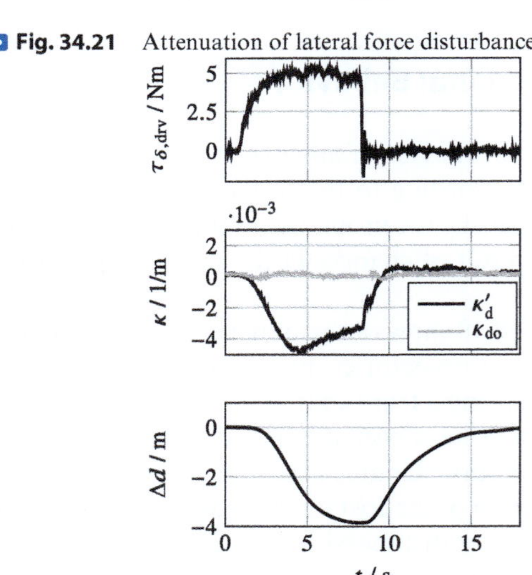

Fig. 34.22 Behavior of the curvature disturbance observer during active driver intervention

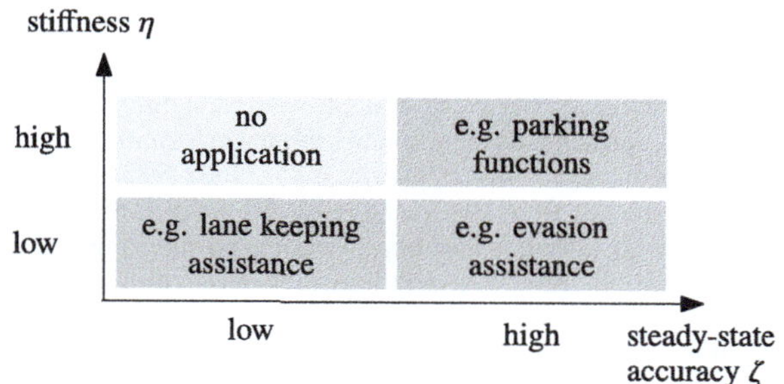

◻ **Fig. 34.23** Reasonable combinations of stiffness and steady-state accuracy for various driver assistance functions inspired by [18]

effected by the parameters η (stiffness) and ζ (steady-state accuracy). ◻ Figure 34.23 shows reasonable combinations of the parameters for different driver assistance functions. For autonomous functions, such as autonomous parking, high precision of the control is required and unwanted interventions by the driver should be compensated. This is achieved by high values for stiffness and steady-state accuracy. Cooperative comfort functions such as lane keeping assistance can be implemented by lower values for η and ζ, since they have comparatively lower demands on the tracking accuracy. Safety functions such as evasion assistance, on the other hand, require good tracking behavior, since the functions are often active only shortly before a possible collision. However, since the driving task is solved jointly between the driver and the vehicle, cooperativeness is also required here. This is achieved by low values for η and high values for ζ.

Apart from shaping the cooperativity, the signals stiffness and steady-state accuracy can also be used to tune the transitional behavior of the function when it is activated or deactivated, e.g., by ramping both signals up or down, respectively. Furthermore, transitions between different assistance systems can be shaped by interpolating η and ζ.

34.7 Conclusion and Outlook

From the authors' point of view, today's driver assistance systems with at least temporary autonomous guidance can only be implemented efficiently by means of a unified and integrated approach for lateral and longitudinal control. Future driver assistance systems with even higher automation levels than those achieved today can only be realized on the basis of such an approach.

The functional architecture proposed here is based on a strict, modular hierarchy that essentially reflects the separation of the driving task into three levels. From a control point of view, the resulting design is a structure with two degrees of freedom, which enables, to a large extent, the independent shaping of the systems's tracking and disturbance attenuation behavior. The tracking behavior is primarily defined by the trajectory planning. Trajectory planning determines, for example, whether a maneuver should be performed comfortably or—because it is an emergency situation—with maximum utilization of the road adhesion coefficient. The underlying controllers, which track the planned trajectories, are identical for all driver assistance systems and for all driving situations. Only the characteristics with respect to steady-state accuracy and with respect to the compensation of external disturbances and driver interventions are adapted to the driving situation. The activation and deactivation of individual functions can also be represented by selectively setting these properties.

The approach has been used successfully for several years in all longitudinal and lateral driver assistance systems industrialized by the BMW Group, thus proving that both autonomous and cooperative functions with both comfort and safety as their principal objective can be realized throughout the entire operating range of vehicle speed and road adhesion utilization. In particular, the ability to continuously realize transitions between manual, cooperative and autonomous driving represents an important innovation that can also be used to combine functions of various automation levels in optional equipment packages. Future vehicles that are operated only autonomously merely constitute a special case in which all features and components that are required for the realization of cooperative behavior could be omitted.

The unification of the functional logic and, to a large extent, of the parameterization of the control structure, too, keeps the calibration effort low. In particular, it was shown that the models used in the controllers must not be highly accurate and can, therefore, be used for many vehicle variants with one parameterization that can be identified with simple driving or test bench experiments.

In addition, the holistic approach will enable merging driver assistance systems for longitudinal and lateral guidance which are still implemented as individual functions today, thus achieving higher levels of automation. From the author's point of view, higher levels of automation also require the behavior at the handling limits to be enhanced. With predictive planning of the trajectories, including the driving dynamics potential, the vehicle can already be guided automatically up to the handling limits in specific emergency situations. However, if the road adhesion coefficient is assumed too high, infeasible trajectories could be planned. In such situations, the controllers of the trajectory and vehicle guidance levels implemented today react virtually like inexperienced drivers and would, for example, further increase the steering angle in case of understeering. Fully autonomous obstacle avoidance maneuvers are, therefore, presently not realizable, yet, as illustrated by the gap in ◻ Fig. 34.1

at the coordinates *autonomous*, *lateral* and *high road adhesion utilization*.

However, the potentials for further development outlined above can be seamlessly integrated into the presented approach, such that the road towards industrialized fully autonomous driving is paved at least from a control engineering point of view.

References

1. Ackermann, J., Blue, P., Bünte, T., Guvenc, L., Kaesbauer, D., Kordt, M., Muhler, M., Odenthal, D.: Robust Control: The Parameter Space Approach. Springer (2002)
2. Ackermann, J., Guldner, J., Sienel, W., Steinhauser, R., Utkin, V.I.: Linear and nonlinear controller design for robust automatic steering. IEEE Trans. Control Syst. Technol. 3(1), 132–143 (1995)
3. Astrom, K.J., Rundqwist, L.: Integrator windup and how to avoid it. In: 1989 American Control Conference, pp. 1693–1698 (1989)
4. Bauer, A., Wollherr, D., Buss, M.: Human-robot collaboration: a survey. Int. J. Humanoid Rob. 05(01), 47–66 (2008)
5. Donges, E.: Driver behavior models. In: Winner, H., Hakuli, S., Lotz, F., Singer, C. (eds.) Handbook of Driver Assistance Systems: Basic Information. Components and Systems for Active Safety and Comfort, pp. 19–33. Springer International Publishing, Cham (2016)
6. Garcia, C.E., Morari, M.: Internal model control. A unifying review and some new results. Ind. Eng. Chem. Process Des. Dev. 21(2), 308–323 (1982)
7. Gasser, T.M., Arzt, C., Ayoubi, M., Bartels, A., Bürkle, L., Eier, J., Flemisch, F., Häcker, D., Hesse, T., Huber, W., et al.: Rechtsfolgen zunehmender fahrzeugautomatisierung. Berichte der Bundesanstalt für Straßenwesen. Unterreihe Fahrzeugtechnik, vol 83 (2012)
8. Horowitz, I.M.: Synthesis of feedback systems. Technical Report (1963)
9. Lewis, F.L., Vrabie, D., Syrmos, V.L.: Optimal Control. John Wiley & Sons (2012)
10. Ohnishi, K.: A new servo method in mechatronics. Trans. Jpn. Soc. Electr. Eng. 107, 83–86 (1987)
11. Rasmussen, J.: Skills, rules, and knowledge; signals, signs, and symbols, and other distinctions in human performance models. IEEE Trans. Syst. Man. Cybern. 3, 257–266 (1983)
12. Rathgeber, C., Winkler, F., Mueller, S.: Collisionfree longitudinal and lateral trajectory planning considering vehicle-dependent potentials. At-Automatisierungstechnik 64(1), 61–76 (2016)
13. Rathgeber, C., Winkler, F., Odenthal, D., Müller, S.: Lateral trajectory tracking control for autonomous vehicles. In: European Control Conference (ECC), pp. 1024–1029. IEEE (2014)
14. Rawlings, J.B.: Tutorial overview of model predictive control. IEEE Control Syst. Mag. 20(3), 38–52 (2000)
15. Riekert, P., Schunck, T.E.: Zur Fahrmechanik des gummibereiften Kraftfahrzeugs. Ingenieur Archiv 11, 210–224 (1940)
16. Sae International.: Taxonomy and definitions for terms related to on-road motor vehicle automated driving systems (2014)
17. Villani, L., De Schutter, J.: Force control. In: Siciliano, B., Khatib, O. (eds.) Handbook of Robotics, Chapter 7, pp. 195–220. Springer, Cham (2016)
18. Walter, M., Nitzsche, N., Odenthal, D., Müller, S.: Lateral vehicle guidance control for autonomous and cooperative driving. In: 2014 European Control Conference (ECC), pp. 2667–2672. IEEE (2014)
19. Werling, M.: Integrated trajectory optimization. In: Winner, H., Hakuli, S., Lotz, F., Singer, C. (eds.) Handbook of Driver Assistance Systems, pp. 1413–1435. Springer International Publishing, Cham (2016)

Open Access This chapter is licensed under the terms of the Creative Commons Attribution-NonCommercial-NoDerivatives 4.0 International License (▶ http://creativecommons.org/licenses/by-nc-nd/4.0/), which permits any noncommercial use, sharing, distribution and reproduction in any medium or format, as long as you give appropriate credit to the original author(s) and the source, provide a link to the Creative Commons license and indicate if you modified the licensed material. You do not have permission under this license to share adapted material derived from this chapter or parts of it.

The images or other third party material in this chapter are included in the chapter's Creative Commons license, unless indicated otherwise in a credit line to the material. If material is not included in the chapter's Creative Commons license and your intended use is not permitted by statutory regulation or exceeds the permitted use, you will need to obtain permission directly from the copyright holder.

Motorcycle Riding Assistance Systems

Raphael Pleß, Andreas Georgi, Jonas Lichtenthäler, Gerald Matschl and Sebastian Will

Contents

© The Author(s) 2026
H. Winner et al. (eds.), *Handbook Assisted and Automated Driving*,
https://doi.org/10.1007/978-3-658-45276-6_35

35.1 Introduction

While the majority of households in Asia own a motorcycle[1] (China 60%, Southeast Asia > 80%, India 47%), these figures are much lower in European countries (with Italy leading the way at 26%) [40]. In the Asian region in particular, motorcycles are an indispensable means of transport due to their low acquisition and maintenance costs, without which a large proportion of the population would be unable to participate in individual transport. In addition to financial considerations, the lower space requirements and the better ecological footprint are advantageous in increasingly densely populated urban centers. These factors are also causing a growing market share in western urban regions. However, especially in less densely populated and wealthier regions, motorcycles—similar to sports cars and classic cars—are used as leisure vehicle. The market for "adventure bikes" in particular is currently experiencing steady growth, so that requirements for use, off the beaten track are also appearing in the specifications of many vehicles. These diverse use cases—from urban transportation and recreational vehicles to sports and competition use, on and off paved roads—place equally diverse demands on motorcycle riding assistance systems. Due to the special dynamic characteristics of single-track vehicles, great efforts have to be made to adapt technologies from the passenger car sector for their use in motorcycles.

These efforts primarily aim at reducing the high number of motorcycle accidents compared to the passenger car population and at alleviating the severe consequences of most accidents. In 2019, just under 20% of all road traffic fatalities in Germany were users of (small) motorcycles. In terms of vehicle population, the risk of being killed on a motorcycle was four times higher than in a passenger car [7]. In addition to the direct benefit of motorcycle riding assistance systems as a safety aid, they can also contribute indirectly to improve safety by increasing rider comfort, for example by reducing stress and distraction so that motorcyclists can better concentrate on—and enjoy—riding.

This chapter deals with the supplementary requirements that motorcycles place on the implementation of assistance systems compared with passenger cars. Based on the fundamentals of slip control systems, the chapter first explains system enhancements that influence braking and acceleration behavior when the vehicle is running straight. This is followed by a description of the enhancements required to implement antilock braking systems (ABS) and traction control systems (TCS) that are suitable for riding in curves. Then, environment sensor-based assistance systems at the path guidance level are discussed that warn (e.g., blind spot assist, rear-end collision warning) or intervene (e.g., adaptive cruise control, emergency brake assist) in critical situations. In both cases, questions arise regarding the triggering of such systems (environmental sensors, trajectory estimation, connectivity, …) as well as the perception and controllability (human–machine interfaces, rider state and skills). The motorcycle-specific extension of ISO26262(-12) is described as the state-of-the-art development and validation process of these systems. For certain development steps, motorcycle riding simulators have also proven themselves useful. Their applicability will be discussed before the chapter ends with an outlook on the future of motorcycle safety systems.

35.2 Basics of Motorcycle Dynamics

35.2.1 Stabilization and Dynamic Equilibrium

Similar to an inverse pendulum, a motorcycle standing vertically is only in an unstable equilibrium. Even small angular deviations at standstill or low speed will lead to capsizing. At increasing speeds this leaning motion is increasingly dampened. Steering and rider motion cause transitions between different equilibrium positions. The equilibrium condition in curved trajectories is the greatest and most obvious difference between passenger cars and motorcycles. While cars build up a roll angle to the outside of the curve due to the centrifugal effect at the vehicle's center of gravity, motorcycles use this same centrifugal effect to generate an equilibrium against the overturning moment resulting from the lateral displacement of the system's center of gravity at roll angles.

During steady-state cornering, the equilibrium position is the so-called theoretical (physically effective) roll angle φ_{th} which, according to Eq. 35.1, is mainly depending on the horizontal lateral acceleration $\ddot{y}$. This again is determined by the square of the set riding speed v and the trajectory curvature κ.

$$\varphi_{th} = \arctan \frac{m \cdot \ddot{y}}{m \cdot g} = \arctan \frac{v^2 \cdot \kappa}{g} \qquad (35.1)$$

As the curvature can only vary within the limits set by the course of the road, the selected velocity dictates the necessary theoretical roll angle of a motorcycle. When

---

1 To ease the flow of reading in this chapter, all forms of motorized single-track vehicles—from e-bikes to mopeds, scooters, light motorcycles, motorcycles and compact two-track tilting vehicles—are referred to as motorcycles. No differentiation is necessary with regard to assistance systems.

riding on the same curves, greater roll angles can only be achieved by riding at higher speeds until a maximum roll angle $\varphi_{th,max}$ is finally reached, which is determined by the maximum lateral friction coefficient of the tires

$$\mu_{lat,max} = \frac{|\ddot{y}|}{g} \tag{35.2}$$

and results to

$$|\varphi_{th}| = \arctan \mu_{lat} \leq \arctan \mu_{lat,max} = |\varphi_{th,max}|. \tag{35.3}$$

With modern tires on a dry road with good grip, this maximum lies above 50°. However, the achieveable roll angle is generally limited by e.g. exhaust system, frame or frame-mounted components (e.g. footrests and foot-controls).

The total roll angle visible from the outside φ_{tot} of the motorcycle (between its plane of symmetry and the vertical) is equal to the theoretical roll angle only assuming infinitely thin tires and a center of gravity located in the plane of symmetry of the motorcycle. ◘ Figure 35.1 shows a motorcycle during steady-state cornering assuming a tire contour radius $r_r \neq 0$ and a rider leaning out of the plane of symmetry. Here, the centrifugal force F_F is in equilibrium with the lateral wheel forces F_Y and the weight force G with the wheel normal forces F_Z.

Neglecting the rider's offset position, it can be seen that the tire contact point moves sideways (tire-scrub-radius l_{rr}) at roll angles $\varphi \neq 0$ due to the tire width. This reduces the effective length of the lever arm between the weight force G and the tire-road contact point. The physically effective roll angle (blue dashed line) is then smaller $|\varphi'|$ than the total roll angle of the motorcycle φ_{tot}, by the amount of the additional roll angle (symbol). The tire-width-related additional roll angle is about 10% of the theoretical roll angle, depending on the tire width (larger additional roll angle) and center of gravity height of the motorcycle (smaller additional roll angle). Especially with so-called "choppers" with very wide tires, low centers of gravity and total roll angles limited by frame-fixed add-on parts, the physically effective roll angle—and therefore the achievable cornering speed or cornering curvature—is severely restricted.

This restriction can be partially eliminated if a movable center of gravity of a rider is taken into account. Depending on the mass ratio between the rider and motorcycle, the system center of gravity can be shifted by a few centimeters. Leaning into the curve ("hanging-off") will increase the physically effective roll angle $|\varphi_{th}|$ at a constant total roll angle $|\varphi_{tot}|$ while leaning out reduces it. While the latter appears disadvantageous in steady-state cornering, the vehicle roll rate can be in-

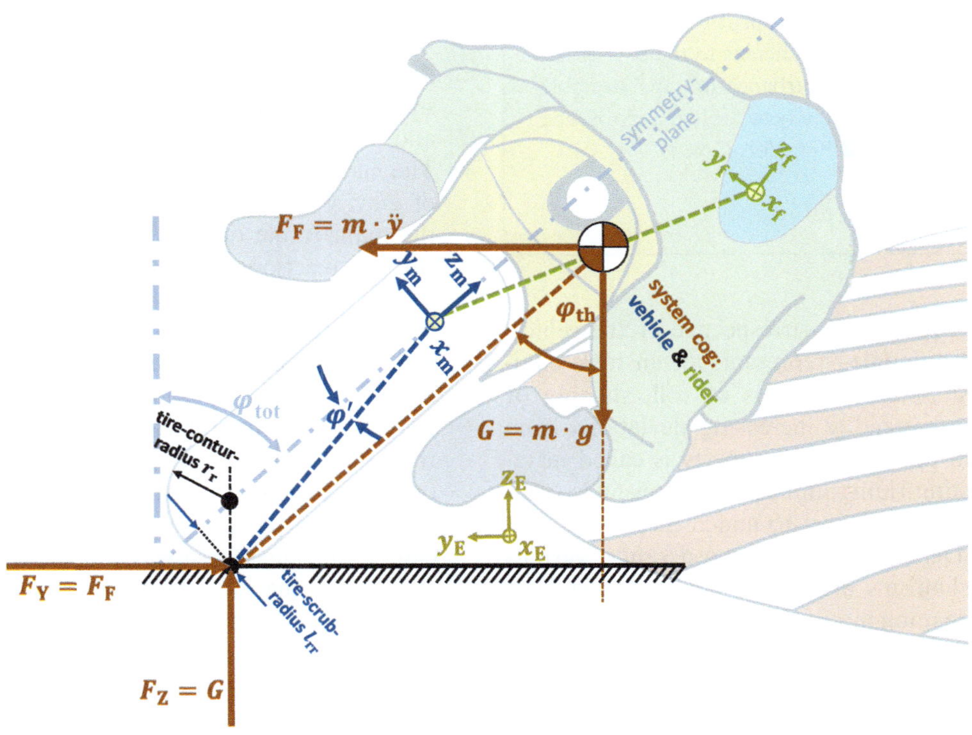

◘ **Fig. 35.1** Equilibrium of forces during constant cornering with tire-induced additional roll angle and rider offset

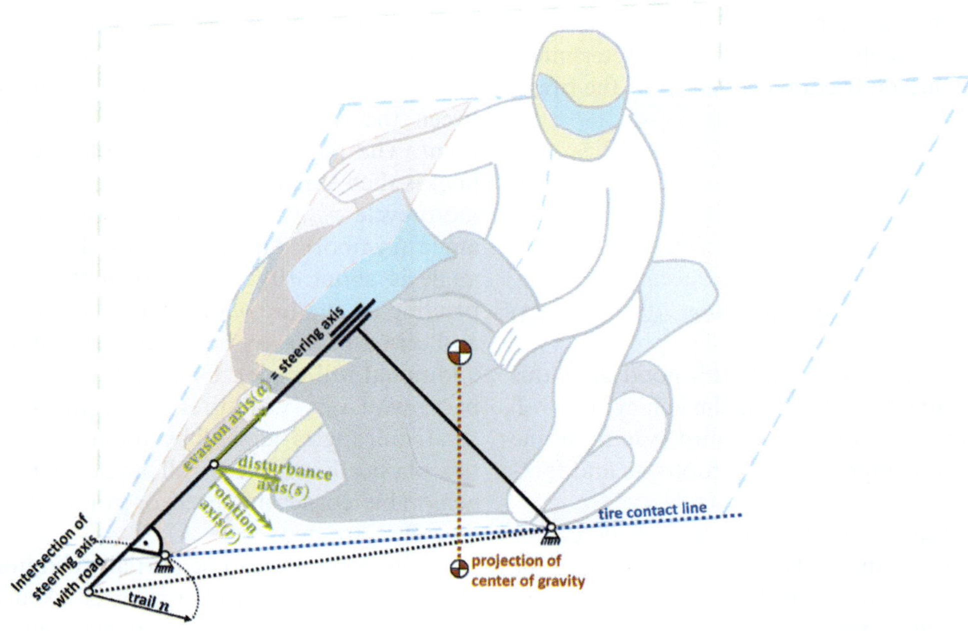

Fig. 35.2 Stabilization by steering angle, weight shift and gyroscopic effect on the front wheel

creased in transient states, which has a stabilizing effect due to the gyroscopic effect of the front wheel.

> Gyroscopic Effect: A body rotating about a principal axis tends to maintain the orientation of this axis of rotation (r). If the rotating body experiences a tilting motion about an axis perpendicular to the axis of rotation (s), the gyroscopic moment causes a turning motion about an axis perpendicular to the axis of rotation and the axis of disturbance (a). the steerable front wheel of a motorcycle behaves like a gyroscope turned on its side (See **Fig. 35.2**, Green).
>
> $$T_a = \Theta_{rr} \cdot \omega_r \cdot \omega_s$$

Every reached equilibrium—irrespective of the vehicle roll and rider lean angle—represents an unstable equilibrium. The smallest deflections of roll, steering or lean angles cause steadily increasing deviations from this equilibrium position; this property is called instability. By adjusting riding inputs, a new, unstable equilibrium can eventually be set to match the aimed vehicle trajectory. Depending on the speed range, various stabilizing mechanisms arise, that prevent or at least slow down the leaning motion:

— At speeds above around 30 km/h, the gyroscopic effect of the rotating front wheel makes a significant contribution to stabilization, as the leaning of the chassis causes the front wheel to turn in the same direction. If the motorcycle leans to the left, a leftward-steering gyroscopic moment is generated. As a result of the inward turning of the front wheel, first the lateral slip force and then the trajectory curvature increase. The centrifugal effect increases proportionally, causing a torque counteracting the leaning motion. Within sufficiently small roll angles, the system thus always strives towards the equilibrium position at $\varphi = 0$. Around this equilibrium, the motorcycle shows permanent oscillations with small amplitudes (no visible deflections of steering and roll angles).

— The mass distribution of the steering system (center of gravity in front of the steering axis), as well as of the main frame (deviation moments), can increase steering reactions resulting from the tilting motion, or the uplifting motion resulting at increasing yaw rates, respectively. This increases the stability of the motorcycle during cornering by preventing capsizing.

— By changing the vehicle geometry (in particular trail, steering head angle, wheelbase), the lever arms of the longitudinal, lateral and normal forces from the front wheel contact point to the steering axis can be influenced in a way that results in increased stability of the motorcycle. Such changes often result in a deterioration in general handling behaviour (▶ Sect. 35.1).

— At low speeds close to standstill, the effects of gyro stabilization, centrifugal effect, deviation torques, etc. are reduced to such an extent that only active riding inputs can stabilize the leaning motion. To achieve stability, the system center of gravity (red in ◻ Fig. 35.2) must always be positioned centrally above the tire contact line (dashed blue).

- In analogy to the inverse (double) pendulum, this balancing process always requires a lateral displacement of the base (here: the tire contact points with the road), since there is no other possibility to support forces between the motorcycle and the environment. The lateral tire displacement is primarily caused by the rolling motion of the steered front wheel. In addition to side slip forces, camber side forces also contribute to the wheels' displacement. Steering and leaning movements enable motorcyclists to control slip and camber angles independently. The sensorimotor skills to perform this complex balancing process are often learned in early childhood on a balance bike or bicycle, so that this balancing process happens unconsciously.

- When the vehicle is at a standstill, a rider can still achieve a lateral displacement of the tire contact point: firstly, due to the transverse motion of the contact point as a result of steering angles of the trailing front wheel; secondly due to the tire-scrub-radius as a result of roll angles. Model calculations show that, depending on the vehicle geometry and assuming an ideal controller, deflections of approx. 4° to 5° roll angles can be compensated and stabilized [41].

35.2.2 Maneuverability and Handling

The dynamic effects described above enable motorcyclists to perform the intended maneuvers. The ability to move the motorcycle along a target trajectory through specific riding inputs is here referred to as maneuverability. Complementary to this, handling refers to the rider workload to perform a particular maneuver. Therefore, a motorcycle has good maneuverability if it offers sufficient inherent stability even at low speeds or allows small turning circles, while it has good handling if the required operating forces (steering torque) are low and the vehicle behaviour is precise and neutral [2, 5, 6].

The execution of a maneuver can be broken down into various subsections. Using the example of a simple cornering maneuver, these are the initiation of the cornering maneuver, the constant cornering maneuver and the righting maneuver.

When initiating a turn, a steering torque directed outward of the curve will cause a side slip force at the front wheel. This results in a torque around the longitudinal axis of the motorcycle and, with it, a roll acceleration towards the desired equilibrium position. This initial leaning motion is supported by the gyroscopic ef-fect at the front wheel and the centrifugal effect at the center of gravity.

Leaning motions of the rider and passenger eventually also produce the desired motorcycle leaning but much more slowly. Since the inertia of the rider and passenger is supported by the motorcycle chassis, leaning towards the inside of the curve causes the motorcycle to tilt in the opposite direction (combined with gyroscopically induced steering). This process depends on the rider-vehicle weight ratio and is increasingly damped at higher velocities. The resulting camber (and side slip) lateral forces of both tires cause the tire contact line to strive towards the outside of the turn, while the inert center of gravity maintains its direction of motion. This creates a lateral offset between the system center of gravity and the tire contact line, which generates the required rolling moment toward the inside of the curve.

As the roll angle increases, camber side forces build up between the tire and the road. The handlebar, which is initially directed towards the outside of the turn, is increasingly rotating in the direction of the turn in order to slow down the rolling motion. This motion comes to an end when the overturning moment induced by the weight force is in equilibrium with the uprising moment induced by lateral wheel forces.

Depending on vehicle geometry, riding speed, etc., both slightly positive or negative steering angles are possible during steady state cornering. During the turn (with the rider and passenger in a neutral position) the contact point between the front tire and the road has a lever arm to the steering axis, so that the longitudinal, lateral and normal wheel forces cause a constant, inwards turning steering torque [31]. Motorcyclists must support this constant steering torque with a countering handlebar torque directed outwards of the turn in order to be able to maintain the reached equilibrium steering angle. Due to this out-turning handlebar torque as well as the steering motion facing in the opposite direction when initiating a turn, the term "counter-steering" is often used in connection with the steering of motorcycles. The magnitude of the in-turning steering torque takes on large values when braking at large roll angles. This is a consequence of the braking force acting on the steering axle via the tire-scrub-radius and the wheel normal force (which increases during braking) acting on the steering axle via the geometric trail. This effect is referred to as brake steering torque and is discussed in detail in [32, 33].

To end the steady state cornering, motorcyclists can bring the motorcycle to an upright position by generating a lifting torque around the roll axis. To do this for the majority of road going motorcycles, the applied counter-torque on the handlebars is reduced, automatically causing the handlebars to turn to the inside of the curve and build up side slip forces. This process can again be supported by active steering or leaning motions. Increasing the riding speed at the exit of the turn increases the centrifugal effect and thus additionally supports the straightening motion.

Compared to passenger cars, special considerations about the maneuverability of motorcycles have to be made:

- The trajectories of the tire contact points differ significantly from the trajectory of the vehicle's center of gravity
- The width of the motorcycle/ rider unit is not constant, but changes due to the roll angle when cornering
- Changing the direction of travel requires roll angles, which are achieved by steering inputs in the opposite direction
- Leaning motions and body tension influence steering behavior.

35.2.3 Dynamic Instabilities

In addition to the stabilizing effects described above, there exist motorcycle-specific dynamic instabilities. When discussing riding assistance systems, however, these have little relevance, which is why reference is made here only to specialist literature on the subject: [1, 6, 36, 37].

35.2.4 Special Requirements for Motorcycle Riding Assistance Systems

Many of the assistance systems used in motorcycles today are derived from systems widely used in the passenger car sector. However, the market launch of new systems for motorcycles happens with a delay of several years. This is the result of lower levels of investment available for such developments in the motorcycle market, and the special requirements that motorcycles place on assistance systems. These can result from, but are not limited to the vehicle geometry, tyre and suspension specification, the inherent instability of a motorcycle and the direct coupling between rider and motorcycle. ◘ Table 35.1 provides a brief overview of some special properties and their effects in the field of motorcycle assistance system development before individual assistance systems are discussed below.

35.3 Stabilization Assistance

In order to guarantee the riding stability of a motorcycle, it is necessary that forces can be transmitted between the tire and the road surface to a sufficient

extent. The so-called "Magic Formula" according to Pacejka [25] and Kamm's circle, which are shown in ◘ Fig. 35.3, are typically used to illustrate this force transmission. The former represents the dependence of a force variable on a slip variable (here: related tire circumferential force F_U and longitudinal slip s). The latter illustrates the maximum possible force that a tire can transmit at a given combination of longitudinal and lateral slip. Its diameter depends on the available coefficient of friction μ which can vary greatly depending on the surface conditions or tire parameters. In steady-state cornering with 40° roll angle and $\mu = 1$ the tire consumes about 84% of its lateral force potential. In this riding situation, just about half (54%) of its circumferential force potential is now still available for acceleration or braking. If however a higher circumferential force is demanded, e.g. by pulling the brake F_U, the tire cannot provide sufficient lateral force F_S. This will lead to the tire sliding towards the outside of the curve.

The assistance systems for vehicle stabilization are assigned to two main categories: the antilock braking system (ABS), which intervenes in braking situations in which the brake slip exceeds a critical value (3rd and 4th quadrant in ◘ Fig. 35.3.) and the traction control system (TCS), which in turn defuses critical ride situations (1st and 2nd quadrant) related to acceleration maneuvres. Basic ABS and TCS can only be used in the middle zone of the Kamm circle (green-yellow), while the curve-compatible systems described later are also effective in the lateral zones (yellow–red). Firstly, the basics of both systems for straight-ahead riding are explained below, before ▶ Sect. 35.3.4 deals with system extensions suitable for cornering.

35.3.1 Anti Lock Braking System

The main task of ABS is to prevent the wheels from locking so that the tire can continue to transmit sufficiently high side forces. A locking front wheel almost inevitably leads to a fall after a short time. This is avoidable by reducing the brake pressure. In addition, most modern ABS prevent the rear wheel from lifting off the ground. (▶ Sect. 35.3.5). The ABS is integrated into the hydraulic line between the brake pump and the brake caliper as shown in ◘ Fig. 35.4. It detects the speed of the wheels via wheel speed sensors and uses these signals to determine the vehicle reference speed (v_{ref}), which serves as the basis for further slip calculation. If the wheel speed (v_w) becomes too low, this is detected as the excessive brake slip (s_B) and the brake pressure is reduced via the brake pressure control. The brake slip results in

$$s_B = \frac{v_{ref} - v_w}{v_{ref}} \qquad (35.4)$$

Table 35.1   Special properties of motorcycles compared to passenger cars and their effect on motorcycle riding assistance systems

Property	Impact	Relevance for assistance
Low vehicle mass	Great influence of the masses of rider, passenger and payload on geometric parameters such as center of gravity position, trail, etc	eCBS, LC, WC, ▶ Sect. 35.3.3, wheel lift-off detection ▶ Sect. 35.3.5
Ratio between height of center of gravity and wheelbase	Lift-off of the tires is possible at reasonably low acceleration and deceleration values. This makes it difficult to calculate a reference speed, which provides the basis for slip control systems	eCBS, LC, WC, ▶ Sect. 35.3.3, wheel lift-off detection ▶ Sect. 35.3.5
Single track	No actuation of yaw moments (e.g. by wheel-selective braking → ESC) possible	Sliding Mitigation ▶ Sect. 35.3.6
Wear-dependent tire contour radius	Wheel speed signals do not directly indicate the riding speed, as they are dependent on the roll angle	Determining the reference velocity ▶ Sect. 35.3.5
Different contour radii front/rear	As the roll angle increases, the dynamic tire radius drops faster at the front wheel than at the rear wheel. This influences the calculation of reference speed and wheel slip	Cornering ABS and TCS ▶ Sect. 35.3.4
Strong coupling between longitudinal, transverse and vertical dynamics	Even slight braking decelerations can cause strong steering reactions when the vehicle is tilted	Cornering ABS and TCS ▶ Sect. 35.3.4
Riding with roll angle	Lowering the system center of gravity, lateral offset of the tire contact points, changing the dynamic rolling radius, changing the ideal brake force distribution, increasing the ground clearance requirements, and much more	Cornering ABS and TCS ▶ Sect. 35.3.4, Assistance with lateral guidance functions ▶ Sect. 35.4.4, Trajectory planning, environment detection and Connectivity ▶ Sect. 35.6
Low construction volume and total weight	Difficult to accommodate sensors, actuators, antennas, control units, etc. Strong protection of components against environmental influences is necessary	Sliding Mitigation ▶ Sect. 35.3.6, Connectivity ▶ Sect. 35.6
Direct rider coupling	Motorcycle users are highly exposed to vehicle vibrations, environmental effects, vehicle reactions, etc	ACC ▶ Sect. 35.4, AEB ▶ Sect. 35.4.2
	Decoupling, e.g. by brake boosters or even steering boosters, does not take place	eCBS, (Cornering) ABS ▶ Sect. 35.3.4
	The rider's seating position influences the motorcycle dynamics. Longitudinal and vertical vehicle movements are transmitted via the upper body to the steering	eCBS, LC, WC, ▶ Sect. 35.3.3, AEB ▶ Sect. 35.4.2
Aerodynamic interaction	The rider and passenger position and posture change the point of air attack and can lead to weight distribution changes at at high speeds	LC, WC ▶ Sect. 35.3.3 wheel lift-off detection ▶ Sect. 35.3.5
Manual wheel selective braking	Riders are free to choose how they distribute braking force between the front and rear wheels, with the exception of models fitted with integral braking systems	Cornering ABS ▶ Sect. 35.3.4, Slip prevention ▶ Sect. 35.3.6

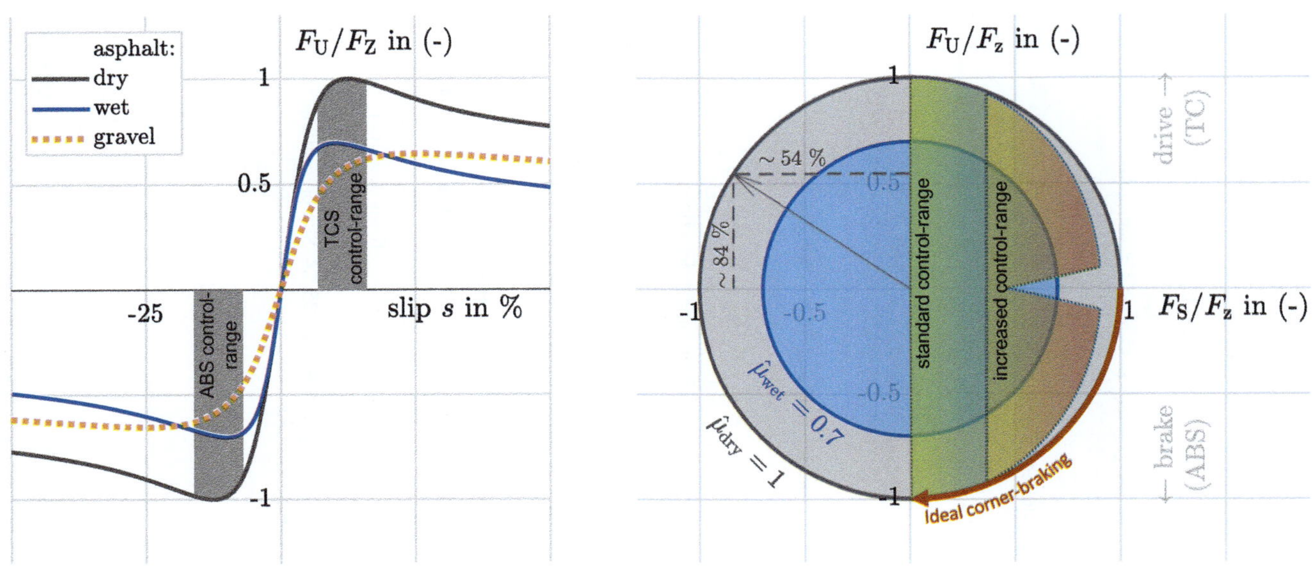

Fig. 35.3 Pacejka's magic formula and Kamm's circle for different adhesion limits and application ranges of the wheel control systems

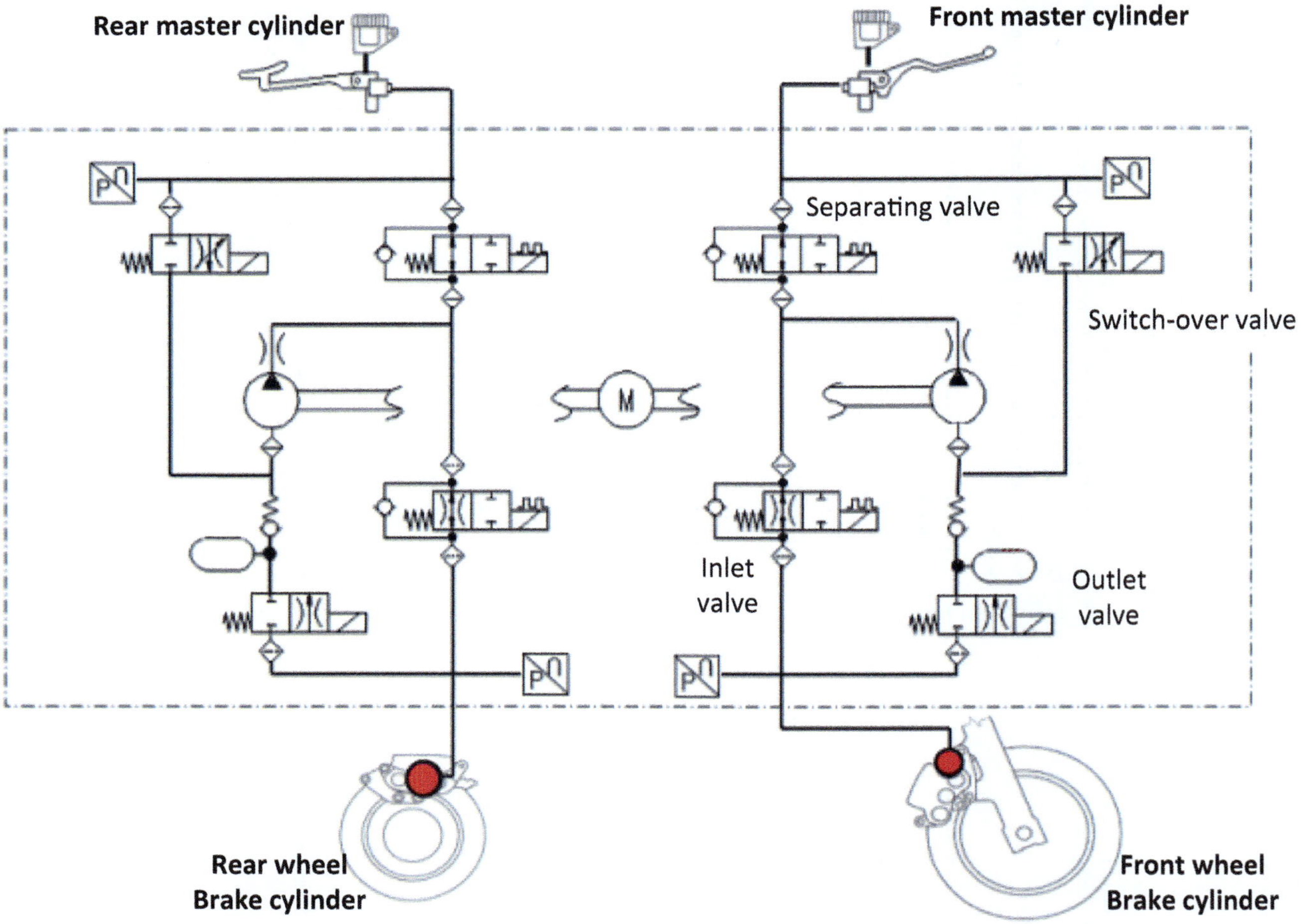

Fig. 35.4 Hydraulic circuit diagram of a motorcycle ABS

Once the target slip is reached again, the brake pressure is increased again to brake the vehicle further.

Determining the reference speed (v_{ref}) presents a particular challenge for two-wheeled vehicles, since only two wheels are available and at least one of them is subject to slippage in critical riding situations, while the other may be lifted at the same time. For example, in the case of a so-called stoppie, the rear wheel lifts off the ground during heavy braking. As a result, the rear wheel speed no longer corresponds to the vehicle speed. Only the remaining front wheel speed signal is available to both estimate the vehicle speed and the wheel slip at the same time. Since neither can be reliably determined in such a situation, the ABS uses other indicators in addition to slip, such as wheel deceleration, to detect wheel instabilities.

Today, dual-channel systems are established on the market that intervene on both the front and rear brakes. In the low-price segment, single-channel systems are also offered which only act on the front-wheel brake. The fitment of ABS on motorcycles is mandatory in the EU since 2016 for new vehicles with an engine capacity of 125 cc or more.

35.3.2 Traction Control System

Considering the high specific performance[2] of modern motorcycles, traction control is an important addition to the now established brake control systems. The primary aim of traction control systems is to prevent the rear wheel from spinning uncontrollably. This is needed on the one hand to support riders when accelerating, especially on roads with changing and reduced friction values. On the other hand, it helps maintaining the lateral force potential at the rear wheel. Especially when cornering, lateral slides with the risk of a 'highside' type accident[3] must be prevented. In most cases, traction control is implemented on the engine control unit. Today, this is typically networked via CAN-BUS with the system components shown in ◘ Fig. 35.5.

Depending on the manufacturer, the control algorithm of the traction control system is located on the engine control unit (ECU) or the ABS control unit. In the latter case, the time delay for detecting the riding condition is shorter because the ABS measures and processes the wheel speed sensors. However, more time is needed to set the torque desired by the traction control system. If in contrast the traction control system is implemented on the ECU, more time is required to detect the riding state, since the measured wheel speeds have to be sent from the ABS to the ECU. On the other hand, the ride torque is converted more quickly. In both cases, the reaction time of the overall system is approx. 50 … 160 ms. To determine the time and intensity of a control intervention, the control unit receives the signals from the wheel speed sensors. The current slip value is determined from the reference and rear wheel speed. For this purpose, vehicle-specific parameters—including characteristics of the wheel-tire pairings approved for the respective vehicle—are stored in the control unit. Since tires from different manufacturers differ slightly in terms of rolling radii and contours and are also subject to manufacturing tolerances and increasing wear in operation, deviations from the basic data are automatically adapted in defined riding conditions by comparing the wheel speeds. If the ride slip determined in this way exceeds an acceptable level for maintaining riding stability, the control system intervenes by reducing the engine torque. The aim is always to keep the slip within the stable range of the µ-slip curve (◘ Fig. 35.3, left). Different slip thresholds are stored for different speeds.

Due to the different tire dimensions on the front and rear wheels, the dynamic rolling radius of both wheels does not change in the same proportion when the vehicle is at an angle. As a result, the system detects "false slip" so that, even without lean angle sensors, a control intervention occurs earlier when accelerating out of curves than when riding straight ahead. However, a compromise must be made between safety and sporty design. For this reason, many motorcycles also offer different traction controller parametrizations, which can usually be selected via the instrument cluster. Depending on the situation and the rider's preference, it can intervene earlier to ensure maximum safety (e.g. rain mode) or later to achieve maximum performance without power reduction by the traction control (e.g. race mode). Depending on the system, the engine torque can be decreased by adjusting the ignition angle, interrupting the fuel supply or—if available—by the electrically controlled throttle valve.

As can be seen in ◘ Fig. 35.3, for example, significantly more drive slip is required on gravel than on asphalt to transmit the same power. The road-specific slip thresholds of a traction control system are therefore often not suitable for off-road use. For this reason, additional parametrizations are usually applied for off-road motorcycles, which take into account the special slip characteristics of loose surfaces such as sand and gravel by means of higher threshold values.

There also exist traction controls on the market with low functional requirements and low complexity. Some of these are even offered as retrofit solutions for racing purposes and are often limited to monitoring the rear wheel alone. Based on engine speed, gear,

---

2   Engine power/vehicle mass.

3   Lateral rollover of the motorcycle as a result of a sudden increase in the side slip force after reduction of the longitudinal tire slip at large vehicle slip angles.

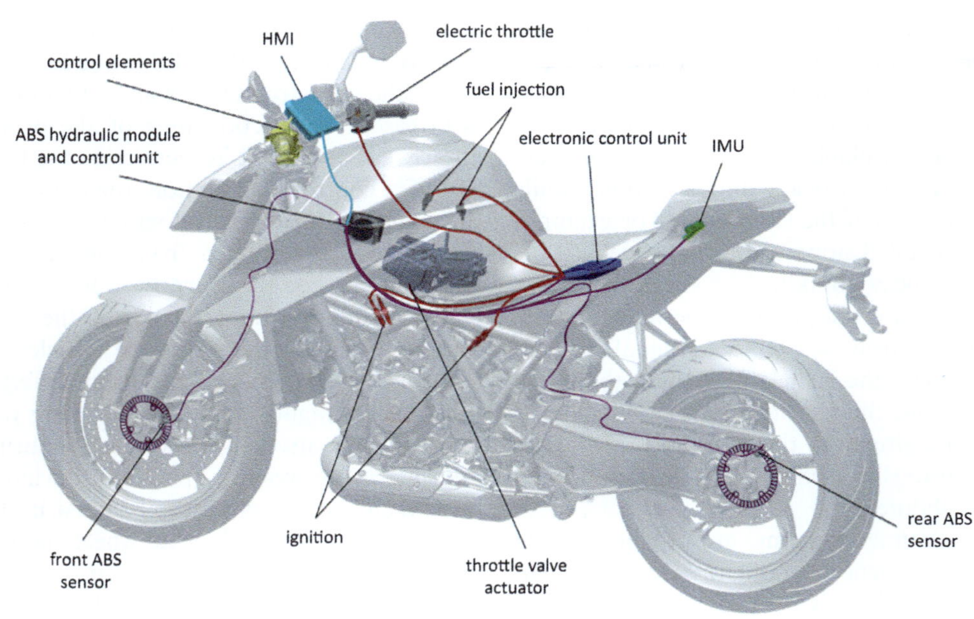

◘ Fig. 35.5   System overview ABS/TCS of a KTM 1290 SuperDuke

throttle position and engine speed increase, the stored algorithm detects an unnaturally strong increase in rear wheel speed and intervenes via the engine management system.

35.3.3 System Enhancements of Wheel Control Systems

35.3.3.1 Electronic Combined Braking System (eCBS)

In higher-quality braking systems, integral or partial integral braking systems are used which can actively build up braking pressure at both wheel brakes or only at the rear wheel. However, not all motorcycles that are technically capable of this function make use of it, as it can—among other things—restrict off-road capability. An automatic actuation of the brakes is therefore not always active but may only be utilized for certain functions, such as adaptive cruise control (cf. ▶ Sect. 35.4). The Bosch ABS 10. ◘ Figure 35.4 ME hydraulic module shown schematically in ◘ Fig. 35.4 consists of a total of eight solenoid valves, four of which are used for the front wheel brake circuit and four for the rear wheel brake circuit. In addition, two pressure sensors, a low-pressure accumulator and a hydraulic pump are installed for each brake circuit. A separate control unit (ECU—Electronic Control Unit) takes over the control task. The two pumps are powered jointly by one electric motor. When the rider actuates the handbrake lever, the pressure is transmitted hydraulically to the front brake caliper; at the same time, a sensor measures the pressure increase and transmits the information to the ECU. The pump motor is controlled according to predefined look-up-tables, operating conditions or other parameters. To actively build up pressure at the rear wheel, the separating valve is closed and the change-over valve is opened. This allows the pump to move the brake fluid from the reservoir to the rear caliper and build up pressure. If riders simultaneously operate the brake pedal with their foot, the change-over valve is closed again when wheel brake pressure is reached and the separating valve is opened again, so that riders once again have direct access from the brake pedal to the rear brake. In systems without integral brake function, the cut-off and changeover valves are omitted, so that only two valves are required per wheel and the pump is only used to return brake fluid to the reservoir.

35.3.3.2 Engine Drag Torque Control (MSR + PreMSR)

On modern motorcycles that have an electrically controlled throttle ("ride-by-wire"), the MSR function can be used to stabilize the vehicle even in situations in which the throttle grip is fully closed by the rider. With the help of the wheel speed sensors, the system detects excessive brake slip that can occur due to the engine drag torque or immediately after downshifting. In response, the engine torque is increased to reduce rear wheel slip to a non-critical level. The main advantage of an MSR is in the low-friction range, where an anti-hopping clutch[4] can't reliably disengage. In vehicles with an inertial measurement unit ("IMU", ▶ Sect. 35.3.4), this function can also prevent the rear wheel from lateral slides which are the result of excessive engine drag torques at large roll angles. This is achieved by actively increasing the engine torque when shifting to a lower gear.

35.3.3.3 LaunchControl (LC)

The launch-control allows maximum acceleration from a standstill. It helps riders to keep the engine speed at a level optimized for rapid acceleration, so that they only have to concentrate on releasing the clutch. Once the vehicle has launched, the system limits the maximum acceleration to that at which the front wheel just maintains contact with the ground. This ensures that the vehicle both remains stable and achieves the maximum physical acceleration. If the throttle is closed, the vehicle is shifted into a certain gear or a certain lean angle is reached, the LC is deactivated and the regular traction control takes over.

35.3.3.4 Slip Adjuster (SA)

In order to take the different road surfaces and tires used on the racetrack into account, many sport motorcycles are equipped with a slip adjuster. This allows riders to adjust the target slip in small increments with the simple press of a button. By this, riders can react to decreasing grip due to worn tires during a race or track riding sessions. Also, hard or soft rubber compounds can have a different target slip in order to be able to transmit maximum power.

Slip adjusters are also available for off-road use, allowing the traction control to be adapted to the wide range of possible surfaces such as scree or sand. By adding an IMU, a slip adjuster can be used to implement vehicle slip angle control (adjustment of the side slip at the rear wheel).

35.3.3.5 Wheelie Control (WC)

Motorcycles equipped with an IMU and traction control can also employ WC. The IMU is used to determine the current pitch angle and to control the acceleration to reach a set target angle. For this purpose, a wheelie must be reliably detected to prevent an incline or pillion operation from being interpreted as a pitch angle (▶ Sect. 35.3.5). The function ultimately serves less to ensure riding stability (keep both tires in contact with the ground) but to increase acceleration.

35.3.3.6 Hill Hold Control (HHC)

Hill Hold Control supports riders when starting uphill. This function actively builds up the brake pressure in the rear brake caliper to prevent the vehicle from rolling backwards. Due to the necessary active pressure build-up, HHC can only be implemented in conjunction with (partial) integral ABS. If an engine torque is detected that is sufficient for starting on the current incline, the brake is released. This allows the rider to concentrate on starting off. The system can also be used for the reverse case, holding a vehicle on a declining road. This type of system can also be used as a 'hand brake' function in some systems, being applied and released by the rider via a specific application of brake pressure when stationary and the vehicle is running.

35.3.4 Sensors and Control of Cornering ABS/TC

The central requirement for slip control in curves is to maintain the lateral force reserves while simultaneously aiming for maximum deceleration (cornering ABS) or acceleration (cornering TCS). Particularly at low friction values, physical limits are quickly reached. With increasing lean angle, the tires' transmissible circumferential force decreases, since an ever-greater part of the tires' total transmissible force is already used up by the side force, as the example in ◘ Fig. 35.3. shows on the Kamm's circle. For this reason, the current lean angle must be determined during cornering, e.g. in order to adapt the control strategy of the ABS to the special features of corner braking (▶ Sect. 35.3.4). Roll-dependent slip thresholds enable more sensitive control including consideration of the false slip mentioned in (▶ Sect. 35.3.2).

There are various inertial measurement units (IMUs) on the market that use acceleration and angular rate sensors to measure the motion of the vehicle. IMUs are offered in various forms. In the entry-level segment, IMUs with three sensors are used (two acceleration (x,y) and one angular rate sensor (z)). Typically, only the roll angle is determined with these sensors. However, for a robust calculation of the pitch or roll angle in the dynamic range, an angular velocity signal is required for each direction of motion. For this reason, sensors that measure all six degrees of freedom of the motorcycle frame are used in the mid- to high-price segment. They have three acceleration and three rotation rate sensors [28]. Highly dynamic controls such as wheelie control are only possible with such 6D IMUs. The IMU itself provides measured values that are referenced to the sensor coordinate system. The conversion to the vehicle coordinate system, e.g. for recording the theoretical roll angle, is usually performed in the ABS control unit [17].

---

4 Mechanical system for limiting the torque that can be transmitted by the clutch during engine overrun.

Due to the coupling of steer and roll dynamics, it is important not only to maintain the adhesion reserves but also to control the inwards-turning brake-steer-torque (◘ Fig. 35.2). This must always be countered by the rider's handlebar torque in order to maintain the target trajectory. If the adhesion limit is exceeded, the slip angle at the front wheel increases, causing the yaw rate of the vehicle as well as the trajectory curvature to decrease. At the onset of an incipient wheel lockup and the resulting control action, the brake force on the front wheel is reduced. Motorcyclists are usually not able to reduce their handlebar torque acting to the outside of the curve at the same speed as the braking force is reduced. This results in a steering angle that is directed towards the outside of the turn. When the control action is completed, the maximum brake force is again applied to the front wheel, and with it the strongly inwards-turning steer torque builds up again. Motorcyclists have typically reduced their handlebar torque at this point, so the handlebars turn inwards again. In repetition of these effects, the handlebar begins to oscillate. If the amplitude of this oscillation is sufficient, a rapid reduction of the roll angle (and associated large roll rates) is observed. This is followed by a yaw and roll oscillation of the vehicle that happens throughout the entire further course of braking. In real road traffic, this could result in the vehicle leaving its lane. The steer and roll angle oscillations are therefore caused by the brake-steer-torque in combination with the kinematic instability of the vehicle and the control of the trajectory by the rider.

The brake pressure gradients and the maximum brake pressure level are limited as a function of the roll angle at the front and rear wheels. On the one hand, this prevents dynamic front wheel over-braking, which is particularly critical when cornering. On the other hand, it helps to control brake pitching and to give the rider a little more time to compensate for the brake-steer-torque, especially right at the start of braking. One to four pressure sensors are installed in the better-equipped ABS variants for this purpose. If only one pressure sensor is used, only the brake pressure at the front brake caliper is monitored. With four pressure sensors, the pressure at both the master and slave cylinders for the front and rear brakes can be measured. This means that the pressure gradient can be taken into account and the pressure increase at the brake can be better limited depending on the situation.

The control of ABS and TCS suitable for cornering requires that the demanded circumferential forces of both tires (due to throttle and brake commands) always remain small enough, so that a force reserve to the edge of the Kamm's circle is maintained. This ensures that the lateral force potential is never exceeded, thus preventing falls caused by lateral slides. The reserve is necessary, for example, to counteract minor fluctuations in the coefficient of friction. Therefore, the theoretically possible ideal deceleration during cornering (red arrow in ◘ Fig. 35.3) is not achieved in reality. With regard to real road use, however, accident data shows that the gain in stability during braking with cornering-ABS is of higher importance than the associated loss of maximum deceleration. Practical experience from development shows that the uplifting motion during braking in turns with cornering-ABS matches the deceleration very well. Thanks to the improved stability, riders can even achieve higher mean decelerations during emergency braking at a high lean angle than with conventional braking systems.

35.3.5 Wheel Lift-Off Detection

In contrast to passenger cars, the brake and engine performance of motorcycles, combined with their short wheelbase and high center of gravity, further limits acceleration. Furthermore, the center of gravity will vary considerably in height due to the load and the mass of the rider and any passengers. The front to rear weight distribution changes with luggage, which often sits directly above the rear wheel. A variation of the system center of gravity height by ± 10 mm resulted in a braking distance variation of $\pm 0,5$ m ($\pm 1,3\%$, according to measurements with KTM 790 DUKE from 100 km/h initial speed).

Braking with a sufficiently high coefficient of friction can cause the rear wheel to lift off the ground, which leads to a rollover if the brake pressure is not reduced. ABS can prevent a rollover even by just measuring the wheel speeds while braking. When the braked rear wheel lifts off, its speed decreases significantly more compared to the vehicle's reference speed, or does not follow the vehicle's deceleration—depending on whether the rear wheel is braked as well or not. This effect is used to detect a lifting rear wheel. In the case of curve **A** in ◘ Fig. 35.6. the rear wheel brake was applied. As the speed of the rear wheel falls far below the reference speed, it is concluded that the rear wheel is losing grip and is considered as over-braked. If in contrast only the front wheel brake is used for deceleration, curve **B** shows that the rear wheel maintains faster than the reference speed and doesn't follow the vehicle's deceleration. It can be concluded again that the rear wheel is no longer in contact with the road. A lifting rear wheel can usually be detected better if its brake is applied (curve **A**), which is an advantage of vehicles with eCBS.

The latest generation of ABS also use information from the IMU and can thus deliver a better compromise of deceleration and stability. Inaccurate lift-off detection can sometimes lead to very large pitch angles close to a possible brake-rollover, tempting riders to re-

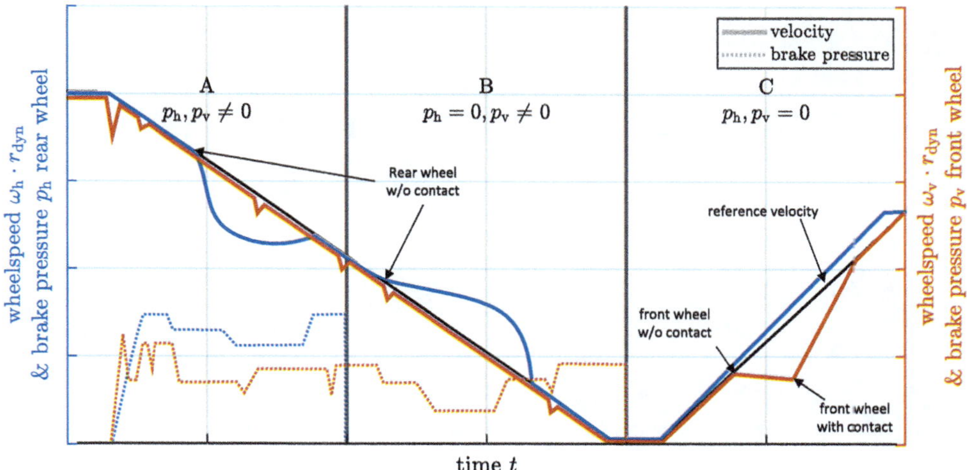

Fig. 35.6 Detection of lifting wheels from their speed signals

lease the brakes prematurely. In emergency situations where riders are not prepared for braking, this can lead to a significantly longer braking distance compared to an ABS that is optimized for riding stability instead of maximum performance. In turn, front wheel lift ('wheelies') can occur during acceleration. With the aid of the wheel speed sensors, the lifting front wheel can also be detected, as the lifted, non-driven front wheel inevitably slows down compared to the rear wheel. With the help of the IMU, a moderate wheelie can be tolerated to avoid constant lifting and lowering of the front wheel and to ensure smooth acceleration. In some cases, the wheelie is even intended ▶ Sect. 35.3.3. But even with simple traction controls that do not have IMU-based wheelie detection, a control is triggered by the front wheel slowing down (red line, course **C** in ■ Fig. 35.6) because the system recognizes this as rear wheel slip and reduces the engine torque.

Similar to the brake pressure gradient, vehicle-specific parameters are also set with regard to the wheelie in order to prevent excessive torque increases and to avoid a sudden and too fast wheelie.

35.3.6 Mitigating Lateral Sliding

The lean angle-dependent wheel control systems described above can prevent over-braking of both wheels or excessive drive-slip at the rear wheel, even at high lean angles, within the limits of physics. This significantly reduces the risk of crashes when cornering.

However, unexpected friction value jumps due to gravel, crushed stones or spilled oil that occur over longer distances during cornering can become critical. Such low friction areas lead to vast lateral slides of the front wheel and, consequently, often to a fall. In a research project of Robert Bosch GmbH, a prototype system was developed that detects these riding situations and uses a recoil actuator (gas pressure accumulator) to provide the missing lateral force [38].

As in the case of passenger cars, essential variables for detecting incipient skidding are the wheels' slip angles. These are calculated by means of a rigid transformation to the wheel contact points from the vehicle slip angle, which is estimated based on a 6D IMU. For this purpose, the existing roll and pitch angle estimation of the MSC algorithm [19] is used, plus an additional vehicle slip angle estimation to improve detection. A trigger criterion combining thresholds for the slip angles, their temporal changes and the roll angle, detects incipient slippage and triggers the intervention by opening the gas accumulator.

The practical functional verification was performed by riding over a one-meter-wide gravel patch with a roll angle of 40° and 50 km/h. The tests were carried out with active MSC. The lateral slide is provoked by a short application of the front brake at the gravel patch. The maneuver without intervention (top of ■ Fig. 35.7), clearly shows how the front wheel collapses and the roll angle increases. Without a support frame, this maneuver would have clearly resulted in a fall. With active system intervention (Fig. below) on the other hand, the missing side force is compensated by the recoil force of the actuator and cornering can be continued with the same roll angle. Tests without a support frame have also confirmed the functional effectiveness. For possible implementation in production vehicles, further technical challenges such as the given weight and size restrictions, durability, crash safety and exclusion of hazards from the actuator must be solved.

Fig. 35.7 BOSCH sliding mitigation

35.4 Assistance Systems at Path Guidance Level

35.4.1 Adaptive Cruise Control (ACC)

Adaptive cruise control (ACC, also known as Active Cruise Control at BMW Motorrad) is a functional extension of the cruise control system, which is considered state of the art. It controls the distance to the vehicle in front by regulating the engine and brake torques (▶ Chap. 32). The function was first announced for motorcycles for model year 2021 by the manufacturers BMW Motorrad, Ducati and KTM and was developed together with Robert Bosch GmbH.

The ACC function covers a set speed range from 30 to 160 km/h and detects ambient information using a mid-range radar (MRR). Control is performed on moving as well as stationary objects.[5] ACC exclusively takes over the longitudinal control of the motorcycle and can be overridden by riders at any time, by accelerating or braking. Braking by the rider leads to deactivation of the system.

From the rider's point of view, ACC on a motorcycle behaves very similarly to ACC in a passenger car that is based exclusively on radar and uses manual transmission. However, there are characteristics of the motorcycle that have a direct impact on the longitudinal controller and especially on the evaluation of the environment data acquisition. Due to the low system mass, high engine drag torque and high drag coefficient compared to passenger cars, motorcycles are subject to higher deceleration values even without the use of the brakes. It follows that in ACC mode on motorcycles—assuming the same quality of object detection—deceleration by braking is required less frequently. This makes for a more comfortable riding experience for riders, since the need to apply the brakes even for small target decelerations is eliminated. Since current motorcycle integral braking systems are optimized for setting larger decelerations, they do not allow sufficiently precise control of the brake pressure at small pressure levels. At the same time, even low brake pressures are clearly noticeable to motorcyclists. The brakes are applied as soon as the other means of deceleration (engine drag torque and riding resistance) are not sufficient to resolve the existing situation. The pressure build-up rate of the brake as well as the maximum value are limited so that braking remains in the comfort range. AEB is employed in an emergency brake situation (▶ Sect. 35.4.2).

Regarding the HMI concept, the differences to the implementation in the passenger car are minor. Although the design options differ due to the smaller space available for accommodating a control, existing control elements of the cruise control can be used and expanded to include a distance setting. In the BMW Motorrad example (■ Fig. 35.8), a button has been introduced in the direct vicinity of the cruise control operation, which is used to cycle through the distance settings (■ Fig. 35.10d). The status of the function is displayed in the cockpit of the vehicle in the same way as in passenger cars (■ Fig. 35.10c).

Another feature of the motorcycle is the quick shifter function, which enables manual shifting with-

---

5   Stopped objects are objects that were detected as moving by the environment sensor system in the past. In comparison, stationary objects were not detected as moving in the past.

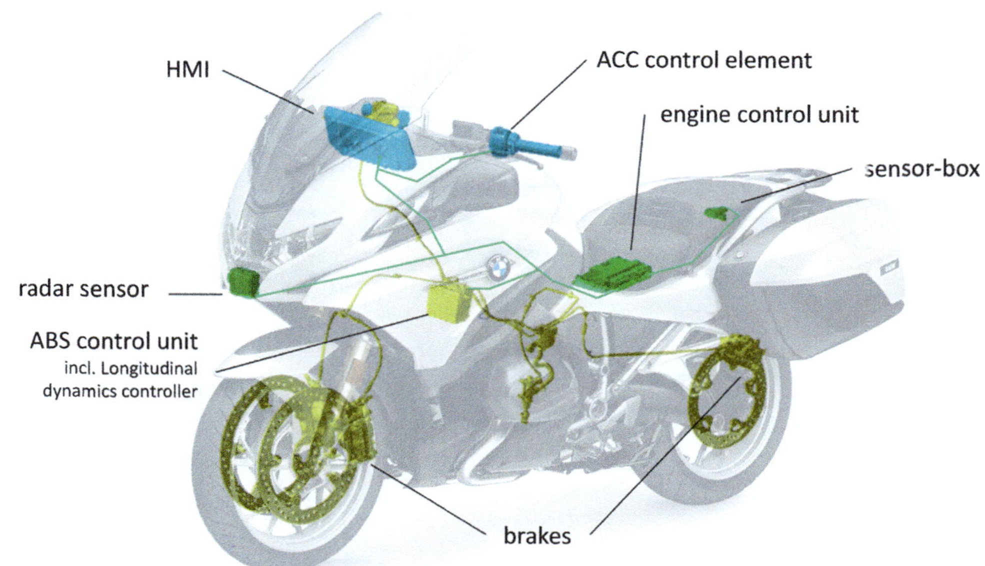

◻ Fig. 35.8 System components of the motorcycle ACC of a BMW R1250 RT

out using the clutch. The "Shift Assistant Pro" from BMW Motorrad allows upshifts as well as downshifts. Unlike conventional cruise control systems, situations that necessitate a gear change occur more frequently in ACC operation. Thus, there is a need to perform gear changes with the "shift assistant" function on the one hand and to enable manual gear changes using the clutch on the other. The interaction between the ACC function and the "shift assistant" function is very small, since the engine torque control takes place in the engine control unit and the torque requested by the ACC function can be treated in the same way as the torque request of the rider. In the case of manual shifting using the clutch, the engine torque requested by the ACC function must be suppressed as soon as the clutch actuation is detected. After clutch engagement, the converted engine torque is matched to the requested torque of the ACC function. The resumption of the engine torque must be limited in gradient so as not to generate uncomfortable longitudinal jerk and at the same time be sufficiently dynamic so as not to give riders the impression that the system is switched off.

Further differences to passenger cars arise in the integration of the radar component into the overall vehicle. In passenger cars, a radar component is usually integrated into the front apron. This results in a distance to the road of less than one meter. The lowest point at which a radar component can be integrated into a motorcycle that is also body-suspended is above the tire at maximum front wheel deflection.

Furthermore, the installation of a component in the motorcycle places high demands on the hardware. The component must be qualified for more demanding vibration profiles compared to passenger car applications due to the lack of vibration decoupling between the engine and chassis. Since the possible locations for a radar integration in the front area of the motorcycle are limited and typically occupied by other control units (e.g. the instrument cluster), challenges can arise regarding the electromagnetic compatibility.

In contrast to the fender-mounted integration common in passenger cars, naked bikes and motorcycles with handlebar-mounted front fairings only allow an integration of the radar component on the steering assembly (e.g. BMW R18 Transcontinental).

The rotational degree of freedom resulting from the handlebar-mounted installation does allow movements of the radar sensor of approx. $\pm 30°$ around the yaw axis, but in real riding conditions at speeds above 30km/h in steady-state cornering, significantly smaller steering angles result. The initial steering angles when initiating a turn (▶ Sect. 35.2.2) also show only small amplitudes. The angular error in azimuth (angular measurement in a spherical co-ordinate system) is considered small, as it is below 1° which results in less than 2m in Y-direction at a distance of 100 m in front of the motorcycle. In addition, this value is oscillating around 0° and on average equal to zero. The initial detection distance is almost identical in the comparison to both integration variants. Differences arise in the domain of object stability.

The rolling dynamics of single-track vehicles lead to further interactions in object detection. In the motorcycle, the curve capability (▶ Chap. 32) depends not only on the azimuth opening angle and the corresponding range, but—with increasing roll angles—on the ele-

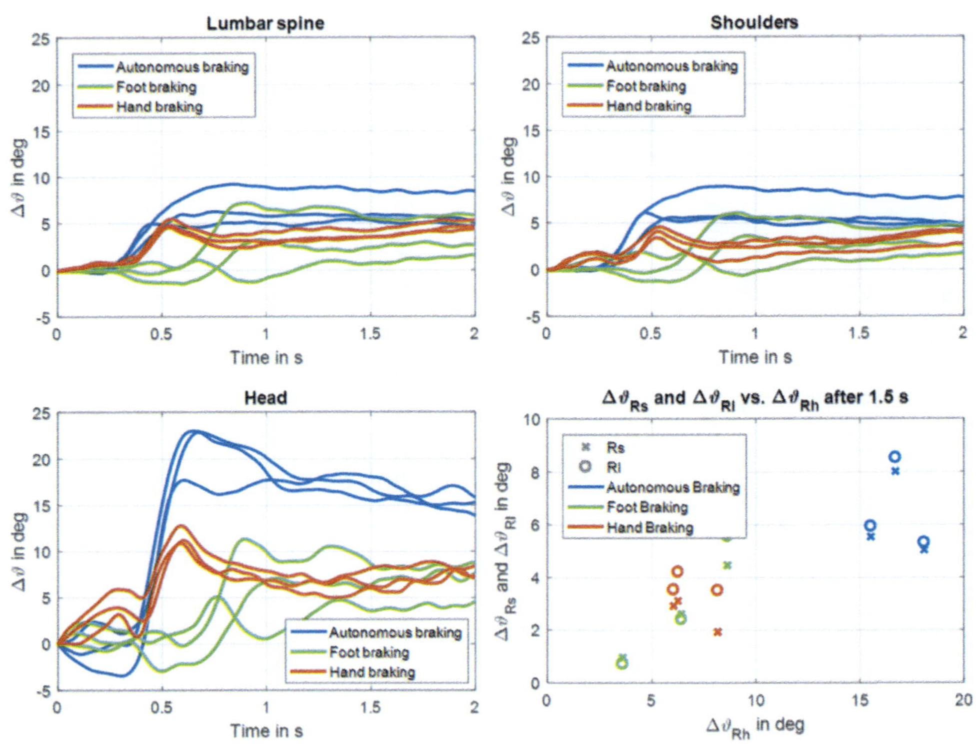

Fig. 35.9 Timeseries of different body difference-angles $\Delta\vartheta$ in the sagittal plane (Rh—Head, Rs—Shoulders, Rl—Lumbar Spine) during autonomous (blue) and manually applied brakes (read—front wheel braking, green—pure rear wheel braking)

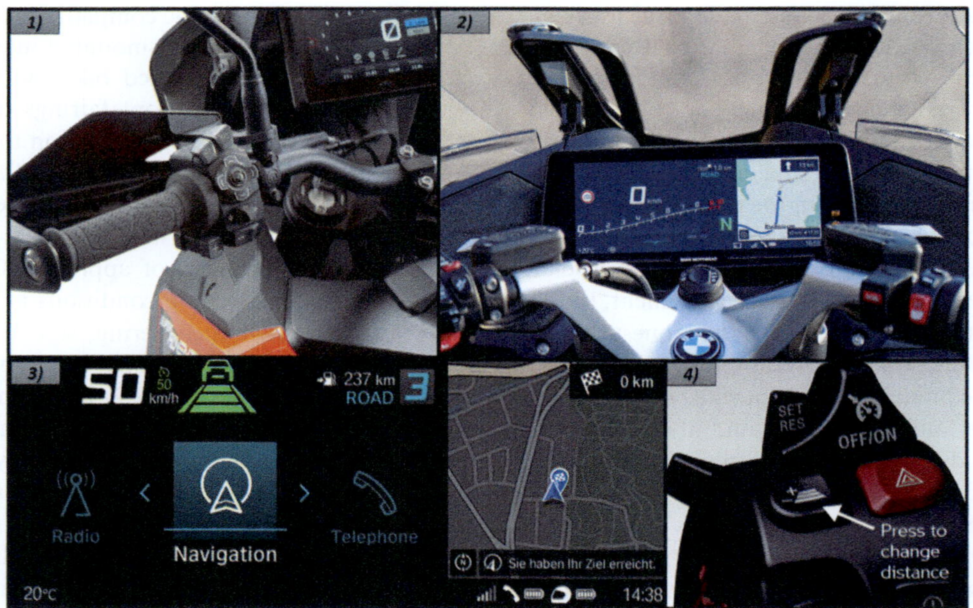

Fig. 35.10 HMI concepts of modern motorcycles. (1) KTM SuperAdventure with 6-way switchcube and additional pedal buttons (2) BMW R1250RT with split-screen, keyless-go and audio system (3) BOSCH motorcycle HMI; the left split-screen shows the board-menu and system information, the right split-screen navigation and infotainment (4) Controls of the ACC function of BMW Motorrad

vation opening angle as well. According to the comparison tables in ▶ Chap. 34 (Comparison of radar components), the elevation aperture angle is approx. 13° (MRR) with an azimuth aperture angle of approx. 12° in the far range and 60° in the near range. Due to the same magnitude of the aperture angles in the far range, there are no negative influences of the roll angle on the effective azimuth aperture angle. However, the smaller the elevation aperture angle becomes compared to the azimuth aperture angle, the more pronounced is the reduction of the effective azimuth aperture angle with roll angle in the direction of the curve. Thus, especially with tight curves and high roll angles at low speeds, target losses occur that would not occur with two-track vehicles. A curve controller function can be used to counteract this system weakness. Limiting the roll angle in ACC mode by reducing the vehicle speed, limits the negative effect of the roll angle on the field of view (FOV). At the same time, the function ensures that in the event of a target loss, the vehicle does not immediately accelerate to the set the target speed.

In passenger cars, the course curvature can be calculated and fused via the steering angle, the yaw rate, the lateral acceleration, and the wheel speed differences (▶ Chap. 34). When used in a single-track vehicle, an estimation of the current curvature can only be drawn via the yaw rate and the roll angle. It should be noted here that the IMU is installed in a body-fixed manner and the influence of the roll angle is taken into account. When riding straight ahead, the self-stabilization due to the steering-roll coupling (▶ Sect. 35.2.1) leads to small oscillations around the roll angle zero position, which results in a measured lateral acceleration and yaw rate. These influences must be taken into account when calculating the course curvature. The use of a steering angle sensor to determine the course curvature on a single-track vehicle is conceivable, but not state of the art at the time of printing of this edition.

35.4.2 Autonomous Emergency Brake (AEB)

The abovementioned ACC is developed as a comfort function, to follow a leading vehicle in a smooth matter. The used radar sensors and integral braking systems however also provide the necessary building blocks to map an autonomous emergency braking function. In addition to investigating the protective potential and the necessary triggering algorithms of motorcycle AEB, ongoing research and development activities are focusing in particular on their controllability and customer acceptance of such systems.

We assume that the maximum permissible decelerations that are not generated by riders themselves may only be set if they are able to stabilize the vehicle during the maneuver. This is because motorcyclists will only be able to support their own inertia and that of any passenger via upper arms and abdominal muscles if they are under high (body) tension and concentrating on the emergency braking situation. However, this is only possible if both hands are on the handlebars—which complies with public regulations, but does not always coincide with real practice ("expected misuse" ▶ Sect. 35.7). In a situation requiring emergency brake intervention, it must be assumed that the readiness to brake is not initially given. Thus, before a maximum deceleration can be set, a transition to a ready-to-brake state must be achieved.

Three key questions arise from this:

1. How is it possible to ensure that a motorcyclist transitions to a ready-to-brake state as quickly as possible?
2. What decelerations are already acceptable during this transition phase while ensuring riding stability and user acceptance?
3. How is it possible to detect the end of the transition phase, from which a deceleration with the maximum allowed value can be set?

To investigate these questions, participant studies were conducted for the first time starting in 2018, in which unprepared participants were placed in an artificial emergency braking situation. In the trials, the emergency braking was triggered by means of a remote control, while at the same time a deceleration of the dummy target EVITA (▶ Chap. 12) generated a plausibility of the emergency braking scenario for the participants [22, 23, 24].

The transition to the ready-to-brake state was initiated by partial braking, as it is also common in the passenger car sector. This allows a reduction in speed even before the transition phase is completed. Both the maximum deceleration intensity (0–7 m/s^2) and the character of the partial braking (impulse braking, braking ramp, continuous braking) were varied. The test results prove that partial braking even up to 5 m/s^2 deceleration is suitable in a motorcycle to initiate a transition to a ready-to-brake state. In addition, the controllability and tolerability of deceleration values far above the maximum applied in motorcycle ACC was verified.

◼ Figure 35.9 shows body-measurements from a rider on an autonomously braking motorcycle while performing a braking maneuver on a closed track. It can be seen that only small angular changes can be observed at the back (top left plot) and at the shoulder level (top right plot). Moreover, they build up slowly when only the foot brake is applied. Especially in the pitching movement of the head $\Delta\vartheta_{Rh}$ (bottom left plot), large dynamic deflections can be observed. The data allows a clear clustering between autonomous and manually triggered braking. Thus, the measurement of the upper body movement appears to be a suitable means to evaluate the rider state with respect to the above-mentioned brake readiness. However, subsequent studies show that shoulder movement is more suitable as an indicator compared to head movement due to its reproducibility [24].

With the design recommendations determined, a subsequent participant study identified states that indicate the subjects' readiness to brake. In particular, the pitch rates of the upper body and head showed characteristics that allow a distinction to be made between states in which a rider is ready to fully brake and states where he is not.

The subjective acceptance of an AEB was investigated in a complementary simulator study [21]. For this purpose, subjects were confronted with an emergency braking situation while performing an operating task (e.g., navigation system) or riding with no hands. It was shown that the acceptance of the system benefits from a warning preceding the braking intervention.

35.4.3 Side View Assist (SVA)

Back in 2015, BMW Motorrad was the first manufacturer to equip a two-wheeler with Side View Assist (SVA). The environment detection is based on the ultrasound technology established in passenger cars, which is usually used in parking scenarios as a distance warning, but also in some models as Side View Assist. The range of use is limited due to a range of only approx. 5m at relative speeds of up to 10km/h in the speed range from 25 up to 80km/h is designed for use in urban environments.

Greater ranges and relative speeds are made possible by the use of radar technology. The first motorcycle models such as the Ducati Multistrada V4 will be equipped with mid-range radar sensors from Bosch from the model year 2021. Due to the limited installation space available, only one radar will be installed centrally in the rear. To implement an SVA, aperture angles of approx. 150° are required. As a result, the antenna pattern and signal processing chain differ significantly from functions acting in the direction of travel, such as ACC or AEB.

Similar to passenger cars, LEDs are on or in the mirror signal when a vehicle enters the blind spot. If a lane change is initiated in this situation by activating the direction indicator, a warning is issued on the respective side. Such systems only find user acceptance if there is a very low false trigger rate. For this reason, two-wheel-specific characteristics such as the change in lean angle during lane changes were taken into account in the algorithms at an early stage of development and adjusted via endurance tests.

Initial independent tests show that Side View Assist aids riders with all-round visibility on a motorcycle [30].

35.4.4 The Future of Active Longitudinal and Lateral Assistance

The use of rider assistance systems in motorcycles has always followed the trend of adopting passenger car technology within varying degrees of time lag. For an ACC function, this would mean extending the functionality to include route control—the automatic adaptation of speed to traffic signs, curves or roundabouts—and/or extending the speed range to include stop-and-go operation. Due to decreasing self-stabilization with decreasing vehicle speed, the implementation of near-stop ACC control is not feasible without applying additional artificial stabilizing technologies. In prototypes and small series, this is made possible, for example, by the use of gyroscopes [15] or the combination of steer-by-wire and electrically actuated front wheel [11]. The use of a camera or a lidar system would enable an extension of the functional scope of the rider assistance systems in both lateral and longitudinal directions. In order to implement assistance functions for lateral guidance, a steering angle sensor and steering actuator are also required.

The route control already established in passenger cars as part of the ACC function (▶ Sect. 32.10.3) can also be implemented in motorcycles. This is based on the assumption that a target trajectory with a steady, continuous curvature can be defined based on the given lane's course at every point. The assumption is easily confirmed for passenger cars, where the lane width corresponds closely to the vehicle width. Due to the smaller ratio of vehicle width to lane width, a motorcycle can travel with a larger lateral deviation from the center of the lane without leaving it compared to a passenger car. A variety of possible curve trajectories results from different degrees of corner cutting. Therefore, depending on the selected trajectory, there is a greater variance of curve speeds even without changing the demand for lateral acceleration (i.e. roll angle).

As explained in ▶ Sect. 35.2.1, changes in the course angle are the result of changes in the roll angle. The buildup of these however, requires initial steering inputs in the opposite direction to the desired course. This effect is particularly pronounced in case of rapidly changing curvatures, such as those required for evasive maneuvers. As a result, the paths covered by the tires in the process differ fundamentally from those of a passenger car. Therefore, existing lateral control strategies from passenger cars cannot be transferred to a motorcycle without adaptation [10].

An essential step in the development of new motorcycle assistance systems at the path guidance level is to consider the overall vehicle dynamics. Self-riding prototype motorcycles have already provided evidence that, with appropriate modeling, even riderless riding would be possible. However, these models will rather be used to develop predictive assistance systems that not only monitor current riding conditions, but can also warn riders of a future critical riding situation before it occurs [16].

35.5   Human Machine Interface (HMI)

In principle, HMI concepts in the area of rider assistance on motorcycles fulfill the same purposes as in passenger cars or other vehicles (see ▶ Chap. 26). Feedback is provided on the system status (e.g., activity of an assistance system, mode, error status), settings of the assistance systems may be changed (e.g., distance when using ACC), and are used to warn the rider (e.g., Forward Collision Warning (FCW)).

The main differences between motorcycles and passenger cars in terms of user-friendly HMI design are the vehicle geometry, ergonomics, and available display size [26]. Control elements are usually located at the motorcycle handlebar (◘ Fig. 35.10a), where especially the left hand sometimes encounters a large number of individual buttons and switches. Visual information is often limited to the presentation in the instrument cluster, which typically offers much less space than the display panels known from passenger cars. Furthermore, the motorcycle's instrument cluster is located close to or outside the edge of the rider's peripheral field of view. Nowadays, modern and large displays can also show map-based navigation as shown for example in ◘ Fig. 35.10b.

Weather influences such as precipitation or glare additionally impede the visual perception of information on the dashboard. In addition to the instrument cluster, optical information can be provided by the use of additional LEDs. The placement of such LEDs in the mirror arms is utilized e.g. for warnings of a blind spot assistant. This enables visual guidance in the direction of the potential danger (see SVA ▶ Sect. 35.4.3). Acoustic information, which is widely used in the passenger car sector and is particularly effective for warnings, is difficult to implement in motorcycles, as there is no closed passenger compartment on the motorcycle and often a lack of installation space for loudspeakers. Furthermore, a motorcyclist is subject to loud ambient noise levels and noise insulation due to wearing a helmet [3, 20]. In-helmet sound presentation requires a reliable communication between the motorcycle and helmet. Based on initial findings, haptic information seems potentially suitable for conveying warnings to motorcyclists [13, 34]. However, these are not yet available in production vehicles. One reason for this is that verification of the perceptibility of different haptic stimuli on the motorcycle is still pending. On the one hand, this is challenging if riders have no contact with the relevant vehicle component (e.g., one hand is not always on the handlebar) or, on the other hand, haptic feedback is superimposed by engine vibrations or chassis excitations. A type of haptic feedback in the broader sense is the so-called brake jerk, which was used in the research of the emergency brake assistant (▶ Sect. 35.4.2). Here, a short braking impulse is triggered by the assistance system. Current research is concerned with the strength and also the buildup of the braking pressure in order to direct the attention of the rider to the riding task without endangering him or her through the braking intervention.

Complementing motorcycle-fixed HMI concepts with rider-fixed HMI solutions (e.g., wearable devices such as protective gloves with vibration feedback for warnings) requires reliable and secure communication between the motorcycle and the external HMI. For now, there is still a lack of uniform interfaces and definition of the liability issue in the event of system failures.

35.6   Connectivity

The inclusion of motorcycles in connected mobility offers great safety potential for motorcyclists [4]. The near absence of passive protection exposes motorcyclists to a higher risk of injury. Especially in accident scenarios where motorcyclists have no possibility for acute accident avoidance, connected assistance systems could increase safety in the future. It should be noted that in some cases the data from motorcycles must be processed differently than that of other road users. On a highway, for example, messages from multiple, particularly slow cars within a small distance can indicate a traffic jam. Based on a temporally and spatially limited number of messages, the system can display a traffic jam message to other approaching vehicles. However, many countries permit motorcyclists filtering through a traffic jam. A motorcycle filtering through such a congestion passage must not cause the congestion message to be canceled.

35.6.1   Localization

Safety-relevant C-ITS (Connected—Intelligent Transport Systems) applications place high demands on positioning accuracy. Passenger cars are localized by sensor data fusion of different measured variables. Predominantly, GNSS data and inertial measurements are merged in a so-called dead reckoning [35]. Knowing the current driving condition, steering angle, etc., it is also possible to estimate a preview of the vehicle trajectory with an accuracy of a few meters. In motorcycles, riding with roll angle represents an additional degree of freedom. Furthermore, the acquisition of steering inputs (steering angle or steering torque) is not common and not very helpful for more precise predictions about the position and future trajectory of a motorcycle. Algorithms for the localization of motorcycles are currently under development and, according to the state of the art, do not yet achieve the accuracy offered in the passenger car sector.

35.6.2 Antennas

In terms of antenna performance, further differences between the use of C-ITS on a motorcycle compared to a passenger car become apparent. In the case of a passenger car, it is obvious to use a roof antenna for a 360°-radio coverage. However, panorama and sunroofs as well as roof rails have a negative impact on signal strength and must be taken into account in the design. Alternative approaches, such as placing the antennas in the bumper, cause problems with regard to pedestrian protection (affecting the impact area). Experience has shown that the transmission range in the 5.9 GHz band deteriorates drastically if the antennas are installed in the passenger compartment. The lack of a passenger compartment on a motorcycle initially leads one to expect an advantage. This is correct in the first instance, but a closer look reveals other challenges. Once again, an influence due to the different riding dynamics of a motorcycle becomes apparent. In order to ensure good transmission performance when riding with roll angle, antennas with omni-directional characteristics must be used. On the contrary, it is sufficient in passenger cars to use antennas with a hemispherical characteristic (▶ Chap. 23). In addition, a wide variety of designs exist in the motorcycle sector, so it is not possible to make a generally valid statement about the ideal positioning of the antenna of a V2X system on a motorcycle. Research results from the Connected Motorcycle Consortium [4] show that the use of two antennas is advantageous. If one antenna is installed in the front and one in the rear of the motorcycle, interference can be reduced (antenna diversity). The system is also less susceptible to interference from the motorcycle itself, rider, passenger or luggage.

35.7 Development and Safeguarding Methods

Operational safety (OS) and functional safety (FS) (see ▶ Chap. 6) address the safety of customer functions, realized by electric/electronic (E/E) solutions. The focus

of functional safety is the safety of the intended function[6] [14], which includes the intended use as well as the expected misuse [29]. Functional safety according to ISO 26262 covers safety in the event of malfunction.

With the increasing introduction of actively intervening rider assistance systems in motorcycles (▶ Sect. 35.4), new challenges are arising to ensure operational safety. The transfer of knowledge from the passenger car sector is more difficult due to greater interactions between the rider and the vehicle. Automated emergency braking with low rider attention and without body tension in a passenger car can result in the rider having to be restrained by a seat belt. On a motorcycle, the same deceleration with inattentive riders could even lead to separation between the rider and the vehicle. This results in different requirements for operational safety.

Uniform, motorcycle-specific methods of ensuring safety in use have had limited availability to date. It makes sense to adopt methods from the passenger car sector and, if necessary, to adapt them specifically to motorcycles. One example is the method for HMI assessment with regard to the distraction effect during intended use [39]. Similar procedures could be used for future methods to check the operational safety, which will become increasingly necessary.

35.7.1 Extension of ISO26262 for Motorcycles

Since the revision of ISO26262 and the official publication of its second edition, the scope of the standard has been extended to include "series production road vehicles, excluding mopeds". The necessary changes and additions have been integrated directly into the individual chapters in the "Trucks and busses" section, while a separate addition has been created for motorcycles: ISO26262, Part 12.

ISO26262 was originally created specifically with the requirements from the passenger car industry in mind, which means that some requirements and processes cannot be directly applied to motorcycles.

The assessment of S, E, C in the Hazard Analysis and Risk Assessment (▶ Chap. 6) using the assessment criteria defined in ISO26262, Part 3, consistently produces assessment results that are one level higher than the assessment in the passenger car. Therefore, this in-

---

6   The intended function corresponds to the specified function in the normal state. This is contrasted with the function in the event of an error (malfunction).

termediate result (Motorcycle Integrity Level, MSIL) is determined and then reduced one level to obtain an ASIL rating. This ensures that the established rating scheme can be used. In addition, this procedure offers the advantage that the known ASIL can continue to be used when communicating with suppliers, authorities, etc.

For the evaluation of controllability and also for verification/validation tests, it is sometimes not possible to conduct a participant study for motorcycles because crashes and injuries must be expected. In this case, the evaluation can be done by "expert riders" and a panel of experts who can assess the technical impact on the vehicle (Controllability Classification Panel, CCP).

Another difference arises from the typical size of developing departments of motorcycle manufacturers compared to passenger car manufacturers. While independent reviews can usually be carried out by other business units at large corporations, this is not possible at typical motorcycle manufacturers with only one business unit. Here, the highest Independence Level I3 is achieved by independent departments with an independent approval authority.

In the area of component and function development (definition of safety goals, concepts, etc.), the motorcycle industry follows the same established principles and processes as for other road vehicles. Nevertheless, it is often not possible to transfer functions already developed and established in passenger cars to motorcycles without change. This is for example prevented by the high power-to-weight ratio of the vehicles: the maximum brake torque on the driven rear wheel is usually lower than the maximum engine torque. Therefore, a safety concept based on over-braking the drivetrain in the event of a fault cannot be applied. On the other hand, the technical requirements are also different. For example, only two wheel-speeds are available, and stronger roll and pitch motions have to be taken into account. At the same time, a significantly smaller installation space is available for components and sensors. The purely technical implementation of electronic rider assistance systems for motorcycles, taking into account performance and functional safety, therefore remains a challenge.

Moreover, it must be taken into account that motorcyclists differ to a large extent in terms of their riding skills, ergonomics, and attitudes toward the use of assistance systems.

35.7.2 Motorcycle Riding Simulators

While driving simulators for cars have been used for decades in safety research, motorcycle riding simulators have so far been less strongly represented in the field of research and development. However, the advantages of riding simulation (▶ Chap. 8) have recently led to increased applications here as well. Depending on the problem to be solved, different fidelity levels of motorcycle riding simulators are available, which vary greatly in terms of their system complexity [9].

Static motorcycle riding simulators are characterized by their simple design and their ease of riding for participants. Although the lack of motion feedback does not allow for a realistic riding impression, simple perception or HMI studies, for example, can still provide valid results.

Dynamic motorcycle riding simulators utilize platforms for motion representation. This increases system complexity, but allows for more sophisticated study designs, where a high fidelity of the riding experience is important. Recent studies verify the usability of such high-fidelity dynamic motorcycle simulators for example for comparing handling characteristics of different chassis configurations [18]. In addition, highly critical riding situations, such as the unexpected intervention of an emergency brake assistant during single-handed riding, can be investigated in order to e.g. obtain data on stability reserves, triggering algorithms, or user acceptance early in the development process [21].

To date, however, only a few dynamic motorcycle riding simulators are known that offer a level of immersion that would be described as "true to reality" by everyday riders. The inevitable differences in perception between real and simulated riding are countered by training study participants on the simulator in advance to the study itself. In this training, the study participants become accustomed to the specific riding behavior of the simulator and learn to deal with possible misperceptions. Compared to car simulators, special challenges arise in order to achieve a high system fidelity, which are described in ◘ Table 35.2.

◘ Figure 35.11 shows the DESMORI motorcycle riding simulator of WIVW GmbH as an example of a dynamic motorcycle riding simulator as it is commonly used today in the field of research and development. It is based on a hydraulic hexapod located in the middle of a cylindrical screen with 4.5 m diameter. This provides a horizontal field of view of 220°. Alternatively, the simulator can be operated with a head-mounted display. Sound is presented via headphones in the helmet or body transducers attached to the helmet. The system detects both steering and leaning inputs from the rider and can apply longitudinal forces to the rider's upper body via a rope towing mechanism.

35.8 Conclusion and Outlook

The technologies and methods presented in this chapter have greatly improved motorcycle safety in recent years. The most significant role is played by the wheel control

■ Table 35.2   Special challenges of motorcycle riding simulation

Motion simulation
To ensure that participants perceive the simulator as a motorcycle that is as true to reality as possible, it is important to generate correct vestibular and proprioceptive stimuli using motion cueing and force feedback methods (▶ Chap. 8)

Roll angle	In real cornering, a moment equilibrium results from gravitational force and centrifugal force. However, simulator platforms cannot represent this centrifugal force or can only do so for a short time. A 1:1 representation of roll angles is therefore not possible. Rather, it seems reasonable to set the platform roll angle to zero in stationary cornering in order to do justice to the realistically resulting force vector. However, this results in deviations in the representation of transient roll dynamics. Platform roll angles are always superimposed by image rotations to represent the total roll angle
Instability	The system inherent instability of a motorcycle, especially at low speeds, is difficult to replicate by fixed coupling of the motorcycle chassis to a motion platform due to the signal propagation times of the underlying data acquisition, simulation, and actuator control. A pivoting coupling of the motorcycle chassis to a motion platform could create a "natural" instability, but is not compatible with realistic motion cueing, that would also provide platform roll angles of $\varphi_{\text{Motion}} \neq 0$ in steady-state cornering
Standstill	Until now, standstill has only been possible on motorcycle riding simulators by means of artificial stabilization measures. For this purpose, the vehicle dynamic models are fixed in the vertical plane by artificial stiffnesses and are only released again above a certain speed that can be controlled by the rider. This can result in a riding behavior that is perceived as unnatural when starting off
Steering	Even without taking into account the required control algorithms, the representation of steering dynamics is a major challenge. The required actuator must be characterized by zero backlash, high dynamics, high stall torque and low friction and often represents one of the largest investments of a motorcycle simulator
Wind load	Motorcyclists are directly exposed to wind loads and other environmental influences, each of which has a major impact on e.g. speed perception. Simulating these wind loads is either uneconomical (e.g. wind tunnel) or requires compromises in the type or strength of the representation. For example, the wind load can be imitated by a cable mechanism attached to the study participant, by dynamically increasing or decreasing the distance between seat and handlebar, or by a high power fan
Longitudinal dynamics	In passenger car simulation, tilt coordination is used regularly (▶ Chap. 8), which makes gravity perceptible as constant longitudinal or lateral acceleration after tilting movements below the perception threshold were performed. The large accelerations, which are achieved even with low-power motorcycles under usual riding conditions, can hardly be realized by this kind of motion cueing
Steer-roll coupling	In the operation of dynamic and static simulators, different preferences for steering characteristics can be found. For example, study participants on a static simulator often prefer the perceived "positive-steering," whereas on a dynamic simulator, "counter-steering" receives more approval from the subjects. (▶ Sect. 35.2.2)

Visualization
Motorcyclists see their surroundings virtually unobstructed. Fairing parts block the view of surrounding details at most temporarily, but quickly release it again depending on the rider's position and head rotation

Projection	The realization of a 360° field view on the basis of projectors or monitors requires high investments. Regularly, for example, there is no floor projection in the near field, even though this is expected to improve speed perception. In addition, dynamic motorcycle simulators are usually not operated in closed cabins, so that study participants can reference (orientate) themselves in their (real) environment regardless of platform movements. This may affect the perception of platform motion
VR/AR glasses	By wearing VR goggles, subjects can be freed from this optical reference, providing additional freedom in motion cueing. Platform and image roll angles can be freely combined, and changes in body position and head rotation can be well mapped. However, there are still general challenges regarding the display quality (image resolution, refresh rate, latencies). In addition, issues arise, for example, for HMI and ergonomics evaluations that require vision on the real components (switches, buttons, displays, etc.)

Rider input acquisition
Motorcyclists can influence lateral dynamics by steering and leaning. Depending on the riding situation, a motorcycle reacts very differently to both inputs. At the same time, the handlebar and chassis also provide feedback to the rider

Steering	Steering represents the dominant input variable for the lateral dynamics of a motorcycle. Until today, steering motors and their control represent a major challenge in the development of motorcycle riding simulators. The sensitivity of motorcycle steering known from real riding cannot be achieved in the virtual environment according to the state of the art. If the study design and dynamic requirements allow it, passive handlebars are sometimes used, or the handlebars are even firmly braced, which significantly simplifies measuring the handlebar torque. The lack of steering angles is hardly perceived in typical riding situations on rural roads or highways due to its small amplitudes during real riding. However, in urban environments, e.g. during turning maneuvers, a recognizable steering angle is indispensable to avoid errors in perception

(continued)

◼ Table 35.2 (continued)

Leaning	Leaning movements of a motorcyclist influence the stationary equilibrium during cornering and can be used to change the trajectory. Measuring it as an input variable in motorcycle riding simulators is not very common and is implemented in very different ways. Sometimes only the footrest pressure is measured, sometimes a marker placed on the rider's back is located by image processing. The DESMORI simulator of WIVW GmbH (◼ Fig. 35.11) uses a force-based measurement method that makes rider-induced roll torques measurable independently of platform movements. This opens up additional application possibilities, such as balancing the motorcycle at a standstill by pressing the feet on the motion platform

Ergonomics
The riding experience on a motorcycle—like on a motorcycle riding simulator—also depends on the vehicle ergonomics. It influences, among other things, the connection of the rider to the vehicle, their body tension and the actuation of controls

Vehicle types	Dynamic motorcycle riding simulators often use a single chassis that is an integral part of the simulator. This limits the extent to which ergonomics can be adapted between different studies. Changing the chassis is often impractical due to the peripheral measurement and control equipment. Typically, deviations in ergonomics are accepted in order to be able to avoid cumbersome retrofitting and the associated costs, if the focus of the study permits this

Study participants
Without study participants, almost every simulator loses its justification. It is therefore important to design and operate a simulator in such a way that it is perceived and used by the study participants as a true-to-life replica of the real system. However, this is not always possible without limitations

Expectation	Study participants vary in their ability to cope with deviations between reality and simulation. Some can quickly adjust to the specifics of simulator riding and ignore errors in motion or image representation, while others do not manage this adjustment. In particular, not all motorcycle riders are aware of the dynamic effects that they utilize to control the motorcycle in real riding. A rider being particularly conscious and concentrated while using the simulator can therefore even be counterproductive if relying on a wrong understanding of the system, which makes its control more difficult

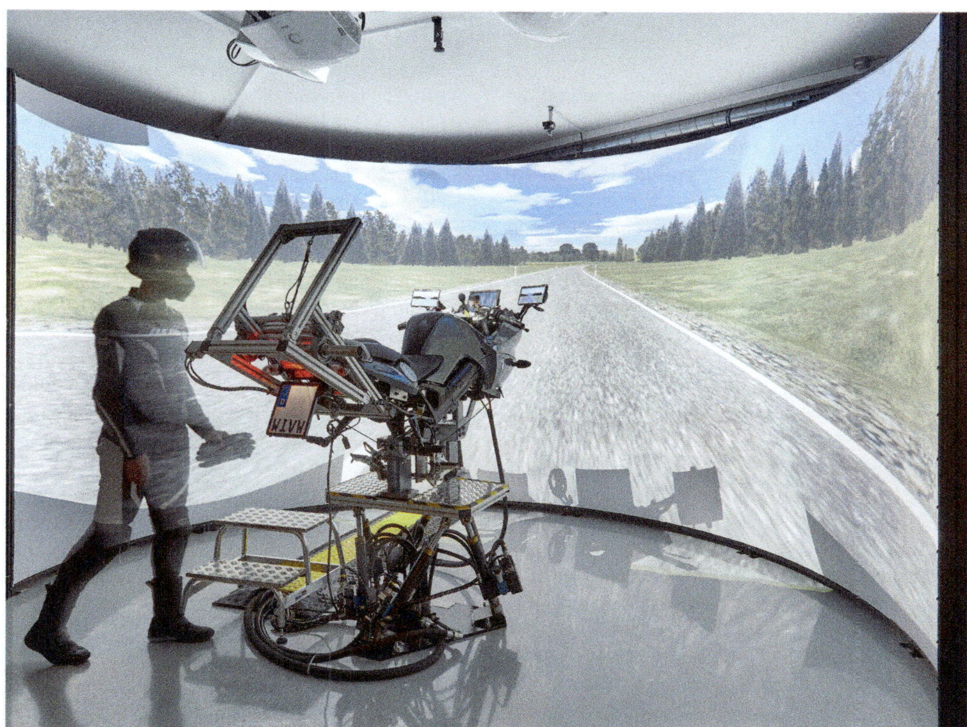

◼ Fig. 35.11   The DESMORI motorcycle riding simulator

systems, which now function robustly over the entire dynamic range relevant to everyday use, even in curves or on unpaved tracks. The systems are capable of maintaining riding stability under adverse environmental conditions or when the rider is overstrained (e.g. during emergency braking). The next step towards higher motorcycle safety now lies in the development of actively intervening rider assistance systems, as these allow earlier intervention compared to wheel control systems. This chapter presented the basics of environment sensing and active interventions in longitudinal dynamics. The resulting challenges, e.g. in the area of controllability, will be further increased by active interventions in lateral assistance systems. In return, assistance functions that can actively intervene in both longitudinal and lateral dynamics and ideally communicate with their environment will prevent a large number of accidents in the future that are often still considered unavoidable today.

References

1. Bayer, B.: Das Pendeln und Flattern von Krafträdern. Ausgabe 4 von Forschungshefte Zweiradsicherheit (1987)
2. Bocciolone, M., et al.: Experimental identification of kinematic coupled effects between rider and motorcycle (2007)
3. Brown, C.H., Gordon, M.S.: Motorcycle helmet noise and active noise reduction. Open Acoust. J. 4(1) (2011)
4. Connected Motorcycle Consortium: CMC basic specification—system specification (2020)
5. Cheli, F., et al.: Rider's movements influence on the lateral dynamic of a sport motorbike (2011)
6. Cossalter, V.: Motorcycle Dynamics. Padova, IT (2006)
7. Deutsches Statistisches Bundesamt: Kraftrad-und Fahrradunfälle im Straßenverkehr 2019 (2020)
8. Guth, S.: Absicherungsmethode von Anzeigekonzepten zur Darstellung fahrfremder Informationen mittels eines Motorrad-Fahrsimulators. Technische Universität Darmstadt (2017)
9. Hammer, T., Merkel, M., Pleß, R., Will, S., Neukum, A.: Anwendungsmöglichkeiten von Motorrad Fahrsimulatoren. BASt FE 82.0700/2017 (bislang unveröffentlicht) (2019)
10. Hans, S., Köbe, M., Prokop, G.: Warum Automatisierung die Zukunft der Motorradsicherheit ist!. Forschungshefte Zweiradsicherheit Nr. 18 (2018)
11. ▶ https://www.honda.de/motorcycles/experience-honda/news-and-events/2017-01-27-honda-moto-riding-assist.html. Last accessed 18 Nov 2021
12. Hoffmann, J., Winner, H.: EVITA—Das Prüfverfahren zur Beurteilung von Antikollisionssystemen. Handbuch Fahrerassistenzsysteme, 2. Aufl. Springer Fachmedien. Wiesbaden (2015)
13. Huth, V., Biral, F., Martín, Ó., Lot, R.: Comparison of two warning concepts of an intelligent curve warning system for motorcyclists in a simulator study. Accid. Anal. Prev. 44, 118–125 (2012)
14. International Organization for Standardization: ISO/PAS 21448 road vehicles-safety of the intended functionality (2019)
15. Kim, D.K.Y:. Dynamically balanced flywheel. Patent Application US20130233100A1 (2013)
16. Köbe, M., Hans, S., Prokop, G.: Von der Automatisierung zur Assistenz–Intervenieren, bevor es kritisch wird. Forschungshefte Zweiradsicherheit Nr. 18 (2018)
17. Lemejda, M., Willig, R.: A new inertial sensor unit for dynamic stabilizing systems of powered two wheelers. In: Proceedings of the 9th International Motorcycle Conference, pp. 66–84 (2012)
18. Massaro, M., Cossalter, V., Sadauckas, J., Lot, R.: Using simulators for the assessment of handling of motorcycles. In: Proceedings of the Bicycle and Motorcycle Dynamics Conference (2016)
19. Matschl, G., Mörbe, M., Gröger, C.: Motorcycle stability control—MSC the next step into safety solutions for motorcycles. In: Proceedings of the 10th International Motorcycle Conference, pp. 128–154 (2014)
20. McKnight, A.J., McKnight, A.S.: The effects of motorcycle helmets upon seeing and hearing. Accid. Anal. Prev. 27(4), 493–501 (1995)
21. Merkel, M., Pleß, R., et al: Automatische Notbremssysteme für Motorräder. BASt FE 82.0661/2015 (bislang unveröffentlicht) (2019)
22. Merkel, N., Pleß, R., Winner, H., Hammer, T., Schneider, N., Will, S.: Tolerability of unexpected autonomous emergency braking maneuvers on motorcycles—a methodology for experimental investigation. In: International Technical Conference and exhibition on the Enhanced Safety of Vehicles. Eindhoven, NL (2019)
23. Merkel, N., Winner, H.: Measures for the evaluation of riders' adaption to the changing vehicle state during autonomous emergency braking maneuvers on motorcycles. In: Symposium on the Dynamics and Control of Single Track Vehicles. Padova, IT (2019)
24. Merkel, N., Winner, H.: Characteristic rider reactions to autonomous emergency braking maneuvers on motorcycles. In: 13th International Motorcycle Conference. Cologne, GER (2020)
25. Pacejka, H.B.: Tire and Vehicle Dynamics, 3rd edn. Butterworth-Heinemann, Oxford (2012)
26. Pieve, M., Tesauri, F., Spadoni, A.: Mitigation accident risk in powered two wheelers domain: improving effectiveness of human machine interface collision avoidance system in two wheelers. Human Syst. Interaction. Catania, IT (2009)
27. Pleß, R., Büttner, A., Merkel, N., Winner, H., Will, S., Hammer, T.: The influence of rider motion on motorcycles and riding simulators. In: 12th International Motorcycle Conference, Cologne, GER (2018)
28. Rexroth: Inertialsensor MM7.10, Technical customer documentation: ▶ https://www.boschrexroth.com/documents/12605/25209290/TCD_MM7.10_V5.pdf. Last accessed 18 Nov 2021
29. Schnieder, L., Hosse, R.S.: Leitfaden Safety of the Intended Functionality. Springer Fachmedien, Wiesbaden (2019)
30. Schönherr, M., Grelaud, M., Hirano, A.: Side view assist—the world's first rider assistance system for two-wheelers. SAE Int. J. Vehicle Dyn. Stability NVH 1(1), 38–43 (2017)
31. Schröter, K., Wallisch, M., Weidele, A., Winner, H.: Bremslenkmomentoptimierte Kurvenbremsung von Motorrädern. Automobiltechnische Zeitschrift (ATZ) Special Motorrad (2013)
32. Schröter, K.: Brake steer torque optimized corner braking of motorcycles. Dissertation, TU Darmstadt (2017)
33. Schröter, K., Pleß, R., Seiniger, P.: Fahrdynamikregelsysteme für Motorräder. In: Handbuch Fahrerassistenzsysteme: Grundlagen, Komponenten und Systeme für aktive Sicherheit und Komfort. Springer Fachmedien Wiesbaden (2015)
34. Sevarin, A., Will, S., Rößer, J., Mikschofsky, N., Menato, L., Hammer, T., Schneider, N., Mark, C.: Assessment of visual and haptic HMI concepts for hazard warning of powered two-wheeler riders. Proceedings of the 13th International Motorcycle Conference. Cologne, GER (2020)
35. Steinhardt, N., Leinen, S.: Datenfusion für die präzise Lokalisierung. In: Handbuch Fahrerassistenzsysteme: Grundlagen, Komponenten und Systeme für aktive Sicherheit und Komfort. Springer Fachmedien Wiesbaden (2015)
36. Stoffregen, J.: Motorradtechnik. Vieweg+Teubner Verlag, Wiesbaden (2012)

35. Tanelli, M., Corno, M., Savaresi, S.M.: Modelling, Simulation and Control of Two-Wheeled Vehicles. John Wiley & Sons, Ltd, Chichester, UK (2014)

38. Wahl, A., Klews, M., Georgi, A.: Novel motorcycle safety system preventing lateral sliding in curves. Automobiltechnische Zeitschrift (ATZ) 9/2018 (2018)

39. Will, S., Hammer, T., Rothe, N., Matschl, G.: Development of an assessment method for powered two-wheeler human machine interfaces. In: Proceedings of the 13th International Motorcycle Conference (2020)

40. WorldAtlas: Countries with the highest motorbike usage, abrufbar unter ► https://www.worldatlas.com/articles/countries-that-ride-motorbikes.html. Last access 18 Nov 2021

41. Zhang, Y., Li, J., Yi, J., Song, D.: Balance control and analysis of stationary riderless motorcycles. In: IEEE International Conference on Robotics and Automation, pp. 3018–3023 (2011)

Open Access This chapter is licensed under the terms of the Creative Commons Attribution-NonCommercial-NoDerivatives 4.0 International License (► http://creativecommons.org/licenses/by-nc-nd/4.0/), which permits any noncommercial use, sharing, distribution and reproduction in any medium or format, as long as you give appropriate credit to the original author(s) and the source, provide a link to the Creative Commons license and indicate if you modified the licensed material. You do not have permission under this license to share adapted material derived from this chapter or parts of it.

The images or other third party material in this chapter are included in the chapter's Creative Commons license, unless indicated otherwise in a credit line to the material. If material is not included in the chapter's Creative Commons license and your intended use is not permitted by statutory regulation or exceeds the permitted use, you will need to obtain permission directly from the copyright holder.

Driver Assistance Systems in Commercial Vehicles

Christian Ballarin, Felix Manuel Reisgys, Ingo Scherhaufer and Christoph Tresp

Contents

© The Author(s) 2026
H. Winner et al. (eds.), *Handbook Assisted and Automated Driving*,
https://doi.org/10.1007/978-3-658-45276-6_36

36.1 Overview

36.1.1 Motivation

The day-to-day work of professional drivers entails numerous challenges such as time pressure, heavy traffic, fatigue and difficulty finding adequate parking, as well as unfavorable weather conditions. At the same time, the slightest lapse in attention can have serious consequences, especially for vulnerable road users such as pedestrians or cyclists. Because of this great responsibility, measures for increasing road safety focus on heavy goods vehicles in particular. Driver assistance systems offer great potential in this respect, especially in commercial vehicles. As the regulatory activities as part of the European Commission's e-safety initiatives are already making clear, driver assistance systems are also seen as an important component of efforts to increase road safety by both legislators and the society.

Technological progress in the area of driver assistance systems is also part of the reason that DEKRA's *Verkehrssicherheitsreport Lkw* [Road Safety Report—Trucks] [9] documents that the frequency of accidents involving trucks per vehicle kilometer in Germany has fallen by more than 70% since 1970. In addition, the number of road users seriously injured or killed in truck accidents decreased by more than 36% and 40% respectively between 1992 and 2018, while freight traffic volume increased by 84.9% in the same period.

The traffic volumes accounted for by trucks will continue to increase in the future. For Germany, the Federal Ministry of Transport's *Verkehrsprognose 2030* [2030 Traffic Forecast] [3] anticipates an increase of 39%. A study conducted by IFMO [15] also assumes an annual increase of 2% in the EU, which corresponds to an increase of 50% within 20 years.

In addition to the increase in transport volume, growing cost pressure as a result of rising fuel prices and toll charges represents a challenge for road haulage companies. Truck procurement costs are also increasing due to new, demanding legal requirements. This demonstrated by the introduction of the Euro VI emissions level in 2014, for example.

The number of mandatory safety systems is likewise increasing. Due to the high kinetic energy of commercial vehicles, passive safety measures have only limited potential, especially for protecting other road users. First-generation active safety systems such as ABS or ESC are already mandatory for trucks larger than 3.5 metric tons. In addition, initial driver assistance systems such as AEBS (Advanced Emergency Brake System) and LDWS (Lane Departure Warning System) have been mandatory since November 2015 [11].

An improvement in road safety, thanks to the market penetration of these systems, is also noticeable in the statistics. Up to 2019, commercial vehicles registered since 2015 caused only 27% of the rear-end collisions in Germany, while they accounted for more than 75% of the long-distance fleet. Furthermore, a general reduction in accidents involving trucks can also be seen: For 2018, federal statistics [24] show 31,803 such accidents involving personal injuries, which corresponds to a reduction of 1.3% compared with the previous year. The number of fatalities from truck accidents fell by 59.6% between 1992 and 2018, while transport capacity doubled (252.3 to 502.2 billion ton-kilometers) [25]. A more precise analysis for goods vehicles larger than 8t is provided by the main accident patterns with personal injuries according to accident type and location in ◉ ◘ Fig. 36.1.

36.1.2 History

The history of safety systems in commercial vehicles dates back to 1981, when the first series ABS System was presented to the public. A decisive milestone in the area of vehicle automation was the PROMETHEUS research project that, in a consortium comprising automobile manufacturers, suppliers and universities, drove forward the use of driver assistance systems to a considerable extent. From the year 2000, first driver assistance systems to analyze the vehicle's surroundings, such as the Telligent distance control ACC system or Lane Keeping Assist for Mercedes-Benz Trucks emerged from the research project. Radar technology was already available in heavy-duty trucks shortly after its introduction in the luxury class of passenger cars, while the first camera was used even four years before application in passenger cars.

In the "Promote Chauffeur" project from 1998 onward, there was also a focus on interconnectivity between trucks, which grew in importance because of on-board electrical systems, mobile communications and the use of GPS data. This involved two tractor/semi-trailer combinations a short distance apart being connected with the aid of "electronic drawbars". The vehicles were coordinated with the aid of infrared signals on the trailer of the leading vehicle and cameras on the following truck.

Nearly all assistance systems that are either industry standard or legally required today were already available much earlier in series production trucks. ◉ ◘ Figure 36.2 shows a selection of safety systems available from Mercedes-Benz and their introduction as systems required by law.

2014 saw systems meeting SAE Level 2 presented in commercial vehicles for the first time in the Mercedes-Benz Actros with "Highway Pilot" concept truck and its North American equivalent, the "Freightliner Inspiration Truck". The "Highway Pilot" combines the

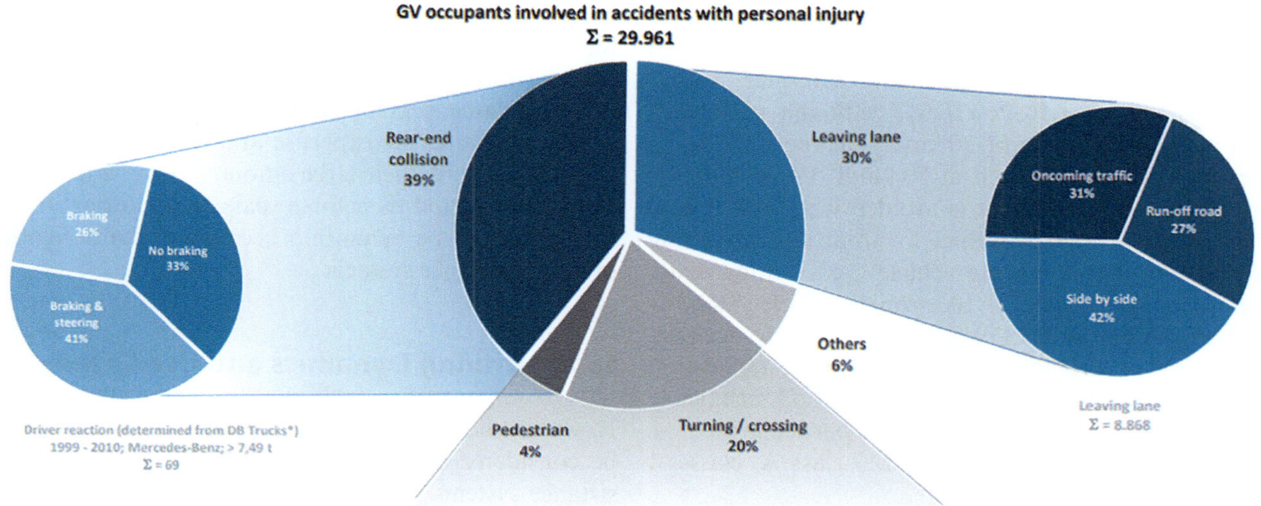

◼ Fig. 36.1   Accident statistics for goods road transport vehicles > 8 t in 2018

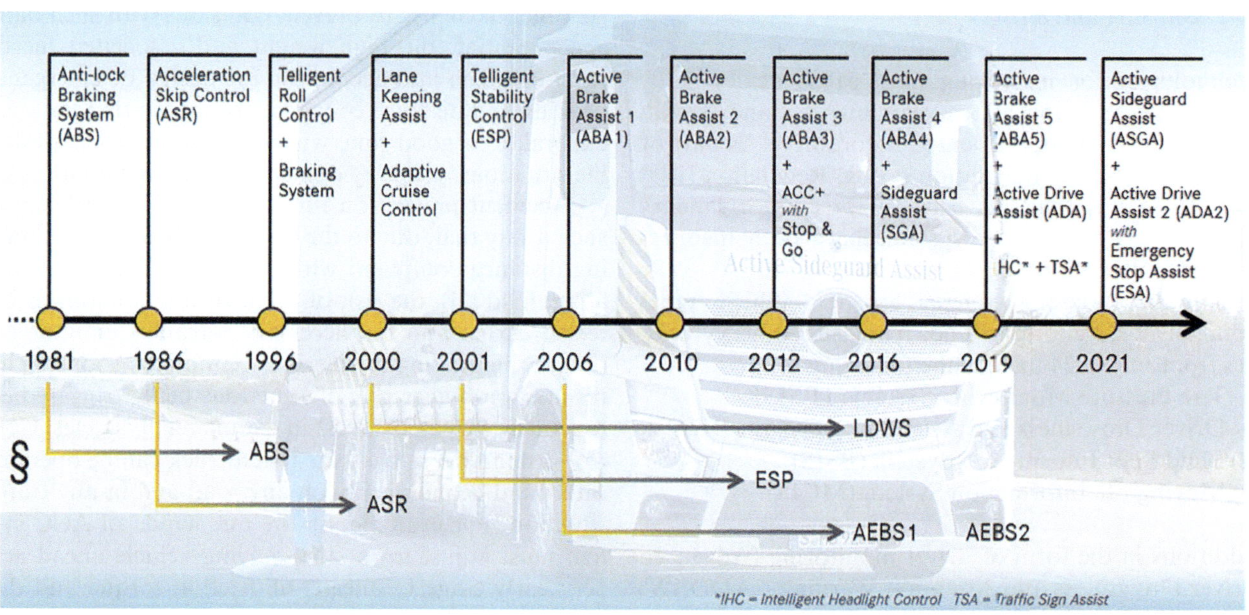

◼ Fig. 36.2   Safety roadmap for Mercedes-Benz trucks

functions of the familiar ACC and Lane Keeping Assist systems, supplemented by interventions in the steering. As a result, simultaneous longitudinal and lateral guidance of the vehicle is assured. On this basis, it was possible to bring partially automated driving (SAE Level 2) into series production for the first time in 2019, in a commercial vehicle with Active Drive Assist in combination with Active Power Steering (APS).

36.1.3 Legislation

The first driver assistance systems were described by lawmakers in 2009 as part of General Safety Regulation 661/2009/EC [11]. Thus, all electronic driving stability control systems (including ESP or ESC) initially became mandatory for all vehicles, including trucks, from November 2014. The positive effect of such systems is demonstrated on the market and by several investigations, e.g. in the Dekra Safety Report [9]. The significant benefit of these systems is indisputable, especially for heavy touring coaches and trucks and their trailers as well.

Lane Departure Warning Systems (LDWS) as well as Advanced Emergency Braking Systems (AEBS) have been specified for all newly registered commercial vehicles since 11/2015. This concerns buses with more than nine seats as well as trucks with a permissible gross vehicle weight exceeding 3.5 metric tons. For the second level of AEBS, the requirement for stationary obstacles was tightened from 11/2018 with the stipulation that a speed reduction of 20 km/h must be achieved in the event of an approach at 80 km/h. Proof must be provided that an accident is avoided when the vehicle drives towards another vehicle in front that is moving at 10 km/h. The following vehicle types are currently exempted from EVSC (Electronic Vehicle Stability Control), AEBS and LDWS because they are not relevant to the occurrence of accidents and the application of these technologies is not yet mature and technically feasible for all-wheel drive vehicles in particular:

- City buses—vehicle class Mz/M3 class A, classes I and II
- All-terrain trucks—vehicle class N2/N3 category G
- Special-purpose vehicles (SPVs)
- Vehicles with more than three axles
- Semitrailer trucks with a maximum GVW of between 3.5 t and 8 t.

Technologies for improving road safety reviewed by the EU Commission at ten-year intervals and regulations adapted or supplemented accordingly. As part of the General Safety Regulation (GSR, Regulation (EU) 2019/2144) [12] that has been in force since January 2020, further mandatory systems have been incorporated into the catalog.

Thus, mandatory systems for new vehicle types from July 2022 and for all newly registered heavy vehicles from July 2024 include the following:

- Tire Pressure Monitoring System (TPMS)
- Driver Drowsiness and Attention Warning (DDAW)
- Blind Spot Information System (BSIS)
- Moving-off Information System (MOIS).

Additions in the form of driver monitoring (Advanced Driver Drowsiness and Attention Warning—ADDAW) and an electronic data recorder (EDR) are already planned for 2026.

36.2 Differences Between Trucks and Passenger Cars

36.2.1 Drivers and Utilization Time

With driving times of nine hours per day [10], driving a commercial vehicle is more demanding than operating an ordinary car, which is used for less than one hour per day on average [14]. The tendency to suffer lapses in concentration resulting from long working and driving times leads to a higher risk of an accident that can be reduced by monitoring and warning drivers. Thus, the use of warning systems that monitor the driver's concentration, driving performance or compliance with statutory rest periods plays an important part in preventing human error.

Furthermore, the expertise and experience of drivers as a result of their large amounts of time that they spend driving and their three years of vocational training [13] are to be taken into account in the design of driving assistance systems.

36.2.2 Driving Dynamics and Use Cases

However, the physical properties of a truck must also be considered during the development of driving assistance systems. At a typical speed of 80 km/h on the freeway, the high mass of a commercial vehicle weighing 40 metric tons means it has the same kinetic energy as a passenger car that weighs 1.5 metric tons traveling at a less realistic speed of more than 460 km/h (almost six times as high). To prevent collisions with such damage potential, the high weight and associated inertia must be taken into account in the design of emergency braking and distance control assistants so that the vehicle brakes in good time without causing any undesirable situations, property damage or personal injury.

Accident prevention functions must be designed in such a way that, due to the significantly increased braking distance compared with a car (+50% to 70% with a 40 t load [1]), the risk of a hazardous situation is detected earlier and the necessary warnings or intervention in the braking or steering are initiated sooner. This results in a higher risk of functions being triggered erroneously (false positives) in commercial vehicles. Control systems need to ensure that a truck can be operated safely and economically on any road and in any traffic situation. For example, on the one hand, an ACC system must adjust for a slow-moving vehicle ahead at a very early stage by means of friction torque and displays a possible overtaking recommendation in order to prevent increased brake wear. On the other hand, adjustments to maintain the safety distance when vehicles cut into the same lane or overtake must be performed very smoothly to avoid triggering a sequence of vehicles dropping back in these typical situations on the freeway.

By using the actuators of a commercial vehicle, it is possible to determine the total weight of a tractor/semitrailer combination from the electronic brake system. This procedure is, however, relatively imprecise. It is not currently possible to reliably determine the height of the center of gravity or the friction values between the tires and the road before an active intervention. Another piece of information that is important while driving in a curve is the kink angle between the towing vehicle and the trailer.

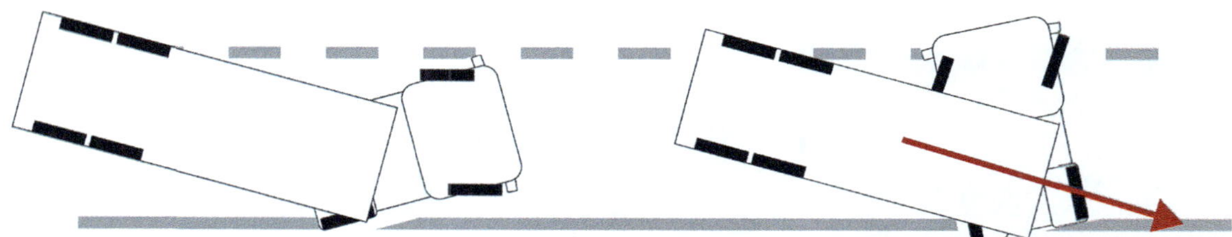

□ Fig. 36.3   Schematic diagram of jack-knifing

Moreover, the selection of possible evasive and braking maneuvers is further restricted by an optional trailer. A truck with a trailer has a larger tractrix curve. Depending on the scenario, pedestrians, stationary obstacles or other road users could be injured or damaged by the trailer. Another point is jack-knifing, whereby the towing vehicle and trailer fold together. This is usually caused by surfaces that are slippery in places coupled with heavy braking, strong crosswinds, a driving maneuver on a hill or an unevenly distributed load in the trailer, which the driver may not be able to influence depending on the load (e.g. transportation of liquids or animals) [6]. You can find an illustration in ◉ □ Fig. 36.3.

36.2.3 Field of View and Sensor System

In addition to the weight, the vehicle geometry is decisive for the design of driver assistance systems. For example, to configure turning assistants, sensors must be positioned such that all of the driver's possible blind spots are covered and objects of different sizes can be correctly detected. It is particularly important to cover the zone in front of the truck as, unlike in a car, a wider blind spot is caused in this area by the driver's seat position [7] (see ◉ □ Fig. 36.4).

In addition, the different activities of the vehicles will create further variance in the requirements for the sensor system. For example, a construction vehicle has different ground clearance and therefore a different height. As a result, the position of the entire sensor system shifts, which means that its functionality must be validated again. In the worst-case scenario, additional climbing bars and wider access ladders can cause reflections for the radar sensor system and thus lead to undesirable behavior. Moreover, the sensor system needs to work with a higher degree of dirt buildup on the windshield or front end. Another aspect is the pitch of the driver's cab. This is of interest in that the forward-looking camera is attached to the inside of the windshield inside the towing vehicle. This results in a different perception of the environment for the camera, which could lead to a false alarm or an alarm not being triggered.

To enable identification of all types of traffic situations, vehicle types, trailers, vehicle bodies, people, bicycles, electric scooters, obstacles on the roadway (e.g. a pallet or lost load) and animals, all-round visibility from the towing vehicle to the semitrailer or trailer with a range of 250 m to the front, 100 m to the side (e.g. at intersections) and up to 30–50 m to the rear (including trailer or semitrailer) must be ensured. At the same time, pedestrians must be detected as far as 60 cm in front of the vehicle in the close-up range. In the case of double or triple trailers, such as those used in South America, for example, it must be possible to monitor a vehicle length of more than 50 m. In addition, the chassis can vary by up to 1 m in height depending on the model designation, meaning that radar sensors need to cope with different installation heights, for example. It must also be possible to parameterize camera systems that are generally installed behind the windshield (in the area cleared by the windshield wipers) with an installation variance of 1–3 m above ground while ensuring the same quality of identification in all cases (e.g. people on the road, lane markings on bends and traffic signs on gantries).

36.3 System and Safety Strategies

36.3.1 Design of Driver Assistance Systems for Commercial Vehicles

When designing an assistance system, the desired or required applications (use cases) are initially used as a basis. A subsequent system design used to identify the "borderline cases", i.e. those cases where the system reaches its limits and possibly no longer works reliably.

In today's systems, these borderline cases usually arise from major differences in the speeds of vehicles, short range and restricted field of view for sensor systems, unknown friction values, total masses and positions of the center of gravity as well as from particular driver behaviors. Since the accident prevention systems include interventions that pose a danger to traffic (emergency braking with deceleration >6 m/s^2), high

Fig. 36.4   Installation position and field of view of the front camera of a Mercedes-Benz Actros

confidence in the evaluation of the underlying situation is required.

In addition, the following aspects must be taken into account as part of the system design for commercial vehicles:

- Environment recognition using different sensor technologies (range, field of view, resolution, accuracy, robustness)
- Variance in sensor installation positions in the variety of model variants (tractor/semitrailer combination versus construction vehicle)
- System operation and displays (visual, auditory, haptic)
- Secured driver monitoring (e.g. steering wheel and driver's seat)
- Electronic control of the service, parking and continuous brake systems (engine brake, retarder)
- Electronic control of the drivetrain (engine, transmission, coupling)
- Electronic actuation of the steering
- Override options for the driver (e.g. using the brake or accelerator pedal)
- Information or warnings for following traffic
- For buses: protection for standing passengers
- Secured real-time-capable data communication
- High computing power in the sensors and central processing units
- Avoidance of false positive reactions with the full intervention strength for the given use cases
- Data backup for accident reconstruction and product liability.

36.3.2 System Architecture

To enable assistance systems be realized economically, architectures, components and sensors based on the passenger car market are used in commercial vehicles. ◉ ◘ Figure 36.5 illustrates the connection between the architectures used at Mercedes-Benz Cars and Truck.

Using a high-resolution state-of-the-art radar sensor system, objects can be measured very accurately in terms of distance (travel time measurement) and speed (frequency shift), while camera systems determine road markings, object classifications (e.g. persons, vehicle types etc.), light sources and brightness with great precision. Thanks to the combination of beam sensors and sensors with optical image processing, strengths and weaknesses of the respective technology can be combined on the basis of different physical measuring methods such that robust detection can be achieved independently of light or weather influences, for example. The combined field of view is shown in ◉ ◘ Fig. 36.6.

The sensor data is processed in a central computer by means of object, track and lane data fusion so that relevant information to control the engine, brake, steering and light functions as well as to issue required HMI outputs for informing and warning the driver being are computed. A corresponding hardware topology can be seen in ◉ ◘ Fig. 36.7. All data must be transmitted in a fail-safe way by means of time and information redundancy, for example, so that a data error or a hacker attack cannot result in malfunctions. In particular, access to the steering and brakes must never result in a safety–

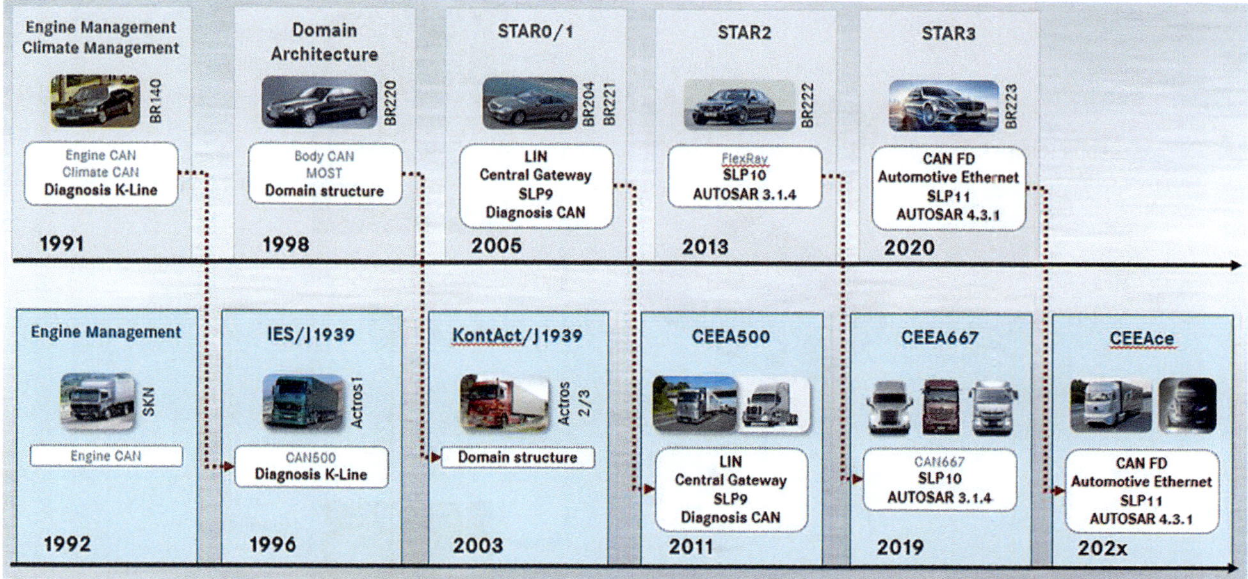

Fig. 36.5   Evolution of E/E architecture at Daimler truck

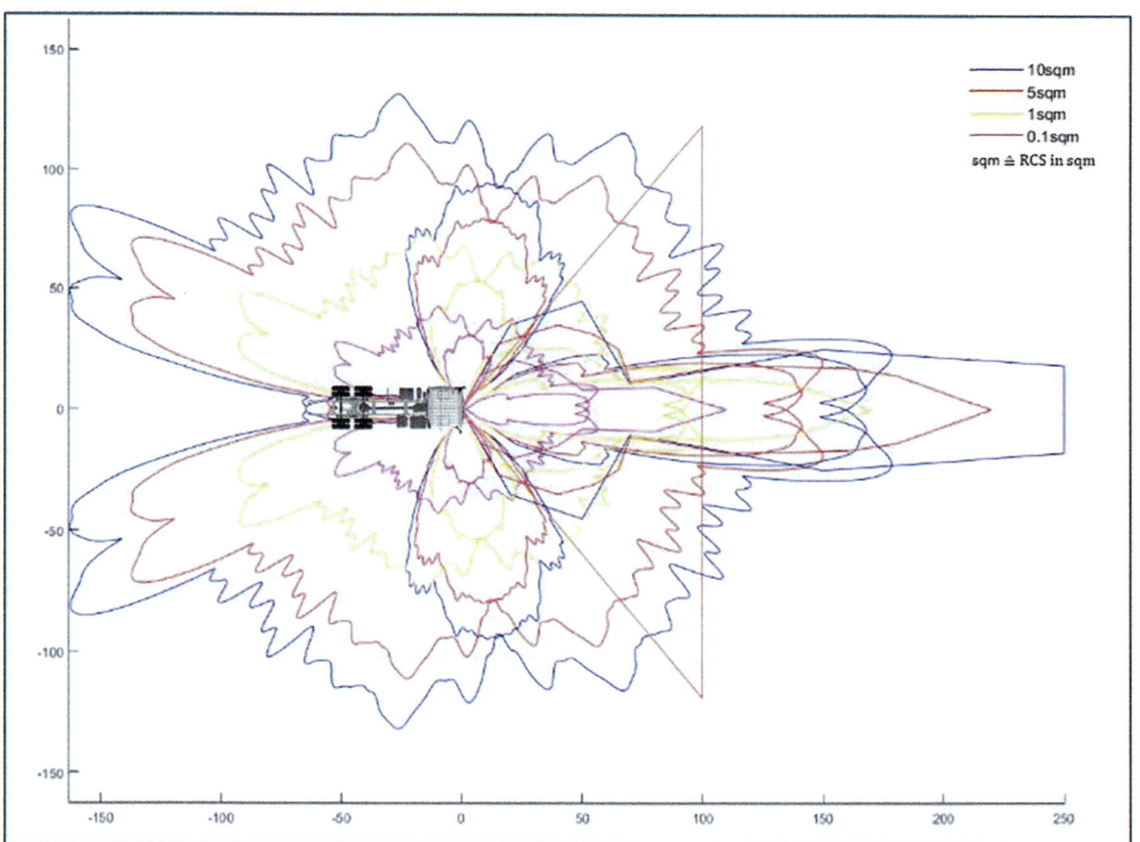

Fig. 36.6   Environment detection in a commercial vehicle

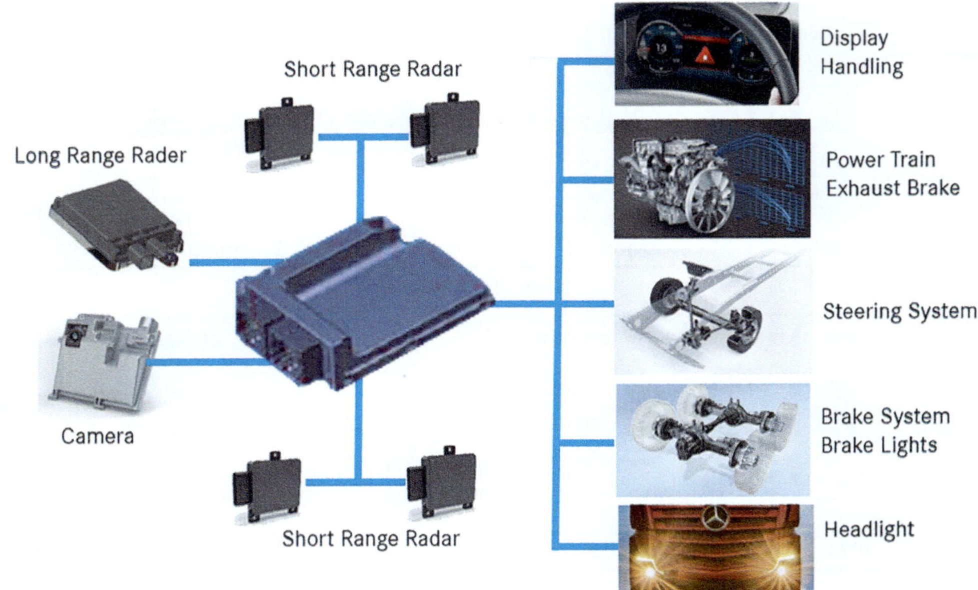

◘ Fig. 36.7 Hardware topology in the Mercedes-Benz Actros

critical situation. To that end, the steering torques can be limited and deceleration can take place only in a limited or cascaded manner, for example. Comfort control for an ACC system can be limited to a maximum of 3 m/s^2 in stop-and-go traffic and to 2 m/s^2 while following a vehicle for example.

The human–machine interface (HMI) is also of enormous importance for driver assistance systems in commercial vehicles. On the one hand, system status, system intervention and warnings for all functions must be clear, intuitive and understandable. On the other hand, drivers must never be distracted to ensure that they give their undivided attention to road traffic. Drivers always bear responsibility and must also be able to override the system at any time in accordance with the Vienna Convention on Road Traffic. With the General Safety Regulation (GSR) [12], it should no longer be possible to deactivate emergency functions that are intervening via a direct switch or pushbutton. In addition, an unwitting driver reaction must never cause to override a system intervention, e.g. in a critical accident situation. In the case of rear-end collisions in particular, it has been shown that drivers apply too little braking force in many cases. For this reason, the emergency braking system is not overridden by an actuation of the brake pedal and implements maximum full-stop braking even if the driver is already operating the brake pedal. By contrast, any warning or brake intervention can be overruled via kickdown on the accelerator pedal.

It is necessary to always optimize the system design such that a high degree of customer acceptance is achieved, regardless of whether used off-road, on a construction site, in city or intercity traffic, on expressways or on freeways. This means that the control systems must operate the motor vehicle just as well and as economically as very good professional drivers could themselves. In contrast, safety systems must act exactly where humans fail and avoid an accident as far as possible at the critical moment. However, to ensure that the systems are accepted, they must run inconspicuously in the background during normal driving and avoid false positive interventions.

An assistance system that constantly sounds the alarm without obvious cause or intervenes inconveniently in events on the road will never be accepted and therefore make no contribution to ensuring safe and accident-free traffic.

36.4 Driver Assistance Systems in Series Use

36.4.1 Information and Warning Functions

36.4.1.1 Lane Departure Warning Systems

A Lane Departure Warning System (LDWS) is an image-processing safety system that helps the driver with staying in lane and issues a warning in good time when it detects the vehicle straying from its lane. The system ensures to maintain the minimum distances from the vehicle to the lane markings on the road. This function intended primarily to provide support on long, monotonous routes.

A digital camera behind the windshield records the vehicle position relative to the left and right lane markings.

◉ ▣ Figure 36.8 shows a possible depiction from Mercedes-Benz on the digital display on the instrument cluster as HMI interface after the function is being triggered by the left lane marking.

A powerful microcomputer that, depending on the imager, continuously evaluates 23,000 to 30,000 pixels is connected downstream of the digital camera. The subsequent evaluation logic ultimately selects the corresponding line candidate. On the basis of training data, the performance of this assistance system can be continuously improved by means of neural networks. In addition, cameras with special filters are often used for this so that the sensitivity to light is increased. A clear marking should be present in order to ensure that the camera detects the lanes flawlessly. The evaluation algorithm is able to predict short, temporary interruptions for a brief period. Whereas daylight is sufficient for detection, the headlamps of commercial vehicles need be switched on in the dark.

As soon as the vehicle undershoots the minimum distance from the lane marking or even crosses the lane marking without actuating a turn signal indicator or initiating a braking procedure, the driver is alerted by a warning signal that lasts for approximately 3 s. This may be relayed to the driver acoustically, e.g. as a rattling sound from the left or right inside speaker; haptically, e.g. by means of vibrations in the driver's seat; or purely visually. As a rule, a combination of these options is used to call the driver's attention to the possible hazardous situation.

The usage of such a system is recommended on good roads with clearly visible lane markings at speeds above 60 km/h, since frequent lane changes are part of normal driving in urban areas. The exact boundary conditions as to when a warning must be triggered are defined in the EU regulation [26].

36.4.1.2   Attention Assist

The Attention Assist is based on the LDWS and uses the extent to which the driver is staying in lane combined with driver activities (actuating turn signals/the accelerator pedal/the retarder or initiating engine braking etc.) and steering to determine a driver's attention level (see ◉ ▣ Fig. 36.9). The LDWS continuously records the position of the vehicle between the lane markings. The Attention Assist assumes that as drivers' alertness decreases, so does his or her ability to keep the vehicle "cleanly" in lane.

The driver activities used to calculate a "monotony value" that assumes that drivers' attention levels falls as driving time increases. However, if drivers exhibit active driving behavior, e.g. by actuating turn signals/the accelerator pedal/the retarder or initiating engine braking, this will have a positive effect on the "monotony value" accordingly.

The Attention Assist automatically activates every time the vehicle is restarted. It records all system-relevant data in the first 15 min, given that the vehicle be traveling on a road with identifiable lane markings at a speed greater than 60 km/h. If the limits stored in the system regarding attention are reached after a defined

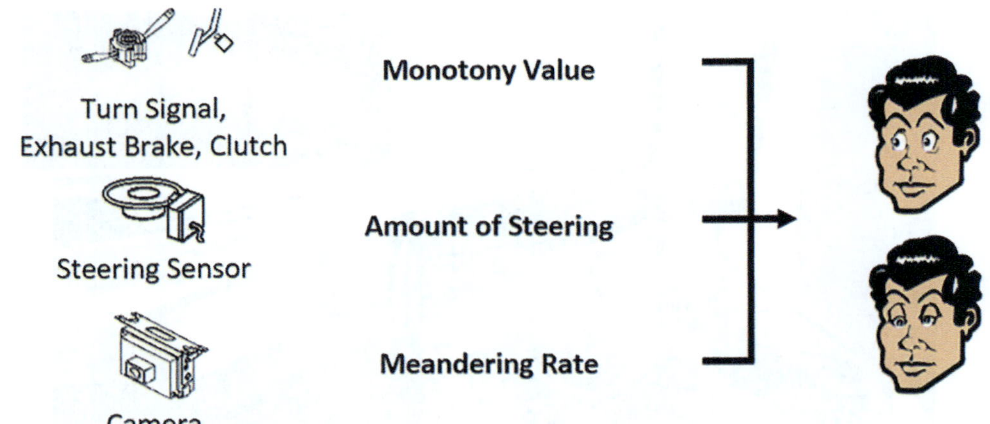

Fig. 36.9 Inputs for the attention assist

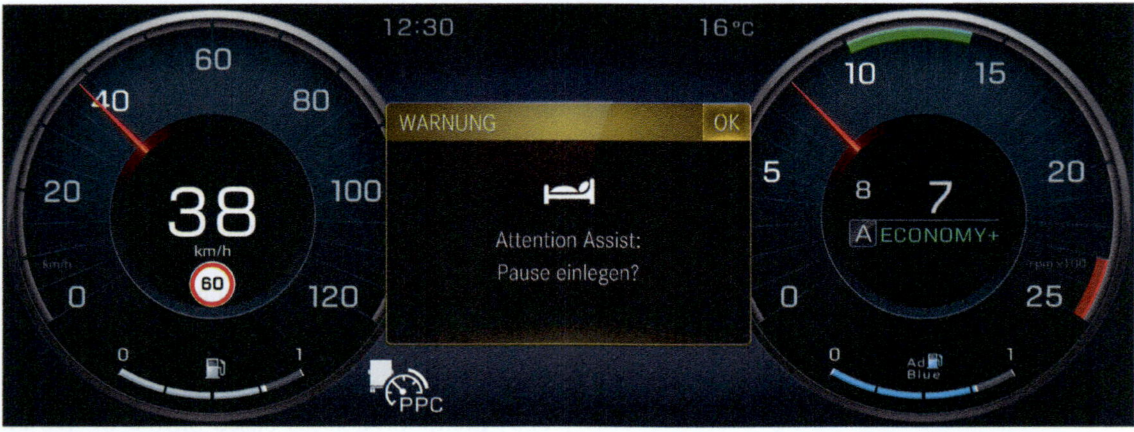

Fig. 36.10 "Please take a break" warning display on the instrument cluster

distance, drivers are warned via an optical signal (see ◉ **Fig. 36.10**) and a brief acoustic signal on the instrument cluster.

The basis for the development of the Attention Assist is the number of accidents that are attributable to driver fatigue. Germany's Federal Office for Roads and Traffic (BASt) states in its report [21] entitled *Erfassung der Fahrermüdigkeit* [Assessment of driver fatigue] that a considerable 10% to 20% of road traffic accidents are attributable to tiredness while driving ◉ **Fig. 36.10**. With the two regulation amendments Driver Drowsiness and Attention Warning (DDAW) [12] and Advanced Distraction Recognition (ADR) [12] EU, such systems will become mandatory in the coming years.

With Advanced Distraction Recognition, an additional camera (Driver Monitoring Camera, DMC) on or in the instrument panel analyzes driver's eyelid movement and line of sight, monitoring and warning the driver.

36.4.1.3 Traffic Sign Recognition

The Traffic Sign Recognition (TSR) supports drivers by displaying traffic signs relevant to commercial vehicles such as speed limits, overtaking bans, height, width or weight restrictions as well as additional information for commercial vehicles. The information generated from camera data that is processed via classification scripts as well as map data. Country-specific regulations, such as maximum speeds on freeways, are also taken into account as part of this. Multiple picture processing steps are necessary for analyzing a camera image for traffic signs. First of all, the system searches for shapes commonly used for traffic signs, such as circles, triangles or diamonds. During this step, all other image areas not required for this function are filtered out. If it is certain that the image section contains a traffic sign, a statistical analysis is carried out with the aim of determining which traffic sign this is. In addition, the position of the traffic sign is tracked.

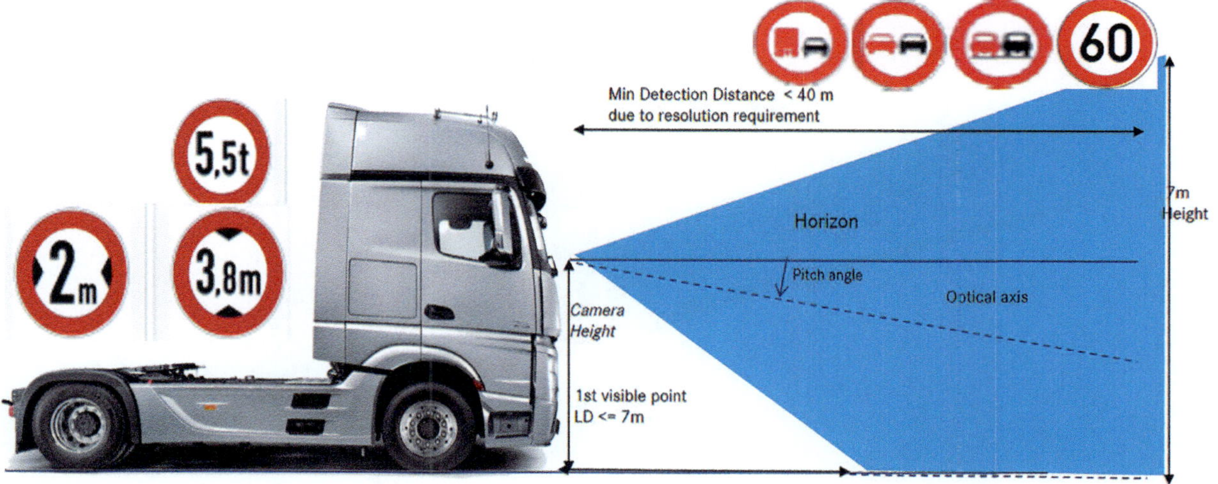

◘ Fig. 36.11 Overview of the field of view of a traffic sign recognition and possible recognizable traffic signs

The TSR intended primarily to inform the driver and draw his or her attention to prohibitions or speed limits. In combination with the ACC, automatic adjustment of the set speed is also a convenient extension of the active use of speed limiters (◘ Fig. 36.11).

36.4.1.4 High-Beam Assist

The High-beam Assist controls the automatic activation and dimming of the headlamps and thus provides optimum illumination over a long range as well as when the vehicle is turning off a road or driving around a bend. Using a camera, light sources from vehicles' rear lamps, headlamps and street lighting, e.g. within a range up to 200 m in urban areas, detected and the high beam is automatically turned on or dimmed. If the national approval regulations permit it, the roof headlamp is also controlled.

In addition, both the steering wheel angle and the turn signal position are processed and, as a result, the front fog lamps on the relevant side of the truck are activated to provide ideal illumination in the direction of travel on bends and when turning. This can significantly reduce accident risks that are caused by obstacles being detected too late or oncoming traffic being dazzled because the driver has forgotten to dim the lights.

36.4.2 Intervening Emergency Functions

36.4.2.1 Advanced Emergency Braking Systems

Advanced Emergency Braking Systems serve to support drivers in situations in which they are at risk of colliding with a moving or stationary vehicle in front as a result of distraction or inattention, for example. The objective of such systems is to prevent collisions in longitudinal traffic.

Since November 2015, the EU Commission has defined the Advanced Emergency Brake System (AEBS) as mandatory equipment for all newly produced trucks. With the aid of radar signals on the front of the truck, the AEBS measures the distance and speed differential to the vehicle ahead and calculates the "time to collision" (TTC); see also [28]. If a possible collision becomes relevant due to the TTC, the system will warn the driver 1.4 s before the necessary full braking, thus giving the driver the opportunity to prevent a collision by means of suitable braking or steering maneuvers. If the driver does not react to the warning signal, emergency braking is carried out by the AEBS system in order to avoid a collision or mitigate the impact. Active Brake Assist 5 (ABA 5) is a Mercedes-Benz-specific extension of such a system and detects or classifies the object with the aid of radar and camera systems (see ◉ ◘ Fig. 36.12). Merging this object data means that the system can carry out emergency braking to avoid stationary and moving vehicles. The system also reacts to moving pedestrians by initiating emergency braking.

ABA 5 has a three-stage principle for vehicle warnings. If the system detects a vehicle that is moving, has stopped or is stationary, a visual and acoustic warning will be issued to alert the driver first. In the process, the radio and hands-free system will be muted so that the driver is aware of the situation. If the driver does not respond adequately despite the visual and acoustic warning, the system will initiate partial braking at 3 m/s^2 in the next step, which corresponds to approximately 50% of the maximum deceleration. In the third warning stage, the system performs autonomous maximum full braking in the event of an imminent collision, which brings the vehicle to a standstill within the system limits of the truck if necessary. At the same time, the traffic behind is made aware of the critical situation by automatic activation of the hazard warning lights.

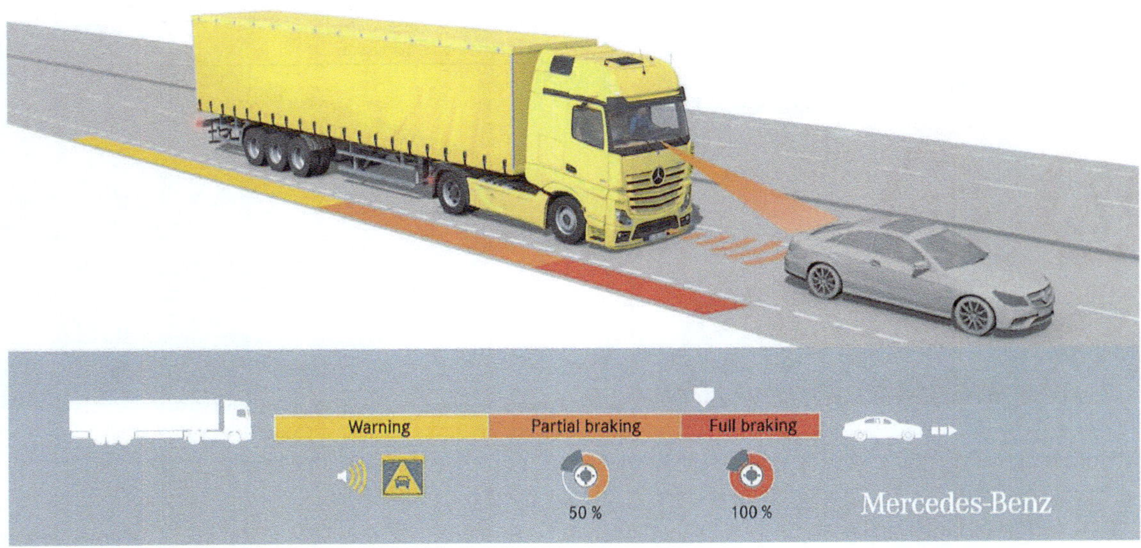

◘ Fig. 36.12 Cascade of functions for ABA 5: warning|partial braking|full braking

At standstill, the vehicle is automatically secured using the parking brake and "Emergency braking in progress" is displayed on the instrument cluster. The vehicle will remain secured at standstill until the driver wishes to remove the vehicle from the danger zone by pressing the acceleration pedal. At that point, the hazard warning lights will also be switched off again.

ABA 5 also supports drivers by means of a two-stage warning principle when a collision with a person who is crossing, approaching or walking in the same lane is imminent. At first, the system responds with partial braking (50% of the maximum braking power) and then, if necessary, with full braking (see ◉ ◘ Fig. 36.13).

◘ Fig. 36.13 ABA 5 traffic situations with moving pedestrians

36.4.2.2 Sideguard Assist

The areas that are not visible or are difficult to see are significantly larger for heavy commercial vehicles than for passenger cars due to their own size. This blind spot is a problem during turning and changing lane especially. In particular, vulnerable road users such as pedestrians or cyclists, who are particularly at risk in the event of a collision, are not always aware that truck drivers may not be able to see them if they are next to the vehicle.

Sideguard Assists supports drivers in situations with restricted visibility. The aim here is to prevent critical situations involving the "collision with vulnerable road users during turning off" and "collision during lane change" major accident types without creating new hazards in road traffic in the process.

Such systems monitor the vehicle's surroundings, especially on the front passenger side and directly in front of the vehicle, using a radar sensor system to detect obstacles in the blind spot (see ◉ ◘ Fig. 36.14). If an obstacle is detected, a multi-level warning system is used. If there is no direct risk for a collision, drivers will be informed of the detected obstacle at the first warning level. This first warning level is intended to bring drivers to observe their surroundings more carefully and to take this into account in subsequent activities. From a technical point of view, the first warning level can be implemented as an optical signal in the form of a warning lamp, for example. If a situation intensifies such that it poses imminent danger, the second warning level will become active. This second warning level should trigger an immediate driver reaction in order to prevent the imminent collision. For technical implementation, it is possible to alter the optical signal from the first warning level by changing the color, turning on additional warning lamps or having the warning lamps flash, for example. An acoustic warning signal can also be added. Both optical and acoustic signals are usually issued in a targeted manner to aid the driver's situational awareness. For example, the acoustic warning signal sounds from the speakers on the side of the vehicle where the hazardous situation is occurring.

In particular, Active Sideguard Assist (ASGA) from Mercedes-Benz Truck offers the opportunity to defuse critical situations via an active brake intervention. This brake intervention takes place at the same time as or after the second warning level and slows the vehicle down to a standstill. Drivers can override the emergency braking that is initiated by pressing the acceleration pedal. Such a situation is shown in ◉ ◘ Fig. 36.15. The person driving the commercial vehicle overlooks a cyclist in the blind spot when turning right. ASGA detects the cyclist and warns the driver of the commercial vehicle. If the driver continues with the turning maneuver, ASGA will initiate emergency braking to prevent a collision.

36.4.2.3 Emergency Stop Assist

Emergency Stop Assists (ESA) bring the vehicle to a standstill independently and safely when drivers are no longer able to, e.g. due to an emergency. If no driver activity is detected over a longer period of time, an optical and acoustic warning is issued requesting the driver to put his or her hands on the steering wheel (see ◉ ◘ Fig. 36.16). If, after 60 s and even after multiple warnings, a driver does not interact with the steering wheel, the accelerator pedal, the brakes or with vehicle systems via the steering wheel buttons, for example, the system will safely slow the truck down to a standstill in its lane and warn the following traffic using hazard warning lights.

◘ **Fig. 36.14** Sideguard assist radar sensor system

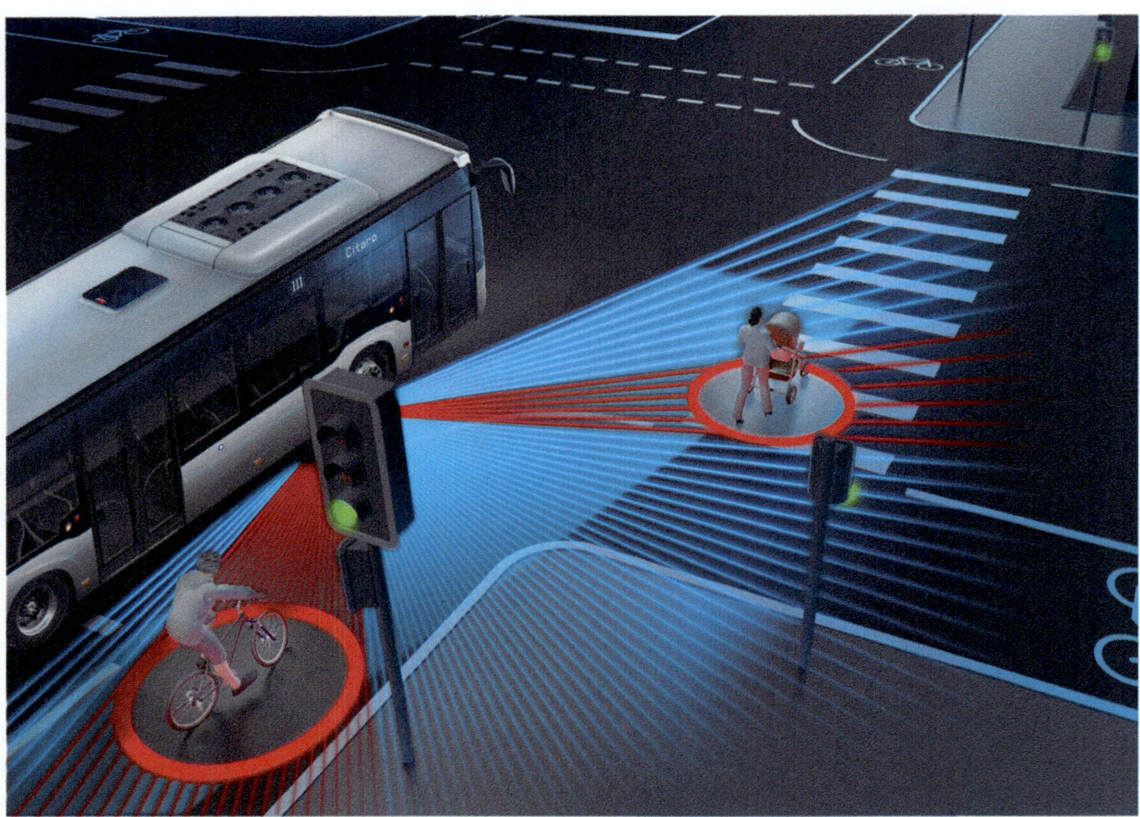

Fig. 36.15 Critical traffic situation for potential sideguard assist intervention while turning right

Fig. 36.16 Traffic situation for emergency stop assist intervention

36.4.3 **Continuously Automating Functions**

36.4.3.1 **Adaptive Cruise Control**

Adaptive Cruise Control (ACC) extends the simple cruise control that is used to control speed with the addition of distance control. Once this system is switched on and set, it continuously monitors the distance. A typical scenario is shown in ◉ ◘ Fig. 36.17. In order to drive the vehicle safely in the best possible way, the system is controlled by deceleration (limited to 2 m/s^2, or 3 m/s^2 in stop-and-go traffic) and acceleration (limited to 1 m/s^2) as required. Detection of vehicles in front is possible up to 200 m and an ego speed of 0 to 89 km/h (up to 100 km/h for touring coaches, up to 120 km/h in some regions). The stop-and-go function is used in traffic jams. If necessary, the vehicle brought to a standstill using the brakes and then accelerated again. However, if a standstill lasts for longer than 2 s, the driver must permit the system to move the vehicle off by pressing a button or actuating the accelerator pedal. If the time at a standstill is less than 2 s, ACC can accelerate the vehicle again independently and without any intervention from the driver. At standstill, the parking brake is automatically engaged and the ACC function deactivated when the driver's seat is left or the driver's door is opened. The range for the target time gap can be set to any of five levels between 1.5 s and 3.6 s depending on the driver's input.

To ensure that ACC operated with minimum fuel consumption in commercial vehicles, alongside a satellite-based positioning system and precise digital road maps, the vehicle can accelerate predictively, make active use of roll phases and automatically adapt the gear selection to the topography and route events. This involves predictive distance control using the vehicle's kinetic energy as often as possible and, for example, beginning a coasting phase at an early stage prior to steep descents. The truck's momentum thereby helps it over the crest of a hill, and the downhill gradient is used to speed the truck up to the set speed once more. This also minimizes the use of the service brake. In addition, torque management predictively optimizes gear selection and shift timing points for driving at the optimum engine operating points for fuel consumption on all routes and in all load conditions.

36.4.3.2 **Lane Keeping Assist**

This system designed to actively guide the vehicle back to its lane in the event that it strays. As a basic principle, this function is linked to a LDWS. This means that intervention from the lane keeping assist is possible only with an active LDWS. To enable the system to intervene, the lanes must be clearly identified. If the vehicle crosses the lane markings without actuating a turn signal lamp and without the driver intervening, the driver will be warned and the vehicle will be guided back to its lane and straightened up for maintaining this course. This process is subdivided into two steps:

1. Steering intervention to guide the vehicle back to its lane
2. Center guiding is established (lateral control).

The warnings issued are acoustic. After every intervention by this system, Lane Keeping Assist remains inactive for 10 s, meaning that the driver needs to take over within this time.

36.4.3.3 **Active Drive Assist**

Active Drive Assist is the first partially automated system as per the SAE definition of Level 2 that designed for freeways and similar roads. The system can take over longitudinal and lateral guidance throughout the entire vehicle speed range of 0–89 km/h (up to 100 km/h for touring coaches, up to 120 km/h in some regions). For this purpose, the ACC (with stop-and-go function), Lane Keeping Assist and ESA are combined.

◘ **Fig. 36.17** Traffic situation with active adaptive cruise control and display on the HMI

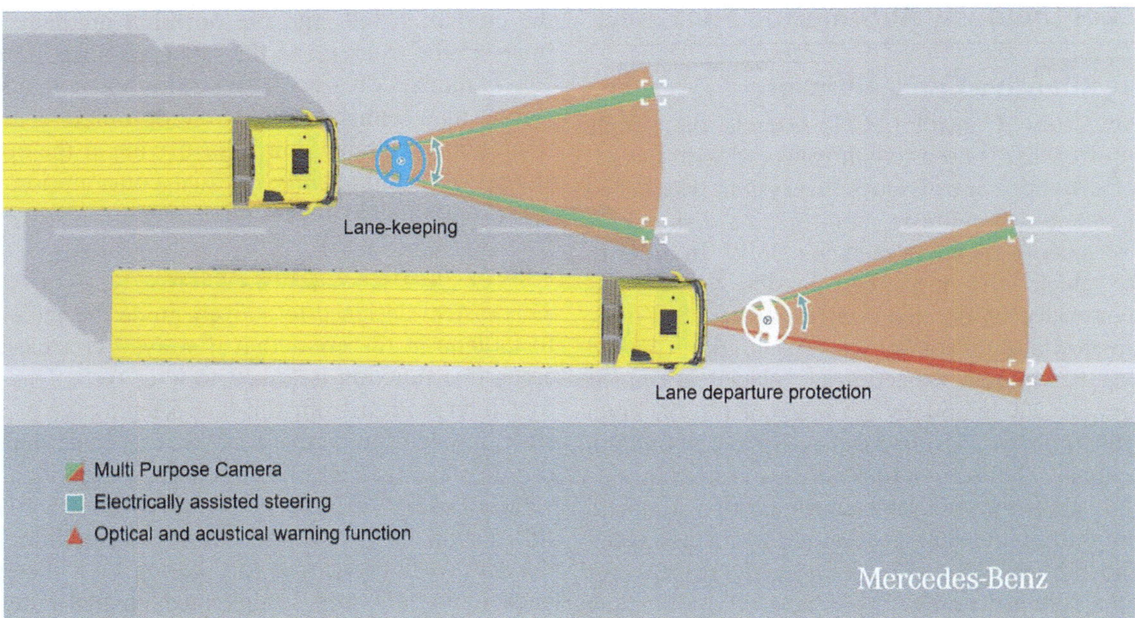

Fig. 36.18 Function principle for lateral control of active drive assist

Active Drive Assist helps drivers to maintain a safe distance from the vehicle in front. Lane Keeping Assist and the Servotwin electrohydraulic steering system are responsible for actively guiding the truck laterally in the lane (see ◉ **Fig. 36.18**). and also taking care that the truck follows the street. The driver can define via a menu if the system will keep the truck exactly on the center of the street lanes or if it orientates a little bit more left or right. With that the driver can configure the system similar to the own driving style. Another feature is driver monitoring. If there is no driver activity for an extended period of up to 60 s, an optical and acoustic warning will be triggered and the vehicle slowed down to a standstill (ESA). It is also possible to switch off the system.

36.5 Testing Procedures

Essentially, the testing processes and procedures applied in the commercial vehicle segment are based on familiar approaches from the passenger car segment (see ▶ Chaps. 9 to 13). In addition to providing a general overview, this section will therefore address special considerations for commercial vehicles.

36.5.1 Testing and approval processes

V-model [22] and the ISO 26262 [16] standard based on it are also important references for the development and testing processes in the area of driver assistance. Since driver assistance systems consist of multiple components, it is essential to coordinate responsibilities for the testing of components at the respective test stages between the OEM and suppliers. Thus lower test levels are frequently taken over by the supplier, while the OEM is responsible for integrating and testing the overall system.

To issue binding approvals for vehicle tests and the start of production (SOP), there is a separate approval process that also picks up on some of the existing test stages. An example of such a process is shown in ◉ **Fig. 36.19**. Once all component tests are passed, an approval for proving ground tests is issued. If this test stage is also successfully passed, an approval for field tests and later customer trials can be issued as well. As part of the field tests for the latest generation of Active Brake Assist from Daimler Truck AG, driving data for a total of 5 million km was recorded and analyzed. All noticeable problems are collected and reported back to

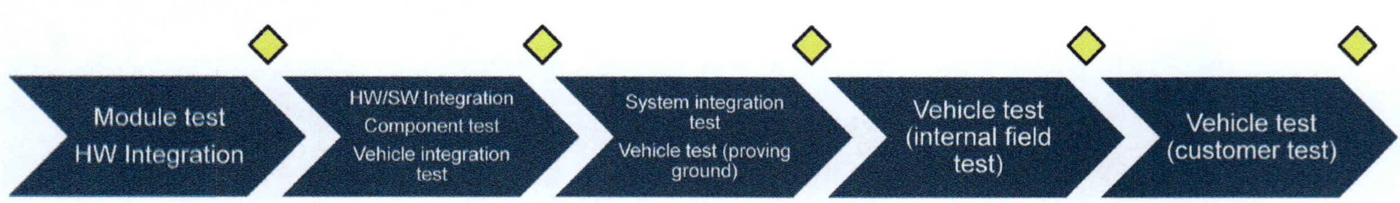

Fig. 36.19 Release process for driver assistance systems of commercial vehicles

the development team so that any necessary functional modification can be implemented.

36.5.2 Test Objectives and Test Environments

Test goals during the development of driver assistance systems are derived from functional safety (ISO 26262) [16] as well as SOTIF (Safety of the Intended Functionality, ISO/PAS 21,448) [17]. In this process, different test case derivation methods are used depending on the function or system to be tested. The test cases generated are assigned to test levels or test environments and are documented in a general test case catalog.

36.5.2.1 Test Environments

The following test environments are being used in developing ADAS for series production:

- XiL (X-in-the-Loop): Deterministic test cases or stochastic scenario-based sampling [18] is used for the unit test (SiL), module test (SiL), hardware integration test (HiL), component test (HiL) and system integration test (HiL) [19].
- Open-loop resimulation: Use of recorded data for efficient regression testing of the software for sensor or functional algorithms. Closed-loop can also be represented to a limited extent [2].
- Proving ground: deterministic test cases, repeatedly tested in the vehicle. Also EMC and rough-road test.
- Field test/trial: specified or random route, global objectives for route to be covered (region/type of route, number of test kilometers), KPIs derived from driving data.
- In addition, there are test environments especially for diagnostic tests, on-board electrical system test and network test.

36.5.3 Special Considerations for Commercial Vehicles

In addition to increased requirements in terms of the availability, reliability and robustness of a driver assistance system for commercial vehicles that are ensured through an effective validation process, the following two aspects must be taken into account in testing:

- Inclusion of customers: Because a commercial vehicle is an item of work equipment that is used by professional drivers over a long period, it is essential that the system is user-friendly. Customer trials allow to collect driving data and driver feedback for the respective use cases.

- Global variants: For driver assistance systems in commercial vehicles that developed for a global market, the diversity of the applicable markets must be taken into account in the testing process. While this can still take place centrally for proving ground tests, it is essential to carry out field tests locally in each region. Thus, data is recorded in five different regions around the world during the development of ADAS for Daimler Truck AG. Global driving data from customers are analyzed even after SOP continuously to gain relevant insights for development.

36.6 Future Developments

36.6.1 Vision Zero

Zero fatalities by 2050—that is the goal of Vision Zero. Germany's government made this vision an integral part of its program. It is also to be incorporated into the German Road Traffic Regulations (StVO) as a guiding principle. The next milestone on the journey is cutting the number of fatalities and severe injuries to half by 2030 compared to 2020. With their safety and assistance systems, vehicle manufacturers can also make an important contribution to achieving these goals.

Safety systems are playing an increasing role in the context of consumer protection, too. The European New Car Assessment Program (Euro NCAP) is a program for consumer protection that evaluates the safety of vehicles. Euro NCAP enables you to find out about vehicle safety quickly and in detail. So far, the ratings have been carried out for passenger cars; an Euro NCAP rating for commercial vehicles is in preparation [27]. In addition to various crash scenarios, the rating also covers the most important driver assistance systems. The rating of the assistance systems is determined by means of tests on the assistance technologies that prevent accidents and help to reduce the extent of injuries. For commercial vehicles, various systems for urban traffic (pedestrians, cyclists, maneuvering, intersections) and freeway traffic (AEB, driver attentiveness) could play a role.

Thus the challenges for future driver assistance systems consist of mastering more complex traffic and accident scenarios. If the active brake assist reacts to precisely one object that is traveling or has stopped in a defined lane in front on the freeway, there is a large number of possible configurations at an intersection in an urban area. Two or more vehicles traveling at considerably different speeds can react to a collision course with entirely different maneuvers, such as braking to adjust speed, accelerating or evading. Added to this are restricted fields of view and a lack of information on the other vehicle's intended change of direction.

All these circumstances represent additional challenges for sensor systems as well as their signal and data processing that are not covered adequately at present. Interesting technological developments that can meet these requirements in the future are addressed below.

36.6.2 Sensors and Applications

The dominant sensors in commercial vehicles today are radar and camera sensors. Ultrasonic sensors play a subordinate role due to their very limited scope of application in the immediate vicinity and their susceptibility to malfunctions under the tough operating conditions of a truck.

While radar sensors are particularly well suited to detecting moving objects and their longitudinal speed, camera systems are good for detecting stationary objects, their physical size and lateral speed. The following chapters will look at future developments of these sensors as well as other interesting technologies and their possible application in commercial vehicles.

36.6.2.1 Camera Sensors

Cameras as sensors can basically be divided into three groups. The first group comprises the "smart" cameras that have integrated or external image processing and extract information from the recorded image, such as vehicles or pedestrians, in a targeted manner. The second group includes cameras that allow recordings and evaluations in respect of optical wavelengths. The third group comprises cameras that are able to obtain three-dimensional information from the image recorded. In addition to an improvement in performance characteristics such as resolution and field of view (FoV), the great potential for future applications is attributable to thermal cameras and three-dimensional cameras in particular.

Thermal cameras use thermographic sensors that are sensitive in the infrared range of the spectrum. The image data obtained can be used to deduce the temperature of the recorded object. As a result, objects that are only partially visible, such as a person standing between parked cars, can also be detected more easily and reliably than with classic camera technology. A significantly improved ability to see at night and in poor weather represents another advantage. Eventually, many accidents occur at night or in the event of poor visibility (◘ Fig. 36.20).

The three-dimensional cameras include stereo cameras that can be used to determine distances via the principle of triangulation, as well as cameras that work with an optical lens in accordance with the time-of-flight (ToF) principle. ToF cameras have an additional active light source in the non-visible wavelength range and thus make it possible to obtain information on the distance to an object through the travel time of the reflected light. They also enable detection of free spaces, which is crucial for some new applications, such as support for safely launching a truck and not least autonomous driving.

36.6.2.2 NLOS Systems

One interesting application is non-line-of-sight (NLOS) technology, on which research is being conducted in combination with radar and optical sensors.

In NLOS imaging, a diffusely reflecting surface is scanned with a laser, whereby the reflected light is re-

◘ **Fig. 36.20** Comparison of a multi purpose camera with a thermal camera; camera: AdaSky 2020; picture: Daimler Truck AG 2020

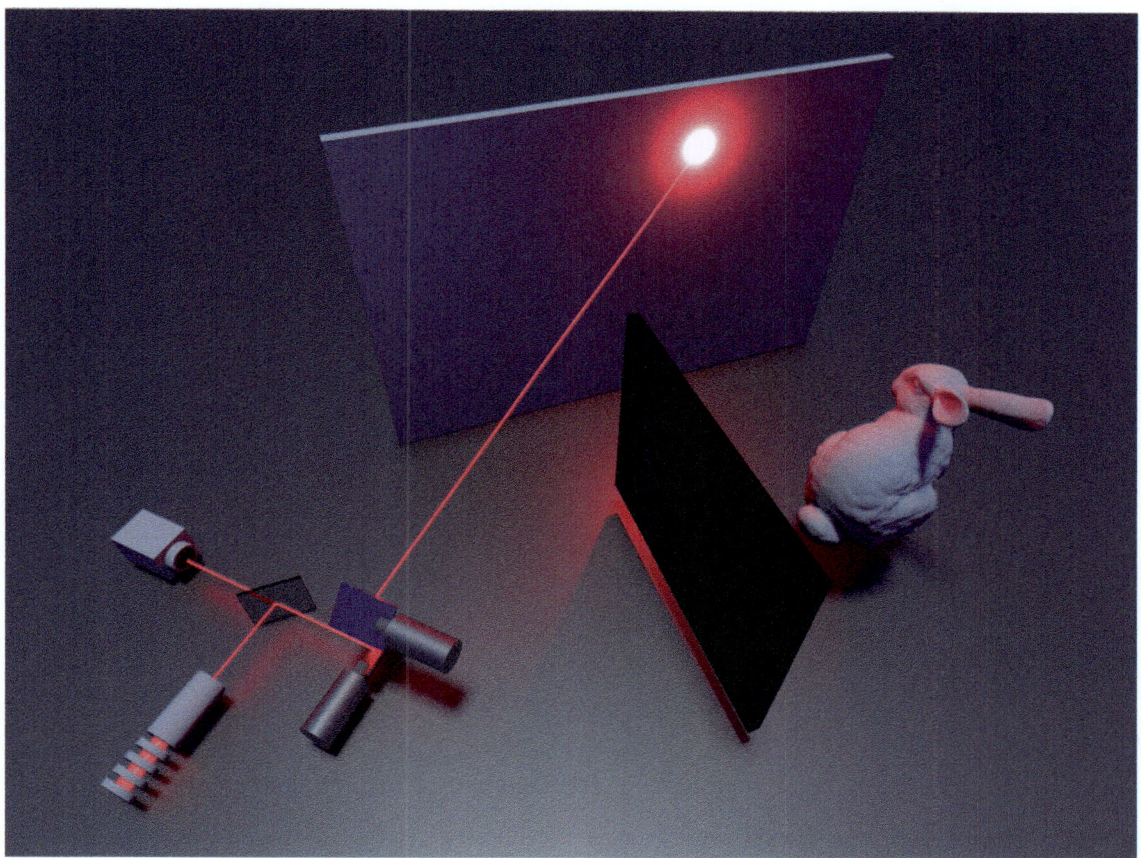

◘ Fig. 36.21   Functional principle NLOS imaging [20]

directed from there to a concealed object. Via the same path, parts of the light reflected by the object then find their way back into the camera. The light indirectly reflected from the object that is captured can be used to reconstruct an image of the concealed object by means of algorithms. What makes this method particularly interesting is that it can be used effectively with current lidar scanners. However, with today's high-performance computers, it still takes several minutes to generate such an image. For practical use, the algorithm would have to be further optimized and the processing power in the vehicle significantly increased (◘ Fig. 36.21).

The principle of NLOS radar technology is very similar. This too works with reflections, in this case from radar waves, and also uses state-of-the-art technologies such as Doppler radar and GPUs. Unlike the lidar-based technology, the required information from the radar signals can already be calculated virtually in real time today (◘ Fig. 36.22).

36.6.2.3 Lidar

Lidar (light detection and ranging) technology is very closely related to radar technology, but also to the time-of-flight principle of camera technology. In con-

trast to radar, lidar simply uses a different part of the electromagnetic spectrum. Thereby, light pulses are emitted by a sensor, reflected by an object and captured again by a detector. The distance to the object can then be calculated using the time difference between transmission and reception.

Systems available today already include imaging lidars that enable high-precision three-dimensional mapping of the object space with a high resolution. The high-resolution point clouds of such imaging lidars can be used for surveying surrounding areas and in helping autonomous vehicles to navigate (◘ Figs. 36.23 and 36.24).

Lidar sensors are still very expensive compared with conventional environment sensors such as cameras or radar. Due to the moving parts required (rotating sensor head, MEMS mirror) such high-performance lidar systems can't be used in the tough conditions that a commercial vehicle is driving in. Lidar systems without moving parts, known as solid-state lidars, currently achieve only short ranges, which limits their possible applications in commercial vehicles. In particular lidar systems will be used in the context of autonomous driving.

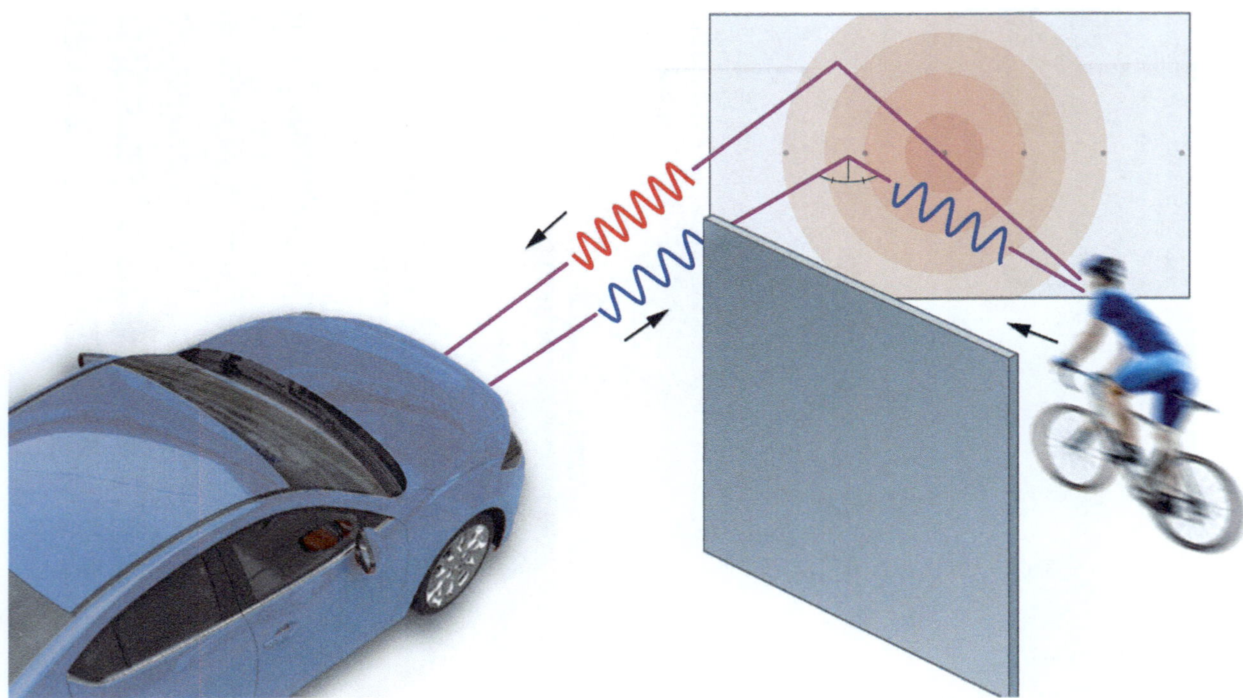

Fig. 36.22 Functional principle NLOS radar [23]

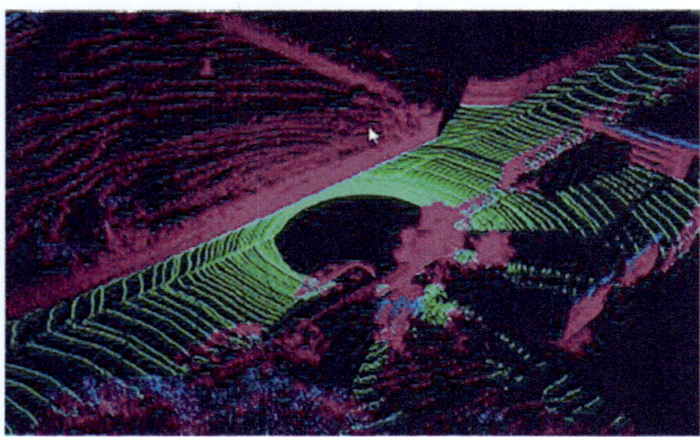

Fig. 36.23 Example of a lidar point cloud; Daimler Truck AG: own measurement

36.6.2.4 Other Innovative Sensors and Sensor Technologies

In addition to the conventional sensors, sensors for automotive use are emerging that can be used to cover other aspects of driver assistance.

These include biometric sensors that can be used in the implementation of biometric person recognition systems. One possible automotive application could be the automatic opening of the vehicle with the aid of an outward-facing camera.

Contactless measurement of heart rate and breathing rate by means of radar and optoelectronic sensor technologies (e.g. cameras) is particularly inter-esting for commercial vehicles, because fatigue and stressful situations that put strain on the heart are recurring causes of accidents involving professional drivers. Monitoring the driver's vital signs, visual behavior and general attentiveness could reduce the risk of such accidents in road traffic. The development of automated driving also increases the need to monitor the driver's condition. Even convenience applications could be geared to the needs of the occupants.

Today, methods of this kind are found in research projects or stationary applications such as identity checks at the airport and are not yet sufficiently reliable to detect stress or physical impairments in the vehicle environment. To further increase the reliability of driver status detection, a large amount of information from redundant sensors could be linked and evaluated using artificial intelligence (**Fig. 36.25**).

Quantum sensors are particularly interesting for use in automobiles due to their extremely high sensitivity compared with today's conventional sensors.

One great advantage of the quantum sensor compared with the quantum computer is the required temperature range, which, in principle, enables use in natural environments. However, the quantum sensor is still at the research stage, with the materials and the miniaturization of the sensor in particular representing challenges. One interesting application for quantum sensors from today's perspective lies in high-precision inertial sensors for GPS-free navigation. With them, the highly sensitive quantum sensors can become one of the tech-

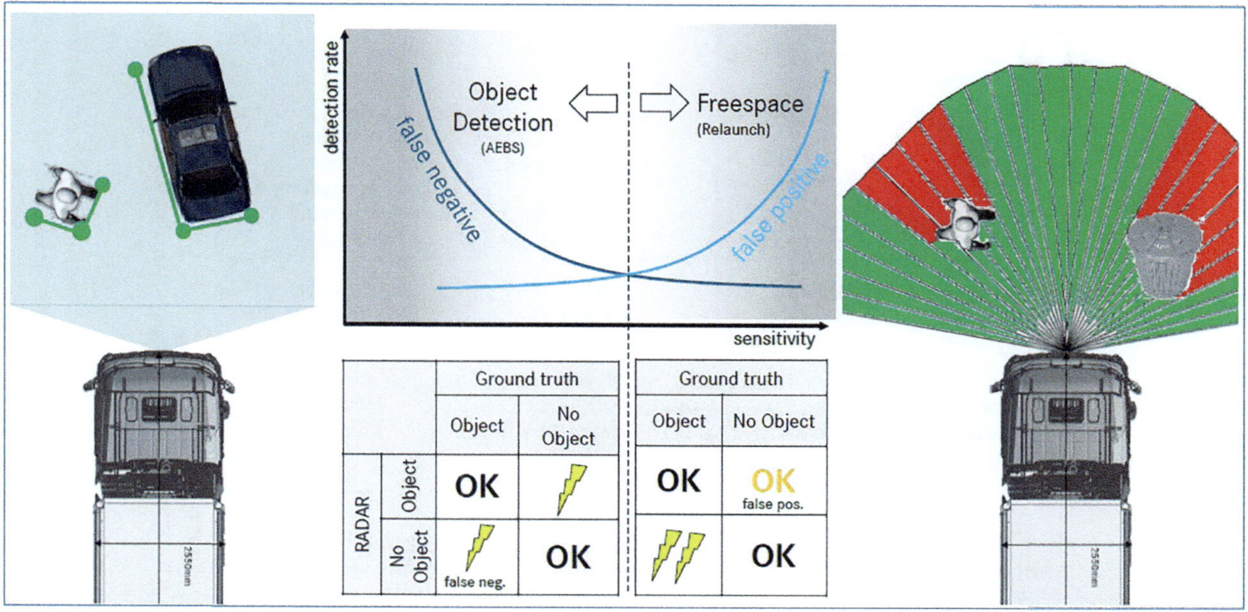

		Ground truth		Ground truth	
		Object	No Object	Object	No Object
RADAR	Object	OK	⚡	OK	OK false pos.
RADAR	No Object	⚡ false neg.	OK	⚡⚡	OK

Fig. 36.24 Sensitivities in object detection and freespace detection with radar

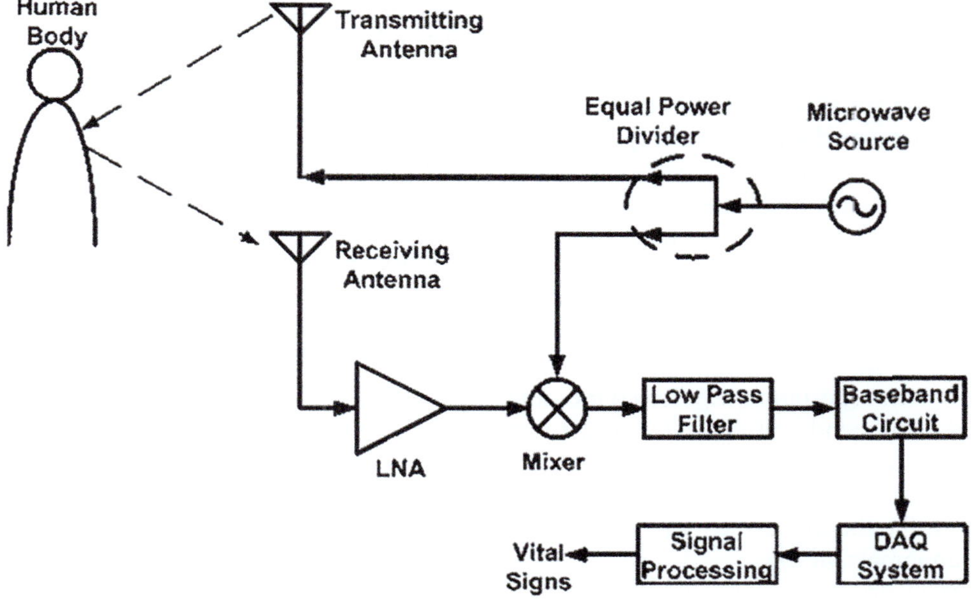

Fig. 36.25 Block diagram of a doppler radar for vital sign recording [8] (modified)

nological keys to GPS-free fully autonomous driving (◻ Fig. 36.26).

Machine learning (ML) is one of the most important elements for future driving control and driver assistance systems. With ML, data can be trained in terms of its statistical properties and therefore used for regression predictions and the classification of detected objects. One special characteristic of ML in this respect is that it can be used to process time series and high-dimensional data structures in a far shorter time than is possible with conventional rules-based approaches.

In particular, the challenges presented by the use of ML lie in user acceptance and the validation of functions based upon it. Because, unlike rules-based approaches, ML-based results are not guaranteed to be 100% correct, but rather represent a statistical estimation with a defined degree of confidence. Another challenge arises from data access and data utilization,

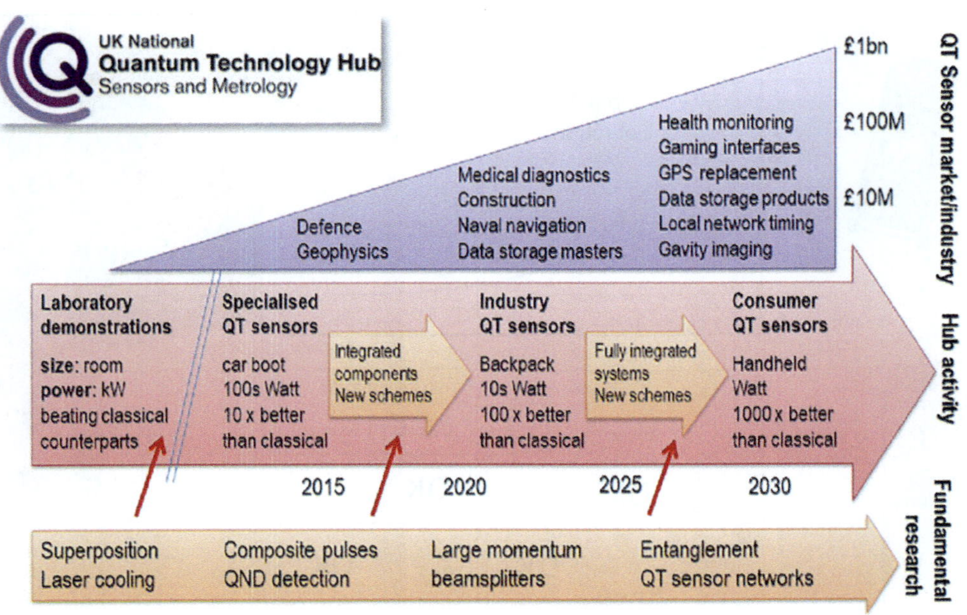

Fig. 36.26 10–years forecast on market development in quantum sensor systems [5]

which must meet the stringent legal requirements for handling data.

36.6.2.5 Other Applications in Driver Assistance and Active Safety

The availability of the technologies described above gives reason to expect many other useful safety and assistance systems in commercial vehicles in the future.

In particular, great importance is to be attached for the detection of pedestrians and cyclists, referred to as vulnerable road users, or VRUs. Urban populations are growing and the bicycle is becoming increasingly popular as a means of transportation, e.g. for courier and delivery services. This trend will inevitably lead to an increase in serious accidents with vulnerable road users that must be counteracted. In this area, new technologies, such as lidar, the thermal camera or NLOS imaging, could play an important role.

Another area in driver assistance is complex traffic situations such as those that arise at intersections. In contrast to the emergency braking assist, a variety of configurations are possible at intersections and turnings, and it is necessary to look out for a large number of objects with different directions of travel and degrees of maneuverability. This task can be completed fully only with sophisticated and powerful algorithms and connected sensors across the involved vehicles.

Another area of action in driver assistance, especially for commercial vehicles with trailers, is safe maneuvering. This is because a driver does not have a direct view in various situations when driving with a trailer. In contrast to passenger cars, driving with a trailer is a standard case in road freight transport (semitrailer truck). For a sensory rear view, interconnected systems must be established between vehicle and trailer offering a standardized broadband data interface. This allows sensor data from the trailer to be processed in the towing vehicle and thus generate an all-round view that enables supporting maneuvering functions.

Since 90% of the serious accidents involving commercial vehicles are attributable to human error, drivers and their condition and attentiveness play a major role in the occurrence of accidents. In the future, effective attention monitoring systems must be developed that completely obviate interventions by an emergency braking assist by restoring the driver's attention in advance of the collision. Visual interior and driver monitoring is assigned an important role here, but contactless heart and breathing rate measurement and the "electronic nose" based on quantum sensors, e.g. for determining alcohol levels from the driver's breath, can also make a contribution (**Fig. 36.27**).

36.6.3 Fully Automated Driving

Emission-free and fully automated driving are among the greatest challenges facing the automotive industry in the 2020s. While the transition to emission-free drive systems is highly motivated by sociopolitical factors, fully automated driving promises a new future for a transport sector stricken by a driver shortage and low margins. Fully automated driving is to be understood

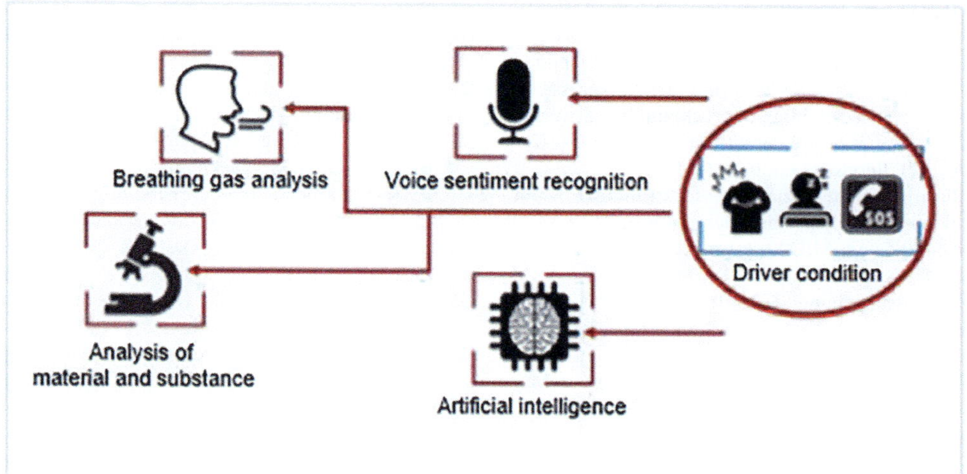

Fig. 36.27 Driver status detection by means of sensors/data fusion

as the driverless operation of vehicles in a specific part of a transport chain at the beginning and end of which is a technically and economically reasonable handover to the preceding or subsequent process step.

The VDA (German Association of the Automotive Industry) divides the automation of vehicles into five levels that do not differentiate between passenger cars and commercial vehicles. While people are responsible for driving the vehicle at all times at levels 0–2, levels 3–5 either partially or fully relieve them of this responsibility. A grouping that is suitable for use in practice with commercial vehicles can be found in the expert opinion of the Board of Academic Advisers to Germany's Federal Minister of Transport and Digital Infrastructure. Here, the automation levels are subdivided into two groups in view of the future technical implementation possibilities.

Group A comprises vehicles that continue to be operated by capable drivers who can activate and deactivate automated driving as well as complete the journey if automated driving reaches its limits. Therefore, the psychological aspects of the driver's role are of particular importance in the expert opinion [4].

Vehicles in group B drive a (possibly limited) route from the beginning to the end with no driver. An operator is required instead of the driver to operate these vehicles. Therefore, the expected operator models are being analyzed with regard to their tasks and obligations [4].

While the driver assistance systems described in the preceding chapters all correlate with Group A, the following chapters will address also the systems and technologies for vehicles in group B.

References

1. ADAC: Bremswege im Vergleich (2019). ► https://www.adac.de/rund-ums-fahrzeug/tests/autotest/bremswege-vergleich/. Last accessed 9 Nov 2021
2. Bach, J., Holzäpfel, M., Otten, S., Sax, E.: Reactive-replay approach for verification and validation of closed-loop control systems in early development. SAE Technical Paper Series. WCX™ 17: SAE World Congress Experience (2017)
3. BMVI: Verkehrsverflechtungsprognose 2030 (2014). ► https://www.bmvi.de/SharedDocs/DE/Artikel/G/verkehrsverflechtungs-prognose-2030. Last accessed 9 Nov 2021
4. BMVI: Automatisiertes Fahren im Straßenverkehr—Herausforderungen für die zukünftige Verkehrspolitik. Gutachten des Wissenschaftlichen Beirats beim Bundesminister für Verkehr und digitale Infrastruktur. In: Zeitschrift für Straßenverkehrstechnik, Issue 8 and 9 (2017)
5. Bongs, K., Boyer, V., Cruise, M., Freise, A., Holynski, M., et al.: The UK National Quantum Technologies Hub in Sensors and Metrology (Keynote Paper). SPIE Photonics Europe, Brussels (2016)
6. Bouteldja, M., Cerezo, V.: Jackknifing warning for articulated vehicles based on a detection and prediction system. In: International Conference on Road Safety and Simulation, vol. 3, pp. 14–16 (2011)
7. Cornelis, S., Todts, W.: Eliminating truck blind spots—a matter of (direct) vision (2016). ► https://www.transportenvironment.org/discover/eliminating-truck-blind-spots-matter-direct-vision/. Last accessed 8 Nov 2021
8. Dalal, H., Basu, A., Abegaonkar, M.P.: Remote sensing of vital sign of human body with radio frequency. CSI Trans. ICT **5**, 161–166 (2016)
9. DEKRA: Verkehrssicherheitsreport 2018 Güterverkehr (2018). ► https://www.dekra-roadsafety.com/media/de/dekra-evs-report-2018-de-final.pdf. Last accessed 8 Nov 2021
10. Europäisches Parlament: Verordnung (EG) Nr. 561/2006 des Europäischen parlaments und des rates. EG 561/2006 (2006)
11. Europäisches Parlament: Verordnung (EG) Nr. 661/2006 des Europäischen parlaments und des rates. EG 661/2009 (2009)

12. Europäisches Parlament: Verordnung (Eu) 2019/2144 des Europäischen parlaments und des rates. EU 2019/2144 (2019)
13. Frühauf, N., Roth, J.J., Schygulla, M.: Aus-und Weiterbildung von Lkw-und Busfahrern zur Verbesserung der Verkehrssicherheit: training and advanced training of truck and bus drivers for improving traffic safety. Berichte der Bundesanstalt für Straßenwesen Reihe M (197). Bergisch Gladbach (2011)
14. infas Institut für angewandte Sozialwissenschaft GmbH: Mobilität in Deutschland 2017—Ergebnisbericht. ► http://www.mobilitaet-in-deutschland.de/pdf/MiD2017_Ergebnisbericht.pdf. Last accessed 8 Nov 2021
15. Institut für Mobilitätsforschung: Zukunft der Mobilität. Szenarien für das Jahr 2030. BMW Verl. (ifmo-Studien) second update, 1st edn. Munich (2010)
16. ISO 26262–2: Road vehicles—functional safety—part 2: Management of functional safety (2018)
17. ISO/PAS 21448: Road vehicles—safety of the intended functionality (2019)
18. Jesenski, S., Stellet, J.E., Branz, W., Zollner, J.M.: Simulation-based methods for validation of automated driving: a model-based analysis and an overview about methods for implementation. In: 2019 IEEE Intelligent Transportation Systems Conference—ITSC. Auckland (2019)
19. von Neumann-Cosel, K.: Virtual test drive. Diss. Technische Universität München, Munich (2014)
20. O'Toole, M., Lindell, D.B., Wetzstein, G.: Confocal non-line-of-sight imaging based on the light-cone transform. Nature (2018)
21. Platho, C., Pietrek, A., Kolrep, H.: Berichte der BASt—Erfassung der Fahrermüdigkeit. HFC Human-Factors-Consult GmbH, Berlin (2021)
22. Sax, E. (Hrsg.): Automatisiertes Testen Eingebetteter Systeme in der Automobilindustrie, 1st edn. Hanser Verlag, Munich (2008)
23. Scheiner, N., Kraus, F., Wei, F., Mannan, F., Phan, B., Appenrodt, N., Ritter, W., Dickmann, J., Dietmayer, K., Sick, B., Heide, F.: Seeing around street corners: non-line-of-sight detection and tracking in-the-wild using doppler radar. In: CVPR (2020)
24. Statistisches Bundesamt (Destatis): Verkehrsunfälle—Series 8 (7) (2020)
25. Statistisches Bundesamt (Destatis): Unfälle von Güterkraftfahrzeugen im Straßenverkehr 2019 (2020)
26. UNECE 130: Uniform provisions concerning the approval of motor vehicles with regard to the Lane Departure Warning System (LDWS) (2014). ► http://data.europa.eu/eli/reg/2014/130/oj. Last accessed 8 Nov 2021
27. Euro NCAP: Euro NCAP Mobilizes New Partners to Promote Safer and Cleaner Mobility in Europe (2020). ► https://www.euroncap.com/de/presse/pressemitteilungen/euro-ncap-mobilizes-new-partners-to-promote-safer-and-cleaner-mobility-in-europe/. Last accessed 10 Nov 2021
28. Winner, H.: Grundlagen von Frontkollisionsschutzsystemen In: Winner, H., Hakuli, S., Lotz, F., Singer, C. (eds.) Handbuch Fahrerassistenzsysteme, 3rd edn, ATZ/MTZ-Fachbuch ©. Springer Fachmedien Wiesbaden, pp. 893–013 (2015). ► https://doi.org/10.1007/978-3-658-05734-3_46

Open Access This chapter is licensed under the terms of the Creative Commons Attribution-NonCommercial-NoDerivatives 4.0 International License (► http://creativecommons.org/licenses/by-nc-nd/4.0/), which permits any noncommercial use, sharing, distribution and reproduction in any medium or format, as long as you give appropriate credit to the original author(s) and the source, provide a link to the Creative Commons license and indicate if you modified the licensed material. You do not have permission under this license to share adapted material derived from this chapter or parts of it.

The images or other third party material in this chapter are included in the chapter's Creative Commons license, unless indicated otherwise in a credit line to the material. If material is not included in the chapter's Creative Commons license and your intended use is not permitted by statutory regulation or exceeds the permitted use, you will need to obtain permission directly from the copyright holder.

Support of Driving Functions by Digital Maps

Roland Homeier, Roland Jentsch, Katharina Jülge, Benjamin Lippelt, and Werner Pöchmüller

Contents

© The Author(s) 2026
H. Winner et al. (eds.), *Handbook Assisted and Automated Driving*,
https://doi.org/10.1007/978-3-658-45276-6_37

37.1 Introduction

This section gives an overview of the different roles of the map. The individual map-based functions and the necessary map information are explained further in the course of the chapter.

37.1.1 Maps in Navigation

The oldest map-based application in vehicles is navigation. While navigation devices used to be reserved for a small group of vehicle customers, they can now be found in many vehicles and are also available offboard in the form of smartphone apps, personal navigation devices (PND) and web-based services. Digital maps are provided free of charge in a simple version (e.g. from Apple, Google or the community-based map service Open Street Map). Navigation has become commonplace. Commercial providers such as TomTom and HERE offer more up-to-date maps with more comprehensive content.

While navigation in the vehicle was originally limited to embedded systems, the navigation functionality is now found separately from the hardware on various systems. An example is smartphone navigation and navigation on a backend, the result of which is transmitted to the vehicle via an over-the-air interface. A mixed form is hybrid navigation, in which it is not necessarily clear to the end user where a specific function is located—although this information is not required at all, as explained below. The advantages of processing on the backend are better performance and scalability. However, this method is not very robust against dead spots. The onboard variant performs better here.

In the case of purely onboard variants, a data carrier is used for storing the map. These used to be optical data carriers, later SD cards or flash memory. A map update was published approximately every 3 months, and it was the driver's responsibility to buy a new data carrier and carry out an update. If you didn't do that, the map gradually became outdated and the navigation could no longer perform its tasks in the usual quality.

With backend-only solutions, the map is located exclusively on a server. It's easier to keep it up-to-date there. In hybrid variants, flash memory is used for storing maps locally in order to be able to tide over interruptions in the connection to the backend.

37.1.2 Maps in Driver Assistance—and Highly Automated Systems

In driver assistance systems, the map can play a crucial role as an additional sensor, especially when it comes to foresight of the vehicle. This role has been spurred on by the current development of partially and fully automated systems. Here, sensors do not always provide the forecast required for location determination and driving maneuver planning, and not all relevant information is provided by fully automatically processed sensor information from an individual vehicle.

For the contribution of the map, it is important to ensure that it is up-to-date: the update cycles of 3 months known from onboard navigation are too long. Rather, it is necessary to be updated to the day, sometimes even to the hour or minute. This can only be achieved through a hybrid or backend solution. The boundary between a "static" map and dynamic traffic information is also blurred: these differ only in their update frequency.

At least according to the current state of the art, the map as a sensor is also relevant to safety for some functions. This means, among other things, that the decision about an update can no longer be left to the driver. There are also redundancy concepts in which sensor information is checked for plausibility against map information in order to achieve the combined confidence required for safety.

While the quality of map content for navigation purposes was still purely a convenience feature, it is essential for automated driving that the confidence of map elements is known and stored together with the purely functional description. Based on the quality, an application can decide whether a certain function is actually offered in the vehicle or not. There may be regional differences as well as differences depending on the map provider.

37.1.3 Layered Map Model

When modeling the map, it has proven beneficial to proceed in a layered model approach and to group map content logically. Suitable criteria for such a grouping are:

- Different data sources: Map content can come from different providers. A layer then contains content from a provider. As part of their product management, commercial data suppliers can combine content that they in turn obtain from various sources in one layer.
- Different use cases: Map content for a specific purpose (e.g. entering a destination) is combined in one layer.
- Relevance for functional safety: Map content can be grouped according to different safety relevance in order to enable a tailor-made and minimally complex development of the software with which it is processed in the vehicle. According to current practice, this is not yet the case and requires a uniform classification according to safety aspects.

The NDS format (see ▶ Chap. 22) implements such a layer model. A georeferencing method is required so that map content from different layers can refer to one another. This can be static, by referencing a map element using an appropriate primary key, or dynamic, using a referencing method such as OpenLR [1] or AGORA C [2]. Dynamic referencing methods enable provider-independent referencing.

The layered model implies that maps for navigation and driver assistance purposes do not have to be separated, but rather can be presented in an integrated manner. This avoids redundant storage of map content that serves multiple applications. Certain basic structures (e.g. for representing geometry) only need to be defined once.

37.2 Navigation

In this section, the classic navigation functionality is recapitulated and an insight into recent developments is given.

37.2.1 Navigation Functions

37.2.1.1 Route Guidance

The classic main task of a navigation system is to guide the user to a geographic destination. Sensors for position determination and digitized road data in the map are available as input variables; the road data is a digital representation of the real-world road network.

Based on this input data, after user input, the driver is given visual and acoustic information with which the vehicle can be guided to the destination. This is done via voice output about the route to be taken, via display instruments in the navigation device (map display and/or symbol display) or via additional visual information in the instrument cluster (usually symbol displays).

The navigation consists of the following software components; the reference to the relevant content in the NDS is given in each case:

- Positioning for location determination with the data available from the sensors. The Routing Building Block is required for this (see ▶ Sect. 22.4.1.2).
- Destination input to describe the destination by the user. The building blocks Name (▶ Sect. 22.4.1.4), FTS (▶ Sect. 22.4.2.2) and SLI (▶ Sect. 22.4.2.1) are required for this.
- Route calculation to determine the path from the current location to the entered destination (route) as a path search on the road data graph. The routing building block is required for this (▶ Sect. 22.4.1.2).
- Route guidance to guide the driver along the route using visual and acoustic information. The routing building block is required for this (▶ Sect. 22.4.1.2).
- Map display to display the geographic map with the current location, route and additional information. The Basic Map Display Building Block (▶ Sect. 22.4.1.5) and the Advanced Map Display Building Block (▶ Sect. 22.4.3) are required for this.
- Dynamization to include and take into account environmental events (e.g. fog, icy roads) and traffic information (e.g. traffic jams, road closures) in the current route. The Traffic Information Building Block (▶ Sect. 22.4.1.6) is required for this (◘ Fig. 37.1).

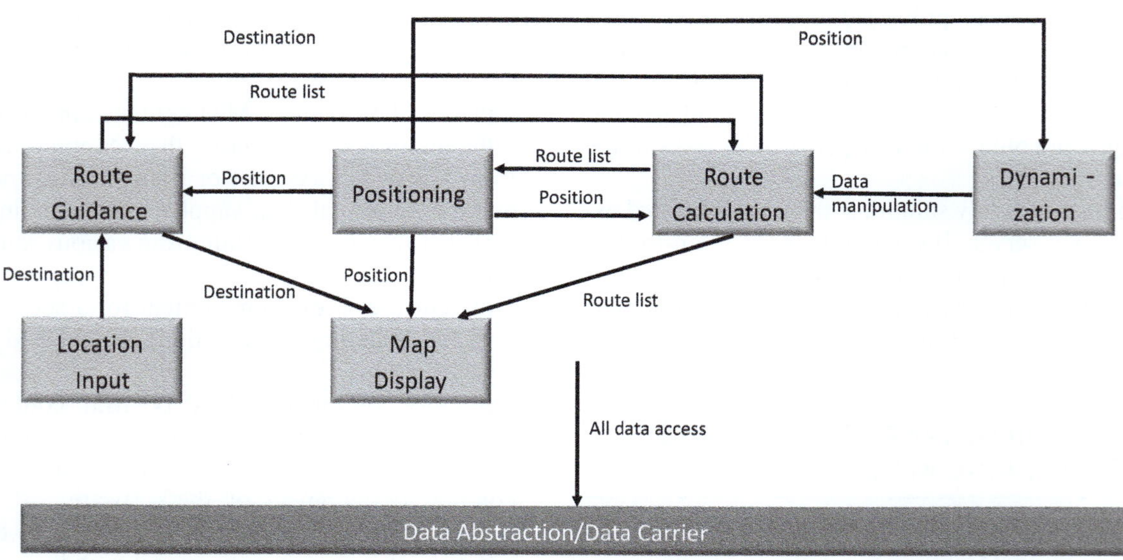

◘ **Fig. 37.1**   Software modules of navigation

37.2.1.2 Learning Navigation

The features of destination input and route calculation can be supplemented with learning functions to relieve the driver and offer a more comfortable experience. The behavior of the driver is stored beyond a single trip. Aspects of data security and data protection must be taken into account during storage.

When entering a destination, the system remembers which destinations are selected most frequently and prioritizes them by excluding rarer destinations early and giving priority to more frequent destinations in the visual display. Causal chains can also be traced. For example, certain hotels near airports are very frequently visited.

During route calculation, previous routes are remembered and given priority over heuristically calculated routes. Here, "secret paths" come into their own, which only people familiar with the area know and which would not result from a path search on the road data graph using fixed map features.

Both learning methods are based on the age and frequency of the information and have the property of unlearning. For example, older routes that have not been used recently are removed from memory. The reason could be that people have changed their place of residence and the route is no longer required. Learning can be limited to a single vehicle, or crowd-sourced data can be used to model the behavior of a larger group. The latter produces useful results more quickly, but is less individually tailored to a person.

37.2.1.3 Navigation for Electric Vehicles

Another useful feature is the addition of electric vehicle features to navigation. This already creates a first bridge to driver assistance systems: the navigation benefits both the driver and the battery management of the vehicle.

The range of the battery is included in the route calculation. To do this, the corresponding consumption must be known as part of the road data graph. For example, gradients are included in this feature, but also traffic jam information. Certain information about the vehicle itself is also required, such as how high the consumption is depending on the engine type and what other consumers are currently in the vehicle, in particular heating or air conditioning.

The result is on the one hand the area that can be reached by the vehicle without further charging as a radius around the vehicle, and on the other hand an optimal route to the selected destination, which takes stations for necessary charging processes into account during the journey (◘ Fig. 37.2).

37.3 Cloud Services

The following subchapter provides a brief overview of cloud backends and the applications running on them, the cloud services. Finally, some topics relevant to the automotive world are briefly presented.

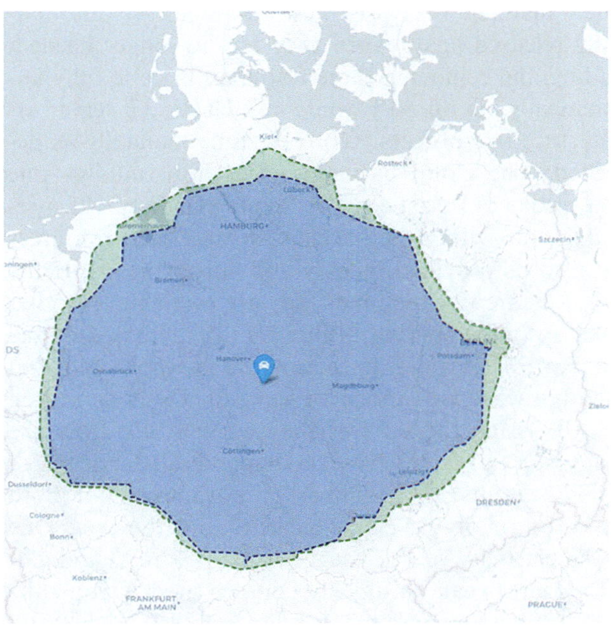

◘ **Fig. 37.2**   Reachable area of an electric vehicle according to battery capacity (left) of 80% (yellow) and 100% (green) and according to driving style (right) with 145 km/h (blue) and 100 km/h (green)

37.3.1 Properties of a Cloud Backend

There is an important difference between a classic IT server architecture and a cloud backend architecture. A server usually refers to a computer that is often very powerful and can be addressed via its address in the network. If the server's resources become scarce, a server scales vertically. This means that the server itself gets larger hard drives, more powerful CPUs or more memory, for example. This usually requires manual intervention. If a server can no longer scale vertically, it must be exchanged for a larger model. In this context, changes to the application architecture are often necessary. This can be associated with downtimes, extensive migration processes and the like. Procuring a server often involves high investment costs that have to be amortized over the course of operation. If a server fails, downtimes and sometimes high downtime costs can occur. An IT server architecture can also contain several computers from the start to which the load is distributed with a load balancer. However, these must be kept permanently available for this purpose and can also reach the limits described.

With a cloud backend, on the other hand, a specific computer or a specific resource is typically not accessed directly. The cloud backend provides its services via standard mechanisms in the network (broad network access). Incoming requests to the cloud backend are typically distributed across multiple resources via a load balancer. If the load on the cloud backend increases, additional resources are automatically added so that the requests can continue to be processed in a timely manner. Likewise, these resources are automatically released again when they are no longer needed. Adding and removing these resources is done fully automatically via an API, while in a classic IT server architecture this process has to be done manually as described. The cloud backend scales horizontally—one also speaks of "elasticity" or "rapid elasticity". If a resource fails within the backend, replacement resources can also be provided quickly and automatically. If the resources are not required, they are available for other tasks (resource pooling). Billing is usually based on usage. For this purpose, the actual use of resources is frequently measured (measured service). Users can often quickly assign the services they want to use from the cloud backend themselves (on-demand self-service).

This results in a number of advantages for both the operator of the cloud backend and the user. Due to the elasticity of the backend, the failure of individual resources can be quickly compensated. Due to the fact that new resources can be integrated quickly and automatically via APIs, cloud services can be designed to be particularly fail-safe. This increases reliability and availability. It is also possible to react quickly to peak loads. All this contributes to increased performance.

Another advantage of a cloud backend compared to local installations is that it can be easily updated. Bug fixes and technical updates only have to be provided in the backend. Large cloud backends receive many thousands of updates every day. This would be unthinkable for local installations. In addition, cloud backends can usually be set up quickly, can grow fast and are usually independent of the hardware on which they run.

Cloud backends can be hosted in the data centers of the respective companies and are then available to the company exclusively for its purposes. In this case one speaks of a "private cloud". In this case, however, the company usually has to bear the entire investment in the cloud itself. A private cloud is particularly suitable for very sensitive data and business processes.

An increasing trend are so-called "public clouds" as offered by companies such as Amazon, Google and Microsoft. In this case, a cloud provider offers the general public to outsource business processes to the backend of the cloud provider. The range of public clouds is often very wide and many different services are offered. This starts with classic virtual machines and extends to storage solutions, data analysis, artificial intelligence offerings and cognitive services such as image recognition and voice control. As a rule, private clouds cannot offer this range. Payment for customers is usually based on usage. Costs for the operation of data centers etc. are eliminated. Public cloud providers usually have a global offering. That means they run data centers all over the world and you can decide which region of the world you want the backend to be used in. In this way, the cloud service that is to be offered can be operated particularly close to future customers. In this way, latencies can be significantly reduced. Incidentally, "public" does not mean that the outsourced processes and data in a public cloud are accessible to everyone. Of course, these can be protected using cyber security mechanisms so that only authorized persons have access to them.

If private clouds and public clouds are used together by a company, this is referred to as a "hybrid cloud".

37.3.2 Technical Realization of a Cloud Backend

Technically, cloud backends are often structured in at least three service layers: infrastructure services, platform services and software services.

The infrastructure services or cloud foundation—often also called infrastructure as a service (IaaS)—are the lowest layer of a cloud backend. In principle, they provide the infrastructure components of a data center, such as computers, network and storage media as a service that can be booked. In addition to the hardware mentioned, this can already include cloud services such as storage mechanisms and messaging services. In the context of IaaS, however, the resources mentioned must typically be managed by the users themselves as far as possible, while this is provided and maintained centrally in a classic data center. The computing capacity is purely virtual—one speaks here of virtual machines. A well-known product of this type is the Elastic Compute Cloud (EC2) from Amazon. The procurement and operation of computers, hardware failure scenarios, security updates from the operating system and the like are taken over by the IaaS provider.

The second layer of a cloud backend is the cloud platform. It is also known as Platform as a Service (PaaS). Within a cloud platform, the user can provide cloud services. The platform service ensures that the necessary resources to run the application are provided. The user does not have to worry about managing the computing capacity. The user does not need to know the underlying operating system either, since the necessary interfaces are provided via the platform service. For example, a car manufacturer could offer a secure API to access vehicle data in a public cloud. In this case, he would essentially have to take care of the development and provision of this cloud service in the public cloud.

On the basis of a platform, a wide variety of software services (Software as a Service, SaaS) can be offered. As with other applications, there are no limits to the variety of services. For example, a car manufacturer could offer an application for keeping a logbook as SaaS.

37.3.3 Virtualization and Containers

In order to be able to scale flexibly horizontally, different forms of virtualization are used within cloud backends. A well-known example that has long been used in classic data centers is a virtual machine: This represents a software-emulated form of hardware—for example an $\times 86$ computer—which is abstracted from the actual hardware. An operating system is usually installed within a virtual machine. The operating system running in the virtual machine only "sees" the virtual machine and can only access the resources allocated to the virtual machine. Nevertheless, it behaves like an operating system running on real hardware. On this basis, applications can be run on a virtual machine in the same way as on an actual machine. The virtual machine runs as a guest on a host computer. The software that creates and runs the virtual machine is called a hypervisor. A hypervisor can run directly on the host hardware—this is common for cloud backends—or on a host operating system running the hypervisor alongside other applications.

Virtual machines can be used in many different ways. An obvious example is installing multiple virtual machines on powerful hardware. The virtual machines then share the hardware and performance of their host computer. Various virtual machines can be used for different purposes; for example, they can be rented to different customers. This idea is similar to the resource pooling approach of a cloud infrastructure. However, this is not uncommonly practiced on automotive computers. For example, one virtual machine can be used for infotainment applications, while another is used for connectivity functions that are to be isolated from other parts of the system for security reasons.

Another virtualization concept that has meanwhile become very popular are containers. While virtual machines, in principle, completely virtualize a computer, virtualization of a container takes place at the operating system level. A separate operating system does not have to be installed in a container. Applications running in a container do not "see" emulated hardware, but—to put it somewhat simply—the actual API of the operating system. Containers are very common in cloud backends.

The container concept is not new either. It comes from the Unix world. FreeBSD has jails, which are very similar to containers. Containers can be configured in such a way that the application running in them can only "see" a certain part of the operating system's API. This increases security because applications in containers cannot access potentially critical areas of a system in this way. In addition, applications can bring the libraries and frameworks they need with them in a container. These do not necessarily have to be present on the host system. Using an application in a container is therefore often very easy. Here it becomes clear why the term "container" is appropriate: Similar to a standard container in transport, a container has only a few external interfaces and can have almost any content—from a simple script to a complex application. Containers can therefore also be easily distributed. Depending on the container framework used, there is, for example, the option of providing containers via central registries. In this way, a container can quickly be offered to a very large community, for example an open source community or employees in a large company. Before the development of a solution is started, the central registry can be searched to see whether there is already a solution to the problem. The containers can then often be run directly in their own environment. This results in a very fast deployment time. Rapid deployment times are essential in order to be successful in the fast-paced software business.

In addition to the advantages presented, which result from the easy handling of containers, there are also advantages in terms of runtime. They start and stop very quickly, which has advantages in scale-out cloud environments. Containers share the kernel and resources of the operating system they run on. This makes them more efficient than virtual machines and also cause little more overhead compared to native applications that run without a container.

Running a container is often far less complex than running a virtual machine. There are therefore cloud providers who only provide applications in the form of containers. Containers are also used on automotive computers. For example, applications that are exposed to the Internet and thus pose a potential security risk can be deployed in a container that has very limited access to the rest of the system. This approach represents an increase in security and requires very few resources on the vehicle computer. Another conceivable approach is the distribution of applications to fundamentally different automotive computers, for example such different series or even different manufacturers. Even the use of applications that were originally developed for a completely different environment is conceivable with containers.

37.3.4 Technical Characteristics of a Cloud Service

The architecture of cloud services or cloud applications must meet the requirements of the cloud platform on which they run. Here they differ greatly from applications that run on classic automotive computers or control units. The classic applications are often closely linked to the hardware and the operating system (strong coupling) on which they run. A control unit often only serves one purpose, the application is completely accommodated on this control unit. In addition, they are often implemented as state machines. They are often expected to behave deterministically, e.g. sending an alive message in a fixed time frame.

Cloud services, on the other hand, follow the principle of loose coupling. The need for this results from the properties of a cloud system. This includes horizontal scaling and frequent updating of individual components. In addition, not all components are immediately available in a cloud backend and sometimes have to be started first. Another design principle is that of the distributed system. If a system scales horizontally, additional CPU cores may have to be integrated, which automatically creates a distributed system. This is one of the reasons why cloud services are often designed to be stateless. Because the cloud service does not have to hold any states, scaling is significantly simplified. Different requests from a client to a cloud service typically do not always reach the same service instance of the cloud service. This is also partly due to the concept of horizontal scaling. If such a service were to be implemented in a stateful manner, it would have to be ensured that this state is always available in every instance of the cloud service or that a client always reaches the same service instance. The administrative effort for such a solution is very high, which is why it is often dispensed with. A possible solution is to keep the state client-side. This leads to a significantly simplified design of the cloud service, but also means that the client has to repeatedly transmit certain information. A good balance is needed here.

Cloud services are often set up as microservices. This is an "isolated, cooperative and autonomous service that has exactly one task" [3]. The microservice masters this one small task particularly well. A microservice follows an idea from the Unix philosophy that goes back to Ken Thompson: "Make each program do one thing well" (Raymond). Microservices support loose coupling. They communicate with each other via interfaces. Different microservices can be implemented with different technologies and programming languages. For example, a computationally intensive service can be developed in C++, one with heavy use of artificial intelligence in Python, and a messaging service in Erlang. Innovations, such as a new framework, can first be introduced in a microservice without influencing others. All can communicate with each other via programming language-independent interfaces. A more complex application can be created by composing several microservices. In addition to the various advantages of the design and implementation, there are also advantages in the development work: Due to its simplicity, the throughput time of the microservice is very short. The term throughput time comes from the lean philosophies and describes the time from the customer's request to the fulfillment of the request. Short throughput times are desirable, among other things, because they allow quick feedback from the customer and thus also a quick reaction to possible change requests. Microservices allow the technical implementation of this philosophy. As a rule, a microservice is only managed by a small team consisting of no more than 5 to 7 employees. Conway's law states that system designs often follow the communication structures of the enterprise. If the service remains the responsibility of a small team, it is also secured by the organization of the company. There are also advantages when deploying and operating a microservice.

There are a number of reasons for implementing applications as cloud services in the automotive environment. An example of this is the ability to update cloud services very frequently. Despite the increasing number of over-the-air updates, updating software directly in the vehicle is still relatively time-consuming—

think of the many millions of vehicles that may need an update. Keeping a cloud service up to date is much easier. It should be mentioned at this point that the APIs often change due to the high update frequency in the cloud. While in the automotive industry vehicle systems and thus their interfaces can be in the field for up to 15 years, cloud APIs are sometimes stable for less than a year. The challenge in this example is to keep the API stable in the direction of the vehicle. This must be taken into account in the architecture of the system and the design of the API. One possibility is the use of a digital twin.

37.3.5 Internet of Things

The "Internet of Things" (IoT) describes the possibility that physical devices—the "things"—can communicate via the Internet. Such a "thing" can be, for example, sensors, actuators, but also an entire car. A possible added value of this approach is the processing and combination of data from very different devices, which was previously not possible in this form. There are many definitions to differentiate IoT from the mere networking of devices. One possible requirement is that an IoT solution must meet the criteria "Aware", "Autonomous" and "Actionable". This means that an IoT device—the "thing"—can "sense" data via a sensor and automatically transmit this data to an IoT application. The IoT application can then react to this, e.g. by using machine learning methods to generate knowledge from the data and in this way make decisions or support the decision-making process.

37.3.5.1 IoT Hub

An IoT hub is where IoT devices connect to the cloud. It is part of a backend offering and provides cloud services for IoT devices to connect and interact with the backend. Other cloud services in the backend communicate with the IoT Hub via standardized backend communication mechanisms such as an AMQP message queue. The IoT devices, on the other hand, can communicate with the IoT Hub via a variety of protocols, such as HTTP or AMQP, but also via special IoT protocols such as MQTT, LwM2M or CoAP. The special features of an IoT device are taken into account, for example the fact that the device cannot always be online. From the point of view of the backend, however, a large number of different devices can be addressed via uniform interfaces. The backend does not need to know which protocol the corresponding IoT device supports; this is implemented transparently via the IoT Hub. All major public cloud providers offer such a service, and there are also various open source projects for IoT hubs. A well-known open source project is e.g. Eclipse Hono [4].

37.3.5.2 Digital Twin

An IoT device can be represented in a backend, for example, via a digital twin. A digital twin mirrors a physical device in the cloud. It contains the most up-to-date representation of the data measured by sensors in the IoT device. However, a Digital Twin goes beyond the mere provision of this data and it is a complete, digital model of the represented object. The model contains all information about the properties, data and functions of the device. Cloud applications can access it like a normal cloud service. For example, the trivial digital twin of a car could know the charge level of the battery for the electric drive, the range and the tire pressure. Based on these values and the route that is also known from the routing cloud service, a cloud service can suggest an optimal charging stop to the driver and indicate that the tire pressure should also be checked. The digital twin of an engine would be much more complex, not only knowing the condition of wearing parts, but also being able to transmit certain data anonymously to the manufacturer so that the design can be optimized using large amounts of data. A digital twin is also a tried and tested means of dealing with frequently changing cloud services and their APIs. While the digital twin keeps the interface to the vehicle stable, the interfaces to cloud services can be adjusted regularly. An exemplary framework for developing digital twins is Eclipse Ditto [5] and the associated Eclipse Vorto [6] for formal modeling of the twins.

37.3.6 Edge Computing

Edge computing can be seen as a further development of the IoT. Here, computing capacity and also cloud functions are shifted from central data centers more to the edge of the cloud. For example, hardware from a cloud provider is installed directly in the data center of an automobile plant. Certain backend offers that otherwise only run in the cloud provider's data center are then available directly on site and can be used. Cloud services can thus be rolled out directly to the local backend. Nothing changes for the cloud service itself. The advantages are obvious: on the one hand, the latencies are much lower, on the other hand, not all data has to be transmitted to the cloud provider's data centers, which can also be necessary in certain cases for data protection reasons. In this context, there are many possible applications, especially for Industry 4.0 applications in automotive production.

In addition to being installed in a data center, edge computing is also relevant on vehicle computers. In addition to the classic tasks from the IoT environment, a much higher level of data processing can also be carried out on the edge. A camera from a driver assistance system may serve as an example. In principle, the camera could send its raw data, namely the captured images, directly to a cloud service in the backend, which then processes the images. However, this creates latencies and a very large amount of data that has to be transmitted and stored. Certain image processing steps can also be carried out on the Edge directly in the camera or in a connected vehicle computer. The raw data is processed directly in the vehicle and, for example, only the extracted feature vectors are sent to the backend. In this way, enormous bandwidth is saved during transmission, and data protection is also satisfied because no images are transmitted, only feature vectors. It is also conceivable that a vehicle manufacturer configures the cameras in its vehicle fleet in such a way that specific images are searched for—e.g. traffic signs—and only this data is then used further, e.g. to train AI models in a targeted manner. If other data becomes interesting after a while, a corresponding command or software update can be sent to the vehicle fleet and from then on other image data will be processed. Another advantage is the fast reaction; for example, if a sensor detects a problem in a security-relevant system, an edge computing system can react directly locally and does not have to first contact a cloud backend.

Cloud services often run in containers in the backend. In principle, edge computing also allows containers to be run on a vehicle computer. It is therefore possible to run cloud services directly in the vehicle. This usually requires a full operating system on the vehicle computer, for example a variant of Linux, and a number of libraries so that the container with the cloud service can also run in the vehicle. Classic control units that work with microcontrollers often do not have these options. Edge computing opens up completely new possibilities here in connection with modern vehicle computers and containers.

37.3.7 Cloud-Services in Practice

The use of cloud services brings a whole range of advantages. However, their use also requires a rethink compared to the way in which classic control units have been developed up to now. Cloud services are almost completely decoupled from the hardware and require powerful software APIs, while in the case of ECUs the hardware used often has a major impact on the system design. In addition, cloud services play to their strengths primarily through powerful scaling mechanisms. High update frequencies can only be achieved if the changes only have to be made in a few places. Providers of cloud services will therefore be interested in setting up a platform whose cloud service offering can be used by as many customers as possible. Other fundamental properties of cloud services such as resource pooling and usage-based billing are particularly viable

if the cloud service is set up uniformly. In practice, this means that a cloud service has to work with many different variants of a control unit or a vehicle computer, i.e. across the boundaries of a control unit family, a vehicle series and even across manufacturer boundaries. Therefore, ECU-specific implementations of cloud services should be avoided.

Uniformly set up cloud platforms also hold the possibility of completely new business models. Let's think of a vehicle that uses its sensors to recognize the characteristics of the road and the environment in which it is currently driving. If as many vehicles as possible not only use this data in the vehicle, but also make it available to a cloud service in the form of feature vectors, for example, this cloud service can automatically create a digital road map from the data obtained. While the actual sensors in the vehicle only use the data in the context of local assistance systems, a completely new product is being created here that can be marketed as such. Customers who want to remove sensors or assistance systems are now joined by those who are interested in digital maps. A new cloud service is emerging that offers new customers a digital map. Thanks to the ability to quickly allocate this cloud service when needed, new customers can make use of this cloud service very quickly. This is an essential feature of a platform: customers can come together and exchange goods. Platforms live from network effects. A positive network effect is, for example: the more vehicles that contribute data to the digital map, the higher the quality of the map, and the more customers will decide in favor of the digital map product. On the other hand, a negative network effect would be that although more and more vehicles are providing data, the cloud service cannot process this data quickly enough due to the amount or the variety. The users of the "digital map" product may then no longer receive up-to-date maps and will turn away from the product. This small example shows how important uniformly designed cloud services are and why one should refrain from designing services that are too specific.

37.4 Support Through Smartphone Connection in the Car

Smartphone integration in automobiles means the wired or wireless connection of a smartphone to the infotainment system of an automobile. This enables the user to interact with the smartphone, e.g. by displaying the smartphone screen on the on-board display and operating it using control buttons via the head unit or the steering wheel remote control, for example. This allows the user to use many smartphone apps in the vehicle thanks to Apple Carplay, Android Auto and other systems.

The reasons are both the use as a replacement for the built-in navigation and as a telephone and multimedia center as well as for a wide variety of social media services. In addition, there are lower costs and a faster update cycle (both for the hardware and the software) than for the systems installed in the vehicle. Remote apps from the vehicle manufacturer on the smartphone enable various infotainment functions to be queried, set and controlled.

In principle, the use of smartphones for map updates is conceivable. Currently this is more interesting for navigation; it remains to be seen whether there are also applications for automated driving. This subchapter provides an introduction to the basics.

37.4.1 Motivation for Smartphone Integration in Automobiles

The forecast for sales of smartphones worldwide shows a clear upward trend: while around 300 million smartphones were sold in 2010, by 2020 there will be around 1.3 billion. The number of smartphone users worldwide rose to 3.5 billion. However, according to paragraph 23 of the German Road Traffic Act, the driver may not operate a smartphone while driving if it has to be picked up or held. There are comparable rules in other EU countries and worldwide. The use of smartphones for navigation in the car is therefore also limited. A possible alternative is the use of a holder for the smartphone or control via the integrated voice control. Solutions have been developed so that the user can access their smartphone while driving without violating the applicable legislation or reducing safety while driving.

However, smartphone integration is also relevant for integrated vehicle information systems. Manufacturer-specific remote apps (on the smartphone and/or in a secure domain of the infotainment system) access vehicle data via a software interface and sometimes allow vehicle functions and settings to be controlled. The range of these services is extensive. It goes from reading out the fuel level and fuel consumption to location information of the vehicle and remote input of the destination for the built-in navigation system to the transmission of service messages and workshop appointments as well as opening and locking the vehicle doors. Updating the software of the various control units in the vehicle is also an important application. Features such as remote parking are planned for the future. Further integration of vehicle functionality into the smartphone ecosystem is also under development. With the CarKey Interface, Apple is integrating a standard interface for car manufacturers from iOS 13.4, which enables the user to use a smartphone or SmartWatch via Near Field Communication (NFC) as a replacement for the car key.

Access to current map data is an important area of application for driver information systems, as these become outdated within a certain period of time. Although the new construction of the transport infrastructure in the industrialized nations is low (less than 1% per year for motorways, see for example the transport investment report in Germany [7]), specific details such as speed limits and even lane openings change much more frequently and are partly integrated in electronic traffic control systems. For driver assistance systems that use map data, it is essential that the data is up to date and enriched with dynamic data such as traffic reports, weather information and traffic and congestion forecasts. The smartphone integration and thus the use of their fast Internet connection (LTE and 5G) is therefore an important part of these systems and a replacement for Traffic Message Channel (TMC) and Digital Audio Broadcasting (DAB)+.

In addition to smartphone integration, many manufacturers rely on their own telematics box in the car for communication. One reason for this is the eCall emergency call system. It is mandatory from 2018 for new vehicle models to be approved. A data connection is then part of the standard equipment.

37.4.2 Possibilities of Smartphone Integration

Today, due to the practically uniform equipment of smartphones with USB or Lightning connections, the previously manufacturer-specific charging and data interfaces via docking stations have become obsolete, and only a few manufacturers still offer them. Special storage places with a corresponding USB connection or radio-based QI charging technology (inductive energy transfer) are increasingly being used for this. The audio connection as the basic functionality of the smartphone integration is often taken over by a Bluetooth connection. This enables access to the audio channel of the smartphone, so that an output on the loudspeakers integrated in the car is permitted. In some cases, however, the USB and Lightning interfaces are also used for audio output.

37.4.2.1 Basic Functions

The simplest type of smartphone integration (apart from looping in the audio signal via the AUX input) is the support of the various Bluetooth profiles (such as the hands-free profile for telephony in the car or audio streaming), which are integrated into the vehicle manufacturers' infotainment systems. Although rudimentary operation from the vehicle is possible, this type of integration is now being replaced by other types, since neither the display of the smartphone is mirrored nor the wide range of smartphone apps can be used. In addition, the usable functionality is determined by the infotainment system in the vehicle. Since the vehicle manufacturers offer practically no continuous updates and functional extensions of the infotainment systems after the purchase of a vehicle, the systems become outdated very quickly.

With Apple CarPlay and Android Auto, more modern types of smartphone integration are available, which allow the use of smartphone apps on the vehicle's central display via appropriate APIs and interfaces. The smartphone apps that can be used are limited because they have to be adapted for the interfaces and in some cases approved by the manufacturers of the smartphone operating systems. The vehicle infotainment systems must also integrate the appropriate hardware and software interfaces in order to use their functions. For example, Apple's wireless CarPlay uses Bluetooth for pairing the smartphone, but the WiFi interface for data transmission because of the higher bandwidth and QI for wireless charging of the smartphone. With Apple's CarKey as a vehicle key replacement, NFC is also used. The complex software interfaces ensure that only a few vehicles currently allow the smartphone apps to be operated using hardware buttons (e.g. steering wheel buttons and rotary pushbuttons), a touchscreen in the infotainment display and voice control.

37.4.2.2 Vehicle Manufacturer-Specific Interfaces

In addition to the normal smartphone apps, there are also vehicle-specific functions that enable information to be retrieved from the vehicle and vehicle functions to be controlled via smartphone. To do this, either the interfaces of the corresponding software system are used (in Apple CarPlay, for example, the air conditioning, logbook, payment [8]), which the vehicle manufacturer must support, or special proprietary smartphone apps from the manufacturer are used, which access the vehicles via the network. For security reasons, gateways or proxies are used in the vehicles. The term "proxy" refers to the part of the head unit software that can communicate with the apps on the smartphone. The information from the smartphone is provided in a format that is compatible with the head unit. In addition, secure data communication often takes place via special back-end servers from the vehicle manufacturers.

In addition to the normal smartphone apps, there are also vehicle-specific functions that enable information to be retrieved from the vehicle and vehicle functions to be controlled via smartphone. To do this, either the interfaces of the corresponding software system are used (in Apple CarPlay, for example, the air conditioning, logbook, payment [8]), which the vehicle manufac-

turer must support, or special proprietary smartphone apps from the manufacturer are used, which access the vehicles via the network. For security reasons, gateways or proxies are used in vehicles. The term "proxy" refers to the part of the head unit software that can communicate with the apps on the smartphone. The information from the smartphone is provided in a format that is compatible with the head unit. In addition, secure data communication often takes place via special backend servers from the vehicle manufacturers.

The aim of manufacturer-specific apps is to create a smartphone application that is specific to the "look and feel" of the car manufacturer and allows communication with the car manufacturer's head unit or other vehicle systems. The background to this approach is that applications from third-party manufacturers do not have direct access to the infotainment system, as the car manufacturers only allow selected contractual partners full access. Third-party manufacturers can design compatible multi-apps themselves using the APIs offered by some automakers. One of the main disadvantages is that there is currently no generalized solution for integrating an app into automobiles from different car manufacturers. A modification of the app is therefore often necessary for compatibility. In addition, the user interface is limited to the specified functionality and the expansion of the interface requires new apps or software updates in the automobile. The maintenance effort is quite high in these cases with little flexibility. Despite the disadvantages, car manufacturers currently give preference to such solutions.

37.4.2.3 Future of Smartphone Connectivity

The future and further development of smartphone integration depends, among other things, on which technologies are developed to connect the car to the mobile network. Three technologies can be differentiated [9]:

Embedded solutions: The connection to the mobile network and all the functionalities provided are realized by systems integrated in the automobile.

Tethering solutions: In order to use mobile phone-dependent functionalities, a connection to a mobile phone that is used as a modem is necessary. Modem and hotspot/access point solutions via Bluetooth and WiFi are available as connection types.

Integrated solutions: Smartphone functionalities—especially apps—are mirrored in the automobile.

None of these solutions should be seen as an exclusive solution. Most automakers are developing strategies using multiple of these connectivity solutions for different market segments (e.g. embedded solutions for compact class models and tethering solutions for sub-compact class models). In addition, different technologies are used depending on the application. Embedded solutions are preferred for safety aspects, while integrated solutions are used for infotainment aspects. Future developments such as car-sharing apps and electromobility, but also data protection and security requirements will further increase the range of developments.

37.5 Functional Safety

37.5.1 Safety of the Map-Based Function

According to ISO26262, safety–critical applications such as for automated driving, including those on static and dynamic map data, must be free of unacceptable risks. Protection against manipulation (security) that has already taken place is a prerequisite. Risks in the processing of map data arise as a result of systematic hardware or software failures and random hardware failures. This can result in faulty map data at a subsequent interface. Such erroneous or inconsistent map data must be reliably detected. In order to achieve this, further consistency checks are carried out in addition to the usual checks such as CRC or hashes. A full check should be carried out based on the map specification, for example by using code generators. Other verifiable details, such as freedom from overlapping and consistency of the map data used, should always be checked completely before use.

ISO26262 stipulates extensive test coverage. Since any combination of attributes can be found on the same street element in the map, testing the processing software is correspondingly complex. Test automation is indispensable here. Other measures that ISO26262 prescribes, such as failure mode and effects analysis and fault tree analysis, are correspondingly complex to carry out due to the complexity of the map-based function.

Further challenges are that a map-based function does not have a fixed time behavior, as is often assumed when safe-guarding safety-related software. Depending on the geographic density of the map information, processing can be faster or slower. New concepts are required here that cannot be directly derived from the previous applications of navigation and driver assistance systems.

Map data must be current at all times to be accurate. Outdated map data should be viewed in the same way as incorrect map data and shall not be used.

37.5.2 Safety-Relevant Attributes and Profiles

First of all, all profiles related to the topology, geometry and properties of the roads are relevant to safety.

Furthermore, profiles for the permitted speed, the presence of bridges, tunnels and road signs are relevant.

37.6 Electronic Horizon

37.6.1 Definition

The preview of map information in front of the current vehicle position is referred to as the electronic horizon. An absolute position (e.g., GPS position) from the vehicle sensors is required for this. This position is assigned to an element of the road data graph from the map, i.e., an edge. In the simplest case, the perpendicular is dropped from the GPS position to all edges and the edge is selected for which the distance is minimal. In addition, the offset (the distance from the beginning of this element) is determined by calculating the length from the starting point of the edge to the dropped perpendicular foot. This corresponds to the method of positioning. Now all possible paths on the road data graph are calculated up to a previously defined length. The path that the vehicle is most likely to take is called the Most Probable Path (MPP), and all other paths are called Sub Paths. Now the relative distances of all existing map elements to the vehicle can be calculated on this graph. The result is converted into bus messages and distributed in the vehicle. Various control units can now receive the map information and use it for their respective vehicle application.

This electronic horizon can be added as another feature to navigation, especially since some functions overlap. Alternatively, it can be implemented as a standalone application. It should be noted that with a certain classification in terms of functional safety, corresponding requirements must be met, which limits the use of existing software. Usually, navigation software is developed as QM software, which means that it cannot be reused for an automated system.

ADASIS has established itself as the interface protocol. For driver assistance systems there is the protocol definition ADASIS v2; for systems for automated driving, the protocol definition ADASIS v3. The basic principle of the ADASIS interface protocol is that the so-called ADAS Horizon Provider (AHP) breaks down the data on the sender side into data telegrams and makes them available on the vehicle bus. On the receiver side, the ADAS Horizon Reconstructor (AHR) reassembles the electronic horizon from the data telegrams (see ◘ Fig. 37.3).

The electronic horizon implies transmission over a vehicle bus; this is what the ADASIS standard is designed for. If the provision of maps and the driving function are located on the same control unit or if an

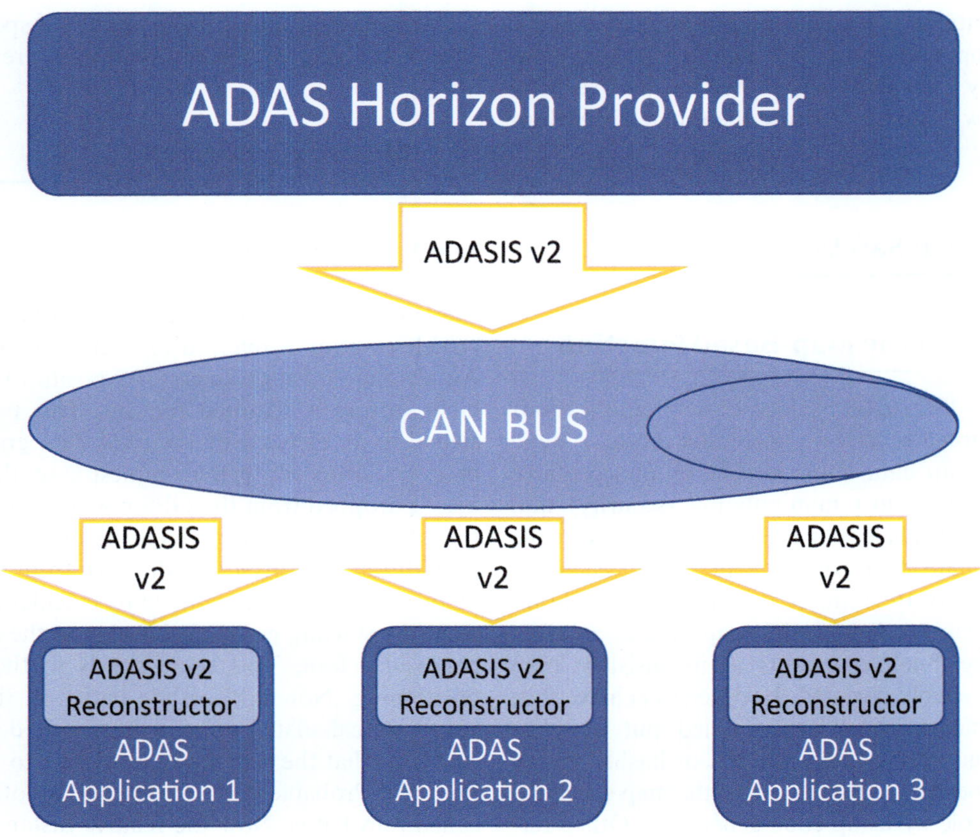

◘ Fig. 37.3 Basic principle of the interface protocol ADASIS v2 (*Source* ADASIS v2 protocol)

IDC (Information Domain Computer) is used (see ▶ Sect. 37.7.2.3), then interprocess communication or a functional interface in the software is more appropriate. But telegrams are not required. The ADASIS standard provides the basis for this, in that a standardization of the functional interface of the AHR is also envisaged. The communication of map data across the electronic horizon is by definition unidirectional. The application has no way of making specific requests but is supplied with the data defined in the system specification.

37.6.2 Electronic Horizon for Driver Assistance Systems

The term assistance function is broad and not used uniformly: It ranges from functions with an informative character—sometimes the navigation itself is regarded as an assistance function—to safety-relevant functions with intervention in vehicle guidance.

Four classes of assistance functions can be distinguished:

1. Functions that increase driving safety (e.g., electronic stability control, brake assistant)
2. Functions that increase driving comfort (e.g., park pilot)
3. Functions to reduce fuel/energy consumption (e.g., gear selection recommendation, coasting assistance)
4. Functions to increase driving performance (e.g., deactivation of the air conditioning when accelerating).

The navigation can play a decisive part in one of these assistance functions. In the following explanations, a distinction is made between navigation-generated and navigation-supported assistance functions.

Navigation-generated assistance functions are generated and provided by the navigation itself. Examples of such assistance functions are:

— the "congestion ahead" warning, which allows the driver to take the next exit and avoid the traffic jam or at least approach the end of the traffic jam at an adjusted speed;
— the curve warning, which warns the driver if the speed is too high for the curve ahead;
— the dangerous spot warning, which warns the driver of dangerous spots (e.g. accident black spots, kindergartens/schools).

With the navigation-supported assistance functions, the navigation functions as a sensor for other assistance functions that are typically implemented on other control units. The navigation provides the situation at the

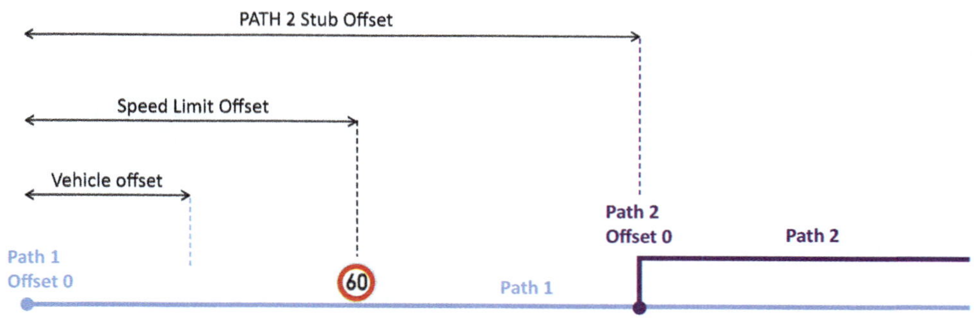

PATH 2 Stub Offset

Speed Limit Offset

Vehicle offset

Path 2 Offset 0   Path 2

Path 1 Offset 0   60   Path 1

◘ Fig. 37.4 ADASIS position encoding (*Source* ADASIS v2 protocol)

current position, the route with active route guidance and the road network ahead (also called ADAS horizon or electronic horizon) via a standardized interface (ADASIS) [10]. The data is broken down into data telegrams on the sender side via the AHP and made available on the vehicle bus. On the receiver side, the AHR reassembles the ADAS horizon from the data telegrams. Thanks to standardization, the assistance functions are independent of the navigation system used. Examples of navigation-supported assistance functions are:

- adaptive light control for better illumination of curves and intersections,
- Fuel-saving driving style thanks to anticipatory gear selection of the automatic transmission according to the route profile.

Thanks to new assistance functions, new sensors—such as video cameras and radar—are establishing themselves in the vehicle, from which navigation benefits. The lane-precise position determination is possible, for example, by evaluating the camera image. Current information,such as traffic sign recognition or the resolution of special situations such as a construction site, can be used in the future for improved driving recommendations or warnings.

So far, the CAN bus has been used as a communication channel in this context. Due to the increase in data volumes, there is a change to Ethernet. The electronic horizon contains information on the road network ahead, whereby only those roads are relevant that can be reached from the vehicle's current position and that have a certain probability of actually being driven on. The electronic horizon for driver assistance is usually generated by a navigation system.

Typically, the look-ahead length of the electronic horizon is around 500 m to 6 km, depending on the requirements of the assistance functions used. Because of the way information is encoded, the look-ahead length.

- when using the typical length grid of 1 m
- is limited to about 8 km for electronic horizon content.

In addition to typical information from a digital map that is used for navigation systemsVsuch as road class and speed limits \special data, in particular gradients and curvatures, are also transmitted in the electronic horizon. Special requirements are also placed on the quality of the location accuracy of information in the electronic horizon; For this reason, this data is usually offered by map data suppliers as so-called ADAS data and marked accordingly in the digital map.

The use of the electronic horizon for assistance functions also requires special properties with regard to the accuracy of the positioning of the navigation system. Only very good localization provides a precise and stable electronic horizon: In this context, only navigation systems that are permanently installed in the vehicle and use the vehicle sensors to optimize the position calculations are used (◘ Fig. 37.4).

37.6.3 Electronic Horizon for Automated Driving

For the purpose of highly automated driving, the ADASIS standard has been further developed and comes into use in version 3. While ADASIS v2 still requires a CAN bus and the standard specifies the actual messages on the bus, only data types are specified in ADASIS v3. The transport layer and the type of data bus are not specified by the standard. It is assumed that the bandwidth of CAN is insufficient. It is to be expected and that Ethernet will prevail here.

37.6.3.1 Automated Driving

Automated driving means systems as described in Level 3 of the definition of the Federal Highway Research Institute. In certain situations, the vehicle takes over the driving task, i.e., the longitudinal and lateral guidance of the vehicle, completely. The definition of these situations is up to the manufacturer of the system. The system must independently recognize which situations it can no longer handle and hand over the driving task in good time or assume a risk-minimal state (e.g. emergency stop on the hard shoulder) (◘ Fig. 37.5).

Map data is used in this context, and the representation as an electronic horizon makes sense, since the

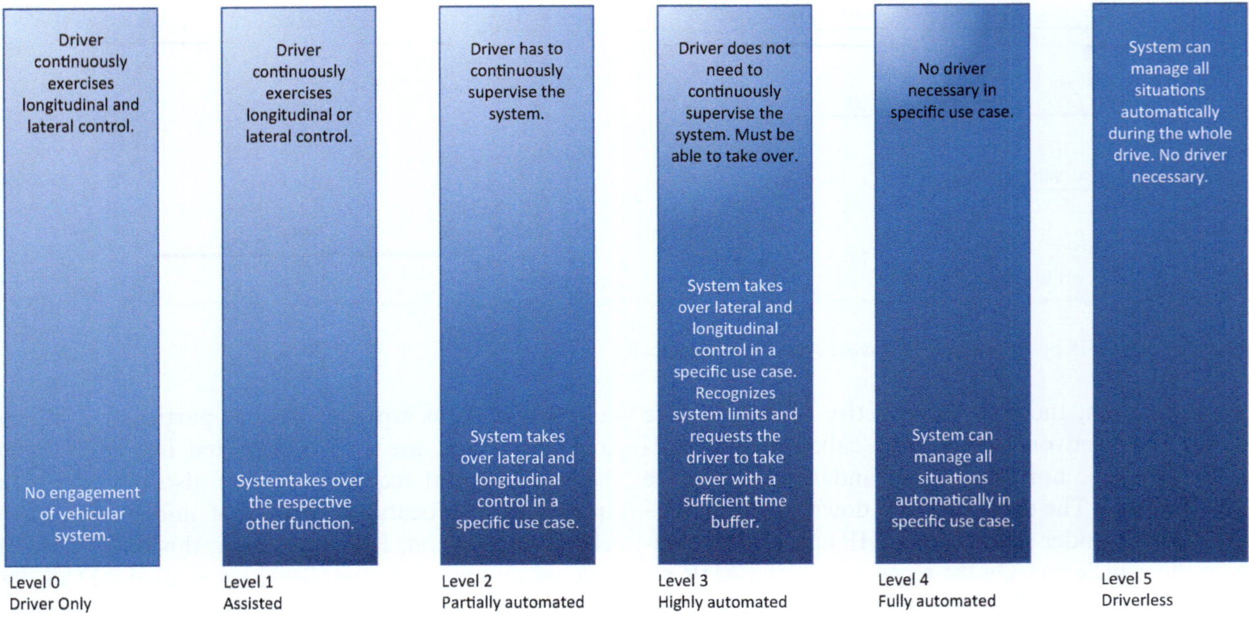

Fig. 37.5 Levels of automation (Quelle: VDA)

properties of the horizon mentioned (unidirectional flow of information, definition in a standard) are compatible with the required system design. Further fields of application are conceivable in levels 4 and 5, but the map model described could be insufficient there and the system partitioning not suitable for using further geographic information from the traffic infrastructure and from surrounding vehicles.

The transfer of responsibility from the driver to the system results in the functional safety requirements described. The driver is initially absent as a "fallback" solution for the system. If they need to take over again, they cannot do so in a few milliseconds. The system therefore needs its own mechanisms to achieve a risk-minimal state. Some of the derived requirements are also assigned to the map or the software that processes them. For example, an emergency trajectory is calculated by the electronic stability control, which in turn is based on a representation of the lane on the map.

37.6.3.2 Use-Cases for Map Data

The modeling of the map for automated driving is based on the map for navigation. In particular, in addition to ADASIS v3, the NDS standard has been expanded for this purpose. The basic structures of a map (geometry, topology, attributes, relationships between the various map elements) have been adopted. Automated driving requires greater geometric accuracy and more detail. For example, it is not enough to know how many lanes a road consists of. Information about the lane boundary and the geometry of the lane is also required. The meaning of this information becomes clear

when you look at what tasks a self-driving car has to perform and what traffic rules it has to obey. In the road traffic regulations of many countries, a solid line at the lane boundary means that this line may not be crossed. This also implies a ban on overtaking. In Germany, yellow lane markings take precedence over white ones, as they represent temporary lanes in construction sites.

A map can thus be divided into use cases for navigation and automated driving. For example, map display and destination input continue to serve navigation, while lane information benefits automated driving. The street graph is required for both.

The use cases for maps for automated driving are divided into localization and planning. For this purpose, the map can be further subdivided. Lane information and 3D objects are required for localization; for planning also lane information including their topology as well as traffic lights and signs are required.

Localization

Localization means determining the position of the vehicle. The vehicle needs to know which road it is on (this is the commonality with navigation), what lane within that road, and where exactly within that lane. A localization can be absolute (as a GPS position on the map) or relative (as an angle and distance to map elements).

Map matching is performed using the GPS position for absolute localization. The accuracy depends on the map used and the quality of the sensor that records the GPS position. If both are high, this means localization is an even simpler task for an algorithm than positioning in a navigation system, which has to work with imprecise input data: Map and sensor errors do not have to be compensated for in the calculation.

For relative localization, the vehicle position is determined using 3D objects—known as landmarks—which can be detected from the vehicle using video, lidar and radar sensors. The landmarks are represented in the map and enable the position to be determined by aligning the landmarks recognized by the vehicle with them (see ◘ Fig. 37.6). It is conceivable that an absolute localization can also be carried out with landmarks. However, this requires consistently high positional accuracy. For example, due to tectonic plate displacement, an existing map can get an offset on all absolutely recorded coordinates. On the other hand, absolute localization with high accuracy is not absolutely necessary for the driving task. In order to guide a vehicle, it is sufficient to know its position with high precision relative to the road on which it is traveling.

Planning

Planning includes the desired guidance on a route calculated by a navigation system in the street graph (strategic planning), a lane selection and a trajectory calculation within this lane and through intersections (tactical planning). Added to this is the avoidance of collisions with obstacles (reactive planning). All this must be done in compliance with all traffic rules. As you can easily see, this is a very complex task. The various parts of the planning differ in their foresight. Strategic planning takes place over periods ranging from minutes to hours and days, tactical planning rather over seconds. Reactive planning takes place within milliseconds and is a task that is also performed by assistance systems in manually operated vehicles (see ◘ Fig. 37.7).

Not all subtasks are implemented with map support. The map is used in its navigation form in strategic planning. Here, the classic navigation map can be enriched with additional attributes, e.g., information about which sections of the route are basically suitable for automated driving and which are not. This allows a route to be optimized for automated driving. Tactical planning relies on dedicated map elements for automated driving. Reactive planning takes place exclusively on the basis of sensor input.

It is to be expected that the first systems in series operation will be limited to certain, highly classified roads in order to make the driving task easier for the system. For this purpose, the information on how a street is classified is taken from the map.

In order to be able to drive safely within a lane, in addition to the geometry, information is also required on where a lane begins and where it ends. For example, if several lanes merge, the vehicle must plan a maneuver in good time to change to a lane that will continue. Collisions with vehicles are to be avoided.

Another use case is a safe stop in situations in which the system cannot continue driving the vehicle. First it will try to hand over the task of driving to the driver. If there is no takeover, a possible reaction is an emergency stop on the hard shoulder. The map provides information on which sections of the route have a hard shoulder. If none is available, this can be a criterion for preventing the system activation or canceling it in good time.

Driving in construction sites represents a particular challenge. This requires very up-to-date information about lanes within the construction site. The first systems could exclude driving within construction sites. To do this, however, they must at least know where construction sites are located and arrange for a takeover or, if this fails, an emergency stop in good time.

Last but not least, numerous traffic regulations that vary from country to country must be observed. The first thing that comes from the map is the information about the country in which you are actually located. In addition, speed limits with areas of validity, e.g., according to vehicle type, time or weather conditions are listed. These can be used to validate and sup-

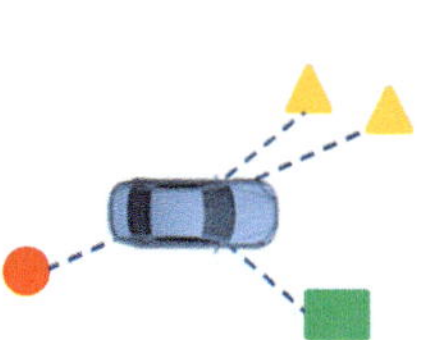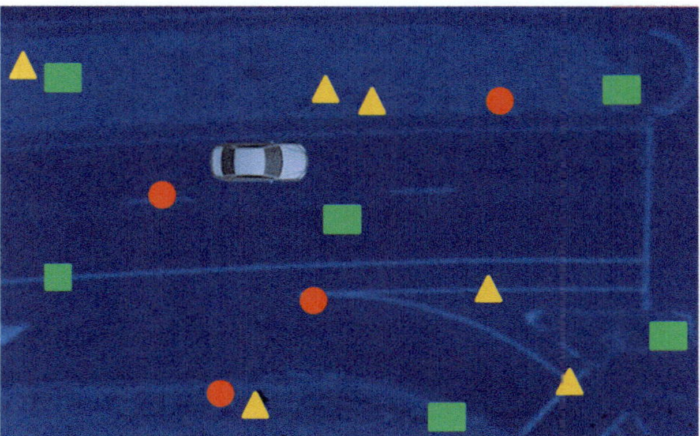

◘ **Fig. 37.6**   Landmarks in sensor view (left) and fitted to a map (right)

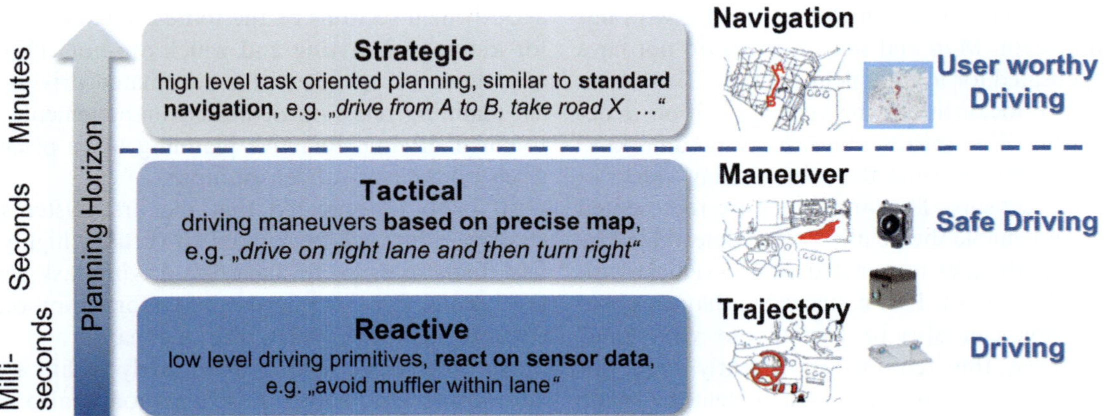

◻ Fig. 37.7   Different levels of planning

plement camera-based traffic sign recognition, e.g., if a sign is currently occluded. Variable message signs and traffic lights are also shown on the map, but without their current status. This must be captured by cameras or, if possible, by communication with the respective infrastructure. Traffic infrastructure can also be important for implicit traffic rules. There are countries like the USA or Australia, where traffic lights at intersections automatically mean that you are not allowed to do a U-turn. The information in the map is also relevant for this, so that it can be taken into account in strategic planning.

37.6.4 Relevant Map Content

Roughly classified, the following map content is required for planning:
- Region affiliation in order to be able to select and implement country-specific traffic rules
- Vehicle-specific restrictions in order to be able to adapt the driving strategy to the vehicle type. There are different restrictions for trucks than for cars.
- Traffic lights
- Structural conditions
- Road class
- Physical separation of lanes from oncoming traffic
- Traffic information
- Variable information as represented by variable traffic signs.

Some of this content is also relevant for navigation systems and its use is already known. Higher accuracy is usually required for automated driving functions, and there is a higher demand for quality. Mistakes here can not only be annoying, but also dangerous.

Some map content (e.g., traffic information, variable speed limits) changes faster and needs to be updated much more frequently. In this case, we speak of dynamic data, which is processed separately from the static data and has a shorter validity for the system.

Further information on digital map data can be found in ▶ Chap. 22.

37.6.5 Repesentation in ADASIS Version 3

Some information such as map data, region affiliation, map matching status, units, etc. are transferred from the AHP to the AHR as global data. The most probable path and sub-paths (generally referred to as path below) are formed.

In order to achieve the higher accuracy required for automated driving functions, the accuracy resolution has been increased from 1 m to 1 cm compared to version 2. The vehicle position and all properties are mapped by distance in centimeters from the start position of the path.

In ADASIS v3, geometries of lanes and their markings as well as landmarks were introduced for the first time. Lanes are represented with absolute coordinates, the geometries of landmarks relative to an absolute coordinate.

37.6.6 Incorporation of the Navigation Route into the Horizon

This section deals specifically with the connection of a classic navigation route (strategic planning) to the other planning tasks (tactical and reactive planning) in a system for highly automated driving.

As already described, automated driving implies functional safety requirements. This prevents navigation systems that were developed before these requirements were defined from being used without further ado. Functional safety cannot be "added" to software that has already been developed.

In order to still be able to use existing navigation software in strategic planning, to which it corresponds functionally, it is desirable to assign safety requirements primarily to tactical and reactive planning. Since reactive planning does not use a map, the mapping of map-based, safety-related functions to tactical planning remains.

You can picture it like this: A level 3 system, for example, cannot perform any driving maneuvers that involve crossing water on ferries. This is also recorded in the system specification and made known to the end user. At the level of strategic planning, this means avoiding ferries (which, by the way, have long been a feature of navigation maps) when automated over manual driving is preferred. Avoiding ferry connections is implemented in route calculation (which corresponds to strategic planning), although of course this is not always successful if there is no overland route to a destination. In case the safety requirement is assigned to strategic planning, no route should be announced at all. When assigning safety requirements to tactical planning, there is a route, and the system gives control back to the driver or stops in good time before reaching the ferry terminal. At the level of strategic planning, in an example erroneous case, it is announced that the entire route, including the ferry, can be driven automatically. The fact that this is ultimately not the case is a loss of comfort, but not a danger.

On the other hand, tactical planning without strategic planning is out of the question, even for simple systems: For example, when driving through motorway junctions, the motorway can be changed without changing lanes.

The information about which lanes are recommended at decision points for following the route can be easily accommodated in the electronic horizon. In the map data model, lanes reference elements of the road data graph. At exits there are at least two road elements to choose from. Lane elements are assigned to these road elements, which can be identified as being on the route.

37.6.7 Map Provision in the Vehicle

As already mentioned, the decision to update map content for safety-related functions cannot be left to the driver. Backend-based solutions are therefore used, with intermediate storage in local flash memory in the control unit if necessary. The function of providing maps precedes horizon generation.

While it is common with classic navigation systems to have the entire, product-specific data area available locally (e.g., Europe), this makes little sense for automated driving functions. The map data becomes outdated too quickly for this. The trend is towards keeping only the (slightly expanded) vehicle environment available in the vehicle, updating it if necessary and deleting or overwriting it if it is not used for a longer period of time. An almost continuous connection to a backend is therefore essential.

Consistency must be ensured when providing maps locally: Map data is usually partitioned in geographical areas (e.g., in tiles in the NDS). Data from different partitions must match and cannot belong to different versions of the map. Depending on the application, this property can be relevant to safety. On the other hand, the availability of map data per se is not relevant to safety, since in the worst case this an availability problem for the application. The contents of the map itself must also be safeguarded: end-to-end monitoring methods from communication technology are used here. Data integrity is ensured by hashes, e.g., a CRC (Cyclic Redundancy Code). In addition, there are security requirements that are a prerequisite to safety. The data is signed for this purpose. Other security measures that do not come from safety considerations concern the restriction of use to an authorized group of people. This is done through encryption.

37.7 ECUs for Navigation and Horizon

A modern vehicle has 30–100 electronic control units, which interact in a complex electrical/electronic architecture (hereinafter referred to as E/E architecture). This paragraph discusses which control units in a vehicle can carry the navigation function or the electronic horizon.

37.7.1 E/E Architecture in Vehicles

In the past, up until about the 1970s, vehicles featured an all-electric architecture. The first electronic control devices found their way into vehicles around the 1970s and 1980s, e.g. for controlling fuel injection into a carburetor. The use of microcontrollers, which can be flexibly programmed using software, and digital buses have opened up completely new possibilities for controlling vehicle functions. Many functions, such as electronic stability control, only became possible through the use of microelectronics. For this reason, equipping vehicles with electronic control units has developed rapidly since 1990 (◘ Fig. 37.8).

In the beginning, isolated electronic control units with little networking for the realization of dedicated functions dominated, but with an increasing number of control units and ever stronger networking, a domain-oriented E/E architecture developed. Here, control units that are largely locally networked with one

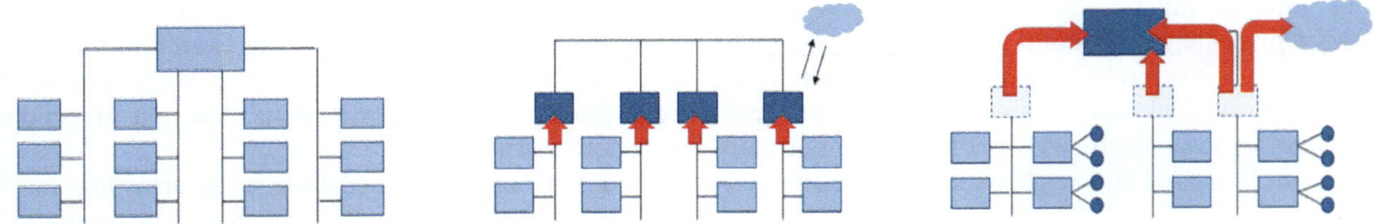

Fig. 37.8 E/E architecture types with dedicated ECUs connected via a gateway (left), domain-oriented architetecture (center) and zone architecture with central computers (right)

another are organized in a domain. E/E domains of today's vehicles are the information/infotainment domain (includes radio, navigation, infotainment in the cockpit with central controls for the driver), the chassis domain (control units for various assistance systems, such as ABS, ESP, ACC, but also for passive safety such as belt tensioners and airbags), the drive train domain (control units for drive and transmission) and the body domain (control units for lighting, central networking, access to the vehicle, air conditioning). In the meantime, the centralization of the control units has progressed so far that the first vehicle manufacturers are beginning to implement central computer architectures in which 1–3 very complex central computers represent extensive functionality in the vehicle. The actuators and sensors distributed throughout the vehicle are connected to the central computers via so-called zone control units. The reduced number of control units leads to less networking complexity and thus a reduction in weight with increased functionality. Figure 1.9 outlines the E/E architecture development.

37.7.2 ECUs for Carrying the Horizon Function

Before placing the electronic horizon on an electronic control unit, the requirements for this must be clarified. The horizon function requires a positioning signal (GNSS positioning), map data, which should be as up to date as possible, and, if possible, information about a currently active navigation route. Depending on the application, the electronic horizon must also meet functional safety requirements (e.g., ASIL A, ASIL B when used in automated driving), which affects the control unit to be selected.

37.7.2.1 Dedicated ECU

For use in automated driving, it is advisable to choose a separate ECU for the electronic horizon. This solution results in relatively high component and interconnectivity costs. However, as long as the equipment rate of vehicles is relatively low in the initial phase of the introduction of automated driving functions, this can be a suitable solution. In connection with networking with a server via a telematics device, current horizon data can always be requested and the memory requirement for holding the map data remains low. Functional safety requirements in the area of ASIL A and ASIL B can be met by a separate control unit, in contrast to today's navigation device, which is usually developed as a QM control unit. However, a disadvantage of this solution is that synchronizing the electronic horizon with an active route on a separate navigation controller is not trivial. On the one hand, this is because they can have different map data. On the other hand, the driver can influence the route in the navigation device via settings, while this is not usually the case with a dedicated ECU. Therefore, the route of the navigation device and the horizon generated in the dedicated ECU can show differences and lead to inconsistent behavior for the driver. In the area of heavy commercial vehicles, dedicated Horizon control units are a common solution today, as some of these do not have navigation devices and the dedicated Horizon ECU is part of the powertrain.

37.7.2.2 Information Domain Computer (IDC)

The former radio/navigation device has meanwhile developed into a very complex domain control device. The first vehicle manufacturers are developing this into a central information domain computer. Since the navigation function is on this control unit, it usually has its own positioning sensors and has a large amount of memory and computing power, it is also very well suited as a carrier for the electronic horizon. The Horizon solutions that have been available in passenger cars since 2010 are implemented in almost all cases on this control unit. The advantage lies in the fact that the navigation function and horizon are operated from a common map base and the active route guidance of the navigation is used to define the horizon. In the past, however, the map basis for the navigation was not up-to-date and the control units did not meet any ASIL requirements. However, the vehicle manufacturers are increasingly making demands on the functional safety of IDCs (up to ASIL B) and their connectivity via a telematics control unit can also be used to keep the map database up to date. Today, the first IDCs are being deployed that enable ASIL requirements up to ASIL B, and maps for fulfilling functions such as speed information and automated driving can be updated in short cycles via mobile phone connections or are even downloaded dynamically from a server and are therefore always up to date.

37.7.2.3 Computers for Driver Assistance and Automated Driving

Today, many vehicles have so-called "Advanced Driver Assistance Systems" (ADAS) and attempts are being made to develop automated driving (= Automated Driving, AD) to the point where it is ready for series production. In the past, each ADAS function had its own control unit, some of which were closely interconnected. As with the IDC, ADAS systems use domain centralization in order to integrate as many ADAS functions as possible, including AD, on as many central computers as possible. Since ADAS functions usually have to meet ASIL requirements from ASIL A to ASIL D, these control units are designed accordingly. They are therefore suitable as carriers for Horizon solutions with ASIL requirements. Furthermore, these control units have an environment sensor system for environment detection and environment interpretation, which can be used for location based on landmarks. This also makes the ADAS/AD computer attractive as a carrier of the horizon function. On the other hand, there is the separation from the navigation, which is on the IDC. This means that no common map basis can be used and the synchronization of the horizon with an active route also requires additional measures. The way to update the horizon on the ADAS/AD computer must also be done via connecting with a telematics control unit to a backend that has up-to-date map data.

37.7.2.4 Telematic Control Unit

Another possibility is to provide the horizon function on a telematics control unit. The main advantage is that updating the map or requesting up-to-the-minute horizon data is very easy, since the telematics control unit provides access from the vehicle to the backend using mobile communications. This means that data for updating the map does not first have to be distributed to another control unit via the vehicle network. Due to the eCall functionality, many telematics devices also have a location sensor. However, this is offset by the fact that telematics control units today generally do not meet ASIL requirements and here, too, no common database is formed with the navigation map. Synchronizing the horizon with an active route is also complex to implement, such as placing the horizon function on a separate control device and ADAS/HAD computer. Telematics devices have relatively little computing power and memory—with the exception of telem-

atics devices in the commercial vehicle sector. However, it is foreseeable that telematics devices will also have to meet functional safety requirements to a greater extent in the future, for example in cooperative systems using vehicle-to-vehicle communication.

37.7.3 Summary

The most common solution today is to generate an electronic horizon on an IDC or its predecessor, an infotainment control unit that also contains the navigation function and the digital map. However, these are exclusively solutions to support comfort, ADAS and energy saving functions that do not put ASIL requirements on the horizon. Solutions on separate control units and telematics devices are less common. ADAS/ AD computers could also provide a horizon solution in the future. In the end, a vehicle manufacturer has to decide which solution is the most economical option given its boundary conditions, requirements and equipment rates. There is much to suggest that implementation on the IDC in conjunction with the navigation system in the passenger car sector and implementation on a telematics control unit in the heavy commercial vehicle sector continue to be particularly useful designs. With the distribution of future HAD functions in vehicles, the ADAS/AD computer also offers itself as a useful, future carrier for the horizon function.

References

1. OpenLR Association. ► http://www.openlr.org/. Last accessed 31 Jan 2022
2. ISO 17572-3:2015: Intelligent transport systems (ITS)—location referencing for geographic databases—part 3: dynamic location references (dynamic profile). ► https://www.iso.org/standard/63402.html. Last accessed 31 Jan 2022
3. Takai, D.: Architektur für Websysteme. Carl Hanser Verlag (2017)
4. Eclipse Hono project. ► https://www.eclipse.org/hono/. Last accessed 31 Jan 2022
5. Eclipse Ditto project. ► https://www.eclipse.org/ditto/. Last accessed 31 Jan 2022
6. Eclipse Vorto project. ► https://www.eclipse.org/vorto/. Last accessed 31 Jan 2022
7. Bundesministerium für Digitales und Verkehr: Verkehrsinvestitionsbericht für das Berichtsjahr (2019). ► https://www.bmvi.de/SharedDocs/DE/Publikationen/G/verkehrsinvestitionsbericht-2019.html. Last accessed 31 Jan 2022
8. Apple Car Play. ► https://developer.apple.com/carplay/. Last accessed 31 Jan 2022
9. mAutomotive: Connecting cars: The technology roadmap. ► https://www.gsma.com/iot/wp-content/uploads/2012/04/gsma-connectingcarsthetechnologyroadmapv2.pdf. Last accessed 31 Jan 2022
10. ADASIS Forum: ADASIS v3 protocol specification v3.1.0 (2018)

Open Access This chapter is licensed under the terms of the Creative Commons Attribution-NonCommercial-NoDerivatives 4.0 International License (▶ http://creativecommons.org/licenses/by-nc-nd/4.0/), which permits any noncommercial use, sharing, distribution and reproduction in any medium or format, as long as you give appropriate credit to the original author(s) and the source, provide a link to the Creative Commons license and indicate if you modified the licensed material. You do not have permission under this license to share adapted material derived from this chapter or parts of it.

The images or other third party material in this chapter are included in the chapter's Creative Commons license, unless indicated otherwise in a credit line to the material. If material is not included in the chapter's Creative Commons license and your intended use is not permitted by statutory regulation or exceeds the permitted use, you will need to obtain permission directly from the copyright holder.

Automated Driving

Human Factors: Level 3+

Klaus Bengler, Johanna Josten and Claus Marberger

Contents

© The Author(s) 2026
H. Winner et al. (eds.), *Handbook Assisted and Automated Driving*,
https://doi.org/10.1007/978-3-658-45276-6_38

38.1 Introduction

Following the logic of SAE J3016 [1], automation Level 3 will be reached as technology advances. At this level, all aspects of the dynamic driving task are transferred to the automation feature and the user is not obliged to permanently monitor the system's performance any more. However, it is assumed that the user remains receptive in order to be able to take over the driving task when prompted by the vehicle. If the takeover does not occur, the vehicle will initiate a failure mitigation strategy in order to reach a state of reduced risk. For limited operational design domains (ODD) this high degree of automation can be achieved very soon. For example, it is quite conceivable that a traffic jam assistance function can operate with very high technical reliability in the low-speed range over a long period of time.

Thus, while in the case of Level 2 and Level 1 the driver's attention is permanently required, it may be averted in the case of Level 3. Homans et al. [2] were able to show that users may easily overestimate the technical capabilities of, vehicles, operating at automation Level 3. According to the mental model of many users it is difficult to assess the limitations of automation and to recognize the correct allocation of tasks between the user and the system. Therefore, the success of high automation levels depends on a complex interplay of technical and human factors. Above all, it must be possible for the user to realistically assess the functional range and functional limitations so that an appropriate level of trust can develop in the automation and that it will be used as intended by the manufacturer.

38.2 Control Transitions as a Safety-Relevant Aspect of Level 3 Functions

Vehicle automation shifts responsibility for and control over the driving task from humans to a technical function. In level 3 [1] vehicles, the user can divert his attention from the driving task in addition to withdrawing from the active execution of longitudinal and lateral guidance. While the initiation of such a shift of control toward automation is at the sole responsibility of the driver, who is free to activate the function, the shift of control back to the driver can be initiated either by the driver himself or by the system. In particular the latter case poses a challenge for system design and a key difference to lower levels of automation. It must be ensured that the user, who acts as a fallback level for the system in the case of level 3 automation, can perform this task safely and as comfortably as possible despite being distracted from the traffic situation, last but not least for reasons of system acceptance. To this end, however, the function must be able to continue to operate for a certain period of time after detecting and indicating the need for a control transition back to the user [1]. Only this will allow the user to safely change roles from passenger in level 3 to supervisor or driver. Based on the permissible diversion of attention from the driving task and resultant requirements for the design of control transitions, a clearly specified area of operation (ODD) represents therefore one probable prerequisite for level 3 functionalities.

A sufficient timeframe from the takeover request to the handover of control towards the driver is necessary for level 3 automation, as the driver may have to refocus his attention from a non-driving related activity back to the driving context. In addition to establishing a readiness to take over, for example by returning the hands to the steering wheel, particularly all relevant aspects of the current traffic situation must be processed for an appropriate takeover of vehicle control to occur [3–5]. In research, the temporal analysis of the driver's takeover behavior is often supplemented with an assessment of the resulting takeover quality to conclude the successful return to the diving task in the current scenario [6–8].

The knowledge established by research on the sequence of control transitions (see ◻ Figs. 38.1 and 38.2 as a conceptual framework) can be used in the design of Level 3 functions for a definition of a necessary duration of system-initiated takeover situations: The minimum time in seconds that should be provided to the user for a safe takeover of vehicle control in Level 3 is in most instances located in the upper single-digit range (for a review Eriksson and Stanton; Seppelt and Victor [9, 10]). However, establishing a sufficient timeframe for the safe design of control transitions to the user is complicated by the fact that a variety of factors influence both the takeover time required by the user and the resulting takeover quality [5]. Influencing factors can be attributed to the user and his condition, such as age, system expertise, or the current engagement in non-driving related activities, but also to the functional design, such as characteristics of the takeover request, and to the current situation of use, including the complexity of the driving scenario [8, 11–15].

As soon as one or more conditions for the ODD cease to apply, the user will again be needed as a supervisor of a level 2 system or as a driver (level 0 or 1) after an appropriate premonition period. Even if Level 3 functions can be used on defined sections of the journey, it will always be necessary to drive the vehicle at a lower automation level or even manually before and after Level 3 appropriate sections. The transitions between these sections must be designed with great care. Furthermore, the user must at all times be aware of the

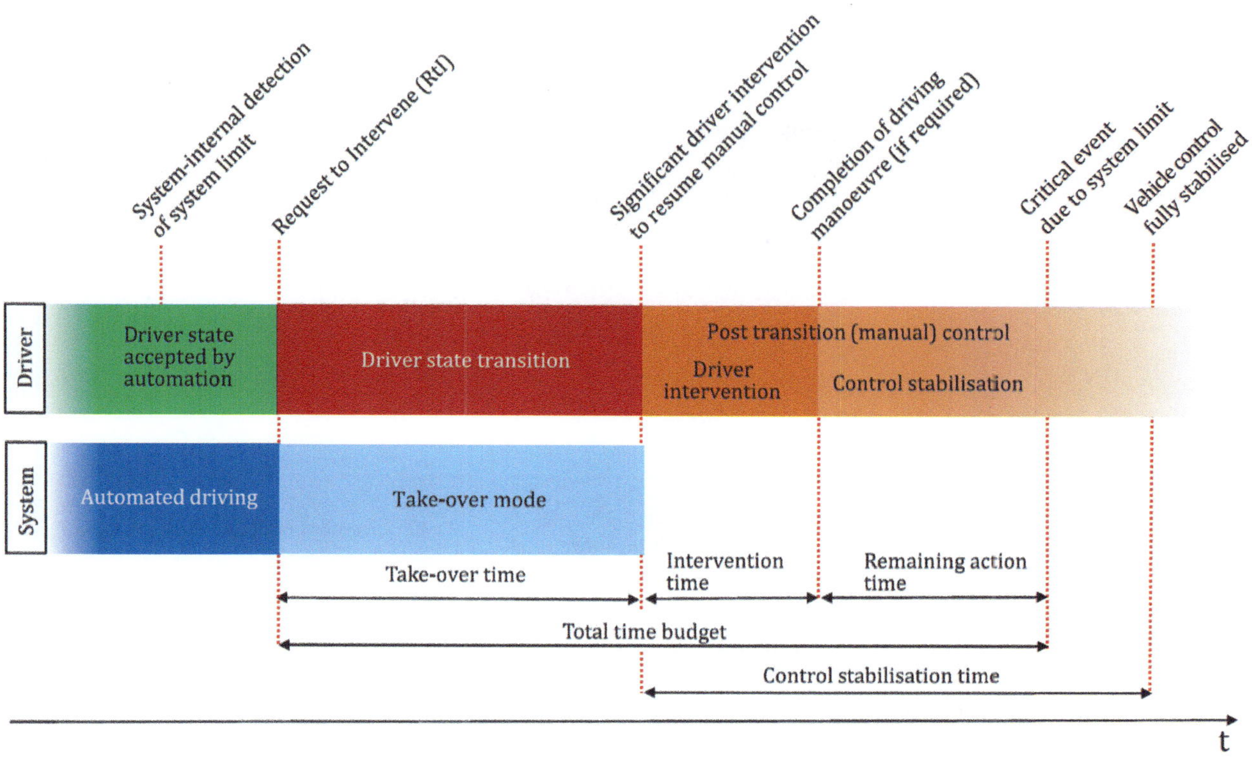

Fig. 38.1 System-initiated transition process from SAE L3 automated mode to manual driving according to ISO TR21959-1:2018

responsibilities currently assigned to the automation. The number of transitions and the duration of continuous automation of one level will be a notable factor for the acceptance of Level 3 vehicles (Figs. 38.1 and 38.2).

Last but not least, the expectation of and experience with takeover situations may also influence the extent to which drivers remain willing to relinquish control to the automated function or to engage in non-driving related activities [16]. One factor that has a significant influence on the transfer of control toward automation and on the reliance upon automation is the trust that users place in the function with regard to the safe and continuous execution of the driving task [17].

38.3 Trust and Level 3 Automation

Trust is considered an important influence on the interaction with automation [18, 19] and is defined as "the attitude that an agent will help achieve an individual's goal in a situation characterized by uncertainty and vulnerability" [19]. Following this definition, the degree of trust in automation becomes especially relevant for user behavior with increasing complexity. Here, the user's understanding of the system is often not sufficient to make a reliable prediction about automation behavior in the current situation.

Due to the large number of scenarios covered by the ODD of a Level 3 automated vehicle, an exact understanding of the function cannot be assumed in all cases, even with increasing user experience. The user's trust is therefore particularly relevant for higher levels of automation with a correspondingly more extensive ODD. Initial trust is, amongst others, defined by available information about a function, while the congruency between the expected automation behavior and the behavior experienced during use determines, for example, the extent of a possible loss of trust in the function [3, 16, 18–22]. Thus, trust in automation changes dynamically, depending on the experience with a specific system and the experienced system behavior, sometimes even within the course of a single interaction [18, 20, 22].

Similar to takeover performance, trust in a specific function depends on many influencing factors which can be attributed to the user, to the function itself, as well as to the environment. In their model of influences on trust in automation, authors in [18] distinguish between dispositional trust, situational trust, and initial and dynamically learned trust. Amongst others, age, attention capacity, culture, expertise, and system complexity are cited as determinants of trust in automation.

As illustrated by Lee and See [19], the so-called calibrated trust, i.e. trust adapted to the capabilities of the automation, represents the optimum for safe interac-

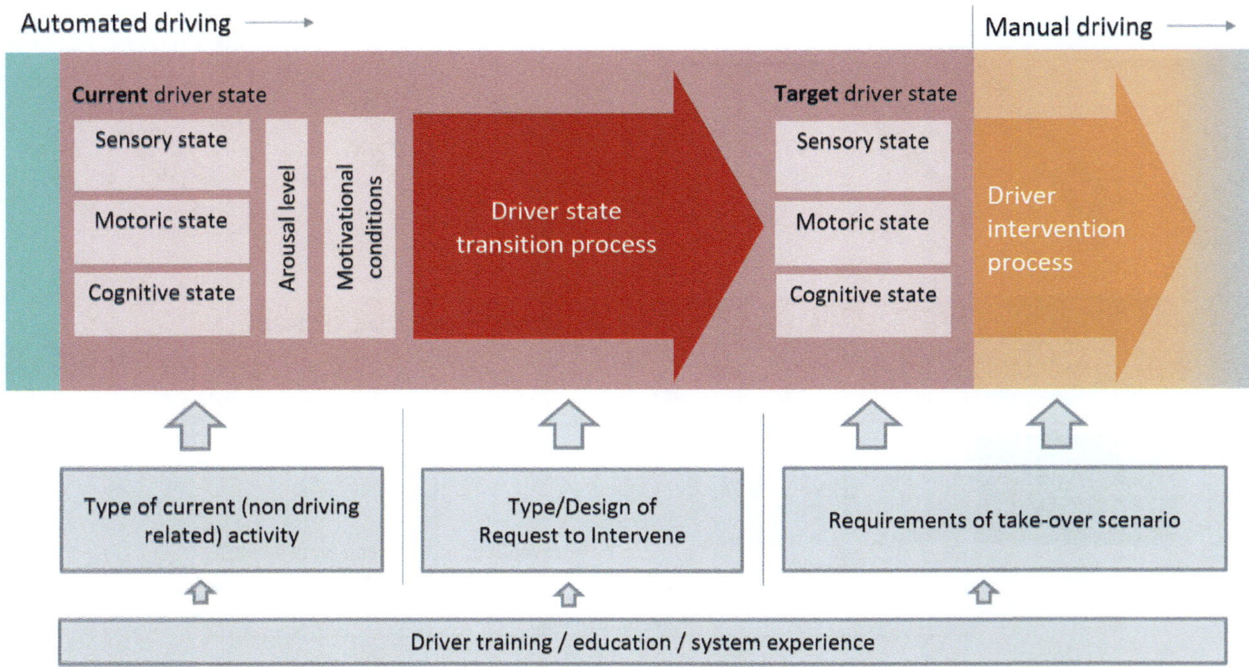

Fig. 38.2 Conceptual model of the driver state transition process according to Marberger et al. (2018)

tion. On the one hand, an underestimation of a function's capabilities has to be avoided, which may result in a less frequent or less comfortable use, e.g., by not making use of the offer to divert attention from traffic when using Level 3 automation. On the other hand, excessive trust (overtrust) also has negative consequences for the interaction with automation [24]. For example, Payre et al. [25] demonstrated a negative impact of high levels of trust on takeover performance.

Especially the latter case, i.e., excessive trust in automation, results in safety-related consequences for the interaction with level 3 functions. For trust to be calibrated to level 3 automation, it is therefore particularly necessary for the user to be fundamentally aware of the possibility of a system-initiated change in automation level, up to and including a complete takeover of vehicle control. The expectation regarding such takeover situations has a relevant influence on user behavior in interaction with automation, as described earlier [16]. In particular, high trust causes a stronger engagement in non-driving related activities and thus a lower attribution of attention towards traffic events [22]. From the research of Albert et al. [23], it is known that while users engage in non-driving related activities, glances to the driving scene can hardly be expected. This is a clear indication of user's trust in the automated function, but it is also a clear indication that level 3 functions must be implemented with the highest technical reliability, as automation errors would not be detected by the user.

In contrast to the negative effect of high trust on system-initiated transitions, research has also shown that individuals with low trust are more likely to override automation even if a driver intervention would not have been necessary [22]. However, driver-initiated interventions into vehicle guidance should rather be avoided, especially with respect to an often initially reduced quality in driving performance after automation use. Due to the dynamic nature of trust, the design of level 3 automation, including, but not limited to, sufficient information for the user about current and expected system states, offers especially great potential with regard to an appropriately calibrated, learned trust.

38.3.1 Interior

Once users are able to turn away from the driving task, there are extensive opportunities to engage in non-driving related activities or adopt non-driving postures. Studies by Yang et al. [26] describe what postures and seat settings occur in conjunction with higher levels of automation. The research by [27] indicates that recline angles of approximately 30° and up to 60° for sleeping and resting are desired by users of a level 3 automation.

These studies allow estimating the adoption process from atypical initial conditions. For example, authors in show that subjects need significantly more time to straighten up into the driving position and grip the steering wheel starting from a recline angle of 24°. In detail, Kerschbaum [28] investigated on which movement paths and trajectories the hands are moved to finally grasp the steering wheel in the takeover case.

Various desires to engage in non-driving related activities are associated with Level 3 [29]. From an ergonomic perspective, these non-driving related activities also require a new arrangement of displays and storage options in the interior.

Furthermore, suitable digital human models for the design of future vehicle interiors can be derived based on the posture data.

38.3.2 Aspects of Comfort and Discomfort in L3+ Automation

A major goal in the development of automated driving systems is to increase the experienced ride comfort. This applies especially to higher levels of automation, since in those scenarios the user does not permanently serve as a fallback-ready driver at system limits. Instead, the user can be regarded as a consumer of a mobility service provided by the vehicle. But what does "comfort" mean and how can it be interpreted in the context of automated driving?

There is no uniform understanding of the construct "comfort" due to the wide range of application domains. However, a core element seems to be "the pleasant state or relaxed feeling of a person in response to his or her environment" (cf. [30, 31]). According to this understanding, comfort is considered a psychological construct that refers to the subjective evaluation of external impact factors. Since the reference conditions for evaluating comfort differs between individuals (e.g., due to different expectations), the perceived comfort can significantly vary between different individuals despite identical environmental conditions [32, 33]. The scientific community also agrees that detrimental factors on comfort draw human attention more easily, which implies that the reduction of "discomfort" is essential to establish comfort [34]. However, it remains debatable whether eliminating sources of discomfort is sufficient to generate a comfortable motion experience [35, 36] using the example of seat comfort). In any case, human factors researchers agree that in the context of automated vehicles, not only somatosensory effects of vehicle movement need to be considered but also the psychological effects of automation on comfort perception [37, 38]. Therefore, this chapter will address various aspects of "ride comfort" [39] that are relevant for describing the automated driving experience at higher levels.

38.3.3 Somatosensory Effects of Vehicle Motion

One of the most obvious approaches to increase ride comfort is to minimize the forces acting on the user. In the context of automated driving there are possibilities to accomplish this goal on several levels of the driving task (e.g., according to Michon [40]):

1. Strategic level: If automated vehicle guidance also includes route selection, the expected level of accelerations acting on the passengers can be estimated by pre-assessing the route characteristics. Information on speed limits, road curvature, the type of the road surface and traffic conditions allows optimizing route selection according to comfort-oriented criteria (e.g. [41]).

2. Tactical level: At this level, the forces acting on the passengers are influenced by the choice and execution of necessary driving maneuvers (e.g., starting/accelerating, decelerating/stopping, changing lanes) as well as the general speed selection [42]. A comfort-oriented design of these driving style-related parameters is typically based on keeping maximum longitudinal and lateral accelerations as low as possible [37, 43] or keeping acceleration changes (jerks) as infrequent and small as possible [44, 45]. Unpleasant vertical accelerations and jerks, caused by locally occurring unevenness of the road surface (e.g. potholes), can also be reduced by early adjustment of the driving trajectory.

3. Operational level: Comfort-oriented design of longitudinal and lateral control also plays an important role at the basic level of the driving task, such as keeping the vehicle on the road. Disturbing effects on motion comfort may arise from non-optimal steering control algorithms (e.g., frequent changes of the steering direction to follow the target trajectory). At this level, the requirement is to maintain harmonic trajectories with small acceleration and jerk values as well as sufficient longitudinal distance to other road users (cf. [44]).

38.4 Travel Progress

Minimizing the vehicle accelerations (as described above) may increase the automated ride comfort only in a range where no significant negative effects on travel progress are perceived. Summala [46] emphasizes "travel progress" among other factors in his "multiple comfort zone" model. This also explains why Lange et al. [47] conclude that a comfortable automated driving style should not necessarily be significantly less dynamic than manual driving. Chauffeur drivers (which often serve as a reference for comfortable driving) are typically required to not only drive smoothly, but also to drive as time efficient as possible. This includes anticipatory selection of the optimal lane to avoid following slower road users.

38.5 Subjective Feeling of Safety

An essential precondition for generating ride comfort in highly automated vehicles is the subjective experience of safety (e.g. [48]). Even if objective safety of the highly automated driving feature is guaranteed by the manufacturer, the subjective experience may differ. Summala's [46] comfort model clearly includes safety-related requirements resulting from adequate speed selection, maintenance of sufficient distances to prevent collision of objects and compliance with traffic rules.

38.5.1 Feeling of Control

By transferring the driving task to an automated system, the driver's role changes from an active controller to a passive passenger. Giving away control authority can be perceived with negative or positive feelings by individuals—from 'loss of control' to 'relief from burden'. The different perspective has already been investigated for the passenger's role in manually driven vehicles (e.g., [49, 50]). In this context, the individual assessment of releasing control does not only depend on the quality of automated vehicle control, but also on the desirability of the replaced driving task for manual control [51]. In this sense, many drivers will likely welcome highly automated driving features in slow-moving Stop and Go traffic, whereas the same feature would lose its desirability for use on rural roads. Moreover, the design of the human–machine interface can have a significant impact on how loss of control is perceived by users. The retained level of perceived control also depends on how easily, safely, and conveniently control transitions can be initiated by the user. According to the SAE J3016 [1], Level 4 features may even deny or delay user-initiated requests for control for safety reasons. It remains to be seen how this system behavior is assessed by users in terms of ride comfort.

Uncertainties regarding the current system state ("mode awareness") and the associated uncertainty regarding the current user role should be avoided as far as possible to prevent so-called "automation surprises" (cf. [52]). Automation surprises refer to unpleasant driving situations that arise when actual system performance limits are not expected by the user. In order

to counteract this risk, ergonomically well-designed information displays with explicit references to the user's tasks can be helpful to mitigate discomfort caused by feelings of being out of control [53].

38.6 Social Influence

Another facet of ride comfort results from the perceived impact of automation on the social environment or the perceived feedback of the social environment on the use of vehicle automation. The social environment can refer to other passengers inside the vehicle or to other road users interacting with the automated driving vehicle in the traffic environment. Especially when the automated vehicle does not externally indicate its current mode, the social environment may easily mistake automated driving behavior for human control. In case automation shows significant deviations from typical human behavior, this may be negatively attributed to the person sitting in the driver's seat. For example, specific characteristics of automation such as lower driving speed, extensive safety distances or hesitant initiation of lane changes could be attributed to an "incompetent driver". Similarly, an automated driving strategy that is not designed to be cooperative could be attributed to an "unfriendly driver". Socially induced discomfort is not only triggered by explicit negative feedback from the social environment, the mere idea, that this could be the case is sufficient for generating psychological discomfort. This is also reflected in the numerous technology acceptance models, in which the "expected" or "suspected" effect on the personal environment plays a major role in determining the intention to use (see [54]). However, the influence of the social environment can also have a positive impact on the perceived comfort. A well performing automated driving system represents a hedonic product quality that goes beyond mere usability [55]. Through the (public) use of an innovative driving feature, a desirable message about one's own self-image (including social status) can be communicated to the external environment.

38.6.1 Influence of Non-driving Related Activities

In higher levels of automation (SAE L3 +) users will be allowed to turn away from the monitoring task and focus their attention on other activities during the automated journey. If users engage in non-driving related activities, the perceived driving comfort is also significantly influenced by the type of non-driving activity. For example, Festner [38] was able to show in an on-road study that users prefer a more defensive driving style when engaged in non-driving related tasks. Ride comfort during visual-manual tasks is primarily affected by factors perceived by the vestibular system, especially when high levels of motor skills are required. In contrast, ride comfort during auditive tasks (allowing the user to look outside) is also affected by visually perceptible factors such as lateral offset in lane position or low distances to other road objects [38].

38.6.2 Motion Sickness

One facet of ride discomfort that is often associated with non-driving related activities is the development of motion sickness or kinetosis. Based on survey data, authors in [56] estimate that about 25% of the population will develop symptoms of motion sickness when reading in automated vehicles. Taking into account all other activities that passengers would like to carry out during an automated ride, this would result in 6–12% of the population who will encounter a "medium to large problem" "at some point" with regard to motion sickness. The most cited and widely accepted theory on kinetosis is based on the notion of sensory conflict. It goes back to Reason and Brand [57] and postulates that motion sickness results from conflicts in processing different motion related stimuli. The conflict may arise in the central nervous system due to several reasons:

- The sensory stimuli perceived by the visual, vestibular, and proprioceptive sub-systems of the central nervous system do not lead to identical assessment of self-movement in a given environment.
- Rotatory and translatory motion stimuli, which are detected by different subsystems of the vestibular organ, do not lead to identical estimation of the body orientation.

The conflicts are mainly due to individual expectations which motion stimuli are consistent to each other and how an assessment of one's own body position in space can be derived from this. These expectations are formed within the individual learning histories of humans when interacting with the natural environment. Higher levels of driving automation could increase the problem due to the following aspects:

- Critical oscillations in longitudinal and lateral dynamics could occur more frequently in automated driving scenarios due to technical limitations of the automation feature (e.g., less performance in anticipating other road user's behavior, more uncertainties in the detection and interpretation of a driving situation).
- The possibility to engage in visual non-driving related tasks cause conflicts in the processing of visual and vestibular motion cues.

- The change from the active driver role to the passive observer role makes the prediction of future vehicle acceleration more difficult.
- Higher level automation is often associated with new interior concepts that allow seating positions not aligned with the driving orientation.

Based on the available knowledge about the root causes, countermeasures can be developed and evaluated to specifically counteract the development of motion sickness. Authors in [58], for example, recommend optimizing the predictability of the vehicle's movement and to avoid incongruent signals for perception of self-motion. Measures that aim to maintain visibility of the driving situation ahead should help to mitigate motion sickness in automated vehicles. Bohrmann [59] has demonstrated the positive effect of reclined seating in several user studies. The modified body posture has an impact on the stimulation of the vestibular organ during vehicle maneuvers and reduces unintentional head movements by utilizing the headrest. In addition, slightly reclined seating positions (backrest angle approx. 40°) or supine positions (backrest angle approx. 60°) are generally preferred by passengers for comfort-related reasons [60].

38.7 User Acceptance of Automation Functions (L1–L3) up to Acceptance of New Mobility Offers (L4)

Gold et al. [3] show that the acceptance of highly automated driving functions can vary greatly before future users have had the opportunity to gain experience with them and can change again after initial user experiences. While older users of automation would like to maintain their mobility even in complex driving situations and would use it as a highly extended assistance, younger users focus on the benefits of non-driving related activities. This results in quite different requirements for the reliability of the technical function in different traffic situations and the distribution of roles between humans and vehicles.

Damböck et al. [61] were already able to show the importance of visualizing the automation function and its future maneuver plans/intentions using the H-mode concept as an example. They investigated different displays in the contact-analog head-up display in connection with haptic feedback.

Even in the case of highly reliable automation functions, users must be given the opportunity to infer current and future functionality from the various displays of HMI. This also applies to automation level 4.

Danner et al. [62] show that predicting the availability of automation on the planned driving route is an important HMI function for building mental models, system understanding and user acceptance.

38.7.1 Summary/Outlook

A prerequisite for Level 3 functionalities is the clear definition of ODDs. However, as soon as one or more conditions for the ODD cease to apply, humans will be needed as drivers or supervisors of lower automation levels (Level 2–Level 0).

Even if Level 3 functionalities can be used on certain road sections, it will always be necessary to drive the vehicle in a lower level or even manually before and after. The transitions between these levels must be designed with great care. The number of transitions and the duration of continuous automation within one level will be a noteworthy factor for acceptance.

Since Level 3 will continue to be used in mixed traffic with manually controlled vehicles for a long time, users will have to repeatedly clarify situations that arise from interactions with other vehicles when turning in/away, in order to maintain the flow of traffic. The general topic of ride comfort in automated driving is especially relevant for higher automation levels when user intervention is expected to occur less frequently. Non-driving related tasks will generate new user requirements not only for the design of the vehicle interior, but also for the design of the automated driving style.

It is quite conceivable to use Level 3 and higher in spatially restricted areas under completely different conditions. Applications in terminals, logistics centers, pedestrian zones or on agricultural land show that continuous efficient operation—mostly in the low speed range—is feasible. Again, it is evident that automated vehicles need to communicate to their users and their environment/interation partners through appropriate HMI components [63].

References

1. SAEJ3016: Taxonomy and definitions for terms related to driving automation systems for on-road motor vehicles, J3016. SAE International/ISO, 400 Commonwealth Drive, Warrendale, PA, United States (2021) and (2018). ▶ https://www.sae.org/standards/content/j3016_201806/
2. Homans, H., Radlmayr, J., Bengler, K.: Levels of driving automation from a user's perspective: How are the levels represented in the user's mental model? In: Human Interaction and Emerging Technologies (IHIET 2019). Springer, Cham (2019)
3. Gold, C., Körber, M., Hohenberger, C., Lechner, D., Bengler, K.: Trust in automation—before and after the experience of takeover scenarios in a highly automated vehicle. Proc. Manuf. 3, 3025–3032 (2015)
4. Lu, Z., Coster, X., de Winter, J.: How much time to drivers need to obtain situation awareness? A laboratory-based study of automated driving. Appl. Ergon. 60, 293–304 (2017)
5. Zeeb, K., Buchner, A., Schrauf, M.: What determines the take-over time? An Integrated model approach of driver take-over after automated driving. Accid. Anal. Prev. 78, 212–221 (2015)
6. Happee, R., Gold, C., Radlmayer, J., Hergeth, S., Bengler, K.: Take-over performance in evasive manoeuvres. Accid. Anal. Prev. 106, 211–222 (2017)
7. Merat, N., Jamson, A.H., Lai, F.C.H., Daly, M., Carsten, O.M.J.: Transition to manual: driver behavior when resuming control from a highly automated vehicle. Transp. Res. Part F 27, 274–282 (2014)
8. Zeeb, K., Buchner, A., Schrauf, M.: Is take-over time all that matters? The impact of visual-cognitive load on driver take-over quality after conditionally automated driving. Accid. Anal. Prev. 92, 230–239 (2016)
9. Eriksson, A., Stanton, N.: Takeover times in highly automated vehicles: Noncritical transitions to and from manual control. Hum. Factors 59(4), 689–705 (2017)
10. Seppelt, B.D. Victor, T.W.: Potential solutions to human factors challenges in road vehicle automation. In: Meyer, G., Beiker, S. (eds.) Road Vehicle Automation, vol. 3, pp. 131–148. Lecture Notes in Mobility (2016). ▶ https://doi.org/10.1007/978-3-319-40503-2_11
11. Gold, C., Körber, M., Lechner, D., Bengler, K.: Taking over control from highly automated vehicles in complex traffic situations: the role of traffic density. Hum. Factors 58, 642–652 (2016)
12. Körber, M., Gold, C., Lechner, D., Bengler, K.: The influence of age on the take-over of vehicle control in highly automated driving. Transp. Res. Part F 39, 19–32 (2016)
13. Merat, N., Jamson, A.H., Lai, F.C.H., Carsten, O.M.J.: Highly automated driving, secondary task performance, and driver state. Hum. Factors 54(5), 762–771 (2012)
14. Petermann-Stock, I., Hackenberg, L., Muhr, T., Josten, J., Eckstein, L.: Bitte übernehmen Sie das Fahren! Ein multimodaler Vergleich von Übernahmestrategien. In Intelligente Transport-und Verkehrssysteme und-dienste Niedersachsen e.V. (Ed.), AAET: Beiträge zum 16. Braunschweiger Symposium, pp. 345–369. ITS Niedersachen e.V Radlmayr 2014, Braunschweig (2015)
15. Radlmayr, J., Gold, C., Lorenz, L., Farid, M., Bengler, K.: How traffic situation and non-driving related tasks affect the take-over quality in highly automated driving. In: Proceedings of the Human Factors and Ergonomics Society, 58th Annual Meeting, pp. 2063–2067 (2014)
16. Pöhler, G., Heine, T., Deml, B.: Einfluss der Erwartungshaltung auf das Übernahmeverhalten in automatischer Fahrzeugführung. 11. Workshop Fahrerassistenzsysteme und automatisiertes Fahren (FAS 2017), Walting (2017)
17. Körber, M.: Individual differences in human-automation interaction: a driver-centered perspective on the introduction of automated vehicles. Doctoral dissertation, Universitätsbibliothek der TU München (2018)
18. Hoff, K.A., Bashir, M.: Trust in automation: integrating empirical evidence on factors that influence trust. Hum. Factors 57(3), 407–434 (2015)
19. Lee, J.D., See, K.A.: Trust in automation: designing for appropriate reliance. Hum. Factors 46(1), 50–80 (2004)
20. Beggiato, M., Krems, J.F.: The evolution of mental model, trust and acceptance of adaptive cruise control in relation to initial information. Transp. Res. Part F 18, 47–57 (2013)
21. Hergeth, S., Lorenz, L., Krems, J.F.: Prior familiarization with takeover requests affects drivers' takeover performance and automation trust. Hum. Factors 59(3), 457–470 (2017)

22. Körber, M., Baseler, E., Bengler, K.: Introduction matters: manipulating trust in automation and reliance in automated driving. Appl. Ergon. **66**, 18–31 (2018). ► https://doi.org/10.1016/j.apergo.2017.07.006

23. Albert, M., Lange, A., Schmidt, A., Wimmer, M., Bengler, K.: Automated driving–assessment of interaction concepts under real driving conditions. Procedia Manuf. **3**, 2832–2839 (2015)

24. Parasuraman, R., Riley, V.: Humans and automation: use, misuse, disuse, abuse. Hum. Factors **39**, 230–253 (1997)

25. Payre, W., Cestac, J., Delhomme, P.: Fully automated driving: impact of trust and practice on manual control recovery. Hum. Factors **58**(2), 229–241 (2016)

26. Yang, Y., Klinkner, J.N., Bengler, K.: How will the driver sit in an automated vehicle? The qualitative and quantitative descriptions of non-driving postures (NDPs) when non-driving-related-tasks (NDRTs) are conducted. In: Congress of the International Ergonomics Association, pp. 409–420. Springer, Cham (2018a)

27. Yang, Y., Gerlicher, M., Bengler, K.: How does relaxing posture influence take-over performance in an automated vehicle? In: Proceedings of the Human Factors and Ergonomics Society Annual Meeting (HFES). Sage Publication Ltd (2018b)

28. Kerschbaum, P.: Design for automation: the steering wheel in highly automated cars. Technical University of Munich (2018). ► https://mediatum.ub.tum.de/1356752

29. Wandtner, B., Schömig, N., Schmidt, G.: Secondary task engagement and disengagement in the context of highly automated driving. Transport. Res. F: Traffic Psychol. Behav. **58**, 253–263 (2018)

30. Vink, P., Hallbeck, S.: Comfort and discomfort studies demonstrate the need for a new model. Elsevier (2012)

31. Bellem, H., Thiel, B., Schrauf, M., Krems, J.F.: Comfort in automated driving: an analysis of preferences for different automated driving styles and their dependence on personality traits. Transport. Res. F: Traffic Psychol. Behav. **55**, 90–100 (2018)

32. Krist, R.: Modellierung des Sitzkomforts: Eine experimentelle Studie (Modeling sit comfort: an experimental study), Katholischen Universität Eichstätt, Germany. Doctoral dissertation, Doctoral thesis (1994)

33. Carsten, O., Martens, M.H.: How can humans understand their automated cars? HMI principles, problems and solutions. Cogn. Technol. Work **21**(1), 3–20 (2019)

34. Bubb, H., Bengler, K., Grünen, R.E., Vollrath, M.: Automobilergonomie. Springer-Verlag (2015)

35. Hertzberg, H.T.E.: The human buttocks in sitting: pressures, patterns, and palliatives. SAE Technical Paper (No. 720005) (1972)

36. Zhang, L., Helander, M.G., Drury, C.G.: Identifying factors of comfort and discomfort in sitting. Hum. Factors **38**(3), 377–389 (1996)

37. Elbanhawi, M., Simic, M., Jazar, R.: In the passenger seat: investigating ride comfort measures in autonomous cars. IEEE Intell. Transp. Syst. Mag. **7**(3), 4–17 (2015)

38. Festner, M.: Objektivierte Bewertung des Fahrstils auf Basis der Komfortwahrnehmung bei hochautomatisiertem Fahren in Abhängigkeit fahrfremder Tätigkeiten: Grundlegende Zusammenhänge zur komfortorientierten Auslegung eines hochautomatisierten Fahrstils. Doctoral dissertation, Universität Duisburg-Essen (2019)

39. Da Silva, M.C.G.: Measurements of comfort in vehicles. Measur. Sci. Technol. **13**(6) (2002)

40. Michon, J.A.: A critical view of driver behavior models: what do we know, what should we do? In: Human Behavior and Traffic Safety, pp. 485–524. Springer, Boston, MA (1985)

41. Li, Z., Kolmanovsky, I.V., Atkins, E.M., Lu, J., Filev, D.P., Bai, Y.: Road disturbance estimation and cloud-aided comfort-based route planning. IEEE Trans. Cybern. **47**(11), 3879–3891 (2017). ► https://doi.org/10.1109/TCYB.2016.2587673

42. Hartwich, F., Beggiato, M., Dettmann, A., Krems, J.F. (Hg.): Drive me comfortable. Individual customized automated driving styles for younger and older drivers, 8. VDI-Tagung. Der Fahrer im 21. Jahrhundert. Braunschweig. 10–11 Nov 2015

43. Freyer, J.: Vernetzung von Fahrerassistenzsystemen zur Verbesserung des Spurwechselverhaltens von ACC. Dissertation. Unter Mitarbeit von Berthold Färber. Göttingen: Cuvillier Verlag (Audi Dissertationsreihe, Band 9) (2008)

44. Bellem, H.: Analysis of driving style preference in automated driving. Dissertation. Technische Universität Chemnitz, Chemnitz (2017)

45. Dovgan, E., Tušar, T., Javorski, M., Filipič, B.: Discovering comfortable driving strategies using simulation-based multiobjective optimization. Informatica **36**(3) (2012)

46. Summala, H.: Towards understanding motivational and emotional factors in driver behaviour: Comfort through satisficing. In Modelling Driver Behaviour in Automotive Environments, pp. 189–207. Springer, London (2007)

47. Lange, A., Maas, M., Albert, M., Siedersberger, K.H., Bengler, K.: Automatisiertes Fahren—So komfortabel wir möglich, so dynamisch wie nötig. Vestibuläre Zustandsrückmeldung beim automatisierten Fahren. In: VDI (Hg.): 30. VDI/VW-Gemeinschafts-tagung. Fahrerassistenz und Integrierte Sicherheit. Wolfsburg, 14–15 Oct 2014. VDI Verlag GmbH, VDI/VW, pp. 215–228 (2014)

48. Bellem, H., Schönenberg, T., Krems, J.F., Schrauf, M.: Objective metrics of comfort: developing a driving style for highly automated vehicles. Transport. Res. F: Traffic Psychol. Behav. **41**, 45–54 (2016)

49. Schönhammer, R.: Zur Psychologie der Beifahrersituation. In: Deutschen Gesellschaft für Psychologie (Hg.): Bericht über 38. Kongreß der Deutschen Gesellschaft für Psychologie. Trier. Göttingen, Hogrefe, pp. 321–322 (1992)

50. Ellinghaus, D., Schlag, B.: Beifahrer. Eine Untersuchung über die psychologischen und soziologischen Aspekte des Zusammenspiels von Fahrer und Beifahrer. Köln, Hannover (2001)

51. Engelbrecht, A.: Fahrkomfort und Fahrspaß bei Einsatz von Fahrerassistenzsystemen. Disserta Verlag (2013)

52. Sarter, N.B., Woods, D.D., Billings, C.E.: Automation surprises. In: Handbook of Human Factors and Ergonomics, vol. 2, pp. 1926–1943 (1997)

53. Flemisch, F., Schieben, A., Schoemig, N., Strauss, M., Lueke, S., Heyden, A.: Design of human computer interfaces for highly automated vehicles in the EU-Project HAVEit. In: International Conference on Universal Access in Human-Computer Interaction, pp. 270–279. Springer, Berlin, Heidelberg (2011)

54. Chang, A.: UTAUT and UTAUT 2: a review and agenda for future research. The Winners **13**(2), 10–114 (2012)

55. Hassenzahl, M.: The hedonic/pragmatic model of user experience. Towards UX manifesto **10** (2007)

56. Sivak, M., Schoettle, B.: Motion sickness in self-driving vehicles. University of Michigan, Ann Arbor, Transportation Research Institute (2015)

57. Reason, J.T., Brand, J.J.: Motion Sickness. Academic Press (1975)

58. Diels, C., Bos, J.E.: Design guidelines to minimise self-driving carsickness. In: 7th International Conference on Automotive User Interfaces and Interactive Vehicular Applications, Nottingham, UK (2015)

59. Bohrmann, D., Bengler, K.: Reclined posture for enabling autonomous driving. In: International Conference on Human Systems

Engineering and Design: Future Trends and Applications, pp. 169–175. Springer, Cham (2019)

60. Bohrmann, D., Koch, T., Maier, C., Just, W., Bengler, K.: Motion comfort–human factors of automated driving. In: Proceedings of the 29th Aachen Colloquium of Sustainable Mobility, pp. 1697–1708 (2020)

61. Damböck, D., Weissgerber, T., Kienle, M., Bengler, K.: Evaluation of a contact analog head-up display for highly automated driving. In: Advances in Human Aspects of Road and Rail Transportation. CRC Press (2012)

62. Danner, S., Hecht, T., Steidl, B., Bengler, K.: Why is the automation not available and when can i use it? In: Black N.L., Neumann W.P., Noy I. (Hrsg.) Proceedings of the 21st Congress of the International Ergonomics Association (IEA 2021). Springer International Publishing (2021)

63. Bengler, K., Rettenmaier, M., Fritz, N., Feierle, A.: From HMI to HMIs: towards an HMI framework for automated driving. Information **11**(2), 61 (2020). ▶ https://doi.org/10.3390/info11020061

Open Access This chapter is licensed under the terms of the Creative Commons Attribution-NonCommercial-NoDerivatives 4.0 International License (▶ http://creativecommons.org/licenses/by-nc-nd/4.0/), which permits any noncommercial use, sharing, distribution and reproduction in any medium or format, as long as you give appropriate credit to the original author(s) and the source, provide a link to the Creative Commons license and indicate if you modified the licensed material. You do not have permission under this license to share adapted material derived from this chapter or parts of it.

The images or other third party material in this chapter are included in the chapter's Creative Commons license, unless indicated otherwise in a credit line to the material. If material is not included in the chapter's Creative Commons license and your intended use is not permitted by statutory regulation or exceeds the permitted use, you will need to obtain permission directly from the copyright holder.

Architecture Views for Automated Driving Systems

Marcus Nolte and Markus Maurer

Contents

© The Author(s) 2026
H. Winner et al. (eds.), *Handbook Assisted and Automated Driving*,
https://doi.org/10.1007/978-3-658-45276-6_39

39.1 Introduction

Current discussions about the potential of automated driving systems often revolve around questions of safety. As, e.g., pointed out in ▶ Chap. 40 the key question of "how safe is safe enough" is still largely unanswered as far as a wide acceptance of automated driving systems is concerned. The introduction of automated vehicles into an environment which is uncontrollable for the developers creates several challenges. First and foremost, the systems must be able to behave safely in an (almost) open system context—almost anything can happen at any time. Secondly, the system context can change over time. It cannot be ensured that all properties of the targeted operational design domain (ODD) persist over the system's entire life cycle. Finally, the real world is a source of uncertainty which is for example caused by interactions with other agents whose behavior cannot be predicted unambiguously.

If drivers are fully relieved from control and supervisory tasks, additional consequences arise. In contrast to SAE Level 1 or 2 systems [60], it is not sufficient anymore to formulate the required system performance as a function of human capabilities to intervene in case of an error. The system must be capable of handling critical events on its own when control would be handed back to human drivers for Level 1 or 2 automation.

This does not only imply the necessity for robust sensors, actuators and algorithms for the realization of the system's functionality: Writing a suitable system specification comes with its own particular challenges when considering the (almost) open system context. With respect to the aforementioned influences of uncertainty, it needs to be assumed that sets of requirements which are formulated during design and development are inherently incomplete (cf. [46]). This poses new challenges for established development processes in the automotive domain (cf. ▶ Chap. 45) on the one hand, and new requirements for the work products, such as system architectures, produced in those processes on the other hand.

In this context, system architectures are established means to dealing with complexity during the design and development of technical systems [69]. Architectures are (sometimes more, sometimes less) abstract representation of a system of interest. Among other things, they allow a structured decomposition of an entire system into sets of sub-systems [77], the definition of interfaces, data flow, and system behavior on different levels or, in other words, in different *views*. As architectures abstract from the concrete physical system design, the formulation of architecture(s) (or rather architecture viewpoints and views) is an economic way of iteratively discussing requirements and system design (cf. ▶ Chap. 45) without having to build possibly expensive physical prototypes.

The abstract nature of architectures also facilitates communication between people involved in the design and development. The realization of modern vehicle systems requires interdisciplinary teams, each of whose disciplines can favor their own established architectures (e.g., functional, hardware, software, etc.). In this respect, it is required to be able to look at a system from different perspectives and to respect a variety of concerns, such that all involved stakeholders can agree on a common understanding of the system of interest that should be developed [69]. Particularly for complex and highly safety-critical systems, such as automated driving systems, this is crucial, as this common understanding can ensure that the system architecture efficiently supports a safety argument.

The design of architectures and architecture views is, of course, not a new challenge that is particular to the field of automated driving. On the contrary, the design of functional and technical architectures as a basis for traceable safety goals is already demanded by [27], which is the gold standard for functional safety for road vehicle E/E-systems (c.f. ▶ Chap. 6). However, with respect to designing and realizing safe vehicle behavior for an almost open world create new challenges which are not covered by [27] alone. Additional approaches are required to design architectures that enable traceability from the specification of the desired vehicle behavior to technical architectures, and implementation details. The arising questions are not mere questions of *functional* safety. Moreover, questions regarding a definition of safe vehicle behavior with respect to ISO/DIS 21448 (Safety of the Intended Functionality—SOTIF[1]) need to be addressed, which [72] refers to as *behavioral safety*.

With regard to the contents of this chapter, a particular challenge arises regarding state of the art of architecture views for automated driving systems: On the one hand, the topic has been widely addressed within the domains of vehicle electronics and even automated driving. These approaches have often been driven from direct practical needs—architectures have been designed from a bottom-up-perspective, often in a pen-and-paper fashion, without leveraging formal description languages such as SysML. At the same time, (Model-Based) Systems Engineering ((MB)SE) has been established as a field of its own (cf. [30, 31]). Here, approaches have been developed, motivated from a top-down perspective. That is, the approaches were designed from the need of creating universal (and by that domain-independent) conventions for the design of system architectures. One of the tools that have been developed in this context are architecture frameworks [31] which purpose is to achieve a

---

1   ISO/DIS 21448 focuses on the safe *specification* of the intended functionality. That is, the standard deals with safety in the absence of systematic and random E/E-faults which are addressed by [27], as well as with safety under foreseeable misuse.

holistic system description by addressing different stakeholder concerns in sets of architecture views. Literature shows a variety of different architecture frameworks, which can be extremely generic, as they address broader scopes than the description of automated driving systems.

Both fields have developed in parallel which results in gaps regarding terminology and ways how architectures are described. On the one hand, there are the already mentioned broad and generic architecture frameworks (cf. ▶ Sect. 39.5) which come with extensive ontologies for the description of system and/or enterprise architectures. On the other hand, tailoring generic frameworks toward the needs and established concepts in the domain of automated driving (e.g., regarding the description of expected behavior, or more general questions with respect to behavioral safety) is subject to ongoing development and research efforts (▶ Sect. 39.4). Closing these gaps between the application domain and generic (MB)SE methods has great potential to provide significant benefit to a safety argument for and with that the verification and validation of automated driving systems.

For this reason, this chapter argues for the application of architecture frameworks for automated driving systems as a key contribution to requirement traceability—as a way to enable consistent design when facing an (almost) system open context. This chapter also provides an overview over established terminology with respect to system architecture frameworks (cf. ▶ Sect. 39.3) based on ISO/IEC/IEEE 42010. Neither the general approaches from the domain of (model-based) Systems Engineering nor the specific solutions from the domain of automated driving can be covered exhaustively in the scope of this chapter. For this reason, this chapter presents a selection of relevant contributions from both domains which illustrate the advantages of creating architecture views (and viewpoints). At the same time, the chapter will point out the potential that results from combining more formal (MB)SE approaches with domain-specific knowledge.

Concrete processes for designing architecture frameworks for SAE Level 3+ systems, the required views and viewpoints are subject to intense research and development at the time of writing this chapter. Specific recommendations with respect to *what* aspects an architecture framework for these systems must cover can only partially be given. The readers should be presented with the basic principles of architecture views and viewpoints, as well as questions that are relevant when it comes to actually designing views and viewpoints. Unfortunately, this means that ready-made solutions and recipes cannot be provided. However, to conclude the chapter, an example will be given, accompanied by a variety of stakeholders and their concerns and some examples how these concerns can be addressed in different architecture views for an SAE 3+ system.

39.2 Motivation: Interdisciplinary Challenges for Automated Driving Systems Safety

As mentioned in the introduction, safety can be seen as a driver for formulating extensive architecture descriptions for complex systems. A key aspect that is often mentioned in this context is emergence. This is due to the fact that complex systems consist of a high number of components that interact over a variety of interfaces. Emergent properties are then referred to properties that cannot be traced back to the properties of the individual sub-systems. Instead, emergent properties manifest only in the interplay of all components [15] (often referred to by the Aristotelian phrase *the whole is not the same as the sum of its parts*).

Under safety aspects, it is crucial that emergent effects do not cause a composed system to show unsafe behavior despite all of its sub-systems have been approved as safe. Emergence in this sense implies that, without proper methods to establish traceability, safety of a composed system cannot be proven by proving safety of each individual component. Considering emergent properties, safety of sub-systems is only a necessary, not a sufficient condition.

Controlling effects that are caused by emergence requires effective interdisciplinary collaboration during design and development. This fact is a key issue of safety engineering that has, e.g., been highlighted by [9] and [34, p. 92].

The statements by [34] basically follow Dijkstra's [18] argumentation with respect to the "separation of concerns" for looking at complex (scientific) contexts. According to [18], it is not beneficial to try "tackling [...] various aspects simultaneously" [18] when looking at a "program".[2] Instead, it would be more helpful to focus on single aspects that are most relevant for solving a given problem at first—while not ignoring but rather prioritizing on other concerns.

Considering the arguments of Koopman and Wagner [34] and Dijkstra [18], two core aspects of safety engineering can be summarized: A suitable system description must be able to reflect all relevant aspects of the complex overall system. Relevance here is determined with respect to the stakeholders from different disciplines. Relevant aspects must include all required information so that the stakeholders from different disciplines can communicate efficiently and agree on neuralgic points of the system description. In such a way, each individual discipline can contribute its competencies effectively. As safety is an emergent property that can only be attributed to the entire system, it is the task of safety engineering to ensure

---

2 Although the term program can be used interchangeably with "software system" in this context.

that the separation of concerns does not lead to an unsafe system.

With regard to the separation of concerns, literature discusses a variety of approaches from a safety engineering perspective. A popular approach in this context is scenario-driven development. First ideas behind the paradigm of scenario-driven development are described and evaluated by [73] (cited according to [8]) in an interdisciplinary design and development setting.

The basic idea behind scenario-based development approaches is to use scenarios as a common basis to facilitate communication between different stakeholders (customers, designers, developers, etc.) during design and development. These scenarios should be described in a language[3] accessible to all stakeholders to enable efficient and accurate communication[4]. Hence, scenarios can be helpful along the entire development process. For additional information regarding scenario-based development for automated vehicles, see, e.g., [8, 67]. Following these references, this chapter will introduce a scenario-based example (▶ Sect. 39.6) to illustrate the argumentation.

The requirements and use cases which designers and developers can formulate by leveraging scenario descriptions typically serve as drivers for system architectures. However, as already mentioned, many disciplines that are involved in the design and development processes have their own established forms of architecture descriptions: Software developers are familiar with using UML for describing the structure and behavior of software components. A popular additional description language in the systems engineering domain is SysML, which is based on UML 2.0. Control engineers might model views on their systems as block diagrams, electrical engineers may use schematics for describing electrical components, and so on and so forth.

With these different perspectives, it is not only important to be able to communicate via common scenarios. In an interdisciplinary setting, for tracing possible emergent properties, it is crucial to assess the impact of architecture decisions not only within individual architecture views but also for the entire system—this is especially true for safety-critical applications such as vehicle automation systems. While scenarios can serve as a first approach for an efficient separation of concerns, they do not directly contribute to keeping track of emergent effects during the design and/or development of a system. Additional methods are needed. This is where the definition of architecture frameworks can provide valuable assistance, as they provide means to establishing traceability within and throughout different architecture views.

Architecture frameworks are concerned with modeling system architectures under the influence of different stakeholder concerns (cf. ▶ Sect. 39.3). Different views address different concerns; traceability can be established via so-called correspondences (cf. ▶ Sect. 39.3). Traceability allows to trace design (and development) decisions from initial use-cases to concrete hardware and/or software solutions. Architecture frameworks are popular tools in the Systems Engineering domain so that their description has been standardized in [31]. In addition to ISO/IEC/IEEE 42010, there are also standards for particular enterprise architecture frameworks [25, 26, 29].

For the sake of completeness, it should be mentioned that in the context of enterprise and systems architecture frameworks, the boundary of the system of interest is of particular importance. For this chapter, in the context of architecture frameworks for automated vehicle systems, we draw the system boundary around a vehicle that is equipped with an automated driving system (ADS-equipped vehicle, according to [60]). Of course, this system interacts with a number of other systems, including the enterprise that builds and operates the ADS-equipped vehicle, the traffic system in general, possible systems for V2X-communication, etc. Hence the ADS-equipped vehicle constitutes a sub-system in a larger system of systems, which can again be described in different views [10, p. 1058]. However, as the fundamental concepts for applying different views to describe a system are the same for larger systems of systems and individual systems, this chapter restricts the focus to the ADS-equipped vehicle as an example.

After introducing the required terminology (▶ Sect. 39.3) and giving a perspective from the field of automated driving (▶ Sect. 39.4), this chapter will also discuss more generic architecture frameworks from a (model-based-)systems-engineering perspective. We will focus on two aspects in this context: On the one hand, we will show that domain knowledge is extremely important when the applicability of architecture frameworks to a particular problem shall be given. On the other hand, we will argue that the generic approaches from a systems-engineering perspective have great potential to contribute to designing and developing safe automated driving systems.

39.3 Terminology

This section presents an overview of the most important terms and concepts from ISO/IEC/IEEE 42010 with respect to architecture frameworks. The central term, as it is the subject of all architecture-related processes and work products is the term *system*. Literature shows many different definitions of what a system actually is, often based on the original, Greek term which describes "a whole that is compounded of parts" [77]. It may be

---

3   This cannot only include natural language descriptions but also formal description languages such as UML [73].

4   According to [73, p. 43 f.], this also requires common taxonomies and glossaries.

seen as a self-referential or recursive definition, as even something that is composed can again be part of a larger composition. That is, a system can itself be part of what Systems Engineers refer to as a *system of systems*. In practice, this means that the term *system* should always be accompanied with a description of the intended *system boundary* which separates the system that is the subject of discussion from other parts of a system of systems, that are not intended to be focused on. For this chapter, we already defined the system boundary around the vehicle that is equipped with an automated driving system. Typical sub-systems in this context would be control units that contribute to the functionality of the automated driving system. To allow a clear and distinct differentiation of a system that is in focus of, e.g., an architecture description, ISO/IEC/IEEE 42010 introduces the term *system of interest*.

As mentioned in the introduction, different disciplines that contribute to system design and development may have their own preferred models for describing architectures. Common to all these models is that they at least describe a system of interest as a composition of system elements (i.e., parts) and the relationship between these parts [54, 69].

ISO/IEC/IEEE 42010 defines the term *architecture* as a collection of "fundamental concepts or properties of a system in its environment embodied in its elements, relationships, and in the principles of its design and evolution" [31, p. 2].

In the standard, architecture is seen as a "holistic conception of that system's fundamental properties" [31, p. 6]. With that, an architecture should define all required rules and concepts that are found relevant for the system of interest. The idea of an *architecture* being a holistic concept is intentionally separated from a more colloquial use of the term in combinations such as "functional architecture", "software architecture", or "physical architecture". According to [31], an architecture describes the whole system of interest while *architecture views* (definition see below) can be used to address individual stakeholder concerns, such as the description of a *functional view*.

The central artifact that defines the architecture of a system of interest (e.g., by means of documents or models) is what ISO/IEC/IEEE 42010 refers to as the *architecture description*. Within this artifact, the so-called *architecture framework* defines the "conventions, principles and practices for the description of architectures established within a specific domain of application and/or community of stakeholders" [31, p. 2]. ◘ Figure 39.1 gives an overview over the conceptual relations between the architecture framework and related concepts that are presented by ISO/IEC/IEEE 42010.

A stakeholder is defined as an "individual, team, organization or classes thereof, having an interest in a system" [31, p. 2]. With respect to an automated driving system, this could, among others, include customers, system architects, developers, but also legislators, organizations for technical supervision, or society as a whole. The particular "interest in a system [that is] relevant to one or more of its stakeholders" [31, p. 2] is called *concern*.

In order to properly frame and address concerns within an architecture framework, [31] introduces the concepts of viewpoints and views: In this context, a viewpoint is defined as a "work product establishing the conventions for the construction, interpretation and use of architecture views to frame specific system concerns" [31, p. 2]. Viewpoint descriptions can hence define *model kinds* and description languages. In other words, a viewpoint is a meta model for the creation of architecture views.

A view then is a "work product expressing the architecture of a system from the perspective of specific system concerns" [31, p. 2]—it instantiates the models defined in its corresponding viewpoint. According to ISO/IEC/IEEE 42010, one viewpoint governs one view that means that the concerns framed by one viewpoint are addressed by exactly one view. In addition, the standard defines that a view needs to express the *entire* system of interest. As can be seen from the literature (cf. ► Sect. 39.5), these demands of ISO/IEC/IEEE 42010 are not always met in practice—and there may be good reasons for deviating from these demands, e.g., for providing a better overview about single aspects of the systems in a view (cf. ► Sect. 39.3).

The relations between individual elements of an architecture description are defined via *correspondence rules* and *correspondences* [31, p. 7]. The distinction between correspondences and correspondence rules follow the same rationale as the differentiation between a view and a viewpoint: Correspondence rules formulate *how* correspondences can be formulated, correspondences can then establish a concrete relation between individual elements in an architecture description. In this way, the formulation of correspondences is an important method to ensure consistency between architecture views. Hence, they contribute to keeping track of possible emerging effects and present an effective way of establishing traceability of stakeholder concerns within and across different architecture views.

Example: Correspondences and Correspondence Rules

Even without defining which specific architecture views are required for the realization of automated driving systems, it is safe to assume that a driver assistance system requires sensors, software for processing sensor data, as well as control units that serve as a platform to execute the software.

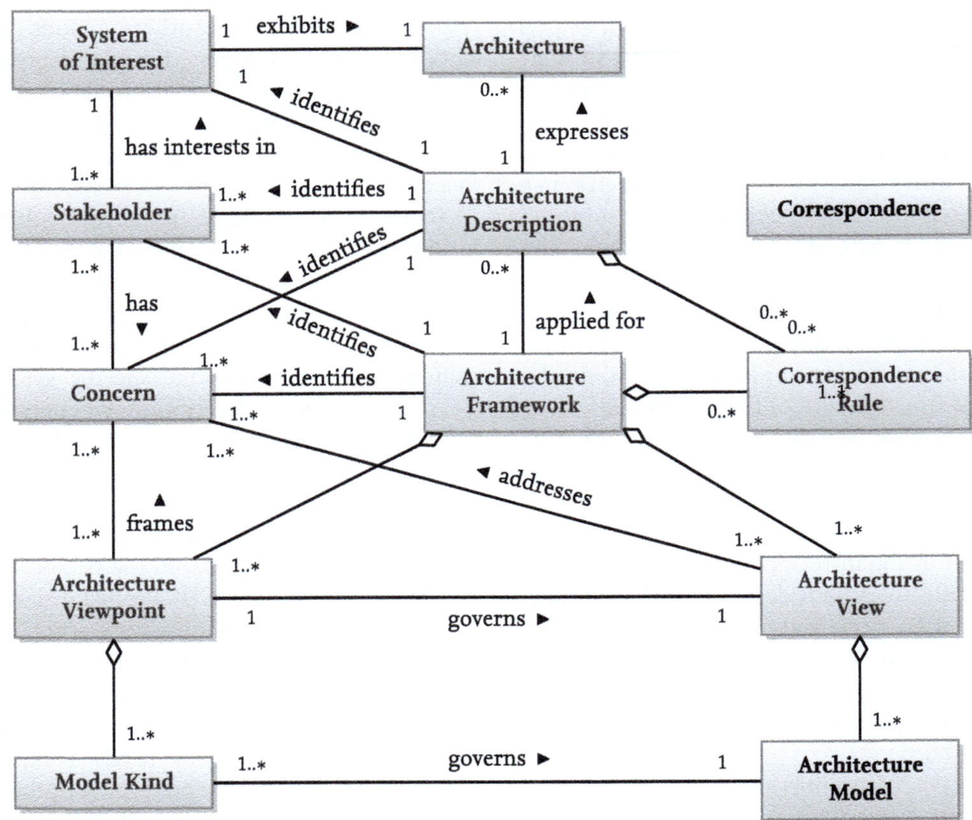

Fig. 39.1 Fundamental concepts around the architecture framework according to [31, p. 10] in a class diagram (UML notation with cardinalities; filled diamonds are UML aggregating relations)

▶ Chapter 32, e.g., gives an overview about the "functional modules" of an adaptive cruise-control system (ACC). We will use the explanations given in that chapter to illustrate the concept of correspondences and correspondence rules.

Correspondence rules can be very general. They can be contained in a meta model or can be written down in natural language. An example (altered from [31, p. 24]) could be phrased as follows:

KR1 – *Each element in a software view must be executed on at least one element in a hardware view.*

KR2 – *Correspondence rules KR1 will be expressed by the following relation:*

The general relation between correspondence rules and correspondences is illustrated in ◘ Fig. 39.2. The figure illustrates a hardware- and a software-centric viewpoint with examples for different model kinds that can be used to create views. Arrows between the views indicate correspondences between the software and the hardware view in terms of an instantiation of software components on different processors.

In the example of the adaptive cruise control system, there are several hardware platforms: a radar sensor, the so-called "intelligent drive controller" (cf. ▶ Chap. 32,

will be called "ACC control unit"), as well as the control unit for the brake control system. Regarding software components, the system contains components for managing logic system states (state manager), as well as a speed and acceleration controller.

In this case, according to ▶ Chap. 32, three correspondences that adhere to correspondence rules **KR1** and **KR2** could be written in natural language as

K1 – State Manager executes on ACC Control Unit

K2 – Speed Controller executes on ACC Control Unit

K3 – Acceleration Controller executes on Brake Control System

◘ Figure 39.3 shows these correspondences in a diagram that resembles a UML deployment diagram in the example of the ACC system. Dashed arrows depict the assignment of software components in the ACC system to different physical control units.

Regarding the traceability of concerns and requirements, these correspondences allow to derive the fact that data which is exchanged between the speed and acceleration controller are exchanged over a CAN bus connection. What could introduce unwanted communication latency. From a control-theoretic perspective, this could

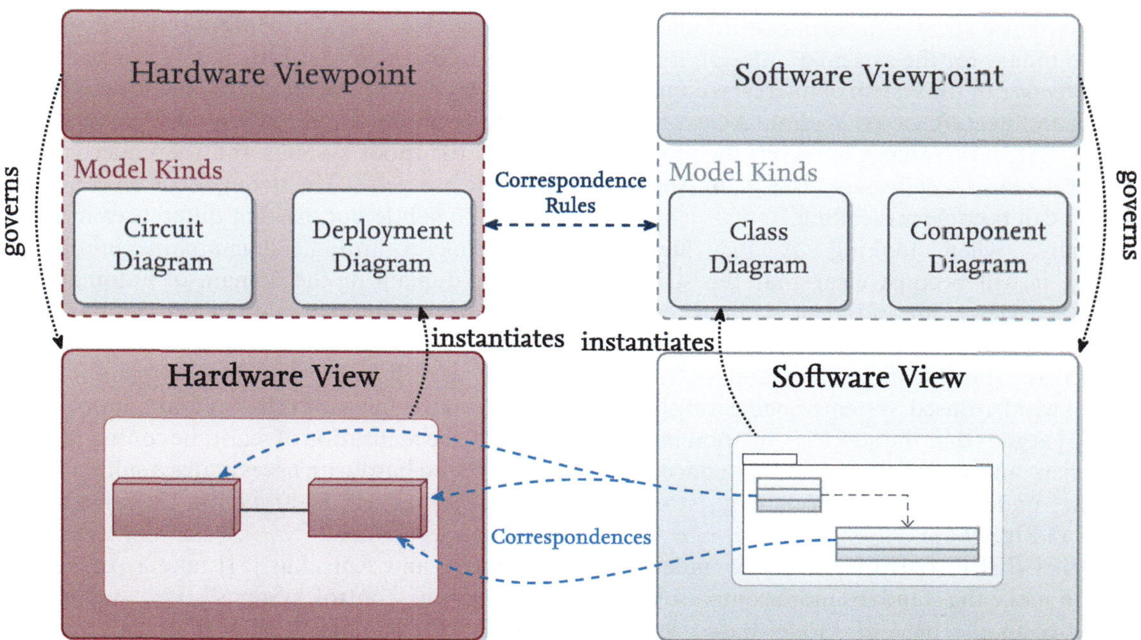

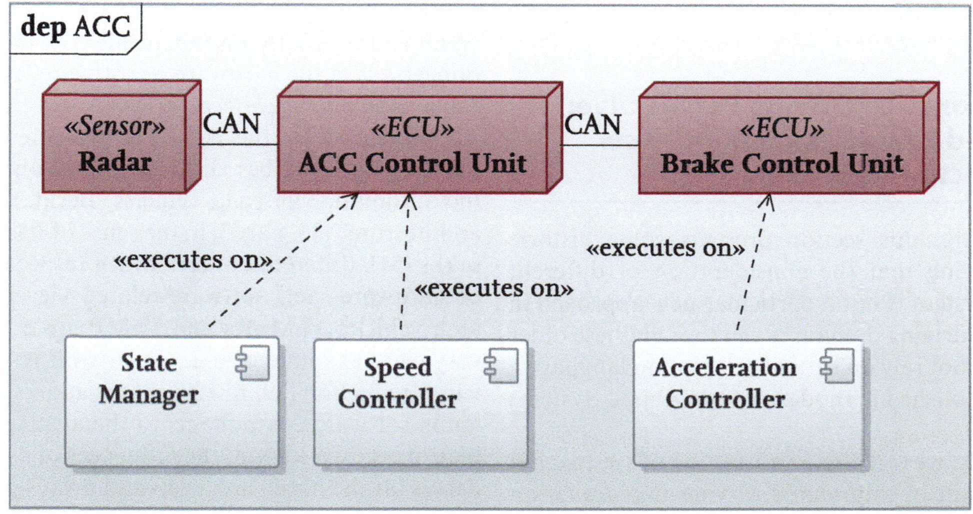

Fig. 39.2 Visualization of the relations between architecture viewpoints, -views, correspondences and correspondence rules according to [36, p. 20]

Fig. 39.3 Modeled correspondences at the example of an excerpt from an ACC system—assuming that the control units communicate via a CAN bus interface

lead to unstable control loops in the worst case. Without noting the mentioned correspondences, this instability would be an emergent effect that might not be encountered when each controller is tested on its own, but which would only be observable in the integrated system.

Thoughts on IEEE/IEC 42010:2011

▶ Section 39.5 presents a number of architecture frameworks that have been developed from a Systems Engineering perspective. Some of them strongly rely on ISO/IEC/IEEE 42010, but deviate explicitly from the ontology showed in Fig. 39.1 at the same time.

It is indisputable that the norm provides an important and consistent frame for describing system architectures and can serve as an important guide in the field of Systems Engineering for practitioners from other disciplines. The norm underlines the importance of respecting the different concerns that different stakeholders can have in a system of interest by standardizing the concepts of viewpoints and views. It stresses the importance of documenting the architecture design process and contributes to the coherent and traceable design of architectures.

The standard also stresses the explicit formulation of the system boundary for the system of interest, it argues that the *architecture* of the system of interest is comprised of all existing architecture views. Without a clear formulation of that system boundary, however, the more colloquial use of the term *architecture* (as, e.g., in "software architecture") can become misleading.

Additionally, when looking at the literature (▶ Sect. 39.5), it will become clear that the strict 1:1 relation between viewpoint and view is almost entirely ignored in current architecture frameworks. This has been identified as a shortcoming of the standard from the perspective of Model-Based Systems Engineering [37]. In this sense, [37] argues that the strict 1:1 relation impedes flexibility when addressing cross-cutting concerns in multiple views under different aspects (e.g., structure, behavior, cf. ◻ Fig. 39.6).

The ISO/IEC/IEEE 42010 is currently under revision[5]—also to make the standard more compatible with current practice in the field of Model-Based Systems Engineering [37]. Amongst others, the strict 1:1 relation and the fact that a view must present the entire system of interest, will be softened [23, personal communication]. A detailed preview of the basic ideas behind the revision of the standard is given by [37].

39.4 An Automated Driving Perspective: Selected Examples for Multi-view Architectures

In the following, this section presents some historic examples showing that the consideration of different views onto a system is not a particular new approach in the automated driving domain. Even though these older approaches do not rely on formal description languages, as they are established in modern Model-Based Systems Engineering.

Additionally, we will discuss a selection of approaches from the domain of automated driving that touch on more formal aspects of Systems Engineering, while only partially defining explicit architecture frameworks.

39.4.1 Origins of Different Architecture Views for Automated Vehicles

Historically, the development of automated driving systems is closely related to fields as robotics and control engineering [2, 17]. These origins impact how system architectures for automated driving systems have been considered:

On the one hand, the concept of architecture is, even still in recent publications, closely connected to early concepts of robot control architectures[6]. This creates a close conceptual connection between systems architectures for robot systems and the underlying control-theoretic problems. The strong focus on control systems has led to subtle but existing differences in the applied terminology regarding system architectures between the robotics domain or the domain of automated driving, respectively, and the domain of (Model-Based) Systems Engineering.

Taking different views onto a system is established practice in the fields of robotics and automated driving. After all, the realization of real-time control algorithms in software and hardware necessitates a suitable separation of concerns, as will be argued in the remainder of this section.

Passino and Antsaklis [51] take a perspective from "autonomous control systems", i.e., systems that can operate with minimal human intervention under high uncertainty. They argue the need for different abstraction levels for the description of such systems, particularly with respect to the large degree of uncertainty that these systems need to deal with. As a central contribution to the required abstraction levels, [51] introduce an architecture view that is independent of the implementation details at the hardware or software level which they call a *functional architecture*.[7]

Influenced by the concepts presented by [51] and others, [39] describes different views on a system for the automation of road vehicles. Besides a "functional architecture" [39, part 3] in the sense of Passino and Antsaklis [51], different concerns with respect to the design of hardware- and software-related views for the research vehicles VaMoRs and VaMP are elaborated. One very clear example is, e.g., given with respect to bandwidth reduction [39, p. 129] that is addressed when assigning software components to the available computing platforms. In addition, the influences of communication delays (at the hardware level) and delays caused by processing times (software level) on the stability of the lateral and longitudinal vehicle dynamics controllers (which is a functional requirement) [39, p. 84] are discussed. While different views are introduced, other integral parts of an architecture framework such as explicit stakeholder definitions, viewpoints, correspondence rules and correspondences are not elaborated on.

In summary, the described references introduce three typical types of views in the context of automated vehicles:

---

5   At the time of writing this chapter in March 2022, no public draft was available.

6   Mataric [38] defines an architecture as a "principled way of organizing a control system" [38, p. 1].

7   Although it should be noted that [51] understanding of a functional architecture as a concept is slightly different from the understanding that is dominant in modern (Model-Based) Systems Engineering.

- a *functional* or *logical view* for describing system elements and their interfaces in a hardware- and software-independent way that is derived from the abstract description of the tasks that the system should fulfill,
- a *software view* for describing implemented software components and the exchanged information (such as interfaces and their datatypes),
- and a *hardware view* for describing computers and/or control units on which software components are executed, as well as the description of bus interfaces for the communication between hardware devices.

Maurer [39] applies UML-based state charts and class diagrams for describing software parts of the system. Other than that, pen-and-paper-style block diagrams are used for displaying the above mentioned views.

39.4.2   A Selection of Recent Approaches for the Formulation of Architecture Views for Automated Driving Systems

Apart from the references mentioned above, [62] also argue for the design of functional, hardware and software views in a broader vehicle engineering context. [62] relate the explicit formulation of these views to traceability requirements in terms of functional safety (▶ Chap. 6) for road vehicles. In this respect, ISO 26262 addresses the need for establishing traceability for safety goals to technical, as well as hardware- and software safety requirements in the technical safety concept. However, as a standard, ISO 26262 makes concrete demands how this traceability should be achieved. For vehicle electronic systems, this yields a strong argument for an explicit formulation of correspondence rules and correspondences between different viewpoints and views, respectively.

However, as argued in the introduction, requirements with respect to the safe behavior of automated vehicles extend the scope of safety considerations beyond established functional safety practices, while demands for traceability of requirements with respect to the required system behavior must be met, as well. For this reason, this section will provide an overview of approaches from the automated driving domain which address questions of traceability either implicitly or via the explicit formulation of viewpoints, views, correspondence rules, and correspondences.

A concept primarily focused on questions of functional safety is presented in the form of *Digital Dependability Identities* (DDI) [56]. DDIs are based on the Open Dependability Exchange meta model [50] and serve for the description of safety analyses for cyber-physical systems—i.e., technical systems that interact with their environment. The meta model has been explicitly built for establishing traceability across architectural views

and allows DDIs to be used for a traceable assurance argumentation for safety-critical automotive systems. DDIs provide means to model the assurance argument in Goal Structuring Notation (GSN) [75] and have, e.g., been used to model a cooperative adaptive cruise controller for a truck platooning system.

DDIs can be used to link GSN elements in an assurance case such as claims, strategies, assumptions, justifications and solutions/evidence to process artifacts (e.g., hazard and risk assessments, functional architecture views, hardware and software views or even component failure trees). Artifacts, such as the results of safety analyses, serve as evidence in the GSN model and can be traced to safety goals or elements in the supported architecture views. With this, e.g., the existence or the redundant realization of certain functional elements can be traced back to the fulfillment of safety requirements or the related safety strategies, respectively.

Although the concept of architecture views or the creation of an architecture framework is not mentioned explicitly, the meta model behind the DDIs can be interpreted as an important part of an architecture framework definition: Models, e.g., for the safety argumentation or failure tree analyses are model kinds that could be assigned to a *safety* viewpoint in an architecture framework. The meta model also establishes correspondence rules. Instances of the correspondence rules can then be used to formulate correspondences between elements in the different views that can be described by the DDIs.

In summary, DDIs cover important aspects of a formalized assurance case according to ISO 26262. With respect to the challenges of behavioral safety (cf. ▶ Sect. 39.4) the DDIs currently lack model kinds and views to establish correspondences between the behavior of the system in a scenario, the respective behavioral requirements, and the respective elements in functional or logical architecture views.

Bach et al. [5] present a taxonomy for classifying E/E features in modern vehicle electronics systems. The authors structure the functions in the systems' processing chains, the realization of functions in software components, and the assignment of software components to control units in a generalized system overview. This overview is comprised of a *logical*, a *software*, and a *hardware architecture* [5]. Architecture views or viewpoints are not mentioned explicitly, although the realization and instantiation relations discussed in the paper could be seen as correspondences between different architecture views.

Similar views are addressed in a viewpoint-based system model presented by [63]. The viewpoints frame concerns such as (functional) safety, cyber security or availability [63]. The resulting system model is graph- and component-based. A component is the smallest, atomic, element in a view. Nodes in the graph that describes the system model represent *components*, edges in the graph can represent activation patterns or

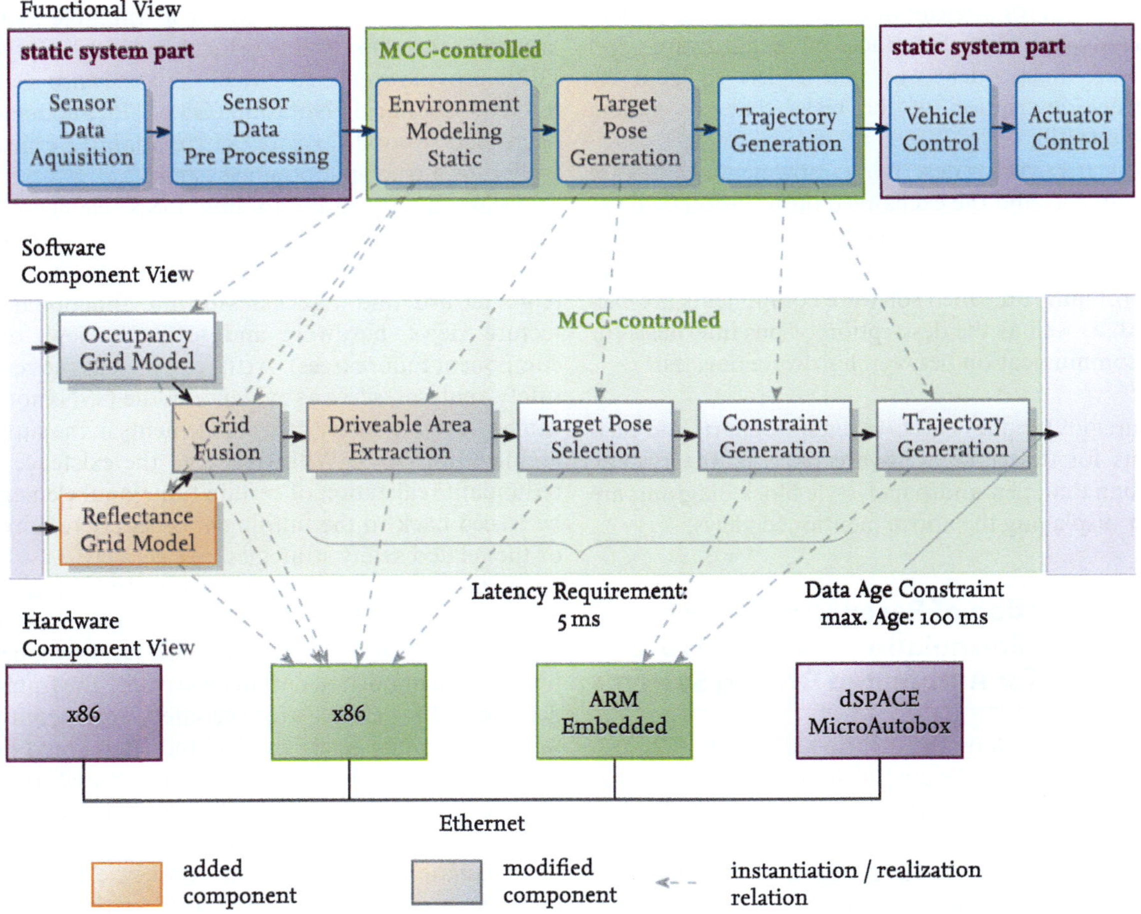

Fig. 39.4 Visualization of an excerpt from a system model based on [63]: Functional, software, and hardware view. Dashed arrows depict the realization of functional elements by software components, and the instantiation of software components on control units, respectively. Additionally, different types of timing requirements are displayed that can be annotated in a timing view. MCC: *Multi Change Controller*, a central operating system component that manages system configurations and software updates

communication interfaces between the components. The system model can express a *functional*, a *communication*, a *software*, a *hardware* and a *thread* or *timing* view, respectively.

Relations between the different views are modeled as graph transformations, i.e., functions for transforming nodes and edges from one view to another. Requirements can be annotated to single or groups of components and their interfaces, which facilitates traceability. ■ Figure 39.4 shows an example from a use case for automated driving in a static environment.

■ Figure 39.4 also provides an example of how correspondences between views can be leveraged for traceability: The dashed arrows show which software components realize which functional elements. In this context, the function for modeling the static environment is realized in two distinct software components that realize discrete occupancy grid maps and one component that fuses those resulting grid maps. These software components are executed on an x86-based controller, while the software components responsible for trajectory plan-

ning are executed on an ARM-based embedded PC. Within the processing chain, requirements for processing times can be defined. These requirements must or should be adhered to (depending on required data integrity), even when including communication overhead over an Ethernet interface.

The resulting system model was designed without relating to the concepts of an architecture framework or ISO/IEC/IEEE 42010. The system model is based on a custom-made description language, without relying on existing description languages, such as SysML. Hence, there are significant differences in the presented approaches and established practice in the Systems Engineering domain. However, the graph transformations presented by [63] define explicit correspondence rules for the applied model kinds in each viewpoint. Also, aspects of behavioral safety are at least partially considered, even if the original focus of the system model was more tailored toward functional safety.

Concerns that address behavioral safety are discussed explicitly by [7]. Starting from an example scenario (ego

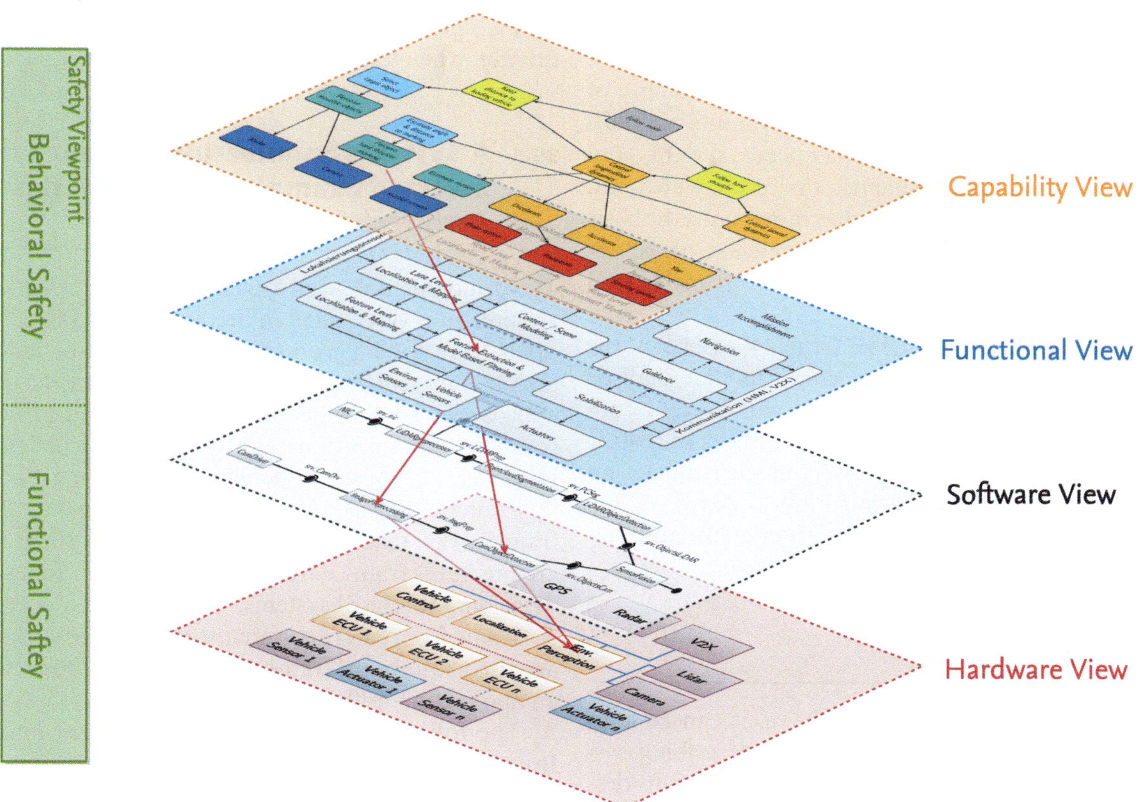

Fig. 39.5 Views according to [7]—red arrows depict correspondences between views, *safety* serves as an example for an orthogonal viewpoint

vehicle approaching a pedestrian crossing), situation-dependent safety goals are derived. Based on these safety goals, the paper discusses how the system behavior can be broken down to functional safety requirements, technical performance requirements and measurable performance metrics in the context of multiple architecture views. The role of correspondences and correspondence rules is emphasized in this context, partially based on [63]. The approach demonstrates a possible way of traceably formulating safety goals on a behavioral level and connecting them to concretely measurable technical measurements. A view for formulating system "skills and abilities" based on [39, 53, 57, 65] serves as a connecting element between the abstract requirements for system behavior to established functional, logical, hardware- and software viewpoints. Orthogonal to the viewpoints that are the basis for building the aforementioned views, [7] propose a safety-centric viewpoint to capture safety goals and requirements (cf. **Fig. 39.5**).

The presented approach does, however, only present basic concepts. The proposed viewpoints and views are not modeled formally. Additionally, the modeled scenario is only informally described. Hence, there is no formal model of use cases, or, in contrast to [56], for safety goals.

Based on the approaches presented in [7, 63], a view centric architecture is presented in the context of the research project UNICAR*agil* [33]. In contrast to [7], different views, requirements, and guarantees for elements (functional elements, software-based services, hardware components, etc.) are modeled within a "configuration tool" [33]. While this implies a form of formal modeling, there is no particular focus on modeling safety goals at a behavioral level. Complementary approaches that aim at safety-related issues [22] and additional user needs (e.g., for physically impaired users) [64] are discussed in the context of the research project [22], as well. These considerations, while addressing functional architecture views and system capabilities are, however, not integrated in a coherent architecture framework.

Carré [14] proposes a "method and architecture for the safety management [of automated vehicles]" [14, p. 73]. Based on an established MBSE-based design process (ARCADIA [14, 59]) discusses functional, logical, and physical viewpoints (physical viewpoints summarize hardware- and software views in ARCADIA).

Carré [14] presents some details from functional architecture views without explicitly discussing the concerns that should be addressed by these views. Only an implementation-centric viewpoint is discussed in detail with respect to a micro-service-based *software* view. Decisions and argumentation during the design of the presented architecture views are not mentioned, except for broad concerns such as "safety".

Şahin et al. [61] argue for an architecture framework for "dynamically configurable vehicles". They define nine viewpoints, mainly based on the so-called "MBSE Grid" presented in ▶ Sect. 39.5. A core concern behind the creation of the framework is to establish traceability of requirements. Requirement-centric viewpoints are defined to explicitly capture stakeholder needs, system-, and component-related requirements. For each viewpoint, the authors propose one SysML diagram type as the basis for defining views on an automated vehicle. Correspondences as means to establish requirements are mentioned but not exhaustively discussed.

Apart from taking the MBSE Grid [44] as a basis, the authors do not mention further concerns that shall be framed by the chosen viewpoints. Domain-specific concerns are not mentioned in detail. The authors only state to consider the role of "scenarios" and their impact on their architecture framework in future work [61, p. 30]. Concerns with respect to (behavioral) safety of automated driving systems are not mentioned at all.

39.4.3 Summary

Considering the references from above, it can be concluded that non-formal approaches for describing system architecture are (up to now) more common to describe automated driving systems [5, 7]. The mentioned approaches also either lack explicit descriptions of the systems' target behavior [56, 61], the explicit traceability of requirements [14] or the expressiveness of a formal SysML-based (or comparable) description languages [7, 63].

Viewpoints, or views, respectively, that are considered by all authors are *functional* or *logical*, *hardware*, and *software* views. At the same time, additional important viewpoints or concretely formulated views are introduced. These include

- views for modeling *scenarios*,
- views for modeling situation-specific *capabilities* of an automated driving system,
- views for modeling *requirements* (e.g., for system behavior in a scenario or the timing behavior of software components).

As a whole, the presented approaches consider important domain-specific aspects *and* important aspects from (model-based) Systems Engineering. At the same time, it becomes apparent, that a consistent application of formal MBSE-methods under explicit domain knowledge regarding safety of automated driving systems is still missing.

For this reason, this chapter will discuss several more generic approaches from the model-based system engineering domain which are tailored toward an application in vehicle engineering on the one hand. On the other hand, the next section will also give a short look into more generic MBSE architecture frameworks that could yield a significant contribution to the field of automated driving, when tailored appropriately.

39.5 A Systems Engineering Perspective: Selected Examples for (MB)SE-Based Architecture Frameworks

The fundamental approaches for structuring system architectures often cited in literature (e.g., [24, 31, 69]) often have a background from the field of enterprise architecture (such as [19, 55, 76]). This means that the approaches have originally been developed to describe complex organizations as socio-technical systems. However, the basic principles that are, e.g., listed in [31] remain the same. Even if enterprise architecture frameworks tend to be more complex than system architecture frameworks that are more focused on describing technical systems.

Apart from enterprise architecture frameworks, there are also a number of frameworks that are tailored toward an application in the automotive industry.

The following section gives some insight into both: Broad and generic, as well as specialized frameworks from an automotive context. The latter frameworks are mainly considered to show which viewpoints have been addressed from a systems-engineering perspective for the automotive domain and to point out which aspects are missing for the description of automated driving systems.

39.5.1 Generic Architecture Frameworks

As discussed in ▶ Sect. 39.2, there are several generic (enterprise) architecture frameworks that have emerged from the developments in the field of (model-based) Systems Engineering. Depending on the focus of the individual frameworks, the design of the considered viewpoints may vary, while fundamental concepts are well established. These fundamental concepts mainly relate to the separation of (stakeholder) concerns discussed in ▶ Sect. 39.2 in the context of designing and developing complex systems.

In general, a recurring pattern in Systems Engineering is the differentiation between *problem space* and *solution space*, or the description of the *problem* and the *solution* domain, respectively.

The description of the problem domain is mainly concerned with understanding the system that is supposed to be developed. (Customer) needs and requirements are captured here to ensure that the designers and developers get an exact understanding of what a customer needs and wants [70]. This also includes taking a black-box view on the system for the most part, focusing on requirements elicitation by discussing the system's interaction with its environment. A white-box perspective is taken to start describing the system's architecture, e.g., starting with the processes to describe (possible candidates for) *functional* or *logical* architecture views [41].

Following activities in the problem domain, in the *solution* domain, the functional and logical views are taken as a basis to develop concrete realization candidates by means of hardware- and software views [41].

In this context, [44] separate the solution and the problem domain into four aspects for describing "requirements", "behavior", "structure", and "parameters", according to [20]. Each cell of the resulting grid can be seen as a possible viewpoint that a generic architecture framework should address (cf. ◘ Fig. 39.6).

Grid representations as pictured in ◘ Fig. 39.6 (sometimes even shown in three-dimensions, or with an added time axis) are common to many generic architecture frameworks (as can be seen in [37]).

A modern and generic architecture framework is the *Unified Architecture Framework* (UAF) [48] (standardized as ISO 19540:2022 [26]). It is an enterprise architecture framework and by this aims at the description, e.g., of an enterprise's missions and structures. It is based on the concepts of older architecture frameworks, often with military background (e.g., the NATO Architecture Framework, the Department of Defense Architecture Framework (USA), or the Ministry of Defense Architecture Framework (GB)). The Unified Architecture Framework, however, wants to be a generic framework that suits industrial, governmental and military organizations [49]. The chosen grid-structure allows the description of over 70 viewpoints—which illustrates the complexity of generic architecture frameworks.

With respect to the MBSE-Grid according to [44], the Unified Architecture Frameworks with its high number of different viewpoints is a toolbox. Not all viewpoints need to be used when applying the Unified Architecture Framework for a specific enterprise architecture description. As long as all relevant stakeholder concerns can be addressed with the set of viewpoints that is chosen eventually. However, it should be mentioned in the context of this chapter, that the Unified Architecture Framework in its current version does not possess any viewpoints for explicitly addressing stakeholder concerns regarding safety. A possible reason is that the Unified Architecture Framework in its current version is a classic enterprise architecture framework whose main focus is the description of organizations where security concerns may be more prominent than safety concerns. Currently, the modeling of technical system has not been a particular focus of the framework. This shall be changed in the upcoming versions of the UAF profile, which puts more emphasis on modeling technical systems in an organizational context [43].

An architecture framework that follows the opposite approach compared to the Unified Architecture Framework is the *Systems Architecture Framework* (SAF).

It is developed under the roof of the German Society for Systems Engineering (GfSE) [1]. The Systems Architecture Framework and the Unified Architecture Framework both base their viewpoint models on SysML. The Systems Architecture Framework has a strong focus on modeling technical systems. However, a system-of-systems approach is taken, so that the concepts in the Systems Architecture Framework can be connected to corresponding concepts in larger enterprise architecture frameworks, such as UAF [35].

The Systems Architecture Framework has the potential to address more than 40 viewpoints which are ordered along domains and aspects, so that its organization is comparable to the unified architecture framework. The transition between the problem and solution domain inside the Systems Architecture Framework is provided between the so-called *Operational Domain* and *Functional Domain* [35]. In this sense, the Operational Domain, e.g., provides concepts in an "Operational-Structure" and an "Operational-Behavior" viewpoint, that allows to capture and understand a system's structure and behavior of a system in an operational context, i.e., in the systems interaction with its environment. Additional viewpoints should frame safety and security concerns in the future.

An important concept that is picked up in the Unified and Systems Architecture Frameworks' ontologies is the concept of *capabilities*. The formulation of enterprise or system capabilities is related to the problem domain. Capabilities describe the potential of a system or an enterprise to achieve an outcome with a certain performance[8]. With respect to automated driving systems, this terminology can provide a crucial connection, e.g., to [32][9] which defines several abstract capabilities that an automated driving system needs for behaving safely in its Operational Design Domain. At the same time, the terminology around capabilities is also familiar, if not fully compatible, yet, to the already mentioned concepts presented by [7, 39, 47, 57] who define "skill and abilities" that an automated driving system needs to show safe behavior in certain scenarios. However, in this respect, compatibility of the terminology in [7, 47, 57] to established Systems Engineering terminology is yet to be established.

Summary

Considering the more general architecture frameworks with a solid MBSE-background above, it becomes clear that they are powerful tools for modeling system architectures. At the same time, these frameworks

---

8   This is a capability definition according to [71] that is widely compatible with Systems Engineering literature.

9   This technical report presents a concept titled "Safety by Design" for automated vehicles and deals with a structured design, as well as verification and validation of automated driving systems.

			Pillar			
			Requirements	Behavior	Structure	Parametrics
Layer of Abstraction	Problem Domain	Black Box	Stakeholder Needs	Use Cases	System Context	Measurement of Effectiveness (MoE)
		White Box	System Requirements	functional Analysis	logical Subsystem Communication	MoE for Subsystems
	Solution Domain		Component Requirements	Component Behavior	Component Assembly	Component Parameters

Fig. 39.6 MBSE-Grid according to [44], which can be recognized in the organization of viewpoints in a variety of general (enterprise) architecture frameworks

are highly complex, as they consist of high numbers of general viewpoints. This means that tailoring the generic frameworks to a specific application domain requires high proficiency with MBSE methods and processes. For this reason, a concrete recommendation how to approach the tailoring of generic architecture frameworks to the field of automated driving cannot be given—also because this is subject to current R&D efforts. A few important steps that should be considered when using or tailoring a generic architecture framework, without any claim of completeness could, however, be listed as (not necessarily to be performed in the exact order):

- definition of stakeholders who may have concerns with respect to the potential system of interest,
- eliciting concerns of those stakeholders,
- definition of the system boundary[10],
- assigning concerns to the viewpoints in the generic framework,
- modeling views with the model kinds defined by the architecture framework for the respective viewpoints.

39.5.2 MBSE-Related Architecture Frameworks for Vehicle-Engineering Applications

Regarding applications in the automotive industry, [13] propose the "Automotive Architecture Framework" (AAF). They discuss five viewpoints: a functional, a technical, an information, a driver/vehicle operations, and a value net viewpoint. Additionally, several optional and orthogonal viewpoints, such as safety and security, are proposed.

The *functional* viewpoint is comprised of models that describe logical interaction between system components. The *technical* viewpoint is concerned with the (geometric) description of physical components within the vehicle as a system. The *driver/vehicle operations* viewpoint serves to describe the interfaces, dependencies and the interaction between drivers and the vehicle, and the interaction with the vehicle's environment. The *value net* viewpoint allows to model typical supply chains and value-creation processes in the automotive industry. It allows to model relations between OEMs and suppliers and is concerned with stakeholder role descriptions in the supply chain as well as the required activities to creating customer value.

These four viewpoints are considered as parallel viewpoints that are required for a complete system description [13]. The mentioned optional and orthogonal viewpoints should serve for the framing of concerns that are deemed less important by the authors. Questions with regard to safety and traceability are not discussed in detail.

---

10 For an automated driving system, it can, e.g., be important if the system should work on its own or whether it should be subject to a technical supervision instance. Also if the automated driving system should be able to use information from traffic infrastructure or other systems that are part of the traffic system, this would need to be reflected in the formulation of the system boundary.

The framework presented by [13] can be considered to be a hybrid between an enterprise and a systems architecture framework, as it consists of viewpoints that frame concerns from both perspectives.

Góngora et al. [21] argue for the creation of architecture frameworks in the automotive domain from an OEM perspective, which is Renault in this case. They base their argument on the growing complexity of automotive E/E-systems by the example of safety concerns. From their perspective, many work products demanded by ISO 26262 show that established development processes before the release of the standard were lacking the necessary formalism to describe safety-critical systems [21, p. 242] (and that this lack is oftentimes still apparent today). For this reason, they suggest an MBSE-based approach to architecture design, also by defining an architecture framework based on the demands of ISO 26262.

The architecture frameworks proposed by [21] consists of four viewpoints: Three parallel viewpoints describe technical aspects (*constructional viewpoints*), functional aspects (*functional viewpoint*), and operating conditions of the vehicle (*operational viewpoint*). These viewpoints shall frame concerns with respect design, development, and operation of a vehicle. An orthogonal *requirements* viewpoint serves for describing requirements.

The operational viewpoint provides models for creating views that describe use cases and operational scenarios.

The functional viewpoint allows to define views for a coarse- (functional architecture) and fine- (functional breakdown structure) granular description of functional system elements.

Eventually, the technical viewpoint allows to define structural views for defining hardware- and software components, as well as a view for allocating requirements to such components, as, e.g., described in ▶ Sect. 39.4.2.

In addition to Renault [21], Volvo has published studies [52] for the definition of automotive architecture frameworks. Once more, the authors emphasize the need for consistent architecture descriptions in the context of an interdisciplinary development of safety-critical, complex vehicle electronics systems. By consistency, the authors refer to the consistency between the initially designed and the finished vehicle at the end of the production process. They see architecture frameworks as crucial tools in this context that allow the formulation of correspondence rules as well as the documentation of architecture decisions. The latter is important, as the authors argue, as different stakeholders would often have the need to adjust architectures due to concerns coming up during development. While the authors present interesting insights into their motivation, the actual viewpoints they present are more focused on the entire supply chain and will thus be not discussed in detail. The authors explicitly consider automated driving systems

as one possible area of application. However, safety concerns are not addressed beyond classic functional safety.

Finally, [16] propose an architecture framework consisting of six viewpoints, framing eight concerns of eight stakeholder groups, based on [21]. The authors name concerns such as *functionality*, *traceability*, and *safety*. These concerns are framed by a *functional*, a *requirements* and an *implementation* viewpoint.

As for the other presented frameworks, the *functional* viewpoint serves for modeling the abstract relations and interaction between functional system elements. According to [16], apart from functionality, the functional viewpoint frames concerns such as dependability, cost, and maintainability. Views and the suggested model kinds for modeling these views are the same as suggested by [21]. Compared to the previous approaches, [16] provides the only summary of explicit correspondence rules. These correspondence rules are established to assign requirements to functional elements to make requirements traceable.

The implementation viewpoint addresses the same concerns as the functional viewpoints and frames *safety* as an additional concern. The viewpoint defines models for a hardware-, a software and a topology view and defines correspondence rules for defining instantiation relations to enable the definition of which software and hardware element(s) realize(s) which functional element(s).

As the implementation viewpoint is the only one that frames safety as a concern, it can be assumed that safety is only understood as *functional* safety in the sense of ISO 26262, rather than as *behavioral safety* as discussed above.

In summary, compared to [13, 16, 21, 52] keeps a closer connection to ISO/IEC/IEEE 42010—particularly with respect to explicitly mentioning correspondence rules.

39.5.3 Summary

Despite that the authors of the four architecture frameworks above emphasize similar challenges and despite the fact that the architecture frameworks are partially based on one another, it can be summarized that the descriptions of [13, 16, 21, 52] only share partial common ground. In their particular design, there are massive differences how which concerns are framed.

Common aspects, that can be summarized, however, are the *separation of concerns* as a means to *cope with complexity*, the importance of *interdisciplinary* work during the design and development of vehicle systems and maybe also the discussion of *potential conflicts* when it comes to establishing formal modeling methods in established automotive development processes.

		Broy et al. 2008		Gongóra et al. 2013		Dajsuren 2019
View-points	5 +6	functional technical [...] interaction information value net security safety quality energy vibration weight	4	functional technical requirements operational	6	functional implementation requirements information deployment feature
Model Kinds	2	petri nets ladder logic	7	block diagrams use case diagrams sequence charts state charts activity diagrams requirement diagrams	9	block diagrams use case diagrams sequence charts state charts activity diagrams requirement diagrams feature diagrams process model
Corres-pondence-rules	-	not named explicitly	3	requirm. ▸ function requirm. ▸ component function ▸ component	8	requirm. ▸ all others function ▸ requirm. component ▸ function deployment ▸ implementation

Fig. 39.7 Comparison of the relevant viewpoints, model kinds, and correspondence rules of the discussed vehicle-engineering-related architecture frameworks. –Orange shade: common elements, grey font: optional viewpoints; model kinds are only listed if explicitly mentioned

Figure 39.7 summarizes the similarities of the frameworks that [13, 16, 21] present. The figure contains all viewpoints that the authors mention, also those that have been omitted in the summaries above.

In the following, further common ground and differences with respect to the handling of stakeholders, concerns, and the corresponding viewpoints will be discussed.

39.5.3.1 Stakeholder and Concerns

Concrete stakeholder groups and their concerns are only mentioned explicitly by [16, 52].

Pelliccione et al. [52] phrase stakeholder concerns as questions[11], which is established practice in the literature. Further discussions in this chapter will adopt this practice. Dajsuren [16], on the contrary, follows the example of ISO/IEC/IEEE 42010 and limits the formulation of concerns to the formulation of keywords. Experience shows that a combination of both approaches can be helpful. Concrete questions can imply the need for concrete proposed solutions, while a formulation of keywords can be easier to start off with.

39.5.3.2 Viewpoints

All discussed architecture frameworks choose a broad formulation of viewpoints, specializing the viewpoints in multiple views per viewpoint. All three frameworks that are summarized in Fig. 39.7 use a *functional* and a *technical* viewpoint (which is called a *implementation* viewpoint in [16]). The latter contains *hardware-* and *software-*related views, such that the already mentioned established views (▸ Sect. 39.4) can be modeled.

39.5.4 Conclusion

The references mentioned in this subsection show several challenges when it comes to defining architecture frameworks. General recommendations for an effective and efficient design of architecture frameworks for automated driving systems are hard to give. Neither taking a top-down-systems-engineering perspective, nor focusing on the domain-specific knowledge with respect to the design and development of automated driving systems helps on its own. The presented architecture frameworks that have been designed from an automotive perspective cannot cover important aspects of behavioral safety.

This emphasizes the need of leveraging both, i.e., the specific domain knowledge in the field of automated

---

11 This includes questions such as: "How can system behavior be modeled?".

vehicles *and* the powerful tools provided by MBSE, when suitable architecture frameworks for automated driving systems shall be designed. What possible approaches could look like and what are suitable generic architecture frameworks that are candidates for tailoring, are still open research questions.

Currently, or at least in the scope of this chapter, it is impossible to give appropriate answers to these questions. What is possible, though, is to show some of the stakeholders and concerns that are relevant for the design of an architecture framework for automated driving systems and, by doing so, to give a first starting point for future work on closing the mentioned gaps.

39.6 Example Scenario and Approach in an Automated Driving Context

From the previous discussions in this chapter, it becomes clear that the definition of architecture views and entire architecture frameworks is established practice for describing systems in the Systems Engineering domain.

While there is often no direct reference to a formal definition of architecture frameworks in the field of automated driving (▶ Sect. 39.4), the application of different views on automated driving systems is established practice, as well. At the same time, there is a number of publications from the field of cyber-physical systems that aim at dealing with the uncertainty that an open context creates for autonomous systems, by leveraging formal methods from the Systems Engineering domain [3, 40] on the one hand and by leveraging knowledge representation on the other hand [14, 66].

The explicit connection between the application in the domain of automated driving and the formal domain of Systems Engineering has not been fully established, yet. Even if the concepts presented by [14] can be seen as an important step into that direction.

At the end of this chapter, we would like to provide a (mostly non-formal) example of how a possible approach for formulating architecture views for ensuring traceable safe system behavior can look like. A scenario as well as a collection of stakeholder concerns shall serve as a starting point for establishing architecture views that can address these concerns. A vehicle which is equipped with an automated driving system (ADS-equipped vehicle, according to SAE taxonomy [60]) serves as the system of interest.

At the same time, this chapter can, of course, not give a complete list of all relevant concerns and stakeholders. The given example serves illustrative purposes. With respect to iterative development processes as presented in ▶ Chap. 45, the given examples could at least be seen as a basis for future contributions to the architecture definition for automated driving systems. For this purpose, the presented stakeholders and their concerns should, however, be extended, e.g., by performing further analysis of relevant stakeholders as well as conducting stakeholder interviews.

39.6.1 Example Scenario

The following scenario is a variant of the one presented in [7]: The ADS-equipped vehicle is approaching a row of side-parked vehicles. A pedestrian is occluded by the parked vehicles. The pedestrian could step onto the road. An oncoming vehicle is driving in the left lane.

This simple scenario alone can illustrate the most aspects of uncertainty discussed in the introduction of this chapter. Particularly, two distinct types of uncertainty are relevant here. On the one hand, there is a lack of knowledge about the state of the environment inside of the occluded areas. Theoretically, each gap between two parked vehicles could allow a pedestrian to potentially step into the path of the ego vehicle. On the other hand, considering the predicted behavior of the oncoming vehicle, its motion is, at least theoretically, only bound by the laws of physics. Without relying on the driver's appropriate behavior, the vehicle *could* enter the ego lane at any time. Considering that the ego vehicle is not standing still, the ego vehicle needs to deal with a permanent theoretical collision probability with other traffic participants. Combined with the possibility that other traffic participants can suffer physical harm in such a collision, technically speaking (also with respect to ISO 26262-terminology), the actors in the scenario are subject to a certain risk.

From this perspective, human drivers are subject to a comparable risk as the automated driving system—as such uncertainty is the cause of an inherent risk that comes with moving in public traffic [46]. Differences may emerge in terms of how this risk is handled, e.g., by assuming which actions of other traffic participants are more likely than others (e.g., neglecting the possibility that the oncoming vehicle enters the ego lane, as such a behavior is considered unlikely). In consequence, human drivers might pass such scenarios without adapting their speed—accepting residual risks in favor for the need of mobility.

In a way human drivers constantly perform such trade-offs, even if only implicitly. From a technical perspective, when designing automated driving systems, such trade-offs should not be made implicitly—at least when it comes to a transparent acknowledgment of the risks that the technical system poses to other traffic participants (cf. also ▶ Chap. 40). Reflecting on stakeholder needs and concerns (such as safety or mobility) can be a way to force such trade-offs to be made explicit when formulating requirements. An architecture framework can then allow for a traceable transformation of stakeholder concerns into concrete requirements for system- and safety requirements. By providing traceability up into the development of software realizing the system's functionality, an architecture framework also provides a basis for tracing the aforementioned trade-off to actual aspects of the concrete implementation of run time algorithms.

39.6.2 Relevant Stakeholders and Stakeholder Groups

An important group of stakeholders are probably the *users* of the system. The same is true of all persons that interact with the system—in the given example, this includes possible pedestrians as well as possible occupants of the oncoming vehicle. These examples can be summarized in the stakeholder group of *traffic participants*. A further generalization of the needs relevant to *users* and other *traffic participants* (such as safety and mobility) yields the entire *society* as a relevant stakeholder group.

Possible groups directly concern with the design and development of automated driving systems are, e.g., *system engineers*, *safety engineers*, and *developers*.

The goals behind the development processes and architectures that are applied in the design and development of technical systems are always also linked to enabling a sound safety assurance argument for these systems. In this respect, being on the receiving end of the assurance process, *regulatory authorities* and finally also *lawyers* can be named as possible stakeholder groups that have an interest in an automated driving system.

This list is by no means complete. Additionally, as this chapter is written from a technical background, we can only take a technical standpoint for the following discussions. Hence, we will focus on the discussion of possible concerns that developers and engineers may have.

39.6.3 Motivating an Architecture Framework for Automated Driving Systems

Given these preconditions, we can formulate several concerns that an architecture framework (or in its instance, an architecture) should be able to frame and address:

❓ Concerns

- How can safe system behavior be described and generated in a traceable fashion?
- How can safety and mobility be traded off during the development of a technical system that operates in an open context?

Example concerns that can be derived from the system's interaction in the open context and that are directly

related to dealing with uncertainty could be phrased as follows:

❓ Concerns

- Which capabilities does the system need for successfully mastering the dynamic driving task under uncertainty?
- Which functionality does the system need to possess for successfully mastering the dynamic driving task under uncertainty?
- How are the required functions be related to the required capabilities?

While the last three concerns can be addressed in individual views, further example concerns require multiple views to be addressed:

❓ Concerns

- How does the timing behavior of an instantiated software component influence system-level behavior?
- What control quality can be achieved if latency requirements are violated?
- How can requirements with respect to the system's capabilities be monitored at run time?
- How can system-level behavior be monitored at run time?

All these concerns are related to aspects of the automated driving systems (control engineering, timing behavior, vehicle behavior) that cannot be addressed by a single view. In this respect, they can serve as an illustrative example for the separation of concerns for which a number of views are required, while only establishing traceability between the views can contribute to avoiding unwanted effects caused by emergent properties.

Concerns with Respect to Functional Architecture Views for Automated Driving Systems

A practical example for concrete architecture concerns for automated driving systems, rather than more general system concerns, has been discussed at a workshop with participants from the Karlsruher Institute of Technology (KIT), the RWTH Aachen and the TU Braunschweig[12].

An extract of discussed questions can be given as follows:

---

12 Workshop "Functional System Architecture" in the context of the DFG SPP 1835 "Cooperatively Interacting Automobiles" (Braunschweig, March 3rd, 2017)—Prof. Dr.-Ing. Christoph Stiller (KIT), Prof. Dr.-Ing. Markus Maurer (TU Braunschweig), Maximilian Naumann (KIT), Eugen Altendorf (RWTH Aachen), Susanne Ernst, Gerrit Bagschik, Marcus Nolte (all TU Braunschweig).

? Concerns

- How does functional safety influence a functional architecture?
- How does a functional architecture consider robustness?
- How is self-perception addressed in a functional architecture?
- How is redundancy considered in a functional architecture?
- Where can concepts of performance monitoring be addressed in a functional architecture?

The workshop was focused on functional architectures (or more correctly functional views, given the background of this chapter) for automated driving systems. However, not all of the concerns above can be addressed by a functional view alone. A good example for this is provided by the question regarding functional safety and redundancy. As functional safety is concerned with controlling random or systematic E/E-faults caused at a hardware or software level. At the same time, a functional (or logical) architecture view specifically abstracts from concrete hardware- and software solutions. Hence, such views do not necessarily address concerns related to functional safety: A functional element can be realized by a number of redundant software components and can be executed redundantly on a number of different control units, to fulfill integrity levels demanded by a safety goal or the derived functional safety requirements according to ISO 26262.

Questions of functional safety are hence also good examples of concerns that cannot be addressed in a single architecture view. Just as it is true for questions regarding safe system-level vehicle behavior.

At the same time, these specific concerns (e.g., with respect to monitoring) show how concerns can shape the appearance of an architecture view. Depending on the concerns that shall be addressed, even functional views for automated driving systems can look differently from one application to another.

39.6.3.1 Examples for Architecture-Related Concerns in the Scenario

Concrete architecture concerns with respect to safety aspects can also be derived from the example scenario. According to [7, 56], one can ask:

? Concerns

- How can safety analysis results be represented within the architecture?
- How can hazards be represented within the architecture?
- How can safety goals be represented within the architecture?

In the context of safety-related concerns, a hazard relevant in the presented scenario can be phrased as follows:

H-1 – Collision with pedestrian stepping onto the road.

This hazard could be addressed by the following safety goal:

SG-1 – Avoidance of collisions with pedestrians entering the ego lane.

Which could be addressed by the following functional safety requirement:

FSR-1 – The vehicle must approach occluded areas with appropriate speed.

Regarding the fulfillment of the example safety goal by approaching occluded areas with *appropriate* speed, a number of situation specific requirements arise, as the question of what *appropriate* means is highly context-dependent. This, e.g., includes sensor ranges, maximum available deceleration, or how uncertainty that is caused by occlusions is handled in general.

Some concrete concerns in this specific context could be formulated as

? Concerns

- How can the given scenario be formally modeled?
- Which capabilities are required at what performance level to fulfill the safety goal with sufficient confidence?
- How can the abstract safety goal be tested by situation-specific metrics?
- Which aspects of the situation determine how an "*appropriate speed*" can be quantified?
- Which development decisions (e.g., acceptable risks) have an impact on the quantification of the safety goal?
- How can the safety goal be monitored at run time?
- Why are which probabilities assumed for jaywalking pedestrians?
- Which failure rates are acceptable for fulfilling the safety goal with the demanded integrity?

This is, of course, only a loose collection of concerns. The collection that shall illustrate that the scenario can again trigger concerns in different contexts for different stakeholders. A suitable architecture framework for automated driving systems should answer such concerns in a way that such situation-dependent concerns can be traced back over different views to the corresponding safety goals.

39.6.4 Examples for Architecture Views

This chapter alone cannot provide a fully tailored architecture framework for automated driving systems that completely addresses all of the aforementioned concerns. Instead, some of those concerns shall be taken as a basis to formulate a few example views that could be relevant in the context of such an architecture framework. The examples will, by no means, be complete, but should serve as an illustration of the principles that can be used to construct additional views. The approach is closely related to [7, 47].

Requirements with Respect to Vehicle Behavior in the Given Scenario

To address concerns with respect to the description of the system's desired target behavior (e.g., in a *behavior viewpoint*), views can be modeled in a variety of ways. Scenario-based approaches are currently in the focus of R&D efforts when it comes to describing and/or verifying and validating the behavior of automated driving systems on different levels of abstraction. From a design perspective, a coarse description of the scenarios[13] and the desired target behavior for the ego vehicle often serve as a starting point.

Scenarios can be seen as parts of *use cases* [68]. Hence, the fundamental elements (or actors) of a (functional) scenario can be modeled in a use case diagram (cf. ◘ Fig. 39.9). The modeled use case is even more abstract than the scenarios it contains. This means that, for this example, the number of parked vehicles could vary between the scenarios contained in the use case. Theoretically, it would also be possible to model some scenarios that contain an occluded pedestrian and some that do not. Hence the use case is only a first, extremely abstract description to gain an overview over possible scenarios, and to formulate first requirements for the behavior of the automated driving system. In this context, it should be mentioned that the challenges when modeling architecture views do not lie in the formally *correct* modeling—as there are tools that can support here—but rather in formulating sufficiently expressive models. In this example, decisions must be made, if using multiplicities for the actors in the use case should be used or if they should not—depending on the stakeholder concerns, this decision may vary.

The use case diagram shown in ◘ Fig. 39.9 contains the infrastructure elements and objects in the scenario which can be found in ◘ Fig. 39.8 as actors. This includes the ego vehicle, the parked vehicle, the oncoming vehicle as well as the infrastructure elements, such as lane markings. The relations of these elements to a description of the Operational Design Domain (ODD) are indicated in the diagram, as well. With these relations, corre-

spondences can be established between requirements, elements of the system architecture, and the actors of the use case[14], including the elements of the ODD description.

Dynamic interactions between the actors of the use case, and by that concerns with respect to the actual behavior, e.g., of the ego vehicle, cannot be modeled in a use case and hence require separate views. Model kinds which allow this kind of behavior modeling are sequence (cf. ◘ Fig. 39.9) or activity diagrams. Such diagrams could, e.g., also be generated from other formal scenario description languages (e.g., [11]).

Finally, it should be mentioned that the same doubts with respect to scalability than can apply to general scenario-based approaches in the field of automated driving, can also apply to the approach of modeling the behavior of automated driving systems in architecture views. A discussion of possible issues follows this example.

Required Capabilities[15]

Having a scenario description, requirements for the desired target behavior of the vehicle can be formulated and structured. An approach of modeling properties or potentials that a system needs to safely master a scenario is the concept of so-called skill- and ability- (in English often subsumed as capabilities) graphs introduced in [45, 58] (cf. left-hand side of ◘ Fig. 39.10).[16] The root node of the graph represents the capability of executing a *maneuver* and, hence, the capability of showing a certain system-level behavior.

According to the definition, each node in the graph is combined with quality metrics. The semantic relations of the edges in the graph are dependency relations: A system capability is only available with full performance, if the subsequent capabilities are available with full performance.

For the given scenario, the desired target behavior should consist of a reduction of the *reference speed* when approaching the parked vehicles. A corresponding excerpt from a skill/ability graph is shown on the left-hand side of ◘ Fig. 39.10. In order to be capable of choosing an appropriate reference speed, the automated driving system must be capable of detecting stationary, as

---

13 Often in terms of *abstract* [67] or *functional* [6] scenarios.

14 The term actor does not only subsume actual people, but all elements that cannot be changed by the system of interest [42].

15 As already discussed shortly, this term is not used consistently between the domains of Systems Engineering and automated driving. In this chapter, the term capability is more related to current SAE-definitions with respect to "behavioral competencies" [4], not in the more abstract understanding of the term in the Systems Engineering domain.

16 The skill graph is intentionally not modeled formally in UML- or SysML-compatible concepts here to illustrate that there still are inconsistencies to be resolved with respect to terminology around the concept of capabilities.

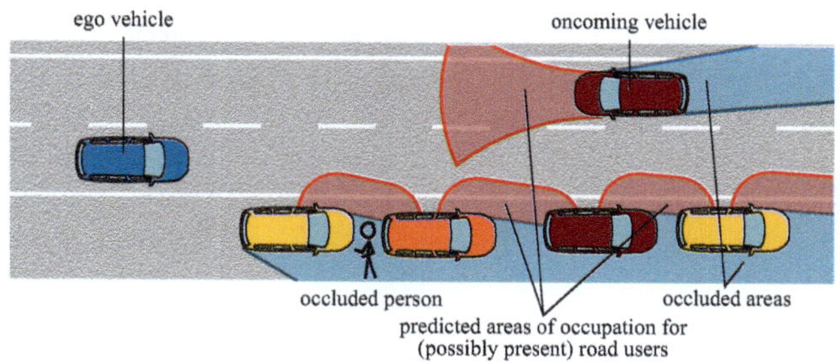

Fig. 39.8 Example scenario for displaying some of the described sources of uncertainty

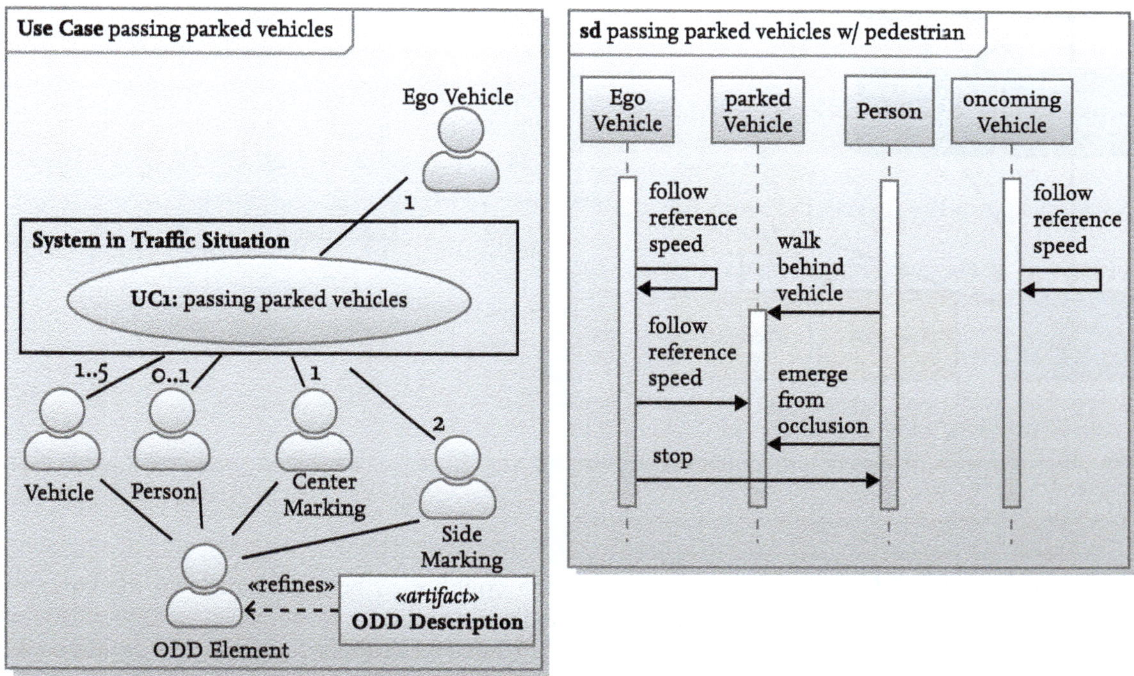

Fig. 39.9 Use case (left) and sequence diagram (right) for scenario description—parallel regions and irrelevant actions (e.g., lateral maneuvers) are omitted in the sequence chart for the sake of simplified visualization

well as dynamic objects in order to be capable of identifying relevant occluded areas.

By structuring the required abilities and skills in the scenario, we also gain a possibility of structuring system requirements, e.g., by a structured decomposition of safety goal. This is sketched on the right-hand side of ■ Fig. 39.10 by means of a SysML requirements diagram. The SysML stereotype «requirement» can, for example, be sub-typed for an application in the domains of functional or behavioral safety. The diagram shows explicitly that the functional safety requirement FSR-1 has been derived from a safety goal and that there are correspondences between these requirements and related skills and abilities (dashed arrows). Looking at the required abilities and skills for a given scenario description and a number of assumptions, coarse functional requirements can be formulated which can then be constrained by corresponding performance requirements.

Functional/Logical View

The definition of functional architecture views is, as mentioned above, intended to create an implementation-independent understanding of the system of interest. A function is constituted by an input-output relationship. Functional elements represent system elements that relate at least one input and one output by means of a function [74]. In this context, the exact software and hardware components which realize a function are irrelevant. This implementation-independent perspective yields structures that are long-term stable with respect to design and development processes [74]. While the concrete hardware (e.g., control units, sensors, etc.) or implementation-specific details in the applied algorithms may change, without having an impact on the functional view onto the system of interest. Functional views are often depicted in activity or (internal) block description diagrams (cf. ■ Fig. 39.11).

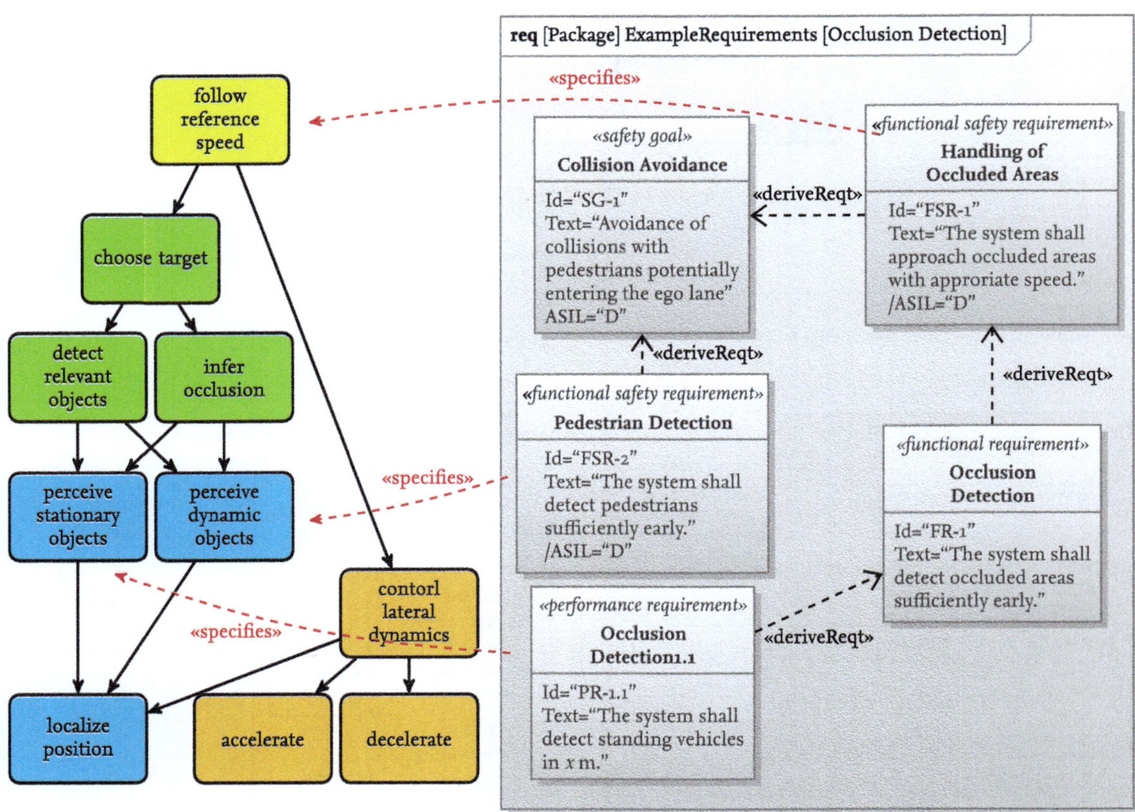

Fig. 39.10 Excerpt from an "skill graph" (left) and requirement diagram (right) for displaying and structuring behavior-related requirements for the ego vehicle

The terms *functional* and *logical* architecture (view) are often used synonymously [12]. When both concepts are differentiated from each other, logical views are often closer related to abstract physical architectures [74]. That means that a logical architecture view abstracts from concrete technical components. Different technological aspects may remain visible, while the concrete choice of physical components is not required. Consider an abstract LiDAR sensor as an example. While such a component may define the technology that a physical component requires, it does not define that the LiDAR sensor must be provided by supplier X with a range of Y m. Hence, the logical view remains independent of a concrete technical implementation. If functional and logical views are explicitly differentiated, functional views are often modeled in activity diagrams, while logical views are more often modeled with (internal) block diagrams.

Figure 39.11 (left-hand side) is illustrated according to [74] and is thus modeled in a internal block description diagram. The displayed functions contribute to the realization of the abilities and skills that are concerned with the detection of stationary and dynamic objects, as well as with the recognition of occluded areas. Functional elements, their interfaces, and the direction of data flow are shown. The relation of requirements to a func-

tion could, e.g., be established via the corresponding skills and abilities.

Under traceability aspects, the functional safety requirement FSR-1 can be assigned to the functional element *pedestrian classification*. The function inherits the demanded integrity level (in this example, this is an ASIL D, cf. ▶ Chap. 6) of the related functional safety requirement.

Software View

Based on the functional view, more implementation-specific software views can be created that describe how functions are realized. As an example, Fig. 39.11 shows, apart from the functional view, a software view with a component diagram. The software components realize, as indicated by arrows in the figure, the function for pedestrian classification, as well as the interface that serves as an input to the respective function. Lines with solid black circles at their ends indicate interfaces that are provided by a component. Unfilled half circles indicate that a component requires the respective interface.

In this example, we assume that sensor data representing dynamic objects is published via an Ethernet interface. This requires additional software components in terms of a driver component (NIC_drv) that makes the bus accessible for the processing components, and a

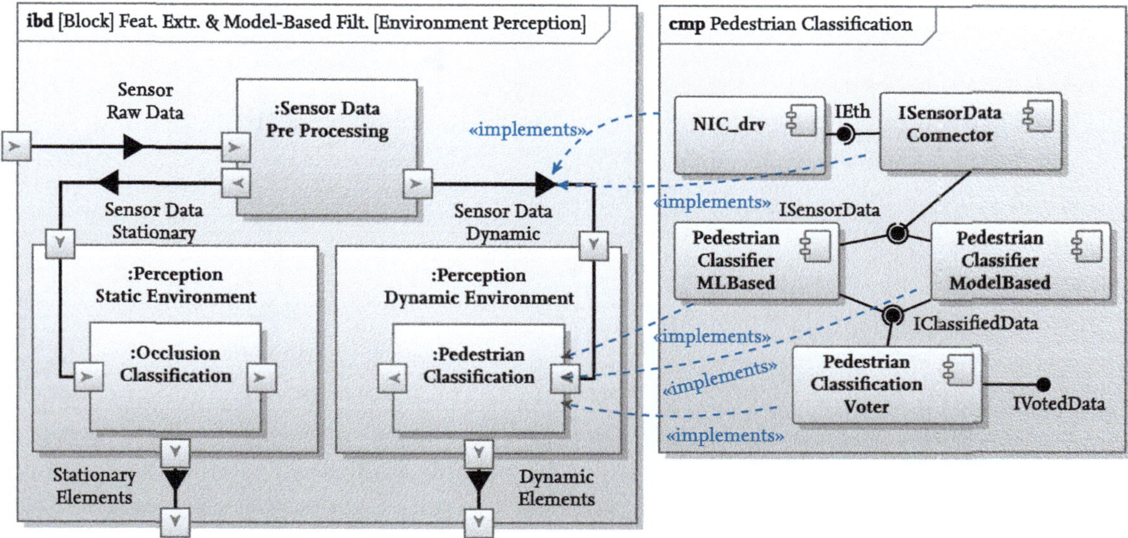

Fig. 39.11 Example of functional elements and interfaces (left) as well as the corresponding software components that realize the different functions (right)

converter component (ISensorDataConnector) that can transform Ethernet data (e.g., raw TCP frames) into an interface that is understandable for the processing components (ISensorData is such a supposed interface for sensor data).

Other diagram types that are typically used for modeling software-centric views are class diagrams. These diagrams cannot only model the properties of software components in more detail, but they can, of course also be used to further specify the interfaces that the components use for communication (e.g., by defining the datatypes in the interfaces).

Figure 39.11 also relates to the concept of traceable requirements. As the function for pedestrian detection is assigned with a functional safety requirements that has a ASIL-D integrity level, the function is realized by two redundant software components[17] and a corresponding voter component.[18]

Hardware View

To conclude the example, Fig. 39.12 shows correspondences between the software view described above (left-hand side) and a possible hardware- or physical view (right-hand side). The hardware view can, e.g., be modeled in an internal block definition diagram

or a deployment diagrams. In the case of the internal block diagram shown in Fig. 39.12, the blocks depict hardware components, i.e., control units and/or communication buses. The relations between the blocks for the hardware components and the block for the aforementioned Ethernet interface show that the hardware component all communicate via Ethernet. For the sake of a clearer visualization, the blocks are depicted without additional properties, although data such as IP addresses or communication ports could be modeled as such properties.

This example also shows that the software components which already imply a redundant realization of the ASIL-D functionality are additionally instantiated on redundant control units, as well (dashed arrows). A deployment diagram as an alternative model kind (cf. Fig. 39.3) could visualize the particular configuration of the Ethernet bus, and its participants in a marginally simpler fashion—while also providing the possibility to model further details such additional routers, switches, or subnet configurations.

39.6.4.1   Summary

The views and the viewpoints that govern the views that are mentioned in this example do not constitute a complete architecture framework. The examples shall provide an impression of how the architecture views that have been mentioned in the course of this chapter can be modeled. The formalization of such models comes with the benefit of automatically establishing correspondence rules. Applying these correspondence rules to formulate correspondences across views greatly contributes to concerns with respect to traceability of requirements and/or design decisions.

---

17 In this case, the function is realized by a machine-learning-based (MLBased) and model-based (ModelBased) classification algorithms, respectively.

18 For a real application, it must be argued, of course, why the implementation of the ASIL-D function with the shown components actually reaches an ASIL-D integrity level. Such rationale can again be modeled by comments or links to the documentation in an architecture view.

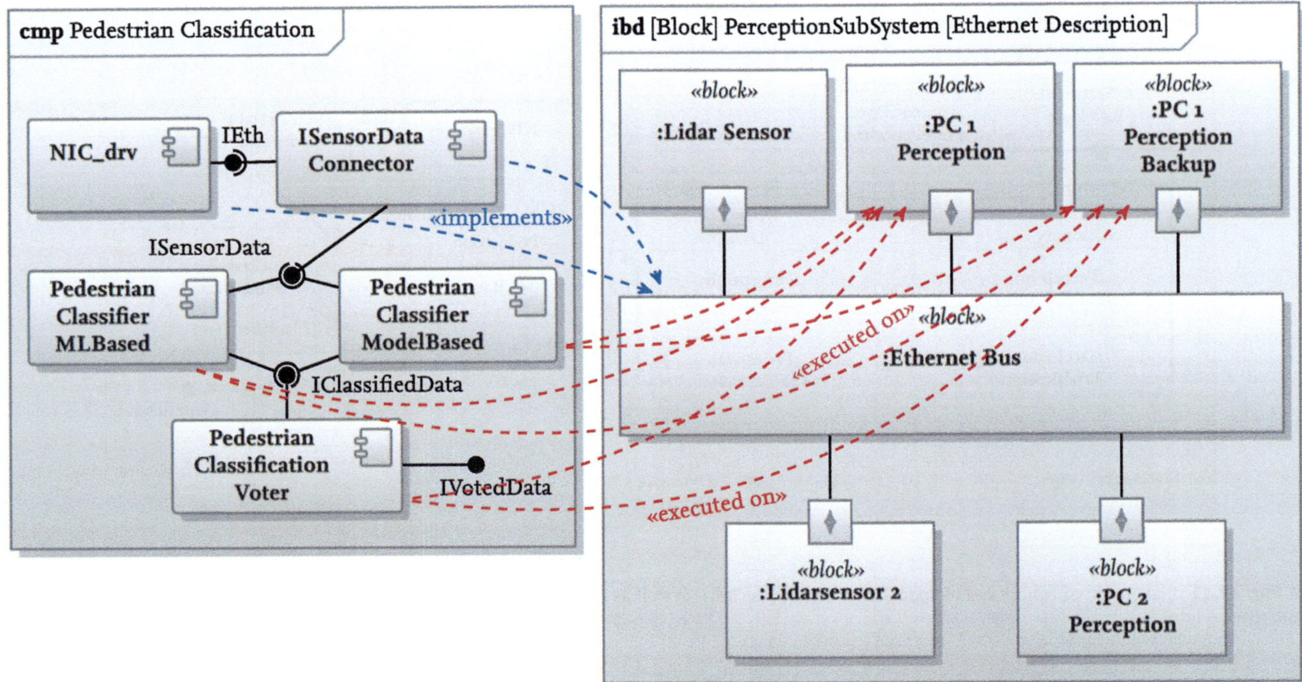

Fig. 39.12 Software- (left) and hardware components (right) with redundant mapping from software- to hardware components

As it has already been mentioned, doubts with respect to the scalability of the presented approach are valid, as such doubts can already be raised in the context of scenario-based development. Nonetheless, scenario-based approaches are currently treated as an important tool in the domain of automated driving. Many publications argue that scenario-based approaches can present a suitable tool for handling the complexity of the real world. The scalability from a selection of scenarios to a sufficiently complete description of the desired target behavior for automated driving systems[19] is, however, subject to current research and remains to be proven. Approaches as presented in this chapter can contribute to solutions, as formal models enable consistency checks of the modeled concepts. What no formal method can do for systems that need to act in an open context is to provide *completeness* guarantees that all required concepts have been modeled.

Eventually, the efforts put into modeling must always be discussed in terms of their benefits. Models should thus only be as complex as required to answer to stakeholder concerns—which implies a great deal of communication with all involved stakeholders or stakeholder groups.

39.7 Conclusions

This chapter has presented an overview of approaches for designing architecture views for automated driving systems from different perspectives. In summary, it can be said that the methods of modern model-based systems engineering can provide important contributions to the realization of safe automated driving systems. At the same time, there is still a gap between the domains of formal model based systems engineering and automated driving, which can be attributed, at least partially, to terminology which uses the same concepts in different semantic meanings (e.g., *capabilities* or *functional architecture*). The reason for this can be traced back to the different directions from which the fields have developed— i.e., from the fields of cognitive control systems on the one hand and from the fields of software engineering and enterprise architecture on the other hand.

Closing these gaps and combining the established formal MBSE methods with the domain knowledge in the field of automated driving presents a promising approach for modeling architecture views for automated driving systems.

Acknowledgements We would like to thank all colleagues from the projects"Controlling Concurrent Change", UNICAR *agil*, as well as the working group "Architecture" from the project VVMethods. They have contributed to the contents of this chapter as partners in countless discussions. Finally, we would like to thank

---

19 Suitable in terms of an argument that the unknown unknown scenarios [28] and the missing requirements caused by these scenarios, fall under an *acceptable* residual risk that is exposed by an automated driving system.

Klaus Dietmayer for his thorough review of the chapter, particularly with respect to the input for the final version of ▶ Sect. 39.6.

References

1. Ackva, S.: Die operational domain im system architecture framework. In: Tag des Systems Engineering (TdSE). Gesellschaft für Systems Engineering (GfSE) e.V. online (2020)
2. Albus, J.P., Huang, H.M., Messina, E.R., Murphy, K., Juberts, M., Lacaze, A., Balakirsky, S.B., Shneier, M.O., Hong, T.H., Scott, H.A., Proctor, F.M., Shackleford, W.P., Michaloski, J.L., Wavering, A.J., Kramer, T., Dagalakis, N., Rippey, W.G., Stouffer, K.A., Legowik, S.: 4D/RCS Version 2.0: A Reference Model Architecture for Unmanned Vehicle Systems (NIST IR 6910). National Institute of Standards and Technology, Gaithersburg, MD (2002)
3. Aniculaesei, A., Grieser, J., Rausch, A., Rehfeldt, K., Warnecke, T.: Toward a holistic software systems engineering approach for dependable autonomous systems. In: 2018 IEEE/ACM 1st International Workshop on Software Engineering for AI in Autonomous Systems (SEFAIAS), pp. 23–30 (2018)
4. Automated Vehicle Safety Consortium (AVSC): AVSC Best Practice for Evaluation of Behavioral Competencies for Automated Driving System Dedicated Vehicles (ADS-DVs). AVSC00008202111 (2021)
5. Bach, J., Otten, S., Sax, E.: A taxonomy and systematic approach for automotive system architectures—from functional chains to functional networks. In: Proceedings of the 3rd International Conference on Vehicle Technology and Intelligent Transport Systems, pp. 90–101. SCITEPRESS—Science and Technology Publications, Porto, Portugal (2017)
6. Bagschik, G., Menzel, T., Maurer, A.R.M.: Szenarien Für Entwicklung, Absicherung Und Test von Automatisierten Fahrzeugen. In: 11 Workshop Fahrerassistenzsysteme Und Automatisiertes Fahren, pp. 125–135 (2017)
7. Bagschik, G., Nolte, M., Ernst, S., Maurer, M.: A system's perspective towards an architecture framework for safe automated vehicles. In: IEEE International Conference on Intelligent Transportation Systems (ITSC), pp. 2438–2445. Maui, HI, USA (2018)
8. Bagschik, G.: Systematischer Einsatz von Szenarien für die Absicherung automatisierter Fahrzeuge am Beispiel deutscher Autobahnen. PhD thesis. TU Braunschweig, Braunschweig (2020)
9. Behere, S., Törngren, M.: Systems engineering and architecting for intelligent autonomous systems. In: Watzenig, D., Horn, M. (eds.) Automated Driving, pp. 313–351. Springer International Publishing, Cham (2017)
10. Bengler, K., Dietmayer, K., Eckstein, L., Stiller, C., Winner, H.: Fahrerassistenzsysteme und Automatisiertes Fahren. In: Pischinger, S., Seiffert, U. (eds.) Vieweg Handbuch Kraftfahrzeugtechnik, pp. 1009–1072. ATZ/MTZ-Fachbuch. Wiesbaden, Springer Fachmedien (2021)
11. Bock, F., Sippl, C., Heinz, A., Lauer, C., German, R.: Advantageous usage of textual domain-specific languages for scenario-driven development of automated driving functions. In: IEEE International Systems Conference (SysCon), pp. 1–8 (2019)
12. Borky, J.M., Bradley, T.H.: Designing in a logical/functional viewpoint. In: Effective Model-Based Systems Engineering, pp. 153–216. Springer International Publishing, Cham (2019)
13. Broy, M., Gleirscher, M., Merenda, P., Wild, D., Kluge, P., Krenzer, W.: Toward a holistic and standardized automotive architecture description. Computer **42**(12), 98–101 (2009)
14. Carré, M.: Autonomic Framework for Safety Management in the Autonomous Vehicle. Université de Pau et des Pays de l'Adour (2019)
15. Checkland, P.: Systems Thinking, Systems Practice, p. 330. J. Wiley, Chichester [Sussex], NY (1981)
16. Dajsuren, Y.: Defining architecture framework for automotive systems. In: Dajsuren, Y., van den Brand, M. (eds.) Automotive Systems and Software Engineering, pp. 141–168. Springer International Publishing, Cham (2019)
17. Dickmanns, E.D.: Dynamic Vision for Perception and Control of Motion, p. 492. Springer, London (2007)
18. Dijkstra, E.W.: On the role of scientific thought (1974). In: Selected Writings on Computing: A Personal Perspective, pp. 60–66. Springer, New York (1982)
19. Fong, E.N., Goldfine, A.H.: Information management directions: the integration challenge. ACM Sigmod Record **18**(4), 40–43 (1989)
20. Friedenthal, S., Moore, A., Steiner, R.: Model-based systems engineering. In: A Practical Guide to SysML, 3rd edn, pp. 15–29. Elsevier (2015)
21. Góngora, H.G.C., Gaudré, T., Tucci-Piergiovanni, S.: Towards an architectural design framework for automotive systems development. In: Aiguier, M., Caseau, Y., Krob, D., Rauzy, A. (eds.) Complex Systems Design & Management, pp. 241–258. Springer, Berlin, Heidelberg (2013)
22. Graubohm, R., Stolte, T., Bagschik, G., Maurer, M.: Towards efficient hazard identification in the concept phase of driverless vehicle development. In: IEEE Intelligent Vehicles Symposium (IV), pp. 1297–1304 (2020)
23. Hilliard, R.: Question on Views and Viewpoints in ISO/IEC/IEEE 42010 (2020)
24. Holt, J., Perry, S.: Architectures and architectural frameworks with MBSE. In: SysML for Systems Engineering: A Model-Based Approach, pp. 425–466 (2013)
25. ISO 15704:2019: Enterprise Modelling and Architecture—Requirements for Enterprise-Referencing Architectures and Methodologies (2019)
26. ISO 19540:2022: Information Technology—Object Management Group Unified Architecture Framework (OMG UAF) (2022)
27. ISO 26262:2018: Road Vehicles—Functional Safety (Version 2) (2018)
28. ISO/DIS 21448:2021: Road Vehicles—Safety of the Intended Functionality (SOTIF). (Version 2) (2021)
29. ISO/IEC 10746:2009: Information Technology—Open Distributed Processing—Reference Model (2019)
30. ISO/IEC/IEEE 15288:2015: International Standard—Systems and Software Engineering—System Life Cycle Processes (2015)
31. ISO/IEC/IEEE 42010:2011: Systems and Software Engineering—Architecture Description (2011)
32. ISO/TR 4804:2020: Road Vehicles—Safety and Cybersecurity for Automated Driving Systems—Design, Verification and Validation (2019)
33. Kampmann, A., Alrifaee, B., Kohout, M., Wüstenberg, A., Woopen, T., Nolte, M., Eckstein, L., Kowalewski, S.: A dynamic service-oriented software architecture for highly automated vehicles. In: IEEE Intelligent Transportation Systems Conference (ITSC), pp. 2101–2108 (2019)
34. Koopman, P., Wagner, M.: Autonomous vehicle safety: an interdisciplinary challenge. IEEE Intell. Transp. Syst. Mag. **9**(1), 90–96 (2017)
35. Leute, M., Haarer, A., Ackva, S., Lalitsch-Schneider, C., Andres, M., Husung, S., Malecki, P.: Modellbasierter Bärentango mit dem system architecture framework. In: Tag des Systems Engineering (TdSE). Gesellschaft für Systems Engineering (GfSE) e.V. online (2021)

36. Malavolta, I.: Software Architecture Modeling by Reuse, Composition and Customization. PhD thesis. Universita di L'Aquila, L'Aquila (2012)
37. Martin, J.N.: Overview of the revised standard on architecture description—ISO/IEC 42010. In: INCOSE International Symposium, vol. 31(1), pp. 1363–1376 (2021)
38. Mataric, M.: Behavior-based control: main properties and implications. In: Proceedings of the IEEE International Conference on Robotics and Automation, Workshop on Architectures for Intelligent Control Systems, pp. 46–54 (1992)
39. Maurer, M.: Flexible Automatisierung von Straßenfahrzeugen Mit Rechnersehen. PhD thesis. University der Bundeswehr München (2000)
40. Mauritz, M.: Engineering of safe autonomous vehicles through seamless integration of system development and system operation. PhD thesis. Technische Universität Clausthal, Clausthal (2019)
41. Mhenni, F., Choley, J.Y., Penas, O., Plateaux, R., Hammadi, M.: A SysML-based methodology for mechatronic systems architectural design. Adv. Eng. Inf. 28(3), 218–231 (2014)
42. Miles, R., Hamilton, K.: Learning UML 2.0. O'Reilly Media Inc, Sebastopol, CA (2006)
43. Morkevicius, A.: Fragerunde zum UAF im Anschluss an einen Vortrag. Misc. persönliche Kommunikation (2021)
44. Morkevicius, A., Aleksandraviciene, A., Mazeika, D., Bisikirskiene, L., Strolia, Z.: MBSE grid: a simplified SysML-based approach for modeling complex systems. In: INCOSE International Symposium, vol. 27(1), pp. 136–150 (2017)
45. Nolte, M., Bagschik, G., Jatzkowski, I., Stolte, T., Reschka, A., Maurer, M.: Towards a skill- and ability-based development process for self-aware automated road vehicles. In: IEEE International Conference on Intelligent Transportation Systems (ITSC), pp. 739–744. Yokohama, Japan (2017)
46. Nolte, M., Ernst, S., Richelmann, J., Maurer, M.: Representing the unknown—impact of uncertainty on the interaction between decision making and trajectory generation. In: IEEE International Conference on Intelligent Transportation Systems (ITSC), pp. 2412–2418. Maui, HI, USA (2018)
47. Nolte, M., Rose, M., Stolte, T., Maurer, M.: Model predictive control based trajectory generation for autonomous vehicles—an architectural approach. In: IEEE Intelligent Vehicles Symposium (IV), pp. 798–805. Los Angeles, CA, USA (2017)
48. Object Management Group (OMG): Unified Architecture Framework Specification Version 1.1 (2020)
49. Object Management Group (OMG): UAF Wiki (2021). ▶ https://www.omgwiki.org/uaf/doku.php. Last accessed on 11 July 2021
50. Papadopoulos, Y., Sorokos, I., Reich, J.: DEIS White Paper—ODE Profile V2 (2019)
51. Passino, K.M., Antsaklis, P.J.: An introduction to intelligent and autonomous control. In: Antsaklis, P.J., Passino, K.M. (eds.), pp. 191–214. Kluwer Academic Publishers, Norwell, MA, USA (1993)
52. Pelliccione, P., Knauss, E., Heldal, R., Ågren, S.M., Mallozzi, P., Alminger, A., Borgentun, D.: Automotive architecture framework: the experience of volvo cars. J. Syst. Archit. 77, 83–100 (2017)
53. Pellkofer, M.: Verhaltensentscheidung für autonome Fahrzeuge mit Blickrichtungssteuerung. PhD thesis. Universität der Bundeswehr München, München (2003)
54. Perry, D.E., Wolf, A.L.: Foundations for the study of software architecture. ACM SIGSOFT Softw. Eng. Notes 17(4), 40–52 (1992)
55. PRISM: Dispersion and Interconnection: Approaches to Distributed Systems Architecture. Cambridge, MA (1986)
56. Reich, J., Zeller, M., Schneider, D.: Automated evidence analysis of safety arguments using digital dependability identities. In: Romanovsky, A., Troubitsyna, E., Bitsch, F. (eds.) Computer Safety, Reliability, and Security, pp. 254–268. Lecture Notes in Computer Science. Springer International Publishing, Cham (2019)
57. Reschka, A.: Fertigkeiten-und Fähigkeitengraphen als Grundlage des sicheren Betriebs von automatisierten Fahrzeugen im öffentlichen Straßenverkehr in städtischer Umgebung. PhD thesis. TU Braunschweig (2017)
58. Reschka, A., Bagschik, G., Ulbrich, S., Nolte, M., Maurer, M.: Ability and skill graphs for system modeling, online monitoring, and decision support for vehicle guidance systems. In: IEEE Intelligent Vehicles Symposium (IV), pp. 933–939 (2015)
59. Roques, P.: MBSE with the ARCADIA method and the Capella tool. In: 8th European Congress on Embedded Real Time Software and Systems (ERTS 2016). Toulouse, France (2016)
60. SAE: Taxonomy and Definitions for Terms Related to Driving Automation Systems for On-Road Motor Vehicles (J3016 Ground Vehicle Standard), Version 202106 (2021)
61. Şahin, T., Raulf, C., Kızgın, V., Huth, T., Vietor, T.: A cross-domain system architecture model of dynamically configurable autonomous vehicles. In: Bargende, M., Reuss, H.C., Wagner, A. (eds.) Proceedings of the 21 Internationales Stuttgarter Symposium, pp. 13–31. Springer Fachmedien, Wiesbaden (2021)
62. Schäuffele, J., Zurawka, T., Carey, R.: Automotive Software Engineering: Principles, Processes, Methods, and Tools (SAE-R 361), p. 385. SAE International, Warrendale, PA (2005)
63. Schlatow, J., Nolte, M., Möstl, M., Jatzkowski, I., Ernst, R., Maurer, M.: Towards model-based integration of component-based automotive software systems. In: Annual Conference of the IEEE Industrial Electronics Society (IECON17), pp. 8425–8432. Beijing, China (2017)
64. Schräder, T., Stolte, T., Jatzkowski, I., Graubohm, R., Nolte, M., Maurer, M.: Compensating for the absence of a required accompanying person: a draft of a functional system architecture for an automated vehicle. In: IEEE International Intelligent Transportation Systems Conference (ITSC), pp. 2264–2271. Indianapolis, IN, USA (2021)
65. Siedersberger, K.H.: Komponenten zur automatischen Fahrzeugführung in sehenden (semi-)autonomen Fahrzeugen. PhD thesis. University der Bundeswehr München (2003)
66. Sifakis, J.: Autonomous systems—an architectural characterization. In: Boreale, M., Corradini, F., Loreti, M., Pugliese, R. (eds.) Models, Languages, and Tools for Concurrent and Distributed Programming: Essays Dedicated to Rocco De Nicola on the Occasion of his 65th Birthday. Lecture Notes in Computer Science, pp. 388–410. Springer International Publishing, Cham (2019)
67. Sippl, C.S.: Identifikation relevanter Verkehrssituationen für die szenarienbasierte Entwicklung automatisierter Fahrfunktionen. PhD thesis. Friedrich-Alexander-Universität Erlangen-Nürnberg (FAU) (2020)
68. Ulbrich, S., Reschka, A., Menzel, T., Schuldt, F., Maurer, M.: Defining and substantiating the terms scene, situation and scenario for automated driving. In: IEEE International Annual Conference on Intelligent Transportation Systems (ITSC), pp. 982–988. IEEE, Las Palmas, Spanien (2015)
69. Vogel, O., Arnold, I., Chughtai, A., Ihler, E., Kehrer, T., Mehlig, U., Zdun, U.: Software-Architektur: Grundlagen–Konzepte–Praxis, 2nd edn, p. 556. Springer Spektrum
70. Walden, D.D., Roedler, G.J., Forsberg, K., Hamelin, R.D., Shortell, T.M. (eds.): Systems Engineering Handbook: A Guide for System Life Cycle Processes and Activities, 4th edn., p. 290. Wiley, Hoboken, New Jersey (2015)
71. Wasson, C.S.: System Engineering Analysis, Design, and Development: Concepts, Principles, and Practices, p. 881. John Wiley & Sons Inc, Hoboken, New Jersey (2015)

72. Waymo: On the Road to Fully Self-Driving, p. 43. Mountain View (2021)
73. Weidenhaupt, K., Pohl, K., Jarke, M., Haumer, P.: Scenarios in system development: current practice. IEEE Softw. **15**(2), 34–45 (1998)
74. Weilkiens, T., Lamm, J.G., Roth, S., Walker, M.: Model-Based System Architecture: Weilkiens/Model-Based System Architecture. John Wiley & Sons, Inc, Hoboken, NJ, USA (2015)
75. Working Group, T. A. C.: Goal Structuring Notation Community Standard Version 2 The Assurance Case (2018)
76. Zachman, J.A.: A framework for information systems architecture. IBM Syst. J. **26**(3), 276–292 (1987)
77. Zuccaro, C., Rambo, J., Hüffer, H., Fritz, J., Dorsch, T.: Systems engineering die. In: Geisreiter, M. (ed.) @Klammer in der technischen Entwicklung, 2nd edn, p. 239. Gesellschaft für Systems Engineering e.V. (2019)

Open Access This chapter is licensed under the terms of the Creative Commons Attribution-NonCommercial-NoDerivatives 4.0 International License (▶ http://creativecommons.org/licenses/by-nc-nd/4.0/), which permits any noncommercial use, sharing, distribution and reproduction in any medium or format, as long as you give appropriate credit to the original author(s) and the source, provide a link to the Creative Commons license and indicate if you modified the licensed material. You do not have permission under this license to share adapted material derived from this chapter or parts of it.

The images or other third party material in this chapter are included in the chapter's Creative Commons license, unless indicated otherwise in a credit line to the material. If material is not included in the chapter's Creative Commons license and your intended use is not permitted by statutory regulation or exceeds the permitted use, you will need to obtain permission directly from the copyright holder.

Safety and Risk—Why Their Definitions Matter

Nayel Fabian Salem, Sophie Le Page, Jason Millar, Philipp Junietz, Marcus Nolte, Robert Graubohm, and Markus Maurer

Contents

© The Author(s) 2026
H. Winner et al. (eds.), *Handbook Assisted and Automated Driving*,
https://doi.org/10.1007/978-3-658-45276-6_40

40.1 Introduction

The acceptance of automated road vehicles (amongst others) depends on their level of safety [1]. With regard to this generally accepted claim, it has been historically neglected that the discussion about the safety of automated vehicles requires further differentiation.

Surprisingly, compared to the research on automated driving in general, safety assurance of automated vehicles (cf. ▶ Chap. 46) is a relatively novel discipline. In 2004, the first DARPA Grand Challenge focused on automated driving and was held in the United States. Competing teams were challenged to design an automated automobile capable of navigating a 240 km course through the Mojave Desert. No team completed the course; the most successful entry completed a mere 11.78 km of the challenge. Even though no winner was declared, and even though automated driving technology has been a research topic in academia and industry since at least the 1980s [2–4], the 2004 Challenge is often cited as a landmark event in the evolution of automated driving technology. This is, in part, because a few years later in 2009 some of the participants joined Google to launch its self-driving car program. Google's project is significant because it captured the imagination of the public, sparking a public discourse around the goals for the safety of self-driving cars. Google's self-driving project was, and still is, hailed as an opportunity to significantly improve road safety by eliminating the human driver. Perhaps the most common statistic that one finds quoted alongside such claims is that over 90% of US road accidents are caused at least in part by driver error [5, 6]. Thus, since self-driving cars eliminate errors from the human driver, it is sometimes wrongly concluded that self-driving cars could eliminate up to 90% of road accidents. In light of such safety expectations it is not surprising that self-driving cars have received so much attention in the 13 years since Google launched its project. However, in the research community of automated vehicles the issue of safety assurance was underestimated for a long time. An example can be given by how safety was treated on the IEEE Intelligent Vehicles Conference in 2016 in Gothenburg. Some researchers suggested that safety was mostly an issue for series development of automated vehicles [7].

One important contribution to the acknowledgment of safety assurance for automated vehicles (especially in the German research policy) was made with the release of the book "Autonomous Driving: Technical, Legal and Social Aspects" [8] by giving recommendations for research demands in the field of automated vehicle safety. Publications from, for example, Schnieder and Schnieder [9], Reschka et al. [10], Bagschik et al. [11], Wachenfeld [12], and others [13–17] further highlighted the importance of safety in the field of automated driving. From a German perspective, these works set the stage for research on safety assurance of automated vehicles [18]. This trend was successfully followed by the initiation of publicly funded projects such as PEGASUS[1] in 2016. Subsequently, the PEGASUS project family was established and still contributes to the field of safety assurance research in Germany with projects such as "Verification and Validation Methods for automated vehicles Level 4 and 5"[2] (VVM) or "SET Level".[3]

In 2017 the German ethics commission "Automated and Connected Driving" elaborates that "the primary purpose of partly and fully automated transport systems is to improve safety for all road users" [19]. Consequently, the priority of safety in favor of utility was defined for all levels of automated driving as constraints of the technology. Whereas the statement of the ethics commission seems to be clear, implications for the development of automated driving systems are still unclear. Grunwald [1] pointed out that the trade-off between safety and utility is a key concern for the public acceptance of automated driving. For driver assistance systems (SAE Level 0–2 [20]) it was assumed that the driver finally takes the responsibility for the vehicle's behavior and hence also handles the trade-offs involved in participating in road traffic. However, automated driving systems (SAE Level 3+ [20]) will be required to perform this task on their own, which is a new challenge for the development and operation of such systems. The novelty of this challenge leads to ongoing questions. How safe do technical agents have to be that weigh more than one ton and move in public traffic with a speed of 30 m/s and more? Which mistakes are they allowed to make without losing general acceptance of automated driving? How does the paradigm of "vision zero" and the sustained claim of research and industrial institutions of the feasibility of zero fatalities and severe accidents in road traffic influence public expectations towards automated driving? Which significance for the acceptance of automated vehicles does the common language understanding of safety as the "freedom from harm or danger" (Merriam-Webster Dictionary) [21] have, as it implies the complete absence of risks that are associated with "safe" technical systems? Please note, that for the sake of simplified public communication we do not distinguish between automated driving systems and automated vehicles in the context of this chapter. We acknowledge that the definition of the system boundary influences aspects that should be considered for system safety. However, for the fundamental discussion of automated driving safety we intentionally neglect this issue.

---

1   ▶ https://www.pegasusprojekt.de/en/home.
2   ▶ https://www.vvm-projekt.de/en/.
3   ▶ https://setlevel.de/en.

In contrast to the definition given in the Merriam-Webster Dictionary, safety engineering standards such as ISO 26262:2018 and IEC 61508:2010 define safety as the "absence of unreasonable risk" or "freedom from unacceptable risk". The semantic difference between the common definition of 'safety' and those definitions provided by engineering standards depend on what is considered "unreasonable". In safety engineering standards a clear definition of unreasonable risk is not provided. However, the guidelines in these standards in terms of arguing for safety always rely on the definition of unreasonable risk. Based on this vagueness it becomes clear that the meaning and use of the term 'safety' is of fundamental importance for assuring and hence arguing for the safety of an automated driving system (SAE Level 4+). The public acceptance of a provided argumentation of safety will finally have a major influence on the acceptance of the technology [1]. It is already acknowledged in the standardization of "safety of the intended functionality" of automated road vehicles [22] that meeting a risk acceptance criterion is required in order to deploy these systems. However, the definition of such publicly accepted criteria is still an open challenge [23] (cf. ▶ Chap. 6). Additionally, even the metrics for measuring safety are not established yet, which however is a necessary step before defining acceptance criteria.

One could ask why the fundamental discussion of safety and risk is relevant now and why it was successfully neglected (rightly or wrongly) for driver assistance systems. Unfortunately, current road traffic involves a number of fatalities and severe injuries. However, this fact seems to be accepted by the public as an inherent risk of road traffic that can be reduced but not completely eliminated [24, 25]. Such an inherent risk being accepted is far from obvious. It is the result of ongoing negotiations between the objectives of road safety and mobility of involved stakeholders. For conventional road vehicles in Germany there has been a clearly decreasing amount of accidents over the past decades [14]. However, it is unclear how this evaluation will be conducted for automated vehicles (SAE Level 3+) as a part of road traffic [17].

The aim of this chapter is to point out the lack of a common understanding of the term 'safety' where automated vehicles are concerned and to give examples of the different meanings and uses of the term from the perspectives of automotive development engineers, regulators, legislators, advocacy groups, and the public. Please note that this analysis shall not represent any of the included stakeholder groups exhaustively. Thus, we seek to provide a description of the problem based on the selected examples rather than attempting to propose solutions that harmonize the term 'safety'. Furthermore, we will provide guidance for a more cohesive discussion of safety across disciplines and stakeholder groups. One could ask why differences in the meanings and uses of 'safety' matter after all, since the more pressing question, at least in the automotive engineering community, has been "How safe is safe enough?" for some time now. We argue that the inability to provide an answer to this question by merely focusing on the engineering aspects of the problem is also based on the different answers to the question across all relevant stakeholders of: "What does safety (of automated vehicles) even mean?" Hence, the focus of this chapter is to provide evidence for the plethora of possible answers to the latter question.

Discussing safety engineering in the context of automated driving and pointing out its associated risks is often (mis)interpreted as an impeding activity. On the contrary, in this chapter we contribute to the goal of enabling sustainable acceptance of automated vehicles as participants of road traffic. Hence, we focus on showing differences in the meaning and use of the term safety between different stakeholders in ▶ Sect. 40.2. Based on our analysis we pose questions that could help facilitate an explicit and more detailed distinction between aspects of risk considered by different stakeholders in ▶ Sect. 40.3.

40.2 The Many Meanings and Uses of the Term 'Safety'

In the current landscape of standards for the safety assurance of automated driving systems (cf. ▶ Chap. 46) the term 'safety' is more specifically considered by focusing on aspects of functional safety and safety of the intended functionality (SOTIF) (cf. ▶ Chap. 6). Definitions for these specific aspects of safety are provided by ISO 26262:2018 (functional safety) and ISO/DIS 21448:2022 (SOTIF). Functional safety thereby is defined as the "absence of unreasonable risk due to hazards caused by malfunctioning behavior of E/E systems" [26] whereas SOTIF is referred to as the "absence of unreasonable risk due to hazards resulting from functional insufficiencies of the intended functionality or its implementation" [22].

Taking a more general perspective, the term 'safety' tends to refer to a state in which there is an absence of unreasonable risk [22–29]. Applying that generalized meaning of safety to automated vehicles (SAE Level 3+), a safe vehicle would be a vehicle that poses no unreasonable risk. Notice, however, the degree of ambiguity and vagueness included in that general sense of the term: it is unclear who, precisely, is otherwise at risk. It could be drivers, passengers, pedestrians who are exposed to risk. It is unclear what amount of risk would be considered unreasonable for any particular hazard. It is further unclear what specific hazards are bundled in the concept of safety. Are the hazards of physical nature? Are they associated with psychic or even environmental aspects? These details matter since to design a "safe" vehicle, or to designate a vehicle as "safe", one must agree on what, specifically, is meant by "safe".

It is important to note from the outset that safety critical functionality as a term encompasses specific risks. Determining what is understood as "safe" thus requires to clarify the kind of risks being discussed, what would be a reasonable, or unreasonable amount of risk, and who ends up being safe once all that is sorted out. This is no easy task as we will discuss in this section. There are many ways that the terms "safe/safety" are currently deployed in the context of automated vehicles. Currently, there is no consensus among stakeholders, be they engineers, regulators, legislators, advocacy groups, or the public, what would constitute a safe automated vehicle. As such, different stakeholders could mean fundamentally different things when they raise concerns or make claims about the safety of automated vehicles. This is a problem for all these stakeholders because their diverging use of the concept of "safety" could set up uneven expectations of how automated vehicles will behave. Engineers using separate definitions of safety to design vehicles could result in vehicles that behave very differently from one another, and differently from how other stakeholders, the public included, expect them to behave. Ultimately, mismatched expectations of this sort could impact the actual or perceived trustworthiness of automated vehicles, possibly leading to lower trust among regulators, legislators, advocacy groups, and the public and thus could lead to problems related to acceptance.

In this section, we analyze how different stakeholders understand and conceptualize safety by breaking down safety statements into their associated risks. Since engineering frameworks (in industry or research) for designing safe automated vehicles are explicit about associated risks in their definitions of safety, we summarize their use of the term 'safety' into the following safety statements: "A safe automated vehicle poses no [insert risk]". In contrast, other stakeholder groups such as legislators, regulators, advocacy groups, and the public media are less explicit about the risks associated with their safety statements about AVs. Therefore, we found that it would not improve the clarity of this section to include similar safety statements for other stakeholder perspectives. Please note that these chosen stakeholder perspectives are not necessarily distinctive but enable us to highlight their differences. Hence, our summaries of safety statement shall not represent the perspective of a stakeholder group exhaustively but give an example of how the term 'safety' is used sometimes.

In order to elaborate the many meanings and uses of the term 'safety' we first analyze several automotive safety engineering frameworks provided by standards or proposed in the research community

(▶ Sect. 40.2.1). With our analysis we provide examples of how the concept of safety is used among automotive engineers as one relevant stakeholder group directly influencing the design of automated vehicles. We then survey several sources to gain some understanding of how "safety" is used among other stakeholder groups (▶ Sect. 40.2.2), which directly or indirectly influence, or are influenced by, the design of automated vehicles such as regulators, legislators, corporations, advocacy groups (e.g. Mothers Against Drunk Driving (MADD)), and public media. What we find in our analysis is that safety and its associated risks are talked about differently between stakeholder groups, which may lead to mismatched expectations of what would constitute a safe vehicle. A synthesis of all possibly considered risks is not in the scope of this chapter as it would require a more thorough review of existing literature as well as qualitative and quantitative studies on relevant stakeholder perspectives. With the identification of a diverging meaning and use of the term 'safety' we aim for contributing to a more explicit discourse of risks without assuming consensus on what constitutes safe automated driving.

40.2.1 How the Term 'Safety' is Used in Automotive Safety Engineering Frameworks

In this chapter, we provide indications to support the hypothesis of mismatched meanings and uses of the term 'safety' for automated vehicles with regard to the risks that are found to be relevant for different stakeholder groups. In this section, we focus on different understandings of safety in the automotive industry and engineering research community. First, we give an example of how safety is interpreted in normative automotive safety engineering frameworks (▶ Sect. 40.2.1.1). Additionally, we show how suggested safety engineering frameworks that were published by corporate and academic research groups use the term "safety" differently. As two examples of such suggested frameworks, we elaborate how safe automated vehicles could be understood by analyzing formal safety verification approaches (▶ Sect. 40.2.1.2) and how one could include human values as a part of the definition of safety following an approach of value-sensitive development (▶ Sect. 40.2.1.3).

40.2.1.1 Established Automotive Safety Engineering Frameworks

Functional Safety and SOTIF

For the development of driver assistance systems, standards for functional safety (e.g. ISO 26262:2018)

and SOTIF (e.g. ISO 21448:2022) provide guidelines to assure these different aspects of safety. In order to analyze the meaning and use of the term 'safety' in both of these safety engineering frameworks, we provide an example in the scope of functional safety. Since the terminology used for SOTIF in the ISO/DIS 21448:2022 often times refers back to definitions from functional safety, we focus on analyzing terminology provided in the ISO 26262:2018.

The ISO 26262:2018 is an international series of standards that provides a reference framework for the automotive functional safety lifecycle. In the standard, there is a very particular understanding of safety and risk. According to the ISO 26262:2018 risk is a "combination of the probability of harm and the severity of that harm" (which is compatible with Bayesian risk [30]). Harm is referred to as "physical injury or damage of the health of persons". In order to quantify physical injuries, the ISO 26262:2018 suggests using for example the abbreviated injury scale (AIS). However, the exact extent of harm is hard to determine [31]. Hence, severity is defined as an "estimate of the extent of harm to one or more individuals that can occur in a potentially hazardous event". A hazardous event is considered as the "combination of a hazard and an operational situation". In order to separate any possible harm from the harm that is considered within the scope of the ISO 26262:2018, a hazard is defined as a "potential source of harm caused by malfunctioning behaviour of the item". The item would in this case be an electric or electronic subsystem, which is potentially part of a larger system. Furthermore, the probability of harm was mentioned in the definition of risk. This probability is constituted by the exposure to the operational situation and the probability of a system failure. The exposure of a hazardous event refers to the "state of being in an operational situation", which again is defined as a "scenario that can occur during a vehicle's life". With these terms being introduced in the ISO 26262:2018-1 [26], there is a clear focus on physical injury as the kind of risk being addressed by the standard.

In order to develop functionally safe systems based on this terminology, the ISO 26262:2018 demands to determine automotive safety integrity levels (ASIL) (cf. ▶ Chap. 6) based on a classification of identified hazards and a subsequent risk assessment. The parameters of the risk assessment are the probability of a "state of being in an operational situation" (exposure), the "estimate of the extent of harm to one or more individuals that can occur in a potentially hazardous event" (severity), and the potential for the "ability to avoid a specified harm [...] through timely reactions of the persons involved" (controllability). After identifying a hazard, it can be assessed by using these three parameters. The system design subsequently has to fulfill the derived goals with an ASIL that is based on this classification.

As an illustration of a hazard analysis and risk assessment in the concept phase of the safety lifecycle and its implications for the underlying meaning of safety, we provide an example of a safety analysis of an automated driving system that shall be operated in SAE Level 4. As the focus of this example is not the exact result of the safety analysis, but rather the workflow, we generously simplify multiple process steps.

For the operational design domain, we assume rural roads with a maximum allowed speed of 100 km/h. One hazard could be that the vehicle leaves the road and crashes into one of the neighboring trees. A possible reason could be a malfunction that leads to the complete loss of the steering function. An operational situation of driving along a curved road sequence with trees on both sides of the road has a high probability (exposure class E4) as it is highly probable (more than 10% of average operation time in the selected target market [31]) for rural roads to have neighboring trees. For the severity of the hazard, we estimate at least life threatening injuries (severity class S3) since even after executing an emergency braking maneuver, the impact speed for a frontal collision with a neighboring tree would be more than 40 km/h (suggested minimum speed for S3 by SAE J2980 [32]). As there are no persons involved in the operational situation (the system is operating on SAE Level 4) that could contribute to the controllability of the hazardous event we have to assess it as uncontrollable (controllability class C3). Based on this example of a hazard analysis, we take the recommendation from the ISO 26262:2018 and derive a required ASIL D for the preservation of the steering function. In the ISO 26262:2018 it is suggested that for example for an ASIL D function, a random hardware failure target value of less than 10^{-8} h^{-1} has to be met. This target value would be applied to hardware components in later steps of the safety lifecycle. However, with respect to the definition of safety as the absence of unreasonable risk this step is not straightforward. What is implicitly involved in determining an ASIL is first of all a risk assessment of the classified hazard. Afterwards, this estimated risk is compared with an acceptance criterion. If the risk involved in the analyzed hazard is higher than required by the risk acceptance criterion, measures to reduce the risk of the hazard are taken. These measures are specified using safety integrity levels and are applied to software and hardware components in subsequent phases of the safety lifecycle.

Based on our example, we argue that the underlying meaning of safety in the ISO 26262:2018 for this specific hazard and functionality is that the system is safe, if among others the design goal of a failure rate of (in this example) less than 10^{-8} h^{-1} is met, because it is defined as reasonable by a risk acceptance criterion. However, why is the design goal not 10^{-12} h^{-1}? Since functional safety has a legacy of systems to refer to, its risk

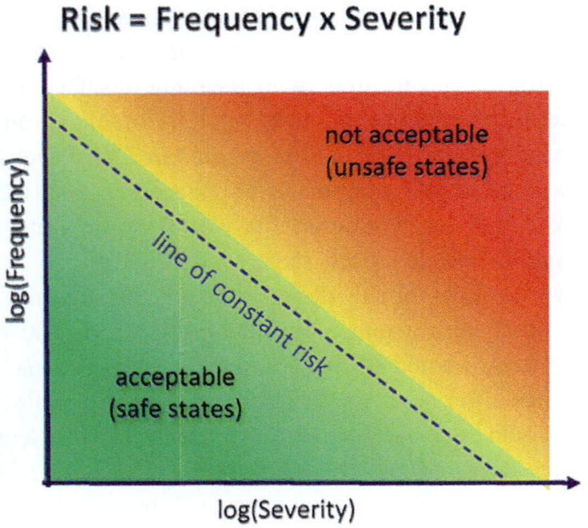

Fig. 40.1 Illustration of risk

acceptance criterion is based on a mix of proven-in-use observations and assumptions based on accident research. For the assurance of automated vehicle safety especially considering SOTIF on the other hand, it is not clear yet, which risk acceptance criterion should be chosen.

In the following section (▶ Sect. 40.2.1.1.2), we present in more detail how quantitative risk assessment is implicitly performed in the established safety engineering frameworks, explain differences in possible risk acceptance criteria such as ALARP, MEM and GAMAB, and elaborate what considerations could be included in choosing one over the other acceptance criterion.

Quantitative Risk Assessment[4]

The usual quantitative risk definition "risk equals frequency times severity" is illustrated in ▪ Fig. 40.1. Due to the double logarithmic scale, a line represents constant risk.

Considering the occurrence of unintended events (frequency) and extent of damage (severity), ideally two areas are defined. In the green area, the system is in a safe state; the corresponding risk is accepted. In the red area, the system is in an unsafe state; the corresponding risk is not accepted. The borderline between these two areas is probably not sharp; there can be a kind of transition area in between.

Although this simple definition is very useful for many questions in technology and insurance industry, it neglects aspects like aversion against high severity, lack of controllability, and personal benefit, which are relevant for risk perception and acceptance by individuals and society. Intensive research on risk percep-

tion and risk acceptance started in the second half of the last century. Different authors analyzed risk acceptance and risk–benefit constellations in various studies (Douglas [33], Crouch [34], Gibson [35] Grunwald [1], Kinchin [36], Kletz [37], Starr [38, 39], and Webb [40]). Fritzsche discussed risk acceptance relating to voluntary nature of exposure based on the above-mentioned studies [41]. Slovic concludes similar results in a more recent, updated publication [42]. It is interesting that both authors conclude similar risk numbers despite the major gap of several decades. The reason might be that risk perception studies reached their peak in the 1970s with the introduction of nuclear power. Junietz et al. [43] suggest correcting the numbers with a factor derived from the change in mortality rate.

For voluntary activities, Fritzsche [41] found that the willingness to accept risks is nearly unlimited, depending on the experienced personal benefit. We can see this by the example of high-risk sport or other leisure activities, e.g. free climbing or motorcycling. Job-related activities are important for a deeper understanding of the subject. Acceptance is relatively well investigated in this field and there is a common understanding of accepted individual mortality risk in the order of 10^{-5} per person and year, for example by professional associations and insurance companies, on the one hand. On the other hand, job-related risks are useful to bridge the gap between voluntary and involuntary risks.

Fritzsche [41] found that for involuntary risks, e.g. death of passengers due to a train or airplane crash, the acceptance level is an order of magnitude lower than for job-related risks. Moreover, acceptance decreases another order of magnitude, if the risk is caused by major technology, e.g. chemical industry or nuclear power generation. Beside the fact that the experienced personal benefit of those technologies is low (at least from a subjective point of view), the low degree of self-determination or rather controllability by individuals plays an important role for the low acceptance level as well as the potentially high number of mortalities (severity). Nevertheless, the summarized studies show that it is generally possible to deal with risk, risk perception, and acceptance in a quantitative manner.

To implement safety requirements based on risk acceptance, several concepts have been developed in different application areas. As there is a relationship between railway and road traffic, it is useful to refer to the CENELEC safety standard EN 50126. The development of this standard has been started in the 1990s where safety requirements based on quantitative risk analysis have been implemented and ALARP, MEM, and GAMAB have been introduced as principles for risk acceptance. Those principles are briefly explained in the following sections.

---

4   This section is largely adopted from Junietz [17].

☐ Table 40.1 Accidents on German highways [47]

Severity	ISO 26262 severity level	Average distance between two accidents of this level (km)	Accident rate per driven distance
Fatal	S3	660×10^6	1.52×10^{-9}
Severe injuries	S2	53.2×10^6	1.88×10^{-8}
Injuries	S1	12.5×10^6	8.00×10^{-8}
W/O injuries	S0	7.5×10^6	1.33×10^{-7}

As Low as Reasonably Practicable (ALARP)

ALARP tries to assess what is technically feasible considering economic sense and socially acceptable. Between the two regions of generally unaccepted and broadly accepted risk, there is a tolerance range where risk is undertaken only if a benefit is desired and where each risk must be made as low as reasonably practicable.

Derivation of risk figures is not directly applicable as EN 50126 failed to give certain values for generally unaccepted and broadly accepted risk. However, other authors, for example Risk & Reliability Associates, deliver both values [44]: The two key levels seem to be located around road death statistics (about 10^{-4} per person and year) and the chances of being struck by lightning (about 10^{-7} per person and year). If something is more dangerous than driving a car, the risk is unacceptable. If something is less dangerous than being struck by lightning, it is not required to be reduced further. In the range between these two figures, cost benefit studies are appropriate to reduce the risk to as low as reasonably practicable. Especially this lower ALARP limit corresponds very well with the acceptance criterion for major technology risks shown in ☐ Fig. 40.3.

Minimum Endogenous Mortality (MEM)

MEM is based upon age- and gender-specific mortality rates [45]. Although the absolute values of the mortality rates change with birth cohort, they show a typical development over age as well as a significant minimum at an age of about 10 years, which is about $5 \cdot 10^{-5}$ per person and year. The related mortality at an age of 10 years is defined as "minimum endogenous mortality".[5] The MEM principle demands that a new system does not significantly contribute to the existing minimum endogenous mortality. EN 50126 specifies that the individual risk due to a certain technical system must not exceed 1/20th of the minimum endogenous mortality, considering that people are normally exposed to the risk of several technical systems. This means that the acceptable individual risk of a certain technical system is $2.5 \cdot 10^{-6}$ per person and year, when

using latest mortality rates as a basis (EN 50126 uses mortality rates from the 1980s). This value corresponds very well with the acceptance criterion for involuntary risks shown in ☐ Fig. 40.3.

Globalement Au Moins Aussi Bon (GAMAB)[6]

GAMAB, (or GAME globalement au moins équivalent), requires, unlike MEM, the existence of a reference system with—currently—accepted residual risks. According to GAMAB, residual risks caused by a new system must not exceed those of the reference system. That is, a new system must offer a level of risk generally at least as good as the one offered by any equivalent existing system. Hence, it is necessary to identify the risk of an equivalent existing system.

Looking for the acceptable risk of automated driving (SAE Level 3+) in a certain application area according to GAMAB, the current risk of an equivalent existing system in the same application area needs to be identified. To derive acceptance requirements for a controlled-access highway pilot, the current risk on German highways (so called "Autobahn") during manual driving is analyzed. ☐ Table 40.1 shows average distances between two accidents referring to severity levels according to ISO 26262 [46]. However, for other Level 3+ systems, there might be no reference system as today's vehicle fleet does not cover the use-case.

☐ Figure 40.2 shows the observed accident rates on German highways versus severity levels according to the ISO 26262. Comparing risks of the different severity categories requires weighting of the different levels. However, there is no standardized way (see also Baum [48], Hydén [49], and Wachenfeld [12]). The lines of constant risk in ☐ Fig. 40.2 assume that the difference between adjacent severity levels is one order of magnitude, or in other words that an accident with fatalities is assessed with a ten times higher severity than an accident with severe injuries. However, that does not mean that a fatal accident is definitely ten times worse. The factor of ten fits to the current accident frequency, which could change in the future e.g. due to enhanced passive safety. However, this assumption allows defin-

---

5   Although the mortality statistics does not differentiate between endogenous and exogenous causes for the death.

6   English: generally at least as good as.

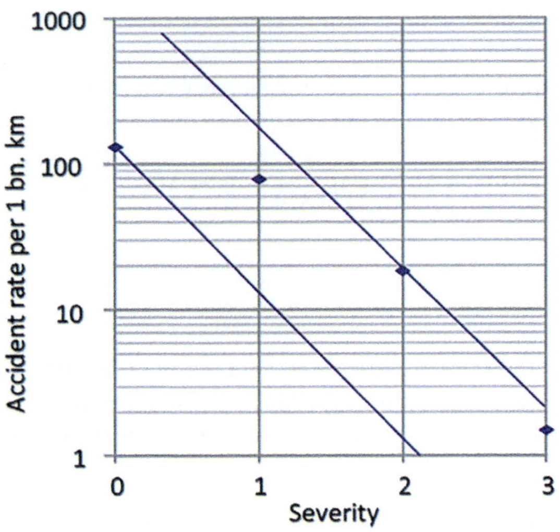

◘ Fig. 40.2   Quantitative accident risk on German highways [47, 50]

ing a band of constant risk in current traffic, which can be used as reference. As already discussed, the risk will not be accepted above the upper envelope. Beyond the lower envelope, the risk might be accepted. Between both lines, there is a transition area. In accordance with the ALARP principle, this is a tolerance zone where risk is undertaken if a benefit is desired and where each risk must be made as low as reasonably practicable.

In ◘ Fig. 40.3, the results of the application of the different risk acceptance principles are displayed related to the risk acceptance limits of the different exposions explained previously. The different approaches for the mortality risk are summarized. On the one hand, it shows that the application of different risk acceptance principles delivers comparable and consistent results. On the other hand, it demonstrates that we have to deal with a relatively broad range of applicable acceptance criteria.

Considering the impact of voluntary exposure, different groups of users have to be distinguished, for example users of automated driving systems and other traffic participants. Finally, a comparison with other technologies—especially other traffic systems and technologies that deliver a high personal benefit—seems to be useful. Additionally, a decrease in total mortality risk is expected in the future, following the trend in the last decades and centuries. Therefore, risk acceptance might change over time.

Summary

In this part of our analysis, we gave examples of normative frameworks in automotive safety engineering and examined how they describe and address safety. There seems to be an agreement in frameworks like functional safety (ISO 26262:2018) and SOTIF (ISO/DIS 21448:2022) that the risks associated with a safe system are related to physical injuries. As we will do for the other frameworks and perspectives we consider in this chapter, we summarize the respective meaning of safety for automated vehicles:

❯ A safe automated vehicle poses no unreasonable physical harm to persons involved in an operational situation; unreasonable is defined as an agreed upon maximum probability of specific severities of harm.

Additionally, we pointed out that safety engineering involves including retrospective acceptance criteria like for example a number of acceptable injuries. Such criteria are used as part of the assessment of prospective estimations in terms of exposure, severity and controllability. Within this process, there is also the task of changing the scope in the analysis. Risk acceptance criteria refer to the entire operational environment on a macroscopic level. For example, the

40

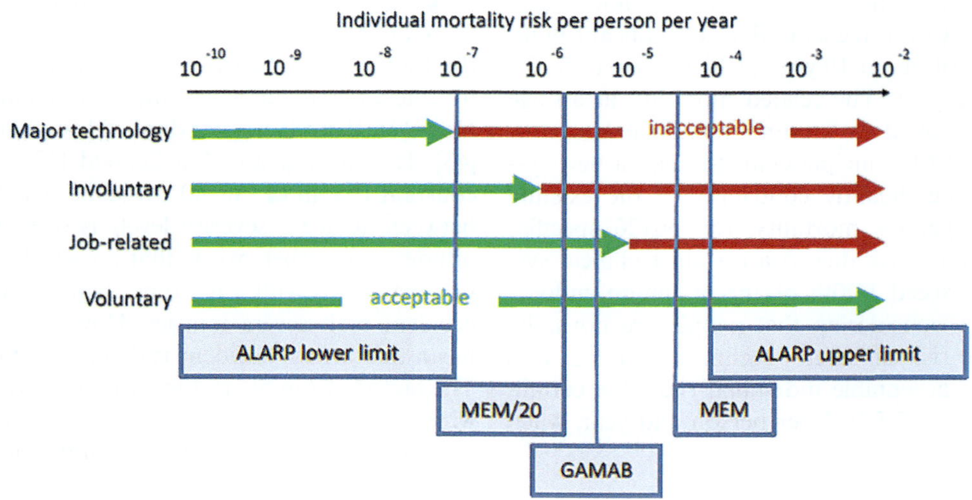

◘ Fig. 40.3   Application of different risk acceptance principles to highway accidents, translated from Steininger and Wech [51], based on Fritzsche [41], GAMAB is based on the risk on German controlled-access highways and MEM is based on the current mortality tables

German Ethics Commission on automated and connected driving raised the goal of a positive risk balance [19] and generally refers to the guiding principle of avoiding accidents. However, it is still unclear how to account for this goal in every considered scenario [52] on a microscopic level. What the positive risk balance already defines is the reference for what makes an automated vehicle safe. A positive risk balance or "safety" is thus achieved when an automated vehicle introduces less risk with respect to accidents than human drivers.

40.2.1.2 Frameworks for Formal Safety Assurance

As automated driving introduces new challenges for safety assurance in an open traffic context (cf. ▶ Chap. 46) several possible solutions have been proposed in the research community. We present one group of approaches for assuring safety in an open traffic context in this section. The concept of formally proving safety is generally realized by agreeing on basic rules and implement their formal representations as part of the safety assurance process. Approaches that realize such a safety assurance strategy were presented for example by Shalev-Shwartz et al. [53] with Responsibility Sensitive Safety (RSS), Nistér et al. [54] with Safety Force Fields (SFF) or Althoff [55] with the utilization of reachable set analysis. For an in-depth presentation of the approaches, we refer the reader to the respective cited sources. In this section, we focus on the underlying meaning of safety and thus only provide an overview of the selected approaches. As an illustration, we will furthermore provide an example of how the interpretation of formally provable safety is manifested in a scenario using the RSS framework.

In their introduction of Safety Force Fields (SFF) Nistér et al. [54] limit their notion of safety to the avoidance of colliding with obstacles. By conceptually separating the nominal behavior planner from a distinct safety behavior planner that could rely on Safety Force Fields, they argue that a vehicle would never contribute to an unsafe situation. Given this condition, Nistér et al. [54] define: "If all actors do what they are required, no collisions can occur." Hence, this use of the term 'safety' is clearly limited by the acceptance of the assumption, that a safe vehicle is one that never contributes to an unsafe situation.

Althoff [55] relies on a similar notion of safety: "A major constraint for the trajectory planner is that the generated trajectories have to be safe, i.e. static and dynamic obstacles must not be hit." However, in his approach Althoff [55] proposes to use stochastic models (reachable sets) to proof that no collision is possible in a given situation. When comparing the notion of safety from Althoff's work [55] to Nistér et al. [54], it

becomes clear that both offer closed solutions for handling safety in an open context by relying on necessary assumptions for their proven models. Making assumptions in order to formally prove safety (or any other property for that matter) cannot be attributed to their specific approaches but is an inherent challenge of using closed mathematical models in an open world.

Shalev-Shwartz et al. [53] introduce Responsibility Sensitive Safety (RSS) and acknowledge necessary trade-offs for automated vehicles acting in an open traffic context. In comparison to established safety engineering frameworks Shalev-Shwartz et al. [53] propose that instead of identifying particular hazards and quantifying the associated physical risks, five "common sense" rules can be formulated and proven mathematically. There are multiple implications of this strategy with respect to the meaning of safety. First, there is an implicit representation of the risks that are considered by RSS. High-level rules try to generalize different kinds of risks to different stakeholders. Therefore, the traceability of stakeholder concerns and development assumptions to their implementation in the behavior of the automated vehicle in road traffic is limited. Second, RSS requires stakeholders to agree on the parameters (just like the assumptions made by Nistér et al. [54] and Althoff [55]) that are used in the formal proof of such kind of safety. Finally, when automated vehicles would be deployed adhering to RSS, then the argument by Shalev-Shwartz et al. [53] suggests that in case of an accident where an automated vehicle behaves compliant with the defined RSS parameters, the responsibility would not be ascribed to the vehicle or respectively the car manufacturer. In contrast to established automotive safety engineering frameworks, there is no explicit integration of risk acceptance criteria. We show what this means for the understanding of safety using RSS by giving a simplified example.

In ◼ Fig. 40.4 Error! Reference source not found. the blue automated vehicle drives from left to right and passes a row of parked vehicles. Oncoming traffic is represented by the red vehicle and added to limit the degrees of freedom of the blue vehicle to the longitudinal speed. Designing a vehicle for such scenarios includes trading of mobility and risk [25]. Merely obeying the speed limit could mean that the mobility of the passengers is as high as the legal traffic code allows. However, by including the uncertainty about a child potentially running on the street between the parked cars and onto the road makes it difficult to argue merely for legal responsibility. In Germany, the argumentation would be even more challenging as the "principle of legitimate expectations" [translated by Salem, Vertrauensgrundsatz] does not apply to children and persons, who are obviously impaired with respect to acting in or perceiving their environment. The uncertainty in the open traffic context leads to an inherent risk of

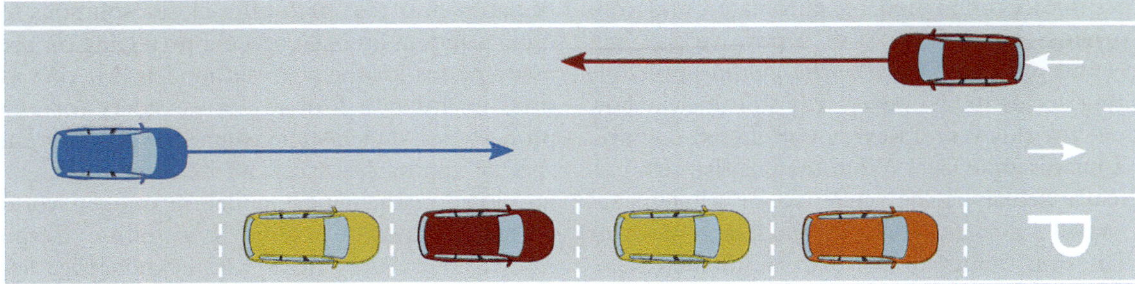

◘ Fig. 40.4 Example of a scenario where the automated vehicle passes a row parked vehicles with possibly occluded pedestrians

participating in road traffic. In the scenario, this uncertainty is represented by the potentially crossing child behind one of the parked vehicles.

RSS proposes a framework for engineers and regulators to agree on a parametrization (in this scenario implying a certain speed) that is associated with an accepted residual risk. RSS acknowledges that trade-offs have to be made due to uncertainties in the open context. However, a consequence of applying the RSS framework to this scenario implies that there is a speed that would exempt the blue vehicle from any (legal) responsibility for an unavoidable collision with a crossing child. Hence, RSS assumes that the residual risk in traffic resulting from the chosen parameters is accepted by all stakeholders.

Even though each of the selected approaches towards formal safety assurance suggest different frameworks to prove system safety, they generally rely on a similar meaning of safety. Implicit trade-offs with respect to the risks considered are a fundamental assumption that approaches such as RSS, safety force fields or reachability analysis make by their very nature of attempting a formal proof of safety. Based on our analysis, we suggest an explicit summary of how the term 'safety' is used in these frameworks:

> A safe automated vehicle is one that poses no risk of violating a formally proven model; the validity of the model is based on agreed assumptions.

This poses challenges for drawing conclusions for our initial questions. Who is at risk? What kinds of risks are addressed? With respect to which reference is the system safe? However, with regard to mismatched expectations between stakeholders it seems important to be explicit about assumptions and trade-offs that are made with respect to safety [23].

40.2.1.3 Value-Sensitive Development Framework

While formal methods for proving safety considered in the context of this chapter do not offer explicit trade-offs, in this section we introduce a framework for value-sensitive development that acknowledges the importance of making trade-offs with respect to safety and risk in an explicit manner. From the perspective of value-sensitive development, explicitly weighing values is important because the risks that are associated with the meaning of safety otherwise are unclear to the relevant stakeholders. Hence, value-sensitive development includes stakeholder values explicitly and provides methods to design for safety, while acknowledging risks of violating stakeholder values when making trade-offs. As the term 'value' is associated with terminological discussions, we do not attempt to define the term universally but we rather show how the term is used within a value-sensitive development framework.

A common definition of the term 'value', which is also referred to by Ulbrich [56] in his work on value-sensitive behavior planning, is given by Kluckhohn [57].

> A value is a conception, explicit or implicit, distinctive of an individual or characteristic of a group, of the desirable which influences the selection from available modes, means, and ends of action. [57]

What remains unclear is, whether the 'conception' of a value has an absolute nature or if it can also be expressed relatively as a result of a weighing process. Albert [58] elaborates the differences concerning the terminology of a "value" from a psychological, sociological and cultural science perspective and proposes a more detailed definition.

> A value as a measure with an adaptable weight and arbitrarily in relation with further measures alike constitutes a value pattern that reduces complexity for stakeholders and determines actions within situational possibilities variously direct. (translated by Salem, based on Albert [58])

In the context of value-sensitive development this definition of values accounts for the need of weighing stakeholder values depending on the situational context as it differentiates between value patterns and values. Hence, based on the definition by Kluckhohn [57] and Albert [58], values can be understood as objectives desired by stakeholders that can be weighed against each other in a given context.

After having discussed the interpretation of the term 'value' in the context of value-sensitive development, we will focus on its relation to the term 'safety' and 'risk'. Value-sensitive development, just like established automotive safety engineering frameworks (▶ Sect. 40.2.1.1), interprets safety as the absence of unreasonable risk. However, the argumentation for reasonable design decisions is not enabled by fulfilling an abstract risk acceptance criterion but by explicitly documenting trade-offs between stakeholder values and formally representing them throughout the product lifecycle of an automated vehicle. In contrast to for example the work by Thornton [59] and Ulbrich [56] safety is therefore not understood as a value but as a result of weighing the violations of stakeholder values.

> A safe automated vehicle is one that poses no unreasonable risk of violating stakeholder objectives (or values).

As a result of associating safety with a reasonable risk of violating stakeholder values, risks can not only be related to physical harm (cf. ▶ Sect. 40.2.1.1) as stakeholders could desire more than physical wellbeing. This notion of safety including risks beyond physical harms is also in line with the kinds of risks included in the concept of safety in the UL 4600, and ISO 31000:2018. Risk as "a combination of the probability to occurrence of a loss event and the severity of that loss event" is defined in the UL 4600. A loss is further defined as "human death, human injury, animal death, animal injury, property damage, environmental damage, or other substantive adverse outcome". With regard to the losses that should be considered when designing safe automated vehicles, UL 4600 enables the discussion of what could be understood as "substantive adverse outcome". The example of including financial cost caused by a failure already indicates that risks beyond physical nature could be included in the safety case.

ISO 31000 further defines risk as the "effect of uncertainty on objectives". This definition enables an even broader consideration of risks that should be addressed in a safety case for automated vehicle. However, note that ISO 31000 is originally intended as a guideline for risk management in organizations, where the assessment of, for example, fatal injuries plays a subordinate role. We apply this meaning of safety to the context of automated vehicles in order to show a potential mismatch within organization's risk management as they introduce this technology.

As a result, in value-sensitive development the meaning of the term safety is associated with risks related to the objectives of stakeholders such as mobility, physical wellbeing, or legal compliance. Because these values or objectives are represented explicitly in the design process it is possible to trace them to the behavior of an automated vehicle. Explicitness of values enables stakeholders to evaluate, whether the behavior could be considered safe from their perspective. However, stakeholder values are highly context dependent. That is, whereas in the scope of one scenario a value trade-off could be considered safe, it might be found unsafe within the scope of the entire operating environment when considering (possibly additional) stakeholders and their values. Hence, a clear definition of the scope is necessary when making trade-offs between values.

In order to describe challenges of integrating the notions of values and safety and since value-sensitive development is not an established framework for handling safety, we introduce essential activities that are part of a value-sensitive development framework (◘ Fig. 40.5) and afterwards reflect the respective meaning and use of the term 'safety'.

The foundation of value-sensitive development is the acknowledgement of stakeholder values that shall be represented in the behavior of an automated driving system. IEEE Std 7000™-2021 further elaborates how values can be elicited and explicitly considered as part of the design process of any product or service. In the context of value-sensitive development the public discourse should be performed throughout the lifecycle of an automated vehicle. Initially, the public debate serves as a basis for the stakeholder related analysis of values and, if possible, the elicitation of context specific requirements. As a prerequisite of releasing automated vehicles will be the definition of acceptance criteria, which will likely be documented as part of standards or laws, this explicit exchange of values has to be performed before the release of the technology. Furthermore, acceptance criteria are volatile. They can change due to specific events as it was observable for nuclear power plants e.g. in Germany, and might change over time as automated vehicle technology matures.

Throughout the design and operation of the automated vehicle, stakeholder values build the reference for validation of the implemented behavior. In order to integrate stakeholder values into the development process, Graubohm et al. [60] present a value-sensitive design [61] inspired approach for the concept phase of an automated vehicle. The iterative nature of value-sensitive design is explicitly built into the process model and facilitates the adjustment of value trade-offs. How these trade-offs are performed will be highly dependent on the defined acceptance criteria and possibly even company culture. Hence, a traceable and transparent documentation of assumptions and design decisions is required in order to be able to argue for sufficient safety. A framework for performing value trade-offs is for example described in the IEEE Std 7000™-2021 Standard.

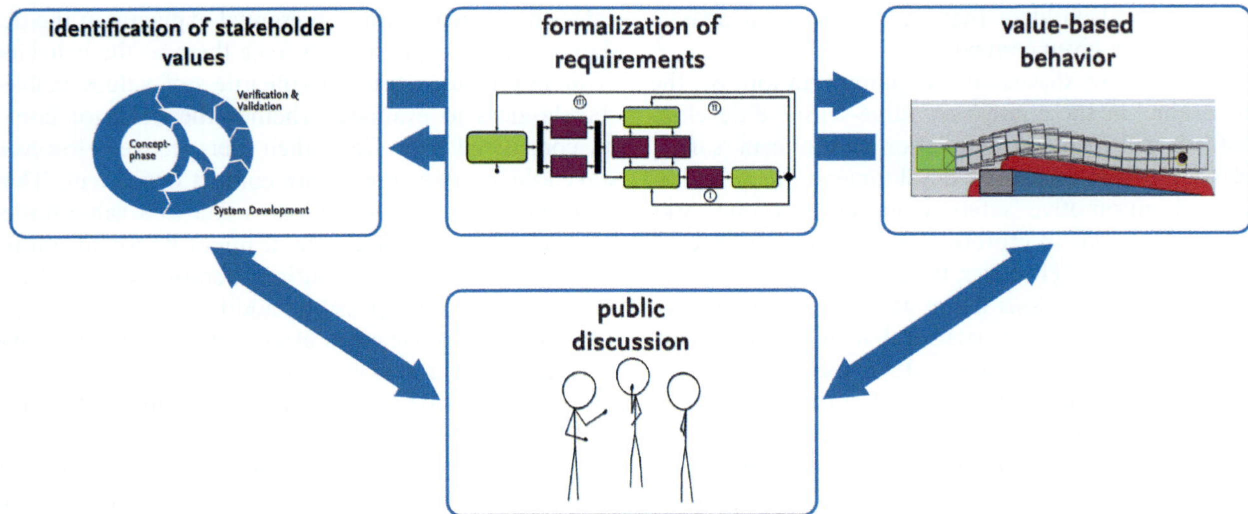

Fig. 40.5 General framework for value-sensitive development of automated vehicles

Since values tend to be abstract and difficult to generally weigh against each other without a given context [62, 63], Salem et al. [64] and Beck et al. [65] introduce the semantic norm behavior analysis and phenomenon signal model as methods to contribute to a context specific behavior specification. The authors use a scenario-based analysis to define concepts and rules for a given scenario and finally validate the formally inferred target behavior against the elicited values, or in their specific approach, behavioral norms. With respect to weighing values, Millar et al. [63] perform a context-specific assessment of values and necessary trade-offs within stakeholder workshops. Hence, the elicitation of values, trade-offs and their respective integration in a formal behavior specification could for example be performed by combining the aforementioned approaches. The scalability of such a proceeding, especially in the context of assuring safety of automated vehicles, remains to be analyzed.

Another aspect of value-sensitive development is the implementation of the specified vehicle behavior. For the tactical formulation of weights and constraints within a Model Predictive Control framework, Nolte et al. [66] present an architectural approach that enables a traceable integration of target behavior in the design of an automated vehicle. In order to assess this implemented behavior with regard to reasonable risk—or violation of stakeholder values—a transparent communication of the safety argument and the included value trade-offs will be required. Finally, all of the aforementioned approaches can contribute to an overarching value-sensitive development framework.

While giving a brief overview of a value-sensitive perspective on safety, we first elaborated a respective meaning and use of the term safety and subsequently discussed possible challenges in the implementation of a value-sensitive development framework. These challenges include aspects like an uncertain scalability of the overall approach as values and especially value trade-offs are context-dependent. Furthermore, integrating the public in debates on acceptance criteria regarding values can lead to development uncertainties for companies as they require very rigorous and obligatory agreements especially regarding safety. An additional obstacle with respect to integrating explicit value trade-offs in a notion of safety is the necessary transparency in a safety argument. Since there is not only technical but especially legal and ethical considerations included in such trade-offs it is a foreseeable challenge for the communication of companies developing automated vehicles.

However, the uncertainty involved in assuring safety of automated vehicles will inherently result in scenarios in which developers will be required to argue for specific design decisions and assumptions that lead to the observed behavior (cf. example in ▶ Sect. 40.2.1.2). Thus, a value-sensitive development approach can extend an argumentation for safe behavior (especially in edge cases).

40.2.1.4 Summary

In this section we discussed the meaning and use of the term safety in three engineering frameworks for designing safe automated vehicles. First, we showed how established frameworks for assuring functional safety and SOTIF take an accepted risk criterion that is set as a design goal, which are expressed as safety integrity levels in the case of functional safety. Current risk acceptance criteria used or those that could be used in the future focus on defining a level of risk in terms of the probability of physical harm that is considered reasonable. Thus, the assurance of a certain safety integrity, which is based on a classification of the probability and severity of physical harm associated with specific hazards, is sufficient to classify a system as 'safe' in the scope of functional safety.

Next, we gave examples of formal methods for safety assurance and showed what they mean by "safety". Based on the approach of proving safety mathematically, there are multiple assumptions being made by for example the frameworks of Safety Force Fields [54], Reachable Set Analysis [55] and Responsibility Sensitive Safety [53]. These assumptions need to be agreed upon as being reasonable in order to deem a system as being safe. However, we elaborated that such approaches do neither account for the consideration of risk acceptance criteria nor an explicit documentation of trade-offs that are involved in specifying safe behavior of an automated vehicle.

Finally, we introduced the framework of value-sensitive development, where safety is a state in which stakeholder values are being at reasonable risk. By adding further categories of risk into the notion of safety that extend the focus on physical wellbeing to additional values such as mobility, a value-sensitive development framework acknowledges the importance of value trade-offs when specifying safe behavior. We also highlighted challenges like uncertainty regarding scalability, transparent risk communication, and the challenge of agreeing on acceptance criteria while accounting for multiple stakeholder values.

40.2.2 How the Term 'Safety' is Used by Other Stakeholders

In the previous section we analyzed different uses of 'safety' in the automotive safety engineering community, focusing on 'safety' as a concept that acknowledges the existence of residual risk. According to the analysis so far, the risks included in each definition can differ from one definition to another, leading to a different meaning of 'safety' in each case. We extend this analysis here and survey several credible and influential sources to gain an understanding of how the term 'safety' is used in relation to automated vehicles among four other stakeholder groups: regulators and legislators, corporations, advocacy groups and public media. Note that we include examples of statements from these groups and do not claim to comprehensively represent all relevant perspectives. We organize and analyze how the term 'safety' is used by a stakeholder group (at least partially) because the perspective may have an impact on the type of risks included in a safety claim about automated vehicles.

What we find in our analysis in this section is that the definitions of 'safety' are sometimes stated explic-itly, but are also often implied, along with the associated risks. Sorting out what is meant in any safety claim related to automated vehicles thus requires some further guesswork or analysis. What we also find in our analysis is that safety is talked about differently when comparing statements from one stakeholder group to another group. However, while safety statements differ within a single group, there are also several similarities in how safety is talked about within a group.

40.2.2.1 Legislators and Regulators

In 1997, the Road Traffic Safety Bill founded on Vision Zero was passed by a large majority in the Swedish parliament. The Vision Zero policy sets out aspirational goals in that "eventually no one will be killed or seriously injured within the road transport system", while "long-term disabling injuries, and non-injury accidents are more or less outside the scope of the Vision" [67]. In this goal, The Vision Zero policy is making a safety claim, and is explicit about the associated risks that are included and excluded in the claim. We can understand that the policy is making a safety claim that accounts for the risk of any serious injury or lethal accidents and neglects risks associated with less serious or non-injury accidents. We can also understand that any probability of occurrence of a serious injury or lethal accident is unreasonable in Vision Zero's safety claim.

More recently in 2020, a project and report was undertaken by The Expert Panel of Connected and Autonomous Vehicles and Shared Mobility, which was assembled by The Canadian Council of Academies upon request by the federal institution that leads the Industry, Science and Economic Development portfolio in Canada. In the report, The Expert Panel presents road safety policy recommendations based on the statement that "fully autonomous vehicles could further reduce injuries and fatalities by eliminating human error in driving, which is a factor in over 90% of collisions" [68]. This statement in part draws on a widely cited statistic from the United States National Highway Traffic Safety Administration [6], which is further elaborated on NHTSA's website page discussing the safety benefits of AVs: "Automated vehicles' potential to save lives and reduce injuries is rooted in one critical and tragic fact: 94% of serious crashes are due to human error [in driving]. Automated vehicles have the potential to remove human error [in driving] from the crash equation" [69]. In this statement from NHTSA, which informs policy recommendations and statements such as those made by The Expert Panel, NHTSA is making a safety claim that is different from Vision Zero's claim. What we can understand from NHTSA's safety claim is that it addresses risks of serious injury or lethal accidents caused by human error, whereas Vision Zero does not focus on whether accidents are caused by human error. We can also understand that NHTSA's safety claim considers severe or lethal accidents with a probability of occurrence that is lower than a referenced probability for manually driven vehicles, whereas Vision Zero's claim addresses severe or lethal accidents with no probability of occurring.

NHTSA also mentions risks of injuries and crashes in its mission to "Save lives, prevent injuries, reduce vehicle-related crashes". However, in other major statements by NHTSA, it is not clear what NHTSA means by the term safety. For example, during a keynote speech at the Automated Vehicles Symposium, NHTSA's Deputy Administrator stated that the development of automated vehicles will "promote safety" without further clarifying the term safety [70]. Distinguishing NHTSA's safety claim from others' safety claims is thus not straightforward in this case.

In Germany the traffic law requires automated vehicles to transition into a minimal risk condition in order to ensure the highest possible traffic safety for passengers, other traffic participants, and third parties in a given situation. However, it is not further specified, which risks should be minimized exactly. Thus, we assume that, based on the terminology used in several paragraphs of the German traffic law (e.g. StVG sec. 40.1e), the considered risks account for physical injuries due to accidents. Further challenges in the context of minimal risk conditions become evident by comparing the use of the term 'safety' between a legal and a technical perspective. The ISO/TR 4804:2020 describes a minimal risk condition as a condition of an automated vehicle that can be transitioned into by executing a minimal risk maneuver. Given the potential mismatch of a condition with an estimated minimal risk and the condition actually associated with a minimal risk, based on ex post data analysis [71], the definition of appropriate minimal risk maneuvers becomes even more challenging. Thus, fulfilling the legal requirement of transitioning into a minimal risk condition requires a clarification of how the definition of such a condition would be assessed retrospectively.

40.2.2.2 Corporate Statements

Corporations from the automotive industry have made several claims about the expected safety of their vehicles. For example, Tesla's Vehicle Safety Report states that "That's why Tesla vehicles are engineered to be the safest cars in the world… [and] have achieved the lowest overall probability of injury" [72]. From this statement, Tesla is making a safety claim which includes the explicit risk of road injuries. We can also guess implicitly that Tesla's safety claim is associated with the probability of a road injury occurring that is lower than a referenced probability of occurrence.

Other corporations such as Waymo are more explicit in their Safety Report about what they mean by the term safety. In the Glossary of Waymo's Safety Report, safety is defined as "Freedom from those conditions that can cause death, injury, occupational illness, damage to or loss of equipment or property, or damage to the environment" [73]. The risks associated with Waymo's safety claim are explicit and account for the risk of death or injury which are similar to the risks addressed in the previous safety claims that we have reviewed so far. Furthermore, we can understand that from Waymo's use of the word 'freedom' in their safety claim, their claim includes risks with no probability of occurrence, which is similar to Vision Zero's focus on the absence of risk in their safety claim. The risks associated with Waymo's safety claim differs from the previous claims in that the risks are associated with "conditions" (such as accidents) that can cause death or injury, thus capturing a broader list of risks from the previous claims which are associated with just accidents. The risks associated with Waymo's safety claim also include other risks than physical injuries, such as occupational illness (possibly referring to health issues beyond injuries), damage to equipment or property (possibly from the result of a collision between vehicles or between vehicles and the surrounding infrastructure), and damage to the environment (such as perhaps pollution caused by the accumulation of vehicle exhaust systems on roads). Overall, we can understand that Waymo is including further risks in their safety claim associated with AVs compared to the previous safety claims we have reviewed so far.

Volvo Group on the other hand is not explicit about what they mean by the term safety. For example, in a column on Volvo's website, The Safety Director at Volvo Group wrote about whether automated vehicles will improve safety without any further clarification of what the term safety means [74]. Distinguishing Volvo Group's safety claim from others' safety claims is thus not straightforward.

40.2.2.3 Advocacy Groups

Advocacy groups such as Mothers Against Drunk Driving (MADD) Canada support the development of automated vehicles to improve road safety. The President of MADD Canada communicated this support stating that automated vehicles will "eventually eliminate impaired driving completely" and that "many lives will be saved due to the reduction in impaired driving and driver error" [75]. The implicit risk included in this safety claim is likely that impaired driving can lead to severe or lethal accidents. We can also understand that the risk of impaired driving is associated with a probability of occurrence that is reduced compared to a referenced probability. Similar to NHTSA's safety claim above, the safety claim from Mothers Against Drunk Driving includes the risk of an accident associated with driver error, given that impaired driving can be understood as a type of human error in driving.

Other organizations such as End Distracted Driving (EndDD) have talked about the potential of automated vehicle features to improve road safety. In an article on their website, the organization makes the safety claim that "autonomous features will still be there to protect all of us […] against distracted driving" [76]. In a second article, the organization also states that automated driving features could "improve their vehicles' ability to detect other cars in the road" as well as "protect [distracted] pedestrians from themselves" and "spot pedestrians and cyclists from a few yards away, predict their trajectory, sound an alarm and, if necessary, activate the brakes if you get too close" [77]. The implicit risks associated with the safety statement in the first article is likely that, similar to the statements reviewed above from Mothers Against Drunk Driving about impaired driving, distracted driving can lead to severe or lethal accidents. Furthermore, the safety statement in the second article seems to include an implicit risk associated with the limits of human driver reaction time to compensate for other distracted drivers or distracted pedestrians.

Advocacy groups such as the National Safety Council (NSC) also spoke about distracted driving in their statements about the potential of automated vehicles to improve road safety. In a Testimony before the Senate Appropriations Committee Transportation, Housing and Urban Development, and Related Agencies Subcommittee, Deborah A.P. Hersman, the former President and CEO of NSC and the former Chief Safety Officer at Waymo, stated that automated driving features would "allow a vehicle to communicate with a red light to compensate for a fatigued driver, stop a car to prevent a collision with a pedestrian if a driver fails to detect him or her, and prevent or mitigate collisions between vehicles" [78]. From this statement, we can understand that the risks associated with safety include risks of collisions (either with a pedestrian or with another vehicle) associated with distracted or fatigued driving.

The organization Let's Talk about Autonomous Driving (LTAD) further represents a diverse set of communities which include Mothers Against Drunk Driving and the National Safety Council, as well as the Foundation for Senior Living, the Foundation for Blind Children, and Students Against Destructive Decisions. On LTAD's website, the organization states that these communities share the belief that "autonomous driving cars can save lives" [79]. In an article on LTAD's website about the links between automated vehicles and safety/stress, the organization further talks about the potential safety benefits of automated vehicles "to reducing stress, road rage, and reckless driving"

[80]. The organization further cites evidence that suggests that the stress of driving in traffic can harm a person's psychic well-being. Thus the organization's safety statement about stress, road rage, and reckless driving addresses risks of psychic harms.

40.2.2.4 Public Media

The public media such as Wired and Canadian Broadcasting Corporation (CBC) are releasing stories about safety concerns associated with automated vehicles. In a story released by Wired about the links between automated vehicles and safety/health concerns, Peter Norton, a transportation historian at the University of Virginia says "Why do we care about safety? Because we care about human health" [81]. Norton also warns that the automated future will result in less walking, more sprawl, and—without dedication to electric propulsion—a huge spike in pollution. The article goes on to cite evidence that suggests sedentary behavior and pollution associated with driving are linked to health risks of cancer and heart disease. The safety concerns associated with AVs in this article differs from the previous statements we have reviewed, which tend to focus on the safety benefits of AVs in removing human drivers from the driving task. While human driven cars kill people in crashes, the article states that cars driven on roads also lead to sedentary behavior and pollution, and that depending on how AVs are designed there could be an increase in cars being driven on roads. Overall, from the statements made in this article we can understand that the risks associated with the safety of AVs include risks of cancer and heart disease due to sedentary behavior and pollution.

Another story released by the CBC talks about the links between automated vehicles and safety/security concerns, where the term security tends to be related to the term safety in that security addresses the risk of harm due to intentional criminal acts. In the CBC article, Sebastian Fischmeister, a Professor in the Department of Electrical and Computer Engineering at Waterloo University in Canada, says "Now that all these vehicles are connected, you have the risk of someone sitting somewhere and through the internet and through connectivity reaching out to all vehicles of a particular manufacturer," and that "That's why safety and security are the key challenges" [82]. Here we understand that Fischmeister is including risks of hackers in their concept of safety, suggesting that automated vehicles are increasingly being built with more software components which can be hacked and controlled. A further implied risk made in Fischmeister's statement may be the risk of loss of control of the vehicle to terrorist hackers, and a hacker's potential ability to control hundreds or thousands of vehicles all at once from the comfort of their home.

40.2.2.5 Summary

Focusing on "safety" as a concept that includes a bundle of risks, we break down safety claims associated with automated vehicles that are made by different stakeholders. We find that stakeholders such as legislators and regulators, corporations, advocacy groups, and public media talk about safety differently given that, when the risks were made explicit, there are different risks addressed among stakeholder safety claims. We find that legislators and regulators tend to talk about the risk of serious injury or fatal accidents when talking about automated vehicle safety, where their claims differ depending on whether their claims account for accidents in general or accidents associated with driver error, and depending on whether their claims consider accidents with a lower probability of occurring (compared to a referenced probability) or accidents with no probability of occurring. In contrast, we find advocacy groups tend to talk about the risk of accidents associated with impaired, fatigued, or distracted driving. Notably we also found an organization of advocacy groups that spoke about the risks of psychic harms and stress associated with driving in their claim about automated vehicle safety. Furthermore, we find that some corporations cover a broader list of explicit risks in their safety claims, such as occupational illness, damage of equipment or property, and damage to the environment caused by "conditions", while other vehicle corporations were not explicit about the risks associated with their safety claims, requiring further guesswork or analysis to distinguish their claims from others. Finally, we find that the public media covered broader risks associated with automated vehicle safety concerns, including the risk of heart disease and cancer due to sedentary behavior and pollution, as well as the risk of loss of control due to hackers taking over control of the vehicles remotely.

Based on the uses of the term 'safety' by multiple stakeholders and our analysis of safety engineering frameworks in ▶ Sect. 40.2.1 we summarize implications with respect to our hypothesis of a diverging meaning and use of the term 'safety'.

40.2.3 Conclusions About the Meanings and Uses of the Term 'Safety'

In the previous section we analyzed how the term 'safety' is used in selected safety engineering frameworks and how other stakeholders use the term. We focused our analysis on the relation between different kinds of risk being considered when raising safety claims or concerns.

For the stakeholder group of engineers we elaborated how safety is traditionally handled as the absence of unreasonable physical harm and how existing standards provide guidelines on how to assure safety of automated vehicles within this defined understanding of related risks. We also pointed out challenges that emerge from the shift from assisted to automated driving with regard to safety measures and quantitative risk assessments. These challenges include for example the absence of the measure of controllability, the issue of adhering to a positive risk balance given differently severe accidents, and more generally the determination of quantitative risk acceptance criteria for the new technology of automated vehicles (SAE Level 3+).

Additionally, we showed examples of approaches for safety engineering from corporate and academic research that use the term 'safety' differently. On the one hand there are approaches that address safety as a formally provable property of an automated vehicle. Whereas formal proofs provide (economically) scalable frameworks, they require fundamental assumptions on reasonable parameters chosen that implicitly might lead to safe systems. However, the relation between a safe vehicle and the risks that are associated with its operation remain implicit and hence unclear.

On the other hand we presented an approach for value-sensitive development in order to give an example of how safety can also be understood as a concept that accounts for stakeholder objectives explicitly. Based on such an explicit consideration of stakeholder values, the risks that are associated with an automated vehicle can be evaluated and explicitly included in the design of these systems.

Other stakeholder groups use the term 'safety' and are not always explicit about the risks they include in their understanding of a safe automated vehicle. Furthermore, the statements we collected as part of this book chapter do not mean to represent a consistent view within each stakeholder group. However, we support our claim that safety is used differently across different stakeholders. In addition to the kinds of risk that are included, which range from physical harm to for example damage of equipment or property, we elaborated that there is a plethora of further considerations that go into the classification of a 'safe' automated vehicle. For example, one question could be if automated vehicles lead to less accidents in road traffic than human drivers in general or if they perform better than impaired drivers. This question becomes even more demanding by comparing automated vehicles with performant and attentive drivers.

Given the provided evidence for different meanings and uses of the term 'safety', in the following section we provide a set of questions about aspects of risk that could be asked when facing safety claims or concerns from different stakeholders. These questions can be used to facilitate an explicit and more detailed distinction between meanings of safety and associated risks and can further support an open discussion among stakeholder groups and disciplines about approaches for handling safety and risk communication for AVs.

40.3 Questions About the Risks Associated with the Term 'Safety'

In the previous section we gave examples of different meanings and uses of the term 'safety'. We found that different stakeholders include different risks when they raise concerns about the safety of automated vehicles or address these concerns with respective claims in a safety engineering framework.

In this section we pose questions about the risks associated with the term 'safety' in order to breakdown the risks into aspects that can be compared. These aspects of risk that we found are as follows: category, scope, target, time, reference, and stakeholder-dependence. The questions we pose to consider these aspects of risk can facilitate future discussions among stakeholder groups and disciplines about how safety statements and associated risks differ, and about approaches for handling safety communication for automated vehicles.

The first question that can be raised to break down a safety concern or claim is:

❯ Which objective is at risk? (category)

This aspect specifies the category of the considered risks. Given the definitions of functional safety (ISO 26262:2018) and safety of the intended functionality (ISO 21448:2022), we elaborated that the objective at risk in these meanings of safety is the physical well-being of persons involved in an operational situation. Statements from Waymo extend this meaning to further objectives such as occupational illness and damage of property. These extensions of the risks to be considered when using the term 'safety' can be generalized as defined for risk management in the ISO 31000:2018. Here, any objective under uncertainty is included as a category of risk that can be managed.

In order to explicitly state the scope when using the term 'safety', one can raise the question:

❯ What scope is considered with the risks? (scope)

Again starting from established safety engineering frameworks for functional safety and SOTIF, hazards and their associated risks are explicitly considered at the vehicle level [22, 26]. Hence, the system boundary is drawn between the vehicle and its immediate environment in an operational situation or scenario [52]. Statements in the public press like those from Peter Norton in Wired suggest to consider risks on a societal level in order to include health risks caused by for example less walking due to automated driving. Based on this dis-

tinction, it becomes clear that different risks can be considered depending on the system boundary or the scope.

Furthermore, the entity which is at risk can be made explicit by asking the question:

❯ Who is at risk? (target)

The answer to this question also depends on the scope at which risks are included in a meaning of safety. For example in a scenario where only the immediate vicinity of the vehicle is considered, passengers and other road users could be considered as affected by different risks. In the ISO 26262:2018 possible targets of harm explicitly defined as "the driver, passengers or persons in the vicinity of the vehicle's exterior". With regard to a larger scope there could also be companies at risk or specific parts of society that do not necessarily interact with automated vehicles directly. When introducing RSS, Shalev-Shwartz et al. [53] write: "The second area of risk lies with lack of scalability." We attribute the kind of entities concerned with such a risk as being companies for which scalability is an objective.

In our analysis we elaborated that there can be differences between considering risks as an outcome of deploying the system and estimating possible risks and defining design goals. Hence, we ask the question:

❯ Are the risks judged retrospectively or prospectively? (time)

The goal of achieving a positive risk balance, which was raised as a requirements by the German ethics commission, and translating that into design goals for an unreasonable amount of risk was for example discussed by Favaró [83]. This process has plenty of challenges as elaborated by Favaró. One of these challenges is changing the perspective towards the deployment of automated vehicles regarding the aspect of time. A positive risk balance could for example be interpreted as the reduction of fatalities in road traffic for each year individually after the technology is released. Also one could tolerate a relative increase in fatalities in the first year but manage the balance with a stepwise market penetration as Wachenfeld et al. [15] suggested and as it is practiced in aviation domain [17].

Even though we focus our analysis on the meanings and uses of the term 'safety' and do not attempt to define an amount of risk that could be reasonable for automated vehicles, the following question helps identifying different kinds of risks:

❯ With respect to which reference does the system pose risks? (reference)

Whether or not an automated vehicle is reasonably safe requires a clarification of how "reasonable" is defined.

Quantitative risk acceptance criteria take different reference values of risks below which an automated vehicle could be considered reasonably safe. These can for example be the probability of being struck by a lightning or of being involved in an accident with a conventional road vehicle. Advocacy groups like Mothers Against Drunk Driving (MADD) or End Distracted Driving (EndDD) define the goal of reducing accidents due to different sorts of impaired driving. Also, statements from for example NHTSA [69] suggest to compare automated vehicles to human drivers and assess the impact of mitigating human error for the safety of road traffic. A more rigorous reference is given by Vision Zero with the goal of completely eliminating fatalities and severe accidents in road traffic. Waymo's definition of safety—comparable to Vision Zero—suggests a complete freedom of harmful conditions. Therefore, in both cases the absence of all risk emerging from automated driving could also be understood as a possible reference to evaluate the safety of automated vehicles.

As we compared perspectives from different stakeholders on what constitutes the safety of a system with respect to the included risks we found that there are risks that only apply to some group of stakeholders and those that can be applied to all of them. The following question helps to make this difference explicit:

> Are the risks based on a perspective from a group of stakeholders or do they apply to all stakeholders? (stakeholder-dependence)

For example an aspect of risk that was only implicitly addressed in this analysis so far is concerned with factors of risk perception. Statements from Waymo considering occupational illness suggest that perceived safety can play a major role for the acceptance of automated driving. The presented framework for value-sensitive development can facilitate including such considerations by accounting for risk perception in the process of weighing stakeholder values. On the other hand, established safety engineering frameworks and formal safety assurance methods rely on mathematical metrics to assess different kinds of risk. These metrics are especially necessary to find a common ground for different stakeholder groups with otherwise very heterogeneous subjective perceptions of risk. However, this does not imply that a seemingly more objective "expert-based risk assessment" is superior to aspects of for example public risk perception as also pointed out by Slovic [42].

40.4 Limitations and Recommendations

In the previous chapter we analyzed different stakeholder perspectives on the safety of automated vehicles and how they use the term 'safety'. We elaborated that there are different kinds of risk being considered by those stakeholder groups. Our analysis is limited in so far that we can neither claim to have considered all relevant stakeholder groups nor that those included are being comprehensively represented. We deem this limitation to be acceptable as it does not compromise our claim that there are different meanings and uses of the term 'safety'.

In order to support future research on this issue we give recommendations on how this initial analysis could be complemented. From an engineering perspective we only included examples of automotive safety engineering frameworks. This selection and thereby also our classification of three kinds of frameworks could be extended by analyzing further standardization work such as the SAE J3016:2021, SAE J3018:2019, ISO/TR 4804:2020, or IEEE 7000-2021. Additionally, further research in the field of automated vehicle safety could be included such as work from Koopman et al. [84], Wachenfeld [12], Fraade-Blanar et al. [85], or Stolte et al. [86]. We also did not include perspectives of stakeholder groups from other parts for the world such as Asia. Thus, there are other influential and reliable sources whose safety claims and associated risks could be included in future analyses. Regarding experience from other domains such as aviation or rail, we found differences with respect to the issue of an open traffic context and the coping mechanisms that can be put in place in order to assure safety. For example the existence of air traffic control enables different safety argumentation strategies compared automated driving. Hence, we did not focus on notions of safety in these domains but we acknowledge the potential of including considerations about safety from these and further fields (like robotics) in the automotive domain.

Generally, we follow the recommendation from Homann [87, 88], who advocated a public discourse of risks in the context of driver assistance systems in 2002, and support his approach in the domain of automated vehicles. Given Homann's argumentation, we further recommend to intensify efforts of an open discourse of risks associated with automated driving *before* the release of respective systems. This becomes increasingly important as first automated driving systems with SAE Level 3 have recently been certified in Germany [89].

Burton et al. [90] present an analysis of potential gaps in safety assurance from an interdisciplinary perspective and offer suggestions for the practical realization of their findings. Thus, Burton et al. [90] recommend to thoroughly define the operational design domain of the automated driving system in order to provide a basis of the system specification. Furthermore, an iterative refinement of safety assurance activities in parallel to the system design are required in order to identify safety-critical scenarios and system limitations as soon as possible and define necessary measures like limiting the operational design domain or adapting the system design (cf. SOTIF [22]). Finally, Burton et al. [90] suggest that in liability cases, there should be an interdisciplinary evaluation where assumptions and decisions in the design process are assessed.

40.5 Summary and Outlook

With 'safety' and 'risk' being two terms of major importance in the field of automated driving this chapter provides a selection of meanings and uses of these terms that are found either in safety engineering frameworks or in statements of different stakeholder groups. The analysis showed that there are fundamental differences in the kinds of risks that are addressed in various safety claims and concerns.

In section ▶ Sect. 40.1 we argued that the shift from assisted to automated driving implies new challenges for safety assurance. We discussed the role of uncertainty in the design of automated vehicles and the participation in road traffic. As a result of these uncertainties in combination with stakeholder objectives, we referred to the concept of inherent risk. Based on this motivation of why safety and risk as concepts should be considered differently for automated instead of assisted driving, we raised the concern that there are different meanings and uses of these terms among stakeholders and this might lead to mismatched expectations of the safety of automated vehicles.

Hence, we gave examples of safety engineering frameworks that are either used in the automotive industry or proposed by corporate or academic research groups. Additionally, we gathered statements from different stakeholder groups. Based on an analysis of these selected frameworks and statements, we concluded that there are in fact different meanings and uses of the term 'safety' and these differences highly depend on the kinds of risk that are considered (▶ Sect. 40.2).

As a refinement of the claim of mismatched meanings and uses of the term 'safety', we provided a set of questions to further break down a safety concern into the included aspects of risk (▶ Sect. 40.3). These questions help to clarify and compare the details of the different risks associated with the meanings and uses of the term 'safety' that we found in our analysis of safety engineering frameworks and stakeholder concerns.

In order to support future research we gave recommendations to extend this initial analysis by including further work addressing safety terminology of automated vehicles. An additional challenge, which was not part of this chapter, lies in answering the question of "How safe is safe enough?" This question can be raised on the societal level but also on the vehicle level.

With automated vehicles being required to be capable of transitioning into a minimal risk condition for example by the German traffic law, clarity about aspects of risk that should be considered and the applied metrics and criteria for automated driving is important as well. As an answer to this question depends on a common understanding of different meanings and uses of the term 'safety' we suggest a more transparent communication of assumptions that are made when raising safety claims or concerns.

Lastly, in a context of rapid technological change with urgent societal and environmental implications, it is essential that trust remains the bedrock of societies, communities, economies, and sustainable development [91]. While the focus of further research has been on addressing issues which affect the release and acceptance of automated vehicles, further research is also needed to realize, identify, and address related issues which affect the perceived or actual trustworthiness of automated vehicles and associated actors and processes, which can lead to lower trust among stakeholders such as companies or developers and the public.

Acknowledgements For their financial support of this work we thank the German "Federal Ministry of Economic Affairs and Climate Action" in the context of the project "Verification and Validation Methods for Automated Vehicles Level 4 and 5", the "Daimler and Benz Foundation" in the context of the project "Value-based decision-making", and the German "Federal Ministry of Education and Research" for funding the project "Automatisiertes Fahren im Mischverkehr". We further thank Till Menzel and Torben Stolte for their valuable feedback on this chapter. Special thanks go to Hermann Winner, who has fundamentally contributed to the discussions that where part of developing this work.

References

1. Grunwald, A.: Societal risk constellations for autonomous driving. Analysis, historical context and assessment. In: Autonomous Driving, pp. 641–663. Springer (2016)
2. Tsugawa, S., Jeschke, S., Shladover, S.E.: A review of truck platooning projects for energy savings. IEEE Trans. Intell. Veh. **1**(1), 68–77 (2016). ▶ https://doi.org/10.1109/TIV.2016.2577499
3. Shladover, S.E.: PATH at 20—history and major milestones. IEEE Trans. Intell. Transp. Syst. **8**(4), 584–592 (2007). ▶ https://doi.org/10.1109/TITS.2007.903052
4. Dickmanns, E.D., Zapp, A.: Autonomous high speed road vehicle guidance by computer vision 1. In: IFAC Proceedings Volumes, 10th Triennial IFAC Congress on Automatic Control—1987 Volume IV, Munich, Germany, 27–31 July 20, pp. 221–226 (1987). ▶ https://doi.org/10.1016/S1474-6670(17)55320-3
5. Smith, B.W.: Human error as a cause of vehicle crashes. The Center for Internet and Society, Stanford Law School (2013). ▶ http://cyberlaw.stanford.edu/blog/2013/12/human-error-cause-vehicle-crashes. Accessed 13 Dec 2021

6. Critical Reasons for Crashes Investigated in the National Motor Vehicle Crash Causation Survey (No. DOT HS 812 115). NHTSA's National Center for Statistics and Analysis, United States, DC (2015)

7. Bagschik, G.: Personal communication (2016)

8. Maurer, M., Gerdes, J.C., Lenz, B., Winner, H.: (eds.): Autonomous Driving. Springer Berlin Heidelberg, Berlin, Heidelberg (2016). ▸ https://doi.org/10.1007/978-3-662-48847-8

9. Schnieder, E., Schnieder, L.: Verkehrssicherheit. Springer, Berlin, Heidelberg (2013). ▸ https://doi.org/10.1007/978-3-540-71033-2

10. Reschka, A., Bagschik, G., Maurer, M.: Towards a system-wide functional safety concept for automated road vehicles. In: Automotive Systems Engineering II, pp. 123–145. Springer (2018)

11. Bagschik, G., Reschka, A., Stolte, T., Maurer, M.: Identification of Potential Hazardous Events for an Unmanned Protective Vehicle (2018). arXiv preprint ▸ arXiv:1804.08728

12. Wachenfeld, W.H.K.: How Stochastic Can Help to Introduce Automated Driving. Technische Universität Darmstadt, Germany, Darmstadt (2017)

13. Stolte, T., Bagschik, G., Reschka, A., Maurer, M.: Hazard analysis and risk assessment for an automated unmanned protective vehicle. In: 2017 IEEE Intelligent Vehicles Symposium (IV). Presented at the 2017 IEEE Intelligent Vehicles Symposium (IV), pp. 1848–1855 (2017). ▸ https://doi.org/10.1109/IVS.2017.7995974

14. Kühn, M., Hannawald, L.: Verkehrssicherheit und Potenziale von Fahrerassistenzsystemen. Handbuch Fahrerassistenzsysteme: Grundlagen, Komponenten und Systeme für aktive Sicherheit und Komfort (2015)

15. Wachenfeld, W., Winner, H.: The new role of road testing for the safety validation of automated vehicles. In: Automated Driving: Safer and More Efficient Future Driving (2017)

16. Junietz, P., Bonakdar, F., Klamann, B., Winner, H.: Criticality metric for the safety validation of automated driving using model predictive trajectory optimization. In: 2018 21st International Conference on Intelligent Transportation Systems (ITSC). Presented at the 2018 21st International Conference on Intelligent Transportation Systems (ITSC), pp. 60–65 (2018). ▸ https://doi.org/10.1109/ITSC.2018.8569326

17. Junietz, P.M.: Microscopic and Macroscopic Risk Metrics for the Safety Validation of Automated Driving (Dissertation). Technische Universität, Darmstadt (2019). ▸ https://doi.org/10.25534/tuprints-00009282

18. Gasser, T.M., Schmidt, E.A., Bengler, K., Chiellino, U., Diederichs, F., Eckstein, L., Flemisch, F., Fraedrich, E., Fuchs, E., Gustke, M., Hoyer, R., Hüttinger, M., Jipp, M., Köster, F., Kühn, M., Lenz, B., Lotz-Keens, C., Maurer, M., Meurer, M., Meuresch, S., Müller, N., Reitter, C., Reschka, A., Riegelhuth, G., Ritter, J., Siedersberger, K.-H., Stankowitz, W., Trimpop, R., Zeeb, E.: Bericht zum Forschungsbedarf. Runder Tisch Automatisiertes Fahren—AG Forschung. Federal Ministry of Transport and Digital Infrastructure (2015)

19. Di Fabio, U., Broy, M., Jungo Brüngger, R., Eichhorn, U., Grunwald, A., Heckmann, D., Hilgendorf, E., Kagermann, H., Losinger, A., Lutz-Bachmann, M., Lütge, C., Markl, A., Müller, K., Nehm, K.: Ethics Commission—Automated and Connected Driving. Federal Ministry of Transport and Digital Infrastructure (2017)

20. SAE: J3016: Taxonomy and Definitions for Terms Related to On-Road Motor Vehicle Automated Driving Systems. SAE International (2021)

21. Safety: In Merriam-Webster.com Dictionary (n.d.). ▸ https://www.merriam-webster.com/dictionary/safety. Accessed 13 Oct 2021

22. International Organization for Standardization: ISO 21448 Road vehicles—Safety of the intended functionality (2022)

23. Shladover, S.E.: Framework for safety assurance of automated driving systems: unresolved issues. In: Gehalten auf der 27th ITS World Congress, Hamburg, Germany (2021)

24. Maurer, M.: Hochautomatisiertes und vollautomatisiertes Fahren. In: Deutscher Verkehrsgerichtstag—Deutsche Akademie für Verkehrswissenschaft (Ed.), 56. Deutscher Verkehrsgerichtstag 2018. Presented at the 56. Deutscher Verkehrsgerichtstag 2018, pp. 43–57. Luchterhand Verlag (2018)

25. Nolte, M., Ernst, S., Richelmann, J., Maurer, M.: Representing the unknown—impact of uncertainty on the interaction between decision making and trajectory generation. In: 2018 21st International Conference on Intelligent Transportation Systems (ITSC). Presented at the 2018 21st International Conference on Intelligent Transportation Systems (ITSC), pp. 2412–2418 (2018). ▸ https://doi.org/10.1109/ITSC.2018.8569490

26. International Organization for Standardization: ISO 26262 Road Vehicles—Functional Safety, Part 1: Vocabulary (2018)

27. International Electrotechnical Commission: IEC 61508-1, Functional Safety of Electrical/Electronic/Programmable Electronic Safety-Related Systems—Part 1: General Requirements (2010)

28. Deutsches Institut für Normung: DIN EN ISO 12100, Sicherheit von Maschinen—Allgemeine Gestaltungsleitsätze—Risikobeurteilung und Risikominderung (2011)

29. International Organization for Standardization, International Electrotechnical Commission: ISO/IEC Guide 51, Safety Aspects—Guidelines for Their Inclusion in Standards (1999)

30. Koch, K.-R.: Einführung in die Bayes-Statistik. Springer Berlin Heidelberg, Berlin, Heidelberg (2000). ▸ https://doi.org/10.1007/978-3-642-56970-8

31. International Organization for Standardization: ISO 26262 Road Vehicles—Functional Safety—Part 3: Concept Phase (2018)

32. Society of Automotive Engineers: SAE J2980—Considerations for ISO 26262 ASIL Hazard Classification (2018)

33. Douglas, M., Wildavsky, A.: How can we know the risks we face? Why risk selection is a social process. Risk Anal. **2**, 49–58 (1982). ▸ https://doi.org/10.1111/j.1539-6924.1982.tb01365.x

34. Crouch, E.A.C., Wilson, R.: Risk/Benefit Analysis. Ballinger Publishing Company, Cambridge, MA (1982)

35. Gibson, S.B.: The use of quantitative risk criteria in hazard analysis. J. Occup. Accid. **1**, 85–94 (1976). ▸ https://doi.org/10.1016/0376-6349(76)90009-2

36. Kinchin, G.H.: Design criteria, concepts and features important to safety and licensing. ANS, In: Proceedings on the International Meeting on Fast Reactor Safety Technology, WA, United States (1979)

37. Kletz, T.A.: Hazard analysis, its application to risks to the public at large. Reliab. Eng. **3**, 325–338 (1978)

38. Starr, C.: Social benefit versus technological risk. Science **165**, 1232–1238 (1969)

39. Starr, C.: Benefit-cost relationships in socio-technical systems. In: Environmental Aspects of Nuclear Power Stations (1971)

40. Webb, G.A.M., McLean, A.S.: Insignificant Levels of Dose: A Practical Suggestion for Decision Making. National Radiological Protection Board (1977)

41. Fritzsche, A.F.: Wie sicher leben wir? Risikobeurteilung und -bewältigung in unserer Gesellschaft. Verlag TÜV Rheinland (1986)

42. Slovic, P.: Perception of risk. Science **236**, 280–285 (1987). ▸ https://doi.org/10.1126/science.3563507

43. Junietz, P., Steininger, U., Winner, H.: Macroscopic safety requirements for highly automated driving. Transp. Res. Rec. **2673**, 1–10 (2019). ▸ https://doi.org/10.1177/0361198119827910

44. Robinson, R.M., Anderson, K.J., Meiers, S.J., Browning, R.W., Alam, T.: Risk and Reliability: An Introductory Text. Risk & Reliability Associates Pty Limited (1997)

45. DESTATIS Statistisches Bundesamt: Sterbetafeln und Lebenserwartung [WWW Document] (2017). ▸ https://www.destatis.de/DE/Themen/Gesellschaft-Umwelt/Bevoelkerung/Sterbefaelle-Lebenserwartung/Methoden/sterbetafeln-lebenserwartung.html. Accessed 05 Dec 2021

46. International Organization for Standardization: ISO 26262 Road vehicles—Functional Safety—Part 4: Product Development at the System Level (2018)

47. Steininger, U., Schöner, H.-P., Schiementz, M., Mazzega, J.: Validation of Assisted and Automated Driving Systems. Crash. Tech., München (2016)

48. Baum, H., Kranz, T., Westerkamp, U.: Volkswirtschaftliche Kosten durch Straßenverkehrsunfälle in Deutschland (2010)

49. Hydén, C.: The Development of a Method for Traffic Safety Evaluation: The Swedish Traffic Conflicts Technique. Bulletin Lund Institute of Technology, Department (1987)

50. Schöner, H.-P.: Challenges and approaches for testing of highly automated vehicles. In: Langheim, J. (ed.) Energy Consumption and Autonomous Driving, Lecture Notes in Mobility, pp. 101–109. Springer International Publishing, Cham (2016). ► https://doi.org/10.1007/978-3-319-19818-7_11

51. Steininger, U., Wech, L.: Wie sicher ist sicher genug? Sicherheit und Risiko zwischen Wunsch und Wirklichkeit. VDI-Berichte (2013)

52. Ulbrich, S., Menzel, T., Reschka, A., Schuldt, F., Maurer, M.: Defining and substantiating the terms scene, situation, and scenario for automated driving. In: 2015 IEEE 18th International Conference on Intelligent Transportation Systems. Presented at the 2015 IEEE 18th International Conference on Intelligent Transportation Systems (ITSC 2015), pp. 982–988. IEEE, Gran Canaria, Spain (2015). ► https://doi.org/10.1109/ITSC.2015.164

53. Shalev-Shwartz, S., Shammah, S., Shashua, A.: On a Formal Model of Safe and Scalable Self-driving Cars (2018). arXiv preprint ► arXiv:1708.06374

54. Nistér, D., Lee, H.-L., Ng, J., Wang, Y.: The safety force field. NVIDIA White Paper (2019)

55. Althoff, M.: Reachability Analysis and Its Application to the Safety Assessment of Autonomous Cars. Technische Universität München, Germany, Munich (2010)

56. Ulbrich, S.M.: Towards Tactical Lane Change Behavior Planning for Automated Vehicles. TU Braunschweig, Braunschweig (2018). ► https://doi.org/10.24355/dbbs.084-201809171100-0

57. Kluckhohn, C.: In: Parsons, T., Shils, E.A. (eds.) Toward a General Theory of Action, pp. 388–433. Harvard University Press (2013). ► https://doi.org/10.4159/harvard.9780674863507.c8

58. Albert, E.: Zur Wertekonzeption in den Sozialwissenschaften—Potenzial, fachliche Prägungen und mögliche Blockaden eines strategisch benutzten Begriffes (Online Publication). Sociology in Switzerland: Contributions to General Sociological Theory, Zurich (2008)

59. Thornton, S.M.: Autonomous Vehicle Motion Planning with Ethical Considerations. Stanford University, Stanford, CA (2018)

60. Graubohm, R., Schräder, T., Maurer, M.: Value sensitive design in the development of driverless vehicles: a case study on an autonomous family vehicle. In: Proc. Des. Soc.: Des. Conf., vol. 1, pp. 907–916 (2020). ► https://doi.org/10.1017/dsd.2020.140

61. Friedman, B.: Human Values and the Design of Computer Technology. Cambridge University Press (1997)

62. Christian Gerdes, J., Thornton, S.M., Millar, J.: Designing automated vehicles around human values. In: Meyer, G., Beiker, S. (Trans.) Road Vehicle Automation 6, pp. 39–48. Springer International Publishing, Cham (2019)

63. Millar, J., Paz, D., Thornton, S.M., Parisi, C., Gerdes, J.C.: A framework for addressing ethical considerations in the engineering of automated vehicles (and other technologies). In: Proc. Des. Soc.: Des. Conf., vol. 1, pp. 1485–1494 (2020). ► https://doi.org/10.1017/dsd.2020.78

64. Salem, N.F., Haber, V., Beck, H.N., Rauschenbach, M., Nolte, M., Reich, J., Stolte, T., Maurer, M.: Semantische Normverhaltensanalyse—ein Beitrag zur durchgängigen Formalisierung von Verhaltensnormen für das automatisierte Fahren (2021). ► https://www.ifr.ing.tu-bs.de/static/files/forschung/papers/download_pdf.php?id=1174-11

65. Beck, H.N., Salem, N.F., Haber, V., Rauschenbach, M., Reich, J.: Phänomen-Signal-Modell: Formalismus, Graph und Anwendung (2021). arXiv preprint ► arXiv:2108.00252 [physics]

66. Nolte, M., Rose, M., Stolte, T., Maurer, M.: Model predictive control based trajectory generation for autonomous vehicles—an architectural approach. In: 2017 IEEE Intelligent Vehicles Symposium (IV). Presented at the 2017 IEEE Intelligent Vehicles Symposium (IV), pp. 798–805. IEEE, Los Angeles, CA, USA (2017). ► https://doi.org/10.1109/IVS.2017.7995814

67. Tingvall, C., Haworth, N.: Vision Zero-An ethical approach to safety and mobility. In: 6th ITE International Conference Road Safety & Traffic Enforcement: Beyond 2000 (1999)

68. CCA (Council of Canadian Academies): Choosing Canada's Automotive Future, Ottawa (ON). The Expert Panel on Connected and Autonomous Vehicles and Shared Mobility, Council of Canadian Academies (2021)

69. Automated Vehicles for Safety: NHTSA (n.d.). ► https://www.nhtsa.gov/technology-innovation/automated-vehicles-safety. Accessed 03 Nov 2021

70. Owens, J.: Automated Vehicles Symposium. NHTSA (2020). ► https://www.nhtsa.gov/speeches-presentations/automated-vehicles-symposium. Accessed 04 Nov 2021

71. de Gelder, E., Elrofai, H., Saberi, A.K., Paardekooper, J.-P., Op den Camp, O., de Schutter, B.: Risk quantification for automated driving systems in real-world driving scenarios. IEEE Access 9, 168953–168970 (2021). ► https://doi.org/10.1109/ACCESS.2021.3136585

72. Tesla: Tesla Vehicle Safety Report (2021). ► https://www.tesla.com/VehicleSafetyReport. Accessed 03 Nov 2021

73. Waymo: Safety Report and Whitepapers. Waymo (2021). ► https://waymo.com/safety/. Accessed 07 Nov 2021

74. Kronberg, P.: So, Will Autonomous Vehicles Improve Safety? (2016). ► https://www.volvogroup.com/en/news-and-media/columns/columns/traffic-safety/will-autonomous-vehicles-improve-safety.html. Accessed 04 Nov 2021

75. MADD: MADD Canada Supports Development of Automated Vehicles to Improve Road Safety (2016)

76. EndDD (End Distracted Driving): Are We Ready for Autonomous Car Features? | End Distracted Driving. EndDD (2019, October, 02). ► https://www.enddd.org/end-distracted-driving/are-we-ready-for-autonomous-car-features/. Accessed 07 Nov 2021

77. Madia, E.: Protecting Pedestrians from Themselves in the Age of Distraction | End Distracted Driving. EndDD (2018, June 10). ► https://www.enddd.org/end-distracted-driving/distracted-driving-updates/protecting-pedestrians-from-themselves-in-the-age-of-distraction/. Accessed 07 Nov 2021

78. NSC (National Safety Council): Automated & Self-Driving Vehicle Revolution: What is the Role of Government?—National Safety Council (2016). ► https://www.nsc.org/company/speeches-testimony/testimony-the-automated-self-driving-vehicle-revolution. Accessed 07 Nov 2021

79. LTAD (Let's Talk about Autonomous Driving): Partners and Community | Let's Talk Autonomous Driving. ► https://ltad.com/partners.html. Accessed 07 Nov 2021

80. LTAD (Let's Talk about Autonomous Driving): Can Fully Autonomous Vehicles Help Drive Away Stress? | Let's Talk Autonomous Driving. ► https://ltad.com/news/can-fully-autonomous-vehicles-help-drive-away-stress.html. Accessed 07 Nov 2021

81. Stockton, N.: "So, Self-Driving Cars Could Make Humans Unhealthier Than Ever. Wired. ► https://www.wired.com/story/self-driving-cars-obesity-pollution-public-health/. Accessed 05 Nov 2021

82. Groleau, C.: Engineering Researchers at UW Look at Safety and Security of Automated Vehicles | CBC News. CBC (2021, October, 03). ► https://www.cbc.ca/news/canada/kitchener-waterloo/university-of-waterloo-automates-vehicles-magna-international-1.6196110. Accessed 05 Nov 2021

83. Favaro, F.: Exploring the Relationship Between "Positive Risk Balance" and "Absence of Unreasonable Risk" (2021). arXiv preprint ▶ arXiv:2110.10566

84. Koopman, P., Wagner, M.: Toward a Framework for Highly Automated Vehicle Safety Validation (SAE Technical Paper No. 2018-01-1071). SAE International, Warrendale, PA (2018). ▶ https://doi.org/10.4271/2018-01-1071

85. Fraade-Blanar, L., Blumenthal, M.S., Anderson, J.M., Kalra, N.: Measuring Automated Vehicle Safety: Forging a Framework (2018)

86. Stolte, T., Ackermann, S., Graubohm, R., Jatzkowski, I., Klamann, B., Winner, H., Maurer, M.: A taxonomy to unify fault tolerance regimes for automotive systems: defining fail-operational, fail-degraded, and fail-safe. IEEE Trans. Intell. Veh. 1–1 (2021). ▶ https://doi.org/10.1109/TIV.2021.3129933

87. Homann, K.: Wirtschaft und gesellschaftliche Akzeptanz: Fahrerassistenzsysteme auf dem Prüfstand (2002)

88. Homann, K.: Wirtschaft und gesellschaftliche Akzeptanz: Fahrerassistenzsysteme auf dem Prüfstand. In: Maurer, M., Stiller, C. (eds.) Fahrerassistenzsysteme mit maschineller Wahrnehmung, pp. 239–244. Springer-Verlag, Berlin/Heidelberg (2005). ▶ https://doi.org/10.1007/3-540-27137-6_11

89. Wilkes, W.: Mercedes Beats Tesla to Hands-Free Driving on the Autobahn. bloomberg.com (2021). ▶ https://www.bloomberg.com/news/articles/2021-12-09/mercedes-beats-tesla-to-hands-free-driving-on-german-autobahn. Accessed 13 Dec 2021

90. Burton, S., Habli, I., Lawton, T., McDermid, J., Morgan, P., Porter, Z.: Mind the gaps: assuring the safety of autonomous systems from an engineering, ethical, and legal perspective. Artif. Intell. **279**, 103201 (2020). ▶ https://doi.org/10.1016/j.artint.2019.103201

91. High-Level Expert Group on Artificial Intelligence set up by the European Commission: Ethics Guidelines for Trustworthy AI (2019). ▶ https://digital-strategy.ec.europa.eu/en/library/ethics-guidelines-trustworthy-ai. Accessed 13 Dec 21

40

Open Access This chapter is licensed under the terms of the Creative Commons Attribution-NonCommercial-NoDerivatives 4.0 International License (▶ http://creativecommons.org/licenses/by-nc-nd/4.0/), which permits any noncommercial use, sharing, distribution and reproduction in any medium or format, as long as you give appropriate credit to the original author(s) and the source, provide a link to the Creative Commons license and indicate if you modified the licensed material. You do not have permission under this license to share adapted material derived from this chapter or parts of it.

The images or other third party material in this chapter are included in the chapter's Creative Commons license, unless indicated otherwise in a credit line to the material. If material is not included in the chapter's Creative Commons license and your intended use is not permitted by statutory regulation or exceeds the permitted use, you will need to obtain permission directly from the copyright holder.

Tactical Safety for Autonomous Vehicles on Highways

Hans-Peter Schöner and Jacobo Antona-Makoshi

Contents

© The Author(s) 2026
H. Winner et al. (eds.), *Handbook Assisted and Automated Driving*,
https://doi.org/10.1007/978-3-658-45276-6_41

41.1 The Concept of Tactical Safety

When people bid farewell to friends or family leaving on a long journey by car, they commonly say "Drive carefully!". What does this safety-related notion mean in detail, and how should it be adapted to Automated or Autonomous Driving? It definitively implies more than just obeying specific traffic rules. It is a behavior guideline, leading to a cautious and cooperative driving style. Some examples of this kind of human driver behavior follow:

- Looking far ahead, *anticipating* possible risks (in addition to recognizing acute prevailing danger) in order to adapt speed, longitudinal distance, and also lateral position in the lane;
- Changing behavior to the safe side based on faint clues ("the sixth sense");
- Understanding performance limits (especially the own) in unusual situations and knowing the conditions for such unusual situations being likely to occur;
- Showing cooperative and fair behavior in order to reduce risk in many difficult or unusual situations ("roadmanship");
- Adopting general rules intelligently for specific situations.

The common idea behind all of these examples is that safety does not only result from the driver's *reaction* on a given critical situation but also from how the driver's *own behavior* influences the likelihood of occurrence and the level of risk of such situations. *Tactical Safety* refers to such behavior, in particular to the change from *normal* behavior due to a recognition of an *unusual* situation. In contrast to some other approaches in the literature (e.g., Weast [28]), the Tactical Safety concept is based on the recognition, that there is not one single set of suitable behavior parameters (e.g., suitable speed, safety distances) for all conditions; safe and at the same time efficient behavior in traffic must adapt itself to the conditions with its potential dangers (Zhao et al. [32], Mattas et al. [12], Schöner et al. [19]).

This document establishes a technical approach to *Tactical Safety*, and how such behavior can be implemented and verified in autonomous vehicles, with a focus on highway driving.

41.2 Precautious Driving as a Complement to Collision Avoidance

Exposure to, *Controllability* of, and *Severity* of accident situations are well established as the three main contributing (multiplying) factors for quantified risk management for both aircraft operation and general complex systems, as illustrated by the "System Safety Handbook" by the US Federal Aviation Association [5], and by the "System reliability theory" by Rausand and Høyland [16], respectively.

Similarly, the risk management methods of IEC 61508 [7] and ISO 26262 [9] deduct the quality requirements of electronic architectures and components based on the *risk* of traffic situations. Here the risk level is derived from a combination (summation in a logarithmic scale) of *Exposure*, *Controllability*, and *Severity* classes.

Exposure can be measured as the occurrence rate E of a challenging situation, typically in relation to operating time or (in transportation systems) to driven distance. *Severity* S of a resulting accident is classified by accident statistics [17] and accounts also for the number of involved (injured or killed) persons. With *controllability* being measured as the probability C to master a challenging situation without damage, the applicable factor for calculating its risk contribution is $(1-C)$ (Schöner [20]). As a result, the formula to calculate the risk contribution R of a single situation is

$$R = E\,(1 - C)\,S \tag{41.1}$$

- E: exposure to an accident-prone situation (by outer conditions or own behavior);
- C: controllability (preventability of a damage) of an acute accident-prone situation;
- S: severity of an accident, if damage could not be avoided.

The overall risk results from the integration over possible scenarios using relevant parameter distributions within these scenarios.

During the development of safety systems in cars, the focus of *Passive Safety* has been on reducing the severity of accidents, while *Active Safety Systems* have mainly focused on improving the controllability aspect; it has still been the task of the driver to avoid the exposure to hazardous (i.e., potentially dangerous) situations (see ◘ Fig. 41.1). The essentially new option of reducing overall risks with Autonomous Drive Systems (compared to a mature driver assistance system) is applying cautious and cooperative driving skills in order to reduce the exposure to dangerous situations: look far ahead, drive precautiously and predictively, know your own limits. By doing this, an autonomous vehicle might avoid getting into collision-prone situations with significant severity which are not (or less likely) controllable for the system. The term *"Tactical Safety"* has been recently proposed for such precautious safety measures in traffic (i.e., with positive influence on risk exposure) [29] by the UK CertiCAV project. Since the term perfectly indicates how such safety measures provide a complement to Passive Safety and Active Safety, it is adopted in this book chapter.

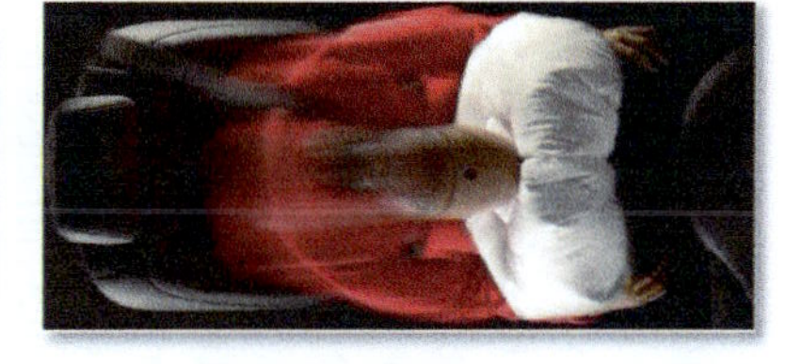

Fig. 41.1   Three factors of risk management in driving scenarios (based on [20]), (*) term *tactical safety* proposed by CertiCAV [28]

Since most driver assistance systems are designed to not interfere with the driver unnecessarily, the intervention of the system thus was limited in most cases to situations when a collision is getting almost unavoidable. For this reason, there has been a predominant focus on very intense interventions (emergency braking, evasive steering) of the driver assistance system, striving to increase the controllability C of the situation in the last possible moment. Autonomous vehicles, however, have a chance to perform automatic vehicle reactions on changes in traffic situations in a much earlier state. This allows to intervene in a more comfortable or even proactive way, or use cooperative actions in order to resolve a traffic conflict in an early stage. All this finally has the effect of reducing the exposure E to critical traffic situations, including the need for intense emergency interventions.

When issuing a driver's license, the assessment of human ability to drive safely and in a responsible manner (Schöner [23]) real traffic does not explicitly look at last second accident-avoidance skills: the focus is primarily on the driver's ability to look ahead, anticipate and avoid dangerous situations, to drive carefully and in a predictable way for others, and to judge and act according to both safety and traffic flow (Schöner [21]). While safe handling of crash-imminent situations is in fact one important feature of autonomous vehicles, anticipative strategies, including the use of adequate safety margins, must adequately be considered before releasing such vehicles into traffic on public roads (Kitajima [10], MobilEye [12], Ploeg [15], Peng [14], Schöner [22]). The Standard ISO/PAS 21448 *Safety of the intended functionality* [8] requires to present solutions for all foreseeable and preventable dangerous situations. In the sense of this standard, the term "preventable dangers" should be considered as a combination of the dangerous situation being either "controllable" (i.e., react fast enough and adequate to master the situation) or "proactively avoidable" (i.e., reduce the exposure probability to sufficiently low values).

Up to now, the test specifications for verification and validation of autonomous vehicles had little focus on the aspect of reducing exposure E to collision-prone situations. This entire chapter has the intention to provide an approach for such testing: The final ▶ Sect. 41.7 of this chapter describes test cases which become significant through the combination of *all three* risk factors.

Some collision scenarios have clear limits of controllability and at the same time high possible severity; this is explained in Sects. 41.3 and 41.4. Such situations need a focus on avoidance by tactical behavior (i.e., proactively reducing exposure) in order to reduce the overall risk; such scenarios are the focus in ▶ Sect. 41.7.2.

Situations with very low probability of severe collisions (low risk factor S) were neither in the main focus of common testing scenarios. However, if these scenarios have a very high occurrence rate E, like merging situations on highways, a flawless and traffic-flow preserving behavior in those situations has also a considerable impact on collision risk. Unusually strong braking or over-cautious behavior might cause surprise, irritation or impatience of other traffic participants, and in this way increase the probability for subsequent collisions. On the other hand, cooperative behavior might be mandatory to reduce the risk of situations like forced merging situations. Test cases for adequate behavior under such high exposure conditions (see ▶ Sect. 41.7.1) are necessary and should be part of a comprehensive safety assurance of Autonomous Vehicles.

Other scenarios are beyond the scope of normal behavior design of Autonomous Vehicles, since they depend on extreme and/or rare outer conditions, which might nevertheless have severe consequences when they happen. Even if they are not to be expected to happen in the normal service life of a single autonomous vehicle, they are significant when a large number of vehicles gets involved. To expect the "unexpected" and to be prepared for severe foreseeable conditions is required by the Standard ISO/PAS 21448 [8]. ▶ Section 41.7.3 proposes several of those severe foreseeable hazard conditions and corresponding test cases and outlines expected behavior and associated pass/fail criteria.

From the discussion of collision avoidance scenarios with limited controllability at highway speeds in ▶ Sect. 41.5, it becomes evident that precautious measures are needed in order to avoid coming into dangerous situations. It turns out that the residual risk depends significantly on
a. the probability of a *free road ahead* and
b. a change in driving behavior when this *free road ahead* cannot be guaranteed.

Occluded dangers on the road ahead would not even be detectable by a perfect long-distance sensor, but *Cooperative Sensing* of dangers on the road ahead by a sufficient number of traffic participants can provide a significant risk reduction; a concept is outlined in ▶ Sect. 41.6. The last subset of test cases in ▶ Sect. 41.7.4 thus proposes a method for the verification of a successful participation in such Cooperative Sensing. The test cases should help to substantiate applicable safety concepts within the already advertised "*Dynamic Map Content*" (HERE [6]) or "*Hazard Warnings*" (TomTom [26]), which are based on the open *SensorIS interface specification* (SensorIS [25]), or the joint *Safety Related Traffic Information Ecosystem* [3], promoted and in preparation in Europe.

41.3 Main Contributing Risk Parameters

When searching systematically for scenarios with a high-risk contribution, it makes sense to first identify the parameters which have the highest influence on the three separate factors *Severity*, *Controllability*, and *Exposure*.

The main contributing parameter for *Severity* is the energy involved in a collision. Since collision energy relates to the square of the speed difference of the collision partners, the relative speed is the main contributing parameter in fast highway driving scenarios. Several vehicles traveling with the same speed pose only a small risk if compared to situations when a fast-driving vehicle interferes with a standing object or even an oncoming car. Lateral intrusions into vulnerable parts of the vehicle may also strongly influence severity, but for a given accident scenario speed difference is the most important parameter.

The main contributing parameter to *Controllability* is the available time to perceive, judge, and perform adequate control actions. Surprises by unexpected emerging conditions with a too short available reaction time, lead to wrong or insufficient control actions, and with increasing speed, the inertia requires additional time for braking or steering actions. With longer available reaction time most situations become controllable by a defect-free system which is well designed for its operational domain (ensured by ISO-26262). But it has to be noted, that the borderline for sufficiently good controllability varies with scenario conditions which influence reaction time (e.g., via weather-dependent sensing quality) and feasible level of reaction (e.g., via road friction coefficient).

Safety distance rules and speed limits intend to influence behavior positively; not following the traffic rules and continuously driving in an inappropriate driving style causes increased exposure to dangers. But this general misbehavior should be precluded for autonomous vehicles. Apart from this, the main contributing parameters to *Exposure* are behavior deficits based on false judgement of situations and too late recognition of (un-usually) dangerous conditions, which require additional safety distances or further speed reduction. In some cases, uncooperative actions have also an important impact on the exposure to accident-prone situations.

For a rough assessment of the risk of a given scenario, it is helpful to represent the possible situations in a coordinate system using as axes the two main contributing parameters *Relative Speed* and *Available Reaction Time*. It is important to note, that both severity and controllability are monotonic functions of relative speed and available reaction time, respectively. Thus, using this coordinate system allows to sort situations by the expected risk of a given situation (when actual exposure is irrelevant, constant or unknown). This also reduces testing efforts, because test cases for verifying controllability for a specific scenario type can be placed along a limit line; successful tests along this line provide also evidence for situations for lower relative speed and higher available reaction time. The effect of these parameters on the risk factors is demonstrated by color codes in ◘ Fig. 41.2.

The severity (intensity of red color) increases strongly with relative speed, however, controllability (intensity of green color) can reduce or even cancel out the risk for large values of available reaction time. Traffic rules also contribute to risk reduction by eliminating exposure to most risks caused by too short reaction times (by requiring a temporal safety distance, amber) and by too large speed differences (by a speed limit, blue). The resulting risk based on the stepwise multiplication of the three factors is shown in ◘ Fig. 41.3 (the intensity of red color depicts the residual risk).

The left diagram of figure 41.3 shows a white area in the lower left, which is caused by little severity although at the same time it shows bad controllability because a minimum reaction time is not available. At larger relative speed, however, the situation switches from good controllability (dense green) to high severity (dense red) with small changes in available reaction time.

The limit line for placing borderline controllability tests is indicated by a red dashed-dotted line; success-

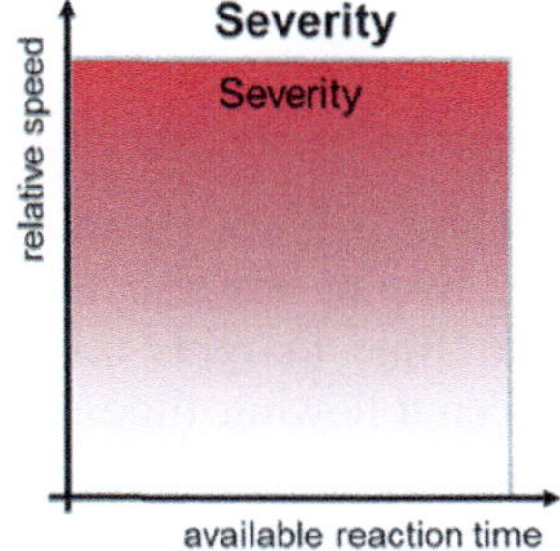

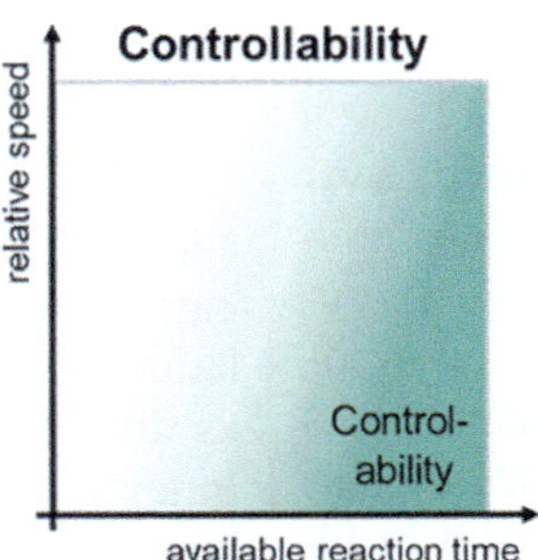

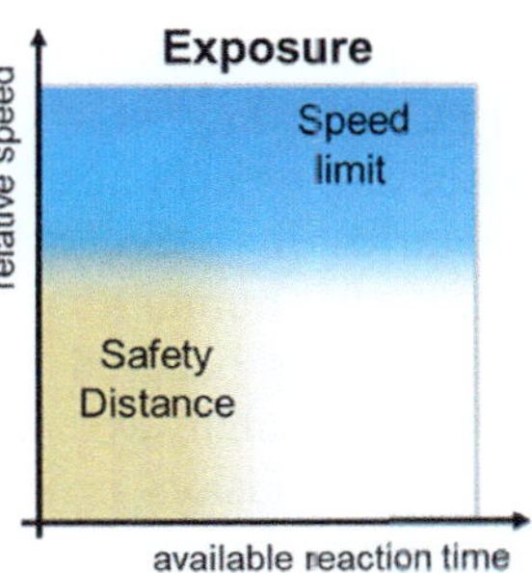

◘ **Fig. 41.2** Effect of main contributing parameters *relative speed* and *available reaction time* on the risk factors severity, controllability and exposure. *Graphics* © H.P. Schöner

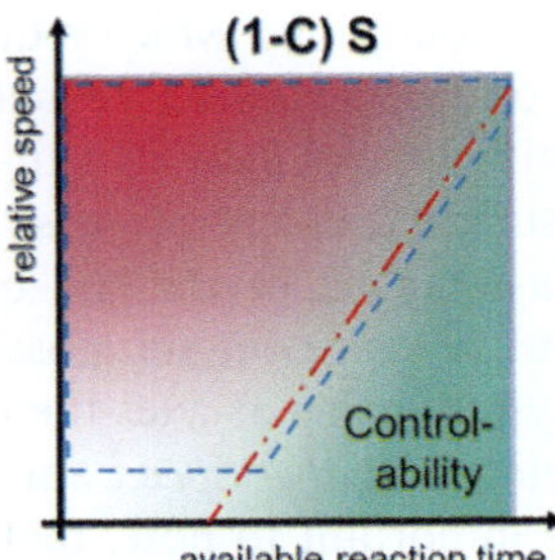

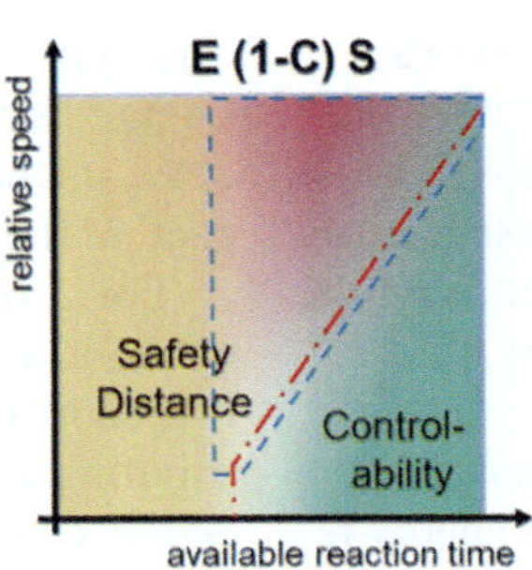

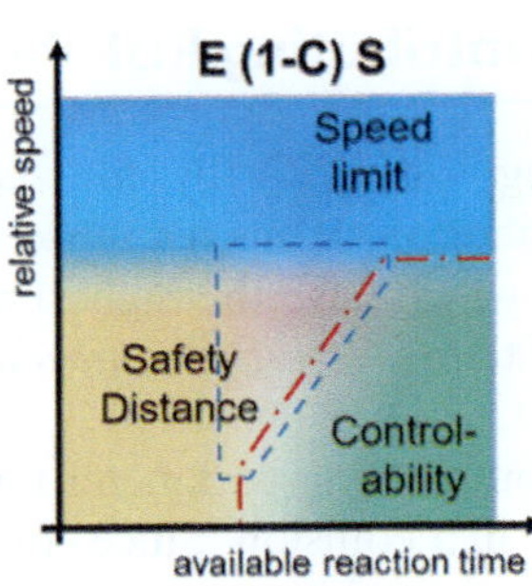

Fig. 41.3 Generic risk assessment on the basis of main contributing parameters: red dashed-dotted line: limit for placing borderline controllability test cases, blue dashed-line limited area: zone with need for *tactical safety*, including its verification. *Graphics* © H.P. Schöner

ful test outcomes along this line indicate controllability as well for other situations of the same scenario category with higher available reaction times and lower relative speed. The position of this limit line needs to allow for foreseeable cases of unobeyed traffic rules: too low safety distances and speeding above the prescribed speed limit. The ranges which still remain red after factoring in controllability (indicating high severity and low controllability) show the need for prevention by precautious behavior, also called *Tactical Safety*; these areas are marked with a blue dashed line. Note that this area overlaps with the limit line for controllability test cases. And further note that the region of controllability, especially depends on environmental scenario conditions influencing reaction time and feasible level of reaction.

The middle chart in **■** Fig. 41.3 shows qualitatively, that there opens a controllability gap above a given safety distance, increasing with higher relative speeds ! inherently based on the properties of the main contributing risk parameters. A lower speed limit might decrease the residual risk (right chart). But depending on the properties of some scenarios, a larger than normal safety distance or a further reduced speed is still necessary to limit the residual risk in the center of the last chart: a situation-dependent increased safety distance and reduced speed, whenever collision scenarios with unusually high relative speeds become more likely or less controllable (e.g., caused by low friction coefficients) than under usually safe conditions.

It turns out, that with commonly used safety distances the described controllability gap opens up at least for scenarios including slow moving or even standing and oncoming objects at driving speeds over 50 km/h (see scenarios in **▶** Sect. 41.7.2). This controllability gap exists for human-driven as well as autonomous vehicles, and is believed to cause many of the existing fatalities on highways. In accident statistics the reason for such collisions is usually identified as "non-appropriate speed". In order to reduce this residual risk, *Passive Safety* and *Active Safety* must be complemented by *Tactical Safety*.

41.4 Risk Assessment for Highway Following Task

The qualitative results of **▶** Sect. 41.3 are quantitatively evaluated in this section for one specific hazardous scenario. For this purpose, we consider a set of situations of a car which follows with constant speed (v_F) behind another vehicle driven by an inattentive driver; this vehicle collides with the standing end of a traffic jam, in the worst case without any previous braking. This "*Inattentive Driver*" scenario is estimated to happen about once per month on German highways (Sat1NRW [18]) based on expert judgements, although it might have been resolved by an evasive steering maneuver in other (typically undocumented) cases; often an evasion is not possible because of surrounding traffic, and evasive steering during an emergency brake situation is typically beyond the capabilities of a normal driver.

According to its distance keeping strategy, the car follows with a safety distance of a certain constant following time gap (t_F). Assuming that the driver or the control software of the vehicle has an effective reaction time (t_R) for initiating an emergency braking maneuver until it reaches a constant deceleration. The effective time gap available to decelerate down to zero speed in this situation is the time margin (t_A):

$$t_A = t_F - t_R \tag{41.2}$$

The resulting distance available for braking (d_A) between the two vehicles at the instant when the constant deceleration would start is calculated as

$$d_A = t_A \cdot v_F = (t_F - t_R) \cdot v_F \tag{41.3}$$
$$\text{(no braking reaction of preceding vehicle)}$$

and increases linearly with following speed. **■** Figure 41.4 shows a graphic representation of this relation, for different time margins (effective time gap as difference between following time gap and reaction time). In the case that the inattentive driver still brakes, but too late to avoid a crash, the time margin (t_A) is increased by the "time to collision"-value (TTC_B) of the crashing vehicle when its brakes are activated:

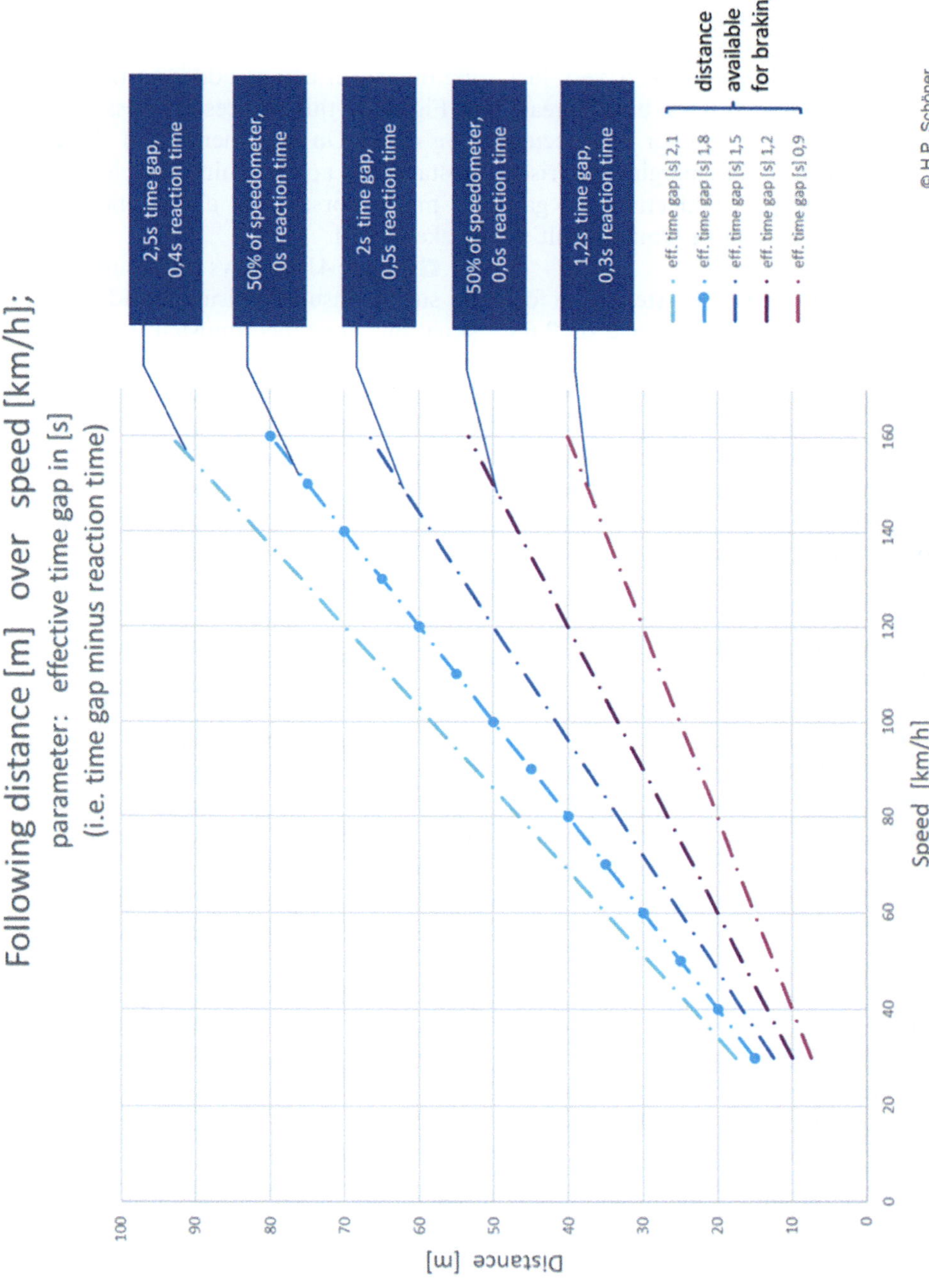

Fig. 41.4   The effect of time gap and reaction time on available braking distance in the "*Inattentive Driver*" scenario

$$d_A = t_A \cdot v_F = (t_F - t_R + \mathrm{TTC_B}) \cdot v_F$$

$$\text{(braking starts at } \mathrm{TTC_B} \text{ before crash).} \tag{41.4}$$

The proposed safety distance for highway traveling in many countries is either 2 or 1.8 s (the latter being equivalent to 50% of the indicated km/h speedometer value in meters; in Germany, steeply progressing monetary fines may be imposed below 0.9 s) [2]. However, in practice 60% of the drivers use distances of below 1.8 s, and still 20% drive with a very reduced safety distance below 0.9 s (Krajewski [11], Schöner [19]). A very fast human reaction time may be low as 0.6 s; but the reaction time varies widely, typically 1.25 s for unexpected routine situations and up to 1.5 s for emerging surprises are published by Abendroth [1]. The effective time gap available for the braking process alone would result in 1.2 s and even very much lower.

In Adaptive Cruise Control (ACC) systems, the following distances can be set within the range of 2.6 s down to 1.0 s according to Winner and Schopper [30]. In combination with a reaction time of ACC/Autonomous Drive systems in the range of 0.4–0.5 s (Winner [31]) the effective time gap may be assumed between 2.1 s (upper limit) down to as short as 0.6 s. It is important to note that the depicted lines in the chart do not even cover the shortest effective time gaps of practical highway driving.

In the considered simplified scenario the following car starts braking at the beginning of the effective time gap with a constant deceleration a_B. Emergency braking decelerations of human drivers typically perform with 4 m/s^2 and may reach up to 8 m/s^2 (Bussgeldkatalog [3]); Emergency Brake Systems (EBS) - at their strongest level - typically starting with 6 m/s^2 and possibly ramping up to the physical limit of 10 m/s^2 (Winner [31]). The theoretical braking distance d_B to bring the vehicle down to a full stop from the initial speed v_F can be calculated as

$$d_B = v_F^2 / (2 \cdot a_B). \tag{41.5}$$

The resulting distance needed to avoid a collision is shown as curved solid line in ◘ Fig. 41.5.

The curves for a very good human driver performance (1.2 s effective time gap and 6 m/s^2 deceleration) intersect at about 50 km/h, more or less at the same speed as a fast-reacting ACC system set at a rather short following distance and braking at its physical limit (0.9 s effective time gap and 8.5 m/s^2 average deceleration). As a result, for speeds significantly higher than 50 km/h the given scenario leads to severe collisions if an evasive maneuver is not possible.

◘ Table 41.1 gives an overview of the residual speed of the following car at the time when it collides with the vehicle of the inattentive driver. Even with relatively large time gaps the collision speed easily exceeds the

values up to which passive safety systems can provide injury-free protection (ca. 30 km/h).

The results for the *"Inattentive Driver"* scenario are similar to other scenarios when the following car has to react on standing objects on the road which are suddenly revealed, e.g., by an evasive steering maneuver of a preceding car (with roughly a vehicle's length longer available braking distance). If the inattentive driver of the front vehicle does react, but too late to avoid a collision, the TTC at the start of his emergency braking maneuver can be added to the effective time gap in ◘ Fig. 41.5; this reduces the missing distance and collision speed. On the other side, if there is not a standing obstacle but an oncoming vehicle, the conditions could be much worse, even if the oncoming vehicle would brake as well.

◘ Figure 41.6 shows the resulting safety assessment of such unusual collision avoidance scenarios. These situations are called unusual, because standing objects on the road - in combination with inattentive drivers (or with a preceding vehicle which only reveals sight on the object in a very last moment) are so rare that a single driver may never experience these situations. Nevertheless, these situations are frequent enough that in combination with their severity they become a significant safety case for a busy highway system. It is important to note, that one essential condition for such situations to happen is the presence of standing (or slow-moving) objects on the road-in combination with at least one additional negative condition. Since the additional condition "Inattentive driver" is hard to perceive by sensors, the presence of standing or slow-moving objects on a highway needs to be considered as a hazard by itself and should trigger additional caution - at least if the following distance to the preceding car is the only available safety reference.

When driving at a certain highway speed, the combination of traveling speed and (effective) safety distance can be considered *safe* in the upper left (green) part of ◘ Fig. 41.6, but clearly *unsafe* in the lower right (red) part of the figure. However, there is a wide (yellow hashed) region in which the assessment of safety depends on the presence of standing (or slow-moving) objects. If there is a method to know with a sufficiently high probability that the road ahead (in front of your preceding car) is free from such objects, the yellow hashed region can be assumed to be *safe* as well. If, however, there is no knowledge or even negative evidence about the free drivability of the road ahead, the yellow hashed region has to be considered *potentially unsafe*. The tactical safety consequence in behavior would be to increase the safety distance or to reduce traveling speed when driving on such *potentially unsafe* highway sections. In more colloquial terms: if you do not definitively *know* whether the road ahead is

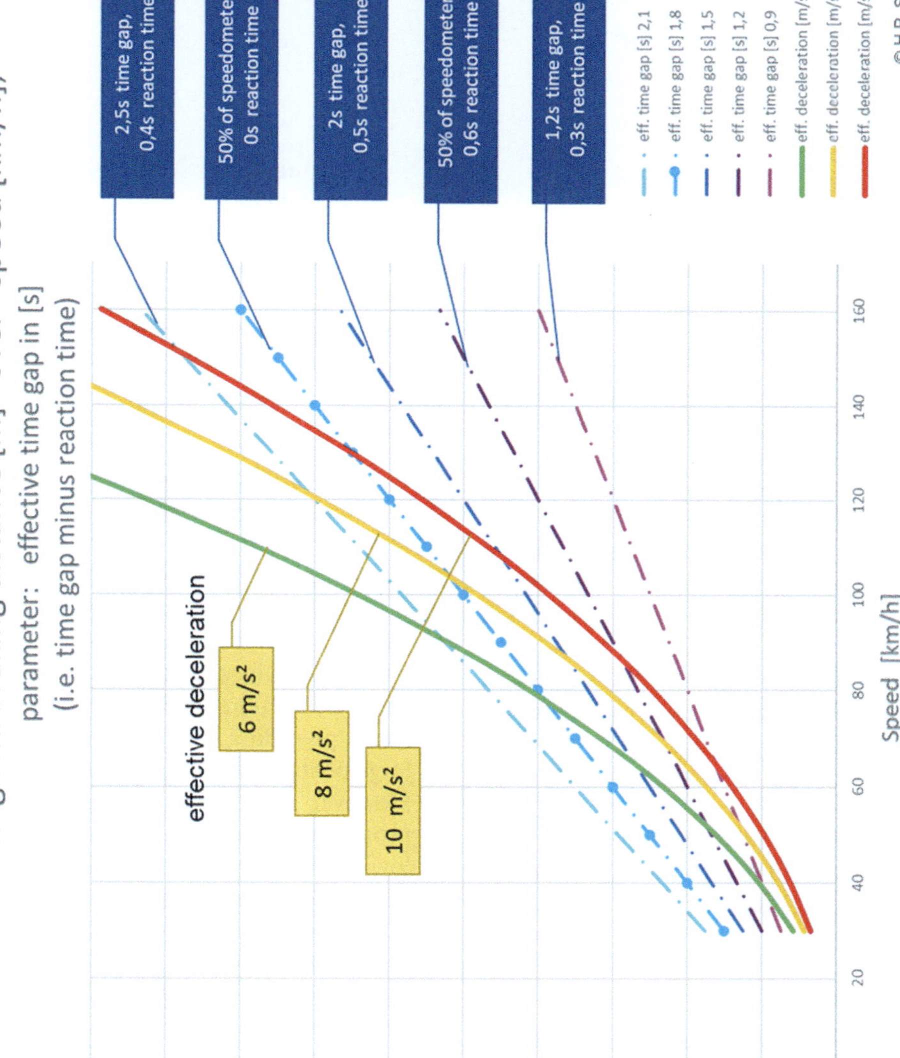

Fig. 41.5   The influence of effective deceleration on braking distance in the "*Inattentive Driver*" scenario

☐ Table 41.1 The influence of initial speed, time gap, reaction time and effective deceleration on residual speed at the time of collision in the (no braking) *"Inattentive Driver"* scenario

Initial speed (km/h)	Following time gap (s)	Reaction time (s)	Effective deceleration (m/s²)	Residual speed (km/h)
90	2	0.4	6	43.3
90	2	0.8	8	43.3
90	2	1.2	10	54.0
90	1.6	0.4	6	58.6
90	1.6	0.8	8	62.9
90	1.6	1.2	10	74.2
120	2	0.4	6	78.1
120	2	0.8	8	78.1
120	2	1.2	10	86.5
120	1.6	0.4	6	90.4
120	1.6	0.8	8	94.2
120	1.6	1.2	10	104.6

free, you should drive with a combination of speed and safety distance that you are sure to handle even unusual surprising situations.

In order to assess generally the severity of a collision resulting from a wrong estimation of the safety distance, the following relation should be noted: the final speed at a crash into an assumed standing vehicle ahead depends only on the missing final deceleration of the following vehicle; it is not a function of *initial speed*, but only of missing safety distance and assumed braking deceleration for the safety distance. From kinematics theory, it is a square root relation, as depicted in ☐ Fig. 41.7.

This finding can be used for further discussions and conclusions, using the SAE accident severity classification S0–S3 (SAE J2980 [17]), which is roughly equivalent to the following rules of thumb for passenger car accidents:

— A residual speed of up to 30 km/h (8 m/s) at a front collision can be expected to lead to an outcome without injury for the passenger in most cases, provided the availability of state-of-the-art passive safety systems.

— A residual speed of more than 60 km/h (16 m/s), however, might go beyond the verified capabilities of modern passive safety systems; the outcome is not predictable and may lead to severe injury or even death of the passengers.

In short, a missing safety distance of 15 m (equivalent to the added length of three vehicles) leads to a medium risk if the applied safety distance is based on low braking deceleration, but it is likely to lead to severe risks when safety distances are designed "to the point", i.e., based on high emergency braking decelerations. The short safety distances based on large emergency brake decelerations seem to provide the same safety margins than the longer safety distances based on less deceleration, but their robustness to miss-judgments is much worse.

41.5 Implications for Autonomous Vehicle Testing Scenarios

The main results which we can take away from the discussion in ▶ Sect. 41.4 are the following:

— The common method of keeping a constant time gap as a safety distance is sufficient to provide enough reaction time for initiating similar actions like a preceding car; this is safe enough for a very large portion of everyday driving.

— However, this comes to its practicable limit on approach to an unusual dangerous situation if the following car cannot perform the same evasive maneuver as the preceding vehicle, or should not share the same fate as the preceding vehicle (like crashing into an object after reacting too late).

— Human drivers use additional hints to perceive danger on the road ahead beyond just detecting and following the action of the preceding vehicle: they observe the traffic further ahead, recognize distant brake lights, or other (blinking yellow, red, or blue) warning signals, react on human or technical warning signs or warning messages. And they consequently change their following behavior by increasing safety distance or by reducing their speed in order to cope with unusual highway situations: they apply precautious driving, or "tactical safety". Human drivers who fail to do so are prone to be involved in severe accidents.

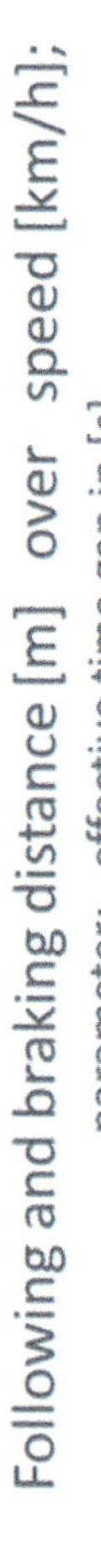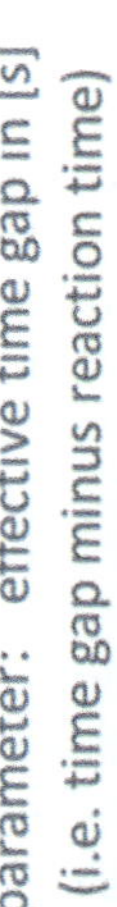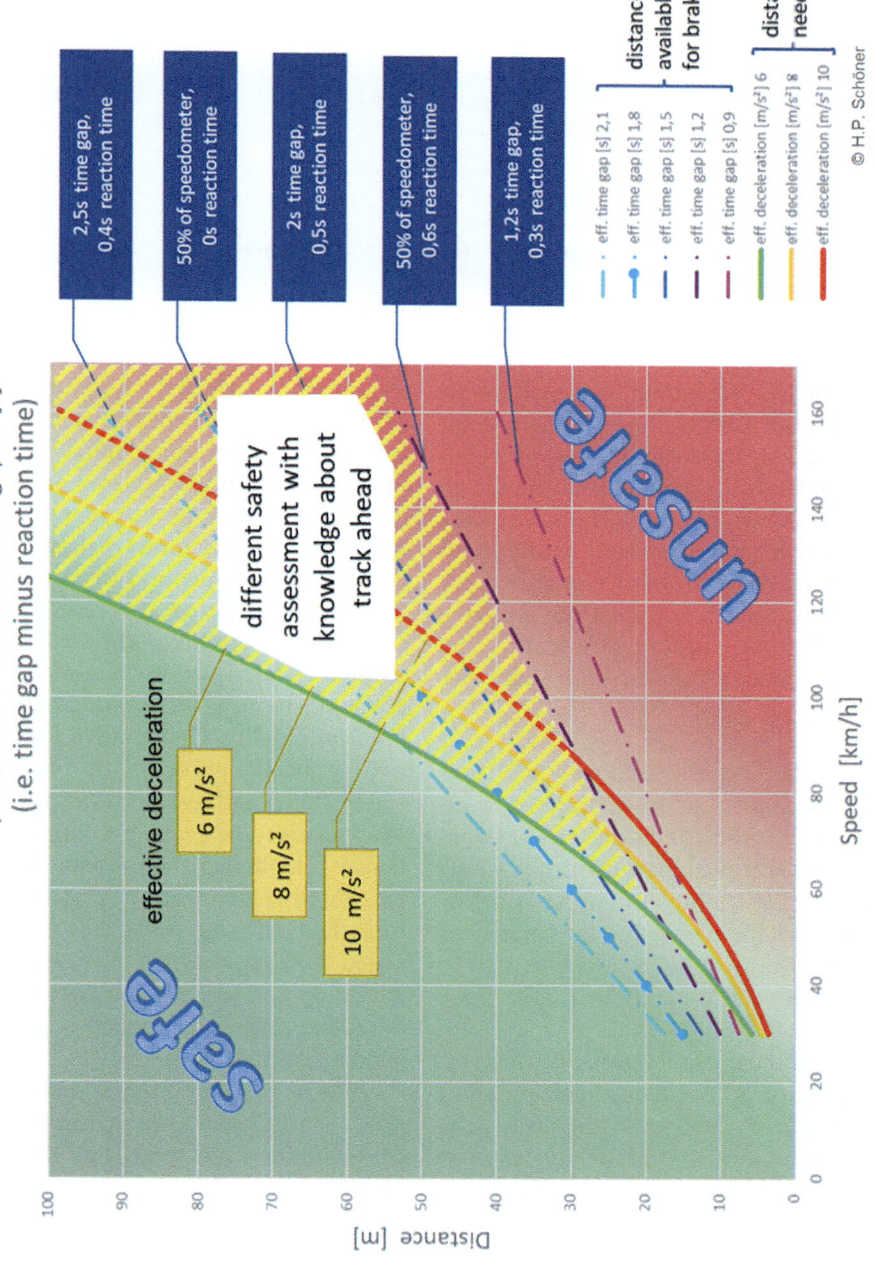

Fig. 41.6 Safety assessment of car following in unusual collision avoidance scenarios, based on the *"Inattentive Driver"* scenario

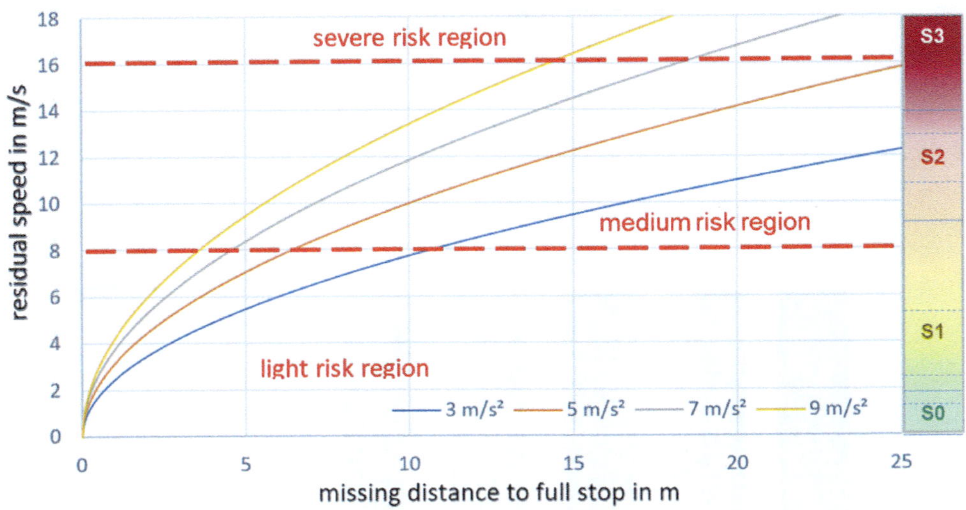

Fig. 41.7 Effect of a too short safety distance on collision speed, and finally on collision severity according to SAE J2980 [17]

- Autonomous vehicles, including state-of-the-art adaptive cruise control systems, using only a common constant time gap following strategy are neither able to cope safely with such unusual situations on highways. Even worse: if set to a short following time gap at high driving speed, several autonomous vehicles would all crash when following an inattentive human driver who does not react on approach to a standing object (e.g., vehicle in a traffic jam, lost cargo) or an oncoming vehicle and collides violently.

In these unusual cases, high severity and very poor controllability come in combination. In addition, the exposure to such unusual traffic conditions is not negligible: collisions of inattentive human drivers crashing into the standing end of a traffic jam happen roughly once a month on German highways, and the severity of such collisions is high. Many of such collisions lead to three and more involved vehicles, causing a large share of deaths on highways.

The consequence of these observations is, that verification and validation of autonomous vehicles which are designed for high speed (≫50 km/h) highway driving should include to check, if and how they can cope with these unusual critical situations (see test cases in ▶ Sect. 41.7.2). Weast [28] pointed out that the definition of parameter values for the physics-based formulas for safe driving behavior is a matter of assumptions. The established parameter values (like recommended safety distances) for human driving are based on the assumption of "normal" driving conditions, and asking for human common sense and good judgement ("roadmanship") to cope with unusual conditions. For autonomous vehicles, we propose the incorporation of a technical concept for such Tactical Safety Behavior: change the recommended behavior parameter values early and smoothly based on indicators for possibly dangerous situations in order to prevent getting into uncontrollable situations (Schöner et al. [23]).

One obvious method to identify such unusual conditions would be to use recently generated information about the state of the road ahead in order to activate a different (more cautious than usual) following behavior, when a free road cannot be expected with a high probability. Cooperative Sensing of the road status by traffic participants, providing crowd-sourced safety status and hazard warnings in a Real-Time HD-Map seems to be a viable solution: the higher the traffic density, the higher the probability of an occluded danger, but also the more vehicles are ahead with a chance to issue a warning. Test cases for this concept are proposed in ▶ Sect. 41.7.4; they require to fulfill several parameters in order to be effective. With such specific parameters, the effect on the residual risk can be calculated quantitatively.

A generally accepted specification of the Cooperative Sensing test case parameters should give clear goals for the development of the Real-Time HD-Map safety status and hazard warning function in order to provide a dependable basis for Tactical Safety behavior, and in such a way fill a significant gap in the safety concept of Autonomous Highway Driving at speeds clearly above 50 km/h. Especially for highway driving, low-speed traffic conditions, and traffic jams (although being everyday conditions) must be considered as safety hazards—not only icy roads and accidents which have happened already.

41.6 A Concept for Dependable Cooperative Sensing

Although there is a wide consensus that communication must play an important role for the realization of safe autonomous traffic, there are basically no concepts with specific allocated tasks for the communication in the safety argument. In UN regulation 157 [32] which allows ALKS systems on highways with speeds up to 130km/h, communication is mentioned as a possibility to issue warning messages, but there is no specific requirement assigned to communication. And with the lack of these concepts, testing scenarios for verification and validation of the safety impact of communication are not yet defined and implemented.

But the technical foundation is available: Standards have been developed for transmitting messages about relevant events and specific dangers via communication networks, like SensorIS [25]. Map providers are working on hazard warning concepts which are able to broadcast traffic conditions within a real-time HD-map, e.g., accident locations, slippery road sections and bad weather conditions; the information takes only a few seconds from first observations to public availability (e.g., HERE [6], TomTom [[25]]). The political will to implement such a safety-related traffic information ecosystem has also been established (e.g., ERTICO [3]).

One important reason for the missing role of communication in safety concepts is that the information coming in by traditional communication is considered to be *not dependable*:

1. In contrast to information produced by in-vehicle sensors, there is no way to request a specific information about a certain region of interest; relevant data is coming in stochastically and cannot be guaranteed to be complete. And with a certain small probability it could be completely wrong or misleading information because of unreliable sources, intentional misuse or other security-related reasons. The absence of a hazard-warning cannot be taken as evidence for the absence of a hazard, since a failure of communication would have the same effect.

2. Even if data is available, it might come in too late for fast and strong Active Safety actions, like emergency braking or evasive steering, or might be outdated; there is always the need for checking the timing and intensity of action based on vehicle-internal sensors.

In the context of Tactical Safety, however, information via communication can be used in a different way: in contrast to Active Safety, Tactical Safety does not act based on acute last-second information, but it can make use of information from a larger time frame.

Thus, it poses less requirements on latency and precision.

Nevertheless, it needs to fulfill certain quality requirements. As we know from many crowd-sourced data, these can be quite reliable when they are abundant, coming from different sources and their statistical data properties are evaluated for consistency. Cooperative sensing needs to support this quality assurance.

In addition, at least for driving on high-speed highways, turning around the logic of emergency warnings will improve the situation. Instead of waiting for a warning in order to drive more carefully than what is normal for highway driving, an autonomous traffic information system should be set up in a safety-oriented way: the "default standard" should be sufficiently large safety distances even for the worst-case conditions, and only if a dependable information is available about a free road ahead, then driving with "normal" safety distances and speeds becomes acceptable. This is the type of logic which safety-related systems are using in all other implementations.

The most important point for this goal is a consequent step

- from reporting traffic hazards stochastically "without any obligation"
- to a dependable information about road status with indicated actuality and known completeness,

at least for highways with operating speed for AVs above 60 km/h. An implementation is possible with a sufficiently large number of participants jointly providing cooperative sensing, for example in the following way.

Since the most important factor for a high severity accident (▶ Sect. 41.2) is a large speed difference, an easy and probably most significant indicator for the risk status of the road ahead is the speed at which vehicles are driving. In addition, almost any relevant incident on the road will be reflected by unusual or changing driving speed. This data is available in high volume and good coverage at any traffic data provider, deducted from existing navigation systems or road sensors. However, the precision, time and location reference of this data can be significantly improved by direct provision from the vehicle control systems of every participant; in addition, the knowledge of traffic speed for each driving lane separately would significantly improve its usability. Own speed and driven lane are robust state-of-the-art state values in modern driver assistance and navigation systems.

This data becomes dependable information if supporting meta data can be provided with it by a service provider, especially how many data reports are available per road section and in a defined recent time interval. The information *"road traffic is flowing in lane i at the indicated speed"* can be trusted above a certain number of consistent reports in the last few minutes (e.g., the last 3 min). But also the knowledge that there are none or very few data points is an evaluable information that can be used, in conjunction with the knowledge of the own vehicle's unobstructed sensing horizon, to deduct suitable or possibly dangerous driving behavior. The communication serves as a redundant, diverse and far-reaching source for the speed of the traffic ahead.

As an example, a possible representation of the traffic flow ahead is depicted in ◘ Fig. 41.8, showing one driving direction of a three-lane highway: for each highway section of a suitable length the prevailing speed is provided, most important being the slowest speed, but also average and maximum speed can be relevant to interpret a changing or chaotic traffic situation. It might be sufficient to provide this in speed intervals. Paramount is the actuality of the information and the data density in time and space, for example by providing the number of speed reports per section in a gliding recent time interval.

In addition to the basic data from vehicles reporting their own speed and lane, which supply the information of actuality and completeness by their regular data flow, further reports about information derived from vehicle sensors (see SensorIS [25]) can enrich the data. This may include speed and location of non-reporting vehicles or objects, but also categories of other traffic events, environmental hazards, or inconsistency warning messages (see test cases in ▶ Sect. 41.7.4). Again, this data has to be provided with the number of reports in a gliding recent time interval.

What kind of functions can be deducted from this information? Here are some examples:

— Every participating user gains information about the traffic speed ahead and can deduct (in its own responsibility) a speed profile without the need for heavy braking. The user can do this independently from occlusions and from the behavior of its direct predecessor and thus becomes more robust against possibly incorrect actions of a single traffic participant.

— Every user has also an indication how old and how robust this information is and has a basis to decide for tactical speed reduction/safety gap increase because of missing or insufficient information, in conjunction with its own sensing horizon.

— Road management authorities can use this information to adapt dynamic speed limit signs.

— It would give authorities more freedom to specify the values for safety-relevant parameters in RSS (Weast [28]): the safe driving rules can be defined dependent on a specific availability or quality of look-ahead data. This can resolve trade-off situations in the selection of parameter values between traffic flow and safety for unusual situations.

For the concept presented above, a lot of questions still need to be addressed, discussed, and harmonized: Are the indicated attributes sufficient for these functions? Which precision and resolution are needed? What is a practical gliding time window? How long should the lane sections be to be practical, e.g., for not missing a standing object, because reporting vehicles drive around? Which minimum requirements should there be on the number of reporting cars, length of sections, precision of information, turn-around time from sensing to map, etc.? And there might be other good solutions for dependable cooperative sensing of road status in addition to this proposal.

But this example should provide evidence, that cooperative sensing can substantially contribute towards a safety case of communication when it provides some new ingredients and features. It should be considered to *require* that autonomous vehicles - and maybe new ADAS functions as well - must participate in such cooperative sensing, in order to assure a sufficiently large number of participants and a fast traffic penetration even for a slow introduction of autonomous vehicles. Tactical Safety actions based on such information would not only significantly reduce the risk for severe collisions of autonomous vehicles in unusual traffic sit-

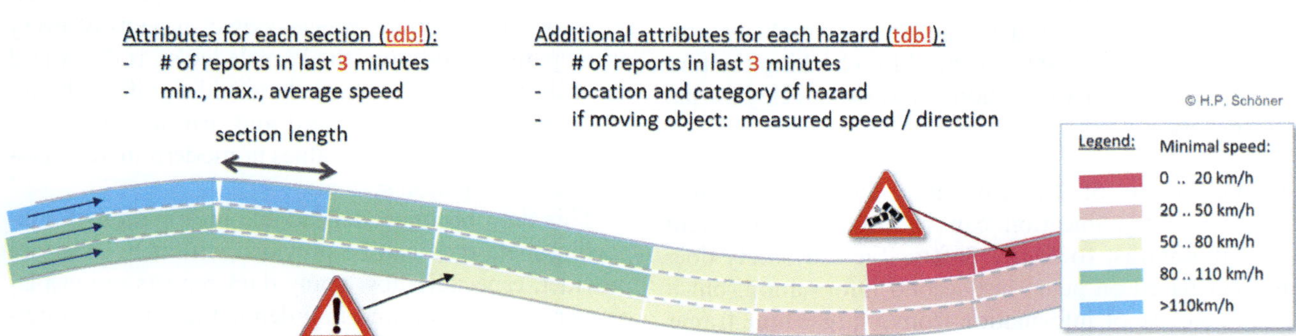

◘ **Fig. 41.8**   Sections of a three-lane highway with indicated actuality and known completeness of speed information and specific other hazards

uations but also could be issued as a driver counseling function for human driven cars.

41.7 Challenging Highway Scenarios

In this section, a set of challenging highway scenarios is collected which have a chance to avoid uncontrollable situations by cooperative or precautious actions. These should be considered as one part (especially in addition to acute collision avoidance scenarios) for the safety assessment of autonomous vehicles before large scale introduction of automatic highway driving.

Some of the interaction scenarios call for a definition of borderline cases (also called "limit situations"); i.e., scenarios with a specified set of parameters at the limit of endangering behavior or collision avoidance. They need an international agreement on some parameters, because there is no "natural" limit for outer conditions. A proposal for an initial set of such parameters has been published in an online document (Schöner [22]).

In addition to the interaction scenarios, several severe foreseeable hazard scenarios and scenarios for long-range object detection are included here, because they are relevant for the overall safety assessment of driving in (highway) traffic scenarios.

Table 41.2 gives an overview of the highway scenario categories which are included in the set. The four categories are characterized by different expected occurrence and expected severity of a possibly resulting collision. Because of these different conditions there are also different goals for the controllability of the scenarios.

1. Difficult Traffic Situations

These are foreseeable traffic situations with some challenges for driving skills, which happen every day; cooperative and predictive behavior helps in reducing the occurrence of danger and collisions out of these situations. An example of this category is "merging at lane closures" on highways. The passing criteria for those situations are: flawless behavior, i.e., do not endanger yourself and others and enable smooth traffic flow.

2. Extraordinary Traffic Situations

These are foreseeable traffic situations, which might happen once per week or less in the traffic statistics of a region. They are unusual for a single vehicle but need to be considered for the traffic community in total. The "inattentive driver" scenario is one example of this category. The passing criteria for those situations, depending on specific parameters, are: early hazard perception and behavior adaption, no collision/at least no injury.

3. Severe Foreseeable Hazard Situations

These hazardous situations are rare and unpreventable, but foreseeable situations, although not foreseeable in detail. Because of their probably severe effects, they have to be considered in a safety assessment. An example of this category is coping with "oncoming traffic" on a one-way highway. The passing criteria for these situations are showing a reasonable mitigation effort and avoiding severe collisions.

4. Long-Range Sensing and Occluded Situations

These situations are needed for high-speed driving on highways with a safety-relevant sensing horizon beyond the practical limits of the own sensors; risk reduction is based on use of communication and cooperative vehicle interaction. An example is verifying adequate "reactions on warnings via communication". Passing criteria here are: showing evidence of basic cooperative capabilities in traffic, suitable supervision of these capabilities, and reasonable reaction in case of faults in the cooperative system.

The scenarios in those different categories are shortly described in the next sections. A more detailed discussion of these scenarios is provided in (Schöner [22]), including:

- purpose of the scenario,
- layout of the scenario,

Table 41.2 Categories of challenging highway scenarios with different exposure, severity, and controllability levels

Category	Occurrence / Exposure	Expected Severity	Foreseeable	Preventable	Goal for Controllability
Difficult traffic situations	~ every day	low - medium	yes	yes	**flawless behaviour**
Extraordinary traffic situations	~ once per week or month	high	yes	limited	**avoidance, no injury**
Severe foreseeable hazard situations	rare or very rare	probably high	yes, but not in detail	no	**mitigation**
Long range sensing & occluded situations	~ every day	possibly high	yes	largely by communication	**verify the basic function**

- key parameters with need for an international harmonization,
- passing criteria.

41.7.1 Difficult Traffic Situations

» Drive flawlessly in difficult everyday situations.

These are foreseeable traffic situations (see ⬛ Fig. 41.9) with some challenges for driving skills, which happen every day; cooperative and predictive behavior helps in reducing the occurrence of danger and collisions out of these situations. Such danger can occur through unexpected behavior of a vehicle because it is surprising for surrounding human drivers with respect to timing and intensity of the action. Since everyday scenarios happen in extremely high numbers, unusual behavior will inevitably lead to a significant number of accidents in traffic, although the single unusual behavior (like a strong braking reaction) should be manageable by most traffic participants under nearly any conditions.

- **Driving towards a blocked road**: When driving towards a known end of a traffic jam or a known object on the road, the vehicle should show a deceleration profile which does not lead to even stronger braking reactions of following traffic. The vehicle should show that it is not only relying on the environmental sensors but also on information provided by a "Live Map" or C2X communication, given this information is available. Independent from the speed of the preceding car the vehicle should reduce its own speed on early approach in order not

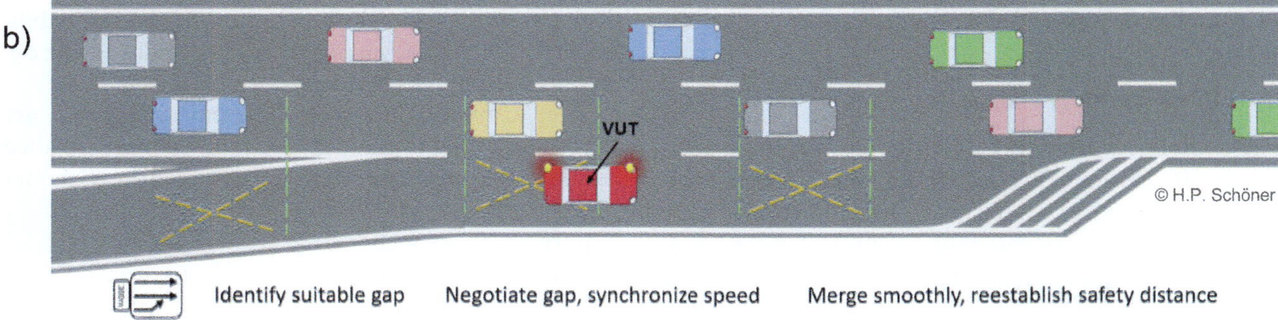

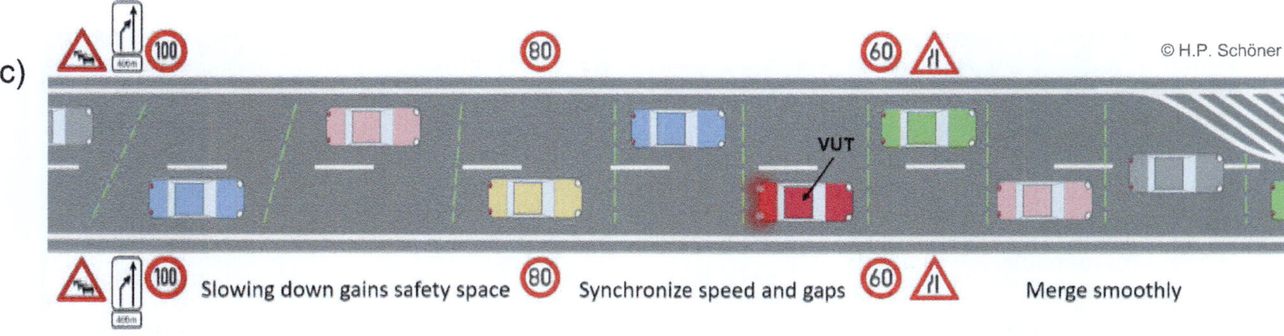

⬛ **Fig. 41.9**   Everyday difficult situations: **a** driving towards a blocked road, **b** forced merging, **c** cooperative merging, **d** emergency vehicle

to brake too hard during final approach, even if the preceding car reacts very late.

- **Forced merging**: This scenario is a required lane change, as seen from the lane-changing vehicle, when one lane in front is ending or blocked, under dense traffic conditions. Changing lanes is especially challenging, if the own lane is ending in a short distance (be it an acceleration lane, a regular merging zone of two lanes, or because of lane closure due to road work). This scenario should verify whether the vehicle is able to identify in an early stage its place for merging into the parallel traffic, adjusting speed and position for this lane change, and perform it safely without forcing other vehicles to brake or steer vehemently.

- **Cooperative merging**: In contrast to the previous scenario, for this one the vehicle under test is in the receiving lane. The vehicle should show cooperative behavior by early and adequate reaction on lane change indicators (turn signal, body move). Goal is the avoidance of too severe reactions which can propagate with increasing effect to the following traffic, thus impairing safety.

- **Emergency vehicle**: an emergency vehicle in service is approaching from the rear with horn and warning lights on. Verify the early recognition of and reaction on the approaching emergency vehicle. Passing criteria is slowing down/stopping and providing space for passing.

The general passing criteria for the behavior against other participants in these difficult traffic scenarios should be, not to endanger anybody or to irritate the other participants, and not to hinder traffic flow more than necessary. Important for the limit cases is the maximum to be expected speed of other traffic participants; without a speed limit no reasonable safety assurance can be provided. For example, the limit for expected speed should be 20 km/h (this value to be discussed and decided among international experts) more than the prevailing speed limit; emergency vehicles (at least when driving faster than this limit) need to be equipped with a radio warning system.

41.7.2 Extraordinary Traffic Situations

» Know your limits in extraordinary situations.

Extraordinary traffic situations are foreseeable traffic situations, which might happen once per week or less in traffic statistics. They are unusual for a single vehicle, but need to be considered for the traffic community in total (see ■ Fig. 41.10).

- **Following an inattentive driver**: Verify the performance of the vehicle under test in case of an inattentive preceding driver who crashes violently without braking into the end of a traffic jam. Inattentive drivers in unexpected traffic jam scenarios are in the range of 1–10%. If this extraordinary situation happens, it always leads to multi-vehicle collisions because safety distances which are used in practice by human drivers do not provide enough reserves for this case, see ► Sect. 41.4. Validate the safety distance strategy with respect to this collision type under maximum speed conditions for the ODD of the autonomous vehicle; verify that the ve-

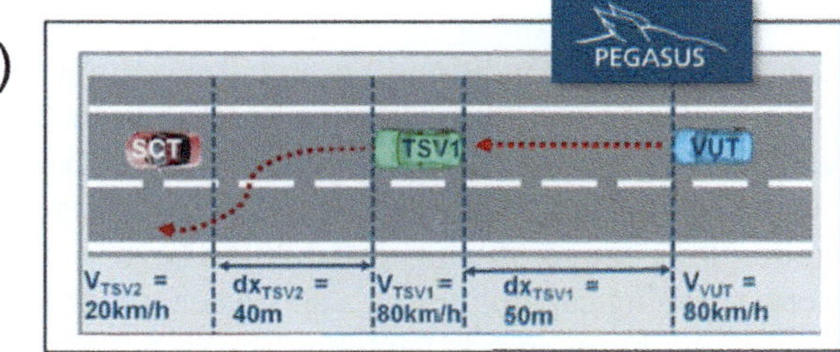

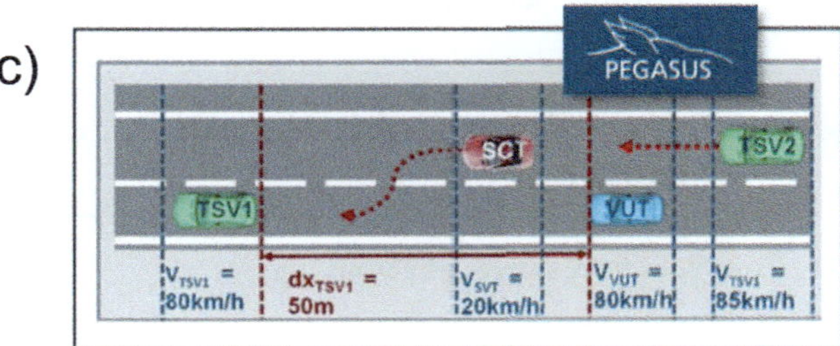

■ **Fig. 41.10** Extraordinary traffic situations: **a** following an inattentive driver, **b** cut-out with obstacle on the road, **c** cut-in with heavy braking

hicle considers hereby its ability to reliably judge the traffic speed in front of a preceding vehicle. For this scenario, a borderline constellation for "no collision"/"collision without injury" has to be established. For several following vehicles compare the performance of the second and third vehicle to the first vehicle; perform an assessment of the convoy stability of this severe braking scenario.

— **Cut-out with obstacle on the road**: Cut-out of the preceding vehicle gives a view onto a blocked road with a standing obstacle (a car or any other object). Verify the performance of the vehicle's collision avoidance under tight timing conditions when changing from vehicle following to collision avoidance with respect to a newly appearing object. Validate whether the following distance strategy of the vehicle is suited for such extraordinary traffic situations under maximum speed conditions for the ODD of the autonomous vehicle.

— **Cut-in with heavy braking**: Cut-in of a vehicle with high differential speed and/or heavy braking in front (challenging the vehicle on the receiving lane). Verify the fast and adequate reaction on a vehicle which violates the normal safety margins of the vehicle under test by this close cutting-in and full braking. Verification of short reaction time and adequate braking. The scenario simulates rowdily take-over maneuvers. One important aspect to be defined is the maximum speed to be expected for the cutting-in vehicle, and the maximum deceleration when it brakes just in front of the vehicle under test. Anticipation of the cut-in based on understanding of the traffic situation and early braking as a tactical safety measure reduces risk of a collision.

The passing criteria for the scenarios depend on the specific parameters; they range from "no collision" up to "mitigation of the collision" (no injury expected under the specified conditions, i.e., crash speed not higher than, for example, 30 km/h (this value has to be discussed and decided among international experts)). Avoiding collisions with practicable safety distance requires long-range sensing capabilities beyond the inherent sensing capabilities of a single vehicle, at least at normal highway speeds (>60 km/h); see ▶ Sect. 41.7.4 for the long-range sensing test scenarios.

41.7.3 Severe Foreseeable Hazard Situations

» Cope with rare, surely dangerous failures.

Severe foreseeable hazard situations are rare and unpreventable, generally (but often not in detail) foreseeable situations with possibly severe implications

(▪ Fig. 41.11). Because of their high severity, they have to be considered in a safety assessment. In some of the perceivable situations, a collision might be unavoidable, and there is not the perfect "expected behavior" which works for all of them.

— **Cut-out with oncoming vehicle**: Cut-out of preceding vehicle gives view onto an oncoming third vehicle in the own lane; defines borderline challenge of the following vehicle; since the oncoming vehicle is endangering (and heavily against the rules), an outcome without collision cannot be guaranteed. Demonstrate the strategy of the vehicle under test for mitigating a collision with an oncoming vehicle on the own lane. Verify reaction time, braking, and steering strategy. Vehicle should move out of the lane or at least to the side of the lane as much as possible (provide enough passing space), bringing speed down by braking; and sending out warning signals.

— **Human warning signs**: Verify the recognition of and reaction on an unequipped human making a warning sign. This could as well be a police officer or traffic controller giving signals. The warning sign has to be defined; for example, a person with a waving arm is standing next to the lane. Passing criteria is to reduce speed to a very low value, and come to a full stop when the person is stepping into the driving corridor.

— **Blind stopping procedure**: Emulation of failure of the complete set of environment sensors (like in a sudden heavy thunderstorm or dust storm); verify the avoidance/mitigation of a severe collision by the vehicle's strategy. Implement a sudden shut-off of all sensor signals, performed with maximum speed on a winding road (parameters according to ODD); goal is to come to a full stop based on knowledge about the track before loss of sensor data. Passing criteria is: staying on the drivable road, and not braking harder than following vehicles can expect and cope with.

— **Failure to take over by driver**: The driver does not react on a take-over request of the vehicle. Verify the vehicle's strategy in the case that the driver does not respond (in time) to a take-over request. Slowing down, increasing the urgency of the take-over request, turning on warning signals (via lights and communication), and finally stopping at a safe location outside of the lane; if the latter is not possible in a reasonable distance, stopping even in the lane is acceptable.

— **Defensive fall-back mode**: After an unexpected, sudden take-over request the driver might not be able to judge all situations adequately and perform wrong (un-reflected, possibly dangerous) actions. Verify adequate provisions of the vehicle for this case. An adequate fall-back mode hinders the driver to perform stupid or dangerous actions, providing

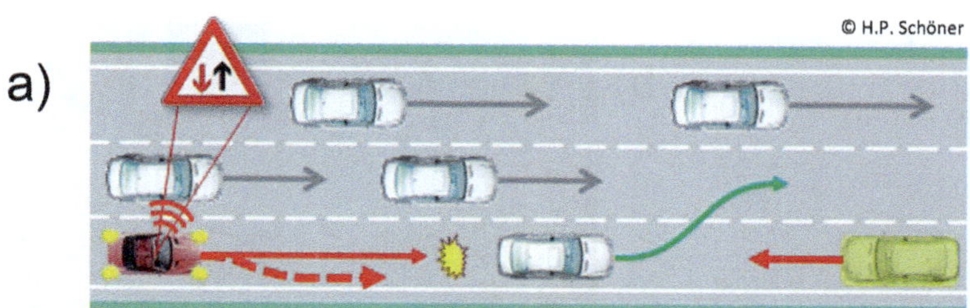

a)

b)

c)

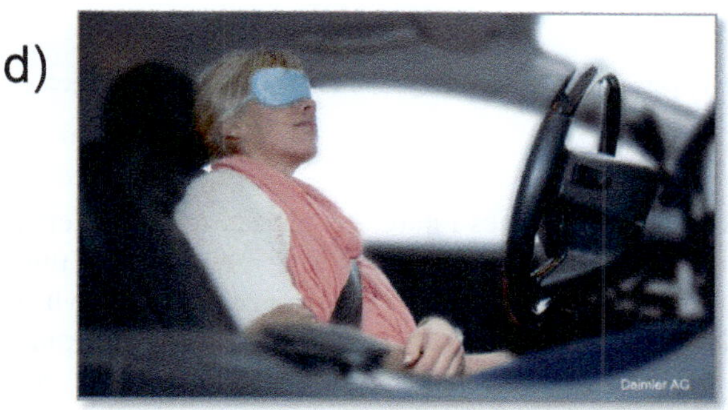

d)

e)

Fig. 41.11 Severe foreseeable hazard situations: **a** cut-out with oncoming vehicle, **b** human warning signs, **c** blind stopping procedure, **d** failure to take over by driver, **e** defensive fall-back mode

a "Protector" function (active control functions are handed over to the driver, but protective features are still fully active).

The passing criteria for all these hazard situations are: show a reasonable mitigation effort.

41.7.4 Long-Range Sensing and Occluded Situations

» Cooperate in long-range sensing.

Long-range sensing situations are needed for high-speed driving on highways with a safety-relevant sensing horizon beyond the range or accessibility of the own sensors; under such conditions risk reduction relies on the use of communication and cooperative ve-

hicle interaction. The vehicle has to prove that it is able to know and understand its own sensing limitations, and it has to be able to judge the availability of adequate cooperative information. Besides this, the vehicle has to show that it can actively participate in the collective effort to increase safety, and how it reacts in traffic when the safety-relevant information is unavailable or unreliable.

- **Awareness of sensing horizon**: Verify awareness for maximum sensing distance. The vehicle should show that it is aware of a reduced detection range, and that it has measures to react on this fact. Detect a small object (limit case) and lane markings under different lighting and weather conditions. Show awareness of a reduced detection range, and perform reactions on this fact.
- **Cooperative long-range detection**: Show that the vehicle is able to collect relevant data for its own and adjacent lanes on the road in order to provide warn-

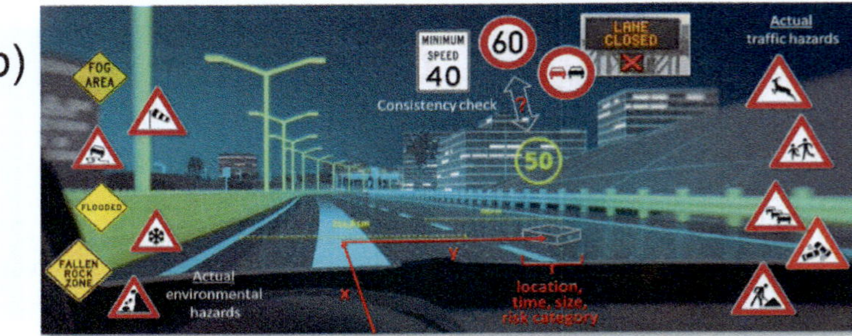

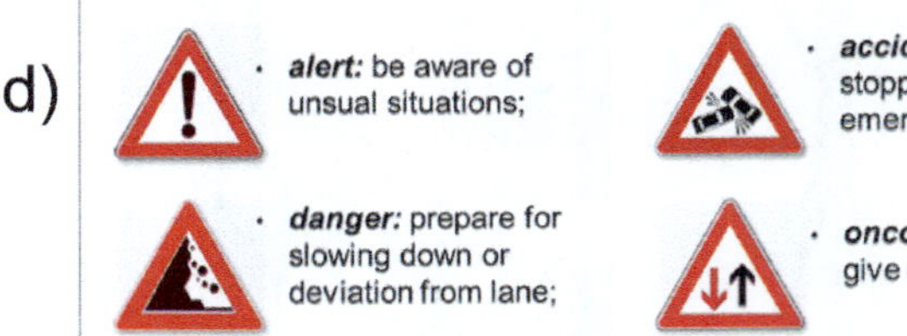

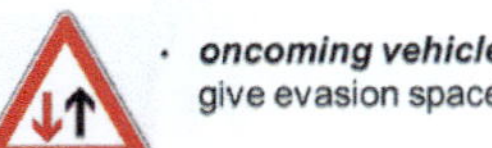

◻ Fig. 41.12   Long-range sensing situations: **a** awareness of sensing horizon, **b** cooperative long-range detection, **c** unusual hazardous object on the road, **d** reaction on warnings via communication

ing messages about road status for a real-time map (road status data base in the cloud). The vehicle can distinguish different danger classes of obstacles in/ near the own or adjacent lanes, and it is able to provide a classified warning message. Turn-around time of a warning message from detection to occurrence of the warning in the real-time map is smaller than 1 min {this value has to be agreed upon internationally, because it determines the residual risk of not providing an up-to-date real-time map}.

- **Reaction on loss of electronic horizon**: If (at high speed) the safe detection distance of the vehicle relies on knowledge of the real-time map (like "road ahead is free with a high probability", but it cannot verify this fact with own sensors), the vehicle has to show how it reacts on the loss of this affirmative real-time information. The vehicle changes after a tolerated time without information of 1 min {this value has to be defined internationally} its following strategy: The vehicle should assume, that the road ahead might no longer be free. The behavior change with respect to following distance and speed is in accordance to the results of scenarios, which assess the additional risk of driving without this information.

- **Reaction on warnings via communication**: Check warning reception and behavior adaption of the vehicle (incl. driver take-over request) when communication (real-time information) issues a warning of

different categories related to traffic, road and environmental conditions (e.g., *alert*: be aware of unusual situations; *danger*: prepare for slowing down or deviation from lane; *accident*: prepare for stopping, expect emergency vehicles; *oncoming vehicle*: give evasion space) (◻ Fig. 41.12).

Passing criteria for long-range sensing situations are: showing evidence of basic cooperative capabilities in traffic, suitable supervision of these capabilities and reasonable reaction in case of faults in the cooperative system.

41.8 Tactical Safety Supplements Passive and Active Safety

One important goal of the design of technical systems is to reach a certificate as a "safe product". This goal tends to bias the mindset of vehicle designers towards the idea that autonomous vehicles are either safe or unsafe, and that *controllability* performance tests can make a clear cut between safe and unsafe. This approach to safety works for a well-defined set of use cases within a restricted operational domain.

For a larger operational domain with traffic and environmental conditions that may reach the controllability limits, however, a residual risk will always remain. Therefore, the management of the system's controllability limitations becomes necessary. By adapting the driving behavior in a way that reduces the *exposure* to uncontrollable dangerous conditions, the risk can be limited to sufficiently low values, without degrading the performance of the system under normal conditions. Test cases for verifying such *Tactical Safety* behavior shall address:

- a clear understanding and supervision of the system's limitations,
- the recognition of unusual conditions causing approach to these limitations,
- an early and smooth reaction on those conditions.

The set of scenario tests proposed in ▶ Sect. 41.7 is not necessarily complete for a scenario-based safety assessment, but it covers some important aspects of highway traffic situations and possible severe hazard conditions. The scenario-based safety assessment has to be complemented with other kinds of safety issues, for example technical vehicle integrity, weather conditions, etc. The proposed scenario set and the details of description of the scenarios need to be adapted based on discussions in standardization working groups.

Although this chapter focusses on highway scenarios, the concept of *Tactical Safety* can be applied in a similar way to other operational driving domains. The principal objective *"stay consciously away from hazards which you might not control"* stays valid. The main contributing risk parameters for exposure, controllability and severity need to be adapted and extended, and new scenarios and hazards must be added. In city driving domains, the prevailing vulnerable road users open further severity hazards, and the severity levels caused by high differential speed need to be interpreted differently. In cross traffic and turning situations, controllability is still dominated by available reaction time, but this does no longer depend on longitudinal safety distances alone. In fact, the principles compiled in this chapter need to be further extended and elaborated for such new domains.

Tactical Safety is emerging as a relatively new field that is complementary to the well consolidated *Passive* and *Active Safety* fields. As increasingly complex automated and autonomous systems meant for complex operational domains penetrate the market, the field of *Tactical Safety* can be expected to evolve and gain relevance. In the views of the authors of this chapter, the success of researchers, autonomous driving systems developers and safety policy makers in evolving this field in the right direction will be critical to reach the full potential safety benefits of autonomous driving.

Acknowledgements The research effort that led to the contents of this chapter was supported during the years 2019 to 2021 by the Ministry of Economy, Trade and Industry of Japan, through the SAKURA project. Sou Kitajima (skita@jari.or.jp), a senior manager from the Japan Automobile Research Institute (JARI) and currently responsible for the SAKURA project is sincerely acknowledged for supporting this research and its publication.In the meantime, many of the challenging highway scenarios mentioned in ▶ section 41.7 have been covered by the general performance requirements in EU Regulation 2022-1426 [31], Annex II. In addition, in the same regulation the term "tactical functions" has been defined in Article 2.6 as control actions of the 'Dynamic Driving Task' DDT with a time constant in the range of seconds (in contrast to "operational functions" in the millisecond range). This is in good accordance to the term and concept of "Tactical Safety" described in this chapter.

References

1. Abendroth, B., Bruder, R.: Capabilities of humans for vehicle guidance. In: Winner et al. (ed.) Handbook of Driver Assistance Systems. Springer International Publishing, Cham (2016)
2. Bussgeldkatalog (2025). ▶ www.bussgeldkatalog.org/abstand/
3. ERCITO et al.: Safety related traffic information ecosystem: Data for road safety (2023). ▶ https://www.dataforroadsafety.eu/
4. EU Regulation 2022-1426.: Type-approval of the automated driving system (ADS) of fully automated vehicles (2022).**150** (2020). ▶ https://eur-lex.europa.eu/legal-content/EN/TXT/?uri=CE-LEX%3A32022R1426
5. FAA: System Safety Handbook. Federal Aviation Association, Washington, DC (2000)
6. HERE: Dynamic Map Content (2025). ▶ https://www.here.com/platform/dynamic-map-content
7. IEC 61508: Functional Safety of Electrical/Electronic/Programmable Electronic Safety-Related Systems. International Electrical Commission (2010)
8. ISO/PAS 21448: Road Vehicles-Safety of the Intended Functionality. International Standardization Organization (2019)
9. ISO 26262: Road Vehicles-Functional Safety. International Standardization Organization (2018)
10. Kitajima, S., Takayama, S., Uchida, N., Yamazaki, K.: Development of a new test center for the evaluation of safety related performance of automated vehicles in Japan. In: 14th International Symposium on Advanced Vehicle Control, AVEC'18, Beijing (2018)
11. Krajewski, R., et al.: The HighD Dataset: a drone dataset of naturalistic vehicle trajectories on German highways for validation of highly automated driving systems. In: 21st IEEE International Conference on Intelligent Transportation Systems (2018)
12. Mattas, K., et al.: Fuzzy surrogate safety metrics for real-time assessment of rear-end collision risk. Accid. Anal. Prev. **148** (2020)
13. MobilEye: Implementing the RSS Model on NHTSA Pre-Crash Scenarios (2018). ▶ www.mobileye.com/responsibility-sensitive-safety/rss_on_nhtsa.pdf
14. Peng, H.: MCity ABC Test: A Concept to Assess the Safety Performance of Highly Automated Vehicles (2019). mcity.umich.edu/conducting-the-mcity-abc-test-a-testing-method-for-highly-automated-vehicles/
15. Ploeg, J., de Gelder, E., Slavik, M., Querner, E., Webster, T., de Boer, N.: Scenario-based safety assessment framework for automated vehicles. In: Proceedings of the 16th ITS Asia-Pacific Forum 2018, pp. 713–726 (2018)

16. Rausand, M., Høyland, A.: System Reliability Theory: Models, Statistical Methods, and Applications. Wiley (2000)

17. SAE J2980: Considerations for ISO 26262 ASIL Hazard Classification. Society of Automotive Engineers (2018)

18. Sat1NRW: Deathly Accident on Highway A2 (2019). ► https://www.sat1nrw.de/aktuell/toedlicher-unfall-auf-a2-192864/

19. Schöner, H.P., Pretto, P., Sodnik, J., Kaluza, B., Komavec, M., Varesanovic, D., Chouchane, H., Antona-Makoshi, J.: A safety score for the assessment of driving style. Traffic Inj. Prev. (2021). ► https://doi.org/10.1080/15389588.2021.1904508

20. Schöner, H.P.: Challenges and approaches for testing of highly automated vehicles. In: 3rd CESA Automotive Electronics Congress, Paris (2014)

21. Schöner, H.P.: How good is good enough in autonomous driving? In: Langheim, J. (ed.) Electronic Components and Systems for Automotive Applications. Lecture Notes in Mobility. Springer, Cham (2019)

22. Schöner, H.P.: Challenging Highway Scenarios Beyond Collision Avoidance for Autonomous Vehicle Certification. Research Gate (2020). ► https://doi.org/10.13140/RG.2.2.29355.05926

23. Schöner, H.P., Antona-Makoshi J.: Testing for tactical safety of autonomous vehicles. 30th Aachen Colloquium Sustainable Mobility (2021)

24. Schöner, H.P.: Assuring Responsible Driving of Autonomous Vehicles. 33rd IEEE Intelligent Vehicles Symposium, Aachen (2022)

25. SensorIS: SensorIS_specification_v1.6.0 public (2024) ► http://sensoris.org/presentations/

26. Tomtom. Hazard Warnings (2025). ► https://www.tomtom.com/products/hazard-warnings/

27. UN Regulation 157 Amend.4, UNECE, 2023-03-07, E/ECE/TRANS/505/Rev.3/Add.156/Amend.4 (2023)150 (2020).

28. Weast, J.: Metrics, methods and assumptions: the state of AV safety assurance. In: Autonomous Vehicle Test and Development 2020, Virtual Conference (2020)

29. Webster, T.: HAV assurance requirements, goals, and strategy. In: CertiCAV Engagement Session 2 on Nov. 12, 2020 (2020). ► https://cp.catapult.org.uk/event/certicav-safety-assurance/, ► https://youtu.be/TQUoPmyqPrM?t=2286

30. Winner, H., Schopper, M.: Adaptive cruise control. In: Winner et al. (ed.) Handbook of Driver Assistance Systems. Springer International Publishing, Cham (2016)

31. Winner, H.: Fundamentals of collision protection systems. In: Winner et al. (ed.) Handbook of Driver Assistance Systems. Springer International Publishing, Cham (2016)

32. Zhao, C., Li, L., Pei, X., Li, Z., Wang, F., Wu, X.: A comparative study of state-of-the-art driving strategies for autonomous vehicles. Accid. Anal. Prev. **150** (2020). ► https://doi.org/10.1016/j.aap.2020.105937

Open Access This chapter is licensed under the terms of the Creative Commons Attribution-NonCommercial-NoDerivatives 4.0 International License (► http://creativecommons.org/licenses/by-nc-nd/4.0/), which permits any noncommercial use, sharing, distribution and reproduction in any medium or format, as long as you give appropriate credit to the original author(s) and the source, provide a link to the Creative Commons license and indicate if you modified the licensed material. You do not have permission under this license to share adapted material derived from this chapter or parts of it.

The images or other third party material in this chapter are included in the chapter's Creative Commons license, unless indicated otherwise in a credit line to the material. If material is not included in the chapter's Creative Commons license and your intended use is not permitted by statutory regulation or exceeds the permitted use, you will need to obtain permission directly from the copyright holder.

Decision-Making for Automated Driving

Piotr Spieker, Johannes Fischer, and Christoph Stiller

Contents

© The Author(s) 2026
H. Winner et al. (eds.), *Handbook Assisted and Automated Driving*,
https://doi.org/10.1007/978-3-658-45276-6_42

42.1 Introduction

The ability to make decisions is one of the most important properties that define higher organisms. Behavioral decision-making also plays a key role for automated vehicles, significantly shaping comfort, efficiency, and safety.

Human drivers structure their vehicle guidance primarily not in the form of trajectories, but through maneuvers. These abstract the actual course of movement to a qualitative description. Although there have been approaches to maneuver free trajectory planning, the separation of the modules *decision-making* and *trajectory planning* has also been established in the architecture of automatic vehicles. On the one hand, this separation of tasks reduces the complexity of motion planning. On the other hand, the separation also makes sense in order to generate explainable behavior that is transparent for passengers and other road users. This book also takes such a division of tasks into account by assigning separate chapters.

For many years, the decision-making was predominantly implemented as state machines, which have already enabled amazing handling sequences and seemingly *intelligent behavior* in robotics. After initial successes in automatic driving, however, they proved to be complex and difficult to understand and in handling safety assessment as the variety of situations increased. Accordingly, a variety of alternative approaches exist today, ranging from graph-based planning to probabilistic methods and learning techniques.

In the remainder of this chapter, various planning methods from the literature will first be presented. The focus is on methods that have already been used for automated driving. Rule-based methods have been used in this environment due to their good traceability. Graph search methods, on the other hand, transform motion planning into a graph search. Probabilistic methods modeled as MDP or POMDP take into account all observations and uncertainties continuously perceived during the trip. More recently, learning methods—often using deep neural networks—have also been trained for this application area. The subsequent section describes architectural aspects of behavioral decision-making. Finally, an arbitrator concept that has been implemented in real test vehicles is presented as an application example.

42.2 Methods for Decision-Making

42.2.1 Rule-Based Methods

Rule-based methods enjoy great popularity in the context of driver assistance systems [1, 5, 60, 95]. Their advantages include the ease of use, numerous freely available software frameworks [17, 27, 74, 86, 87], direct integration into MATLAB [55], or even the standardization in the Unified Modeling Language [89]. They can be divided into classical state-based methods like finite state machines and decision trees, and behavior-based methods like behavior trees and arbitration graphs.

42.2.1.1 Finite State Machines

Finite state machines, more precisely deterministic finite automata, originated in hardware design [91] and found their way into theoretical computer science since the 1940s/50s [31]. There, they are still used today in the field of formal languages and complexity theory, among others [90]. On the application side, they are used today in hardware and software design [91], robotics [77], and in the field of driver assistance systems [1, 94]. Their popularity in practical applications is partly due to the fact that they are intuitive and easy to understand and at the same time very simple to implement.

For a practical introduction to the theory of finite (state) automata, please refer to [90]. This section summarizes the most important contents from that and from [31].

Finite state machines can be described either in the Mealy or Moore model. Since both can be easily transformed into each other, in the following, the latter is presented w.l.o.g. A Moore automaton consists of

- a finite set of states S,
- a finite set of input symbols Σ (also events),
- a finite set of output symbols Δ (also actions),
- a state transition function $\delta : S \times \Sigma \to S$,
- an output function $\lambda : S \to \Delta$,
- an initial state $s_0 \in S$, and
- a finite set of final states $F \subseteq S$.

Starting at the initial state s_0, a finite state machine changes to a new state $s_{i+1} = \delta(s_i, e_i)$ at each event $e_i \in \Sigma$, which is usually triggered from outside. Remaining in the same state $s_{i+1} = s_i$ is also possible. Upon (re)entering a state s_i, the state machine outputs $a_i = \lambda(s_i)$.

In the graphical representation according to the UML standard [89], finite state machines are represented by directed graphs. The nodes represent the states and the edges, which are labeled with corresponding input symbols, represent the state transitions. The state names are shown in their headings and the corresponding outputs as further labels. A small filled circle marks the initial state with an arrow, while final states are marked with small filled circles surrounded by a ring. In extensions such as Hierarchical State Machines, individual states may in turn be state machines themselves, or may also contain concurrent state machines.

In robotics, events are usually generated as the result of a situation interpretation, while outputs realize the

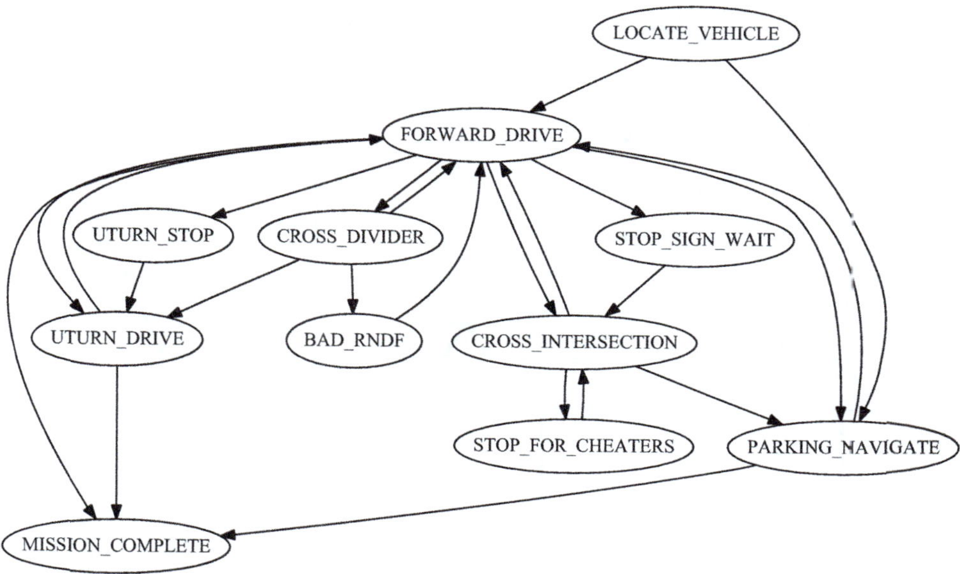

◘ Fig. 42.1   State machine used by Team Junior to participate in the 2007 DARPA Urban Challenge [60]

intended actions as input values for the subsequent executing layer. ◘ Figure 42.1 shows a simplified representation of a state machine by Team Junior of Stanford University [60], as used similarly by almost all finalists of the DARPA Urban Challenge. Here, the states cover tactical driving maneuvers, such as forward driving, crossing an intersection, or parking. For clarity, the two states *Escape* and *Traffic Jam* have been omitted because they are reachable from almost all other states. Likewise, events, actions, the output function, and initial and final states have been omitted. Directed edges represent possible transitions.

Finite state machines are a good choice for reasonably small behavioral problems due to their simple implementation and intuitive representation. However, they scale poorly for more complex systems, since the number of possible state transitions grows quadratically with the number of states in the worst case. Hierarchical state machines can at least limit this growth to manually defined hierarchical levels. Another disadvantage is the poor modifiability of finite state machines, since when adding or removing states, many of the already existing states and state transitions may have to be considered and adapted as well.

Moreover, state transitions in finite state machines essentially resemble jump instructions (similar to the "GoTo command" outlawed in programming) which make it difficult to understand, analyze, or verify source code. Thus, it is hardly possible to trace back why a certain state is currently active without collecting an event history and tracing it step by step in a time-consuming manner. In software development, "GoTo commands" have been avoided at least since the introduction of Structured Programming Languages [20] and Edsger

W. Dijkstra's open letter "Go To Statement Considered Harmful" [22].

Finally, when representing state machines of complex systems, clarity quickly suffers due to the large number of state transitions. This can already be seen from the fact that Team Junior omit two strongly interconnected states in the example of ◘ Fig. 42.1 to preserve clarity.

42.2.1.2 Decision Trees

Decision trees in their original form were defined as recursive structures for describing classification rules [61, 71].

Formally, they are directed ordered trees whose roots and inner nodes represent conditions in the form of discrete or even Boolean decision variables χ_i and whose leaves represent the resulting decisions $f(\chi_1, \ldots, \chi_n)$. The recursive evaluation starts at the root node and, depending on the value of its decision variable $\chi_i = k$, is continued at the k-th child node until a leaf node and thus a decision is reached.

In practical implementation, decision trees can simply be realized as a hierarchical concatenation of if/else branches and therefore enjoy great popularity in reasonably small decision problems.

In order to use decision trees in automated driving, a suitable state-space representation is usually chosen, which is then adequately subdivided with the decision variables [4]. ◘ Figure 42.2 shows an example of two decision trees as they were used—in combination with finite state automata—by BMW for highly automated driving on highways and an emergency stop assistant [3]. Here, two separate decision trees were designed for lateral and longitudinal driving, whose decision variables are derived from the situation interpretation, for example, as

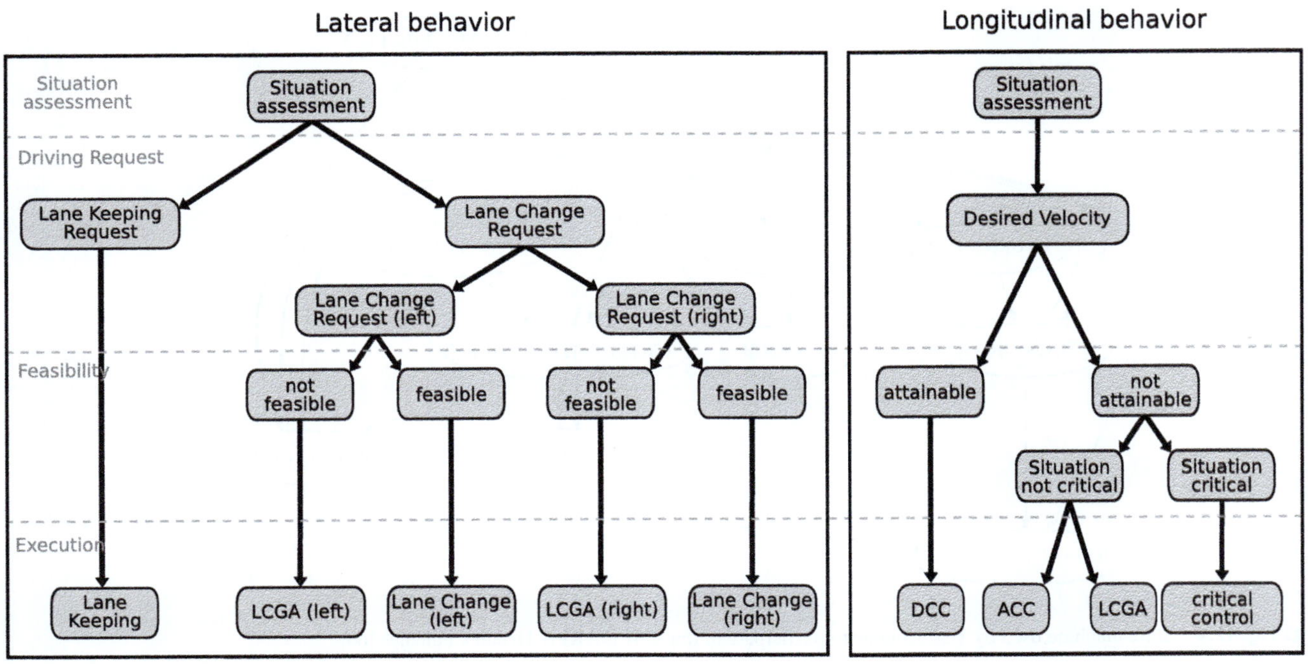

Fig. 42.2 Decision trees for highly automated driving on highways and an emergency stop assistant [3]

a lane change request. Finally, the leaves generate target maneuvers and constraints for the downstream trajectory planning—which is also separated in lateral and longitudinal components.

Decision trees, like finite state machines, are easy to implement and intuitive to grasp. Moreover, they are modular and extensible, so that subtrees can be designed, developed, and added independently of the rest of the tree. However, it may prove to be a disadvantage that the leaves do not return any (success) status or other performance criteria, so that the result of a behavior cannot directly influence the further behavior selection.

42.2.1.3 Behavior Trees

Behavior trees can be represented by connected circle-free undirected graphs whose leaves describe possible behaviors as well as conditions, while the inner nodes define the selection mechanism. The decision-making is not event based, but timer based. Moreover, the action nodes have direct interfaces to the perception and actuation modules, while their state is propagated through the tree.

Historically, behavior trees originated in computer game development [37] and have only been used in robotics since 2012 [6, 63]. To the authors' knowledge, the use in driver assistance systems or automatic vehicles is so far limited to only one simulative work, which indicates the superiority of behavior trees over finite state machines by investigating scalability using software metrics [64].

However, since behavior trees generalize many other architectures such as (Hierarchical) Finite State Machines, Decision Trees, and Subsumption [16] and

offer many advantages, this method will be presented here as well.

In particular, it convinces by modularity, hierarchical structure, reusability of components, responsiveness, and interpretability [18]. In addition, tools have been developed to formally analyze behavior trees or even to synthesize them automatically.

Compared to finite state machines, one of the main advantages of behavior trees is that individual behaviors can be (re)used within a higher order behavior without having to specify what relation they have to the other behaviors [6].

A detailed overview and introduction to behavior trees is provided by both the survey study by Iovino et al. [37] and the reference work by Michele Colledanchise and Petter Ögren [18], each of whom have themselves made significant contributions to the formalization and application of behavior trees.

Behavior trees are characterized by a strict functional separation between decision-making and behavior execution. The structure of the tree determines the decision-making by means of so-called *control flow nodes*, i.e., how and in which order the tree and finally the leaf nodes, the so-called *execution nodes*, are evaluated.

The tree is evaluated sequentially with a defined frequency starting from the root node. In the process, an evaluated node indicates via its return value whether it was completed successfully, has failed, or is still being executed.

Depending on the return value, the parent control node decides whether further child nodes are evaluated or whether the node itself returns its status. Essential *control*

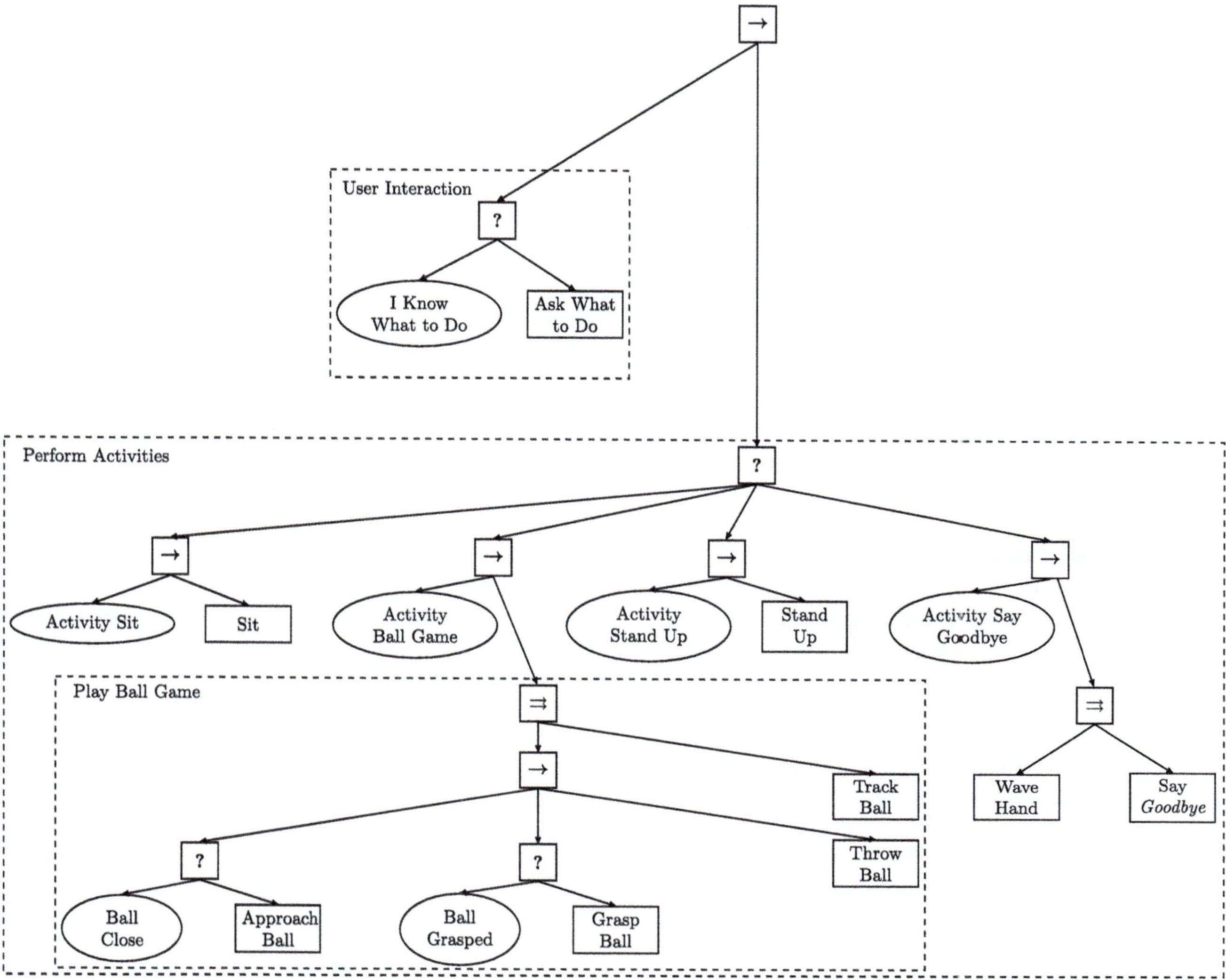

Fig. 42.3   Behavior tree for an interactive humanoid robot (adapted from [18])

flow nodes include nodes for implementing sequences, fallback structures, and concurrency. For example, a sequence node evaluates its children in given order until one returns that it has failed or continues to execute. Thus, behaviors that depend on the successful execution of their predecessor behaviors can be combined.

The *execution nodes*, in turn, describe the individual behavior options of a behavior tree. Here, an additional distinction is made between condition and action nodes, which decouples the preconditions of a behavior from the actual execution. When evaluated, the condition nodes use the return value to indicate whether a particular condition is fulfilled without influencing the environment. The action nodes, on the other hand, execute the underlying behavior when called and return the status of this behavior via the return value.

Figure 42.3 shows an example behavior tree, for an interactive humanoid entertainment robot (simplified example from [18]). Sequence nodes are marked with an arrow ($\rightarrow$), fallback nodes with a question mark (?), and parallel nodes with a double arrow ($\Rightarrow$). Round leaves represent the condition nodes and square leaves represent the action nodes. The robot's task is to sit down, stand up, play ball, or say goodbye according to the operator's request. The *Ask What to Do* action takes over the user interaction, while the other actions execute the activity selected by the user. For this purpose, each action is linked directly or indirectly via a sequence node to its corresponding condition node, which returns whether this action should be executed. For example, the action *Stand Up* depends on the condition *Activity Stand Up*. Furthermore, the dashed outlines show how the modular hierarchical structure of the tree allows a subdivision into subtasks. The activity *Play Ball Game* can be modeled as a single action, for example, or again as a behavior tree, as in this example.

In the context of automatic driving, action nodes could realize driving maneuvers such as following, changing lanes, or parking, and they would need to be linked to appropriate condition nodes to design a safe system.

Behavior trees are characterized by greater flexibility and at the same time more compact modeling than, for example, finite state machines. In addition, the functionality and the selection mechanism of a behavior tree can be intuitively grasped in the graphical representation, even during online operation.

However, since in practice each precondition is often modeled as a separate leaf node, behavior trees can become very extensive in their representation. Due to the decoupling of preconditions and the execution of a behavior, the safety of the system also depends significantly on the arrangement of the nodes in the tree.

Both drawbacks are addressed by the arbitration graphs described below.

42.2.1.4 Arbitration Graphs

Arbitration graphs, although they can be viewed as a specialization of behavior trees, emerged independently in the context of robot soccer [50]. They combine key ideas of the subsumption principle [14], knowledge-based architectures such as belief-desire-intention [72], and programming paradigms such as object-oriented programming [78]. Thus, arbitration graphs are characterized by a behavior-based structure that combines atomic so-called behavior blocks or options hierarchically in a directed graph by means of arbitrators. In contrast to behavior trees, the preconditions and the actual execution of a behavior are linked in one and the same component, the behavior block. This ensures that behavior options— regardless of their arrangement in the graph— can only be selected if they are currently applicable and, above all, safe in the given situation.

The preconditions of a behavior block are determined by the so-called invocation and commitment condition. The *invocation condition* indicates whether the behavior can be started safely in the current situation. If a behavior block is already active, the *commitment condition* returns whether the behavior can still be continued. Thus, for example, a behavior block for lane changing can help to ensure that a lane change is not interrupted (by selecting a different behavior) in an unsafe state, but is either completed successfully or aborted in a controlled manner. The *gain control* and *lose control* functions are called when the behavior block becomes active or inactive again. They can be used, among other things, for initialization and memory management of a behavior block. Finally, the *get command* function generates the actual commands of the behavior.

An arbitrator contains an ordered list of behavior blocks and selects from these options by means of a selection logic. The priority, sequence, random, and cost arbitrators have been presented as fundamental arbitrator types [50, 65]. The priority arbitrator chooses the applicable behavior option with highest a priori set priority. A sequence arbitrator calls its options in a given order, and the random arbitrator chooses randomly among its applicable options. Finally, to allow for a wider range of selection mechanisms, a cost arbitrator chooses the most favorable behavior at the moment. In order to do so, cost functions have to be added to the behavior blocks to estimate their cost in a given situation. To avoid oscillations in the selection by cost, the cost function can be additionally embedded into a hysteresis function.

Arbitrators can also be combined with other arbitrators and/or behavior blocks to form more complex arbitration graphs. For this purpose, arbitrators adopt the interfaces defined for behavior blocks (by inheritance), as illustrated by the simplified class diagram in ◘ Fig. 42.4. The invocation and commitment conditions of an arbitrator depend exclusively on the invocation and verification conditions of its options as well as its selection logic. For example, the invocation condition of a priority arbitrator is true as long as at least one of its options is currently applicable, i.e., either the invocation condition of one of its options is true or, if an option is already active, the commitment condition of this active option is true.

The decision process, i.e., the evaluation of an arbitration graph, starts at the root and is continued recursively up to the leaf nodes, depending on the arbitrator types and the invocation or commitment conditions of the behavior blocks. The command of the selected behavior is then returned as the result of this recursion throughout the tree. The application example in ▶ Sect. 42.4 introduces arbitration graphs in more detail in the context of urban automated driving.

Thanks to their strict modularity, arbitration graphs are characterized by high flexibility and good extensibility. Behavior blocks or even entire parts of an arbitration graph can be designed, added to the graph, or removed again without having to necessarily adapt other parts of the graph. They are also intuitively understandable and comprehensible in design as well as in operation. At the same time, they have a more compact representation compared to behavior trees.

The returning of the command values via consistent interfaces of the graph components also reduces error-proneness and facilitates code analysis.

Finally, the strict coupling of preconditions with the behavior planning increases the safety of the overall system—after all, this means that a behavior option that has not explicitly signaled to be reasonable and safe to execute in the current situation is never chosen. However, these constraints have the consequence that arbitration graphs are less flexible and efficient than optimized behavior trees.

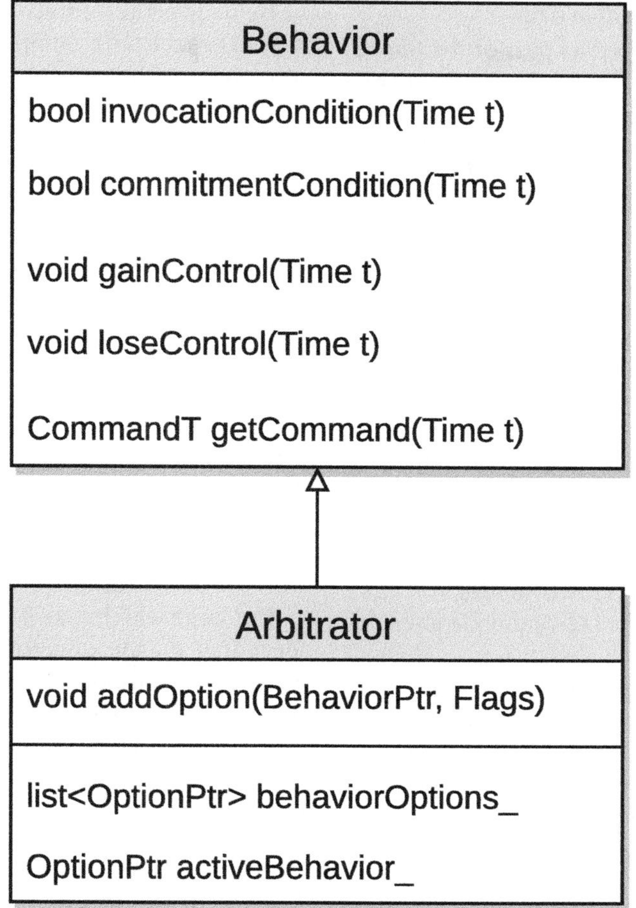

Fig. 42.4 Arbitrators inherit the interface of the behavior blocks to enable hierarchical chaining to create arbitration graphs

42.2.2 Graph Search Methods

Up to now, graphs have been used to link states or behavior blocks with each other. In the methods described below, the behavior sequence itself is represented by a path in a search graph. These methods are mostly used for path planning, i.e., motion planning in static environments, where speed is irrelevant. In principle, however, they can also be used for trajectory planning in dynamic environments and—using appropriate constraints and cost functions—for decision-making. Good introductions that form the basis for this section are provided by LaValle [51], Karaman and Frazzoli [41], and Gammell and Strub [25].

The planning problem of these graph search methods is first defined as a search problem: In the state space $\mathbf{X} \subseteq \mathbb{R}^n$, a trajectory, i.e., sequence of states $\mathbf{x} : \{0, 1, \ldots, T\} \to \mathbf{X}_{\text{free}}$, is sought. It should connect the initial state $\mathbf{x}(0) \in \mathbf{X}_{\text{free}}$ with the target region $\mathbf{X}_{\text{goal}} \subset \mathbf{X}_{\text{free}}$ admissibly, i.e., over a collision-free and, if necessary, kinodynamically constrained space $\mathbf{X}_{\text{free}} := \mathbf{X} \backslash \mathbf{X}_{\text{invalid}}$, and optimally with respect to a non-negative cost function c.

For this purpose, the usually continuous search space $\mathbf{X}$ can either be discretized in advance in order to determine an optimal trajectory within this graph, for example, by means of A* search. Or, it can be sampled online during the search by means of sampling-based methods, such as Probabilistic Roadmaps (PRM*) or Rapidly exploring Random Trees (RRT*). Searching in pre-generated graphs is easier to implement in practice and is guaranteed to return an optimal solution, if it exists, or return an error. However, in many real applications, it is difficult to specify a suitable a priori discretization, or even impossible because the search space is unbounded. Moreover, a discretization that is too sparse may prevent a solution from being found at all, while a sampling that is too dense comes with heavy computational load and consequently large latency. For this reason, sampling-based approaches have recently become very popular, even though they can only provide convergence guarantees in the probabilistic sense.

Finally, to apply graph search methods to decision problems, such as maneuver planning of automated vehicles, it may be helpful to combine them with process algebra or temporal logic [25].

42.2.2.1 Dijkstra and A*-Algorithm

The A* algorithm was developed as part of the Shakey project and has been one of the most popular algorithms for shortest path search in graphs since its publication in 1968 [29]. It essentially extends Dijkstra's method [21] by a so-called heuristic, whereby the search is directed and thus considerably more efficient.

For both methods, the search space must be discretized a priori in the form of a weighted graph $(\mathbf{N}, E)$. It consists of admissible nodes $\mathbf{N} \subseteq \mathbf{X}_{\text{free}}$, a start node $\mathbf{x}(0) \in \mathbf{N}$, a nonempty set of goal nodes $\mathbf{G} \subseteq \mathbf{X}_{\text{goal}}$, and edges E, whose weights $c : E \to \mathbb{R}$ represent the cost of this trajectory part. The search starts with the expansion of the starting node, with the cost to reach this node set to $g(\mathbf{x}(0)) = 0$. For all other nodes, the cost of the shortest known path to reach the respective node is initialized to infinity. The list of "open" nodes, those that are eligible for expansion in the next step, is initialized as an empty set. Similarly, the list of "closed" nodes, for which the cost to reach these from the starting node is known exactly, is initialized empty.

As the search proceeds, the algorithm successively expands selected nodes n_{expand} from the "open" list. The expansion of a node n_{expand} includes the following steps:

- Each of its child nodes n_{child}, which is not yet on the closed list, is added to the list of "open" nodes.
- For each child node n_{child}, it is investigated whether the path via n_{expand} forms a shorter path than the previously known shortest path. This is true if $g(n_{\text{expand}}) + c((n_{\text{expand}}, n_{\text{child}})) < g(n_{\text{child}})$ holds. In this case, $g(n_{\text{child}})$ is reduced to this value and n_{expand} is marked as the predecessor of n_{child}.
- The expanded node n_{expand} is added to the "closed" list and removed from the "open" list.

After expansion of a node n, the costs from the starting node to n are known exactly in $g(n)$. Thus, the best path to reach n is determined by backtracking the respective predecessors. Hence, once a target node has been expanded, the search is completed.

The key to an effective search lies in the selection of the respective node to be expanded. The Dijkstra algorithm works depth prioritized, i.e., it always first expands those nodes that can be reached from the start node with the least number of steps.

The A* algorithm introduces a so-called *heuristic* $h(n)$, which forms a lower bound on the path from n to the target. In the case of geometric path lengths, a possible heuristic is given, for example, by Euclidean distance. With this, the A* algorithm now underestimates the cost of a path from the start node, via node n to one of the goal nodes. For this purpose, the cost to reach the node from the start node $g(n)$ is supplemented by an estimate of the remaining cost to reach the goal $\hat{h}(n)$:

$$f(n) = g(n) + \hat{h}(n).$$

By its general formulation, the A* algorithm can be applied to arbitrarily complex search spaces $\mathbf{X}$, cost functions $c(e)$, and heuristics $h(n)$.

Provided that the heuristic $h(n)$ forms a lower bound on the cost of n to the goal, $h(n)$ is called *admissible*. Then it can be proved that the A*-algorithm is complete, i.e., if a solution exists, the algorithm is guaranteed to find a solution. If no solution exists, this is revealed in finite time by the fact that the "open" list is empty. Compared to randomized methods, this is an important advantage, because if the latter do not find a solution, usually no statement can be made about whether no solution exists or merely the number of iterations was too small to find it.

If the heuristic additionally satisfies the inequality $h(n) \leq c(n, n_{\text{child}}) + h(n_{\text{child}})$ for any node n and one of its children n_{child}, it is called *monotonic*. In this case, when expanding a node, the cost estimate can never decrease. As a result, the algorithm becomes optimal, i.e., it always finds the best solution.

For large search spaces, the A* algorithm is slow and, depending on the discretization, yields only piecewise continuous trajectories. The better a heuristic approximates the real costs, the faster the algorithm converges, i.e., the fewer nodes have to be expanded. Accordingly, in practice, often heuristics are used that estimate the costs as accurately as possible, but satisfy one or even both of the required properties only approximately. In this case, the A* algorithm quickly finds a good but not necessarily the optimal solution.

42.2.2.2 Probabilistic Roadmaps

The Probabilistic Roadmap Method [44] first scans the search space with a randomized sample in its so-called learning phase and then searches for the shortest path within the generated "roadmap" in the evaluation phase. Due to their very efficient evaluation phase, they are particularly suitable for motion planning of robots with many degrees of freedom operating in a static environment, but have also been used for vehicle-like multiagent problems [84].

In the learning phase, admissible states $\mathbf{x} \in \mathbf{X}_{\text{free}}$ are drawn first. These are combined with the next possible already existing node n to form an undirected circle-free graph. A simple efficient motion planner, called the local planner, is used to check beforehand whether a path between the two states $\mathbf{x}$ and n can be computed. The states of the drawn sample thus form the nodes of the resulting graph, while the paths determined by the local planner represent its edges.

In the evaluation phase, which may be executed several times, the roadmap is used to determine the shortest path between two states of the robot. For this purpose, the start and end states s and e are first connected to their closest nodes $\tilde{s}$ and $\tilde{e}$ on the roadmap. A subsequent graph search is used to determine the optimal path and the corresponding state sequence.

Depending on the implementation, the paths are not explicitly stored in the roadmap to save memory, so that the local planner must be executed again in the last step to generate a continuous trajectory.

PRMs, available in their original form [43] or the further development PRM* [41], are probabilistically complete, converge asymptotically toward the optimum with increasing number of nodes, and are computationally efficient. PRMs are designed to allow multiple queries in the same roadmap. This provides significant runtime advantages in static or low-changing environments, as the roadmap only needs to be created once. However, in highly dynamic environments or applications where the search space is not limited, the creation of an a priori roadmap becomes expensive for multiple search queries at different times.

42.2.2.3   Rapidly Exploring Random Tree

Rapidly exploring Random Trees (RRT) were originally proposed by LaValle [52, 53] and some important extensions, such as RRT*, were proposed by Karaman and Frazzoli [41]. They have been successfully used in numerous highly complex applications and continue to enjoy considerable attention today. Among others, MIT's DARPA Urban Challenge team used an RRT-based trajectory planner [48], Karaman et al. used an Anytime-RRT* to plan trajectories for a forklift [42], and Boston Dynamics used so-called reachability-guided RRTs for motion planning over rough terrain of their four-legged robots, LittleDog and BigDog [76].

In their simplest form, RRTs iteratively build a tree of admissible partial trajectories. For this, the tree is initialized with the initial state $\mathbf{x}(0)$ as the root node and then iteratively states are drawn from the admissible search space $\mathbf{X}_{\mathrm{free}}$. In each iteration step, for such a random state $\mathbf{x}_{\mathrm{rand}}$ the closest existing node $\mathbf{x}_{\mathrm{nearest}}$ is searched for. A so-called steering function, which approximates the system dynamics, now determines a new node $\mathbf{x}_{\mathrm{new}} \approx \mathbf{x}_{\mathrm{rand}}$ based on $\mathbf{x}_{\mathrm{nearest}}$ and the targeted $\mathbf{x}_{\mathrm{rand}}$. $\mathbf{x}_{\mathrm{nearest}}$ may differ from $\mathbf{x}_{\mathrm{rand}}$, for example, in the case of non-holonomic systems. If an admissible, i.e., collision-free, trajectory from $\mathbf{x}_{\mathrm{nearest}}$ to $\mathbf{x}_{\mathrm{new}}$ can be determined, the new node $\mathbf{x}_{\mathrm{new}}$ and the edge $(\mathbf{x}_{\mathrm{nearest}}, \mathbf{x}_{\mathrm{new}})$ are added to the tree. ◘ Figure 42.5 shows the 2D projection of an explored RRT tree including the solution for five-dimensional trajectory planning of a kinodynamic vehicle [52].

Since in RRTs the search ends directly as soon as a node of the tree reaches the goal region, the search is fast, but usually yields a suboptimal trajectory. The RRT* extension invests slightly more computation time (but with the same complexity) to optimize the tree during the search. For this, on the one hand, the new node x_{new} is not immediately connected to the next node of the tree, but all nodes in the closer neighborhood of x_{new} are examined. The new parent node x_{near} is selected out of these neighboring nodes, to the one x_{near} which provides the most favorable admissible trajectory from x_0 to x_{new}. Additionally, the neighboring nodes are examined to see if they would also have better paths via x_{new} than via their previous parent node. In this case, x_{new} is marked as the new parent node by so-called "rewiring". As a result, the resulting tree contains the edges with stronger orientation away from the start node and toward the goal node, respectively, as can be seen in ◘ Fig. 42.6.

RRTs are probabilistically complete, computationally efficient, and (in the RRT* variant) converge probabilistically toward the optimal trajectory. Moreover, thanks to the steering function, they do not require an exact connection between a new random state and the tree, making them more suitable for systems with kinematic or kinodynamic constraints. With a few adjustments, the RRT* method can also be used efficiently as an anytime method in online applications [42], i.e., after any computation time at least one solution is known, which is subsequently improved and becomes asymptotically optimal. Since RRT* works probabilistically, the trajectories generated in practice are inevitably suboptimal (due to finite computation time), jerky, and occasionally even zigzag. Therefore, in the context of driver assistance systems, a subsequent local trajectory planner is necessary, for example, by using a model predictive controller. Some works, such as Theta*-RRT* [66], combine the RRT* method with discrete search methods or with linear-quadratic controllers [69] to counteract such deficits. Due to the large variety of variants in the RRT family, it is therefore worthwhile to search for the most suitable method depending on the application [25].

42.2.3   Methods with Probabilistic State and Observation Models

In the previous sections, methods for planning problems with deterministic system dynamics were presented. In contrast, in this section, methods are considered which model the state evolution and observations of the environment probabilistically. This corresponds to the requirements of real traffic situations, in which behavioral decisions often have to be made under imperfect information about other road users and the environment. A suitable model for probabilistic planning tasks is the Markov decision problem (MDP).

In this (usually discrete-time) model, the system is in a state s_t at any time. This state contains all information about the agent (usually the ego vehicle; only in central planning problems all controlled vehicles) and the environment (relevant road information and other road users). At any point in time, the agent makes a decision for an action a_t, which affects the system state s_t. The successor state s_{t+1} results stochastically via the transition model $\mathcal{T}(s_{t+1}|s_t, a_t)$ from the current system state

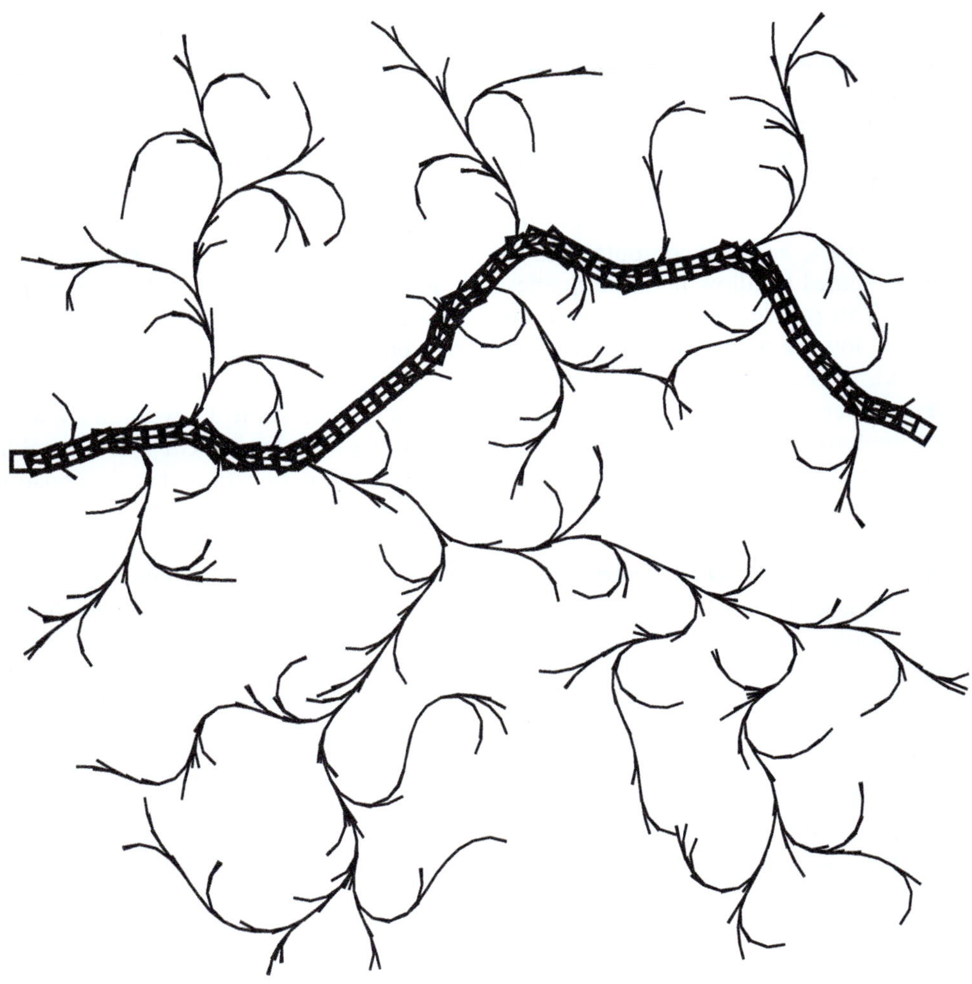

Fig. 42.5   RRT tree for five-dimensional trajectory planning of a kinodynamic vehicle [52]

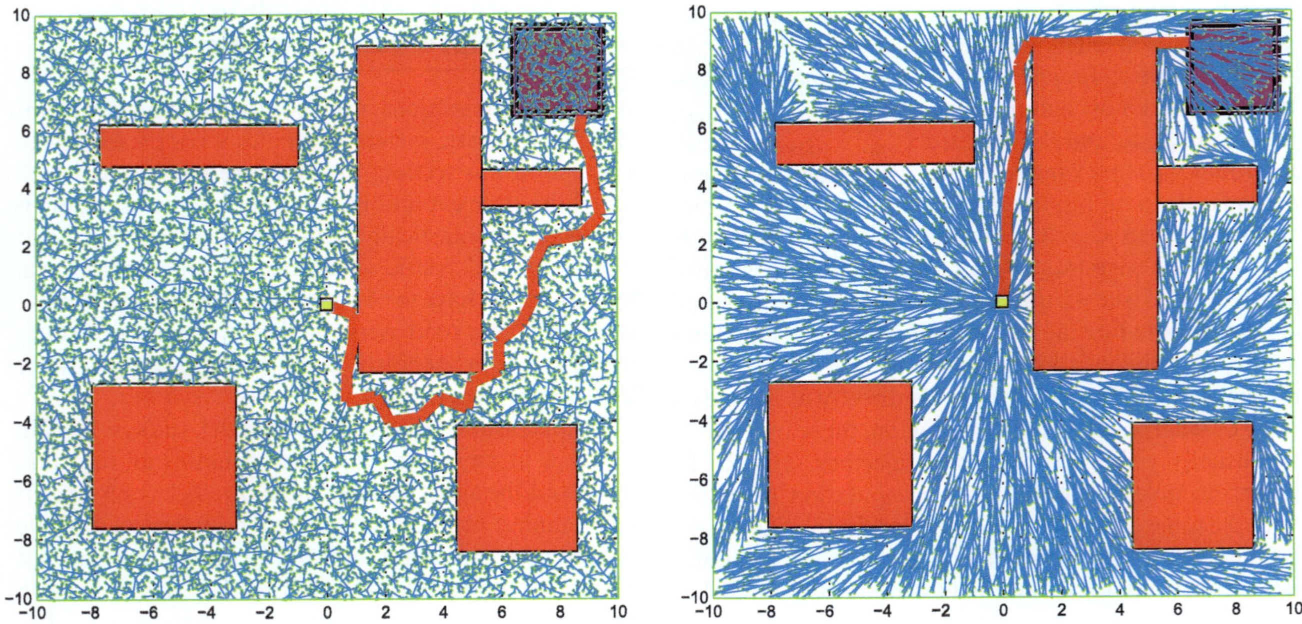

Fig. 42.6   Comparison of RRT (left) and RRT* (right) for the same problem and sampling [41]

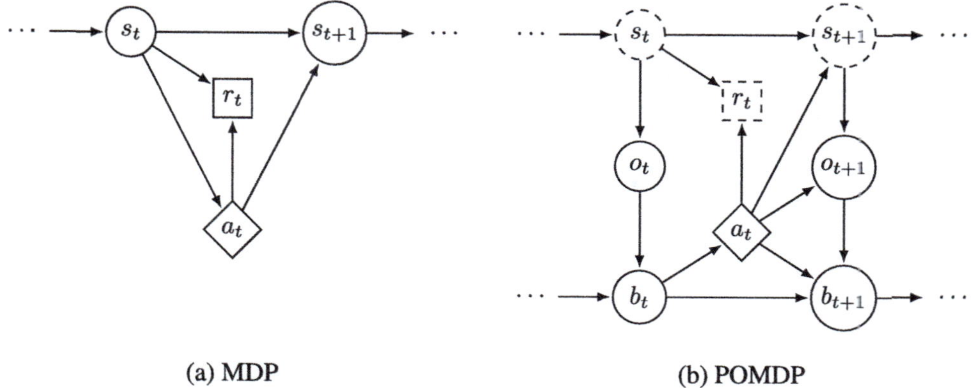

Fig. 42.7 Probabilistic graphical models of MDP and POMDP. Dashed nodes are not directly observable

and the chosen action. After each action, the agent also receives a reward $\mathcal{R}(s_t, a_t)$ depending on the state and the chosen action.

A *policy* π denotes a control strategy in an MDP, i.e., a mapping that assigns to each state an action to be performed. The objective is to find the policy π that maximizes the cumulative reward $\mathbb{E}_{s \sim d}[V^\pi(s)]$ expected over the initial state distribution d.

Here

$$V^\pi(s) = \mathbb{E}\left[\sum_{t=0}^{\infty} \gamma^t \mathcal{R}(s_t, \pi(s_t)) | s_0 = s\right]$$

denotes the *value function* of policy π, with future rewards discounted by a factor $\gamma < 1$ at each step. Overall, a Markov decision problem is thus defined by the tuple $(\mathcal{S}, \mathcal{A}, \mathcal{T}, \mathcal{R}, \gamma, d)$, where $\mathcal{S}$ denotes the state space and $\mathcal{A}$ the action space. **Figure** 42.7a shows a Bayesian network representing the conditional dependencies in an MDP.

MDPs are structurally similar to the finite state machines described in ▶ Sect. 42.2.1. In contrast to these, however, the state space of an MDP can be infinite. In real problems, it is often even continuous and high dimensional. Furthermore, the state transitions of an MDP are in general probabilistic, whereas they are deterministic in state machines. While MDPs are targeted for optimal planning (proactive), state machines are more often used to select a behavior based on events (reactive). The optimal policy of an MDP can be found, for example, using dynamic programming algorithms.

MDPs allow to model planning problems where the state transitions are not deterministic. An example is the consideration of a stochastic prediction of other road users. To realistically model a driving situation, however, additional uncertainties in the system state have to be considered that cannot be modeled with MDPs. For example, due to sensor uncertainties, the localization of both other road users and the ego vehicle is always sub-

ject to uncertainty. Furthermore, the planned route and the cooperative behavior of other road users are usually unknown. Finally, **Fig.** 42.8 also illustrates uncertainties in the existence of other road users, which frequently occurs in traffic scenes due to occlusions.

Such uncertainties can be modeled as partially observable Markov decision problems (POMDPs). In this extension of MDPs, it is assumed that the system state cannot be observed directly. Instead, at each time step, the agent receives a stochastic observation o_t from an observation model $\mathcal{Z}(o_t | s_t, a_{t-1})$, whose distribution depends on the unknown state s_t and the latest action a_{t-1}. The observation model allows the agent to infer the unknown state from the observations.

The probability distribution of the current state using all information known up to the corresponding time t is called the *belief* b_t. To define a POMDP, in addition to the elements of an MDP, one must specify the observation space O, the observation model $\mathcal{Z}$, and the initial belief b_0. The initial belief can be used to specify prior knowledge about the problem definition. If there is no prior knowledge, a uniform distribution over the unobservable states can be assumed. **Figure** 42.7b shows the conditional dependencies in a POMDP.

Using a Bayesian filter, the belief b_t at time t can be determined recursively via the state transition model and the observation model. Starting with the initial belief b_0, the chosen actions and obtained observations are used to update

$$b_{t+1}(s_{t+1}) = p(s_{t+1}|b_t, a_t, o_{t+1}) \propto \mathcal{Z}(o_{t+1}|s_{t+1}, a_t) \int_S \mathcal{T}(s_{t+1}|s_t, a_t) b_t(s_t) ds_t.$$

$$(42.1)$$

In practice, particle filters are often used for this purpose, for which a generative state transition model is sufficient. However, the observation model must be explicitly stated to reweight the particles according to the probability of the latest observation. Since the state itself is not directly observable, the belief is the basis of decision-making. Thus, in POMDPs, a policy maps beliefs to actions.

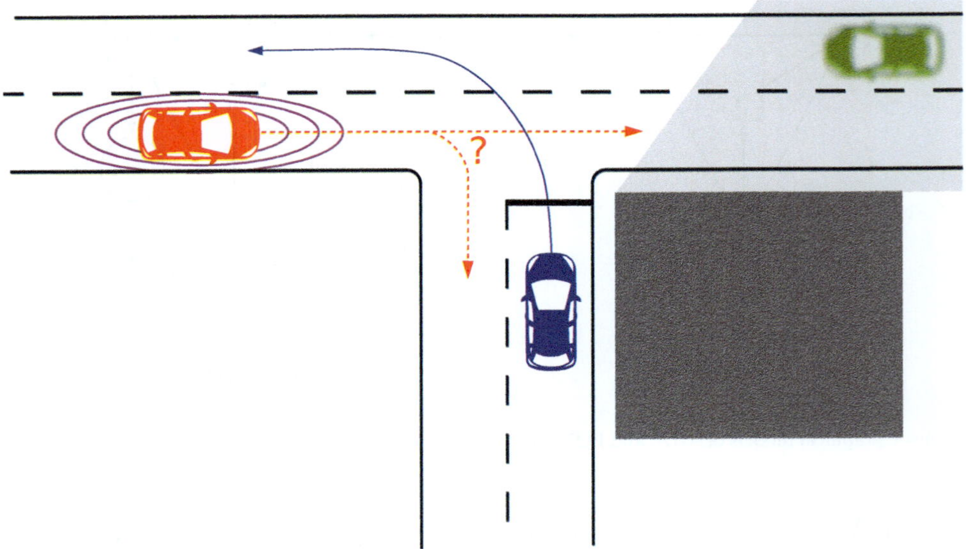

Fig. 42.8 Typical uncertainties in automated driving. The blue ego vehicle does not know the future route of the red vehicle. Furthermore, the position estimate of the red vehicle is uncertain. Due to the occlusion, it is additionally unknown whether a vehicle is coming from the right

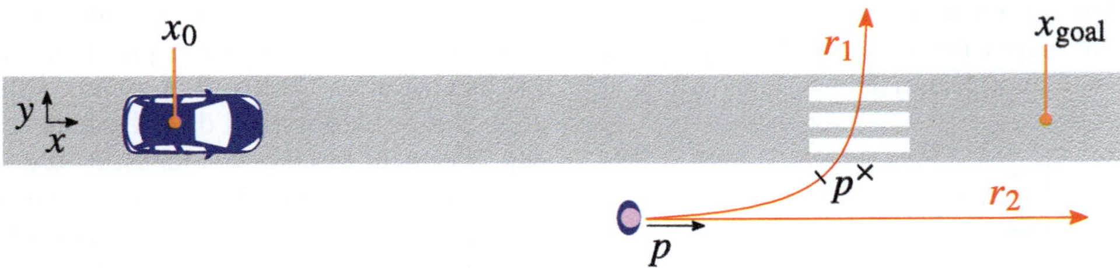

Fig. 42.9 Modeling of a POMDP using the example of a crosswalk. The pedestrian's route is not directly observable. If she wants to cross the crosswalk, the pedestrian stops briefly at position $p^\times$ before crossing

POMDPs can be used to model a much larger class of problems than MDPs. In particular, the above situations related to uncertain system states can be addressed. A major strength is that the optimal policy implicitly collects information needed to optimally solve the problem. However, this comes at the cost of a high algorithmic complexity in solving a POMDP [67].

There are different methods to solve POMDPs. While offline algorithms aim at determining a global solution which then only has to be evaluated in the current belief at runtime, online algorithms continually compute a solution only for the current belief in each step.

In the following, this section discusses how the behavioral decision problem in typical automated driving scenarios can be modeled using POMDPs. A good overview and further reading on solution methods for POMDPs can be found in [47]. Current real-time capable algorithms are presented in [15, 46, 80]. These algorithms are based on Monte Carlo tree search and are able to determine the best action for the current belief online. In addition to the following example, the works cited in the remainder of this paper describe approaches for concrete modeling of various scenarios.

Crosswalk example: State space

As an example, we will show how driving behavior at a crosswalk can be modeled. The scenario considered is shown in **Fig. 42.9**. The vehicle starts at position x_0 with velocity v_0 and has to reach position x_goal. Reversing is not possible and the maximum velocity is v_max. In addition, the vehicle has to watch out for a pedestrian who will either cross the crosswalk (path r_1) or continue walking along the road (path r_2). To summarize the state of this scenario, the vehicle is described by longitudinal position x and velocity v_x and the pedestrian is described by her future route r, her longitudinal position p along this route, and her velocity v_p. Thus, the state has five dimensions $s = (x, v_x, r, p, v_p)$ and the state space can be expressed as the Cartesian product

$$S = [x_0, x_\text{goal}] \times [0\tfrac{\text{m}}{\text{s}}, v_\text{max}] \times \{r_1, r_2\} \times [p_\text{min}, p_\text{max}] \times [0\tfrac{\text{m}}{\text{s}}, 2\tfrac{\text{m}}{\text{s}}].$$

42.2.3.1 State Transition Model

The state transition model describes the system dynamics. It defines the probability distribution of the successor state for a given state and a chosen action. Hence, the first question is how to choose the state space and the action space.

The state space must contain all information about the entire environment at a point in time that are necessary for decision-making, in particular, the physical state of all road users involved. Static knowledge about the road topology and geometry can be assumed as background knowledge and does not have to be included in the (dynamic) state.

It is convenient to model vehicles in a path-oriented Frenet coordinate system, i.e., to use one coordinate for the longitudinal motion along a reference path and a second coordinate for the lateral deviation from the reference path [11, 33, 35]. Bouton et al. [10] instead use a global coordinate system, which makes it necessary to additionally include the orientation of the vehicle in the state [10].

For longitudinal motion, position and velocity are usually chosen as states and acceleration as action, both for the ego vehicle and for the other vehicles. For lateral motion, different approaches are conceivable, depending on the exact problem. In intersection scenarios, it can be assumed that vehicles do not change lanes and therefore neglect lateral states [12, 35]. In this case, the state only contains the traveled lane for the identification of collision points between the paths.

The usual way to consider lane changes is to include the lateral position in the state and to use the lateral velocity as the action. The assumption behind this is that (especially at higher speeds) a desired lateral velocity can be built up quickly by steering [33, 81].

It is also possible to use different lateral dynamics for the ego vehicle than for the other vehicles. For example, sometimes only lane changes of the ego vehicle are considered, while other vehicles remain on their initial lane [33].

If pedestrians are to be considered, a path is usually assumed for them, along which they move, for example, when crossing a pedestrian crossing. The position and velocity of pedestrians along this path are then added to the state [12].

The most interesting part of the state modeling is the internal, latent state of the other road users. This hidden state is strongly dependent on the problem and encodes unknown properties of the other road users, which can only be inferred indirectly via observations. For example, the intention or destination of the other road users can be modeled as a hidden state if there are multiple route options [35]. Another possibility is to consider different motion models or parameters of motion models as latent state, or interaction models that cause cooperative or selfish behavior [10, 81].

Another common application of POMDPs is the modeling of occlusions by static or dynamic obstacles [11, 32]. Here, the appearance of new agents at the boundaries of the visible region must be modeled by a probability distribution, e.g., a uniform distribution on the relevant hidden region. This distribution then enters the optimization as part of the belief.

The action space defines the possible decisions that the agent can make in each step and is closely related to the state space. In most cases, the agent is simply the ego vehicle, and only for this a decision can be made. The behavior of the other road users must then be provided as reasonable as possible by the state transition model. An exception here are central planning problems, where the behavior of all agents is centrally planned. In this case, an action includes the behavior of all centrally planned road users.

Unfortunately, the vast majority of algorithms for POMDPs only allow the use of a discrete action space. Since the algorithmic complexity usually increases exponentially with the number of actions, it is important to use an action space as small as possible. In order to describe the longitudinal behavior sufficiently well, for example, four acceleration values are selected which cause moderate acceleration, driving with constant speed, moderate braking, and strong braking, respectively [10, 35]. If lateral motion is also considered, three actions can be used that cause leftward motion, rightward motion, and no lateral motion, respectively [33, 81]. The action space of the agent then consists of the Cartesian product of the lateral and longitudinal actions. To reduce the number of actions again, one can make additional assumptions, e.g., that heavy braking will not occur at the same time as a lane change or that lateral and longitudinal actions are generally mutually exclusive [32].

In the state transition model, it is defined how an action transfers the current state into the successor state. This state transition is generally stochastic, but can of course also be deterministic. For the longitudinal or lateral dynamics of the ego vehicle, a simple second- (or first-)order motion equation is integrated, where the selected action represents the input as is assumed as constant for the duration of the time step. Additionally, noise can be added to the input to conservatively compensate the simplified modeling of the system dynamics.

Different motion models can be used for other vehicles. The simplest option is to use a model with constant speed or constant acceleration or a randomly selected acceleration. More realistic driving behavior is obtained by using improved motion models such as the Intelligent Driver Model (IDM) [45] or by considering inter-vehicle interactions [35].

A constant velocity motion is typically chosen as the motion model for pedestrians, where the sometimes abrupt movements of pedestrians are compensated by a large amount of uncertainty [11].

The latent state is often assumed to be constant, since the intentions and driving behavior of other road users are assumed not to change. However, especially for pedestrians, sudden changes in behavior are to be expected.

To account for this, it is a good idea to allow the latent state to change by randomly selecting a different state with a certain probability.

Crosswalk example: State transition model

Discrete acceleration values are used to model the action space

$$\mathcal{A} = \{-4\tfrac{m}{s^2}, -2\tfrac{m}{s^2}, 0\tfrac{m}{s^2}, 2\tfrac{m}{s^2}\}.$$

In each time step, only those accelerations a are available, where the velocity v'_x remains non-negative. With the time discretization Δt, the state transition to the successor state $s' = (x', v'_x, r', p', v'_p)$ is modeled as follows:

$$x' = x + v_x \Delta t + \tfrac{1}{2} a \Delta t^2,$$

$$v'_x = \min(v_{\max}, v_x + a\Delta t),$$

$$r' = r,$$

$$p' = p + v_p \Delta t,$$

$$v'_p = \frac{p^\times - p'}{\Delta t}, \quad \text{if } r = r_1 \wedge 0\text{m} < p^\times - p' \le 2\tfrac{m}{s}\Delta t,$$

$$v'_p = \max(0\tfrac{m}{s}, \min(2\tfrac{m}{s}, v_p + \delta v\tfrac{m}{s})), \quad \text{else,}$$

where the terms δv in each time step are independently sampled from a standard normal distribution. The specified dynamics for v'_p causes the pedestrian to stop in any case at position $p^\times$ in front of the crosswalk if she wants to cross it (see ◻ Fig. 42.9). Apart from that, her velocity is randomly increased or decreased within $[0\tfrac{m}{s}, 2\tfrac{m}{s}]$. The route of the pedestrian is assumed to be constant. These equations overall define the stochastic state transition model $\mathcal{T}(s'|s, a)$.

42.2.3.2 Observation Model

At each time step, the agent receives an observation of the environment, which can be used to draw conclusions about the state in the form of the belief. Since localization and sensor inaccuracies are usually of minor importance for the planning problem, the positions and velocities of visible road users are often assumed to be fully observable [32, 81] or to have only a small uncertainty [10]. In the case of occlusions, obtained observations are within the limits of the visible range [13].

The observations allow to implicitly draw conclusions about the latent states of the other road users. The observation model defines the probability distribution of the received observation for a given state and action. Given the observations, Bayes' theorem can be used to determine the belief according to Eq. (42.1). For instance, the uncertainty in the route choice of another vehicle can be reduced by the fact that the observed poses are significantly more probable for one of the possible routes than for the other route options [35]. Similarly, certain observations are more likely for an aggressive driving behavior than for a cooperative driving behavior [9].

Crosswalk example: Observation model

From the vehicle perspective, the vehicle state (x, v_x) can be assumed to be known. In addition, the position of the pedestrian in Cartesian coordinates (x, y) can be measured with unbiased normally distributed uncertainty $\Sigma = \begin{pmatrix} \sigma^2 & 0 \\ 0 & \sigma^2 \end{pmatrix}$. However, we assume that nothing is known about the pedestrian's route and her velocity cannot be measured. The observation space thus can be summarized as

$$O = [x_0, x_{\text{goal}}] \times [0\tfrac{m}{s}, v_{\max}] \times \mathbb{R}^2$$

and the observation model as $o = (x, v_x, \text{cart}(p, r) + \delta p)$ with measurement noise $\delta p \sim \mathcal{N}([0, 0], \Sigma)$. Here, the function cart transforms the pedestrian position p along the route r into Cartesian coordinates $\text{cart}(p, r) = (x(p, r), y(p, r)) \in \mathbb{R}^2$. Additionally, the measurement uncertainty

$$\sigma^2 = \left\| \text{cart}(p, r) - \begin{pmatrix} x \\ 0 \end{pmatrix} \right\|_2^2$$

is assumed to be distance dependent.

42.2.3.3 Reward Function

The optimal policy of a POMDP maximizes the expected discounted cumulated reward. Therefore, the desired driving behavior of the ego vehicle is implicitly defined by the reward function. Usually, the reward function is composed of different (sometimes opposing) terms that are weighted against each other. Typical is the penalization of quadratic deviation from a desired velocity. Alternatively, overshooting the desired velocity can be penalized quadratically, while undershooting is only penalized linearly [34]. In order to achieve efficient driving, a deviation constant velocity lane following is also frequently penalized quadratically, i.e., longitudinal and possibly lateral acceleration [33]. Convenience can be further enhanced by penalizing the longitudinal jerk and lateral acceleration, which results in action selection becoming more consistent in time [35]. Maneuver safety is predominantly achieved by heavily penalizing collisions [10, 34]. An alternative to this is to allow only those actions at each time step that have been verified as collision free by an independent safety module [81]. To this end, it must be ensured that at least one action is always safe, for instance, emergency braking.

At last, there are problem-specific terms in the reward function. In order to safely pass through an intersection or crosswalk, a reward for reaching a target region is often used [11, 13]. Similarly, in scenarios where a lane change is required, reaching the target lane can be rewarded and deviating from the center of the lane can be penalized [33]. Since the rewards are discounted over time, the agent is implicitly induced to reach the destination as quickly as possible. In the context of occlusions, an increase in the field of view can be rewarded to make the agent position itself optimally before the occlusion [32]. Another way to explicitly reward a gain in information is to penalize high entropy over the unobservable states [23].

Crosswalk example: Reward model

The reward function encodes the desired behavior. First and foremost, collisions with the pedestrian must be avoided. Furthermore, comfortable driving and quickly reaching the destination is desired. A possible reward function to realize this behavior is therefore

$$\mathcal{R}(s, a) = \begin{cases} -100, & \text{if collision,} \\ 10, & \text{if } x + v_x \Delta t \geq x_{\text{goal}}, \\ -a^2 - 1, & \text{else.} \end{cases}$$

Reaching the goal region or a colliding terminates the scenario with a high or a negative reward, respectively. In all other cases, high acceleration values are penalized, and constant step costs cause reaching the goal as quickly as possible.

42.2.4 Reinforcement Learning Methods

The ideas and procedures of the previous section are well suited to solve problems for which one has a sufficiently good model description. Consequently, the quality of the obtained optimal policy depends also on the quality of the state transition and observation models. Sometimes, however, it can be difficult to find realistic models that can still be computed efficiently at the same time.

In such cases, reinforcement learning methods are suitable. To use them, the problem is also modeled as MDP or POMDP, but no knowledge of the state transition and observation models is assumed, only the possibility of interacting with the environment, either in a simulation or in the real world. The ideas underlying reinforcement learning also originate in dynamic programming.

However, due to the unknown environment model, the problem can only be solved statistically. For this purpose, a sample is used that originates from the agent's interaction with the environment and consists of trajectories and the corresponding rewards received. To obtain the sample, the agent first moves exploratively and with growing experience more and more determined in the environment. The accumulated experience is then used to iteratively learn the optimal behavior in the form of a policy [83].

In the literature, different algorithms for reinforcement learning are used. One class of algorithms directly optimizes the policy (so-called "actor" methods) [82]. Other methods are based on estimating the *Q-function* of a policy π

$$Q^\pi(s, a) = \mathcal{R}(s, a) + \gamma \mathbb{E}_{s' \sim \mathcal{T}(\cdot \mid s, a)} \left[V^\pi(s') \right],$$

which for a state s reflects the expected cumulative reward if action a is chosen in the first step and policy π is followed thereafter ("critic" methods). A Q-function immediately yields an improved policy by selecting the action with the largest Q-value in each state. This insight forms the basis for the well-known Q-learning, which iteratively learns the Q-function of the optimal policy from experience of the form (s, a, s') with successor states $s' \sim \mathcal{T}(\cdot \mid s, a)$ via the update rule

$$Q(s, a) \leftarrow (1-\alpha)Q(s, a) + \alpha \left(\mathcal{R}(s, a) + \gamma \max_{a'} Q(s', a') \right),$$

where α is the so-called learning rate [92]. Finally, there are also many methods that combine both ("actor-critic" methods) [58].

Apart from simple tasks, the Q-function and the policy cannot be reproduced exactly, such that in practice they are parameterized using function approximators. The use of neural networks has recently proven to be particularly successful in this regard [59]. In the following, we will describe how certain aspects in automated driving can be modeled and which approaches and special requirements arise in the architecture of neural networks in this application.

42.2.4.1 Modeling

The state and action space are modeled similarly as described in the previous section. It is often assumed that one can observe the positions and velocities of all vehicles (i.e., their physical states) within a certain sensor range. Some work additionally models unobservable, internal states of the other vehicles, but usually without explicitly inferring a belief over them. Instead, the belief can be implicitly captured with recurrent neural network architectures [88]. Alternatively, a k-Markov approximation of the POMDP can be used, where the last k observations are concatenated as input to the neural network [12, 30]. However, it is also possible to explicitly estimate the belief and append it to the network input [9].

In the case of the action space, in addition to modeling it with concrete acceleration values, there is also the possibility of defining different maneuver options as actions. In an intersection scenario, for example, the maneuver classes *stopping before the intersection, entering the intersection carefully*, and *crossing the intersection* can be distinguished [38, 39, 88]. Similarly, on a motorway, entering any gap in traffic can be considered as a maneuver [56]. The trajectory is then optimized based on the selected maneuver by a lower level trajectory planning system.

The reward function can also be defined using essentially the same criteria as described in the previous section. In addition, risk-sensitive behavior can be achieved in the process by penalizing risky states, using heuristics to evaluate criticality [40]. Alternatively, algorithms can be used that learn a distribution over the cumulative reward instead of just the Q-function, which corresponds to the expected cumulative reward. This allows risk measures such as conditional value at risk to be evaluated, which optimizes the behavior for the riskiest α-quantile [8, 19].

42.2.4.2 Safety

In reinforcement learning, special attention must be paid to safety verification. In contrast to planning algorithms, at runtime only execution of the optimal policy takes place. Due to function approximation or situations not represented in the training data, it can happen that an action that appears optimal is actually unsafe. The safety of the selected action must therefore be verified separately.

One approach to achieve this is to train the neural network normally at first and only allow safe actions to be executed after training [57]. However, it is better to take safety verification into account already during training by including only safe actions when calculating the expected rewards [24]. In contrast, the approach of penalizing the selection of unsafe actions during training via the reward function is not sufficient on its own, since this does not rule out the possibility that they will nevertheless be selected later. Moreover, the magnitude of the punishment terms then automatically introduces a trade-off between reward received and safety, although the latter should always be guaranteed.

42.2.4.3 Neural Network Architecture

The neural networks most commonly used as function approximators are often comparatively small in reinforcement learning. In particular, for a compact state representation, only a few hidden layers with a three-digit number of neurons are used. For a discrete action space, the Q-function can be learned such that each action is represented by one network output. If continuous actions are modeled, an "actor-critic" architecture is suitable, which learns the optimal policy in a second neural network in addition to the Q-function [28, 54]. However, if the actions are to represent entire maneuvers, the number of available actions may be variable. This is the case, for example, if the maneuvers each represent a lane change into a different traffic gap in highway driving. In this case, a vectorized maneuver representation can be added as an additional network input, so that the network can then be evaluated for all possible maneuvers sequentially to find the best action [56].

The formulation of the state as a network input also plays an important role. If the physical states of the individual vehicles are simply concatenated in a vector, the problem arises that, firstly, this vector has a variable length (depending on the number of vehicles), and secondly, it depends on the (arbitrarily chosen) order of the vehicles. This problem can be circumvented by using a learned dynamic input representation [36]. This involves learning a neural layer, which is then applied separately to each vehicle. Summing the individual results allows for variable length as well as being invariant regarding the vehicle order [36]. An equivalent behavior can also be achieved by using convolutional neural networks with a specific choice of the convolution kernel parameters [30]. Another possibility is to parameterize the state in the form of a grid where the cells describe the position and occupied cells contain the respective vehicle velocities [38, 39].

42.3   Architectures for Decision-Making

Contrary to occasional claims and demands, there is no such thing as absolute safety in either human-guided or automated road traffic. However, by suitably linking powerful modules, architectures can be formed that offer a high level of safety, so that automatic vehicles can be expected to have significantly fewer accidents than we accept today in human-guided traffic.

The following sources of uncertainty must be taken into account and designed in a fail-safe manner, i.e., in the event of a fault in a component, the overall system must switch to a minimum-risk state on its own:

Hardware Faults In the automotive sector, safety requirements are determined according to ISO/DIS 26262 and described by various SILs (safety integrity levels). This makes it possible to design fault-tolerant systems that provide sufficient functional safety against hardware faults. However, this article places the consideration of functional safety behind behavioral safety. Interested readers are referred to the corresponding standard [73] and to the UNICARagil project [79].

Perception Uncertainty Measurement uncertainty is inherent in sensors. It is subject to a variety of influences, ranging from weather conditions to the properties of the measured object to the degradation of sensors. Accordingly, it makes sense to require sensors to be self-monitoring, i.e., to assign uncertainty measures to their individual measurements that describe the range of possible deviations. Of particular importance is the uncertainty description of existence and class information about road users. In this context, especially hidden areas have to be described explicitly, see ◘ Fig. 42.8.

Prediction Uncertainty By its very nature, future traffic behavior can only be predicted with uncertainty. Various authors introduce as a requirement that behavior of automated vehicles is acceptable if they are thereby not considered to cause accidents in the sense of the respective laws (blame free) [75]. Obviously, such a requirement is necessary, but hardly sufficient

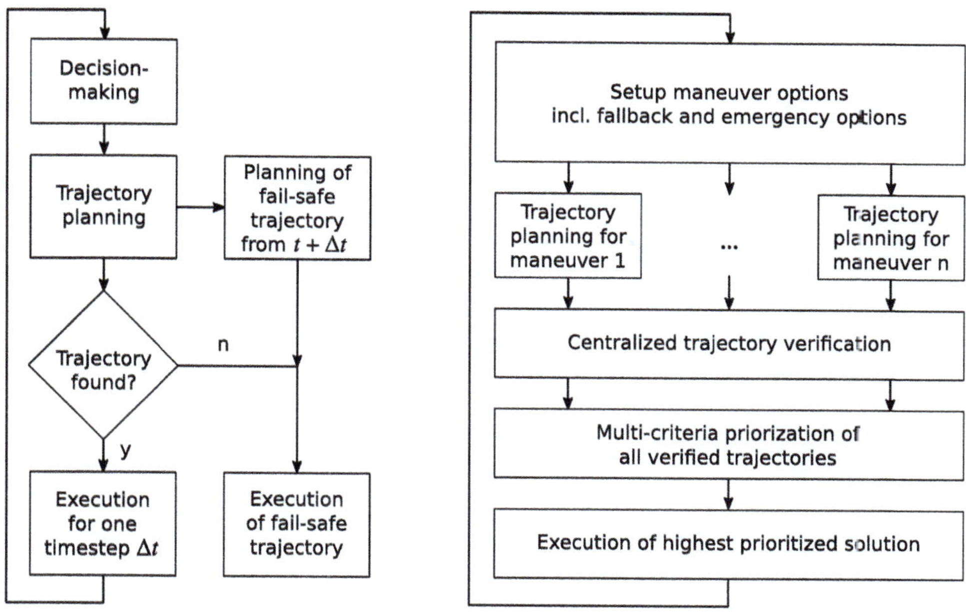

Fig. 42.10 Architectures for safe behavior. Left: Sequential maneuver and trajectory planning with a fallback trajectory. Right: Planning of alternative maneuvers and trajectories with central verification and decision-making

for automatic vehicles due to the frequency of incompliant behavior in road traffic. Rather, an appropriately tolerant behavior toward rule violations by others is necessary. While automatic—as well as human-guided—vehicles cannot protect themselves against all erratic driving maneuvers in their environment, tolerant behavior can at least be realized with respect to moderate rule violations. Therefore, a probabilistic prediction of the behavior of all road users can be taken into account in the behavior generation. In case the scene develops as expected, comfortable and speedy trajectories are driven, but these are planned in such a way that, in the case of unexpected events, they can be transferred into—then possibly uncomfortable—safe alternative trajectories [85].

In the first implementations of automatic driving, the main aim was to cover as much distance as possible automatically. The corresponding *sense-plan-act* structure was very simple. It proceeded bottom-up from perception to action. First, as much information about the driving environment as possible was collected. On this basis, the best possible planning and control was then carried out.

An essential property of safe architectures for automated driving is their awareness about available alternative options for the unexpected case that a planned trajectory appears increasingly uncertain in the course of its execution. Beyond the assessment of the risk of the currently planned trajectory, behavioral alternatives must therefore be planned through. Accordingly, today's architectures are top-down oriented.

Figure 42.10 shows on the left a fail-safe architecture going back to [68], in which a fallback trajectory is planned for each planned trajectory, which is identical to the planned trajectory for one time step of the planning, but then transfers the vehicle to a risk-minimal state. As long as a safe trajectory is found at each time step, it is executed. If a sufficiently reliable trajectory can no longer be planned in the main branch, the fallback trajectory is used.

The right side of Fig. 42.10 shows an alternative architecture in which any number of maneuvers with their trajectories are planned out. It is also permissible that no satisfactory trajectory can be determined in individual maneuver options. A central verification evaluates all trajectories and finally the best trajectory is selected and executed. This can be multi-criteria and adapted to the current driving situation. Various planning methods can be used, ranging from deep neural networks and MPC methods to simple stop and emergency stop maneuvers. Thus, in case of unexpected events in the traffic as well as in the own vehicle, it is possible to react safely in an appropriately graduated manner. In the following section, an arbitrator concept is presented based on this architecture.

42.4 Experiments with Arbitration Graphs

The concept of arbitration graphs was transferred from robot soccer to automated driving in urban traffic at the Karlsruhe Institute of Technology in 2020 [65]. The goal was to use an arbitration graph for decision-making at

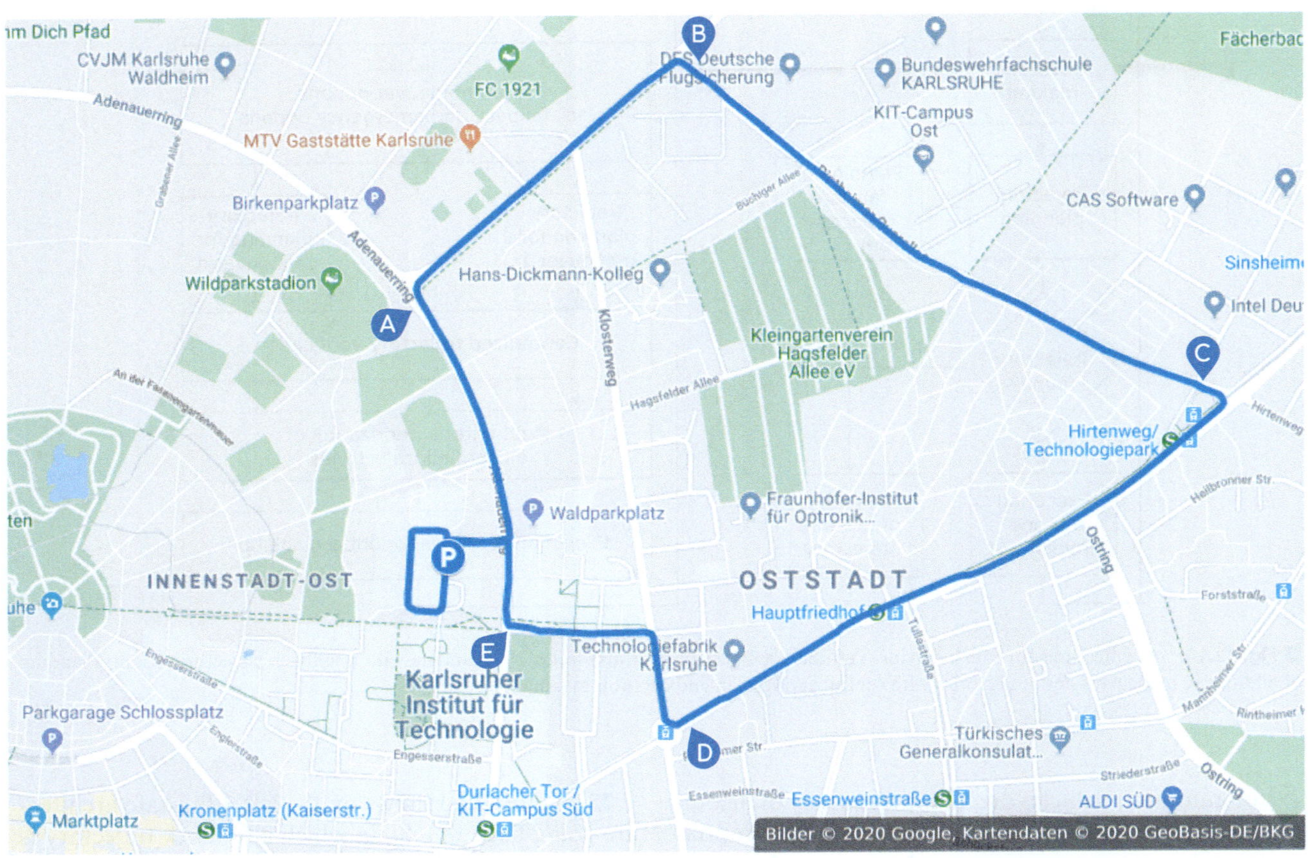

Fig. 42.11 Test track along the "Test Area Autonomous Driving Baden-Württemberg"

a maneuver level in order to drive the test route shown in Fig. 42.11, first in simulation and then in an automated manner with the test vehicle Bertha. For this purpose, basic driving maneuvers were designed as behavior blocks and linked to an overall system with arbitrators.

42.4.1 Test Track

The route stretches over 5.7km, mostly across the "Test Area Autonomous Driving Baden-Württemberg", and includes both urban and rural road-like elements with speed limits of 30, 50, and $60\frac{km}{h}$. As an example, intersection B with the planned route is shown in Fig. 42.12a. At this point, turning is especially a challenge as the crossing bicycle traffic has right of way. Subsequently, a lane change is necessary to continue following the route in a southeasterly direction. Figure 42.12b shows the traffic circle D at Karl-Wilhelms-Platz, known from the KITTI dataset [26], where the route turns north.

42.4.2 Environment Model

In this implementation, each behavior block has access to a central environment model that contains an abstract representation of the static and dynamic environment. For this purpose, the environment model continuously aggregates the preprocessed and predicted sensor data, aligns it to a highly accurate planning map [70], and transforms it into Frenet coordinates [93]. Specifically, the environmental model consists of

- the current ego motion state,
- the detected dynamic objects,
- open-loop predictions of the objects,
- the Lanelet map, and
- the planned route.

42.4.3 Maneuver Representation

In automated driving, different representations are available for the command of a behavior, i.e., the intended maneuver. In structured environments, a corridor-based representation in the form of constraints is suitable for subsequent trajectory planning. These are mainly defined by the lane boundaries (also called driving corridor), traffic rules, predicted objects as well as the selected homotopy class. In unstructured environments, such as parking lots, on the other hand, an additional reference path is necessary, or even a complete target trajectory is more suitable. Since there

(a)　　　　　　　　　　　　　　　(b)

Fig. 42.12 Excerpts from the "Test Area Autonomous Driving Baden-Württemberg". **a** Intersection B at Rintheimer-Querallee with crossing bike lane and subsequent lane change. **b** Roundabout D at Karl-Wilhelms-Platz

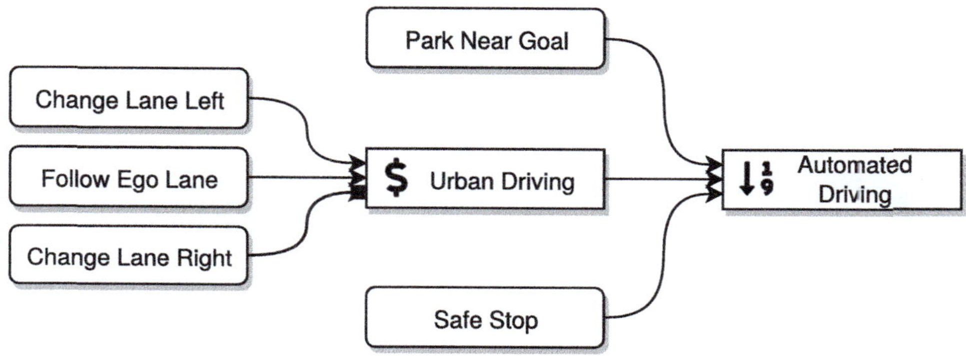

Fig. 42.13 Arbitration graph with basic behavior blocks for automated urban driving

are good reasons for both variants, a task-independent meta-representation was chosen as output of the *get command* function, which supports both corridor-based and trajectory-based representations.

42.4.4 Behavior Blocks

For the first field test, the basic behavior blocks have been developed, which are sufficient in principle to drive the selected test route automatically. Further behavior blocks will be developed at a later stage, mainly to optimize the behavior in challenging scenarios. The basic behavior blocks cover following the ego lane, lane change maneuvers, parking, or an emergency stop maneuver and are illustrated in ● Figs. 42.13 and 42.14.

In the FollowEgoLane behavior block, the invocation and commitment conditions check whether the ego vehicle is located within a lane of the route at all. If this is the case, the subsequent trip can be performed in an ACC-like approach. For this purpose, the *get command* function generates a driving corridor from the ego lane boundaries, adds any stop lines or the vehicles ahead,

and returns this target maneuver to the calling arbitrator. The stop lines cover a variety of situations with longitudinal behavior, such as crosswalks, intersections with traffic lights, and giveway situations. They can be released either already in the preprocessing of the environment model, by the behavior blocks themselves or manually (by a passenger or remote control). A subsequent trajectory planner, similarly modeled as in [95], finally executes the corridor-based maneuvers.

A lane change is only possible if the vehicle is in a lane of the roadway, this lane has an adjacent lane that may be changed to, and there is a sufficiently large gap at this point for safe lane changing. These conditions are checked in the call condition of the ChangeLane behavior block. In addition, a lane change should not be interrupted in the middle of the maneuver by other behaviors, but should either be finished successfully or aborted in a controlled manner. Therefore, the commitment condition is true until the ego vehicle has either fully reached the target lane or got back to the original lane. Finally, for the corridor-based maneuver representation, a corridor is generated that transitions from the ego lane to the target lane. The *get command* function returns this corri-

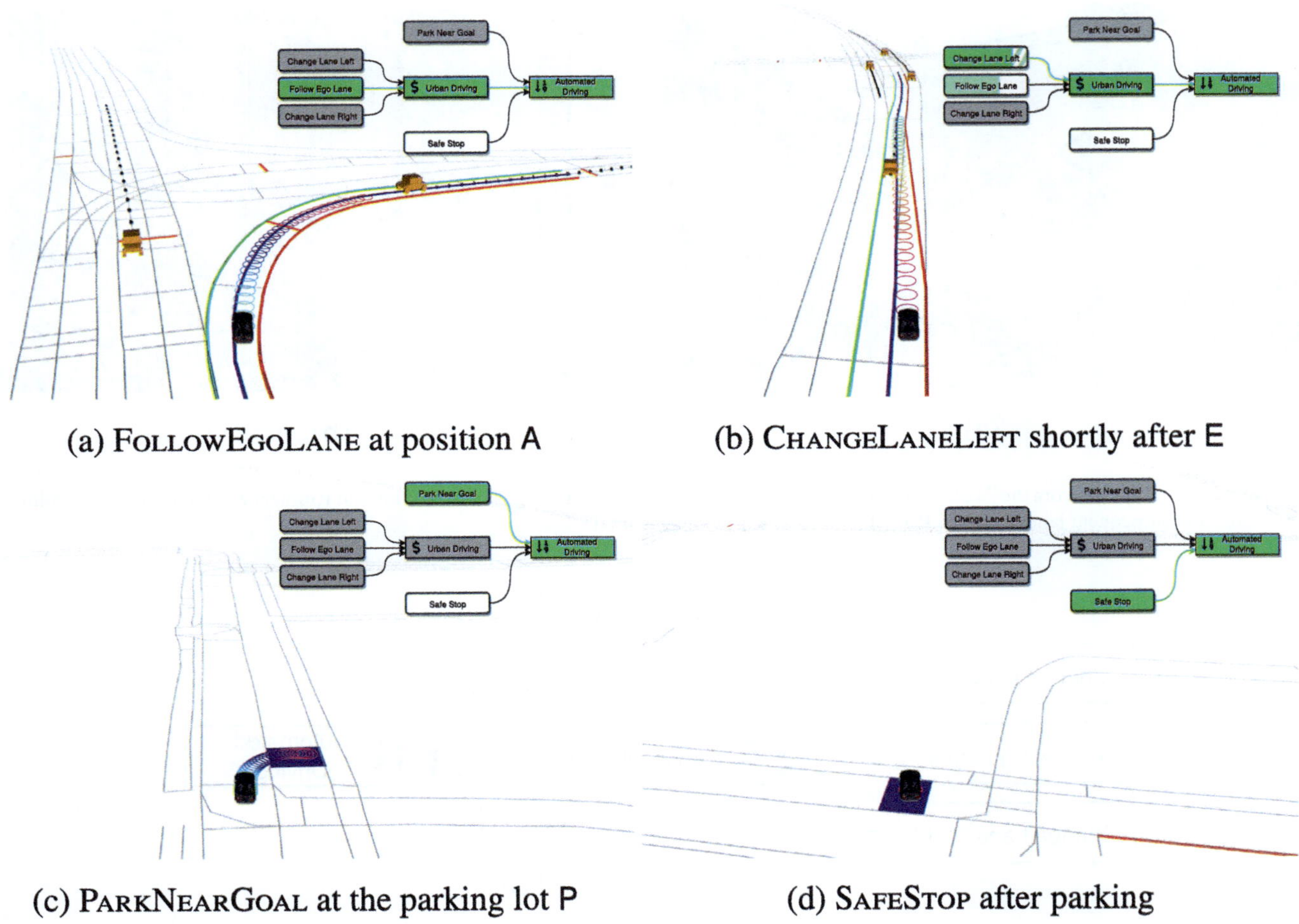

(a) FollowEgoLane at position A

(b) ChangeLaneLeft shortly after E

(c) ParkNearGoal at the parking lot P

(d) SafeStop after parking

Fig. 42.14 Extracts from the simulation results, each with the state of the arbitration graph. Behavior blocks or arbitrators with a gray background are currently not applicable. The green background marks the active behavior option, with the bar plot for ChangeLaneLeft and FollowEgoLane representing the expected benefits, i.e., the negative normalized costs of these options

dor, all objects to be considered and potential stop lines. The latter are important, for example, for an anticipatory behavior when changing to a turning lane. It is needed to reduce the velocity during the lane change already, when approaching a crosswalk.

The behavior block ParkNearGoal is supposed to maneuver the ego vehicle into a selected parking space at the end of the route. Thus, the invocation condition is only true when the vehicle has almost come to a standstill near the parking lot. The commitment condition subsequently remains true until the vehicle has come to a sufficiently accurate stop again at the parking position. For maneuver planning of the parking behavior, a hybrid curvature trajectory is determined based on [7] and an LQR-based longitudinal planning in the *get command* function.

At the start of a journey, a fully automated vehicle must also maneuver out of its parking space or garage. However, since the test vehicle Bertha can only be switched to automatic mode when it is in slow rolling motion and thus already on the track, such a parking-out behavior was not implemented.

The emergency stop maneuver SafeStop is intended to bring the vehicle to a safe standstill as a fallback level if no other behavior is applicable or if behavior blocks fail. SafeStop is thus intended to be applicable at all times, such that the invocation and commitment conditions always return true. In a first and simplest implementation, the *get command* function implements full braking. However, in order to implement more comfortable and less risky emergency stopping maneuvers, further approaches should be taken into account. For example those that combine convex optimization with reachable set analysis [68] and thus generate emergency stopping maneuvers that are provably safe and at the same time as comfortable as possible.

42.4.5 Cost Arbitrator

In automated driving, different driving maneuvers do not always have clear priority compared to each other. When designing the decision-making module, it is impossible to determine in advance whether, for example, a lane change is always preferable to follow lane behavior, as long as it is applicable. Therefore, so-called cost arbitrators have been presented for the use of arbitration graphs in automated driving [65]. In principle, they can be based on arbitrary cost functions and are thus also suitable for other application areas.

In this implementation, the cost function considers long-term routing costs and penalizes short-term maneuver changes in order to finally decide whether a possible lane change is also reasonable. For this purpose, the cost function for maneuver choice J is determined as follows:

$$J = -\hat{v} + n \cdot J_{\text{LCRoute}} + J_{\text{LC}}. \tag{42.2}$$

Thus, it rewards the medium-term expected average speed of the maneuver $\hat{v}$, penalizes the required number of lane changes n that would be necessary to follow the route after this maneuver, and penalizes maneuvers that themselves perform lane changes with J_{LC}. First heuristic analyses in simulation have shown that such a cost function is robust in its parameterization, so that the illustrative as well as appropriate values of $J_{\text{LCRoute}} = 10\frac{\text{km}}{\text{h}}$ and $J_{\text{LC}} = 5\frac{\text{km}}{\text{h}}$ were chosen.

42.4.6 Full Arbitration Graph

The behavior blocks are now to be transformed into a suitable arbitration graph as shown in ◘ Fig. 42.13. First, the corridor-based maneuvers FollowEgoLane and ChangeLane are combined with the presented cost arbitrator to form UrbanDriving. ChangeLane is instantiated twice—once for the lane change to the left and once to the right. Since the ParkNearGoal parking behavior only becomes applicable near the goal parking lot anyway, it is added to the top-level arbitrator AutomatedDriving with higher priority than UrbanDriving. Finally, the SafeStop behavior is added as the lowest priority option of the AutomatedDriving arbitrator to constitute the fallback layer.

42.4.7 Results

The presented arbitration graph was first evaluated in the CoInCar simulation environment [62]. The trip lasts a total of 9:40 min, with the follow lane maneuver FollowEgoLane active for 91% of the time. This is expected, because FollowEgoLane also integrates longitudinal behavior for different types of stop lines. At points A, C, D, and E lane changes to the right, i.e., ChangeLaneRight, are performed; ChangeLaneLeft, on the other hand, is performed at B and several times in a row at point E. At the end of the route, ParkNearGoal moves the vehicle to the designated parking space P on the KIT campus. Since in the end the drive is completed and thus none of

the other behavior options is applicable anymore, Safe-Stop prevents undefined behavior and keeps the vehicle at a standstill.

◘ Figure 42.14 shows various situations and the state of the arbitration graph as they occurred in the simulation. On the turning lane at location A in ◘ Fig. 42.14a, neither a lane change nor parking is applicable and is therefore grayed out in the visualization. This is because, first, there are no adjacent lanes here to switch to and, second, the target position is too far away to start a parking maneuver. UrbanDriving is nevertheless applicable, because the FollowEgoLane option is applicable. Therefore, AutomatedDriving chooses its higher priority option UrbanDriving, which in turn chooses FollowEgoLane and calls its *get command* function so that the ego vehicle executes a lane following maneuver.

In ◘ Fig. 42.14b, i.e., shortly after location E, the vehicle is still not close to the goal parking lot. In addition, there is no right adjacent lane, but there is a left adjacent lane. Furthermore, since the route is about to turn left onto the KIT campus, ChangeLaneLeft has lower costs than FollowEgoLane or, as indicated by the green bars in the figure, a higher benefit. AutomatedDriving thus chooses the UrbanDriving option again in this situation, and it in turn chooses the lower cost ChangeLaneLeft option.

In ◘ Fig. 42.14c, the ego vehicle is shown during the parking maneuver. Since it came to a stop just before the end of the route in the follow lane mode, ParkNearGoal was applicable and executed by AutomatedDriving as the highest priority behavior option. Subsequently, the ego vehicle left the route in parking mode, such that now none of the corridor-based maneuvers were applicable.

Finally, as soon as the ego vehicle comes to a standstill in the parking space, ParkNearGoal is also no longer applicable, as can be seen in ◘ Fig. 42.14d. Consequently, the fallback option SafeStop is the only behavior that can still be executed, is selected by AutomatedDriving, and thus ensures that the vehicle remains in standstill.

42.4.8 Summary

In summary, arbitration graphs have proven to be a particularly suitable method for decision-making at maneuver level. Due to the consequent modularity and decoupling of the individual behavior blocks, they are manageable, and the entire arbitration graph is very easy to maintain. In addition, the decision-making can be easily traced and thus examined due to its compact hierarchical representation including costs and invocation conditions. By integrating the preconditions of a behavior into the invocation and commitment condition, it is ensured that only behaviors can be invoked that are actually app-

Fig. 42.15   Arbitration graphs have replaced finite state machines on Bertha

licable in the current situation. If one also employs verification methods such as reachable set analysis [2] in these precondition functions, only guaranteed safe behaviors would be executed.

Based on these good simulation results and positive experience with arbitration graphs, the approach was transferred to the real test vehicle Bertha (see Fig. 42.15) in the spring of 2020. In the summer of 2020, the arbitration graph finally replaced the previously used finite state machine. It has been used since then successfully in many real test drives on the road. For safety reasons, however, in some challenging situations, such as crossing bike lanes with bicycles having right of way or merging by means of a zipper procedure, manual release of the stop lines is used. This makes it possible to solve these situations semi-automatically without having to interrupt the automated mode of the vehicle until sufficiently reliable perception and planning procedures have been developed at the institute for these situations as well. The presented arbitration architecture considerably simplifies such a subsequent extension of the decision-making by its modular design.

As already indicated, the arbitration approach will be extended by verification methods based on reachable set analysis. However, since this requires planned trajectories, an MPCC-based trajectory planner like [49] will be integrated into the behavior blocks for follow and change lane behaviors, thus transferring all maneuver options into a trajectory-based maneuver representation. In addition, further behaviors will be developed, especially for interactive maneuvers and the anticipatory handling of occlusions and uncertainties.

42.5 Conclusion

Linking the internal environment representation and trajectory planning, behavior planning plays a central role. The resulting driving strategy must process the predicted development of the scene and make the essential decisions that a human driver must also make continuously. The downstream trajectory planning then essentially implements the made decision in an efficient and convenient way.

In this chapter, essential methods of decision-making were presented. Due to their clarity and widespread use, rule-based methods were introduced, in which the behavior strategy can be represented in the form of a graph whose nodes form behavior modules. In contrast, in the subsequent graph search methods, the decision-making is made by searching a motion graph. For this, an objective function is defined, and the optimal behavior is determined as a minimizing sequence of states. However, these methods are primarily suitable for planning in an environment whose future evolution is known or at least can be well predicted. In contrast, the probabilistic methods presented next can be used when the further evolution of the scene has to be modeled stochastically due to insufficient knowledge or when essential parameters of the scene (e.g., future behavior of other road users) are unknown or highly uncertain. Using a probabilistic model of the dynamics and the observations, a policy is then determined that optimizes the objective criterion. Reinforcement learning, presented next, uses (real or simulated) data instead of explicit dynamics and observation models to learn optimal behavior from collected experience. Learning methods are nowadays used with increasing success, but a major drawback is the enormous demand for training data and the particular challenges in validating safety for learning methods.

A central role in terms of safety is determined by the architecture of behavioral decision-making. Although absolute safety in road traffic will remain unattainable, the choice and implementation of a suitable architecture enables the realization of automated vehicles that are resilient to a wide range of errors, from perception to the traffic violations of others. Finally, in light of these architectural considerations, a concept for behavioral decision-making based on the arbitration graphs presented in ▶ Sect. 42.2.1.4 has been described and is already successfully used in practice. This architecture increases traceability, scalability, and safety. At the same time, the different planning strategies presented in ▶ Sect. 42.2 can be used as behavior blocks within the arbitration graph, combined by suitable arbitrators.

References

1. Aeberhard, M., et al.: Experience, results and lessons learned from automated driving on Germany's highways. IEEE Intell. Transp. Syst. Mag. **7**(1) (2015). ▶ https://doi.org/10.1109/MITS.2014.2360306
2. Althoff, M., Magdici, S.: Set-based prediction of traffic participants on arbitrary road networks. IEEE Trans. Intell. Veh. **1**(2) (2016). ▶ https://doi.org/10.1109/TIV.2016.2622920
3. Ardelt, M., Waldmann, P.: Hybrides Steuerungs- und Regelungskonzept für das hochautomatisierte Fahren auf Autobahnen. In: at – Automatisierungstechnik **59**(12) (2011). ▶ https://doi.org/10.1524/auto.2011.0961
4. Ardelt, M., et al.: Strategic decision-making process in advanced driver assistance systems. In: IFAC Proceedings Volumes, 6th IFAC Symposium on Advances in Automotive Control, vol. 43, no. 7 (2010). ▶ https://doi.org/10.3182/20100712-3-DE-2013.00006
5. Bacha, A., et al.: Odin: team Victortango's entry in the Darpa urban challenge. J. Field Robot. **25**(8) (2008). ▶ https://doi.org/10.1002/rob.20248
6. Bagnell, J.A., et al.: An integrated system for autonomous robotics manipulation. In: 2012 IEEE/RSJ International Conference on Intelligent Robots and Systems (2012). ▶ https://doi.org/10.1109/IROS.2012.6385888
7. Banzhaf, H., et al.: From footprints to beliefprints: motion planning under uncertainty for maneuvering automated vehicles in dense scenarios. In: International Conference on Intelligent Transportation Systems. IEEE (2018). ▶ https://doi.org/10.1109/ITSC.2018.8569897
8. Bernhard, J., Pollok, S., Knoll, A.: Addressing inherent uncertainty: risk-sensitive behavior generation for automated driving using distributional reinforcement learning. In: 2019 IEEE Intelligent Vehicles Symposium (IV), pp. 2148–2155 (2009). ▶ https://doi.org/10.1109/IVS.2019.8813791
9. Bouton, M., et al.: Cooperation-aware reinforcement learning for merging in dense traffic. In: 2019 IEEE Intelligent Transportation Systems Conference (ITSC), pp. 3441–3447 (2019). ▶ https://doi.org/10.1109/ITSC.2019.8916924
10. Bouton, M., Cosgun, A., Kochenderfer, M.J.: Belief state planning for autonomously navigating urban intersections. IEEE Intelligent Vehicles Symposium (IV), Los Angeles, CA, USA (2017) (2017)
11. Bouton, M., et al.: Scalable decision making with sensor occlusions for autonomous driving. In: IEEE International Conference on Robotics and Automation, ICRA. Brisbane, Australia (2018)
12. Bouton, M., et al.: Utility decomposition with deep corrections for scalable planning under uncertainty. In: Proceedings of the 17th International Conference on Autonomous Agents and Multi-agent Systems, pp. 462–469. AAMAS '18. International Foundation for Autonomous Agents and Multiagent Systems, Richland, SC (2018)
13. Brechtel, S., Gindele, T., Dillmann, R.: Probabilistic decision-making under uncertainty for autonomous driving using continuous POMDPs. In: 2014 IEEE 17th International Conference on Intelligent Transportation Systems (ITSC), pp. 392–399 (2014)
14. Brooks, R.: A robust layered control system for a mobile robot. IEEE J. Robot. Autom. **21** (1986). ▶ https://doi.org/10.1109/JRA.1986.1087032
15. Cai, P., et al.: HyP-DESPOT: a hybrid parallel algorithm for online planning under uncertainty. Int. J. Robot. Res. 027836492093707 (2020). ISSN: 0278-3649, 1741–3176. ▶ https://doi.org/10.1177/0278364920937074
16. Colledanchise, M., Ögren, P.: How behavior trees modularize hybrid control systems and generalize sequential behavior compositions, the subsumption architecture, and decision trees. In: IEEE Trans. Robot. **33**(2) (2017). ▶ https://doi.org/10.1109/TRO.2016.2633567
17. Colledanchise, M., Faconti, D.: BehaviorTree.CPP. Version 3.5.4. Fundació Eurecat, Barcelona, Spain. ▶ https://behaviortree.dev
18. Colledanchise, M., Ögren, P.: Behavior Trees in Robotics and AI: an Introduction. CRC Press Boca Raton, FL, USA (2018). ▶ https://doi.org/10.1201/9780429489105
19. Dabney, W.: et al. Implicit quantile networks for distributional reinforcement learning. In: Dy, J., Krause, A. (eds.) Proceedings of the 35th International Conference on Machine Learning, vol. 80. Proceedings of Machine Learning Research. PMLR, pp. 1096–1105 (2018)
20. Dahl, O.J., Dijkstra, E.W., Hoare, C.A.R. (eds.): Structured Programming, 234 pp. Academic Press Ltd., London, United Kingdom (1972)
21. Dijkstra, E.W.: A note on two problems in connexion with graphs. Numer. Math. **1**(1) (1959)
22. Dijkstra, E.W.: Letters to the editor: go to statement considered harmful. Commun. ACM **11**(3) (1968). ▶ https://doi.org/10.1145/362929.362947
23. Fischer, J., Tas, O.S.: Information particle filter tree: an online algorithm for POMDPs with belief-based rewards on continuous domains. In: Daumé, H. (ed.): Proceedings of the 37th International Conference on Machine Learning. III and Aarti Singh. Proceedings of Machine Learning Research. PMLR, vol. 119, pp. 3177–3187 (2020)
24. Fischer, J., et al.: Sampling-based inverse reinforcement learning algorithms with safety constraints. In: 2021 IEEE/RSJ International Conference on Intelligent Robots and Systems (IROS), pp. 791–798 (2021). ▶ https://doi.org/10.1109/IROS51168.2021.9636672

25. Gammell, J.D., Strub, M.P.: Asymptotically optimal sampling-based motion planning methods. In: Annual Review of Control, Robotics, and Autonomous Systems, vol. 4, no. 1 (2021). ▶ https://doi.org/10.1146/annurev-control-061920-093753

26. Geiger, A., et al.: Vision meets robotics: The KITTI dataset. Int. J. Robot. Res. (IJRR) (2013). ▶ https://doi.org/10.1177/0278364913491297

27. Gresyk, A.: High-Performance Hierarchical Finite State Machine. Version 1.8.0. (2021). https://hfsm.dev. Visited on 18 Jan 2021

28. Haarnoja, T., et al.: Soft actor-critic: off-policy maximum entropy deep reinforcement learning with a stochastic actor. In: Dy, J., Krause, A. (eds.) Proceedings of the 35th International Conference on Machine Learning, vol. 80. Proceedings of Machine Learning Research. PMLR, pp. 1861–1870 (2018)

29. Hart, P.E., Nilsson, N.J., Raphael, B.: A formal basis for the heuristic determination of minimum cost paths. IEEE Trans. Syst. Sci. Cybern. 4(2) (1968). ▶ https://doi.org/10.1109/TSSC.1968.300136

30. Hoel, C.J., Wolff, K., Laine, L.: Automated speed and lane change decision making using deep reinforcement learning. In: 2018 21st International Conference on Intelligent Transportation Systems (ITSC), pp. 2148–2155 (2018). ▶ https://doi.org/10.1109/ITSC.2018.8569568

31. Hopcroft, J.E., Motwani, R., Ullman, J.D.: Introduction to Automata Theory, Languages, and Computation. 3rd edn, 535 pp. Pearson Addison Wesley, Boston, MA, USA (2007). ▶ https://doi.org/10.1145/568438.568455

32. Hubmann, C., et al.: A POMDP maneuver planner for occlusions in urban scenarios. In: 2019 IEEE Intelligent Vehicles Symposium (IV), pp. 2172–2179 (2019). ▶ https://doi.org/10.1109/IVS.2019.8814179

33. Hubmann, C., et al.: A belief state planner for interactive merge maneuvers in congested traffic. In: IEEE International Conference on Intelligent Transportation Systems (ITSC) (2018)

34. Hubmann, C., et al.: Automated driving in uncertain environments: planning with interaction and uncertain maneuver prediction. IEEE Trans. Intell. Veh. (2018). ISSN: 2379-8904, 2379-8858. ▶ https://doi.org/10.1109/TIV.2017.2788208

35. Hubmann, C., et al.: Decision making for autonomous driving considering interaction and uncertain prediction of surrounding vehicles. In: Intelligent Vehicles Symposium (IV), pp. 1671–1678. IEEE (2017)

36. Huegle, M., et al.: Dynamic input for deep reinforcement learning in autonomous driving. In: 2019 IEEE/RSJ International Conference on Intelligent Robots and Systems (IROS), pp. 7566–7573. IEEE, Macau, China (2019). ISBN: 978-1-72814-004-9. ▶ https://doi.org/10.1109/IROS40897.2019.8968560

37. Iovino, M., et al.: A Survey of Behavior Trees in Robotics and AI. Version 2. May 13 (2020). arXiv: arxiv: 2005.05842 [cs]. ▶ http://arxiv.org/abs/2005.05842

38. Isele, D., Nakhaei, A., Fujimura, K.: Safe reinforcement learning on autonomous vehicles. In: 2018 IEEE/RSJ International Conference on Intelligent Robots and Systems (IROS) (2018). ▶ https://doi.org/10.1109/IROS.2018.8593420

39. Isele, D., et al.: Navigating occluded intersections with autonomous vehicles using deep reinforcement learning. In: 2018 IEEE International Conference on Robotics and Automation (ICRA), pp. 2034–2039 (2018). ▶ https://doi.org/10.1109/ICRA.2018.8461233

40. Kamran, D., et al.: Risk-aware high-level decisions for automated driving at occluded intersections with reinforcement learning. In: 2020 IEEE Intelligent Vehicles Symposium (IV), pp. 1205–1212 (2020). ▶ https://doi.org/10.1109/IV47402.2020.9304606

41. Karaman, S., Frazzoli, E.: Sampling-based algorithms for optimal motion planning. Int. J. Robot. Res. 30(7) (2011). ▶ https://doi.org/10.1177/0278364911406761

42. Karaman, S., et al.: Anytime motion planning using the RRT*. In: International Conference on Robotics and Automation. IEEE (2011). ▶ https://doi.org/10.1109/ICRA.2011.5980479

43. Kavraki, L.E., Kolountzakis, M.N., Latombe, J.: Analysis of probabilistic roadmaps for path planning. IEEE Trans. Robot. Autom. 14(1) (1998). ▶ https://doi.org/10.1109/70.660866

44. Kavraki, L.E., et al.: Probabilistic roadmaps for path planning in high-dimensional configuration spaces. IEEE Trans. Robot. Autom. 12(4) (1996). ▶ https://doi.org/10.1109/70.508439

45. Kesting, A., Treiber, M., Helbing, D.: General lane-changing model mobil for car-following models. Transp. Res. Record J. Transp. Res. Board, 86–94 (2007). ISSN: 0361-1981, 2169-4052. ▶ https://doi.org/10.3141/1999-10

46. Klimenko, D., Song, J., Kurniawati, H: TAPIR: a software toolkit for approximating and adapting POMDP solutions online. In: Proceedings of the Australasian Conference on Robotics and Automation, Melbourne, Australia, vol. 24 (2014)

47. Kochenderfer, M.: Decision Making Under Uncertainty—Theory and Application, 1st edn. MIT Press (2015). ISBN: 0-262-02925-1 978-0-262-02925-4

48. Kuwata, Y., et al.: Real-time motion planning with applications to autonomous urban driving. IEEE Trans. Control Syst. Technol. 17(5) (2009). ▶ https://doi.org/10.1109/TCST.2008.2012116

49. Lam, D., Manzie, C., Good, M.: Model predictive contouring control. In: Conference on Decision and Control (CDC). IEEE (2010). ▶ https://doi.org/10.1109/CDC.2010.5717042

50. Lauer, M., et al.: Cognitive concepts in autonomous soccer playing robots. Cogn. Syst. Res. 11(3) (2010). ▶ https://doi.org/10.1016/j.cogsys.2009.12.003

51. LaValle, S.M.: Planning Algorithms. Cambridge University Press, New York, NY, USA (2006)

52. LaValle, S.M.: Rapidly-Exploring Random Trees: A New Tool for Path Planning. Computer Science Dept., Iowa State University, Ames, IA, USA (1998)

53. LaValle, S.M., Kuffner, J.J.: Randomized kinodynamic planning. Int. J. Robot. Res. 20(5) (2001). ▶ https://doi.org/10.1177/02783640122067453

54. Lillicrap, T.P., et al.: Continuous control with deep reinforcement learning. In: Bengio, Y., LeCun, Y. (eds.) ICLR (2016)

55. MATLAB Stateflow. The MathWorks, Inc. Natick, MA, USA (2020). ▶ https://www.mathworks.com/products/stateflow.html

56. Mirchevska, B., et al.: Amortized Q-learning with model-based action proposals for autonomous driving on highways. In: IEEE International Conference on Robotics and Automation, pp. 1028–1035. IEEE, Xi'an, China (2021). ▶ https://doi.org/10.1109/ICRA48506.2021.9560777

57. Mirchevska, B., et al.: High-level decision making for safe and reasonable autonomous lane changing using reinforcement learning. In: 2018 21st International Conference on Intelligent Transportation Systems (ITSC), pp. 2156–2162. IEEE, Maui, HI (2018). ISBN: 978-1-72810-321-1 978-1-72810-323-5. ▶ https://doi.org/10.1109/ITSC.2018.8569448

58. Mnih, V., et al.: Asynchronous methods for deep reinforcement learning. In: Balcan, M.F., Weinberger, K.Q. (eds.) Proceedings of the 33rd International Conference on Machine Learning, Vol. 48. Proceedings of Machine Learning Research. PMLR, pp. 1928–1937, New York, New York, USA (2016)

59. Mnih, V., et al.: Human-level control through deep reinforcement learning. Nature 518(7540), 529–533 (2015). ISSN: 0028-0836, 1476-4687. ▶ https://doi.org/10.1038/nature14236

60. Montemerlo, M., et al.: Junior: the stanford entry in the urban challenge. J. Field Robot. **25**(9) (2008). ▶ https://doi.org/10.1002/rob.20258

61. Moret, B.M.E.: Decision trees and diagrams. ACM Comput. Surv. **14**(4) (1982). ▶ https://doi.org/10.1145/356893.356898

62. Naumann, M., et al.: CoInCar-Sim: an open-source simulation framework for cooperatively interacting automobiles. IEEE Intell. Veh. Symp. (2018). ▶ https://doi.org/10.1109/IVS.2018.8500405

63. Ögren, P.: Increasing modularity of uav control systems using computer game behavior trees. In: AIAA Guidance, Navigation, and Control Conference. American Institute of Aeronautics and Astronautics, Minneapolis, MN, USA (2012). ▶ https://doi.org/10.2514/6.2012-4458

64. Olsson, M.: Behavior trees for decision-making in autonomous driving. KTH, School of Computer Science and Communication (CSC), Stockholm, Sweden (2016). ▶ http://urn.kb.se/resolve?urn=urn:nbn:se:kth:diva-183060

65. Orzechowski, P.F., Burger, C., Lauer, M.: Decision-making for automated vehicles using a hierarchical behavior-based arbitration scheme. In: Intelligent Vehicles Symposium. IEEE, Las Vegas, NV, USA (2020). ▶ https://doi.org/10.1109/IV47402.2020.9304723

66. Palmieri, L., Koenig, S., Arras, K.O.: RRT-based nonholonomic motion planning using any-angle path biasing. In: International Conference on Robotics and Automation. IEEE (2016). ▶ https://doi.org/10.1109/ICRA.2016.7487439

67. Papadimitriou, C.H., Tsitsiklis, J.N.: The complexity of markov decision processes. Math. Oper. Res. **12**(3), 441–450 (1987). ISSN: 0364765X, 15265471

68. Pek, C., Althoff, M.: Computationally efficient fail-safe trajectory planning for self-driving vehicles using convex optimization. In: International Conference on Intelligent Transportation Systems. IEEE (2018). ▶ https://doi.org/10.1109/ITSC.2018.8569425

69. Perez, A., et al.: LQR-RRT*: optimal sampling-based motion planning with automatically derived extension heuristics. In: International Conference on Robotics and Automation. IEEE (2012). ▶ https://doi.org/10.1109/ICRA.2012.6225177

70. Poggenhans, F., et al.: Lanelet2: a high-definition map framework for the future of automated driving. In: International Conference on Intelligent Transportation Systems. IEEE (2018). ▶ https://doi.org/10.1109/ITSC.2018.8569929

71. Quinlan, J.R.: Decision trees and decision-making. IEEE Trans. Syst. Man, Cybern. **20**(2) (1990). ▶ https://doi.org/10.1109/21.52545

72. Rao, A.S., Georgeff, M.P.: An abstract architecture for rational agents. In: Proceedings of the Third International Conference on Principles of Knowledge Representation and Reasoning. KR'92. Morgan Kaufmann Publishers Inc., San Mateo, CA, USA (1992)

73. Road Vehicles—Functional Safety: ISO 26262. International Organization for Standardization, Geneva, Switzerland (2018). https://www.iso.org/cms/render/live/en/sites/isoorg/contents/data/standard/06/83/68383.html. Visited on 19 Apr 2021

74. Schäling, B.: The Boost C++ Libraries, 2nd edn, 570 pp. XML Press, Laguna Hills, CA, USA (2014)

75. Shalev-Shwartz, S., Shammah, S., Shashua, A.: On a Formal Model of Safe and Scalable Self-driving Cars (2017). arxiv: 1708.06374 [cs, stat]. visited on 04 Dec 2018

76. Shkolnik, A., et al.: Bounding on rough terrain with the LittleDog Robot. Int. J. Robot. Res. **30**(2) (2011). ▶ https://doi.org/10.1177/0278364910388315

77. Siciliano, B., Khatib, O. (eds.): Springer Handbook of Robotics. Springer Handbooks. Springer International Publishing (2016). ▶ https://doi.org/10.1007/978-3-319-32552-1

78. Stefik, M., Bobrow, D.G.: Object-oriented programming: themes and variations. AI Mag. **6**(4) (1985). ▶ https://doi.org/10.1609/aimag.v6i4.508

79. Stolte, T., et al.: Towards Safety Concepts for Automated Vehicles by the Example of the Project UNICARagil (2020). ▶ https://doi.org/10.24355/dbbs.084-202011171557-0

80. Sunberg, Z., Kochenderfer, M.: Online algorithms for POMDPs with continuous state, action, and observation spaces. In: Proceedings of the Twenty-Eighth International Conference on Automated Planning and Scheduling, pp. 259–263. AAAI (2018)

81. Sunberg, Z.N., Ho, C.J., Kochenderfer, M.J.: The value of inferring the internal state of traffic participants for autonomous freeway driving. In: American Control Conference (ACC). Seattle (2017)

82. Sutton, R.S., et al.: Policy gradient methods for reinforcement learning with function approximation. In: Solla, S., Leen, T., Müller, K. (eds.) Advances in Neural Information Processing Systems, vol. 12, pp. 1057–1063. MIT Press (2000)

83. Sutton, R.S., Barto, A.: Reinforcement learning: an introduction. In: Adaptive Computation and Machine Learning, 2nd edn. The MIT Press, Cambridge. MA London (2018). ISBN: 978-0-262-03924-6

84. Svestka, P., Overmars, M.H.: Coordinated motion planning for multiple car-like robots using probabilistic roadmaps. In: International Conference on Robotics and Automation, vol. 2. IEEE (1995). ▶ https://doi.org/10.1109/ROBOT.1995.525508

85. Taş, O.S., Stiller, C.: Limited visibility and uncertainty aware motion planning for automated driving. In: 2018 IEEE Intelligent Vehicles Symposium (IV) (2018). ▶ https://doi.org/10.1109/IVS.2018.8500369

86. The State Machine Framework. Version 5.15. Espoo, Finland: The Qt Company Ltd., (2020). ▶ https://doc.qt.io/qt-5/statemachine-api.html

87. TinyFSM. Version 0.3.2. Zürich, Switzerland: Digital Integrity GmbH (2018). ▶ https://digint.ch/tinyfsm/

88. Tram, T., et al.: Learning negotiating behavior between cars in intersections using deep Q-learning. In: 2018 21st International Conference on Intelligent Transportation Systems (ITSC), pp. 3169–3174 (2018). ▶ https://doi.org/10.1109/ITSC.2018.8569316

89. Unified Modeling Language. Standard 2.5.1. Milford, MA, USA: Object Management Group, Inc (2017). ▶ https://www.omg.org/spec/UML/2.5.1

90. Vossen, G., Witt, K.-U.: Grundkurs Theoretische Informatik: Eine anwendungsbezogene Einführung - Für Studierende in allen Informatik-Studiengängen, 6th ed. Springer, Wiesbaden, Germany (2016). ▶ https://doi.org/10.1007/978-3-8348-2202-4

91. Wagner, F., et al.: Modeling Software with Finite State Machines: A Practical Approach, 391 pp. CRC Press, Boca Raton, FL, USA (2006)

92. Watkins, C.J.C.H., Dayan, P.: Q-learning. Mach. Learn. **8**(3), 279–292 (1992). issn: 1573-0565. ▶ https://doi.org/10.1007/BF00992698

93. Werling, M., et al.: Optimal trajectory generation for dynamic street scenarios in a Frenét frame. In: International Conference on Robotics and Automation. IEEE (2010). ▶ https://doi.org/10.1109/ROBOT.2010.5509799

94. Ziegler, J., et al.: Making bertha drive—an autonomous journey on a historic route. IEEE Intell. Transp. Syst. Mag. **6**(2) (2014). ▶ https://doi.org/10.1109/MITS.2014.2306552

95. Ziegler, j., et al.: Trajectory planning for bertha—a local, continuous method. In: Intelligent Vehicles Symposium. IEEE (2014). ▶ https://doi.org/10.1109/IVS.2014.6856581

Open Access This chapter is licensed under the terms of the Creative Commons Attribution-NonCommercial-NoDerivatives 4.0 International License (▶ http://creativecommons.org/licenses/by-nc-nd/4.0/), which permits any noncommercial use, sharing, distribution and reproduction in any medium or format, as long as you give appropriate credit to the original author(s) and the source, provide a link to the Creative Commons license and indicate if you modified the licensed material. You do not have permission under this license to share adapted material derived from this chapter or parts of it.

The images or other third party material in this chapter are included in the chapter's Creative Commons license, unless indicated otherwise in a credit line to the material. If material is not included in the chapter's Creative Commons license and your intended use is not permitted by statutory regulation or exceeds the permitted use, you will need to obtain permission directly from the copyright holder.

Optimale Trajektorien

Moritz Werling

Contents

© The Author(s) 2026
H. Winner et al. (eds.), *Handbook Assisted and Automated Driving*,
https://doi.org/10.1007/978-3-658-45276-6_43

43.1 Introduction

Advanced collision avoidance systems, lane keeping support, traffic jam assistance, and remote valet parking; they all operate on the actuators to relieve the drivers of the lateral and/or longitudinal vehicle control or make it safer for them. Well-defined tasks, such as staying in the middle of the marked lane while following the vehicle ahead, can be handled by a set-point controller [15]. And yet standard automated parking maneuvers already require a calculated trajectory[1] that must be adapted to the available parking space. As systems need to cover more and more situations, the number of degrees of freedom increase, which makes a trajectory parameterization very complex, especially when the vehicle must take numerous obstacles into account. This calls for a systematic approach based on mathematical optimization (as opposed to heuristic approaches such as potential field and elastic bands methods, see, e.g., [6, 28], with their inherent limitations, cf. [27]. In the chapter at hand, we address real-time trajectory optimization,[2] a task that an automated vehicle faces when it travels through its environment, also referred to as motion planning in robotics [29, 30]. The focus will be on methods that engage with the longitudinal and lateral vehicle movement. However, the results can be transferred to novel warning systems that can also benefit from an optimal trajectory prediction, see, e.g., [11]. Generally speaking, a trajectory optimization method is sought that can handle both structured (e.g., streets) and unstructured environments (parking lots), one that works among cluttered static obstacles and in moving traffic as well as exhibits a natural, human-like, anticipatory driving behavior. Using more technical terms, the method should be easy to implement, parameterize, adapt, scale well with the number of vehicle states and the length of the optimization horizon, incorporate nonlinear, high fidelity vehicle models, combine the lateral and longitudinal motion, be complete,[3] allow for both grid maps and object lists representations (with predicted future poses) of the obstacles, be numerically stable, and transparent in its convergence behavior (if applicable). Also, the calculation effort must be low to allow for short optimization cycles on (low performance) electronic control units so that the vehicle can quickly react to sudden changes in the environment. Unfortuna-

tely, there is no such single method that has all these properties. And, most likely, there will never be one. However, different optimization methods can be combined to get as close as possible to the above requirements. The next section therefore gives a closer look into the basic principles of trajectory optimization and their application.

43.2 Dynamic Optimization

When engineers speak about optimization, they usually refer to static optimization, in which the optimization variables p are finite, also called parameters (e.g., finding the most efficient operating point of an engine). Then optimal refers to some well-defined optimization criterion, usually the minimization of a cost function $J(p)$ (e.g., fuel consumption per hour). Trajectory optimization is different in that the optimization variables are functions $x(t)$ of an independent variable t, usually time. It is also called dynamic optimization or infinite-dimensional optimization. Evaluating $x(t)$ therefore requires a cost functional (a "function of a function"), which quantifies the "quality" of the trajectory $x(t)$ by a scalar value. Due to the vehicular focus, a special case will be considered, one that requires the trajectory $x(t)$ to be consistent with some dynamical system model which has an input u. Without such model, the optimization cannot incorporate the inherent properties and physical limitations of the vehicle. This special case of dynamic optimization is called an optimal control problem (e.g., [31]).

43.2.1 Optimal Control Problem

A fairly general formulation of the optimal control problem (OCP) reads:

Minimize the cost functional

$$J(\boldsymbol{u}(t)) = \int_{t_0}^{t_f} l(\boldsymbol{x}(t), \boldsymbol{u}(t), t)\,\mathrm{d}t \;+\; V(\boldsymbol{x}(t_f), t_f)$$

$$(43.1a)$$

subject to the system dynamics

$$\dot{\boldsymbol{x}}(t) = \boldsymbol{f}(\boldsymbol{x}(t), \boldsymbol{u}(t), t), \quad \boldsymbol{x}(t_0) = \boldsymbol{x}_0 \qquad (43.1b)$$

as well as the equality and inequality constraints

$$\boldsymbol{g}(\boldsymbol{x}(t_f), t_f) = \boldsymbol{0} \qquad (43.1c)$$
$$\boldsymbol{h}(\boldsymbol{x}(t), \boldsymbol{u}(t), t) \leq \boldsymbol{0}, \quad \forall t \in [t_0, t_f]\,. \qquad (43.1d)$$

In other words, for our (possibly nonlinear and time variant) system with state $\boldsymbol{x} \in \mathbb{R}^n$ and input $\boldsymbol{u} \in \mathbb{R}^m$ we seek on the interval $t \in [t_0, t_f]$ the input trajectory $\boldsymbol{u}(t)$ that minimizes the cost functional J while steering (in

---

1   More precisely: a path, which does not have any time dependency.

2   Notice, that in control theory the term trajectory planning usually implies that there is no feedback of the actual system states on the trajectory. The dynamical system is then only stabilized by a downstream trajectory tracking controller, which is not always advisable. We therefore use the term trajectory optimization instead to be independent of the utilized stabilization concept.

3   A complete algorithm always finds the solution if it exists.

43

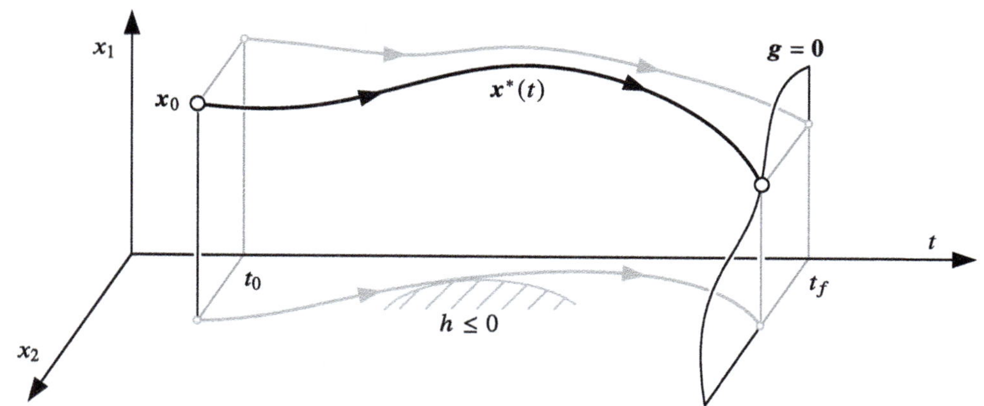

Fig. 43.1 Example of an optimal trajectory $x^*(t) \in \mathbb{R}^2$ with end constraints $g = 0$ at a fixed end time t_f and inequality constraint $h \leq 0$ for x_2

the truest sense of the word) the system from its initial state x_0 to an end state $x(t_f)$, so that the equality and inequality constraints are fulfilled at all times. The optimal input trajectory is usually denoted by $u^*(t)$ and the resulting state trajectory by $x^*(t)$, see **Fig. 43.1**. Notice that J comprises integral costs l and end point costs V and is not only a functional of the input $u(t)$ but also of the states $x(t)$. Furthermore, if the final time t_f is not given, it becomes part of the optimization, so that the length of the trajectory will also be optimized.

43.2.2 Problem Formulation for DAS and Automated Driving

Coming back to the automotive application, equation (43.1b) of the previous section describes the dynamics of the vehicle. This state space model also includes the planar motion, either relative to some reference such as the lane center, see ▶ Sect. 43.3.1.2, or relative to a stationary origin, see ▶ Sect. 43.3.3.4. Undesired vehicle motion such as deviations from the lane center, detours, dangerous vehicle states (e.g., large slip angles), or uncomfortable jerks, e.g., caused by hectic steering actions will be penalized in the cost functional (43.1a). The free space prediction between the generally moving obstacles can be described by the inequality constraints (43.1d). And the end constraints (43.1c) can be utilized to require that the optimized vehicle trajectory will be aligned to the road at the end of the optimization interval. As will be explained in ▶ Sect. 43.5, this final state plays an important role for the stability of the "replanning" algorithm, therefore special costs $V(x(t_f), t_f)$ can be introduced in (43.1a).

43.3 Solving the Optimal Control Problem

All known approaches to the OCP can be assigned to one of the following three principles, see, e.g., [8].

43.3.1 Approach I: Calculus of Variations

The classical approach to the OCP is *calculus of variations*, which delivers valuable insight in the solution.

43.3.1.1 Theoretical Background: Hamilton Equations

Static optimization problems can be tackled by *differential calculus*. It is well known that the first derivative of a function $J(p)$ is equal to zero at a minimum (or any other stationary point). For multivariate problems we can write $\nabla J(p) = 0$ which leads to a set of (algebraic) equations that the optimum p^* must satisfy. The extension to problems with equality constraints requires the method of so-called *Lagrange-multipliers*, yielding first order necessary conditions for optimality. Analogously for dynamic optimization, variational calculus requires that the first variation of the functional $J(u(t))$ vanishes for the optimal control function u^* which is often written as $\delta J(u(t)) = 0$. In order to incorporate the system dynamics in the OCP, which constitute (differential) equality constraints, we can also apply the Lagrange-multiplier method. This yields a set of differential equations, the so-called *Hamilton equations*:

$$\dot{x} = f(x, u, t) \tag{43.2a}$$

$$\dot{\lambda} = -\frac{\partial l}{\partial x} - \left[\frac{\partial f}{\partial x}\right]^{\mathrm{T}} \lambda \tag{43.2b}$$

$$0 = \frac{\partial l}{\partial u} + \left[\frac{\partial f}{\partial u}\right]^{\mathrm{T}} \lambda \tag{43.2c}$$

They are *first order necessary conditions* for our OCP (leaving out the inequality constraints (43.1d)). The function $\lambda(t)$ constitutes the Lagrange-multipliers, here called *co-states*. Besides the initial condition

$$x(t_0) = x_0 \tag{43.3}$$

the optimal trajectory must fulfill the (algebraic) *transversality conditions*, depending on whether the end state $\boldsymbol{x}(t_f)$ is constrained by (43.1c) and/or the final time t_f is given, see, e.g., [31]. The simplest condition requires a fixed end state $\boldsymbol{x}_f$ at a given end time t_f so that

$$\boldsymbol{x}(t_f) = \boldsymbol{x}_f \tag{43.4}$$

Either way, this results in a *boundary value problem*, which in general needs to be solved numerically. This so-called *indirect approach* is very accurate but, from experience, not as flexible as the direct approach that we will introduce in ▶ Sect. 43.3.2. However, for simple OCPs the resultant boundary value problem can be solved analytically, leading to fast computable optimal trajectory primitives with broad applications (see ▶ Sect. 43.4).

43.3.1.2 Example Application: Automated Lane Change

We will now apply the described method to the generation of optimal lane change primitives. The lateral motion across the road can be modeled as a triple integrator system with states $\boldsymbol{x} = [x_1, x_2, x_3]^{\mathrm{T}}$, namely the position, the lateral velocity, and the lateral acceleration, respectively, all within the reference frame of some curve, see ◻ Fig. 43.2.

Then, the system dynamics are described by

$$\dot{\boldsymbol{x}} = \boldsymbol{f}(\boldsymbol{x}, u) = [x_2, x_3, u]^{\mathrm{T}} \tag{43.5}$$

where u represents the lateral jerk, which is the third derivative of the lateral position. We will now seek for the optimal system input $u^*(t)$ that transfers the integrator system from its initial state

$$\boldsymbol{x}(0) = \boldsymbol{x}_0 \tag{43.6}$$

to a given end state

$$\boldsymbol{x}(t_f) = \boldsymbol{0} , \tag{43.7}$$

at a given end time t_f. Among all trajectories we seek for the one that minimizes the integral of the jerk-square, that is

$$J = \int_0^{t_f} \frac{1}{2} u(t)^2 \mathrm{d}t \tag{43.8}$$

so that the movement feels most pleasant to the passengers. Notice that the final cost here has no influence on the solution due to the fixed end state, so that we can set $V = 0$ in (43.1a).

With $\boldsymbol{\lambda} = [\lambda_1, \lambda_2, \lambda_3]^{\mathrm{T}}$, evaluating the so-called control equation (43.2c) we obtain

$$0 = u + \lambda_3 \quad \Rightarrow \quad u = -\lambda_3 . \tag{43.9}$$

The co-state equation (43.2b) yields

$$\dot{\boldsymbol{\lambda}} = \begin{bmatrix} \dot{\lambda}_1 \\ \dot{\lambda}_2 \\ \dot{\lambda}_3 \end{bmatrix} = \begin{bmatrix} 0 \\ -\lambda_1 \\ -\lambda_2 \end{bmatrix}$$

and therefore, we get

$$\dot{\lambda}_1 = 0 \quad \Rightarrow \quad \lambda_1(t) = -c_1$$
$$\dot{\lambda}_2 = -\lambda_1 \Rightarrow \lambda_2(t) = c_1 t + c_2$$
$$\dot{\lambda}_3 = -\lambda_2 \Rightarrow \lambda_3(t) = -\frac{1}{2}c_1 t^2 - c_2 t - c_3 , \tag{43.10}$$

With λ_3 from (43.10) it can be seen from (43.9) that the optimal control input function is a third order polynomial with yet unknown integration constants c_1, c_2, and c_3. Substituting $u(t)$ in the state equation (43.5) we find the optimal trajectory to

$$\dot{x}_3 = \lambda_3 \quad \Rightarrow \quad x_3(t) = \frac{1}{6}c_1 t^3 + \frac{1}{2}c_2 t^2 + c_3 t + c_4 \tag{43.11}$$

$$\dot{x}_2 = x_3 \Rightarrow x_2(t) = \frac{1}{24}c_1 t^4 + \frac{1}{6}c_2 t^3 + \frac{1}{2}c_3 t^2 + c_4 t + c_5 \tag{43.12}$$

$$\dot{x}_1 = x_2 \Rightarrow x_1(t) = \frac{1}{120}c_1 t^5 + \frac{1}{24}c_2 t^4 + \frac{1}{6}c_3 t^3 + \frac{1}{2}c_4 t^2 + c_5 t + c_6 \tag{43.13}$$

with additional integration constants c_4, c_5, and c_6. To comply with the initial state (43.6) and end constraint (43.7) the integration constants need to be chosen accordingly. As (43.10)-(43.13) are linear with respect to the constants this can be done by simple linear algebra, leading to the trajectory (a) in ◻ Fig. 43.2.

Similarly, we can find the solution to the OCP with a free end state (and a free end time). In this case we require the end state to be as close to (and as soon at) the reference as possible by setting

$$V(\boldsymbol{x}(t_f)) = k_1 x_1(t_f)^2 + k_2 x_2(t_f)^2 + k_3 x_3(t_f)^2, \qquad k_i > 0$$
$$(V(\boldsymbol{x}(t_f), t_f) = k_1 x_1(t_f)^2 + k_2 x_2(t_f)^2 + k_3 x_3(t_f)^2 + k_t t_f, \qquad k_i, k_t > 0)$$

The new transversality conditions will only lead to a different end state and end time, see (b) and (c) in ◻ Fig 43.2. However, the optimal function class, that is the 5th-order polynomial, will stay the same.

43.3.1.3 Further Readings

The previous calculations can be analogously carried out for the longitudinal movement, which leads to a very comfortable braking characteristic [22]. The longitudinal and lateral movements can also be combined by a regular local and temporal sampling of target states across and along the street. This leads to a reactive algorithm [49], which prevents collisions with static and moving obstacles widely used in academia and industry. The variational approach was generalized to problems with input constraints, known as *Pontryagin's minimum principle*. As for the vehicular application, it yields the shortest path connecting two poses of a vehicle with a limited turn

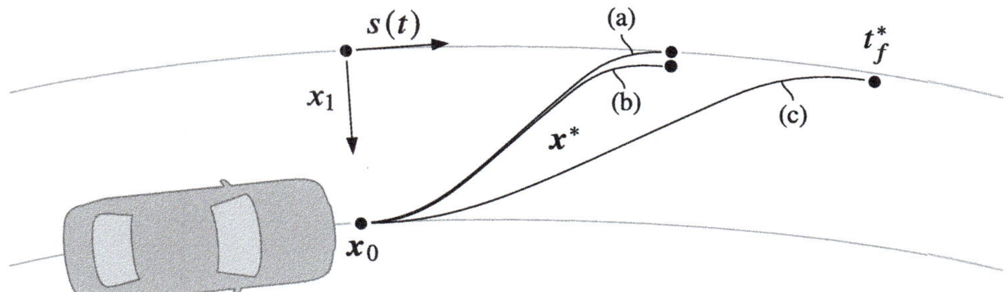

Fig. 43.2 Optimal lane changes for **a** a given end time and end state; **b** a given end time and a free end state; **c** a free end time and a free end state

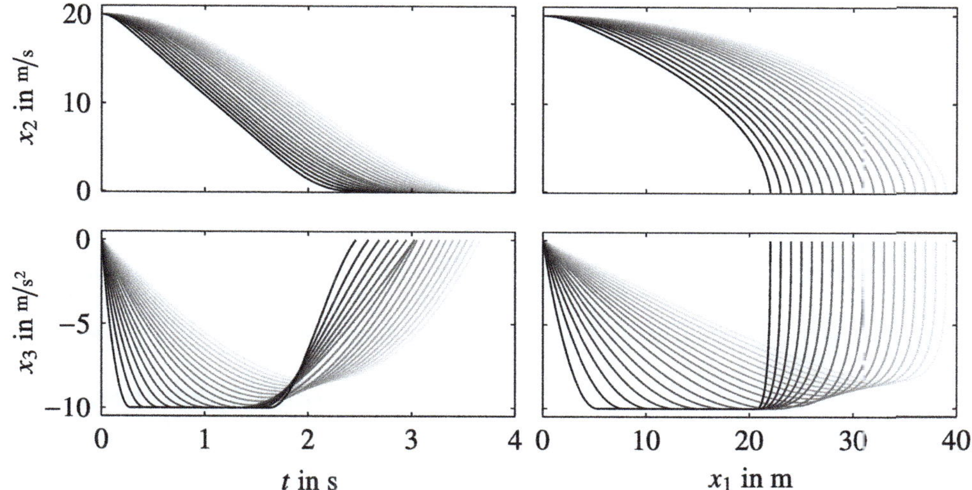

Fig. 43.3 Minimal jerk-square minimal time stopping trajectories with acceleration constraint with $a \geq -10$m/s^2

radius [5, 10, 43]. Even more involved is finding variational solutions to problems with state constraints such as the optimal braking application in ◻ Fig. 43.3 from [47]. Numerical solutions to the first order necessary conditions can be found by so-called indirect methods (see, e.g., [17]), which provide very accurate results in general. Contrary to direct methods, as described in the sequel, they need to determine initial conditions for the co-states, which makes the application challenging for an automotive online application.

43.3.2 **Approach II: Direct Optimization Techniques**

Direct optimization is probably the most widely explored approach in *model-predictive control* (MPC), see ▶ Sect. 43.5. It approximates the dynamic optimization problem of the OCP as a static one, since the latter can be efficiently solved by well-established numerical solvers. The essence of this approach is a *finite-dimensional parameterization*, typically of the input, sometimes the state, or the output trajectory.

43.3.2.1 **Theoretical Background: Finite Parameterization Approximation**

We will introduce a very common method for nonlinear MPC called single shooting. In a first step, we chose a finite-dimensional parameterization of the input

$$u(t) = \psi(t, \bar{u})$$

such as a piecewise constant interpolation (see ◻ Fig. 43.4), a polynomial, or a spline. The input trajectory is therefore fully described by the finite parameter vector $\bar{u}$. The system dynamics (43.1b) now read

$$\dot{x}(t) = f(x(t), \psi(t, \bar{u}), t), \quad x(t_0) = x_0 .$$

This constitutes an initial value problem, which can be solved by an ordinary differential equations solver. We denote the resultant trajectory by

$$x(t) = \phi(t, \bar{u}).$$

Furthermore, it is standard practice that the inequality constraints are only required to hold at N discrete equidistant points in time t_i, $i = 1, \ldots, N$, so that the number of inequality constraints is also finite.

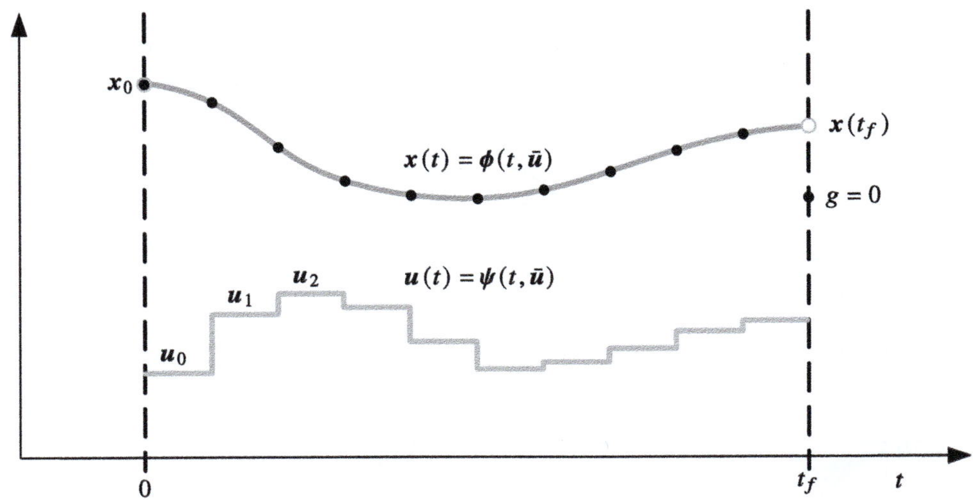

◘ Fig. 43.4   Finite parameterization of the input and sampling of the constraints [37]

For a fixed optimization horizon, the OCP is transformed to the following static optimization problem:
Minimize the cost function

$$J(\bar{u}, t) = \int_{t_0}^{t_f} l(\phi(t, \bar{u}), \psi(t, \bar{u}), t)\, dt + V(\phi(t_f, \bar{u}))$$

(43.14a)

subject to the equality and inequality constraints

$$g(\phi(t_f, \bar{u}), t_f) = 0 \tag{43.14b}$$
$$h(\phi(t_i, \bar{u}), \psi(t_i, \bar{u})) \le 0, \quad i = 0, \dots, N . \tag{43.14c}$$

Loosely speaking, for a first guess $\bar{u}_0$ the system will be simulated in a "single shot" for $t \in [t_0, t_f]$ starting from x_0. Then, the total costs J are evaluate as well as the constraints g and h in (43.14a), (43.14b), and (43.14c), respectively. These values are then fed back to a numerical solver, which repeats this procedure by a variation of $\bar{u}$ to conclude how to modify the parameter so that J gets smaller without violating $g = 0$ and $h \le 0$. When the solution does not significantly change any more or a certain number of iterations has been reached, the optimization will be terminated.

For a nonlinear system model this approach is also referred to as *nonlinear programming* (NP). It has been successfully demonstrated, e.g., in [48]. However, the computational demand is very high for today's electronic control units, since the NP needs to be broken down to a sequence of *quadratic programs* (QP) at real-time. In order to reduce the computational effort to a few milliseconds, [7, 12], for instance, approximate the formulation of trajectory optimization directly as a QP

$$J = \bar{u}^{\mathsf{T}} H \bar{u} + F \bar{u} \tag{43.15a}$$
$$A_c \bar{u} \le b_c \tag{43.15b}$$

with suitably sized matrices H, F, A_c, b_c. For the OCP (43.1), this can be accomplished by a modeling of time-discrete, linear system dynamics (43.18), linear state (43.20), (43.21) and input constraints (43.22), as well as quadratic costs (43.23), as shown in the sequel.

43.3.2.2 Example Application: Emergency Obstacle Avoidance

We now sketch the approach from [21], which has the potential to cover a wide range of series applications. In doing so, we focus on the lateral motion along a reference curve Γ such as a lane center.

We assume in a first step that the longitudinal profile $v(t)$ is given from an upstream algorithm. The difference in orientation $\theta - \theta_r$, see ◘ Fig. 43.5, is typically smaller than 20° [46] and d_r is small with respect to the reference curve's radius $r_r = \frac{1}{\kappa_r}$. We therefore model $\sin(\alpha) \approx \alpha$ and $\cos(\alpha) \approx 1$. Furthermore $\frac{d_r}{r_r} = d_r \kappa_r \ll 1$, which give us for the system state $x = [d_r, \theta, \kappa, \theta_r, \kappa_r]^{\mathsf{T}}$ the following linear time-variant model

$$\dot{x}(t) = A_C(t)x(t) + B_C(t)u(t) + E_C(t)z(t), \ x(t_0) = x_0 \tag{43.16}$$

where

$$A_C(t) = \begin{bmatrix} 0 & v(t) & 0 & -v(t) & 0 \\ 0 & 0 & v(t) & 0 & 0 \\ 0 & 0 & 0 & 0 & 0 \\ 0 & 0 & 0 & 0 & v(t) \\ 0 & 0 & 0 & 0 & 0 \end{bmatrix},$$

$$B_C(t) = \begin{bmatrix} 0, 0, 1, 0, 0 \end{bmatrix}^{\mathsf{T}}, \quad E_C(t) = \begin{bmatrix} 0, 0, 0, 0, 1 \end{bmatrix}^{\mathsf{T}},$$

Notice that the state has been extended by the instantly driven curvature κ of the reference point, see ◘ Fig. 43.5, to ensure continuous curvatures, as well as the reference value κ_r and its integral θ_r of Γ. The latter simulates curvature influence as a known disturbance.

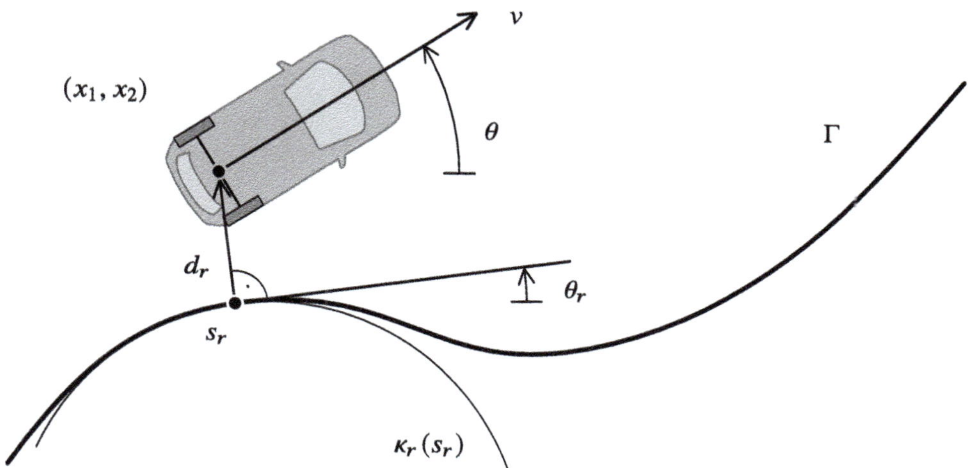

Fig. 43.5 Kinematic vehicle model with respect to a given reference curve Γ

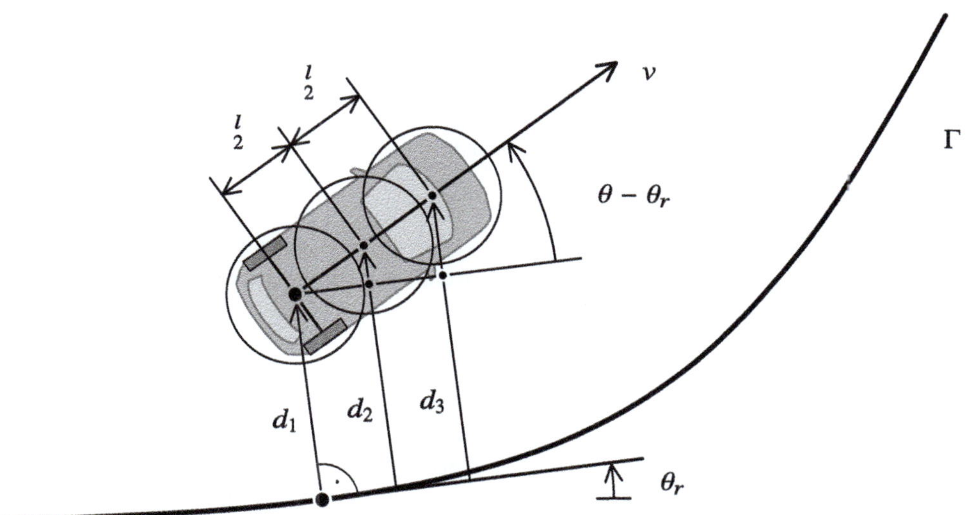

Fig. 43.6 Visualization of the defined system outputs

To minimize the number of constraints, we define the system output based on the idea of approximating the vehicle shape by several circles [53], in our case three, see **Fig. 43.6**. With $l_1 = 0, l_2 = \frac{l}{2}, l_3 = l$, we define the circles' center positions with respect to the reference curve Γ by

$$d_i = d_r + l_i \sin(\theta - \theta_r) \approx d + l_i(\theta - \theta_r), \quad i = 1, 2, 3. \quad (43.17)$$

Next, we follow the idea of a finite parameterization of the input and transform the continuous-time vehicle prediction model (43.16) into its equivalent time-discrete form in one step. Assuming the system matrix $A_C(t)$ as well as the system input $u(t)$ and the system disturbance $z(t)$ to be constant within each discretization interval k with length T_s, we can use the *Laplace transformation*, see [21] for details, to get the time-discrete, time-varying system model

$$x(k + 1) = A(k)x(k) + B(k)u(k) + E(k)z(k),$$
$$y(k) = C(k)x(k), \qquad x(0) = x_0. \quad (43.18)$$

As we want to minimize the discretization error caused by the system disturbance $z(k)$ and keep $z(k)$ constant within each discretization interval we define

$$z(k) = \frac{\kappa_r(k + 1) - \kappa_r(k)}{T_s} \quad (43.19)$$

approximating the reference curvature as a forward first order hold.

In the third step, we define upper and lower limits for the system output based on the perceived environment and the vehicle dynamics.

As depicted in **Fig. 43.6**, the system output (43.17) can be interpreted as a normal distance between the reference curve Γ and the center position of each circle. Due

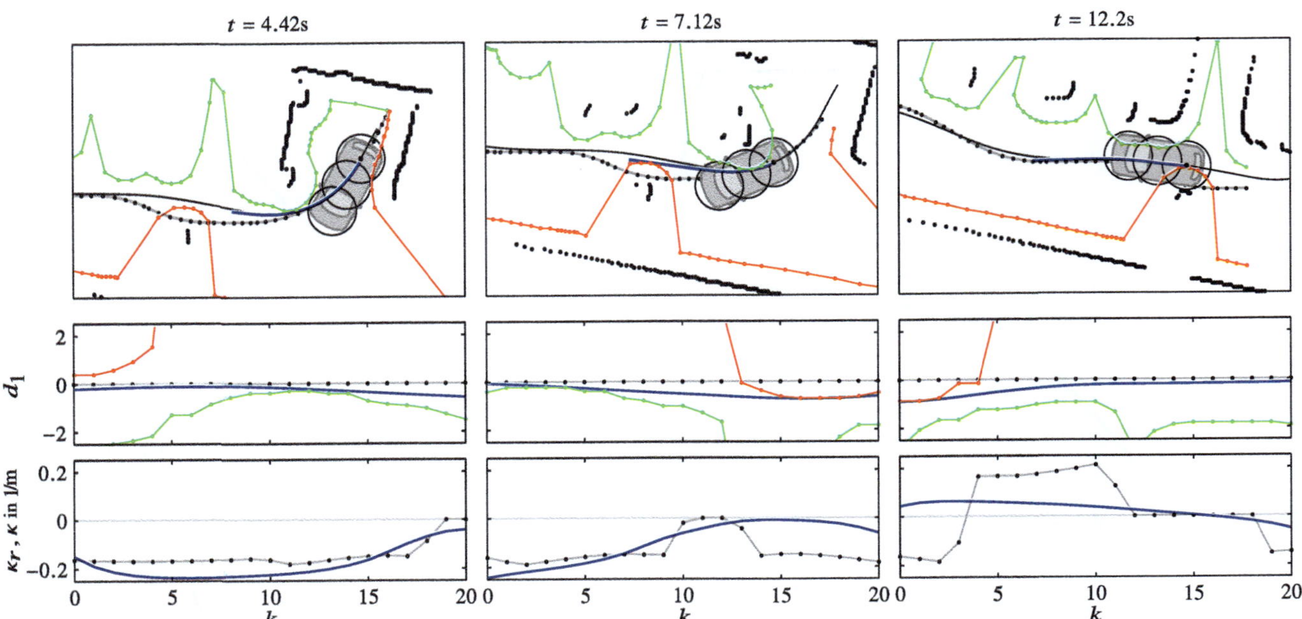

Fig. 43.7 Bird's eye view of the moving vehicle for the parking scenario at three consecutive instants of time with its correspondent results for the lateral position for the rear axle d_1 and the curvature κ, where the blue bold line represents the optimized trajectory, the grey, black dotted line a reference curve and the red and green dotted lines the left and right boundaries

to the presence of moving traffic the maximum admissible deviation, that is how far each circle can be shifted form the center to either side before intersecting an object, obstacle or the lane markings, is not only dependent on the vehicle's position s_r along the reference curve Γ, but also on the time step k, namely

$$d_{i,\min}\left(s_r(k),k\right) \le d_i(k) \le d_{i,\max}\left(s_r(k),k\right), \quad i = 1,2,3, \quad (43.20)$$

where $s_r(k)$ is given by the integration of the desired future velocity profile.

Also, we would like to account for the limited turning circle by introducing a fixed time-invariant upper $\kappa_{\max,\delta}$ and lower $\kappa_{\min,\delta}$ for the κ signal in the system output. Additionally, to account for the total wheel traction, we combine these bounds by a time-varying term which is dependent on $v(k)$ so that

$$\underbrace{\max\left(\kappa_{\min,\delta},\kappa_{\min,\mu}(k)\right)}_{\kappa_{\min}} \le \kappa(k) \le \underbrace{\min\left(\kappa_{\max,\delta},\kappa_{\max,\mu}(k)\right)}_{\kappa_{\max}}, \quad (43.21)$$

where $\kappa_{\min,\mu}(k) = \kappa_{\min}(v(k),\mu)$ and $\kappa_{\max,\mu}(k) = \kappa_{\max}(v(k),\mu)$ are derived using the relations stated by the so-called *Kamm's circle* based on an the estimated friction parameter μ.

Lastly, as the steering actuator can only supply a certain maximum power, the vehicle's steering rate is limited. We therefore confine the curvature's rate of change, that is the system input, by box constraints

$$u_{\min} \le u(k) \le u_{\max}. \quad (43.22)$$

In the last step of the problem modeling, we define a quadratic running cost function of the form

$$l\left(x(k),u(k)\right) = w_d d_r^2 + w_\theta[\theta - \theta_r]^2 + w_\kappa[\kappa - \kappa_r]^2 + w_u u^2 \quad (43.23)$$

with $w_d, w_\theta, w_\kappa, w_u > 0$. The first three terms penalize deviations from the reference curve Γ whereas the last term encourages a smoothing behavior. In some straightforward but laborious transformations we substitute the system state in the cost function and the output constraint equation and finally get the aspired quadratic program (43.14c), which needs to be updated and solved as the vehicle drives along the reference curve. As can be seen in the parking scenario in Fig. 43.7, the vehicle stays close to the reference but smooths out its curvature without violating the lateral position and curvature constraints.

43.3.2.3 Further Readings

Single shooting has been intensively studied for many vehicular applications such by [12, 16, 24, 26, 38, 50]. Nevertheless, many other numerical methods exist such as *multi-shooting* [4] and *collocation* [23] that might gain in importance in the future when dealing with instable vehicle models in challenging driving conditions. Instead of SQP techniques so-called *interior point methods* (IP) can be applied, see [36]. As an alternative to a finite parameterization of the input, for flat systems [44] the flat output can be parameterized instead [25], which has been successfully demonstrated in complex inner-city scenarios by [51].

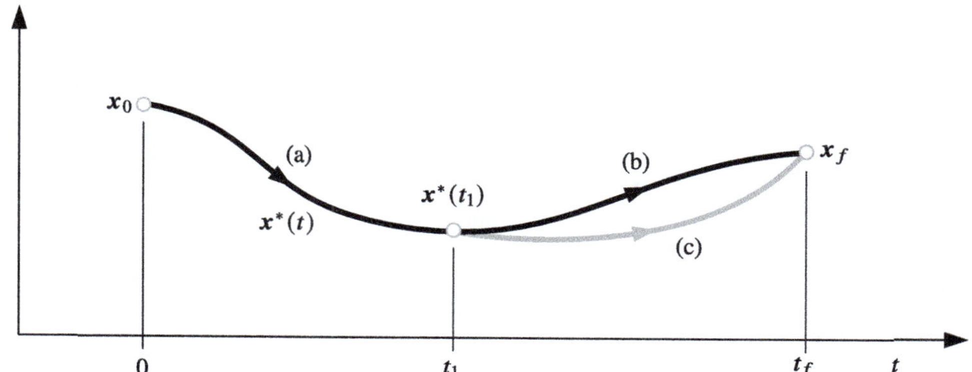

Fig. 43.8   Illustration of Bellman's Principle of Optimality

43.3.3 Approach III: Dynamic Programming

Certain tasks such as parking in several moves or finding the way through multiple moving obstacles are *combinatorial (non-convex) problems*. They cannot be tackled by (local) *direct optimization methods* as the latter rely on an initial solution. However, *dynamic programming* is a principle from [1] that significantly reduces the computational burden of these combinatorial problems, so that the *global optimum* can be efficiently obtained. It is therefore no surprise that it can be found in numerous (discrete) optimization algorithms.

43.3.3.1 Theoretical Backgorund: Bellman's Principle of Optimality

Richard *Bellman's principle of optimality* states:

Definition 1 An *optimal policy* has the property that whatever the initial state and initial decision are, the remaining decisions must constitute an optimal policy with regard to the state resulting from the first decision."

In other words, an optimal trajectory is composed of optimal sub-trajectories. This can be understood by looking at **Fig. 43.8**, which shows the optimal trajectory $x^*(t)$ in black connecting x_0 with x_f. The sub-trajectory (b), bringing $x^*(t_1)$ to x_f, must be optimal, too. Otherwise, there would be another sub-trajectory (c) (grey), that would lead to a better total trajectory (a)+(c), which is contrary to the optimal trajectory $x^*(t)$ comprising (a)+(b).

The largest practical benefit from this principle is received when time-discretizing the OCP. We therefore consider the time-discrete process

$$x(k+1) = f(x(k), u(k), k), \quad k = 0, \ldots, k_f - 1 \quad (43.24)$$

and seek for the optimal steering sequence $u^*(k)$ that minimizes the cost function

$$J = \sum_{k=0}^{k_f-1} l(x(k), u(k), k) . \quad (43.25)$$

Notice, that constraints can be easily incorporated in (43.25) by setting the costs $l = \infty$ when the equality or inequality equations are violated. A basic element of dynamic programming is memorizing the costs of sub-trajectories. Therefore, we define the so-called *cost-to-go*

$$G = \sum_{\kappa=k}^{k_f-1} l(x(\kappa), u(\kappa), \kappa)$$

(notice the difference between κ and k), which incur when going from an intermediate state $x(k)$ all the way to k_f.

At this point, it should be noticed that the minimal cost-to-go, denoted by G^*, is nearly as good as the solution u^* itself, as will be seen later. The minimal cost-to-go can be found in *Bellman's Recursion Formula*

$$G^*(x(k), k) = \min_{u(k)} \{ l(x(k), u(k), k) + G^*(f(x(k), u(k), k), k+1) \},$$

which can be derived from the principle of optimality in a few steps. In words, it relates the minimal cost-to-go of the state $x(k)$ at the kth step, namely $G^*(x(k), k)$, to the subsequent minimal cost-to-go $G^*(f(x(k), u(k), k), k+1)$ in consideration of the best choice of the possible inputs $u(k)$ with its associated stage cost $l(x(k), u(k), k)$. The formula is used to break down the multi-staged OCP into simpler single-staged optimizations, which numerous algorithms exploit to their advantage.

43.3.3.2 Dynamic Programming in a Temporal Decision Process

We assume that the system input can only take discrete values from a given set, which depends on the current state and the current time, that is $u(k) \in \mathcal{U}(x(k), k)$. The set ensures that the input will transfer the system from one

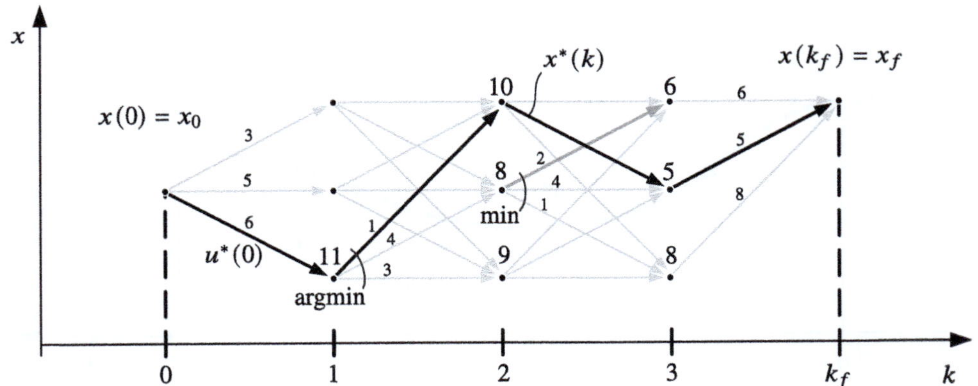

Fig. 43.9   Discrete decision process with the optimal trajectory

discrete state $x(k) \in \mathcal{X}(k)$ to the next. We therefore get a multi-stage decision process, such as the simple example in ◘ Fig. 43.9 with three discrete states in each time step (except for the goal state x_f).

Even though the naive approach of evaluating all $3^3 = 27$ possibilities would here be feasible, this is clearly prohibitive for realistically sized problems due to an exponential runtime of $\mathcal{O}(m^{k_f})$, where m is the number of state transitions and k_f the optimization horizon. However, we can apply the recursion formula, start with the last stage $k = k_{f-1}$, and work our way back to the first step. In doing so we calculate the optimal cost-to-go of every state of each stage by contemplating all possible transients and memorize it. This way it can be accessed when evaluating the optimal cost-to-go of the previous stage. Every stage can be done in $\mathcal{O}(mn)$, where m and n equals the number of its inputs and states. This adds up to a total runtime of $\mathcal{O}(mnk_f)$, which is only linear in the length of the optimization horizon. Algorithm 6 summarizes the algorithm.

Algorithm 1 Dynamic Programming

1: $G^*(x_f, k_f) \leftarrow 0$
2: **for** $k = k_f - 1$ **to** 0 **do**
3:     **for all** $x \in \mathcal{X}$ **do**
4:         $G^*(x(k), k) \quad = \quad \min\limits_{u(k)} \{ l(x(k), u(k), k) \ +$
   $G^*(f(x(k), u(k), k), k+1) \}$
5:     **end for**
6: **end for**

The optimal input and state sequence $u^*(k)$ and $x^*(k)$ can now be found by an efficient forward search starting at x_0 and alternating

$$u^*(k) = \underset{u(k)}{\mathrm{argmin}} \{ l(x(k), u(k), k) + G^*(f(x(k), u(k), k), k+1) \} \quad (43.26)$$

and (43.24) for every stage to the end, see black arrows in ◘ Fig. 43.9.

43.3.3.3 Example Application: Optimal Overtaking

A simplified problem setup and its solution can be found in ◘ Fig. 43.10. Here, the lateral motion is optimized, so that the car can proceed at a constant speed without getting too close to the other vehicles. The system state comprises the lateral position and velocity, hence $x(k) = [d(k), v_{\mathrm{lat}}(k)]^T$, which is discretized with $\Delta d = 0.5$m and $\Delta v_{\mathrm{lat}} = 0.5$m/s at $t_k = k \cdot \Delta t$ with $k = 1, \ldots, 10$ and $\Delta t = 1.0$s . Furthermore, the input set $\mathcal{U}$, i.e., all lateral acceleration profiles, is chosen as a first order polynomial with -10.0m/s$^2 < u(t) < 10.0$m/s^2. And lastly, the stage costs are defined as

$$l(x(k), u(k), k) = \int_{t_k}^{t_k + \Delta t} u(t)^2 \mathrm{d}t + k [d(k) - d_{\mathrm{nearest}}(k)]^2 + C_{\mathrm{collision}}(x(k), u(k), k)$$

with $k = 1.0$, where $d_{\mathrm{nearest}}(k)$ denotes the lateral coordinate of the lane center closest to $d(k)$. The cost term $C_{\mathrm{collision}}(x(k), u(k), k)$ equals to zero if the vehicle does not collide within the interval $[t_k, t_k + \Delta t]$ and else equals to infinity. Due to the few system states with their coarse discretization, the (global) solution is found within a few milliseconds.

43.3.3.4 Further Readings

The presented iteration scheme is especially suited for structured dynamic environments as shown by [33, 52], as well as [19]. As can be seen in ◘ Fig. 43.12, for natural, human like trajectories it is most advantageous to align the sampling along and across the road course. The costs usually penalize not only jerky, accelerant movements, and the approximation to obstacles, but also deviations from the desired road center. As opposed to that, in unstructured environments, such as parking lots, there is no preferred direction, so that the sampling uniformly covers the $[x, y]$ plane as shown in ◘ Fig. 43.11. As static obstacles dominate the problem, temporal aspects are usually neglected meaning that it is not important at what time a location is visited by the vehicle (see

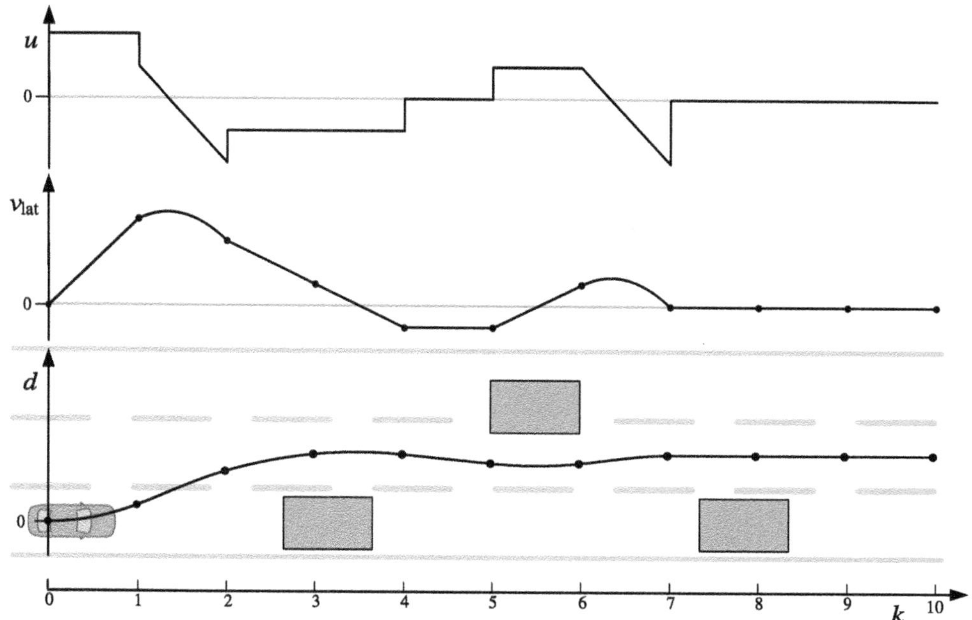

□ Fig. 43.10   Dynamic programming example

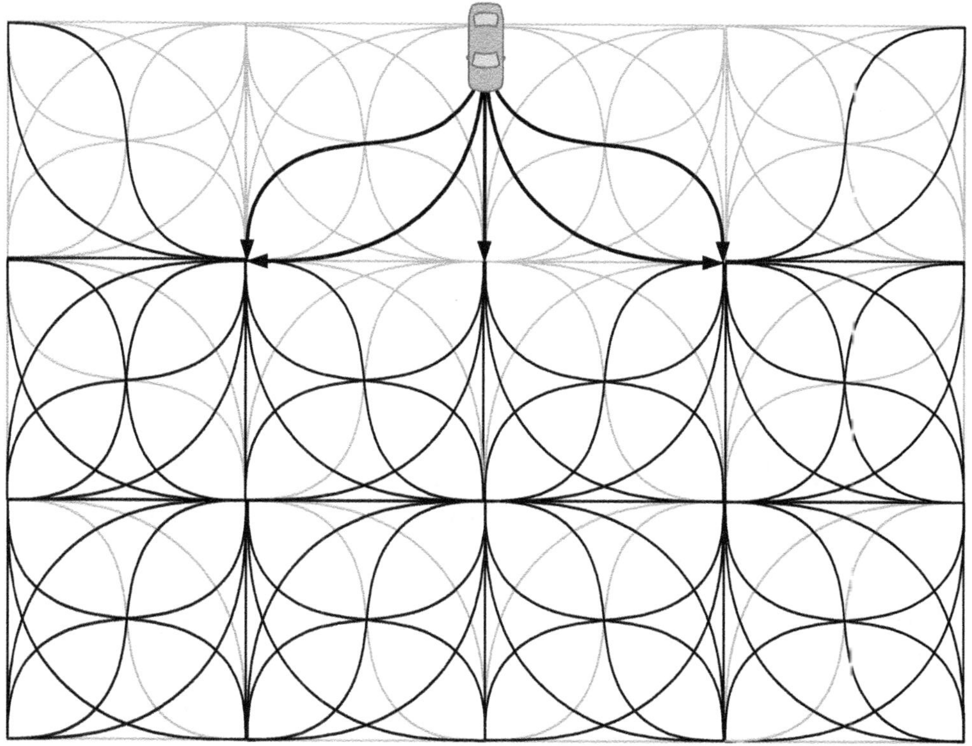

□ Fig. 43.11   State lattice for unstructured environments (illustration based on [34])

[9, 32, 35, 41, 54]). This drastically simplifies the problem, however, the iteration cannot be applied anymore due to the loss of the processing sequence given by k (the decision graph becomes cyclic). Therefore, a substitute order needs to be chosen, one that still leads to the optimum: Investigating the state (expanding the node) with the lowest cost-to-come leads to *Dijkstra's algorithm*, unlike considering the state with the lowest (under)estimate on the total cost, which results in A^* *algorithm*, an informed search. The latter is especially

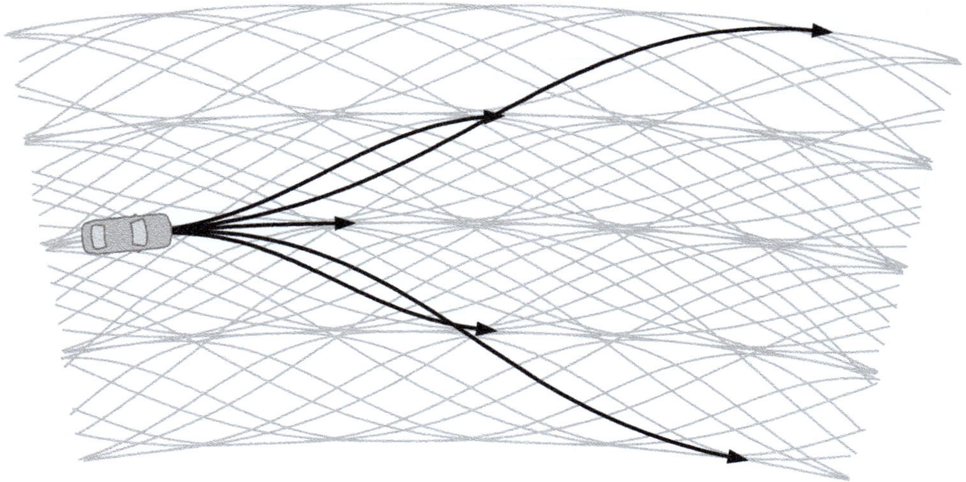

Fig. 43.12 State lattice for structured environments (illustration based on [33])

Table 43.1 Comparison and combination of the approaches

Approach	Many states	Continuous states	Global optimum	Long horizon
DP	⊖	⊖	⊕	⊕
DO	⊕	⊕	⊖	⊖
DP + DO	⊕	⊕	⊕	⊕
DP + DO + IO	⊕	⊕	⊕	⊕⊕

DP: Dynamic Programming   DO: Direct Optimization   IO: Indirect Optimization

efficient in unstructured environments, as the costs are dominated by the covered distance, which can be well estimated by the Euclidean distance (the so-called heuristic), see [2].

43.4 Comparison of the Approaches

When we compare dynamic programming with direct optimization, we realize that the two approaches possess orthogonal capabilities, see table below. Dynamic programming suffers the so-called *curse of dimension* [1] meaning that the approach does not scale well with the number of states. Its application is therefore limit to models with few system states (<3-4) with a coarse discretization. Direct optimization, however, can incorporate system models with numerous and continuous system states (>4) and is therefore able to directly feed the system input $u(t)$. In turn, direct optimization cannot deal with arbitrary cost functionals due to numerical limitations (local convergence) of the underlying solver. Also, its runtime usually grows exponentially with the number of optimization variables so that the length of the optimization horizon is restricted to a few seconds. As opposed to that, dynamic programming can handle arbitrary costs and will always lead to the global optimum. Furthermore, dynamic programming scales comparably well with the length of the optimization horizon (cf. the linear runtime of the dynamic programming iteration scheme in ▶ Sect. 43.3.3.2). For complex, farsighted trajectory optimization tasks, the two approaches need to be combined. Dynamic programming will then yield only a "rough long-term plan", which either serves as the reference trajectory (see, e.g., [13, 14, 20]) and/or provide an initial guess [25] for the locally working direct optimization method. The latter will take a detailed model of the vehicle into account and improve the dynamic programming solution on a reduced optimization horizon to a feasible trajectory. The optimal state trajectory is either forwarded to a low-level feedback steering/acceleration controller or the optimal input trajectory is directly fed to the vehicle actuators. Closed form solutions from the calculus of variations are thereby often used to speedup dynamic programming. This can be in the form of a heuristic for an informed search (see, e.g., [54]) or a so-called analytical expansions [9], both of which, roughly speaking, approximate the remaining trajectory and therefore extend the computable optimization horizon, see **Table 43.1**.

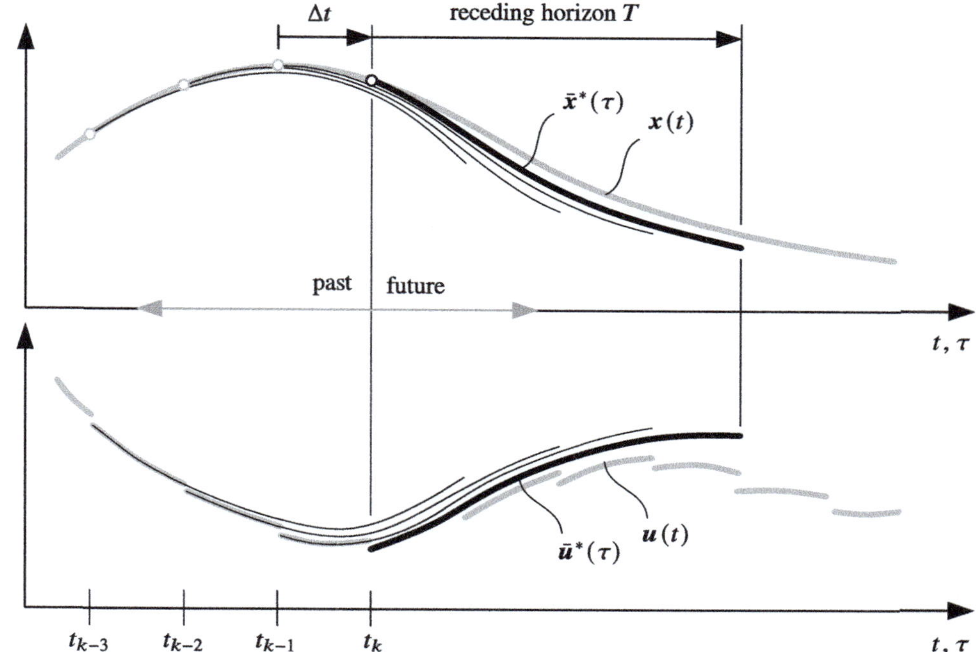

◘ Fig. 43.13   Receding horizon optimization with optimization length T, cycle time Δt, current time step t_k, predicted and actual trajectory $\bar{x}^*$, x predicted and actual system input $\bar{u}^*$, u

43.5   Receding Horizon Optimization

The receding horizon approach is the gist of *model-predictive control* (MPC, see, e.g., [42]), which makes a numerical optimization practical for closed loop control. Therein, in each step t_k, the OCP is solved on a finite horizon T, which calculates the optimal open-loop trajectories $\bar{x}^*(\tau)$ over $\tau \in [t_k, t_k + T]$, see ◘ Fig. 43.13. Only the first part of the optimal control $\bar{u}^*(\tau)$ is implemented on Δt. Right in time the new solution is available for the OCP that has been shifted by Δt. In classical MPC, at each t_k the current plant state is fed back as the new initial state of the OCP. Altogether, this leads to a closed control-loop that anticipates future events, such as input and state saturation, and takes control actions accordingly.

This procedure is completely compatible with trajectory optimization for vehicles. Even more, its replanning mechanism can innately take the limited sensor range and predictability of the other traffic participants into account, which can fundamentally change the OCP from one optimization step to the other. Furthermore, the approach leaves additional degrees of freedom, which can be used to increase the overall robustness of the closed loop system. Firstly, the prediction model may not only include the plant but also the underlying fast low-level feedback or feedforward controllers, which are intended to simplify the resultant optimization model (43.1b). And secondly, the initial state of the OCP does not necessarily have to be the actual vehicle state but

can also be the optimal trajectory of the last step sampled at the current time—or a combination of both. This works as long as the fed-back desired states of the optimal trajectory are tracked by an above-mentioned low-level feedback controller. However, as soon as components of the current vehicle state are used, that show up in the optimization constraints, so-called *slack variables* need to be introduced in the optimization. They transform the otherwise hard inequalities into *soft constraints* as they are referred to in the MPC lingo. This is required as otherwise the slightest disturbance or model uncertainty would ultimately lead to an initial state that the OCP with hard constraints does not have a solution for. MPC theory offers even more. It also deals with *stability* issues of the closed-loop system such as in [18]. More precisely, even for time-invariant system dynamics, constraints, and costs (typical of conventional control problems), the OCP solution changes over time due to the receding horizon, as indicated by ◘ Fig. 43.13. In the worst case the differences of the consecutive solutions build up and the system destabilizes. Well-established MPC-schemes therefore propose an augmentation of the cost functional and constraints, e.g., by a special terminal constraint (*invariant set* [3]) and cost (*control Lyapunov function* [18]) in order to guarantee stability. These schemes can also be transferred to the automotive application. Even more interesting for collision avoidance are the *permanent feasibility* guarantees, which can prevent the sub-optimal short-sighted solutions from "dead ends" or "surprises" from the environment [39, 40, 45].

43.6 Conclusion

Owing to the increase in processing power, computationally intensive optimization algorithms can be executed in real-time in many industrial areas. As automotive electronic control units follow this trend, it is only a matter of time until receding horizon control techniques will emerge in production cars. As has been shown in this chapter, these techniques are most suitable for solving complex trajectory optimization tasks for novel driver assistance systems and automated driving. Based on well-known, elaborated principles we can combine different optimization techniques and implement fast, powerful algorithms with no need to reinvent the wheel. To surmount the complex integration of numerous safety and comfort functions to a complete driver-friendly unit, an integrated trajectory optimization module is required. Ideally, such a module covers the superset of all emerging use cases. Conventional functions such as emergency braking and lane keeping will then only be special cases within the algorithm.

References

1. Bellman, R.: The Theory of Dynamic Programming. RAND-P-550. Rand Corporation Santa Monica CA (1954)
2. Bertsekas, D.: Dynamic Programming and Optimal Control. Athena Scientific Belmont, MA (1995)
3. Blanchini, F.: Survey paper: set invariance in control. Automatica (J. IFAC) **35**(11), 1747–1767 (1999)
4. Bock, H., Diehl, M., Leineweber, D., Schlöer, J.: A direct multiple shooting method for real-time optimization of nonlinear dae processes. In: Allgöwer, F., Zheng, A., Byrnes, C. (eds.) Nonlinear Model Predictive Control, Progress in Systems and Control Theory, vol. 26, pp. 245–267. Birkhäuser Basel (2000)
5. Boissonnat, J.D., Cerezo, A., Leblond, J.: A note on shortest paths in the plane subject to a constraint on the derivative of the curvature. Report 2160, INRIA (1994)
6. Brandt, T.: A predictive potential field concept for shared vehicle guidance. Ph.D. thesis, Universität Paderborn (2008)
7. Carvalho, A., Gao, Y., Gray, A., Tseng, H., Borrelli, F.: Predictive control of an autonomous ground vehicle using an iterative linearization approach. In: Intelligent Transportation Systems-(ITSC), 2013 16th International IEEE Conference on, pp. 2335–2340 (2013)
8. Diehl, M., Bock, H., Diedam, H., Wieber, P.: Fast direct multiple shooting algorithms for optimal robot control. Fast Motions Biomech. Robot. **340**, 65–93 (2006)
9. Dolgov, D., Thrun, S., Montemerlo, M., Diebel, J.: Path planning for autonomous vehicles in unknown semi-structured environments. Int. J. Robot. Res. **29**(5), 485–501 (2010)
10. Dubins, L.: On curves of minimal length with a constraint on average curvature and with prescribed initial and terminal positions and tangents. Amer. J. Mathem. **79**(3), 497–516 (1957)
11. Eichhorn, A., Werling, M., Zahn, P., Schramm, D.: Maneuver prediction at intersections using cost-to-go gradients. In: International Conference on Intelligent Transportation Systems. IEEE (2013)
12. Falcone, P., Borrelli, F., Asgari, J., Tseng, H.E., Hrovat, D.: Predictive active steering control for autonomous vehicle systems. IEEE Trans. Control Syst. Technol. **15**(3), 566–580 (2007)
13. Ferguson, D., Howard, T., Likhachev, M.: Motion planning in urban environments: Part I. In: International Conference on Intelligent Robots and Systems, pp. 1063–1069. IEEE/RSJ (2008)
14. Ferguson, D., Howard, T., Likhachev, M.: Motion planning in urban environments: Part II. In: International Conference on Intelligent Robots and Systems, pp. 1070–1076. IEEE/RSJ (2008)
15. Gayko, J.: Lane keeping support. In: Winner, H., Hakuli, S., Wolf, G. (eds.) Handbuch Fahrerassistenzsysteme, pp. 554–561. Vieweg + Teubner Verlag, Wiesbaden (2012)
16. Gerdts, M., Karrenberg, S., Müller-Beßler, B., Stock, G.: Generating locally optimal trajectories for an automatically driven car. Optim. Eng. **10**(4), 439–463 (2009)
17. Graichen, K.: Nichtlineare modellprädiktive regelung basierend auf fixpunktiterationen. at - Automatisierungstechnik **60**(8), 442–451 (2012)
18. Grüne, L., Pannek, J.: Nonlinear Model Predictive Control: Theory and Algorithms. Springer (2011). ▶ http://books.google.de/books?id=kBnz6OFxO8wC
19. Gu, T., Dolan, J.: On-road motion planning for autonomous vehicles. In: Intelligent Robotics and Applications, pp. 588–597. Springer (2012)
20. Gu, T., Snider, J., Dolan, J.M., Lee, J.: Focused trajectory planning for autonomous on-road driving. In: Intelligent Vehicles Symposium, 2013 IEEE, pp. 547–552. IEEE (2013)
21. Gutjahr, B., Gröll, L., Werling, M.: Lateral vehicle trajectory optimization using constrained linear time-varying mpc. IEEE Trans. Intelli. Transp. Syst. **18**(6), 1586–1595 (2016)
22. Gutjahr, B., Werling, M.: Automatic collision avoidance during parking and maneuvering—an optimal control approach. In: Intelligent Vehicles Symposium Proceedings, 2014 IEEE, pp. 636–641. IEEE (2014)
23. Hargraves, C.R., Paris, S.W.: Direct trajectory optimization using nonlinear programming and collocation. J. Guidance Control Dyn. **10**(4), 338–342 (1987)
24. Howard, T., Kelly, A.: Optimal rough terrain trajectory generation for wheeled mobile robots. Int. J. Robot. Res. **26**(2), 141–166 (2007)
25. Kang, K.: Online optimal obstacle avoidance for rotary-wing autonomous unmanned aerial vehicles. Ph.D. thesis (2012)
26. Kelly, A., Nagy, B.: Reactive nonholonomic trajectory generation via parametric optimal control. Int. J. Robot. Res. **22**(7–8), 583–601 (2003)
27. Koren, Y., Borenstein, J.: Potential field methods and their inherent limitations for mobile robot navigation. In: Robotics and Automation, 1991. Proceedings, 1991 IEEE International Conference on, pp. 1398–1404. IEEE (1991)
28. Krogh, B.: A generalized potential field approach to obstacle avoidance control. In: International Robotics Research Conference. Bethlehem, Pennsylvania (1984)
29. Latombe, J.: Robot Motion Planning. Springer (1990)
30. LaValle, S.: Planning Algorithms. Cambridge University Press (2006)
31. Lewis, F., Syrmos, V.: Optimal Control. John Wiley & Sons, Inc. (1995). ▶ http://de.scribd.com/doc/49575744/Opti-Control-Lewis
32. Likhachev, M., Ferguson, D.: Planning long dynamically feasible maneuvers for autonomous vehicles. Int. J. Robot. Res. **28**(8), 933–945 (2009)
33. McNaughton, M.: Parallel algorithms for real-time motion planning. Ph.D. thesis, Carnegie Mellon University (2011)
34. McNaughton, M., Urmson, C., Dolan, J., Lee, J.: Motion planning for autonomous driving with a conformal spatiotemporal lattice. In: International Conference on Robotics and Automation, pp. 4889–4895. IEEE (2011)

35. Montemerlo, M., Becker, J., Bhat, S., Dahlkamp, H., Dolgov, D., Ettinger, S., Haehnel, D., Hilden, T., Hoffmann, G., Huhnke, B., et al.: Junior: the Stanford entry in the urban challenge. J. Field Robot. **25**(9) (2008)

36. Nocedal, J., Wright, S.: Numerical Optimization. Springer Science + Business Media (2006)

37. Papageorgiou, M.: Optimierung: Statische, dynamische, stochastische Verfahren. Springer (2012)

38. Park, J., Kim, D., Yoon, Y., Kim, H., Yi, K.: Obstacle avoidance of autonomous vehicles based on model predictive control. Proc. Inst. Mech. Eng. Part D J. Autom. Eng. **223**(12), 1499–1516 (2009)

39. Pek, C., Althoff, M.: Fail-safe motion planning for online verification of autonomous vehicles using convex optimization. IEEE Trans. Robot. (2020)

40. Pek, C., Manzinger, S., Koschi, M., Althoff, M.: Using online verification to prevent autonomous vehicles from causing accidents. Nature Mach. Intell. **2**(9), 518–528 (2020)

41. Pivtoraiko, M., Knepper, R., Kelly, A.: Differentially constrained mobile robot motion planning in state lattices. J. Field Robot. **26**(3), 308–333 (2009)

42. Rawlings, J.: Tutorial overview of model predictive control. Control Syst. IEEE **20**(3), 38–52 (2000)

43. Reeds, J., Shepp, L.: Optimal paths for a car that goes both forwards and backwards. Pacific J. Mathem. **145**(2), 367–393 (1990)

44. Rouchon, P., Fliess, M., Lévine, J., Martin, P.: Flatness, motion planning and trailer systems. In: Conference on Decision and Control, pp. 2700–2705. IEEE (1993)

45. Shalev-Shwartz, S., Shammah, S., Shashua, A.: On a formal model of safe and scalable self-driving cars. arXiv preprint arXiv:1708.06374 (2017)

46. Turri, V., Carvalho, A., Tseng, H., Johansson, K.H., Borrelli, F.: Linear model predictive control for lane keeping and obstacle avoidance on low curvature roads. In: 2013 16th International IEEE Conference on Intelligent Transportation Systems: Intelligent Transportation Systems for All Modes, ITSC 2013; The Hague; Netherlands; 6 October 2013 through 9 October 2013, pp. 378–383. IEEE (2013)

47. Werling, M.: Optimale aktive Fahreingriffe: für Sicherheits- und Komfortsysteme in Fahrzeugen. Walter de Gruyter GmbH & Co KG (2017)

48. Werling, M., Liccardo, D.: Automatic collision avoidance using model-predictive online optimization. In: Conference on Decision and Control, pp. 6309–6314. IEEE (2012)

49. Werling, M., Ziegler, J., Kammel, S., Thrun, S.: Optimal trajectory generation for dynamic street scenarios in a frenet frame. In: International Conference on Robotics and Automation, pp. 987–993. IEEE (2010)

50. Yoon, Y., Shin, J., Kim, H., Park, Y., Sastry, S.: Model-predictive active steering and obstacle avoidance for autonomous ground vehicles. Control Eng. Pract. **17**(7), 741–750 (2009)

51. Ziegler, J., Bender, P., Dang, T., Stiller, C.: Trajectory planning for bertha - a local, continuous method. In: Intelligent Vehicles Symposium Proceedings, 2014 IEEE, pp. 450–457. IEEE (2014)

52. Ziegler, J., Stiller, C.: Spatiotemporal state lattices for fast trajectory planning in dynamic on-road driving scenarios. In: International Conference on Intelligent Robots and Systems, pp. 1879–1884. IEEE/RSJ (2009)

53. Ziegler, J., Stiller, C.: Fast collision checking for intelligent vehicle motion planning. In: Intelligent Vehicles Symposium, pp. 518–522. IEEE (2010)

54. Ziegler, J., Werling, M., Schröder, J.: Navigating car-like robots in unstructured environments using an obstacle sensitive cost function. In: Intelligent Vehicles Symposium, pp. 787–791. IEEE (2008)

Open Access This chapter is licensed under the terms of the Creative Commons Attribution-NonCommercial-NoDerivatives 4.0 International License (► http://creativecommons.org/licenses/by-nc-nd/4.0/), which permits any noncommercial use, sharing, distribution and reproduction in any medium or format, as long as you give appropriate credit to the original author(s) and the source, provide a link to the Creative Commons license and indicate if you modified the licensed material. You do not have permission under this license to share adapted material derived from this chapter or parts of it.

The images or other third party material in this chapter are included in the chapter's Creative Commons license, unless indicated otherwise in a credit line to the material. If material is not included in the chapter's Creative Commons license and your intended use is not permitted by statutory regulation or exceeds the permitted use, you will need to obtain permission directly from the copyright holder.

AI for Automated Driving

Eike Rehder and Marius Cordts

Contents

© The Author(s) 2026
H. Winner et al. (eds.), *Handbook Assisted and Automated Driving*,
https://doi.org/10.1007/978-3-658-45276-6_44

44.1 Introduction to Machine Learning

Machine learning describes the ability of a machine to learn how to perform a task from data without being explicitly programmed to do so [65]. This does not mean that machine learning requires no programming at all. Instead, it uses parameterizable models with a structure that *implicitly* allows to solve the task at hand. Then, the actual solution is obtained by a learning algorithm. In this, the parameters of the model are adjusted such that the task is solved on given data. The algorithm thus *learns* the desired processing.

44.1.1 Learning as an Optimization Problem

Machine learning serves as a tool in data processing. Its task is to generate specific output values from input values provided to the algorithm [8]. For this, most machine learning approaches model a mapping $f(\circ)$ from input $\mathbf{x}$ to output $\mathbf{y}$ controlled by a set of parameters $\boldsymbol{\Theta}$,

$$f(\mathbf{x}, \boldsymbol{\Theta}) \mapsto \mathbf{y} . \tag{44.1}$$

This mapping is often referred to as the machine learning *model*.

The task of a learning algorithm is to adjust the parameters $\boldsymbol{\Theta}$ such that $f(\mathbf{x}, \boldsymbol{\Theta})$ produces the desired output. This is where the data comes into play. For the actual learning, it consists of pairs $(\mathbf{x}, \mathbf{y}')$ of known input values $\mathbf{x}$ and corresponding desired output values $\mathbf{y}'$. Sets of such pairs $\{(\mathbf{x}_1, \mathbf{y}'_1), (\mathbf{x}_2, \mathbf{y}'_2), \ldots (\mathbf{x}_N, \mathbf{y}'_N)\}$ make up so called *data sets*. Methods in which the desired output values $\mathbf{y}'$ are provided from the outside are called *supervised learning*, whereas in *semi-* or *unsupervised learning*, no such values are available. In the following, we will mostly consider supervised methods as they are used in the majority of applications. Many of the presented concepts, however, can also be transferred to unsupervised learning.

In the process of learning, the data is used to find suitable parameters $\boldsymbol{\Theta}$ for the model f. The task is to find the parameter set $\boldsymbol{\Theta}^*$ for which the mapping generates the desired output from given input values, i.e.

$$f(\mathbf{x}_i, \boldsymbol{\Theta}^*) = \mathbf{y}'_i, \quad \forall i \in 1, 2, \ldots, N . \tag{44.2}$$

Depending on the choice of f and the available data, this task may not be solvable exactly. Instead, finding the parameter set $\boldsymbol{\Theta}^*$ is treated as an optimization problem. For this, a loss function $\mathcal{L}(\mathbf{y}, \mathbf{y}')$ is defined. This loss function measures the deviation between the model's output $\mathbf{y}$ and the desired values $\mathbf{y}'$. This loss should be minimized for all examples in the dataset. Substitution of $\mathbf{y}$ with $f(\mathbf{x}, \boldsymbol{\Theta})$ yields

$$\boldsymbol{\Theta}^* = \arg\min_{\boldsymbol{\Theta}} \sum_{i=1}^{N} \mathcal{L}(f(\mathbf{x}_i, \boldsymbol{\Theta}), \mathbf{y}'_i) . \tag{44.3}$$

The optimization problem (44.3) can be solved through various approaches, depending on loss function $\mathcal{L}$, model f as well as the provided data. While for some of these, closed form solutions exist, most methods employ iterative algorithms to find $\boldsymbol{\Theta}^*$, called *training*. The data consequently is referred to as *training data set*.

Once the optimal parameter set $\boldsymbol{\Theta}^*$ is found in training, it can be used in $f(\mathbf{x}, \boldsymbol{\Theta}^*)$ to generate outputs on previously unseen data. For this, (44.1) is evaluated using the fixed parameter set $\boldsymbol{\Theta}^*$ to generate output values $\mathbf{y}$ from the provided input data $\mathbf{x}$. This process is called *inference*. A model's *generalization* describes its ability to produce correct outputs from new data.

While machine learning models exist that do not rely on optimization for training, the majority of techniques employed in today's automated vehicles uses this technique. Therefore, in the following, we focus on optimization-based learning.

44.1.2 Machine Learning Methods

Machine learning is used to automatically extract information from data. Different categories of such methods exist. The task of *classification* is to produce discrete decisions about every data point, e.g. to distinguish a vehicle from a pedestrian. In *regression*, on the other hand, the model should produce continuous values, for example, a traffic participant's velocity. Furthermore, methods for automatic data analytics or data mining exist, e.g. to identify hidden patterns in accident statistics.

In automated vehicles, classification and regression are most common. They are used to interpret the traffic situation around the vehicle. Whether a model serves as classification or regression is often only determined by the loss $\mathcal{L}$ it has been trained with as well as the final processing of its output. Therefore, for the following explanations, we only consider classification models. The presented concepts, however, also apply to regression.

An intuitive classification model is a linear classifier. In this, decisions are modeled as a data point's position relative to a hyper plane, $\boldsymbol{\Theta}$. Here, the model's parameters $\boldsymbol{\Theta}$ represent the hyper plane's normal vector. By adding a 1 as an additional dimension to $\mathbf{x}$, then called homogeneous coordinates, a shift of the plane from the origin can be realized. The inner product of an input $\mathbf{x}$ with the plane's parameter vector, $\boldsymbol{\Theta}^\top \mathbf{x}$, results in a single scalar. Its sign reflects a binary decision whether the point lies on the left or right side of the plane, i.e.

$$y = \operatorname{sign}(\boldsymbol{\Theta}^\top \mathbf{x}) . \tag{44.4}$$

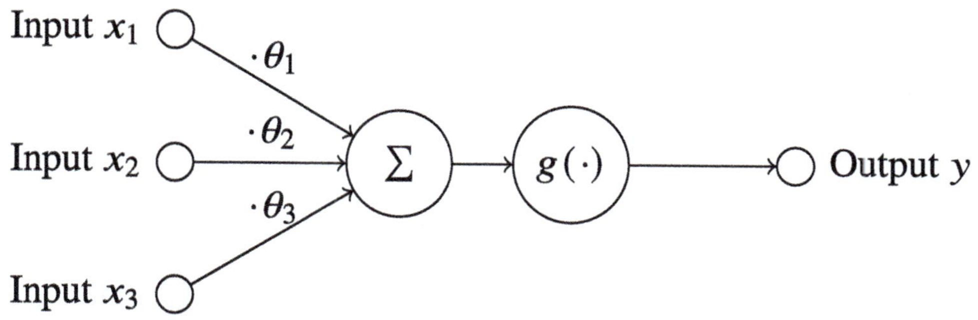

Fig. 44.1 An artificial neuron according to (44.5). All inputs x_i are multiplied with θ_i and summed. Then, a non-linear function g is applied to generate the neuron's final output y

The norm $|\Theta^{\top}\mathbf{x}|$ represents the input's distance to the decision plane. Instead of the sign, a logistic function with values $[-1, 1]$ can be used to represent the decision, e.g. the hyperbolic tangent. This special class of decisions is called *logistic regression*. To optimize the parameters of this decision, often the complement of the product of output and desired value is minimized, i.e. $\mathcal{L}(\mathbf{y}, \mathbf{y}')=1-\mathbf{y}\cdot\mathbf{y}'$ [8]. Are the two signs identical, the loss is decreased, and increased otherwise. Minimizing this loss therefore prefers identical signs for $\mathbf{y}$ and $\mathbf{y}'$.

The hyper plane's parameters and their decision quality are mostly determined by the position of the input values. The further they lie away from the decision boundary, the less sensitive the decision is to small variations in the input data. This reasoning gave rise to Support Vector Machines (SVM). For these classifiers, the decision plane's distance to data points of the separate classes, the so-called support vectors, is maximized [16]. Additional non-linear transformations of these points also allow for arbitrary shaped decision boundaries. A widespread example of such transformations is kernel functions.

Such simple classifiers often display poor learning capacities. More powerful models can be constructed from combinations of multiple individual models. For example, a set of linear classifiers can be combined to represent arbitrary decisions. In *Boosting*, models like the one in (44.4) are cascaded such that every stage is trained to correct errors of the previous classifiers. Furthermore, multiple models can be combined in majority decisions in so-called *Ensembles*.

Another class of machine learning models is the so-called *artificial neural networks (ANNs)*. Inspired by biological information processing in neurons, an artificial neuron combines input, excitation, and output in a single building block, see **Fig. 44.1**. The individual components of the input vector $\mathbf{x}$ are weighted with corresponding parameters Θ and then summed. The result is fed to a non-linear activation function g to compute the artificial neuron's output

$$y = g(\Theta^{\top}\mathbf{x}) . \tag{44.5}$$

The artificial neuron resembles the simple linear decision explained in (44.4), but replaces the sign function by arbitrary non-linearities. Instead of a single decision function, an artificial neural network consists of a multitude of individual neurons. For this, these are combined to so called layers in which many such neurons process the same input in parallel. A layer's output can then serve as the input to a next layer to construct complex network architectures. The resulting models can approximate arbitrary non-linear functions [46]. In practice, the processing steps of an artificial neural network are treated as a compute graph, as shown in **Fig. 44.2**. The optimization of the parameters can then be modeled as an alternating propagation of information forward and backward through the graph [78]. This allows efficient training of even highly complex architectures.

In many applications, data should be analyzed for local patterns independent of their exact location within an input sequence. However, the neuron described by Eq. (44.5) associates one specific weight from Θ for every entry in the input vector $\mathbf{x}$. Because there is a connection of every input entry to the neuron, it is called *fully connected*. A shift of a pattern within the input vector will result in a completely different output. To compensate for this, a neuron can be applied to only parts of the vector. It can then be evaluated on different positions within the data and, thus, identify local patterns. **Figure 44.3a** displays such a neural network. The input consists of four values while the subsequent neurons only make use of three of those. It can therefore be placed at two different positions within the vector. This results in two individual neurons that share their weights. The two neuron's output can be combined in the subsequent network processing.

A special class of weight sharing in networks is so-called *convolutional neural networks (CNNs)*. In these, an input grid is convolved with a filter mask [61]. The result of this discrete convolution represents neurons with shared weights, evaluated on all possible positions in the input grid (see **Fig. 44.3b**). As a result the number of learnable parameters is reduced drastically. Furthermore, the network inherently can detect local patterns in a grid while being robust against translation of these

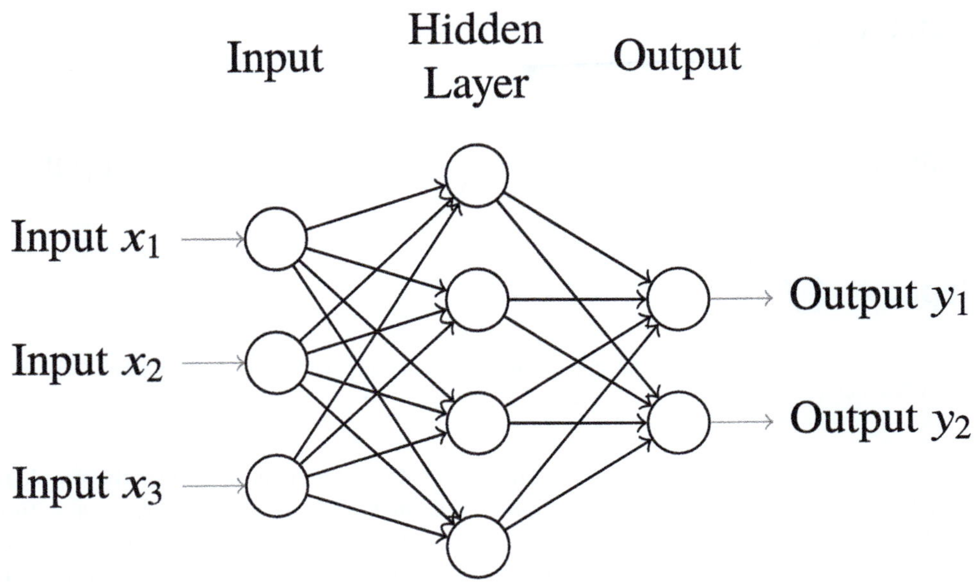

Fig. 44.2 A simple artificial neural network. Every circle in hidden and output layer represents an artificial neuron as shown in ▶ Fig. 44.1

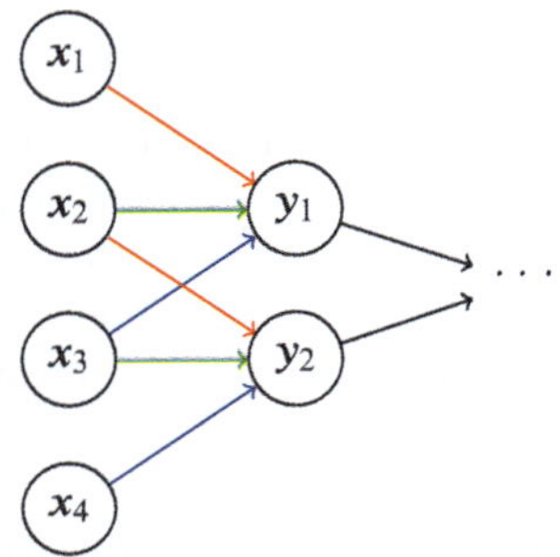

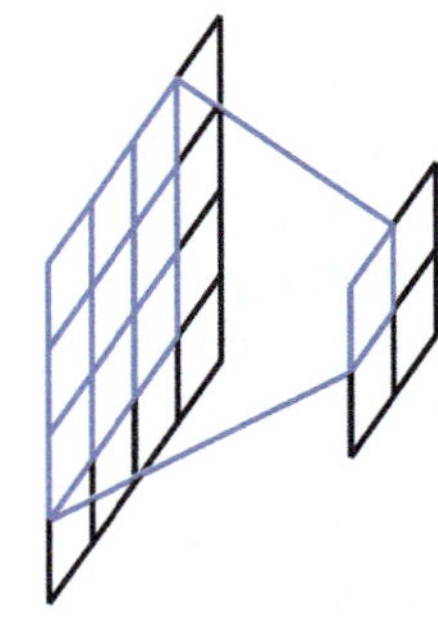

(a) A sparsely connected network with shared weights. Both neurons y_1 and y_2 use the same parameters (visualized with same color of connections) and are, thus, functionally identical, but operate on different pieces of the input data.

(b) In a convolutional network the output is computed from a local neighborhood. In this example, a 3×3 window of input cells contributes to a neuron's output. By moving it over the input grid, the neuron can be evaluated at different positions.

Fig. 44.3 When neurons with the same weights are evaluated at different positions, they can be used to identify local patterns. In regular grids, this can be computed efficiently using discrete convolutions

patterns. This is especially useful in sensor processing. For example, this allows to detect an object in an image regardless of its exact position therein. Consequently, today's CNNs are able to classify image content on a human level [43].

44.1.3 History of Machine Learning in Automated Driving

Machine learning algorithms can perform tasks without being explicitly programmed to do so. Consequently, they find their application wherever explicit programming is challenging. In vehicle automation, this is especially true for sensor data processing where the desired output can be specified but hardly be modeled explicitly. As an example, machine learning plays a significant role in image processing since the 1980s [68]. Due to the great variety in visual appearances of objects, it is simply infeasible to find suitable explicit models. A learning method on the other hand can identify the relevant features and optimal representations by itself.

One of the earliest machine learning models used in vehicle automation was artificial neural networks (ANNs). Already in 1989, a vehicle was controlled directly based on image data by a neural network [68]. With the development of ever more powerful learning techniques, they found their way from research into driver's assistance systems in series production. As an example, ANNs, SVMs, and boosting were combined to realize camera-based pedestrian safety systems [25, 26, 96]. Justified by their performance, learning algorithms are ubiquitous in today's vehicle development.

With recent advances in data processing, the trend in machine learning goes into the direction of deep neural networks [58, 62]. In contrast to early ANNs that consisted of only one or two layers, today's networks feature tens to hundreds of these. This results in millions of trainable parameters that can only be efficiently trained through the availability of large amounts of data and parallel compute. Today, powerful convolutional architectures are at the core of many vehicle automation systems. Due to their superior performance, they increasingly replaced all other learning methods. Therefore, in the following, we will focus on this class of machine learning systems.

44.2   Development Cycle

The development of a machine learning system goes hand in hand with the construction of a suitable database. To this end, data is created, the algorithm is trained, its performance is assessed, and from this additional needs in terms of data are identified. By iterating through this cycle, a powerful model, i.e. a neural network, is created (cf. ▶ Sect. 44.1). This model is deployed to the vehicle to interpret its environment or generate a desired behavior, cf. ▶ Sect. 44.3. In contrast to classic software development, this model is not designed by hand, e.g. from physical or logical relationships. Instead, it is obtained through optimization, using the collected training database. The engineering expertise therefore lies less in the model itself but rather in the development cycle that generates it. Ultimately, this boils down to effective data processing.

Typically, the development cycle consists of multiple steps on which we will shine light on in the following: data collection, annotation, training, and test, as shown in ◘ Fig. 44.4. These steps are usually run through iteratively until the desired results are achieved by the model [38]. Besides this central cycle, various smaller loops and hybrid development forms exist depending on the application.

44.2.1   Data Collection

A successful application of machine learning methods requires the availability of suitable data as this forms the basis for the training algorithm that learns the mapping from input to output values from this data. This puts high demands on quantity and quality of the data since a network can only learn relationships that are covered by it.

For development, this means that data collection should cover a product's operational design domain (ODD) as much as possible [29]. Should, for example, a traffic sign detection system be used worldwide, the data has to be collected from various countries around the globe. For an image-based lane detection that should function day and night, the different weather conditions must be covered in the data acquisition. To ensure that a network generalizes to the real world, and not only to laboratory conditions, a high variance and diversity of data is required [37]. A pedestrian detection system, for example, will only work well if samples of pedestrians in various scenes and appearances were collected in development.

To collect data for machine learning in driver's assistance systems, vehicles are equipped with special recording hardware. Additionally, data from customer vehicles may be used. This allows a fast and efficient collection of rare events. To reduce the amount of data collected from the many sensors of today's vehicles, special data filters are employed that extract relevant events depending on the application. These filters can be employed directly in-vehicle [95] or in a dedicated filtering step during ingest to a data center [40]. Apart from real-world recordings, data can be created in simulation or computer rendering [76, 80] as well as augmentation of real scenes [24, 69]. This allows a targeted and systematic generation of rare or difficult cases. Often a combination of these approaches is employed during development of a machine learning system.

In practice, it is infeasible to predefine the exact required data quantities and properties upfront. Instead, the database needs to be adjusted continuously with respect to the current state of development [38]. Therefore, testing as the last stage of the development cycle provides valuable insights and allows targeted data collection. If a system shows unexpected behavior in certain situations, more of these can be added to the database. From this, the neural network will learn the desired behavior and the feedback loop is closed.

44.2.2   Annotation

Once data has been collected, it has to be prepared for the actual training (cf. ▶ Sect. 44.2.3). For supervised learning (cf. ▶ Sect. 44.1.1), the training data requires pairs of known inputs **x** together with the corresponding desired outputs **y**′. Creating these desired outputs, also called *ground truth*, is the goal of data annotation. Depending on the application, this is done either manually or automatically, while hybrid forms may exist.

In manual annotation [15, 62, 79], humans create the desired outputs for every single input sample by hand. For example, for an object detection system, all objects like vehicles or pedestrians need to be marked within a

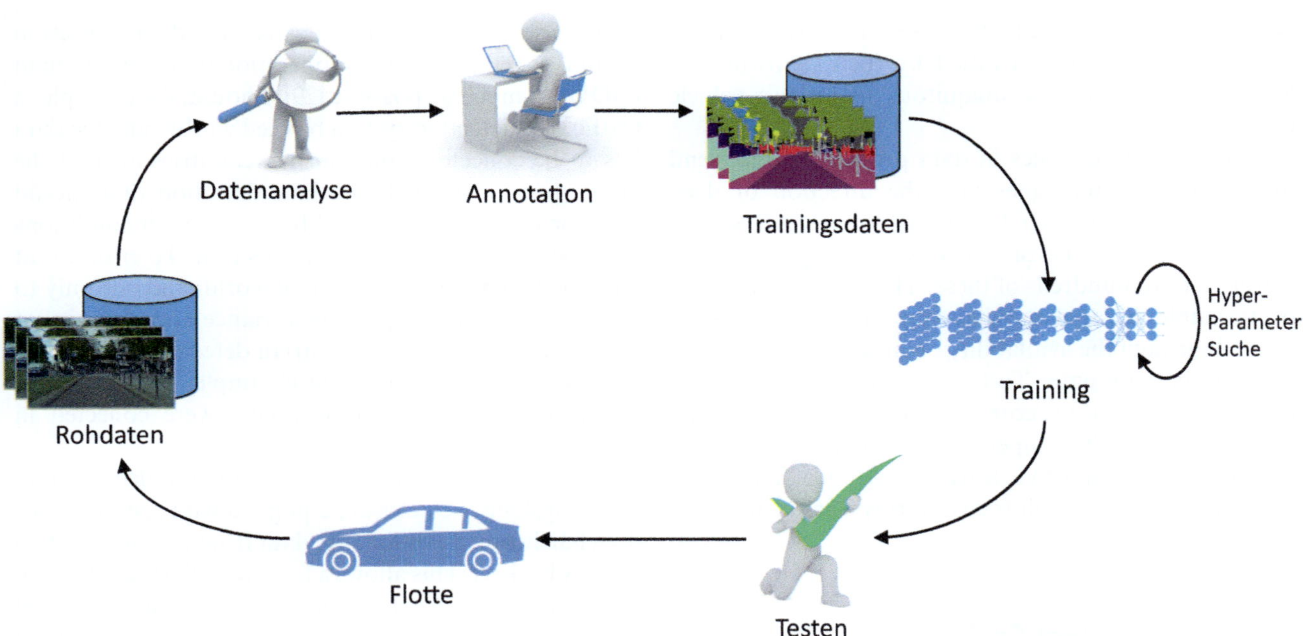

▫ Fig. 44.4   Development cycle of a machine learning system for automated vehicles. Raw data is collected from a fleet of vehicles and ingested into a data center. Depending on the application, the data is analyzed and relevant samples are extracted. These samples are annotated, either manually or automatically, to create the pairs of input and desired output values for training. The actual training often includes hyperparameter tuning. The final model is then tested and rolled out to the vehicle fleet

scene, as shown in ▫ Fig. 44.5. Typically, the annotation of an entire dataset is performed by many human labelers in parallel [94]. As these annotations ultimately define a network's performance, quality control has to be put in place [49]. Moreover, the desired output values have to be specified as precisely as possible [15]. If an annotator should mark passenger cars and trucks, it should be defined how to proceed for vans.

Many annotation tasks can also be fully [22, 100] or at least partly automated [1, 97]. If, for example, a system for prediction of future motion should be designed, the actual motion can be retrieved in hindsight from recorded scenes automatically [45, 95]. For some tasks, reference sensors can be used to generate annotations, e.g. a stereo camera to learn to predict the distance of objects using a monoscopic camera [36] or classified scenes from a camera to train the same classification for laser [67] or short range sensors [66].

44.2.3 Training

The goal of training is to solve the optimization problem (44.3) for the previously collected and annotated dataset to find the model's parameters. For every machine learning system, there is the risk of *overfitting* [21]. In these cases, the model only works well on training data or even perfectly solves (44.3). This can trivially be achie-

ved by memorizing the training data. In the real world, the model shows much worse performance, the network doesn't sufficiently *generalize*. Do identify such issues during training, parts of the training data are set aside and not used for training but for validation [15, 35]. At training time, the network's performance is assessed on this validation data periodically. Once this performance saturates or even deteriorates, the training is terminated. In addition to this, overfitting can be mitigated by using smaller models with fewer parameters or other regularization techniques [30, 58].

Artificial neural networks are trained using gradient descent methods that require large amounts of parallel compute capacities due to typically large models and data sets. Such compute is often provided via graphics processing units (GPUs). Additionally, the neural network development is driven by empirical results that require experiments for different network architectures, loss functions, regularizations, and other hyperparameters [38]. Again, the validation data is used for such hyperparameter evaluation and optimization. This process can also be executed automatically which requires vast compute capacities. In extreme cases, even the entire network architecture can be sought for like this [102].

Fig. 44.5 Manually annotated objects in the Cityscapes Dataset [15], used as training data for 3D vehicle detection [32]

44.2.4 Testing

Machine learning systems for automated driving are usually tested according to a multi-stage strategy [99]. Initially, a model's performance can be assessed in terms of relevant metrics using a test data set. Since the model is created using the training data and the validation data is used to oversee the training, the test data needs to be strictly separate. To obtain meaningful results, the test data may be recorded at different locations, times or situation configurations than the training and validation data. Nevertheless, the same requirements for diversity and quality hold true for the test data. To evaluate a model's performance, its output for the test data is compared to the annotated desired output and assessed quantitatively. As a second testing stage, the model is deployed together with the entire system including pre- and post-processing, either in a *software in the loop* or *hardware in the loop* setup. These tests focus less on the actual performance and more on the interplay between the different components. Finally, the entire system is rolled out to test vehicles and examined in real-world application.

Every testing stage can give valuable insights into a system's deficiencies. These findings define the requirements for the data collection for the next development phase and the cycle starts over.

44.3 Vehicle Applications

The flexibility of machine learning methods allows a wide field of applications in automated driving. Some exemplary use cases are presented in the following. These, however, are not intended as an exhaustive list and only shall give an insight into recent developments.

44.3.1 Semantic Segmentation

Environment perception of automated vehicles often demands a full semantic understanding of sensor raw data. The goal of this semantic segmentation is to assign a distinct semantic class to every single measurement, e.g. every pixel in a camera image, every lidar reflection, or every radar measurement. This semantic class comes from a set of previously defined categories, e.g. vehicle, pedestrian, road, vegetation, or sky. With this, the entire scene around a vehicle is interpreted. A successful classification facilitates the grouping of neighboring measurements of the same class into clusters, hence the term *semantic segmentation*. Some examples for different sensors are displayed in Fig. 44.6.

Semantic segmentation allows a detailed analysis of the surrounding scene. It provides important information on road layout and can distinguish drivable space from other surfaces such as sidewalk or terrain. Knowledge on locations of vegetation can help identify occlusions, e.g. if an object expected from the map might be covered by foliage. Detection of infrastructure elements such as buildings or poles can guide localization or odometry. Even the position of the sky can be informative to suppress noisy distance measurements.

Early works on semantic segmentation commonly used an unsupervised clustering to reduce the amount of data to process, e.g. *superpixels* in image processing. These were then classified using machine learning models such as SVMs [31]. Additionally, graphical models such as conditional random fields (CRFs) were used as regularization [57]. Today, this task is solved almost exclusively with deep neural networks, for which *fully convolutional networks (FCNs)* were the groundbreaking work [86]. This class of networks generates one output for every pixel of an input image, in this case the semantic class of that pixel. These networks consist of a feature extractor,

(a) Camera: real-time capable FCN [14].

(b) Lidar: LiLaNet. Source: [67]

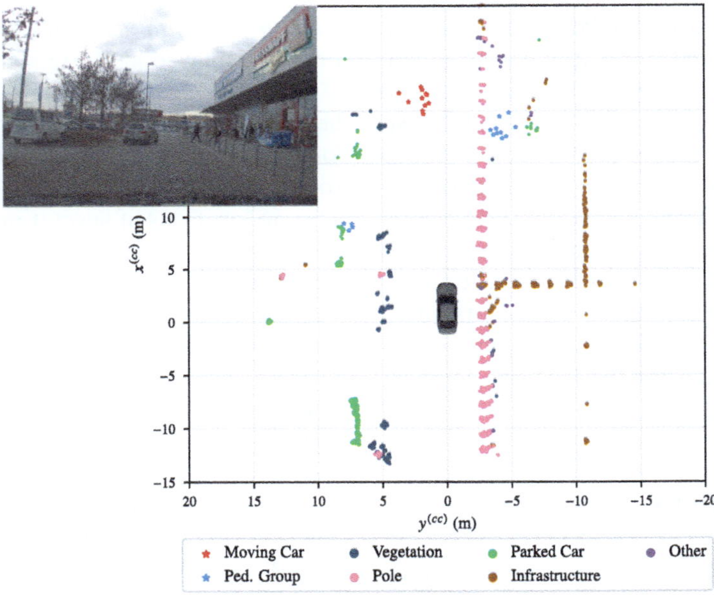

(c) Radar. Source: [84] (modifiziert)

Fig. 44.6 Examples of semantic segmentation generated with deep neural networks for different sensor modalities. Semantic classes are visualized using colormapping

also called *backbone* or *encoder*, and a *head* or *decoder*. The backbone compresses raw measurement data and extracts relevant features of the scene. At the same time, it reduces the input resolution drastically, e.g. by a factor of 32 [44, 88, 89]. The head's task is then to project these low resolution features back to the original size and finally produce a classification for the input pixels. For this, intermediate layers are extracted from the backbone in FCNs, then interpolated linearly to the same resolution and combined for classification. The DeepLab model [13] uses so called *atrous* convolutions for interpolation. This special class of convolutions uses sparsely populated filter masks so that larger filter sizes can be realized with the same amount of parameters as in regular convolutions. Nevertheless, they can be trained the same as the rest of the network and can be computed efficiently. The U-Net architecture [77], initially designed for medical image processing, mirrors the backbone in the head with upsampling instead of downsampling, where connections are drawn between same size resolution layers in backbone and head.

The downside of semantic segmentation is that it provides no information on instances of objects of the same class. Adjacent cars, for example, will be combined in the same segment which complicates downstream processing such as tracking or maneuver planning. However, since semantic segmentation and segmentation of instance (cf. ▶ Sect. 44.3.2) are related methods and representations, it's natural to combine them into a single network. The task of segmenting countable objects as individual instances and non-countable regions as class segments is termed *panoptic segmentation* [56].

Methods for semantic segmentation are compared in public benchmarks. Especially relevant are PASCAL VOC 2012 [27] for general images and Cityscapes [15] with a focus on road scenes. The latter also provides a comparison of methods for panoptic segmentation.

44.3.2 Object Detection

In addition to general scene understanding, detection of other traffic participants is of crucial importance for an automated vehicle. As opposed to semantic segmentation, here, the focus lies on the perception of individual object instances. As these might be moving traffic participants, they have to be handled explicitly in order to interact with them and avoid collision.

Machine learning is the established method for object detection from sensor measurements. Interpretation of raw sensor data requires recognition of diverse and highly variable patterns. As this can hardly be modeled explicitly, learning algorithms provide the preferred solution. Hand engineered approaches exist that can reduce raw data to expressive features [17, 28, 51, 96] but today, most of them have been replaced by deep neural networks that perform object detection directly from raw data.

These detectors mostly consist of two components. First, a deep neural network extracts and compresses features of the input data. This backbone serves as the basis for detection. A second network, the head, is attached to the backbone and interprets the features to perform object detection. This head is the actual detector. As both parts are neural networks, they can be trained jointly for optimal detection performance.

The detection network's construction from two individual parts allows a modular configuration of these components. This means that a given detector head can be combined with different backbones and vice versa. Since only the backbone performs feature extraction from sensor data, a detector can easily be transferred between different sensor modalities. If, for example, an architecture exists for image processing and another one that works on lidar data, both can serve as the backbone for the same detector head [71, 87]. This concept also allows an extension of the backbone with heads for various different tasks, e.g. simultaneous semantic and instance segmentation [56]. With this, the feature extraction of the backbone has to be computed only once for all tasks which can greatly reduce inference time.

A neural network can be trained as an object detection system by designing a suitable output representation. All detectors have in common that they predict some existence confidence measure that reflects whether an input is considered an object or not. This confidence often includes class attributes to provide information on what kind of object was detected, e.g. a cyclist or pedestrian. Furthermore, the object is localized in space to determine its location. Commonly, this is represented using geometric primitives, often rectangles in two or cuboids in three dimensions, so called *bounding boxes*. In image processing, these rectangles can easily be described by four values, e.g. the position of two corners or center and extent of the box as shown in ◘ Fig. 44.7. These parameters can serve as additional outputs of the network [63, 71, 75]. More complex architectures can even perform pixel-wise segmentation of objects, see ◘ Fig. 44.8 [42, 93]. Compared to image processing, object detection in three dimensions comes with greater variety [4]. For example, detections can be represented as a cuboid with nine degrees of freedom [33]. For practical reasons, often only parts of these are actually used [41]. In analogy to pixel-wise segmentation, it is also possible to reconstruct the complete 3D shape objects [59].

The modularity of neural networks allows for flexible adjustment of detectors to the individual needs of a system. For best detection performance, two stage detectors are used. Here, a first stage suggests possible object candidates while a second stage classifies them in detail [75]. The improved accuracy comes at the price of higher run-

Fig. 44.7 Object detection from camera images. *Source* [93]

Fig. 44.8 Instance segmentation from camera images. *Source* [93]

time. In contrast to that, single stage detectors are less accurate but much faster [63, 71, 92], An extension, e.g. with class hierarchies or additional attributes, is possible for all architectures [42, 72].

An overview over recent methods and their properties such as accuracy, runtime, or input data requirements can be gained from scientific benchmarks. Especially MS COCO [62] compares methods for detection as well as instance segmentation. KITTI provides a benchmark more focused on the automated driving domain [35] for detection, while Cityscapes [15] assesses instance segmentation and 3D object detection [32]. As of today, all these benchmarks are led by machine learning systems proving them an indispensable tool for object detection.

44.3.3 Prediction

In automated driving, planning one's own actions requires the prediction of other traffic participants' motion. Once they are detected as described in ▶ Sect. 44.3.2 and tracked over longer periods of time, assumptions about their future behavior can be made. This prediction is often modeled as an extrapolation of the current dynamics [5, 83]. However, a traffic participant's motion is also influenced by more complex factors such as scene geometry, intentions, or interactions. Especially the last two aspects are diverse to a degree that they can't be modeled explicitly. The human on the other hand demonstrates that reasonable predictions can be learned [82].

The advantage of a learned prediction is that it can include information that can hardly be integrated into physics- and dynamics-based models. For example, interactions of multiple actors in a scene can be learned without the need to understand the exact mechanics of it [2]. Also, intentions play a vital role in prediction. Machine learning methods can infer these directly from raw sensor data or track information [54, 74]. As all motion is bound to geometric constraints of the scene, e.g. road boundaries, these should also be incorporated into prediction. This can be realized by explicit detection of relevant features [10] or directly inferred from other representations, e.g. topological or occupancy maps [74, 81].

Instead of predicting motion of individual traffic participants, machine learning methods can also reason about them implicitly. From occupancy maps that also contain information on object dynamics, neural networks can be used to predict future occupancy maps [45]. This implicit model no longer requires any detection of individual actors. However, the model still has all relevant information available to reason about even most complex interactions. In analogy to detection from raw data, prediction can also be performed directly from sensor measurements. Neural networks have been used to process multiple subsequent laser scans in order to simultaneously detect and predict other traffic participants [64]. As the sensor data is not preprocessed, the network automatically learns features to include intentions, interactions and scene geometry into prediction.

In contrast to sensor data processing, no benchmark was yet established as standard for prediction. As potential candidates, Argoverse, Lyft, or nuScenes provide the possibility to evaluate prediction methods [11, 12, 47]. However, since it is a fundamental requirement for successful automation of driving, increased activity is to be expected.

44.3.4 Behavior Planning

In the same way future behavior of others can be predicted with machine learning, it can be used to control one's own behavior. In behavior generation, the automated vehicle plans its own motion to follow its mission while avoiding collisions. The mission may encode reaching a destination or following a specific route while adhering to traffic rules.

An automated vehicle should not only plan its path through traffic but also its velocity profile along this path. These two factors may also impact each other. Furthermore, the planning itself is subject to constraints such as vehicle dynamics or spacial restrictions through scene geometry and other traffic participants. All these constraints result a complex combinatorial problem for decision-making. In classic programming, this is often simplified by path-velocity-decomposition [53] or local solutions [6, 101].

Machine learning allows to solve these problems implicitly. For this, different approaches exist. One such solution is the planning of a trajectory through traffic. In classical methods, graph search algorithms are employed by treating discrete positions in the world as nodes in a graph. Then, an optimal path through this graph is found and, thus, a trajectory is generated. The search algorithms are parameterized through cost functions for nodes or transitions between them that should be optimized for. When the graph search is formulated as a neural network [85, 91], the cost function can be learned from examples [73]. Instead of planning a trajectory, it can also be directly proposed from sensor processing [48].

The execution of a planned trajectory requires a controller that generates signals to actuate the vehicle accordingly. Sensorimotor control presents an alternative. In this, a closed control loop is constructed directly from raw sensor data to the actuators of the vehicle. Thus, the controller is replaced by a neural network that sets throttle or steering from sensor measurements. The design of such a system, however, requires careful optimization. The automated vehicle has to learn by trying out actions and observing their consequences. From these, it receives a feedback in form of a cost function. If, for example, the vehicle follows the road the cost will be small while a collision results in high cost. The learning algorithm tries to control the vehicle to minimize the cost. Such a system is called *reinforcement learning*. Since failure is a fundamental part of the feedback, a system cannot be trained in real-world traffic. Instead, it has to be done in simulation [7, 9]. As a result, the challenge now becomes to transfer the learned system from there to reality. To close this domain gap, real-world data can be augmented to give the impression of a real feedback loop [3].

The demand for comparison in behavior generation poses a severe challenge in research. It is not trivial to assess the performance of such a behavior generator. Depending on the point of view, different metrics can be used, e.g. safety, comfort, efficiency, or even impact on others. This already points to the next issue. Since the own behavior influences that of others, it is hard to conduct standardized reference experiments. The INTER-ACTION data set provides the possibility to evaluate decisions in recorded real-world scenarios [98]. Naturally, this does not include interactions. Currently, these can only be assessed in simulation [23].

44.4   Dealing with Unknowns

Even though deep neural networks deliver impressive performance, not all challenges are solved to this day. In order for an automated vehicle to navigate autonomously through real-world traffic, it has to be able to interpret any given situation correctly. This is especially challenging since learning algorithms are taught only on a finite data set. In reality, they will most likely be confronted with situations that were not included in that data. In most cases, these situations can be explained through redundant sensor modalities and algorithms. Whenever this is not the case, an automated system has to assess its own limitations.

From the perspective of system design and safety, two separate aspects of self-assessment play a role. First, a learned component should be able to estimate the uncertainty in its own decisions and the probability of errors. Second, a safe operation should be ensured even in presence of previously unknown and unexpected situations. Even though these concepts are related, they differ in a central detail. While self-assessment of decision uncertainty can be learned from finite data just as in training, the detection of unknowns requires an understanding of not understanding.

tors that stem from the measurement process, e.g. sensor noise. The epistemic part relates to variations in the model parameters itself. As these are estimated from training, this type of uncertainty can be reduced with more data. The aleatoric uncertainty is a property of the sensor system and remains unaffected by this.

Modern machine learning methods tend to overestimate their own confidence [39]. Classifiers, for example, produce pseudo-probabilities for how likely a given sample belongs to the predicted class. If a classifier could correctly assess its uncertainty, a computed pseudo-probability would be equal to the rate of correct decisions with that confidence value. Of all predictions with a pseudo-probability output of 80 %, 80 % should be correct. This property is called *calibration*. In reality, a neural network tends to produce the extreme points, i.e. either 0 % or 100 %, even if the error rates lie far from it [18, 19]. As an example, the confidence values displayed in ◘ Fig. 44.9 do not indicate erroneous classification of the truck at the left image border. To overcome this problem, networks have to be extended for improved self-assessment. In classification, this can be achieved by smoothing confidences for the predicted classes [39]. Furthermore, ensembles can mitigate the issue as explained in ▶ Sect. 44.1.2, as multiple models can complement each other [60]. ◘ Figure 44.10 shows how calibration can be improved from such ensembles. Alternatively, if only a single model is available, it can be perturbed during inference. Often, minimal noise is sufficient to change uncertain decisions while others remain unaffected [34, 55]. If a decision varies in the presence of perturbations, it can be considered uncertain.

Self-assessment of neural networks remains an active field of research. While methods for classification or segmentation have reached production quality, the uncertainty estimation lags behind. To overcome this, multiple sensor modalities or processing algorithms can be fused. However, in order for fusion systems to show their full potential, system self-assessment will be required.

44.4.1   Uncertainty Estimation

An automated vehicle is controlled by individual specialized systems that build up on one another. In order for this processing chain to operate safely, every piece of the chain has to work faultlessly. If errors go undetected, they will propagate through the chain and lead to system failure. Therefore, self-assessment of every component is crucial. For example, a radar sensor should provide uncertainties for a distance measurement of a lead vehicle so that these can be incorporated in object tracking for adaptive cruise control.

Uncertainties in neural networks are attributed to different sources. These are distinguished in *aleatoric* and *epistemic*. The aleatoric uncertainty describes all fac-

44.4.2   Detection of Unknowns

Machine learning methods are developed on finite data sets. This means that in deployment, such a system will encounter situations that it was not trained for. Nevertheless, it should still be able to maintain a safe system state. Here, a distinction has to be made between generalization and recognition of actual unknown situations. While for generalization, known concepts can be applied to previously unseen data, this is not true for unknowns. If, for example, a network was trained to detect vehicles, it should also work for new types or makes of vehicles. If, on the other hand, it is confronted with an animal, it simply *cannot* classify this correctly. Recognition of such situations is desired.

(a) Erroneous semantic segmentation of a road scene

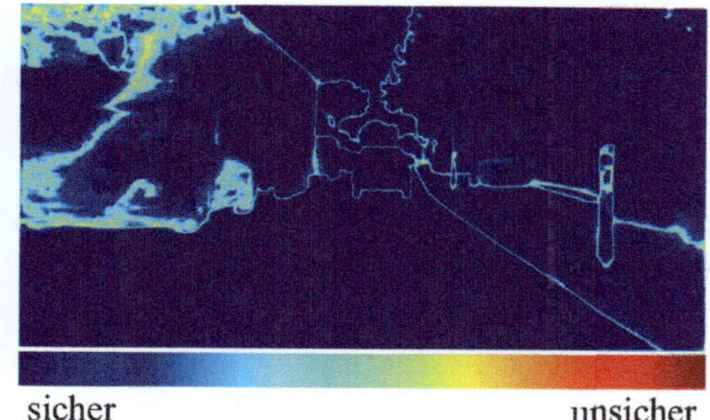

(b) Predicted decision uncertainty.

Fig. 44.9   Neural networks tend to be overconfident. *Source* [52]

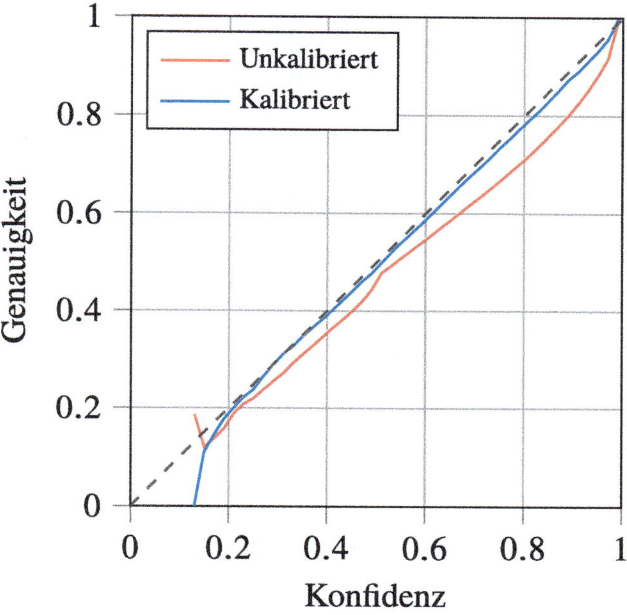

Fig. 44.10   Calibration of two different neural networks. An uncalibrated network is overconfident while the calibrated version outputs realistic confidences

(a) The semantic segmentation detects the lawn mower as a vehicle, even though none was included in the training base.

(b) The networks self-assessment highlights the lawn mower and small structures as uncertain.

Fig. 44.11   Decision uncertainty in presence of unknowns. *Source* [52]

The crucial difference between generalization and detection of unknowns lies in the containment of the decision space. For all known patterns, a similar sample can be found. This is no longer true for completely unknown patterns. Instead, the opposite is true: situations are unknown if they cannot be matched to something known.

Exclusion is one way to approach unknowns in the design of learning systems. An example is small objects on the road, e.g. lost cargo [70]. A model that segments the road should detect that these resemble *no* road without the need of explicit classification. If a closed set of possible decisions exists, the possibility to predict *none of them* can be used to represent unknowns [20]. Alternatively, a learning system can also be trained to incorporate uncertainties in its decision. Such a system can detect unknown without sacrificing generalization [52]. ◘ Figure 44.11 demonstrates this advantage.

Unfortunately, already the definition of unknowns is challenging. When a system for lane detection is trained exclusively on asphalt roads, how should it react to cobblestone streets? Obviously, in this case, the detection of such a road is justified. On the other hand, it was never included in the database and, therefore, it would also be correct to classify it as unknown. Many such examples can be constructed for various applications in decision-making. Decision hierarchies pose a valid solution to resolve this ambiguity. In the example of road surface types, levels of ground, drivable space and road can be defined. This way, a cobble stone street could be classified as drivable, even if it is not recognized as a road. Such constructs are especially helpful in object detection, as all moving objects need to be considered in behavior planning, regardless of their exact semantic class [72]. An example for such a hierarchy is shown in ◘ Fig. 44.12.

While machine learning algorithms demonstrate impressive performance in classification, they have no understanding of their own limitations. In practice, this means that results have to be verified in post-processing. A system that has knowledge of its own insufficiencies could contribute valuable information to such a post-processing or even completely replace it.

44.5 Outlook

Machine learning allows to generate implicit models about the vehicle environment or behavior from data. This technique can solve even the most complex tasks with comparably little effort. Especially deep neural networks stand out for their exceptional performance. Fueled by the availability of large amounts of data and efficient processing units, they increasingly replace other methods in automated driving. Today, deep neural networks are the first choice in development of systems for environment perception, prediction, or behavior generation.

Yet, many challenges remain. A reliable uncertainty estimation and detection of unknowns are active fields of research. The implicitness of models poses a severe challenge on interpretability of a network's decision-making. This, however, is especially desired for failure cases. At the same time, prior knowledge from physical relationships are hard to integrate explicitly. The training of neural networks requires full differentiability even up to the input data. This can be used to create targeted perturbations to the inputs, so called *adversarial attacks*. Here, instead of optimizing the network's parameters, modifications on the input are learned to introduce large errors to the network's processing [90]. Finally, established standards and norms on certification and functional safety, e.g. the ISO26262 [50], cannot be applied directly to machine learning.

In light of the pace of machine learning development in the recent years, it is expected that this will shape the future of automated driving. Considering its exceptional performance, solutions for interpretability and certifications are expected to enable the large-scale application of machine learning.

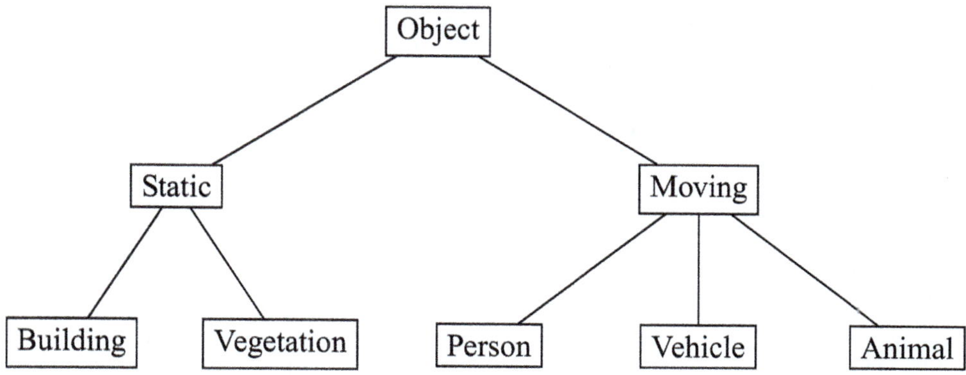

Fig. 44.12   A class hierarchy allows to compromise between decision certainty and fine-grained semantic understanding

References

1. Acuna, D., Ling, H., Kar, A., Fidler, S.: Efficient interactive annotation of segmentation datasets with polygon-rnn++. In: Proceedings of the IEEE Conference on Computer Vision and Pattern Recognition (2018)

2. Alahi, A., Goel, K., Ramanathan, V., Robicquet, A., Fei-Fei, L., Savarese, S.: Social LSTM: Human Trajectory Prediction in Crowded Spaces. In: IEEE Conference on Computer Vision and Pattern Recognition (CVPR), pp. 961–971. Las Vegas, NV, USA (2016)

3. Amini, A., Gilitschenski, I., Phillips, J., Moseyko, J., Banerjee, R., Karaman, S., Rus, D.: Learning robust control policies for end-to-end autonomous driving from data-driven simulation. IEEE Robot. Autom. Lett. **5**(2), 1143–1150 (2020)

4. Arnold, E., Al-Jarrah, O.Y., Dianati, M., Fallah, S., Oxtoby, D., Mouzakitis, A.: A survey on 3d object detection methods for autonomous driving applications. IEEE Trans. Intell. Transp. Syst. **20**(10), 3782–3795 (2019)

5. Barth, A., Franke, U.: Where will the oncoming vehicle be the next second? In: 2008 IEEE Intelligent Vehicles Symposium, pp. 1068–1073. IEEE (2008)

6. Bender, P., Taş, Ö.Ş., Ziegler, J., Stiller, C.: The combinatorial aspect of motion planning: Maneuver variants in structured environments. In: 2015 IEEE Intelligent Vehicles Symposium (IV), pp. 1386–1392. IEEE (2015)

7. Bewley, A., Rigley, J., Liu, Y., Hawke, J., Shen, R., Lam, V.D., Kendall, A.: Learning to drive from simulation without real world labels. In: 2019 International Conference on Robotics and Automation (ICRA), pp. 4818–4824. IEEE (2019)

8. Bishop, C.M.: Machine Learning and Pattern Recognition. Springer, Heidelberg, Germany (2006)

9. Bojarski, M., Del Testa, D., Dworakowski, D., Firner, B., Flepp, B., Goyal, P., Jackel, L.D., Monfort, M., Muller, U., Zhang, J., et al.: End to end learning for self-driving cars. arXiv preprint arXiv:1604.07315 (2016)

10. Bonnin, S., Weisswange, T.H., Kummert, F., Schmuedderich, J.: Pedestrian crossing prediction using multiple context-based models. In: IEEE International Conference on Intelligent Transportation Systems (ITSC). Qingdao, China (2014). ▶ https://doi.org/10.1109/itsc.2014.6957720

11. Caesar, H., Bankiti, V., Lang, A.H., Vora, S., Liong, V.E., Xu, Q., Krishnan, A., Pan, Y., Baldan, G., Beijbom, O.: Nuscenes: A multimodal dataset for autonomous driving. In: Proceedings of the IEEE/CVF Conference on Computer Vision and Pattern Recognition, pp. 11621–11631 (2020)

12. Chang, M.F., Lambert, J., Sangkloy, P., Singh, J., Bak, S., Hartnett, A., Wang, D., Carr, P., Lucey, S., Ramanan, D., et al.: Argoverse: 3d tracking and forecasting with rich maps. In: Proceedings of the IEEE Conference on Computer Vision and Pattern Recognition, pp. 8748–8757 (2019)

13. Chen, L.C., Papandreou, G., Kokkinos, I., Murphy, K.P., Yuille, A.L.: Semantic image segmentation with deep convolutional nets and fully connected CRFs. In: International Conference on Learning Representations (2015)

14. Cordts, M.: Understanding Cityscapes: Efficient Urban Semantic Scene Understanding. Ph.D. thesis, Technische Universität, Darmstadt (2017)

15. Cordts, M., Omran, M., Ramos, S., Rehfeld, T., Enzweiler, M., Benenson, R., Franke, U., Roth, S., Schiele, B.: The cityscapes dataset for semantic urban scene understanding. In: IEEE Conference on Computer Vision and Pattern Recognition (CVPR), pp. 3213–3223. Las Vegas, NV, USA (2016)

16. Cortes, C., Vapnik, V.: Support-vector networks. Mach. Learn. **20**(3), 273–297 (1995)

17. Dalal, N., Triggs, B.: Histograms of oriented gradients for human detection. In: 2005 IEEE Computer Society Conference on Computer Vision and Pattern Recognition (CVPR'05), vol. 1, pp. 886–893. IEEE (2005)

18. Dawid, A.P.: The well-calibrated bayesian. J. Amer. Stat. Assoc. **77**(379), 605–610 (1982)

19. DeGroot, M.H., Fienberg, S.E.: The comparison and evaluation of forecasters. J. Royal Stati. Soc. Series D (The Statistician) **32**(1–2), 12–22 (1983)

20. DeVries, T., Taylor, G.W.: Learning confidence for out-of-distribution detection in neural networks. arXiv preprint arXiv:1802.04865 (2018)

21. Dietterich, T.: Overfitting and undercomputing in machine learning. ACM comput. Surv. (CSUR) **27**(3), 326–327 (1995)

22. Doersch, C., Zisserman, A.: Multi-task self-supervised visual learning. In: Proceedings of the IEEE International Conference on Computer Vision, pp. 2070–2079 (2017). ► https://doi.org/10.1109/ICCV.2017.226

23. Dosovitskiy, A., Ros, G., Codevilla, F., Lopez, A., Koltun, V.: Carla: An open urban driving simulator. arXiv preprint arXiv:1711.03938 (2017)

24. Dwibedi, D., Misra, I., Hebert, M.: Cut, paste and learn: surprisingly easy synthesis for instance detection. In: Proceedings of the IEEE International Conference on Computer Vision, pp. 1310–1319 (2017). ► https://doi.org/10.1109/ICCV.2017.146

25. Enzweiler, M., Gavrila, D.M.: Monocular pedestrian detection: survey and experiments. IEEE Trans. Pattern Anal. Mach. Intell. **31**(12), 2179–2195 (2008)

26. Enzweiler, M., Gavrila, D.M.: A multilevel mixture-of-experts framework for pedestrian classification. IEEE Trans. Image Process. **20**(10), 2967–2979 (2011)

27. Everingham, M., Eslami, S., Van Gool, L., Williams, C., Winn, J., Zisserman, A.: The PASCAL visual object classes challenge: A retrospective. Int. J. Comput. Vis. **111**(1), 98–136 (2014)

28. Felzenszwalb, P.F., Girshick, R.B., McAllester, D., Ramanan, D.: Object detection with discriminatively trained part-based models. IEEE Trans. Pattern Anal. Mach. Intell. **32**(9), 1627–1645 (2009)

29. Fratrik, F.: How many operational design domains, objects, and events? In: SafeAI@AAAI (2019)

30. Friedman, J., Hastie, T., Tibshirani, R.: The Elements of Statistical Learning, vol. 1. Springer, New York, NY, USA (2001)

31. Fulkerson, B., Vedaldi, A., Soatto, S.: Class segmentation and object localization with superpixel neighborhoods. In: Proceedings of the IEEE International Conference on Computer Vision, pp. 670–677 (2009)

32. Gählert, N., Jourdan, N., Cordts, M., Franke, U., Denzler, J.: Cityscapes 3D: Dataset and benchmark for 9 DoF vehicle detection. In: Proceedings of the IEEE Conference on Computer Vision And Pattern Recognition Workshop on Scalability in Autonomous Driving (2020)

33. Gählert, N., Wan, J.J., Jourdan, N., Finkbeiner, J., Franke, U., Denzler, J.: Single-shot 3d detection of vehicles from monocular rgb images via geometry constrained keypoints in real-time. arXiv preprint arXiv:2006.13084 (2020)

34. Gal, Y., Ghahramani, Z.: Dropout as a bayesian approximation: Representing model uncertainty in deep learning. In: International Conference on Machine Learning, pp. 1050–1059 (2016)

35. Geiger, A., Lenz, P., Urtasun, R.: Are we ready for autonomous driving? The KITTI vision benchmark suite. In: IEEE Conference on Computer Vision and Pattern Recognition (CVPR), pp. 3354–3361. Providence, RI, USA (2012)

36. Godard, C., Mac Aodha, O., Brostow, G.J.: Unsupervised monocular depth estimation with left-right consistency. In: Proceedings of the IEEE Conference on Computer Vision and Pattern Recognition (2017)

37. Gong, Z., Zhong, P., Hu, W.: Diversity in machine learning. IEEE Access **7**, 64323–64350 (2019)

38. Goodfellow, I., Bengio, Y., Courville, A.: Deep Learning. MIT Press (2016)

39. Guo, C., Pleiss, G., Sun, Y., Weinberger, K.Q.: On calibration of modern neural networks. In: International Conference on Machine Learning, pp. 1321–1330 (2017)

40. Haussmann, E., Fenzi, M., Chitta, K., Ivanecky, J., Xu, H., Roy, D., Mittel, A., Koumchatzky, N., Farabet, C., Alvarez, J.M.: Scalable active learning for object detection. In: IEEE Intelligent Vehicles Symposium (IV) (2020)

41. He, C., Zeng, H., Huang, J., Hua, X.S., Zhang, L.: Structure aware single-stage 3d object detection from point cloud. In: Proceedings of the IEEE/CVF Conference on Computer Vision and Pattern Recognition, pp. 11873–11882 (2020)

44

42. He, K., Gkioxari, G., Dollár, P., Girshick, R.: Mask r-cnn. In: Proceedings of the IEEE International Conference on Computer Vision, pp. 2961–2969 (2017)

43. He, K., Zhang, X., Ren, S., Sun, J.: Deep residual learning for image recognition. In: Proceedings of the IEEE Conference on Computer Vision and Pattern Recognition, pp. 770–778 (2016)

44. He, K., Zhang, X., Ren, S., Sun, J.: Deep residual learning for image recognition. In: IEEE Conference on Computer Vision and Pattern Recognition (CVPR), pp. 770–778 (2016). ▶ https://doi.org/10.1109/CVPR.2016.90

45. Hoermann, S., Bach, M., Dietmayer, K.: Dynamic occupancy grid prediction for urban autonomous driving: a deep learning approach with fully automatic labeling. In: IEEE International Conference on Robotics and Automation (ICRA), pp. 2056–2063. Brisbane, Australia (2018)

46. Hornik, K.: Approximation capabilities of multilayer feedforward networks. Neural Networks **4**(2), 251–257 (1991)

47. Houston, J., Zuidhof, G., Bergamini, L., Ye, Y., Jain, A., Omari, S., Iglovikov, V., Ondruska, P.: One thousand and one hours: Self-driving motion prediction dataset. ▶ https://level-5.global/level5/data/ (2020)

48. Hubschneider, C., Bauer, A., Doll, J., Weber, M., Klemm, S., Kuhnt, F., Zöllner, J.M.: Integrating end-to-end learned steering into probabilistic autonomous driving. In: 2017 IEEE 20th International Conference on Intelligent Transportation Systems (ITSC), pp. 1–7. IEEE (2017)

49. Ipeirotis, P.G., Provost, F., Wang, J.: Quality management on amazon mechanical turk. In: Proceedings of the ACM SIGKDD Workshop on Human Computation, pp. 64–67 (2010)

50. ISO 26262:2018: Road vehicles – Functional safety. Standard, International Organization for Standardization (2018)

51. Jones, M., Viola, P.: Fast multi-view face detection. Mitsubishi Electric Research Lab TR-20003-96 **3**(14), 2 (2003)

52. Jourdan, N., Rehder, E., Franke, U.: Identification of uncertainty in artificial neural networks. In: Proceedings of the 13th Uni-DAS eV Workshop Fahrerassistenz und automatisiertes Fahren, vol. 2 (2020)

53. Kant, K., Zucker, S.W.: Toward efficient trajectory planning: the path-velocity decomposition. Int. J. Robot. Res. **5**(3), 72–89 (1986)

54. Karasev, V., Ayvaci, A., Heisele, B., Soatto, S.: Intent-aware long-term prediction of pedestrian motion. In: IEEE International Conference on Robotics and Automation (ICRA), pp. 2543–2549. Stockholm, Sweden (2016)

55. Kendall, A., Badrinarayanan, V., Cipolla, R.: Bayesian segnet: model uncertainty in deep convolutional encoder-decoder architectures for scene understanding. arXiv preprint arXiv:1511.02680 (2015)

56. Kirillov, A., He, K., Girshick, R., Rother, C., Dollár, P.: Panoptic segmentation. In: IEEE Conference on Computer Vision and Pattern Recognition (CVPR) (2019)

57. Kohli, P., Ladický, L., Torr, P.: Robust higher order potentials for enforcing label consistency. Int. J. Comput. Vis. **82**(3), 302–324 (2009)

58. Krizhevsky, A., Sutskever, I., Hinton, G.E.: Imagenet Classification With Deep Convolutional Neural Networks. In: Advances in Neural Information Processing Systems (NeurIPS), pp. 1097–1105. Lake Tahoe, NV, USA (2012)

59. Ku, J., Pon, A.D., Waslander, S.L.: Monocular 3d object detection leveraging accurate proposals and shape reconstruction. In: Proceedings of the IEEE Conference on Computer Vision and Pattern Recognition, pp. 11867–11876 (2019)

60. Lakshminarayanan, B., Pritzel, A., Blundell, C.: Simple and scalable predictive uncertainty estimation using deep ensembles. In: Advances in Neural Information Processing Systems, pp. 6402–6413 (2017)

61. LeCun, Y., Bottou, L., Bengio, Y., Haffner, P.: Gradient-based learning applied to document recognition. Proc. IEEE **86**(11), 2278–2324 (1998)

62. Lin, T.Y., Maire, M., Belongie, S., Bourdev, L., Girshick, R.B., Hays, J., Perona, P., Ramanan, D., Zitnick, C.L., Dollár, P.: Microsoft COCO: common objects in context. In: European Conference on Computer Vision, pp. 740–755 (2014)

63. Liu, W., Anguelov, D., Erhan, D., Szegedy, C., Reed, S., Fu, C.Y., Berg, A.C.: Ssd: Single shot multibox detector. In: European Conference on Computer Vision, pp. 21–37. Springer (2016)

64. Luo, W., Yang, B., Urtasun, R.: Fast and Furious: real time end-to-end 3D detection, tracking and motion forecasting with a single convolutional net. In: IEEE Conference on Computer Vision and Pattern Recognition (CVPR), pp. 3569–3577. Salt Lake City, UT, USA (2018)

65. Mitchell, T.M., et al.: Machine Learning (1997)

66. Nava, M., Guzzi, J., Chavez-Garcia, R.O., Gambardella, L.M., Giusti, A.: Learning long-range perception using self-supervision from short-range sensors and odometry. IEEE Robot. Autom. Lett. **4**(2), 1279–1286 (2019). ▶ https://doi.org/10.1109/LRA.2019.2894849

67. Piewak, F., Pinggera, P., Schafer, M., Peter, D., Schwarz, B., Schneider, N., Enzweiler, M., Pfeiffer, D., Zollner, M.: Boosting lidar-based semantic labeling by cross-modal training data generation. In: Proceedings of the European Conference on Computer Vision (ECCV) (2018)

68. Pomerleau, D.A.: Alvinn: An autonomous land vehicle in a neural network. In: Advances in Neural Information Processing Systems, pp. 305–313 (1989)

69. Qi, X., Chen, Q., Jia, J., Koltun, V.: Semi-parametric image synthesis. In: Proceedings of the IEEE Conference on Computer Vision and Pattern Recognition (CVPR) (2018)

70. Ramos, S., Gehrig, S., Pinggera, P., Franke, U., Rother, C.: Detecting unexpected obstacles for self-driving cars: Fusing deep learning and geometric modeling. In: 2017 IEEE Intelligent Vehicles Symposium (IV), pp. 1025–1032. IEEE (2017)

71. Redmon, J., Divvala, S., Girshick, R., Farhadi, A.: You only look once: Unified, real-time object detection. In: Proceedings of the IEEE Conference on Computer Vision and Pattern Recognition, pp. 779–788 (2016)

72. Redmon, J., Farhadi, A.: Yolo9000: better, faster, stronger. In: Proceedings of the IEEE Conference on Computer Vision and Pattern Recognition, pp. 7263–7271 (2017)

73. Rehder, E., Quehl, J., Stiller, C.: Driving like a human: Imitation learning for path planning using convolutional neural networks. In: International Conference on Robotics and Automation Workshops (2017)

74. Rehder, E., Wirth, F., Lauer, M., Stiller, C.: Pedestrian Prediction by Planning using Deep Neural Networks. In: IEEE International Conference on Robotics and Automation (ICRA), pp. 1–5. Brisbane, Australia (2018)

75. Ren, S., He, K., Girshick, R., Sun, J.: Faster r-cnn: towards real-time object detection with region proposal networks. IEEE Trans. Pattern Anal. Mach. Intell. **39**(6), 1137–1149 (2016)

76. Richter, S.R., Vineet, V., Roth, S., Koltun, V.: Playing for data: ground truth from computer games. In: European Conference on Computer Vision, pp. 102–118 (2016)

77. Ronneberger, O., Fischer, P., Brox, T.: U-net: Convolutional networks for biomedical image segmentation. In: Medical Image Computing and Computer-Assisted Intervention—MICCAI 2015, pp. 234–241 (2015)

78. Rumelhart, D.E., Hinton, G.E., Williams, R.J.: Learning representations by back-propagating errors. Nature **323**(6088), 533–536 (1986)

79. Russakovsky, O., Deng, J., Su, H., Krause, J., Satheesh, S., Ma, S., Huang, Z., Karpathy, A., Khosla, A., Bernstein, M., Berg, A.C., Fei-Fei, L.: ImageNet large scale visual recognition challenge. Int. J. Comput. Vis. **115**(3), 211–252 (2015)

80. Sakaridis, C., Dai, D., Van Gool, L.: Semantic foggy scene understanding with synthetic data. Int. J. Comput. Vis. **126**(9), 973–992 (2018). ▶ https://doi.org/10.1007/s11263-018-1072-8

81. Salzmann, T., Ivanovic, B., Chakravarty, P., Pavone, M.: Trajectron++: Multi-agent generative trajectory forecasting with heterogeneous data for control. arXiv preprint arXiv:2001.03093 (2020)

82. Schmidt, S., Färber, B., Pérez Grassi, A.: Geht er oder geht er nicht?–Ein FAS zur Vorhersage von Fußgängerabsichten. In: Workshop Fahrerassistenzsysteme, pp. 2–4. Walting, Germany (2008)

83. Schneider, N., Gavrila, D.M.: Pedestrian path prediction with recursive Bayesian filters: a comparative study. In: German Conference on Pattern Recognition (GCPR), pp. 174–183. Saarbrücken, Germany (2013)

84. Schumann, O.: Machine learning applied to radar data: classification and semantic instance segmentation of moving road users. Ph.D. thesis, Universitätsbibliothek Dortmund (2021)

85. Shankar, T., Dwivedy, S.K., Guha, P.: Reinforcement Learning via Recurrent Convolutional Neural Networks, pp. 2592–2597 (2016)

86. Shelhamer, E., Long, J., Darrell, T.: Fully convolutional networks for semantic segmentation. IEEE Trans. Pattern Anal. Mach. Intell. **39**(4), 640–651 (2017)

87. Simony, M., Milzy, S., Amendey, K., Gross, H.M.: Complexyolo: an euler-region-proposal for real-time 3d object detection on point clouds. In: Proceedings of the European Conference on Computer Vision (ECCV) (2018)

88. Simonyan, K., Zisserman, A.: Very deep convolutional networks for large-scale image recognition. In: International Conference on Learning Representations (2015)

89. Szegedy, C., Liu, W., Jia, Y., Sermanet, P., Reed, S., Anguelov, D., Erhan, D., Vanhoucke, V., Rabinovich, A.: Going deeper with convolutions. In: Proceedings of the IEEE Conference on Computer Vision and Pattern Recognition, pp. 1–9 (2015)

90. Szegedy, C., Zaremba, W., Sutskever, I., Bruna, J., Erhan, D., Goodfellow, I., Fergus, R.: Intriguing properties of neural networks. arXiv preprint arXiv:1312.6199 (2013)

91. Tamar, A., Wu, Y., Thomas, G., Levine, S., Abbeel, P.: Value Iteration Networks. In: Advances in Neural Information Processing Systems (NeurIPS), pp. 2154–2162. Barcelona, Spain (2016)

92. Tan, M., Pang, R., Le, Q.V.: Efficientdet: Scalable and efficient object detection. In: Proceedings of the IEEE/CVF Conference on Computer Vision and Pattern Recognition, pp. 10781–10790 (2020)

93. Uhrig, J., Rehder, E., Fröhlich, B., Franke, U., Brox, T.: Box2pix: Single-shot instance segmentation by assigning pixels to object boxes. In: 2018 IEEE Intelligent Vehicles Symposium (IV), pp. 292–299. IEEE (2018)

94. Welinder, P., Perona, P.: Online crowdsourcing: Rating annotators and obtaining cost-effective labels. In: IEEE Computer Society Conference on Computer Vision and Pattern Recognition - Workshops, pp. 25–32 (2010). ▶ https://doi.org/10.1109/CVPRW.2010.5543189

95. Wirthmüller, F., Klimke, M., Schlechtriemen, J., Hipp, J., Reichert, M.: A fleet learning architecture for enhanced behavior predictions during challenging external conditions. In: IEEE Symposium on Computational Intelligence in Vehicles and Transportation Systems (IEEE CIVTS) (2020)

44

96. Wohler, C., Anlauf, J.K.: An adaptable time-delay neural-network algorithm for image sequence analysis. IEEE Tran. Neural Networks **10**(6), 1531–1536 (1999)

97. Xie, J., Kiefel, M., Sun, M.T., Geiger, A.: Semantic instance annotation of urban scenes by 3D to 2D label transfer. In: Proceedings of the IEEE Conference on Computer Vision and Pattern Recognition, pp. 3688–3697 (2016)

98. Zhan, W., Sun, L., Wang, D., Shi, H., Clausse, A., Naumann, M., Kummerle, J., Konigshof, H., Stiller, C., de La Fortelle, A., et al.: Interaction dataset: an international, adversarial and cooperative motion dataset in interactive driving scenarios with semantic maps. arXiv preprint arXiv:1910.03088 (2019)

99. Zhang, J.M., Harman, M., Ma, L., Liu, Y.: Machine learning testing: survey, landscapes and horizons. IEEE Trans. Software Eng. (2020). ▶ https://doi.org/10.1109/TSE.2019.2962027

100. Zhou, T., Brown, M., Snavely, N., Lowe, D.G.: Unsupervised learning of depth and ego-motion from video. In: Proceedings of the IEEE Conference on Computer Vision and Pattern Recognition, pp. 1851–1858 (2017)

101. Ziegler, J., Bender, P., Schreiber, M., Lategahn, H., Strauss, T., Stiller, C., Dang, T., Franke, U., Appenrodt, N., Keller, C.G., et al.: Making Bertha drive-an autonomous journey on a historic route. IEEE Intell. Transp. Syst. Mag. **6**(2), 8–20 (2014)

102. Zoph, B., Le, Q.V.: Neural architecture search with reinforcement learning. In: International Conference on Learning Representations (2017)

Open Access This chapter is licensed under the terms of the Creative Commons Attribution-NonCommercial-NoDerivatives 4.0 International License (▶ http://creativecommons.org/licenses/by-nc-nd/4.0/), which permits any noncommercial use, sharing, distribution and reproduction in any medium or format, as long as you give appropriate credit to the original author(s) and the source, provide a link to the Creative Commons license and indicate if you modified the licensed material. You do not have permission under this license to share adapted material derived from this chapter or parts of it.

The images or other third party material in this chapter are included in the chapter's Creative Commons license, unless indicated otherwise in a credit line to the material. If material is not included in the chapter's Creative Commons license and your intended use is not permitted by statutory regulation or exceeds the permitted use, you will need to obtain permission directly from the copyright holder.

Special Demands of Automated Driving on the Design Process

Robert Graubohm and Markus Maurer

Contents

© The Author(s) 2026
H. Winner et al. (eds.), *Handbook Assisted and Automated Driving*,
https://doi.org/10.1007/978-3-658-45276-6_45

45.1 Introduction

The (temporary) release of human drivers from control and monitoring tasks presents a disruptive innovation in vehicle development for public road transport. Accordingly, developers face great challenges—including the creation and application of structured and systematic process models. Compared with the development of conventional road vehicles equipped only with assistance functions, new systems and processes are needed to meet the demands of higher levels of automation. In particular, solutions must be identified that enable the use of automated driving functions and vehicles (SAE Level 3+ [26]) in an open world. To date, the development of driver assistance systems mainly relies on humans as a fallback in demanding and unforeseen situations, which significantly limits the applicability of established design methods when it comes to automated driving.

Defining a sound set of system requirements in terms of a functional specification can be seen as a crucial part of a design process. While the abstract description of the intended functional scope may seem trivial, the comprehensive requirements analysis is a complex problem that is difficult to master in an early development stage. In particular, first decisions on intended fallback strategies and safety mechanisms are already made in the course of initial requirements definitions; these decisions have a significant impact on system behavior in unforeseen situations and when encountering system limitations.

The central challenge for the introduction of novel systems for automated driving is the approval of what has been developed, both by those responsible at manufacturing companies and by potentially required third-party auditors. The prerequisite to this is a convincing safety argumentation, which includes proof that relevant laws and standards are respected (see ▶ Chap. 46). However, a proof of the complete absence of any residual risks appears to be impossible in the course of the release of automated driving functions [22], at least as long as the vehicles operate in the traffic system present today. Therefore, a convincing safety concept that identifies, names, and addresses risks in the best possible way becomes essential in the context of driverless vehicles. The basis of the safety concept and thus the foundation of the release argumentation is created as part of the system design process, which is explained in more detail in this chapter. First, ▶ Sect. 45.2 deals with the definition of the term *design* in the context of mechatronic system development. Subsequently, ▶ Sect. 45.3 explains particular challenges that arise during related development phases in the use case of automated driving. Based on identified challenges, a reference development process for automated driving functions is presented and discussed in ▶ Sect. 45.4. Fi-

nally, an introduction to the example application of the structured design model in the completed research project aFAS follows in ▶ Sect. 45.5.

45.2 Design of Mechatronic Systems

In literature, the *design* of mechatronic systems is described variously but not principally contradictorily. Therefore, definitions used in technical literature and standards serve at this point to derive a consistent definition for this chapter and, on this basis, to place the design in a complete development process.

A frequently quoted description of design as part of product development comes from the theory of axiomatic design. In axiomatic design, the *design* definition describes a mapping of solutions and requirements [37, pp. 2ff]. "Design is an interplay between what we want to achieve and how we want to achieve it" [37, p. 3]. The mapping process covers four domains: customer domain, functional domain, physical domain, and process domain. Functional requirements, design parameters, and finally process variables are derived from customer needs or attributes. Functional requirements are defined as "a minimum set of independent requirements that completely characterizes the functional needs of the product […]" [37, p. 14]. Along with functional requirements, constraints are also captured as "bounds on acceptable solutions" [37, p. 14]. In this context, the challenge of translating customer needs into functional requirements in a *solution-neutral* environment is described to avoid results similar to existing products, "forestalling creative thinking by introducing personal bias" [37, p. 15].

The guideline VDI 2206 [38] adopts the definition of the term *design* in the context of mechatronic system development from explanations about the synthesis step by Haberfellner et al.:

> It consists of "intuiting" the whole solution concept, of discerning or "finding" the requisite solution elements, and of intellectually assembling and combining these elements into a viable whole [13, p. 211].

Hence, the design is basically an early step in the overall development of a product which, contrary to other definitions in mechanical engineering, also includes the conceptual stage and leads to a concretization of requirements in the technical system [38, Sect. 3.1.2]. Regarding the result of a completed design, the guideline states:

> The result of system design is the assured concept of a mechatronic system. This is understood as meaning the solution established in principle and checked by verification and validation [38, p. 8].

In accordance with the definitions presented above, *design* should be understood as the concretization of an abstract development goal (usually based on an identified customer benefit) through concrete technical solution approaches. The resulting solution should be based on a comprehensive set of functional as well as non-functional requirements and already validated by concept studies, prototypes, or samples. The following sections additionally explain that successful system design in the context of automated driving also depends on the specification of a preliminary but robust safety concept. The final specification and production planning of a series product, as well as its validation and release for production and sales, are steps in the product development process that follow the design stage.

Thereby, positioning *design* in the phases of the safety lifecycle according to ISO 26262 (see ▶ Chap. 6) is challenging: The objective of a sound functional concept is shared with the concept phase according to ISO 26262, which is addressed in Part 3 of the standard [16, Part 3]. In the course of designing electrical and electronic vehicle systems, special attention should therefore be paid to the process requirements and work products of the concept phase. However, as soon as solutions are already prototyped and tested for concept validation, later phases of the ISO 26262 safety lifecycle (development at the system level, at the hardware level, and at the software level) are also run through in early stages as part of the design process.

45.3 Challenges in Concept Specifications in the Field of Automated Driving

The development of elaborate mechatronic systems, such as automated vehicles and driving functions, presents substantial challenges that result in particular from the complexity of the technical systems and the diversity of situations in their operational domains. The early consideration and addressing of these challenges already at the design stage should be considered of particular importance for a successful concept specification. For example, guideline VDI 2206 states that the problem of interactions between a large number of components that influence the behavior of the overall system is to be "already taken into account in the early phase of design" [38, p. 4].

Thus, considering emergent properties of a system under design is key, i.e., properties "of a system of elements that cannot be derived from the properties of the individual elements" [28, p. 97]. The concept of emergent properties of a system, resulting from intended or unintended interaction of components, is seen as an essential feature of systems engineering [5, p. 5]. One key emergent property of systems is their safety [21, p. 64] a property that is also of highest relevance in the context of automated driving. Accordingly, a concept specification is expected to create a basis for the development item to develop emergent properties or emergent system behavior in a desired manner.

Another aspect of the complexity of mechatronic development projects is the interdisciplinary nature of the product and its conceptual design. The goal of mechatronic system design is explained, for example, by guideline VDI 2206 as establishing "a cross-domain solution concept", which highlights the need for interdisciplinary collaboration and communication [38, Sect. 3.1.2]. It is is argued that systematic coordination of domains in early development phases avoids cost- and time-intensive iterations that result from a sequential procedure without coordination [38, Sect. 2.4]. The cross-domain character of a development of mechatronic systems referenced in the guideline refers—in reference to definitions from literature—to the interaction of the disciplines of information technology, mechanical engineering, and electrical engineering [38, Sect. 2.1].

For the design of automated vehicles and driving functions, the relevant disciplines need to be extended even further. At SAE Level 3+, the responsibility for executing the driving task is at least temporarily delegated to the system [26]. Thus, ergonomics (cf. ▶ Chap. 38) and law (cf. ▶ Chaps. 3 and 4) are examples of disciplines that must additionally be taken into account early in the concept specification.

The short innovation cycles in key technologies for automating the driving task are both an opportunity and a challenge within the design of future vehicle systems. Progressive developments and improvements in the sensors, controllers, and algorithms used create a basis for ambitious novel functional scopes. However, this evolution of employed technology simultaneously manifests the need for predictions about anticipated series readiness as well as availability and cost levels within concept specifications of automated driving. The validity of these predictions must be continuously assessed in the course of a progressing development process. Furthermore, the popular use of innovative software development approaches (e.g., machine learning) can have a significant impact on structuring of subsequent phases of overall system development, which must already be considered at the time of concept specification. Functions that are based (partially) on machine learning require, on the one hand, the availability or preparation of exhaustive training data and, on the other hand, additional test efforts in order to check the robustness and safety of the software components usually not interpretable by humans.

The aspects of high system and development complexity, the interdisciplinary character of a concept specification, and the advancing evolution of available technology lead to the particular importance of the requirements definition for the design of automated vehicles and driving functions. In general, the design of a technical system is often based on the concretization of abstract requirements that already exist initially. For example, guideline VDI 2206 describes that a development order, in which the objective is specified and described in the form of requirements, forms the starting point of the system design [38, Sect. 3.1.2]. In the field of automated driving, however, diverse and often conflicting stakeholder interests exist that need to be harmonized in the course of concept specification. Driverless vehicles should meet the expectations of their users and achieve the highest levels of safety, but they must ultimately also be tolerated and accepted by other traffic participants and society [31, p. 159]. Accordingly, even abstract system requirements are challenged in the course of a design and often have to be revised.

The particular placement of the concept phase and concept specification in the automotive domain is also reflected in the guideline VDI 2221: In examples of various design processes, the clarification of the problem or the task is described for automotive manufacturers and suppliers as part of a comprehensive phase of concept development [39]. In the field of automated driving, the challenging definition, concretization, and revision of requirements actually represents a main aspect of the concept phase. Safety requirements, whose role in the design process is explained in more detail in ▶ Sect. 45.3.2, exemplify this fact. Regardless of the degree of abstraction, a safety concept can only be generated on the basis of previously defined functional scopes. At the same time, however, a safety concept has a considerable influence on the requirements list of the overall system: New requirements may need to be generated or adaptations of existing requirements can be provoked under safety aspects.

Within an explanation of the V-model XT, which was introduced as a process model for system development, it is stated that new functional or non-functional requirements can result from safety analyses, which supplement the requirements defined in a specification document [15, p. 192]. While the explanations demonstrate the dynamic nature of a requirements list in large-scale system development projects, at least through safety considerations; they nevertheless do not illustrate the recursive and iterative nature of the actual requirements definition for the specification of automated driving functions. As previously explained, safety analyses in the concept phase actually have a direct impact on the concrete specification of a driverless application and may require adjustments to any previously documented functionality requirement. The lay-

ers of abstraction for requirements engineering introduced by Dick et al. [5, pp. 11ff]. illustrate the continuous expansion and concretization of requirements through the entire design process as well. However, the layers also suggest a sequential definition of requirements as long as there are no explicit design changes. In contrast, the ISO 9241-210 standard for human-centered design for interactive systems [17] portrays that the context of use of a system must be iteratively reanalyzed on the basis of generated solution concepts, which leads to new requirements and, if necessary, new solution concepts. The standard's recommendations for design activities thus appear particularly well transferable for safety–critical applications. In this context, Witte et al. [41] state that the evaluation of the context of use should be continued over the entire lifecycle of a product in order to be able to identify the need for adaptations at later points in time as well.

45.3.1 Uncertainties During Requirements Definition

As previously described, for various reasons, requirements definition during the design of automated vehicles and driving functions takes place iteratively in the course of a comprehensive concept phase. In addition to the central challenge of systematically conducting requirements analyses and thereby creating sound and traceable lists of requirements, uncertainties are also of great importance in the field of automated driving. The previous section already mentioned the need to deal with technological uncertainties about performance and costs achievable in the future. In addition, automated vehicles and driving functions are examples of disruptive innovative products for which willingness to pay and market acceptance are subject to major uncertainties.

An important uncertainty during the concept phase is whether in complex systems the definition of specific requirements actually leads to a desired emergent behavior. Documenting the system design in different architectural views is one approach to managing existing complexity (see ▶ Chap. 39). However, it is also possible that a preliminary system architecture is designed in accordance with the existing requirements, but that the product actually created deviates from the conceived solution (cf. explanations on architecture drift and architecture erosion in ▶ Chap. 39).

In the application of automating the driving task for public road traffic, handling of uncertainties due to possible faults in the developed technical system also plays a particular role. Faults can typically contribute to a significant deviation of the vehicle behavior in operation from that intended during concept specification. In the context of road traffic, such deviations often lead to violations of rules and can cause considerable hazards. In general, a distinction must be made between dealing with systematic faults in a technical system and dealing with random hardware faults: Random hardware faults are based on a failure of at least one hardware component caused by chance. Systematic faults occur in particular due to human errors or misjudgments during development and deficits during production (cf. [28, p. 30], [16, Part 1, 3.119 and 3.165]).

Both types of faults are directly related to the handling of uncertainties during requirements definition. Systematic faults arise in part already during the concept specification. Incomplete or ambiguous requirements, for example, can result in faults in the developed system. The possibility of incomplete requirements can partly be attributed to further uncertainties present during the design process. For instance, there is uncertainty about possible characteristics and probabilities of occurrence of individual operating scenarios in an intended operational environment. Even in highly restricted applications, there is an infinite number of distinguishable situations in the real world that an automated vehicle can encounter and that it must master in principle. There is additional uncertainty whether other road users will behave differently in interaction with driverless vehicles than they do in road traffic today.

If the unpredictability of certain scenarios in connection with performance limitations or foreseeable misuse of individual vehicle functions leads to unsafe behavior, the term *residual risks* is used in accordance with the ISO/DIS 21448 standard [19]. The basic principle is that residual risks must be evaluated and are only acceptable up to a certain level (cf. ▶ Sect. 6.3). While residual risks can and should be reduced with particular effort in the course of a development, they can never be completely eliminated, as there are inherent risks associated with the use of automated vehicles and driving functions in the public traffic environment [24]. The requirements definition can in part influence inherent risk levels by modifying functional scopes and the intended area of operation. However, the remaining risks and their causal uncertainties should be disclosed accompanying the design process to provide a basis for public debate on acceptable risk levels [14, 22].

While systematic faults in a developed technical system are therefore not the sole cause of existing risks, they do lead to a direct increase in residual risks if they result in unsafe behavior in certain situations. Accordingly, a process should always be structured in such a way that through systematic procedures and sufficient examination of work products, systematic faults are avoided or identified and eliminated through testing and system adaptation. A common approach to avoid systematic faults in the context of system design is model-based systems engineering. A central as-

pect of the approach is ensuring that requirements are addressed by functions to be implemented by linking them in models (cf. [42]).

In addition to avoiding systematic faults, clear limits should be assigned to random faults in the course of the requirements definition, since otherwise arbitrarily high failure rates of the overall system are possible in principle. If it is determined in a hazard analysis that certain possible failures lead to significant risks, thresholds are expressed by automotive safety integrity levels (ASIL) according to ISO 26262. If specific failure modes occur more frequently than allowed by the integrity level, the use of the developed system in the intended operational environment poses unreasonable risks [16, Part 5, 9]. Within ISO 26262, ASIL classifications also serve to prescribe a necessary scope of processes and development methods to avoid systematic faults (see ▶ Sect. 6.2.2.1).

45.3.2   Concept Phase as the Basis of Product Safety Argumentation

It already became apparent in the previous sections that safety aspects play an important role starting in the initial phases of the design of automated vehicles and driving functions. Since the definition of necessary safety strategies and safety mechanisms can have significant implications for the system design, it must be carried out at an early stage on the basis of an evaluation of potential hazards. These hazards essentially result from the intended functional scope and the limits of the operational environment. Whether the development of certain systems is pursued further thus also depends on whether a system-wide safety concept that adequately addresses potential hazards appears feasible.

Therefore, the goal of driverless operation of road vehicles is also associated with a development of mechatronic components (sensors, actuators, and control units) and special algorithms that do not directly serve the fulfillment of the intended function. These elements rather serve the argumentation of sufficient safety in unknown scenarios or under degradation conditions. However, the previously mentioned feasibility of safety strategies depends not only on technical aspects but also on economic aspects. For example, anticipated costs of the hardware components required to realize an intended functionality at an acceptable safety level could in fact lead to a significant adaptation of the development objective.

In the course of system design, therefore, not only concrete technical solution approaches for functionalities are identified, but also solution approaches for safety strategies. The effectiveness and feasibility of individual strategies can already be validated with studies, prototypes, or samples. The design of automated vehicles and driving functions also includes considerations of how safety, as an emergent property of the conceived system, is eventually to be evaluated and argued.

In this respect, the early documentation of a safety concept forms the basis for a comprehensive safety argumentation supported by test results, which is required for a series release of the function or the vehicle. Such a series release is always carried out by decision-makers of the manufacturing company, but may additionally be required to be carried out by third-party auditors prior to market launch, for example within the European Union. Essential contributions of the system design to the safety argumentation are the documentation of

- legal and normative requirements,
- a comprehensive system description that includes a definition of the operational domain, envisioned safety mechanisms, and a preliminary system architecture,
- applied procedures for hazard identification and their results,
- identified causes of risk and existing uncertainties about unknown scenarios as well as targeted upper limits for residual risks,
- strategies developed to avoid unreasonable risks (expressed in safety requirements),
- evidence of the effectiveness of introduced safety strategies,
- qualitative and quantitative residual risks already known during concept specification, and
- necessary procedures and methods for ensuring conformity with requirements during implementation in a series development as well as through testing.

In summary, ▶ Sect. 45.3 discussed that functional concepts and scopes of automated driving functions must be designed to serve multiple critical aspects. Central examples are

- a sound safety concept,
- customer benefit (including, in particular, subjective benefits),
- marketability (cost/benefit and communication aspects), and
- integration in a vehicle concept (especially technological aspects: computing power, energy requirements, packaging, …).

After all, compared to many other systems, vehicles that move automatically in public spaces have very diverse stakeholder groups: In addition to the expectations of passengers, in particular the expectations of other traffic participants—including pedestrians and cyclists—must be systematically taken into account during the design process. Accordingly, sound concept specifications must be acceptable to both users and other road users, while at the same time providing a societally tolerable level of safety.

45.4 Reference Development Process for Automated Vehicles

A reference development process makes an important contribution to overcoming the concept specification challenges in a design of automated vehicles and driving functions explained in ▶ Sect. 45.3. Orientation toward a structured development process helps to ensure early and systematic consideration of the previously listed crucial aspects. The process structure presented in this chapter is thus intended to serve as a reference for organizing and monitoring ongoing development projects that are in early development stages; in an industrial context, this would typically be research and advanced development.

45.4.1 V-Model as a Cross-Domain Process Model

An established reference for the development procedure for complex systems is the so-called V-model. Essentially, the process model describes the combination of a top-down decomposition with a bottom-up integration, which is illustrated in the form of the letter V. The model originated in software development projects of the U.S. national aeronautics and space administration (NASA) [8].

Guideline VDI 2206 introduces the V-model as a macro-cycle of mechatronic system development, which features the phases system design, domain-specific design, and system integration [38, Sect. 3.1.2].

» The V model describes the generic procedure for designing mechatronic systems, which is to be given a more distinct form from case to case [38, p. 29].

In addition, the phases of the macro-cycle are flanked by the process module modeling and model analysis in the sense of model-based system development (see ◘ Fig. 45.1).

In addition to the potential to deal with the complexity of mechatronic systems through a rigorous top-down approach to defining the basic structure, the guideline also emphasizes the need to iterate through the macro-cycle represented by the V-model.

» A complex mechatronic product is generally not produced within one macro-cycle. Rather, a number of cycles are required.

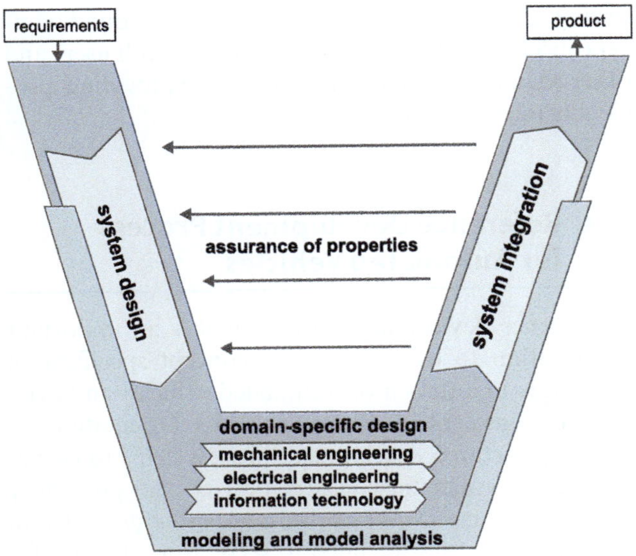

requirements

product

system design

system integration

assurance of properties

domain-specific design

mechanical engineering

electrical engineering

information technology

modeling and model analysis

Fig. 45.1 V-model as a macro-cycle [38, p. 29]

In a first cycle, for example, the product is functionally specified, first operating principles and/or solution elements are selected and roughly dimensioned, checked for consistency in the system context and realized in an exemplary form. [38, p. 30]

The possibility of multiple cycle runs is illustrated within guideline VDI 2206 by nested V-models, in which the product maturity increases with each cycle (cf. [38, p. 31]). However, the macro-cycle shares the highly simplified assumption with many other versions of the V-model that the system requirements are available as input data for system development from the beginning (see ▪ Fig. 45.1). In this respect, this process model does not place any particular focus on reproducing a systematic requirements definition at different levels of abstraction. Graphically, there is also no emphasis on immediate feedback for adapting the functional specification as soon as significant design conflicts arise that can be considered unresolvable (cf. ▶ Sect. 45.4.3).

Dick et al. define different layers of abstraction for requirements and traceable links between the layers within the top-down-bottom-up methodology of the V-model [5, pp. 11ff]. First, requirements are concretized and extended by the design process. Then, in the course of qualifying, the design is tested against the requirements at various layers of abstraction. The authors argue that a structured requirements engineering process forms the basis of a traceable linkage of requirements, which is the condition for impact analyses as part of the change management. As already described in ▶ Sect. 45.3, however, Dick et al. present a rather sequential procedure for requirements specifica-

tion and likewise do not depict a dynamically iterative process.

A concrete extension of the macro-cycle *V-model* reflecting the special demands of systems engineering is presented by Eigner et al. [7]. In particular, the integration of model-based systems engineering approaches into the process model is further elaborated. In addition to advantages in dealing with existing complexity in the design of mechatronic systems, the authors also describe the support of requirements specification through early modeling of the system description in semi-formal notation such as the systems modeling language (SysML).

In a paper on autonomous system design, Sifakis expresses various points of criticism of the V-model [34, Sect. 2.3.2]. The author shares the view that the assumption depicted in the model that all system requirements are initially available in clearly formulated form is not realistic. Furthermore, Sifakis criticizes the notion of a consistent top-down system development, since in reality it is much more likely that existing systems would be modified incrementally. Finally, the paper questions the practice of verification after completed implementation as depicted in the V-model. Sifakis cites these deficits as the reason "the V-model has been abandoned in modern software engineering in favor of the so-called Agile methodologies" [34, p. 10].

When presenting the V-model as a reference process for automotive software engineering, Schäuffele and Zurawka also list the limitations of the process model as a representation of a real development procedure, [32, Sect. 1.5] which have already been mentioned above. The authors emphasize that in development projects, user requirements are not fully understood at the start, but that the V-model implicitly assumes this to be the case. Schäuffele and Zurawka likewise share the view that "the reality of development is characterized by incremental and iterative procedures, forcing developers to repeat some steps or even all steps of the entire V-Model many times" [32, p. 32].

Although it becomes clear that the V-model does not provide a realistic process representation for software-intensive development objects such as automated vehicles, it is still of essential importance for illustrating the general procedure and the phases of a system development. For example, both the automotive standard ISO 26262 [16, Part 1, Fig. 1] and the annex to ISO/DIS 21448 [19, Fig. A.15] define process steps using a V-model. In addition to the V-model, other established process models for product development exist in various engineering fields. Eigner [6] provides a comprehensive overview of the history and form of various process models in mechatronics and its subdisciplines.

45.4.2 Requirements for a Domain-Specific Reference Process

The introduction of a domain-specific reference process makes a significant contribution to illustrating the special demands of automated driving on the design process. For this purpose, this section will first describe important requirements for a dedicated process model. The requirements are explained in detail below, supported by related work and additional literature, and listed in summary at the end of this section.

Haberfellner et al. present "four basic principles" for systems engineering process models [13, pp. 27ff]. The authors state that it is appropriate

- to proceed from the general to the detailed,
- to consistently look for alternatives,
- to divide the system development into temporally distinguishable process phases, and
- to apply a formal procedural guideline in solving problems.

Haberfellner et al. further argue that the proposal to establish the "from the general to the detail" approach—i.e., a top-down principle—is closely related to further requirements for a process model [13, pp. 54f]. On the one hand, the top-down procedure can be concretized by defining phases as a macroscopic strategy for process structuring. On the other hand, a strict top-down system specification enables the also recommended systematic generation and evaluation of solution variants.

Wilmsen et al. [40] define a process for specifying requirements for reference processes in large companies. The authors define the necessary process characteristics on the basis of stakeholder requirements, with various users of the process (designers, project managers, etc.) representing important stakeholders. Wilmsen et al. apply their method to automotive predevelopment and name a high degree of detailing, variability in the work packages, and enabling an agile way of working through iterative and flexible process models as examples of important requirements in this field.

In their work, Wilmsen et al. [40] also draw on the results of a survey of experts from large German companies in the automotive industry presented by Pfeffer et al. [27]. Based on the assumption that many development projects of conventional vehicle systems to date are carried out following the V-model, Pfeffer et al. identify challenges with regard to process models for the development of automated vehicles and driving functions. On average, the respondents clearly agreed that development processes and process models in the automotive industry would change considerably within the next decade. As a frequently expressed expectation, the authors mention the adaptation of existing processes to integrate agile methods, e.g., to establish shorter development cycles, especially for supplied components.

The importance of a systematic requirements definition for a mechatronic system to be developed has already been highlighted in the previous sections. Automated vehicles and driving functions represent complex development goals whose requirements analysis drives the early development phase. The goal of a process description should therefore also be the consideration and concrete structuring of this development phase. With regard to the procedure of a systematic definition of necessary safety goals, the safety standard ISO 26262 [16] already makes important contributions (cf. ▶ Chap. 6). The safety lifecycle of ISO 26262 includes a concept phase, which precedes the concretization of the system under development and specifies implementation-independent safety requirements. In this context, the concept phase has several steps such as a functional specification of the system (item definition), a hazard analysis, and a safety concept specification. However, the concept phase is not mapped in the V-model-based reference process used in ISO 26262 but is described by a preceding block that does not allow any conclusions about its structure or stages.

Cross-domain process descriptions—such as guideline VDI 2206—cannot provide a concrete reference at this point either, since they do not sufficiently consider the requirements of safety–critical systems. A concrete example is the indispensable need to perform a hazard analysis in the course of designing an electronic vehicle system. Although guideline VDI 2206 uses the development of an electromechanical vehicle brake as an accompanying application example, the guideline does not go into more detail about this dimension of safety consideration.

In this context, Sexton et al. [33] describe that during early development activities of an electronic vehicle system, an informal process of safety consideration accompanies the V-model. The process is characterized by a cyclic concatenation of early hazard and safety analyses with design decisions. In particular, safety requirements from potential hazards continuously influence the system architecture decisions during the concept phase (cf. ▶ Sect. 45.3.2).

Reschka [29] adopts the structure of the development process reflected in ISO 26262 for the design of a vehicle guidance system and thus demonstrates the suitability of the process phases of the general automotive safety standard for the use case of automated driving. In addition, however, Reschka also proposes in particular an extension of the specifications for the concept phase in the form of a structured process for the systematic generation of an item definition for automated driving functions.

In summary, a domain-specific process model for the design of automated vehicles and driving functions should therefore

- differentiate between process phases,
- establish a top-down principle,
- enable the comparison of different solutions,
- reflect the agile nature of early development,
- represent the process phase of requirements definition, and
- take into account the state of the art (e.g., the reference process of ISO 26262).

45.4.3   Systematic Design of Automated Vehicles and Driving Functions

In this section, a reference process is presented that describes a systematic design in the area of automated driving and takes into account the previously listed requirements of a process description. The model presented is based on the process description of the pre-development of an automatic emergency brake as an assistance system for road vehicles [23]. A detailed description of the background of the development project is given by Rieken et al. in a previous version of this handbook [30]. Among other things, their article points out the specific benefits of frontal collision avoidance by assistance systems and explains the enabling technological development in perception systems. In addition, the still existing uncertainty with respect to intervention decisions (i.e., interventions of a technical system in the longitudinal vehicle guidance) and the challenge of a sound safety concept for assistance systems, especially with regard to ergonomic aspects, are discussed in more detail.

◻ Figure 45.2 shows the reference process developed on this basis for a systematic design of automated vehicles and driving functions. In its form presented here, the reference process was first explained in [11]. The adaptation of the process model to reflect the challenges of automated driving is based on findings from the research project aFAS, which is presented as an application example in ▶ Sect. 45.5.

The process model has a circular form with two possible loops and is thus intended to particularly emphasize the iterative character of the design. The iteration via the inner loop is called *item refinement* and means a highly resource-efficient revision of the functional specification at an early stage of development. The steps of the inner loop are characterized by discussions about the benefits, feasibility, and consequences of the intended functional scope with experts from different disciplines. The triggering of an inner iteration is possible at any time within the first development phases, as soon as unresolvable conflicts arise due to design decisions. For the first illustrated development phase of the systematic definition of functional technology-independent requirements, the term *concept phase* is chosen in accordance with the automotive standard ISO 26262. The other designations *system design*, *domain-specific design*, *system integration*, and *verification and validation* are based on the phases of the reference process of guideline VDI 2206 (cf. ▶ Sect. 45.4.1).

Specifically, the process model describes the beginning of the concept phase as the collection and documentation of ideas, use cases, etc., which essentially result from the motivation to offer the greatest possible customer benefit. The core benefit of an automated vehicle is the provision of mobility, but users will also benefit from other aspects such as comfort, efficiency, or safety. The ideas and use cases captured represent a central objective in the further course of the design. Restrictions to the functional scope may be decided later for different reasons and often require an adjustment of the intended use cases. This is expressed in the process model by the circular shape and in both loops.

Following the collection of use cases, the intended functional specification is captured abstractly in an item definition, which also addresses aspects such as the preliminary structure and expected operational environments of the system to be developed. This forms the basis of a subsequent hazard analysis. In this, starting from possible operating situations, hazards due to vehicle behavior are identified and a risk assessment is performed. A safety concept suitable for the intended functional scope must then be developed. In the course of the concept phase, a functional safety concept is created that describes safety mechanisms and safety strategies for reducing identified risks to an acceptable level through implementation-independent functional safety requirements (cf. ▶ Sect. 6.2.2.2).

The relevance of early safety analyses depicted in the process model follows the principle of *safety-by-design*: Hazards resulting from an intended functional scope and context of use are identified early in the design process and actively avoided as far as possible by adapting a preliminary system design or conceiving and implementing suitable mechanisms. Safety-by-design thus differs from design approaches which, on the basis of good professional practice, assume sufficient safety throughout the development phase, which ultimately is proven by testing the system. The technical report ISO/TR 4804 [18] emphasizes that the consistent application of safety-by-design principles makes key contributions to the safety of automated vehicles.

The systematic design represents the condition stated in ▶ Sect. 45.3.2 that, in the field of automated driving, the development of certain functional characteristics is only pursued if the implementation of a system-wide safety concept that adequately addresses potential hazards appears feasible. Otherwise, the func-

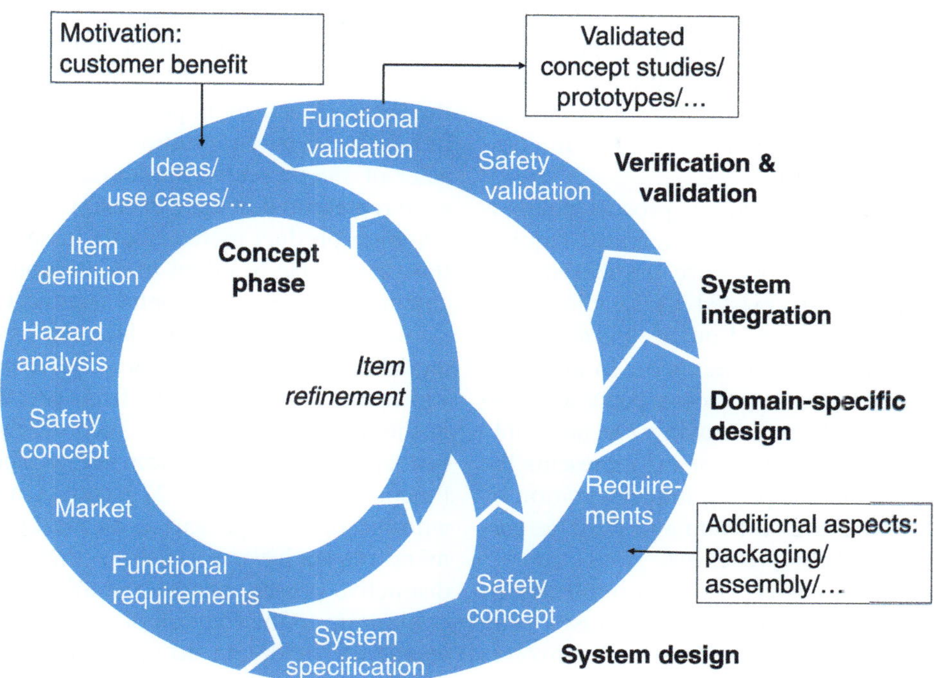

Fig. 45.2 Systematic design of automated vehicles and driving functions following [11]

tional specification must be adapted to form a sound safety concept through implementable safety strategies. For example, the operational environment can be restricted or use cases and thus functional scopes can be discarded. In the process model, this is a key aspect of the *item refinement*: If no satisfactory safety concept is available, the functional objective should be adjusted instead of unnecessarily putting resources into the further realization of concepts that must be assumed unreleasable. An equivalent of this mechanism within the design of automotive functions can also be found in the safety standard ISO/DIS 21448. During the development and as part of a safety analysis, the standard suggests a modification of the functional system specification or a restriction of applications if risks above an acceptable level are identified and cannot be further reduced [19, Clause 8].

With the subsequent keyword *market*, the process model encourages a critical review whether the intended functional scopes—in particular also under the aspect of potential restrictions due to the safety concept—represent a marketable concept that the developers consider suitable for further concretization and implementation. Up to this point, only limited resources have gone into the mainly conceptual work. The actual creation of concept studies and prototypes, as well as their evaluation in extensive tests, requires a multiple of time, financial resources, and other resources and should only be done for promising concepts.

As a product of the concept phase, a sound concept is expressed by the specification of functional requirements. In addition to the specification of intended functional scopes, all safety requirements for functional components from the previously developed safety concept are also included at this point. As already explained in ▶ Sect. 45.2, the functional requirements should be formulated as solution-neutral as possible—i.e., technology-independent and implementation-unspecific—in order to avoid unnecessarily restricting the solution space.

The system specification then takes place based on the functional requirements within the system design process phase. This is the first time that concrete technology decisions are necessary, which are incorporated into a technical system architecture to be implemented in hardware and software (cf. ▶ Chap. 39). The selection of which control units, sensors, and actuators are to be used and how they are to be linked together then forms the basis for the further specification of the safety concept. For automated vehicles and driving functions, the definition of a technical safety concept that specifies the necessary implementation of functional safety requirements in concrete mechatronic components (cf. ▶ Chap. 6) is a central and at the same time very challenging step in the system design process.

At this point, the process model provides another option for an item refinement. This possibility of a late concept adaptation becomes necessary in particular because available technological capabilities can be compared with the functional requirements here for the first time. A simple example of this would be requirements that demand a performance from the environment sensor technology (e.g., in terms of detection field or range) that cannot be achieved by any known sensor technology. In this case, it would not make sense to continue with the creation of a concept study; instead, the requirements must be adapted on the basis of a new iteration of the concept phase.

In the model, the system design process phase is concluded with the specification of technical requirements. These requirements concretize the concept description, which previously existed as functional requirements, based on the carried out technical specification and subsequent safety assessment. In addition, other—possibly automotive-specific—aspects are also taken into account, such as packaging and assembly constraints, etc. The list of requirements represents the starting point for a domain-specific implementation of the system in hardware and software and must be correspondingly sound and comprehensive.

The mechatronic subsystems are then implemented by specialized developers in their own processes using the technical system description. This usually involves the parallel implementation of individual components, which necessitates a subsequent integration into an overall system. Following the implementation and integration of all hardware and software components of the system under development, system tests can be performed in the course of verification and validation. Here, the process model distinguishes in particular between the validation of the system-wide safety concept and the functional validation.

Within the scope of safety validation, it is checked whether the safety concept has been implemented as intended and whether its effectiveness in terms of hazard avoidance appears sufficient in the intended context of use according to tests. Functional validation, on the other hand, describes empirical measures for checking whether the concept study or prototype created sufficiently serves the customer benefits initially documented as motivation. Findings from the system tests and, in particular, identified deficits of the system created in a previous iteration serve as an important source of information for a next iteration in the process model in order to finally create an assured concept for a potential series product in accordance with the design definition of VDI 2206 (cf. ▶ Sect. 45.2).

In general, it should be noted that the circular shape of the process model presented emphasizes the possible reuse of already generated artifacts in the course of both the inner and outer iterations beyond the end product of a previous run. Specifically, large parts of the safety analysis, conceived mechanisms and system architectures, functional requirements, etc. can be reused or only slightly revised and extended over several iterations of a concept study. This aspect is shared by the described systematic design of automated vehicles and driving functions with the spiral model of software development according to Boehm [3]. However, the presented process model complements the aspects of continuous requirements revision and feedback to the initial objectives, in particular via the mechanism of item refinement. In addition, the highly iterative character of the presented model basically serves the principle represented in agile development methods of generating a product incrementally in short iterations and thus being able to adapt to continuous changes in the circumstances of a development project [2].

According to Cooper [4], the systematic design model explained here is a descriptive process model in which process observations from practice are reflected in the sense of a reference. Accordingly, the illustrated systematic design of automated vehicles and driving functions (cf. ◨ Fig. 45.2) is not a normative model with the aim of representing an ideal process. Although many normative models share the aspect of defining distinctive process phases, the arbitrary number of iterations up to the creation of a validated concept in the course of the design do not represent an ideal guideline. For this, a process without iterations would probably have to be aimed for, but this is not realistic in practice.

Extensions and further developments of the design model are conceivable from various perspectives. As already explained in ▶ Sect. 45.3, the development of automated vehicles and driving functions is characterized by a large number of potentially conflicting stakeholder interests. The safety of systems under development is a central aspect here and leads to a concentration of the presented process model on safety aspects already in the early design stage. However, a societal tolerance for an achievable level of safety must be verified throughout the design process—also by explicitly disclosing existing residual risks. Systematic identification and consideration of stakeholder groups is suitable for determining further important constraints for the concept specification beyond safety aspects. For this purpose, methods that aim to consider human values in the course of system development can make important contributions. An approach to combine the value-oriented development paradigm of *value sensitive design* according to Friedman et al. [9] with the process structure from ◨ Fig. 45.2 is presented in [10].

45.5   Application Example aFAS

The research project aFAS (automatic driverless safety vehicle for motorway roadworks), which was successfully completed in 2018 with a public demonstration in flowing traffic, served as an important source of information for the further advancement of the reference process originally created for the design of assistance systems. The current process model has been presented in Sect. 45.4.3 and is applied in this section using the example of the development of a prototype driverless protective vehicle carried out in the aFAS project.

As described in the process model, the motivation for the development of an automated protective vehicle in the aFAS project was also a specific customer benefit: employee safety. For human drivers of a protective vehicle that announces roadworks, there is a real risk of life-threatening injuries due to accidents caused by carelessness of the drivers passing by. The research project therefore aimed to develop a system that automates slow driving on the hard shoulder of highways and thus makes driverless operation in public road traffic possible.

Based on this objective, different use cases were concretized, such as regular manual operation of the protective vehicle for the traveling between work sites and the depot without system interventions as well as automated following of a work truck as the leading vehicle at different distances. The intended functional scopes were subsequently documented in an item definition together with preliminary assumptions on the functional system structure and other constraints. Ohl et al. [25] and Stolte et al. [36] explain essential contents of this artifact, which is a living document during the course of a design. At this point, figures illustrating the operating modes (◘ Fig. 45.3) and the functional system structure (◘ Fig. 45.4) are included, which served as central elements of the function specification in the item definition and for further functional development.

With the assumptions about functional scopes, system structure, and the operational setting documented in an item definition, the processing of the next task described in the process model began within the research project: performing a hazard analysis. Structure, contents, and results of the hazard analysis for the automated protective vehicle are described by Stolte et al. [35]. Of central importance here is that the causes of hazards to be investigated in the context of driverless operation do not lie exclusively in potential faults and failures of electronic and electrical system components, which are the focus of the ISO 26262 safety standard (cf. ▶ Chap. 6). Instead, ergonomic aspects—such as possible mode confusion for vehicle operators—were also investigated within the safety analysis of the development project. In addition, functional insufficiencies of the system were evaluated as possible causes of hazards. Examples include inadequate recognition of lane markings or the leading vehicle by the vehicle guidance system. In this respect, a procedure was implemented in the hazard identification that is now described and required in the ISO/DIS 21448 standard.

In principle, a hazard analysis is concluded initially by defining safety goals that address all identified risks. Before this has been achieved in the course of the hazard analysis for the automated protective vehicle, an item refinement became necessary, which is represented in the process model by the inner loop. Stolte et al. [35] describe that originally a use case was intended to be implemented where the automated protective vehicle is able to pass a stationary obstacle on the shoulder of the highway by entering the adjacent traffic lane. During safety considerations, a major hazard potential of such a maneuver sequence in flowing traffic was determined. Mechanisms to mitigate the resulting risks were not feasible within the scope of the project. In particular, this special case would have made it impossible for a safety concept to generally prohibit the automatic vehicle guidance system from crossing the hard shoulder marking and entering the traffic lane. In this respect, there was a significant design conflict that had to be resolved by an iteration within the concept phase. Within this iteration, the previously described particularly challenging use case was excluded for the time being and the item definition was subsequently adapted accordingly. With the adjusted functional scopes, it was then possible to conclude the hazard analysis and risk assessment by specifying 17 safety goals with the corresponding required integrity levels (see [35]).

As illustrated in the process model, the next step of the design in the research project was the preparation of a functional safety concept. For this purpose, the safety goals were recorded and concretized by developing safety mechanisms and further safety strate-

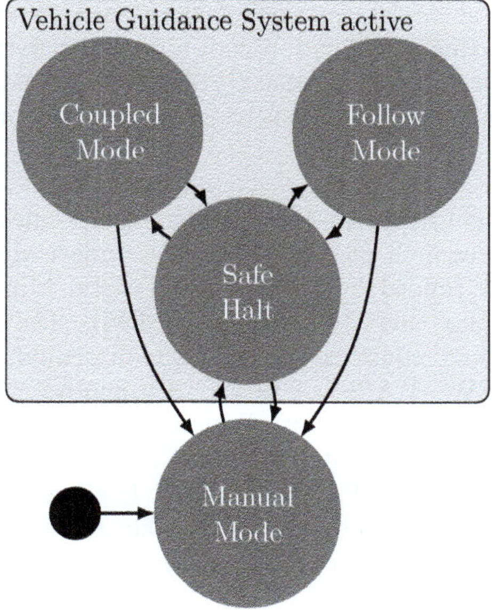

◘ **Fig. 45.3** Operating modes of the automated protective vehicle in the aFAS project [36]

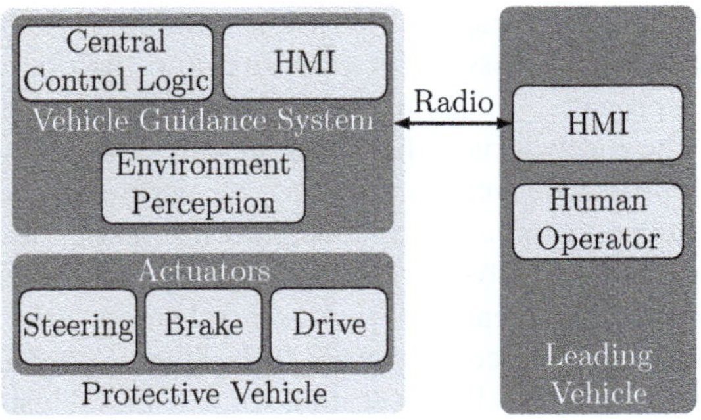

Central Control Logic | HMI

Vehicle Guidance System

Environment Perception

Actuators

Steering   Brake   Drive

Protective Vehicle

Radio

HMI

Human Operator

Leading Vehicle

Fig. 45.4   Functional components of the overall system in the aFAS project [12]

gies. The defined goals and strategies were then used to specify (as far as possible solution-independent) functional safety requirements for the individual components shown in Fig. 45.4. In reality, however, some decisions already made in the project about the structure and capabilities of individual control units were also included at this point. The specific contents of the safety concept created in the aFAS project are explained in [12].

Due to its nature as a research project, market analyses, which are listed as the next instance in the process model, were not of great importance. However, the realization of the automation system with limited project resources and the successful demonstration without the presence of a person in the protective vehicle received considerable attention following the completion of the project; this was due to the project presenting a solution to a safety problem of international relevance [20].

The composition of the functional requirements, which describe the safety concept, the functional scopes, the system structure, etc., also formed the conclusion of the concept phase in the aFAS project. The system design was then concretized by participating industry partners, and within this process, the safety concept was translated into technical requirements. The same applies to the implementation and integration of the individual system components.

Finally, the tests required to verify and validate the function and safety of the prototype for a demonstration were carried out in different settings. Moreover, specific faults were injected to investigate the effect of safety mechanisms [1]. First, the protective vehicle controlled by the automated vehicle guidance system drove only on a closed test track, then on an idle newly constructed highway, and finally in moving traffic on the section of highway designated for the demonstration. During the tests, vehicle operators of the protective vehicle had means to intervene in the vehicle guidance system. During the actual demonstration, however, the person was no longer in the protective vehicle but in the leading vehicle and only had the option of stopping the protective vehicle by radio command.

45.6   Conclusion

The application of suitable methods and processes makes a significant contribution to the success of the development of automated vehicles and driving functions, already in the early phases. In this chapter, the constraints of a design in the context of automated driving were explained. For this purpose, the definition of design within the development of complex mechatronic systems was addressed first, followed by a discussion of the special challenges involved in carrying out a design of automated vehicles and driving functions. Overcoming various identified challenges can be supported by structuring the design process. Accordingly, a process model for systematic design in automated driving was presented and explained. In addition to a reference for distinguishable phases, the reference process also establishes mechanisms to avoid unnecessary inefficiencies by continuously questioning the functional specification and further design decisions. Finally, the successful implementation of the design of an automated vehicle guidance system for a protective vehicle from the research project aFAS was described, which made major contributions to the current version of the process model.

Acknowledgements   This chapter incorporates results of the research project aFAS. The project aFAS was partially funded by the German Federal Ministry of Economics and Technology (BMWi). The project consortium consisted of MAN (consortium leader), ZF TRW, WABCO, Bosch Automotive Steering, Technische Universität Braunschweig, Hochschule Karlsruhe, Hessen Mobil—Road and Traffic Management, and BASt—Federal Highway Research Institute. The authors would like to thank their current or former colleagues Andreas Reschka, Gerrit Bagschik, Torben Stolte, and Markus Steimle for their work on the development of an automated protective vehicle in the aFAS project during their employment at the institute. Their contributions form the basis of the further development of the process model for systematic design in automated driving explained in this chapter.

The research on the design of automated vehicles and driving functions was partly accomplished within the project "UNICARagil" (FKZ 16EMO0285). We acknowledge the financial support for the project by the German Federal Ministry of Education and Research (BMBF).

References

1. Bagschik, G., Stolte, T., Steimle, M., Maurer, M.: Funktionale Sicherheit für das automatisch fahrerlos fahrende Absicherungsfahrzeug. Presented at the aFAS Schlusspräsentation, House of Logistics & Mobility, Frankfurt am Main, 20 June 2018 (2018)

2. Beck, K., Beedle, M., van Bennekum, A., Cockburn, A., Cunningham, W., Fowler, M., Grenning, J., Highsmith, J., Hunt, A., Jeffries, R., Kern, J., Marick, B., Martin, R.C., Mellor, S., Schwaber, K., Sutherland, J., Thomas, D.: Manifesto for Agile Software Development (2001). ▶ http://agilemanifesto.org/. Accessed 5 May 2021

3. Boehm, B.W.: A spiral model of software development and enhancement. Comput. **21**(5), 61–72 (1988). ▶ https://doi.org/10.1109/2.59

4. Cooper, R.G.: A process model for industrial new product development. IEEE Trans. Eng. Manag. EM-**30**(1), 2–11 (1983). ▶ https://doi.org/10.1109/tem.1983.6448637

5. Dick, J., Hull, E., Jackson, K.: Requirements Engineering, 4th edn. Springer, Cham (2017). ▶ https://doi.org/10.1007/978-3-319-61073-3

6. Eigner, M.: Überblick Disziplin-spezifische und -übergreifende Vorgehensmodelle. In: Eigner, M., Roubanov, D., Zafirov, R. (eds.) Modellbasierte virtuelle Produktentwicklung, pp. 15–52. Springer Vieweg, Berlin, Heidelberg (2014). ▶ https://doi.org/10.1007/978-3-662-43816-9_2

7. Eigner, M., Dickopf, T., Apostolov, H.: The evolution of the V-model: from VDI 2206 to a system engineering based approach for developing cybertronic systems. In: Ríos, J., Bernard, A., Bouras, A., Foufou, S. (eds.) Product Lifecycle Management and the Industry of the Future. 14th IFIP WG 5.1 International Conference PLM 2017, Seville, 10–12 July 2017. IFIP Advances in Information and Communication Technology, vol. 517, pp. 382–393. Springer, Cham (2017). ▶ https://doi.org/10.1007/978-3-319-72905-3_34

8. Forsberg, K., Mooz, H.: The relationship of system engineering to the project cycle. In: First Annual Symposium of the National Council on Systems Engineering, Chattanooga, 21–23 Oct 1991. INCOSE International Symposium, vol. 1, pp. 57–65. Wiley, Hoboken (1991). ▶ https://doi.org/10.1002/j.2334-5837.1991.tb01484.x

9. Friedman, B., Kahn, P.H., Borning, A., Huldtgren, A.: Value sensitive design and information systems. In: Doorn, N., Schuurbiers, D., van de Poel, I., Gorman, M.E. (eds.) Early Engagement and New Technologies: Opening Up the Laboratory, Philosophy of Engineering and Technology, vol. 16, pp. 55–95. Springer, Dordrecht (2013). ▶ https://doi.org/10.1007/978-94-007-7844-3_4

10. Graubohm, R., Schräder, T., Maurer, M.: Value sensitive design in the development of driverless vehicles: a case study on an autonomous family vehicle. In: Marjanović, D., Štorga, M., Pavković, N., Bojčetić, N. (eds.) 16th International Design Conference DESIGN 2020, Cavtat, 26–29 Oct 2020. Proceedings of the Design Society: DESIGN Conference, vol. 1, pp. 907–916. Cambridge University Press, Cambridge (2020). ▶ https://doi.org/10.1017/dsd.2020.140

11. Graubohm, R., Stolte, T., Bagschik, G., Reschka, A., Maurer, M.: Systematic design of automated driving functions considering functional safety aspects. In: 8. Tagung Fahrerassistenz, Lehrstuhl für Fahrzeugtechnik mit TÜV SÜD Akademie, Munich, 22–23 Nov 2017 (2017)

12. Graubohm, R., Stolte, T., Bagschik, G., Steimle, M., Maurer, M.: Functional safety concept generation within the process of preliminary design of automated driving functions at the example of an unmanned protective vehicle. In: Wartzack, S., Schleich, B., Gonçalves, M.G., Eisenbart, B., Völkl, H. (eds.) 22nd International Conference on Engineering Design (ICED19), Delft, 5–8 Aug 2019. Proceedings of the Design Society: International Conference on Engineering Design, vol. 1 no. 1, pp. 2863–2872. Cambridge University Press, Cambridge (2019). ▶ https://doi.org/10.1017/dsi.2019.293

13. Haberfellner, R., de Weck, O., Fricke, E., Vössner, S.: Systems Engineering—Fundamentals and Applications. Birkhäuser, Cham (2019). ▶ https://doi.org/10.1007/978-3-030-13431-0

14. Homann, K.: Wirtschaft und gesellschaftliche Akzeptanz: Fahrerassistenzsysteme auf dem Prüfstand. In: Maurer, M., Stiller, C. (eds.) Fahrerassistenzsysteme mit maschineller Wahrnehmung, pp. 239–244. Springer, Berlin, Heidelberg (2005). ▶ https://doi.org/10.1007/3-540-27137-6_11

15. Höppner, S.: AG-/AN-Schnittstelle—Schwerpunkt Ausschreibungen/Vertragswesen. In: Das V-Modell XT. eXamen.press, pp. 173–274. Springer, Berlin, Heidelberg (2008). ▶ https://doi.org/10.1007/978-3-540-30250-6_3

16. International Organization for Standardization: ISO 26262: Road Vehicles—Functional Safety. ISO, Geneva (2018)

17. International Organization for Standardization: ISO 9241-210: Ergonomics of Human-System Interaction—Part 210: Human-Centred Design for Interactive Systems. ISO, Geneva (2019)

18. International Organization for Standardization: ISO/TR 4804: Road Vehicles—Safety and Cybersecurity for Automated Driving Systems—Design, Verification and Validation. ISO, Geneva (2020)

19. International Organization for Standardization: ISO/DIS 21448: Road Vehicles—Safety of the Intended Functionality. ISO, Geneva (2021)

20. Jentzsch, G.: MAN aFAS Receives the First Truck Innovation Award at the IAA Commercial Vehicles (2018). ▶ https://press.mantruckandbus.com/corporate/man-afas-receives-the-first-truck-innovation-award-at-the-iaa-commercial-vehicles-2018/. Accessed 29 Dec 2021

21. Leveson, N.G.: Engineering a Safer World—Systems Thinking Applied to Safety. MIT Press, Cambridge (2011)

22. Maurer, M.: Hochautomatisiertes und vollautomatisiertes Fahren. In: Deutscher Verkehrsgerichtstag—Deutsche Akademie für Verkehrswissenschaft (ed) 56. Deutscher Verkehrsgerichtstag 2018, Goslar, 24–26 Jan 2018, pp. 43–57. Luchterhand, Cologne (2018)

23. Maurer, M., Wörsdörfer, K.-F.: Unfallschwereminderung durch Fahrerassistenzsysteme mit maschineller Wahrnehmung—Potentiale und Risiken. Presented at the Seminar Fahrerassistenzsysteme und aktive Sicherheit, Haus der Technik, Essen, 20 Nov 2002 (2002)

24. Nolte, M., Jatzkowski, I., Ernst, S., Maurer, M.: Supporting Safe Decision Making Through Holistic System-Level Representations and Monitoring—A Summary and Taxonomy of Self-Representation Concepts for Automated Vehicles (2020). ▶ https://arxiv.org/abs/2007.13807. Accessed 5 May 2021

25. Ohl, S., Maurer, M., Häusler, K., Holldorb, C.: Autonomes Fahren im Straßenbetriebsdienst auf Autobahnen. In: Intelligente Transport- und Verkehrssysteme und -Dienste Niedersachsen e.V. (ed.) AAET 2012—Automatisierungssysteme, As-

sistenzsysteme und eingebettete Systeme für Transportmittel. Beiträge zum gleichnamigen 13. Braunschweiger Symposium, Braunschweig, 8–9 Feb 2012, pp. 252–272. ITS Mobility, Braunschweig (2012)

26. On-Road Automated Driving (ORAD) Committee: SAE J3016: Taxonomy and Definitions for Terms Related to Driving Automation Systems for On-Road Motor Vehicles. SAE, Warrendale (2021). ▶ https://doi.org/10.4271/J3016_202104

27. Pfeffer, R., Basedow, G.N., Thiesen, N.R., Spadinger, M., Albers, A., Sax, E.: Automated driving—challenges for the automotive industry in product development with focus on process models and organizational structure. In: 2019 IEEE International Systems Conference (SysCon), Orlando, 8–11 Apr 2019 (2019). ▶ https://doi.org/10.1109/SYSCON.2019.8836779

28. Rausand, M., Haugen, S.: Risk Assessment: Theory, Methods, and Applications, 2nd edn. Wiley, Hoboken (2020). ▶ https://doi.org/10.1002/9781119377351

29. Reschka, A.: Fertigkeiten- und Fähigkeitengraphen als Grundlage des sicheren Betriebs von automatisierten Fahrzeugen im öffentlichen Straßenverkehr in städtischer Umgebung. Dissertation, Technische Universität Braunschweig (2017). ▶ https://doi.org/10.24355/dbbs.084-201707280929

30. Rieken, J., Reschka, A., Maurer, M.: Development process of forward collision prevention systems. In: Winner, H., Hakuli, S., Lotz, F., Singer, C. (eds.) Handbook of Driver Assistance Systems, pp. 1177–1206. Springer, Cham (2016). ▶ https://doi.org/10.1007/978-3-319-12352-3_48

31. Ross, H.-L.: Safety for Future Transport and Mobility. Springer, Cham (2021). ▶ https://doi.org/10.1007/978-3-030-54883-4

32. Schäuffele, J., Zurawka, T.: Automotive Software Engineering—Principles, Processes, Methods, and Tools, 2nd edn. SAE, Warrendale (2016). ▶ https://doi.org/10.4271/r-432

33. Sexton, D., Priore, A., Botham, J.: Effective functional safety concept generation in the context of ISO 26262. SAE Int. J. Passeng. Cars – Electron. Electr. Syst. 7(1), 95–102 (2014). ▶ https://doi.org/10.4271/2014-01-0207

34. Sifakis, J.: System design in the era of IoT—meeting the autonomy challenge. In: Bliudze, S., Bensalem, S. (eds.) Proceedings of the 1st International Workshop on Methods and Tools for Rigorous System Design (MeTRiD 2018), Thessaloniki, 15 Apr 2018. Electronic Proceedings in Theoretical Computer Science, vol. 272, pp. 1–22. Open Publishing Association, Waterloo (2018). ▶ https://doi.org/10.4204/EPTCS.272.1

35. Stolte, T., Bagschik, G., Reschka, A., Maurer, M.: Hazard analysis and risk assessment for an automated unmanned protective vehicle. In: 2017 IEEE Intelligent Vehicles Symposium (IV), Los Angeles, 11–14 June 2017, pp. 1848–1855 (2017). ▶ https://doi.org/10.1109/IVS.2017.7995974

36. Stolte, T., Reschka, A., Bagschik, G., Maurer, M.: Towards automated driving: unmanned protective vehicle for highway hard shoulder road works. In: 2015 IEEE 18th International Conference on Intelligent Transportation Systems, Gran Canaria, 15–18 Sept 2015, pp. 672–677 (2015). ▶ https://doi.org/10.1109/ITSC.2015.115

37. Suh, N.P.: Axiomatic Design—Advances and Applications. Oxford University Press, New York (2001)

38. Verein Deutscher Ingenieure: VDI 2206: Design Methodology for Mechatronic Systems. Beuth, Berlin (2004)

39. Verein Deutscher Ingenieure: VDI 2221 Part 2: Design of Technical Products and Systems—Configuration of Individual Product Design Processes. Beuth, Berlin (2019)

40. Wilmsen, M., Gericke, K., Jäckle, M., Albers, A.: Method for the identification of requirements for designing reference processes. In: Marjanović, D., Štorga, M., Pavković, N., Bojčetić, N. (eds.) 16th International Design Conference DESIGN 2020, Cavtat, 26–29 Oct 2020. Proceedings of the Design Society: DESIGN Conference, vol. 1, pp. 1175–1184. Cambridge University Press, Cambridge (2020). ▶ https://doi.org/10.1017/dsd.2020.301

41. Witte, T.E.F., Hasbach, J., Schwarz, J., Nitsch, V.: Towards iteration by design: an interaction design concept for safety critical systems. In: Sottilare, R.A., Schwarz, J. (eds.) Adaptive Instructional Systems. 2nd International Conference AIS 2020, Copenhagen, 19–24 July 2020. Lecture Notes in Computer Science, vol. 12214, pp. 228–241. Springer, Cham (2020). ▶ https://doi.org/10.1007/978-3-030-50788-6_17

42. Wymore, A.W.: Model-Based Systems Engineering: An Introduction to the Mathematical Theory of Discrete Systems and to the Tricotyledon Theory of System Design. CRC Press, Boca Raton (1993)

Open Access This chapter is licensed under the terms of the Creative Commons Attribution-NonCommercial-NoDerivatives 4.0 International License (▶ http://creativecommons.org/licenses/by-nc-nd/4.0/), which permits any noncommercial use, sharing, distribution and reproduction in any medium or format, as long as you give appropriate credit to the original author(s) and the source, provide a link to the Creative Commons license and indicate if you modified the licensed material. You do not have permission under this license to share adapted material derived from this chapter or parts of it.

The images or other third party material in this chapter are included in the chapter's Creative Commons license, unless indicated otherwise in a credit line to the material. If material is not included in the chapter's Creative Commons license and your intended use is not permitted by statutory regulation or exceeds the permitted use, you will need to obtain permission directly from the copyright holder.

Test Concepts for the Safeguarding of Automated Driving

Lutz Eckstein and Hermann Winner

Contents

© The Author(s) 2026
H. Winner et al. (eds.), *Handbook Assisted and Automated Driving*,
https://doi.org/10.1007/978-3-658-45276-6_46

46.1 Introduction and Scope

For every product, a minimum level of assurance of functional quality is required before it is introduced into the market, whether this is due to legal requirements, the avoidance of manufacturer liability issues, or to establish or maintain a positive reputation. In countries that are transposing the rules defined at the UN-ECE level into national law, two decision-making milestones are particularly relevant: firstly, the manufacturer's release, which implicitly assumes that the approval can be fulfilled, and secondly, the actual approval of the product for a market (homologation). Approval is granted by one or more natural persons who run the risk of being prosecuted if this approval was not carried out in accordance with the state of the art. From the company's point of view, the release confirms that the intended functionality, reliability and safety are given.

The term safeguarding is used here in a similar way as in [1, 2] as the process that precedes a release and produces a safety argumentation. This process includes many aspects from specification to market monitoring. ▶ Chapter 6 presents key aspects of it. This chapter focuses on selected test concepts for validating the safety of the automated driving function.

In the case of newly developed products retrospective field experience is principally lacking. Therefore, the release decision is always based on a prognosis, which is based on the results of the quality assurance accompanying the development, mostly from virtual and real tests. These tests are on the one hand derived from the product requirements. On the other hand, they are based on experience with other products, especially with predecessor products. In the case of "grown" products, this creates a solid base of test sets whose statement of validity is high, since test results and field experience over previous product generations can be compared with each other. Furthermore, quality assurance processes can be optimized in this way because the effort required for testing is a significant cost factor even for mass products such as automobiles, but is nevertheless not free of errors, as recalls reveal again and again. Nevertheless, this is balanced by the fundamentally high reliability of the products, which is reflected in the steadily increasing lifetime of the vehicles in operation and the low proportion of accidents caused by technical defects. This provides a mature basis for the further development of the next generations of motor vehicles.

Automated driving at level 3 and above introduces for the first time a function in the motor vehicle that not only performs the driving task but also takes responsibility for it within the operational design domain (ODD). This fundamentally changes the prerequisites for safeguarding, since the driver is no longer available as responsible monitoring and fallback instance like in automation levels 1 and 2. Although the function of automation levels 3 or 4 will be very similar to level 2 in most use cases, the fundamental difference lies in the driver role. All continuously acting driver assistance systems (Level 1 or 2) emphasize the controllability of system limits and failures by the driver. This leads to the limitation of the system's actions (e.g., max. accelerations, max. steering torque), which are sufficient for comfortably driving, but often insufficient to handle a challenging situation, so that drivers are experiencing these limits from time to time. In contrast, intervening assistance systems such as the Automatic Emergency Brake (AEB) are designed in a compromise between effectiveness and controllability, so that the required intervention is delayed or even absent in unclear situations, cf. ▶ Chap. 6.

Automated driving functions from level 3 and above on the other hand, must react in any case and, if necessary, with maximum possible driving dynamics. If, as at Level 3, a human driver is present to continue driving after a sufficiently long takeover, this does not make safeguarding any easier, because maneuver support by human driver cannot be relied on—he/she does not have the function of a fallback instance as at Levels 1 and 2. In addition, new challenges are added with the design of the takeover and the safeguarding of takeover readiness. The distribution of functions between human and machine at Level 3 allows for an expansion of the Operational Design Domain (ODD), but it also potentially leads to mode confusion, provided Level 2 functionality is offered alongside Level 3 [3].

These special level 3 challenges will not be discussed below. The basic function of automated driving can also be safeguarded at level 3 without any significant differences from the higher levels. In contrast, type and scope of the safeguarding, and even the entire process, depend strongly on the selected ODD.

In addition to the driving function, which is handed over by the human driver to the automated vehicle guidance system, the human is eliminated as an "energy source" in automated driving, which is why braking and steering systems must also have "energy" redundancies. This is discussed in detail in ▶ Chaps. 24 and 25 and therefore the safeguarding and testing requirements for this are not considered in this chapter. Also, motion control per se hardly raises any new questions for safeguarding, so that the focus of safeguarding automated driving is on perception (environment detection and representation) and cognition (situation interpretation and behavior generation). To this end, we first discuss the potential and limitations of driving tests as a testing concept, including the statistical basis for a safety argumentation (▶ Sect. 46.2). Scenario-based testing is presented in ▶ Sect. 46.3 as an alternative or at least as a supplement. The challenges for practical implementation are also highlighted. In the outlook (▶ Sect. 46.5), an approach for the future is briefly touched upon, which takes a process-oriented look at the testing challenge.

46.2 Potential and Limits of Test Drives

46.2.1 Role of Test Drives in Safeguarding of Motor Vehicles

Test drives traditionally play a major role in evaluating new vehicles and new functions. For this purpose, mainly near-production prototypes and pre-production vehicles are used. These vehicles can be driven under more difficult conditions by test drivers, whether in the driving dynamics limit mainly on test tracks or in environments with extreme ambient or road conditions (e.g., hot-land or winter testing). Most of the testing is carried out under "normal" conditions, as so-called near-customer test drives (in German: Kundennahe Fahrerprobung, KFE). This can be supplemented with pilot studies of pre-production vehicles in selected customer hands, consisting of "willing to test" customers and/or company employees, but always after a (first) successful KFE campaign.

One goal of conducting test drives is to stimulate vehicle functions, i.e., to challenge in situations that are far more complex than the test scenarios on test tracks. The other goal is the subjective and objective evaluation of innovative functions during these drives. Trained testers in particular can provide valuable feedback from the driving impressions, especially if comfort impressions are also to be collected. Data processing in the vehicle is equipped to the extent that a sufficient amount of data can be recorded and objectively evaluated. Statistical evaluations can be used for this purpose, providing a frequency distribution on predefined criteria. These can also be data-intensive collections, ranging from data read in from a ring buffer in response to a trigger to complete raw data collections that allow post-processing in the development lab, e.g., using Replay2Sim concepts[1] as open-loop simulation. This allows to use the tests multiple times, e.g., if only for perception software is changed. However, it is not sufficient to just "replay" the data, because a test requires criteria for (automated) evaluation.

Depending on the application, it is sufficient to use certain metrics to draw on the result of functional pro-

cessing, e.g., one or more criticality measures. In the field of perception, efforts are made to add a ground truth (GT) to the recorded data. In selected test series, this can be formed with reference sensor technology, e.g., dynamic objects equipped with RTK-based measurement technology and/or high-precision maps of the environment. If such a reference cannot be formed, a reference environment representation is added to the recordings of the vehicle under test in post-processing (called annotation or labeling). Using measures such as retrospective state filtering, AI-assisted classification and human classification, the goal is to provide the lowest-error comparative reference against which the perceptual software results can be compared. For economic motivation, an attempt is made to reduce the amount of data, predominantly by omitting "uninteresting" sections.

The "data treasures" from test drives can also serve in part as training data for data-driven AI algorithms for perceptual purposes, in which case the data sets must meet further requirements. If the sensor hardware or arrangement is changed, which already includes changing the mounting position, the data set loses significant value. It can still be used to check the robustness of algorithms with regard to the changes. However, it is no longer representative for the "system-under-test" (SuT).

Test drives are considered the "real" testing compared to the test track, component and virtual testing variants. However, this reality often forgets that the vehicles are, after all, only slightly representative of the variety of later production vehicles, either because of the large number of configurations that can be ordered or because of possible component tolerance combinations of the individual components. Although boundary prototypes can be tested for known critical tolerances, this only applies to a few preselected tolerance dimensions for reasons of expense.

In the design of near-customer test drives, the focus is on the length of the route, since the costs depend on this. For assistance systems, distances between 100,000 km and 10 million km are common, cf. [4], depending on whether only one system variant ("carry over") or a new development with a large scope of changes is to be tested. In addition to the cost of the testing effort, the test drives put pressure on the development time budget. Therefore, attempts are made to combine the requirements as well as possible, mostly based on experience. These are:

- Drive through the entire range of use as far as possible (and in the process identify the difficult sections for follow-up tests).
- Cover operating conditions such as weather, time of day and seasons as completely as possible.
- Experience traffic situations in the greatest variety (from lonely night driving to vacation traffic jams).

---

1   The data required for the function are recorded in real runs and are replayed for simulation with existing or new function software.

◼ Table 46.1   Accident frequency and mean driving distance between two accidents in the same category (rounded numerical values)

Accident consequences	Accidents/victims per year	Distance/accident
Property damage (all roads)	2.35 million	320.000 km
Personal injury (all roads)	300.000	2.5 million km
Fatalities (all roads)	3.000	>250 million km
Personal injury (freeway only)	20.000	12.5 million km
Fatalities (freeway only)	356	>700 million km

As the test distance increases, a "saturation of knowledge" occurs, i.e., the anomalies still to be observed are in most cases already known, so that a termination criterion can be found about them, since a sufficient prognosis for functional quality and safety can be derived from the experience gained up to that point. The tests of successor systems and variants can draw on this wealth of experience and actually only have to check whether the system behaves as known, for which significantly shorter distances are required, provided that this result is confirmed.

46.2.2 Safety Proof of Automated Driving based on Test Drives

46.2.2.1 Target Values

If a long distance is driven for testing, the idea is to put this distance and/or the driving time required for it in relation to the accident occurrence in order to estimate how safely driving is done with this driving function. However, one quickly comes up against limits here. For one thing, the test drives only cover a certain, albeit broad, range of situations. So there are always gaps in coverage that cannot be ruled out without field experience that goes far beyond the test routes. Without a previous system, knowledge of exposure in accident-prone situations is lacking, which means that the testing profile cannot be adapted to it in advance. On the other hand, it is difficult to find and statistically prove criteria for safe (automated) driving, which will be briefly derived in the following.

The criteria for a macroscopic consideration are discussed in detail in ▶ Chap. 40, which is why only one criterion is adopted here: the at least equal safety to today's road traffic.

The criterion sounds simple at first and also socially acceptable. In the application of the criterion, it becomes more complex. Accident statistics can be used as a measure of current safety. It provides values for the number of accidents recorded by the police, which in turn are broken down into accidents with property damage,[2] accidents with personal injury and fatal accidents. Furthermore, accident casualty figures can be broken down by injury class to road category. These can be converted, as shown in Table. 46.1, into a mean distance between two accidents.[3] The total annual mileage of all motor vehicles in Germany is 750 billion km for all roads and 250 billion km for highways, cf. [5], (figures rounded).

The average distance between two (police-registered) accidents of 320,000 km appears to be easily verifiable as a target with a driving distance of several million km, but accident damage varies greatly, so that this figure alone is not very informative about the risk to health and life. Simply because of the constitutional special protection of integrity and life, it must be verified whether safety is given in this respect. The average distance between two accidents with personal injury is in the range of the distance of test drives for new assistance systems. However, this is not sufficient for a statistically reliable statement, as will be shown in the following section.

Even if the prognosis could be made that the number of accidents with personal injury is below the reference value, this still does not allow a forecast of the number of accidents with fatalities. While there are about 100 accidents with personal injury per fatality in Germany (all roads), there are only 56 for the autobahn. The ratio for serious injuries varies between 4.6 (all roads) and 3.4 (autobahn), i.e., lower than for fatalities by factors of 21 and 16. In principle, extrapolation with these factors allows a prediction of fatality risk, but these empirical values are only valid for road traffic with driving by humans as known today. For example, the operation of driverless shuttle vehicles (cf. ▶ Chap. 49) is different since accidents on the one hand lead to lower injury severities because of the lower speed, but become statistically "conspicuous" in severe accidents

---

2   A distinction between "serious accidents with property damage" and "other property damage accidents", as listed in the statistics [5] is not made hereafter, since accidents with serious property damage can also fall under the second category if the cause of the accident is neither a misdemeanor nor a felony.

3   The mean distance between two fatal accidents is estimated downward via number of fatalities per mile driven.

because of the higher number of passengers and reduced restraint protection. Therefore, the mentioned extrapolation can only be applied with great caution with a corresponding safety margin. However, this caution can also lead to a false unfavorable prognosis of safety, as the following example shows.

A system twice as safe as the reference would record about half the number of accidents on which the reference is based in a test section designed for the number of fatalities. Extrapolating to the number of fatalities with a safety factor of two (versus the ground truth not yet known) extrapolates the number of fatalities to the same level as the reference. This would mean that it would no longer be possible to prove that the safety with regard to fatal accidents is higher than that of the reference. The consequence: The test would have to be extended so that the safety factor would be reduced, whereby the distance would then fall into the range of a multiple of the reference distance between two accidents with fatal consequences. Then the proof of the comparative number of accidents with fatalities would also be within reach, as it is in [6, 7] discussed in detail, with the consequence that the test distance requirement rises into the range of several billion km. Only a system that is safer by an order of magnitude compared to the reference would have a chance to be "free-tested" with significantly shorter distances, although it is difficult to predict whether the number of accidents caused by other road users would decrease in the same way, because in "only" 55–60% of accidents, the vehicle involved is considered to be the main cause. If the number of accidents caused by other road users remained constant, a reduction below 40–45% the reference would not be possible at all. The only way out would be to count only the accidents caused, but the number of accidents not caused must still be considered, because the total number must not increase. Here, too, a statistically robust proof is only possible over a very large distance, which would have to exceed one billion km in most cases.

46.2.2.2 Distance Factor for the Design of Test Drives

If the reference distance between two accidents of a certain accident category is itself already long, it must be increased for a proof by a distance factor that allows a statistically robust proof. Three assumptions are made for this purpose:

- The confidence level (1—probability of error); 95% is usually chosen;
- The stochastic process; here the Poisson distribution is assumed, since this requires only one parameter (expected value) and is often used in the modeling of accidents (model assumption: all accidents occur independently);

- The ratio of the accident expected value versus the reference.

In [6] and in more detail in [7] results are presented for different parameter combinations of confidence level and expected value ratio. This is shortened here to two examples.

With 95% confidence for the proof of a higher safety, at least three times the reference distance must be achieved if this is completed without any event (accident in the category under consideration). For this, the safety must be about six times higher, so that with 50% probability this case also occurs.

If the safety is "only" twice as high, about ten times the distance is required to make the proof statement at the 95% level with the expected ratio of 5 to 10 accidents.

46.2.2.3 Conclusion on Statistical Proof Through Safety Assurance Test Drives

Proving safety via test drives requires distances in the order of one or more billion km, which are not economically viable and would also significantly extend the development time. In optimistic cases, where the accident frequency is at least one order of magnitude lower than the reference, extrapolating accident numbers from one or more lower severity categories, the proof could be provided with mileages up to two orders of magnitude lower a robust prediction of higher safety. However, the number of accidents caused by other road users would also have to be significantly lower than today. In turn, the operational design domain (ODD) would have to be so favorable that the ability to avoid accidents in the event of behavioral errors by other road users compared to humans increases significantly, or the behavior of other road users changes in a way that reduces accidents due to the presence of an automated vehicle. There are no reliable forecasts for any of these three factors, which is why the conclusion remains that it is not currently possible to *prove the* safety of automated driving before the introduction of this technology. According to the above considerations, a fictitious free-testing of a "highway pilot" would require about 10 billion km, which would result in 2 million vehicle-years of driving, assuming an average annual driving distance of about 5000 km per vehicle on highways. This figure -or other figures obtained with the same calculation- shows the factual impossibility, but remains relevant for observation in the field. Until then, the use of automated (Level 3+) driving remains in a "probationary" state. As discussed in [7] in detail, this "probationary period" does not allow us to conclude that automated driving "on probation" must be assessed as a higher risk. For one thing, even proving a higher risk requires very long distance, and for another, field experience can be rapidly fed back to improve safety. Another helpful factor is the "dilution of risk", since in the introduction phase the fleet under observation forms only a very small group in road traffic and thus the collision probability would be still negligible due to the low exposure to other road users.

46.2.3 Silent Testing

A more economically attractive approach to enable large scale testing is in what is known as silent testing, or in [8] 'Virtual Assessment of Automation in Field Operation' (VAAFO). The idea is also based on the idea of running a virtual shadow instance (also known as shadow mode) in the vehicle in parallel with the current vehicle system, which generates setpoints for the vehicle control, but which are not executed by the actuator system. Tesla [9] reports the use of this technology for the further development of assistance systems. In greater detail, Wang looks at [10] the possibilities of silent testing and suggests possible system designs.

Provided that the computer architecture supports silent testing, at least one additional behavior planning process is performed on the basis of the equipped perception sensor technology, the result of which allows conclusions to be drawn about the intended future movement, e.g., via a planned trajectory. If this deviates from the real trajectory of the reference vehicle guidance (human or automation predecessor system), the shadow instance "continues" on the virtual trajectory, which is why the environment mapping is consequently transformed to the virtual position. If the deviations between virtual and real trajectory are large, this is a cause for situation evaluation, e.g., whether the virtual trajectory is to be evaluated critically with respect to a collision danger. But also the behavior in real critical situations can be evaluated. The goal of silent testing is to record potentially critical driving situations and to record the data required for retrospective consideration, which can be forwarded to the developers telematically, for example. These can be used for future scenarios for testing, as described in the subsequent section, or for a safety assessment if an accident probability can be derived from the data.

The main difficulty of this concept lies in the divergence of the virtual and the real instance. For example, if the virtual instance were to follow a vehicle ahead, but overtake the real one, the virtual vehicle would soon get no valid information about the virtual object ahead, while in the real world the planner would be given different planning tasks than the virtual instance. Therefore, the virtual instance must always be reset to the current state and thus restarted. This can be done according to certain criteria or in fixed time steps. But each reset bears the risk of resetting the shadow instance just at the beginning of a wrong action and thus overlooking the error. Since behavior planning requires little computing power compared to perception and environment representation, several shadow instances can be operated simultaneously with staggered "life cycles", which defuses the resetting problem. However, it is not completely solved with it also, since with resetting the memory of the algorithm is overwritten. For this reason, the algorithm development of software that is to be suitable for silent testing must include appropriate concepts that take this problem into account.

Silent testing is particularly suitable for qualifying updates in the field when a large number of vehicles with suitable equipment is available. Even in the case of new functions or new areas of application, this method can gather valuable insights. However, it quickly loses the ability to derive a safety prognosis from it because the reference system differs more from the system under test and dilutes the argumentation. For example, common situations with human drivers, such as tailgating, are already planned differently from the approach with machine-based vehicle guidance. And thus the behavior under these conditions is not comparable with the behavior in real operation.

Silent-testing will complement the safeguarding process as an important monitoring tool from the field and, in particular, can significantly reduce the effort required to update software. For a safety argumentation, however, it is only one building block on which alone no safety proof can be made. However, this methodology can provide insights from practice about challenging scenarios and their manifestation as well as frequency, which in turn are of great value for scenario-based testing, which is presented in the next section.

46.3 Concept of Scenario-Based Testing

46.3.1 Idea and Genesis of Traffic Scenarios

The idea of abstracting and parameterizing a challenging traffic situation in the form of a reusable scenario or "script" can be traced back to the evaluation of new system concepts in driving simulators. For example, in 1997, the parameterizable scenario of an interfering vehicle entering the vehicle's path was used in two versions in the dynamic driving simulator of Daimler-Benz AG in Berlin in order to compare driving performance with sidesticks as alternative controls to steering wheel and pedals (cf. ◘ Fig. 46.1).

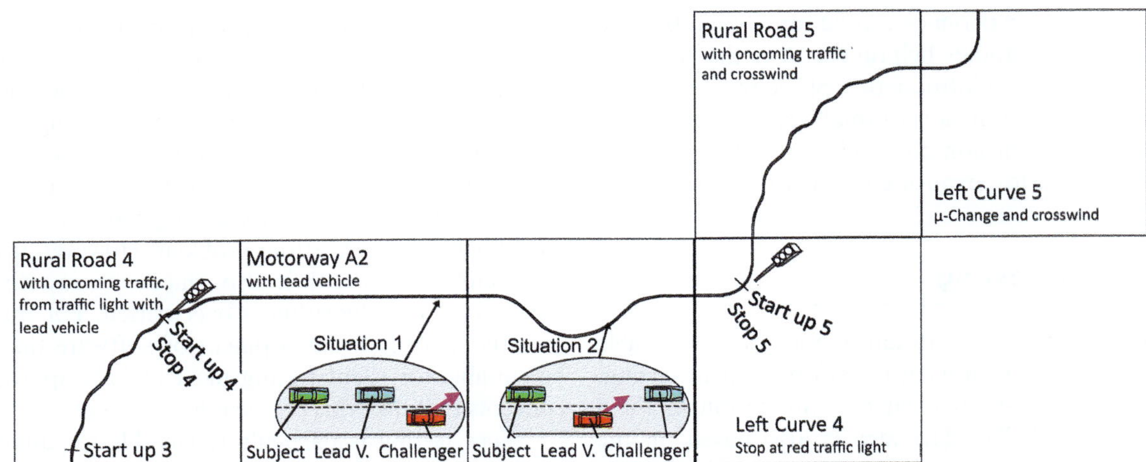

Fig. 46.1   Route with two challenging traffic situations in the dynamic driving simulator of Daimler-Benz AG [11]

While the initial situation and the logical sequence were the same for all participating subjects, the further course and the outcome depended on the reaction behavior of the human and the driving performance with the respective control concept.

These traffic scenarios were initially derived on the basis of accident data analyses since accident data provide statistically reliable information about situations where drivers obviously have a special need for support. Since official accident statistics are not very detailed, automobile manufacturers began more than fifty years ago to collect accident data in detail on their own to better understand causes and effects [12]. Building on this, the German Federal Highway Research Institute (BASt), together with the Research Committee on Automotive Technology (FAT), initiated GIDAS (German In-Depth Accident Study) in 1999, the most comprehensive accident research project in Germany, and created a common format and procedures. [13]. For a long time now, accident data analysis has not only provided evidence for further improving passive vehicle safety, but also valuable indications for potential driver assistance needs, i.e., for active safety measures.

For example, it has been found that in many cases the human driver is not able to use the full deceleration potential of the vehicle, because the operation of the brake pedal is too weak. This led to the invention of the brake assistant [14], which has been further optimized by the additional use of environment sensors and, in particular, long-range radar. The effectiveness of the brake assistant was demonstrated in Daimler AG's dynamic driving simulator using a scenario derived from GIDAS data [15]. In this specific test, the following scenario was formulated and implemented:

» The test person drove through a town at 50 km/h. The time to collision was determined in relation to a possible collision. At a fixed time in relation to a possible collision (time to collision-controlled), a child approaching from the left ran onto the road in front of the subjects' vehicle. The child was previously hidden at the roadside by parked vehicles. The situation was parameterized such that the accident could be prevented or mitigated with a good braking response by the driver. …

The programmed sequence depicting the scenario ensures that the required reaction in order to avoid a collision is identical for all test subjects, which is often realized via the so-called Time-To-Collision (TTC). Each test person thus has the same time budget to avert the collision. If a test person reacts very late, he or she would have to decelerate the vehicle more strongly, with the brake assistant exerting its supporting effect. If they were to act very early, for example when a collision warning is given, a correspondingly lower vehicle deceleration would be sufficient to avoid an accident.

The description of the routes and the scenarios was harmonized early on with the goal of reusability, especially in driving simulators [16]. In 2018, as part of the PEGASUS project, VIRES handed over the harmonization of OpenDRIVE and OpenSCENARIO to ASAM e.V. [17].

Inspired by the 100-Car Study [18] with the so-called **Field Operational Test**s (FOT), another methodology for the evaluation of innovative driver assistance systems was established, which was systematically used by several vehicle manufacturers. While the 100-Car Study focused on the behavior of drivers, in particular to understand the distracting effect of the then emerging cell phones in relation to "accepted" secondary tasks such as "radio tuning", the EU project euroFOT, for example, focused on the interaction of drivers with different driver assistance systems as well as their effectiveness [19]. Matlab-based analysis of the continuous recording of video and measurement data first identified points in time when, for example, a high level of deceleration or a small distance to the vehicle in front was measured. These sections were analyzed by an interdisciplinary team to evaluate the potential event. Against the background of the FOT analyses, it was obvious to no longer use only accidents as a source for relevant scenarios in tests in the driving simulator, but also events from FOT that did not lead to an accident.

Against the background of the infinite variety of challenging situations in road traffic, it was proposed in [20] for the first time to systematically extract these action-relevant scenarios from driving test data and to abstract them by suitable parameters. It was also suggested to collect and use these scenarios across companies, which was finally realized in the project PEGASUS [21] project described below.

46.3.2 Systematic Collection and Use of Traffic Scenarios

The core idea in [20] described above, which was introduced into the conception of the PEGASUS-project, was to perform the majority of the statistical proof required under ▶ Sect. 46.2.2 based on a growing collection of relevant traffic scenarios in the virtual world and to combine the tool of simulation with other established evaluation methods.

In the first step, the automated driving function to be safeguarded is "confronted" in the virtual world with a large number of scenarios extracted from a growing amount of real driving data, i.e., it must react to the behavior of other road users, such as a cut-in vehicle, in order to avoid an accident. In the second step, according to [20] such scenarios, which are not adequately handled by the automated driving function in simulation, can first be analyzed by experts in the driving simulator in order to narrow down possible causes

for the "conspicuous" functional behavior. If it was unclear whether human drivers would be able to cope with this challenging scenario, this could be checked with a representative group of test subjects to generate a reference. In the third step, tests on a test site (controlled field) are planned in order to be able to analyze the real function and, if necessary, to optimize it appropriately. Finally, it is inevitable to specifically evaluate the driving function in real road traffic within the framework of a FOT, i.e., with a focus on those conditions that have proven to be particularly challenging.

Another basic idea of the "cycle of critical situations" shown in ■ Fig. 46.2 was to share these scenarios which have been abstracted from concrete events and experiences and thus anonymized, and to jointly establish an accepted state of the art. The still desirable goal is to establish a cooperative database of relevant scenarios CODAS (COoperative Database for Automated Systems) analogous to GIDAS, based on a growing, verifiable number of kilometers driven by several vehicle manufacturers, suppliers and other organizations.

For this purpose, different sources should be used to extract parameterizable scenarios from factual events and real observations. ■ Figure 46.3 shows the concept of extracting relevant scenarios from different data sources and their use in different instances of the development process.

Scenarios resulting from accident reconstruction on the one hand and critical traffic situations recorded in the course of field and road tests prior to the market launch (SOP) of a product on the other are integrated into the symbolically represented database. In addition, real traffic data is also used, as an increasing proportion of vehicles on the market comprise extensive environmental sensor technology as well as powerful data processing and communication with the backend of the respective vehicle manufacturer. Real traffic data is of particular importance, as not only new scenarios may be continuously extracted from it, but the frequency of

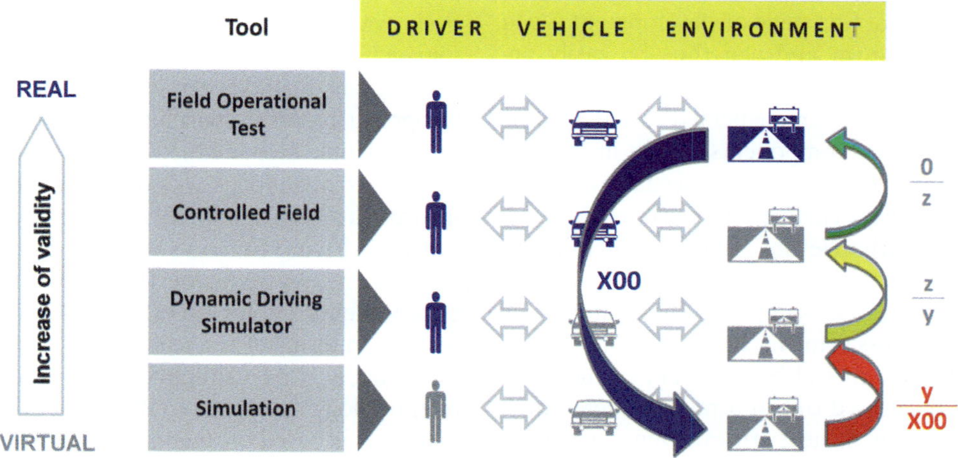

■ **Fig. 46.2** Core idea of the "cycle of critical situations" [20]

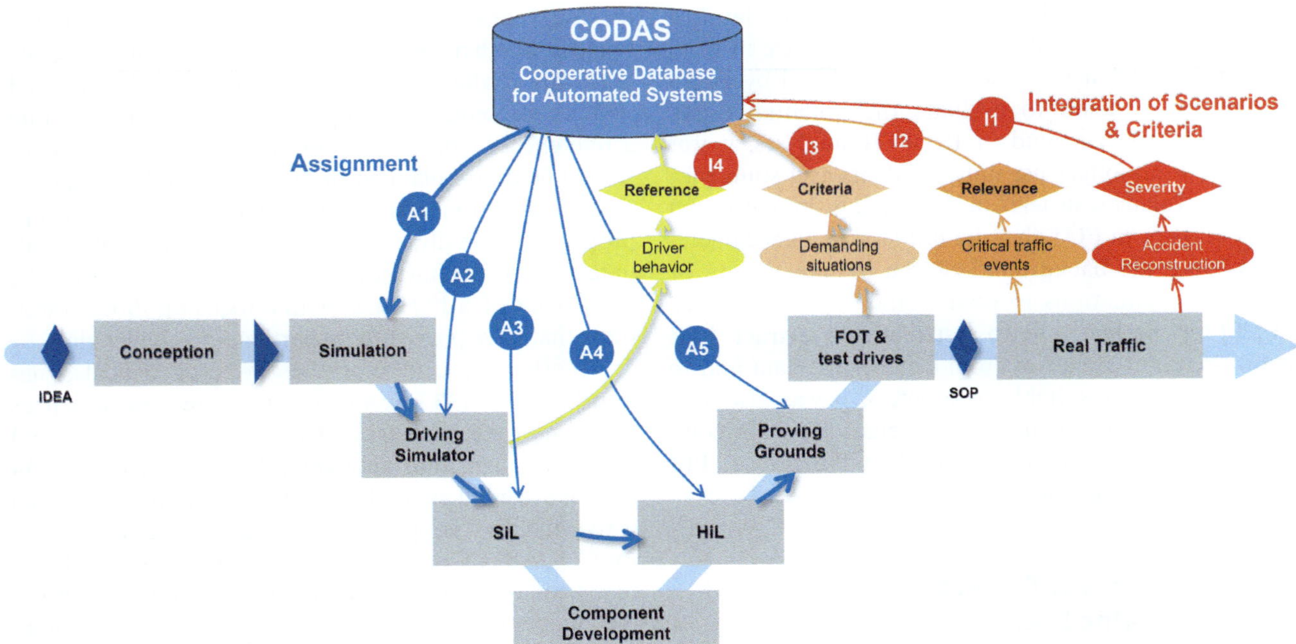

▫ Fig. 46.3   Extraction of scenarios from multiple sources and use in the development and safeguarding process, based on [21]

occurrence can also be determined as a key determinant of their relevance.

The scenarios collected in the database are made usable by playouts A1 to A5 in different instances of the development process, which is shown in ▫ Fig. 46.3 in a simplified form based on the V model. The use of scenarios relevant to a driving function in traffic simulation initially enables macroscopic evaluation of the potential effectiveness of new concepts and systems for assisted and automated driving in terms of traffic safety and traffic efficiency. If scenarios are used in static or dynamic driving simulators, it is possible on the one hand to make the acceptance of the tactical and dynamic behavior of automated driving functions subjectively assessable by humans, and on the other hand to objectively compare the performance of the function directly with the performance of human drivers. While the driving function at this stage of development is usually in algorithmic form and can thus be easily and comprehensibly adapted, in subsequent instances of the development process it is implemented in the form of software and hardware components, which in turn can be evaluated on the basis of relevant scenarios. Finally, such scenarios that have emerged in the course of development as particularly relevant for the evaluation of a driving function can also be implemented on a closed test site, as was done, for example, in the course of developing a steering assistant in 2000 [22]. The tests on public roads, which will still be necessary in the future but which can be carried out in a more targeted manner, potentially result in new critical events which are abstracted and made usable in the form of scenarios so that the **cycle of relevant scenarios is** closed.

46.3.3 Main Features of the PEGASUS Project

The PEGASUS research project (Project for the Establishment of Generally Accepted Quality Criteria, Tools and Methods as well as Scenarios and Situations for the Release of Highly Automated Driving Functions) pursued in the years 2016–2019 the goal of developing a manufacturer-independent approach to safeguarding highly automated driving functions. The SAE Level 3 function of the highway chauffeur was chosen as an application example in order to develop a safety argumentation for the manufacturer-side release of such a driving function [23].

To this end, a continuous tool chain has been developed in the project that includes simulation methods and testing in real environments in order to be able to objectively evaluate the respective development stage of an automated driving function. It was also taken into account that the new methods can be integrated into existing development processes, which is why the definition of harmonized interfaces and metrics also played a major role.

A particular focus was on simulation-based testing in order to gain comprehensive knowledge of an automated driving function early on in the development process, thereby reducing the scope of testing in real environments and for overall validation to a level that is realistic in terms of time and cost.

In ▫ Fig. 46.4 a simplified illustration of the overall methodology developed in the PEGASUS project is depicted. It is based on the use of scenarios derived from different data sources and from expert knowledge (data processing).

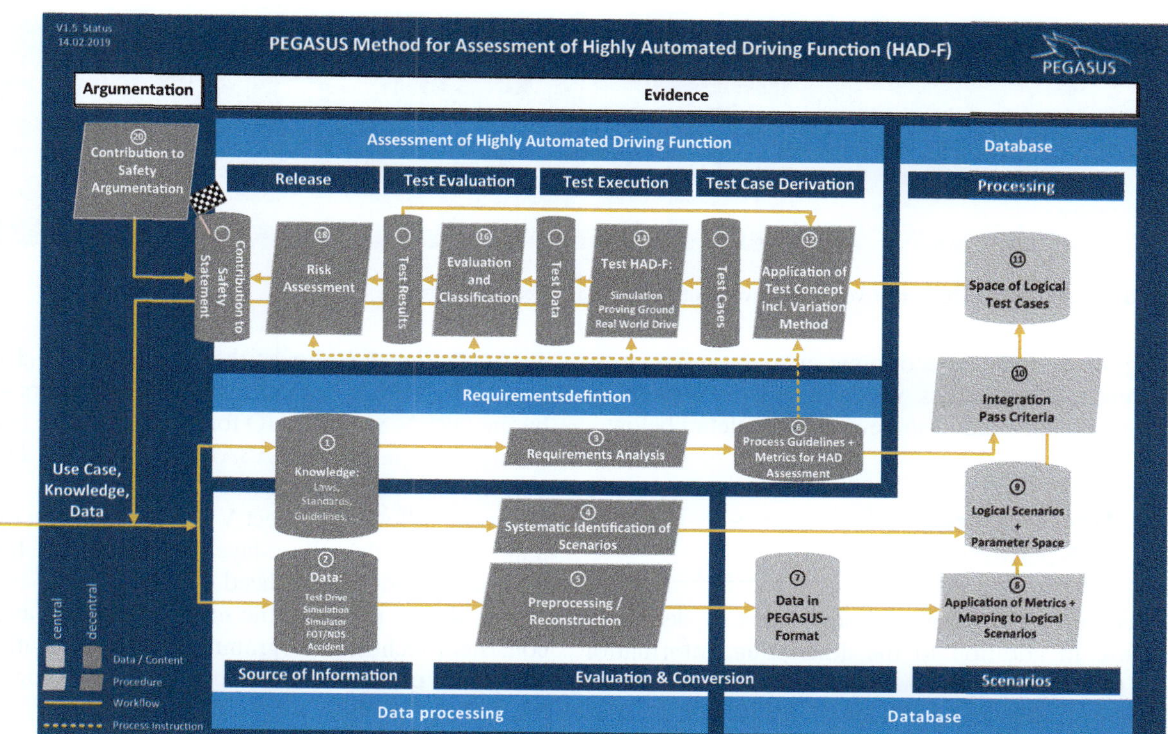

Fig. 46.4 Illustration of the overall methodology for safeguarding automated driving functions researched in the PEGASUS project [23]

Scenarios are extracted from the data provided by contributing companies in a standardized format and stored in the database of logical scenarios with the corresponding parameter spaces shown on the right below (Database). To evaluate the performance of a driving function in these scenarios, pass/fail criteria have been developed based on an analysis of the requirements placed on the automated driving function.

At the top in ▪ Fig. 46.4 the basic evaluation process is shown, which in the first step requires the derivation of the relevant test cases. A test case comprises the scenario, the test method and the criteria on the basis of which an automated driving function is evaluated. The execution of the tests is followed by the evaluation and classification of the test results. Finally, all results are incorporated into an overall risk assessment, which is an important building block in the safety argumentation required for approval.

To complement the test drive data provided by some companies in the PEGASUS project, drone- or UAV-based traffic detection was tapped as another data source [24]. Trajectories of multiple vehicles without occlusions can be extracted simultaneously from the images taken by one aircraft (cf. ▪ Fig. 46.5). The traffic environment can then be analyzed for each of these vehicles and potentially relevant traffic scenarios can be abstracted from them. At the same time, data protection challenges are circumvented because neither faces nor license plates of vehicles are visible in the aerial images.

Special focus of the development of evaluation methods addressed the efficient use of the scenarios on the one hand in the simulation, and on the other hand on test sites [23]. Due to the limited validity of current simulation models, cf. ▶ Sect. 46.3.6.2, it is imperative to test in particular the perception of an automated vehicle in relevant scenarios in a real driving environment that at the same time excludes public road traffic, ref. ▶ Chap. 13. Therefore, in recent years, some universities have created specialized test tracks such as Mcity,[4] Asta Zero[5] and the ATC,[6] and companies also invest large sums in suitable test environments at home and abroad.[7]

With the PEGASUS project, an essential cornerstone for scenario-based testing was laid and discussed internationally in the so-called PEGASUS Expert Workshops. In the meantime, the concept of scenario-based safeguarding is being researched and applied

---

4   University of Michigan test site, see ▶ https://mcity.umich.edu/our-work/mcity-test-facility/ (retrieval data for each of the links?).

5   Test site at Chalmers University in Gothenburg, see ▶ https://www.astazero.com/

6   Aldenhoven Testing Center of RWTH Aachen University, see ▶ https://www.aldenhoven-testing-center.de/de/

7   For example, the Zala Zone in Hungary (▶ http://zalazone.hu/) or the Mercedes-Benz Immendingen Test and Technology Center (▶ https://group.mercedes-benz.com/innovation/specials/immendingen.html?r=dai).

◼ Fig. 46.5   UAV-based detection of vehicles for trajectory and scenario determination [24]

in all globally relevant economic areas by research institutions and companies, which is why the concept of scenarios is explained in somewhat more detail below.

46.3.4 Description and Relevance of Scenarios

This section first presents possible ways of **describing** scenarios. In addition to the **description**, information on the **relevance of the scenarios** is also required. In the simplest case, **relevance** includes information about how often events were observed. The information of relevance is required in order to take frequent scenarios into account when selecting scenarios for safeguarding an automated driving function, while extremely rare events e may be excluded for good reason if necessary.

46.3.4.1 Scenario Description

The idea and the description of traffic scenarios have their origin in driving simulation, as shown in ▶ Sect. 46.3.1, where a distinction was initially made between the definition of the static driving environment in the form of road geometry, lanes, traffic signs and rules, boundary development, etc., and the description of dynamic traffic. While the behavior of the traffic participants and especially of the so-called interfering vehicle in the scenario under investigation was programmed as a logical sequence of events similar to the script of a movie, stochastic traffic models were used to give, for example, highway traffic on the oncoming lanes a realistic appearance.

In the context of development and testing of assisted and automated driving functions, in [25] suitable definitions of the terms scenario, scene, and situation have been published. The structuring in the form of static and dynamic description levels has been concretized in [26] and called the 4-layer model.

In the PEGASUS project, this layer model was initially extended to five layers [27] and, based on a suggestion from John Maddox at AVS 2018, supplemented by a sixth layer that, in addition to physical realities, includes digital information that a connected vehicle can already use today [28].

As ◼ Fig. 46.6 shows, the static layers 1–3 are described by OPEN DRIVE, the dynamic layers 4 and 5 by the OPEN SCENARIO format, which has been harmonized in parallel by ASAM e.V. The 6th layer has played practically no role in PEGASUS yet.

In the follow-up project VVM (Verification and Validation Methods), the 6-layer model (6LM) has been optimized and standardized with regard to its application also to urban traffic scenarios as well as a more consistent technical programming implementation [30]. The main changes are briefly explained below on the basis of ◼ Fig. 46.7.

On the one hand, Layer 1 still includes the road geometry, but also the temporally invariant traffic signs and locations of traffic signal systems. However, the current switching status of the traffic signal system is now assigned to Layer 6, which, as before, also includes other digital information that is exchanged via vehicle-to-infrastructure communication (V2X), for example. Layer 2 now focuses on structures and objects away from the roadway that could, for example, influence the perception of a system at an intersection or even interfere with localization due to the reflection of electromagnetic waves. This simplifies the modeling of the layers in the form of scenarios.

With regard to the **nomenclature of scenarios** a distinction is made between different levels of abstraction. First, a scenario can be described verbally, which is referred to as a "**functional scenario**". If the scenario is implemented in programming terms and parameters are assigned to this logical flow control, this is referred to as a **logical scenario.** If concrete values are assigned to the parameters, such as the distance and longitudinal speed of an approaching vehicle relative to the "ownship", this is referred to as a **concrete scenario**. This can correspond to a reconstructed accident, for example. By a reasonable variation of the parameters, as suggested in [20], an "event space" can be generated from an event, which theoretically represents a higher number of kilometers driven. The nomenclature of scenarios, like the definition of automation levels, is in a process of discussion and optimization that has not yet led to a commonly supported standard.

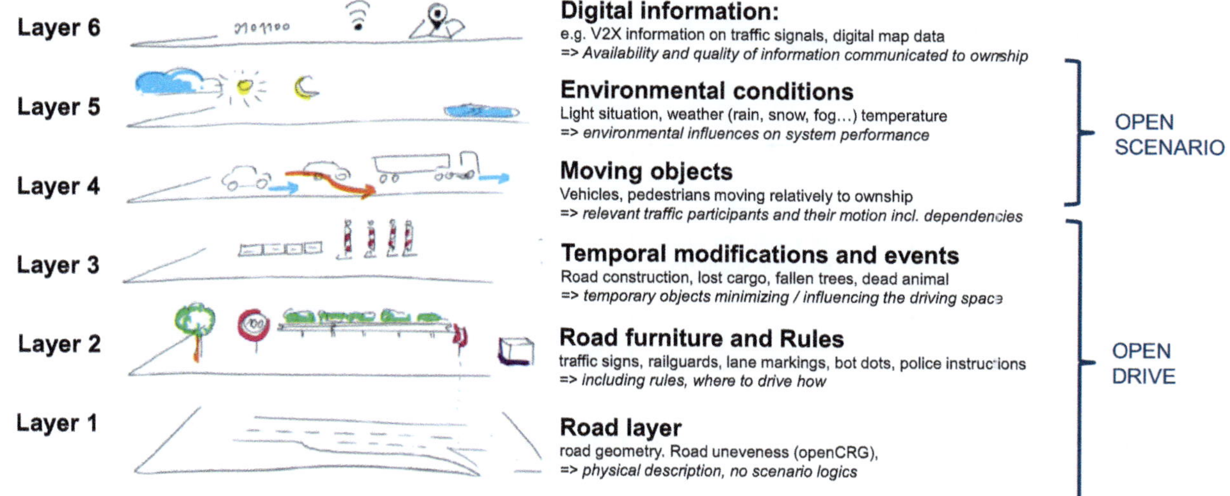

Fig. 46.6 Description of the 6 layers of the scenario model from PEGASUS [29]

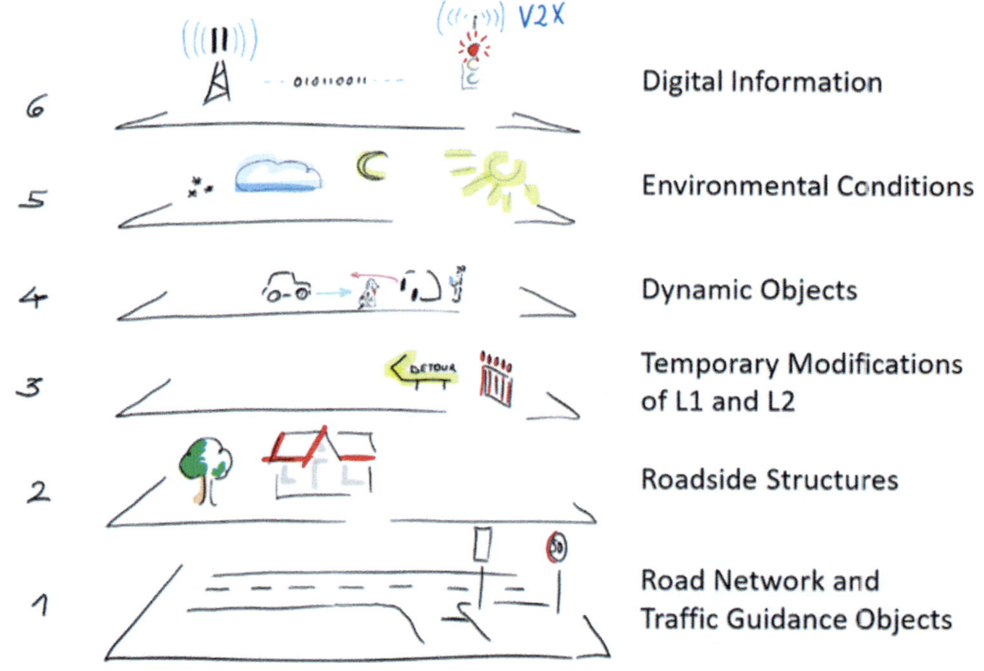

Fig. 46.7 Scenario layer model agreed in the VVM (verification and validation methods) project [30]

46.3.4.2 Relevance of the Scenarios

The **relevance** of a scenario initially plays an important role in the question of whether it should be used in the development process for evaluation and optimization. It is necessary to differentiate extremely rare scenarios from those that occur very frequently and thus absolutely have to be taken into account. Chris Urmson presented an illustrative example in numerous lectures: one of Google's vehicles encountered a wheelchair user who was chasing a family of ducks on the road and therefore had to brake.[8]

For each scenario and the associated parameter configurations, which are divided into classes, it is important to note the **frequency** as an essential criterion of relevance. For example, the basic scenario "cut-in from the right" generally occurs frequently on the highway, while concrete manifestations such as the cut-in of a motorcycle with a TTC of less than one second is al-

---

8   ▶ https://www.mercurynews.com/2016/04/05/google-self-driving-car-brakes-for-wheelchair-lady-chasing-duck-with-broom/,   accessed 3/28/2022.

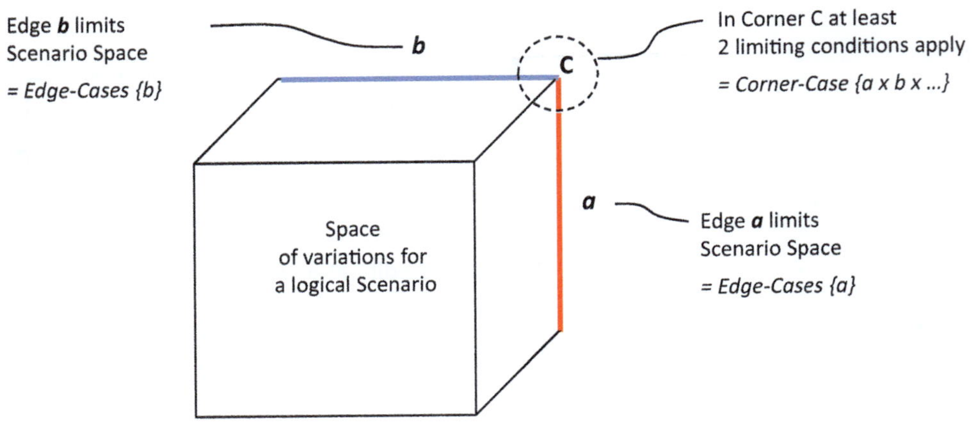

Fig. 46.8 Illustrative explanation of the terms edge case and corner case using a cube that symbolizes the space of possible manifestations of a scenario

ready observed much less frequently. Particularly efficient compared to the evaluation of measurement data from a vehicle is the external observation from an bird's eye perspective by UAVs,[9] because they can cover several hundred meters of road without occlusions. Networked road side units,[10] which create a digital traffic twin [31], provide an extensive database, since, for example, the movement and interaction of over 50,000 vehicles per day can be observed at a busy intersection.

While on the one hand there are scenarios that occur practically every time a person drives a car in less demanding manifestations, there are on the other hand also very rare situations in traffic. If a driver is overwhelmed by the demands resulting from a challenging situation, this rare event is reflected in the accident statistics. In the meantime, the terms **edge case** and **corner case** have become established to describe these very rare constellations, which are also used in communications technology. The difference can be easily explained using the following graphic (**Fig. 46.8**).

In the case of an **edge case**, only one of the edges shown in **Fig. 46.8**, which limits the space of possible expressions of a scenario, takes on an extreme value. This can be due to any of the six layers of the scenario, e.g., glare from a low sun on layer 5 or a fallen tree on layer 3. If another extreme manifestation of at least one other factor is added, e.g., a moose as an unusual road user on layer 4, the edge-shaped boundaries of the scenario space form an intersection or "corner" and thus a corner **case**.

In the course of safeguarding an automated driving function, the question consequently arises as to which of these edge cases or corner cases should be included

in the process. It is easy to imagine that there are extremely rare situations that a technical system will not be able to handle. The question of whether it is permissible, in analogy to passive safety, to compensate for a questionable performance of an automated driving function in a very rare scenario by a good performance in many other, very frequent scenarios, has so far been little discussed and is unanswered.

46.3.5   Human as a Reference

The **reference** comes into play when evaluating the performance of an automated driving function with regard to coping with a scenario: if the function to be safeguarded cannot avoid a collision in a demanding manifestation of a scenario, the question arises as to whether human drivers could have succeeded in doing so. In accident-based scenarios it is obvious that a human being was not able to master the specific scenario, otherwise the accident would not have occurred. For the most part, however, little is known about the individual who drove the vehicle: How capable was the person? What were the individual performance limits? Was he or she fatigued or distracted at the time of the accident? This ultimately leads to the question of whether, for example, 100 other people could have avoided the accident or—on the contrary—would also have had an accident.

If one first considers a concrete scenario as such, it strictly speaking cannot be titled as critical. Rather, the degree of difficulty can be described by quantifying the reaction required to avoid an accident. In the simplest case, this can be described by different combinations of reaction time and reaction strength, as was made clear in the description of the driving simulator test for the brake assistant, cf. ▶ Sect. 46.3.1.

The description of the required action to master the challenge resulting from a scenario becomes more complex in the case of multiple response options. However,

---

9   Unmanned Areal Vehicle, colloquially also called drone.
10 RSU: measuring equipment installed along the road for sensory detection of traffic, often using multiple sensor principles such as camera, radar and lidar.

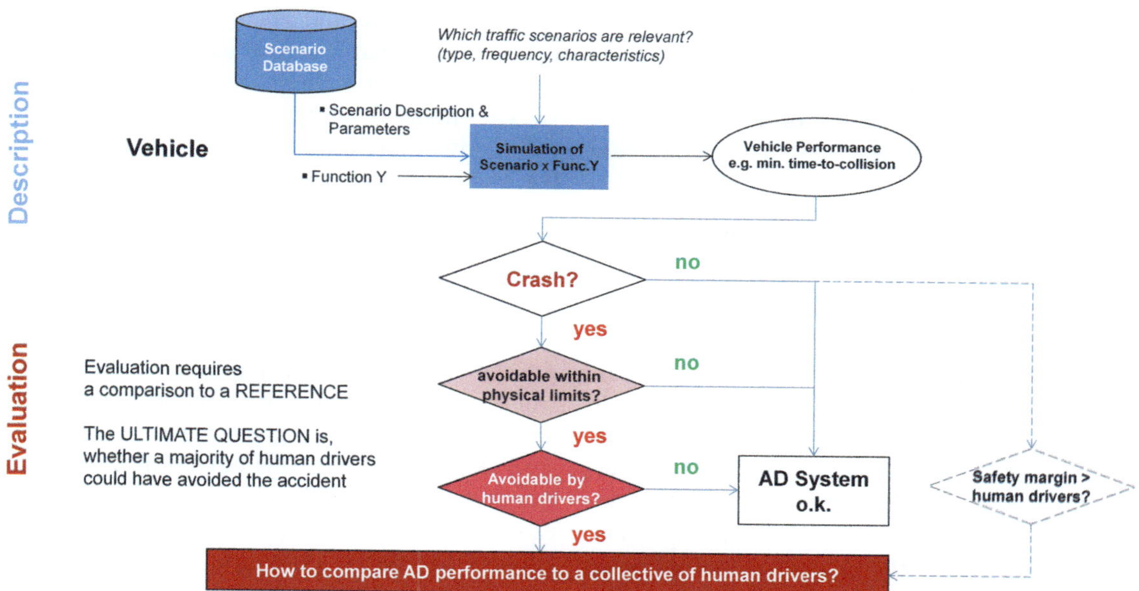

◘ Fig. 46.9 Multi-level assessment of coping with specific scenarios, discussed at the safety assurance breakout session at the 2018 automated vehicle symposium

an evaluation of criticality is only possible when the action requirements of a concrete scenario are compared with the capabilities of a human being or a machine—similar to an examination in studies. The exam as such cannot be critical, but it can be demanding. Only the solution of the tasks can be evaluated. It becomes "critical" from the perspective of an exam participant if he or she is poorly prepared, whereas the same person would be able to hand in the correct solution before the end of the exam period if well prepared.

Ideally, the human or the machine succeeds in avoiding a potential accident by fully exploiting the physical limits of driving. However, automated driving functions—like humans—require time for perception, cognition and action. It should therefore not be expected that an automated driving function will actually avoid every accident that is avoidable in terms of driving physics.

In terms of **perception,** the technical system has the advantage of being able to perceive 360 degrees of the environment at all times, but it also has the disadvantage of not being able to match the perceptual performance of a concentrated human being at great distances with the sensors currently in use. In terms of **cognition,** anticipating the future behavior of other road users is a key technical challenge. While humans learn continuously, the learning process of a technical system needs to be more structured in order to have a finite number of known and validated software states in the field. Finally, the dynamics of an **action are** limited on the part of both humans and machines. Advantages of the technical system are fatigue-free and reproducible actions and simultaneity of longitudinal and lateral control.

Because of these similar limitations of humans and machines, it seems advisable to distinguish between several steps when evaluating the objective performance of both in a concrete scenario (cf. ◘ Fig. 46.9).

If the automated driving function can avoid a collision, there is usually no need for further investigation in the course of safeguarding a system. If an accident occurs, the first question is whether it could have been physically avoided—if not, this cannot be expected from an automated driving function. If an accident could have been avoided within the physical limits of driving, it is advisable to seek a comparison with the performance of human drivers. If a physically avoidable accident can't be avoided by most human drivers, it may be accepted that an automated driving function also reaches its performance limits.

Consequently, it is more than obvious to operationalize the comparison with human drivers in a suitable way already during system development. One approach is to think of the automated driving function as a driver. WAYMO, for example, explicitly speaks of the "WAYMO driver" and publishes with [32] a study that aims to show the superiority of their robotic driver. Japanese car manufacturers also rely on the human driver as a reference [33] and are actively introducing this concept into ISO standardization.

There is also another reason why humans may represent the ultimate benchmark with regard to evaluating the performance of an automated driving function: on the one hand, it is necessary to convince a society and its representatives, including the responsible authorities, before an automated driving function is introduced that it drives at least as safely as an average driving human, although the discussion about the required level of safety of automated driving functions has not yet been conclusively conducted either nationally or internationally. Secondly, in the event of an accident, the court will ultimately ask whether a human driver could have avoided the accident. Therefore, automotive manufacturers are already preparing for a possibly necessary defense of their systems in court by carrying out scientifically based investigations in the course of system development, cf. also ▶ Sect. 40.2.2 in the chapter on safety and risk.

In order to efficiently compare the performance of an automated driving function with that of humans in all instances of the development process, a mathematical description or modeling of the relevant characteristics of human vehicle drivers can be helpful. The **modeling of driver behavior** has a long tradition. In his dissertation published in 1991 [34], Bösch meticulously systematized the approaches, which were already very diverse at that time. In the meantime, numerous other control models have been developed, but very few of them with the aim of serving as a reference for the performance of an automated driving function. This is due to the challenge, already described by Bösch, to determine the parameters of control-engineering human models, which is why so-called human performance modeling (HPM) has so far only dealt with quantitative driver models to a limited extent. Therefore, in the following, exemplary driver models are discussed, in whose development the authors have participated.

With the goal of prospective effectiveness evaluation, Rösener et al. in [35] modeled the performance of driver collectives using analysis of drone data and complementary driving simulator experiments to compare the performance of automated driving functions with this reference. The advantage of the bird's-eye view in ◼ Fig. 46.5 lies in the large image section of approximately 400 m in length, which enables an evaluation of both the tactical and reactive behavior of human drivers. Within the framework of a test in the driving simulator, the behavior of a representative test subject collective can be examined in a statistically robust manner in order to also capture the dispersion of human performance capabilities. The evaluation methodology also takes into account how the frequency of the relevant scenarios changes depending on the penetration of traffic by the respective automated driving function.

A very versatile driver model was presented in [36] and published on GitHub. The motivation of the model was initially to represent the most realistic possible surrounding traffic in traffic simulations. However, it also proved suitable for provoking a traffic situation that is challenging for the automated vehicle by deliberately manipulating the modeled behavior, e.g. as a virtual driver of a disturbing vehicle in a scenario (cf. ◼ Fig. 46.1). Finally, the model can also be used to rep-

resent the tactical and reactive behavior of real driver collectives, so that it can be used as a reference for automated driving functions in offline simulations.

The model described in [36] presented is based on the modular architecture shown in ◼ Fig. 46.10. On the one hand, this results from the levels of the driving task navigation, guidance (tactical behavior) and stabilization (includes reactive behavior), on the other hand from the steps of human information processing, namely perception, cognition and action. It is not necessary to use and parameterize all building blocks to use the model, which distinguishes it from known control-engineering models with numerous parameters to approximate the observed behavior.

The parameters of this human model vary to some extent depending on the driving task. Furthermore, the extent to which there are cultural or market-specific differences should be examined. Suitable processes and methods must therefore be developed for parameterization. In [37] simpler modeling approaches that could be sufficiently validated for defined scenarios were chosen. With regard to parameterization, especially of the tactical behavior, the UAV-based trajectories presented in ▶ Sect. 46.3.4.2 proved to be useful, since the behavior of road users, e.g., in the form of lane changes, can be extracted very well from the aerial images.

Nevertheless, there is a need for further research on the modeling of human performance as a reference, both in terms of the methods used to determine the parameters and the resulting validity, as well as potential differences between cultures and markets.

46.3.6 Implementation Challenges

The scenario-based testing approach attempts to transfer the relevant knowledge about the driving task in the specified ODD into a test catalog in such a way that the test results allow a reliable argumentation about the achieved safety level. This requires representative completeness, i.e., the potential gaps must be able to be estimated as small enough so that the resulting risks remain below the accepted residual risk. This tempts to make the scenario catalog extensive and dense in the choice of parameters. As a result, the economic advantages over pure test driving would be lost, since simulation also costs time and resources. It is therefore not sufficient to deductively develop generic scenarios, as illustrated in the following section.

In contrast, there is the challenge of incompleteness in inductive scenario generation, cf. ▶ Sect. 46.3.6.2. If a valid and presumably thus also comprehensive test scenario catalog is available, the challenge of valid test execution remains. As described in ▶ Chap. 13 (ID 0303), tests can be performed reproducibly on test sites with real vehicles without risking hazards for oc-

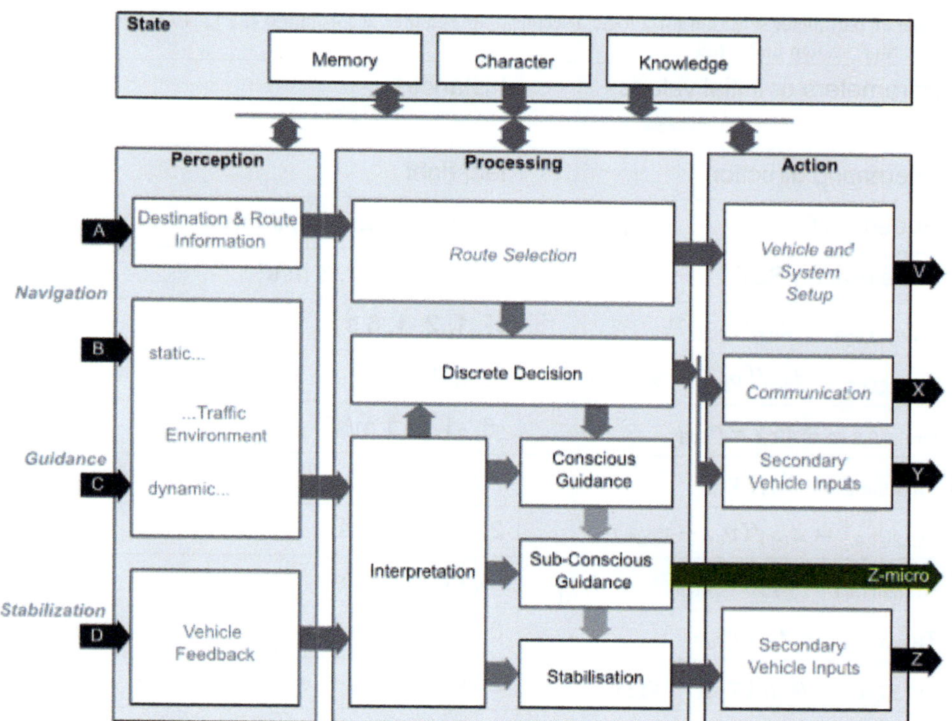

□ Fig. 46.10   Functional architecture of the agent-based human model according to. [36]

cupants and road users. Likewise, the chapter also describes the effort that must be expended for such tests, which is why at best only a small selected subset of the required tests can be performed on the proving ground. Therefore, it is inevitable to perform the test series mainly in virtual test environments, as it has been researched in the PEGASUS project in its main features. Already the tests on proving grounds struggle with the question whether the test is validly chosen and executed. This question arises even more when digital models are used for the vehicle, the functional components, and the environment. This will be discussed in ▶ Sect. 46.3.6.3.

46.3.6.1 Combinatoric Problem of Deductive Generation of Scenarios

Deductive generation of scenarios starts from a typical traffic situation that leads to a functional scenario, cf. ▶ Sect. 46.3.4.1. Here, the example of a situation with a possible lane change for the use case "highway pilot up to higher speeds" will be used. This makes it possible to evaluate whether the automated vehicle guidance system correctly assesses the kinematics of the situation, derives the correct maneuver decision, and performs the maneuver in such a way that no relevant hazard arises.

As clear as this functional scenario may seem, there are many parameters to consider, as the □ Table 46.2 shows.

The choice of parameters on time basis (τ_{HW}, τ_{tC}) avoids a much larger number of combinations with completely unrealistic initial states that would be generated if, for example, independent parameter sets for distance, velocity, and acceleration were used for the object vehicles, which would have to represent approximately the same space as in the □ Table 46.2 would have to be mapped. This alone would save about 3 orders of magnitude! Nevertheless, there are still many combinations which could and must be sorted out, because what does such a number of combinations mean?

Assuming real-time execution of the scenarios with an assumed average duration of 10 s results in a test time of $3.5 \cdot 10^{13}$s $\approx 10^{10}$h $\approx 10^6$a. Assuming a mean distance of 300 m per scenario, this comes to $1.05 \cdot 10^{13}$km, i.e., about 1,000 times more than would have to be assumed for statistical test drives, cf. ▶ Sect. 46.2.2.3. Nevertheless, criticism can still be made that test gaps cannot be excluded. The above parameter sets can be justified as plausible, both in terms of the range of values and the discretization gradation. However, they have representative coverage only if the result space is "flat", i.e., the results of the test do not have a high local sensitivity to the change of the parameters. This is often not the case, especially in collision scenarios with intersecting objects. If the discretization of a crossing scenario is too coarse, it could happen that the collision is excluded from the initial conditions alone, because in one case the crossing

◻ **Table 46.2** Example of parameter selection for a logical scenario lane change according to ◻ Fig. 46.11, parameters for initializing the vehicles assigned in color; SuT: system under test

i		Parameters or initial values	Instances	Parameter number n_i
1		Alternating direction	left, right	2
2		Speed SuT	10, 15, 20, 25, 30, 35 m/s	6
3		Acceleration SuT	-5, -3, -1, 0, 1, 3 m/s²	6
4		$\tau_{HW,0,f,SuT} = d_{0,f}/v_{SuT}$ [11]	½, 1, 2, 4, 6 s	5
5		$\tau_{tC,0,f,SuT} = d_{0,f}/(v_{SuT} - v_{0,f})$ [12]	2, 4, 8, $\propto$ s	4
6		$a_{rel,0,f,\_SuT} = a_{0,f} - a_{SuT}$	-5, -1, 0, 1 m/s²	4
7		$\tau_{HW,Sut,0,r} = d_{0,r}/v_{0,r}$	½, 1, 2, 4, 6 s	5
8		$\tau_{tC,SuT,0,r} = d_{0,r}/(v_{0r} - v_{SuT})$	2, 4, 8, $\propto$, -8 s	5
9		$a_{rel,SuT,0,r} = a_{SuT} - a_{0,r}$	-3, -1, 0, 1 m/s²	4
10		$\tau_{HW,1,f,SuT} = d_{1,f}/v_{SuT}$	0, ½, 1, 2, 4 s	5
11		$\tau_{tC,1,f,SuT} = d_{1,f}/(v_{SuT} - v_{1,f})$	2, 4, 8, $\propto$, -8, -4 s	6
12		$a_{rel,1,f\_Sut} = a_{1,f} - a_{SuT}$	-3, -1, 0, 1 m/s²	4
13		$\tau_{HW,1,r,0,f} = d_{1,r}/v_{1,r}$	½, 1, 2, 4, 6 s	5
14		$\tau_{tC,1,r,1,f} = d_{1,r}/(v_{1,f} - v_{1,r})$	2, 4, 8, $\propto$, -8 s	5
15		$a_{rel,1,r.1,f} = a_{1,r} - a_{S1,f}$	-3, -1, 0, 1, 3 m/s²	5
16		More lanes	1 left, 0, 1 right	3
17		Curve curvature	$\frac{-1}{100}, \frac{-1}{1000}, 0, \frac{1}{1000}, \frac{1}{100}$ $\frac{1}{m}$	5
18		Visibility condition	Without impairment, impairment due to rain/spray, fog	3
19		Brightness	Dark, cloudy bright, sunny	3
20		Pavement grip	High, reduced, low (slippery)	3
		Full factorial combinatorics K	$S_N = \prod_{i=1}^{N} n_i = \prod_{i=1}^{20} n_i$	$3{,}5 \cdot 10^{12}$

would still be in front of the object and in the next parameter value it would already be behind it and thus the ability to avoid collisions would not be put to the test. Statistical methods according to the Monte Carlo principle offer the way out to systematically avoid such discretization artifacts, but generate higher test numbers for it, if the same minimum discretization density is to be guaranteed.

A very common way to reduce the number of combinations is to use the concept "t-wise", cf. [17]. Instead of representing all possibilities (corresponding to N-wise), only $t < N$ parameters are completely combined, namely the t parameters with the highest number of parameters. This results in a saving of the factorization of all combinations not covered under the t parameters, which are listed with Index j in the ascending order of the parameter numbers up to $N - t$ are sorted:

$$\frac{S_t}{S_N} = \prod_{j=1}^{N-t} n_j \tag{46.1}$$

In the above example, with $t = 10$ a reduction by the factor $2.07 \cdot 10^5$ to $S_{10} = 1.7 \cdot 10^7$ would be achieved. As shown in [38], a balance between discretization density and total number of combinations can be achieved with the choice of t a balance between discretization density and total number of combinations can be achieved. With the exemplarily calculated number of S_{10} one now reaches a range which allows a systematic filtering according to exclusion criteria with limited (computational) effort in order to exploit the remaining reduction potential.

46.3.6.2 Incompleteness of the Inductive Generation of Scenarios

The inductive generation of scenarios is based on events and thus on concrete sequences of scenarios that are available from experience, e.g., from accident databases. Based on the assumption that machine-based vehicle guidance also has to face similar challenges as humans, scenarios can be generated from accident reconstruction. The advantage of this approach is its representativeness in practice; the disadvantage is the difficult transferability if, as may be expected, driving behavior differs significantly from that of humans. Furthermore, the accident database lacks information on accidents avoided by active avoidance decisions, e.g., when other road users disregard the right of way.

Field-operational tests or natural-driving studies can partially compensate for this deficiency, but even these have only a small and thus poorly representative number of near-accident cases, thus demonstrating the need for the "cycle of relevant scenarios" described in ▶ Sect. 3.1 for the continuous improvement of this knowledge.

Another challenge of inductive scenario generation is not to overestimate the significance of accidents that have occurred, because in a certain sense each accident is always unique and will never occur exactly again. Therefore, in a scenario generated from an accident sequence, the parameters must also be varied in a small range, which is then equivalent to a "local" deductive generation, albeit with previously strongly limited parameter space.

46.3.6.3 Validity of Virtual Testing

In the following, virtual tests are all tests that represent essential components of the system-under-test with software and hardware models and are based on an at least partially digitally represented environmental simulation. The range of virtual tests thus includes model-in-the-loop (MiL), software-in-the-loop (SiL), hardware-in-the-loop (HiL) and vehicle-in-the-loop (ViL), whereby the former can only be used for testing in the early stages of development.

The ultimately always high number of necessary tests forces the test execution as virtual tests, on the one hand because of the high possible test automation level and on the other hand because of the higher parallelization potential. Furthermore, they have the advantage of easier reproducibility compared to real test drives. Virtual tests can offer further advantages in that faults can be injected or the robustness to component or functional tolerances can be tested without having to make complex changes to the vehicle and its

equipment. The software in the system-under-test can be taken over for a virtual test largely unchanged. It is therefore easier to subject it to different test stimuli in the virtual environment than in the real environment.

However, the statements of the virtual tests can only be used for validation if the underlying digital models are valid for the respective test objective. Since models always have deviations from reality, it must be ensured that the result of the test with these models is not changed, let alone that the passing of the test is caused by the model deviation. An error in the other direction (false failure) is not critical for the safety statement, but can result in great expense, e.g., by the fact that the test must be repeated in real tests. Therefore, a validity statement must be determined prior to test design. The following questions should be answered:

- What is the test objective?
- Which phenomena must be correctly mapped or modeled for this test objective?
- According to which test criteria is the test success determined?
- What deviations from model to reality affect the evaluation with these criteria?
- What series of tests are required for validation to identify possible deviations?
- How sensitive are the test criteria to model deviations?
- Can the possible deviations be mapped by targeted parameter scattering without questioning the test result in case of failure?

The last point addresses the problem of false failures mentioned above.

The modeling of the vehicle dynamics of a motor vehicle has now reached a high level of maturity and abundant experience about the validity ranges of the models, although a well-founded and quantitative validity statement is often not available for this area either, cf. [39, 40].

Thus, the major challenge of modeling for automated driving concerns the perceptual models. It is true that the software functions for filtering, tracking, or sensor data fusion can be largely integrated one-to-one into the virtual test environment. Much more difficult is the modeling of the raw signals, e.g., the pixel brightness information of the camera, for lidar the time spectra, and for radar the frequency spectra. Although techniques based on ray tracing methods, among others, have been established for this purpose, they do not succeed in comprehensively representing all potentially relevant phenomena. To structure the multitude of phenomena and their chains of effects and causes alone is an enormous challenge, which is being addressed with a method PERCOLLECT (Perception Sensor Collaborative Effect and Cause Tree) [41] for camera, radar, lidar and ultrasonic sensor technology and has

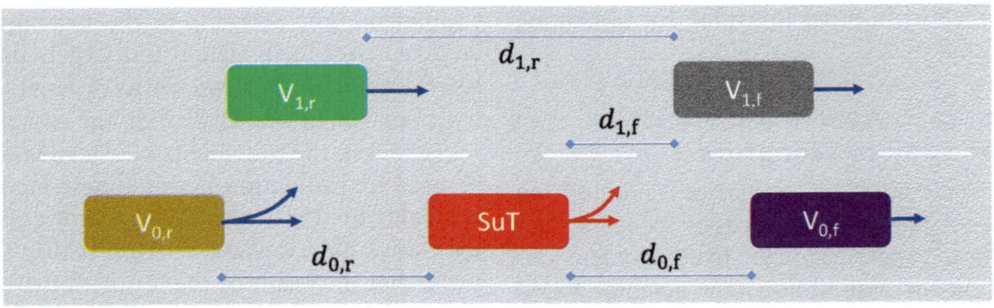

Fig. 46.11 Lane change scenario with the objects involved (blue) and the subject vehicle (red)

been published on GitHub. Further, [41] also proposes a method for finding out which of the underlying effects and causes of the phenomena should be considered in the modeling (Cause, Effect, and Phenomenon Relevance Analysis (CEPRA) for Sensor Modeling).

For many test objectives of automated driving, however, less stringent requirements are placed on fidelity, e.g., for testing behavioral algorithms. Therefore, perceptual models are used that "abbreviate" the signal flow from the ground truth environmental model of the simulation environment to the environmental representation as the result of the perceptual chain. Instead of the raw signals mentioned above, intermediate quantities such as detections or straight object data are generated. The loss of fidelity is compensated by the significantly reduced computation time and simpler parameterization. Despite immense progress in modeling in recent years, we are still far from a target state in which the model specifies which properties the subsequently developed component must have in order to be considered a valid component, just as it is the case today in the field of mechanical products that the CAD component specifies the scale. For some years to come, modeling will lag behind development, and the model will have to be repeatedly revalidated on the real component.

46.3.7 Approaches to the Implementation Challenges

46.3.7.1 Classification and Scenario Generation in Analogy to Accident Reconstruction

One challenge of the approach of extracting relevant scenarios exclusively from actually observed events such as accidents or near-collisions lies in its incompleteness. Only after the utilization of several billion kilometers one could one argue that at least regularly occurring events were covered by the extracted scenarios.

One in [42] proposed approach is to identify and discretize relevant characteristics on each level of the 5-level model in order to then cover the theoretically possible traffic situation space sufficiently well by combining all factors and their gradations. This leads on the one hand to an extremely high number of concrete scenarios, but also to the fact that the resulting situations have to be classified afterwards.

In order to avoid the necessity of classifying the generated sequences ex-post in the form of scenarios, an alternative approach has been conceived, based on **accident reconstruction** [43]. In the analysis of an accident, **backward calculations** are performed from the accident event to determine possible initial positions of the vehicles. Since the automated vehicle, unlike a vehicle involved in an accident, is not necessarily stationary, print fat of the vehicles involved in the scenario are considered instead of absolute trajectories. In **Fig. 46.12** a longitudinal traffic scenario is shown in analogy to the longitudinal traffic accident type, and in analogy to the collision type, a passenger car scenario is shown in the absence of a reaction of the automated driving function in the rear-end collision type under 0° with 25% overlap and damage to the front of the vehicle.

Analogous to the accident reconstruction, the question arises as to how the potential opponent of the accident drawn in **Fig. 46.12** in red could have moved relative to the automated vehicle before the impending collision. **Figure 46.13** shows some of the relative movements of the potential collision partner that are possible in principle. Accordingly, it could be a vehicle cutting in from the left or right as well as a decelerating vehicle in front.

The requirements for the perception and cognition of the automated vehicle are clearly different for the paths shown as examples. Therefore, it seems obvious to focus on those trajectories, i.e., paths with relative speeds, which should be particularly challenging due to the selected sensor concept, in the course of the development-accompanying validation. For this purpose, the degree of difficulty for perception can be systematically increased by adding sight-obscuring objects. In the same way, the demands on cognition and action can be further increased by considering action-constraining objects.

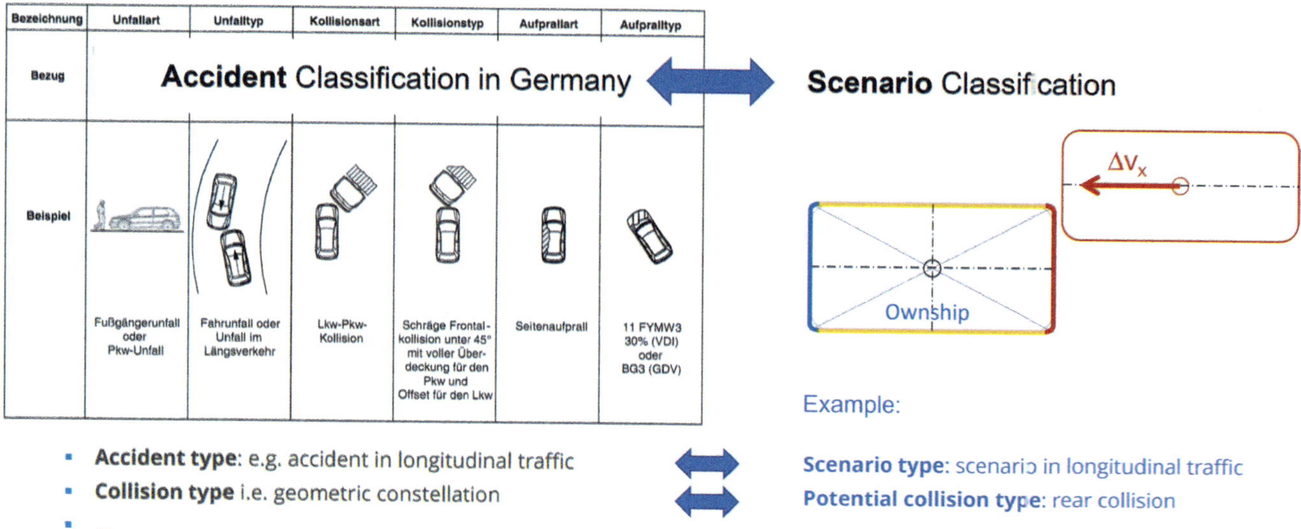

Fig. 46.12 Classification of a scenario analogous to the established accident data analysis

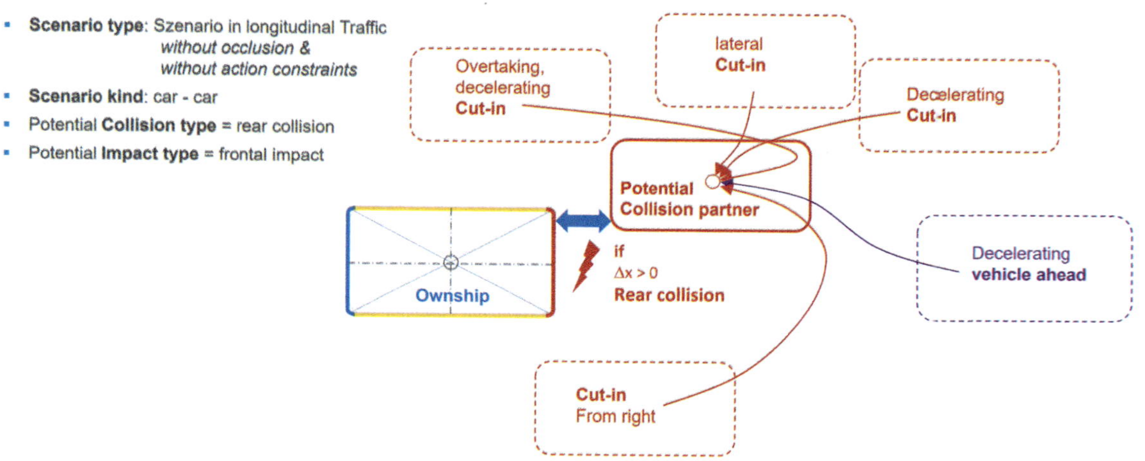

Fig. 46.13 Exemplary "reconstruction" of the scenario in longitudinal traffic with possible paths of the potential collision partner

This consideration, which is based on accident reconstruction, also leads to the finding documented in [37] that no fundamentally new scenarios will arise as a result of automated driving: neither the potential type of collision deviates from the known accident event, nor the space of possible trajectories along which the potential collision object could have moved relatively towards the automated vehicle.

This results in the following scenarios for driving on a freeway as described in [28], cf. left graph in **Fig. 46.14**. The abstracted process for deriving these scenarios is shown on the right in the same figure [43].

On the basis of this reasoning, a clear distinction can also be made between the **effectiveness** of the function to be demonstrated and the **automation risk** to be minimized. **Figure 46.15** asks in the first step who or what induced the need for an accident-preventing action—if it is the highly automated function (HAF) itself, then this can be understood as an original automation risk, analogous to the **accident causer** in accident reconstruction. If, on the other hand, it is not the function but other road users or a factor at one of the other levels of the scenario that constitutes the cause of the need for action, the **effectiveness** of the automated driving function is assessed. The latter was the focus of the PEGASUS project.

The differentiation contained in **Fig. 46.15** between the automated driving function as a potential cause of an accident and the "challenged entitiy", which is the need for action caused by other road users or factors on another level of the 6LM, is also reflected in the safety argumentation of the Japanese car manu-

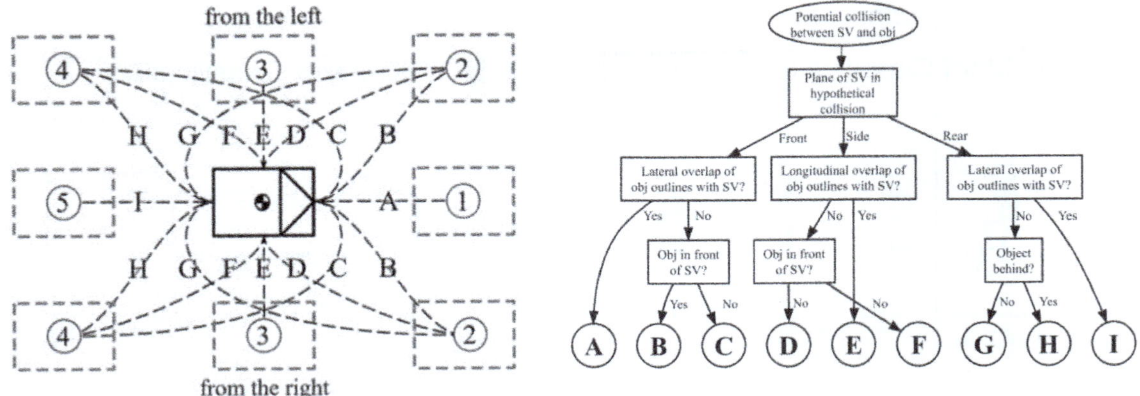

◻ Fig. 46.14　Basic scenarios on a highway represented by relative paths and initial positions of potential collision partners

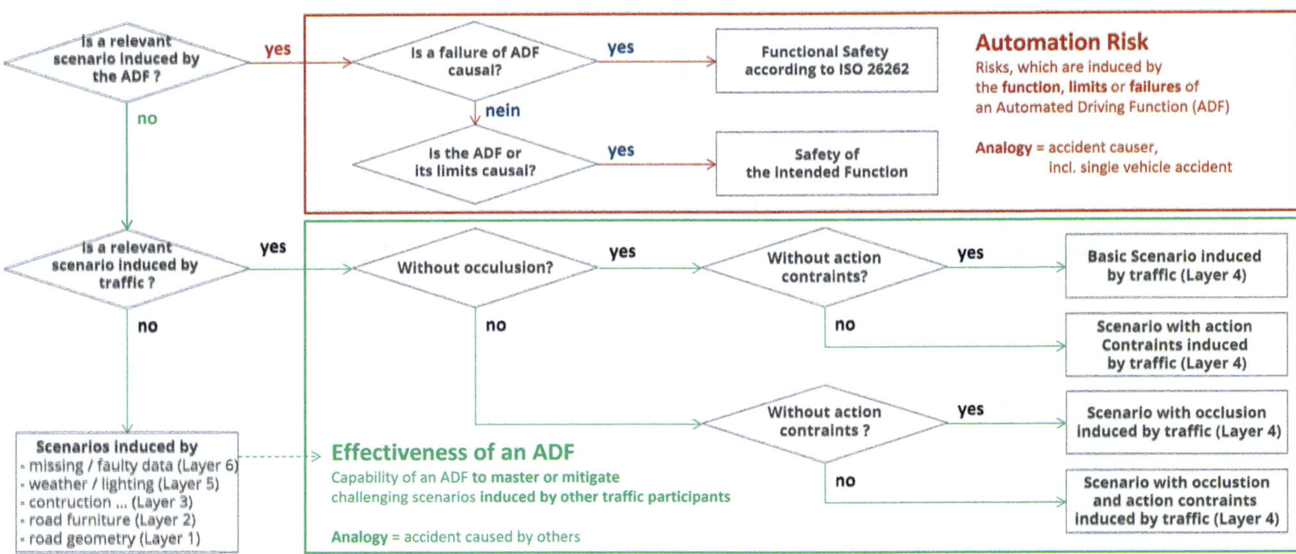

◻ Fig. 46.15　Delineation of effectiveness and automation risk based on scenario classification

facturers [44]. This makes the claim that an automated driving function must not cause any unjustifiable risk, thus focusing on limiting the automation risk. It seems obvious that both approaches or perspectives are necessary. This also opens up the possibility of comparing the **benefit** of a function in terms of effectiveness with the **risk potentially introduced by** this function.

The approach presented here can be used to generate specific scenarios that place particular demands on the individual system components. The approach also pursued in Japan of defining dedicated scenarios for machine perception, cognition and reaction, can also be understood as a **decomposition** of the automated driving function, and has apparent advantages.

46.3.7.2　Functional Decomposition

A high potential for controlling the test variety is the decomposition of the chain of effects from the information acquisition process to the action. Thus, instead of performing a series of tests with complete system-under-test under multiple variation, so-called particular tests are performed along the processing chain, where only the parameters affecting these subsystems need to be varied. A partitioning, developed from the accident analysis scheme, is presented in [38], cf. ◻ Fig. 46.16. Example calculations on different logical scenarios prove a reduction potential of about one order of magnitude, although the difficulty should not be concealed that the pass and fail criteria of particular

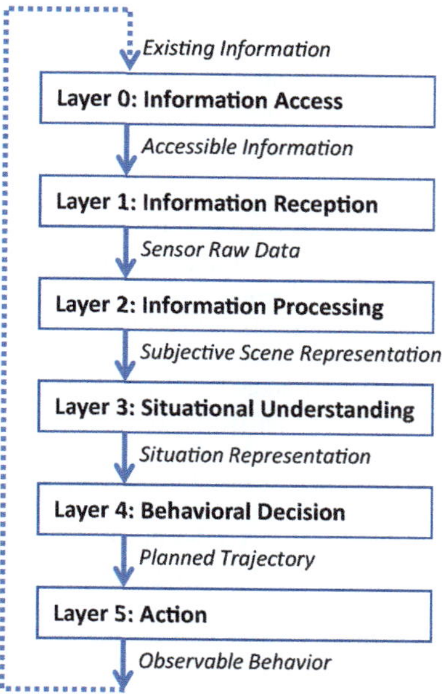

Fig. 46.16 Functional decomposition of the processing chain of automated driving according to [38]

tests have to be proved as sufficient for the prediction of pass or fail in full tests. In return, it has the advantage that the cause of failure can be found more easily in addition to the test result.

46.4 Conclusion

For automated driving, there are various approaches for assuring safety. What they all have in common is that they lack the prior knowledge from many years of practice, i.e., the statement of how good the test coverage is with these approaches. Test drives for statistical proof of safety turn out to be too costly and time consuming. Feedback from field use, whether via silent testing in predecessor systems or from field observation during a "trial" introduction, can provide much of this missing prior knowledge. This needs to be further developed with today's knowledge of relevant scenarios into efficient scenario-based test concepts so that safety validation does not prevent the introduction of automated driving functions.

Analogous to the established continuous collection and analysis of accidents, the aim is to systematically record critical traffic situations and convert them into scenarios that can be used in various simulation and evaluation methods along the development process. The foundations for this approach have been laid by the PEGASUS project and similar projects, such as SAKURA in Japan. However, operationalization, harmonization and standardization are not yet complete. Since simulation plays an essential role in the validation of automated driving functions, the validity of simula-

tion models is of great importance. Efficient simulation of the virtually infinite number of scenarios requires optimized test variation and decomposition concepts, which are currently subjects of research and development.

The evaluation of the performance of a driving function requires a suitable metric or reference. Although humans are considered as a reference in many countries, there are so far only few approaches to represent the tactical and reactive behavior of a representative collective in such a way that it is already possible in simulation to directly compare the performance of an automated driving function with that of humans.

Finally, the release of an automated driving function requires a robust safety argumentation that is based both on findings from simulation and on results from driving tests on closed test sites and in real road traffic. The development and standardization of such safety argumentation is the subject of ongoing projects and international discussions, which is why these could not be addressed in depth in this chapter.

46.5 Outlook

As explained in previous sections, many challenges for safeguarding still await an appropriate solution. In most cases, there is a balancing act between, on the one hand, a very high level of effort and, on the other hand, the remaining uncertainty as to whether the prognosis is valid for safe operation. There is simply a lack of experience available in other product areas. From the authors' point of view, this leads to three conclusions:

A one-off product qualification carried out before introducing an automated driving function into the market cannot guarantee the necessary safety. A maturing phase of the product in operation, which can be compared to the growth in experience of a human being, is required. The introduction strategy has to reflect this finding by having an introduction phase on "trial" and by successively determining the achieved safety in more detail during this phase. A method for this is presented in [45].

The operational design domain (ODD) must reflect this uncertainty about the achieved level of safety by designing the boundary conditions for the operational design domain to mitigate hazards and avoid potential excessive demands on the automated vehicle guidance system. Thereafter, a gradual expansion of the operational envelope may take place, provided that the achieved level of knowledge about safety in the proven operational envelope is sufficiently high.

The maturation of the product (automatic vehicle guidance) must be accompanied by a maturation of the associated safeguarding process. The quality of the safeguarding process can be made measurable through constant feedback of experiences in use. In particular, surprising events are helpful in revealing gaps or weaknesses in the previous process, especially in the test area. If an undesirable behavior occurs, it should be clarified,

— whether this has been previously considered in the development and testing process,

— if not, this case must at least be added into the test suite,

— if known, it should be questioned why the occurrence of this event was not prevented, e.g., whether the test had deficiencies in its execution.

The trend of the surprise rate per operational distance, which is to be determined continuously, can be used to extrapolate how large the remaining risk is, although this also requires assumptions that still have to be validated, cf. [46]. Qualification of the assurance process instead of product qualification has long been established in the field of software development, but does not (yet) play a significant role for approval and release in the automotive sector. The test cases discussed today for an approval test appear to be of a symbolic nature from a statistical point of view.

All three conclusions imply a protracted period of time, which is needed for responsible introduction and expansion. From this point of view, it becomes understandable why the optimistic, in parts capital market-oriented promises of the last decade could not be kept so far.

References

1. Weitzel, D.A., Winner, H., Cao, P., Geyer, S., Lotz, F., Sefati, M.: In: Bundesanstalt für Straßenwesen (Hrsg.) Absicherungsstrategien für Fahrerassistenzsysteme mit Umfeldwahrnehmung (translated: Safeguarding strategies for driver assistance systems with environment perception): Bericht zum Forschungsprojekt FE 82.0546/2012. 1. Auflage. Bergisch Gladbach, Deutschland: Wirtschaftsverlag NW Verlag für neue Wissenschaft, 2014 (Berichte der Bundesanstalt für Straßenwesen 98) (2014)

2. Bagschik, G.: Systematischer Einsatz von Szenarien für die Absicherung automatisierter Fahrzeuge am Beispiel deutscher Autobahnen, Dissertation Technische Universität Braunschweig (2022)

3. Winner, H., Merkel, N.: Mode-Confusion und Inkompatibilitäten in der Migrationsphase des automatisierten Fahrens (translated: Mode-confusion and incompatibilities in the migration phase of automated driving). In: Winner H, Bruder R (Hrsg.) (Wie) wollen wir automatisiert fahren? 8. Darmstädter Kolloquium Mensch&Fahrzeug, Darmstadt, 7.+8. März 2017, pp. 53–64 (2017). ► https://tuprints.ulb.tu-darmstadt.de/5672/1/Mensch%20und%20Fahrzeug%20Tagungsband%202017.pdf

4. Fach, M., Baumann, F., Breuer, J., May, A.: Bewertung der Beherrschbarkeit von Aktiven Sicherheits- und Fahrerassistenzsystemen an den Funktionsgrenzen (translated: Assessment of the controllability of active safety and driver assistance systems at the functional limits). In: 26. VDI/VW-Gemeinschaftstagung Fahrerassistenz und Integrierte Sicherheit, 6./7. Oktober 2010 in Wolfsburg (2010)

5. DESTATIS, Federal Statistical Office: ► https://www.destatis.de/DE/Themen/Gesellschaft-Umwelt/Verkehrsunfaelle/Publikationen/Downloads-Verkehrsunfaelle/verkehrsunfaelle-zeitreihen-pdf-5462403.pdf?__blob=publicationFile. Accessed January 2022

6. Winner, H.: Quo vadis, ADAS? In: Winner, H., Hakuli, S., Lotz, F., Singer, C. (eds.) Handbook Driver Assistance Systems, pp. 1557–1584. Springer Reference (2016) ► https://doi.org/10.1007/978-3-319-12352-3 © Springer Switzerland 2015

7. Wachenfeld, W.: How Stochastic can Help to Introduce Automated Driving, Dissertation Technische Universität Darmstadt (2017). ► http://tuprints.ulb.tu-darmstadt.de/5949

8. Wachenfeld, W., Winner, H.: Virtual assessment of automation in field operation—a new runtime validation method. In: 10. Workshop Fahrerassistenzsysteme, pp. 161–170 (2015). ISBN: 978-3-00-050746-5. ► https://www.uni-das.de/images/pdf/fas-workshop/2015/FAS2015_Tagungsband.pdf. Accessed January 2022

9. Tesla: What is Shadow Mode Tesla Autonomy (2019). ► https://www.youtube.com/watch?v=SAceTxSelTI. Accessed Jan 2022

10. Wang, C.: Silent testing for safety validation of automated driving in field operation. Dissertation Technical University of Darmstadt (2021). ► https://doi.org/10.26083/tuprints-00019819

11. Eckstein, L.: Entwicklung und Überprüfung eines Bedienkonzepts und von Algorithmen zum Fahren eines Kraftfahrzeugs mit aktiven Sidesticks (translated: Development and verification of an operating concept and algorithms for driving a motor vehicle with active sidesticks) Dissertation Universität Stuttgart, VDI Fortschritt-Berichte Reihe 12, No. 471 (2000)

12. Buerkle, H., Feese, J.: 50 Years of Mercedes-Benz accident research—ready for the next level. In: The 26th International Technical Conference on the Enhanced Safety of Vehicles (ESV) Eindhoven, Netherlands, June 10–13, 2019, Paper ID: 19-0178 (2019)

13. German In-Depth Accident Study (GIDAS): ► https://www.bast.de/EN/Automotive_Engineering/Subjects/gidas.html

14. Frank, P., Scheible, G., Reichelt, W.: Verfahren und Vorrichtung zum Abbremsen eines Fahrzeugs (translated: Method and device for braking a vehicle) Patent disclosure document DE19822860A (1998)

15. Unselt, T., Breuer, J., Eckstein, L.: Fußgängerschutz durch Bremsassistenz (translated: Pedestrian protection through brake assistance). TÜV Akademie GmbH: Conference Active safety through driver assistance. Garching (2004)

16. Dupuis M, Grezlikowski H (2006) OpenDRIVE - An Open Standard for the Description of Roads in Driving Simulations, in: M. Dupuis: Proceedings of the Driving Simulation Conference DSC Europe 2006 , p. 25–35.

17. All-Electronics: ASAM adopts standards for highly automated driving (2018). ► https://www.all-electronics.de/automotive-transportation/asam-open-drive-automatisiertes-fahren.html. Accessed 12 Sept 2018

18. Neale, V.L., Dingus, T.A., Klauer, S.G., Sudweeks, J., Goodman, M.: An Overview of the 100 Car Naturalistic Study and Findings National Highway Traffic Safety Administration United States Paper Number 05-0400 (2005)

19. Benmimoun, M., Zlocki, A., Ljung Aust, M., Faber, F., Saint Pierre, G.: Safety analysis method for assessing the impacts of advanced driver assistance systems within the European large scale field test "euroFOT". In: 8th ITS European Congress, Lyon, 06–09 June 2011 (2011)

20. Eckstein L, Zlocki A (2013) Safety Potential of ADAS - Combined Methods for an Effective Evaluation, 23rd International Technical Conference on the Enhanced Safety of Vehicles (ESV), Seoul, Republic of Korea, 27–30 May 2013.

21. PEGASUS Project Office: Website (2016). ► https://www.pega-susprojekt.de/de/about-PEGASUS. Accessed April 2022

22. Barthel, R.: Ergonomische Betrachtung von Lenkassistenzfunktionen zur Fahrerunterstützung in kritischen Fahrsituationen (translated: Ergonomic Consideration of Steering Assistance Functions for Driver Support in Critical Driving Situations). Dissertation Technical University of Darmstadt (2003)

23. PEGASUS Final Report Overall Project (2020). ► https://www.pegasusprojekt.de/de/lectures-publications

24. Krajewski, R., Bock, J., Kloeker, L., Eckstein, L.: The highD Dataset: a drone dataset of naturalistic vehicle trajectories on german highways for validation of highly automated driving systems. In: 2018 21st International Conference on Intelligent Transportation Systems (ITSC), 2018, pp. 2118–2125 (2018). ► https://doi.org/10.1109/ITSC.2018.8569552.

25. Geyer, S., Baltzer, M., Frank, B., Hakuli, S., Kauer, M., Kienle, M., Meier, S., Weißgerber, T., Bengler, K., Bruder, R., Flemisch, F., Winner, H.: Concept and development of a unified ontology for generating test and use-case catalogues for assisted and automated vehicle guidance. In: IET ITS, 2014, vol. 8, no. 3, pp. 183–189 (2014)

26. Schuldt, F., Saust, F., Lichte, B., Maurer, M., Scholz, S.: Effiziente systematische Testgenerierung für Fahrerassistenzsysteme in virtuellen Umgebungen (translated: Efficient systematic test generation for driver assistance systems in virtual environments). Institute of Control Engineering, TU Braunschweig (2013). ► https://publikationsserver.tu-braunschweig.de/receive/dbbs_mods_00052570

27. Bagschik, G., Menzel, T., Maurer, M.: Ontology based scene creation for the development of automated vehicles. In: 2018 IEEE Intelligent Vehicle Symposium, 2018, pp. 1813–1820 (2018)

28. Bock, J., Krajewski, R., Eckstein, L., Klimke, J., Sauerbier, J., Zlocki, A.: Data Basis for Scenario-Based Validation of HAD on Highways. Aachen Colloquium Automobile and Engine Technology, Aachen (2018)

29. Eckstein, L.: The PEGASUS Project—Results and Future Prospects. Plenary presentation at Automated Vehicle Symposium (AVS) 16th July 2019, San Francisco (2019)

30. Scholtes, M., Westhofen, L., Turner, L.R., Lotto, K., Schuldes, M., Weber, H., Wagener, N., Neurohr, C., Bollmann, H., Körtke, F., Hiller, J.H.M., Bock, J., Eckstein, L.: 6-layer model for a structured description and categorization of urban traffic and environment (2020). arXiv: 2012.06319 [cs.OH]

31. Kloeker, L., Kloeker, A., Thomsen, F., Erraji, A., Eckstein, L.: How to build a highly accurate digital twin - intelligent infrastructure in the corridor for new mobility. Aachen Colloq. Sustain. Mobil. 4–6(10), 2021 (2021)

32. Scanlon, J.M., Kusano, K.D., Daniel, T., Alderson, C.O.A., Victor, T.: Waymo simulated driving behavior in reconstructed fatal crashes within an autonomous vehicle operating domain. Accid. Anal. Prev. (2021). ► https://doi.org/10.1016/j.aap.2021.106454

33. Taniguchi, S., Ozawa, K., Kitahara, E., et al.: Automated Driving Safety Evaluation Framework Ver.1.0. Japan Automobile Manufacturers Association, Inc (JAMA), AD Safety Assurance Expert Group (2020). ► https://www.jama.or.jp/english/publications/environmental.html. Accessed 13 May 2022

34. Bösch, P.: Der Fahrer als Regler (translated: The driver as a controller). Dissertation Vienna University of Technology (1991). ► https://repositum.tuwien.at/handle/20.500.12708/13454

35. Rösener, C., Sauerbier, J., Zlocki, A., Eckstein, L., Hennecke, F., Kemper, D., Oeser, M.: Potenzieller gesellschaftlicher Nutzen durch zunehmende Fahrzeugautomatisierung (translated: Potential societal benefits from increasing vehicle automation), Berichte der Bundesanstalt für Straßenwesen, Reihe F: Fahrzeugtechnik (128) (2019). ► https://bast.opus.hbz-nrw.de/files/2140/F128_barrierefr+ELBA+PDF+neu.pdf

36. Becker, D., Klimke, J., Eckstein, L.: System Design of an Agent Model for the Closed-Loop Simulation of Relevant Scenarios in the Development of ADS. ATZelectronics, vol. 16 (2021)

37. Rösener, C.: A traffic-based method for safety impact assessment of road vehicle automation. Dissertation RWTH Aachen University, Aachen, fka GmbH (2020). ISBN: 978-3-946019-39-8. ► https://publications.rwth-aachen.de/record/801589/files/801589.pdf

38. Amersbach, C.T.: Functional decomposition approach—reducing the safety validation effort for highly automated driving. Dissertation, Technical University of Darmstadt (2020). ► https://doi.org/10.25534/tuprints-00011520

39. Viehof, M.: Objektive Qualitätsbewertung von Fahrdynamiksimulationen durch statistische Validierung (translated: Objective quality assessment of vehicle dynamics simulations by statistical validation). Dissertation Technical University of Darmstadt (2018). ► http://tuprints.ulb.tu-darmstadt.de/7457

40. Danquah, B., Riedmaier, S., Lienkamp, M.: Potential of statistical model verification, validation and uncertainty quantification in automotive vehicle dynamics simulations: a review. In: Vehicle System Dynamics, pp. 1–30 (2020). ► https://doi.org/10.1080/00423114.2020.1854317

41. Linnhoff, C., Rosenberger, P., Schmidt, S., Elster, L., Stark, R., Winner, H.: Towards serious perception sensor simulation for safety validation of automated driving—a collaborative method to specify sensor models. In: 2021 IEEE International Intelligent Transportation Systems Conference (ITSC), 2021, pp. 2688–2695 (2021). ► https://doi.org/10.1109/ITSC48978.2021.9564661

42. Schuldt, F.: Ein Beitrag für den methodischen Test von automatisierten Fahrfunktionen mit Hilfe von virtuellen Umgebungen (translated: contribution for the methodical testing of automated driving functions using virtual environments). Diessertation Technical University of Braunschweig (2017). ► https://publikationsserver.tu-braunschweig.de/servlets/MCRFileNodeServlet/dbbs_derivate_00043536/Diss_Schuldt_Fabian.pdf

43. Weber, H., Bock, J., Klimke, J., Roesener, C., Hiller, J., Krajewski, R., Zlocki, A., Eckstein, L.: A framework for definition of logical scenarios for safety assurance of automated driving. In: Traffic Injury Prevention (2019). ISSN: 1538-9588

44. Antona-Makoshi, J., Uchida, N., Kitahara, E., Ozawa, K., Tanaguchi, S.: A Safety assurance process for automated driving systems, Paper ID AP-TP2329. In: 26th ITS World Congress, Singapore (2019)

45. Wachenfeld, W.: How stochastic can help to introduce automated driving. Ph.D. thesis, Darmstadt University of Technology (2017). ► http://tuprints.ulb.tu-darmstadt.de/5949

46. Winner, H., Wachenfeld, W., Junietz, P.: Validation and introduction of automated driving. In: Winner, H., Prokop, G., Maurer, M. (eds.) Automotive Systems Engineering II, pp. 177–196. Springer International Publishing, Cham (2018). ISBN: 978-3-319-61605-6

47. Eckstein, L.: Safety Assurance—Developing and Assessing Automated Driving, in Automated Vehicles Symposium (AVS)—Breakout Session 6: Safety Assurance, San Francisco 2016 (2016)

48. Grindal, M., Offutt, J., Andler, S.F.: Combination testing strategies: a survey, In: Software Testing, Verification and Reliability, vol. 3, Issues 15, pp. 167–199 (2005)

49. Rosenberger, P., Linnhoff, C.: PerCollECT—Perception Sensor Collaborative Effect and Cause Tree. GitHub Repository (2022). ► https://github.com/PerCollECT

50. Krajewski, R., Bock, J., Kloeker, L., Eckstein, L.: The highD Dataset: a drone dataset of naturalistic vehicle trajectories on German highways for validation of highly automated driving systems. In: 2018 21st International Conference on Intelligent Transportation Systems (ITSC), pp. 2118–2125 (2018). ► https://doi.org/10.1109/ITSC.2018.8569552

Open Access This chapter is licensed under the terms of the Creative Commons Attribution-NonCommercial-NoDerivatives 4.0 International License (▶ http://creativecommons.org/licenses/by-nc-nd/4.0/), which permits any noncommercial use, sharing, distribution and reproduction in any medium or format, as long as you give appropriate credit to the original author(s) and the source, provide a link to the Creative Commons license and indicate if you modified the licensed material. You do not have permission under this license to share adapted material derived from this chapter or parts of it.

The images or other third party material in this chapter are included in the chapter's Creative Commons license, unless indicated otherwise in a credit line to the material. If material is not included in the chapter's Creative Commons license and your intended use is not permitted by statutory regulation or exceeds the permitted use, you will need to obtain permission directly from the copyright holder.

Maintainability and Updateability of ADAS Software

Mischa Möstl, Johannes Schlatow, Kai-Björn Gemlau, and Rolf Ernst

Contents

© The Author(s) 2026
H. Winner et al. (eds.), *Handbook Assisted and Automated Driving*,
https://doi.org/10.1007/978-3-658-45276-6_47

47.1 Introduction

Software is one of the central building blocks of every modern vehicle. While this trend already began in classic vehicle technology, in the field of autonomous driving (Level 4+) almost all functionality is implemented in software. The resulting systems come with new requirements, with flexibility, in particular the ability to change software after deployment, as an important requirement. At the same time, supposedly outdated core aspects of vehicle development such as functional safety and the closely related topic of real-time capability are not losing their importance at all. On the contrary, the absence of a human driver in level 4+ autonomous driving means that correct software timing is indispensable for safe operation, and any unsafe behavior must at least be detected (fail-safe) or even corrected (fail-operational). Thus, while updates often address functional requirements, they are inextricably linked to non-functional requirements.

The need to update software in the vehicle has different causes and also occurs in different degrees of difficulty. First, there are updates of existing software elements. These include, for example, function and security patches, which are intended to ensure continued software operation. The complexity here is initially limited, since the previously running system is already known and does not change its functionality.

Upgrades, on the other hand, add new functions to an existing system and are typically intended to generate additional revenue for the vehicle manufacturer. In the simplest case, this may be an optional convenience feature that the customer can unlock through a pay-to-use model. Due to the importance of software, however, vehicles are increasingly developed as universal hardware/software platforms. Those platforms shall allow adding functionality, which does not yet exist at market launch. In the field of automated driving, upgrades play a special role in the Operational Design Domain (ODD) concept. The ODD specifies operating conditions for which a vehicle was developed and under which safe operation can be ensured [4]. These include intended traffic situations (e.g., highway or downtown traffic), but also environmental conditions such as weather conditions (temperature, rain or snow, lighting conditions) or road surface characteristics (condition and presence of pavement markings, paved roads, dirt roads). Outside the ODD, operation is either not possible, or only possible under specific restrictions (e.g., greatly reduced speed). A vehicle can thus initially be developed for a limited ODD, with the models required for operation being part of the initial software state. Upgrades allow to extend the range of vehicle applications at a later point in time, and to import optimized models for the operation of the vehicles. For example, a fleet of vehicles could be optimized for operation in a specific major city.

Further steps toward flexibility include automated variant management and virtualized or multi-tenant platforms. While, in the first case, a manufacturer wants to develop functionality only once in order to be able to use it on different vehicle series and configurations as independently of the platform as possible, the second case aims to provide a uniform software platform for several customers. In a variant-rich system, a function upgrade must therefore be rolled out to many different vehicles, while an upgrade of a virtualization platform potentially affects many customers.

Overall, update intervals are shortened compared to today's vehicles. In consequence, upgrades are no longer installed during workshop visits but are rolled out "over-the-air". Vehicles are equipped with a mobile communications interface by means of which new software can be made available, or installed directly, by the vehicle manufacturer. Another step is live upgrading, which changes the system at runtime. As safe vehicle operation must be guaranteed even while the software is being updated, live upgrades place the highest demands with respect to safety.

A major challenge arises from the growing system complexity of hardware and software. In order to meet the performance requirements of data-intensive algorithms, vehicles must provide hardware/software platforms that were previously used in high-performance computing. They include heterogeneous multicore processors with high-performance memory architectures, including caches and virtualized address spaces. To provide the necessary bandwidths for communication, Ethernet-based backbones are integrated into vehicles, while the vehicles are interconnected with their environment using wireless communication.

Non-functional timing requirements increase the effort for verification and validation. Because of the low timing predictability of high-performance hardware architectures, it is difficult to predict the effect of an update on the timing behavior of an entire system. Since formal analysis methods for such platforms provide too pessimistic results and meaningful tests are very costly, the costs for updates increase. The situation is even more problematic when variants are taken into account, which lead to an explosion in complexity for the required test cases, if the procedure remains unchanged. Updates therefore become an integration problem with respect to non-functional requirements and testing or verification costs can no longer be considered as one-off development costs. Popular agile development methods are a particular challenge, as they usually produce new software versions more frequently.

In order to enable a high degree of flexibility with regard to subsequent updates, update effects, their test and verification must be taken into account from the outset, in the development process as well as in the system architecture. This includes, among other things:

- uniform interfaces for the abstraction of a concrete hardware/software platform, such that, once created, a software component can be used on many platforms.
- a well-defined representation of updates and their boundary conditions (software manifests).
- a certain degree of platform-independent software timing (e.g., through programming with timing constructs).
- runtime mechanisms for monitoring and safeguarding the timing behavior of the software.

Maintenance and update of Advanced Driver Assistance Systems (ADAS) will play a special role in the future development of vehicle systems, insofar as high software complexity and dynamics, with extensive requirements for functional safety, meet an equally growing complexity of the hardware and software architecture. This chapter will therefore focus on emerging non-functional dependencies and their control.

After an overview of the state of the art in the following section, proposals for the integration and verification of ADAS will be developed that draw on existing software standards and meet the above requirements. The methods presented have been developed in various projects, and a good part of them have already been adopted in standardization, such as the concept of Logical Execution Time (LET), which is now used in series development. The proposals should therefore be seen as a snapshot of an active method development. The chapter closes with a conclusion.

47.2   State of the Art

The state of the art for basic software, which includes operating systems and communication stacks, is largely covered by the family of AUTOSAR standards [2]. They are divided into three groups: The first group corresponds to the original AUTOSAR standard and, for the first time, provided a uniform basis for distributed software systems in vehicles. In this standard, all components are statically configured at design time. This part of AUTOSAR is now known as the *AUTOSAR Classic Platform*. It is in continuous use, primarily for applications with high timing and functional safety requirements, and for classic functions implemented on microcontrollers. The *AUTOSAR Adaptive Platform* has been added, which is particularly suitable for complex software with high demands on computing power. This includes ADAS of newer generations, that e.g., provide longitudinal and/or lateral vehicle guidance or process a multitude of sensor data. Common elements are described in the *AUTOSAR Foundation*. To realize maintainability and updateability of AUTOSAR software in the vehicle, AUTOSAR provides different methods

for the two platforms, Classic and Adaptive. They are described in the following section. For both platforms, application software is only installed—monitoring of the function and non-functional requirements of an update are not provided. For this purpose, other mechanisms must be used that are only partially present in the AUTOSAR standard. ▶ Section 47.2.2 provides an overview of applicable AUTOSAR mechanisms and the state of research in this area.

47.2.1   Updates

The *Update and Configuration Management (UCM)*, which can update one or more AUTOSAR systems, was already introduced for the AUTOSAR Adaptive platform with Release 17–10. UCM is primarily responsible for package management of both executable program files and required data, such as engine maps. How an update must proceed is described in so-called manifest files for an update. These manifests describe the various configuration aspects for the application software and the AUTOSAR basic software on the target system of the update. The requirements placed on an AUTOSAR implementation stipulate a "package management", that provides cryptographic measures to ensure the authenticity as well as the integrity of an update. In concrete terms, only software and data from known and trusted sources should be imported. Especially in the case of over-the-air updates, the communication paths must be secured separately and independently of UCM. For example, a separate application is required that addresses the interface of UCM.

UCM can also be used to address the update path for AUTOSAR Classic systems by emulating a flashing adapter in software on an AUTOSAR Adaptive system. The respective *Diagnostic Protocol Data Unit Application Programming Interface (D-PDU API)* implementation forwards the binary data passed to the flashing adapter and manages an update session as well as the control of transport and network layer during the update. For updating systems that do not comply with the AUTOSAR standard, ISO 22900-2:2017 [9] is used. However, since the ISO standard is very broad, it is permitted for AUTOSAR conformance to implement only a subset of the ISO standard. On AUTOSAR Classic systems, an update process replaces software, data and configuration at the granularity of AUTOSAR software clusters.

Both AUTOSAR mechanisms and, in particular, ISO 22900-2:2017 only solve the delivery problem (so-called deployment), but not the integration problem. For example, an update could change scheduling parameters, which can also have an impact on unchanged applications on a system. If functional as well as non-functional parameters of unmodified applications must not change, this severely restricts the design space

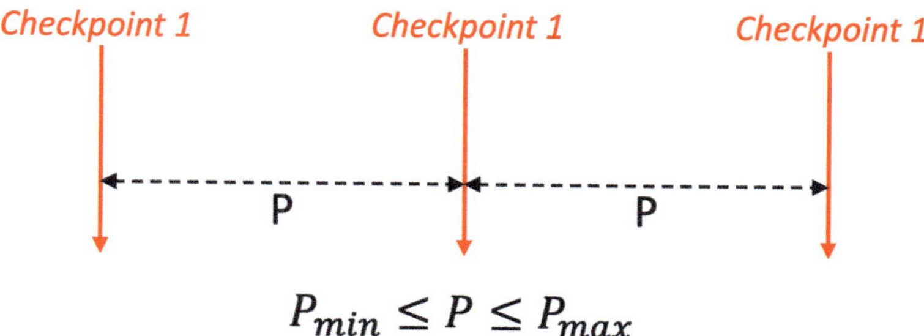

$$P_{min} \leq P \leq P_{max}$$

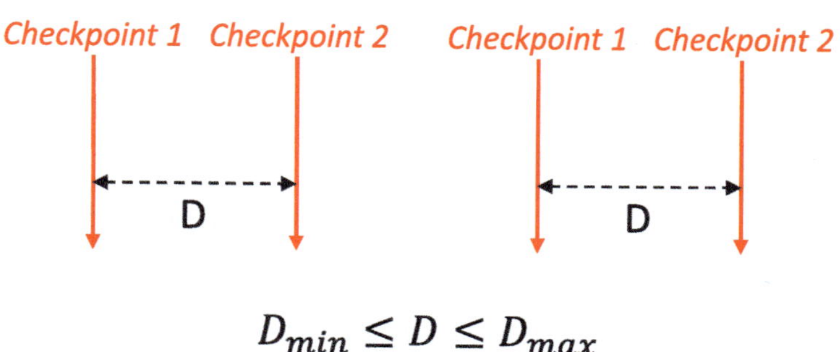

$$D_{min} \leq D \leq D_{max}$$

Fig. 47.1 AUTOSAR watchdog manager: timing of checkpoints with *alive supervision* and *deadline supervision*, P = period, D = deadline

for updates. If variations are allowed, quality assurance and (operational) security tests must guarantee that all specified parameters stay in their bounds. This in turn can entail a high testing effort, since the corresponding dependencies of an update must be known and further side effects must be excluded. However, this integration task becomes increasingly difficult to solve in advance as the number of applications, as well as their complexity, grows. Monitoring applications at runtime can be a mechanism to catch side effects at runtime and prevent malfunctions.

47.2.2 Monitoring

Depending on the AUTOSAR platform variant, the AUTOSAR standard family provides various mechanisms for observing and monitoring a system at runtime. This monitoring is referred to as *System Health Monitoring* across AUTOSAR platforms. For the AUTOSAR Classic platform, the *Watchdog Manager* [1] is a basic technology that is often used in systems with functional safety requirements. Watchdogs are timers that trigger a specified reaction when a defined anomalous situation occurs. Typically, this response is a reset of the CPU, a SoC compound, or an entire Electronic Control Unit (ECU). However, other corrective responses can also be implemented. These hardware watchdogs are managed,

for example, by the AUTOSAR watchdog manager. In this context, the manager represents an intermediate software layer that can enable various monitoring strategies. The three basic ones are

- Deadline Supervision
- Alive Supervision
- Logical Supervision

For this purpose, special calls of watchdog manager functions are introduced into the application or into other basic software, which are called *checkpoints*. The watchdog manager can now monitor whether checkpoints are executed with a certain periodicity (alive supervision), or whether the time between two checkpoints is less than a given barrier (deadline supervision). **Figure** 47.1 illustrates the difference between the two approaches. In case of *alive supervision*, the time between each two executions of the same checkpoint is compared. *Deadline supervision*, on the other hand, compares the distance between two different checkpoints and is therefore also suitable for aperiodic executions, but it is more complicated to map this to simple watchdog timers. According to the AUTOSAR specification of the watchdog manager, a failure of the second checkpoint would therefore not be recognized as a deadline violation.

Logical supervision, in contrast, aims at a correct logical sequence of different checkpoints. If this logical

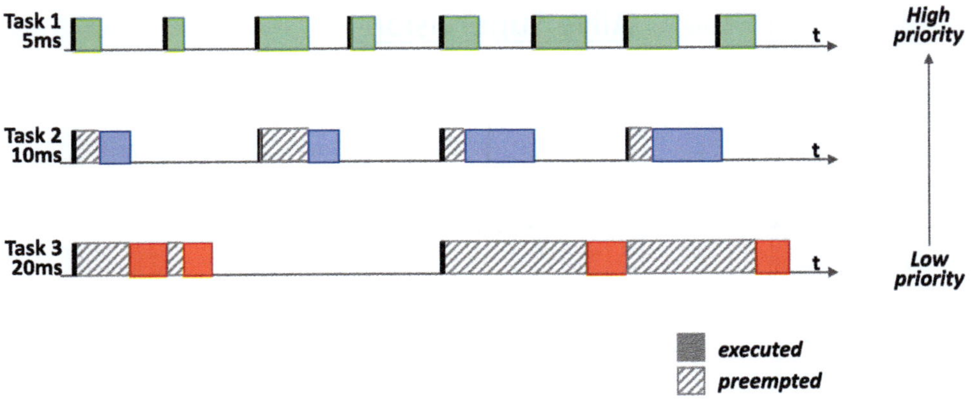

Fig. 47.2   Example real-time system with several tasks of different priority

sequence is not adhered to, the watchdog reaction is also triggered. The AUTOSAR watchdog manager or the implementer is free to either trigger a reset directly, or simply not reset the hardware watchdog's count value, which also results in a reset.

Platform and health management of the AUTOSAR Adaptive platform works according to a similar principle with the difference that, in addition to the pure reset, error handling routines can also be called in software.

Monitoring in distributed systems is usually supported by *heartbeat* mechanisms. Heartbeats are periodic signals sent by a system to be monitored, usually over a network. A monitoring system can detect a failure from the absence of these signals and trigger appropriate responses. Heartbeats, which are primarily used in network applications, thus follow the same monitoring principle as watchdogs, which outsource the monitoring task to specialized hardware timers.

Health monitoring explained here is used specifically to monitor the processes in the software system in order to intercept recognizable anomalies in the function and runtime behavior of the software. It is not sufficient to monitor functional correctness, i.e., it does not replace functional testing. However, it is very useful for monitoring the non-functional influence of updates on other parts of a software system, which are not detectable in the test of an updated software element, because the effects only arise during integration. While such influences already exist in classical software systems, they will be much more difficult to control in future systems. The main reasons are, on the one hand, the already described high complexity of hardware and software, which leads to correspondingly complex interferences, and, on the other hand, the growing agility made possible by the AUTOSAR Adaptive platform. Whereas in the AUTOSAR Classic platform a change entailed the adaptation of the static configuration and software integration, which ended in a retest of the ECU, individual software components can now be exchanged in an existing software state. Therefore, the following section will address the resulting integration problem in an AUTOSAR Adaptive platform.

47.3   Integration and Verification of Non-functional Requirements

Typical non-functional requirements relate to the timing of a function realization or parts thereof. A latency requirement, for example, can be used to specify an upper bound for the time span between two events. Since functions respectively their implementations in software share the platform with other software, a further non-functional requirement for *limited interference* often arises. The goal of this requirement is to prevent a function from being affected by the resource consumption of other software. Since complete prevention of interference is usually associated with a greater loss of efficiency, designers typically settle for bounded interference, i.e., an interference that does not impact functionality.

This becomes clear when looking at a simplified real-time system in **Fig. 47.2**, such as can be implemented with AUTOSAR Classic. In this system, three periodic tasks are executed with different priorities. Since only one execution can occupy the processor at any time, lower priority tasks are interrupted and suspended if a higher priority task requests processor time. If we consider Task2 or Task3 in this example, it becomes clear that the latency (the time between activation and completion of an execution) depends on the computation time of Task1 (or all higher-priority tasks together). In practice, analytical methods are limited to simple systems, since they depend on many factors such as processor type, code base, tool availability, etc. Similarly, such analytical methods require that the runtime behavior of all tasks—i.e., the pure execution time of the code without interruptions due to scheduling and activation timing—is precisely known.

In recent systems, however, dynamic call structures and often event-driven chains, where a task is activated by a predecessor task, are increasingly encountered. Such event-driven chains complement the fixed cyclic/periodic task activation of established software generations. Those systems are then often not designed as *hard real-time* systems, but as *firm* or *soft real-time* systems. That is, a violation of a deadline does not necessarily have catastrophic consequences but can be tolerated in many cases, as long as the violation does not occur too frequently or in combination with other anomalies. When integrating a wide variety of functional implementations on a platform, the problem now arises of securing them against each other. Since current platforms with high computing power and correspondingly complex networking and memory structures exhibit a strongly state-dependent timing behavior, guarantees for computing time and maximum interference can only be given according to the state of the art under very conservative assumptions. However, such conservative assumptions mean that much of the gain in computing power over simple architectures is lost. Nonetheless, if timing constraints are violated very rarely, it is possible to introduce a safeguard for the function that is activated only in these rare cases. A timing violation thus corresponds to the occurrence of an error, with subsequent error handling. The error handling can lead, in the simplest case, to an abort of the function, or to an application-specific countermeasure, like skipping a single calculation, or switching to a fallback level with changed functionality. For the detection of a response time violation, the aforementioned monitoring techniques are suitable. For their employment, however, a consistent collection of application-specific requirements is necessary.

In case of timing requirements, data timing, i.e., the time when data are processed or produced, increasingly comes into focus, whereas the timing of software execution recedes to the background.

47.3.1 Decoupling of Data and Timing

Current ADASs are characterized by functions with cause-effect chains that are distributed across several ECUs. In this context, the distribution and provision of data in the vehicle is of growing importance. The application, or the function implementations, that provide a specific sample or information are increasingly interchangeable. This also results in greater flexibility in function design and, above all, in function implementations: The location where data is provided becomes secondary and, for some functions, may even be located in a cloud- or edge-server outside the vehicle. The focus is on the data itself and its quality. As a consequence, modern ADAS become data-centric systems, which should be specifically considered in the design. In the event of an update, the entire set of cause-effect chains must be validated,

which can lead to enormous costs without the appropriate interfaces and guarantees.

In addition, the applications that provide certain data are often not known at the time of the initial design, either because a design is still developing, or because concrete implementations are only added in later update cycles, or existing functionality is expanded in the process. It is important to note, however, that data version management and data flow timing can be a decisive factor for the quality of the function, because even small changes to the system—for example, an update of a third software component—can influence the timing of a software component under consideration. For this very reason, increased robustness against such types of perturbations is advantageous and necessary if updateability is required for a system or its components. Therefore, three aspects must be considered:

- separation of data transport from data processing, which also allows flexible subscription and provisioning of data,
- identification and management of different versions of a data object,
- definition and application of mechanisms for specifying and fixing application timing and the associated data flows.

Mechanisms for decoupling interfaces

A middleware is used to decouple individual applications from the underlying platform. It provides means for both communication and execution abstraction. Communication between different applications is implemented via data objects to which applications are given access rights (publish, subscribe). The actual distribution/transmission of the samples (which publisher sends to which subscribers) is the task of the middleware and remains transparent for the applications.

Differences arise from the configuration, which can be either static or dynamic using service discovery. The AUTOSAR Classic platform uses a static configuration. Here, the Virtual Functional Bus (VFB) is used at design time to connect applications without knowledge of the platform. In the synthesis process, this is used to create a mapping of the software components to the hardware that satisfies the required constraints, e.g., with regard to timing. In contrast, the AUTOSAR Adaptive platform allows the use of a service-oriented middleware, such as SOME/IP or the Data Distribution Service (DDS). This enables, in addition to a static configuration, a linking of publishers and subscribers at runtime. Here, an application is regarded as a service that has certain requirements for input data and provides output data on the other side. The middleware can now link corresponding services at runtime.

SOME/IP is also defined in the AUTOSAR Classic platform as an interface to the AUTOSAR Adaptive

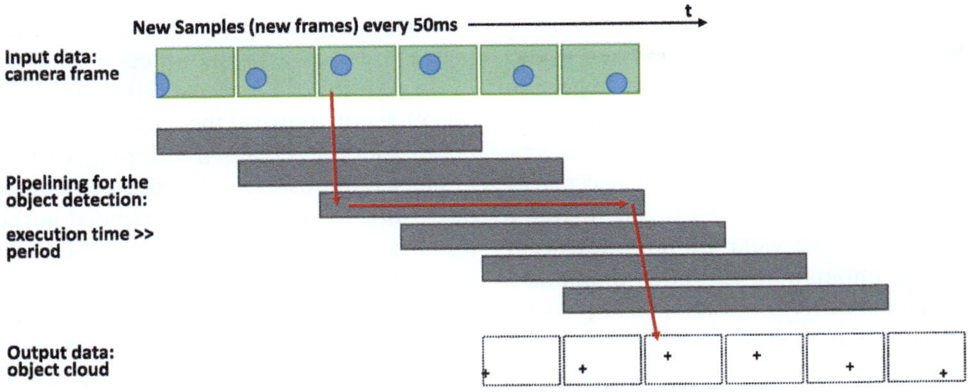

Fig. 47.3 Management of multiple samples (versions) of the same data object with pipelining

platform. DDS, on the other hand, is widespread in the field of industrial automation and represents the middleware for the Robot Operating System (ROS) 2. Differences arise in the possible Quality-of-Service parameters that an application can negotiate with the middleware. While SOME/IP itself does not provide any QoS mechanisms and outsources this either to the application or to the underlying communication protocol (UDP/IP or TCP/IP), DDS already knows a wider range of QoS parameters, which are also matched during service discovery. Services can thus only be linked if not only the data type at the interfaces match but also the quality requirements, such as the planned frequency with which new data is provided.

Platform separation or partitioning is used to divide and strictly separate resources such as memory and CPU between different applications or groups of applications. Virtualization using hypervisor architectures aims to make such separation transparent to the application layer as well as to provide a unified execution platform abstraction for the application. Middleware and hypervisor architectures thus achieve a "functional" separation of platform and function. Interfaces, both for communication between different applications and for execution, are thus platform independent, which increases portability and flexibility during the application development. At the same time, isolation mechanisms can specifically limit many errors (e.g., due to memory corruption). However, there are two main challenges. First, stronger isolation or platform abstraction typically induces higher runtime overhead. Secondly, the platform timing (of both execution and communication) influences the function or the quality of the result. Hence, further mechanisms are necessary, which decouple the temporal behavior from the execution/communication.

Mechanisms for decoupling data versions
In addition to the complex timing behavior of current platforms, there is a second effect that is determined by the type of calculation. While in classic automotive systems the execution time needed to produce the next output sample is generally shorter than the period of the calculation activation, the extensive calculations in driver assistance systems form multi-stage flow graphs that are processed according to the assembly line principle (pipelining). Typical for this pipelining is that the period with which sensor data is provided at the input is shorter than the latency through the overall computation. This difference has an important impact on age and versioning of data samples in a system. While in a classical system there is only one version of a data object at a time, which is overwritten by the next execution, in pipelining there are multiple versions with different ages at the same time. An example is shown in **Fig. 47.3. While new sensor data is provided by the camera every 50ms, processing a camera image and subsequent decision-making and action takes significantly longer time. In this case, the middleware must ensure to provide the correct version of the camera image to an object-detection job. A version of the data object (also called a sample) is uniquely identifiable (e.g., via a sequence number) and is stored in a so-called history (buffer) until it is no longer needed. **Figure 47.3 is still a simplifying representation, since the execution time of each object detection is constant and no linking of different input data (data fusion) takes place. Changes in the flow of a pipeline, as caused by variations in computation time or by updates, lead to a potential change of the involved data samples, so that an older or younger sample of a data object is used. This can, for example, result in incorrect assignments of camera and radar data in a sensor fusion or in trajectory calculations being performed on outdated track data.

Mechanisms for decoupling the response time
Shared resources as well as data dependencies between applications can cause them to affect each other's timing. While these dependencies are at least identifiable a priori in a statically configured system, they represent a major challenge for dynamic systems in which updates, but also changes to the platform, constantly change the possible

effects of dependencies. In this context, an influence in the timing behavior can not only result from obvious misbehavior of an application (e.g., a software bug), but can also be the result of a changed schedule (e.g., an update in another part of the system).

Thus, different motivations exist for decoupling the timing behavior in a system. On the one hand, this can be the containment of timing errors. This is a reactive approach that monitors the timing behavior of different applications at run time and compares the observation with a target behavior. The target behavior must be defined at design time and is used to protect other applications from the effects of misbehavior. The goal is therefore to prevent error propagation. One step further are mechanisms that allow the data flow to be specified via a logical timing behavior already at design time and enforce this at runtime. This is to permit, beyond the pure error handling, also a decoupling of the function from the run time behavior in order to obtain a higher degree of platform independence.

A timing failure is, for example, a response time of a task in the system that exceeds the specified deadline [3]. The causes of timing failures are of different nature. For example, serious runtime errors can be caused by hardware defects that result in an overload of interrupts or bus messages. These can prevent a task from meeting its planned maximum execution time. Especially with the introduction of high-performance and less time predictable platforms, however, another group of error causes is gaining importance, that already occur at design time. Since it is almost impossible to analytically determine a formally provable worst case timing at design time, function and integration timing must be measured or even estimated. Depending on the criticality of the software, an appropriate effort for determining timing behavior must be made, which also varies the quality of the estimated worst cases. It is therefore very likely that software with low criticality will deviate from its specified timing behavior at runtime while the probability decreases as criticality increases.

Mechanisms to safeguard timing behavior therefore rely on detecting such misbehavior and preventing propagation. One possibility is a purely time-driven scheduling [11]. Such Time Division Multiple Access (TDMA) scheduling is used, for example, in the central driver assistance unit (in German short "zFAS") proposed by Audi [12]. Software is grouped into partitions. The CPU is assigned exclusively to one partition during its time slot, isolating different partitions from each other. However, an isolation of different software within the partition is not achieved. At the same time, a strict TDMA schedule is not load-preserving, i.e., if a partition does not use its time slot, this computing time cannot be used elsewhere. In order to achieve high utilization of the hardware, complex static partitioning is necessary, which in the end offers little flexibility. For the development of future systems with highly dynamic behavior, more flexible methods therefore become much more attractive.

As an alternative to strict TDMA scheduling, the AUTOSAR Classic platform therefore supports various mechanisms for error detection and containment, by means of the AUTOSAR OS Timing Protection. These techniques, originally from the OSEK time standard, monitor the execution of each task using timers. Firstly, the inter-arrival time, i.e., the time between two successive activations of a task, can be monitored here and compared with a predefined lower bound. The task can thus not induce more system load than planned. On the other hand, the pure execution time of the task can be measured with the help of a watchdog timer. Assuming that the measuments are correct, the task does not interfere more than planned with other low priority tasks. Finally, AUTOSAR OS also allows critical sections (areas in which other interrupts and scheduling are deactivated) to be protected in order to prevent errors from blocking the entire system or to prevent a low-priority task from influencing higher-priority tasks more than planned.

AUTOSAR Adaptive systems differ greatly from AUTOSAR Classic systems in terms of call structures and scheduling. This has direct impact on the necessary mechanisms for timing isolation in AUTOSAR Adaptive. While AUTOSAR Classic has a fixed task structure with predominantly periodic calls and fixed communication relationships, e.g., to implement basic control tasks, AUTOSAR Adaptive has a service-oriented architecture. A service can be called sporadically. Typically, processing pipelines are implemented in this way, where, for example, a sensor produces data periodically at the beginning, but this data is then passed on between the various services in a processing pipeline on an event-driven basis. A service thus calls its subsequent service as soon as it has processed the data and thus completed its subtask in the pipeline. In doing so, it may call other services during its processing, resulting in multi-level client-server relationships. Service-oriented software architectures thus lead to branched call chains with correspondingly complex timing behavior.

With the sporadic server mechanism, POSIX provides a concept for budget-based scheduling, which was also implemented in Linux [6]. The sporadic server ensures that a service always has a specified minimum amount of computing time available in a certain interval and at the same time cannot take up more computing time than planned. During its execution, the service gradually consumes its budget, which is replenished at a later point in time. This time is called the replenishment period. When the budget is completely used up, the priority of the service is lowered so that it cannot further interfere with other services. Conceptually, the sporadic server thus allows to model interference by a sporadic service as periodic, again.

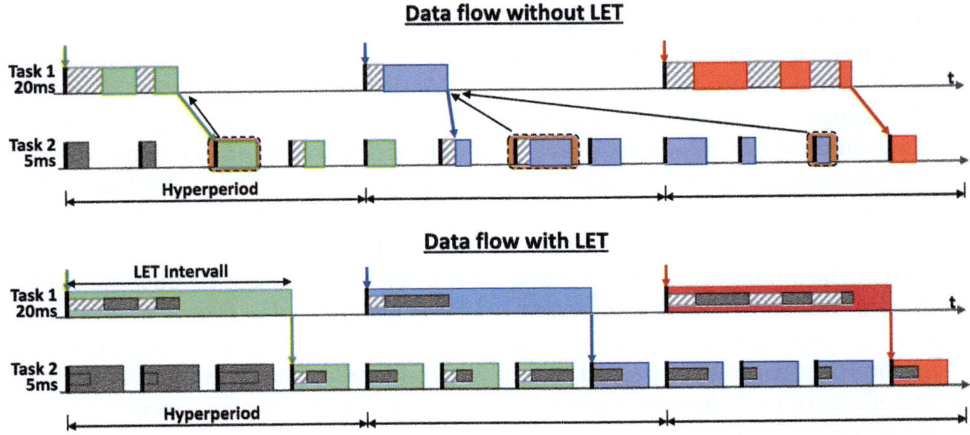

Fig. 47.4 Deterministic data flow with LET (Logical Execution Time)

Mechanisms for deterministic data flow

AUTOSAR Classic and AUTOSAR Adaptive are no alternative architectures, but will be used in combination for a longer period of time. At the interfaces between the two software architectures with their different activation and communication semantics, a breaking point arises which can easily lead to the uncontrolled loss of an unambiguous (deterministic) data flow and thus to a faulty system function. At this point, there have been recent developments with regard to the use of time-constrained programming paradigms that could provide a systematic way to solve the problem. In the following, we will first give a brief insight into the communication in AUTOSAR Classic and show how data flow determinism is established there using a new approach. Initially, the method called LET was used only in single ECUs, where LET was used to achieve non-blocking communication between multiple computational cores. Then we show an extension of LET to a System-Level Logical Execution Time (SL-LET), which allows the method to be applied to distributed systems including pipelining. LET is part of the AUTOSAR Classic Timing Extensions, SL-LET is currently in AUTOSAR standardization. Finally, SL-LET provides the necessary properties to form an interface with deterministic data flow between AUTOSAR Classic and AUTOSAR Adaptive that can be monitored during operation.

All the mechanisms presented so far have in common that they support monitoring an upper bound for the response time (worst-case response time, WCRT). However, this is only a first step. Decisive for the behavior of a function is not only the compliance with an upper time bound, but also the data flow. An example of this is given in the upper part of **Fig. 47.4**.

Here, two tasks with different periods are considered, where the 20ms task produces data that is read by the 5ms task. Such a periodic multirate system is typical for the implementation of control tasks in an AUTOSAR Classic system. In this case, the data flow describes, which job of the 5ms task reads from which job of the 20ms task. A job is defined as the instantiation (execution) of a task. In the classic automotive software convention, a task reads input data as soon as its job starts execution and writes the results as soon as the job is finished. However, this behavior is subject to jitter (e.g., due to scheduling and variable execution time), which causes the response time of a task to vary between the best case and the worst case. Looking at the third job of the 5ms task in **Fig. 47.4** (highlighted in orange in each case), we see that due to this jitter, some input data is from the same hyperperiod, but older data is read in the last hyperperiod.

The data flow problem becomes obvious if a task must set several input data in relation, as it is for example the case with sensor fusion. Here, the quality of the sensor fusion is strongly influenced by how old the respective sensor data are (i.e., whether they were recorded at the same time or not). This behavior poses a major problem for the updateability of the software. Thus, the data flow itself can change due to an indirect influence. For example, if a third task is modified (even though it has nothing to do with the current function), this can affect the runtime behavior of the two tasks under consideration, even if no upper bound is violated in the process. Furthermore, since most functions are distributed across multiple ECUs, these effects are not limited to local scheduling, but are also affected by changes in communication latency.

This is where mechanisms such as LET [10] come in, which place an additional logical abstraction over the timing behavior. Here, a job no longer reads and writes its data at the limits of its physical execution time, but at the limits of a predefined LET interval. This behavior can be implemented with little effort using a middleware. The response time is thus constant and independent of scheduling. This logical envelope is shown in the lower part of **Fig. 47.4** and leads to a deterministic data flow. For AUTOSAR Classic systems, the necessary mechanisms for specification by means of LET are standardized in the timing extensions. In addition to the deterministic data flow, this has other important advantages. On the one hand, LET explicitly establishes robustness against changes in the response time. For example, if in ◘ Fig. 47.4 the last execution of the 20ms task (marked in red) in the current system would be a worst-case behavior, the LET interval can also be chosen somewhat larger. Thus, the data flow does not change even if the execution time becomes longer due to an update, as long as it is not larger than the LET interval. On the other hand, compliance with the LET specification can be checked very easily at runtime by means of monitoring. For the determination of an LET interval, it is irrelevant whether it is chosen on the basis of a formal worst-case analysis (e.g., systems with hard real-time conditions) or whether only a worst-case observed by measurements in a prototypical system serves as a basis. Similar to the interface specification in a middleware, LET thus allows a building block principle for the specification of timing behavior, which is also called composability. An important purpose of this composable timing is the reuse of software in different model variants and on different hardware platforms. LET systems thus allow a certain degree of platform independence, since the same specification is usable in different products or implementations.

With regard to modern ADAS, pipelining and the distribution of functions across multiple ECUs play a major role, as already described. Here, however, the classic LET concept reaches its limits. On the one hand, the processing time due to the overall calculation of a pipeline, as well as the communication between distributed systems, is often significantly greater than the period with which sensor data is recorded at the input. On the other hand, the time bases in a distributed system are subject to a permanent drift and can only be synchronized with sufficient accuracy. However, LET only allows an interval length less than or equal to the period and requires a uniform time base. This is necessary because LET does not support versioning of the data, which can cause unwanted changes in the data flow when LET intervals overlap in the implementation. In order to be able to use the mentioned advantages of LET, with SL-LET [7] an extension was created that unites these aspects. In addition to including synchronization inaccuracies between distributed time bases, SL-LET allows, under certain conditions, an interval length larger than the period. The key here is that there are no data dependencies between the now overlapping SL-LET intervals. This generally follows the pipelining approach and already implicitly applies to different use cases. For example, SL-LET is applicable with very little effort to distributed communication, where data is communicated in the form of messages (message passing). Here, the communication channel works like a pipeline in which two successive messages are transported and processed independently of each other. Similarly, SL-LET is applicable to applications that span

both the AUTOSAR Classic and the AUTOSAR Adaptive platforms. Here, SL-LET intervals can cover entire pipeline segments and, for example, enforce a deterministic data flow at the interface between the two platforms. This example is already sketched in ◘ Fig. 47.3 by considering the gray blocks of the pipeline execution as SL-LET intervals.

SL-LET inherits the central properties of LET. Thus, SL-LET allows both to specify a monitorable timing behavior and, through non-blocking synchronization, to implement efficient communication that includes version management of data objects.

In collaboration between industry and research [5, 8], LET has now been established in industrial series development, initially in the field of drive technology. SL-LET is at the AUTOSAR standardization stage.

47.3.2 Runtime Monitoring as a Protection Mechanism

In the previous sections, we explained that decoupling timing and data simplifies integration. Especially for well-tested components, side effects are thus greatly reduced during integration. Nevertheless, certain (emergent) effects only arise through integration. For example, even an inconspicuous update of an existing software component can lead to the use of more CPU cores or to a higher single core load. Decoupling mechanisms, such as the aforementioned LET, can localize the resulting effect, since these mechanisms create deterministic data flow and timing, while being robust to changes in the component's internal timing. Nevertheless, the changed software component usually also influences the scheduling and can thus cause race conditions in other components. These race conditions occur when a function depends on the order of arrival of data at its input. Although these race conditions are defects of the influenced software component, detection is difficult, since testing never covers all temporal variations. Optimally, one would be able to prove the absence of such defects by formal methods, which is hardly practicable with the complexity of the used software libraries in modern ADAS. The only remaining way out is deterministic, time-controlled or partitioned scheduling. But even this approach entails a great deal of configuration complexity, and does not fit the variable workload generated by data-dependent algorithms. This is exactly where runtime monitoring comes in, as it can be used as a safety mechanism to catch rare and/or emergent errors.

The term *monitoring* has different connotations at different levels of a system. In this chapter, we consider non-functional aspects of function realization, e.g., observing and potentially intervening in the timing of the system. The scientific literature has made several contributions

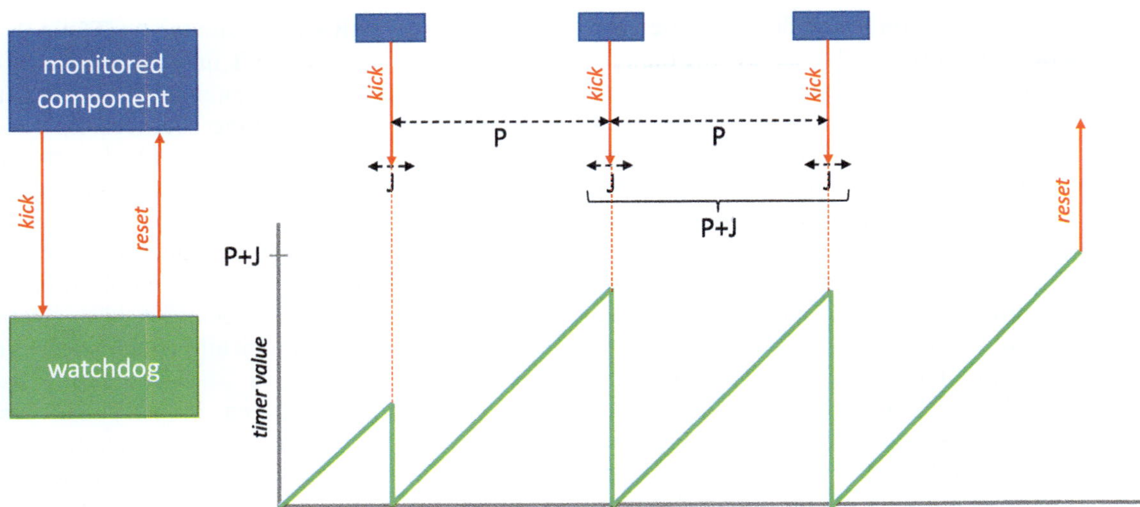

◻ Fig. 47.5   Functionality and timing of watchdog monitoring

specifically on observing and influencing classical task-based systems. Either the activation patterns (such as a periodicity) in [14], or the combination of activation patterns and worst-case execution time [13] are observed and, in case of non-conformance with a stored specification, the load is shifted backwards in time. Overload cases are thus avoided, as well as those case where tasks can sensitively interfere with each other's timing, thus safeguarding the previously mentioned property of *bounded interference*. This type of monitoring is called *protective monitoring* because it protects the system and other applications from misbehavior (out-of-spec behavior). These mechanisms assume that (a) a deviation from a specification can be detected and (b) appropriate actions can be taken. If countermeasures are omitted or if there are no countermeasures for a certain form of misbehavior, the result is purely *detective monitoring*. In this case, monitoring only serves to collect data and thus has no direct feedback to the running system. However, this information is still usable in the safeguarding process to detect whether, for example, an update in the system changes the behavior in an unacceptable way.

The watchdog and/or heartbeat mechanisms mentioned at the beginning can be understood neither as *protective* nor as *detective*. Rather, they result in *reactive monitoring* because, although they do not prevent the misbehavior, they can react within certain time limits.

This classification becomes clearer if we additionally include the aspect of implementation. Monitoring mechanisms can be implemented either *asynchronously* or *synchronously*. For example, a watchdog is an asynchronous implementation because it observes the behavior of the monitored subsystem independently, from the outside. The mechanisms for monitoring activation patterns, on the other hand, must be implemented synchronously, since misbehavior cannot be prevented otherwise. Although both mechanisms aim at protection, their implementation (asynchronous or synchronous) results in a different quality (reactive or protective).

In practice, the implementation of protective monitoring is difficult because the synchronous implementation additionally affects the timing of a monitored subsystem. Moreover, even with protective monitoring, we cannot provide arbitrary guarantees and achieve safeguards. This is due to uncertainties in the implementation of the function, the platform itself, and external factors (environmental influences). In order to meet safety standards, a certain independence of the protection mechanisms is also necessary, which is more feasible for reactive monitoring with asynchronous implementation.

The simple example of a watchdog timer will suffice to explain the protection provided by monitoring. ◻ Figure 47.5 illustrates how the watchdog works and the resulting guarantees it provides. The watchdog's internal timer is reset periodically. Small time deviations of the periodic kick signal are usually modeled as jitter, i.e., this signal can be triggered earlier or later than the ideal periodic signal with period P by the value of jitter J. If this jitter can be determined, it results in a threshold value of $P + J$ for triggering the watchdog. This threshold value then also determines the maximum response time of the watchdog. In addition, the recovery time should not be ignored. In case of a reset, this is the time from the triggering of the watchdog to the recovery of the function.

The watchdog mechanism can already be used to monitor some functions, but modern ADAS require an extended view. As already mentioned, data-dependent algorithms and complex hardware platforms are used here. Determining a WCRT is often difficult because a WCRT only appears in rare constellations. Therefore, one tries to develop these systems as *firm real-time*, which means that occasional violations of the timing behavior are allowed.

Especially in image data processing, where a stream of camera images is continuously processed at a certain frame rate, the absence of a single frame does not necessarily cause the function to fail. This aspect was considered in [15] using the example of distributed processing chains in Autoware.Auto, which is built on ROS 2. These processing chains consist of several software components distributed across different ECUs. The software components communicate via a DDS middleware. In this work, it was shown that the latencies of event chains can be monitored across different software components and ECUs with reasonable execution costs while tolerating occasional errors. Furthermore, in contrast to the AUTOSAR watchdog manager, the absence of events is also monitored.

When using such monitoring mechanisms, two additional questions arise in practice. On the one hand, it must be decided which reactions should be performed when a monitoring mechanism is triggered. A complete reset of the CPU or the ECU is not very practical for modern ADAS due to the long start-up time. On the other hand, the question arises which events must be monitored or where the corresponding checkpoints should be implemented in the monitored software components. Especially when using foreign and/or complex software libraries, the implementation of checkpoints is associated with extra effort.

Yet, data-centric application systems help in checkpointing. Since data-centric middlewares already denote component interactions and their timing, the monitoring of just the data timing should be sufficient to observe the functionally relevant processing latencies in the system.

The correct recovery strategy cannot be addressed in general, i.e., independently of the function itself. Rather, error handling should be the application's obligation. In the example of image data processing, an additional transient reduction of resolution is a possible reaction to observed performance problems. It is important that there are clear specifications with regard to which guarantees a platform is able to provide as part of the integration and which error cases can nevertheless occur and must be handled by the application. In other words, the function must provide for appropriate error handling.

47.4 Conclusion

Updates are of crucial importance for modern driver assistance systems. The basic possibility of carrying out software updates in the vehicle over-the-air (OTA), i.e., without a workshop visit, is already industrial practice for some types of software components. Due to the high demands on functional safety, updates for driver assistance cannot be regarded as isolated updates or additions to functionality, but require inclusion of non-functional effects such as timing, which can influence not only the modified function itself, but also other functions. In contrast to classical software in vehicles, data versioning and pipelining play a central role in driver assistance systems with mutual impact on the well-known effects of time-variant cause-effect chains.

Seamless and platform-independent integration therefore requires coordination between applications (functions) and platform. Here, mechanisms for decoupling can be applied to mitigate interference on other functions, to monitor execution and to increase robustness against change of platform properties.

References

1. AUTOSAR GbR: AUTOSAR—Specification of Watchdog Manager, R20-11 edn (2020). ▶ https://www.autosar.org/fileadmin/user_upload/standards/classic/20-11/AUTOSAR_SWS_WatchdogManager.pdf
2. AUTOSAR GbR: AUTOSAR—Standards, R20-11 edn (2020). ▶ https://www.autosar.org/standards
3. Bertrand, D., Faucou, S., Trinquet, Y.: An analysis of the AUTOSAR OS timing protection mechanism. In: IEEE Conference on Emerging Technologies Factory Automation (ETFA) (2009). ▶ https://doi.org/10.1109/ETFA.2009.5347159
4. Colwell, I., Phan, B., Saleem, S., Salay, R., Czarnecki, K.: An automated vehicle safety concept based on runtime restriction of the operational design domain. In: 2018 IEEE Intelligent Vehicles Symposium (IV), pp. 1910–1917. IEEE (2018)
5. Ernst, R., Kuntz, S., Quinton, S., Simons, M.: The logical execution time paradigm: new perspectives for multicore systems (Dagstuhl seminar 18092). Dagstuhl Rep. **8**, 122–149 (2018)
6. Faggioli, D., Mancina, A., Checconi, F., Lipari, G.: Design and implementation of a posix compliant sporadic server for the Linux kernel. In: Proceedings of the 10th Real-Time Linux workshop, pp. 65–80 (2008)
7. Gemlau, K.B., Köhler, L., Ernst, R.: A platform programming paradigm for heterogeneous systems integration. In: Proceedings of the IEEE, pp. 1–22 (2020). ▶ https://doi.org/10.1109/JPROC.2020.3035874
8. Hennig, J., von Hasseln, H., Mohammad, H., Resmerita, S., Lukesch, S., Naderlinger, A.: Towards parallelizing legacy embedded control software using the LET programming paradigm. In: 2016 IEEE Real-Time and Embedded Technology and Applications Symposium (RTAS), pp. 1–1. IEEE Computer Society (2016)
9. ISO22900: ISO 26262:2018 road vehicles—modular vehicle communication interface (MVCI)—part 2: diagnostic protocol data unit application programming interface (D-PDU API). Standard, International Organization for Standardization, Geneva, CH (2017)
10. Kirsch, C.M., Sokolova, A.: The logical execution time paradigm. In: Advances in Real-Time Systems, pp. 103–120. Springer (2012)
11. Kopetz, H., Bauer, G.: The time-triggered architecture. Proc. IEEE **91**(1), 112–126 (2003)
12. Lang, M.: Integration of Adas into one central platform controller. Auto Tech Rev. **5**(7), 32–35 (2016)
13. Neukirchner, M., Axer, P., Michaels, T., Ernst, R.: Monitoring of workload arrival functions for mixed-criticality

systems. In: Proceedings of Real-Time Systems Symposium (RTSS) (2013)

14. Neukirchner, M., Michaels, T., Axer, P., Quinton, S., Ernst, R.: Monitoring arbitrary activation patterns in real-time systems. In: Proceedings of IEEE Real-Time Systems Symposium (RTSS) (2012)

15. Peeck, J., Schlatow, J., Ernst, R.: Online latency monitoring of time-sensitive event chains in safety-critical applications. In: 2021 Design, Automation and Test in Europe Conference and Exhibition (DATE), pp. 539–542. IEEE (2021)

Open Access This chapter is licensed under the terms of the Creative Commons Attribution-NonCommercial-NoDerivatives 4.0 International License (▶ http://creativecommons.org/licenses/by-nc-nd/4.0/), which permits any noncommercial use, sharing, distribution and reproduction in any medium or format, as long as you give appropriate credit to the original author(s) and the source, provide a link to the Creative Commons license and indicate if you modified the licensed material. You do not have permission under this license to share adapted material derived from this chapter or parts of it.

The images or other third party material in this chapter are included in the chapter's Creative Commons license, unless indicated otherwise in a credit line to the material. If material is not included in the chapter's Creative Commons license and your intended use is not permitted by statutory regulation or exceeds the permitted use, you will need to obtain permission directly from the copyright holder.

ODD Freight Transportation

Axel Gern, Michael Haag, Jens Kotte, Kai Furmans,
and Peter Vaughan Schmidt

Contents

© The Author(s) 2026
H. Winner et al. (eds.), *Handbook Assisted and Automated Driving*,
https://doi.org/10.1007/978-3-658-45276-6_48

48.1 Importance of Road Freight Transport

The volume of road freight transport has been closely linked to economic growth in recent decades. However, as a result of a transformation toward a service economy, growth in road freight transport in the USA has become detached from economic growth; in Germany and Europe, this detachment has only just begun [1]. As a result of this separation, a further increase in the volume of road freight transport must be anticipated. Although the haulage capacity of rail and inland water transportation of goods will increase as well, a rise in road freight transport has also been forecast (see ◘ Fig. 48.1).

In order to be able to classify the various freight transportation applications, it is essential to understand how freight traffic works. This is explained below.

48.2 How Does Freight Transport Work?

In contrast to passenger transport, freight transportation is designed to be handled consistently from the starting point to the end point of the transport process using transport equipment. Today, only comparatively short distances are bridged by humans alone.

Historically, freight transportation has evolved from point-to-point transport with a correspondingly low transport frequency to transport within a networked transportation system. Efficient point-to-point transports are handled as a full truck load, whereby the capacity of the transport vehicle cannot necessarily be utilized in terms of volume or mass, as this is not always possible within the time frame for transport.

Transport networks are being set up as transport requirements are increasingly shifting toward smaller quantities which are transported at a higher frequency. Vehicles travel on fixed or variable routes and the transport units are typically handled once or more at hubs. This results in a large consolidation with simultaneous high transport frequency (see [3], section A1.2.4). The loading and unloading typically use large load carrier levels, such as pallets, pallet cages, or colli, that are handled with forklifts.

These concepts have also become established in parcel and postal services, however, individual parcels are directly unloaded, sorted, and reloaded in this case.

A transport network (see ◘ Fig. 48.2) consists of depots in which the shipments collected by individual vehicles are registered and pre-sorted or in which the shipments delivered overnight are distributed to the delivery vehicles, as well as hubs at which large shipment volume are collected and sorted centrally to then be re-distributed back to the depots (this time for delivery). ◘ Figure 48.3 schematically shows the coverage of a country by such networks. On the right-hand side, only transportation chains that run through at least one hub are shown; on the left-hand side, shipments from depot to depot are also included which can be useful for large transport volumes.

Using a parcel service as an example, the route of a parcel or a large load carrier[1] through the transportation network generally runs as follows (see ◘ Fig. 48.4):

Collection: – Packages are collected from their origin as part of preparation. Depending on the volume of shipments, this is done with parcel distribution vehicles, in medium-sized trucks, often with high volumes by collecting and replacing semitrailers, fully enclosed swap bodies, or even by full trucks.

Sorting at the depot: – At the first hub, sorting is carried out for the next transportation step and the shipment is loaded into the vehicles that will drive to the first hub.

Depot-to-hub transportation: – Transportation from depot to a feeder hub or regular hub, typically in full loads (exception: last truck of the day).

Sorting at the hub: – At the hubs, large shipping volumes are consolidated and sorted mechanically. Shipments are then loaded onto vehicles that drive to another hub or directly to a depot.

Hub-to-hub transportation: – Depending on the structure of the network, transport from hub to hub is also necessary. This is all the more necessary the larger the area covered.

Hub-to-depot transportation: – Parcels are transported to the delivery depot.

Delivery: – At the delivery depot, parcels are loaded onto the parcel distribution vehicles in order to then deliver them to the destination address.

Loading and unloading vehicles at hubs and depots may be supported mechanically or the entire process might even be automated. However, loading at the origin and unloading at the final recipient are both still heavily reliant on humans loading and unloading either manually or with tools—typically forklifts in case of large load carriers.

Comprehensively, we can see that transportation logistics is a very complex issue, which is characterized, among other things, by the heterogeneity of the transportation equipment as well as the load and rapid changes in the goods to be transported.

48.3 Motivation for Automating Freight Transport

In addition to increasing road safety, which plays a central role in all types of mobility, economic aspects are of great importance, in particular in freight transportation,

---

1   Such as pallets, pallet cages, colli.

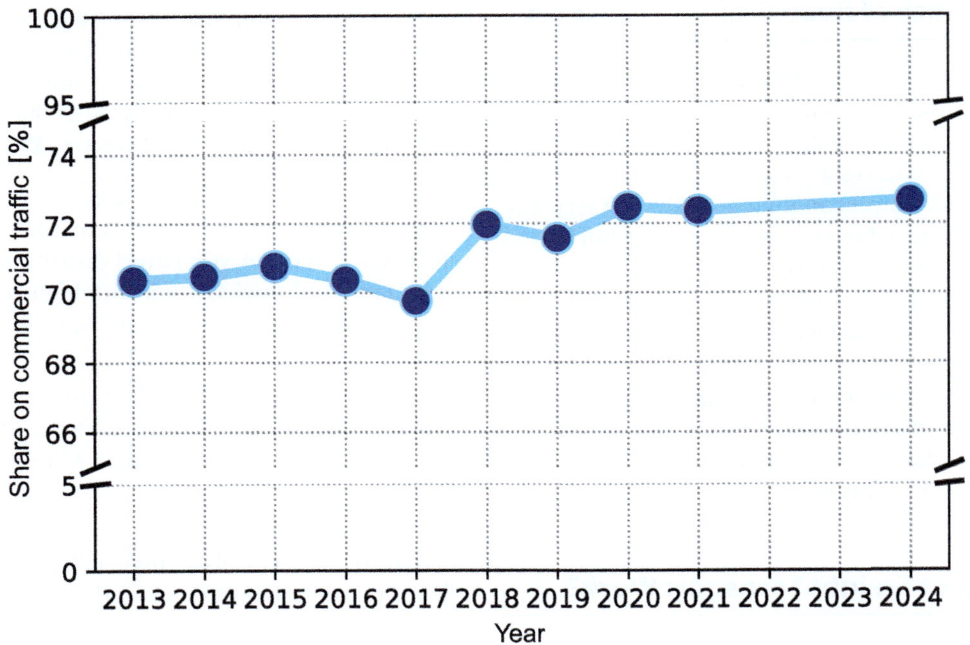

Fig. 48.1   Proportion of trucks in freight transport haulage in Germany (according to modal split) (own illustration based on [2])

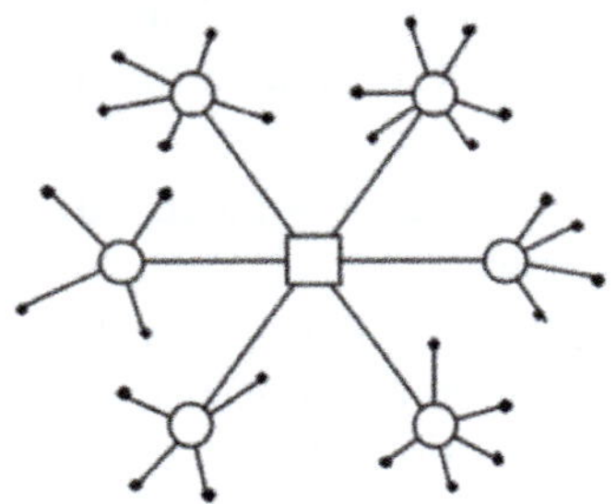

a Hub-and-Spoke-Net

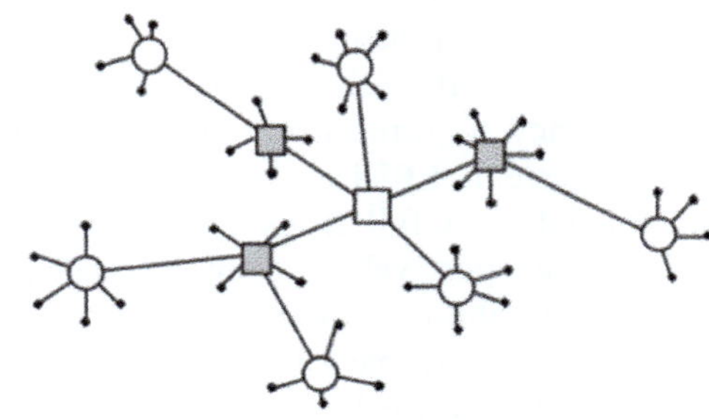

b Net with Feeder-Hubs

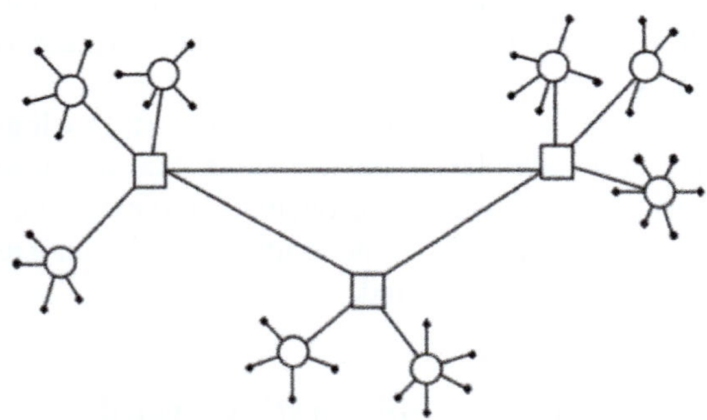

c Multi-Hub-Net z

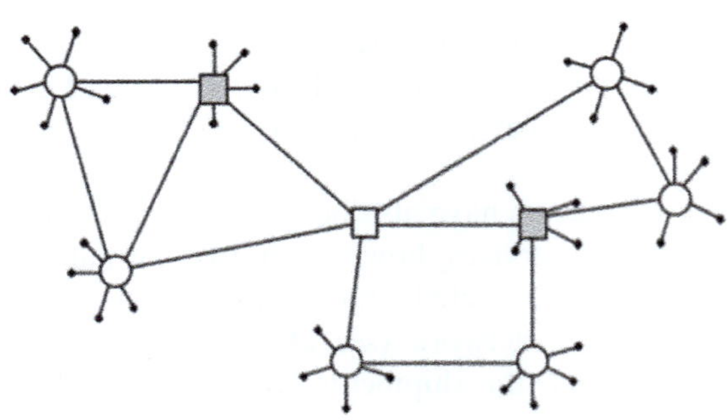

d Mixed structure

Fig. 48.2   Haulage network structures (from [3])

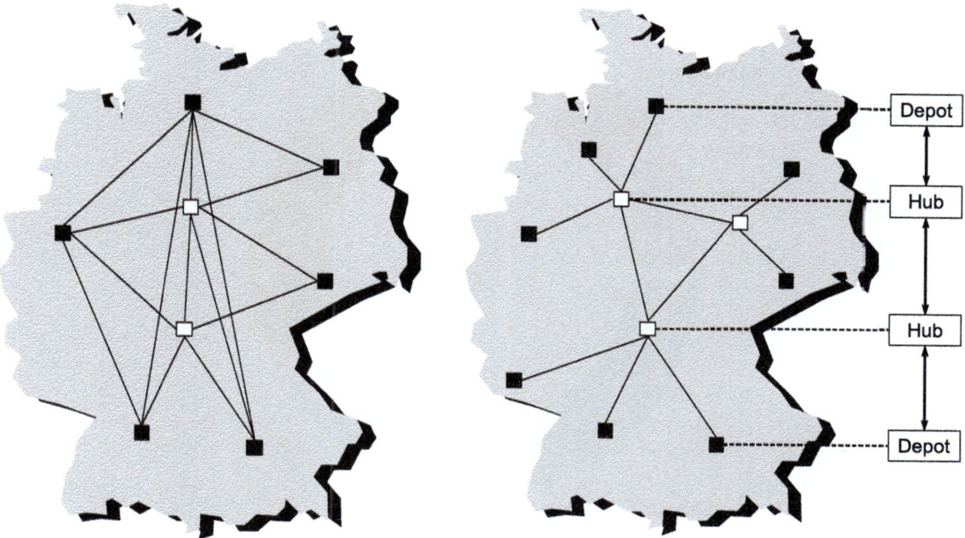

Fig. 48.3 Schematic illustration of the geographical coverage of a country by a multi-level transportation network run by a parcel service provider (from [4])

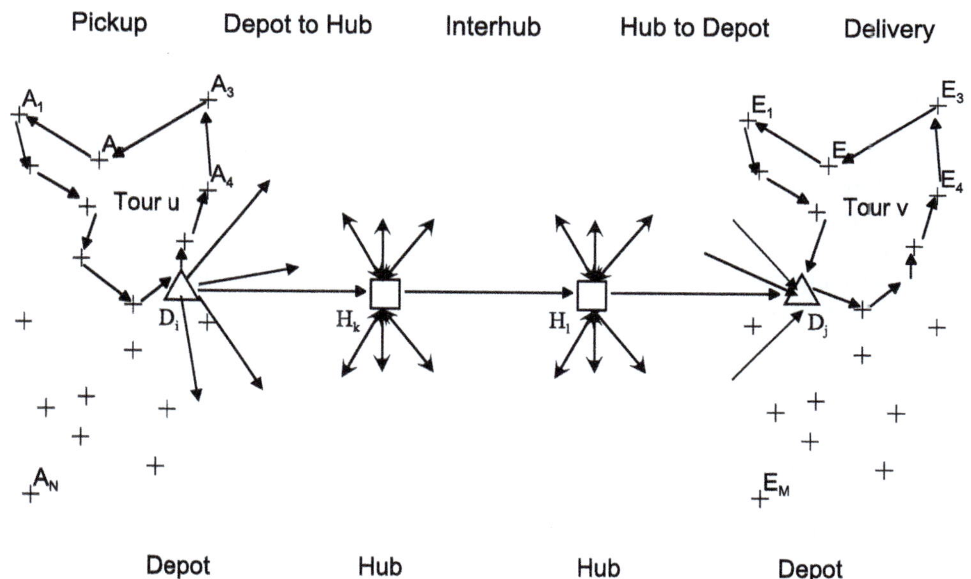

Fig. 48.4 Procedure for a shipment from sender A_n to recipient E_m using a multi-level transportation network run by a parcel service provider

and here especially the total cost of ownership (TCO); not forgetting the contribution to protecting the climate. In this section, we will look at these factors in more detail.

48.3.1 Economic Factors

The logistics market generated a total revenue of USD 9.6 trillion in 2018 and contributed around 12% of the global Gross Domestic Product (GDP). The proportion of the trucking industry is around 43% with an estimated annual revenue of around USD 4.1 trillion [5]. The trucking industry is experiencing strong growth and is expected to generate projected sales of USD 5.5 trillion by 2027 [6]. In particular, in the United States of America road freight transportation is of considerable economic importance, with revenue of USD 791.7 billion in a transportation sector that is worth USD 1 trillion [7]. This is due to the advantages of truck transport, such as flexibility, costs, and speed, compared with other modes of transport, such as railway transportation.

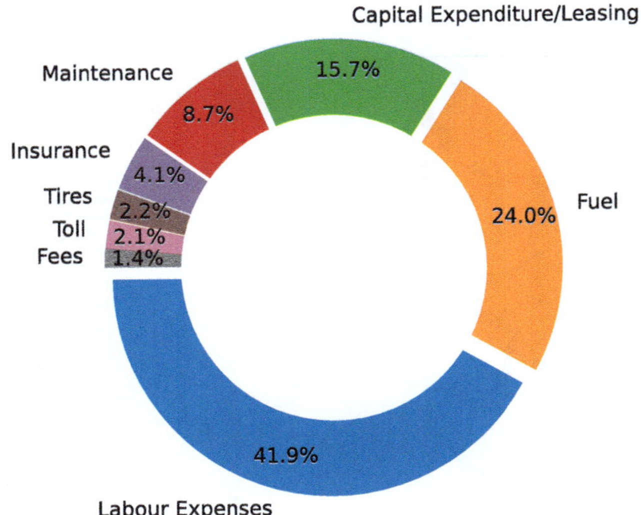

Fig. 48.5 Relative composition of the total cost of ownership for a truck in the USA [8]

48.3.1.1 Lower Total Cost of Ownership

There is intense competition in the commercial vehicle sector, which puts the players involved under great cost pressure. Enterprises in the shipping and transportation industry generally make investment decisions for new vehicle procurement on the basis of economic determinants. Acquisition expenses and operating costs incurred over the usage period form the most important decision-making basis for vehicle procurement. The consideration of the total cost of ownership allows all costs incurred over the acquisition and use phase to be recorded in their entirety.

Figure 48.5 shows the total cost of ownership in the USA per mile resulting from the operation of a truck according to the various cost items for 2019. According to the American Transportation Research Institute, monetary compensation of truck drivers accounts for around 42% of operating costs per mile, which represents the highest relative cost component of the total usage costs. This corresponds to absolute wage costs (salary and supplements) of USD 0.693 per mile. Fuel costs are 24%, and the investment costs or leasing fees amount to around 16% of the total cost of ownership [8].

When considering the relative proportion of the driver costs in relation to the total cost of ownership, it is clear that it is the fastest growing cost component. In 2012, the share of the truck driver costs was still 33%. This fact can be explained by a surplus in an excess demand for drivers and increased salaries for truck drivers (see ▶ Sect. 48.3.1.3).

By automating long-distance haulage, it is expected that the labor costs for the driver on medium hauls can be largely compensated for. Instead, truck drivers could be deployed on the significantly more complex first and last mile (see ▶ Sect. 48.3.1.3).

48.3.1.2 Restrictive Driving Times and Rest Periods

In addition to drivers compensation, another important economic aspect is the regulation for driving times and rest periods. ▶ Table 48.1 provides a condensed and non-exhaustive overview of the driving times and rest periods applicable in the United States and Germany. Labor regulations are intended to ensure safety in road traffic plus the health and safety of the drivers. The table shows that both daily and weekly driving time restrictions apply to drivers. After the permitted driving time is reached, rest periods must be taken during which the driver is not allowed to drive the truck.

Analyzing existing data sets with regard to the real use of trucks shows that the actual vehicle utilization is pretty low. Taking into account the applicable breaks in driving time and rest periods, a truck is at a standstill or parked around 50% of the total time. Only 34% of the total usage time is spent on the actual task of transporting goods. Furthermore, trucks are stuck waiting 9% of the time, while the remaining 6% of the total time is characterized by loading and unloading processes.

On the other hand, highly automated trucks (SAE level 4; also referred to as autonomous trucks in the following, as they must make decisions independently), ref. to Chap. 2, can be available for the transportation of goods for around 22 h per day [11]. As a result, the individual vehicle utilization can be increased accordingly and more freight capacity is available for goods transport overall. This can be illustrated by the transport of goods from the west coast to the east coast of the United States (distance of around 2,500 miles). Taking into account the driving times and rest periods for truck drivers, a truck takes about 4 days for this route (assuming one driver per truck). Using highly automated trucks,

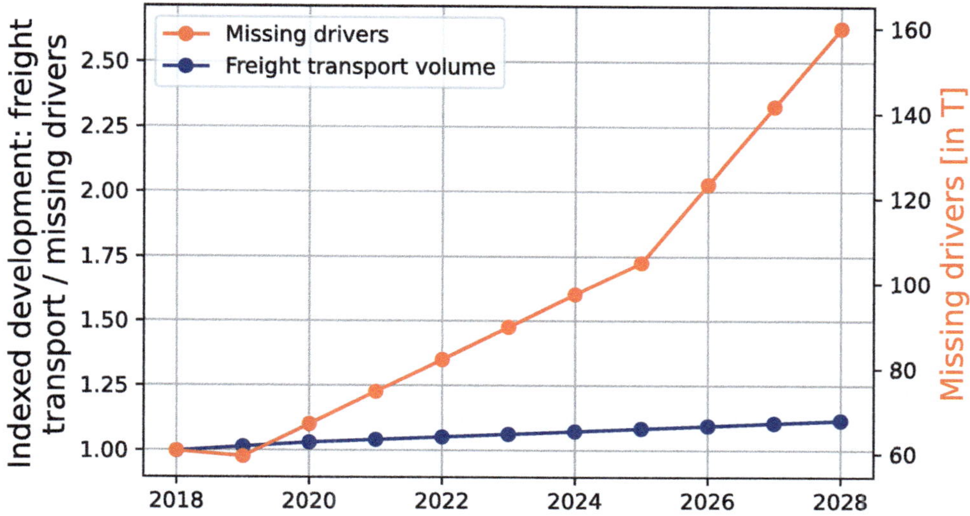

Table 48.1 Overview of daily driving and rest periods in Germany [9] and the United States [10]

		Germany	USA
Driving times	Daily	Max. 9 hrs	Max. 11 hrs
		Twice weekly, max. 10 hrs	
	Weekly	Max. 56 hrs	Max. 60 hrs (7 days)
		Max. 90 hrs in two consecutive weeks	Max. 70 hrs (8 days)
Break in driving time		Min. 45 min. after 4.5 hrs at the wheel It is possible to split the time: 15/30 min	30 min after 8 hrs at the wheel
Rest periods	Daily	Min. 11 hrs	Min. 10 hrs
		Reduction to 9 hrs possible (over 2 weeks)	Can be split

Fig. 48.6 Indicated development of road freight traffic and forecast of driver shortage in the USA by 2028 [14, 18]

this journey can be reduced to around 36 h, which is less than half the conventional driving time [12]. In addition, completely new business models are possible thanks to the significant reduction in goods transport times. It can also be assumed that transport activities will increasingly be shifted to overnight delivery, as the driver's sleep pattern on the medium-haul routes does not have to be taken into account in logistics chain planning. Drivers then take over the regional distribution of the goods at the end of the supply chain.

48.3.1.3 Increasing Shortage of Drivers with Simultaneous Increase in Transport Volume

Another factor is the continuous growth in freight transportation, which comes along with an increasing shortage of truck drivers to transport the goods. A general definition of labor shortages can be formulated as "lack of supply in relation to demand, taking into account prevailing wages and conditions" [13].

In the media, there is talk of high job turnover and a driver shortage, especially among long-distance drivers in the USA. The main reason for the lack of drivers on the labor market is the high average age of the workforce employed in this sector as well as the retirement age. The average age of drivers in 2018 was 46 [14]. Other reasons include poor working conditions with long working hours, long absences and work-related illnesses such as chronic exhaustion and lack of sleep [15–17].

In 2018, the driver shortage summed up to a lack of approximately 60,800 drivers. An extrapolation of the trend shows a shortage of around 100,000 drivers in 2023 and around 160,000 drivers in 2028 (see Fig. 48.6). This corresponds to a percentage growth in the driver resource deficit of around 163% between 2018 and 2028. 110,000 new drivers are required every year in the USA to cover the demand for qualified drivers [14].

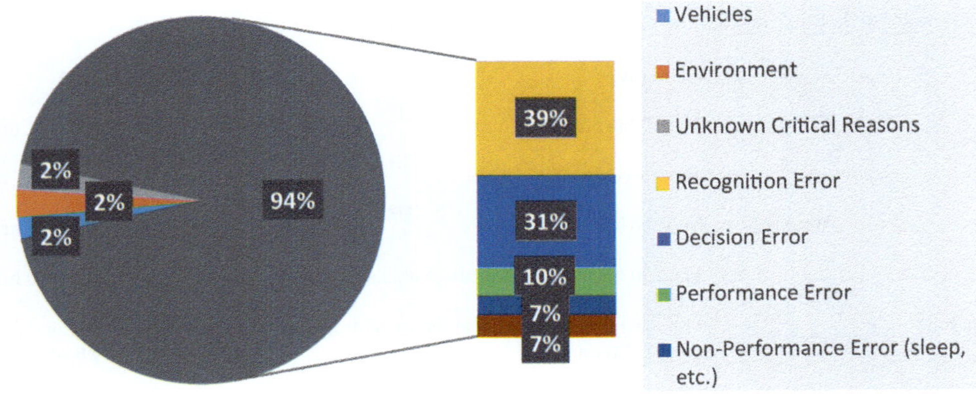

Fig. 48.7 Causes for incidents in US road traffic [19]

Table 48.2 Truck miles covered and accidents on interstates in the USA in 2017 [20, 21]

Truck miles/year on interstates	96,544 million Miles
Accidents involving trucks (>3,5t)	75,500
Of which killed	505
Of which severely injured	2,700
Of which suffered minor injuries	13,500
Of which caused material damage	58,800

An additional driver for the automation of long-haul truck transportation is the steadily increasing volume of freight transportation. The increasing freight volume is intensifying demand on the driver labor market, further exacerbating the driver shortage. **Figure 48.6** shows rising freight transport and a forecast up to 2028. Growth of around 14% between 2020 and 2028 is assumed for road freight transport [18].

Automating trucks on long-distance routes can increase the attractiveness of the truck driver's job profile. This means that the monotonous journeys on long-distance routes are no longer necessary and drivers can take on the more demanding first and last mile sections in which truck automation reaches its limits.

48.3.2 The Factor of Traffic Safety

The automation of trucks offers an opportunity to significantly increase road safety and thus reduce the number of traffic incidents and fatalities. According to the NHTSA, 94% of all traffic incidents in the United States are caused by human driving errors [19]. **Figure 48.7** shows a breakdown of the causes of road traffic accidents. It is clear that a lack of attention or distraction are the main causes in 39% of the incidents caused by drivers in the USA. Inappropriate handling, which can be described as driving too fast as well as misinterpretation of other road users' behavior, causes 31% of traffic incidents. Other reasons for traffic incidents are insufficient vehicle control (10%) and tiredness/falling asleep at the wheel (7%).

Table 48.2 shows the mileage of trucks on US interstate highways as well as the associated incidents involving trucks in 2017. In total, the total truck mileage on US interstates is around 152.6 billion km per year. There are more than 75,000 incidents with direct truck involvement. In this context, there are incidents involving injuries with 500 fatalities and almost 3,000 people suffering serious injuries.

Considering road traffic incident statistics, it can be assumed that highly automated systems will lead to a reduction in the number of incidents. This hypothesis is based on the assumption that automating the task of driving can reduce human error as the main cause of traffic incidents. Automated systems do not become inattentive and cannot be distracted or become tired. Furthermore, they do not take any increased risks but adapt their driving behavior to the current conditions (road traffic regulations, taking into account the speed and distance regulations, light and weather conditions, etc.). However, it is a major challenge to also statistically validate the safety gains to be anticipated from vehicle automation—which are based on assumptions (please see Chap. 46).

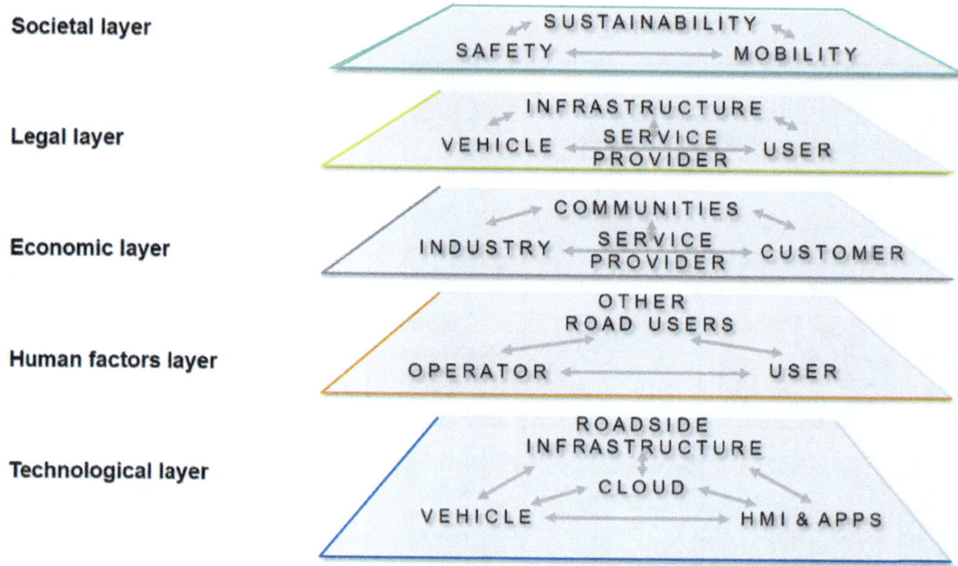

Fig. 48.8 Challenges for autonomous driving [24]

48.3.3 Ecological Factors

The emission of greenhouse gases contributes to the acceleration of climate change. The transport sector accounts for around 20% of global CO_2 emissions [22]. Motorized road transportation accounts for 78% of the CO_2 emitted in the transport sector, with around 44% attributable to individual car transport and 34% to commercial vehicles [23]. The transport sector must also play its part in tackling the climate crisis and reduce the consumption of fossil fuels and the associated emissions.

Initial studies have shown that automating the task of driving can reduce fuel consumption by up to 10% [11]. This can be achieved by optimizing the driving task and more efficient control of the truck. In addition, an increasing number of automated vehicles contributes to improved traffic flow and less congestion overall in the medium-to-long term.

48.4 Fundamental Challenges

Highly automated systems offer the potential to solve many of the previously described challenges of future mobility or to be at least part of the solution. However, the introduction of a new technology to the market usually comes with new challenges. This is also the case for highly automated vehicles. The fundamental challenges of automated driving can be structured in a 5-level model [24] (**Fig. 48.8**).

At the highest level, this model summarizes the societal challenges. It is particularly important to show that the new technology is an advantage for society. One challenge at this level is, for example, to define what level of safety is accepted by society and how this can be determined for automated vehicles.

If these challenges can be resolved, the framework for business models, user behavior, and the technology itself must be defined for the next level down, the regulatory level. For example, at this level, governments must look at how the potential of new innovations needs to be reconciled with protecting the public from the risks of developing technologies.

All economic challenges are summarized at the next level down. In order for a new technology to enter the market sustainably, it must deliver economic benefits to various stakeholders. In the case of automated driving, these could be the manufacturers of the vehicles, service providers, and customers.

The human factors are considered at the next level down. People assume various roles in automated driving. A human might have the role of user, operator, or other traffic user. At this level, it must be clarified how automated vehicles interact with other traffic users.

The lowest level deals with challenges and technical solutions. As a result of the complexity of automated vehicles, it must be clarified, among other things, how the architecture (function, software, and hardware) of the vehicles must be designed efficiently and sustainably. It is also important to develop new methods for validating the vehicles. Many of the challenges are also linked across levels. This can be illustrated with the example of sensor selection. At the technical level, it is important to select a system that can accurately perceive the environment in a wide variety of weather conditions. However, legal regulations (for example, approved frequency ranges, radiation intensities, and data protection) and the costs of the systems must also be considered during the selec-

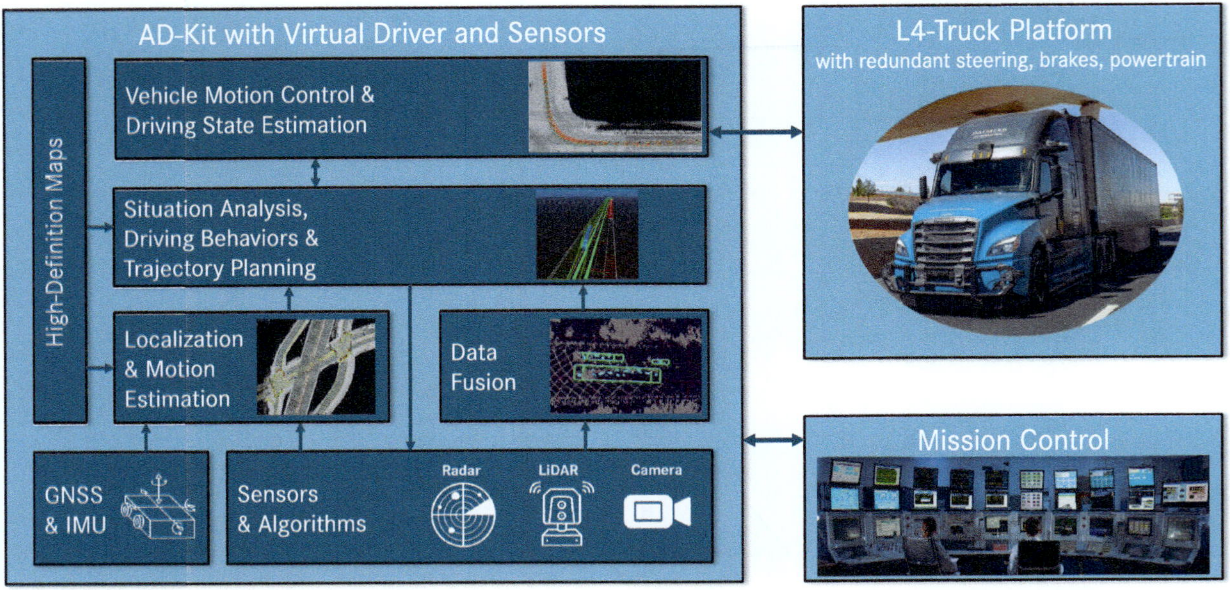

◘ Fig. 48.9   System architecture of a highly automated truck [26]

tion process. For example, if technical implementation is only possible with a more cost-intensive sensor, the challenge on how to maintain the economic advantage despite the higher system costs must be solved at the economic level.

In the following, the focus will be on the technical challenges and solutions.

48.5   The Structure of a Highly Automated Truck

48.5.1   System Architecture

The basic structure of robots for different domains is very similar, regardless of whether it is a vacuuming robot in a private environment or an autonomous truck. Traditionally, the architecture can be divided into three levels [25]:

1. Sense: The robot perceives the environment.
2. Plan: The robot understands the traffic situation and plans its behavior.
3. Act: The robot acts according to the planned behavior.

Applied to autonomous trucks, this means: The vehicle performs all three of these tasks continuously and repetitively in real time. Like a human driver, it is able to recognize a wide range of different objects (for example cars and motorcycles), derive and understand the current traffic situation and derive and carry out the necessary action.

◘ Figure 48.9 illustrates the system architecture of a highly automated truck. This consists of a redundant vehicle platform, the Autonomous Driving Kit (AD kit), and a Mission Control Center (backend). The individual components of the AD kit are the so-called Virtual Driver Software (VD) and the required sensor and computer hardware.

48.5.1.1   The Virtual Driver

The AD kit maps the virtual driver that replaces the function of a human truck driver on certain routes as long as a truck is operated within the approved Operational Design Domain (ODD). The ODD describes the conditions for which a specific automated vehicle is designed. The ODD covers the specific operating conditions in which the automated driving system must operate correctly in terms of roadway infrastructure, speed range, other road users, lighting conditions (day and/or night), weather conditions, and other operating restrictions [27]. The exact specification of the ODD, but also its limits and its robust detection, are key to the development and release of a highly automated truck.

The AD kit system consists of three subsystems: sensors, compute platform, and virtual driver software. The sensor system is responsible for the perception of a vehicle's surroundings. The sensors are selected in such a way that they can detect the entire environment around the vehicle seamlessly and in certain areas, even redundantly. ◘ Figure 48.10 shows the sensor equipment based on a Freightliner Cascadia from Torc Robotics. It consists of a large number of sensors and includes various radars (radio detection and ranging), lidars (light detection and ranging) and camera systems. Lidars and radars are active sensors. Cameras receive the natural light of the surroundings. Each sensor type has its own speci-

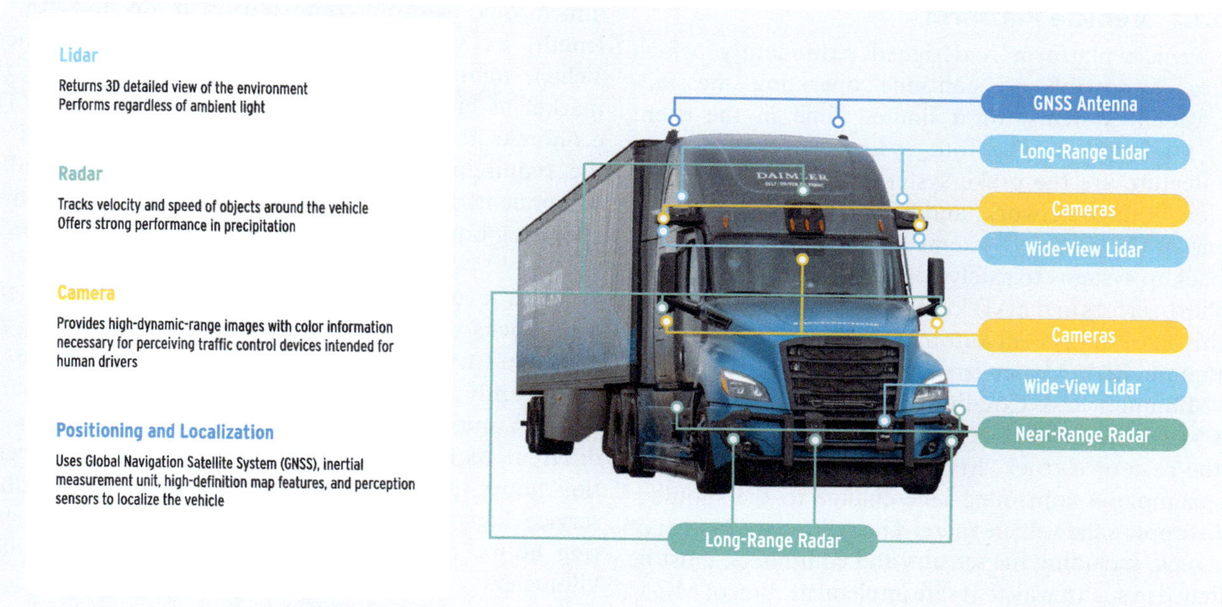

Fig. 48.10 Torc Robotics sensor equipment as an example [28]

fic advantages and disadvantages. Radars, for example, allow the direct calculation of the relative velocity of other road users by utilizing the Doppler effect. In addition, they are significantly less sensitive to weather than other sensor types. However, they are typically not as accurate in terms of angular resolution. Cameras, on the other hand, are high-resolution and allow the classification of objects. However, a direct determination of the distance to other objects using a single camera is only possible by making assumptions (for example on lane level). The various sensor modalities also work in different frequency ranges, which means that they have different sensitivities to environmental conditions such as humidity, rain, and fog. The objective of the sensor configuration is also to use the specific advantages and compensate for their disadvantages in order to achieve highly available and robust system behavior.

The compute platform is home to the virtual driver software. It must provide corresponding connections for the large number of sensors and computing power for the complex algorithms.

The virtual driver software represents the actual intelligence of the system. The data from the sensor system represents the input information for the virtual driver. From these, the environment perception component calculates all relevant environment information, such as the position and size of other road users, the lanes, the position of crash barriers, traffic signs, and traffic lights. At the same time, the environmental information is used to determine the position of your own truck on the digital map (called localization). The digital map contains further information that cannot necessarily be determined directly or only using the compute resources extensively. These include elements, such as traffic regulations, such as unprotected turns' or stop lines at intersections that are not marked. The map information and the results of the environment detection are then represented in the so-called environment model. This is the basis for predicting the behavior of other road users and the evolution of the traffic situation. Decisions on safe behavior are made tactically, strategically, and reactively in different time horizons in line with the levels [25]. There are individual patterns of behavior, for example for intersections, lane changes, merging as well as cyclists or pedestrians. Motion planning is responsible for planning a safe, collision-free, and convenient vehicle trajectory. The central output is a safe path. The vehicle control system uses this information to control the actuators of the truck.

There are several concepts in algorithmic implementation. Among others, these can be distinguished by utilizing pre-recorded and post-processed knowledge in the form of a high-resolution digital map (for example to detect the correct traffic situation) or detecting the situation in real time by the sensors. Furthermore, the systems can be classified whether classic algorithms for decision-making are used or whether the system is trained using machine learning. The latter approach includes learning the complete functional chain from recognition to the correct vehicle response (end-to-end learning, for example, see [29]).

48.5.1.2 Vehicle Platform

The vehicle platform is designed redundantly, which makes it possible to continue operating the fail-operational systems for a limited time in the event of a malfunction. The four systems with redundant architecture are the brake system, the steering system, the low-voltage network, and network communication. If one of the primary systems fails, the vehicle can use its backup systems to safely control the truck. The same applies to errors in the AD kit, for example, sensor failure. In this case, the system automatically executes so-called Minimum Risk Maneuvers (MRM) in order to reach the Minimum Risk Condition (MRC) [27, 30]. MRM and MRC are defined as part of the safety assessment. In the case of a truck driving on the highway, this is, for example, a controlled lane change to the shoulder and stopping the vehicle there. The redundant system of the truck, including the sensors and computers, must be designed in such a way that safe implementation of MRM and MRC is possible even if individual components fail.

48.5.1.3 Mission Control Center

There are several requirements for the operation of a highly automated truck. The objective is to monitor all automated trucks at all times with the ability to intervene if necessary. For this purpose, it is mandatory that the truck is constantly connected to the operator's backend.

- The Mission Control Center (MCC) is responsible for monitoring and controlling the transportation process. Relevant information is exchanged with the truck via the backend. The MCC represents the interface between the operator and its logistics system. For example, the MCC provides the truck with the mission information that it needs to complete the transportation task.
- Continuous tracking of the vehicles and monitoring of the vehicle's state of health. In the event of a fault or breakdown, the vehicle control center is informed so that appropriate action can be taken.
- Constant changes to the infrastructure as a result of construction or roadworks make it necessary to update the map information continuously and keep the vehicles up-to-date. For this purpose, the autonomous truck constantly sends information to the backend. Updated maps are uploaded to the truck.
- If the truck gets 'stuck' in a situation, i.e., it is unable to find its own solution for the current situation, an operator can help get the vehicle back on track (Teleguidance).

48.5.2 Challenges in Automating a Truck

Cars and trucks differ significantly in terms of weight (up to 2.5 t compared with up to 40 t), the vehicle dimensions (4–6 m compared with up to 20 m in length) as well as the required braking distances. Vehicle manufacturers position passenger cars on the market as emotionally charged consumer goods. The economic and technical design of a car is based on the requirements placed on a vehicle. The installed components are developed on the basis of the required operating hours and the mileage throughout the service life.

On the other hand, commercial vehicles, in this case trucks, are capital assets that form the link for cross-industry value chains. They are subjected to significantly higher loads than cars, loads which the vehicles must withstand. This inevitably results in different requirements in terms of service life, operating hours, and mileage. Trucks are designed with a service life of at least 10 years, 25,000–30,000 operating hours, and a mileage of more than 1.3 million kilometers.

Automated cars (self-driving cars or SDC) and automated trucks (SDT or self-driving trucks) differ greatly also in terms of vehicle automation, ref. to Table (◘ Table 48.3). The SDC use case typically focuses primarily on ride-hailing vehicles in densely populated urban areas with high demand. The urban use case provides a high level of traffic complexity as a result of the diverse motion patterns of other road users such as pedestrians and cyclists. SDTs, on the other hand, are used on selected routes along the more structured highway environment. Simpler road geometry and less dynamic objects make this use case seem easier to master. Structurally separated direction lanes and thus only road users traveling in one direction support the simpler structure of the traffic to be taken into consideration.

However, the higher inherent speed of the truck—up to 75 miles per hour in the USA—means that objects must be detected at a larger distance in order to take the longer braking distances into account. Visibility also plays a decisive role when discussing use cases such as merging onto the highway. The limited acceleration capacity compared with a passenger car makes it necessary to plan over a larger planning horizon. This poses a major challenge, especially for the sensor system.

The sensors of a passenger car can be positioned so that they allow for complete 360° all-around visibility and thus seamless monitoring of the traffic area around the vehicle. In the case of a truck, providing complete all-round visibility requires the installation of sensors on the trailer. This is difficult for the logistics chain to achieve, as, for example, a ratio tractor to trailer of 1 to 3 can be expected in the USA. In this respect, the focus is on installing sensors only on the tractor. Consequently, environment perception must also be optimized for the correct tracking of objects in the trailer's blind spot .

Table 48.3 Comparison of highly automated cars and trucks

	Passenger Vehicles (RoboTaxis)	Autonomous Trucks (Hub2Hub)
Use case	Urban (70 km/h)	Highway (110 km/h)
Weight and Length	2 t/6 m	40 t/up to 20 m
Sensor requirements	Detection and classification of highly complex objects	Extended detection based on speed and mass
Road users	Unpredictability (pedestrians, cyclists)	Structured traffic, incidents with lost cargo

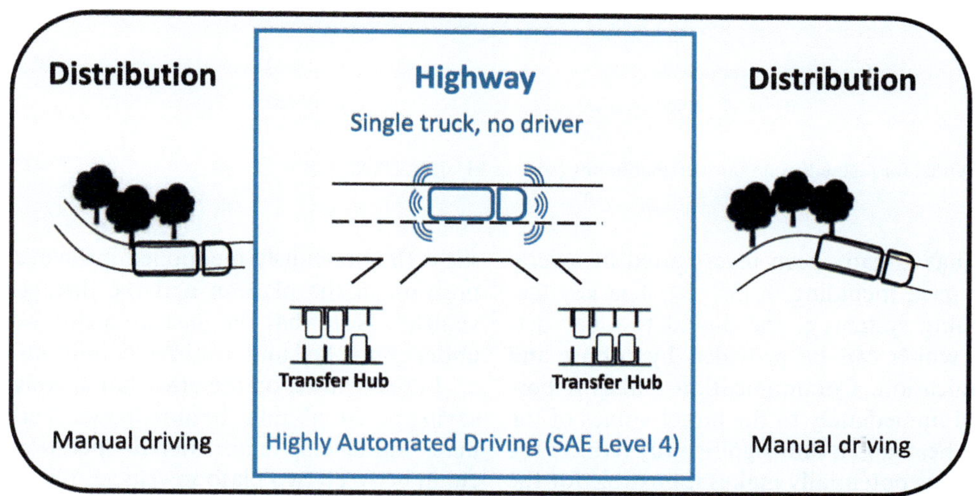

Fig. 48.11 Operational design domain for the hub-to-hub use case [26]

48.6 "Middle Mile" Case Study

There is great potential in the automation of freight transportation as has become clear in the previous sections. But it also poses a number of challenges. In contrast to chaotic inner-city traffic, one of the focal points for freight transport is the automation of the middle mile with the more structured highway environment. In this case, the technological objectives and validation of continuously safe behavior of the automated vehicle appear to be achievable earlier on, if the ODD is chosen correctly.

48.6.1 Introduction to the Hub-to-Hub Concept

In logistics, the so-called 'middle mile' refers to the link in the supply chain in which the goods are transported from a warehouse, a distribution center, or a transshipment point over a large distance to another similar location in the logistics network. This includes processes to reach the hub, hub-to-hub transports, and hub-to-depot processes (see **Fig. 48.4**). In this section, hubs are defined as such transshipment centers. **Figure 48.11** illustrates this use case.

The driver must be removed from the cab in order to leverage the potential described above. For the so-called hub-to-hub concept, this means that the goods are transported by a human driver to a hub near the highway or that the network's hubs already have highway connections. The highly automated truck then drives from the hub to the highway and the long distance along the highway to the target hub. At that point, a human driver resumes the further journey and the distribution of the goods.

As indicated in ▶ Sect. 48.5.2, automation on the highway is also a major challenge. However, it is significantly smaller than tackling inner-city traffic with a truck.

48.6.2 Adding Platooning

An expansion of the SDT system to include robust and safe communication between several autonomous trucks offers the potential to convert several vehicles into a convoy or platoon. Therefore the distances between the vehicles can be reduced and it allows for better coordination of the driving strategies. Potentially, there are positive effects on fuel consumption [31, 32] and traffic flow [33, 34], which have been explained in more detail below.

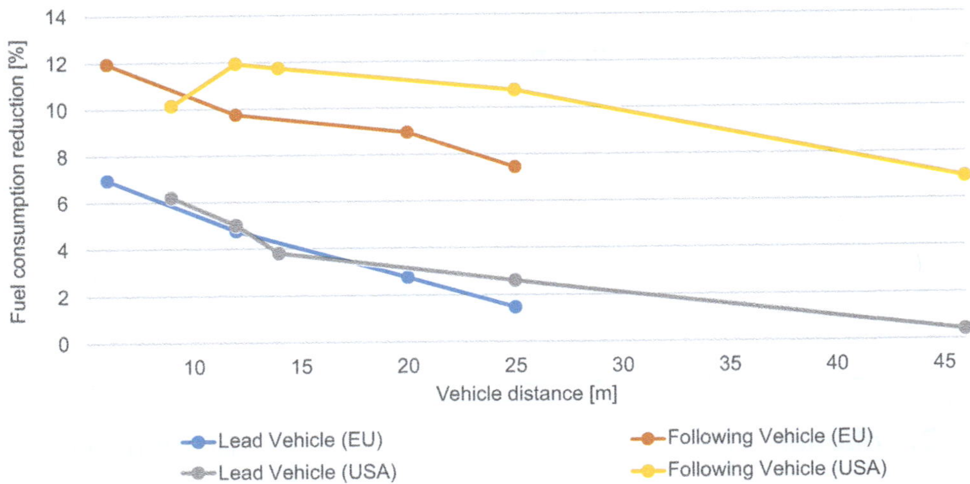

◘ Fig. 48.12   Potential for fuel reduction through platooning (*sources* EU: [31], USA: [32])

Platooning has already been investigated in several projects in the past, including in [35–37]. The key feature of platooning systems is the digital tow bar between vehicles, which can be provided by robust and secure communication. Communication makes it possible to respond immediately to the target values of the vehicle driving ahead and it thus significantly reduces the response time. This potentially makes it possible for the distances between the vehicles to be significantly reduced. This advantage can be easily explained using a braking maneuver. The vehicle convoy is informed immediately via communication if the system or the driver of the vehicle ahead sets negative acceleration. As a result, all vehicles in the convoy can respond with minimal latency. Without direct communication, it would only be possible to respond once the sensors of the following vehicle had detected the vehicle's reaction.

The potential to reduce the distance between vehicles is a decisive advantage of platooning systems. This makes it possible to optimize traffic flow [33], and to reduce fuel consumption thanks to aerodynamic advantages [34]. The shorter distances lead to a reduction in the headwind pressure at the front and the negative pressure at the rear of the vehicle. The potential depends on the position in the platoon, the distance between the vehicles, and the traffic context.

The resulting fuel consumption advantages are shown in ◘ Fig. 48.12. The graph shows measurements on test tracks in the USA and Germany for different distances. When evaluating the graph, note that the tests were carried out at a speed of 80 km/h in Europe and at 105 km/h (65 mph) in the USA. However, the influence of speed on the relative consumption advantages in the speed range between 80 and 105 km/h can be neglected as a result of studies [38], since the differences are also influenced by other factors, such as the different vehicle shapes and configurations as well as the system design. The results show the previously mentioned influence of the vehicle position in the platoon and the distance between the vehicles. Note that the measurements were carried out under controlled and idealized conditions on test tracks.

In real traffic, on the other hand, considerably lower saving potentials must be anticipated. This is the result of many influencing factors in real operation. Vehicles that cut in between two platoon vehicles play a key role here. In these situations, a safety distance to the vehicle cutting in must be established. Merging vehicles are particularly common when driving onto or off a highway. Coupling and de-coupling the platoon is energy-inefficient. This inefficiency must be offset by the advantages of platoon driving in order for the overall journey to provide an efficiency advantage. In addition, significant optimizations in trucks' aerodynamics have been achieved in recent years [39], which are expected to lead to a reduction in the actually achievable savings. At the same time, it must be ensured that the engines of the successor vehicles are cooled correctly, as these may tend to overheat given the reduced distance [40].

In addition, the braking potential of trucks varies greatly and depends significantly on the load. In order to prevent rear-end collisions within the platoon, the order of the vehicles in the platoon and the safety distance must be determined in relation to the braking potential of all vehicles. This requires robust identification of the relevant parameters.

In principle, a platoon can be set up for each automation level 1 to 5, but considerable potential can only be exploited at level 4, i.e., if the driver can be removed from at least one cab. The basis for this is a fully automated truck with AD kit equipment as described above, in order to ensure that coupling and de-coupling of the platoon works robustly and safely. It also seems impossible to drive an entire platoon onto a highway without appropriate structural measures. Pre-assembled platoons are

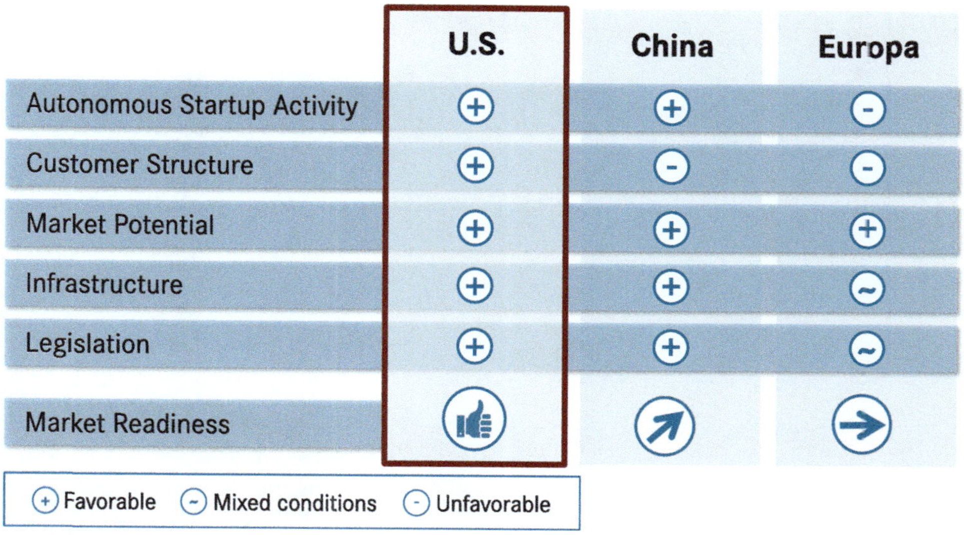

	U.S.	China	Europa
Autonomous Startup Activity	+	+	−
Customer Structure	+	−	−
Market Potential	+	+	+
Infrastructure	+	+	~
Legislation	+	+	~
Market Readiness	👍	↗	→

+ Favorable   ~ Mixed conditions   − Unfavorable

Fig. 48.13 Comparison of autonomous truck markets [26]

only possible if there are larger transportation quantities with the same origin and destination. In practice, this occurs less frequently, which reduces the frequency of this application. It also limits the flexibility of transportation companies.

From these observations, the conclusion is, that platoons do not provide a quicker introduction of automated trucks as originally hoped, but instead present a number of new and complex challenges. Platooning can be a useful extension as soon as a high number of highly automated trucks are on the road. However, in order to be effective, it is imperative that trucks can build platoons on an ad hoc basis and do not have to be predefined. This requires a cross-manufacturer definition of the requirements and interfaces.

48.6.3 Choice of Region and Initial Market

In addition to the technical restrictions arising for operation from the ODD, there are other regulatory and economic conditions for the introduction of highly automated trucks in various regions and markets. ▪ Figure 48.13 gives an overview of the three current main markets and their assessment based on various criteria. Note that as of 2021, a driverless truck cannot be certified in any of the regions.

In addition to the ODD, topics such as traffic density, the length of the relevant routes, but also the prevailing relative vehicle speeds must be taken into account. In the USA, for example, trucks drive on highways within speed ranges similar to those of passenger cars. At 55 to 75 mph, this is significantly higher than Europe's 80 km/h, but it leads to relatively harmonized traffic. In Germany, on the other hand, relative speeds can easily exceed 100 km/h. When the road infrastructure in the two regions is taken into account, it becomes clear that the lane and road widths in the USA—in terms of highways—are significantly larger than in many European countries. The distances covered in the USA are also significantly longer. In Europe, transportation routes are predominantly over shorter distances and within narrow national borders overall. As already mentioned in ▶ Sect. 48.3.1.1, wage costs for truck drivers are very high, especially in the USA. Taking into account economic, legal and technical aspects, the USA therefore appears to be the preferred initial market for the introduction of highly automated trucks.

Looking specifically at the use case of hub-to-hub driving in the USA, it becomes clear that three main routes from east to west in the southern US states, the so-called 'Sunbelt', are particularly suitable (see ▪ Fig. 48.14). Significant proportions of goods are transported along these routes. At the same time, the weather conditions are more favorable here, as it is dry for a very large part of the year and there is only very little rainfall. Snow and icy conditions occur only in very few parts and very rarely. These conditions promote a faster introduction of highly automated trucks.

As already described in ▶ Sect. 48.5.2, the development of a highly automated truck is technologically very challenging. The objective is to minimize complexity by choosing the right ODD. Key is to focus on the structured surroundings of the highway as much as possible. The correct choice of hubs, including their proximity to the highway, minimal complexity of the route to and from the highway to the hub, as well as the design of the on-ramps, play important roles.

For the successful introduction of highly automated driving in this ODD, validation must be provided (Please see Chap. 46) that the system responds reliably in all

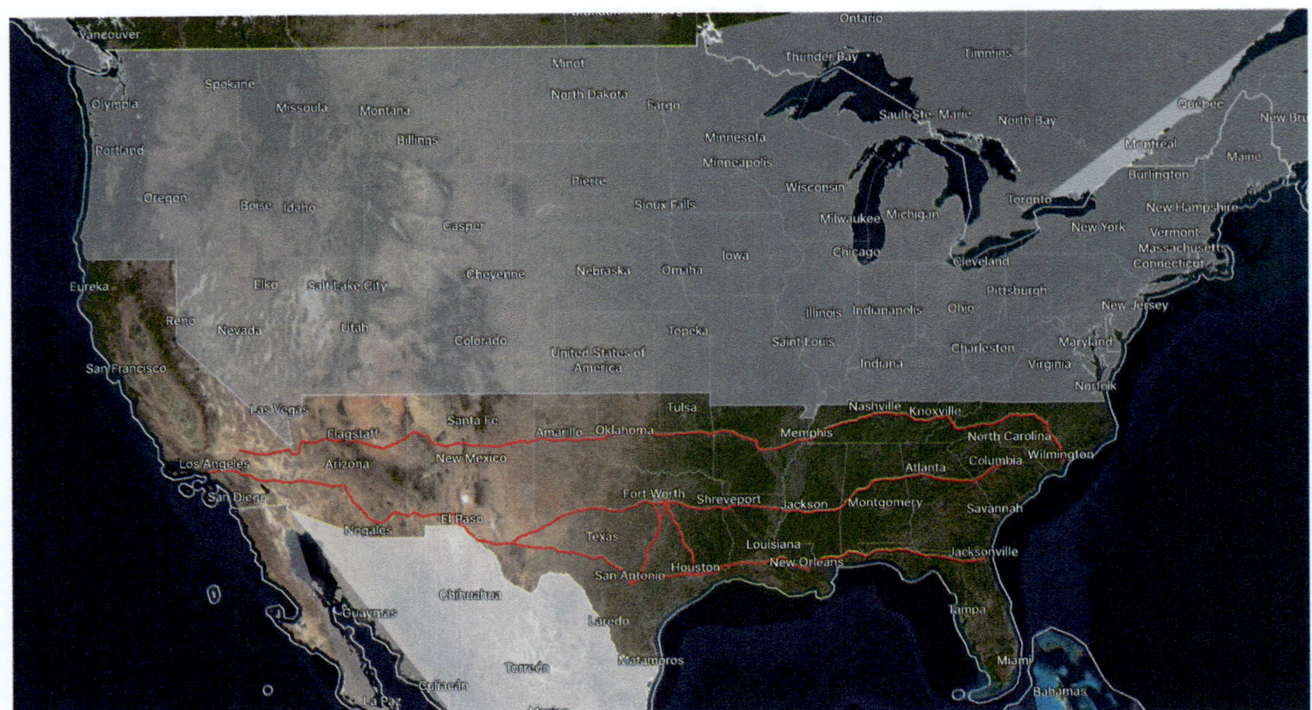

■ **Fig. 48.14** The Sunbelt in the south of the USA with interstates I10, I20 and I40

■ **Fig. 48.15** Autocargo for the transfer to a public parcel box storage facility (*Source* Project Unicaragil (▶ www.unicaragil.de))

situations. Thereafter there will be only a few restrictions in the expansion of the use cases and the ODD.

48.7 Case Study "Last Mile Delivery" Preparation and Follow-up

Preparation and follow-up pose different challenges than the "middle mile". These challenges are discussed using parcel delivery as an example, as this part of the process possesses the most difficult requirements. The following tasks must be completed for successful, autonomous delivery of a package:

Identifying the destination – Not all regions have address systems with a system for city/town—street—house number that support determining the destination. Systems such as *what3words*[2] represent a solution for narrowing down the destination options, but such a system must be established first on a broader scale. If the shipment is to be handed over to a waiting person, who can move toward a delivery vehicle, the destination has been sufficiently determined. However, if a transfer station ("letterbox 2.0") is to be used, this must be detected and a docking maneuver must be carried out. The destination must be reachable at ground level, which is why, from today's point of view, a transfer station is required for multi-story buildings.

Drive to destination – Driving from the current location to the destination requires autonomous driving in municipal, urban, and rural areas. Not only roads are used, but depending on the size of the vehicle, also sidewalks and pedestrian crosswalks. Conceptually, these vehicles function like the highly automated trucks described above. However, due to size, weight, and cost, such comprehensive sensor equipment cannot be implemented. At the same time, these vehicles, which have been specially developed for the application, also drive relatively slow. If the package must be delivered to a transfer station, a position must be reached that allows a transfer. Tight spatial conditions at the transfer point result in high requirements for the mobility of the vehicle and thus for the chassis kinematics.

Transfer of the shipment – Two concepts are currently being pursued for the hand-over of the consignment:

Handover to a person – If the shipment is handed over to a person, it is sufficient to grant access to the vehicle's load compartment, while making sure, that only designated parcels are available. In this case, the arrival of the vehicle must be synchronized with the actual presence of the recipient. For this purpose, this person must be located near the destination and must be notified when the delivery vehicle has reached the destination position.

Transfer to a transfer station – On the other hand, if the shipment is handed over to a transfer station, a handling device is required in the vehicle or in the statio-

nary transfer station. The shipment is stored in the transfer station until it is picked-up by the recipient.

Security against theft – When transferring the shipment, it must be ensured that it is only handed over to authorized persons. This is done by authorizing the recipient at the vehicle or at the transfer station.

At present, several research and pilot projects are examining two fundamental approaches to automating preparation and follow-up based on autonomous driving.

48.7.1 Example of AutoCargo

One variant consists of large vehicles that transport many shipments simultaneously, driving to a series of destinations and independently transferring the shipments to transfer stations (see ◘ Fig. 48.15).

As part of the UNICARagil project funded by the BMBF, German universities leading in the field of automated driving have teamed up with selected industry specialists to rethink automated vehicles and their architecture. Based on the latest research findings on automated and networked driving as well as electromobility, the project will develop disruptive modular architectures made up of hardware and software components for driverless vehicle concepts (please see Chap. 51).

The autonomous delivery vehicle drives at moderate speed on public roads and, if permitted, on private roads, and thus flows with the traffic. In the case of deliveries during off-peak hours (for example, at night), the vehicle could also be driven more slowly to increase safety and, if necessary, reduce the noise level.

Thanks to the dynamic modules with a steering angle of up to 90°C, the vehicle has great mobility to navigate up close to the necessary transfer stations. In these prototypes, the transfer is carried out by a robot arm with seven degrees of freedom, which can reach all packages in the load compartment. It picks up the package to be delivered with the aid of a suction gripper and places it through the vehicle opening on an extended platform of the transfer station. The packages are placed in an approximate sequence of the scheduled stops in an exchangeable container at the depot so that they can be delivered during the trip. For space efficiency reasons, shelves were not installed as the heterogeneity of the packages is estimated to be so high that it would generate too much empty space with standardized shelf height and depth. It must be anticipated that the position of the packages can change throughout the journey, as there is no stable stacking pattern at all times. Therefore, it must be possible to identify the packages in the load compartment and determine their position using sensors inside the vehicle so that the transfer can take place using the robot arm. After positioning, the vehicle communicates with the transfer station

---

2   See ▸ https://what3words.com/.

Fig. 48.16   Overview of efeu-Campus (left) and vehicle (right), (*source* Project efeu-Campus (▶ https://efeucampus-bruchsal.de))

to enable the hand-over. At the same time, the area is secured to rule out theft of the parcel during the transfer.

Packages (for example, returns) can also be accepted in the same way, however, managing the packages within the load compartment poses additional challenges in this case. A comparatively small, separate area in the vehicle is planned for the start.

The challenge with this concept lies in the necessary infrastructure on the recipient side, which must be established.

48.7.2   Example of the Efeu-Campus

The efeu-Campus project is funded by the European Regional Development Fund (ERDF) for Innovation and Energy Transition, as well as by the state of Baden-Württemberg. The facility is located in Bruchsal. Among other things, the aim is to show how a low-emission residential district of the future can be created with innovative transport concepts for the "last mile" (▪ Fig. 48.16).

The efeu-Campus' approach is to bundle the deliveries for a district at a central transfer point, the so-called QuartiersDepot. Here, the transport of shipments within the district is separated from transport via the large-scale transport network. At this point, storage and, if necessary, support for the loading and unloading of non-standardized goods is available. Small-scale vehicles with limited speed and capacity travel within the district. At the request of the residents, vehicles bring shipments from the central transfer point to the destination where the person removes the shipments from the vehicle after authorization has been given. Standardized transport containers are used. Vice versa, residents can also initiate transport from the place of residence to the transfer point and the subsequent onward transport. In the district, residents can decide individually whether they want synchronous delivery, i.e., with a personal hand-over, or asynchronous delivery and collection. In this case, a transfer station is required where the transportation box is handed over.

The vehicles drive on the streets and sidewalks of the district and are subject to the separate requirements of the district. Navigation is based on a map of the district generated with the help of LiDAR sensors. Vehicles avoid other road users and people in particular as long as space allows for it. Otherwise, they stop and wait until the situation is resolved, for instance by remote control. In addition, the vehicle has a safety architecture similar to that of indoor robots in order to stop safely in the event of a predicted collision.

48.8   Summary and Outlook

This section shows that automation in freight transportation has great potential. Practical implementation essentially depends on four factors:

- Regulatory premise for the use of autonomous driving
- Provision of infrastructure for autonomous driving
- Technological progress and future developments
- Acceptance of autonomous driving in the society.

At the moment, the various applications are being intensively researched and developed. It remains exciting to see the introduction and the future potential of autonomous freight transportation.

References

1. OECD: Decoupling the Environmental Impacts of Transport from Economic Growth. OECD (2006). https://doi.org/10.1787/9789264027138-en, https://www.oecd-ilibrary.org/content/publication/9789264027138-en
2. Statista: Anteil der Lkw an der Transportleistung im Güterverkehr in Deutschland in den Jahren von 2013 bis 2024 (2022). https://de.statista.com/statistik/daten/studie/12195/umfrage/anteil-der-lkw-am-gueterverkehr-in-deutschland/#professional
3. Arnold, D., Kuehn, A., Isermann, H., Tempelmeier, H., Furmans, K.: Handbuch Logistik. Springer (2007)
4. Blunck, S.: Modellierung und Optimierung von Hub-and-Spoke-Netzen mit beschränkter Sortierkapazität. KIT Scientific Publishing (2005)
5. Maiden, T.: How big is the logistics industry? (2020). https://www.freightwaves.com/news/how-big-is-the-logistics-industry
6. Analysts, G.I.: Global Freight Trucking Industry (2020). https://www.researchandmarkets.com/reports/4663018/freight-trucking-market-global-industry
7. Associations, A.T.: Trucking moved 11.84 billion tons of freight in 2019 (2019). https://www.trucking.org/news-insights/trucking-moved-1184-billion-tons-freight-2019
8. Williams, N., Murray, D.: American Transportation Research Insitute (2020)
9. Für Güterverkehr, B.: Fahrpersonalrecht (2021). https://www.bag.bund.de/DE/Themen/RechtsentwicklungRechtsvorschriften/Rechtsvorschriften/Fahrpersonalrecht/fahrpersonalrecht_node.html
10. Administration, F.M.C.S.: Summary of hours of service regulations (2021). https://www.fmcsa.dot.gov/regulations/hours-service/summary-hours-service-regulations
11. TuSimple. U.S. Securities and Exchange Commission Registration Statement (2021). https://www.sec.gov/Archives/edgar/data/1823593/000119312521091150/d909743ds1.htm
12. Reuters: TuSimple to demo self-driving trucks in Q4 2021 (2021). https://www.reuters.com/business/autos-transportation/tusimple-demo-self-driving-trucks-q4-source-2021-04-15/
13. Richardson, S.: What Is a Skill Shortage? ERIC (2007)
14. Costello, B., Karickhoff, A.: Truck driver shortage analysis 2019. Technical report, American Trucking Associations (2019)
15. Häkkänen, H., Summala, H.: Accid. Anal. Prev. **33**(2), 187 (2001)
16. Beilock, R.: Accid. Anal. Prev. **27**(1), 33 (1995)
17. Adams-Guppy, J., Guppy, A.: Ergonomics **46**(8), 763 (2003)
18. B. of Transportation Statistics: Weight of shipment by transportation mode (2019). https://www.bts.dot.gov/browse-statistical-products-and-data/freight-facts-and-figures/weight-shipments-transportation-mode
19. N.H.T.S. Administration: US Department of Transportation, Washington, DC, pp. 1–3 (2018)
20. F.H. Administration: Highway statistics 2019 (2020). https://www.fhwa.dot.gov/policyinformation/statistics/2019/vm1.cfm
21. N.H.T.S. Administration: Large trucks crash stats (2018). https://crashstats.nhtsa.dot.gov/api/public/viewpublication/812663
22. Commission, E., Centre, J.R., Olivier, J., Guizzardi, D., Schaaf, E., Solazzo, E., Crippa, M., Vignati, E., Banja, M., Muntean, M., Grassi, G., Monforti-Ferrario, F., Rossi, S.: GHG Emissions of All World : 2021 Report (Publications Office, 2021). https://doi.org/10.2760/074804
23. McBain, S., Teter, J.: Tracking transport 2021. Technical report, IEA (2021). www.iea.org/reports/tracking-transport-2021
24. VDI-Handlungsempfehlung Dezember 2019–Automatisiertes und autonomes Fahren (2019)
25. Siciliano, B., Khatib, O., Kröger, T.: Springer Handbook of Robotics, vol. 200. Springer (2008)
26. Daimler Truck AG (2022)
27. S. International: Taxonomy and definitions for terms related to driving automation systems for on-road motor vehicles (2021). https://doi.org/10.4271/J3016_202104
28. Torc Robotics: Innovating safety and efficiency. Technical report (2022). https://torc.ai/wp-content/uploads/2022/01/Torc-Robotics-VSSA.pdf
29. Bojarski, M., Testa, D., Dworakowski, D., Firner, B., Flepp, B., Goyal, P., Jackel, L., Monfort, M., Muller, U., Zhang, J., Zhang, X., Zhao, J., Zieba, K. (2016) arXiv preprint arXiv:1604.07316
30. Wood, M., Robbel, P., Maass, M., Tebbens, R.D., Meijs, M., Harb, M., Schlicht, P.: Safety first for automated driving. Technical report, Aptiv, Audi, BMW, Baidu, Continental Teves, Daimler, FCA, HERE, Infineon Technologies, Intel, Volkswagen (2019)
31. Davila, A.: Report on fuel consumption: SARTRE (2013)
32. McAuliffe, B., Raeesi, A., Lammert, M., Smith, P., Hoffman, M., Bevly, D.: In: SAE Technical Paper Series. SAE International400 Commonwealth Drive, Warrendale, PA, United States (2020), SAE Technical Paper Series.https://doi.org/10.4271/2020-01-0679
33. Kotte, J., Huang, Q., Zlocki, A.: In: 19th ITS World Congress, ed. by ITS. Wien (2012)
34. Shladover, S.: Introduction to Truck Platooning (2017)
35. Bonnet, C.: Chauffeur 2 final presentation (07.05.2003). http://www.itsforum.gr.jp/Public/E4Meetings/P01/fremont5_2_2.pdf
36. Robinson, T., Coelingh, E.: Operating Platoons On Public Motorways: An Introduction to the SARTRE Platooning Programme (2010). https://www.researchgate.net/profile/erik-coelingh/publication/268300380_operating_platoons_on_public_motorways_an_introduction_to_the_sartre_platooning_programme
37. KONVOI: Verbundprojekt KONVOI: Entwicklung und Untersuchung des Einsatzes von elektronisch gekoppelten LKW-KONVOIs: Abschlussbericht
38. McAuliffe, B.R.R., Croken, M., Ahmadi-Baloutaki, M.B., Raeesi, A.: Fuel-economy testing of a three-vehicle truck platooning system. https://doi.org/10.4224/23001922
39. Georgiev, P., Cornelis, S.: US truck fuel efficiency standards: costs and benefits compared. Tech. rep., Transport&Environment (2018). https://www.transportenvironment.org/wp-content/uploads/2021/07/2018%2001%2010%20US%20truck%20standards%20FINAL.pdf
40. Zhang, C., Lammert, M.P.: Impact to Cooling Airflow from Truck Platooning. Technical report, National Renewable Energy Lab.(NREL), Golden, CO (United States) (2020)

Open Access This chapter is licensed under the terms of the Creative Commons Attribution-NonCommercial-NoDerivatives 4.0 International License (▶ http://creativecommons.org/licenses/by-nc-nd/4.0/), which permits any noncommercial use, sharing, distribution and reproduction in any medium or format, as long as you give appropriate credit to the original author(s) and the source, provide a link to the Creative Commons license and indicate if you modified the licensed material. You do not have permission under this license to share adapted material derived from this chapter or parts of it.

The images or other third party material in this chapter are included in the chapter's Creative Commons license, unless indicated otherwise in a credit line to the material. If material is not included in the chapter's Creative Commons license and your intended use is not permitted by statutory regulation or exceeds the permitted use, you will need to obtain permission directly from the copyright holder.

Driverless Shuttles as a Supplement to Local Public Transport

Timo Woopen and Lutz Eckstein

Contents

© The Author(s) 2026
H. Winner et al. (eds.), *Handbook Assisted and Automated Driving*,
https://doi.org/10.1007/978-3-658-45276-6_49

49.1 Introduction

This chapter introduces the special case of driverless shuttles, which supplement local public transportation with a flexible component. To this end, the specific operational design domain of low-speed automated driving (LSAD) is examined in more detail and analyzed at various layers.

> **Operational Design Domain (ODD)**
>
> Operating conditions for which a particular automated driving system or feature is designed to function, including, but not limited to, environmental, geographic, and time-of-day constraints and/or the required presence or absence of certain traffic or roadway features [1].

The ODD describes the general conditions under which a vehicle can move in road traffic and to which driving functions are to be designed accordingly. The restrictions of an ODD are often selected in such a way that technically controllable situations for vehicle automation are delimited, enabling early testing and roll-out of functionalities. It thus provides the framework both for development and validation as well as for approval of the respective automated driving functions.

> The following examples are given in SAE J2016 [1] for specifying an ODD:
> - An *ADS feature* is designed to *operate* a *vehicle* only on fully access-controlled freeways in low-speed traffic, under fair weather conditions and optimal road maintenance conditions (e.g., good lane markings and not under construction).
> - An *ADS-dedicated vehicle* is designed to *operate* only within a geographically defined military base, and only during daylight at speeds not to exceed 25 mph.
> - An *ADS-dedicated* commercial truck is designed to pick up parts from a geofenced sea port and deliver them via a specific route to a distribution center located 30 miles away. The *vehicle's ODD* is limited to day-time operation within the specified sea port and the specific roads that constitute the prescribed route between the sea port and the distribution center.

These examples show that the constraints of an ODD have a significant impact on the subsequent use of the specific automated vehicle. An ODD can start with "simple" conditions, such as a specified route in specified environmental and ambient conditions, and extend to mastering an entire route network in a large city at day time and night time and under all weather conditions. The requirements for mastering these ODDs imposed on automated vehicles vary enormously.

The ODD may define several parameters. Among others, these include

- maximum permissible vehicle speed;
- operation venue;
- road conditions;
- environment conditions, such as structurally separated lanes;
- limitations to permitted surrounding traffic;
- operating time; and
- weather conditions.

In order to ensure a uniformly standardized description of the ODD, the Association for Standardization of Automation and Measuring Systems (ASAM e.V.) is striving for standardization in the form of a uniform, machine-readable description format for operational design domains [2]. Such a format enables the integration of a vehicle-specific ODD into different simulation environments to facilitate the development and validation processes of automated driving functions.

The ODD therefore takes a decisive role in driver assistance systems and automated driving functions of automation levels one to four according to SAE definition [1]. While level 0 assistance functions are not ODD restricted and provide situational support to the driver when needed, level 1 and 2 systems are actively enabled by the user in certain speed ranges and under prescribed environmental conditions. Level 3 and 4 systems, by definition, may only operate in their specific ODD and take over the driving task within that ODD. Only Level 5 systems are not restricted by an ODD according to SAE and can operate in all conditions that a driver can also control.

The following chapters will now introduce the specific ODD of driverless shuttles as well as their respective specifics. For this purpose, an overview of current shuttle concepts, pilot tests, and research activities is given before the operational design domain LSAD is specifically classified.

49.2 State of the Art

Driverless shuttles of simple characteristics are in commercial production since 2014 [3] and a limited number of them are used in pilot tests in cities and municipalities in many countries. In general, they can be characterized as low-speed automated vehicles that operate on predefined routes. For their development and use, international standardization committees are working on the definition of performance (speeds, object recognition) and system requirements (data recording) as well as requirements for test procedures to establish the automation systems and vehicles [4]. Partly, the concepts and application areas significantly differ. In this chapter, the best-known shuttle concepts are presented first, before pilot applications and their specific ODD are summarized. In addition, an overview of current research projects in the field of driverless shuttle concepts is given.

Driverless shuttles travel on very short (often only a few hundred meters) and predefined routes in pilot tests. These routes are mostly mapped in high detail and selected to exclude challenging scenarios, such as left turns on busy intersections. The vehicles have highly accurate satellite-based localization mechanisms and thus mostly move along a previously defined path. In most cases, independent and predictive behavior and trajectory planning takes place only for longitudinal dynamic maneuvers. For the time being, a situation-dependent re-planning of the path or trajectory is still omitted. Nevertheless, these functionalities are researched and improved in parallel developments. It is therefore to be expected that driverless shuttles may also include further functionalities in the future.

49.2.1 Available Shuttle Concepts

Looking at the various vehicles and prototypes present in public road traffic as well as the pilot applications carried out worldwide, it is noticeable that a large proportion of the applications rely on the vehicles of only a few manufacturers. This chapter compares current concepts and highlights commonalities as well as differences. The resulting findings provide information on the specific ODD for low-speed automated vehicles.

Basically, the concepts can be divided into two categories: those vehicles with an integrated driver's workspace and those that do not provide direct vehicle control for a human on board. These vehicles usually only have attendants on board as supervisors for the automated driving function.

> An attendant accompanies the journey of a driverless shuttle on board of the vehicle with the task of monitoring the automated ride and, if necessary, intervening in the execution via a previously defined interface. In most cases, a stopping maneuver is initiated and there is no direct intervention in the vehicle control. In addition, the attendant can control the vehicles outside of the automated driving function or ODD using suitable actuators, which are usually in the form of a gaming console controler or a stick.

By eliminating the driver's workplace, these concepts offer more efficient space and higher capacities. The vehicles can act autonomously within their specific ODD and are therefore mostly assigned to automation level 4 (automated driving). ◘ Table 49.1 provides an overview of shuttle vehicles deployed on public roads in 2021 and their specific characteristics.

The shuttle concepts listed above have common features in terms of maximum speeds and passenger capacity. All vehicles move at a maximum speed of 60 km/h and carry between 10 and 22 passengers. While the specified speed range is comparable to that of conventional buses in public transport, the vessel size, defining the maximum number of passengers, differs from currently used city buses. The implementation of the automated driving function varies between the different vehicles and demonstration sites. Vehicles that pursue simple approaches to automation merely drive along predefined paths on which only the driving speed is selected. Few vehicles already have the capability to reschedule path and trajectory in real time, for example, to avoid a vehicle parked in its own driving corri-

◘ **Table 49.1**   Overview on shuttle concepts used on public roads in 2021

Manufacturer	Type	Origin	Driver workplace	Max. speed (km/h)	Max. number of passengers
EasyMile	EZ10	France	No	40	12–15
Navya	Autonom® Shuttle Evo	France	No	25	15
Local Motors	Olli	USA	No	40	12
2getthere	Shuttle	Netherlands	No	60	22
The Moove	Mover	Germany	Yes	60	15
Apollo	Minibus	China	No	40	14
May Mobility	Shuttle	USA	Yes	40	
IAV	Heat Shuttle	Germany	No	50	10
Coast Autonomous	P-1	USA	No	40	14
Transdev, Lohr	I-Crystal	France	No	40	16

a

b

Fig. 49.1 Shuttle vehicles of Navya and EasyMile

dor. The challenges and opportunities of these vehicles are discussed in more detail in ▶ Sect. 49.3. For comparison, ▪ Fig. 49.1 shows the mostly used vehicles of the manufacturers Navya and EasyMile.

49.2.2 Pilot Projects

Shuttle vehicles are used in pilot projects of various scopes and volumes around the world. Regular operation without an attendant in mixed traffic, i.e., together with human road users, has not yet been established. The degree of integration into the overall public transportation network varies from simple point-to-point connections to dedicated lines covered by driverless shuttle vehicles. An overview of different activities in Europe, the USA, and Asia is given in [5–9]. In most cases, the above-mentioned vehicles from Easy Mile and Navya are used, both of which have already been operating on public roads since 2015 by means of individual permits. Furthermore, the pilot projects mostly operate in a much lower speed range than the vehicles allow according to the data sheet. Thus, most projects in Europe operate between 6 and 26 km/h on relatively short routes of up to about 2 km [8]. The vehicles move along predefined routes and paths, guided by GNSS. They thus travel on predefined "virtual tracks" and react to dynamic obstacles by perceiving them with the installed sensor technology (cf. ▶ Sect. 3.5), usually only in the form of a speed adjustment. To some extent, these limitations occur due to the admission bases for automated vehicles in the respective countries. Regulatory aspects linked to the ODD LSAD are discussed in more detail in ▶ Sect. 3.2.

49.2.3 Research Activities

The large number of pilot projects mentioned above demonstrate the great interest in the use of driverless shuttles to supplement the existing public transport ser-

vice. However, the limitations described, such as low speeds, fixed and short routes and trajectories, and reaction only through deceleration, also indicate that driverless shuttle systems have not yet reached a sufficient level of maturity for regular operation without an attendant. For this reason, in addition to the pilot projects presented, research projects are being launched and conducted on automated driving functions that go well beyond this. These projects furthermore focus on very different topics, such as

- innovative vehicle concepts and structures;
- fault-tolerant electrics and electronics architectures;
- upgrade- und update-capable software for
  - perception,
  - planning,
  - control;
- connection to cloud and infrastructure;
- usage and design of control rooms/centers;
- safety assurance.

The research project UNICAR*agil* [10] combines many of these efforts in one project and is therefore described here as a representative research project for the usage of the ODD LSAD. The goal of the project is to research and exemplarily implement novel architectures for agile automated vehicle concepts. ▪ Figure 49.2 illustrates the overall concept of the project. Based on a developed modular driving platform with innovative, mechatronically independent dynamics modules, each of which combines the tasks of steering the drive and the brake and enables steering angles of up to 90° per wheel. Four different vehicle versions in two different sizes are being realized for different use-cases. One of these versions is the "autoSHUTTLE", which has similar characteristics to the shuttle concepts presented in ▶ Sect. 2.1. The modular overall system spots human beings as users at the center of the development. The vehicles are connected to each other and the cloud. In addition to environment information perceived by the

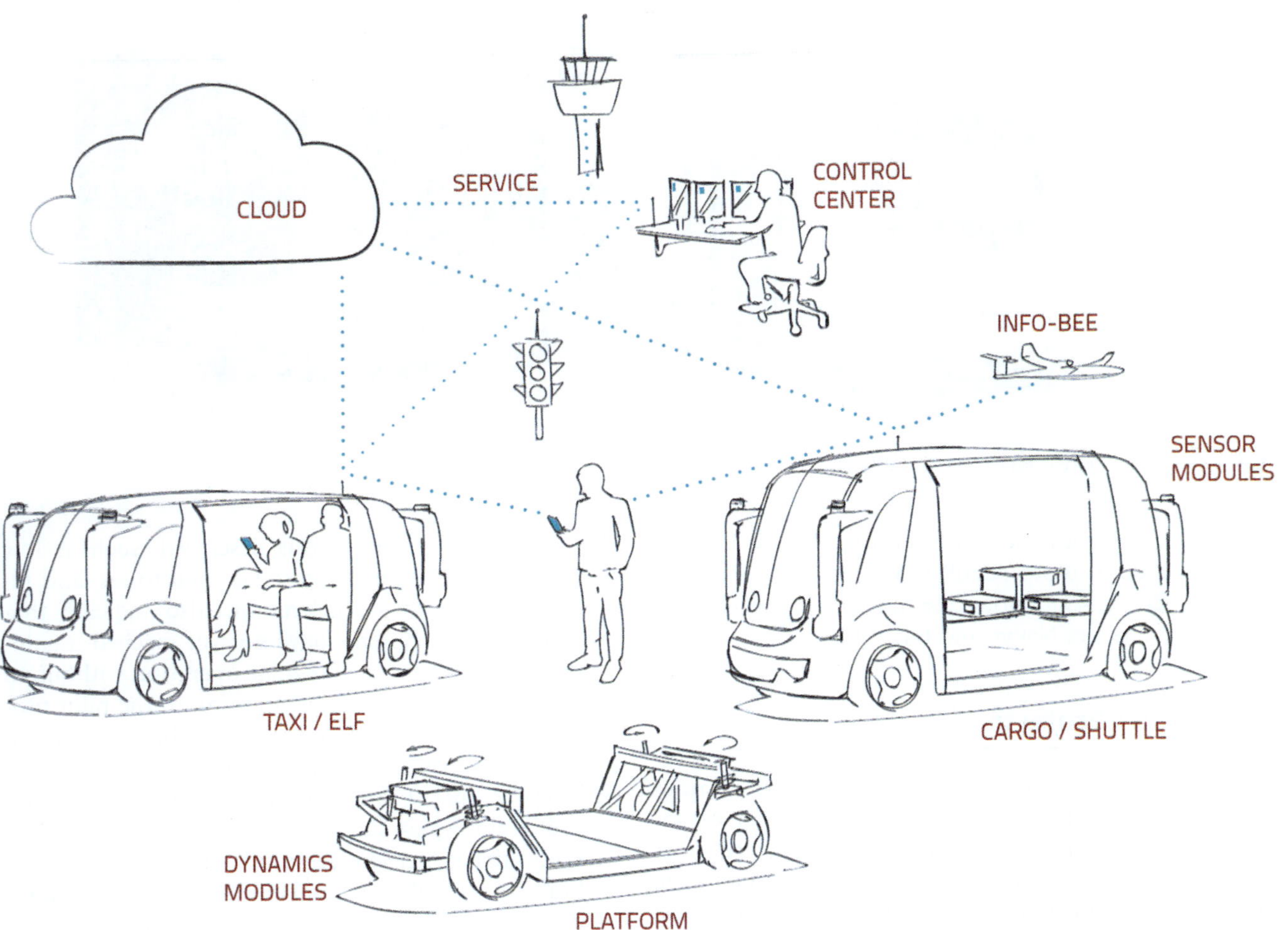

Fig. 49.2   Overall system sketch of the research project UNICAR*agil* [10]

vehicle, the cloud can also access data from intelligent transient infrastructure sensors, the info bee, to process information collectively and make it available to the vehicle fleet. The control center accompanies the journey of all vehicles in the fleet and can take over decisions in situations that driverless vehicles cannot or may not make independently.

The project continues to develop a novel E/E architecture that is oriented towards human information processing and at the same time provides the basis for an update- and upgradeable service-oriented software architecture (ASOA). An additional unique feature is the consistent consideration of safety and security aspects beginning in the early concept phase.

49.3   Positioning the Operational Design Domain

Driverless shuttles are a special application of automated vehicles. They are used in local public transport and, compared to conventional services, promise greater individuality and a demand-oriented use of

vehicles in local public transport with reduced costs in the medium term. The small vessel size for 4–20 passengers (► Sect. 2.1) differs from the usual vessel sizes of other local public transport vehicles like buses. Automated shuttles could thus make an important contribution to first- or last-mile mobility or flexibly connect rural areas to urban centers in the future.

> Driverless shuttles will not replace the classic public transport system with its large vessel sizes, such as buses, trams, or subways. Rather, driverless shuttles represent an extension of public transport system that may efficiently and economically connect less frequented routes.

In order to systematically present the relevant characteristics and challenges of the operational design domain driverless shuttles operate in, these are structured using the 5-Layer-Model (**Fig. 49.3**) [11, 12]. Firstly, the model is generally explained.

The model addresses the societal goals and challenges on the top layer. In terms of sustainable mobility, the aim is not only to make traffic safer and more time

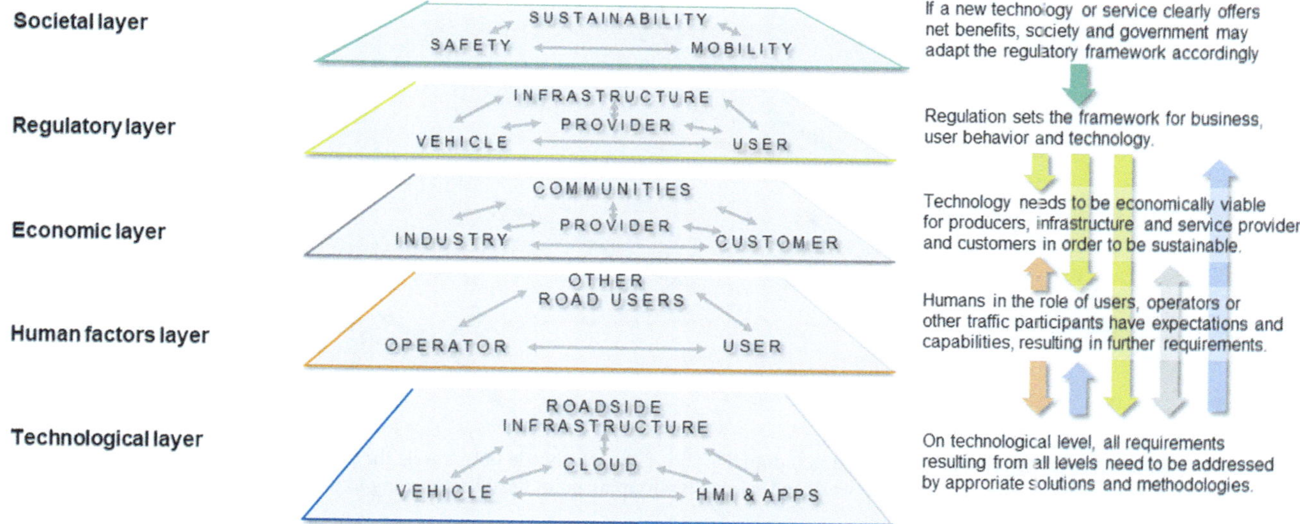

◘ **Fig. 49.3**   5-Layer-model to structure mobility-challenges

efficient, but also to make mobility accessible to all social groups. The regulatory layer is deliberately arranged below the societal layer because a society can adapt the regulatory framework. On the one hand, the layer addresses vehicle admission, and on the other hand, it reflects requirements for vehicle owners, fleet operators, and safety personnel as well as liability issues. The economic layer arranged below is also influenced by the regulatory framework and targets both economic and business aspects, since finally functioning business models are needed to deploy driverless shuttles on a larger scale. The ergonomic layer focuses on human beings, both as users and as stewards or pilots, as well as road users who interact with driverless shuttles in road traffic. For example, ergonomic design rules for the interaction between user and vehicle or app as well as between controller and control room are addressed here. Additionally, factors determining acceptance are identified. At the lowest layer, the technical layer, the challenges and requirements from the layers above must be addressed in the form of suitable concepts and technologies. These technologies are to be developed using efficient methods and need to be effectively secured. The specific characteristics of these layer with respect to driverless shuttles will be discussed in the following chapters in detail.

Before taking a detailed look at the challenges and interactions between the various layers, the relationship between the ODD LSAD and the ODD taxi, described in ▶ Chap. 51, will be highlighted. Basically, both chapters describe the operational design domain for automated driving systems that may supplement local public transport. ◘ Figure 49.4 illustrates the main differences between the two operational design domains [12].

Compared to the ODD of an automated cab or taxi, driverless shuttles can efficiently be used in much more limited areas. Driverless shuttles, like automated cabs in the first instance, travel a selected route network that can be specifically limited in terms of the resulting technical requirements. The routes driven by driverless shuttles can be selected in advance and thus adapted to the respective degree of maturity or the capabilities of the automation systems. In addition, structural adjustments are conceivable, such as equipping all traffic signals with communication modules, so that the feasibility of the automated driving functions can be specifically influenced and favored.

Automated cabs, on the other hand, travel on a route network that is ideally freely selectable, i.e., significantly more diverse, since the user is free to choose his or her location and destination. As for automated shuttles, there will be restrictions in real operation regarding the route network to be served, as well. However, for a reasonable operation of automated cabs, the route network must be considered much broader than for driverless shuttles. The two use-cases also differ in the available stopping options. While cabs should be able to approach any destination point, shuttles continue to move within a defined stop-network, only the "approach" of stops is designed according to demand. The number of stops within a public transport network can be increased compared to the current coverage, since the relatively small shuttles require far less space and are more maneuverable than conventional buses. Further core differences arising are listed as follows:

- Users: As a complement to public transport, shuttles are designed for multiple passengers with different destinations, whereas a cab is booked and used by one or more passengers for a single destination.
- Vessel size: While driverless shuttles carry between 6 and 25 passengers, cab vehicles are generally designed for about 4 passengers.

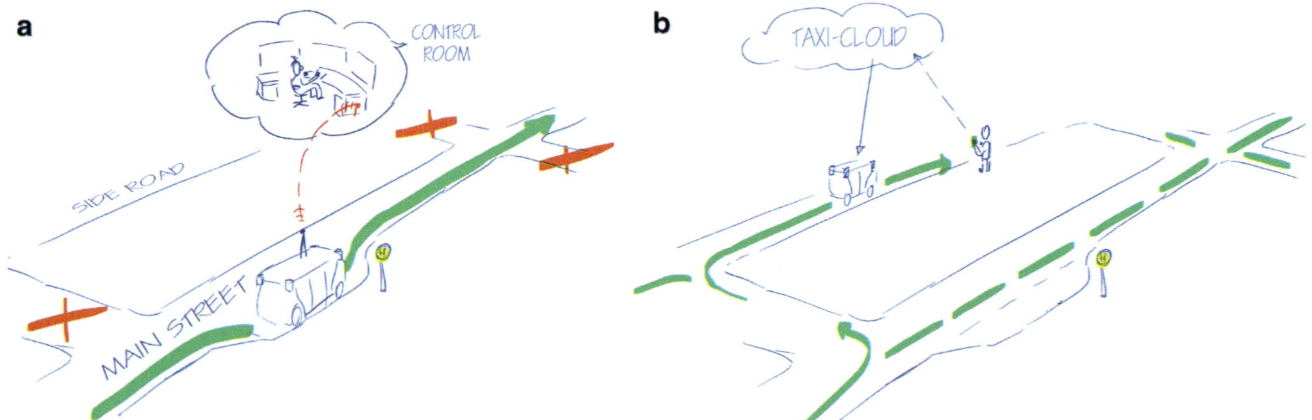

Fig. 49.4 Comparison of the shuttle/LSAD and taxi domains [12]. Left: the shuttle only travels the predefined route and stops at defined stops. Right: the cab travels an extended route network and selects the stopping point flexibly according to the customer's wishes

- Fleet operation: In the first stages of expansion, driverless shuttles move according to a defined route schedule, while cabs operate flexibly and on demand.
- Interior design: The design of shuttles is largely based on public transport and can be characterized as demand-responsive, since the number of passengers remains variable. Cabs can be designed to comfortably transport two to four passengers.

With further development of the vehicles and the automation functions, the ODD LSAD can also be expanded. In the future, it is conceivable that shuttle vehicles will operate more flexibly outside of a fixed schedule and in larger numbers, but still on defined routes. In the long term, the fixed route and stop-network can also be eliminated, and shuttles can thus stop anywhere in the city for access and egress demand driven [12]. Accordingly, in this long-term future picture, the ODD LSAD and cab are clearly converging.

The individual aspects, special characteristics, and possible extensions of the ODD LSAD will be examined in detail in the next chapters.

49.3.1   Societal Layer

It is important to evaluate the effectiveness of driverless shuttles in terms of road safety and sustainability on the societal layer. Furthermore, on this basis, one need to contrast the predicted benefits with the potential risks to create acceptance in society and build trust in automated vehicles without drivers with the ability to intervene.

Built on a simulation-based scenario calculation, EU [13] showed that the operation of automated vehicles and consequently driverless shuttles (starting from a penetration rate of 50%) achieves a 27% reduction in the number of road accidents.

Regarding sustainability, it is not only about energy or transport efficiency, but also about access to mobility at reasonable costs. Rural regions, in particular, suffer from poor public transport connections, as an economic consideration does not allow for close or even demand-driven scheduling of conventionally controlled buses. Automating the vehicles offers the possibility of a demand-driven, flexible deployment of several, relatively small vehicles. The connection can thus be improved and the efficiency (by avoiding empty runs) as well as the comfort (through better public transport connections for citizens) can be increased. Better availability of public transport services using driverless shuttles can also help to increase the acceptance of this new vehicle category. Only the ODD LSAD offers the opportunity to improve the mobility offer and thus increase the acceptance of road users in this way.

Driverless shuttles are already operating in various pilot applications. These pilots not only target the technological development, but also aim to build trust by making the technology accessible to the mainstream. However, the highly restricted ODD in many pilot tests, in which shuttles operate at very low speeds on fixed routes, also results in new challenges on the societal layer. On the one hand, people can experience the technology directly, but they can also learn about the maturity-related limitations of the systems and the reasons for manual control by the steward. This requires trained personnel to accompany rides and transparently educate interested passengers about the respective limitations of shuttle automation. Given the very limited capabilities, this is necessary to gradually build trust in automation. With transparent presentation of the capabilities of the systems, but also their limitations, user trust can potentially be increased [14]. On the other hand, current pilot projects offer some risks in terms of acceptance by other road users. Many pilot applications still operate in very low-speed ranges, often below 20 km/h. If the vehicles move in mixed traffic, i.e., on a road together with other road users such as buses, delivery vehicles, and passenger cars, driverless shuttles post a potential traffic obstacle that can affect the traffic flow in a negative way. Less critical in this regard are pilot tests in dedicated lanes or in areas not used for public transport at all, such as campus areas. Wintersberger et al. [15] compared user acceptance, trust, and subjective time perception between current driverless shuttles (corresponding to the ODD described here) and manually controlled group cabs in a field study. The results not only indicate a positive influence due to the use of the systems, but also the reservation due to the still low speed as mentioned above.

Another challenge is the perceived change in the employment structure. Increasing vehicle automation is causing growing job-loss concern in some parts of society, even though transit agencies and carriers have in fact been struggling with an acute shortage of qualified personnel for several years. Moreover, the ODD LSAD only complements existing public transportation. In metropolitan areas and on high-frequency routes, large-volume trams or buses will continue to be used to efficiently transport large numbers of people. Furthermore, the expansion of the mobility system will not only help in creating new jobs, e.g., in control rooms, but, of course, also in the development, safety assurance, and production of these vehicles.

49.3.2 Regulatory Layer

The approval of driverless shuttles poses the greatest challenge on the regulatory layer. Since the vehicles usually do not have a driver's workplace, they have no direct mechanical intervention option for braking or steering. Safe operation of the vehicles in automated driving mode must therefore be ensured by other means (cf. ▶ Sect. 49.3.5). Current pilot applications do operate driverless shuttles on public roads in some scenarios, but are currently still implemented via special permits. These exemptions require both trained safety personnel (operators) and a detailed description of the ODD. Type approval of an automated vehicle is a necessary prerequisite for regular operation on public roads. State-of-the-art concepts are often assigned to class M2[1] (cf. [16]), which, however, still refers to conventionally controlled vehicles. Furthermore, vehicles' safety testing is not easily adoptable to automated systems (▶ Chap. 46). For these reasons, the Operational Design Domain takes a crucial role in current projects. By restricting the ODD, e.g., by operating at daylight,

boundary conditions that facilitate these exemptions are created.

For further establishment of driverless shuttles, it will be necessary to establish technical regulations and may thus be reasonable to create a separate vehicle class. At the time of writing, the German government has passed an act that will allow automated vehicles to be admitted to SAE level 4 for the first time in the world [17]. Core components include regulations on technical requirements for vehicles with autonomous driving functions, tests, and procedures for issuing operating permits, and regulations relating to duties of the person involved in operation.

The text of the law itself stipulates that automated vehicles may operate without a safety driver and must be monitored by a "technical supervision", which may order deactivation of the driving function or enable exceptional driving maneuvers, if necessary. The law regulates the responsibility of this technical supervision. In order to be able to operate driverless, the vehicles must also always be able to "put themselves into the minimum risk state if necessary" [17].

The use of driverless shuttles lends itself to the establishment of control rooms with the inclusion of a technical supervisor, since several vehicles can be "operated" by just one operator in the future. The control room could thus also be combined with fleet management services, which are responsible for the distribution and routing of the vehicles. In addition, a service control center can take on the important role of direct communication with passengers (cf. ▶ Sect. 49.3.4). Further information on the design and use of control centers can be found in ▶ Chap. 50. Additional information on the regulatory basis for the development of assistance systems and automated driving functions can be found in ▶ Chap. 3. ▶ Chapter 4 also lists the general conditions for the type approval of vehicles with assistance systems and automated driving functions.

49.3.3 Economic Layer

The economic layer plays an indirect role for the operational design domain but is of great importance in the case of the specific application of driverless shuttles from the operator's and user's perspective. Driverless shuttles can only be used meaningfully as a supplement to established mobility modules if they are both profitable for the operator and attractive for the user.

State-of-the-art technology currently is a early stage to evaluate concrete experiences. However, studies like [18] certainly show the potential for increasing mobility connection, i.e., the quality of connection, in rural regions through driverless shuttles. Concluding this, a possible passenger increase for local public transport

---

1   Class M2 describes "raft vehicles with more than eight seats in addition to the driver's seat and with a total mass not exceeding 5 tons, regardless of whether these vehicles have standing room" [16].

operators seems possible, which may make the application of low-speed automated shuttles economically attractive.

Current pilot tests offer the service of a driverless shuttle mostly free of charge. However, the actual costs incurred in an operational system depend heavily on the operating costs and maintenance costs of the driverless shuttle fleet and its interaction with the rest of the conventional public transport network.

Municipalities and cities have another role to play. Based on the extensive sensor technology of vehicles and infrastructure, they can additionally act as providers of data and information and provide additional information and services not only to the transport operators and their automated vehicles, but also to logistics service providers. The drivable routes defined in the ODD LSAD offer cities and municipalities the opportunity to initiate infrastructural measures that support automated driving in the early phase. In this way, they can become the central provider of a so-called digital twin of traffic and benefit economically by merging the data of their own traffic infrastructure with that of connected automated vehicle fleets in real time, thus providing valuable information for logistics and traffic.

49.3.4 Human Factors Layer

From a passenger's perspective, first and foremost interaction and acceptance play a decisive role, as they influence the users' willingness to pay for the service on the economic layer. The interaction begins with the submission of the ride request, usually by using a provider-specific app that informs the passenger about the planned time, the stop, and the cost of the ride. Once the shuttle arrives at the stop, the passenger authenticates himself or herself with the smartphone, thereby gaining access to the vehicle. The interaction between the passenger and the driverless shuttle requires little information, as in the simplest case it can be designed as follows:

1. Booking of the trip: Desired start and time of boarding, destination.
2. Entry: Authentication to open the door.
3. Stop request.
4. (Exit): Opening the door.
5. Emergency communication: Establishing a voice connection with the control room.

■ Figure 49.5 illustrates an example of a "passenger story" derived in the UNICAR*agil* project for the use case of a driverless shuttle [19]. Based on the various interactions during a shuttle ride, technical requirements linking the ergonomic and technical layers can be derived.

Travelers will be able to perform most of the interaction via their own smartphone. However, since not every traveler has these capabilities, shuttles often provide for additional options for interaction in the interior. The results from [20] suggest that a large pro-

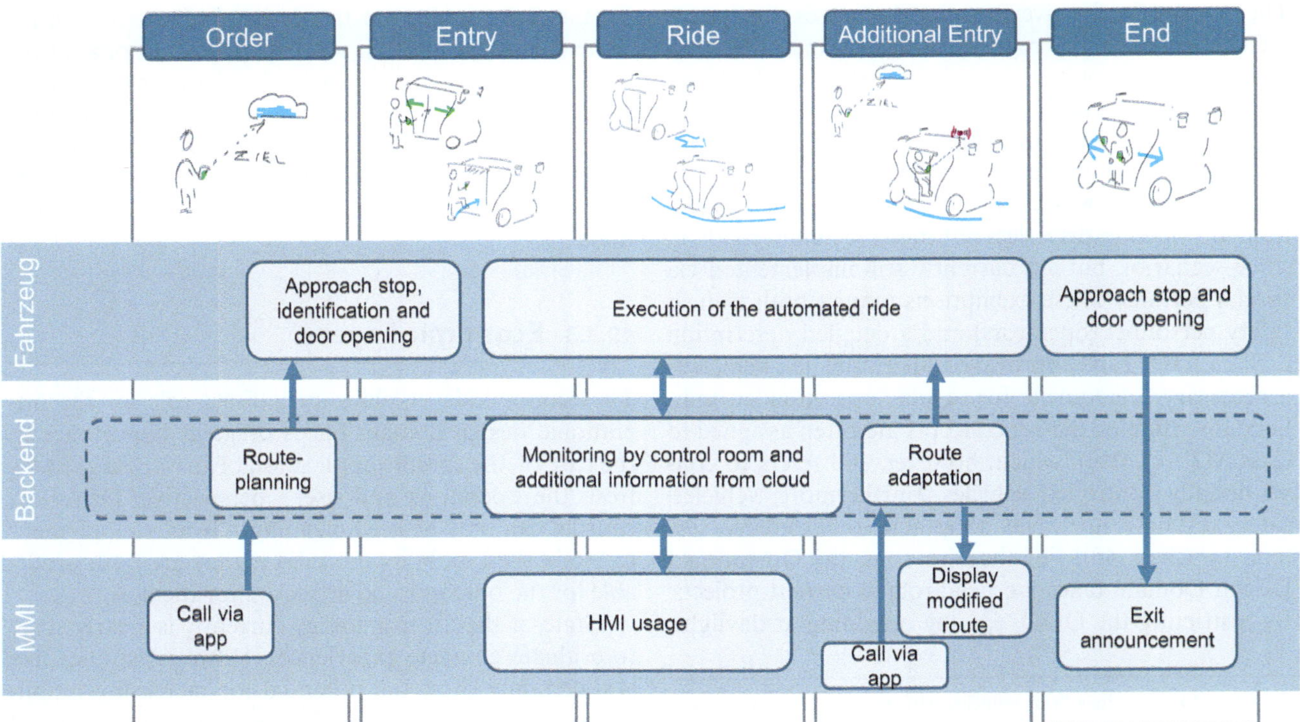

■ **Fig. 49.5** Simplified passenger story for deriving technical requirements. The passenger story describes the course of a typical trip with a driverless shuttle from booking the trip to disembarking [19]

◘ Fig. 49.6   Examples of shuttle interiors (left: interior of a Navya EVO Shuttle, right: UNICAR*agil* concept of a flexible shuttle interior)

portion of passengers would like to receive additional information about the next stop, the current vehicle position, or further details such as the status of the automated journey. For this reason, various interaction options are provided in state-of-the-art shuttle concepts. Often, a central interaction display is combined with several additional display screens. Via the central display, route changes, such as changing one's own destination or spontaneous stop requests, can be expressed. ◘ Figure 49.6 shows examples of the interior concepts of various shuttle implementations.

If shuttle vehicles are not accompanied by a safety driver or steward, communication with the control room or another service point is an essential interaction channel between passengers and the vehicle or its operator. Some possible situations arising during a trip may result in passengers' uncertainty and willingness to contact a "real" person in charge. In driverless vehicles, this functionality will be realized by a direct voice connection with the control room or service center. Thus, one person can accompany several vehicles at the same time.

Automation offers people confined to wheelchairs the ability to travel in traffic. This requires an electric wheelchair ramp or lift that automatically extends when the vehicle comes to a planned stop. A minimum requirement for driverless shuttles is therefore a barrier-free design at least equal to that of today's public transportation. Automation also offers the ability for individual to travel in traffic for people confined to a wheelchair. The prerequisite for this is an electric wheelchair ramp or lift that automatically extends as the stop approaches.

As long as driverless shuttles are operated on predefined and fixed routes with a permanent stop-network, the entry height into the vehicles, i.e., the distance from roadway to ground level, can be selected similarly to current low-floor buses, since entry is usually proceeded from the curb. However, if the ODD is expanded for future developments and shuttles can board or alight passengers at any location, special attention must be paid to boarding height in vehicle design. Thus, by eliminating boarding from the curb, the floor height must be adjusted to the extent that ergonomic boarding can still be ensured. Alternatively, appropriate access aids, such as retractable ramps, must be provided.

At the ergonomic layer, it is also necessary to describe the interaction of the operator in the control room with the driverless shuttles, as described in ► Sect. 49.3.1. As in aviation, economical operation is only feasible if one operator is responsible for several vehicles. This constraint requires a system-ergonomic design of the control room to keep the load-related stress on humans as low as possible. A detailed consideration of the control room can be found in ► Chap. 50.

Finally, the design of the interaction of automated vehicles, including their behavior with other road users, plays a decisive role. It is both a question of how courteously shuttles should behave toward others and how they communicate with other road users. In terms of communication, driverless shuttles can interact directly with other road users via external human–machine interaction (e-HMI) interfaces, such as monitors or microphones.

Concerning vehicle behavior, it stands to reason that a shuttle can display planned driving behavior in a way that can be easily interpreted by external road users. In addition to using the indicators, it seems reasonable to design the parameterization of a maneuver or trajectory planner (cf. ► Chap. 43) for this indirect communication. However, these methods for external communication, i.e., to other direct road users, are not specific to the ODD LSAD, but can also be applied to both vehicles equipped with driver assistance systems and automated vehicles, respectively. The tactical behavior of automated vehicles is defined in research projects such as "Autoknigge" [21], and by consortia in the form of rules. The most prominent example is currently the "Responsibility-Sensitive Safety" (RSS) model based on human driving behavior [22]. Further information is summarized in ► Chap. 40.

In addition to system ergonomics, new possibilities are also emerging with regard to anthropometry. By eliminating the driver's workplace, novel interior concepts can be realized in driverless shuttles. To expand capacity, standing room in the vehicles is conceivable and has already been implemented in current concepts. This allows a variety of possibilities for positioning seating and standing areas, which at the same time require relatively high vehicles to realize upright boarding and standing for the majority of passengers. In [23], Köhler et al. investigated how interior concepts and seat positioning in automated vehicles can be realized and evaluated.

Driverless shuttles are often structurally designed to move bidirectionally, i.e., equally in both directions of travel. In combination with the novel seating concepts described above, a potential risk of experiencing symptoms of motion sickness, kinetosis [24], arises for many travelers. However, studies indicate that these can be reduced or even avoided by foreseeable vehicle behavior [25].

49.3.5 Technical Layer

In technical perspective, driverless shuttles differ significantly from conventional minibuses, buses, or passenger cars available today. As already discussed, most shuttle concepts dispense with a dedicated driver's workplace. The resulting geometric freedom allows vehicles to be designed with greater interior capacity, which therefore can be designed for bidirectional traffic, optionally. Yet, the elimination of the possibility of direct human intervention creates additional technical challenges that need to be solved. One example is the creation of necessary redundancies and machine self-awareness in order to avoid individual errors and to be able to transfer the driverless vehicles to a risk-minimal state independently. The automation of the driving function requires different sensor principles for a robust environment perception, as well as powerful control units for processing the sensor data, planning the maneuvers, and controlling the actuators. The positioning of the required sensors and the electronic and electric connection of the required control units are important constraints for the design and conception of the vehicle onboard network architecture.

49.3.5.1 Geometric Characteristics

Driverless shuttles are designed to complement public transport. The ODD of the vehicles limits their area of operation both geographically to a defined area or even to defined routes and to a specified speed range. These limits also play a role in the geometric design of the vehicles.

The vehicles are battery-electric. By eliminating the combustion engine and integrating the battery into the vehicle floor, the vehicles can be designed compactly and offer maximum usable interior volume, from the front to the rear of the vehicle. Individual wheel drives, e.g., using wheel hub motors, enable further space gains in the interior. Combined with by-wire steering and brake actuators, which are also wheel-specific, this results in so-called dynamics modules (also known as corner modules) in which each wheel can be controlled individually and independently. The resulting steerability of all four wheels enables the shuttle to move agilely in road traffic. The ability to select the instantaneous center of the vehicle's motion relatively freely during steering maneuvers also makes it possible to realize trajectories that prevent passengers from feeling motion sickness, for example.

Compared to conventional buses, shuttles have a relatively low moving load. The safety of passengers in the event of an accident is therefore of increased importance for driverless shuttles. The vehicles must provide deformation and safety zones to protect passengers in the event of a collision. In addition, new restraint systems need to be developed to secure passengers in various sitting and, if necessary, standing positions. Initial developments in this regard are already emerging for automated cab vehicles [26], as well as elsewhere. The main challenge in ODD LSAD in the future will be to ensure safety for standing passengers.

49.3.5.2 Perception Characteristics

While the driver's view is an important design criterion for conventional vehicles, it is the positioning of the sensor system for automated vehicles. The sensors percept the environment through various physical sensor principles (cf. ▶ Chaps. 15, 16, 17 and 18) and provide the input data for the automation function. The sensor system is to be positioned in such a way that it can ideally perceive the vehicle's surroundings without any blind spots. The choice of sensor system is highly dependent on the respective ODD. If, for example, the vehicle only drives in good environmental conditions (here: without rain and snow) and daylight, low speeds on secured routes without other road users, the sensor equipment can be relatively simple. However, if bidirectional driving in mixed traffic and at higher speed range is desired, a more complex perception concept is necessary.

In general, there are various options for installing sensors on the vehicle. In the usual way, as in current passenger cars, sensors are integrated into the vehicles' design. Cameras are located behind the windshield or radar sensors in the radiator grille, for example. This has the advantage of an unaffected design as well as an included cleaning mechanism for the camera by the windshield wiper but offers the disadvantage of a high occlusion of the sensor's viewing angles, especially with

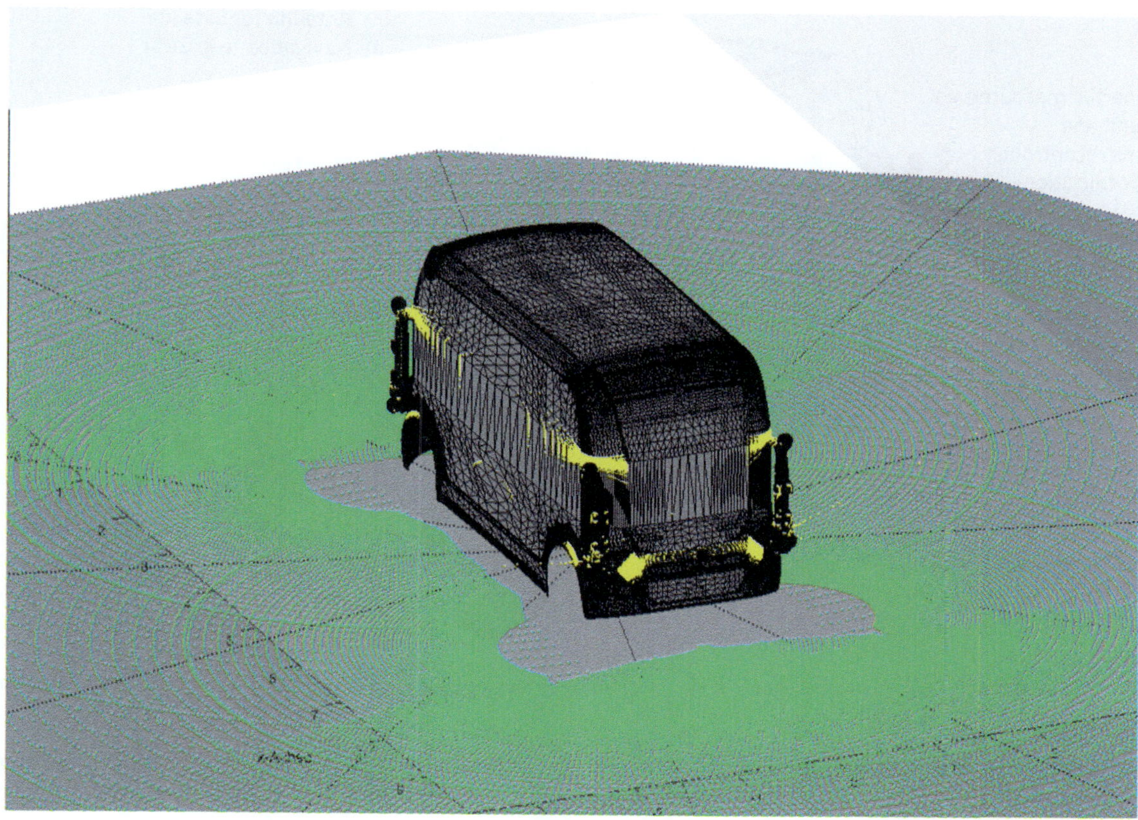

◘ Fig. 49.7 Simulation of the ground coverage of the sensors used from external sensor modules when using four modules, each with a 360° LiDAR, four cameras, and two radar sensors

360° sensor technology. For this reason, some concepts introduce the usage of externally installed sensor technology to optimize viewing areas. ◘ Figure 49.7 shows an example of the simulated sensor coverage when using external sensor modules. ◘ Figure 49.8 additionally shows an example of the sensors and computing units installed in the autoSHUTTLE of the UNICAR*agil* project.

Especially for automated shuttles, the local restriction of the ODD offers the possibility to access external information by using connectivity. Additionally, it is possible to act in a coordinated way with other road users that not necessarily belong to the same fleet. At neuralgic points of the previously determined routes, a useful supplement can be established through infrastructure-based sensor technology, which provides additional information about objects, for example, in side streets that are difficult to see from the vehicles' point of view. Compared to other ODDs, such as the ODD defined for cabs, the ODD LSAD is particularly well suited for the economically reasonable integration of infrastructure sensor technology, since the route network can be well defined and relevant locations for the use of infrastructure sensor technology can be examined in advance.

In the simplest case, a traffic light signal system communicates directly with the vehicles and provides the information on their timing. In addition to this, a digital twin of the traffic is computed either by the urban control center or in the provider-specific cloud. This "collective environment model" can serve as an additional source of information for the shuttle [27]. Especially in the case of visibility restrictions, e.g., due to occlusion, this information generated outside the vehicle has a particular value, as it enables a predictive and comfortable driving style. The external environment model can be fed by infrastructure-based sensor data, so-called Road-Side Units (RSU), as well as by automated air vehicles, whose potential contribution is being evaluated for the first time in the UNICAR*agil* project. Various communication mechanisms can be considered for communication (cf. ► Chap. 23). For direct, low-latency vehicle-to-vehicle communication of small data packets, the automotive W-LAN IEEE 802.11p is the most suitable. Mobile communication standards such as 5G also strive for low-latency transmission so that diversely redundant communication can be established.

The post-processing of sensor data does not notably differ from that carried out in other automated vehicles, see Sect. 49.4 for a more detailed compilation.

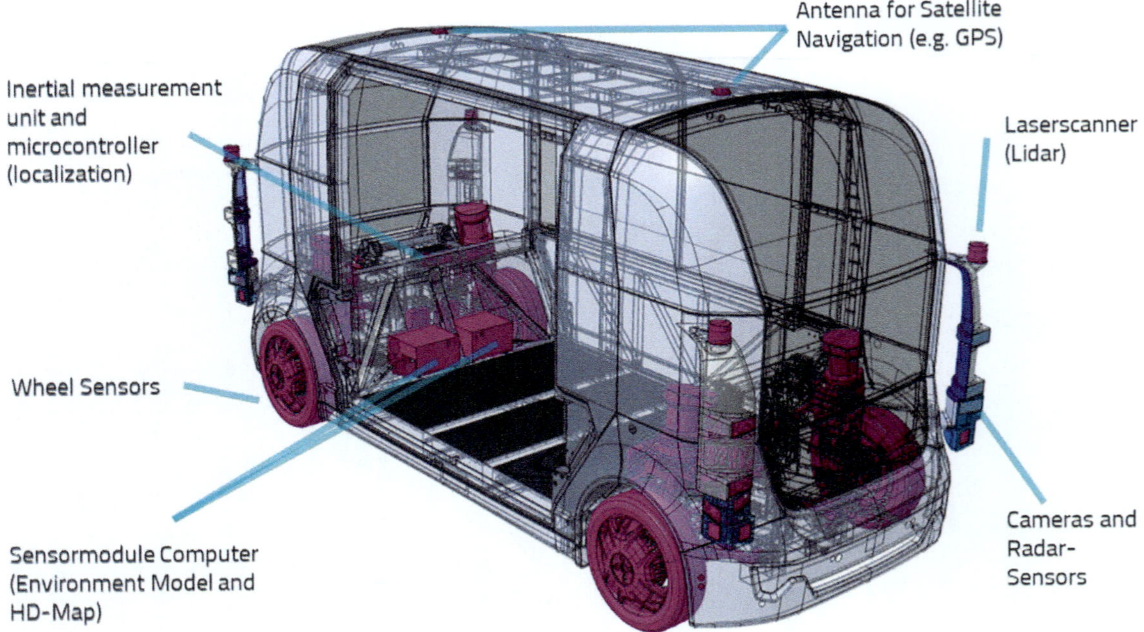

▫ Fig. 49.8 Sensor technology in a driverless shuttle using the example of the open autoSHUTTLE-concept from the UNICAR*agil* project

49.3.5.3 Behavioral Decision and Trajectory Planning

However, special attention in ODD LSAD should be paid to behavioral decision-making and trajectory planning. Since driverless shuttles transport passengers of different ages and physical conditions, both sitting and standing in the vehicles, the behavior of the automated driving function should always adapt to the given situation to maximize comfort for all passengers. This exemplarily includes planning trajectories with reduced lateral and longitudinal forces as far as possible. Furthermore, the interconnection of vehicles with each other or the cloud may also offer significant advantages for behavioral decision-making mechanisms. Not only can the journeys of different vehicles be coordinated with each other, but also behavioral decisions in traffic situations, respectively. Coordinated decision-making currently is the subject of DFG-funded fundamental research [28].

By transferring large amounts of data to the cloud, automated driving functions can also be improved in the long term by retrospectively evaluating the behavioral decisions, similar to what a human driver does. In this way, a collective memory can be realized so that automated vehicles, exposed to an unfamiliar intersection scenario, can benefit by the experience of many other vehicles. This concept is addressed for the first time in the UNICAR*agil* project and is initially implemented and evaluated in simulation [27].

For driverless shuttles in operation without an accompanying person with the possibility to intervene, safety mechanisms also play an essential role. Additional functions in the processing chain provide redundant information with which the vehicle guidance function can always be transferred to a risk-minimal state, even in the event of a fault. In detail, this means that, depending on the situation and the error, the vehicle continues to drive a few more meters to identify a safe place to stop. An emergency stop at a busy intersection or railroad crossing must be avoided in all cases. Given situations where automation cannot make a behavioral decision because, for example, traffic rules would have to be disregarded by crossing a solid line driving around an incorrectly parked vehicle in the second row, a human operator in a control room takes over the maneuver decision and enables the vehicle to move on with its planned trajectory. If the decision is not available in time, the shuttle stops autonomously and waits for the maneuver decision. From standstill, a person from the control room could take over control via tele-operation in exceptional cases (cf. ▶ Chap. 50).

49.4 Summary and Outlook

In summary, the ODD LSAD in which driverless shuttles are operated can be described as an operational design domain that is well containable and therefore is highly relevant for the current state of the art. By defining drivable routes, speeds, and environmental conditions, it is suitable for testing automated driverless vehicles at an early stage and at the same time demonstrating the benefits of automated vehicles to the interested public. The containment of the ODD allows state-of-the-art automated vehicles to operate on a pilot basis without a safety driver, but only with stewards on board.

In the medium to long term, it will be possible to expand the ODD LSAD so that driverless shuttle vehicles can travel larger route networks, are not dependent on predefined stops, and do not need to have a steward on board anymore. Driverless shuttles will then supplement local public transportation in areas where low-capacity vehicles at high frequency significantly improve the mobility of the population. This will include first- and last-mile transfers to mobility hubs where passengers can connect to various high-volume modes of transportation.

Functions such as safe stopping and connectivity to a control center or technical supervision play a crucial role in enabling these vehicle concepts. At the same time, boundary conditions are being created to make driverless shuttles type-approvable, possibly as a separate vehicle class. Furthermore, there is enormous potential for connecting intelligent infrastructure within the ODD LSAD such as the connection with traffic signals via vehicle-to-infrastructure communication or integrating infrastructure sensors to extend the vehicle's own sensor visibility. In the future, collective cloud functions, such as a collective memory, will also be able to play a central role in shaping driverless shuttle services within the ODD LSAD.

References

1. SAE International, On-Road Automated Vehicles Standards Committee: Taxonomy and Definitions for Terms Related to Driving Automation Systems for On-Road Motor Vehicles. Document no. J3016_201806. Revision June 2018. ▶ https://www.sae.org/works/documentHome.do?comtID=TEVAVS&docID=-J3016_201806&inputPage=dOcDeTaIlS. Accessed 07 March 2022

2. Association for Standardization of Automation and Measuring Systems (ASAM) e.V.: OpenODD: Project Proposal, version 1.0 (2021). ▶ https://www.asam.net/project-detail/asam-openodd/. Accessed 02 April 2021

3. Maisto, M.: Induct Now Selling Navia, First Self-Driving Commercial Vehicle. eWeek (2014). ▶ https://www.eweek.com/innovation/induct-now-selling-navia-first-self-driving-commercial-vehicle/. Accessed 11 July 2021

4. International Standardization Organization (ISO): ISO/DIS 22737/Intelligent Transport Systems—Low-Speed Automated Driving (LSAD) Systems for Predefined Routes—Performance Requirements, System Requirements and Performance Test Procedures. DRAFT International Standard (2020). ▶ https://www.iso.org/standard/73767.html. Accessed 07 March 2022

5. Iclodean, C., Cordos, N., Varga, B.O.: Autonomous shuttle bus for public transportation: a review. Energies **13**(11) (2020, June)

6. Haque, A.M., Brakewood, C.: A synthesis and comparison of American automated shuttle pilot projects. Case Stud. Transp. Policy **8**(3), 928–937 (2020, September)

7. Verband Deutscher Verkehrsunternehmen: Autonome Shuttle-Bus-Projekte in Deutschland (2021). ▶ https://www.vdv.de/liste-autonome-shuttle-bus-projekte.aspx. Accessed 08 April 2021

8. Hagenzieker, M., Boersma, R., Nuñez Velasco, P., Ozturker, M., Zubin, I., Heikoop, D.: Automated Buses in Europe: An Inventory of Pilots (2021). ▶ https://research.tudelft.nl/en/publications/automated-buses-in-europe-an-inventory-of-pilots-version-10. Accessed 07 March 2022

9. Ainsalu, J., Arffman, V., Bellone, M., Ellner, M., Haapamäki, T., Haavisto, N., Josefson, E., Ismailogullari, A., Lee, B., Madland, O., Madžulis, R., Müür, J., Mäkinen, S., Nousiainen, V., Pilli-Sihvola, E., Rutanen, E., Sahala, S., Schønfeldt, B., Smolnicki, P.M., Soe, R.-M., Sääski, J., Szymańska, M., Vaskinn, I., Åman, M.: State of the art of automated buses. Sustainability **10**(9) (2018, August)

10. Woopen, T., Lampe, B., Böddeker, T., Eckstein, L., Kampmann, A., Alrifaee, B., Kowalewski, S., Moormann, D., Stolte, T., Jatzkowski, I., Maurer, M., Möstl, M., Ernst, R., Ackermann, S., Amersbach, C., Leinen, S., Winner, H., Püllen, D., Katzenbeisser, S., Becker, M., Stiller, C., Furmans, K., Bengler, K., Diermeyer, F., Lienkamp, M., Keilhoff, D., Reuss, H., Buchholz, M., Dietmayer, K., Lategahn, H., Siepenkötter, N., Elbs, M., v. Hinüber, E., Dupuis, M., Hecker, C.: UNICAR*agil*—Disruptive Modular Architectures for Agile, Automated Vehicle Concepts, 27. Aachener Kolloquium, Aachen (2018, October). ▶ https://www.unicaragil.de/images/publications/2018-07-23_UNICARagil_AachenerKolloquium_final.pdf. Accessed 07 March 2022

11. Zlocki, A., Eckstein, L.: Challenges and research needs on automated driving. In: 3. International VDI Conference Automated Driving, Düsseldorf (2016, June 29)

12. Dietmayer, K., Eckstein, L., Form, T., Flemisch, F., Friedrich, R., Fulger, D., Gasser, T., Harms, K., Hense, B., Müller, N., Schäfer, C., Zlocki, A.: Automatisiertes und autonomes Fahren, VDI-Handlungsempfehlung (2019, December). ▶ https://www.vdi.de/ueber-uns/presse/publikationen/details/vdi-handlungsempfehlung-automatisiertes-und-autonomes-fahren. Accessed 07 March 2022

13. Rösener, C.: A Traffic-based Method for Safety Impact Assessment of Road Vehicle Automation, Dissertation, Aachen (2020, July 07). ▶ https://publications.rwth-aachen.de/record/801589. Accessed 07 March 2022

14. Lee, J.D., See, K.A.: Trust in automation: designing for appropriate reliance. Hum. Factors: J. Hum. Factors Ergon. Soc. **46**(1), 50–80 (2004)

15. Wintersberger, P., Frison, A.-K., Thang, I., Riener, A.: Autonome Shuttlebusse im ÖPNV Analysen und Bewertungen zum Fallbeispiel Bad Birnbach aus technischer, gesellschaftlicher und planerischer Sicht, Ch. 6, pp. 95–113 (2020). ▶ https://doi.org/10.1007/978-3-662-59406-3. Accessed 07 March 2022

16. EU: Verordnung (EU) 2018/858 des Europäischen Parlaments und des Rates vom 30. Mai 2018 über die Genehmigung und die Marktüberwachung von Kraftfahrzeugen und Kraftfahrzeuganhängern sowie von Systemen, Bauteilen und selbstständigen technischen Einheiten für diese Fahrzeuge, zur Änderung der Verordnungen (EG) Nr. 715/2007 und (EG) Nr. 595/2009 und zur Aufhebung der Richtlinie 2007/46/EG (Text von Bedeutung für den EWR.), May 2018 (2018)

17. Bundesministerium für Verkehr und digitale Infrastruktur: Entwurf eines Gesetzes zur Änderung des Straßenverkehrsgesetzes und des Pflichtversicherungsgesetzes–Gesetz zum autonomen Fahren, Bearbeitungsstand (2021, February 08). ▶ https://www.bmvi.de/SharedDocs/DE/Artikel/DG/gesetz-zum-autonomen-fahren.html

18. Jürgens, L.: Autonome Shuttlebusse im ÖPNV Analysen und Bewertungen zum Fallbeispiel Bad Birnbach aus technischer, gesellschaftlicher und planerischer Sicht, Ch. 4, S. 39–54 (2020). ▶ https://doi.org/10.1007/978-3-662-59406-3. Accessed 07 March 2022

19. UNICAR*agil* consortium, autoSHUTTLE user-story, You-Tube-Video: ▶ https://www.youtube.com/watch?v=gJ24HGEU-QJc&t=1s. Accessed 17 July 2021

20. Herzberger, N.D., Schwalm, M., Reske, M., Woopen, T., Eckstein, L.: Ergebnisse einer Befragung von 619 Personen in Deutschland im Rahmen des Projekts UNICAR*agil* (2019). ▶ https://www.unicaragil.de/images/downloads/2019-01-01-18NHE022-Mobilitaetskonzepte-der-Zukunft.pdf. Accessed 07 March 2022

21. Eckstein, L., Kowalewski, S., Rumpe, S.: AutoKnigge—Modellierung, Bewertung und Absicherung von Verhalten für Kooperativ Interagierende Automobile, Deutsche Forschungsgemeinschaft (DFG)—project number 273237627 (2015). ▶ https://gepris.dfg.de/gepris/projekt/273237627. Accessed 07 March 2022

22. Mobileye: Responsibility-Sensitive Safety: A Mathematical Model for Automated Vehicle Safety (2021). ▶ https://www.mobileye.com/responsibility-sensitive-safety. Accessed 17 July 2021

23. Köhler, A.-L., Prinz, F., Wang, L., Becker, J., Voß, G., Ladwig, S., Eckstein, L.: Wie fahren wir autonom? Nutzerbedürfnisse für Innenraumkonzepte sowie Anforderungen für den Insassenschutz. 28. Aachener Kolloquium 2019, Aachen, Oktober 2019 (2019). ▶ https://www.unicaragil.de/images/downloads/Publication_2019-10-08_Koehler_et_al._2019_UNICAR_OSCCAR_AC_Colloquium.pdf. Accessed 07 March 2022

24. Bohrmann, D., Koch, T., Maier, C., Just, W., Bengler, K.: Motion comfort—human factors of automated driving. In: 29th Aachen Colloquium of Sustainable Mobility 2020, Aachen, 2nd edn., vol. 29, pp. 1697–1708 (2020, October)

25. Ladwig, S., Voß, G.M.I., Wolligandt, C., Arling, V.: Passenger psychology and automated driving—the impact of novel interior concepts on human perception. Symposium auf der 62. Tagung experimentell arbeitender Psychologen, Jena, Germany (2020). ▶ https://www.researchgate.net/publication/344553886_Passenger_psychology_and_automated_driving_-_The_impact_of_novel_interior_concepts_on_human_perception. Accessed 11 July 2021

26. Zoox Representation: Keeping every rider safe (2021). ▶ https://zoox.com/vehicle/. Accessed 10 April 2021

27. Lampe, B., Woopen, T., Eckstein, L.: Collective Driving—Cloud Services for Automated Vehicles in UNICAR*agil*. 28. Aachener Kolloquium 2019, Aachen (2019, October). ▶ https://www.unicaragil.de/images/publications/Publication_2019-10-08_Kollektives_Fahren_AachenerKolloquium_v2.pdf. Accessed 07 March 2022

28. Stiller, C.: Schwerpunktprogramm 1835—Kooperativ interagierende Automobile, Deutsche Forschungsgemeinschaft (DFG), 14.07.2021 (2015). ▶ https://gepris.dfg.de/gepris/projekt/255645177?context=projekt&task=showDetail&id=255645177&. Accessed 07 March 2022

Open Access This chapter is licensed under the terms of the Creative Commons Attribution-NonCommercial-NoDerivatives 4.0 International License (▶ http://creativecommons.org/licenses/by-nc-nd/4.0/), which permits any noncommercial use, sharing, distribution and reproduction in any medium or format, as long as you give appropriate credit to the original author(s) and the source, provide a link to the Creative Commons license and indicate if you modified the licensed material. You do not have permission under this license to share adapted material derived from this chapter or parts of it.

The images or other third party material in this chapter are included in the chapter's Creative Commons license, unless indicated otherwise in a credit line to the material. If material is not included in the chapter's Creative Commons license and your intended use is not permitted by statutory regulation or exceeds the permitted use, you will need to obtain permission directly from the copyright holder.

ODD Taxi

Frank Diermeyer, Dominik Raudszus, and Marc Wimmershoff

Contents

© The Author(s) 2026
H. Winner et al. (eds.), *Handbook Assisted and Automated Driving,*
https://doi.org/10.1007/978-3-658-45276-6_50

50.1 Introduction

In addition to the operational domains of freight transport and shuttle, the operational design domain (ODD) taxi is worthwhile to be discussed. As with the automated shuttle, the focus is on passenger transportation in short-distance traffic. In contrast to the shuttle, on- and deboarding does not take place at fixed stops but, similar to a conventional taxi, flexibly at freely selectable locations (◘ Fig. 50.1).

The traffic area to be covered in this use case is accordingly larger and more diverse than in the case of the ODD shuttle with predefined routes in scheduled traffic. This degree of freedom poses additional challenges and also limits the solutions that can be implemented. This has far-reaching implications for the technical concept of automated taxi fleets and thus justifies separate consideration. For example, concepts that rely on roadside units (RSU) to ensure traffic safety are usually not economically feasible for an automated taxi service.

Taxis contribute to personal, independent mobility in an urban or regional transportation concept. The need for individual mobility is opposed to limited resources such as traffic space, energy requirements, and costs. With automation, costs may be reduced in the medium term and thus the conflict of goals can be partly resolved. The required traffic space and energy demand are hardly influenced positively by the automation of taxis. In contrast, more favorable transportation fees may increase the attractiveness and the automated taxi could be used more often and could be chosen more often instead of bus and train. If the trend toward sharing of transport does not continue, the use of automated taxis could lead to an increase in the number of journeys and ultimately result in increased demands for traffic space and energy. In the course of the digitization of transportation, additional modes of operation are going to be offered that make sharing or pooling rides easy and convenient. Consequently, the conflicting goals of the need for individual mobility and limited resources can be gradually resolved in the future through the combination of automation and digitization.

The goal for vehicles of the ODD taxi is to implement individual mobility in a comfortable, accessible, and resource-saving way.

In the following, the topic of automated taxis is considered and discussed in detail. First, the most common terms in this context are defined and the common types of transport are described. A market overview in ▶ Sect. 50.3 shows the best-known players in the field of automated taxis and ride brokers. Sections 50.4 and 50.5 are dedicated to the technology of taxi automation. First, the specific challenges and requirements of ODD taxi are collected and explained in detail. Then, the current and future available technical solutions are presented and described. In ▶ Sect. 50.6, the impact of

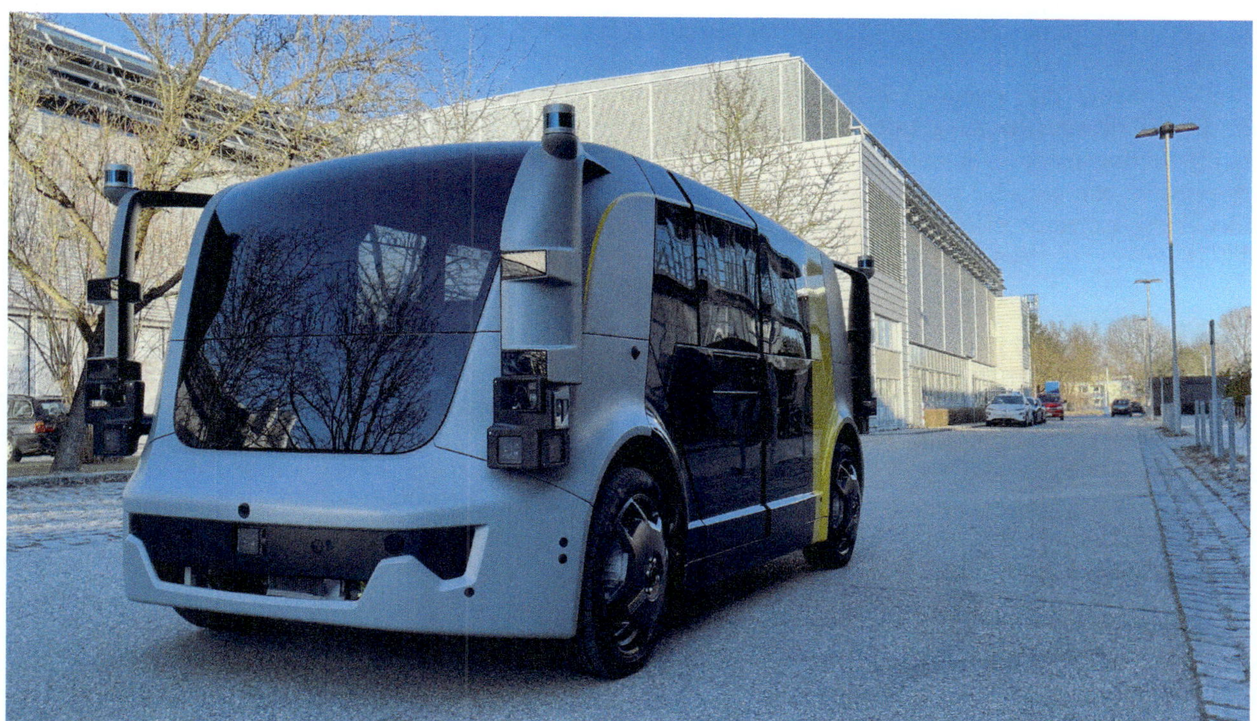

◘ **Fig. 50.1**   Automated taxi autoTAXI; developed as part of the UNICARagil project funded by the German Federal Ministry of Education and Research (BMBF)

the highly automated vehicle of the ODD taxi is highlighted and discussed in several dimensions. The outlook in ▶ Sect. 50.7 concludes the topic.

50.2 Types of Transport and Definitions of Terms

Different service concepts can be defined within the ODD taxi. In literature, the nomenclature of transport types is used inconsistently. Usually the terms "ride-hailing" or "ride-sourcing" [1] refer to exclusive on-demand passenger transport. In the case of combined rides, the terms "ride-pooling" and "ride-sharing" are often used synonymously [2] even though the term "ride-sharing" originally refers to the transport of one or more passengers by a private driver [3], which is coordinated by an agency for arranging rides.

Whereas in ride-hailing operations the route is determined individually by the passenger, who thus reaches his or her destination as quickly as possible, in ride-pooling or sharing operations the journeys of several passengers are combined and served by the same vehicle. In this way, the travel time for individual passengers may increase compared to ride-hailing, but from a global perspective, efficiency increases due to higher utilization of the fleet, resulting in lower traffic volumes and lower costs for the individual passenger.

In the context of this chapter, the term "on-demand passenger transportation" is generally used, as well as the term "ride-hailing" for the special case of exclusive transportation and the term "ride-sharing" for the case of shared transportation.

In Germany, the amendment of the Passenger Transportation Act (PBefG) created a legal framework for ride-sharing services for the first time [4]. Previous services were only possible with temporary exemptions under the so-called experimentation clause (§ 2 Abs. 7 PBefG).

50.3 Market Survey

Ride-sharing organizations are taking advantage of recent technological advances to bring drivers and passengers together through a customer-to-customer business model. Consequently, this type of company acts more as a platform that allows those who own a car to offer taxi rides whenever they want. By using this platform, sophisticated algorithms bring drivers and passengers together so that the social good can be better maximized. In this respect, ride-sharing works much like a taxi service. One important difference, however, is that ride-sharing companies are subject to different regulatory and licensing requirements than traditional taxi companies, which are country-specific but generally not as stringent as those for traditional taxi companies. This usually gives ride-sharing companies a price advantage over taxi companies, which is why the ride-sharing business model is often referred to as a "disruptive" technology.

With autonomous taxi, the business model of ride-sharing platforms changes from a customer-to-customer model to a business-to-customer model, as the driver no longer exists as a "customer" of the platform. For example, an automated taxi provider might decide to design, manufacture, and operate an automated taxi fleet, but could rely on existing ride-matching platforms to handle the end-customer relationship. Another example would be that the company decides to also develop the ride-matching platform itself, which would mean that they interact directly with the end customer.

In November 2019, Uber, DiDi Chuxing and Lyft were the global leaders in the ride-hailing business [39]. DiDi Chuxing operated in China, while Lyft and Uber operated in North America and Europe. ◘ Figure 50.2 provides an overview of the ride-sharing market share per company in 2019.

To get a better global overview of the main players, ◘ Fig. 50.3 shows selected major players and their main areas of influence in the global ride-sharing market on a map.

In 2020, the global ride-sharing market size was USD 75.39 billion and it is projected to grow to USD 117 billion in 2021 [5]. Note that this projected growth of more than 50% is also related to COVID-19, but the rapid growth of the market is also driven by other key factors: consumers, especially younger adults, want to avoid the high fixed costs of owning a vehicle. Furthermore, ride-sharing is expected to become especially prevalent in cities where owning a car is not only costly but also less practical due to traffic congestion and limited parking.

The large impact area of ride-sharing has been enabled by the widespread use of smartphones. Mobility apps are particularly popular in India and China, so mobility services in regions like China will see large revenue streams in perspective. In regions, where public transport is well funded and attractive, ride-sharing will not play a significant role. Europe definitely includes regions where the market for urban mobility platforms combining individual and shared mobility options can play a noteworthy role.

As explained in ▶ Sect. 11.6.3, the driver represents the largest cost position in non-autonomous ride-sharing services, accounting for 80% of the total cost per kilometer. By replacing the driver with automation and connectivity, fully automated vehicles will, in perspective, drastically reduce the cost of a ride while increasing the addressable market, which is estimated to

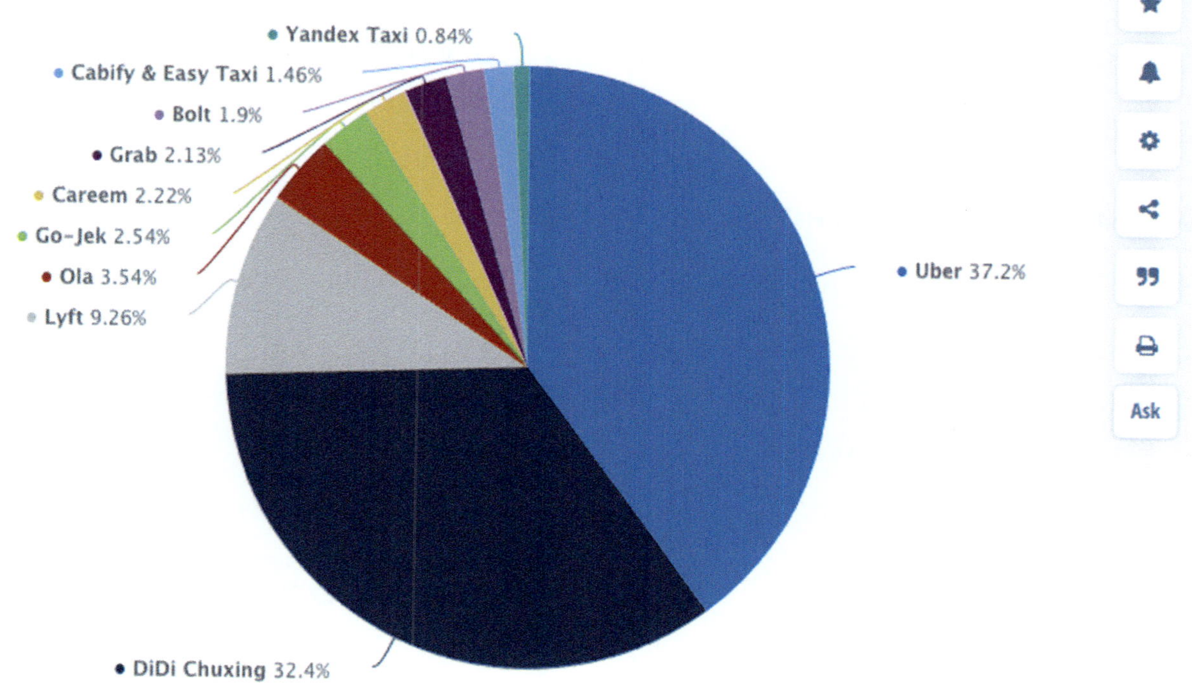

Fig. 50.2 Ride-sharing market share per company in 2019

grow to more than $1 trillion USD by 2030 [6]. Therefore, existing ride-sharing companies are interested in removing the driver to gain access to this large market.

In addition to existing ride-sharing companies, there are startups that aim to develop an automated ride-sharing service without having an existing ride-sharing service with human drivers. These companies differ in how the automated driving stack is integrated into the vehicle platform: Startups like Cruise and Zoox aim to develop an automated vehicle platform from scratch, while companies like Waymo integrate their automated driving stack on top of customized but existing vehicle platforms. Both approaches have their advantages and disadvantages: for example, a custom-developed automated vehicle platform offers early and visually appealing integration of sensors into the vehicle design. The passenger compartment can still be designed specifically for the use case and a bidirectional driving capability of the vehicle can be provided to avoid complicated driving maneuvers such as multi-point turns. The disadvantages of developing a purpose-built vehicle platform include dramatically increased development effort and expenses, and greater challenges in the vehicle approval process due to significant differences from human-controlled vehicles (e.g., no steering wheel or driver pedals).

50.4 Taxi-Specific Requirements

50.4.1 Unlimited Number of Scenarios Makes Release More Complex

The requirement to operate automated vehicles in a large area such as a city results in a significantly larger number of different scenarios compared to the operation of individual defined lines. Since line operation is mostly conducted along main traffic routes, area operation extends the ODD primarily in the direction of secondary routes. Therefore, traffic-calmed paths, play streets, and dead-end streets are increasingly encountered. The challenges of path planning and vehicle control precision are increased by the fact that traffic lanes are narrower, curve radii are smaller, and therefore distances from surrounding objects become shorter. In addition, occlusions caused by parked vehicles within a short distance of the ego vehicle make it difficult to move forward smoothly and quickly.

With a finite number of routes, it is possible to modify complex road sections so that they are more manageable for the automation. Structural modifications can so bridge gaps in the ODD. This approach is not feasible for the operation of driverless taxis. The number of road sections without clear structural in-

Ride-Hailing pioneer Uber is facing increased competition from companies worldwide

Deep dive: shared mobility

Selected key players in the Ride Hailing market worldwide

Note: applications are not exhaustive and serve only as examples
Source: Statista Mobility Market Outlook 2020

Fig. 50.3 Global overview of the main players and their main areas of influence

formation such as road markings or curbs is too high. Adapting the environment to the capabilities of the automation system does not seem economically feasible. Instead, the automated vehicle itself must have the ability to navigate safely in unstructured environments.

In summary, it can be stated that in the case of the ODD taxi, the capabilities of the system must be more comprehensive, as well as the traffic constellations that occur are more diverse than in the case of the ODD shuttle. In combination, this leads to a significant higher number of relevant scenarios, which again greatly increases the effort required to release a corresponding traffic system.

50.4.2 Potentials and Limits of Roadside Units

In addition to structural measures to reduce complexity, stationary devices may also provide supplementary information that supports the automation in its behav-

ioral decision-making. These roadside units (RSUs) can sense and transmit information on the topology of the road or intersection, as well as data of the current status of the traffic signal system or sensor data such as live video streams of the intersection from a bird's-eye view. Both acquisition and commissioning as well as operation and maintenance incur high costs for each RSU. The attempt in Germany to reduce investment costs by integrating RSUs as part of the regular renewals and updates of the traffic infrastructure leads to delays due to the long operating lifetimes of existing traffic signal systems and makes it difficult to implement them quickly throughout the country. From the perspective of individual cities, a complete deployment in larger areas therefore does not appear to be economically feasible. Cities and municipalities, as operators of road and traffic infrastructure, are also hindered in their investments by the lack of reliable refinancing. This results in the requirement for the automated taxis to ensure safe travel even without the use of roadside units.

50.4.3 Up-to-Date and Comprehensive Map

Static environment information can be provided together with the dynamic information by RSUs specific to the currently relevant environment. In contrast, conventional approaches keep the static part of the information in the form of detailed map material in the vehicle. Due to the limited usability of RSUs in the taxi use case, it is not possible to rely on these memory-saving methods. The driverless taxi depends on keeping detailed map material in the vehicle with correspondingly high demands on the required storage space.

These maps, often referred to as HD maps, contain precise geometries of the roadway and surroundings as well as semantic information such as topology of the road with lanes and turn-by-turn directions, speed limits, and other driving rules.

In addition, it is state of the art to use maps for self-positioning (localization) [7]. In one method, for example, georeferenced LiDAR point clouds are stored for this purpose [8]. The current sensor signals are compared with the stored data and, based on the highest match, the current vehicle position and orientation can be determined precisely. Another, data-saving method integrates the positions and optical features of only prominent elements, also called landmarks, into the map. If these features are detected in the camera data, the vehicle is able to verify or even precisely determine its own position based on the relative position to these landmarks.

Regardless of the type of maps used, the amount of data to be provided for the map information is proportional to the area to be covered by the respective ODD. The requirements imposed by the map material extend not only to the amount of storage space in the vehicle, but also to the amount of data to be transmitted as a result of updating the maps. In particular, data from route closures and road works must be kept up to date from a larger area for this application than when operating on dedicated lines.

50.4.4 Dealing with Construction Sites

An automated taxi will also encounter road works that may temporarily change the drivable space or the traffic flow. Therefore, the automated vehicle must be able to navigate in such construction sites. This can happen either with the help of remote assistance and/or fully automated by the automated driving function.

In both cases, however, the construction site must be identified first. Then, either remote assistance or instructions need to be requested, or construction site navigation may be triggered in the automated software stack.

In the case of using remote assistance, an operator in a control room may provide instructions to the auto-

mated vehicle that the software can validate on board. For example, the human operator could suggest a path through a diverted lane that is validated on board before the automated vehicle confirms and travels it [9].

If fully automated navigation in construction sites is used, additional requirements need to be considered. The roadwork site not only needs to be recognized, but the automated driving stack has to interpret it. In doing so, the automated driving software cannot rely completely on its high-resolution maps, since, for example, lanes may be shifted, traffic routing may be changed, and/or lane markings may be invalid or non-existent.

The prediction of other road users also needs to be adjusted in construction sites—for example, a crosswalk might be moved to a different location due to the construction site, which would change pedestrian behavior. Moreover, construction workers move differently in a construction site than ordinary pedestrians.

Another challenge near construction sites is that construction workers or other involved persons may direct traffic, e.g., with the use of temporary stop signs or hand signals.

After an automated vehicle passed a construction site, it should share the collected data with the backend servers or other vehicles in the fleet so that this data can be considered for future route planning [10].

50.4.5 Passenger Interaction

In addition to automating the driving task, driverless passenger transport vehicles also require technical solutions for interacting with passengers. In existing non-automated hailing and pooling services, the driver interacts with the customer in a variety of ways in addition to the driving task. In addition to general communication with passengers, these include, for example, authenticating passengers (do the names and number of passengers match the booking?), opening and closing the door or the luggage compartment if necessary, and responding to stop requests or emergencies. Pre-ride and post-ride operations are already widely supported by non-automated ride-sharing and ride-hailing providers by technical solutions, especially smartphone apps. In the following, the requirements are described for an automated hailing or pooling service regarding passenger interactions [11] along the customer experience from booking to arrival at the destination:

- **Booking**: Customer data and customer request (number of passengers, start and destination, preferences) are recorded. The customer is provided with the information relevant for the booking decision (e.g., place and time of the start of the trip, destination of the trip, costs).

— **Entry**: The customer needs to receive information to identify the vehicle. For this purpose, the vehicle must have a static or dynamic label for identification.

Access to the vehicle can be regulated by an authentication process to prevent unauthorized use as well as confusion. After successful authentication, the door can be opened automatically. Before the vehicle starts moving, it needs to be ensured that the passengers are ready to leave and have fastened their seat belts.

— **During the ride**:
   - Assistance may be provided to the customer, e.g., in the case of questions about the trip or how to interact with the vehicle.
   - Personalization options can be provided (e.g., air conditioning, music).
   - The customer should be able to adjust the destination or end the trip prematurely.
   - The user must be able to report emergencies or initiate rescue operations via suitable user interfaces.
   - The interior space and, in particular, the well-being of the user needs to be monitored so that appropriate actions can be taken in the event of a medical emergency, for example.
   - The customer should be informed about the course of the journey.
— **Exit**: The customer has to be informed when the destination is reached. When the vehicle has reached a safe stopping position, the door shall be opened automatically or manually by the passenger.

After the customer has left the vehicle, navigation instructions may also be provided, if the exit position is not identical with the requested destination.

▶ Section 50.5 discusses various approaches for solving the described requirements.

50.4.6 Fleet Management

Compared to regular services with fixed routes and schedules and the private use of automated vehicles, the use of automated vehicles in fleets for demand-driven pooling or hailing operations is highly complex in terms of fleet management. Fleet management facilitates to consider a large number of requirements, some of which conflict with each other, and on this basis to centrally specify the assignment of passengers to vehicles and the routes of the entire fleet (e.g., within the operating area in a city). In this way, the efficiency and utilization of the fleet can be increased. Thus, fleet management has a significant impact on the profitability of operations. The underlying requirements are summarized below.

From the customer's point of view, the following essential requirements can be named in a simplified way:
— The start of the journey or the arrival at the destination should deviate as little as possible from the start or arrival time requested by the customer.
— The walking distance to be covered by the customer to the place of the start of the journey or from the place of arrival of the vehicle to the customer's destination should be as short as possible.
— The travel time should be as short as possible.
— The price should be as low as possible.

To fulfill the customer's wishes in terms of location and time in a cost-efficient manner, the aim must be to achieve the highest possible utilization of the vehicles (in the pooling case) and to avoid empty runs (both in the pooling and in the hailing case). This can be achieved, for example, by positioning the vehicles according to their demand.

In addition to the requirements from the customer's point of view, operational constraints need to be considered in the fleet management. The fleet can only be operated within its ODD. This results in restrictions both in terms of the routes that can be driven and in terms of other environmental conditions (e.g., weather conditions) in which fleet operation is possible. In addition, regulatory requirements may have to be complied with. For example, it may be necessary not to drive on certain roads due to specifications by the municipality. Another influencing factor is the need to schedule breaks for charging the high-voltage battery or refueling, as well as for maintenance and cleaning work. With regard to the business case, the productive time of the vehicles needs to be maximized.

50.4.7 Maneuver Planner

Automated taxis should be able to pick up and drop off passengers as close as possible to their desired location. This means that the vehicle may have to enter a dead-end street, for example. Here, bidirectional vehicle concepts have the advantage of not having to perform complicated turning maneuvers to turn the vehicle around. The bidirectional vehicle can enter the dead-end street, drop off or pick up the passenger, and drive in the opposite direction to exit.

Furthermore, when dropping off or picking up passengers, it is desirable to bring the vehicle as close to the curb as possible, both to achieve a comfortable entry/exit for the passenger and to allow space for traffic behind the automated taxi to pass. In such scenarios, four-wheel steering allows the vehicle to move closer to the curb while maintaining the length of the parking space.

Additionally, the automated taxi must pick up and drop off passengers in parking lots, which generally have more unstructured traffic. Pedestrians could suddenly cross the roadway, people loading and unloading their vehicles, temporarily backing up, and out of a parking space. Stores could receive delivery from a large truck, for example, that needs to be maneuvered to the loading dock. Social, nonverbal interactions between road users in parking lots present another challenge for automated driving software, such as when a pedestrian waves his hand to indicate he will not cross in front of the vehicle.

50.5 Technical Solutions

50.5.1 Passenger App

Existing solutions for customer interactions use smartphone apps. Trip-independent user data such as name or payment data are stored as a user profile. The process for booking a trip differs between providers only in details. The main steps are summarized below:

- The customer enters the starting point and destination as an address or by selecting them on a map.
- The app displays the boarding and exit positions, which may differ from the start and destination points requested by the customer. In addition, the customer is shown the walking distance from the desired starting point to the boarding position and from the exit position to the desired destination point.
- In addition, the app displays the estimated time of arrival of the vehicle at the starting point and the destination point (◘ Fig. 50.4).

- The cost of the trip is displayed (multiple options if necessary).
- The customer confirms the booking of the trip.
- Information identifying the vehicle, such as a vehicle number, is displayed in the app.
- Until the vehicle arrives at the boarding position, the customer is informed about the estimated time of arrival via the app.

The interactions in the app for booking automated vehicles are essentially identical to ride-hailing offerings operated by drivers. For example, Lyft [12] offers an automated ride-hailing service that can be booked through the Lyft app. Similarly, the service Waymo One [13], which offers ride-hailing exclusively with automated vehicles, is no different from non-automated providers in terms of app interactions. Some providers additionally offer booking via the Internet browser. In contrast to line-based transport, both the departure time and the starting point and destination can be customized by individually entering the customer's request using the app or browser.

50.5.2 Screens in the Vehicle and Exterior/HMI

As soon as the customer encounters the vehicle, the vehicle HMI provides another option for interaction in addition to the app. The first step is to identify the vehicle from the outside. Particularly, if there are several vehicles from the same provider near the boarding

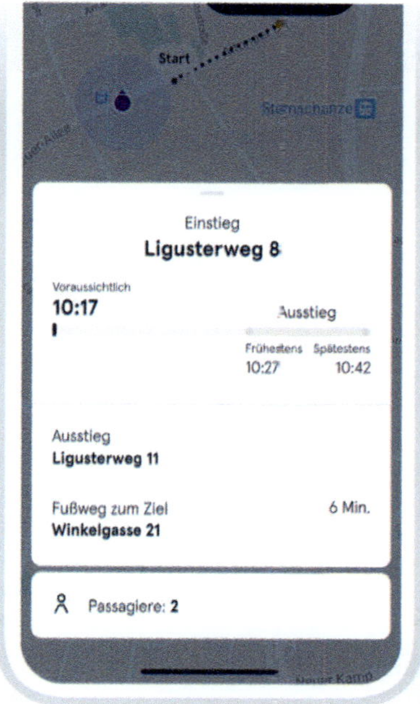

◘ **Fig. 50.4** Exemplary representation of the trip booking and the information on departure and arrival time

position, the customer needs to be able to identify the correct vehicle quickly and clearly. Displays on the exterior of the vehicle are particularly suitable for this.

Currently used vehicles are generally based on series-production passenger cars that are equipped with additional sensor technology and computing hardware. Simultaneously work is being carried out on vehicle concepts for so-called special-purpose vehicles (SPV), which are being developed specifically on the basis of the requirements for MaaS use. One example is the vehicle concept Cruise Origin [14] which has screens to dynamically display a vehicle number on the side of the vehicle. In the vehicles of the Waymo One service, for example, two letters that the customer can specify individually via the app are displayed on a screen behind the windshield.

In the Cruise Origin vehicle concept, additional authentication takes place before boarding by means of a PIN pad attached to the door.

During the journey, passengers are provided with information via screens in the vehicle. Depending on the concept, passenger-specific screens are used, which also allow settings to be made using touch functionality (Zoox [15]) or shared screens that provide information to several passengers at the same time (Cruise Origin).

50.5.3   Control Room/Vehicle Telematics

In control rooms, people control and monitor complex technical systems. Control rooms can be found in aviation, in rail traffic as well as in road traffic, e.g., for monitoring traffic on highways [16, 17]. Control rooms will also play a central role in the operation of automated vehicle fleets. In the control room, the correct and reliable performance of the individual vehicles is monitored. In the event of deviations from the normal state, further procedures are initiated to maintain the operation of the mobility system and ensure safety, cf. ◘ Fig. 50.5. In addition, the control room provides interfaces that enable sovereign intervention by the police, fire department, and rescue services. In detail, the tasks of the control room can be assigned to three categories:

— Monitoring of the vehicle status.
— Resolve technical troubles remotely.
— Communication with passengers and emergency services.

A prerequisite for the function of a control room is communication with the vehicles. Automated test routines regularly evaluate data streams from the vehicle and infrastructure sensors and provide an up-to-date

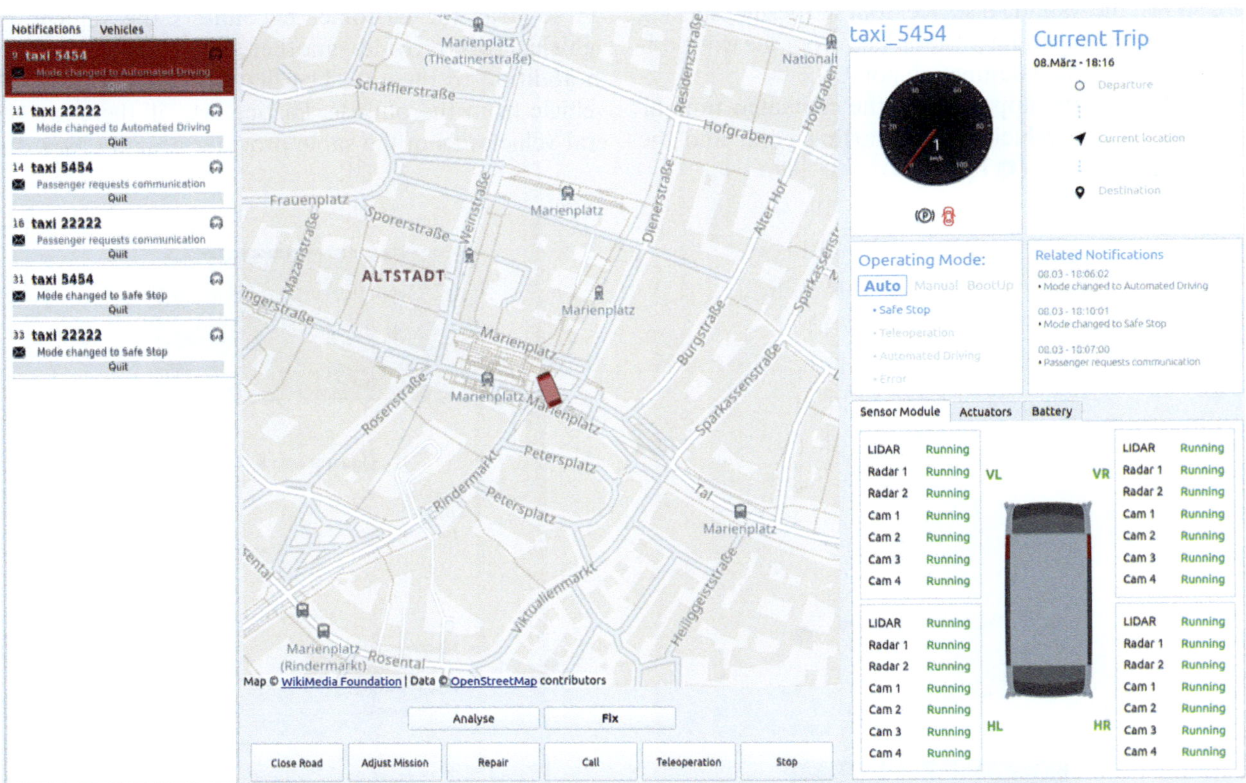

◘ **Fig. 50.5**   Control room user interface

status of the currently active vehicles. In addition to the position and speed of the automated vehicle, the status also includes the number of occupied seats and the currently planned route with stops.

Any faults that occur can be detected in the vehicle and transmitted to the control room. They may also be identified directly in the control room. If the situation appears to be unclear, other data sources such as traffic cameras and infrastructure sensors are evaluated. Based on this, escalation operations are processed and coordinated in the control room. The range of operations spans from communication with passengers or passers-by, remote support of the automated system by a human operator, temporary takeover of the driving function by a human operator, or up to dispatching service personnel.

Areas with little or no cellular reception, such as underpasses, tunnels, or streets with tall houses and/or metal facades, pose a particular challenge. One solution is to avoid these areas and block them in the map data for navigation. Another option, for smaller, limited areas, is to estimate the communication downtime and check for contact re-establishment accordingly. For example, if the planned route is known to pass through a tunnel with limited cellular reception, the control room can actively check to see if the vehicle will exit the tunnel within a suitable time and resume transmitting its status messages. If the feedback fails to arrive beyond the estimated time, procedures are initiated to check the vehicle's status, such as evaluating traffic cameras or dispatching service personnel. With this approach, the operational area can be expanded without reducing safety or service quality in the event of an incident.

With good cellular reception, many solutions are available [18–22] which allow technical incidents to be resolved remotely. The German amended Road Traffic Act (StVG) takes this development into account by introducing the concept of technical supervision. Currently, the options for action by technical supervision listed in the StVG are limited to actions such as deactivating the vehicle or enabling maneuvers. Continuous human monitoring or teleoperation, on the other hand, is explicitly not intended. It can be assumed that with further findings from research and operation, additional interaction options for the technical supervision will find their way into the StVG. Depending on their characteristics, the interaction concepts are divided into the areas of remote support/teleassistance and teleoperation [23]. The common feature of these technical solutions is that the human interacts with the automated vehicle from a distance, taking over partial or complete behavioral decisions and thus responsibility [18–22]. For this purpose, video streams from the vehicle cameras and, in some cases, other sensor signals are transmitted to the human operator [20], see ◘ Fig. 50.6. Relying on his extensive prior knowledge and specific training, the human operator is able to interpret the specific situation more precisely and make decisions for further action.

At teleassistance, the automated system is mostly supported in the area of environment perception. Objects can be unambiguously classified by the human operator, ghost objects can be removed, and objects that

◘ Fig. 50.6   Teleoperation interface with display of multiple video streams

can be driven over can be marked as such. The area marked as traversable can also be specified and extended if necessary. Based on the updated environment model, the technical system takes over the further steps of the driving task such as behavior planning and control. As soon as the vehicle has left the unclear or restricted area, the automated vehicle control system takes over the driving task again completely.

If the trajectory is specified by the human operator or the vehicle is directly controlled by him, this is called teleoperation. The term teleoperation spans a large field of applications defined by a wide range of capabilities of the teleoperated vehicle. In some cases, the release of a path proposed by the automated system and its monitoring during execution is already referred to as teleoperation. In this use case, high demands are placed on the local intelligence in the vehicle. At the other end of the range of definitions direct control is meant, in which the accelerator and brake pedal positions and the steering wheel angle are specified by the human operator. Here, only minor skills of the machine controller in the vehicle are required. The requirements for the bandwidth and latency of communication between the control room and the vehicle are inversely proportional to the requirements for the capabilities of the automated system. For example, monitoring of automated driving at low speeds is possible even with higher latencies of image transmission. Direct control, on the other hand, requires significantly lower latency and higher image quality of the video streams, which are also reflected in higher bandwidth requirements. As well intermediate stages can be found in the area of teleoperation such as speed specification similar to cruise control or point specifications from which the final trajectory is derived in the vehicle. The development of suitable human–machine interaction concepts is currently the subject of many research projects [24–26].

Control rooms with teleassistance and teleoperation are used in all domains of automated transport. Due to the largest number of (unknown) scenarios, the number of deployments will be highest in the automated taxi fleet domain. Effective approaches to solving technical incidents remotely increase the availability and thus the reliability of the transport system. Efficient operations and the choice of the appropriate method of remote support mean that more and more vehicles can be serviced by fewer and fewer people. This reduces the running costs of control rooms and contributes to the always safe operation of automated vehicle fleets.

Just as with the driving functions, situations may arise in passenger interactions that cannot be handled fully automatically by functions in the vehicle and require remote intervention by an operator.

This requires a communication channel between operator and passengers inside the vehicle. Access to vehicle data and vehicle functions is also required. In this way, the operator is able to evaluate the situation and initiate appropriate actions. These can consist of providing information to the customer (e.g., PIN to open the door), remote access to vehicle functions (e.g., opening the door by the operator), or initiating further processes (e.g., sending a replacement vehicle, in case of a technical defect).

50.5.4 Fleet Management: Demand Prediction

In taxi fleet operations, fleet management has a significant impact on primary evaluation variables such as operating costs, resource consumption, and mobility service quality. Controllable variables of the fleet management are the dispatching strategy, vehicle distribution, and fleet size. The operation of automated taxi fleets offers the possibility to optimally adjust the controllable variables at any time. The current demand, including the start and end points of the trips, is recorded centrally and serves as input variables for ongoing optimization. In addition, sophisticated prediction algorithms are used to predict the future development of customer demand [1] with high spatial and temporal resolution. Historical data and other relevant information such as the current weather forecast, dates of cultural events such as concerts as well as sporting events and trade fairs are used [27]. Based on longer term forecasts, the minimum fleet size is determined with which the defined service quality is achieved. As a result, more vehicles may be put into active service or taxis are ordered back to the depot, if necessary. Forecasts over a shorter period of time and with a high spatial resolution are used to ideally reposition the vehicles after the completed trip so that the journey for the next operation is as short as possible in terms of time and space.

In summary, it can be stated that the solution modules shown point to a realization of automated taxis in the near future. Key areas of research and development in the near future will be in the areas of increasing reliability, reducing homologation costs and configuring control rooms.

50.6 Impact of Highly Automated Vehicle ODD Taxi

50.6.1 Societal Layer

With regard to the impact of mobility on society, the high consumption of resources, local emissions, and the high land requirements of individual transport are increasingly coming into focus.

In the European Commission Mobility Strategy and Action Plan [28], automated on-demand services are an important component of a seamless multi-modal mobility system. By linking them with other modalities, such as rail, an alternative to the privately used car can be created for many people.

However, some transport researchers also see the risk of an increase in individual transport in urban areas [29] if the share of passenger kilometers were to shift from public transport to on-demand mobility services due to falling costs and greater convenience of automated vehicles.

Especially in major European cities, traffic is increasingly regulated. Examples are the congestion charge in London [30] or car-free zones in Vienna [31], Amsterdam, or Hamburg.

An additional offer of on-demand mobility is especially needed in rural areas, due to the generally poorer public transport connections. Nevertheless, economic efficiency will be lower compared to urban driving due to low population density and thus limited demand.

The expected cost reduction of on-demand mobility services compared to non-automated ride-hailing or taxi services will result in new mobility options for elderly and physically impaired people in the long run.

50.6.2 Legal Layer

The legal framework is provided on the one hand by legislation on automated driving and on the other hand by passenger transport law. Laws and rules may be adapted, if the potential societal benefits appear to be significantly greater than potential risks associated with the legalization of a new technology.

In Germany, the legal basis for driverless operation of automated vehicles is provided by the Autonomous Vehicle Approval and Operation Ordinance (AFGBV) [32] based on the amended Road Traffic Act (StVG). In addition to technical requirements for the vehicle, the AFGBV also regulates the approval of the operating area. It also defines the role of technical supervision, which is responsible for monitoring the vehicle and—if necessary—initiating the minimum risk condition.

With the amendment of the Passenger Transportation Act in 2021 (PBefG) [33], Germany created the basis for ride-sharing services, which were previously only possible with the help of time-limited special permits. For example, the PBefG provides for the transport form of bundled on-demand transport. Furthermore, it is up to the municipalities to define the concrete framework conditions for the operators of mobility services. Among other things, the PBefG stipulates that the licensing authority must specify a quota for the share of pooled trips. The PBefG also sets the goal of achieving accessibility in local public transport by 2022. It is to be expected that the requirement of accessibility will also be imposed on on-demand passenger transport.

50.6.3 Economic Layer

The legal layer defines, which business models may be chosen in order to make a new technology commercially viable. When considering the economic potential of automated taxi services, the following cost factors should be considered:

- vehicle costs,
- the cost of licensing,
- insurance costs,
- maintenance costs,
- cleaning costs,
- fuel costs,
- parking costs as well as
- costs for remote support such as technical supervision, teleassistance, and/or operation.

In addition to these costs, the utilization of the automated taxi and the transport fee that can be achieved also influence profitability.

The largest cost factor of a conventional ride-hailing service today is the driver, which indicates the potential of automated taxi. It is estimated that the driver represents the largest single cost in non-automated ride-hailing services, accounting for 80% of the total cost per kilometer.

However, there will still be humans involved in an automated taxi ride, for example, in teleoperation or vehicle maintenance. The goal is to reduce human involvement per vehicle as much as possible while still ensuring a safe and comfortable ride.

Looking at the cost breakdown for automated taxi, it is clear that there are three key challenges to making an automated taxi service viable: first, maximizing the number of vehicles that a human can oversee for remote assistance. For example, if there were one teleoperation person per automated taxi, the cost savings would be marginal or even negative due to the expensive technology. Therefore, the automated taxi company must maximize the number of vehicles that one teleoperation person can supervise.

Second, the utilization rate of automated taxis needs to be optimized. The utilization rate of today's human-operated taxis in a region like San Francisco is about 50% [34], which means that the taxi carries a passenger only 50% of the time. Potential ways that increase the utilization rate for automated taxis include a ride-pooling approach or using the automated vehicles during idle times for other services such as delivery services. Furthermore, the automated taxi can collect data during the journey. The use or commercialization of these data also offers added economic value. Examples include the detection of free parking spaces, map updates, and the collection of information on road works.

Third, the cost of the automated platform itself, including sensors and computing power, is a non-trivially quantifiable cost position today. The cost of computing power is falling rapidly, by an order of magnitude every 4–5 years, and the cost of sensors such as LiDAR, radar, and vision is also expected to fall significantly. However, today's sensor architectures of automated vehicles have multi-modal 360-degree sensor fields of view around the vehicle, which requires a large number of sensors. This makes it a challenge to drastically reduce the cost of sensors.

When pricing automated taxi services, the cost-based determination of an offer price should be contrasted with a consideration of the benefits for the customer. Humans can make different use of the time they have been investing in driving themselves. For example, the passenger in the automated taxi can already process emails or documents during the ride, while in the self-driven vehicle he must devote his cognitive attention to the driving task. In addition, the passenger can be provided with other paid services in the interior, such as music offerings or an Internet connection, so that this can have a positive effect on the profitability of automated taxi.

It is not uncommon for a new mobility service to be initially offered free of charge in order to obtain feedback from as many users as possible [35, 36]. Of course, this is only possible on a temporary basis, as high costs are incurred, especially during the introductory phase. Thus, in the shuttle model region of Upper Franconia [35] one route each with automated shuttles was implemented in the cities of Hof, Kronach, and Rehau and operated according to a fixed schedule.

50.6.4 User Layer

The user layer is positioned between the economic layer and the technical layer, since the user's willingness to pay is directly influenced by the perceived benefits of the technology or function.

The stated goal is that trips in an automated taxi will be cheaper for customers than trips in a taxi with a human driver, so that users change their behavior. Although journeys in (automated) buses or by streetcar or suburban railway are still cheaper, these are only offered on a limited number of defined routes and at fixed times. Individual travel by car is a completely different matter: the costs are lower than for automated taxis if the annual mileage is high. Also, the starting point and destination as well as the time of the trip can be determined individually. However, disadvantages of

driving a vehicle are the preoccupation with the driving task, which is perceived as exhausting especially at peak times in gridlocked or congested traffic, as well as the search for a parking space. In summary, from the user's point of view, the application profile for the automated taxis may be attractive, mainly because of the high gain in comfort combined with low additional costs compared to alternative mobility offers.

The targeted user group of automated taxis is therefore made up of people who have previously used their own vehicle. Less desirable from a traffic point of view are users who switch from public transport to the automated taxi. The switch from large shared transport vehicles to several small, individually used transport vehicles would increase traffic density. In contrast, there is a net total benefit if ride-sharing is used intensively. Because of the very comfortable ride-sharing of the automated taxis, former private car users could overcompensate the disadvantage caused by the former public transport users.

Ride-sharing needs to be made easy to use with the help of apps that are optimized in terms of system ergonomics. In addition to a new form of mobility, ride-sharing also initiates a social interaction that is new for many people. People are used to meeting familiar people in small vehicles and meeting strangers in large vehicles such as buses or trains. The unfamiliar situation of riding with unknown people in a small vessel, combined with the absence of a human taxi driver during the introductory phase, could lead to limited acceptance. Analyzing this and deriving recommendations for action is the subject of current research [37].

The absence of a driver also has a very practical impact on the services offered by the mobility system. Passengers who depend on individual assistance require additional services and solutions for journeys in automated vehicles. For example, people with reading difficulties or visual impairments can be addressed with voice instructions and control. Technical solutions for boarding and taking along wheelchairs are also available. In the future, individual support can only be fully guaranteed by an accompanying person, so that existing mobility services will continue to have their justification.

50.6.5 Technical Layer

On the technical layer, all requirements resulting from the layers above need to be considered in order to develop suitable technical solutions. This is not only true for the vehicle itself. The use of automated taxis on a large scale in inner-city traffic will most likely have an impact on infrastructure. For example, depots will be required for storage, maintenance, cleaning, and charging/fueling of the vehicles. If the automated fleet consists of electric vehicles, the depots will require a high-capacity connection to the power grid for charging.

In addition, automated taxis in large numbers in inner cities have the potential to reduce the number of private vehicles used, resulting in less demand for parking spaces and thus reducing parking search traffic. Studies [38] show that this traffic is of a non-negligible magnitude. Ride-pooling can also reduce traffic in downtowns, especially if the number of trips does not increase at the same time.

Automated taxis use the mobile network primarily to increase the reliability and user experience of the transport service, while safety needs to be ensured without the mandatory need for a mobile connection. Consequently, a very good expansion of the mobile network is required, especially in terms of reliability and data throughput.

The integration of automated taxi fleets into the mobility concept of municipalities and cities still involves unresolved issues. In particular, the design and networking of the provider-specific control rooms poses various challenges, also with regard to the legal requirements for technical supervision.

Technical supervision is an integral part of the operating concept of driverless vehicles and is currently designed individually by the manufacturer or provider—thus the resulting solutions are largely incompatible with each other. The technical solution to ensure efficient and safe operation is a prerequisite for approval for public road traffic. It should be emphasized that this forms the core competence of the mobility system provider and thus represents one of the key competitive differentiating factors. The operation of vehicles from several providers in a joint technical control room currently does not appear to be feasible, neither technically nor in terms of corporate policy, so that the combination of the provider-specific control rooms with a higher level, municipal control room appears to be the most likely solution.

It is also unclear which instance or organization will operate the technical supervision. Three variants are currently being discussed. The technical supervision may be integrated into the control room of the municipal traffic control and also operated by it. The second variant envisages an independent operator of the technical supervision. Predestined for this appear providers who already offer services in the area of monitoring aviation and train traffic or in connection with telecommunications. On the one hand, this operator is instructed by the manufacturer or mobility provider in the safety concept of the mobility system, and on the other hand, the operator provides the municipal control room with the necessary status information. In the last variant, the manufacturer operates the technical

supervision itself as an extension of his business field. The UNICARagil concept envisages a municipal control room as the superordinate and coordinating instance for several provider-specific control rooms.

Another degree of freedom arises in the centralized or decentralized mode of operation of the technical supervision. Technical or regulatory requirements may necessitate a technical supervisory authority close to the region. These may be requirements for low latency of data transmission or legal requirements that presuppose local knowledge on the part of the personnel. This would make it impossible to implement the decentralized approach, which envisages a few, large, and thus cost-efficient control rooms. These requirements also determine whether technical supervision can be relocated to other (low-wage) countries. In this case, it is then also necessary to evaluate customer acceptance.

In pilot projects, only vehicles from one manufacturer are currently being used and technical supervision is being implemented on site. Different approaches are already being used for operation. In the near future, cross-regional supervision by one manufacturer seems to be more obvious than cross-manufacturer operating concepts. For the latter, dedicated standardization procedures of several stakeholders would be required or the standard-setting specification of a market leader. To the knowledge of the authors, no approaches to this have been published so far.

50.7 Outlook

Currently, Waymo offers its passenger transportation service with automated driverless taxis Waymo One [13] in the cities of Phoenix, Arizona, and San Francisco, California. Apollo Go, the automated taxi service from provider Baidu, is currently available in five major Chinese cities, including parts of Beijing and Shanghai [40]. It is expected that offerings from other vendors such as Zoox and Cruise will follow. Further cities will gradually be opened up and more and more people will be able to use the services of automated taxi. It remains to be seen at what point in time and under which conditions the ODD taxi services may be operated economically.

References

1. Liu, T, Krishnakumari, P., Cats, O.: Exploring Demand Patterns of a Ride-Sourcing Service using Spatial and Temporal Clustering (2019). ▶ https://doi.org/10.1109/MTITS.2019.8883312
2. Kaddoura, I., Schlenther, T.: The impact of trip density on the fleet size and pooling rate of ride-hailing services: a simulation study. Procedia Comput. Sci. **184**, 674–679 (2021). ISSN: 1877-0509. ▶ https://doi.org/10.1016/j.procs.2021.03.084

3. Karaenke, P., Schiffer, M., Waldherr, S.: The Customer is Always Right: Customer-Centered Pooling for Ride-Hailing Systems (2021). arXiv preprint ▶ arXiv:abs/2107.01161

4. ▶ https://www.eurofound.europa.eu/data/platform-economy/initiatives/amendment-to-passenger-transport-act-in-germany

5. Ride-sharing market size worldwide in 2020 and 2021 (2021). ▶ https://www.statista.com/statistics/1155981/ride-sharing-market-size-worldwide/

6. Size of the global market for autonomous vehicles in 2030, by segment (2021). ▶ https://www.statista.com/statistics/875069/av-global-market-size-by-segment/

7. Levinson, J., Thrun, S.: Robust vehicle localization in urban environments using probabilistic maps. In: IEEE International Conference on Robotics and Automation (ICRA), pp. 4372–4378 (2010)

8. Elhousni, M., Huang, X.: A survey on 3D LiDAR localization for autonomous vehicles. In: IEEE Intelligent Vehicles Symposium (IV), pp. 1879–1884 (2020). ▶ https://doi.org/10.1109/IV47402.2020.9304812

9. Offizielles Video zum Thema TeleGuidance des Start Ups Zoox (2020). ▶ https://www.youtube.com/watch?v=NKQHuutVx78

10. Lampe, B., Woopen, T., Eckstein, L.: Collective driving - cloud services for automated vehicles in UNICARagil, 28. Aachener Kolloquium, Aachen (2019). ▶ https://doi.org/10.18154/RWTH-2019-10061

11. Kim, S., Chang, J.J.E., Park, H.H., Song, S.U., Cha, C.B., Kim, J.W., Kang, N.: Autonomous taxi service design and user experience. Int. J. Hum.-Comput. Interact. **36**(5), 429–448 (2020). ▶ https://doi.org/10.1080/10447318.2019.1653556

12. Kirsten Korosec, Argo, Ford to launch self-driving vehicles on Lyft's ride-hailing app (2021). ▶ https://techcrunch.com/2021/07/21/argo-ford-to-launch-self-driving-vehicles-on-lyfts-ride-hailing-app/

13. ▶ https://waymo.com/waymo-one/

14. ▶ https://www.getcruise.com/technology

15. Sean O'Kane, Zoox unveils a self-driving car that could become Amazon's first robotaxi (2020). ▶ https://www.theverge.com/2020/12/14/22173971/zoox-amazon-robotaxi-self-driving-autonomous-vehicle-ride-hailing

16. Wikipedia: Flugverkehrskontrolle (2022). ▶ https://de.wikipedia.org/w/index.php?title=Flugverkehrskontrolle&oldid=222428895

17. Wikipedia: Transportleitung (2022). ▶ https://de.wikipedia.org/w/index.php?title=Transportleitung&oldid=217567673

18. Feiler, J., Diermeyer, F.: The perception modification concept to free the path of an automated vehicle remotely. In: Proceedings of the 7th International Conference on Vehicle Technology and Intelligent Transport Systems—VEHITS, pp. 405–412 (2021). ISBN: 978-989-758-513-5; ISSN: 2184-495X. ▶ https://doi.org/10.5220/0010433304050412

19. Phantom Auto: Teleoperation safety solution for autonomous vehicles (2019). ▶ https://phantom.auto/

20. Hosseini, A., Lienkamp, M.: Enhancing telepresence during the teleoperation of road vehicles using HMD-based mixed reality. In: IEEE Intelligent Vehicles Symposium (IV), pp. 1366–1373 (2016)

21. Liu, R., Kwak, D., Devarakonda, S., Bekris, K., Iftode, L.: Investigating remote driving over the lTE network. In: Proceedings of the 9th International Conference on Automotive User Interfaces and Interactive Vehicular Applications, pp. 264–269 (2017)

22. Schimpe, A., Diermeyer, F.: Steer with me: a predictive, potential field-based control approach for semi-autonomous, teleoperated road vehicles. In: IEEE 23rd International Conference on Intelligent Transportation Systems (ITSC), pp. 1–6 (2020). ▶ https://doi.org/10.1109/ITSC45102.2020.9294702

23. SAE J3016_202104: Taxonomy and Definitions for Terms Related to Driving Automation Systems for On-Road Motor Vehicles (2020)

24. UNICARagil (2022). ▶ https://www.unicaragil.de/de/

25. M Cube: Wies'n Shuttle (2022). ▶ https://www.mcube-cluster.de/projects/wiesnshuttle/

26. REmote Driving Operation—REDO (2022). ▶ https://www.vinnova.se/en/p/remote-driving-operation---redo

27. Wittmann, M., Neuner, L., Lienkamp, M.: A predictive fleet management strategy for on-demand mobility services: a case study in Munich. Electronics (2020). ▶ https://doi.org/10.3390/electronics9061021

28. Sustainable and Smart Mobility Strategy—Putting European Transport on Track for the Future (2020) ▶ https://eur-lex.europa.eu/legal-content/EN/TXT/?uri=CELEX:52020DC0789 [just one source]

29. Overtoom, I., Correia, G., Huang, Y., Verbraeck, A.: Assessing the impacts of shared autonomous vehicles on congestion and curb use: a traffic simulation study in The Hague, Netherlands. Int. J. Transp. Sci. Technol. **9**(3), 195–206 (2020). ISSN: 2046-0430. ▶ https://doi.org/10.1016/j.ijtst.2020.03.009

30. Green, C.P., Heywood, J.S., Paniagua, M.N.: Did the London congestion charge reduce pollution? Reg. Sci. Urban Econ. **84**, 103573 (2020). ISSN: 0166-0462. ▶ https://doi.org/10.1016/j.regsciurbeco.2020.103573

31. Wiener City wird noch vor der Wien-Wahl autofrei (2020). ▶ https://www.vienna.at/wiener-city-wird-noch-vor-der-wien-wahl-autofrei/6649029

32. Entwurf eines Gesetzes zur Änderung des Straßenverkehrsgesetzes und des Pflichtversicherungsgesetzes—Gesetz zum autonomen Fahren (2021). ▶ https://www.bmvi.de/SharedDocs/DE/Anlage/Gesetze/Gesetze-19/gesetz-aenderung-strassenverkehrsgesetz-pflichtversicherungsgesetz-autonomes-fahren.pdf?__blob=publicationFile

33. Personenbeförderungsgesetz (PBefG) (2021). ▶ https://www.gesetze-im-internet.de/pbefg/PBefG.pdf

34. Rayle, L., Shaheen, S., Chan, N., Dai, D., Cervero, R.: App-Based On-Demand Ride Services: Comparing Taxi and Ridesourcing Trips and User Characteristics in San Francisco (Working paper) (2014). ▶ https://www.its.dot.gov/itspac/dec2014/ridesourcingwhitepaper_nov2014.pdf

35. Shuttle Modellregion Oberfranken (2021). ▶ https://www.shuttle-modellregion-oberfranken.de/

36. KelRide (2021). ▶ https://kelride.com/

37. König, A., Brandies, A., Schnieder, L., Meike, J., Dotzauer, M.: Zukunftsszenarien autonomer Fahrzeuge—Nutzungsorientierte Gestaltung eines individuell abrufbaren Personentransportsystems (2016)

38. Searching for Parking Costs Americans $73 Billion a Year (2017). ▶ https://inrix.com/press-releases/parking-pain-us/

39. Leading ride-hailing operators worldwide as of November 2019, based on market share (2021). ▶ https://www.statista.com/statistics/1156066/leading-ride-hailing-operators-worldwide-by-market-share/

40. Erl, J.: Baidu schlägt Waymo: Apollo Go größter autonom fahrender Taxi-Dienst (2021). ▶ https://mixed.de/baidu-schlaegt-waymo-apollo-go-groesster-autonom-fahrender-taxi-dienst/

Open Access This chapter is licensed under the terms of the Creative Commons Attribution-NonCommercial-NoDerivatives 4.0 International License (▶ http://creativecommons.org/licenses/by-nc-nd/4.0/), which permits any noncommercial use, sharing, distribution and reproduction in any medium or format, as long as you give appropriate credit to the original author(s) and the source, provide a link to the Creative Commons license and indicate if you modified the licensed material. You do not have permission under this license to share adapted material derived from this chapter or parts of it.

The images or other third party material in this chapter are included in the chapter's Creative Commons license, unless indicated otherwise in a credit line to the material. If material is not included in the chapter's Creative Commons license and your intended use is not permitted by statutory regulation or exceeds the permitted use, you will need to obtain permission directly from the copyright holder.

Driver State Monitoring in Automated Driving

Claus Marberger and Dietrich Manstetten

Contents

This article elaborates on the challenges of driver state management in the context of partially and conditionally automated driving.

© The Author(s) 2026
H. Winner et al. (eds.), *Handbook Assisted and Automated Driving*,
https://doi.org/10.1007/978-3-658-45276-6_51

51.1 The Driver's Role in Automated Driving

The vision of automated driving often refers to the idea of a safe, fast, and comfortable form of mobility that no longer requires a human driver (e.g., Mercedes-Benz Group [1]). At the same time, it is currently not foreseeable how rapidly and to what extent this vision turns into reality. For current driving automation systems on public roads, the user still plays a specific, safety-relevant role. This fact is also reflected in the established taxonomies on the classification of vehicle automation (e.g., according to SAE J3016 [2]). ◘ Table 51.1 shows an overview of the assigned driver roles for partial and conditional automation levels which this chapter focuses on.

While automation level 2 requires a driver, who is fully involved in the driving task and ready to intervene at any time, the driver at automation level 3 should remain sufficiently alert to detect explicit takeover requests and other obvious system limitations. This understanding of roles is also reflected in the current legal requirements in Germany (see §1 and §23 StVO and §1b paragraph 1 of the StVG). For a deeper understanding of the legal aspects of changing driver duties at different levels of driving automation, please refer to ► Chap. 3 of this book.

51.2 Known Challenges

This section describes individual findings from psychological research showing that maintaining a suitable driver state in the context of automated driving (levels 2 and 3) poses certain challenges.

51.2.1 Vigilance

In partial driving automation (SAE level 2), the driver may be physically decoupled from the longitudinal and lateral aspects of vehicle control. At the same time, he/she retains full responsibility for system monitoring in order to be able to intervene in a timely manner in the event of an error or in situations that are not controlled by the system. Since partially automated functions will only be introduced to the market with sufficient reliability (due to customer acceptance), errors and necessary user interventions may occur only rarely. There is a rich body of research on human performance in supervising automated systems (see research on "vigilance") which consistently indicates a sharp decline in the detection of errors after only a few minutes (Mackworth [4], for an overview see Warm et al. [5]). Humans are not particularly eager to carry out the monitoring task, and if they do, they perceive the task as being monotonous and mentally exhausting (Warm et al. [5], Greenlee [6]), which in turn leads to task-related fatigue (Körber et al. [7]). In order to combat monotony and boredom humans may lean towards other, subjectively more rewarding activities and, thus, shift their attention away from the monitoring task (Carsten and Martens [8], Llaneras et al. [9]). This effect is particularly reinforced when users rely excessively on L2 automation (Banks [10]). The more reliable an automated system is, the more critical are behavioral adaptations of the users in case of surprising system limits. Hancock [11] summarizes this dilemma by concluding "if you build systems where people are rarely required to respond, they will rarely respond when required" (Hancock [11], S. 453). In addition to these critical findings

◘ **Table 51.1** The role of the driver in partial and conditional automation according to SAE (excerpt from [3]). For clarification of the abbreviations ODD (operational design domain) and OEDR (object and event detection and response), please refer to the SAE standard J3016 [2]

SAE L2 ("partial driving automation")	SAE L3 ("conditional driving automation")
Determine when activation or deactivation of the system is appropriate	Determine when activation or deactivation of the automated driving system is appropriate
Execute the OEDR by monitoring the driving environment and responding if necessary (e.g., emergency vehicles coming)	Does not need to execute the longitudinal, lateral driving tasks and the monitoring of the environment for operational decisions in the ODD
Constantly supervise the dynamic driving task executed by the system. Although the driver may be disengaged from the physical aspects of driving, he/she must be fully engaged mentally with the driving task and shall immediately intervene when required by the environment or by the system (no transition demand by the system, just warning in case of misuse or failure)	Shall remain sufficiently vigilant as to acknowledge the transition demand and acknowledge vehicle warnings, mechanical failure, or emergency vehicles (increased lead time compared to level 2)
The driver shall not perform secondary activities which will hamper him in intervening immediately when required	May turn his attention away from the complete dynamic driving task in the ODD but can only perform secondary activities with appropriate reaction times. It would be beneficial if the vehicle displays were used for secondary activities

(mainly originating from laboratory research), the effects of real systems in the field must also be taken into account. Gershon et al. [12] find that users of a partially automated driving feature switch frequently and freely between different automation levels in real traffic—not only in the context of critical driving situations.

51.2.2 Fatigue

As long as people perform a safety-relevant function in automated driving, fatigue remains an important issue—as it does in manual driving (Schömig et al. [13], Vogelpohl et al. [14]). Partially and highly automated driving features that require the driver as fallback at system limits place high demands on the user to stay awake. Generally, fatigue refers to a feeling of being extremely tired, usually because of hard work or exercise (Oxford Learner's Dictionaries [15]). According to the underlying root causes, May and Baldwin [16] differentiate between three types of fatigue: sleep-related fatigue due to sleep deprivation, active fatigue due to too high task load, and passive fatigue due to too low task demand/monotony (see vigilance). In a highly automated ride, passive fatigue may emerge after about 20 min (see Jarosch [17]). At the same time, however, fatigue can also be counteracted by suitable non-driving-related activities (at SAE level 3) (see Jarosch [17], Weinbeer [18]). Although sleep-related fatigue is not a phenomenon specifically caused by automation, it may trigger unintended use of the automation feature for some users. Severe drowsiness and sleep states are not compatible with the general requirements for L2 and L3 automation and should therefore be avoided.

51.2.3 Limited Time Budget for Transition Demands

In contrast to partial driving automation, automation features at SAE L3 are defined by the fact that they must provide the driver a sufficient "time budget" to safely resume vehicle control. A large number of user studies have been conducted to analyze how large this time window must be (for an overview, see Zhang et al. [19] or Jarosch et al. [20]). In this context, the evaluation criteria should not only refer to the temporal dimension, but especially to the quality of the transition. In order to make sure that a large proportion of drivers (e.g., 90%) is able to handle a challenging takeover situation, the recommended takeover time budget extends to no less than 8–10 s (see Damböck [21], Peter-

mann-Stock et al. [22] or Naujoks et al. [23]). Providing such large time budgets is a technical challenge for the development of highly automated systems.

51.2.4 Effects of Inappropriate Mental Models

Building adequate mental models regarding the performance of automation and the associated role of the driver is a crucial factor for the safe use of automated systems. After all, the product-specific system limits and the intended use of an automation function do not always correspond to the initial expectations of users. In particular, highly reliable partial driving automation may falsely be perceived as a higher level of automation that no longer requires permanent supervision (Banks et al. [10], Boos et al. [24]). When critical driving situations are developing with active partial automation, it is often not easy for users to recognize whether a corrective intervention is needed or whether automation can handle the situation on its own. If the mental model about the system performance does not correspond with the actual design of the system, so-called "automation surprises" may arise (Sarter et al. [25]). Overreliance on partial driving automation may even prevent drivers to intervene (sufficiently), even though they are paying full visual attention to the critical traffic situation (see Viktor et al. [26]).

Automated driving functions may likely exist at different levels within one vehicle, which poses another challenge on the human user: there must be a clear understanding, which tasks are currently being taken over by the system and what the remaining role of the user is. An incorrect assessment of the current operating mode, the so-called "mode confusion", can easily lead to safety–critical situations, especially if the performance of the current automation level is overestimated (Wilson et al. [27], Kurpiers et al. [28]).

Automated driving functions take over certain aspects of the driving task and may trigger behavioral adjustments that go far beyond the intended use as described by the manufacturer (see Banks et al. [10] or Morando et al. [29]), especially when there is great subjective benefit of the behavioral adaptation. For example, if partial automation makes drivers feel safer to interact with mobile electronic devices, this behavior (not intended by the manufacturer) may manifest itself over time. The Federal Highway Research Institute in Germany (BASt) further distinguishes between two types of misuse: (1) unintended use that is caused by a faulty mental model and (2) unintended use despite a correct mental model (Marberger [30]). For reasons of product liability, manufacturers of automated driving functions are obliged to take foreseeable misuse into account for system design and marketing.

51.3 Solution Approaches

Due to the phenomena described above, manufacturers of automated driving functions can not only refer to the legal duties of the user but must meet additional requirements in the product design. For example, basic and binding design requirements for the human–machine interface are described in the type approval regulations for specific functions (e.g., in UN ECE regulations R79 [31] or R157 [32]). In addition, when designing and marketing systems, manufacturers should adhere to relevant development standards such as the RESPONSE Code of Practice ([33]) or the international standard ISO/PAS 21448:2019 ([34]) which also includes an analysis of possible undesirable driver states.

51.3.1 Strategies for Partial Driving Automation (SAE L2)

At the level of partial driving automation, various approaches have been explored in the scientific debate on how to technically support the maintenance of a desirable driver state (for an overview see Cabrall et al. [35] or Mueller et al. [36]).

The goal of **proactive strategies** (see Mueller et al. [36]) is to design the automation feature or the conditions of use in such a way that the driver cannot leave the control loop in the first place or at least remains motivated not to do so. This should implicitly make it clear to users that partial automation merely supports the task of vehicle control and does not replace it (as is the case from SAE level 3 onward).

Examples of this strategy are as follows:

- Cooperative concepts for vehicle guidance have been proposed by which the lateral control of the vehicle is based on the combined inputs of automation and driver. By involving the driver in a continuous sensorimotor sub-task of vehicle guidance, awareness of the driving situation should be maintained as much as possible. For an overview of this principle, please refer to the relevant literature on "shared control" (Abbink et al. [37], Flemisch et al. [38]).
- Since critical vigilance reduction only sets in after a certain duration of the monitoring task, the maximum time window for SAE L2 driving could be limited accordingly (Parasuraman et al. [39]).

- In order to facilitate the monitoring task, HMI concepts were proposed that give feedback about the current level of system reliability, and hence the need to supervise system performance. By modulating the user's task load, continuous monitoring should be made easier and more acceptable to users (Beller et al. [40]).

In addition to the proactive strategies, **system-based monitoring of the current driver state** (as a **reactive strategy**) addresses those cases in which the driver does not seem to be sufficiently involved in the driving or monitoring task anymore. These technical devices record and evaluate (reactively) suitable behavioral indicators and trigger a cascade of system interventions (e.g., warnings, takeover request, feature deactivation) when predetermined thresholds are exceeded (Mueller et al. [36]). Currently available systems for partial automation include some forms of driver state monitoring, but they differ with respect to which indicators are monitored and by the design of the system intervention. So-called "hands-on" systems measure the extent of the driver's involvement in lateral vehicle control and, if minimum requirements are violated, a warning and intervention cascade is triggered (for example, in accordance with UN ECE R79 [31]). However, other driver state monitoring approaches are also in place which focus on head orientation and gaze direction of the driver (e.g., utilized in Cadillac Super Cruise or Nissan ProPilot). As with the proactive strategies, warnings and sanctions (up to feature deactivation) are intended to motivate the driver to use the system as intended and thus ensure an appropriate driver state. The positive effect of individual measures on driver behavior can be demonstrated in respective user studies (e.g., Llaneras et al. [41]). However, the final assessment of whether these measures keep users of a partially automated driving feature "sufficiently engaged" is subject of current discussions for regulation of functions at SAE Level 2.

Downgrading the automation level in case of detected misuse can be seen as a form of **adaptive automation**. In adaptive automation (for an overview see Sheridan and Parasuraman [42]), the scope and type of assistance adapts according to predetermined criteria, such as the currently detected driver state. Alternatively, the automation level could also be increased as long as an inappropriate driver state is detected. For example, Cabrall et al. [43] compared two scenarios how to deal with visual distraction: (1) automated activation of steering support in manual driving (increase of the automation level from SAE L0 to L1) and (2) request to resume manual control in partially automated driving (decrease of the automation level from SAE L2 to

L0). In their study, the first scenario was preferred over the second one. However, in real life, a driver-adaptive increase to higher automation levels is often not feasible for technical reasons.

51.3.2 Strategies for Conditional Driving Automation (SAE L3)

An essential user requirement in conditional driving automation (SAE L3) is related to being "receptive" for explicit requests to intervene and for general events that obviously prevent driving automation to safely control the vehicle. This requirement differs significantly from those in SAE Level 2 as SAE L3 does not require the driver to continuously monitor the driving situation and may even allow the user to engage in visually distracting, non-driving-related activities. Various strategies can be pursued in order to keep the user in fallback-ready state. Here are some examples for **proactive strategies**:

— In order to support the perceptibility of an "RtI" (Request to Intervene) signals, other visual and acoustic signals in the vehicle interior should be temporarily blocked (see UN ECE R157 [32]).
— Physical readiness to resume control can be supported by limiting strong changes in the seating position during conditionally automated driving.
— Since vehicles with conditionally automated driving features usually also offer lower levels of automation (in other situations), it is particularly important to clearly communicate the specific driver task for the current mode (Flemisch et al. [44]).
— In order to prevent the various forms of fatigue (see ▶ Sect. 51.2), non-driving activities are particularly suitable in the context of conditionally automated driving (SAE L3). They should be easy to interrupt and be perceived by the drivers as useful and attractive (see Sprung [45] or Weinbeer et al. [18]).

Technical means for driver state monitoring also play an important role in maintaining driving readiness for a fallback-ready user as these systems can trigger system interventions in case critical behavior is detected.

51.4 Toward the Assessment of Driver Readiness

The term "driver readiness" or "driver availability" is often used in discussions about driver state requirements not only in the context of conditional driving automation, but also for partial driving automation. A reference document on the development of UN regulations for automated driving (UN ECE [3]) mentions functions for detecting driver availability for SAE L2 to L4. In the ISO technical report 21959-2:2020 "Human Performance and State in the Context of Automated Driving" [46], the concept of "driver availability" or "driver readiness" is introduced (as synonyms) for the assessment of driver states for partial and conditional automation. In this chapter, we regard both terms as synonyms for a concept that relates the current driver state with the predicted human takeover performance at specific system limits (see Marberger et al. [47]). The current driver state (bottom-up analysis) and the requirements of the upcoming driving situation (top-down analysis) play an essential role in assessing the readiness of the driver. In the following sections, the interaction of the driver state with situational requirements will be elaborated in more detail (independent of specific and binding design requirements, e.g., based on UN ECE regulations). ▫ Figure 51.1 shows a model for assessing and maintaining driver readiness in automated driving, which serves as a basis for the following sections.

51.4.1 Demands of Takeover Situations

The required state of the driver during automated driving basically depends on the requirements of potential takeover situations resulting from a specific system design. Key questions here include the following:
— How easily can the driver detect the need for intervention? Will there be system-initiated warnings such as a "Request to Intervene" or does the driver have to recognize the need for intervention him/herself?
— How large is the time budget between the detectability of a critical situation and the last possible moment of intervention? Which safety buffers can the automation and the vehicle environment offer?
— Which aspects of the driving task must the driver take over? The monitoring task or, in addition, lateral and/or longitudinal control?
— How complex is it to resume the driving task? Does surrounding traffic have to be taken into account?

The assessment of such questions is usually done in the context of a hazard and risk analysis (see ISO/PAS 21448 [34]) for a concrete automation function and is therefore not only dependent on the assigned SAE level. Different designs of a function at SAE level 2 can therefore lead to different requirements for the driver state. Potential takeover situations in a conditionally automated drive differ significantly from those that may occur in partially automated driving, as has been mentioned several times. In the context of conditional automation, the current driving situation can further be differentiated whether the system is operating within its system limits ("steady-state operation") or whether the user is expected to resume control following a Request to Intervene.

51.4.2 Indicators for Estimating Driver Readiness

According to this model, understanding the abstract state "driver readiness" can be inferred from the states of associated sub-constructs. Although these are still not yet directly measurable, they usually correspond to the constructs, which are considered in the literature as relevant aspects of the driver state (e.g., "situational awareness" or "receptivity"—see ▶ Sect. 51.4.3. In the end, the assessment of the driver's state is based on the analysis of measurable driver state data and behavioral indicators (such as head orientation or hand position on the steering wheel), which provide estimates of the higher, more abstract constructs. In the following, different types of driver state indicators and their classification into specific classes are described as examples:

Indicators can be classified on a continuum as to how directly or indirectly they are associated with the corresponding driver state component. ◘ Figure 51.1 provides an overview of different classes of indicators for estimating driver readiness.

- For example, measures describing the **physical state of specific body features** of the user are considered to be rather direct indicators. A prominent example is the camera-based recording of gaze direction or eyelid closure behavior (Rauch et al. [48], Mueller et al. [36]). According to the eye-mind hypothesis (Just and Carpenter [49]), the analysis of gaze direction represents a widely used basis for assessing visual attention. In principle, the current state or state change over time can also be recorded and analyzed for many other body features, such as the ears, skin, feet, or voice.
- Another class of rather direct indicators refers to the behavior of users after the presentation of a target signal **(stimulus response performance)**. A typical example is the analysis of the driver response after a request to intervene, e.g., in terms of type of behavior and latency.
- In addition to the non-invasive methods for collecting driver state data, there is also the possibility that users provide direct and explicit information about their **self-assessed driver state**. Although these methods are typically only used in a research context, they could also be part of an interactive product design.

In addition to rather direct ways of inferring a driver's condition, several categories of indirect indicators can also be distinguished. In contrast to the detection of driver states in manual driving, however, these indirect indicators cannot be inferred from the quality of manual vehicle guidance (such as current systems for detecting fatigue or distraction) (see Hecht et al. [50] or Manstetten et al. [51]).

- Indirect information on the state of the driver can be obtained, for example, from the **user's interaction with on-board (or integrated) systems**. Relevant parameters can be: time and duration of user input, recognition of non-driving activity and classification, task difficulty, required human resources or expected stress level.
- While information about **general user characteristics** does not directly refer to the current user state, it can be used to better interpret other indicators. The knowledge of when and how a specific user has taken over vehicle control in the past is helpful to better classify current behavior. When collecting and storing user-related data, particular attention must be paid to stay compliant with data protection regulations.
- Another class of indirect indicators refers to the **temporal and spatial characteristics of the current situation**. For example, what is the noise level in the interior? Are there any other passengers inside the vehicle? Who is talking to whom? What time is it? Is the view of the traffic situation restricted? Answers to these questions can facilitate the interpretation of direct indicators or serve as a fallback level if direct indicators are (currently) not available.

51.4.3 Components of Driver Readiness

▶ Section 51.4.2 described various indicators that are hypothesized to influence the level of driver readiness. Individual indicators can be interpreted via one or more intermediate levels, where they address key aspects of driver readiness that cannot be measured directly, such as "perceptual readiness" or "fatigue". Ideally, individual indicators are combined on the basis of a model. This facilitates, for example, the interpretation of a single direct variable by considering another indirect variable as context. Selecting a broad set of indicators is also recommended due to limited availability of single signals under real-life conditions: less precise but robust features can serve as fallback levels for the primary indicators. This being said, highly informative, single driver state indicators (such as leaving the driver's seat) may directly determine a critical driver state. In the following sections, individual components of driver readiness will be explained in more detail.

51.4.3.1 Receptivity

In the context of automated driving, the term "receptivity" was introduced by SAE as a specific requirement for a user state in the J3016 standard. In this standard, "receptivity" is defined as an "aspect of human consciousness that refers to a person's ability to focus his/her attention in a reliable and appropriate way in response to a signal" (SAE J3016, Chap. 3.23 [2]). The signal must be able to attract the driver's attention. In this way, the requirement clearly differs from a permanently attentive (vigilant) driver state, in which relevant signals must be derived from a person's interpretation of the situation. The ability of the sensory organs to process the stimulus is a necessary precondition for the readiness to perceive a signal (e.g., a request to intervene).

Accordingly, examples of impaired receptivity are
- An optical signal cannot be perceived visually due to closed eyes or continuous gazes to other areas.
- An acoustic signal cannot be perceived due to deafness or because it is masked by acoustic noise (not controllable in level).
- A haptic signal (e.g., vibration) cannot be perceived due to the lack of contact between the body and the actuator.
- A signal may no longer be perceivable due to an impaired form of consciousness such as sleep or a medical emergency.

In order to facilitate the perceptibility of a signal, a multimodal design is generally recommended (see Naujoks et al. [52]). Visual output should be combined with acoustic or haptic signals. Additionally, driver state monitoring may include different types of indicators (see ◘ Fig. 51.1), such as
- the detection of the eye condition (e.g., eyelid opening and spatial orientation of the pupil);
- the type and latency of the expected user response to a signal;
- the analysis of ambient conditions (e.g., sun glare, other sources of noise in the vehicle).

As a theoretical construct, receptivity cannot be measured directly, but is ideally operationalized via several indicators. The monitoring of the indicators can provide evidence of impaired receptivity, but it does not ensure receptivity of a specific user in a specific situation.

51.4.3.2 Situational Awareness

Another concept often used in this context is "situational awareness" (Endsley [53]). It refers to the cognitive state of being accurately aware of one's environment by perceiving the objects in the environment that are relevant for a task, understanding their meaning, and being able to predict their future state for a sufficiently long period of time. In the context of multiple automation levels in one vehicle, awareness of the current operating mode (e.g., partial, conditional, or high automation) plays a particularly important role, since this determines the specific requirements on the user state. As a construct, situational awareness clearly goes beyond receptivity, since the adequate prediction of a situation depends on further cognitive processes. Available knowledge about a system's behavior in similar situations, which can be retrieved from stored past experiences in long-term memory, plays an essential role here. Due to the strongly cognitive character of the construct, situation awareness is difficult to capture by methods of driver observation. Nevertheless, individual behavioral indicators can point toward limited situation awareness during automated driving or in a takeover situation, such as
- The user visually turns away from the driving scene and/or engages in non-driving-related activities.
- The driver's gaze is focused on a single area inside or outside the vehicle for an extended period of time, hindering comprehensive monitoring of the surrounding traffic and the current state of the automation feature. A decrease in horizontal gaze dispersion may also indicate increased cognitive load from non-driving-related activities (see Wang et al. [54]).
- After a request to intervene, the user does not show the expected reaction (e.g., change of glance behavior or body position). When evaluating the behavior, however, it would be necessary to differentiate whether the user deliberately suppresses the expected reaction (e.g., in order to test system limits) or whether the reaction is unintentional.
- The driver takes control of the vehicle without having observed the traffic environment sufficiently (e.g., missing glances to rear-view mirror before initiating a lane change maneuver).
- After a takeover, the driver initiates a driving maneuver that leads to a potentially dangerous situation.

51.4.3.3 Motor Readiness

Motor readiness refers to the driver's current ability to physically operate the vehicle controls or the effort required to create a physical condition suitable for safely resuming vehicle control. In today's vehicle concepts, control elements primarily include the steering wheel, the pedals, and other driving-relevant controls. Restrictions on the motor readiness can be assumed, for example, if
- both hands are off the steering wheel,
- one hand or both hands hold an object,
- a hand-held object cannot easily be put away,
- the current seat position makes it difficult to reach the pedals and steering wheel,
- the driver shows postures unsuitable for immediate resumption of vehicle control,
- the driver leaves the driver's seat.

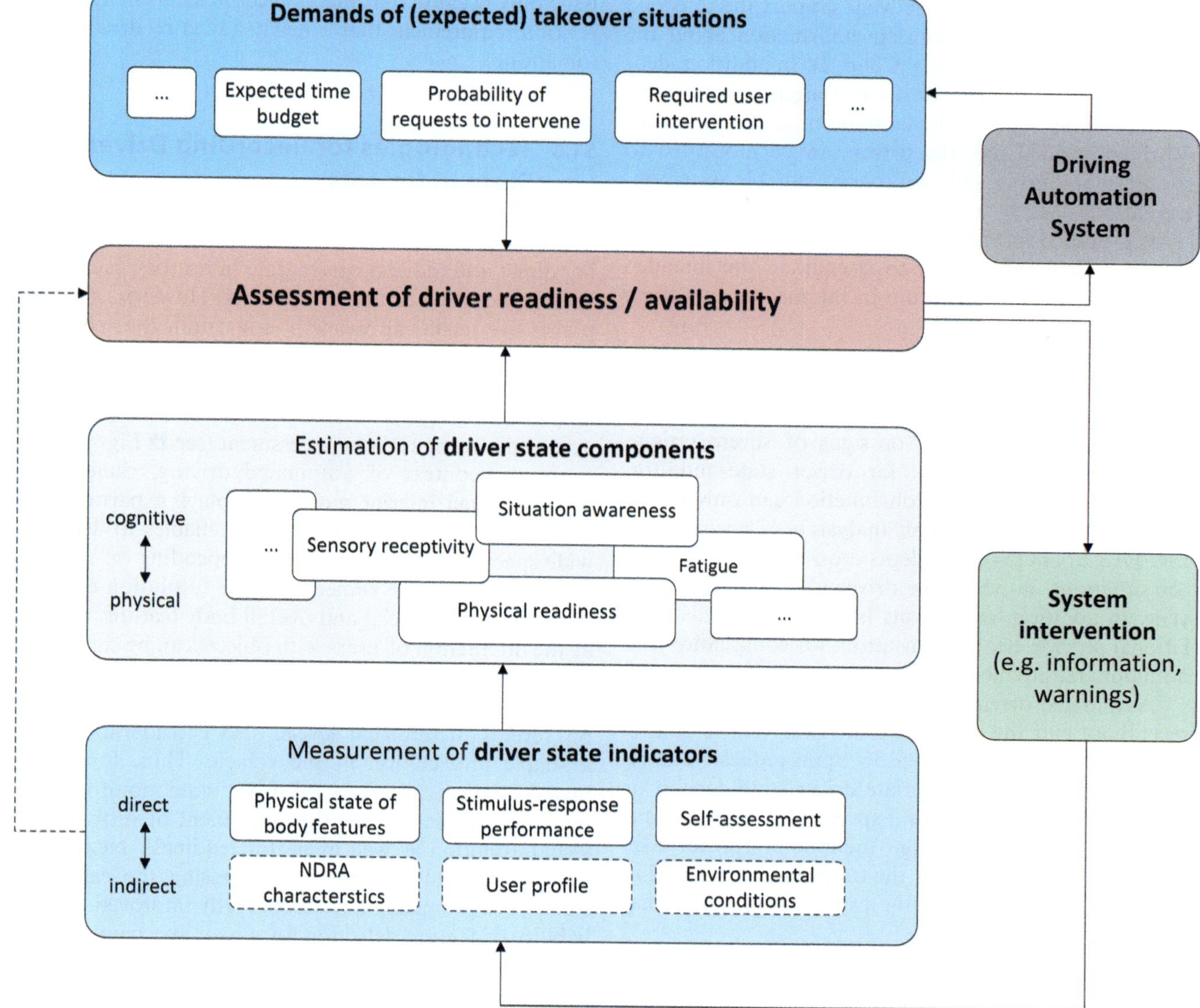

◻ Fig. 51.1   Model for assessing and maintaining driver readiness (or driver availability) in automated driving

Individual requirements of the motor readiness to take over can be checked in principle by suitable characteristics of interior sensors (see ▶ Chap. 50). Minimum requirements are already formulated at the regulatory level (e.g., UN ECE R79 [31] or R157 [32]).

51.4.3.4   Fatigue

The terms fatigue, drowsiness, and sleepiness are often not clearly distinguished from each other in the literature and estimated by the same metrics, such as those based on the degree of eye opening or eyelid closure. In particular, the metric PERCLOS (proportion of time the eyelid is closed more than 80%) has been described as a robust indicator for assessing sleepiness (e.g., Wierwille et al. [55]; Sato et al. [56] or Jarosch [17]). Reference measures based on EEG, such as the analysis of so-called alpha spindles (cf. Frey [57]), may be used in a laboratory context, but do not play a role in current production vehicles due to the complex and error-prone procedure.

As with the estimation of other driver states, it is advantageous to integrate multiple indicators of drowsiness in a meaningful way, such as considering time of day or driving duration. However, indirect indicators of drowsiness based on the analysis of manual control can no longer be used in an automated driving context (Manstetten et al. [51]).

51.5   Designing Driver State Monitoring Systems

In principle, driving automation systems should be designed in a way that the driver's task is communicated clearly and unambiguously when using these systems. The design of the human–machine interaction in the vehicle is supposed to proactively support the intended use and to provide appropriate information about the remaining driver task (see ▶ Chap. 2). In addition, dedicated technical devices for driver state monitoring can be utilized: as soon as driver state indicators point toward unintended use, the driver can be motivated to re-engage in his/her assigned role by suitable system interventions (e.g., by a scheme of escalating warning levels as proposed in Mueller et al. [36]). Driver state monitoring systems are meant to strengthen the intended use of an automation feature in the medium and long terms.

At the level of partial automation, system interventions are typically based on indicators such as long glances away from the driving scenery, prolonged hands-off wheel episodes, or signs of severe fatigue. The specific requirements for driver state monitoring for a specific automation function can only be derived from a hazard and risk analysis (e.g., according to ISO/PAS 21448 [34]) and depend on the specific takeover situations, in which the driver is expected to intervene. In addition, regulations (such as UN ECE regulations) provide clear specifications for compliance with minimum requirements in many countries of the world.

Automated driving on SAE L3 requires permanent receptivity and the ability to take over vehicle control within a sufficient time window upon request. The requirements for the appropriate driver state depend on the operating conditions of a specific automation function, such as the speed range, the time budget for takeovers, or the complexity of the traffic environment. The minimum requirement of the expected user state may be

- the driver is in a sitting position which is generally suitable for resuming vehicle control;
- the driver is generally awake and responsive.

The first UN ECE regulation for the SAE L3 function "Automated Lane Keeping System" (UN ECE R157 [32]) describes specific requirements for driver state monitoring in the context of automated driving up to 60 km/h in motorway-like environments. This regulation also describes how the driver's ability to override the automated driving system can depend on an assessment of the current user state: for example, if visual distraction from the traffic scene is detected, the thresholds for deactivating the system by user intervention should be increased.

It may seem surprising that health-related user data has not been mentioned so far in assessing driver readiness. A medical emergency such as a sudden loss of consciousness due to a hypoglycemic shock would clearly prevent driver readiness. The detection of medical emergencies or even signs of intoxication should therefore be considered in the context of automated driving as well. This chapter does not elaborate on this issue only because the detection of medical emergencies is not a requirement that is newly added by driving automation.

51.6 Technologies for Recording Driver State Indicators

There are a variety of technical approaches for recording direct and indirect driver state indicators, each with specific advantages and drawbacks. However, a comprehensive technical review is not within the scope of this chapter. In the following sections, a number of approaches for estimating driver states are outlined as examples and roughly assessed in terms of their potential according to the authors' assessment (see ◘ Fig. 51.2).

In the context of automated driving, **camera-based driver and interior monitoring** plays a particularly important role, since this approach enables to detect a wide spectrum of body features. Depending on the detection range of the camera and the resolution or type of image sensor, facial and overall body features as well as the interaction of users with objects can be captured. The software-based interpretation of the features in terms of driver state components typically takes place on several hierarchical levels, also considering other available information in the vehicle. Thus, for automated driving, camera-based driver state monitoring is regarded as essential for the assessment of drowsiness, visual attention as well as motor readiness. New technologies and algorithms for processing the captured signals are promising and have greatly improved the reliability of feature detection for a wide spectrum of human faces/bodies even under challenging lighting conditions. Automated driving features at SAE Levels 2 and 3 that allow hands-off operation typically require camera-based monitoring solutions for detecting relevant driver state indicators.

In order to determine whether specific body parts of a user are in contact with an object in the vehicle, dedicated sensors can be used that can register **human touch or force**. Capacitive and resistive sensors are effectively used in the context of automated driving, for example, for the evaluation of hand position at the steering wheel or seating position.

The detection of **user input on vehicle-related controls** can be a reliable source of information about the user's state: is there any user input (e.g., steering intervention) at all? How intense is the user input or how frequently does a particular user input occur? The user's action may be especially diagnostic if it can be associated with triggering events (such as a request to intervene). This requires a comprehensive integration of signals at the level of driver state monitoring. Relevant user inputs may not only be related to primary vehicle controls (like steering wheel, pedals), but also to the control of non-driving-relevant systems in the vehicle (e.g., in-vehicle infotainment or integrated mobile devices).

Another method to infer the user's state in the vehicle interior refers to the **analysis of acoustic information** (although suitable technologies have not been installed in vehicles on a large scale so far). Apart from interpreting the intensity and direction of a sound source (e.g., "driver is listening to loud music and singing"), the semantic interpretation of nonverbal signals could provide additional cues about the driver's state (e.g., changes in spoken language due to alcohol consumption or detection of emotional states such as anger and aggression). The computer-based analysis of nonverbal voice features as well as the correct interpretation of verbal information can be a valuable complement to the visual analysis of the interior.

Although the collection of physiological data by means of external measurement techniques (e.g., derivation of EEG, ECG, or EMG) is widely used in a research context, they do not play a significant role in practical applications so far for various reasons. In principle, however, it is conceivable that data from **wearable sensors** (such as from fitness wristbands) could be made available to the vehicle's user state monitoring system. Thus, physiological parameters such as heart rate, respiratory rate, or galvanic skin response could be considered in the overall assessment of the user's state. The inclusion of external sensors does not only require standardized data interfaces but also comes with high requirements for availability and validity of the data—especially for safety-relevant applications. As of now, these pre-conditions are hard to meet, which is why wearable sensor technology is not (yet) considered to a significant extent in driver state monitoring.

Further research is also needed for the **prediction of driver states** on the basis of measurable indicators. The prediction of a driver state is particularly helpful if an intervention in the current user state is no longer feasible (for example, sending a request to intervene to a sleeping user). Compared to the estimation of the current driver state, forecasts can naturally only be modeled with (even) greater uncertainty. Apart from general assumptions about the development of certain user states, data about individual driver characteristics (e.g., from the usage history or from a usage profile) can also be used for prediction.

Even if the identification of the current user as well as the storage of personal data does not pose a technical challenge, the legal requirements for data protection must always be considered. Currently, personalized predictive driver state estimation does not yet play a significant role in the field of vehicle automation.

51.7 Summary and Outlook

In this book chapter, the role of driver state monitoring in partial and conditional driving automation is examined in more detail. Driver state monitoring appears to be an important measure for supporting the driver in using an automated driving system as intended.

The theoretical construct "driver readiness" serves as a (possible) framework model that includes the state of the driver on the basis of measurable criteria in relation to the demands of the driving situation. Several sections of this article point toward the need to integrate information from several indicators, since single driver state criteria are often not sufficiently diagnostic to infer a complex (latent) driver state. It is also evident that individual technologies for driver state detection differ by the number of use cases they can address. In this regard, the camera-based observation of the vehicle interior stands out from alternative methods as it covers a broad range of applications.

The specific requirements for driver state monitoring in automated driving are not (yet) clearly defined for many future functional features and therefore cannot be described in this article. Organizations for standardization of processes and products (such as ISO or SAE) as well as interest groups in the context of the regulation of automated driving functions are currently working intensively on this topic and will further specify the role of driver state monitoring in automated vehicles.

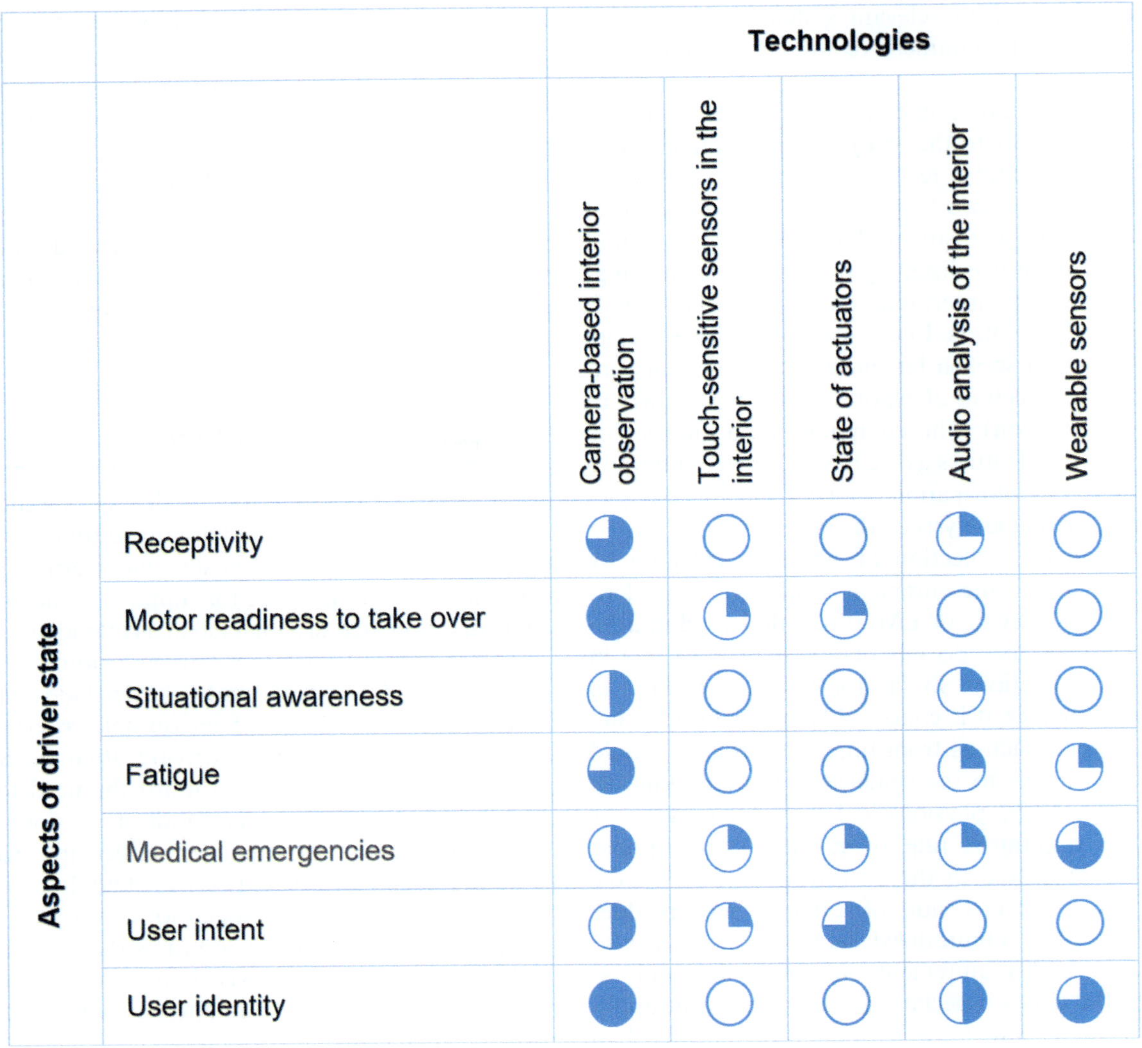

Technologies

Aspects of driver state	Camera-based interior observation	Touch-sensitive sensors in the interior	State of actuators	Audio analysis of the interior	Wearable sensors
Receptivity					
Motor readiness to take over					
Situational awareness					
Fatigue					
Medical emergencies					
User intent					
User identity					

Fig. 51.2 Assessment of the potential of selected technological approaches for estimating specific driver state aspects (filled circle corresponds to "very large potential", empty circle corresponds to "little/no potential")

References

1. Mercedes-Benz Group: Autonomous Driving (o. J.). ► https://www.daimler.com/innovation/product-innovation/autonomous-driving/. Accessed 14 July 2021
2. SAE International: Taxonomy and Definitions for Terms Related to Driving Automation Systems for On-Road Motor Vehicles J3016. SAE International, Warrendale, PA, USA (2018). ► https://doi.org/10.4271/J3016_201806
3. United Nations Economic Commission for Europe: Reference Document with Definitions of Automated Driving under WP.29 and the General Principles for Developing a UN Regulation on Automated Vehicles (2018, April 23). ► https://unece.org/fileadmin/DAM/trans/main/wp29/wp29resolutions/ECE-TRANS-WP29-1140e.pdf
4. Mackworth, N.H.: The breakdown of vigilance during prolonged visual search. Q. J. Exp. Psychol. **1**(1), 6–21 (1948)
5. Warm, J.S., Parasuraman, R., Matthews, G.: Vigilance requires hard mental work and is stressful. Hum. Factors: J. Hum. Factors Ergon. Soc. **50**(3), 433–441 (2008). ► https://doi.org/10.1518/001872008X312152
6. Greenlee, E.T., DeLucia, P.R., Newton, D.C.: Driver vigilance in automated vehicles: hazard detection failures are a matter of time. Hum. Factors: J. Hum. Factors Ergon. Soc. **60**(4), 465–476 (2018). ► https://doi.org/10.1177/0018720818761711
7. Körber, M., Cingel, A., Zimmermann, M., Bengler, K.: Vigilance decrement and passive fatigue caused by monotony in automated driving. Procedia Manuf. **3**, 2403–2409 (2015). ► https://doi.org/10.1016/j.promfg.2015.07.499
8. Carsten, O., Martens, M.H.: How can humans understand their automated cars? HMI principles, problems and solutions. Cogn. Technol. Work **21**(1), 3–20 (2019). ► https://doi.org/10.1007/s10111-018-0484-0
9. Llaneras, R.E., Salinger, J., Green, C.A.: Human factors issues associated with limited ability autonomous driving systems: drivers' allocation of visual attention to the forward roadway. In: Proceedings of the 7th International Driving Symposium on Human Factors in Driver Assessment, Training, and Vehicle Design: Driving Assessment 2013, The Sagamore on Lake George, Bolton Landing, New York, USA, June 17–20, 2013 (S. 92–98). The University of Iowa, Public Policy Center, Iowa City, Iowa (2013). ► http://pubs.lib.uiowa.edu/drive/article/id/27368/

10. Banks, V.A., Eriksson, A., O'Donoghue, J., Stanton, N.A.: Is partially automated driving a bad idea? Observations from an on-road study. Appl. Ergon. **68**, 138–145 (2018). ► https://doi.org/10.1016/j.apergo.2017.11.010

11. Hancock, P.A.: Automation: how much is too much? Ergonomics **57**(3), 449–454 (2014). ► https://doi.org/10.1080/00140139.2013.816375

12. Gershon, P., Seaman, S., Mehler, B., Reimer, B., Coughlin, J.: Driver behavior and the use of automation in real-world driving. Accid. Anal. Prev. **158**, 106217 (2021). ► https://doi.org/10.1016/j.aap.2021.106217

13. Schömig, N., Hargutt, V., Neukum, A., Petermann-Stock, I., Othersen, I.: The interaction between highly automated driving and the development of drowsiness. Procedia Manuf. **3**, 6652–6659 (2015). ► https://doi.org/10.1016/j.promfg.2015.11.005

14. Vogelpohl, T., Kühn, M., Hummel, T., Vollrath, M.: Asleep at the automated wheel—sleepiness and fatigue during highly automated driving. Accid. Anal. Prev. **126**, 70–84 (2019). ► https://doi.org/10.1016/j.aap.2018.03.013

15. Oxford Learner's Dictionaries: Fatigue. ► https://www.oxfordlearnersdictionaries.com/definition/american_english/fatigue. Accessed 11 March 2022

16. May, J.F., Baldwin, C.L.: Driver fatigue: the importance of identifying causal factors of fatigue when considering detection and countermeasure technologies. Transport. Res. F: Traffic Psychol. Behav. **12**(3), 218–224 (2009). ► https://doi.org/10.1016/j.trf.2008.11.005

17. Jarosch, O.: Non-driving-related tasks in conditional driving automation (Dissertation). Technische Universität München, München (2020, January 7)

18. Weinbeer, V., Muhr, T., Bengler, K.: Automated driving: the potential of non-driving-related tasks to manage driver drowsiness. In: Bagnara, S., Tartaglia, R., Alboline, S., Thomas, A., Fujita, Y. (eds.) Proceedings of the 20th Congress of the International Ergonomics Association (IEA 2018) (Bd. VI: Transport Ergonomics and Human Factors (TEHF), Aerospace Human Factors and Ergonomics). Springer International Publishing, Cham (2019)

19. Zhang, B., de Winter, J., Varotto, S., Happee, R., Martens, M.: Determinants of take-over time from automated driving: a meta-analysis of 129 studies. Transport. Res. F: Traffic Psychol. Behav. **64**, 285–307 (2019). ► https://doi.org/10.1016/j.trf.2019.04.020

20. Jarosch, O., Naujoks, F., Wandtner, B., Gold, C., Marberger, C., Weidl, G., Schrauf, M.: The Impact of Non-driving Related Tasks on Take-over Performance in Conditionally Automated Driving—A Review of the Empirical Evidence. 9. Tagung Automatisiertes Fahren, München (2019)

21. Damböck, D.: Automationseffekte im Fahrzeug—von der Reaktion zur Übernahme (Dissertation). Technische Universität München (2013, Juli 16)

22. Petermann-Stock, I., Hackenberg, L., Muhr, T., Mergl, C.: Wie lange braucht der Fahrer? Eine Analyse zu Übernahmezeiten aus verschiedenen Nebentätigkeiten während einer hochautomatisierten Staufahrt. 6. Tagung Fahrerassistenzsysteme. Der Weg zum automatischen Fahren (2013)

23. Naujoks, F., Purucker, C., Wiedemann, K., Marberger, C.: Noncritical state transitions during conditionally automated driving on German freeways: effects of non-driving related tasks on takeover time and takeover quality. Hum. Factors: J. Hum. Factors Ergon. Soc. **61**(4), 596–613 (2019). ► https://doi.org/10.1177/0018720818824002

24. Boos, A., Feldhutter, A., Schwiebacher, J., Bengler, K.: Mode errors and intentional violations in visual monitoring of level 2 driving automation. In: 2020 IEEE 23rd International Conference on Intelligent Transportation Systems (ITSC), S. 1–7. IEEE, Rhodes, Greece (2020). ► https://ieeexplore.ieee.org/document/9294690/

25. Sarter, N.B., Woods, Billings: Automation surprises. In: Handbook of Human Factors and Ergonomics, 2. Aufl., S. 1926–1943. Wiley, New York (1997)

26. Victor, T.W., Tivesten, E., Gustavsson, P., Johansson, J., Sangberg, F., Ljung Aust, M.: Automation expectation mismatch: incorrect prediction despite eyes on threat and hands on wheel. Hum. Factors: J. Hum. Factors Ergon. Soc. **60**(8), 1095–1116 (2018). ► https://doi.org/10.1177/0018720818788164

27. Wilson, K.M., Yang, S., Roady, T., Kuo, J., Lenné, M.G.: Driver trust and mode confusion in an on-road study of level-2 automated vehicle technology. Saf. Sci. **130**, 104845 (2020). ► https://doi.org/10.1016/j.ssci.2020.104845

28. Kurpiers, C., Biebl, B., Mejia Hernandez, J., Raisch, F.: Mode awareness and automated driving—what is it and how can it be measured? Information **11**(5), 277 (2020). ► https://doi.org/10.3390/info11050277

29. Morando, A., Gershon, P., Mehler, B., Reimer, B.: Driver-Initiated Tesla Autopilot Disengagements in Naturalistic Driving, S. 57–65. ACM (2020). ► https://doi.org/10.1145/3409120.3410644

30. Marberger, C.: Nutzerseitiger Fehlgebrauch von Fahrerassistenzsystemen: Bericht zum Forschungsprojekt FE 82.275/2004. Bundesanstalt für Straßenwesen, Bergisch Gladbach (2007)

31. United Nations Economic Commission for Europe: Uniform provisions concerning the approval of vehicles with regard to steering equipment, Regulation No. 79 (2018)

32. United Nations Economic Commission for Europe: Uniform provisions concerning the approval of vehicles with regards to Automated Lane Keeping Systems, Regulation No. 157 (2021)

33. RESPONSE 3 Consortium: Code of Practice for the Design and Evaluation of ADAS. RESPONSE 3: A PReVENT Project (2009). ► https://www.acea.auto/files/20090831_Code_of_Practice_ADAS.pdf

34. International Organization for Standardization: Road Vehicles—Safety of the Intended Functionality. ISO/PAS 21448:2019 (2019). ► https://www.iso.org/standard/70939.html

35. Cabrall, C.D.D., Eriksson, A., Dreger, F., Happee, R., de Winter, J.: How to keep drivers engaged while supervising driving automation? A literature survey and categorisation of six solution areas. Theor. Issues Ergon. Sci. **20**(3), 332–365 (2019). ► https://doi.org/10.1080/1463922X.2018.1528484

36. Mueller, A.S., Reagan, I.J., Cicchino, J.B.: Addressing driver disengagement and proper system use: human factors recommendations for level 2 driving automation design. J. Cogn. Eng. Decis. Mak. 155534342098312 (2021). ► https://doi.org/10.1177/1555343420983126

37. Abbink, D.A., Mulder, M., Boer, E.R.: Haptic shared control: smoothly shifting control authority? Cogn. Technol. Work **14**(1), 19–28 (2012). ► https://doi.org/10.1007/s10111-011-0192-5

38. Flemisch, F., Abbink, D., Itoh, M., Pacaux-Lemoine, M.-P., Weßel, G.: Shared control is the sharp end of cooperation: towards a common framework of joint action, shared control and human machine cooperation. IFAC-PapersOnLine **49**(19), 72–77 (2016). ► https://doi.org/10.1016/j.ifacol.2016.10.464

39. Parasuraman, R., Mouloua, M., Molloy, R.: Effects of adaptive task allocation on monitoring of automated systems. Hum. Factors: J. Hum. Factors Ergon. Soc. **38**(4), 665–679 (1996). ► https://doi.org/10.1518/001872096778827279

40. Beller, J., Heesen, M., Vollrath, M.: Improving the driver-automation interaction: an approach using automation uncertainty. Hum. Factors: J. Hum. Factors Ergon. Soc. **55**(6), 1130–1141 (2013). ► https://doi.org/10.1177/0018720813482327

41. Llaneras, R.E., Cannon, B.R., Green, C.A.: Strategies to assist drivers in remaining attentive while under partially automated driving: verification of human-machine interface concepts.

Transp. Res. Rec.: J. Transp. Res. Board **2663**(1), 20–26 (2017). ► https://doi.org/10.3141/2663-03

42. Sheridan, T.B., Parasuraman, R.: Human-automation interaction. Rev. Hum. Factors Ergon. **1**(1), 89–129 (2005). ► https://doi.org/10.1518/155723405783703082

43. Cabrall, C.D.D., Janssen, N.M., de Winter, J.C.F.: Adaptive automation: automatically (dis)engaging automation during visually distracted driving. PeerJ Comput. Sci. **4**, e166 (2018). ► https://doi.org/10.7717/peerj-cs.166

44. Flemisch, F., Kaussner, A., Petermann, I., Schieben, A., Schömig, N.: HAVEit Deliverable D.33.6: Validation of Concept on Optimum Task Repartition (2011)

45. Sprung, A.J.: Auswirkung von Müdigkeit auf die Fahrleistung und Präventionsmöglichkeiten—SAE L3. Gehalten auf der TANGO Abschlusspräsentation (2020, September). ► https://tangoversuch1.files.wordpress.com/2020/10/04_auswirkung_von_muedigkeit_und_praevention_sae_l3_v3.pdf

46. International Organization for Standardization: Road vehicles—human performance and state in the context of automated driving—part 2: considerations in designing experiments to investigate transition processes (ISO Technical Report No. 21959-2:2020) (2020). ► https://www.iso.org/standard/73928.html

47. Marberger, C., Mielenz, H., Naujoks, F., Radlmayr, J., Bengler, K., Wandtner, B.: Understanding and applying the concept of "Driver Availability" in automated driving. In: Stanton, N.A. (Hrsg.) Advances in Human Aspects of Transportation, Bd. 597, S. 595–605. Springer International Publishing, Cham (2018). ► https://doi.org/10.1007/978-3-319-60441-1_58

48. Rauch, N., Kaussner, A., Krüger, H.-P., Boverie, S., Flemisch, F.: The importance of driver state assessment within highly automated vehicles. In: Gehalten auf der 16th ITS World Congress and Exhibition on Intelligent Transport Systems and Services, Stockholm, Sweden (2009)

49. Just, M.A., Carpenter, P.A.: A theory of reading: from eye fixations to comprehension. Psychol. Rev. **87**(4), 329–354 (1980)

50. Hecht, T., Feldhütter, A., Radlmayr, J., Nakano, Y., Miki, Y., Henle, C., Bengler, K.: A review of driver state monitoring systems in the context of automated driving. In: Bagnara, S., Tartaglia, R., Albolino, S., Alexander, T., Fujita, Y. (Hrsg.) Proceedings of the 20th Congress of the International Ergonomics Association (IEA 2018), Bd. 823, S. 398–408. Springer International Publishing, Cham (2019). ► https://doi.org/10.1007/978-3-319-96074-6_43

51. Manstetten, D., Beruscha, F., Bieg, H.-J., Kobiela, F., Korthauer, A., Krautter, W., Marberger, C.: The evolution of driver monitoring systems: a shortened story on past, current and future approaches how cars acquire knowledge about the driver's state. In: 22nd International Conference on Human-Computer Interaction with Mobile Devices and Services, S. 1–6. ACM (2020). ► https://doi.org/10.1145/3406324.3425896. 28 June 2022

52. Naujoks, F., Wiedemann, K., Schömig, N., Hergeth, S., Keinath, A.: Towards guidelines and verification methods for automated vehicle HMIs. Transport. Res. F: Traffic Psychol. Behav. **60**, 121–136 (2019). ► https://doi.org/10.1016/j.trf.2018.10.012

53. Endsley, M.R.: Automation and situation awareness. In: Parasuraman, R., Mouloua, M. (eds.) Automation and Human Performance: Theory and Applications, pp. 163–181. CRC Press (1996)

54. Wang, Y., Reimer, B., Dobres, J., Mehler, B.: The sensitivity of different methodologies for characterizing drivers' gaze concentration under increased cognitive demand. Transport. Res. F: Traffic Psychol. Behav. **26**, 227–237 (2014). ► https://doi.org/10.1016/j.trf.2014.08.003

55. Wierwille, W., Wreggit, S.S., Kirn, C.L., Ellsworth, L.A., Fairbanks, R.J.: Research on vehicle-based driver status/performance monitoring; development, validation, and refinement of algorithms for detection of driver drowsiness. Final report (No. HS-808 247). Virginia Polytechnic Institute and State University, Blacksburg, Washington DC (1994). ► https://rosap.ntl.bts.gov/view/dot/2578. 28 June 2022

56. Sato, T., Takeda, Y., Akamatsu, M., Kitazaki, S.: Evaluation of driver drowsiness while using automated driving systems on driving simulator, test course and public roads. In: Krömker, H. (Hrsg.) HCI in Mobility, Transport, and Automotive Systems. Driving Behavior, Urban and Smart Mobility, Bd. 12213, S. 72–85. Springer International Publishing, Cham (2020). ► https://doi.org/10.1007/978-3-030-50537-0_7

57. Frey, A.: Zum Fahrerzustand beim automatisierten Fahren: Objektive Messung von Müdigkeit und ihre Einflussfaktoren (2021)

Open Access This chapter is licensed under the terms of the Creative Commons Attribution-NonCommercial-NoDerivatives 4.0 International License (► http://creativecommons.org/licenses/by-nc-nd/4.0/), which permits any noncommercial use, sharing, distribution and reproduction in any medium or format, as long as you give appropriate credit to the original author(s) and the source, provide a link to the Creative Commons license and indicate if you modified the licensed material. You do not have permission under this license to share adapted material derived from this chapter or parts of it.

The images or other third party material in this chapter are included in the chapter's Creative Commons license, unless indicated otherwise in a credit line to the material. If material is not included in the chapter's Creative Commons license and your intended use is not permitted by statutory regulation or exceeds the permitted use, you will need to obtain permission directly from the copyright holder.

Supplementary Information

© The Editor(s) (if applicable) and The Author(s) 2026
H. Winner et al. (eds *Handbook Assisted and Automated Driving*, https://doi.org/10.1007/978-3-658-45276-6

Index

The manufacturer's authorised representative in the EU is Springer
Nature Customer Service Centre GmbH, Europaplatz 3, 69115 Heidelberg,
Germany. If you have any concerns regarding our products, please
contact ProductSafety@springernature.com

Printed and bound by CPI Group (UK) Ltd, Croydon, CR0 4YY
03/07/2026
02156410-0001